WITHDRAWN
UTSA LIBRARIES

RENEWALS 458-4574

Techniques in AQUATIC TOXICOLOGY

Volume 2

Techniques in AQUATIC TOXICOLOGY

Volume 2

Edited by

Gary K. Ostrander, Ph.D.

Johns Hopkins University
Baltimore, Maryland

Boca Raton London New York Singapore

A CRC title, part of the Taylor & Francis imprint, a member of the Taylor & Francis Group, the academic division of T&F Informa plc.

Library of Congress Cataloging-in-Publication Data

Catalog record is available from the Library of Congress

This book contains information obtained from authentic and highly regarded sources. Reprinted material is quoted with permission, and sources are indicated. A wide variety of references are listed. Reasonable efforts have been made to publish reliable data and information, but the author and the publisher cannot assume responsibility for the validity of all materials or for the consequences of their use.

Neither this book nor any part may be reproduced or transmitted in any form or by any means, electronic or mechanical, including photocopying, microfilming, and recording, or by any information storage or retrieval system, without prior permission in writing from the publisher.

All rights reserved. Authorization to photocopy items for internal or personal use, or the personal or internal use of specific clients, may be granted by CRC Press, provided that $1.50 per page photocopied is paid directly to Copyright Clearance Center, 222 Rosewood Drive, Danvers, MA 01923 USA. The fee code for users of the Transactional Reporting Service is ISBN 1-56670-664-5/05/$0.00+$1.50. The fee is subject to change without notice. For organizations that have been granted a photocopy license by the CCC, a separate system of payment has been arranged.

The consent of CRC Press does not extend to copying for general distribution, for promotion, for creating new works, or for resale. Specific permission must be obtained in writing from CRC Press for such copying.
Direct all inquiries to CRC Press, 2000 N.W. Corporate Blvd., Boca Raton, Florida 33431.

Trademark Notice: Product or corporate names may be trademarks or registered trademarks, and are used only for identification and explanation, without intent to infringe.

Visit the CRC Press Web site at www.crcpress.com

© 2005 by CRC Press
No claim to original U.S. Government works
International Standard Book Number 1-56670-664-5
Library of Congress Card Number
Printed in the United States of America 1 2 3 4 5 6 7 8 9 0
Printed on acid-free paper

Dedication

Books such as this are usually dedicated to family members, colleagues, collaborators, or mentors who have played a meaningful role in the author's life and career. Sometimes, however, some of the most significant people in our lives are outside the circle of our family or our discipline.

I dedicate this volume to the folks at Johnson Orchards in Yakima, Washington, and in particular to Roy Johnson, Jr. ("Mister Johnson"), Donna Johnson, and Eric Johnson. I gratefully acknowledge that for many years I always had employment, whether it be for a summer or for a few hours over a weekend home from college. More importantly, I am most appreciative for their always believing in me and supporting me, no matter what objective I was pursuing. This has been true for nearly 30 years.

Preface

The initial volume in this series was published in 1996 and was well received by the scientific community. In fact, it has been most gratifying to observe the volume entering a second printing and the positive feedback I continue to receive from colleagues around the world. While the original intent was to follow with additional volumes every 2 to 3 years (wishful thinking!), other commitments have dictated a more modest approach. Nonetheless, herein I am pleased to present 39 additional techniques chapters.

As with *Volume 1*, this volume contains a blend of established and recently developed techniques that have the potential to significantly impact the expansive field of aquatic toxicology. I have divided the chapters into four broad sections to include techniques for assessment of toxicity in whole organisms, cellular and subcellular toxicity, identification and assessment of contaminants in aquatic ecosystems, and I conclude with a general techniques section, techniques for aquatic toxicologists, which contains chapters that could be of value for anyone working in this field.

Each of the individual chapters covers a specific procedure in detail. A brief *Introduction* serves to highlight the technique and the *Materials* section provides a very detailed list of what is needed to conduct the procedure(s). The *Procedures* sections are written so that the procedure can be easily followed and reproduced by a technician, graduate student, or someone with a basic knowledge of the field. In the *Results and Discussion* section, I have asked the contributors to provide and describe typical results, as well as anomalous results, false positives, artifacts, etc. In some instances, the contributors have provided data from their recently published work. Alternatively, some contributors have provided the data and discussed the previously unpublished experiments. Each chapter concludes with a list of pertinent *References* and a few also include appendices as necessary.

In addition to my review, all the chapters in this volume were reviewed by at least one individual with appropriate subject matter expertise. I am most grateful to the following scientists for providing thoughtful commentaries on one or more (usually more!) of the manuscripts contained herein: Gary Atchison, Keith Cheng, Dominic M. Di Toro, Craig Downs, Damjana Drobne, Howard Fairbrother, Jeffrey P. Fisher, Douglas J. Fort, Marc M. Greenberg, Mary Haasch, William E. Hawkins, James P. Hickey, James N. Huckins, David Janz, J. McHugh Law, Lawrence LeBlanc, Paul McCauley, Mark S. Myers, Dana Peterson, Taylor Reynolds, Colleen S. Sinclair, Terry W. Snell, Larry G. Talent, and Rebecca Van Beneden.

Finally, I express my sincere appreciation to the contributors for taking the time to distill their techniques into a consistent, easily accessible format. I believe that this consistency, more than anything else, has contributed to the success and the utility of this series.

About the Editor

Gary K. Ostrander received a B.S. degree in biology from Seattle University in 1980, an M.S. degree in biology from Illinois State University in 1982, and his Ph.D. degree in 1986 from the College of Ocean and Fisheries Sciences at the University of Washington, where he specialized in aquatic toxicology. He was an NIH postdoctoral fellow in the Department of Pathology, School of Medicine, at the University of Washington from 1986 to 1989 and also served as a staff scientist at the Pacific Northwest Research Foundation from 1986 to 1990.

In 1990, he joined the Department of Zoology at Oklahoma State University as an assistant professor and was tenured and promoted to the rank of associate professor in 1993. He assumed a dual role as director of the Environmental Institute and associate dean of the Graduate College in 1995.

Dr. Ostrander joined the faculty at Johns Hopkins University in 1996 and is currently the associate provost for research and chair of the Graduate Board at Johns Hopkins University. He holds his academic appointments in the Department of Biology in the School of Arts and Sciences and in the Department of Comparative Medicine in the School of Medicine.

Dr. Ostrander has authored over 80 technical papers and book chapters, edited 4 books, and has written a field guide. His primary research interest has been elucidating mechanisms of chemical carcinogenesis for which he employed aquatic, rodent, and human models. A second aspect of his work has been laboratory and field studies focused on understanding the mechanisms behind the worldwide decline of coral reef ecosystems.

Contributors

Merrin S. Adams
CSIRO Energy Technology
Bangor, New South Wales, Australia

Katie M. Anderson
Pacific Northwest Research Institute
Seattle, Washington

Mary R. Arkoosh
National Oceanic and Atmospheric Administration
Seattle, Washington

David H. Baldwin
National Marine Fisheries Service
Seattle, Washington

C. Basslear
University of Guam Marine Laboratory
Mangilao, Guam

Ronny Blust
University of Antwerp
Antwerp, Belgium

J.L. Bolton
National Oceanic and Atmospheric Administration
Seattle, Washington

Doranne J. Borsay Horowitz
U.S. Environmental Protection Agency
Narragansett, Rhode Island

Daryle Boyd
National Marine Fisheries Service
Seattle, Washington

R.H. Boyer
National Oceanic and Atmospheric Administration
Seattle, Washington

Deborah Boylen
National Oceanic and Atmospheric Administration
Seattle, Washington

Sandra K. Brewer
U.S. Army Corps of Engineers
Rock Island, Illinois

Kay Briggs
George Mason University
Manassas, Virginia

D.W. Brown
National Oceanic and Atmospheric Administration
Seattle, Washington

D.G. Burrows
National Oceanic and Atmospheric Administration
Seattle, Washington

Jon Buzitis
National Marine Fisheries Service
Seattle, Washington

Thomas Capo
University of Miami
Miami, Florida

Wenlin Chen
Syngenta
Greensboro, North Carolina

Laura Coiro
U.S. Environmental Protection Agency
Narragansett, Rhode Island

Tracy K. Collier
National Oceanic and Atmospheric Administration
Seattle, Washington

James M. Conder
University of North Texas
Denton, Texas

D.E. Conners
University of Georgia
Athens, Georgia

Simon C. Courtenay
University of New Brunswick
St. John, New Brunswick, Canada

Wim De Coen
University of Antwerp
Antwerp, Belgium

Karel A.C. DeSchamphelaere
Ghent University
Ghent, Belgium

Richard T. Di Giulio
Duke University
Durham, North Carolina

Aaron G. Downs
Yale University
New Haven, Connecticut

Craig A. Downs
EnVirtue Biotechnologies, Inc.
Winchester, Virginia

Damjana Drobne
University of Ljubljana
Ljubljana, Slovenia

Samo Drobne
University of Ljubljana
Ljubljana, Slovenia

Monique G. Dubé
National Water Research Institute
Environment Canada
Saskatoon, Canada

Jim Ferretti
U.S. Environmental Protection Agency
Edison, New Jersey

Douglas J. Fort
Fort Environmental Laboratories, Inc.
Stillwater, Oklahoma

Natasha M. Franklin
McMaster University
Hamilton, Ontario, Canada

Pier Francesco Ghetti
University Ca' Foscari of Venice
Venice, Italy

Annamaria Volpi Ghirardini
University Ca' Foscari of Venice
Venice, Italy

Naomi K. Gilman
Pacific Northwest Research Institute
Seattle, Washington

Karen L. Gormley
University of New Brunswick
St. John, New Brunswick, Canada

Virginia M. Green
Pacific Northwest Research Institute
Seattle, Washington

Torsten Hahn
Technical University of Braunschweig
Braunschweig, Germany

Dagobert G. Heijerick
Ghent University
Ghent, Belgium
and
EURAS
Zwijnaarde, Belgium

Caren C. Helbing
University of Victoria
Victoria, British Columbia, Canada

David P. Herman
National Oceanic and Atmospheric Administration
Seattle, Washington

James P. Hickey
U.S. Geological Survey/Great Lakes Science Center
Ann Arbor, Michigan

J. Hoguet
College of Charleston
Charleston, South Carolina

Lawrence Hufnagle, Jr.
National Marine Fisheries Service
Seattle, Washington

Rebecca E.M. Ibey
University of New Brunswick
St. John, New Brunswick, Canada

Colin R. Janssen
Ghent University
Ghent, Belgium

Lyndal L. Johnson
National Oceanic and Atmospheric Administration
Seattle, Washington

Robert B. Jonas
George Mason University
Manassas, Virginia

Andrew S. Kane
University of Maryland
College Park, Maryland

Ioanna Katsiadaki
Centre for Environment, Fisheries, and Aquaculture Science
Weymouth, Dorset, United Kingdom

E.T. Knobbe
Sciperio Inc.
Stillwater, Oklahoma

Tomoko Koda
National Institute for Environmental Studies
Tsukuba, Ibaraki, Japan

Margaret M. Krahn
National Marine Fisheries Service
Seattle, Washington

Leslie Kubin
National Marine Fisheries Service
Seattle, Washington

Thomas W. La Point
University of North Texas
Denton, Texas

Roman Lanno
Ohio State University
Columbus, Ohio

J.M. Law
North Carolina State University
Raleigh, North Carolina

James M. Lazorchak
U.S. Environmental Protection Agency
Cincinnati, Ohio

D.W. Lehmann
North Carolina State University
Raleigh, North Carolina

J.F. Levine
North Carolina State University
Raleigh, North Carolina

Chiara Losso
University Ca' Foscari of Venice
Venice, Italy

Deborah L. MacLatchy
University of New Brunswick
St. John, New Brunswick, Canada

Donald C. Malins
Pacific Northwest Research Institute
Seattle, Washington

Anne McElroy
Marine Science Institute
Stony Brook University
Stony Brook, New York

James P. Meador
National Marine Fisheries Service
Seattle, Washington

Masatoshi Morita
National Institute for Environmental Studies
Tsukuba, Ibaraki, Japan

Diane Nacci
U.S. Environmental Protection Agency
Narragansett, Rhode Island

Michael C. Newman
Virginia Institute of Marine Science
Gloucester Point, Virginia

Michelle B. Norris
University of Georgia
Athens, Georgia

Alessandra Arizzi Novelli
University Ca' Foscari of Venice
Venice, Italy

Gary K. Ostrander
Johns Hopkins University
Baltimore, Maryland

R.W. Pearce
National Oceanic and Atmospheric Administration
Seattle, Washington

Esther C. Peters
Tetra Tech, Inc.
Fairfax, Virginia

Kathy L. Price
Cooperative Oxford Laboratory
Oxford, Maryland

Robert H. Richmond
Kewalo Marine Laboratory
Honolulu, Hawaii

Amy H. Ringwood
Marine Resources Research Institute
Charleston, South Carolina

L.A. Ringwood
Wake Forest University
Winston-Salem, North Carolina

Robert L. Rogers
Fort Environmental Laboratories, Inc.
Stillwater, Oklahoma

Jeanette M. Rotchell
University of Sussex
Falmer, Brighton, United Kingdom

Michael H. Salazar
Applied Biomonitoring
Kirkland, Washington

Sandra M. Salazar
Applied Biomonitoring
Kirkland, Washington

James D. Salierno
University of Maryland
College Park, Maryland

Yelena Sapozhnikova
University of California
Riverside, California

Daniel Schlenk
University of California
Riverside, California

Nathaniel L. Scholz
National Marine Fisheries Service
Seattle, Washington

Ralf Schulz
University Koblenz-Landau
Landau, Germany

Rainie L. Sharpe
University of New Brunswick
St. John, New Brunswick, Canada

Kevin S. Shaughnessy
University of New Brunswick
St. John, New Brunswick, Canada

Colleen S. Sinclair
Towson University
Towson, Maryland

C.A. Sloan
National Oceanic and Atmospheric Administration
Seattle, Washington

Mark E. Smith
SoBran, Inc.
Cincinnati, Ohio

Roel Smolders
University of Antwerp
Antwerp, Belgium

Shane Snyder
Southern Nevada Water Authority
Henderson, Nevada

Frank C. Sommers
National Marine Fisheries Service
Seattle, Washington

Yoshihiro Soya
Tsuruga Institute of Biotechnology
Toyobo Co. Ltd.
Toyo-cho, Tsuruga, Fukui, Japan

Coral L. Stafford
National Oceanic and Atmospheric Administration
Seattle, Washington

Jennifer L. Stauber
CSIRO Energy Technology
Bangor, New South Wales, Australia

John J. Stegeman
Woods Hole Oceanographic Institution
Woods Hole, Massachusetts

Henry H. Tabak
U.S. Environmental Protection Agency
Cincinnati, Ohio

Karen L. Tilbury
National Marine Fisheries Service
Seattle, Washington

Nigel L. Turner
Cranfield University
Silsoe, Bedfordshire, United Kingdom

Vivek P. Utgikar
University of Idaho
Idaho Falls, Idaho

Glen J. Van Der Kraak
University of Guelph
Guelph, Ontario, Canada

Nik Veldhoen
University of Victoria
Victoria, British Columbia, Canada

Deena M. Wassenberg
Duke University
Durham, North Carolina

Jason B. Wells
ILS, Inc.
Atlanta, Georgia

Richard N. Winn
University of Georgia
Athens, Georgia

Robert J. Wolotira
National Oceanic and Atmospheric Administration
Seattle, Washington

Cheryl M. Woodley
National Oceanic and Atmospheric Administration
Charleston, South Carolina

Gladys K. Yanagida
National Marine Fisheries Service
Seattle, Washington

Gina M. Ylitalo
National Marine Fisheries Service
Seattle, Washington

Yuan Zhao
Virginia Institute of Marine Science
Gloucester Point, Virginia

Contents

Section III: Techniques for identification and assessment of contaminants in aquatic ecosystems

section one

Techniques for assessment of toxicity in whole organisms

chapter one

Integrative measures of toxicant exposure in zebra fish (Danio rerio) *at different levels of biological organization*

Roel Smolders, Wim De Coen, and Ronny Blust
University of Antwerp

Contents

Introduction

When aquatic organisms are exposed to pollutants, a cascade of biological events occurs if the exposure concentration is high enough and/or the exposure duration is long enough.[1–3] The general biochemical basis of stress responses was first described by Seleye[4] and is usually referred to as the general adaptation syndrome (GAS). This concept distinguishes three phases in response to stress: (1) primary alterations at the biochemical level, (2) secondary responses on a physiological level, and (3) tertiary effects on the whole organism level.

While a great deal of effort is directed towards eliciting the effects of pollutants on the first and second phases of the GAS, often referred to as biomarkers, there is still a profound need for a detailed description of effects at the whole organism level of biological organization.[1,5–7] Not only is the ecological relevance of effects at the whole organism level of biological organization, like condition, growth, or survival, often easier to establish than responses at lower levels, but also the integrative character of these

1-56670-664-5/05/$0.00+$1.50
© 2005 by CRC Press

measurements is one of the essential advantages of whole organism measurements.[8–10] These advantages promote the use of whole organism-based measurements, while a major difficulty with these measures is that there is often no identification of cause or mechanism of toxicity.

Though individual chemicals can have an impact on specific biochemical pathways, it is often difficult to extrapolate these effects to higher levels of biological organization, especially in the case of complex exposure scenarios, like field exposure or whole effluent toxicity.[10–12] Thus, although effects at the whole organism level in many cases lack the descriptive power to differentiate among the impact of specific toxicants, they give an integrative and holistic overview of the culminate effect of the disturbance of processes at lower levels of biological organization.

Moreover, effects on condition, growth, and survival, measured at the individual level, are an essential step to translate responses to higher levels of biological organization, like reproduction, population, and community effects.[9,13–16] There is, however, no "right" level of biological organization at which the effects of pollution should be studied and endpoints should be evaluated as part of a continuum of effects. Responses at a molecular or a cellular level of biological organization can provide detailed information on how chemicals interact with specific target sites and molecular pathways, but often provide little information concerning the ecological consequences of the effects. Studies on populations and communities do incorporate a high level of ecological relevance, since the well-being of populations and communities is the ultimate protection goal of ecotoxicology, but do not provide sufficient information on the eventual causes of the effects.[2,3,7,12,17]

What is required, is an integrated approach, where an understanding of the mechanistic basis of the stress responses in individuals is used to predict and interpret the penultimate effects at the population level. It should be obvious that the description of effects at the individual level is essential as a bridge between the mechanistic specificity of biochemical and physiological processes and the ecological relevance of population studies.

In order to obtain this holistic and integrative overview of how toxicants have an impact on organisms, the link should be made between different levels of biological organization and a "currency" has to be identified to extrapolate from one level to the next.[10,12,13,18,19] Energy budgets have been proposed as a very useful example of such a currency, not only because energy budgets can be determined at different levels of biological organization but also because they can provide a causal relationship between different levels, potentially relating cellular effects to growth or reproduction.[1,7,14,20]

Data that were previously presented by Smolders et al.[16,21] and unpublished data on the impact of different effluent concentrations on zebra fish (*Danio rerio*) will be used as an example on how endpoints describing cellular energy budgets, condition and reproduction are quantified, how data can be interpreted, and how these data can be combined to provide a holistic assessment of the impact at different levels of biological organization. All these data should be viewed within the framework of the metabolic cost hypothesis.[10,12,19,22] An active organism has a certain amount of energy available to it through its normal feeding, which will be used for storage in energy reserves, and will subsequently be used for investment in somatic growth, maintenance, or reproduction. If, however, a stressor impacts the fish, a shift in energy allocation will occur. Instead of the earlier balance between growth, maintenance, and reproduction, more energy will be allocated towards maintenance, leaving less energy available for growth and reproduction (Figure 1.1). Hence, from a theoretical point of view, stressed fish will show lower energy reserves, poorer condition, reduced growth, and impaired reproduction due to a decreased availability of energy.

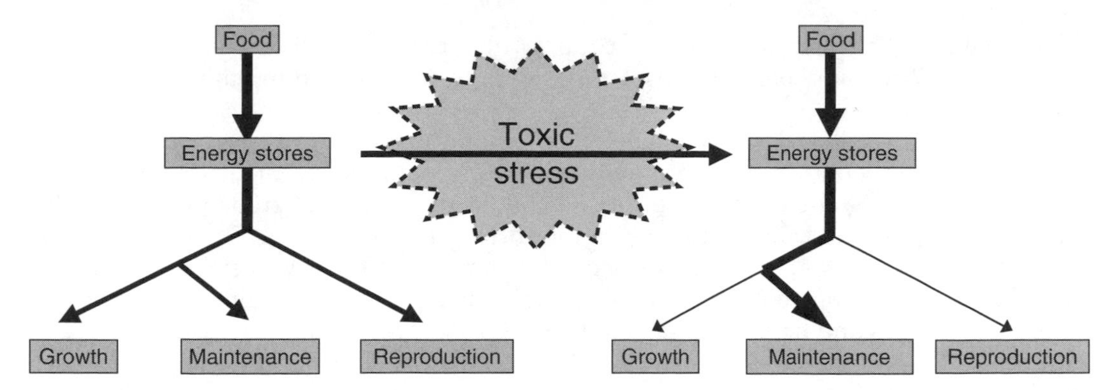

Figure 1.1 Outline of the metabolic cost hypothesis; in the presence of a toxic stressor, an increased amount of energy will be directed towards the basal maintenance of exposed organisms, leaving less energy available for growth and reproduction.

Also, indirect effects of pollutant exposure can be taken into account using this metabolic cost hypothesis. Even when organisms are not directly impacted by pollutant exposure, there may be other trophic levels that are impacted, leading to changes in food availability and trophic relations. Through a trophic cascade, this reduced food availability or change in food quality will eventually also have an impact on the energy availability of target organisms.[12,22–25]

In this chapter, we will present a number of different measurements, all focussed on determining the integrative effect of pollutant exposure at different levels of biological organization, illustrating the metabolic cost hypothesis. Briefly, methods will be presented that measure the energy reserves at a cellular level of biological organization, condition at the whole organism level, and reproduction as a final, ecologically relevant endpoint. These endpoints can be linked by one single currency, energy budgets, and we will try to illustrate how effects are interrelated.

Procedures and materials required

General zebra fish maintenance

Zebra fish *Danio rerio* (Hamilton) is recommended as test species in a number of ecotoxicological test protocols, e.g., different Organization for Economic Co-operation and Development (OECD) and International Organization for Standardization (ISO) guidelines. The advantages of zebra fish as bioassay organism include its small size, robustness, short life cycle, and the fact that under laboratory conditions it can be induced to breed all year round. Development from the fertilized egg to full reproductive maturity takes only 3–4 months. Additionally, this relatively short generation time makes zebra fish suitable for partial and full life cycle tests to evaluate the effects of chemicals on growth, development, and reproduction of fishes.[26–29]

Zebra fish are available at most pet stores and with commercial fish suppliers throughout the world. For general maintenance procedures, please refer to "The Zebrafish Book. A Guide for the Laboratory Use of Zebra fish (*Danio rerio*)"* by

* This book is available online through the Zebrafish Information Network at http://zfin.org/zf_info/zfbook/zfbk.html.

Westerfield.[26] This standard work deals with most of the aspects of feeding, rearing, and breeding of zebra fish and provides detailed information on general test procedures.

For the experiments described in this chapter, we started the test with adult zebra fish with an average length of 37.7 mm (range 33–42 mm) and an average weight of 0.52 g (range 0.31–0.81 g). The test was performed following the OECD guideline 204 for fish toxicity tests and the USEPA guidelines for chronic toxicity testing.[27,28] Fish were fed commercial fish feed at a ratio of 2% of the maximum average body weight, set at 0.93 g. For optimal spawning conditions, water temperature should be around 25°C. The test lasted for 28 days and data were gathered weekly.

Cellular energy reserves

Toxicants can have an effect on the feeding rate, biomass conversion efficiency, and energy requirements of aquatic organisms.[1,30,31] All these different aspects will have an effect on the energy budget at a cellular level of biological organization. Carbohydrates, lipids, and proteins are the major energy reserves present in aquatic organisms, and thus according to the metabolic cost hypothesis, these reserves will be impacted by toxicant exposure. Methods to determine glycogen (as a main source of carbohydrate), lipid, and protein reserves have been developed and are easily performed using basic laboratory equipment.

Glycogen reserves: The analysis is based on the method by Roe and Dailey.[32]

- Pipette 200 μl of homogenate into an 1.5-ml Eppendorf tube and add 50 μl (1 *N*) perchloric acid (PCA).
- As preliminary measurements, determine the proper dilution of homogenate so that the measurements are within the range of the calibration curve (see the following).
- Vortex for 30 s and subsequently incubate on ice for 30 min.
- After incubation, centrifuge for 3 min at 10,000*g*.
- Transfer 200 μl of the supernatant into 14-ml test tubes (make dilutions if necessary) and add 1 ml of Anthrone reagent.
- For Anthrone reagent, measure 100 ml of concentrated sulfuric acid and place in a 250-ml beaker under a fume hood. Add 0.2 g reagent grade Anthrone, mix by stirring and cool on ice for 1–2 h.
- Make a calibration curve of at least five concentrations between 0 and 1000 μg/ml glycogen dissolved in 0.2 *N* PCA. Also add 200 μl of the calibration standards to 14-ml test tubes and treat equal to samples.
- Transfer 200 μl of the supernatant and calibration standards to 14-ml test tubes (make dilutions if necessary) and add 1 ml of Anthrone reagent.
- Incubate the test tubes at 95°C (in a water bath) for 30 min.
- Allow to cool down, transfer 200 μl to microtiter plates (perform measurements in triplicate) and read in microtiter plate reader at 620 nm.

Lipid reserves: The determination of lipids is based on the chloroform–methanol method described by Bligh and Dyer,[33] adapted for microplate reader. The extraction procedure can be performed in 1.5-ml test tubes and measurements are done in a microtiter plate reader.

- Take 250 μl of homogenate. Again, the proper dilution of homogenate needs to be determined first for the measurements to be within the range of the calibration curve.
- Add 500 μl of chloroform (reagent grade) and vortex for 5 s.
- Add 500 μl methanol and 250 μl water and vortex for 5 s.
- Centrifuge the test tubes for 3 min at 3000*g* at 4°C.
- After centrifugation, there will be three distinct fractions in the test tubes. The upper part is the methanol–water fraction, then there is a small layer containing the tissue homogenate residue, and the lower part is the chloroform fraction that contains the dissolved lipids.
- Remove the two first layers (pipette the methanol–water fraction and discard, remove the tissue residue layer with a pipette-tip), and transfer a 100-μl subsample of the third, chloroform fraction to a glass test tube.
- Make a calibration curve of at least five concentrations between 0 and 3000 μg/ml tripalmitin dissolved in chloroform. Also add 100 μl of the calibration curve to glass test tubes and treat equal to samples.
- Add 500 μl sulfuric acid to the 100 μl chloroform–lipid sample in the glass test tubes and incubate for 30 min in an oven at 200°C. Be careful to use only test tube racks and test tubes that can resist 200°C.
- Be extremely careful when taking the samples out of the oven because of the concentrated sulfuric acid at 200°C. Always use protective gloves and safety goggles.
- Let the samples cool down, first in air, later on ice. When samples have cooled down, add 500 μl of water. Again be careful because adding water to concentrated sulfuric acid may cause splashes. As always, make sure you wear proper protection. If sample absorption is higher than the range of the calibration curve, add more water.
- Transfer 200 μl of sample (in triplicate) to a 96-well microtiter plate, also incorporating the calibration curve in triplicate. Read in a microtiter plate reader at 340 nm.
- Recalculate lipid content of samples based on the known concentrations of the calibration curve.

Protein reserves: The procedure is based on the Bradford method.[34] It is a very simple quantification procedure that uses Coomassie brilliant blue dye:

- Add 50 μl of sample to a 1.5-ml test tube and add 950 μl of 1 *M* NaOH.
- Vortex for 5 s and incubate for 30 min at 60°C.
- After incubation, cool on ice, vortex again and transfer 200 μl of sample (in triplicate) to a 96-well microtiter plate.
- Make a calibration curve of at least five concentrations between 0 and 1000 μg/ml bovine serum albumin (BSA) dissolved in 0.2 *N* NaOH. Also add 200 μl of all concentrations of the calibration curve to microtiter plates and treat equal to samples.
- Subsequently add 50 μl of Coomassie brilliant blue dye reagent.
- Absorption measurements are done in a microtiter plate reader at a wavelength of 595 nm.
- Protein content is recalculated to micrograms per milliliter by means of the calibration curve.

Note: It is essential that for every measurement of energy reserves, the sample preparation is accompanied by a calibration curve. This calibration curve is needed to recalculate from the absorption data to the energy reserve concentrations (in μg/ml). Since calibration curves may slightly vary due to small differences in measurement procedures, the use of an average calibration curve is not advisable.

Calculation of energy budgets

Changes in body composition were expressed as changes in energy budget (EB_x) and were calculated using the following formula:[18,19,21]

$$EB_x = [((T_x - T_{x-1})(Y_x - Y_{x-1})/2) + (T_x - T_{x-1})(Y_{x-1} - Y_0)]/T_x \quad (1.1)$$

where T_x is the exposure time x, Y_x the composition (glycogen, lipid, and protein content) at time x, and $x - 1$ is the previous measurement time. Data on energy reserves and condition for Y_0 and T_0 need to be measured in the same batch of zebra fish before the start of the exposure. This approach allows the quantification of changes in the energy budget in zebra fish between different exposure regimes and periods. Since energy budgets reflect all physiological changes the exposed organisms had to make to survive during the entire exposure period, they are more relevant than the mere absolute levels of energy reserves. Whole body energy budgets can be calculated by summing the energetic values for the different reserves, using an enthalpy of combustion of 17 kJ/g for glycogen, 39.5 kJ/g for lipids, and 24 kJ/g for proteins.[21,35]

Condition indices

Since condition indices and growth describe key processes in individual aquatic organisms, these indices are a good basis for bridging the gap between cellular and population levels of biological organization. The most common type of condition indices is ratios between morphological features of fish.[36–38] Because length and weight are often routinely measured, measures of growth and condition can provide basic information about the general well-being of organisms. Length and weight should be determined for a relatively large number of fish (20–50 individuals) because the more data are available, the more accurate the condition determination will be. In the experiment presented here, length was measured for 50 fish per aquarium on a plastic covered sheet of millimeter paper, which was kept moist. Total length was determined up to 1 mm. Weight was determined up to 0.01 g using a simple analytical balance. In a review paper, Bolger and Connolly[39] identified as many as eight forms of condition index that have been used in fisheries research. However, only two frequently used indices describing condition in fish are described and illustrated:

Fulton's condition factor (FCF):

$$FCF = WL^{-3} \quad (1.2)$$

where W is the total body weight (in g) and L is the total length (in mm). Sometimes it may be more interesting to use the standard length (length of the fish from head to the base of the tail fin). This can be more accurate since the shape of the tail fin may vary significantly, thus skewing results. The main problem associated with FCF is that the formula assumes isometric growth (growth with unchanged body proportions), an assumption often violated when adult fish are used.

Relative condition factor (RCF):

$$\mathrm{RCF} = W/(aL^b) \tag{1.3}$$

The parameters a and b are determined from a control (unstressed) population and are determined by log W = log $a + b$ log W (i.e., a is the regression intercept and b is the regression slope). This procedure automatically leads to the result that the condition of the control population is 1 and the RCF of the exposed populations is a fraction of this value (hence, relative condition factor). A problem encountered with the RCF is that there needs to be a clear reference population to determine a and b. For laboratory experiments, this is usually not a problem, but for field experiments the choice is much less straightforward.

Other condition indices have been proposed, some using additional parameters like height of the fish.[38] Though these indices are reported to be less variable and more accurate, the need to measure both length and height will cause additional stress to the fish. Handling has been reported to cause severe stress in fish, including reduced growth, increased sensitivity to infections,[40,41] and delayed reproduction[42,43] and should therefore be minimized.

With the recent developments in image analysis and the declining cost of digital cameras, the analysis of digital pictures of exposed fish provides a new and very sensitive opportunity to measure subtle changes in growth, condition, and more complex aspects of development, including truss analysis[44] and fluctuating asymmetry.[45]

Reproduction

There are a number of methods to determine the reproductive potential of zebra fish. A method that has proven very efficient in our experiments is the marbling of small cages.[16,26] Briefly, four males (longer, slimmer, and more yellow especially on the belly) and two females (plumper and more silvery) are transferred from the exposure aquaria to smaller breeding containers. The bottom of every spawning container is covered with marbles, so any eggs spawned are protected from predation from the parents. Apart from the marbling technique, many varieties of mesh mating cages with pull-out dividers have been devised. Though these cages usually are custom-made, they should be large enough to enable the fish to swim freely and to prevent the eggs from being eaten by the parents. Zebra fish breed photoperiodically and normally produce eggs in the morning, shortly after sunrise.[26,46,47] The parent fish are left in the cages overnight, they are removed from the spawning containers and the marbles are carefully removed and cleaned. If spawning occurred, eggs will be visible at the bottom of the cage, and the number of eggs can be counted to quantify egg production (expressed as average number of eggs per female).

Endpoints that should be measured if a proper and accurate evaluation of effects of toxicants on reproduction is desired, should include the following:

- *Number of female fish spawning*: By individually caging females, this endpoint can be quantified. As we will illustrate later, the number of female fish spawning is related to energy budget of fish. However, there is a relatively high natural variability in zebra fish spawning, so the use of controls, replicates, and possibly repetition of experiments must be encouraged.

- *Number of eggs spawned per fish*: When parent fish are removed from the small spawning containers and marbles or cages are removed, eggs can be observed by eyes without the aid of any optical devices at the bottom of the containers as small spheres of about 0.5–1 mm diameter.
- *Hatching*: If the eggs are left in the small spawning tanks after quantification of spawning, the percentage hatching can also be quantified. Hatchlings can be observed after 3–4 days as little black juveniles (1–1.5 mm long).
- Since juvenile life stages are generally considered to be the most sensitive in ecotoxicological testing, juvenile growth, development rate, occurrence of malformations, and survival may be the very sensitive additional endpoints, which are beyond the scope of this chapter; details on rearing and maintaining juvenile zebra fish have extensively been discussed by Westerfield.[26]

Results and discussion

The data that are used here to illustrate the methods outlined in the previous section are extracted from Smolders et al.[16,21] and unpublished data. These data describe the impact of different concentrations of an industrial effluent on the energy budget, condition, and reproduction of zebra fish. As an example, we will compare responses of exposure to the control and 100% effluent aquarium.

Figure 1.2 illustrates how the total energy budget of zebra fish depends on both effluent concentration and exposure duration. The effects of effluent-exposure on the cellular energetics are expressed as energy budgets, instead of simply comparing energy content. As already indicated before, the presentation of energetics through energy budgets represents an integration over time. Because pollutant exposure is a continuous process, and the aim of our experiment is to correlate the cellular energy data with other integrative measures like condition, growth, or reproduction, the integrated approach is a

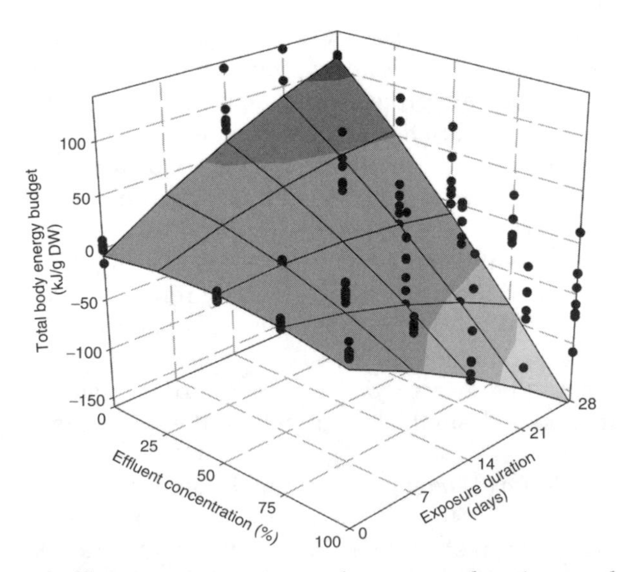

Figure 1.2 The effect of effluent concentration and exposure duration on the total body energy budget of effluent-exposed zebra fish (quadratic smoothening, $n = 156$; $R^2 = 0.531$).

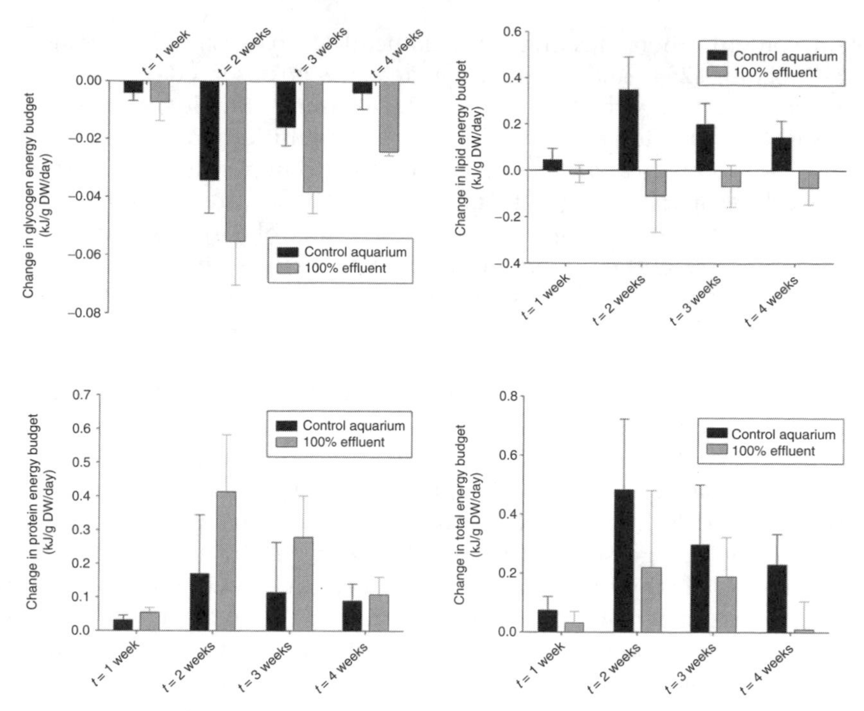

Figure 1.3 Overview of changes in energy budgets in control and exposed zebra fish ($n = 8$, mean $\pm$ SD). Complete data and discussion are presented in Smolders et al. (2003).[21]

good reflection of the continuous stress that is posed upon the fish throughout the experiment.

Figure 1.3 represents the total energy budget of effluent-exposed fish through time, and is the sum of the three major energetic components in fish, i.e., carbohydrates, lipids, and proteins. Though especially the total energy budget is relevant to extrapolate effects at a cellular level to effects at higher levels of biological organization, it also pays to look at the individual components. Our data showed that especially lipid budgets were related to effluent-exposure in a dose-dependent manner. Effluent-exposure caused a rapid and dose-dependent depletion of lipid energy budgets. The depletion of lipids by pollutant exposure has been documented for, among others, carp fingerlings,[48] rainbow trout,[49] and eel[50] chronically exposed to different pollutants. Also glycogen (as the primary source of carbohydrates) was significantly affected by effluent-exposure. Since glycogen is a rapidly available source of energy, it was not surprising, however, that this source of reserves was depleted by effluent-exposure. There are many examples in the literature where pollutant stress depleted glycogen levels in fish.[51,52]

Finally, protein levels were significantly different in fish exposed to effluents compared to the control. Even though protein is a prominent source of energy in fish, stress preferably causes depletion of glycogen and lipid reserves instead of protein.[1,51,53] We observed a significant increase in protein content in the second and third weeks of exposure to different effluent concentrations, but the levels restored to normal in the fourth week. An increase in protein content after pollutant exposure has also been observed by a number of other authors. For example, Wall and Crivello[54] found an increased microsomal protein content when starving juvenile winter flounder (*Pleuronectes americanus*) for 2 weeks, Brumley et al.[55] reported a 1.5-fold increase in liver protein content when injecting sand flathead (*Platycephalus bassensis*) with up to 400 mg/kg Arochlor-1254, and a similar increase was observed in the liver of carp (*Cyprinus carpio*) injected with the herbicide 2,4-Diamin.[56] Smolders et al.[21] formulated the hypothesis that low to intermediate levels of pollution trigger increased protein synthesis (e.g., for detoxification processes and other defense mechanisms) when other sources of readily available energy like glycogen and lipids are still sufficiently present. On a more theoretical basis, Chapman[57] discussed this phenomenon using a hormetic concentration–response curve, where hormesis consists of a stimulatory response on a given endpoint (in our case whole body protein content) of 30–60% above the controls and comprises a general biological phenomenon that may represent overcompensation to an alteration in homeostasis.

Figure 1.4 shows the difference in length–weight relationship between the control and 100% effluent aquarium after 28 days of exposure. For fish of the same length, fish in the control aquarium are significantly heavier than fish exposed to the undiluted effluent. Equation (1.3) gives the values for a and b for the fish population in the control aquarium, which were used to determine the RCF. The advantage of using condition indices over simply determining changes in length or weight can be illustrated by the results presented in Figure 1.4. There will always be differences in growth within distinct populations when a large number of fish are used for testing. Even though fish in the control aquarium are of the same age and have the same life history, there is an intrinsic natural variability in length and weight. While this variability is natural, it hampers finding significant differences in average length among exposure aquaria due to high standard deviations. If, however, data are being rescaled based on length, which basically is what condition indices represent, much more subtle effects can be detected. Furthermore, since condition indices are a population response, they can also be used for populations with an uneven age distribution. As already mentioned, a problem often faced when using RCF is

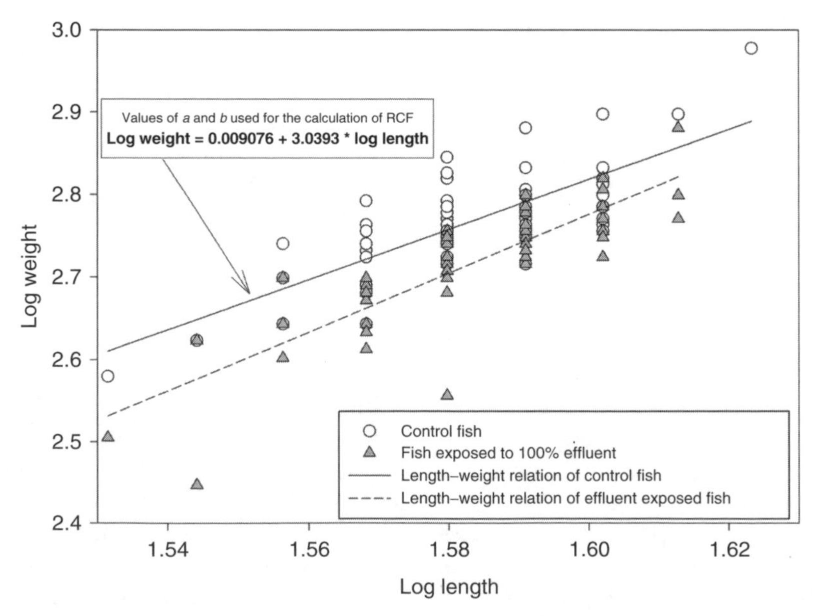

Figure 1.4 Length–weight for control and effluent-exposed zebra fish. The parameters of the length–weight relationship for the control fish are used to calculate the RCF.

the lack of a clearly defined reference population. While this is not a problem for laboratory experiments, it may be an important issue for field studies. Among different locations, food availability may not be comparable, thus making comparison among different locations difficult. Since in many rivers, pollution and eutrophication are linked, increased nutrient levels may be an important factor determining the growth performance of different food sources (e.g., macroinvertebrates and algae), masking the pollutant-related effects. Increased growth and/or condition of fish downstream point sources of pollution has been reported earlier, downstream bleached kraft mill effluents in particular. For example, Hodson et al.[58] reported elevated lipid levels and condition in white sucker (*Catostomus commeroni*) downstream BKME, and linked this with increased nutrient levels and richer food sources. Also Adams et al.[13] and Gibbons et al.[24] found increased energy storage, growth, and condition in redbreast sunfish (*Lepomis auritus*) and spoonhead sculpin (*Cottus ricei*) caught downstream bleached kraft pulp mills and contributed this improved performance to eutrophication and increased food sources associated with the mill effluent. In a comparison of 51 mills in Quebec, Langlois and Dubud[59] reported that if significant differences in the condition of fish were observed downstream pulp and paper mill effluents, in nine out of ten cases, fish exposed at the effluent discharge show a significantly better condition than fish at an unperturbed reference site. Food availability might have been one of the confounding factors, since they also reported that effluent discharge areas had a significantly higher abundance of macroinvertebrates, a very important food source for most fish species, compared to reference sites.

Also Fulton's condition factor (FCF) can be used to express the effects of effluent-exposure on the well-being of the fish. The interpretation of the FCF is not without danger; however, since the FCF requires isometric growth, an assumption that is often violated. However, RCF does not allow making a distinction between improved condition in the control aquarium and reduced condition in the 100% effluent aquarium. Figure 1.5 clearly indicates that the differences in condition among treatments are mainly due to an increased condition of fish in the control aquarium. FCF increases as the exposure period

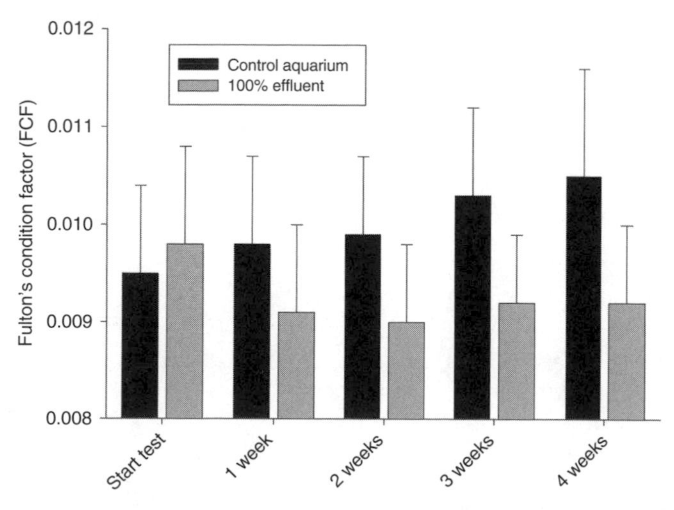

Figure 1.5 The effect of effluent-exposure on Fulton's condition factor (FCF; mean $\pm$ SD, $n = 50$). Note that this index requires isometric growth, an assumption frequently violated for adult fish.

increases, while fish in the 100% effluent aquarium have a reduced FCF compared to that at the start of the test, but no further decrease afterwards.

The ultimate level of concern in ecotoxicology is the stability and persistence of populations and communities of organisms. Therefore, spawning (the number of eggs produced) and hatching (the number of eggs leading to juveniles after fertilization) as endpoints of reproduction should be the penultimate result of all toxicological investigations. In addition, since the fecundity of fish often seems to be limited by the energetic cost of reproduction, spawning and hatching are key concepts within the bioenergetics approach.[60] Though spawning and hatching in zebra fish are relatively easy to quantify, the interpretation of the data is less straightforward. One of the main disadvantages of spawning and hatching as toxicological endpoints is that reproduction provides an ecologically relevant description of the effects of stress but does not, by itself, provide information on potential causes.

In the experiment presented here, there was a significant reduction of spawning and hatching in effluent-exposed fish (Table 1.1). However, this reduction was not caused by a reduction of the average number of eggs spawned per female, but by a reduction of spawning females. In other words, if fish in the 100% effluent aquarium spawned, they produced the same number of eggs as fish in the control aquarium, but not as many female fish in the 100% effluent aquarium actually made it to spawning. Again, this can be explained within the framework of bioenergetics. Fish will relocate energy towards maintenance, growth, and reproduction based on energy availability. If under toxic stress, however, female fish will not be able to relocate the necessary amount of energy towards reproduction, spawning frequency will reduce and the length of the reproductive cycle will increase. Within our effluent toxicity example, this was illustrated by the significant correlation between RCF of the mother fish and number of eggs spawned ($R^2 = 0.778$, $n = 20$, $p < 0.001$). Also multiple regressions describing the relationship between glycogen, protein, and lipid budgets and reproduction was significant ($R^2 = 0.410$, $n = 20$, $p < 0.034$),

Table 1.1 Effects of effluent-exposure on spawning and hatching of zebra fish (*D. rerio*)

Aquarium	Time (weeks)	No. cages	Spawning (% of cages)	Spawning (total No. eggs)	No. eggs/spawning female	Percentage of eggs hatching	No. juveniles/cage
Control	1	3	67	165	82.5	32.5	65
	2	4	75	485	161.7	55.5	304
	3	5	80	663	165.8	25.9	169
	4	5	60	589	196.3	73.8	371
50% effluent	1	3	67	328	164	26.9	70
	2	4	75	573	191	21.7	179
	3	5	40	275	137.5	32.7	72
	4	5	20	296	296	58.5	173
75% effluent	1	3	67	249	124.5	44.4	170
	2	5	40	173	86.5	32.3	66
	3	5	40	291	145.5	38.2	112
	4	5	60	601	200	33.3	210
100% effluent	1	3	33	51	51	58.8	30
	2	5	40	219	109.5	37.9	121
	3	5	40	105	52.5	47.9	75
	4	5	20	12	12	33.3	4

Table 1.2 Multiple linear regression describing the relationship between changes in energy budgets and integrated condition indices in effluent-exposed zebra fish (*D. rerio*). Data are given as average ($\pm$ SD), $n = 20$

	Intercept	Glycogen	Lipids	Protein	R^2	p-level
FCF ($\times$1000)	9.596 (0.091)	0.000066 (0.00029)	0.00016 (0.00003)	−0.000028 (0.000057)	0.643	$p < 0.001$
	$p < 0.001$	$p = 0.822$	$p < 0.001$	$p = 0.634$		
RCF ($\times$1000)	993.12 (8.20)	0.091 (0.026)	0.0105 (0.0029)	−0.0013 (0.0051)	0.795	$p < 0.001$
	$p < 0.001$	$p = 0.003$	$p = 0.002$	$p = 0.803$		
Reproduction	100.42 (13.89)	0.034 (0.044)	0.0113 (0.0049)	−0.0049 (0.0087)	0.410	$p < 0.0339$
	$p < 0.001$	$p = 0.455$	$p = 0.035$	$p = 0.579$		

indicating that indeed a reduced energy budget was causing a reduction in reproduction (Table 1.2). Based on this information, we concluded that, in this case, the effect of effluent-exposure on reproduction was not caused by a direct toxic effect but by an indirect reduction in energy availability for reproduction. The observation that there was no significant reduction in hatching also indicated that there was no direct toxic effect of the effluent on the reproductive process. There are, however, several examples where reduced spawning and hatching in aquatic organisms is caused by direct toxic action of chemicals, so caution in the interpretation of data is necessary. Many toxicants are reported to directly influence spawning and hatching, often through a delay of spawning and a reduction of hatching percentage. Though we did not quantify these endpoints in the example presented earlier, the time until hatching appears to be a very sensitive endpoint when exposing zebra fish[61–64] or Japanese medaka (*Oryzias latipes*)[65,66] to a wide variety of pollutants.

The aim of this chapter was to illustrate how relatively simple measures of energy budgets, condition indices, and reproduction can be combined within the framework of bioenergetics. The methods presented do not require high-tech equipment, are relatively easy and fast to perform but give a good overview of the potential effects of stress on fish. The main idea is that combining cellular energy budgets (fast and sensitive yet ecologically questionable endpoints) with reproduction through individual endpoints like growth and condition will increase the ecological relevance of toxicity testing and provide an integrative and holistic overview of effects. The results presented show that the use of different levels of biological organization within one test and one test species are not only useful to quantify effects but can also show how disturbance at one level of biological organization can influence other levels of organization.

References

1. Giesy, J.P. and Graney, R.L., Recent developments in and intercomparisons of acute and chronic bioassays and bioindicators, *Hydrobiologia*, 188/189, 21–60, 1989.
2. Munkittrick, K.R. and McCarty, L.S., An integrated approach to aquatic ecosystem health: top-down, bottom-up or middle out? *J. Aquat. Ecosyst. Health*, 4, 77–90, 1995.
3. Adams, S.M., Establishing causality between environmental stressors and effects on aquatic ecosystems, *Human Ecol. Risk Assess.*, 9, 17–35, 2003.
4. Seleye, H., Stress and the general adaptation syndrome, *Br. Med. J.*, 1, 384–392, 1950.

5. Wedemeyer, G.A. and McLeay, D.J., Methods for determining the tolerance of fishes to environmental stressors, in *Stress and Fish*, Pickering, A.D., Ed., Academic Press, London, 1981, pp. 247–275 (367 pp.).
6. Beitinger, T.L. and McCauley, R.W., Whole-animal physiological processes for the assessment of stress in fishes, *J. Gr. Lakes Res.*, 16, 542–575, 1990.
7. Maltby, L., Studying stress: the importance of organism-level responses, *Ecol. Appli.*, 9, 431–440, 1999.
8. De Kruijf, H.A.M., Extrapolation through hierarchical levels, *Comp. Biochem Phys. C.*, 100, 291–299, 1991.
9. Kooijman, S.A.L.M. and Bedaux, J.M., Analysis of toxicity tests on fish growth, *Water Res.*, 30, 1633–1644, 1996.
10. Smolders, R., Bervoets, L., Wepener, V. and Blust, R., A conceptual framework for using mussels as biomonitors in whole effluent toxicity, *Human Ecol. Risk Assess.*, 9, 741–760, 2003.
11. Schreck, C.B., Physiological, behavioral, and performance indicators of stress, *Am. Fish. Soc. Symp.*, 8, 29–37, 1990.
12. Calow, P., Physiological costs of combating chemical toxicants: ecological implications, *Comp. Biochem. Phys. C.*, 100, 3–6, 1991.
13. Adams, S.M., Crumby, W.D., Greeley, M.S., Ryon, M.G. and Schilling, E.M., Relationships between physiological and fish population responses in a contaminated stream, *Environ. Toxicol. Chem.*, 11, 1549–1557, 1992.
14. Beyers, D.W., Rice, J.A., Clements, W.H. and Henry, C.J., Estimating physiological cost of chemical exposure: integrating energetics and stress to quantify toxic effects in fish, *Can. J. Fish. Aquat. Sci.*, 56, 814–822, 1999.
15. Beyers, D.W., Rice, J.A. and Clements, W.H., Estimating biological significance of chemical exposure to fish using a bioenergetics-based stressor-response model, *Can. J. Fish. Aquat. Sci.*, 56, 823–829, 1999.
16. Smolders, R., Bervoets, L., De Boeck, G. and Blust, R., Integrated condition indices as a measure of whole effluent toxicity in zebrafish (*Danio rerio*), *Environ. Toxicol. Chem.*, 21, 87–93, 2002.
17. Clements, W.H., Integrating effects of contaminants across levels of biological organization: an overview, *J. Aquat. Ecosys. Stress Recov.*, 7, 113–116, 2000.
18. De Coen, W.M., Janssen, C.R. and Giesy, J.P., Biomarker applications in ecotoxicology: bridging the gap between toxicology and ecology, in *New Microbiotests for Routine Toxicity Screening and Biomonitoring*, Persoone, G., Janssen, C. and De Coen, W.M., Eds., Kluwer Academic, Dordrecht, 2000, 210 pp.
19. De Coen, W.M. and Janssen, C.R., Cellular energy allocation: a new methodology to assess the energy budget of toxicant-stressed *Daphnia* populations, *J. Aquat. Ecosys. Stress Recov.*, 6, 43–55, 1997.
20. Adams, S.M. and Greeley, M.S., Ecotoxicological indicators of water quality: using multi-response indicators to assess the health of aquatic ecosystems, *Water Air Soil Pollut.*, 123, 103–115, 2000.
21. Smolders, R., De Boeck, G. and Blust, R., Changes in cellular energy budget as a measure of whole effluent toxicity in zebrafish (*Danio rerio*), *Environ. Toxicol. Chem.*, 22, 890–899, 2003.
22. Calow, P. and Sibly, R.M., A physiological basis of population processes: ecotoxicological implications. *Funct. Ecol.*, 4, 283–288, 1990.
23. Barreire Lozano, R. and Pratt, J.R., Interaction of toxicants and communities: the role of nutrients, *Environ. Toxicol. Chem.*, 13, 361–368, 1994.
24. Gibbons, W.N., Munkittrick, K.R. and Taylor, W.D., Monitoring aquatic environments receiving industrial effluents using small fish species 1: response of spoonhead sculpin (*Cottus ricei*) downstream of a bleached-kraft pulp mill, *Environ. Toxicol. Chem.*, 17, 2227–2237, 1998.
25. Stuijfzand, S.C., Helms, M., Kraak, M.H.S. and Admiraal, W., Interacting effects of toxicants and organic matter on the midge *Chironomus riparius* in polluted river water, *Ecotox. Environ. Safe.*, 46, 351–356, 2000.

26. Westerfield, M., *The Zebrafish Book. Guide for the Laboratory Use of Zebrafish* (Danio rerio), 3rd ed., University of Oregon Press, Eugene, OR, 1995 (http://zfin.org/zf_info/zfbook/zfbk.html).
27. Organization for Economic Cooperation and Development, Fish, Prolonged Toxicity Test: 14-day Study, OECD Guideline 204, Paris, France, 1993.
28. U.S. Environmental Protection Agency, *Short-term Methods for Estimating the Chronic Toxicity of Effluents and Receiving Waters to Freshwater Organisms*, 3rd ed., EPA 600/4-91/002, Cincinnati, OH, 1994.
29. Maack, G. and Segner, H., Morphological development of the gonads in zebrafish, *J. Fish Biol.*, 62, 895–906, 2003.
30. Widdows, J. and Donkin, P., Role of physiological energetics in ecotoxicology, *Comp. Biochem. Phys. C.*, 100, 69–75, 1991.
31. Smolders, R., Bervoets, L. and Blust, R., Transplanted zebramussels (*Dreissena polymorpha*) as active biomonitors in an effluent-dominated river, *Environ. Toxicol. Chem.*, 21, 1889–1896, 2002.
32. Roe, J.H. and Dailey, R.E., Determination of glycogen with the anthrone reagent, *Anal. Biochem.*, 15, 245–250, 1966.
33. Bligh, E.G. and Dyer, W.J., A rapid method of total lipid extraction and purification, *Can. J. Biochem. Phys.*, 37, 911–917, 1959.
34. Bradford, M.M., A rapid and sensitive method for the quantitation of microgram quantities of protein utilizing the principle of protein-dye binding, *Anal. Biochem.*, 72, 249–254, 1976.
35. Jobling, M., *Fish Bioenergetics*, Chapman & Hall, London, 1994, 309 pp.
36. Goede, R.W. and Barton, B.A., Organismic indices and an autopsy-based assessment as indicators of health and condition of fish, *Am. Fish. Soc. Symp.*, 8, 93–108, 1990.
37. Lambert, Y. and Dutil, J.-D., Can simple condition indices be used to monitor and quantify seasonal changes in the energy reserves of Atlantic cod (*Gadus morhua*)? *Can. J. Fish. Aquat. Sci.*, 54 (Suppl. 1), 104–112, 1997.
38. Jones, P.E., Petrell, R.S. and Pauly, D., Using modified length–weight relationships to assess the condition of fish, *Aquacult. Eng.*, 20, 261–276, 2000.
39. Bolger, T. and Connolly, P.L., The selection of suitable indexes for the measurement and analysis of fish condition, *J. Fish Biol.*, 34, 171–182, 1989.
40. Saeij, J.P.J., Verburg-van Kemenade L.B.M., Van Muiswinkel, W.B. and Wiegertjes, G.F., Daily handling stress reduces resistance of carp to *Trypanoplasma borreli*: *in vitro* modulatory effects of cortisol on leukocyte function and apoptosis, *Dev. Comp. Immunol.*, 27, 233–245, 2003.
41. Davis, K.B., Griffin, B.R. and Gray, W.L., Effect of handling stress on susceptibility of channel catfish *Ictalurus punctatus* to *Icthyophthirius multifiliis* and channel catfish virus infection, *Aquaculture*, 214, 55–66, 2002.
42. Cleary, J.J., Pankhurst, N.W. and Battaglene, S.C., The effect of capture and handling stress on plasma steroid levels and gonadal condition in wild and farmed snapper *Pagrus auratus* (Sparidae), *J. World Aquacult. Soc.*, 31, 558–569, 2000.
43. Schreck, C.B., Contreras-Sanchez, W. and Fitzpatrick, M.S., Effects of stress on fish reproduction, gamete quality, and progeny, *Aquaculture*, 197, 3–24, 2001.
44. Fitzgerald, D.G., Nanson, J.W., Todd, T.N. and Davis, B.M., Application of truss analysis for the quantification of changes in fish condition, *J. Aquat. Ecosyst. Stress Recov.*, 9, 115–125, 2002.
45. Gronkjaer, P. and Sand, M.K., Fluctuating asymmetry and nutritional condition of Baltic cod (*Gadus morhua*) larvae, *Mar. Biol.*, 143, 191–197, 2003.
46. Eaton, R.C. and Farley, R.D., Spawning cycle and egg production of zebrafish, *Brachydanio rerio*, in the laboratory, *Copeia*, 1, 195–204, 1974.
47. Laale, H.W., The biology and use of zebrafish *Brachydanio rerio* in fisheries research. A literature review, *J. Fish Biol.*, 10, 121–173, 1977.
48. Palackova, J., Pravda, D., Fasaic, K. and Celchovska, O., Sublethal effects on cadmium on carp (*Cyprinus carpio*) fingerlings, in *Sublethal and Chronic Effects of Pollutants on Freshwater Fish*, Muller, R. and Lloyd, R., Eds., Fishing News Books, London, pp. 53–61, 1994 (371 pp.).
49. Handy, R.D., Sims, D.W., Giles, A., Campbell, H.A. and Musonda, M.M., Metabolic trade-off between locomotion and detoxification for maintenance of blood chemistry and growth

parameters by rainbow trout (*Oncorhynchus mykiss*) during chronic dietary exposure to copper, *Aquat. Toxicol.*, 47, 23–41, 1999.
50. Sancho, E., Ferrando, M.D. and Andreu, E., Effects of sublethal exposure to a pesticide on levels of energetic compounds in *Anguilla anguilla*, *J. Environ. Sci. Health*, 33B, 411–424, 1998.
51. Heath, A.G., *Water Pollution and Fish Physiology*, CRC Press, Boca Raton, 1987, 243 pp.
52. Omoregie, E., Ufodike, E.B.C. and Onwuliri, C.O.E., Effect of petroleum effluent pollution on carbohydrate reserves of the Nile tilapia, *Oreochromis niloticus* (L.), *Discov. Innovat.*, 12, 26–29, 2000.
53. McKee, M.J. and Knowles, C.O., Protein, nucleic acid and adenylate levels in *Daphnia magna* during chronic exposure to chlordecone, *Environ. Pollut.*, 42, 335–351, 1986.
54. Wall, K.L. and Crivello, J., Effects of starvation on liver microsomal P450 activity in juvenile *Pleunonectes americanus*, *Comp. Biochem. Phys. C.*, 123, 273–277, 1999.
55. Brumley, C.M., Haritos, V.S., Ahokas, J.T. and Holdway, D.A., Validation of biomarkers of marine pollution exposure in sand flathead using aroclor-1254, *Aquat. Toxicol.*, 31, 249–262, 1995.
56. Oruc, E.O. and Uner, N., Effects of 2,4-Diamine on some parameters of protein and carbohydrate metabolisms in the serum, muscle and liver of *Cyprinus carpio*, *Environ. Pollut.*, 105, 267–272, 1999.
57. Chapman, P.M., Whole effluent toxicity testing — usefulness, level of protection and risk assessment, *Environ. Toxicol. Chem.*, 19, 3–13, 2000.
58. Hodson, P.V., McWhirter, M., Ralph, K., Gray, B., Thivierge, D., Carey, J.H., Vanderkraak, G., Whittle, D.M. and Levesque, M.C., Effects of bleached kraft mill effluent on fish in the St-Maurice river, Quebec, *Environ. Toxicol. Chem.*, 11, 1635–1651, 1992.
59. Langlois, C. and Dubud, N., Pulp and Paper Environmental Effects Monitoring (EEM). Results Synthesis for the 47 Cycle 1 Studies Conducted in Quebec, Environmental Canada, Montreal, 1999, 123 pp.
60. Lyons, D.O. and Dunne, J.J., Reproductive costs to male and female worm pipefish, *J. Fish Biol.*, 62, 767–773, 2003.
61. Todd, N.E. and Van Leeuwen, M., Effects of Sevin (carbaryl insecticide) on early life stages of zebrafish (*Danio rerio*), *Ecotox. Environ. Safe.*, 53, 267–272, 2002.
62. Dave, G. and Xiu, R.Q., Toxicity of mercury, copper, nickel, lead, and cobalt to embryos and larvae of zebrafish, *Brachydanio rerio*, *Arch. Environ. Con. Tox.*, 21, 126–134, 1991.
63. Örn, S., Andersson, P.L., Forlin, L., Tysklind, M. and Norrgren, L., The impact on reproduction of an orally administered mixture of selected PCBs in zebrafish (*Danio rerio*), *Arch. Environ. Con. Tox.*, 35, 52–57, 1998.
64. Roex, E.W.M., Giovannangelo, M. and Van Gestel, C.A.M., Reproductive impairment in the zebrafish, *Danio rerio*, upon chronic exposure to 1,2,3-trichlorobenzene, *Ecotox. Environ. Safe.*, 48, 196–201, 2001.
65. Nirmala, K., Oshima, Y., Lee, R., Imada, N., Honjo, T. and Kobayashi, K., Transgenerational toxicity of tributyltin and its combined effects with polychlorinated biphenyls on reproductive processes in Japanese medaka (*Oryzias latipes*), *Environ. Toxicol. Chem.*, 18, 717–721, 1999.
66. Villalobos, S.A., Papoulias, D.M., Meadows, J., Blankenship, A.L., Pastva, S.D., Kannan, K., Hinton, D.E., Tillitt, D.E. and Giesy, J.P., Toxic responses of medaka, d-rR strain, to polychlorinated naphthalene mixtures after embryonic exposure by in ovo nanoinjection: a partial life-cycle assessment *Environ. Toxicol. Chem.*, 19, 432–440, 2000.

chapter two

Use of disease challenge assay to assess immunotoxicity of xenobiotics in fish

Mary R. Arkoosh, Deborah Boylen, Coral L. Stafford, Lyndal L. Johnson, and Tracy K. Collier
National Oceanic and Atmospheric Administration

Contents

Introduction

A properly functioning immune system is critical in maintaining the fitness or health of an organism. The immune system is a complex network that involves regulation by both the nervous and endocrine systems, which allows an individual to respond to or fight against an invading parasite or foreign material (antigens). An improperly functioning immune system may respond to "self" as foreign causing autoimmune disease or hypersensitivity or, the opposite can occur, leading to immunosuppression. Foreign chemicals, or "xenobiotics" have the potential of deregulating a healthy immune system. Xenobiotics can shift the system from operating at a healthy homeostatic level to being either

1-56670-664-5/05/$0.00+$1.50
© 2005 by CRC Press

hyper- or hyposensitive. Both results may have drastic consequences on the individual organism and at the population level as well.

An important tool for examining the status of the immune system is disease challenge studies, also referred to as host resistance challenge studies. In our target species, juvenile chinook salmon (*Oncorhynchus tshawytscha*), this technique provides an opportunity to determine if xenobiotic exposure can alter the ability of fish to respond immunologically to bacteria known to be harmful or pathogenic to the species in its natural environment. We will present a study that has been published in part[1,2] whereby we examined the effects of various xenobiotics on the disease susceptibility of juvenile chinook salmon. We will discuss the number of variables that need to be considered when planning a disease challenge experiment involving xenobiotics and how to address them.

Materials required

Equipment:

- Low temperature incubator: model 815 (Precision)
- Digital oscillating orbital shaker (Thermolyne)
- UV/Vis spectrophotometer: model Du 530 (Beckman)
- Tissue culture enclosure
- Biological safety cabinet
- Ultrapure water system
- Autoclave
- Magnetic stir plate
- pH meter
- Analytical balance
- Cryogenic can
- Adjustable micropipettes (10, 200, 1000 μl)
- Vortex mixer
- Compound microscope with oil immersion magnification

Supplies:

- 15-ml graduated tubes with screw caps and conical bottoms
- Sterile serological pipettes (10, 5, 1 ml)
- Micropipette tips (10, 200, 1000 μl)
- Latex gloves
- 1-cc tuberculin syringes
- 100 × 15 mm sterile Petri plates
- 1-μl sterile disposable inoculating loops
- Erlenmeyer culture flasks
- Weighing boats
- Cheese cloth
- Necropsy instruments
- Alcohol burner
- Frosted microscope slides
- Paper towels
- Spray bottle

- Ethanol
- Novobiocin sensitivity discs
- Vibriostatic sensitivity discs 0/129 (2,4-diamino-6,7-diisopropyl pterdine phosphate)
- Necropsy cutting board
- Metomidate hydrochloride (Wildlife Laboratories)
- Bacteriological media
  Tryptic soy agar (TSA)
  Tryptic soy broth (TSB)
  Sodium chloride (NaCl)
- *Vibrio anguillarum* media (VAM)
  Sorbitol
  Yeast extract
  Bile salts
  Cresol red
  Bromthymol blue
  Bacteriological agar
  Ampicillin
- Bacteriological diagnostic test kits
  Oxidase Test kit (bioMérieux Vitek)
  Rapid agglutination test: Mono-Va 50 Tests (Bionor)
  Gram stain kit (Sigma)
- Aquaculture supplies
  450-l plus 2400-l circular aquaculture tanks
  Fishnets
  1.5-in.^{2} air diffuser for each tank
  Fish food (BioOregon)
  Aquaculture disinfectant: I-O-Safe
  Buckets

Xenobiotics:

- Polycyclic aromatic hydrocarbon (PAH) model mixture (analytical grade, Sigma)
  Fluoranthene
  Pyrene
  Benz[*a*]anthracene
  Chrysene
  Benzo[*b*]fluoranthene
  Benzo[*k*]fluoranthene
  Benzo[*a*]pyrene
  Indeno[1,2,3-*cd*]pyrene
  Dibenz[*a,h*]anthracene
  Benzo[*g,h,i*]perylene
- Hexachlorobutadiene (HCBD; Sigma)
- Polychlorinated biphenyl (PCB) mixture: Aroclor 1254 (AccuStandard)
- 7,12-Dimethylbenz[*a*]anthracene (DMBA; Sigma)
- Chlorinated-enriched Hylebos Waterway sediment extract (CHWSE)

Procedure

Contaminants

We are interested in examining the effect of contaminants that are representative of the type of chemicals found in a contaminated Puget Sound estuary, on the health of juvenile salmon. Accordingly, the following five contaminant solutions were chosen and prepared for the pathogen challenge study: (1) a mixture of organic contaminants extracted from contaminated sediment collected from the Hylebos Waterway (CHWSE) located in Puget Sound, WA. We used an extraction method that enriched for chlorinated butadiene-compounds inclusive of HCBD; (2) a solution of HCBD, which we determined to be a marker chemical for the Hylebos Waterway[3]; (3) a model mixture of PAHs prepared to represent high molecular weight PAHs (3–5 rings) found in the Hylebos Waterway sediment; (4) a PAH compound, DMBA, shown previously to suppress immune responses in juvenile chinook salmon[4]; (5) a commercially acquired PCB mixture (Aroclor 1254), similar to the mixture of PCBs found in the Duwamish and Hylebos waterways and shown previously to suppress immune responses in juvenile chinook salmon.[4] The sum concentrations and compositions of the chemicals and chemical mixtures are listed in a previous study.[2]

Caution must always be used when handling chemicals and contaminated sediments. This paper describes procedures that may involve hazardous materials but does not purport to address all of the safety issues involved. Collection and use of sediments with unknown chemical contamination and preparation of chemical test mixtures may involve substantial risk to personal safety and health. It is highly recommended that appropriate precautions be taken to minimize contact with test chemicals and sediments. Laboratory personnel should consult their facility's Chemical Hygiene Plan for specific policies and emergency procedures. In addition, current Material Safety Data Sheets (MSDS) should be consulted for all known chemicals used and/or suspected contaminants. Further information can be obtained from the U.S. Occupational Safety and Health Administration (OSHA)[5] and the Environmental Protection Agency (EPA).[6,7]

Contaminants in field-collected samples may include mutagens, carcinogens, and other potentially toxic compounds. When working with sediments with unknown contaminants, and/or with known chemical hazards, it is essential to minimize worker contact by employing appropriate safety equipment and procedures. This includes the use of appropriate gloves, laboratory coats or protective suits, safety goggles, face shields, and respirators where needed. Samples and chemical test mixtures should be handled and prepared in a ventilated safety hood, and used materials disposed of in an appropriate manner. Laboratory personnel should be trained in proper practices for handling, using, and disposing of all chemicals used in the procedures described here. It is the laboratory's responsibility to comply with federal, state, and local regulations governing waste management and hazardous materials disposal. Further information is available in "The Waste Management Manual for Laboratory Personnel"[8] and from EPA.

Fish

Fish must be "healthy" so that confounding factors, such as a pre-existing disease, are not present. Inspections performed by state fish health inspectors and observations by the hatchery managers revealed no evidence or signs of the principal salmonid diseases in salmon (9–12 g) used in these studies. A fish health inspection[9] examines for the following salmonid pathogens: *Listonella anguillarum, Renibacterium salmoninarum, Yersinia ruckeri,* infectious hematopoietic necrosis virus, infectious pancreatic necrosis virus, and viral hemorrhagic septicemia virus. If these pathogens are identified in the fish, the investigator may consider using a different source of fish for the experiment.

Juvenile chinook salmon were collected and transported to the laboratory for use in generating the lethal dose (LD)–response curve relationship for DMBA, Aroclor 1254, CHWSE, and HCBD solutions. Care must be taken so that undue stress does not occur during transport. To this end, the salmon were placed in 95-l coolers equipped with air stones to maintain proper oxygen levels during transport. Upon arrival, the fish were immediately transferred to 2400-l circular fiberglass tanks with fresh dechlorinated water. The juvenile salmon were slowly acclimated from freshwater to seawater over a 5-day period. Juvenile salmon were allowed to acclimate for a minimum 2-week period in seawater (9–12°C, 30–32 ppt salinity) prior to the beginning of the LD tests. The juvenile salmon were fed 3% of their total body weight per day with Biodiet Grower Peletized Feed (2.5 mm, BioOregon Incorporated, Warrenton, OR).

Generation of the LD–response curves for DMBA, Aroclor 1254, CHWSE, and HCBD solutions

"Measurement of lethality is precise, quantal, and unequivocal and is, therefore, useful in its own right if only to suggest the level and magnitude of the potency of the substance."[10] Prior to examining an effect of a chemical on the immune response of fish, it is important to determine its LD–response curve with the species of interest. The contaminant may be delivered in a number of ways (i.e., diet, injection, or contaminated sediment using a MESOCOM approach for exposure). However, for initial studies, we recommend using injection delivery of the contaminant. This ensures that each fish receives a specific amount of chemical, and that you are not introducing other variables that come into play when using a diet or MESOCOM approach to contaminant delivery. For example, in a diet study, each fish may not eat equivalent amounts of diet thereby creating a range of contaminant exposure in the fish. Also, it may be unclear how a diet or sediment exposure may translate to contaminant body burden in the fish.

Our LD–response curves were determined by exposing 10 juvenile chinook salmon to each of the doses that consisted of: 50, 200, 400, 600, and 800 g sediment equivalent per kilogram of fish for CHWSE; 0.1, 1, 5, 30, 60, 100, 500, and 1000 mg kg^{-1} of fish for HCBD; 1, 10, 50, 100, 250, and 500 mg kg^{-1} of fish for Aroclor 1254; and 15, 25, 35, 50, 65, 75, and 88 mg kg^{-1} of fish for DMBA. Control fish were also exposed to the acetone/emulphor carrier solution. The fish were anesthetized with 1.5 mg of metomidate hydrochloride per liter (Wildlife Laboratories, Fort Collins, CO), weighed, measured, and injected intraperitoneally (i.p.) with 1 $\mu l\ g^{-1}$ body weight of the contaminant. The salmon were held in 450-l tanks with running seawater and were observed for mortality over a 96-h period. This information generated for the PAH compound, DMBA, was also used to determine the dose of the PAH model mixture to be used in the disease challenge study.

Contaminant exposure and pathogen challenge

In Trial 1, fish were exposed by i.p. injection to sublethal doses of either CHWSE, HCBD or the PAH model mixture. In Trial 2, fish were injected with one dose of either Aroclor 1254 or DMBA. Doses of the contaminants for the *L. anguillarum* challenge were chosen based on their LD–response curve, targeting doses that would allow for a substantial but sublethal dose. In both trials, control fish were injected with the acetone/emulphor carrier solution. One week after contaminant injection, fish were exposed to either *L. anguillarum* or to seawater diluent alone (see the following). In Trial 1, the juvenile salmon were anesthetized with 1.5 mg l^{-1} of metomidate hydrochloride, individually weighed, and then injected i.p. with 2 μl of inoculum per gram body weight of fish. Fish were injected with either HCBD, the PAH model mixture, CHWSE, or acetone/emulphor carrier solution. Concentrations of the test solutions were 10% of the LD_{50} of HCBD (20 mg of HCBD per kilogram of fish), 10% of the LD_{50} of the PAH model mixture (6.3 mg of the model PAH mixture per kilogram of fish), or 41% of the LD_{30} of CHWSE (307 g sediment equivalent per kilogram of fish). Since we were only able to achieve 30% mortality in the fish exposed to a very high concentration of CHWSE, we determined the LD_{30} for this contaminant instead of the LD_{50} (see the section "Results and discussion"). The groups of fish injected with each contaminant were kept separate in individual tanks. Thirty fish were randomly assigned to six tanks for each of the four treatments. Three tanks of each treatment were challenged with *L. anguillarum*. The remaining three tanks in the treatment received only the TSB diluent. This latter unexposed group provided an estimate of background mortality.

In Trial 2, juvenile salmon were anesthetized as described earlier for Trial l, weighed and injected i.p. with 2 μl of inoculum per gram body weight of fish. Fish were injected with DMBA, Aroclor 1254, or acetone/emulphor carrier solution. Concentrations of the DMBA and Aroclor 1254 solutions were 20% of the LD_{50}. Therefore, the concentration of DMBA used was 13 mg of the PAH per kilogram of fish. The concentration of Aroclor 1254 used was 54 mg of the PCB mixture per kilogram of fish. The groups of fish injected with each contaminant were kept separate in individual tanks. Fifteen fish were randomly assigned to six tanks for each of the three treatments. Three tanks of each treatment were challenged with *L. anguillarum*. The remaining three tanks in the treatment received only the TSB diluent. This latter unexposed group provided an estimate of background mortality.

Infection of salmon with L. anguillarum

A number of factors need to be addressed when deciding which pathogen to use in the disease challenge. For example: Is the fish likely to come into contact with the pathogen in its natural environment? Is the pathogen easily grown in culture? Are there techniques for identification of the pathogen in exposed fish? For our disease challenges, we have chosen a Gram-negative bacteria *L. anguillarum*. Relative to some pathogenic bacterial species a lot is known about the epizootiology of this predominately marine pathogen.[11] *L. anguillarum* causes vibriosis that can occur in preacute, acute, and chronic forms in salmonids[12] and has been reported to have a major effect on over 48 species of marine fish. As discussed in the following, this pathogen can be grown to log phase within 24 h and both presumptive and confirmatory tests are available for identification of the pathogen. All of the above qualities make this pathogen an invaluable tool in disease challenges.

A lyophilized preparation of *L. anguillarum* (strain 1575), a gift from Biomed, Bellevue, WA, was kept at −70°C until use. The bacteria were rehydrated with 2 ml of TSB supplemented with 0.5% NaCl,[13] then the purity of the culture was determined by Gram stain, cell morphology, and agglutination with the rapid agglutination test (Bionar, Norway). To ensure further that the rehydrated preparation was free of contaminating bacteria, an aliquot was cultured onto a TSA plate supplemented with

0.5% NaCl for 24 h at 25°C. At the same time, 2 ml of the rehydrated preparation was placed into 100 ml of TSB supplemented with 0.5% NaCl for production of a stock culture. After 18 h of growth at 25°C on a shaker, the stock culture was mixed 1:1 (v/v) with a glycerol and saline buffer. The stock culture of *L. anguillarum* was aliquoted into cryovials (2 ml each) and stored at −70°C for future use.

A growth curve of *L. anguillarum* (strain 1575) at 25°C was determined to ensure that at the time of challenging the salmon with bacteria, the bacteria would be close to the peak of their exponential growth phase. In brief, 2 ml of the stock culture was placed into 500 ml TSB supplemented with 0.5% NaCl and placed on a shaker at 25°C. Two milliliters of aliquots were removed every hour from the 500 ml of bacterial suspension, and the turbidity of the culture determined with a UV–Vis recording spectrophotometer (Shimadzu Scientific Instrument, Columbia, MD) at 525 nm wavelength until just after the beginning of the stationary phase. During the exponential growth phase, the bacteria divide steadily at a constant rate and the population is uniform in terms of chemical composition of cells, metabolic activity, and other physiological characteristics.[14] The stationary phase of bacterial growth may be defined as the period of time after the exponential phase, in which the growth of the bacteria slows down or is no longer exponential either because a required nutrient is limiting or because the metabolic products become inhibitory to bacterial growth.[15]

Prior to conducting the disease challenge, a lethal concentration (LC) curve needs to be generated with the pathogen and the species of interest. Once the curve is generated, the low, medium, and high concentrations of the pathogen should be selected for testing host resistance.[16] If the contaminant is immunoenhancing, an increased resistance to the pathogen should be observed at the higher concentrations. If the chemical is immunosuppressive, an increase in disease susceptibility should be observed in the lower concentration. However, if the compound is immunosuppressive, care must be taken not to use too high or too low of a concentration of the pathogen. If the pathogen concentration is too low, contaminant-exposed fish may have the ability to fight the low number of bacteria in the same manner as the noncontaminant-exposed (control) fish. If the concentration of the pathogen is too high, it may overwhelm even the healthiest immune system and a difference in susceptibility will not be observed between the two treatment groups. Therefore, an LC–response curve for juvenile chinook salmon was determined with *L. anguillarum* (strain) 1575. Bacterial concentration was quantitated turbidimetrically at 525 nm, until an optical density of 1.7 was achieved, corresponding to the period near the end of the exponential growth phase. Seven log 10 dilutions (10^{-1}, 10^{-2}, 10^{-3}, 10^{-4}, 10^{-5}, 10^{-7}, and 0 ml bacterial culture per milliliter seawater) of the stock growth culture of *L. anguillarum* were used to determine the LC–response curve.

Bacterial plate counts of the stock culture were performed to determine the concentration of live *L. anguillarum*, to which the fish are exposed (Figure 2.1). To calculate the concentration of *L. anguillarum*, or colony forming units (cfu) per milliliter growing in a suspension culture, 10-fold serial dilutions were made of the bacteria and then plated onto TSA supplemented with 3% NaCl. Dilutions were made using eight sterile 15-ml graduated tubes with screw caps and conical bottoms filled with 9 ml of sterile TSB supplemented with 1.5% NaCl. With a sterile pipette, 1 ml of *L. anguillarum* suspension culture is transferred to the first tube, creating a dilution factor of 1:10. This dilution is mixed thoroughly by vortexing. With a sterile pipette 1 ml of the first dilution is transferred to the second tube creating a dilution factor of 1:100. This process was repeated until all successive serial dilutions are made. Each of the last three serial dilutions (10^{-6}, 10^{-7}, and 10^{-8}) are plated on three replicate Petri plates containing TSA supplemented

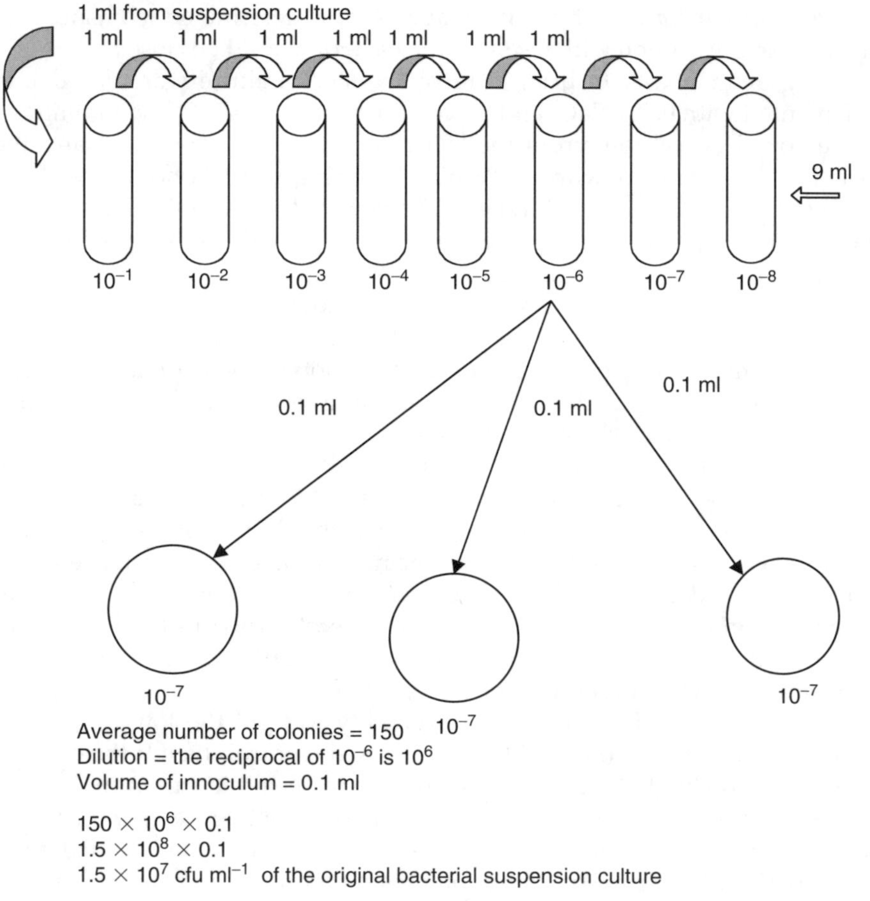

Figure 2.1 Plate count method for determining the number of colony forming units per milliliter of *L. anguillarum* in the stock solution.

with 3% NaCl. The center of the plates are inoculated with 0.1 ml of the dilution and a thin uniform layer of the inoculum is spread across the entire plate using a glass "hockey stick" dipped in 95% ethanol, flamed, and cooled on the edge of the agar plate. This procedure is repeated for each plate. The plates are labeled and incubated at 25°C for 48 h. Countable plates contain 20–200 colonies and the average numbers of colonies for the three plates of each dilution are used to calculate the bacterial concentration (cfu ml^{-1}) in the stock culture.

Twenty salmon, approximately 10 g each, were placed in 7.6-l buckets containing 4 l (1 l for every 50 g of fish) of the various concentrations of bacteria diluted in filtered seawater, or into a control bucket containing 4 l seawater. The fish were exposed to the bacteria under static conditions for 1 h with aeration. Following each challenge, the fish were placed in 450-l tanks with flow-through sand-filtered and UV-treated seawater. Mortalities were collected and tabulated twice a day for a minimum of 7 days (168 h). The salmon were not fed during the experimental period.

Water treatment

When dealing with contaminants and pathogens, it is imperative to ensure that both the influent and effluent are contaminant and pathogen free. Our influent was sand-filtered and UV-treated seawater. Before the effluent was released into Yaquina Bay it was treated with charcoal and chlorine to prevent the introduction of chemicals and surviving bacteria, respectively.

Confirmation of pathogen-induced mortality

Necropsies should be performed on all mortalities to ensure that the dead fish have been infected with *L. anguillarum*. First, dead fish are sprayed with 75% ethanol. A small incision is made into the ventral abdomen with sterile scissors taking care not to damage any of the internal organs. With a sterile scalpel blade, the swim bladder and internal organs are pushed aside to expose the kidney. A sterile loop is inserted into the kidney and then aseptically struck onto a TSA plate supplemented with 3.0% NaCl. After 48 h of incubation at 25°C, colonies of *L. anguillarum* appear on the plate as shiny cream-colored raised and round mounds (0.5 × 1.5 μm; Figure 2.2).[11] Presumptive tests were initially performed on bacterial colony growth on the TSA plates. Presumptive identification of

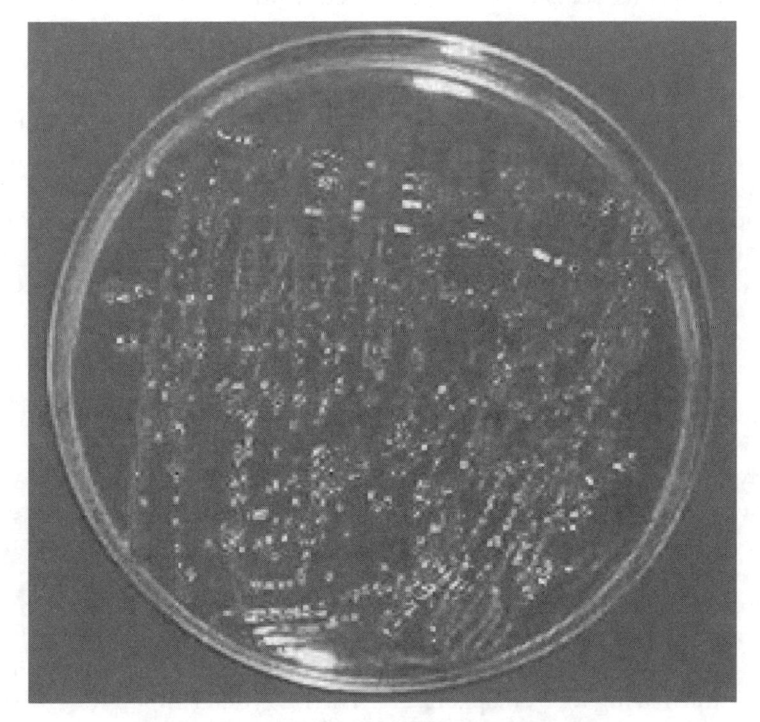

Figure 2.2 Colony morphology of *L. anguillarum* grown on TSA.

L. anguillarum infection was determined by inhibition of bacterial growth by novobiocin and the vibriostatic agent 0/129 (2,4-diamino-6,7-diisopropyl pterdine phosphate) on TSA (Figure 2.3A) and growth on VAM (Figure 2.3B). *V. anguillarum* media is a selective media for *Listonella* species and used to presumptively identify *L. anguillarum. L. anguillarum* on VAM produces a bright yellow, round, flat colony with a yellow halo after 48 h of incubation at 25°C.[17] Also as part of the presumptive identification of *L. anguillarum*, bacterial colonies were Gram stained and examined for the presence of cytochrome oxidase with a dry slide oxidase test (Figure 2.4). Bacteria presumptively identified as *L. anguillarum* were confirmed by a commercially available rapid agglutination test kit (Bionor; Figure 2.5).

Generalized linear models: Statistical analyses

Generalized linear modeling (GLM) was used to determine the LD_{30} for CHWSE and the LD_{50} for Aroclor 1254, HCBD, and DMBA solutions and the confidence limits for these values.[18] Statistical significance between treatments in the pathogen challenge studies was also assessed using GLM. We assumed that the number of survivors and mortalities in both assays follow a binomial distribution.[18,19] It has been determined that the most appropriate way to analyze the binomially distributed response data generated in toxicity tests is by using GLM.[18]

The analyses were performed with the GLMStat computer application.[20] To define a GLM it is important to identify the error structure and the link function that relates the linear predictor to the expected survival/mortality probabilities.[21] The logistic GLM was used for the analysis of data. For this model, the error structure is binomial and the linear predictor was related to the expected value of the datum by the logit link function.

Specifically for the pathogen challenge study, we used the logistic model to evaluate if survival/mortality of fish treated with a contaminant and fish treated only with the acetone/emulphor carrier solution were significantly different ($P_{\alpha} \leq 0.05$) beginning at 7 days post challenge. This analysis was conducted after correction for background mortalities. To correct for background deaths, the number of mortalities of a particular treatment group not treated with *L. anguillarum* were subtracted from the mortalities of that treatment group exposed to the bacteria. If required, analyses were conducted beyond day 7. Experiments were continued until mortalities began to level off at an asymptote in at least one treatment group.

Results and discussion

The cumulative 96-h LD curves of salmon given various doses of DMBA (Figure 2.6), Aroclor 1254 (Figure 2.7), CHWSE (Figure 2.8), or HCBD (Figure 2.9), were determined. The 96-h LD_{50} was determined for DMBA, Aroclor 1254, and HCBD to ensure the use of sublethal dosages of the test solutions for use in the pathogen challenge experiments. Since we were only able to achieve 30% mortality in the fish exposed to a very high concentration of CHWSE, we determined the LD_{30} for this contaminant. The LD_{50} determined for Aroclor 1254 was 270 mg kg^{-1} of salmon. The upper and lower 95% confidence limits were 500 and 100 mg kg^{-1}, respectively. The LD_{50} determined for DMBA was 63 mg kg^{-1} of salmon. The upper and lower 95% confidence limits were 73 and 56 mg kg^{-1},

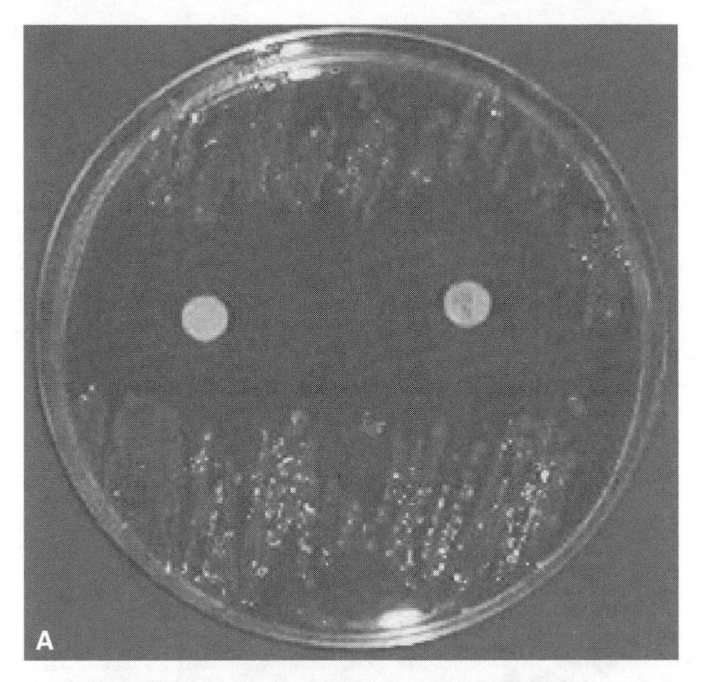

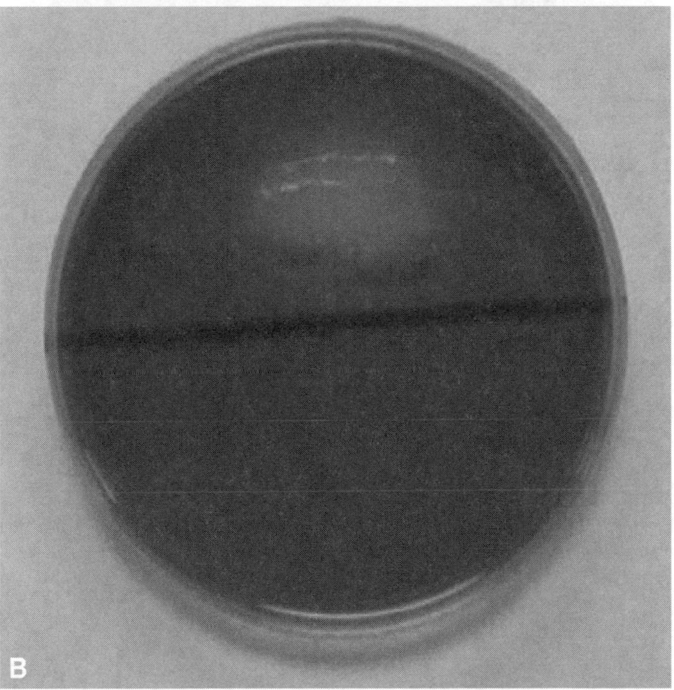

Figure 2.3 Two presumptive tests for *L. anguillarum*. (A) Zones of inhibition (or no bacterial growth) of *L. anguillarum* around the novobiocin and the vibriostatic agent 0/129 disks. The disks absorb water from the agar allowing the antibiotic to dissolve. The antibiotic migrates through the agar. There will be no growth in the areas where the antibiotic is at inhibitory concentrations for the bacteria on the plate.[21] (B) Growth of *L. anguillarum* on VAM. The media contains bile salts, a high concentration of NaCl and is at a high pH. These three factors select mostly for *Listonella* species. The color change (green to yellow) is due to the ability of *L. anguillarum* to ferment sorbitol.[17]

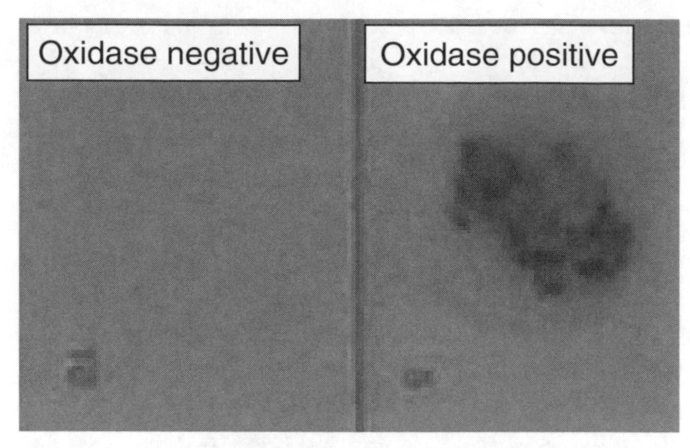

Figure 2.4 The presence or absence of cytochrome can be detected with an oxidase test that is commercially available. The enzyme is present if the bacteria are capable of oxidizing a reagent present on the test strip into a blue colored product. If a blue product forms, the bacteria is considered "oxidase positive." *L. anguillarum* is oxidase positive.

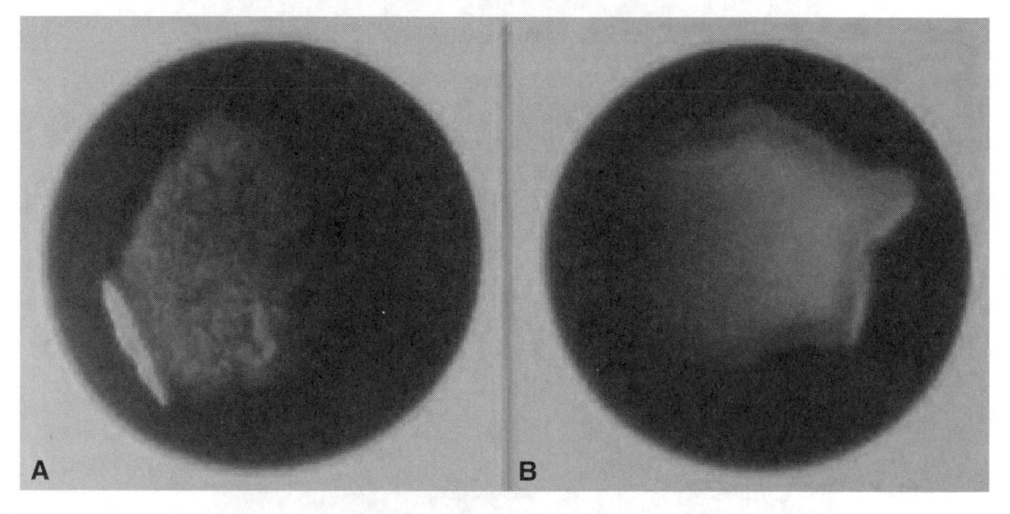

Figure 2.5 Agglutination of bacterial colonies from TSA with a monoclonal antibody to *L. anguillarum* is considered to be a confirmatory test for the identification of *L. anguillarum*. (A) A positive reaction or agglutination with the antibody. (B) The negative control shows no agglutination with the specific monoclonal antibody.

respectively. The LD_{50} for HCBD was 200 mg HCBD per kilogram of fish. The upper and lower 95% confidence limits were 450 and 130 HCBD per kilogram of fish, respectively. The LD_{30} determined for CHWSE was 741 g CHWSE equivalent per kilogram of fish. The upper and lower 95% confidence limits were 1500 and 540 g CHWSE equivalent per kilogram of fish, respectively.

The peak of the exponential growth phase of *L. anguillarum* grown in 500 ml of TSB supplemented with 1.5% NaCl and slowly agitated at 25°C occurred at approximately 15 h with an optical density of 1.9 (Figure 2.10). In subsequent challenge experiments,

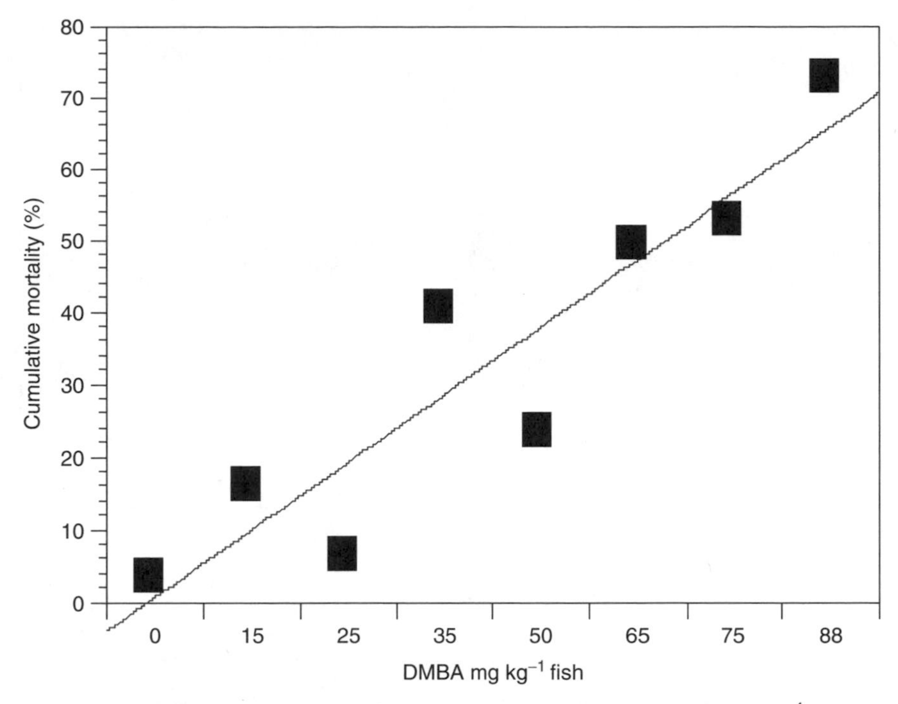

Figure 2.6 The cumulative 96 mortality of salmon given various doses of DMBA.[4] (From Arkoosh, M.R., Clemons, E., Huffman, P., Kagley, A., Casillas, E., Adams, N., Sanborn, H.R., Collier, T.K. and Stein, J., *J. Aquat. Anim. Health*, 13, 257–268, 2001. With permission.)

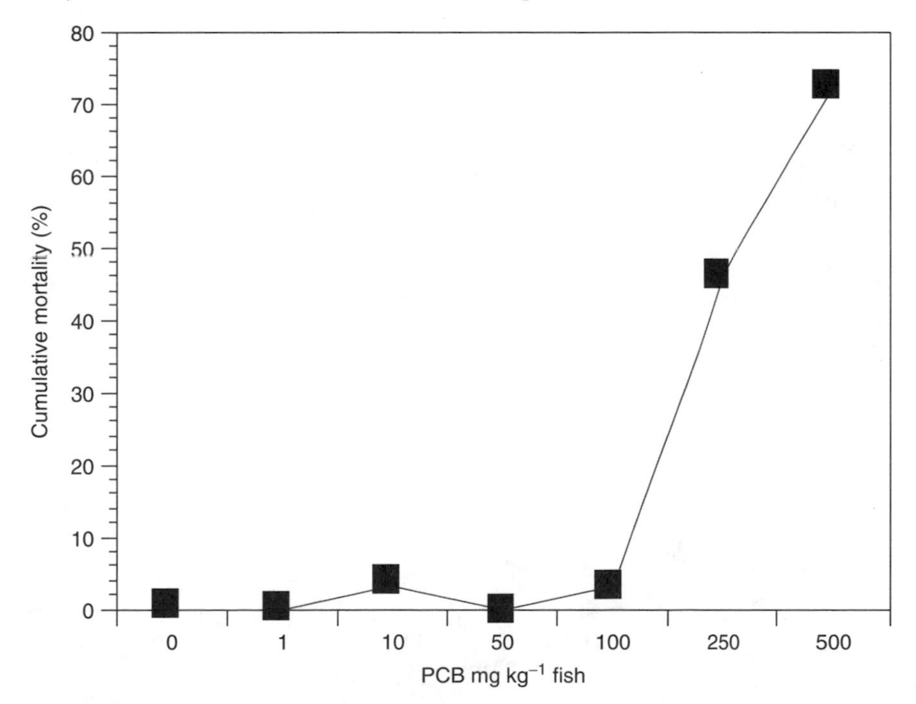

Figure 2.7 The cumulative 96 mortality of salmon given various doses of Aroclor 1254.[4] (From Arkoosh, M.R., Clemons, E., Huffman, P., Kagley, A., Casillas, E., Adams, N., Sanborn, H.R., Collier, T.K. and Stein, J., *J. Aquat. Anim. Health*, 13, 257–268, 2001. With permission.)

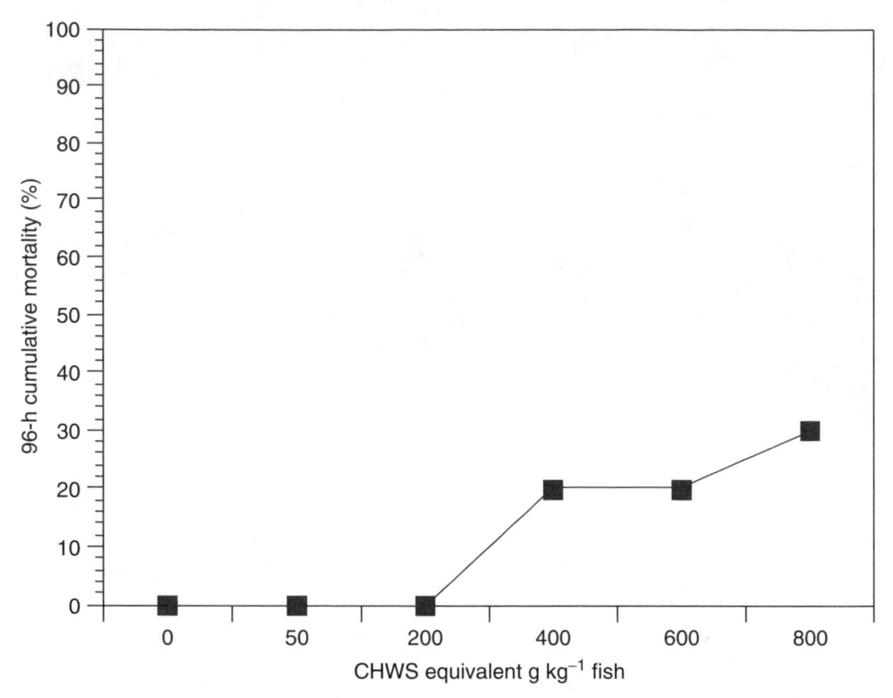

Figure 2.8 The cumulative 96 mortality of salmon given various doses of CHWSE.[2] (From Arkoosh, M.R., Clemons, E., Huffman, P., Kagley, A., Casillas, E., Adams, N., Sanborn, H.R., Collier, T.K. and Stein, J., *J. Aquat. Anim. Health*, 13, 257–268, 2001. With permission.)

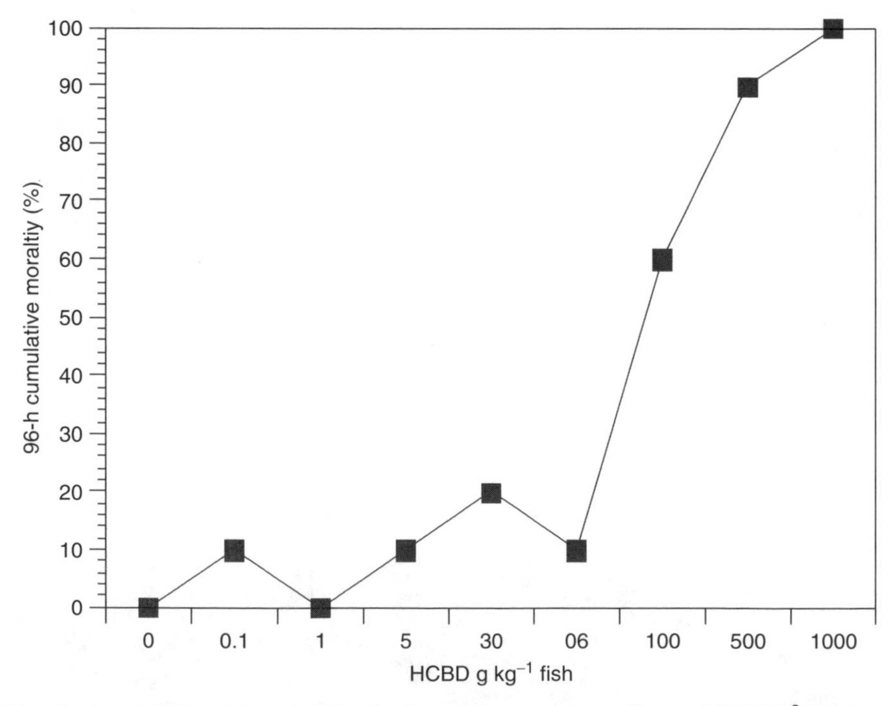

Figure 2.9 The cumulative 96 mortality of salmon given various doses of HCBD.[2] (From Arkoosh, M.R., Clemons, E., Huffman, P., Kagley, A., Casillas, E., Adams, N., Sanborn, H.R., Collier, T.K. and Stein, J., *J. Aquat. Anim. Health*, 13, 257–268, 2001. With permission.)

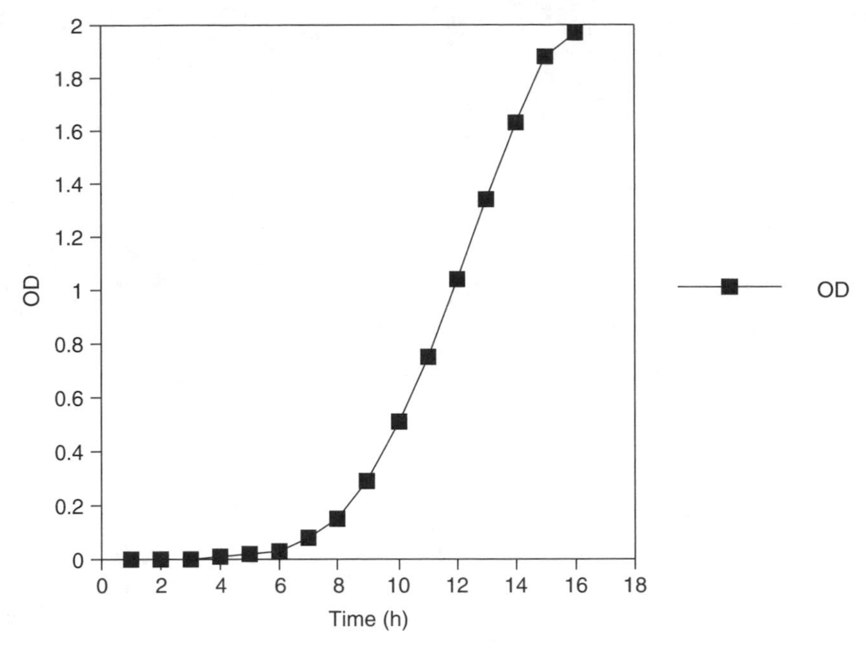

Figure 2.10 Growth curve of *L. anguillarum* measured as optical density (OD) at 525 nm (λ). The peak of the exponential growth phase of the bacteria occurs after 15 h.[1] (From Arkoosh, M.R., Casillas, E., Huffman, P., Clemons, E., Evered, J., Stein, J.E. and Varanasi, U., *Trans. Am. Fish. Soc.*, 127, 360–374, 1998. With permission.)

L. anguillarum cultures were grown to an optical density of approximately 1.7 prior to preparation of bacterial culture dilutions used for juvenile salmon exposures. The 168-h (7-day) LD–response curve for juvenile salmon exposed to *L. anguillarum* was determined by a logit regression analysis (Figure 2.11).

The percent cumulative mortality in Trials 1 and 2 of the different groups of juvenile chinook salmon exposed to *L. anguillarum* is shown in Figures 2.12 and 2.13, respectively. These figures represent the net cumulative mortality attributed to exposure to the bacteria after subtracting background mortality observed in juvenile chinook salmon that received chemical contaminants but were not exposed to bacteria. Background mortality at the end of the experiments was very low. Specifically, background mortality in Trial 1 at day 7 for the various treatments was the following: acetone/emulphor carrier solution (1.4%), CHWSE (1.2%), HCBD (2.7%), and the model mixture of PAH (0%). Background mortality in Trial 2 at day 9 for the various treatments was the following: acetone/emulphor carrier solution (6.6%), Aroclor 1254 (2.3%), and DMBA (6.7%). Statistical testing was performed beginning with 7 days post exposure data to determine treatment differences.[1] If statistical differences were not noted at 7 days post exposure, data obtained on the following days were also examined.

In Trial 1, the net cumulative mortality of juvenile chinook salmon exposed to *L. anguillarum* after receiving either CHWSE, HCBD, or the model mixture of PAHs ranged from 28% to 31% compared to the 16% observed in the acetone/emulphor control group at 7 days post bacterial challenge (Figure 2.12). Although in Trial 2, a significant difference in the net cumulative mortality between juvenile chinook salmon

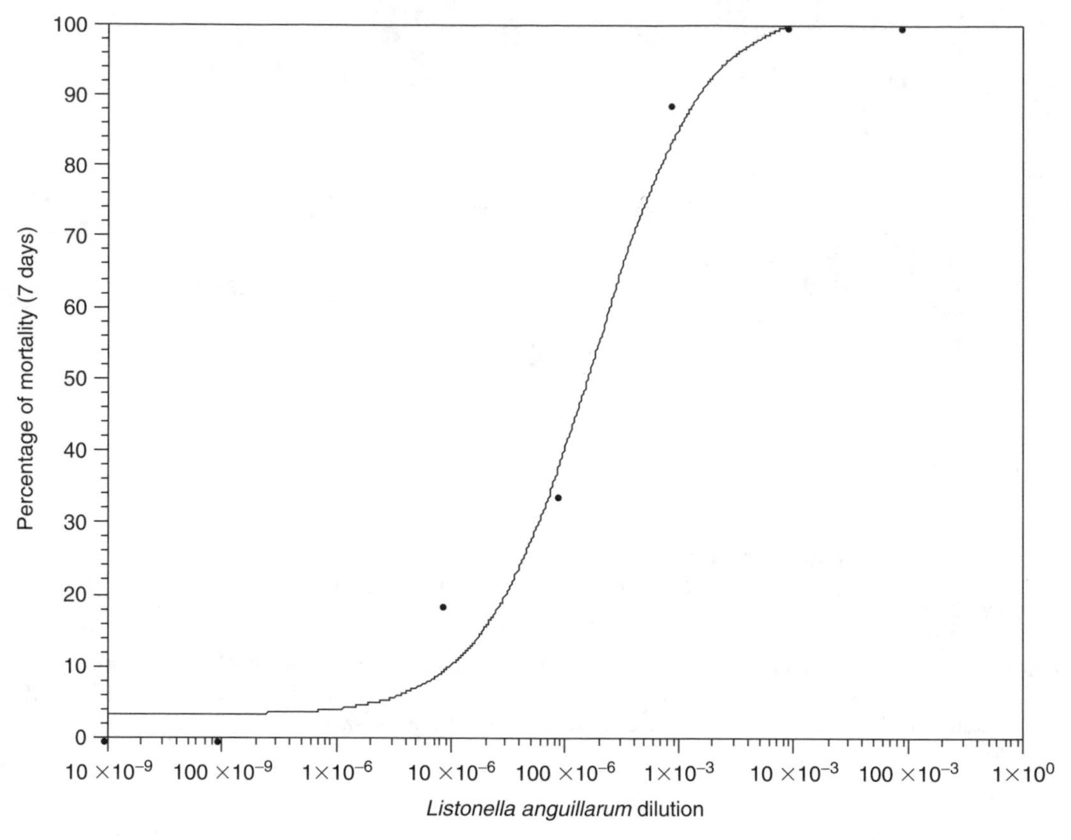

Figure 2.11 LD–response curve of juvenile chinook salmon at 7 days (168 h) post exposure to *L. anguillarum*.[1] (From Arkoosh, M.R., Casillas, E., Huffman, P., Clemons, E., Evered, J., Stein, J.E. and Varanasi, U., *Trans. Am. Fish. Soc.*, 127, 360–374, 1998. With permission.)

exposed to Aroclor 1254 and acetone/emulphor was detected on day 8, a difference between DMBA and acetone/emulphor was not detected until day 9 after challenge. The net cumulative mortality of juvenile chinook salmon exposed to the bacteria after receiving either DMBA or Aroclor 1254 ranged from 46% to 49% compared to the 25% observed in the acetone/emulphor control group at 9 days post challenge (Figure 2.13). The cumulative mortality was significantly higher in fish exposed to CHWSE, DMBA, Aroclor 1254, the PAH model mixture, or HCBD relative to fish receiving only the acetone/emulphor carrier.

Therefore, juvenile chinook salmon exposed to contaminants associated with urban estuaries in Puget Sound, such as the Hylebos and Duwamish waterways, exhibited a higher susceptibility to mortality induced by the marine pathogen *L. anguillarum*, than did the pathogen exposed juvenile chinook salmon treated only with the carrier acetone/emulphor. The contaminants tested represent specific subsets of predominant estuarine chemical pollutants. The chlorinated hydrocarbons tested were characterized by HCBD, PCBs, and CHWSE (which is composed primarily of HCBD-like compounds) and the aromatic hydrocarbons were characterized by the model mixture of PAHs and DMBA. Results obtained using the disease challenge assay described here, together with our previous studies,[4,22] support the hypothesis that chemical contaminant exposure

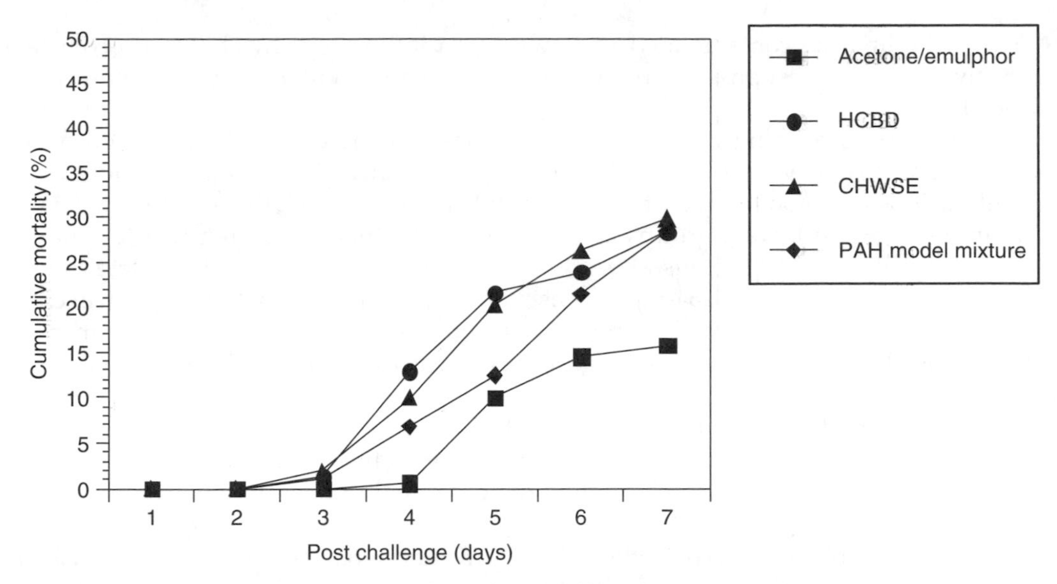

Figure 2.12 Percent cumulative mortality of juvenile chinook salmon injected with either HCBD, a model mixture of PAHs, a chlorinated-enriched sediment extract form the Hylebos Waterway, or the carrier control (acetone/emulphor) after exposure to 6×10^{-5} ml bacterial solution per milliliter seawater. The cumulative mortalities are corrected for mortalities observed in juvenile chinook salmon injected with either HCBD, a model mixture of PAHs, a chlorinated-enriched sediment extract form the Hylebos Waterway, or the carrier control (acetone/emulphor) but not exposed to *L. anguillarum*.[1] (From Arkoosh, M.R., Casillas, E., Huffman, P., Clemons, E., Evered, J., Stein, J.E. and Varanasi, U., *Trans. Am. Fish. Soc.*, 127, 360–374, 1998. With permission.)

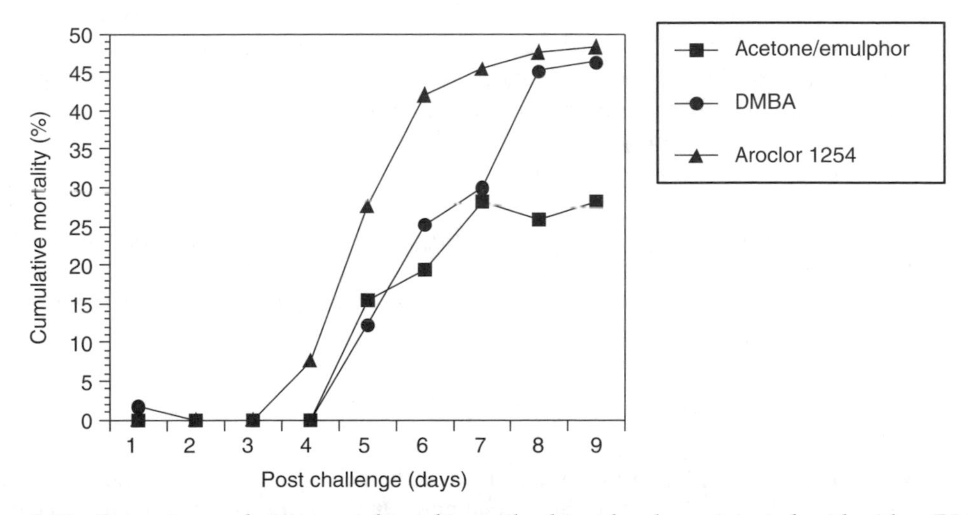

Figure 2.13 Percent cumulative mortality of juvenile chinook salmon injected with either DMBA, Aroclor 1254, or the carrier control (acetone/emulphor) after exposure to 6×10^{-5} ml bacterial solution per milliliter seawater. The cumulative mortalities are corrected for mortalities observed in juvenile chinook salmon injected with either DMBA, Aroclor 1254, or the carrier control (acetone/emulphor) but not exposed to *L. anguillarum*.[1] (From Arkoosh, M.R., Casillas, E., Huffman, P., Clemons, E., Evered, J., Stein, J.E. and Varanasi, U., *Trans. Am. Fish. Soc.*, 127, 360–374, 1998. With permission.)

of juvenile salmon in contaminated waterways can influence their ability to produce a protective immune response such that their survival potential may be significantly reduced.

Observations spanning the entire temporal scale of the experimental period in the present study provide useful data in evaluating differences in disease susceptibility of juvenile chinook salmon exposed to contaminants. This is especially apparent in Trial 2. Differences between DMBA groups and the control acetone/emulphor-injected groups were not significant until 9 days after exposure, whereas we were able to determine a difference between Aroclor 1254-injected fish and the control acetone/emulphor-injected fish after 8 days of exposure. Therefore, it is important to monitor the experiment to determine if it should be extend beyond the classical 7-day period in order to observe differences.

In conclusion, disease challenge assays in fish are useful in determining the effects of xenobiotics on their immune response. A number of variables need to be considered prior to performing the assay. They include but are not limited to the following:

- Which xenobiotic to expose the fish to, why, and what dose to expose the fish at? What method of exposure to use? How long after exposure to the contaminant prior to the disease challenge?
- Which pathogen to use in the challenge, how to challenge the fish with the pathogen (bath or injection) and what concentration to use. What confirmatory and presumptive tests are available?
- The temporal scale of the disease challenge assay needs to be determined.
- The fish need to be kept at a constant temperature, salinity, and flow since these variables have the ability to influence disease progression.
- Appropriate controls need to be used for quality assurance. For example, in our study, we used two types of controls. One control included juvenile chinook salmon, which did not receive bacteria, and the second control included juvenile chinook salmon, which received only the carrier acetone/emulphor. This way it is possible to determine how many fish might be dying from the procedure itself and not from *L. anguillarum* exposure.

If the xenobiotics under study are found to impair the ability of the fish to fight the pathogen, more specific tests can be used to determine which functional aspect (i.e., adaptive and/or innate immunity), or cell types (i.e., B cell, T cell, and/or macrophage) of the immune system is targeted by the xenobiotic.

References

1. Arkoosh, M.R., Casillas, E., Huffman, P., Clemons, E., Evered, J., Stein, J.E. and Varanasi, U., Increased susceptibility of juvenile chinook salmon (*Oncorhynchus tshawytscha*) from a contaminated estuary to the pathogen *Vibrio anguillarum*, *Trans. Am. Fish. Soc.*, 127, 360–374, 1998.
2. Arkoosh, M.R., Clemons, E., Huffman, P., Kagley, A., Casillas, E., Adams, N., Sanborn, H.R., Collier, T.K. and Stein, J., Increased susceptibility of juvenile chinook salmon to vibriosis after exposure to chlorinated and aromatic compounds found in contaminated urban estuaries, *J. Aquat. Anim. Health*, 13, 257–268, 2001.
3. Collier, T.K., Johnson, L.L., Myers, M.S., Stehr, C.M., Krahn, M.M. and Stein, J.E., Hylebos Fish Injury Study. Round II Part 3. Exposure of Juvenile Chinook Salmon to Chemical Contaminants Specific to the Hylebos Waterway: Tissue Concentrations and Biochemical Responses.

Overview of Commencement Bay Natural Resource Damage Assessment, http://www.darcnw.noaa.gov/nrda.htm, 1994.

4. Arkoosh, M.R., Clemons, E., Myers, M. and Casillas, E., Suppression of B-cell mediated immunity in juvenile chinook salmon (*Oncorhynchus tshawytscha*) after exposure to either a polycyclic aromatic hydrocarbon or to polychlorinated biphenyls, *Immunopharmacol. Immunotoxicol.*, 16, 293–314, 1994.
5. U.S. Occupational Safety and Health Administration, Regulations: Hazard Communication Standard 29 CFR 1910.1200.
6. U.S. Environmental Protection Agency, Occupational Health and Safety Staff/Office of Administration, Occupational Health and Safety Manual, Transmittal 1440, 1986.
7. U.S. Environmental Protection Agency, Pesticides and Toxic Substances Enforcement Division, Toxic Substances Control Act Inspection Manual, 1990.
8. American Chemical Society, Department of Government Relations and Science Policy, The Waste Management Manual for Laboratory Personnel.
9. United States Fish and Wildlife (USFW) Service Manual, Part 713, Fish Health Operations, 1995.
10. Doull, J. and Bruce, M.C., Origin and scope of toxicology, in *Casarett and Doull's Toxicology. The Science of Poisons*, Klaassen, C.D., Amdur, M.O. and Doull, J., Eds., Macmillan, New York, 1986, pp. 3–10 (974 pp.).
11. Austin, B. and Austin, D.A., Vibrionaceae representatives, in *Bacterial Fish Pathogens. Disease in Farmed and Wild fish*, Springer-Praxis, pp. 266–307 (384 pp.).
12. Noga, E.J., Vibriosis (salt water furunculosis, *Vibrio* infection, Hirta disease), in *Fish Disease Diagnosis and Treatment*, Mosby, St. Louis, MO, 1996, pp. 149–151 (367 pp.).
13. Beacham, T.D. and Evelyn, T.P.T., Population variation in resistance of pink salmon to vibriosis and furunculosis, *J. Aquat. Anim. Health*, 4, 168–173, 1992.
14. Pleczar, M., Jr. and Reid, R., Reproduction and growth, in *Microbiology*, McGraw-Hill, New York, 1972, pp. 123–138 (948 pp.).
15. Davis, B.D, Dulbecco, R., Eisen, H.N. and Ginsberg, H.S., Bacterial nutrition and growth, in *Microbiology*, Harper & Row, New York, 1980, pp. 60–72 (1355 pp.).
16. White, K.L., Jr., Specific immune function assays, in *Principles and Practice of Immunotoxicology*, Miller, K., Turk, J. and Nicklin, S., Eds., Blackwell Scientific Publications, Cambridge, MA, 1992, pp. 304–323 (379 pp.).
17. Alsina, M., Martfnex-Picado, J., Jofre, J. and Blanch, A.R., A medium for presumptive identification of *Vibrio anguillarum*, *Appl. Environ. Microbiol.*, 60, 1681–1683, 1994.
18. Kerr, D.R. and Meador, J.P., Hazard risk assessment. Modeling dose response using generalized linear models, *Environ. Toxicol. Chem.*, 15, 395–401, 1996.
19. Hildén, M. and Hirvi, J.P., Survival of larval perch, *Perca fluviatilis* L., under different combinations of acidity and duration of acid conditions, analyzed with a generalized linear model, *J. Fish Biol.*, 30, 667–677, 1987.
20. Beath, K.J., *GLMStat User Manual*, Version 1.5, 1995.
21. Baker, R.J. and Nedler, J.A., Susceptibility of chinook salmon *Oncorhynchus tshawytscha* (Walbaum), and rainbow trout, *Salmo gairdneri* Richardson, to infection with *Vibrio anguillarum* following sublethal copper exposure, *J. Fish Dis.*, 6, 267–275, 1978.
22. Arkoosh, M.R., Casillas, E., Clemons, E., McCain, B. and Varanasi, U., Suppression of immunological memory in juvenile chinook salmon (*Oncorhynchus tshawytscha*) from an urban estuary, *Fish Shellfish Immunol.*, 1, 261–277, 1991.
23. Barry, A.L. and Thornsberry, C., Susceptibility testing: diffusion test procedures, in *Manual of Clinical Microbiology*, American Society of Microbiology, Washington, D.C., 1980, pp. 463–477 (1044 pp.).

chapter three

Enhanced frog embryo teratogenesis assay: Xenopus *model using* Xenopus tropicalis

Douglas J. Fort and Robert L. Rogers
Fort Environmental Laboratories, Inc.

Contents

1-56670-664-5/05/$0.00+$1.50
© 2005 by CRC Press

Introduction

The expense and time-consuming nature of conventional testing has warranted the development, validation, and eventual widespread use of alternative, nonmammalian model systems. Frog embryo teratogenesis assay—*Xenopus* (FETAX), a 4-day, nonmammalian, whole embryo bioassay designed to evaluate potential teratogenic hazard by direct chemical compounds and complex environmental mixture screening, and used as a tool to evaluate toxicological mechanisms of action, was developed to evaluate organ system malformation during early embryo and larval development.[1–3] In its original format, FETAX validation studies have been performed using the model as a screening assay using both pure chemical compounds and complex environmental mixtures (wastewater effluent, surface water, sediments, and groundwater).[4–12] Because *Xenopus* lack many metabolic enzyme systems, including the mixed-function oxidase (MFO) system, through the first 96 h of development, an exogenous metabolic activation system (MAS) was developed[13] and evaluated[14–21] using both Aroclor 1254- and isoniazid-induced rat liver microsomes. Currently, no standardized and well-validated alternative models exist for screening in teratogenesis. Today, only the limb micromass culture and the rodent whole embryo culture are used as an *in vitro* test of developmental toxicity.

An *ad hoc* Interagency Coordinating Committee on the Validation of Alternative Methods (ICCVAM) was established in 1994 by National Institute of Environmental Health Sciences (NIEHS) to develop a report recommending criteria and processes for validation and regulatory acceptance of toxicological testing methods that would be useful to federal agencies and the scientific community. In late 1998, ICCVAM initiated an evaluation of FETAX.[22] The ultimate objective of the ICCVAM evaluation was to determine applications in which FETAX may be successfully used and to determine areas of use that require further validation. Ultimately, the outcome of the ICCVAM evaluation prioritized further research needs and validation efforts to maximize understanding of how FETAX can be most effectively used as an alternative test method. From the ICCVAM, one overwhelming recommendation was brought forth.[22] This suggestion was to strongly consider replacing the currently used *Xenopus laevis* with a species with greater potential, *Xenopus tropicalis*.

The comparative advantages of using *X. tropicalis* over *X. laevis* are summarized in Table 3.1. FETAX can effectively be conducted in *ca.* 2-days-using of *X. tropicalis* as opposed to 4-days-using with *X. laevis*.[23] The increased rate of development of *X. tropicalis* is presently being evaluated, however, as it may have an impact on the sensitivity of the test. Many species that develop at cooler temperatures and, thus, develop more slowly than those at higher temperatures, are more sensitive, presumably due to longer developmental windows (i.e., larger exposure targets). Since FETAX uses an exogenous MAS consisting of rat liver microsomes, the increased temperature may increase the effectiveness and efficiency of the MAS. *Xenopus tropicalis*, being diploid, also represents a cytogenetic genetic advantage over *X. laevis*, which is oligotetraploid.

The smaller size of *X. tropicalis* compared to *X. laevis* represents practical advantages for the assay. Less test material is required for the smaller organism. More organisms can

Table 3.1 Comparison of *X. laevis* and *X. tropicalis* attributes as a test species

Consideration	*X. laevis*	*X. tropicalis*
Rate of development (time to complete FETAX test)	4-days	2-days[a]
Culture temperature	23.5 ± 0.5°C	26.5 ± 0.5°C[a]
Ploidy	Oligotetraploid	Diploid[a]
Egg/larvae size (culture density)	Larger	Smaller[a]
Waste production (NH_3)	Moderate	Negligible[a]
Clutch size	1500 per female	2500 per female[a]
Transgenic capacity	Moderate to good	Good[a]
Time to sexual maturity	1.5–2 years	4–6 months[a]
Consistency in long-term developmental kinetics	Moderate	Relatively tight[a]
Capacity to establish test battery with longer-term assay (including reproductive and life cycle assessments)	Low to moderate	Relatively high[a]
Chromosomes/genome size	36/3.1 × 10^9 bp	20/1.7 × 10^9 bp[a]
Disease susceptibility	Low to moderate[a]	Moderate
Capacity to develop inbred lines	Little	Great[a]
Literature available	Large database[a]	Emerging database

[a] Represent a significant advantage as a test species.

be accommodated in each replicated treatment, thus offering a density advantage over *X. laevis*. *Xenopus laevis* produce a greater quantity of waste products during the assay than *X. tropicalis* (*ca.* 1.5 mg/l NH_3-N versus <0.5 mg/l NH_3-N).[24] Increased clutch size also creates a significant production advantage for the FETAX model. One of the most exciting advantages is the capacity to create transgenic lines of *X. tropicalis* and the use of cDNA microarray technology. In addition to the aforementioned anticipated advantages of *X. tropicalis*, the shorter life cycle allows more expedient and practicable reproductive toxicity studies and endocrine evaluation of the thyroid axis during metamorphosis. It is also possible to create inbred strains of *X. tropicalis*, which will likely decrease the genetic variability observed in *X. laevis*. Developing inbred lines of *X. laevis* is difficult at best. In the present chapter, we describe a modified FETAX method using *X. tropicalis* and present results of studies comparing the sensitivity and practicability of using *X. tropicalis* in this developmental toxicity model system.

Materials and methods

Equipment and supplies

The equipment, chemicals, and consumable supplies required for performing FETAX are provided in Table 3.2.

Reagents

The following solutions were used to perform FETAX. Reagents that were used in solutions are presented in Table 3.2.

1. *FETAX solution*: A reconstituted water medium[4] was used as a negative control and as a diluent for preparing aqueous solutions of test chemicals by dissolving

Table 3.2 Equipment, chemicals, and supplies used to perform FETAX

Item	Model/CAS number	Source	Catalog number	Chemical purity (%)
Incubator, 23 ± 0.5°C	307C	Fisher	11-679-25C	—
Ohaus balance	E10640	Fisher	02-112-4	—
Zoom microscope	Stereomaster	Fisher	12-562-1	—
Denver pH meter	9340.1	Fisher	02-226-83	—
YSI DO meter	52	Fisher	13-298-00	—
Fume hood	47	Fisher	16-500-10	—
Corning stirrer/hot plate	PC-620	Fisher	11-497-8A	—
Nikon digital camera	Coolpix 5000	B&H Photo	NICP5000J	—
Rena submersible heater	RH100	Aquatic Eco	RH100	—
Sodium chloride	7647-14-5	Sigma	S9888	99.0
Sodium bicarbonate	144-55-8	Sigma	S8875	99.0
Magnesium sulfate	7487-88-9	Sigma	M7506	99.0
Calcium sulfate dihydrate	10101-41-4	Sigma	C3771	99.0
Potassium chloride	7447-40-7	Sigma	P3911	99.0
Calcium chloride dihydrate	10035-04-8	Sigma	C5080	96.0
Chorionic gonadotropin	9002-61-3	Sigma	CG5	—
L-Cysteine	52-90-4	Sigma	C7352	98.0
Streptomycin sulfate	3810-74-0	Sigma	S6501	98.0
Penicillin-G	69-57-8	Sigma	PENNA	—
Cyclophosphamide	6055-19-2	Sigma	C0768	98.0
Acetic hydrazide	1068-57-1	Sigma	A3031	90.0
6-AN	329-89-5	Sigma	A0630	99.0
NADP, B-	1184-16-3	Sigma	N0505	99.0
NADPH, B-	2646-71-1	Sigma	N1630	95.0
Glucose-6-phosphate monopotassium salt	103192-55-8	Sigma	G6526	99.0
Glucose-6-phosphate dehydrogenase	9001-40-5	Sigma	G7878	—
DMSO	67-68-5	Sigma	154938	99.9
3-Aminobenzoic acid ethyl ester (MS-222)	886-86-2	Sigma	A5040	—
Sodium hydrosulfite	7775-14-6	Sigma	S1256	85.3
Formaldehyde solution, 37% (w/w)	50-00-0	Sigma	F1635	95.0
Petri dishes, 60 mm	—	Fisher	08-757-13A	—
Petri dishes, 100 mm	—	Fisher	08-757-12	—
Transfer pipettes, 3 ml	—	Fisher	13-711-7	—
Tuberculin syringes, 1 ml w/26 g $\frac{3}{4}$ needles	—	Fisher	14-823-2E	—
Syringes, 5 ml	—	Fisher	14-823-35	—
Needles, 18 g $1\frac{1}{2}$	—	Fisher	14-826-5D	—

0.625 g NaCl, 0.096 g $NaHCO_3$, 0.075 g $MgSO_4$, 0.06 g $CaSO_4{\cdot}2\ H_2O$, 0.03 g KCl, and 0.02 g $CaCl_2{\cdot}2\ H_2O$ per liter deionized (DI) water.

2. *1% Dimethyl sulfoxide* (DMSO) (v/v) *in FETAX solution* water was used as a negative control and as a diluent for preparing aqueous solutions of test chem-

Table 3.3 FETAX culturing conditions for *X. tropicalis*

Parameter	Condition
Test type	Static renewal
Temperature	26.5 ± 0.5°C
Light quality	Ambient laboratory illumination
Light intensity	10–20 uE/m^2/s (50–100 ft-c)
Photoperiod	12 h light; 12 h dark
Test chamber size	25 ml
Test solution volume	10 ml per replicate
Renewal of test solutions	Daily
Age of test organisms	Small cell blastulae (stages 8 and 9)
Replicates per treatment	4 per negative control; 2 per positive control (reference toxicants); 2 per test substance
Test organisms per chamber	20
Test organisms per treatment	80 per negative control; 40 per positive control (reference toxicants); 40 per test substance
Test vessel randomization	Randomization of tray placement in incubator
Feeding regime	None, prefeeding stage
Cleaning	Siphoned daily, prior to treatment renewal
Aeration	None
Dilution water	FETAX solution prepared using E-PURE® deionized water and reagent grade chemicals
Test duration	Developmental stage 46 (*ca.* 48 h)
Endpoints	Survival, malformation, growth (length)
Test acceptability	Control mortality/malformation rates ≤10%

icals that are not readily dissolvable in water. Ten milliliters of DMSO were added to FETAX solution water and diluted to 1 l.

3. *6-Aminonicotinamide* (6-AN)[2] reference toxicant stock solution (at 2500 mg/l) was used as a positive control to induce mortality. The stock was prepared by dissolving 2.5 g 6-AN per liter FETAX solution water and storing at room temperature to prevent the formation of crystals.
4. *6-Aminonicotinamide* (6-AN)[2] reference toxicant stock solution (at 5.5 mg/l) was used as a positive control to induce effect (malformations). The stock was prepared by adding 2.2 ml of 2500 mg/l 6-AN stock solution to FETAX solution water to make 1 l of stock solution. The solution was stored at 4.0°C.
5. *Human chorionic gonadotropin* (hCG) *stock A solution* (1000 units/ml) was prepared by injecting 5.0 ml of 0.9% saline solution into a 5000 unit vial of hCG, using a 10-cc syringe and shaking until dissolved. The hCG stock A was used to artificially induce maturation and amplexus.
6. *Human chorionic gonadotropin* (hCG) *stock B solution* (150 units/ml) was prepared by adding 0.15 ml of hCG stock A solution to 0.85 ml of 0.9% saline solution using a 1.0-cc syringe and shaking until mixed. The hCG stock B was used to artificially induce maturation and amplexus.
7. L-*Cysteine solution* (2.0%) was prepared by dissolving 2.0 g of the chemical per 100 ml of FETAX solution water and used to remove jelly coat from fertilized eggs.
8. *3-Aminobenzoic acid ethyl ester* (MS-222) stock solution (at 100 mg/l) was prepared by dissolving 0.1 g MS-222 in FETAX solution water and diluted to 1 l. MS-222 was used to anesthetize test specimens at test takedown.

9. *Formalin solution* (3.0%) was used to preserve test specimens indefinitely. Formalin solution was prepared by mixing 40.5 ml of formaldehyde solution (37.0%) in 900 ml of FETAX solution water. The pH was adjusted to 7.0 using 0.1 *N* hydrochloric acid (HCl) solution. FETAX solution water was used to bring the reagent volume to 1 l. The formalin reagent was stored at room temperature.
10. *Penicillin–streptomycin stock solution* was used as an antibiotic to prevent bacterial growth and is prepared by mixing 10,000 units/ml penicillin G and 10,000 units/ml streptomycin sulfate into 100 ml FETAX solution water. Solution was stored at 4.0°C.
11. *FETAX antibiotics* (FAB) *solution* water was used in place of FETAX solution water in activated (MAS) FETAX. Ten milliliters of penicillin and streptomycin stock solution was added to FETAX solution water and diluted to 1 l. The solution was stored at 4.0°C. In the event a carrier solvent (1% v/v DMSO) was used as a diluent of the test material, a FAB and 1% v/v DMSO should be prepared and stored at 4.0°C.
12. *Acetic hydrazide* (AH) *stock solution* (at 3000 mg/l) was used as a reference toxicant by dissolving 3.0 g AH in FAB solution water and diluting to 1 l. The AH solution was stored at 4.0°C.
13. *Cyclophosphamide* (CP) *stock solution* (at 4000 mg/l) was used as a reference toxicant by dissolving 4.0 g CP in FAB solution water and diluting to 1 l. The stock solution was stored at 4.0°C.
14. *Metabolic activation system* (MAS): Rat liver microsomes. Microsomes were stored in freezer.
15. *MAS NADPH generator*: Dissolved 16.8 mg NADPH, 264 mg NADP, and 3.7 g glucose-6-phosphate in 33.6 ml FAB solution water and aliquots were separated into three scintillation vials. The generator was labeled and stored in freezer.
16. *MAS NADPH dehydrogenase*: Injected 5.0 ml of FAB solution water into a 500-unit vial of glucose-6-phosphate dehydrogenase, swirling to dissolve. The MAS enzyme was transferred to a scintillation vial, labeled, and stored in freezer.

Test system

Animal husbandry, breeding, and embryo collection were performed as described in American Society for Testing and Materials (ASTM) E1439-98.[2] *Xenopus tropicalis* embryos were harvested from adult breeding stock originally obtained from *Xenopus* I (Dexter, MI). Guidelines provided by Nieuwkoop and Faber[25] were used to stage the embryos. Both Nieuwkoop and Faber's[25] guidelines and the *Atlas of Abnormalities*[3] were used to help determine normal embryo and cleavage patterns.

Adult husbandry

Sexually mature *X. tropicalis* were separated by sex and housed in 50-l plastic tubs (6–8 animals per tub) containing 20 l of aged (dechlorinated) tap water. Temperature in each tub was maintained at 26 $\pm$ 1°C with the use of submersible heaters. The frogs were fed Salmon Starter (#4 pellet) from Zeigler Brothers (Gardners, PA), three times per week (Monday, Wednesday, and Friday), 3–4 h prior to cleaning the tubs. Feeding amounts varied between tubs based on number, size, and sex of the frogs. It was estimated that females received 40–50 mg and males received 25–30 mg per animal per feeding event.

Adult breeding

Adult female and male frogs were chosen based on their readiness for breeding. This was determined by physical appearances. Indicators used to determine female readiness were the appearance of full ovaries (sides of the abdomen appeared swollen from dorsal view) and protruding cloaca (small tail-like papillae located at the posterior end). Male frogs chosen for breeding possessed darkened nuptial pads on the palms of their forelimbs and/or light mating strips on the underside of their forelimbs.

Each male and female frog chosen for breeding was separated and administered a priming dose of 0.1 ml of hCG stock B (solution 6) subcutaneously in the dorsal lymph sac. Approximately 48 h later, immediately prior to introduction to the breeding chambers, each frog was given a booster injection of 0.1 ml of hCG stock A (solution 5). Each male and female was then paired and placed in 5 l breeding chambers containing approximately 3 l of FETAX solution (solution 1). Aquarium pumps were attached to each chamber to aerate the water. Each chamber was then covered and placed in a quiet, dark location overnight.

Embryo collection

Xenopus tropicalis embryos were collected by carefully pouring off as much water from the breeding chamber as possible, without losing eggs, and adding 500 ml of 2% (w/v) L-cysteine solution (pH 8.1) (solution 7). The eggs were gently swirled in the solution for 2–3 min to remove the jelly coat. The cysteine solution was then poured off and the remaining eggs were immediately rinsed with FETAX solution water until there was no cysteine odor (5–6 times). The de-jellied embryos were then divided into 100 mm Petri dishes containing FETAX solution water using 3-ml transfer pipettes. Unfertilized and necrotic eggs were discarded, leaving only normal-appearing fertilized embryos.[25] When developing embryos reached mid-to-late blastula stage (stages 8 and 9), they were transferred to 60-mm Petri dishes containing the test material.

Test substances

FETAX is designed to identify toxicity in environmental samples, both aqueous and solid phase, and aqueous solutions of chemical compounds.[2] Before initiating FETAX, environmental samples were analytically characterized by measuring pH, dissolved oxygen, conductivity, hardness, alkalinity, ammonia-nitrogen, and residual chlorine (residual oxidants). Only dissolved oxygen was adjusted by aeration if below 4.0 mg/l. Stock solutions made from chemical compounds were analyzed for pH and dissolved oxygen and adjusted if pH was outside the 6.0–9.0 range and dissolved oxygen was below 4.0 mg/l. Temperature, pH, and dissolved oxygen were monitored daily throughout the test.

Test design

The culture conditions used in the FETAX adapted for *X. tropicalis* are described in Table 3.3.

FETAX overview

The basic experimental designs may be considered to be the screen, the range finding or limit test, and the definitive concentration response assay. The FETAX screen was used to test material at 100% concentration, such as environmental samples. The FETAX range test was used to estimate the 96-h LC_{50} and EC_{50} and help define the definitive test concentrations. The FETAX definitive test was used to test materials at several concentrations to more precisely determine an LC_{50} and an EC_{50}. These materials included environmental samples and/or chemical compounds. The definitive assay was also modified to evaluate developmental toxicity for human health hazard assessment by incorporating an exogenous MAS. The MAS was composed of microsomes induced from rat livers and a nicotinamide adenine dinucleotide generator system, which simulated mammalian metabolism.

Microsome preparation

Rat liver microsomes were prepared using post-isolation mixes of phenobarbital-, β-naphthoflavone-, and isoniazid-induced preparations.[13] Microsomal P-450 activity was inferred by measuring the *N*-demethylation of aminopyrine (phenobarbital and β-naphthoflavone preparations) or *N*-nitrosodimethylamine (isoniazid preparation) to formaldehyde.[26] Post-isolation mixed microsome sets were prepared by mixing equal activities of each of the three differently induced microsomes. Post-isolation mixed MAS preparations have performed the best in FETAX validation studies and represent a broad spectrum of MFO activity.[27,28] Protein was determined by the BioRad method (BioRad, Richmond, CA).[29] Specific aliquots of the mixed microsomes were pretreated with carbon monoxide (CO) to inhibit MFO activity. Inhibition with CO was achieved by chemically reducing designated aliquots with dithionite and subsequent bubbling of CO gas through the reconstituted microsomes for 3 min.[13]

FETAX screen procedure for aqueous solutions

Sixty-millimeter Petri dishes were set up to test each treatment in duplicate (additional replicates may be tested, if desired). Four replicates of FETAX solution (solution 1) were used as the control. Also, four additional control replicates of 1% (v/v) DMSO in FETAX solution (solution 2) were used if test treatments were insoluble in water. Two replicates each of 5.5 mg/l and 2500 mg/l of 6-AN (solutions 3 and 4) were set as reference toxicants. Blastula stage embryos (*ca*. stages 8 and 9) were then added to each test dish containing treatments and controls. Loaded Petri dishes were then placed in an incubator with temperature maintained at 26.5 $\pm$ 0.5°C. The test was renewed daily and mortality and stage data were collected. When at least 90% of the control larvae reached stage 46, the test was terminated by anesthetizing the specimens in 100 mg/l MS-222 (solution 8) and fixing in 3% (v/v) formalin (solution 9). Data endpoints collected from the FETAX screen were percent mortality, percent malformation, and statistical differences in growth compared to the control.

FETAX range and definitive procedures for aqueous solutions

The FETAX range and definitive tests were performed in a manner similar to that performed for the screen, with only the treatment setup being different. Instead of screening different samples at concentrations of 100%, treatments in the range and definitive tests consisted of one sample or chemical tested at several concentrations. Data collected from the FETAX range test were used to better define concentrations selected for the FETAX definitive test. Data endpoints collected from the FETAX definitive test were percent mortality, percent malformation, LC_{50}, EC_{50}, and minimum concentration to inhibit growth (MCIG).

FETAX definitive procedure with MAS for aqueous solutions

The FETAX definitive MAS test consisted of two current definitive tests, one without MAS (unactivated) and one with MAS (activated). Enough 60-mm Petri dishes and stages 8 and 9 embryos were set up to test each of the following two sets:

1. Unactivated (without MAS) set:
   - Four replicates of FETAX solution as a control (solution 1)
   - Four replicates of FAB solution water without MAS (solution 11)
   - Four replicates of 1% (v/v) DMSO in FAB solution water (if required) (solution 11)
   - Two replicates of 6-AN solution at 5.5 mg/l concentration (solution 3)
   - Two replicates of 6-AN solution at 2500 mg/l concentration (solution 4)
   - Two or more replicates of each sample concentration to be tested
2. Activated (with MAS) set:
   - Four replicates of FAB solution water with MAS or 1% (v/v) DMSO in FAB solution water with MAS (if required) (solutions 11, 14–16)
   - Two replicates of 3000 mg/l AH with the CO-treated MAS and the MAS generator system (solutions 12, 14–16)
   - Two replicates of 4000 mg/l CP with the CO-treated MAS and the MAS generator system (solutions 13–16)
   - Two replicates of 3000 mg/l AH with the MAS and NADPH generator system (solutions 12, 14–16)
   - Two replicates of 4000 mg/l CP with the MAS and NADPH generator system (solutions 13–16)
   - Two or more replicates of each sample concentration to be tested

Twenty milliliters (10 ml per replicate) of each test treatment was placed in its own Erlenmeyer flask. Forty milliliters of reagent was used for the controls with four replicates. Microsomes (0.2 ml per 20 ml of solution) (solution 14) were added to each flask of the activated (MAS) set only, with the exception of unactivated AH and CP. Before adding microsomes to AH and CP, the microsomes inactivated were first treated by adding 5–6 granules of sodium hydrosulfite to *ca.* 0.5 ml microsomes and then bubbling CO gas through the microsomes for 3 min at a rate of one bubble every 1–2 s. MAS NADPH generator (192 μl) (solution 15) and 63 μl of MAS NADPH dehydrogenase (solution 16) were added to each activated (MAS) flask and the contents mixed. Ten milliliters aliquots of solution from each flask were transferred to appropriately labeled Petri dishes. Blastula stage embryos (*ca.* stages 8 and 9) were then added to each test dish and dishes were placed in an incubator with temperature maintained at 26.5 ± 0.5°C.

FETAX scoring

After FETAX test takedown, larval specimens were anesthetized using a 100-mg/l solution of MS-222 (solution 8) and then preserved in 3% (v/v) formalin (solution 9). Specimens were then observed using a dissecting microscope and scored. Scoring data included total number of specimens per treatment, number of survivors per treatment, number of survivors per treatment with malformations, and a specific description of the malformations induced.

FETAX digitizing

After preservation, the larvae in each Petri dish were photographed using a digital camera and then electronically digitized using a personal computer and SigmaScan® Pro digitizing software (SPSS, Chicago, IL). Whole body length measurements, from snout to tail, were made of each specimen from each treatment.

Statistical analysis

Trimmed Spearman–Karber analysis was used to determine the median lethal concentration (LC_{50}), the EC_{50} (the concentration inducing gross terata in 50% of the surviving larvae), and their respective 95% fiducial intervals. A teratogenic index (TI) value was calculated by dividing the LC_{50} by the EC_{50} (malformation). The TI value represents the separation in the lethal and malformation-inducing concentration–response curves. Increasing TI values typically indicate increasing teratogenic potential.[5,21] Head-to-tail length was measured as an indicator of embryo growth using SigmaScan® Pro digitizing software (SPSS, Corte Madera, CA) and a personal computer. Concentrations inducing growth inhibition (MCIG) were calculated by ANOVA or ranked ANOVA [Bonferroni *t*-test, $P < 0.05$ (parametric data sets) or Dunn's test, $P < 0.05$ (nonparametric data sets)] using SPSS.

Results

Results of concurrent FETAX tests using *X. laevis* and *X. tropicalis* are presented in Table 3.4.

Control results

In 24 separate tests performed with *X. tropicalis*, the control mortality and malformation was 3.2 ± 0.8% and 4.8 ± 0.7%, respectively (present chapter and Reference 24). In either species, the incidences of mortality and malformation in the 5.5 mg/l 6-AN positive control treatment ranged from 0% to 15% and 47.5% to 82.7%, respectively. In either species at both culture temperatures, the incidences of mortality and malformation for the 2500 mg/l 6-AN positive control treatment ranged from 50% to 75% and 100%,

Table 3.4 Comparative sensitivity of *X. laevis* and *X. tropicalis* in FETAX model

Test substance [CAS No.]	Species	Temperature (°C) (± 0.5)	Exogenous MAS[a]	Unit	LC_{50}[b] (CI)	EC_{50}[c] (CI)	TI[d] (CI)	MCIG[e] (% LC_{50})	Reference
Acetylhydrazide [1068-57-1]	*X. laevis*	23.5	No	mg/l	9850.0 (9800.0–10150.0)	53.5 (48.2–58.8)	184.1 (164.0–204.2)	50.0 (50.0)	
	X. tropicalis	26.5	No	mg/l	8325.0 (8150.0–8500.0)	71.1 (64.1–78.2)	117.1 (102.0–132.2)	0.1 (9.1)	
	X. laevis	23.5	Yes	mg/l	8325.0 (8000.0–8650.0)	72.5 (64.0–81.0)	115.2 (102.1–128.3)	40.0 (0.5)	
	X. tropicalis	26.5	Yes	mg/l	4050.0 (3850.0–4250.0)	84.7 (80.5–88.9)	47.8 (33.5–56.5)	35.0 (0.9)	
6-AN [329-89-5]	*X. laevis*	23.5	No	mg/l	2264.6 (2249.1–2285.1)	4.7 (6.0–7.5)	481.8	100.0 (4.4)	24
	X. tropicalis	26.5	No	mg/l	2315.6 (2205.5–2420.7)	8.4 (8.1–8.7)	275.7	100.0 (4.3)	
Atrazine [1912-24-9]	*X. laevis*	23.5	No	mg/l	24.5 (19.3–31.1)	4.1 (3.0–5.7)	6.0	5.0 (20.4)	24
	X. tropicalis	26.5	No	mg/l	31.8 (29.8–33.8)	7.2 (5.2–9.2)	4.4	5.0 (15.7)	
Copper [7758-99-8]	*X. laevis*	23.5	No	mg/l	0.4 (0.3–0.4)	0.3 (0.2–0.5)	1.3	0.2 (50.0)	24
	X. tropicalis	26.5	No	mg/l	0.3 (0.2–0.3)	0.1 (0.1–0.2)	3.0	0.1 (33.3)	
Cyclophosphamide [6055-19-2]	*X. laevis*	23.5	No	mg/l	8150.0 (7500.0–8800.0)	6325.0 (6000.0–6650.0)	1.3 (1.0–1.6)	10,000.0 (>100.0)	
	X. tropicalis	26.5	No	mg/l	6350.0 (6100.0–6600.0)	5150.0 (5200.0–6450.0)	1.0 (0.9–1.1)	8000.0 (>100.0)	
	X. laevis	23.5	Yes	mg/l	1.1 (0.9–1.4)	0.4 (0.2–0.5)	2.8 (2.5–3.1)	50.0 (0.6)	
	X. tropicalis	26.5	Yes	mg/l	0.2 (0.1–0.3)	0.04 (0.02–0.06)	5.0 (4.0–6.0)	0.01 (5.0)	
Ethanol [64-17-5]	*X. laevis*	23.5	No	% (w/v)	1.0 (0.8–1.1)	0.7 (0.6–0.8)	1.4	1.0 (100.0)	24
	X. tropicalis	26.5	No	% (w/v)	1.6 (1.5–1.7)	1.0 (0.9–1.1)	1.6	1.3 (81.3)	
Methotrexate [60388-53-6]	*X. laevis*	23.5	No	mg/l	508.0 (475.0–543.0)	22.0 (21.0–24.0)	23.1 (20.1–26.1)	10.0 (2.0)	
	X. tropicalis	26.5	No	mg/l	322.5 (298.5–346.5)	39.3 (30.1–48.4)			
						8.2 (7.1–9.3) 5.0 (1.6)			
Semicarbazide [563-41-7]	*X. laevis*	23.5	No	mg/l	6956.0 (6642.0–7270.0)	9.5 (8.2–11.0)	734.2	7.5 (0.1)	24
	X. tropicalis	26.5	No	mg/l	7386.0 (7000.4–7768.6)	12.5 (11.4–13.8)	590.9	20.0 (0.3)	
Complex environmental mixture (Bermuda pond water-BARP)[7]	*X. laevis*	23.5	No	% (v/v)	32.4 (28.0–36.0)	10.2 (9.0–11.5)	3.2 (2.8–3.6)	6.1 (18.8)	
	X. tropicalis	26.5	No	% (v/v)	27.2 (23.1–32.3)	13.9 (10.3–17.5)	2.0 (1.5–25.)	12.5 (46.0)	

[a] Rat liver metabolic activation system.

[b] Median lethal concentration with respective 95% fiducial interval in parenthesis.

[c] Median teratogenic concentration with respective 95% fiducial interval in parenthesis.

[d] Teratogenic Index = LC_{50}/EC_{50}.

[e] Minimum concentration inhibiting growth (% of LC_{50} = $MCIG/LC_{50}$).

respectively. The performance results met the requirements established in ASTM E1439-98 and compared favorably to historical *X. laevis* tests.[2]

6-Aminonicotinamide

The stage 46 LC_{50}, EC_{50} (malformation), and TI values for 6-AN are presented in Table 3.4. The 6-AN[24] induced gut and cardiac miscoiling and abdominal edema in *X. tropicalis* at concentrations $\geq$10 mg/l. 6-AN induced notochord lesions, craniofacial mal-development, mouth abnormalities, rupture of the pigmented retina, and microcephaly at concentrations $\geq$10 mg/l in *X. tropicalis*. 6-AN induced gut and cardiac miscoiling, abnormal mouth development, and pericardial edema in *X. laevis* at concentrations $\geq$10 mg/l. 6-AN induced notochord lesions, craniofacial mal-development, abnormal pigmented retina formation, and microcephaly at concentrations $\geq$10 mg/l in *X. laevis*.

Ethanol

The stage 46 LC_{50}, EC_{50} (malformation), and TI values for ethanol are presented in Table 3.4. Ethanol[24] induced craniofacial dysmorphology, microphthalmia, and abnormal mouth development in *X. tropicalis* at concentrations $\geq$0.5% (w/v). Ethanol induced miscoiling of the gut, visceral hemorrhage and edema, and microcephaly at concentrations $\geq$1% in *X. tropicalis*. Ethanol induced abnormal gut development, craniofacial dysmorphogenesis, and mouth malformations in *X. laevis* at concentrations $\geq$0.5% (w/v). Ethanol induced abnormal myotome development, microphthalmia, microcephaly, and visceral hemorrhage and edema at concentrations $\geq$1% (w/v) in *X. laevis*.

Semicarbazide

The stage 46 LC_{50}, EC_{50} (malformation), and TI values for semicarbazide are presented in Table 3.4. Semicarbazide[26] induced notochord lesions, gut miscoiling, craniofacial and mouth mal-development, and microphthalmia at concentrations $\geq$10 mg/l in *X. tropicalis*. Semicarbazide induced other eye malformations, including rupture of the pigmented retina, at concentrations $\geq$100 mg/l in *X. tropicalis*. Semicarbazide induced gut malformations, craniofacial and mouth dysmorphogenesis, and notochord lesions in *X. laevis* at concentrations $\geq$5 mg/l. Semicarbazide induced eye malformations (ruptured pigmented retina and microphthalmia) and microcephaly at concentrations $\geq$100 mg/l in *X. laevis*.

Copper

The stage 46 LC_{50}, EC_{50} (malformation), and TI values for copper are presented in Table 3.4. Copper[24] induced gut miscoiling, craniofacial dysmorphogenesis, microcephaly, and mouth defects in *X. tropicalis* at concentrations $\geq$0.05 mg/l. Notochord lesions, abnormal coiling of the heart, cardiovascular malformations, microphthalmia, rupture of the pigmented retina, and abdominal edema at concentrations $\geq$0.1 mg/l in *X. tropicalis*. Copper induced gut miscoiling, abdominal and ophthalmic edema, craniofacial dysmorphogenesis, microphthalmia, rupture of the pigmented retina, and hypognathia in *X. laevis* at concentrations $\geq$0.05 mg/l. Microcephaly, hydrocephalus, and cardiovascular malformations were induced at concentrations $\geq$0.25 mg/l in *X. laevis*.

Atrazine

The stage 46 LC_{50}, EC_{50} (malformation), and TI values for atrazine are presented in Table 3.4. Atrazine[24] induced abnormal development of the mouth and craniofacial region in *X. tropicalis* at concentrations ≥1 mg/l. Atrazine induced notochord lesions, tail flexure, microphthalmia, rupture of the pigmented retina, and microcephaly at concentrations ≥10 mg/l in *X. tropicalis*. Atrazine induced notochord lesions, tail flexure, craniofacial defects, microphthalmia, lens and pigmented retina malformations, and abnormal development of the mouth in *X. laevis* at concentrations ≥5 mg/l. Microcephaly, visceral hemorrhage, and cardiovascular malformation were noted at concentrations ≥25 mg/l in *X. laevis*.

Methotrexate

The LC_{50}, EC_{50} (malformation), TI, and MCIG values for tests of methotrexate are presented in Table 3.4. Methotrexate induced gut and cardiac miscoiling and abdominal edema in *X. tropicalis* at concentrations ≥10 mg/l. Methotrexate induced notochord lesions, craniofacial mal-development, microophthalmia, and microcephaly at concentrations ≥10 mg/l in both species.

Cyclophosphamide

The LC_{50}, EC_{50} (malformation), TI, and MCIG values for tests of cyclophosphamide are presented in Table 3.4. Microophthalmia, mouth malformations, and gut miscoiling were induced in *X. tropicalis* as the result of exposure to ≥0.02 mg/l activated cyclophosphamide. In addition, concentrations ≥0.05 mg/l also induced rupturing of the pigmented retina, abnormal development of the lens, and microcephaly. These abnormalities were observed in markedly greater severity with increasing concentration of activated cyclophosphamide and were consistent with malformations induced in *X. laevis*.

Acetylhydrazide

The LC_{50}, EC_{50} (malformation), TI, and MCIG values for tests of acetylhydrazide are presented in Table 3.4. Gut miscoiling and notochord lesions were induced in *X. tropicalis* as the results of exposure to ≥20 mg/l acetylhydrazide. In addition, concentrations ≥150 mg/l produced abnormal development of the mouth, eye malformations, visceral edema, craniofacial defects, and abnormal heart development. These abnormalities dramatically increased in severity with increasing test concentration and were consistent with malformations induced in *X. laevis*.

Bermuda pond sediment extract (complex mixture)

The LC_{50}, EC_{50} (malformation), TI, and MCIG values for tests of pond sediment extracts from Bermuda are presented in Table 3.4. Microophthalmia, mouth malformations (including hypognathia), craniofacial defects, and cephalic malformations were induced in *X. tropicalis* as the results of exposure to these samples. These abnormalities dramatically increased in severity with increasing test concentration and were consistent with malformations induced in *X. laevis*.

Discussion

Results from these studies indicated that *X. laevis* and *X. tropicalis* generally responded similarly to the test compounds evaluated based on embryo lethality, the gross malformations observed, and embryo–larval growth. From a quantitative standpoint, some variability in the developmental endpoints was observed when comparing results for *X. tropicalis* to *X. laevis* results. These results were not necessarily unexpected. Since *X. tropicalis* is typically cultured at greater temperatures ($26.5 \pm 0.5°C$) than *X. laevis* ($23 \pm 0.5°C$), the rate of embryonic development is dependent on temperature, lethality in short-term alternative developmental toxicity test systems is influenced by exposure length, and malformation is specifically affected by exposure at critical periods of development, some endpoint variability was not unexpected. However, marked differences between the results in *X. tropicalis* and *X. laevis* were not found in this study. The magnitude of difference found is likely to be test compound dependent.

The teratogenic potential of chemicals assayed with FETAX has been routinely assessed based on TI values, growth endpoints, and types and severity of induced terata. In general, TI values <1.5 indicate low teratogenic potential as little or no separation exists between the concentrations that induce malformation without embryo lethality and those concentrations causing lethal effects. Thus, greater TI values represent a greater separation in malformation and lethal response curves and, thus, a greater potential for embryos to be malformed in the absence of lethality. Other endpoints, including types and severity of abnormalities and growth inhibition, are also considered in evaluating teratogenic hazard. Each of the compounds tested was determined to have teratogenic potential with varying degrees of potency. Regardless of species, the teratogenic potential (separation between lethal- and malformation-inducing concentration ranges) of the test materials evaluated in the present study was relatively similar: The teratogenic potency (concentration at which malformations are observed) and growth inhibiting potential (concentration that inhibit growth relative to the lethal concentration) were also similar regardless of species. The growth inhibiting potential tracked directly with the trends in teratogenic potential.[24] The same similarity in trends was identified during the validation of FETAX with *X. laevis*.[4–21,27,28,30,31,32]

In addition to the aforementioned anticipated advantages of *X. tropicalis*, the shorter life cycle allows more expedient and practicable reproductive toxicity studies and endocrine evaluation of the thyroid axis during metamorphosis. It is also possible to create inbred strains of *X. tropicalis*, which will likely decrease the genetic variability observed in *X. laevis*. Developing inbred lines of *X. laevis* is difficult at best. The only negative aspect of using *X. tropicalis* at this point is anecdotal evidence that this species may be more prone to diseases than *X. laevis*. More work will be required to determine the significance of the disease sensitivity issue.

Conclusions

In summary, results from the previous studies[23,24] suggested that *X. tropicalis* is an acceptable alternative species for *X. laevis* in FETAX. The revised alternative methods provided in this manuscript should enable interested investigators to conduct FETAX using *X. tropicalis*.

References

1. Dumont, J.N., Schultz, T.W., Buchanan, M. and Kai, G., Frog embryo teratogenesis assay—*Xenopus*—a short-term assay applicable to complex mixtures, in *Symposium on the Application of Short-term Bioassays in the Analysis of Complex Mixtures III*, Waters, M.D., Sandhu, S.S., Lewtas, J., Claxton, L., Chernoff, N. and Nesnow, S., Eds., Plenum Press, New York, 1983, pp. 393–405.
2. American Society for Testing and Materials, Standard guide for conducting the frog embryo teratogenesis assay—*Xenopus* (FETAX), in *Annual Book of ASTM Standards*, Vol. 11.05, ASTM, Philadelphia, PA, 1998, pp. 826–836.
3. Bantle, J.A., Dumont, J.N., Finch, R.A., Linder, G. and Fort, D.J., *Atlas of Abnormalities: A Guide for the Performance of FETAX*, 2nd ed., Oklahoma State University Press, Stillwater, OK, 1998.
4. Dawson, D.A. and Bantle, J.A., Development of a reconstituted water medium and initial validation of FETAX, *J. Appl. Toxicol.*, 7, 237–244, 1987.
5. Dawson, D.A., Fort, D.J., Newell, D.L. and Bantle, J.A., Developmental toxicity testing with FETAX: evaluation of five compounds, *Drug Chem. Toxicol.*, 12, 67–75, 1989.
6. Bantle, J.A., Burton, D.T., Dawson, D.A., Dumont, J.N., Finch, R.A., Fort, D.J., Linder, G., Rayburn, J.R., Buchwalter, D. and Maurice, M.A., Initial interlaboratory validation study of FETAX: phase I testing, *J. Appl. Toxicol.*, 14, 213–223, 1994.
7. Bantle, J.A., Fort, D.J., Rayburn, J.R., DeYoung, D.J. and Bush, S.J., Further validation of FETAX: evaluation of the developmental toxicity of five known mammalian teratogens and non-teratogens, *Drug Chem. Toxicol.*, 13, 267–282, 1990.
8. Bantle, J.A., Burton, D.T., Dawson, D.A., Dumont, J.N., Finch, R.A., Fort, D.J., Linder, G., Rayburn, J.R., Buchwalter, D. and Maurice, M.A., Initial interlaboratory validation study of FETAX: Phase I testing, *J. Appl. Toxicol.*, 14 (3), 212–223, 1994.
9. Bantle, J.A., Burton, D.T., Dawson, D.A., Dumont, J.A., Finch, R.A., Fort, D.J., Linder, G., Rayburn, J.R., Buchwalter, D., Gaudet Hull, A.M., Maurice, M.A. and Turley, S.D., FETAX interlaboratory validation study: phase II testing, *Environ. Toxicol. Chem.*, 13 (10), 1629–1637, 1994.
10. Bantle, J.A., Finch, R.A., Burton, D.T., Fort, D.J., Dawson, D.A., Linder, G., Rayburn, J.R., Hull, M., Kumsher-King, M., Gaudet-Hull, A.M. and Turley, S.D., FETAX interlaboratory validation study: Phase III part 1 testing, *J. Appl. Toxicol.*, 16, 517–528, 1996.
11. Fort, D.J., Stover, E.L., Bantle, J.A., Rayburn, J.R., Hull, M.A., Finch, R.A., Burton, D.T., Turley, S.D., Dawson, D.A., Linder, G., Buchwalter, D., Dumont, J.N., Kumsher-King, M. and Gaudet-Hull, A.M., Phase III interlaboratory study of FETAX, part 2: interlaboratory validation of an exogenous metabolic activation system for frog embryo teratogenesis as *Xenopus* (FETAX), *Drug Chem. Toxicol.*, 21 (1), 1–14, 1998.
12. Bantle, J.A., Finch, R.A., Fort, D.J., Stover, E.L., Hull, M., Kumsher-King, M. and Gaudet-Hull, A.M., Phase III interlaboratory study of FETAX, part 3: FETAX validation using compounds with and without an exogenous metabolic activation system, *J. Appl. Toxicol.*, 19 (6), 447–472, 1999.
13. Fort, D.J., Dawson, D.A. and Bantle, J.A., Development of a metabolic activation system for the frog embryo teratogenesis assay—*Xenopus* (FETAX), *Teratogen Carcinogen Mutagen*, 8, 251–263, 1988.
14. Fort, D.J., James, B.L. and Bantle, J.A., Evaluation of the developmental toxicity of five compounds with the frog embryo teratogenesis assay—*Xenopus* (FETAX) and a metabolic activation system, *J. Appl. Toxicol.*, 9, 377–388, 1989.
15. Fort, D.J. and Bantle, J.A., Analysis of the mechanism of isoniazid-induced developmental toxicity with frog embryo teratogenesis assay—*Xenopus* (FETAX), *Teratogen Carcinogen Mutagen*, 10, 463–476, 1990.
16. Fort, D.J., Rayburn, J.R. and Bantle, J.A., Evaluation of acetaminophen-induced developmental toxicity using FETAX, *Drug Chem. Toxicol.*, 15, 329–350, 1992.
17. Fort, D.J., Stover, E.L., Rayburn, J.R., Hull, M. and Bantle, J.A., Evaluation of the developmental toxicity of trichloroethylene and detoxification metabolites using *Xenopus*, *Teratogen Carcinogen Mutagen* 13, 35–45, 1993.

18. Fort, D.J., Propst, T.L. and Stover, E.L., Evaluation of the developmental toxicity of 4-bromobenzene using frog embryo teratogenesis assay—*Xenopus*: possible mechanisms of action, *Teratogen Carcinogen Mutagen*, 16, 307–315, 1996.
19. Fort, D.J., Stover, E.L., Propst, T., Hull, M.A. and Bantle, J.A., Evaluation of the developmental toxicity of theophylline, dimethyluric acid, and methylxanthine metabolites using *Xenopus*, *Drug Chem. Toxicol.*, 19, 267–278, 1996.
20. Propst, T.L., Fort, D.J., Stover, E.L., Schrock, B. and Bantle, J.A., Evaluation of the developmental toxicity of benzo(*a*)pyrene and 2-acetylaminofluorene using *Xenopus*: modes of biotransformation, *Drug Chem. Toxicol.*, 20, 45–61, 1997.
21. Fort, D.J., Stover, E.L., Farmer, D.R. and Lemen, J.K., Assessing the predictive validity of frog embryo teratogenesis assay—*Xenopus* (FETAX), *Teratogen Carcinogen Mutagen*, 20, 87–98, 2000.
22. Interagency Coordinating Committee on the Validation of Alternative Methods (ICCVAM), Minutes of the Expert Panel Meeting on the Frog Embryo Teratogenesis Assay—*Xenopus* (FETAX): A Proposed Screening Method for Identifying the Developmental Toxicity Potential of Chemicals and Environmental Samples, May 2000, pp. 1–4.
23. Song, M.O., Fort, D.J., McLaughlin, D.L., Rogers, R.L., Thomas, J.H., Buzzard, B.O., Noll, A.M. and Myers, N.K., Evaluation of *Xenopus tropicalis* as an alternative test organism for frog embryo teratogenesis assay—*Xenopus* (FETAX), *Drug Chem. Toxicol.*, 26 (3), 177–189, 2003.
24. Fort, D.J., Rogers, R.L., Thomas, J.H., Buzzard, B.O., Noll, A.M. and Spaulding, C.D., The comparative sensitivity of *Xenopus tropicalis* and *Xenopus laevis* as test species for the FETAX model *J. Appl. Toxicol.*
25. Nieuwkoop, P.D. and Faber, J., *Normal Tables of* Xenopus laevis (*Daudin*), Garland Publishing, New York, 1975.
26. Lucier, G., McDaniel, O., Burbaker, P. and Klien, R., Effects of methylmercury chloride on rat liver microsomal enzymes, *Chem. Biol. Interact.*, 4, 265–280, 1971.
27. Fort, D.J., Rogers, R.l., Stover, E.L. and Finch, R.A., Optimization of an exogenous metabolic activation system for FETAX: part II—post-isolation rat liver microsome mixtures, *Drug Chem. Toxicol.*, 24, 103–116, 2001.
28. Fort, D.J., Rogers, R.L., Paul, R.R., Stover, E.L. and Finch, R.A., Optimization of an exogenous metabolic activation system for FETAX: part II—post-isolation rat liver microsome mixtures, *Drug Chem. Toxicol.*, 24, 117–128, 2001.
29. Bradford, M.M., A rapid sensitive method for the quantification of microgram quantities of protein utilizing the principle of protein-dye binding, *Anal. Biochem.*, 72, 248–254, 1976.
30. Fort, D.J., Stover, E.L., Bantle, J.A. and Finch, R.A., Evaluation of the developmental toxicity of thalidomide using frog embryo teratogenesis assay—*Xenopus* (FETAX): biotransformation and detoxification, *Teratogen Carcinogen Mutagen*, 20, 35–47, 2000.
31. Dresser, T.H., Rivera, E.R., Hoffmann, F.J. and Finch, R.A., Teratogenic assessment of four solvents using the frog embryo teratogenesis assay—*Xenopus* (FETAX), *J. Appl. Toxicol.*, 12, 49–56, 1992.
32. Sunderman, F.W., Jr., Plowman, M.C. and Hopfer, S.M., Embryotoxicity and teratogenicity of cadmium chloride in *Xenopus laevis*, assayed by the FETAX procedure, *Ann. Clin. Lab. Sci.*, 21, 381–391, 1991.

chapter four

A short-term mummichog (Fundulus heteroclitus) *bioassay to assess endocrine responses to hormone-active compounds and mixtures*

Deborah L. MacLatchy, Karen L. Gormley, Rebecca E.M. Ibey, Rainie L. Sharpe, and Kevin S. Shaughnessy
University of New Brunswick
Simon C. Courtenay
Gulf Fisheries Centre, Fisheries and Oceans Canada
Monique G. Dubé
National Water Research Institute, Environment Canada
Glen J. Van Der Kraak
University of Guelph

Contents

1-56670-664-5/05/$0.00+$1.50
© 2005 by CRC Press

Introduction

There has been intense international interest in the development, validation, and standardization of laboratory fish tests to examine the responses of fish to endocrine disrupting substances (EDSs) and hormone-active effluents.[1–4] It is doubtful that a single surrogate species can be used to extrapolate results from laboratory testing to effects in wild species, as fish exposed to the same effluent source can vary in their endocrine and reproductive fitness responses.[5,6] Development of partial and full life cycle fish tests for EDS testing has primarily focussed on fathead minnow, *Pimephales promelas*[3,4] and other freshwater species, e.g., Japanese medaka (*Oryzias latipes*).[7] Less effort has been applied to estuarine and marine species, although efforts in sheepshead minnow[8] and eelpout[9] have yielded viable protocols. The application of full life cycle methodology for EDS testing can be costly and occurs in a timeframe not suited to testing purposes.[4] Therefore, there is a need to validate and characterize short-term, whole organism laboratory bioassays to identify responses suggestive of whole organism responses in the wild.

Extensive effort has been placed on the development of physiological tools, e.g., receptor,[10] vitellogenin,[11] plasma steroid,[12,13] and gonadal *in vitro*[14] assays. Additional advances in gene array[15] and molecular biology[15,16] techniques have also helped elucidate mechanisms of action, determine species differences, and define exposures. In total, studies measuring a variety of endpoints on numerous species have shown that although endocrine systems during vertebrate evolution have been relatively conserved,[17] significant species differences in receptor binding,[18] hormone-mediated responses,[6] and reproductive status[5] exist even among fish exposed to EDSs or hormone-active effluents.

To predict whole organism or population-level adverse health effects, exposure protocols must be designed to supply mechanism-specific information that provides opportunities to extrapolate across species and to use laboratory species as surrogates for wild fish. Here, we describe a short-term gonadal recrudescence bioassay for the estuarine killifish or mummichog (*Fundulus heteroclitus*), as well as the techniques used to assess changes in reproductive and endocrine status (including organ size, fecundity, plasma steroid and vitellogenin levels, and *in vitro* gonadal steroid production). The protocol has been validated with exposures to model compounds, representative of major routes by which EDSs exert their effects (anti/estrogens and anti/androgens).[2,19] The protocol has also been used to investigate causality at a Canadian pulp mill by determining the waste stream source and thereafter identifying hormone-active contaminants.[20–22] The mummichog is a good candidate for bioassay development as it is the numerically dominant fish species in salt marshes along the eastern coast of North America, and much is known about its reproductive biology in the wild[23,24] and the laboratory.[25,26] In addition, mummichog were chosen for bioassay development as their natural range, size, and adaptability make them suitable for laboratory, artificial stream, and field studies, thus allowing extrapolation across study designs and enhancing the application of results to "real world" contaminant effects.

Materials required

Most of the chemicals, materials, and equipment can be purchased from any scientific or aquatic systems suppliers. Specific catalog numbers are included where relevant to specific techniques. Hagen® materials can be purchased from most local pet stores. Where helpful, sample recipes for relevant volumes are included. Animal care protocols should be approved by institutional animal care committees prior to beginning.

Mummichog field collection

Equipment required:

- Fishing license
- Seine net (approximately 6 m × 2 m, 0.6 cm mesh; e.g., Aquatic Eco-Systems, Apopka, FL, USA, catalog #HDS4)
- Chest waders
- Life jackets
- Buckets (approximately 20 l capacity)
- Aquarium dip nets
- Battery powered aquarium aerators (e.g., Aquatic Eco-Systems DC5)
- Air tubing
- Air stones
- Large insulated transport containers [approximately 150 gal (700 l) capacity; e.g., Aquatic Eco-Systems B2300]
- Plastic or galvanized minnow traps (43 cm L × 23 cm W; e.g., Aquatic Eco-Systems MT2)
- Floating rope
- Dog food (generic, approximately 1.5 cm diameter; cat food is too small and turns to mush quickly)
- Film canisters (optional)

Animal husbandry

Equipment required:

- Dechlorinated fresh water
- Salt water (true salt water or artificial, e.g., Coralife Instant Ocean™)
- Fish food (standard commercial trout pellets; crushed variety)
- Hagen® Nutrafin® floating pellets for cichlids (Rolf C. Hagen Inc. International, Montreal, QC, Canada)
- Standard glass aquaria [e.g., 50 gal (190 l) minimum], larger stock tanks or water tables as available

- Multiparameter environmental meter capable of measuring dissolved oxygen (DO), temperature, pH, conductivity, and salinity (e.g., YSI 556 Multi-Probe System, Yellow Springs Instruments, Yellow Springs, OH, USA)
- Aquarium-size mechanical, chemical, and biological filtration filters (e.g., Hagen® AquaClear 500 System) for static systems
- Aquarium dip nets
- Air tubing
- Air stones
- Air supply (battery powered, standard electrical aquarium air pumps, or in-building system)
- Water quality test kits (e.g., Hagen® test kits)
- Water conditioner for biological systems (e.g., Hagen® Cycle™)

Artificial regression and recrudescence for year-round supply of fish

Equipment required:

- Standard glass aquaria [30–50 gal (114–190 l) capacity], larger stock tanks or water tables as available
- Filters (e.g., Hagen® AquaClear 200 suitable for 150 l aquaria)
- Flow-through or recirculation water chilling system for fresh and salt water suitable for holding tanks at 4°C or cold room set at 4°C for static system

Note: Chillers, pumps, and filters for chilling and/or recirculation systems are based on flow rates and volumes; please contact a supplier, such as Aquatic Eco-systems, to determine products suitable for your needs.

- Water quality test kits (e.g., Hagen® test kits)
- Multiparameter environmental meter
- Lights on timer device; either for whole room or in sectioned off area of room containing aquaria (various models are available at local hardware stores or from dealers, e.g., Intermatic, Energy Federation Incorporated, Westborough, MA, USA)
- Fish food, as described for animal husbandry

Short-term bioassay

Equipment required:

- Multiple glass aquaria [9 gal (34 l) minimum capacity]
- Dechlorinated fresh water
- Salt water
- Submersible fish tank filters (e.g., Hagen® Marina Jet Flow Corner Filter 10894)
- Filter media (e.g., Hagen® Poly Filter Wool A1031 and Living World Filter Box Carbon A1332)
- Large standard plastic garbage cans
- Air tubing
- Air stones
- Air supply (aerators or in-building)
- Dip nets

- Buckets [4–5 gal (16–20 l) capacity]
- Fish food (standard commercial trout pellets; crushed variety)
- Multiparameter environmental meter

Fish sampling

Equipment required:

- Bucket [approximately 2.5 gal (10 l) capacity]
- Air tubing
- Air stones
- Air supply (aerators or in-building)
- Latex or vinyl gloves
- Safety goggles
- Lab coat
- Aquarium dip nets
- Paper towel (cut into sizes approximately 2 to 3 × size of fish; 1-$\frac{1}{2}$ times the number of fish to be sampled)
- Needles (e.g., Beckton-Dickinson 26G3/8 or 25G5/8)
- Syringes (e.g., Beckton-Dickinson 1-cc syringe)
- Microfuge tubes (polypropylene, 1.5 ml)
- Microfuge tube racks (commercially purchased, or can be made easily by making suitable-size holes in sheets of Styrofoam)
- Borosilicate glass test tubes (12 mm × 75 mm)
- Test tube racks for 12 mm × 75 mm tubes
- Waterproof markers for labeling
- Ice
- Ice tray [dishpan or kitty litter tray or equivalent (46 × 76 cm)]
- Sharps container
- Refrigerated centrifuge (capable of at least 2500*g* and holding 1.5 ml Microfuge tubes; e.g., IEC Centra GP8R, with 316 rotor and 5827/5862 adaptor for 1.5 ml Microfuge tubes, Fisher Scientific, Nepean, ON, Canada, catalog #05-112-120 plus accessories; or IEC 3592 microcentrifuge, Fisher Scientific 05-112-114D)
- Freezer (−80°C)
- Electronic balance(s) (± 0.001 g sensitivity)
- Dissection kit (scalpel, forceps, ruler)
- pH meter
- Magnetic stir plate
- Stir bars

Reagents required:

- Anaesthetic: tricaine methanesulfonate (TMS)–water
  - TMS (Syndel International, Vancouver, BC, Canada, catalog #18323)
  - Salt water
  - Dechlorinated fresh water

Dissolve TMS in combined salt and fresh water at 0.05–0.1 g/l. *Note*: TMS in fresh water should be buffered but does not need to be if used in salt or partial salt water as described here. Please check the Material Safety Data Sheet (MSDS) for TMS, a potential carcinogen in humans.

- Heparin solution
  - 10 mg heparin sodium salt (Sigma-Aldrich H-0777, Sigma-Aldrich Corp., St. Louis, MO, USA)
  - 10 ml double-distilled water (ddH_2O)

Dissolve heparin in ddH_2O to obtain a concentration of 1 mg/ml; can be stored at 4°C for 2–3 days.

- NaCl solution (150 m*M*)
  - 0.876 g NaCl (Sigma-Aldrich A-9625)
  - 100 ml ddH_2O

Combine NaCl and ddH_2O; can be stored at 4°C for approximately 10–12 months.

- Aprotinin solution (1 KIU/μl)
  - NaCl solution (150 m*M*)
  - 1 mg aprotinin (Sigma-Aldrich A-1153; 4.3 TIU/vial where 1300 KIU = 1 TIU)

Add 1 mg aprotinin to 559 μl NaCl solution (150 m*M*) to produce a solution of 10 KIU aprotinin/μl. Store at 4°C. This solution is diluted further NaCl using (150 m*M*) to obtain an appropriate volume of 1 KIU aprotinin/μl. Store at 4°C.

- 1 m*M* hydrochloric acid (HCl) solution
- 1 m*M* sodium hydroxide (NaOH) solution
- Medium 199 buffer (M199)
- One bottle (final concentration 11.0 g/l) of M199 containing Hank's salts without bicarbonate (Sigma-Aldrich M-0393)
  - 6.0 g HEPES sodium salt (Sigma-Aldrich H-3784)
  - 0.35 g $NaHCO_3$
  - 0.1 g streptomycin sulfate (Sigma-Aldrich S-9137)
  - 1.0 g bovine serum albumin (BSA; Sigma-Aldrich A-7888)
  - 1l ddH_2O

Dissolve powdered M199 in 900 ml ddH_2O and stir gently with magnetic stir plate and stir bars until dissolved. Rinse original package with small amount of ddH_2O to remove all traces of powder. Add HEPES, $NaHCO_3$, and streptomycin and stir. Determine pH with a pH meter; adjust pH to 7.4 using either 1 m*M* HCl or 1 m*M* NaOH. Make up final volume to 1 l using ddH_2O. Add BSA when ready to use. M199 must be used within 24 h following addition of BSA. It can be stored for a limited time without BSA added (<1 week).

In vitro *incubations*

Equipment required:

- Glass Petri dishes (100 mm × 10 mm)
- Ice

- Ice tray [dishpan or kitty litter tray or equivalent (46 × 76 cm)]
- Scalpel
- Magnetic stir plate
- Stir bars
- Electronic balance (± 0.001 g sensitivity)
- 24-well culture plate (e.g., tissue culture treated non-pyrogenic polystyrene, Fisher Scientific C5003473) and/or 12 mm × 75 mm borosilicate glass test tubes
- Racks for 12 mm × 75 mm test tubes (if used)
- Waterproof markers for labeling
- Refrigerator (4°C)
- Repeater pipette (capable of dispensing 1000 μl
- 1000 μl pipette with disposable tips
- Incubator capable of holding at 18°C
- Scintillation vials (glass, 7 ml)
- Freezer (–20°C)

Reagents required:

- M199 buffer
- M199 + IBMX solution
  - 1 l M199 buffer
  - 0.222 g IBMX (3-isobutyl-1-methylxanthine; Sigma-Aldrich I-5879)

Add IBMX to M199 buffer and stir to dissolve using magnetic stir plate and stir bars.

- hCG stock solution
  - hCG (chorionic gonadotropin lyophilized powder; Sigma-Aldrich C-0434)
  - Ethanol (EtOH; Sigma-Aldrich E-7023)

Add hCG to EtOH to produce a 20 IU hCG per 5 μl EtOH solution.

Classification of follicle stage

Equipment required:

- Standard dissecting scope
- Dissecting forceps, blunt-pointed
- Glass Petri dishes (100 mm × 10 mm)

Reagents required:

- M199 or Bouin's solution (Sigma-Aldrich HT101128)

Bouin's contains picric acid; please check the MSDS.

- Plastic scintillation vials (20 ml, Fisher Scientific, 03 337 11B)

Plasma steroid extractions

Equipment required:

- Fume hood
- Borosilicate test tubes (16 mm × 150 mm)
- Racks for 16 mm × 150 mm test tubes
- Waterproof markers for labeling
- Glass scintillation vials (7 ml)
- Metal pan (approximately 20 cm wide × 30 cm long × 5 cm deep)
- pH meter
- Magnetic stir plate/hot plate combination
- Stir bars
- Pipette (50–200 μl tips)
- Repeater pipette capable of dispensing 500 μl
- Vortex mixer
- Freezer (–20°C)
- Thermos for liquid nitrogen (if this method of freezing is chosen)

Reagents required:

- Acetone (99.5% ACS reagent)
- Dry ice (approximately 1 kg)

Alternately, liquid nitrogen may be used instead of an acetone/dry ice bath.

- Diethyl ether
- ddH_2O
- 1 m*M* HCl
- 1 m*M* NaOH
- Phosgel buffer
  - 5.75 g Na_2HPO_4 (Sigma-Aldrich S9763)
  - 1.28 g $NaH_2PO_4 \cdot H_2O$ (Sigma-Aldrich S9638)
  - 1.0 g gelatin (Sigma-Aldrich G2500)
  - 0.1 g thimerosal (Sigma-Aldrich T5125)
  - 1 l ddH_2O

Dissolve Na_2HPO_4, $NaH_2PO_4 \cdot H_2O$, gelatin, and thimerosal in ddH_2O and heat to 45–50°C to dissolve the gelatin. Adjust pH to 7.6 (using 1 m*M* HCl or 1 m*M* NaOH) if required. Store at 4°C for up to 1 week.

Radioimmunoassay

Equipment required:

- Facilities licensed for radioactive use
- Borosilicate test tubes (12 mm × 75 mm) = assay tubes
- Assay tube rack for assay tubes (Fisher Scientific 14-809-22)

- Borosilicate test tubes (16 mm × 150 mm)
- Racks for 16 mm × 150 mm test tubes
- Waterproof markers for labeling
- Microfuge tubes (polypropylene, 1.5 ml)
- Test tube rack
- Pipette(s) and tips capable of 100, 200, and 1000 μl
- Vortex mixer (for test tube volumes)
- Refrigerated centrifuge (capable of holding at 2500*g* with buckets for multiple 12 mm × 75 mm glass test tubes; e.g., IEC Centra GP8R, with 316 rotor and 37-place 5737 adaptor; Fisher Scientific 05-112-120 plus accessories)
- Incubator capable of holding at 18°C
- Graduated cylinder (100 ml)
- Plastic disposable beakers with lids, 250 ml or 8 oz (e.g., Fisher Scientific 02-544-125)
- Fume hood
- Repeater pipette (5–200 μl)
- Paper towels or Kimwipes® (Fisher Scientific S47299)
- Plastic scintillation vials (7 ml, e.g., HDPE scintillation vials, Fisher Scientific 03-337-20)
- Scintillation counter (e.g., LS 6500, Beckman Coulter, Mississauga, ON, Canada) with counting racks for 7 ml scintillation vials
- pH meter
- Magnetic stir plate/hot plate combination
- Stir bars
- Electronic balance (± 0.001 g sensitivity)

Reagents required:

- Phosgel buffer (as described in the section "Plasma steroid extractions")
- Steroid standards
  - Estradiol (E_2; Sigma-Aldrich E8875)
  - Testosterone (T; Sigma-Aldrich T1500)
  - Ethanol (100%)

Both hormones are stored at a working concentration of 1000 ng/ml in ethanol (EtOH) at −20°C. Other hormones can be measured depending on commercial availability of standards, antibodies (Ab), and tracers.

- Steroid Ab (e.g., Medicorp, Montreal, QC, Canada or Steraloids, Newport, RI, USA)
  - E_2-Ab
  - T-Ab

Our Ab is received from Medicorp in 1 ml aliquots. Each aliquot is diluted in approximately 9.2 ml of phosgel (= stock dilution). This stock is then separated into 0.5 ml aliquots in 1.5 ml Microfuge tubes and stored at −20°C. The dilution required to provide 50% binding (= working dilution) is established by serial dilution of the stock followed by radioimmunoassay (RIA). The working dilution is specific to each shipment of Ab and must always be titered and assessed for binding prior to doing RIAs with samples. For 11-ketotestosterone, a steroid synthesized in male fish, Ab cannot be purchased from commercial suppliers. Details for Ab production are too detailed to describe here.

- Steroid tracers (Amersham International, Buckinghamshire, UK)
  - 2,4,6,7-^{3}H-17β-estradiol (E_2*) [product TRK322, specific activity (s.a.) = 3.22 TBq/mmol, 87 Ci/mmol]
  - 1,2,6,7-^{3}H-testosterone (T*) (product TRK921, s.a. = 3.52 TBq/mmol, 95 Ci/mmol)
  - Ethanol (100%)

Stock steroid tracers are diluted (100 μl of tracer in 10 ml of ethanol) and stored in glass scintillation vials at −20°C until used in the assays.

- Inter-assay (IA) samples

Samples of a known concentration of steroid (pooled plasma or standard) must be run concurrently with each assay. Pooled plasma is obtained by bleeding a number of fish, centrifuging blood, combining resultant plasma, and redistributing into aliquots for freezing at −20°C. Alternately, as described for production of standards, a specific concentration of steroid can be made in phosgel, aliquoted, and frozen at −20°C. It is important to make up a large number of IA samples (in the 100s) to make them available over a large number of assays.

- Scintillation cocktail (e.g., ScintiSafe Econo 1 Sigma-Aldrich SX20-5)
- Charcoal solution
  - 0.5 g activated charcoal (Sigma-Aldrich C4386)
  - 0.05 g dextran T70 (Sigma-Aldrich D1537)
  - 100 ml phosgel

Add charcoal and dextran to phosgel and stir continually on magnetic stir plate. Store at 4°C for 2–3 days.

- Ice
- Ice tray (as previously described)

Vitellogenin assay

Equipment required:

- 1 l bottles with lids (e.g., Fisher Scientific 06-414-1D)
- Amber bottle (at least 500 ml) (e.g., Fisher Scientific 06-423-2C)
- Graduated cylinders (10, 100, 1000 ml)
- 12 mm × 100 mm test tubes
- Electronic balance (± 0.001 g sensitivity)
- Repeater pipette
- 8-Channel attachment for repeater pipette and appropriate tips (25–125 μl)
- Solution reservoirs for multi-tip pipette

- 96-well EIA/RIA polystyrene assay plates (e.g., Corning 3590, Fisher Scientific 07-200-35)
- 100 ml glass beakers
- Microfuge tubes (polypropylene, 1.5 ml)
- Pipette for 2–20 μl and disposable tips
- Incubator capable of maintaining 37°C
- Plastic wrap (e.g., Saranwrap®)
- Shaker plate capable of 0–500 rpm (e.g., Gyrotory Shaker, Model GZ; New Brunswick Scientific Co. Edison, NJ, USA)
- Vortex mixer
- Spectrophotometer capable of reading 96-well plates at 490 nm
- Magnetic stir plate
- Stir bars

Reagents required:

- ddH_2O
- 5 *M* H_2SO_4
- HCl (1 m*M*)
- NaOH (1 m*M*)
- $NaHCO_3$ (baking soda is acceptable)
- Sodium bicarbonate buffer (SBB) solution
  - 4.20 g $NaHCO_3$
  - 5.0 mg gentamycin sulfate (Fisher Scientific BP918-1)
  - 1 l ddH_2O

Stir on magnetic stirrer until all reagents are dissolved. Adjust pH to 9.6 with HCl or NaOH; may be stored up to 1 year at 4°C.

- Tris buffered saline — Tween (TBS-T) 10× solution
- 12.1 g Tris–HCl (Tris[hydroxymethyl]aminomethane hydrochloride; Sigma-Aldrich T-3253)
- 87.7 g NaCl
- 10 ml Tween 20 (1%) (enzyme grade polyoxyethylene 20-sorbitan monolaurate; Fisher Scientific BP337-500)
- 50 mg gentamycin sulfate (Fisher Scientific BP918-1)
- 1 l ddH_2O

Stir on magnetic stirrer until all reagents are dissolved. Adjust pH to 7.5 (with HCl or NaOH); may be stored up to 1 year at 4°C.

- TBS-T working solution
  - 100 ml TBS-T 10 × solution
  - 900 ml ddH_2O

TBS-T 10× solution is diluted with ddH_2O to create TBS-T working solution. Stir on magnetic stirrer until all reagents are dissolved. Adjust pH to 7.5 as previously described; may be stored up to 1 year at 4°C.

- TBS-T-BSA solution
  - 100 ml TBS-T working solution
  - 0.5 g BSA 98% ELISA grade (Sigma-Aldrich A-7030)

Note: BSA must be ELISA grade. Stir on magnetic stirrer until all reagents are dissolved. Adjust pH to 7.5 and the solution may be stored up to 3 weeks at 4°C (watch for flocculents).

- Ammonium acetate solution
  - 0.385 g ammonium acetate 98% (Sigma-Aldrich A-7330)
  - 100 ml ddH_2O

Stir on magnetic stirrer until all reagents are dissolved. Combine ammonium acetate and ddH_2O.

- Citric acid solution
  - 0.525 g anhydrous citric acid (Sigma-Aldrich C-0759)
  - 50 ml ddH_2O

Stir on magnetic stirrer until all reagents are dissolved. Combine citric acid and ddH_2O.

- Ammonium acetate–citric acid (AACA) solution
  - 100 ml ammonium acetate solution
  - citric acid solution

Adjust the pH of the ammonium acetate solution to 5.0 with the citric acid solution, while stirring on magnetic stir plate. Store at 4°C for 4 months in amber bottle.

- *o*-1,2-phenylenediamine (OPD) solution
  - 16 ml of AACA solution
  - 8 μl of 30% hydrogen peroxide (Sigma-Aldrich H-1009)
  - 8 mg OPD (Sigma-Aldrich P-1526) powder

Mix well on magnetic stir plate in 100 ml beaker. This is enough for 1 plate. Prepare immediately before use. (*Do not* make it up beforehand and store it.)

- Vitellogenin stock solution

Vitellogenin protein for standards for our work was prepared by Nancy Denslow, University of Florida, FL, USA, according to the methodology described by Denslow et al.[11] Currently, there is no commercial supplier for mummichog vitellogenin standards. Nancy Denslow may be contracted to create vitellogenin standards through the Molecular Biomarkers Core Facility, associated with the Interdisciplinary Center for Biotechnology Research at the University of Florida. The facility may be accessed through: www.biotech.ufl.edu/MolecularBiomarkers. Store 50 μl aliquots at −80°C.

- Vitellogenin coating solution
  - Vitellogenin stock solution (concentration will vary with batch)
  - SBB

Vitellogenin must be added to SBB to coat the wall of the plate. We have found that 16.5 ng vitellogenin per well is a good concentration for this solution and 150 μl of vitellogenin coating solution is required per well. Therefore, a solution of 0.11 ng/μl vitellogenin is used. For a 96-well plate, 18.8 ml of vitellogenin coating solution is required. (For example, assuming 0.60 mg/ml vitellogenin stock solution, we need 3.45 μl of 0.60 mg/ml solution in 18.8 ml of SBB to create the coating solution for 1 plate.) Vortex vitellogenin aliquot prior to use. Mix well on a magnetic stir plate in a small beaker.

- Primary Ab specific for mummichog vitellogenin aliquot

Our primary Ab was developed by Dr. C. Rice at Clemson University as described by MacLatchy et al.[2] It is now possible to purchase monoclonal Ab specific for mummichog vitellogenin from Euromedex, a company based in France (http://www.euromedex.com/New_t.html#) and EnBio Tec Laboratories in Japan (http:/www.enbiotec.co.jp/en/product/index.html). Ab should be diluted with TBS-T-BSA to aliquots of 50 μl and stored at −80°C.

- Primary Ab solution
  - TBS-T-BSA
  - Primary Ab aliquot

Each batch of antiserum will contain different levels of Ab and must be diluted to a working concentration. A dilution factor will need to be determined for each individual batch of Ab. (For example, our laboratory uses a 4250× dilution factor. The Ab stock required for 1 plate is the total volume of Ab solution needed divided by the dilution factor, 2.36 μl of antiserum in 10.0 ml of TBS-T-BSA.) Vortex primary Ab aliquot prior to use. Mix solution on magnetic stir plate in a small beaker.

- Secondary Ab aliquot
  - Peroxidase conjugate — goat anti-mouse IgG (whole molecule, Sigma-Aldrich A-6154)
  - TBS-T-BSA

The Ab should be initially diluted to 5× with TBS-T-BSA and frozen at −80°C in aliquots of 100 μl until use.

- Secondary Ab solution
  - Secondary Ab aliquot
  - TBS-T-BSA

Each batch requires different dilutions; optimal standard curves have been achieved with dilutions in the range of 100 to 2000×. For 1 plate, 18 ml of secondary Ab solution is needed. For a 200× dilution, the secondary Ab aliquots (diluted 5× prior to storage) must be further diluted by 40×. For example, 450 μl of the secondary Ab aliquots are pipetted into 18 ml of TBS-T-BSA to achieve the desired dilution. Vortex secondary Ab aliquot prior to use. Mix solution on magnetic stir plate in a small beaker.

- Vitellogenin IA
  - Vitellogenin stock solution
  - TBS-T-BSA

Vitellogenin stock solution is added to TBS-T-BSA to create a known concentration of vitellogenin (e.g., 18.75 ng/50 μl), which is pipetted into aliquots of at least 100 μl and stored at −80°C.

Procedures and protocols

Mummichog field collection

The mummichog is a euryhaline estuarine fish species found along the Eastern North American coast from northern New Brunswick to northern Florida. They are most easily located in shallow areas within tidal creeks where the tide flows over eelgrass.[27] Mummichog complete their entire life cycle within the estuaries.[28,29] Reproductive males have yellow bellies and a dark spot towards the rear of the dorsal fin. Females are generally larger and rounder, with dark vertical stripes on their bodies.

Seine netting

1. Once an appropriate fishing location is identified, a fishing license must be obtained for the area. The time of day for collection should also be considered to optimize catch success: mummichog tend to move closer to shore during rising tide and away during falling tide. However, it is not uncommon to find mummichog at the shoreline throughout the entire tidal cycle.
2. An appropriate area is located within the estuary where it is safe to pull the seine net through the water with minimal obstructions. Mummichog prefer soft or muddy bottoms as opposed to rocky substrates.
3. The seine is stretched perpendicular to the shoreline by one person, while a second person holds the net close to shore (chest waders and life jackets are recommended). An adequate maximum depth at the furthest distance from the shoreline is approximately 1 m; however, mummichog can also be found in shallower waters.
4. The net is pulled through the water parallel to the shoreline. It is imperative that the lead line be kept along the substrate throughout the duration of the tow to avoid fish loss. Each person should move through the water at the same rate, while maintaining a "U-shape" in the net. An appropriate tow distance is approximately 20–30 m; however, this will vary depending on environment and fish abundance.
5. To complete the tow, the person closest to shore slows, while the other person swings toward the shoreline. The loop is closed and the net is dragged up onto the shore, pulling the fish within the net completely out of the water.
6. The fish are quickly removed from the net and placed in a bucket (either by hand or with a dip net). It may be necessary to have multiple buckets available depending on the catch size, as it is imperative not to overcrowd the fish. Battery-powered aerators should be used to prevent low levels of DO.

7. Fish are sorted by size and by species (by-catch should be released as soon as possible). Small mummichog of less than 60 mm standard length should be released as they are too small to be used in the bioassay. Suitable mummichog are transported to the laboratory in a large well-aerated transport container.

Minnow trapping

1. Individual minnow traps are attached to floating rope, and spaced by approximately 3 m. The user may attach any number of traps to reasonable lengths of rope. A single buoy should be attached to one end of the rope to aid in trap positioning and recovery.
2. Dog food may be used to draw mummichog to the minnow traps (food may be placed directly in each trap or enclosed within an empty film canister containing various small holes).
3. Traps are placed in the water at the desired sites at an approximate depth of 30–100 cm. Traps should be used during low tide to ensure that fluctuating tides do not result in traps being exposed to air. In addition, a length of rope should be attached to the shoreline (above the maximum tide level) to prevent loss of the traps and ease of retrieval.
4. Traps should remain undisturbed for an appropriate length of time relative to the abundance of fish in the area (1–24 h). *Note*: the minnow trapping method may be used at any time during the tide cycle; however, caution must be taken so that traps are not exposed to air with the changing tides.
5. The rope is pulled toward the shore to retrieve the traps. Each trap is removed from the water one at a time to minimize air exposure and the fish are emptied into an aerated bucket of water.
6. Fish are sorted and transported as previously described.

Animal husbandry

Mummichog are easily maintained in the laboratory. Fish may be held in large glass aquaria filtered with high-quality standard aquarium filters (static system), in large flow-through stock tanks or in a flow-through system in a water table. We have found that flow rates of about 200 ml/min provide adequate renewal of the water. It is recommended that filters also be installed on flow-through system to optimize fish health. Stocking densities should be as low as possible; we have found that a good benchmark for fish density is 20–30 g biomass/l of water, roughly 2–3 fish/l in a static system and 30–50 g biomass/l of water, roughly 3–5 fish/l in a flow-through system.

1. Fish are held at ambient light and temperature, 15–20 ppt salinity, and DO of greater than 85%. Water temperature, salinity, and DO should be monitored regularly using a multiparameter environmental meter. If any of the parameters are out of a normal range, a partial or complete water change should be done for static systems, or a partial water change with adjustment of the freshwater and saltwater flow rates for flow-through systems.
2. Filter components are regularly cleaned and/or replaced. When cleaning filters, changing water or establishing new tanks, a water conditioner for biological

systems should be used according to instructions provided. These practices will help maintain fish health.

3. Other water quality testing (ammonia, nitrates, nitrites, pH, etc.) should be carried out periodically using standard test kits. Abnormal levels in any of the parameters should be corrected with a partial water change and cleaning or replacement of the filter components.
4. Overall fish health should be observed daily and any symptoms of fish disease, such as abnormal behavior, changes in activity levels, presence of physical peculiarities (color, skin, and fin condition), or external parasites, should be noted. Any sick or dead fish should be removed immediately. Sick fish should be treated in an isolation tank with a treatment specific to their ailment and not returned to the stock tanks.
5. The fish should be fed daily with standard commercial crushed trout pellets at approximately 1–3% body weight, and this diet should be supplemented regularly (every 2–3 days) with cichlid pellets (1–3%). Food consumption should be monitored and the amount of food given should be adjusted accordingly. Uneaten food should be removed regularly along with any waste products.

Artificial regression and recrudescence for year-round supply of fish

It is well established that reproductive cycles in mummichog can be artificially manipulated to induce spawning.[2,26] During April–September, laboratory fish captured from the wild can be used as spawning cycles are maintained in laboratory on 2–4 week cycles. Bioassay exposures are optimal when fish are sampled a few days to a week prior to the full moon during recrudescence (period of gonadal maturation). Availability of pre-spawning fish year-round is essential if studies are to be done outside the natural spawning cycles. Fish can be artificially regressed and recrudesced by the following protocol.

1. Male and female mummichog are separated into tanks by sex. Separation of sexes during regression yields pre-spawning fish that have not initiated spawning behaviors and gamete release.
2. Fish can be kept in filtered aquaria with maximum density of 20–30 g biomass/l water. For 114-l aquaria, we use one or two AquaClear 200 filters. Filters should be cleaned as needed and water quality parameters measured as previously described in "Animal husbandry." Alternately, fish can be kept in flow-through systems. Salinity should be maintained at 16 ppt and DO $>$ 85%.
3. Once acclimated (usually 1 week), photoperiod and water temperature are adjusted to 8-h L:16-h D and 4°C over a period of a week. Water temperature can be maintained by placing static aquaria in a cold room. Flow-through systems can be maintained at 4°C by using readily available in-line commercial chillers (contact Aquatic Eco-systems or other suppliers for examples suited to particular water flow needs).
4. Fish should be fed to satiation daily (approximately 1% body weight/day). *Note*: fish feeding rates decrease in cold temperatures and they can be fed every other day depending on consumption.

5. After 8 weeks of cold exposure fish should be completely regressed. A good external indicator of regression is the fading of male spawning colors (yellow ventral surface).
6. Following 8 weeks of cold exposure, day length and temperature should be increased gradually to 16-h L:8-h D and 18–20°C, respectively, over a period of a week.
7. Gonadosomatic index (GSI) should be significantly increased by 2 weeks; hormone levels should be equivalent to pre-spawning fish in the wild or laboratory by 4 weeks[2]). Noticeable yellowing of male ventral surfaces at 2 weeks identifies progressing gonadal recrudescence.
8. Once fish have recrudesced, they can be used in the short-term bioassay as per the following descriptions.

Short-term bioassay

The following protocol is a description of a static exposure with daily renewal. However, flow-through exposures can be done using diluter systems.[3] These systems are costly but ensure constant exposure concentrations and minimize exposure to metabolic by-products and wastes. Flow-through systems are recommended when available.

1. Glass aquaria are filled with 16 ppt water (equal parts of de-chlorinated fresh water and salt water). Aquaria are aerated and filtered using corner filters during the acclimation period (at least 1 week).
2. Mummichog with a minimum standard length of 70 mm are selected, weighed (0.01 g), and randomly allocated to each aquarium (6–8 fish/12–16 l of 16 ppt water). Fish smaller than 70 mm may not provide adequate volumes of blood for steroid and vitellogenin analysis. Sex ratio should be 1:1 in all tanks. Fish can be kept at ambient photoperiod in spring/summer or on 16-h L:8-h D.
3. Fish are fed approximately 3% body weight/day with standard crushed commercial trout pellets and maintained under static conditions with a natural spring/summer photoperiod throughout the acclimation period.
4. Water quality parameters should be measured daily with the multiparameter environmental meter. DO should be >75%, temperature between 10°C and 20°C, and salinity 16 ppt. Filter media may be changed as necessary during the acclimation period.
5. Following acclimation, filters are removed and aquaria are randomly allocated to treatment groups (each group consisting of a minimum of four separate tanks). Treatments may include whole effluents (e.g., pulp mill effluent), model steroid compounds (e.g., ethynyl estradiol), or other compounds of interest.
   (a) If the chemicals used for the exposure are mixed in a solvent, such as methanol or ethanol, equivalent amounts of the solvent must be added to the control groups. We have done trials that show small solvent volumes (35 μl in 16 l $= 35 \times 10^{-6}$ l/16 l $= 2.2 \times 10^{-4}$% solvent) have no effect on the reproductive endpoints measured in this bioassay (unpublished data). An extra control group in which no solvents are administered may be added to the experimental design if it is necessary to determine the effects of the solvent.
6. Static water is completely replenished daily. Water should be allowed to come to ambient temperature in large plastic garbage cans prior to use. To safely perform

water changes, fish are removed with a dip net (one tank at a time, separate nets per treatment) and placed into a bucket containing the pre-determined volume of 16 ppt water (e.g., 16 l). Aquaria are then emptied, and fish are replaced in the aquaria with the new water.

7. Each tank receives the designated treatment following water renewal. It is recommended that the treatments be administered for each tank immediately following the water change in order to minimize the duration of time that the fish are unexposed. *Note*: when effluents are used in place of water during water changes (e.g., 100%, 50%, 1%, etc.), dilute and use as exchange water accordingly.
8. During the exposure period water quality measurements are taken (temperature, DO, conductivity, salinity), and fish are fed as described for acclimation period. Fish should be allowed at least 1 h for food consumption prior to water renewal.
9. Repeat steps 6–8 every 24 h for 7–15 days. Length of bioassay is dependent on potency of material being tested; we have found that 7-day exposures are adequate for demonstrating endocrine effects for many EDSs and pulp mill effluents.[2,19,21]
10. Throughout the exposure period, fish should be monitored daily for overall health status (e.g., presence or absence of parasites), abnormal behaviors, and physical appearance (color, skin, and fin condition). Any dead or sick fish should be removed immediately. Sick fish should be treated for any infections but must not be returned to the exposure tanks following removal.
11. On the final day, fish are sampled as described in the section "Fish sampling".

Fish sampling

1. Prior to sampling, needles and syringes must be heparinized (usually a day or more before sampling date). Heparin solution is drawn into and emptied from each assembled needle-and-syringe pair 2 to 3× (approximately 1 ml each draw). *Note*: no liquid is to be left in the syringe or needle, and the plunger should be used repeatedly to clear out any of the remaining liquid.
2. All required Microfuge tubes (for blood when sampling and later for plasma after centrifugation) should be prelabeled using a waterproof marker prior to the sampling day. These Microfuge tubes can be laid out in racks in order of sampling to enhance organization on the sampling day. As well, test tubes for holding gonads following dissections can be prelabeled and placed in racks.
3. Sampling with a group of people in an assembly line works best given the number of fish (e.g., 8 fish × 4 replicate tanks × 5 treatments = 160 fish) to be sampled. Preprinted index cards with spaces for all necessary information (treatment group, fish number, sex, length, total body weight, liver weight, and gonad weight) can be used and kept with each fish during sampling to avoid confusion. We will generally have two bleeders (one for males, one for females), as well as one or two dissectors for each sex. An additional person acting as "gopher" to get fish and run other errands is helpful to have on hand. We usually line up the two teams across from each other on opposite sides of a long table, with enough balances, dissecting tools, etc., for each person or for easy sharing. In step 2, we will generally separate tubes for male and female fish to organize the assembly lines.

4. Fish are sampled one tank at a time. Fish are anesthetized by placement in bucket(s) of 8 l of aerated 16 ppt water containing TMS–water. Gloves must be worn by all people in contact with TMS–water or fish placed in TMS–water. Lab coats and safety goggles are also recommended (refer to MSDS or equivalent information provided by supplier).
5. To simplify sampling we find it easiest to separate fish by sex at this time by placing female and male fish in different buckets. Three to four fish can be anesthetized at one time.
6. Fish are anesthetized until they are no longer swimming but opercular movement is still evident. Fish are removed one at a time for sampling.
7. Anesthetized fish are bled while held in a piece of paper towel (allows a firm grip on the fish). The fish is held ventral side up and the needle is inserted posterior to the pelvic fins and anus, until it hits the vertebrae (caudal puncture). The needle is just slightly withdrawn away from the vertebrae. Suction is then applied on the syringe plunger until blood begins to enter the needle and syringe; usually 100–300 μl of blood can be collected. Blood samples are then ejected, without the needle attached to prevent hemolysis, into a 1.5-ml Microfuge tube. To the Microfuge tube, the aprotinin solution (a protease inhibitor) must be added at this time at a concentration of 10 KIU/ml blood (e.g., 2 μl KIU solution/200 μl blood). Syringes and needles are disposed of into a biohazard/sharps container. Microfuge tubes are kept on ice until centrifugation. If kept cold, blood can be held prior to centrifuging for a few hours until all fish are sampled.
8. Fish are measured (standard length; rostrum to peduncle; ± 1 mm) and weighed (± 0.01 g).
9. Fish are euthanized by spinal severance with scissors or scalpel.
10. Fish are dissected; liver and gonadal tissue are removed and weighed separately (± 0.001 g). If multiple balances are used among dissectors, balances need to be calibrated prior to sampling. Gonadal tissue is placed into the prelabeled borosilicate glass test tubes containing 1 ml medium 199 buffer and kept on ice for *in vitro* assays.
11. Fish carcasses are disposed of properly according to local regulations.
12. Plasma is centrifuged (15–20 min at 2400*g* at 4°C) to separate the plasma from the red blood cells. Plasma is then transferred to new, clean, prelabeled 1.5 ml Microfuge tubes and stored at −80°C until analysis. We separate plasma into two Microfuge tubes at this point, one for subsequent steroid analysis and one for subsequent vitellogenin analysis. For steroids, freeze >50 μl plasma. For vitellogenin, freeze 10–20 μl.
13. Transfer data from index cards to an electronic spreadsheet. Columns for fish number, treatment, sex, weight, length, gonad weight, liver weight, GSI (gonad weight/body weight × 100), liver-somatic index (liver weight/body weight × 100), condition factor (weight/length^3 × 100), and comments are typical values recorded and calculated at this point.

In vitro *incubations*

The following method is based on the protocol of McMaster et al.[14]; included here are the directions applicable to mummichog. However, McMaster et al.[14] is a detailed protocol, and it is advised that it be reviewed for additional information regarding methodology for measuring *in vitro* gonadal steroid production in fish.

1. The protocol for gonadal *in vitro* incubations should be begun as soon as sampling is finished for all the fish, or even part-way through sampling if there are available people. Gondal tissue must be kept on ice until the incubations are started. Gonadal tissue is removed from one test tube at a time and placed into a glass Petri dish on ice containing M199. Tissue from each fish is individually processed. For testes, and ovaries with no distinct follicles easily separated by the naked eye, tissue is divided with a scalpel, and 18–25 mg of tissue is placed into individual wells each containing 1 ml M199 in a 24-well cell culture plate. Alternately, borosilicate glass tubes (12 mm × 75 mm) may be used for tissue incubations. If ovaries have prematurational follicles (<1.25 μm in diameter), the follicles should be separated and counted, and 10–15 follicles (consistent within experiment) placed in each well (24-well culture plates should be used if follicles are counted). Maturational follicles and mature eggs will probably also be present; however, steroid production of T and E_2 are generally low in these samples (unpublished data). The number of wells that can be processed per fish is dependent on the amount of gonadal tissue available, up to 6–8 wells/fish. The mass of the tissue (or number of follicles) and fish identification number must be recorded for each well. The tissue in each well should be in two pieces to maximize exposed surface area. Tissue should be kept on ice at all times. When the plate is full, store in a refrigerator at 4°C until step 2. Repeat for all gonadal samples.
2. When all gonadal tissue has been allocated, the media in each well is drawn off using a 1000-μl pipette. Each well then receives 1 ml of M199 + IBMX solution. We have shown that IBMX, a phosphodiesterase inhibitor that stimulates steroidogenesis by prolonging the presence of cAMP (3′,5′-cyclic adenosine monophosphate), enhances steroid production in mummichog incubations.[2] We use steroid production in the M199 + IBMX samples as our "basal" production levels. Within the 6–8 wells of gonadal tissue or follicles for each fish, half of the wells (3–4) are designated with this basal treatment label and require no further additions. The remaining wells are designated stimulated treatment and receive 5 μl hCG stock solution. HCG is an analog of gonadotropin hormone, and stimulates cAMP mobilization of cholesterol into the steroidogenic pathway. The basal and stimulated samples are then placed into an incubator at 18°C for 24 h. For additional precursor incubation options refer to the work of McMaster et al.[14]
3. After incubation, media is drawn off plates using a 1000-μl pipette using a separate tip for each well. Media from each well is then placed into appropriately labeled glass scintillation vials, capped, and frozen at −20°C until analyzed by RIA.

Classification of follicle stage

Follicles produce different levels of steroids at different stages (e.g., pre-vitellogenic, vitellogenic, maturational),[14] therefore effects on plasma steroids or *in vitro* steroid production due to treatment can be difficult to tease apart from stage-dependent changes. As in other fish, mummichog generally have high levels of circulating estradiol (and T) when undergoing vitellogenesis.[26,30] Steroid levels, particularly estradiol, drop off during maturation; the maturation-inducing progesterones then become important.[31] Therefore, treatment effects causing changes in gonadal maturation can potentially be confounding when plasma steroid levels are looked at in isolation from developmental stage of the follicles.

It is, however, possible to classify follicular development stage and to determine if there are treatment effects. This can be done precisely by histology[32] or electron microscopy,[33] but the methodology is too detailed to relay here. It is also possible to detect treatment differences in fecundity and mature egg size in mummichog using image analysis.[34]

In a less precise fashion, one can separate and count under a dissecting scope different follicle classes. This can be done on sampling day, using follicles in M199 or at a later date with follicles preserved in Bouin's solution. Counting mature follicles and eggs also provides a fecundity estimate for each fish, i.e., an estimate of the number of mature eggs to be released during the next spawn. It is possible to count either all the (mature) follicles in the ovaries, or to count follicles in only part of the gonad (e.g., the left ovary).

1. On the sampling day, if follicles are to be sized immediately, place the ovary (or part thereof) of one fish in a glass Petri dish with a small amount of M199 to keep the ovary moist. If follicles are to be sized at a later date, store the ovary from each fish separately, in a plastic scintillation vial, with a generous amount of Bouin's solution to completely cover the ovary. At the time of staging for stored follicles, place the ovary of one fish in a glass Petri dish with a small amount of Bouin's solution.
2. Using two pairs of forceps, gently separate the follicles from each other and the connective tissue. Separate opaque and translucent (yellow) follicles. Maturational follicles and mature eggs are generally >1.3 mm in diameter (this can vary depending on fish stock[33]) and are translucent, pale yellow, and have lipid droplets at one pole. Mature eggs are generally loose and 1.7–1.8 mm in diameter.
3. Count the number of maturational follicles and mature eggs, as well as the opaque (previtellogenic and vitellogenic) follicles. The number of maturational follicles and mature eggs represents a fecundity estimate for each fish. The relative proportions of opaque follicles, translucent follicles, and mature eggs provide an estimate of staging or maturity for the gonad and may be useful for discriminating between differences in hormonal levels due to reproductive stage and differences due to treatment effects.

Plasma steroid extractions

Steroid hormones are bound to steroid binding proteins in the plasma, and they must be separated before they can be assayed using RIA. Ether extraction based on the methods of McMaster et al.[12] using an acetone dry ice bath (or liquid nitrogen) is a relatively quick, inexpensive, and simple method to achieve separation of the protein–hormone complexes.

Spiking samples (with cold or radiolabeled steroid) can be done to validate recovery and, therefore, the extraction procedure. The phosgel solution should be at 4°C and should be made on Day 1 for use on Day 2. Following extraction, phosgel-reconstituted extracted steroids can be stored at −20°C for up to a year.

Day 1:

1. Plasma samples are thawed.
2. 16 mm × 150 mm test tubes are labeled as per the samples to be extracted.
3. Equal volumes (200, 100, 50 μl; dependent on the amount available in the samples) of plasma are pipetted into the test tubes.
4. Using a repeater pipette, 500 μl of ddH_2O are added to each tube.

Note: The following steps must be performed under the fume hood.

5. Ether (5 ml) is added to each of six tubes.
6. Each tube is vortexed for 20 s and allowed to settle.
7. Each tube is vortexed again and allowed to settle.
8. An acetone/dry ice bath is set up by placing 2–3 pieces (5–10 cm diameter, 3–4 cm thick) of dry ice in a metal pan and adding a volume of acetone to have a pool of approximately 4 cm depth. *Note*: it helps to tip the metal pan so as to be working only in one corner. Alternately, the tube may be immersed into a thermos of liquid nitrogen.
9. The tubes are slowly placed in acetone (liquid nitrogen) to freeze the aqueous fraction. Do all six (one or two at a time).
10. One tube at a time, the edges are thawed at the solid phase level (using your hands) to ensure all the solid phase goes to the bottom.
11. The tube just thawed is re-frozen.
12. The ether phase is decanted into respective, labeled glass scintillation vials.
13. The vials are left in the fume hood to evaporate the ether overnight. Alternatively, the ether can be evaporated under N_2 to shorten the evaporation time.

Day 2:

14. When the scintillation vials are dry, 1 ml of phosgel buffer is added.
15. The reconstituted samples are frozen at −20°C until use. The reconstituted samples should be left at room temperature for 3 h prior to freezing to ensure complete reconstitution of the steroids into the phosgel.

Radioimmunoassay

RIAs are used to measure the concentration of steroids in the blood (plasma protocol) or the biosynthetic capacity of the gonads (*in vitro* protocol). The assay is based on competitive binding by a fixed concentration of steroid tracer (E_2^* or T*) with unlabeled steroid (E_2 or T) in standards or in samples. Competition is for a limited and constant number of binding sites available on the steroid Ab (E2-Ab and T-Ab), such that only 50% of total steroid concentration will be bound by the Ab.

There are a number of quality controls required to ensure precision and accuracy within and between RIAs in a laboratory. These parameters must be established for each laboratory. The following is a summary of RIA performance characteristics described in detail by McMaster et al.[12] In the initial stages of establishing an RIA, parallelism (presence of interfering compounds) should be examined by analyzing serial dilutions of media or plasma; details of linearity should be provided. Similar to parallelism, specificity (cross-reactivity of the Ab) must be examined by comparing the ability of related hormones to bind the specific Ab. Accuracy is simply determined by spiking samples with known concentrations of unlabeled steroid and doing regression analysis to determine the slope of the line when steroid added is plotted against steroid measured. A slope different than 1 indicates the presence of interfering compounds and the RIA must be reassessed. Precision (inter- and intra-assay variabilities) must be considered by running a sample from a common pool of plasma or media in each assay (IA) and by replicating this IA several times within the same assay (intra-assay variability). An IA variability >15% is unacceptable, as is an intra-assay >10%. The IA calculation is the standard deviation divided by the mean of your IA samples from multiple assays. The values used in the similar intra-assay calculation come from 3–5 samples of the common pool run in the same assay. Upper and lower detection limits must be determined prior to interpreting a data set, and there is no universal means by which to establish these limits. The lower limit can be taken to be the 0 steroid standard value minus 2 standard deviations,[12] and this value should be reported with the data. The upper limit can be established as the steroid concentration that gives less than 10% of total binding. If a sample falls within this upper limit, it should be diluted to give approximately 50% binding and re-assayed.

The RIA described here has a number of modifications from the original assay described by McMaster et al.[12] Optimization was required for mummichog due to the small blood volumes and sometimes low steroid production levels obtained from these small-bodied fish. We increase the number of standard replicates (from 3 to 5) to allow for a more precise standard curve in the 0–12.5 ng/tube region of the curve. Binding of the "0" standard is set at 50% rather than the broader acceptable range of 35–50% described by McMaster et al.[12] Adding Ab prior to the addition of tracer gives increased opportunity for endogenous hormone to bind to the Ab. The volume of extracted sample used in the assay is always 200 μl and the incubation time is 24 h versus the less conservative parameters (50–200 μl and 3–24 h, respectively) described by McMaster et al.[12]

The phosgel and charcoal solutions should always be at 4°C. A little extra of each solution should be made to ensure there is always enough.

Day 1:

1. Samples (plasma or *in vitro*) and IA are thawed at room temperature.
2. Steroid standards are made up as follows:
   (a) Stock steroid (100 μl of 1000 ng/ml) is diluted with 24.9 ml of phosgel buffer (Table 4.1, row 1) in a plastic beaker to produce a final steroid concentration of 800 pg/tube.
   (b) Serial dilutions are performed in 16 mm × 150 mm test tubes to prepare subsequent standards: final steroid concentrations of 400, 200, 100, 50, 25, 12.5, 6.25, 3.125, 1.56 pg/tube (Table 4.1, rows 2–10). Each standard should be vortexed prior to serially diluting.
3. Assay tubes (12 × 75 mm) are placed in racks. Respective standards, samples and the IA and intra-assay sample(s), and phosgel are added as indicated in Table 4.2, using 200-μl pipettes and separate pipette tips for each different addition.
4. Aliquots of the desired Ab (E_2-Ab or T-Ab) are thawed and phosgel is added to make the working Ab dilutions. *Note*: Ab for all tubes = 200 μl × (No. standard tubes + No. samples + 2 IA).

Table 4.1 Steroid standard preparation by serial dilution for the RIA

Initial [steroid]	Serial dilution: volume of initial [steroid] to be combined with volume of phosgel to yield final [steroid]	Volume of phosgel to be added (ml)	Final [steroid] (pg/tube)	Standard No. (std #)
STOCK (1000 ng/ml)	100 μl	24.9	800	1
800 pg/tube	1 ml	1	400	2
400 pg/tube	1 ml	1	200	3
200 pg/tube	1 ml	1	100	4
100 pg/tube	1 ml	1	50	5
50 pg/tube	1 ml	1	25	6
25 pg/tube	1 ml	1	12.5	7
12.5 pg/tube	1 ml	1	6.25	8
6.25 pg/tube	1 ml	1	3.125	9
3.125 pg/tube	1 ml	1	1.56	10

Table 4.2 Assay tube contents for the RIA

Tube No.	Contents	Phosgel (μl)	Standard (μl)	Unknown (μl)
1,2,3	NSB[a]	400	—	—
4,5,6,7,8	0	200	—	—
9,10,11,12,13	1.56	—	200 (std #10)	—
14,15,16,17,18	3.125	—	200 (std #9)	—
19,20,21,22,23	6.25	—	200 (std #8)	—
24,25,26,27,28	12.5	—	200 (std #7)	—
29,30,31	25	—	200 (std #6)	—
32,33,34	50	—	200 (std #5)	—
35,36,37	100	—	200 (std #4)	—
38,39,40	200	—	200 (std #3)	—
41,42,43	400	—	200 (std #2)	—
44,45,46	800	—	200 (std #1)	—
47,48	Unknown 1	—	—	200
—	Unknown 2 to x (including some intra-assay controls)	—	—	200
145,146[b]	Unknown x	—	—	200
147,148	IA[c]	—	—	IA 200

[a] Non-specific binding (NSB).

[b] Unknown samples up to the maximum capacity of the centrifuge (minus spaces for the intra-assay controls).

[c] IA aliquots: always included as the last two tubes in each assay.

Note: Ensure all steps from this point forward are being performed in a radioactivity-licensed area.

5. The incubator is turned on and set to 18°C.
6. The working steroid tracer solution is made by adding an appropriate amount of phosgel to stock steroid tracer.
7. The tracer dilution is verified.

   (a) Two scintillation vials containing 200 μl of working steroid tracer solution + 600 μl phosgel + 5 ml scintillation cocktail are counted in the scintillation counter and the total counts should be approximately 5000 counts per minute (cpm).
8. Working Ab solution (200 μl) is added using a repeater pipette to every assay tube except tubes 1, 2, and 3 as these tubes are used to calculate non-specific binding (NSB).
   (a) NSB accounts for the presence of tracer not directly related to binding by the Ab (possible contamination or background counts). The mean value of the counts in these tubes is later subtracted from the counts per minute of the standards and tracer in subsequent calculations.
9. Steroid tracer (200 μl) is added to every assay tube using a repeater pipette.
10. Tubes are incubated at 18°C for 24 h.
11. Total counts reference (TCR) tubes are made by adding 200 μl of tracer and 600 μl of phosgel to each of three scintillation vials. These tubes are capped and incubated with the other tubes in the assay.

Day 2:

12. Centrifuge is turned on (4°C).
13. At the end of the incubation period, the tubes are placed in ice/ice water for exactly 10 min.
14. After 10 min in ice-cold water, 200 μl of charcoal is added to each tube (not the TCRs) using a repeater pipette. Time from the addition of charcoal to the last assay tube to the beginning of the spin (step 16) should be 10 min exactly. *Note*: the charcoal solution must be stirred with a magnetic stirrer at all times during additions so that the charcoal is suspended evenly in the solution.
15. Tubes are wiped with paper towels or Kimwipes®, vortexed, and loaded into the centrifuge.
16. Tubes are spun at 4°C for 12 min at 2400–2500*g*.
17. The liquid phase is decanted from the tubes into the scintillation vials, by quickly pouring the contents of the tubes into the vials, being careful not to disturb the charcoal pellet. Scintillation vials can be ordered in bulk to save money. It is recommended that in preliminary scintillation vial orders, the vials be ordered with the re-usable cardboard racks included to provide racks for holding of vials during decanting. Set up the vials for decanting and counting in the order of the assay: TCR, NSB, standards, samples, IA.
18. Scintillation cocktail (5 ml) is added to each scintillation tube, including the TCRs. A preset volume pump attached to the scintillation fluid container makes dispensing easier.
19. Scintillation tubes are capped and vortexed.
20. Scintillation tubes are loaded into the counter, using a counting program appropriate to the radiolabeled ^{3}H steroid isotope.
21. Waste materials are disposed of in a manner consistent with local radioactive garbage regulations.

Data analysis

The raw data (cpm) produced by the scintillation counter can be analyzed using commercially available programs or programs associated with the specific scintillation counter used.

Presently, we are using templates set-up using Microsoft Excel® and SPSS SigmaPlot® version 8.0 or higher. The pharmacology software in SigmaPlot (Simple Ligand Binding using the 4-parameter logistic curve fit) is used to define the parameters of the standard curve, and these parameters are then inserted into an Excel worksheet that is set up to calculate the concentration of steroid hormones in the original samples. The first step required is subtracting the NSB values from the standard curve counts and sample counts. The concentrations in the samples are then calculated as follows.

Example calculations

A typical standard curve data set might give the following parameters for the equation to the 4-par logistic curve:

$$y = D + \frac{A - D}{1 + 10^{(x - \log C)^B}} \tag{4.1}$$

$A = 1973.69, \quad B = 1.12, \quad C = 30.6300 \ (\log C = 1.4862), \quad D = 18.2951$

These values are calculated by SigmaPlot® using the standard curve data generated by the scintillation counts and the Simple Ligand Binding function. The 4-par logistic curve equation [Equation (4.1)] can solve for x (the sample steroid concentration) using the counts per minute of the sample (y) in the Microsoft Excel® spreadsheet. For example, if $y = 1750$ cpm and this value is entered into Equation (4.1) as solved for x, one would find $x = 4.93$ pg/tube. The units are pg/tube at this point because this value reflects the concentration of steroids in that particular scintillation vial of the RIA. To calculate back to the original sample one must factor in the volume of plasma used in the steroid extraction (typically 30–100 μl), the volume of phosgel used to reconstitute the steroids after the extraction (typically 1 ml) and the volume of sample used in the RIA (typically 200 μl). Therefore, if 100 μl of plasma was extracted and reconstituted in 1 ml of phosgel, then:

$$\begin{aligned}[\text{steroid}] &= \frac{\text{steroid content in tube}}{\text{volume of sample used}} \times \frac{\text{reconstituted plasma volume}}{\text{volume of plasma extracted}} \\ &= \frac{24.93\ \text{pg/tube}}{200\ \mu\text{l}} \times \frac{1000\ \mu\text{l}}{100\ \mu\text{l}} \\ &= 1.25\ \text{pg}/\mu\text{l}\ (= 1.25\ \text{ng/ml})\end{aligned} \tag{4.2}$$

The respective calculations for an *in vitro* RIA are slightly different. For example, assume a gonadal tissue mass of 23.6 mg used in the incubation, 1000 μl of M199 used in the incubation period of 24 h and a volume of incubation media used in the RIA of 200 μl/tube. If the steroid content of one tube in the assay is 57 pg/tube, the calculations are as follows:

$$\begin{aligned}&[\text{steroid}]\ (\text{produced by gonad } \textit{in vitro}) \\ &\quad = \frac{\text{steroid content in tube}}{\text{volume of sample used}} \times \frac{\text{media volume in incubation}}{\text{gonadal tissue mass used in incubation}} \times \frac{1}{\text{time}} \\ &\quad = \frac{123\ \text{pg/tube}}{200\ \mu\text{l/tube}} \times \frac{1000\ \mu\text{l}}{23.6\ \text{mg}} \times \frac{1}{24\ \text{h}} \\ &\quad = 1.09\ \text{pg produced/mg testes/h}\end{aligned} \tag{4.3}$$

Vitellogenin assay

This assay measures vitellogenin levels in plasma. Vitellogenin is a precursor to the egg yolk protein and is synthesized in the liver under estrogenic control. While both males and females carry the gene for this protein, under normal circumstances male fish do not synthesize vitellogenin. However, when exposed to substances with estrogenic properties, males will begin to synthesize the protein. Vitellogenin induction has been explored as a biomarker for several fish species.[11]

The assay to detect vitellogenin works through a chain of reactions. The plates are precoated with vitellogenin. The plasma samples and a primary Ab are then added; this Ab binds to the vitellogenin in the samples and to the vitellogenin that is coating the walls. A secondary Ab is subsequently added, which binds to the primary Ab. The secondary Ab is conjugated to an enzyme, which will act upon a substrate (OPD) to produce a characteristic color. The more enzyme present, the more color that will be developed. If there is a low level of vitellogenin in the plasma, relatively more Ab will bind to the vitellogenin on the side of the wall of the plate and more color will be produced.

1. Ensure there is enough of each of the buffer solutions and that all pH values are correct. Also ensure that the solutions are not past their expiry dates.

Day 1:

2. Vitellogenin coating solution (150 μl) is added to each well with a repeater pipette. The plate is covered tightly with plastic wrap and incubated for 3 h at 37°C.
3. The standards, samples, IA, and primary Ab are prepared during this incubation.
   (a) Vitellogenin standards are made up by diluting with TBS-T-BSA stock vitellogenin to a final concentration of 75 ng/50 μl in a test tube. Serial dilutions are performed to prepare subsequent standards with final concentrations of 37.5, 18.75, 9.4, 4.69, 2.34, 1.17, 0.59, 0.29, and 0.15 ng/50 μl. Each standard should be vortexed prior to the subsequent dilution. The initial dilution will vary according to the concentration of the stock solution. [For example, 5 μl vitellogenin stock solution (assuming 0.60 mg/ml) is pipetted into 2000 μl of TBS-T-BSA to create the required 75 ng/50 μl.] Subsequent dilutions are 1:1 volumes of previous standard and TBS-T-BSA as with the RIA standard dilutions.
   (b) Our laboratory has found that laboratory-held mummichog plasma must be diluted further in order to have vitellogenin levels low enough to compare to the standard curve. We have found that a 1001× dilution works best for these fish. This dilution may be accomplished by pipetting 2 μl of plasma into 2000 μl TBS-T-BSA. For field samples, we have found the best dilution to be much lower, about 501×, which may be achieved by pipetting 4 μl into 2000 μl of TBS-T-BSA. The dilution that results in the most plasma vitellogenin samples within the limits of the standard curve is dependent upon the type of plasma samples that are being analyzed and should be tested for the best dilution each time samples are to be analyzed from a specific experiment.

(i) IA: Thaw aliquot
(ii) Ab: Thaw and prepare as described in reagents section

4. Plates are removed from the incubator and excess vitellogenin solution is shaken off into the sink.
5. The plate is washed by adding 300 μl of TBS-T going up and down the plate with the multi-tip repeater pipette (three runs of 100 μl) followed by shaking the solution off into the sink. This is repeated five times.
6. TBS-T-BSA (200 μl) is added to every well.
7. The plate is incubated for 30 min at 37°C under plastic wrap. This time is crucial, so that the BSA builds up but does not coat over the vitellogenin. The solution is discarded afterwards, and the plate is not washed.
8. Primary Ab, standards, samples, and IA are added to the appropriate wells as described in Table 4.3.
9. The plate is covered with plastic wrap and incubated on a desktop overnight.

Day 2:

10. The plate is rinsed five times with TBS-T as in step 5.
11. Secondary Ab solution (150 μl) is added to every well.
12. The plate is incubated under plastic wrap for 2 h at 37°C.
13. OPD solution is prepared when the 2-h incubation period is nearly over.
14. The plate is removed from the incubator and rinsed five times with TBS-T (as in step 5).
15. OPD solution (150 μl) is added to each well with a repeater pipette, and the plate is covered with plastic wrap. The plate is allowed to incubate at room temperature in a dark drawer for 30 min.
16. The spectrophotometer is turned on and set at 490 nm.
17. The plate is removed from the drawer. Some of the wells should have yellow coloration.
18. H_2SO_4 (50 μl) is added to each well to denature proteins and stop the enzymatic production of color.
19. The plate is placed on the shaker for 10 min at 100 rpm.

Table 4.3 Suggested assignment of wells within plate for vitellogenin assay

	1	2	3	4	5	6	7	8	9	10	11	12
A	Buf	Buf	Buf	Buf	9.4	9.4	9.4	9.4	S9	S9	S17	S17
B	Ab	Ab	Ab	Ab	18.75	18.75	18.75	18.75	S10	S10	S18	S18
C	0.15	0.15	0.15	0.15	37.5	37.5	37.5	37.5	S11	S11	S19	S19
D	0.29	0.29	0.29	0.29	75	75	75	75	S12	S12	S20	S20
E	0.59	0.59	0.59	0.59	S1	S1	S5	S5	S13	S13	S21	S21
F	1.17	1.17	1.17	1.17	S2	S2	S6	S6	S14	S14	S22	S22
G	2.34	2.34	2.34	2.34	S3	S3	S7	S7	S15	S15	S23	S23
H	4.69	4.69	4.69	4.69	S4	S4	S8	S8	S16	S16	IA	IA

Buf: Buffer only, 150 μl of TBS-T-BSA. Ab: Ab only, 50 μl of TBS-T-BSA, plus 100 μl of antiserum. 0.15–75: standards from serial dilution, add 50 μl plus 100 μl antiserum. S1–S23: samples, 50 μl of samples plus 100 μl antiserum. IA: inter-assay, 50 μl of IA plus 100 μl antiserum.

20. The plate is read on the spectrophotometer at 490 nm.
21. The acid is neutralized by shaking $NaHCO_3$ over the plate so that it enters each well. The plate is then discarded.

Data analysis

The data are analyzed by comparison to a standard curve. This curve is generated from the absorbance values ($A_{490\,nm}$) of the vitellogenin standards. The data are transformed prior to construction of the standard curve by taking the log of the concentrations of the vitellogenin standards, thus providing the x-values for the curve. Secondly, the $A_{490\,nm}$ values for each vitellogenin standard concentration are converted to a binding over origin binding (B/B_o) or relative maximum absorbance value. This is done by subtracting the mean absorbance values of the blank (or buffer-only) wells (which removes the amount of color production that is not a direct result of the OPD acting upon secondary Ab bound to the walls of the plate) and then setting the absorbance of the Ab-only wells (less the absorbance of the blank wells) to 100% or maximum absorbance, and determining the percentage of maximum absorbance for each standard concentration. These values become the y-values, and a standard curve such as Figure 4.1 may be plotted.

It is imperative that the standard curve produces a reliable regression, as it will be used to determine the concentration of vitellogenin in the plasma samples. Therefore, we recommend that an effective standard curve is one that has an r^2 of at least 0.95. When new reagents are acquired or the assay has not been used recently, the standard curve should be performed to ensure the assay is working correctly (i.e., there has been no

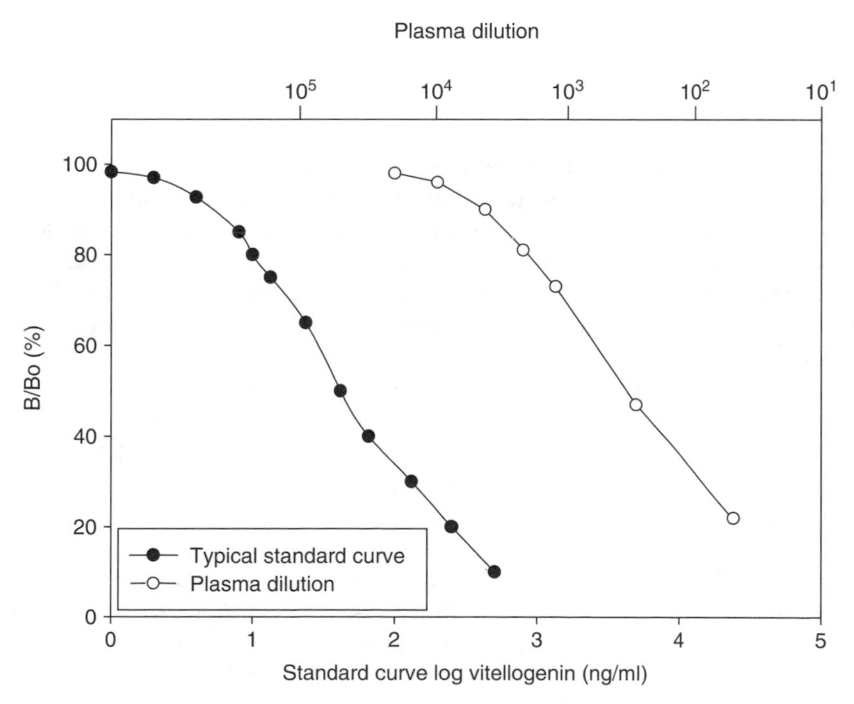

Figure 4.1 Lower x-axis: typical binding displacement curve (–●–) obtained with vitellogenin standards. Upper x-axis: Parallel binding displacement curve (–○–) obtained with serial dilution of plasma from a male mummichog exposed to waterborne ethynyl estradiol.

degradation of Ab or other reagents). In addition, parallelism of multiple standard curves should be confirmed using analysis of covariance (ANCOVA) to ensure that there is no significant difference among the slopes of multiple standard curves.

It is important that the diluted plasma samples fall within the standard curve of purified vitellogenin. It is recommended that a serial dilution of a plasma sample be tested prior to running samples, in order to show linearity of the plasma dilution (Figure 4.1), as well as to determine the dilution of the plasma that lies within the vertical slope of the standard curve. IA variability should be assessed with every assay by including an IA sample as described earlier. Please refer to the work of Denslow et al.[11] for further details on assessing assay quality.

Once the regression on the standard curve values has been determined to be acceptable, the sample concentrations can be calculated. Two replicate $A_{490\,\mathrm{nm}}$ values for each plasma sample are averaged, and the relative absorbance for each sample as a percentage of the maximum absorbance is determined as described earlier. If this value is >95% of the maximum absorbance, the concentration of vitellogenin is considered to be below detectable limits. If this value is below 95% of the maximum absorbance, then the log concentration of vitellogenin in the sample is calculated from the parameters of the regression equation. This value must then be inverse-log transformed to give the concentration of vitellogenin (ng/50 μl as per the standard curve) within the well. To convert this concentration (ng/50 μl) to the actual concentration of vitellogenin in the sample, this must be multiplied by the plasma dilution factor. This will give a concentration in nanograms per milliliter that may be converted to more appropriate units if necessary (e.g., mg/ml). We have used both Corel QuattroPro® and Microsoft Excel® to create templates to extrapolate the vitellogenin concentrations for the samples from the standard curve.

Statistics

Differences in mean gonad and mean liver size among treatments can be tested with an ANCOVA with total fish weight as the co-variate of organ weight. Condition factor can be tested in a similar way, with weight as the dependent variable and standard length as the co-variate. When the p-value of the interaction term is >0.1, treatment and covariate are considered as the only possible sources of variation in subsequent analyses.[35] Scatterplots, with regression, of log covariate (y-axis) against the log dependent value (x-axis) can be given; however, it is often easier to indicate differences by listing in a table the morphological variables as the index or calculated value (Table 4.4).

Statistical test(s) involved in the data analysis (e.g., steroid hormones and vitellogenin) are dependent on outliers, normality, and variability. Before performing any tests of significance, the data can be examined for outliers using a test such as the Dixon test for outliers.[36] Given a normal distribution, a nested analysis of variance (ANOVA) can be performed with fish as the units of replication to determine presence or absence of tank effects and to test for treatment differences. If the ANOVA identifies treatment differences, Tukey's, Dunn's, or Dunnett's *post-hoc* tests can be used to determine where these treatment differences exist. A decision should be made *a priori* if there is an interest in comparing all treatments (Tukey's or Dunn's multiple comparison test) or in comparing each treatment to the control (Dunnett's multiple comparison test). For the former, Tukey's test is used if sample sizes are equal, Dunn's test if they are not.[37]

Table 4.4 Effects of 7 days of exposure of female (F) and male (M) mummichog to final effluent (0–100%) collected from a Canadian paper mill. Values are means (standard errors of the means). Differing superscript letters indicate groups that are different

Sex	Effluent concentration	Weight (g)	Gonadosomatic index, GSI (%)	Liver-somatic index, LSI (%)	*In vitro* Testosterone (T) production (pg/mg tissue/h)	Maturational follicles and mature eggs (*N*)	Plasma vitellogenin (mg/ml)
F	0	13.0 (0.662)	10.4 (1.35)[a]	5.75 (0.244)		156.5 (20.3)[a]	3.14 (0.189)[a]
	1	12.6 (0.332)	9.58 (0.758)[a]	5.58 (0.202)		143.7 (11.4)[a]	2.12 (0.231)[ab]
	5	12.9 (0.281)	8.38 (1.01)[a]	5.18 (0.376)		117 (13.3)[a]	2.54 (0.456)[ac]
	15	12.6 (0.624)	7.86 (0.888)[a]	6.088 (0.290)		91.0 (12.0)[b]	2.78 (0.329)[a]
	30	11.5 (0.721)	6.89 (0.864)[a]	5.04 (0.313)		103.3 (13.0)[a]	1.56 (0.274)[bc]
	50	12.6 (0.625)	6.48 (0.749)[a]	5.32 (0.473)		82.9 (10.4)[b]	1.47 (0.129)[bc]
	100	13.6 (0.682)	6.07 (0.800)[b]	5.118 (0.285)		68.9 (8.64)[b]	1.24 (0.165)[b]
	p-value	0.067	0.011	0.222		<0.001	<0.001
M	0	15.4 (0.581)	1.65 (0.226)[a]	2.52 (0.262)	1.25 (0.115)[a]		1.51 (0.134)
	1	13.8 (0.644)	1.39 (0.118)[a]	2.46 (0.226)	1.06 (0.0900)[ab]		1.71 (0.540)
	5	15.3 (0.501)	1.35 (0.0808)[a]	2.43 (0.138)	0.978 (0.136)[ab]		1.43 (0.157)
	15	15.3 (0.373)	1.25 (0.105)[a]	2.27 (0.142)	0.943 (0.160)[b]		1.58 (0.482)
	30	14.3 (0.499)	1.21 (0.0115)[a]	2.45 (0.145)	0.767 (0.110)[b]		1.69 (0.132)
	50	14.5 (0.563)	1.23 (0.0843)[a]	2.98 (0.137)	0.570 (0.0496)[c]		1.56 (0.142)
	100	14.7 (0.291)	0.981 (0.0706)[b]	2.62 (0.141)	0.436 (0.0675)[c]		1.39 (0.278)
	p-value	0.257	0.022	0.265	<0.001		0.988

If tank effects are present as determined by the nested ANOVA, the unit of replication becomes the tank rather than the individual fish (with an associated decrease in statistical power). RIA data are often non-normally distributed. Performing a log-transformation will typically permit use of parametric tests; however, if the transformation is not effective or desirable, the equivalent non-parametric test (e.g., Kruskal–Wallis ANOVA on ranks) can be performed.[37]

Any statistical software program with these various tests can be used. Examples of programs we have used include SAS/STAT (SAS, Cary, NC, USA) and Systat (Systat Software, Richmond, CA, USA) and SPSS (SPSS, Chicago, IL, USA).

Results and discussion

As an example of the type of date resulting from the mummichog bioassay, mummichog were exposed to final effluent from a Canadian paper mill to determine fish responses at environmentally relevant and greater effluent concentrations. Pre-spawning fish were exposed in static systems for 7 days in February 2001 following use of the artificial regression-recrudescence protocol. In brief, 3 male and 3 female mummichog were weighed prior to allocation into 32.4 l filtered aquaria containing 16 l of 16 ppt water at 18–20°C and >85% DO. Fish were acclimated for 1 week in the exposure aquaria. Beginning on Day 1, fish were exposed in replicates of four aquaria to the following concentrations of paper mill effluent: 0%, 1%, 5%, 15%, 30%, 50%, and 100%. The receiving environment normally receives 1–5% dilutions of the effluent. Water and effluent changes were made daily. Following exposure, fish were sampled as previously described. *In vitro* incubations were done with testes tissue only. The number of maturational follicles and mature eggs were counted; opaque (pre-maturational) follicles made up only 5% of the total follicles counted.

Female ($p < 0.11$) and male ($p < 0.022$) GSIs were significantly reduced at 100% effluent (Table 4.4). There was no effect on liver-somatic index (LSI) (Table 4.4) or condition factor (data not shown) in either males or females. Female plasma T was depressed at 30–100% ($p < 0.001$), while male T was depressed at 100% ($p < 0.001$) (Figure 4.2). Responses for

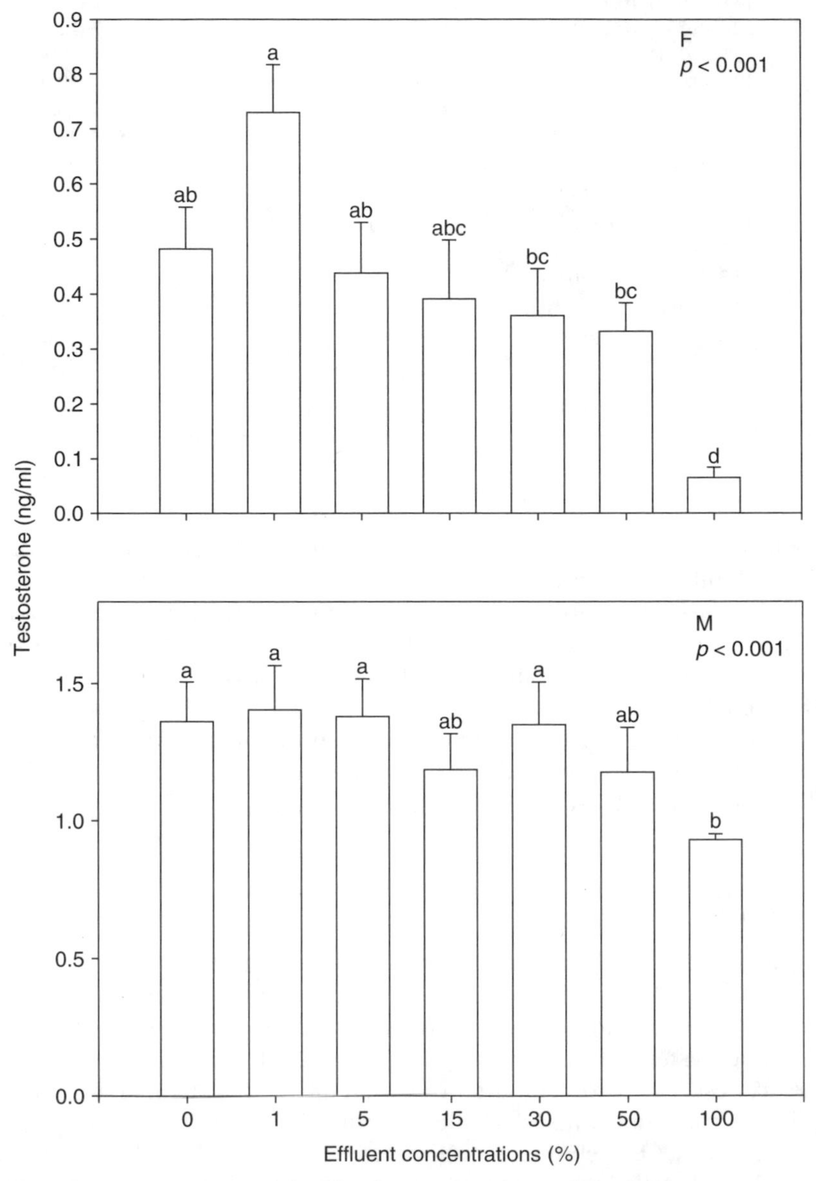

Figure 4.2 Plasma testosterone (T) levels of female (F; top) and male (M; bottom) mummichog exposed to effluent from a Canadian paper mill for 7 days in the bioassay. Values are mean (+standard error of the mean). Significant differences are indicated by differing letters. Plasma hormone levels are routinely presented in graphical form. Figures are especially suitable for clearly representing dose–response data.

plasma estradiol (females) and 11-ketotestosterone (males) paralleled those of plasma T (data not shown). *In vitro* incubations of testes pieces were decreased at 50% and 100% ($p < 0.001$) (Table 4.4).

These results are typical of those obtained when using the bioassay to assess effects of hormone-active effluents (or EDSs). Depressions in T production in males have been observed in males exposed to environmentally relevant concentrations of anti/androgens,[19] high concentrations of anti/estrogens,[2] and pulp mill effluents.[20,21] In males, *in vitro* depressions in T production were affected at lower effluent concentrations that plasma T (Table 4.4, Figure 4.2). Observed treatment differences in gonadal steroid production are often correlated with changes in circulating steroid levels. Gonadal production is often altered at lower exposure concentrations than plasma steroid levels.[2,19] Pulp mill effluent effects, and effects of other EDSs, are often manifested at the level of the gonad[38] through effects on steroid production.[39,40] The *in vitro* incubation assay is a relatively simple method of determining effects at the gonadal level. Additional methodologies using precursor additions can be used to identify enzymes within the steroidogenic pathway affected by exposure.[14]

The number of maturing eggs (maturational follicles plus mature eggs) was decreased starting at 15–30% ($p < 0.001$; Table 4.4). This is the first time in our bioassay in which we have detected significant changes in follicle development during the 7-day bioassay. At 100% effluent, male testes size was also reduced. In females, high endocrine toxicity at 15–30% was observed; the lack of follicular development correlates with reduced plasma steroid levels. The number of mature follicles/eggs is in the same range as those previously found in field studies,[33,34] as well as our own fecundity values for pair-breeding laboratory stocks (R. Ibey and D. MacLatchy, unpublished). Whether follicular development was delayed by direct effects of the effluent on follicular cell growth and differentiation, or indirectly by effects on endocrine parameters, is unknown. Mummichog follicular development is under hypothalamo-pituitary control,[41] and pulp mill effluents are known to interfere with the pituitary–gonadal axis (decreased gonadotropin level, decreased gonadotropin response at gonads) in fish.[42] Estradiol is also required in fish to direct ovarian growth and differentiation[43]; depressions in estradiol production may have caused changes in paracrine or autocrine control. Further work is required to fully understand the control and function of endocrine systems in fish. Without such understanding, our ability to understand EDS effects is limited.

No effects on plasma vitellogenin levels were found in males (Table 4.4). Induction of vitellogenin production in (male) fish is a commonly measured indicator of exposure to an estrogenic compound or effluent.[44] Pulp and paper mill effluents are not normally associated with estrogenic activity[45] although constituents of pulp mill effluent have been shown to be estrogenic.[46] This is because final effluent chemistry can be quite distinct from the chemistry of the component waste streams that make up the final effluent. It has previously been established that mummichog do induce vitellogenin when exposed to estrogen mimics (ethynyl estradiol[2]) and sewage.[6] Females in this study had depressed vitellogenin levels (Table 4.4), most likely due to depressed estradiol levels (data not shown). We have obtained similar results in female mummichog exposed to methyl T.[19] In juvenile fathead minnow, high concentrations (>75 μg/l) of a pharmaceutical antiestrogen (ZM 189,154) caused depressions in whole-body homogenate concentrations of vitellogenin.[47] It is important to note that many EDSs are able to interact with both estrogenic and androgenic receptors,[48] and a vitellogenic response to EDSs indicates interaction with just one receptor-mediated pathway.

In sum, these results are typical of those observed when mummichog are exposed to hormone-active compounds or effluents. Taken together, the endpoints are able to identify whole organism responses (plasma steroids and vitellogenin) and a population-level effect (reduced fecundity). As well, one mechanism by which the effluent manifests its effects is decreased steroidogenesis. The effluent does not appear to be estrogenic (vitellogenin data). This paper mill data reflects pulp mill effluent studies using this bioassay in that the effects resemble responses found in fish exposed to model anti/androgenic compounds rather than model estrogenic compounds.[45]

Acknowledgments

A number of graduate and undergraduate students in the Canadian Rivers Institute are thanked for their significant contributions to the development of this mummichog bioassay. They include: A. Belyea, J. Beyea, M. Beyea, K. Costain, C. Gilman, J. Ings, F. Leusch, C. MacNeil, A. Smitheram, and L. Vallis. Drs. Nancy Denslow (University of Florida) and Charlie Rice (Clemson University) are thanked for their assistance in the development of the mummichog vitellogenin assay. Collaborations with Dr. Mark Hewitt (Environment Canada) have been fundamental in developing the use of the bioassay for investigation of cause protocols. This work has been funded by a Health Canada/Environment Canada Toxic Substances Research Initiative grant to G. Van Der Kraak (PI), D. MacLatchy, and S. Courtenay; Natural Sciences and Engineering Research Council of Canada grants to D. MacLatchy; and a Network of Centres of Excellence (Canadian Water Network) grant to D. MacLatchy (K. Munkittrick, PI).

References

1. Organization for Economic Development, Final Report from the OECD Expert Consultation Meeting, London, UK, October 28–29, 1998, Report 9906, Environmental Health and Safety Division, Paris, France, 1999.
2. MacLatchy, D.L., Courtenay, S.C., Rice, C.D. and Van Der Kraak, G.J., Development of a short-term reproductive endocrine bioassay using steroid hormone and vitellogenin endpoints in the estuarine mummichog, *Fundulus heteroclitus*, *Environ. Toxicol. Chem.*, 22, 996–1008, 2003.
3. Parrot, J.L. and Wood, C.S., Fathead minnow life cycle tests for detection of endocrine-disrupting substances in effluents, *Water Qual. Res. J. Can.*, 37, 651–667, 2002.
4. Ankley, G.T., Hensen, K.M., Kahl, M.D., Korte, J.J. and Makynen, E.A., Description and evaluation of a short-term reproduction test with the fathead minnow (*Pimephales promelas*), *Environ. Toxicol. Chem.*, 20, 1276–1290, 2001.
5. Munkittrick, K.R., Van Der Kraak, G.J., McMaster, M.E. and Portt, C.B., Reproductive dysfunction and MFO activity in three species of fish exposed to bleached kraft mill effluent at Jackfish Bay, Lake Superior, *Water Pollut. Res. J. Can.*, 27, 439–446, 1992.
6. McArdle, M., Elskus A., McElroy, A., Larsen, B., Benson, W. and Schlenk, D., Estrogenic and CYP1A response of mummichogs and sunshine bass to sewage effluent, *Mar. Environ. Res.*, 50, 175–179, 2000.
7. Gray, M.A., Teather, K.L. and Metcalfe, C.D., Reproductive success and behavior of Japanese medaka (*Oryzias latipes*) exposed to 4-*tert*-octylphenol, *Environ. Toxicol. Chem.*, 18, 2587–2594, 1999.
8. Zillioux, E.J., Johnson, I.C., Kiparissis, Y., Metcalfe, C.D., Wheat, J.V., Ward, S.G. and Liu, H., The sheepshead minnow as an *in vivo* model for endocrine disruption in marine teleosts: a partial life cycle test with 17β-ethynyl estradiol, *Environ. Toxicol. Chem.*, 20, 1968–1978, 2001.

9. Rasmussen, T.H., Andreassen, T.K., Pedersen, S.N., Van der Ven, L.T.M., Bjerregaard, P. and Korssgaard, B., Effects of waterborne exposure of octylphenol and oestrogen on pregnant viviparous eelpout (*Zoarces viviparous*) and her embryos *in ovario*, *J. Exp. Biol.*, 205, 3857–3876, 2002.
10. Balaguer, P., Joyeux, A., Denison, M.S., Vincent, R., Gillesby, B.E. and Zacharewski, T., Assessing the estrogenic and dioxin-like activity of chemicals and complex mixtures using *in vitro* recombinant receptor-reporter gene assays, *Can. J. Physiol. Pharmacol.*, 74, 216–222, 1996.
11. Denslow, N.D., Chow, M.C., Kroll, K.J. and Green, L., Vitellogenin as a biomarker of exposure for estrogen or estrogen mimics, *Ecotoxicology* 8, 385–398, 1999.
12. McMaster, M.E., Munkittrick, K.R. and Van Der Kraak, G.J., Protocol for measuring circulating levels of gonadal sex steroids in fish, *Can. Tech. Rep. Fish. Aquat. Sci.*, 1836, 1–29, 1992.
13. McMaster, M., Jardine, J., Ankley, G., Benson, B., Greeley, M., Gross, T., Guillette, L., MacLatchy, D., Martel, P., Van Der Kraak, G. and Munkittrick, K., An interlaboratory study on the use of steroid hormones in examining endocrine disruption, *Environ. Toxicol. Chem.*, 20, 2081–2087, 2001.
14. McMaster, M.E., Munkittrick, K.R., Jardine, J.J., Robinson, R.D. and Van Der Kraak, G.J., Protocol for measuring *in vitro* steroid production by fish gonadal tissue, *Can. Tech. Rep. Fish. Aquat. Sci.*, 1961, 1–78, 1995.
15. Larkin, P., Folmar, L.C., Hemmer, M.J., Poston, A.J. and Denslow, N.D., Expression profiling of estrogenic compounds using a sheepshead minnow cDNA macroarray, *Environ. Health Perspect.*, 111, 839–846, 2003.
16. Safe, S., Modulation of gene expression and endocrine response pathways by 2,3,7,8-tetrachlorodibenzo-*p*-dioxin, *Pharmacol. Ther.*, 67, 247–281, 1995.
17. Lister, A.L. and Van Der Kraak, G.J., Endocrine disruption: why is it so complicated? *Water Qual. Res. J. Can.*, 36, 175–190, 2001.
18. Wells, K. and Van Der Kraak, G., Differential binding of endogenous steroids and chemicals to androgen receptors in rainbow trout and goldfish, *Environ. Toxicol. Chem.*, 19, 2059–2065, 2000.
19. Sharpe, R.L., MacLatchy, D.L., Courtenay, S.C. and Van Der Kraak, G.J., Effects of a model androgen (methyl testosterone) and a model anti-androgen (cyproterone acetate) on reproductive endocrine endpoints in a short-term adult mummichog (*Fundulus heteroclitus*) bioassay, *Aquatic Toxicol.*, 67, 203–215, 2004.
20. Dubé, M.G. and MacLatchy, D.L., Identification and treatment of a waste stream at a bleached kraft pulp mill that depresses a sex steroid in the mummichog (*Fundulus heteroclitus*), *Environ. Toxicol. Chem.*, 20, 985–995, 2001.
21. Hewitt, M.L., Smythe, S.A., Dubé, M.G., Gilman, C.I. and MacLatchy, D.L., Isolation of compounds from bleached kraft mill recovery condensates associated with reduced levels of circulating testosterone in mummichog (*Fundulus heteroclitus*), *Environ. Toxicol. Chem.*, 21, 1359–1367, 2002.
22. Hewitt, M.L., Dubé, M.G., Culp, J.M., MacLatchy, D.L. and Munkittrick, K.R., A proposed framework for investigation of cause for environmental effects monitoring, *Human Ecol. Risk Assess.*, 9, 195–212, 2003.
23. Taylor, M.H., Environmental and endocrine influences on reproduction of *Fundulus heteroclitus*, *Amer. Zool.*, 26, 159–171, 1986.
24. Cochran, R.C., Zabludoff, S.D., Paynter, K.T., DiMichele, L. and Palmer, R.E., Serum hormone levels associated with spawning activity in the mummichog, *Fundulus heteroclitus*, *Gen. Comp. Endocrinol.*, 70, 345–354, 1988.
25. Day, J.R. and Taylor, M.H., Photoperiod and temperature interaction in the seasonal reproduction of female mummichogs, *Trans. Am. Fish. Soc.*, 113, 452–457, 1984.
26. Shimizu, A., Reproductive cycles in a reared strain of the mummichog, a daily spawner, *J. Fish Biol.*, 51, 724–737, 1997.
27. Nelson, D., Irlandi, E., Settle, L., Monaco, M. and Coston-Clements, L., Distribution and Abundance of Fishes and Invertebrates in Southeast Estuaries, NOAA/NOS Strategic Environmental Assessments Division, Silver Spring, MD, USA, 1991, 221 pp.

28. Scott, W.B. and Scott, M.G., Atlantic fishes of Canada, *Can. Bull. Fish. Aquat. Sci.*, 219, 1–731, 1988.
29. Stone, S., Lowery, T., Field, J., Williams, C., Nelson, D. and Jury, S., Distribution and Abundance of Fishes and Invertebrates in Mid-Atlantic Estuaries, NOAA/NOS Strategic Environmental Assessments Division, Silver Spring, MD, USA, 1994, 280 pp.
30. MacLatchy, D.L., Dubé, M.G., Kerin, B. and Leusch, F.D.L., Species selection for understanding reproductive endocrine effects of xenobiotics on fish: *Fundulus heteroclitus* use in E. Canada, in *Proceedings of the Sixth International Symposium on the Reproductive Physiology of Fish*, Norberg, B., Kjesbu, O.S., Taranger, G.L., Andersson, E. and Stefansson, S.O., Eds., University of Bergen, Bergen, Norway, 1999, p. 376 (499 pp.).
31. Greeley, M.S., Jr., Calder, D.R., Taylor, M.H., Hols, H. and Wallace, R.A., Oocyte maturation in the mummichog *Fundulus heteroclitus*: effects of steroids on germinal vesicle breakdown of intact follicles *in vitro*, *Gen. Comp. Endocrinol.*, 62, 281–289, 1986.
32. Yasutake, W.T. and Wales, J.H., Microscopic Anatomy of Salmonids: an Atlas, Resource Publication 150, Fish and Wildlife Service, US Department of the Interior, Washington, D.C., 1983, 190 pp.
33. Selman, K. and Wallace, R.A., Gametogenesis in *Fundulus heteroclitus*, *Amer. Zool.*, 26, 173–192, 1986.
34. Leblanc, J., Couillard, C.M. and Brêthes, J.-C.F. Modifications of the reproductive period in mummichog (*Fundulus heteroclitus*) living downstream from a bleached kraft pulp mill in the Miramichi Estuary, New Brunswick, Canada, *Can. J. Fish. Aquat. Sci.*, 54, 2564–2573, 1997.
35. Environment Canada, Fish monitoring, fish Survey, Section 5.1. Technical Guidance Document for Pulp and Paper Environmental Effects Monitoring, Ottawa, Canada, 1997, 33 pp.
36. Kanji, G.K., *100 Statistical Tests*, SAGE Publications, London, UK, 1993, 224 pp.
37. Zar, J.H., *Biostatistical Analysis*, 4th ed., Prentice-Hall, New Jersey, 1999, 929 pp.
38. Van Der Kraak, G.J., Munkittrick, K.R., McMaster, M.E. and MacLatchy, D.L., A comparison of bleached kraft mill effluent, 17β-estradiol, and β-sitosterol effects on reproductive function in fish, in: *Principles and Processes for Evaluating Endocrine Disruption in Wildlife*, Kendall, R.J., Dickerson, R.L., Giesy, J.P. and Suk, W.P., Eds., SETAC Press North America, Pensacola, FL, 1998, pp. 249–265 (515 pp.).
39. Leusch, F.D.L. and MacLatchy, D.L., Implants of β-sitosterol impede cholesterol transfer across the mitochondrial membrane isolated from gonads of male goldfish (*Carassius auratus*), *Gen. Comp. Endocrinol.* 134, 255–263, 2003.
40. McMaster, M.E., Van Der Kraak, G.J. and Munkittrick, K.R., Exposure to bleached kraft pulp mill effluent reduces the steroid biosynthetic capacity of white sucker ovarian follicles, *Comp. Biochem. Physiol.*, C112, 169–172, 1995.
41. Brown, C.L., Grau, E.G. and Stetson, M.H., Functional specificity of gonadotropin and thyrotropin in *Fundulus heteroclitus*, *Gen. Comp. Endocrinol.*, 58, 252–258, 1985.
42. Van Der Kraak, G.J., Munkittrick, K.R., McMaster, M.E., Portt C.B. and Chang, J.P., Exposure to bleached kraft pulp mill effluent disrupts the pituitary-gonadal axis of white sucker at multiple sites, *Tox. Appl. Pharmacol.*, 115, 224–233, 1992.
43. Srivastava, R.K. and Van Der Kraak, G.J., Regulation of DNA synthesis in goldfish ovarian follicles by hormones and growth factors, *J. Exp. Zool.*, 270, 263–272, 1994.
44. Routledge, E.J., Sheahan, D., Desbrow, C., Brighty, G.C., Waldock M. and Sumpter, J.P., Identification of estrogenic chemicals in STW effluent. 2. *In vivo* responses in trout and roach, *Environ. Sci. Technol.*, 32, 1559–1565, 1998.
45. MacLatchy, D.L., Dubé, M.G., Hewitt, M.L., Courtenay, S.C. and Van Der Kraak, G.J., Development of a fish bioassay to test for hormonally-active contaminants in pulp mill effluents, in *Proceedings of the Fifth International Conference on Fate And Effects of Pulp Mill Effluents*, DEStech Publications, Lancaster, PA, USA, 410–419, 2004.
46. Zacharewski, T.R., Berhane, K., Gilleby, B.E. and Burnison, B.K., Detection of estrogen and dioxin-like activity in pulp and paper mill black liquor and effluent using *in vitro* recombinant receptor/reporter bioassays, *Environ. Sci. Technol.*, 29, 2140–2146, 1995.

47. Panter, G.H., Hutchinson, T.H., Längem R., Lye, C.M., Sumpter, J.P., Zerulla, M. and Tyler, C.R., Utility of a juvenile fathead minnow screening assay for detecting (anti-) estrogenic substances, *Environ. Toxicol. Chem.*, 21, 319–326, 2002.
48. Sohoni, P. and Sumpter, J.P., Several environmental oestrogens are also anti-androgens, *J. Endocrinol.*, 158, 327–339, 1998.

chapter five

*Conducting dose–response feeding studies with salmonids: Growth as an endpoint**

James P. Meador, Frank C. Sommers, and Leslie Kubin
National Marine Fisheries Service

Robert J. Wolotira
National Oceanic and Atmospheric Administration

Contents

* Reference to a company or product does not imply endorsement by the U.S. Department of Commerce to the exclusion of others that may be suitable.

1-56670-664-5/05/$0.00+$1.50
© 2005 by CRC Press

Introduction

This chapter provides details on the procedures for conducting toxicant dose–response feeding studies with salmonids. The main focus in our laboratory is to assess the effects of various toxicants that juvenile salmonids are exposed to as they migrate through urban estuaries. Many of the procedures outlined here are general and may be applied to other fish species. This chapter will primarily discuss how growth is evaluated; however, once the fish are dosed for a specific time period, most other desired biological response, such as immunochallenge, physiological function, behavioral assays, or reproductive abnormalities may also be assessed.

Feeding toxicants to fish is an ideal way to mimic environmentally realistic exposure. Injecting toxicants in large doses can overwhelm physiological systems producing results that may never be observed in fish from contaminated sites. Information on contaminant concentrations in stomach contents and prey species can be used to design dose–response studies where single toxicants or mixtures can be applied to food and fed for a specified period of time. Although it is difficult to accurately mimic the exposure fish experience in the field, the results from controlled laboratory studies with defined doses can add greatly to ecological risk assessments, critical body residue determinations, and bioaccumulation evaluations. These types of studies are also valuable for the researcher interested in physiological mechanisms of response from field-like exposures.

Materials

Water system

Fiberglass tanks for holding fish and conducting experiments (e.g., 1.3 and 2 m diameter), polyvinyl chloride (PVC) water supply lines, valves, flexible tubing, and PVC drain lines; biofilter (surface area determined by biomass loading), nitrifying bacteria to seed biofilter, biosolids filter (capacity determined by biomass loading), charcoal filter (size determined by flow rate), submersible pumps, water chiller (size determined by flow rate and degrees cooling), water heater (size determined by flow rate and degrees heating), and ultraviolet (UV) sterilizer (wattage determined by flow rate); airstones, Tygon tubing, and air blower (oil free).

Fish husbandry/experimental

Egg transfer containers, coolers, egg incubation trays, automatic fish feeders, fish nets (large and small), iodophor for the eggs, disinfectant for fish nets and floor mats, and several clean 20-l buckets; balances (e.g., basic toploader for weighing food and a high capacity balance for weighing fish in water); water chemistry test kits (e.g., ammonia, nitrite, alkalinity), oxygen meter, flow meters, pH meter, salinometer, water level sensor, and thermometer (continuous/recording).

Fish food and dosing

Fish food (commercial food for the early life stages, low fat pellets for the exposure phase, and food with antibiotics for disease outbreaks); pipettes, covering the range of 10–100 μl, 100 μl to 1 ml, 1–10 ml, Erlenmeyer flasks with glass stopper, and a large spoon; balance (high precision for weighing test chemicals); tricaine (MS-222), dichloromethane ($MeCl_2$), fume hood, stainless steel bowls, trays/tubs (e.g., polycarbonate, stainless steel), aluminum foil, and polyethylene tubs for food storage.

Procedures

A. Source of fish

Fish for experimentation can be obtained from many sources and usually at any desired life stage. They can be obtained from hatcheries as eggs and raised in the laboratory to the desired life stage for testing or collected from the field. We recommend obtaining eggs (as gametes or at the eyed stage) and culturing fish to ensure high quality juveniles for testing and total control over all stages. We routinely raise 6000 juveniles from eggs, which is not difficult, if water quality parameters are kept within prescribed limits (see Appendix I). We obtain our eggs from the University of Washington hatchery and in the past have acquired fish in the smolting stage from the Soos Creek Hatchery in Auburn, WA. Salmonids at different stages of development can be purchased from various sources (e.g., Trout Lodge, P.O. Box 1290, Sumner, WA. 98390, 253.863.0446; AquaSeed, 2301 NE Blakeley Street, Suite 102, Seattle, WA 98105, 206.527.6696, info@aquaseed.com).

B. Raising fish from eggs

1. Obtain the desired number of high quality eggs from a reputable source that can certify the eggs are disease free. If the eggs are to be transferred to another location, a permit may be required from the appropriate agency (e.g., State Department of Fisheries).
2. The transportation of eggs requires egg containers. In our laboratory, we use sections of 4-in. PVC pipe with screens attached to one end, a cooler, and clean cotton towels. For transport, eggs are placed in the containers, which are then stacked in a cooler. Wet towels are placed on all sides of the egg containers. Eggs are then transported to the laboratory and placed in an incubation rack.
3. The total number of eggs received can be estimated by the following method. The number of eggs in a small volume is determined (e.g., 100-ml beaker). The total volume is then determined with a suitable container (e.g., 1-l glass Pyrex beaker). This total volume and the actual count of eggs/ml are multiplied to determine the total number of eggs in culture. Estimating the total number of eggs can also be accomplished by weighing; however, determination by volume is more convenient for dispensing eggs to the incubation trays.

4. The eggs are then placed in an incubation chamber that has been disinfected with an appropriate disinfectant (e.g., iodophor). The incubation tray should be mounted on supports over a water reservoir tank (e.g., 2-m-diameter fiberglass tank). When additional water (makeup water) is needed to keep a constant volume, it should be passed through an activated charcoal filter to remove contaminants and chlorine and a UV sterilizer to reduce pathogen exposure. Water in the reservoir tank is recirculated through a chiller or heater, biofilter, and UV sterilizer. This water is then pumped to the top of the incubator by submersible pump and allowed to flow over all trays holding eggs. Two pumps are used for redundancy and should be plugged into separate electrical circuits. A recirculating water system is desirable for this phase to control water quality parameters; however, freshwater can be supplied from diverted natural sources (e.g., lake or stream) or from a municipal system and treated.
5. The density of eggs in the hatching trays should be low enough to allow a complete flow of water over the eggs, as well as sufficient access to remove dead eggs. A flow of 1 l/min should be more than sufficient for a large number of eggs. It is important that the oxygen content of the water exiting the incubation trays is maintained at 100% saturation. If less than 100%, the flow rate will have to be increased. The front and back of the incubator are covered with black plastic to protect the eggs from exposure to light. The temperature of the water should be optimized for this life stage. This is approximately 10°C for chinook salmon (*Oncorhynchus tshawytscha*).
6. Environmental parameters (see Appendix I).
7. The incubation chamber should be checked daily for dead eggs or larvae, which must be removed to reduce fungal infection.
8. Alevins (newly hatched salmon with yolk sac attached) should remain in the incubator until the yolk sac is almost completely absorbed or when larvae are swimming freely around the trays. Fish can then be gently transferred to a holding tank (e.g., the 2-m water reservoir tank).

C. *Raising salmon fry to presmolt juveniles*

1. A loading density not to exceed 20–30 larvae per liter of water is recommended. The characteristics of the water source will have to be taken into consideration for the loading density. A flow-thorough system will support more fish than a recirculating system.
2. Environmental parameters (see Appendix I).
3. Feeding the larvae is initiated at an appropriate time post hatch. For chinook salmon, this occurs when fish begin spending more time in the water column instead of the tank bottom. Commercially prepared fish food specially formulated for salmon larvae is recommended. This food should be introduced every 0.5–1 h via automatic feeders, at the rate and pellet size recommended by the feed manufacturer. A 2-m diameter fiberglass tank with autofeeder is shown in Figure 5.1.
4. When fish reach an average size of approximately 1–2 g, they may be fed 3–4 times/day by hand. The size of the fish food should be increased according to the schedule presented in Table 5.1, which is approximate for chinook salmon held at 9–11°C. Additional details on pellet sizes and feeding rates can be obtained from

Figure 5.1 Large fiberglass tanks (2 m diameter) used for rearing juvenile chinook. Automatic feeder shown above each tank.

Table 5.1 Example of feeding schedule for juvenile chinook salmon

Stage	Size of fish	Size of pellet	Percentage of bw/day
1	Larvae to 4 g	Micro pellet	2.8–2.0 as fish increase
2	4–6 g	1.2 mm	1.7
3	6–25 g	1.5 mm	1.6–1.8
4	25 to 80+ g	2.0 mm	1.4–1.5

manufacturers (e.g., EWOS Canada, www.EWOS.ca). The schedule in Table 5.1 showing pellet size and rate of feeding may be different for other species of fish in culture and should be adjusted according to feed manufacturer specifications.

5. As soon as the fish weigh approximately 1 g, a subsample of 100 fish should be weighed individually to determine their mean weight. After this, all juvenile fish can be weighed in large batches to determine the total biomass (see Appendix II). The total biomass in grams and the average weight of the sample will be used to determine the total number of fish in culture. This value can be used to determine feeding rates.
6. Once the total number of fish and biomass are established, the amount of food to be dispensed can be determined. For juvenile chinook salmon, we dispense food at a rate of 1.5–2.0 % body weight (bw)/day. An appropriate amount of food is weighed and can be given to the fish in 2 or 3 feedings within normal working hours (e.g., at 10 am and 4 pm).
7. A sample of 100 fish should be weighed once per week to determine the average weight of all fish in the population. This average weight is used to adjust the amount of food given to the entire population.

8. When juvenile fish weigh approximately 2 g, they are switched to a fish pellet that contains a total lipid content that mimics the prey for wild fish (see Appendix IV.A).

D. Introducing smolts to seawater

If experiments are conducted in seawater, the fish must be acclimated to high salinity water. Several parameters may be used to determine when smolts are ready for entry into seawater.[1] These include gill ATPase, endocrine hormones (e.g., thyroxin), body lipid, blood sodium, or skin coloration. The following procedures are described for juvenile chinook salmon. This protocol may need to be adjusted for other salmon species.

1. When the juveniles weigh approximately 2–4 g, they should be moved into larger tanks (e.g., 2-m diameter) at a density not to exceed approximately one fish per liter. At this time, the fish should remain in the freshwater system and allowed a few days after the transfer to acclimate. In our laboratory, we keep up to 2000 fish in 2-m-diameter tanks, which hold approximately 2000 l of water. For each tank, we maintain a flow of 40–50 l/min, which is well within the prescribed limits of 0.6 l/min/kg at 10°C.[2] It has been our experience that fish at a density of one fish per liter need relatively high flows, otherwise dissolved oxygen levels will decline rapidly.
2. Environmental parameters (see Appendix I).
3. For juvenile chinook salmon, it has been demonstrated that the physiological parameters for making the transition to seawater are optimized during the new moon.[3] When our young of the year chinook salmon are larger than 5 g and their skin coloration is silvery, we introduce them to seawater during the new moon in May, June, or July. During this time, the salinity is raised from 0 to 7 parts per thousand (ppt) on the first day. After this, the salinity is increased by 5 ppt/day until full strength seawater is attained. After achieving full strength seawater, the fish are given a minimum of 2 weeks to acclimate before handling.
4. Once in seawater, these fish are fed 1.5% of their body weight per day with the low fat fish food at rate of 5–6 times per week. The food is generally delivered in two or three equal batches over the day to total the 1.5% bw ration.

E. Distribution of fish to experimental tanks

1. Before fish are distributed, the experimental tanks should be thoroughly cleaned with acid and disinfectant. All fouling organisms need to be removed from the tanks, and all water delivery lines and drains need to be checked for obstructions. In our laboratory, we use a dilute solution (1 *M*) of muriatic or hydrochloric acid to remove fouling organisms from the tanks, which is applied twice. After this treatment, we apply a solution of quaternary ammonium for 15 min to disinfect the tanks. This disinfectant comes with directions for concentration and time. These cleaners can be purchased from vendors specializing in aquaculture supplies. After cleaning the tanks, the entire system should be operated for at least 24 h to flush out any remaining

acid and disinfectant. After this, the tanks should be drained, refilled, and allowed to operate normally for at least 48 h before fish are added.

2. Within 1 week prior to distribution of fish to experimental tanks, a subsample of fish (e.g., 100 fish) should be selected at random from the common pool of fish to determine the mean and standard deviation (SD) of their weight. This information will be used to select the size range of fish to be included in the experiment. For growth studies, it is ideal to keep the SD of the mean weight for each experimental tank relatively low and uniform across tanks.
3. For growth studies, a size range of fish that produces a relatively low SD should be selected before hand. In our experiments, we strive for a coefficient of variation (CV = SD/mean * 100) of approximately 10–15%. This range in weights is subject to change to meet experimental needs and open to interpretation at the time of distribution. One major consideration is the number of fish available and the expected rejection rate. For example, in one of our experiments, we selected a weight range of 8–10 g, which produced a mean fish weight for each tank of approximately 9 g with SD = 0.5–0.6 (= CV of 7%). In our experience, a CV of 10–15% requires twice as many fish as needed to run the experiment due to a rejection rate of approximately 50%. Another strategy is to distribute all fish randomly to the experimental tanks; however, this will likely lead to greater difficulty in data interpretation because of the larger variance for mean weight.
4. Fish are weighed individually to determine their weight (Appendix II). If a weight range has been selected as described in step 3, each fish will be accepted or rejected for the experiment based on its weight. Two clean buckets with approximately 15 l of clean water from the system should be close by. Fish that will be used in the experiment go into one bucket and rejected fish into the other. Each bucket will need an airstone to keep the oxygen saturation level between 70% and 100%, and the temperature should not rise more than 1°C from the system temperature.
5. The number of fish added to each experimental tank will vary depending on the species selected, tank volume, and flow rate of water. In our experiments, we place approximately 75 fish in each 1.3-m-diameter tank. This has proven to be an optimal number of fish for these tanks based on consideration of biomass/liter, behavioral interactions, and water quality. The density for salmonids of this size is approximately 1.5 g of fish/l. Figure 5.2 shows an array of 1.3-m-diameter fiberglass tanks. This facility has 32 experimental tanks.
6. The number of experimental tanks for each treatment should be determined by a power analysis. See Appendix III for statistical methods. The order of filling of the tanks with fish will be determined with a random number generator. Because selecting fish from the common pool of fish is often not random, fish need to be added to the experimental tank in batches. When dip-netting fish from the common pool, often the smaller and slower fish are caught more easily and these should be randomly distributed among the tanks (see Appendix III.B). Fish are usually added in three equal batches. If 75 are to be added, approximately 25 fish will be added to each tank based on the order determined with the random number generator. After this, the next 25 should be added in the same order as the first round of additions.
7. Once all the fish are distributed to the experimental tanks, an analysis of variance (ANOVA) and *post-hoc* test is performed to determine if the mean

Figure 5.2 Picture shows an array of 1.3-m-diameter (500 l) fiberglass tanks. Total array is 32 tanks. Note the air supply lines, drain stand pipes, spray bars for water delivery (inside the tanks), overhead lighting, and the tank with mesh cover.

fish weight in any tank is significantly different from that in any other tank. If a mean fish weight of a tank is significantly different, fish may be removed or new fish added until no significant differences are observed. This can also be accomplished in an iterative fashion by examining the ANOVA for each round of 25 fish that are added to tanks. If any of the tanks exhibit deviation at any stage, corrective measures (e.g., adding heavier or lighter fish within the selected range) can be taken to ensure that a non-significant mean weight will be obtained. It is also important that the variance in fish weights among tanks is uniform. A test to examine the homogeneity of variance (e.g., Bartlett's) should be performed.

8. After all fish are distributed to the experimental tanks and the mean weights for each tank are deemed non-significant, then a treatment designation for each tank will be chosen with a random number generator such that replicates from a given treatment are scattered through the complex of tanks. This procedure will minimize the effects of any potential environmental gradients that may occur in the physical space containing the experimental tanks (see Appendix III.B).
9. Any fish that dies between the time of distribution and the start of the experiment should be replaced. Because dead fish will assume the same osmotic pressure as their environment and dehydrate or superhydrate, depending on the salinity, they should placed in a physiologically isotonic solution (0.17 M or approximately 1% salt solution) for 2 h, removed, blotted of excess water, and weighed. A fish of comparable weight should then be removed from the database for that tank and the weight of the replacement fish inserted in its place.
10. After distribution to the experimental tanks, the fish should be allowed to acclimate for 7–10 days before feeding with experimental food. During this

period, the fish are fed unadulterated fish pellets that will be used in the experiment.

11. Tank covers are highly recommended to keep fish from jumping out and to exclude predators. We use a soft fiber mesh material (white) attached to two curved pieces of flexible 1-in. tubing that allows easy access to the fish and the ability to add food without lifting the cover (Figure 5.2). The covers should also be disinfected and dried thoroughly before fish are added to the tanks.

F. *Conducting the experiment*

1. After distribution of the fish to experimental tanks is completed and an acceptable mortality has been achieved (e.g., <0.2%/day), fish will be fed contaminated fish pellets at rate to be determined (e.g., 2.0% bw/day). The total amount fed per tank is based on this rate and the total biomass for that particular tank. See Appendix IV.B for information on preparing toxicants-dosed fish pellets. Fish are fed 5 or 6 days/week. Appendix I.D contains details on feeding schedule and increasing the food ration.
2. In our laboratory, juvenile chinook salmon are usually fed 2.0% bw/day to achieve a sufficient rate of growth for testing the hypothesis of toxicant effects on growth. A feeding rate of 1% bw/day for juvenile salmonids at their optimum temperature (approximately 10–11°C) is very close to a maintenance dose that would produce very little growth. For juvenile chinook salmon in our system under our usual conditions, the growth efficiency is approximately 20%, i.e., 1 g of dry food produces approximately 1 g of fish biomass based on wet weight ($\approx$0.2 g of fish dry weight).
3. Environmental parameters (see Appendix I).
4. The fish in each tank are fed contaminated pellets that have been dosed with toxicants for a period of time appropriate for the study. Considerations for the dosing period include the growth rate and life cycle of the fish, the amount of dosed food that can be made in one batch, and the types of biological responses measured. In our laboratory, we commonly feed dosed food for 45 days and unadulterated food for several months after the dosing period.
5. After a predetermined time period (e.g., 45 days), the fish in each tank are weighed. Each individual fish in a tank is weighed by one of the methods described in Appendix II. The data are generally recorded and saved in an Excel spreadsheet then exported to a statistical program for analysis (see Appendix III). If the experiment continues, the total biomass obtained for each tank should be used to adjust the amount of food given for the next feeding cycle. Each tank is then given a unique amount of food based on the total biomass obtained from the weighing. See Appendix I.D for information on increasing the weekly food ration.
6. After a predetermined time (e.g., another 45 days or day 90 of the experiment), the fish in each tank can be weighed again in the manner described in Appendix II and at regular time intervals, if so desired.
7. This experimental design also allows several other chemical and biological responses to be quantified. If sufficient fish are available from each tank, samples for whole-body chemistry, blood for physiological parameters,

internal organs for biomarkers or morphological indices (e.g., hepatosomatic index) can be taken. A select number of dosed fish may also be removed for separate experiments, such as a disease challenge or behavioral tests.

8. At a minimum, the fish food should be analyzed for the added toxicants to determine the actual exposure concentrations for experimental fish. If resources permit, a broad analytical scan of several potential contaminants [e.g., polycyclic aromatic hydrocarbons (PAHs), polychlorinated biphenyls (PCBs), phthalates, and metals] should be conducted for one or two samples of the control food to determine background contamination.
9. For some experimental designs, it may be desirable to determine the toxicant concentrations in fish tissue (e.g., whole-body, liver, gill) or metabolites in bile. Our laboratory routinely analyzes whole-body fish to determine bioaccumulation of toxicants from fish food and for correlation analysis with biological responses. Fish are starved for 24 h before whole-body or bile samples are taken. Stomach contents and the alimentary canal should be removed during sampling if correlations are to be generated for fish responses, or they may be left in place if the total concentration is desired for assessing exposure to those species that prey on salmon.

G. *Quality control/assurance*

The details of a quality assurance program are to be decided by the laboratory conducting the experiments and the intended use of the data. Examples of quality assurance plans and guides can be found on various government websites.[4] It is highly recommended that a database program (e.g., Filemaker Pro™ or MS Access™) be used to keep track of feeding regimes, mortalities, and fish removed for various samples or other experiments. Chain of custody numbers and forms are desirable for every sample, even if this research is not conducted under the auspices of any environmental statute. For these types of studies, standard practice for quality control includes calibration of balances and measuring instruments (e.g., thermometers, oxygen meters), feed quality (low contaminants, correct nutrient mix), and associated controls for analytical chemistry. Periodic checks on food allocation should also be made to ensure correct dosage amounts.

All results for water quality measurements, adjusted water quality parameters, fish weights, feeding schedules and amounts, mortalities, abnormalities, corrective actions taken, and any observational notes are recorded in a laboratory notebook by the scientist in charge of fish husbandry. These notes will be useful for assessing problems and should be reported with the results of each experiment.

Discussion/considerations

The preceding has been a discussion on how to set up and conduct growth studies with juvenile salmonids exposed to toxicants in their food. Some of these techniques and analytical tools will be appropriate for other species of fish; however, several aspects of the species life history will need to be considered.

Growth is just one biological response that can be assessed with this design. Once the fish are dosed, they can be used in a multitude of other studies designed to assay for effects. If growth is the main concern, the researcher may want to delve into the mechanism responsible for the response. For example, decreased growth may be due to a physiological response, and even though the treatment fish are eating just as much as the control fish, their metabolism may be altered producing a negative effect on growth. A reduction in growth may also be caused by the fish becoming lethargic and not ingesting the same amount of food as the control fish or rejecting food because of reduced palatability. These should all be part of the routine suite of behavioral observations for this type of experiment.

In our experiments, we consider fish weight to be a more important variable than fish length for assessing growth impacts due to toxicant exposure. This is supported by a few studies noting that fish weight was more affected than fish length for salmonids exposed to toxicants.[5,6] Measuring length is important, however, for determining the condition factor (K) (see Appendix III.C), which can provide general information about the health of the fish. In general, when fish are stressed, the condition factor usually declines indicating that a fish will be lighter for a given length.

There are several excellent books on fish aquaculture, systems engineering, and fish diseases that should be consulted for more details and specific situations.[7–11]

Acknowledgments

Many thanks to Paul Plesha, Mark Tagal, and Casimir Rice for advice on fish husbandry and aquaculture and for helping out with system maintenance and fish care when needed. A very special thanks to Dr. Karl Shearer for technical advice regarding fish growth and food formulation and for his expert review of this manuscript.

Appendix I Environmental conditions

Freshwater or seawater water can be supplied to fish with a recirculating or flow-through system. This delivery system should have a charcoal filter in line to remove contaminants and sufficient UV sterilizers to keep the water pathogen free. For both flow-through and recirculating water systems, a biosolids filter is highly recommended to remove wastes and excess food. A biofilter is highly recommended for recirculating systems to provide sufficient surface area for nitrifying bacteria to keep ammonia and nitrite levels low. It is highly recommended to seed the biofilter with a bottle of nitrifying bacteria, which can be purchased from any aquarium or pond store. A flow-through system is recommended for the toxicant exposure phase because recirculating systems are difficult to maintain and it is a challenge to keep fish in different treatments from being cross-contaminated. Figure 5.3 shows the biosolids filter, air blower, and UV sterilizers for the experimental facility in Figure 5.2.

I.A Water quality parameters

Water quality (pH, ammonia, nitrite, temperature, hardness, oxygen content, and alkalinity) should be checked frequently. Recirculating systems will require more frequent analysis of water quality parameters than flow-through systems. Additionally, some parameters are more variable in freshwater systems (e.g., pH, hardness, alkalinity) than in seawater systems. Parameters that are likely to change quickly and can cause acute stress in fish (e.g., pH, temperature, ammonia, nitrite) should be checked several times per week and other

Figure 5.3 Picture shows a drum filter to remove solid material from the water line, two UV light sterilizers (mounted on wall), and an air blower for supplying air to each tank. The large UV unit contains 8 tubes (40 W each) and the small unit under the plywood contains two tubes (200 W each). These components are for the tank array in Figure 5.2.

parameters (alkalinity and hardness) may be checked less frequently (e.g., once or twice per week). All values should be recorded in a notebook or computer spreadsheet. Water temperature should be maintained at an appropriate level for the species of choice and it should be checked several times per day, preferably on a continuous basis.

Many of these parameters can be determined with electrodes (pH, oxygen content) or colorimetric test kits (hardness, ammonia, nitrite, alkalinity). Corrective measures must be taken if any of the parameters are outside the accepted range. The values in Table 5.2 are for juvenile chinook salmon, different values for each parameter may be required for other species. Any corrections made to water quality should always be done in a cautious manner to minimize any large changes in the parameter being optimized. The parameter being changed should be measured within 1 h after corrective action is taken and every hour thereafter until the parameter value has stabilized. Oxygen content should be monitored continuously until optimized. If a flow-through seawater system is being used for delivery, it is the discretion of the researcher to determine how to optimize the parameter (e.g., increase flow if oxygen content is low, addition of a charcoal filter if ammonia is high, use chillers to reduce temperature).

I.B Water flow rate

The flow of water will depend on the biomass in the tank. For the incubation trays, a flow sufficient to keep the eggs at the proper temperature should be maintained. For the small salmonids (<10 g), Heen et al.[7] recommend 0.3–1.0 l/min/kg fish, for temperatures found in Table 5.2. In our laboratory, we have found that a flow rate of approximately 2–4 l/min for the 1.3-m-diameter tanks (500 l) with 75 fish is adequate to maintain proper water

Table 5.2 Example of water quality parameters for juvenile chinook salmon in a freshwater recirculating system

Parameter	Minimum limit	Maximum limit	Optimum value
Ammonia	0 ppm	0.5 ppm	<0.1 ppm
Nitrite	0 ppm	0.5 ppm	<0.1 ppm
pH	6.5	8.5	7.8
Hardness	50 ppm	200 ppm	125–150 ppm
Alkalinity	50 ppm	150 ppm	75–100 ppm
Temperature	8°C	12°C	10–11°C
Oxygen (saturation)	70%	100%	95–100%

quality. For our larger 2-m tanks (2000 l) with up to 2000 juveniles (4–10 g), we maintain a flow rate of at least 40 l/min to maintain the oxygen content above 85–90% saturation. In each of our tanks, we have a "spray bar" connected to the inflow hose. This is a section of PVC pipe (2 cm diameter by 30–50 cm length) with approximately 8–10 holes (0.6–1 cm) drilled in a line. Water exits the spray bar at a high rate into the tank, which helps aerate the water (Figure 5.2). The jets of water are set at about 45° to the water surface and help to create a current in the tank that the fish orient to and swim against. As a fail safe measure, a sufficient number of airstones should be placed in each tank to ensure that the oxygen content does not fall below 70% saturation at any time, especially in the event that the recirculating or flow-through delivery system fails (Figure 5.2). This should be tested periodically by shutting off the water flow to a tank with fish and recording the change in oxygen content. Corrective measures may be needed (e.g., more air stones or a more powerful blower).

I.C Light

If the tanks are in a building, the light will be artificial. It is important to have adequate light in the building and bulbs that reflect the natural spectrum as close as possible. The daily light schedule should match that of ambient daylight. Ideally, the light fixtures will be on timers and rheostats that will gradually increase and decrease the light level.

I.D Feeding fish and increasing food ration

Fish in culture and those in the experiment are fed 5–6 times/week. Generally, an appropriate amount of food will be weighed and given to the fish in 2–3 feedings over the day. Personnel feeding the fish should take all the necessary precautions to avoid contact with the food containing toxicants. The amount of food placed in each tank is usually the same for a 1-week period and is based on the total biomass and the selected feeding rate (e.g., 2.0% bw/day). At the beginning of the second week, the amount of food distributed to each tank is increased to reflect an increase in fish biomass. The rate fed will be the same 2.0% bw/day, and the new total biomass is based on a growth curve projection from previous studies or actual data. The actual data can be obtained from total biomass weights that are determined from a separate "growth control" tank of fish that is treated identically as the experimental tanks but fed control food (solvent treated). Weekly or daily growth increments and increased ration amounts may also be calculated with equations similar to those used in financial applications to determine compound growth.

I.E General husbandry

Each tank should be inspected daily for dead fish and checked for properly working spray bars, airstones, and drains. If any abnormalities are observed, they should be corrected immediately. Dead fish are removed daily and recorded for each tank and the data used in the statistical analysis. Once per week, excess food and fish waste should be removed from the bottom of each tank. This can be accomplished by gently breaking the waste loose for suction down the drain or by a separate device that will siphon or suck the material out of the tank. If the fish appear lethargic or mortalities increase independent of toxicant treatment, a fish health specialist should be consulted immediately. The specialist may recommend fish food with antibiotics, which can be purchased from feed manufacturers, or possibly a formalin dip if external parasites are evident. There are several references (e.g., Stoskopf[8]) with additional information on treatments for fish diseases. It is also recommended to have floor mats (specially designed) with disinfectant at the entrance to the fish holding/experiment area to reduce the transfer of pathogens. In addition, materials that come in contact with the fish (e.g., buckets and nets) should also be disinfected after each use.

The size of the experimental tank is also an important consideration. We have found the 1.3-m-diameter circular tank to be ideal because of its volume and the ability to allow a circular current for the fish to swim against. In the past, we conducted experiments in 0.7-m square tanks (approximately 200 l) that were shown in a paired comparison test (unpublished) to cause increased mortality in juvenile chinook salmon. This increased mortality was likely due to increased stress caused by crowding and the lack of a circular current.

I.F Miscellaneous

The building that houses the tanks should be secured such that predators and pests (e.g., cats, otters, rats) cannot enter. If there is evidence of such animals in the building, they should be trapped and removed. Sensors to detect system malfunctions are highly recommended. In our laboratory, temperature probes are connected to a computer that can notify personnel by auto dialer and pager if values occur outside a preset range. We also have flow meters that connect to our security system, which is monitored continuously. Any interruptions in flow are detected by the security monitoring company, which can initiate a call to the appropriate personnel.

Appendix II Method of weighing fish

Fish can be weighed as a group for determining total biomass, as individuals when distributing to experimental tanks, or when assessing the frequency distribution of their individual weight. Weighing individual fish can be done by two different methods. One method is to narcotize the fish and weigh in a tared plastic container on a top-loading balance. Another method of weighing is by water displacement, which is less stressful on the fish and minimizes handling.

For the first method, a narcotizing agent, such as tricaine methanesulfonate (also known as MS-222), is used to immobilize the fish. For juvenile chinook salmon, we have used a solution of 125 mg/l, which takes approximately 5 min to affect the fish. This narcotizing agent is difficult to work with because it suppresses the respiratory system and fish can die if exposed too long. Research has also shown that MS-222 may adversely affect physiological parameters[12] and possibly the toxicological response of the fish.

For the water displacement method, we use a Sartorius 4000 balance that weighs items up to 50 kg with a 0.1-g accuracy. A 20-l bucket with approximately 15 l of water is placed on the balance and tared to 0.0 weight. Fish are netted and placed in the bucket as a group or individually to obtain their weight. The net with fish is allowed to drip for a short period of time (e.g., 3–5 s) such that very few, if any, drips fall from the net. This will ensure relatively accurate weight of each fish being added to the container of water. At the beginning of each day when fish are weighed, each scientist weighs one fish 3 times to assess variability in their technique. This information can be used as a quality control guide for fish weighing.

For all weighing methods that hold fish in buckets, airstones must be present to keep oxygen content between 70% and 100% saturation and fish biomass kept below 20 g/l. Oxygen content should be checked frequently with an oxygen meter to ensure adequate levels (70–100% saturation) for fish respiration. Fish are generally held in the buckets for no longer than 20 min. When this time limit is reached, the bucket with fish is gently poured into the holding tank from where they originated or a suitably prepared tank for long-term holding or experimentation.

Appendix III Statistical design and analysis

III.A Determining the number of experimental tanks

The first step is to determine how many treatment concentrations are to be tested. This usually requires some previous information regarding the expected response for a given exposure concentration, such as previous studies in your laboratory or literature values. If this type of information is not available, a range-finding assay can be conducted to provide a crude estimate of the concentration needed to produce an adverse effect. This can be accomplished with one or two replicates for each order of magnitude increase in exposure concentration (e.g., 0.1, 1, 10, 100 μg/g of food). This design will likely produce results with no response for the lower concentrations and a 100% response for one or more of the higher treatment concentrations. Once an adverse exposure concentration range is determined, a more refined experiment can be designed with smaller differences in exposure concentrations (e.g., 2- or 3-fold increases) that will likely to provide partial responses for one or two of the exposure concentrations (e.g., 25% reduction in growth). Statistically, it is advantageous to use the same factor difference between exposure concentrations, the same number of replicates per treatment, and to have partial responses. The number of fish (sampling unit) per replicate (experimental unit) will depend on various parameters for the test species, such as accepted biomass per liter, behavioral interactions, the number of desired samples, and long-term goals.

To determine how many replicates are needed for each treatment, a discussion regarding the "power of the test" is relevant. It should be noted that independent replicates are required for hypothesis testing and inferential statistics. It is often tempting to have only one or two tanks (experimental units) and measure some response in several individuals and call these "replicates" (i.e., the sample size (n) becomes all measured values). Using data from several individuals from one tank would be pseudoreplication because they are not truly independent.[13] For testing the hypothesis that there are no differences among treatment means ($\mu_1 = \mu_2 = \mu_3 = \mu_n$), each tank would be a replicate and one data-point (e.g., mean weight) from that tank would be used for the analysis.

Power is the ability to avoid false negatives or type II errors. A false negative is equivalent to finding no effect when in fact one actually exists. Also, a false negative is accepting the null hypothesis (H_0: $\mu_1 = \mu_2 = \mu_3 \cdots$,) of equal treatment means when you should reject the null hypothesis and is determined by β, the type II error. In many experiments, β is high, meaning there is a high probability that the null hypothesis will not be rejected. A power analysis is highly dependent on the experimental variance, which may be higher or lower than estimated for a proposed experiment. The variance that determines power is the variance among tank means within treatments. The actual power analysis cannot be conducted until after the experiment is performed. Because the variability between replicates cannot be known before the experiment is conducted, maximizing the number of replicates in the experimental design is advantageous to assure sufficient power of the test. An example of the expected power for a typical experiment follows.

Example:
Treatments (k) = 9, replicates (n) = 5, total tanks (N) = 45, $\alpha = 0.05$, $v_1 = 8$ degrees of freedom (df) = $k - 1$, $v_2 = 36$ df = $k(n - 1)$, σ^2 (mean square error) is estimated by $s^2 = 1.4\,g^2$ (from previous experiment); δ = minimum detectable difference = 3 g (a value based on previous work to maximize the power of the test):

$$\phi = \sqrt{\frac{n\delta^2}{2ks^2}} = 1.33.$$

Once the quantity ϕ (phi) is calculated, the power of the test is then computed for a specific difference. From a statistical table (e.g., Figure B.1h in Reference 14), this ϕ value equates a power of 0.7. This is equivalent to saying that there will be a 30% chance of committing a type II error in this analysis. Compared to other biological experiments, this power is acceptable.[15] If only 4 replicates were used, the power would drop to 0.67 and it would increase to 0.72 for 6 replicates.

To determine the number of replicates for each treatment, the following equation for independent samples can be used[16];

$$N = \frac{(Z_\alpha + Z_\beta)^2 2\sigma^2}{\delta^2}$$

Z_α and Z_β are the probability associated with one tail of the normal distribution. For a two-tailed probability, use $Z_{\alpha/2}$. For example, using a type I α error rate of 0.05 and a type II β error rate of 0.2 (20%; which equals a power of 80%), $Z_{0.05/2} = 1.96$ and $Z_{0.2} - 0.85$ for a two-tailed test. As suggested by Steel and Torrie,[16] because the variance is often not known, it would be advantageous to define the minimum detectable difference in terms of the SD. If the SD is set to 1, $2\sigma^2/\delta^2 = 2$, our result for a two-tailed test would be $N = (1.96 + 0.85)^2 * 2$ or a sample size of 15.8 (round up to 16). For most experiments, this is far too many replicates for the available resources. By adjusting the minimum detectable difference, type I or type II error rate, or the experimental variance, the number of replicates needed for avoiding type II errors can be reduced.

There is nothing sacred about using $\alpha = 0.05$ for determining statistical significance. Because α and β are inversely related for a given sample size, choosing an α of 0.10 would decrease β and increase the power of the test. For many situations, a higher α value is desirable to assure environmental protection (see an in-depth discussion by Peterman[15]).

By weighing several fish from each tank, we may have much more information that may be used to determine the power of the test; however, a statistician should be consulted if this approach is used. By taking multiple samples per tank (e.g., 15 fish), we may be able to consider the power analysis in terms of composite sampling. Recent research has provided formulae for calculating power when composites are used.[17] According to the method provided by Edland and van Belle,[17] the power for our example above would be approximately 0.88, or there would be a 12% chance of making a type II error. This would be for a smaller minimum detectable difference (= 0.6 g). Predictions for power using the same design but with 4 replicates would yield a power of 0.80. Six replicates would boost the power to 0.95.

III.B Randomization of fish and treatments

It is crucial that each replicate in the experiment is treated identically. Because gradients exist, a simple way to minimize the impact of all uncontrolled factors on the dependent variable is to randomize individuals over treatments and experimental units. When designing an experiment, there are many physicochemical and biological factors, such as light, temperature, salinity, food ration, behavior, and organism interaction that must be considered. These factors are either controlled, i.e., included in the experimental design, or their influence is minimized.

The experimental units in the study are the tanks, and these can affect the dependent variable in a variety of ways. The most obvious sources of tank effect are location and the order that the tanks are filled with fish. For example, tanks closest to an exterior wall may experience greater temperature fluctuations, light levels, or disturbance than the interior tanks. Also, the first tanks filled may contain fish that are less robust than fish placed later in the tanks because the "weaker" ones may be captured more easily from the larger pool of fish. To avoid confounding these effects with treatment effects, the order that the tanks are filled should be done randomly with respect to location.

As described in step 6 of the section "Distribution of fish to experimental tanks," we add fish in three groups of 25 to the experiment tanks. A recent study that examined the statistics of various random allocation schemes of organisms to experimental chambers (6 treatments, 4 replicates, 10 fish per tank) found that the best scheme was to randomly add each individual to tanks over all treatments.[18] The authors also concluded that a scheme that randomly allocates all fish or half of the total to each tank over all treatments also produces acceptable statistical results. Our scheme for allocation is similar to the latter one recommended by the authors and is more appropriate for experiments with large numbers of fish.

After fish allocation is completed, the next step is to randomize treatments (and replicates) among the experimental units (tanks). The treatments must be randomized in regard to the order the tanks were filled because of the potential for interactive effects of the toxicant with some factor related to the order of tank filling. Each tank is assigned to a treatment in a random fashion until all treatments and replicates are determined.

The order by which fish and treatments are randomized is accomplished with a random number generator. These can be obtained from statistical tables (e.g., statistical tables of Rohlf and Sokal[19]) or from a spreadsheet (e.g., Microsoft Excel).

III.C Analyzing the results

While it might be tempting to weigh only a subsample of all fish in a tank, there is usually such high variability among fish weights that a relatively large proportion of the total would be required for sufficient representation. This, in addition to occasional non-random sampling, argues for weighing the entire population of each tank.

While we do not measure length in the beginning of the experiment, it is useful to determine length at the end. Using length and weight, a condition factor (K) can be calculated that will provide information on the general health of the fish. Comparing K values among treatments can be an important endpoint for assessing toxicant impacts.

$$\text{condition factor } (K) = \frac{\text{weight}}{\text{length}^3} * 100$$

Weight is in grams and length is in centimeters. For most salmonids, a value of 1.0 is considered healthy.[20] See Reference 20 for a discussion on condition factor for other fish species.

Several types of analyses would be appropriate for analyzing the results of a growth study. The two basic approaches are ANOVA and regression analysis.

ANOVA

The first step is to test the assumptions of the ANOVA. (That is: Do these data come from a normal distribution and are the variances homogeneous?) References on basic statistical analysis[14,16] will have information on how to determine if the data are suitable for ANOVA and, if not, how data transformations may help. If the fish are weighed just once, a univariate ANOVA consisting of treatment concentrations (independent variable) and fish weight (dependent variable) will suffice. A repeated measures ANOVA would be statistically appropriate if fish weights are taken at several time points from the same fish. *Post hoc* tests, such as Scheffe's or Student–Newman–Keuls, can be run to determine which pairs of treatments are significantly different from each other. Dunnett's test, which generates p-valves for comparisons involving the control to each treatment, may be a more appropriate *post hoc* test.

The ANOVA can be run on the tank weight means or growth rate constants, if several weights are obtained at several time points. Growth rate can be determined with the equation

$$W_t = W_0 * e^{k_g t}$$

where W_t and W_0 are tissue weights at time t and time 0, k_g is the growth rate constant in units of d^{-1}, and t is the specified time interval.

Regression analysis

Many different types of growth curves are possible[20]. For many, standard regression analysis can be performed and their slopes tested for significant differences. All intercepts will be the same because the experiment will have started with no significant differences in fish weight.

Growth and mortality data can be analyzed by generalized linear models (GLM) to produce LCp, ECp, LRp, and ERp values. LCp and ECp statistics are point estimates that define the lethal and sublethal (i.e., growth) exposure concentration for a given proportion of the population. LRp and ERp values are also point estimates for lethal and sublethal responses that are based on tissue concentrations (residue). The point estimate for the degree of growth inhibition (e.g., EC10, ER25) can be determined with equations provided by Bailer and Oris.[21] Growth, which can be exponential, would be modeled as:

$$\text{net growth } (g) = e^{(a+bx)}$$

where x is the fish food or whole-body tissue concentration, and a and b are coefficients determined by GLM using a log link function and a gamma error distribution. The confidence intervals for these growth ECp and ERp values can be determined with the delta method.[22] If mortality occurs, GLM equations can also be used to generate point estimates[23] (e.g., LC10, LR50), which are more accurate than standard probit methods.

Appendix IV Fish food

IV.A Fish food for experimentation

It is desirable to keep the total fat content of food given to experimental fish low and close to that found in wild fish and their prey. A diet that is artificially high in fat will produce fish with excess body fat[24] that may lead to physiological abnormalities. This is also important for toxicity evaluation because body fat can affect the toxic response for hydrophobic contaminants.[25,26] These studies identify an inverse correlation between the toxicity response and lipid content (higher lipid reduces the response), implying that studies with laboratory fish fed high fat commercial diets may underestimate the toxic response. This is especially important for studies that attempt to mimic the toxic response in juvenile salmonids in an estuary. As these fish transit from freshwater to seawater (smoltification) their whole-body lipid content falls to low levels (5–10% dry weight).[27–29]

Juvenile salmonids generally prey on invertebrates[30] that contain variable amounts of lipids. The lipid content for polychaetes and molluscs are generally in the range of 5–10% dry weight[30,31]; however arthropods (insects and crustaceans) can be more variable and substantially higher for some species.[30] Commercially prepared fish food for salmonids is usually very high in total fat content (15–25%), which causes the lipid content of juvenile fish to increase beyond that found in wild fish.[24,29] Additionally, commercially prepared feeds can contain relatively high levels of contaminants, such as PCBs, DDTs, and PAHs.[32,33] In order to mimic as closely as possible the food that a wild fish would be obtaining, it may be necessary to formulate your own fish pellets or contract a laboratory or wholesaler to make it to your specifications. For studies using juvenile salmonids, a lipid content of approximately 7–10% dry weight is desirable to mimic natural prey.

Prepared fish food can be purchased from one of several manufacturers. The basic ingredients (fish meal, vitamins, and binder) can also be obtained from some of these manufacturers, and pellets can be made in the laboratory with the appropriate machinery. We prefer to use cod liver oil purchased from a health food store because we have found it to be very low in contaminants, such as PCBs. We have found that by making our own fish food, the total fat content can be minimized (e.g., 7–10% of dry weight) to mimic natural prey levels.

When juvenile fish weigh approximately 2 g, they should be switched to low fat pellets to allow their body lipid composition to reflect their diet before they are fed dosed food. This will continue until the experiment begins, at which time fish will be fed with the same low fat food that has toxicants added. It would be advantageous to sample a few fish occasionally from the common pool of fish and conduct a proximate analysis[27,30] to specifically determine lipid content. Ideally, experimental fish that are in the presmolting or smolting phase should have whole-body lipid levels in the 5–10% (dry weight) range. Lipid content in the fish pellets should be maintained above 6.5% (dry weight) to assure adequate energy and growth for juvenile chinook salmon.[30] If the experiment is conducted past the smolt phase, the lipid content of the feed should be increased to match the life stage and trophic level at which the fish would be naturally feeding.[30]

IV.B Dosing the fish pellets

It is crucial that all preparations of fish food are treated identically to avoid confounding factors. The only differences among treatments should be the amount of toxicant being tested. Ideally, all food should be dosed in one batch to avoid batch-to-batch variation. Testing the dosed food before the experiment starts is required. A simple test with several fish is required to ensure there are no problems with pellet size or palatability of the food, especially at the high doses. Depending on the toxicant of interest, fish pellets with high concentrations may be rejected by fish.

The main consideration for dosing fish pellets is how to get the toxicants on the food. Toxic elements (metals) may be dissolved in acidic water and added to the dry ingredients during formulation of the fish pellets. One important aspect for testing the toxicity of an element is the species (form) that is added to food (e.g., inorganic versus alkylated, metal valence, redox state). For organic toxicants, non-polar solvents are often required to dissolve the compound to assure uniform amounts on the pellets. Some of the solvents used by researchers to add organic toxicants to fish food include fish oil, dichloromethane (methylene chloride, $MeCl_2$), petroleum ether, acetone, and hexane.

Methylene chloride is a popular solvent because it is non-polar. It is advantageous over a more polar solvent, such as acetone, which may soften the fish pellets and cause them to disintegrate. Disadvantages of using $MeCl_2$ include its toxicity to humans and emissions as it volatilizes. Other solvents may be used, depending on the toxicant of interest. Fish oil may be preferred for some contaminants; however, non-polar solvents may be necessary for the more hydrophobic test compounds. We have found that $MeCl_2$ works very well for PAHs and PCBs because the more hydrophobic congeners in these groups are difficult to get into solution. At least one study has looked at the effects on fish growth from treating fish pellets with $MeCl_2$.[34] These authors found no statistical difference in fish growth when comparing $MeCl_2$ treated pellets with unadulterated pellets.

In our laboratory, we have noticed that the moisture content of the fish pellet can change during the dosing procedure, which may be important for comparing growth between treatments and controls (solvent control and regular control, if used). Because the amount fed per day is based on a percentage body weight of fish and a consistent dry weight to wet weight ratio for the fish pellets, each batch of pellets should be checked for moisture content. If differences are found, adjustments in the amount fed should be made so a consistent amount of food (based on dry weight) is given to each group of fish.

The first step in preparing dosed fish pellets is to make a concentrated stock solution of the selected toxicant. For example, if dosing fish pellets with Aroclor 1254 (a mixture of PCB congeners), a stock solution of 4000 μg/ml would be reasonable. Based on the density of Aroclor 1254 (1.5 g/ml) it would take 0.133 ml in 50 ml of an appropriate solvent, such as $MeCl_2$ to make a stock solution of 4000 μg/ml. The stock solution is kept in an appropriate sized Erlenmeyer flask (with glass stopper) in the dark. The same approach can be taken with toxicants that occur as powders, which are weighed out on a balance and added to the $MeCl_2$ solution.

One approach for dosing the fish pellets with organic toxicants is to add an appropriate amount of food to a stainless steel bowl. In our studies, we routinely dose a batch of 3500 g of fish food, which is enough to feed 150 fish (10 g each) at 2% bw/day for 45 days. After the food is weighed out and placed into the bowl, an appropriate amount of stock solution is added to a 4-l bottle of $MeCl_2$. For example, to generate a PCB dose of 1.5 μg/g on the fish pellets, one would pipette 1.31 ml of the stock solution (see earlier) into the 4-l bottle. The solution is shaken, allowed to equilibrate for at least 10 min, and then the entire bottle is poured onto the pellets in the stainless bowl. The size of the bowl should be large enough to hold the entire batch of food but allow the solution to completely cover the pellets.

This mixture is stirred with a large metal spoon for 1 min initially, for 1 min after a 1-h period, and then for 1 min on the following day. Progressing from low to high doses and rinsing the spoon with $MeCl_2$ between bowls will reduce cross-contamination. All bowls are left under a fume hood and the solution is allowed to evaporate for approximately 1 week, which is usually sufficient for dryness. After this time, the fish pellets are placed into a large plastic or stainless steel tub lined with aluminum foil and allowed to dry for another 24 h. Once dried, the pellets are placed into a plastic tub with a tight-fitting lid and kept in a freezer at −20°C until needed to feed fish. A comparison of our nominal concentrations for this procedure to the analytical results for fish food and tissue concentrations have shown a high degree of success for this method.

References

1. Wedemeyer, G.A., Saunders, R.L. and Clarke, W.G., Environmental factors affecting smoltification and early marine survival of anadromous salmonids, *Mar. Fish. Rev.*, 42, 1–14, 1980.
2. Wallace, J., Environmental considerations, in: *Salmon Aquaculture*, Heen, K., Monahan, R.L. and Utter, F., Eds., vol. 5, Fishing News Books, London, 1993, pp. 127–143 (278 pp.).
3. Grau, E.G., Specker, J.L., Nishioka, R.S. and Bern, H.A., Factors determining the occurrence of the surge in thyroid activity in salmon during smoltification, *Aquaculture* 28, 48–57, 1982.
4. US EPA, Guidance on Quality Assurance Project Plan, US EPA QA/G-5, 2002, 58 pp. plus appendices.
5. Moles, A. and Rice, S.D., Effects of crude oil and naphthalene on growth, caloric content, and fat content of pink salmon in seawater, *Trans. Amer. Fish. Soc.*, 112, 205–211, 1983.
6. Heintz, R.A., Rice, S.D., Wertheimer, A.C., Bradshaw, R.F., Thrower, F.P., Joyce, J.E. and Short, J.W., Delayed effects on growth and marine survival of pink salmon *Oncorhynchus gorbuscha*

after exposure to crude oil during embryonic development, *Mar. Ecol. Prog. Ser.*, 208, 205–216, 2000.
7. Heen, K., Monahan, R.L. and Utter, F., *Salmon Aquaculture*, Fishing News Books, London, 1993, 278 pp.
8. Stoskopf, M.K., *Fish Medicine*, WB Saunders, Philadelphia, 1993, 882 pp.
9. Wheaton, F.W., *Aquacultural Engineering*, Krieger Publishing, Malabar, 1993, 708 pp.
10. Roberts, R.J. and Shepard C.J., *Handbook of Trout and Salmon Diseases*, 3rd ed., Fishing News Books, London, 1997, 179 pp.
11. Huguenin, J.E. and Colt, J., *Design and Operating Guide for Aquaculture Seawater Systems*, Elsevier Science, Amsterdam, 2002, 332 pp.
12. Fagerlund, U.H.M., McBride, J.R. and Williams, I.V., Stress and tolerance, in: *Physiological Ecology of Pacific Salmon*, Groot, C., Margolis, L. and Clarke, W.C., Eds., Chapter 8, UBC Press, Vancouver, 1995, pp. 459–503 (510 pp.).
13. Hurlbert, S.H., Pseudoreplication and the design of ecological field experiments, *Ecol. Monogr.* 54, 187–211, 1984.
14. Zar, J.H., *Biostatistical Analysis*, 2nd ed., Prentice-Hall, Englewood Cliffs, NJ, 1984, 718 pp.
15. Peterman, R.M., Statistical power analysis can improve fisheries research and management, *Can. J. Fish Aquat. Sci.*, 47, 2–15, 1990.
16. Steel, G.D. and Torrie, J.H., *Principals and Procedures of Statistics. A Biometrical Approach*, 2nd ed., McGraw-Hill, New York, 1980, 633 pp.
17. Edland, S.D. and van Belle, G., Decreased sampling costs and improved accuracy with composite sampling, in: *Environmental Statistics, Assessment, and Forecasting*, Cothern, C.R. and Ross, N.P., Eds., Lewis Publishers, Boca Raton, 1994, pp. 29–55.
18. Davis, R.B., Bailer, A.J. and Oris, J.T., Effects of organism allocation on toxicity test results, *Environ. Toxicol. Chem.*, 17, 928–931, 1998.
19. Rohlf, F.J. and Sokal, R.R., *Statistical Tables*, W.H. Freeman, San Francisco, 1969, 253 pp.
20. Weatherley, A.H, *Growth and Ecology of Fish Populations*, Academic Press, London, 1972, 293 pp.
21. Bailer, A.J. and Oris, J.T., Estimating inhibition concentrations for different response scales using generalized linear models, *Environ. Toxicol. Chem.*, 16, 1554–1559, 1997.
22. Seber, G.A.F., *The Estimation of Animal Abundance and Related Parameters*, Charles Griffen, Bucks, England, 1982, 654 pp.
23. Kerr, D. and Meador, J.P., Modeling dose–response with generalized linear models, *Environ. Toxicol. Chem.*, 15, 395–401, 1996.
24. Shearer, K.D., Silverstein, J.T. and Dickhoff, W.W., Control of growth and adiposity of juvenile chinook salmon (*Oncorhynchus tshawytscha*), *Aquaculture*, 157, 311–323, 1997.
25. Lassiter, R.R. and Hallam, T.G., Survival of the fattest: implications for acute effects of lipophilic chemicals on aquatic populations, *Environ. Toxicol. Chem.*, 9, 585–595, 1990.
26. van Wezel, A.P. and Opperhuizen, A., Narcosis due to environmental pollutants in aquatic organisms: residue-based toxicity, mechanisms, and membrane burdens, *Crit. Rev. Toxicol.*, 25, 255–279, 1995.
27. Shearer, K.D., Factors affecting the proximate composition of cultured fishes with emphasis on salmonids, *Aquaculture* 119, 63–88, 1994.
28. Brett, J.R., Energetics, in: *Physiological Ecology of Pacific Salmon*, Groot, C., Margolis, L. and Clarke, W.C., Eds., Chapter 1, UBC Press, Vancouver, 1995, pp. 3–68 (510 pp.).
29. Meador, J.P., Collier, T.K. and Stein, J.E., Use of tissue and sediment based threshold concentrations of polychlorinated biphenyls (PCBs) to protect juvenile salmonids listed under the U.S. Endangered Species Act, *Aquatic Conserv. Mar. Freshwater Ecosys.*, 12, 493–516, 2002.
30. Higgs, D.A., MacDonald, J.S., Levings, C.D. and Dosanjh, B.S., Nutrition and feeding habits in relation to life history stage, in: *Physiological Ecology of Pacific Salmon*, Groot, C., Margolis, L. and Clarke, W.C., Eds., Chapter 4, UBC Press, Vancouver, 1995, pp. 161–315 (510 pp.).
31. Boese, B.L. and Lee, H., II. Synthesis of Methods to Predict Bioaccumulation of Sediment Pollutants, US EPA ERL-N Contribution N232, 1992.

32. Easton, M.D.L., Luszniak, D. and Von der Geest, E., Preliminary examination of contaminant loadings in farmed salmon, wild salmon and commercial salmon feed, *Chemosphere* 46, 1053–1074, 2002.
33. Jacobs, M.N., Covaci A. and Schepens, P., Investigation of selected persistent organic pollutants in farmed Atlantic salmon (*Salmo salar*), salmon aquaculture feed, and fish oil components of the feed, *Environ. Sci. Technol.*, 36, 2797–2805, 2002.
34. Wang, S.Y., Lum, J.L., Carls, M.G. and Rice, S.D., Relationship between growth and total nucleic acids in juvenile pink salmon, *Oncorhynchus gorbuscha*, fed crude oil contaminated food, *Can. Jour. Fish. Aquat. Sci.*, 50, 996–1001, 1993.

chapter six

Field experiments with caged bivalves to assess chronic exposure and toxicity

Michael H. Salazar and Sandra M. Salazar
Applied Biomonitoring

Contents

1-56670-664-5/05/$0.00+$1.50
© 2005 by CRC Press

Introduction

This chapter reviews the use of field experiments with caged bivalves as an evolving technique in aquatic toxicology to assess chronic exposure and toxicity. The caged bivalve approach has become more feasible, practical, and routine as a result of experience and expertise gained in a number of developing paradigms, including (1) ecological risk assessment, (2) tissue residue effects approaches, and (3) bivalve biomarkers. Ecological risk assessment has helped focus a more equal emphasis on the importance of characterizing exposure and effects, tissue residue effects approaches have facilitated the utility of tissue chemistry data from laboratory and field studies, and bivalve biomarkers have helped characterize and understand the biochemical processes linking exposure and effects.[1] While there are no "ideal" indicator organisms, bivalves have many attributes necessary for effective sentinels of exposure and effects in environmental monitoring and assessment programs in aquatic toxicology.[2]

Nearly all clinical measurements (e.g., histological, biochemical, and physiological) used in laboratory tests are applicable to field testing, but their limited application in the field is due to a lack of understanding of the utility and knowledge in the flexibility of these diagnostic tools.[3] We believe that the same is true for caged bivalves. While caged and indigenous bivalves have been used extensively in "Mussel Watch" programs to quantify chemical exposure, they have been used less frequently in effects-based monitoring programs. Nevertheless, caged bivalves are equally suited to characterizing exposure and effects. Caging facilitates exposure and effects measurements. Caging also facilitates the utility of mussels as monitoring and assessment tools for several reasons. These include a well-defined exposure period, comparisons between beginning- and end-of-test measurements, and experimental control of size range, exposure history, and genetic makeup of the test animals. Therefore, field experiments using caged bivalves utilize many characteristics of traditional laboratory bioassays in terms of experimental control and many characteristics of field monitoring in terms of environmental realism.[4] Strategic deployment of caged bivalves along suspected chemical gradients can help identify the source of chemicals in the water column and the bioavailability of sediment-associated chemicals. Transplant studies have been used successfully to identify the fine vertical structure of bioaccumulation and associated bioeffects in the water column when the cages were separated by as little as 2 m vertical distance.[5–8]

McCarty[9] proposed developing a single bioassay methodology that includes an integration of tissue chemistry and effects measurements. Bivalves are well suited for this approach because they integrate external chemical exposure as they filter the water for food. With respect to the methods described here, the most important attribute may be their ability to be easily collected, measured, caged, and transplanted.[4] Since the gill is both a respiratory and food-collecting organ, chemical exposure in the field includes both waterborne and dietary exposure pathways, an element often missing from laboratory exposures. Parrish et al.[10] suggest that the real variable in risk assessment is exposure rather than toxicity, and that shifting from laboratory tests to field testing could reduce the uncertainty in the assessment. We believe that the variability in laboratory measurements of exposure and effects are inexorably linked, and that field experiments are necessary to bridge the gap and establish links with more traditional assessment methods.

The inclusion of more sophisticated measurements, such as biomarkers, in field experiments further increases the ability to bridge gaps between exposure and effects.

Many biochemical methods developed for marine, estuarine, and freshwater bivalves can be performed quickly and inexpensively. Some of those developed for reproduction (e.g., the vitellin assay) give an useful estimate of reproductive condition in a relatively short period of time.[11–13] These biochemical measurements be supplemented with other observations (e.g., sex reversal which is a more direct method of demonstrating possible endocrine disruption) to produce more conclusive information on effects.[14]

Aquatic toxicology would be enhanced by using a more ecological risk assessment-based approach that includes field experiments. The methods described here are important to aquatic toxicology because they provide a standardized field approach to implement long-term, chronic testing.[4] We believe that most toxicity and bioaccumulation tests conducted in the laboratory[15–18] can be conducted in the field, and that field experiments are necessary to increase the utility of the data in terms of predicting effects and establishing causality. The purpose of this chapter is to advance the knowledge of practitioners, and promote field experiments with caged bivalves as a useful, cost-effective, and practical diagnostic tool in aquatic toxicology.

Test organisms and materials required

Recommended test organisms

Recommended test species include (but are not restricted to) those shown in Table 6.1. These species have been used more than any other bivalve species but because of the similarity in bivalve shape and structure, almost any bivalve is suited to caging and transplantation.

Source of test organisms

Most marine bivalves are bottom-dwellers that can be easily collected in shallow water by wading or by divers or dredging in deeper water. Some marine bivalves are routinely cultured, and culturing facilities can generally provide test animals of uniform size. Obtaining bivalves from a culturing facility eliminates uncertainties associated with age, genetics, and previous exposure history. Many freshwater bivalves are becoming more difficult to collect in large numbers because of their threatened or endangered status. Regardless of whether marine or freshwater bivalves are being collected, a permit will be necessary for collection and transplantation because of concerns regarding introduction of exotic species.[4]

Table 6.1 Species most commonly used in field testing

Marine bivalves		Freshwater bivalves
(1) Family Mytilidae	(3) Family Ostreidae	(1) Family Unionidae
Mytilus californianus	*Crassostrea gigas*	*Elliptio complanata*
Mytilus edulis	*Crassostrea virginica*	*Pyganodon grandis*
Mytilus galloprovincialis	*Ostrea edulis*	(2) Family Corbiculidae
Mytilus trossulus	(4) Family Tellinidae	*Corbicula fluminea*
Perna perna	*Macoma balthica*	(3) Family Sphaeridae
Perna viridis	*Macoma nasuta*	*Sphaerium simile*
(2) Family Pectinidae	(5) Family Veneridae	(4) Family Dreissinidae
	Mercenaria mercenaria	*Dreissena polymorpha*

It is recommended to sample a number of surrogates at the beginning of the test to establish their condition and to quantify the concentration of chemicals in their tissues to assess their chemical exposure history. It is usually preferable to transplant bivalves to test areas as soon as possible after collection because laboratory holding can also induce stress to test animals. The spawning period should be avoided because of potential effects on bioaccumulation and growth. If it is necessary to hold bivalves for extended periods in the laboratory, they must be fed, even if unfiltered water is used in the holding tanks.[4]

Materials required

The supplies and materials shown in Table 6.2 may be required to conduct a field experiment with caged bivalves. The quantity of each item depends on the number of bivalves used and the study design.

Table 6.2 Materials and supplies commonly used for field testing with caged bivalves

Field supplies	*Decontamination supplies*
Aluminumfoil	Brush, scrub
Buckets, 1-gal	Biodegradable cleaning solution
Buckets, 5-gal	Distilled water
Cable ties, 14-in.	Weigh pans
Cable ties, 4-in.	*Data sheets and forms*
Cable ties, 8-in.	Chain of custody forms
Calipers, field-plastic	Electronic data sheets
Clip boards	Hard copy data sheets
Compartmentalized trays	Extra hard copy data sheets
Cutting boards (plastic/glass)	*Sample jars*
Deployment line	Sample jars, supplied by analytical lab
Distribution rack	Sample jar labels
Extension cords	Tamper-proof tape
Forceps, plastic	*Electronic gear*
Gloves, surgical	Balance, electronic
Ice chests	Calipers, electronic
Labels for mesh bags	Computer, portable
Mesh bags	Data transfer hardware
Trays, flat w/grid	*In-situ* temperature monitoring devices
Paper towels	Surge protectors
Pliers, cross-cut (small and large)	Power strips
Predator mesh	*Miscellaneous*
Scalpels/knives (stainless/ceramic)	Tables
Scissors	Chairs
Tarp	Bottle to stretch bags
Trash bags	Identification tags for cages
Tubs, bus	Tape, duct
Weights/anchors, cinder blocks	Tape, electrical
Ziploc® polybags	Marking pens, fine and thick point

Procedures

Data set

Controlled experiments with caged bivalves under field conditions facilitate the simultaneous collection of data to help characterize chemical exposure and associated biological effects in the same organism and at the same time. The procedures outlined here are designed for collecting measurements of bivalve shell lengths, whole animal wet weights, tissues, and shell weights. These field measurements are used to characterize effects due to exposure at the deployment site. The tissue chemistry data, reported by an analytical laboratory, are used to characterize exposure conditions. If *in-situ* temperature monitors are used, these data can be used to help characterize other factors that may be affecting bivalve growth and response.

Exposure chambers

Although it is possible to mark individuals, cages with individual compartments are generally recommended for field studies with caged bivalves to minimize potential interferences, standardize the procedures, and minimize the time associated with handling individuals. For long-term exposures to characterize sediment, compartments may not be appropriate. The basic concept behind the cage design is to maximize water flow to the test animals, while containing test animals within the cage and ensuring that each individual has the same water exposure. Cages can be flexible material with compartments attached to a rigid frame (i.e., mesh bags attached to a PVC frame or heavy plastic mesh), rigid with fixed compartments (i.e., plastic trays or wire baskets with internal divisions), or a benthic cage where test animals are held in contained sediment without compartments (Figure 6.1). The separation of test animals into individual compartments allows equal exposure to each bivalve. Compartmentalization also facilitates tracking individuals throughout the test and eliminates the need to mark or notch individuals.[4] Although cages without compartments are not recommended for most water column exposures because of the potential for clumping of individuals and uneven exposure conditions, a "bucket" or benthic cage is appropriate for long-term exposures where the bivalves must have the ability to migrate within the surficial sediments.[19,20]

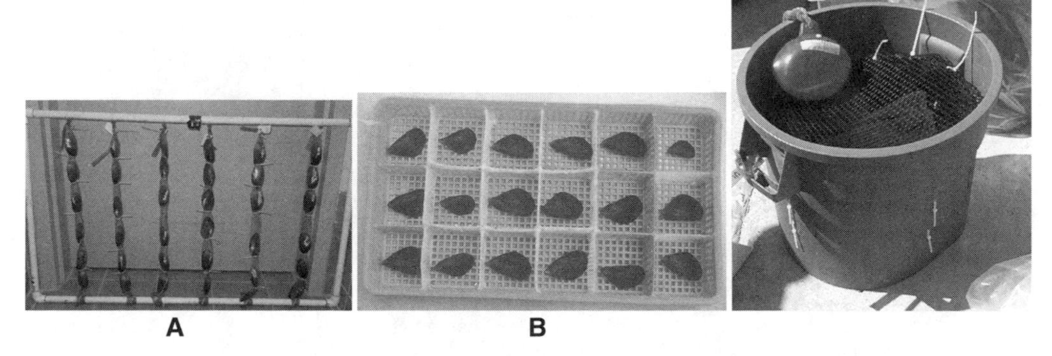

Figure 6.1 Cage examples: (A) compartmentalized flexible mesh; (B) compartmentalized rigid; (C) non-compartmentalized benthic.

Flexible mesh bags and rigid PVC cage construction

The dimensions of the flexible mesh bags and the PVC frames depend on the species, size of the individual test organisms at the start of the test, expected growth, and the number of organisms per cage. Oyster culch netting, similar to that used in bivalve aquaculture, is recommended to make the mesh bags. This netting is available in many diameters and mesh sizes, comes in tubular form, and can be cut to length as needed. An approximate 6-in. diameter mesh material is recommended for smaller smoothed-shelled species like mussels and clams because there is less mesh at the point of constriction. For larger bivalves with rough shells and irregular shapes, such as oysters, a tubing of larger diameter should be used. Individual compartments are created by separating the bivalves within the mesh bags with a plastic cable tie or other restricting device. Sufficient space should be provided in each compartment to allow test animals to grow during the exposure period. The mesh bag should be long enough to accommodate the desired number of bivalves per bag plus sufficient material to allow secure attachment to the PVC frame. Approximately 30 cm of mesh netting on either end of the bag is generally sufficient for attachment to a PVC frame constructed from 1.9 cm diameter material.[4]

For most applications, schedule 40 PVC pipe (3/4 in diameter) is sufficient for frame construction (Figure 6.2). Practical frame sizes range from 30 cm × 30 cm to about 20 cm × 100 cm. To minimize crowding and possible damage to bivalve shells, the PVC frame should be approximately 5 cm longer and wider than the space occupied by the bivalves once placed in the mesh bag. To remove any water soluble and/or volatile chemicals, the frames should be soaked, preferably in flowing fresh or seawater, for at least 24 hours after construction. Number each cage with an indelible marker to facilitate attachment of mesh bags.

Negative buoyancy can be achieved by drilling the frame approximately every 25 cm. with a hole between 1/8 and 3/16 in. to allow water to enter the pipe and remove trapped air. The corners of the frame should not be drilled to eliminate the potential for weakening the overall structure of the frame. Frames that are not drilled may tend to float.

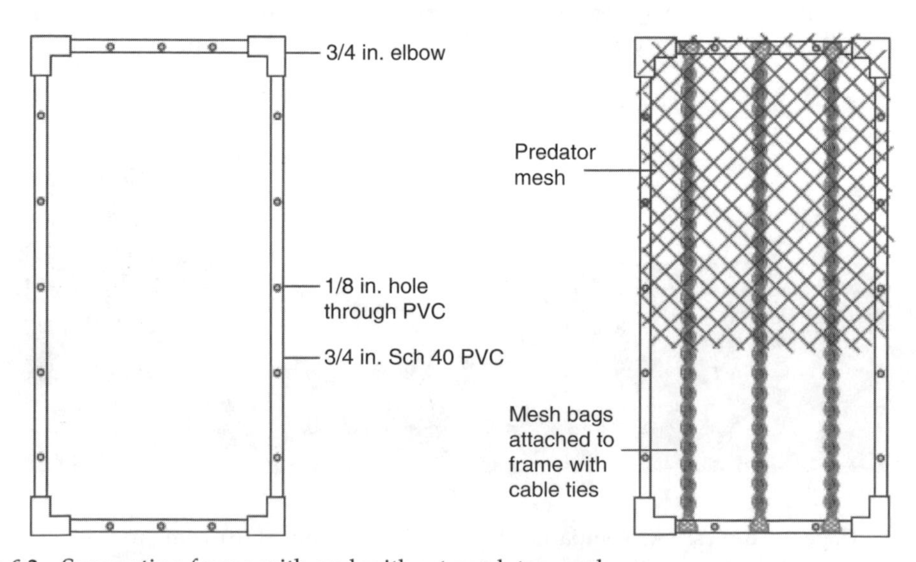

Figure 6.2 Supporting frame with and without predator mesh.

Because each PVC frame will hold one or more mesh bags containing bivalves, each mesh bag must be labeled with an identification tag. Durable plastic or other inert material is recommended. For consistency and ease in identifying the beginning of a bag, it is recommended that the bag be knotted approximately 23 cm from the end and the tag be attached near the knot with a plastic cable tie. Tags should include both the cage number and the bag number, i.e., a label of 9–3 indicates that this is the third bag assigned to Cage #9.

Benthic cage

The overall design of the benthic cage (Figure 6.3) includes a solid outer container and an inner mesh chamber held in place with plastic cable ties.[19,20] Mussels are confined by this mesh insert; there are no individual compartments. The bottom of the mesh chamber is lined with biodegradable newspaper, and the mesh chamber is then filled with clean sand. The newspaper helps retain the clean sand during set up and deployment. Mussels are distributed evenly on top of the clean sand in the benthic chamber before deployment. The benthic cage is buried in sediment with only the top 15 cm exposed. Sediment trapped by the cage will be retained by the solid outer walls.

Conducting the tests

Pre-sort

To minimize the potential for size differences among cages at the start of the test and to achieve a more even size distribution, the bivalves should be pre-sorted into size groups, with each size group in its own container. Sorting can be based on either shell length, as determined with vernier calipers with a measurement accuracy of 0.1 mm, or whole-animal wet-weight, as determined with an analytical balance with a measurement accuracy of 0.01 g. For most species, the pre-sort is based on shell length, with size groups in 1-mm increments (e.g., 50, 51, 52 mm, etc.). Measure the shell length and place the individual in the appropriately labeled bucket or holding container. The bivalves should be kept moist and cool by using wet ice and moist paper towels as necessary. They should

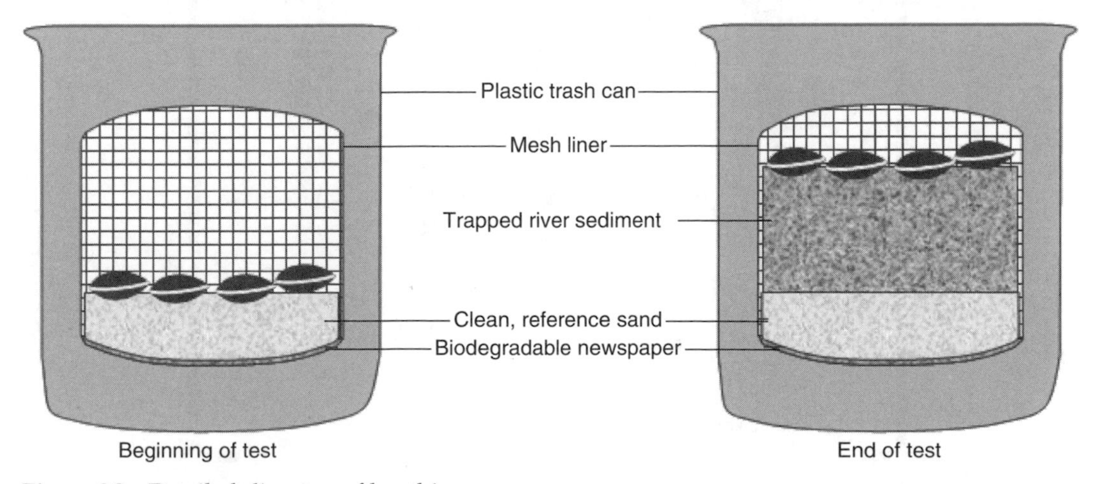

Figure 6.3 Detailed diagram of benthic cage.

be handled as carefully, gently, and quickly as possible so that animals are not unnecessarily stressed. Water should not be placed in the holding containers during the pre-sort because it could lead to oxygen-deficient conditions. The bivalves should be kept out of direct sunlight to eliminate heat stress. The bivalves can be held in an ice chest with wet ice if air temperatures are excessively warm.[4]

After the pre-sort, count the number of bivalves in each size category. To achieve the goal of starting the test with bivalves of a similar size, identify the smallest series that contains the number of bivalves required to initiate the test. The overall target is a 5- to 10-mm range for the size of bivalves used (i.e., 20–25, 33–38, 36–46 mm), except for very large freshwater bivalves where it may be very difficult to obtain a sufficient number in a narrow size ranges. Once the test specimens have been identified, wet ice and paper toweling can be used to keep them wet and cool. The bivalves can be placed in a flow-through system, or other aerated holding device, if measurement and distribution will not occur until a later time.

Set-up for distribution

A PVC distribution rack (Figure 6.4) is used to hold the mesh bags to facilitate distribution of measured mussels. Stretch the mesh material with your hand or an object approximately so that the bivalve will easily fall within the bag during the distribution process. To ensure an even distribution of mussels in each size group across all stations,[4] attach all bags with the "–1" designation to the distribution rack (i.e., for

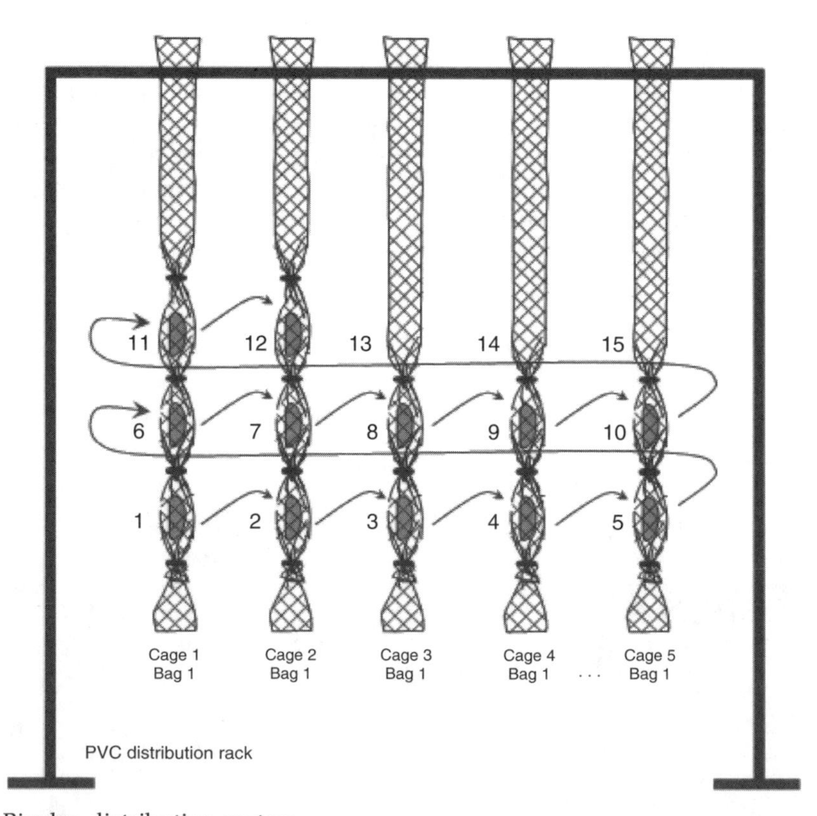

Figure 6.4 Bivalve distribution system.

the first round of distribution, attach bags that are labeled 1-1, 2-1, 3-1, 4-1, 5-1, etc.). The bags should be hung from the distribution rack in cage number sequence. Once the "-1" bags are filled, the second round of distribution will utilize all bags with the "-2" designation.[4]

Measurement and distribution

Starting with either the smallest or largest size group identified in the pre-sort, place all bivalves within the size group into a tray or tub containing water. The bivalves need to be held in water before measurement to eliminate air between the valves. Bivalves must be completely submerged and flat on the bottom prior to measurement. Bivalves that float on the surface or sit upright on the bottom may have air trapped between their valves. These individuals should not be used because the air will bias the whole-animal wet-weights, because air weighs less than water. Transfer these suspect individuals to a separate container containing water. If left undisturbed for a brief period (e.g., 5–10 min), they will likely purge the trapped air. Water temperature should be maintained as close as possible (approximately $\pm$ 5°C) to temperatures at the deployment site or lower. Plastic bags containing wet ice can be placed in the tub with the bivalves to maintain desired water temperature. A rapid change in water temperature could induce spawning in adult organisms that have ripe gametes.

Shortly after placement in the tub with water, the bivalves will begin to respire, and their shells will be slightly agape (approximately 2–3 mm). Most species will tightly close their shells upon light physical stimulation. Bivalves that do not completely close their shells upon movement or light physical stimulation (i.e., agitation of the water around the bivalves or lightly tapping the shell) should not be used. Prior to making the final measurements, examine the exterior of the shell. Individuals that have broken shells or holes in their shells should not be used.

Randomly select one specimen from the holding tray. Using a paper towel, blot excess water from exterior of the shell, measure its shell length with a caliper, and weigh with an analytical balance. The data can be recorded electronically and/or manually. Once the specimen is measured and weighed, place it into the first mesh bag on the distribution rack.

Affix a 4-in. plastic cable tie around the mesh material above this individual. The cable tie should be adjusted so that it is tight enough to prevent the animal from passing through, but loose enough so that the cable tie can be moved if necessary. Do not constrict the mesh to the point that it might restrict valves from opening. There should be enough slack in the mesh to allow movement and growth of the individual during the test. Randomly select another specimen from the tray and make desired measurements. Place this individual into the second mesh bag on the distribution rack, affixing a cable tie as previously described. Repeat this process until one individual has been placed into each mesh bag. Continue adding bivalves, one at a time to the mesh bags, completing one row before another is started, until each mesh bag contains the desired number of individuals. When all of the bags on the distribution rack have been filled, remove the bags, knot or cable tie the open end, leaving a tail length of approximately 25 mm. Place the completed bags into a cooler lined with ice and moist paper towels. Repeat the above process until all the mesh bags are filled.[4]

This distribution process is also recommended for bivalves to be deployed in the benthic cage even though the bivalves will be removed from the mesh bags before deployment. It is not necessary to separate bivalves with plastic cable ties during the distribution process if the benthic cages are used. The mesh bags facilitate transportation to the deployment sites, and identification of individuals assigned to a given benthic cage.

Once the bivalves have been distributed to all mesh bags, sort the bags by cage number (i.e., the first number on the label). Either attach the bags to the PVC frames, as described below, or place them in uncontaminated flowing water (either laboratory or field) until ready to attach bags to the PVC cages or distribute the mussels in the benthic cage.

Measuring and distributing bivalves to be used for baseline (beginning-of-test) tissue chemistry

Bivalves to be used for baseline tissue chemistry should be identified during the measurement and distribution of test specimens to ensure similar sizes. This can be accomplished by randomly assigning a cage number for each baseline tissue chemistry sample. However, instead of distributing the bivalves for baseline tissue chemistry to mesh bags, they can be distributed to rigid, compartmentalized boxes, which eliminates the need to remove individuals from the mesh bags once the distribution process is completed. The individuals should be placed into the compartmentalized boxes in order (i.e., with the first individual measured placed into compartment #1, the second into compartment #2, etc.). At the end of the distribution process, the tissues are removed for chemical analysis.

Although it is possible to measure chemicals in individual bivalves and this approach could provide more data on individual variability, the cost of chemical analysis of individuals is usually prohibitive. Furthermore, pooling provides a more integrated sample that reduces individual variability. The methods described here assume that tissues will be pooled to create composites for chemical analyses. The number of individuals necessary for the composite will depend on the (1) the size of the soft tissue mass and (2) the amount of tissue required by the analytical laboratory for the desired analyte(s).

Attachment of mesh bags to PVC frame

A set of mesh bags attached to a PVC frame constitutes a cage (Figure 6.2). The mesh bags can be attached to the PVC frame by (1) knotting the tail ends of the mesh directly to the PVC, or (2) using cable ties to firmly attach mesh to the PVC frame. There should be sufficient slack in the mesh bag after it is attached so that the bivalves are able to open their shells and filter, but not so much slack that the bivalves hang more than about 5 cm below the plane of the cage. If the mesh is stretched too tightly, it will restrict valve opening, filtration, and growth. If the bag is too loose, it will tend to get tangled in the deployment array. If necessary, slide the cable ties to increase or decrease the space between individuals without compromising the space available for each individual (i.e., do not decrease the space between animals so that there is insufficient space for them to open their valves during respiration). A temperature-recording device can be attached to the frame at this time. If desired the cage can be enclosed within a heavy duty plastic mesh envelope, with a mesh size appropriate to exclude predators (e.g., approximately 1.25–2.5 cm).[4]

Water column deployment

For water column deployments, cages can be either suspended from a fixed mooring, such as a floating pier or piling, or they can be suspended within the water column by attaching them to a line that has an anchor on one end and a surface or subsurface buoy attached to the other end. Factors that should be considered during the deployment of cages for surface water assessments include change in tidal height (i.e., to ensure the cages are at the desired depth during both low and high tides), slope of bottom material (i.e., to ensure the cages do not slide down a steep slope during the exposure period), and boating and recreational activities in the vicinity of the cages (i.e., to avoid cages being removed by or tangled within propellers). Water column deployments are primarily used to assess chemicals in the water column, but this approach can also be used to assess chemicals associated with sediments if the cage is suspended so that it is close to the bottom (e.g., $\approx$1 m).

Fixed bottom deployment

PVC cages can be deployed directly on top of the sediments or set at a fixed distance from the bottom by attaching legs to the cage and pushing the legs into the sediments to hold the cage in place. Depending on the species of bivalve used, bivalves in cages deployed directly on top of the sediments can be used to assess chemicals associated with both sediments and the water column. If the goal is to assess subsurface sediments, it is recommended to place the cages directly on top of the sediments. The bivalves gain exposure to chemicals in the sediments as the sediments infiltrate the mesh material or as the foot rakes the sediment surface. Attach sufficient weights or anchors to ensure the cages remain at the desired position. Rebar can also be bent into a "U" and pushed over the cage into the sediment to secure the cage in position. Iron rebar should be coated with appropriate rubberized coatings or covered by plastic bags to prevent potential metal exposure.[4]

Benthic cage deployments

Prior to deploying the benthic cage, remove bivalves from their mesh bags and place them on the clean sand. Affix the flexible mesh top to the inner mesh with cable ties, so the bivalves cannot escape during transport to the bottom. Benthic cage deployment is best accomplished with divers. The diver will need to dig a hole in the bottom substrate, set the benthic cage in the hole, place sediment around the outside of the benthic cage, and secure the cage to the bottom with either weights or rebar.[19,20]

Deployment period

The deployment period is a function of the type of chemicals being accumulated. A minimum 30-day deployment period is recommended; a preferred deployment period is 60–90 days. A period of less than 30 days is not recommended unless the chemicals of concern are low molecular weight organic compounds, such as some PAHs. Equilibrium for most other chemicals, such as metals and high molecular weight organic compounds, is generally achieved in marine and freshwater bivalves within a period of approximately 60–90 days. A test period of 60–90 days helps ensure reaching chemical equilibrium and provides sufficient time for adverse effects to manifest themselves. The benthic cage is recommended for long-term exposures where more subtle effects, such as endocrine disruption and sex reversal, are the measurement endpoints.[19,20]

Retrieval and end-of-test measurements, collection and preparation of bivalve tissues for chemical analysis

The same care in maintaining moisture and water temperatures should be followed at the end of the experiment as in the beginning. It is critical to retain the order of bivalves during the end-of-test measurements, so the measurements can be paired with beginning of test values. Compartmentalized trays are used to hold the bivalves during the end-of-test measurements. Otherwise, procedures are virtually identical to those employed at the beginning except that tissues must be removed from all the animals exposed in the field. Additional details on these procedures can be found elsewhere.[4] Specific procedures for collection, preparation, and preservation of bivalve tissues are often provided by project requirements or the analytical laboratory but additional guidance is also provided.[4]

Results and discussion

Three case studies are presented to demonstrate that the caged bivalve methodology can identify the fine structure of exposure and effects along suspected chemical gradients. The rationale is, if the methodology is useful for identifying differences over these small spatial and temporal scales, it will also be useful over larger scales. These studies also show that the methodology is versatile and robust. The San Diego Bay study is important because it was the first study, in which these specific methods were used to demonstrate spatial differences in exposure and effects over a small vertical scale (3 m). It represents the initial development and field testing of the method and application of a long-term record (3 years) at several different sites that facilitated development of tissue residue effects relationships. The Port Valdez study is important because it represents the deepest transplant (70 m) we conducted at that time. It was used to demonstrate near-bottom gradients in PAH tissue burdens horizontally (kilometer scale) and vertically (2 m). The Port Alice study is important because it represents near-surface horizontal (kilometer scale) and vertical (2 m) gradients in growth rates. Mussel survival was high in all three tests with exposure periods from 56 to 84 days.

Regarding the relevance to ecological risk assessment, exposure and effects were measured in all three studies and the San Diego Bay results were used as part of the U.S. Navy's ecological risk assessment for utilization of TBT antifouling coatings. Since the San Diego Bay study provided data over a series of years, it was most applicable for determining tissue residue effects, and these relationships have been discussed elsewhere.[6] Biomarkers were also measured in subsequent San Diego Bay studies that established links with the growth endpoint.[21,22] Biomarkers and tissue residue effects approaches will not be emphasized in this chapter because they have been addressed extensively elsewhere. It should be remembered, however, that in its simplest form the caged bivalve methodology could be considered as a platform for almost any clinical measurement. This combination of caged bivalves, tissue residue effects approaches, and biomarkers is what makes the approach so potentially powerful.

In each study, exposure was characterized by measuring bioaccumulation of chemicals of concern and effects by measuring survival and growth. Three different species of marine mussels in the genus *Mytilus* were used (*Mytilus galloprovincialis*, *M. trossulus*, and *M. edulis*), and in each study, bioaccumulation and growth were used to assess exposure and effects over relatively small spatial scales of either 2 or 3 m vertical distance in areas suspected of physical and chemical stratification. All cages were deployed with floats to maintain a fixed position relative to the surface of the water column or bottom sediment, unlike natural populations that are attached to fixed substrates where water exposures would change with tidal cycle. A general description of each of these studies is provided in Table 6.3.

San Diego Bay, California

The San Diego Bay study is important because it shows the ability to quantify exposure and effects over a small spatial scale (3 m). These relationships were quantified over an extended temporal scale (3 years).[5] Concentrations of TBT in seawater were also measured to complete the exposure-dose–response relationships. These San Diego Bay data led to our development of the exposure-dose–response paradigm.[6] These data were also important from the ecological risk assessment perspective because they showed that ship hulls floating on the surface were the source of TBT in mussel tissues and not the contaminated sediment.[23]

Statistically significant differences were found in seawater TBT concentrations, tissue TBT concentrations, and mussel growth rates when the site 1 m below the surface was compared with the site 1 m above the bottom (sites were only 3 m apart) (Figure 6.5). The data show a precipitous decline in seawater TBT concentrations (exposure) that coincides with restrictions on the use of TBT antifouling coatings (Figure 6.5A). This decline was also associated with a general decrease in TBT concentrations in mussel tissue (Figure 6.5B), although confounding factors, such as adverse effects, of weekly measurements in the first four tests made data interpretation more difficult. Excessive handling during these measurements reduced mussel growth rates and probably TBT accumulation as well. When length and weight measurements were made every other week, both growth rates and TBT accumulation increased. The important lesson here is that measurement stress can affect both laboratory and field experiments. Another apparent anomaly appears in the growth rate data where growth rates generally increase at both sites as TBT in water and tissues decreases. In Test 8, however, growth rates decrease significantly, and this was attributed to lower water temperatures near 14°C. Water temperatures above 20°C also reduced mussel growth rates. This corroborated results from other studies suggesting that 20°C was near optimum for mussel growth.[6,23]

Port Valdez, Alaska

The Port Valdez study is important because it shows the ability to quantify total PAHs in mussel tissues (dose) over a small vertical spatial scale (2 m) and a larger horizontal spatial scale (1 km). The main purpose of this test was to demonstrate if the caged mussels would survive, grow, and accumulate chemicals of concern; the test was considered successful from that perspective. Differences were found in PAH bioaccumulation and mussel growth rates among sites. At the time, this caged mussel study was the best example of how a gradient design could be used to demonstrate that the source of PAHs in mussel tissues was the Ballast Water Treatment Facility diffuser, which discharged at a depth of 70 m. This is shown by the decreasing gradient of PAHs in mussel tissues (Figure 6.6) with distance away from the diffuser.[7]

The gradients shown for each of the three monitoring depths (i.e., 5, 7, 9 m above the bottom) are very similar, except for the concentration of total PAHs at 1200 m. At 9 m above the bottom, the cage was lost and no data are available, at 7 m above the

Table 6.3 Summary of case study details

Year	Location	*Mytilus* species	Size range (mm)	Exposure duration (days)	Number of mussels	Percent survival	Depths/ distances	Type of gradient
1987–1990	San Diego Bay, CA	*M. galloprovincialis*	10–12	84	300	96	1 m below surface, 1-m above bottom	Vertical
1997	Port Valdez, AK	*M. trossulus*	30–36	56	2100	97	5, 7, and 9 m above bottom, 200 m intervals from diffuser	Vertical and horizontal
1997	Port Alice, B.C., Canada	*M. edulis*	14–21	68	1620	95	2, 4, and 6 m below surface, 0.3, 3, and 10 km from diffuser	Vertical and horizontal

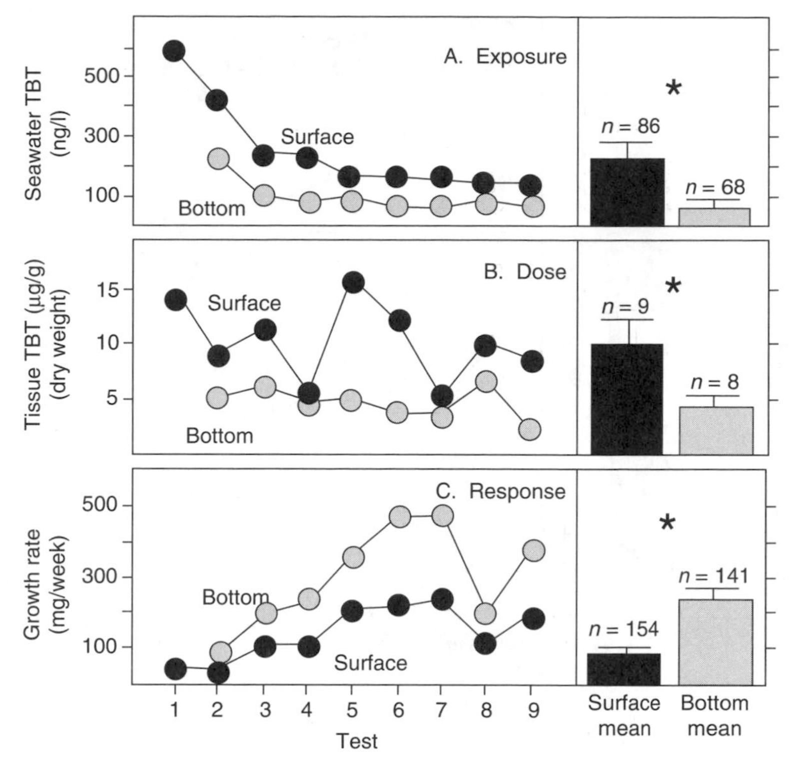

Figure 6.5 Examples of fine structure in (A) chemical exposure, (B) internal dose, and (C) growth response at two sites separated by 3 m vertical distance.

bottom the data point fell along the regression line, and at 5 m above the bottom the value was higher than expected. These anomalies may preclude a more extensive interpretation. Without the data point for 9 m above the bottom, it was difficult to hypothesize whether the regression should be more like that for 5- or 7-m above the bottom, as well as the significance of the regression. When using all the data, the poorest relationship is found at 5 m above the bottom. For the 5-m depth, it is possible that the total PAH concentration in mussel tissues from the 1200 m site is either an outlier or representing another source. Chemical fingerprinting suggests that it is another source. If only the data between 200 and 1000 m are used, the significance of the regression is much better.[7]

Port Alice, B.C. Canada

The Port Alice study shows the ability to detect differences in effects among sites along a suspected chemical gradient even when the chemical of concern has not been identified.[8] Environment Canada has based its Environmental Effects Monitoring program on effects because they are the ultimate concern. In addition, scientists still have not identified the causal factor for those effects measured near pulp and paper mills over the last 25 years. While most of these effects have been measured in fish, we have recently demonstrated similar effects through elevated vitellin production that suggest effects on reproduction and endocrine disruption in freshwater bivalves. The main purpose of the Port Alice test was to quantify effects in caged mussels and determine if possible tracers or causative

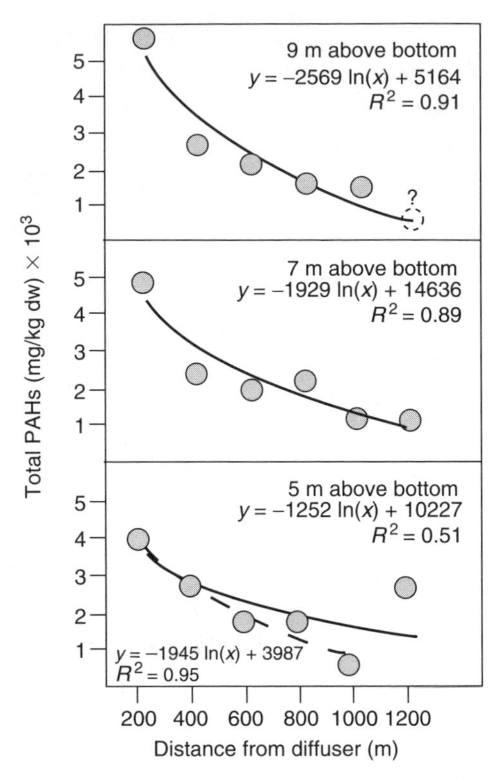

Figure 6.6 Examples of fine structure and decreasing chemical gradient with distance at 5-, 7-, and 9-m above the bottom in a depth of 70 m.

agents responsible for these effects could be identified. Correlations were found for growth effects and the plant sterol campesterol, but campesterol may not be the causative agent because it has not shown significant adverse effects in other studies. Effects were shown along the suspected chemical gradient, as well as differences among sites (Table 6.4).

Table 6.4 summarizes results of statistical comparisons for the three most important growth rate metrics at pooled Stations 1 and 2, 3 and 4, and 5 and 6 at distances of 0.3, 3, and 10 km from the diffuser, respectively. The lack of continuity in statistical results for these three metrics suggests that there is some uncertainty regarding the spatial influence of the pulp mill effluent. Statistical analyses of weight and length growth rate suggest that growth was reduced only at a distance of approximately 0.3 km from the mill. End-of-test tissue weights, however, suggest that growth was reduced up to a distance of 3 km from the mill. Growth rates were correlated with spent sulphite liquor, dissolved oxygen, temperature, and distance from the diffuser. The relationship between growth rate and distance from the diffuser and depth are shown in Figure 6.7. The best relationship between plant sterols and distance from the effluent diffuser was for campesterol. Not only was there a gradient of decreasing concentrations of campesterol with distance from the diffuser, but there was also a statistically significant relationship between the decreasing concentration of campesterol in mussel tissues and the increasing mussel growth rates. Although these relationships do not establish causality, the data suggest that campesterol could be a factor affecting mussel growth. Even if it does not adversely affect growth, it might be useful as a tracer for pulp and paper mill effluents.[8]

Table 6.4 Statistical comparisons among pooled stations at distances of 0.3, 3, and 10 km from the diffuser (statistically similar values are italicized)

	Pooled stations		
	1 and 2	3 and 4	5 and 6
Weight growth rate (mg/week)	218	*236*	*248*
Length growth rate (mm/week)	1.22	*1.28*	*1.32*
EOT tissue weight (g, wet)	*0.68*	*0.69*	0.78
EOT percent lipid (%)	*1.00*	*1.14*	1.48
EOT percent water (%)	*79.5*	*78.4*	*77.5*

As with the Port Valdez study, not all questions were answered, but the advantage of an experimental field approach is the ability to improve the experimental design to answer the questions being asked and develop new questions and new approaches. This is important to aquatic toxicology in the sense of hypothesis development and hypothesis testing and is the core of the caged mussel methodology.

Summary and conclusions

We believe that field experiments using caged mussels can enhance modeling and assessments in aquatic toxicology because of the ability to characterize exposure and effects over

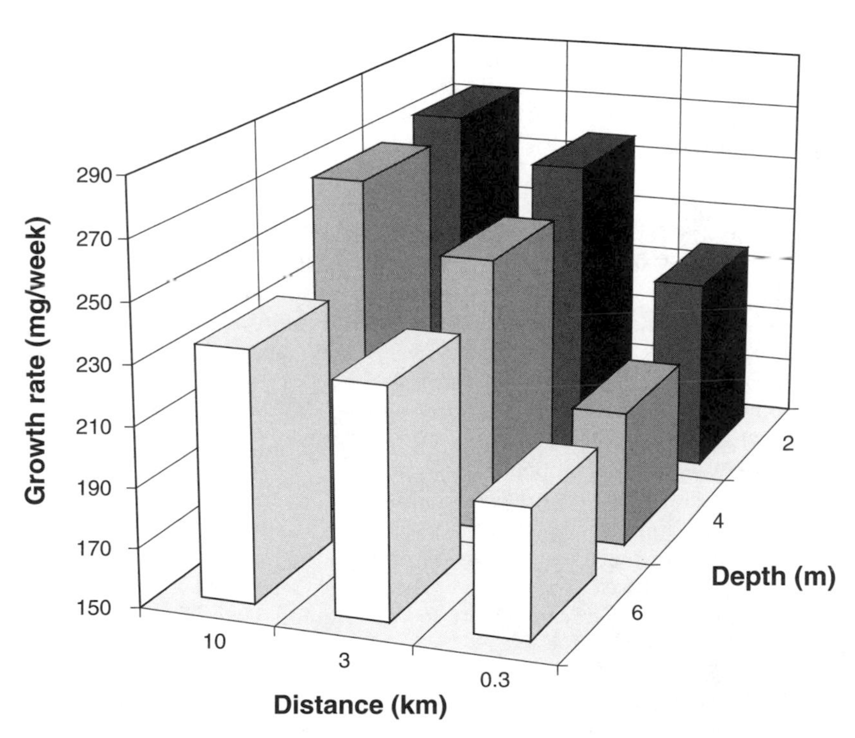

Figure 6.7 Examples of fine structure and increasing growth gradient with distance at 2-, 4-, and 6-m below the surface.

space and time. Caging facilitates the utility of both tissue residue effects approaches and bivalve biomarkers. The field experiments account for the importance of receiving water because the tests are conducted under environmentally realistic conditions that include all possible exposure pathways. It is this ability to characterize exposure and effects that makes these field experiments consistent with ecological risk assessment-based monitoring and a weight of evidence approach. The examples provided here demonstrate how caged bivalves can be used to characterize the fine structure of exposure and effects in horizontal and vertical chemical gradients. This approach can also help establish causal relationships between exposure, dose, and response, particularly when paired with tissue residue effects approaches and bivalve biomarkers. While there are no perfect monitoring and assessment tools, field experiments with caged bivalves possess many characteristics of a practical and useful methodology. In each of the examples provided here, field experiments using caged bivalves provided important information that could not have been obtained through traditional field observations or laboratory toxicity tests. More importantly, the experimental approach is consistent with a basic tenet of aquatic toxicology, which includes hypothesis testing and hypothesis development.

References

1. U.S. EPA, Guidelines for Ecological Risk Assessment, EPA/630/R-95/002F, Final. Risk Assessment Forum, U.S. Environmental Protection Agency, Washington, D.C., April 1998.
2. Widdows, J. and Donkin, P., Mussels and environmental contaminants: bioaccumulation and physiological aspects, in *The Mussel* Mytilus*: Ecology, Physiology, Genetics and Culture*, Gosling, E., Ed., Elsevier, Amsterdam, 1992, pp. 383–424.
3. Versteeg, D.J., Graney, R.L. and Giesy, J.P., Field utilization of clinical measures for the assessment of xenobiotic stress in aquatic organisms, in *Aquatic Toxicology and Hazard Assessment*, Adams, W.J., Chapman, G.A. and Landis, W.G., Eds., vol. 10, ASTM STP 971, American Society for Testing and Materials, Philadelphia, PA, 1988, pp. 289–306.
4. ASTM, E-2122, Standard guide for conducting *in-situ* field bioassays with marine, estuarine and freshwater bivalves, in *2001 Annual Book of ASTM Standards*, American Society for Testing and Materials (ASTM), Conshohocken, PA, 2001, pp. 1546–1575.
5. Salazar, M.H. and Salazar, S.M., Assessing site-specific effects of TBT contamination with mussel growth rates, *Mar. Env. Res.*, 32 (1–4), 131–150, 1991.
6. Salazar, M.H. and Salazar, S.M., Using caged bivalves as part of an exposure-dose–response triad to support an integrated risk assessment strategy, in *Proceedings, Ecological Risk Assessment: A Meeting of Policy and Science*, de Peyster, A. and Day, K., Eds., SETAC Special Publication, SETAC Press, Pensacola, FL, 1998, pp. 167–192.
7. Applied Biomonitoring, Final report, Caged Mussel Pilot Study, Port Valdez, Alaska, 1997. Kirkland, Washington, Report to Regional Citizens' Advisory Council, RCAC Contract Number 631.1.97, 1999, 96 pp. plus appendices.
8. Applied Biomonitoring, Final report, Caged Mussel Pilot Study, Port Alice Mill, EEM Program, Regional Manuscript Report: MS 00-01, 2000, 85 pp. plus appendices.
9. McCarty, L.S., Toxicant body residues implications for aquatic bioassays with some organic chemicals, in *Aquatic Toxicology and Hazard Assessment*, Mayes, M.A. and Barron, M.G., Eds., vol. 14, ASTM STP 971, American Society for Testing and Materials, Philadelphia, PA, 1991, pp. 183–192.
10. Parrish, P.R., Dickson, K.L., Hamelink, J.L., Kimerle, R.A., Macek, K.J., Mayer, F.L., Jr., and Mount, D.I., Aquatic toxicology: ten years in review and a look at the future, in *Aquatic Toxicology and Hazard Assessment*, Adams, W.J., Chapman, G.A. and Landis, W.G., Eds., vol. 10, ASTM STP 971, American Society for Testing and Materials, Philadelphia, PA, 1988, pp. 7–25.

11. Blaise, C., Gagné, F., Pellerin, J. and Hansen, P.-D., Determination of vitellogenin-like properties in *Mya arenaria* hemolymph (Saguenay Fjord, Canada): a potential biomarker for endocrine disruption, *Environ. Toxicol.*, 14(5), 455–465, 1999.
12. Gagne, F., Blaise, C., Salazar, M., Salazar, S. and Hansen, P.D., Evaluation of estrogenic effects of municipal effluents to the freshwater mussel *Elliptio complanata*, *Comp. Biochem. Physiol.*, C., 128, 213–225, 2001.
13. Gagne, F., Blaise, C., Aoyama, I., Luo, R., Gagnon, C., Couillard, Y. and Salazar, M., Biomarker study of a municipal effluent dispersion plume in two species of freshwater mussels, *Environ. Toxicol.*, 17, 149–159, 2002.
14. Blaise, C., Gagne, F., Salazar, M., Salazar, S., Trottier, S. and Hansen, P.-D., Experimentally-induced feminisation of freshwater mussels after long-term exposure to a municipal effluent, *Fresenius Env. Bull.*, 12 (8), 865–870, 2003.
15. ASTM E 724-94, Standard guide for conducting static acute toxicity tests starting with embryos of four species of saltwater bivalve molluscs, in *1998 Annual Book of ASTM Standards*, 1998, pp. 192–209.
16. ASTM E 729-96, Standard guide for conducting acute toxicity tests on test materials with fishes, macroinvertebrates, and amphibians, in: *1998 Annual Book of ASTM Standards*, 1998, pp. 218–238.
17. ASTM E 1022-94, Standard guide for conducting bioconcentration tests with fishes and saltwater bivalve molluscs, in *1998 Annual Book of ASTM Standards*, 1998, pp. 218–238.
18. ASTM E 1688-97, Standard guide for determination of the bioaccumulation of sediment-associated contaminants by benthic invertebrates, in *1998 Annual Book of ASTM Standards*, 1998, pp. 1075–1124.
19. Salazar, M.H., Salazar, S.M., Gagne, F., Blaise, C. and Trottier, S., Developing a benthic cage for long-term *in-situ* tests with freshwater and marine bivalves, in *Proceedings of the 29th Annual Aquatic Toxicity Workshop, Canadian Technical Report of Fisheries and Aquatic Sciences 2438*, vol. 62, Whistler, British Columbia, 2002, pp. 34–42.
20. Salazar, M.H., Salazar, S.M., Gagne, F., Blaise, C. and Trottier, S., An *in-situ* benthic cage to characterize long-term organochlorine exposure and estrogenic effects, *Organohalogen Compounds* 62, 440–443, 2003.
21. Anderson, J.W., Jones, J.M., Steinert, S., Sanders, B., Means, J., McMillin, D., Vu, T. and Tukey, R., Correlation of CYP1A1 induction, as measured by the P450 RGS biomarker assay, with high molecular weight PAHs in mussels deployed at various sites in San Diego Bay in 1993 and 1995, *Mar. Environ. Res.*, 48, 389–405, 1999.
22. Steinert, S.A., Strieb-Montee, R., Leather, J.M. and Chadwick, D.B., DNA damage in mussels at sites in San Diego Bay, *Mutat. Res.*, 399, 65–85, 1998.
23. Salazar, M.H. and Salazar, S.M., Mussels as bioindicators: effects of TBT on survival, bioaccumulation and growth under natural conditions, in *Tributyltin: Environmental Fate and Effects*, Champ, M.A. and Seligman, P.F., Eds., Chapman & Hall, London, 1996, pp. 305–330.

chapter seven

Application of computer microscopy for histopathology in isopod toxicity studies

Damjana Drobne and Samo Drobne
University of Ljubljana

Contents

Introduction

Histopathological changes are integrators of the cumulative effects of alterations in physiological and biochemical systems in an organism exposed to a natural or anthropogenic stress.[1,2] The primary advantage of histopathology is that it permits the visual localization of injury to cells and tissues in multiple organs as it existed just prior to sacrifice and fixation of the tissue. In contrast to tissue homogenates, this approach provides a window to understanding the functional architecture of cells, tissues, and organs. The use of histopathology as a research tool also permits characterization of the essential biological features of the animal being investigated for its sexual, histological, and reproductive status, all variables that may also influence the outcome of experimental manipulations. Furthermore, histopathological examination can be used as an adjunct method to diagnose certain infectious diseases that may be linked to environmental or anthropogenic stressors.

1-56670-664-5/05/$0.00+$1.50
© 2005 by CRC Press

In mammals and in fishes, histopathological examination is widely recognized as a reliable method for disease diagnosis and for assessing acute and chronic effects of exposure to toxicants at the cellular and organ levels. This is not the case in invertebrates. One reason for this lies in the fact that knowledge and experience of fundamental invertebrate biology, pathology, and toxicology is far behind that of mammals and fishes.

For an accurate histological interpretation of sections, the tissue must be properly fixed, processed, and stained, and the examiner must be familiar with the range of normal morphological variations due to sex, reproductive and nutritional status, age, and season. In recent years, histopathology has reached a breakthrough due to the availability of computer-assisted tools for microscopy.

The aim of computer microscopy is the accurate mapping and quantification of biological tissue whether it is of individual cells, groups of cells, or of entire histological regions.[3] In computer microscopy, the user selects those features of the tissue that are to be studied and then makes boundary contours and/or sequences of isolated points. The contours and points are coded for identification purposes. Line lengths and closed areas are measured by tracing and standard computer numerical integration techniques, respectively.

In our work, computer-assisted morphological mapping was employed for estimation of the histological characteristics of the digestive gland epithelium. We determined the morphometric parameters and lipid surface density on serial sections of digestive glands (hepatopancreas) of the terrestrial isopod *Porcellio scaber* (Isopoda, Crustacea). Isopods are among the most popular organisms in terrestrial, as well as aquatic, toxicology.[4] A variety of biomarkers have already been developed to be used in isopod toxicity studies. We established some standardized criteria for histopathological assessment of the state of the organism, and an example is given of a terrestrial isopod *P. scaber* exposed to metal- and pesticide-contaminated food.

Materials required

Tissue preparation

The terrestrial isopods, *P. scaber* (Latreille, 1809) (Isopoda, Crustacea), were collected under concrete blocks and pieces of decayed wood lying on earth in the garden. We followed the toxicity testing protocol as proposed by Drobne and Hopkin.[5] Before dissection, sex, moult stage, presence or absence of marsupium, and weight of each animal were determined. Sex was determined by the characteristic shape of the male's first and second pleopods, which differ from the other pleopods (Figure 7.1b). In females, all pleopods have a similar shape. Moult stage could be determined by milky white deposits on the ventral part of the pereon, indicating the premoult stage (Figure 7.1b).[6] The absence of deposits is a characteristic of the intermoult and postmoult animals. The postmoult stage is difficult to determine antimortem. One characteristic of this stage is an empty gut. In the toxicity studies, only the intermoult animals, as determined postmortem, were included.

After exposure to contaminated food, the animals were dissected and the digestive system was isolated (from 9 to 10.30 a.m.). The digestive system of *P. scaber* is composed of a short stomach, a gut, and four blind ending digestive gland tubes (Figure 7.1c). The animals were decapitated, and the last two pereonies were cut off (Figures 7.1a,b). The gut and four digestive gland tubes were pulled out by tweezers and immediately dipped into fixative. The hepatopancreas is best preserved if it is pulled out together with the stomach and the gut. In such a case, the digestive juices remain within the lumen of the tube, and the least mechanical damage is caused to the gland epithelium. If the digestive gland

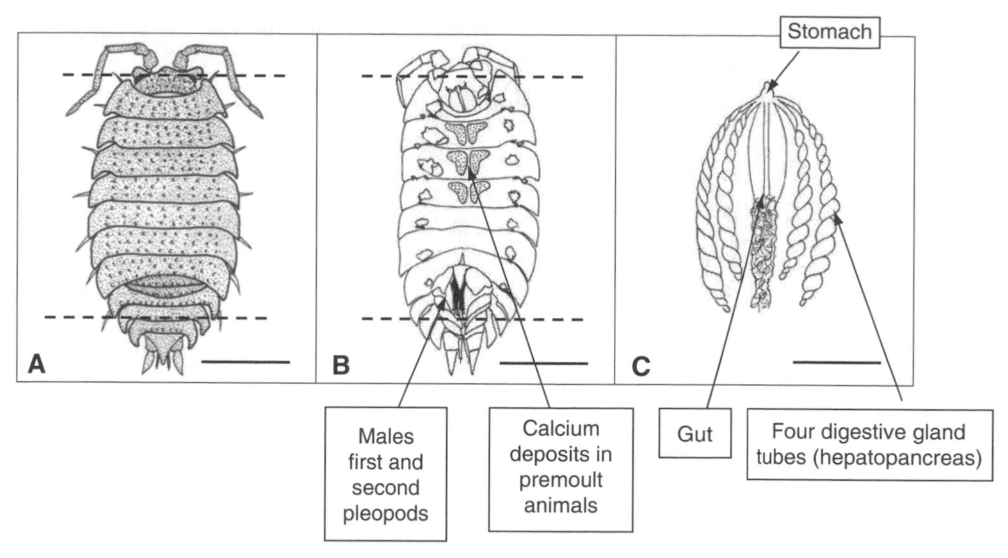

Figure 7.1 (a) Dorsal site of *P. scaber*. (b) Ventral site of *P. scaber*. Broken lines indicate parts of body that are cut off before the digestive system is isolated. (c) The isolated digestive system: stomach, gut, and four digestive gland tubes. Scale bar is 2.5 mm.

tubes are not pulled out together with the gut, it is not recommended to pull them out tube by tube, but to open the lateral parts of the body, in order to gently remove the ventral site and only then pull out the gland tubes. In such cases, the most anterior part of the tube is damaged, but the rest is satisfactorily preserved. If the gland tube is stretched during the isolation, it can be further processed; however, some artifacts are expected, for example, a changed luminal area. After isolation and before fixation, the color and the shape of the gland tubes are described, as well as the amount of digesta (Tables 7.1a and 7.2a).

Procedures

The isolated digestive gland tubes were fixed in Carnoy-B fixative, a mixture of glacial acetic acid (10%), absolute ethanol (60%), and chloroform (30%) for 3 h at room temperature. Fixative was washed out with absolute ethanol for 2 h. Then the tissue was transferred to xylene (3 × 15 min) and subsequently embedded in the melted paraplast at 58°C. Samples were oriented parallel to the edges of the model and incubated in the melted paraplast for 12 h. The paraplast was allowed to harden for 2 days at room temperature. Then 8-μm sections (Reichert-Joung 2040 rotary microtome) of the entire tube were cut (approximately 600–700 sections per gland tube). All sections were stained with eosin for 30 min, then briefly rinsed in 70% alcohol (2 s) and in 96% alcohol (2 s), and dehydrated in 96% ethanol (3 × 2 min) and xylene (2 × 5–10 min). In staining only with eosin, no differentiation of the dye is needed and thus the procedure is highly repeatable. This is of significant importance for comparative, computer-based objective analyses of histological images.

Manual contouring of the inner and the outer epithelial surface of a serial section of one gland tube was performed to calculate the average epithelial area and epithelial thickness on a section (computer program for image analyses, KS-400 Kontron Electronic,

Table 7.1 Description of the digestive system of a control, intermoult *P. scaber*

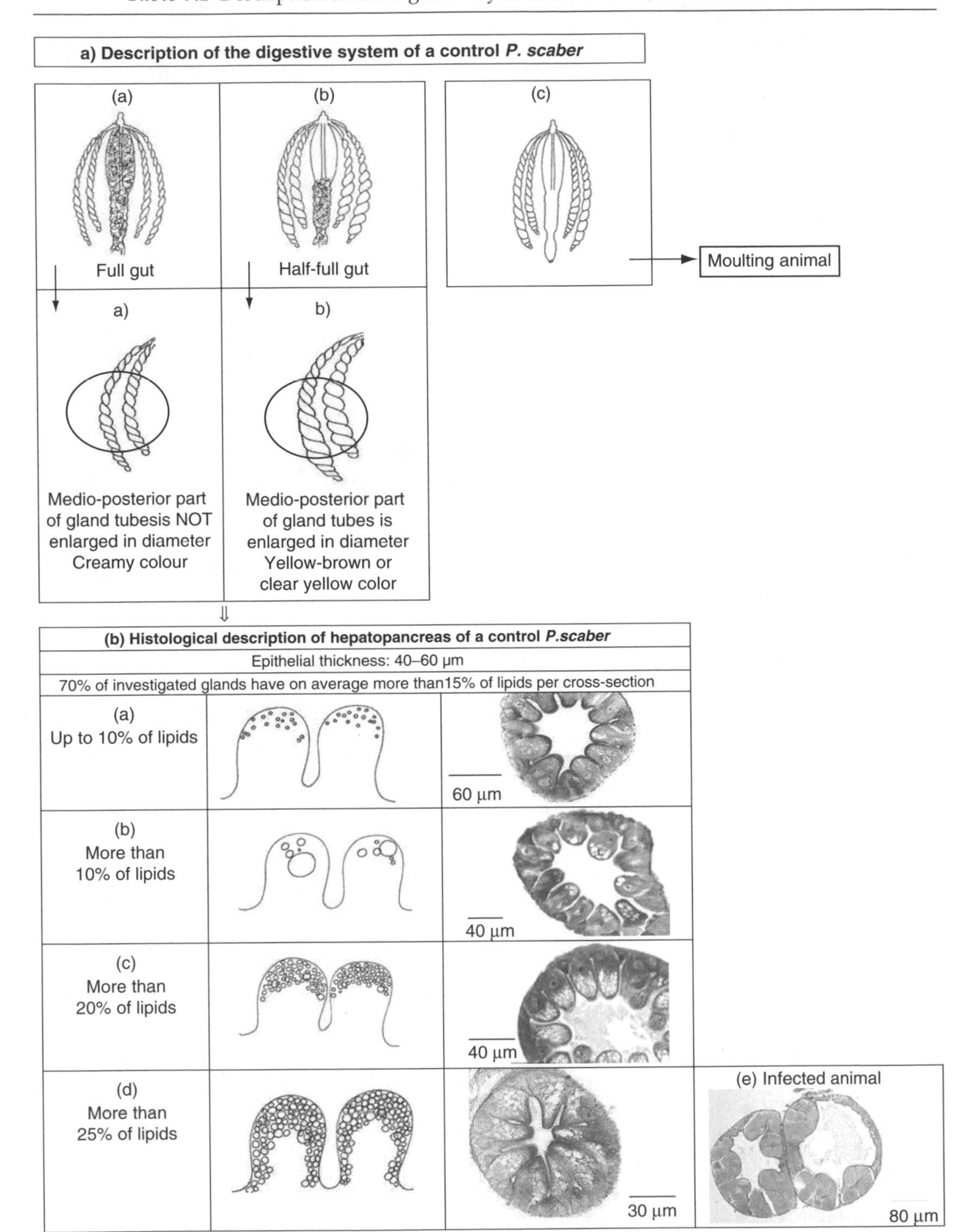

Table 7.2 Description of the digestive system of a stressed *P. scaber*

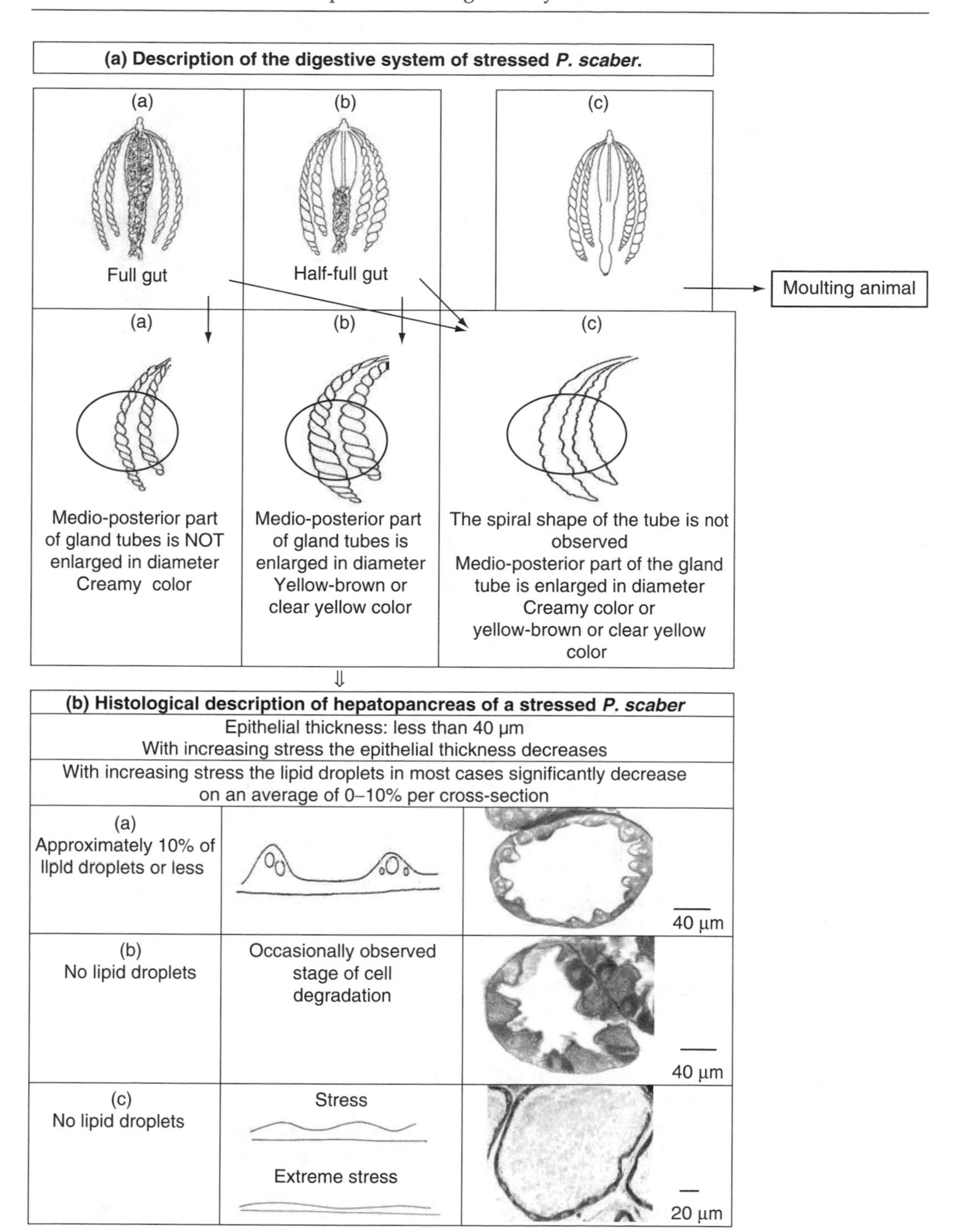

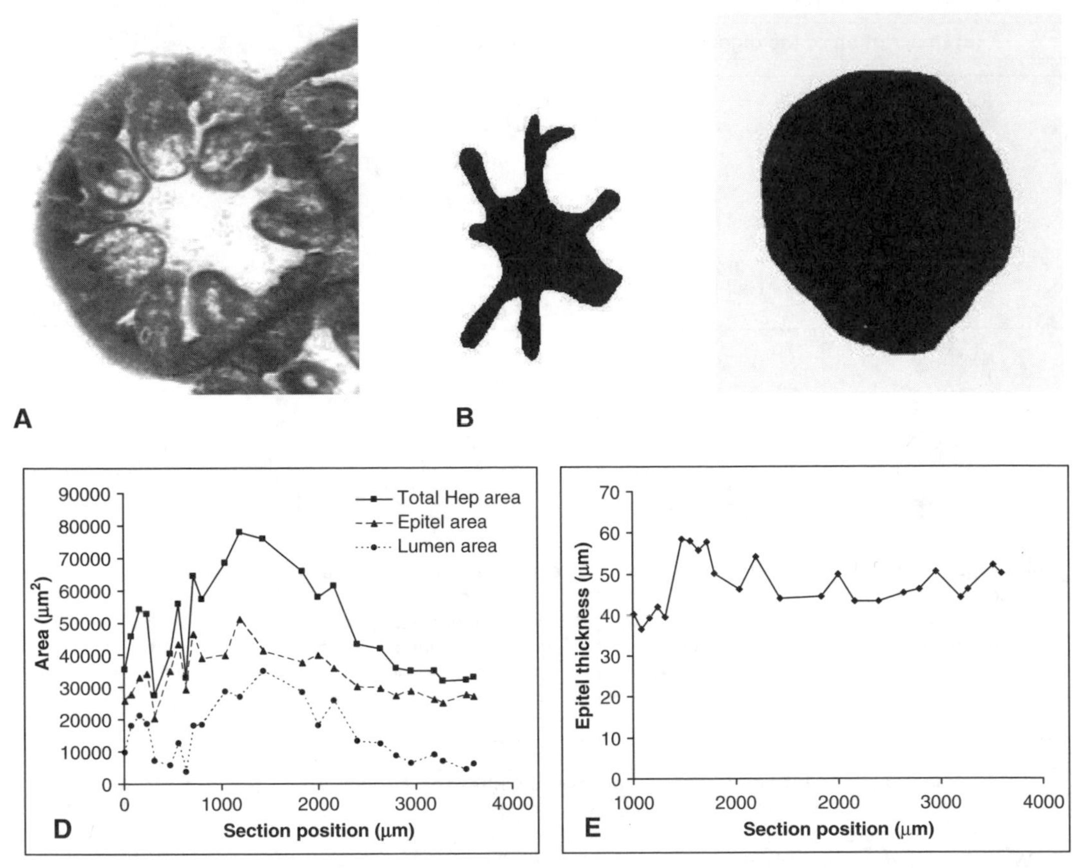

Figure 7.2 An example (Animal 387) of determining the lumen, epithelial and total tube area on cross-sections along a gland tube. See also http://www.fgg.uni-lj.si/~/sdrobne/DDrobne/CRC_Histopathology/. (a) The image of a histological section of hepatopancreas. (b) A contour of the inner epithelial area. (c) A contour of the outer epithelial area. (d) Lumen, epithelial, and total gland tube areas on serial sections along one gland tube. (e) Epithelial thickness on serial sections along the same gland tube.

Germany, Figure 7.2). Manual contouring of the inner surface was necessarily performed, since, using the computer program, the borders between the apical epithelial surfaces usually cannot be distinguished from the lumen that is filled with digestive juices. In some control animals, contouring of all serial sections was performed (http://www.fgg.uni-lj.si/~/sdrobne/DDrobne/CRC_Histopathology/). The morphological data of all serial sections were used to evaluate statistically the number of sections needed to be analyzed along the gland tube to obtain a consistent morphometric description of the tube. It was calculated that the image analyses of each 25th section along the gland tube provide the same morphometric data that cumulative analysis of all serial sections would yield.

On the selected section, the lipid surface density ($SD_{lipid} = S_{lipid}/S_{epithelium}$, SD, surface density; S, surface) was also determined. In this case, the surface is the epithelial area on a cross-section. The median lipid surface density along a tube was calculated. An example of a report on lipid surface density determination (Animal 387) that was generated in "Mathematica" and "Excel" is shown in Table 7.3.

Results and discussion

Histological characteristics of unstressed, intermoult digestive glands of P. scaber

Intermoult *P. scaber* have either full or half-full guts. Those with half-full guts have brown-yellow or clear yellow glands with an increased diameter in the medio-posterior part. The glands of animals with full guts are of uniform diameter and are of white or white-yellow color. Microscopically, the digestive gland epithelium is composed of dome-shaped cells protruding into the lumen of a gland tube. They are filled with lipid droplets, but the quantity of lipid droplets varies among individuals. In the majority of animals, the cell cytoplasm is composed of up to 25–30% lipids (Table 7.3). In more than 300 animals investigated, none were found to possess digestive glands totally free of intracytoplasmic lipid droplets. The droplets are situated in the apical part of the cytoplasm or are distributed throughout the entire cytoplasm. The distribution of lipids in the cytoplasm along the entire gland tube does not vary significantly (Table 7.3).

Irrespective of the fullness of the gut, the color of the glands, and the diameter of the medio-posterior part of the gland, the average epithelial thickness along a gland tube is no less than 40 μm. This indicates that the epithelial thickness does not change as the diameter of the gland tube changes during the digestive cycle.[7] The epithelial thickness for that reason is a more reliable measure of stress than the epithelial area, expressed as a percentage of the cellular area on a cross-section. However, in severely stressed animals (long duration of exposure and higher doses of chemicals), the epithelial area as a fraction of tube area is also significantly reduced (Table 7.2b). This was previously shown by Odendaal and Reinecke.[8]

In a group of unstressed animals some have empty guts; these are probably post-moult animals and were therefore not included in our study. Also, in the unstressed group, some animals have larger and softer glands than expected. A microscopic investigation of these glands revealed that they are infected by intracellular bacteria.[9,10] In the infected individuals, the epithelial cell cytoplasm is filled with vacuoles of different sizes, which contain bacteria (Table 7.1b). The infected animals were also excluded from this investigation.

Macroscopic and histological characteristics of the digestive system of stressed P. scaber

As in unstressed *P. scaber*, the animals fed with metal- and pesticide-contaminated food have either full or half-full guts. The majority of glands have a similar macroscopic appearance to those in the unstressed group. However, some animals have large and soft glands, and the spiral shape of the tube is not seen.

In comparison to control animals, the average microscopic thickness of the digestive gland epithelium is less than 40 μm, and the cell cytoplasm usually contains less than 10% lipid droplets. With increasing stress, the percentage of cytoplasm occupied by lipid droplets drops to under the detectable limit. The epithelial cells have a pyramidal shape

Table 7.3 An example of a report on lipid droplets (Animal 387) generated in Mathematica and Excel

Relative statistics: evaluation of the quantity of lipid droplets

Animal:	387

m =	558	mm =	475	d =	10	k =	0.90
n =	750	nn =	475	dout =	4	level =	1.35

	Lipids
N_total=	24
N_bad=	2
Avg =	13.9 %
StDev =	3.0 %
Min =	7.1 %
Max =	18.9 %
D1 =	10.2 %
D9 =	17.2 %

Section: 387-009-400

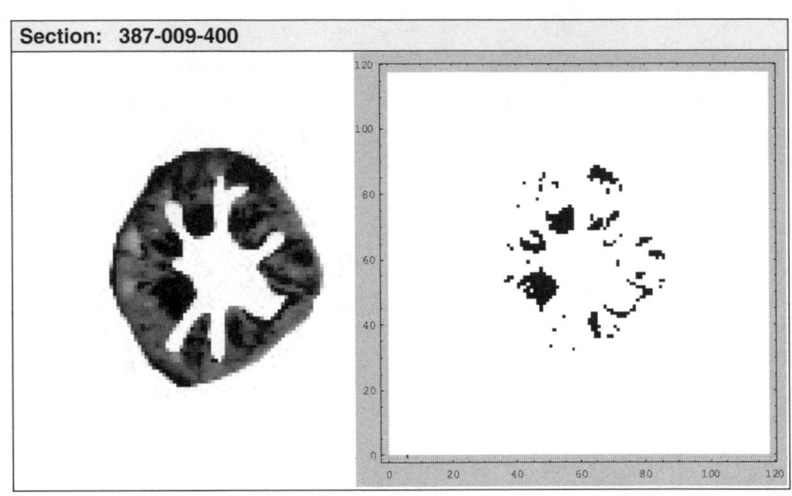

Section	Lipids
1	
10	0.127888
20	0.102136
30	0.115739
40	0.146259
60	0.163593
70	0.086154
80	0.138809
90	0.147927
100	0.165459
130	0.172546
150	0.148345
180	0.165293
230	0.123077
250	0.070901
270	0.158668
300	0.138303
330	0.145473
350	0.174803
370	0.158543
400	0.189398
410	
440	0.124360
450	0.102524

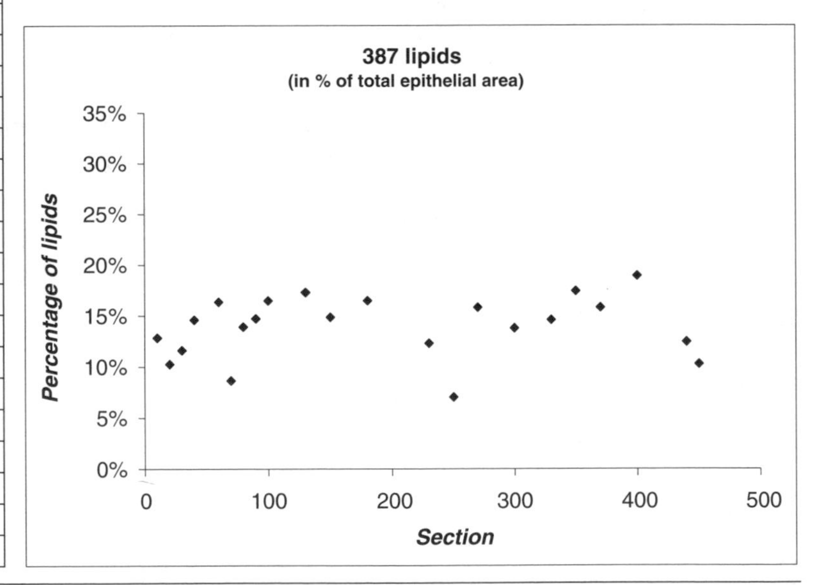

Note: N_total is the total number of the analyzed sections, N_bad is the number of bad (not analyzed) sections, Avg is the average of lipids in the analyzed sections, StDev is the standard deviation, Min and Max are the minimum and maximum of lipids, and D1 and D2 are the first and the ninth decile of values in the analyzed section. Parameters m, n, mm, nn, d, and dout are used in previous phases of preparation and alignment of section images. Parameter k defines the percentage of central pixels that were analyzed (sometimes there occur some errors in the margins of the outlined sections), and level is an empirically defined level of lipid droplets in the section.

(Table 7.1b). In some animals, degradation of epithelium was observed over regionally extensive areas all along the tube (Table 7.2b).[11] With increased stress, the flattened regions become larger and the pyramid-shaped cells are fewer. The overall change consisted of an absence of lipid droplets and the extreme attenuation of the digestive gland epithelium, which were correlated with the duration of exposure to the stressor and with the dose of the chemical.

Distinguishing histological characteristics of unstressed and stressed digestive glands of P. scaber

An adult intermoult *P. scaber* (40–60 mg wet weight) is fit if the average epithelial thickness of a gland tube is usually more than 40 μm, the cells are dome shaped and filled with lipids along the entire gland tube.

The histopathological characteristics of the digestive glands of an organism exposed to stress are as follows: the average epithelial thickness along a gland tube is less than 40 μm, up to 10% of the epithelial cell cytoplasm is composed of lipids in some cells only, there are large areas with pyramid-shaped cells, areas with flattened cells, and/or areas with degrading cells. The epithelial thinning and reduction of lipids in cells show a dose response.

It is difficult if not impossible to distinguish on the basis of histological parameters whether an organism is fit or slightly stressed. Characteristics of the "borderline" group include an epithelial thickness of less than or around 40 μm, regions in the tube where up to 25% of lipids are observed and also regions with no lipids in the cells, and flattened regions, as well as regions with degrading cells. However, dome-shaped cells prevail. We suggest that in animals exposed to low and moderate levels of stress, the judgment whether histological examination indicates the effect of a toxicant or not should be based on observations of more digestive glands of organisms exposed to the same stress. Based on our experiences, 5–8 organisms per experimental group would give reliable information on the effect of a stressor.

Histopathology in isopod toxicity studies

In the protocol suggested for histopathological studies on isopods, the two most serious limitations of histopathology are satisfactorily reduced. The tissue fixation, processing, and staining procedures are highly reproducible. A range of normal morphological variations due to sex, reproductive and nutritional status, and season were thoroughly examined before the histopathological criteria were selected.

An additional advantage of using the digestive glands of isopods in toxicity studies is that each gland tube can be used for a different type of analysis; for example, histopathology, analyses of biochemical biomarkers, analyses of energy reserves (lipids, proteins, glycogen), etc.

It is concluded that in analogy with mammalian and fish toxicology, the histopathology of invertebrates deserves a place in the toxicological toolbox (Table 7.4). Various computer-assisted approaches in pathology can equally be applied to the isopod tissue, which will in turn increase its value in environmental toxicity studies.

Table 7.4 Characteristics of histopathology in toxicity studies with isopods

Animals	Terrestrial and aquatic isopods
Equipment	Standard equipment for a histological laboratory and computer-assisted light microscopy
	Digestive gland histopathology
Characteristics	Applicability
Easy to perform	Moderately
Reliable	Yes, when a large data base on histological responses of selected species is established
Repeatable	Yes
Fast screening	Yes, using reference light micrographs for comparison
Sensitive	Moderately
Early warning	No
Ecologically relevant	When applied with other lower and higher level biomarkers
Laboratory validated	For Cd, Zn, Hg, organophosphorus pesticide diazinon

References

1. Myers, M.S. and Fournie, J.W., Histopathological biomarkers as integrators of anthropogenic and environmental stressors, in *Biological Indicators of Aquatic Ecosystem Stress*, Adams, S.M., Ed., American Fisheries Society, Bethesda, MD, 2002, p. 221.
2. Wester, P.W., van der Ven, L.T.M., Vethaak, A.D., Grinwis, G.C.M. and Vos, J.G., Aquatic toxicology: opportunities for enhancement through histopathology, *Environ. Toxicol. Phar.*, 11, 289–295, 2002.
3. Glaser, J.R. and Glaser, E.M., Stereology, morphometry, and mapping: the whole is greater than the sum of its parts, J. *Chem. Neuroanat.*, 20, 115–126, 2000.
4. Drobne, D., Terrestrial isopods — a good choice for toxicity testing of pollutants in the terrestrial environment, *Environ. Toxicol. Chem.*, 16, 1159–1164, 1997.
5. Drobne, D. and Hopkin, S.P., The toxicity of zinc to terrestrial isopods in a standard laboratory test, *Ecotox. Environ. Safe.*, 31, 1–6, 1995.
6. Zidar, P., Drobne, D. and Štrus, J., Determination of moult stages of *Porcellio scaber* (Isopoda) for routine use, *Crustaceana*, 71, 646–654, 1998.
7. Hames, C.A.C. and Hopkin, S.P., The structure and function of the digestive-system of terrestrial isopods, J. *Zool.*, 217, 599–627, 1989.
8. Odendaal, J.P. and Reinecke, A.J., Quantifying histopathological alterations in the hepatopancreas of the woodlouse *Porcellio laevis* (Isopoda) as a biomarker of cadmium exposure, *Ecotox. Environ. Safe.*, 56, 319–325, 1989.
9. Drobne, D., Štrus, J., Žnidaršič, N. and Zidar, P., Morphological description of bacterial infection of digestive glands in the terrestrial isopod *Porcellio scaber* (Isopoda, Crustacea), J. *Invertebr. Pathol.*, 73, 113–119, 1999.
10. Kostanjšek, R., Štrus, J., Drobne, D. and Avguštin, G., Candidatus *Rhabdochlamydia porcellionis*, an intracellular bacterium from hepatopancreas of the terrestrial isopod *Porcellio scaber* (Crustacea: Isopoda), *Int. J. Syst. Evol. Microbiol.* 54, 543–549, 2004.
11. Drobne, D. and Štrus, J., The effect of Zn on the digestive gland epithelium of *Porcellio scaber* (Isopoda, Crustacea), *Pflügers. Arch.*, 431, 247–248, 1996.

chapter eight

Sperm cell and embryo toxicity tests using the sea urchin Paracentrotus lividus *(LmK)*

Annamaria Volpi Ghirardini, Alessandra Arizzi Novelli, Chiara Losso, and Pier Francesco Ghetti
University Ca' Foscari of Venice

Contents

1-56670-664-5/05/$0.00+$1.50
© 2005 by CRC Press

Introduction

A major requirement in environmental health assessment and monitoring of coastal marine environments is the availability of "laboratory biological instruments" allowing reliable evaluation of sublethal toxicity endpoints for single pollutants and environmental matrices. In this field, the sea urchin broadly meets the above-mentioned requirements, allowing evaluation and integration of a multiple set of sublethal endpoints in early life stages, and thus has great importance for population health, although conducted on a single-species basis. In fact, key-events evaluated in sea urchin bioassays include: (a) reproductive success, (b) offspring quality following gamete exposure, (c) larval development, and (d) cytogenetic anomalies. Any change in these events is evaluated by following exposure to single toxicant or polluted environmental matrix of sea urchin gametes or embryos, which are easy to obtain in a good state of health of adult organisms during their reproductive period.

There are many advantages with the utilization of sea urchin bioassays. They include: the availability of detailed basic information on echinoid biology, the year-round availability of animals, the possibility of using organisms at the top of evolutive scale of invertebrates, of having available multiple life stages together with short exposure times, sensitive responses and multiple test endpoints. As regard the test itself, it has low laboratory costs, it is highly reproducible, it has a cosmopolitan test system, and there is also a wealth literature available.[1]

This work describes the procedure developed in our laboratories for performing two toxicity bioassays: the first based on reproductive success as endpoint (sperm cell toxicity test) and the second on larval development (embryo toxicity test), using the sea urchin *Paracentrotus lividus* Lmk. The procedure, derived from the original protocol by Dinnel et al.,[2] was developed following the auto-ecological characteristics of *P. lividus* and in harmony with US Environmental Protection Agency standard procedures using the species *Arbacia punctulata* and *Strongylocentrotus purpuratus*.[3,4] *P. lividus* is the most abundant sea urchin species along northern Atlantic and Mediterranean coasts, with a long reproductive period, from October to June.[5]

Toxicity tests using the early life stages of *P. lividus* are demonstrated to be reliable and useful tools in assessing the toxicity of xenobiotic compounds, which are very dangerous for aquatic life. The contemporary use of these two toxicity tests combines very important toxicological information, for a predictive view of the effects that dangerous pollutants may have on populations in coastal marine environments.[6] Moreover, our research group proposes bioassays with *P. lividus* as methods for quality assessment and monitoring of marine coasts and transitional environments in particular, the Lagoon of Venice. The suitability of bioassays for assessing lagoonal sediments has been tested by their application at sites in the Lagoon of Venice typified by differing kinds and levels of pollution. Elutriates were chosen to assess the potential effects of pollutants that are made available in the water column as a consequence of sediment resuspension (dredging, fishing gear, etc.). Both tests were effective in discriminating several different pollution/bioavailability situations, although their combined use showed higher efficacy in discriminating between stations and also periods.[7]

A preliminary investigation with pore waters (by extraction with centrifugation) was also carried out. Preliminary results showed that both bioassays were able to highlight differences in the toxicity of sediment samples from varying contamination sites (unpublished data).

Materials required

Animal collection and culture

- PET vessel (30 l) for animal transfer preferably in an insulated transport case
- Sampling sack
- Portable aeration systems
- Glass aquarium (250 l) (thermostated, or in a climate-controlled chamber)
- Reconstituted seawater prepared with a commercial salt for aquaria
- Aeration system, using filtered compressed air and equipped with air lines, tubes, and air stones
- Filter systems equipped with filtering materials and blast holes
- Temperature (T), pH, and oxygen meters

Dilution water

- Analytical balance.
- Dilution water tank to prepare reconstituted water.
- Deionized water purified by the Milli-RO system/Milli-Q system (Millipore, Bedford, MA, USA).
- Reconstituted seawater prepared by adding sea salts reported in Table 8.1 following the ASTM formula[8] in due amounts and order list, to 890 ml of distilled water (Milli-RO) to reach a salinity value of 34%.

Reagents

- Reference toxicant: copper standard solution for atomic adsorption (1000 mg/l) in nitric acid 0.5 M/l (Baker, Deventer, Holland)
- KCl reagent grade

Table 8.1 Salts used for preparing reconstituted seawater following the formula ASTM[8]

Compound	Amount (mg)
NaF	3
$SrCl_2\ 6H_2O$	20
H_3BO_3	30
KBr	100
KCl	700
$CaCl_2\ 2H_2O$	1470
Na_2SO_4	4000
$MgCl_2\ 6H_2O$	10,780
NaCl	23,500
$Na_2SiO_3\ H_2O$	20
$NaHCO_3$	200

- Formaldehyde (37%) buffered with sodium tetraborate
- Glacial acetic acid 10% reagent grade
- Kits for NH_3 and NO_3 analysis
- Chloride acid
- Acetone

Test apparatus and equipment

- Small syringe
- Parafilm
- Latex gloves
- Glass vessels
- Pasteur pipettes
- Eppendorf vessels
- Ice bucket
- Pipettes (1, 10 ml)
- 10-ml graduated cylinders
- Neubauer hemacytometer counting chamber
- Slides
- Count register (at least two places)
- Dissecting microscope, at 10× and at 40× for determination of number of eggs and sperm cells, respectively
- Fume hood
- Gauze or sieve (200 μm)
- Beakers (100, 500, 1000 ml)
- Flasks (100, 250, 500 ml)
- Sterile polystyrene 6-well micro-plates with lids (Iwaki Brand, Asahi Techno Glass Corporation, Tokyo, Japan) as test chambers
- Thermostatic bath (at least 45 × 45 cm) and/or thermostatic room
- Invert microscope for counting eggs and embryos (optional)
- pH meter.

Procedures

Animal collection and culture

Collection: Adult sea urchins are collected in marine coastal zones at depths from 1 to 4 m, far from sources of large-scale industrial and domestic pollution.

Collection procedure: A variable number of adult sea urchins (we suggest 30–70, at least 4 cm in diameter) are collected by divers during the natural reproductive period, in order to avoid any stress to the organisms. The adults should be handled carefully and gently, so we suggest that divers should be experienced.

Animal transfer: A 30-l PET vessel in an insulated, aerated transport case, partially filled with seawater from the sampling site, is used to transfer animals to the laboratory.

Animal culture: A maximum of 20–30 animals for each 100 l of water is stored in one 250-l glass aquarium containing aerated seawater from the sampling site, kept in a conditioned room (18 ± 0.5°C) with a natural photoperiod. Animals already used for experiments are kept in an identical aquarium and then taken back to the sea. The temperature of seawater is gradually changed until the culture temperature was reached.

The aquaria must be provided with both aeration and filter systems; the later may be composed, for example, of a series of two filtering units, each of which is used to filter about 50 l of seawater. These units are made of 1-l PET pierced columns and filled with various filtering materials and blast holes. The first column is filled with a porous substrate that supplies bacteria of a suitable substratum for their growth; the second column contains granular activated carbon. Each column is kept topped up by a marine pH stabilizer. A small pump (e.g., Mod. Zippy 50-Ocean Fish, Prodac International, Cittadella, PD, Italy) is applied to each column, for continuous re-circulation and filtering of seawater.

Sea urchins are fed every 2–3 days with macroalgae (e.g., *Ulva* sp.) and molluscs (e.g., *Mytilus* sp.), collected at the same sampling site.

Sea urchins are kept at mean salinity 35 $\pm$ 1‰, pH 7.8–8.2, and oxygen at saturation level, respectively. Fecal pellets must be removed every 2 days and the seawater partly replaced by filtered artificial seawater (e.g., Ocean Fish, Prodac International).

Sea urchins can be used for tests after their gradual acclimatization to culturing conditions. Our experience suggests that 1 week of acclimatization is usually sufficient, when differences in water temperature between sampling site and culturing conditions were at most 8–10°C. Acclimatization must be longer, up to 3 weeks, when differences in temperature are higher (about 15°C).

Values of T, pH, DO, O_2, NH_3, and NO_3 are periodically checked, and any dead organisms are removed. If a culturing step is not possible, tests can also be performed with animals as soon as they are collected; but this approach, in our opinion, may not guarantee high data reproducibility (see the following). Some of our data also suggest that organisms are more sensitive as soon as they are collected with respect to cultured ones: EC_{50} data obtained with the reference toxicant (copper) showed values of 28 (26–30) μg/l and 41 (40–43) μg/l with animals just collected, but 46 (44–49) μg/l and 60 (58–62) μg/l, after 1 week of acclimatization, respectively.

Pre-test phases

The test procedure has been developed with reference to the protocol proposed by Dinnel et al.[2] according to the auto-ecological characteristics of *P. lividus*. It is strongly suggested to execute both toxicity tests at the same time in order to use the same pool of gametes. Step-by-step procedure is reported as follows:

Gamete emission: Sea urchins are induced to spawn by injecting 1 ml of 0.5–1 *M* KCl solution into the coelom through the peristome; the animals are then allowed to spawn for about 30 min in 50 ml of artificial seawater in a glass vessel. We suggest inducing spawning at least 9 animals, in order to obtain a minimum of 3 males and 3 females with a good emission. If no good emission is achieved, or there are not enough males and females, other animals must be induced to spawn.

Male gamete collection: The sperm cells (lactescent emission) obtained from a minimum of 3 males are put together in a seawater volume ranging from 50 to 100 ml, depending on the abundance of emission, washing the gonadic plate of each male with the 5-ml pipette. This wet collection is suitable if the test is performed within 1 h of collection. Otherwise, it is advisable to collect the pool of sperm cells dry with a Pasteur pipette in an Eppendorf vessel, and keep it in a covered ice bucket or refrigerator (4°C). However, before cell density is determined, the dry pool needs to be diluted in 50 ml of artificial seawater in a glass vessel, for more reliable evaluation of the sperm cell concentration.

Determination of sperm cell density: The density (no. sperm cells/l) of this pre-diluted sperm suspension is determined by adding a 0.1-ml subsample to 1 ml of glacial acetic acid in a 10-ml graduated cylinder, brought to volume with artificial seawater (dilution factor, df = 100). After the mixing of the suspension by cylinder inversion and Pasteur pipette, one drop is added to each of the two counting places of the Neubauer hemacytometer counting chamber. The count is performed, after 15 min waiting, under a dissecting microscope at 40×. Sperm cell density is determined by applying the formula:

$$[\text{sperm/ml}] = [(\text{df})\ (\text{sperm counted})\ (\text{hemacytometric conversion factor})/\text{no. squares}]\ \text{mm}^3/\text{ml}$$

$$[\text{sperm/ml}] = [(100)\ (\text{sperm counted})\ (4000)/400] \times 1000$$

Sperm/egg ratio: The optimal value for *P. lividus* is 20000:1. In order to keep this ratio constant, sperm density must be adjusted suitably (4×10^7) by applying the formula:

$$\text{df} = [(\text{sperm/ml})/\text{no. sperm desired}] \times 0.1 = [(\text{sperm/ml})/4 \times 10^7] \times 0.1$$

where 0.1 is the sperm volume in ml to add to 10 ml test solution.

The results of a previous set of experiments, in which four different sperm/egg ratios were used (5000:1, 10000:1, 20000:1, 30000:1), showed that, at a sperm/egg ratio of 5000:1, fertilization success was low (39 ± 3.5%), whereas a high percentage of fertilized eggs resulted (88 ± 4.24%, 92 ± 3.46%, 95 ± 1.15%) at higher ratios, respectively. Consequently, a sperm/egg ratio of 20000:1 was chosen as the most suitable, to ensure higher constancy in the percentage of fertilization in controls (Figure 8.1).

Egg collection: The eggs (orange emission), obtained by washing the gonadic plates of minimum of 3 females with the 5-ml pipette, are filtered through a 200-μm gauze and put together in a large beaker (500 ml) containing artificial seawater, in order to obtain a prediluted egg suspension.

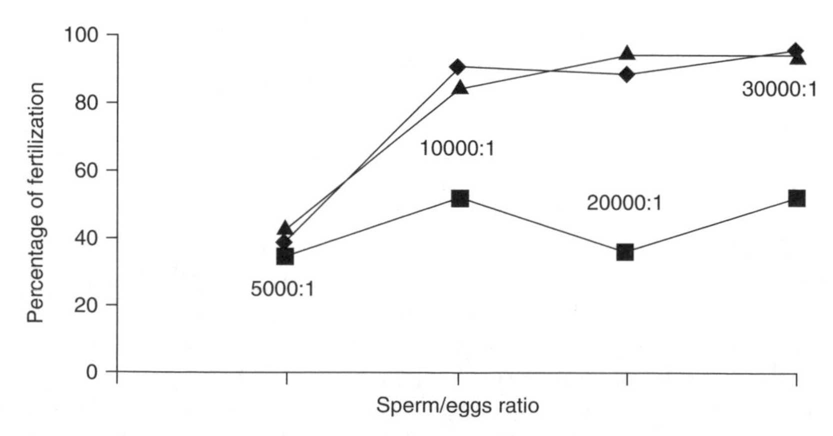

Figure 8.1 Results of sperm cell toxicity test conducted at different sperm/egg ratios (5000:1, 10000:1, 20000:1, and 30000:1). Execution conditions: gametes pool from 3 males; ◆, eggs from first female; ■, eggs from second female; ▲, eggs from third female.

Determination of egg density: Egg suspension density is determined by adding 0.1 ml of the pre-diluted suspension of eggs in a 10-ml graduated cylinder and counting 1 ml of this diluted suspension using a slide (a plankton counting camera with vertical bars facilitates counting), under a dissecting microscope at 10×.

In order to keep the sperm/egg ratio constant, egg density must be standardized to 2000/ml by applying the formula:

$$df = (\text{no. eggs}/\text{no. eggs desired}) \times 1 = (\text{no. eggs}/2000) \times 1$$

where 1 is the egg volume in ml to add to 10 ml test solution.

Some experiments showed that, using the eggs obtained from 3 females separately, the success of fertilization was closely linked to differences in egg quality, less evident when the eggs were put together in a single pool (Figure 8.1).

Sperm cell test procedure

Test procedure conditions: Temperature 18 $\pm$ 0.5°C; salinity 35 $\pm$ 2‰ $O_2 > 40\%$; pH 7.8–8.2.

Execution: 0.1 ml of adjusted sperm suspension is exposed to 10-ml aliquots of test solution in test chambers, and left to incubate in a thermostatic bath at 18°C for 60 min. It is recommended to shake gently and constantly the flask by hand during the transfer of the 0.1 ml of sperm in each well.

After 1 h of exposure, 1 ml of standardized egg suspension is added directly to the solution, in which the sperm cells are exposed. Like the sperm, the flask containing the egg suspension must be gently and constantly shaken by hand during the transfer of the 1 ml of eggs to each well. The number of eggs added to each well containing sperm cells exposed to toxicant solution is about 2000. A period of 20 min is allowed to pass, to ensure that fertilization has occurred.

End of test: The test is stopped adding 1 ml of concentrated buffered formalin, which also allows samples to be preserved for the test lecture from few days up to 1 month.

Endpoint: The endpoint considered in this test is fertilization success/failure, based on observation of the presence of a fertilization membrane surrounding the eggs (Figure 8.2).

Counting eggs: The percentage of fertilization in each treatment is determined by counting 200 eggs at 10× using a plankton counting camera. For correct evaluation, each suspension must be gently mixed by hand in order to re-suspend the eggs uniformly, in a subsample of 1 ml. When using an invert microscope, count directly by observing the whole well randomly.

Repeated counts of 50, 100, 200, and 300 eggs showed that determination of the fertilization percentage is more reliable when the number of counted eggs was at least 200. Counts of 50 and 100 eggs provided coefficients of variation, respectively, of 11.6%

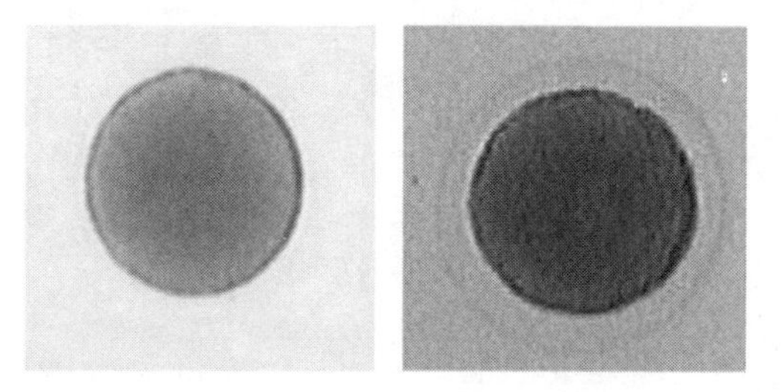

Figure 8.2 Fertilized and unfertilized eggs in sperm cell toxicity test (10×).

and 14%, whereas when the number of counted eggs was increased, the coefficient of variation decreased to 4% and 3% for 200 and 300 eggs, respectively (Figure 8.3).

Embryo toxicity test procedure

Test procedure conditions: Temperature 18 ± 0.5°C; salinity 35 ± 2‰ $O_2 > 40\%$; pH 7.8–8.2.

Zygote achievement: Adjusted sperm and egg suspensions are put together at a sperm/egg ratio of 10:1. For example, prepare at least 100 ml of egg suspension in a beaker, add 10 ml of sperm suspension, and put in the thermostatic bath or room.

Wait for fertilization: A period of 20 min is allowed to pass, to ensure that fertilization has occurred.

Test procedure execution: The test is performed by adding 1 ml of fertilized egg suspension to 10-ml aliquots of test solution contained in test chambers, incubated in a dark thermostatic room or bath at 18°C for 72 h. The number of embryos in the 10-ml test solution is about 2000 (in each well). Normally, zygotes develop in embryos after 48 h, but the time chosen for the test guarantees that all zygotes reach the embryo stage in the negative control.

End of test: The test is stopped by adding 1 ml of concentrated buffered formalin, which also allows the embryos to be preserved for the test lecture from a few days up to 1 month.

Endpoint: The endpoint considered in this test is larval (pluteus) development success/failure. Test results may be evaluated following two different approaches, corresponding to different levels of in-depth examination of anomalies in development.

The first level is a simple, quick discrimination between the normal and anomalous development of each (pluteus) larva (considering all anomalies, including malformations, delays, and blockages at pre-larval stages together). This procedure, which is also currently used in standard U.S. EPA tests, gives the percentage of effect and allows calculating EC_{50} and NOEC. Most data reported in the section "Sensitivity toward pure substances" (and related tables) are of this kind.

The second level requires more accurate assessment of larval anomalies, which may be distinguished into at least four categories:

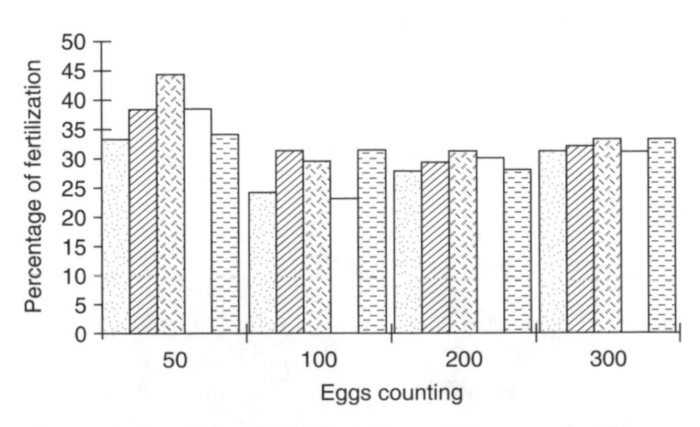

Figure 8.3 Results of repeated counts of 50, 100, 200, and 300 eggs for the sperm cell toxicity test. Data are referred to an experiment using the reference toxicant (60 μg/l test concentration, 1 replicate). Execution conditions: pools of gametes from 3 males and 3 females.

(1) "Malformed plutei," including larvae that are developed but show some malformations (e.g., defects of skeleton and/or digestive apparatus) (Figure 8.4B).
(2) Prisms, including the stage that is normally reached at about the 40th h but abnormal after 72 h (Figure 8.4C).
(3) "Retarded plutei," including larvae that are late, even after 72 h of development time (Figure 8.4D).
(4) Phases blocked before differentiation in larvae (i.e., pre-gastrula stages, gastrula) (Figure 8.4E).

Such careful examination of results after an experiment better discriminates only embryotoxic effects from teratogenic ones. A case study for heavy metals is reported at the end of the section "Sensitivity toward pure substances."

Counting: The percentage of plutei with normal development in each treatment is determined by counting 100 larvae. Each suspension must be gently mixed by hand in order to re-suspend larvae uniformly in a subsample of 1 ml. When using an invert microscope, count directly by observing the whole well randomly.

Repeated counts of 50, 100, 200, and 300 larvae showed more reliable counts when the number of plutei is at least 100: counts of 50 plutei provided larger coefficients of variation (6.9%) than counts of 100 or more larvae (2.6–2.8%) (Figure 8.5).

Preparation of test solutions

Preparation of solutions to perform tests with a pure toxicant (copper was chosen also for its use as a reference toxicant) and with elutriate from a contaminated sediment as a complex matrix, is briefly described here. Results are reported in the following.

The copper solution is prepared by adding 1 ml of the concentrated standard solution (1000 mg/l) to 100 ml of double-distilled water (Milli-Q system, Millipore, Bedford, MA, USA). The solution at concentration 10 mg/l is then diluted using artificial seawater (35‰, 18°C, pH 8) to prepare the toxicity test solutions. The number of reference toxicant

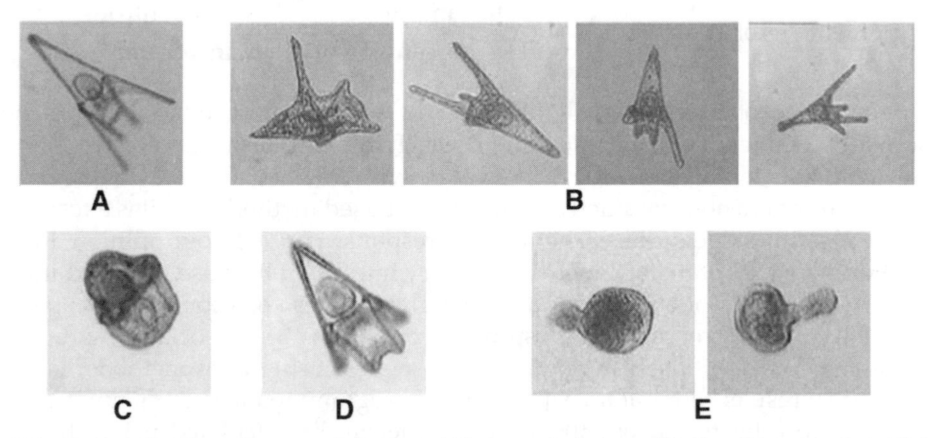

Figure 8.4 Anomalies in embryo toxicity test: (A) normal pluteus; (B) "malformed plutei," including developed larvae that are developed but show some malformations (e.g., defects of skeleton and/or digestive apparatus); (C) prisms; (D) "retarded plutei"; (E) phases blocked before differentiation in larvae (in this example exogastrulas) (10×).

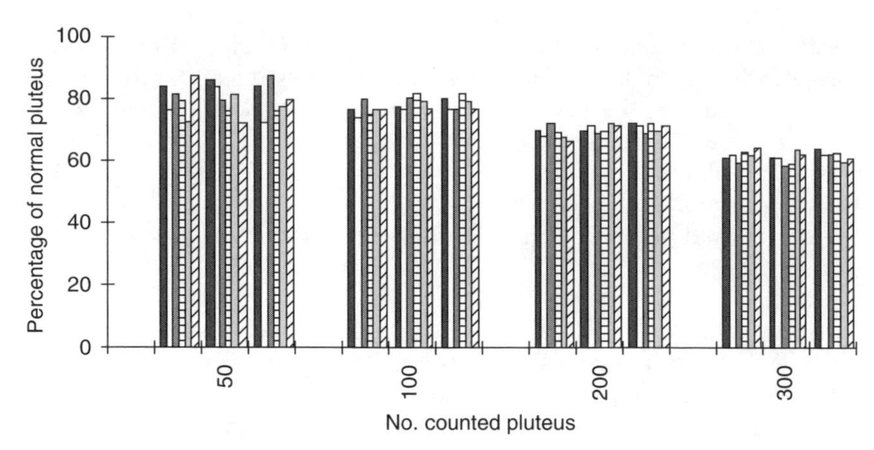

Figure 8.5 Results of repeated counts of 50, 100, 200, and 300 embryos for the embryo toxicity test. Data are referred to a negative control (3 replicates). Execution conditions: pool of gametes from 3 males and 3 females.

concentrations is usually at least 6, chosen in logarithmic scale (24, 36, 48, 60, 72, 84 μg/l) plus the negative control, and the test is performed with 3 replicates.

Elutriates are usually prepared performing dilutions at 6‰, 12%, 25%, 50%, 75%, and 100%, using artificial seawater. Three experimental replicates are used for each dilution and for negative controls.

Data analysis

Toxicity data obtained from a single concentration of pure toxicant or environmental sample may be expressed as the percentage of effect, taking into account the effect of the control, according to Abbott's formula[9]:

$$\text{percentage of effect} = \frac{100 \times \left(\begin{array}{c}\text{percentage of effect}\\ \text{in sample}\end{array} - \begin{array}{c}\text{percentage of effect}\\ \text{in control}\end{array}\right)}{100 - \text{percentage of effect in control}}$$

Percentage of effect obtained from all following dilutions can be used to calculate the EC_{50} and the NOEC values. Two statistical methods, Trimmed Spearman–Karber[10] and Probit,[9] are available to calculate the EC_{50} value with 95% confidence limits. Trimmed Spearman–Karber is a non-parametric test, the most used method; Probit is a test allowing to calculate also the slope of the concentration–response curve. In our opinion, the slopes as calculated by Probit statistical method are very important because they add important information on toxicity of a compound. The higher values correspond to steeper slopes meaning that small increases in exposure concentrations are associated with large increases in observed responses. In any case, we suggest using both methods, in order to verify statistical results better. If an experiment is programmed in order to investigate low concentrations producing effects, the NOEC value can be calculated using the Dunnett program.[11] In Table 8.2, an example with an inorganic toxicant (copper) is reported, with EC_{50} values calculated with the two statistical methods, and the NOEC. Table 8.3 gives an example of spermiotoxicity determination (as percentage of effect and EC_{50}) of environ-

Table 8.2 Example of determination of EC_{50} (with two statistical methods) and NOEC starting from experimental data of sperm cell toxicity test with a pure substance (copper). All data are expressed as μg/l

Copper (Cu^{2+})	Percentage of unfertilized eggs			
	1 replicate	2 replicates	3 replicates	Mean SD
Control	10	14	13	12
24	15	19	23	19
36	26	27	32	21
48	42	31	35	36
60	49	32	43	41
72	76	72	75	74
84	81	85	92	86
	EC_{50} trimmed Spearman–Karber method			62.42 (59.46–65.53) μg/l
	EC_{50} Probit method			64 (60–68) μg/l, slope = 7.73
	NOEC Dunnett program			24 μg/l

Table 8.3 Examples of spermiotoxicity determination (as percentage of effect and EC_{50}) of environmental samples (elutriates): sample n.1, sample with a low toxicity; sample n.2, sample with a high toxicity; n.c., not calculable

	Percentage of unfertilized eggs			
Sample n.1, percentage of sample	1 replicate	2 replicates	3 replicates	Mean
0 (control)	7	9	8	8
6	6	8		
12	10	7	11	9
25	11	10	10	10
50	10	12	11	11
75	15	14	14	14
100	22	25	20	22
	Percentage of Effect (Abbott's formula)			15
	EC_{50}			n.c.
Sample n.2, percentage of sample	1 replicate	2 replicates	3 replicates	Mean
0 (control)	10	8	8	9
6	15	17		
12	28	28	27	28
25	42	50	44	45
50	54	55	55	55
75	85	88	80	84
100	100	100	100	100
	Percentage of Effect (Abbott's formula)			100
	EC_{50} (Spearman-Karber)			33 (29–37)

mental samples (elutriates). Percentage of effect ranks less toxic samples whereas, EC_{50} allows to better discriminating most toxic ones.[7]

Quality assurance/quality control

A good quality assurance (QA)/quality control (QC) program in toxicological experiments requires various phases, including: (a) biological materials (adult quality used for obtaining gametes, negative and positive controls necessary to ascertain gamete quality and to accept or reject experimental data); (b) experimental conditions (constancy of conditions, cleaning of equipment); (c) toxicant characteristics (water solubility and possible used carrier, stability, volatility, ability to modify media, e.g., pH values); (d) abiotic matrices for testing (sampling methods ensuring representative samples of environmental media, sample storage, standardization of procedures to obtain test matrices). In this context, only some guidelines pertaining to points *a* and *b* are reported in the following because of the extent of topics in points *c* and *d*.

Adult quality

Adult quality can be periodically checked by morphometrical–physiological parameters, such as the gonadal index (G.I.) and percentage of maturity. These parameters are closely related to the maturity and health of the sea urchin population used for bioassays and can reveal possible reproductive anomalies. Our experience highlighted the importance of this check, also when the sampling site was changed, sacrificing some adult specimens for information on their morphometrical–physiological characteristics, in order to ascertain the state of the animals before conducting toxicity bioassays.

According to Fenaux[12] the G.I. for sea urchin is calculated as follows:

$$\text{G.I.} = \frac{\text{humid weight of gonads (ml)}}{\text{total humid weight of adult}} \times 100$$

Each adult organism is first weighed using a technical balance and the weight expressed as gram per liter. Through an incision in the peristomial membrane, all five gonads are gently extracted (using a spoon) and the sex recorded. In a 10-ml graduate cylinder containing 5 ml of seawater, the gonads are then added and the weight calculated in milliliter (the water volume displaced by the gonads). In any case the literature also reports other methods for calculating G.I.[5,13]

The percentage of maturity is calculated as the rate between the number of spawning animals during each bioassay and the total number of animals injected with KCl. For *P. lividus* at a stage of good maturation, the G.I. ranges from 5 to 9.[5] The percentage of maturity in males is never lower than 50% but is very variable in females, depending on seasons.[12]

Negative controls

Negative controls are necessary to ascertain gamete quality and to accept or reject experimental data. In all kinds of experiments, first of all artificial seawater must be used as a negative control. The acceptability of test results is fixed both at a fertilization rate and at a percentage of normal plutei of $\geq$70% in all negative controls.[8]

For experiments performed with environmental samples (e.g., elutriates and pore water extracted from sediment), an unpolluted natural or artificial sediment (control sediment) is recommended, in order to take into account also the possible "matrix effect." The experiment excludes a matrix effect when no statistically significant differences are found with tests using the control sediment.

Control chart with a reference toxicant (positive control)

In examining the intralaboratory precision of a method, a reference toxicant control chart must be generated, reporting EC_{50} values as they are obtained by new experiments.[14] This control chart includes both valid and not-valid EC_{50} data considered for the precision of the methods used and intralaboratory reproducibility calculations, and also reveals the sensitivity of the test organisms each time a new test (e.g., with environmental samples) is performed.

For both toxicity bioassays with *P. lividus*, the iterative use of the reference toxicant (copper) and the building up of the control chart yielded information on method precision (considering the minimum number of variables as one operator and one/few batch of organisms) and on intralaboratory reproducibility in several years, with different operators and a lot of batches of organisms from different sampling sites (spatial–temporal variability).[6,14] Copper was selected because it answers to most requisites for a good reference toxicant.[15]

Equipment cleaning

At the end of each test, equipment used to prepare and store water, and to prepare gamete dilution and positive control solutions are rinsed with water and left in 10% chloride acid overnight in separate tanks; this measure is strongly recommended in order to avoid any cross-contamination between the equipment used for biological materials and that used for the reference toxicant or other contaminated material. The materials are then rinsed at least six times with Milli-RO water and the same with Milli-Q water. The glass vessels in which environmental sample dilutions are prepared are washed with detergent, rinsed with water, washed with an organic solvent, such as acetone, and then the normal cleaning procedure is followed.

Results and discussion

Quality assurance/quality control

Adult quality

The G.I., calculated according to Fenaux,[12] found for *P. lividus* of the north Adriatic Sea during the period 1998–2001, revealed that the species generally shows good maturation during the period February–June. In 1999, the mean G.I. value was 7.1 ± 3 ($n = 48$, one sampling site), and in 2001, in three different sites of the Gulf of Venice, the mean G.I. values were 5.4 ± 2.2 ($n = 38$), 4.7 ± 2.3 ($n = 46$), and 5.0 ± 2.7 ($n = 35$) (unpublished data). Anomalous spawning was found in one site of the Gulf of Venice in 1997, with gametes very sensitive to the reference toxicant: the G.I. values were 2.2 ± 2 ($n = 150$) in January and 3.4 ± 2.4 ($n = 10$) in February, lower than those from a reference site (7.7 ± 1.8, $n = 115$ in January).

Negative controls

For each set of experiments (with pure substances or environmental samples), the data for negative controls are expressed by averaging all tests performed with diluted water. For example, in all experiments with organotin compounds, controls showed $95 \pm 2\%$ of fertilized eggs and $83 \pm 5\%$ of normal plutei,[6] in experiments with heavy metals, controls showed $88 \pm 5\%$ of fertilized eggs and $84 \pm 3\%$ of normal plutei.[16]

Control chart with copper as reference toxicant

The control charts of the embryo toxicity and sperm cell toxicity tests are reported in Figures 8.6(a) and (b), respectively. Detailed indications for the building up of the control chart with increasing numbers of variables are reported for the sperm cell toxicity test by Volpi Ghirardini and Arizzi Novelli.[14] After many years, and considering both the natural variability and the high number of introduced variables (i.e., sampling sites, operators, organism batches), the acceptability range considered for EC_{50} is 39–71 and 50–86 μg/l for the sperm cell and embryo toxicity tests, respectively. When EC_{50} values fall outside this range, all data produced using the same pool of gametes should be rejected or considered with due caution.

Intralaboratory variability obtained on the sperm cell toxicity test ($CV = 16\%$, $n = 46$) was lower than the intralaboratory variability range reported for the Atlantic species *Arbacia punctulata* (23–48%)[17]; data for the embryo toxicity test were comparable with those reported by Phillips et al.[18] ($CV = 12\%$, $n = 30$, for *P. lividus* versus $CV = 20\%$ for *S. purpuratus*) using the same reference toxicant and similar methodological procedures. Moreover, the NOEC values (copper) calculated for both sperm cell (years 1998–1999) and embryo toxicity tests (years 1997–1999) yielded means $\pm$ SD of 0.032 ± 0.008 mg/l ($n = 12$) and 0.037 ± 0.015 mg/l ($n = 9$), respectively. The mean NOEC/EC_{50} $\pm$ SD ratios were, 0.60 ± 0.12 and 0.47 ± 0.29, in good agreement with those of 0.53 ± 0.13 and 0.46 ± 0.093 ($n = 6$, calculated) found for *S. purpuratus*[19] for the sperm cell and the embryo toxicity tests, respectively.

Sensitivity toward pure substances

In order to investigate the sensitivity and discriminatory ability of both tests, various compounds, such as anionic and non-ionic surfactants (only with the sperm cell toxicity test),[19] organotin compounds,[6] heavy metals,[16] and sulphides and ammonia[20,21] were studied in our laboratory.

As regards surfactants, the investigated compounds were aromatic and aliphatic surfactants of anionic (linear alkylbenzene sulfonates, LAS) and non-ionic (alcohol polyethoxylates, AE, nonylphenol polyethoxylates, NPE) types, and their aerobic biodegradation products, sulfophenylcarboxylates (SPC), polyethylene glycols (PEG), carboxylated polyethylene glycols (PEGC), carboxylated AE (AEC), and nonylphenol (NP). Results are reported in Table 8.4. The sperm cell toxicity test showed good discriminatory ability among all tested surfactants and their biodegradation products: EC_{50} values differed by about 4 orders of magnitude, ranging from 0.06 to >200 mg/l. This feature allows the test to record significantly different toxicity responses, even when differences in molecular structure are slight. The toxicity of anionic surfactants depends on the length of the alkyl chain, and that of non-ionic surfactants is due to their length and branching. Much lower toxicity was shown by aerobic biodegradation products, in comparison with that of their parent compounds, with the exception of NP. In this study, the sperm cell test

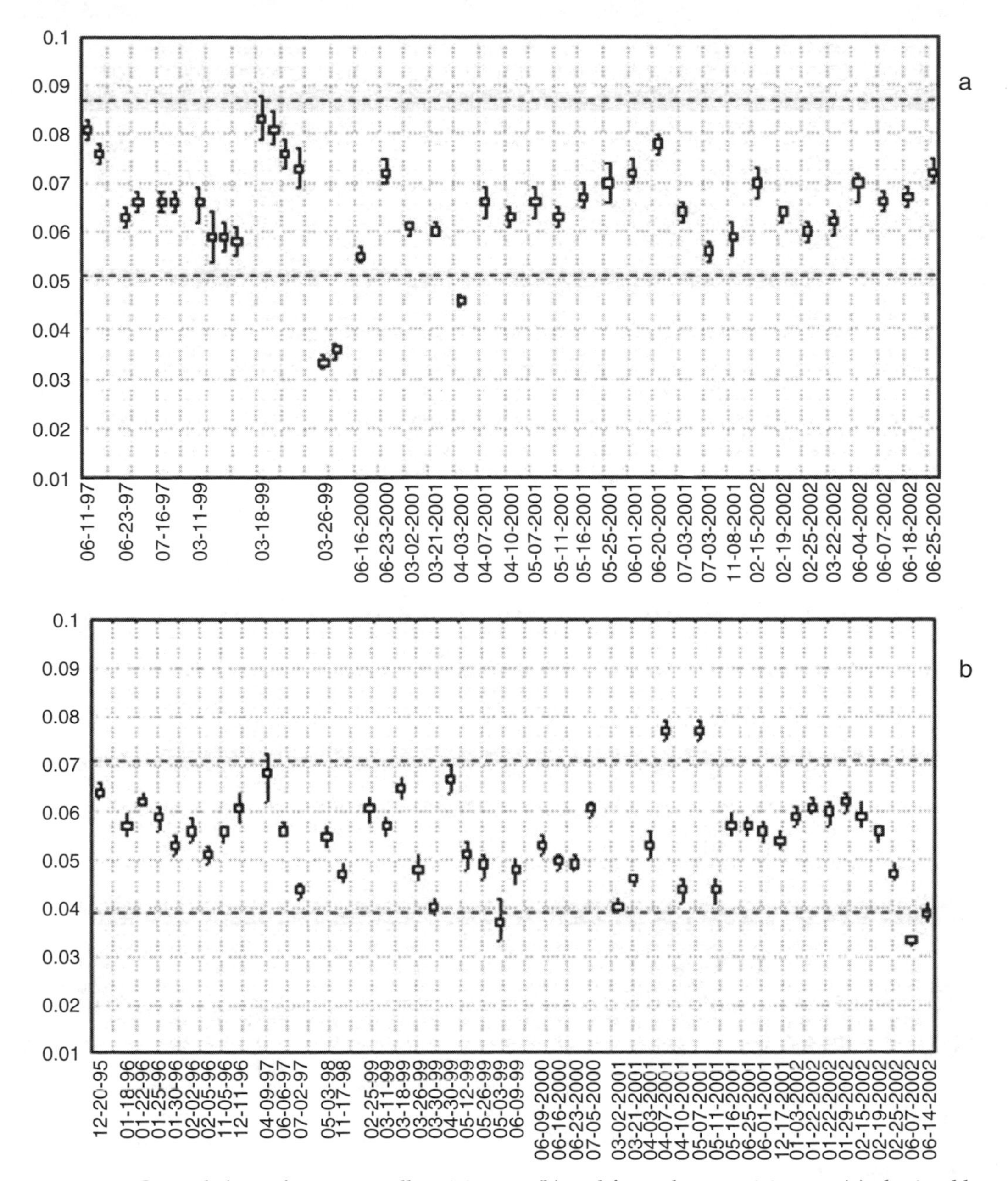

Figure 8.6 Control charts for sperm cell toxicity test (b) and for embryo toxicity test (a) obtained by repeated experiments with the reference toxicant (copper). Dotted lines: acceptability ranges. In ordinate the EC50 values in mg/L are reported.

demonstrated its predictive value in determining the toxicity of xenobiotic compounds with similar molecular structures.

As regards organotin compounds, EC_{50} and NOEC values are listed in Table 8.5. Comparisons between sperm cell and embryo assay results highlighted the fact that the latter has greater sensitivity toward triphenyl, trimethyl, and triethyl compounds, which were instead more toxic for embryos. Comparable sensitivity between sperm and

Table 8.4 Toxicity values (EC_{50} with 95% confidence limits) of anionic and nonionic surfactants for sperm cell toxicity test with *P. lividus*: euAV, average number of ethoxy units; acAV, average length of alkyl chain

Compound		Sperm cell toxicity test, EC_{50}
C_9-LAS	Nonylbenzene sulfonate	2.71 (2.64–2.77)
$C_{11.6}$-LAS	C_{10}–C_{13} alkilbenzene sulfonate mixture	1.12 (1.10–1.14)
C_4-SPC	2-sulfophenyl butirrate	>200
C_{11}-SPC	11-sulfophenyl undecanoate	21.3 (21.1–21.5)
NPE	Nonylphenol polyethoxylate with euAV = 10	1.94 (1.91–1.97)
NP	Nonylphenol	0.27 (0.27–0.28)
L-C_{12}AE	Dodecanol polyethoxylate with euAV = 9	0.94 (0.87–1.01)
L-$C_{13.7}$AE	Linear $C_{12, 14, 16, 18}$ AE blend with acAV = 13.7 and euAV = 10	0.062 (0.060–0.064)
O-$C_{13.6}$AE	Oxo-C12-C15 (46% linear, 54% mono-branched) blend with acAV = 13.6 and euAV = 7	0.12 (0.11–0.12)
B-C_{12}AE	Monobranched C12AE with euAV = 5	4.03 (4.01–4.05)
M-C_{13}AE	Multibranched AE blend with acAV = 13 and euAV = 8	0.92 (0.89–0.95)
PEGs 200, 400, 3000	PEG blend with euAV = 4.5, 9, 68	>200
E_4DC	Carboxylated PEG with euAV = 4	>200
AEC-3.5; AEC-5.5	Carboxylated AE blend with euAV = 3.5, 5.5	2.98 (2.87–3.10); 4.41 (4.33–4.50)

Table 8.5 Toxicity values (NOEC and EC_{50} with 95% confidence limits) of tributyltin and triphenyltin for sperm cell and embryo toxicity tests with *P. Lividus*. Data expressed in μg/l (n.c., not calculable)

Compounds	Sperm cell toxicity test		Embryo toxicity test	
	NOEC	EC_{50}	NOEC	EC_{50}
Tributhyltin chloride[a]	1.3	5.8 (5.5–6.2)	1.3	2.5 (2.4–2.6)
Tributhyltin oxide[a]	n.c.	3.0 (2.7–3.3)	1.8	2.6 (2.5–2.7)
Triphenyltin acetate[a]	n.c.	18.5 (17.8–19.3)	0.5	1.17 (1.11–1.24)
Triphenyltin hydroxide[a]	11	16.5 (15.8–17.2)	0.4	1.11 (1.04–1.18)
Trimethyltin chloride[b]	25,000	30,360 (29,040–31,730)	n.c.	139 (129.1–149.6)
Triethyltin bromide[b]	n.c.	728.7 (699.7–758.9)	n.c.	16.9 (15.8–18.1)

[a] Data from Reference 6.

[b] Unpublished data.

embryos was observed for tributyltin compounds. Investigation of larval anomalies highlighted the fact that damage caused by triorganotins is highly concentration-dependent, and that the most sensitive stages correspond to the crucial phases of differentiation (gastrula and prisma). Data for tributyltin compounds compared well with literature values for early life stages of other marine organisms. NOEC values for the sea urchin embryotoxicity test ranged from 0.4 to 1.8 μg/l and are thus not far from chronic toxicity data in the literature, in spite of very different exposure times (72 h versus several weeks).

Despite the high number of researches on the effects of tributyltin compounds, there is scarce comparable information about sea urchin early life stages sensitivity towards them. Our results concerning embryo toxicity are in agreement with those obtained for tributyltin oxide by Ozretic et al.[22] for *P. lividus*; our EC_{50} values are close to those reported by Ozretic et al. ($EC_{50} = 2\ \mu g/l$), although they used a different methodology. The proportional relationship between the damage rates occurring during differentiation with increasing toxicant concentration were confirmed by Ozretic et al.[22] and Marin et al.[23] Other authors investigated the toxicity of TBTO with other echinoid species, but it is very difficult to compare the results.[24,25] The only toxicity data concerning effects on sea urchin sperm cells are reported by Ringwood[26] for tributyltin chloride towards *Echinometra mathaei* (1-h $EC_{50} = 1.8\ \mu g/l$).

For improved information about the comparative sensitivity of sea urchin bioassays to heavy metals, which are one of the most important causes of contamination in the ecosystem of the Lagoon of Venice, the toxicity of As^{3+}, Cd^{2+}, Cr^{3+}, Ni^{2+}, Pb^{2+}, Cu^{2+}, Zn^{2+}, and Hg^{2+} was investigated. Toxicity values, expressed as EC_{50} and NOEC, are listed in Table 8.6, together with the EC_{50} literature range for other echinoids. Both tests were able to discriminate heavy metal toxicity, showing differences in EC_{50} values of 3 and 2 orders of magnitude for sperm and embryo toxicity tests, respectively. The latter test was particularly sensitive to Cd^{2+}, Ni^{2+}, Pb^{2+}, and Zn^{2+}, whereas both tests showed similar sensitivity toward Cu^{2+}, Cr^{3+}, and As^{3+}. Mercury was shown to be the most toxic of all the heavy metals examined: EC_{50} values were of the same order of magnitude, although the sperm cell test was about three times more sensitive. However, the general higher sensitivity of the embryo toxicity test with respect to the sperm cell test was also confirmed with these pollutants, and is still clearer when the NOEC values are examined. Moreover, for both heavy metals and organotin compounds, the quality and degree of defects in developmental stages depend on increasing concentration.

For all these compounds, toxicity tests using the early life stages of *P. lividus* were demonstrated to be reliable and useful tools to assess the toxicity of xenobiotic compounds, which are very dangerous for aquatic life. The combined use of these

Table 8.6 Toxicity values (NOEC and EC_{50} with 95% confidence limits) of heavy metals for sperm cell and embryo toxicity tests with *P. lividus*. EC_{50} range from literature are reported, regarding the following sea urchins species: *P. Lividus*,[27–30] *Arbacia punctulata*,[31–34] *Arbacia spatuligera*,[31] *Echinometra mathaei*,[26,31] *S. droebachiensis*,[31,35] *S. purpuratus*,[18,31,35] *S. franciscanus*,[31,35] *Dendraster excentricus*,[35] *Lytechinus variegatus*.[31] Data from Reference 16, expressed as mg/l

Metal	Sperm cell toxicity test			Embryo toxicity test		
	NOEC	EC_{50}	EC_{50} range from the literature	NOEC	EC_{50}	EC_{50} range from the literature
As^{3+}	1.9	3.9 (3.7–4.2)	No data as EC_{50}	0.1	1.9 (1.8–2.1)	No data as EC_{50}
Cd^{2+}	1.6	8.4 (7.3–9.7)	8–38[33–35]	0.01	0.2 (0.2–0.3)	0.5–11.24[29,32,35]
Cr^{3+}	0.7	3.2 (3.0–3.5)	20.7 and 341.8[31]	0.2	3.1 (2.4–3.8)	No data as EC_{50}
Ni^{2+}	0.5	5.1 (4.7–5.6)	No data as EC_{50}	0.05	0.32 (0.28–0.35)	No data as EC_{50}
Pb^{2+}	0.5	16.2 (15.6–16.8)	1.3–19[34,35]	0.003	0.07 (0.06–0.08)	0.04–0.54[26,28,34,36]
Cu^{2+}	0.03	0.07 (0.05–0.07)	0.01–0.06[26,31,34,35]	0.04	0.06 (0.05–0.07)	0.02–0.11[28–30,32]
Zn^{2+}	0.1	0.21 (0.20–0.23)	0.028–0.38[31,33–35]	0.01	0.049 (0.045–0.053)	0.02–0.58[17,30,35]
Hg^{2+}	0.01	0.017 (0.016–0.017)	0.0014[32]	0.01	0.05 (0.04–0.05)	0.008–0.044[27–29,32]

two toxicity tests supplies very important information, for a predictive view of some effects, which dangerous pollutants may have on populations in coastal marine environments.

Sulfide and ammonia are considered possible confounding factors when toxicity bioassays are used for monitoring purposes in environmental matrices (e.g., elutriates and pore water extracted from sediments). To improve the very scarce information on the sensitivity of both sea urchin methods to sulfide and ammonia, experiments with *P. lividus* were performed in the same aerobic conditions used for testing environmental samples. Toxicity data for sulfide and ammonia expressed as EC_{50} and NOEC are given in Table 8.7. Both bioassays are very sensitive toward sulfide, their EC_{50} values being comparable. In the sperm cell toxicity test, increasing sulfide concentrations corresponded to an increase in the toxic effect (i.e., an increase in unfertilized eggs). In the embryo toxicity test, although sulfide was almost completely oxidized during the 72-h period, increasing sulfide concentrations corresponded to an increase in the sum of anomalies, meaning that a sufficient sulfide concentration was present in the delicate phases of larval development to determine toxic effects. The embryotoxicity data are in agreement with previous literature toxicity data on other species of echinoids.[20]

As regards ammonia, results confirmed that for sea urchin embryos, the toxicity of ammonia is strictly pH dependent; the higher is the pH of the medium, the lower is the concentration of total ammonia required to reach 50% of embryotoxic effect (due to increased undissociated form). The sperm cell toxicity test was less sensitive to total ammonia than the embryo toxicity test by 1 order of magnitude.

Toxicity tests using the early life stages of *P. lividus* are demonstrated to be reliable, useful tools in assessing the toxicity of xenobiotic compounds, which are very dangerous for aquatic life. Both tests demonstrated high intralaboratory reproducibility, considering data obtained using the reference toxicant over many years of experimentation. Experiments with pure compounds confirmed the high sensitivity of both tests; the sperm cell test has sensitivity comparable to that of an acute sensitive test, whereas the embryo toxicity test shows sensitivity close to that shown in chronic toxicity tests. The combined use of these two toxicity tests yields very important toxicological information, for a predictive view of some effects, which hazardous pollutants may have on populations in coastal marine environments.

Table 8.7 Toxicity values (NOEC and EC_{50} with 95% confidence limits) of sulfide and ammonia for sperm cell and embryo toxicity tests with *P. lividus*. The only EC_{50} data available in literature regarded the sea urchins *S. purpuratus*.[36,37] Data from References 20 and 21, expressed as mg/l

mg/l	Sperm cell toxicity test		Embryo toxicity test		
	NOEC	EC_{50}	NOEC	EC_{50}	EC_{50} range from the literature
Total sulfide	0.1	1.2 (1.1–1.3)	0.1	0.4 (0.4–0.5)	0.19[32]
Total ammonia					
pH 7.7	—	—	1.0	5.70 (5.3–6.1)	7.2[33]
pH 8.0	—	20.7 (18.0–23.9)	0.5	4.24 (3.9–4.6)	2.98[33]
pH 8.3	—	—	0.1	3.10 (2.9–3.3)	1.38[33]

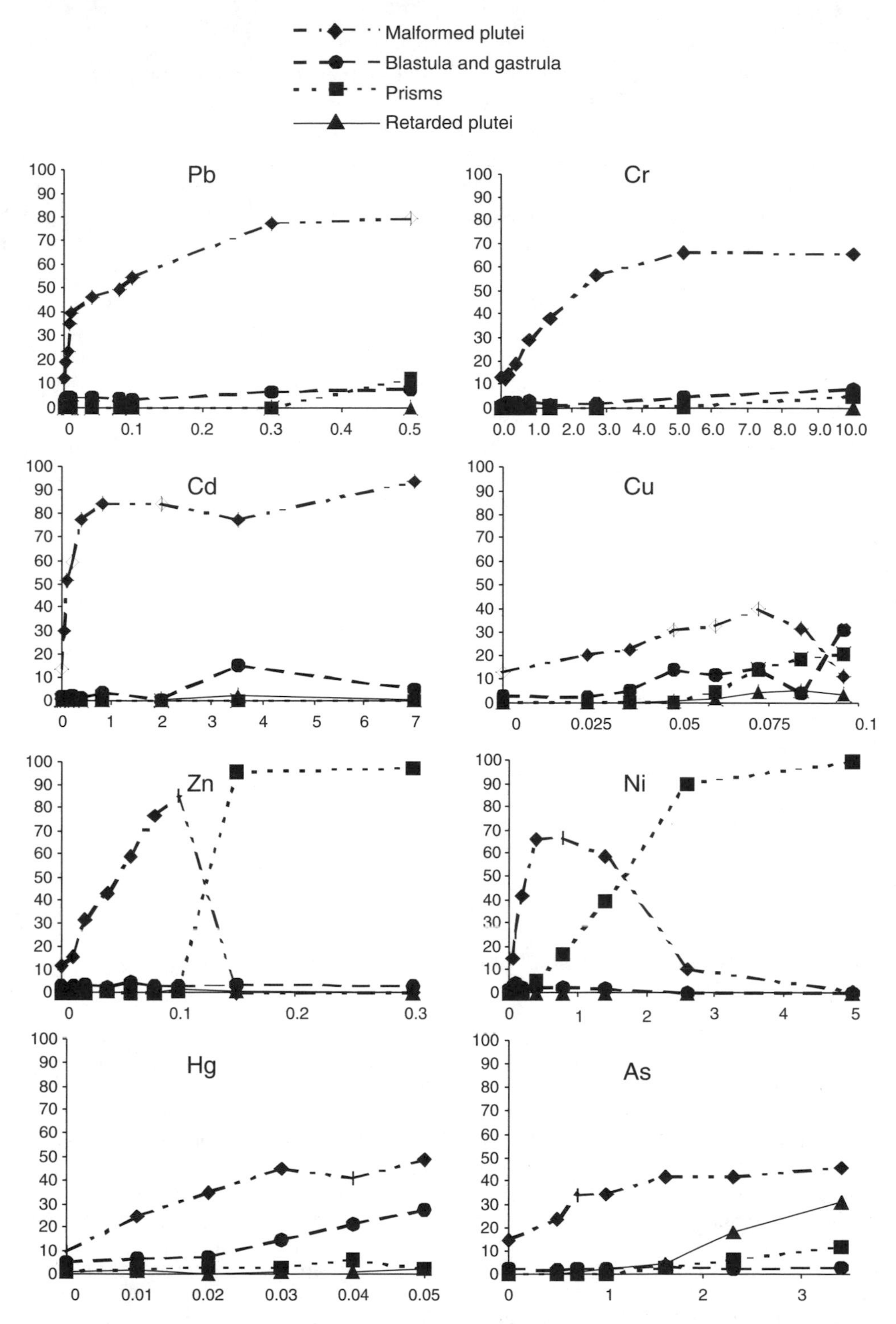

Figure 8.7 Quality and degree of defects in the developmental stages of embryos of *P. lividus* exposed to heavy metals. In abscissa concentrations in milligrams per liter are reported. In ordinate the percentage of effects is reported.

Determination of anomalies in embryo development

Careful examination of anomalies in embryo development after an experiment allows us to better discriminate the kinds of effects produced by each toxicant or environmental matrix. Determination of larval anomalies may yield many indications on the toxicity of pure substances and/or environmental samples tested. An example of the quality and degree of defects caused in developmental stages by pure substances is shown in Figure 8.7 and concerns the study carried out on heavy metals.[16] Generally, the increased concentrations of heavy metals mainly caused larval malformations, particularly evident for Pb^{2+}, Cr^{3+}, and Cd^{2+}, which showed similar trends. For Ni^{2+} and Zn^{2+}, increasing concentrations gave rise to irregular development, which led to an increase of prisms. Malformed prisms were found with 2.6–5 mg/l of Ni^{2+} and 0.15–0.3 mg/l of Zn^{2+}. For Hg^{2+} and As^{3+}, increasing concentrations caused a linear increase in malformed plutei, blastula, and gastrula stages from 0.02 mg/l for Hg^{2+} and in prisms and retarded plutei from 1.6 and 2.3 mg/l for As^{3+}. Cu^{2+} showed a different pattern in metal toxicity, with a decrease in malformed plutei and an increase in prisms, at about 0.072 mg/l and at the blastula and gastrula stages at 0.048 mg/l.

Acknowledgments

The authors are particularly grateful to D. Tagliapietra, C. Pantani, G. Pessa, and M. Picone. Gabriel Walton revised the English text.

References

1. Dinnel, P.A., Pagano G. and Oshida, P.S., A sea urchin test system for environmental monitoring, in *Echinoderm Biology*, Balkema, A.A., Ed., Rotterdam/Brookfield, 1988, pp. 611–619 (ISBN 90 6191 7557).
2. Dinnel, P.A., Link J.M. and Stober, Q.J., Improved methodology for a sea urchin sperm cell bioassay for marine waters, *Arch. Environ. Contam. Toxicol.*, 16, 23–32, 1987.
3. U.S. EPA, Earl-standard Operating Procedure Conducting the Sea Urchin Larval Development Test, U.S. Environmental Protection Agency Research Laboratory, Narragansett, RI, 1990, pp. 135–140.
4. U.S. EPA, Earl-standard Operating Procedure Conducting the Sea Urchin *Arbacia punctulata* fertilization test, U.S. Environmental Protection Agency Research Laboratory, Narragansett, RI, 1991, pp. 125–131.
5. Byrne, M., Annual reproductive cycles of the commercial sea urchin *Paracentrotus lividus* from an exposed intertidal and a sheltered habitat on the west coast of Ireland, *Mar. Biol.*, 104, 275–289, 1990.
6. Arizzi Novelli, A., Argese E., Tagliapietra D., Bettiol C. and Volpi Ghirardini, A., Toxicity of tributyltin and triphenyltin towards early life stages of *Paracentrotus lividus* (Echinodermata: Echinoidea), *Environ. Toxicol. Chem.*, 21 (4), 859–864, 2002.
7. Volpi Ghirardini, A., Arizzi Novelli, A., Losso, C. and Ghetti, P.F., Sea urchin toxicity bioassays for sediment quality assessment in the Lagoon of Venice (Italy), *Chem. Ecol.*, 19 (2/3), 99–111, 2003.
8. ASTM, Standard Guide for Conducting Static Acute Toxicity Tests with Echinoid Embryos, E 1563-98, 1998.
9. Finney, D.J., *Probit Analysis*, Cambridge University Press, London, UK, 1971.

10. Hamilton, M.A., Russo R.C. and Thurston, R.V., Trimmed Spearman–Karber method for estimating median lethal concentrations in toxicity bioassays, *Environ. Sci. Technol.*, 12, 714–720, 1978.
11. U.S. EPA, Short-term Methods for Estimating the Chronic Toxicity of Effluents and Receiving Waters to West Coast Marine and Estuarine Organisms, U.S. EPA, 600/R-95/136, Cincinnati, OH, US, 1995.
12. Fenaux, L., Maturation des gonades et cycle saisonnier des larves chez *A. lixula, P. lividus* et *P. microtuberculatus* (Echinoides) à Villefranche-sur-mer, *Vie et milieu*, 13, 1–52, 1968.
13. Guettaf, M. and San Martin, G.A., Etude de la variabilitè de l'indice gonadique de l'oursin comestible *Paracentrotus lividus* (Echinodermata: Echinoidea) en Mediterranee Nord-Occidentale, *Vie Milieu*, 45 (2), 129–137, 1995.
14. Volpi Ghirardini, A. and Arizzi Novelli, A., A sperm cell toxicity test procedure for the Mediterranean species *Paracentrotus lividus* (Echinodermata: Echinoidea), *Environ. Technol.*, 22, 439–445, 2001.
15. Lee, D.R., Reference toxicants in quality control of aquatic bioassays, in *Aquatic Invertebrate Bioassays*, Buikema, A.L., Jr. and Cairns, J., Jr., Eds., ASTM STP 715, American Society for Testing and Materials, Philadelphia, 1980, pp. 188–199.
16. Arizzi Novelli, A., Losso, C., Ghetti, P.F. and Volpi Ghirardini, A., Toxicity of heavy metals using sperm cell and embryo toxicity bioassays with *Paracentrotus lividus* (Echinodermata: Echinoidea): comparisons with exposure concentrations in the Lagoon of Venice (Italy), *Environ. Toxicol. Chem.*, 22 (6), 1295–1301, 2003.
17. Chapman, G.A., Sea urchin sperm cell test, in *Fundamentals of Aquatic Toxicology: Effects, Environmental Fate and Risk Assessment*, 2nd ed., Rand, G.M., Ed., vol. 6, 1995, pp. 189–205.
18. Phillips, B.M., Anderson, B.S. and Hunt, Y.W., Spatial and temporal variation in results of purple urchin (*Strongylocentrotus purpuratus*) toxicity tests with zinc, *Environ. Toxicol. Chem.*, 17 (3), 453–459, 1998.
19. Volpi Ghirardini, A., Arizzi Novelli, A., Likar, B., Poiana, G., Ghetti, P.F. and Marcomini, A., Sperm cell toxicity test using sea urchin *Paracentrotus lividus* Lamarck (Echinodermata: Echinoidea): sensitivity and discriminatory ability towards anionic and nonionic surfactants, *Environ. Toxicol. Chem.*, 20 (3), 644–651, 2001.
20. Losso, C., Arizzi Novelli, A., Picone, M., Volpi Ghirardini, A., Ghetti, P.F., Rudello, D. and Ugo, P., Sulphide as confounding factor for toxicity bioassays with the sea urchin *Paracentrotus lividus*: comparison with chemical analysis data, *Environ. Toxicol. Chem.*, 23 (2): 396–401, 2004.
21. Arizzi Novelli, A., Picone, M., Losso, C. and Volpi Ghirardini, A., Ammonia as confounding factor in toxicity tests with the sea urchin *Paracentrotus lividus* (Lmk), *Toxicol. Environ. Chem.* 85(4/6): 183–190, 2003.
22. Ozretic, B., Petrovic, S. and Krajnovic-Ozretic, M., Toxicity of TBT-based paint leachates on the embryonic development of the sea urchin *Paracentrotus lividus* Lam, *Chemosphere*, 37, 1109–1118, 1998.
23. Marin, M.G., Moschino, V., Cima, F. and Celli, C., Embryotoxicity of butyltin compounds to the sea urchin *Paracentrotus lividus*, *Mar. Environ. Res.*, 50, 231–235, 2000.
24. Kobayashi, N. and Okamura, H., Effects of new antifouling compounds on the development of sea urchin, *Mar. Poll. Bull.*, 44: 748–754, 2002.
25. Hamoutene, D., Rahimtula, A. and Payne, J., Development of a new biochemical assays for assessing toxicity in invertebrate and fish sperm, *Wat. Res.*, 16, 4049–4053, 2000.
26. Ringwood, A.H., Comparative sensitivity of gametes and early developmental stages of a sea urchin species (*Echinometra mathaei*) and a bivalve species (*Isognomon californicum*) during metal exposure, *Arch. Environ. Contam. Toxicol.*, 22, 288–295, 1992.
27. Warnau, M., Iaccarino, M., De Biase, A., Temara, A., Jangoux, M., Dubois, P. and Pagano, G., Spermiotoxicity and embryotoxicity of heavy metals in the echinoid *Paracentrotus lividus*, *Environ. Toxicol. Chem.* 15, 1931–1936, 1996.

28. His, E., Heyvang, I., Geffard, O., De Montaudouin, X., A comparison between oyster (*Crassostrea gigas*) and sea urchin (*Paracentrotus lividus*) larval bioassays for toxicological studies. *Water Res.*, 33, 1706–1718, 1999.
29. Fernandez, N., Beiras, R. Combined toxicity of dissolved mercury with copper, lead and cadmium on embryogenesis and early larval growth of the *Paracentrotus lividus* sea-urchin. *Ecotoxicology*, 10, 263–271, 2001.
30. Radenac, G., Fichet, D. and Miramand, P., Bioaccumulation and toxicity of four dissolved metals in *Paracentrotus lividus* sea-urchin embryo, *Mar. Environ. Res.*, 51, 151–166, 2001.
31. Larrain, A., Riveros, A., Silva, J. and Bay-Schmith, E., Toxicity of metals and pesticides using the sperm cell bioassay with the sea urchin *Arbacia spatuligera*, *Bull. Environ. Contam. Toxicol.*, 62, 749–757, 1999.
32. Carr, S.R., Long, E.R., Windom, H.L., Chapman, D.C., Thursby, G., Sloane, G.M. and Wolfe, D.A., Sediment quality assessment studies of Tampa Bay, Florida, *Environ. Toxicol. Chem.*, 15, 1218–1231, 1996.
33. Burgess, R.M., Schweitzer, K.A., McKinney, R.A. and Phelps, D.K., Contaminated marine sediments, water column and interstitial toxic effects, *Environ. Toxicol. Chem.*, 12, 127–138, 1993.
34. Nacci, D., Jackim, E. and Walsh, R., Comparative evaluation of three rapid marine toxicity tests: sea urchin early embryo growth test, sea urchin sperm cell toxicity test and Microtox, *Environ. Toxicol. Chem.*, 5, 521–525, 1986.
35. Dinnel, P.A., Link, J.M., Stober, Q.J., Letourneau, M.W. and Roberts, W.E., Comparative sensitivity of sea urchin early embryo growth test, sea urchin sperm cell toxicity test and Microtox, *Environ. Toxicol. Chem.*, 5, 521–525, 1989.
36. Knezovich, J.P., Steichen, D.J., Jelinski, J.A. and Anderson, S.L., Sulfide tolerance of four marine species used to evaluated sediment and pore-water toxicity, *Bull. Environ. Contam. Toxicol.*, 57, 450–457, 1996.
37. Greenstein, D.J., Alzadjali, S., Bay, S.M., in *Southern California Coastal Water Research Project Annual Report 1995–96*, Allen, M.J., Francisco, C. and Hallock, D., Eds., Westminster, CA.

chapter nine

Assessment of metal toxicity to sulfate-reducing bacteria through metal concentration methods

Vivek P. Utgikar
University of Idaho

Henry H. Tabak
U.S. Environmental Protection Agency

Contents

Introduction

This is a method for the quantification of the toxic effects of a heavy metal on the activity of the sulfate-reducing bacteria (SRB). SRB have tremendous potential as remediating agents for the treatment of acid mine drainages or other metal-contaminated, sulfate-rich streams through the reactions shown in the following[1]:

1-56670-664-5/05/$0.00+$1.50
© 2005 by CRC Press

$$\text{organic matter} + SO_4^{2-} \rightarrow HS^- + HCO_3^- \tag{9.1}$$

$$Me^{2+} \text{ (metal ion)} + HS^- \rightarrow MeS \text{ (metal sulfide)} \downarrow + H^+ \tag{9.2}$$

The overall effect of the two reactions is the reduction in sulfate concentration, acidity, and dissolved metal concentrations. Heavy metals present in the waste streams can impact the bacterial community adversely through various mechanisms, including enzyme deactivation, protein denaturation, and exclusion of essential cations.[2,3] Quantification of the toxic effects is vital for the design and operation of an SRB-based treatment process for remediating these waste streams.[4] The technique presented in this chapter allows the determination of the toxic concentration — defined as the minimum metal concentration at which complete cessation of the sulfate reduction activity takes place, and an EC_{50} concentration — effective concentration 50, the initial metal concentration at which the sulfate reduction activity is reduced by 50% over that of a culture not exposed to the metal. The toxic concentration of the metal is determined by exposing the SRB culture to progressively higher concentrations of the metal under investigations and monitoring the activity of the culture. An active culture generates sulfide according to reaction (9.1), which then combines with the ferrous ion present in the system to form a black precipitate of ferrous sulfide. This blackening of the culture serves as the visual indicator of the activity. The concentration at which the characteristic blackening is not observed is the toxic concentration of the metal. If the metal concentration is lower than the toxic concentration, then the SRB culture retains its activity and sulfide generated due to the bacterial metabolism reduces the metal concentration to zero through precipitation as metal sulfide. Consequently, monitoring the metal concentration with time allows one to quantify the sulfate-reduction activity. The SRB culture is exposed to different concentration levels (lower than the toxic concentration) and the sulfate-reduction activities are determined as a function of the initial toxicant metal concentration. The activity of the control culture not exposed to the toxicant metal is monitored through the ferrous ion concentration measurements.

Materials required

SRB culture

Analytical instruments

Inductively coupled plasma (ICP) emission spectrometer, Perkin-Elmer Optima 3300 DV or equivalent
pH meter, Accumet AR25 (Fisher Scientific, catalog #13-636-25) or equivalent

General purpose laboratory equipment

Aluminum Seal Crimper/Decrimper, Fisher Scientific, catalog #03-375-24A/24AC
Anaerobic chamber system, Model 855 AC, Plas Labs, Lansing, Michigan
Analytical balance, Fisher Scientific, catalog #02-112-8
Autoclave, Barnstead/Thermolyne 215560
Drying oven, Fisher Isotemp* Model 506G Economy Lab Oven
General purpose refrigerator, Fisher Scientific, catalog #97-920-1
New Brunswick I2400 incubating shaker, Fisher Scientific, catalog #14-728-2
Magnetic stirrer, Barnstead/Thermolyne Fisher Scientific, catalog #11-496-30 to 32

Model 225 Benchtop centrifuge with rotor and tubes, Fisher Scientific, catalog #04-978-50, 05-111-10, 05-111-18
Muffle furnace, Thermolyne F114215

Reagents

Ammonium chloride, Fisher Scientific, catalog #A661-500
Argon gas, prepurified grade, for ICP analysis
Calcium chloride, Fisher Scientific, catalog #C79-500
Copper sulfate, Fisher Scientific, catalog #C495-500
Distilled deionized (DI) water
Ferrous sulfate, Fisher Scientific, catalog #I146-500
Hydrochloric acid, trace metal, Fisher Scientific, catalog #A508-500
Magnesium sulfate, Fisher Scientific, catalog #M63-500
Nitric acid, trace metal, Fisher Scientific, catalog #A509-500
Nitrogen, hydrogen, and carbon dioxide gases as needed for the anaerobic chamber
Potassium phosphate monobasic, Fisher Scientific, catalog #P285-500
Resazurin, Aldrich, catalog #19930-3
Sodium acetate, Fisher Scientific, catalog #S210-500
Sodium citrate, Fisher Scientific, catalog #S279-500
Sodium hydroxide, Fisher Scientific, catalog #S318-500
Sodium sulfate, Fisher Scientific, catalog #S421-500
Yeast extract, Fisher Scientific, catalog #BP1422-100
Zinc sulfate, Fisher Scientific, catalog #Z68-500

Glassware/miscellaneous supplies

0.2-μm Gelman Nylon Acrodisc Syringe Filter, Fisher Scientific, catalog #4550
Desiccator
Disposable needle, B-D 305194, Fisher Scientific
Disposable syringe, B-D 309603, Fisher Scientific
Falcon Blue Max Jr. graduated tubes (for ICP analysis), Fisher Scientific, catalog #14-959-49B
Fisherbrand Finnpipette pipetters, 100–1000 μl and 1–5 ml, Fisher Scientific, catalog #14-386-74 and 21-377-244
Fisherbrand Finntip pipet tips, Finntip 1000 and Finntip 5 ml
Fisherbrand graduated polypropylene tube (for sample storage), Fisher Scientific, catalog #14-375-150
Flexible tubing (for vacuum filtration, sparging)
Gas regulators, for gases used with anaerobic chamber and ICP
Glass beakers, from 50 ml to 4 l capacity
Glass fiber filter disks, Gelman type AE or Whatman grade 934AH
Gooch crucible and crucible holder
Gray butyl stoppers for serum bottles, Fisher Scientific, catalog #06-447G
Metal tubing (for gas connections)
Oxygen trap, Alltech Associate part #4831
Stir bars, Fisher Scientific, catalog #14-511-59
Suction flask
Volumetric flasks (for dilutions), 250 ml

Weighing dishes
Wheaton aluminum seals, Fisher Scientific, catalog #06-406-14B
Wheaton serum bottles, 125 ml, Fisher Scientific, catalog #06-406K
Wide-bore pipets, Kimble #37005 or equivalent

Procedures

Preparation of the nutrient medium

The high pH recommended for the most of the nutrient media[5] for the cultivation of SRB would result in the precipitation of the heavy metal as hydroxide. Therefore, it is recommended to adjust the pH to a lower value where the heavy metal would remain in the dissolved state. Further, the phosphate concentration also needs to be reduced to prevent metal-phosphate precipitation. The composition of the nutrient medium is shown in Table 9.1.

One liter of nutrient medium is prepared by starting with slightly less than 1 l (ca. 950 ml) of distilled, DI water in a 1.5- or 2-l glass beaker. Quantities of the components 1, 3, 4, 5, 6, and 9 shown in Table 9.1 are weighed using the analytical balance and added to the water while stirring. Components 2, 7, and 8 are added through concentrated stock solutions having the following concentrations: KH_2PO_4 5000 mg/l; $CaCl_2 \cdot 2H_2O$ 6000 mg/l; $MgSO_4 \cdot 7H_2O$ 6000 mg/l. The stock solution concentrations are tenfold higher than the component concentrations in the nutrient medium, and the desired component concentration is achieved by adding 10 ml of the stock solution to the beaker. The order of addition of the components has no effect on the medium; however, it is recommended that the component 9 (ferrous sulfate) be added at the end. One milliliter of 0.1% (w/v) indicator resazurin solution is also added per liter of the nutrient medium. The nutrient medium volume is adjusted to 1 l by adding the required amount of DI water after all the components are dissolved in water. The pH of the medium is adjusted to 6.6 ± 0.1, by addition of dilute hydrochloric acid or sodium hydroxide solutions. This nutrient medium is used for the cultivation of control cultures — cultures that are not exposed to the toxicant heavy metal. The media containing the toxicant heavy metal are obtained by spiking the control medium with varying amounts of stock solutions of the heavy metal sulfate solution. For example, spiking 200 ml of the control media with 0.1, 0.2,

Table 9.1 Nutrient medium composition

No.	Component	Concentration (mg/l)
1	Ammonium chloride (NH_4Cl)	1000
2	Potassium phosphate monobasic (KH_2PO_4)	50
3	Sodium acetate (CH_3COONa)	6000
4	Sodium sulfate (Na_2SO_4)	4500
5	Yeast extract	1000
6	Sodium citrate ($Na_3C_6H_5O_7 \cdot 2H_2O$)	300
7	Calcium chloride ($CaCl_2 \cdot 2H_2O$)	60
8	Magnesium sulfate ($MgSO_4 \cdot 7H_2O$)	60
9	Ferrous sulfate ($FeSO_4 \cdot 5H_2O$)	500

0.4, and 0.8 ml of 5000 mg/l of Zn^{2+} solution (as zinc sulfate) yields nutrient media containing 2.5, 5, 10, and 20 mg/l (ppm) of Zn^{2+}. The 200 ml of nutrient medium at each metal concentration and the control medium are divided into two aliquots of 100 ml each and dispensed into the 125 ml serum bottle. The serum bottles are stoppered with the butyl rubber stoppers and sealed using the aluminum crimp seals. These are then autoclaved at 121°C for 15 min for sterilization. The medium shown in the table is specific to acetate-utilizing SRB, as the organic substrate (electron donor) in the medium is acetate. The media for other SRB can be similarly designed to prevent precipitation of the heavy metal studied. The high concentration of the ferrous ion in the nutrient medium serves to trap the biogenic sulfide and thus is essential for measuring the SRB activity.

Removal of dissolved sulfide from SRB culture

It is presumed that an active SRB culture is available for toxicity assessment. The culture may be a pure strain procured from an organization, such as ATCC (American-type culture collection), and activated according to the procedure suggested, or it may be a mixed culture obtained from a bioreactor. The dissolved sulfide (hydrogen sulfide) associated with an active culture needs to be removed from the system as it will cause metal precipitation and confound results. The SRB culture obtained above is subjected to centrifugation at 2500*g* for 10 min to form a pellet. The clear supernate containing nutrient salts and sulfide is discarded. The pellet is resuspended in a volume of deaerated DI water equal to the original culture volume and again subjected to centrifugation at 2500*g* for 10 min. The deaerated DI water is obtained by purging oxygen-free nitrogen (nitrogen stripped off the residual oxygen by passing it through the oxygen trap) gas through it for 30 min prior to use. The clear supernate left after the settling of the SRB culture is discarded. This wash procedure is repeated twice to remove all the traces of dissolved sulfide. The washed culture is resuspended in a smaller quantity of deaerated DI water to obtain a concentrated suspension and used for the inoculation of the nutrient media. The culture should be washed preferably just prior to the inoculation and used without delay. However, it can be stored in the refrigerator at 4°C for several hours if needed.

The SRB concentration in the toxicity experiments is approximately 100 mg/l on a dry basis. Thus ca. 100 mg SRB culture is needed per liter of the nutrient medium prepared as described earlier. The volume of the SRB culture activated or withdrawn from the bioreactor must contain at least as much SRB biomass on a dry basis. For example, if the bioreactor SRB concentration is 2000 mg/l, at least 50 ml need to be withdrawn for the experiment. The actual volume withdrawn should be preferably between 150 and 200 ml to account for any losses, as well as the determination of biomass concentration as described in the following. The final concentrated suspension after the removal of dissolved sulfide should have the biomass concentration ca. 10,000 mg/l, so that the addition of 1 ml of this suspension to 100 ml nutrient medium results in the SRB concentration of 100 mg/l without any appreciable change in volume.

Determination of biomass concentration

The suspended biomass concentration in the washed concentrate is determined using the standard method 2540 for the measurement of biosolids concentration.[6]

A measured volume of the sample is filtered through a preweighed glass fiber filter in the Gooch crucible using the suction flask. The filtered solids are rinsed with DI water. The filter-crucible assembly is dried in the drying oven for at least 1 h at 103–105°C, transferred to the desiccator, and reweighed after cooling. The difference between the weights divided by the sample volume yields the suspended solids concentration. The volatile suspended solids are obtained by igniting the residue after drying in a muffle furnace at a temperature of 550°C. The weight loss divided by the sample volume yields the volatile solids concentration.

Inoculation and incubation of the cultures

The inoculation of the cooled media is carried out in the anaerobic chamber. The details on the setup and operation of the anaerobic chamber are available in the manual supplied by the vendor and are not repeated here. The serum bottles are opened inside the anaerobic chamber, spiked with the washed SRB stock culture seed, and resealed. As mentioned earlier, the final concentration of the biomass in the seeded serum bottles should be approximately 100 mg/l. This is achieved by spiking an appropriate volume of the concentrated, washed culture. As described earlier, spiking 1 ml of washed culture having a suspended biomass concentration of 10,000 mg/l into 100 ml nutrient medium yields the desired concentration. Serum bottles should be set up at least in duplicate and preferably in triplicate for incubation at each metal concentration. The seeded serum bottles are removed from the chamber and incubated in the shaker-incubator at ca. 150 rpm to ensure well-mixed conditions. The temperature of incubation should be the same as the temperature of activation of the pure strain or the bioreactor temperature from which the SRB culture is derived.

Sampling and analysis

The toxic concentration of the heavy metal under investigation is determined from visual observation of blackening of the serum bottles. Samples are withdrawn at specified time interval (7-day, 10-day, etc.) for analysis of metal concentration and quantification of the sulfate reduction activity as mentioned in the section "Introduction." The concentration data are used to obtain the toxic concentration and EC_{50} values (defined later) as discussed in the following section. A sample is obtained from the serum bottle using a B-D disposable needle attached to a B-D disposable syringe through a 0.20-μm syringe filter. The sample is acidified to a concentration of 2% nitric acid for storage at 4°C in a graduated polypropylene tube prior to the analysis. The metal concentrations in the sample are determined by ICP method as described in the standard methods.[6] Copper and zinc are determined at 324.75 and 213.8 nm, respectively. The standard operating procedures for the instrument are specific to the instrument, and the manufacturer's instruction should be followed for analysis. Two percent nitric acid is used as a diluent to adjust the metal concentration in the sample to within the range spanned by the calibration standards.

Results and discussion

Toxic concentrations

An acetate-utilizing mixed culture of SRB was exposed to copper and zinc in one series of experiments. The source for the mixed SRB culture was anaerobic digester sludge of a municipal wastewater treatment plant, and the SRB culture derived from this source was maintained at 35°C in a master culture reactor in the laboratory.[7] Consequently, the serum bottles were also incubated at 35°C. The toxic concentrations obtained in these experiments are shown in Table 9.2.

As seen from Table 9.2, the toxic concentration of copper was found to be 18.1 mg/l in the first set and 12 mg/l for the second set. The corresponding values for zinc were 23.4 and 19.4 mg/l, respectively. The discrepancy between the two toxic concentrations for each metal obtained in the two different serum bottle sets is attributed to the fact that the initial metal concentration is a discrete variable in the experiments. In the first set of experiments, the highest initial concentrations at which sulfate reduction activity (blackening of the inoculated bottles) was observed were 9 mg/l of copper and 10.6 mg/l of zinc. Sulfate reduction was not observed in the serum bottles containing the next higher initial concentrations of metals, which are the values shown in Table 9.2. The initial metal concentration ranges were reduced in the second set, such that the highest initial metal concentrations in the second set were 16 mg/l for copper and 20 mg/l for zinc. Further, serum bottles were set up at intermediate initial concentrations (12 mg/l for copper and 14.5 mg/l for zinc) to reduce the uncertainty in the determination of toxic concentrations. Sulfate reduction activity was observed in all serum bottles with initial concentrations less than 12 mg/l of copper and 19.4 mg/l (ca. 20 mg/l) of zinc. It can be concluded that the toxic concentrations of copper and zinc for this culture were 12 and 20 mg/l, respectively.

EC_{50}

The metal ions present in the system are consumed by the biogenic sulfide according to Equation (9.2). The toxicant metal ions, as well as the ferrous ion, present in the nutrient medium acted as sinks for the biogenic sulfide. The difference between the initial and final total metal ion concentrations (on a molar basis) was attributed to precipitation of the metal ion as sulfide, and thus served as a measure of the SRB activity. The dissolved ion concentrations were measured at the end of 7 days in tube set 2. The metabolic activity of the SRB can be correlated to ionic concentrations only when an excess of metal ions

Table 9.2 Toxic concentrations of copper and zinc

Serum bottle set	Toxic concentration, mg/l	
	Cu	Zn
1	18.1	23.4
2	12	19.4

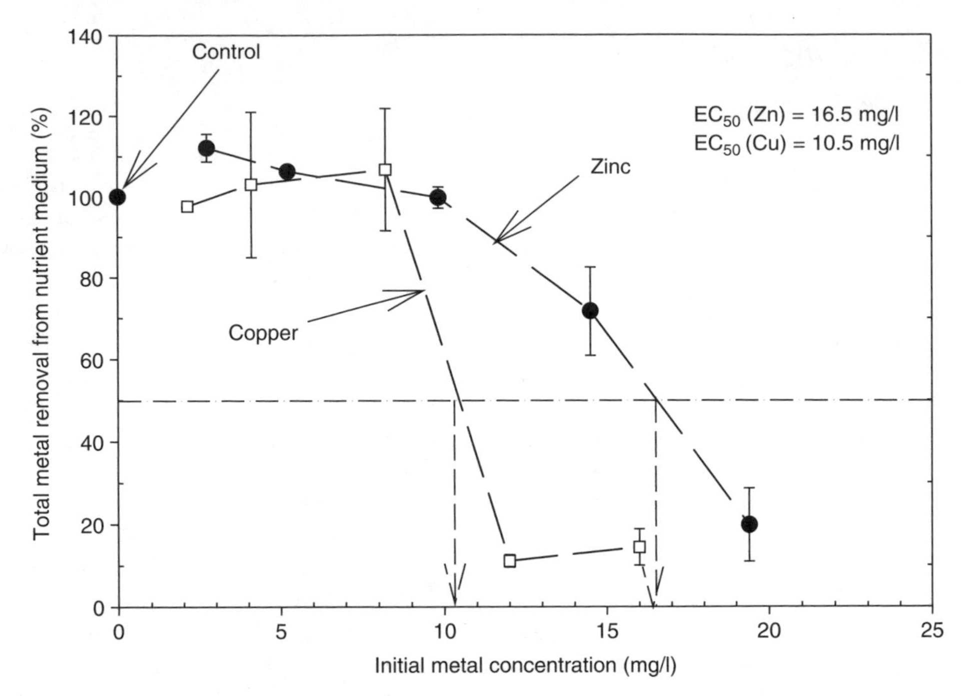

Figure 9.1 EC_{50} of zinc and copper.

(provided in this study through high ferrous ion concentrations) is present in the system. If the experiment is continued for a prolonged period of time, then all the metal ions will be consumed, and the measurements will be indistinguishable across the initial metal ion concentrations. Figure 9.1 shows the amounts of metal ions removed from the solution as a function of initial metal ion concentrations. The amount of metal removed is normalized with respect to the amount of metal removed in the control tubes and plotted as a function of initial metal concentration in the figure. The initial concentration of the metal ion at which the total metal removal was 50% of the control was defined as EC_{50} (7 days), and the values were found to be 10.5 and 16.5 mg/l for copper and zinc, respectively.

EC_{50} is an effective parameter that can be used to compare and rank different metal ions with respect to their toxicity to SRB. It can also be used to compare the toxicity of a toxicant (metal ions in this case) to SRB to the toxicity of the same toxicant to other microorganisms.

References

1. Dvorak, D.H., Hedin, R.S., Edenborn, H.M. and McIntire, P.E., Treatment of metal contaminated water using bacterial sulfate reduction, *Biotechnol. Bioeng.*, 40, 609–616, 1992.
2. Mazidji, C.N., Koopman, B., Bitton, G. and Neita, D., Distinction between heavy metal and organic toxicity using EDTA chelation and microbial assays, *Environ. Toxicol. Water Quality: An Int. J.*, 7, 339–353, 1992.
3. Mosey, F.E. and Hughes, D.A., The toxicity of heavy metal ions to anaerobic digestions, *Water Pollut. Contr.*, 74, 18–39, 1975.

4. Utgikar, V.P., Tabak, H.H., Haines, J.R. and Govind, R., Quantification of toxic and inhibitory impact of copper and zinc on mixed cultures of sulfate-reducing bacteria, *Biotechnol. Bioeng.*, 82, 306–312, 2003.
5. Atlas, R.M., *Handbook of Microbiological Media*, Park, L.C., Ed., CRC Press, Boca Raton, FL, 1993.
6. American Public Health Association, American Water Works Association, Water Environment Federation, *Standard Methods for the Examination of Water and Wastewater*, 20th ed., American Public Health Association, Washington, D.C., 1998.
7. Utgikar, V.P., Chen, B.-Y., Chaudhary, N., Tabak, H.H., Haines, J.R. and Govind, R., Acute toxicity of heavy metals to acetate-utilizing mixed cultures of sulfate-reducing bacteria: EC100 and EC50, *Environ. Toxicol. Chem.*, 20, 2662–2669, 2001.

section two

Techniques for measurement of cellular and subcellular toxicity

chapter ten

Cellular diagnostics and its application to aquatic and marine toxicology

Craig A. Downs
EnVirtue Biotechnologies, Inc.

Contents

Introduction

Theory

Biomarkers indicate the status or condition of a specific biological property. In principle, biomarkers are generally classified into three categories: biomarkers of exposure, biomarkers of effect, and biomarkers of susceptibility.[1–3] Biomarkers of exposure indicate that the organism is or has been exposed to a specific class of xenobiotics or environmental conditions. Biomarkers of effect are indicative of alterations in molecular and cellular processes of an organism as a result of a xenobiotic or environmental exposure. Biomarkers of susceptibility reflect the risk for which an organism may acquire an altered physiological condition. The major difficulty in the application of biomarkers within toxicology is

1-56670-664-5/05/$0.00+$1.50
© 2005 by CRC Press

defining and linking different biomarkers within and between categories into an integrated system that can (1) identify a stressor or stressors affecting an organism, (2) distinguish between different physiological conditions, (3) elucidate the mechanism of toxicity as a result of the interaction, and (4) predict the fate of the interaction.[2]

Cellular diagnostics is a systematic approach to defining and integrating biomarkers of exposure, effect, and susceptibility based on their functionality within a cell and how alterations in the behavior of a single cellular parameter or set of cellular parameters (biomarkers) may affect overall cellular operation or performance. The cell is a dynamic system comprised of both macro- and microstructures and processes. Many of these sub-cellular processes are key metabolic pathways and cellular structural components that are essential for maintaining cellular operations and homeostasis. These cellular metabolic pathways and structural operations can be divided into categories or sub-systems of cellular integrity and function, which can be further defined by discrete parameters (Table 10.1). The behavior of these components and processes defines the physiological condition of the cell. Hence, changes in the behavior of the cellular components/processes are, in effect, changes in the physiological condition of the cell.

Adapting the concept of cellular diagnostics to aquatic toxicological assessments first requires defining cellular toxicity. The cellular diagnostic approach posits that changes in *cellular integrity* and *homeostatic responses* are fundamental end-points in discriminating whether a xenobiotic or environmental factor is affecting the cell. The terms *cellular integrity* and *homeostatic responses* can be partially and stipulatively defined by the some of the categories in Table 10.1. It should be noted that these *definiens* of cellular integrity and homeostatic responses are used because they are measurable by the present technology and are well understood. As our understanding and technology of cell biology evolves, so too will the categories and parameters that define cellular integrity and homeostatic response.

Each category is defined by a functional process and the physical components that are essential to that process. For example, protein metabolic condition can be defined, in part, by many of the RNA, protein, and enzymatic components (sub-systems) that play a role in protein synthesis, protein maturation, and protein degradation (Figure 10.1). To use the category of protein metabolic condition as an example, the three sub-systems of protein metabolic condition can be defined and measured by the following parameters:

Protein synthesis:

- Aminoacyl-tRNA synthase
- eIF1 (translation initiation factor 1; stress inducible A121/SUI1)
- p67 (modulator of eIF2 activation)

Protein chaperoning:

- Hsp72 B′ (inducible Hsp70 cytosolic homolog)
- Grp75 (inducible Hsp70 mitochondrial homolog)
- Grp78 (Hsp70 endoplasmic reticulum homolog)
- Hsp60
- Hsp90α (chaperonin)

Protein degradation:

- Ubiquitin (free and conjugated)
- Ubiquitin activase (E1)
- Ubiquitin ligase (E2)
- Lon (mitochondrial-localized protease)

Table 10.1 Categories of cellular integrity and homeostasis

Genomic integrity:
The ability of the genomic process to maintain a functional state. Some parameters that are used as an index for this condition include DNA damage products (e.g., xenobiotic adducted to DNA, oxidized DNA, and DNA abasic sites), DNA repair systems (e.g., DNA glycosylases), occurrence of single and double strand breaks, chromosomal response, and mutation

Protein metabolic condition:
The process of protein synthesis, protein maturation, and protein degradation. Parameters that are sensitive to changes in this category include p67, concentration and activation states of various Hsp90s, Hsp60, grp78, ubiquitin, LON proteases, and specific ubiquitin ligases

Xenobiotic detoxification:
The process of preventing or reducing the adverse (toxic) effects of exposure to a xenobiotic. Some parameters of the detoxification condition of the cell include concentration and activity of various cytochrome P450 monooxygenase, glutathione-*s*-transferases, and members of the ABC protein family (P-glycoprotein 180/MDR)

Metabolic integrity:
The process of a cell in maintaining a differentiated state from its environment and is the product of sub-processes or "metabolic" pathways. Parameters include the concentration and activity of specific enzymes (e.g., aconitase, PEPCase, ferrochelatase) and discrete organic products (e.g., pyruvic acid, ATP)

Oxidative damage and response:
The process of maintaining a viable condition in an oxygen-laden environment. Parameters include changes in the concentration of oxidative damage products (e.g., protein carbonyl groups, oxidized DNA, aldehyde) and anti-oxidant enzymes and solutes (e.g., catalase and sorbitol)

Immuno-competence:
The process of defending against microbial invasion. Parameters include anti-microbial peptides (*e.g.*, defensins), sterols, sulfonated glycolids, and polyphenols

Autophagy:
The process of removing organelles and other cellular structures by self-digestion. This includes sub-processes, such as eradication of mitochondria, via lysosomal digestion and micronuclei deletion. Parameters that reflect autophagy state include dye retention of lysosomes (e.g., neutral red assay), activation state of Apg16p, Apg12p, and ceramide content

Endocrine competence:
The ability of the cell to communicate with other cells to coordinate phenotypic expression and tissue function. This includes processes of differentiation, inflammation, and stress tolerance

Membrane integrity:
The purpose of a membrane is to act as a selectively permeable barrier. Components and activities that can be used as an index for membrane condition include lipid peroxide and peroxidation breakdown products, sterol content and composition, fatty acid composition, and integral and peripheral membrane protein complexes

These molecular entities were selected as cellular diagnostic parameters for protein metabolic condition because changes in the concentration or activity of these entities may reflect a change in that sub-system. The other categories that define cellular integrity and homeostatic response are similarly structured (Table 10.1).

Shifts in the steady-state rates or levels of these parameters are indicative of a shift in the equilibrium of these sub-systems (Figure 10.2). These shifts contribute to defining the

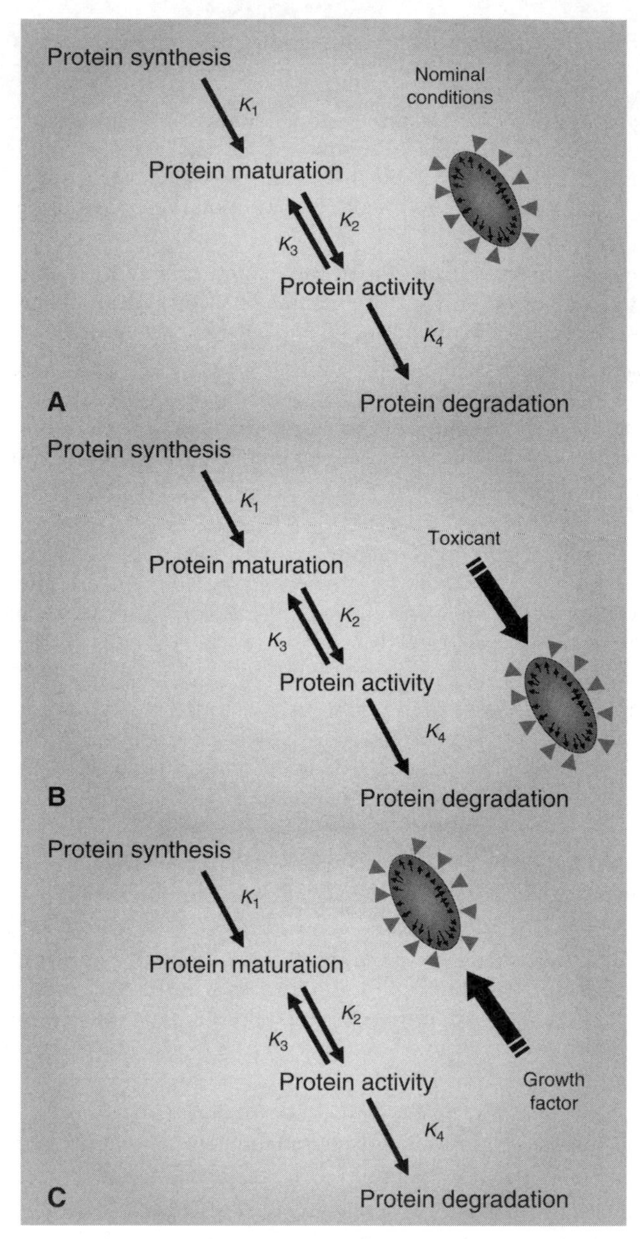

Figure 10.1 Detecting shifts in the kinetics of cellular sub-system equilibria. A primary objective of cellular diagnostics is to detect and understand the consequences of shifts in the steady-state rates of specific cellular sub-systems (e.g., Table 10.1). For example, in protein metabolic condition, there are four primary rates (K_1, K_2, K_3, K_4). Under nominal conditions (Panel A), an equilibrium is formed among the four rates. Introduction of a toxicant that either damages proteins directly, and/or up-regulates the production of specific proteins can alter the steady-state rates that were formed under nominal conditions (Panel B). The final result is a shift in the steady-state rates. Toxicants are not the only factor that can alter steady-state rates. Hormones, pheromones, cellular growth, and differentiation factors can alter the regulation of the physiological condition of the cell, possibly resulting in an increase in protein synthesis and maturation, but not degradation (Panel C). By measuring parameters of protein metabolic condition, one can determine if there has been a change in equilibrium kinetics, and in what direction the change occurred.

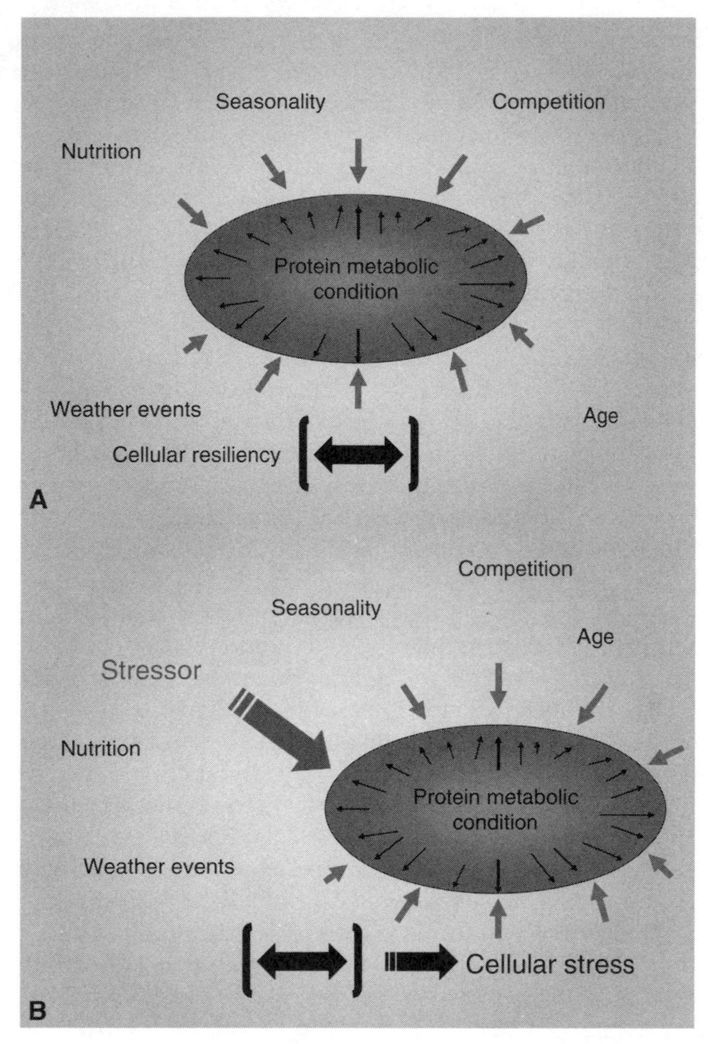

Figure 10.2 Natural variation, cellular resiliency, and cellular pathology. The steady-state rates of a cellular sub-system (e.g., protein metabolic condition) have a dynamic range. Internal and external factors shape the equilibrium of the cellular sub-system. Examples of internal factors include gene regulation, gene/protein structure (mutations), and availability of ligands and substrates. Examples of external factors include temperature, nutritional status, and age of the organism. Under most conditions in the life of the organism, the steady-state rates usually occur within a specific range. The cell has a specific range, in which the steady-state rates can exist and not incur an altered phenotype. This range is known as the range of cellular resiliency or the cellular stress capacity (Panel A). If a "force" factor affects the steady-state rates and so that an altered cellular condition emerges, then this shift in steady-state rates can be categorized as a stressed condition (Panel B). A stressed condition does not necessarily imply a pathological condition. Designation of a pathological condition requires detecting and associating decreases in aspects of cellular performance, such as a decrease free energy levels (e.g., ATP) or a decreases in the reducing/oxidation capacity (e.g., glutathione, NADH) of the cell.

cellular condition. Deviation of the behavior of a specific parameter from the reference is an altered state. An altered state is defined as a pathology or "diseased condition" for the individual (or population) only if that phenotype is associated with condition that adversely affects *performance*.[4] Performance is defined by qualitative states of reproductive fitness, metabolic condition, immuno-competence, genomic integrity, and other similar aspects (Table 10.1). Hence, a pathology or diseased condition may be defined by an abnormal vital function involving any structure, process, or system, whether it is for a cell, individual, or population. In the context of toxicology and cellular diagnostics, a diseased state is any condition where there is deterioration in cellular integrity, an inability or reduction to function, or a reduction in homeostatic capacity.

Cellular diagnostics is analogous in its methodology to clinical and engineering diagnostics, both in theory and practice. Three elements or "concepts" are inherent to the process of diagnosis, which include (1) mastery of the literature (scholarship) and mastery of technique, (2) observation (data) and history of the subject, and (3) the reasoning and interpretative process.[5] The criteria established in all three elements of the diagnostic process must first be met in order to achieve the goals of cellular diagnostics; to detect changes in the subject and to relate or designate an understanding of what those changes mean in terms of the quality of cellular "health."

The first element of the diagnostic process requires a thorough knowledge of the principles and facts of biochemistry, cell physiology, and physiology, and includes mastering the literature pertaining to each of the cellular parameters that are to be examined.[6] For example, "Hsp70" (heat-shock protein 70) is not just "Hsp70"; there are constitutive and inducible isoforms and isoforms that have distinctive sub-cellular localizations that possess slightly different functional properties.[7] Knowing the functional role of each isoform is necessary for not only deriving an accurate interpretation/diagnosis of what the data means, but also to design the appropriate tests and assays in a test array. The other aspect of this element is achieving a level of skill and competence in performing the assay protocols and in handling of the specimen. Additionally, it requires the ability to modify or optimize these standardized protocols to account for the nuances inherent to the sample matrix. For example, to successfully (i.e., an acceptable level of confidence in the data produced) conduct a SDS-polyacrylamide gel electrophoresis (PAGE)/western blot immuno-assay for protein carbonyl residues on blood serum from sea otters and on polyp tissue of *Porites lobata* (a species of hard coral), different ingredients in the buffers as well as different PAGE running conditions are required for each species. An understanding of the matrix composition of the sample is crucial for optimizing assay protocols to provide an acceptable level of confidence in the accuracy of the data produced.

The second element of the diagnostic process is composed of two types of observations: *anamnesis* and test evaluation. *Anamnesis* or "history" is the process of recognizing and understanding the context in which the samples will be analyzed. Anamnesis is the act of taking into account a patient's history. In the context of cellular diagnostics and toxicology, it is the appreciation of an organism's life history, the organism's current state of development, the motives underlying the investigation, and the justification for sampling and testing specific tissue type(s) or the whole organism. Cellular diagnostics is based on the premise that there is a hierarchical relationship between cellular level behavior and phenomena that occur at the tissue, organismal, and population levels.[8] The toxicological investigation may be focusing on the question as to why an exposure to a particular xenobiotic decreases the fecundity of that organism or it may be focused on why mortality occurs after several weeks of exposure to a xenobiotic within a certain concentration range. The motivation of the investigation should provide a historical and conceptual context in which the cellular diagnostic approach can operate. This is no different than the patient history that is essential to an accurate and confident medical diagnosis by the clinician.

The second component of this element is the *test evaluation*: this includes not only processing and assaying the samples but also experimental design and statistical analysis. The act of collecting, processing, and assaying the samples is the most visible component of cellular diagnostics. Designing the assay arrays and the competent execution of the assay protocols is only one aspect of the *test* evaluation. Appropriate experimental design and rigorous statistical methods are required to ensure validity of the data, and that the data can be used as legitimate premises to infer logical conclusions. Hence, not only must there be a level of confidence in the accuracy and precision of the data, but also there has to be a level of confidence in discriminating the differences between populations (reference versus treatments). Only after acquiring an acceptable level of confidence in both aspects of the test evaluation process can the clinician move to the third element of cellular diagnostics, the interpretative process.

The third element of the diagnostic process, reasoning and interpretation, can be divided into two different methodologies; mathematical theory-based methodology and an empirically driven methodology.[9] The mathematical-theory methodology focuses upon symbolic logic and probability theory. Similar to recent developments in bioinformatics, the mathematical theory-based methodology uses methods of Bayesian analysis, neural networks, and fuzzy sets to create algorithms for pattern recognition and discrimination of biomarker data and to test (correlate) for relationships between biomarker pattern and qualitative categories of cellular conditions (e.g., cellular resiliency verusus cellular stress versus cellular degeneration). The empirically driven methodology is a five-step process similar to that described by Ledley and Lusted (1959).[9] The first step is establishing the question being asked concerning the cellular condition. For example: Does exposure to a pesticide induce oxidative stress and to what extent? The second step lists the facts established from the test evaluation and the anamnesis. The third step is ranking or prioritizing the significance of the facts based on values perceived by the diagnostician. For example, the diagnostician may grade the accumulation of hydroxynonenal (HNE) in a pesticide-exposed population to be of great importance because of HNE's known role in cellular pathology and as a key indicator of oxidative damage. The fourth step is creating categories that qualify a cellular condition. In the context of oxidative stress, categories, such as "nominal," "responding to an oxidative stress, but not exhibiting damage," and "exhibiting cellular damage," are crude but effective designations that provide a means to differentiate levels of physiological or process decline. The fifth step is the inclusionary and exclusionary categorizations of the data with the appropriate cellular condition. These two methods (mathematical versus empirical) are not exclusive of one another but are rarely used in conjunction, probably because of the lack of available exposure and formal training to these two methods. Recent work with *in silico* cellular modeling using biomarker data from field-based studies may provide an effective solution in combining the two methodologies for a more robust diagnosis.[2]

Materials required

Coral nubbins for experimental application

- Hammer or mallet
- Steel or titanium leather punch

- Metal, glass, plastic, or Teflon bolt with a head 1–2 cm in diameter
- Marine epoxy or biological glue

Sample preparation and assay protocols

Equipment:

- Ceramic or stainless steel mortar and pestle
- Microcentrifuge
- Vortex mixer
- pH meter
- Analytical balance
- Hot plate
- Container to boil water
- PAGE set
- Wet or semi-dry western transfer apparatus
- 96-well dot blotter or 48-well slot blotter
- Incubation chambers
- Rocker
- Micropipetters (10, 200, and 1000 μl)
- Multi-channel (10–200 μl) pipetter (8 or 12 tips placements)
- Scanner or camera imaging system
- Liquid nitrogen dewar
- Styrofoam box with lid (1 l)
- Computer

Supplies:

- Locking microcentrifuge tubes
- Cryovials (1.8–4 ml)
- Zinc-based spatulas
- Pyrex-based bottle (250–1000 ml)
- Micropipette tips (20, 200, and 1000 μl)
- Graduated cylinders (25 and 500 ml)
- PVDF membrane or nitrocellulose membrane with a pore size 0.2 μm or less
- Flat-plated forceps
- Microcentrifuge tube holders
- Whatman No. 5 filter paper

Reagents:

- Liquid nitrogen
- Dry ice
- Distilled water or distilled/deionized water
- Trizma base (ultra-pure grade)
- Sodium chloride
- Sodium dodecyl sulfate (ultra-pure grade)

  - *Note*: Use SDS from a bottle that has *not* been opened for more than 45 days. SDS can oxidize over time reducing its properties and compromise the sample preparation.
- Dithiothreitol (DDT) or B-mercaptoethanol (BME)
  - *Note*: DDT should be kept in the refrigerator and in a desiccator. When in solution, it should have a "sweet" smell to it when wafted; if it has a "sour" smell to it, the DDT is oxidized and its capacity as a thiol reductant is severely reduced.
  - BME has a shelf life of about 60 days after initially opened and will undergo considerable oxidation, reducing its effectiveness as a thiol reductant.
- Disodium ethylenediaminetetraacetate (EDTA)
  - *Note*: EDTA will dissolve into solution if it is titrated with 5 *M* NaOH, final pH of the solution should be 8.0
- Desferoximine methylate
- Phenylmethylsulfonyl fluoride (PMSF)
- Aprotinin
- Leupeptin
- Bestatin
- Pepstatin A
- α-amino-caproic acid
- Salicylic acid
- Polyvinylpolypyrrolidone (PVPP)
- Dimethyl sulfoxide (DMSO)
- Coomassie blue RR 250
- Methyl alcohol (methanol)
- Glacial acetic acid
- Non-fat dry milk
- Bovine serum albumin
- Tween-20

Procedures

Cellular diagnostics is a methodology that can be applied to almost any given situation in aquatic and marine toxicology. The important question is: Is its application in a given investigation appropriately and correctly applied? One of the pillars of toxicology is the investigation of a dose response of an organism or cell to exposure of a xenobiotic. In keeping with the theme of this book, a strategy is described for using cellular diagnostics to examine the serial dose effects of a single xenobiotic to the hard coral species, *P. porites*. The procedures needed to execute such an experimental design are also included.

Step 1: preparing the organism and experimental design

P. porites has a boulder-like coral structure that includes finger-like extensions from the main body. Unlike most branching coral species (e.g., *Acroporids* and *Stylophora*), the coral tissue extends well into the coralline skeleton. Tissue with an abundance of zooxanthallae (the algal symbiont found in coral) can go as deep as 11 mm into the skeleton, and Coomassie blue or Ponceau red staining suggests coral tissue devoid of obvious zooxanthallae may extend 1–2 mm deeper. Coral plugs should be 1–2 cm in diameter, depending on how much tissue is needed for assaying and should also be 2–4 mm deeper than the zooxanthallae-tissue zone. A steel leather punch is one of the best and most inexpensive tools for creating the initial coral plug. Steel leather punches can be obtained from any local hardware store. Care should be taken to ensure that the leather punch is rust free, and washed with soap (dish soap is best) to remove any oils or silicon grease before use. For the most part, the skeleton of most *Porites* species can easily be cored without much damage to the edge of the leather punch. Extended use or denser skeletons can damage the edge of the leather punch. Jagged or bent edges on a leather punch should be sharpened or, if beyond repair, the leather punch should be discarded.

Once the coral plug is made, it can be affixed to an appropriate mounting fixture, such as the top of a clam shell, or better, the top of a plastic or Teflon-coated bolt with a 1- to 2-cm diameter head is preferable (Figure 10.3). Using a bolt-like fixture will allow for higher densities of coral plugs to be placed in the same chamber, as well as ease of manipulating the plugs between chambers. The nature of the xenobiotic should be taken into consideration. If using a fuel/oil or other types of polycyclic aromatic hydrocarbons (PAHs) as toxicants, Teflon-coated or solid-Teflon fixtures should be used because PAHs have a habit of adhering to ''non-stick'' materials, altering the working concentrations of the xenobiotic in the dosing chambers. Care should also be taken in choosing the type of Teflon used. PFE or PFA Teflon is best and PTFE and related Teflons should be avoided because of their susceptibility to degradation in saline environments. There are several types of adhesives that can be used to affix the coral plugs to the fixture. Most workers use a marine epoxy (suggest Mr. Sticky's Underwater Glue [Fairbanks, California, USA] for its low toxicity) or a biological glue. Biological glues include byssal-thread glue or collagen-based glues that are commonly used in the make-up industries that are saline resistant. Other glues that work well are medical glues, such as Pros-Aide (ADM Tronics), used in surgery or dentistry. It is important not to use standard marine epoxies because their potential toxicity, especially prolonged exposures to ultra-violet light.

The act of creating a coral core induces a stress response. Cellular damage products, such as lipid peroxides, malondialdehyde, hydroxynonenals, and protein carbonyl

Figure 10.3 Coral plugs. Corals plugs can be mounted on an appropriate fixture so that they can be grown at high density and be easily handled. On the left, *P. porites* coral plug in the branching form. On the right, *P. porites* coral plug in a non-branching form. Photos courtesy of Tom Capo.

residues may take as long as 7–14 days to clear from the cellular system and return to basal levels. Chaperonins (e.g., Hsp60) and other stress-associated proteins may take as long as 21 days to return to basal levels. It is suggested that coral plugs be allowed to recuperate and acclimate for at least 2 months before being used in an experiment. Culture conditions like water flow, temperature, light, and feeding methods will not be discussed here, but the U.S. Coral Disease and Health Consortium can be used as a good resource for this information due to the participation of coral culturing experts from both commercial and academic sectors (http://www.coral.noaa.gov/coral_disease/cdhc.shtml).

The dosing system itself may be a significant source of artifact. Dosing corals in static chambers for even 24 h can produce significant artifact. Most coral species require a specific rate of fluid flow over them. In some coral species, the optimal flow rate is almost negligible, while in other species, the optimal rate may be as high as 4 ml min^{-1}. *P. porites* requires a high flow rate, and low flow rates can result in decreased growth rates, increased ammonia secretion, and a significant increase in many stress and growth proteins within 48 h. Ammonia/nitrogen toxicity is an important factor to consider, especially with static dosing systems. Frequent water changes (e.g., every 12 h) or a water flow-through system can control ammonia artifact effects. Light concentration is another significant source of artifact. Indoor ambient lighting is relatively low, even if it seems bright to our eyes (compare 50 μmol of photons m^{-1} s^{-1} in most laboratories versus 2100 μmol of photons m^{-1} s^{-1} outside at noon on a clear day). The rate of photosynthesis is dependent on the concentration rate of light, as are all of its potentially adverse by-products, such as active oxygen species. For photosynthetic organisms, many herbicides and toxicants are light dependent for acute toxicity (e.g., methyl violagen, atrazine, diuron). At least 500 μmol of photons m^{-1} s^{-1} should be used during an experiment. Light quality is also another factor. Like plants, coral zooxanthallae require certain wavelengths of light more than other wavelengths to have optimal photosynthesis. The quality of light emitted by a sulfur lamp or a standard fluorescent lamp is very different compared to the spectrum composition of natural light. Two options in resolving the issue of quality and quantity of light are to use natural light and alter the quantity of light by means of neutral density filters or to use a solar simulator generator. Though solar simulator generators can produce some of the most consistent light conditions from one experiment to another, their expense makes their use prohibitive.

One of the best systems for coral dosing is a recirculating system (Figure 10.4). The system can be placed on a standard laboratory cart that has two benches. On the top bench sits an insulated dosing chamber that is deep enough so that the coral core is submerged at least 4 cm below the surface. On one end of the chamber is a flow-in port, while on the opposite end of the chamber is the flow-exit port; seawater flow enters from the flow-in port and exists from the flow-exit port. On the bottom bench sits a 10- to 20-l reservoir. A variable flow-rate pump will pump water from the reservoir to the chamber. Water flows back to the reservoir from the dosing chamber via the flow-exit port by gravity. Shut-off valves on the tubing that interfaces between the reservoir and the dosing chamber can be installed to make it easy to switch out the reservoir at appropriate time intervals, depending on the length of the exposure. The top part or lid of the chamber can be left open to allow for lighting. A coil connected to a recirculating chiller/heater can be placed in the reservoir or dosing chamber to control for temperature.

Coral plug samples that are collected for testing should be rinsed free of excess mucus with fresh seawater, and quickly "toweled" or dried with a fiber-free towel for excess seawater. Excess seawater with the samples can dilute the final concentration of the

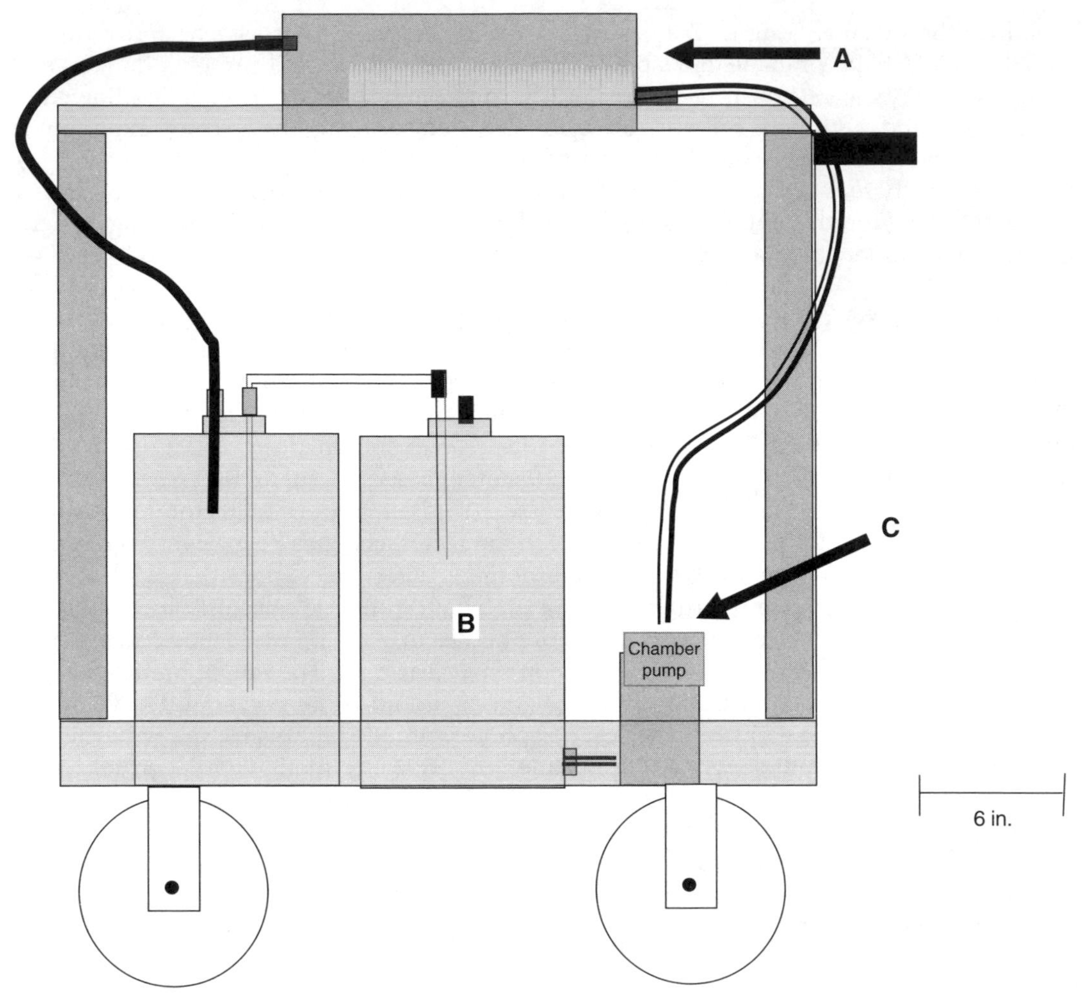

Figure 10.4 Dosing system for corals. Static systems can produce significant artifact, especially if the exposure to the xenobiotic is longer than 24 h. The entire system can be placed on a two-shelf cart. On the top shelf is the dosing chamber, a chamber about 2 l in volume, large enough so that the top of the coral plug is covered by at least 6 cm of water. The top of the dosing chamber is open to the air (light transmission). Water is pumped into the chamber from a set of 25-l reservoirs by a variable-flow pump that sits on the bottom shelf of the cart. Reservoirs can be changed out at designated time points (8–12 h) with fresh solution of seawater/xenobiotic.

sample during sample preparation and can cause artifact when running the sample on a SDS-PAGE. Coral should be quickly frozen in liquid nitrogen or dry ice and kept at –80°C or colder until sample preparation.

Sample preparation, quality control validation, and ELISA protocol

Tests

The protocols described in the following focus only on enzyme-linked immuno-sorbent assays (ELISA). There are a number of non-ELISA tests that can be included in a standard cellular diagnostic assay arrays. For example, spectrophotometric assays, such as those for lipid peroxide, glutathione, and for measuring specific enzyme activities (e.g., esterase, EROD), can readily be included in an array of tests with some modification of the sample preparation protocols. Other innovative assay protocols, such as in this volume by Dr. Kouda and co-authors (Chapter 15), PCR-based assays, COMET assay, and HPLC-based assays, can readily be included in a cellular diagnostic assay array.

Sample preparation

Pre-chill the mortar and pestle with liquid nitrogen equivalent to two volumes of the mortar. Add half or all of a coral plug sample to the mortar. Immediately half-fill the mortar with liquid nitrogen, and after about 5 s begin grinding the sample with the pestle, making sure that the sample never thaws. The sample should be ground to a consistency between that of table salt and fine flour. The spatula and cryovial are pre-chilled in dry ice. Use the spatula to transfer the sample powder to the cryovial. The sample should be kept at –80°C or lower. Thawing and refreezing will cause the sample to solidify, making it impossible to use in further sample preparation steps unless it is reground.

The denaturing buffer should be made within 4 h of use and consists of 2% SDS, 50 m*M* Tris–HCl (pH 6.8), 25 m*M* DDT, 10 m*M* EDTA, 0.001 m*M* sorbitol, 4% PVPP (w/v), 0.005 m*M* salicylic acid, 1% (v/v) DMSO, 50 μ*M* desferoximine methylate, 0.04 m*M* Bestatin, 0.001 m*M* E-64, 2 m*M* PMSF, 2 m*M* benzamidine, 5 μ*M* a-amino-caproic acid, and 1 μg/100 μl pepstatin A. Transfer 50–100 mg of sample from the cryovial to the locking-cap microcentrifuge. Add 1400 μl of denaturing buffer to the microcentrifuge tube, lock the cap to the tube, and vigorously vortex the sample for at least 10 s. Incubate the tube in a 90°C water bath for 3 min, vortex the tube for 10 s, and then reincubate at 90°C for 3 min. Remove the tube and allow it to cool at room temperature for at least 5 min. Centrifuge the tube for 15 min at least 13,000*g*. Three phases will be evident after centrifugation; the bottom phase is the insoluble phase consisting predominantly of coralline skeleton and PVPP (Figure 10.5). The middle phase should be transparent, brownish in color, and free of a whitish film that characterizes the top phase. The top

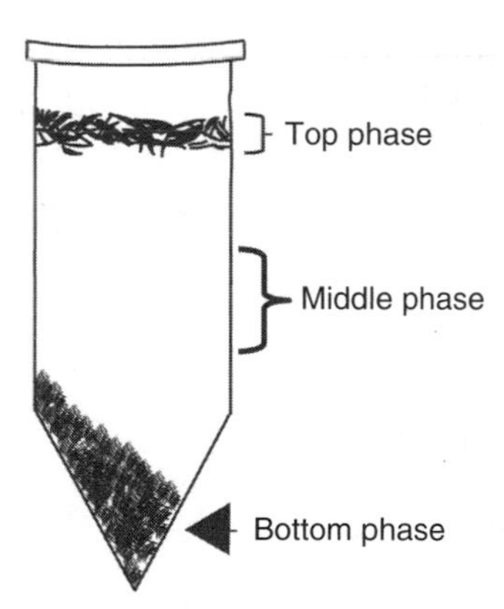

Figure 10.5 Sample preparation. Three phases are formed after centrifugation of the sample. The bottom phase is insoluble debris, primarily calcium carbonate skeleton, and insoluble PVPP. The middle phase should be relatively transparent and should be absent of residual mucus. It should be non-viscous. The top phase should be somewhat viscous and sticky, whitish in color. The demarcation between the middle and top phases is not always apparent, and care should be taken that when aspirating the middle phase, there should be no contamination from the top-phase matrix.

phase is a very viscous matrix composed predominantly of cross-linked polysaccharides. Aspirate 200–300 μl of the middle phase; be careful not to collect any supernatant that may be contaminated with the whitish matrix, which is the top phase. Deposit the middle-phase supernatant into a new tube.

Determining the protein concentration of the sample is necessary to equalize the sample for loading onto SDS-PAGE/western blotting assays, ELISA, and for normalizing the data when calculating the quantity of the parameter in the sample. Protein concentration is one of the most common parameters assayed, but unfortunately, it is one of the least understood assays. Besides sample preparation, assaying for protein concentration is a potential source of artifact and variation that can affect all subsequent assays. There are a number of good sources that explain both the method and theory of a number of different protein concentration assays that should be studied.[10] Many components of a denaturing buffer, as well as compounds inherent to the sample, may potentially interfere with a particular protein concentration assay. For example, chlorophylls and carotenoids in plant and photosynthetic material (e.g., coral, some species of gastropods and mollusks) can interfere with accurate measurements using Coomassie-based spectrophotometric assays. For corals, Ghosh et al.'s (1988)[11] protein concentration assay is a preferred method, because it is amendable to high-throughput, does not require a protein precipitation step to exclude interfering substances produced by the zooxanthallae, and is not affected by the harsh reagents found in the denaturing buffer recipe described in this chapter.[11] Using Whatman No. 5 filter paper rather than using a microtiter plate is a better medium for a blotting platform. Using SDS denatured bovine serum albumin or oval albumin is acceptable as protein concentration standards, though the highest concentration standard that should be used is 2.5 mg ml^{-1}. Sample protein concentrations above 2.5 mg ml^{-1} are suspect because the ratio of SDS to protein does not ensure adequate binding of the SDS to protein, which can lead to protein aggregation and precipitation as a result of freeze/thaws and during the heat-denaturation step of the sample preparation.

Apply 1 μl of sample in triplicate to the filter paper (5 or 10 cm diameter). Samples and calibration curves can be designated on the paper using a pencil (No. 2). Allow the samples to completely dry, either air dry for 10 min or with a hair dryer. Incubate the filter paper on a rocker in the stain solution for at least 10 min. Pour off the stain solution, and then apply 4–5 washes of the destain solution, incubating 5–8 min for each wash. The filter paper untouched by sample or calibrant should be white and not have any remnants of blue stain (Figure 10.6).

Stain solution:

- 200 ml of distilled water
- 50 ml of glacial acetic acid
- 250 ml of methanol
- 4 g of Coomassie blue RR 250

Destain solution:

- 800 ml distilled water
- 200 ml of glacial acetic acid
- 1000 ml of methanol

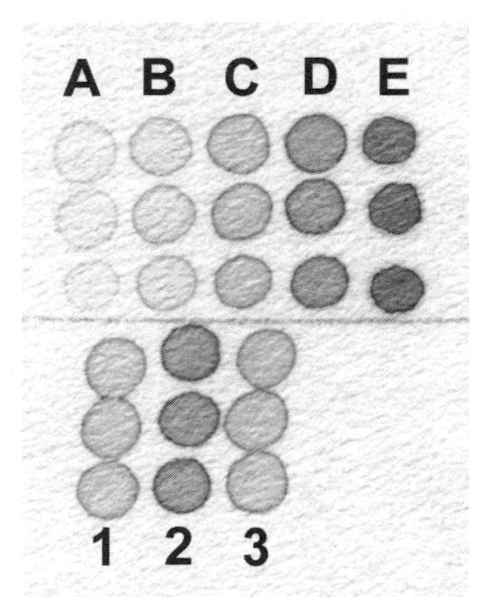

Figure 10.6 Protein concentration assay. Results of protein concentration assay using a modified method of Ghosh et al. (1988).[11] Concentration calibrants are plated on the Whatman No. 5 filter paper as in columns A–E. Column A = 0.125 $\mu g\ \mu l^{-1}$. Column B = 0.250 $\mu g\ \mu l^{-1}$. Column C = 0.5 $\mu g\ \mu l^{-1}$. Column D = 1.0 $\mu g\ \mu l^{-1}$. Column E = 2.0 $\mu g\ \mu l^{-1}$. Samples done in triplicate indicated in columns 1–3. All calibrants and samples are plated (hence assayed) on the filter paper in triplicate. Assay paper can be scanned by a scanner to create a digital image. Optical densities for each replicate can be determined using a densitometry program, such as NIH Image developed at the U.S. National Institutes of Health and available on the Internet at http://rsb.info.nih.gov/nih-image/.

Sample validation

Quality control step 1: Confidence that the assays are measuring the appropriate cellular marker is usually required for most scientific publications, government reports, or litigious discussions. SDS-PAGE/western blot analysis is the minimum level of quality control/quality assurance. Western blot analysis will demonstrate that (1) preparation of the sample was satisfactory, and that (2) the antibody for the set of samples can be validly used for ELISA. Criteria for confidence assurance are:

- There is no "smearing" of the sample lane
- Band(s) are present at the expected migration rate
- Absence of non-specific cross-reactivity

Aliquots of 100 μl from four random samples should be pooled, the protein concentration determined for the pooled sample and applied to SDS-PAGE. Concentration of acrylamide and other gel contents are specific to the type of parameter being examined. For example, assaying of small heat-shock proteins and metallothioneins will require specific reductants in both the denaturing buffer and in the separating gel. Adding reductants, such as TCEP (pH neutral) at 1 m*M* in the separating gel, will ensure thiol reduction as the thiol-rich proteins move through the gel. Altering the percent of acrylamide in the separating gel can enhance the resolution in seeing differences, depending on the molecular weight of the target protein. Finally, it is important to ensure that the gel has SDS in both the stacking and separating solution. Many pre-cast gels lack SDS, which can affect the migration rate of many of the target proteins.

Contents of gels are transferred to activated PVDF or nitrocellulose membrane using either a semi-dry or wet transfer system. PVDF blots are incubated in a blocking buffer and then transferred to the primary antibody solution for an allotted time at a specific temperature. Blots are washed four times with a wash buffer, incubated with an appropriate conjugated secondary antibody for 1 h, washed four times with a wash buffer, and then developed using an appropriate detection system. Replicates of the sample blots should also be assayed with only the conjugated secondary antibody with no exposure to a primary antibody in order to determine level of artifact caused by the conjugated secondary antibody (non-specific cross-reactivity).

Blocking solution:

- 1× Tris buffered saline (pH 8.4); 500 ml; 15 m*M* NaCl, 50 m*M* Trizma base; titrate pH of buffer with 3 *N* HCl to 8.4
- 30 g of non-fat dry milk

Wash solution:

- 2× Tris buffered saline (pH 8.4).
  - Tween-20, 0.01% (v/v). If the Tween is old (shelf life of 45 days), it may undergo oxidation, causing an increase in non-specific cross-reactivity.

Primary and secondary antibody solution:

- 2× Tris buffered Saline (pH 8.4)
- Pre-determine titer for optimal specificity

Blots can be developed either colorimetrically, chemiluminescently, or fluorimetrically. There are a number of good references that describe standard SDS-PAGE/western blotting procedures in detail.[10,12]

The purified protein of the target parameter should be run on the lane next to the sample to ensure that the migration rates are the same. Using only molecular weight standards is a practical means of approximating that the band is the right migration rate, but there is more confidence in using the purified protein or a sample that you know has the protein, because you can account for the migration nuances of the protein running under specific buffer and gel conditions.

There are five types of results that can be observed on a blot when working with coral or other invertebrates that house an algal symbiont and plant samples. The five types of results are:

1. Band at the expected migration position with no other non-specific cross-reactive bands (Figure 10.7A).
2. Band is *not* at the expected migration position, but bands can be observed at different migration positions. There is no confidence that the antibodies, the sample preparation, or the running conditions are optimal for ELISA (Figure 10.7B).
3. Band at the expected migration position, but with bands at unexpected migration positions. For example, if you are assaying for invertebrate manganese superoxide dismutase, you should see a band around 25 Ku (standard international designation for migration of proteins for gel electrophoresis). You may also see

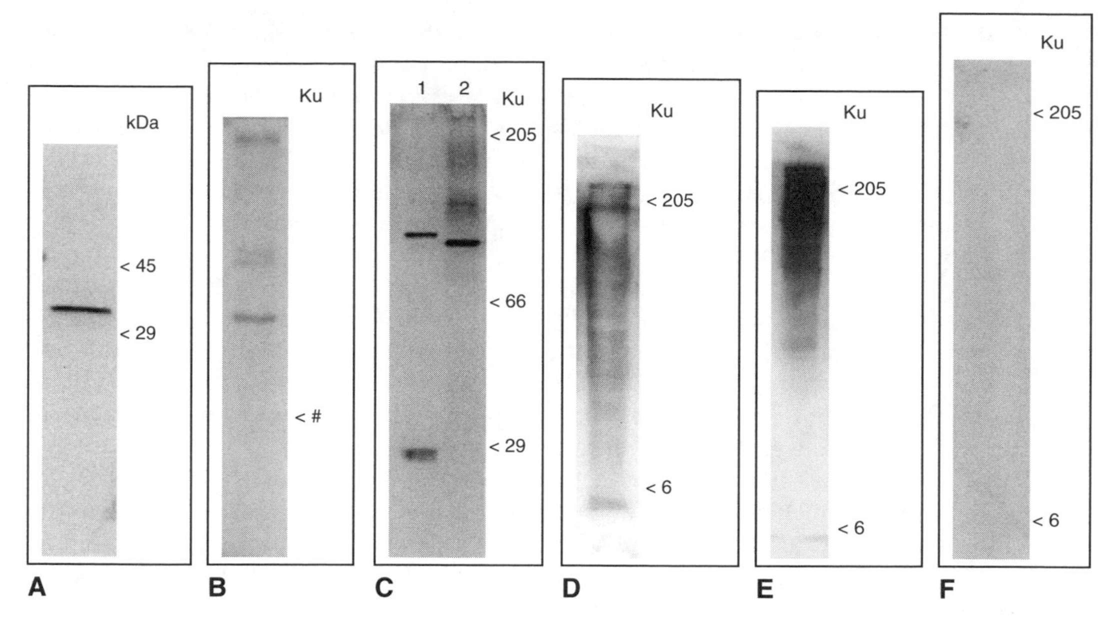

Figure 10.7 Quality control step 1: possible results for immuno-assaying with western blotting. Panel A: acceptable result of a sample assayed for heme oxygenase, which should have a band around 32 Ku. Panel B: unacceptable result of a sample assayed for glutathione-*s*-transferase, which should have a band at the position designated by #. Panel C: unacceptable result of a sample assayed for chloroplast small heat-shock protein (sHsp). In lane 1, the monomer of the chloroplast sHsp is detected at the 29 Ku migration distance, but there are also higher-molecular weight bands. The sample in lane 1 was made up in sample buffer with 2% SDS, 10 m*M* thiol reductant, and was not heat-denatured. In lane 2, the sample was made up in sample buffer with 0.5% SDS, no thiol reductant, and was not heat-denatured. Notice that the monomer is not present. Panel D: coral sample (*Montastrea annularis*) that was prepared in a sample buffer lacking PVPP and exhibits streaking as a result of extensive Maillard product contamination. Panel E: coral sample (*M. faveloata*) had extensive mucus contamination that was not removed during sampling. Panel F: coral sample that was assayed using only the HRP conjugated secondary antibody to ensure that the secondary antibody was not associated with any non-specific cross-reactivity.

bands around 40 Ku, and above the 190 Ku markers. There may be something wrong with the sample preparation, with the antibodies, or with the running conditions of the gel. For example, many enzymes exist in a multimeric structures and require harsh denaturing and reducing conditions to break the enzyme complex apart into its separate components. These slower migrating bands may be the non-denatured, multimeric structures, as in the case of homogenization buffer lacking SDS and run on a native gel. In any case, there is no confidence in the sample or conditions to conduct an ELISA (Figure 10.7C).

4. There is a dark smear going the entire length of the sample lane. This is most likely due to extensive Maillard product formation. Polyphenols from the algae are cross-linking proteins and carbohydrates. Increasing the concentration of PVPP or adding PVP (0.5% w/v maximum) may rectify this artifact. Additionally, Na tetraborate (5 m*M* maximum) may further help reduce the level of Maillard product formation. Nonetheless, there is no confidence that the samples can be subject to an ELISA (Figure 10.7D).

5. There is a dark smear running from the top of the lane to about the 66 Ku marker. This artifact is most likely the result of high polysaccharide contamination. When aspirating the middle phase during sample preparation, some of the top phase was also aspirated as part of the sample. Multiple centrifugations/aspirations, subjecting the sample to a polysulfone microcentrifugation filter, or increasing the level of DMSO or adding L-cysteine (1 m*M* maximum) may help to reduce this artifact. There is no confidence that the samples can be subject to an ELISA under their present state (Figure 10.7E).

When assaying a sample blot with only the secondary antibody, the blot should be clear. Any bands or smears appearing on the blot can mean that the secondary antibody used for both western assay and the ELISA is non-specifically cross-reacting with some epitope in the sample (Figure 10.7F). Fab fragments conjugated with the appropriate reporter enzyme usually give the best results under most situations.[13]

Quality control step 2 and ELISA optimization: The pooled sample from step 1 is serially diluted eightfold, the highest concentration is 6.667 ng μl^{-1}. A 16-fold serial dilution of the calibration standard is also created. The calibration standard is the purified protein or antigen being assayed. The concentration of the calibration standard should be known and designated either in moles or grams of the antigen. Both the sample serial dilution and the calibrant serial dilution should be plated on a 384-well microtiter plate or on a membrane dot or slot blotter. Because most labs can affordably access a dot or slot blotter, the protocol will focus exclusively on using a dot/slot blotter for carrying out the ELISA. If using a Bio-Rad or other brand of dot/slot blotter, activate the PVDF membrane in methanol and then equilibrate the membrane in 1× TBS for 1 min. If using a nitrocellulose membrane, equilibrate the membrane for 1 min in 1× TBS. Quickly place the membrane within the manifold of the dot/slot blotter and assemble the dot/slot blotter. Apply the serial dilutions of both the sample and the calibrant in triplicate on the dot/slot blotter. Contrary to most manufacturers' instructions, do not incubate the serial dilutions on the dot/slot blotter for 20 min or longer. Once the samples and calibrants are loaded onto the blotter, suctioning the samples through the membrane should be done immediately. A hand-held vacuum, such as the Nalgene hand pump, is one of the best instruments for creating a vacuum for a dot/slot blotter. You can control the vacuum pressure much more readily than with a motorized pump, as well as maintain better control in releasing the vacuum. A slow release of the vacuum pressure is best because it prevents backwash of the samples back across the membrane; a major source of artifact for dot/slot blotters. Immediately transfer the membrane to the blocking solution and carry out the development of the blot using the same protocols as done in "quality control step 1."

Once the blot is developed, scan or image-capture the blot. NIH Image is an easy to use and free optical density program that can be used on PC or Macintosh computers. Determine the optical density of each point and then plot the data. A trend line can be fitted using either a liner or polynomial (second order) regression, and the equation and R^2 for the trend lines are determined.[13] Lack of linearity (or curvilinearity), such as a low R^2 value, is an indication of artifact.[13] Sample loading, primary and secondary antibody titers, as well as incubation, and wash conditions can all be optimized during this step. Intra-specific variation for each triplicate concentration point should be less than 15%. Finally, the calibrant standard curve should be optimized with the sample loading, so that the signal of the sample falls somewhere in the middle of the calibrant standard curve. To increase the probability that all the samples will fall within the calibrant standard curve, create a serial dilution of a sample from the reference treatment and a serial dilution of a sample from the highest concentration from the xenobiotic-exposure treatment.

A "third quality control step" is expensive, training intensive and time consuming but bestows the highest confidence that the antibody is truly detecting the correct protein or antigen in the samples. This step entails the construction of an immuno-column using the primary antibody, purification of the immuno-column elutant, extensive fractionation protocols if there is more than one protein eluted from the columns, and Edman degradation or Mass spectrometry peptide sequencing. This step is perceived as optional only because of the expense, extensive training, and accessibility to expensive supplies and equipment that is required to conduct the quality control step. Protocols for this step can be found in a number of good resources.[10]

ELISA: All samples and calibrants should be plated onto the dot/slot blotter in triplicate. Optimally, a single calibrant curve should be plated to the left, middle, and the right sides of the dot/slot blotter (Figure 10.8). Sample replicates can be plated in a similar fashion, or plated close together. The reason for dispersing the calibrant curves throughout the microtiter plate is to account for the potential of edge effects or differential staining from one side of the blot to the other. Data should be calculated as mole or gram per nanogram of total soluble protein (normalization). For example, when measuring Hsp60, the data should be calculated as moles of Hsp60 per nanogram total soluble protein or grams of Hsp60 per nanogram total soluble protein.

Statistical analysis of data

The first step of the analysis is to determine if the level of a specific parameter in any of the treatments are significantly different from the control population. Data should be first tested for normality and equal variance. If the data are normally distributed, apply a one-way ANOVA; but if the data are not normally distributed, consider the assumptions of other types of analyses, such as a one-way ANOVA on Ranks (e.g., WELCH ANOVA). If treatment differences are found, apply the appropriate planned comparison test[14] (Figure 10.9).

Canonical correlation analysis (CCA) has been used in the past as a heuristic tool to illustrate how biomarkers can be used to discriminate among xenobiotics and

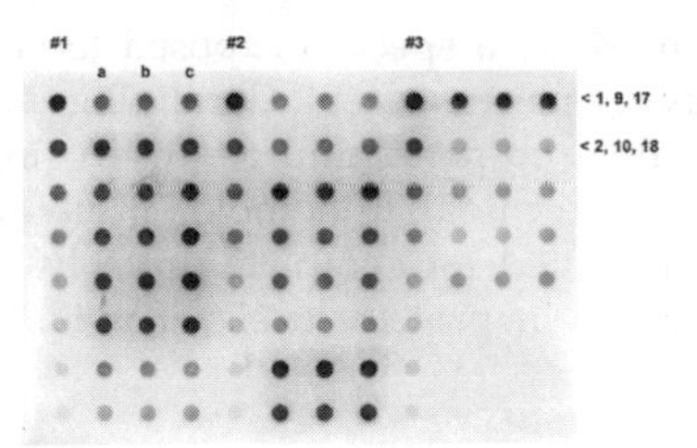

Figure 10.8 ELISA's result using a 96-well dot blotter. # = calibrant standard curve. The calibrant with the highest concentration in plated in row 1, the calibrant with the lowest concentration is plated in row 8. Optimally, replicates of the standard curve should be evenly distributed across the dot blotter. In row 1, triplicate assays of samples 1, 9, and 17. In row 2, triplicate assays of samples 2, 10, and 19. In column a, replicate one of samples 1, 2, 3, etc., are plated. In column b, replicate two of samples 1, 2, 3, etc., are plated. In column c, replicate three of samples 1, 2, 3, etc., are plated. Plate represents data for the chloroplast small heat-shock protein in samples of the hard coral, *M. annularis*.

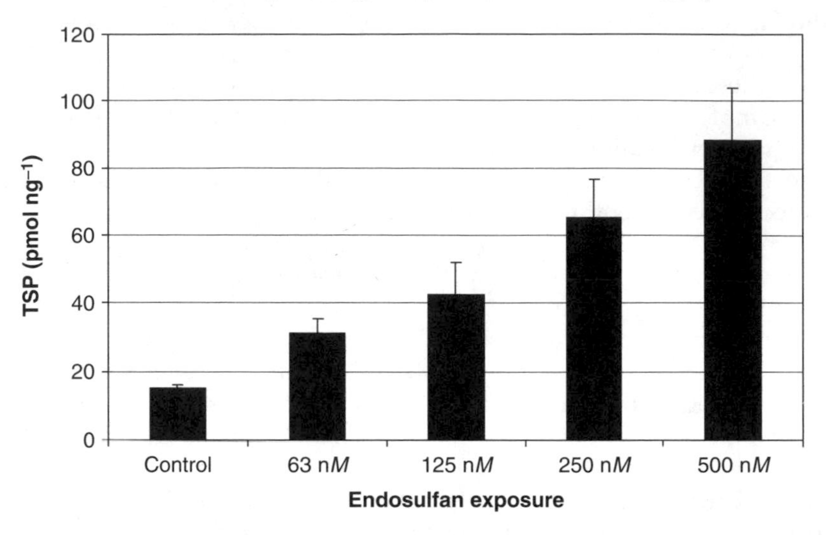

Normality test: Passed ($P > 0.050$)
Equal variance test: Passed ($P = 0.057$)

One-way ANOVA

Source of variation	**DF**	**SS**	**MS**	***F***	***P***
Between groups	4	11619.839	2904.960	34.222	<0.001
Residual	15	1273.281	84.885		
Total	91	2893.120			

α=0.05

Mutiple comparisons versus control group (Holm-Sidak method):
Ovcaerall signifince level = 0.05

Comparsion	Diff. of means	t	Unadjusted *P*	Critical level	Significant?
Col 1 vs. Col 5	67.814	10.409	0.000	0.013	Yes
Col 1 vs. Col 4	50.314	7.723	0.000	0.017	Yes
Col 1 vs. Col 3	27.564	4.231	0.001	0.025	Yes
Col 1 vs. Col 2	16.377	2.514	0.024	0.050	Yes

Figure 10.9 *P. lobata* (tropical hard coral species) exposed to endosulfan (a pesticide) in a dose response experimental design. $N = 4$ per treatment. Coral plugs were exposed to differing concentrations of endosulfan for 12 h, then prepared and assayed using the protocol described in this chapter. Top panel is the data for the cnidarian homolog of ubiquitin ligase (E2), an enzyme that conjugated ubiquitin to a protein slated for complete proteolysis. Bottom panel is the statistical method used to test for differences in protein levels of ubiquitin ligase between the different treatments and the reference control.

environmental stressors.[15,16] CCA is an eigen analysis method that reveals the basic relationships between two matrices, in this case those of xenobiotic treatments and biomarker data. The CCA provides an objective statistical tool for determining which biomarker (or suite of biomarkers) best reveals the presence or response of a particular xenobiotic and environmental stressor. In the case of discriminating the effects of two or more xenobiotics/environmental exposures, two assumptions of CCA are made that result from the experimental design used: (1) stressor gradients are independent and linear and (2) biomarker responses are linear (Figure 10.10).

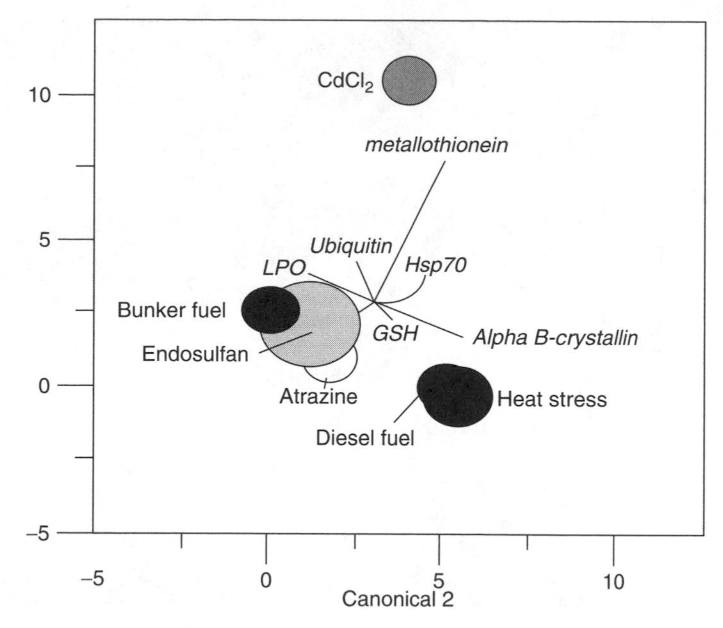

Figure 10.10 Canonical centroid plot of standardized biomarker responses in an organism exposed to a pesticide, an herbicide, a heavy metal, heat stress, and two different types of fuel. Original variates were biomarker levels expressed as a percentage of the control value in each experiment. Circles show the 95% confidence intervals around the distribution centroid of each stressor. Biplot rays radiating from the grand mean show directions of original biomarker responses in canonical space. Metallothionein, which responds to heavy metal stress, distinguished the cadmium chloride treatments from all others; αB-crystallin distinguished diesel fuel and heat stress versus the others.

CCA can discriminate between the effects of the different concentration doses using the same protocol as above, but CCA can also differentiate the patterns of system resiliency and system degeneration. Instead of randomly choosing or compiling all the biomarker data into a single matrix, biomarkers from only a specific cellular diagnostic category are included into the matrix. For example, only parameters of protein metabolic condition are included into one matrix, while the other matrix is the reference and xenobiotic treatments compose the second matrix (Figure 10.11).

Results and discussion

Case study: hexachlorobenzene and corals

The objective of the investigation is to begin to answer the question: "What effect does an exposure to a hexachlorobenzene (HCB) have on coral cellular physiology?" This inquiry concerning the mechanisms of toxicity can be further distilled into three questions:

- Does exposure to a HCB produce an oxidative stress?
- Does exposure to HCB affect heme metabolism?
- Is light a factor for HCB toxicity and pathology?

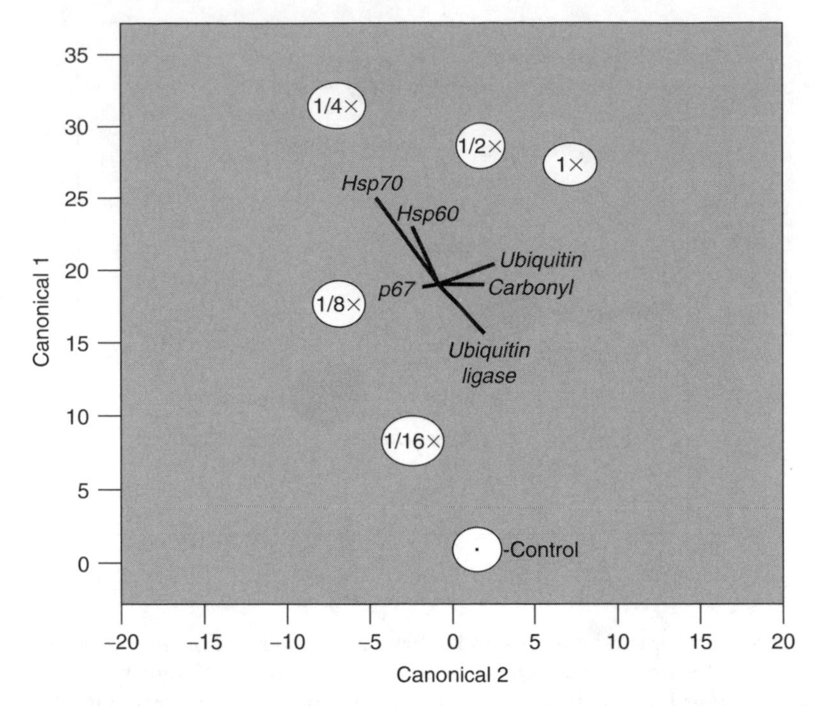

Figure 10.11 Canonical centroid plot using six parameters of protein metabolic condition in liver cells exposed to the different concentrations of the hepatotoxicant, thioacetimide. Exposure to thioacetimide was 24 h. Concentration of thioacetimide is: 1 m*M* at centroid plot 1×, 500 μ*M* at centroid plot 1/2×, 250 μ*M* at centroid plot 1/4×, 125 μ*M* at centroid plot 1/8×, and 63 μ*M* at centroid plot 1/16×. Biplot rays radiating from the grand mean show directions of original biomarker responses in canonical space. Notice that protein oxidative damage product (protein carbonyl) and a marker for protein degradation (ubiquitin) define differences between the 1× centroid and the control, indicating that at 1 m*M* thioacetimide induces protein oxidation (oxidative stress) and increase in the rate of protein degradation.

These three questions have particular relevance because an examination of the mammalian, vertebrate, and invertebrate literature indicates that oxidative stress and disruption of heme synthesis are critical aspects of HCB toxicity.[17] In this case study, we will only consider a 24-h exposure of 750 ng l^{-1} of HCB.

Coral nubbins the size of 1 cm^2 of *P. porites* were cultured in the laboratory under 800 μmol m^{-1} s^{-1} photosynthetic active radiation (PAR) for two months (12-h night-and-day cycles) and supplemented with *Artemia* on a tri-weekly basis. Coral nubbins came from a single coral colony from 6-m depth.

Corals nubbins were transplanted to a recirculating dosing system and allowed to acclimate for 1 week. Light levels and feeding schedule were unchanged, and there was an 85% volume complete water change-out of the recirculating dosing system every 12 h. Water temperature range was 25–27°C. A single coral nubbin was placed in a single recirculating dosing system; $N = 6$ per treatment, there was a total of 18 independent recirculating dosing units and 18 coral nubbins. There were three treatments: (1) control with light at 800 μmol m^{-1} s^{-1} PAR, (2) 750 ng l^{-1} of HCB at 800 μmol m^{-1} s^{-1} PAR, and (3) 750 ng l^{-1} of HCB at 50 μmol m^{-1} s^{-1} PAR. Exposure began 30 min before the beginning of the light period and ended 30 min before the end of the dark period.

Corals were collected and prepared as described earlier, with slight modifications to accommodate an assay that measures protein carbonyl.[18,19] The criteria for using the biomarkers, measured in this study, were partly constrained by technology and available resources. The cellular parameters included in this study are evolutionarily conserved, thus making it easy to produce and validate ELISAs for this species. Assays for biomarkers that are not genetically constrained, such as GSH and protein carbonyl, had to be modified to account for qualities inherent to coral sample matrix. As in any situation, resources are limited and assaying for 30–40 different parameters can be cost and time prohibitive. We measured seven parameters that are particularly relevant to HBC toxicity in both vertebrate and invertebrate species.[17] These seven parameters can be categorized into the cellular sub-systems of "oxidative damage and response" and porphyrin metabolism (Table 10.2).

Protein carbonyl formation increased in response to HCB exposure, and the rate of lesion formation was exacerbated by high-light. MnSOD was unchanged by exposure to HCB in both light conditions, suggesting that oxidative damage was not occurring in the mitochondria, or the ability to respond to damage was somehow hindered. In contrast, cytosolic Cu/ZnSOD showed an expression pattern similar to protein carbonyl, suggesting that the oxidative damage/response was, in part, localized to the cytosol, but that light was again an exacerbating factor. Catalase levels significantly decreased with exposure to HCB under both light regimes. This decrease in catalase can be interpreted as: (1) the ability of the cell's oxidative stress capacity is greatly reduced, especially its ability to deal with hydrogen peroxide; and (2) a mechanism is present that is specifically altering catalase steady-state levels. From these markers, we can conclude that the HCB exposure induces a significant oxidative stress, and that it is exacerbated by light. The pattern of the anti-oxidant enzymes, by themselves, does not confer a clear understanding of why this pattern exists. From a prognostic perspective, an increase in protein lesions by HCB suggests membrane integrity and even genomic integrity may be jeopardized by oxidative damage, suggesting further investigation into the mechanism of HCB-associated oxidative stress toxicity.

Scholarship of the literature surrounding HCB biochemistry and toxicology can help elucidate the pattern observed in the oxidative damage and response category. A classic

Table 10.2 Example data for diagnostic interpretation[a]

Cellular parameter	Control	High light HCB	Low light HCB
Oxidative damage and response			
Protein carbonyl (pmol mg^{-1} TSP)	147 ± 8	$1631 \pm 42^{A,B}$	$465 \pm 27^{A,B}$
Cu/Zn superoxide dismutase (fmol ng^{-1} TSP)	67 ± 8	$368 \pm 16^{A,B}$	$128 \pm 22^{A,B}$
Mn superoxide dismutase (fmol ng^{-1} TSP)	598 ± 120	628 ± 92	580 ± 39
Catalase (fmol ng^{-1} TSP)	142 ± 7	15 ± 3^{A}	22 ± 6^{A}
Porphyrin metabolism (subset of metabolic condition)			
Ferrochelatase (fmol ng^{-1} TSP)	66 ± 6	7 ± 2^{A}	12 ± 4^{A}
PPO IX oxidase (fmol ng^{-1} TSP)	32 ± 3	58 ± 3^{A}	62 ± 4^{A}
Protoporphyrin species (nmol ng^{-1} TSP)	28 ± 7	1855 ± 351^{A}	1695 ± 178^{A}

[a] Data from an experiment where corals were exposed to hexachlorobenzene for 24 h. $N = 5$ per treatment. All parameters except for protoporphyrin and protein carbonyl are for cnidarian homologs. Under these assay conditions, it is impossible to distinguish between dinoflagellate and cnidarian protein carbonyl and protoporphyrin. All parameters were analyzed by one-way ANOVA. Data are given as means and standard errors. Superscript "A" indicates that the treatments are significantly different from the control. Superscript "B" indicates that the treatments are not only significantly different from the control, but are significantly different from one another.

paper by Cam and Nigigisyan[20] showed that mammals exposed to HCB resulted in a pathology known as *porphyria cutanea tarda symptomatica* (PCTS). The more obvious manifestation of HCB exposure is a dark red hue discoloration of urine (described as a port wine color) and severe skin lesions when the individual is exposed to sunlight.[17,21] Experimentation of HCB in rats demonstrated that HCB produced porphyria as a result of interference in the pathway of heme metabolism.[22] These studies were done, in part, to help elucidate the toxicological mechanism associated with an acquired outbreak of porphyria in more than 3000 individuals in Turkey between 1955 and 1959, which was finally traced to the consumption of wheat contaminated with HCB.[21,23] Since the 1970s, studies investigating the toxicological mechanisms of HCB have been able to demonstrate that HCB inhibits a number of enzymes in the porphyrin synthesis pathway, producing an accumulation of protoporphyrin and porphyrinogen species.[21,24] These porphyria species can readily absorb light, react with divalent oxygen, and produce singlet oxygen, a noxious reactive oxygen species (ROS).[25]

This literature can aid in explaining *why* we see an increase in protein oxidation lesions, especially the increase of these lesions associated with light.[26,27] In fact, the phenomenon of light-induced ROS from a porphyrin-like compound has been capitalized on as a new and effective treatment of certain cancers (e.g., skin cancer, Barlett's esophageal cancer, colon cancer): hypericin-based photo-dynamic therapy.[28] The decrease in catalase can be explained by two separate but synergistic mechanisms. Catalase is a porphyrin-dependent enzyme whose accumulation and activity is known to decrease as a result of porphyrin synthesis interference.[29,30] In addition, catalase is extremely susceptible to oxidation and inhibition by both singlet oxygen and superoxide.[31,32] Both mechanisms are relevant to this situation and most likely acting to affect catalase levels.

To confirm that a porphyrin-based toxicological mechanism is occurring as a consequence of HCB exposure, three parameters central to porphyrin metabolism (a subset of the category, metabolic condition) were assayed: ferrochelatase, protoporphyrinogen oxidase (PPO) IX, and total porphyrin content. PPO IX is the second to the last enzyme in heme synthesis and converts PPO IX into protoporphyrin.[33] Ferrochelatase is the last enzyme of heme synthesis and insets the iron into the protoporphyrin to form heme.[33] Ferrochelatase levels decreased significantly in both HCB treatments, most likely resulting in decreased heme synthesis and increased protoporphyrin accumulation. Ferrochelatase is an iron-sulfur protein, and its activity and structure are sensitive to changes in redox state of its sub-cellular environment.[34] Ferrochelatase is the "failure point" of heme synthesis in response to a number of toxins and is not surprising that it would be adversely affected by HCB.[34] PPO levels significantly increased, probably as a result for cellular requirement for metalloporphyrins (e.g., heme). Increase of PPO with a decrease in ferrochelatase suggests that protoporphyrin species may be hyper-accumulating as a result of this defect in the heme synthesis pathway. A microplate fluorimetric method was used to detect total concentration of uroporphyrin, coproporphyrin, and protoporphyrin in the coral sample.[34] If enzymes in the porphyrin synthesis pathway are inhibited, porphyrin synthesis intermediates will accumulate.[35] Measurement of total porphyrin species (actually porphyrinogen and non-metal porphyrins) revealed that HCB in both the light treatments resulted in over a 20-fold increase compared to controls, supporting the argument that (1) HCB interferes with the porphyrin/heme synthesis pathway in corals, and (2) the presence of non-metal porphyrins and porphyrinogens is likely a primary source of oxidative stress.

From these seven biomarkers, we were able to address the three questions posed at the beginning of this case study, and more importantly, we established a foundation for the mechanism of HCB toxicity in corals; HCB exposure causes an oxidative stress that is exacerbated by light as a result of heme/porphyrin synthesis interference. In addition, the cellular diagnostic approach can immediately provided a basis for further investigation into the nature of HCB toxicity in corals. HCB causes an oxidative stress, and we know that it adversely affects protein function and steady-state levels. Does it also cause DNA lesions, resulting in increased mutation rates? Does this oxidative stress translate into a reduction in reproductive fitness? Is there oxidation of lipid membranes with HCB exposure? Is this lipid peroxidation associated with accumulation of malondialdehyde and hydroxynonenal-known mutagens and enzyme inhibitors? Examining biomarkers in the categories of genetic integrity and membrane integrity can provide answers to these questions.

A principal goal of any kind of systematic biomarker approach is to link behavior occurring at the molecular and cellular level to the population level and, perhaps, even to the community and ecosystem levels.[36] Predicting population responses and *ecological forecasting* is an esteemed aspiration of biomarker research.[37] Approaching this problem with a thorough understanding of biochemistry and cellular physiology is an essential prerequisite for accurate predictions. For example, if we know that HCB exposure causes or at least increases the risk for oxidative damage in corals, and that it also diminishes aspects of the cell's anti-oxidant enzyme defense, we could prognosticate that a HCB exposure would increase the risk of a coral bleaching event in corals experiencing conditions conducive to coral bleaching. This prognosis is further based on the knowledge that oxidative stress is a major mechanism for many forms of environmentally associated coral bleaching.[19] Hippocrates aphorism of "he will manage the cure best who has foreseen what is to happen from the present state of matters"[38] best encapsulates the central tenet of cellular diagnostics and prognostics: to determine the present and future physiological condition of the cell through an understanding of the cell's functional components and processes.

Acknowledgments

I sincerely thank Cheryl Woodley, John Fauth, Charles Robinson, and Michael Moore for discussion concerning biomarkers and its application to ecotoxicological issues; Gerald Marks and Stig Thunell for introducing me to the world of porphyria; and Bruce N. Ames for continued support and patience. I also wish to thank Tom Capo for the photographs in Figure 10.3, V. Dean Downs for providing the design schematic in Figure 10.4, Aaron Downs for providing the graphics in Figure 10.5, John Fauth for the canonical plots in Figures 10.10 and 10.11, and Richard Owen for inspiring the case study of HCB. Finally, I would like to thank U.S. NOAA's Center for Coastal Environmental Health and Biomolecular Research for working with me in developing and validating early aspects of molecular biomarker systems and cellular diagnostics.

References

1. Depledge, M.H., Amaral-Mendes, J.J., Daniel, B., Halbrook, R.S., Kloepper-Sams, P., Moore, M.N. and Peakall, D.P., The conceptual basis of the biomarker approach, in: *Biomarkers —*

Research and Application in the Assessment of Environmental Health, Peakall, D.G. and Shugart, L.R., Eds., Springer, Berlin, 1993, pp. 15–29.

2. Moore, M.N., Biocomplexity: the post-genome challenger in ecotoxicology, *Aquat. Toxicol.*, 59, 1–15, 2002.
3. Decaprio, A.P., Biomarkers: coming of age for environmental health and risk assessment, *Environ. Sci. Tech.*, 31, 1837–1848, 1997.
4. Downs, C.A., Shigenaka, G., Fauth, J.E., Robinson, C.E. and Huang, A., Cellular physiological assessment of bivalves after chronic exposure to spilled Exxon Valdez crude oil using a novel molecular diagnostic biotechnology, *Environ. Sci. Tech.*, 36, 2987–2993, 2002.
5. Clendening, L. and Hashinger, E.H., *Methods of Diagnosis*, Mosby, St. Louis, MO, 1947.
6. Aristotole, ~350 B.C.E. *Posterior Analytics*, Book 1. Translated G.R.G. Mure.
7. Bukua, B. and Horwich, A.L., The Hsp70 and Hsp60 chaperone machines, *Cell* 92, 351–366, 1998.
8. Allen, T.F.H. and Starr, T.B., *Hierarchy: Perspectives for Ecological Complexity*, The University Chicago Press, Chicago, 1982, pp. 209–216.
9. Ledley, R.S. and Lusted, L.B., Reasoning foundations of medical diagnosis, *Science* 130, 9–21, 1959.
10. Walker, J.M., Ed., *The Protein Protocols Handbook*, Humana Press, Totowa, NJ, 1996, pp. 3–50.
11. Ghosh, S., Gepstein S., Heikkila, J.J. and Dumbroff, B.G., Use of a scanning densitometer or an ELISA plate reader for measurement of nanogram amounts of protein in crude extracts from biological tissue, *Anal. Biochem.*, 169, 227–233, 1988.
12. Malik, V.S. and Lillehoj, E.P., Eds., *Antibody Techniques*, Academic Press, San Diego, CA, 1994.
13. Crowther, J.R., *The ELISA Guidebook*, Humana Press, Totowa, NJ, 2001.
14. Sokal, R.R. and Rohlf, F.J., *Biometry*, W.H. Freeman, New York, 1995.
15. Gad, S. and Weil, C.S., *Statistics and Experimental Design for Toxicologists*, Telford Press, Caldwell, NJ, 1986, pp. 1–29.
16. Downs, C.A., Dillon, R.T., Jr., Fauth, J.E. and Woodley, C.M., A molecular biomarkers system for assessing the health of gastropods (*Ilyanassa obsoleta*) exposed to natural and anthropogenic stressors, *J. Exp. Mar. Biol. Ecol.*, 259, 189–214, 2001.
17. Marks, G.S., Zelt, D.T. and Cole, S.P., Alterations in the heme biosynthesis pathway as an index of exposure to toxins, *Can. J. Physiol. Pharmacol.*, 60, 1017–1026, 1982.
18. Downs, C.A., Mueller, E., Phillips, S., Fauth, J.E. and Woodley, C.M., A molecular biomarker system for assessing the health of coral during heat stress, *Mar. Biotechnol.*, 2, 533–544, 2000.
19. Downs, C.A., Fauth, J.E., Halas, J.C., Dustan, P., Bemiss, J. and Woodley, C.M., Oxidative stress and seasonal coral bleaching, *Free. Radical Biol. Med.*, 33, 533–543, 2003.
20. Cam, C. and Nigigisyan, G., Acquired toxic porphyria cutanea tarda due to hexachlorobenzene, *J. Am. Med. Assoc.*, 183, 88–91, 1963.
21. Elder, G.H., Porphyria caused by hexachlorobenzene and other polyhalogenated aromatic hydrocarbons, in: *Handbook of Experimental Pharmacology*, vol. 44, De Matteis, F. and Aldridge, W.N., Eds., Springer, Berlin, 1978, pp. 157–200.
22. San Martin De Viale, L.C., Rios De Molina, M.D.C., Wainstock De Calmanovici, R. and Tomio, J.M., Experimental porphyria produced in rats by hexachlorobenzene. IV. Studies on step-wise decarboxylation of uroporphyrinogen and phyriaporphyrinogen *in vivo* and *in vitro* in several tissues, in *Porphyrins in Human Disease*, Doss, M., Ed., Karger, Basel, 1976, pp. 445–452.
23. Schmid, R., Cutaneous porphyria in Turkey, *N. Engl. J. Med.*, 263, 1960.
24. Marks, G.S., Exposure to toxic agents: the heme biosysnthetic pathway and hemoproteins as indicator, *Crit. Rev. Toxicol.*, 15, 151–179, 1985.
25. Salet, C., Moreno, G., Richelli, F. and Bernardi, P., Singlet oxygen produced by photodynamic action causes inactivation of the mitochondrial permeability transition pore, *J. Biol. Chem.*, 272, 21938–21943, 1997.
26. Burnnett, J.W. and Pathak, M.A., Effect of light upon porphyrin metabolism of rats. Study of porphyrin metabolism of normal rats and rats with hexachlorobenzene-induced porphyria, *Arch. Dermatol.*, 251, 314–322, 1964.

27. Thunell, S. and Harper, P., Porphyrins, porphyrin metabolism, porphyries. III. Diagnosis, care and monitoring in porphyria cutanea tarda — suggestions for a handling programme, *Scand. J. Clin. Lab. Invest.*, 60, 561–580, 2000.
28. Batlle, A.M., Porphyrins, porphyrias, cancer, and photodynamic therapy, *J. Photochem. Photobiol. B.*, 20, 5–22, 1993.
29. Haeger-Aronsen, B., Experimental disturbance of porphyrin metabolism and of liver catalase activity in guinea pigs and rabbits, *Acta Pharmacol. Toxicol.*, 21, 105–115, 1964.
30. Teschke, R., Boelsen, Landmann, H. and Goerz, G., Effect of hexachlorobenzene on the activities of hepatic alcohol metabolizing enzymes, *Biochem. Pharmacol.*, 32, 1745–1751, 1983.
31. Lledias, F., Rangel, P. and Hansberg, W., Oxidation of catalase by singlet oxygen, *J. Biol. Chem.*, 273, 10630–10637, 1998.
32. Shimizu, N., Kobayaski, K. and Hayashi, K., The reaction of superoxide radical with catalase: mechanism of the inhibition of catalase by superoxide, *J. Biol. Chem.*, 259, 4414–4418, 1984.
33. Thunell, S., Porphyrins, porphyrin metabolism and porphyries. I. Update, *Scand. J. Clin. Lab. Invest.*, 60, 509–540, 2000.
34. Dailey, H.A., Dailey, T.A., Wu, C.K., Medlock, A.E., Wang, F.K., Rose, J.P. and Wang, B.C., Ferrochelatase at the millennium: structures, mechanisms, and [2Fe-2S] clusters, *Cell. Mol. Life Sci.*, 57, 1909–1926, 2000.
35. Grandchamp, B., Deybach, J.C., Grelier, M., De Verneuil, H. and Nordmand, Y., Studies of porphyrin synthesis in fibroblasts of patients with congenital erythropoietic porphyria and one patient with homozygous coporphyruria, *Biochim. Biophys. Acta*, 620, 577–586, 1980.
36. Downs, C.A. Fauth, J.E. and Woodley, C.M., Assessing the health of grass shrimp (*Palaeomonetes pugio*) exposed to natural and anthropogenic stressors: a molecular biomarker system, *Mar. Biotechnol.*, 3, 380–397, 2001.
37. Fauth, J.E., Downs, C.A., Halas, J.C., Dustan, P. and Woodley, C.M., Mid-range prediction of coral bleaching: a molecular diagnostic system approach, in *Ecological Forecasting: New Tools for Coastal and Ecosystem Management*, Valette-Silver, N.J. and Scavia, D., Eds., National Oceanic and Atmospheric Administration Technical Memorandum NOC NCCOS 1, 2004, pp. 5–12 (116 pp.).
38. Hippocrates of Chios, ~450 B.C.E. Book of Prognostics.

chapter eleven

A non-destructive technique to measure cytochrome P4501A enzyme activity in living embryos of the estuarine fish Fundulus heteroclitus

Diane Nacci and Laura Coiro
U.S. Environmental Protection Agency

Deena M. Wassenberg and Richard T. Di Giulio
Duke University

Contents

1-56670-664-5/05/$0.00+$1.50
© 2005 by CRC Press

Introduction

An embryonic and larval bioassay using the estuarine fish, *Fundulus heteroclitus*, was developed to detect and measure the relative potency of bioavailable compounds that act through the aryl hydrocarbon receptor (AhR). AhR agonists, including poly-halogenated dioxins/furans, and some poly-aromatic hydrocarbons (PAHs), and polychlorinated biphenyls (PCBs), are widely distributed globally and potentially toxic to the wide array of vertebrate species possessing this signal transduction pathway.[1,2] Exposure to this class of contaminants activates this toxicological pathway, resulting in induction of cytochrome P4501A (CYP1A). CYP1A induction is assessed typically as increased mRNA or protein concentration or enzyme activity, often measured as ethoxyresorufin-*O*-deethylase (EROD) activity.[3–5] In addition, CYP1A induction may be causally related to the early life stage effects produced by AhR agonists in fishes,[6,7] although probably not involved in all toxic effects produced by these compounds. Therefore, EROD activity is useful as an indicator of exposure and some adverse effects produced by this ubiquitous and important class of toxic contaminants.

In the embryonic and larval fish bioassay described here, EROD activity is measured using a novel *in ovo* non-invasive fluorescent technique applied to individual embryos. Because this endpoint is measured non-destructively, embryos for which EROD activity has been measured can also be assessed for traditional toxicity endpoints, including survival, development, and growth. Thus, this bioassay provides a simple, safe, and small volume whole animal method, which is highly responsive to AhR agonists. This *in ovo* method is applicable for use with single compounds and mixtures, as well as environmental media, such as water, sediment, sediment pore waters, and organic extracts. Therefore, the assay can be used to detect bioavailable contamination by compounds that act through the AhR, rank toxic potency of environmental media, and further understanding of the mechanisms of toxicity for compounds and environmental mixtures. In addition, lethal and sub-lethal effects of contaminants on fish early life stages can be included in species-specific population models used to project ecological effects of contamination.[8] A summary of procedures is shown in Figure 11.1, and a detailed description of materials and methods based on published methods[9,10] follows.

Materials required

- Water purification system — biologically pure deionized water (DI), e.g., Millipore Super-Q®, or equivalent
- Air pump for oil free air supply
- Air lines, plastic or Pasteur pipettes, or air stones
- pH and dissolved oxygen (non-stirring probe) meters, for routine physical and chemical measurements
- Light box for counting and observing embryos and juvenile culture animals
- Refractometer or salinometer for determining salinity
- Thermometers, glass or electronic, laboratory grade for measuring water temperatures
- Thermometers, bulb-thermograph or electronic-chart type for continuously recording temperatures
- Rinse bottles for DI and seawater
- Temperature-controlled room or chamber (23°C) for maintenance of *Artemia* and fish cultures

Embryo exposure (1–6 days post-fertilization)

- Collect fertilized eggs from parent tanks
- Transfer individual eggs into exposure vials
- Add contaminant and EROD substrate to exposure vials
- Maintain in sealed vials for 5 days
- Observe microscopically for developmental progress and abnormalities

Embryo development (7 days post-fertilization -- hatching)

- Transfer eggs into uncontaminated sea water
- Observe microscopically for developmental progress and abnormalities
- Renew sea water on alternate days
- Quantify EROD activity using fluorescence microscopy/photometry (days 7–10 post-fertilization)
- Monitor daily for hatch (~14 days post-fertilization), feed hatchlings *Artemia*

Measurement endpoints on individual fish

- Embryonic survival
- Embryonic development: rate and normality
- Embryonic EROD fluorescence
- Hatch age
- Larval survival
- Larval length (7 days post-fertilization)
- Histological lesions (fixed samples)

Figure 11.1 Protocol for measuring cytochrome P4501A enzyme EROD activity in a standardized embryonic and larval fish bioassay with the estuarine fish *F. heteroclitus*.

- Adult *F. heteroclitus* for use as brood stock
- Aluminum minnow traps (20 cm diameter) for collecting adult fish
- Squid-bait for minnow traps
- Wooden stakes-hold minnow traps from drifting with tide
- Large (100 l) coolers for transport of field-collected adult fish to the laboratory
- Newly hatched Great Salt Lake *Artemia* nauplii for feeding larval and juvenile fish
- Pellet fish food (e.g., Biodry 1000®, Bio Vita FF Pellet Food, Bio-Oregon, Warrenton, OR) for feeding adult fish
- Flake fish food (e.g., TETRA Standard Mix SM-80®, Tetra Werke Baensch, Melle, Germany) for feeding adult fish
- Frozen adult *Artemia* for feeding adult fish
- Egg collection chambers constructed as two separable polyvinyl chloride pipe rings: one covered with large nylon mesh screening upon which adults spawn, and a second inner ring with fine screening that retains eggs and protects them from consumption (Figure 11.2)
- Crystallization dishes (20 cm diameter) for holding embryos
- Transparent plastic wrap for covering vials and dishes

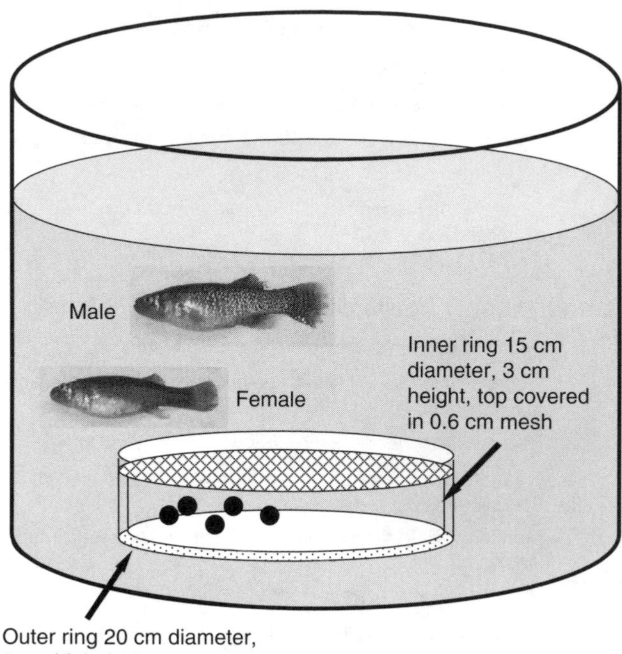

Figure 11.2 Collection of fertilized eggs from reproductively active *F. heteroclitus* into chambers made of readily available materials: large mesh nylon screening as top layer (spawning substrate), large-diameter polyvinyl chloride pipe for framework, and fine mesh nylon screening as bottom layer (egg retention).

- Natural or artificial sea water, 20–30 ppt dependent on local salinity of parental fish collection
- Glass scintillation vials (20 ml, e.g., ASTM Specifications E438 Type 1 Class A borosilicate) with plastic caps for embryo exposure and embryonic and larval maintenance
- Solvents, including acetone and dimethyl sulfoxide, ACS reagent grade or better
- Enzyme substrate: ethoxyresorufin (ER; Molecular Probes, Eugene, OR, USA) dissolved in dimethyl sulfoxide, and stored frozen prior to use
- Forceps and Pasteur pipets for transferring embryos
- Dissecting microscope for observation of embryos
- Depression microscope slides with coverslips
- Fluorescence microscope, equipped with G 365/LP 420 (UV excitation) and LP 510/LP 590 (or rhodamine) excitation/emission filters
- Microscope photometer to quantify fluorescence, or microscope image capture and analysis software
- Buffered formalin (10%) to preserve embryos and larvae after testing

Procedures

Spawning stock

Adult *F. heteroclitus* used as brood stock are collected using baited traps from Atlantic coast estuaries. Adult fish can also be acquired from commercial sources, or from young fish raised to maturity in the laboratory. Adult fish can be maintained in natural seawater, flow-through or artificial, re-circulated aerated systems. Typically, the fish are maintained at salinities that reflect collection conditions, i.e., from 20 to 30 ppt, and a photoperiod of 14-h light:10-h dark, and 23 ± 3°C. To maintain spawning condition, adults are fed commercial fish food *ad libitum*.

Egg collection

Spawning chambers are placed into tanks containing mature adult fish on nights preceding the full and new moons (Figure 11.2). The following mornings, eggs deposited into the lower portion of the chambers are rinsed with uncontaminated seawater, and gently washed into collection dishes. Up to about 200 eggs are selected randomly for each test. Each egg is transferred by forceps into a 20-ml glass scintillation vials, containing 7.5 ml filtered seawater at room temperature. These vials are covered loosely with transparent wrap and incubated at 23°C, under standard laboratory fluorescent lights, cycling 12 h "on/off."

Embryo-larval exposure and development

Embryo-larval exposure and assessment[9,11] is described briefly here. On the day following collection, each egg is reviewed under a dissecting microscope for development. Undeveloped or abnormal eggs are discarded (with percentage of viable eggs noted). Chemical exposure to embryos (aged 2 days post-fertilization) begins by introducing solvent (acetone) or toxicant in 0.1–1 μl volume contained in 2.5 ml seawater (i.e., final solvent concentration $\leq$ 0.01%). Seawater also includes ER dissolved in dimethyl sulfoxide to produce a final ER concentration of 21 μg/l, and final solvent concentration of 0.001%. Vials are capped, incubated at 23°C, and gently rolled or rotated daily. At 7 days post-fertilization, after 5 days static exposure, embryos are transferred (by forceps or Pasteur pipette) into clean vials containing uncontaminated seawater. These vials are left uncapped and renewed with fresh seawater every other day. Developmental progress is reviewed by observation using a dissecting microscope, and abnormalities are evaluated as described[12,13] (Figure 11.3).

After initiation of spontaneous hatching (beginning about 11 days post-fertilization using these laboratory conditions), eggs are checked every day. Beginning on the day of hatching, larvae are fed 24-h-old *Artemia* nauplii on a daily basis. Seawater is renewed every other day throughout larval development. On 7 days post-hatching, larvae are fixed by transfer into buffered formalin (10%) and preserved for future histological examination.

In ovo *EROD assay*

Embryos are observed for EROD activity using fluorescence microscopy (Figure 11.4). Although EROD can be measured in embryos as young as 3 days post-fertilization, embryos aged 7–10 days are observed typically.[9] Embryos of this age are convenient; bladders are large enough to find easily, and manipulations do not prompt premature hatching. To measure EROD, individual embryos are placed onto glass depression slides containing a small volume of seawater and covered with a glass cover slip. Embryos are

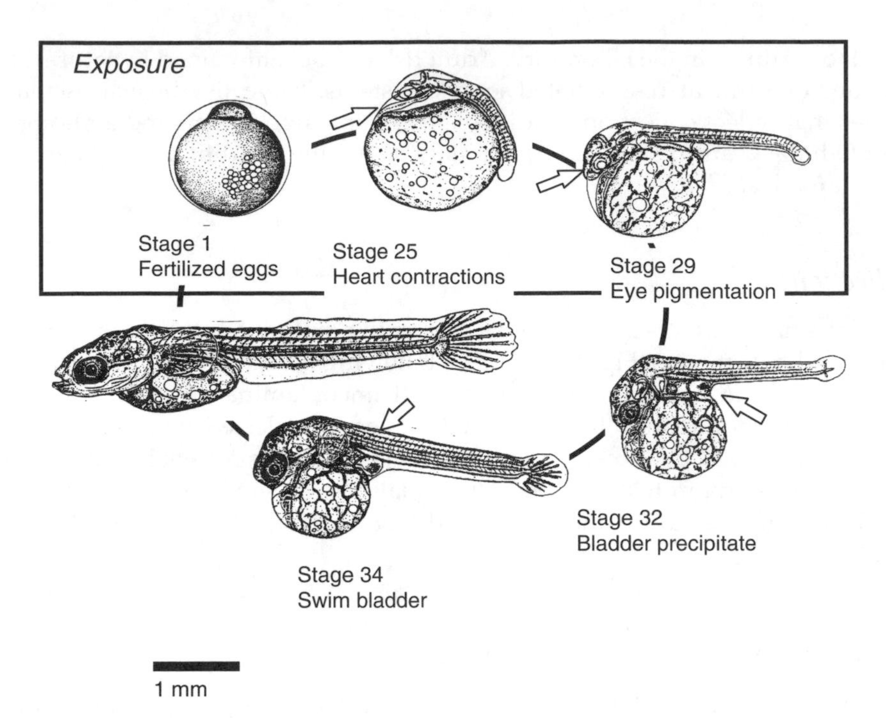

Figure 11.3 Embryonic development rate and features of estuarine fish *F. heteroclitus*. (Adapted from Armstrong, P.B. and Child, J.S., *Biol. Bull.*, 128 (2), 143–168, 1965.)

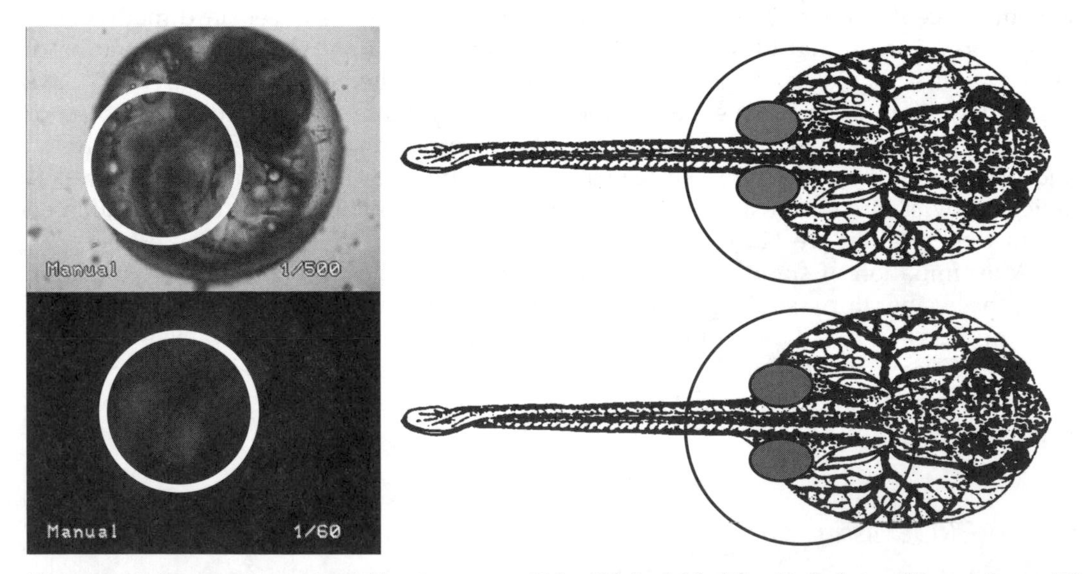

Figure 11.4 **(see color insert following page 464)** Bilobed bladder in *F. heteroclitus* embryo (10 days post-fertilization) fluoresces blue (420 nm) under UV excitation (365 nm); bladders in embryos exposed to aryl hydrocarbon agonists and the EROD substrate (ethoxyresorufin) accumulate resorufin and fluoresce red (590 nm) under green (510 nm) excitation.

observed using a fluorescent microscope, equipped with G 365/LP 420 (UV excitation) and LP 510/LP 590 (or rhodamine) excitation/emission filters. (Appropriate procedures for fluorescence microscope use should be followed. For example, fluorescence lamp should be warmed up and adjusted to produce stable, maximal output prior to test measurements.) Gentle pressure on the cover slip is used to rotate the embryo and localize the large, bilobed urinary bladder using low power (2.5×) magnification. Under UV excitation, the bladder is blue in color. This feature (and 10× magnification) allows easy localization, and orientation prior to fluorescence measurements (Figure 11.4).

EROD quantification

Once the bladder is located, the rhodamine filter set is used to observe the fluorescent product of the EROD reaction, resorufin. Red resorufin fluorescence, indicative of EROD activity, is quantified using image analysis or a photometer system operated as an attachment for the standard fluorescent microscope. For the latter method, fluorescent intensity is quantified using a standardized pinhole size or area of measurement, i.e., a pinhole aperture size 8 equivalent to 0.10 mm^2, centered over a bladder lobe. Fluorescence can be quantified as signal per unit area (V/mm^2) by normalizing to pinhole area. Alternatively, relative fluorescence can be reported by dividing measurements by average fluorescence values in control embryos.

For image analysis techniques, quantification of EROD intensity is based on captured images. Briefly, fluorescence intensity is quantified as pixel density in a small area over one lobe of the bladder, normalized by subtraction of background fluorescence and exposure time. While many data analysis programs can be used, image quantification has been accomplished using IP Lab Software (Scanalytics, Fairfax, VA).

Data management, storage and analysis

Test data include a unique test identification number, parental source of embryos, and date of fertilization for the batch of eggs tested. Daily logs include incubation temperature, observations by individual of developmental features, and presence and severity of lesions or abnormalities. Other useful endpoints include post-fertilization and post-hatching age of mortality, and age at hatching. In addition, length at hatching and histological lesions can be recorded after test termination. Typically, data are entered into spreadsheet formats and, 100% error-checked before statistical analysis by individual or treatment group.

Bioassays include 100–300 embryos in treatment groups of 20–40 embryos. To characterize the toxicity of a compound or substance replicate bioassays (2–3 per compound) are often conducted using different batches of embryos from the same parental stocks. Statistical procedures to develop concentration-dependent responses of EROD and health endpoints are described fully elsewhere.[11] Briefly, quantitative models are constructed using a non-linear least-squares regression procedure.[14] Survival response is described using this equation: $R_0 \times \phi\ [(\log(EC_{20}) - \log(C))/\phi + 0.8416]$. In this equation, C is the tested concentration and the estimated parameters are: R_0, the predicted control response; F, the cumulative area under the standard, normal distribution; s, the standard deviation of the normal distribution. A similar model is used to describe EROD responsiveness: $(R_m - R_0) \times \phi\ [(\log(EC_{20}) - \log(C))/\phi + 0.8416] + R_m$. The parameters are defined as for the previous model, with the addition of R_m, the predicted maximal response. To make statistical comparisons between these concentration-dependent responses, data can be fit to logistic regression models as described elsewhere.[15] Variances of log-transformed EROD responses can be tested for normality using the Shapiro–Wilks analysis (univariate procedure) and if normal, can be tested to determine differences between treatments, e.g., using one-way analysis of variance followed by Duncan's multiple range test.

Troubleshooting

Poor condition of parental fish may produce embryos with decreased survival compared with eggs produced by parents in good condition. Toxic substances may be introduced by contaminants in dilution water or solvents, glassware, sample hardware, and testing equipment. Be sure that all materials that are in contact with test materials or animals are thoroughly clean, but rinsed free of any cleaning products, and then soaked in seawater. Optionally, glassware can be prepared using acid soaking (e.g., up to 1 h in 5% HCl) and then heating (250°C for 6 h).

Results and discussion

To compare results across methods and to evaluate the sensitivity, specificity, and applicability of the *in ovo* EROD method, tests have been conducted using single compounds and complex environmental mixtures. Specifically, positive controls include strong AhR agonists, such as "dioxin," 2,3,7,8-tetrachlorodibenzo-*p*-dioxin (TCDD), and dioxin-like compounds, i.e., 3,3′,4,4′,5-pentachlorobiphenyl (PCB126), and PAHs, including benzo(*a*)pyrene (BaP) and 3 methyl cholanthrene (3MC). Another PAH that has been used as a negative control is fluoranthene (FLU), a poor AhR agonist.[3] In addition to single chemicals, environmental samples have also been tested using this method. These samples include marine and estuarine waters and sediment-derived samples, i.e., pore waters and organic extracts. Examples of results from these tests are provided to demonstrate test attributes and uses.

Comparison with standard EROD method

Increased EROD activity has been shown to be a sensitive response of *F. heteroclitus* embryos exposed to AhR agonists.[16] Typically, EROD activity is measured *in vitro* using microsomal fractions of tissues or whole embryos.[4] To compare *in ovo* and embryonic *in vitro* EROD results directly, embryos from reference site populations were exposed to PCB126, then tested using the *in ovo* method, or sacrificed and measured using the *in vitro* method. For this purpose, *in vitro* enzyme activity was measured using fish embryonic microsomes prepared as described[17] and modified for embryonic *F. heteroclitus*[15] and is described fully elsewhere.[18] Results using both methods were correlated, showing significant increases at comparable concentrations (Figure 11.5).

Associations between EROD and developmental effects

Developmental anomalies are shown (Figure 11.6) that typify exposures to AhR agonists at concentrations that increase EROD activity. Characteristic developmental embryonic

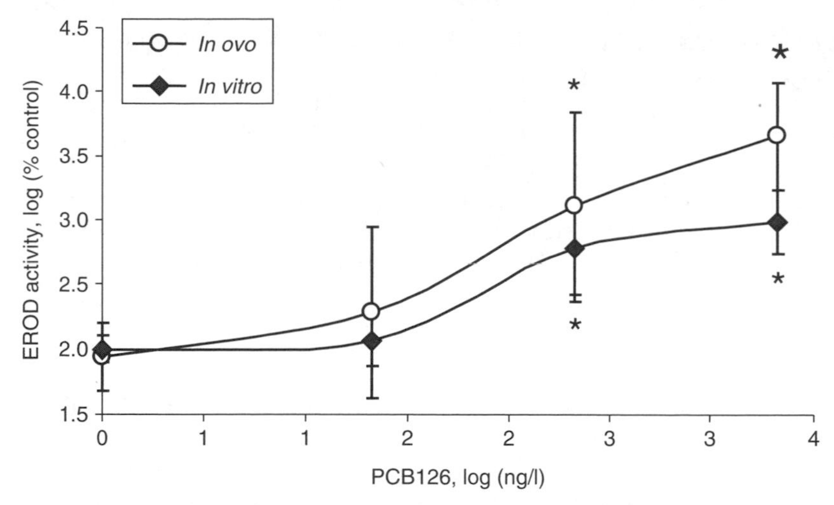

Figure 11.5 EROD activity in embryonic *F. heteroclitus* measured using microsomal *in vitro* production or *in ovo* fluorescence (mean ± SD), three determinations per treatment with PCB126. Significant differences from respective controls as determined using analysis of variance procedures are indicated (*).

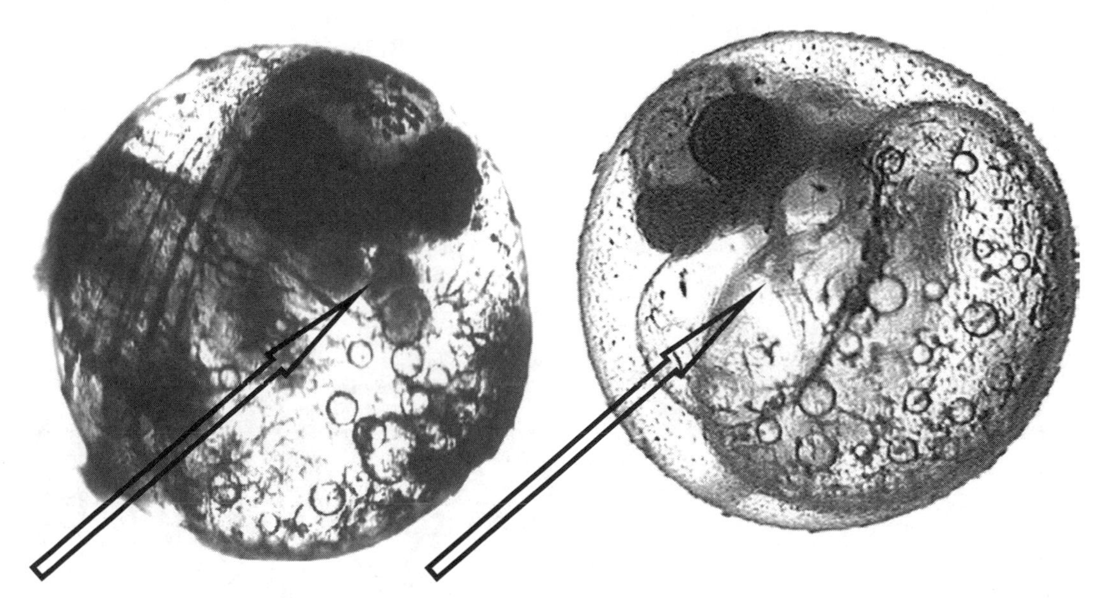

Figure 11.6 Photomicrograph of *F. heteroclitus* embryos (10 days post-fertilization). Arrows indicate normal heart development (left photo), and pericardial edema and "tube heart" (right photo) in an embryo exposed to an aryl hydrocarbon receptor agonist, PCB126.

anomalies are consistent with descriptions of "blue sac" like syndrome observed following exposures to AhR agonists of early life stages of *F. heteroclitus*[19,20] and other fish species.[21] These abnormalities included circulatory lesions, such as pericardial edema and "tube heart," and tail hemorrhages.[7,19,22] While characteristically observed, this syndrome of lesions is not uniquely diagnostic of AhR agonist compounds.[19]

To explore the relationship between EROD induction and developmental deformities, reference site embryos were exposed to a range of PCB126 concentrations, and assayed for *in ovo* EROD activity on day 7 of development.[23] The same embryos were assessed for deformities on day 10 post-fertilization. The frequency and severity of deformities were correlated with EROD fluorescence (Figure 11.7), consistent with a mechanistic association between EROD and health effects. In addition, this relationship suggests that concentrations of test substances that produce EROD responses are predictive of concentrations that produce adverse effects on fish development and early life stage survival.

Method sensitivity and specificity

Table 11.1 provides a comparison of the sensitivity of the *in ovo* EROD method with some other published methods used to quantify the potency of AhR agonists. Test endpoints selected for comparison were related to CYP1A induction, i.e., EROD activity and AhR response via reporter gene-linked activation. Although species/tissues, exposure durations and test parameters differ among test systems, nominal exposure concentrations of TCDD and PCB126 producing responses using the *in ovo* EROD method are 1–2 orders of magnitude lower than those reported in *in vitro* systems. This limited comparison suggests that the *F. heteroclitus in ovo* method is relatively sensitive to AhR agonist effects.

In ovo EROD activity has been measured in *F. heteroclitus* embryos using chemicals that vary in their potency as AhR agonists, such as PCB126, 3MC, BaP, and FLU.[9] Test data and modeled response relationships show that embryonic EROD fluorescence is specific for AhR agonists (Figure 11.8). In the order of their potency to produce EROD fluorescence, EC_{50} values for PCB126, 3MC, and BaP were 74, 123, and 7112 ng/l,

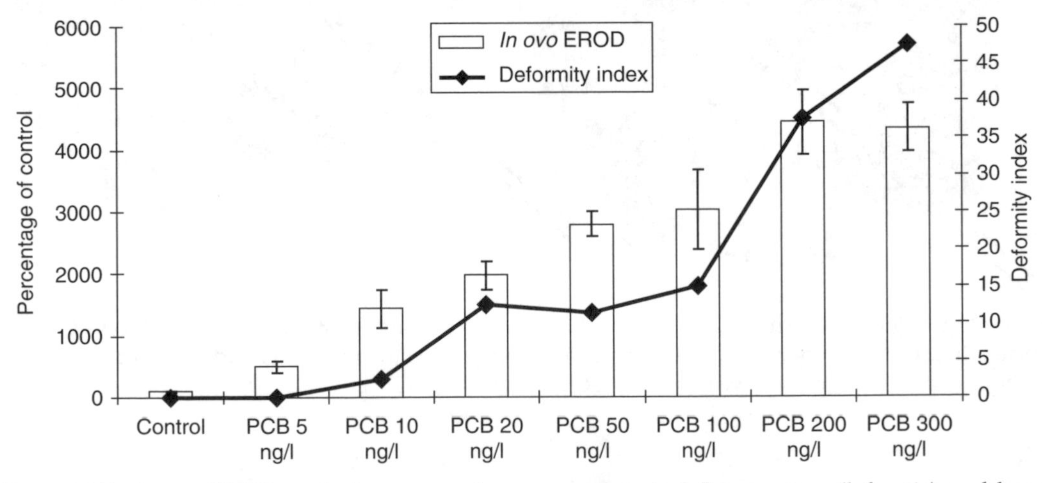

Figure 11.7 *In ovo* EROD analysis expressed as percent control fluorescence (left axis) and heart deformity assessment (right axis) of PCB126 exposed-embryos. Embryos were dosed on the day of spawning (day 0), EROD was assessed on day 7 post-fertilization, and deformities were assessed on day 10 post-fertilization. Heart deformities were scored for severity between 0 and 5 (no deformity to most severe). A deformity index was calculated by dividing the sum of the scores for a treatment group by the total possible score ($5 \times n$) and multiplying the quotient by 100.

Table 11.1 Comparative responses to aryl hydrocarbon agonists, TCDD and PCB126, measured or estimated using cell culture or *F. heteroclitus* embryo systems

Endpoint	Test population	TCDD EC_{50} (ng/l)	PCB126 EC_{50} (ng/l)	References
EROD, in ovo	*F. heteroclitus*, reference population	0.2[a]	36	(11)
EROD, in vitro	HepG2 (human)	32.2		(30)
	H4IIE (rat)	6.4	86	(29)
	PLHC-1 (fish)	41.9	121	(31)
	RTL (trout)	2.0	75	(32, 33)
Reporter gene, in vitro	HepG2 (human)	112.7		(34)
	101L (human)	32.2	3263	(35)
	H4IIE-Luc (rat)	1.9	94	(29)
	RTL 2.0 (trout)	20.6		(36)

[a] Estimated for TCDD using measured values for PCB126 and toxic equivalency in fish for PCB126 (i.e., 0.005).[37]

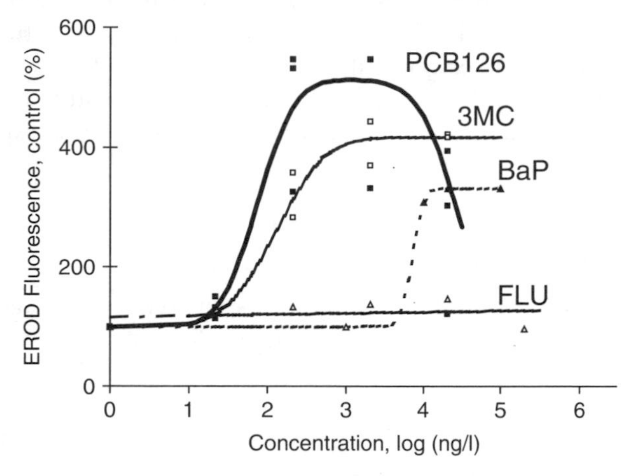

Figure 11.8 AhR-agonist specificity and concentration-responsiveness of EROD fluorescence in 10-day post-fertilization *F. heteroclitus*. Modeled response curves and mean (± SD) responses for replicated tests of strong AhR agonists, PCB126 (triangles), 3MC (squares), BaP (circles), and FLU, a poor AhR agonist (diamonds). (Adapted from Nacci, D., Coiro, L., Kuhn, A., Champlin, D., Munns, W., Jr., Specker, J. and Cooper, K., *Environ. Toxicol. Chem.*, 17, 2481–2486, 1998. With permission.)

respectively.[9] Consistent with expectations for non-AhR-mediated effects, no significant increase in EROD fluorescence was detected for FLU (<20,000 ng/l). Taken together, these results demonstrate that embryonic EROD fluorescence is a specific indicator of AhR-mediated effects, and that fluorescence results are consistent with other methods to assess similar responses.

Testing mixtures and environmental samples

The *in ovo* EROD assay has also been used to explore how environmentally relevant mixtures affect EROD activity. For example, EROD activity can be inhibited by combinations of chemicals acting through the AhR and other toxicological pathways.[24] To test the effects of chemical combinations, embryos were exposed to BaP, and PCB126, agonists alone in and in combination.[23] BaP and PCB126 both induced EROD activity, but when dosed in combination, resultant EROD measurements were less than those that would be predicted if the compounds were strictly additive (Figure 11.9).

Because the effects of chemical combinations cannot be predicted accurately, it is often desirable to test effects directly of environmental samples composed of chemical mixtures. Whole sediment testing is generally unacceptable because adhering sediment obscures the observation of *in ovo* EROD fluorescence. However, the *in ovo* EROD method has been used to test estuarine sediment-associated samples, including pore waters[10] and extracts.[25] Typically, results include positive and negative control exposures using chemicals, such as 3MC and FLU, respectively, and tested materials derived from sediments that range in severity and categories of chemical contamination (Figure 11.10). In the example shown, pore waters and organic extracts were prepared from sediments collected from a relatively uncontaminated reference site and one highly contaminated reference site with a complex mixture of contaminants, including TCDD. Test media from the highly contaminated sediment, both the pore water (tested as 71% of the exposure media) and organic extract (tested as 0.02% of the exposure media) produced strong *in ovo* EROD responses. Testing pore waters and organic extracts from the same sediment provides a strategy to estimate the bioavailability of AhR agonists in sediments. However, caution must be applied to the interpretation of results because exposure to complex mixtures often produces non-additive effects (as demonstrated in the previous example).

Investigating toxicity mechanisms

Relative to other species, *F. heteroclitus* is highly responsive to AhR agonists during early embryonic development.[9,16,22] However, some populations of *F. heteroclitus* indigenous to

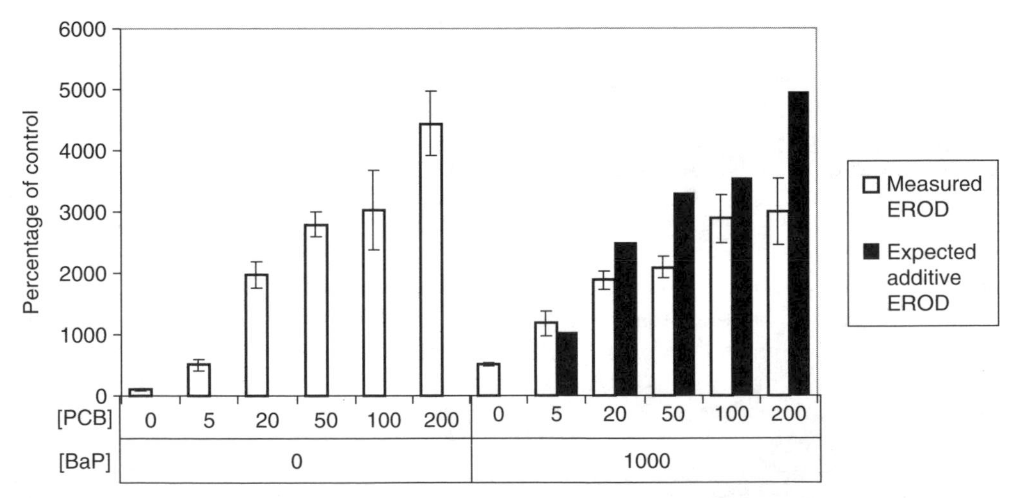

Figure 11.9 *In ovo* EROD activity (% solvent control, mean and standard error) for embryos treated with BaP (1000 ng/l, white bar), and a range of concentrations of PCB126 alone (white bars) or in combination with BaP (white bars), and expected additive EROD fluorescence (black bars). (Adapted from Wassenberg, D.M. and Di Giulio, R.T., Teratogenesis in *Fundulus heteroclitus* embryos exposed to a creosote-contaminated sediment extract and CYP1A inhibitors, *Mar. Env. Res.*, 54, 279–283, 2002. With permission.)

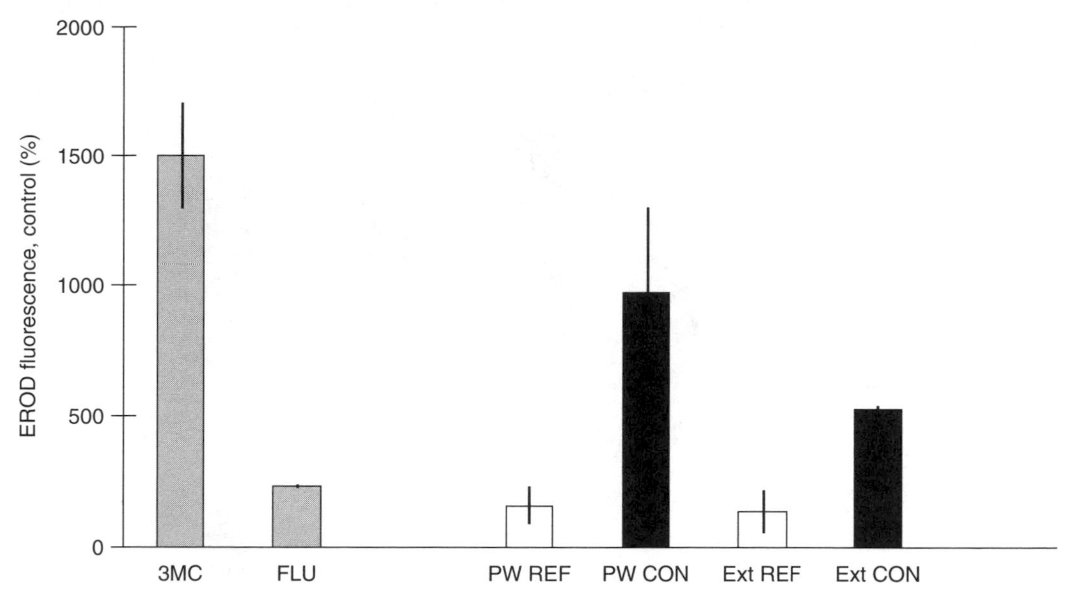

Figure 11.10 *In ovo* EROD responses of *F. heteroclitus* embryos exposed to positive control, 3MC (light bar), negative control, FLU (light bar), and sediment-derived materials: pore waters (PW) and organic extracts (EXT) from a reference (REF white bars) site or a site contaminated with complex mixtures of AhR agonists and other contaminants (CON dark bars).

contaminated sites are highly resistant to the effects of AhR agonists.[9,10,26] This tolerance is demonstrated as poor CYP1A inducibility, and resistance to the health effects associated with exposure to AhR agonists.[8,9] As shown, *in ovo* EROD fluorescence and embryonic deformities have been shown to increase when embryos from reference site populations are exposed to AhR agonists. However, embryos from populations of *F. heteroclitus* that have been characterized as tolerant to AhR agonists are relatively unresponsive to 3MC exposure (Figure 11.11).

These tolerant populations are indigenous to sites that are highly contaminated with complex mixtures that include PAHs (VA population) and PCBs (MA population). Similarities and differences between tolerant populations of *F. heteroclitus* are suggested by their intergenerational responses to 3MC. Specifically, F1 and F2 embryos from the population indigenous to the PCB-contaminated are unresponsive to 3MC exposure. However, while F1 embryos from the PAH-contaminated site are unresponsive to 3MC, their progeny (F2 embryos) are as sensitive to 3MC as those from a local reference population. Although the specific biochemical mechanism for reduced sensitivity in *F. heteroclitus* is not known (although see Reference 27), these differences suggest tolerance that has evolved in independent populations (like these) may have different mechanistic bases.[28] These examples demonstrate how the *in ovo* EROD method can be used in conjunction with other endpoints to elucidate mechanisms of toxicity and tolerance to contaminants that act as AhR agonists.

Summary

A non-destructive indicator of cytochrome P450A1 embryonic enzyme activity in fish has been developed that provides information relevant to exposure and effects of environmental contaminants. This information complements information obtained

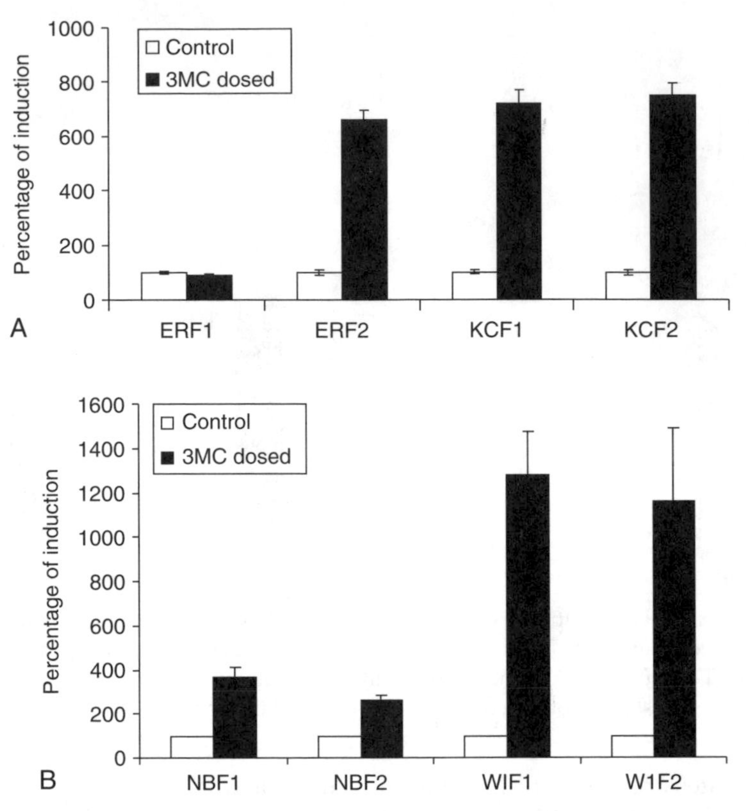

Figure 11.11 *In ovo* EROD activity in response to AhR agonist exposure in populations of *F. heteroclitus* that display tolerance to local contaminants in comparison to local reference populations: (a) from Virginia, tolerant (creosote-contaminated Elizabeth River, ER) embryos and reference (King's Creek, KC) embryos (F1 and F2 generations) exposed to 3MC (268 ng/l) or solvent control (From Meyer, J.N., Nacci, D.E. and Di Giulio, R.T., *Toxicol. Sci.*, 68, 69–81, 2002. With permission.); (b) from Massachusetts, tolerant (PCB-contaminated New Bedford, NB) and reference (West Island, WI) embryos (F1 and F2 generations) exposed to 3MC (2.68 μg/l) or solvent control.

using traditional fish embryonic and larval bioassays. In addition, because measurements using this technique are made non-invasively, they can be used to examine mechanisms linking AhR-mediated toxicity and long-term effects of environmental contaminants on individuals.

This method is based on the accumulation in the embryonic fish bladder of the fluorescent product of the EROD enzyme, indicative of cytochrome P450A1 enzyme activity. Results derived from this method are specific to AhR agonists and are correlated quantitatively with standard methods to measure EROD activity in fish embryos and larvae. This technique can be used diagnostically to identify bioavailable classes of contaminants in environmental media. However, complex mixtures may inhibit EROD responses, masking the occurrence of some toxic chemicals.

While many fish species are potentially suitable, this technique has been optimized using an estuarine fish species, *F. heteroclitus*, indigenous to estuaries along the east coast of the US. Adults of this species are easy to collect and can be maintained in the laboratory in fertile condition year round, producing many eggs per female on a semi-lunar cycle. Relative to other species, *F. heteroclitus* is highly responsive to AhR agonists, although populations vary in their responsiveness to contaminants in manners that are adaptive to their home grounds.[10,11] This intra-specific variation in responsiveness (reflected as variation in *in ovo* EROD fluorescence[10,11]) makes this species and this measurement endpoint useful for the investigation of mechanisms of toxicity for an important class of environmental contaminants.

The methods described here were developed specifically for *F. heteroclitus*. However, with minor modifications, they are appropriate for other fish species (not shown). Simple experimental procedures using positive and negative control exposures provided as examples here can be used to validate the utility of these methods for other fish species.

Acknowledgments

This is contribution number AED-04-016 of the U.S. EPA ORD NHEERL Atlantic Ecology Division. Although the research described in this contribution has been funded by the U.S. EPA, it has not been subjected to Agency-level review. Therefore, it does not necessarily reflect the views of the agency. Mention of trade names, products, or services does not constitute endorsement or recommendation for use. Support also includes NIEHS Superfund Basic Research Program (P42 ES10356 to R.T.D.) and U.S. EPA STAR fellowship (to D.M.W.).

References

1. Safe, S., Polychlorinated biphenyls (PCBs): environmental impact, biochemical and toxic responses, and implications for risk assessment, *Crit. Rev. Toxicol.*, 24, 87–149, 1994.
2. Hahn, M.E., The aryl hydrocarbon receptor: a comparative perspective, *Comp. Biochem. Physiol.*, 121C, 23–53, 1998a.
3. Stegeman, J.J. and Hahn, M.E., Biochemistry and molecular biology of monooxygenases: current perspectives on forms, function, and regulation of cytochrome P450 in aquatic species, in: *Aquatic Toxicology, Molecular, Biochemical and Cellular Perspectives*, Eds., Malins, D.C. and Ostrander, G.K., Eds., CRC Press, Boca Raton, 1994, pp. 87–206.
4. Whyte, J.J., Jung, R.E., Scmitt, C.J. and Tillitt, D.E., Ethoxyresorufin-*O*-deethylase (EROD) activity in fish as a biomarker of chemical exposure, *Crit. Rev. Toxicol.*, 30 (4), 347–569, 2000.
5. Hahn, M.E., Biomarkers and bioassays for detecting dioxin-like compounds in the marine environment, *Sci. Total Environ.*, 289, 49–69, 2002.
6. Cantrell, S.M., Lutz, L.H., Tillit, D.E. and Hnnink, M., Embryotoxicity of the 2,3,7,8-tetrachlorodibenzo-*p*-dioxin (TCDD): the embryonic vasculature is a physiological target for TCDD-induced DNA damage and apoptotic cell death in medaka (*Orizias latipes*), *Toxicol. Appl. Pharmacol.*, 141, 23–34, 1996.
7. Guiney, P.D., Smolowitz, R.M., Peterson, R.E. and Stegeman, J.J., Correlation of 2,3,7,8-tetrachlorodibenzo-*p*-dioxin induction of cytochrome P4501A in vascular endothelium with toxicity in early lifestages of Lake trout, *Toxicol. Appl. Pharmacol.*, 143, 256–273, 1997.
8. Nacci, D., Gleason, T., Gutjahr-Gobell, R., Huber, M. and Munns, W.R., Jr., Effects of environmental stressors on wildlife populations, in: *Coastal and Estuarine Risk Assessment: Risk on the Edge*, Newman, M.C., Ed., CRC Press/Lewis Publishers, Washington, D.C., 2002.
9. Nacci, D., Coiro, L., Kuhn, A., Champlin, D., Munns, W., Jr., Specker, J. and Cooper, K., A nondestructive indicator of EROD activity in embryonic fish, *Environ. Toxicol. Chem.*, 17, 2481–2486, 1998.
10. Meyer, J.N., Nacci, D.E. and Di Giulio, R.T., Cytochrome P4501A (CYP1A) in killifish *Fundulus heteroclitus*: heritability in altered expression and relationship to survival in contaminated sediments, *Toxicol. Sci.*, 68, 69–81, 2002.

11. Nacci, D.E., Coiro, L., Champlin, D., Jayaraman, S., McKinney, R., Gleason, T.R., Munns, W.R., Jr., Specker, J.L. and Cooper, K.R., Adaptation of wild populations of the estuarine fish *Fundulus heteroclitus* to persistent environmental contaminants, *Mar. Biol.*, 134, 9–17, 1999.
12. Armstrong, P.B. and Child, J.S., Stages in the normal development of *F. heteroclitus, Biol. Bull.*, 128 (2), 143–168, 1965.
13. Middaugh, D.P. and Whiting, D.D., Responses of embryonic and larval inland silversides, *Menidia beryllina*, to No. 2 fuel oil and oil dispersants in seawater, *Arch. Environ. Contam. Toxicol.*, 29, 535–539, 1995.
14. Bruce, R.D. and Versteeg, D.J., A statistical procedure for modeling continuous toxicity data, *Environ. Toxicol. Chem.*, 11, 1485–1494, 1992.
15. Oris, J.T. and Bailer, A.J., Equivalence of concentration–response relationships in toxicological studies: testing and implications for potency estimates, *Environ. Toxicol. Chem.*, 16, 2204–2209, 1997.
16. Binder, R.L. and Stegeman, J.J., Microsomal electron transport and xenobiotic monooxygenase during embryonic period of development in the killifish, *F. heteroclitus, Toxicol. Appl. Pharmacol.*. 73, 432–443, 1984.
17. Wisk, J.D. and Cooper, K.R., The stage specific toxicity of 2,3,7,8-tetrachlorodibenzo-*p*-dioxin in embryos of the Japanese medaka (*Oryzias latipes*), *Environ. Toxicol. Chem.*, 9, 1159–1169, 1990.
18. Nacci, D.E., Retinoid Homeostasis of an Oviparous Fish, Ph.D. Thesis, University of Rhode Island, Kingston, 2000, 270 pp.
19. Prince, R. and Cooper K.R., Comparisons of the effects of 2,3,7,8-tetrachlorodibenzo-*p*-dioxin on chemically-impacted and non-impacted subpopulations of *F. heteroclitus*. I. TCDD toxicity, *Environ. Toxicol. Chem.*, 14 (4), 579–588, 1995.
20. Prince, R., Comparisons of the Effects of 2,3,7,8-tetrachlorodibenzo-*p*-dioxin on chemically impacted and Non-impacted Subpopulations of *Fundulus heteroclitus*, Ph.D. Thesis, Graduate School-New Brunswick, Rutgers, The State University of New Jersey and the Graduate School of Biomedical Sciences, Robert Wood Johnson Medical School, Piscataway, NJ, USA, 1993.
21. Walker, M.K. and Peterson, R.E., Potencies of polychlorinated dibenzo-*p*-dioxin, bibenzofuran and biphenyl congeners, relative to 2,3,7,8-tetrachlorodibenzo-*p*-dioxin, for producing early life stage mortality in rainbow trout (*Oncorhynchus mykiss*), *Aquat. Toxicol.*, 21, 219–238, 1991.
22. Toomey, B.H., Bello, S., Hahn, M.E., Cantrell, S., Wright, P., Tillit, D. and DiGiulio, R.T., TCDD induces apoptotic cell, death and cytochrome P4501A expression in developing *Fundulus heteroclitus* embryos, *Aquat. Toxicol.*, 53 (2), 127–138, 2001.
23. Wassenberg, D.M., Swails, E.E. and Di Giulio, R.T., Effects of single and combined exposures to benzo(*a*)pyrene and 3,3′,4,4′,5-pentachlorophenol on EROD activity and development in *Fundulus heteroclitus, Mar. Environ. Res.*, 54, 279–283, 2002.
24. Willett, K.L., Wassenberg, D., Lienesch, L., Reichert, W. and Di Giulio, R.T., *In vivo* and *in vitro* inhibition of CYP1A-dependent activity in *Fundulus heteroclitus* by the polynuclear aromatic hydrocarbon fluoranthene, *Toxicol. Appl. Pharmacol.*, 177, 264–271, 2001.
25. Wassenberg, D.M. and Di Giulio, R.T., Teratogenesis in *Fundulus heteroclitus* embryos exposed to a creosote-contaminated sediment extract and CYP1A inhibitors, *Mar. Env. Res.*, 58, 163–168, 2004.
26. Nacci, D., Coiro, L., Champlin, D., Jayaraman, S. and McKinney, R., Predicting responsiveness to contaminants in wild populations of the estuarine fish *Fundulus heteroclitus*, *Environ. Toxicol. Chem.*, 21 (7), 1525–1532, 2002.
27. Hahn, M.E., Mechanisms of innate and acquired resistance to dioxin-like compounds, *Rev. Toxicol.*, 2, 395–443, 1998.
28. Van Veld, P. and Nacci, D.E., Chemical tolerance: acclimation and adaptations to chemical stress, in: *The Toxicology of Fishes*, DiGiulio, R.T. and Hinton, D.E., Eds., Taylor & Francis, Washington, 2004.
29. Sanderson, J.T., Aarts, J.M.M.J.G., Brouwer, A., Froese, K.L., Denison, M.S. and Giesy, J.P., Comparison of Ah receptor-mediated luciferase and ethoxyresorufin-*O*-deethylase induction in H4IIE cells: implications for their use as bioanalytical tools for the detection of polyhalogenated aromatic hydrocarbons, *Toxicol. Appl. Pharmacol.*, 137, 316–325, 1996.

30. Wiebel, F.J., Wegenke, M. and Kiefer, F., Bioassay for determining 2,3,7,8-tetrachlorodobenzo-*p*-dioxin equivalents (TEs) in human hepatoma HepG2 cells, *Toxicol. Lett.*, 88, 335–338, 1996.
31. Hahn, M.E., Woodward, B.L., Stegeman, J.J. and Kennedy, S.W., Rapid assessment of induced cytochrome P4501A (CYP1A) protein and catalytic activity in fish hepatoma cells grown in multiwell plates: response to TCDD, TCDF, and two planar PCBs, *Environ. Toxicol. Chem.*, 15, 582–591, 1996.
32. Clemons, J.H., van den Heuvel, M.R., Stegeman, J.J., Dixon, D.G. and Bols, N.C., Comparison of toxic equivalent factors for selected dioxin and furan congeners derived using fish and mammalian liver cell lines, *Canad. J. Fish. Aquat. Sci.*, 51, 1577–1584, 1994.
33. Clemons, J.H., Lee, L.E.J., Myers, C.R., Dixon, D.G. and Bols, N.C., Cytochrome P4501A1 induction by polychlorinated biphenyls (PCbs) in liver cell lines from rat and trout and the derivation of toxic equivalency factors, *Canad. J. Fish. Aquat. Sci.*, 53, 1177–1185, 1996.
34. Postlind, H., Vu, T.P., Tukey, R.H. and Quattrochi, L.C., Response of human CYP1A-luciferase plasmids to 2,3,7,8-tetradichlorodibenzo-*p*-dioxin and polycyclic aromatic hydrocarbons, *Toxicol. Appl. Pharmacol.*, 118, 255–262, 1993.
35. Anderson, J.W., Rossi, S.S., Tukey, R.H., Vu, T. and Quattrochi, L.C., A biomarker, P450 RGS, for assessing the induction potential of environmental samples, *Environ. Toxicol. Chem.*, 14 (7), 1159–1169, 1995.
36. Richter, C.A., Tieber, V.L., Denison, M.S. and Giesy, J.P., An *in vitro* rainbow trout cell bioassay for aryl hydrocarbon receptor-mediated toxins, *Environ. Toxicol. Chem.*, 16 (3), 543–550, 1997.
37. van den Berg, M., Birnbaum, L., Bosveld, B.T.C., Brunstrom, B., Cook, P., Feeley, M., Giesy, J.P., Hanberg, A., Hasagawa, R., Kennedy, S.W., Kubiak, T., Larsen, J.C., Leeuwen, F.X.R., Liem, A.K.D., Nolt, C., Petersen, R.E., Poellinger, L., Safe, S., Shrenk, D., Tillit, D., Tysklind, M., Younes, M., Waern, F. and Zacharewski, T., Toxic equivalency factors (TEFs) for PCBs, PCDDs, and PCDFs for human and wildlife, *Environ. Health Perspect.*, 106, 775–792, 1998.

chapter twelve

Determination of lipid classes and lipid content in tissues of aquatic organisms using a thin layer chromatography/flame ionization detection (TLC/FID) microlipid method

Gina M. Ylitalo, Gladys K. Yanagida, Lawrence Hufnagle, Jr., and Margaret M. Krahn
National Marine Fisheries Service

Contents

Introduction

Lipid content of aquatic animals is used frequently to normalize the concentrations of lipophilic contaminants.[1] In partitioning studies of lipophilic contaminants among various tissues in aquatic organisms, concentrations of these compounds are often

1-56670-664-5/05/$0.00+$1.50
© 2005 by CRC Press

lipid-normalized to elucidate how contaminants partition among the various tissues of an organism. This adjustment is also used when examining differences in contaminant levels among different species, as well as for modeling biomagnification of lipophilic contaminants in food webs and examining biota-sediment accumulation factors.

Tissues of marine biota consist of many different classes of lipids that differ significantly in polarity and may require specific solvents for their extraction. For example, blubber of cetaceans is comprised primarily of neutral lipids, such as triglycerides (TG) or wax esters,[2–5] whereas brain samples of these animals contain a much wider suite of lipids, including cholesterol and more polar lipid classes (e.g., phospholipids).[4] Solvents that have been used for lipid extractions range from non-polar/semi-polar solvents, such as hydrocarbon solvents (e.g., hexane, pentane) and chloroform, to polar solvents, such as methanol for the extraction of more polar lipids (e.g., membrane-bound phospholipids). Differences in lipid class composition can impact lipid normalization and, therefore, interpretation of results. For example, Ewald and Larsson[6] observed that fish lipids with high phospholipid content accumulated less PCB77 than did fish lipids with low phospholipid content. A review on the interactions among lipids and lipophilic contaminants in fish can be found in Elskus *et al.*[1] Kawai *et al.*[4] and Tilbury *et al.*[7] found lower lipid-normalized PCB concentrations in marine mammal brain compared to other tissues. The majority of lipids in brain are comprised of phospholipids and cholesterol rather than the neutral lipids that make up most of the total lipid in other tissues. Therefore, a lipid method that is capable of determining "total lipids," as well as lipid class distribution, can be very valuable in understanding the differences in organochlorine (OC) concentrations among various tissues, species, and trophic levels.

A number of methods are used to extract and quantitate lipid content in a wide array of marine biota tissues[8–11] The most widely accepted gravimetric method, developed by Bligh and Dyer,[12] employs a chloroform/methanol solvent system to extract both neutral and polar lipids from a wide array of matrices. Although the most frequently used methods for lipid quantitation are based on gravimetric techniques, these analyses provide no information about the classes of lipids present in a tissue. A method has been developed to quantitate lipids using thin layer chromatography (TLC) coupled with flame ionization detection (FID), TLC/FID, system (see review in Reference 9). This method has been used to measure concentrations of lipid classes and calculate percent total lipid in phytoplankton, shrimp, bivalves, fish, and marine mammals.[5,13–15] Although this instrument system does have some limitations (e.g., time-consuming setup and calibration of instrument, variability of lipid quantitation due to sample spotting technique, a small range of linearity of the FID response), it provides rapid quantitation of lipid classes and lipid content in one analysis, including low-lipid tissues, as well as samples too small for good gravimetric quantitation.

Over the past decade, non-destructive collection of tissue samples (e.g., whale blubber biopsy, fish blood) from marine animals, especially populations of species that are threatened or endangered, has increased for biological and ecological studies. The tissue samples collected via biopsy are used to determine, for example, contaminant levels and profiles, stable isotope ratios and biomarkers.[15–18] However, due to the small size (<0.30 g) of some biopsy samples, only a selected number of analyses (e.g., contaminant and genetic analyses) can be completed. For contaminant analyses of small (<1.0 g) tissue samples, especially low-lipid containing samples, it is important to develop a single method that is capable of efficiently extracting both lipophilic contaminants, as well as lipids. Recently, we have developed a method that simultaneously extracts lipophilic OCs and lipids in tissue samples of marine organisms that range in size from 0.2 to 3.0 g using an accelerated solvent extraction (ASE) method (see Chapter 35)[19]. With this method, the majority of the sample extract (>90% by volume) is used for OC analyses [e.g., high-performance liquid chromatography with photodiode array detection (HPLC/PDA) or gas chromatography/mass spectrometry (GC/MS)]. The remaining portion of the sample extract is then used to determine percent lipid and profiles of lipid classes using TLC with FID. In this chapter, the modifications made to determine lipids by ASE are compared to two previous analytical methods (ASE dichloromethane extraction/gravimetric quantitation and pentane/hexane extraction with TLC/FID analyses) to demonstrate the advantages and limitations of each lipid method.

Materials required

Tissue extractions

See Chapter 35[19] for details.

Lipid analyses

Equipment:

- Iatroscan MK-5®, Bioscan; Washington, D.C.
- Baxter Temp Con Oven
- A Dell Dimension 8300 computer or equivalent computer, which can run Microsoft Windows XP Pro®, has at least 2 GB hard disk space and 384 MB of RAM and is not a hard disk configured to be FAT 16
- Waters Empower® data acquisition system (Waters; Milford, MA)

Supplies:

- 100-ml Pyrex glass graduated cylinder (Fisher Scientific, Pittsburgh, PA, catalog #08-552E)
- 25-ml Pyrex glass graduated cylinder (Fisher Scientific, Pittsburgh, PA, catalog #08-552C)
- Chromarods®, type S-III (Shell-USA, Fredericksburg, VA, catalog #3248)
- Chromarod holder SD-5 (Shell-USA, Fredericksburg, PA, catalog #5321)
- Chromarod spotting guide (Shell-USA, Fredericksburg, VA, catalog #5231)
- TLC development tank, 6 in. wide × 7.5 in. high × 1.25 in. deep (Shell-USA, Fredericksburg, PA, catalog #3201)
- 32 in.-diameter Whatman Grade No. 1 filter paper, cut to 5.75 in. wide × 7.35 in. high (VWR Scientific, West Chester, PA, catalog #78450-229)
- Hamilton syringe 1 μl (Fisher Scientific, Pittsburgh, PA, catalog #14813112)

Reagents:

- Cholesterol (Sigma Chemical, St. Louis, MO, catalog #C-8667)
- Oleic acid (Sigma Chemical, St. Louis, MO, catalog #O-1008)
- L-α-phosphatidylcholine (Sigma Chemical, St. Louis, MO, catalog #P-3556)
- Triolein (Sigma Chemical, St. Louis, MO, catalog #T-7140)
- Lauryl stearate (Nucheck Prep, Elysian, MN, catalog #WE-1304)

- Diethyl ether (Fisher Scientific, Pittsburgh, PA, catalog #E197-1)
- Formic acid (88%) (Fisher Scientific, Pittsburgh, PA, catalog #A-118P-100)
- Dichloromethane, pesticide grade (VWR Scientific, West Chester, PA, catalog #BJ300-4)
- Ultra high-purity hydrogen, 5.0 grade (PraxAir, Danbury, CT)

Procedures

TLC/FID calibration standards

Stock solutions of lauryl stearate (10 mg/ml), triolein (10 mg/ml), oleic acid (10 mg/ml), cholesterol (2.0 mg/ml), and L-α-phosphatidylcholine (2.0 mg/ml) were prepared in dichloromethane and stored at −20°C. Four lipid calibration standards for the Iatroscan were prepared in dichloromethane and stored in a −20°C freezer. Calibration standard #1 contained the following compounds: lauryl stearate (10.0 mg/ml), triolein (10.0 mg/ml), oleic acid (10.0 mg/ml), cholesterol (2.0 mg/ml), and L-α-phosphatidylcholine (2.0 mg/ml). Calibration standard #2 contained the following: lauryl stearate (5.0 mg/ml), triolein (5.0 mg/ml), oleic acid (5.0 mg/ml), cholesterol (1.0 mg/ml), and L-α-phosphatidylcholine (1.0 mg/ml). Calibration standard #3 contained the following: lauryl stearate (2.5 mg/ml), triolein (2.5 mg/ml), oleic acid (2.5 mg/ml), cholesterol (0.50 mg/ml), and L-α-phosphatidylcholine (0.50 mg/ml). Calibration standard #4 contained the following: lauryl stearate (1.25 mg/ml), triolein (1.25 mg/ml), oleic acid (1.25 mg/ml), cholesterol (0.25 mg/ml), and L-α-phosphatidylcholine (0.25 mg/ml). We recommend making new calibration solutions every 4–6 months to ensure that the standards have not decomposed or oxidized.

Microlipid extraction of samples

Prior to use, all tubes and glass transfer pipettes are rinsed three times with acetone to remove any potential contaminants. As a safety precaution, all glassware rinsing, tissue extraction, and lipid spotting procedures are performed in fume hoods by personnel wearing nitrile gloves and safety glasses.

Lipids and OCs were extracted from tissues using an ASE as described by Sloan *et al.* in Chapter 35 of this volume. For HPLC/PDA and TLC/FID analyses, the amount of tissue extracted ranges from 0.20 to 5.0 g, depending on the expected lipid and water content. Briefly, a tissue sample was weighed (to the nearest 0.01 g) in a tared 10-oz jar. Sodium sulfate (15 cc) and magnesium sulfate (15 cc) were added sequentially to each sample jar and mixed thoroughly using a clean metal spatula. Each sample mixture was transferred to a 33-ml ASE extraction cell and an OC surrogate standard (see Chapter 25 of Reference 20 by Ylitalo *et al.*) was added to each cell. Lipids and OCs were extracted using dichloromethane (see Chapter 35 of Reference 19 for details on ASE extraction parameters).

After the ASE extracted a sample set, each 60-ml collection tube was weighed to the nearest 0.01 g and the weight was recorded. Using a solvent-rinsed 9-in. glass pipette, a 1-ml aliquot of each sample extract was transferred to a 2-ml GC vial, capped and stored at −20°C until TLC/FID lipid analysis. Each 60-ml collection tube was recapped and the remaining sample extract was weighed and recorded. For low-lipid tissues (<10%), each 1-ml lipid extract portion was reduced in volume to 100-μl using a gentle stream of ultra-pure nitrogen. By concentrating the low-lipid tissue extracts to 100 μl, we are fairly confident that we can quantitate the lipid classes within the linear response ranges of the FID. The remaining sample extract was cleaned up and analyzed for OC contaminants by HPLC/PDA or GC/MS (see Chapter 25 of this volume).

Separation of lipid classes

Silica gel chromatography rods (type S-III Chromarods) were used for separation of each lipid class in the tissue extracts. The Chromarods are quite sensitive to humidity,[9] so we suggest storing the racks of Chromarods in a 60°C oven or desiccator until needed. Ten Chromarods were placed into a SD-5 Chromarod holder, and each Chromarod was blank scanned (scanned from top to bottom at a speed of 30 s/scan) three times by the Iatroscan MK-5 TLC/FID to remove any organic compounds that were present and to "activate" the Chromarod. While the Chromarods were blank scanned, a TLC development tank containing hexane/diethyl ether/formic acid [60:10:0.02 (v/v/v)] with a rectangular piece of filter paper (5.75 in. × 7.35 in., to help saturate the tank) was set up in a laboratory hood. The development tank is performed in a fume hood of a temperature-regulated laboratory (temperature range 20–25°C) to prevent shifts in retention times of the lipid classes. To ensure that the development tank was completely saturated, we added the filter paper and developing solvents to the tank at least 30 min prior to placing the first set of Chromarods in the tank. After blanking the Chromarods, the rod holder was removed from the Iatroscan and placed onto a spotting guide. A 1-μl aliquot of sample extract was carefully spotted onto each Chromarod near the base of the Chromarod using a 1-μl Hamilton syringe. The spotting technique was one of the most difficult techniques to master. For example, if the sample extract was spotted over a large area of the Chromarod, we have found that the concentration of phospholipids is 20–50% less than the phospholipid concentration in the same extract spotted over a smaller region. In order to provide better separation of the various lipid classes, we try to keep the sample spot as small as possible to avoid spreading at the origin of the Chromarod. Although we have not used a semi-automatic sample spotter (available from Shell-USA, Fredericksburg, VA, model SES 3202/IS-02), we have been told by colleagues that sample spots made with the spotter are more consistent and smaller compared to the spots delivered with a 1-μl syringe. After spotting each Chromarod with the sample extract, the rod holder was placed in a 60°C oven for 2 min to evaporate the sample solvent. Alternately, the sample solvent can be rapidly evaporated using a hair dryer. The rod holder was removed from the oven and placed in the chromatography development tank for 24 min in order to separate the various lipid classes (e.g., wax esters, TG, free fatty acids, cholesterol, phospholipids) in each sample extract. After 24 min in the development tank, the Chromarod holder was removed immediately from the tank and placed in a 65°C oven for 5 min to evaporate any remaining solvent. The rack should not stay in the tank any longer than 24 min because the separation of lipid classes can be altered with additional time in the development tank.

TLC/FID analysis

The FID was operated with hydrogen and airflow rates of 160 and 2000 ml/min, respectively. We have found that if the hydrogen flow on the Iatroscan was changed (e.g., hydrogen flow dial is inadvertently bumped or turned), the calibration curves of the lipid classes must be recalculated. Each Chromarod was scanned at 30 s/scan, which

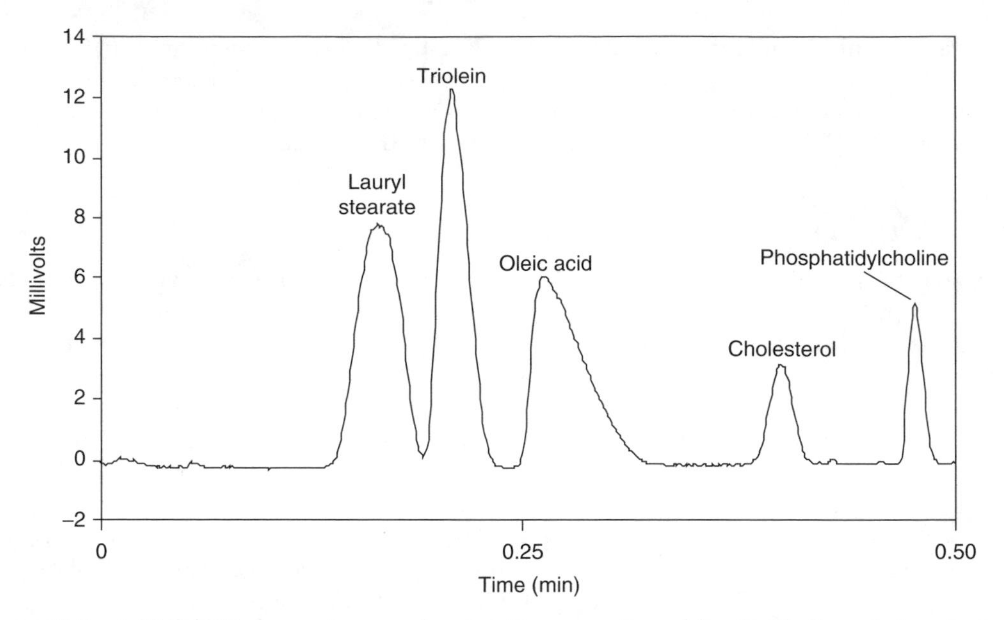

Figure 12.1 A TLC/FID chromatogram showing the separation of five lipids [lauryl stearate (wax ester), triolein (TG), oleic acid (free fatty acid) cholesterol, and L-α-phosphatidylcholine (phospholipid)] contained in a calibration standard.

allowed for complete combustion of all components on the rod. The Chromarods were then passed through the hydrogen flame to ionize the sample. The ionization caused changes in electric current flowing through the FID. The Iatroscan MK-5 was interfaced with a Dell Dimension 8300 computer and data were collected and reprocessed using the Waters Empower data acquisition system. The currents were recorded by the Empower data acquisition system. Each sample chromatogram was automatically integrated using the Empower software system. The integration of each sample peak was checked, reintegrated (if needed), and each sample report with corresponding chromatogram was printed.

Quality assurance

Each extraction sample set consisted of 11–14 field samples, a method blank, and a National Institute of Standards and Technology (NIST) standard reference material (i.e., SRM 1974b, SRM 1945, SRM 1946). To determine if the extraction solvents, glassware, or other supplies were free of lipid contaminants, a solvent blank was analyzed with every sample set. To monitor the accuracy of our extraction method, an NIST standard reference material was analyzed with each sample set and results met laboratory criterian.[21] The percent lipid concentrations of the SRM 1945, SRM 1946, and SRM 1974b determined by our TLC/FID method were compared to "total extractable organic" values published by NIST.

External standard curves for each lipid class were created every 3 weeks. To determine if the Iatroscan TLC/FID was operating properly, a calibration standard was analyzed with each Chromarod rack that consisted of 10 Chromarods. If the concentration of a lipid in the calibration standard was not within $\pm$ 15% of the known concentration of that compound, the field samples were reanalyzed. The limits of detection for each of the five lipids contained in the calibration standards were the following: lauryl stearate

(0.20 mg/ml), triolein (0.20 mg/ml), oleic acid (0.30 mg/ml), cholesterol (0.25 mg/ml), and L-α-phosphatidylcholine (0.25 mg/ml).

Calculations of lipid class concentrations and percent lipid

The Empower data acquisition software integrates the area under each peak. Once the calibration curves for the lipid classes are completed, the software automatically calculates the concentration of each lipid class. These data are then used to calculate the concentration of each lipid class per gram sample extracted. For example, to calculate the concentration of TG in a tissue sample, Equation (12.1) was used:

$$\mu\text{g TG/g sample} = (\text{amount of TG in sample} * \text{volume } (\mu\text{l}) \text{ of sample extract}) / \text{sample weight (g)} * \text{volume } (\mu\text{l}) \text{ of extract spotted on Chromarod}) \quad (12.1)$$

This concentration was then converted to g TG/kg samples extracted using Equation (12.2):

$$\text{g TG/kg sample} = \mu\text{g TG}/(\text{g sample extracted} * 0.001) \quad (12.2)$$

Percent total lipids were calculated by summing the concentrations (g of lipid class/kg sample) of the five lipid classes and multiplying this sum by 0.1. A four-point linear external calibration curve for each lipid class was used for quantitation. Duplicate TLC/FID analyses were performed for each sample extract, and the mean value was reported.

Results and discussion

Lipid classes are readily separated on the Chromarods and are rapidly quantitated using the Iatroscan TLC/FID system. In Figure 12.1, five lipids [lauryl stearate (wax ester), triolein (TG), oleic acid (free fatty acid), cholesterol, and L-α-phosphatidylcholine (phospholipid)] contained in a calibration standard were separated using a hexane/diethyl ether/formic acid development system. With this non-polar/semi-polar solvent system, the neutral lipids migrate up to the Chromarod further than with the polar lipids. Similar to our findings, other studies that determined lipid content of marine organisms and fish meal using Iatroscan TLC/FID system reported good separation and quantitation of various classes of lipids on the Chromarods.[11,13,14]

We compared the lipid content of various tissues of marine organisms determined by three different methods to evaluate the best method to extract and quantitate lipids. The three methods were the following: (1) dichloromethane ASE extraction/gravimetric analyses[20]; (2) pentane/hexane extraction with TLC/FID analysis (old method);[5] and (3) dichloromethane ASE extraction with TLC/FID analysis (new method). The percent lipid values of the NIST SRMs and various tissues of marine biota determined by these methods are reported in Table 12.1. The mean lipid values of the three SRMs determined by the ASE/gravimetric method or by either of the Iatroscan methods are comparable or lower than the mean values published by NIST. We also found that, although the percent lipid values determined by TLC/FID were comparable to or lower than the values measured gravimetrically, the lipid concentrations quantitated by these two methods were correlated. For example, the lipid values of various SRMs and marine tissues determined by the new method were correlated with the lipid concentrations measured gravimetrically ($r^2 = 0.990$, $P < 0.0001$). Similarly, Delbeke *et al.*[13] found that the lipid concentrations determined by TLC/FID of tissues of various species of marine biota were correlated but were approximately half as great as those determined by the gravimetric method. Because the gravimetric method measures lipids, as well as other exogenous materials extracted from a tissue, it is likely that the lipid content determined gravimetrically is overestimated due to interferences from non-lipid substances. In contrast, non-lipid compounds do not substantially affect the TLC/FID lipid measurements. Therefore, our preliminary findings indicate that lipid concentrations can be readily measured using the new ASE/TLC/FID method.

To evaluate the extraction efficiency of the two solvent systems used to extract lipids and then quantitated by TLC/FID, we compared the lipid values in three SRMs and various marine biota tissues extracted with dichloromethane (new method) or pentane/hexane (old method). In the present study, it should be noted that the SRMs and fish tissue samples were homogenates, whereas the blubber of the California sea lion were individual subsamples. We found that the lipid values were highly correlated ($r^2 = 0.951$, $P < 0.0001$) between the two systems but did not find any consistent trends. For example, the dichloromethane-extracted blubber samples (SRM 1945 and blubber of sea lion) contained comparable or higher percent lipid values than did the same samples extracted with pentane/hexane, whereas the dichloromethane-extracted rockfish liver samples (14965 and 14968) had lipid values that were comparable or lower than the lipid concentrations of the same liver samples extracted by the pentane/hexane method (Table 12.1). Because a limited number of samples (<50 samples) have been analyzed by both the old and new methods, the results presented in this chapter should be considered primarily qualitative.

Data on lipid classes measured in tissues of marine biota can provide information on quality of the samples that are being analyzed for chemical contaminants. In a contaminant study on eastern North Pacific gray whales, Krahn *et al.*[5] found that the blubber samples of "decomposed" whales (decomposition state rated by field personnel) contained lower proportions of TG and higher proportions of free fatty acids, cholesterol, and phospholipids than the blubber of "fresh dead" whales. Because free fatty acids are generally attributable to hydrolysis of TG, increase in this lipid class is an indication of decomposition. In fact, whales showing advanced decomposition had higher levels of free fatty acids than were found in moderately decomposed whales. In the present study, blubber of the California sea lions ($n = 3$) contained high proportions of TG (~95%) and low proportions of phospholipids (~5%). These results are consistent with other studies that show that blubber of "healthy" marine mammals is comprised primarily of neutral lipids (e.g., TG, wax esters).[4,5,7,15] The lipid class profiles (no free fatty acids or cholesterol) of the California sea lion blubber samples (data not shown) indicated that these samples were stored properly. Over the past few years, we have determined the lipid profiles of a number of blubber samples of stranded marine mammals prior to OC analyses to determine if the tissue quality may be compromised. In the future, we will continue to screen blubber samples for quality using ASE with TLC/FID.

The TLC/FID analyses of the lipid extracts provide information on lipid classes, as well as percent lipid on a wide array of marine biota tissues, including low-lipid containing samples, such as rockfish muscle samples, and low-weight blubber biopsy samples. Data on lipid classes can provide valuable information on the quality of the tissues being analyzed. In addition, both contaminant and lipid analyses can be conducted on the same dichloromethane ASE extract — saving analyses time and reducing costs.

We recommend extraction of marine biota tissues using the dichloromethane ASE method — the lipid values of the NIST SRMs determined by the new method (ASE with dichloromethane) are more similar to the NIST published values than the lipid values measured by the old method (pentane/hexane). The advantages of lipid analyses by the Iatroscan TLC/FID are: (1) data on lipid classes are obtained that can be used to

Table 12.1 Percent lipid values measured gravimetrically and by TLC with FID in various NIST SRMs and tissues of marine biota

Sample number	Sample type	NIST published values	Iatroscan TLC/FID		Gravimetric GC/MS method, dichloromethane
			Pentane/hexane[a]	Dichloromethane[b]	
SRM 1974b ($n = 5$)	Blue mussel homogenate	0.64 ± 0.13	0.32 ± 0.046[c]	0.20 ± 0.038	0.36 ± 0.03
SRM 1945 ($n = 15$)	Whale blubber homogenate	74.29 ± 0.45	57 ± 4.9	63 ± 7.6[d]	74 ± 2.7
SRM 1946 ($n = 5$)	Fish muscle homogenate	10.17 ± 0.48	7.2 ± 0.21	5.6 ± 0.34	10 ± 3.4
CSL 4951	CA sea lion blubber		39	56	Not determined
CSL 4955	CA sea lion blubber		50	49	Not determined
CSL 4914	CA sea lion blubber		48	54	Not determined
SK-XM2	Coho salmon muscle		5.0	4.1	4.7
NQ-XM9	Coho salmon muscle		3.2	3.9	5.3
14965	Rockfish liver		4.2	4.1	7.6
14968	Rockfish liver		13	8.3	13
PO-PHW10	Pacific herring whole body		Not determined	5.3	6.0
CH-PHW07	Pacific herring whole body		Not determined	1.3	3.0
02SM-PHW04	Pacific herring whole body		4.4	2.6	6.6
1572	Rockfish muscle		Not determined	0.55	0.94
1583	Rockfish muscle		Not determined	0.19	0.27

[a] Old method.

[b] New method.

[c] $n = 6$.

[d] $n = 5$.

determine the quality of certain tissue samples (e.g., blubber); (2) overestimation of percent lipid due to extraction of exogenous substances is eliminated; and (3) lipid content can be determined for low-lipid tissues and small tissue samples. However, there are some limitations: (1) the initial setup is more costly than for gravimetric quantitation; (2) calibration of the TLC/FID is fairly time consuming compared to gravimetric analyses; and (3) sample spotting technique is difficult to master. We recommend a performance-based quality assurance program for lipid analyses. For example, we include a method blank and an SRM or control material, for which there are published or certified percent lipid values, with each sample set to determine how well the dichloromethane ASE with TLC/FID system is operating.

Acknowledgments

We appreciate the technical assistance or advice of Cheryl Krone, Don Brown, David Herman, John Stein, and Tracy Collier. We thank Sandra O'Neill, James West, Greg Lippert, and Steve Quinnell of the Washington State Department of Fish and Wildlife for the collection of the fish tissue samples as part of the Puget Sound Ambient Monitoring Program (PSAMP). We thank Frances Gulland and Denise Greig from The Marine Mammal Center in Sausalito, CA, for the collection of the California sea lion blubber samples. We also appreciate the careful review of the manuscript by James Meador and Jon Buzitis.

References

1. Elskus, A.A., Collier, T.K. and Monosson, E., Interactions among lipids and persistent organic pollutants (POPs) in fish, in *The Biochemistry and Molecular Biology of Fishes*, Mommsen, T.P. and Moon, T.W., Eds., Elsevier, Amsterdam (in press).
2. Litchfield, C., Greenberg, A.J., Caldwell, D.K., Caldwell, M.C., Sipos, J.C. and Ackman, R.G., Comparative lipid patterns in acoustical and nonacoustical fatty tissues of dolphins, porpoises and toothed whales, *Comp. Biochem. Physiol.*, 50B, 591–597, 1975.
3. Lockyer, C.H., McConnell, L.C. and Waters, T.D., The biochemical composition of fin whale blubber, *Can. J. Zool.*, 62, 2553–2562, 1984.
4. Kawai, S., Fukushima, M., Miyazaki, N. and Tatsukawa, R., Relationship between lipid composition and organochlorine levels in the tissues of striped dolphin, *Mar. Pollut. Bull.*, 19 (3), 129–133, 1988.
5. Krahn, M.M., Ylitalo, G.M., Burrows, D.G., Calambokidis, J., Moore, S.E., Gosho, M., Gearin, P., Plesha, P.D., Brownell, R.L., Jr., Blokhin, S.A., Tilbury, K.L., Rowles, T. and Stein, J.E., Organochlorine contaminant concentrations and lipid profiles in eastern North Pacific gray whales (*Eschrichtius robustus*), J. *Cetacean Res. Manage.*, 3 (1), 19–29, 2001.
6. Ewald, G. and Larsson, P., Partitioning of 14C-labelled 2,2′,4,4′-tetrachlorbiphenyl between water and fish lipids, *Environ. Toxicol. Chem.*, 13 (10), 1577–1580, 1994.
7. Tilbury, K.L., Stein, J.E., Meador, J.P., Krone, C.A. and Chan, S.-L., Chemical contaminants in harbor porpoise (*Phocoena phocoena*) from the North Atlantic Coast: tissue concentrations and intra- and inter-organ distribution, *Chemosphere*, 34 (9/10), 2159–2181, 1997.
8. Nilsson, W.B., Gauglitz, E.J., Jr., Hudson, J.K., Stout, V.F. and Spinelli, J., Fractionation of menhaden oil ethyl esters using supercritical fluid CO_2, *J. Am. Oil Chem. Soc.*, 65, 109–117, 1988.
9. Shantha, N.C., Thin-layer chromatography-flame ionization detection Iatroscan system, *J. Chromatogr.*, 624, 21–35, 1992.
10. Perkins, E.G., Ed., *Analyses of Fats, Oils, and Derivatives*, AOCS Press, Champaign, IL, 1993, 438 pp.

11. Johnson, R.B. and Barnett, H.J., Determination of fat content in fish feed by supercritical fluid extraction and subsequent lipid classification of extract by thin layer chromatography-flame ionization detection, *Aquaculture*, 216 (1–4), 263–282, 2003.
12. Bligh, E.G. and Dyer, W.J., A rapid method of total lipid extraction and purification, *Can. J. Biochem. Physiol.*, 37 (8), 911–917, 1959.
13. Delbeke, K., Teklemariam, T., de la Cruz, E. and Sorgeloos, P., Reducing variability in pollution data: the use of lipid classes for normalization of pollution data in marine biota, *Intern. J. Environ. Anal. Chem.*, 58, 147–162, 1995.
14. Bergen, B.J., Nelson, W.G., Quinn, J.G. and Jayaraman, S., Relationships among total lipid, lipid classes, and polychlorinated biphenyl concentrations in two indigenous populations of ribbed mussels (*Geukensia demissa*) over an annual cycle, *Environ. Toxicol. Chem.*, 20 (3), 575–581, 2001.
15. Ylitalo, G.M., Matkin, C.O., Buzitis, J., Krahn, M.M., Jones, L.L., Rowles, T. and Stein, J.E., Influence of life-history parameters on organochlorine concentrations in free-ranging killer whales (*Orcinus orca*) from Prince William Sound, AK, *Sci. Total Environ.*, 281, 183–203, 2001.
16. Todd, S., Ostrom, P., Lien, J. and Abrajano, J., Use of biopsy samples of humpback whale (*Megaptera novaeangliae*) skin for stable isotope ($d^{13}C$) determination, *J. Northw. Atl. Fish Sci.*, 22, 71–76, 1997.
17. Fossi, M.C., Casini, S. and Marsili, L., Nondestructive biomarkers of exposure to endocrine disrupting chemicals in endangered species of wildlife, *Chemosphere*, 39 (8), 1273–1285, 1999.
18. Hobbs, K.E., Muir, D.C.G., Michaud, R., Beland, P., Letcher, R.J. and Norstrom, R.J., PCBs and organochlorine pesticides in blubber biopsies from free-ranging St. Lawrence River Estuary beluga whales (*Delphinapterus leucas*), 1994–1998, *Environ. Pollut.*, 122, 291–302, 2003.
19. Sloan, C.A., Brown, D.W., Pearce, R.W., Boyer, R.M., Botton, J.L., Burrows, D.G. Herman, D.P. and Krahn, M.M., Determining aromatic hydrocarbons and chlorinated hydrocarbons in sediments and tissues using accelerated solvent extraction and gas chromatography/mass spectrometry, in Techniques in Aquatic Toxicology, vol. 2, Ostrander, G.K., Ed., CRC Press, Boca Raton, FL (Chapter 35) (in press).
20. Ylitalo, G.M., Buzitis, J., Boyd, D., Herman, D.P., Tilbury, K.L. and Krahn, M.M., Improvements to high-performance liquid chromatography/photodiode array detection (HPLC/PDA) method that measures dioxin-like polychlorinated biphenyls and other selected organochlorines in marine biota, in Techniques in Aquatic Toxicology, vol. 2, Ostrander, G.K., Ed., CRC Press, Boca Raton, FL (Chapter 25) (in press).
21. Wise, S.A., Schantz, M.M., Koster, B.J., Demiralp, R., Mackey, E.A., Greenberg, R.R., Burow, M., Ostapczuk, P. and Lillestolen, T.I., Development of frozen whale blubber and liver reference materials for the measurement of organic and inorganic contaminants, *Fresenius J. Anal. Chem.*, 345, 270–277, 1993.

chapter thirteen

Larval molting hormone synthesis and imaginal disc development in the midge Chironomus riparius *as tools for assessing the endocrine modulating potential of chemicals in aquatic insects*

Torsten Hahn
Technical University of Braunschweig

Contents

Introduction

In metazoans, hormones play a crucial role in control and fine-tuning of physiological processes. Such diverse and complex functions as energy metabolism, cellular homeostasis, reproduction and development, behavioral patterns, or phenotypic characteristics are governed by endocrine regulation.[1] Thus, endocrine disruption, the impairment of one or more of these processes by exogenous compounds, may well have uncalculable consequences in the living environment, including human health. Consequently, in the last decade,

1-56670-664-5/05/$0.00+$1.50
© 2005 by CRC Press

considerable research effort has been directed towards this issue.[2–4] Most of these studies focussed on vertebrate hormone systems, and although invertebrate species represent about 95% of all animals, effects in invertebrates have never been thoroughly investigated.[5–8] Probable reasons are that invertebrate hormone systems are much more divergent and much less examined (and understood) than those of vertebrates, and that the high degree of conservation within vertebrate hormone systems makes these better animal models with respect to possible impacts of endocrine-disrupting chemicals on human health.

In selecting a suitable test endpoint for detecting endocrine disruption, it is important to consider the multitude of levels of interaction, in which a chemical may impact hormonal function in an organism. These may comprise production, metabolism, release, transport, receptor interaction, or metabolism of natural hormones.[5] The present chapter describes a method that uses a combined *in vivo* and *in vitro* approach for detecting endocrine modulation of development in aquatic larvae of the non-biting midge *Chironomus riparius* Meigen. The test endpoints, *in vitro* synthesis of ecdysteroidal molting hormones by prothoracic glands and imaginal disc development *in vivo*, represent different levels of biological organization, but both functions are clearly correlated with endocrine regulation.

The choice of *C. riparius* as a test organism is based on a variety of factors that make working with these animals advantageous. First, as insects, their endocrinology is well understood when compared to other invertebrate taxa. Furthermore, chironomids are widely used as test organisms in ecotoxicology, culturing is easy and standard methods have been developed.[9,10] Chironomids are sexually reproducing animals (in contrast, e.g., to daphnids) and their larvae are of a size that allows dissection of hormone-secreting glands. Finally, particularly in the older literature, a wealth of information is available on developmental biology and endocrinology of this species.

Wülker and Götz[11] have suggested a method for exact determination of sex and developmental stage of fourth-instar larvae, which will be summarized in the following. King et al.[12] first described the analysis of molting hormone production by *in vitro* incubation of prothoracic glands and, since then, it has been applied in a huge variety of problems, including neuroendocrine control of ecdysteroid synthesis[13] and the effects of synthetic insect growth regulators.[14]

Materials required

A prerequisite for the experiments described in this chapter is a healthy laboratory culture of *C. riparus* or related *Chironomus* species (e.g. *Camptochironomus* (*Chironomus*) *tentans*). Methods for setting up and maintaining a culture have been described, e.g., in References 9 and 10.

General:

- A set of 0.2 μl to 5 ml pipettes
- 25, 100, and 300 μl and 1.5 ml repeater pipettes
- Tubes, tube racks, volumetric flasks (10 and 100 ml)

For staging:

- Microscope or stereomicroscope equipped with bottom illuminator
- Glass slides
- A set of tweezers, including spring steel and self-locking tweezers
- Vials or dishes to collect the staged animals

For dissection:

- Stereomicroscope
- Fiber optic illuminator with dual light pipes
- Very fine curved micro-scissors
- Precision tweezers Dumont No. 5 (available through laboratory suppliers; be sure to have some spare tweezers available because they are very sensitive to inadvertent damage)
- Tweezer set, including spring steel and self-locking tweezers
- Insect minute pins
- Ice pack for cooling the dissection dish from below
- Paper tissue
- Ice-cold water for anesthesia
- Cold 70% EtOH
- Pasteur pipettes
- Sterile glass Petri dishes, approximately 50 mm diameter (the use of plastics should be avoided because several plastic additives are suspected to possess endocrine activity)
- Autoclaved 8 × 40-mm GC sample vials (Merck Eurolab, or comparable product) (avoid autoclaving the caps unless you are sure that they are resistant)
- Water bath or electronic heating block
- Beadle's ringer solution: 7.5 g/l NaCl, 0.35 g/l KCl, 0.21 g/l $CaCl_2$, as given in Reference 15, autoclaved or sterile filtered
- Cannon's ringer solution as given in Reference 16, see Table 13.1 for details
- Finally, for a quiet hand, I have found it helpful to renounce the morning cup of coffee when dissections are planned

For ecdysteroid extraction:

- Methanol HPLC grade or better
- Table top centrifuge (1000 *g*)
- Ultrasonic bath
- Nitrogen
- Water bath or heating block
- Evaporator manifold (this can easily be self-made using plastic tube and injection needles). Alternatively a vacuum concentrator can be used ("Speed-Vac" or similar product)

For radioimmunoassay:

- Access to a working place where handling and storage of substances with activities of up to 50 μCi tritium is allowed
- Tritium-labeled ecdysone (New England Nuclear, #NET621)
- Unlabeled ecdysone standard (Sigma-Aldrich, #E9004)
- Ecdysone-specific antiserum (Trifolio, Lahnau, Germany, DBL-1)
- Normal sheep serum (ICN Biomedicals, #642952)
- Sodium azide
- RIA buffer: 100 m*M* boric-acid/sodium tetraborate, 75 m*M* NaCl, pH 8.4
- Scintillation fluid
- Refrigerated centrifuge capable of a speed of at least 5000 *g* and capacity for 24 or more tubes

- Ice-cold saturated ammonium sulfate solution
- Refrigerator
- Scintillation counter

Procedure

1. *Staging and sexing of the developmental phase of 4th larval instars*

A method for exact determination of sex and developmental stage of fourth-instar chironomid larvae was described by Wülker and Götz.[11] By examining the larval imaginal discs they characterized nine morphologically distinct phases of equal duration, considering developmental characteristics of gonads, genital and thoracic imaginal discs. For the method presented here, only the genital imaginal discs were used for determination of the developmental stages. This was found to provide sufficiently detailed information to classify a single larva within a reasonable amount of time.

Single specimens are collected from the culture vessels containing larvae from a synchronized culture (i.e., having left their egg masses within 24 h or less), usually on day 11 or day 12 after hatching (*C. riparius* at 20°C). For sexing and staging, larvae are blotted on tissue and placed dorsal surface down on a microscope slide. Another slide is carefully placed on top to keep the larva in place. Using a dissection microscope (light from below) at 50-fold magnification, the genital imaginal discs in the 8th and 9th abdominal segment are easily observable. No symptoms were found to indicate that intact larvae were hurt by this procedure. However, in a few cases, the larvae did not stay intact and hemolymph leaked out of the body. These larvae are moribund and should be excluded from any further experimentation. A detailed description of the developmental characteristics of the single phases is given in Figure 13.1 (from Reference 11, with permission).

After the determination, larvae are collected in appropriate vessels, one for each sex and stage. For the experiments of substance-induced alteration of development, larvae in phases 5 and 6 should be chosen, as these larvae usually do not enter pupal stage within the test duration of 48 h. For dissection and *in vitro* incubation of prothoracic glands, animals in phase 9 were most suitable because molting-hormone production in insects reaches its maximum just before molting begins (Figure 13.2, and Reference 17).

2. In vivo *development*

Single larvae to be used in these experiments are transferred into 20-ml test vessels that contain 2 ml medium to which the desired test substance was added, and are exposed for 48 h under rearing conditions, but without aeration, food, or sediment. After this time, the larvae are removed from the vessels and staged again as described earlier. For each larva, the individual 48-h development is then computed by subtracting the larval phases' value at the start of the experiment from the value determined after 48 h of exposure.

3. *Dissection of larvae and* in vitro *incubation of prothoracic glands*

Larvae are sexed and staged as described earlier, and males and females only from the 9th (= last) larval stage are used for dissection. It was long known that ecdysteroid release by the prothoracic glands reaches its maximum activity directly prior to the molting process. This was also established for chironomids by Laufer et al.,[17] who determined the whole-body ecdysteroid titer in relation to days of larval age after hatching. In

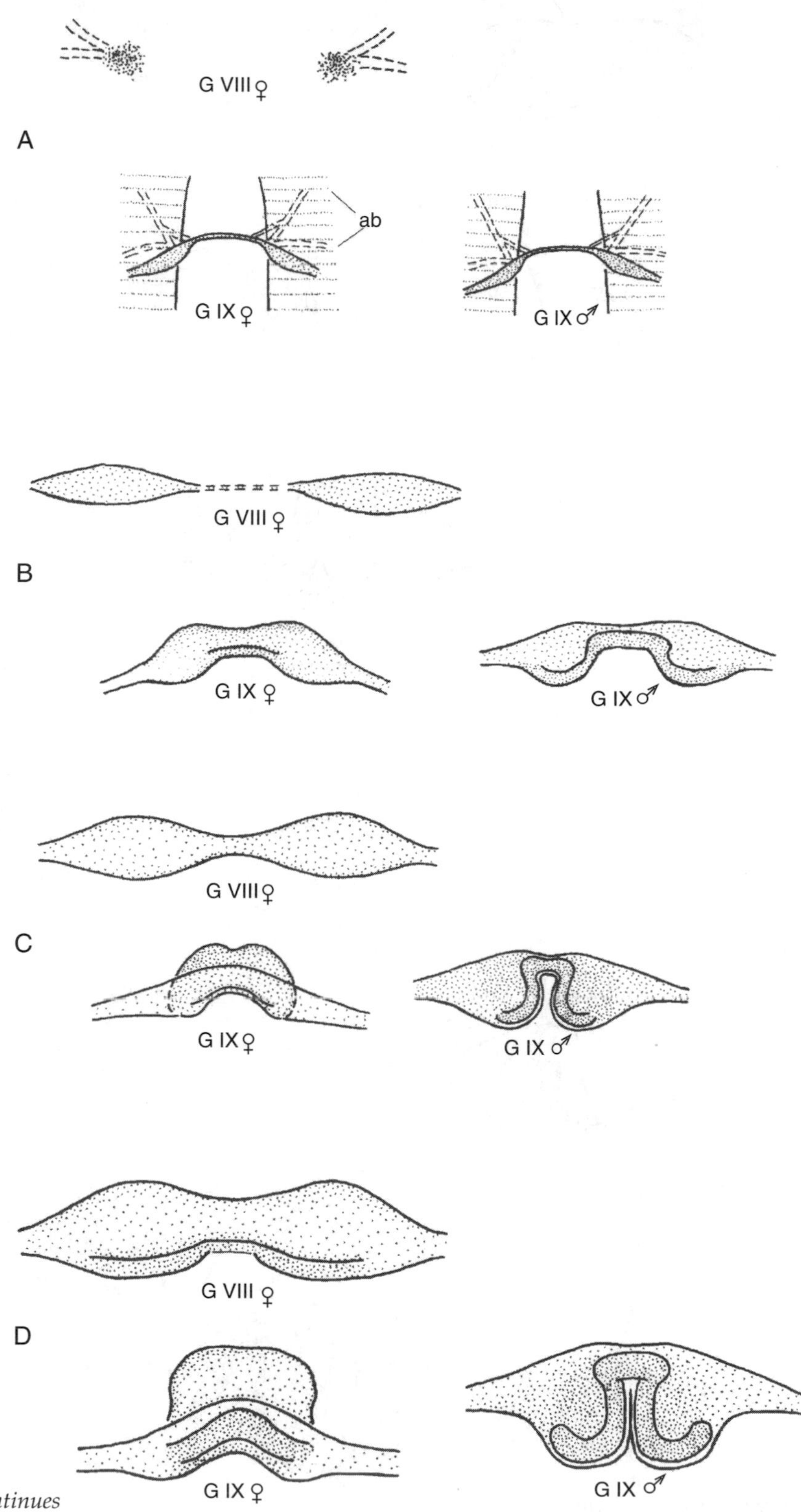

Figure 13.1 *A–D Continues*

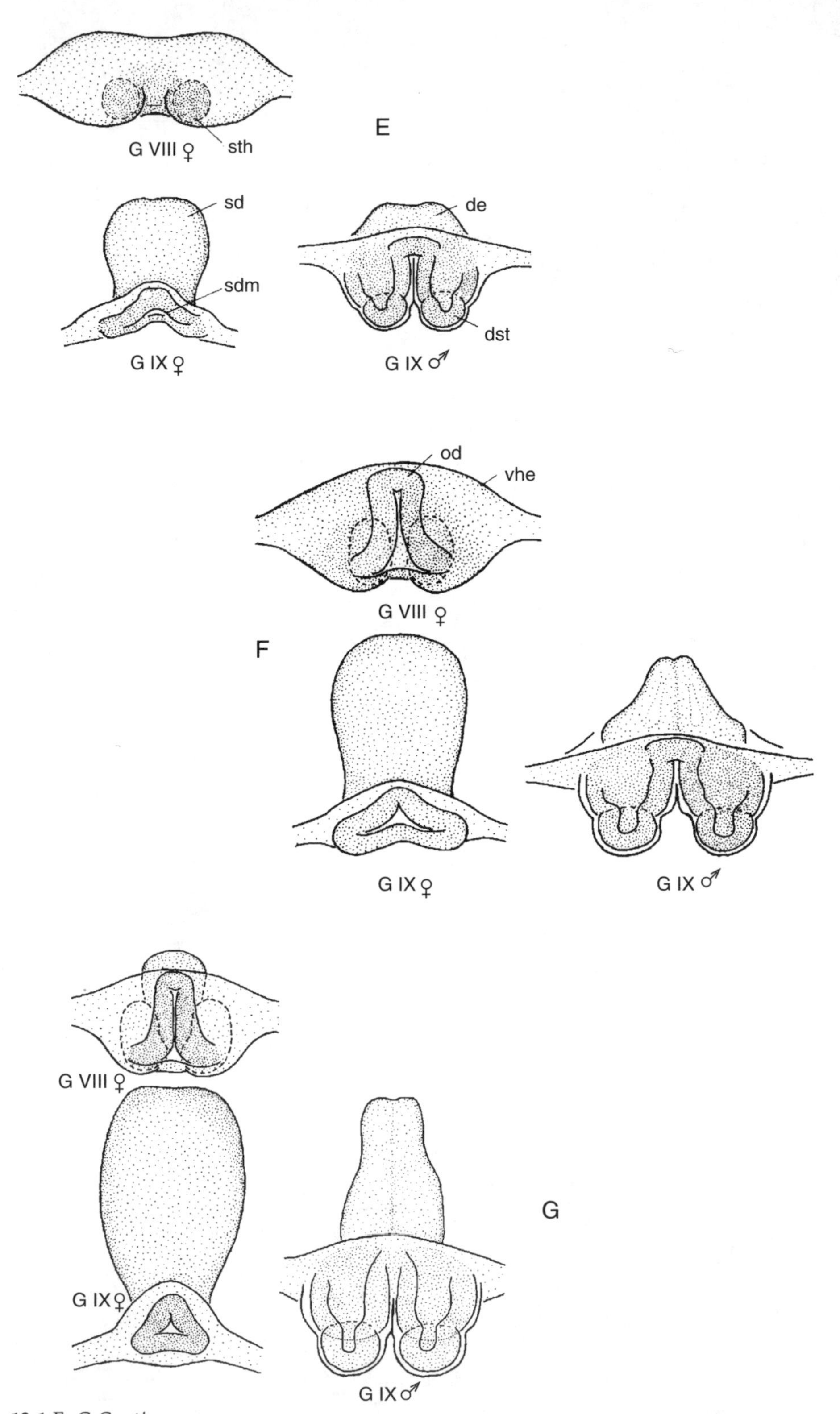

Figure 13.1 *E–G Continues*

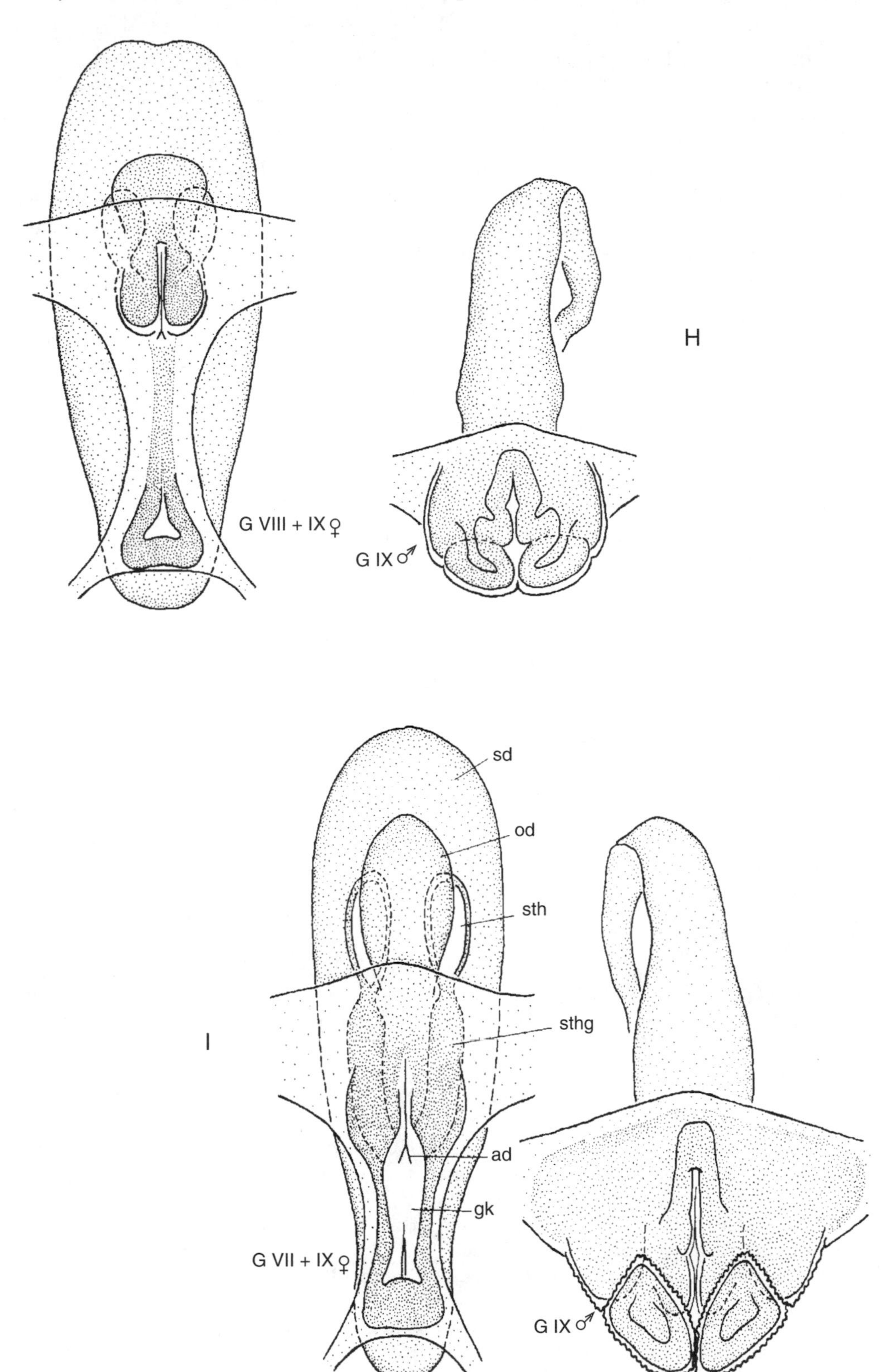

Figure 13.1 *H and I Continues*

preparation for the present experiments, tests of ecdysteroid titers in hemolymph reverified these results and additionally confirmed that 9th-phase midges display the highest molting hormone content (Figure 13.2).

Late larvae in phases 8 and 9 can, without any microscopical observation, easily be recognized by a thick white "collar" in the first thoracic segment, but for an unequivocal determination, sexing and staging by imaginal disc characters is necessary. The collar is formed by the pupal respiration organ, which has already developed under the larval cuticle. Because late-fourth-instar larvae already contain the developing or fully developed pupae, they are often referred to as pharate pupae. However, before final ecdysis these animals are still true larvae, in overall physical state and in their response to their environment.[18]

Staged animals are immersed in ice-cold water for anesthesia until they stop moving. (I have no experience with CO_2 anesthesia but this also might be an option.) They are then briefly washed in cold 70% ethanol and blotted dry on tissue. The dissection dish is placed on an ice pack for continuous cooling and filled with ice-cold Beadle's ringer solution. Under a stereomicroscope the larva is fixed ventral surface down with a minute pin, which is inserted directly between the head capsule and the first thoracic segment. This is crucial because the head capsule itself is too brittle and may break if the larva should move, and the thoracic segments contain the central nervous system and adhering neuroendocrine glands, which should not be harmed. An overview of the organization of the chironomid retrocerebral system[19,20] is provided in Figure 13.3. For

---

Figure 13.1 Overview of the developmental phases in 4th-stage *Chironomus* larvae, depending on the differentiation grade of the genital imaginal discs. (Reproduced from Wülker, W. and Götz, P., *Z. Morphol. Tiere*, 62, 363–388, 1968. With permission.) (A) Phase 1: G VIII ♀, in some cases, groups of enlarged hypodermis cells at the typical site. G IX ♀ and ♂, hypodermal depressions in the region of the terminal ampullae, linked together by a narrow hypodermal ridge. Sexes cannot be differentiated. (B) Phase 2: G VIII ♀, the two parts of the anlage connected by a narrow, often incomplete band. G IX ♀ and G IX ♂, anlage parts larger and connected by a broad bridge. Sexes cannot be differentiated. (C) Phase 3: G VIII ♀, anlage parts oval, with a wide connecting strip. G IX ♀, anlage expanded and fused medially; mucus gland sometimes visible as transverse slit. G IX ♂, anlage spectacles-shaped; medial and posterior edge reinforced by a ridge. (D) Phase 4: G VIII ♀, anlage has formed a unit, with indented oral margin. G IX ♀, mucus gland growing out like a tongue; ratio of tongue length to minimal width (L:W) < 1.2. G IX ♂, caudal parts of the anlage medially apposed to one another over a large area. (E) Phase 5: G VIII ♀, anlage a wide oval; spermathecal anlagen emerging dorsally as round bulges. G IX ♂, mucus gland, L:W = 1.2–2.0. G IX ♂, ejaculatory duct growing out at the front edge, as a two-peaked hillock; in the back part dististyli already stand out slightly. (F) Phase 6: G VIII ♀, anlage of the oviduct extends to the front edge of the hypodermal infolding, but not beyond. G IX ♀, mucus gland, L:W = 2.0–3.0. G IX ♂, in ventral view, the anlage of the ejaculatory duct appears about half as long as the hypopygal anlage; dististyli clearly demarcated. (G) Phase 7: G VIII ♀, oviduct anlage projects beyond the hypodermal indentation. G IX ♀, front edge of mucus gland extends to the same level as G VIII. G IX ♂, anlage of the ejaculatory duct is at least as long as the hypopygal anlage. (H) Phase 8: G VIII ♀, an elongated oval. G IX ♀, front edge of mucus gland extends far beyond G VIII; the regions around the openings (G IX and G VIII) are beginning to fuse; apodeme already visible. G IX ♂, anlage of ejaculatory duct twice as long as hypopygal anlage; dististyli form tongue-like inward projections. (I) Phase 9: G VIII ♀ and G IX ♀, in the fusion region of G VIII and G IX a forked apodeme and complete genital chamber have formed. Spermathecae have long excretory ducts. G IX ♂, Hypopyge in conspicuously curled pupal sheath. *Key to lettering*: ab, suspension bands; ad, apodeme; de, ejaculatory duct; dst, dististylus of the male genital anlage; G VIII, GIX, genital imaginal disks of the 8th and 9th abdominal segments; gk, genital chamber; od, oviduct; sd, mucus gland; sdm, opening of mucus gland; sth, spermatheca; sthg, spermathecal duct; vhe, front edge of the hypodermal infolding.

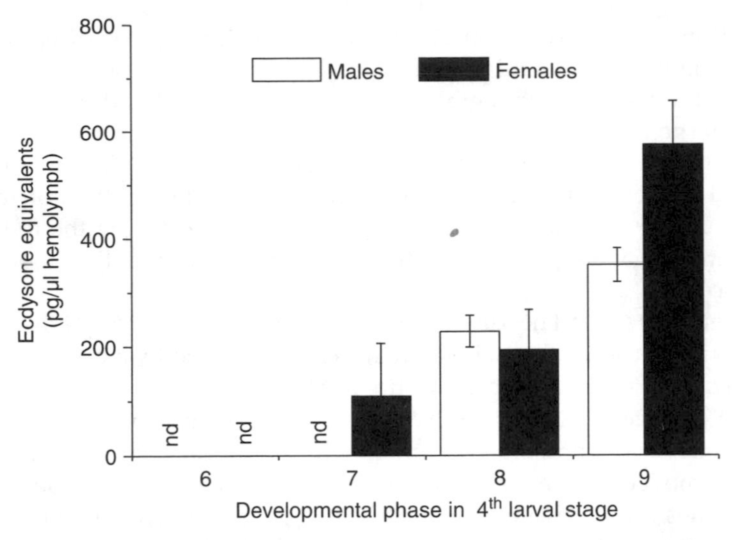

Figure 13.2 RIA-analyzed hemolymph-ecdysteroid concentration in relation to larval developmental phase. Between 1 and 2 μl larval hemolymph, measured precisely (± 0.05 μl), were collected, extracted, and RIA-analyzed as described in the text. Columns represent the means of data from 4 to 8 larvae ± S.E.; n.d., not detectable.

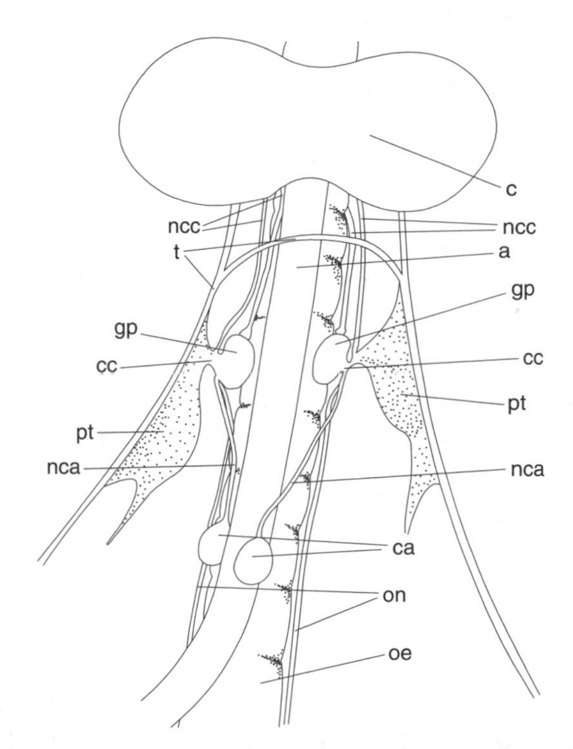

Figure 13.3 Overview of the organization of the retrocerebral glands in *Chironomus* larvae. (Lettering in accordance with Possompès, B., *Arch. Zool. Exp. Gen.*, 89, 201–364, 1953; Credland, P., Scales, D. and Philipps, A., *Zoomorphologie*, 92, 65–75, 1979.) *Key to lettering*: a, aorta; c, brain; ca, corpus allatum; cc, corpus cardiacum; gp, post-cerebral gland; nca, nervus corporis allatum; ncc, nervus corporis cardiacum; oe, oesophagus; on, oesophageal nerve; pt, peritracheal tissue; t, trachea.

further dissection, the first or second abdominal segment is gripped with a pair of fine tweezers, and the larva is carefully elongated. The whole abdomen can now be severed directly behind the last thoracic segment. Usually, the midgut and parts of both salivary glands then emerge from the opened body cavity. The midgut is *carefully* pulled out until the transition to the oesophagus appears, where it is severed. The salivary glands are also pulled out and the salivary ducts are severed. Be careful not to damage the glands; the tissues are very sticky and may cause problems during the following dissection steps. If removal of the glands is difficult in this stage of the dissection, it is better to leave them in place.

To fix the tissue, a second needle is now placed in the ventral part of the opened thorax, so that the dorsal skin of the thoracic segments can be opened by a proximad longitudinal cut up to the recess of the first needle. Be sure not to make the cut too deep, so that only the integument but no deeper tissues are hurt. The fine white structures of the paired pupal respiratory organs may hamper the view inside the thoracic segments. However, they should not be removed, as the main trachea at their bases lies close to the almost invisible peritracheal tissues (the equivalent to prothoracic glands, the source of the ecdysteroids), which it is important not to damage. If possible, the silky-white transversal trachea directly behind the brain should be cleaved to facilitate the opening of the segments. After removal of the head capsule, the opened thoracic segments are carefully flushed with Beadle's ringer using a Pasteur pipette to remove residues of hemolymph. If salivary glands have not been removed yet, they can now easily be accessed to complete the dissection. Tissues are then placed carefully in a sterile Petri dish containing Cannon's medium for an acclimation period of 30–60 min at room temperature. With some practice, a single dissection can be performed in about 5 min.

After acclimation single tissues are transferred to autoclaved 8 × 40-mm sample vials filled with 80 μl Cannon's medium, in which the appropriate amounts of test substance had previously been dissolved. We have always used pure ethanol as a carrier solvent and not exceeded a solvent concentration of 0.01%, but if one wishes, other combinations remain to be tested. The vials are closed and incubated at 30°C in a water bath for 24 h.

Preliminary experiments on *in vitro* ecdysteroidogenesis revealed that Cannon's medium was indeed best suited for chironomid tissue. Attempts to use Beadle's ringer solution, Grace's insect medium, and Medium 199, pure or supplemented with 25 m*M* HEPES, 1.73 m*M* $CaCl_2$, and 1% (w/v) Ficoll 400 (chemicals from Sigma-Aldrich, Taufkirchen, Germany)[21] resulted in low or undetectable ecdysteroid concentrations after 24 h of incubation. This might be due to the relatively high osmolality of the commercial media (around 400 mos*M*/kg). For Cannon's medium, a value of merely 260 mos*M*/kg was found (Table 13.1). This is supported by the findings of Scholz and Zerbst-Boroffka,[22] who measured an osmolality of not more than 180 mos*M*/kg in *C. plumosus* hemolymph.

4. *Ecdysteroid extraction*

At the end of the incubation period, 320 μl cold methanol are added to each vessel to give a final concentration of 80% (v/v). Vials are then briefly vortexed and the tissues are removed. Samples can now be stored at –20°C or directly processed.

For ecdysteroid extraction, the vials are vortexed and sonicated in an ultrasonic bath for 2 min and then centrifuged at room temperature at 1000 *g* for 5 min to remove any solid tissue residues. The supernatant is transferred to a reaction tube, and the solid residues extracted once with 80% aqueous methanol as above. The supernatants are

Table 13.1 Cannon's modified insect medium,[a] as described in Ringborg and Rydlander[16]

Na_2HPO_4	356	L-Arginine	140	Thiamin–HCl	0.004
$MgCl_2 \times 6\ H_2O$	93.5	DL-Lysine	250	Riboflavin	0.004
KCl	82	L-Histidine	500	Nicotinic acid	0.004
NaCl	420	L-Aspartate	70	Pantothenic acid	0.004
$Na_2SO_4 \times 10H_2O$	2000	L-Asparagic acid	70	Biotin	0.004
$CaCl_2$	42	L-Glutamate	120	Folic acid	0.004
		L-Glutamic acid	120	Inositol	0.004
		Glycine	130	Choline	0.004
Glucose	140	DL-Serine	220		
Fructose	80	L-Alanine	45		
Sucrose	80	L-Proline	70	Cholesterol	6
Trehalose	1000	L-Tyrosine	10	Penicillin	100 U/ml
		DL-Threonine	70	Streptomycin	0.1 mg/ml
		DL-Methionine	180	Phenol red[b]	20
Malic acid	134	L-Phenylalanine	30		
α-Ketoglutaric acid	74	DL-Valine	40	pH	7.2
Succinate	12	DL-Isoleucine	20	Osmolality[c]	260 mos*M*/kg
Fumarate	11	DL-Leucine	30		
		L-Tryptophan	20		
		L-Cystine	5		
		L-Cysteine–HCl	16		

[a] All concentrations as mg/300 ml water (cell culture grade).

[b] Omitted.

[c] Measured.

combined and evaporated to dryness. I have found that, with respect to speed, evaporation at 55°C under a stream of nitrogen is superior to the use of a vacuum centrifuge (Speed-Vac). The dry residues are resuspended in 300 μl RIA buffer (vortex, ultrasonic bath 2 min), and two aliquots of 100 μl are used for RIA analysis. (The volumes given in this chapter have in our hands yielded precise results; however, transferring a method from one laboratory to another often causes weird problems, so method optimization may be necessary in preparation for the RIA.)

5. *Radioimmunoassay*

A RIA for the determination of ecdysteroids was first introduced by Borst and O'Connor,[23] and since then it has been applied and methodically improved by many researchers, too numerous to be reviewed in detail in this volume. For a general treatise on theory and practice of immunoassays, the reader may refer to, e.g., References 24 or 25. In brief, the procedure applied here is as follows: an antiserum, raised against ecdysone or 20-hydroxyecdysone, is added to a solution containing known amount of radioactive-labeled ecdysone and an unknown amount of unlabeled ecdysone. The antiserum will now competitively bind both the labeled and the unlabeled ecdysone in a ratio that reflects the ratio of the ligands' concentration. After incubation the antibodies are precipitated by centrifugation in the presence of 50% ammonium sulfate, and the supernatant is removed. Radioactivity originating from labeled ecdysone bound to the pelleted antibodies can now be determined by liquid scintillation counting. Results are quantified by a standard curve, in which known amounts of "cold" ecdysone are analyzed instead of unknown samples.

Antiserum DBL-1 can be purchased from Trifolio GmbH, Lahnau, Germany. This antiserum was originally produced by the group lead by Prof. J. Koolman in Marburg, Germany, and displays high affinity to ecdysone and its active metabolite 20-hydroxyecdysone. A detailed characterization can be found in References 26 and 27. The freeze-dried material is resuspended in 0.5 ml bidistilled water with 0.05% sodium azide added as a preservative. From this stock a 1:1000 working solution in RIA buffer with 5% (v/v) normal sheep serum (ICN Biomedicals GmbH, Eschwege, Germany) is freshly prepared for every assay. (If, for any reasons, you wish to replace the sheep serum, be sure not to use any bovine protein source in an assay with DBL-1.)

Tritium-labeled ecdysone (α-[22, 23-^{3}H(N)]-ecdysone) was obtained from New England Nuclear, Zaventam, Belgium. Specific activity of the batch used in our laboratory was 53 Ci/mmol. A working solution with an activity of 0.1 μCi/ml was prepared in RIA buffer as follows: from the original package of 50 μCi in a volume of 500 μl ethanol, 25 μl were pipetted to a volume of 25 ml RIA buffer. A volume of 25 μl (= 0.0025 μCi) yielded an activity of approximately 2700 counts/min, when 1.5 ml of Wallac HiSafe2 scintillation liquid were added and samples counted in a Wallac 1409 scintillation counter (Wallac, Freiburg, Germany).

Pure ecdysone was obtained from Sigma-Aldrich, Taufkirchen, Germany. An amount of 1 mg ecdysone is weighed into a 10 ml volumetric flask and then dissolved in pure methanol to give a stock solution of 100 μg/ml and stored at -20°C. Because it was difficult to precisely match the desired nominal weight of 1 mg, it was necessary to approximate the weight as near as possible and then to read the exact weight from the balance's display and to use this for further calculations. Stock solutions in our laboratory were between 95 and 106 μg/ml. From this stock, a working solution of (about) 50 pg ecdysone per microliter RIA buffer is prepared by dissolving 50 μl in 100 ml RIA buffer.

Following the scheme given in Table 13.2, an ecdysone standard curve is pipetted into 2-ml reaction tubes (Eppendorf SafeLock). Non-specific binding (NSB) is the amount of labeled ecdysone that unspecifically (i.e., not by antibodies) binds to the protein matrix in the reaction mixture. This should generally not exceed 10% of the maximum bound radioactivity (B_0).[26] There are two ways of preparing NSB samples, either by omitting antiserum so that the mixture will only contain the sheep serum as protein matrix, or by adding a large amount (50× the highest concentration applied in the standard curve) of unlabeled ecdysone, which will readily prevent antibodies from binding significant amounts of labeled ecdysone.[26] Under the assay conditions described here, I have found no qualitative differences between these possibilities, but usually applied the latter one.

Volumes of 100 μl unknown sample, NSBs, or standards, 100 μl radioactive tracer solution (0.1 μCi/ml), and 100 μl of 1:1000 DBL-1 antibody dilution are combined in this order in 2-ml reaction tubes (Eppendorf SafeLock), and after mixing the samples are incubated overnight (>12 h) at 4°C. Incubation is stopped by adding 300 μl ice-cold saturated ammonium sulfate solution to each tube to yield an ammonium sulfate concentration of 50% (tests using polyethylene glycol 6000 instead of ammonium sulfate[24,25] were not successful in this assay). Samples should be mixed well soon after addition of the ammonium sulfate. For all the above steps it is extremely helpful to use a repeater pipette or a suitable multichannel pipette. Samples are then placed on ice for 30 min and centrifuged 12 min at 5000 *g* or more and 4°C to precipitate the protein-bound ligands, both labeled and unlabeled. In the centrifuge, make sure to orientate all tubes in such a way that the position of the almost invisible pellet is still known after removing

Table 13.2 Preparation of a standard curve for quantification of ecdysteroids by radioimmunoassay

Unlabeled ecdysone per well (pg)	Ratio ecdysone standard[a] to RIA buffer (μl), to give a sample volume of 100 μl	Suggested number of replicates
NSB	—[b]	6
Maximum amount of bound radioactivity (B_0)	0.0/100	6
12.5	0.25/99.75	2
25	0.5/99.5	2
50	1/99	2
100	2/98	2
200	4/96	2
400	8/92	2
800	16/84	2
1600	32/68	2
3200	64/36	2
5000	100/0.0	2

[a] 50 pg/μl.

[b] For the preparation of NSB samples, see text.

the tubes (e.g., hinge up). When the centrifuge's capacity is too small for precipitating all samples at one time, temperature control is crucial because centrifugation temperature may rise when several batches of tubes are processed. This could severely affect RIA results.

After centrifugation the supernatants are removed by vacuum aspiration using a Pasteur pipette or a smoothed flat-bottomed injection needle. While doing so, be careful not to injure the pellet. The collected supernatants must then be discarded as radioactive waste. To speed up the procedure, I have found that a washing step with 50% saturated ammonium sulfate solution, as prescribed by Gelman et al.,[28] can be omitted without loss of accuracy when the liquid phase is aspirated twice; after removal of the bulk of liquid allow each tube to stand for 2 min, during which the remainder of the viscous ammonium sulfate solution will collect at the tubes' bottom, then aspirate again.

To the pelleted antigen-antibody complexes 25 μl cell-culture-grade water and a suitable volume of scintillation liquid are added. We have used 1.5 ml of Wallac HiSafe2 scintillation liquid, Wallac, Freiburg, Germany. For other products it may be necessary to optimize the added volume. Though time consuming, each single tube should now be thoroughly vortexed. A white "flash" in the tube during vortexing indicates that water and scintillation fluid have formed a homogeneous suspension.

The tubes are placed in a scintillation counter and antibody-bound radioactivity is determined as counts per minute. Studies to determine the optimal counting time have revealed that 2 min are enough to achieve an adequate accuracy under the conditions given here. Mean NSB activity, which should not exceed 10% of the mean B_0 value, is subtracted from all samples and standards. Results are expressed as percent radioactivity bound (B/B_0), and concentration is plotted on a logarithmic scale. A sigmoidal regression model, e.g., of the form

$$f(x) = \frac{A_1 - A_2}{1 + \left(\frac{x}{x_0}\right)^p} + A_2$$

can be used for analysis, with A_1 and A_2 determining the initial and final y-values, respectively, x_0 determining the concentration where $B/B_0 = 50\%$, and p determining the curve slope at x_0. This regression can be generated with any suitable statistical package for personal computer. Generally, quantifiable results are in a range of B/B_0 between 20% and 70%,[26] which in this assay corresponded to 100–1000 pg ecdysone. Inter-assay stability in terms of x_0 values was found to display a coefficient of variation of 7.6%, and the correlation coefficient of each single standard curve was always better than 99%. Because of the cross-reactivity of the antiserum with 20-hydroxyecdysone and other free ecdysteroids,[27] results should be expressed as picograms ecdysone-equivalents synthesized per hour. For the above procedure, a recovery rate of $99.8 \pm 9.4\%$ ($n = 10$) was determined by extracting known amounts of ecdysone dissolved in Cannon's medium. Duplicate values that differ by more than 10% from their means should be rejected as a criterion of assay validity.

In our laboratory, the above method has proven to provide accurate results with acceptable efforts in experimental time and handling. However, if a higher throughput of samples is planned, the methods described in Reference 28 may provide useful hints for further optimization.

Results and discussion

In a combined experiment, both endpoints were investigated for the effects of the endocrine-disrupting biocide tributyltin oxide.[29] As described above, male and female ecdysteroid production was monitored in tissues exposed to 50, 500, and 5000 ng/l tributyltin oxide (as Sn). In a parallel experiment, single larvae in phases 5 or 6 were exposed to 10, 50, 200, and 1000 ng/l tributyltin oxide, and the development within 48 h was observed. The results are presented in Figures 13.4 and 13.5. These concentrations were in the sublethal range, as under the same test conditions the mean lethal concentration was estimated to be 25 μg/l.[29]

This experiment revealed that mean ecdysteroid-synthetic rates were different between males and females from the control groups (Figure 13.4), which is consistent with earlier findings by Laufer et al.[17] and with the results presented in Figure 13.2. Furthermore, in treated males, the ecdysteroidogenic activity was significantly increased in the 500-ng/l exposure, but in the other two test concentrations no significant differences from the control were found. In females a different effect occurred (Figure 13.4) because in all three groups ecdysteroid production was significantly lowered, by about 50% compared to the control group. In both sexes, therefore, the influence of tributyltin oxide was not dose dependent. A hypothesis focussing particularly on endocrine disruptors has recently raised the possibility that dose–response curves in the endocrine system might sometimes differ from traditional paradigms in toxicology, due to the possible involvement of various mechanisms of action and multiple target sites.[30]

The *in vivo* endpoint, larval development over a 48-h exposure period, also revealed a sex-specific influence of tributyltin oxide (Figure 13.5). In both males and females, the proportions of larvae in distinct developmental phases at the end of the experiment exhibited significantly different patterns, depending on TBTO dose. In general, exposed

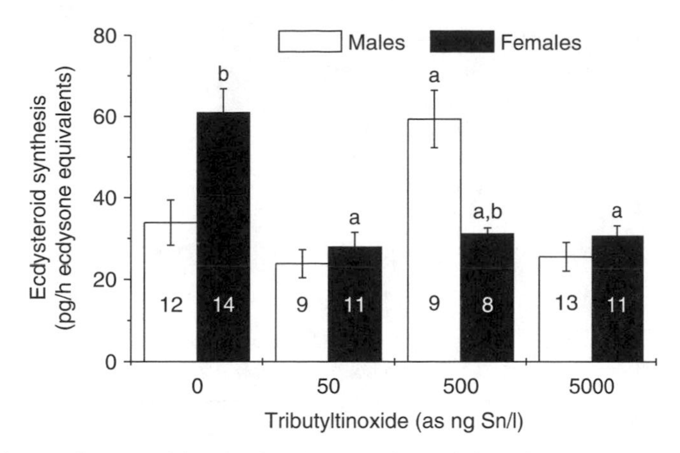

Figure 13.4 *In vitro* ecdysteroid biosynthesis in male and female *C. riparius* larvae exposed to tributyltin oxide. Numbers in columns show number of analyzed animals, error bars indicate ± S.E.; a, significantly different from the respective control; b, significantly different from the respective male group (Scheffé's test, $p < 0.05$). (From Hahn, T. and Schulz, R., *Environ. Toxicol. Chem.*, 21, 1052–1057, 2002. With permission.)

males tended to develop faster, i.e., within 48 h most of the larvae reached a more advanced developmental phase than in the control. TBTO-treated females, in contrast, clearly showed retarded development relative to the controls. Again a significant difference between the sexes was also observed in the untreated control. This probably reflects protandry, which is typical of chironomidae; that is, male emergence starts before that of females.[11,18] This might enhance mating success in species with short-lived adults.[18] In tributyltin oxide-exposed animals, this natural asynchrony in larval development increased, but whether this is of ecological relevance or not is difficult to decide from these results alone.

However, the results from the *in vitro* and *in vivo* experiments hint at a connection between larval development and molting-hormone production, as in females lowered ecdysteroid synthesis coincided with slower development. In males a partly opposite pattern was present. Exposed male larvae developed faster, which was consistent with the elevated ecdysteroid synthesis *in vitro* found in the 500-ng/l group but not in the groups exposed to 50 and 5,000 ng/l. It cannot be judged from these results whether the effects are connected, either directly or indirectly (or not at all). Several modes of action, either alone or in combination, may be responsible for these effects,[29] and for a clearer elucidation of the mechanism(s), more detailed experiments have to be carried out.

However, rather than to study target–site interactions, it is the aim of this combined *in vitro* and *in vivo* approach to provide a screening tool for the detection of a chemical's endocrine modulating potential by combining two levels of biological organization. In this context, the preparation and incubation of almost complete thoracic segments, containing the intact neural complex of cerebral and subesophageal ganglia, corpora allata, corpora cardiaca, and prothoracic glands, is advantageous for this method. Thus, multiple substance-induced modulations in a variety of neuroendocrine organs and functions within the complex hierarchy of the hormone system can be visualized in only two endpoints, ecdysteroid synthesis and imaginal disc development.

Depending on the size of the midge culture and the number of larvae in a suitable 4th stage larval phase available simultaneously, the *in vivo* tests on imaginal disc

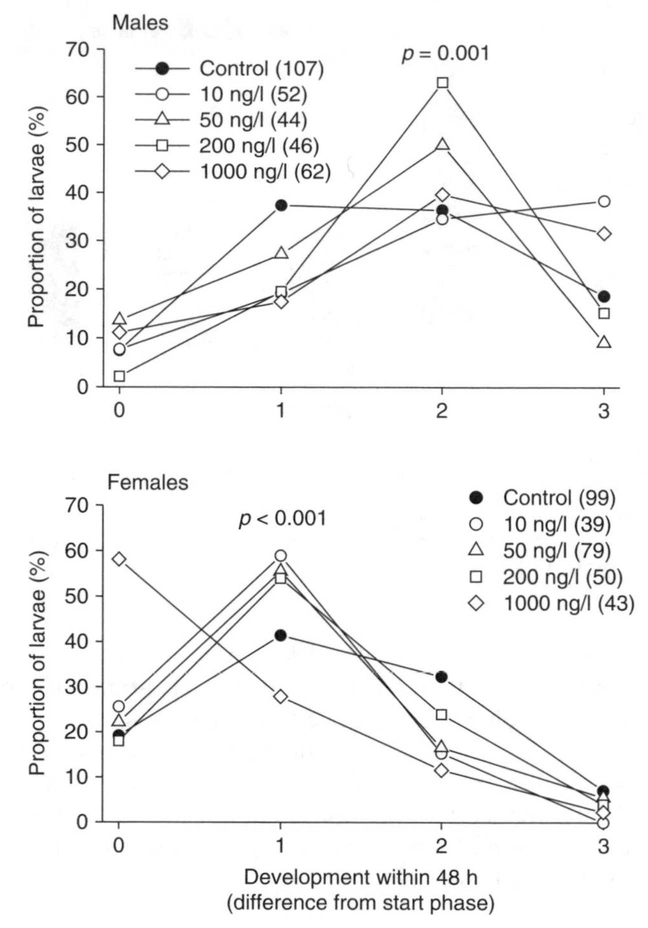

Figure 13.5 Development of fourth larval instar *C. riparius* within a 48-h period of exposure to different concentrations of tributyltin oxide. Numbers in parentheses show number of analyzed animals. *p*-Values indicate statistical significance when distributions of larvae in developmental phases (given as the numerical difference of phase at start and end of the experiment) were different between the experimental groups, due to tributyltin oxide treatment (contingency table tests). (From Hahn, T. and Schulz, R., *Environ. Toxicol. Chem.*, 21, 1052–1057, 2002. With permission.)

development can be performed in relatively short time. Also a sufficient number of *in vitro* ecdysteroid synthesis experiments can be performed in a reasonable time period, although sample extraction and radioimmunoassay are time-consuming steps, and successful dissections require prior training of a researcher's surgical skills.

A disadvantage is certainly the unavoidable remaining subjectivity connected with the microscopic identification of the larvae's developmental stage. Minimizing this subjectivity to ensure reliable and consistent work through several experiments requires a certain degree of practice, which can be acquired only by previous training. If larvae are to be staged by more than one person, all persons involved in this work should harmonize with their definitions of diagnosis of imaginal disc differentiation. This is crucial because the drawings of Wülker and Götz[11] (Figure 13.1) only provide snapshots of a continuous process of development, and the persons involved should agree in staging intermediate appearances. Therefore, repeatability controls (e.g., staging of the same larvae by different persons) should be carried out at regular intervals.

The combined methods described earlier are focussed on interaction of xenobiotics with ecdysteroid metabolism. However, the *in vitro* part does not include ecdysone receptor interaction of the test substance, which should not be ignored as a further mode of action. For this endpoint, another test for hormonal activity of xenobiotics in insects has recently been applied by Dinan et al.,[31] who were able to detect agonistic and antagonistic effects of a wide array of tested chemicals using an ecdysteroid-responsive cell line in a microplate assay.

Furthermore, the focus of experiments dealing with insect endocrine disruption could be extended to another class of molting hormones, the juvenile hormones. Using the experimental processes described in this chapter as a model, it should be possible to establish a similar method for measurements on *in vitro* juvenile hormone production, e.g., using a radiochemical assay[32] or an antibody-based method.[33]

References

1. Spindler, K.D., *Vergleichende Endokrinologie,* Georg Thieme Verlag, Stuttgart, New York, 1997, 215 pp.
2. Ankley, G., Mihaich, E., Stahl, R., Tillitt, D., Colborn, T., McMaster, S., Miller, R., Bantle, J., Campbell, P., Denslow, N., Dickerson, R., Folmar, L., Fry, M., Giesy, J., Gray, L.E., Guiney, P., Hutchinson, T., Kennedy, S., Kramer, V., Leblanc, G., Mayes, M., Nimrod, A., Patino, R., Peterson, R., Purdy, R., Ringer, R., Thomas, P., Touart, L., Van Der Krakk, G. and Zacharewski, T., Overview of a workshop on screening methods for detecting potential (anti-) estrogenic/androgenic chemicals in wildlife (Kansas City, MO, USA, March 1997), *Environ. Toxicol. Chem.*, 17, 68–87, 1998.
3. Crews, D., Wellingham, E. and Skipper, J.K., Endocrine disruptors: present issues, future directions, *Quart. Rev. Biol.*, 75, 243–260, 2000.
4. McLachlan, J.A., Environmental signaling: what embryos and evolution tells us about endocrine disrupting chemicals, *Endocr. Rev.*, 22, 319–341, 2001.
5. DeFur, P.L., Crane, M., Ingersoll, C. and Tattersfield, L., Eds., *Endocrine Disruption in Invertebrates: Endocrinology, Testing, and Assessment*, Society of Environmental Toxicology and Chemistry, Pensacola, FL, 1999, 303 pp.
6. Lafont, R., The endocrinology of invertebrates, *Ecotoxicology*, 9, 41–57, 2000.
7. Hutchinson, T.A., Reproductive and developmental effects of endocrine disrupters in invertebrates: *in vitro* and *in vivo* approaches, *Toxicol. Lett.*, 131, 75–81, 2002.
8. Oehlmann, J. and Schulte-Oehlmann, U., Endocrine disruption in invertebrates, *Pure Appl. Chem.*, 75, 2207–2218, 2003.
9. ASTM, Standard test methods for measuring the toxicity of sediment-associated contaminants with fresh-water invertebrates, E1706-95b, in *Annual Book of ASTM Standards*, vol. 11.05, Philadelphia, PA, 1999.
10. OECD Draft Document, *OECD Guidelines for the Testing of Chemicals*, Proposal for a new guideline 219, Sediment-water chironomid toxicity test using spiked water, 2001.
11. Wülker, W. and Götz, P., Die Verwendung der Imaginalscheiben zur Bestimmung des Entwicklungszustandes von *Chironomus*-Larven (Diptera), *Z. Morphol. Tiere*, 62, 363–388, 1968.
12. King, D.S., Bollenbacher, W.E., Borst, D.W., Vedeckis, W.V., O'Connor, J.D., Ittycheriah, P.I. and Gilbert, L.I., The secretion of α-ecdysone by the prothoracic glands of *Manduca sexta in vitro*, *Proc. Natl. Acad. Sci. USA*, 71, 793–796, 1974.
13. Gilbert, L.I., Rybczynski, R., Song, Q., Mizoguchi, A., Morreale, R., Smith, W.A., Matubayashi, H., Shionoya, M., Nagata, S. and Kataoka, H., Dynamic regulation of prothoracic gland

ecdysteroidogenesis: *Manduca sexta* recombinant prothoracicotropic hormone and brain extracts have identical effects, *Insect Biochem. Mol. Biol.*, 30, 1079–1089, 2000.
14. Oberlander, H. and Silhacek, D.I., Mode of action of insect growth regulators in lepidopteran tissue culture, *Pestic. Sci.*, 54, 300–322, 1998.
15. Laufer, H. and Wilson, M.A., Hormonal control of gene activity as revealed by puffing of salivary gland chromosomes in dipteran larvae, in *A Student's Guide to Experiments in General and Comparative Endocrinology*, Peter, R.F. and Gorbman, A., Eds., Prentice-Hall, New Jersey, 1970, pp. 185–200 (210 pp.).
16. Ringborg, U. and Rydlander, L., Nucleolar-derived ribonucleic acid in chromosomes, nuclear sap, and cytoplasm of *Chironomus tentans* salivary gland cells, *J. Cell. Biol.*, 51, 355–368, 1971.
17. Laufer, H., Vafopoulou-Mandalos, X. and Deak, P., Ecdysteroid titres in *Chironomus* and their relation to haemoglobins and vitellogenins, *Insect Biochem.*, 16, 281–285, 1986.
18. Armitage P.D., Cranston P.S. and Pinder L.C.V., Eds., *The Chironomidae. Biology and Ecology of Non-biting Midges*, Chapman & Hall, London, 1995, 570 pp.
19. Possompès, B., Recherches expérimentales sur le déterminisme de la metamorphose de *Calliphora erythrocephala* Meig, *Arch. Zool. Exp. Gen.*, 89, 201–364, 1953.
20. Credland, P., Scales, D. and Philipps, A., A description of the retrocerebral complex in the adult midge, *Chironomus riparius* (Diptera: Chironomidae), and a comparison with that of the larva, *Zoomorphologie*, 92, 65–75, 1979.
21. Oeh, U., Lorenz, M.W. and Hoffmann, K.H., Ecdysteroid release by the prothoracic gland of *Gryllus bimaculatus* (Ensifera: Gryllidae) during larval-adult development, *J. Insect Physiol.*, 44, 941–946, 1998.
22. Scholz, F. and Zerbst-Boroffka, I., Environmental hypoxia affects ionic and osmotic regulation in freshwater-midge larvae, *J. Insect Physiol.*, 44, 427–436, 1998.
23. Borst, D.W. and O'Connor, J.D., Arthropod molting hormone: radioimmunoassay, *Science*, 178, 418–419, 1972.
24. Law, B., *Immunoassay. A Practical Guide*, Taylor & Francis, London, 1996, 222 pp.
25. Gosling, J.P., Ed., *Immunoassay: A Practical Approach*, Oxford University Press, Oxford, 2000, 304 pp.
26. Reum, L. and Koolman, J., Radioimmunoassay of ecdysteroids, in *Ecdysone: From Chemistry to Mode of Action*, Koolman, J., Ed., Thieme Medical Publishers, Stuttgart, New York, 1989, pp. 131–143 (481 pp.).
27. Reum, L., Haustein, D. and Koolman, J., Immunoabsorbtions as a means for the purification of low molecular weight compounds: isolation of ecdysteroids from insects, *Z. Naturforsch.*, 36C, 790–797, 1981.
28. Gelman, D.B., Khalidi, A A. and Loeb, M.J., Improved techniques for the rapid radioimmunoassay of ecdysteroids and other metabolites, *Invert. Reprod. Develop.*, 32, 127–129, 1996.
29. Hahn, T. and Schulz, R., Ecdysteroid synthesis and imaginal disc development in the midge *Chironomus riparius* as biomarkers for endocrine effects of tributyltin, *Environ. Toxicol. Chem.*, 21, 1052–1057, 2002.
30. Dickerson, R.L., Brouwer, A., Gray, L.E., Grothe, D.R., Peterson, R.E., Sheehan, D.M., Sills-McMurry, C. and Wiedow, M.A., Dose–response relationships, in *Principles and Processes for Evaluating Endocrine Disruption in Wildlife*, Kendall, R.J., Dickerson, R.L., Giesy, J.P. and Suk, W.P., Eds., Society of Environmental Toxicology and Chemistry, Pensacola, FL, 1998, pp. 69–96 (303 pp.).
31. Dinan, L., Bourne, P., Whiting, P., Dhadialla, T. S. and Hutchinson, T.H., Screening of environmental contaminants for ecdysteroid agonist and antagonist activity using the *Drosophila melanogaster* Bii cell *in vitro* assay, *Environ. Toxicol. Chem.*, 20, 2038–2046, 2001.
32. Yagi, K.J. and Tobe, S.S., The radiochemical assay for juvenile hormone biosynthesis in insects: problems and solutions, *J. Insect Physiol.*, 47, 1227–1234, 2001.
33. Goodman, W.G., Coy, D C., Baker, F C., Xu, L. and Toong, Y.C., Development and application of a radioimmunoassay for the juvenile hormones, *Insect. Biochem.*, 20, 357–364, 1990.

chapter fourteen

The electro-olfactogram: An in vivo *measure of peripheral olfactory function and sublethal neurotoxicity in fish*

David H. Baldwin and Nathaniel L. Scholz
National Marine Fisheries Service

Contents

1-56670-664-5/05/$0.00+$1.50
© 2005 by CRC Press

Disclaimer:

Reference to a company or product does not imply endorsement by the US Department of Commerce to the exclusion of others that may be suitable.

Introduction

There is an emerging realization that environmental contaminants can have important sublethal impacts on fish and other aquatic organisms.[1] While traditional measures of acute mortality (i.e., lethal concentrations, or LC_{50}s) may be useful for predicting responses to accidental chemical spills or similar scenarios that release unusually high concentrations of a contaminant into fish habitats, most forms of pollution in natural systems do not result in acute fish kills. This is particularly true for stormwater runoff and similar kinds of non-point source pollution. Consequently, a current challenge in the field of aquatic toxicology is to develop and refine sublethal endpoints for fish. Such endpoints are most useful for the purposes of natural resource management if they can ultimately be related to survival, reproduction, or other life history parameters that are important for the dynamics of natural populations.

It is important that experimental measures of sublethal toxicity be sufficiently sensitive to detect changes in physiological function in response to low level, environmentally realistic contaminant exposures. The approach should also be quantitative, technically sound, and reproducible. Finally, measures of physiological function should be matched, to the extent possible, to the known or suspected mechanism of action for the contaminant (e.g., reproductive measures for endocrine disruptors or neurobehavioral measures for neurotoxicants).

Olfaction is an important sensory system that underlies several critical behaviors in fish. For example, olfaction plays a key role in predator detection and avoidance,[2,3] kin recognition,[4] reproduction,[5] and navigation.[6] Consequently, chemicals that damage the olfactory system have the potential to reduce the survival and/or reproductive success of individual animals. The peripheral portion of the olfactory system in fish (reviewed in Reference 7) consists of a pair of olfactory rosettes, each located within a chamber on either side of the rostrum of the fish. A single rosette from a juvenile coho salmon (*Oncorhynchus kisutch*) is shown in Figure 14.1A. The surface of the rosette is covered to a large extent by a sensory epithelium (Figure 14.1B) containing, among other cell types, olfactory receptor neurons (ORNs). These cells are in direct contact with the surrounding environment, and thus they are vulnerable to the neurotoxic effects of dissolved contaminants. The ORNs are responsible for detecting odorant molecules in the surrounding aquatic environment. The odorants bind to membrane-associated receptor proteins in the apical cilia or microvilli of ORNs. Odor binding results in a second messenger cascade that ultimately leads to the propagation of action potentials in the individual ORN axons projecting to the olfactory regions of the brain. The overall process of olfactory signal transduction has been described in considerable detail (reviewed in Reference 8). Some of the principal components are highlighted in Figure 14.1C. Contaminants that interfere with any part of this process could potentially impair peripheral olfactory function and, by extension, olfactory-mediated behaviors crucial for survival or reproduction.

The electro-olfactogram (EOG) is an established technique for measuring peripheral olfactory function in fish[9] and other vertebrates (reviewed in Reference 10). The odor-evoked EOG is an extracellular field potential that consists of a large, negative voltage

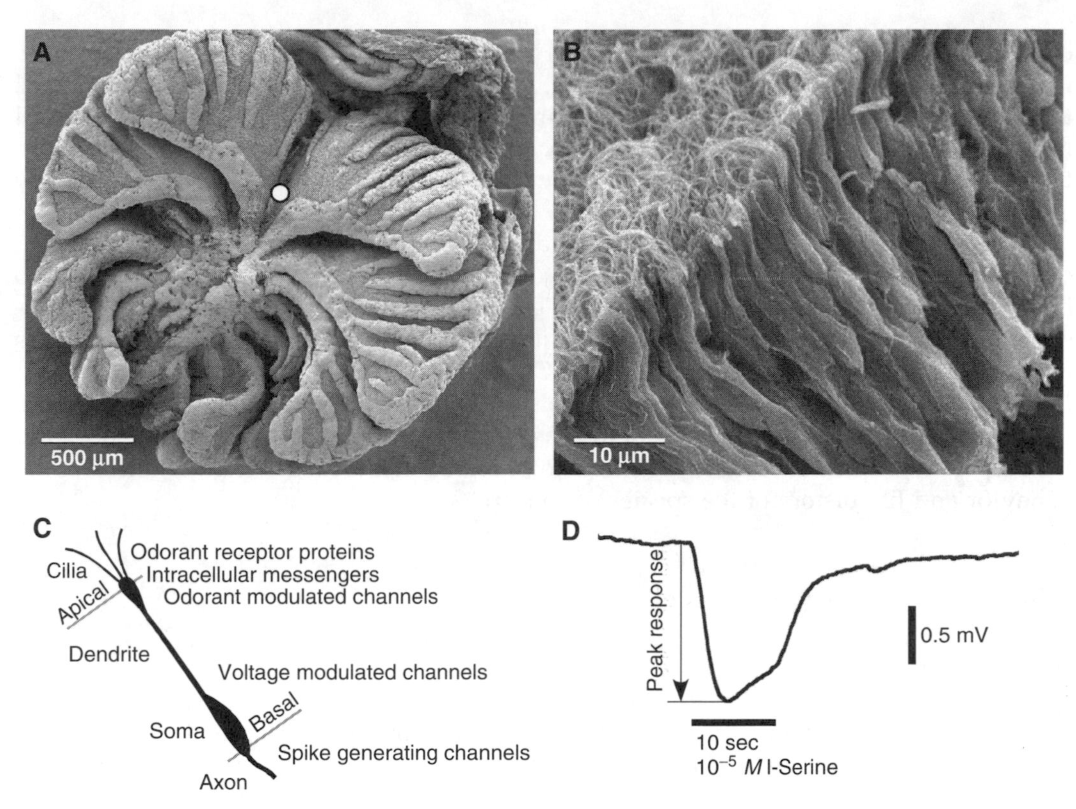

Figure 14.1 Features of the coho salmon peripheral olfactory system. (A) Scanning electron micrograph showing an entire olfactory rosette from a juvenile coho salmon. Each of the lamellae (major folds) is covered in an epithelium that includes regions of sensory neurons. The open circle denotes the location and approximate size of the tip of a standard recording microelectrode. (B) Scanning electron micrograph showing a cross-section of the sensory epithelium along a single lamella. In the upper left is the apical surface containing the cilia and microvilli of the ORNs. The dendrites connect the apical (or ciliated) ends of the ORNs with the somata, which are evident in the center of the epithelium. The axons of the ORNs emerge from the basal surface at the lower right and eventually produce the olfactory nerve (not seen). (C) Schematic of a single ORN showing the major anatomical compartments of the cell (labeled to the left). Major biochemical components of the olfactory transduction process are labeled to the right. Gray lines indicate the apical and basal surfaces of the epithelium. (D) Typical odor-evoked EOG obtained from a juvenile coho salmon. The horizontal bar indicates a 10-s switch from clean background water to water containing 10^{-5} *M* L-serine. The L-serine evoked EOG consists of a negative deflection in the transepithelial voltage potential. This includes a transient peak followed by a decay towards a tonic plateau that persists for the duration of the odor pulse. Scanning electron micrographs are courtesy of Carla Stehr, Northwest Fisheries Science Center.

transient measured with an electrode positioned near the surface of the sensory epithelium (dot in Figure 14.1A). The amplitude of the EOG reflects the summed electrical response of many ORNs as they bind to dissolved odorants.[7] An example of an EOG is shown in Figure 14.1D. The EOG is a robust and direct measure of ORN function in the intact animal. Because of this, the technique has been widely used to investigate the fundamental mechanisms of olfactory signal transduction (reviewed in Reference 10). However, while the EOG was recognized early as a potential tool for monitoring the effects of contaminant exposure on olfactory function in fish,[9,11] it has been used only infrequently over the past two decades (e.g., References 12–18). This may be due, in part, to a lack of well-described, consistent experimental procedures. Also, neurophysiological recording methods are somewhat complex from a technical standpoint. Advances in technology over the last two decades, however, have reduced much of this complexity and made EOG recordings more tractable.

This chapter provides a detailed and updated description of standard EOG recording methods for fish. Experiments recently performed in our laboratory[17,18] will be used to highlight the application of the technique. This research investigated the effects of dissolved copper on the olfactory function of juvenile coho salmon (*O. kisutch*). Copper is used here only as an example, and the EOG approach should apply equally well to investigations involving other neurotoxicants and other fish species. For toxicology, the overall aim of the EOG recording method is to detect contaminant-induced changes in the responses of primary sensory neurons to odorants that are significant in terms of the behavior and life history of the species of concern.

Materials required

Equipment

- Fiber-optic Illuminator
- Stereoscopic zoom microscope (SMZ645, Nikon Instruments, Melville, NY, USA)
- Boom stand for microscope (SMS6B, Diagnostic Instruments, Sterling Heights, MI, USA)
- Vibration isolation table with magnetic top (Newport, Irvine, CA, USA)
- Air compressor or air cylinder and regulator to pressurize the vibration isolation table

This equipment collectively provides a magnified view of a stable, illuminated surface upon which to position the animal and the micromanipulators.

- Aquarium chiller (PowerCooler 1/5 HP Chiller, CustomSeaLife Inc., San Marcos, CA, USA)
- Aquarium pump (Mag-Drive 7, EG Danner Mfg, Central Islip, NY, USA)
- Water reservoir (a large plastic garbage will work)
- Fish holder (see the following)

This equipment provides the life-support system that maintains the fish during the experiments. The system used in our laboratory consists of a fish holder made from Plexiglas sheets forming a watertight container with a "V" cross-section to hold the fish upright. Small pieces of foam alongside the fish help support it within the holder. The inflow at one end of the holder includes a mouthpiece made from a Y-adaptor fitting normally used with Tygon tubing. The aquarium pump recirculates water between the aquarium chiller and the water reservoir to maintain a supply of chilled water. Fish requiring a temperature above ambient will need a heater instead. A PVC T-adaptor and valve divert some of the output of the aquarium pump to the fish holder for the continuous perfusion of the gills with chilled, oxygenated water containing anesthetic via the mouthpiece. For experiments that expose the fish to contaminants while in the holder (via a separate perfusion of the olfactory chamber described in the following), the outflow of the holder is collected in a separate container and disposed of in a manner appropriate for the contaminant. This requires periodically refilling the water reservoir. In situations where the fish is exposed prior to being placed in the holder, and therefore no exposure will occur in the holder, the outflow can be routed back to the water reservoir for recirculation.

- Perfusion system (ValveLink8 pinch valve, Automate Scientific, San Francisco, CA, USA)

The perfusion system consists of a set of eight pinch valves connected to a micromanifold. A system is also available with 16 valves. Solenoid valves can be used instead, but pinch valves allow the silicone tubing (the part in contact with the solution) to be replaced if desired. This is routinely done between experiments in our laboratory to reduce contamination from previous experiments. Switching between the valves is accomplished manually using controls on the perfusion system or by external signals (e.g., from a computer) connected to the digital inputs of the valve controller.

- Thermoelectric chiller (Z-max prototype kit, Tellurex, Traverse City, MI, USA)
- Aluminum plate (3 × 3 in., 1/2-in. thick)
- Thermistor
- Ohmmeter

This equipment is used to chill the solution as it flows from the micromanifold to the olfactory rosette. For fish requiring a temperature greater than ambient, a thermoelectric device can be configured to heat instead of cool. The compact nature of this system allows it to be placed on the vibration isolation table next to the fish holder, thus reducing the length of the delivery tube and minimizing any warming that may occur before the solution reaches the rosette. Also, with this configuration, the individual solutions upstream of the micromanifold do not have to be chilled. The system constructed in our laboratory includes an aluminum plate with a serpentine groove (several back-and-forth channels connected by hairpin turns) milled into one side. A length of tubing lies within the groove with short lengths of the tubing emerging from two holes in the side of the plate. The entire groove is packed with heat sink grease to maximize thermal conduction from the plate to the tubing. The plate is securely clamped, with a good layer of heat sink grease, to the face of the thermoelectric chiller. The temperature of the solution flowing out of the aluminum plate is measured using an ohmmeter connected to a thermistor sealed into a T-adapter with aquarium sealant. This positions the thermistor in the solution as it flows through the tubing. The thermistor reading (resistance) is used to set the controller of the thermoelectric chiller to maintain the desired solution temperature. If the controller lacks feedback input from the thermistor (as does the unit in our laboratory), the thermoelectric chiller needs to be turned off when there is no flow through the perfusion system. Otherwise, the thermoelectric device will continue chilling and may ultimately freeze the solution within the aluminum plate.

- Electrode puller (P-30, Sutter Instrument, Novato, CA, USA)
- Microelectrode holders (MEH1S, WPI, Sarasota, FL, USA) mounted to Plexiglas rods
- Micromanipulators (MM-3, Narashige International, East Meadow, NY, USA)

- Magnetic bases for micromanipulators (M9, WPI, Sarasota, FL, USA)
- AC/DC differential amplifier (Model 3000, A-M Systems, Carlsborg, WA, USA)
- Oscilloscope (TDS224, Tektronics, Wilsonville, OR)

This is standard electrophysiological equipment for making and using microelectrodes. The large diameter (low impedance) of the microelectrode tips that are used for recording EOGs (see the section "Procedures") obviates the need for a headstage pre-amplifier sometimes used in electrophysiology.

- Computer (desktop with a PCI slot, PowerMac G4, Apple Computer, Cupertino, CA, USA)
- Data acquisition hardware (6035E, National Instruments, Austin, TX, USA)
- Data acquisition software (LabVIEW, National Instruments, Austin, TX, USA)
- Data analysis and graphing software (KaleidaGraph, Synergy Software, Reading, PA, USA)

The computerized data acquisition system serves to both control the perfusion system and to record the odor-evoked EOG. This requires hardware with at least one analog input and eight digital outputs and software that can simultaneously collect analog input and change digital outputs. This can be accomplished with a variety of hardware/software combinations that are commercially available. The system used in our laboratory (listed earlier) includes a custom LabVIEW program (available upon request from the authors) that allows for data collection to be largely automated (see the section "Results and discussion"). Data analysis can also be performed using a variety of software programs that are commercially available. Our laboratory uses a second custom LabVIEW program (also available upon request from the authors) to measure the waveform and amplitude of odor-evoked EOGs. Conventional commercial software is used for graphics, statistics, and other standard analyses.

Reagents

- Agar (Sigma, St. Louis, MO, USA)
- MS-222 stock solution

5 g/l MS-222 (tricaine methane sulfonate, Sigma, St. Louis, MO, USA) and 1 g/l sodium carbonate (as a pH buffer) in distilled water

- Physiological saline (for freshwater-phase coho salmon)

140 m*M* NaCl, 10 m*M* KCl, 1.8 m*M* $CaCl_2$, 2.0 m*M* $MgCl_2$, and 5 m*M* HEPES in distilled water (pH to 7.4 with NaOH)

- Gallamine triethiodide solution

300 mg/l gallamine triethiodide (Sigma, St. Louis, MO, USA) in physiological saline

- Microelectrode saline

140 m*M* NaCl, 10 m*M* KCl, 1.8 m*M* $CaCl_2$, and 2.0 m*M* $MgCl_2$ in distilled water

- Microelectrode holder filling solution

3 *M* KCl in distilled water

- Odorant stock solutions

Prepare as appropriate for the desired odorants (including a carrier if one is needed). For the experiments shown here, the solutions consisted of taurocholic acid (TCA, 10^{-2} *M*), L-serine (10^{-2} *M*), and an amino acid mixture (L-arginine, L-aspartic acid, L-leucine, and L-serine each at 10^{-2} *M*) all in distilled water.

- Contaminant stock solution

Prepare as appropriate for the contaminant being studied. For the data shown here, the solution consisted of 3.4 g/l copper chloride (Sigma, St. Louis, MO, USA) in distilled water (pH to 3.0 with HCl).

Supplies

- Microelectrode glass (borosilicate 1 mm OD/0.75 mm ID w/filament, TW100F-3, WPI, Sarasota, FL, USA)
- Metal ground electrode (a hypodermic needle connected to ground will work)
- Syringes (having a variety of sizes is useful)
- Hypodermic needles (having a variety of sizes is useful)
- Microfil syringe tip (MF34G-5, WPI, Sarasota, FL, USA) for filling microelectrode holders
- Dissecting tools, including scalpels, small scissors, and small forceps
- PVC piping and valves
- Tygon tubing and fittings
- Silicone tubing
- Electrical wiring and connectors

Procedures

Fish

The data presented in this chapter are from experiments on juvenile coho salmon reared at the Northwest Fisheries Science Center in 2400-l fiberglass tanks supplied by a filtered, recirculating water system (11–13°C, pH 7.1, buffered to 120 ppm total hardness as $CaCO_3$). Fish were fed commercial salmon pellets (Bio-Oregon, Warrenton, OR, USA). Whatever the method of animal husbandry and handling it is important to avoid introducing any chemicals into the tanks housing the fish that might impair the olfactory system. This might include, for example, contaminants in the source water for the hatchery or metals originating from certain kinds of plumbing. Also, fish rearing facilities occasionally use algicides or fungicides to control the growth of algae in the aquaculture system or fungi on the fish. While these chemicals may not have any discernible effects on the overt behavior of fish in the rearing system, they may nevertheless impair the olfactory system. This may lead to uncontrolled variation in EOG recordings among all fish in a given study. Avoiding this requires knowing, and perhaps modifying, the protocols used for animal husbandry.

While the procedures in the following are based on specific studies by our laboratory and others,[9,12,17] they can be adapted to a range of fish species, sizes, and ages. Different sized fish might require changes to the dose and duration of anesthesia, physical modifications of the fish holder and tubing, and changes to the temperature and/or flow rate over the gill and rosette. For the experiments presented here, fish ($n = 77$) were age 1 + with an average size ($\pm$ 1 SD) of 22.7 $\pm$ 0.4 cm and 143 $\pm$ 8 g.

Solutions

Prepare concentrated stock solutions (odorants and contaminant) each week in the appropriate carrier (e.g., distilled water for amino acids) and store appropriately. Prepare odorant stimulus solutions daily by diluting the stock into the background water (filtered, dechlorinated municipal water for the experiments shown in this chapter). Prepare contaminant exposure solutions as needed based on the experimental protocol. For the examples presented in this chapter, copper solutions were made daily by diluting copper chloride stock into background water. The selection of olfactory stimuli (odorants) will depend on the fish species of interest. Ideally, they should be naturally occurring olfactory cues with relevance to olfactory-mediated behaviors. In the case of salmon, the rationale for selecting four amino acids and TCA as odorants will be discussed.

Exposure prior to electrophysiological recording

Odor-evoked EOG recordings are typically stable for several hours.[17] Exposures lasting longer than this should be conducted separately, before preparing the fish for electrophysiological recording. Since EOG responses are recorded from fish after the exposure, the exposure protocol can vary (e.g., flow-through versus static renewal) without requiring changes to the EOG procedure itself.

Microelectrode preparation

Microelectrodes should be prepared each day. Adjust the electrode puller to make glass microelectrodes with gentle tapers to ~100-μm diameter tips. An easy approach is to pull electrodes with relatively fine tips and manually break the tips under a dissecting microscope to the desired diameter. Gently heat a 2% agar-saline solution (0.1 g agar in 5 ml of electrode saline) until the agar dissolves completely. Remove from the heat and load the solution into a small syringe connected to narrow diameter Tygon tubing that snugly fits over the microelectrode glass. The agar solution will begin to cool, so this needs to be done quickly. Using the syringe/tubing, carefully fill the microelectrodes from the rear (non-tip) end until solution comes out of the tip. Make sure that there are no gaps or air bubbles in the electrode. Gently remove excess agar from the tip of the microelectrode. Fill several microelectrodes in the approximately 5 min before the solution cools. Microelectrodes can be stored for use later in the day.

Microelectrodes are fitted to holders and then connected to the electronics prior to use in recording. Fill a microelectrode holder with 3 *M* KCl using a microfil needle attached to a syringe. Gently insert a microelectrode halfway into a holder and tighten the screw to seal the o-ring. Make sure that there are no air bubbles trapped in the holder. Mount a second microelectrode in the same way. Attach each electrode holder to a micromanipulator and temporarily secure the micromanipulators via their magnetic bases on the vibration isolation table away from the fish holder. If desired, the electrical properties of the microelectrodes can be easily tested at this point by inserting them both into a beaker of physiological saline. The method for microelectrode testing is not described here, since it will be specific to the amplifier and should be described in the associated manual.

Fish preparation

A schematic diagram and photograph of a coho salmon with the perfusion and electrophysiological recording systems are shown in Figure 14.2. Anesthetize a fish with MS-222 (10 ml of MS-222 stock per liter of water, 50 mg/l final concentration) for approximately 20 min. The fish should be unresponsive yet still show some opercular movement. Inject the paralytic gallamine triethiodide (1 ml of paralytic solution per kg body mass, 0.3 mg/kg final concentration) into several locations in the muscles along the side of the fish. The anesthesia and paralytic are necessary to prevent movement during recordings. We tested several other procedures for anesthetizing fish, including chilling, and observed no differences in terms of the odor-evoked EOGs. Thus, we use MS-222 as a standard anesthetic. Also, any effect of the anesthetic/paralytic will occur in both treatment and control fish.

Place the fish into a Plexiglas holder on the vibration isolation table with a mouthpiece (an appropriately sized Y-adaptor for the Tygon tubing works) inserted into the

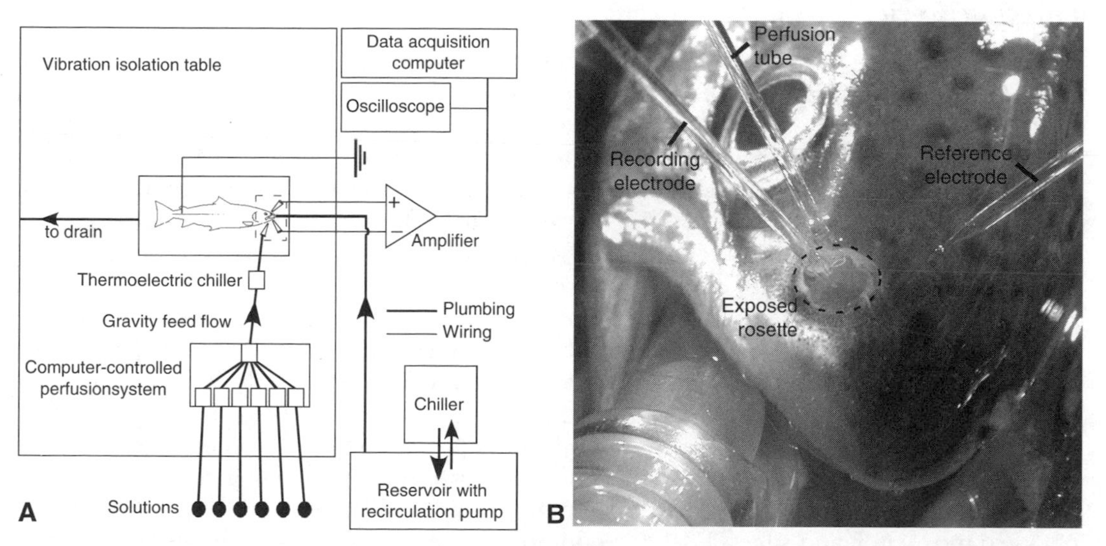

Figure 14.2 Electophysiological recording system used to measure odor-evoked EOGs from the sensory epithelium of juvenile coho salmon. (A) Schematic diagram showing the major components of the apparatus. The dashed box denotes the area shown in more detail in (B). (B) Photograph showing the rostrum of a coho salmon, the mouthpiece providing water to the gills, and the positioning of the perfusion tube and glass microelectrodes. See the sections "Materials required" and "Procedures" for more information. Figure adapted from Reference 17.

mouth, just anterior to the gill arches. Deliver chilled (12°C), oxygenated water containing MS-222 (50 mg/l) through the mouthpiece at a flow of 120 ml/min. To improve water flow across the gills, clip a piece of each operculum and hold the remaining portion open with a piece of foam. Place a wet paper towel over the body of the fish (and in contact with the water) to keep it moist. For larger fish, add ice to keep animals cool.

Remove the skin overlying the olfactory chamber to expose the olfactory rosette. Care must be taken to minimize bleeding, particularly from the blood vessels near the rosette. Excessive blood loss will deprive the rosette of circulation and render the olfactory epithelium essentially unresponsive to odorant stimulation. In coho salmon, this is readily apparent as a change in the color of the rosette from pink to pale white.

Electrophysiological recording

With the aid of a stereomicroscope, position the recording microelectrode along the midline of the rosette at the base of the large, posterior-most lamella (see Figure 14.1A) using a micromanipulator. Place the reference microelectrode in the skin of the rostrum, dorsal to the rosette (see Figure 14.2B). The two microelectrodes and the fish need to remain stable to avoid movement artifacts in the recording. Insert a separate grounded metal electrode (a hypodermic needle will work) into the muscle near the tail. Using a differential amplifier, amplify (500×) and filter (100 Hz low pass) the differential signal from the two microelectrodes (voltage) and display the signal on an oscilloscope. The output of the amplifier should also be connected to the computerized data acquisition system for recording.

Direct delivery of odorants and contaminants to the olfactory epithelium

During recordings, deliver a continuous flow of chilled water (12°C) to the exposed rosette at a rate of 5 ml/min through a perfusion tube that terminates in the olfactory chamber. Since gravity drives the perfusion flow, the length and diameter of the tubing and height of the storage syringes relative to the table will determine (and limit) the rate. A microelectrode glass mounted to a micromanipulator can be used as the terminal end of the perfusion tube (Figure 14.2B). Fill the individual storage syringes of the perfusion system with the desired solutions (e.g., water only, water plus odorant, water plus contaminant, or water plus odorant and contaminant). Opening a pinch valve of the perfusion system allows the selected solution to flow (via gravity feed) to the rosette. One line to the manifold should be dedicated to background water alone. The remaining lines can vary depending on the experimental design. The pinch valves allow for automated, rapid (40 ms) switching between the different solutions. Pass the single output from the manifold through a thermoelectric chiller (see the section "Materials required") to chill the solution (12°C) just prior to delivery to the rosette. Using this approach, a perfusion source can be selected from as many as eight different solutions (one line for background water and seven lines for odorant or contaminant solutions) by setting the appropriate digital output on the computerized data acquisition system. If necessary, a line can be changed during a recording to deliver a different solution as long as the line is carefully flushed of the previous solution. Use the data acquisition system to record the electrical response from the rosette, while simultaneously switching the perfusion system to produce odor pulses (see the following).

Testing procedure

Following the placement of electrodes, allow the fish to acclimate for approximately 15 min before testing. EOG responses are elicited by briefly switching from background water to an odorant-containing solution. Blank (or control) responses are obtained by switching between background water delivered through the dedicated line and background water from a line normally used for odorant-containing solutions. Blank pulses should be delivered routinely to test for any responses that may reflect mechanical switching artifacts or residual odorants in the lines or micromanifold.

In our laboratory, a computerized system allows for EOGs to be evoked by multiple odor pulses that can be easily varied in terms of stimulus solution, pulse duration, or pulse interval. At the onset of any new experiment, typical responses to a standard set of odor pulses should be determined. This is best done empirically, since species, size, flow rate, and other experimental parameters can influence odor-evoked responses from the olfactory epithelium of fish (see Figures 14.4–14.6). Once the optimal conditions for delivering stimuli have been determined, the computerized system can reproducibly present the odor pulses to multiple fish within and between exposure groups. For brief exposures (one to a few hours), measure baseline responses to the odor pulses, and then change the solution perfusing the rosette to a separate line containing the contaminant. Wait for the duration of the exposure, switch back to clean background water, and then obtain a second set of responses to the same set of olfactory stimuli (Figure 14.3). For experiments in which fish are exposed for longer durations, it is typically not possible to compare pre-exposure versus post-exposure EOG responses of individual animals. In these situations, only post-exposure EOG responses can be collected. These data are then compared to the responses obtained from control (unexposed) fish.[18]

Data analysis

Quantify the EOG responses by measuring the peak negative amplitude relative to the pre-odorant baseline (Figure 14.1D). Since the EOG represents the summed activity of the ORNs,[7,10] the peak amplitude is presumably proportional to the electrical impulses (action potentials) that are transmitted to the olfactory forebrain via the olfactory nerve. EOG responses to an odor pulse can be reported in absolute terms (mV) or as a relative response. Relative responses may be intra-animal (pre-contaminant versus post-contaminant exposures) or inter-animal (exposed versus unexposed treatment groups). EOG responses should not be expressed relative to responses to a standard odor pulse (e.g., 10^{-5} *M* L-serine), since the responses to the standard pulse may also change with contaminant exposure. It is relatively straightforward to collect contaminant dose–response data by monitoring the EOG responses of fish in different treatment groups. Several methods exist for determining effects thresholds from these types of data. As an example, a regression-based approach is presented in the section "Results and discussion".

Determining an appropriate pulse duration

For sufficiently long pulse durations, a typical EOG consists of a negative phasic peak that decays slowly to a tonic plateau (see Figures 14.1D and 14.4A). If the pulse duration is too short, there will be insufficient time for the phasic peak to develop, and the true peak

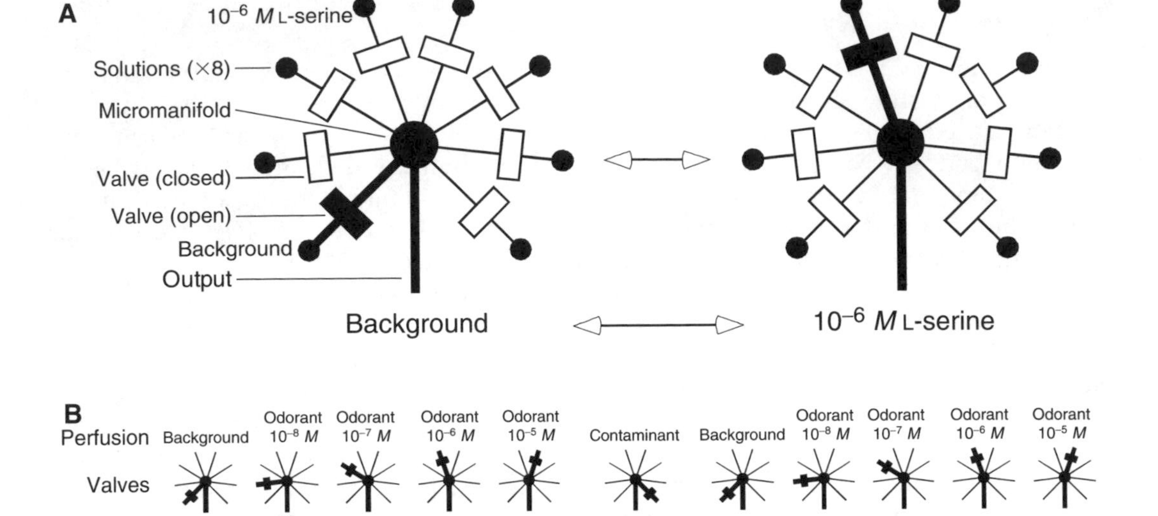

Figure 14.3 Schematic diagrams illustrating the perfusion apparatus and an example of a solution delivery sequence used to expose the olfactory epithelium to a contaminant and to record odor-evoked EOGs pre-exposure and post-exposure. (A) A schematic representation of the perfusion system showing the switch between solutions. Each solution container (a large syringe or glass bottle) is connected to a single micromanifold with a silicone tube that passes through a pinch valve. Switching from one open pinch valve to another changes the solution flowing out of the manifold to the olfactory rosette. As an example, a switch from the background solution to a solution containing background plus 10^{-6} *M* L-serine is illustrated. The double arrow denotes a switch back to background water, which terminates the odor pulse. (B) A schematic representation of the delivery of multiple odor pulses and a dissolved contaminant. Although four separate odor pulses are shown here as an example, the number and combination of odor pulses may vary for different experiments. In the present example, odor pulses are 10 s long (black bars) and separated by 120-s washes with background water (light gray bars). Each stimulus is represented by a different valve and line to the manifold. Here, a single odorant is presented in an ascending order of four concentrations. Diagonal white gaps denote breaks in the time bar. The four pulses are used to gauge the pre-exposure response, and then the perfusion is switched to a line containing a dissolved contaminant (note the different valve). After a brief exposure period (30 min), the perfusion sequence for the four odor pulses is repeated to determine the post-exposure olfactory response. The EOGs shown are responses to a mixture of four amino acids (see the section "Materials required") before and after a 30-min exposure to 10 μg/l copper (see the section "Results and discussion"). The horizontal bars below each post-exposure EOG shows the peak amplitude of the corresponding pre-exposure EOG for that odorant.

amplitude of the olfactory response to the odorant will not be recorded. The kinetics of the phasic peak will depend on the properties of the delivery system (e.g., flow rate, diffusion at the leading edge) and on the neurophysiological properties of the responsive ORNs. Figure 14.4A shows EOG recordings from a single fish using various pulse durations. Pooled data from several juvenile salmon (Figure 14.4B) empirically show that pulse durations of 10 s are sufficient to reach maximum amplitude given the fish and stimulus delivery apparatus used during these experiments. Shorter pulses produce smaller peaks

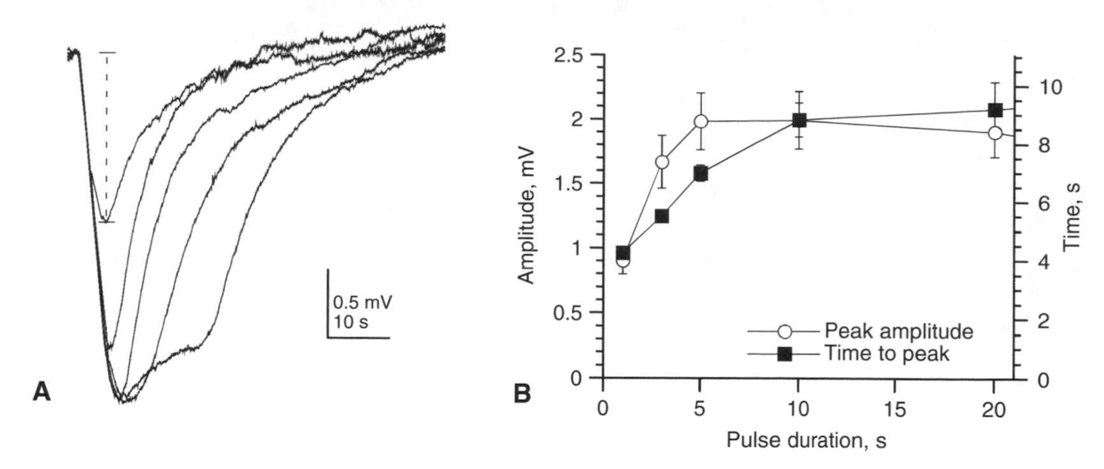

Figure 14.4 Varying the duration of odorant pulses changes the waveform and amplitude of evoked EOGs. (A) Superimposed EOG responses to 10^{-5} *M* L-serine pulses delivered with durations of 1, 3, 5, 10, and 20 s (obtained from the same fish). The dashed vertical line indicates the peak amplitude (as measured from the baseline) for the EOG evoked by a 1-s pulse. (B) Double *Y*-axis plot showing the effect of pulse duration on peak amplitude and time to peak (mean $\pm$ 1 standard error, $n = 4$ fish). Time to peak was measured from the initial deflection from the baseline to the peak of the EOG. Figure adapted from Reference 17.

and/or reduce the time to peak. Longer pulse durations do not increase peak amplitude, but they do increase the duration of ORN activity and thus the potential for sensory adaptation of the olfactory responses.

Determining an appropriate pulse interval

If two successive odor pulses are presented too close together, the amplitude of the second pulse will be less than the first (even if the stimulus pulses are the same). This is a form of sensory adaptation, and it limits the minimum rate at which a set of odor pulses should be presented. Figure 14.5A shows the consequences of varying the interval between two pulses (10^{-5} *M* L-serine) on the EOG responses of a single fish. When odor pulses are presented in sequence, the amplitude of the second pulse increases with the duration of the interpulse interval. The pooled data from several fish (Figure 14.5B) shows that the responses of the olfactory epithelium to identical odorant pulses are equivalent if the pulses are separated by at least 120 s. Therefore, in this specific example, individual odor pulses should be separated by at least a 2-min interval.

Selecting odorants and concentrations

Fish use distinct ORNs to detect different classes of odor molecules that vary in terms of their molecular structure.[7] These include bile salts (e.g., TCA), amino acids (e.g., L-serine), and steroid hormones (e.g., prostaglandins). For example, amino acids have been shown to activate at least four non-overlapping populations of ORNs.[19,20] The ability of specific odorants to evoke EOGs can be relatively species specific (e.g., the prostaglandins[21]) or generalized across species (e.g., bile salts and amino acids[7]). Since individual contaminants could potentially have different effects on different populations of ORNs, EOG-based

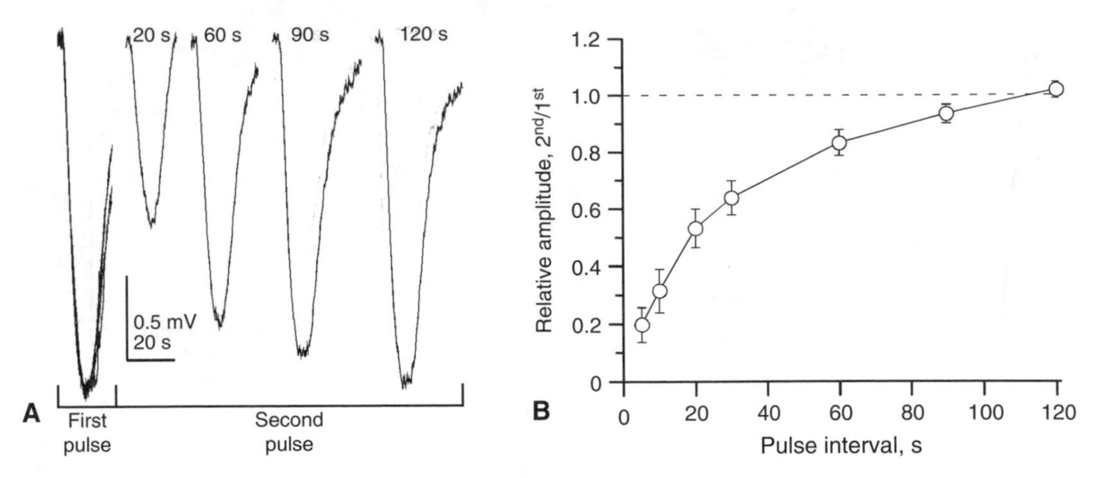

Figure 14.5 Varying the relative timing of two odor pulses changes the amplitude of the evoked EOG for the second pulse. (A) Traces showing pairs of 10^{-5} M L-serine pulses (each 10-s long) presented with inter-pulse intervals of 20, 60, 90, and 120 s (obtained from a single fish). For clarity, the intervening data have been deleted and only the evoked EOGs are shown. The first pulses (four total) are superimposed and the second pulses have been arbitrarily positioned horizontally and aligned vertically by their initial downward deflections. (B) Pooled data (mean $\pm$ 1 standard error, $n = 4$ fish) showing the effect of pulse interval on the amplitude of the second pulse relative to the first pulse. Figure adapted from Reference 17.

experiments should, to the extent possible, use multiple and dissimilar odorants as olfactory stimuli. The bile salt TCA and four amino acids serve as examples here since they are known to stimulate distinct olfactory pathways,[19,20] and they do not appear to be species specific.[7] Ultimately, of course, the choice of odorants will depend, in part, on the species being studied and on the olfactory-mediated behaviors of interest.

Odorants can be presented individually or as mixtures. Using odor pulses with a single odorant will only measure the effect of contaminant exposure on the population of ORNs sensitive to that odorant. Using mixtures of odorants will stimulate multiple populations of ORNs and mean that any effect of contaminant exposure on the EOG represents the combined effect on multiple populations of ORNs. Whether the contaminant exposure had an effect on any specific populations of ORNs cannot be discerned. In fact, a population of ORNs stimulated by the mixture may have been unaffected by the exposure. However, since the EOG response represents the summed activity of the multiple independent pathways, the mixture will produce relatively large EOG responses while limiting the amount of stimulation to any one population of ORNs, thus reducing the chance of saturation or adaptation. This makes the EOG responses easier to resolve relative to background noise and blank responses while at the same time reducing the chance of saturation or adaptation of the ORNs.

In addition to selecting odorants, consideration should be given to the concentrations at which the odorants are presented to the sensory epithelium. Again, initial range finding experiments are recommended here as there may be differences in olfactory sensitivity due to differences in species, age, etc. Figure 14.6A shows EOG responses to several concentrations of a single odorant (L-serine). Figure 14.6B shows a range of concentrations of several odorants producing EOG responses above the noise in the measurements of amplitude and the responses to blank pulses, but below concentrations that saturate the response. Using concentrations in this range will produce robust measurements and

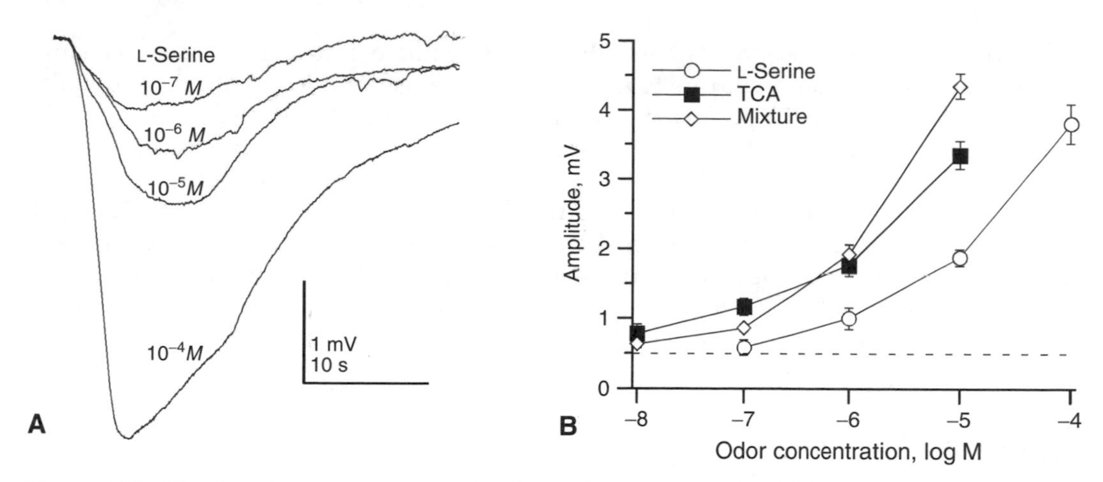

Figure 14.6 Varying the concentration of odor pulses changes the amplitude of evoked EOGs. (A) The amplitudes of EOGs evoked in response to 10 s L-serine pulses increase with increasing odorant concentration (10^{-7}, 10^{-6}, 10^{-5}, and 10^{-4} *M*). EOGs from a single fish have been superimposed. (B) Dose–response data (mean $\pm$ 1 standard error, $n = 6$ fish) for three odorants (L-serine, TCA, and the amino acid mixture). The dotted line shows the mean response to a blank odor pulse (background water; $n = 4$ fish). Figure adapted from Reference 17.

reduce the likelihood of changes in the sensory physiology of the ORNs, such as adaptation. For the amino acids, these concentrations may also reflect actual levels seen in the natural environment.[22] In summary, the effect of contaminant exposure on the olfactory response should ideally be evaluated for multiple odorants over a range of stimulus concentrations. The use of a multi-line perfusion system, such as the one described earlier, makes these types of experiments more tractable.

Determining recording stability

Preliminary experiments should be performed to test the stability of the EOG response over time. Experimental recordings may last for a few hours if they involve a pre-exposure series of odor pulses followed by a delivery of contaminant, and then a post-exposure odorant presentation. Changes in the condition of the fish (rundown) or the properties of the microelectrodes may lead to non-specific changes in EOG amplitude. Prior to the experiments presented here, preliminary recordings were performed to show that EOG responses to repeated pulses of 10^{-5} *M* L-serine do not significantly change over a 2-h recording interval (linear regression, $n = 6$, data not shown). This stable recording interval is typically sufficient to conduct most kinds of experiments.

Sources of artifacts

Several sources of artifacts can cause non-olfactory signals to be recorded. These include a high level of electrical noise, or an unstable recording baseline.[10] Electrical artifacts may also originate from the operation of the valves. These will appear as transient spikes coinciding with the opening/closing of the valves. Improving the electrical isolation between the wiring to the valve controller and the wiring from the amplifier can reduce these artifacts. However, they usually do not interfere with the measurement of EOGs.

Any movement of the rosette relative to the recording microelectrode will cause a change in the baseline. This can occur due to movement of the fish or microelectrode. It can also occur due to changes in the perfusion flow rate that will arise if there are differences in the heights of the solutions in the supply syringes. Problems with the microelectrodes, such as blockage with tissue or an air bubble, will lead to increased noise in the recordings. Make extra microelectrodes, so that a faulty microelectrode can be changed if needed. Finally, there will almost always be a slight cross-contamination between the lines and micromanifold when multiple solutions are used. To gauge the magnitude of this artifact, periodically conduct tests with blank stimulus solutions. A small response is likely even when switching between two lines that each contain source water.

Results and discussion

Copper exposure inhibits EOG responses

Dissolved copper is a common non-point source pollutant in fish habitats. Recent measurements of the effect of copper on the odor-evoked EOGs in coho salmon[17,18] will be used as an example of EOG recordings from contaminant-exposed fish. Figure 14.7 shows EOG responses obtained from a single fish before and after a 30-min perfusion of 10 μg/l copper chloride over the rosette. Seven odor pulses of 10-s duration were presented at 2-min intervals in the following sequence: 10^{-5} *M* L-serine, 10^{-6} *M* TCA, 10^{-8} *M* amino acid mixture, 10^{-7} *M* mixture, 10^{-6} *M* mixture, 10^{-5} *M* mixture, and 10^{-5} *M* L-serine. The perfusion was then changed to the copper-containing solution for 30 min. After the exposure interval, the perfusion was changed back to background water, and the series of seven odor pulses was repeated. A schematic illustration of the solution delivery is shown in Figure 14.3B. For each set of odor pulses, only the first L-serine odor pulse is shown in Figure 14.7. For data analysis, the data (post/pre-exposure ratios) from the two responses were averaged. Whereas the effect of copper on the responses to L-serine and TCA are point estimates (i.e., inhibition of an olfactory response to an individual odorant at a single concentration), the EOGs evoked by the amino acid mixture captures the effects of copper on multiple receptor populations over three log units of stimulus intensity. The effect of copper was quantified by averaging the reductions measured for the four mixture concentrations. For the fish shown in Figure 14.7, the 30-min exposure to 10 μg/l copper reduced the L-serine response by 57%, the TCA response by 67%, and the response to the amino acid mixture by 35% (all relative to the pre-exposure EOG amplitudes for the same fish). EOGs recorded from fish exposed to copper for 96 h show reduced EOGs similar to those seen in Figure 14.7.[18]

For brief exposure durations, such as the example above, the fish can serve as both the treatment and a control. Comparing post-exposure EOG responses to pre-exposure responses in the same fish provides some control over the biological variability between fish in the recordings. A separate control group exposed to uncontaminated water rather than copper should still be tested. Recordings from this unexposed treatment group represent an additional control for any effects of time and/or artifacts in the apparatus.

Estimating thresholds for sublethal copper neurotoxicity

The EOG recording method is particularly useful for determining sublethal thresholds for copper-induced neurotoxicity. This is true for fish exposed to copper for relatively short

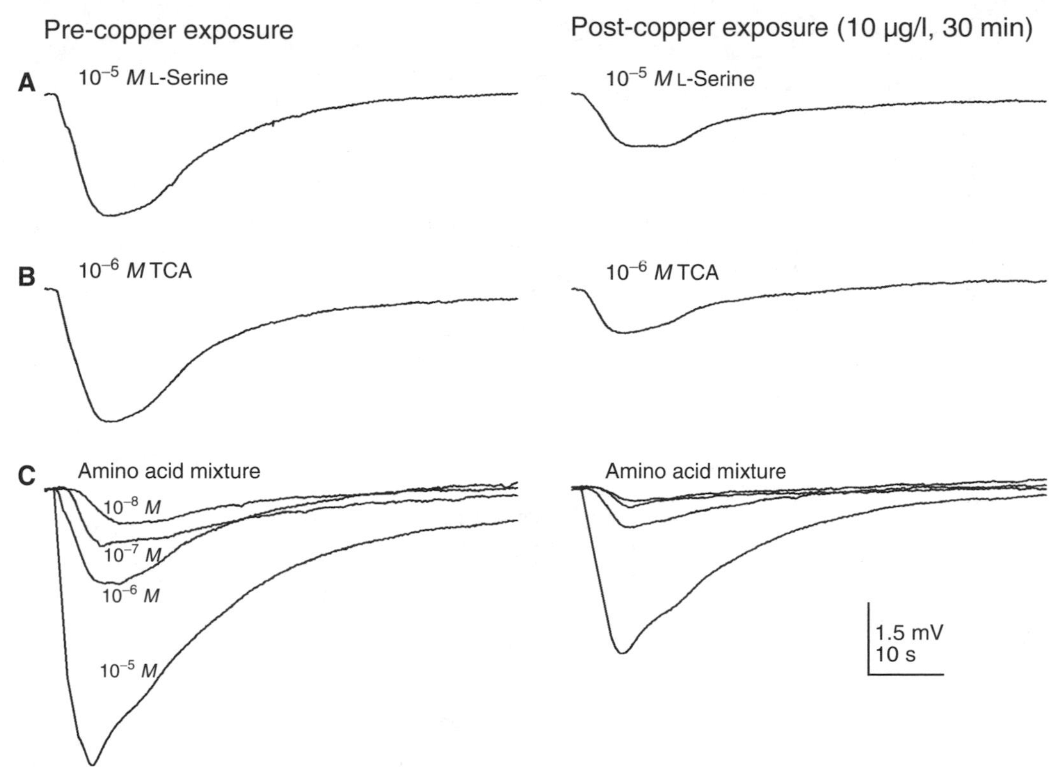

Figure 14.7 Short-term copper exposure diminishes the responsiveness of the olfactory epithelium to natural odorants. EOGs obtained from the same fish before and after copper exposure are shown. A 30-min exposure to 10 μg/l copper reduced the EOG evoked by 10^{-5} *M* L-serine by 57% (A) and the response to 10^{-6} *M* TCA by 67% (B). Similarly, the EOG responses to all four concentrations of the amino acid mixture (C) were reduced. The average reduction in the responses to the mixtures was 35%. Not shown are EOG responses to blank pulses, which were also reduced, and by a similar percentage. Figure adapted from Reference 17.

intervals (30 min) while on the recording apparatus (as shown here and in Reference 17), and also for animals exposed for longer intervals (i.e., days) in separate tanks (e.g., Reference 18). For the short-term experiment shown here, where copper was delivered directly to the olfactory chamber for 30 min, EOG responses were collected from six different exposure groups (see the section "Procedures" and Figure 14.7). These consisted of an unexposed (control) group and fish exposed to copper at nominal concentrations of 1, 2, 5, 10, and 20 μg/l ($n = 6$ fish per group). During the exposure interval for the control group, the perfusion was switched from the dedicated background water line to the line normally used to deliver copper-containing solutions, but this time used to deliver background water instead. This was done, in part, to monitor for any residual copper contamination in the delivery line.

For the purposes of demonstration, only data from the EOGs evoked by L-serine are shown here. Notably, the inhibitory effects of copper on the olfactory responses to other odorants are similar to those for L-serine.[17,18] The inhibitory dose–response relationship for the effect of short-term (30 min) copper exposure on the EOG is shown in Figure 14.8. Notably, there was a slight reduction in EOG responses to the odorants (expressed as the post-exposure/pre-exposure ratio) for the control fish

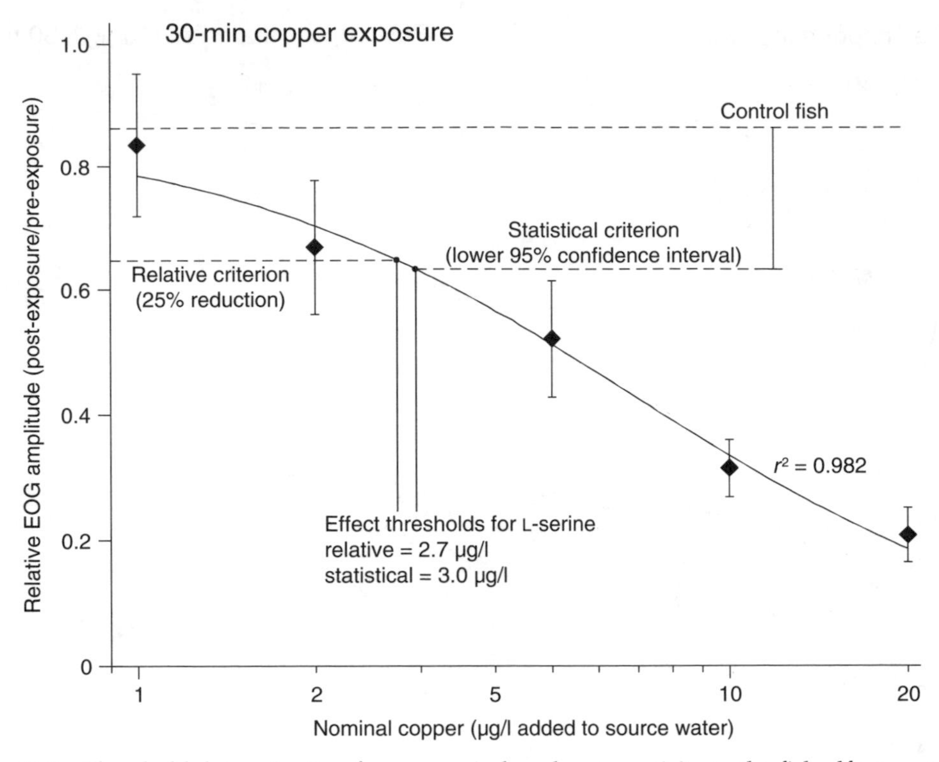

Figure 14.8 Threshold determination for copper-induced neurotoxicity to the fish olfactory epithelium. Data were obtained from six treatment groups (control and five copper exposures; $n = 6$ fish per group) exposed for 30 min. For each exposure group, the mean (± 1 standard error) relative EOG amplitude in response to 10^{-5} *M* L-serine is shown. The upper dashed line indicates the mean reduction in EOG responses seen in control fish (a reduction discussed in the text). A dashed line indicates the relative threshold criterion (a 25% reduction relative to controls). The vertical line in the upper right shows the lower limit of the 95% confidence interval for the control fish. A dashed line indicates the resulting statistical criterion. See the section "Results and discussion" for the equation used to fit the data. Filled circles indicate the relative and statistical threshold concentrations of 2.7 and 3.0 μg/l (respectively) for copper's effect on L-serine olfactory sensitivity. Note that these threshold values are nominal concentrations above background, or an increase from the approximately 3 μg/l copper present in the source water for the NWFSC hatchery. Adapted from Reference 17.

(0.84 ± 0.09, mean $\pm$ 1 standard error). This intra-animal reduction in evoked EOGs was not statistically significant (one-group *t*-test, hypothetical mean $= 1$, $p > 0.05$). However, it does indicate the potential for slight residual contamination in the lines used to deliver copper to the olfactory chamber. Care should therefore be taken to flush the lines between experiments.

Several methods have previously been used to describe toxicological effects thresholds (e.g., lowest observable effect concentration, inhibition concentration, benchmark dose). In this example, the effect threshold for a 30-min copper exposure was estimated by fitting the data for L-serine with a sigmoid logistic model:

$$y = m/[1 + (x/k)^n]$$

where m is the maximum relative EOG amplitude (fixed at the control mean of 0.84), y the relative EOG amplitude, x the copper concentration, k the copper concentration at half-maximum relative EOG amplitude (IC_{50}), n is the slope.

Figure 14.8 shows that the model is a good fit for the range of responses to L-serine. The values ($\pm$ standard error) for k and n were 6.8 ± 0.6 and 1.2 ± 0.1, respectively. Similar fits with the model were found for the copper-induced reductions in EOGs evoked by the other odorants (TCA and the amino acid mixture; Reference 17). The threshold concentration was then determined based on the nominal concentration at which the model crosses a criterion level. The criterion was set to a relative departure of 75% of the control mean (or a 25% reduction relative to controls), a level very close to a statistical departure based on the lower 95% confidence interval for the control group (Figure 14.8). Using this approach, the nominal threshold concentration for copper yielding 25% inhibition (IC_{25}) was found to be 2.7 μg/l for L-serine. This inhibitory threshold is very similar to the relative thresholds for TCA (2.3 μg/l) and the amino acid mixture (3.0 μg/l).[17]

Additional applications of EOG recordings to aquatic toxicology

The case study presented earlier used the EOG recording method to determine thresholds for copper-induced reductions in the olfactory response of coho salmon. The application of the EOG technique to other fish species, other contaminants, and other natural odorants has already been discussed. The EOG can also be used to monitor the sublethal effects of dissolved contaminants in terms of time-to-effect and time-to-recovery. For contaminants that impact the sensory epithelium relatively quickly (e.g., copper[17]), an advantage of the EOG technique is that olfactory function can be measured in a single fish before, during, and after the exposure of the olfactory rosette to track both time-to-effect and time-to-recovery over the short term. Additionally, due to the importance of olfactory-mediated behaviors mentioned earlier, effects of contaminant exposure observed in EOG studies can be used to infer a contaminant's effect on higher endpoints, such as survival or reproduction, where data may be lacking or difficult to obtain. When experiments on a contaminant's effect on higher endpoints are undertaken, aspects of the experimental design, e.g., the exposure concentrations to test, can be guided by the results from EOG studies. The EOG recording technique, therefore, can serve both as a standalone measure of the sublethal impact of contaminant exposure on peripheral olfactory physiology and as a link between contaminant-induced disruptions of physiological function and disruptions of behavior, reproduction, or survival.

Acknowledgments

The authors wish to thank Jason Sandahl and Jana Labenia for their assistance in collecting the data presented here; Cathy Laetz, Julann Spromberg, and an anonymous reviewer for comments on this manuscript; Brad Gadberry for the care and maintenance of coho salmon; and Carla Stehr for the electron micrographs in Figure 14.1.

References

1. Peterson, C.H., Rice, S.D., Short, J.W., Esler, D., Bodkin, J.L., Ballachey, B.E. and Irons, D.B., Long-term ecosystem response to the Exxon Valdez oil spill, *Science*, 302, 2082–2086, 2003.

2. Brown, G.E. and Smith, R.J., Conspecific skin extracts elicit antipredator responses in juvenile rainbow trout (*Oncorhynchus mykiss*), *Can. J. Zool.*, 75, 1916–1922, 1997.
3. Hiroven, H., Ranta, E., Piironen, J., Laurila, A. and Peuhkuri, N., Behavioural responses of naive Arctic charr young to chemical cues from salmonid and non-salmonid fish, *Oikos*, 88, 191–199, 2000.
4. Quinn, T.P. and Busack, C.A., Chemosensory recognition of siblings in juvenile coho salmon (*Oncorhynchus kisutch*), *Anim. Behav.*, 33, 51–56, 1985.
5. Moore, A. and Waring, C.P., Electrophysiological and endocrinological evidence that F-series prostaglandins function as priming pheromones in mature male Atlantic salmon (*Salmo salar* parr), *J. Exp. Biol.*, 199, 2307–2316, 1996.
6. Wisby, W.J. and Hasler, A.D., Effect of occlusion on migrating silver salmon (*Oncorhynchus kisutch*), *J. Fish Res. Board Can.*, 11, 472–478, 1954.
7. Hara, T.J., Mechanisms of olfaction, in: *Fish Chemoreception*, Hara, T.J., Ed., Chapman & Hall, London, 1992, pp. 150–170.
8. Schild, D. and Restrepo, D., Transduction mechanisms in vertebrate olfactory receptor cells, *Physiol. Rev.*, 78, 429–466, 1998.
9. Evans, R.E. and Hara, T.J., The characteristics of the electro-olfactogram (EOG): its loss and recovery following olfactory nerve section in rainbow trout (*Salmo gairdneri*), *Brain Res.*, 330, 65–75, 1985.
10. Scott, J.W. and Scott-Johnson, P.E., The electroolfactogram: a review of its history and uses, *Microsc. Res. Tech.*, 58, 152–160, 2002.
11. Caprio, J., *The underwater EOG: A Tool for Studying the Effects of Pollutants on the Oolfactory Receptors of Fish*, Presented at Conference Workshop on Chemoreception in Studies of Marine Pollution, Leangkollen, Norway, 13 July 1980, 1983.
12. Baatrup, E., Døving, K.B. and Winberg, S., Differential effects of mercurial compounds on the electroolfactogram (EOG) of salmon (*Salmo salar* L.), *Ecotoxicol. Environ. Saf.*, 20, 269–276, 1990.
13. Bjerselius, R., Winberg, S., Winberg, Y. and Zeipel, K., Ca^{2+} protects olfactory receptor function against acute Cu(II) toxicity in Atlantic salmon, *Aquat. Toxicol.*, 25, 125–138, 1993.
14. Moore, A. and Waring, C.P., Sublethal effects of the pesticide diazinon on olfactory function in mature male Atlantic salmon parr, *J. Fish Biol.*, 48, 758–775, 1996.
15. Waring, C.P. and Moore, A., Sublethal effects of a carbamate pesticide on pheromonal mediated endocrine function in mature male Atlantic salmon (*Salmo salar* L.) parr, *Fish Physiol. Biochem.*, 17, 203–211, 1997.
16. Moore, A. and Lower, N., The impact of two pesticides on olfactory-mediated endocrine function in mature male Atlantic salmon (*Salmo salar* L.) parr, *Comp. Biochem. Physiol. Biochem. Mol. Biol.*, 129, 269–276, 2001.
17. Baldwin, D.H., Sandahl, J.F., Labenia, J.S. and Scholz, N.L., Sublethal effects of copper on coho salmon: impacts on nonoverlapping receptor pathways in the peripheral olfactory nervous system, *Environ. Toxicol. Chem.*, 22, 2266–2274, 2003.
18. Sandahl, J.F., Baldwin, D.H., Jenkins, J.J. and Scholz, N.L., Odor-evoked field potentials as indicators of sublethal neurotoxicity in juvenile coho salmon (*Oncorynchus kisutch*) exposed to copper, chlorpyrifos, or esfenvalerate, *Can. J. Fish Aquat. Sci.* 61, 404–413, 2004.
19. Sveinsson, T. and Hara, T.J., Multiple olfactory receptors for amino acids in Arctic char (*Salvelinus alpinus*) evidenced by cross-adaptation experiments, *Comp. Biochem. Physiol. A.*, 97, 289–293, 1990.
20. Kang, J. and Caprio, J., Electro-olfactogram and multiunit olfactory receptor responses to complex mixtures of amino acids in the channel catfish, *Ictalurus punctatus*, *J. Gen. Physiol.*, 98, 699–721, 1991.
21. Laberge, F. and Hara, T.J., Behavioural and electrophysiological responses to F-prostaglandins, putative spawning pheromones, in three salmonid fishes, *J. Fish Biol.*, 62, 206–221, 2003.
22. Shoji, T., Ueda, H., Ohgami, T., Sakamoto, T., Katsuragi, Y., Yamauchi, K. and Kurihara, K., Amino acids dissolved in stream water as possible home stream odorants for masu salmon, *Chem. Senses*, 25, 533–540, 2000.

chapter fifteen

Enzyme-linked immunosorbent assay for screening estrogen receptor binding activity

Tomoko Koda and Masatoshi Morita
National Institute for Environmental Studies

Yoshihiro Soya
Tsuruga Institute of Biotechnology, Toyobo Co. Ltd.

Contents

Introduction

Endocrine disrupter chemicals (EDCs) include those that mimic estrogen (xenoestrogens) and cause reproductive problems.[1] Xenoestrogens can produce adverse effects by binding to the estrogen receptor (ER).[2] In addition, they may also alter hormone metabolism, affect various neural centers or the pituitary, and modify serum hormone binding proteins.[2] We need an efficient screening strategy for xenoestrogens because a large number of chemicals are commercially produced and widely used, and this type of information is not generally available.

1-56670-664-5/05/$0.00+$1.50
© 2005 by CRC Press

A final assessment of endocrine disrupting ability must be based on whole animal studies[3–5]; however, alternative tests, such as high through-put assays of receptor-dependent responses, can be used to prioritize chemicals before conducting time-consuming and expensive *in vivo* assays. Receptor binding assays and other *in vitro* tests, including cell proliferation assays, expression of specific proteins, yeast two-hybrid assays, and reporter gene assays, are examples of these high through-put methods.[1,2,6–13] We considered that from these tests, a receptor binding assay was a rapid and cost-effective screening technique. Therefore, we establish an enzyme-linked immunosorbent assay (ELISA) for assessing the binding ability of chemicals to an ER.[14,15] Our competitive ELISA used estrogen labeled with horseradish peroxidase, as opposed to radioisotopes, was safe and very easy to perform.

Materials required

ELISA kit for EDCs

- Human ERα[16]
- Anti-human ERα mouse IgG (monoclonal antibody); the antibody was fixed to microplate by physical absorption
- Unlabeled ligand, which was usually 17β-estradiol
- Horseradish peroxidase labeled 17β-estradiol
- Substrate was 3,3′,5,5′-Tetramethylbenzidine (Cas No. 54827-17-7)
- 1N sulfuric acid
- Buffer solutions: aqueous buffer (10 m*M* phosphate-buffered saline, pH 7.2) and washing solution (50 m*M* disodium phosphate, 150 m*M* NaCl, 0.05% [w/v] Tween-20)
- Microplate (96 wells)

The materials listed above are available as a kit (Ligand Screening System, ERα, Code No. ERA-101, Toyobo Co., Biochemical Department, 2-2-8 Dojimahama, kita-ku, Osaka, 530-8230, Japan. Tel.:+81-6-6348-3786, fax:+81-6-6348-3833, e-mail: order lifescience@bio.toyobo.co.jp).

Equipment

- Immuno Wash MODEL 1575 (Bio-Rad Laboratories, Arakawa, Tokyo, Japan)
- Microplate reader (Benchmark, Bio-Rad Laboratories, Arakawa, Tokyo, Japan)
- Incubator

Chemicals for analysis

Chemicals for analysis were purchased from the following sources: bisphenol A (BPA), 3-*tert*-butylphenol (*m*-*t*-BP), 4-*n*-pentylphenol (*p*-*n*-PeP), 4-*tert*-pentylphenol (*p*-*t*-PeP), 4-octylphenol (*p*-*n*-OP), *p*-*t*-OP, and 4-*n*-nonylphenol (*p*-*n*-NP) from the Wako Pure Chemical Industries, Osaka, Japan; diethylstilbestrol (DES) from the Sigma Chemical Co., St.Louis, MO, USA; 4-nonylphenol (*p*-NP) from the Kanto Chemical Co., Tokyo, Japan; 4-*tert*-butylphenol (*p*-*t*-BP) from Nacalai Tesque, Kyoto, Japan. The 4-*n*-nonylphenol consisted of only *n*-isomer, the 4-nonylphenol consisted of an undefined mixture of isomers.

Procedures

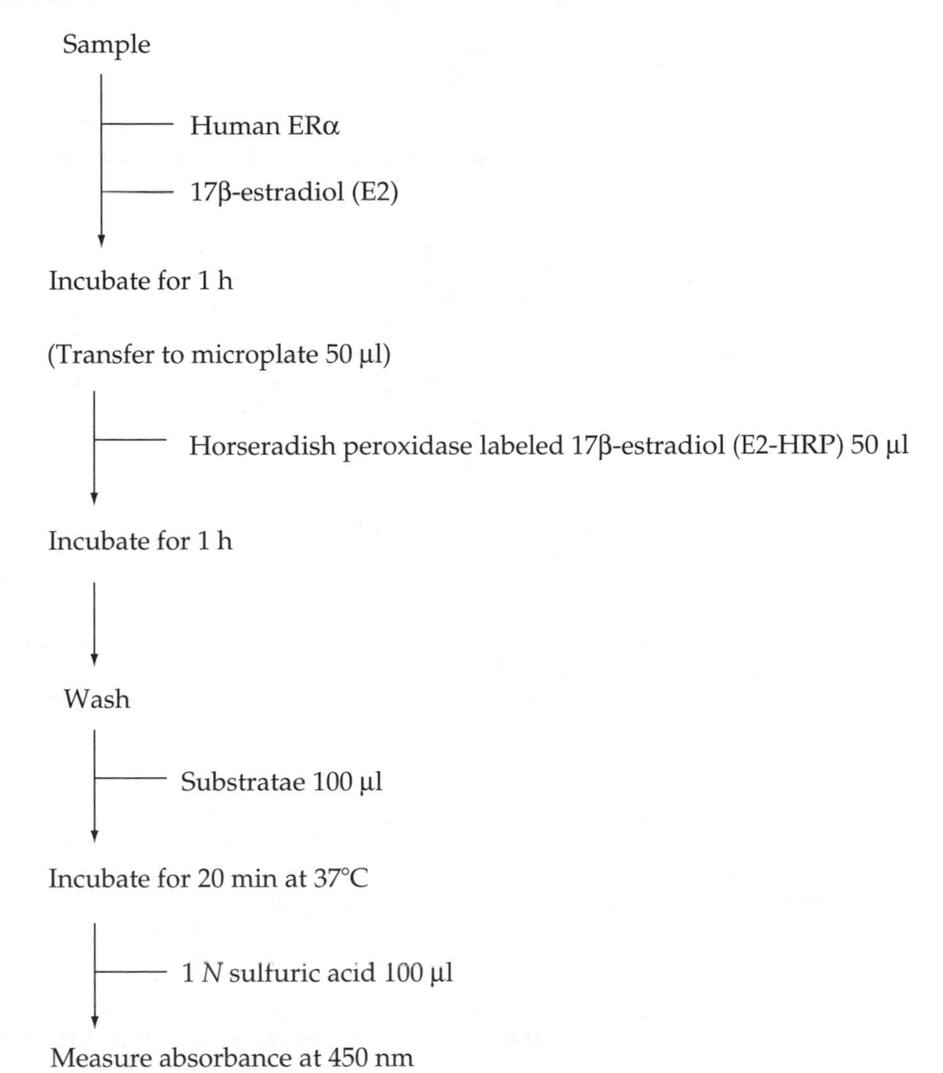

The method involves two steps of competitive reactions. The first step is the competition between the ligand (E2) and test chemicals on the ERα. When test chemicals have affinities for ERα, the concentration of the free ligand in the medium increases depending upon the binding affinities of test chemicals for ERα. If a test chemical does not have any binding affinity for ERα, the ligand in the medium is completely absorbed by the ERα. The second step is the competition between the free ligand, which was obtained at the first step reaction, and E2-HRP on the anti-E2 monoclonal antibody. When the free E2 exists in the medium, the competition against E2-HRP occurs on the antibody. Thus, color development by HRP decreases depending upon the concentration of the free ligand. If the free ligand in the medium is absent, the color development by HRP occurs fully. A scheme of the assay is shown in Figure 15.1.

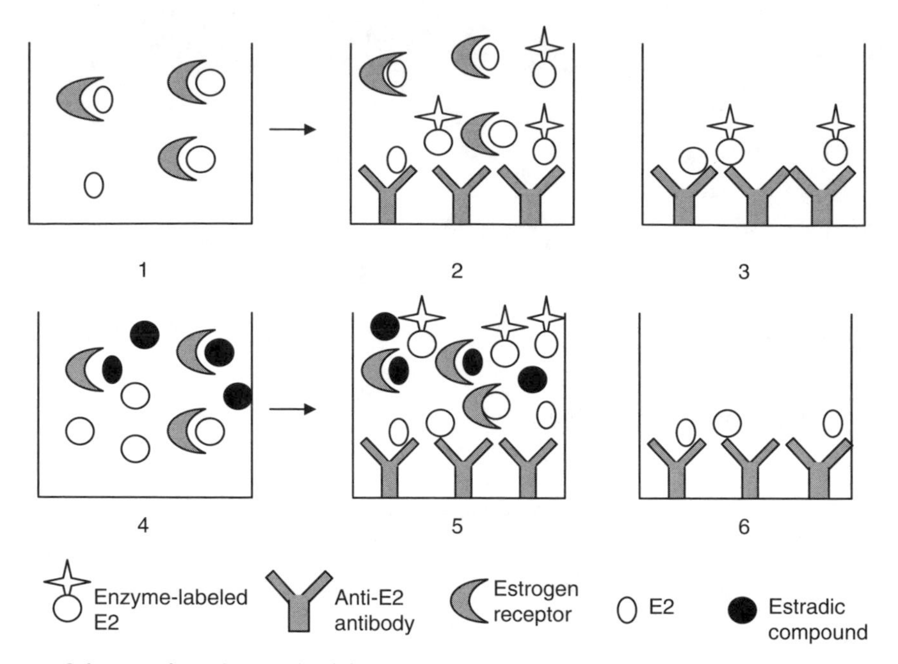

Figure 15.1 Scheme of ELISA method for measuring ERα binding activity of chemicals. Wells 1–3 show binding of 17β-estradiol to antibody and receptor. Wells 4–6 show how an estrogenic compound decreased the binding of enzyme-labeled estradiol to the anti-estradiol antibody on the microplate. Wells 2–3 and 5–6 were coated with anti-estradiol antibody. Well 1 (or 4) was different from wells 2–3 (or 5–6). Estradiol, compounds and ERs were mixed in well 1 (or 4), and they were transferred in well 2 (or 5), and then enzyme-labeled estradiol was added. After incubated, the plate was washed before the substrate was added. The compound competed with 17β-estradiol for binding to the ER, and the free 17β-estradiol then bound to the antibody and decreased the amount of the enzyme-labeled estradiol available to bind to the anti-estradiol antibody on the microplate. The antibody dose not bind other estrogenic compounds apart from estradiol. The enzyme-labeled estradiol does not bind to the receptor.

We usually used three concentrations of test chemicals in a range that ideally included 50% inhibition of each chemical. Test chemicals were dissolved in dimethylsulfoxide and diluted in aqueous buffer to a dimethylsulfoxide concentration of 1%. This sample solution (30 μl), receptor solution (20 μl), and E2 solution (30 μl) were added together to wells in the reaction plates (96-well polystyrene plates). The reaction plates were incubated at 4°C for 1 h, and 50 μl of the reaction solution and 50 μl of E2-HRP were added onto antibody plates that were coated with the E2 antibody, and incubated at 4°C for 1 h. Each well was washed three times with 150 μl of washing solution using an Immuno Wash MODEL 1575. Washing solution was removed from each well, 100 μl of substrate solution that contained 3,3′,5,5′-tetramethylbenzidine was added to the well and the antibody plates were incubated at 37°C for 20 min in the absence of light. Sulfuric acid (1 *N*, 100 μl) was added to each well to stop the reaction and the absorbance of each antibody plate well was measured at 450 and 655 nm using a microplate reader

Dimethylsulfoxide (1% in aqueous buffer) and DES (300 n*M*) were used as negative and positive controls, respectively.

Calculation of binding affinity

ERα inhibition was calculated based on the following equation:

$$\text{inhibition } (\alpha) = ([B] - [S])/([B] - [P])$$

where $[B]$ is absorbance measured in the absence of the test chemical, $[S]$ is absorbance measured in the presence of the test chemical, and $[P]$ is absorbance measured in the presence of the positive control chemical (DES).

Calculation of receptor binding index

ERα index was calculated based on the following equation:

$$\text{ER}\alpha \text{ index} = (1 - \alpha) \times (S \times 3/8 - R \times \alpha)/\alpha$$

where α is inhibition, S is sample concentration (nM), 3/8 is the volume correction factor, and R is the receptor concentration (nM). ERα index, which is calculated using the above equation, is approximately equal to the dissociation constant, and we use ERα index as an approximation for Kd.

Calculation of IC_{50}s

Regression was made of inhibition (α) on sample concentrations, and IC_{50} (concentrations that cause 50% inhibition of binding of E2 to ERα) was calculated from the regression curve. When α α zwas lower than 0.5, IC_{50} was calculated using the equation:

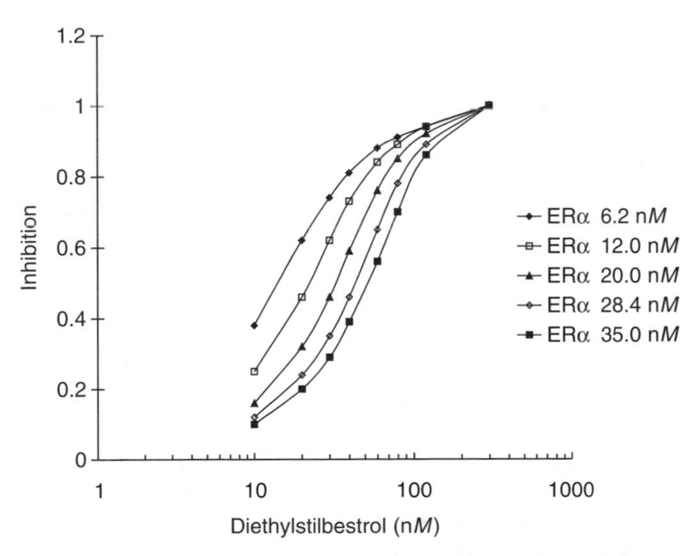

Figure 15.2 Inhibition curves in the presence of 6.2, 12.0, 20.0, 28.4, and 35.0 nM ERα. Ligand concentration was 12.5 nM, reaction temperature was 37°C, 17β-estradiol antibody concentration was 62.5 ng/well, 17β-estradiol-horseradish peroxidase was diluted 1/1000.

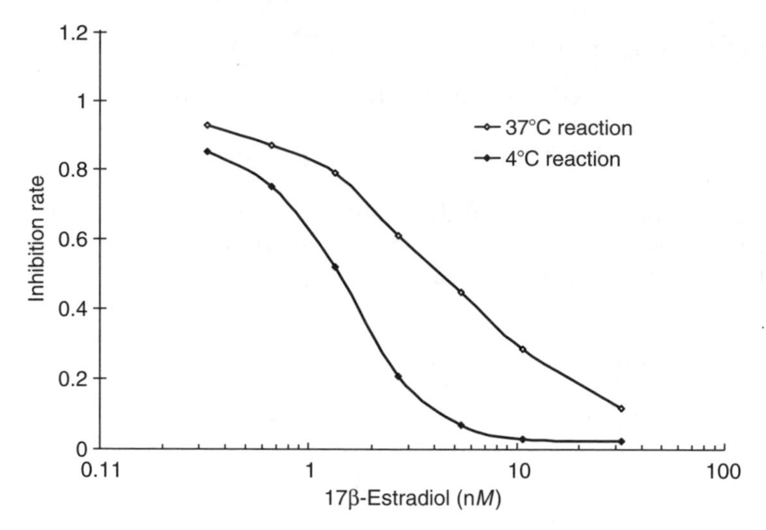

Figure 15.3 Inhibition curves for the 17β-estradiol and 17β-estradiol-horseradish peroxidase reaction for 17β-estradiol antibody at 4°C and 37°C. 17β-Estradiol antibody concentration was 62.5 ng/well and 17ββ-estradiol-horseradish peroxidase was diluted 1/1000. The inhibition rate was set at 1 when the concentration of 17β-estradiol was 0.

$$IC_{50}\ (\mathrm{n}M) = (\mathrm{ER}\alpha\ \mathrm{index} + R \times 0.5) \times (8/3)$$

Calculation of relative binding affinity

The relative binding affinity (RBA) was calculated based on the equation:

$$\mathrm{RBA} = IC_{50}\ \mathrm{of\ DES}/IC_{50}\ \mathrm{of\ test\ chemical}$$

Results and discussion

Factors that affected results

- Concentrations of ERα

When the concentrations of ERα at the first step varied between 6.2 and 35.0 n*M*, the binding rate of DES to ERα (the competition rate against E2) differed, as shown in Figure 15.2. When the ERα concentration was decreased, the quantity of the test chemical required to produce the same degree of inhibition also decreased. The concentration of the ligand (E2) was 12.5 n*M*, the reaction temperature for the first step reaction was 37°C, HRP-E2 was diluted 1/1000, and the concentration of the anti-E2 antibody was 62.5 ng/well.

- Temperature of the reaction between E2 and HRP-E2

The reaction rate depends on reaction temperature, as shown in Figure 15.3. When reaction temperature decreased, the reaction of E2 or HRP-E2 with coated E2 antibody also decreased. Since we anticipated that the measured concentration range would also decrease, we compared reaction temperatures of 4°C and 37°C. The measured concentration range was decreased from 1.5 approximately 20 n*M* at 37°C to 0.5 approxi-

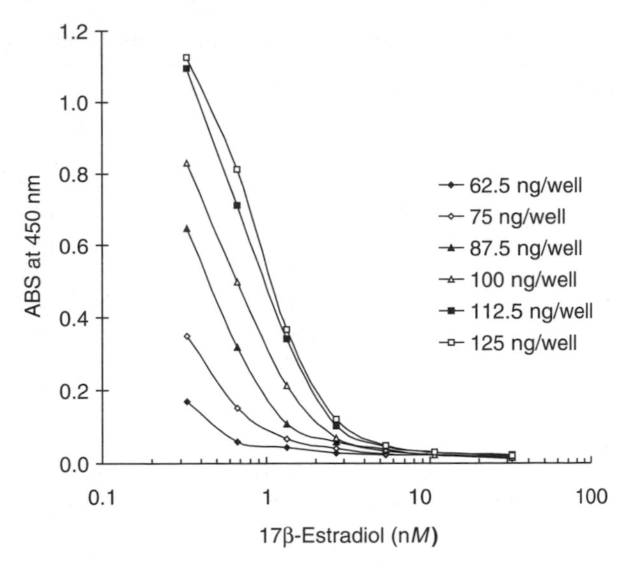

Figure 15.4 17β-Estradiol and 17β-estradiol-horseradish peroxidase reaction for 17β-estradiol antibody. 17β-Estradiol antibody concentrations were 62.5, 75, 87.5, 100, and 112.5 ng/well. 17β-Estradiol-horseradish peroxidase was diluted 1/1000.

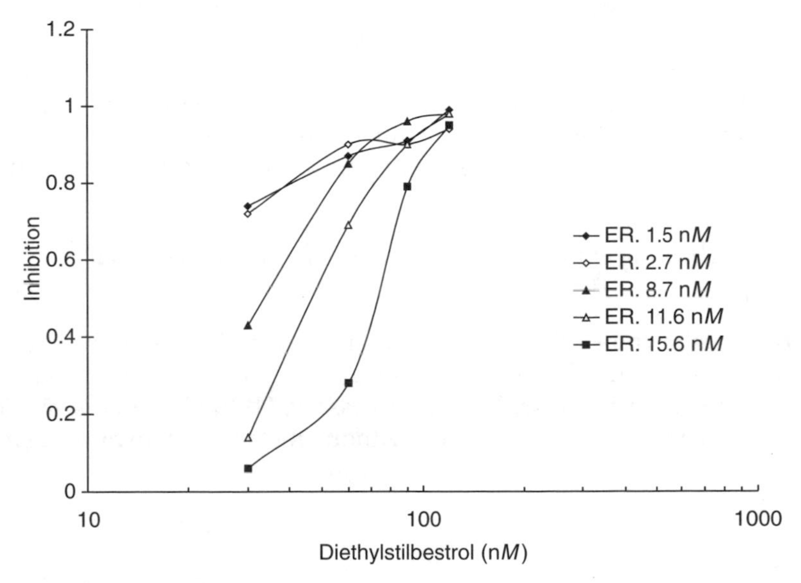

Figure 15.5 Inhibition curves in the presence of 1.5, 2.7, 8.7, 11.6 and 15.6 n*M* ERα. Ligand concentration was 4.2 n*M*, reaction temperature was 4°C, 17β-estradiol antibody was 125 ng/well and 17β-estradiol-horseradish peroxidase was diluted 1/1000.

mately 3 n*M* at 4°C. HRP-E2 was diluted 1/1000 and the concentration of E2 antibody was 62.5 ng/well.

- Concentration of the E2 antibody

The reaction rate (absorbance of 450 nm) between E2 and HRP-E2 depended on the concentration of the E2 antibody, as shown in Figure 15.4. When the solid-phase antibody

Table 15.1 IC_{50} and RBA values of each chemical[a]

Compound	IC_{50} (n*M*)	RBA
DES	8	1.13E + 02
BPA	883	1.00E + 00
m-*t*-BP	28,776	3.07E − 02
p-*t*-BP	31,651	2.79E − 02
p-*n*-PeP	27,487	3.21E − 02
p-*t*-PeP	9757	9.04E − 02
p-*n*-OP	32,792	2.69E − 02
p-*t*-OP	2832	3.12E − 01
p-*n*-NP	25,625	3.44E − 02
p-NP	699	1.26E + 00

[a] DES, diethylstilbestrol; BPA, bisphenol A; *m*-*t*-BP, 3-*tert*-butylphenol; *p*-*t*-BP, 4-*tert*-butylphenol; *p*-*n*-PeP, 4-*n*-pentylphenol; *p*-*t*-PeP, 4-*tert*-pentylphenol; *p*-*n*-OP, 4-*n*-octylphenol; *p*-*t*-OP, 4-*tert*-octylphenol; *p*-*n*-NP, 4-*n*-nonylphenol; and *p*-NP, 4-nonylphenol.

concentration was increased, absorbance also increased. HRP-E2 was diluted 1/1000 and the reaction temperature was 4°C.

Finally, the receptor concentration was optimized while other parameters were constant (Figure 15.5). When the receptor concentration was low, the same inhibition could be achieved with less test chemical. Thus, conditions can be optimized to require less test chemical and reduce cost. We usually performed the assay with 12 n*M* of ERα, 4.2 n*M* of E2, 4°C at the reaction, 125 ng/well of E2 antibody, and HRP-E2 diluted 1/1000.

ERα binding abilities of endocrine disrupting chemicals assessed using this method

IC_{50}s and RBA are shown in Table 15.1. The advantages of this method are: (1) it does not use radioisotopes, (2) many test compounds can be screened simultaneously in a short time at low cost, (3) specific apparatus is unnecessary, (4) the method is simple and easy to use, and (5) it can be applied with or without metabolic activation using rat liver S9 mix.

References

1. Soto, A.M., Sonnenschein, C., Chung, K.L., Fernandez, M.F., Olea, N. and Serrano, F.O., The e-screen assay as a tool to identify estrogens: an update on estrogenic environmental pollutants, *Environ. Health Perspect.*, 103 (Suppl. 7), 113–122, 1995.
2. Blair, R.M., Fang, H., Branham, W.S., Hass, B.S., Dial, S.L., Moland, C.L., Tong, W., Shi, L., Perkins, R. and Sheehan, D.M., The estrogen receptor binding affinities of 188 natural and xenochemicals: structural diversity of ligands, *Toxicol. Sci.*, 54, 138–153, 2000.
3. Alworth, L.C., Howdeshell, K.L., Ruhlen, R.L., Day, J.K., Lubahn, D.B., Huang, T.H.-M., Besch-Williford, C.L. and vom-Saal, F.S., Uterine responsiveness to estradiol and DNA methylation are altered by fetal exposure to diethylstilbestrol and methoxychlor in CD-1 mice: effects of low versus high doses, *Toxicol. Appl. Pharmacol.*, 183, 10–22, 2002.

4. Cunny, H.C., Mayes, B.A., Rosica, K.A., Trutter, J.A. and Miller, J.P.V., Subchronic toxicity (90-day) study with para-nonylphenol in rats, *Regul. Toxicol. Pharmacol.*, 26, 172–178, 1997.
5. Law, S.C., Carey, S.A., Ferrell, J.M., Bodman, G.J. and Cooper, R.L., Estrogenic activity of octylphenol, nonylphenol, bisphenol A and methoxychlor in rats, *Toxicol. Sci.*, 54, 154–167, 2000.
6. Fang, H., Tong, W., Perkins, R., Soto, A.M., Prechtl, N.V. and Sheehan, D.M., Quantitative comparisons of in vitro assays for estrogenic activities, *Environ. Health Perspect.*, 108, 723–729, 2000.
7. Andersen, H.R., Andersen, A.-M., Arnold, S.F., Autrup, H., Barfoed, M., Beresford, N.A., Bjerregaard, P., Christiansen, L.B., Gissel, B., Hummel, R., Jorgensen, E.B., Korsgaard, B., Guevel, R.L., Leffers, H., McLachlan, J., Moller, A., Nielsen, J.B., Olea, N., Oles-Karasko, A., Pakdel, F., Pedersen, K.L., Perez, P., Skakkeboek, N.E., Sonnenschein, C., Soto, A.M., Sumpter, J.P., Thorpe, S.M. and Grandjean, P., Comparison of short-term estrogenicity tests for identification of hormone-disrupting chemicals, *Environ. Health Perspect.*, 107 (Suppl. 1), 89–108, 1999.
8. Perez, P., Pulgar, R., Olea-Serrano, F., Villalobos, M., Rivas, A., Metzler, M., Pedraza, V. and Olea, N., The estrogenicity of bisphenol A-related diphenylalkanes with various substituents at the central carbon and the hydroxy groups, *Environ. Health Perspect.*, 106, 167–174, 1998.
9. Soto, A.M., Justicia, H., Eray, J.W. and Sonnenschein, C., *p*-Nonylphenol: an estrogenic xenobiotic released from "modified" polystyrene, *Environ. Health Perspect.*, 92, 167–173, 1991.
10. Korach, K.S. and McLachlan, J.A., Techniques for detection of estrogenicity, *Environ. Health Perspect.*, 103 (Suppl. 7), 5–8, 1995.
11. Gaido, K.W., Leonard, L.S., Lovell, S., Gould, J.C., Babai, D., Porter, C.J. and McDonnell, D.P., Evaluation of chemicals with endocrine modulating activity in a yeast-based steroid hormone receptor gene transcription assay, *Toxicol. Appl. Pharmacol.*, 143, 205–212, 1997.
12. Shelby, M.D., Newbold, R.R., Tully, D.B., Chae, K. and Davis, V.L., Assessing environmental chemicals for estrogenicity using a combination of in vitro and in vivo, *Environ. Health Perspect.*, 104, 1296–1300, 1996.
13. Coldham, N.G., Dave, M., Sivapathasundaram, S., McDonnell, D P., Connor, C. and Sauser, M.J., Evaluation of a recombinant yeast cell estrogen screening assay, *Environ. Health Perspect.*, 105, 734–742, 1997.
14. Koda, T., Soya, Y., Negishi, H., Shiraishi, F. and Morita, M., Improvement of a sensitive enzyme-linked immunosorbent assay for screening estrogen receptor binding activity, *Environ. Toxicol. Chem.*, 21, 2536–2541, 2002.
15. Morohoshi, K., Shiraishi, F., Oshima, Y., Koda, T., Nakajima, N., Edmonds, J.S. and Morita, M., Synthesis and estrogenic activity of bisphenol A mono- and di-beta-D-glucopyranosides, plant metabolites of bisphenol A, *Environ. Toxicol. Chem.*, 22, 2275–2279, 2003.
16. Green, S., Walter, P., Kumar, V., Krust, A., Bornert, J.-M., Argos, P. and Chambon, P., Human oestrogen receptor cDNA: sequence, expression and homology to v-erb-A, *Nature*, 320, 134–139, 1986.

chapter sixteen

Lysosomal destabilization assays for estuarine organisms

A.H. Ringwood
University of North Carolina at Charlotte
D.E. Conners
University of Georgia
J. Hoguet
College of Charleston
L.A. Ringwood
Wake Forest University

Contents

Introduction

Estuarine ecosystems are subject to increased stress associated with human population growth, in some cases nearly explosive, in coastal areas of the United States.[1,2] Estuaries provide essential functions, including serving as nursery grounds, for a variety of shellfish and fish species.[3] There is no question that organisms in these habitats are being exposed to contaminants, and bioaccumulation can be readily documented.[4] While acute toxicity incidents (e.g., fish kills, depauperate communities) are highly visible occurrences, it is more difficult to determine the impacts associated with chronic or recurring exposures to sublethal levels of contaminants. In some cases, compensatory mechanisms may function to sequester, detoxify, or ameliorate the effects of stressors, so exposures do not always translate into adverse effects. In other cases, individual stressors or combinations of stressors may cause chronic stress that can compromise basic physiological functions and long-term population sustainability. Sensitive biomarker tools should facilitate our ability to recognize when habitat conditions adversely affect biotic integrity before the effects are irreversible or very expensive to remedy. Cellular biomarker responses are valuable for identifying when conditions have exceeded compensatory mechanisms and the individuals and populations are experiencing chronic stress, which, if unmitigated, may progress to severe effects at the ecosystem level.

1-56670-664-5/05/$0.00+$1.50
© 2005 by CRC Press

Lysosomes are intracellular organelles that are involved in many essential functions, including membrane turnover, nutrition, and cellular defense.[5,6] The internal acidic environment of lysosomes, integral for the optimal activity of acid hydrolases, is maintained by a membrane-bound, ATPase-dependent proton pump. Lysosomes sequester metals and other contaminants, which make them prone to oxidative damage, etc., that can lead to membrane destabilization. Contaminants can also impair lysosomal function by disruption of the proton pump. Either mode of action can lead to the impairment of vital functions and cell death.[7–10]

Lysosomal function assays of hepatic or blood cells are regarded as valuable indicators of pollutant-induced injury, and there is a substantial body of literature validating that environmental pollutants cause destabilization of lysosomes (e.g., see References 11–15). Neutral red (NR) techniques for assessing lysosomal destabilization have been used successfully in fish and invertebrate taxa, and have been incorporated into major European programs (BIOMAR, Black Sea Mussel Watch). In this chapter, the methods that we developed and our experiences with a lysosomal destabilization assay based on NR retention in three common estuarine species (oysters, *Crassostrea virginica*; grass shrimp, *Palaemonetes pugio*; and mummichogs, *Fundulus heteroclitus*) are described.

NR retention assays are relatively simply assays for assessing lysosomal stability. The most common sources of cells for the assay are blood cells or hemocytes and hepatic cells. Hepatic tissues are especially rich in lysosomes. Since liver or hepatic tissues are a common site of accumulation and detoxification of contaminants, they are particularly relevant for identifying sublethal impacts of contaminants. Moreover, hepatic preparations typically yield a large number of cells and are the only real option for small organisms or those that are difficult to bleed. Cells incubated in NR accumulate the lipophilic dye in the lysosomes, where it is trapped by protonization. In healthy cells, NR is taken up and retained in stable lysosomes for extended periods; whereas in damaged cells the NR leaks out of lysosomes and into the cytoplasm, and are referred to as cells with destabilized lysosomes. Therefore, the presence of NR in the cytoplasm reflects the efflux of lysosomal contents into the cytosol, which ultimately causes cell death.[13] We use a single-time-point assay, in which the endpoint for a site or treatment is based on microscopic examinations after a set incubation period (e.g., 60 min) of cell preparations from multiple individuals in which cells are scored as stabilized or destabilized, and the percentages of cells that are destabilized are determined. Other investigators use a NR retention time endpoint, a multiple-time-point assay in which individual cell preparations are examined repeatedly, and the time duration required until approximately 50% of the cells are destabilized cells is determined. The initial processing steps for both approaches are the same; only the endpoints are slightly different.

Materials required

- Ultrapure deionized water should be used for all buffers, and all chemicals should be cell biology grade.
- Calcium magnesium free saline (CMFS), used for all species; 20 m*M* HEPES, 360 m*M* NaCl, 12.5 m*M* KCl, and 5 m*M* tetrasodium EDTA
  - 4.766 g HEPES, 20.00 g NaCl, 0.932 g KCl, and 1.901 g EDTA in 995 ml DI H_2O
  - Adjust pH to 7.35–7.40 with 6 *N* NaOH
  - Adjust final volume to 1000 ml and filter through a 0.45-μm screen
  - Check pH and salinity just prior to use (salinity should be approximately 25‰)
  - Store refrigerated up to 5 days

- Magnesium free saline (MFS), used for shrimp and fish; 20 m*M* HEPES, 480 m*M* NaCl, 12.5 m*M* KCl, and 5.0 m*M* $CaCl_2$
  - 0.477 g HEPES, 2.661 g NaCl, 0.093 g KCl, and 0.055 g $CaCl_2$ in 99.5 ml DI H_2O
  - Adjust pH to 7.50–7.55 with 6 *N* NaOH
  - Adjust final volume to 100 ml with DI H_2O and filter through a 0.45-μm screen
  - Check pH and salinity just prior to use (salinity should be approximately 25‰)
  - Store refrigerated up to 5 days.
- Trypsin (bovine) (1.0 mg/ml), prepare just prior to use
- Collagenase (Type IV) (1.0 mg/ml), prepare just prior to use
- NR dye, Prepare just prior to use.
  - Make a 1° stock solution by adding 4 mg of NR powder to 1 ml of DMSO
  - Make a 2° stock solution (NR concentration of 0.04 mg/ml) by adding 100 μl of the 1° stock solution to 9.90 ml CMFS; wrap in foil to protect from light and keep at room temperature
- Hepatic or hepatopancreatic tissues (small piece, approximately 2–4 mm square, cleaned of extraneous tissues, and minced)
- 24-well cell culture plates
- Nylon mesh
- Reciprocating shaker
- Microcentrifuge tubes (2 ml)
- Pasteur pipettes
- Pipetter and tips
- Centrifuge
- Compound microscope
- pH meter
- Refractometer

Procedures

While we have more experience with oysters, like other investigators, we have been able to successfully modify the assay for other species, including marsh mussels (*Geukensia demissa*), as well as grass shrimp and mummichogs. In general, these methods have been readily applicable to a wide range of organisms with fairly minor modifications, a valuable attribute for cellular response assays, so these techniques should be transferable to other mollusks, crustaceans, and fish species. As part of our routine sampling protocols, the animals are kept cool (not cold) in aerated water collected on site from the field to the laboratory. We have also conducted assays with oysters that have been shipped cool (e.g., approximately 5–10°C) or held under refrigeration overnight; we have conducted holding studies and found that oysters can be kept cool in the refrigerator for up to 3 days before significant increases in lysosomal destabilization are observed; we routinely keep our analysis time within 2 days, preferably within 1 day of collection.

The basic protocols used for oysters are summarized in Figure 16.1 and described in more detail in the following. Only minor modifications were required for shrimp and fish, indicated with a numerical superscript in Figure 16.1, and described in the following:

1. The first series of steps involves dissection of hepatopancreatic (oysters and shrimp) or liver (fish) tissues and chemical dissociation of the tissues. Since in oysters the gonadal tissues are closely associated with the hepatopancreatic tissues, it is important to trim away as much gonadal tissue as possible and

Oyster lysosomal destabilization assay

1. Dissect out a small piece of hepatopancreas tissue (2–4 mm square), and rinse with CMFS; mince into smallpieces, rinse again, and place tissues into a 24-well plate (each with 600 μl cold CMFS).[a]
- Shake on a reciprocating shaker (100–120 rpm for 20 min).
- [b]Add 400 μl trypsin (in CMFS) to each sample, and shake for 20 min.
- Keep cool through all steps. Conduct initial dissection steps on chilled microscope slides placed on a Petri dish filled with ice; keep plates cool (e.g., the plate can be placed in a plastic container on top of a cold ice pack covered with a towel as shown).

2. Shear samples with a glass pipette, and transfer to a microcentrifuge tube/filter apparatus (e.g., place a clean square of nylon mesh with a trimmed pipette tip in the tube to form a filter apparatus as shown).
- Centrifuge at 200–225 g, 5 min (15°C).
- Remove filter, discard supernatant, and resuspend cells in 1 ml CMFS.
- Perform 1–2 rinses with CMFS, i.e., centrifuge at 200–225 g, 5 min (15°C).
- Resuspend cells in 50–300 μl CMFS, depending on size of cell pellet.

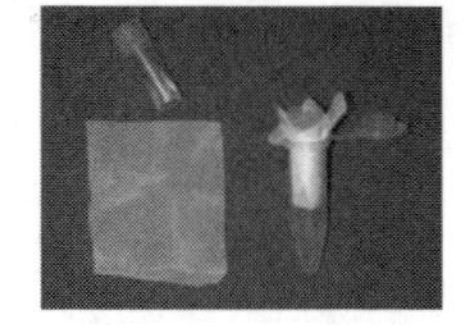

3. Add NR solution (2^0 stock, 0.04 mg/ml) to cells using a volume equal to that of the final volume, and mix gently with a plastic pipette tip. Stagger the time of addition of NR to each tube by 1–2 min to allow reading time (typically 20–30 samples are processed at the same time).
- Store in a light protected chamber at room temperature.

4. Score the cells using a compound microscope after 60 min[c] (e.g., mix gently and transfer an aliquot of cells to a microscope slide; this step may be conducted for a set of slides by placing themin a dark humidified chamber approximately 5–15min before the reading time).
- Score cells (≥50 from each individual preparation) using a 40× lens, as either dyepresent in the lysosome (stable) or dye present in the cytosol (destabilized). Score only hepatic cells that contain lysosome
- Calculate percentage of cells with destabilized lysosomes for each individual.

Figure 16.1 Annotated flow diagram of the lysosomal destabilization assay with oysters. The superscripts indicate the steps that must be modified for use with grass shrimp and mummichogs as follows: [a]Use 500 μl CMFS for shrimp and fish; [b]add 500 μl collagenase in MFS for shrimp and fish; [c]score shrimp and fish cells after 90 min.

rinse well (i.e., by pipetting the tissues on the cold microscope slide) with CMFS to remove extraneous gametes. For the fish tissues, extensive rinsing is required to remove excess blood cells. For all species, CMFS is used to initiate dissociation of cell junctions (while the CMFS solution described here was suitable for these three estuarine species, it may be necessary to adjust the osmolality, primarily by adjusting the NaCl concentrations, for species from higher or lower salinity habitats). Then for oysters, trypsin is used to further facilitate dissociation. However, shrimp and fish tissues have more collagen in their connective tissues, so collagenase rather than trypsin is used. The activity of collagenase requires calcium and also has a somewhat higher pH requirement, so it is made up in an MFS, pH 7.5, which contains the necessary calcium.

2. The next series of steps involves physical dissociation of the tissues and rinsing of the cellular preparation. The cells should be gently sheared as they are drawn in and out of a glass Pasteur pipette, and then transferred to a microcentrifuge tube/

filter apparatus. The entire apparatus is then gently centrifuged (approximately 500*g*) and the cells are pulled through the nylon screening during centrifugation to complete the dissociation. The cells are then rinsed and resuspended in CMFS. A microcentrifuge placed in a cooled chamber can be used, as well as a refrigerated centrifuge.

3. An equal volume of the NR solution (2° stock, 0.04 mg/ml) is then added to each cell suspension and then the cells should be mixed gently with a pipetter (vortexing is not recommended). NR is a light-sensitive vital dye, so it is important to minimize light exposure through all steps.
4. After the incubation period in NR (60 min for oysters; 90 min for grass shrimp and mummichogs), the cells should be scored as either stable (NR contained within lysosomes) or destabilized (NR leaking out of enlarged and damaged lysosomes and diffusing into the cytoplasm) (Figure 16.2). For fish and shrimp, a slightly longer incubation period (90 min) is required because the initial uptake of NR into the lysosomes is somewhat slower. Since these are mixed cell preparations, it is important to establish rules or criteria for scoring the cells. The rules that we typically use are:

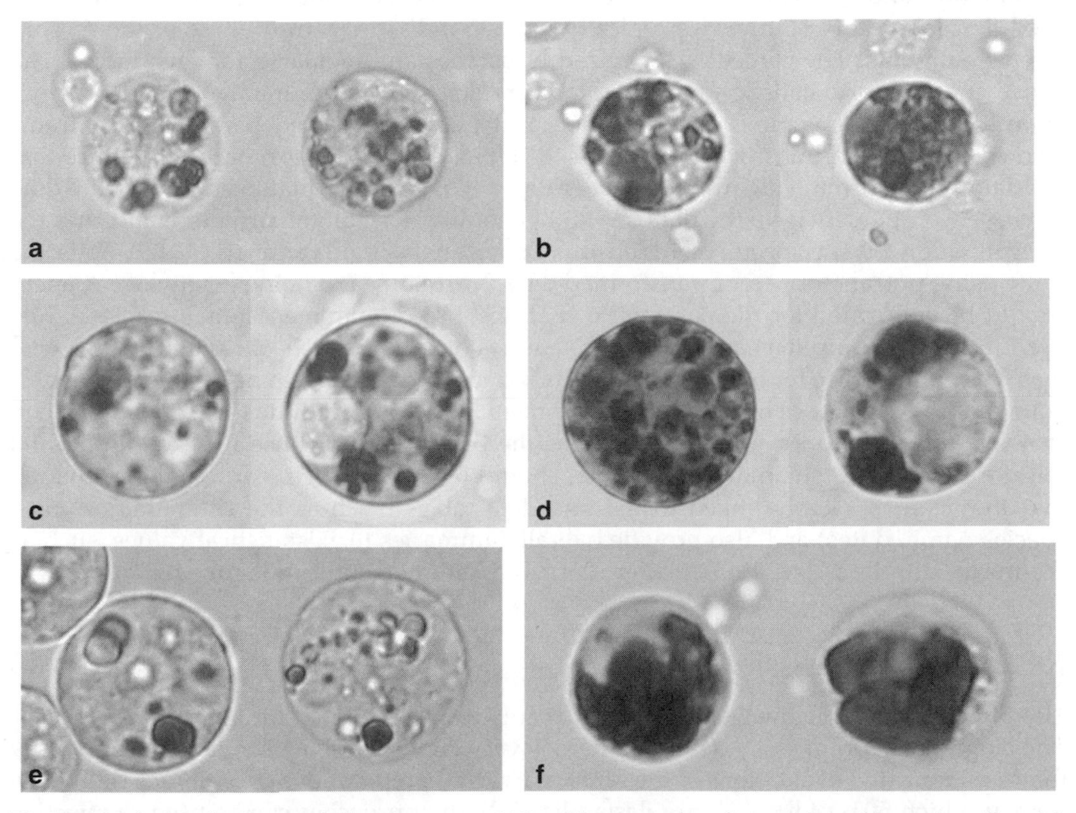

Figure 16.2 **(see color insert following page 464)** Photographs of examples of hepatic cells with stable and destabilized lysosomes for oysters, grass shrimp, and mummichogs: (a) oyster hepatopancreas cells scored as stable; (b) oyster hepatopancreas cells scored as destabilized; (c) grass shrimp hepatopancreas cells scored as stable; (d) grass shrimp hepatopancreas cells scored as destabilized; (e) fish liver cells scored as stable; (f) fish liver cells scored as destabilized.

- Score only those cells within a certain size range (oysters, 25–40 μm; shrimp, 60–75 μm; mummichogs, 35–50 μm).
- Score only those hepatic cells that accumulate NR with distinct lysosomes (the preparations sometimes also contain extraneous egg or yolk cells that will appear almost solid red especially during reproductive periods). Since the cells are still in CMFS at this stage, the cells will remain in a somewhat rounded state, so it is important to focus up and down a little with the microscope to observe the different planes of each cell.
- Score at least 50 cells from each individual preparation.
- Score conservatively. Most cells are readily scored as stable or destabilized. However, there may be some cases where there is greater uncertainty, but we typically score the somewhat marginal cells as stable. This assay is a microscopic assay, in which the live cells are evaluated by a reader, so there is a level of subjectivity. Reader training is critical, and routine QA is required. At least one sample from each preparation should be reviewed by a second reader. A library of photographic images can also be maintained for reference.

Using the single-time-point assay, studies with field-collected animals are typically based on as many animals as possible, preferably 20 individuals, but as few as 10 have been used. For laboratory studies, we typically use 3–5 individuals from three or more replicated treatments. The percent destabilized lysosomes are calculated for each individual, and then summary statistics are generated for each site (mean, median, standard deviation, quartiles). It is often recommended that percentage-based parameters be transformed (e.g., arcsin transformation) prior to statistical analyses, primarily to ensure that the data set is normally distributed. Generally, we have found that even though the data are based on percentages, the data are still generally normally distributed, and that the interpretation of site-specific comparisons using analysis of variance (ANOVA) is the same using untransformed or transformed data. Normality and homogeneity of variances should be confirmed for the parametric analyses, or non-parametric analyses should be used if these assumptions are violated. Examples of the spreadsheet formats (e.g., Excel) that we routinely use for raw data include information on organism height, length, sex, or gonadal index (an index of gonadal development based on 1 for little to no gonad present, to 4 for fully developed gonad), as well as the biomarker response for each individual (Figure 16.3a). The "summary" data tables should include enough redundant information that they can be clearly linked to the raw data tables (such as the site name or code, species name, dates) and also provide overall summaries (e.g., statistical values, such as the mean, standard deviation, median, 25th and 75th percentiles) (Figure 16.3b).

Results and discussion

The NR assay is a relatively simple assay that is readily adapted for a variety of species. The single-time-point assay described here is very similar to the NR retention time assay that uses repeated evaluations of the same individual preparations over time to define the time at which 50% of the cells are destabilized. Both approaches can be used as effective biomarkers of contaminant exposure. In our laboratory, we were faced with processing resident animals or those from caged deployments from multiple sites over a limited window of time, so the retention time approach was logistically difficult and would allow only a small sample size. After a careful review of the existing data on mussels, there was

(a) Example of raw data table for the lysosomal destabilization assay.

Site	Species	Sampling date	Animal #	Height (cm)	Length (cm)	Gonadal index	Percentage of lysosomal destablization	Lysosomal analysis date	QA code
CIAAAW00	crasvirg	2/9/2000	1	9.3	4.1	4	24.53	2/10/2000	
CIAAAW00	crasvirg	2/9/2000	2	9.0	4.4	3	26.56	2/10/2000	
CIAAAW00	crasvirg	2/9/2000	3	9.1	3.5	3	29.63	2/10/2000	
CIAAAW00	crasvirg	2/9/2000	4	8.6	3.6	4	26.23	2/10/2000	
CIAAAW00	crasvirg	2/9/2000	5	9.4	3.8	3	25.97	2/10/2000	
CIMOSW00	crasvirg	4/12/2000	1	6.7	2.6	4	47.17	4/13/2000	
CIMOSW00	crasvirg	4/12/2000	2	2.9	2.2	2	39.22	4/13/2000	
CIMOSW00	crasvirg	4/12/2000	3	2.9	2.3	4	49.06	4/13/2000	
CIMOSW00	crasvirg	4/12/2000	4	5.2	2.2	3	63.64	4/13/2000	
CIMOSW00	crasvirg	4/12/2000	5	6.9	2.5	4	36.67	4/13/2000	

(b) Example of summary data table for the lysosomal destabilization assay.

Site	Species	Assay	#Animals	Mean (%) Lysosomal Destabilization	STD	Median (%) Lysosomal Destabilization	25%	75%	QA code
CIAAAW00	crasvirg	Lyso	5	26.58	1.87	26.2	25.61	27.33	
CIMOSW00	crasvirg	Lyso	5	47.15	10.59	47.1	38.58	52.71	

Figure 16.3 Examples of raw data and summary tables used for the lysosomal destabilization assay for oysters for two sites. Site and species codes are followed by ancillary data, such as information on size, gonadal condition, etc.

a consistent general pattern that those from polluted sites had NR retention times of less than 60 min, often less than 30 min, whereas those from reference sites were often stable for hours. After some preliminary laboratory studies, we decided that a single-time-point of 60 min should provide good discrimination between treatments and sites. By using this approach, we were then able to substantially increase our sample size. Furthermore, it has been our experience that in most cases, cells harvested from oysters from control conditions or clean reference sites have very low destabilization rates for many hours (good morning-preparations can often be used for afternoon demonstrations or laboratory exercises); those exposed to contaminants often begin to show significant destabilization within 30 min.

The primary difference between the processing of the different species was that CMFS and trypsin are used as the primary means of chemical dissociation, whereas with the shrimp and fish, CMFS and collagenase are used due to the higher collagen content of their tissues. In general, oysters and mussels are the easiest to process because both have substantial amounts of hepatopancreatic tissue, and generate high quality cell preparations. Grass shrimp were the hardest to process because they have smaller hepatopancreases. With grass shrimp, there were sometimes problems with consistency in acquiring high quality cell preparations, as some preparations were very sticky and cell recoveries were sometimes very low while most preparations were of very high quality. Mummichogs also have very small livers, but generally enough cells could be obtained for the assay. Tissue pieces had to be carefully rinsed to avoid contamination by excess blood cells. The NR incubation period for grass shrimp and mummichogs is a little longer than that used for mollusks. In general, more work is needed with both of these species to continue to refine cell-handling protocols, but these studies have served to establish the feasibility of the general approaches.

The most challenging aspect of any biomarker is developing a framework for interpretation. To do this effectively requires a sound basis for interpreting cellular data, including expected values and an appreciation of the potential variation, as well as the sources of the variation. In the biomedical context, this is analogous to defining the normal range of responses. With a good baseline knowledge of what is normal or what is expected, cellular assays are used as diagnostic tools in medical applications, as warning signals of early disease conditions, for prognosis, and evaluating the effectiveness of remedies. Our overall goal has been to apply this kind of framework to estuarine organisms as a means of characterizing organismal health and habitat quality. In the medical arena, it is sometimes necessary to define different normal ranges based on age, sex, etc. In estuarine habitats, species-specific responses may be affected by habitat differences, such as salinity regimes, seasonal differences, etc., as well as age and sex. Some physiological processes, such as overall metabolic rate and growth of oysters, are known to be affected by a variety of natural habitat factors,[16] which can make it difficult to distinguish contaminant stress from differences associated with natural factors. Therefore, one of the goals of our biomarker studies has been to identify essential cellular functions that are not readily affected by "natural stressors" (e.g., have less variation over broad environmental regimes) but are affected by contaminants. Ultimately, our ability to correctly interpret the responses is based on the robustness of the normal range criteria. In human medicine, the normal ranges are based on an extensive database from a wide range of sources; and likewise we draw on both laboratory studies and field studies from a wide range of habitats to delineate the "normal" responses that are characteristic of a healthy estuarine organism.

Examples of some of the ways that we have tried to address these important interpretation issues, as well as some of the kinds of data generated, are presented, particularly with regard to distinguishing contaminant effects from potential salinity effects. Laboratory studies were used to determine the effects of salinity, a variable that can range from nearly fresh water to full-strength seawater within an estuary. Oysters are typically found in low, as well as high, salinity regimes. Laboratory studies were conducted with oysters and shrimp (and also marsh mussels), in which animals were collected from a reference, mid-range salinity site (e.g., approximately 25‰), returned to the laboratory, and immediately placed in one of three treatments: 10‰, 20‰, or 30‰ seawater, with no acclimation. Then the lysosomal destabilization responses were evaluated over time, e.g., within 24 h in some cases and for extended periods up to 2 weeks. The results of these kinds of studies have consistently indicated that lysosomal destabilization rates are not different between any of the treatments in both oysters and grass shrimp (Figures 16.4a and b); results with marsh mussels are virtually the same as that observed with oysters.

Similar studies have been conducted with pollutants (e.g., Cu, benzo-*a*-pyrene), in which oysters and shrimp were exposed to a range of concentrations and sampled after various time points (from 18 h to 14 days). The patterns with both oysters and shrimp were very different from those observed for the salinity studies. After only 18 h, there were significant increases in oyster lysosomal destabilization rates that persisted over time, in a dose–dependent manner (Figure 16.4c). Significant elevations in lysosomal destabilization rates were also observed in 7-day exposures in grass shrimp (Figure 16.4d). Numerous laboratory studies have supported the model that lysosomal destabilization in

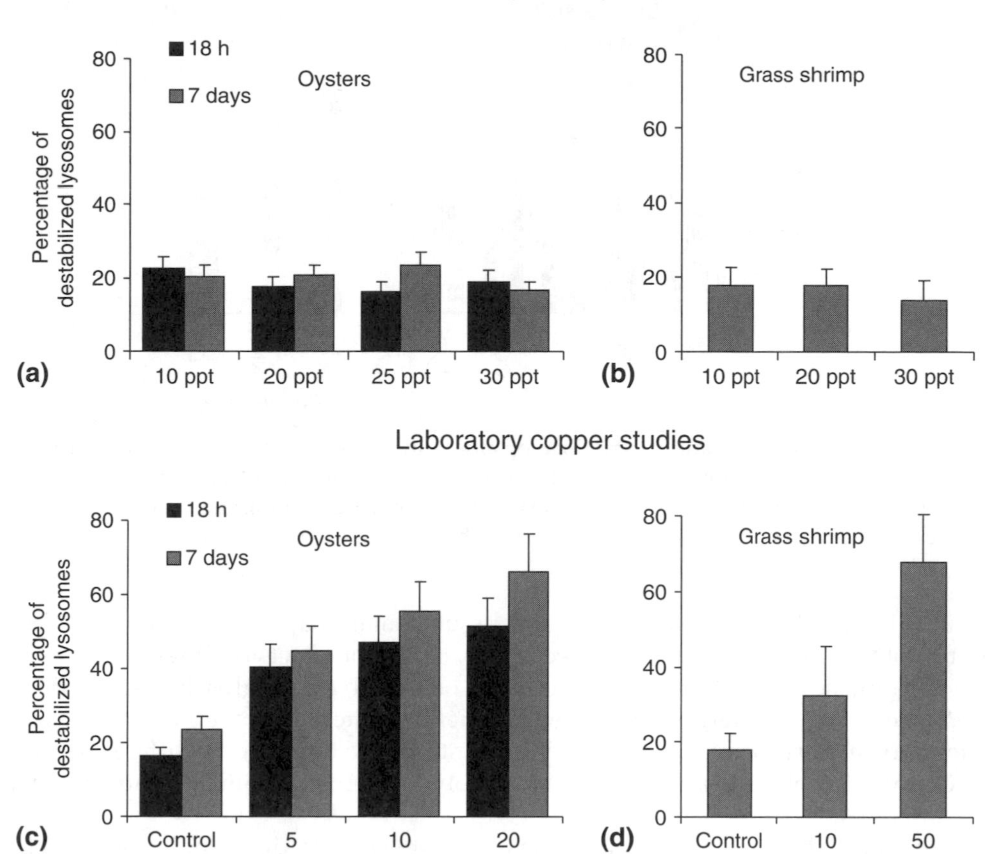

Figure 16.4 Results of laboratory studies regarding the effects of salinity on lysosomal destabilization in (a) oysters (modified from data published in Ringwood et al., 1998) and (b) grass shrimp. For the salinity studies, animals collected from 20 to 25 ppt (‰) salinity regimes were placed in lower and higher salinities, and lysosomal destabilization was evaluated over time. Results of laboratory studies regarding the effects of copper exposures on lysosomal destabilization in (c) oysters (modified from data published in Ringwood et al., 1998) and (d) grass shrimp. All Cu treatments were significantly higher than controls. Data are means + standard deviations.

mollusks and grass shrimp is not significantly affected by different salinity conditions but increases dramatically when animals are exposed to toxic chemicals.

Extensive field studies with both caged oysters deployed *in situ* and organisms collected from resident populations have also been conducted. Lysosomal destabilization rates of organisms from reference sites remain low over broad salinity regimes; this is consistent with the results of the laboratory studies. Also consistent with the laboratory studies, oysters from polluted sites have significantly higher lysosomal destabilization rates as shown in Figure 16.5; similar results have been observed in oysters collected from polluted sites in Galveston Bay, Texas.[17] Therefore with oysters, field data further support the model that lysosomal destabilization is not significantly affected by salinity regimes but increases in response to contaminants. More details of field and laboratory work have been published and include additional issues, e.g., lysosomal responses were

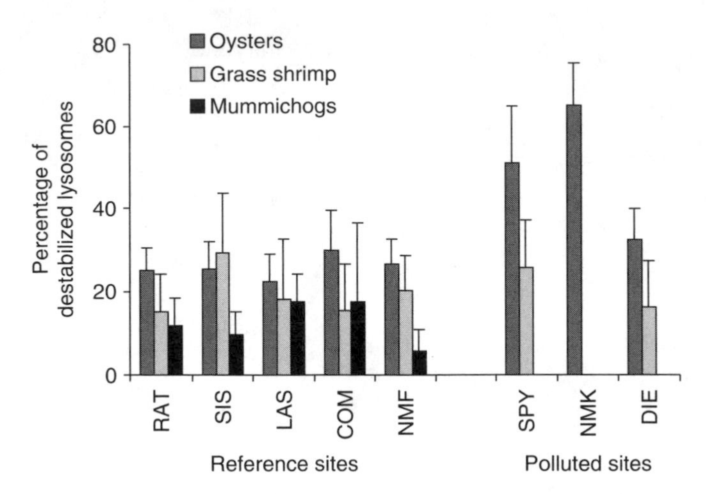

Figure 16.5 Results of lysosomal destabilization assays in field-collected oysters, grass shrimp, and mummichogs during the summer of 2000. The first five sites were classified as reference sites. The last three sites were polluted sites from Charleston Harbor, SC; the missing data for shrimp and mummichogs were due to very low collection numbers of these species in polluted systems.

significantly correlated with sediment and tissue contaminants.[15,17–19] We have seen no evidence of significant differences based on sex or size or effects of different temperature regimes. Some seasonal differences have been observed, e.g., higher lysosomal destabilization rates in the winter, but the rates are highly related to tissue metal levels, which also tend to be more bioavailable during the winter; since there are no differences at some very clean reference sites, the assay may truly reflect contaminant loads more than seasonal issues.[20]

With grass shrimp and mummichogs, the site-specific differences have been less clear-cut, e.g., organisms from polluted sites do not consistently have higher destabilization rates. Currently, there are only limited data for these species, in part, because we often had difficulty collecting shrimp and fish at polluted sites, in spite of repeated visits, different sampling techniques, etc., we caught only a few or no organisms. It could reflect that the animals have already died from the conditions or that they are mobile and have left the system. For those collected, there is generally not an easy way to know if the animals have been exposed for short or extended periods of time. Certainly, one of the advantages of working with sessile organisms like oysters and mussels is that they can not escape exposure to contaminants when they are present, so their responses reflect the integration of exposure and detoxification responses.

Overall, this body of work conducted since 1995 reinforces the value of the lysosomal assay as a valuable indicator of contaminant exposure. We have worked extensively at a range of contaminated sites, but have also focused on working at as many reference or uncontaminated sites as possible to be able to identify and characterize the variation associated with the normal ranges of responses. Using a much larger data set based on the kinds of data illustrated in the figures, we have begun to define normal ranges, as well as toxicologically significant levels, for various biomarker responses, including lysosomal destabilization. The single-time-point percentage endpoint results are highly repeatable between assays, so our interpretation can be based on a larger data base and not have to rely on specific assay controls or a limited number of reference sites. Summaries of data (means and standard deviations) can be used for estimating normal ranges for oysters, grass shrimp, and mummichogs for lysosomal destabilization. Based on data for field-collected organisms during the summers of 1999 and 2000, compiled from 10 to 17 reference sites and >200 individuals per species, the normal ranges (means and standard deviations) were:

Oysters: 25.21 $\pm$ 4.08
Grass shrimp: 20.92 $\pm$ 10.33
Mummichogs: 16.97 $\pm$ 16.61

Overall, oyster lysosomal destabilization rates <30% are associated with reference or control conditions, e.g., normal responses of healthy oysters. These levels are believed to represent the normal background levels of lysosomal destabilization, and probably reflect a combination of low normal turnover rates of hepatic cells, as well as some low level damage associated with the assay itself. For grass shrimp and mummichogs, these levels may be a little lower, around 20%; although the variation was greater for shrimp and fish, and more work is needed for these species.

Therefore, we have some confidence with the oysters that based on the data from contaminated sites and laboratory exposure studies, as well as those under reference conditions, various toxicity criteria can be established. For example, consider the following criteria:

Oyster	Normal range	Concern	Stress
Lysosomal destablization	<15–30%	30–40%	>40%

The "Normal range" defines the optimum conditions and indicates that there is no evidence of stress. The "Stress" levels are believed to indicate that homeostatic and detoxification mechanisms have been overwhelmed, and that the oysters are significantly stressed. From our experience, these levels are consistently observed at the most polluted field sites or at the highest contaminant exposures. The "Concern" values represent levels that are somewhat outside the normal range limits, and should be regarded as indicating that the animals are experiencing some exposures or stressful conditions. These are levels that are often observed at moderately polluted sites. We hypothesize that lysosomal destabilization levels at or above the 40% stress level would represent serious damage that would translate into significant organismal effects. It is not difficult to comprehend that organisms with more than 40% dysfunctional hepatic cells would experience significant physiological impairment or death. Recent studies regarding relationships between gamete viability and lysosomal destabilization have supported this hypothesis, e.g., very poor rates of normal development are observed with gametes from parents whose lysosomal destabilization rates are >40%.[21] Oyster levels in both field and laboratory studies have gone as high as 60–70%, which could represent a threshold for acute toxicity. The laboratory studies with shrimp also support this. In the shrimp studies shown in Figure 16.4b, there was also a 100-ppb treatment, and the sole shrimp that survived had 70.4% destabilized lysosomes. An important question is whether or not the levels that we have defined as levels of concern reflect reversible effects if conditions improve.

It is important to consider how issues regarding statistical and biological significance may be relevant for interpretation of biomarker studies. For example, in some specific cases, control or reference conditions may be sufficiently low and have a low variance term, so that statistical differences may be observed at levels less than the 30% normal criteria. Based on our broader experience at reference sites, lysosomal destabilization rates of <30% are not believed to represent biologically significant perturbations even though the results may be statistically significant. Therefore, we prefer to use a combination of biologically and statistically significant requirements for identifying significant toxicological effects, e.g., >30% and statistically significant from reference conditions.

Finally, as with any biomedical tool, the techniques, utility, and interpretation of biomarker responses will continue to be refined and characterized. One of the obvious future directions that are currently being explored is the development of an image analysis system for the lysosomal destabilization assays. Photographic analyses and records would provide an archival record of the assay and allow more rigorous quality control for scoring the cells. However, there are some important issues that need to be considered. For example, since the cells are held in CMFS and remain somewhat rounded and three-dimensional, it is difficult to appropriately evaluate cells by still photography. Real-time video scans of the slides with up-and-down focus would be better but can require more time (and requires higher equipment funds) than is typically needed for analyzing an individual slide; a trained reader typically can read a slide in 1–2 min, and since the assessments are with live cells, it is desirable to process the slides as quickly as possible. There have been some discussions about using the quantity of NR in cells as an index of destabilization, but it is important to realize that the distribution of NR is more important than the quantity. Cells with lots of lysosomes may accumulate a lot of NR, but as long as it is confined to the lysosomes, the functional integrity is maintained.

While there are a number of valuable reasons for using hepatic preparations, we like a number of investigators have used these techniques with circulating hemocytes and blood cells. Circulating hemocytes of bivalves and crustaceans are particularly relevant immunologically. Bivalve hemocytes may be sampled either destructively via intracardial puncture or nondestructively by withdrawing hemolymph from the adductor muscle. Nondestructive sampling techniques (i.e., the organism is not killed) have a variety of advantages, including the allowance for repeated sampling on a single individual and sampling of rare or threatened species.[22] Good candidates for these approaches include the freshwater Unionid mussels, some of which are among the most imperiled fauna in North America.[23] Moreover, adapting these techniques for fresh water species should serve to improve our ability to monitor and manage those groups, as well as estuarine species.

In summary NR techniques are relatively simple techniques for evaluating lysosomal function and organismal health that can be used in a variety of organisms. The ability to interpret the results requires a substantial effort under non-polluted conditions in order to build a sound database for defining criteria associated with toxicological effects. It should be appreciated that these ranges represent a combination of "scientific art" that draws on knowledge of mechanisms and experience, as well as statistical rigor and lots of samples. More data, from both laboratory and field studies, are needed for grass shrimp and mummichogs. With oysters, the considerable laboratory and field efforts have enabled us to conduct assessments and begin to interpret the results with a high degree of certainty (it has been used recently for oil spill, superfund, and other habitat assessments). It has a number of valuable attributes, including relatively inexpensive, rapid, and is believed to provide high value information regarding organismal health. Broader use of cellular biomarkers should improve our ability to monitor and manage important estuarine species, and protect these important habitats.

Acknowledgments

We thank C.J. Kepper, M. Gielazyn, and B. Ward for their assistance with the field work and various aspects of these studies. This research was supported in part by a grant from NOAA/UNH Cooperative Institute for Coastal and Estuarine Environmental Technology (CICEET), NOAA Grant Number NA87OR0512.

References

1. Culliton, T.J., Warren, M.A., Goodspeed, T.R., Remer, D.G., Blackwell, C.M. and McDonough, J.J., III, 50 Years of Population Change along the Nation's Coasts, 1960–2010, The Second Report of a Coastal Trends Series, OCRA/NOS/NOAA, Silver Springs, MD, 1990, p. 41.
2. Cohen, J.E., Small, C., Mellinger, A., Gallup, J. and Sachs, J., Estimates of coastal populations, *Science*, 278, 1211–1212, 1997.
3. Beck, M.W., Heck, K.L., Jr., Able, K.W., Childers, D.L., Eggleston, D.B., Gillanders, B.M., Halpern, B., Hays, C.G., Hoshino, K., Minello, T.J., Orth, R.J., Sheridan, P.F. and Weinstein, M.P., The identification, conservation, and management of estuarine and marine nurseries for fish and invertebrates, *Bioscience* 51, 633–641, 2001.
4. Rainbow, P.S. and Phillips, D.J.H., Cosmopolitan biomonitors of trace metals, *Mar. Pollut. Bull.*, 26, 593–601, 1993.
5. Moore, M.N., Cellular responses to pollutants, *Mar. Pollut. Bull.*, 16, 134, 1985.
6. Moore, M.N., Reactions of molluscan lysosomes as biomarkers of pollutant-induced cell injury, in: *Contaminants in the Environment: A Multidisciplinary Assessment of Risks to Man and Other Organisms*, Lewis Publishers, Boca Raton, FL 1985, pp. 111–123.
7. Lowe, D.M. and Fossato, V.U., The influence of environmental contaminants on lysosomal activity in the digestive cells of mussels (*Mytilus galloprovincialis*) from the Venice Lagoon, *Aquat. Toxicol.*, 48, 75–85, 2000.
8. Ohkuma, S., Moriyama, Y. and Takano, T., Identification and characterization of a proton pump on lysosomes by fluoroscein isothiocyanate-dextran fluorescence, *Proc. Nat. Acad. Sci. USA*, 79, 2758–2762, 1982.
9. Lowe, D.M., Fossato, V.U. and Depledge, M.H., Contaminant-induced lysosomal membrane damage in blood cells of mussels *Mytilus galloprovincialis* from the Venice Lagoon: an in vitro study, *Mar. Ecol. Prog. Ser.*, 129, 189–196, 1995.
10. Viarengo, A. and Nott, J.A., Mechanisms of heavy metal cation homeostasis in marine invertebrates, *Comparative Biochem. Physiol. C.*, 104, 355–372, 1993.
11. Lowe, D.M., Moore, M.N. and Evans, B.M., Contaminant impact on interactions of molecular probes with lysosomes in living hepatocytes from dab *Limanda limanda*, *Mar. Ecol. Prog. Ser.*, 91, 135–140, 1992.
12. Krishnakumar, P.K., Casillas, E. and Varanasi, U., Effect of environmental contaminants on the health of *Mytilus edulis* from Puget Sound, Washington, USA. 1. Cytochemical measures of lysosomal responses in the digestive cells using automatic image analysis, *Mar. Ecol. Prog. Ser.*, 106, 249–261, 1994.
13. Lowe, D.M., Soverchia, C. and Moore, M.N., Lysosomal membrane responses in the blood and digestive cells of mussels experimentally exposed to fluoranthene, *Aquat. Toxicol.*, 33, 105–112, 1995.
14. Regoli, F., Lysosomal responses as a sensitive stress index in biomonitoring heavy metal pollution, *Mar. Ecol. Prog. Ser.*, 84, 63–69, 1992.
15. Ringwood, A.H., Conners, D.E. and Hoguet, J., Effects of natural and anthropogenic stressors on lysosomal destabilization in oyster *Crassostrea virginica*, *Mar. Ecol. Prog. Ser.*, 166, 163–171, 1998.
16. Kennedy, V.S., Newell, R.I.E. and Eble, A.F., *The Eastern Oyster* Crassostrea virginica, Maryland Sea Grant Books, College Park, MD, 1996, p. 734.

17. Hwang, H.-M., Wade, T.L. and Sericano, J.L., Relationship between lysosomal membrane destabilization and chemical body burden in eastern oysters (*Crassostrea virginica*) from Galveston Bay, Texas, USA, *Environ. Toxicol. Chem.*, 21, 1268–1271, 2002.
18. Ringwood, A.H., Conners, D.E., Keppler, C.J. and DiNovo, A.A., Biomarker studies with juvenile oysters (*Crassostrea virginica*) deployed *in-situ*, *Biomarkers* 4, 400–414, 1999.
19. Ringwood, A.R., Conners, D.E. and Keppler, C.J., Cellular responses of oysters, *Crassostrea virginica*, to metal-contaminated sediments, *Mar. Environ. Res.*, 48, 427–437, 1999.
20. Ringwood, A.H., Hoguet, J. and Keppler, C.J., Seasonal variation in lysosomal destabilization in oysters, *Crassostrea virginica, Mar. Environ. Res.*, 54, 793–797, 2002.
21. Ringwood, A.H., Hoguet, J., Keppler, C. and Gielazyn, M., Linkages between cellular biomarker responses and reproductive success in oysters, *Mar. Environ. Res.*, 2004, 58: 151–155.
22. Fossi, M.C., Nondestructive biomarkers in ecotoxicology, *Environ. Health Perspect.*, 102 (Suppl.), 49–54, 1994.
23. Ricciardi, A. and Rasmussen, J.B., Extinction rates of North American freshwater fauna, *Conserv. Biol.*, 13, 1220–1222, 1999.

chapter seventeen

IMCOMP-P: An assay for coral immuno-competence

Craig A. Downs
EnVirtue Biotechnologies, Inc.
Aaron G. Downs
Yale University
Robert B. Jonas and Kay Marano-Briggs
George Mason University
Thomas Capo
University of Miami
Cheryl M. Woodley
National Oceanic and Atmospheric Administration

Contents

Introduction

Corals have recently been afflicted with epizootic outbreaks in the Caribbean, the Great Barrier Reef, and in the Indian Ocean.[1,2] These disease outbreaks have caused mortality rates as high as 90% on some reefs.[1] Only recently have some of the pathogens been identified.[1,3,4] *Vibrio shiloi* or similar species/strains[5] can cause bleaching or rapid tissue lysis in corals. Recent evidence indicates that *Serratia marcescens* causes white pox on corals.[4] Some evidence suggests that environmental factors may alter pathogen physiology, inducing a more infectious or pathogenic state[3] or alternatively that environmental conditions compromise coral defense mechanisms, rendering them more susceptible to infection. Unfortunately, little is known of coral defense systems (immunology), other than having allorecognition[6] and phagocytic cells.[7,8] Microbial pathology of coral disease is an issue that

1-56670-664-5/05/$0.00+$1.50
© 2005 by CRC Press

continues to baffle scientists and resource managers. Is the disease occurring because of an introduction of a novel pathogen into the environment, addition of abiotic factors that induce pathogenicity (e.g., increased iron availability or increased temperature), or factors influencing a decrease in cnidarian immuno-competence? Resolving this issue is paramount in effectively understanding and managing coral disease outbreaks.

Innate immunity and antimicrobial peptides

Innate immunity is an ancestral defense system found in all multicellular organisms (including plant and animal phyla) and is considered the first line of defense against invading microorganisms. In contrast to the vertebrate's adaptive immunity with its complex network of humoral and cellular responses, the innate immune system, in general, is characterized by three types of processes: (1) phagocytosis, (2) initiation of proteolytic cascades, or (3) synthesis of antimicrobial peptides (AMPs).[9]

Due to their rapid production and diffusible nature, AMPs compose one of the most important elements in lower invertebrate defense systems.[10] The discovery of the first AMPs, gramicidin and tyrocidine,[11] catalyzed a whole discipline of research that has made an impressive progress in the past 60 years. Antimicrobial peptides are distributed throughout the entire animal kingdom. Some AMPs are as small as three amino-acid residues long, while others are as large as 45 kDa.[12] AMPs are diverse in structure and function[13] with some AMPs permeabilizing cell membranes or blocking ion channels, while others inhibit metabolic electron transport or protein chaperoning.[14,15]

Anti-microbial peptides have distinct secondary structures that lend potency to their function and are commonly categorized into four groups according to their secondary structure (Table 17.1).[16] AMP structures often contain a region of cationic (i.e., positively charged) amino acids. This cationic region allows the AMP to bind to negatively charged microbial membranes, providing a simple but effective targeting system. In non-polar solvents, such as microbial membranes, AMPs often form separate hydrophobic and hydrophilic domains, making AMPs *amphipathic* proteins.

There are several mechanistic models that describe how AMPs function.[16] The exact mechanism varies from AMP to AMP depending on the mode of antimicrobial action. The barrel-and-stave model used to describe the activity of some AMPs consists of the AMPs aggregating into barrel-like structures within the microbial membrane to form pores in the membrane (Figure 17.1, Panels A–C).[10,16] These barrel structures will increase in size as more AMPs bind to the membrane, increasing both the frequency and size of pores. These pores begin a process of diffusion of vital intercellular components from the microbe and culminate in cell death. The aggregate channel model, which describes the activity of AMPs, such as histatin 5, is similar in that the AMPs cause the formation of small transient pores in the membrane.[16] However, the function of these pores is not to let vital components out as in the barrel-and-stave model. Instead, the pores facilitate the invasion of the microbes by the

Table 17.1 Common classification of antimicrobial peptides according to structure (Adapted from Olano, C.T. and Bigger, C.H., *J. Invert. Pathol.*, 76, 176–184, 2000.)

Group	Structure
I	Linear peptide with an α helical structure
II	Linear peptide with high representation of specific amino acids
III	Peptides with loops
IV	Peptides that contain β-strands or other structural restraints

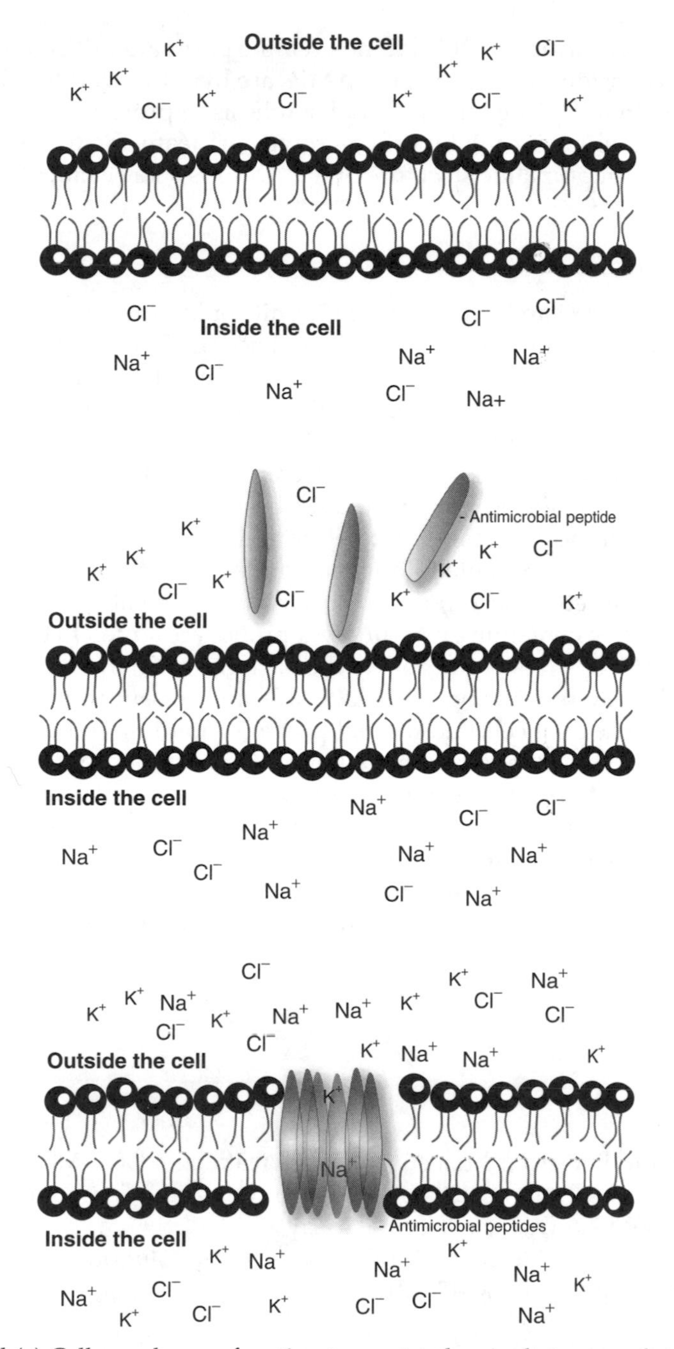

Figure 17.1 Panel (a) Cell membranes function to create a barrier between solutes outside the cell and solutes inside the cell. There are cellular pumps in the cell membrane that maintain this difference in solute composition and concentration on either side. Panel (b) AMPs (anti-microbial peptides) are produced by the coral and significantly contribute to the protein composition in coral mucus. AMPs can also be found within the cell, and within the extracellular space that exists between coral cells. Panel (c) Some types of AMPs can aggregate together and "punch" holes in the bacterial cell membrane, but not in the coral or zooxanthellae membranes. These AMPs create channels across the bacterial cell membrane.

AMP causing the pore formation. It is unknown if the pores are utilized by other AMPs to enter the cell. Once inside the microbe, the AMPs are hypothesized to inhibit a variety of cellular processes, from DNA expression and synthesis to protein folding to intercellular signaling events, resulting in inhibition of growth or death for the microbe.[16] In both of these models, the amphipathic structure of the AMPs is a key element to their function.

IMCOMP-P assay

To advance scientific research into coral immunity and coral epidemiology, we have developed an easy to use, inexpensive, quick, and accessible assay that is functional and quantitative. This assay measures one aspect of coral innate immunity, permeability of bacteria by a protein fraction of coral tissue: IMCOMP-P (immuno-competence-permeability). This assay is based on two fluorescent DNA probes, each having different accessibilities to microbial DNA. The total soluble protein fraction is extracted from the coral sample. The protein fraction is precipitated with acetone, and resolubilized in a non-denaturing buffer. This homogenate is filtered through a 10,000 molecular-weight cut-off filter, and the collected filtrate contains an abundance of putative coral AMPs. This filtrate is added to a normalized concentration of bacteria, and incubated for 30 min at room temperature. A solution containing two different probes are added to the bacterial culture. The green probe (SYTO 9®) is permeable to the bacterial membrane and can readily bind bacterial DNA, thereby fluorescing. Only if AMPs are present in the coral filtrate, thus permeabilizing the bacterial cell, can the red probe (propidium iodide) pass through the bacterial membrane. Once the red probe is within the bacterial cell, it will preferentially bind to the bacterial DNA and displace the DNA binding of the green probe, thereby quenching the green probe's fluorescent abilities. Thus, a decrease in the fluorescence of the green probe indicates the presence and action of AMPs (Figure 17.2 Panels A–D). This assay can be adapted to any bacterial, fungal, or algal species and requires only altering the bacterial growth conditions and using a different DNA binding fluorescent probe.

Materials required

Reagents:

- Propidium iodide from Molecular Probes (catalog #3566, Molecular Probes, OR, USA). A stock solution of 20 m*M* propidium iodide (Component A) in dimethyl sulfoxide (DMSO) was used for the assay
  - SYTO 9® from Molecular Probes as a stock solution of 3.34 m*M* in DMSO (Component B; catalog #7012)
  - DMSO
  - Guanidine HCL, EM Sciences
  - Trizma base, EM Sciences
  - EDTA, EM Sciences
  - Deferoxamine mesylate, Sigma-Aldrich
  - Phenylmethylsulfonyl fluoride (PMSF), Sigma-Aldrich
  - Acetone, EM Sciences
  - Sorbitol, Sigma-Aldrich
  - Millipore Microcon diafiltration tube with a molecular weight cutoff of 10,000 Da

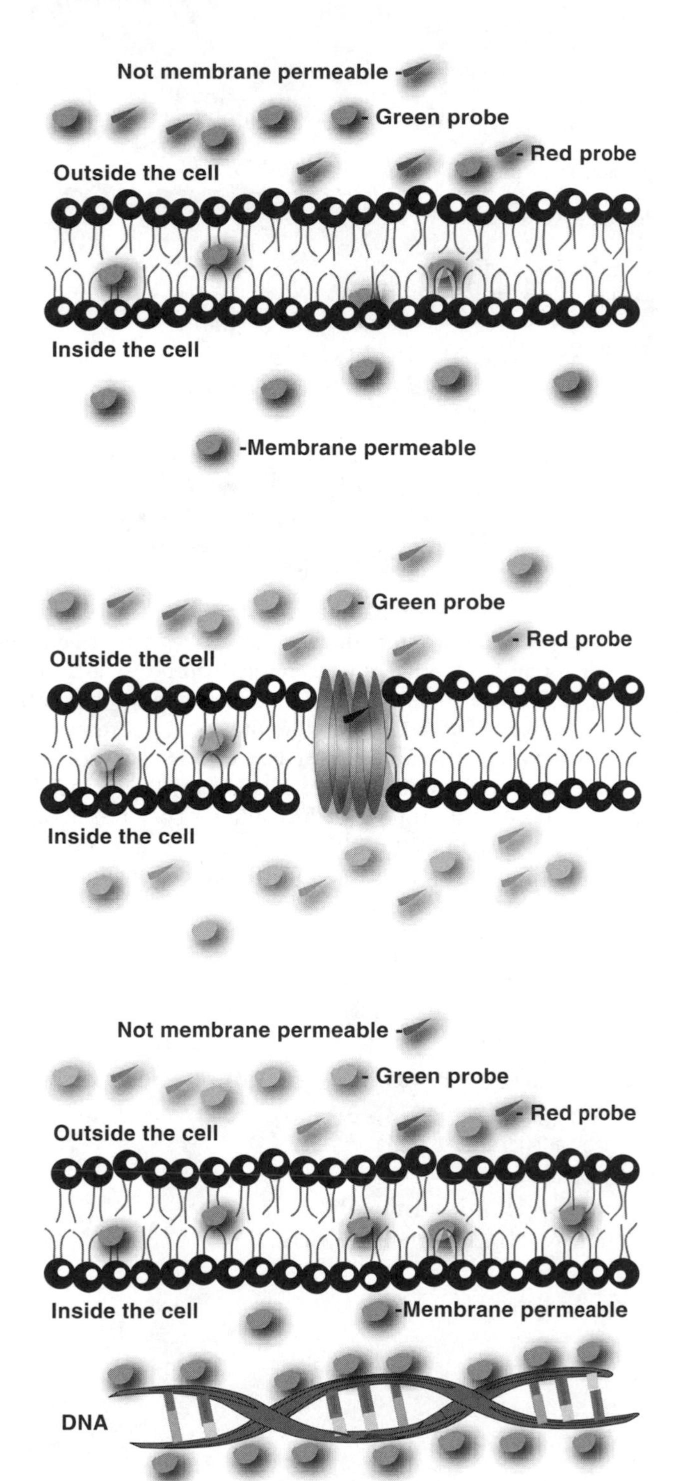

Figure 17.2 *Continues*

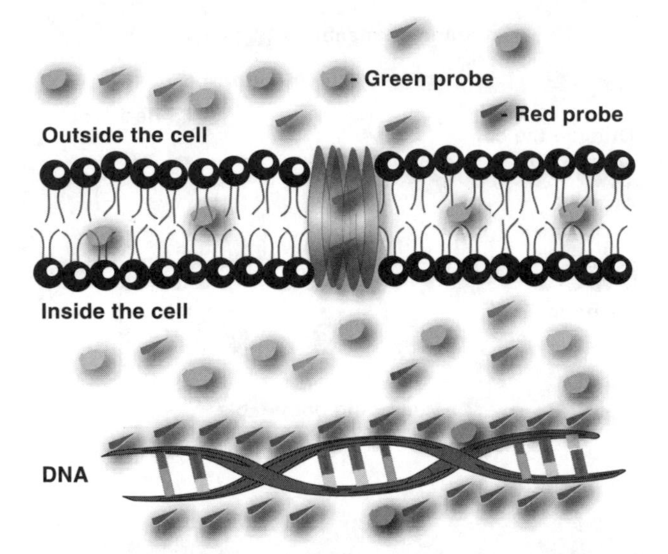

Figure 17.2 *Continued* Panel (a) Two different fluorescent probes are added to bacteria grown in culture, a green fluorescent probe and a red fluorescent probe. The green fluorescent probe is permeable to bacterial cell membranes; the red fluorescent probe is not. These two probes will *only* fluoresce when bound to DNA. Panel (b) The green fluorescent probe crosses the bacterial cell membrane and binds to the bacterial DNA. Once bound, the green fluorescent probe will emit light, or fluoresce. The red probe, which cannot bind to the bacterial DNA, does not fluoresce. Panel (c) Extract from healthy corals that have an abundance of AMPs will permeabilize the bacterial cell membrane, allowing the red fluorescent probe to diffuse into the bacterial cell. Panel (d) Once inside the cell, the red fluorescent probe will exclude the green fluorescent probe from binding to the bacterial DNA. This will do two things: (1) decrease the fluorescence emitted by the green fluorescent probe, and (2) once bound to the DNA, the red fluorescent probe will fluoresce.

- Tryptic soy broth
- Isopropanol
- *S. marcescens* cultures (American Type Culture Collection; www.atcc.org)
- *Supplies*:
- Sterile culture test tubes (8 ml)
- Sterile tooth picks
- Clear-bottom 96-well microtiter plates
- Micropipette tips (20, 200, and 1000 μl)
- Micropipettors (10, 200, and 1000 μl)
- Microcentrifuge tubes

Equipment:

- Vortex mixer
- pH meter
- Analytical balance
- Microcentrifuge tubes
- Temperature-controlled shaker
- Microplate Fluoresence Reader

Method

Sample preparation:

1. All coral samples should be ground frozen to a fine powder in a mortar and pestle using liquid nitrogen. About 200 mg of frozen coral powder is placed in a locking microcentrifuge tube.
2. To each tube, 1.4 ml of denaturating buffer 140 (DB140) containing 4 *M* guanidine HCl, 10 m*M* EDTA, 50 u*M* deferoxamine mesylate, and 1 u*M* PMSF were added. Tubes were vortexed for 30 s and then incubated at 40°C for 20 min, vortexed every 5 min.
3. Tubes are then centrifuged at 14,000 *g*, for 15 min.
4. After centrifugation, 300 ul of the middle phase, with care to avoid all liposulfopolysaccharides, are aspirated and placed in a new tube.
5. Acetone that had been chilled at –80°C for over 1 h is added to the aspirant at a 4:1 ratio (v/v; acetone/aspirant).
6. Vortex the tubes for 20 s, and then incubate the tubes at –80°C for 1 h.
7. After the incubation, the tube is given a quick vortex then centrifuged at room temperature for 15 min at 14,000 *g*.
8. The acetone fraction is aspirated and discarded, and the pellet air-dried for 15 min at 36°C.
9. The dry pellet is then solubilized in a buffer containing 50 m*M* Tris/HCl (pH 8.5) and 50 u*M* sorbitol. Solubilization of the pellet may take between 10 min to over an hour, but the pellet must be completely solubilized. Mixing with a pipettor may accelerate the rate of solubilization.
10. Tubes are centrifuged for 5 min at 14,000 *g*.
11. The supernatant is transferred into a Millipore Microcon diafiltration tube with a molecular weight cutoff of 10,000 Da.
12. Filter tubes with sample are centrifuged at 6000 *g* for 8 min. The filtrate was collected into a new tube.
13. Determine the protein concentration of the filtrate. We suggest the protein concentration method of Ghosh et al. 1988 (see Chapter 10, this volume), though other methods may be used because the acetone precipitation step should have removed all interfering carotenoids.
14. Samples can be stored at –20°C for 1 week.

Microbial conditions

Bacteria used in the IMCOMP assay should be growing exponentially in tryptic soy broth. A growth curve must be generated to determine the range of OD_{640} that correlates to exponential growth. To generate the growth curve, incubate a 2-ml culture of *S. marcescens* in tryptic soy broth as below. Determine and plot OD_{640} every 2 h. The exponential fraction of the resulting sigmoid curve will determine the OD_{640} range for exponential growth. Use only bacteria within this OD_{640} range to generate the OD_{640} 0.157 stock suspension.

- To a sterile culture tube with a sterile stopper, add 2 ml of tryptic soy broth.
- With a sterile toothpick, dab a single colony of *S. marcescens* that had been plated and cultured on standard microbial agar media.

- Drop the toothpick into the culture tube with the tryptic soy broth, stopper the tube.
- Place tube in a temperature-controlled shaker (temperature set at 36°C) and incubate for at least 12 h until cloudy.
- After the culture has reached an appropriate amount of turbidity, generate a 1:10 dilution by taking 100 ul of culture and adding it to 900 ul of tryptic soy broth. Determine OD_{640}.
- Calculate the necessary volume of culture required to make sufficient 0.157 OD_{640} stock suspension for assay, i.e., 2 ml of 0.785 OD_{640} is required to generate 10 ml of 0.157 OD_{640}. Continue incubation and OD_{640} determination until the required OD_{640} is cultured.
- Centrifuge the exponentially growing cells at 10,000 rpm for 10 min.
- Decant the tryptic soy broth, resuspend the pellet in 1× PBS and diluted to 0.157 OD_{640} to generate a stock suspension of *S. marcescens* for the propidium iodide/SYTO 9 competition assay (see the following). A standard curve for this assay also requires dead *S. marcescens* cells. To generate the dead cell stock solution, *S. marcescens* cells were spun down at 10,000 rpm for 10 min and resuspended in 70% isopropanol for 1 h. The isopropanol-killed *S. marcescens* were then pelleted again, resuspended in water, and diluted to 0.157 OD_{640}.

Standard live/dead curves

- Mix stock solutions of live and dead *S. marcescens* in ratios of 100:0, 90:10, 50:50, 10:90, and 0:100 of live and dead bacteria.
- Aliquot 35 ul of each ratio into a 96-well plate. To account for possible artifact of "plate" effects, aliquot a replicate of the live dead curve in columns 1, 5, and 9 on the 96-well plate.
- Add 20 ul of water to each sample well to bring the final OD_{640} to 0.1.
- Add 55 ul of the fluorescent stock solution, incubating the cells on a platform shaker in the dark for 20 min. Take special care not to expose the sample to light before reading fluorescence to avoid bleaching of the fluorophores.
- Read the microtite plate using Flx800 Microplate Fluorescence Reader (Bio-Tek Instruments) using KCJunior software (v. 1.31.2, Bio-Tek Instruments). Excitation was 485 nm and emission was 528 nm for Component A and 485 nm excitation and 620 nm emission for Component B (Figure 17.3).

Experimental curves

In a 96-well plate, add 35 ul of live *S. marcescens* ($OD_{640} = 0.157$), 15 ul of water, and 5 ul of coral extract was added to each well, for a final $OD_{640} = 0.1$. To account for sample variation, each coral extract and bacteria sample must be repeated on the plate in triplicate. To minimize variability due to plate positioning, the placement of each sample should be randomized as much as possible. Finally, inclusion of a standard live/dead curve, in triplicate, on each experimental plate will minimize the variability when comparing results from plate to plate.

Incubate the samples on a platform shaker at 30°C for 1 h to allow cells to recover from the stress of sample preparation. Remove the samples from the shaker, and add 55 ul of 0.6% fluorescent stock solution to each well. Incubate the plate on a rocker at room

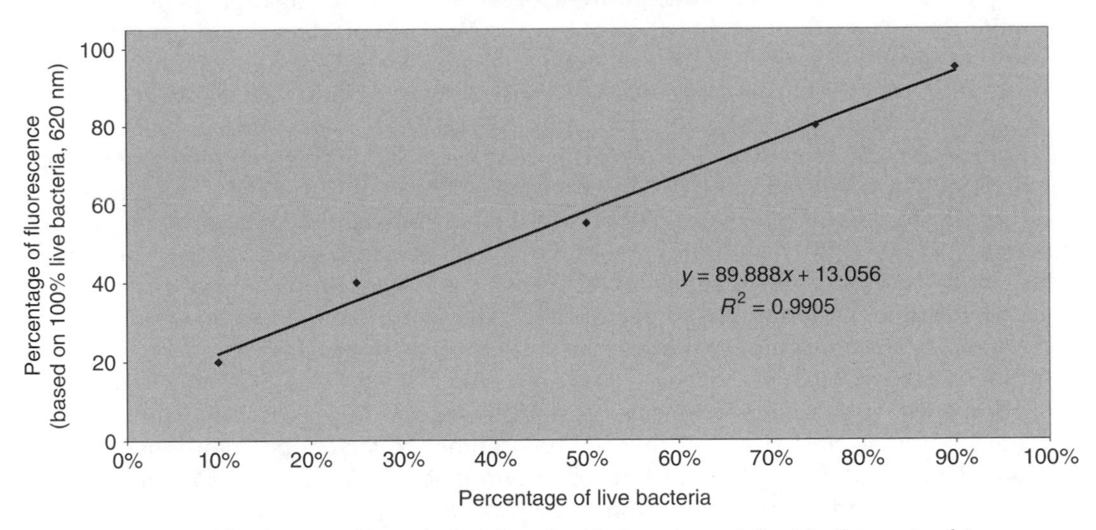

Figure 17.3 Live/death curve. Bacteria incubated with isopropanol for 1 h, then mixed in proportions to live bacteria. Mixed bacteria are then incubated with the probe solution, incubated for 15 min at room temperature. Bacteria are then plated on a 96-well microtiter plate, and read on a fluorescent plate reader at the appropriate excitation/emission wavelengths.

temperature in the dark for 20 min. Taking care to avoid light exposure, read the sample fluorescence as stated above.

Results and discussion

IMCOMP-P was developed to address certain deficiencies in discerning the nature and processes of coral pathology. Reports of disease outbreaks in corals have increased in the past ten years.[17] The Caribbean is notorious for extensive outbreaks of black-band and white-plague disease.[1] In the Indian Ocean, there are increasing reports of a "red plague" of an unknown etiology that can kill coral tissue along a front at a rate of 10–15 cm a day, and as fast at 35 cm a day (Dr. R. Jeyabaskaran, personal communication). Are these recent outbreaks the result of the introduction of opportunistic pathogens, changes in the environment that affect pathogen physiology, or changes in host immunity? Changes in host immunity can be the result of host age, nutritional status, seasonal physiology (i.e., spawning seasons), cost of intra- and inter-species competition, and environmental pressures. Because of the alarming declines of coral reefs in the Caribbean and around the world and of the increase of anthropogenic factors associated with these declines (i.e., agricultural run-off, water-shed alterations, residential and urban development), understanding the relationship between environmental factors and coral health is becoming vitally important.

Invertebrate physiology relies solely on innate immunity as a means of protection against microorganisms. IMCOMP-P is one of several technologies we have developed to better define cnidarian immunology. IMCOMP-P assesses the antimicrobial activity of small antimicrobial peptides or compounds that have a molecular weight smaller than 10,000 Da. In the format described in this chapter, IMCOMP-P cannot determine the source of the antimicrobial compounds (host versus dinoflagellate versus resident microbial community). Many of the antimicrobial peptides and algal phenolic secondary compounds can cause permeablization of the microbial cell membranes, either by altering membrane–protein channel pores or creating channels.[16] IMCOMP-P capitalizes on the characteristics of two fluorescent probes that have different affinity constants to DNA and different membrane permeability constants. SYTO 9® is permeable to bacterial cell membranes, while propidium iodine (PI) is not permeable to the cell membrane. If the integrity of the cell membrane is compromised, PI can enter the cell and compete for binding of DNA with SYTO 9®. Because PI has a higher binding affinity, it can outcompete SYTO 9® for DNA binding sites. Hence, SYTO 9® is excluded from binding to DNA, and the fluorescence signal of SYTO 9® will decrease proportionally with the permeabilization of bacterial cell membrane. Therefore, an increase in antimicrobial membrane permeabilization peptides or compounds will decrease the fluorescence of SYTO 9® fluorescence.

To test the applicability of this assay to different species of coral and preliminarily investigate the possible effects of relevant environmental factors on coral immunity, we examined the treatment effects of hydrogen peroxide on *Porites porites*, endosulfan on *Pavona gigantea*, and Irgarol 1051 on *Madracis mirabilis*.

Rates of oxidative damage increase in coral experiencing elevated sea-surface temperatures, hypo-salinity, high-light stress, and high ultraviolet light exposure. Exposure to exogenous hydrogen peroxide (H_2O_2) can induce intracellular oxidative damage because of the permeability of H_2O_2 to membrane lipid bi-layers.[18] Oxidative stress shifts the equilibrium of a cell's protein metabolic condition, by denaturing enzymes and proteins, decreasing transcriptional and translational efficiency, and altering primary amino-acid structure through mutagenesis.[19] Data in Figure 17.4 supports the hypothesis that increased oxidative stress decreases immuno-competence, either by decreasing antimicrobial compound production and/or decreasing exocytosis rates.

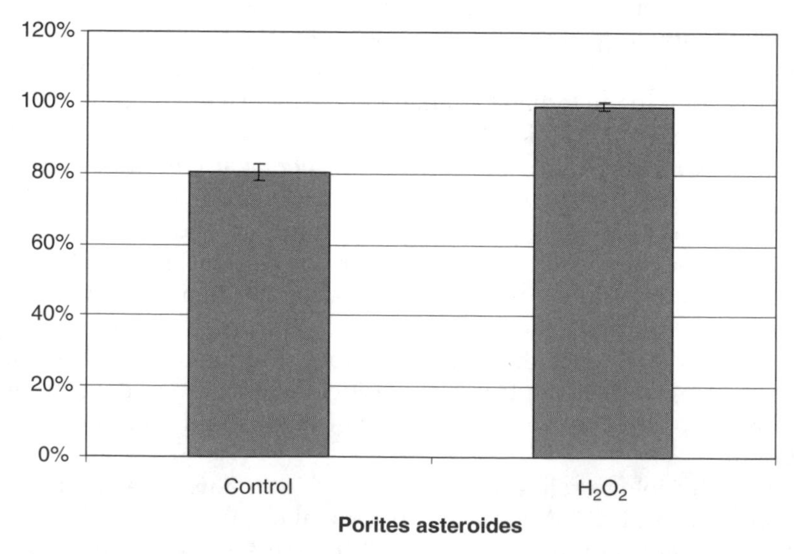

Figure 17.4 The percent of bacteria alive after a 1-h incubation with coral protein extract from *Porites porites* exposed to hydrogen peroxide for 8 h and controls. Exposure to H_2O_2 significantly decreased AMP activity, potentially making the coral more susceptible to infection ($p < 0.05$). $N = 3$ per treatment.

Endosulfan is an agricultural pesticide used world-wide, especially in districts whose watersheds spill into regions of coral reefs (i.e., Great Barrier Reef, Belizian Barrier Reef, Florida Keys). Exposure to endosulfan decreased the antimicrobial activity of corals (Figure 17.5). Endosulfan inhibits cholinesterase activity and induces massive oxidative damage in invertebrates.[20] The half-life of endosulfan is longer in marine environments than in aquatic environments.[21] Exposure of 100 u*M* endosulfan in *Pavona* and in *Montastrea annularis* induces extensive oxidative stress, decreasing protein metabolic condition, metabolic homeostasis, and increasing destabilization of genomic integrity [Downs et al., manuscript in preparation (*Pavona*); Downs et al., preliminary results (*Montastraea*)]. Endosulfan could be reducing immuno-competence via the oxidative stress mechanism as seen in *Porites* exposed H_2O_2, or it could also be reducing immuno-competence as an inhibitor. Endosulfan is a sulfonated chlorinated phenol, which can be oxidized either metabolically or by photolysis to create a chlorinated phenol with an active thiol group. A number of AMP classes have abundant cysteine residues whose thiols are functional groups. Thiolated endosulfan could bind to these thiol groups on the AMPs, thereby inhibiting activity.

Irgarol 1051 is an herbicide component used in boat antifoulant paints. Work from Owens and colleagues demonstrated that 100 ppb Irgarol in seawater can significantly reduce net photosynthesis of coral dinoflagellate symbiont after an 8-h exposure.[22,23] Further studies into the effects of Irgarol on both host and dinoflagellate symbiont cellular physiology show that an 8-h exposure to 10 ppb Irgarol can significantly reduce host levels of catalase and ferrochelatase, indicating a decrease in antioxidant capacity, as well as possible decrease in host immuno-competence (Owens et al., in preparation). Because of increasing concentrations of Irgarol in waters associated with recreational marinas near or on coral reefs, Irgarol may be considered a possible risk factor for compromising coral immuno-competence. However, in these preliminary experiments, there was no significant difference in antimicrobial activity between control and Irgarol exposed coral in the IMCOMP-P assay (Figure 17.6). This finding should not however be interpreted as Irgarol 1051 having no effect on coral immunity, but rather that IMCOMP-P does not assay all coral immune system components and other components may indeed be affected by this compound.

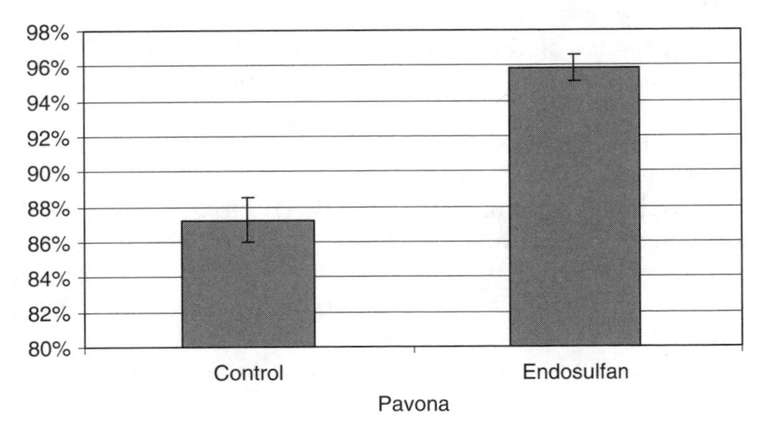

Figure 17.5 The percent of bacteria alive after a 1-h incubation with coral protein extract from *P. gigantea* exposed to an 8-h exposure of 100 u*M* endosulfan. Exposure to endosulfan significantly decrease AMP activity ($p < 0.05$). $N = 4$ per treatment.

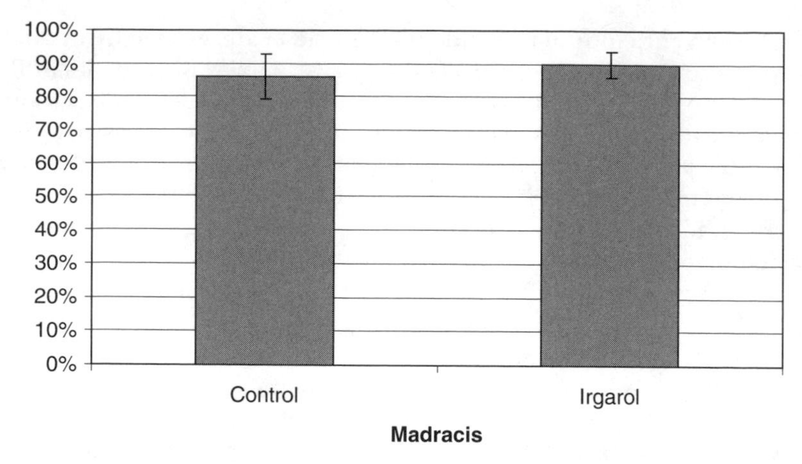

Figure 17.6 The percent of bacteria alive after a 1-h incubation with coral protein extract from *Madracis* controls and exposed to 10 ppb Irgarol 1051 for 24 h. There was no significant difference between control and coral exposed to 10 ppb Irgarol 1051 for 24 h. $N = 3$ per treatment.

Aspects of this assay and its protocol can easily be modified to meet end-user's needs. Instead of acetone precipitation, an acid or salt precipitation could be used to ensure better precipitation of more hydrophilic, short polypeptides that might remain soluble in a partial acetone environment. Care must be taken though, with a trichloroacetic acid or trifluoroacetic acid precipitation because of potential acid hydrolysis of polypeptides. For many samples, the final protein concentration may be too dilute to properly determine; hence, samples can be concentrated via vacuum centrifugation or lyophilization. We used extracts of whole coral tissue for antimicrobial activity. It may be possible to use just the mucus produced from the coral to determine the level of antimicrobial activity in coral, thereby avoiding tissue damage created from the biopsy, as well as alleviating the need to obtain collection permits.

Acknowledgments

This research is a product of the multi-institutional U.S. Coral Disease and Health Consortium. We would like to thank Richard Owen and Lucy Buxton of the Bermuda Biological Station for Research for the *Madracis* samples.

This publication does not constitute an endorsement of any commercial product or intend to be an opinion beyond scientific or other results obtained by the National Oceanic and Atmospheric Administration (NOAA). No reference shall be made to NOAA, or this publication furnished by NOAA, to any advertising or sales promotion that would indicate or imply that NOAA recommends or endorses any proprietary product mentioned herein, or which has as its purpose an interest to cause the advertised product to be used or purchased because of this publication.

References

1. Richardson, L.L., Coral diseases: What is really known? *Trends Ecol. Evol.*, 13, 438–443, 1998.

2. Al-Moghrabi, S.M., Unusual black band disease outbreak in the northern tip of the Gulf of Aqaba (Jordan), *Coral Reefs*, 19, 330–331, 2001.
3. Kushmaro, A., Loya Y, Fine, M. and Rosenberg, E., Bacterial infection and coral bleaching, *Nature*, 380, 396, 1996.
4. Patterson, K.L., Porter, J.W., Ritchie, K.B., Polson, S.W., Mueller, E., Peters, E.C., Santavy, D.L. and Smith, G.W., The etiology of white pox, a lethal disease of the Caribbean elkhorn coral, *Acropora palmata*, *Proc. Natl. Acad. Sci. USA*, 99, 8725–8730, 2002.
5. Ben-Haim, Y., Zicherman-Keren, M. and Rosenberg, E., Temperature-regulated bleaching and lysis of the coral *Pocillopora damicornis* by the novel pathogen *Vibrio coralliilyticus*, *Appl. Environ. Microbiol.*, 69, 4236–4242, 2003.
6. Lang, J.C. and Chornesky, E.A., Competition between scleractinian reef corals — a review of mechanisms and effects, in: *Ecosystems of the World*, Dubinsky, Z., Ed., vol. 25, Elsevier, Amsterdam, 1990, pp. 209–252.
7. Olano, C.T. and Bigger, C.H., Phagocytic activities of the Gorgonian coral *Swiftia exserta*, *J. Invert. Pathol.*, 76, 176–184, 2000.
8. Bosch, T.C.G. and David C.N., Immunocompetence in *Hydra*: epithelial cells recognize self-nonself and react against it, *J. Exp. Zool.*, 238, 225–234, 1986.
9. Kimberly, D. and Beutler, B., The evolution and genetics of innate immunity, *Nat. Rev. Genet.*, 2, 256–267, 2001.
10. Kamysz, W., Okroj, M. and Lukasiak, J., Novel properties of antimicrobial peptides, *Acta Biochim. Polo.*, 50, 461–469, 2003.
11. Hotchkiss, R.D. and Dubos, R.J., Chemical properties of bactericidal substances isolated from cultures of a soil bacillus, *J. Biol. Chem.*, 132, 793–794, 1940.
12. Sarmasik, A., Antimicrobial peptides: a potential therapeutic alternative for the treatment of fish diseases, *Turk. J. Biol.*, 26, 201–207, 2002.
13. Boman, H.G., Peptide antibiotics and their role in innate immunity, *Annu. Rev. Immunol.*, 13, 61–92, 1995.
14. Zasloff, M., Antibiotic peptides as mediators of innate immunity, *Curr. Opin. Immunol.*, 4, 3–7, 1992.
15. Zasloff, M., Antimicrobial peptides of multicellular organisms, *Nature*, 415, 389–395, 2002.
16. Van't Hof, W., Veerman, E.C.I., Helmerhorst, E.J. and Amerongen, A.V.N., Antimicrobial peptides: properties and applicability, *Biol. Chem.*, 382, 597–619, 2001.
17. Woodley, C.M., Bruckner, A.W., Galloway, S.B., McLaughlin, S.M., Downs, C.A., Fauth, J.E., Shotts E.B. and Lide, K.L., *Coral Disease and Health: A National Research Plan*, National Oceanic and Atmospheric Administration, Silver Spring, MD, 2003, 66 pp.
18. Downs, C.A., Fauth, J.E., Halas, J.C., Dustan, P., Bemiss, J.A. and Woodley, C.M., Oxidative stress and seasonal coral bleaching, *Free Rad. Biol. Med.*, 33, 533–543, 2002.
19. Halliwell, B. and Gutteridge, J.M.C., *Free Radicals in Biology and Medicine*, Oxford University Press, Oxford, 1999.
20. Downs, C.A., Dillon, R.T., Jr., Fauth, J.E. and Woodley, C.M., A molecular biomarker system for assessing the health of gastropods (*Illyanassa obsoleta*) exposed to natural and anthropogenic stressors, *J. Exp. Mar. Biol. Ecol.*, 259, 189–214, 2001.
21. U.S. EPA, Toxicological Profile for Endosulfan, Draft for public comment, U.S. Department of Health and Human Services, 1999.
22. Owen R., Knap A.H., Toaspern M. and Carbery K., Inhibition of coral photosynthesis by the antifouling herbicide Irgarol 1051, *Mar. Poll. Bull.*, 44, 623–632, 2002.
23. Owen R., Knap, A.H., Ostrander N. and Carbery K., Comparative acute toxicity of herbicides to photosynthesis of coral zooxanthellae, *Bull. Environ. Cont. Toxicol.*, 70, 541–548, 2003.

chapter eighteen

Monitoring gene expression in Rana catesbeiana *tadpoles using a tail fin biopsy technique and its application to the detection of environmental endocrine disruptor effects in wildlife species*

Nik Veldhoen and Caren C. Helbing
University of Victoria

Contents

Introduction

Wildlife species are being exposed to an ever-increasing number of chemical compounds that are generated as byproducts of human activity.[1–4] These chemicals encompass a large group of environmental pollutants that include pharmaceuticals, pesticides, herbicides, and industrially relevant chemicals. Many of these compounds, while not overtly toxic, have the potential or have been demonstrated to interfere with the normal growth and development of a variety of species and could directly interfere with regulatory endocrine systems.[3–6] Exposure to such chemical contaminants during critical life stages could result

1-56670-664-5/05/$0.00+$1.50
© 2005 by CRC Press

in permanent deleterious effects to the reproductive potential[7] or physiological fitness[8,9] of species at risk. Of great concern is the increasing usage of these potential endocrine disrupting chemicals (EDCs) and a concomitant rise in their concentrations in surface and groundwater sources within the environment.[10–13]

With the number of chemicals registered for commercial use in the United States alone estimated to exceed 70,000 (Toxic Substances Control Act Inventory; U.S. E.P.A., www.epa.gov/opptintr/opptdiv), a toxicity assessment method must be applicable to a rapid, high throughput experimental format and allow for sample collection in both laboratory and field settings. While mortality and morphometric analyses have proven useful in the characterization of toxic exposure to chemical contaminants, such endpoints are limited in their sensitivity and processivity. In addition, design of non-lethal, minimally invasive procedures for sample collection and assessment of contaminant exposure is warranted for endangered or threatened species.[14–18] The techniques employed must also provide an increased predictive ability and be amenable to regulatory standardization. To this end, we have developed a biopsy procedure combined with genetic biomarker analysis that can be performed on live animals. This technique takes advantage of advances made in the stabilization of tissue samples prior to experimental processing and allows for sample collection in the field using an animal capture-and-release methodology. The use of the biopsy procedure standardizes the amount of tissue to be analyzed, thereby helping to minimize sampling variance and can be applied to a high throughput screening process. The introduction of genetic biomarkers as an analytical endpoint allows for early detection of chemical contaminant exposure effects prior to overt morphometric changes during development.[5] Comparison of the resulting gene expression information from test versus reference sampling locations can help in the determination of whether individual species have been exposed to potentially deleterious substances.

To demonstrate the utility of the biopsy procedure, we have collected tissue samples from North American bullfrog, *Rana catesbeiana*, tadpoles undergoing natural metamorphosis and from premetamorphic tadpoles exposed to exogenous thyroid hormone (TH). In addition, the effects of exposure to the potential EDC, acetochlor, on TH-induced metamorphosis of live animals were assessed. Biopsies were obtained from tadpole tail fins under three different scenarios: from the dorsal tail region of live animals, from tadpoles euthanized immediately after experimental exposures, and from cultured tadpole tail tips. Steady-state levels of TRβ mRNA, a sensitive and early biomarker of the TH-induced metamorphic genetic program,[5,17] were determined using reverse transcription followed by real-time quantitative polymerase chain reaction (QPCR) analysis. Such genetic assessment in conjunction with morphometric and population analyses can contribute to the identification of potentially hazardous chemical contaminants in the environment that disrupt normal developmental processes. The utility of the biopsy procedure across a variety of wildlife species is also discussed.

Materials required

- *R. catesbeiana* (Taylor–Kollros (TK) Stages VI–VIII, collected locally)
- Sterile disposable 2-mm dermal punch (catalog #162-33-31, The Stevens Company, www.stevens.ca)
- Extra fine-point curved dissecting forceps (catalog #25607-890, VWR International, www.vwr.com)
- 20 cm × 20 cm plastic Plexiglas plate
- Capture net

- Paper towel
- Protective latex gloves
- 1.5-ml sterile safe-lock plastic sample tubes (catalog #0540225, Fisher Scientific, www.fisherscientific.com)
- Tissue preservative (RNAlater, catalog #7021, Ambion, www.ambion.com)
- Tissue homogenization reagent (TRIzol, catalog #15596-018, Invitrogen Life Technologies, www.invitrogen.com)
- Retsch MM300 mixer mill (catalog #85110, QIAGEN, www1.qiagen.com)
- 2 × 24 position shaker racks (catalog #69998, QIAGEN, www1.qiagen.com)
- 3-mm tungsten-carbide homogenization beads (catalog #69997, QIAGEN, www1.qiagen.com)
- Glycogen (catalog #901-393, Roche Molecular Biochemicals, www.biochem.roche.com)
- pd(N)$_6$ random hexamer oligonucleotide (catalog #27216601, Amersham Biosciences, www1.amershambiosciences.com)
- Moloney murine leukemia virus (MMLV) RNase H$^-$ reverse transcription kit (Superscript II reverse transcriptase, catalog #18064-014, Invitrogen Life Technologies, www.invitrogen.com)
- Ribonuclease inhibitor (catalog #15518-012, Invitrogen Life Technologies, www.invitrogen.com)
- Thermopolymerase (Platinum Taq DNA polymerase, catalog #10966-034, Invitrogen Life Technologies, www.invitrogen.com)
- Deoxynucleotide set (catalog #10297-018, Invitrogen Life Technologies, www.invitrogen.com)
- SYBR Green I fluorescent DNA-binding dye (catalog #S-7563, Molecular Probes, www.probes.com)
- ROX passive reference dye (catalog #600536, Stratagene, www.stratagene.com)
- MX4000 quantitative real-time thermocycler system (catalog #401260, Stratagene, www.stratagene.com)
- Sequence-specific DNA primer sets (see Table 18.1)

Procedures

Tissue biopsy

Tissue biopsies are taken from the dorsal tail fin of whole animals or from the fin region of dissected and cultured tail tips using a dermal punch that isolates a 2-mm diameter tissue sample. For amphibian organ culture conditions, see Reference 19. During collection of

Table 18.1 Primer sequences and target *R. catesbeiana* genes used in DNA amplification

Gene target	Genbank accession #	Amplicon size (bp)	Primer sequences
TRβ	L27344	538	UP 5′-AGCAGCATGTCAGGGTAC-3′ DN 5′-TGAAGGCTTCTAAGTCCA-3′
16S ribosomal RNA	M57527	533	UP 5′-AGAAGGAACTCGGCAAAT-3′ DN 5′-CCAACATCGAGGTCGTAA-3′
ribosomal protein L8	AY452063	270	UP 5′-CAGGGGACAGAGAAAAGGTG-3′ DN 5′-TGAGCTTTCTTGCCACAG-3′

tissue samples from live tadpoles, the biopsy is taken quickly in order to minimize stress to the animals. Tadpoles are placed briefly on a paper towel positioned on top of a plastic Plexiglas biopsy support plate. Each live animal must be gently immobilized body-first into the palm of the hand to prevent sudden movement that could result in tearing of the biopsy region. Figure 18.1A This is accomplished by one investigator immobilizing the animal, while a second investigator retrieves the biopsy sample. Figure 18.1B Up to four biopsies can be obtained from the dorsal fin of a live premetamorphic (TK Stages VI–VIII; Reference 20) *R. catesbeiana* tadpole allowing for time-course studies with a repeated measures design (Figure 18.1C). Due to the membranous nature of the tadpole tail fin sampled, biopsies collected as described do not cause any observable discomfort for live animals or affect their swimming behavior. Tissue biopsy collection from dissected and cultured tail samples are performed in a disposable plastic weigh-boat. Approximately two fin biopsies are available from a cultured 2-cm tail tip (Figure 18.1D). All collected samples are immediately immersed in the RNA preservative, RNAlater and stored at 4°C for up to a few months prior to RNA isolation.

Tissue homogenization

Homogenization of the biopsies was performed using the TRIzol reagent as described by the manufacturer. Mechanical disruption of tadpole tail fin tissue utilized 100 μl TRIzol reagent, a 3-mm diameter tungsten-carbide bead, and safe-lock Eppendorf tubes in a

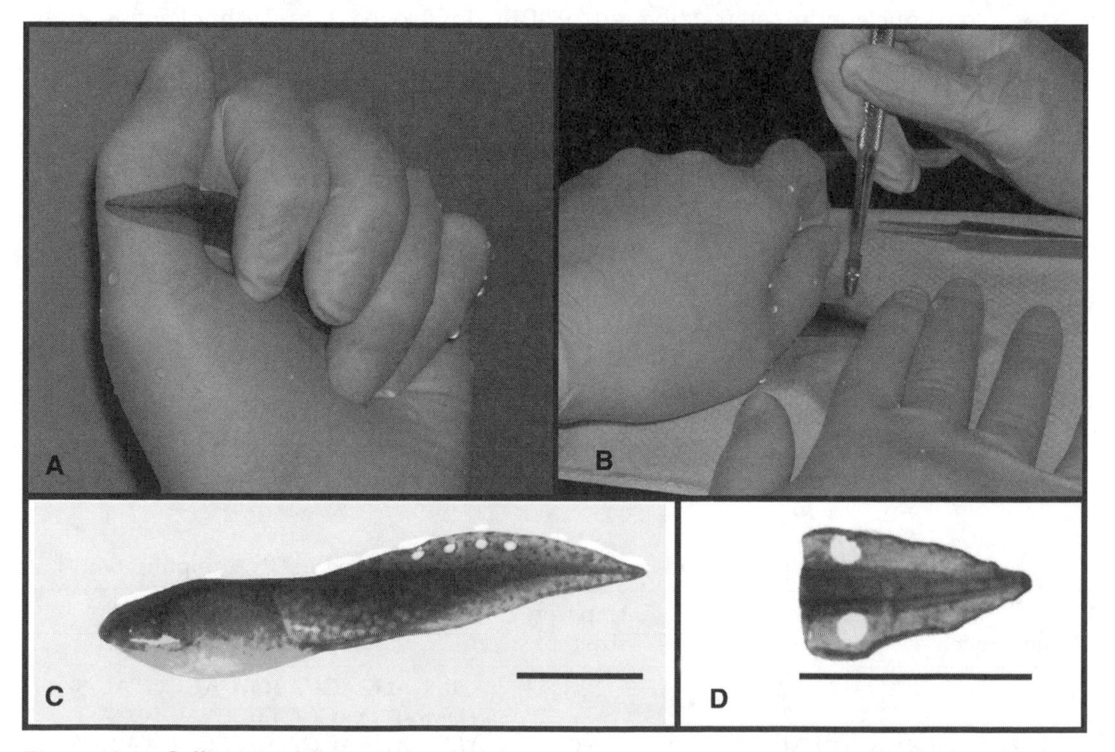

Figure 18.1 Collection of *R. catesbeiana* tail fin tissue using a biopsy procedure. Tissue for genetic analyses can be collected from live animals handled in an appropriate manner (A and B). The location of sampled tissue is shown within the dorsal tail region of live tadpoles (C) or from cultured tadpole tail tips (D). Bars in C and D represent a 2-cm length.

Retsch MM300 mixer mill at 20 Hz for 6 min. Mixing chambers are rotated 180° halfway through the homogenization procedure. Following phase separation, 20 μg glycogen was added as a nucleic acid carrier. Isolated RNA was subsequently resuspended in 10 μl of diethyl pyrocarbonate (DEPC)-treated RNase-free water and stored at −70°C. RNA is very sensitive to conditions that promote degradation. To maintain the integrity of RNA samples, ensure that proper laboratory methods that minimize potential RNase contamination are employed during the isolation procedure.[21] All plasticware used, such as pipette tips and sample tubes, should be guaranteed RNase-free.

Preparation of cDNA

A typical RNA yield was 1 μg per tail fin biopsy, and all of this RNA was annealed with 500 ng $pd(N)_6$ random hexamer oligonucleotide at 65°C for 10 min followed by a quick cool-down on ice. RNA was converted to cDNA using 200 units of MMLV RNase H^- Superscript II reverse transcriptase as described in the manufacturer's recommended protocol. The 20-μl reaction was incubated at 42°C for 2 h and diluted 20-fold prior to DNA amplification.

DNA amplification

All DNA amplification primers were designed with Primer Premier Version 4.1 software (Premier Biosoft International, www.premierbiosoft.com) and synthesized by AlphaDNA (www.alphadna.com) (Table 18.1). Amplification of specific cDNA targets was performed using real-time QPCR performed on an MX4000 system as described by the manufacturer. One major advantage of QPCR is a non-reliance on collection of endpoint expression data. Information associated with a given DNA amplification reaction is collected during each thermocycle, and differences in gene expression can be determined using the highly sensitive early phase of the amplification process. All reagents are prepared in distilled water, which has been autoclaved. The 15 μl QPCR reaction contained 10 m*M* Tris–HCl pH 8.2 at 20°C, 50 m*M* KCl, 3 m*M* $MgCl_2$, 0.01% Tween-20, 0.8% glycerol, 40,000-fold dilution of SYBR Green I, 200 μ*M* dNTPs, 83.3 n*M* ROX reference dye, 10 pmol of each primer, 2 μl of diluted total cDNA, and 1 unit of Platinum Taq DNA polymerase.[5] The thermocycle program for all target DNA sequences under investigation included a denaturation step at 95°C (9 min); 40 cycles of 95°C (15 s), 55°C (30 s), 72°C (45 s), and a final elongation step at 72°C (7 min). Cycle threshold (Ct) data obtained from QPCR reactions were compared to standard curves of Ct versus DNA copy numbers that are generated for each individual target DNA using known amounts of plasmid containing the amplicon of interest. Data obtained from quadruplicate QPCR reactions performed for each sample are averaged and normalized to the invariant ribosomal protein L8 gene control (Table 18.1). The choice of a given normalizer gene product is crucial for correct interpretation of QPCR-derived data and must be supported by statistical determination of its invariant nature throughout the sample population. A trend or fluctuation in the expression of the normalizer could indicate a treatment or exposure-related effect. In general, the chosen normalizer gene should also display an amplification profile within the same dynamic range of detection to that of the target genes of interest. Controls that lack cDNA template (non-specific amplification control) or thermopolymerase enzyme (no amplification control) are included to ensure the specificity of target DNA amplification. In addition, QPCR reaction products that utilize newly designed primer pairs are initially separated on an agarose gel and visualized with ethidium bromide. This is done to ensure that the SYBR dye-based signal derived during QPCR is specific for the target amplicon under investigation and is distinct from noise created through primer dimer effects or spurious non-specific DNA amplification. An alternative DNA amplification method that does not require use of a real-time QPCR system can also be employed.[17] This earlier study used endpoint collected data from standard DNA amplification reactions separated by agarose gel electrophoresis and visualized using ethidium bromide staining. Densitometric values for each amplicon band were quantified from captured digital images, and relative gene expression determined using 16S ribosomal RNA to normalize expression data between samples following the DNA amplification procedure (Table 18.1).

Results and discussion

TH action plays a crucial role in growth, development, and homeostasis of humans and other vertebrates. Of particular importance is the association of TH with brain development and normal cognitive function.[22,23] Exposure to chemical contaminants has been associated with changes in TH levels in human populations[24] and is also linked to neurotoxicity and modulation of gene expression in brain tissue of laboratory animals.[25,26] With respect to wildlife populations, amphibians represent a recognized sentinel species for the detection of EDC-associated effects due to their aquatic larval life stage and a dependence on hormone activity for metamorphosis.[27–30] TH plays a critical role in regulating the tissue remodeling events that occur during metamorphosis, and it is this process that could potentially prove a sensitive target for EDC action.[31–33] TH acts primarily through nuclear TH receptors (TRα and TRβ) that initiate tissue-specific genetic programs by the activation and repression of specific genes.[34–36] It is remarkable that this single hormone signal can induce such diverse tissue fates as resorption of the larval tadpole tail (involving cell apoptosis) and development of the limbs (involving cell proliferation). In addition to TH, the actions of other hormones, such as estrogen and corticosterone, also modulate this critical developmental period making metamorphosis a prime target for endocrine disruption.[37–39] The number of developmentally abnormal frogs observed from the wild and population declines suggests that some form of disruption in gene expression programs crucial for normal development may exist in certain amphibian populations.[40,41]

To test the potential of the biopsy procedure to detect changes in gene expression associated with amphibian development, the steady-state levels of TRβ mRNA in tail tissue were investigated during natural metamorphosis of *R. catesbeiana* tadpoles. This represents a period of increased TH-dependent gene expression and marked changes in target tissues that result in gross anatomic and physiologic changes culminating in the metamorphosis to a juvenile froglet. The steady-state levels of TRβ transcript increased approximately 9-fold during the prometamorphic phase of development up until metamorphic climax (TK Stages XIV–XXII) (Figure 18.2). Increased TRβ gene expression correlates with increased levels of endogenous TH during metamorphosis.[36,42] A similar upregulation in TRβ gene expression during metamorphosis has been reported previously for a number of amphibian species that undergo metamorphosis.[19,35,43]

Induction of precocious metamorphosis can occur following exposure of premetamorphic tadpoles to exogenous 3,3′,5-triiodothyronine (T_3).[44] At this stage of development, tadpoles are functionally athyroid (Regard et al., 1978), yet are highly sensitive to TH exposure[44] thus providing a useful physiological baseline from which to study both

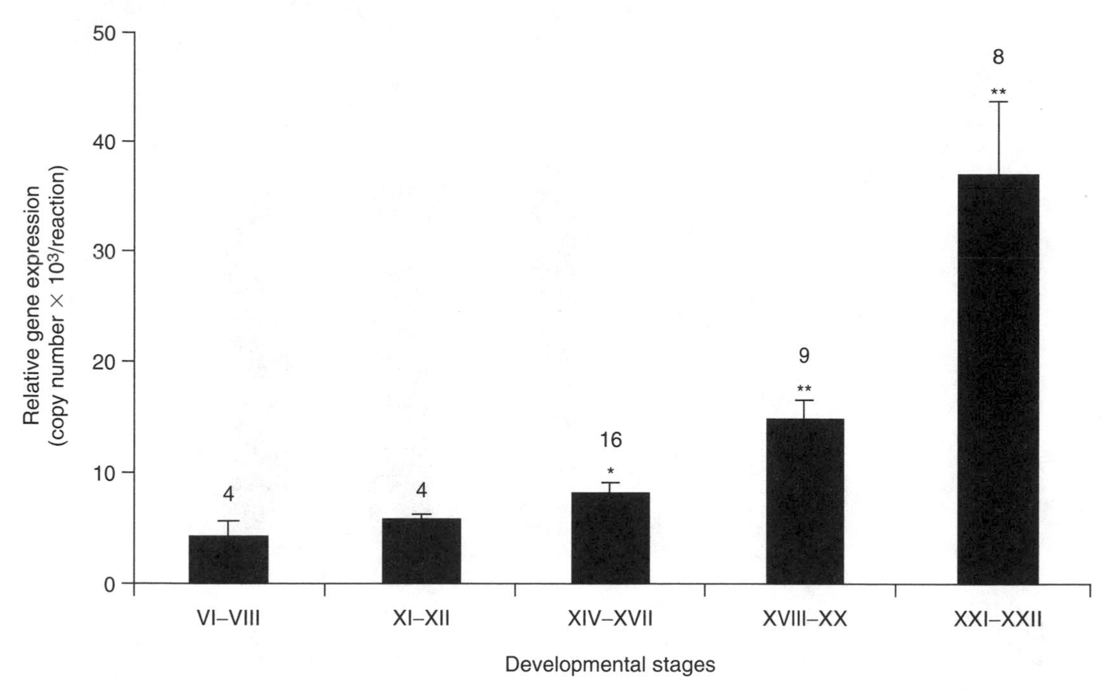

Figure 18.2 Expression of the TRβ gene in tadpole tail fin tissue during natural metamorphosis of *R. catesbeiana*. Biopsy samples were taken from the tail fin of euthanized tadpoles, which had reached the indicated developmental stage. The level of TRβ gene expression was determined using QPCR analysis and normalized across samples using the invariant ribosomal L8 gene expression. The means and standard errors are shown and the number of animals assessed per developmentally staged group is indicated above each bar. Significant differences compared to TK Stages VI–VIII are denoted by a single asterisk ($p < 0.05$) or a double asterisk ($p < 0.001$). Determination of standard errors and one-way analysis of variance (Tukey–Kramer multiple comparisons test) for significance were carried out using InStat V3.01 (GraphPad Software, www.graphpad.com).

hormone-dependent development and the effects of potential EDCs. Using this hormone inducible system, TH-dependent changes in TRβ mRNA expression were assessed in tail fin tissue obtained from T_3-exposed whole animals that were sacrificed immediately prior to the biopsy (Figure 18.3A) and from dissected tail tips cultured in the presence of T_3 (Figure 18.3B). A significant increase in hormone-dependent TRβ mRNA expression is detected within 24 h for both intact animals exposed to hormone and with the isolated tail tissue. A similar upregulation in TRβ expression is also observed in tail fin samples collected from live tadpoles following a 24-h exposure to T_3 (Figure 18.4).

In order to test whether this method is capable of detecting a disruption of gene expression associated with EDC exposure during precocious metamorphosis, we treated tadpoles with the environmentally relevant dose of 10 n*M* acetochlor[10] in the presence or absence of 100 n*M* T_3 for 24 h (Figure 18.4). This pre-emergent herbicide has been shown to modulate TH-dependent gene expression and accelerate precocious metamorphosis in *Xenopus laevis*.[5] Acetochlor alone has no effect on TRβ mRNA expression in premetamorphic tadpole tails; however, the combination of T_3 with acetochlor results in a statistically significant elevation of the T_3-induced response. This observation in *R. catesbeiana*, in addition to effects on mRNA expression and morphometric endpoints observed in *X. laevis*, identifies acetochlor as a potential environmental EDC that targets

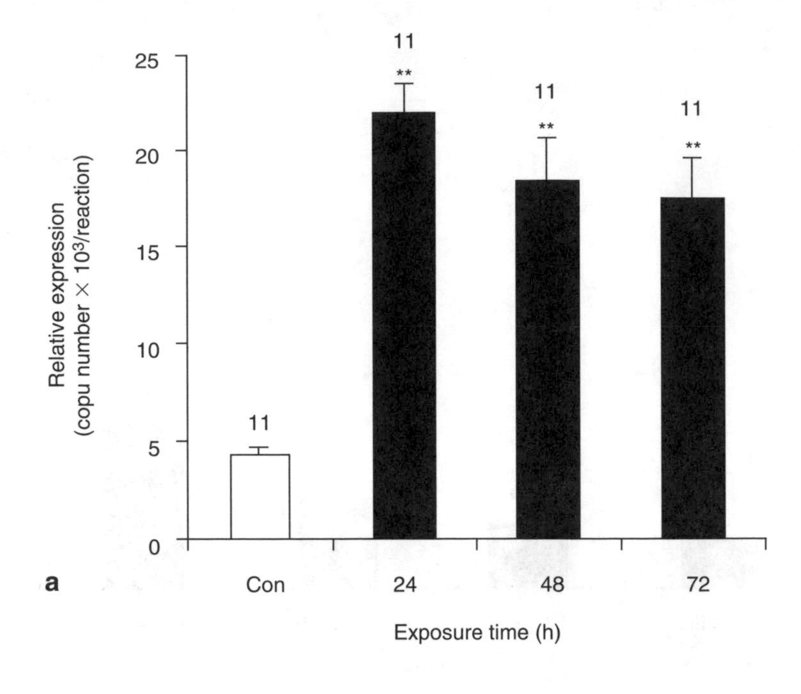

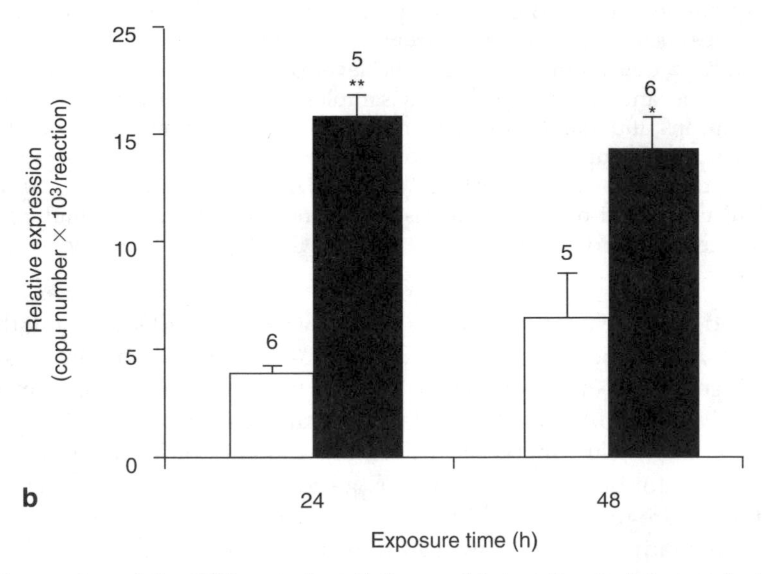

Figure 18.3 Expression of the TRβ gene in tail tissue of intact *R. catesbeiana* tadpoles or isolated cultured tail tips following exposure to exogenous TH. Biopsy samples were collected from the tail fin of tadpoles treated with 100 n*M* T_3 and euthanized immediately following treatment (a) or from cultured tail tissue exposed to 10 n*M* T_3 (b) for the noted time period. Control exposures included treatment with the DMSO vehicle (white bars). The level of TRβ mRNA expression was determined using QPCR analysis and normalized across samples using the invariant ribosomal L8 gene expression. The means and standard errors are shown and the number of animals used per time point is indicated above each bar. Significant differences compared with the vehicle control (Con) are denoted by a single asterisk ($p < 0.05$) or a double asterisk ($p < 0.001$). Statistical analyses were performed as described in Figure 18.2.

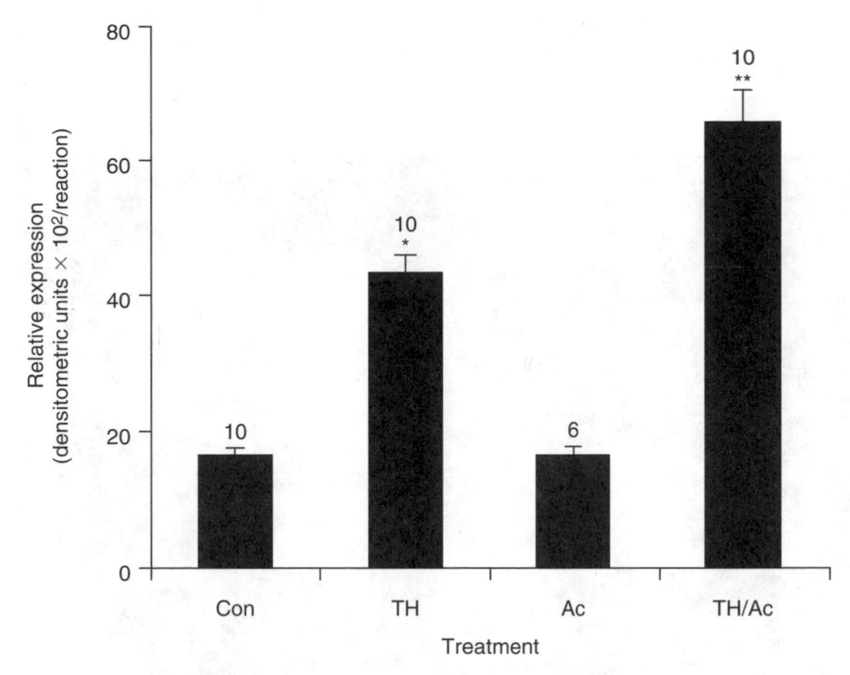

Figure 18.4 Identification of EDC action of the herbicide acetochlor in live *R. catesbeiana* tadpoles. Animals were acclimatized to 20°C for 2 days prior to treatment with 100 n*M* T_3 (TH), 10 n*M* acetochlor (Ac), or a combination of both (TH/Ac) for 24 h. Tissue biopsy samples were collected from tadpole tail tips. The level of TRβ mRNA expression was assessed using standard PCR analysis and normalized across samples using the invariant expression of the 16S ribosomal RNA gene. The means and standard errors are shown and the number of animals used per treatment is indicated above each bar. A single asterisk denotes a significant difference when compared with the control (Con) and acetochlor (Ac) treatments ($p < 0.001$). A double asterisk shows a significant difference compared with all other treatments (Con, $p < 0.001$; Ac, $p < 0.001$; TH, $p < 0.01$). Statistical analyses were performed as described in Figure 18.2. (From Veldhoen, N. and Helbing, C.C., *Environ. Toxicol. Chem.*, 20 (12), 2704–2708, 2001. With permission.)

TH-associated gene expression during frog development. This gene-screen could be extended to evaluate the effects of potential EDCs that target other hormone-dependent developmental processes in amphibians (e.g., testosterone and estrogen in sex determination and the contribution of corticosterone and retinoids during metamorphosis). It is important to note that significant acetochlor-associated effects on TRβ gene expression following exposure to the herbicide are observed prior to any measurable changes in morphological endpoints.[5] Thus, the use of biopsy sampling in combination with genetic biomarkers may allow for greater predictive capability and earlier assessment of EDC action in wildlife species compared with established methods that rely solely on morphometric or mortality endpoints.

While invaluable for repeated measurement-based analyses within a controlled laboratory setting, the biopsy procedure can also be easily applied to the collection of samples from the field in an attempt to address the effects of acute and chronic environmental contaminant exposure on the health of wildlife species. This becomes particularly important in the context of exposure to complex mixtures of contaminants (e.g., effluents) and the contribution of environmental factors to the generation and/or propagation of bioactive chemical components. The equipment required is highly portable (Figure 18.5),

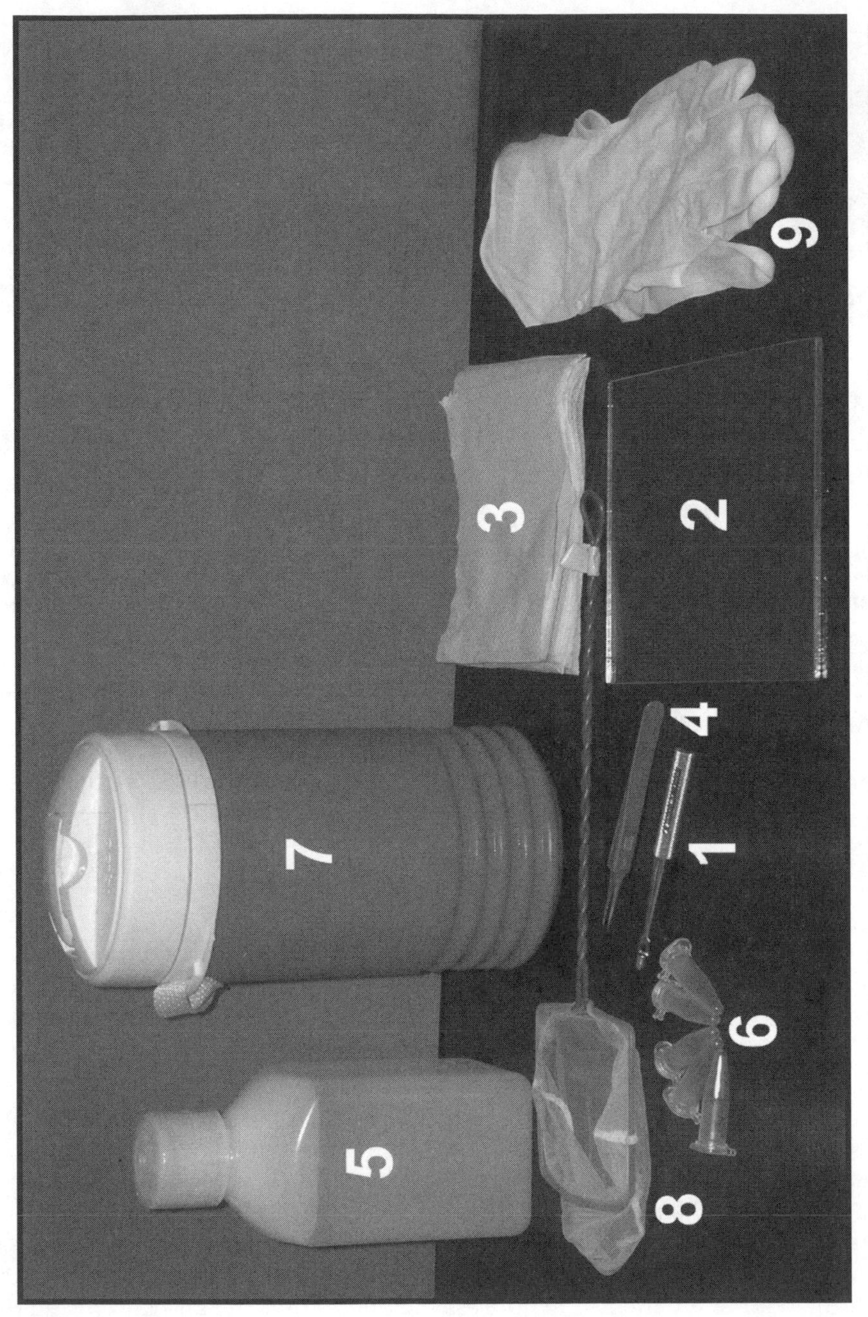

Figure 18.5 Equipment required for field collection of tissue samples. The portable kit includes: a dermal punch biopsy tool (1), a Plexiglas biopsy support (2), paper towel (3), fine-tip forceps (4), RNA preservative (5), plastic storage tubes (6), cooler container (7), capture net (8), and latex gloves (9).

and samples retrieved in this manner are stable at 4°C for several months prior to RNA isolation. It is estimated that a single 2-mm biopsy of tadpole tail fin yields enough RNA to perform 200 QPCR reactions allowing for repeated measurement of targeted gene expression. In addition, genetic material obtained from the biopsy sampling procedure can be used, in combination with recently developed amplified RNA (aRNA) techniques and DNA array technology, to obtain extensive multi-gene expression profiles of the effects of a suspected EDC on the growth and development of species at risk.[5,45–48]

It is also conceivable that this gene-screen procedure could be adapted for any species of interest by selecting the appropriate biopsy instrument for the size and location of tissue to be sampled and identifying appropriate genetic biomarkers for use in QPCR or DNA array-based assessments. Tissue samples could be obtained from wildlife species as diverse as rodents (ear), waterfowl (foot webbing), fish (fin), bats (ear or wing membrane), and amphibians (tail fin and foot webbing). Indeed, biopsy collection of skin and blubber tissue samples from marine mammals is well established.[14,15,49,50] By examining a number of different animal species that occupy both terrestrial and aquatic environments, a greater profile of the potential impact of chemical contamination within an ecosystem may be generated.

References

1. Vos, J.G., Dybing, E., Greim, H.A., Ladefoged, O., Lambre, C., Tarazona, J.V., Brandt, I. and Vethaak, A.D., Health effects of endocrine-disrupting chemicals on wildlife, with special reference to the European situation, *Crit. Rev. Toxicol.*, 30 (1), 71–133, 2000.
2. Cooper, R.L. and Kavlock, R.J., Endocrine disruptors and reproductive development: a weight-of-evidence overview, J. *Endocrinol.*, 152 (2), 159–166, 1997.
3. Tanabe, S., Contamination and toxic effects of persistent endocrine disrupters in marine mammals and birds, *Mar. Pollut. Bull.*, 45 (1–12), 69–77, 2002.
4. McLachlan, J.A., Environmental signaling: what embryos and evolution teach us about endocrine disrupting chemicals, *Endocrinol. Rev.*, 22 (3), 319–341, 2001.
5. Crump, D., Werry, K., Veldhoen, N., Van Aggelen, G. and Helbing, C.C., Exposure to the herbicide acetochlor alters thyroid hormone-dependent gene expression and metamorphosis in *Xenopus laevis, Environ. Health Perspect.*, 110 (12), 1199–1205, 2002.
6. Hayes, T., Haston, K., Tsui, M., Hoang, A., Haeffele, C. and Vonk, A., Atrazine-induced hermaphroditism at 0.1 ppb in American leopard frogs (*Rana pipiens*): laboratory and field evidence, *Environ. Health Perspect.*, 111 (4), 568–575, 2003.
7. Guillette, L.J., Jr. and Gunderson, M.P., Alterations in development of reproductive and endocrine systems of wildlife populations exposed to endocrine-disrupting contaminants, *Reproduction*, 122 (6), 857–864, 2001.
8. Grasman, K.A., Fox, G.A., Scanlon, P.F. and Ludwig, J.P., Organochlorine-associated immunosuppression in prefledgling Caspian terns and herring gulls from the Great Lakes: an ecoepidemiological study, *Environ. Health Perspect.*, 104 (Suppl. 4), 829–842, 1996.
9. Luebke, R.W., Hodson, P.V., Faisal, M., Ross, P.S., Grasman, K.A. and Zelikoff, J., Aquatic pollution-induced immunotoxicity in wildlife species, *Fundam. Appl. Toxicol.*, 37 (1), 1–15, 1997.
10. Kolpin, K.W., Nations, B.K., Goolsby, D.A. and Thurman, E.M., Acetochlor in the hydrologic system in the midwestern United States, *Environ. Sci. Tech.*, 30, 1459–1464, 1996.
11. Barbash, J.E., Thelin, G.P., Kolpin, D.W. and Gilliom, R.J., Distribution of Major Herbicides in Ground Water of the United States, 98-4245, U.S. Geological Survey, Sacramento, CA, 1999.
12. Scribner, E.A., Battaglin, W.A., Goolsby, D.A. and Thurman, E.M., Changes in herbicide concentrations in Midwestern streams in relation to changes in use, 1989–1998, *Sci. Total Environ.*, 248, 255–263, 2000.

13. National Agricultural Statistics Service, Agricultural Chemical Usage: 2000 Field Crops Summary, United States Department of Agriculture, Washington, D.C., 2001, 136 pp.
14. Newman, J.W., Vedder, J.M., Jarman, W.M. and Chang, R.R., A method for the determination of environmental contaminants in living marine mammals using microscale samples of blubber and blood, *Chemosphere*, 29 (4), 671–681, 1994.
15. Fossi, M.C., Casini, S. and Marsili, L., Nondestructive biomarkers of exposure to endocrine disrupting chemicals in endangered species of wildlife, *Chemosphere*, 39 (8), 1273–1285, 1999.
16. Simms, W. and Ross, P.S., Vitamin A physiology and its application as a biomarker of contaminant-related toxicity in marine mammals: a review, *Toxicol. Ind. Health*, 16 (7/8), 291–302, 2001.
17. Veldhoen, N. and Helbing, C.C., Detection of environmental endocrine-disruptor effects on gene expression in live *Rana catesbeiana* tadpoles using a tail fin biopsy technique, *Environ. Toxicol. Chem.*, 20 (12), 2704–2708, 2001.
18. Champoux, L., Rodrigue, J., Desgranges, J.L., Trudeau, S., Hontela, A., Boily, M. and Spear, P., Assessment of contamination and biomarker responses in two species of herons on the St. Lawrence River, *Environ. Monit. Assess.*, 79 (2), 193–215, 2002.
19. Helbing, C.C., Gergely, G. and Atkinson, B.G., Sequential up-regulation of thyroid hormone β receptor, ornithine transcarbamylase and carbamyl phosphate synthetase mRNAs in the liver of *Rana catesbeiana* tadpoles during spontaneous and thyroid hormone-induced metamorphosis, *Dev. Genet.*, 13, 289–301, 1992.
20. Taylor, A.C. and Kollros, J.J., Stages in the normal development of *Rana pipiens* larvae, *Anat. Rec.*, 94, 7–24, 1946.
21. Ausubel, F.M., Brent, R., Kingston, R.E., Moore, D.D., Seidman, J.G., Smith, J.A. and Struhl, K., Eds., *Current Protocols in Molecular Biology*, Chandra, V.B., Series Ed., John Wiley & Sons, Brooklyn, NY, 1999.
22. Sohmer, H. and Freeman, S., The importance of thyroid hormone for auditory development in the fetus and neonate, *Audiol. Neurootol.*, 1 (3), 137–147, 1996.
23. Zoeller, T.R., Dowling, A.L., Herzig, C.T., Iannacone, E.A., Gauger, K.J. and Bansal, R., Thyroid hormone, brain development, and the environment, *Environ. Health Perspect.*, 110 (Suppl. 3), 355–361, 2002.
24. Persky, V., Turyk, M., Anderson, H.A., Hanrahan, L.P., Falk, C., Steenport, D.N., Chatterton, R., Jr. and Freels, S., The effects of PCB exposure and fish consumption on endogenous hormones, *Environ. Health Perspect.*, 109 (12), 1275–1283, 2001.
25. LoPachin, R.M., Lehning, E.J., Opanashuk, L.A. and Jortner, B.S., Rate of neurotoxicant exposure determines morphologic manifestations of distal axonopathy, *Toxicol. Appl. Pharmacol.*, 167 (2), 75–86, 2000.
26. Zoeller, R.T. and Crofton, K.M., Thyroid hormone action in fetal brain development and potential for disruption by environmental chemicals, *Neurotoxicology*, 21 (6), 935–945, 2000.
27. Cooke, A.S., Tadpoles as indicators of harmful levels of pollution in the field, *Environ. Pollut. (Series A)*, 25, 123–133, 1981.
28. LeBlanc, G.A. and Bain, L.J., Chronic toxicity of environmental contaminants: sentinels and biomarkers, *Environ. Health Perspect.*, 105 (Suppl, 1), 65–80, 1997.
29. Lutz, I. and Kloas, W., Amphibians as a model to study endocrine disruptors. I. Environmental pollution and estrogen receptor binding, *Sci. Total Environ.*, 225, 49–57, 1999.
30. Kloas, W., Lutz, I. and Einspanier, R., Amphibians as a model to study endocrine disruptors. II. Estrogenic activity of environmental chemicals *in vitro* and *in vivo*, *Sci. Total Environ.*, 225, 59–68, 1999.
31. Brucker-Davis, F., Effects of environmental synthetic chemicals on thyroid function, *Thyroid*, 8 (9), 827–856, 1998.
32. Brouwer, A., Morse, D.C., Lans, M.C., Schuur, A.G., Murk, A.J., Klasson-Wehler, E., Bergman, A. and Visser, T.J., Interactions of persistent environmental organohalogens with the thyroid hormone system: mechanisms and possible consequences for animal and human health, *Toxicol. Ind. Health*, 14 (1/2), 59–84, 1998.

33. Rolland, R.M., A review of chemically-induced alterations in thyroid and vitamin A status from field studies of wildlife and fish, J. *Wildl. Dis.*, 36 (4), 615–635, 2000.
34. Yaoita, Y., Shi, Y-B. and Brown, D., *Xenopus laevis* alpha and beta thyroid hormone receptors, *Proc. Natl. Acad. Sci. USA*, 87, 7090–7094, 1990.
35. Yaoita, Y. and Brown, D.D., A correlation of thyroid hormone receptor gene expression with amphibian metamorphosis, *Genes Dev.*, 4 (11), 1917–1924, 1990.
36. Eliceiri, B. and Brown, D., Quantitation of endogenous thyroid hormone receptors α and β during embryogenesis and metamorphosis in *Xenopus laevis, J. Biol. Chem.*, 269, 24459–24465, 1994.
37. Galton, V.A., Mechanisms underlying the acceleration of thyroid hormone-induced tadpole metamorphosis by corticosterone, *Endocrinology*, 127 (6), 2997–3002, 1990.
38. Rabelo, E.M., Baker, B.S. and Tata, J.R., Interplay between thyroid hormone and estrogen in modulating expression of their receptor and vitellogenin genes during *Xenopus* metamorphosis, *Mech. Dev.*, 45 (1), 49–57, 1994.
39. Hayes, T.B., Interdependence of corticosterone and thyroid hormones in larval toads (*Bufo boreas*). I. Thyroid hormone-dependent and independent effects of corticosterone on growth and development, *JM. Exp. Zool.*, 271 (2), 95–102, 1995.
40. Houlahan, J.E., Findlay, C.S., Schmidt, B.R., Meyer, A.H. and Kuzmin, S.L., Quantitative evidence for global amphibian population declines, *Nature*, 404 (6779), 752–755, 2000.
41. Alford, R.A., Dixon, P.M. and Pechmann, J.H., Global amphibian population declines, *Nature*, 412 (6846), 499–500, 2001.
42. Yaoita, Y., Shi, Y.-B. and Brown, D.D., *Xenopus laevis* alpha and beta thyroid hormone receptors, *Proc. Natl. Acad. Sci. USA*, 87, 7090–7094, 1990.
43. Howe, C.M., Berrill, M., Pauli, B.D., Helbing, C.C., Werry, K. and Veldhoen, N., Toxicity of glyphosate-based pesticides to four North American frog species, *Environ. Toxicol. Chem.* 23 (3), 1928–1938, 2004.
44. Shi, Y.-B., *Amphibian Metamorphosis: From Morphology to Molecular Biology*, Wiley, New York, 2000, 288 pp.
45. Dent, G.W., O'Dell, D.M. and Eberwine, J.H., Gene expression profiling in the amygdala: an approach to examine the molecular substrates of mammalian behavior, *Physiol. Behav.*, 73 (5), 841–847, 2001.
46. Helbing, C.C., Werry, K., Crump, D., Domanski, D., Veldhoen, N. and Bailey, C.M., Expression profiles of novel thyroid hormone-responsive genes and proteins in the tail of *Xenopus laevis* tadpoles undergoing precocious metamorphosis, *Mol. Endocrinol.*, 17 (7), 1395–1409, 2003.
47. Veldhoen, N., Crump, D., Werry, K. and Helbing, C., Distinctive gene profiles occur at key points during natural metamorphosis in the *Xenopus laevis* tadpole tail, *Dev. Dyn.*, 225, 457–468, 2002.
48. Frog MAGEX DNA array, ViagenX Biotech Inc., www.viagenx.com.
49. Hobbs, K.E., Muir, D.C., Michaud, R., Beland, P., Letcher, R.J. and Norstrom, R.J., PCBs and organochlorine pesticides in blubber biopsies from free-ranging St. Lawrence River Estuary beluga whales (*Delphinapterus leucas*), 1994–1998, *Environ. Pollut.*, 122 (2), 291–302, 2003.
50. Ylitalo, G.M., Matkin, C.O., Buzitis, J., Krahn, M.M., Jones, L.L., Rowles, T. and Stein, J.E., Influence of life-history parameters on organochlorine concentrations in free-ranging killer whales (*Orcinus orca*) from Prince William Sound, AK, *Sci. Total Environ.*, 281 (1–3), 183–203, 2001.
51. Regard, E., Taurog, A. and Nakashima, T., Plasma thyroxine and triiodothyronine levels in spontaneously metamorphosing Rana catesbeiana tadpoles and in adult anuran amphibia, *Endocrinology*, 102 (3), 674–684, 1978.

section three

Techniques for identification and assessment of contaminants in aquatic ecosystems

chapter nineteen

Coral reproduction and recruitment as tools for studying the ecotoxicology of coral reef ecosystems

Robert H. Richmond
Kewalo Marine Laboratory

Contents

Introduction

Aquatic toxicology is a discipline that deals with the effects of chemicals and potential pollutants on individual organisms, usually with the intent of applying such data to understanding effects on populations and community structure and function. Aquatic organisms have a number of different life-history stages through which they must successfully pass, in order to become part of a population, and hence, assays that deal with only one stage, such as the adult form, may miss critical effects and have limited predictive value.

Many benthic aquatic organisms in both the protostome and deuterostome phyla, including Porifera, Cnidaria, Annelida, Crustacea, Echinodermata, and Mollusca, spawn gametes (eggs and sperm) into the water column, with subsequent external fertilization and development into planktonic larval stages, which eventually settle on specific substrata prior to metamorphosis into the juvenile form. These organisms respond to

1-56670-664-5/05/$0.00+$1.50
© 2005 by CRC Press

chemical cues for synchronization of spawning events among conspecifics, for mediation of egg–sperm interactions and for metamorphic induction in response to particular substrata, aggregations of adults or preferred prey.[1] Since so many steps in the life cycles of marine organisms are chemically mediated, the lack of a measurable effect of a xenobiotic on adult organisms at a range of concentrations does not necessarily mean the chemical is safe from a population or community perspective if it interferes with chemical cues at earlier life-history stages. Additionally, 100% successful fertilization of eggs and development of larvae in a bioassay with a subsequent 0% recruitment rate has the same overall effect on the population as 100% mortality at the fertilization stage. Finally, the duration of the observation period following exposure of an early life-history stage of an organism is important, as initial survival does not necessarily mean growth, development and reproductive ability will be unaffected later on.

Scleractinian corals provide an excellent set of organisms for use in aquatic toxicologic studies in tropical marine environments. Many species are simultaneous hermaphrodites that release combined egg–sperm packets during highly predictable annual spawning events. The gametes are easy to collect and manipulate, the large numbers available lend themselves to experiments with large sample sizes and hence rigorous statistical testing, and the larvae that result from successful fertilizations often have specific cues necessary for settlement and metamorphosis, allowing for multiple stage testing. Additionally, specific genetic crosses can be controlled and as a result, the confounding effects of genetic variability among individuals can be controlled. Genetic lines can be reared for later exposure and determination of biomarker expression. Gametes can be collected from the same coral colonies over several years if care is taken in collecting and maintaining "donor" corals.

Materials required

Material	Source	Comments
Gravid corals; at least two different colonies of the same species	Local coral reef communities	See (Richmond and Hunter, 1990; Richmond, 1997) Reference 2, for timing and location. Collecting permits is usually necessary
Sea water system	Field station, aquarium, or laboratory; can be open, closed, or recirculating	Corals have specific requirements for water quality, light, pH, temperature, and water motion
Basins/containers large enough to hold individual colonies	Aquarium store, hardware store, department store	Containers should be conditioned prior to use; glass is preferable for some experiments; plastic basins should be soaked to remove plasticizers
Plankton netting (Nytex) 105-μm mesh	Aquatic Ecosystems catalog #M105 (www.aquaticeco.com)	The netting is used to make sieves to collect gamete bundles. Coral eggs typically range in size from ca. 180 to 360 μm. A mesh size of 105 μm works for most species

Material	Source	Comments
2-in. PVC pipe	Hardware stores	Good diameter for making sieves for separating egg–sperm clusters. Nytex is attached to end of PVC pipe with hot melt glue
Hot melt glue and glue gun	Hardware or department stores	Used to attach plankton netting to PVC pipe sections for sieves
250-ml specimen cups, beakers or equivalent	VWR, Scientific Products or most laboratory suppliers	Glassware is preferable over plastic
Pasteur pipettes and bulbs	VWR, Scientific Products, etc.	Both 5- and 9-in. pipettes are useful
1-l glass beakers, fleakers or jars	Most labware suppliers	1200-ml vessels allow for raising batches of 1000 larvae at a density of 1 larva/ml
Air pumps (aquarium type or system), lines and airstones	Any pet shop or store with pet department	Need gentle aeration to keep floating eggs from getting trapped in the surface tension of the water
Hemocytometer	VWR, Scientific Products, etc.	Necessary for determining sperm density if the separation technique is used

Procedures

Selection of coral species

Coral species are selected based on their reproductive behavior and timing, growth rate, growth form, sensitivity to environmental parameters, cultivation characteristics, and biogeographic distribution patterns. Our laboratory has been able to routinely raise larvae from 13 species of corals that represent a variety of growth forms, habitats, and range of sensitivities to stress. *Pocillopora damicornis*, a brooding species, is widely distributed across the Pacific Ocean and produces viable planula larvae monthly, throughout the year, in Micronesia and Hawaii.[3] These larvae contain a full complement of zooxanthellae upon release from the parent colony. As such, this coral is good for studies requiring competent larvae but not for experiments to test the effects of chemical compounds on fertilization, larval development, or acquisition of symbiotic zooxanthellae. *P. damicornis* is a non-perforate species, meaning tissue is relatively easy to remove from the carbonate exoskeleton, which is a helpful characteristic for assays requiring subsequent colony tissue analysis. This species also releases relatively low quantities of mucus and is easy to maintain and transplant.

Several broadcast spawning acroporids (e.g., *Acropora humilis, A. tenuis, A. valida, A. danae, A. surculosa,* and *A. wardii*) and favids (*Goniastrea retiformis* and *Leptoria phrygia)* have proven to be useful in fertilization, development, symbiotic algal uptake, and recruitment bioassays. Species in these two groups, along with corals in the family Poritidae, are among the most commonly studied coral species. Since our research aim is to develop protocols that are widely applicable, we continue to add coral species that have appropriate biogeographic distribution patterns.

Collection of corals

Corals are collected from the field several days prior to predicted planulation or spawning. For the brooding species *P. damicornis*, colonies are collected 2–3 days after the new moon, as this species has been found to planulate within several days of the lunar first quarter, each month of the year, in Micronesia.[3] Timing is different in Hawaii and Australia, demonstrating the need for accessing data on the reproductive timing of local populations.

Spawning species are collected 2 days after the June–August full moons on Guam (Richmond and Hunter, 1990),[1] and in March–May in Palau. In all cases, corals are checked in the field for the presence of gametes each month leading up to spawning. Generally, if a coral is ripe, the eggs will be pigmented (pink, red, or orange) and are easily seen on the colony side of a broken branch or fragment (Figure 19.1). If no color is observed, the branch tip or fragment is brought back to the laboratory for examination under a compound microscope. For the acroporid and favid corals studied on Guam to date, eggs can be observed developing up to 6 months prior to spawning (January), appearing as white growths along the mesenteries. As spawning approaches, the eggs increase in size and eventually become pigmented. In the species that are simultaneous hermaphrodites (containing both eggs and sperm within the same colony and polyp), sperm begin to develop within a month or two of spawning, after eggs have already formed.

When ripe colonies are found, they are gently removed from the substratum with a hammer and chisel, and kept submerged until they are transferred to coolers filled with seawater and quickly transported to running seawater tables or aquaria. It is important to collect at least two colonies of each species, preferably some distance apart (to increase the probability of the two being genetically different) to allow for outcrossing in fertilization assays. Experiments have demonstrated self-fertilization rates for the simultaneously hermaphroditic corals are low or non-existent.[1,4]

Collection of gametes and larvae

The acroporid corals proven most appropriate for toxicology studies are all simultaneous hermaphrodites that release combined egg–sperm bundles[2] (Richmond and Hunter, 1990;

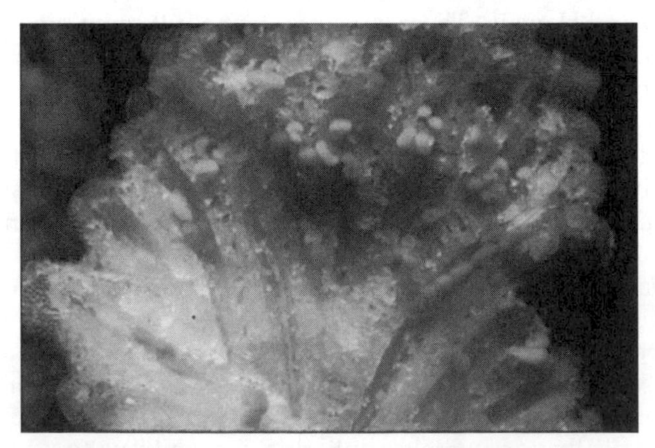

Figure 19.1 **(see color insert following page 464)** Cross-section of coral colony with pink eggs and white spermaries.

Figure 19.2 **(see color insert)** Combined egg–sperm clusters.

Figure 19.3 **(see color insert)** Floating gamete bundles collected in sieves.

Figure 19.2). Upon release, these float to the surface and can be collected on sieves with a mesh size of 105 μm or, alternatively, collected individually with a Pasteur pipette and placed intact into a 250-ml container with UV-sterilized/filtered seawater (Figure 19.3).

Two different systems have been developed, depending upon the experiments that will be performed. In fertilization studies, or those testing the effects of water-soluble pollutants, the clusters are washed with 0.45 μm (Millipore) UV-sterilized/filtered seawater, with the sperm being collected as the wash. The eggs are re-suspended in filtered/sterilized seawater, and fertilized with sperm from another colony of the same species, at a sperm concentration of ca. 10^5 sperm/ml.[4] Outcrossing has been found to be necessary as self-fertilization rates are relatively low. Higher concentrations of sperm often result in polyspermy and non-viable embryos. Fertilized embryos are placed into 1- and 2-l glass or plastic culture flasks with aeration, to provide a degree of water motion. Densities of eggs and developing larvae should not exceed 1/ml of seawater. Water is changed every 12 h during the first 48 h by siphoning from the bottom of the vessel, and daily afterwards until the larvae are ready to be settled onto appropriate substrata. Antibiotics are not necessary and handling of larvae should be minimized.

In studies using larval development, recruitment, zooxanthellae acquisition, and survivorship, we have simplified the fertilization procedure, and simply added the appropriate number of gamete clusters from two different colonies of the same species to 1 l of filtered seawater to achieve the proper sperm density. For example, for most acroporids, 50–100 gamete clusters from each of two colonies yield a sperm density of ca. 10^5 sperm/ml, and yield between 800 and 1200 larvae, with over 90% fertilization success. For the favid corals, crossing 150–200 clusters from each of two colonies in 1 l of seawater has been effective. The "cluster" technique makes a number of experiments logistically feasible at high levels of replication, and hence statistical rigor.

The planula larvae of the brooding coral *P. damicornis* are approximately 1 mm in diameter. Adult colonies measuring 15–20 cm in diameter are placed into 3- to 9-l vessels that overflow into containers constructed from plastic beaker bases with walls of 80-μm Nytex plankton netting. The corals are kept under continuous flow conditions of 1 l/min, and the buoyant larvae, which float out of the bowls, are collected in the mesh of the cups. This species has been found to planulate at night, so collectors are checked each morning for the presence of larvae.

Bioassays

Fertilization bioassays address the first critical, chemically-mediated step in the reproductive process and can be performed in a variety of vessels and a using a range of concentrations based on the question being asked and the characteristics of the substance being tested. When gamete quantity is limited, 15-ml prewashed glass scintillation vials are used with ca. 100 eggs (50 each from two different colonies; 5–8 clusters from each, for a variety of acroporid and favid corals) per vial, in 10 ml of UV-sterilized/filtered seawater, with a sperm density of ca. 10^5 sperm/ml. With the expectation that outcrossing will occur when using the "cluster" technique, developing larvae will be of two different genotypes. If follow-up experiments are to be performed on the larvae that form from the fertilization assays, using 100 eggs collected from a single colony and fertilized with sperm separated from a conspecific colony will presumably yield larvae of a single genotypic cross, and may reduce variability in experimental results.

Experiments are scored by viewing samples under a dissecting microscope for the number of eggs fertilized, the number of embryos reaching the planula larval stage versus the number of eggs/embryos that are non-viable.[5] A watch glass or other concave dish makes counting easier. Careful and limited handling allows for the larvae that do form to be reared further for recruitment and algal acquisition assays.

Recruitment assays

Coral larvae from spawning species become competent to settle between 18 and 72 h following fertilization, depending on species and egg size.[1] The smaller eggs (from favids) develop cilia and become competent more quickly than the larger *Acropora* eggs. The brooded larvae of *P. damicornis* are fairly non-specific and will settle on a variety of substrata pre-conditioned with bacterial/diatomaceous films. The *Acropora* and *Goniastrea* larvae are more selective, and our previous experiments indicate that several species are highly specific, settling only on particular species of crustose coralline algae (*Hydrolithon reinboldii*).

For recruitment bioassays, chemical effects of potential pollutants are tested by exposing preferred substrata to the xenobiotic, rinsing the substrata, then placing them into 250-ml beakers with clean, filtered seawater or by exposing larvae to the chemical and providing them with untreated substrata. A known quantity of larvae is added to each beaker (50–100 per replicate), and the substrata scored daily under a dissecting microscope for 5 days for larvae that have settled (come into contact with the substrata), metamorphosed (cemented to the surface and displaying evidence of calcification; Figure 19.4), remained swimming or died. Alternatively, larvae are exposed (generally for 12–24 h) and subsequently added to beakers with the appropriate untreated substrata. Our research has demonstrated that pollutants can interfere with metamorphic inducers associated with the substrata and/or with the inducer receptor of the larvae. It is important to address both possibilities. While fertilization and embryological development, stages in corals are particularly sensitive to water-soluble chemicals; lipophilic substances appear to have a greater effect on larval recruitment.

In assays designed to study the effects of chemicals on zooxanthellae uptake, competent larvae are placed into basins or aquaria containing cleaned coral rubble in 0.45 μm (Millipore) filtered seawater. Pieces are checked daily for the presence of settled larvae, and if present, fragments are transferred to separate containers for exposure to chemicals and the addition of zooxanthellae. Zooxanthellae for use in these assays can come from pieces of donor colonies of the same species placed in the container, from cultured lines, from zooxanthellae centrifuged from tissue preparations of conspecific colonies, or from zooxanthellae collected and filtered from "induced expulsions." One known stress response of corals is the breakdown of the animal–algal symbiosis. A by-product of cyanide bioassays performed to determine the effects of cyanide fishing on reefs was the discovery that exposure of coral branches to cyanide at 0.1 g/l for 5 min resulted in free-swimming zooxanthellae evacuating the host cells and tissue within hours of exposure. Since zooxanthellae appear to have the same cyanide-resistant respiratory pathway found in other plants, clean zooxanthellae free of coral tissue residue can be acquired this way that have found to be functional and capable of colonizing coral recruits.

For bioassays using adult corals, colonies are cultivated from larvae to a size of approximately 5–10 cm in diameter. Depending on the question, corals are either

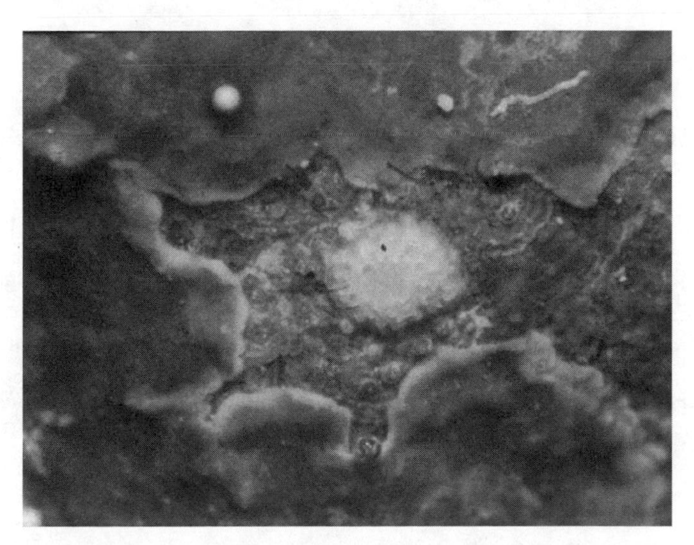

Figure 19.4 **(see color insert)** Coral larval recruit, settled and metamorphosed.

"pulsed" and transferred to grow-out tanks with flowing seawater or maintained under test conditions in a closed system with air stones. We use separate containers for each coral, 10 corals per concentration (for replication, avoiding pseudoreplication) in a randomized block design. In pulse experiments, after exposure, tagged, treated and control corals are placed into a flowing seawater tank with a flushing rate of 21/min and observed for signs of bleaching, tissue loss, and death for a period of 30 days.

A great advantage of using corals is the ability to control genetic variability by using larvae from a single cohort. A single colony can be repeatedly "harvested" for gametes and larvae over a period of years with careful handling. For spawning species, once gametes are released and collected, colonies are tagged and transplanted back into the field, by cementing them in a marked area using a mixture of 7 parts cement to 1 part plaster of paris. Previous experiments have demonstrated that for several types of stress assays, genetic considerations are important, as variability among coral colonies across different genotypes was greater than the effect measured within a single genotype.

In conclusion, corals have proven to be valuable tools for ecotoxicological studies. Their reproductive behavior provides opportunities for studying effects of chemicals on cueing between conspecific colonies during spawning events, for studies of egg–sperm recognition, fertilization, embryological development, metamorphic induction, and acquisition of symbiotic zooxanthellae. The ability to rear corals of known genotypes in large numbers allows for statistically rigorous testing and for monitoring specific sites over time. Recent advances in the development of biomarkers of exposure in corals hold promise for determining the effects of pollutants at sublethal levels, and for measuring responses to mitigative measures. Such tools are proving useful to help resource managers address the effects of human activities on coral reef ecosystems, presently under threat world-wide.

Acknowledgments

This research was supported by grants from the US EPA STAR program and the NOAA COP/CRES program. I thank Yimnang Golbuu, Steven Victor, Walter Kelley, Wendy Chen, Aja Reyes, Jack Idechong, Sarah Leota, and Teina Rongo for their assistance with the bioassays and coral reproductive experiments.

References

1. Richmond, R.H., Reproduction and recruitment in corals: critical links in the persistence of reefs, in *Life and Death of Coral Reefs*, Birkeland, C.E., Ed., Chapman & Hall, New York, pp. 175–197, 1997.
2. Harrison, P.L. and Wallace, C.C., Coral reproduction, in *Ecosystems of the World: Coral Reefs*, Dubinsky, Z., Ed., Elsevier, Amsterdam, pp. 133–208, 1990.
3. Richmond, R.H. and Jokiel, P.L., Lunar periodicity in larva release in the reef coral *Pocillopora damicornis* at Enewetak and Hawaii, *Bull. Mar. Sci.*, 34 (2), 280–287, 1984.
4. Heyward, A.J. and Babcock, R., Self-and cross fertilization in scleractinian corals, *Mar. Biol.*, 90, 191–195, 1986.
5. Richmond, R.H., Effects of coastal runoff on coral reproduction, in *Global Aspects of Coral Reefs: Health, Hazards and History*, Ginsburg, R.N., Ed., University of Miami, pp. 360–364, 1994b.
6. Richmond, R.H. and Hunter, C.L., Reproduction and recruitment of corals: Comparisons among the Caribbean, the tropical Pacific, and the Red Sea, Mar. Ecol. Prog. Ser. 60, 185–203, 1990.

chapter twenty

Using the stickleback to monitor androgens and anti-androgens in the aquatic environment

Ioanna Katsiadaki
Centre for Environment, Fisheries, and Aquaculture Science (CEFAS)

Contents

Note: Dr. Katsiadaki is a British Crown employee and the chapter was written as part of her official duties at the CEFAS Weymouth Laboratory, and therefore the British Crown copyright for the chapter cannot be assigned. CRC Press is, however, hereby authorized to include the article in the volume and in any reprintings or reproduction of the book in any form or media, including the sale of reprints, providing it is understood that British Crown copyright and British Crown user rights are reserved. The rights granted by this authority do not extend to authorizing third parties to reproduce British Crown copyright material. Any such applications received by the publisher should be addressed to CEFAS.

1-56670-664-5/05/$0.00+$1.50
© 2005 by CRC Press

Introduction

Over the past few years there has been increasing evidence of the hormone-like effects of environmental chemicals, such as pesticides and industrial chemicals, in both wildlife and humans. These so-called endocrine disruptive chemicals (EDCs) mainly act by mimicking or antagonizing the effect of the endogenous hormones estradiol and testosterone but may also disrupt the synthesis and metabolism of endogenous hormones and/or their receptors.[1] A large number of compounds have been reported to possess endocrine modulating activity. These include natural products, pesticides, fungicides and insecticides, medical drugs, and commercial and/or industrial chemicals. Although a causal relationship between exposure to these substances and human/wildlife reproductive health has not been fully established, the characterization of the adverse effects of EDCs, at environmentally relevant concentrations, is very desirable.

Much research appears to have been done already on the role of estrogenic xenobiotics. The most widely used biomarker for estrogenic exposure is the presence of the female specific protein, vitellogenin in the plasma, liver or whole-body homogenates of male fish.[2–6] The role of androgenic xenobiotics has not been studied in nearly so much detail, despite the fact that there is an increasing concern for clinical implications of these chemicals in humans.[7] One of the clearest observations of androgenicity in the aquatic environment has been made in female mosquito fish (*Gambusia* sp.), living downstream of kraft mill effluent discharges.[8–10] While the detection of androgenic compounds in the environment is presently restricted to pulp mill effluents (PMEs) and sewage treatment works without secondary treatment,[11] compounds with anti-androgenic activity appear to be more widespread. At the moment, the only official *in vivo* test able to provide information on the androgenic or anti-androgenic effect of suspected chemicals is the Hershberger castrated male rat assay.[12]

The three-spined stickleback (*Gasterosteus aculeatus*) offers a great potential for the assessment of reproductive disturbances caused by androgenic xenobiotics due to its pronounced androgen-dependent male secondary sexual characters that present during its breeding season (late spring and summer). These characters include development of nuptial coloration, kidney hypertrophy, territorial and nest-building behavior. The kidney hypertrophies under the control of androgens to produce a "glue" protein that is used to build the nest out of algae, plant material, sand, and detritus. This glue protein was first characterized by Jakobsson et al.[13] and was given the name spiggin from the name of the stickleback in Swedish, the spigg. Spiggin is assembled from three subunits in the urinary bladder[14] and is deposited on suitable nest material by contractions of the urinary bladder. The production of the glue protein by the male stickleback has potential as a biomarker for androgenic and anti-androgenic xenobiotics because (a) it is well established that is androgen-dependent,[15–18] and (b) it has easily measurable response parameters that include kidney weight changes (nephrosomatic index), histological changes on the height of the epithelial cells,[17,19,20] spiggin changes,[21,22] and spiggin mRNA changes.[14]

To date, spiggin is the only androgen-induced protein that has been isolated from fish. Recent research at Centre for Environment, Fisheries and Aquaculture Science (CEFAS) has firmly established that female stickleback kidneys can produce spiggin in response to several model androgens, such as 17α-methyltestosterone (17α-MT), dihydrotestosterone (DHT), testosterone (T), and 11-keto-testosterone (11-KT), added to the ambient water. As we show in this chapter, spiggin production is inhibited in males (and in 17α-MT-stimulated females) by flutamide (FL; a well-known anti-androgenic drug that is used in the treatment of prostate cancer.[23] A number of other xenobiotics with suspected anti-androgenic activity have been tested and most showed clear inhibition of spiggin production.

Materials required

Husbandry

Glass aquaria of various sizes for maintaining fish stock and experimental populations; de-chlorinated freshwater and/or seawater supply; multi-channel peristaltic pumps; air (O_2) supply; tubing (high quality Portex or similar polypropylene); pipe work fittings; flow meters; oxygen and pH meters; equipment for determining water hardness; thermometers; under-gravel bacteria filters; coarse gravel (approximately 5 mm); controlled day-length light; siphoning tube; nets; various sized beakers; Pasteur pipettes.

Histological processing

Dissection kit; glass vials (50 or 20 ml); 10% neutrally buffered formalin (BDH); industrial methylated spirit (IMS); xylene or the safer alternative Histosolve™ solvent (Thermo electron/Shandon); ethanol (reagent grade); cassettes (5 × 3 cm uni-cassette, Tissue Tek); paraffin wax (W1 formula wax, RA Lamb or similar with a melting point of 57–58°C); Historesin (Leica or Taab, UK, comes with extensive manufacturers instructions to use); molds; vacuum infiltration processor (VIP; Tissue TEK VIP 2000 or similar); tissue embedding center (Tissue TEK II or similar); microtome (Shandon Finesse E or similar); standard cryostat; disposable microtome blades (80 × 0 mm, RA Lamb, Catalog E53.37/S35); stainless steel knife; routine histological stains (see Tables 20.2 and 20.3); histology glass slides and cover slips; xylene-based polymer mountant (DPX or similar); standard microscope equipped with a graticule or an eyepiece micrometer; microscope (Nikon E800 or similar) linked to camera (DXM 1200F or similar) and PC; image analysis software (Lucia G LIM screen measurement or VIDS or other similar software).

ELISA for spiggin

Polystyrene plates (high protein-binding, flat-bottomed, Costar® EIA/RIA 96-well); polypropylene plates (low-binding, round-bottomed); plate covers (Mylar sealing tape from Sigma); plate washer (e.g., MRW, 8-channel plate washer made by Dynex Technologies, The Microtiter® Company); plate reader (e.g., MRX microplate reader made by the same company as the washer).

Coating buffer (0.05 *M* sodium bicarbonate–carbonate, pH 9.6); plate wash buffer (0.1 *M* sodium phosphate (di-basic salt, 0.072 *M*, mono-basic salt, 0.028 *M*), 0.14 *M* sodium chloride, 27 m*M* potassium chloride, 0.05% Tween-20); assay buffer (same as plate wash buffer plus 0.1% bovine serum albumin (Sigma) and 0.15 m*M* sodium azide); second antibody (affinity isolated anti-rabbit IgG whole molecule conjugated to alkaline phosphatase (Catalog A-3937, Sigma Immuno-Chemicals); alkaline phosphatase substrate (1 mg/ml 4-nitrophenyl phosphate (*p*-NPP) in 0.2 *M* Tris buffer (SIGMA FAST™ tablets). There are also specific reagents (that have to be produced in-house). These include polyclonal antisera against spiggin and a spiggin standard.

Procedures

Fish collection and husbandry

Three-spined sticklebacks are ubiquitous in the whole of the North hemisphere and can be found in almost all aquatic habitats from freshwater until full seawater. However, they spend the best part of their lifetime in freshwater systems where they reproduce, and it is relatively easy to catch them with simple netting devices or traps.

Although juvenile fish can also be used as a model for endocrine disruption research, the proposed tests are designed to use adult fish that are 8–18 months old and weigh at least 0.8 g. Fish used in the experiments should be disease-free and have no obvious signs of parasitism. The fish can be maintained in the laboratory in either freshwater or brackish or even full seawater. However, it is preferable that they are kept in brackish water at 10°C and short photoperiod, 8 h light and 16 h dark (*L*:*D* 8:16) until used for the experiments and in some cases during the exposure period. The fish are fed *ad libitum* with a variety of fresh or frozen food [mosquito larvae, bloodworm, *Daphnia*, brine shrimp (*Artemia*), etc.] once daily throughout the exposure period. Stock populations can be fed on dried food (TetraMin) also but not exclusively. Dry food however, whenever used, should be estrogen free.

Stock populations can be kept at a density of 2–10 fish per litre of water, while the experimental tanks should not exceed 1 fish/l. The duration of the proposed tests is 21 days.

Exposure to test compounds

Test compounds can be delivered in a number of ways, such as injections (IM, IP), implantation, and via the food. However, fish will readily absorb compounds added to the water, and this is the preferred method of exposure. Water-borne exposure to androgenic and anti-androgenic compounds/effluents can be accomplished using two systems, namely, continuous flow and semi-static. The former requires the use of peristaltic pumps that can be adjusted to deliver the desired flow rate/concentration of test compound to the water inflow. The semi-static system that we use at CEFAS involves a 70% water change (by siphoning out the water slowly and refilling the aquaria) and addition of fresh test solution(s) every 48 h. For establishment of actual (as opposed to nominal concentrations) of the compounds, water samples (1–2 l) are taken at least once a week into dark glass bottles when using a continuous flow system to verify the concentration of the test compound(s) in the water. When using a semi-static system of exposure, the frequency should be increased to include water samples just prior to and immediately after the water change in order to establish the variation of test compound concentration in the 48-h period. Extraction and then measurement of the tested chemicals in the water can be achieved by a number of methods that are outside the scope of this chapter.

Light microscopy

At the end of the exposure period, the fish are humanly killed and their weight recorded. The kidneys and gonads should be dissected out and immediately placed in 10% neutrally buffered formalin. Kidney dissection allows better fixation and embedding. It is advisable that the gonads are processed along with the kidneys to allow determination of the reproductive status of the fish and to allow easy and unmistakable identification of fish gender.

If whole fish sections are preferred, then an incision should be made across the ventral lining to allow penetration of fixative. The ratio of tissue: formalin should be 1:10 (w/v). Fixed dissected tissues (or whole fish) should then be placed in perforated plastic cassettes and processed using a VIP as described in Table 20.1. The times and solutions can be varied to specifically suit the size of the samples. If using a rotary processor or immersion technique, duration of steps need to be extended. Samples can be placed into 70% IMS before processing begins. Solutions described as "sitting" solutions (stations 1 and 14) are called this, as these are the stages immediately before and after the processing schedule commencement. The processor is normally run overnight will finish as set on the processors timer. In the case of our VIP, station 14 will not empty out at the end of processing, and cassettes can be allowed to stand until removal for embedding. Station 1 may be replaced by 10% neutral buffered formalin to aid fixation or to speed up sample processing. Thermo electron/Shandon Histosolve™ is used as a safer alternative to xylene. Following impregnation, tissues are ready to be embedded. The lids of the cassettes are removed, and molten paraffin (58°C) is poured into the cassette. Paraffin is allowed to solidify around and within the tissue forming blocks, which are then rapidly cooled over a cold plate.

Although paraffin wax is a standard embedding material, resin embedding allows for a more detailed picture of renal histology and is therefore recommended. For this, fixation and dehydration steps are the same as for paraffin wax embedding, followed by the steps pre-infiltration, infiltration, polymerization, and embedding, which are described in great detail in the product instructions.

Both paraffin and resin sections are cut at 5 μm using a standard microtome and disposable blades or a cryostat and stainless steel blade, respectively. The sections are mounted on glass slides and stained.

Lee's methylene blue-basic fuchsin stain is an inexpensive, one-step simple stain that offers great detail in the renal epithelium. Table 20.2 describes the staining schedule. The

Table 20.1 Fish tissue processing schedule for VIP

Station No.	Solution	Duration (min)
1	70% I.M.S.	35 (sitting solution)
2	90% I.M.S.	35
3	100% I.M.S.	35
4	100% I.M.S.	35
5	100% I.M.S.	35
6	100% I.M.S.	35
7	Histosolve solvent	35
8	Histosolve solvent	35
9	Histosolve solvent	35
10	Histosolve solvent	35
11	W1 paraffin wax at 60°C	45
12	W1 paraffin wax at 60°C	45
13	W1 paraffin wax at 60°C	45
14	W1 paraffin wax at 60°C	45 (sitting solution)

Table 20.2 Staining schedule for Lee's methylene blue-basic fuchsin (500 ml)

Step No.	Procedure	Chemicals
1	Dissolve stain in 100 ml water	0.128 g methylene blue
2	Dissolve stain in 100 ml water	0.128 basic fuchsin
3	Dissolve chemicals in 175 ml water	0.52 g NaH_2PO_4 $2H_2O$, 2.52 g Na_2 HPO_4 $2H_2O$
4	Combine solutions (steps 1–3)	
5	Add ethanol and mix	125 ml ethanol
6	Filter stain (can be used for 4 days)	
7	Immerse slides into stain and agitate gently for 10–15 s	
8	Rinse in de-ionized water (DW)	
9	Air dry	
10	Mount under cover slip	

stain is useful for 4 days when kept at 4°C and can be used for both paraffin and resin sections. Standard hematoxylin–eosin staining schedule can also be used.

When greater detail is desired, then resin sections and periodic acid Schiffs–Mallory trichrome (PAS) schedule can be used. This stain highlights the presence of carbohydrates in purple red (and since spiggin is a glycoprotein, it can be clearly seen in renal sections). Details are provided in Tables 20.3 and 20.4.

In each renal section, the height of the cells that line the secondary proximal kidney epithelium height (KEH) should be measured under a microscope equipped with a video camera and linked to a PC with image analysis software. If such a system is not available, then a standard microscope equipped with a graticule or an eyepiece micrometer can be

Table 20.3 Staining schedule for Schiffs–Mallory trichrome schedule (resin sections)

Step No.	Procedure/solution	Duration
1	Immerse in 5% (w/v) periodic acid	4.5 min
2	Immerse in DW	Wash 5×
3	Immerse in Schiffs reagent (Table 20.4)	60 min
4	Place under running tap water	10 min
5	Immerse in 1% (w/v) acid fuchsin	1 min
6	Wash with DW	30 s
7	Wash with DW	30 s
8	Immerse in 1% (w/v) phosphomolybdic acid	1 min
9	Wash with DW	10 s
10	Immerse in Mallory trichrome stain (Table 20.4)	15 s
11	Wash with DW	10 s
12	Immerse in 90% (v/v) IMS	5 s
13	Immerse in 100% (v/v) IMS	5 s
14	Immerse in 100% (v/v) IMS	5 s
15	Immerse in 1:1 IMS/Citroclear	5 s
16	Immerse in Citroclear	5 s
17	Immerse in Citroclear	5 s
18	Air dry, mount DPX and dry at 40°C	Overnight

Table 20.4 Preparation of special solutions for Schiffs–Mallory trichrome schedule

Step No Schiffs 200 ml	Procedure	Chemicals
1	Bring to boiling point	Distilled water (200 ml)
2	Add stain (water is just off boiling point)	1 g pararosaniline
3	Add chemical (temperature 50°C)	2 g $K_2O_2S_2$
4	Add acid (room temperature)	2 ml HCl (35%)
5	Add charcoal and leave overnight in stoppered flask	2 g activated charcoal
6	Filter through Whatman No. 1 filter paper. Stain can be used several times and is stable at 4°C for several weeks. Discharge when pink	
Mallory trichrome 200 ml		
1	Dissolve stain in distilled water	1 g aniline blue
2	Add stain	2 g orange G
3	Add stain	2 g oxalic acid
4	Mix and keep at 4°C until use	

used. From each fish, ten renal sections should be prepared, and the KEH of 40 secondary proximal convoluted tubules measured. The measurements should include all tubules of the second proximal convoluted segment until the pre-set number of 40 is reached.

ELISA for spiggin

In CEFAS, we have developed a high throughput ELISA for the androgen-induced protein spiggin, which has the capacity of measuring the protein titers in whole kidney homogenates. Both the production of antisera and the ELISA procedure have been described in detail previously.[21]

ELISA procedure

A large number of hypertrophied kidneys and urinary bladders from breeding males were pooled and dissolved in the strong urea solution described earlier (1:5 w/v). This was divided into 100-ml aliquots, which were frozen and given an arbitrary value of 10,000 units/ml. The coating material for the ELISA plates was prepared solely from nest glue. A large pool of nest-derived material was collected, treated with the strong urea buffer and frozen in aliquots of 100 μl. The procedure was finalized as follows. All kidneys to be assayed should be mixed with 200 μl of spiggin buffer and heated at 70°C for 30 min. The spiggin content of 20 kidneys can be determined, without replication, in one plate. Assays are carried out over 48 h. On day 1, the nest-derived material is diluted 1:1000 with coating buffer, and 100 μl is dispensed into every well of the appropriate number of high protein-binding plates. The plates are then sealed, wrapped in moistened paper and a plastic bag and stored at 4°C overnight. Also on day 1, 135 μl of assay buffer is added to all wells of an appropriate number of low protein-binding plates (i.e., 1 plate/20 samples). A vial of spiggin standard is thawed out and diluted 1:10 (1000 units/ml) and 1:50 (200 units/ml) in assay buffer. Each of the wells in columns 1 and 5 and wells A–D of column 9 receives 15 μl of each sample. Wells E and F of column 9 receive 15 μl of the high standard, and wells G and H receive 15 ml of the low standard. Using a multipipette set to 15 μl, the contents of the wells in column 1 are mixed thoroughly by pipetting up and down 20 times; 15 μl are then transferred to column 2 and the procedure repeated up to column 4. The final 15 μl are discarded and the pipette tips changed. This complete procedure is then in turn carried out on the samples/standards that had been added to the wells in columns 5 and 9. The result is a set of 10-fold dilutions of all samples and standards.

Rabbit anti-spiggin serum is diluted 1:10,000 in assay buffer and 65 μl are added to all wells. The plates are then sealed, wrapped in moistened paper and a plastic bag, and stored at 4°C overnight.

On the morning of day 2, the high protein-binding plates are washed three times with 200 μl of wash buffer. A multipipette is used to transfer 150 μl of the contents of the wells of the low protein-binding plate into the corresponding wells of the high protein-binding plate. The transfer starts with column 12 and work backwards. Pipette tips are discarded after every four columns so as to avoid contamination between samples. The plates are resealed and wrapped and left to incubate for 4–6 h at room temperature. After this period, they are washed three times with wash buffer followed by addition of 150 μl of second antibody, diluted 1/15,000 in assay buffer. The plates are resealed, wrapped, and incubated overnight at 4°C.

On the morning of day 3, the plates are washed three times with distilled water followed by the addition of 150 μl of *p*-NPP. Color development is measured within an hour with the microplate reader set at a wavelength of 405 nm. The number of spiggin units/kidney is then divided by the weight of fish to give spiggin units/g of body weight.

The validation procedure of the developed ELISA for spiggin in CEFAS involved bisection of 160 kidneys in approximately two equal parts (which were weighed to the nearest mg), one of which was used for the immunoassay and the other half for light microscopy. Estimates of intra- and inter-assay variations, which measure the precision and reproducibility of the assay, respectively, were determined by repeatedly measuring the same kidney extracts and calculated as coefficients of variation (CV). Values were analyzed by ANOVA with a *post-hoc* Duncan's test during statistical analysis. Data were logarithmically transformed prior to analysis.

At the moment there is no available commercial kit for an immunoassay. However, it is anticipated that due to the recent advances in the molecular characterization of the protein, recombinant spiggin and antibodies will soon become commercially available.

Results and discussion

Besides the fact that the numbers of tests that measure estrogenic action are increasing,[3–6,24–28] to date there has been no reliable *in vivo* screening test system for environmental androgens and/or antiandrogens in the aquatic environment.

The experiments that took place in CEFAS were the first to demonstrate kidney hypertrophy and spiggin production in intact males (not gonadectomized) and females exposed to a number of androgens added to the water. A series of both natural and synthetic androgens have been tested in our lab and successfully induced spiggin production in female sticklebacks. However, the majority of the data we have produced (about 15 experiments) are on 17α-MT and DHT.

17α-MT was proved to be very efficient at inducing kidney hypertrophy in female sticklebacks in a dose response manner (Figure 20.1). The increase in the kidney epithelium

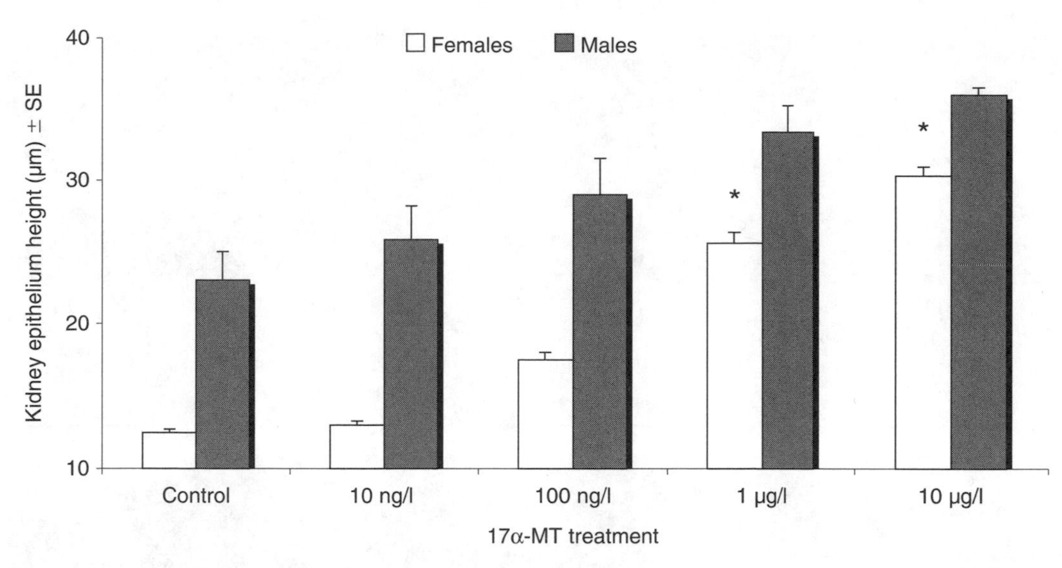

Figure 20.1 KEH of sticklebacks treated with 17α-MT. Asterisk indicates significant difference from control group ($p < 0.01$).

height (on average about 13 μm in female fish) between the different treatment groups was statistically significant ($p < 0.01$) for the highest 17α-MT doses used (1 and 10 μg/l).

The novel structural glycoprotein, spiggin, which was first characterized by Jakobsson et al. (1999),[13] is localized in the tubules of the secondary proximal segment, as shown by staining of androgen-treated female kidney sections with PAS (Figure 20.2, Plates (b)–(f)). A few females demonstrated small amounts of a PAS-positive material in their kidney tubules upon treatment with 17α-MT at 100 ng l/l. The increase in the KEH in this group, however, was not statistically significant. No PAS-positive material was detected when using lower concentrations of 17α-MT. Therefore, the lowest observed effect concentration (LOEC) for 17α-MT based on histological sections after a 3-week exposure was determined as 1 μg/l of ambient water and the no observed effect concentration (NOEC) as 100 ng/l.

Although the histological changes in the kidney provided a dose–response curve for 17α-MT (based on the height of the kidney cells) and proved to be a valid bioassay, the processing of a large number of samples for microscopy was time consuming. The development of an immunoassay based on the androgen-induced protein, spiggin, was expected to provide a much faster and specific bioassay. Our validation method, which consisted of comparing the KEH of 160 fish that received different doses of 17α-MT and/or FL (as measured on histological sections of half of the kidney) with the amount of spiggin units (as measured in the other half of the kidney), resulted in an excellent correlation ($r^2 = 0.93$, Figure 20.3). We observed no false positive or negative samples and parallelism in the ELISA assay were good. The spiggin assay was completed within 3 days of sacrificing the fish, while the histological method took a further 4 weeks. Spiggin concentrations showed a 10^5-fold variation — as opposed to a 4-fold difference in KEHs. This highlighted the differences between the different doses of 17α-MT much more clearly than the histological method. The excellent correlation between the histological assay and the developed ELISA for spiggin provided solid evidence of a fully functional novel bioassay for androgens. The precision of the assay, determined by calculating the intra-assay CV of repeated measures of samples within the same assay, was less than 9%.

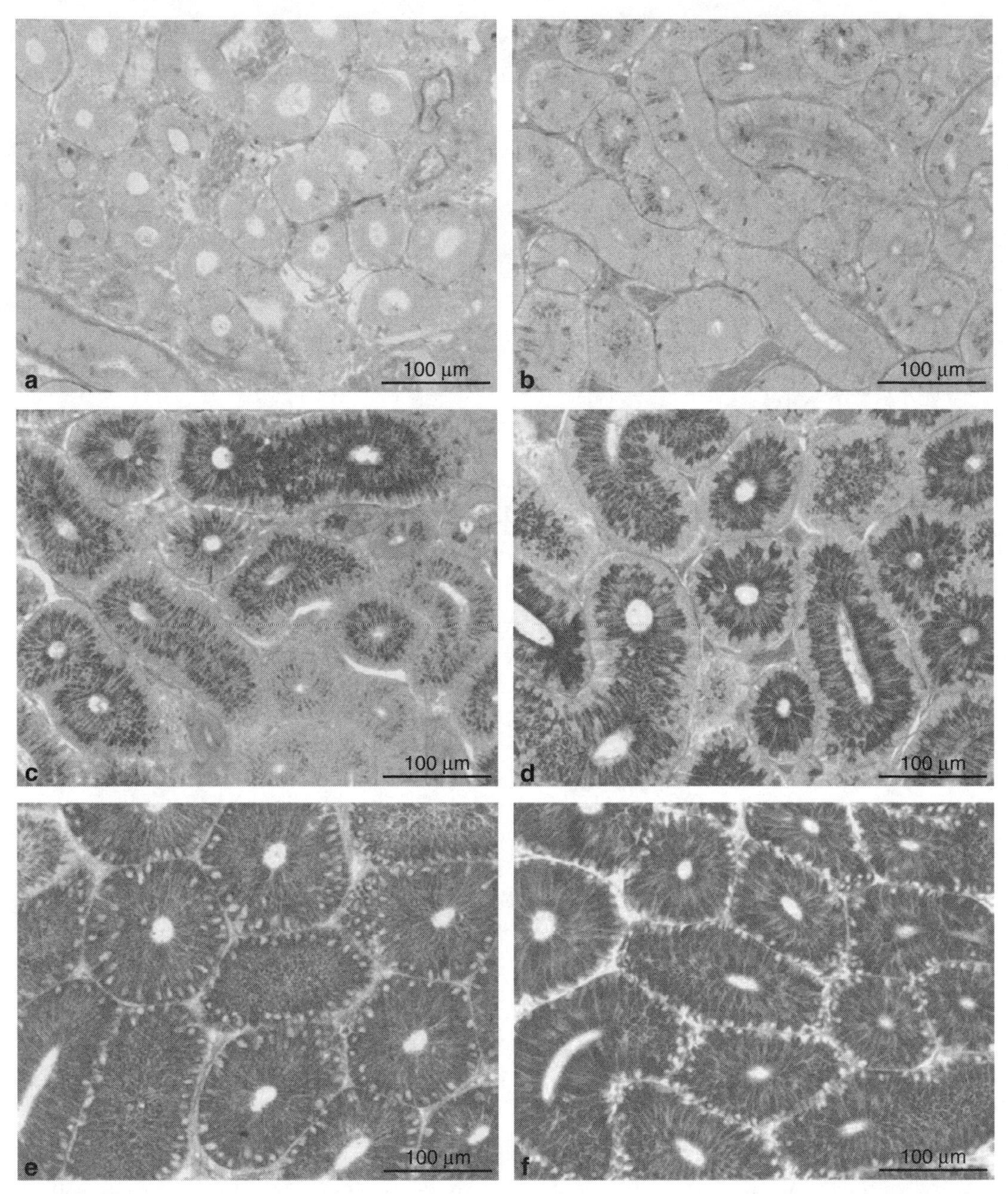

Figure 20.2 **(see color insert following page 464)** Renal histology of 17α-MT-treated female sticklebacks (a–e) and a naturally breeding male (f). Plates are at the same power and stained with periodic acid-Schiff (PAS). (a) control fish (KEH = 13.5 μm); (b) 100 ng/l 17α-MT (KEH = 18 μm); (c) 1 μg/l 17α-MT (KEH = 24.2 μm); (d) 10 μg/l 17α-MT (KEH = 29.7 μm); (e) 500 μg/l 17α-MT (KEH = 36.2 μm); (f) Breeding male (KEH = 34.5 μm).

The reproducibility of the assay, calculated as the inter-assay CV, was determined by measuring the same samples in four separate assays. The average inter-assay CVs of samples were less than 13%.

With the aim of applying the assay in the field of ecotoxicology, investigations into the potency of other model androgens, such as DHT, on the kidney hypertrophy in the three-spined stickleback started. Figure 20.4 displays the results of a time course of action experiment (1–5 weeks) using DHT to induce spiggin in female kidneys. Unfortunately, the fish present in the tanks with a DHT concentration at 5 μg/l developed a bacterial infection and died after day 14. This dose had nevertheless caused significant ($p < 0.01$) induction of spiggin at 7 and 14 days. One female fish showed spiggin induction at the

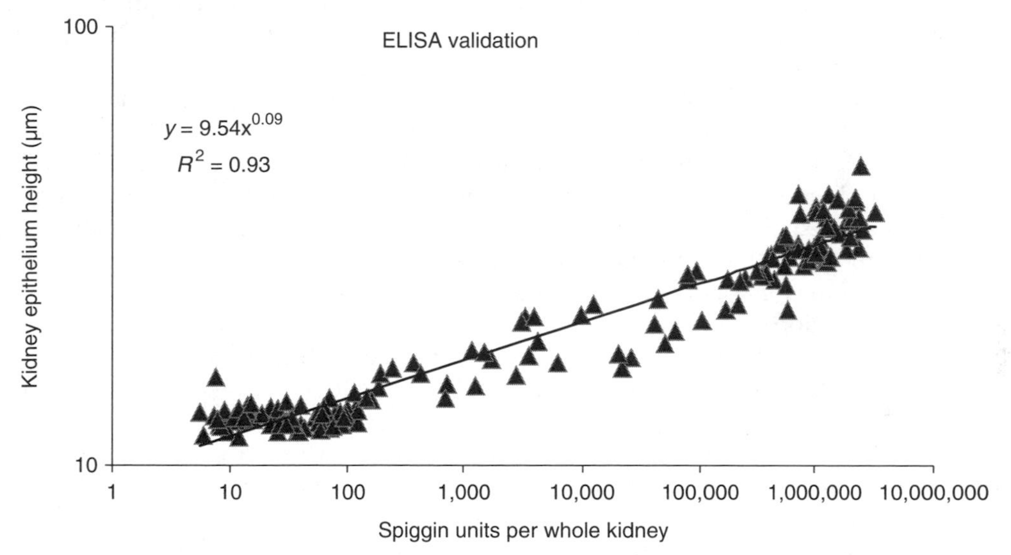

Figure 20.3 Comparison of the developed ELISA for spiggin with the KEHs obtained with histological examination.

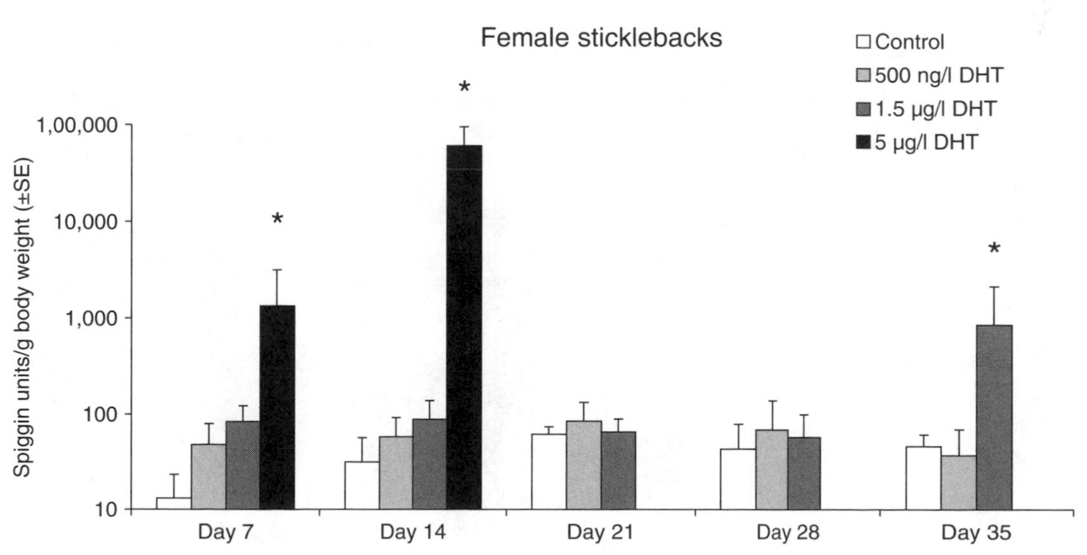

Figure 20.4 Time course of action of DHT as a model androgen. The asterisk denotes significant difference from control ($p < 0.01$).

dose of 1.5 μg/l, but only after 35 days exposure. No spiggin-positive females were observed at the dose of 500 ng/l.

The dose response experiment with DHT used the model androgens at the range 1–5 μg/l. Fish were exposed for either 3 or 5 weeks (Figure 20.5). Upon statistical analysis of the results, the NOEC and LOEC for a 3-week exposure to DHT were 2 and 3 μg/l, respectively ($p < 0.05$). However, after 5 weeks of exposure, statistical analysis ($p < 0.05$) indicated that the 5-week NOEC and LOEC for DHT were 1 and 2 μg/l, respectively.

Our results suggested that 17α-MT is slightly more potent than DHT. A differential potency of androgens, however, is not always related to a higher affinity for the relevant receptor; it could reflect a different degree of binding to steroid serum binding proteins (not mediated via the receptor).

During the main dose response experiment with DHT, the determination of the NOEC and LOEC was very accurate, and the differences between the DHT doses were very narrow. The only paradox in the construction of a dose–response curve for DHT was the apparently higher spiggin levels at 2 μg/l in comparison to the 3 μg/l during the 5-week exposure test. This difference, however, was non-significant. Several biological factors (i.e., other than the androgen dose used) — for instance size, age, social interaction, and infection — may influence the total amounts of spiggin produced in the kidney upon androgen stimulation of very similar doses. Another possible explanation could lie in the presence of natural androgens excreted by the males into the aquaria water. Indeed, in the group treated with 2 μg/l, there were more males present (11 males) than in the 3-μg/l group (8 males).

A laboratory exposure of female sticklebacks to several dilutions of Swedish PME for a period of 3 weeks took place in collaboration with Stockholm University. After treatment with PME, the kidneys were excised and divided into two parts, one of which was studied with light microscopy and the other assayed with the spiggin immunoassay. Both assays indicated significant kidney stimulation, showing on average 83,000 spiggin units per kidney (Figure 20.6) and a KEH of 27 μm. Regression analysis of the data obtained

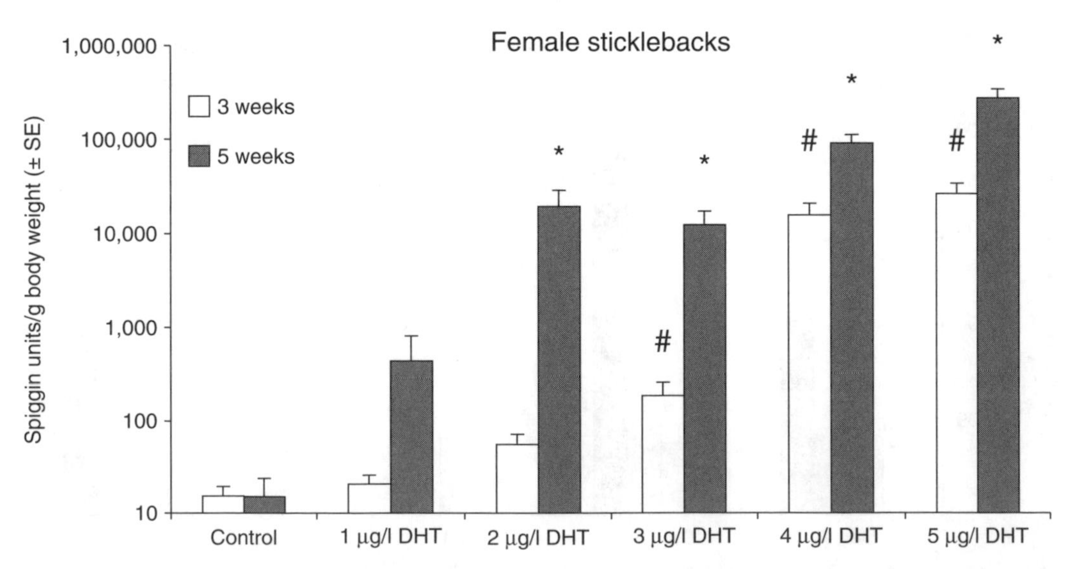

Figure 20.5 Dose–response curve for DHT after 3 and 5 weeks exposure. Significant inductions ($p < 0.05$) of spiggin in relation to the control groups is denoted by # (3 weeks) and * (5 weeks).

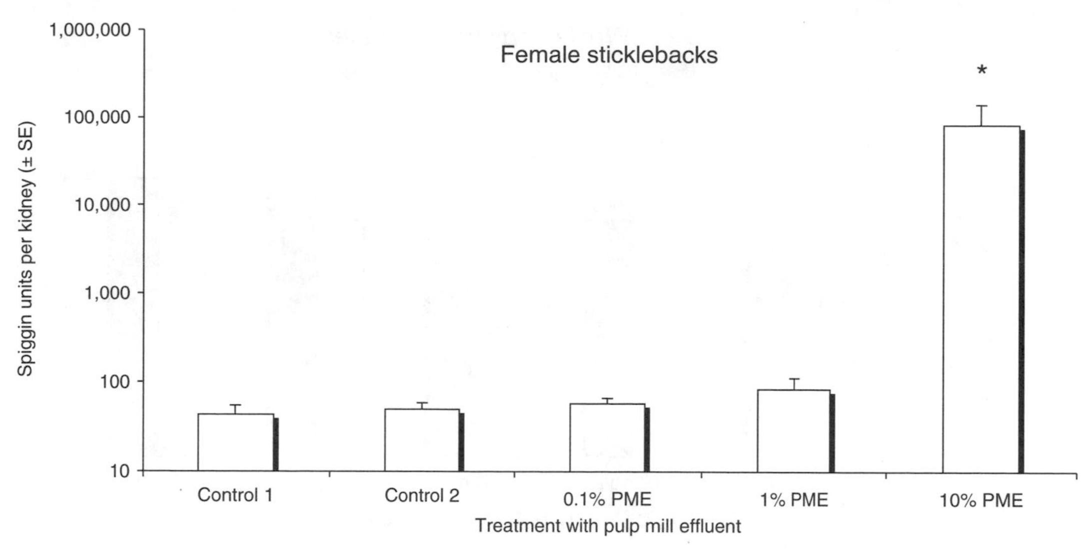

Figure 20.6 Female stickleback exposure to a well-known androgenic effluent. Significantly elevated spiggin denoted by asterisk ($p < 0.05$).

from the two assays revealed an excellent coefficient of correlation ($r^2 = 0.96$). The masculinizing effect of PME on female mosquito fish has been reported long ago,[8] but we have demonstrated for first time the androgenicity of this type of effluent in the stickleback.

Since the dissection of kidneys creates a potential bottleneck in what is an otherwise rapid method for assessing *in vivo* androgenicity of compounds, we investigated whether spiggin could be usefully measured in whole-body homogenates. The sticklebacks were first treated with different doses of 17α-MT for 3 weeks, then humanly killed, and their bodies digested with the same urea buffer used for preparing the kidney digest (Figure 20.7). It should be noted that the actual spiggin units induced by the 17α-MT concentrations, shown in Figure 20.7, seem low because the results are expressed in spiggin units per milliliter of homogenate, not in units per kidney. Although the range in response is not as good as when measured in kidneys alone, there is nevertheless a clear dose response effect in female fish. We have used this procedure to measure spiggin in juvenile sticklebacks with a very small body size (<0.1 g), where dissection is impossible, and have confirmed the validity of the method.

With the aim of addressing the problem of environmental anti-androgens, potentially a far more interesting group of xenobiotics than androgens themselves, we designed an exposure system where anti-androgenic activity could be tested. In this system, female sticklebacks are exposed simultaneously to an androgen and an anti-androgen, and the degree of kidney hypertrophy is measured. We have chosen FL at 500 μg/l of ambient water and demonstrated a very clear inhibitory effect (Figure 20.8). At both concentrations of 17α-MT tested (100 ng/l and 1 μg/l) the resulting spiggin units of fish treated with FL were statistically significantly lower ($p < 0.05$) than the protein units of the fish that were treated with 17α-MT only. In more recent experiments, we have used 17α-MT at 500 ng/l and found it to induce on average 90,000 spiggin units/g body weight in female sticklebacks. This lower concentration of 17α-MT allows the FL-induced inhibition of spiggin production to be become evident at lower concentrations of FL (Figure 20.9).

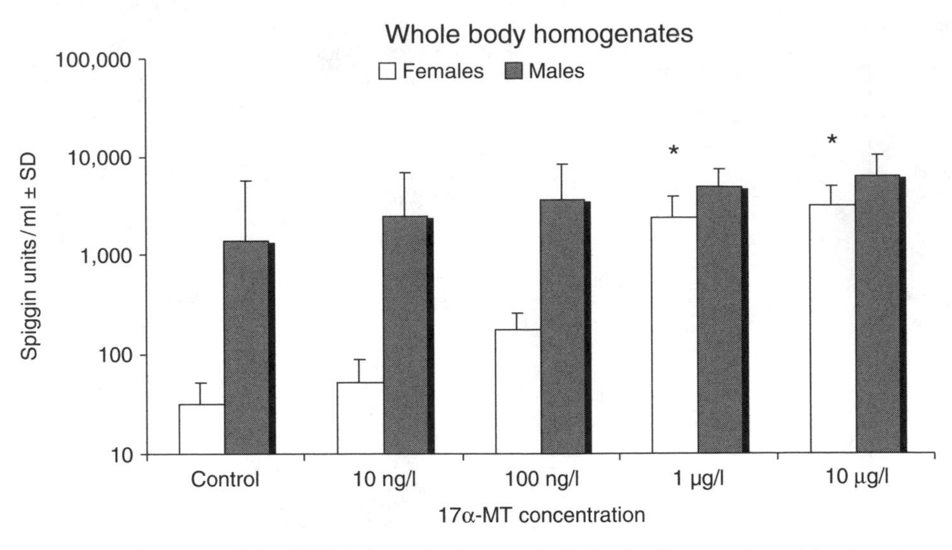

Figure 20.7 Application of the ELISA for spiggin on whole-body digests. Asterisks denote concentrations of 17α-MT that significantly induced spiggin ($p < 0.05$).

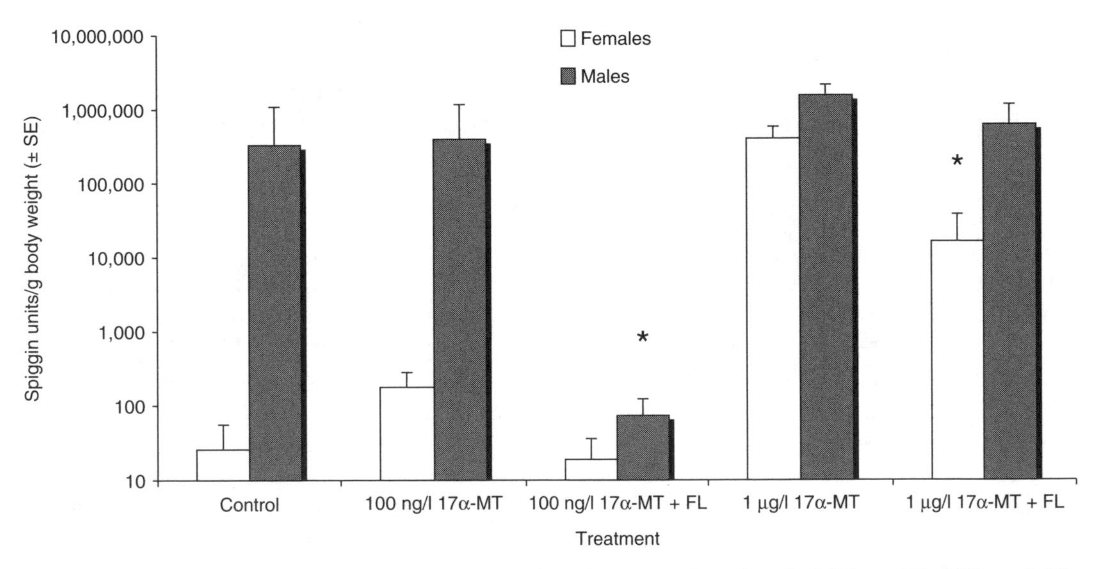

Figure 20.8 Spiggin units per kidney of sticklebacks treated with 17α-MT or 17α-MT and FL at 500 µg/l. Asterisks denote 17α-MT and FL treatments that were significantly different from the 17α-MT only group ($p < 0.05$).

In addition, the dose response experiment with DHT included EE_2 at 20 ng/l and FL at 500 µg/l. Figure 20.10 presents the results; which show that after 3 weeks of exposure, no spiggin-positive females were found in the tanks containing FL (along with 5 µg/l DHT), EE_2, or methanol only, and no spiggin-positive males were found in the FL-treated tanks, only 1 in the EE_2-treated tank (low levels) and 2 in the control tanks. After 5 weeks of exposure, no spiggin-positive females were found in the tanks containing FL (along with 5 µg/l DHT), EE_2, or methanol only, and no spiggin-positive males were found in the FL-treated tanks, only 1 in the EE_2-treated tank (low levels) whilst there were 6 in the

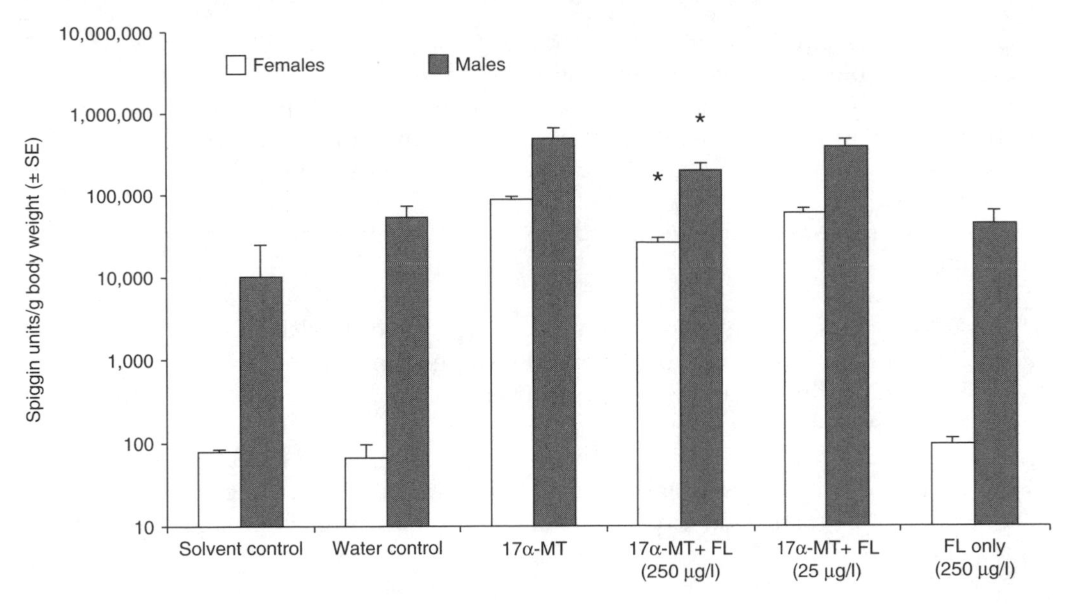

Figure 20.9 Spiggin units per kidney of sticklebacks treated with 17α-MT at 500 ng/l or 17α-MT and FL. Asterisks denote 17α-MT and FL treatments that were significantly different from the 17α-MT only group ($p < 0.05$).

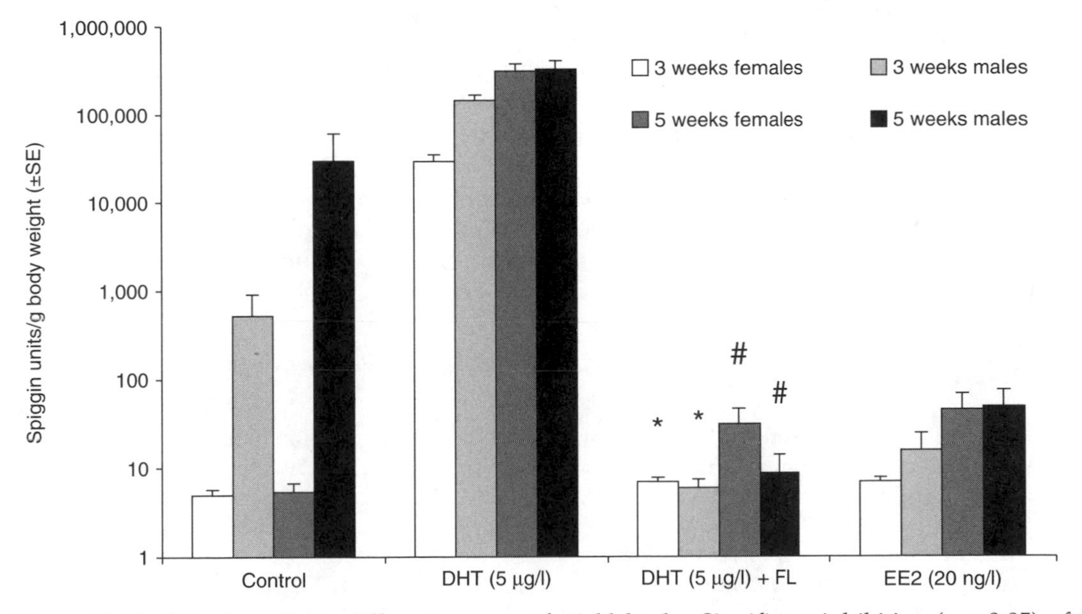

Figure 20.10 Spiggin units in different groups of sticklebacks. Significant inhibition ($p < 0.05$) of spiggin relative to the DHT-treated group is denoted by # (3 weeks) and * (5 weeks).

control tanks. The lack of kidney stimulation in the EE_2-treated groups in comparison to the controls suggests that estrogens antagonize natural androgens. Indeed, in all control male groups used throughout the whole study, kidney stimulation was observed in a significant number of males under laboratory conditions. More recent experiments have confirmed that relatively low concentrations of estrogens (i.e., $EE_2 > 10$ ng/l and E_2 >100 ng/l) are able to inhibit spiggin production in male fish.

More recently, we have run a series of experiments using suspected environmental antiandrogens and have demonstrated that vinclozolin, fenitrothion, linuron, and DDE metabolites all inhibit or significantly reduced spiggin production in the androgen-stimulated female kidney.

The following list summarizes the advantages of the stickleback as a test and sentinel species for EDCs:

1. During the spawning season, the kidneys of male sticklebacks produce substantial amounts of a nest-building protein, spiggin, in response to androgenic stimulation, and this response is highly specific to androgens (i.e., progestogens and estrogens are ineffective).
2. The spiggin can be readily quantified using the validated ELISA (providing a 10^5-fold range in spiggin content of stickleback kidneys) and can be induced in female kidneys by androgen exposure, while this induction is antagonized by anti-androgens.
3. The stickleback is: a recognized Organisation of the Economic Co-operation of Developed countries (OECD) test species (OECD Guideline 210); an extremely ubiquitous species found in all aquatic habitats from full seawater to freshwater; endemic in Europe and North America, unlike other fish used as test-species, it is easily obtained in very large numbers in the field, allowing for *in situ* biomonitoring; easy to maintain in laboratory aquaria (both adults and larvae); has a short life cycle and is readily induced into breeding condition all-year-round by photothermal manipulation; its reproductive behavior is better known than any other fish and its reproductive physiology/endocrinology is well documented; it presents very high egg/fry survival rates (close to 100%).
4. A vitellogenin assay has also been developed at CEFAS, thus the stickleback has the potential to act as a simultaneous biomarker for androgens (spiggin induction in females) and estrogens (vitellogenin production in males), dramatically reducing the number of fish and the costs used for testing.

References

1. Kelce, W.R. and Wilson, E.M., Environmental antiandrogens: developmental effects, molecular mechanisms, and clinical implications, J. *Mol. Med.*, 75, 198–207, 1997.
2. Purdom, C.E., Hardiman, P.A., Bye, V.J., Eno, N.C., Tyler, C.R. and Sumpter, J.P., Estrogenic effects of effluents from sewage treatment works, *Chem. Ecol.*, 8, 275–285, 1994.
3. Sumpter, J.P. and Jobling, S., Vitellogenesis as a biomarker for oestrogenic contamination of the aquatic environment, *Environ. Health Perspect.*, 103 (Suppl. 7), 173–178, 1995.
4. Folmar, L.C., Denslow, N.D., Rao, V., Chow, M., Crain, D.A., Enblom, J., Marcino, J. and Guillette, L.J., Vitellogenin induction and reduced serum testosterone concentrations in feral male carp (*Cyprinus carpio*) captured near a major metropolitan sewage treatment plant, *Environ. Health Perspect.*, 104, 1096–1101, 1996.
5. Allen, Y., Thain, J., Matthiessen, P., Scott, A.P., Haworth, S. and Feist, S., A Survey of Oestrogenic Activity in UK Estuaries and Its Effect on Gonadal Development of the Flounder *Platichthys flesus*, International Council for the Exploitation of the Sea, ICES report, C.M. 1997/U:01, 1997.

6. Denslow, N.D., Chow, R.M., Chow, M.M., Bonomelli, S., Folmar, L.C., Heppell, S.A. and Sullivan, C.V., Development of biomarkers for environmental contaminants affecting fish, in *Chemically Induced Alterations in Functional Development and Reproduction of Fishes*, Proceedings, Wingspread Conference, 1995, SETAC Publication Series, Holland, R.M., Gilbertson, M. and Peterson, R.E., Eds., SETAC Press, Pensacola, FL, 1997, pp. 73–86.
7. Kelce, W.R., Gray, L.E. and Wilson, E.M., Antiandrogens as environmental endocrine disruptors, *Reprod. Fert. Devel.*, 10, 105–111, 1998.
8. Howell, W.M., Black, D.A. and Bortone, S.A., Abnormal expression of secondary sex characters in a population of mosquito fish, *Gambusia affinis holbrooki*: evidence for environmentally-induced masculinization, *Copeia*, 1980 (4), 676–681, 1980.
9. Howell, W.M. and Denton, T.E., Gonopodial morphogenesis in female mosquito fish, *Gambusia affinis affinis*, masculinized by exposure to degradation products from plant sterols, *Environ. Biol. Fish*, 24, 43–51, 1989.
10. Cody, R.P. and Bortone, S.A., Masculinization of mosquito fish as an indicator of exposure to kraft mill effluent, *Bull. Environ. Contam. Toxicol.*, 58, 429–436, 1997.
11. Thomas, K.V., Hurst, M.R., Matthiessen, P., McHugh, M., Smith, A. and Waldock, M.J., An assessment of *in vitro* androgenic activity and the identification of environmental androgens in United Kingdom estuaries, *Environ. Toxicol. Chem.*, 20 (7), 1456–1461, 2002.
12. Hershberger, L.G., Shipley, E.G. and Meyer, R.K., Myotrophic activity of 19-nortestosterone and other steroids determined by modified levator ani muscle method, *Proc. Soc. Exp. Biol. Med.*, 83, 175–180, 1953.
13. Jakobsson, S., Borg, B., Haux, C. and Hyllner, S.J., An 11-ketotestosterone induced kidney-secreted protein: the nest building glue from male three-spined stickleback, *Gasterosteus aculeatus*, *Fish Physiol. Biochem.*, 20, 79–85, 1999.
14. Jones, I., Lindberg, C., Jakobsson, S., Hellqvist, A., Hellman, U., Borg, B. and Olsson, P.-E., Molecular cloning and characterization of spiggin: an androgen-regulated extraorganismal adhesive with structural similarities to Von Willebrand factor-related proteins, *J. Biol. Chem.*, 276, 17857–17863, 2001.
15. De Ruiter, A.J.H. and Mein, C.G., Testosterone-dependent transformation of nephronic tubule cells into serous and mucous gland cells in stickleback kidneys *in vivo* and *in vitro*, *Gen. Comp. Endocrinol.*, 47, 70–83, 1982.
16. Mayer, I., Borg, B. and Schulz, R., Seasonal changes in and effect of castration/androgen replacement on the plasma levels of five androgens in the male three-spined stickleback, *Gasterosteus aculeatus* L., *General Comparative Endocrinol.*, 79, 23–30, 1990.
17. Borg, B., Antonopoulou, E., Andersson, E., Carlberg, T. and Mayer, I., Effectiveness of several androgens in stimulating kidney hypertrophy, a secondary sexual character, in castrated male three-spined sticklebacks, *Gasterosteus aculeatus*, *Can. J. Zool.*, 71, 2327–2329, 1993.
18. Jakobsson, S., Mayer, I., Schulz, R.W., Blankenstein, M.A. and Borg, B., Specific binding of 11-ketotestosterone in an androgen target organ, the kidney of the male three-spined stickleback, *Gasterosteus aculeatus*, *Fish Physiol. Biochem.*, 15, 459–467, 1996.
19. Van Oordt, G.J., Die veranderungen des hodens wahrend des auftretens der sekundaren geschlechtiesmerkmale bei fischen. I. *Gasterosteus pungitious* L, *Arch. Mikroskop. Anatomie Entwicklungsmechanik*, 102, 379–405, 1924.
20. Katsiadaki, I., Scott, A.P. and Matthiessen, P., The use of the three-spined stickleback as a potential biomarker for androgenic xenobiotics, in *Proceedings of the Sixth International Symposium on the Reproductive Physiology of Fish*, Norberg, B., et al., Eds., Bergen, Norway, 4–9 July, 1999, 2000, pp. 359–361.
21. Katsiadaki, I., Scott, A.P., Hurst, M.R., Matthiessen, P. and Mayer, I., Detection of environmental androgens: a novel method based on enzyme-linked immunosorbent assay of spiggin, the stickleback (*Gasterosteus aculeatus*) glue protein. *Environ. Toxicol. Chem.*, 21 (9), 1946–1954, 2002.

22. Katsiadaki, I., Scott, A.P. and Mayer, I., The potential of the three-spined stickleback, *Gasterosteus aculeatus* L., as a combined biomarker for oestrogens and androgens in European waters, *Mar. Environ. Res.*, 54, 725–728, 2002.
23. Singh, S.M., Gauthier, S. and Labrie, F., Androgen receptor antagonists (antiandrogens): structure–activity relationships, *Curr. Med. Chem.*, 7, 211–247, 2000.
24. Allen, Y., Scott, A.P., Matthiessen, P., Haworth, S., Thain, J.E. and Feist, S., Survey of estrogenic activity in United Kingdom estuarine and coastal waters and its effects on gonadal development of the flounder *Platichthys flesus*, *Environ. Toxicol. Chem.*, 18, 1791–1800, 1999.
25. Bowman, C.J. and Denslow, N.D., Development and validation of a species- and gene-specific molecular biomarker: vitellogenin mRNA in Largemouth Bass (*Micropterus salmoides*), *Ecotoxicology*, 8, 399–416, 1999.
26. Denslow, N.D., Chow, M.C., Kroll, K.J. and Green, L., Vitellogenin as a biomarker of exposure for estrogen and estrogen mimics, *Ecotoxicology*, 8, 385–398, 1999.
27. Heppell, S.A., Denslow, N.D., Folmar, L.C. and Sullivan, C.V., Universal assay of vitellogenin as a biomarker for environmental estrogens, *Environ. Health Perspect.*, 103 (Suppl. 7), 9–15, 1995.
28. Tyler, C.R., van der Eerden, B., Jobling, S., Panter, G. and Sumpter, J.P. Measurement of vitellogenin, a biomarker for exposure to oestrogenic chemicals, in a wide variety of cyprinid fish, *J. Comp. Physiol. B.*, 166, 418–426, 1996.

chapter twenty-one

Simple methods for estimating exposure concentrations of pesticide resulting from non-point source applications in agricultural drainage networks

Wenlin Chen
Syngenta

Contents

Introduction

Refinement of quantitative assessment for pesticides and other agricultural chemicals in surface water networks (rivers, streams, lakes, and reservoirs) is being pursued with increased intensity. Recent passage of the Food Quality Protection Act (FQPA) by Congress and the conclusions reached by the US EPA Ecological Committee on FIFRA Risk Assessment Methods (ECOFRAM)[1] have particularly accelerated the effort, requiring that

1-56670-664-5/05/$0.00+$1.50
© 2005 by CRC Press

such assessment be conducted accurately with reasonable certainty. Pesticide occurrence in surface water systems is complicated and highly variable over time and geographical locations due to different use patterns (application timing, rate, and method), weather, and watershed characteristics. Monitoring to directly determine exposure with sufficient frequency on a site-by-site basis is almost prohibitively expensive and is impossible for new compounds. As a result, predictive models with measurable accuracy have become important assessment tools to address ecological, as well as drinking water, concerns.

There have been a large number of process-based models that have been developed to estimate pesticide concentrations in the environment.[2–5] These models are mostly mechanistic in nature, each representing a specific homogeneous environment with well-defined pesticide use patterns and physical processes of environmental fate, weather, and hydrology. For example, the Pesticide Root Zone Model (PRZM)[3] simulates surface runoff, soil erosion, and leaching from a specified homogeneous field, while the Exposure Analysis Modeling System (EXAMS) predicts the fate processes in an aquatic environment (such as a pond or lake).[4] Mechanistic models are most useful in parameter sensitivity analysis and comparative assessment where simulations of different chemicals are conducted under the same environmental conditions. Although mechanistic models adequately describe relatively homogeneous and small plot runoff and microcosm/mesocosm experiments (generally <1 ha), extrapolating these models to large-scale watersheds is not as successful.[6,7] Model performance can be further degraded, when compounding worst-case assumptions on use rate and environmental fate parameter values as required by current regulatory guidelines.[5,7] As a result, predictions may be many orders of magnitude different from the monitoring data.

Mechanistic models are developed and calibrated at a small field scale. However, very little is known as to how a physical or chemical process calibrated on a small field scale of given landscape characteristics may be extrapolated to a much larger-scale watershed, within which many variable small-scale or sub-watersheds/catchments are arranged in a particular pattern or with variable degrees of spatial correlation. Consider a pulse of an environmental chemical resulting from a runoff event in a relative homogeneous environmental segment. The fate and transport processes may be well described deterministically by a mechanistic model. However, as the pulse continues to move through the watershed, experiencing different segments of wide variability in surface hydrology, wetness, topography, soil, and vegetation, the resulting process, and interactions may be more complex than the original model would predict.

The response of the whole watershed is not only dependent on a set of parameter values (even they are distributed) but also on the spatial correlation of all segments (i.e., the physical segment layout and geometry) and their inter-segment linkages that are often poorly understood. It is arguable that such problem of spatial variability can be effectively dealt with probabilistically without considering spatial correlation. Spatial and temporal variability in farming practices, hydrological properties, and coincidence of agricultural applications and storm/runoff events at a given point in time and space can all easily lead to additional complexity in the development of a comprehensive and fully physically based watershed model.

To overcome the complexity and uncertainty of mechanistic models as above, several simple regression approaches have been proposed recently in the development of watershed scale models.[8–11] Larson and Gilliom[8] found that, regardless the time dimension, percentiles of the monitoring data of several major herbicides obtained from the US Geological Survey (USGS) National Water-Quality Assessment (NAWQA) Program could be explained in a significant regression by chemical use intensity and four watershed hydrological characteristics (runoff factor, soil erodibility coefficient, Dunn overland flow, and drainage area). Larson et al.[9] later further extended the regression to include more percentiles derived from the atrazine monitoring data and named the regression as Watershed Regression for Pesticides (WARP) model.

Since the WARP model does not include pesticide environmental fate properties, different WARP models would need to be developed for different pesticides. In 2002, an independent regression approach was developed by Chen et al.[10] using the newly proposed pesticide surface water mobility index (SWMI) and watershed use intensity to explain pesticide percentiles.[10] As SWMI is determined by the organic carbon normalized soil/water sorption coefficient (K_{oc}, ml/g) and soil degradation half-life, the SWMI regression models would be able to differentiate pesticides of different properties. More recently, the SWMI concept was further incorporated into the WARP method to obtain ability of multi-compound predictions.[11] None of the approaches, however, have been able to predict the time series data required for accurate environmental exposure assessments.

Regression models are completely based on monitoring data. While not explicitly concerned with the exact mechanisms involved in the watershed transport process, the removal of *a priori* assumptions in regression models may offer the opportunity to simulate reality more closely as they are distilled past experiences (data). In the next few sections, the SWMI regression approach for distributional concentration (percentile) predictions is reviewed, and the SWMI approach is extended to establish a new regression method for predicting daily time series concentrations. Wherever it is deemed necessary, specific calculation procedures are also included.

Mobility of organic compounds to surface water

Any organic compound is more or less mobile in the environment. The term mobility can be defined as its inherent propensity susceptible to off-site transport by a hydrologic process, such as soil leaching or surface runoff. Traditionally, pesticide mobility has been quantified by the organic carbon normalized soil/water sorption coefficient (K_{oc}, ml/g). Compounds with higher K_{oc} values tend to bind more to soil particles than the lower K_{oc} compounds, thus movement by water is retarded.

The parameter K_{oc} alone, however, is not a full indicator of mobility, as the chance of a pesticide to be potentially transported off-site is also dependent on how quickly or slowly it is degraded, i.e., its persistence in soil. The concept of mobility was thus extended to include degradation half-life ($T_{1/2}$) by forming a combined mobility index to rank soil leaching potential of various chemicals.[12–14] However, only recently has the concept been further developed to benchmark pesticide mobility specifically relevant to surface water transport processes, such as surface runoff and soil erosion.[10]

Recognizing that surface water transport processes like storm-generated runoff events are more transient and distinct intermittent events compared to soil leaching (generally gradual, slow, and continuous), SWMI was derived from the basic equations of field runoff and erosion.[10] A standard environmental scenario was parameterized to benchmark off-site runoff potential of various chemicals with different pairs of K_{oc} and $T_{1/2}$. The resulting SWMI expression in terms of K_{oc} and $T_{1/2}$, was simplified as[10]:

$$\mathrm{SWMI} = \frac{\exp(-3.466/T_{1/2})}{(1 + 0.00348K_{oc})}(1 + 0.00026K_{oc}) \tag{21.1}$$

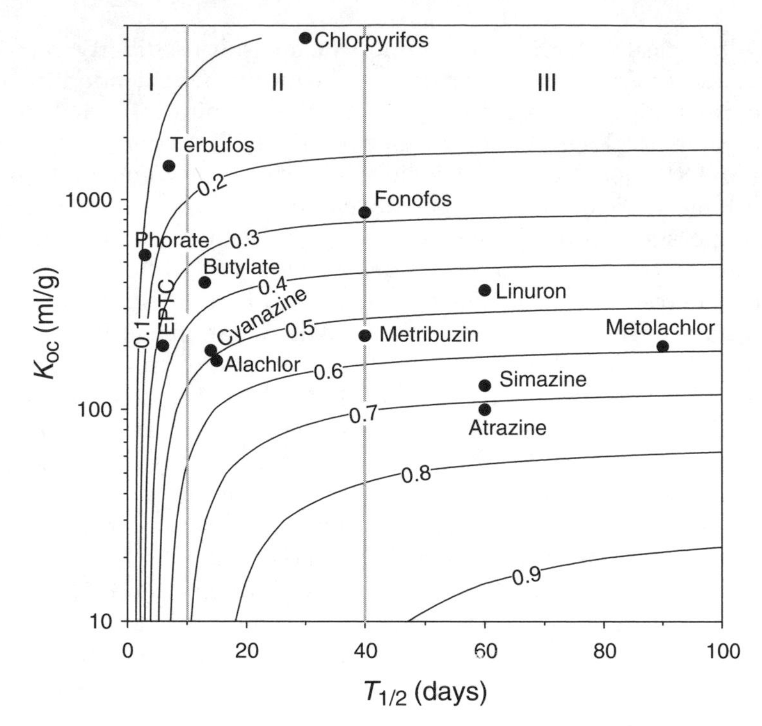

Figure 21.1 Contour plot of the pesticide SWMI as a function of the organic carbon-normalized soil/water sorption partition coefficient (K_{oc}) and degradation half-life ($T_{1/2}$). (From Chen, W. et al., *Environ. Toxicol. Chem.*, 21, 300, 2002. With permission.)

SWMI, as calculated in Equation (21.1), provides a relative mobility scale, with value 0 being the asymptotically least mobile and 1 the asymptotically most mobile and susceptible to transport by the processes of surface water runoff and soil erosion.

Calculating SWMI is straightforward. Paired K_{oc} and $T_{1/2}$ values measured from the same soil in the laboratory or field studies are preferred whenever possible for a given compound. If multiple paired measurements are available for K_{oc} and $T_{1/2}$, calculating SWMI for each individual paired values and averaging should provide a more representative estimate for SWMI.

A contour plot of SWMI as a function of K_{oc} and $T_{1/2}$ is provided in Figure 21.1 for a number of commonly used pesticides. The lines in the plot represent equal values of SWMI. As discussed by Chen et al.,[10] three sensitivity regions in the SWMI plot can be identified to illustrate the differential roles of K_{oc} and $T_{1/2}$ in determining mobility. For Region I ($T_{1/2} < 10$ days), SWMI is predominantly a function of half-life, and it is much less sensitive to K_{oc}. Off-site movement of these short-living compounds is expected to be much less dependent on their soil adsorption but more correlated with the temporal and spatial coincidence of pesticide applications and runoff events during the life span. For region II, half-life is between 10 and 40 days, and the mobility index is sensitive to both parameters. For Region III, compounds are more persistent, and K_{oc} primarily determines SWMI. Such is expected, as runoff is more likely during a longer half-life period.

Predictive regression models using SWMI

With the pesticide mobility index defined in the previous section, one would expect that a regression relationship could be established between SWMI and the corresponding watershed observed pesticide concentrations. Depending on sorting patterns of the observed data, two general types of regression models can be classified. In this chapter, we refer to the first type as the distributional concentration model (DCM) and the second as the time series concentration model (TSCM).

DCM is developed on data points at a given percentile level, which is obtained from sorting a data set in ascending order according to the magnitude of the observed values. Note that a percentile is the rank of a value in a data set as a percentage of that entire data set (but see next section for time-weighted percentiles). For example, the 95th percentile value means that 95% of the data values in a data set are equal to or less than the given value. A 95th percentile DCM, therefore, is the regression based on the 95th percentile data and would predict the possible 95th percentile values.

Unlike DCM that disregards the time dimension, TSCM retains the temporal sequence in the same order of sampling. Time thus is always an independent variable in TSCM regressions. As such, TSCM is able to simulate the fluctuation, trend, and seasonality in the concentration dynamics of a surface water system.

Time-weighted sampling

Most surface water monitoring programs are designed to sample more frequently immediately after field application of pesticides (usually late spring to early summer) when higher runoff is expected. Less frequent samples are collected in the rest of the year. The uneven sample frequency within a calendar year requires the measured pesticide concentrations to be time-weighted to remove bias. By time weighting, the 95th percentile value would mean that over 95% of the total time during an observation period the observed data are equal to or less than the given value.

Assuming each sampling point represents half the time to the preceding sample and half to the following sample as described by Chen et al.,[10] a time weight for each sample can be calculated as:

$$w_i = [(t_{i+1} - t_{i-1})/2] \Big/ \sum_{i=1}^{n} t_i \tag{21.2}$$

where t_i is time point of sample i; n is number of samples during the total sampling period. The weights are then organized by their associated corresponding concentrations in an ascending order. The cumulative time-weighted percentile associated with each concentration is thus determined as:

$$P_i = 100 \sum_{j=1}^{i} w_j \tag{21.3}$$

where P_i corresponds to the time-weighted percentile at concentration C_i of sample i. An exact percentile, for example, the 95th, is estimated by linear interpolation between two adjacent samples. With Equation (21.2), the time-weighted mean concentration (TWMC) is then calculated:

$$\mathrm{TWMC} = \sum_{i=1}^{n} w_i C_i \tag{21.4}$$

Distributional concentration model (percentiles)

Data

The data set used for the SWMI DCM development and evaluation was obtained from the pesticide monitoring program conducted in six drainage basins of Lake Erie tributaries by the Water Quality Laboratory at Heidelberg College, Ohio, USA.[10,15] Sampling locations in each of the six drainage basins are shown in Figure 21.2. We refer this data set as Heidelberg data.

The Heidelberg data set is one of the most comprehensive databases available in the U.S., with recorded pesticide riverine concentrations continuously since 1983. Sampling frequencies in the data set range from 3 samples/day during high runoff season (typically April 15 to August 15) to about 2 samples/month in the rest of each calendar year.[15] A list of the pesticide environmental fate properties, total use, and calculated SWMI values is provided in Table 21.1. Each drainage basin and its area are in Table 21.2. A more detailed description of the monitoring program can be found in Reference 15.

Model development

A power function of SWMI with a linear factor of pesticide watershed use intensity was found to describe the Heidelberg data well at various percentiles for each tributary or all six tributaries combined. Detailed descriptions of the model development are available in Reference 10. The SWMI DCM reads:

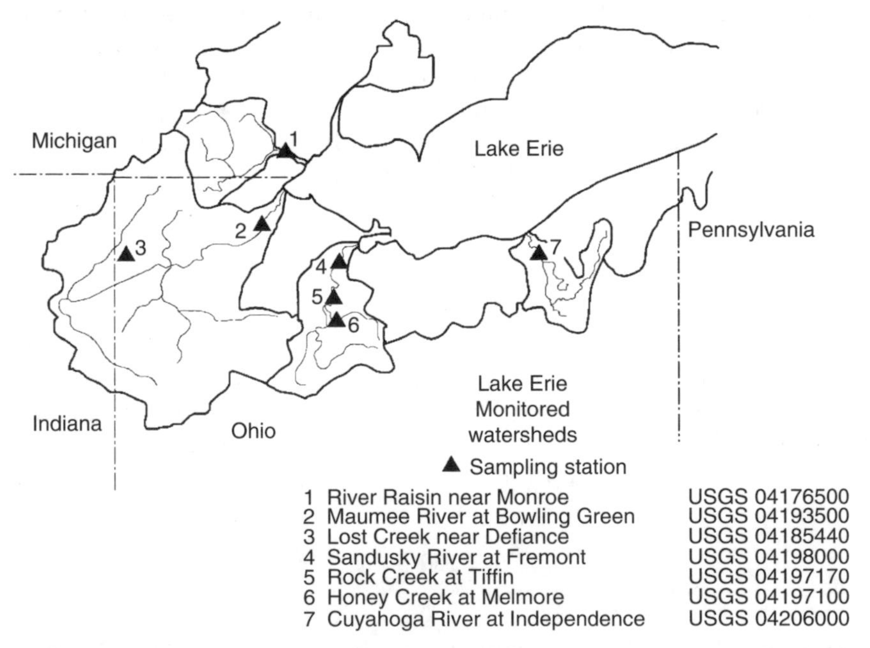

Figure 21.2 Pesticide sampling stations of the agricultural runoff-monitoring program in the Lake Erie tributary drainage basins, Heidelberg College, Ohio, USA. The U.S. Geological Survey (USGS) stream gauging station numbers are shown on the right side of the label. (From Chen, W. et al., *Environ. Toxicol. Chem.*, 21, 301, 2002. With permission.)

Table 21.1 Pesticide information used in the development of the DCM with the SWMI. (Modified from Chen, W. et al., *Environ. Toxicol. Chem.*, 21, 301, 2002. With permission.)

Pesticides	Type	K_{oc} (l/kg)	$T_{1/2}$ (days)	SWMI	Total use in 1986 (kg)	Use intensity[a] (kg/ha)
Alachlor	Herbicide	170	15	0.5207	1,058,896	0.4522
Metolachlor	Herbicide	200	90	0.5969	665,417	0.2842
Atrazine	Herbicide	100	60	0.7184	581,750	0.2485
Cyanazine	Herbicide	190	14	0.4932	206,463	0.0882
Metribuzin	Herbicide	224	40	0.5453	203,405	0.0869
Linuron	Herbicide	370	60	0.4523	110,931	0.0474
Terbufos	Insecticide	1540	7	0.1388	59,265	0.0253
Butylate	Herbicide	400	13	0.3536	50,558	0.0216
Chlorpyrifos	Insecticide	6070	30	0.1038	24,565	0.0105
EPTC	Herbicide	200	6	0.3482	30,757	0.0131
Phorate	Insecticide	540	3	0.1248	25,358	0.0108
Fonofos	Insecticide	870	40	0.2792	19,432	0.0083
Simazine	Herbicide	130	60	0.6718	14,132	0.0060

[a] Based on the total land areas of 2,341,468 ha where the total use data we obtained in the Lake Erie tributary region, 1986.

Table 21.2 Drainage areas of Lake Erie major tributaries, Ohio, and White River, Indiana

River name	Drainage area (ha)
Lost Creek	1,130
Rock Creek	8,800
Honey Creek	38,600
River Raisin	269,900
Sandusky River	324,000
Maumee River	1,639,500
White River	2,938,382

$$C_{P(x)} = \frac{M_{use}}{A_b} \alpha \text{SWMI}^{\beta} \tag{21.5}$$

where $C_{P(x)}$ is the xth percentile concentration, μg/l, or ppb; M_{use} is total annual use of a pesticide in a basin or watershed, kg; A_b is total area of a drainage basin or watershed, ha; α and β are regression constants. Defining $R_b = M_{use}/A_b$, R_b then becomes the annual watershed use intensity, kg/ha. As a power function, the SWMI DCM would predict 0 concentrations for any percentile levels as SWMI approaches 0 (the most immobile compound).

The SWMI DCM was calibrated initially based on data at three concentration levels (TWMC, the 95th percentile, and peak) for the entire monitoring period (1983–1991) and for year 1986 data only as the pesticide use information in the region was only surveyed in that year.[10] The model calibration was further extended to include the 80th, 90th, and 99th percentiles for the 1986 data. Lower percentile regressions are not available as most of the concentrations were non-detectable (recorded as 0). The calibrated 1986 model parameters are summarized in Table 21.3. Results of other calibrations, such as for each individual river or over the entire 9-year's data set (1983–1991), are available in Reference 10.

Model use

Calculation of the SWMI DCM [Equation (21.5)] is simple once the parameter values are determined. It can be easily implemented on spreadsheet programs, such as Excel® (Microsoft Corporation, Washington, USA). When a range of environmental fate (i.e., variable K_{oc} or $T_{1/2}$) or variability in use intensity is expected, the calculation can be done in a probabilistic manner. Three simple steps are summarized below:

1. Selection of a percentile model should be based on the assessment purpose. For example, if the TWMC is of interest, the regression constants for TWMC should be used (Table 21.3).
2. Calculate SWMI using Equation (21.1) if a pair of K_{oc} and $T_{1/2}$ determined from the same soil is given. If multiple paired K_{oc} and $T_{1/2}$ data are available, calculate SWMI per each pair of parameters, then average the SWMI estimates. For multiple unpaired K_{oc} and $T_{1/2}$ data, an independent random sampling from the distributions of the two parameters may be executed to obtain a distribution of DCM estimates (e.g., using Crystal Ball® software, Decisioneering, Denver, Colorado, USA).
3. Use intensity should be estimated always as accurately and as currently as possible. If the total use (M_{use}) in a watershed of interest is not known, it may be estimated by multiplying the label use rate (often available on the product package) by the total treated area expected in the watershed. If a distribution of M_{use} is available, such as through survey, a distribution of the model runs can be carried out accordingly.

Some example runs and cautions for interpreting the model results are provided in the section "Results and discussion."

Time series concentration model (daily)

The occurrence of pesticides in natural water systems is generally seasonal. In addition to seasonality, local transient fluctuations are inevitable, as agricultural practices, weather, and pesticide dissipation processes are all time-dependent, either deterministically or

Table 21.3 Calibrated model parameters for the DCM based on the single year (1986) data of the six Lake Erie agricultural tributary drainage basins, Ohio. (Modified from Chen, W. et al., *Environ. Toxicol. Chem.*, 21, 305, 2002. With permission.)

Parameters	Log(α)	β	R^2	n [a]	S^2 [b]
TWMC	1.253	2.854	0.67	72	0.24
80th percentile	1.558	2.745	0.74	54	0.11
90th percentile	1.937	3.131	0.81	63	0.13
95th percentile	2.093	2.825	0.87	67	0.07
99th percentile	2.293	2.442	0.65	71	0.19
Peak	2.492	2.267	0.53	72	0.26

[a] Number of data points. The different number of data points is due to no samples or values excluded when below limit of detection.

[b] Sample variance of regression residuals.

stochastically. Predictions of the detailed time sequence of pesticide concentrations are challenging and yet critical in assessing exposures of different time periods often important to address eco-toxicological or drinking water concerns. In this section, we extend the SWMI concept to a TSCM for predicting pesticide daily concentrations in rivers and streams. We first establish a specific atrazine TSCM as a base model. The specific TSCM is then generalized for other compounds through SWMI and product use rate in a relationship with atrazine.

Model development

The TSCM to be developed considers three components in a typical pesticide daily concentration sequence: seasonality, long-term trend, and local instant fluctuation (random or flow-caused). Its basic form of the equation, modified after Helsel and Hirsch,[16] reads:

$$\ln (C + 1) = (1 + b_0 \ \exp [b_1 \sin (6.2832t) + b_2 \cos (6.2832t) + b_3 t])\, x^k \qquad (21.6)$$

where C is concentration, μg/l, or ppb; t is time, year; 6.2832 is constant for the periodic function expressed in years; x is the ratio of stream flow rate (m^3/day) to the watershed area (m^2), m/day; b_0, b_1, b_2, b_3, and k are regression constants. The left-hand side of Equation (21.6) is a transformation to ensure valid mathematical operations in case of non-detectable concentration records (i.e., 0's).

A meaningful interpretation for each of the regression constants in Equation (21.6) is: b_1, and b_2 in the sine and cosine terms determine the amplitude and the time of the cycle of the concentration peaks, b_3 represents a global trend (slope), k is a coefficient accounting for the transient fluctuation caused by short-term flow dynamics, and b_0 is an overall adjusting coefficient.

Equation (21.6) is calibrated using the atrazine concentrations of the Heidelberg data set observed at the Maumee River station in a 10-year time span from January 1, 1986 to December 31, 1995. The stream flow data was obtained from the corresponding USGS gauging station (Figure 21.2). Results of the regression are presented in Figure 21.3A (transformed scale) and Figure 21.3B (linear scale). Residual quantile–quantile, 1:1, and percentile plots of the regression are shown in Figures 21.4A, B, and C, respectively.

The resulting atrazine TSCM coefficients are $b_0 = 0.00957$, $b_1 = 1.393$, $b_2 = -6.109$, $b_3 = -0.0173$, and $k = 0.14$. Using the phase shift relationship with b_1 and b_2, one can easily calculate the peak day[16]:

$$\text{peak day} = 58.019 \left[1.5708 - \tan^{-1} \left(\frac{b_2}{b_1} \right) \right] \approx 169 \text{ (or June 18)}$$

which indicates that over the 10 years from 1986 to 1995 there is a global seasonality of atrazine concentrations that peaked around June 18 each year in the Maumee River drainage watershed. Actual peak time may be influenced by the instant local fluctuations caused by stream flow ($k = 0.14$) and other random factors, such as storm events. The negative sign of b_3 suggests a slight long-term decline trend for atrazine in the watershed.

The specific atrazine TSCM can be generalized for other compounds by incorporating two key factors; the chemical environmental fate properties (as represented by SWMI) and annual product use intensity in a watershed (R_b). The assumption is that the calibrated atrazine TSCM reflects the basic structure of the pesticide occurrence patterns in a watershed that are primarily influenced by agricultural seasonality, long-term trend,

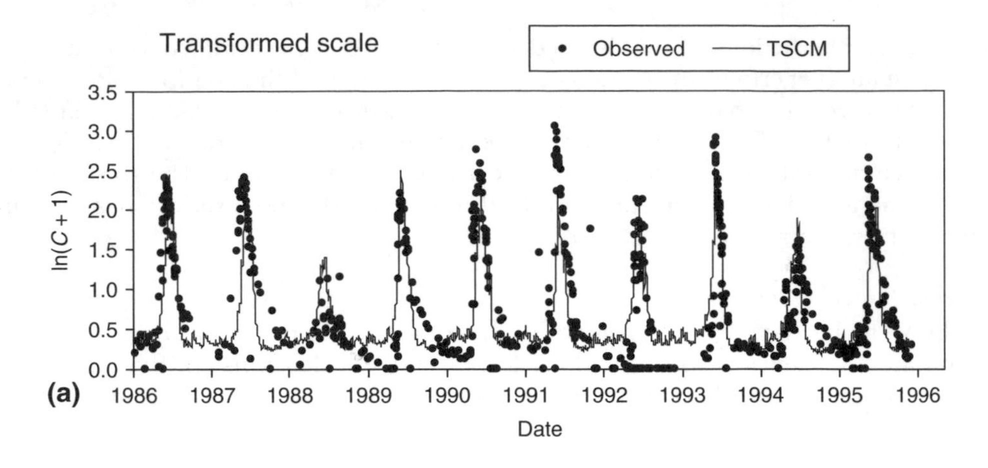

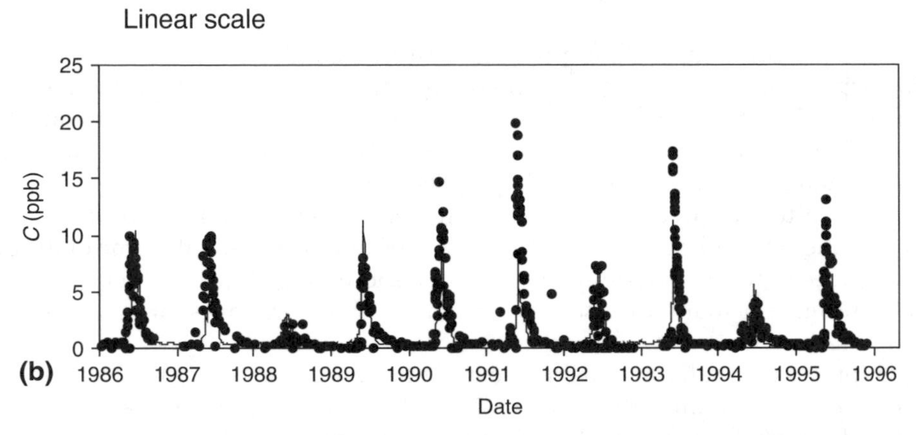

Figure 21. 3 The atrazine TSCM and the observed concentrations of the Heidelberg data set at the Maumee River station, Ohio, USA, in a 10-year time span between January 1, 1986 and December 31, 1995. The resulting model coefficients $b_0 = 0.00957$, $b_1 = 1.393$, $b_2 = -6.109$, $b_3 = -0.0173$, and $k = 0.14$.

and stream flow (an integrated parameter of weather and watershed hydrology). To achieve a generalized TSCM, atrazine is thus used as a standard to which other pesticide concentrations are compared. If the general use pattern and agricultural seasons are similar, the ratio of a pesticide's concentration to atrazine observed at the same time and location would be expected to correlate to the pesticide's environmental fate properties and its use rate in the watershed.

Atrazine as a comparison standard is primarily rationalized by its moderate mobility and persistence in the environment and the fact that it is one of the most popular, widely used, and longest established herbicides in the US agriculture. It can be shown that atrazine concentrations in the Heidelberg data set correlate well with many other compound's observed data. To seek for accuracy and representativeness, however, only the 1986 data from Maumee River, the largest Lake Erie tributary watershed, is processed as the pesticide use information was surveyed only specifically in that year.[10] As examples shown in Figure 21.5, a general good correlation exists between atrazine and

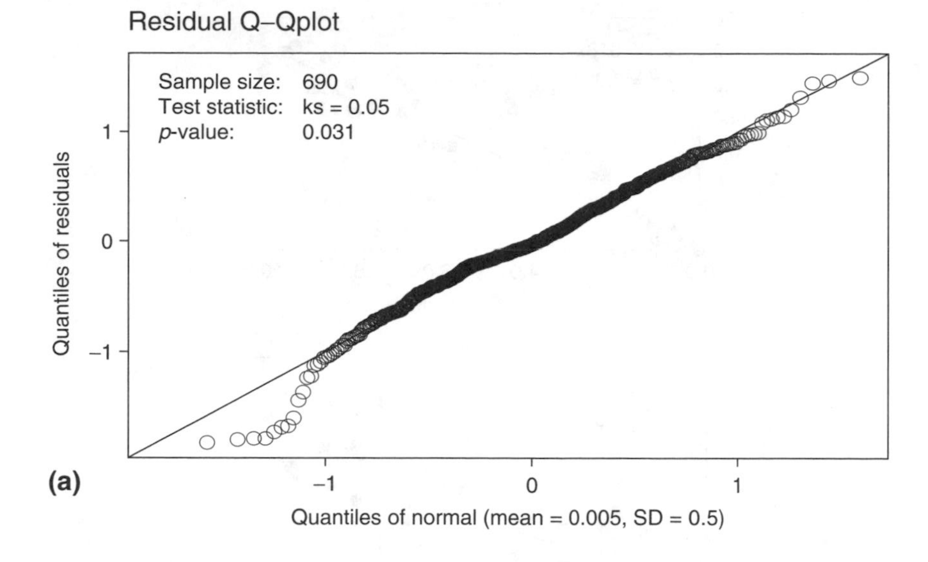

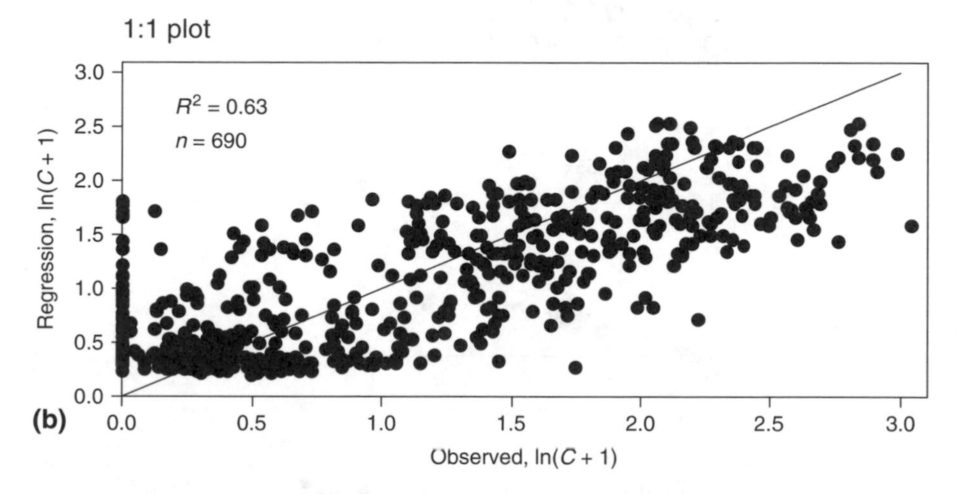

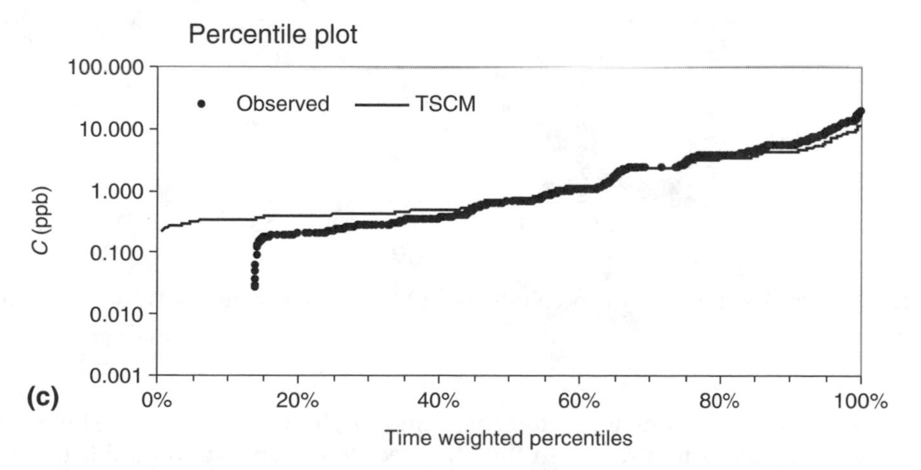

Figure 21.4 Residual quantile–quantile (a), 1:1 (b), and percentile (c) plots of the regression between the atrazine TSCM and the observed concentrations of the Heidelberg data set at the Maumee River station, Ohio, USA. Shown on the Plot A are also the results of the Kolmogorov–Smirnov (ks) goodness of fit test.

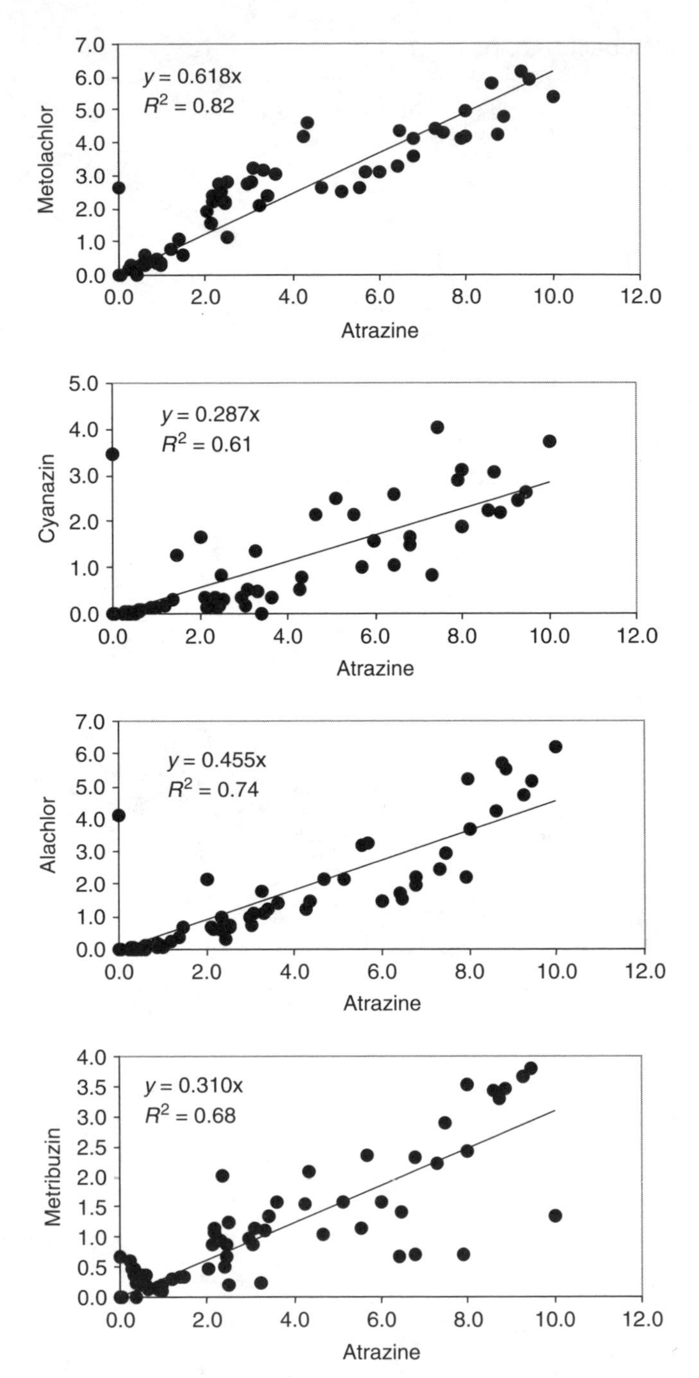

Figure 21.5 Examples of pesticide concentrations (ppb) in relation to atrazine observed in Maumee River, Ohio, 1986.

the concentrations of metribuzin, metolachlor, and alachlor. The resulting slopes of eight pesticide concentrations to atrazine in the Maumee River sub-data set are listed in Table 21.4, together with their SWMI and use intensity values. Pesticides with a majority of

Table 21.4 Ratio (slope) of pesticide concentrations (ppb) to atrazine observed in Maumee River, Ohio, 1986, in relationship with compound environmental fate properties (SWMI) and watershed use intensity (R_b, kg/ha). Example results are also plotted in Figure 21.6

	Simazine	Atrazine	Metribuzin	Alachlor	Metolachlor	Cyanazine	Terbufos	Linuron
SWMI	0.6718	0.7184	0.5453	0.5207	0.5969	0.4932	0.1388	0.4523
Watershed use intensity (Rb, kg/ha)	0.00604	0.248	0.0869	0.452	0.284	0.0882	0.0253	0.0474
Slope to atrazine	0.093	1.000	0.310	0.455	0.618	0.287	0.00106	0.0220

non-detectable concentrations (generally > 80% of the data) and showing little correlation with atrazine were not included in Table 21.4 (mostly insecticides, such as phorate).

As expected, the pesticide–atrazine slope was found to have a good correlation with SWMI and the watershed use intensity R_b, interestingly with a cubic power of SWMI [close to the β value in the DCM; Equation (21.5) and Table 21.3] and square root of R_b (Figure 21.6). This means that less mobile (smaller SWMI) and lower use compounds tend to have lower detectable concentrations. With this simple relation, a generalized SWMI TSCM, expressed in the linear scale, is readily derived from Equation (21.6):

$$C = 5.381\left(\exp\left\{\left(1 + b_0 \exp\begin{bmatrix} b_1 & \sin(6.2832t)+ \\ b_2 & \cos(6.2832t)+ \\ b_3 & t \end{bmatrix}\right)x^k\right\} - 1\right)\sqrt{R_b}\,\mathrm{SWMI}^3 \qquad (21.7)$$

The parameter values in the generalized TSCM [Equation (21.7)] are given previously in the atrazine TSCM calibration. With available compound-specific watershed use intensity, SWMI value, and stream flow data, daily estimates of the likely concentrations in a given year can be easily obtained from Equation (21.7). Validation of the generalized TSCM using independent data sets from different watersheds are presented in the section "Results and discussion."

Model use

Calculation of Equation (21.7) is easy as long as input parameters are prepared properly. A few points, however, are deemed prudent to put the model use into context:

1. Prior to a simulation, potential differences in agricultural use patterns of a test compound versus atrazine need to be understood thoroughly. For example, if a product is used primarily in fall or winter, use of the atrazine-based TSCM will result in wrong predictions of seasonality. However, if the application timing is similar to atrazine, a slight shift of seasonality may be tolerable in an assessment as long as the overall distributions of the estimates are reasonable.

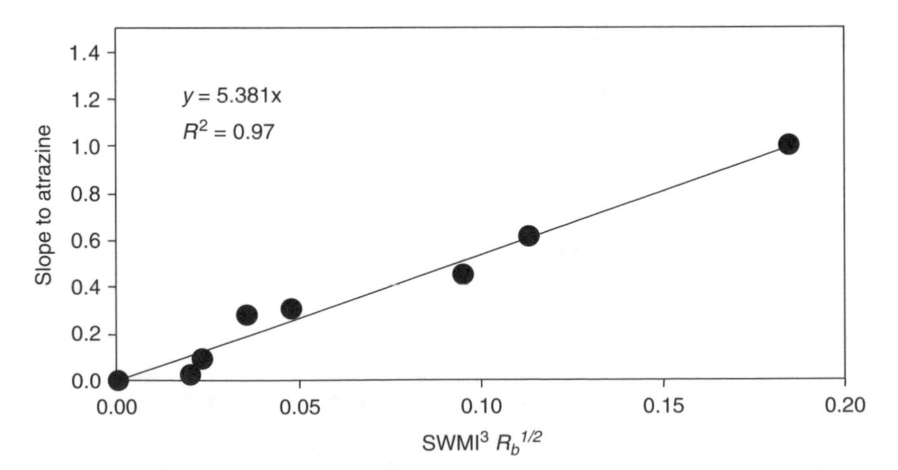

Figure 21.6 Ratio (slope) of pesticide concentrations (ppb) to atrazine observed in Maumee River, Ohio, 1986, in relationship with the compound SWMI and watershed use intensity (R_b, kg/ha).

2. The physical watershed, to which a simulation is relevant, needs to be clearly defined. The TSCM estimated concentrations are only referred to at the watershed exit point. It does not represent spatially averaged predictions of a watershed. As such, the stream flow data should come from the same watershed exit point and the estimate of the annual use intensity (R_b) should be watershed-based (not per political borders).
3. Calculate SWMI and R_b parameters as in the section "Distributional concentration model." Use intensity should be estimated as accurately as possible. If multiple years are simulated, be aware of potential changes in annual use rate.
4. The time unit in TSCM is in years. That is, 1 day is approximately 0.00274 year.

Results and discussions

Distributional concentration model

The SWMI DCM (Equation (21.5]) predicts concentration percentiles in its overall distribution that are likely detected at the exit of a watershed. Limited test of the SMWI DCM on a drainage basin near 3,000,000 ha of a river system (White River, IN, USA) and a 1400 ha of a reservoir basin (Higginsville City Lake, MO, USA) indicates that the model was able to predict TWMC, peak and the 95th percentile concentrations within about 5-fold of the observed data.[10] Comparisons of the model to the observed data from both surface water systems are provided in Figures 21.7 and 21.8.

It should be noted that the response variable was log transformed in the DCM regression. As discussed in Reference 16, inverse transform of the regression would indicate that model predictions are related to the median trend (as oppose to the arithmetic mean) on the original linear scale. For predicting the arithmetic mean, an adjustment factor can be estimated as $\exp(0.5\,S^2)$, where S is the standard deviation of the regression residuals on the natural log scale. As in Table 21.3, the residual variance (S^2) ranges from 0.07 to 0.26 over the developed percentile models, indicating that the adjustment factor to Equation (21.5) will be from 1.04 (the 95th percentiles) to 1.14 (peak). In this chapter, the adjustment factor for the arithmetic mean is not considered.

Consistent with the regression results (Table 21.3), model predictions appear to closely match the observed data around the 95th percentile. Model deviation tends to increase generally as it moves to the lower or higher end in the distribution (Figures 21.7 and 21.8). An explanation could be that data near the 90–95th percentile are most likely generated by "normal" runoff events, while peaks or the higher percentiles are generally associated with extreme events that could not be explained by the chemical environmental fate properties or the use rate alone. Such rare events can be the coincidence of an application and a high runoff-generating storm or snowmelt. Lower percentile data are often dominated with non-detectable records (or 0's as in the Heidelberg data set). Thus, a decent regression at the lower percentiles is hard to obtain, such is the case for data below the 80th percentile. TWMC, however, is a weighted average of the entire distribution. Its "quality" of the model predictions would lie in between those of the peak and the lower portion of the distribution.

For a demonstration, the TWMC DCM in Table 21.3 is executed for a hypothetical compound, assuming a lognormal distribution for each of its three parameters without inter-correlation: $T_{1/2}$ (mean 30 days, SD 15 days), K_{oc} (mean 100 ml/g, SD 50 ml/g), and R_b (mean 0.2 kg/ ha, SD 0.1 kg/ha). The simple Monte Carlo simulation is easily

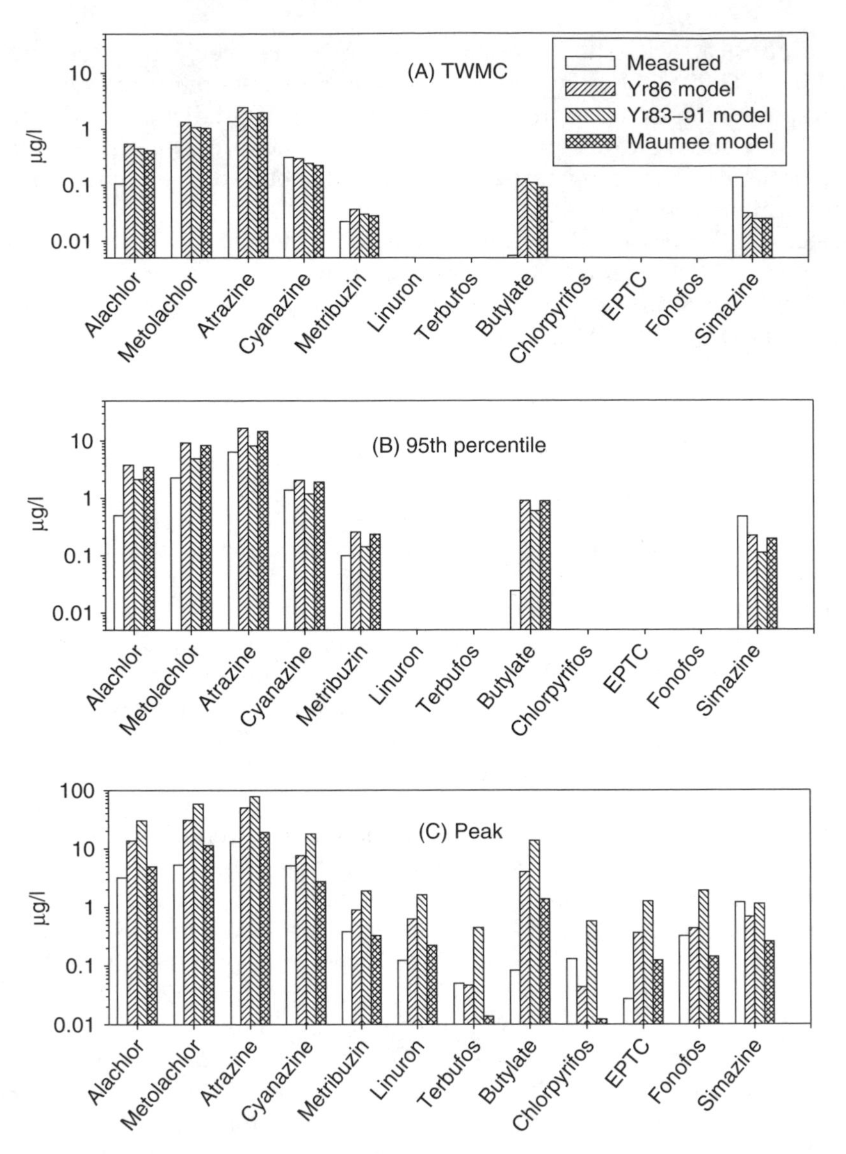

Figure 21.7 Comparison of the DCM predictions to the observed data of 12 pesticides in White River at Hazelton, Indiana, USA, 1991–1996. (From Chen, W. et al., *Environ. Toxicol. Chem.*, 21, 300, 2002. With permission.)

conducted using Excel® with the embedded Crystal Ball software. Results of the estimated TWMC distribution are presented in Figure 21.9 for 1000 runs of repeated random sampling from the three lognormal distributions of the input parameters. The estimated mean TWMC is 1.15 ppb with an SD of 0.7. The Crystal Ball routine also outputs a sensitivity chart that ranks contributions to the total simulated model variance from each parameter (Figure 21.10). For the SWMI DCM, sensitivity is in the order

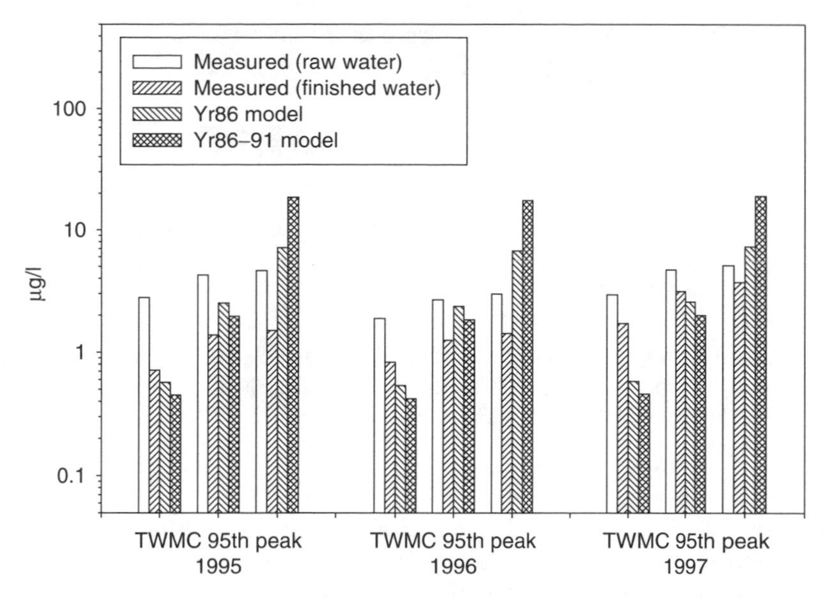

Figure 21.8 Comparison of the DCM predictions to the observed data of atrazine in raw and finish waters at the Higginsville City Lake, Missouri, USA, 1995–1997. (From Chen, W. et al., *Environ. Toxicol. Chem.*, 21, 300, 2002. With permission.)

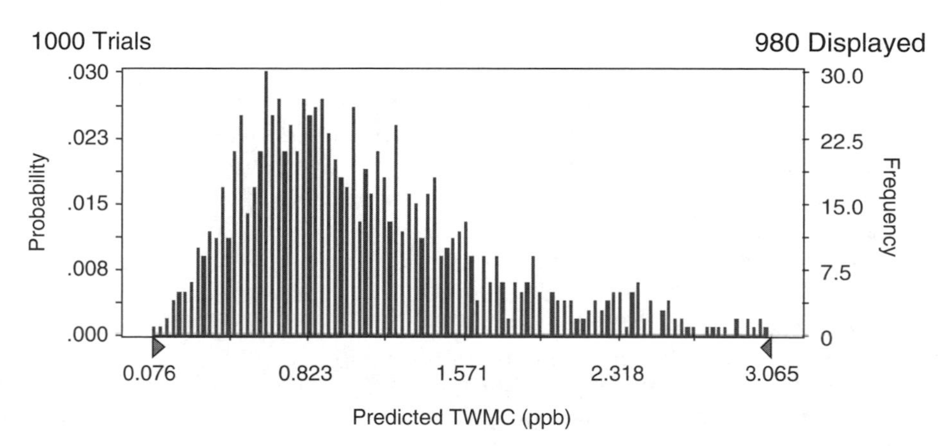

Figure 21.9 Frequency chart of the SWMI DCM for predicting the TWMC, assuming lognormal distribution for half-life (mean 30 days, SD 15 days), K_{oc} (mean 100 ml/g, SD 50 ml/g), and watershed use intensity (mean 0.2 kg/ha, SD 0.1 kg/ha). Simulated by Crystal Ball 2000.2.

$R_b > K_{oc} > T_{1/2}$. The K_{oc} sign is negative, indicating that smaller K_{oc} values mean higher model predictions.

The SWMI DCM is essentially determined by three parameters, the compound use intensity in a watershed (or basin, R_b), soil adsorption property (K_{oc}), and degradation half-life ($T_{1/2}$). The simple form and computational robustness of the model make it an ideal tool for relative risk assessment of different chemicals over large databases. Although some preliminary evaluations indicated no distinct differences in predicting

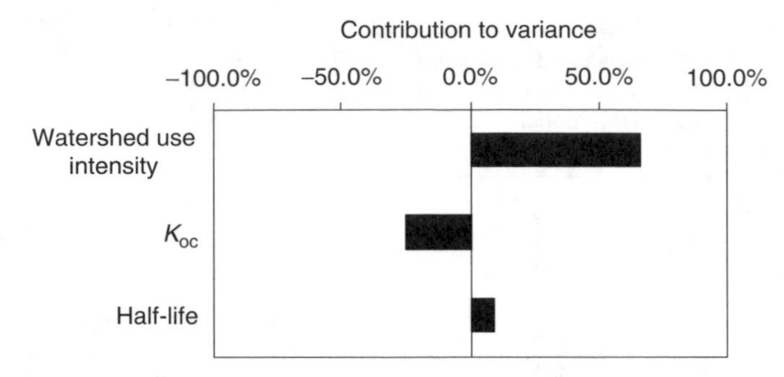

Figure 21.10 Sensitivity chart of the SWMI DCM for predicting the TWMC, assuming lognormal distribution for half-life (mean 30 days, SD 15 days), K_{oc} (mean 100 ml/g, SD 50 ml/g), and basin use intensity (mean 0.2 kg/ha, SD 0.1 kg/ha). Simulated by Crystal Ball 2000.2.

different surface water systems,[10] the model ability to differentiate inherent watershed runoff vulnerability is expected to be limited as no specific watershed hydrological characteristics are considered in the model. However, recognizing the high level of watershed complexity in almost every detailed configuration of the physical mass transport processes, the SWMI DCM is a useful and effective screening tool.

Time series concentration model

As shown in Figures 21.3A and B, a fairly close agreement was achieved between the base atrazine TSCM and the observed data with an R^2 of 0.63 over 690 data points. The regression residuals exhibit approximately a normal distribution with only a slight departure in the lower concentration end (too many non-detectable records) (Figure 21.4A). The residual SD is about 0.5, resulting in an adjustment factor of 1.13 if the regression is intended for predicting the arithmetic mean of $(C + 1)$ on the original linear scale (see the discussion in the section "Distributional concentration model"). However, the adjustment factor was not adopted in the development of the extended TSCM [Equation (21.7)] to retain the lognormal nature in the data.

Figures 21.4B and C further illustrate the model-data agreement by a 1:1 and a percentile plot. The 1:1 plot compares the model to data on a 1-to-1 basis in the order of time, while the percentile plot compares the sorted data and model predictions in an ascending order based on the magnitude of the compared variables. Both comparisons show a reasonable model fit to the majority range of the observed data with some tendency to over-predict small concentrations, while under-estimate higher extremes (e.g., >95th percentile). Some autocorrelation structure of the regression residuals also exists (not shown), but this is not explored further here as the data set has variable sampling intervals.

The atrazine-specific TSCM is extended to a general form through the ratio of a given compound to atrazine that is further related to SWMI and its watershed use intensity [Equation (21.7)]. Model validation results are presented in Figures 21.11–21.17, where the generalized TSCM predictions are compared to eight chemicals with monitoring data available in four different watersheds (White River of Indiana, Sandusky River, Honey Creek, and Rock Creek of Ohio). These watersheds and monitoring data are independent of the Ohio Maumee River watershed, which was used in the model calibration. The eight

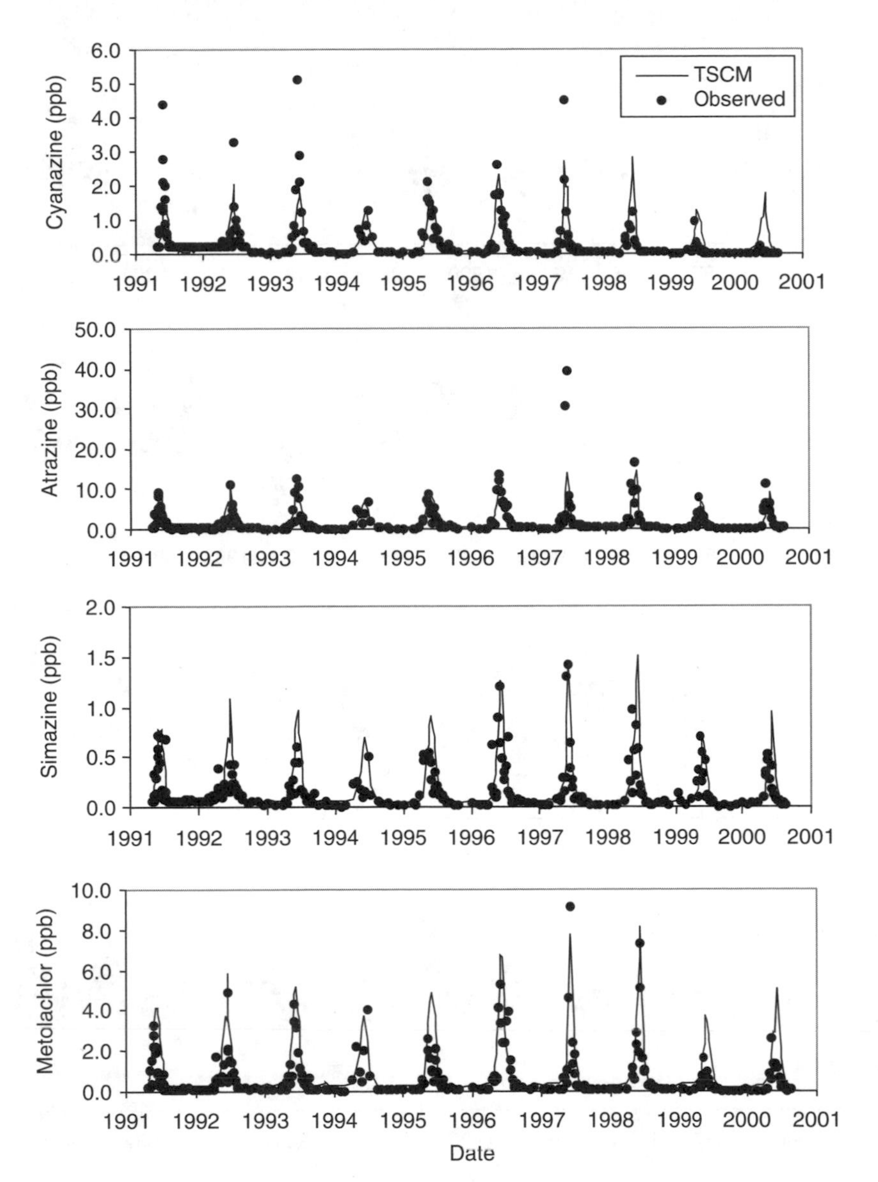

Figure 21.11 Independent predictions of the generalized TSCM compared to the observed herbicides, USGS-NAWQA White River Study Unit at Hazelton, Indiana (May 1, 1991 to August 29, 2000). *Data source*: http://www-dinind.er.usgs.gov/nawqa/wr06000.htm.

chemicals are distributed in the three SWMI regions as shown in Figure 21.1 (Region I: terbufos, phorate; Region II: butylate, cyanazine, fonofos; Region III: atrazine, simazine, metolachlor). Detailed descriptions of these data sets can be found in References 10, 15, and 17.

As shown in Figures 21.11–21.17, the generalized TSCM predicted well the overall trend, seasonality, and to a good extent the local transient fluctuations in all four watersheds for two 10-year time periods: 1986–1995 for the Ohio watersheds; and 1991–2000 for White River, Indiana. The model worked best for the most commonly used triazine and

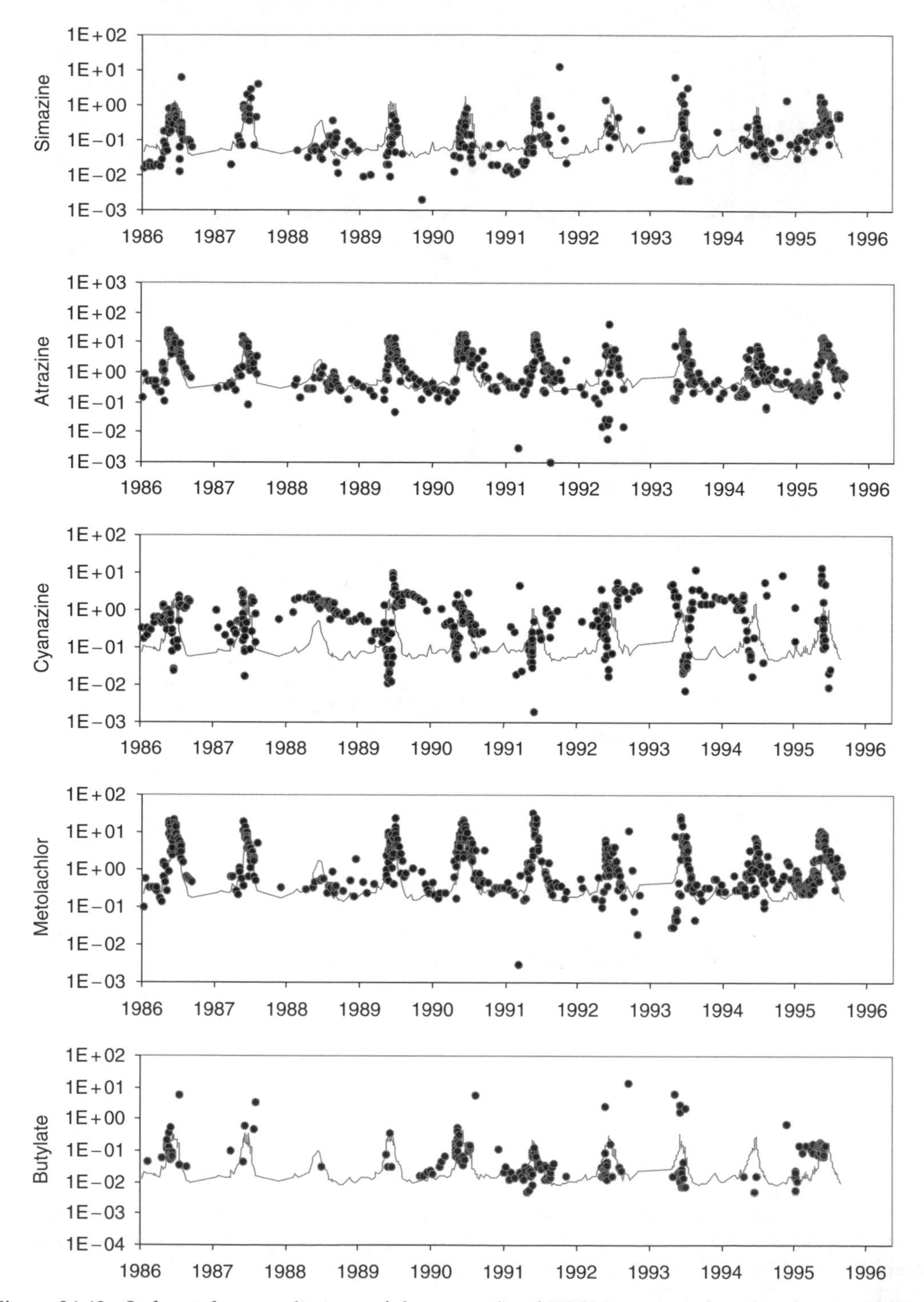

Figure 21.12 Independent predictions of the generalized TSCM compared to the observed herbicides in Sandusky River, Ohio (January 13, 1986 to December 4, 1995). Lines are TSCM predicted and points represent data. Unit: μg/l or ppb.

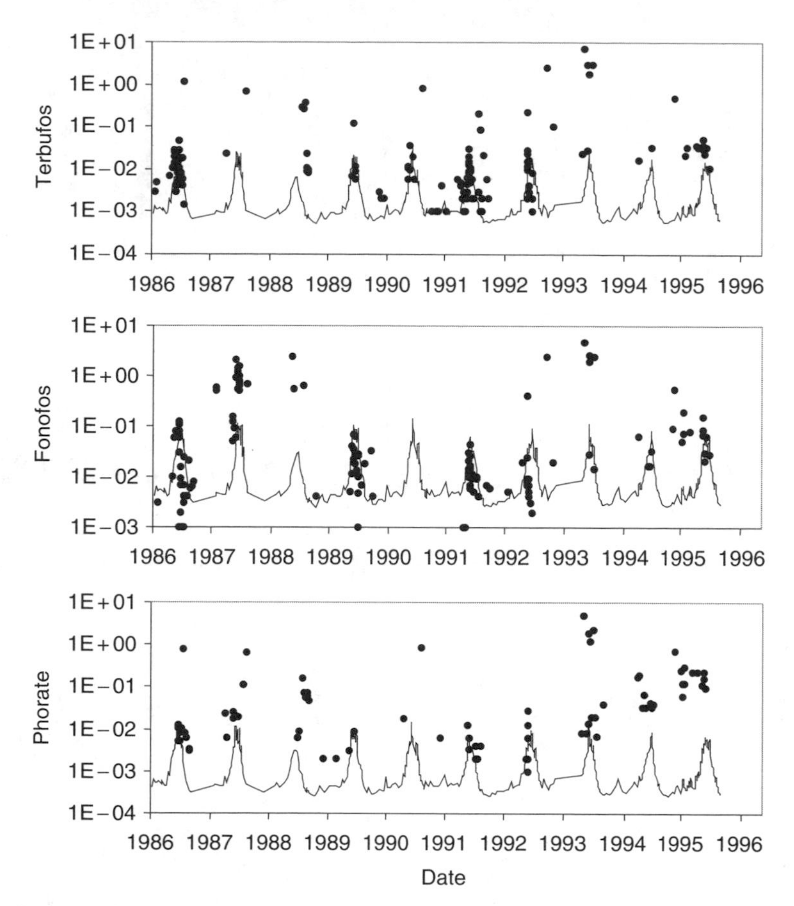

Figure 21.13 Independent predictions of the generalized TSCM compared to the observed insecticides in Sandusky River, Ohio (January 13, 1986 to December 4, 1995). Lines are TSCM predicted and points represent data. Unit: μg/l or ppb.

chloroacetanilide herbicides (atrazine, cyanazine, simazine, and metolachlor), while moderately well for butylate. Although predictions for the insecticides (fonofos, terbufos, and phorate) are not as good as for the herbicides, the model does provide a believable pattern of the occurrence of these chemicals. The less accurate insecticide results can be primarily due to scarce data with a few extreme values, potential inaccurate use rate, and application patterns (timing, frequency, areas of use, etc.) that are significantly different from atrazine.

Differences in model performance were not observed among different watersheds that range in drainage areas from 1130 ha (Rock Creek) to 2,938,382 ha (White River) in this model validation exercise. This is probably attributed to the effective incorporation of the non-linear stream flow term in the model [Equation (21.7)]. Influence of stream flow on chemical mass transport within a watershed is complicated and variable in timing. Straight correlation between daily flow and concentration is often hardly identifiable, at least for the watersheds studied in this work (results not shown). The reason is likely because of the dual roles of stream flow, i.e., mass delivery and dilution. During cropping seasons as new applications of chemicals are made to the watershed, runoff and stream

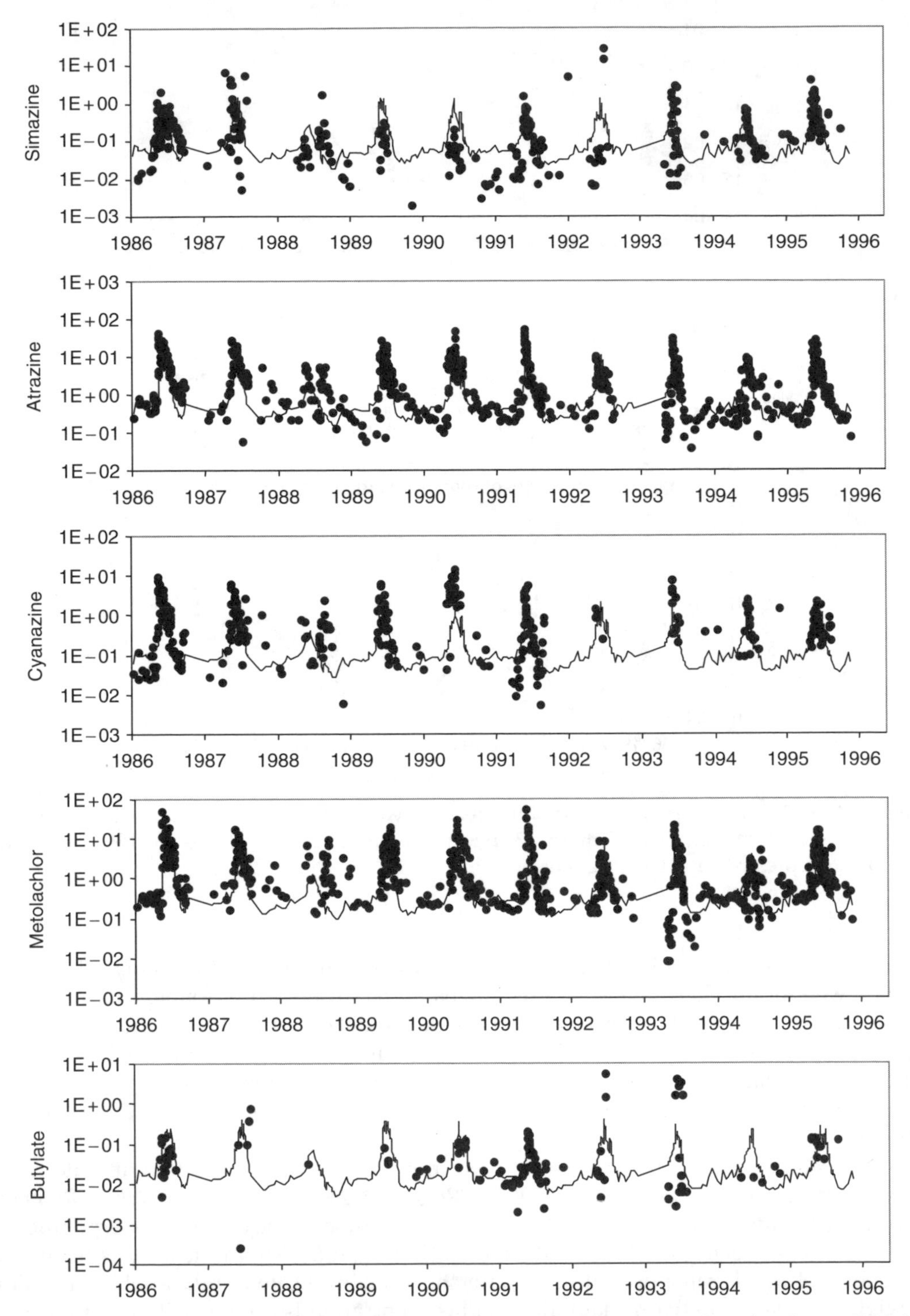

Figure 21.14 Independent predictions of the generalized TSCM compared to the observed herbicides in Honey Creek, Ohio (January 13, 1986 to November 27, 1995). Lines are TSCM predicted and points represent data. Unit: μg/l or ppb.

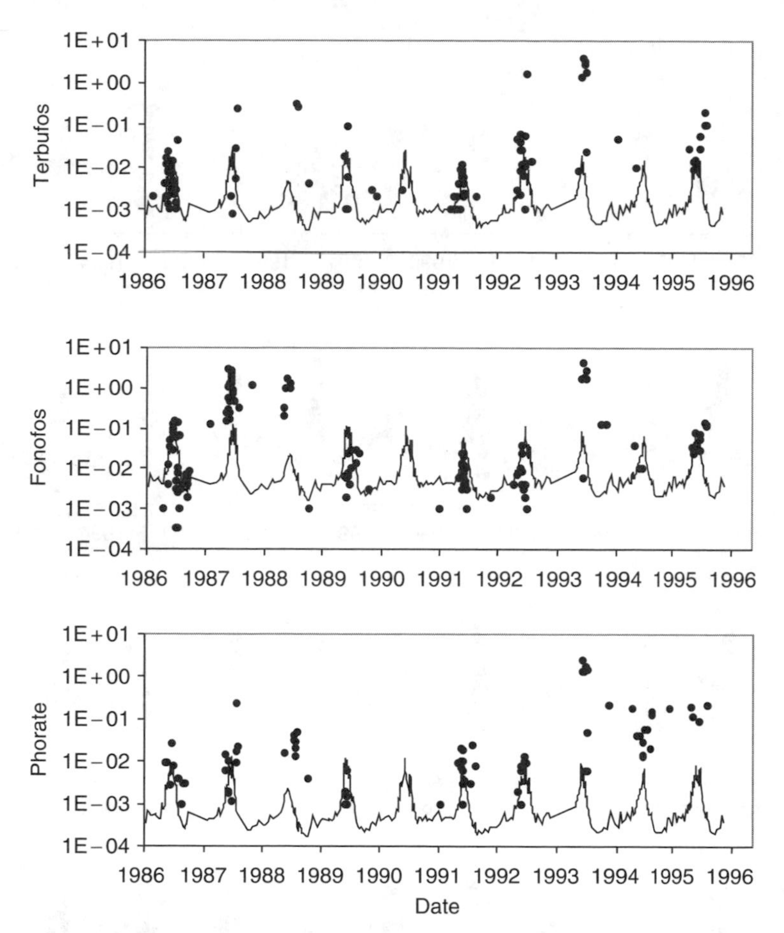

Figure 21.15 Independent predictions of the generalized TSCM compared to the observed insecticides in Honey Creek, Ohio (January 13, 1986 to November 27, 1995). Lines are TSCM predicted and points represent data. Unit: μg/l or ppb.

flow are more likely to play a role of mass delivery, thus they likely positively correlate (albeit noisily) with riverine concentrations. On the other hand, dilution role would be observed during non-cropping seasons, such as winter, when only a small amount of aged residues (if any) in soil is available to runoff.

Beyond the potential inefficiencies of the model inherent structure, a number of data quality factors deserve mentioning. The first is the watershed use intensity that needs to be accurate and watershed-specific for each year. As mentioned previously, only the 1986 pesticide annual use was available for the Ohio watersheds and the 1992–1994 annual average for White River. Assuming constant annual use for every year may not introduce much error for major and well-established herbicide, such as atrazine. However, substantial errors could be generated for herbicides with changing use patterns, and for compounds whose demands are primarily driven by presence of pests (insecticides) and diseases (fungicides). A close examination of the model validation results for the Ohio watersheds (Figures 21.12–21.17) reveals that the model tends to predict more accurately in the first year (i.e., 1986), with model-data correlation (R^2) as high as 0.77. This is

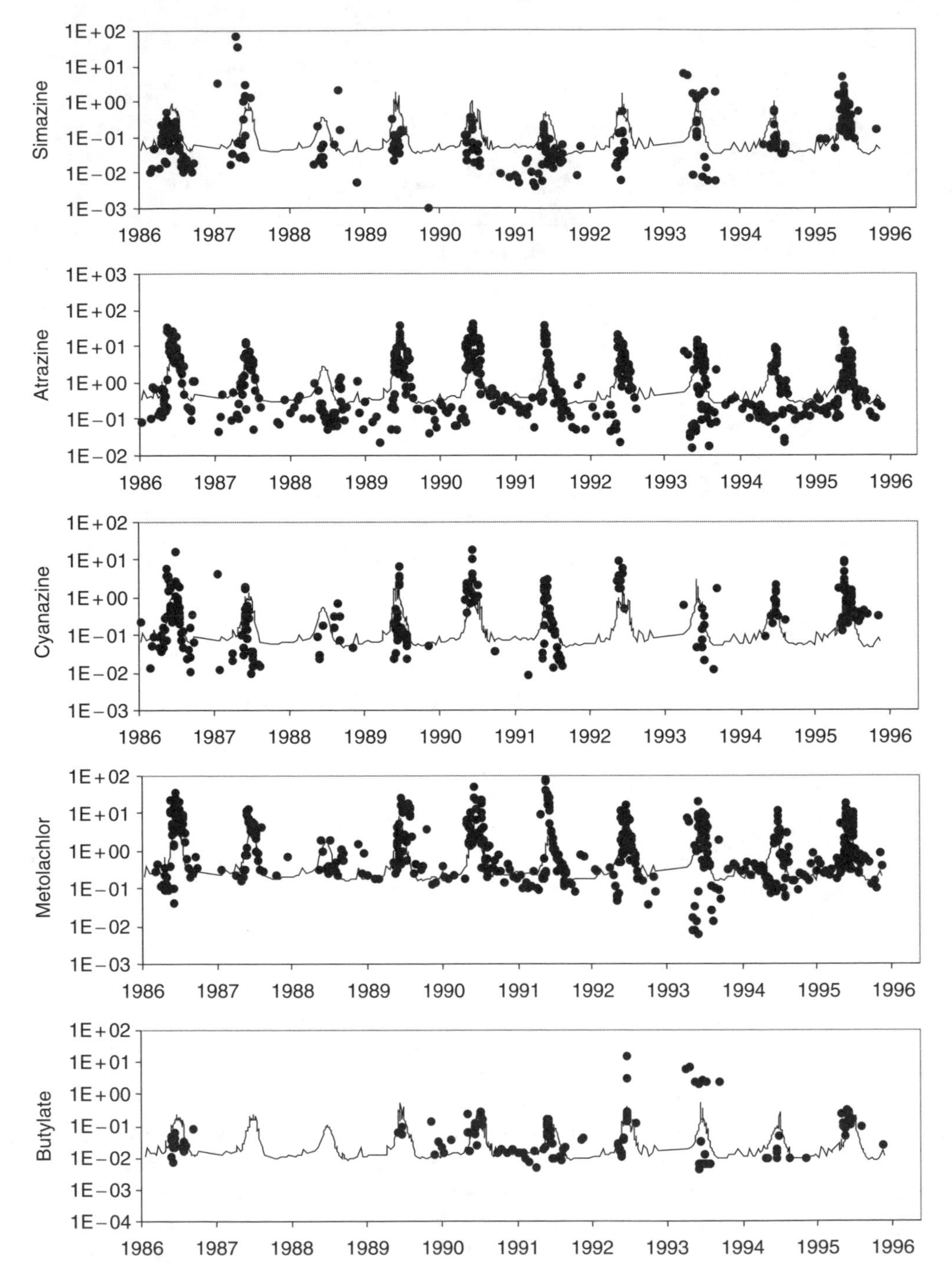

Figure 21.16 Independent predictions of the generalized TSCM compared to the observed herbicides in Rock Creek, Ohio (January 13, 1986 to November 27, 1995). Lines are TSCM predicted and points represent data. Unit: μg/l or ppb.

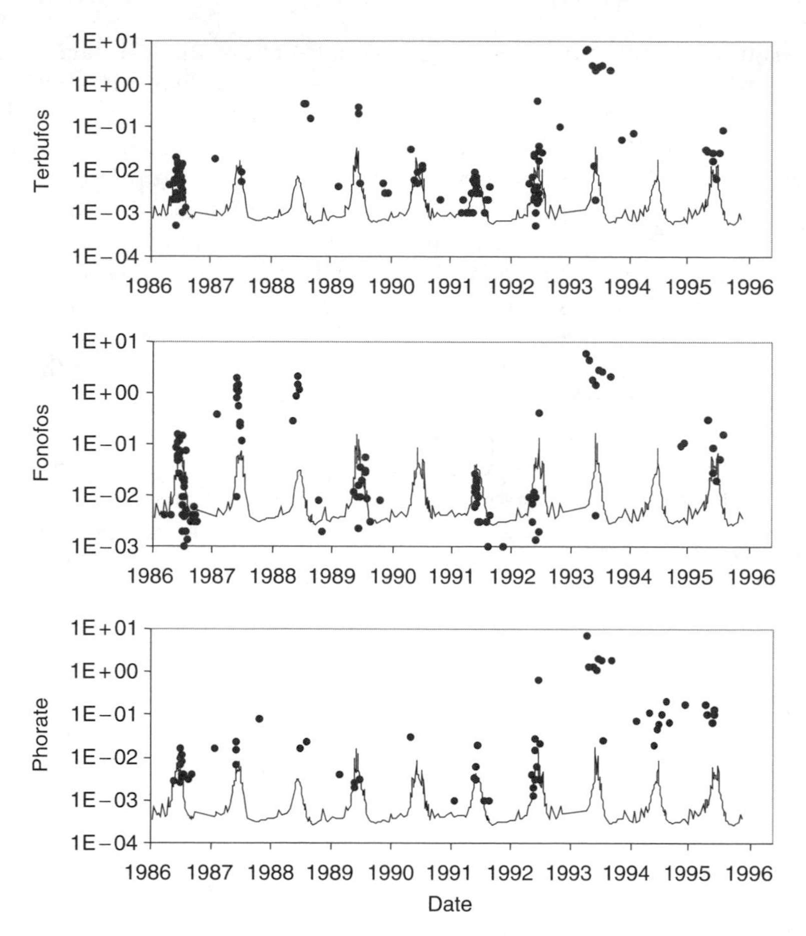

Figure 21.17 Independent predictions of the generalized TSCM compared to the observed insecticides in Rock Creek, Ohio (January 13, 1986 to November 27, 1995). Lines are TSCM predicted and points represent data. Unit: μg/l or ppb.

particularly true for the less used herbicide butylate and all three insecticides, which had no detectable concentrations in some subsequent years after 1986. Similar patterns are not obvious for the White River, as the four herbicides are all commonly used and well established in the region during the monitoring years (except cyanazine after year 2000 due to phase out of the product) (Figure 21.11).

The second data quality factor is the analytical accuracy of low concentrations for infrequently used compounds, such as the scarce insecticide data. This can affect the model calibration. Censorship or data reported only as below laboratory detection limit is another critical aspect that needs to be dealt with properly.[16] Data sets with a large portion that is censored may not be appropriate at all for accurate model calibration. The Heidelberg data set used in this report was not censored. However, the decreased accuracy is noticeable as concentrations approach to 0. While increasing analytical accuracy is a long-term goal, some specifically designed studies targeting at controllable mesoscale watersheds or catchments, where a known amount of such pesticide use is available, would provide accurate data for model calibration.

The environmental fate parameters K_{oc} and half-life are the last data quality factors that deserve attention. In general, short-term (usually 24 h) determined soil adsorption parameters are sufficient in relation to surface runoff processes. Thus, K_{oc} values determined from long-term aged residues may not be necessarily useful for the modeling purpose here. Half-life based on field studies may be more appropriate than the laboratory-determined values if pesticide dissipation due to runoff and leaching are insignificant in the field study. In case of available multiple parameter values, the ones determined from soils that are most similar to the major watershed characteristics should be used.

Summarizing the TSCM generalization from the atrazine-specific version [Equation (21.6)] to a multi-compound model [Equation (21.7)] is of considerable interest. The bridge is to use atrazine as a tracer or surrogate of other compounds so that the general global seasonality and stream flow effects are characterized. Local/specific behavior of an observation in a surface water system would be primarily caused by their specific use patterns and environmental fate properties. The approach is thus general in the sense that a surrogate compound different from atrazine may be selected to simulate different chemical classes in different regions, if the targeted class of chemistry or a geographical region has a distinct use pattern or seasonality.

Acknowledgments

P. Richards provided the Lake Erie drainage basin monitoring data. Beneficial discussions with C.G. Crawford and P. Richards were obtained during the preparation of this chapter, as well as in various previously related works. Comments from colleagues P. Hertl, W. Phelps, S. Chen, P. Hendley, and C. Breckenridge are appreciated.

References

1. Ecological Committee on FIFRA Risk Assessment Methods (ECOFRAM), Draft Reports, U.S. Environmental Protection Agency, Office of Pesticides Programs, Washington, D.C., USA, 1999 (http://www.epa.gov/oppefed1/ecorisk/index.htm).
2. Jones, R.L. and Mangels, G., Review of the validation of models used in Federal Insecticide, Fungicide, and Rodenticide Act environmental exposure assessments, *Environ. Toxicol. Chem.*, 21, 1535, 2002.
3. Mullins, J.A., Carsel, R.F., Scarbrough, J.E. and Ivery, A.M., PRZM-2, A Model for Predicting Pesticide Fate in the Crop Root and Unsaturated Soil Zones: Users Manual for Release 2.0, Technical Report EPA/600/R-93/046, Environmental Research Laboratory, U.S. Environmental Protection Agency, Athens, GA, USA, March, 1993.
4. Burns, L.A., Exposure Analysis Modeling System (EXAMS): User Manual and System Documentation, Technical Report EPA/600/R-00/081, Environmental Research Laboratory, U.S. Environmental Protection Agency, Athens, GA, USA, September, 2000.
5. U.S. Environmental Protection Agency, Guidance for Selecting Input Parameters in Modeling the Environmental Fate and Transport of Pesticides, Version II, U.S. Environmental Protection Agency, Office of Pesticides Programs, Environmental Fate and Effects Division, Washington, D.C., USA, February 28, 2002.
6. Solomon, K.R., Baker, D.B., Richards, R.P., Dixon, K.R., Klaine, S.J., La Point, W., Kendall, R.J., Weisskopf, C.P., Giddings, J.M., Giesy, J.P., Hall, L.W. and Williams, W.M., Ecological risk assessment of atrazine in North American surface waters, *Environ. Toxicol. Chem.*, 15, 31, 1996.
7. Hertl, P., Phelps, W., Gustafson, D.I., Jackson, S.H., Jones R.L., Russell, M.H. and Schocken, M.J., A comparison of US EPA's Tier 1 and 2 index reservoir model estimates to drinking water

reservoir monitoring results in selected US systems in 1999/2000, Presented at the 10th IUPAC International Congress on the Chemistry of Crop Protection, Basle, Switzerland, August 4–9, 2002.

8. Larson, S.J. and Gilliom, R.J., Regression models for estimating herbicide concentrations in U.S. streams from watershed characteristics, *J. Am. Water Resour. Assoc.*, 37, 1349, 2001.
9. Larson, S.J., Crawford, C.G. and Gilliom, R.J., Development and Application of Watershed Regressions for Pesticides (WARP) for Estimating Atrazine Concentration Distributions in Streams, U.S. Geological Survey Water Resources Investigations Report 03–4047, 2004.
10. Chen, W., Hertl, P., Chen, S. and Tierney, D., A pesticide surface water mobility index and its relationship with concentrations in agricultural drainage watersheds, *Environ. Toxicol. Chem.*, 21, 298, 2002.
11. Crawford, C.G., Larson, S.J. and Gilliom, R.J., Development and Application of Watershed Regression for Pesticides (WARP) Methods to Estimate Organophosphate Pesticide Concentration Distributions in Streams, U.S. Geological Survey Water-Resources Investigations Report (Draft), October 23, 2003.
12. Rao, P.S.C., Hornsby, A.C. and Jessup, R.E., Indices for ranking the potential for pesticide contamination of groundwater, *Proc. Soil Crop Sci. Soc. Fla.*, 44, 1, 1985.
13. Jury, W.A., Focht, D.D. and Farmer, W.J., Evaluation of pesticide groundwater pollution potential from standard indices of soil-chemical adsorption and biodegradation, J. *Environ. Qual.*, 16, 422, 1987.
14. Gustafson, D.I., Ground water ubiquity score: a simple method for assessing pesticide leachability, *Environ. Toxicol. Chem.*, 8, 339, 1989.
15. Richards, R.P. and Baker, D.B., Pesticide concentration patterns in agricultural drainage networks in the Lake Erie Basin, *Environ. Toxicol. Chem.*, 12, 13, 1993.
16. Helsel, D.R. and Hirsch, R.M., *Statistical Methods in Water Resources*, Elsevier Science, Amsterdam, 1992, p. 341.
17. Crawford, C.G., Occurrence of Pesticides in the White River, Indiana, 1991–1995, U.S. Geological Survey Fact Sheet 233–95, White River Basin Study, U.S. Geological Survey, 5957 Lakeside Boulevard, Indianapolis, IN, 1995.

chapter twenty-two

Design and analysis of toxicity tests for the development and validation of biotic ligand models for predicting metal bioavailability and toxicity

Karel A.C. DeSchamphelaere and Colin R. Janssen
Ghent University

Dagobert G. Heijerick
Ghent University and EURAS

Contents

1-56670-664-5/05/$0.00+$1.50
© 2005 by CRC Press

Introduction

The biotic ligand model (BLM) is an integrative framework to evaluate and predict bioavailability and toxicity of metals to freshwater organisms.[1–3] It considers both metal complexation and speciation in the solution surrounding the organism and the interactions between metal ions and competing ions at binding sites on the organism–water interface (e.g., epithelial cells of gill tissue).

Some early studies (e.g., References 4–6) led to the formulation of the free-ion activity model[7] and the gill surface interaction model,[8] whose central hypothesis was that metal toxicity was related to the amount of metal bound to toxic action sites on this organism–water interface. Although conceptually those models contained almost all the features of the BLM, the use of the former models for regulatory purposes remained limited.[3] Perhaps, the reason of the current success of the BLM is that, for the first time, a model was able to integrate all state-of-the-science knowledge on complexation, speciation, and interactions at the toxic site of action into a generalized, visually attractive and easy-to-handle computerized framework (Figure 22.1).

Another reason for the increasing success of the BLM is the overwhelming body of evidence that supports the concepts formulated in it. For example, numerous detailed studies have demonstrated the effects of organic matter, pH, and hardness cations on zinc toxicity to fish, crustaceans, and algae.[9–14] Concurrently, increasing physiological evidence has been reported on the protective effects of cations against metal toxicity.[15,16] Additionally, it has been demonstrated that cations like Ca^{2+} and H^{+} decrease the binding of toxic metal cations to fish gills.[17] The derivation of stability constants for the toxic metal cation and competing cations,[17] together with the quantitative relation of metal bound to the fish gill and toxicity, independent of water chemistry,[18] coupled to a geochemical speciation model,[19] resulted in the BLM.[3]

For small invertebrates, the determination of gill-metal concentrations is experimentally difficult to obtain due to the size of the organisms. As a result, the derivation of stability constants for the toxic metal and the competing cations becomes compromised. In this chapter, we present an experimental design to determine those constants directly from toxicity data, and this specifically for the case of zinc toxicity to the crustacean *Daphnia magna*, based on data reported in Reference 9. Special emphasis is also given to the mathematical background of the proposed design, the analyses of the data, and the validation of the BLM with field-collected surface water samples.

Materials required

Supplies

- Polyethylene test cups, 50 ml (TEDECO, Belgium)
- Polyethylene vessels, 2 l, to store stock solutions (#kart614, VWR, Belgium)
- Polyethylene vessels, 5 l, to collect field water samples (#cove5020, VWR, Belgium)
- Volumetric flasks, 200 ml and 2 l (#hirs2800180 and #brnd37265, VWR, Belgium)

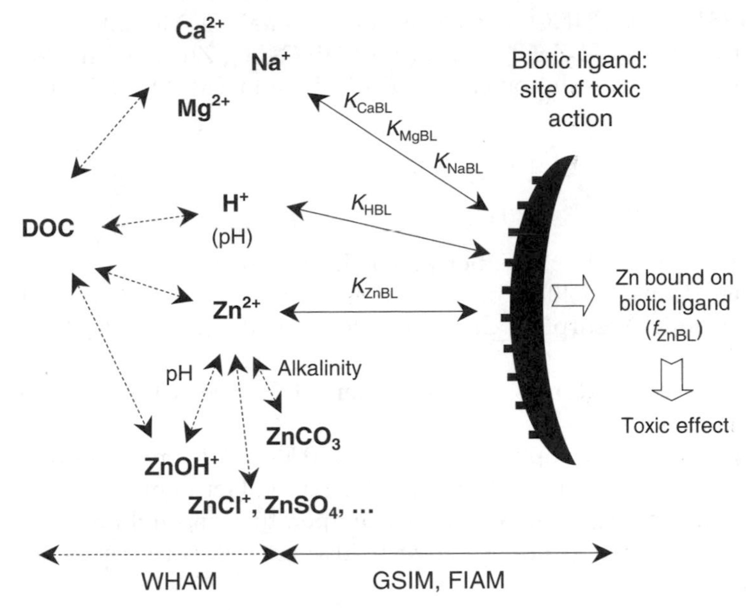

Figure 22.1 Schematic overview of the BLM for metal bioavailability and toxicity (exemplified for Zn, redrawn from Reference 3). Dashed lines represent speciation reactions (modeled by WHAM[19]); solid lines represent binding to the BL (which are assumed to be sites of toxic action). The free zinc ion, i.e., Zn^{2+}, forms complexes with inorganic ligands, such as OH^- (determined by pH), CO_3^{2-} (determined by alkalinity), SO_4^{2-}, and Cl^-. Zn^{2+} and $ZnOH^+$ also form complexes with DOC and Ca^{2+}, Mg^{2+}, and H^+ (also determined by pH) competes with Zn for binding sites on DOC. The speciation of Zn can be calculated with any speciation model (here WHAM[19]) when concentrations of all ligands present in solution and the stability constants of complexes (e.g., Reference 20) are known. Zn^{2+} binds to the BL and the concentration of zinc bound to the BL (or the fraction of BL sites occupied by zinc, f_{ZnBL}) determines the toxic effect. The latter concentration is assumed to be constant for a given effect size (e.g., $f_{ZnBL} = f_{ZnBL}^{50\%}$ at 50% mortality). The formation of complexes reduces Zn^{2+}, resulting in a reduced f_{ZnBL}, resulting in decrease of toxicity. Ca^{2+}, Mg^{2+}, Na^+, and H^+ can compete with copper for BL sites thus decreasing f_{CuBL} and toxicity. This concept was first formulated in the free ion activity model (FIAM[7]) and the gill surface interaction model (GSIM[8]). Binding affinities of Zn^{2+}, Ca^{2+}, Mg^{2+}, Na^+, and H^+ are defined by K_{ZnBL}, K_{CaBL}, K_{MgBL}, K_{NaBL}, and K_{HBL}, respectively.

- Glass beakers, 250 ml and 2 l (#kart1545 and #kart1549, VWR, Belgium)
- 0.45-μm filters, 32 mm diameter, Acrodisc®, polyethersulfon (#pall4655, VWR, Belgium)
- 0.45-μm filters, 142 mm diameter, Supor®, polyethersulfon (#pall60177, VWR, Belgium
- 10-ml syringes
- 10-ml vials for AAS sample storage

Reagents

- $CaCl_2 \cdot 7H_2O$, $MgCl_2$, $MgSO_4$, $NaHCO_3$, KCl, $ZnCl_2$ (pro analyse, Merck)
- pH 4 and pH 7 buffer solutions (Merck), for calibration of pH electrode

- HCl (>35%, w/v), HNO_3 (>69%, w/v) (pro analyse, Merck)
- Standard solution of $ZnCl_2$ in 1% (v/v) HNO_3, 1 g Zn/l (Sigma-Aldrich)
- Certified reference solutions for Zn (TM-25.2 and TMDA-62, Evironment Canada, NWRI)

Equipment model number, source

- Balance, precision 0.1 g (Mettler PJ3600 DeltaRange®)
- pH meter (Consort P407, Turnhout, Belgium)
- Flame Atomic Absorption Spectrophotometer (SpectrAA800, Varian, Mulgrave, Australia)
- Inductively coupled mass spectrometer (ICP, Spectro Analytical Instruments, Kleve, Germany)
- Ion chromatography apparatus (DIONEX2000i/SP, Dionex, Sunnyvale, CA, USA)
- TOC analyzer (TOC-5000, Shimadzu, Duisburg, Germany)
- Facilities for testing under controlled temperature and light cycle
- Automatic pipettes, volumes between 20 μl and 10 ml (Finnpipette® digital, Labsystems)

Software/databases

- Excel® (Microsoft)
- BLM windows version 1.0.0 (free download from www.hydroqual.com/winblm)
- Database of critically selected stability constants, according to Martell et al.[20] (NIST)

Procedure

Mathematical theory

This section gives the mathematical description of the BLM theory and the experimental design that results from it to derive the BLM parameters, as first proposed by De Schamphelaere and Janssen.[1] Although in theory it is applicable to any other toxic cationic metal, zinc is used as an example in all equations in the following. The model can be written for any other metal by replacing zinc by another metal cation in the equations.

The total number of zinc binding sites on the biotic ligand (BL) is called the complexation capacity of the BL (CC_{BL}), which is analogous to a total concentration of any other ligand in the test medium. In the BLM, they are treated as uniformly distributed in the exposure water, i.e., reacting with the entire water volume.[21] A mass balance equation on the BL can be written as:

$$CC_{BL} = [ZnBL] + [CaBL] + [MgBL] + [NaBL] + [HBL] + [BL] \qquad (22.1)$$

where CC_{BL} is the complexation capacity of the BL (mol/l); [ZnBL], [CaBL], [MgBL], [NaBL], and [HBL] are concentrations of cation–BL complexes (mol/l); and [BL] is the concentration of unoccupied BL sites (mol/l). In this reaction, one cation is assumed to react with one ligand site (monodentate binding), and for simplicity, charge balance is not taken into account.

Equilibrium equations for the binding of cations (e.g., Zn^{2+}) to the BL sites can be written as (conditional) stability constant expressions of the form:

$$K_{ZnBL} = \frac{[ZnBL]}{(Zn^{2+}) \times [BL]} \tag{22.2}$$

where K_{ZnBL} is the stability constant for Zn^{2+} binding to BL sites (l/mol) and (Zn^{2+}) is the chemical activity (round brackets) of the free zinc ion (mol/l). It is stressed at this point that this and subsequent equations make use of chemical activities of cations, and not of concentrations of cations. For simplicity, ionic strength corrections of the stability constants that describe cation binding to the BL are not carried out. Similar equations are valid for cations competing with zinc:

$$\begin{aligned} K_{CaBL} &= \frac{[CaBL]}{(Ca^{2+}) \times [BL]}; & K_{MgBL} &= \frac{[MgBL]}{(Mg^{2+}) \times [BL]}; \\ K_{NaBL} &= \frac{[NaBL]}{(Na^{+}) \times [BL]}; & K_{HBL} &= \frac{[HBL]}{(H^{+}) \times [BL]} \end{aligned} \tag{22.3}$$

It is stressed here that this list is not restricted to the cations mentioned here. Neither is it anticipated that all these cations would compete significantly with any toxic metal cation, including zinc. However, these cations are the ones most reported to have significant effects on metal toxicity.[1,9]

The concentration of zinc bound to the BL, which according to the BLM assumptions determines the magnitude of toxic effect, can be expressed as a function of (Zn^{2+}), (Ca^{2+}), (Mg^{2+}), (Na^{+}), and (H^{+}) by combining Equations (22.1)–(22.3):

$$[ZnBL] = \frac{K_{ZnBL} \cdot (Zn^{2+})}{1 + K_{ZnBL} \cdot (Zn^{2+}) + K_{CaBL} \cdot (Ca^{2+}) + K_{MgBL} \cdot (Mg^{2+}) + K_{NaBL} \cdot (Na^{+}) + K_{HBL} \cdot (H^{+})} \cdot CC_{BL} \tag{22.4}$$

Assuming, in accordance with the BLM theory, that the complexation capacity of the BL is independent of the water quality characteristics, the fraction of the total number of zinc binding sites occupied by zinc (f_{CuBL}) equals:

$$\begin{aligned} f_{ZnBL} &= \frac{[ZnBL^{+}]}{CC_{BL}} \\ &= \frac{K_{ZnBL} \cdot (Zn^{2+})}{1 + K_{ZnBL} \cdot (Zn^{2+}) + K_{CaBL} \cdot (Ca^{2+}) + K_{MgBL} \cdot (Mg^{2+}) + K_{NaBL} \cdot (Na^{+}) + K_{HBL} \cdot (H^{+})} \end{aligned} \tag{22.5}$$

This fraction can also be assumed, still in accordance with BLM theory, to determine the magnitude of toxic effect and, therefore, is constant at 50% effect ($f_{ZnBL}^{50\%}$); i.e., independent of the water quality characteristics, as demonstrated recently for copper and nickel.[18,21] As such, Equation (22.5) can be rearranged as:

$$\begin{aligned} EC50_{Zn2+} = &\frac{f_{ZnBL}^{50\%}}{(1 - f_{ZnBL}^{50\%}) \cdot K_{ZnBL}} \\ &\cdot \{1 + K_{CaBL} \cdot (Ca^{2+}) + K_{MgBL} \cdot (Mg^{2+}) + K_{NaBL} \cdot (Na^{+}) + K_{HBL} \cdot (H^{+})\} \end{aligned} \tag{22.6}$$

where $EC50_{Zn2+}$ is the free zinc ion activity resulting in 50% effect or the median effect concentration expressed as free zinc ion activity. The latter equation is not restricted to the EC50 but can be applied to any other point effect estimate (e.g., EC10). In theory, it is applicable to any endpoint (mortality, growth, reproduction) measured and for any exposure duration (short-term, long-term). In this study, the example of a 48-h acute immobilization assay with *D. magna* is presented.[9] Equation (22.6), however, has already been applied successfully to chronic exposures to copper and zinc.[11,22]

The absence of CC_{BL} in Equation (22.6) indicates that measured zinc concentrations on the BL are not essential for the development of a BLM. Consequently, we suggest that BLM parameters can be estimated from toxicity data alone. More importantly, Equation (22.6) clearly shows that, if the BLM concept applies, linear relationships should be observed between $EC50_{Zn2+}$ and the activity of one competing cation when the activity of other competing cations is constant. It thus allows for an explicit test of the BLM theory of cation competition.

How this finding dictates the experimental design to be followed can be understood as follows. If toxicity tests are performed at different concentrations of one competing cation and a constant concentration of all other cations, the linear relation between the activity of this one competing cation and the $EC50_{Zn2+}$ is characterized by a slope and an intercept. The example is given for Ca as competing cation, but similar equations can be written for other cations:

$$\text{intercept}_{\text{Ca}} = \frac{f_{\text{ZnBL}}^{50\%}}{(1 - f_{\text{ZnBL}}^{50\%} \cdot K_{\text{ZnBL}})} \cdot \left\{1 + K_{\text{MgBL}} \cdot \left(\text{Mg}^{2+}\right)_{\text{Ca}} + K_{\text{NaBL}} \cdot (\text{Na}^{+})_{\text{Ca}} + K_{\text{HBL}} \cdot (\text{H}^{+})_{\text{Ca}}\right\} \quad (22.7)$$

and

$$\text{slope}_{\text{Ca}} = \frac{f_{\text{ZnBL}}^{50\%}}{\left(1 - f_{\text{ZnBL}}^{50\%}\right) \cdot K_{\text{ZnBL}}} \cdot K_{\text{CaBL}} \quad (22.8)$$

where $(Mg^{2+})_{Ca}$, $(Na^{+})_{Ca}$, and $(H^{+})_{Ca}$ are the mean activities of other cations in the tests performed at different Ca concentrations. Dividing slope by intercept gives the following ratio (R_{Ca}):

$$\frac{\text{slope}_{\text{Ca}}}{\text{intercept}_{\text{Ca}}} = \frac{K_{\text{CaBL}}}{\left\{1 + K_{\text{MgBL}} \cdot \left(\text{Mg}^{2+}\right)_{\text{Ca}} + K_{\text{NaBL}} \cdot (\text{Na}^{+})_{\text{Ca}} + K_{\text{HBL}}(\text{H}^{+})_{\text{Ca}}\right\}} = R_{\text{Ca}} \quad (22.9)$$

Similar equations can be derived when the $EC50_{Zn2+}$ is determined for varying Mg, Na, and H concentrations. Rearranging these equations into a matrix form, results in:

$$\begin{pmatrix} 1 & -R_{\text{Ca}} \cdot \left(\text{Mg}^{2+}\right)_{\text{Ca}} & -R_{\text{Ca}} \cdot (\text{Na}^{+})_{\text{Ca}} & -R_{\text{Ca}} \cdot (\text{H}^{+})_{\text{Ca}} \\ -R_{\text{Mg}} \cdot \left(\text{Ca}^{2+}\right)_{\text{Mg}} & 1 & -R_{\text{Mg}} \cdot (\text{Na}^{+})_{\text{Mg}} & -R_{\text{Mg}} \cdot (\text{H}^{+})_{\text{Mg}} \\ -R_{\text{Na}} \cdot \left(\text{Ca}^{2+}\right)_{\text{Na}} & -R_{\text{Na}} \cdot \left(\text{Mg}^{2+}\right)_{\text{Na}} & 1 & -R_{\text{Na}} \cdot (\text{H}^{+})_{\text{Na}} \\ -R_{\text{H}} \cdot \left(\text{Ca}^{2+}\right)_{\text{H}} & -R_{\text{H}} \cdot \left(\text{Mg}^{2+}\right)_{\text{H}} & -R_{\text{H}} \cdot (\text{Na}^{+})_{\text{H}} & 1 \end{pmatrix} \cdot \begin{pmatrix} K_{\text{CaBL}} \\ K_{\text{MgBL}} \\ K_{\text{NaBL}} \\ K_{\text{HBL}} \end{pmatrix} = \begin{pmatrix} R_{\text{Ca}} \\ R_{\text{Mg}} \\ R_{\text{Na}} \\ R_{\text{H}} \end{pmatrix} \quad (22.10)$$

The solution of this matrix results in the estimation of the stability constants K_{CaBL}, K_{MgBL}, K_{NaBL}, and K_{HBL}. In practice, the following experimental design is suggested. Complete toxicity tests with Zn should be performed in a base medium, with low concentrations of all cations, and in several test media with increasing concentrations of one competing cation, while other cations are kept constant and as low as possible. The latter is necessary for obtaining reliable slope and intercept estimates. In this study, the individual effect of Ca^{2+}, Mg^{2+}, Na^{+}, and H^{+} activities on the acute toxicity of zinc to *D. magna* is reported.[9]

Steps in the development of a BLM

Figure 22.2 summarizes the different steps that need to be taken for the development and validation of a BLM predicting metal toxicity. The sequence of steps is described in brief in this paragraph, a more detailed explanation of each individual step is given in separate paragraphs in the following. Each step is assigned a number, referring to the different steps depicted in Figure 22.2 and to the separate paragraphs in the section "Toxicity testing, model development and validation."

1. *Experimental design*: Before starting the experiments one needs to decide on several ecotoxicological and bioavailability aspects, for example, which organisms will be used, are acute or chronic effects investigated, which endpoint needs to be modeled (e.g., mortality, growth, reproduction), which competing cations will be investigated (e.g., Ca, Mg, Na).
2. *Testing/measuring*: Once the above-mentioned aspects are addressed, ecotoxicity tests need to be performed to assess the effect of competing cations on the toxic effects of the metal on the chosen endpoint.
3. *Speciation calculation*: A next step is the calculation of speciation of the metal. Not all species are equally toxic, and in general, the free metal ion is considered most toxic. The free metal ion activities are the basis of most BLM and thus need to be determined before BLM parameters can be estimated. Next, the endpoint of interest (e.g., EC50 of mortality) is calculated as free metal ion activity for each test medium (each concentration of competing cation). Care should be taken in choosing the speciation model and the thermodynamic database of stability constants for metal complexes, and in correctly taking into account the complexation properties of dissolved organic carbon (DOC).
4. *Derivation of stability constants for competing cations*: The calculation of these constants is explained in detail in Reference 1 and is based on the solution of matrix equation (22.13).
5. *Estimation of K_{ZnBL} and $f_{ZnBL}^{50\%}$*: The calculation is based on the optimization of the logit-transformed effect versus f_{ZnBL} for varying K_{ZnBL}. Together with the stability constants for competing cations, these parameters can be introduced into the BLM model software (hydroqual). The development of the BLM is now complete.
6. *Validation testing with natural surface waters*: For regulatory application, it is of utmost importance that a BLM can not only predict metal toxicity in synthetic waters but also in natural surface waters. Water samples are taken from natural surface waters varying in important water chemistry characteristics (i.e., relevant for bioavailability and for BLM predictions). Those samples are then spiked and ecotoxicity tests are performed. The endpoint of interest (e.g., EC50) is again calculated.

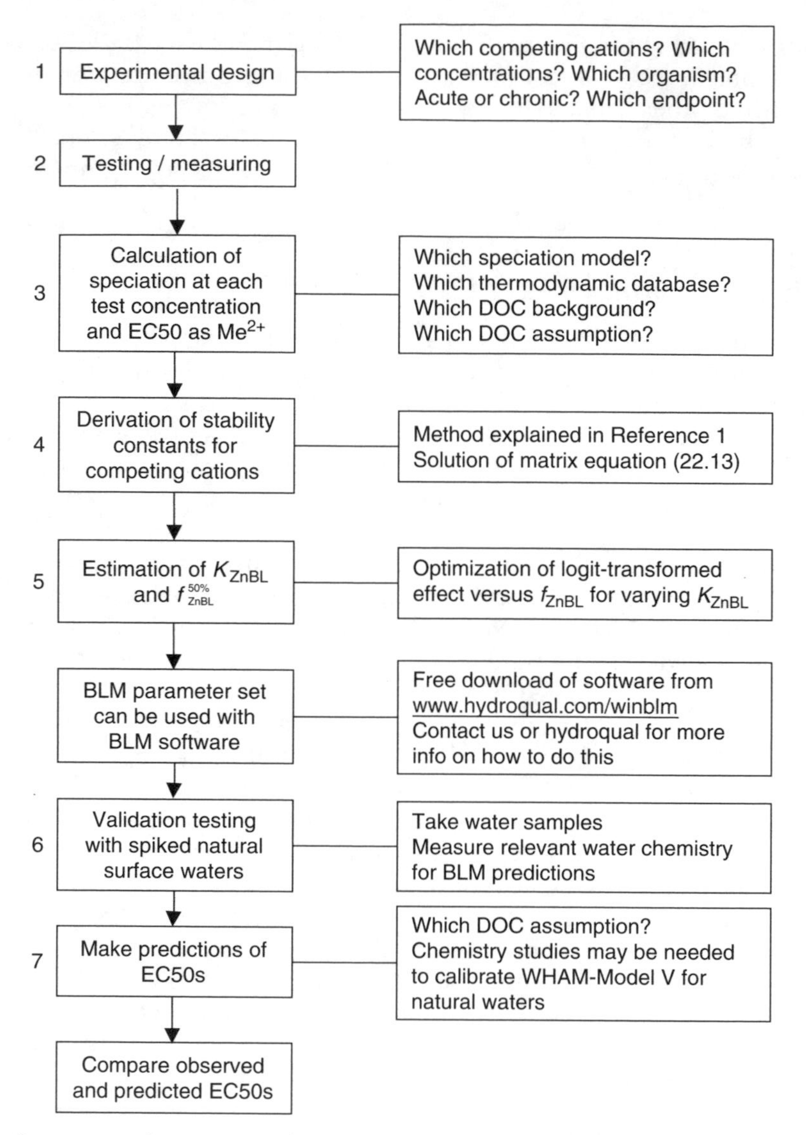

***Figure** 22.2* Summary of sequence of steps in the development of a BLM for predicting metal toxicity, along with questions/critical issues that need to be addressed with each step.

7. *Comparison observed and predicted EC50*: The developed BLM is used to make predictions of EC50 in the natural surface waters. Here it may first be necessary to calibrate the speciation model WHAM-Model V, which calculates complexation of metals to natural DOC, to independently generated chemical speciation data sets. When all this is done, observed and predicted EC50s are compared and the predictive potential of the BLM is evaluated.

The paragraphs below each describe in more detail the above explained procedure with acute zinc toxicity to *D. magna* as an example.

Toxicity testing, model development and validation

Step 1: experimental design

The experimental design is depicted in Figure 22.3. In each test series, the concentration of one cation was varied, while keeping all other cation concentrations as low and as constant as possible. Seven series of bioassays were performed: 2 Ca sets, 2 Mg sets,

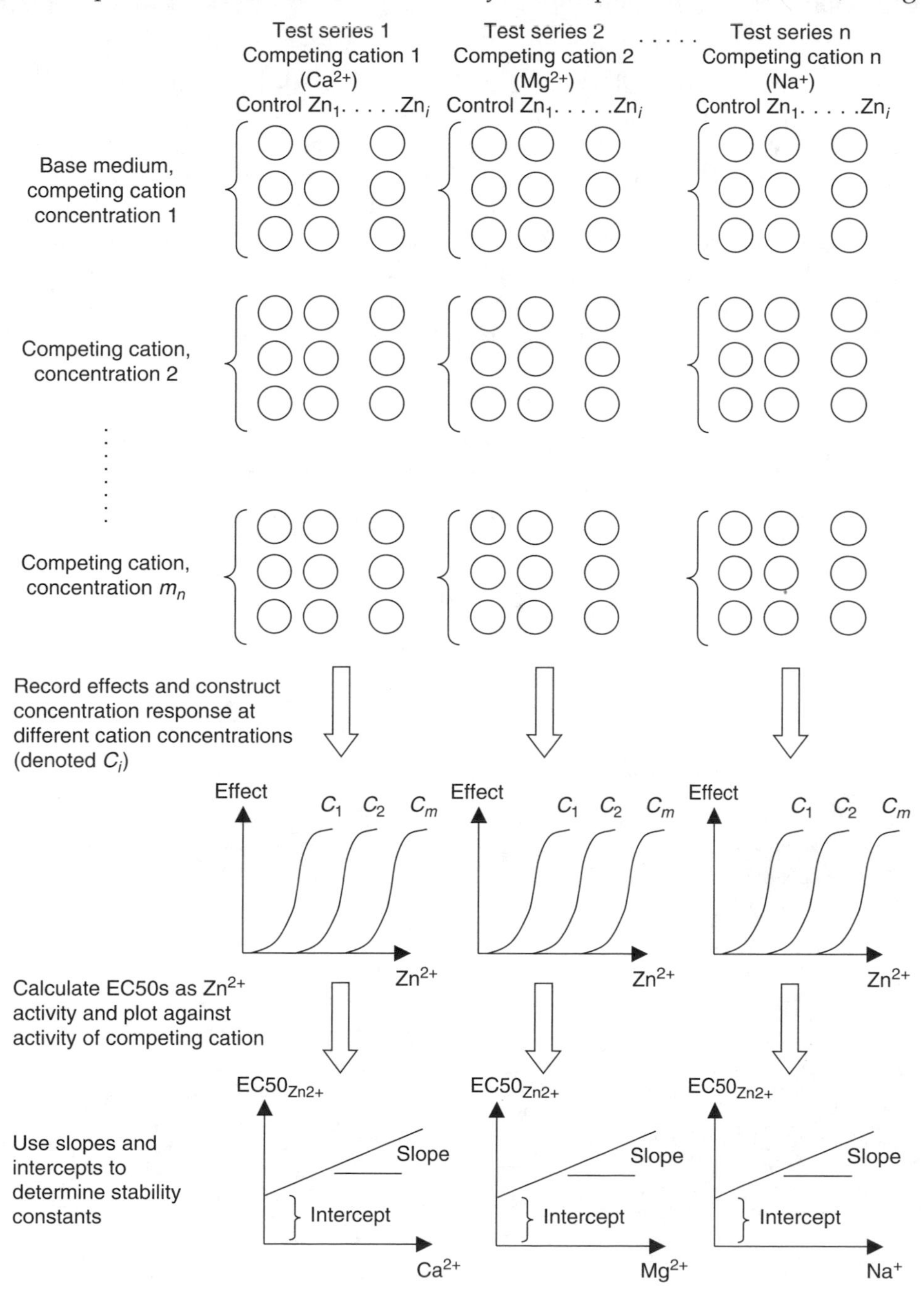

Figure 22.3 Experimental design for estimation of stability constants for competing cations.

1 Na set, 1 K set, and 1 pH set. All test series consisted of at least four bioassays (4 different cation concentrations). All bioassays of one test set were conducted simultaneously to minimize variability due to possible slight changes in *D. magna* culture conditions.

Step 2: toxicity testing/measuring

Preparation of test media

Preparation of test media is illustrated in Figure 22.4. All synthetic test media were prepared by adding appropriate volumes of stock solutions of $CaCl_2$, $MgCl_2$ or $MgSO_4$, NaCl, KCl, and $NaHCO_3$ to a base medium that contained 0.25 m*M* $CaCl_2$, 0.25 m*M* $MgCl_2$, and 0.078 m*M* KCl. Except for the pH sets, these media were adjusted to pH 6.8 by adding 0.078 m*M* $NaHCO_3$. The pH in the pH set was controlled by adding $NaHCO_3$ until the desired pH was reached. Sodium concentrations in all different pH tests were kept constant through addition of NaCl. All test media were prepared in volumetric flasks of 2 l using deionized water as dilution water. For each bioassay, the prepared test medium was then used as the dilution water to make a Zn-concentration series. The

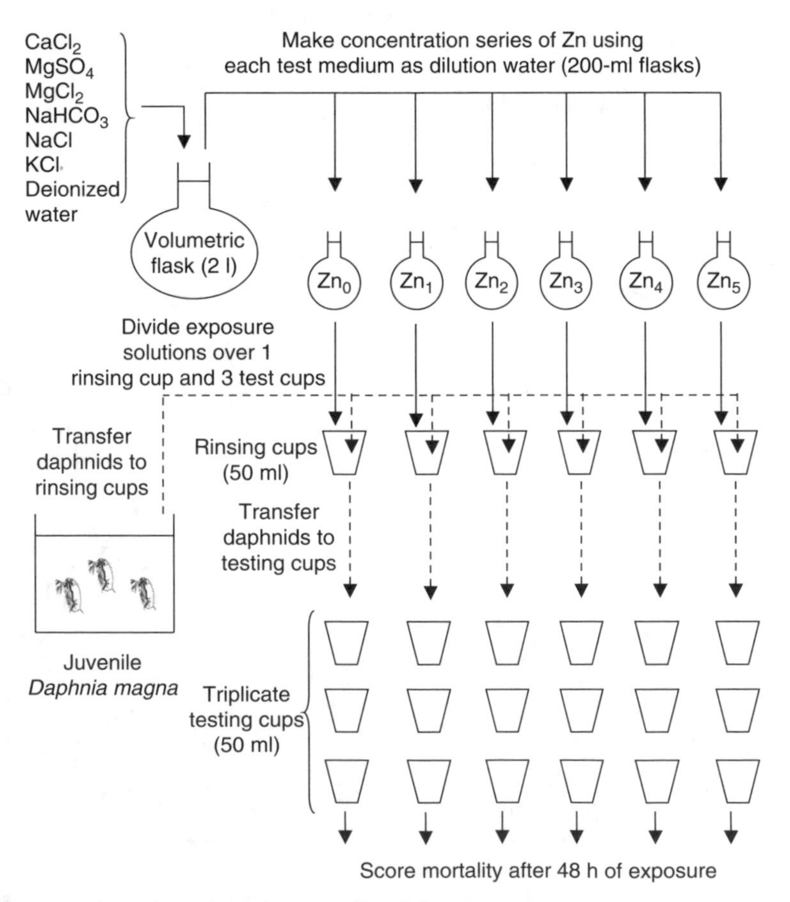

Figure 22.4 Preparation of synthetic test media and test setup.

difference between two consecutive (nominal) zinc concentrations was 0.25 log-units. For each zinc concentration, 200 ml of test solution were prepared in a volumetric flask and divided into 50-ml quantities over four polyethylene cups (three test cups and one rinsing cup). In order to obtain near-equilibrium situations, all media were stored in the test cups at 20°C for 1 day prior to being used in the toxicity tests.

Acute toxicity tests with D. magna

Acute 48-h immobilization assays with juvenile *D. magna* (<24-h old) were performed according to Test guideline 202 of the Organization for Economic Cooperation and Development (OECD, Paris, France).[23] The test organisms originated from a healthy *D. magna* clone (K6), which has been cultured under in M4-medium,[24] at a temperature of 20°C, under a 12-h light:12-h dark cycle, and which is fed a green algal mix of *Pseudokirchneriella subcapitata* and *Chlamydomonas reinhardtii* in a 3:1 ratio. Algae were obtained from the Culture Collection of Algae and Protozoa (CCAP, Windermere, United Kingdom). Juveniles from the first two broods were not be used for testing.

The main testing conditions are summarized in Table 22.1. For each test medium, an acute toxicity assay was conducted consisting of at least six treatments (1 control + 5 zinc concentrations) with a difference of 0.25 log-units between two consecutive zinc concentrations tested. Each treatment was performed with three replicates using 10 organisms per replicate. The number of immobilized juveniles in each cup was recorded after 48 h.

Approximately 40–50 organisms were first transferred to the rinsing cup with a glass pipette in order to dilute the culture medium, which is transferred along with the juvenile daphnids (usually <1 ml), with 50 ml of test medium (Figure 22.4). From this rinsing cup, ten juveniles were then transferred into each of the three replicates. This procedure was repeated for each test concentration.

Chemical measurements

Dissolved zinc levels (0.45 μm filtered) in each test concentration were determined at the beginning of the test using a flame-atomic absorption spectrophotometer (SpectrAA800 with Zeeman background correction, Varian, Mulgrave, Australia). Calibration standards (Sigma-Aldrich, Steinheim, Germany) and a reagent blank were analyzed with every ten samples. Two certified reference samples, TMDA-62 and TM-25.2 (National Water

Table 22.1 Test conditions of the acute toxicity test with *D. magna*

Temperature	20 ± 1°C in a temperature controlled cabinet
Light	12-h light:12-h dark; illumination not exceeding 800 lx
Number of concentrations	Control + 5 concentrations (logarithmically arranged, maximal 1 log-unit apart)
Number of replicates/concentration	3
Number of organisms/replicate	10
Test vessel	Polyethylene cups
Volume of test medium	50 ml
Feeding	None
Test medium	Variable, see Tables 22.4–22.7
Aeration	None

Research Institute, Burlington, ON, Canada) with certified Zn concentrations (mean ± 95% confidence interval) of 110 ± 15.5 μg/l and 24 ± 4.6 μg/l, respectively, were analyzed at the beginning and end of each series of Zn measurements. Measured values were always within 10% of the certified value.

Since earlier measurements had shown that, for synthetic waters, concentrations of other cations (Ca, Mg, Na, K) and anions [Cl, SO_4, dissolved inorganic carbon (DIC)] are always within 10% of the nominal concentrations, nominal concentrations are reported here and are used as input for the speciation calculations.

The pH (pH meter P407, Consort, Turnhout, Belgium) of each test medium was checked before and after the test. pH of the natural water samples was adjusted to the pH measured in the field, just before the start of each test. The pH glass electrode was calibrated daily using pH 4 and pH 7 buffers (Merck, Darmstadt, Germany).

Step 3: calculation of speciation and EC50s

Speciation calculations

For each test medium and for all zinc concentrations tested speciation was calculated using WHAM-V.[19] Before starting speciation calculations of metals in test media, some important issues must be addressed: (1) Which speciation model should be used? (2) Which stability constants for inorganic and/or organic complexes should be chosen? (3) What is the effect of background concentration of DOC in test water?

In principle, the choice of speciation model itself is less critical than the choice of stability constants for complexation reactions (especially those with the toxic metal cation). Using thermodynamic principles, most often-used speciation programs (e.g., MINTEQA2,[25] MINEQL+,[26] WHAM-V[19]) will compute very similar distributions of species provided that the same stability constants are used. Stability constants of metal complexing to chemically defined ligands (e.g., OH^-, CO_3^{2-}, SO_4^{2-}) are available in a number of databases. A particularly good source is the NIST Critical Stability Constants of Metal Complexes Database.[20] The use of standardized constants should definitely be encouraged. In the context of the BLM, WHAM has become an integral part of the BLM software, as it can not only compute inorganic metal speciation but also takes complexation to natural organic matter into account (i.e., humic and fulvic acids).

The advantage of WHAM, however, is that it has an in-built module for complexation to natural organic matter, such as humic and fulvic acids. This model's description of metal complexation with NOM is, compared to interactions with inorganic ligands, much more complicated. Indeed, NOM is an irresolvable mixture of a very large number of compounds varying in their properties, including their ability to bind metal ions. The WHAM-Model V[19,27] is a multiple site model that is able to relatively accurately predict metal and proton binding to NOM. More recently, WHAM-6, an update of WHAM-Model V has become available from the Centre for Ecology and Hydrology (Windermere, United Kingdom).[28] This version has not been included in the BLM software yet.

The advantage of using WHAM is that it can predict speciation both in absence and in the presence of organic matter. Although most often in test media based on deionized water no organic matter is added deliberately, a considerable amount of background DOC may be present. De Schamphelaere and Janssen[1] have measured a background DOC level of ~280 μg/l, and they assumed that 50% reacted as fulvic acid and the other 50% was inert for chemical reactions. Since organic matter usually has a large affinity for metals, the significance of complexing of metals to background DOC should always be considered in speciation calculations. Another important consideration may be the amount of DOC that the organisms themselves introduce release into the test vessels. Although Fish and Morel[29] have demonstrated that daphnids can release important amounts of DOC, De Schamphelaere and Janssen[1] demonstrated that no significant amounts of DOC were introduced in 48-h tests with 10 juveniles of *Daphnia* in 50 ml of test water, the setup that was also used in the present study.

The two issues above are illustrated with two examples for Zn speciation: effect of different inorganic stability constants on Zn speciation and effect of background DOC on zinc speciation. The results of these calculations are given in Table 22.3. Calculations were performed at 20°C, ionic strength of 10 m*M* (as NaCl), DIC = 1 m*M*, Zn = 10 and 100 μ*M*, and at pH levels ranging from 5.5 to 8.5. Calculations were performed for DOC levels of 0 μg/l, 140 μg/l, 500 μg/l, and 1000 mg/l (assumed 100% fulvic acid, performed with BLM windows version 1.0.0.).

The stability constants for inorganic zinc complexes are somewhat different in WHAM (see Reference 19) and in the NIST database (Table 22.2), especially the ones for the $ZnHCO_3$ and the $Zn(OH)_2$ complexes. From Table 22.3, it can be deduced that with the original WHAM constants the calculated Zn^{2+} activity is somewhat lower than with the NIST constants, and that this difference increases with increasing pH levels (up to factor 1.5 at pH 8.5). This clearly illustrates the importance of carefully considering the choice of stability constants whenever speciation calculations are to be carried out. The NIST constants are preferred because they are standardized and they were also used here.

Table 22.3 also demonstrates that for zinc the potential presence of background DOC is not very important at the zinc concentrations typical for toxic effects in the present study (10–100 μ*M*, 654–6540 μg/l). For DOC levels up to 1 mg/l, differences with 0 mg DOC/l assumed are less than factor 1.1 in all cases. Hence, for this study, we pursued with calculations assuming 0 mg DOC/l. However, for studies with other metals and/or with lower concentrations tested, such preliminary calculations should always be carried out before doing calculations for the tests performed.

Calculation of EC50s

Based on recorded immobility percentages at each tested zinc concentration, 48-h EC50s were calculated using the Trimmed Spearman–Karber method.[30] 48-h EC50s were also calculated based on calculated Zn^{2+} activities at each tested zinc concentration. Zn^{2+} activities (and activities of other cations in solution) were calculated using BLM software,

Table 22.2 Stability constants for inorganic zinc complexes according to the thermodynamic databases in NIST[19] and WHAM-Model V[20]

Species	Log *K* (NIST)	Log *K* (WHAM-V)
$ZnOH^+$	5.00	5.04
$Zn(OH)_2$	10.2	11.1
$ZnHCO_3^+$	11.83	13.12
$ZnCO_3$	4.76	4.76
$ZnSO_4$	2.34	2.38
$ZnCl^+$	0.4	0.4

Table 22.3 Effect of choice of thermodynamic database, NIST[20] or WHAM-Model V[19] (see Table 22.2) and of background DOC on Zn speciation at 20°C, ionic strength = 0.01 *M* (as NaCl) and total inorganic carbon = 1 m*M*. Results expressed as μ*M* Zn^{2+} activity. DOC input assumed 100% fulvic acid

pH	Effect of thermodynamic database			
	10 μ*M* total Zn, NIST	100 μ*M* total Zn, NIST	10 μ*M* total Zn, WHAM-V	100 μ*M* total Zn, WHAM-V
5.5	6.44	63.6	6.17	61.0
6	6.36	62.8	5.75	56.8
6.5	6.24	61.7	5.20	51.6
7	6.08	60.4	4.76	47.6
7.5	5.79	57.8	4.43	44.6
8	5.12	51.4	3.92	39.6
8.5	3.73	37.7	2.65	26.9
Effect of background DOC at 1 μM total Zn, calculated with NIST constants				
pH	0 mg DOC/l	0.14 mg DOC/l	0.5 mg DOC/l	1 mg DOC/l
5.5	6.44	6.36	6.16	5.89
6	6.36	6.26	6.01	5.68
6.5	6.24	6.13	5.83	5.44
7	6.08	5.95	5.62	5.19
7.5	5.79	5.65	5.30	4.83
8	5.12	4.99	4.65	4.20
8.5	3.73	3.63	3.37	3.04
Effect of background DOC at 10 μM total Zn, calculated with NIST constants				
pH	0 mg DOC/l	0.14 mg DOC/l	0.5 mg DOC/l	1 mg DOC/l
5.5	63.6	63.3	62.4	61.3
6	62.8	62.4	61.6	60.3
6.5	61.7	61.4	60.5	59.2
7	60.4	60.1	59.1	57.8
7.5	57.8	57.5	56.5	55.2
8	51.4	51.1	50.2	49.0
8.5	37.7	37.5	36.8	35.9

which has WHAM-V incorporated as a geochemical speciation module. In our case, speciation calculations were run assuming no background DOC (see the following) and using standard stability constants for inorganic complexes[20] (Table 22.2).

Effect of Ca, Mg, Na, and pH on zinc toxicity

In Tables 22.4–22.7, next to physico-chemistry of all test media, also the 48-h EC50s are given. An increase of Ca from 0.125 to 5 m*M* resulted in a 10-fold increase of 48-h EC50, i.e., from 4.88 to 45.7 μ*M* Zn (319–2990 μg/l). Mg and Na had a less protective effect than Ca. Mg reduced toxicity only 4-fold between concentrations of 0.21 and 3 m*M*, i.e., 48-h EC50s from 4.53 to 20.4 μ*M* Zn (296–1330 μg/l). Na reduced toxicity only 5-fold between concentrations of 0.077 and 15 m*M*, i.e., 48-h EC50s from 7.31 to 39.8 μ*M* Zn (478–2600 μg/l).

Table 22.4 Water chemistry[a] and speciation in the two Ca test series (separated by dashed lines) and 48-h median effective concentrations of Zn for *D. magna* as dissolved Zn and as Zn^{2+} activity

Ca (m*M*)	Cl^- (m*M*)	$EC50_{dissolved}$ (μ*M*)	Ca^{2+} (*M*)[b]	Mg^{2+} (*M*)[b]	Na^+ (*M*)[b]	$EC50_{Zn2+}$ (*M*)
0.25	0.577	8.69	2.01E – 04	2.01E – 04	7.34E – 05	6.51E – 06
0.5	1.077	11.8	3.92E – 04	1.96E – 04	7.28E – 05	8.72E – 06
1	2.077	21.7	7.50E – 04	1.87E – 04	7.18E – 05	1.44E – 05
3	6.077	33.4	2.03E – 03	1.66E – 04	6.93E – 05	1.98E – 05
4	8.077	39.6	2.56E – 03	1.59E – 04	6.83E – 05	2.12E – 05
0.125	0.327	4.88	1.02E – 04	2.04E – 04	7.37E – 05	3.77E – 06
0.25	0.577	5.79	2.01E – 04	2.01E – 04	7.34E – 05	4.49E – 06
0.5	1.077	9.56	3.92E – 04	1.95E – 04	7.27E – 05	7.23E – 06
1.25	2.577	13	9.19E – 04	1.84E – 04	7.15E – 05	9.65E – 06
2	4.077	22.6	1.32E – 03	1.76E – 04	7.05E – 05	1.49E – 05
2.5	5.077	26.6	1.71E – 03	1.68E – 04	6.95E – 05	1.82E – 05
3.75	7.577	43.4	2.42E – 03	1.61E – 04	6.85E – 05	2.69E – 05
5	10.077	45.7	3.08E – 03	1.54E – 04	6.75E – 05	2.84E – 05

[a] Other chemistry relevant for BLM: pH = 6.8; DOC = 0 mg/l; Mg = 0.25 m*M*; Na = 0.077 m*M*; *K* = 0.077 m*M*; SO_4^{2-} = 0.25 m*M*; DIC = 0.077 m*M*.

[b] Chemical activity of competing cations.

Table 22.5 Water chemistry[a] and speciation in the two Mg test series (separated by dashed lines) and 48-h median effective concentrations of Zn for *D. magna* as dissolved Zn and as Zn^{2+} activity

Mg (m*M*)	Cl^- (mM)	$EC50_{dissolved}$ (μ*M*)	Ca^{2+} (*M*)[b]	Mg^{2+} (*M*)[b]	Na^+ (M)[b]	$EC50_{Zn2+}$ (*M*)
0.25	0.577	8.48	2.02E – 04	2.01E – 04	7.34E – 05	6.31E – 06
0.5	1.077	9.81	1.96E – 04	3.91E – 04	7.28E – 05	7.70E – 06
1	2.077	12.3	1.87E – 04	7.49E – 04	7.18E – 05	8.54E – 06
2	4.077	16.2	1.75E – 04	1.40E – 03	7.04E – 05	1.05E – 05
3	6.077	20.4	1.66E – 04	2.00E – 03	6.92E – 05	1.31E – 05
0.21	0.577	4.53	2.04E – 04	1.71E – 04	7.35E – 05	3.56E – 06
0.41	0.977	4.65	1.98E – 04	3.25E – 04	7.30E – 05	5.68E – 06
0.82	1.797	8.61	1.90E – 04	6.26E – 04	7.21E – 05	6.19E – 06
1.23	2.617	13.3	1.84E – 04	9.07E – 04	7.14E – 05	9.63E – 06
1.65	3.457	15.1	1.79E – 04	1.18E – 03	7.08E – 05	1.07E – 05
2.47	5.097	15.6	1.71E – 04	1.69E – 03	6.98E – 05	1.16E – 05

[a] Other chemistry relevant for BLM: pH = 6.8; DOC = 0 mg/l; Ca = 0.25 m*M*; Na = 0.077 m*M*; *K* = 0.077 m*M*; SO_4^{2-} = 0.25 m*M*; DIC = 0.077 m*M*.

[b] Chemical activity of competing cations.

When pH was increased from 6 to 8, this resulted in a less than 3-fold reduction of toxicity, i.e., 48-h EC50s increased from 8.4 to 22.7 μ*M* Zn (549–1770 μg/l). Although the effect of water hardness (the sum of Ca and Mg effects) on zinc toxicity to freshwater organisms has been investigated by numerous authors (see Reference 31, for a review), the individual effects of Ca, Mg, and Na have not been investigated in detail. Alsop and Wood[14] found that whole body uptake of zinc by the rainbow trout *Oncorhynchus mykiss* was reduced by Ca, Mg, and Na, but that only Ca had a protective effect on acute zinc toxicity. The cation concentrations tested by these authors, was however, much lower (i.e., up to only 1 m*M*) than the levels that were tested in the present study. Perhaps, effects on uptake are detectable at lower cation concentrations than effect on toxicity. This indicates that, if one

Table 22.6 Water chemistry[a] and speciation in the Na test series and 48-h median effective concentrations of Zn for *D. magna* as dissolved Zn and as Zn^{2+} activity

Na (m*M*)	Cl^- (m*M*)	$EC50_{dissolved}$ (μ*M*)	Ca^{2+} (*M*)[b]	Mg^{2+} (*M*)[b]	Na^+ (*M*)[b]	$EC50_{Zn2+}$ (*M*)
0.077	0.577	7.31	2.01E – 04	2.01E – 04	7.34E – 05	5.80E – 06
1	1.5	10.2	1.94E – 04	1.94E – 04	9.44E – 04	7.52E – 06
3	3.5	12.6	1.83E – 04	1.83E – 04	2.78E – 03	8.69E – 06
6	6.5	16.2	1.72E – 04	1.72E – 04	5.46E – 03	1.13E – 05
9	9.5	23.7	1.64E – 04	1.64E – 04	8.07E – 03	1.52E – 05
12	12.5	25.6	1.59E – 04	1.59E – 04	1.07E – 02	1.56E – 05
15	15.5	39.8	1.52E – 04	1.52E – 04	1.31E – 02	1.79E – 05

[a] Other chemistry relevant for BLM: pH = 6.8; DOC = 0 mg/l; Ca = 0.25 m*M*; Mg = 0.25 m*M*; *K* = 0.077 m*M*; SO_4^{2-} = 0.25 m*M*; DIC = 0.077 m*M*.

[b] Chemical activity of competing cations.

Table 22.7 Water chemistry[a] and speciation in the pH test series and 48-h median effective concentrations of Zn for *D. magna* as dissolved Zn and as Zn^{2+} activity

pH	Cl^- (m*M*)	DIC (m*M*)	$EC50_{dissolved}$ (μ*M*)	$EC50_{Zn2+}$ (M)
6	1.737	0.04	8.4	6.40E – 06
6.5	1.687	0.09	11	8.49E – 06
7.5	1.277	0.5	11.4	7.01E – 06
8	0.577	1.2	22.7	9.23E – 06

[a] Other chemistry relevant for BLM: DOC = 0 mg/l; Ca = 0.25 m*M*; Mg = 0.25 m*M*; Na = 1.2 m*M*; *K* = 0.077 m*M*; SO_4^{2-} = 0.25 m*M*.

wants to use toxicity data to quantify protective effects of cations, high enough cation concentrations should be tested.

The strong effect of Ca^{2+}, compared to Mg^{2+} and Na^+, can be understood better when reported interactions between calcium and zinc are considered: studies with *O. mykiss* have shown that elevated calcium concentrations inhibited zinc influx competitively and consequently, it has been postulated that calcium and zinc share a common uptake pathway across the apical membrane of the chloride cells of the gills[15,16] The specific acute toxic mode of action of zinc is the disturbance of calcium uptake by the gills primarily through competitive inhibition, and hence it seems logic that Ca provides more competitive protection against zinc toxicity than Mg and Na.

Whereas the effects of Ca, Mg, and Na can be attributed to competition effects (see the following), the effect of pH is most likely a pure speciation effect. Indeed, at higher pH, more Zn^{2+} is complexed to carbonate and hydroxide complexes, and thus less Zn is left as the free Zn^{2+} ion. This is illustrated by the fact that, when EC50s are expressed as Zn^{2+} activity, no significant relation with pH is observed.

If H^+ would be an important competitor of Zn^{2+} for binding on the BL, an increase of the 48-h EC50 (as free Zn^{2+} activity) with increasing H^+ (decreasing pH) is expected. The absence of this trend indicates that the possible effect of H^+ competition on zinc toxicity to *D. magna* is negligible in the pH range tested and should therefore not be incorporated in the BLM (for this organism and metal). The fact that for other species, a quantitative H^+ competition has been demonstrated,[11,31] illustrates that species-specific differences in competitive interactions may exist. The here-described experimental design easily allows testing competition effects of cations on metal toxicity for various organisms. The H^+ effect will not be considered further in this chapter.

Step 4: estimation of stability constants for competing cations

The relation between activities of Ca^{2+}, Mg^{2+}, and Na^+ and 48-h $EC50_{Zn2+}$ is presented in Figure 22.5. Significant linear relations were observed for Ca ($R^2 = 0.94$, $p < 0.05$), Mg ($R^2 = 0.89$, $p < 0.05$), and Na ($R^2 = 0.98$, $p < 0.05$), and this corroborates with the BLM concept of cation competition according to Equations (22.6)–(22.9). However, since the H^+ competition effect was not significant, the Equations (22.5), (22.6), and (22.10) can be simplified to:

$$f_{ZnBL} = \frac{K_{ZnBL} \cdot (Zn^{2+})}{1 + K_{ZnBL} \cdot (Zn^{2+}) + K_{CaBL} \cdot (Ca^{2+}) + K_{MgBL} \cdot (Mg^{2+}) + K_{NaBL} \cdot (Na^+)} \quad (22.11)$$

$$EC50_{Zn^{2+}} = \frac{f^{50\%}_{ZnBL}}{(1 - f^{50\%}_{ZnBL}) \cdot K_{ZnBL}} \cdot \quad 1 + K_{CaBL} \cdot (Ca^{2+}) + K_{MgBL} \cdot (Mg^{2+}) + K_{NaBL} \cdot (Na^+) \quad 22.12)$$

$$\begin{pmatrix} 1 & -R_{Ca} \cdot (Mg^{2+})_{Ca} & -R_{Ca} \cdot (Na^+)_{Ca} \\ -R_{Mg} \cdot (Ca^{2+})_{Mg} & 1 & -R_{Mg} \cdot (Na^+)_{Mg} \\ -R_{Na} \cdot (Ca^{2+})_{Na} & -R_{Na} \cdot (Mg^{2+})_{Na} & 1 \end{pmatrix} \cdot \begin{pmatrix} K_{CaBL} \\ K_{MgBL} \\ K_{NaBL} \end{pmatrix} = \begin{pmatrix} R_{Ca} \\ R_{Mg} \\ R_{Na} \end{pmatrix} \quad (22.13)$$

It is noted that the equations can easily be adapted for any other case by omitting or adding competing cations to the equations. First, the data presented in Tables 22.4–22.7 were used to solve matrix equation (22.13) (Table 22.8). The matrix equation can be solved on a hand calculator or with the functions "PRODUCTMAT" and "INVERSEMAT" available in Excel (Microsoft). Estimated stability constants are also summarized in Table 22.9, where they are compared to stability constants found for acute zinc toxicity to fish species.[31] The acute BLM for fish did not contain constants for Mg and Na competition, but this is because there were simply no data available for the estimation of these constants. The authors did not explicitly exclude the possibility of Mg or Na competition. The log K_{CaBL} was ~1.5 log-units lower for *D. magna* than for fish and no H^+ competition constant was included for *D. magna*, as opposed to fish. This indicates that species-specific differences may exist in stability constants of competing cations.

Step 5: estimation of Zn binding parameters for the biotic ligand

For the final development of the Zn-BLM for *D. magna*, the K_{ZnBL} and $f_{ZnBL}^{50\%}$ of Equation (22.12) can be optimized as follows, according to the method first proposed by De Schamphelaere and Janssen.[1] It must be noted that the approach presented here is

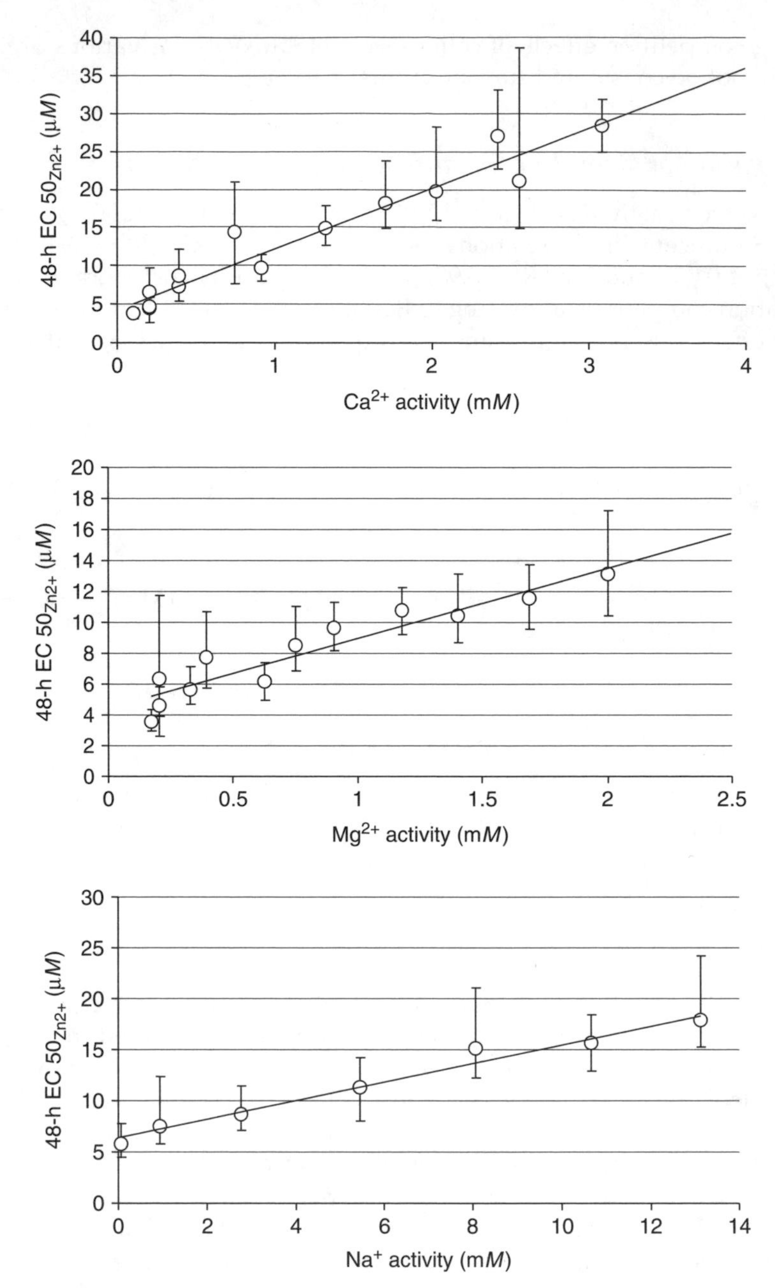

Figure 22.5 Effects of Ca^{2+}, Mg^{2+}, and Na^{+} on the 48-h EC50 (as Zn^{2+} activity) for *D. magna*. Error bars are 95% confidence limits.

an optimization of two parameters to a toxicity data set, and that perhaps other parameters might be found when real binding data of zinc to the BLM were available. Although the authors are aware that this separation of K_{ZnBL} and ${f_{ZnBL}}^{50\%}$ is predominantly a mathematical distinction, they would like to point out that this initial approximation does not influence toxicity predictions. Indeed, it is the ratio

Table 22.8 Estimation of stability constants (K, in M^{-1}) for the competing cations for the acute Zn-BLM for *D. magna* (48-h immobility) derived using data of Tables 22.4–22.7 and Equation (22.13) (NA, not applicable)

$Cation_X$	Slope	Intercept (M)	R_X (M^{-1})	$(Ca^{2+})_X$ (M)	$(Mg^{2+})_X$ (M)	$(Na^+)_X$ (M)	K_{XBL} (M^{-1})	Log K_{XBL}
Ca^{2+}	7.81E – 03	4.49E – 06	1740	NA	1.81E – 04	7.10E – 05	2190	3.34
Mg^{2+}	4.83E – 03	4.66E – 06	940	1.87E – 04	NA	7.17E – 05	1320	3.12
Na^+	9.15E – 04	6.34E – 06	144	1.75e – 04	1.75E – 04	NA	234	2.37

$$\frac{f_{ZnBL}^{50\%}}{(1 - f_{ZnBL}^{50\%}) \cdot K_{ZnBL}}$$

in Equation (22.12) that determines the 48-h $EC50_{Zn2+}$ when competing cation activities are known. As stated by De Schamphelaere and Janssen,[1] this ratio can be regarded as the $EC50_{Zn2+}$ when no cation competition is observed in the test medium and is equal to 3.51 μM ($\pm$ 0.58 μM; 95% CL) ($n = 31$). Nevertheless, below a possible approach is presented to separate K_{ZnBL} and $f_{ZnBL}{}^{50\%}$.

For each treatment (all test media × 5 zinc concentrations per test medium) activities of Ca^{2+}, Mg^{2+}, Na^+, and Zn^{2+} had already been calculated. When a log K_{ZnBL} is now chosen, the f_{ZnBL} can easily be calculated for each treatment. Now, since one of the main assumptions of the BLM is that this f_{ZnBL} is directly related to the toxic effect, a good relation should be observed between f_{ZnBL} and the effect. In practice, the log K_{ZnBL} is varied and for each value, a linear regression is performed on logit (48-h immobility) versus the calculated f_{ZnBL} and the R^2 is determined. The point where the regression line crosses the X-axes is the $f_{ZnBL}{}^{50\%}$. The combination of log K_{ZnBL} and $f_{ZnBL}{}^{50\%}$ that results in the highest R^2 is retained as the final combination to be used in the BLM (see Figure 22.6 for illustration of this method). A log $K_{ZnBL} = 5.31$ with an associated $f_{ZnBL}{}^{50\%} = 0.417$ resulted in the best fit ($R^2 = 0.806$) (Figure 22.5). This means that, when 41.7% of the BL is

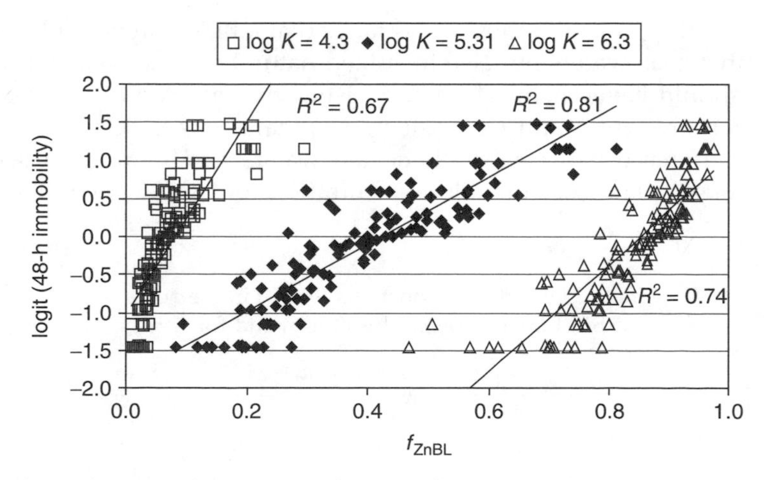

Figure 22.6 Relation between the logit-transformed effect and the fraction of BL sites occupied with Zn for different values of log K_{ZnBL}. For log K_{ZnBL} = 5.31, an optimal R^2 value was obtained. The $f_{ZnBL}{}^{50\%}$ is 0.417 (where logit = 0).

occupied by zinc, 50% effect (immobility) is expected to occur after 48 h of exposure. A log K_{ZnBL} of 5.31 is comparable with stability constants (K_{Zn}) for zinc binding to gills of juvenile rainbow trout.[32,33] Alsop and Wood[32] found a log K_{ZnBL} (as Zn^{2+}) ranging between 5.3 and 5.6. Galvez et al.[33] reported a log K_{ZnBL} of 5.1.

Finally, all BLM parameters for acute zinc toxicity to *D. magna* are reported in Table 22.9. Those values can easily be entered into the BLM modeling software (Hydroqual, Mahwah, NJ, USA), and at this point the model development is complete and the BLM can be used for making EC50 predictions in natural waters.

Step 6: validation testing with spiked natural surface waters

Natural water samples were collected in March 2003 in The Netherlands, Belgium, and France. pH was measured in the field using a portable pH meter. Samples of 5 l arrived in the laboratory the same day of collection and were filtered through a 0.45-μm filter (Gellman, Ann Arbor, MI, USA). Subsamples were taken for analysis of DOC, Ca, Mg, Na, K, SO_4, Cl, and DIC. These parameters were measured as they are required as input for the BLM. For toxicity testing, pH of the natural water samples was adjusted to the pH measured in the field. Samples were subsequently spiked with 5 different zinc concentrations and equilibrated for 2 days at 20°C. This equilibration is of utmost importance, since kinetics of metal complexation to dissolved organic matter may be slow in some cases.[34] Toxicity tests and zinc analysis procedures were similar as those described earlier.

For natural water samples, cations were measured using inductively coupled mass spectrometry (ICP, Spectro Analytical Instruments, Kleve, Germany) and chloride and sulphate concentrations using ion chromatography (DIONEX2000i/SP, Dionex, Sunnyvale, CA, USA). DOC (0.45 μm filtered) and inorganic carbon (IC) were measured using a TOC analyzer based on analytical combustion (TOC-5000, Shimadzu, Duisburg, Germany).

Step 7: comparison of observed and predicted EC50s

In a regulatory context, it is of utmost importance that a BLM can not only predict metal toxicity in synthetic lab waters but also in spiked natural surface waters. Preferably, the tested waters should have chemical characteristics that are representative of the area (stream catchment, region, country, continent, etc.) that needs to be protected.

The main difference between synthetic lab waters and natural surface waters is that the latter may contain considerable concentrations of DOC, which may bind metals

Table 22.9 Stability constants (log *K*) in the acute Zn-BLM for *D. magna* (this study) and for fish[32]

	Daphnia magna	Fish
Ca^{2+}	3.34	4.8
Mg^{2+}	3.12	—
Na^{+}	2.37	—
H^{+}	—	6.7
Zn^{2+}	5.31	5.5

like zinc and reduce their bioavailability. The effect of DOC is that it reduces the free metal ion activity, and if one can accurately estimate the extent to which this occurs, one can also predict the effect on toxicity. Although WHAM-V, the speciation module of the BLM-software, has demonstrated to be a robust tool for predicting metal binding to organic matter, it has been observed that the model may overestimate metal binding to natural organic matter in some cases.[35,36] This can, for example, be solved by assuming that only a fraction of the natural organic matter behaves as active fulvic acid. For copper, accurate toxicity predictions were made using the assumption of 50% active fulvic acid.[35,36] Within the BLM software this can be achieved by using the measured DOC concentration divided by 2 as the DOC input value and by setting the percentage of humic acid content to 0%. However, the default parameters for Zn binding to organic matter in WHAM-V are based on a very limited data set,[27] and until recently, no validation/refinement of these parameters was carried out based on Zn titration data sets with natural water samples. In such cases, before the predictive capacity of the BLM is assessed, a calibration of WHAM-V to organic-metal complexing data sets is mostly preferable. Recently, we have been able to demonstrate that WHAM-V yields the best Zn speciation estimates for natural waters if 61% active fulvic acid is assumed and 39% of the DOC is assumed to be inert.[37] The latter estimate was used for the predictions of 48-h EC50s in all natural waters tested.

Six water samples were taken in three different countries and both running and standing waters were sampled. Table 22.10 gives the water chemistry and the 48-h EC50s of Zn in the six tested waters. 48-h EC50s were between 315 and 3290 μg Zn/l, indicating a 10-fold difference. In Figure 22.6, the predicted 48-h EC50s were plotted against the observed 48-h EC50s. All EC50s were predicted with an error less than factor 2, which should be considered very good, given normal biological variability in replicated toxicity tests.

Table 22.10 Water chemistry and observed 48-h EC50s of zinc for *D. magna* in natural surface waters

Water, type,[a] location, country code	pH	DOC[b] (mg/l)	Ca (m*M*)	Mg (m*M*)	Na (m*M*)	*K* (m*M*)	Cl (m*M*)	SO_4 (m*M*)	DIC[c] (m*M*)	$EC50_{dissolved}$ (μ*M*)	$EC50_{dissolved}$ (μg/l)
Ankeveensche plas, S, Nederhorst-den-Berg, NL	6.77	17.3	0.956	0.269	0.757	0.156	1.322	1.410	0.340	33.3	2180
Ruisseau de St. Martin, R, Bihain, B	6.03	5.37	0.092	0.044	0.272	0.022	0.042	0.423	0.427	4.82	315
Ourthe Orientale, R, Brisy, B	7.34	2.53	0.125	0.139	0.384	0.053	0.010	0.649	0.299	5.41	354
Markermeer, S, Marken, NL	8.03	7.49	1.316	0.577	3.799	0.223	1.135	8.970	2.578	50.3	3290
Rhine, R, Lobith, NL	8.22	2.30	1.522	0.441	2.409	0.129	0.594	6.064	3.195	39.5	2580
Le Voyon, R, Trélon, FR	8.10	4.17	0.925	0.293	0.437	0.034	0.208	0.592	2.527	25.5	1670

[a] R = running water; S = standing water.

[b] Dissolved organic carbon; for input in BLM software, multiply with 0.61 and set humic acid percentage to 0% (set to 0.01% as software does not work with 0%).

[c] Dissolved inorganic carbon; BLM software needs alkalinity as input; DIC can be transformed to alkalinity as explained in the help menu of the BLM software (*www.hydroqual.com/winblm*).

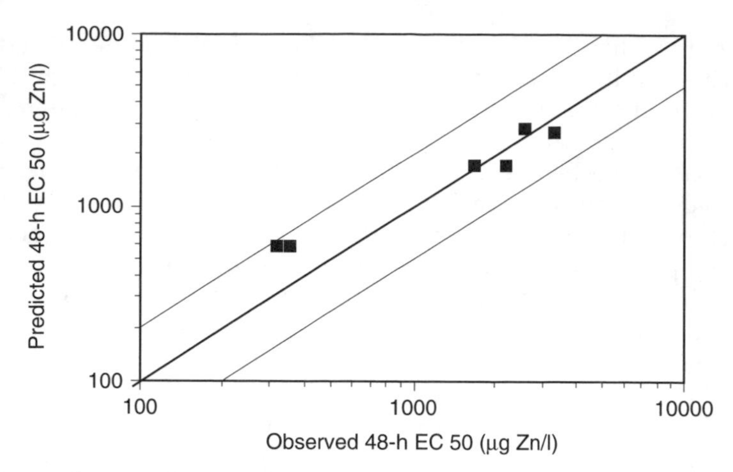

Figure 22.7 Observed versus BLM-predicted 48-h EC50 in six spiked natural surface waters. Bold line, perfect prediction; thin line, factor 2 prediction error.

Summary

The process of developing and validating a BLM for metal toxicity was explained in detail in this chapter. The approach has been originally developed for acute copper toxicity to *D. magna*[1,35] and has been shown a valid alternative of gill-metal binding studies (e.g., Reference 17) for deriving stability constants. The here-shown example for acute zinc toxicity to *D. magna* again indicates that this methodology is very sound and results in a model that yields very accurate toxicity predictions. The methodology can be applied to exposures of any organism to any metal for any exposure duration (acute and chronic). The methodology can yield a BLM in a relative short period of time (<1–2 years) and is thus interesting for the rapid incorporation of bioavailability in criteria setting and risk assessment of metals.

Acknowledgments

The authors wish to thank the International Lead Zinc Research Organization (ILZRO), European Zinc Association (IZA), European Copper Institute (ECI), International Copper Association (ICA), and the Nickel Producers Environmental Research Association (NIPERA). A.C. Karel De Schamphelaere was supported with a Ph. D. grant from Flemish Institute for the Promotion of Scientific and Technological Research in Industry (IWT-Vlaanderen), and additional support was provided by the Ghent University research fund (BOF No. 01110501). We also wish to thank Herbert Allen, Filip Tack, Tao Cheng, Steve Lofts, Virginia Unamuno Rodriguez, Emmy Pequeur, Jill Van Reybrouck, Leen Van Imp, Gisèle Bockstael, Guido Uyttersprot, Marc Vanderborght, and Ria Van Hulle.

References

1. De Schamphelaere, K.A.C. and Janssen, C.R., A biotic ligand model predicting acute copper toxicity for *Daphnia magna*: the effects of calcium, magnesium, sodium, potassium and pH, *Environ. Sci. Technol.*, 36, 48–54, 2002.

2. Di Toro, D., Allen, H., Bergman, H., Meyer, J., Paquin, P. and Santore, R., Biotic ligand model of the acute toxicity of metals. I. Technical basis, *Environ. Toxicol. Chem.*, 20, 2383–2396, 2001.
3. Paquin, P.R., Gorsuch, J.W., Apte, S., Batley, G.E., Bowles, K.C., Campbell, P.G.C., Delos, C.G., Di Toro, D.M., Dwyer, R.I., Galvez, F., Gensemer, R.W., Goss, G.G., Hogstrand, C., Janssen, C.R., McGeer, J.C., Naddy, R.B., Playle, R.C., Santore, R.C., Schneider, U., Stubllefield, W.A., Wood, C.M. and Wu, K.B., The biotic ligand model: a historical overview, *Comparat. Biochem. Physiol. C.*, 133, 3–36, 2002a.
4. Zitko, V., Carson, W.V. and Carson, W.G., Prediction of incipient lethal levels of copper to juvenile Atlantic salmon in the presence of humic acid by cupric ion electrode, *Bull. Environ. Contam. Toxicol.*, 10, 265–271, 1973.
5. Zitko, V. and Carson, W.G., Mechanism of effects of water hardness on lethality of heavy metals to fish, *Chemosphere*, 5, 299–303, 1976.
6. Allen, H.E., Hall, R.H. and Brisbin, T.D., Metal speciation, effects on aquatic toxicity, *Environ. Sci. Technol.*, 14, 441, 1980.
7. Morel, F.M.M., *Principles of Aquatic Chemistry*, John Wiley & Sons, New York, 1983.
8. Pagenkopf, G.K., Gill surface interaction model for trace-metal toxicity to fishes: role of complexation, pH, and water hardness, *Environ. Sci. Technol.*, 17, 342–347, 1983.
9. Heijerick, D.G., De Schamphelaere, K.A.C. and Janssen, C.R., Predicting acute zinc toxicity for *Daphnia magna* as a function of key water chemistry characteristics: development and validation of a biotic ligand model, *Environ. Toxicol. Chem.*, 21, 1309–1315, 2002.
10. Heijerick, D.G., Janssen, C.R. and De Coen, W.M., The combined effects of hardness, pH, and dissolved organic carbon on the chronic toxicity of Zn to *D. magna*: development of a surface response model, *Arch. Environ. Contam. Toxicol.*, 44, 210–217, 2003b.
11. Heijerick, D.G., De Schamphelaere, K.A.C. and Janssen, C.R., Biotic ligand model development predicting Zn toxicity to the alga *Pseudokirchneriella subcapitata*: possibilities and limitations, *Comp. Biochem. Physiol. C.*, 133, 207–218, 2002.
12. Barata, C., Baird, D.J. and Markich, S.J., Influence of genetic and environmental factors on the toxicity of *Daphnia magna* Straus to essential and non-essential metals, *Aquat. Toxicol.*, 42, 115–137, 1998.
13. Paulauskis, J.D. and Winner, R.W., Effects of water hardness and humic acid on zinc toxicity to *Daphnia magna* Straus, *Aquat. Toxicol*,. 12, 273–290, 1988.
14. Alsop, D.H. and Wood, C.M., Influence of waterborne cations on zinc uptake and toxicity in rainbow trout, *Oncorhynchus mykiss*, *Can. J. Fish Aquat. Sci.*, 56, 2112–2119, 1999.
15. Hogstrand, C., Reid, S.D. and Wood, C.M., Ca^{2+} versus Zn^{2+} transport in the gills of freshwater rainbow trout and the cost of adaptation to waterborne Zn^{2+}, *J. Exp. Biol.*, 198, 337–348, 1995.
16. Hogstrand, C., Verbost, P.M., Wendelaar Bonga, S.E. and Wood, C.M., Mechanisms of zinc uptake in gills of freshwater rainbow trout: interplay with calcium transport, *Am. J. Physiol.*, 270, 1141–1147, 1996.
17. Playle, R.C., Dixon, D.G. and Burnison, K., Copper and cadmium binding to fish gills: estimates of metal-gill stability constants and modeling of metal accumulation, *Can. J. Fish. Aquat. Sci.*, 50, 2678–2687, 1993b.
18. MacRae, R.K., Smith, D.E., Swoboda-Colberg, N., Meyer, J.S. Bergman, H.L., The copper binding affinity of rainbow trout (*Oncorynchus Mykiss*) and brook trout (*Salvelinus fontinalis*) gills: implications for assessing bioavailable metal, *Environ. Toxicol. Chem.*, 18, 1180–1189, 1999.
19. Tipping, E., WHAM — a chemical equilibrium model and computer code for waters, sediments, and soils incorporating a discrete site/electrostatic model of ion-binding by humic substances, *Comput. Geosci.*, 20, 973–1023, 1994.
20. Martell, A.E., Smith, R.M. and Motekaitis, R.J., Critical Stability Constants of Metal Complexes Database, Version 4.0, NIST Standard Reference Database 46, National Institute of Standards and Technology, Gaithersburg, MD, USA, 1997.

21. Meyer, J.S., Santore, R.C., Bobbit, J.P., Debrey, L.D., Boese, C.J., Paquin, P.R., Allen, H.E., Bergman, H.L. and Di Toro, D.M., Binding of nickel and copper to fish gills predicts toxicity when water hardness varies, but free ion activity does not, *Environ. Sci. Technol.*, 33, 913–916, 1999.
22. De Schamphelaere, K.A.C. and Janssen, C.R., Development and field validation of a biotic ligand model predicting chronic copper toxicity to *Daphnia magna*, *Environ. Toxicol. Chem.*, 23, 1248–1255.
23. OECD, Guideline for Testing of Chemicals, No. 202, Organisation for Economic Cooperation and Development. Paris, France, 1984.
24. Elendt, B.P. and Bias, W.R., Trace nutrient deficiency in *Daphnia magna* cultured in standard medium for toxicity testing. Effects of the optimization of culture conditions on life history parameters of *D. magna*, *Wat. Res.*, 24, 1157–1167, 1990.
25. Allison, J.D., Brown, D.S. and Novo-Gradac, K.J., MINTEQA2/PRODEFA2, a Geochemical Assessment Model for Environmental Systems: Version 3.0 User's Manual, Environmental Research Laboratory, Office of Research and Development, USEPA, 1991.
26. Schecher, W.D. and McAvoy, D.C., MINEQL+: a software environment for chemical equilibrium modelling, *Comput. Environ. Urban Syst.*, 16, 65–76, 1992.
27. Tipping, E. and Hurley, M.A., A unifying model of cation binding to humic substances, *Geochim. Cosmochim. Acta*, 56, 3627–3641, 1992.
28. Tipping, E., Humic ion-binding Model VI: an improved description of the interactions of protons and metal ions with humic substances, *Aquat. Geochem.*, 4, 3–48, 1998.
29. Fish, W. and Morel, F.M.M., Characterisation of organic copper-complexing agents released by *Daphnia magna*, *Can. J. Fish. Aquat. Sci.*, 40, 1270–1277, 1983.
30. Hamilton, M.A., Russo, R.C. and Thurston, R.V., Trimmed Spearman–Karber method for estimating median lethal concentrations in toxicity bioassays, *Environ. Sci. Technol.*, 11, 714–719, 1977.
31. Santore, R.C., Mathew, R., Paquin, P.R. and Di Toro, D., Application of the biotic ligand model to predicting zinc toxicity to rainbow trout, fathead minnow and *Daphnia magna*, *Comp. Biochem. Physiol. C.*, 133, 271–285, 2002.
32. Alsop, D.H. and Wood, C.M., Kinetic analysis of zinc accumulation in the gills of juvenile rainbow trout: effects of zinc acclimation and implications for biotic ligand modeling, *Environ. Toxicol. Chem.*, 17, 1911–1918.
33. Galvez, F., Webb, N., Hogstrand, C. and Wood, C.M., Zinc binding to the gills of rainbow trout: the effect of long-term exposure to sublethal zinc, *J. Fish Biol.*, 52, 1089–1104, 1998.
34. Ma, H., Kim, S.D., Cha, D.K. and Allen, H.E., Effects of kinetics of complexation by humic acid on toxicity of copper to *Ceriodaphnia dubia*, *Environ. Toxicol. Chem.*, 18, 828–832, 1999.
35. De Schamphelaere, K.A.C., Heijerick, D.G. and Janssen, C.R., Refinement and field validation of a biotic ligand model predicting acute copper toxicity to *Daphnia magna*, *Comp. Biochem. Physiol. C.*, 133, 243–258, 2002.
36. Dwane, G.C. and Tipping, E.. Testing a humic speciation model by titration of copper-amended natural waters, *Environ. Int.*, 24, 609–616, 1998.
37. Cheng, T., De Schamphelaere, K., Lofts, S., Rodriguez, U., Tack, F., Janssen, C.R., Tipping, E. and Allen, H.E., Measurement and computation of zinc binding to neutral dissolved organic matter in European Surface Waters. *Environ. Toxicol. Chem.* (accepted).

chapter twenty-three

Rapid toxicity fingerprinting of polluted waters using lux-marked bacteria

Nigel L. Turner
Cranfield University

Contents

Introduction

The release of toxic chemicals to surface waters in the UK and EU has, in the past, been largely controlled by legislation pertaining to pollutant chemistry alone. Environmental quality standards (EQSs) for controlled substances in the UK, required under the 1976 Control of Dangerous Substances Act, are derived from discharge limits based on total pollutant load.[1] Pollutant bioavailability is accounted for to some extent with this approach (e.g., certain pollutants will have higher discharge limits in hard-water areas); however, actual environmental toxicity (and the resultant ecological implications) are not

1-56670-664-5/05/$0.00+$1.50
© 2005 by CRC Press

implicitly considered. With the adoption of the Water Framework Directive by the EU in 2000, this is now changing. Article 16 of the directive requires all member states to attain "good" ecological status in all surface waters by 2015 (this approach is comparable with the USEPA Toxicity Reduction Evaluation procedures).[1,2] This places increasing importance on defining the *in situ* toxicity of pollutants (and pollutant mixtures) in receiving waters, thus increasing the need for accurate, representative, and innovative methods of toxicity assessment.

Traditional environmental toxicity testing has focused on the use of higher organisms (i.e., macrophytes and animals) as biosensors. However, bacterial assays are being increasingly exploited due to their rapidity, practicality, and versatility.[3] In addition, as bacteria represent the "lowest rung" of the ecological ladder, it can be argued that their inclusion for toxicity testing is essential if ecological impacts on receiving waters are to be accurately determined. A wide range of bacterial assays currently exist; luminescent assays have become especially popular since the commercial introduction of Microtox® in 1981, which uses the naturally luminescent marine bacterium *Vibrio fischeri*.[3,4] Genes from this, and other luminescent species, have been inserted into numerous terrestrial bacteria to create a wide range of "*lux*-marked biosensors."[5]

lux Biosensors have been applied to toxicity testing for a wide range of environmental pollutants.[6–8] However, such biosensors (in common with toxicity assays in general) indicate sample toxicity (and bioavailability) alone, whilst revealing nothing of the chemical nature of the pollutants directly. In other words, there is little mutuality between toxicity biosensors and analytical chemistry. The purpose of the method described here is to bridge the gap between traditional biosensors and chemical analysis, by presenting a technique by which a luminescent biosensor may be used for both pollutant identification and toxicity assessment at the same time. Its primary application is as an environmental screening tool. A related approach has been developed through the use of a battery of catabolic *lux* biosensors, where luminescence is induced by the presence of target chemicals.[9] However, such an approach only works for certain toxicants, is less practical than using just a single biosensor and, most importantly, does not indicate toxicity.

The procedure described here involves a single constitutive *lux* bacterial biosensor, used to kinetically "fingerprint" individual toxicants and complex effluents. This is possible due to the fact that different toxicants elicit highly characteristic light response-curves when the biosensor is exposed to them. Such response-curves are derived by simply measuring luminescence continuously from the point of exposure for a 5-min assay. This technique is relatively inexpensive, rapid, simple to perform, and combines the qualitative potential of chemical analysis with the toxicological information of a biosensor. In addition, the technique can be applied to any pollutant or pollutant mixture that elicits a toxic response in the biosensor.

The methods presented here include a detailed description of the assay procedure, the way in which "reference" toxicant response-curves may be derived, and examples of its application using spiked environmental samples and industrial effluents. Mathematical procedures are demonstrated by which assay results may be interpreted.

Materials required

Biosensor

This research has focused on the use of *Escherichia coli* HB101, containing the full *lux* gene cassette (*lux* CDABE) as a multicopy plasmid insertion (pUCD607).[10] Insertion of the full gene cassette results in the expression of luminescence without the requirement for any exogenous substrate addition. This biosensor was used for toxicity testing because it is robust, simple to use, and shows strong luminescent expression. The biosensor was lyophilized (freeze-dried) at late logarithmic growth stage in 1-ml aliquots and stored at −20°C prior to use.[11,12] These individual aliquots of lyophilized biosensor may be stored for at least 6 months before use. There is no reason why other luminescent constructs (or Microtox®) could not be used in the same way as described here, although minor modifications in protocol may be required depending on the biosensor used (especially with regard to resuscitation procedure). The source of the construct used in this research was the Department of Plant and Soil Science, University of Aberdeen, Scotland.

Equipment

- Temperature-controlled automatic luminometer; BioOrbit 1251 (Turku, Finland): This model uses a carousel pre-loaded with samples in plastic cuvettes. Luminescence of each sample in turn is measured for a pre-determined time (programmed via a computer interface: Multiuse software, version 1.01/April 1991/JN), with individual measurements taken every 2 s. Results are stored in data files for later analysis (they can be transferred into a spreadsheet program, such as Microsoft® Excel).
- Standard PC (with a 486 processor or above) for programming the luminometer.
- BioOrbit integrated injection system (Turku, Finland): enables pre-programmed injection of reagents directly into sample cuvettes during luminescence measurement. This should be fitted with a 100-μl syringe and comes with a programmable interface. The syringe is connected to a reagent source via a Teflon® tube, and the outlet (which leads into the luminometer light chamber) is via a second Teflon tube.
- Glass universal bottles (c. 30 ml): used for re-suspension of freeze-dried bacteria.
- Shaking incubator (at 25°C): used for resuscitation of freeze-dried bacteria.
- Two controlled-temperature chambers/refrigerators to hold biosensor aliquots at 15°C and 5°C, respectively.
- 1.5-ml plastic disposable cuvettes (Clinicon, Petworth, UK): used for holding samples in the luminometer.
- HI 8424 microcomputer pH meter.
- Baird Alpha 4 flame atomic absorption spectrometer (FAAS): used for quantification of heavy metals in environmental samples.
- "Pollution and Process Monitoring" Labtoc UV digestion and infrared TOC detector: for determination of total organic carbon (TOC) in environmental samples.
- Adjustable pipettes covering a range from 0.1 to 1 ml: these should use non-toxic disposable tips.

Chemicals

All chemicals used are analytical grade.

- Potassium chloride (KCl); Aldrich, Gillingham, UK; used for re-suspension of biosensor
- 2,4-dichlorophenol (2,4-DCP); Acros, Geel, Belgium; toxicant

- 3,5-dichlorophenol (3,5-DCP); Aldrich, Gillingham, UK; toxicant
- Sodium arsenite (As); BDH Chemicals, Poole, UK; toxicant
- 2-brom-2-nitro-1,3-propand (bronopol); Aldrich, Gillingham, UK; toxicant
- Copper (II) sulphate (Cu); Fisons, Loughborough, UK; toxicant
- Tetradecyltrimethylammonium bromide (TTAB), Aldrich, Gillingham, UK; toxicant (an anionic surfactant)
- Zinc (II) sulphate (Zn); Fisons, Loughborough, UK; toxicant

Environmental samples for spiking

Water samples were taken from three locations for subsequent spiking: these were kept refrigerated at 5°C in 1-l Duran bottles (without head-space) and were used within 1 week:

- River Don (Seaton Park, Aberdeen, Scotland)
- River Dee (Ballater, Deeside, Scotland)
- Tap water (Aberdeen University)

Toxic environmental effluents

Environmental effluent samples were taken from three locations: these were kept refrigerated at 5°C in 2-l Duran bottles (without head-space) and were used within 2 weeks. Samples were filtered (pore-size of 0.45 μm) prior to use.

- Sample 1: pre-treatment whiskey distillery effluent from the Glenfarclas whisky distillery, Speyside, Scotland
- Sample 2: untreated metal-rich effluent from a galvanizing and electroplating plant, Grampian, Scotland
- Sample 3: untreated effluent from a paper-processing plant, Grampian, Scotland

Procedures

A procedure is described herein by which *lux* bacteria may be used to fingerprint toxic samples by deriving characteristic light-inhibition curves for samples, thus combining toxicity assessment with chemical identification. This technique may be used to identify individual toxicants (or classes of toxicants) by comparing sample response-curves with those of reference toxicants. Because this technique is chemically non-specific, whole complex effluents may also be identified using the same technique (something difficult to achieve by chemical techniques). Moreover, this technique is extremely rapid, repeatable, and simple to perform, making it a potentially valuable tool for environmental screening purposes.

Procedures are described here for deriving series of reference luminescence-inhibition response-curves for specific toxicants (7 in this case). These act as a "database" of known responses against which "unknown" samples may be compared. Toxicant spiked water samples and complex effluents are also characterized in the same way, for comparison with the reference responses-curve series.

The procedures described here are divided into two main sections:

1. Assay protocol: the use of a cellular-injection method for deriving light-inhibition response-curves for a small data set of reference toxicants, spiked water samples, and environmental effluents.
2. Data interpretation:
   (i) the computer-based application of an algorithm for differentiating response-curves;
   (ii) analytical techniques for interpreting output from the algorithm program.

Assay protocol

The following procedure describes the derivation of a series of light-inhibition response-curves for a reference toxicant, in this example Zn, using a range of concentrations. This is done by injection of *lux*-marked bacteria into chemical standards inside a BioOrbit 1251 luminometer, whilst recording resultant luminescence inhibition. This assay series consists of 6 individual assays: 5 concentrations of Zn (from "low" to "high" light inhibition) and a control. This technique can be used to produce a set of reference response-curves for any chemical that elicits a toxic response to the biosensor: the specific reference toxicants characterized will depend upon the research carried out. In this chapter, reference response-curve series were derived for Cu, As, 2,4-DCP, 3,5-DCP, bronopol, and TTAB. These were chosen to represent a range of chemical classes and modes of toxicity, to form a database of characterized toxicity response-curves. Essentially, the same method is used to determine response-curves from experimental samples (spiked water samples and industrial effluents), which represent unknown samples to be tested against the reference database. All reference standards, spiked environmental samples, and industrial effluents were pH adjusted, by addition of HCl_{aq} and NaOH, to a value of 5.5 prior to use. All assays are performed in triplicate.

Resuscitation and control of the biosensor prior to use:

1. Remove a glass vial containing a freeze-dried aliquot of *E. coli* HB101 pUCD607 biosensor from frozen storage (−20°C) and allow to equilibrate with room temperature for half an hour.
2. Pipette 1 ml of a 10-ml, 0.1 *M*, KCl solution from a glass universal bottle into the vial and agitate for 10 s, then pipette the cell suspension back into the universal tube and agitate further using the pipette.
3. Place in a shaking incubator at 25°C for 30 min, then pipette 1-ml aliquots of cell suspension into 3-ml plastic cuvettes and place in a refrigerator at 5°C (repeatability of assays was found to be fairly insensitive to resuscitation time).[12] The cells can be stored at this temperature for up to 2 h prior to use without significantly altering the repeatability of the results (luminescence declines to very low levels at 5°C, indicating reduced metabolic activity).
4. A 1-ml aliquot of cell suspension is used for each assay run, i.e., 6 concentrations tested for a reference series. Therefore, for triplicate replication, 3 aliquots are used. Each aliquot is stored at 5°C prior to use, being raised to 15°C, 15 min prior to use (the incubator block of a Microbics Model 500 luminometer was used for this purpose).

Optimization experiments have shown that following this temperature regime significantly increases experimental repeatability.[12] This is due to the metabolic state of the cells remaining more constant than they are at room temperature. For example, where triplicate cell suspensions were held at either 15°C or 25°C, after 2 h the mean luminescence of the 15°C suspensions had decreased slightly to 80% of the control (SD = 5%), whilst the 25°C suspensions had increased to 270% of the control (SD = 23%). Maintaining cells at 5°C effectively suspends metabolic activity for several hours, although luminescence is too low for experimentation, hence the need to raise the temperature prior to use.

Preparation of chemical standards: All chemical standards used should be made from pre-prepared top standards. Test standards are made in double-deionized water immediately prior to use to represent a toxicity range from low (light inhibition is minimal) to high (light inhibition is considerable). This involves preliminary range-finding experiments (using the techniques described here) to ascertain a desired concentration range for each chemical standard. As response-curves differ in shape between toxicants, it is not necessary (or practical) to produce exact degrees of inhibition for specific reference concentrations. Therefore, a certain degree of subjectivity may be used in deriving reference response-curves. The most important consideration is to ensure that mid-range toxicity is well described. The lowest concentration used for the series of Zn reference response-curves was 0.0065 m*M*, resulting in 17.2% light inhibition after 5 min; the highest was 0.036 m*M*, resulting in 96.8% inhibition.

Aliquots of each standard (900 μl) are aliquoted into individual plastic luminometer cuvettes in triplicate (i.e., $6 \times 3 = 18$ cuvettes in total). These are incubated at 15°C prior to use.

Use of the luminometer and injection system for derivation of light response-curves: The BioOrbit 1251 luminometer is pre-programmed, using the MultiUse software via a computer interface, to measure luminescence of each of the 18 cuvettes in turn as one "assay run." General instructions for programming are given in the luminometer handbook. For each cuvette, luminescence is programmed to be measured "continuously" for 310 s (measurements are actually taken at 2-s intervals) with luminescent cells injected after 10 s, giving an effective assay time of 5 min. It takes around 110 min to process a full set of 18 cuvettes in this way. It was found that with longer assay run times than this repeatability became a problem due to the time that the cells were left standing. The cuvettes are continuously agitated by rotation, whilst in the light chamber of the luminometer.

The luminometer is cooled to 20°C by running tap water through an internal cooling system. This is the lowest temperature that could be reached with this cooling system, although running assays at 15°C would probably be preferable if possible. The injection system is connected to the luminometer via an electrical interface. This enables time of injection to be programmed into the assay run. The injector is programmed via a dedicated controller (which plugs into the injector) to inject a single 100-μl dose of cells, with a 1-s injection time, directly into each cuvette via a narrow Teflon tube fed into the top of the light chamber. This occurs when the cuvette is inside the luminometer light chamber, thus luminescence is recorded from the initial luminescence peak onwards. It has been demonstrated that the force of injection is sufficient to achieve good mixing within the cuvette by the injection of dyes.

Once the luminometer, injector, reagents, and cell suspensions are prepared, the assay series is executed as follows:

1. The 18 toxicant standards (including the three controls) are loaded into the luminometer carousel in random order.

2. The injector syringe is primed with 1 ml of cell suspension. This is loaded into the syringe, via a short Teflon tube, using a built-in piston that automatically re-loads the syringe following each injection.
3. Once everything is primed, the assay run is initiated via the computer interface and may be left to run unattended.
4. On finishing the run, the luminometer readings may be saved as data files to be loaded into a spreadsheet (such as Microsoft Excel).
5. The injection system should be cleaned by flushing with 70% ethanol.

Once the standards/effluents are prepared, this procedure (once set up) involves little actual work, enabling numerous individual samples to be run in a day without too much effort.

The above technique can be used to produce a set of reference curves for any chemical that elicits a toxic response to the biosensor. Reference curves were derived for 7 toxicants for the results presented here: Zn, Cu, As, 2,4-DCP, 3,5-DCP, bronopol, and TTAB. These were chosen to represent a range of chemical classes and modes of toxicity. Mathematical methods for making inter-comparisons between these response-curve series, in order to determine their relative "distinctiveness," are described in the section "Data interpretation."

Derivation of spiked water response-curves: Aliquots of the three water samples described earlier were each spiked with single concentrations of 5 reference toxicants. Concentrations were selected to give mid-range responses based on the reference series as follows: Cu (0.15 m*M*), Zn (0.012 m*M*), As (0.016 m*M*), bronopol (0.016 m*M*), and 2,4-DCP (0.05 m*M*). Response-curves were derived in the same way as described earlier, with single concentrations used rather than a concentration series. For each spiked sample, the corresponding water sample was used as the control.

Derivation of effluent response-curves: Effluent response-curve series (5 concentrations plus 1 control) were derived as for the reference response-curves. This involved range-finding assays to establish a suitable dilution range in each case. Unpolluted controls were unavailable for the three samples, so deionized water was used for the control in each case and was used for making the serial dilutions.

Chemical characterization of effluents: The effluents were characterized with regard to pH, TOC, and heavy metal concentration (Cu, Zn, Cd, and Pb), as these were considered toxicologically relevant for the specific effluents. Toxic concentrations of Cu had previously been detected in whiskey distillery effluent from the same source,[13] and the other metals had previously been found in effluent from the galvanizing and electroplating plant (unpublished data). The paper processing plant effluent was not fully characterized, but previous research had shown it be a highly complex effluent containing significant concentrations of Hg and mixed phenolic compounds.[14]

Data interpretation

Fingerprinting algorithm: A simple algorithm was derived to compare an unknown response-curve with a reference response-curve series. This was implemented as a simple computer program written in Microsoft Visual C++, by which an unknown response-curve could be automatically compared against all available reference response-curves to identify a "best-fit." The algorithm operates as follows:

1. Each 300-s response-curve is simplified by using just 6 time points (t_1, t_2, t_3, etc.) taken at 50-s intervals. This linear time series has been found to give the best identification success in general, although an exponential time series (where time points are biased towards the start of the assay) has also been used successfully.[15] The reason for using a linear series is that there is often a lot of ''noise'' at the start of the assay, resulting in greater error. However, the first 30 s of some toxicant response-curves (e.g., Cu) may show characteristic signatures that may not be captured by a linear series. In practice, it was found that there was greater ''initial'' variability with environmental samples than with laboratory standards, making a linear series preferable overall.
2. L_u is defined as the luminescence value of the unknown toxicant at the first time point. For each of the toxicants in the reference database, there is a luminescent value stored for each concentration (C_1, C_2, C_3, etc.), e.g., for database toxicant 1, these are denoted as: $L_{p1}(C_1)$ for concentration 1, $L_{p1}(C_2)$ for concentration 2, $L_{p1}(C_3)$ for concentration 3, etc.
3. The algorithm determines between which pair of concentrations L_u occurs, as shown in Figure 23.1.
4. The algorithm, based on the simple technique of proportionality, then determines the concentration C_u of the reference toxicant that would have the same toxic response (i.e., percentage luminescence). A linear relationship between concentration and luminescence between each pair of concentrations is assumed. Hence, C_u is given by:

$$C_u = \left(1 - \frac{L_u - L_{p1}(C_2)}{L_{p1}(C_1) - L_{p1}(C_2)}\right)(C_2 - C_1) + C_1$$

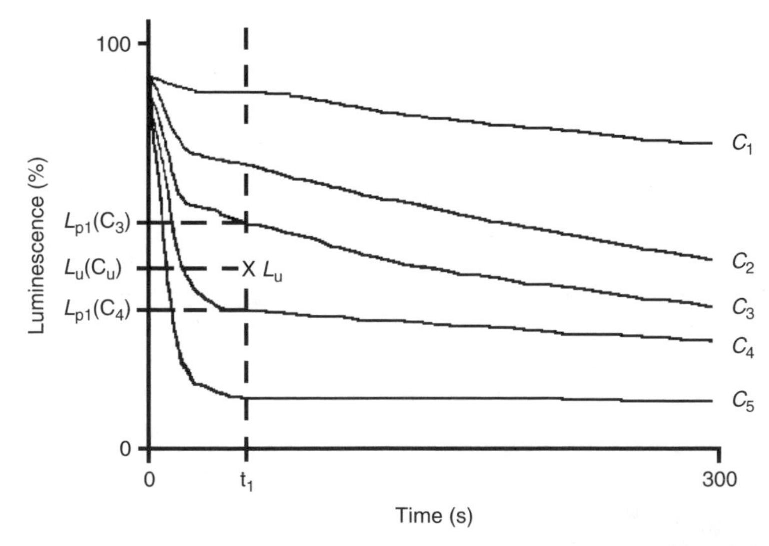

Figure 23.1 Method by which the concentration of an unknown toxicant is predicted: L_u is the luminescence of the unknown toxicant at time t_1. This falls between the luminescence values of concentrations C_3 and C_4 for the reference toxicant (L_{d1}). The predicted concentration (C_u) is determined by the model relative to these two values. (From Turner, N.L. Horsburgh, A., Paton, G.I., Killham, K., Meharg, A.A., Primrose, S., and Strachan N.J.C, *Environ. Toxicol. Chem.*, 20, 2456–2461, 2001. With permission.)

5. The 95% confidence interval for C_u is determined by propagating the confidence interval in L_u through the above equation.
6. These steps are repeated for each reference toxicant for the specific time point, and then the same process is repeated for each of the remaining 5 time points. The result is a temporal series of predicted concentrations and confidence intervals for each reference toxicant, based on the shape of the unknown response-curve. Therefore, if the predicted concentration of a reference toxicant remains consistent over the 6 time points, then it may be considered a likely match for the unknown toxicant.

Interpretation of results

Simply checking the output from the above method visually can indicate roughly which reference toxicants represent a good fit, and which do not. Two different methods for interpreting the computer output are presented, an implicit (yes/no) best-fit approach and a relative "goodness-of-fit" approach. The relative application of these two approaches is dependent on the type of analysis being done.

Best-fit approach: Where the aim of the analysis is to see whether an unknown toxicant response matches a reference toxicant response, one simply determines if the predicted concentration is consistent for each time point (i.e., falls within the 95% confidence intervals derived by applying the algorithm). In addition, the mean predicted concentration can be taken as a quantitative estimate of toxicant bioavailability. Figure 23.2 gives an illustration of this method. In this example, the 0.018-m*M* response for bronopol, when compared against the bronopol reference responses, gives the same predicted concentration within the 95% confidence interval for each time point (as would be expected). When compared against As, however, the predicted concentration is not consistent for all time points. This represents a positive-fit using the best-fit approach for bronopol, but not for As.

Best-fit analysis was used to make comparisons between the response-curves of the reference toxicants to determine how well the different toxicants can be differentiated. The highest and lowest concentrations were excluded in these inter-comparisons, as in many cases they are out of range, e.g., if observed luminescence inhibition is higher than

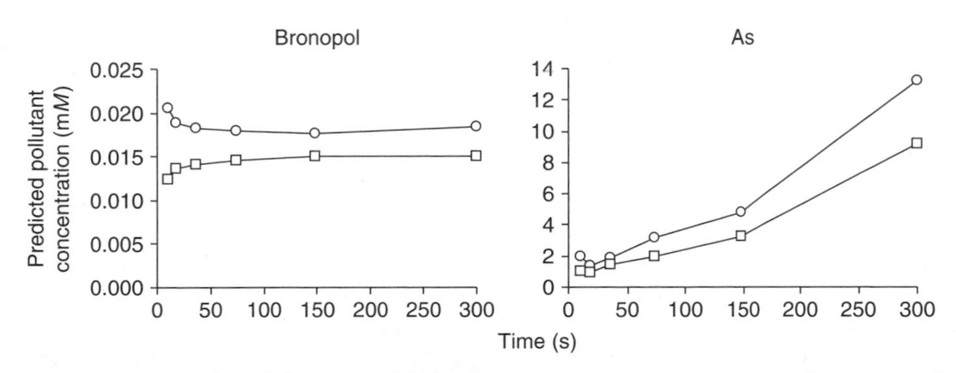

Figure 23.2 Example of model output: 0.018 m*M* bronopol response compared against reference responses for (a) bronopol and (b) As (○ upper 95% CI, □ lower 95% CI). This illustrates a typical positive identification (where predicted concentration is constant with time) versus a negative identification. (From Turner, N.L. Horsburgh, A., Paton, G.I., Killham, K., Meharg, A.A., Primrose, S., and Strachan N.J.C, *Environ. Toxicol. Chem.*, 20, 2456–2461, 2001. With permission.)

for the highest concentration of the reference toxicant, the highest concentration of the reference toxicant is given as the best-estimate, which will be an underestimation. This is because extrapolation beyond the range of the reference series was not done. Where it can be demonstrated that the dose response-curve for a reference toxicant is linear, extrapolation beyond the characterized range could be performed (although this approach is not used here). The best-fit technique was also applied to the identification of the toxicant spikes in environmental samples and to comparison of the industrial effluents.

Relative-fit approach: The above approach is useful when implicit identification is being looked for. However, as the fingerprinting technique is best used as a screening tool, it may often be more useful to determine the relative goodness-of-fit for an unknown toxicant compared against a range of reference toxicants. This may be done as follows:

1. Predicted concentrations (c_i) are obtained, using the algorithm, for comparison of an unknown toxicant to a reference toxicant for each time point. Unless the match is perfect, these 5 concentrations (one for each time point) will vary.
2. The overall predicted concentration (C) is defined as that which results in the smallest summed deviation (F) when subtracted from the individual predicted concentrations (c_1 to c_5). This is done via the following equation, which can be solved using the Microsoft Excel "solver" function to find the minimum value of F by changing C:

$$F = \text{absolute}(c_1 - C) + \text{absolute}(c_2 - C) + \cdots + \text{absolute}(c_5 - C)$$

3. In conclusion, two parameters are derived: the best-estimate concentration of the reference toxicant (C) and a measure of how consistent the predicted concentration is throughout the course of a 5-min assay (the deviation coefficient, F).
4. As the size of F is not independent of the size of C, an unbiased measure of goodness-of-fit is defined as F/C. This F/C value is used to compare goodness-of-fit between different reference toxicants.

This technique was used to demonstrate identification of toxicants in the industrial effluents.

Results and discussion

Reference response series

The temporal response-curves for the 7 reference toxicants are shown in Figure 23.3. Each concentration curve is plotted using the 6 time points required by the model. Clear differences can be seen between the response-curves of the different toxicants, with some more distinctiveness than others. Arsenate, Cu, Zn, bronopol, and TTAB all show responses related to exponential decay, with especially rapid inhibition at the start of the assay for arsenate and bronopol, with the most gradual inhibition for Zn. The response-curves of 2,4-DCP, and to a lesser extent 3,5-DCP, are markedly different from the other toxicants, in that varying degrees of luminescent recovery are seen following initial luminescence inhibition. This effect had been observed in a previous study, where low concentrations of phenolics were shown to cause luminescent stimulation in standard acute assays.[16] The overall shapes of the response-curves shown in Figure 23.3 show good

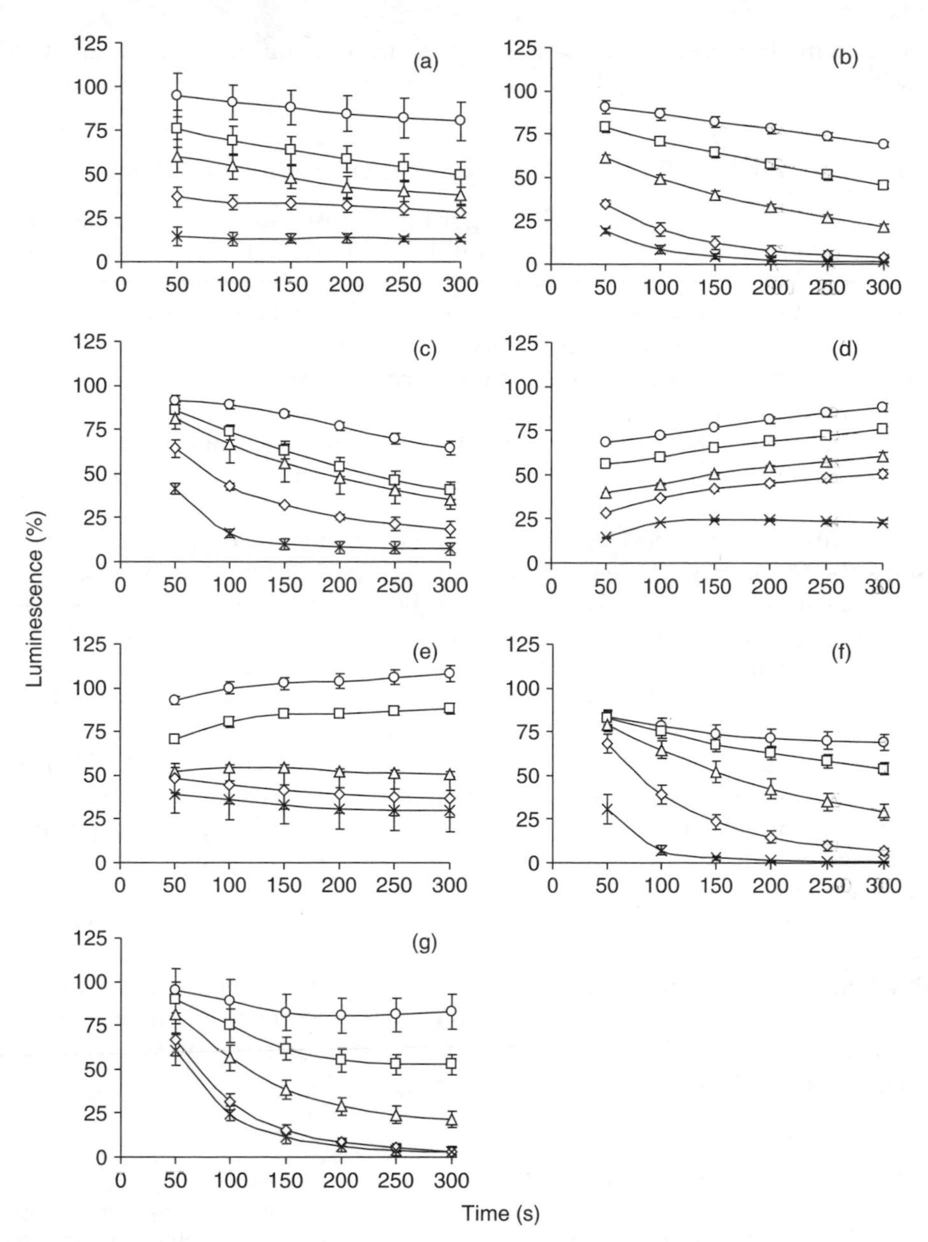

Figure 23.3 Reference toxicant temporal response-curves. Each point on a curve represents the mean value from triplicate data, with 95% confidence interval error bars:

(a) As	(○ 0.3 m*M*,	□ 1.0 m*M*,	△ 2.0 m*M*,	◇ 6.0 m*M*,	× 18.0 m*M*);
(b) Bronopol	(○ 0.004 m*M*,	□ 0.008 m*M*,	△ 0.016 m*M*,	◇ 0.032 m*M*,	× 0.048 m*M*);
(c) Cu	(○ 0.09 m*M*,	□ 0.15 m*M*,	△ 0.18 m*M*,	◇ 0.22 m*M*,	× 0.30 m*M*);
(d) 2,4-DCP	(○ 0.025 m*M*,	□ 0.05 m*M*,	△ 0.10 m*M*,	◇ 0.20 m*M*,	× 0.50 m*M*);
(e) 3,5-DCP	(○ 0.02 m*M*,	□ 0.08 m*M*,	△ 0.12 m*M*,	◇ 0.16 m*M*,	× 0.20 m*M*);
(f) TTAB	(○ 0.05 m*M*,	□ 0.07 m*M*,	△ 0.09 m*M*,	◇ 0.14 m*M*,	× 0.18 m*M*);
(g) Zn	(○ 0.0065 m*M*,	□ 0.009 m*M*,	△ 0.012 m*M*,	◇ 0.024 m*M*,	× 0.036 m*M*).

consistency with earlier assays using different stocks of chemical standards and batches of biosensor.[12]

Fingerprinting algorithm and use of best-fit identification technique

The assay technique and fingerprint algorithm were validated by comparing the reference toxicants to see whether they could be differentiated from each other using the algorithm. This was done by individually comparing the response-curves from each toxicant, as unknown responses, with the reference database responses. The highest and lowest concentrations for each toxicant were not tested as unknowns as many of these are out of range, so only the mid-range concentrations were compared, i.e., the computer program only extrapolates between known values, as an approximately linear dose response outside the measured concentration responses cannot be assumed. This is why each reference response series should cover as full a range of toxic responses as possible.

The results from the best-fit analysis are shown in Table 23.1. The value shown represents the number of "false-positives" for each pair of toxicants (where the concentration response of a given toxicant could not be differentiated from another reference toxicant). The maximum number is 6, as the middle 3 response-curves of "toxicant 1" were compared as unknowns against "toxicant 2" (as a reference toxicant), and vice versa. The model gives slightly different results depending on whether a given toxicant is treated as the unknown test response or the "known" reference response with respect to the comparative toxicant. The most false-positives occurred for TTAB. This is because the TTAB response series was somewhat intermediate between the response series of Cu, Zn, As, and bronopol. Cu differentiation was not as good as might be expected, as the 0.18-m*M* concentration had uncharacteristically high error. Similar results have been published previously, with the difference that an exponential time series was used to describe each curve.[15] This resulted in slightly better identification success, i.e., reduction of false-positives. However, a linear time series is used here as it gave better results overall when considering the spiked samples and the industrial effluents. This is because it puts less emphasis on the start of the assay, where there is greater variability, thus reducing the chance of false-negatives. However, for comparisons between the actual reference responses, where there is no chance of false-negatives arising (as toxicant

Table 23.1 Identification of reference pollutants using the fingerprint algorithm. Values given as the total number of false-positives for each pair of chemicals, where total comparisons for each pollutant pair = 6 (3 comparisons for the first pollutant as an unknown + 3 comparisons for the second pollutant as an unknown)

	Reference toxicants						
	As	Bronopol	Cu	2,4-DCP	3,5-DCP	TTAB	Zn
As	—						
Bronopol	1	—					
Cu	2	2	—				
2,4-DCP	0	0	0	—			
3,5-DCP	1	0	0	0	—		
TTAB	2	2	3	0	0	—	
Zn	0	0	2	0	0	2	—

Table 23.2 Quantification of pollutants spiked in environmental samples

	Predicted concentration (m*M*) in each environmental sample		
Pollutant spike	Tap water	Dee water	Don water
2.0 m*M* As	2.7 (± 0.9)	2.3 (± 0.6)	3.7 (± 0.9)
0.016 m*M* bronopol	0.014 (± 0.001)	0.012 (± 0.002)	0.014 (± 0.001)
0.15 m*M* Cu	0.09 (± 0.01)	0.11 (± 0.01)	0.10 (± 0.01)
0.05 m*M* 2,4-DCP	0.07 (± 0.01)	0.05 (± 0.01)	0.07 (± 0.01)

response-curves are compared against themselves), potential for differentiation is slightly reduced using a linear time series.

Identification of toxicant spikes

Single-concentration response-curves were derived for 2.0 m*M* As, 0.016 m*M* bronopol, 0.15 m*M* Cu, and 0.05 m*M* 2,4-DCP spikes in three environmental samples (tap water, River Don water and River Dee water). Figure 23.4 shows the response-curves for the three environmental samples; it can be seen that there is strong similarity between them. The results from the best-fit analysis showed that all the spikes were successfully identified (at the 95% test limit), in all three samples, with no false-positives. This demonstrates that in simple aquatic media, this technique is highly effective. However, as Cu, bronopol, and As responses are very similar, this degree of success may not necessarily always be repeatable. Table 23.2 shows that the predicted concentrations for the toxicant spikes were similar to the actual concentrations for all 4 toxicants, with

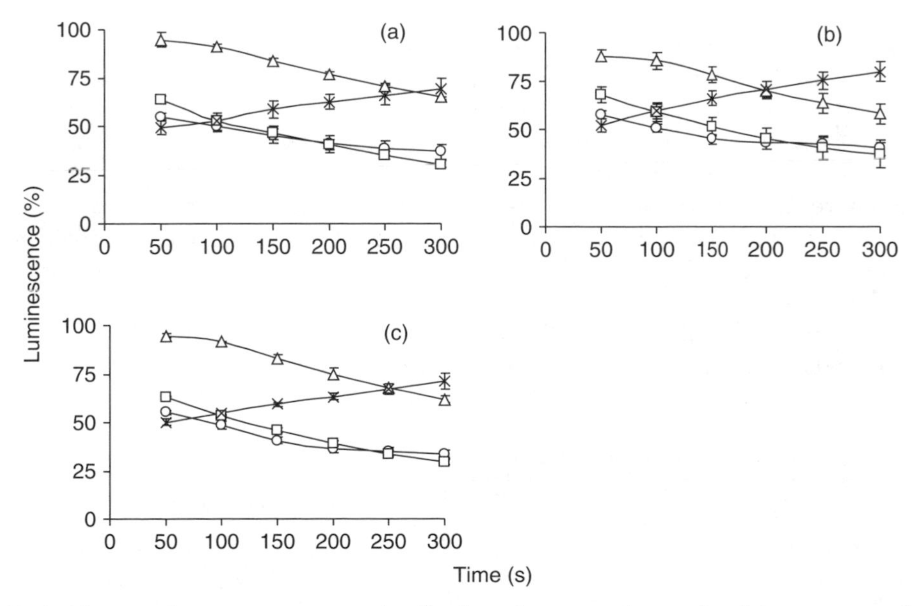

Figure 23.4 Temporal response-curves of spiked environmental samples: (a) tap water, (b) river Dee water, and (c) river Don water, spiked with single concentrations of toxicants: ○ 2.0 m*M* As; □ 0.016 m*M* bronopol; △ 0.15 m*M* Cu; × 0.05 m*M* 2,4-DCP. Each point on a curve represents the mean value from triplicate data, with 95% confidence interval error bars.

As and 2,4-DCP slightly overestimated, and Cu and bronopol slightly underestimated. This indicates that the bioavailability of the toxicants in the samples was approximately the same as for the reference standards. In the presence of significant concentrations of humic acids or dissolved matter, the predicted concentration would be expected to be significantly less as the bioavailable fraction would likely be reduced. The small differences in predicted concentrations, however, may be due in part to differences in water sample chemistry.

Characterization and differentiation of industrial effluents

Using FAAS, the concentrations of Cu, Zn, Cd, and Pb were quantified in the three effluents, as were TOC and pH. These results are shown in Table 23.3. No Cd or Pb was found in any of the effluents. These metals were considered to represent the likely main toxic constituents of the whiskey distillery and galvanizing plant effluents, although the toxic components of the paper processing plant were not identified.

All three effluents were found to be highly toxic to the biosensor and required considerable dilution to bring them "in range." The highest concentrations of the three effluents (expressed as percentage dilution of the raw effluent) were 25% for the whiskey distillery, 0.5% for the galvanizing and electroplating plant, and 0.1% for the paper processing plant. Figure 23.5 shows the response-curve series for each of these effluents. The difference in response-curve shapes between all three effluents is clear, indicating that complex effluents may result in characteristic kinetic signatures. When the algorithm was used to make direct comparisons between the effluent response-curves, there were no false-positives, i.e., all three effluents were differentiated at the 95% test level. The approach used was the same as for the comparisons between the reference toxicants: the middle three response-curves from each effluent were compared against the response-curves of the other two effluents.

The effluent response-curves were next compared with the reference response-curves. As complex effluents were involved and not single toxicants, the "relative-fit" approach was used to compare the effluents with the reference toxicants. This was because the best-fit approach is suitable where an exact fit with a given toxicant is looked for. The relative-fit approach (using F/C values as a measure of goodness-of-fit) is suitable where the relative similarity of an unknown response with respect to a range of toxicants is required, and not implicit identification. F/C values were derived for each concentration of each effluent for comparison with all seven reference pollutants. The average F/C values for each effluent $\times$ reference toxicant comparison (being the average of the three effluent dilutions tested) are shown in Figure 23.6 (error bars = 95% confidence intervals). The F/C values for 2,4-DCP are not shown for any of the effluents, as

Table 23.3 Chemical characterization of environmental effluents

Effluent	Chemical parameter			
	pH	Cu concentration (m*M*)	Zn concentration (m*M*)	TOC (m*M*)
Whiskey distillery (pre-treatment)	6.1	1.05	0.03	3.0
Galvanizing and electroplating plant	6.3	0.66	1.65	2.2
Paper processing plant	3.3	None detectable	None detectable	3.6

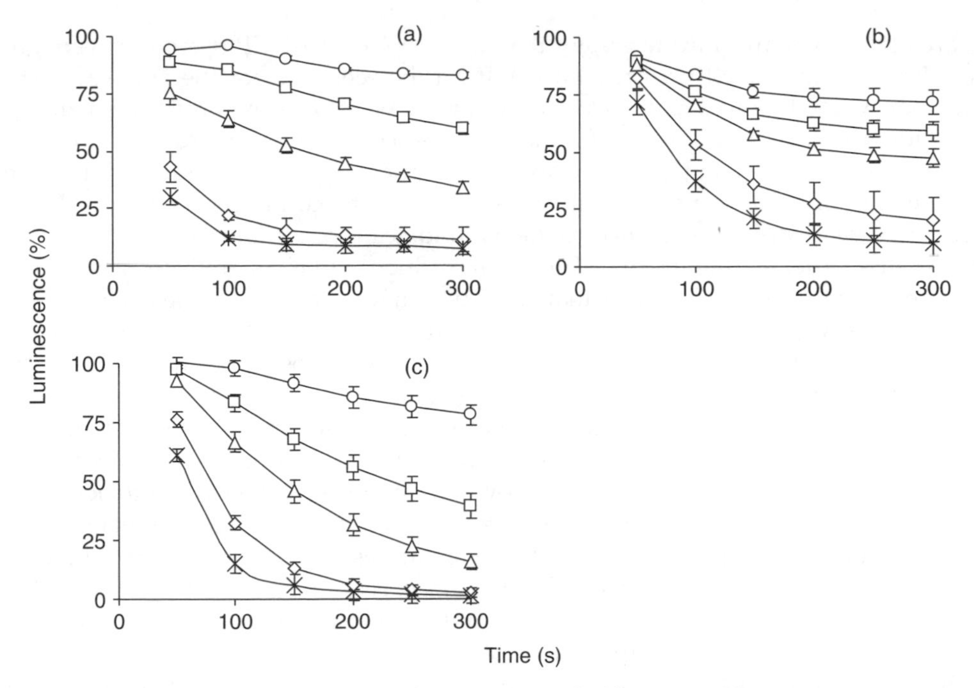

Figure 23.5 Temporal response-curves of environmental effluents. Each point on a curve represents the mean value from triplicate data, with 95% confidence interval error bars. Each curve represents a different percentage dilution of the raw effluent:

(a) Whiskey distillery	(○ 5%,	□ 10%,	△ 18%,	◇ 25%,	× 50%);
(b) galvanizing plant	(○ 0.2%,	□ 0.25%,	△ 0.3%,	◇ 0.35%,	× 0.5%);
(c) paper processing plant	(○ 0.04%,	□ 0.06%,	△ 0.07%,	◇ 0.08%,	× 0.1%).

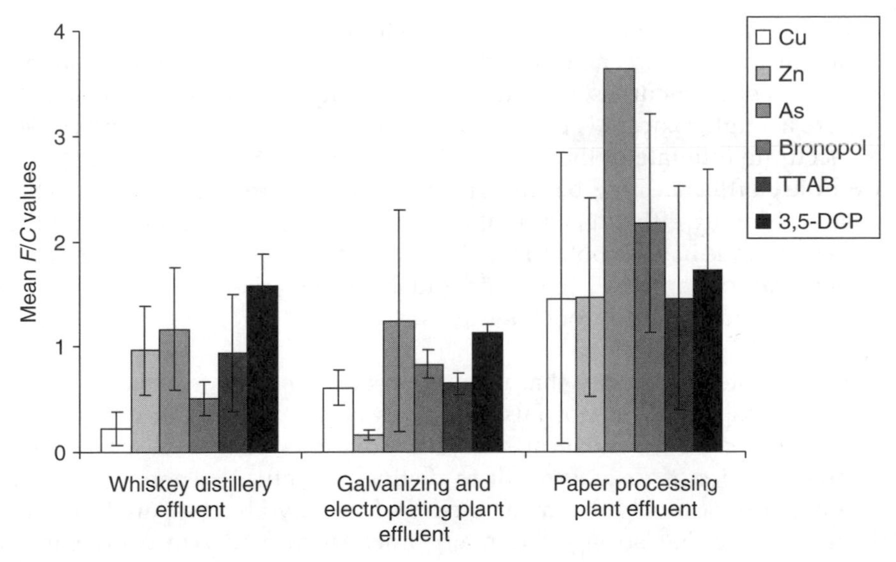

Figure 23.6 Relative similarity of reference toxicant response-curves to effluent response-curves for three industrial effluents. For each effluent, the mean F/C value (from the five effluent dilutions tested) is shown for each reference toxicant. The error bars represent the 95% confidence intervals of the F/C values for each of these dilutions. A small mean F/C value indicates a strong degree of similarity between the effluent and the reference response-curves (a possible identification). The 2,4-DCP comparisons are not included, as the mean F/C values were too great in all cases. The error bars for As are not included due to their size (95% CI $= \pm 2.65$).

they are an order of magnitude greater than the rest (i.e., 2,4-DCP responses were greatly dissimilar from all the effluent responses). It can be seen that for the whiskey distillery effluent Cu was identified as the closest fit; for the electroplating and galvanizing plant Zn was identified as the closest fit; for the paper processing plant there were no close fits.

It is not surprising that there were no close fits for the paper processing plant from the reference toxicants, as none of them were expected to contribute significantly to the toxicity. Cu was correctly identified as the main toxicant present in the whiskey distillery effluent. This indicates that the high TOC of the effluent did not significantly influence the signature of the toxicant. The fact that the Cu signature did not appear altered by the presence of Zn was possibly due to differences in speciation of the two metals, as Zn has been shown to be considerably more toxic to the biosensor than Cu (EC50 of Cu = 0.150 m*M*; EC50 of Zn = 0.07 m*M*).[12] The success in identifying the primary toxicant in a complex effluent was also demonstrated for the galvanizing and electroplating plant effluent, where Zn was identified as the primary toxicant (although Cu was present as well, Zn would be expected to contribute overwhelmingly to the overall toxicity at their respective concentrations). This validates the use of the technique as a screening tool for toxic ingredients in complex effluents. However, successful "identification" of a toxicant should not be taken as implicit proof of its presence; it is an indication of what is likely to be a cause of toxicity or, conversely, what is unlikely (i.e., a screening tool). This can guide subsequent actions, such as chemical analysis.

Overall, this technique represents a reasonably simple and inexpensive screening process using a single bacterial biosensor. Although it does not have the accuracy of chemical analysis, it represents a way in which a simple biosensor may be applied to both effluent characterization and toxicity assessment at the same time. Specific advantages of this technique are:

1. As with any standard biosensor, it reports on the bioavailable fraction of any toxicant that has a toxic effect on the bacterium.
2. Effluents may be screened to identify likely toxicants based on the shape of their toxicity response-curves, or to discard toxicants that are a poor match.
3. Where a single toxicant is present, and compared with reference response-curves, an accurate estimate of the bioavailable concentration may be derived.
4. Complex effluents may be fingerprinted by their response-curves. This can be used to identify effluents without the need to chemically characterize them.
5. This technique has the potential to be adapted for on-line use.[17] This would allow ephemeral toxicants (i.e., highly degradable or volatile organic compounds) to be characterized in flow-through systems.

It should be noted, however, that using a greater range of toxicants will invariably result in an increasing amount of false-positives as this technique does not have the discriminatory potential of chemical analysis. Toxicants can be clearly differentiated where different modes of toxicity result in distinct signatures. Therefore, if chemically distinct groups of toxicants with similar modes of toxicity are compared, differentiation potential may not be that strong. The main practical difficulty to overcome with this technique is controlling between-sample variability, although if the timing and temperature are closely controlled, variability can be kept acceptably low. An important consideration when performing a large number of assays is to ensure that all assays are performed using vials from the same freeze-dried batch. This is because subtle differences in biosensor response may be seen between batches. Quality control procedures ensure that differences between batches are minimal.[12] Overall, this methodology is relatively simple to follow with no major pitfalls. As assays can be performed rapidly, and results obtained instantly (light inhibition is presented on the computer screen during the assay), any mistakes made can be easily identified and rectified.

In conclusion, this is a versatile and novel technique that may be adapted to a wide range of exploratory environmental problems, especially for screening prior to chemical analysis, and for identification of complex effluents.

Acknowledgment

This research was supported by the Natural Environment Research Council (NERC) and AZUR Environmental, and supervised by Dr. Graeme Paton and Prof. Ken Killham of the University of Aberdeen, Scotland. Dr. Norval Strachan (University of Aberdeen) wrote the algorithm program and helped formulate the mathematical techniques used for this research.

References

1. National Society of Clean Air and Environmental Protection (NSCA), *Pollution Handbook 2003*, NSCA, Brighton, 2003.
2. Lanz, K. and Scheuer, S., *Handbook on EU Water Policy*, Hontelez, J., Ed., European Environmental Bureau, 2001 (http://www.eeb.org/activities/water/).
3. Kaiser, K.L.E., Correlation of *Vibrio fischeri* bacteria test data with bioassay data for other organisms, *Environ. Health Perspect.*, 106, 583–591, 1998.
4. Steinberg, S.M., Poziomek, E.J., Engelmann, W.H. and Rogers, K.R., A review of environmental applications of bioluminescence measurements, *Chemosphere*, 30, 2155–2197, 1995.
5. Kado, C.I., Lux and other reporter genes, in *Microbial Ecology: Principles, Methods and Applications*, Levin, M.A., Seidel, R.J. and Rogul, M., Eds., McGraw-Hill, New York, 1992, pp. 371–392.
6. Boyd, E., Killham, K. and Meharg, A.A., Toxicity of mono-, di- and tri-chlorophenols to *lux*-marked terrestrial bacteria, *Burkholderia* species *Rasc c2* and *Pseudomonas fluorescens*, *Chemosphere*, 43, 157–166, 2001.
7. Strachan, G., Preston, S., Maciel, H., Porter, A.J.R. and Paton, G.I., Use of bacterial biosensors to interpret the toxicity and mixture toxicity of herbicides in freshwater, *Water Res.*, 35, 3490–3495, 2001.
8. McGrath, S.P., Knight, B., Killham, K., Preston, S. and Paton, G.I., Assessment of the toxicity of soils amended with sewage sludge using a chemical speciation technique and a *lux*-based biosensor, *Environ. Toxicol. Chem.*, 18, 659–663, 1999.
9. Ben-Israel, O., Ben-Israel, H. and Ulitzer, S., Identification and quantification of toxic chemicals by use of *Escherichia coli* carrying *lux* genes fused to stress promoters, *Appl. Environ. Microbiol.*, 64, 4346–4352, 1998.
10. Shaw, J.J. and Kado, C.I., Development of a *Vibrio* bioluminescence gene-set to monitor phytopathogenic bacteria during the ongoing disease process in a non-disruptive manner, *Biotechnology*, 4, 560–564, 1986.
11. Paton, G.I., *The Development and Application of a Bioassay using Lux-marked Microorganisms to Assess Terrestrial Ecotoxicity*, Ph.D. thesis, University of Aberdeen, Scotland, 1995, 310 pp.
12. Turner, N.L., *Toxicity Fingerprinting of Pollutants and Environmental Effluents Using* lux *Bacterial Biosensors*, Ph.D. thesis, University of Aberdeen, Scotland, 2001, 275 pp.
13. Paton, G.I., Palmer, G., Kindness, A., Campbell, C., Glover, L.A. and Killham, K., Use of luminescence-marked bacteria to assess copper bioavailability in malt whisky distillery effluent, *Chemosphere*, 31, 3217–3224, 1995.

14. Brown, J.S., Rattray, E.A.S., Paton, G.I., Reid, G., Caffoor, I. and Killham, K., Comparative assessment of the toxicity of a papermill effluent by respirometry and a luminescence-based bacterial assay, *Chemosphere*, 32, 1553–1561, 1996.
15. Turner, N.L., Horsburgh, A., Paton, G.I., Killham, K., Meharg, A.A., Primrose, S. and Strachan, N.J.C., A novel toxicity fingerprinting method for pollutant identification using *lux*-marked biosensors, *Environ. Toxicol. Chem.*, 20, 2456–2461, 2001.
16. Sinclair, G.M., Paton, G.I., Meharg, A.A. and Killham, K., *Lux*-biosensor assessment of pH effects on microbial sorption and toxicity of chlorophenols, *FEMS Microbiol. Lett.*, 174, 273–278, 1999.
17. Horsburgh, A., Turner, N.L., Mardlin, D. and Killham, K., On-line fingerprinting of environmental effluents as *lux*-marked biosensors, *Biosen. Bioelectron.*, 17, 495–501, 2001.

chapter twenty-four

Aquatic in situ *bioassays to detect agricultural non-point source pesticide pollution: A link between laboratory and field*

Ralf Schulz
University Koblenz-Landau

Contents

1-56670-664-5/05/$0.00+$1.50
© 2005 by CRC Press

Introduction

Active biomonitoring, which includes all methods that insert organisms under controlled conditions into the site to be monitored, has been developed as a link between laboratory, microcosm, and mesocosm studies on the one hand and passive biomonitoring in the field on the other hand. In ecotoxicology, active biomonitoring is often referred to as the use of *in situ* bioassays. Those *in situ* bioassays may be defined in the context of this book section as the experimental measurement of the responses of test organisms directly exposed to agricultural non-point source pesticide pollution in the aquatic environment over a defined time period. The results of *in situ* bioassays are thus considerably more relevant to the natural situation than those of laboratory experiments, especially with respect to the contamination scenario.[1] Whereas chemical analyses in the field provide abiotic data, the results of the *in situ* bioassay are based on a toxicological response[2] and, accordingly, substantially more informative regarding the protection of animal and plant communities.

The toxicological and ecological relevance are two important requirements that need to be fulfilled when bioassays should be used to assess non-point source pesticide pollution:

- There must be a clear relationship between the pesticide contamination and the response measured in the bioassay. To establish this toxicological relationship is particularly difficult, when runoff is the main route of pesticide entry, as various factors change at the same time (turbidity, flow, nutrients, pesticides). However, this is an intrinsic problem of any sort of ecotoxicological assessment of runoff-related pesticide pollution. The detection and assessment of spray drift is in comparison much easier, as no other environmental factors change at the same time.
- The response in the bioassay must directly or indirectly reflect responses of the same species (or the whole community) in the field. Otherwise, *in situ* bioassays rather serve as a measurement tool to detect unfavorable environmental conditions, which may eventually be identified as pesticide pollution but not necessarily as an indicator of ecologically relevant effects on the population or community level.

The responses of the test organisms used in *in situ* bioassays may be various lethal and sublethal ecotoxicological endpoints measured at suborganismic, organismic, or even the population level, see the section "Procedures" for further details. There are several important considerations for the selection of suitable test organisms, such as reproducibility, sensitivity, and ecological relevance, see the section "Material required" for further details. The time period used for testing are among other factors dependent on the expected exposure scenario and the test organisms used, see the section "Procedures" for further details. The subsequent section "Results and discussion" summarizes the outcome of the case study. The case study results are finally interpreted focusing on the two already mentioned and crucial aspects toxicological and ecological relevance.

Material required

Exposure system

As can be seen from Table 24.1, most *in situ* exposure systems are fairly simple self-made pieces of equipment. Plastic or PVC tubes with the ends closed off either by mesh or by lids have often been used. If lids are used to close off the ends, additional mesh covered

Table 24.1 Examples for characteristics of *in situ* exposure systems

Test organism	Exposure boxes			Openings for water exchange		Access to sediment	Reference
	Type/material	Size	Fixation	Number and size	Mesh size; material		
Mussel (*Mytilus trossulus*)	Bags of flexible vexar mesh tubing	1.5 m long, divided in 10 compartments	Nylon lines, buoyed weights	Completely mesh-covered	Not clear	No	8
Cladoceran (*C. dubia*)	Clear cellulose acetate butyrate sediment core liner tubes	12.7 cm, Ø 5.1 cm, ends closed with lids	In 1-cm nylon mesh holding bags fixed with stakes	2 side openings, 2.5 × 6 cm	149-μm polypropylene microporous filter mesh	No	9
Cladoceran (*D. magna*), zooplankton	Polypropylene beakers	50 ml	Plastic frame and weight	2 side and 1 bottom, Ø 0.2 cm	50 μm; nylon, thermal glue from Elis-Taiwan	Yes, through bottom opening	4
Amphipod (*Chaetocorophium* cf. *lucasi*)	PVC pipe	10 cm, Ø 10 cm	Placed in 2 cm deep depression in sediment	2 side, 8 × 10 cm	440 μm; plastic screen, epoxy glued	Yes, through bottom opening	10
Amphipod (*P. nigroculus*)	Polyethylene jars	10 cm, Ø 9 cm	Metal stakes	2 side, 6 × 8 cm	1 mm stainless steel mesh	No	11
Amphipod (*G. pulex*)	PVC cages	5 cm, Ø 5 cm	In 2.5-cm mesh holding baskets fixed with bricks	Both ends	1 mm	No	6
Amphipod (*H. azteca*), dipteran (*Chironomus tentans*)	Clear plastic tubes	15.2 cm, Ø 6.35 cm, ends closed with lids	In PE mesh dunk holding bags fixed with stakes	2 side openings, 4.5 × 9 cm	149-μm polypropylene woven screen cloth, silicone glued	Yes, through one opening	1
Amphipod (*H. azteca*), dipteran (*C. tentans*)	Clear cellulose acetate butyrate sediment core liner tubes	12.7 cm, Ø 5.1 cm, ends closed with lids	In PE mesh holding bags fixed with bricks	2 side openings, 3 × 5 cm	Polypropylene screen	Yes, through one opening	12
Amphipod (*G. pulex*), caddisfly (*L. lunatus*)	Plastic boxes	40 × 17 × 16 cm	Metal stakes	1 front and 1 rear	1 mm, polypropylene screen, thermal glued	No	7
Crayfish (*Procambarus* spp.)	Conical cages	50.8 × 50.8 × 15.2 cm	Not clear	Completely mesh-covered	Not clear	Yes, embedded into sediment	13

Continues

Table 24.1 Continued

Test organism	Exposure boxes			Openings for water exchange		Access to sediment	Reference
	Type/material	Size	Fixation	Number and size	Mesh size; material		
Dipteran (*C. riparius*)	PVC pipe	Whole water column, Ø 6.8 cm	Pipe driven 10–15 cm into sediment + metal stake	No	Sediment exposure only, uncontaminated water used	Yes, pipe driven into sediment	5
Dipteran (*C. tentans*), oligochaet (*Lumbriculus variegatus*)	Plastic core tube	46 cm, Ø 8 cm	Tube driven 15 cm into sediment	Several side holes, Ø 2.5 cm	224 μm; Nitex®, silicon glued	Yes, tube driven into sediment	3
Leopard frog (*Rana pipiens*), green frog (*R. clamitans*)	Nylon mesh cage	20 cm, Ø 15 cm	Wooden dowels	Completely mesh-covered	500 μm; Nitex Nylon	No	14
Rainbow trout (*Oncorhynchus mykiss*)	Cylindrical nylon mesh cages	38 cm, Ø 38 cm	Three fiberglass rods	Completely mesh-covered	6.3 mm; Nylon	No	15

openings have been placed into the side of the tube. There are no general guidelines with regard to the size of the bioassays or the material provided within the bioassays as food, shelter, or substrate. Both depend greatly on the ecological requirements of the test organisms used. Their density in the bioassay should not be higher than densities commonly found in the field. For frequently used *in situ* test organisms, such as amphipods, cladocerans, or chironomids, several papers focusing on methodological aspects do exist.[1,3–6] For case-bearing caddis flies, it is important to add case-building material to the bioassays.[7] If the emergence of semiaquatic insects is considered as an endpoint, it is required to make sure that the exposure containers do extent over the water surface even at varying water levels (see the section "Case study" for further details).

The mesh size used to cover the openings of the bioassay containers also depends mainly on the size of the test organisms; however, there are some other considerations that need to be taken into account. Too large mesh sizes might allow other organisms (e.g., predators or individuals of the same species) to enter the containers and affect the test organisms' survival or the assessment of results. Too small mesh sizes may reduce the flow within the containers, which, once again, may affect the test organisms, specifically if typical stream organisms are to be used. Moreover, the water exchange within the system or the access of potentially contaminated suspended particles may be reduced, and thus the exposure within the container may be different from the surrounding, which would be a crucial disadvantage of any bioassay design.

As the turbidity of agricultural surface waters varies greatly, clogging of the mesh is another problem that may arise over time and that has the potential to greatly affect the overall-performance of the bioassays. Specifically for studies in running waters, it is thus recommended to ensure sufficient flow of water through the bioassay containers over the duration of the exposure. Many studies have used nylon or polypropylene screens (Table 24.1); however, stainless steel mesh, which is available in various mesh sizes, has been proven to be a very robust and UV-resistant alternative.[11]

The fixation of the bioassay containers needs to be sufficient to ensure flood resistance, e.g., during runoff events. Depending on the purpose of the study, it is also important to make sure that the test organisms are in close contact with contaminated sediments. Various studies have suggested design options, such as exposure to the sediment through one of the mesh-covered openings[12] or through an open end of the bioassay tube directly inserted into the sediment.[3] The latter option in combination with the use of an uncontaminated water layer placed on top of the contaminated sediment may even serve as a sediment-only exposure system.[5]

Type, number, and source of organisms

Generally, the test organisms should be sensitive enough to indicate any potential pesticide effect, but tough enough to survive the exposure procedure at high rates under control conditions. As can also be seen from Table 24.2, the number of individuals used per bioassay ranges from 2 for fish[15] to 100 for mussels[8] or amphipods.[16] The main criteria affecting the decision about numbers of test organisms might be availability, ethical considerations, statistical, and methodological requirements, i.e., if a certain amount of tissue is necessary to measure specific biomarkers.

Many authors have used lab-cultured organisms during their studies (Table 24.2) as they offer the advantage of better reproducibility of the method in other labs and better comparability of results. However, for crayfish and fish, commercial suppliers may be

Table 24.2 Examples for the design of *in situ* exposure studies

Test organism	Site selection	Origin of test organisms	Number of replicates per site	Individuals per replicate size or age	Test duration/ feeding	Endpoints	Statistics	Reference
Mussel (*M. trossulus*)	1 reference + 12 contaminated	Mussel farm, farm holding 1 day	2 or 3	100, 5–6 cm	82 days	Bioaccumulation	Toxicity-normalization and ranking	8
Cladoceran (*C. dubia*)	1 reference + 1 contaminated	Lab culture	4	10, <24-h old	48 h/No	Survival	Not clear	9
Cladoceran (*D. magna*), Zooplankton	2 reference + 6 contaminated	Lab cultures	4–6	5	48 h/No	Survival	Kruskal-Wallis; Mann-Whitney	4
Amphipod (*Chaetocorophium* cf. *lucasi*)	1 contaminated	Field/lab holding 4 days	3 (per treatment)	20, 2–4 mm	10 days	Survival	ANOVA, Dunnett	10
Amphipod (*P. nigroculus*)	2 contaminated, internal control	Field, lab holding 2 days	4	10, juvenile	3–7 days/ butterspoon leaf	Survival, 6 exposure periods	ANOVA-Fisher PLSD	11
Amphipod (*G. pulex*)	24 reference + 15 contaminated	Field/lab holding 10 days	17–18	30, adult	6 days/alder leaf	Feeding rate, survival	ANOVA, *t* test	6
Amphipod (*G. pulex*)	1 contaminated	Field	Not clear	100, adult	11 weeks	Feeding rate and survival, weekly control	No testing	16
Amphipod (*H. azteca*)	3 contaminated	Lab cultures	4	10, <2 weeks old	7 or 18 days, ground rabbit chow	Survival	*T* test	1

Dipteran (*C. tentans*)	1 reference + 2 contaminated	Lab cultures	>4	10, 2nd instar	TetraFin®	Survival	Not clear	12
Amphipod (*G. pulex*), caddisfly (*L. lunatus*)	1 reference + 3 contaminated, internal control	Field, control samples	4	30, adults 30, 4th instar	11 weeks	Survival, weekly control	ANOVA-Fisher PLSD	7
Crayfish (*Procambarus* spp.)	1 reference + 1 contaminated	Commercial supplier, lab holding 1 week	3	10, 6–9 cm	4 days	Survival	Fishers exact test	13
Oligochaet (*L. variegatus*)	1 reference + 2 contaminated	Lab culture	5 or 6	40, adults	10 days/No	Survival. growth	ANOVA-Tukey	3
Leopard frog (*R. pipiens*), green frog (*R. clamitans*)	2 reference + 4 contaminated	Lab culture	4	10 fertilized eggs	2–3 weeks/ boiled lettuce	Hatching success, growth	Nested ANOVA	14
Rainbow trout (*O. mykiss*)	1 reference + 1 contaminated or 1 reference + 2 contaminated	Fish hatchery/ lab holding >4 weeks	1	2–5, 50–150 g weight	2–8 days/No	Biomarker gene expression	ANOVA-Tukey, *t* test	15

indicated; and for species not easy to culture in the lab, such as *Gammarus pulex* (Amphipoda), catches from the field were commonly used (Table 24.2). However, in these cases, a pre-holding period in the lab is often performed in order to adapt the organisms to experimental conditions and to account, as good as possible, for any effects of unwanted and unknown pre-exposure of the organisms at the site, from which they were obtained. Differences in sensitivity of lab-cultures and field catches to the same toxicant levels have been shown, e.g., for amphipods[17] or chironomids.[18] However, the reaction of test organisms present in the field and obtained directly from there may be more relevant for extrapolations of results to the target habitat under study (ecological relevance). If field samples of test organisms are the preferred option, they are ideally collected from uncontaminated control sites within the study catchment, which reflect similar environmental conditions.

Additional measurements

Electronic meters, photometric or spectrometric methods are commonly used to monitor general water-quality parameters (see also Reference 19). Suitable sampling methods for parameters that are directly related to runoff and thus need to be measured using an event-triggered design, such as turbidity, flow, and pesticide levels, have been described by Liess and Schulz.[20] The analysis of pesticide compounds generally follows standardized methods described in detail elsewhere.[19] These methods, however, depend on the equipment available and may be modified to account for matrix effects, e.g., from organic sediment fractions.[21]

Case study

Adult *Gammarus pulex* L. (Crustacea: Amphipoda; carapace length > 6 mm) were used for the case study on *in situ* exposure. The amphipods were obtained from the control sites used during this study. Prior to use, *G. pulex* specimens were kept for 7–9 days in the laboratory at 20 ± 1°C in an aerated 50-l tank and fed with alder (*Alnus glutinosa*) leaves.

Plant containers (40 × 17 × 15 cm with an open top) made from green PVC were used as exposure boxes. The front and rear walls of the boxes were made of white nylon netting (1-mm mesh) to allow water flow through the boxes. The top of each container was also covered by 1-mm nylon mesh attached with Velcro tap to allow easy access. The boxes were supported by Styrofoam swimmers (30 × 5 × 5 cm) attached to the outside of each sidewall; they thus floated with the upper third above the water surface. Two nylon ropes (1 m long; 2 mm diameter) were connected to the front end of each box and the other end of the ropes was fixed to two metal stakes (1 m long; 10 mm diameter) hammered into the sediment. As the boxes were flexibly connected to the metal stakes via the rope, they were always floating at the water surface independent of changes in the actual water level during runoff events. In smaller and shallower water bodies, these exposure boxes may be directly attached to the sediment using metal stakes.

An initial experiment carried out during periods with high turbidity due to rainfall-related edge-of-field runoff indicated that the sediment accumulation rates in the boxes may be quite high and thus affect the survival of the amphipods. To overcome this problem, the design of the boxes was modified by replacing the basement of the boxes also with 1-mm nylon mesh. The aim was to reduce the sediment accumulation rates within the boxes by allowing the sediment to percolate through the bottom mesh. Both box types were compared during exposure periods with runoff (see the section "Results and discussion" for details). The modified exposure box design was used in all subsequent experiments.

Sampling of amphipod populations occurring in the stream itself was conducted using conventional Surber sampling (area: 0.062 m^2; 1-mm mesh). At each site 4 independent samples were taken over a stretch of 50 m.

Total suspended solids (TSS) were measured using a HTS1 turbidity meter (Dr. Lange, Duesseldorf, Germany). To calibrate the turbidity measurements as described by Gippel,[22] certain samples were filtered through pre-weighed Whatman (Springfield Mill, UK) GF/F (0.45-μm pore size) glass microfiber filters and dried at 60°C for 48 h. The filter paper was re-weighed to determine TSS. Physico-chemical water parameters were measured with test kits from Macherey & Nagel, Dueren, Germany or electronic meters from Wissenschaftliche Technische Werkstaetten, Weilheim, Germany.

Analysis for the persistent organochlorine insecticide lindane and the currently used insecticides Fenvalerate (pyrethroid) and parathion-ethyl (organophosphate) was done at the Institute for Ecological Chemistry and Waste Analysis, TU Braunschweig. All water samples were filtered through pre-weighed Whatman GF/F (0.45-μm pore size) glass microfiber filters prior solid-phase extraction of 1000-ml samples with C_{18}-columns (J.T. Baker, Griesheim, Germany). Pesticides from solid particles were extracted with acetone. The measurements were made with GC/ECD and confirmed with GC/MS, with the following quantification limits: lindane and parathion-ethyl 0.01 μg/l for water and 1 μg/kg for particles; Fenvalerate, 0.05 μg/l for water and 5 μg/kg for particles. Details of the sampling methods and pesticide analysis are given by Liess et al.[21]

Procedures

Study objectives

A clear definition of the objectives of a planned *in situ* bioassay study and the data set to be generated are strongly recommended. This will help to identify implications for the test organism, exposure system and site selection, the experimental design, endpoints used, additional measurements required, as well as for the assessment of the toxicological and ecological validity and the statistical data analysis. Some of these aspects have already been discussed in the previous section, the others will follow below.

Site selection

Most importantly, the site or sites used for a particular study should be suitable to reflect the environmental stressor of interest. If agricultural non-point source pesticide pollution is concerned, sites will be most likely situated in upstream headwater sections of running water catchments or in ponds situated in areas of arable land use. Further downstream sections have the disadvantage of representing areas with multiple exposures, specifically if urban runoff, treated wastewater from urban settlements or industry enters the water bodies as well. It may thus be very difficult to separate the effect of non-point source pesticide pollution from other contaminants. If possible, it is also recommended to make sure that non-point source exposure is likely to happen in the area under study, either via spray drift during application or as a result of rainfall-induced runoff events. The presence of erosion rills leading from the agricultural fields into surface waters may be an indicator of edge-of-field runoff.[23]

Generally, it is of great importance to carefully select control or reference sites for each study (Table 24.2). Ideally, the control site is situated in the same catchment in order to show similar environmental conditions. With regard to non-point source pollution, and more specific to runoff, an ideal control site would receive also edge of field runoff, including flow changes, turbidity, and nutrients, but no pesticide contamination. It is quite obvious that such a combination of characteristics of a control site is almost impossible to find. Some investigations have therefore used the same study site as an internal control site over time.[7] Runoff events occurring before commencement of pesticide application in the area may be used as a control to represent the effect of runoff without pesticide exposure, and subsequent events are then used to represent the effects of pesticide-contaminated runoff. For this type of study it is, however, suggested to monitor real exposure in samples taken during all runoff events in order to verify the assumptions.[24] Another option to try and understand or rule out some of the parameters always associated with runoff is to test, e.g., the potential effects of turbidity in the laboratory, as it was done with the South African freshwater amphipod *Paramelita nigroculus*.[11]

Finally, two further aspects need to be considered in the site selection process. On the one hand, it is necessary to regain the exposure systems to be able to retrieve the organisms, which may sometimes be a problem in large uniform or rapidly changing habitats, such as estuaries. Geographical positioning systems (GPS) may be a helpful tool here. On the other hand, the sites need to be sufficiently remote to avoid damage due to vandalism, which has been considered as an important threat by various authors.[10]

Experimental design

Information from other studies about the number of replicate exposure boxes and the number of organisms used per replicate are summarized in Table 24.2. As mentioned before, the density in the exposure boxes should not exceed densities observed for the same species in the field. DeWitt et al.[10] used slightly less than half of the average amphipod population density at the field site in their *in situ* bioassays. There are examples of using two or more species together in the same exposure boxes,[7,25] which may be a valuable to gain more information, given that interference or competition between the species can be ruled out as a factor. As can be seen from Table 24.2, many studies used at least three replicate containers at each site, which may be considered as the very minimum from a statistical point of view.

It is worth to note that inter-laboratory comparisons that have been undertaken with *in situ* bioassays using *G. pulex* and metalliferous effluents indicated striking differences in the results.[2] Unfortunately, there is not much data available to address this aspect further.

The exposure time used depends on the entry routes, i.e., the number and timing of the expected exposure events (runoff or spray drift), on the life cycle of the test organism, and on other aspects, such as the expected control mortality (ideally $\geq$80%). Survival of *G. pulex* over exposure periods of 1 week usually was greater than 80%.[7] For *Hyalella azteca*, survival rates of more than 80% have been observed over time periods of up to 4 weeks.[1]

However, other environmental factors may also influence the survival of the test organisms. Matthiesen et al.[16] reported that transient ice cover killed 53% of *G. pulex* exposed over a 7-day period. Differences between exposure in sun and shade may have an effect,[26] specifically if the photo-induced toxicity, e.g., of polycyclic aromatic hydrocarbons (PAHs) is concerned.[9] As will be pointed out in the case study, suspended particles that accumulate in the exposure boxes may have a great impact on the test organism survival. Tucker and Burton[12] attributed mortalities detected using *H. azteca* exposed *in situ* to agricultural runoff to increased turbidities.

If tube exposure chambers are used that contain sediment-dwelling organisms and that are open at the bottom in order to allow exposure to contaminated sediment, the retrieval of these system needs to follow a procedure that ensures recovery of the test organisms.[3,5] The tube may be first gently loosened from the surrounding sediment by pushing it to a steep angle (relative to the sediment) and rotating it slowly in a circular fashion. With the tube tilted at approximately 45°, a thin plastic or metal disc may be slid over the bottom, while the tube is still in the sediment. With the disc held in place, the test chamber may then be pulled from the sediment.[3]

As a general advice, it is suggested to consult existing guidelines for conducting toxicity tests with the test species considered for *in situ* testing or to have a look at guidelines for similar species, prior to *in situ* testing. Some examples for general guidance are, e.g., References 27–30.

Toxicological endpoints

Survival is the most frequently used test endpoint used in *in situ* exposure studies (Table 24.2). A great advantage is that survival is an endpoint that can be measured in a rather objective way even by different investigators, although the disappearance of test organisms may cause problems if cannibalistic behavior cannot be ruled out. Most small invertebrates will also decay quickly, after they died, and they may thus not be found some days later. On the other hand, mortality is a relatively insensitive test endpoint, and other sub-lethal endpoints, such as behavior, reproduction, development, or sub-organismic biomarkers, may be more relevant, specifically as they do also have potential to indicate effects of endocrine disrupting chemicals.[31] Feeding rate (scope for growth) on pre-conditioned leaf discs has been successfully established in ecotoxicology as a functional sub-lethal biomarker for *in situ* exposed *G. pulex*.[6]

Most studies measure the response of the test organisms only once at the end of the *in situ* exposure period (Table 24.2). Amphipods exposed over a period of 11 weeks in an agricultural headwater stream in south England were controlled weekly and dead animals were replaced.[16] Exposure over the same time period was performed in a stream system receiving insecticide-contaminated runoff in Germany, during which the decrease in the number of test organisms was observed without replacement of dead animals.[7]

Additional measurements

Depending on the objectives of the study, a variety of additional variables are necessary to monitor. A general site characterization aiming at all potential factors affecting the survival of the test organisms and the performance of the bioassay exposure experiment thus needs to be performed. This characterization should include the upstream areas and their catchment, if running waters are concerned. Data on the historic use of the site may be very helpful in the interpretation of results from stagnant water bodies. General water-quality parameters, such as nutrient levels, hardness, temperature, pH, and oxygen levels, will complete the characterization of the study sites.

In the context of studies on non-point source pesticide pollution, information about the timing and location of pesticide application, the application rates, and compounds used will greatly facilitate the study design and interpretation of results. This information may be obtained from the farmers. In order to assess the probability and occurrence of edge-of-field runoff events, rainfall data are required. Data about wind direction and wind speed are relevant for spray drift studies.

As many other factors, such as flow rate, turbidity, other contaminant or nutrient concentrations, may vary during runoff and may also affect the test organism response, these factors need to be monitored. Sampling for this purpose should be triggered by the runoff event, and automatic samplers are thus often used for this purpose.[20] However, passive runoff samplers,[23] suspended particle samplers,[32] and high-water level samplers[33] have been developed and used as cheaper and simpler alternatives. All those devices were specifically designed to detect transient pesticide peaks during edge-of-field runoff conditions. Being the main factor of interest for the *in situ* studies discussed in this chapter, the pesticide exposure needs to be characterized, as detailed as possible, with regard to both peak concentrations and exposure duration. Suitable sampling methods have been described by Liess and Schulz,[20] and the analysis of pesticide compounds generally follows standardized methods.[19] Contaminant concentrations may also be measured in the exposed test organisms in order to assess bioaccumulation.[34]

Data analysis and reporting

Analysis of variance (ANOVA) has been most commonly used to evaluate results from *in situ* exposure studies (Table 24.2). In cases where only pairwise comparisons are needed, *t* tests are frequently performed. However, all these methods may require transformation of the data in order to fulfill the criteria of normal distribution, or non-parametric alternatives may be indicated. If multiple comparisons are performed, a Bonferroni correction should be applied to control for type I statistical errors. Comparisons of various treatments with the same control may be undertaken using Dunnett's *post-hoc* test as part of the ANOVA procedure. There are various standard books that cover statistical procedures in great detail (e.g., Reference 35).

In order to account for the toxicological relevance of the bioassay response, a clear reflection of the cause and effect relationship is of key importance in evaluating *in situ* studies. It should include the temporal, and if relevant, also the spatial relation of exposure and response. Moreover, it should assess the likelihood of the response to be a true result of the observed exposure in the light of known toxicity data and other confounding parameters present. Finally, the ecological relevance of the results should be addressed through comparison of the bioassay response with effects on the population, community, or ecosystem level in the field.

Case study

The present pilot study was designed to answer (1) as to what extent *in situ* bioassays with *G. pulex* are suitable for the detection of short-term (about 1 h) runoff-related pesticide inputs, and (2) whether the response in bioassays is related to the responses of the same species in the stream. The first question was addressed by comparing the bioassay response at two potentially contaminated sites with two uncontaminated control sites. The second question was approached by comparing bioassay responses with in-stream abundance data at the same site. In addition, some methodological aspects of confounding factors, such as elevated sediment accumulation in the exposure boxes or extensive macrophyte growth, were addressed. The assessment and interpretation of the results are finally presented in the section "Results and discussion."

The case study was carried out in 1997 at the Fuhse river system about 20 km west of Braunschweig in northern Germany (N 52° 1′; E 10° 27′; 150 m above sea level). An upstream (region 1; headwater) and a downstream (region 2) regions of the river system were chosen for the study, making sure that arable land was the main type of land-use (Table 24.3). Smaller headwater streams in the upstream region have agricultural non-point source pollution as the main anthropogenic impact. Some of the downstream reaches receive high nutrient loads, which in combination with lower water levels and higher light intensities in summer lead to very dense coverages with submerged aquatic macrophytes. In each region (regions 1 and 2), a control site (Controls 1 and 2), and an exposed site (Exposed 1 and 2) were selected. Sites protected from direct entry of edge-of-field runoff, e.g., by vegetated buffer strips, were assumed to serve as control sites, while sites with a high abundance of erosion rills were considered as exposed sites. Contamination with pesticides was, however, confirmed with event-triggered sampling and analysis of residues. The slopes in the catchment area vary between 1% and 2%. It is intensively cultivated (sugar beets, winter barley, and winter wheat). The most common soil types in the area are loess loam and clayey marl.

At each site, four boxes containing 30 adult *G. pulex* L. (Amphipoda; carapace length $>$ 6 mm) were installed in the stream. The density of test organisms in the boxes corresponded to the densities commonly observed on plant structures in the field.[24] In each box, 100 g sand, two stones (*ca.* 5 × 5 × 3 cm), four water-parsnip plants (*Berula erecta*), and four alder (*A. glutinosa*) leaves provided food and shelter. Amphipods pre-kept in the laboratory (see the section "Material required") were introduced into the boxes and retrieved after 7 days of exposure during all experiments. Survival of exposed amphipods was used as the toxicological endpoint. Three experiments were conducted in total:

- An initial experiment conducted during two subsequent exposure periods (7 days each) in April compared the survival of amphipods in differently designed boxes in order to assess the effects of accumulation of suspended particles in the boxes during runoff events with elevated turbidity. The site Exposed 1 was used for this experiment; however, no insecticide contamination occurred, as the exposure periods finished before commencement of insecticide spraying.

Table 24.3 Mean (± SE; $n = 3$ measurements taken in May) of morphological and structural parameters at the four study sites

Parameter	Upstream sites Control 1	Downstream sites Exposed 1	Control 2	Exposed 2
Width (m)	1.5 ± 0.3	1.3 ± 0.4	4.1 ± 0.8	4.9 ± 1.0
Depth (m)	0.23 ± 0.04	0.25 ± 0.09	0.77 ± 0.32	0.72 ± 0.19
Current velocity (m/s)	0.21 ± 0.01	0.26 ± 0.02	0.42 ± 0.08	0.51 ± 0.12
Arable land (%)	95 ± 6	97 ± 5	83 ± 14	89 ± 15

- A second experiment (the main experiment) aimed to exemplarily assess potential pesticide effects. It consisted of two subsequent exposure periods (7 days each) in May and all four sites (Control 1, Exposed 1, Control 2, and Exposed 2) were used during this experiment. During the first week, no runoff occurred, while a 17-mm storm rainfall event led to edge-of-field runoff from arable land, previously treated with insecticides (the pyrethroid Fenvalerate and the organophosphate parathion-ethyl) as part of the usual farming program.
- The third experiment was conducted in August only at the two downstream sites (Control 2 and Exposed 2). This experiment aimed to exemplify the influence of aquatic macrophytes and related low oxygen concentrations in summer on the bioassay results.

In order to address the ecological validity of the bioassay approach, the densities of *G. pulex* populations at the four study sites were quantified using conventional Surber sampling at monthly intervals between April and August. This survey was done to complement the second exposure experiment. The results of this survey are not the main subject of this chapter; however, they were included, where it is necessary to assess the ecological validity of the *in situ* bioassays.

Various additional measurements of water-quality parameters, pesticide exposure, and site characteristics were undertaken. The deposition rates of sediments in the exposure boxes are based on measurements of volume of sediment layer before its removal at the end of the exposure period. The modified exposure box design that was used during all subsequent exposure experiments lead to reduced and similar sediment accumulation rates in the boxes at all sites.

Water samples for pesticide analysis were taken manually as 1-h composite samples at all four sites using 1-l amber glass bottles. Samples were obtained during elevated water levels resulting from the runoff event that occurred during the second exposure period in May. The particles suspended in the water were accumulated continuously at all four sites using suspended particle samplers,[32] from which they were collected and analyzed at 7-day intervals.

For a description of the abundances in the field, both species were collected at the four study sites at monthly intervals, with a Surber sampler (area: 0.062 m^2; 1-mm mesh). At each site, four independent samples were taken over a stream length of 50 m.

Differences between the mean mortality rates (*in situ* bioassays) or mean densities (field samples) were tested for significance ($p \leq 0.05$) using one-way ANOVA for multiple comparisons or t test for pairwise comparisons. Data were transformed using $\ln(x + 1)$ to satisfy the assumptions of the tests. A Bonferroni correction was applied to control for type I statistical errors, which assessed statistical significance with $\alpha = 0.012$.

Results and discussion

General

As shown in Table 24.4, important water-quality parameters, measured in May and August, are comparable between sites within acceptable limits. Orthophosphate levels showed a tendency to increase from May to August; however, this increase was most pronounced at the generally nutrient-richer downstream sites Control 2 and Exposed 2. At both the sites, the nitrite and nitrate levels were also higher in summer. As a likely

Table 24.4 Mean (± SE; $n = 3$) of water-quality parameters measured during the day and aquatic macrophyte coverage at the four study sites in May (week interval without runoff) and in August

Parameter	Control 1		Exposed 1		Control 2		Exposed 2	
	May	August	May	August	May	August	May	August
Nitrite (mg/l)	0.05 ± 0.01	0.05 ± 0.01	0.05 ± 0.01	0.05 ± 0.02	0.04 ± 0.01	0.05 ± 0.01	0.04 ± 0.01	0.05 ± 0.01
Ammonium (mg/l)	0.5 ± 0.1	0.3 ± 0.1	0.2 ± 0.1	0.2 ± 0.1	0.7 ± 0.2	1.1 ± 0.2	0.9 ± 0.2	1.9 ± 0.5
Nitrate (mg/l)	15 ± 8	14 ± 7	19 ± 4	17 ± 8	9 ± 3	15 ± 4	16 ± 5	35 ± 5
Orthophosphate (mg/l)	0.04 ± 0.03	0.08 ± 0.04	0.05 ± 0.03	0.05 ± 0.05	0.12 ± 0.03	0.17 ± 0.11	0.09 ± 0.07	0.28 ± 0.13
Hardness (mg $CaCO_3$/l)	491 ± 16	484 ± 25	460 ± 21	481 ± 24	369 ± 49	446 ± 25	490 ± 17	417 ± 65
pH	7.9 ± 0.6	7.8 ± 0.5	8.0 ± 0.1	8.1 ± 0.4	8.2 ± 0.2	8.0 ± 0.6	7.9 ± 0.4	8.4 ± 0.7
Oxygen (mg/l)	9.1 ± 1.2	8.4 ± 1.1	9.4 ± 1.5	8.5 ± 1.2	10.2 ± 1.3	7.5 ± 1.0	9.8 ± 0.5	4.8 ± 0.7
Temperature (°C)	10.5 ± 2.2	19.7 ± 3.1	11.1 ± 2.4	20.1 ± 1.7	13.7 ± 2.6	23.4 ± 2.6	13.8 ± 1.1	23.8 ± 1.9
Macrophyte coverage (%)	28 ± 12	25 ± 9	37 ± 15	45 ± 14	21 ± 15	27 ± 19	28 ± 17	89 ± 11[a]

[a] Mainly fennel pondweed (*P. pectinatus*).

result of the higher nutrient levels,[36] the macrophyte coverage of the streambed with funnel pondweed (*Potamogeton pectinatus*) strongly increased at the site Exposed 2 from 28 ± 17% in May to 89 ± 11% in August. Moreover, the oxygen content measured during the day decreased from 9.8 ± 0.5 to 4.8 ± 0.7 mg/l. Low oxygen levels associated with high macrophyte densities are a known feature of nutrient-rich surface waters in summer.[37] Decreased oxygen contents in August were also observed at the other sites; however, they were generally at least as high as 7.5 ± 1.0 mg/l.

First experiment on exposure box design

The initial methodological exposure experiment conducted in April compared two different design options: exposure boxes with a closed PVC bottom (box design 1) and a modified design with the bottom of the box replaced by 1-mm nylon mesh (box design 2). As expected, there is no difference in the survival of *G. pulex* between the two box design options during a no-runoff exposure period with clear water at TSS levels below 20 mg/l (Figure 24.1). The total accumulation of sediment in individual boxes averaged at 0.73 ± 0.22 l/week ($n = 4$) during no-runoff conditions. During the following week interval a rainfall-related edge-of-field runoff event occurred leading to an increased TSS level of 856 ± 49 mg/l ($n = 3$). As a result, the survival of *G. pulex* in box design 1 was significantly reduced ($p < 0.01$) in comparison with box design 2. Although the survival in the modified box design 2 with a mesh bottom was lower during runoff (about 80%) than during no-runoff time intervals (>90%), the differences were not significant (Figure 24.1). The sediment accumulation rate during runoff was 3.76 ± 0.71 l/week in box design

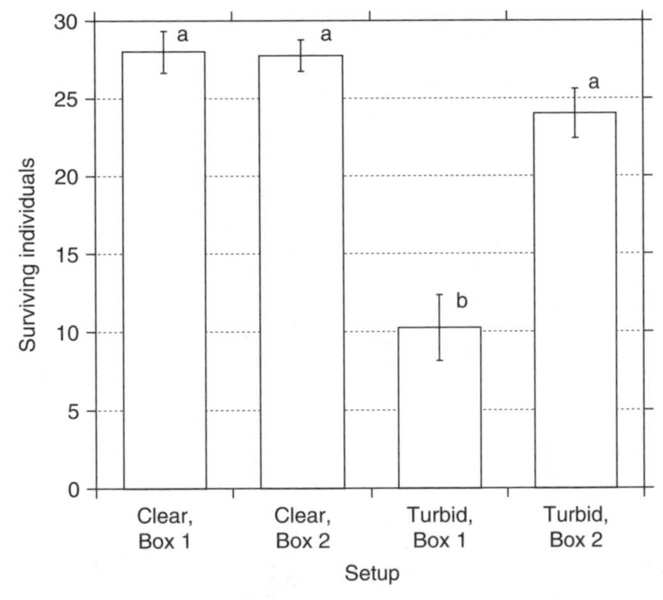

Figure 24.1 Mean (± SE; $n = 4$) survival of *G. pulex* exposed *in situ* in two different exposure box versions at the upstream site Exposed 1 during a no-runoff week interval with clear water and during a subsequent week interval with runoff-related increases in turbidity and sediment accumulation rates in April. Box 1 had a non-pervious PVC bottom, while the modified box 2 had a pervious 1-mm nylon mesh as the bottom. Different letters indicate significant (ANOVA; $p < 0.05$) differences in survival rate between the setups.

1 and 1.32 ± 0.69 l/week in box design 2 ($n = 4$). It is concluded that the modified box design 2 with a mesh bottom is more suitable during runoff studies with potential for high TSS levels. These boxes were thus used for the rest of the study, during which relatively low and similar sediment accumulation rates were observed in the boxes (<1.41 ± 0.95 l/week) at all sites. However, the results also suggest that elevated turbidity levels may affect the survival rates of *G. pulex*, which has already been reported from another study.[38] In contrast, a recent study using the amphipod *P. nigroculus*[11] demonstrated that the 7-day laboratory exposure to TSS levels as high as 1500 mg/l did not affect survival. *In situ* exposed *Ceriodaphnia dubia* showed a lower mortality during elevated turbidity rates in a stream where photo-induced PHAs were the main source of toxicity[9]; however, turbidity levels in this study were generally lower than during agricultural storm runoff in the present study.

Second experiment on pesticide effects: toxicological and ecological evaluation

Pesticide and TSS levels greatly increased during the second exposure interval in May, during which a rainfall-related edge-of-field runoff event occurred. Parathion-ethyl and Fenvalerate were detected at 47.5 and 35.8 μg/kg, respectively, in suspended particles, and at 2.1 and 0.3 μg/l, respectively in water samples taken from the upstream site Exposed 1 (Table 24.5). A much lower parathion-ethyl level of 2.5 μg/kg was detected at the downstream site Exposed 2 during runoff. Slightly elevated parathion-ethyl and background lindane levels were observed during no runoff at site Exposed 1 and during runoff at site Control 2. Compared with other studies the observed particle-associated pesticide levels were rather low, while the water-associated pesticide levels were comparably high.[39] Baughman et al.,[40] e.g., detected effects on *in situ* exposed estuarine shrimps *Palaemonetes pugio* during transient Fenvalerate peaks of 0.11 μg/l.

The survival of *in situ* exposed *G. pulex* was significantly decreased ($p < 0.0001$) to about 4 individuals (*ca.* 12%) during the week interval with runoff at the site Exposed 1 (Figure 24.2), at which elevated insecticide levels were observed (Table 24.5). No significant differences in survival were observed between no-runoff and runoff conditions at both control sites. Survival was also reduced to about 55% during runoff at the downstream site Exposed 2; however, differences were not significant ($p = 0.073$), most likely

Table 24.5 Pesticide levels detected in suspended particles and TSS at the four study sites in May during a week interval without runoff and during a subsequent week interval with edge-of-field runoff due to a 17-mm rainfall event (ND = not detected)

	Control 1		Exposed 1		Control 2		Exposed 2	
Parameter	No runoff	Runoff	No runoff	Runoff[a]	No runoff	Runoff	No runoff	Runoff
Lindane (μg/kg)	ND	ND	0.1	ND	ND	0.2	ND	ND
Parathion-ethyl (μg/kg)	ND	ND	ND	47.5	ND	0.3	ND	2.5
Fenvalerate (μg/kg)	ND	ND	ND	35.8	ND	ND	ND	ND
TSS (mg/l)	15	933	19	818	145	441	166	397

[a] Water samples taken during runoff contained 2.1 μg/l parathion-ethyl and 0.3 μg/l Fenvalerate.

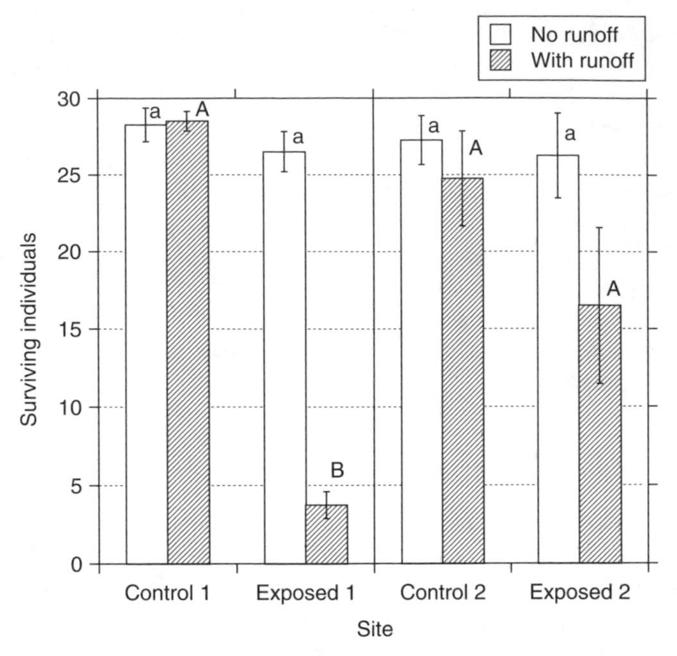

Figure 24.2 Mean (± SE; $n = 4$) survival of *G. pulex* exposed *in situ* during a no-runoff week interval and during a subsequent week interval with runoff in May at all four study sites. Different letters indicate significant (*t* test; $p < 0.05$) differences in survival rate between separate pairwise comparisons of treatments at the control and the respective exposed site for each time interval.

because of the generally higher variation in survival rates that was observed at this site. An increased number of replicates at both downstream sites might have been an option to overcome this problem of a lack in statistical power.

For a toxicological evaluation of the results, the stressor pesticide contamination should be evaluated against other environmental variables. Water-quality parameters did not differ between sites (Table 24.4). A higher rate of sediment deposition resulting from increased TSS levels during runoff (Table 24.5) can sometimes be a cause of increased invertebrate mortality,[41,42] but in the present study, increased TSS levels were observed at the exposed and the respective control sites, which had neither strongly increased insecticide contamination (Table 24.5) nor greater mortality (Figure 24.2). This indicates that in the present study the pesticide contamination and not the water-quality parameters or sediment caused the observed mortalities. This conclusion is corroborated by other *in situ* and survey-based investigations from another agricultural catchment nearby, in which hydraulic and water-quality parameters were tested separately from insecticides and excluded as the responsible factor for effects on macroinvertebrates.[7,24]

As part of the toxicological evaluation, it can be concluded that the *in situ* bioassay with *G. pulex* used in this study serves as a valuable tool for a toxicology-based detection of short-term insecticide input events.

In order to evaluate the presented bioassay results in the ecological context, they need to be compared with the abundance dynamic of the same species in the field. In the bioassays, *G. pulex* exhibited a clear decrease in the number of adult individuals when insecticides were present at the contaminated sites, while no or only a minor decrease occurred at the control site. In the field samples, however, the number of individuals of *G. pulex* present at these times remained almost the same. The abundance before and after the runoff event in May was 674 ± 443 individuals/m^2 and 649 ± 488 individuals/m^2. Overall increased the abundance of *G. pulex* from April to August at all four study sites.

While the bioassay indicated a significant decline of adult *G. pulex* at the contaminated sites, there was a general increase in the density of *G. pulex* in the field. The adult individuals confined to the exposure boxes are thus presumably exposed to higher concentrations and, accordingly, exhibit higher mortalities. In evaluating pesticide contaminations, and in extrapolating the results of *G. pulex* bioassays to the field, the possibility of an overestimation of toxicity must be considered. Even if the results of the *G. pulex* bioassay were not directly transferable to field populations of *G. pulex*, they can be used as a very helpful (measurement) tool to confirm that insecticide input events have occurred and to estimate their effects on other macroinvertebrate species. On the other hand, as shown in a previous study using *G. pulex* and the caddis fly *Limnephilus lunatus*,[7] the difficulty to observe an exposure–effect relationship in the field samples might be due to the elevated coefficients of variance. However, for *L. lunatus* even slightly higher differences in variance had not resulted in differences in measurable mortality.

Third experiment on the influence of extensive macrophyte coverage

As mentioned earlier, extensive coverage of funnel pondweed (*P. pectinatus*) was observed as a result of the combination of high light intensity, temperature, and strongly elevated nutrient levels at the downstream site Exposed 2 in summer. The extensive macrophyte coverage is a well-known phenomenon in the downstream stretches of the Fuhse river system, and mechanical weed control is usually performed during the summer months to avoid flooding of areas adjacent to the river during extensive rainfall events.

The survival of *G. pulex* at the site Exposed 2 in summer, which had an almost complete macrophyte coverage of 89 ± 11%, was significantly ($p < 0.01$) reduced (Figure 24.3). This is most likely due to the low oxygen levels (Table 24.4) that were observed as a result of the high macrophyte coverage. Considering the similarity of all other measured water-quality parameters, the oxygen level and reduced flow rates are assumed to be responsible for the observed increased mortality. It has been reported that *G. pulex* reacts susceptible to reduced oxygen levels.[43] Furthermore, *G. pulex* may be replaced in the field by the isopod *Asellus aquaticus*, which is less sensitive to low oxygen levels,[44] a phenomenon that can also be observed in downstream stretches of the river Fuhse.

Conclusion

Overall, these results emphasize that it is of high importance during *in situ* exposure studies to understand and/or eliminate confounding factors, such as macrophyte coverage or turbidity. Otherwise, a direct cause–effect relationship between contamination and bioassay response might be difficult to establish. However, the results also show that bioassays are a powerful tool for the detection and ecotoxicological assessment of agricultural non-point source pesticide pollution, given that a sufficient understanding of the ecological validity of the bioassay responses is available.

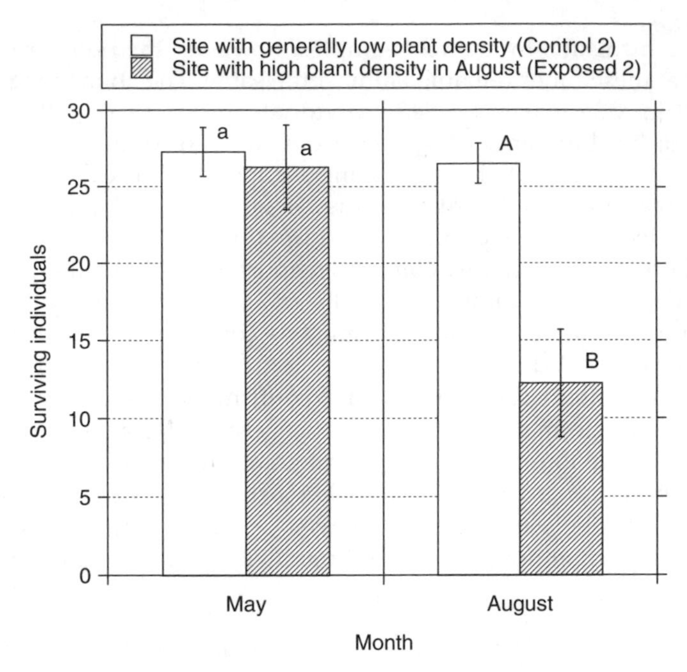

Figure 24.3 Mean ($\pm$ SE; $n = 4$) survival of *G. pulex* exposed *in situ* during a week interval each in May and August at site Control 2 with low macrophyte coverage and at site Exposed 2 with high macrophyte coverage in August. Different letters indicate significant (t test; $p < 0.05$) differences in survival rate between separate pairwise comparisons of sites in each May and August.

References

1. Chappie, D.J. and Burton, G.A., Jr., Optimization of *in situ* bioassays with *Hyalella azteca* and *Chironomus tentans*, *Environ. Toxicol. Chem.*, 16, 559–564, 1997.
2. Crane, M. and Maltby, L., The lethal and sublethal responses of *Gammarus pulex* to stress: sensitivity and sources of variation in an *in situ* bioassay, *Environ. Toxicol. Chem.*, 10, 1331–1339, 1991.
3. Sibley, P.K., Benoit, D.A., Balcer, M.D., Philipps, G.L., West, C.W., Hoke, R.A. and Ankley, G.T., *In situ* bioassay chamber for assessment of sediment toxicity and bioaccumulation using benthic invertebrates, *Environ. Toxicol. Chem.*, 18, 2325–2336, 1999.
4. Pereira, A.M.M., Mortagua, A., Soares, V.M., Goncalves, F. and Ribeiro, R., Test chambers and test procedures for *in situ* toxicity testing with zooplankton, *Environ. Toxicol. Chem.*, 18, 1956–1964, 1999.
5. Crane, M., Higman, M., Olsen, T., Simpson, P., Callaghan, A., Fisher, T. and Kheir, R., An *in situ* system for exposing aquatic invertebrates to contaminated sediments, *Environ. Toxicol. Chem.*, 19, 2000.
6. Maltby, L., Clayton, S.A., Wood, R.M. and McLoughlin, N., Evaluation of the *Gammarus pulex in situ* feeding assay as a biomonitor of water quality: robustness, responsiveness, and relevance, *Environ. Toxicol. Chem.*, 21, 361–368, 2002.
7. Schulz, R. and Liess, M., Validity and ecological relevance of an active *in situ* bioassay using *Gammarus pulex* and *Limnephilus lunatus*, *Environ. Toxicol. Chem.*, 18, 2243–2250, 1999.
8. Salazar, M.H., Duncan, P.B., Salazar, S.M. and Rose, K.A., *In situ* bioassays using transplanted mussels. II. Assessing contaminated sediments at a superfund site in Puget sound, in *Environmental Toxicology and Risk Assessment*, Hughes, J.S., Biddinger, G.R. and Mones, E., Eds., American Society for Testing and Materials, Philadelphia, 1995, pp. 242–263.

9. Ireland, D.S., Burton, G.A., Jr. and Hess, G.G., *In situ* toxicity evaluations of turbidity and photoinduction of polycyclic aromatic hydrocarbons, *Environ. Toxicol. Chem.*, 15, 574–581, 1996.
10. DeWitt, T.H., Hickey, C.W., Morrisey, D.J., Nipper, M.G., Roper, D.S., Williamson, R.B., van Dam, L. and Williams, E.K., Do amphipods have the same concentration-response to contaminated sediment *in situ* as *in vitro*? *Environ. Toxicol. Chem.*, 18, 1026–1037, 1999.
11. Schulz, R., Using a freshwater amphipod *in situ* bioassay as a sensitive tool to detect pesticide effects in the field, *Environ. Toxicol. Chem.*, 22, 1172–1176, 2003.
12. Tucker, K.A. and Burton, G.A., Assessment of nonpoint-source runoff in a stream using *in situ* and laboratory approaches, *Environ. Toxicol. Chem.*, 18, 2797–2803, 1999.
13. Schlenk, D., Huggett, D.B., Allgood, J., Bennett, E., Rimoldi, J., Beeler, A.B., Block, D., Holder, A.W., Hovinga, R. and Bedient, P., Toxicity of fipronil and its degradation products to *Procambarus* sp.: field and laboratory studies, *Arch. Environ. Contam. Toxicol.*, 41, 325–332, 2001.
14. Harris, M.L., Bishop, C.A., Struger, J., Ripley, B. and Bogart, J.P., The functional integrity of northern leopard frog (*Rana pipiens*) and green frog (*Rana clamitans*) populations in orchard wetlands. II. Effects of pesticides and eutrophic conditions on early life stage development, *Environ. Toxicol. Chem.*, 17, 1351–1363, 1998.
15. McClain, J.S., Oris, J.T., BurtonG.A., Jr. and Lattier, D., Laboratory and field validation of multiple molecular biomarkers of contaminant exposure in rainbow trout (*Oncorhynchus mykiss*), *Environ. Toxicol. Chem.*, 22, 361–370, 2003.
16. Matthiesen, P., Sheahan, D., Harrison, R., Kirby, M., Rycroft, R., Turnbull, A., Volkner, C. and Williams, R., Use of a *Gammarus pulex* bioassay to measure the effects of transient carbofuran runoff from farmland, *Ecotox. Environ. Safety*, 30, 111–119, 1995.
17. Maltby, L. and Crane, M., Responses of *Gammarus pulex* (Amphipoda, Crustacea) to metalliferous effluents — identification of toxic components and the importance of interpopulation variation, *Environ. Pollut.*, 84, 45–52, 1994.
18. Hoffman, E.R. and Fisher, S., Comparison of a field and laboratory-derived population of *Chironomus riparius* (Diptera, Chironomidae) — biochemical and fitness evidence for population divergence, J. *Econ. Ent.*, 87, 318–325, 1994.
19. Keith, L.H., *Compilation of EPA's Sampling and Analysis Methods*, CRC Press, Boca Raton, 1996, pp. 1–1696.
20. Liess, M. and Schulz, R., Sampling methods in surface waters, in *Handbook of Water Analysis*, Nollet, L.M.L., Ed., Marcel Dekker, New York, 2000, pp. 1–24.
21. Liess, M., Schulz, R., Liess, M.H.-D., Rother, B. and Kreuzig, R., Determination of insecticide contamination in agricultural headwater streams, *Water Res.*, 33, 239–247, 1999.
22. Gippel, C.J., Potential of turbidity monitoring for measuring the transport of suspended solids in streams, *Hydrol. Processes*, 9, 83–97, 1995.
23. Schulz, R., Hauschild, M., Ebeling, M., Nanko-Drees, J., Wogram, J. and Liess, M., A qualitative field method for monitoring pesticides in the edge-of-field runoff, *Chemosphere*, 36, 3071–3082, 1998.
24. Schulz, R. and Liess, M., A field study of the effects of agriculturally derived insecticide input on stream macroinvertebrate dynamics, *Aquat. Toxicol.*, 46, 155–176, 1999.
25. Jacobi, G.Z., Containers for observing mortality of benthic macroinvertebrates during antimycin treatment of a stream, *The Progressive Fish Culturist*, 39, 103–104, 1977.
26. Hatch, A.C. and Burton, G.A., Jr., Sediment toxicity and stormwater runoff in a contaminated receiving system: consideration of different bioassays in the laboratory and field, *Chemosphere*, 39, 1001–1017, 1999.
27. Carr, R.S. and Chapman, D.C., Comparison of methods for conducting marine and estuarine sediment porewater toxicity tests-extraction, storage, and handling techniques, *Arch. Environ. Contam. Toxicol.*, 28, 69–77, 1995.
28. OECD, Detailed review paper on aquatic testing methods for pesticides and industrial chemicals, *OECD Environmental Health and Safety Publications*, 11, 1–303, 1998.

29. USEPA, *Methods for Measuring the Acute Toxicity of Effluents and Receiving Waters to Freshwater and Marine Organisms*, 4th ed., United States Environmental Protection Agency, Cincinnati, OH, 1993, p. 198.
30. USEPA, *Methods for Measuring the Toxicity and Bioaccumulation of Sediment-associated Contaminants with Freshwater Invertebrates*, EPA 600/R-94/024, Duluth, MN, 1994.
31. Lagadic, L. and Caquet, T., Invertebrates in testing of environmental chemicals: Are they alternatives? *Environ. Health Persp.*, 106, 593–611, 1998.
32. Liess, M., Schulz, R. and Neumann, M., A method for monitoring pesticides bound to suspended particles in small streams, *Chemosphere*, 32, 1963–1969, 1996.
33. Schulz, R., Peall, S.K.C., Dabrowski, J.M. and Reinecke, A.J., Current-use insecticides, phosphates and suspended solids in the Lourens River, Western Cape, during the first rainfall event of the wet season, *Water SA* 27, 65–70, 2001.
34. Ingersoll, C.G., Ankley, G.T., Benoit, D.A., Brunson, E.L., Burton, G.A., Dwyer, F.J., Hoke, R.A. and Landrum, P.F., Toxicity and bioaccumulation of sediment-associated contaminants using freshwater invertebrates: a review of methods and applications, *Environ. Toxicol. Chem.*, 14, 1885–1894, 1995.
35. Sokal, R.R. and Rohlf, F.J., *Biometry*, W.H. Freeman, San Francisco, 1981, p. 859.
36. Portielje, R. and Roijackers, R.M.M., Primary succession of aquatic macrophytes in experimental ditches in relation to nutrient input, *Aquat. Bot.*, 50, 127–140, 1995.
37. Fisher, S.J. and Willis, D.W., Seasonal dynamics of aquatic fauna and habitat parameters in a perched upper Missouri River wetland, *Wetlands*, 20, 470–478, 2000.
38. Schulz, R. and Liess, M., Runoff-related short-term pesticide input into agricultural streams: measurement by use of an *in situ* bioassay with aquatic macroinvertebrates, *Verh. Ges. Ökol.*, 27, 399–404, 1997.
39. Schulz, R., Field studies on exposure, effects and risk mitigation of aquatic nonpoint-source insecticide pollution — a review, J. *Environ. Qual.*, 33, 419–448, 2004.
40. Baughman, D.S., Moore, D.W. and Scott, G.I., A comparison and evaluation of field and laboratory toxicity tests with Fenvalerate on an estuarine crustacean, *Environ. Toxicol. Chem.*, 8, 417–429, 1989.
41. Culp, J.M., Wrona, F.J. and Davies, R.W., Response of stream benthos and drift to fine sediment deposition versus transport, *Can. J. Zool.*, 64, 1345–1351, 1986.
42. Cooper, C.M., Benthos in Bear Creek, Mississippi: effects of habitat variation and agricultural sediments, J. *Freshwater Ecol.*, 4, 101–113, 1987.
43. Meijering, M.P.D., Lack of oxygen and low pH as limiting factors for *Gammarus* in Hessian brooks and rivers, *Hydrobiologia*, 223, 159–171, 1991.
44. Whitehurst, I.T., The *Gammarus–Asellus* ratio as an index of organic pollution, *Water Res.*, 25, 333–340, 1991.

chapter twenty-five

Improvements to high-performance liquid chromatography/photodiode array detection (HPLC/PDA) method that measures dioxin-like polychlorinated biphenyls and other selected organochlorines in marine biota

Gina M. Ylitalo, Jon Buzitis, Daryle Boyd, David P. Herman, Karen L. Tilbury, and Margaret M. Krahn
National Marine Fisheries Service

Contents

1-56670-664-5/05/$0.00+$1.50
© 2005 by CRC Press

Introduction

Organochlorines (OCs) are widespread chemical contaminants that frequently occur in the marine environment. Many of these compounds [e.g., polychlorinated biphenyls (PCBs), dichlorodiphenyltrichloroethane (DDTs), chlordane] possess certain chemical properties (e.g., stable, low flammability, lipophilic) that not only make them effective pesticides and industrial compounds but also make them highly persistent environmental contaminants.[1–3] Accumulation of certain OCs in marine organisms is linked to various deleterious biological and physiological effects, including reproductive impairment, immune suppression, and pathological lesions.[4–9] Measuring for toxic OCs, such as PCBs and DDTs, in marine organisms is therefore necessary to evaluate the risk to their health, as well as provide a measure of ecosystem quality.

Several methods have been developed to accurately determine trace levels of OCs.[10–18] Because numerous steps are needed to remove interfering compounds and to separate the various OCs before analyses by gas chromatography with electron capture detection (GC/ECD), gas chromatography/mass spectrometry (GC/MS), or high resolution GC/mass spectrometry (HRGC/MS), many of these methods are expensive, labor-intensive, and time-consuming. Due to these limitations, only a small number of samples can be analyzed in a short time span. Analytical methods that could rapidly and accurately measure toxic OCs in marine organisms are valuable, especially methods that could measure a suite of OCs at trace levels.

We have developed a rapid and inexpensive method to measure dioxin-like PCBs and other selected OCs using high-performance liquid chromatography with photodiode array detection (HPLC/PDA).[19] Total PCBs, summed DDTs and summed PCB toxic equivalents (TEQs) were calculated using these concentration data. Although the TEQ concentrations determined by HPLC/PDA are more conservative values compared to those measured by HRGC/MS because they include only dioxin-like PCBs and not other compounds (e.g., polychlorinated dibenzodioxins, polychlorinated dibenzofurans) that could contribute to the TEQ values, this method rapidly provided PCB TEQ concentration data for a wide range of marine biota. Recent modifications have been made to this HPLC/PDA method, including a more rapid extraction technique using an accelerated solvent extractor (ASE) with dichloromethane, as well as further analyses of HPLC/PDA extracts by GC/MS for a larger suite of OCs (e.g., additional PCB congeners, *o,p'*-DDE, oxychlordane, mirex, β-hexachlorocyclohexane). In this chapter, we describe the modifications made to the HPLC/PDA method and report some preliminary OC data for reference materials and killer whale samples that were analyzed using the expanded HPLC/PDA method. The OC levels were measured by the modified HPLC/PDA method in National Institute of Standards and Technology (NIST). Standard reference materials (SRMs) were then compared to the values determined by a more comprehensive method (GC/MS),[20] as well as to the certified values reported by NIST.

Materials required

Tissue extraction

See Reference 20 (Chapter 35 of this volume) for description of the ASE extraction method.

Tissue sample cleanup

Equipment:

- Auto desiccator, Sanpla Dry Keeper (Sanplatec Corp., Osaka, Japan)
- Rotovap with water bath (Buchi Switzerland, Krvr 65/45)
- Vacuum pump, model #0523-VHF-GS82DX (Gast Mfg. Corp., Benton Harbor, MI)
- Concentrator tube heaters with glass-cylinder shroud, wire tube holder, and aluminum inserts bored out to fit 50-ml conical tubes (Kontes, Vineland, NJ, catalog #720000-0000)

Supplies:

- Glass wool, Corning® 3950 (VWR Scientific; West Chester, PA; catalog #32848-003)
- Pyrex® glass funnel (Fisher Scientific; Pittsburgh, PA; catalog #10-346C)
- 16 in. Teflon® rod (Fisher Scientific; Pittsburgh, PA; catalog #14-518-10F)
- Glass cleanup columns, 305 mm long and 9 mm i.d., 20 mm i.d. for solvent reservoir (DJ Glass Factory, San Jose, CA; special-ordered glassware)
- 10-ml conical glass tubes (Fisher Scientific; Pittsburgh, PA; catalog #05-569-2)
- Kontes® 2-l one-neck round bottom flask (Fisher Scientific; Pittsburgh, PA; catalog #K601000-0824)
- 9-in. glass Pasteur pipettes (All World; Lynnwood, WA; catalog #020-03609)
- 50-ml concentrator tubes with caps (Fisher Scientific; Pittsburgh, PA; catalog #0553841a)
- 4-ml amber glass vials with self-sealing Teflon-coated caps and 250 μl glass inserts with springs (Sun SRI; Wilmington, NC; catalog #200-596)

Reagents:

- Acetone, HPLC grade (Fisher Scientific; Pittsburgh, PA; catalog #A929-4)
- Dichloromethane, high purity (Burdick & Jackson; Muskegon, MI; catalog #300-4)
- Concentrated hydrochloric acid, reagent grade (Fisher Scientific; Pittsburgh, PA; catalog #A144-212)
- Hexane, HPLC grade (VWR Scientific; West Chester, PA; catalog #BJGC217-4)
- Methanol, nanograde (VWR Scientific; West Chester, PA; catalog #MK516008)
- Concentrated sulfuric acid, trace metal grade (Fisher Scientific; Pittsburgh, PA; catalog #A510-500)
- Davisil® (WR Grace & Co.) 100–200 mesh silica, grade 634 (Fischer Scientific; Pittsburgh, PA; catalog #S734-1)
- Potassium hydroxide (Fisher Scientific; Pittsburgh, PA; catalog #P250-500)
- Ultra-pure nitrogen gas, Grade 4.8 (PraxAir, Danbury, CT)

Analyses of OCs by HPLC/PDA

Equipment:

- Two Cosmosil® 5-PYE analytical column (4.6 mm × 250 mm; 5 μm particles; Nacalai Tesque, Kyoto Japan; purchased through Phenomenex, Torrance, CA)

- Cosmosil 10-PYE guard column (4.6 mm × 50 mm; 10 μm particles; Nacalai Tesque, Kyoto Japan; purchased through Phenomenex, Torrance, CA)
- Two 6-port Rheodyne® electrically actuated valves (Cotati, CA)
- Model 7950 column chiller (Jones Chromatography, Lakewood, CO)
- Waters 515 isocratic pump (Milford, MA)
- Waters 715 Ultra WISP® autosampler (Milford, MA)
- Waters 996 PDA detector (Milford, MA)
- Empower® data acquisition software (Waters, Milford, MA)
- A Dell Dimension 8300 computer or equivalent computer that can run Microsoft Windows XP Pro®, has at least 2GB hard disk space, with at least 384 MB of RAM and is not a hard disk configured to be FAT 16

Reagents:

- Stock solution of 1,7,8-trichlorodibenzo-*p*-dioxin (1,7,8-Tri CDD), 1,2,3,4-tetrachlorodibenzo-*p*-dioxin (1,2,3,4-TCDD), individual PCB congeners, individual DDTs, and hexachlorobenzene; all at 100 ng/μl in isooctane (AccuStandard; New Haven, CT; see AccuStandard catalog for numbers)
- Ultra-high purity helium gas, grade 5, 99.999% (PraxAir; Danbury, CT)
- Copper, reagent grade, granular, 20–30 mesh (JT Baker; catalog #1720-05)

SRMs

- SRM® 1974b, a blue mussel homogenate (NIST; Gaithersburg, MD)
- SRM 1945, a whale blubber homogenate (NIST; Gaithersburg, MD)
- SRM 1946, a fish tissue homogenate (NIST; Gaithersburg, MD)

Analyses of OCs by GC/MS

Equipment:

- Agilent 7683 autosampler with an on-column injection syringe and needle guide (Agilent Technologies, Wilmington, DE)
- Agilent 6890N gas chromatograph with a cool on-column injection port
- Deactivated fused-silica guard column, 10-m × 0.53-mm (460-2535-10, Agilent Technologies)
- DB-5® (J & W Scientific, Folsom, CA) GC column (60-m × 0.25-mm, 0.25-μm film thickness, 122-5062, Agilent Technologies)
- Agilent 5973N Mass Selective Detector® (Agilent Technologies)
- Computer system with ChemStation® software (Version DA, Agilent Technologies)

Supplies:

- Indicating moisture trap (GMT-4-HP, Agilent Technologies)
- Disposable oxygen trap (803088, Supelco)
- Indicating oxygen trap (4004, Alltech Associates, Deerfield, IL)
- Glass, universal column union (705-0825, Agilent Technologies) sealed with polyimide sealing resin (500-1200, Agilent Technologies)

Reagents:

- Ultra-high purity helium gas, grade 5, 99.999% (PraxAir, Danbury, CT)
- Isooctane, Optima (Fisher Scientific; Pittsburgh, PA; catalog #0301-4)
- Stock solutions of individual PCB congeners for GC calibration standards and GC internal standard (PCB 103), 100 ng/μl in isooctane (AccuStandard; New Haven, CT; see AccuStandard catalog for numbers)
- Stock solutions of individual OC pesticides for GC calibration standards, 100 ng/μl in isooctane (AccuStandard; New Haven, CT; see AccuStandard catalog for numbers)

Procedures

HPLC/PDA standards

A standard solution (10 ng/μl) of each internal standard (1,2,3,4-TCDD and 1,7,8-TriCDD) was prepared in hexane. A single HPLC/PDA calibration solution (0.322 ng/μl except where noted) was prepared in hexane and contained the following compounds (in order of HPLC elution): PCB 200, *o,p′*-DDT, *p,p′*-DDE, 101, 153, 110, 138, 180, 128, 118, 114, *p,p′*-DDT, 170, 105, *o,p′*-DDD, 190, 156, 157, 189, 77, HCB, 1,7,8-TriCDD (HPLC internal standard, 0.800 ng/μl), 126, *p,p′*-DDD, 169, and 1,2,3,4-TriCDD (surrogate standard, 0.800 ng/μl). A second HPLC/PDA calibration solution was prepared in hexane as a 1/4 serial dilution of the above calibration solution.

GC/MS standards

A solution of the GC internal standard (PCB 103, 10 ng/μl) was prepared in isooctane. A series of 11 GC calibration standards (Level 1–Level 11) containing the OCs listed in Table 35.1, Chapter 35,[20] were used for calibration.

Sample extraction and cleanup

Prior to use, all glassware were rinsed three times with acetone to remove any potential contaminants. All glassware rinsing, tissue extraction, and cleanup procedures were performed in fume hoods by personnel wearing nitrile gloves and safety glasses. Because PCBs and other OCs may adhere to glass, all glassware once used were discarded.

OCs and lipids were extracted from tissues using an ASE (see Chapter 35 for description).[20] For HPLC/PDA analyses, the amount of tissue extracted ranged from 0.20 to 5.0 g, depending on the expected lipid and water content. In all cases, the sample mass was adjusted such that the total amount of lipid loaded onto the cleanup column was less than 0.3 g. Each tissue sample was weighed to the nearest 0.01 g in a tared 10-oz jar. Sodium sulfate (15 cc) was added and the sample contents were mixed thoroughly using a solvent-rinsed stainless steel spatula. Next, magnesium sulfate (15 cc) was added to each sample jar and mixed thoroughly using the spatula. Each sample mixture was transferred to a 33-ml ASE extraction cell and the OC surrogate standard (1,2,3,4-TCDD, 10 ng/μl) was added to the top of each sample cell. The OCs and lipids were sequentially extracted at 2000 psi and 100°C with two cell volumes using dichloromethane and the combined extract (~ 50 ml) was collected in a 60-ml collection tube. The extract was thoroughly mixed using a Vortex mixer. Prior to cleanup, a 1-ml aliquot of each sample extract was transferred to a 2-ml GC vial for lipid analysis by thin layer chromatography/flame ionization detection (TLC/FID) (see Chapter 12).[21] The remaining volume of each HPLC/PDA extract was then transferred to a 50-ml concentrator tube and reduced to <1 ml by evaporation with heat.

Sample extracts were cleaned up on gravity flow cleanup columns described by Krahn *et al.*[19] Briefly, each cleanup column was prepared with the following media (listed in order of addition to column): a glass wool plug, 0.4 g neutral silica, 0.5 g KOH-impregnated silica (basic silica),[22] and 2.5 g of 40% (w/w) sulfuric acid-impregnated silica gel. Because preparation of the basic silica gel is time consuming due to extensive glassware setup, it was advantageous to prepare several large batches (approximately 500 g each) at one time and store them in an autodesiccator until needed. Conversely, the acidic silica gel was prepared just prior to use by weighing 40 g of activated silica gel in a tared pre-cleaned 4-oz glass jar and then adding 14.5 ml of sulfuric acid. The jar was shaken by hand for 30 min to ensure that the silica gel was impregnated evenly with the sulfuric acid. The cleanup columns were washed with 20 ml of hexane/dichloromethane (1:1, v/v), and the total sample extract (< 1 ml) was added to the top of each column using a 9 in. glass pipette. The volume of sample extract added to the cleanup column must be <1 ml in order for the surrogate standard (1,2,3,4-TCDD) to be eluted properly. To remove any remaining OCs in the concentrator tube, each tube was rinsed with 1 ml dichloromethane and the rinsate was immediately added to the top of the column. After the sample extract was completely loaded onto the column, the OCs were eluted with 22 ml of hexane/dichloromethane (1:1, v/v) and were collected in a clean, pre-weighed 50-ml concentrator tube. The HPLC internal standard (1,7,8-TriCDD; 10 ng/μl) was then added to each tube and mixed thoroughly. Each sample tube was capped, weighed and the weights recorded. For selected samples, an 8-ml aliquot of the sample extract was transferred to 10-ml tapered glass tube to analyze for a larger suite of OCs by GC/MS. The 50-ml concentrator tube was capped and reweighed. Copper was activated by covering it with concentrated hydrochloric acid, stirring the mixture with a glass rod, and allowing it to stand for 5 min. The copper was then sequentially washed three times each with methanol and dichloromethane and was covered with dichloromethane until use. Approximately 0.50 g activated copper was added to each of the sample tubes to remove any elemental sulfur introduced by the sulfuric acid silica gel and the extract was concentrated to ~1 ml by evaporation with heat. The sample extract was transferred to a GC vial and the solvent volume was further reduced to 100–150 μl, depending on expected OC concentrations, under a stream of ultra-pure nitrogen gas. Using a 9-in. glass pipette, the sample extract was transferred to a 250-μl insert in a 4-ml amber vial.

OC analyses by HPLC/PDA

A detailed description of the HPLC/PDA method can be found in Reference 19. Briefly, the dioxin-like congeners (PCBs 77, 105, 118, 126, 156, 157, 169, 189) were resolved from other selected PCBs (PCBs 110, 128, 138, 153, 170/194, 180, 190, 200) and pesticides (*o,p'*-DDD, *p,p'*-DDD, *p,p'*-DDE, *o,p'*-DDT, *p,p'*-DDT, HCB) by HPLC on two Cosmosil PYE analytical columns, connected in series and cooled to 18°C using a column chiller. The flow rate of the eluent was 0.94 ml/min throughout the analysis. A 50-μl portion of each sample extract was injected onto the PYE columns. For optimum separation of PCB congeners on the HPLC system, we have found that the column temperature and HPLC flow must be adjusted for each new set of PYE columns used.

The OCs were measured by an ultraviolet (UV) PDA detector using the following parameters: start wavelength, 200 nm; end wavelength, 310 nm; spectra/s, 2.0; resolution, 4.8 nm; as described previously in Reference 19. The OCs were identified by comparing their UV spectra (200–310 nm) and retention times to those of reference standards in a library. The analyte purity was confirmed by comparing spectra within a peak to the apex spectrum (see Reference 19 for description). The PDA measurements were performed by integrating peaks recorded at 220 nm (wavelength that is near the maximum UV absorbance for most PCB congeners) for all OCs except *p,p′*-DDE. For this analyte, the peak at 266 nm was integrated. Although the total run time for each sample analysis was 55 min, data were collected for only 40 min to minimize the size of the data files.

OC analysis by GC/MS

Each 8-ml aliquot of sample extract was reduced in volume to approximately 250 μl under a stream of ultra-pure nitrogen gas. The extract was then quantitatively transferred to a 1-ml GC vial containing a 450-μl glass insert to which 15 μl of GC internal standard (PCB 103, 2 ng/μl in isooctane) was previously added. The 10-ml tapered tube was rinsed with an additional 150 μl dichloromethane and the rinsate added to the insert and mixed thoroughly. The combined extract (~415 μl) was mixed thoroughly and exchanged into isooctane by first reducing its volume to approximately 100 μl under nitrogen gas and then adding 120 μl of isooctane to the sample insert. The sample extract was then reduced to a final volume of 100 μl.

The analytes of interest were separated on a 60-m DB-5 capillary column (250 μm film thickness) and analyzed by GC/MS (model Agilent 5972/3) operated in the electron impact (EI) single ion monitoring (SIM) mode (see the section "GC/MS for Quantitating CHs" in Chapter 35, for a comprehensive description of the GC/MS method).[20] The instrument was calibrated using a series of 11 multi-level calibrations standard solutions containing known amounts of 69 different OC compounds (PCBs, pesticides) from which response factors relative to PCB 103 were computed. The relative recovery of the surrogate standard, 1,2,3,4-TCDD was then computed for each sample and used to correct computed sample extract concentration results for losses incurred during sample processing.

Quality assurance

Each extraction sample set consisted of 11–14 field samples and various quality assurance samples. To determine if the extraction reagents, glassware, or other supplies were free of OC contaminants, a method blank was analyzed with every sample set. To evaluate the precision of the method, approximately 10% of the tissue samples were analyzed in duplicate. A NIST SRM (e.g., SRM 1945, SRM 1946, SRM 1974b) was analyzed with each sample set to measure the accuracy of our method set and to show that the laboratory quality assurance criteria were met for all analytes detected in the tissue samples.[23]

Results and discussion

In the mid-1990s, we developed a HPLC/PDA method[19] to rapidly measure concentrations of dioxin-like PCBs and other selected OCs in tissues of marine organisms for various studies.[8,24–28] With this older HPLC/PDA method, OCs were extracted from various marine tissue samples using a rather cumbersome and time-consuming technique — homogenization with sodium sulfate using a Tissuemizer and extraction with pentane/hexane (50:50, v/v). Recently, an ASE was set up in our laboratory, and OCs were extracted with dichloromethane from a wide range of marine tissues. Similar to the old method, the sample extracts obtained by the new method (ASE with dichloromethane) were then cleaned up using gravity flow acidic silica cleanup columns and were analyzed rapidly by HPLC/PDA for dioxin-like PCBs and other selected OCs.[19]

Replicate samples of three NIST SRMs were analyzed by HPLC/PDA, and the OC results were compared to NIST certified or recommended concentrations of these analytes, as well as to those measured by a GC/MS method, in which samples were extracted similarly by ASE with dichloromethane but were cleaned up using HPLC/size exclusion chromatography.[20] In general, the OC data (based on wet weight) obtained by our HPLC/PDA were in good agreement with the NIST concentrations, as well as the GC/MS results (Table 25.1). However, some PCB congeners coeluted on the HPLC system (e.g., PCB 170 coeluted with PCB 194; PCB 87 coeluted with PCB 153). In addition, we found coeluting PCBs present at the retention times for PCBs 118 and 128 in certain SRMs (SRM 1945 and SRM 1946). As a result of these interferences, the concentrations of these analytes measured by HPLC/PDA were higher than those measured by GC/MS.[19] For the blue mussel homogenate (SRM 1974b), neither *p,p′*-DDE nor *o,p′*-DDT could be quantitated by HPLC/PDA due to coelution of these pesticides with non-planar PCBs. Therefore, the ΣDDTs determined for SRM 1974b by HPLC/PDA did not include these values and, consequently, led to underestimated values compared to those determined by GC/MS (see Table 25.1).

We compared the HPLC/PDA-measured concentrations of various OCs in all three NIST SRMs using the two extraction methods, specifically homogenization with Tissuemizers (old method) or with ASE using dichloromethane (new method). The OC results are presented in Figure 25.1. In general, comparable levels of OCs were measured by HPLC/PDA in these homogenate samples extracted by either extraction method, with the largest differences occurring for the more polar DDTs. From a more practical viewpoint, the main difference between these two methods was the amount of time needed to extract a sample set. The old method typically required 12 h to prepare the Tissuemizer probes and extract samples, whereas only 8 h is usually needed to extract a set of 14 samples using the new ASE-based method. In addition, the amount of solvent needed to prepare and extract a set of samples was reduced approximately 50% when using the new ASE extraction method compared to the solvent amount used with the old method. Fewer contaminant peaks have been measured in method blanks extracted by ASE with dichloromethane compared to the method blanks extracted using Tissuemizers with pentane/hexane. Furthermore, because the ASE is an automated instrument that can extract a set of samples unattended, other laboratory tasks can concurrently be completed as the samples are being extracted — saving both time and labor costs.

As a test to determine if there were differences in OC values determined by HPLC/PDA and GC/MS, we analyzed the same killer whale sample extracts that had gone through acidic silica cleanup by both quantitation methods. The OC results for tissues collected from three killer whale tissue samples are shown in Table 25.2. Concentrations

Table 25.1 Mean concentrations (± SD) of selected PCBs and other OCs measured in three standard reference materials (SRMs) determined by HPLC/PDA and GC/MS

		NIST SRM 1974b, blue mussel homogenate			NIST SRM 1945, whale blubber homogenate			NIST SRM 1946, fish muscle homogenate		
		HPLC/PDA ($n = 6$)	GC/MS[a] ($n = 15$)	NIST (certified)	HPLC/PDA ($n = 5$)	GC/MS[a] ($n = 15$)	NIST (certified)	HPLC/PDA ($n = 5$)	GC/MS[a] ($n = 8$)	NIST (certified)
Dioxin-like PCBs	77	0.57 ± 0.034	[b]	**0.56 ± 0.023**[c]	[d]	[b]	[e]	0.39 ± 0.027	[b]	0.327 ± 0.025
	105	4.8 ± 0.29	4.3 ± 0.30	4.00 ± 0.18	33 ± 5.2	29 ± 1.8	30.1 ± 2.3	22 ± 1.3	21 ± 1.7	19.9 ± 0.9
	118	12 ± 1.0	11 ± 0.31	10.3 ± 0.4	110 ± 5.5[f]	85 ± 4.0	74.6 ± 5.1	66 ± 1.3[f]	52 ± 3.4	52.1 ± 1.0
	126	[d]	[b]	[e]	[d]	[b]	[e]	0.36 ± 0.038	[b]	0.380 ± 0.017
	156	0.50 ± 0.088	0.80 ± 0.07	0.718 ± 0.080	9.1 ± 0.92	14 ± 1.9	10.3 ± 1.1	7.0 ± 0.38	9.9 ± 0.63	9.52 ± 0.51
	157	[d]	[b]	**0.236 ± 0.024**[c]	[d]	[b]	[e]	2.0 ± 0.31	[b]	[e]
	169	[d]	[b]	[e]	[d]	[b]	[e]	[d]	[b]	0.106 ± 0.014
	189	[d]	[b]	[e]	[d]	[b]	[e]	1.3 ± 0.089	[b]	[e]
Other PCBs	110	9.2 ± 0.10	9.9 ± 0.30	10.0 ± 0.7	37 ± 14	38 ± 3.9	23.3 ± 4.0	32 ± 0.84	25 ± 1.2	22.8 ± 2.0
	128	2.3 ± 0.32	2.1 ± 0.10	1.79 ± 0.12	56 ± 9.6[f]	27 ± 3.5	23.7 ± 1.7	49 ± 2.3[f]	25 ± 1.4	22.8 ± 1.9
	138	8.4 ± 0.28	13 ± 0.42[g]	9.2 ± 1.4	150 ± 16	190 ± 13[g]	131.5 ± 7.4	110 ± 4.5	170 ± 9.1[g]	115 ± 13
	153	17 ± 1.0	15 ± 0.39[h]	12.3 ± 0.8	290 ± 15	250 ± 15[h]	213 ± 19	180 ± 4.5	200 ± 14[h]	170 ± 9
	170/190	[d]	0.20 ± 0.18	0.269 ± 0.034	92 ± 3.8[i]	54 ± 5.1	40.6 ± 2.6	34 ± 0.89[i]	34 ± 2.0	25.2 ± 2.2
	180	0.88 ± 0.14	1.3 ± 0.11	1.17 ± 0.10	140 ± 8.4	160 ± 14	106.7 ± 5.3	74 ± 1.2	81 ± 4.1	74.4 ± 4.0
Pesticides	*o,p′*-DDD	0.70 ± 0.10	1.3 ± 0.39	1.09 ± 0.16	17 ± 2.8	24 ± 3.3	18.1 ± 2.8	[j]	1.1 ± 0.64	2.2 ± 0.25
	p,p′-DDD	2.6 ± 0.17	4.5 ± 0.48	3.34 ± 0.22	120 ± 8.4	150 ± 14	133 ± 10	7.2 ± 0.19	13 ± 1.5	17.7 ± 2.8
	p,p′-DDE	[j]	4.5 ± 0.28	4.15 ± 0.38	500 ± 37	530 ± 40	445 ± 37	290 ± 4.0	380 ± 27	373 ± 48
	o,p′-DDT	[j]	[d]	**0.894 ± 0.057**[c]	83 ± 19	88 ± 6.1	106 ± 14	26 ± 2.6	18 ± 1.4	**22.3 ± 3.2**[c]

Continues

Table 25.1 Continued

	NIST SRM 1974b, blue mussel homogenate			NIST SRM 1945, whale blubber homogenate			NIST SRM 1946, fish muscle homogenate		
	HPLC/PDA (n = 6)	GC/MS[a] (n = 15)	NIST (certified)	HPLC/PDA (n = 5)	GC/MS[a] (n = 15)	NIST (certified)	HPLC/PDA (n = 5)	GC/MS[a] (n = 8)	NIST (certified)
p,p′-DDT	[d]	0.63 ± 0.26	**0.396 ± 0.096**[c]	220 ± 15	270 ± 33	245 ± 15	33 ± 1.8	41 ± 4.7	37.2 ± 3.5
HCB	[d]	[d]	[e]	31 ± 1.6	27 ± 1.3	32.9 ± 1.7	6.3 ± 0.10	7.5 ± 0.28	7.25 ± 0.83
Summed PCBs (ΣPCBs)	120 ± 5.2[k]	160 ± 4.1[l]	**205 ± 42**[c]	1,900 ± 160[k]	1,800 ± 140[l]	[e]	920 ± 24[k]	930 ± 45[l]	[e]
SummedDDTs (ΣDDTs)	3.3 ± 0.14[m]	11 ± 1.1[n]	[e]	930 ± 58[m]	1,100 ± 76[n]	[e]	360 ± 6.7[m]	460 ± 31[n]	[e]

[a] GC/MS method described in Chapter 35.[20]

[b] Cannot be measured.

[c] Values in bold are NIST reference concentrations.

[d] Not detected.

[e] Not reported.

[f] There were coeluting PCBs that affected the quantitation (increased the value).

[g] Known to coelute with PCBs 166, 163, and 164.

[h] Known to coelute with PCB 132.

[i] On HPLC/PDA system, PCBs 170/194 coelute.

[j] Not quantitated because coeluting PCBs were present.

[k] Summed PCBs (ΣPCBs) were calculated by summing the concentrations of 14 reported congeners, then adding the concentrations of the unreported congeners, which are calculated using the response factor of PCB101 when present, of PCB138 if congener 101 is not present, or the average response factor over all congeners if neither PCB is present.

[l] Summed PCBs (ΣPCBs) were calculated by summing the concentrations of 40 reported congeners (PCBs 17, 18, 28, 31, 33, 44, 49, 52, 66, 70, 74, 82, 87, 95, 99,101/90, 105, 110, 118, 128, 138, 149, 151, 153/132, 156, 158, 170/190, 171, 177, 180, 183, 187, 191, 194, 195, 199, 205, 206, 208, 209).

[m] Summed DDTs (ΣDDTs) were calculated by summing the concentrations of *o,p′*-DDD, *o,p′*-DDT, *p,p′*-DDE (at 266 nm), *p,p′*-DDD, and *p,p′*-DDT.

[n] Summed DDTs (ΣDDTs) were calculated by summing the concentrations of *p,p′*-DDT, *p,p′*-DDE, *p,p′*-DDD, *o,p′*-DDD, *o,p′*-DDE, and *o,p′*-DDT.

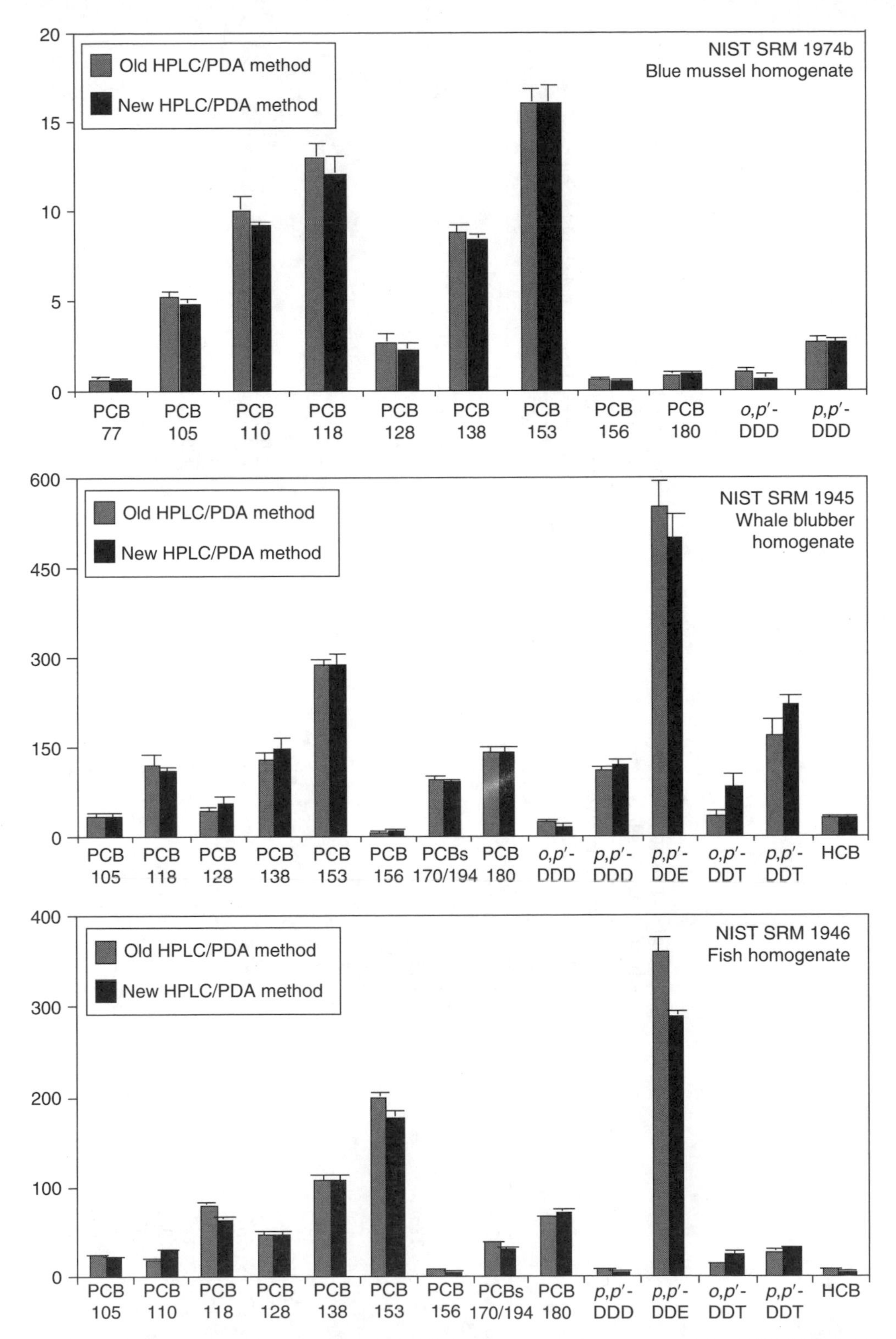

Figure 25.1 OC concentrations for NIST 1974b (a blue mussel homogenate), NIST SRM 1945 (a whale blubber homogenate), and NIST SRM 1946 (a fish muscle homogenate) extracted by two different extraction methods (old method: homogenization with pentane/hexane; new method: ASE with dichloromethane).

Table 25.2 Concentrations (ng/g, wet weight) of selected PCBs and other OCs in the same killer whale sample extracts that had gone through acidic silica cleanup and were measured by HPLC/PDA and GC/MS

		Northern resident, juvenile killer whale whole blood ($n = 1$)		Gulf of Alaska transient, female killer whale blubber biopsy ($n = 1$)		Western Alaska, offshore male killer whale blubber biopsy ($n = 1$)	
		HPLC/PDA	GC/MS	HPLC/PDA	GC/MS	HPLC/PDA	GC/MS
Dioxin-like PCBs	77	<0.056	[a]	<1.9	[a]	<1.1	[a]
	105	1.6	1.3	190	190	44	39
	118	7.7[b]	4.0	1,100[b]	870	190	180
	126	<0.053	[a]	<1.7	[a]	<0.98	[a]
	156	<0.036	0.52	26	57	13	22
	157	<0.033	[a]	20	[a]	5.1	[a]
	169	<0.071	[a]	<2.3	[a]	<1.4	[a]
	189	<0.035	[a]	[c]	[a]	<0.75	[a]
Other PCBs	52	[a]	4.5	[a]	720	[a]	53
	99	[a]	6.5	[a]	2,100	[a]	210
	101	[a]	4.7	[a]	500	[a]	240
	110	[a]	0.64	[a]	31	[a]	21
	128	2.6[b]	1.8	600[b]	430	110[b]	69
	138	8.6	12	3,600	4,200	370	530
	149	[a]	5.7	[a]	1,000	[a]	270
	153	16	17	8,100[b]	6,700	760	700
	170/190[d]	1.3	1.4	900	740	72	70
	180	2.4	3.7	2,300	2,000	160	200
	187	[a]	3.0	[a]	1,100	[a]	180
	194	[d]	0.19	[d]	310	[d]	20

Pesticides	*o,p'*-DDD	<0.098	0.48	230	260	26	23
	p,p'-DDD	2.0	1.8	1,300	1,500	110	150
	o,p'-DDE	[a]	0.80	[a]	240	[a]	46
	p,p'-DDE	110	99	32,000	34,000	4,400	4,600
	o,p'-DDT	<0.12	1.8	1,400	1,400	160	280
	p,p'-DDT	<0.14	0.56	310	840	120	200
	HCB	5.0	4.2	340	240	18	16
	β-HCH	[a]	2.4	[a]	1,500	[a]	17
	Mirex	[a]	[c]	[a]	420	[a]	45
	Oxychlordane	[a]	1.6	[a]	1,500	[a]	44
	Summed PCBs (ΣPCBs)	86[e]	82[f]	34,000[e]	25,000[f]	3,400[e]	3,100[f]
	Summed DDTs (ΣDDTs)	110[g]	100[h]	35,000[g]	38,000[h]	4,800[g]	5,300[h]

[a] Cannot be measured.

[b] There were coeluting PCBs that affected the quantitation (increased the value).

[c] Not quantitated because an interfering compound was present.

[d] On HPLC/PDA system, PCBs 170/194 coelute.

[e] Summed PCBs (ΣPCBs) = sum of concentrations of 14 reported congeners, then adding the concentrations of the unreported congeners, which are calculated using the response factor of PCB 101 when present, of PCB 138 if congener 101 is not present, or the average response factor over all congeners if neither PCB is present.

[f] Summed PCBs (ΣPCBs) = sum of concentrations of 40 reported congeners (PCBs 17, 18, 28, 31, 33, 44, 49, 52, 66, 70, 74, 82, 87, 95, 99, 101/90, 105, 110, 118, 128, 138, 149, 151, 153/132, 156, 158, 170/190, 171, 177, 180, 183, 187, 191, 194, 195, 199, 205, 206, 208, 209).

[g] Summed DDTs (ΣDDTs) = sum of concentrations of *o,p'*-DDD, *o,p'*-DDT, *p,p'*-DDE (at 266 nm), *p,p'*-DDD, and *p,p'*-DDT.

[h] Summed DDTs (ΣDDTs) = sum of concentrations of *p,p'*-DDT, *p,p'*-DDE, *p,p'*-DDD, *o,p'*-DDD, *o,p'*-DDE, and *o,p'*-DDT.

of several OCs measured by HPLC/PDA were, in general, comparable to the GC/MS values. Again, we found that the PCB 128 values determined by HPLC/PDA were higher than those measured by GC/MS and is likely due to coelution with PCB 123 on the Cosmosil PYE column. Conversely, we found that the concentrations of PCB 156 determined by GC/MS in all three killer whale samples were higher than the HPLC/PDA values and is likely due to interference by PCB 171 and/or PCB 202 on the DB-5 GC column. A previous study reported that, although several types of contaminants (e.g., dieldrin, lindane, endosulfan A, endosulfan B, organophosphates) were completely or partially destroyed using acidic silica gel cleanup, many other environmental contaminants (e.g., all PCB congeners, DDTs, HCB) survived this rigorous cleanup technique.[29] Consequently, by analyzing a portion of the HPLC/PDA sample extract using GC/MS, we could successfully acquire OC concentration data for ten additional OC pesticides plus several additional PCB congeners that cannot be measured by the HPLC/PDA method. Because a much larger suite of OCs can be measured in the same sample extracts when combining both the HPLC and GC results, these additional OC data can be used for a wider range of statistical analyses (e.g., principal component analysis).

Dioxin-like PCB congeners and other selected OCs were readily extracted from tissues of marine organisms using the modified HPLC/PDA method. The advantages of the modified method are:

1. the time required to prepare and extract a sample set was reduced by 33%;
2. the amount of solvent used to prepare and extract a set of samples was reduced by approximately 50% compared to the amount needed using the old extraction method;
3. concentrations of dioxin-like PCBs, other PCBs, and selected OCs can be determined on the same sample extract by HPLC/PDA and/or GC/MS.

The only major limitation of the new ASE-based method is the initial cost of the instrument relative to the cost of the Tissuemizer motors and probes.

A performance-based quality assurance program for OC analyses is highly recommended. For that purpose, various marine tissues have been prepared by NIST and other agencies for which there are certified or recommended concentrations of many OCs, including dioxin-like PCB congeners. Analyses of a suitable reference material, method blank and replicate samples provide information on how well the extraction, cleanup, and quantitation methods are operating.

Acknowledgments

We appreciate the technical assistance or advice of Catherine Sloan, Don Brown, Gladys Yanagida, and Richard Boyer. We appreciate the collection of killer whale tissue samples by Craig Matkin of the North Gulf Oceanic Society, Paul Wade of the National Marine Mammal Laboratory in Seattle, WA, Janet Whaley of the NOAA Fisheries Office of Protected Resources in Silver Spring, MD, and Brent Norberg of the NOAA Fisheries Northwest Regional Office in Seattle, WA. We also appreciate the careful review of the manuscript by Jennie Bolton and Don Brown.

References

1. McFarland, V.A. and Clarke, J.U., Environmental occurrence, abundance, and potential toxicity of polychlorinated biphenyl congeners: considerations for a congener-specific analysis, *Environ. Health Perspect.*, 81, 225–239, 1989.
2. Schmidt, C.W., Most unwanted: persistent organic pollutants, *Environ. Health Perspect.*, 107 (1), A18–A23, 1999.
3. de Wit, C.A., An overview of brominated flame retardants in the environment, *Chemosphere*, 46, 583–624, 2002.
4. Myers, M.S., Stehr, C.M., Olson, O.P., Johnson, L.L., McCain, B.B., Chan, S.-L. and Varanasi, U., Relationships between toxicopathic hepatic lesions and exposure to chemical contaminants in English sole (*Pleuronectes vetulus*), starry flounder (*Platichthys stellatus*), and white croaker (*Genyonemus lineatus*) from selected sties on the Pacific Coast, USA, *Environ. Health Perspect.*, 102 (2), 200–215, 1994.
5. Niimi, A.J., PCBs in aquatic organisms, in *Environmental Contaminants in Wildlife: Interpreting Tissue Concentrations*, Beyer, W.N., Heinz, G.H. and Redmon-Norwood, A.W., Eds., Lewis Publishers, Boca Raton, FL, 1996, pp. 117–152.
6. Ross, P.S., de Swart, R.L., Timmerman, H.H., Reijnders, P.J.H., Vos, J.G., van Loveren, H. and Osterhaus, A.D.M.E., Suppression of natural killer cell activity in harbour seals (*Phoca vitulina*) fed Baltic Sea herring, *Aquat. Toxicol.*, 34, 71–84, 1996.
7. Beckmen, K.B., Lowenstine, L.J., Newman, J., Hill, J., Hanni, K. and Gerber, J., Clinical and pathological characterization of northern elephant seal skin disease, *J. Wildlife Dis.*, 33, 438–449, 1997.
8. Johnson, L.L., Sol, S.Y., Ylitalo, G.M., Hom, T., French, B., Olson, O.P. and Collier, T.K., Reproductive injury in English sole (*Pleuronectes vetulus*) from the Hylebos Waterway, Commencement Bay, Washington, J. *Aquat. Ecosystem Stress Rec.*, 6, 289–310, 1999.
9. Arkoosh, M.R. and Collier, T.K., Ecological risk assessment paradigm for salmon: analyzing immune function to evaluate risk, *Human Ecol. Risk Assess.*, 8 (2), 265–276, 2002.
10. Krahn, M.M., Moore, L.K., Bogar, R.G., Wigren, C.A., Chan, S.-L. and Brown, D.W., High-performance liquid chromatographic method for isolating organic contaminants from tissue and sediment extracts, J. *Chromatogr.*, 437, 161–175, 1988.
11. Kuehl, D.W., Butterworth, B.C., Libal, J. and Marquis, P., An isotope dilution high-resolution gas chromatographic-high resolution mass spectrometric method for the determination of coplanar polychlorinated biphenyls: application to fish and marine mammals, *Chemosphere*, 22 (9/10), 849–858, 1991.
12. Kannan, N., Petrick, G., Schultz-Bull, D.E. and Duinker, J.C., Chromatographic techniques in accurate analysis of chlorobiphenyls, *J. Chromatogr.*, 642, 425–434, 1993.
13. Hess, P., de Boer, J., Cofino, W.P., Leonards, P.E.G. and Wells, D.E., Critical review of the analysis of non- and mono-*ortho*-chlorobiphenyls, *J. Chromatogr. A.*, 703, 417–465, 1995.
14. Muir, D.C.G., Ford, C.A., Rosenberg, B., Norstrom, R.J., Simon, M. and Beland, P., Persistent organochlorines in belugas (*Delphinapterus leucas*) from the St. Lawrence River estuary. I. Concentrations and patterns of specific PCBs, chlorinated pesticides and polychlorinated dibenzo-*p*-dioxins and dibenzofurans, *Environ. Pollut.*, 93, 219–234, 1996.
15. Ikonomou, M.G., Fischer, M., He, T., Addison, R.F. and Smith, T., Congener patterns, spatial and temporal trends of polychlorinated diphenyl ethers in biota samples from the Canadian west coast and the Northwest Territories, *Organohalogen Compounds*, 47, 77–80, 2000.
16. Senthil Kumar, K., Kannan, K., Paramasivan, O.N., Shanmugasundaram, V.P., Nakanishi, J. and Masunaga, S., Polychlorinated dibenzo-*p*-dioxins, dibenzofurans, and polychlorinated biphenyls in human tissues, meat, fish and wildlife samples from India, *Environ. Sci. Technol.*, 35, 3448–3455, 2001.
17. Kucklick, J.R., Struntz, W.D.J., Becker, P.R., York, G.W., O'Hara, T.M. and Bohonowych, J.E., Persistent organochlorine pollutants in ringed seals and polar bears collected from northern Alaska, *Sci. Total Environ.*, 287, 45–59, 2002.

18. Hall, A.J., Law, R.J., Wells, D.E., Harwood, J., Ross, H.M., Kennedy, S., Alldrin, C.R., Campbell, L.A., and Pomeroy, P.P., Organochlorine levels in common seals (*Phoca vitulina*) which were victims and survivors of the 1988 phocine distemper epizootic, *Sci. Total Environ.*, 115, 145–162, 1992.
19. Krahn, M.M., Ylitalo, G.M., Buzitis, J., Sloan, C.A., Boyd, D.T., Chan, S.-L. and Varanasi, U., Screening for planar chlorobiphenyl congeners in tissues of marine biota by high-performance liquid chromatography with photodiode array detection, *Chemosphere*, 29 (1), 117–139, 1994.
20. Sloan, C.A., Brown, D.W., Pearce, R.W., Boyer, R.H., Bolton, J.L., Burrows, D.G., Herman, D.P. and Krahn, M.M., Determining aromatic hydrocarbons and chlorinated hydrocarbons in sediments and tissues using accelerated solvent extraction and gas chromatography/mass spectrometry, in *Techniques in Aquatic Toxicology*, vol. 2, Ostrander, G.K., Ed., CRC Press, Boca Raton, FL (Chapter 35) (2005).
21. Ylitalo, G.M., Yanagida, G.K., Hufnagle, L., Jr. and Krahn, M.M., Determination of lipid classes and lipid content in tissues of aquatic organisms using a thin layer chromatography/flame ionization detection (TLC/FID) microlipid method, in *Techniques in Aquatic Toxicology*, vol. 2, Ostrander, G.K., Ed., CRC Press, Boca Raton, FL (Chapter 12) (2005).
22. Lebo, J.A., Zajicek, J.L., May, T.W. and Smith, L.M., Large-scale preparation of potassium hydroxide-modified silica gel adsorbent, *J. Assoc. Off. Anal. Chem.*, 72, 371–373, 1989.
23. Wise, S.A., Schantz, M.M., Koster, B.J., Demiralp, R., Mackey, E.A., Greenberg, R.R., Burow, M., Ostapczuk, P. and Lillestolen, T.I., Development of frozen whale blubber and liver reference materials for the measurement of organic and inorganic contaminants, *Fresunius* J. *Anal. Chem.*, 345, 270–277, 1993.
24. Beckmen, K.B., Ylitalo, G.M., Towell, R.G., Krahn, M.M., O'Hara, T.M. and Blake, J.E., Factors affecting organochlorine contaminant concentrations in milk and blood of northern fur seal (*Callorhinus ursinus*) dams and pups from St. George Island, Alaska, *Sci. Total Environ.*, 231, 183–200, 1999.
25. Ylitalo, G.M., Buzitis, J. and Krahn, M.M., Analyses of tissues of eight marine species from Atlantic and Pacific Coasts for dioxin-like chlorobiphenyls (CBs) and total CBs, *Arch. Environ. Contamin. Toxicol.*, 37, 205–219, 1999.
26. Ylitalo, G.M., Matkin, C.O., Buzitis, J., Krahn, M.M., Jones, L.L., Rowles, T. and Stein, J.E., Influence of life-history parameters on organochlorine concentrations in free-ranging killer whales (*Orcinus orca*) from Prince William Sound, AK, *Sci. Total Environ.*, 281, 183–203, 2001.
27. Loughlin, T.R., Castellini, M.A. and Ylitalo, G., Spatial aspects of organochlorine contamination in northern fur seal tissues, *Mar. Pollut. Bull.*, 44, 1024–1034, 2002.
28. Willcox, M.K., Woodward, L.A., Ylitalo, G.M., Buzitis, J., Atkinson, S. and Li, Q.X., Organochlorines in the free-ranging Hawaiian monk seal (*Monachus schauinslandi*) from French Frigate Shoals, North Pacific Ocean, *Sci. Total Environ.* 322, 81–98, 2004.
29. Bernal, J.L., Nozal, M.J.D. and Jimenez, J.J., Some observations on clean-up procedures using sulphuric acid and Florisil, *J. Chromatogr.*, 607, 303–309, 1992.

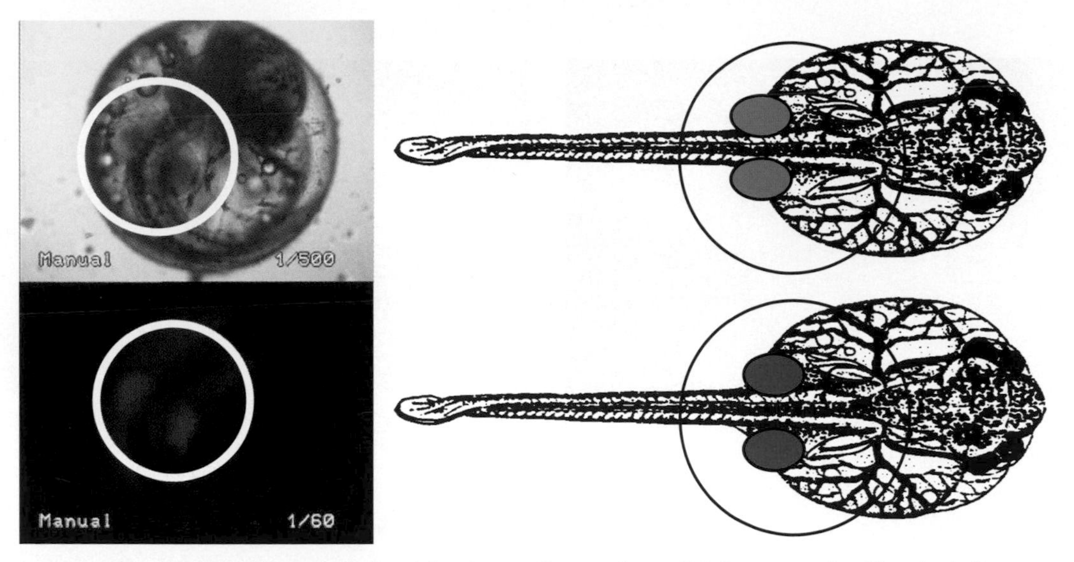

Figure 11.4 Bilobed bladder in *Fundulus heteroclitus* embryo (10 days post-fertilization) fluoresces blue (420 nm) under UV excitation (365 nm); bladders in embryos exposed to aryl hydrocarbon agonists and the EROD substrate (ethoxyresorufin) accumulate resorufin and fluoresce red (590 nm) under green (510 nm) excitation.

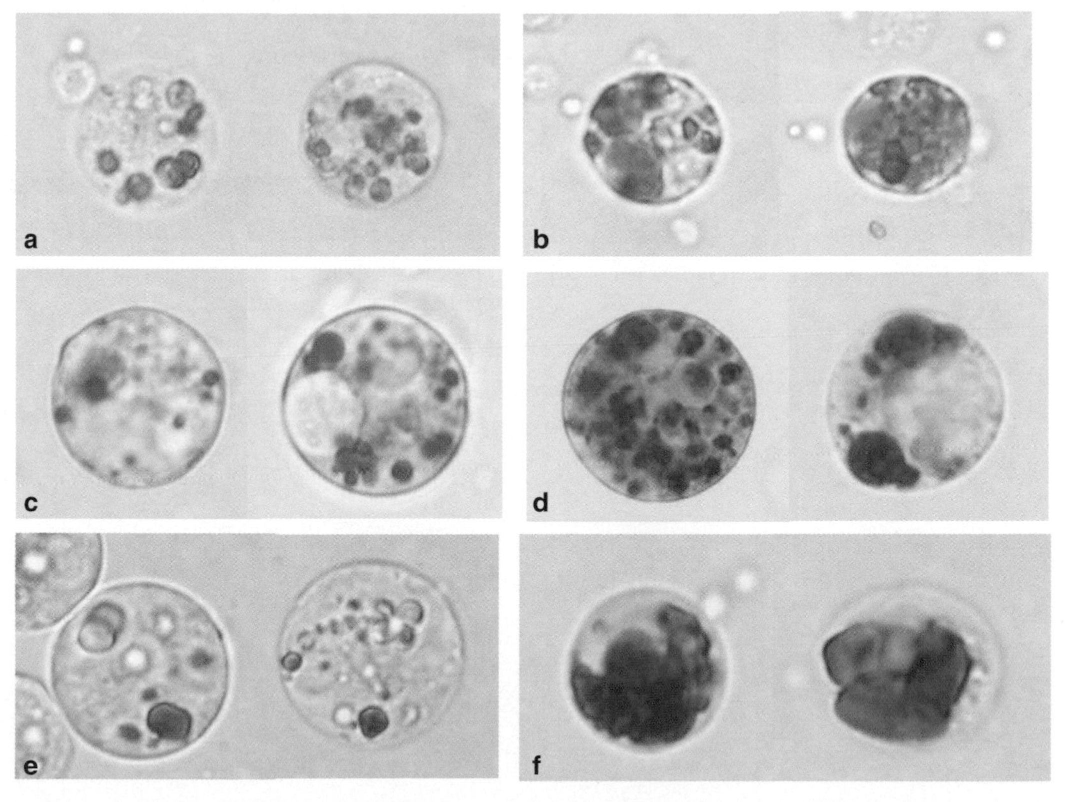

Figure 16.2 Photographs of examples of hepatic cells with stable and destabilized lysosomes for oysters, grass shrimp, and mummichogs: (a) oyster hepatopancreas cells scored as stable; (b) oyster hepatopancreas cells scored as destabilized; (c) grass shrimp hepatopancreas cells scored as stable; (d) grass shrimp hepatopancreas cells scored as destabilized; (e) fish liver cells scored as stable; (f) fish liver cells scored as destabilized.

Figure 19.1 Cross-section of coral colony with pink eggs and white spermaries.

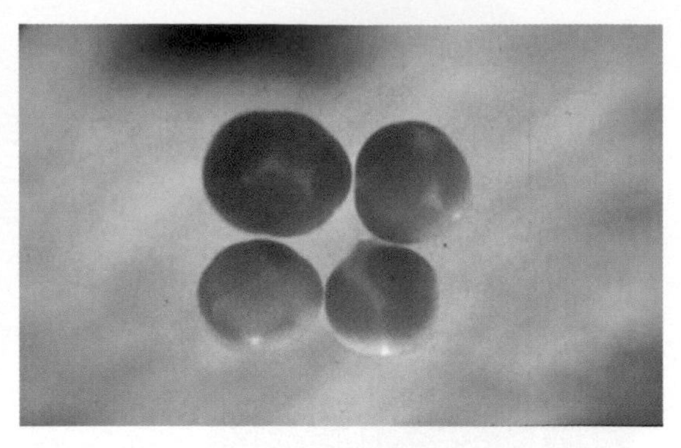

Figure 19.2 Combined egg–sperm clusters.

Figure 19.3 Floating gamete bundles collected in sieves.

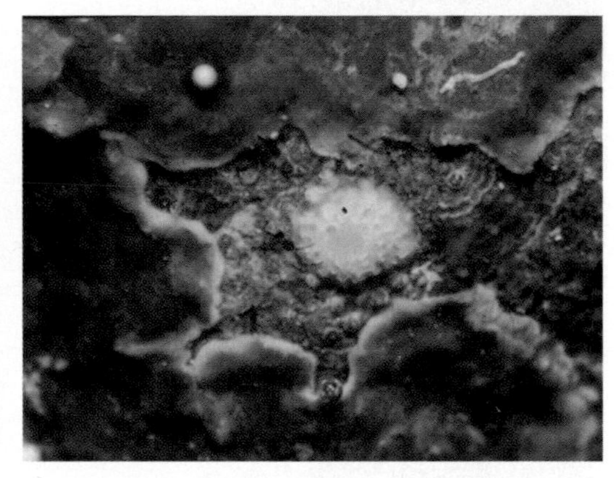

Figure 19.4 Coral larval recruit, settled and metamorphosed.

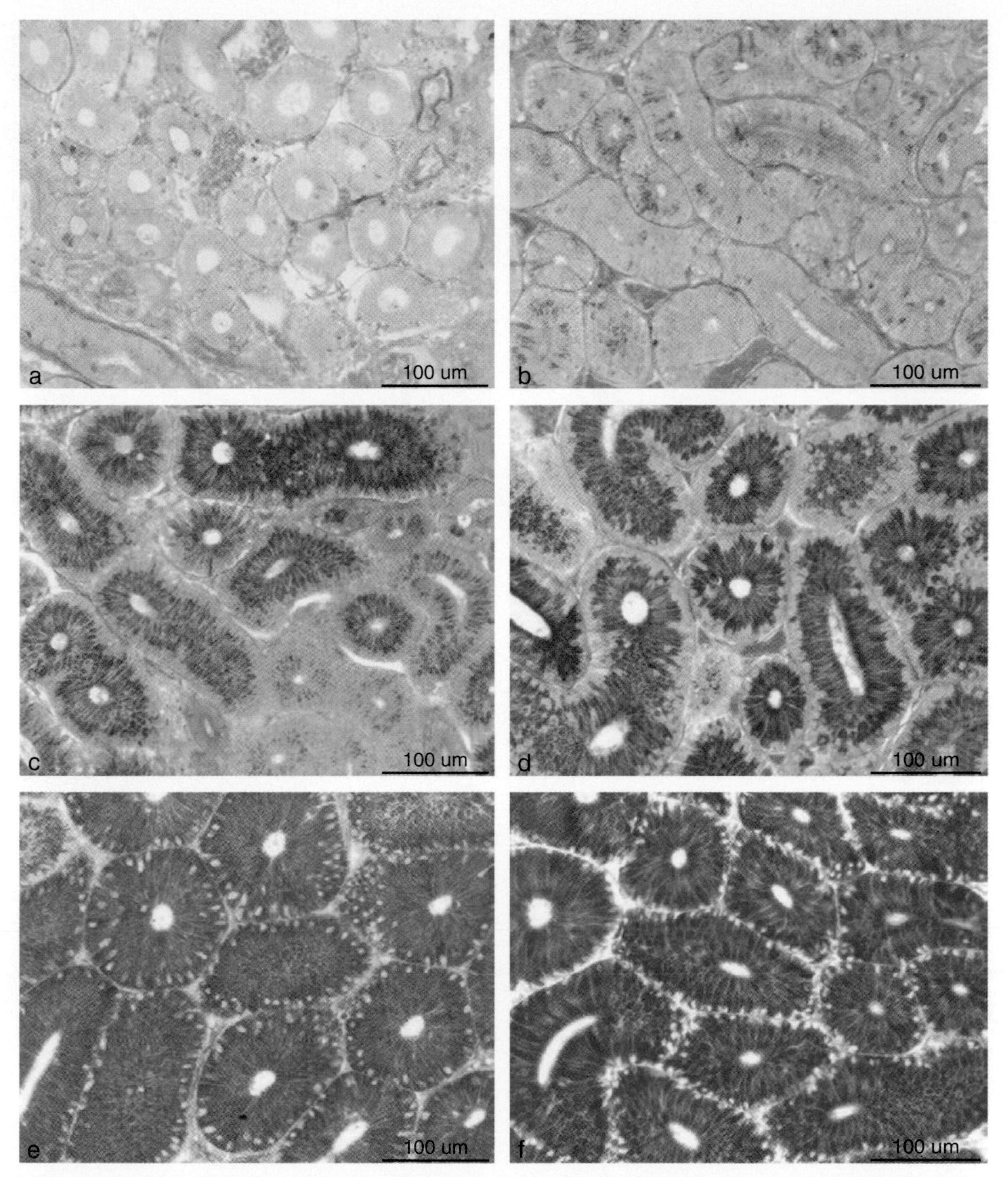

Figure 20.2 Renal histology of 17α-MT-treated female sticklebacks (a–e) and a naturally breeding male (f). Plates are at the same power and stained with periodic acid-Schiff (PAS). (a) control fish (KEH = 13.5 μm); (b) 100 ng/l 17α-MT (KEH = 18 μm); (c) 1 μg/l 17α-MT (KEH = 24.2 μm); (d) 10 μg/l 17α-MT (KEH = 29.7 μm); (e) 500 μg/l 17α-MT (KEH = 36.2 μm); (f) Breeding male (KEH = 34.5 μm).

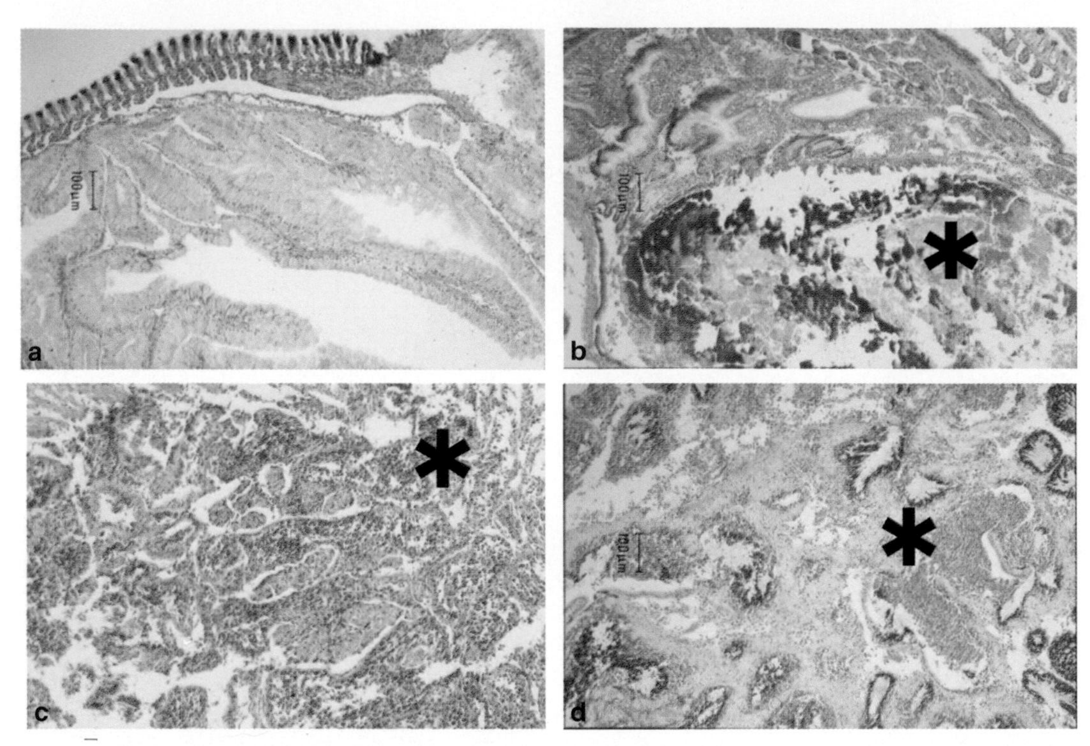

Figure 36.8 Example lesions affecting the health of individual clams and population status in the Gulf of Riga: (a) normal kidney, gill at top; (b) kidney concretion at "*"; (c) germinoma, undifferentiated cells filling gonadal follicles to left of "*"; (d) abnormal thickening of connective tissue among gonadal follicles in male clam to left of "*," infiltration of hemocytes in follicles to right of "*." (Photomicrographs courtesy of E.C. Peters.)

chapter twenty-six

Estrogenic activity measurement in wastewater using in vitro *and* in vivo *methods*

Yelena Sapozhnikova and Daniel Schlenk
University of California
Anne McElroy
State University of New York
Shane Snyder
Southern Nevada Water Authority

Contents

Introduction

Throughout the world wastewater from municipal treatment facilities has been evaluated for the occurrence of estrogenic chemicals. Estrogens have been reported to cause

1-56670-664-5/05/$0.00+$1.50
© 2005 by CRC Press

widespread reproductive dysfunction in aquatic fish species.[1,2] Several studies have shown physiological changes in male fish exposed to estrogens, including intersex conditions.[3,4] Studies initially in the UK have reported that male fish exposed to sewage effluents had increased concentrations of vitellogenin (VTG), a biomarker for environmental estrogens.[5–7]

To date only compounds binding and/or activating the estrogen receptor have been identified from wastewater.[8,9] Since many compounds may elicit estrogenic activity throughout the modulation of endogenous estrogens within organisms, it is likely *in vitro* receptor-driven assays significantly underestimate the estrogenic activity of environmental samples. To gain a better understanding of total estrogenic activity *in vivo* fish VTG and *in vitro* [yeast estrogen screen (YES)] assays were utilized for detecting estrogenic activity in wastewater from New York city. Through a series of chromatographic separations, isolated fractions of varied polarity were evaluated by *in vitro* and *in vivo* assays. Chromatographic steps included Empore SDB-XC extraction disks, Sep Pac Plus C-18 cartridges, and finally C-18 HPLC. All fractions were tested for estrogenic activity by both *in vivo* and *in vitro* assays, and the most active fraction was finally analyzed for unknown compounds using GC/MS/MS and LC/MS/MS.

Estradiol equivalent concentrations (EEQs) were measured in effluent extracts subjected to both assays utilizing standard curves constructed from exposing animals and cells to 17β-estradiol (E2). Identity and, in some cases, concentrations of estrogenic chemicals were determined with GC/MS/MS and LC/MS/MS analysis.

Material required

Animals:

Sexually mature male Japanese medaka (*Oryzias latipes*) (local supplier or other researcher stock).

Equipment:

- +30°C incubator (Precision, Model 815) with orbital platform shaker
- Cell culture cabinet (Precision Scientific Model 52200047)
- −80°C freezer (Revco Legaci Refrigeration system)
- −20°C freezer (Kenmore)
- +4°C Isotemp Chromatography Refrigerator (Fisher Scientific)
- Vortex (Fisher Genie2)
- Plate Reader (Molecular Device, Umax Kinetic microplate)
- HPLC with UV detector (Shimadzu)
- Fraction Collector (Spectra/Chrom CF-1)
- Microcentrifuge (Fisher Scientific, Micro12)
- Rotary Evaporator (Buchi No 14201, Switzerland)
- PH-meter (Orion, model 550A)
- Analytical balance (Mettler Toledo, AB54-S)
- Autoclave (Amsco Eagle, SV-3033)
- Stirrer (Fisher, Model 120S)
- Pipet Dispenser (Gilson)

(See Tables 26.1 and 26.2 for the details about supplies and reagents.)

Table 26.1 Supplies for *in vitro* and *in vivo* assay-guided fractionation

Item	Source	Catalog #
Sterile plates, 100 × 15 mm	Fisher	08–757–12
Disposable sterile centrifuge tubes, 50 ml	Fisher	05–5398
Disposable sterile centrifuge tubes, 15 ml	Fisher	12–565–286A
HPLC column, 25 cm × 4.6 mm × 4 μm Ultrasphere C-18	Beckman	235333
Drierite absorbent	Fisher	07–578–4A
Vacuum manifold Visiprep™ DL	Supelco	57044
Valve Liners to Visiprep™ DL	Supelco	57059
Glass vials 2 ml	Fisher	03–339
Round bottom flasks, 50 ml	Fisher	10–060–4B
Desiccator	Fisher	08–615B
High performance extraction disks SDB-XC, 47 mm, Empore	Empore^{3M}	2240
Assay plate, 96–well, flat bottom	Fisher	12–565–502
Weighing paper	VWR	12578–165
Microcentrifuge tubes, 1.5 ml	Fisher	02–681–320
Disposable inoculation loops/needles	Fisher	13–075–3
5 3/4" Disposable Pasteur Pipets	Fisher	13–678–20A
Sterile glass pipets, 10 ml	Fisher	13–675–71
Disposable sterile filter system, 1 l	Fisher	09–761–42
Glass Microfibre filters GF/C, 1.2 μm, 55 cm	Whatman	1822055
0.45 μm filter	Millipore	GNWPO4700
Brine shrimp	Local supplier	
ELISA Biosense Japanese medaka kit	Biosense	V01013402–096
Microfiltration Assemblies, 90 mm	Fisher	K953755–0090
Flasks, 1 l	Fisher	K953760–000
Flasks, 125 ml	Fisher	K953710–000
Cylinder, 50 ml	Fisher	08–549–17C
Syringe Hamilton, 100 μl	Fisher	14–824–6
Syringe Hamilton, 250 μl	Fisher	14–824–2
Culture tubes	Fisher	14–961–25
Beakers, 1 l	Fisher	02–539P

Procedures

EEQs were calculated for effluent extracts and fractions subjected to YES and VTG assays using constructed dose–response curves. For the YES assay E2 standards in a concentration range from 10^{-4} to 10^{-14} *M* was incubated with yeast suspension for 5 days as described in the following. Each measurement was made in duplicate and a dose–response curve was constructed as E2 concentration versus optical density (OD). E2 recovery was determined using spiked water. The recovery from the YES assay was 69 ± 8%, and the method detection limit (MDL) was 1 ng/l.

For the VTG assay, five replicates per each E2 concentration (from 1 ng/l to 100 μg/l) were applied to construct a concentration–response curve. The control treatment group received only ethanol carrier. Hepatic VTG was measured using the commercially available ELISA Kit for medaka (Biosense, Norway). EEQs were calculated from the linear portion of dose–response curve as previously described.[10–13]

Table 26.2 Reagents for *in vitro* and *in vivo* assay-guided fractionation

Item	Source	Catalog #
Yeast strain (*Saccharomyces cerevisae*)	Provided by Prof. J. Sumpter of Brunel University (United Kingdom)	
Yeast nitrogen base (YNB) without amino acids and ammonium sulfate	Difco	2084620
Ammonium sulfate	Fisher	A702–500
Peptone	Fisher	BP1420–100
Adenine, min 99%	Sigma	A3159
Dextrose anhydrous	Sigma	G-7021
Casamino acids	Bacto(TM)	2007–04–30
Sucrose	IBI Shelton Scientific	IB 37160
Dihydrogen sodium phosphate	Fisher	424390025
Disodium hydrogen phosphate heptahydrate	Fisher	424395000
Potassium chloride	Merck KGaA	PX1405–5
Magnesium sulfate	Merck KGaA	MX0075–1
2-Mercaptoethanol	Merck KGaA	6010
Sodium dodecyl sulfate, 99% min	Fisher	BP 166–100
Agar granulated	Difco(TM)	214530
Copper (II) sulfate	Fisher	197711000
17β-estradiol (E2)	Sigma	E-8875
O-nitro phenyl Β-D-galactopyranoside (ONPG) (store in −20°C)	Sigma	N-1127
Reagent alcohol (ethanol), HPLC grade	Fisher	A995–4
Sodium carbonate	Merck KGaA	SX 0395–1
Methanol, HPLC grade	Fisher	2304/CS
Acetone, HPLC grade	B&J Brand	BJAH0104
Ethyl 3-aminobenzoate methane sulfonic acid salt, 98% (MS222)	Sigma	E1052–50G
Nitrogen gas, 99% pure	Local supplier	

Sample collection

Samples were collected in methanol-rinsed brown glass bottles. One liter sample of wastewater is required for the procedure. Wastewater samples should be maintained at 4°C until filtration within 1–2 days of collection. Portable pumps may be necessary with appropriate tubing length to transport wastewater to the collection bottle.

Wastewater extraction protocol

Removing particles: It is best if water samples are filtered with 1.2 and 0.45 μm filters together within 1–2 days of receiving the samples.

1. Set up a microfiltration apparatus and rinse with methanol. Use one 0.45 μm filter and one 1.2 μm filter to remove the particles from the water.
2. Begin to filter the water through the apparatus by turning on the vacuum. The flow rate does not matter when removing particles. (*Note*: Water containing many particles may clog the filters, causing the water flow to become very slow. Change the filters if needed.)
3. Once all of the water sample has passed through, turn off the vacuum, and disassemble the apparatus. Filters may be thrown away.

Empore disk extraction:

1. Use Empore filter disks to extract estrogens from water samples. (Store Empore filter disks in the desiccator with Drierite absorbents.)
2. Rinse funnel and flask with methanol. Set up microfiltration apparatus as before, using an Empore filter disc. The first vacuum flask will be used as a waste flask to condition the filter disc.
3. Condition the filter disc with 15 ml of acetone. Allow a few drops to pass through the disc, then shut off vacuum and allow the filter to soak for 1 min.
4. After 1 min pass the acetone through by applying the vacuum until the filter disc is dry.
5. Condition the filter with 15 ml of methanol. Allow a few drops to pass through the disc, then shut off the vacuum and allow the disc to soak for 1 min.
6. After 1 min pass the methanol through slowly until the level of methanol is just above the filter disc. It is important that the disc does not dry out. If it does, then repeat the conditioning process.
7. Once the filter disc is conditioned, the flasks must be changed. Carefully pull the stopper out of the waste flask and place it on a clean vacuum flask.
8. Begin to filter the water sample through the apparatus at a rate of 5 ml/min (approximately 1 drop/s). It will take approximately 3 h to filter a 1-l sample. It is important to keep the filter disc wet until all water has passed through.
9. After the water sample has passed though, turn vacuum on and dry filter for 5 min.
10. Remove the filter disc and elute it with methanol (see the following). If it is not possible to elute it right away, store it in a Zip lock bag, and label the bag with the water sample number/name and date of Empore filtration. Store in the −20°C freezer.

Empore disc elution

1. If the Empore filter disc was stored in the freezer, remove it and let it thaw for 15 min.
2. Set up microfiltration apparatus as previously described using a 125-ml vacuum flask.
3. Elute the disc with 2 × 15 ml methanol at a rate of 1 drop/s. It is important to keep the flow rate constant.
4. After methanol has passed through, allow the filter disc to dry with vacuum for a few seconds.
5. Transfer the methanol extract to a round bottom flask or pear-shaped flask for vaporization with a rotary evaporator.
6. Vaporize the extract with rotary evaporation at 65°C until dry. Reconstitute the dry extract in 300–500 μl ethanol, rinsing the walls of the flask, and transfer to a labeled 2-ml vial.
7. Evaporate the extract to 100 μl in a nitrogen stream. The extract is ready now for the YES assay.

Table 26.3 Recipes for preparing media and buffer (1 l)

Ura-Tryp media	YPS media
6.7 g yeast nitrogen base (YNB) w/o amino acids and ammonium sulfate	1% yeast extract (10 g)
5 g ammonium sulfate	0.5% peptone (5 g)
2 ml adenine sulfate (10 mg/ml)[a]	10% sucrose (100 g)
100 ml dextrose (20%)	16.1 g $Na_2HPO_4 \times 7H_2O$ (60 m*M* final)
100 ml casamino acids (5%)	Keep in +4°C refrigerator
800 ml DDI water	
Keep in +4°C refrigerator	*Z-buffer*:
	5.5 g $NaH_2PO_4 \times H_2O$ (40 m*M* final)
Solid media plates:	0.75 g KCl (10 m*M* final)
Add 16 g of agar to 1 l of Ura-Tryp	0.246 g $MgSO_4 \times H_2O$ (1 m*M* final)
Media, autoclave for 20 min, pour on the plates to create 2 mm layer, keep in +4°C refrigerator	2.7 ml B-mercaptoethanol (50 m*M* final)
	Adjust to pH 7.0
	Keep at room temperature
	Filter Sterilize after dissolving

[a] To dissolve adenine sulfate use water bath at 45–50°C.

In vitro *yeast estrogen screen assay*

Day 1

Cell growth on solid media (Table 26.3): Streak yeast cells from previous plate (or glycerol stock, keep yeast stock in −80°C freezer, thaw slowly on ice) on Ura-Tryp media plates and incubate at 30°C for 48 h. (When streaking from the stock, cells require at least 4 cycles of growth on the plates before growing in liquid media.) Single colonies may be observed after 24 h but require 48 h before they can be picked for growth in liquid media.

Day 3

Cell growth in liquid media: Pick one independent colony and inoculate in 50 ml sterile Falcon tube containing 5 ml Ura-Tryp media. Incubate at 30°C in a shaking incubator at 300 rpm for 24 h.

Day 4

Determination of cell density:

1. Vortex the tube containing the yeast.
2. Place 100 μl of the yeast solution on 96-well flat bottom microplate, YPS media in another well on the same plate. Read the plate at 650 nm in a plate reader.
3. Calculate the amount of yeast solution to prepare yeast suspension for YES assay:

$$\text{plate reading} = \text{yeast reading} - \text{media reading}$$
$$\text{special OD(SOD)} = [(\text{plate reading}) + 0.343]/0.3918$$

4. To prepare yeast suspension, use the following equation:

volume (V) of yeast solution to add (ml) = 0.057*10 ml/SOD

Add this amount of yeast solution to (10 − V) ml of YPS media.

5. Vortex, add 100 μl of 1.25 mM $CuSO_4$.

Assay

1. Add 100 μl of E2 of different concentration (10^{-4} to 10^{-14} M), sample extracts and ethanol to 1.5-ml microcentrifuge tubes.
2. Add 700 μl of yeast suspension to each tube.
3. Incubate at 30°C for 5 days with caps open but covered with sterile paper towel.

Day 10

1. After 5 days, centrifuge tubes for 3 min and aspirate supernatant (under hood) leaving pellet intact.
2. Add 200 μl of Z-buffer to each tube. Mix by gentle vortexing.
3. Prepare assay buffer:
   - 10 mg ONPG dissolve in 9.75 ml of Z-buffer, add 125 μl of B-mercaptoethanol,
   - 250 μl 10% SDS, vortex.
4. Add 400 μl of ONPG solution (assay buffer) to the tubes.
5. Incubate (caps closed) at 30°C until color develops (1 h).
6. Add 250 μl of 1 M sodium carbonate to stop the reaction.
7. Centrifuge for 3 min.
8. Withdraw 150 μl from each tube and place on 96-well plate.
9. Measure OD at 405 nm.
10. Subtract blank (ethanol OD from sample OD).
11. Create dose–response curve placing E2 molar concentrations (10^{-4} to 10^{-14} M) on X-axis, and corresponding ODs on Y-axis.
12. Calculate sample EEQs using constructed dose–response curve by linear extrapolation from linear portion of concentration–response curve (R^2 = 0.96).

In vivo *medaka screening bioassay*

Fish cultured as described previously.[12]

1. One-liter glass beakers are filled with 0.8 l water.
2. Add 1 ml of effluent extract at day 1 and day 3 with a complete water renewal at day 3.
3. Spike the control sample with 1 ml of solvent (ethanol).
4. At t = 0, add 3 sexually mature male Japanese medaka (*O. latipes*) to each treatment beaker.
5. After day 6, euthanize the animals in 1 g/l MS222 and remove the liver. Analyze for VTG using the Biosense Japanese medaka ELISA Kit following manufacture's instructions.
6. For the E2 concentration–response curve, carry out exposure with five concentrations of E2 (10 ng/l to 100 μg/l) in ethanol. EEQs are calculated as for the YES assay.

Solid phase extraction fractionation

To differentiate estrogenic compounds by their polarity, fractionation with an ethanol/water gradient on a solid phase extraction (SPE) cartridge can be applied.

1. Set up Sep Pac C-18 plus cartridges in Supelco vacuum manifold.
2. Condition cartridges with methanol allowing a few drops of methanol to pass through to waste vial below the column.
3. Close the valve then condition for 1 min. Pass methanol through, but leave a 1-mm layer on top of the cartridge.
4. Transfer the extract onto the cartridge using a disposable glass pipet.
5. Rinse the pipet with a small amount of ethanol and add it into cartridge.
6. Allow the sample to pass through cartridge.
7. Once the sample level is just above the cartridge surface, close the valve and remove the waste vial from beneath the column. Place a clean vial labeled with the sample number beneath the column.
8. Collect 5 elution fractions of 10 ml: 10%, 25%, 50%, 75%, 100% ethanol/water.
9. Elute the C-18 cartridge with the 5 fractions listed above, placing a clean vial beneath for each fraction.
10. Vaporize fractions to dryness using nitrogen gas.
11. Add 100 μl ethanol to reconstitute.
12. Perform YES and medaka assays.

Fractionation by HPLC

Estrogenic compounds can be further separated by reverse phase HPLC with a shallow methanol/water gradient.

1. Inject 200 μl (injection loop 250 μl) of effluent extract/active fraction onto an HPLC system using a C-18 column (25 cm × 4.6 mm × 4 μm) with flow rate of 1 ml/min with a methanol/water gradient. HPLC elution program:
   - methanol 40% 0–3 min,
   - methanol 40–100% 3–30 min.
2. Collect fractions every 3 min.
3. Assess estrogenic activity of fractions with *in vivo* and *in vitro* assays.

Results and discussion

The technique, described in this chapter, was developed to identify causative agents responsible for estrogenic activity in wastewater collected from two New York city sewage treatment plants (STPs). Wastewater samples were collected in September 2002 from two STPs in Long Island, NY: the Red Hook (RH) Treatment Plant, which is a full secondary treatment plant with 60 million gallons per day (MGD) average capacity; and the New Town Creek (NTC) STP (Advanced Primary, 310 MGD average capacity).

Sample preparation was conducted as shown in Figure 26.1.

The *in vitro* EEQ for Red Hook STP effluent was 22.2 ng/l, whereas it was only 2.4 ng/l for the effluent from New Town Creek STP (Figure 26.2). Therefore, Red Hook

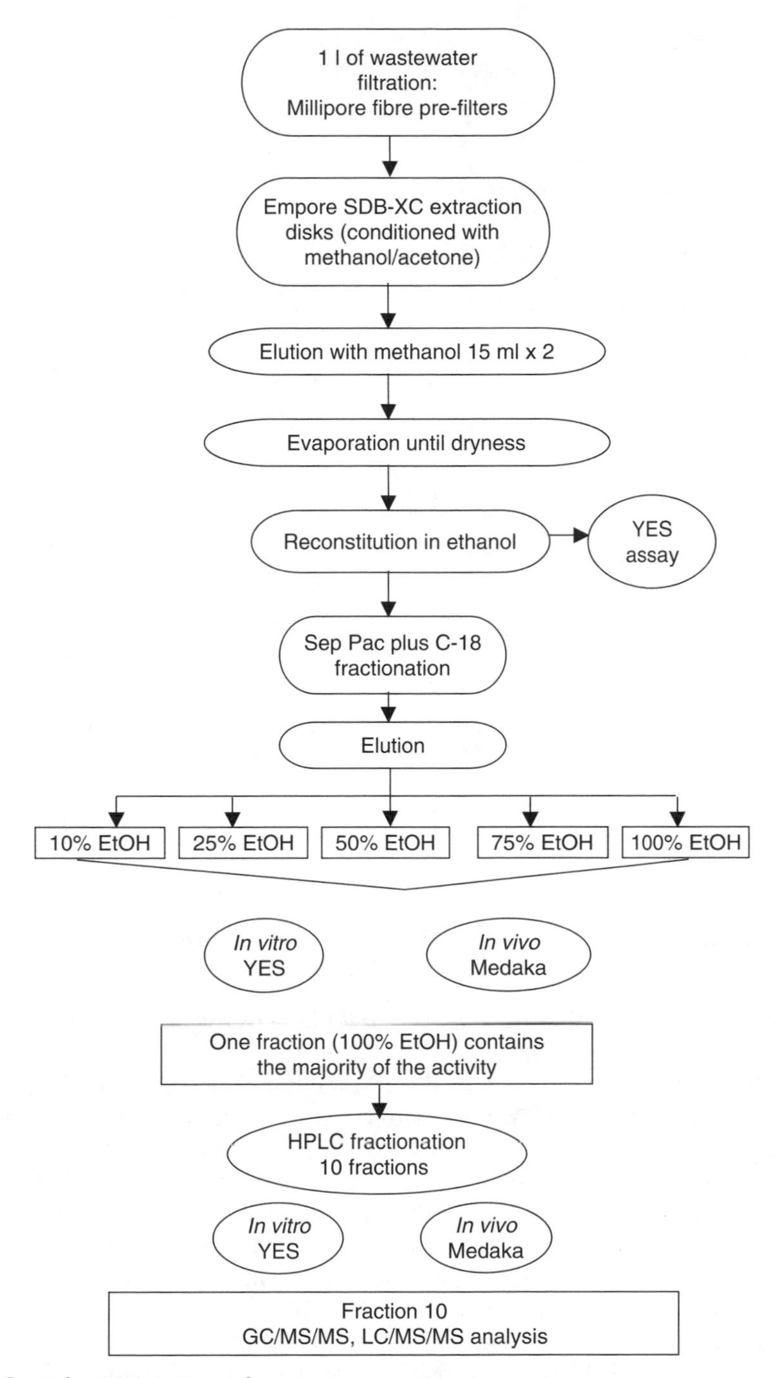

Figure 26.1 Sample preparation scheme.

effluent was used for further study using an ethanol/water fractionation scheme with an SPE C-18 Sep Pac cartridge. All 5 fractions (10%, 25%, 50%, 75%, and 100% ethanol) were assessed for estrogenic activity both by *in vivo* and *in vitro* assays. Both assays showed the highest EEQ in the 100% ethanol fraction (Figure 26.3). The YES assay showed 19.3 ng/l EEQ in the 100% ethanol fraction, but the EEQ measured with the medaka assay was

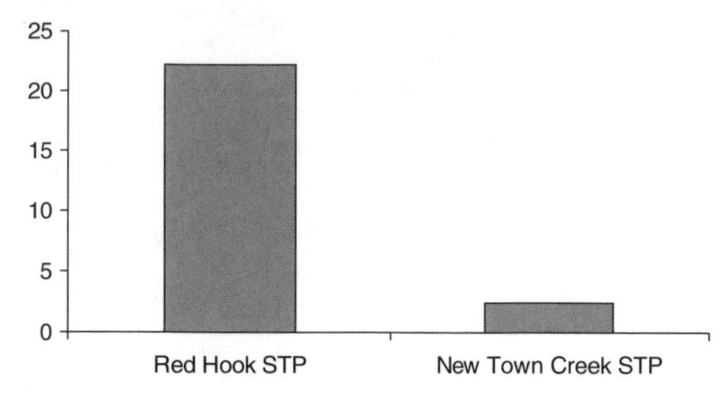

Figure 26.2 EEQs for Red Hook and New Town Creek STPs, ng/l.

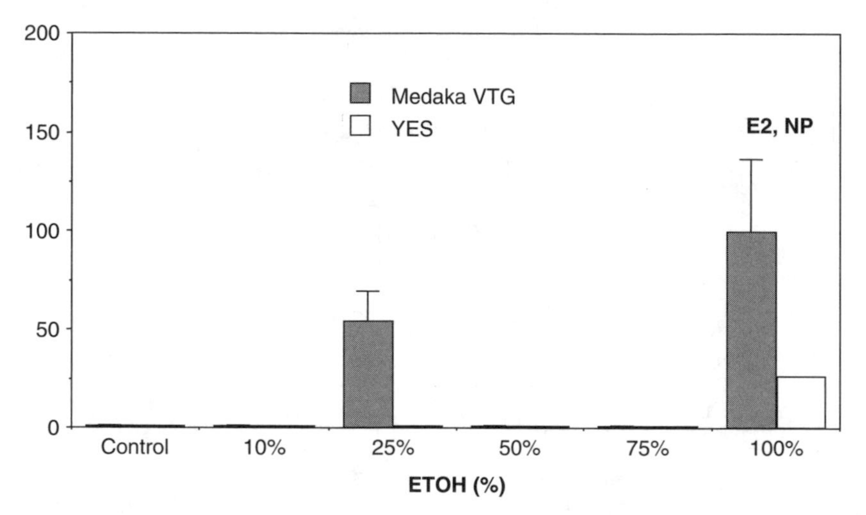

Figure 26.3 EEQs for ethanol fractions from Red Hook effluent, ng/l.

100 ng/l. The *in vivo* assay also detected estrogenic activity of 50 ng/l EEQ in the 25% ethanol fraction, where no YES activity was observed.

With further HPLC fractionation of the 100% ethanol fraction, 10 HPLC fractions of 3-min intervals were collected and measured for estrogenic activity. The YES assay showed 4.3 ng/l in two fractions — 6 and 9 (collected at 15–18 and 24–27 min, respectively) (Figure 26.4). In contrast, *in vivo* assays measured the highest EEQ in fraction 10 (27–30 min) — with 140 ng/l of activity. *In vivo* activity was also observed in fractions 1, 2, and 3 (0–3, 3–6, and 6–9 min, respectively) with activities as high as 80 ng/l EEQ. Fraction 8 (21–24 min) where E2 elutes did not show *in vitro* nor *in vivo* activity. 4-Nonylphenol (0.2 μg/l) was observed in fraction 10.

To identify other estrogenic agents, HPLC fraction 10 was analyzed for unknown compounds using GC/MS/MS and LC/MS/MS as described elsewhere.[14–16] (Table 26.4). Oxybenzone, a chemical used as a sunscreen agent and UV adsorber, and galaxolide used as musk odorants were found at concentrations 19 ng/l and 6 μg/l, respectively. Among other endocrine disruptors, determined in this fraction, triclosan, the antibacterial chemicals, was found in a concentration of 26 ng/l. Triclosan is included in detergents,

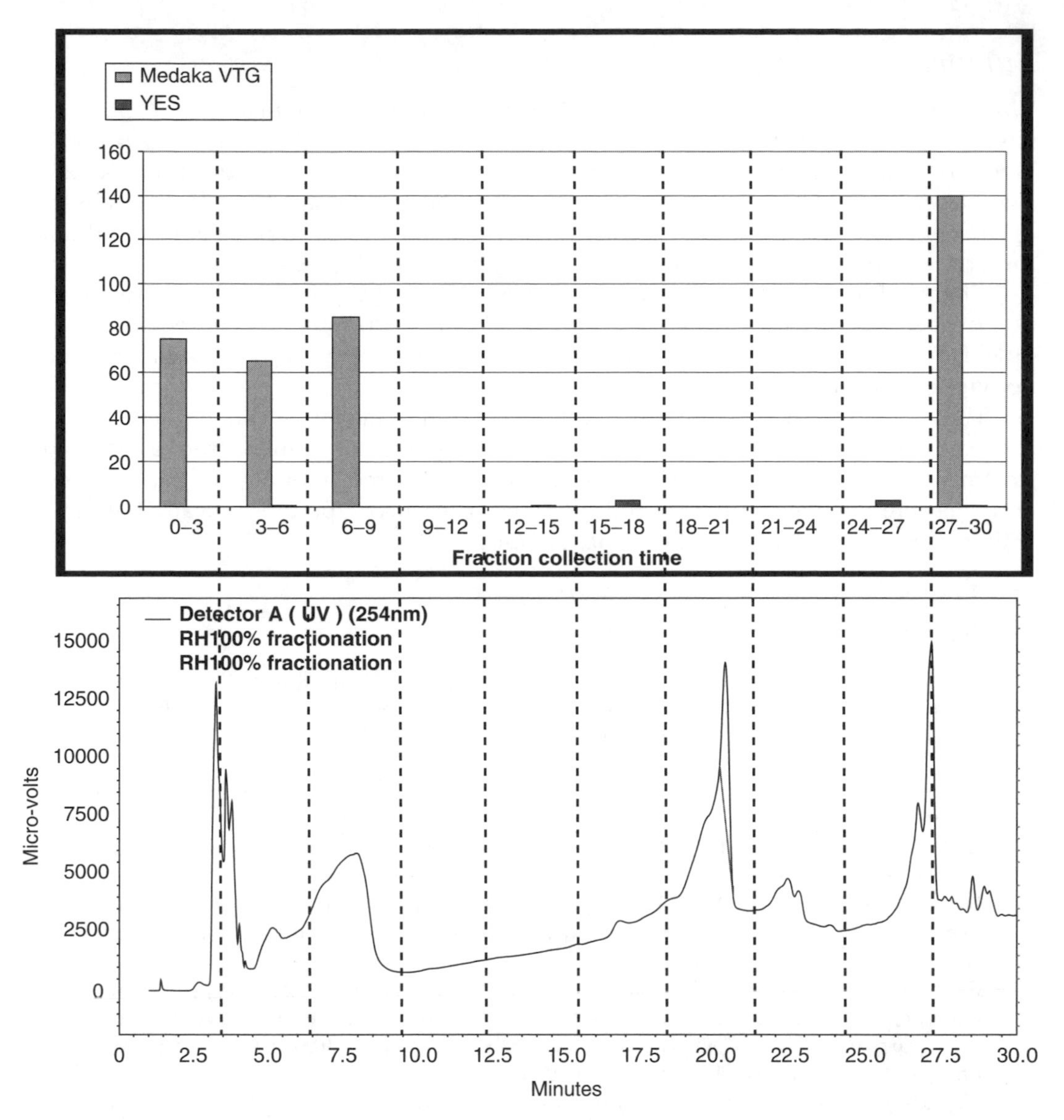

Figure 26.4 *In vitro* and *in vivo* estrogenic activity for 100% ethanol fraction, Red Hook, ng/l.

dish soaps, laundry soaps, deodorants, cosmetics, lotions, creams, and toothpastes and mouthwashes. Oxybenzone and triclosan are shown to be weakly estrogenic.[17,18] Two polyaromatic hydrocarbons (PAHs) pyrene and phenanthrene were detected in concentrations of 5 and 7.5 μg/l, respectively. Other compounds identified but not quantified were: butylated hydroxytoluene (BHT), plasticizers (bisphenol A like), diethyl phthalate, benzophenone, dibutyl phthalate, a TCEP-like fire retardant, triphenylphosphate, formylmethylenetriphenylphosphorane, triphenylphosphine sulfide, nonylphenol isomers, and squalene.

Summary

Wastewater effluents from two New York treatment facilities, Red Hook and New Town Creek, were evaluated for estrogens using *in vitro* and *in vivo* bioassays. EEQs for Red Hook STP showed higher results compared to New Town Creek STP for September 2002. Therefore, the Red Hook effluent extract was further fractionated to identify causative agents. *In vivo* activity was consistently higher than *in vitro*-based calculations indicating the presence of estrogenic agents that were not potent ER-ligands.

Chemical analysis provided with GC/MS/MS and LC/MS/MS showed nonylphenol, oxybenzone, triclosan, galaxolide, phenanthrene, and pyrene in concentrations ranging from 19 ng/l to 7.5 μg/l, and other agents, including plasticizers fire retardant nonylphenol isomers, and squalene.

These results indicate a mixture of estrogenic compounds with varied polarity, which are not potent estrogen receptor ligands within wastewater effluent from a New York treatment facility. Further study is necessary evaluating the amounts of these unknowns and determining the individual contributions of each compounds toward the *in vivo* activity observed in the extract or water sample.

Reference

1. Belfroid, A.C., Van der Horst, A., Vethaak, A.D., Schafer, A.J., Rijs, G.B.J., Wegener, J. and Cofino, W.P., Analysis and occurrence of estrogenic hormones and their glucuronides in surface water and waste water in The Netherlands, *The Science of the Total Environment*, 225, 101–108, 1999.
2. Sumpter, J.P., and Jobling, S., Vitellogenesis as a biomarker for estrogenic contamination of the aquatic environment, *Environ. Health Perspect.*, 103, 173–178, 1995.
3. Jobling, S., Nolan, M., Tyler, C.R., Brighty, G. and Sumpter, J.P., Widespread sexual disruption in wild fish, *Environ. Sci. Technol.*, 32, 2498–2506, 1998.
4. Islinger, M., Yuan, H., Voelkl, A. and Braunbeck, T., Measurement of vitellogenin gene expression by RT-PCR as a tool to identify endocrine disruption in Japanese medaka (*Oryzias latipes*), *Biomarkers*, 7, 80–93, 2002.
5. Folmar, L.C., Denslow, N.D., Rao, V., Chow, M., Crain, D.A., Enblom, J., Marcino, J. and Guilette, L.J., Vitellogenin induction and reduced serum testosterone concentrations in feral male carp (*Cyprinus carpio*) captured near a major metropolitan sewage treatment plant, *Environ. Health Perspect.*, 104, 1096–1101, 1996.
6. Harries, J.E., Sheahan, D.A., Jobling, S., Matthiessen, P., Neall, P., Routledge, E.J., Rycroft, R., Sumpter, J.P. and Tylor, T., A survey of estrogenic activity in United Kingdom inland waters, *Environ. Toxicol. Chem.*, 15, 1993–2002, 1996.
7. Harries, J.E., Sheahan, D.A., Jobling, S., Matthiessen, P., Neall, P., Sumpter, J.P., Tylor, T. and Zaman, N., Estrogenic activity in five United Kingdom rivers detected by measurement of vitellogenesis in caged male trout, *Environ. Toxicol. Chem.*, 16, 534–542, 1997.
8. Harries, J.E., Janbakhsh, A., Jobling, S., Matthiessen, P., Sumpter, J.P. and Tyler, C.R., Estrogenic potency of effluent from two sewage treatment works in the United Kingdom, *Environ. Toxicol. Chem.*, 18, 932–937, 1999.
9. Desbrow, C., Rutledge, E.J., Brighty, G.C., Sumpter, J.P. and Waldock, M., Identification of estrogenic chemicals in STW effeluent. 1. Chemical fractionation and *in vitro* biological screening, *Environ. Sci. Technol.*, 32, 1549–1558, 1998.
10. Legler, J., Zeinstra, L.M., Schuitemaker, F., Lanser, P.H., Bogerd, J., Brouwer, A., Vethaak, A.D., De Voogt, P., Murk, A.J. and Van Der Burg, B., Comparison of *in vivo* and *in vitro* reporter gene assays for short-term screening of estrogenic activity, *Environ. Sci. Technol.*, 36, 4410–4415, 2002.

11. Tilton, F., Benson W. and Schlenk, D., Evaluation of estrogenic activity from a municipal wastewater treatment plant with predominantly domestic input, *Aquat. Toxicol.*, 61, 211–224, 2002.
12. Thompson, S., Tilton, F., Schlenk, D. and Benson, W.H., Comparative vitellogenic responses in three teleost species: extrapolation to *in situ* filed studies, *Mar. Environ. Res.*, 50, 185–189, 2000.
13. Huggett, D., Foran, C., Brooks, B., Weston, J., Peterson, B., Marsh, E., La Point, T. and Schlenk, D., Comparison of *in vitro* and *in vivo* bioassays for estrogenicity in effluent from North American Municipal Wastewater Facilities, *Toxicol. Sci.*, 72, 77–83, 2003.
14. Snyder, S.A., Keith, T.L., Verbrugge, D.A., Snyder, E.M., Gross, T.S., Kannan, K. and Giesy, J.P., Analytical methods for detection of selected estrogenic compounds in aqueous mixtures, *Environ. Sci. Technol.*, 33 (16), 2814–2820, 1999.
15. Snyder, S.A., Villeneuve, D.L., Snyder, E.M. and Giesy, J.P., Identification and quantification of estrogen receptor agonists in wastewater effluents, *Environ. Sci. Technol.*, 35 (18), 3620–3625, 2001.
16. Vanderford, B., Pearson, R., Rexing, D. and Snyder, S., Analysis of endocrine disruptors, pharmaceuticals, and personal care products in water using liquid chromatography/tandem mass spectrometry, *Anal. Chem.*, 75, 6265–6274, 2003.
17. Bruchet, A., Prompsy, C., Filippi, G. and Souali, A., A broad spectrum analytical scheme for the screening of endocrine disruptors (EDs), pharmaceuticals and personal care products in wastewaters and natural waters, *Water Sci. Technol.*, 46 (3), 97–104, 2002.
18. Foran, C.M., Bennett, E.R. and Benson, W.H., Developmental evaluation of a potential nonsteroidal estrogen: Triclosan, *Mar. Environ. Res.*, 50 (1–5), 153–156, 2000.

chapter twenty-seven

A toxicity assessment approach for evaluation of in-situ *bioremediation of PAH contaminated sediments*

Henry H. Tabak, James M. Lazorchak, and Jim Ferretti
U.S. Environmental Protection Agency

Mark E. Smith
SoBran, Inc.

Contents

1-56670-664-5/05/$0.00+$1.50
© 2005 by CRC Press

Introduction

Polycyclic aromatic hydrocarbons (PAHs) represent a group of organic contaminants known for their prevalence and persistence in petroleum-impacted environment, such as groundwater, soils, and sediments. Many high molecular weight PAHs are suspected carcinogens, and the existence of these PAHs in dredged sediments can result in the classification of the sediment as hazardous waste, for which appropriate treatments may be necessary prior to disposal. Unfortunately, *ex-situ* PAH treatment is not cost effective, and natural attenuation/recovery has uncertainties, including possible resuspension, transport, and distribution over wide areas, that can produce substantial risk to human health and to the ecosystems. The widespread contamination of sediments by PAHs has thus created a need for a cost-effective remediation process for restoration of these sediments, and bioremediation is one method that may reduce the risk of sediment associated PAHs.[1–3]

Bioremediation is complicated by three factors: (1) the need to biodegrade a complex mixture of PAHs with varying ring numbers and molecular weight in aged sediments; (2) PAHs in sorbed and aqueous phase sediment must be made available to the degrading microbial consortia; (3) the need to determine whether the cause of toxicity in PAH contaminated sediments is eliminated by the use of selected bioremediation processes.

In recent years, considerable efforts have been made to characterize the impact of PAH bioavailability and the biodegradation of PAH mixtures on bioremediation process and recent advances in our understanding of the first two factors as they relate to sediment bioremediation have been adequately reviewed.[4] Recent studies on sequestration and bioavailability of PAHs in sediments/soils[5–11] indicate the importance of considering the impact of bioavailability in bioremediation strategies for sediments contaminated with PAHs. Studies on sorption and desorption characteristics of PAHs and other hydrophobic compounds in sediments have also been documented in the literature.[12–16] Aerobic biodegradation has been well documented, and PAH degraders have been isolated from geographically diverse PAH contaminated sediments. Recent studies have shown the degradation of lower-ring PAHs to occur under varied redox conditions and sulfate reducing conditions.[17–24] Most of these studies have focused on individual PAH compounds, which were added to the sediment and degraded by enrichments with pure cultures.

Recently, studies on biodegradability, bioavailability, and toxicity of PAHs in aged contaminated marine and estuarine sediments, on the influence of each of these factors on one another and on their impact on bioremediation[25] have helped to clarify these three important factors that complicate *in-situ* bioremediation of PAH contaminated sediments. These studies emphasized the need for determining the baseline and residual toxicity of these sediments before and after biological treatments and physico-chemical sediment manipulation tests in order to optimize the bioremediation strategy for rendering the PAH contaminated sediment sites significantly risk free to human and ecological health.

Recently, advances have been made in the development of toxicity identification and evaluation (TIE) methods for determining the cause(s) of observed acute and sublethal toxicity in contaminated freshwater, interstitial, and marine sediments.[26–32] The TIE methods selectively manipulate the bioavailability of the potential toxicants in sediments. Comparison of the manipulated samples versus non-manipulated samples (baseline) permits a determination of active toxicant(s). These TIE methods serve as a guide for site remediation and are a part of risk assessment for the contaminated sediment sites. The TIE approach comprises of three phases: phase 1, characterization (what classes of toxicants are active? i.e., metals, organic toxicants); phase 2, identification (what specific toxicant is active? i.e., copper, PAHs); phase 3, confirmation (validate findings of phases 1 and 2). The manipulations reported in the literature are comprised of aeration (volatile toxicants), filtration, chelation (metals), solid phase extraction (organic toxicants), graduated pH tests (ammonia), oxidant reduction (chlorine), cation exchange resin addition, acid volatile sulfide addition and base metal reaction (metals), coconut charcoal addition and Ambersorb resin addition (organic toxicants), and zeolite addition (ammonia). Whole sediment manipulations include: (1) addition of toxicants (toxicants interact with materials added to sediment which reduces bioavailability); (2) removal of toxicants (toxicants are removed from the sediment, interstitial water and overlying water); (3) alteration of toxicants (toxicant is altered chemically to a form that is not toxic). The marine toxicity tests utilize:

- amphipods (*Ampelisca abdita*) — mortality, tubular dwelling in sediment;
- mysids (*Americamysis bahia*) — mortality, epibenthic;
- bivalves (*Mercenaria mercenaria*) — mortality and growth, infaunal.

The freshwater toxicity tests utilize:

- amphipods (*Hyalella azteca*) — mortality and growth, epibenthic;
- midges (*Chironomas tentants*) — mortality and growth, surface burrower.

In order to develop an effective strategy for bioremediation of contaminated sediments, it is essential to conduct toxicity tests to measure baseline toxicity, to determine the cause of toxicity, and to evaluate the effectiveness of sediment biotreatment methods in reducing the ecotoxicity. The overall objective of the present research effort was to develop a toxicity assessment approach for the evaluation of *in-situ* bioremediation of PAH contaminated sediments. The studies incorporated the use of: (1) varied biodegradation procedures (under aerobic and sulfate reducing conditions) on sediment samples collected from New York/New Jersey Harbor (NY/NJ H) (Figure 27.1) containing trace quantities of PAHs and from the East River (ER), New York, characterized by high PAH concentration levels; (2) toxicity tests to determine baseline toxicity of these sediment samples before any biological treatment; (3) toxicity tests to determine the residual toxicity of the sediment following the biotreatment procedures and the physico-chemical manipulation tests on those sediment samples to determine the cause of toxicity, the extent of detoxification by the above methods and the effectiveness of the developed biotreatment procedures and sediment manipulation tests in reducing the ecotoxicity and

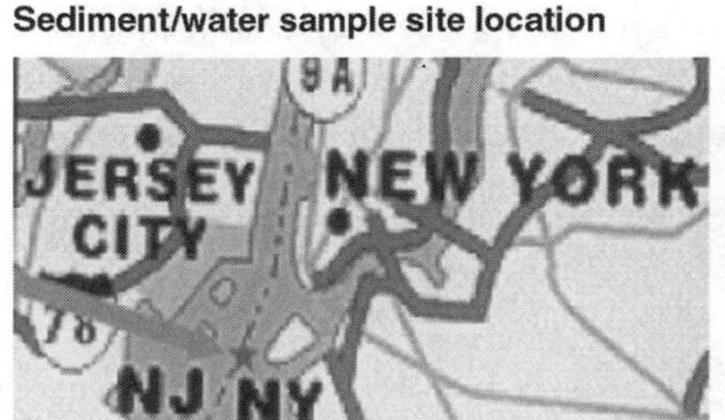

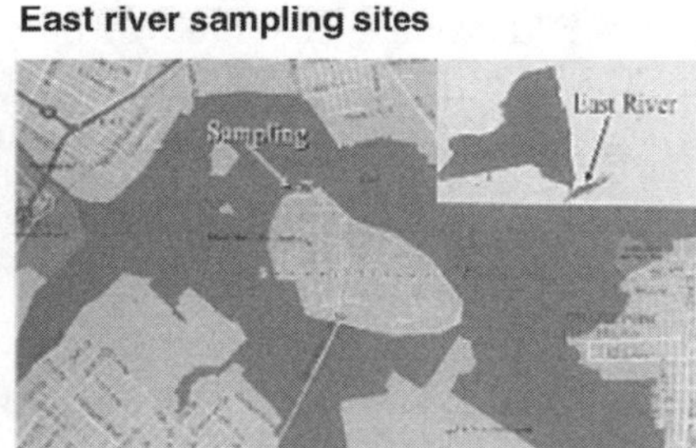

Figure 27.1 Sources of sediment samples.

plant toxicity. The recently published TIE methods served as a background in the development of a toxicity approach for the evaluation of the *in-situ* bioremediation of the PAH contaminated sediments.

The sediment toxicity testing is to provide a measure of biotreatment efficiency based on ecotoxicity values and to relate the reduction of contaminant concentration in sediments to the reduction of ecotoxicity based on biological assay methods. The toxicity methods are to be used to assess how much each biotreatment and sediment manipulation procedure reduces the lethal, sublethal, and bioaccumulative levels of PAH contaminants in sediments.

This chapter emphasizes studies on: (1) biodegradability of PAHs in East River aged sediments under aerobic and sulfate reducing conditions, with and without added co-substrates; (2) toxicity of the untreated and biotreated NY/NJH and East River sediments to aquatic ecosystems and benthic organisms and development of methods to reduce toxicity of untreated sediments. The information gained from the above studies should be useful in developing strategies for the biotreatment of PAH contaminated sediment.

Materials and methods

Characterization of sediment samples

Samples of the sediment from the NY/NJH and from the East River and the respective overlaying water were used in the adsorption/desorption, biodegradation, and toxicity studies. Pebbles, shells, and vegetable matter were removed before storing the sediments and natural water in sealed 20 gal plastic carboys at 4°C. The NY/NJH sediment was air dried in covered shallow pans for 2 weeks, followed by powdering and sieving. The East River sediment was taken directly from the sediment container. Sediment characterization data for the NY/NJH and the East River sediments are presented in Table 27.1. Initial PAH concentration analysis in the NY/NJH and the East River sediments were obtained by extraction with 1:1 methanol/methylene chloride mixture and analyzed by gas chromatography using a HP 5890 Series II GC (Hewlett Packard, Palo Alto, CA) equipped with flame ionization detector, a 7673 automatic sampler, and a PTE-5 fused silica capillary column (J&W. Folsom, CA; 30 m, 0.25 m film thickness). The column temperature started at 100°C for 1 min, increased to 280°C at 4°C/min, then increased to 300°C at 24°C/min, and was held at 300°C for 12 min. The GC/FlD method is similar to the US EPA SW 8100 method for PAH analysis. The 2-fluorobiphenyl was used as the surrogate compound and showed a recovery of 70–80%. The external standard method was used in the analysis. Mass spectroscopic analysis of the chromatography peaks was also performed to confirm the GC results on the individual PAH contaminants. Only trace quantities of PAHs were observed to be present in the NY/NJH sediment, which are shown in Table 27.1. The NY/NJH and East River sediment samples were found to contain 4.18% and 6.0% w/w organic carbon, respectively. Table 27.1 provides characterization of both sediments.

Sources of sediment samples

Figure 27.1 provides the sources of the NY/NJ H and East River sediments used in the toxicity assessment approach. The source of the East River sediment was near the Rikers Island area.

Table 27.1 Sediment characterization

NY/NJH sediment		East river sediment	
Cation exchange capacity	42.9 mEq/100 g	Cation exchange capacity	37.9 mEq/100 g
Organic carbon	4.16% (w/w)	Organic carbon	6.0% (w/w)
pH	7.6	pH	8.3
Total nitrogen	0.279%	Total nitrogen	0.289%
Total sulfur	0.60%	Total sulfur	1.02%
Sulfate	371 ppm	Sulfate	264 ppm
Olsen phosphorus	47 ppm	Olsen phosphorus	15 ppm
Soluble salts	4.25 mmhos/cm	Soluble salts	4.88 mmhos/cm
Metals (ppm)		Metals (ppm)	
Total iron	39,820	Total iron	34,200
Total manganese	894	Total manganese	381
Total copper	86.9	Total copper	321
Total lead	99.1	Total lead	477
Total chromium	59.6	Total chromium	71.3
Total mercury	15.6	Total mercury	11.9
PAHs[c] (ppm)		PAHs (ppm)	
Naphthalene	0.1	Naphthalene (2-ring)	46.7
Phenanthrene	0.2	2-Methyl naphthalene (2-ring)	27.7
Pyrene	0.6	Acenaphthylene (3-ring)	19.9
Benzo(*a*)pyrene	0.2	Acenaphthene (3-ring)	81.1
		Dibenzofuran (3-ring)	5.31
		Fluorene (3-ring)	34.6
		Phenanthrene (3-ring)	190.6
		Anthracene (3-ring)	79.1
		Fluoranthene (4-ring)	97.3
		Pyrene (4-ring)	141.8
		Benzo[*a*]anthracene (4-ring)	70.1
		Chrysene (4-ring)	64.1
		Benzo[*b* + *k*]fluoranthene (5- ring)	67.7
		Benzo[*e*] pyrene (5-ring)	30.9
		Benzo[*a*] pyrene (5-ring)	52.9
		Dibenzo [*a,h*] anthracene (5-ring)	4.4
		Indeno [*1,2,3-c,d*] pyrene (6-ring)	29.8
		Benzo [*g,h,i*] perylene (6-ring)	25.9
Particle size distribution (mm)	Percent	Particle size distribution (mm)	Percent
<2	21	<2	5.0
2–3	7	2–3	1.0
3–5	4	3–5	1.0
5–10	9	5–10	0.5
10–20	8	10–20	1.5
20–45	29.7	20–45	54.2
45–106	15.7	45–106	9.4
106–180	1.8	106–180	10.0
180–250	0.7	180–250	5.7
250–500	1.9	250–500	6.1
0.5–1.00	0.9	0.5–1.00	5.5
>1.00	0.3	>1.00	0.1

Aerobic biodegradation slurry system studies

Experiments were conducted to determine if supplemental nutrients (NH_4Cl, K_2HPO_4) were needed for enhancement of biodegradation of PAHs in East River sediment. In these studies, 963 mg/l NH_4Cl and 280.6 mg/l K_2HPO_4 were added to a set of serum bottles resulting in an N/P ratio of 5:1. A second set of serum bottles were set up as controls without those nutrient supplements. A series of experiments were also conducted to assess possible enhancement of PAH biodegradation by surfactants and co-substrates. Either 0.12 mg/l Triton X-100, 36.4 mg/l ethanol or 48.1 mg/l salicylic acid were added to three sets of serum bottles in these studies, to evaluate their ability to stimulate biodegradative activity of the indigenous microbiota. The 10-g samples of East River sediment and 50 ml of natural overlaying water collected from the same site were added to 100-ml capacity glass serum bottles. The system was buffered by the addition of crushed limestone (2 g) to the liquid. The crushed limestone shells were added to provide pH buffer against the production of sulfuric acid resulting from the oxidation of hydrogen sulfide present in East River sediment. The headspace in the bottles contained an atmosphere of ~75% O_2/~25% N_2 to provide adequate oxygen (DO) in the liquid phase for the duration of the study. The serum bottles sealed with Teflon-faced butyl stoppers and aluminum crimp-seals were placed on the rotating tumblers. In control set of serum bottles, a solution of sodium azide and sodium molybdate (20 m*M* Na_2MoO_4 and 1 mg/l NaN_3) was added to prevent biological activity and growth of microbiota. Samples were analyzed for the 19 contaminant PAHs, pH, and DO at the following time periods: 0, 1, 2, 4, 8, 17, and 24 weeks. Triplicate samples were sacrificed at each sampling event and subjected to analysis. The dissolved oxygen and the oxygen content in the headspace were checked regularly to confirm the availability of oxygen in the aerobic setups throughout the experimental run.

Sulfate reducing conditions slurry system studies

The NY/NJH and East River sediments were shown to have significant concentration of sulfate and hydrogen sulfide, and microbial analyses of the samples illustrated presence and significant activity of the sulfate reducing bacteria (SRB). A series of experiments were conducted to assess the biodegradation of PAHs in East River sediments under sulfate reducing conditions with and without the addition of supplemental nutrients. Studies were undertaken to determine the effect of added organic nutrients and co-substrates (volatile fatty acids, alcohols) and metals (Fe) as electron donors on enhancing the rate of biodegradation of PAHs in sediments slurry systems under sulfate reducing conditions. Experiments reported in this chapter are those without the use of additives and those with either acetate or ethanol addition to the serum bottle setups. For the sulfate reducing conditions, 30 m*M* Na_2SO_4 was added as the electron acceptor and 0.963 g/l NH_4Cl and 0.4209 g/l K_2HPO_4 were added as inorganic N and P supplements. A BOD_5/N/P ratio of 200:5:1 was established in the sediment slurry systems. The natural sediment overlaying water was purged with nitrogen, and the sediment slurry was prepared in an anaerobic hood (Forma Scientific, Marietta, OH) to obtain a pure N_2 headspace. In these studies with co-substrates, either 230 mg/l acetate or 150 mg/l ethanol were added to the serum bottles. The co-substrates were added again two times during the course of the experiment when the substrate was depleted. DO and oxygen in the headspace were measured at each sampling event, and O_2 free condition prevailed throughout the experimental runs. Samples were analyzed for the 19 contaminant PAHs, sulfate, sulfides, DO, and CO_2 at the following time periods: 0, 1, 2, 4, 9, 14, 21, and 29 weeks. Triplicate samples were sacrificed at each sampling event and subjected to analysis. The slurry reactor bottles were provided with hydrocarbon traps and traps to collect CO_2 and H_2S gases when there was an indication of significant production of these gases.

In the aerobic and sulfate reducing studies, whole bottles were sacrificed for PAH analysis at each sampling time. Sediment was separated by centrifugation and extracted for 24 h, using a 50-ml mixture of 50% methanol and 50% dichloromethane. The extract was obtained by filtering the mixture through a 0.1-μm pore nylon membrane and then analyzed using a Hewlett Packard 5890 series II GC/FID for 19 contaminant PAH compounds. Sulfate analysis was performed using a Dionex DX 500 Ion Chromatograph. The O_2, N_2, and CO_2 contents in the headspace were measured using a GC/TCD system. Additives, ethanol, and acetate were measured with GC/FID system, while salicylic acid was measured with HPLC system. DO and pH were measured using an Orion model with 720A pH meter and Corning model 90 DO meter, respectively.

Sediment toxicity tests

Sediment toxicity tests were conducted on the NY/NJH and East River sediments to measure the baseline and residual toxicity of the sediment samples. The tests were undertaken to determine how effective were the developed biotreatment strategies on reducing the ecotoxicity of the PAH contaminated sediments and to provide a measure of biotreatment efficiency based on ecotoxicity values. The objective of running the tests was to relate the reduction of the contaminant concentration in sediments to the reduction of ecotoxicity (lethal, sublethal, or bioaccumulative endpoints) based on biological assay methods.

Four freshwater toxicity tests were as follows:

1. Epibenthic amphipod (*H. azteca*) — mortality and growth tests: a standard 10-day USEPA method,[33] using 100 ml sediment and 175 ml overlaying water, and two 7-day exposure methods (the EMAP method, using 50 ml sediment and 160 ml overlaying water[34]; and a modified volume method, developed by us, that uses 17 ml sediment and 30 ml of overlaying water).
2. A 70-day burrowing worm (*Lumbriculus variegatus*) — mortality and budding test.[35–37]
3. A 7–8-day fish embryo larval survival and teratogenic test with *Pimephales promelas* (fathead minnow embryo-larva, FHM-EL): USEPA[38] method that uses 40 ml sediment and 60 ml overlaying water.
4. A 4-day vascular aquatic plant, *L. minor* (duckweed): frond number and chlorophyl (*a*) test[39–41] that uses 15 ml sediment and 2 ml overlaying water.

Two marine toxicity tests were also used: (1) a marine amphipod *A. abdida*, 10-day mortality test[42] that uses 200 ml sediment and 600 ml overlaying water; (2) a sheepshead minnow, *C. variegatus*, embryo-larval (SHM-EL) sediment mortality test.[43]

The overlaying water for the amphipod and FHM-EL freshwater tests was reformulated using moderately hard water,[33,34] and moderately hard reconstituted water was used in the duckweed test. The reduced volume freshwater amphipod test was developed and used in this study, since existing larger volume requirements of

USEPA standard method exceeded the amounts available from the enhanced biotreatment studies.

To determine the cause of toxicity in these sediments, five sediment manipulations were performed:

1. Two sediment purge procedure (a), where 2–4 volumes of laboratory water were replaced over the sediment in a 24-h period and (b) a thin layer purging method.[44]
2. A sediment dilution procedure, where grade 40 silica sand was mixed with PAH contaminated sediments on a weight/weight basis.
3. A sediment aeration procedure, where sediment samples were aerated by adding 80 ml of sediment (140 g) to a 250-ml glass graduated cylinder and 120 ml of overlaying water followed by aeration for 24–48 h.
4. An Ambersorb treatment procedure, where PAH contaminated sediment samples were treated with two types of organics removal resins, Ambersorb 563 (AS 563) and Ambersorb 572 (AS 572).[45,46]
5. An Amberlite treatment procedure, where Amberlite IRC-718, an inorganic (metal) removal resin was mixed with the metal bearing and PAH contaminated sediments.[27]

Standard operating procedures for sediment toxicity tests

The standard operating procedures (SOPs) of the sediment toxicity tests for the amphipod, *H. azteca*, the worm, *L. variegatus*, the embryo-larva, *P. promelas*, and the plant, *L. minor* (duckweed) are provided in Table 27.2. The SOPs provide the test criteria and the corresponding specifications for each of the criteria. Comparisons of SOPs for *H. azteca* sediment toxicity tests using the standard volume/EMAP method/reduced volume methods are outlined in Table 27.3. Differences in the standard and the miniaturized procedures of sediment toxicity tests for the freshwater amphipod *H. azteca* are shown in Table 27.4. Differences in standard and miniaturized procedures the sediment toxicity tests for freshwater FHM-EL survival (FHM-ELS) and the marine SHM-EL survival SHM-ELS are shown in Table 27.5.

Sediment manipulation methods

To determine the cause of toxicity in the sediments, five sediment manipulations were performed as follows.

Sediment purge procedure

Purging of the sediment consisted of two methods: (1) replacing overlying water in the first 24 h with 4–6 volumes; (2) a thin-layer purging method, where 1.5 l of sediment is placed in a 16-in. wide × 12-in. long × 7-in. deep pan, with 15 l of overlying water added. The overlying water was changed every 24 h for 5 days, and a sediment sample collected for pore water unionized ammonia analysis. The pore water was collected by centrifuging the sediment sample at 2000 rpm for 20 min. Unionized ammonia was measured each day.

Table 27.2 Standard Operating procedures for the Sediment Toxicity Tests

Standard operating procedures for amophipad *H. azfeca* and aqualic worm sediment toxicity tests samples		Standard operating procedures for aquatic worm *L. variegates* sediment toxicity tests	
Test criteria	Specifications	Test criteria	Specifications
Test type	Static-reneval	Test type	Static-renewal
Test duration	10 or 7 days	Test duration	7 days
Temperature	23°C + 1°C or 25°C + 1°C	Temperature	25°C ± 1°C
Phelaperiod	16 h light/8 h dark	Photoperiod	16 h light/8 h dark
Test chamber size	400 or 200 ml	Test chamber size	200 ml
Sediment volume	100 or 40 ml	Sediment volume	40 ml
Overlying water volume	175 or 160 ml daily	Overlying water volume	160 ml
Removal of test solution	*H. azleca* — 7-days old	Renewal of test solution	Daily
Age of test organisms	*L variegalus* — adults	Age of test organisms	Adults
	10 or 20 each species	No. organisms/chamber	10 or 20
No. organisms/chamber	40 or 50 each species	No. replicate/conc.	4
No. replicates/conc.	1.5 or 2.0 ml FFAY*	Organisms/conc.	40 or 80
No. organisms/conc.	Reformulated	Feeding 1.5 or 2.0 ml FFAY*	
Feeding	Moderately hand	Overlying water	Reformulated moderately hard reconstiluted water**
	Grade 40 milica sand	Control sediment	grade 40 silica sand
Control sediment	Mortality and/or growth	Endpoint mortality	and for Growth
Endpoint	>80% survival in the control	Test acceptability	>80% survival in controls
Test acceptability		* = Digested fish flakes/walfar yeast	

* Digested fish flakes walfar yeast
** = Hardness = 80 to 100 mg/l 80–100 mgl
Alkalinity = 60–80 mg/l

Standard operating procedures for *P. propreties* and *C. veginatious* embryo-larval sediment toxicity test samples. (FHM-ELS) (SEM-ELS)		Standard operating procedures for duckweed *Lerravia minor* sediment toxicity tests	
Test criteria	Specifications	Test criteria	Specifications
Test type	Static-renewal	Test type	Static-renewal
Test duration	10 days or 7 days	Test duration	10 or 7 days
Temperature	25°C + 1°C	Temperature	25°C + 1°C
Photoperiod	14 h light/10 h dark	Photoperiod	14 h light/10 h dark
Test chamber size	30 ml	Test chamber size	30 ml
Sediment volume	15 ml	Sediment volume	15 ml

Continues

Table 27.2 Continued

Standard operating procedures for amophipad *H. azfeca* and aqualic worm sediment toxicity tests samples		Standard operating procedures for aquatic worm *L. variegates* sediment toxicity tests	
Test criteria	Specifications	Test criteria	Specifications
Overlying water volume	2 ml	Overlying water volume	2 ml
Removal of test solution	nt 48 h	Removal of test solution	nt 48 h
Age of plants	2 frond plants	Age of plants	2 frond plants
No. 2 frond plants chamber	6	No. 2 frond plants chamber	6
No. replicates chambers conc.	4	No. Replicates chambers/conc.	4
No. plants/conc.	48	No. Plants/conc.	48
Feeding	0.1 ml of 3 nutrient stocks	Feeding	0.1 ml of 3 nutrient stocks
Overlaying water	Moderately hard	Overlaying water	Moderately hard
Control sediment	Reconstituted water*	Control sediment	Reconstituted water*
Endpoint	Grade 40 milica sand +	Endpoint	Grade 40 milica sand + alfalifa
Test acceptability	Alfalifa		
Hardness = 80–100 mg/l	Frond number		
Alkalinity = 60–80 mg/l	Growth as wet wt	Test acceptability	Frond number
			Growth as wet wt. Chlorophyll-*a*
	Number of control fronda		
		Test acceptability	Frond number
	Doubles controls		Growth as wet wt.
			Chlorophyll-*a*
		*Hardness = 80–100 mg/l Alkalinity = 60–80 mg/l	Number of control fronds doubles controls

Sediment dilution procedure

Grade 40 silica sand (wetted by mixing 100 ml of sand with 50 ml overlying water) was used as a dilution substrate. Sediment dilutions were made on a weight/weight basis. For example, 1% East River sediment was prepared by weighing out 1 g of sediment and 99 g of control sand and mixing them. A spoon or spatula was used to completely mix the materials and add them to the appropriate test container, after which the overlying water was added. The diluted sediments were then treated as a standard sediment sample.

Table 27.3 Comparisons of SOPs for *H. azteca* sediment toxicity tests using standard volumes/EMAP methods/reduced volumes[a]

Test criteria	Specifications
Test type	Static-renewal
Test duration	10 days or 7 days/7 days/7 days
Temperature	23 ± 1°C/25 ± 1°C
Photoperiod	25 ± 1°C
Test chamber size	16 h light/8 h dark
Sediment volume	400 ml/200 ml/60 ml
Overlying water volume	100 ml/40 ml/17 ml
Removal of test solution	175 ml/160 ml/30 ml
Age of test organisms	Daily
No. organisms/chamber	7-days old, 24 h age range
No. replicates chambers/conc.	10/20/5
No. organisms/conc.	4/4/4
Feeding	40/80/20
Overlaying Water	2 ml algae/alfalfa
	Reformulated
Control sediment	Moderately hard
Endpoint	Reconstituted water*
Test Acceptability	Grade 40 milica sand

[a] Mortality and/or growth: 80% survival in the controls.

Table 27.4 Differences in Standard and Miniaturized Procedures for *Hyalella azteca*, freshwater amphipod

Test criteria	Specifications	
	Standard test	Miniaturized test
Test temperature	23 ± 1°	± 1°C
Test chamber size	300 ml	60 ml
Sediment volume	100 ml	17 ml
Overlying water volume	175 ml	30 ml
Organisms/chamber	10	10
Replicates	8	4
Test duration	10 days	7 days

Table 27.5 Differences in standard and miniaturized procedures for freshwater FHM-ELS and marine SHM-ELS sediment toxicity tests

Test criteria	Specifications	
	Standard test (ml)	Miniaturized test (ml)
Test chamber size	125	60
Sediment volume	40	17
Overlying water volume	60	30

Sediment aeration procedure

An aeration procedure was developed, to "blow-off" the volatile sulfides and oxidize the remaining sulfides, thus reducing the overall toxicity of these samples. The sediment samples were aerated by adding 80 ml of sediment (140 g) to a 250-ml glass graduated cylinder and then adding 120 ml of overlying water. An air tube was then inserted into this mixture, and a mild aeration started (~100 bubbles/min). The cylinders were placed into the hood, covered, and allowed to aerate overnight. After aeration, the slurry was removed from the cylinder, placed into centrifuge tubes and centrifuged at 2000 rpm for 20 min. After centrifuging, the excess overlying water was discarded, and the sediment samples were collected for use in the sediment toxicity tests. These aerated sediments were used as the 100% samples or diluted with sand as described above. If desired, the overlying water samples can be saved for aqueous testing with any species.

Ambersorb/Amberlite treatment procedures

East River sediment samples were treated with two types of organic contaminants (PAHs) removal resins, AS 563, and AS 572. The resins were mixed in the sediments at the rate of 4% or 8% AS 563 or AS 572 by weight. Each sample was treated as described above in the procedure for sediment aeration.

Amberlite IRC-718 is an inorganic contaminant (metals) removal resin. Data from other researchers indicate that the use of this type of resin can potentially reduce the toxicity associated with a sediment sample. IRC-718 procedure was the same as the Ambersorb except we used 8% by weight IRC-718 only.

Results and discussion

Aerobic biodegradation slurry system studies

A critical factor in the aerobic PAH biodegradation in East River sediment is the availability of a sufficient supply of oxygen to overcome the initial oxygen depletion due to the high content of sulfide in the sediment. Respirometric experimental data indicated that the largest O_2 uptake rate occurred during the first 24 h, and that the O_2 uptake rate was ~3.5 O_2/g sediment/day. In the slurry systems used in the aerobic studies, 10 g sediment would require ~175 ml air to meet the oxygen demand of the first 24-h period. Based on this observation, an O_2 supply protocol was adopted in the aerobic studies to provide an atmosphere ~75% O_2/25% N_2 in the headspace of the serum bottles to provide adequate DO in the liquid phase for the duration of the study. Others have reported similar effect of sulfides in sediment on BOD.[47]

Biodegradation studies with N/P amendments

In the experiments with and without nutrients (NH_4Cl and K_2HPO_4), with sufficient oxygen present and continuous oxygen resupply to avoid oxygen depletion during the course of the experiment, considerable biodegradation of 2-, 3-, 4-, and 5-ring PAHs was observed in sediment slurry bottle systems. Figure 27.2 shows changes in concentration of the 19 contaminant PAHs present in East River sediment under aerobic conditions without inorganic N and P addition over a period of 24 weeks. DO and oxygen content in the headspace were monitored at each sampling event. The existence of aerobic conditions throughout the experiment was confirmed by a final oxygen content of 14% and a DO above 4 mg/l. During a period of 24 weeks, considerable degradation of 2-, 3-, 4-, and 5-ring PAHs was achieved, while PAH concentration in the abiotic controls remained basically unchanged.

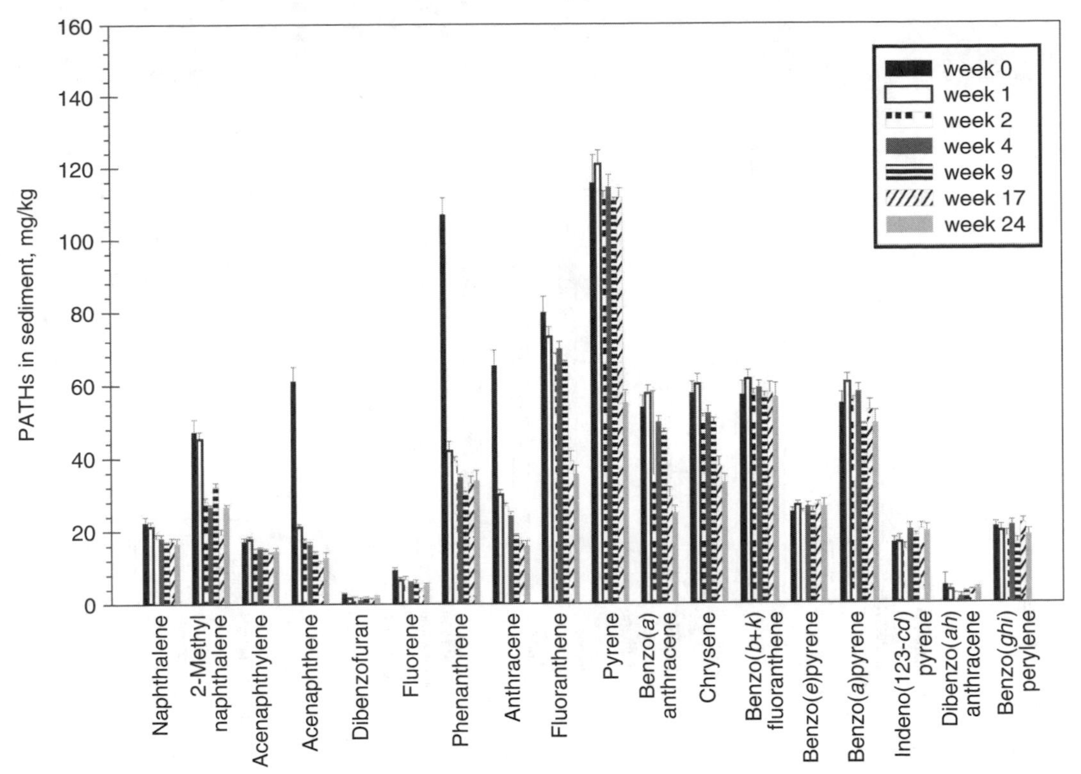

Figure 27.2 PAH concentration in sediment — aerobic, without nutrients.

As seen in Figures 27.2 and 27.3 compounds with lower molecular weight reached lower residual levels in the sediment. The removal of 2-, 3-, and 4-ring compounds was up to 70–80%, 5-ring compounds were removed by just 10–20%, while 6-ring compounds did not exhibit appreciable degradation. The removal of naphthalene, acenaphthylene, dibenzofuran, and fluorene was not as high as that of other lower molecular weight compounds. This may be due to their low initial concentrations in the sediment.

The changes in the concentrations of PAHs over time in the aerobic slurry system with N/P nutrient supplement were very close to those observed in the system without any nutrient addition. No appreciable increase of PAH removal was observed, which suggests that supplying additional N and P to the system did not accelerate biodegradation in the sediment tested. This is probably due to the presence of nitrogen and phosphorus in sufficient quantities in the sediment to support biodegradation.

Biodegradation studies with organic amendments

The experiments with the co-substrates, salicylic acid, and ethanol provided data that showed no considerable improvement of PAHs in the East River sediment over the 27-week incubation period. The addition of the surfactant, Triton X-100 did not appreciably enhanced the biodegradation of those PAHs over the same period. Under all conditions, the residual PAH concentrations were somewhat lower than those observed

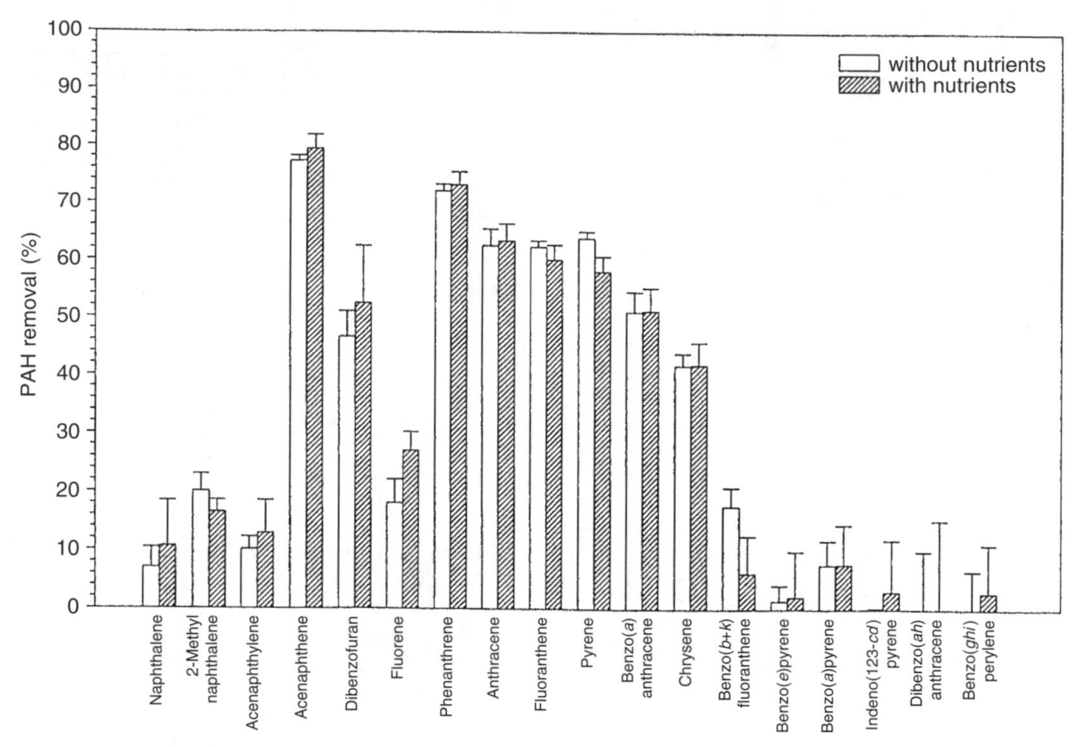

Figure 27.3 PAH Removals — aerobic, without and with nutrients.

in the non-amended systems. Figure 27.4 provides data on biodegradation of anthracene, acenaphthene, fluoranthene, pyrene, and benzo(b + k)fluoranthene in aerobic slurry systems with no additives and with the co-substrates, ethanol, salicylic acid, and surfactant Triton X-100, over a period of 27 weeks. More appreciable losses in PAH concentrations in amended systems were observed, however, after incubation periods longer than 27 weeks as compared to non-amended systems. Aerobic bioslurry biodegradation data for the 19 contaminant PAHs, showing the concentration levels at week 0 and residual concentrations of these PAHs after 54 weeks in milligrams per kilogram of sediment are shown in Tables 27.6–27.9 and in corresponding Figures 27.5–27.8, for systems amended with ethanol, salicylic acid, Triton X-100, and for the non-amended systems, respectively.

Biodegradation studies under sulfate reducing conditions

In the experiments under sulfate reducing conditions, with no co-substrates (ethanol, acetate) added to serum bottles, significant biodegradation of 2- and 3-ring PAHs was observed in sediment slurry bottle systems. In presence of sulfate, PAHs degraded simultaneously with sulfate reduction activity of the SRB, though the extent and rate of biodegradation were much lower then those observed under aerobic conditions. After a lag period of about 2 months, lower-ring PAH compounds up to 3-ring structure exhibited some degree of biodegradation with simultaneous loss of sulfate and significant production of hydrogen sulfide. The generation of biogenic H_2S suggested the activity of SRB in reducing sulfate, as well as SRB population possibly being involved in PAH biodegradation. Figure 27.9 provides a profile of PAH concentrations in sediments at the

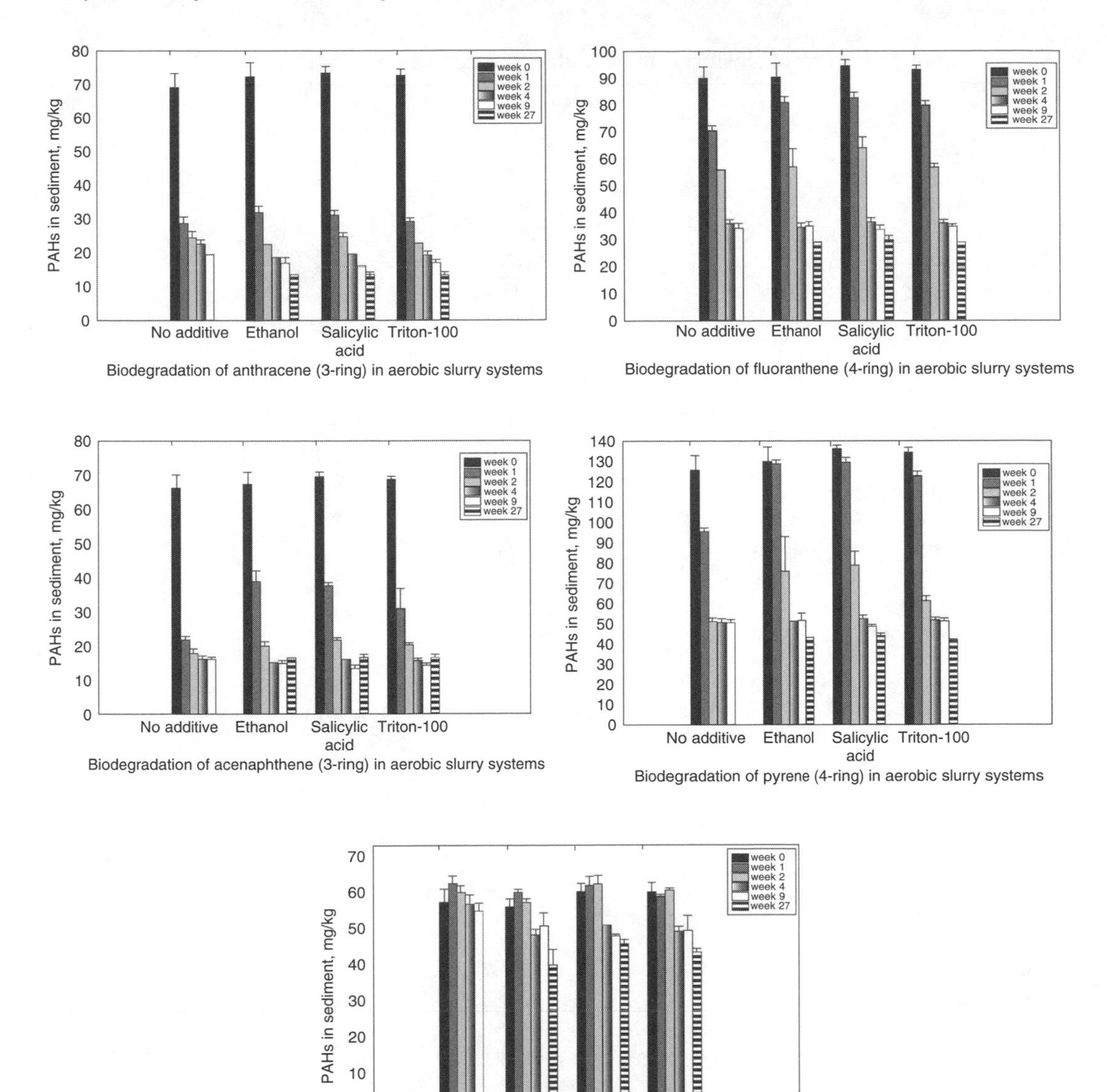

Figure 27.4 Aerobic biodegradation studies — with organic biostimulants.

end of 1, 2, 4, 9, 14, and 19 weeks of incubation. The percentage of removal of PAHs was much lower than under aerobic conditions. PAH levels in control systems with biocides remain the same. Sulfate concentrations were also observed to decrease in the experimental setups. Figure 27.10 provides data on biodegradation of phenanthrene, anthracene, acenaphthene, fluororanthene, pyrene, and benzo(b + k)fluoranthene in sulfate reducing slurry systems with no additives and with the co-substrates, ethanol, and acetate over a period of 21 weeks.

After a lag phase of 2 months, phenanthrene biodegradation was observed to be significant in the slurry systems without co-substrates added. Phenanthrene decreased significantly from 160 to 95 mg/kg in 4 weeks, while biodegradation rate of other

Table 27.6 Aerobic bioslurry biodegradation data in ethanol amended system 3

Aerobic with ethanol	Week 0 (mg/kg)	Week 54 (mg/kg)
Naphthalene	25.456	22.17022
2-Methylnaphthalene	34.54472	37.3621
Acenaphthylene	17.90362	18.84999
Acenaphthene	67.05873	15.33721
Dibenzofuran	4.677455	2.283759
Fluorene	18.46644	4.03039
Phenanthrene	113.5975	30.37519
Anthracene	71.6221	13.37814
Fluoranthene	89.67024	30.36731
Pyrene	130.4363	43.72649
Benzo(*a*)anthracene	53.55876	22.03127
Chrysene	57.38439	31.89784
Benzo(*b* + *k*)fluoranthene	55.74918	44.86067
Benzo(*e*)pyrene	25.78305	23.93589
Benzo(*a*)pyrene	59.2934	42.12915
Indeno(123-*cd*)pyrene	21.49348	17.74094
Dibenzo(*ah*)anthracene	5.088158	6.159342
Benzo(*ghi*)perylene	19.74418	22.4568

Table 27.7 Aerobic bioslurry biodegration data in salicylic amended systems

Aerobic with salicylic acid	Week 0 (mg/kg)	Week 54 (mg/kg)
Naphthalene	26.79728	21.02046
2-Methylnaphthalene	36.41523	36.7819
Acenaphthylene	18.82613	18.11814
Acenaphthene	69.26625	15.57402
Dibenzofuran	5.280595	1.888199
Fluorene	18.64831	3.451252
Phenanthrene	114.8897	31.08614
Anthracene	73.75823	13.27477
Fluoranthene	94.6796	31.38728
Pyrene	136.7311	44.95758
Benzo(*a*)anthracene	57.06599	22.68479
Chrysene	60.7539	32.67509
Benzo(*b* + *k*)fluoranthene	59.82612	45.31577
Benzo(*e*)pyrene	26.21742	24.69259
Benzo(*a*)pyrene	60.08126	42.67976
Indeno(123-*cd*)pyrene	22.86969	18.06649
Dibenzo(*ah*)anthracene	4.337355	5.807288
Benzo(*ghi*)perylene	21.0528	20.1208

molecular PAHs was lower than that of phenanthrene. Loss of sulfate was accompanied by biodegradation of phenanthrene, while in control systems the phenanthrene degradation and sulfate concentration remained unchanged. The data suggest that phenanthrene degradation can be attributed to the activities of SRB existing in the sediment. In the experiments with added co-substrates, acetate, and ethanol, the extent of biodegradation

Table 27.8 Aerobic bioslurry biodegradation data in triton amended systems

Aerobic with triton	Week 0 (mg/kg)	Week 54 (mg/kg)
Naphthalene	25.45247	21.33433
2-Methylnaphthalene	36.29946	37.1525
Acenaphthylene	18.85257	18.29402
Acenaphthene	68.23156	15.26969
Dibenzofuran	4.367361	1.814585
Fluorene	17.87536	3.218282
Phenanthrene	111.7924	30.3522
Anthracene	72.9447	13.48011
Fluoranthene	93.21046	31.25655
Pyrene	134.7868	43.92004
Benzo(*a*)anthracene	56.92604	22.47102
Chrysene	60.33874	31.37765
Benzo(*b* + *k*)fluoranthene	59.79	44.36316
Benzo(*e*)pyrene	27.33923	23.62958
Benzo(*a*)pyrene	60.81231	41.42859
Indeno(123-*cd*)pyrene	22.42689	18.243
Dibenzo(*ah*)anthracene	4.126818	5.820103
Benzo(*ghi*)perylene	19.80722	21.423

Table 27.9 Aerobic bioslurry biodegradation data in non-amended systems

Aerobic without other treatment	Week 0 (mg/kg)	Week 54 (mg/kg)
Naphthalene	27.6845	23.22133
2-Methylnaphthalene	35.95138	26.12385
Acenaphthylene	15.82853	19.24812
Acenaphthene	67.65496	21.92844
Dibenzofuran	3.441659	1.296103
Fluorene	12.91397	4.411246
Phenanthrene	98.38959	29.9047
Anthracene	68.60064	14.64701
Fluoranthene	90.88615	29.45856
Pyrene	133.031	42.16269
Benzo(*a*)anthracene	54.68672	21.8395
Chrysene	58.96554	28.66355
Benzo(*b* + *k*)fluoranthene	57.423	43.88569
Benzo(*e*)pyrene	29.44014	28.22046
Benzo(*a*)pyrene	54.67897	44.81036
Indeno(123-*cd*)pyrene	24.89002	19.42264
Dibenzo(*ah*)anthracene	4.410595	2.527917
Benzo(*ghi*)perylene	20.3554	22.25528

of 2- and 3-ring PAHs was significantly improved, particularly in the later stages of incubation. In the same studies with the addition of co-substrates the biodegradation of phenanthrene was enhanced up to a level comparable to that under aerobic conditions. In experiments with longer incubation times, when ethanol or acetate was added to the system, phenanthrene biodegraded from 120 mg/kg to as low as 20 mg/kg in a period of 4 months, while no obvious lag phase was observed. Higher rate of biodegradation and

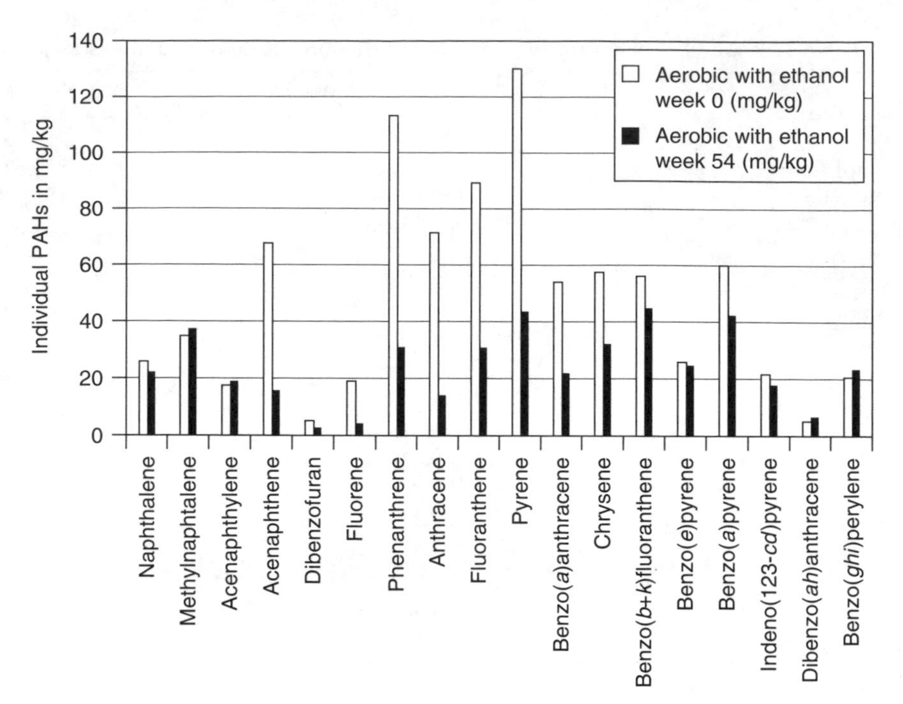

Figure 27.5 Loss of PAHs in aerobic systems amended with ethanol.

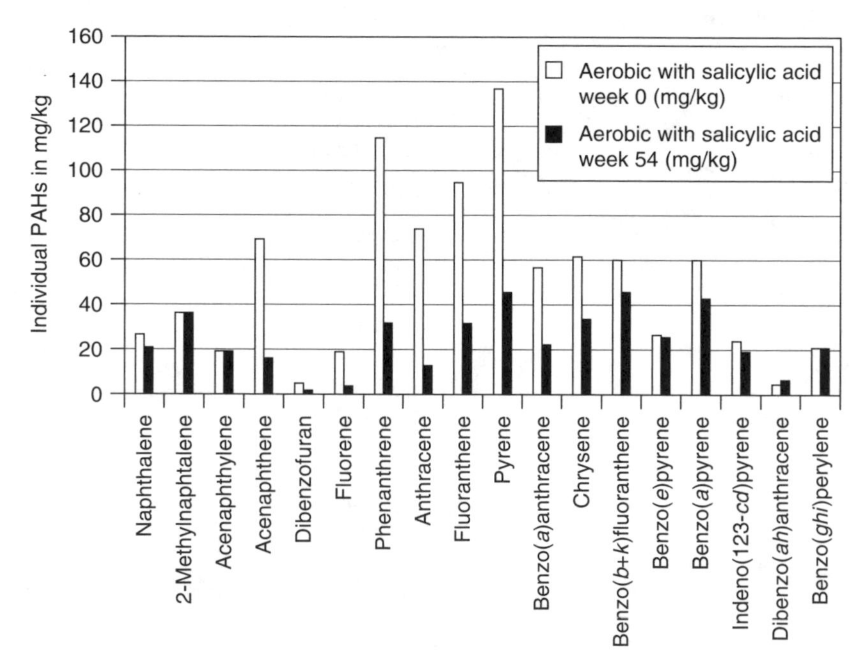

Figure 27.6 Loss of PAHs in aerobic systems amended with salicylic acid.

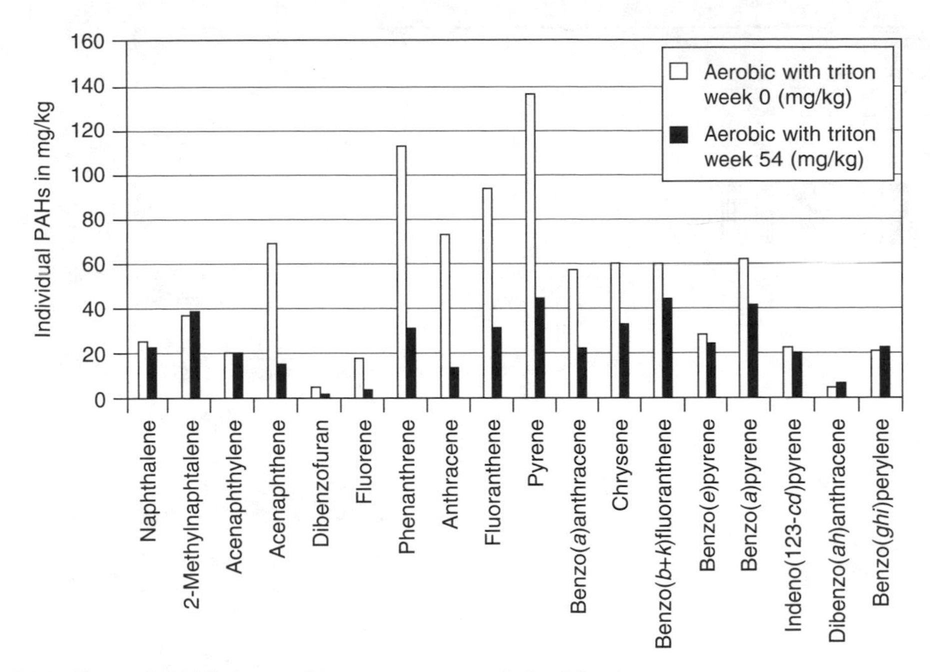

Figure 27.7 Loss of PAHs in aerobic systems amended with triton.

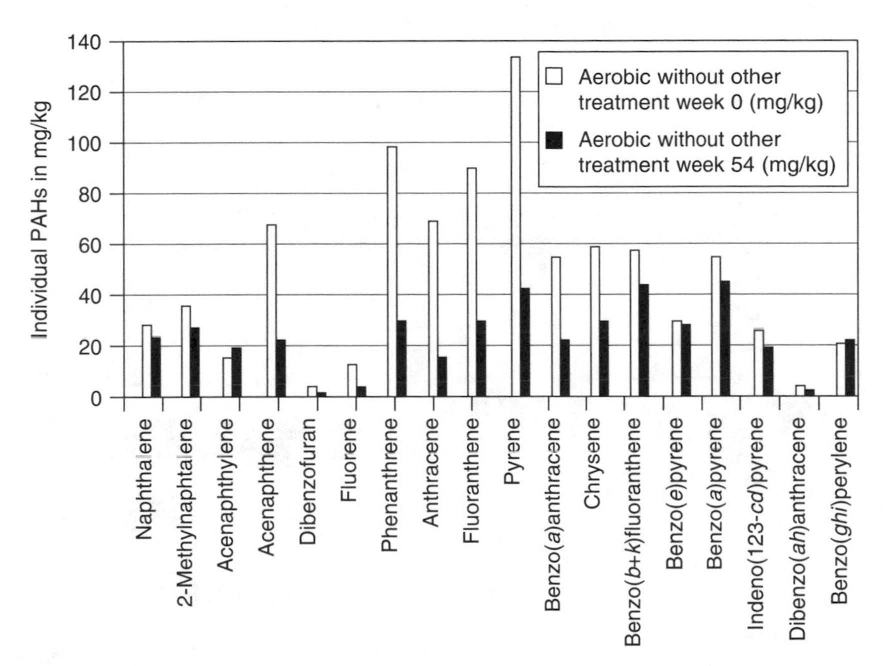

Figure 27.8 Loss of PAHs in non-amended aerobic systems.

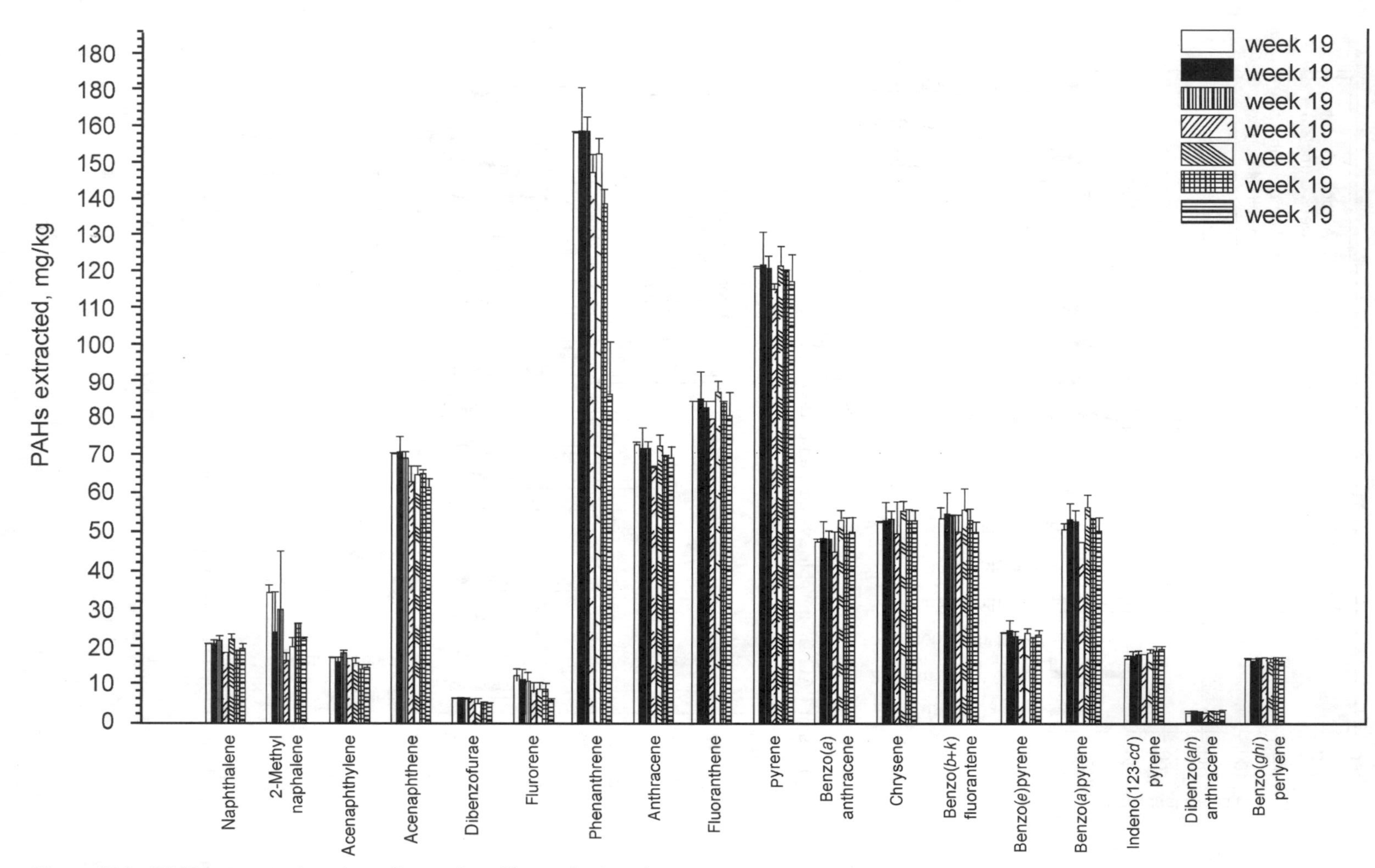

Figure 27.9 PAH concentrations in sediment in sulfate reducing slurry system.

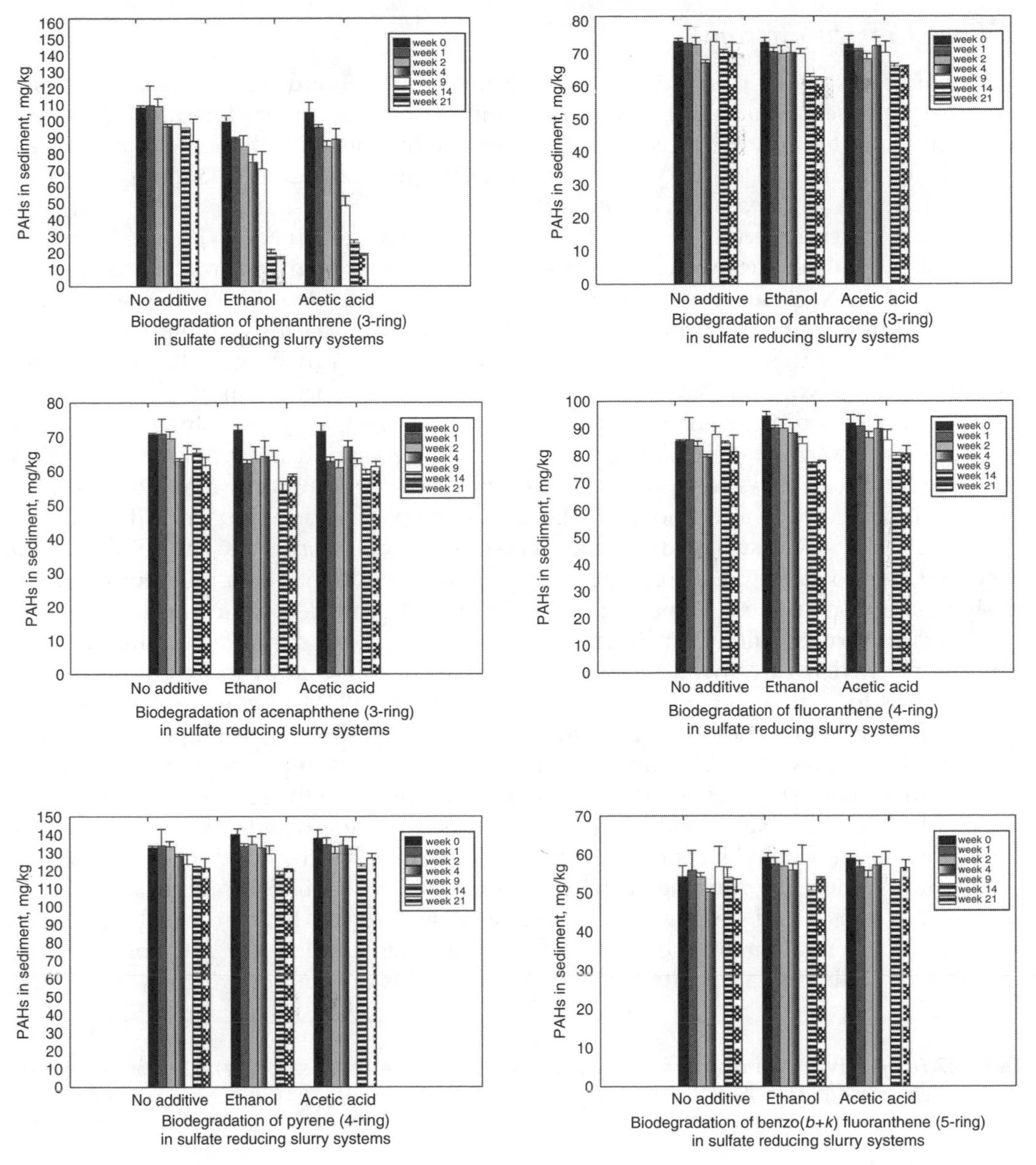

Figure 27.10 Anaerobic biodegradation studies — under sulfate reducing conditions — with organic biostimulants.

lower residuals of phenanthrene were thus achieved in presence of ethanol and acetate. The decrease of sulfate concentration in the systems with co-substrates was greater than that without co-substrates. The enhancement of phenanthrene degradation by ethanol or acetate may be attributed to the amplification of the phenanthrene degrading SRB through the utilization of readily available degradable co-substrates.

Our data on biodegradation of 2- and 3-ring PAHs and East River sediments under sulfate reducing conditions corroborate the findings of others,[21–24] who showed that low molecular weight PAHs can be degraded under these conditions in sediments from heavily contaminated sites.

Sediment toxicity testing

Sediment toxicity tests were conducted with the NY/NJH and East River sediments to measure the baseline and residual toxicity of the sediment samples, determine the source of toxicity, and to reduce the ecotoxicity in these sediments with the use of biotreatment and a series of manipulations, consisting of dilution, purging, aeration, and resin (Ambersorb and Amberlite) treatment procedures.

Table 27.10 summarizes the results of all tests performed on NY/NJH and East River sediments with three freshwater organisms, one freshwater plant, and one marine amphipod. The NY/NJH sediment samples were tested with the freshwater amphipod, *H. azteca*, three times over the course of the study. The marine amphipod, *A. abdita*, when tested with this sediment showed a high level of survival, 95%. Tests performed with the aquatic worm, *L. variegatus*, indicated that this test organism would not survive in the NY/NJH sediment unless the sediment was purged with 2–4 volumes of dilution water. Tests performed with NY/NJH sediment showed significant reduction in survival of the aquatic worms. When sediments were purged with dilution water, no significant reduction in survival was found. FHM-EL tests were performed with NY/NJH sediment and none of the larvae survived. It appears that the aquatic worms and the FHM-EL were very sensitive to salinity and organic and inorganic contaminants in this sediment. The freshwater and marine amphipods gave similar results with respect to survival.

Tests performed with East River sediment (Table 27.10) showed that the sediment was acutely toxic to freshwater and marine organisms, as well as to freshwater plants. Toxicity was found when freshwater aquatic worms and FHM-EL were exposed to NY/NYH sediment in one of three tests performed with the freshwater amphipod. No toxicity was found in the NY/NJH sediments when the marine amphipod was used. This may indicate that the freshwater organisms were more sensitive to the source of toxicity in the sediment. The East River sediment significantly reduced frond (leaves) production (~58.3%) and chlorophyl *a* concentration (~35.4%) in the freshwater duckweed test.

In preparation for running toxicity tests on treated sediment samples, a reduced volume method was developed for the freshwater methods. Because bench scale tests for natural attenuation and enhanced biodegradation studies do not contain large amounts of sediment, a toxicity method that uses less volume or weight was needed.

Table 27.10 Survival data results for *H. azteca* (freshwater amphipods 10-day), *L. variegatus* (freshwater worm), FHM-ElS, SHM-ELS, *L. minor* (duckweed), and *A. abdita* marine amphipod[a]

Sample	*H. azteca* (% survival)	*L. variegatus* (% survival)	FHM-ELS (% survival)	SHM-ELS (% survival)	*L. minor* (% reduction, Frond #/% reduction Chla, mg/g	*A. abdita* (% survival)
NY/NJH	50–100	5–100	0	0–0	# −96%; Chla −41%	95
1% NY/NJH	0–100		80	95	#0; Chla 0	
East River	0–55	0–0	0–0	0–0	# −94 to −58%; Chla −25% to −35.4%	0
1% East River	0–73		0–10	20–80	#0; Chla 0	

[a] NY/NJH, New York/New Jersey Harbor sediments; Chla, chlorophyl *a* in mg/g of dry weight of duckweed.

Table 27.11 shows the results of side-by-side tests performed using a 7-day EMAP[34] that was used in assessing toxicity in sediments collected in a U.S. EPA 1994–1995 Rocky Mountain Regional Environmental Monitoring and Assessment Program (REMAP) and a U.S. EPA Environmental Monitoring and Assessment Program (EMAP) in the mid-Atlantic region in 1994–1998. Table 27.11 shows the results of a number of side-by-side freshwater amphipod tests performed using both of these methods on one East River sediment, two laboratory sediment sand controls, and eight sediments collected from a local project on the Little Miami River, Ohio. All survival results were statistically similar. The miniaturized (or reduced volume) method appears to be very promising, since survivals were not statistically different from existing methods. There were some differences in the growth data. The reduced volume tests showed lower growth results between the two methods. Additional method modifications are currently underway, looking at the feeding regime to see if comparable results can be obtained.

H. azteca toxicity tests with East River sediments using sediment manipulations consisting of sediment dilution, aeration purging, and Ambersorb and Amberlite resin treatment disclosed presence of specific organic and inorganic toxic contaminants in the sediments and provided information on how to significantly reduce the specific contaminant toxicity to the amphipod.

Table 27.12 summarizes the percentage survival results from *H. azteca* sediment toxicity tests with East River contaminated sediments. The East River sediment was

Table 27.11 Comparison tests with *H. azteca* in both control and natural sediments, using the EMAP method and the reduced volume method[a]

	Reduced volume method				7-day EMAP method			
Sample	Survival (%)	C.V. (%)	Grw (μg/sur)	C.V. (%)	Survival (%)	C.V. (%)	Grw (μg/sur)	C.V. (%)
Cnt	100	0	18–42	12.3–45.3	92.5–100	0–6.98	49–89	3.4–29.8
ER	0	0	0	0	0	0	0	0
99-01-99-19	95–100	0–12.8	16–41	22.4–61.3	93.8–97.5	2.6–6.7	48–70	5.1–14.6

[a] All values represent ranges. "Cnt" is the sand control sediment, spiked with 0.5% alfalfa. ER is the East River contaminated sediment. The 99-01-99-19 sediments are sediments collected as part of the Little Miami River project. "Grw" (ug/sur) is the growth in dry weight per surviving individual. C.V. (%) is the percentage of coefficient of variation.

Table 27.12 Results from *H. azteca* sediment toxicity test with East River contaminated sediment. The East River sediment was diluted to 0.1%, 1%, 10%, and 100% for all treatments. The treatments included aeration, and two purging procedures

Sample purged 1 or 2 procedures	Sample aerated	*H. azteca* (% sur)	C.V. (%)
Sand control	24 h	100	0
1% ER P1	24 h	0	0
1% ER P1	24 h	25	40
1% ER P1	24 h	45	22.2
1% ER P1	48 h	25	76.6
Sand control	24 h	100	0
1% ER P2	24 h	35	54.7
1% ER P2	Not aerated	20	141.4

diluted to 1% for all treatments. The treatments performed on different dates, included aeration, no aeration, and two purging procedures. The percentage of survival of *H. azteca* ranged from 0% to 45% as contrasted to 100% survival in the sand controls. Table 27.13 shows the percentage of survival results from *H. azteca* sediment toxicity tests with East River sediment, which was undiluted and diluted to 10%, 1%, and 0.1% for all treatments. The treatment included aeration only procedure and aeration with subsequent addition of 8% AS 572. When only aeration was used, the percentage of survival of *H. azteca* in undiluted, 10% and 1% sediment was shown to be 0%, whereas in the 0.1% diluted sediment, 100% survival was exhibited. The use of aeration combined with the addition of AS 572 resulted in 0%, 20%, 40%, and 95% survival of *H. azteca* in undiluted, 10%, 1%, and 0.1% sediment samples, respectively, as compared to 100% survival in the sand controls. Table 27.14 shows the percentage of survival results from *H. azteca* sediment toxicity tests with East River sediment, which was diluted to 1%, 5%, and 10%. The treatment included aeration and aeration with subsequent addition of 8% Ambarsorb 572 and a mixture of AS 572 and Amberlite IRC-718. When only aeration was used, the percentage of survival of *H. azteca* was shown to be 35%, 25%, and 25% in the 10%, 5%,

Table 27.13 Results from *H. azteca* sediment toxicity tests with East River contaminated sediment. The East River sediment was diluted to 0.1%, 1%, 10%, and 100% for all treatments. The treatments included aeration, and aeration with the subsequent addition of 8% AS 572

Sample	Sample aerated	*H. azteca* (% sur)	C.V. (%)
Sand control	Yes	100	0
0.1% ER	Yes	100	0
1% ER	Yes	0	0
10% ER	Yes	0	0
100% ER	Yes	0	0
0.1% ER + 8% AS 572	Yes	95	10.5
1% ER + 8% AS 572	Yes	40	40.8
10% ER + 8% AS 572	Yes	20	0
100% ER + 8% AS 572	Yes	0	141.4

Table 27.14 Results from *H. azteca* sediment toxicity tests with East River contaminated sediment. The East River sediment was diluted to 1%, 5%, and 10% for all treatments. The treatments included aeration and aeration with subsequent addition of 8% AS 572 and a mixture of 8% AS 572 and 8% Amberlite IRC-718

Sample	Sample aerated	*H. azteca* (% sur)	C.V. (%)
Sand control	Yes	100	0
1% ER	Yes	25	40
5% ER	Yes	25	100.7
10% ER	Yes	35	71.9
Sand control	Yes	100	0
1% ER + 8% AS 572	Yes	90	12.8
5% ER + 8% AS 572	Yes	65	100.7
10% ER + 8% AS 572	Yes	25	64.4
Sand control + 8% IRC 718	Yes	95	10.5
1% ER + 8% AS 572 + 8% IRC-718	Yes	45	42.6
5% ER + 8% AS 572 + 8% IRC-718	Yes	40	57.7

and 1% sediment samples, respectively, as compared to 100% survival in sand controls. When aeration was combined with the addition of AS 572, the percentage of survival of *H. azteca* was 25%, 65%, and 90% in the 10%, 5%, and 1% sediment samples, respectively, as compared to 100% survival in sand controls. When aeration was combined with the addition of a mixture of 8% AS 572 and 8% Amberlite IRC-718, the percentage of survival was 40% and 45% in the 5% and 1% sediment samples, respectively, as compared to 95% survival in the sand control. Table 27.15 shows percentage of survival results of *H. azteca* sediments toxicity tests with East River contaminated sediment diluted to 10% for all treatments. The treatment included aeration with subsequent addition of 8% AS 563 or AS 572, addition of 8% AS 563 with no aeration, and addition of a mixture of 8% AS 572 and 8% Amberlite IRC-718 with aeration. With aeration alone, there was no survival in sediment samples. There were 90% and 85% survival in aerated sediment samples containing 8% AS 563 or AS 5782, respectively, and 0% survival in non-aerated sediment samples containing 8% AS 563. In aerated sediment samples containing a mixture of 8% AS 572 and 8% Amberlite IRC-718, 90% survival was exhibited. Table 27.16 summarizes the results on the percentage of survival of *H. azteca* in aerated or non-aerated, 0.1% and 1% diluted sediments, which were treated with varied combinations of Ambersorb and Amberlite resins and compares these results to a range of 90–100% survival. in sand controls.

Grade 40 silica sand was used as a control in these manipulation toxicity tests. The aeration treatments significantly reduced the concentration levels of dissolved and gaseous hydrogen sulfide in the sediments samples. The treatment of aerated and/or purged sediment samples with AS 563 and AS 552 reduced the concentration of organics, including PAHs in the aqueous phase through formation of bonded organics (PAH)/resin/sediment particle aggregates. Amberlite IRC-718 has significantly adsorbed the metals in the sediment to form metal/resin/sediment particle aggregates. The results on the survival of *H azteca* in sediment samples following the various sediment treatment manipulations illustrate a significant increase in the percentage of survival of the amphipod due to reduction of hydrogen sulfide, metals, and organics, including PAHs in sediment samples and thus implicate them as specific contaminants causing toxicity. Figures 27.11 and 27.12 illustrate percentage of survival of *H. azteca* in the East River sediment samples following aeration, dilution, purging, and Ambersorb and Amberlite resin treatments. The sediment treatment manipulations are listed in an increasing order of effectiveness to raise the percentage of survival of the amphipod, *H. azteca,* in the

Table 27.15 Results of *H. azetca* sediment toxicity tests with East River contaminated sediment. The treatments included aeration, aeration with subsequent addition of 8% AS 563 and AS 572, and addition of 8% AS 563 with no aeration and a mixture of 8% AS 572, and 8% Amberlite IRC-718 with aeration

Sample	Sample aerated	*H. azteca* (% sur)	C.V. (%)
Sand control	Yes	100	0
10% ER	Yes	0	0
10% ER 8% AS 563	No	0	0
10% ER 8% AS 563	Yes	90	12.8
10% ER 8% AS 572	Yes	85	11.8
10% ER + 8% AS 572 + 8% IRC-718	Yes	90	115.5

Table 27.16 *H. azteca* toxicity results of treatments of 0.1% and 1% East River sediments with AS 572 and AS 563, Amberlite IRC-718, with (Y) and without (N) aeration

Sample	Sample aerated	*H. azteca* (% sur)	C.V. (%)
Sand Cnt	Y*	100	0
Sand Cnt 4% AS 563	Y	95	10.5
Sand Cnt 8% AS 563	Y	90	12.8
Sand Cnt 4% AS 572	Y	100	0
Sand Cnt 8% AS 572	Y	90	12.8
Sand control + 8% IRC-718	Y	95	10.5
0.1% ER	Y	100	0
0.1% ER	N*	100	0
0.1% ER + 8% AS 572	Y	95–100	10.5–0
0.1% ER + 8% AS 572	N	100	0
0.1% ER + 8% AS 563	Y	100	0
0.1% ER + 8% IRC-718	N	100	0
0.1% ER + 8% IRC-718	Y	100	0
1% ER	Y	0–45	0
1% ER	N	0	0
1% ER + 8% AS 572	Y	40–90	40.8–12.8
1% ER + 8% AS 572	N	5–35	200–71.9
1% ER + 8% AS 563	Y	40–70	70.7–28.6
1% ER + 8% AS 563	N	45	42.6
1% ER + 8% IRC-718	N	20	115.5
1% ER + 8% IRC-718	Y	75	25.5
1% ER + 8% AS 572 + 8% IRC-718	Y	45–80	42.6–20.4
1% ER + 8% AS 563 + 8% IRC-718	Y	90	12.8

Y = Afrated, * N = Non-Afrated

treated sediments: dilution/aeration/purging > dilution/aeration/purging + AS 563/AS 572 treatment and/or dilution/aeration/purging + Amberlite IRC-718 treatment > dilution/aeration/purging + AS 563/AS 572 + Amberlite IRC-718 treatment.

Table 27.17 illustrates the survival data of *H. azteca* following the aerobic biodegradation treatments of NY/NJH and East River sediments using non-amended samples and samples amended with ethanol, salicylic acid, and the surfactant Triton

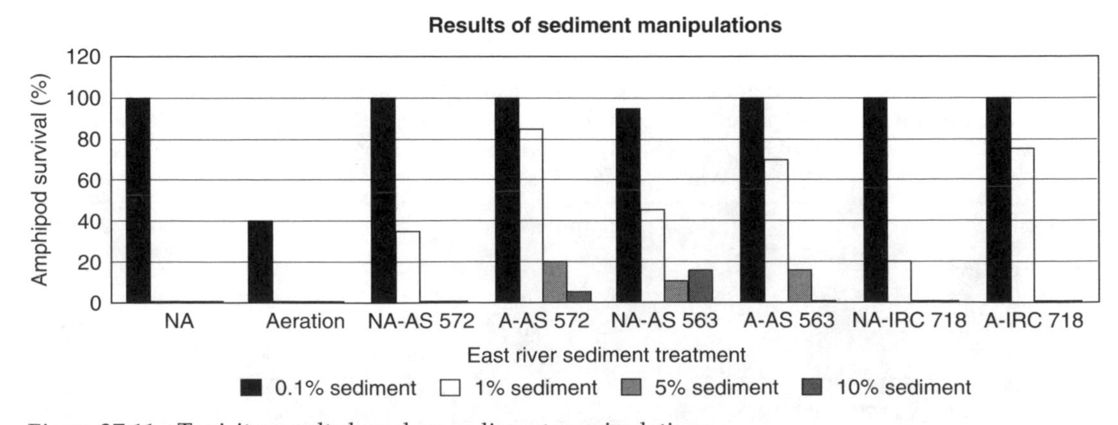

Figure 27.11 Toxicity results based on sediment manipulations.

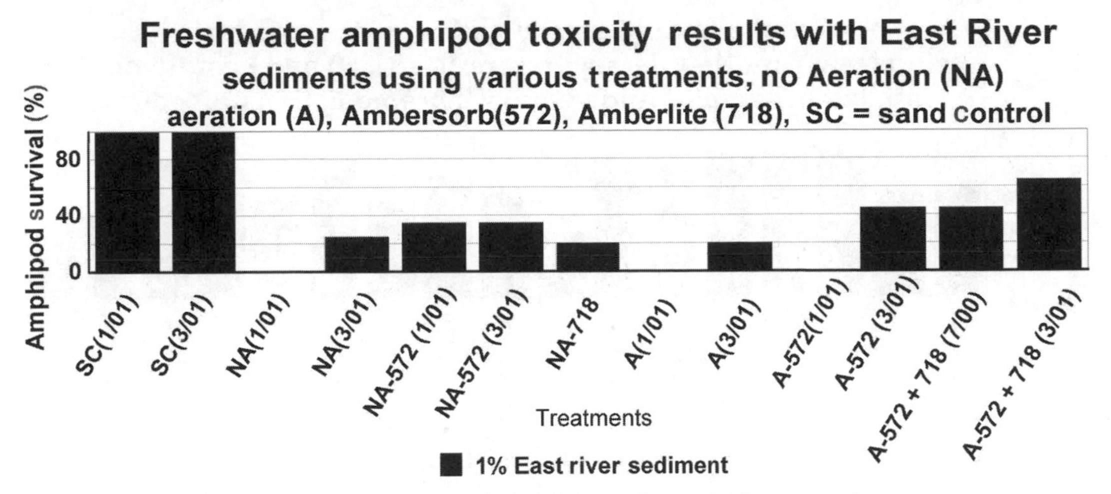

Figure 27.12 Toxicity results based on sediment manipulation without aeration.

Table 27.17 Results from *H. azteca* sediment toxicity tests conducted using East River and NY/NJH sediments treated using an aerobic biodegradation slurry system (bioslurry) and the addition of materials to stimulate the PAH biodegradation activity of the indigenous microbiota in the sediments

Sample	Treatment	*H. azteca* (% sur)	C.V. (%)
Sand control	None	90	22
NY/NJH	Bioslurry	65	15.4
NY/NJH	Bioslurry + ethanol	90	12.8
NY/NJH	Bioslurry + salicylic acid	90	12.8
NY/NJH	Bioslurry + triton	90	12.8
ER 1%	Bioslurry	80	20.4
ER 10%	Bioslurry	85	22.5
ER 100%	Bioslurry	40	40.8
ER 1%	Bioslurry + ethanol	65	29.5
ER 10%	Bioslurry + ethanol	75	33.6
ER 100%	Bioslurry + ethanol	60	47.1
ER 1%	Bioslurry + salicyclic acid	85	22.5
ER 10%	Bioslurry + salicyclic acid	60	47.1
ER 100%	Bioslurry + salicyclic acid	65	52.5

X-100. Figure 27.13 shows the same percentage of survival data as histograms for the varied aerobic biodegradation treatments. The aerobic biodegradation treatments of the NY/NJH and East River sediments have significantly increased the survival of the amphipod, *H. azteca*. The incorporation of the sediment manipulation procedures and the use of the aerobic biodegradation treatments of PAH contaminated sediments was shown to significantly eliminate the cause of toxicity due to the PAHs, metals, and hydrogen sulfide and thus increased the survival of the amphipod, *H. azteca*.

Conclusions

Bench-scale serum bottle sediment slurry studies with aged East River sediment were undertaken to determine the biodegradation of PAH contaminants, with and without the

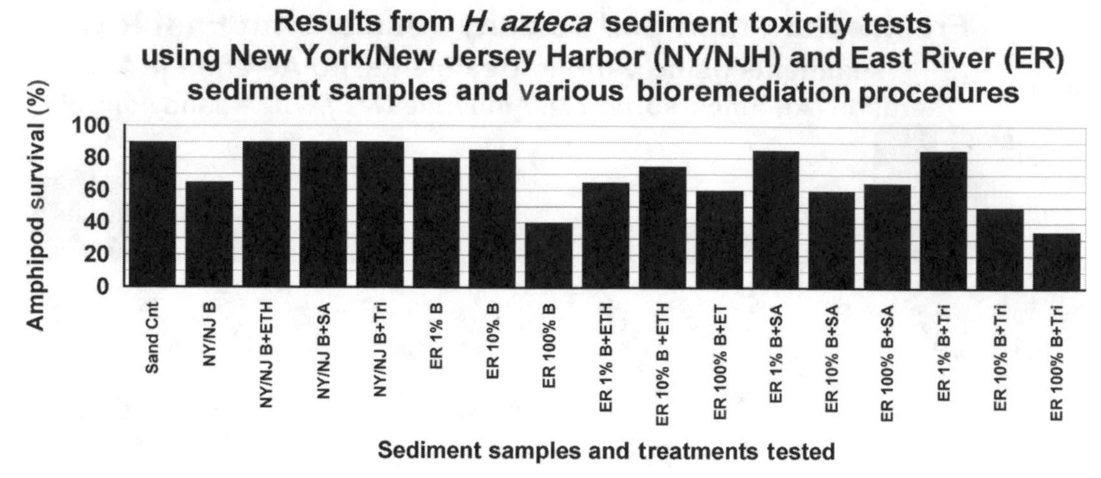

Figure 27.13 Toxicity results based on various biodegradation procedures.

addition of nutrients/co-substrates under both aerobic and sulfate reducing conditions. In the aerobic slurry systems with and without supplemental nutrients (nitrogen and phosphorus) and with pure oxygen in the headspace, appreciable biodegradation (biotransformation) of 2-, 3-, 4-, and 5-ring PAHs occurred within 17 weeks of incubation. Under sulfate reducing conditions, 2- and 3- ring PAHs were biodegraded simultaneously with sulfate reduction, after a lag period of 2 months but the rate of biodegradation was much lower than that under aerobic conditions. The addition of the co-substrates, ethanol, and acetate led to an increase of rate of biodegradation of 2- and 3-ring PAHs, particularly phenanthrene within the same incubation period.

Sediment toxicity tests were undertaken to measure baseline and residual toxicity of the NY/NJH and East River sediments to fresh water and marine amphipods, aquatic worms, FHM-EL, SHM-EL, and a duckweed vascular plant. Tests using five sediment manipulations (dilution, aeration, purging, AS 563 and AS 572, and Amberlite IRC-718 treatments) were preformed to determine the cause of toxicity in the sediments. The East River sediment was shown to be acutely toxic to all the organisms tested, as well as to plants (duckweed). The NY/NJH sediment was slightly toxic to FHM-EL and the aquatic worm *L. variegatus*, but was not toxic to one marine amphipod *A. abiota*. The East River sediment significantly reduced front production (−58.2%) and chlorophyll a levels (−35.4%) in the fresh water duckweed tests. A miniaturized toxicity test, developed by us, shows survival results similar to the old EMAP method for lethality. This miniaturized method could logistically be very desirable for determining toxicity of sediments where only small samples of sediment can be provided. Results from sediment manipulation tests showed that freshwater amphipod, *H. azetca*, survival was improved with aerobic biotreatement and with sediment manipulation procedures consisting of sediment dilution and aeration combined with 8% AS 563 or AS 572 resin treatment and/or 8% Amberlite IRC-718 treatment. The studies revealed that hydrogen sulfide, organics, including PAHs, and metals were factors contributing to East River sediment toxicity.

Studies on bioavailability, biodegradation, and toxicity of NY/NJH and East River sediments stress the importance of understanding the impact of these factors on bioremediation of PAH contaminated sediments and provide an awareness that contaminated sediments with different characteristics may require modification in the biorestoration strategy.

It is essential that the toxicity tests be performed before and after the biotreatments of the contaminated sediments in order to determine the baseline and residual toxicity of these sediments and to determine the cause of toxicity. The toxicity assessment of the contaminated sediments before and after use of biotreatment procedures and sediment manipulation strategies is important for the development of an efficient bioremediation approach for a specific contaminated sediment site.

References

1. Madsen, E.L., Determining *in-situ* biodegradation, *Environ. Sci. Technol.*, 25, 1663–1673, 1991.
2. Alexander, M., Research needs in bioremediation, *Environ. Sci. Technol.*, 25, 1971–1973, 1991.
3. Alexander, M., How toxic are toxic chemicals in soil, *Environ. Sci. Technol.*, 29, 2713–2717, 1995.
4. Hughes, J.B., Beckles, D.M., Chandra, S.D. and Ward, C.H., Utilization of bioremediation processes for the treatment of PAH-contaminated sediments, J. *Ind. Microbiol. Biotechnol.*, 18, 152–160, 1997.
5. White, J.C., Alexander, M., Pignatello, J.J., Enhancing the bioavailability of organic compounds sequestered in soil and aquifer solids, *Environ. Toxicol. Chem.*, 18, 182–187, 1999.
6. Alexander, M, Hatzinger, P.B., Kelsey, J.W., Kottle, B.D. and Nam, K., Sequestration and realistic risk from toxic chemicals remaining after bioremediation, *Ann. NY Acad. Sci.*, 829, 1–5, 1997.
7. Chung, N. and Alexander, M., Effect of concentration on sequestration and bioavailability of two polycyclic aromatic hydrocarbons, *Environ. Sci. Technol.*, 33, 3605–3608, 1999.
8. Hatzinger, P.B. and Alexander, M., Effect of aging of chemicals in soil and their biodegradability and extractability, *Environ. Sci. Technol.*, 29, 537–545, 1995.
9. Tang, J, and Alexander, M., Mild extractability and bioavailability of polycyclic aromatic hydrocarbons in soil, *Environ. Toxicol. Chem.*, 18, 2711–2714, 1999.
10. Pignatello, J.J. and Xing, B., Mechanisms of slow sorption of organic chemicals in natural particles, *Environ. Sci. Technol.*, 30, 1–11, 1996.
11. Luthy, R.G., Aiken, G.R., Brusseau, M.L., Cunningham, S.D., Gschwend, P.M., Pignatello, J.T., Reinhard, M., Traina, S.J., Weber, W.J. and Westall, J.C., Sequestration of hydrophobic organic contaminants by geosorbent, *Environ. Sci. Technol.*, 32, 3341–3347, 1997.
12. Kan, A.T., Fu, G., Chen, W., Ward, C.H. and Tomson, M.D., Irreversible sorption of neutral hydrocarbons to sediments: experimental observations and model predictions, *Environ. Sci. Technol.*, 32, 892–902, 1998.
13. Chiou, C.T., McGroddy, S.E. and Kile, D.E., Partition characteristics of polycyclic aromatic hydrocarbons on soils and sediments, *Environ. Sci. Technol.*, 32, 264–269, 1998.
14. Weber, W.J., Huang, W. and Yu, H., Hysteresis in the sorption and desorption of hydrophobic organic contaminants by soils and sediments. 2. Effects of organic matter heterogeneity, J. *Contam. Hydrol.*, 31, 149–165, 1998.
15. Karrickhoff, S.W., Brown, D.S. and Scott, T.A., Sorption of hydrophobic pollutants in natural sediments, *Water Res.*, 13, 241–248, 1979.
16. Fu, G, Kan, A.T. and Tomson, N.O., Adsorption and desorption hysteresis of PAHs in surface sediment, *Environ. Toxicol. Chem.*, 13, 1559–1567, 1994.
17. Bauer, J.E. and Capone, D.G., Degradation and mineralization of the polycyclic aromatic hydrocarbons anthracene and naphthalene in intertidal marine sediment, *Appl. Environ. Microbiol.*, 50, 81–90, 1985.
18. Bauer, J.E. and Capone, D.G., Effects of co-occurring aromatic hydrocarbons on degradation of individual polycyclic aromatic hydrocarbons in marine sediment slurries, *Appl. Environ. Microbiol.*, 53, 129–136, 1988.

19. Heitkamp, M.A., Freeman, J.P. and Cerniglia, C.E., Naphthalene biodegradation in environmental microcosms: estimates of degradation rates and characterization of metabolites, *Appl. Environ. Microbiol.*, 53, 129–136, 1987.
20. Mihelcic, J.R. and Luthy, R.G., Degradation of polycyclic aromatic hydrocarbon compounds under various redox conditions in soil-water systems, *Appl. Environ. Microbiol.*, 54, 1182–1187, 1988.
21. Zhang, X. and Young, L.Y., Carboxylation as an initial reaction in the anaerobic metabolism of naphthalene and phenanthrene by sulfidogenic consortia, *Appl. Environ. Microbiol.*, 63, 4759–4764, 1997.
22. Coates, J.D., Anderson, R.T. and Lovley. D.R., Oxidation of polycyclic aromatic hydrocarbons under sulfate-reducing conditions, *Appl. Environ. Microbiol.*, 62, 1099–1101, 1995.
23. Coates, J.D., Woodward, J., Allen, J., Philp, P. and Lovely, D.R., Anaerobic degradation of polycyclic aromatic hydrocarbons and alkanes in petroleum-contaminated marine harbor sediments, *Appl. Environ. Microbiol.*, 63, 1099–1101, 1997.
24. Hayes, L.A., Nevin, K.P. and Lovley, D.R., Role of prior exposure on anaerobic degradation of naphthalene and phenanthrene in marine harbor sediments, *Org. Geochem.*, 30, 937–945, 1999.
25. Tabak, H.H., Lazorchak, J.M., Lei, L., Khodadoust, A.P., Antia, J.E., Bagchi, R. and Suidan, M.T., Studies on bioremediation of polycycne aromatic hydrocarbon contaminated sediments: bioavailability, biodegradability and toxicity issues, *Environ. Toxicol. Chem.*, 22 (3), 473–482, 2003.
26. Burgess, R., Charles, J., Kuhn, A., Ho, K., Patton, L. and McGovern, D., Development of a cation exchange methodology for marine toxicity identification (TIE) application, *Environ. Toxicol. Chem.*, 16 (6), 1203–1211, 1997.
27. Burgess, R.M., Cantwell, M.G., Ho, K.T., Kuhn, A., Cook, H. and Rand Pelletier, M.C., Development of a toxicity identification evaluation (TIE) procedure for characterizing metal toxicity in marine sediments, *Environ. Toxicol. Chem.*, 19 (4), 982–991, 2000.
28. Ho, K.T., McKinney, R., Kuhn, A., Pelletier M. and Burgess, R., Identification of acute toxicants in New Bedford Harbor sediments, *Environ. Toxicol. Chem.*, 16 (3), 551–558, 1997.
29. Ho, K.T., Kuhn, A., Pelletier, M., Burgess, R. and Helmstetter, A., Use of *Ulva lactuca* to distinguish ammonia toxicity in marine waters and sediments, *Environ. Toxicol. Chem.*, 18 (2), 207–212, 1999.
30. Ho, K.T., Kuhn, A., Pelletier, M., Mc Gee, F., Burgess, R. and Serbst, J., Sediment toxicity assessment: comparison of standard and new testing designs, *Arch. Environ. Contam. Toxicol.*, 39, 462–468, 2000.
31. Ho, K.T., Burgess, R.M., Pelletier, M., Serbst, J.R., Ryba, S.A., Cantwell, M.G., Kuhn, A. and Raczelowski, P., An overview of toxicant identification in sediments and dredged materials, *Mar. Poll. Bull.*, 44, 286–293, 2002.
32. Pelletier, M.C., Ho, K.T., Cantwell, M., Kuhn-Hines, A., Jayaraman, S. and Burgess, R., Use of *Ulva lactuca* to identify ammonia toxicity in marine and estuarine sediments, *Environ. Toxicol. Chem.*, 20 (12), 2852–2859, 2001.
33. U.S. EPA, Methods for Measuring the Toxicity and Bioaccumulation of Sediment Associated Contaminants with Freshwater Invertebrates, EPA/600/R-94/024, U.S. Environmental Protection Agency, Office of Research and Development, Washington, D.C., USA, 2000.
34. Smith, M.E., Lazorchak, J.M., Herrin, L.E., Brewer-Swartz, S. and Thoeny, W.T., Reformulated reconstituted water for testing with the fresh amphipod, *Hyalella azteca, Environ. Toxicol. Chem.*, 16, 1229–1233, 1997.
35. USEPA, Environmental Monitoring and Assessment Program, Surface Water 1991 Northeast Pilot Field Operations and Training Manual, Tallent-Halsell, N G and Merritt, R.J., U.S. Environmental Protection Agency, Environmental Monitoring Systems Laboratory, Las Vegas, Nevada 89193, 1991.
36. ASTM, Nelson, M.K., Ingersoll, C.G. and Dwyer, F.J., Draft #5 New Standard Guide for Conducting Solid-Phase Toxicity Tests with Freshwater Invertebrates, ASTM Committee E47 on Biological Effects and Environmental Fate, Subcommittee E47.03 on Sediment Toxicity, American Society for Testing and Materials, Philadelphia, PA, 1992.

37. Klemm, D.J. and Lazorchak, J.M., EMAP: Surface Waters and Region 3 Regional Environmental Monitoring and Assessing Program: 1994 Pilot laboratory manual for streams, Cincinnati (OH) U.S. Environmental Protection Agency, Environmental Monitoring Systems Laboratory. Report No. EPA/620/R-94/003, 1994.
38. U.S. EPA, Short-term Methods for Estimating the Chronic Toxicity of Effluents and Receiving Water to Freshwater Organisms, Office of Research and Development, Washington, D.C., EPA/600/4-91/002, July 1994.
39. American Society for Testing and Materials, *Standard Guide for Conducting Static Toxicity with* Lemma gibba *63*, Philadelphia, PA, 1998.
40. American Public Health Association, *Standard Methods for the Examination of Water and Wastewater*, 19th ed., Washington, D.C., 1995.
41. Taraldsen, J.E. and Norberg-King, T.J., New method for determining effluent toxicity using duckweed (*Lemma minor*), *Environ. Toxicol. Chem.*, 9, 761–767, 1990.
42. U.S. EPA, Methods for Assessing, the Toxicity of Sediment Associated Contaminants with Estuarine and Marine Amphipods, EPA600R-94/025, U.S. Environmental Protection Agency, Office of Research and Development, Washington, D.C., 1994.
43. U.S. EPA, Methods for Assessing the Toxicity of Sediment-Associated Contaminants with Estaurine and Marine Amphipods, U.S. Environmental Protection Agency, Office of Research and Development, Washington, D.C., EPA/600/R-94/026, 1994.
44. Ferretti, J.A., Calesso, D.F. and Hermon T.R., Evaluation of methods to remove ammonia interference in marine sediment toxicity tests, *Environ. Toxicol. Chem.*, 18, 201–206, 2000.
45. Kosian, P.A., West, C.W., Pasha, M.S., Cox, J.S., Mount, D.R., Huggett, R.J. and Ankley, G.T., Use of nonpolar resin for the reduction of fluoranthene bioavailability in sediment, *Environ. Toxicol. Chem.*, 18, 201–206, 1999.
46. West, C.W., Kosian, P.A., Mount, D.R., Makynen, E.A. and Ankley, G.T., Amendment of sediments with a carbonaceous resin reduces bioavailability of polycyclic aromatic hydrocarbons, *Environ. Toxicol. Chem.*, 20 (5), 1104–1111, 2000.
47. Jee, V., Beckles, M.C., Ward, C.H. and Hughes, J.B., Aerobic slurry reactor treatment of phenanthrene in contaminated sediment, *Water Res.*, 32 (4), 1231–1239, 1998.

chapter twenty-eight

Application of solid-phase microextraction fibers as biomimetic sampling devices in ecotoxicology

Roman Lanno and Thomas W. La Point
Ohio State University
Jason M. Conder
University of North Texas
Jason B. Wells
ILS, Inc.

Contents

Introduction

A number of passive sampling devices (PSDs) have been used during the past 20 years for environmental monitoring of organic chemicals. Semi-permeable membrane devices (SPMDs) are composed of low-density polyethylene (LDPE) layflat tubing filled with a

1-56670-664-5/05/$0.00+$1.50
© 2005 by CRC Press

known weight of neutral lipid (triolein). LDPE has been shown to effectively mimic the function of bipolar lipid membranes of biological systems in uptake of hydrophobic organic chemicals, and the triolein in SPMDs represents the neutral or storage lipid pool in organisms.[1] SPMDs are the original PSDs used successfully to extract hydrophobic compounds (e.g., PCBs, OCs) from water, sediment, air, and even soil.[1–5] The introduction of solid-phase microextraction (SPME) fibers in 1990[6] provided a new tool for the environmental monitoring of organic chemicals. The SPME comprises a core of fused silica fiber that is coated with a thin polymer phase for the sampling of analytes and subsequent introduction into a GC system via thermal desorption. The polarity of the coating can be varied to optimize the sampling of analytes of varying polarities. For example, a C18 coating (very hydrophobic) is best for extremely hydrophobic organic chemicals (e.g., PAHs, PCBs), while a more polar coating, such as polyacrylate, would be more suitable for detecting more polar compounds (e.g., TNT, herbicides). SPME has been used to distinguish between dissolved and bound fractions of organic pollutants in the environment.[7] A number of other solid-phase extractants (e.g., Tenax beads, C18 disks) and other formats with thin polymer coatings have also recently been reported for use in environmental monitoring.[8]

PSDs have a number of advantages over live organisms for aquatic environmental monitoring, including ease of deployment, low production, and maintenance costs, transportability, and applications to a wide variety of environmental systems (e.g., freshwater, marine, sediments, soils). These devices absorb organic contaminants from the exposure medium following the general pattern of a one-compartment first-order kinetics (1CFOK) model (Figure 28.1) that can be described by the general equation:

$$C_{\mathrm{PSD}}(t) = C_{\mathrm{medium}} \times k_1/k_2 \times (1 - \mathrm{e}^{-k_2 t}) \tag{28.1}$$

where $C_{\mathrm{PSD}}(t)$ is the concentration of the contaminant on the sampler as a function of time, t, C_{medium} is the dissolved contaminant concentration in water, and k_1 and k_2 are uptake and elimination rate constants, respectively. The uptake rate of chemicals by a PSD is a function of its sampling rate that is proportional to the number of volumes sampled per day per unit volume of the device.[9] The time taken to reach a near-equilibrium state (Figure 28.1) is a function of the area/volume (A/V) ratio, with a higher A/V providing a faster sampling rate. For this reason, equilibrium would theoretically be reached much faster in an SPME (e.g., 7 μm PDMS coating) compared to an SPMD (e.g., 100 cm × 2.54 cm containing 1 g of triolein). In practice, this theory is correct, with SPMDs taking weeks to months to reach equilibrium and SPMEs reaching equilibrium in hours, given a chemical of the same log

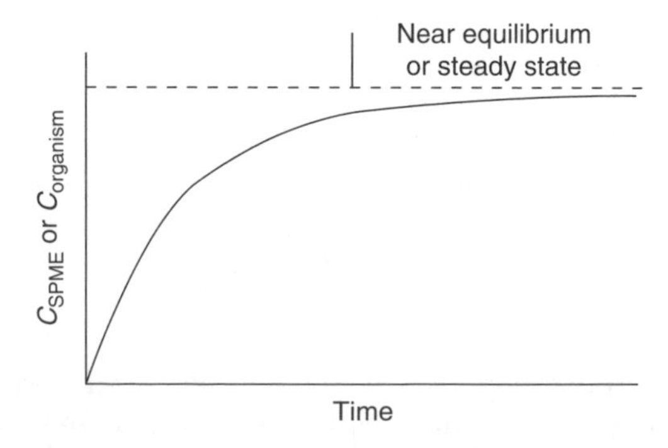

Figure 28.1 Theoretical kinetics of chemical uptake by a PSD (e.g., SPME) or an organism.

K_{ow}, and similar exposure conditions (e.g., water flow). The time taken to reach equilibrium is important for environmental monitoring to help avoid the risks associated with fouling, degradation, and vandalism of the device. Of all the types of PSDs currently commercially available, SPME fibers exhibit the most rapid response time and reach equilibrium in a matter of hours to days for all except the most hydrophobic compounds. PSDs that reach equilibrium during their deployment can be referred to as equilibrium sampling devices (ESDs) and the general theory of ESD application in environmental monitoring is thoroughly described elsewhere.[9,10] It should be noted, however, that SPMDs are not designed to reach equilibrium, and relative to SPMEs, SPMDs have a high sampling matrix volume and high analyte capacity for non-polar organics. Thus SPMDs are "integrative" samplers where relatively large amounts of analytes are sequestered, and equilibrium is generally not approached. On the other hand, SPMEs have very low sampling matrix volumes and capacities, and approach equilibrium much more rapidly than SPMDs. Thus, SPMEs are ESDs and do not truly represent grab samples or point-in-time samples, but exposure times required to reach equilibrium can be very brief relative to SPMDs, which integrate exposure over many days to weeks. The time to equilibrium of SPMEs will vary depending upon phase thickness, exposure conditions, and hydrophobicity of the target molecule.

Another factor to consider during ESD deployment is depletion of the sampling medium. Since uptake of chemicals by ESDs is strictly by passive diffusion, it is imperative that the sampler does not deplete the sampling medium and alter diffusion gradients. This is especially important when water samples with low levels of contaminants are measured in laboratory situations. In field deployments, with a constant renewal of the aqueous medium due to bulk water movement, depletion of the sample medium is usually not as much of an issue. Under conditions of non-depletive biomimetic sampling using SPME, less than 5% depletion of mass balance of sample is suggested so that equilibrium between fiber and solid phases is not disturbed due to altered diffusion gradients.[10,11]

SPME as a biomimetic sampling device

Although the value of SPMEs as a tool for monitoring environmental levels of organic chemicals is presented earlier, their rapid chemical uptake kinetics presents an opportunity not realized by other PSDs for comparison to chemical uptake kinetics by organisms. Bioaccumulation of organic chemicals by aquatic organisms follows the same 1CFOK model as chemical uptake by the SPME. For a given concentration of an organic chemical, both the SPME and the organism will achieve an equilibrium or steady-state concentration, respectively, facilitating a correlation between the two values when different concentrations of the same chemical are compared.[5] In this way, the bioaccumulation potential of a chemical can be predicted from the equilibrium concentration of the chemical in SPME, and the SPME acts as a biomimetic sampling device (BSD). A further extension of this concept is the application of chemical uptake by SPMEs to predict the toxicity of organic chemicals. Chemical kinetics and residues in organisms related to biological responses, such as growth and survival, are termed critical body residues (CBRs)[12,13] and can be compared to kinetics and residues of chemicals in SPMEs in order to make bioavailability comparisons and predict toxicity. Recent research has established such relationships between SPME chemical uptake and CBRs for 1,2,3,4-tetrachlorobenzene and 2,4,5-trichloroaniline in the midge larvae (*Chironomus riparius*),[14] petroleum hydrocarbon mixtures and toxicity to shrimp,[15] pyrene (PYR) and earthworms (*Eisenia fetida*),[15] and 1,2,3,4-tetrachlorobenzene and enchytraeids (*Enchytraeus crypticus*).[16]

The focus of this chapter is the examination of SPME fibers as biomimetic surrogates for estimating the bioavailability of organic chemicals in sediments and soils. SPMEs provide an indirect measure of bioavailability[17] and can be correlated with chemical bioaccumulation by organisms. We discuss two different approaches for the application of SPMEs in environmental matrices. The first method describes the deployment of the SPME in an aqueous suspension of soil while it is contained in the commercially available SPME holder, while the second method describes an approach where a piece of SPME fiber is placed directly in the sediment testing medium during toxicity tests. Each of these approaches has its advantages and disadvantages that will be discussed in the remainder of the chapter.

Validating assumptions for equilibrium sampling with SPME fibers

Prior to conducting chemical determinations in a test system of interest using SPME fibers, it is necessary to determine the time taken to reach equilibrium with respect to the concentration of analyte on the SPME fiber and to validate that sampling in your test system is non-depletive. The theory behind these measurements will not be discussed here and the reader is referred to a detailed treatment of these topics by Mayer et al.[9] Our discussion in this chapter focusses on the practical aspects of conducting these measurements in a laboratory setting.

Determining time taken to reach equilibrium

In order to facilitate a meaningful comparison of chemical levels in a medium or with accumulation by an organism, the chemical concentration on the SPME fiber must be at near-equilibrium conditions with the chemical in the medium (Figure 28.1). This can be achieved by measuring the chemical concentration in the SPME until it is constant, usually involving a number of replicated parallel samples. As an example, for PYR in a soil test system, triplicate samples were taken at 4, 48, and 96 h (Figure 28.2) and showed no increase in PYR concentration over that time period, suggesting that 4 h would be sufficient to achieve equilibrium for PYR determinations in this system. It would appear that in order to determine the kinetic phase of PYR partitioning to the SPME fiber, measurements would need to be made in a geometric time series at durations shorter than 4 h.

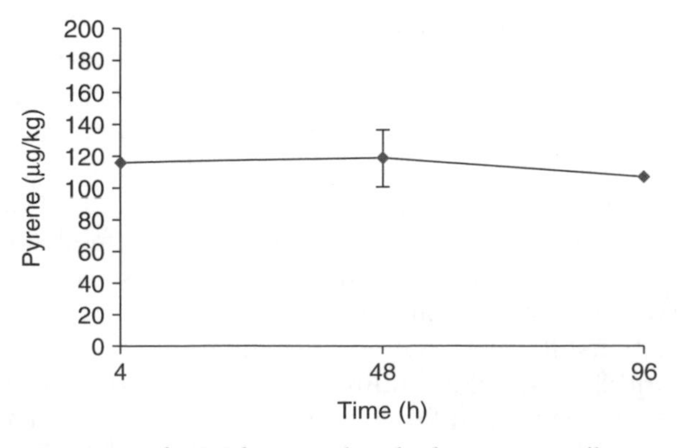

Figure 28.2 SPME exposure time for Webster soil spiked at 1000 mg/kg.

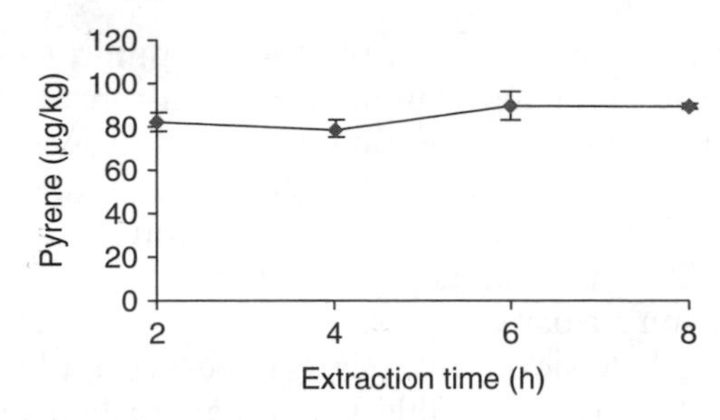

Figure 28.3 Reextraction of Perkins soil spiked at 300 mg/kg.

Validation of sample non-depletion

In order to validate that the SPME is not depleting the available chemical in the sample, resampling of the same sample is required. In the example below (Figure 28.3), PYR concentrations are determined in the same sample for duration of 2 h, followed by two 4-h exposures, followed by an 8-h exposure. The PYR concentration measured by the SPME did not change, suggesting that resampling the soil preparation did not affect the amount of PYR available for partitioning to the SPME, affirming that sampling in this system was indeed non-depletive.

Experiment 1: Determination of phenanthrene residues in soils using SPME fibers exposed in a vial-stir bar system

Procedure

The general experimental design entailed exposing earthworms (*Eisenia fetida*) to phenanthrene (PHE) in artificial soil containing organic matter at two levels (10% or 1% OM, w/w, as peat moss) to establish differences in PHE bioavailability due to differences in OM. The tests were conducted with three nominal concentrations (0.31, 2.05, and 3.10 mmol/kg dry mass soil) of PHE with an acetone solvent control (in triplicate) for each soil OM level, following ASTM guidance for conducting soil toxicity tests with earthworms as detailed in Reference 5.

Solid phase microextraction fiber analysis

The goal of SPME analysis was to determine the potentially bioavailable fraction of PHE in the soil rather than total PHE. In this example, SPMEs were deployed in commercially available SPME holders in order to take advantage of the increased analyte sensitivity due to direct injection into the GC with no solvent extraction. SPME fibers (7 μm polydimethylsiloxane (PDMS) with manual holders, Supelco, Bellefonte, PA) were used to assess uptake of PHE directly from the aqueous phase of soil at the earthworm toxicity test temperature (24 $\pm$ 1°C). For the measurement of analytes in soil, the SPME fiber must be exposed to a soil suspension in water. The soil/water ratio may vary with soil type and analyte concentration. For each SPME determination of PHE in soil in this study, 0.500 g of PHE-spiked freeze-dried artificial soil, 15 ml reagent grade water (RGW), and a Teflon®-coated magnetic stir bar (0.3 cm $\times$ 1.3 cm) were placed in a screw-top amber SPME vial (15-ml headspace with Teflon septum, Supelco). In order to increase sample throughput, a ten-place magnetic stirrer (1200 rpm, IKA) was used with ten sample vials and ten SPME fiber assemblies to obtain steady-state data for PHE concentrations in soil suspensions. A support stand was constructed to hold ten SPME manual holders simultaneously during exposure. The needle of the SPME apparatus was inserted through the Teflon septum of the sample vial when the fiber was deployed. Each vial was aligned on the magnetic stirrer for optimum stirring velocity (~1000 rpm) and SPME fibers were exposed until steady state was achieved. Residue analysis can be accomplished on any GC fitted with an SPME adapter. In this study, this was accomplished using a Tracor 565 GC-FID, megabore fused silica capillary column (DB-5, 30 m × 0.53 mm ID × 1.5 μm, J&W Scientific), 0.75-mm ID SPME-inlet liner (Supelco), JADE septum-less injector with SPME adapter (0.56 mm ID, Alltec). Helium (High Purity, Sooner Airgas) was used as the carrier and makeup gas. The flow rate for the carrier gas was set to 35 cm/s linear velocity and makeup flow rate was set to 45 ml/min. Hydrogen (fuel for FID, High Purity, Sooner Airgas) flow rate was 35 ml/min, and breathing air (oxidant, Grade D, Sooner Airgas) flow rate was 350 ml/min. The temperature program for direct injection GC analysis was: injection port temperature 290°C, detector temperature 300°C, initial oven temperature 160°C (5-min hold) with 35°C/min ramp to 210°C (7-min hold).

It is critical that prior to initial use, each new fiber be conditioned (320°C) for 4 h to remove any adhesive or other contaminants from the fiber as per instructions from the manufacturer. Subsequent thermal desorption and conditioning of the SPME fiber was accomplished by exposing the fiber while inserted into the heated injection port (290°C) of the GC for 5 min. This resulted in adequate desorption followed by blank analyses of each fiber to ensure no PHE carryover problems existed. SPME fiber performance was determined before and after each soil determination by measuring a reference standard solution (1 mg PHE/l RGW). Integration of peaks was done by external calibration using injections from PHE standards with a certified PHE check standard (Chem Service, F81MS). Chromatogram data were collected and analyzed using PeakNet® chromatography software (Version 5.1, Dionex 1999).

One advantage of this method is that SPME fibers attached to the commercially available fiber holders could be reused until their performance degraded and was unsatisfactory. One of the major issues related to the reuse of SPME fibers to determine PHE in soil suspensions was the degradation of the bonded phase of the fibers due to erosion by soil particles during stirring. To monitor fiber performance, fiber apparatus identification and labeling was critical, since all fibers are not identical and degrade at different rates with use. Permanently labeling each SPME injector/holder and very carefully recording the history of use for each fiber accomplished identification. All SPME fibers will erode over time when exposed in an agitated medium, regardless of the exposure medium, due to general use and numbers of collisions with stirred material. This results in degraded performance with repeated use. In media, such as filtered water or air, this is not a big problem. Performance checks were conducted before and after each soil suspension exposure by measuring a reference standard solution (1 mg PHE/l RGW) to ensure that fiber sensitivity had not decreased.

PHE quantification was conducted using external calibration by conventional solvent injections with PHE standards. The quantification can be thought of as the amount of analyte on the GC column. For example, if standards were injected using 1 μl of solution at concentrations of 1, 5, 10, 50, and 100 ng/μl, the resulting mass of analyte on the analytical column would be 1, 5, 10, 50, and 100 ng, respectively. Once the fiber was analyzed, the amount of analyte detected would be that amount on the fiber. The results may be expressed in any number of ways (i.e., amount of *X* on the fiber, amount of *X* per volume of phase on the fiber, the amount of *X* per mass of phase on the fiber, etc.).

For PHE determinations in this study, SPME fiber assemblies with a 7-μm thick PDMS bonded phase were used. SPMEs are available with bonded phases of various thickness, but analyte kinetics are much faster with thinner phases, and this was the only phase thickness that was commercially available bonded to the fiber (phase bonding results in a much more durable fiber).

We constructed a support stand to hold ten SPME manual holders simultaneously during exposure. Alternatively, one could remove the SPME fiber/needle apparatus from the holder and deploy without the need for a support stand. This will only work with SPME fibers for automated systems [SUPELCO-57303] (without the retractable spring included for the operation of manual fibers [SUPELCO-57302]). Residue analysis would require that the fiber apparatus be reinstalled to the holder. Each deployment of the SPME required that the fragile needle must be retracted on insertion and removal from the sampling vial. The SPME apparatus must be firmly supported while being inserted through the Teflon septum of the sample vial. The Teflon septum must be replaced after each use if the sample vials are reused. The fiber was placed near the bottom one-third of the vial where the agitation was the greatest. The fiber was also placed between the vial wall and the center of the vial for greatest agitation potential. It was critical to align each vial on the magnetic stirrer for optimum stirring velocity (~1000 rpm) or just below the vortex point. Creating a vortex will cause cavitation and limit the contact of the fiber with the exposure medium and should be avoided.

Each SPME fiber was exposed to a solution containing PHE until steady state was achieved (5 h). Steady state was determined by exposing SPME fibers to 1 mg PHE/l RGW over a geometric time interval (e.g., 0.5, 1, 2, 4, 8, and 16 h). No differences of PHE uptake existed between the 4, 8, and 16 h exposure ($p < 0.05$).

After the appropriate exposure period, the fiber was retracted into the needle and the device removed from the vial. The bonded phase on the fibers is very fragile and will be stripped if it comes in contact with the vial septum or GC septum or inlet device.

An additional issue related to the reuse of SPME fibers with the commercial SPME holder is water being drawn inside the sheathing needle of the fibers by capillary action during sampling. Fine particles and associated contaminants may also be drawn into the sheathing needle, and contaminants may be thermally desorbed onto the GC column during analysis, possibly compromising the analysis. The debris may also be heat-fixed to the fibers degrading its performance. To reduce this problem, the fiber is removed from the sample vial as described above, exposed, and rinsed with a stream of ultra-high purity (UHP) RGW or placed into a vial with UHP RGW under high stirring velocity for a very short period of time to dislodge any attached particles.

The addition of water into a GC is something to be avoided as when water is heated in the injection port, water expands violently and can damage the GC, injection port, analytical column, or detector. The addition of water may also cause sample analytical problems. To minimize problems caused by water injection, it is necessary to remove the water, and this was accomplished by exposing the fiber and placing an absorptive material (e.g., Kimwipe) at the interface of the needle and plunger sheathed within.

Most of the water will be drawn to the Kimwipe via capillary action. Caution should be used with this technique as the sequestration phase may be eroded easily if the Kimwipe touches the fiber itself.

Immediately upon removal of the fiber from the exposure medium in the sample vial, analytes begin to partition to the air. The rate of loss depends on the fugacity of the analyte, volatility, vapor pressure, and temperature. One way to slow down the loss is to freeze the fiber. Another way is to use a GC septum cored lengthwise to create a "cap." The septum material will have some contaminants on it that will partition to the fiber using this method. The freezing method in addition to capping with a septum is a very good method for extended storage periods. The best way to guard against analyte loss is to desorb (analyze) the fiber immediately after exposure, minimizing analyte loss.

Results and discussion

Comparing SPME measures with biological responses

The application of SPME technology for evaluating dose–response relationships between PHE exposure and earthworm responses allows the examination of the effects of organic matter in artificial soil on PHE bioavailability. Even though total PHE levels were similar in the different soils, there was a dramatic difference in earthworm survival (Table 28.1) with increased organic matter increasing earthworm survival. The effect of organic matter on SPME-measurable PHE can be useful in explaining the observed results. Earthworm body residues are well correlated with SPME-measurable PHE ($r = 0.93$) (Figure 28.4), while no significant correlation between earthworm PHE residues and total PHE levels in the artificial soil was evident.

The determination of PHE in soils using this method that allows the reuse of fibers for multiple determinations provides a cost-effective method for the determination of the potentially bioavailable fraction of PHE in the test soils. The cost of the fibers is ~$60 (US) per fiber, and at least eight bioavailability determinations could be made in a soil matrix without a reduction in sensitivity due to fiber degradation. The design and intended use of SPME fibers was for the determination of total chemical measurements from various media, such as water and air (headspace analysis). However, modifying factors of toxicity, such as dissolved organic matter and suspended particulate matter, reduce the amount of chemical available for SPME measurements in a manner similar to the reduction of chemical bioavailability for organisms.

Table 28.1 Total PHE, SPME-measurable PHE, and mortality of earthworms exposed to PHE in artificial soil varying in organic matter content

Soil OM (%)	Total PHE (mmol/kg)	SPME PHE (mmol/cm^2)	Mortality (%)
1	0	5.7E – 07 (2.1E – 07)	0
1	0.31	2.1E – 06 (2.1E – 06)	97
1	2.05	2.7E – 05 (9.2E – 06)	100
1	3.10	3.1E – 05 (7.0E – 06)	100
10	0	1.2E – 07 (1.4E – 07)	0
10	0.31	6.6E – 07 (4.7E – 07)	0
10	2.05	6.2E – 06 (1.9E – 06)	6.7
10	3.10	1.2E – 05 (5.0E – 06)	68

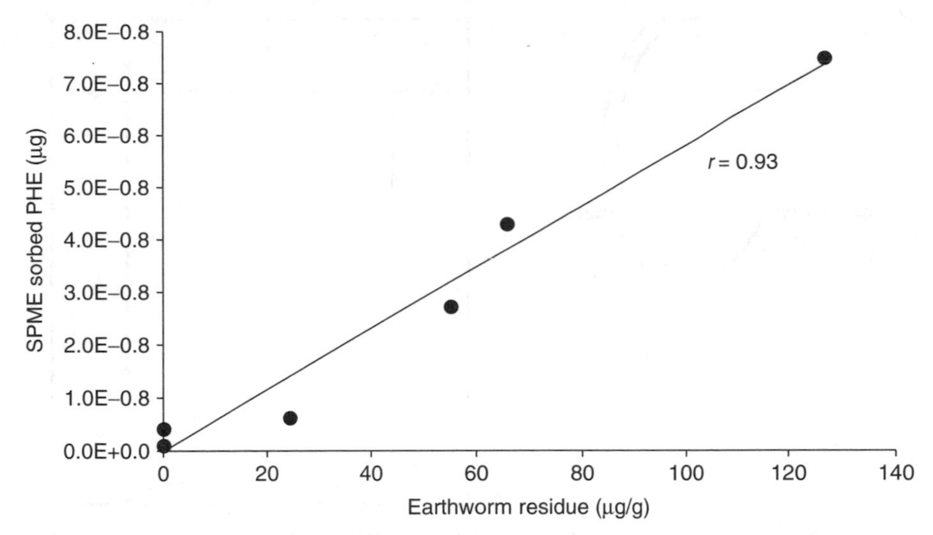

Figure 28.4 Relationship between SPME-measurable PHE and earthworm body residues. Reprinted from Ecotoxicology and Environmental Safety, Vol. 57, Lanno, R.P., Wells, J., Conder, J., Bradham, K., and Basta, N., The bioavailability of chemicals in soil for earthworms, p. 39–47, Copyright 2004, with permission from Elsevier.

Experiment 2: Determination of TNT in sediment using a disposable SPME fiber technique

Procedure

The static, direct-burial, disposable SPME approach for measuring TNT and nitroaromatics (NAs) in sediment is described in detail by Conder et al.[18] Briefly, 85-μm polyacrylate SPME fiber is purchased in bulk lengths, which are cut into 1.00-cm pieces using a double bladed, stainless steel razor blade apparatus. To facilitate handling and retrieval of the thin, transparent fiber, SPMEs are inserted through a stapled, small Teflon disc (Figure 28.5a) or placed in a small (2 cm × 2 cm) 60-μm stainless steel mesh envelope (Figure 28.5b). Chemical uptake by the fiber is independent of the type of holder. Before deployment, the SPME and holder are rinsed with 50:50 HPLC-grade acetonitrile/ultrapure water (Milli-Q Purification System), rinsed with ultrapure water, and allowed to dry at room temperature.

Measurement of NAs by SPMEs can be conducted directly in sediment during organism exposures.[18] SPMEs can be suspended in the overlying water, placed at the sediment/water interface, or buried within the sediment at the desired depth. Exposure time is 48 h, which is sufficient for TNT and its major NA transformation products to reach steady state in static exposures to NA-spiked water at room temperature (23°C). We assume that SPME extractions in sediment and sediment porewater are non-depletive, and that absorption kinetics is similar to that in pure water. Although not as rugged, the stapled Teflon disc holder is less expensive and easier to manipulate than the stainless steel mesh envelope holder, and the disc holder also has the advantage that it can be retrieved easily with a strong magnet.

After exposure, SPMEs are removed from their holders and placed into HPLC autosampler vials. The fibers can be stored (4°C) for up to a week or immediately extracted with 400 μl of 50:50 HPLC-grade acetonitrile/ultrapure water to desorb the compounds from the SPMEs. After desorption (10 min), the SPME is discarded and the solvent analyzed for NAs

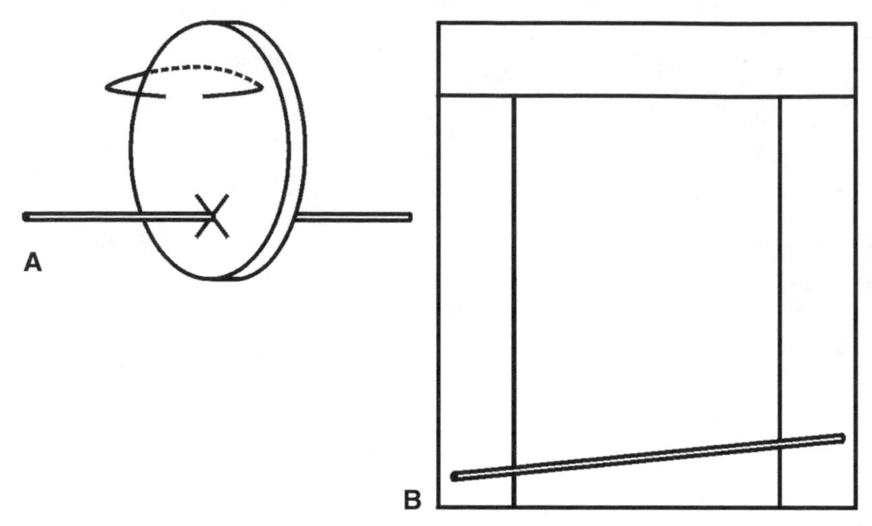

Figure 28.5 Stapled Teflon disc (a) and steel mesh envelope (b) holders for disposable SPMEs.

by HPLC. The 1-cm lengths of polyacrylate SPME fiber contain 0.521 μl polyacrylate; it is best to present SPME data as concentrations (number of molecules absorbed by the polyacrylate divided by volume of the polyacrylate) as in the previous studies.[10,14,19,20]

Routine quality assurance/quality control (QA/QC) procedures include assessment of fiber quality and extraction/handling procedures. It is advantageous to sacrifice 2–3 cm of each bulk SPME order to assess fiber performance. The 1-cm fibers can be exposed to 15-ml spiked water (e.g., 50 nmol TNT/ml) for 48 h and uptake compared to published fiber/solution partition coefficients.[18] In experiments in which disposable SPMEs are used to measure contaminants, extraction procedures should also be assessed. SPMEs can be spiked (coated directly on fiber or placed in their extraction vials) prior to receiving extraction solvent to determine percent recovery of the analytes.

Results and discussion

TNT concentrations in SPMEs were accurate predictors of TNT bioavailability in sediment. Sediments were spiked at the same TNT spiking level, but received five different levels of powdered activated carbon amendments (0–1000 μg carbon/g sediment, dw) to insure differences in bioavailability among the sediments. Each 100-g sediment replicate received 500 ml water and was aged for 14 days.[21] After aging, 10 adult *Tubifex tubifex* (sediment-dwelling oligochaetes) and one SPME were exposed for 48 h. SPMEs were exposed in overlying water at the sediment/water interface in steel mesh envelopes (Figure 28.5b). Because of differences in bioavailability (activated carbon amendments), sediment TNT concentrations determined by the acetonitrile liquid/liquid extraction were not good predictors of *Tubifex* TNT concentrations, which were subject to square root transformation to satisfy assumptions of regression analysis (Figure 28.6a). The regression model was significant; however, the slope was negative, indicating that as sediment TNT concentrations increased, *Tubifex* TNT concentrations decreased. If sediment TNT concentrations are normalized by the nominal amount of activated carbon added, the regression model improves, and the slope becomes positive (Figure 28.6b). However, if the x-axis is replaced by SPME-measurable TNT concentrations (Figure

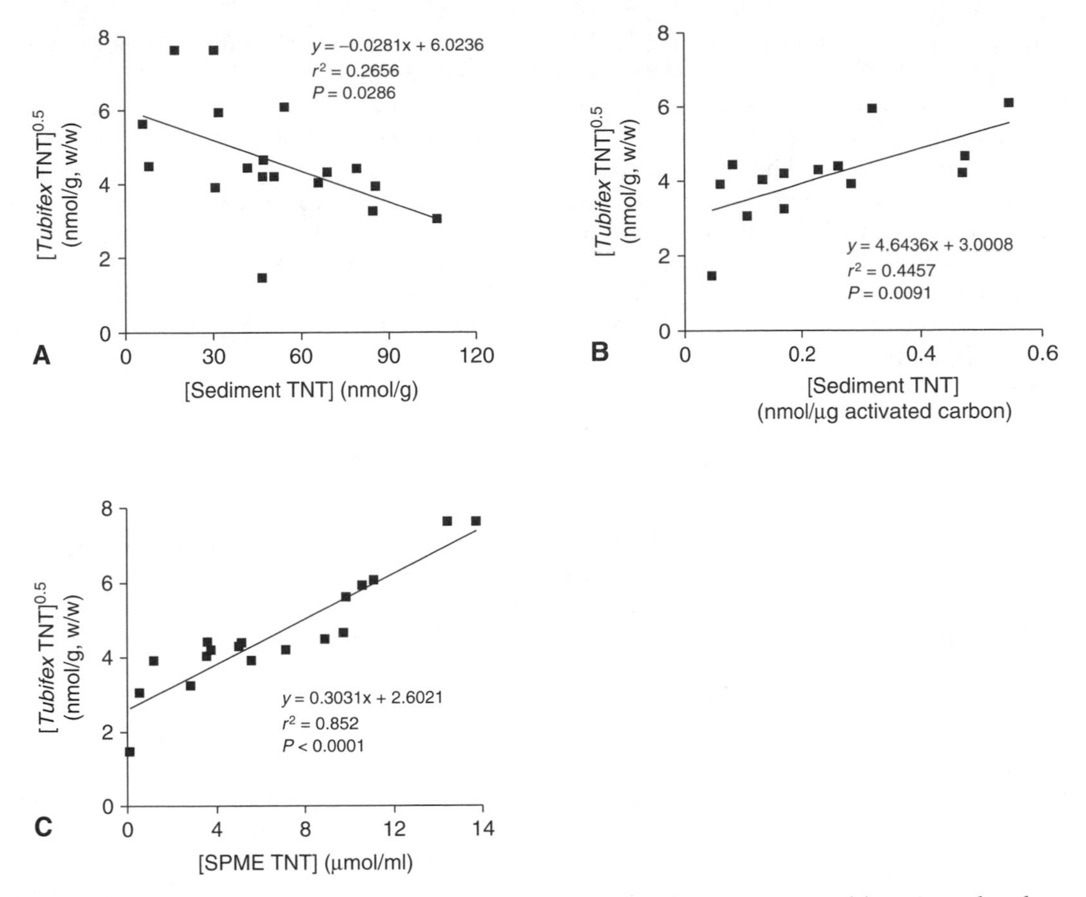

Figure 28.6 *Tubifex* TNT concentrations versus sediment TNT concentrations (a), activated carbon-normalized TNT concentrations (b), and SPME TNT concentrations (c) in TNT-spiked, activated carbon-amended sediment.

28.6c), the coefficient of determination improves from 0.4457 to 0.8520. Thus, SPMEs more accurately predict TNT bioavailability (*Tubifex* TNT concentrations) than either sediment TNT concentrations or sediment TNT concentrations normalized for carbon content.

Aside from the ability to more accurately predict TNT bioavailability in sediment than the liquid–liquid extraction to determine "total" NAs, the disposable SPME technique is slightly less expensive. By using the more rugged stainless steel mesh envelope holders, it may be possible to deploy SPMEs *in situ*. Also, SPMEs may be able to provide matrix-independent measures of toxicity and bioavailability. Median lethal toxicity (LC_{50}s based on SPME concentrations) for *Tubifex* exposed in TNT-spiked water and sediment were within a factor of 1.1.[22]

Even though SPMEs offer some advantages over traditional measures of exposure, they also have some limitations in estimating available compounds. SPMEs cannot mimic dietary uptake in cases where dietary contaminants are metabolized differently or partition to different compartments within an organism compared to chemicals taken up across external surfaces (epidermis or gills). SPMEs cannot simulate the complex compartmentalization of organics among different tissues, organs, and/or biomolecules, nor can they mimic biotransformation due to metabolic reactions with biomolecules, active elimination, or detoxification systems. SPMEs cannot mimic organism movement or other behaviors that alter exposure scenarios. While the static burial method for measuring the relatively hydrophilic NAs (log K_{ow} values ≤ 2) is rapid (48 h), more hydrophobic compounds may need longer times to reach equilibrium in fibers. Time to reach steady state will vary by compound hydrophobicity, agitation method, and fiber type.[23] For example, Leslie et al.[14] found that time to reach equilibrium in static exposures using disposable 2-cm long, 15-μm PDMS fibers took from 10 h to 2 weeks with compounds ranging in log K_{ow} from 3 to 6. If minimization of sampling time is paramount, SPMEs (in holders) can be agitated in a small sediment sample, similar to the technique described previously for soil. We found that SPME exposure time for NAs could be reduced to 24 h with SPMEs exposed to 5–10 g sediment samples in 15-ml vials (topped up with water) in an end-over-end rotating mixer (60 rpm).

Summary and conclusions

Although the application of SPME technology to measure concentrations of organic chemicals in water and air is quite advanced, the application of SPME technology to sediments and soils is in its infancy. The examples of various methods for determining concentrations of PHE in soils and TNT and its metabolites in sediments presented in this chapter are promising but are only initial forays into applying SPMEs in complex environmental matrices. Both the methods discussed in this chapter, the reusable SPME in the commercially available holder and the one-time-only use and disposal of SPME fragments, have advantages and disadvantages. However crude these methods, the results are very promising in being able to detect potentially bioavailable levels of organic contaminants that correlated well with organism uptake and responses. Further research examining relationships between SPME-detectable organic chemicals, bioaccumulation, and organism responses is needed to establish SPMEs as a useful tool for monitoring organic chemicals in the environment.

References

1. Huckins, J.N., Tubergen, M.W. and Manuweera, G.K., Semipermeable membrane devices containing model lipid: a new approach to monitoring the bioavailability of lipophilic contaminants and estimating their bioconcentration potential, *Chemosphere*, 20, 533–552, 1990.
2. Huckins, J.N., Manuweera, G.K., Petty, J.D., Mackay, D. and Lebo, J.A., Lipid containing semipermeable membrane devices for monitoring organic contaminants in water, *Environ. Sci. Technol.*, 27, 2489–2496, 1993.
3. Booij, K., Sleiderink, H.M. and Smedes, F., Calibrating the uptake kinetics of semipermeable membrane devices using exposure standards, *Environ. Toxicol. Chem.*, 17, 1236–1245, 1998.
4. Petty, J.D., Poulton, B.C., Charbonneau, C.S., Huckins, J.N., Jones, S.B., Cameron, J.T. and Prest, H.F., Determination of bioavailable contaminants in the Lower Missouri River following the flood of 1993, *Environ. Sci. Technol.*, 32, 837–842, 1998.
5. Wells, J.B. and Lanno, R.P., Passive sampling devices (PSDs) as biological surrogates for estimating the bioavailability of organic chemicals in soil, in *Environmental Toxicology and Risk Assessment: Science, Policy, and Standardization — Implications for Environmental Decisions*, Greenberg, B.M., Hull, R.N., Roberts M.H., Jr. and Gensemer, R.W., Eds., American Society for Testing and Materials, West Conshohocken, 2001, pp. 253–270 (358 pp.).
6. Arthur, C.L. and Pawliszyn, J., Solid phase microextraction with thermal desorption using fused silica optical fibres, *Anal. Chem.*, 62, 2145–2148, 1990.

7. Poerschmann, J., Zhang, Z., Kopinke, F.-D. and Pawliszyn, J., Solid phase microextraction for determining the distribution of chemicals in aqueous matrices, *Anal. Chem.*, 69, 597–600, 1997.
8. Wilcockson, J.B. and Gobas, F.A.P., Thin-film solid-phase extraction to measure fugacities of organic chemicals with low volatility in biological samples, *Environ. Sci. Technol.*, 35, 1425–1431, 2001.
9. Mayer, P., Tolls, J., Hermens, J.L.M. and Mackay, D., Equilibrium sampling devices, *Environ. Sci. Technol.*, 37, 184A–191A, 2003.
10. Mayer, P., Vaes, W.H.J., Wijnker, F., Legierse, K.C.H.M., Kraaij, R.H., Tolls, J. and Hermens, J.L.M., Sensing dissolved sediment porewater concentrations of persistent and bioaccumulative pollutants using disposable solid-phase microextraction fibers, *Environ. Sci. Technol.*, 34, 5177–5183, 2000.
11. Hermens, J.L.M., Freidig, A.P., Urrestarazu Ramos, E., Vaes, W.H.J., van Loon, W.M.G.M., Verbruggen, E.M.J. and Verhaar, H.J.M., Application of negligible depletion solid-phase extraction (nd-SPE) for estimating bioavailability and bioaccumulation of individual chemicals and mixtures, in *Persistent, Bioaccumulative, and Toxic Chemicals. II. Assessment and New Chemicals*, Lipnick, R.L., Jansson, B., Mackay, D. and Petreas, M., Eds., American Chemical Society, Washington, D.C., 2001, pp. 64–74 (276 pp.).
12. McCarty, L.S. and MacKay, D., Enhancing ecotoxicological modelling and assessment: body residues and modes of toxic action, *Environ. Sci. Technol.*, 27, 1718–1728, 1993.
13. Lanno, R.P. and McCarty, L.S., Worm bioassays: What knowledge can be applied from aquatic toxicity testing? *Soil Biol. Biochem.*, 29, 693–697, 1997.
14. Leslie, H.A., Oosthoek, A.J.P., Busser, J.M., Kraak, M.H.S. and Hermens, J.L.M., Biomimetic solid-phase microextraction to predict body residues and toxicity of chemicals that act by narcosis, *Environ. Toxicol. Chem.*, 21, 229–234, 2002.
15. Parkerton, T.F., Stone, M.A. and Letinski, D.J., Assessing the aquatic toxicity of complex hydrocarbon mixtures using solid phase microextraction, *Toxicol. Lett.*, 112/113, 273–282, 2000.
16. Van der Wal, L., Bioavailability of Organic Contaminants in Soil, Ph.D. thesis, Universiteit Utrecht, Utrecht, Netherlands, 2003, 122 pp.
17. Lanno, R.P., Wells, J., Conder, J., Bradham, K. and Basta, N., The bioavailability of chemicals in soil for earthworms, *Ecotoxicol. Environ. Safety*, 57, 39–47, 2004.
18. Conder, J.M., La Point, T.W., Lotufo, G.R. and Steevens, J.A., Nondestructive, minimal-disturbance, direct-burial solid phase microextraction fiber technique for measuring TNT in sediment, *Env. Sci. Technol.*, 37, 1625–1632, 2003.
19. Verbruggen, E.M.J., Vaes, W.H.J., Parkerton, T.F. and Hermens, J.L.M., Polyacrylate-coated SPME fibers as a tool to simulate body residues and target concentrations of complex organic mixtures for estimation of baseline toxicity, *Environ. Sci. Technol.*, 34, 324–331, 2000.
20. Leslie, H.A., Ter Laak, T.L., Busser, F.J.M., Kraak, M.H.S. and Hermens, J.L.M., Bioconcentration of organic chemicals: Is a solid-phase microextraction fiber a good surrogate for biota? *Environ. Sci. Technol.*, 36, 5399–5404, 2002b.
21. Conder, J.M., Lotufo, G.R., La Point, T.W. and Steevens, J.A., Recommendations for the assessment of TNT toxicity testing in sediment, *Environ. Toxicol. Chem.*, 23, 141–149, 2004.
22. Conder, J.M., Lotufo, G.R., Bowen, A.T., Turner, P.K., La Point, T.W. and Steevens, J.A., Solid phase microextraction fibers for estimating the toxicity of nitroaromatic compounds, *Aquat. Ecosys. Health Manage*, 7, 387–397, 2004.
23. Pawliszyn, J., *Solid Phase Microextraction: Theory and Practice*, Wiley-VCH, New York, 1997, 247 pp.

chapter twenty-nine

Passive dosimeters for measurement of ultraviolet radiation in aquatic environments

C.S. Sinclair
Towson University
R.H. Richmond
University of Hawaii
E.T. Knobbe
Sciperio Inc.
C. Basslear
University of Guam Marine Laboratory
G.K. Ostrander
Johns Hopkins University

Contents

Introduction

Ultraviolet radiation (UVR) is subdivided into UVA (320–400 nm), UVB (280–320 nm), and UVC (220–280 nm). Like visible light (430–790 nm), the longer rays of UVA pass through the atmosphere with little attenuation, while stratospheric ozone absorbs most UVB and all of the dangerous UVC. High-energy rays of UVA and UVB initiate photochemical reactions (e.g., smog generation) and can cause molecular and cellular damage to biological organisms.

Excessive exposure to UVA and UVB may induce the formation of cyclobutane pyrimidine dimers, alkaline-labile lesions, DNA crosslinks or breaks, and lesions in proteins.[1–4] In humans, excessive exposure to UVB leads to skin damage, skin cancer, cataracts, and

1-56670-664-5/05/$0.00+$1.50
© 2005 by CRC Press

immune system suppression.[5] Occurrences of skin lesions, squamous cell carcinomas, and infectious keratoconjunctivitis in domestic animals are attributed to exposure to UVB.[6–10] Once thought to be relatively innocuous, evidence now suggests that UVA exposure may be as damaging as UVB.[11] UVB exposure primarily damages the surface layers of tissue, while UVA penetrates further and causes damage much deeper in human tissues.[12,13]

Despite the partial attenuation of UVR due to reflection at the air/water interface and absorption by dissolved and particulate matter, aquatic organisms are susceptible to UVR induced damage. Calm, clear water allows for deeper penetration of UVR with detrimental biological effects detected at depths of 20 m.[14–20] Sessile organisms (e.g., corals) that lack the ability to move into more sheltered areas and organisms that live suspended in the water (e.g., larval stages of fish and invertebrates, plankton) may be at greatest risk of damage. Many of these at-risk organisms, from Cyanobacteria to teleosts, are protected from excessive UVR exposure by mechanisms, such as melanin pigmentation in amphibian eggs[21,22] and mycosporine-like amino acids in marine organisms,[19,23–26] but limits to this protection may exist. Short-term exposure to increased levels of UVR leads to reduced coral calcification and skeletal growth, decreased photosynthetic activity by zooxanthellae, and reduced survivorship among coral larvae.[27–32] UVR exposure may also contribute to coral bleaching.[29,33–36] Benthic echinoids, *Strongylocentrotus droebachiensis* (green sea urchin) and *Echinarachnius parma* (sand dollar), show decreased survivorship and delayed development in addition to DNA damage when exposed to environmentally relevant levels of UVR.[37] UVB exposure is also reported to reduce activity of bacterioplankton, which consume labile dissolved organic material in the surface layers of the ocean.[38]

Progressive thinning of the protective ozone layer has prompted the initiation of a number of studies on the effects of UVR in aquatic environments. Current technology for measuring UVR in aquatic environments is somewhat limited by cost, electrical power requirements (e.g., batteries), maintenance, deployment depths, and measurable wavelengths. Inherent restrictions of the current technology make it unsuitable for long-term monitoring or sensing in remote locations. Gleason[39] suggests that the ideal instrument for measuring UVR in marine environments would be "small, self-contained, submersible to at least 50 m, able to scan to at least 300 nm, and be within the budget of most research labs". Recently, passive sensors for detection and quantification of UVR and visible light have been developed.[40,41] These passive sensors require no power, can be used in air or water (salt or fresh), are small and self-contained, and have no temperature dependence between 0–50°C making them ideal for many long-term monitoring or remote sensing applications.

Materials required

Dosimeters: Colyott et al.[40] describe the design of the passive dosimeters used throughout this study (Figure 29.1). Briefly, the dosimeters are constructed with a Teflon cap containing an interference filter, an attenuator, and a diffuser. The Delrin base contains the thin-layer aluminum oxide detectors developed by Landauer, Stillwater Crystal Growth Division (formerly Stillwater Sciences), 723 1/2 Eastgate, Stillwater, Oklahoma 74074 (phone: 405–377–5161, www.landauerinc.com/affiliates).[42] A bayonet fitting attaches the base to the cap and provides an air and watertight seal. Dosimeters are 25.4 mm in diameter and 19.1 mm long with a 12.7-mm window through which the radiation passes to reach the detector. Dosimeters are read using either thermoluminescence or optically stimulated luminescence technology.[40,41,43]

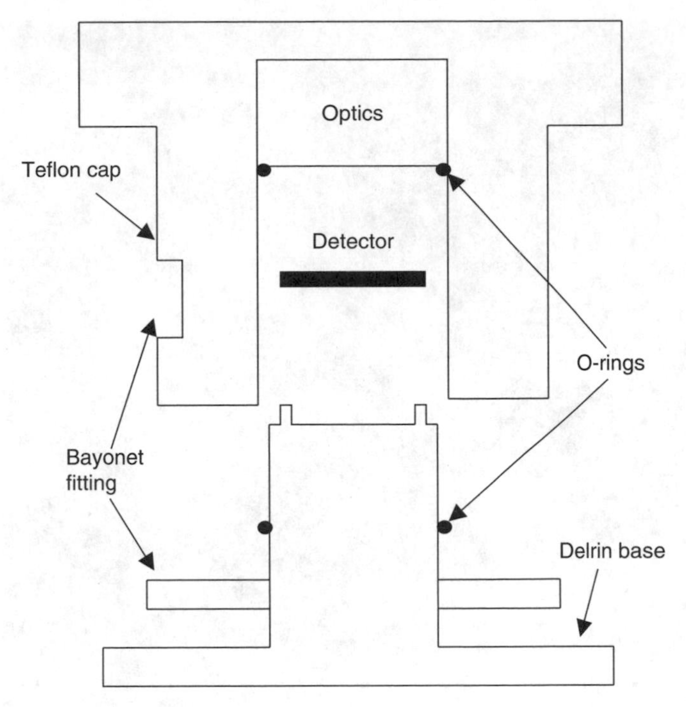

Figure 29.1 Passive dosimeter.

Procedures

Sensor deployment: For the purpose of illustration, we describe a pilot study conducted in a coral reef ecosystem at a remote location. Forty dosimeters were deployed on the edge of a patch reef within Apra Harbor, Guam 13′ 28° N, 144–48° E, on February 19, 2001. Prior to shipping, the manufacturer calibrated the dosimeters to measure UVA or UVB radiation. Dosimeters were fixed in custom-fabricated PVC mounts (Figure 29.2) and attached to the substrate by elastic straps. A bubble level was used to orient the sensors vertically with the collecting window pointed towards the water surface. The most difficult and critical (as supported by the data presented in the following) portion of this process is the stabilizing and leveling of the dosimeters. Even a slight deviation of a few degrees from the plane of the water surface will impact measurements. In shallow water, currents and wave action can have a significant impact. This is especially true for short-term data collection. To this end, it is advisable for dosimeter deployment to be done while on SCUBA if the depths are greater than 1 m. The dosimeters were placed at depths of 3, 9, and 18 m, and exposed for 30 min, 48 h , or 480 h (see Table 29.1). The 30-min exposure was done at a depth of 3 m as the sun approached zenith to achieve optimal penetration.

Climatological data for the period of deployment were obtained from the NOAA website (www.prh.noaa.gov). The availability of such information reinforces the utility of passive dosimetry in remote locations where collection of climatological data on a daily basis would be difficult or impossible. In fact, with appropriate filtering, dosimeters could be left in a remote location without any attention for many months, all the while collecting data. Filtering is pre-set at the factory prior to shipping and can be calibrated based on intended length of deployment. However, it should be pointed out that we have not investigated the potential impact of fouling. It is anticipated that future generations of

Figure 29.2 Passive dosimeter mounted underwater.

Table 29.1 Depth, exposure time, and number of passive dosimeters deployed at each site for UVA or UVB monitoring

Depth (m)	Length of exposure (h)	Number of sensors deployed
3	0.5	2
3	48	3
3	480	3
9	48	3
9	480	3
18	48	3
18	480	3

sensors will be impregnated with anti-fouling compounds and thereby enhance the feasibility of long-term studies. The average air temperature was 33.6°C (range 26.4–39.2°C). Atmospheric conditions ranged from sunny to partly cloudy with no days of complete cloud cover recorded. Average wind was 12.9 kn (range 8.5–15.7 kn). Total precipitation for the duration of sensing was 5.4 cm with a maximum of 2.4 cm falling on one day.

Data retrieval: Upon completion of exposures, the dosimeters were shipped to the manufacturer for data retrieval.

Results and discussion

Cumulative UVA and UVB exposure was determined for three depths, 3, 9, and 18 m, at three exposure times, 30 min (3 m only), 48 h, and 480 h. The results are shown in Figure 29.3 where the average measured dose of UVA or UVB (J/m^2) is plotted against the duration of exposure (h). As expected, the average measured dose decreased with the increase in depth for both UVA and UVB. Average doses of UVA were 2- to 7-fold higher

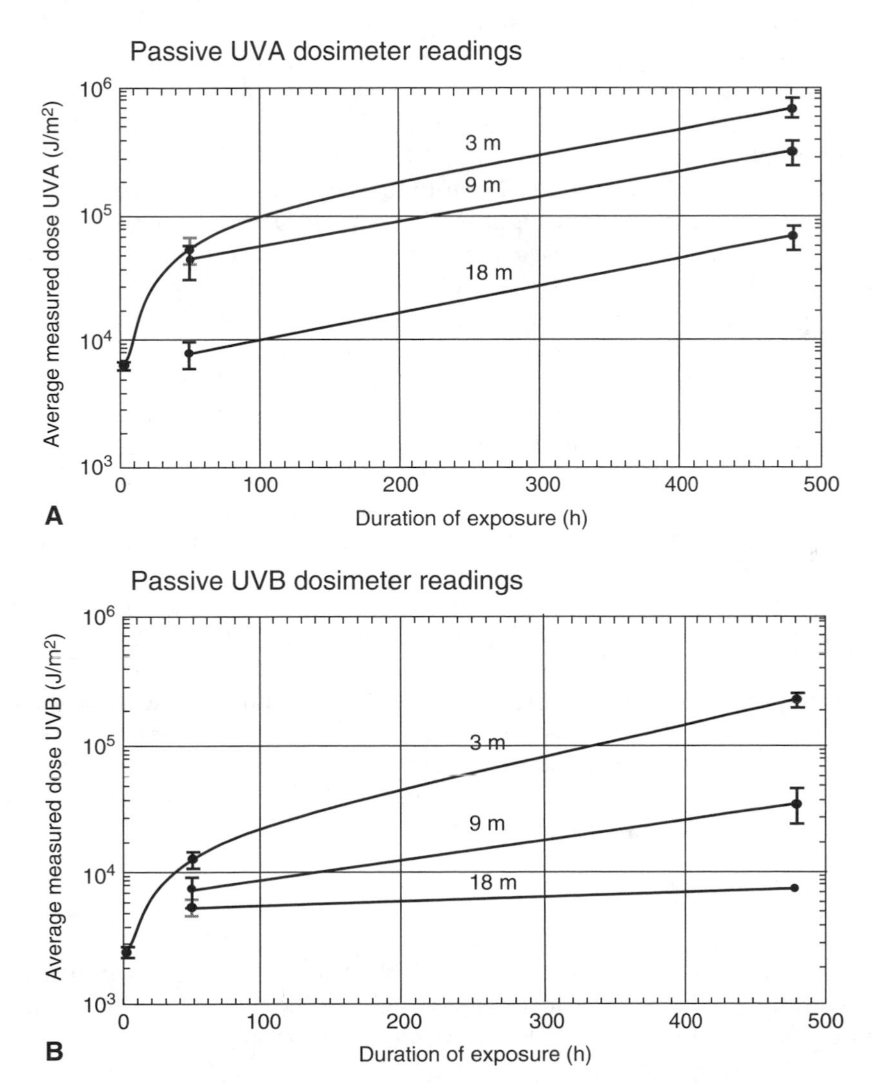

Figure 29.3 Average measured UVR dose as a function of duration of exposure and depth of deployment. Each data point is the mean of measured doses from three separate sensors at the same location with the exception of the 0.5 h time point for both UVA and UVB with a mean calculated from two sensor readings, and 18 m UVB doses calculated from one sensor at 480 h. The error bars represent 1 SD from the mean (A) UVA; (B) UVB.

than UVB levels at the same depths, which is indicative of the higher levels of UVA passing through the atmosphere and the differential attenuation of UVA and UVB as the rays penetrate the water.[44]

The average measured dose of UVA increases almost 10-fold from 48 to 480 h of exposure for all three depths of deployment (Figure 29.3A). An 8-fold difference in average UVA dosage was detected between dosimeters exposed for 30 min and those exposed for 48 h. We would expect to see a much greater difference between these data points. However, the 30-min exposure was performed during the Sun's zenith under clear skies and calm water conditions thereby maximizing the levels of UVA and UVB entering the water. In contrast, the 48 and 480 h exposures include extended periods of partial cloud cover and darkness resulting in a lower cumulative UVA dosage. Decreases in the average measured dose of UVA in relationship to the increase in depth, 1.3-fold for 3–9 m and 6-fold for 3–18 m, suggests that similar levels of UVA rays are able to penetrate to 3 and 9 m. Under clear, calm conditions these results are highly likely.[45]

Average measured UVB exposure for dosimeters placed at a depth of 3 m showed an approximate 8-fold increase from 30 min to 48 h and a 10-fold increase from 48 to 480 h (Figure 29.3B). The 8-fold increase from 30 min to 48 h is identical to the UVA data for the same exposure comparison, which suggests that similar factors might have affected UVB exposure. In contrast, a 5-fold increase and a 1.4-fold increase were measured for 48–480 h UVB exposures at 9 and 18 m, respectively. The dramatic decrease in UVB dosage with the increase of depth is most likely due to increased attenuation of the short UVB rays[44,46,47] and the inability of UVB to penetrate effectively into depths greater than 6 m.

Data from two of the three sensors exposed to 480 h of UVB at 18 m were not used to calculate average dose due to values that suggest a misalignment might have occurred during sensing. Since the UVA dosimeters located in the same areas provided exposure measurements similar to the third UVA dosimeter measurement, we eliminated the possibility that an increase in suspended particles or wave activity is responsible for the unusual UVB readings. This illustrates the importance of properly securing and aligning the dosimeters at the initiation of the sensing period.

Excessive exposure to UVR can lead to reduction in reproductive capacity, impaired larval development, and even death in aquatic organisms. The ability of researchers to perform long-term or remote studies of UVR in aquatic environments has been limited by sensors that require a continuous input of power, are prohibitively expensive, or difficult to use. Recent technological advances have led to the development of passive sensors that function without an active power source. The passive sensors offer researchers a reliable and cost-effective tool for long-term and remote monitoring of UVR and photosynthetically available radiation in aquatic environments.

Acknowledgments

The authors would like to thank A. Lucas at Nextstep Technologies (Stillwater, Oklahoma) for kindly providing the dosimeters used in this study. Logistical support was funded under a grant from the U.S. EPA Star Program (grant R 82-8008).

References

1. Löber, G. and Kittler, L., Selected topics in photochemistry of nucleic acids. Recent results and perspectives, *Photochem. Photobiol.*, 25, 215–233, 1977.

2. Lippke, J.A., Gordon, L.K., Brash, D.E. and Haseltine, W.A., Distribution of UV light-induced damage in a defined sequence of human DNA: detection of alkaline-sensitive lesions at pyrimidine nucleoside-cytidine sequences, *Proc. Nat. Acad. Sci. USA*, 78, 3388–3392, 1981.
3. Peak, M.J. and Peak, J.G., DNA to protein crosslinks and backbone breaks caused by far- and near-ultraviolet and visible radiations in mammalian cells, in *Mechanisms of DNA Damage and Repair. Implications for Carcinogenesis and Risk Assessment*, Simic, M.G., Grossman, L. and Upton, A.C., Eds., Plenum Press, New York, 1986, pp. 103–202 (578 pp.).
4. Shindo, Y. and Hashimoto, T., Time course of changes in antioxidant enzymes in human skin fibroblasts, *J. Dermatol. Sci.*, 14, 225–232, 1997.
5. Acevedo, J. and Nolan, C., Environmental UV Radiation, Research and Development, Brussels, A report prepared for the Commission of the European Communities, Directorate-General XII for Science, 1993.
6. Kopecky, K.E., Pugh, G.W. and Hughes, D.E., Wavelength of ultraviolet radiation that enhances onset of critical infectious bovine keratoconjunctivitis, *Am. J. Vet. Res.*, 41, 1412–1415, 1980.
7. Hargis, A.M., A review of solar induced lesions in domestic animals, *Comp. Cont. Educ. Pract.*, 3, 287–300, 1981.
8. Nikola, K.J., Benjamin, S.A., Angleton, G.M., Saunders, W.J. and Lee, A.C., Ultraviolet radiation, solar dermatosis, and cutaneous neoplasia in Beagle Dogs, *Radiat. Res.*, 129, 11–18, 1992.
9. Teifke, J.P. and Lohr, C.V., Immunohistochemical detection of p53 overexpression in paraffin wax-embedded squamous cell carcinomas of cattle, horses, cats, and dogs, *J. Comp. Pathol.*, 114, 205–210, 1996.
10. Mendez, A., Perez, J., Ruiz-Villamore, E., Garcia, R., Martin, M.P. and Mozos, E., Clinicopathology study of an outbreak of squamous cell carcinoma in sheep, *Vet. Rec.*, 141, 597–600, 1997.
11. Wang, S.Q., Setlow, R., Berwick, M., Polsky, D., Marghoob, A.A., Kopf, A.W. and Bart, R.S., Ultraviolet A and melanoma: a review, *J. Am. Acad. Dermatol.*, 44, 837–846, 2001.
12. Parrish, J.A., Fitzpatrick, T.B., Tanenbaum, L. and Pathak, M.A., Photochemotherapy of psoriasis with oral methoxsalen and long wave ultraviolet light, *New Engl. J. Med.*, 291, 1207–1211, 1974.
13. Robert, M., Bissonauth, V., Ross, G. and Bouabhia, M., Harmful effects of UVA on the structure and barrier function of engineered human cutaneous tissues, *Int. J. Radiat. Biol.*, 75, 317–326, 1999.
14. U.S. Environmental Protection Agency (USEPA), An assessment of the effects of ultraviolet-B radiation on aquatic organisms, in *Assessing the Risks of Trace Gases That Can Modify the Stratosphere*, EPA 400/1-87/001C. EPA, Washington, D.C., 1987.
15. Smith, R.C., Prezelin, B.B., Baker, K.S., Bidigare, R.R., Boucher, N.P., Coley, T., Karentz, D., MacIntyre, S., Matlick, H.A., Menzies, D., Ondrusek, M., Wan, Z. and Waters, K.J., Ozone depletion: ultraviolet radiation and phytoplankton biology in Antarctic waters, *Science*, 255, 952–959, 1992.
16. Scully, N.M. and Lean, D.R.S., The attenuation of ultraviolet radiation in temperate lakes, *Arch. Hydrobiol.*, 43, 135–144, 1994.
17. Häder, D.-P., Photo-ecology and environmental photobiology, in *CRC Handbook of Organic Photochemistry and Photobiology*, Horspool, W.M. and Song, P.-S., Eds., CRC Press, Boca Raton, FL, 1995, pp. 1392–1401 (2500 pp.).
18. Coohill, T.P., Häder, D.-P. and Mitchell, D.L., Environmental ultraviolet photobiology: introduction, *Photochem. Photobiol.*, 64, 401–402, 1996.
19. Shick, J.M., Lesser, M.P. and Jokiel, P.L., Effects of ultraviolet radiation on coral and other reef organisms, *Global Change Biol.*, 2, 527–545, 1996.
20. Booth, C.R., Morrow, J.H., Coohill, T.P., Frederick, J.E., Häder, D.-P., Holm-Hansen, O., Jeffrey, W.H., Mitchell, D.L., Neale, P.J., Sobolev, I., van der Leun, J. and Worrest, R.C., Impacts of solar UVR on aquatic microorganisms, *Photochem. Photobiol.*, 65, 252–269, 1997.
21. Duellman, W.E. and Tureb, L., *Biology of Amphibians*, McGraw-Hill, New York, 1986, 670 pp.
22. Licht, L.E. and Grant, K.P., The effects of ultraviolet radiation on the biology of amphibians, *Am. Zool.*, 37, 137–145, 1997.
23. Shibata, K., Pigments and UV-absorbing substance in corals and blue-green alga living on the Great Barrier Reef, *Plant Cell Physiol.*, 10, 325–335, 1969.

24. Dunlap, W.C., Williams, D.McB., Chalker, B.E. and Banaszak, A.T., Biochemical photoadaptation in vision: UV-absorbing pigments in fish eye tissues, *Comp. Biochem. Physiol.*, 93B, 601–607, 1989.
25. Dunlap, W.C. and Shick, J.M., Ultraviolet radiation absorbing mycosporine-like amino acids in coral reef organisms: a biochemical and environmental perspective, *J. Phycol.*, 34, 418–430, 1998.
26. Cockell, C.S. and Knowland, J., Ultraviolet radiation screening compounds, *Biol. Rev.* (*Camb.*), 74, 311–345, 1999.
27. Jokiel, P.L. and York, R.H., Solar ultraviolet photobiology of the reef coral *Pocillopora damicornis* and symbiotic zooxanthellae, *Bull. Mar. Sci.*, 32, 301–315, 1982.
28. Gleason, D.F., Differential effects of ultraviolet radiation on green and brown morphs of the Caribbean coral *Porites asteroides*, *Limnol. Oceanogr.*, 38, 1452–1463, 1993.
29. Gleason, D.F. and Wellington, G.M., Ultraviolet radiation and coral bleaching, *Nature* (*Lond.*), 365, 836–838, 1993.
30. Masuda, K., Goto, M., Maruyama, T. and Miyachi, S., Adaption of solitary corals and their zooxanthellae to low light and UV radiation, *Mar. Biol.*, 117, 685–691, 1993.
31. Gleason, D.F. and Wellington, G.M., Variation in UVB sensitivity of planula larvae of the coral *Agaricia agaricites* along a depth gradient, *Mar. Biol.*, 123, 693–703, 1995.
32. Shick, J.M., Lesser, M.P., Dunlap, W.C. and Stochaj, W.R., Depth-dependent responses to solar ultraviolet radiation and oxidative stress in the zooxanthellate coral *Acropora microphthalma*, *Mar. Biol.*, 122, 41–51, 1995.
33. Lesser, M.P., Stochaj, W.R., Tapely, D.W. and Shick, J.M., Bleaching in coral reef anthozoans: effects of irradiance, ultraviolet radiation and temperature on the activities of protective enzymes against active oxygen, *Coral Reefs*, 8, 225–232, 1990.
34. Glynn, P.W., Imai, R., Sakai, K., Nakano, Y. and Yamazato, K., Experimental responses of Okinawan (Ryukyu Islands, Japan) reef corals to high sea temperature and UV radiation, *Proc. 7th Int. Coral Reef Symp.* (*Guam*), 1, 27–37, 1992.
35. Reaka-Kudla, M.L., O'Connell, D.S., Regan, J.D. and Wicklund, R.I., Effects of temperature and UV-B on different components of coral reef communities from the Bahamas, in *Proceedings of the Colloquium on Global Aspects of Coral Reefs: Health, Hazards, and History*, Ginsburg, R.N., Ed., Rosenstiel School of Marine and Atmospheric Science, University of Miami, 1993, pp. 126–130 (420 pp.).
36. Shick, J.M., Romaine-Lioud, S., Ferrier-Pages, C. and Gattuso, J.P., Ultraviolet-B radiation stimulates shikimate pathway-dependent accumulation of mycosporine-like amino acids in the coral *Stylophora pistillata* despite decreases in its population of symbiotic dinoflagellates, *Limnol. Oceanogr.*, 44, 1667–1682, 1999.
37. Lesser, M.P. and Barry, T.M., Survivorship, development, and DNA damage in echinoderm embryos and larvae to ultraviolet radiation, *J. Exp. Mar. Biol. Ecol.*, 292, 75–91, 2003.
38. Herndl, G.J., Müller-Niklas, G. and Frick, J., Major role of ultraviolet-B in controlling bacterioplankton growth in the surface layer of the ocean, *Nature*, 261, 717–718, 1993.
39. Gleason, D.F., Ultraviolet radiation and coral communities, in *Ecosystems, Evolution, and Ultraviolet Radiation*, Cockell, C.S. and Blaustein, A.R., Eds., Springer-Verlag, New York, pp. 118–149, 2001 (221 pp.).
40. Colyott, L.E., Akselrod, M.S. and McKeever, S.W.S., An integrating ultraviolet-B dosemeter using phototransferred thermoluminescence from α-Al_2O_3:C, *Radiat. Prot. Dosim.*, 72, 87–94, 1997.
41. Colyott, L.E., McKeever, S.W.S. and Akselrod, M.S., An integrating UVB dosemeter system, *Radiat. Prot. Dosim.*, 85, 309–312, 1999.
42. Akselrod, M.S., McKeever, S.W.S., Moscovitch, M., Emfietzoglou, D., Durham, J.S. and Soares, C.G., A thin-layer α-Al_2O_3:C beta TL detector, *Radiat. Prot. Dosim.*, 66, 105–110, 1996.
43. Walker, F.D., Colyott, L.E., Agersnap-Larsen, N. and McKeever, S.W.S., The wavelength dependence of light-induced fading of thermoluminescence from α-Al_2O_3:C, *Radiat. Meas.*, 26, 711–718, 1996.

44. Kuwahara, V.S., Toda, T., Hamasaki, K., Kikuchi, T. and Taguchi, S., Variability in the relative penetration of ultraviolet radiation to photosynthetically available radiation in temperate coastal waters, Japan, *J. Oceanogr.*, 56, 399–408, 2000.
45. Xenopoulous, M.A. and Schindler, D.W., Physical factors determining ultraviolet radiation flux into ecosystems, in *Ecosystems, Evolution, and Ultraviolet Radiation*, Cockell, C.S. and Blaustein, A.R., Eds., Springer-Verlag, New York, pp. 36–62, 2001 (221 pp.).
46. Smith, R.C. and Baker, K.S., Penetration of UV-B and biologically effective dose-rates in natural waters, *Photochem. Photobiol.*, 29, 311–323, 1979.
47. Dunne, R.P. and Brown, B.E., Penetration of solar UVB radiation in shallow tropical waters and its potential biological effects on coral reefs: results from the central Indian Ocean and Andaman Sea, *Mar. Ecol. Prog. Ser.*, 144, 109–118, 1996.

section four

Techniques for aquatic toxicologists

chapter thirty

Spectral models for assessing exposure of fish to contaminants

Donald C. Malins, Virginia M. Green, Naomi K. Gilman, and Katie M. Anderson
Pacific Northwest Research Institute
John J. Stegeman
Woods Hole Oceanographic Institution

Contents

Introduction

Fourier transform-infrared (FT-IR) spectroscopy is capable of identifying a wide variety of chemical structures on the basis of their unique vibrational and rotational properties.[1–3] DNA has a characteristic "signature" spectrum, the peaks, shoulders, and other spectral properties of which have been identified by spectroscopists as corresponding to specific structures in the DNA molecule.[1] In addition, statistical models of FT-IR spectra have the

1-56670-664-5/05/$0.00+$1.50
© 2005 by CRC Press

remarkable ability to reveal subtle changes in complex cellular structures resulting from various biological and chemical stresses.[3,4] Recent examples include the ability to discriminate, with high sensitivity and specificity, between the DNA of healthy and cancerous prostate tissues, thus providing a basis for predicting the probability of prostate cancer.[5] Furthermore, the unique ability of the FT-IR statistical models to differentiate between diverse groups of tissues was evident when it was shown that primary prostate tumors could be readily distinguished from metastasizing primary tumors. This achievement was the basis for statistical models for predicting which tumors are most likely metastasizing without having to wait for metastatic cells to be detected at distant sites in the body (e.g., the groin), at which point treatment options are limited.[5]

We recognized that the FT-IR statistical models had the potential for studying the effects of toxic chemicals on fish. In 1997, we showed that liver DNA of English sole (*Parophrys vetulus*) from the chemically contaminated Duwamish River (DR) in Seattle, WA was structurally different from that of English sole from the relatively clean, rural environment of Quartermaster Harbor (QMH) in Puget Sound, WA.[4] A subsequent study, conducted in October 2000,[6] showed that the DNA from the gills of English sole from the DR could be readily distinguished from the gill DNA of the same species from QMH. The FT-IR spectral differences between groups were consistent with a marked increase in the sediment contamination (e.g., concentrations of polychlorobiphenyls [PCBs] and aromatic hydrocarbons) and the degree of CYP1A expression in the gills. A logistic regression analysis of the spectral data sets resulted in the development of a *DNA damage index* with high sensitivity and specificity.

The present report illustrates the application of the FT-IR statistics technology to assess differences in the DNA structure of various fish tissues between reference and contaminated environments. The resulting data can be used for assessing the quality of marine environments, toxic effects on fish, and the effectiveness of remediation protocols. Overall, the FT-IR statistics technology is best used in conjunction with other markers of exposure or toxicity, such as CYP1A expression[7–9] and various histological[10] and histochemical indices.[7,11] Although initially applied to fish, this technology has the potential for application to various other aquatic organisms, in addition to a variety of human diseases.

Materials required

Tissues

Groups of fish (preferably sex-matched and not differing significantly in size and mass) are obtained from contaminated and essentially non-contaminated reference sites. Females should be restricted to those with quiescent gonads to minimize the effects of reproductive stage (e.g., suppression of CYP1A by estradiol) on the biomarker data. Each fish should be given a unique identification and carefully weighed and measured. In field studies, fish are kept alive until sacrificed via decapitation aboard the vessel. The desired tissue (e.g., gill, liver) is removed and immediately frozen in liquid nitrogen. Tissues should be maintained in a −80°C freezer until DNA extraction. Prior to freezing, a few milligrams of the tissue are preserved in neutral formalin for histological examination or histochemical determinations or both (e.g., CYP1A).[7,11,12] Otoliths may be removed for subsequent age determinations.[13]

Supplies and equipment for DNA extraction

- Scalpels, forceps, spatulas
- Mortars and pestles
- Liquid nitrogen
- Qiagen Genomic DNA Buffer Set, #19060
- Qiagen Genomic-tip 100/G, #10043
- Falcon 15-ml graduated polypropylene tubes (conical bottom), #352096
- Osmonics Cameo 30N syringe filter, nylon, 5.0 μ, 30 mm, #DDR50T3050
- Roche RNAse A (1 mg ml^{-1}), #109169
- Worthington Proteinase K (20 mg ml^{-1}), #LS004222
- Isopropanol
- Ethanol 70% (ice cold)
- Microcentrifuge tubes 2 ml, polypropylene
- Transfer pipettes 1.5 and 3 ml, disposable, polyethylene
- 50°C water bath
- Refrigerated (4°C) centrifuge
- Microcentrifuge at 4°C
- Optima grade water (Fisher), #W74LC

Supplies and equipment for FT-IR spectral analysis

- FT-IR microscope spectrometer (System 2000, Perkin-Elmer)
- BaF_2 plate, 38.5 mm × 19.5 mm × 2 mm (International Crystal Laboratories, Garfield, NJ)
- Aluminum BaF_2 plate holder (Custom-made for the Pacific Northwest Research Institute, Seattle, WA; See Figure 30.1)
- 0.1–2.5 μl pipetter with tips
- Dissecting microscope

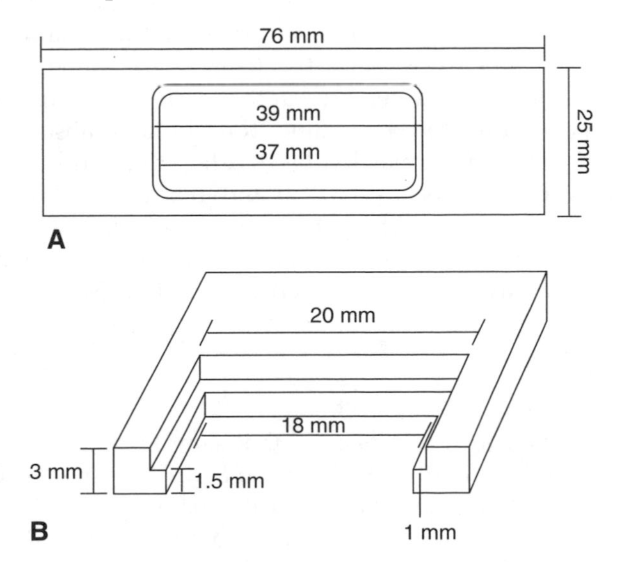

Figure 30.1 (A) Diagram of custom-made aluminum BaF_2 plate holder and (B) cross-section schematic. Dimensions are given; images are not to scale.

Procedures

DNA extraction

Frozen tissue (~100 mg; −80°C) is ground to powder with a mortar and pestle while submerged in liquid nitrogen. DNA (~50 μg) is then extracted from each sample with Qiagen 100/G Genomic-tips (Qiagen, Chatsworth, CA) using the standard Qiagen extraction protocol with the following modification: after elution, the DNA solution (eluate) is passed through a 5.0-μ Cameo 30N filter (Osmonics, Minnetonka, MN) to remove residual resin from the Qiagen Genomic-tip prior to precipitation. After filtration, the Qiagen protocol is resumed. In preparation for FT-IR spectral analysis, the DNA pellet is dissolved in 10–40 μl (depending on size) of optima grade water (Fisher Scientific). The DNA is allowed to dissolve overnight at 4°C. The Qiagen procedure is an ion-exchange system and does not constitute a source for artifactual oxidation of purines during extraction.

FT-IR spectroscopy

A 0.2-μl aliquot of the DNA solution is spotted directly on a BaF_2 plate and allowed to spread, forming an outer ring that contains the DNA. Two separate spots are created for each DNA sample. The spots are allowed to dry. Spotting is repeated until the ring is at least 100 μ wide, the width of the aperture of the System 2000 microscope spectrometer (Perkin-Elmer). The plate is then placed in a lyophilizer for 1 h to completely dry the DNA. Initially, a background energy reading (percent transmittance) is determined from a blank area of the BaF_2 plate. Energy readings are then taken at various points around the ring (Figure 30.2), and the points for spectral determinations are selected where the energy readings are 15–25% less than the background energy (optimally close to 15% less). Ten spectral determinations are made around each of the two rings per sample and the percent transmittance values are converted (Fourier-transformed) into absorbance values (Figure 30.2). The spectral data obtained are saved in a database for subsequent statistical analysis. Using MS Excel, each spectrum is baselined by taking the mean absorbance across 11 wave numbers, centered at the minimum absorbance value between 2000 and 1700 cm^{-1}, and subtracting this value from the total absorbance at each wavenumber. Each spectrum is then normalized by dividing the entire baselined absorbance values by the mean absorbance between 1750 and 1550 cm^{-1}. Baselining and spectral normalization adjust for the optical characteristics of each sample (e.g., related to film thickness). The mean absorbance value of the 20 spectral determinations for each sample is then calculated at each wavenumber between 1750 and 1275 cm^{-1}.

Other considerations

To avoid any batch effects, it is recommended that samples from reference and contaminated sites be randomized during DNA extraction and FT-IR spectroscopy. Tissues may be ground and stored at −80°C prior to DNA extraction. One technician can extract 6–12 samples, using the modified Qiagen protocol, in about 6 h, including a 2-h incubation time. FT-IR analysis should be performed on the extracted samples within a day or two. For long-term storage, we do not recommend that the DNA be kept in water, but rather in the dry state at −80°C.

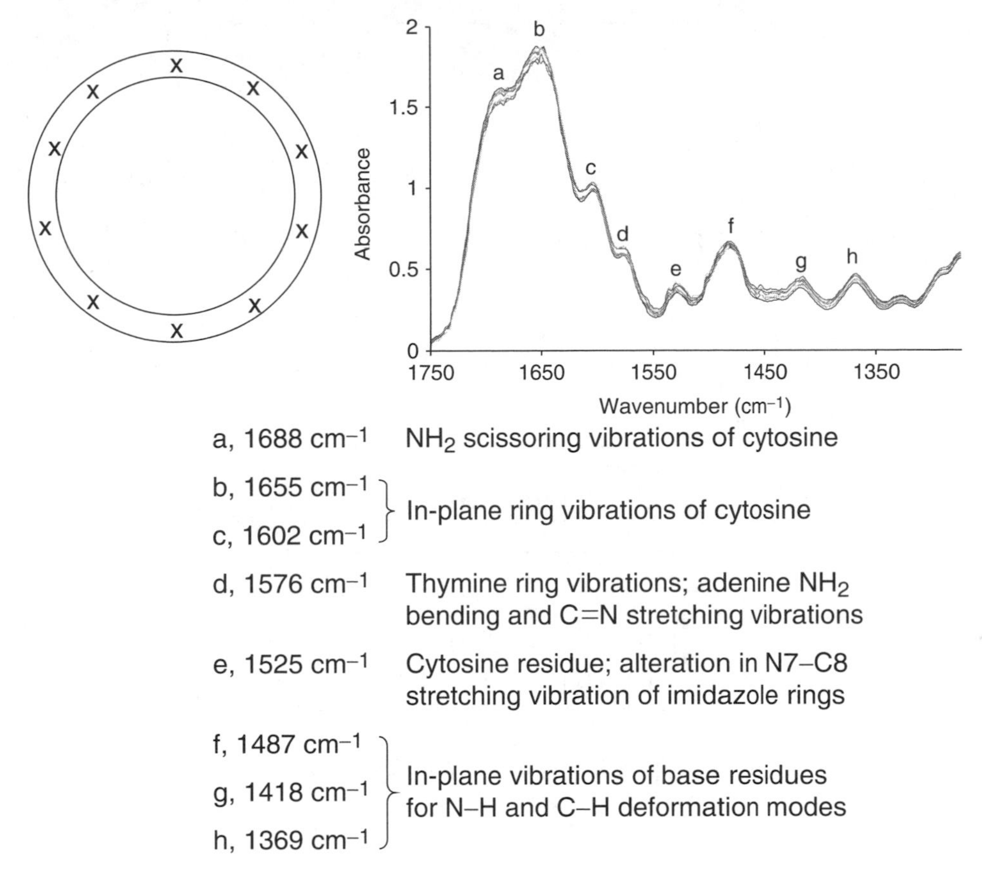

Figure 30.2 A DNA ring with 10 spectral determination points (x), the subsequent DNA spectra obtained, and the wavenumber and structural assignments for designated peaks.

Statistical analyses

FT-IR mean spectra

A mean spectrum is determined for each fish group (i.e., from contaminated and reference sites; Figure 30.3A). A t test is then performed at each wavenumber to establish statistical differences (P values) between the mean spectra (Figure 30.3B). Over the wavenumbers used, spectral regions with $P < 0.05$ are likely to represent real structural differences between groups when they comprise 5% of the spectral range.[3] These structural differences represent alterations in various aspects of the DNA molecule (e.g., as illustrated in Figure 30.2).

Principal components analysis

Statistical model development is accomplished by first conducting principal components analysis (PCA) on the mean spectrum of each individual DNA sample (S-Plus 2000 Professional Release 1, Mathsoft Engineering & Education, Cambridge, MA). PCA entails nearly 1×10^6 correlations between ~1000 independent variables relating to the absorbance, wavenumbers, and other properties of the spectrum.[14] PCA results in 10 principal

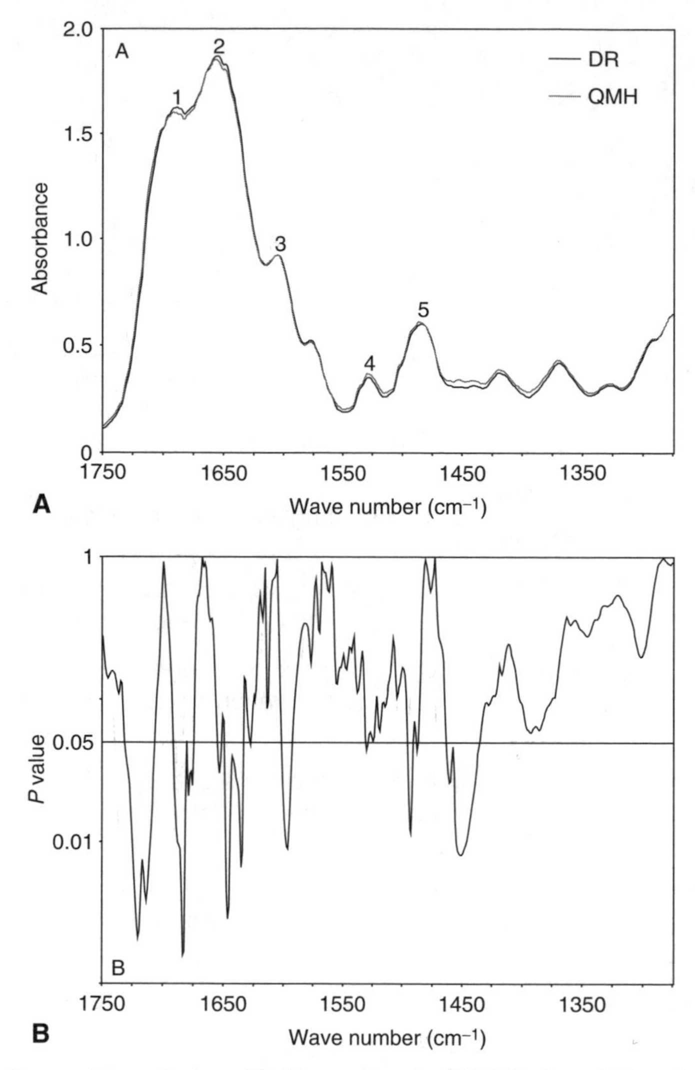

Figure 30.3 (A) Comparison of mean FT-IR spectra of gill DNA from DR and QMH fish; (B) *P* values from a *t* test comparing mean spectra at each wave number. The peak wavenumber designations are as follows: 1, 1688 cm^{-1}; 2, 1655 cm^{-1}; 3, 1602 cm^{-1}; 4, 1525 cm^{-1}; 5, 1487 cm^{-1}. (Modified and reprinted from Malins, D.C., Stegeman, J.J., Anderson, J.W., Johnson, P.M., Gold, J. and Anderson, K.M., *Environ. Health Perspect*. (Perspect., 2004, 112: 511–515.)

component (PC) scores for each sample. Significant differences ($P \leq 0.05$) in each PC score between groups are determined using *t* tests. The PC scores showing the most significant differences between the groups are used to construct two- or three-dimensional PC plots (Figure 30.4). Those PC scores representing fish with similar DNA structures will cluster together and be separated from other clusters reflecting a different DNA structure.

DNA damage index

Logistic regression analysis is performed (SPSS statistical package 10.0, SPSS, Chicago, IL) using a single, significant ($P \leq 0.05$) PC score to establish a DNA damage index. Using a

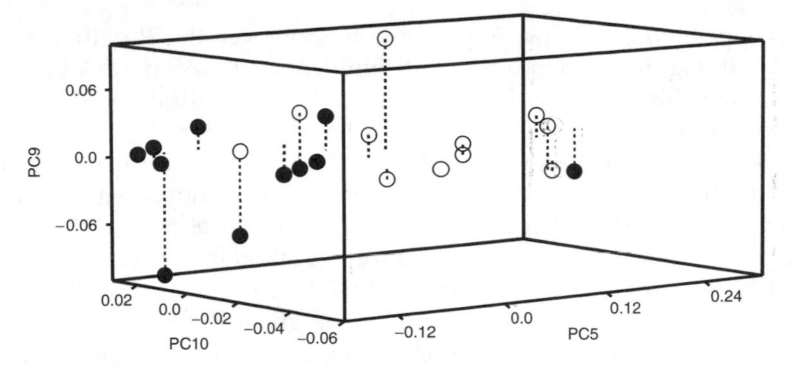

Figure 30.4 Three-dimensional separation of PC scores from the FT-IR spectra of gill DNA from the DR (• ; $n = 11$) and QMH (○ ; $n = 11$) fish. Dotted drop lines represent the distance from the PC9 baseline level of 0. (Reprinted with permission from Malins, D.C., Stegeman, J.J., Anderson, J.W., Johnson, P.M., Gold, J. and Anderson, K.M., *Environ. Health Perspect.* (Perspect., 2004, 112: 511–515.)

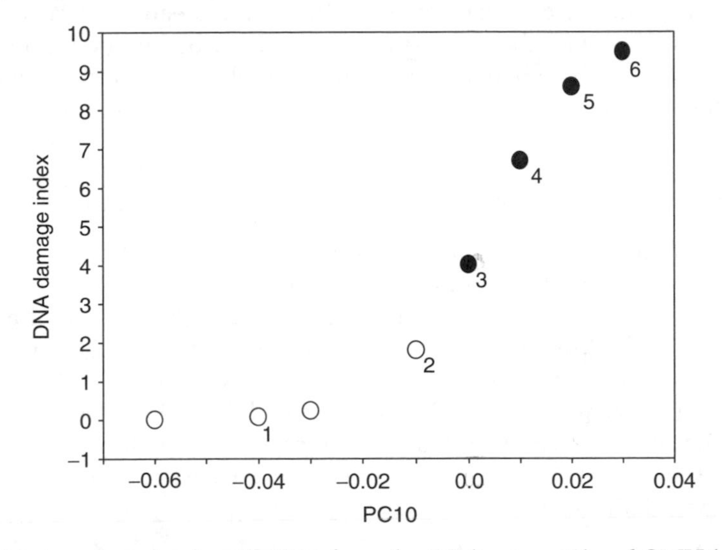

Figure 30.5 DNA damage index for gill DNA from the DR (• ; $n = 11$) and QMH (○ ; $n = 11$) fish. Overlapping points: 1, two QMH; 2, one DR and one QMH; 3, one DR and five QMH; 4, four DR; 5, one DR and one QMH; 6, four DR. (Reprinted with permission from Malins, D.C., Stegeman, J.J., Anderson, J.W., Johnson, P.M., Gold, J. and Anderson, K.M., *Environ. Health Perspect.* (Perspect., 2004, 112: 511–515.)

scale of 1–10, this index is based on the different spectral and structural properties of DNA from each fish group (Figure 30.5) and is a measure of DNA damage that provides a means to discriminate between fish from reference and contaminated sites.[6]

Results and discussion

Comparison of mean DNA spectra

As an example, comparison of the mean FT-IR spectra for the DNA from the DR ($n = 11$) and QMH ($n = 11$) fish is given in Figure 30.3A. Some of the differences in the mean spectra may appear to be almost imperceptible; however the P values at each wavenumber shown in Figure 30.3B indicate that significant differences ($P \leq 0.05$) were found for the five peaks identified in Figure 30.3A. The peak differences occurred at the following wave numbers: 1688, 1655, 1602, 1525, and 1487 cm^{-1}. The structural assignments for these peaks are given in Figure 30.2 and represent an array of differences in the structures of the nucleotide bases between the two DNA groups. Significant differences between the mean DNA spectra for the two fish groups represented 26.5% of the spectral range (differences <5% may occur by chance).[3] Although a broader range of wavenumbers can be used (i.e., from 1750 to 700 cm^{-1}),[4] we have found that the narrower range used here (i.e., from 1750 to 1275 cm^{-1}) has given the most reliable results with studies of fish DNA. This spectral range primarily represents vibrations of the nucleotide bases, thus spectral differences between groups of DNA would include structural changes associated with the genome. Comparison of mean spectra is a procedure for identifying the nature and extent of DNA structural differences (Figure 30.2) in fish from reference and contaminated environments. These comparisons also allow for initial evaluation of whether sufficient differences exist between the group mean spectra to justify further statistical analyses. However, the technique of PCA described in the following paragraph has the ability to discriminate between groups, even when the spectral means show few, or even no differences.

Principal components analysis

PCA is a powerful means of discriminating subtle differences in DNA structures between groups of fish from different environments. Groups of DNA samples representing fish from contaminated and reference environments will cluster in different areas of the PC plots by virtue of their different spectral and structural properties. An example of this discrimination is a three-dimensional projection of PC scores (PC10, PC5, and PC9) representing the gill DNA from the DR (contaminated) and QMH (reference) fish (Figure 30.4). Despite the high degree of separation, the relatively small number of overlapping points between the groups may reflect fish migrations or other aberrations known to exist in natural fish populations.

DNA damage index

In this example with gill DNA, logistic regression analysis was conducted based on PC10, selected for its significance ($P < 0.01$). The resulting sigmoid-like curve shows a distinct separation between the DR and QMH fish groups with 9/11 DR scores falling above ~5.0 on the index and 10/11 of the QMH scores falling below this value. This indicates that the DR fish may have more damage to their gill DNA than the fish from QMH, which is further substantiated by the sediment chemistry and CYP1A data.[6] This results in an 82% probability of correctly identifying a DR sample and a 92% probability of correctly identifying a QMH sample.

The DNA damage index provides a means of quantifying the DNA damage between fish from reference and contaminated environments. This index can be determined for any two fish populations to assess environmentally induced DNA damage using a variety of tissues (e.g., gill, liver, gonads, and kidney). The example described employing fish gill[6] has the added advantage of non-lethality if tissues are obtained via punch biopsies.[15] One attractive application of the FT-IR statistics technology would be to determine the effects of remediation on chemically contaminated aquatic environments. After remediation, DNA samples would be evaluated using the damage index developed for that specific environment, species, and tissue type to determine whether the spectral and structural characteristics had improved to closely match the index values established for the reference site. The usefulness of the DNA damage index has so far been limited to the study of English sole in Puget Sound.[6] We look forward to the application of the FT-IR statistics technology, including the DNA damage index, to other fish and aquatic species, as well as to other environments having different contaminant profiles.

Acknowledgments

We thank Robert Spies and Jordan Gold of Applied Marine Sciences, Inc., 4749 Bennett Dr., Livermore, CA 94550, for fish collections and Nhan Vo for technical assistance. This publication was made possible by the National Institute of Environmental Health Sciences, NIH, grant number P42 ES04696.

References

1. Tsuboi, M., Application of infrared spectroscopy to structure studies of nucleic acids, *Appl. Spectrosc. Rev.*, 3, 45–90, 1969.
2. Parker, F.S., *Applications of Infrared, Raman, and Resonance Raman Spectroscopy in Biochemistry*, Plenum Press, New York, 1983.
3. Malins, D.C., Polissar, N.L., Ostrander, G.K. and Vinson, M.A., Single 8-oxo-guanine and 8-oxo-adenine lesions induce marked changes in the backbone structure of a 25-base DNA strand, *Proc. Natl. Acad. Sci. USA*, 97, 12442–12445, 2000.
4. Malins, D.C., Polissar, N.L. and Gunselman, S.J., Infrared spectral models demonstrate that exposure to environmental chemicals leads to new forms of DNA, *Proc. Natl. Acad. Sci. USA*, 94, 3611–3615, 1997.
5. Malins, D.C., Johnson, P.M., Barker, E.A., Polissar, N.L., Wheeler, T.M. and Anderson, K.M., Cancer-related changes in prostate DNA as men age and early identification of metastasis in primary prostate tumors, *Proc. Natl. Acad. Sci. USA*, 100, 5401–5406, 2003.
6. Malins, D.C., Stegeman, J.J., Anderson, J.W., Johnson, P.M., Gold, J. and Anderson, K.M., Structural changes in gill DNA reveal the effects of contaminants on Puget Sound fish, *Environ. Health Perspect*. Perspect., 112, 511–515, 2004.
7. Woodin, B.R., Smolowitz, R.M. and Stegeman, J.J., Induction of cytochrome P450 1A in the intertidal fish *Anoplarchus purpurescens* by Prudhoe Bay crude oil and environmental induction in fish from Prince William Sound, *Environ. Sci. Technol.*, 31, 1198–1205, 1997.
8. Stegeman, J.J., Schlezinger, J.J., Craddock, J.E. and Tillitt, D.E., Cytochrome P450 1A expression in midwater fishes: potential effects of chemical contaminants in remote oceanic zones, *Environ. Sci. Technol.*, 35, 54–62, 2001.
9. Miller, K., Addison, R. and Bandiera, S., Hepatic CYP1A levels and EROD activity in English sole: biomonitoring of marine contaminants in Vancouver Harbour, *Mar. Environ. Res.*, 57, 37–54, 2004.
10. Moore, M.J. and Myers, M.S., Pathobiology of chemical-associated neoplasia in fish, in *Aquatic Toxicology: Molecular, Biochemical and Cellular Perspectives*, Malins, D.C. and Ostrander, G.K., Eds., Lewis Publishers, Boca Raton, FL, 1994, pp. 327–386.
11. Smolowitz, R., Hahn, M. and Stegeman, J., Immunohistochemical localization of cytochrome P-450IA1 induced by 3,3′,4,4′-tetrachlorobiphenyl and by 2,3,7,8-tetrachlorodibenzoafuran in liver and extrahepatic tissues of the teleost *Stenotomus chrysops* (scup), *Drug Metab. Dispos.*, 19, 113–123, 1991.

12. Van Veld, P.A., Vogelbein, W.K., Cochran, M.K., Goksoyr, A. and Stegeman, J.J., Route-specific cellular expression of cytochrome P4501A (CYP1A) in fish (*Fundulus heteroclitus*) following exposure to aqueous and dietary benzo[*a*]pyrene, *Toxicol. Appl. Pharmacol.*, 142, 348–359, 1997.
13. Secor, D.H., Manual for Otolith Removal and Preparation for Micro Structural Examination, Electric Power Research Institute, Palo Alto, 1991.
14. Timm, N.H., Ed., *Multivariate Analysis*, Brooks/Cole, Monterey, CA, 1975, pp. 528–570.
15. McCormick, S.D., Methods for non-lethal gill biopsy and measurement of Na^+, K^+-ATPase activity, *Can. J. Fish Aquat. Sci.*, 50, 656–658, 1993.

chapter thirty-one

Design and use of a highly responsive and rigidly controllable hypoxia exposure system

D.W. Lehmann, J.F. Levine, and J.M. Law
North Carolina State University

Contents

Introduction

Broad changes in oxygen concentrations are characteristic of aquatic ecosystems impacted by eutrophication. Daily cycles of oxygen levels in estuaries can range from daytime supersaturation reaching as high as 300% to predawn anoxia (0%).[1–3] To test the effects of hypoxia in aquatic animal models, we have designed an exposure system capable of rapidly changing dissolved oxygen (DO) conditions in experimental tanks. The system was assembled from readily available components and allows precise, programmable control of DO concentrations in the laboratory setting.

When dealing with sublethal stressors, complicating factors, such as time to effect and spatial relevance, make field studies impractical. The mobility of free-ranging species, and the dynamic variability of aquatic systems makes clearly characterizing the exposure history of an individual challenging. Exposure to stressors may occur days prior to and miles apart from a fish kill or sampling site.[4–6] Controlled laboratory studies, however, can be

1-56670-664-5/05/$0.00+$1.50
© 2005 by CRC Press

used to document the contribution of individual environmental factors to specific health problems.

Periods of hypoxia or anoxia can have profound effects on aquatic organisms. Anoxia has been associated with the loss of benthic invertebrates in eutrophied ecosystems,[7] as well as fish kills.[8] In addition, hypoxic events have been proposed to play a role in more subtle disease conditions, such as reproductive or developmental abnormalities.[9,10] The spectrum of health effects observed in fish populations, although implied, is poorly understood mechanistically. An example is the association of hypoxic events with epizootic ulcerative syndrome (EUS), a disease which has affected countless Atlantic menhaden (*Brevoortia tyrannus*) along the east coast of the United States from Chesapeake Bay to the estuaries of the Carolinas.[11–13] Originally ascribed to the *Aphanomyces* fungus[14] and more recently to dinoflagellate *Pfiesteria* spp.,[5] the root cause(s) of EUS remain largely unknown.[15–17] Histopathological studies performed in our laboratory on hundreds of specimens frequently showed no evidence of a specific pathogen that could be associated with the disease.[18] Moreover, EUS occurrence in the estuaries of North Carolina follows a seasonal trend with temperature and hypoxic events.[19]

Fish have gained popularity as animal models in aquatic toxicology as recent advances have increased our knowledge of normal physiologic conditions and responses to various stressors.[20] Fish models are also increasingly being used in research leading to information regarding human diseases and genetic and reproductive responses.[21] In this chapter, we describe a laboratory system for examining the response of aquatic species to hypoxia. Experiments with wild caught Atlantic menhaden and laboratory-reared Nile tilapia (*Oreochromis niloticus*) were used to demonstrate the functionality and limits of the system. The tilapia served as a resistant species and the menhaden, based on the association of estuarine hypoxia with ulcerative skin lesions in these fish, served as a susceptible species.

The system described provides a means to investigate responses to hypoxia as a single variable or in combination with other stressors. The use of nitrogen gas to reduce the partial pressure of dissolved gases is not novel in aquatic research. The reduction in oxygen tension can be isolated and studied if the proper mixture of gases is used to emulate atmospheric carbon dioxide, argon, and nitrogen ratios. This method works because fish, unlike other vertebrates, sense O_2 in their environment in place of CO_2.[22–24] The system can readily be used for aquatic organisms other than fish or for complex mesocosm studies. Using O_2 in place of N_2 or component air can also rapidly create hyperoxic experimental conditions. Other advantages this system has over traditional methods include its controllability, monitoring systems, rapid oxygen partial pressure changes, and automated data logging and graphing. Herein, we describe the system and provide some example data from two experiments to demonstrate its use. DO levels for a longer-term hypoxia exposure were based on acute LC_{50} values for the two fish species. We chose a series of endpoints to test the hypothesis that hypoxia and subsequent reperfusion create oxidative cellular damage as a factor in the development of ulcerative skin lesions in fish.[25–28]

Material requirements and setup

A wide range of tank sizes can be used with this system depending on the size and flow rating of the protein skimmer, or foam fractionator, employed and species-specific requirements. Our system includes 260-l fiberglass circular tanks for the 1- and 2-h acute exposures and 855-l dark blue, circular polyethylene tanks for the long-term exposures.

Each tank has a bulkhead located on the bottom (center) and on the sidewall at the bottom. The center bulkhead was for drainage and tank cleaning purposes. The sidewall bulkhead is plumbed into an external Iwaki-Walchem 55RLXT pump (Walchem Corp., Holliston, MA). Two or more tanks should be employed, each as an independent unit, to house control and exposed animals.

The system functions by use of a Neptune Systems (San Jose, CA) Aquacontroller Pro unit connected to laboratory grade pH, DO, temperature, and conductivity probes (Figure 31.1). Data from the probes are monitored by the controller and logged both by the controller and by Aquanotes (Neptune Systems) software on a computer connected directly to the controller via a serial port. The controller can be operated via the computer (local or internet) or directly on the controller via its simple programming language. User-defined settings toggle system components on and off remotely. For example:

$$\begin{aligned} \text{DO} &> 1.0\,\text{mg/l} = \text{on} \\ \text{DO} &< 0.9\,\text{mg/l} = \text{off} \end{aligned} \tag{31.1}$$

When toggled on, the system functions by opening a solenoid and turning on the Iwaki-Walchem feed pump. The skimmer (AE Tech, ETSS Professional 800) is a passive device that forces air (or compressed gases) and water to mix at high rates via a downdraft mechanism. The skimmer must be placed at or above the level of the exposure tank's resting water position to allow for gravity feedback to the tank. This allows efficient and rapid mixing of the exposure tank water with contained gases, in this case pre-purified grade nitrogen.

Standard 300-ft^3 nitrogen tanks with nitrogen-rated regulators are used as a source of nitrogen. (*Note*: Compressed gas tanks should be properly secured to a wall or other stationary structure according to current institutional safety regulations.) The nitrogen gas passes from the tank and regulator through a stainless steel and glass flow meter

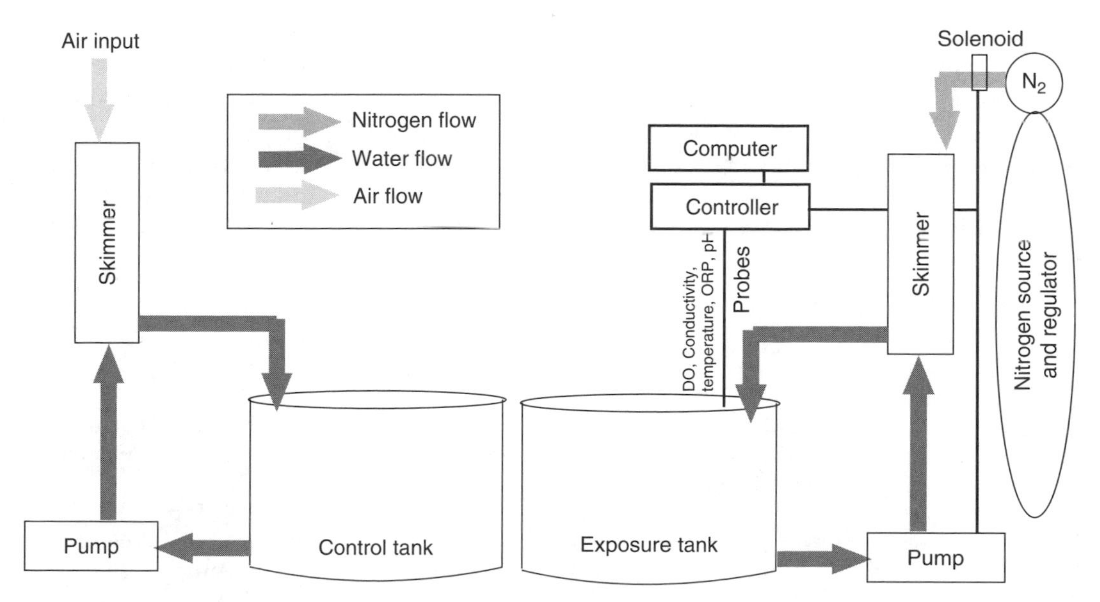

Figure 31.1 DO exposure system schematic showing sequence of controller and direction of laminar fluid and gas flow. Tanks of any size from 150 to 1500 l can be used on this system depending on the capacity and volume rating of the skimmer used.

(Dwyer, model SS-DR12442). Flow meters bracket ranges of gas volume per time so the choice of flow meter model necessarily depends on the volume of nitrogen released per hour into the skimmer. The flow meter has a needle valve for controlling 5–20 standard cubic feet per minute (scfm) input into the system, so that flow rates can be finely adjusted while the regulator is static. The solenoid valve, placed between the regulator on the N_2 source and the skimmer, is turned on and off based on signals from the controller to allow gas to flow to the skimmer.

Signals are relayed from the controller via an x10 control module (www.x10.com). This module codes the signal and passes it along the electrical lines of the building allowing for remote control of x10 appliance modules. The controller specifies the channel, and each module set to that channel will respond with an ON/OFF switch. For example, when the controller reads DO at 1.1 mg/l, it then sends a signal to the unit(s) connected to the solenoid and the feed pump for the skimmer thereby initiating the scrubbing of the exposure tank water that circulates from the tank, through the skimmer, and returns to the tank via PVC plumbing with a reduced oxygen concentration. (Table 31.1)

Water for the menhaden exposures was made with synthetic sea salt (Instant Ocean, Mentor, OH) to 12.5 ppt (19 mS/cm) and allowed to mix for at least 24 h prior to use. Water was mixed between tanks before introducing the randomized fish to eliminate slight differences between water parameters. Each tank was equipped with a 2 mil clear plastic cover secured by elastic lines and clamps to prevent fish from gulping air at the surface during exposure. The clear plastic allowed easy observation of the fish during the study.

Atlantic menhaden were collected from a reference site, the White Oak River, NC. The fish were cast netted, placed into filtered, flow-through tanks and held for a minimum of 2 weeks. Menhaden had a mean fork length of 17.5 cm and mean weight of 71.8 g. Feeding commenced on the second day of holding and continued twice daily with salmon starter crumble (Zeigler Bros., Gardners, PA) ground to a fine consistency. After acclimation, fish were transported to the laboratory in a large, round, enclosed transport tank with heavy aeration and circulation to minimize stress. Temperature was maintained during transport by frozen blocks of water in sealed containers or by aquarium heaters. At the laboratory, fish were acclimated for a further 2 weeks prior to the experiment and remained apparently healthy. In parallel experiments, healthy tilapia with a mean tail length of 19 cm and mean weight of 156 g were randomized into the exposure tanks using aged and dechlorinated tap water. Tilapia were a gracious donation from NC State University Fish Barn.

Table 31.1 Required materials for exposure system setup

Item	Model	Manufacturer
Skimmer	ETSS Pro 800	AE Tech
Pump	55RLXT	Iwaki-Walchem
Solenoid	115 V a/c static off/powered on type	National Welders
Tanks	260 l circular fiberglass	Custom
	855 l circular polyethylene	PolyTank, MN
Regulator	Nitrogen-rated, M1-960-PG	National Welders
Controller	Aquacontroller Pro and lab grade probes	Neptune Systems, CA
Computer	Any Pentium model	Any
PVC pipe	1.5 in. rigid piping	Any
PVC tubing	1 in. flexible tubing	Any

Validation of DO concentrations was performed daily using a handheld YSI-85 portable meter (Yellow Springs, OH). All DO probes had new membranes at the start of each experiment and were calibrated daily. All other probes were calibrated at the start of each experiment.

Procedures

All experiments were performed under protocols approved by the Institutional Animal Care and Use Committee (IACUC), NC State University. Acute exposures were performed in 260-l round fiberglass custom tanks with either menhaden or tilapia. Each tank was prepared as mentioned earlier and contained 5 and 7 randomly selected fish, respectively. Fish were acclimated to the exposure tanks for a minimum of 2 days prior to initiating the experiment. Feeding was halted and tanks were cleaned 24 h prior to initiation. Water quality parameters [ammonia, nitrite, nitrate, and hardness (for tilapia only)] were monitored daily. A separate tank was used for each oxygen saturation level. Menhaden were exposed to 84/6.7 (control), 20/1.59, 15/1.19, 10/0.79, and 5/0.39 (% oxygen saturation/mgL^{-1}) for 1 h in independent tanks. Tilapia were exposed to 82/6.9 (control), 20/1.68,10/0.83,7/0.58, and 3/0.24 (% oxygen saturation/mgL^{-1}) in independent tanks for 2 h.

Exposures were initiated, and a log was kept of DO (% saturation and mg/l), pH, temperature, and mortality at 5-min intervals. Moribund fish, as indicated by uncontrolled swimming behavior or lack of response to physical stimuli, were removed for sampling. At the end of 1 or 2 h of exposure, remaining fish were euthanized and sampled.

Euthanasia was performed by overdose of MS222 (Argent Chemical Laboratories, Redmond, WA) in water from the tank in which the fish was exposed to maintain the oxygen saturation level. Sampling consisted of taking length and weight measurements, drawing blood for clinical pathology, and taking tissue samples for histopathology, oxidative stress, and immune function measurements. Fish were examined for gross abnormalities upon dissection. Blood and spleen samples were taken from all fish for immune function analysis (not covered herein). Samples of heart, liver, anterior kidney, intestine, gonads, gills, and spleen were fixed in 10% neutral buffered formalin for 24–48 h and then held in 70% ethanol for histopathology. Samples of muscle, liver, and blood were placed in 2-ml cryovials and snap-frozen in liquid nitrogen for later analysis of oxidative stress endpoints.

Clinical pathology

To determine blood parameters, we used a Portable Clinical Analyzer (i-Stat Corp., East Windsor, NJ) with expendable cartridges that self-calibrate upon insertion into the unit. The EG7+ cartridge employed displays results from analysis of a few drops of fresh blood for the following parameters: sodium, potassium, ionized calcium, hematocrit, pO_2, and pCO_2. Blood was drawn from the caudal vein of the fish in a syringe free of anticoagulant. Analyses for lactate dehydrogenase (LDH), creatinine kinase (CK), aspartate aminotransferase (AST), glucose, and total protein were performed by the Clinical Pathology Laboratory at the NCSU College of Veterinary Medicine. Blood samples for clinical pathology were taken in heparin-coated syringes, kept on ice, and sent directly for plasma chemistry analysis.

Histopathology

For histopathological examination, tissues were fixed in 10% neutral buffered formalin for 24–48 h, routinely processed by paraffin embedment, sectioned at 5 μm, and stained with hematoxylin and eosin (H&E). Tissue sections were evaluated by a single pathologist and assigned a grade from 0 (no remarkable abnormalities) to 5 (severely lesioned).[29]

Data analysis

Results of clinical pathology and histopathology were analyzed using JMP (SAS, Cary, NC). ANOVA and Dunnett's tests were used for all comparisons using 84% and 80% oxygen saturation levels as controls for menhaden and tilapia, respectively.

Results and discussion

The controlled DO exposure system described here is very efficient and less demanding of personnel for operation than traditional systems. It responds rapidly to computer or controller commands. Monitoring can be performed remotely and data logging is automated, allowing for better control and replication of experiments. DO was reduced to 20% saturation (1.6 mg/l) in 1 h in an 855-l tank, and levels remained steady over a period of 96 h with an SD ± 0.091 mg/l during that time (Figure 31.2). Effective exposure of animals to stressors in toxicologic research hinges on controlled applications, and this system increases the control by reducing variability. Isolation of factors from extraneous

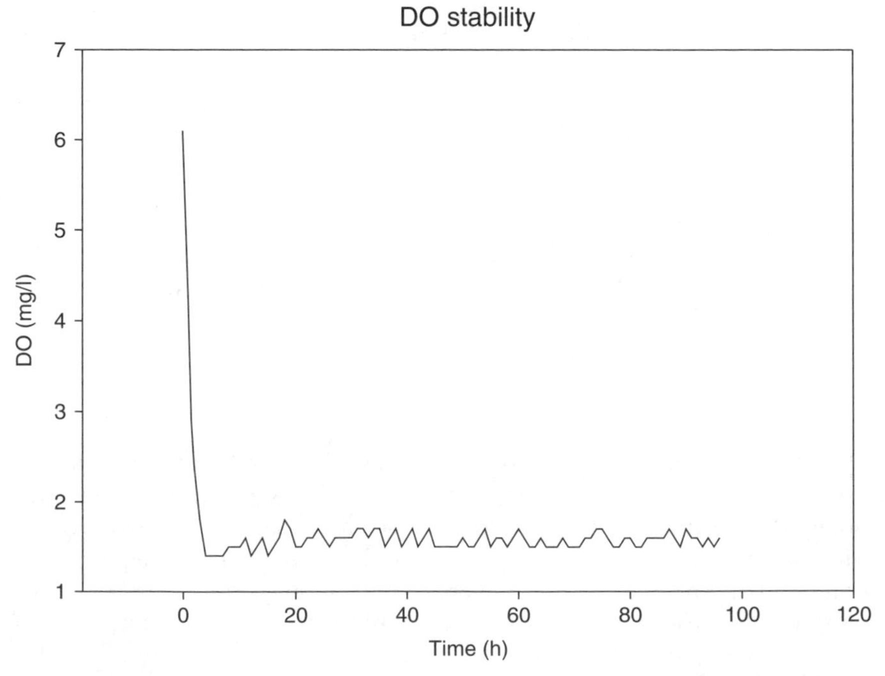

Figure 31.2 DO stability over time in laboratory trials. Trials proceeded for 96 h from initiation time. DO level of 1.6 mg/l (20% oxygen saturation) was held within 0.091 mg/l over the time course after initial oxygen reduction.

inputs is also critical for reproducibility and a realistic determination of the biological effects of each factor.

LC_{50} determination with Atlantic menhaden in these experiments mirrored the previously published data indicating a DO level of 16% saturation (1.2 mg/l) for 1 h as the approximate lethal concentration. Tilapia were highly resistant to challenge by hypoxia evidenced by a moderate response at 3% saturation (0.24 mg/l). Tilapia mortality was low, only 28% over the 2-h exposure period, such that the data generated were insufficient to determine an accurate LC_{50} (Figure 31.3).[6]

Fish showed behavioral changes as a result of the hypoxic stress. Menhaden are a filter feeding, continually active fish. The inability to rest and preserve energy stores is obvious in comparison with tilapia regarding responses to treatment. At 10% saturation, menhaden were obviously stressed and several began to search for the surface to gasp at the air–water interface. At 5% saturation, they were visibly agitated, and all tried to reach the surface of the tank. Tilapia responded to 7% saturation by disengaging territorial behaviors and resting on the bottom and occasionally an individual fish would attempt to gasp from the surface. At 3% saturation, many of the tilapia would intermittently gasp at the surface but then return to the bottom.[30] This response indicates that they have mechanisms for reducing metabolic oxygen demand and are a valid choice as a hypoxia-resistant species.

Results of the acute exposures suggest that the exposure system works very efficiently and mimics environmental hypoxia with minimal additional stress on test subjects. Blood electrolyte changes validated the adverse effect of hypoxia on the test fish, and changed

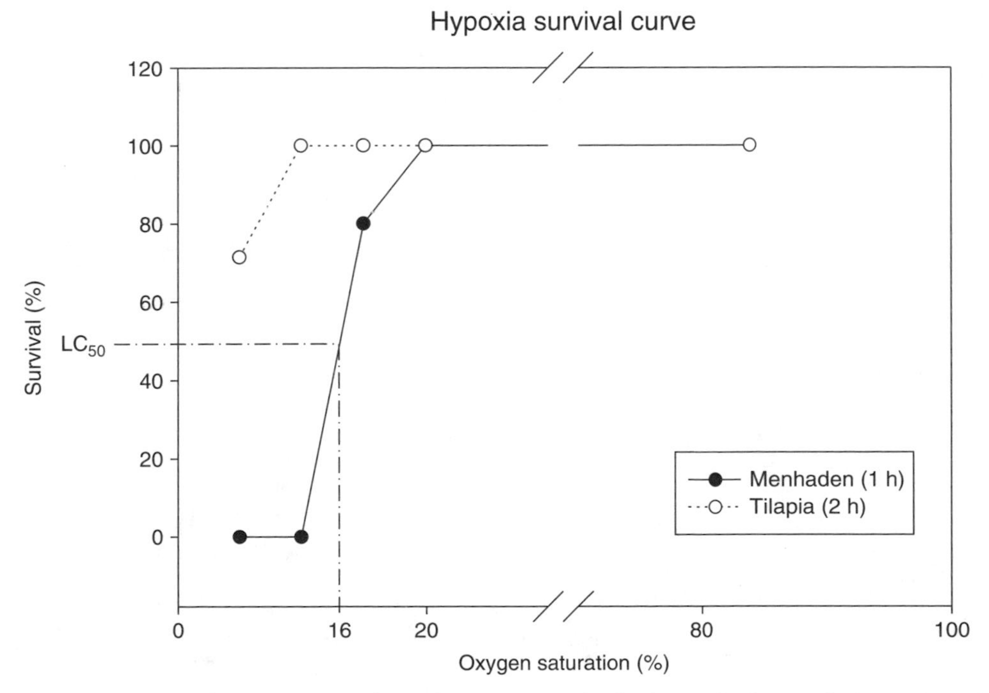

Figure 31.3 Percent survival in acute hypoxia exposures for both menhaden and tilapia (1 and 2 h, respectively). Menhaden displayed an approximate LC_{50} of 16% oxygen saturation (1.2 mg/l), while tilapia proved extremely hardy down to 3% oxygen saturation (0.24 mg/l). $N = 5$ menhaden and 7 tilapia per saturation level.

sharply as the fish passed from a mild to a severe stress state with reduced oxygen tensions. Partial pressure of CO_2 in the blood fell as oxygen saturation decreased in the exposure tank, indicating that O_2 availability is coordinately falling (Figure 31.4). Blood ion concentrations likely shift in response to a depletion of ATP stores and the lack of ability to regenerate those energy stores during a failure in oxidative phosphorylation.[31] A concomitant drop in pH suggests that anaerobic metabolism occurs as a salvage effort. Reduced ATP concentrations would also lead to a failure of the ATPases that maintain homeostasis in the blood. Failure of the sodium–potassium ATPase in many cells of the body and by the ATPases that drive chloride cell function allow for increases in sodium, calcium, and potassium in the blood. Menhaden showed a significant increase in K and Na at 10% saturation. Menhaden also responded with significant increases in ionized Ca (iCa), potassium (K), sodium (Na), and glucose at 5% saturation. Tilapia showed increases in K and iCa at 3% saturation (Figures 31.5 and 31.6).

Histopathology showed only mild parasitism in both treated and control menhaden, a reflection of being wild caught specimens. No significant difference in lesion prevalence was seen between treated and control menhaden. Likewise, no remarkable microscopic lesions were found in the tilapia specimens to suggest that hypoxia alone causes ulcerative skin lesions in fish. This is consistent with our findings from other biomarkers of oxidative stress (not discussed here) using this system. While both menhaden and tilapia showed strong physiologic responses to extremely low oxygen saturation levels, there appeared to be no overt oxidative damage to cells generated from these exposures. This may suggest an indirect or perhaps supplemental role for hypoxia in EUS and is an area in need of further study.

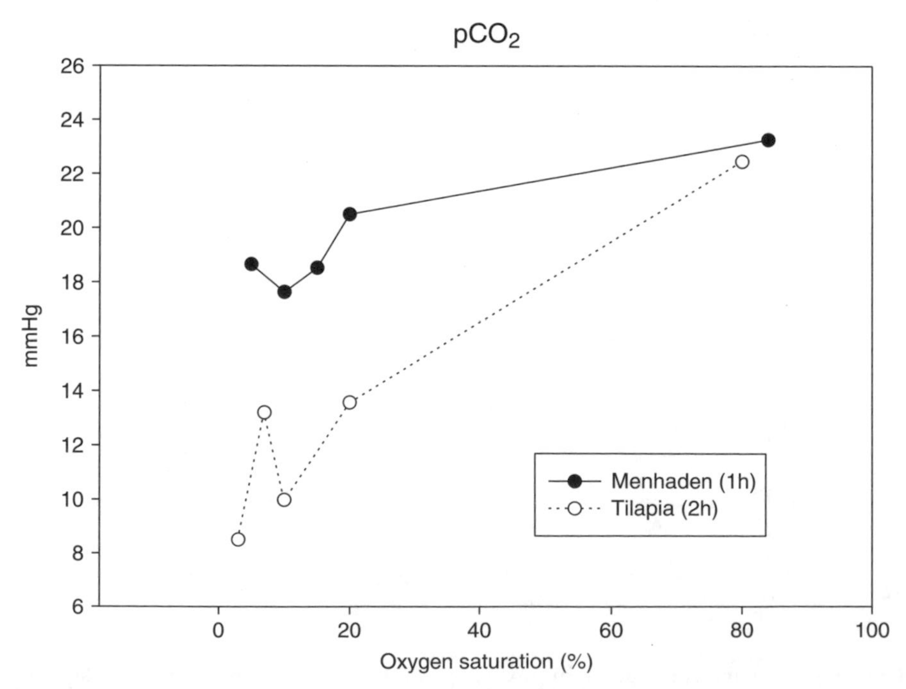

Figure 31.4 Partial pressure of carbon dioxide (pCO_2) in venous blood of exposed menhaden and tilapia. CO_2 concentrations fell as DO levels in the exposure tank decreased. A sharp drop was evident at approximately 10% saturation in both species. $N = 3$.

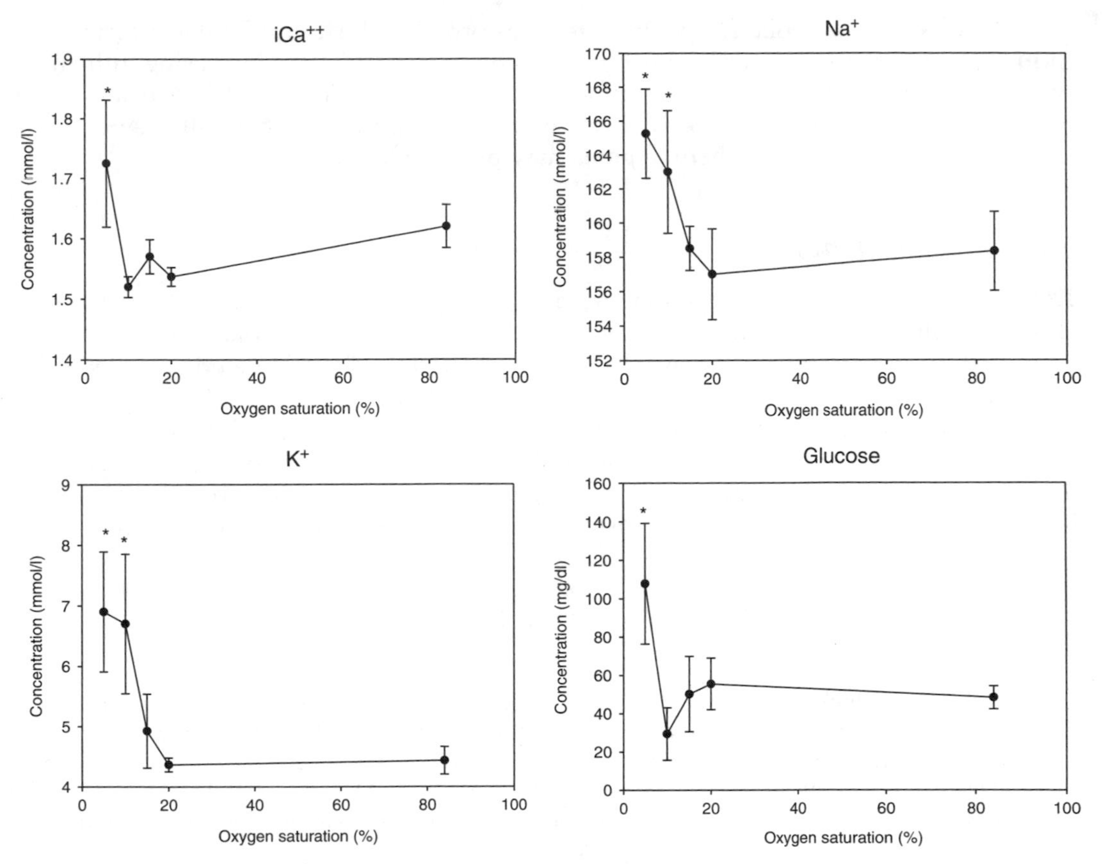

Figure 31.5 Menhaden blood chemistry parameters. Significant changes (*) were noted at critically stressful levels of hypoxia, indicating physiological failure of oxidative phosphorylation or blood ion homeostasis. Na ($p < 0.004$), K ($p < 0.0013$), iCa ($p < 0.003$), and glucose ($p < 0.0016$). Error bars are standard deviation.

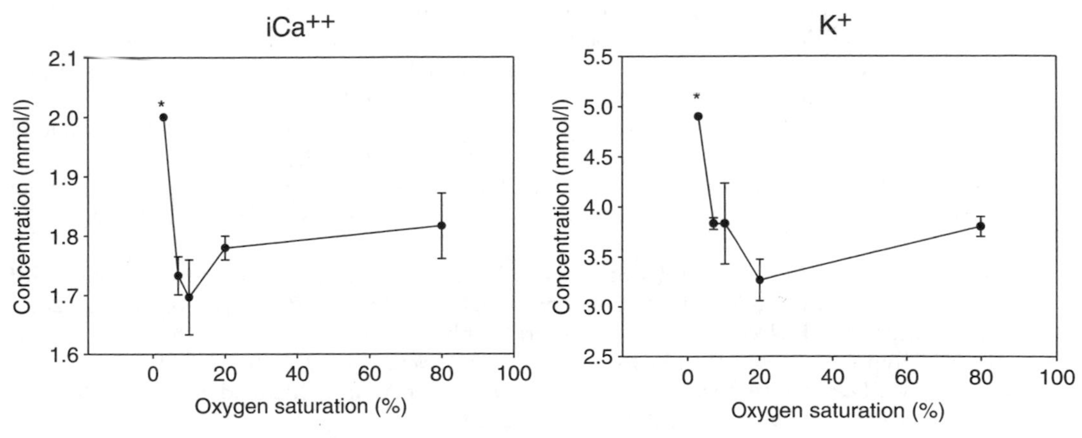

Figure 31.6 Tilapia blood chemistry parameters. Significant changes were seen in fewer ion types as compared to menhaden, indicating less loss of homeostasis and better energy management in tilapia. K ($p < 0.0041$), iCa ($p < 0.0041$). Error bars are standard deviation.

This precisely controlled hypoxia system has proven to be useful for the experimental induction of hypoxic responses in fish in our laboratory. With ever increasing influences of anthropogenic inputs into the nation's watersheds, particularly those resulting in eutrophication, this system is likely to serve broader applications that will answer questions in aquatic toxicology where hypoxia may play a role.

Acknowledgments

Development of this system was supported in part by the North Carolina Department of Environment and Natural Resources, project EW200020; North Carolina Department of Health and Human Services, project OEE 101; and NCSU College of Veterinary Medicine, state appropriated research funds. We sincerely thank the NOAA staff in Beaufort, NC, for their expertise and use of their facilities; J. Overton, L. Ausley, and M. Hale for their support in this project; the members of the Neuse River Rapid Response Team and the Tar/Pamlico Rapid Response Team for valuable technical expertise in our fish sampling efforts; M. Mattmuller for excellent histopathology support; and M. Dykstra and J. Rice for helpful discussions.

References

1. Paerl, H.W., Pinckney, J.L., Fear, J.M. and Peierls, B.L., Ecosystem responses to internal and watershed organic matter loading: consequences for hypoxia in the eutrophying Neuse River Estuary, North Carolina, USA, *Mar. Ecol. Prog. Ser.*, 166, 17–25, 1998.
2. Buzzelli, C.P., Luettich, R.A., Powers, S.P., Peterson, C.H., McNinch, J.E., Pinckney, J.L. and Paerl, H.W., Estimating the spatial extent of bottom-water hypoxia and habitat degradation in a shallow estuary, *Mar. Ecol. Prog. Ser.*, 230, 103–112, 2002.
3. Wu, R., Hypoxia: from molecular responses to ecosystem responses, *Mar. Pollut. Bull.*, 45, 35–45, 2002.
4. Hall, L.W., B.D., Margrey, S.L. and Graves, W.C., A comparison of the avoidance responses of individual and schooling juvenile Atlantic menhaden, *Brevoortia tyrannus* subjected to simultaneous chlorine and delta T conditions, *Toxicol. Environ. Health*, 10 (6), 1017–1026, 1982.
5. Burton, D.T., R.L. and Moore, C.J., Effect of oxygen reduction rate and constant low dissolved oxygen concentrations on two estuarine fish, *Trans. Am. Fish. Soc.*, 109, 552–557, 1980.
6. Miller, D.C., P.S. and Coiro, L., Determination of lethal dissolved oxygen levels for selected marine and estuarine fishes, crustaceans, and a bivalve, *Mar. Biol.*, 140, 287–296, 2002.
7. Diaz, R.J. and R.R., Marine benthic hypoxia: a review of its ecological effects and the behavioral responses of benthic macrofauna, *Oceanogr. Mar. Biol. Ann. Rev.*, 33, 245–303, 1995.
8. Domenici, P., F.R., Steffenson, J.F. and Batty, R.S., The effect of progressive hypoxia on school structure and dynamics in Atlantic herring *Clupea harengus*, *Proc. Roy. Soc. Lond. B.*, 269 (1505), 2103–2111, 2002.
9. Ross, S.D., DA., Kramer, S. and Christensen, B.L., Physiological (antioxidant) responses of estuarine fishes to variability in dissolved oxygen, *Comp. Biochem. Phys. C.*, 130, 289–303, 2001.
10. Baker, S.M. and M.R., Description of metamorphic phases in the oyster *Crassostrea virginica* and effects of hypoxia on metamorphosis, *Mar. Ecol. Prog. Ser.*, 104 (1/2), 91–99, 1994.
11. Dykstra, M., *Pfiesteria piscicida* and ulcerative mycosis of Atlantic Menhaden-current status of understanding, *J. Aquat. Anim. Health*, 12, 18–25, 2000.
12. Noga, E.J., J.S., Dickey, D.W., Daniels, D., Burkholder, J.M. and Stanley, D.W., Determining the Relationship Between Water Quality and Ulcerative Mycosis in Atlantic Menhaden, NCSU, Raleigh, 1993.

13. Dykstra, M.J., L.J. and Noga, E.J., Ulcerative mycosis: a serious menhaden disease of the southeastern coastal fisheries of the United States, *J. Fish Dis.*, 12, 175–178, 1989.
14. Levine, J.F., H.J., Dykstra, M.J., Noga, E.J., Moye, D.W. and Cone, R.S., Epidemiology of ulcerative mycosis in Atlantic Menhaden in the Tar-Pamlico River Estuary, North Carolina, *J. Aquat. Anim. Health*, 2, 162–171, 1990.
15. Pinckney, J.L., P.H., Haugen, E. and Tester, P.A., Responses of phytoplankton and Pfiesteria-like dinoflagellate zoospores to nutrient enrichment in the Neuse River Estuary, North Carolina, USA, *Mar. Ecol. Prog. Ser.*, 192, 65–78, 2000.
16. Dykstra, M.J., L.J., Noga, E.J. and Moye, D.W., Characterization of the *Aphanomyces* species involved with ulcerative mycosis (UM) in Menhaden, *Mycologia*, 78 (4), 664–672, 1986.
17. Noga, E., Skin ulcers in fish: *Pfiesteria* and other etiologies, *Toxicol. Pathol.*, 28 (6), 807–823, 2000.
18. Law, J., Differential diagnosis of ulcerative lesions in fish, *Environ. Health Perspect.*, 109 (Suppl. 5), 681–686, 2001.
19. Pearl, H.W., P.J., Fearm J.M. and Peierls, B.L., Ecosystem responses to internal and watershed organic matter loading: consequences for hypoxia in the eutrophying Neuse River estuary, North Carolina, USA, *Mar. Ecol. Prog. Ser.*, 166, 17–25, 1998.
20. Kelly, S.A., Havrilla, C.M. and Brady, T.C., Oxidative stress in toxicology: established mammalian and emerging piscine model systems, *Environ. Health Perspect.*, 106 (7), 375–384, 1998.
21. Law, J., Issues related to the use of fish models in toxicology pathology: session introduction, *Toxicol. Pathol.*, 31 (Suppl.), 49–52, 2003.
22. K.G., Gas exchange, in *The Physiology of Fishes*, Evans, D., Ed., CRC Press, Boca Raton, FL, 1998, pp. 101–128.
23. M, N., Oxygen-dependant cellular functions — why fishes and their aquatic environment are a prime choice of study, *Comp. Biochem. Phys. A*, 133 (1), 1–16, 2002.
24. Moriwaki, Y., Y.T. and Higashino, K., Enzymes involved in purine metabolism — a review of histochemical localization and functional implications, *Histol. Histopathol.*, 14 (4), 1321–1340, 1999.
25. Levine, J.F., H.J., Dykstra, M.J., Noga, E.J., Moye, D.W. and Cone, R.S., Species distribution of ulcerative lesions on finfish in the Tar-Pamlico River Estuary, North Carolina, *Dis. Aquat. Organ.*, 8, 1–5, 1990.
26. Kurtz, J.C., J.L. and Fisher, W.S., Strategies for evaluating indicators based on guidelines from the Environmental Protection Agency's Office of Research and Development, *Ecol. Ind.*, 1, 49–60, 2001.
27. Noga, E.J., W.J., Levine, J.F., Dykstra, M.J. and Hawkins, J.I I., Dermatological diseases affecting fishes of the Tar-Pamlico Estuary, North Carolina, *Dis. Aquat. Organ.*, 10, 87–92, 1991.
28. Paerl, H.W., P.J., Fear, J.M. and Peierls, B.J., Fish kills and bottom-water hypoxia in the Neuse River and estuary: reply to Burkholder et al., *Mar. Ecol. Prog. Ser.*, 186, 307–309, 1999.
29. Hurty, C.A., B.D., Law, J.M., Sakamoto, K. and Lewbart, G.A., Evaluation of the tissue reactions in the skin and body wall of koi (*Cyprinus carpio*) to five suture materials, *Vet. Record.*, 151, 324–328, 2002.
30. Wannamaker, C.R., JA, Effects of hypoxia on movements and behavior of selected estuarine organisms from the southeastern United States, *J. Exp. Mar. Biol. Ecol.*, 249, 145–163, 2000.
31. Cotran, R.S., V.K., Collins, T. and Robbins, S.L., *Pathologic Basis of Disease*, 6th ed., Robbins, S.L., Ed., Saunders, Philadelphia, 1999.

chapter thirty-two

Fish models in behavioral toxicology: Automated techniques, updates and perspectives*

Andrew S. Kane and James D. Salierno
University of Maryland
Sandra K. Brewer
U.S. Army Corps of Engineers

Contents

Introduction

Since the science of toxicology began thousands of years ago, behavioral endpoints have been used to study the effects of chemicals and drugs on humans and other mammals. In aquatic toxicology, however, the nexus of behavioral sciences with the study of toxicants

* The views expressed herein are of the authors and do not necessarily reflect those of the Federal Government. No endorsement by any Agency of the Federal Government is intended or inferred, including conclusions drawn or use of trade names.

1-56670-664-5/05/$0.00+$1.50
© 2005 by CRC Press

has only become prominent within the last five decades. Behavioral endpoints have been slow to be integrated in aquatic toxicology because, until recently, there was a poor understanding of how alterations in behavior may be related to ecologically relevant issues, such as predation avoidance, prey capture, growth, stress resistance, reproduction, and longevity. Further, the ability to achieve repeatable, quantifiable data from a large number of animals or exposures has been challenging. Recent improvements in computer and video automation have made possible significant progress in the ease, utility, and affordability of obtaining, interpreting, and applying behavioral endpoints in a variety of applications from water quality monitoring to use in toxicity identification evaluation (TIE).[1–9] Consequently, behavioral endpoints in aquatic toxicology are shifting from being met with skepticism by investigators to being received with greater enthusiasm.

One of the first comprehensive reviews on aquatic behavioral toxicology was published by Rand.[10] Over the past 20 years, the field of behavioral toxicology has grown, in part, because of increased interest in the number of species used, endpoints measured, and methods to collect and interpret data. Numerous reviews have traced these advancements in the field of behavioral toxicology.[11–15] However, the recognition of behavioral toxicology as an important tool in aquatic toxicology is most clearly seen in the acceptance of behavioral endpoints in Federal regulations. In 1986, the U.S. government accepted avoidance behavior as legal evidence of injury for Natural Resource Damage Assessments under Proceedings of the Comprehensive Environmental Response, Compensation, and Liability Act of 1980.[16]

This chapter provides updated information with an added perspective on automated systems that evaluate quantitative behavioral endpoints. The chapter also provides a compilation of reference material relating to specific and nonspecific observations of behavioral alterations in fish that will be of use to both experienced and new practitioners of behavioral toxicology. The reader is reminded that much work has also been done with aquatic invertebrates,[17] including crabs,[18,19] daphnids,[20] clams,[21,22] and other animals; however, the volume of such work precludes its inclusion in this chapter.

Why study behavior?

Behavior provides a unique perspective linking the physiology and ecology of an organism and its environment.[15] Behavior is both a sequence of quantifiable actions, operating through the central and peripheral nervous systems,[23] and the cumulative manifestation of genetic, biochemical, and physiologic processes essential to life, such as feeding, reproduction, and predator avoidance. Behavior allows an organism to adjust to external and internal stimuli in order to best meet the challenge of surviving in a changing environment. Conversely, behavior is also the result of adaptations to environmental variables. Thus, behavior is a selective response that is constantly adapting through direct interaction with physical, chemical, social, and physiological aspects of the environment. Selective evolutionary processes have conserved stable behavioral patterns in concert with morphologic and physiologic adaptations. This stability provides the best opportunity for survival and reproductive success by enabling organisms to efficiently exploit resources and define suitable habitats.[15]

Since behavior is not a random process, but rather a highly structured and predictable sequence of activities designed to ensure maximal fitness and survival (i.e., success) of the individual (and species), behavioral endpoints serve as valuable tools to discern and evaluate effects of exposure to environmental stressors. Behavioral endpoints that integrate endogenous and exogenous factors can link biochemical and physiological processes, thus providing insights into individual- and community-level effects of environmental contamination.[24,25] Most importantly, alterations in behavior represent an integrated, whole-organism response. These altered responses, in turn, may be associated with reduced fitness and survival, resulting in adverse consequences at the population level.[26]

Rand[10] stated that behavioral responses most useful in toxicology should be: (1) well-defined endpoints that are practical to measure; (2) well understood relative to environmental factors that cause variation in the response; (3) sensitive to a range of contaminants and adaptable to different species; (4) ecologically relevant. To this list, we add endpoints that should ideally: (5) elucidate different modes of action or chemical classes; (6) be able to ''stand alone'' and be easily incorporated into a suite of assessments; (7) be simple to automate in order to maximize their utility for a broad range of applications; (8) have representation across species (e.g., reproduction, food acquisition) in order to facilitate investigations into the phylogeny and ontogeny of behavior; (9) include a suite of endpoints that focus on innate behavior of sentinel organisms that can be altered in association with stress exposure; and (10) help delineate ecosystem status, i.e., health. Although each of these considerations has merit, often the application of a specific endpoint, or suite of endpoints, is based on the ability to functionally discern exposure-related alterations, using available techniques, with the most appropriate sentinel species.

The application of behavioral endpoints in any toxicity study must also be based on the stressor(s) to be evaluated. Basic knowledge of the compound/toxicant/stressor of interest is necessary. Stress agents of interest should (a) be ''behaviorally toxic,'' (b) have a route of uptake for the aquatic species in question, and (c) structurally resemble a behavioral or neurotoxicant or one of its active metabolites. Of course, toxicants that may be a behavioral or CNS toxicant to mammals may not have similar affects on aquatic animals, and *vice versa*.[10]

The mechanism of action, route of uptake, and behavior of the compound of interest in the aquatic environment must all be understood. In adult fish, gill and gut epithelia are major routes of uptake; however, physiological differences of different life stages need to be taken into consideration. Larval fish utilize skin as a respiratory interface and may uptake more compounds than adults of the same species that utilize gills for respiration. In contrast, for compounds that directly target the gill lamellae, adult fish will have increased sensitivity compared with larvae due to the increased surface area of the gill surface.

Fish models in behavioral testing

To date, there are no standardized species or groups of species used for aquatic behavioral toxicology testing. Different species often have different behavioral and physiological responses to stress and toxicant exposure. Therefore, preliminary observations and assays are required in order to determine the feasibility of a particular species, and if aberrant behavioral patterns can be associated with specific exposure scenarios. It is not unreasonable for preliminary testing with a novel species to take months, and in some cases, years, in order to develop biologically relevant endpoints of exposure.

Fish are ideal sentinels for behavioral assays of various stressors and toxic chemical exposure due to their: (1) constant, direct contact with the aquatic environment where chemical exposure occurs over the entire body surface; (2) ecological relevance in many natural systems[27]; (3) ease of culture; (4) ability to come into reproductive readiness[28]; (5) long history of use in behavioral toxicology. Alterations in fish behavior, particularly in nonmigratory species, can also provide important indices for ecosystem assessment.

Ideally, test organisms should have the following characteristics: (1) high ecological relevance; (2) susceptibility to the stressor(s) in question, both in the field and in the laboratory; (3) have wide geographical distributions; (4) be easy to culture and maintain under laboratory conditions; (5) have relatively high reproductive rates and, should have relatively early maturation and easy fertilization in order to produce sufficient numbers of organisms of the proper age and size for testing; (6) have environmental relevance to the potential exposure (have been exposed to the test contaminant in the wild); (7) have the ability to yield reproducible data under controlled laboratory conditions.

Once the model test species is determined, exposure-related behavioral alterations can be distinguished as fixed action patterns (FAPs) or through specific behaviors discerned from a species' ethogram. FAPs are specific, innate behavioral sequences initiated by specific stimuli that are not a result of gene–environment interactions.[29] These "hard-wired" behaviors are typically under genetic control, may be species specific, go to completion upon initiation, are not regulated through feedback loops, and are not reflexes but complex, coordinated behaviors.[14] As such, alteration(s) of FAPs are good endpoints to include in a suite of behavioral tests.

Quantifiable behavioral changes in chemically exposed fish provide novel information that cannot be gained from traditional toxicological methods, including short-term and sublethal exposure effects, mechanism of effect, interaction with environmental variables, and the potential for mortality.[12,28,30–33] Ecologically relevant behaviors affected by sublethal concentrations include: altered vigilance, startle response, schooling, feeding, prey conspicuousness, migration, and diurnal rhythmic behaviors.[12,34] Changes in behavior may also alter juvenile recruitment, thereby disrupting population demography and community dynamics over time.[26]

Researchers wishing to develop a new fish model for behavioral toxicology must consider the life history and ecology of the species. For example, herring form tight schools in nature, yet if kept solitary in the laboratory, will die after a few days.[35] Other clupeids, such as menhaden (*Brevoortia tyrannus*), can be laboratory-maintained over long periods of time (months to years) in large holding tanks, but when transferred to behavior/exposure arenas, become highly stressed and succumb within 72 h due to sepsis prior to any toxicant/stressor exposure. These examples demonstrate the importance of having insight into the sentinel species' niche, i.e., habitat, diet, foraging strategies, reproductive strategies, home range, and social structures (school versus shoal versus solitary swimmers). These factors will help discern (1) which types of behavior are important for study, (2) appropriate experimental design and exposure parameters, and (3) the ecological importance of the behavior on the life history of the fish. Channel catfish (*Ictalurus punctatus*), brown bullhead

(*Ictalurus nebulosis*), and striped bass (*Morone saxatilus*), for example, may at times share similar geographical habitats but differ greatly in life history characteristics (diet, position in the water column, migration, and social structures). These differences directly translate into differential exposure to chemicals in the environment. Of course, there may also be vast physiological and biochemical differences between species in the metabolism, tissue distribution, and elimination profiles, all of which can alter exposure concentrations at target tissues.

Different species of fish will have different suites of behaviors and adaptive behavior patterns, i.e., responses to stimuli. These patterns may also vary widely under different holding and exposure conditions within individuals of the same species. Therefore, it is critical to carefully document observations of normal baseline behavior under controlled conditions prior to behavioral testing with a chemical or other stress agent. Further, it is important to recognize changes in behavior that are not only associated with controlled, laboratory stress exposure, but also with sub-optimal health. Table 32.1 provides a consolidation of qualitative comments from the literature that indicate behavioral changes associated with a broad variety of different scenarios. This table illustrates the often-nonspecific nature of many different behavioral alterations associated with disease agents, biologicals, sub-optimal water quality, and

Table 32.1 Behavioral alterations observed in fish associated with different stress agents

Host species	Stressor	Behavior/movement comments	References
	Biologicals		
Salmonids	*Aeromonas salmonicida*	Lethargy, inappetence, loss of orientation, abnormal swimming behavior	(37, 38)
Fish	*Bacillus* spp.	Weakness, lethargy	(38)
Cold freshwater fish	Bacterial gill disease	Lethargy, flared opercula, coughing, dyspnea	(39)
Fish	Bacterial gill disease	Lethargy, anorexia, increased respiration	(40, 41)
Fish	Blood flukes	Lethargy, flashing	(39)
Bluegill	40 ppb brevetoxin	Altered ventilatory responses; reversible after 1 h	(42)
Channel catfish	Channel catfish virus disease	Hanging head up in the water, disorientation, corkscrew swimming	(39)
Fish	*Chryseobacterium scopthalmum*	Lethargy	(43, 44)
Coho salmon and farmed trout	*Clostridium botulinum*	Sluggishness, erratic swimming, listlessness, may alternately float and sink before showing temporary rejuvenation	(45)
Warm and cold freshwater fish, cold water marine fish	Columnaris	Dyspnea	(39)
Striped bass	*Corynebacterium aquaticum*	Inappetence, swimming more slowly	(38)

Table 32.1 Continued

Host species	Stressor	Behavior/movement comments	References
Deep angelfish	Deep angelfish disease (herpes-like)	Loss of equilibrium, headstanding	(46)
Catfish	*Edwardsiella ictaluri*	Fish hang listlessly at surface in a head-up-tail down posture, sometimes swimming rapidly in circles; corkscrew spiral swimming, depression	(39, 47)
Lake trout, lake trout × brook trout	Epizootic epitheliotropic disease (herpes virus-like)	Sporadic flashing, corkscrew swimming	(48)
Striped mullet	*Eubacterium tarantellus*	Erratic swimming, loss of equilibrium, spiral swimming; floating at surface and sinking to the bottom repeatedly	(49)
Salmonids	*Exophiala salmonis*	Erratic swimming	(50, 51)
Farmed barramundi	*Flavobacterium johnsoniae*	Listlessness, anorexia	(52)
Coho salmon	*Flexibacter psychrophilus*	Nervous spinning	(53)
Fish	Gill *Cryptobia* infestation	Anorexia	(39)
Cherry salmon	*Hafnia alvei*	Slow swimming	(54)
Salmonids	Infectious hematopoietic necrosis virus	Lethargy, sporadic hyperactivity	(39)
Salmonids	Infectious pancreatic necrosis virus	Corkscrew spiral swimming, whirling	(39)
Fish	*Lactococcus gravieae*	Moribund fish swim erratically just below the surface of the water	(38)
Fish	Motile aeromonads	May swim normally or hang in the water, on their sides	(55)
Fish	Mycobacteria	Listlessness, anorexia, dyspnea, inappetence	(38, 56–59)
Mainly salmonids	*Myxobolus cerebralis*	Whirling or frenzied, tail-chasing behavior, impaired balance	(60)
Fish	*Nocardia*	Inactivity, anorexia	(61)
Bluegill	*Pfiesteria piscicida* (laboratory exposure to non-axenic cultures)	Decreased aggression and social interactions, followed by solitary time spent on or near the bottom. Subsequent sporadic bursts of activity, including tailstanding, bobbing, corkscrewing in place, breaking at the surface, followed by inactivity at a 45° angle in water column, or resting on the bottom prior to morbidity. Strong elevations in cough rate and % movement without notable changes in respiratory rate	(36, 42)
Fish	*Pfiesteria piscicida* cultures	Pigmentation changes, lethargy, episodic hyperactivity and decreased respiration	(62–65)

Table 32.1 Continued

Host species	Stressor	Behavior/movement comments	References
Fish	*Plesiomonas shigelloides*	Inappetence	(66)
Fish	*Piscirickettsia salmonis*	Gathering at the surface of cages, sluggishness, inappetence	(38)
Fish	Protozoan ectoparasites	Dyspnea	(39)
Rio Grande cichlid; zilli cichlid	Rio Grande cichlid rhabdovirus disease	Lethargy	(67)
Salmonids	Salmonid rickettsial septicaemia	Lethargy, swimming near the surface or at the side of the net	(68)
Rabbitfish	*Shewanella putrefaciens*	Lethargy	(69)
Carps; sheatfish; guppy; Northern pike	Spring viremia of carp (*Rhabdovirus carpio*)	Decreased swimming ability	(70)
Fish	*Staphylococcus aureus*	Lethargy	(38)
Fish	*Streptococcus*	Erratic swimming	(71)
Tilapia	*Streptococcus difficilis*	Lethargy, erratic swimming, showing signs of dorsal rigidity	(38)
Farmed Atlantic salmon	Unidentified Gram-negative rod	Lethargy, swimming close to surface, loss of balance	(38)
Warm marine fish	Uronemosis	Dyspnea, hyperactivity, then lethargy	(39)
Rainbow trout	*Vagococcus salmoninarum*	Listless behavior, impaired swimming	(72)
Fish	*Vibrio alginolyticus*	Sluggishness	(73, 74)
Fish	*Vibrio anguillarum*	Anorexia, inactivity	(38, 75)
Sharks	*Vibrio harveyi*	Lethargy, stopped swimming, appearing disorientated	(38)
Fish	*Vibrio salmonicida*	Inappetence, disorganized swimming	(75)
Japanese horse mackerel	*Vibrio trachuri*	Erratic swimming	(38)
Cold freshwater fish (mainly salmonids)	Viral hemorrhagic septicemia	Lethargy, congregating away from the current on the edges of the pond or raceway, looping swimming behavior, darting through the water and spiraling at the bottom of the pond	(39)
Atlantic salmon	*Yersinia intermedia*	Lazy movements, congregating at the surface of the water	(76)
Rainbow trout	*Yersinia ruckeri*	Sluggishness	(77)
	Contaminants		
Rainbow trout	Al	Avoidance behavior(s)	(78)
Rainbow trout	Al	2 minute clips, position holding, slow and burst type swimming	(79)
Atlantic salmon, rainbow trout	Cu and Zn (salmon); zinc sulfate (trout)	Avoidance behavior(s); 53 ppb (salmon); 5.6 ppb (trout)	(80)
Three-spined sticklebacks	BBP (butyl benzyl phthalate)	Shoal choice	(81)
Rainbow trout	Carbaryl	Velocity, school size NNA at 1 fps for 1 min (60 frames)	(82)
Rainbow trout	Carbaryl	Swimming capacity, feeding activity, strikes	(12)
Fathead minnow	Cd	Decreased predator avoidance, 25–375 ppb	(83)

Table 32.1 Continued

Host species	Stressor	Behavior/movement comments	References
Rainbow trout	Cd	Altered dominance, feeding and aggression	(84)
Zebra fish	Cd	Avoidance behavior(s)	(85)
Bluegill	Cd, Cr, Zn	Hyperactivity	(86)
Green sunfish	Chlordane	Avoidance behavior(s) 20, 10, 5 mg l^{-1}	(87)
Rainbow trout	Co	Dominance hierarchy, growth, food intake and coloration	(88)
Rainbow trout	Copper sulfate, dalapon, acrolein, dimethylamine salt of 2,4 D, xylene	Avoidance behavior(s)	(89)
Pink salmon	Crude oil	Avoidance behavior(s) 1.6 mg l^{-1}	(90)
Estuarine fish	Cu	8 ppb olfactory disruption	(91)
Goldfish	Cu	Velocity, TDT, turning angles	(92)
Rainbow trout	Cu	Attraction 460–470 ppb, avoidance 70 ppb	(93)
Salmon	Cu	Altered chemoreception and home stream recognition	(94)
Rainbow trout	Cu and Ni	Attraction 390 ppb (Cu), 6 ppb (Ni), avoidance 4.4 ppb (Cu), 24 ppb (Ni)	(95)
Rainbow trout	Cu, Co	Avoidance behavior(s)	(96)
Atlantic salmon	DDT	Alteration in temperature preference, 5–50 ppb	(97)
Bluegill	DDT	Hyperactivity	(98)
Brook trout	DDT	Biphasic concentration–response relationship for temperature preference and DDT	(99)
Croaker	DDT	Effects on the F_1 generation vibratory/visual stimuli, burst speed	(100)
Goldfish	DDT	Increases in velocity, turns and area occupied	(82)
Brook trout	DDT and methoxychlor analogs	Alteration in temperature preference for methoxychlor analogs	(101)
Mosquito fish	Endrin, toxaphene, parathion	Avoidance behavior(s)	(102)
Mummichog	Environmental MetHg	Prey capture (strikes and captures), predator avoidance, lab and field validation	(103)
Rainbow trout	Heavy metal mix	Avoidance behavior(s)	(104)
Chinook salmon	Kraft mill extract	Avoidance behavior(s) 2.5–10%	(105)
Smelt	Kraft mill extract	Avoidance behavior(s) 0.5%	(106)
Three-spine stickleback	Lead nitrate	Attraction at high concentration, avoidance at low concentration	(107)
Shiners	Malathion	Concentration-dependent decrease in temperature selection	(108)
Medaka	OPs	Vertical path analysis in 1-min clips, velocity, meandering, TDT, smooth versus erratic swimming, 4 fps	(109)

Table 32.1 Continued

Host species	Stressor	Behavior/movement comments	References
Mosquito fish	OPs	Avoidance behavior(s)	(102)
Goldfish	Parathion	Hypoactivity and alteration in angular change	(110)
Mummichog	Pb	Feeding activity and performance, predator avoidance	(111)
Rainbow trout fingerlings	Phenol	Decrease in predator avoidance to adults, 0.5–18 mg l^{-1}	(112)
Minnow	Phenol and *p*-chlorophenol	Avoidance behavior(s)	(113)
Herring	Pulp mill extract	Avoidance behavior(s)	(114)
Rainbow trout	Rotenone	Avoidance behavior(s)	(115)
Roach	2,4,6 trinitrophenol	Attraction	(116)
Three-spined stickleback	Zinc and copper sulfate	Avoidance, "stupefied and motionless"	(107)
	Water quality		
Brook trout	Acidification	Avoidance behavior(s)	(117)
Trout	Acidification	Female reproductive behavior, nest digging	(118)
Carp	Ammonia	Center of gravity of a group of fish/vertical location	(119)
Largemouth bass and mosquito fish	Ammonia	Decreased prey consumption of bass, less effect on mosquito fish	(120)
Fish	Ammonia poisoning	Hyperexcitability, fish often stop feeding	(121)
Fish	Chlorine poisoning	Dyspnea	(39)
Fish	Environmental hypoxia	Fish piping for air, gathering at water inflow, depression	(122, 123
Cold freshwater fish	Hypercarbia	Dyspnea	(39)
Bluegill	Hypoxia	Altered ventilatory and cough responses	(42)
Mullet, menhaden, spot, croaker, pinfish, and mummichog	Hypoxia	Avoidance behavior(s)	(124)
3-spine sticklebacks, minnows, and brown trout	Hypoxia and temperature change	Avoidance behavior(s)	(125)
Fish	Nitrite poisoning	Lethargy, congregating near the water surface	(39)
Fish	Low pH	Hyperactivity, dyspnea, tremors	(126)
Fish	Total suspended solids	Coughing to clear gills	(39)
Fish	Low temperature stress	Inactivity, depression	(39)

contaminants. Further, these collective references suggest the need to provide quantitative data when reporting behavioral alterations to facilitate comparison with other studies or observations.

The requirement to carefully document baseline "normal" behavior should be viewed as strength of behavioral testing. Traditional (LC_{50}) tests do not require stringent documentation of baseline behavior, other than visual observations of whether the test subjects were "healthy," and verification that a minimal amount of mortality (i.e., $\leq$10%)

occurs in the control and treatment group(s). Behavioral toxicology testing, however, allows the control group to subsequently be exposed, if careful documentation on baseline behavior is made. The statistical power of behavioral tests can be greatly improved by using repeated measures analyses, using each animal as its own control. This type of analysis greatly reduces the inherent variability between individuals.

Descriptive behavioral alterations

Behavioral assays provide biologically relevant endpoints to evaluate sublethal exposure effects and may compliment traditional toxicity testing. In order to evaluate behavioral endpoints, specific descriptive observations regarding behavioral alterations in response to low-level stress (deviation from baseline) need to be demonstrated. The degree of alteration that can be experimentally meaningful is typically based on the ability to statistically discriminate differences between treatment and control groups. However, it should be noted that a common misuse of statistics is to find differences between treatment groups solely due to low p-values without empirically observable, exposure-related changes. Factors that can influence p-values include, in addition to sound experimental design, use of proper controls (negative, solvent, and positive), time frame of observation(s), sample size, response precision within treatment groups, and reproducibility between experiments (refer also to ''Preliminary studies'' section, page 578).

Ultimately, the question addressed in behavioral toxicology is: How do alterations in a species' behavior, resulting from sublethal stress exposure, alter individual fitness and have a biologically relevant effect? Even when biologically and statistically significant data are derived from an exposure study, it is typically difficult to extrapolate alterations observed under controlled laboratory conditions to ecologically relevant field scenarios. The answer to bridging this gap lies in the appropriate selection of behavioral endpoints and adding a similarly controlled comparison with the same species under more naturally complex or field exposure conditions (e.g., using predator–prey interactions and avoidance–attractance responses).[26,127,128]

Individual movement and swimming patterns

Avoidance and attractance

When a contaminant triggers a stimulus response, the resulting behavioral reaction (avoidance or attraction) significantly regulates exposure duration of the organism. If the contaminant is perceived by the fish as noxious, the fish responds by avoiding the area containing the chemical. In contrast, if the contaminant triggers an attractance response, the fish will stay in the area, thus increasing exposure duration. Avoidance–attractance responses depend on: (1) the substance activating the receptor; (2) sufficient exposure history of the species to evolve adaptive responses or sufficient experience by the organism to acquire a response to the stimulation; (3) sufficient directional information from the chemical concentration gradient to orient in the proper direction from the chemical plume.[15]

Avoidance and attractance behavior in fish has proven to be an easy and realistic behavioral endpoint of exposure because many contaminants induce avoidance or attractance behavior. The utility of avoidance behavior as an indicator of sublethal toxic exposure has been demonstrated over the past 50 years, and chemically induced avoidance or attractance may significantly alter the distribution and migration patterns of individuals and groups of fish.[129]

Gray[11] demonstrated the avoidance of oil-contaminated water and gas-supersaturated water by free-ranging fish in the field. Avoidance of heavy metals (e.g., cadmium, copper, cobalt, and aluminum) by a variety of freshwater fish has been documented at low environmentally relevant concentrations.[78,85,96,104,130] In addition, fish can actively avoid fluctuations in water quality conditions, such as hypoxia, temperature, acidification, and ammonia.[117,119,124,125] Fish also maintain the ability to avoid anthropogenic compounds released into the environment, including certain pesticides and rotenone.[102,115]

Through the use of elaborate experimental designs, coupled with qualitative and quantitative measures of behavior, avoidance behavior can provide an endpoint that directly correlates to the field. Similar avoidance responses were observed in laboratory and field tests with fathead minnows (*Pimephales promelas*) when exposed to metals characteristic of the Coeur d'Alene River (Idaho, ID) downstream from a large mining extraction operation. Telemetry studies conducted at the confluence of the river with an uncontaminated tributary of the river revealed a similar avoidance of the contaminated water within the concentration range that induced avoidance responses in laboratory studies.[128] Hartwell et al.[127] conducted integrated laboratory and field studies of avoidance and demonstrated that fathead minnows avoided a blend of heavy metals (copper, chromium, arsenic, and selenium) that are typical of effluent from fly ash settling basins of coal-burning electrical plants. Fish avoided a 73.5 $\mu g\ l^{-1}$ mixture of these metals in a natural stream and 34.3 $\mu g\ l^{-1}$ in an artificial stream.

It may appear at first glance that most studies demonstrate avoidance behavior when fish are exposed to contaminants. However, it is difficult to generalize about the avoidance of aquatic contaminants by fish because of the variety of species and experimental designs used to test behavioral responses, as well as variations in the modes and sites of action of the chemicals studied.[95] Beitinger[131] reviewed the published literature on avoidance for over 75 different chemicals. Roughly one-third of the chemicals were avoided, whereas the others either failed to elicit a response or induced inconsistent responses. Many contaminants may cause avoidance reactions but some may attract aquatic organisms: these include detergents,[132] some metals,[30,93,133] and petroleum hydrocarbons.[134,135] Also, different species may have different avoidance responses. Largemouth bass (*Micropterus salmoides*) have been shown to be insensitive to 50 $\mu g\ l^{-1}$ copper sulfate, whereas goldfish (*Carassius auratus*) and channel catfish were attracted to this concentration.[133] Part of the explanation for these apparent conflicting results may be the contaminant-induced alterations of chemosensory systems. Hansen et al.[96,130] found simultaneous alterations in chemosensory-mediated behavior, in the physiologic responsiveness of the olfactory system of chinook salmon (*Oncorhynchus tshawytscha*) and rainbow trout (*O. mykiss*) to L-serine, and evidence of damage to the olfactory tissue responsible for mucosa production and olfactory receptor cells. McNicol and Scherer[136] determined that whitefish (*Coregonus clupeaformis*) avoided cadmium concentrations of 1 $\mu g\ l^{-1}$ and less, and also avoided cadmium at 8 $\mu g\ l^{-1}$ and greater, but showed little response to concentrations between this range.

In recent physiological and behavioral studies (McNichol and Hara, personal communication), electro-olfactorygram (EOG) responses to the lower concentrations were shown to be mediated by the olfactory system. The olfactory system apparently became injured at cadmium concentrations greater than 1 $\mu g\ l^{-1}$, when avoidance responses ceased to occur and EOG responsiveness to L-serine were abolished. The renewed avoidance response to 8 $\mu g\ l^{-1}$ was likely induced by generalized irritation.

Likewise, Hansen et al.[96,130] found that even brief exposure of chinook salmon and rainbow trout to copper (25 μg l^{-1}) was associated with a significant reduction in EOG responses that recovered over several days; however, exposure to higher concentrations (44 μg l^{-1} chinook salmon; 180 μg l^{-1} rainbow trout) abolished behavioral responses. Furthermore, physiologic recording revealed these higher concentrations diminished both the EOG responses recorded on the mucosa and electro-encephalogram (EEG) responses to L-serine recorded from the olfactory tract. Necrosis and reduced density of olfactory receptors were evident injuries to the olfactory epithelium. These studies highlight the versatility and importance of integrating behavioral endpoints into a suite of toxicological studies that include relevant physiological and pathological endpoints helped to elucidate mechanism of action. We refer the reader to the many thorough reviews of the well-studied endpoint of avoidance–attractance behavior.[15,131,137,138]

Swimming patterns

Avoidance behavior is an amalgam of many behaviors that may culminate in a single endpoint, whereas movement analysis is a finer scale technique investigating the components of movement. Neurotoxicity is frequently observed in changes in form, frequency, or posture of swimming movements, with changes often occurring much earlier than mortality.[12,15] Sublethal metal and pesticide exposures have demonstrated alterations in swimming behaviors and serve as models for additional stressors.[79,92,109,139] When bluegill received pulsed doses of the pyrethroid insecticide ES-fenvalerate (0.025 μg l^{-1}), the first indication of toxicity was caudal fin tremors as fish initiated movement.[140] Exposure of rainbow trout to sublethal concentrations of 40 μg l^{-1} malathion resulted in convulsive movements.[139] A review by Little and Finger[12] revealed that the lowest behaviorally effective toxicant concentration that induced changes in swimming behavior of fish ranged from 0.1% to 5.0% of the LC_{50}. When observations were made over time, behavioral changes commonly occurred 75% earlier than the onset of mortality. Development of locomotory responses, frequency of swimming movements, and duration of activity were significantly inhibited before effects on survival or growth were observed in brook trout (*Salvelinus fontinalis*) alevins exposed to aluminum concentrations (300 mg l^{-1}) under acidic conditions ($pH \leq 6.1$).[141]

Movement analysis of individuals and groups of fish continues to be refined as computer technology advances. Swimming responses have been used in automated biomonitoring systems because of their consistent sensitivity to numerous contaminants.[142,143] Studies of fish movement typically involve videography and quantification of movement parameters. Movement endpoints are designed to discern alterations in general swimming patterns in response to stressor exposure. Behavioral endpoints quantified through movement analysis typically include total distance traveled, velocity, acceleration, turning angles and frequency, time spent swimming, as well as horizontal and vertical distributions of individuals.

The measurements of swimming behavior are usually limited to the laboratory. Assessment of fish at contaminated field sites is currently not possible as species-typical responses have not been defined to permit the evaluation of behavioral function except for the most extreme aberrations.[15] In the laboratory, subtle changes that arise from exposure can be confirmed through comparisons with controls or with responses observed during a preexposure period.

Intra- and interspecific interactions

For hazard assessment and environmental regulation, it is important to show a causal linkage with the population in order to provide a predictive index of population-level effects. Recently, behavioral toxicology has focused more on complex behaviors, such as prey capture, predator avoidance, and courtship, and mating. These behaviors maintain high environmental relevance and direct fitness consequences to the individual. Hypoactivity and hyperactivity, as well as deviations in adaptive diurnal rhythmicity, may disrupt feeding and increase vulnerability to predation.[144,145] However, studies on these more complex behaviors are multifaceted and difficult to conduct depending on the amount of ecological realism the researcher wishes to achieve, but the experimental design can also be readily adapted to standard toxicity testing procedures.[146] The advantages, however, can be great and ecosystem effects of toxicant and stressor exposure may be more readily implicated. For example, fish exposed to heavy metals (cobalt, lead, and cadmium) displayed alterations in dominance, feeding behavior, growth, and predator avoidance.[84,111] Faulk et al.[100] demonstrated that the F_1 generation of fish exposed to DDT had deficits in their response to vibratory and visual stimuli, as well as altered swimming behavior. Feeding and prey vulnerability have been used to examine sublethal contamination because predator and prey may be differentially affected by toxicants.[147] Exposure to environmental mercury resulted in alterations in foraging (prey strikes and captures), as well as predator avoidance.[103] Female reproductive behavior and nest digging were found to be disturbed upon exposure to increasing levels of acidification.[118] Alterations in these behaviors can have serious effects to the individual and population of fish exposed and may induce changes in gene flow and demography.[148,149] To date, there are no well-characterized examples of automated systems to detect and evaluate predator–prey interactions. Certainly, this is not to say that the technology is not currently attainable.

Respiratory patterns

Respiration is a rhythmic neuromuscular sequence regulated by an endogenous biofeedback loop, as well as by external environmental stimuli. Acute contaminant exposure can induce reflexive cough and gill purge responses to clear the opercular chamber of the irritant, and can also increase rate and amplitude of the respiratory cycle as the fish adjusts the volume of water in the respiratory stream. As exposure continues, the respiration cycle can become irregular, largely through decreased input, as well as alterations in the endogenous pacemaker. Diamond et al.[3] found that the frequency and amplitude of bluegill opercular rhythms and cough responses were altered following exposure to different contaminants. For example, dieldrin, an organochlorine insecticide, increased ventilatory frequency at concentrations above 24 $\mu g\ l^{-1}$ and caused cough responses and erratic movements. In contrast, zinc at 300 $\mu g\ l^{-1}$ reduced the amplitude of the respiratory response.

A variety of biomonitoring systems have been developed to assess changes in respiratory rate relative to stress exposure. These systems have the great advantage of sensitivity since many waterborne stressors, even at low environmental concentrations, affect gill tissue and respiratory function. Respiratory frequency, depth (volume), and cough frequency can be measured, noninvasively, using physiological signals from restrained sentinel fish. One such system to accomplish this with good repeatability has been described by Shedd et al.[150] Briefly, small flow-though exposure vessels house individual small fish in their respective chambers. Electrical signals generated by the respiratory and body movements of individual fish are detected by electrodes suspended above and below each fish. The signals are amplified, filtered, and analyzed using various algorithms on a personal computer. The muscular electrical output (0.05–1 mV) from each fish is independently amplified by a high-gain, true differential-input, instrumentation amplifier by a factor of 1000. Signal interference by frequencies above 10 Hz is attenuated by low-pass filters. The ventilatory parameters monitored by the computer include ventilatory rate, ventilatory depth (mean signal height), and gill purge (cough) rate.

Since fish are poikilotherms, temperature may also play an important role in determining exposure effects on a given fish species. Efforts at the UM Aquatic Pathobiology Laboratory to validate a respiratory response system, as described above, have recently demonstrated temperature-dependent differences in bluegill (*Lepomis macrochirus*) exposed to brevetoxin at 19°C versus 25°C. Interestingly, respiratory responses (increased ventilatory, cough and "other movement") were altered at 25°C but not at 19°C (Figure 32.1).

As with other behavioral systems, it is essential to properly integrate responses over time in order to achieve a good signal-to-noise ratio. The accuracy of any computer ventilatory parameter can be established by comparing the computer-generated values with concurrent strip chart recorder tracings.[150] Biomonitoring systems that measure fish ventilatory patterns have further application as early warning signals of water quality changes and toxicity.[9]

Social behavior and group dynamics

Toxicology studies typically focus on the exposure of single fish in the laboratory, when in reality, many fishes tend to congregate in groups and interact with many components of their environment. Group living is a basic life history characteristic of many fishes, with 25% of all species forming schools or shoals during their life, and 50% during larval and juvenile stages.[35,151] Pavlov and Kasumyan[151] define a fish school as having all individuals oriented in the same direction, situated at a certain distance from each other, and unitary in all movements (polarized). Shoaling, in contrast, is a simple, spatial aggregation of fish attracted by a stimulus occurring independently of each other with no mutual attraction between individuals (nonpolarized). Schooling and shoaling behaviors are complex social behaviors utilized by a wide diversity of fish species to increase individual fitness and propagate their genes in the population[152] by providing defense from predation, while increasing reproductive, foraging, and migration efficiencies.

These behaviors have predictable structures, shapes, and responses to threats and environmental fluctuations. In addition, these behaviors are intimately tied to, and regulated by, the visual and lateral line systems, and are developed as soon as fish are able to swim and feed.[151] Alterations in school structure and density can be caused by individual differences in motivation, physiology, and abiotic and biotic factors of the environment. It has been demonstrated, through the use of shoal choice experiments and frame capture, that pesticide exposures can alter shoaling and schooling behaviors. Atlantic silversides (*Menidia medidia*) exposed to an acetyl cholinesterase inhibiting insecticide, carbaryl, displayed alterations in parallel orientation and increased distances between fish when compared to controls.[82] In addition, swimming orientation in schools

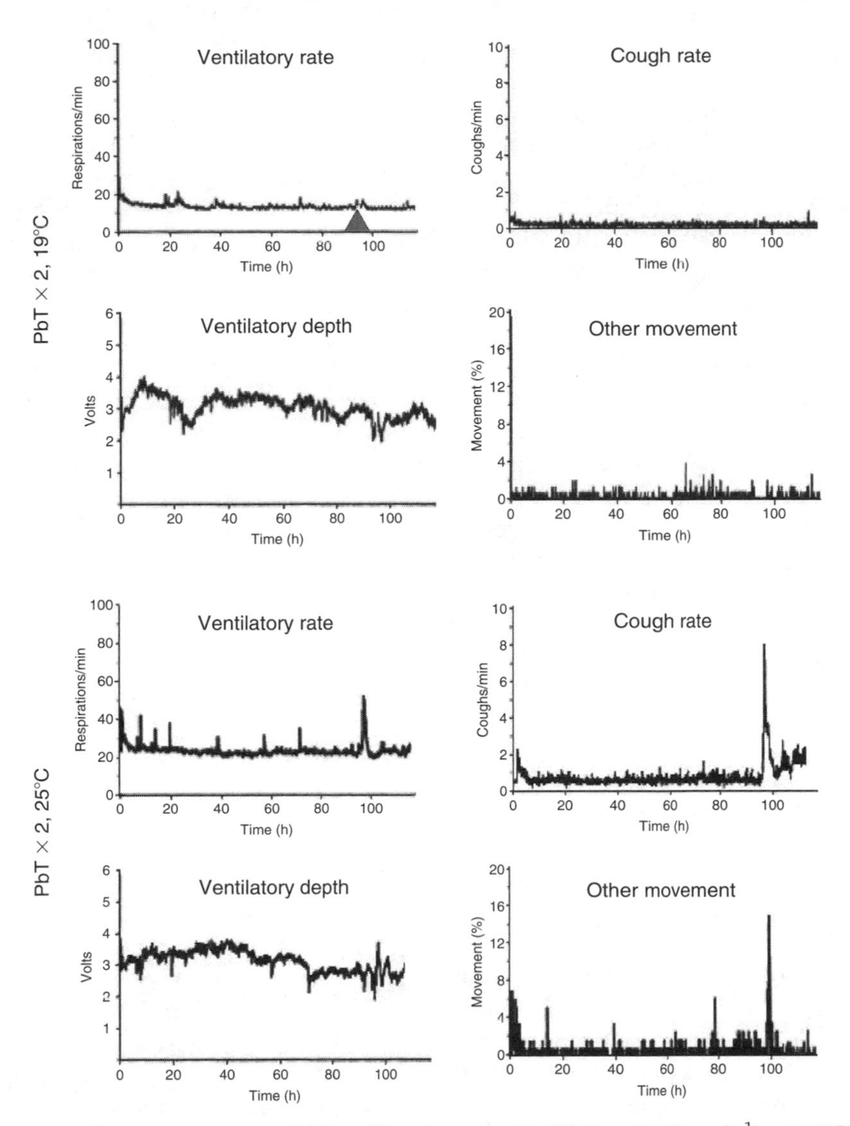

Figure 32.1 Ventilatory responses of bluegill to brevetoxin (PbT × 2, 40 μg l^{-1}) at 19°C and 25°C using a flow-through exposure system as described by Shedd et al.[150] These data illustrate the importance of water quality variables (temperature in this case) in determining the behavioral effects of certain toxicants. The upper panel of four graphs shows respiratory responses of bluegill at 19°C; no significant responses are noted in these data (gray triangle in upper-right graph indicates start of exposure after 96 h of baseline acclimation). At 25°C, bluegill respond to exposure with significantly increased ventilatory rate (but not ventilatory depth), cough rate, and "other movement." This latter category represents data that do not fall into one of the three previous categories, based on the analytical algorithms; it often reflects whole-body movement in the exposure chambers.

of three-spined sticklebacks (*Gasterosteus aculeatus*) was disturbed following exposure to the organotin bis(tributyltin)oxide.[81] Schooling declined following exposure of yearling common carp to 0.05 mg l^{-1} DDT and of fathead minnows to 7.43 mg l^{-1} of the herbicide 2,4-dinitrophenol at a pH of 7.57.[153]

Behavioral analysis systems

Rand[10] extensively reviewed different exposure and tank designs that have historically been used to evaluate fish responses to stress exposures. These systems include both static and flow-through designs, as well as tube, Y-shape, rectangular, square, round, and maze configurations.[80,89,92,105,107,129,154–167] The different exposure designs permit gathering data relevant to avoidance and attraction, ability to detect gradients, orientation to changes in light, sound and temperature, altered performance/stamina and learning. Changes associated with toxicant or stress exposure may be gathered visually, or with electrodes,[168–170] photocells or photoresistors,[171] or videography. Data have been recorded using event recorders, strip chart recorders, polygraphs, and video processors and computers that can integrate signals and generate *x*, *y* coordinate data.

Recent hardware and software updates have been used to develop integrated exposure systems to test suites of behavioral endpoints.[172] These exposure systems are used to investigate the effects of sublethal stressors on fish movement and responses to stimuli. For example, simultaneous video capture from multiple exposure arenas can be used to digitally track movement. Video data can then be used to address questions regarding differences between treatment groups or differences between pre- and postexposure behavior using repeated measures analyses.

Figure 32.2 schematically depicts such a system using video cameras with multiple, dedicated analog video decks. In this system, twelve 10-l exposure arenas were constructed from 14-in. (35.6 cm) diameter polyvinyl chloride (PVC) pipes and end caps. Each arena had two 0.25-in. (0.64 cm) threaded nipples that served as an input and drain. The input was bifurcated to accept both toxicant and dilution water flow lines. Toxicant and dilution water flow were electronically controlled with digital, multichannel peristaltic pumps (Masterflex L/S, Cole-Parmer, Vernon Hills, IL) that were supplied by multiple, aerated 600-l carboys. The exposure arenas were illuminated with 14 h:10 h (light/dark) shadow less fluorescent lighting combined with a computer-controlled dusk and dawn cycle provided by incandescent lights. Lighting was controlled using X-10 computer hardware with an X-Tension (Sand Hill Engineering, Geneva, FL) software interface on a Macintosh operating system. Dusk–dawn lighting systems are very useful in reducing stress in aquatic holding and testing facilities, and can be developed, relatively inexpensively, using other techniques and hardware systems.[173] Twelve color charged-coupled device (CCD) cameras with manual iris and focus control were mounted above the respective arenas and were connected to dedicated VCR decks with time-lapse and dry contact closure capability for recording and computer control. All 12 VCRs were connected to a multiplexer that supported real-time observation (Figure 32.2). VCR recording and stop functions were synchronously activated remotely using X-10 technology.

Analog video data are then digitized in real-time at three frames per second on a Macintosh platform (various hardware and software systems, both Macintosh- and Windows-based are available for this task). Movement data are subsequently imported into a commercial tracking program (Videoscript Professional, version. 2.0©) and converted into *x*, *y* coordinate data. This program uses a custom algorithm designed for fish movement, which identifies and tracks the head of each fish target. The *x*, *y* coordinate data are then analyzed, using proprietary software designed at the Aquatic Pathobiology Laboratory, to obtain the desired behavioral endpoints. There are a variety of ''off the shelf,'' commercially available motion tracking and analysis systems (e.g., *Ethovision*, Noldus Information

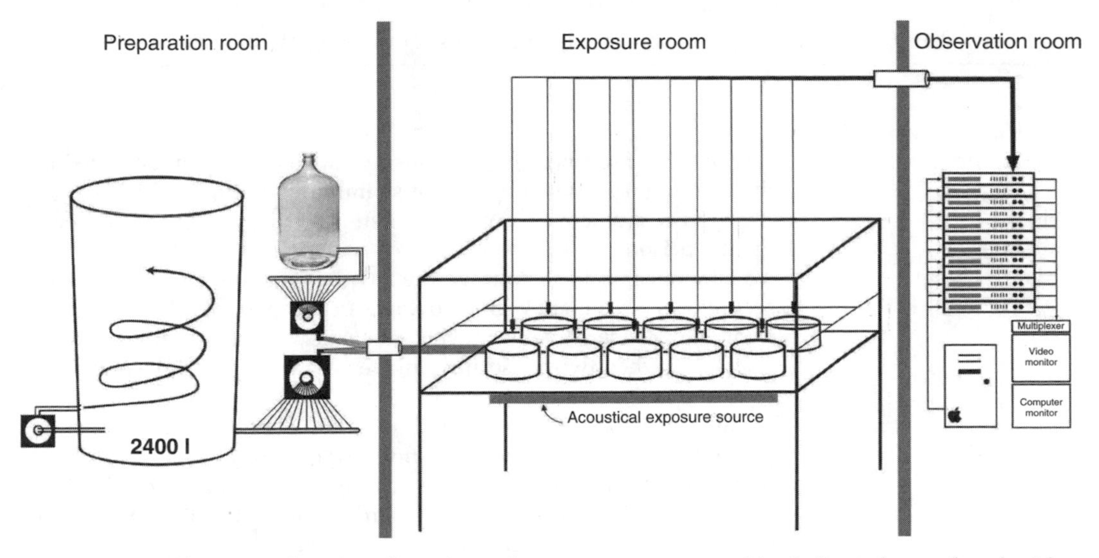

Figure 32.2 Schematic showing flow-through exposure arenas with dedicated, overhead video cameras (shadow less lighting and drainage not shown) in the behavioral toxicology exposure suite at the Aquatic Pathobiology Laboratory, University of Maryland. Water is prepared, temperature adjusted, and delivered from an adjacent "preparation room," while video data are amassed by computer-controlled video decks. Toxicants or aqueous stress agents, and flow-through dilution water, are pumped from the preparation room by computer-controlled peristaltic pumps. An instantaneous acoustic/vibratory stimulus can be provided to discern differences in startle response behavior. In this diagram, an acoustical exposure source is indicated under the platform supporting the exposure arenas. The source in one experiment consisted of three carefully positioned, spring-loaded mousetraps that could be tripped remotely with conjoined pull strings. Video response data are transformed into x, y coordinate data, and relevant endpoints are discerned using our tracking and analytical software. Each of 12 arenas can house individual or multiple fish (up to 120 fish can be monitored simultaneously), or individual arenas can be replaced with multichamber arenas to aid in identifying individual animals that respond (or fail to respond) to different stimuli (see Figure 32.4 describing chambers used to discern startle response with small fish).

Technology; and *Expert Vision*, Motion Analysis Corporation) that can then be customized for particular research requirements.

A major benefit of this behavioral hardware and analysis system (Figure 32.2) is that investigators can take conventional behavioral analyses, which have previously been limited to ranks and counts, and quantify it using computer technology. The behavior and hardware analysis system has potential for greater flexibility in behavioral measurements than commercial behavioral quantification systems, but requires that a programmer be on staff. Current system capabilities and development areas include quantification of a wide range of behaviors, including daily swimming patterns, startle-type responses, avoidance behaviors, and social interactions (Table 32.2). In addition, the system can remotely dose and record up to 12 individual fish or 12 groups of up to 10 individuals simultaneously, for up to 1 h, without the need to mark or tag the animals, thus reducing variance in behavior due to observer or handling disturbances. Finally, the system can be adapted for static and flow-through exposures of environmentally relevant contaminants, and has the ability to be mobile for real-time field assessment.

Table 32.2 Individual and group endpoints for movement analysis

	Definition
Individual endpoints	
Percent movement	The number of seconds the fish satisfies movement criteria divided by the total number of seconds spent swimming, multiplied by 100
Velocity	Average velocity (cm s^{-1}) while the fish is moving during the experimental period
Angular change	The difference (0–180°) between the angular components of two consecutive 1 s movement vectors (degrees per second) divided by the total number of consecutive 1 s movement events. Angular change was only calculated when two consecutive movement vectors met the movement criteria
Path tortuosity (fractal dimension)	Fractal dimension (D) is calculated using the hybrid divider method (Hayward et al., 1989) and is an indicator of path complexity. A series of path generalizations are created at various step sizes, and the data are mathematically drawn on a Richardson plot (log path length versus log step length). D is then calculated as 1 minus the slope of the Richardson plot. $D = 1$ if the path is a straight line and $D = 2$ if it completely fills the 2-dimensional plane
Space allocation	The number of frames fish spend in predefined regions of the exposure arena divided by the total number of frames
Distance from center	Sum of individuals distance from the center of the exposure arena (cm) divided by the total number of frames. This is a measure of how close the fish swims to the walls of the arena
Relative burst frequency	The number of frames that velocity is >3 SD above mean velocity
Startle response	Duration of movement, latency to response, percent response, and burst swimming (response to vibratory/auditory stimulus)
Anti-predator response	Percent fish halting movement, latency to response, percent exhibiting startle-type response, direction of movement (toward or away visual stimuli), and group endpoints (response to overhead "fly-by" of bird silhouette)
Group endpoints	
Interactions	The number of times two fish swim within 0.1 body lengths of each other
Percent shoaling	Number of frames satisfying shoaling criteria divided by the total number of frames
Shoal NNA	Angle of trajectory between two fish in a shoal, must be greater than 45°
Shoal NND	Distance between nearest and second nearest neighbor for each fish in a shoal (minimum three fish)
Percent schooling	Number of frames satisfying schooling criteria divided by the total number of frames
School NNA	Angle of trajectory between two fish in a shoal, must be less than or equal to 45°
School NND	Distance between nearest and second nearest neighbor for each fish in a shoal (minimum three fish)
Percent solitary	Number of frames not satisfying shoaling or schooling criteria divided by the total number of frames
Solitary NND	Distance between nearest and second nearest neighbor for each individual fish not in a shoal or school
Velocity	Speed of fish calculated in centimeters per second

Recent efforts have refined a set of behavioral endpoints in order to investigate the effects of sublethal stressors on fish movement and responses to stimuli. A suite of behavioral endpoints has been developed to evaluate the effect(s) of specific compounds at low levels on fish (Table 32.2). In addition, individual and group models can be utilized with the ability to test different species of small fish. To illustrate the importance of using a suite of exposure endpoints, data from killifish (*Fundulus heteroclitus*) exposed at 25°C to an environmentally relevant concentration (40 μg l^{-1}) of dissolved brevetoxin failed to produce exposure-associated alterations in any nondirected movement parameter. (Brevetoxin is an important biotoxin produced by harmful algae (dinoflagellates) associated with "red tides.") However, this acute, sublethal, low-level exposure significantly altered startle responses in killifish. Conversely, a low anesthetic dose (60 mg l^{-1}) of the common anesthetic and model toxicant, methane tricanesulfonate,[174] was associated with significantly altered movement patterns (Figure 32.3), as well as startle responses.

In addition to exposure-related alterations in movement patterns, changes in startle response parameters can lend valuable insight into changes in the CNS that may have significant environmental consequences. Startle response parameters include, but are not limited to: response frequency, response latency, and average velocity. Startle responses can be elicited using a vibratory stimulus and quantified from small to medium-sized fish using the exposure system described previously in this chapter (Figure 32.2). Alternatively, more precise acoustical tone pips, generated through underwater speakers, can be used as a stimulus with small fish (Figure 32.4). Startle response testing can also yield data sets that can be used for screening large groups of animals.

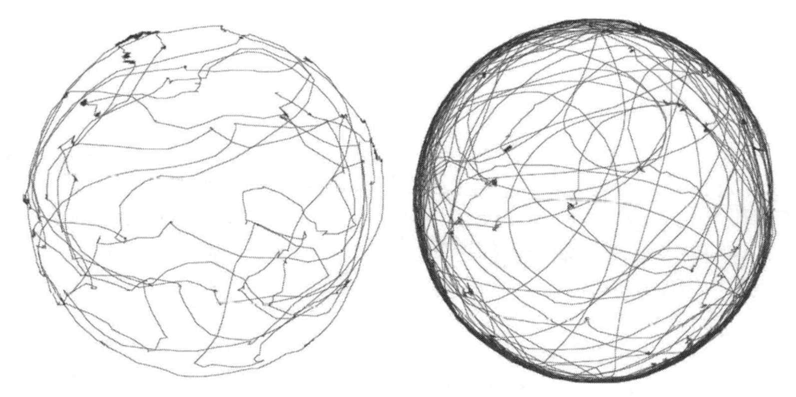

Figure 32.3 Paths of an individual killifish (*F. heteroclitus*) before (left) and after (right) exposure to a sub-anesthetic dose (60 mg l^{-1}) of MS-222. Analysis of these 30-min paths indicate that exposure to low doses of this anesthetic agent causes an increase in percent time in motion (from 12% to 49%) and movement velocity (9.1–11.9 cm s^{-1}), a decrease in path complexity (fractal dimension of 1.082–1.027), and a tendency to swim close to the arena periphery (change in distance from center). All of these endpoints describe quantifiably significant alterations in movement associated with exposure. Functionally, exposed animals tended to increase their speed and stay in motion to compensate for slight loss of equilibrium. The "intoxicated state" was depicted by "hugging" the vessel walls during movement and failing to maintain vigilance (loss of path complexity). MS-222 exposure also significantly altered the startle response of exposed fish such that there was a decrease in the number of responses, a decline in the response duration, and an increased response latency ($p < 0.02$). Startle responses to MS-222 and other chemical stressors were elicited using an instantaneous vibratory stimulus.

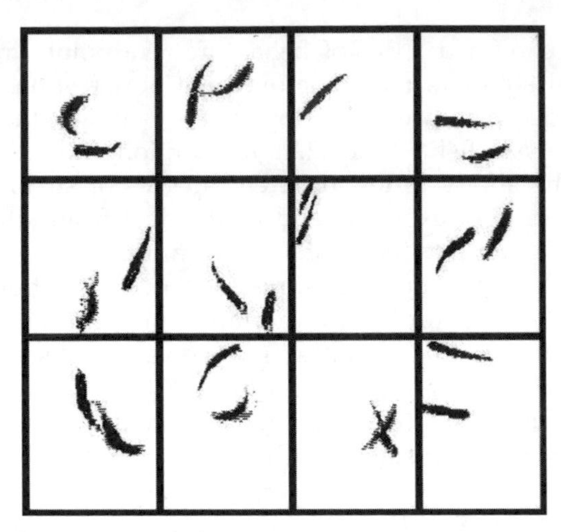

Figure 32.4 An alternative, more precise mechanism to elicit startle response. Video frame captures showing startle response from zebra fish (*Danio rerio*) exposed to instantaneously pulsed, tone pip (400 Hz, 150 db). The stimulus was generated using a tone generator, an amplifier, and a calibrated underwater speaker. Fish were acclimated individually in each of 12, small chambers. Video frames were captured 33 ms prior to the stimulus and 33 ms after the stimulus to visualize the position of the fish prior to and after eliciting startle response. A fish doublet is seen in all chambers where the animal startled in response to the stimulus. All animals startled in this image except for one (top row, 3rd column). In this trial, typical of others at this sound pressure level, 12 out of 12 fish in 8 out of 10 replicate trials startled; 11 out of 12 fish in the other two replicates startled. Note that the acoustic stimulus described here is different from the "acoustical sound delivery board" as illustrated and described in Figure 32.2.

Preliminary studies

When developing or applying automated systems, preliminary studies are essential in order to optimize and validate behavioral hardware and maximize the sensitivity of individual endpoints. For example, it is important to know (1) how a given species will acclimate to the exposure arenas, (2) effects of flow or toxicant infusion relative to distribution within the exposure arena, as well as the sentinel's response to flow (attractance/avoidance to dilution flow), and (3) the optimal duration of observation time intervals. Discerning appropriate observation time intervals (both timing and duration) is critical. Shorter observation time tends to introduce more noise (and a tendency for false positives; type I errors), while longer observation time tends to average out potentially important exposure-related differences (tendency for negatives; type II errors). Figures 32.5 and 32.6 provide sample data indicating changes in movement patterns during an acclimation study, and toxicant distribution visualized in a dye experiment, respectively.

The importance of understanding flow and mixing dynamics in test chambers should not be underestimated.[30] Most studies use steep gradient test systems, in which the chemical concentration increases sharply over a short distance; in most instances, the gradient is unknown to the experimenter but not to the fish. Fish must choose between clean water and contaminated water. When the gradient is steep, it may not mimic what fish actually experience in the field, except under unusual conditions, for example, immediately downstream from an effluent pipe or mine tailing. Kleerekoper et al.[30]

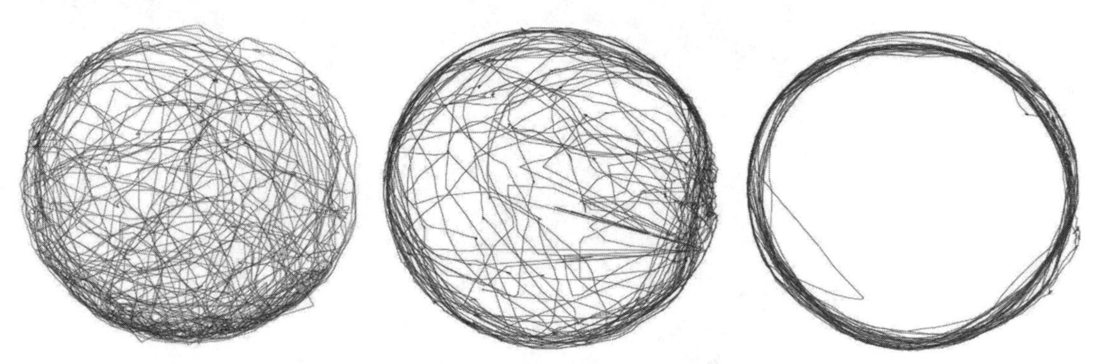

Figure 32.5 Acclimation experiment with killifish showing plotted x, y coordinate data for 30 min of movement from a single animal 1 h after entry into control exposure area (left), after 24 h (center), and after 48 h (right). These plots indicated that, over time, there was a significant, repeatable decrease in path complexity (fractal dimension), a reduction in angular change and time spent in the center of the vessel (swimming close to vessel wall). These data were useful in determining appropriate timing to expose and observe these fish. We found that after 24 h of arena acclimation, animals maintained sufficient movement complexity and repertoire to discern changes in nondirected swimming patterns in response to certain stress exposure scenarios.

showed that when goldfish (*C. auratus*) were attracted to copper concentrations between 11 and 17 $\mu g\ l^{-1}$ under shallow gradient conditions, but when the gradient was steep, fish avoided concentrations as low as 5 $\mu g\ l^{-1}$.[175]

Conclusions

Behavioral toxicology is a useful indicator of sublethal contamination because behavioral endpoints frequently occur below concentrations that are chronically lethal and at lower concentrations than those that affect growth.[140,141] As such, careful selection and design of behavioral tests could produce definable, interpretive endpoints for use in regulatory applications of product registration, damage assessments, formulation of water quality criteria and standards, and understanding changes in fish population dynamics based on anthropogenic effects. Behavioral endpoints could also, through the increased use and decreasing costs, be included in regulation of effluent standards. Although behavioral endpoints may often provide excellent sensitivity, utility, or biologic significance, behavioral responses have historically not been routinely included in standard aquatic toxicity assessment programs.

In developing behavioral test methods for contaminant assessment, behavioral toxicological endpoints must address one or more of three assessment questions: (1) How well does the response measure effect or injury arising from exposure? (2) Does the response aid in the identification of the toxic agent? (3) Does the use of the behavioral response increase the capability of contaminant assessments to predict the ecologic consequences of exposure? About 11 years ago, Little et al.[27] explicitly defined the challenges for behavioral toxicology: "(1) more and better behavioral testing procedures ... must be developed and refined into effective tools which can be readily and clearly applied in contaminant assessments; (2) establishment of the links between behavioral effects observed in the laboratory and ecological effects observed in the field to aid in the development of contaminant assessment that are adequately predictive of population and community response; and (3) behavioral toxicologists must continue to work to dispel the

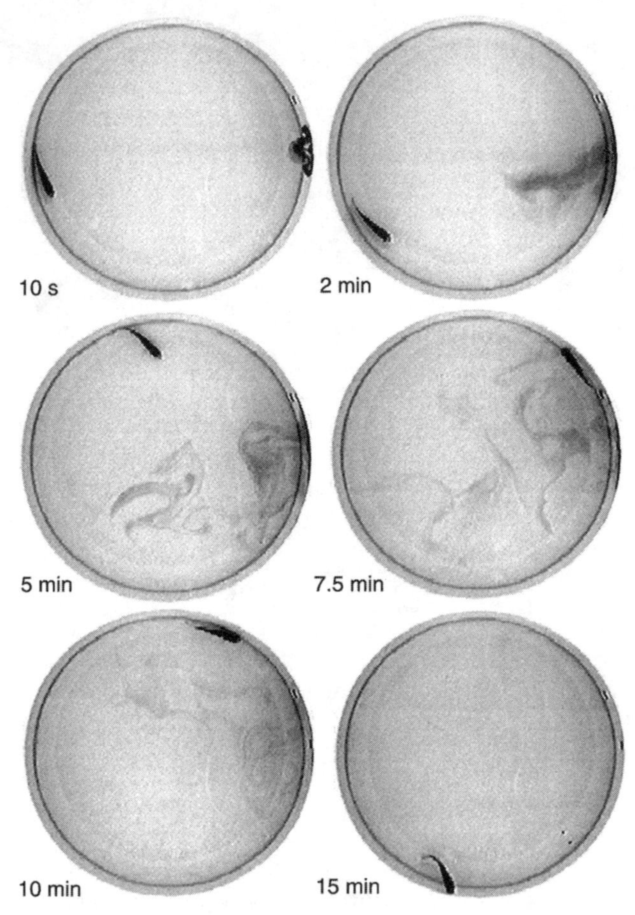

Figure 32.6 Preliminary experiment using nontoxic food-grade dye to discern patterns and timing of "toxicant" dispersion in exposure arenas. A 5-ml bolus of diluted dye was infused into multiple vessels and video frames were captured at 10 s, 2 min, 5 min, 7.5 min, 10 min, and 15 min. These sequential images from a single arena indicate that avoidance/attractance behavior can be observed within the initial 2–5 min of exposure, while the "toxicant" is not yet dispersed throughout the arena and there is still a concentration gradient in the "sector" of the arena where the stressor is pumped. After 15 min the dye is well mixed with the dilution water throughout the vessel. A number of factors will influence the rate of dispersion throughout the vessel, including the flow rate of the infused stress agent, the density and temperature of the mixtures, the number and species of test fish in the arena, and the shape and volume of the arena.

erroneous paradigm prevalent in aquatic toxicology that suggests that behavioral responses are not suitable for inclusion in contaminant evaluation and assessment because they are either unquantifiable, too complex, too variable, extraordinarily difficult to measure, not biologically significant, or lacking in ecologic relevance."

Great strides have been made towards elucidating and developing better testing procedures, in large associated with improvements in videography and the power of personal computers, and with the availability of video equipment to fit almost any budget and work requirement. However, additional work is required to validate the use of many individual and group behaviors as endpoints of sublethal toxicant and stressor exposure.

Today, the relationship between many exposure-related behavioral alterations observed in the laboratory and their relevance to ecological effects observed in the field remains poorly understood. A notable exception is avoidance responses that have been intensively studied in the laboratory and verified in the field through the use of telemetry. As a result, avoidance is the only behavioral response that has legal standing in the United States as evidence of environmental injury under proceedings of the Comprehensive Environmental Response and Compensation Liability Act. Consequently, there is opportunity for researchers to actively contribute to the development of contaminant assessment paradigms that are predictive of population and community response. To accomplish this goal, sound experimental designs, combined with proper statistical analyses, are essential for developing approaches using fish movement for the bioindication of stressors or assessing mechanism of action.

Continued interest and new developments in automated behavioral monitoring and analysis techniques, especially those that can be verified in field trials, are being used to make functional inroads to dispel the erroneous paradigm that "behavioral responses are not suitable for inclusion in contaminant evaluation and assessment."[27] Behavioral responses should be more routinely included and integrated in contaminant evaluation as researchers continue to learn about and develop new automated techniques that allow for relatively inexpensive and rapid collection, processing, and verification of behavioral disruptions associated with contaminant exposure.

The behavioral exposure systems and endpoints described in this chapter can be applied to quantify differences in behavior associated with exposure to metals, organics, pesticides, algal bloom toxins, alterations in water quality, and agricultural waste. Behavioral endpoints also provide a valuable tool for the quantification of differences in reproductive behaviors, predator–prey interactions, behavioral changes across environmental gradients, and differences in swimming behavior between species. Analysis of behavioral responses to a variety of stimuli can provide quantitative measures of neural and mechanical disruption, reflecting biochemical and physiological alterations.[25] These additional tools support toxicological investigation with endpoints other than traditional LC_{50}'s, and may aid in determining new no observed effect levels (NOELs) and lowest observed effect levels (LOELs), as well as investigating the environmental relevance of various low-level exposures. Behavioral toxicology may ultimately provide sensitive, noninvasive, and broadly applicable endpoints for the description of integrated, whole animal responses associated with exposure to a wide variety of stressors.

Acknowledgments

We thank the U.S. Environmental Protection Agency—Science to Achieve Results Program (R828224-010), and the State of Maryland, for their support to develop and utilize the behavioral toxicology suite at the UM Aquatic Pathobiology Center. We gratefully acknowledge Edward Little, Geoffrey Gipson, Aaron DeLonay, Tim Molteno, Tom Shedd, and Colin Hunter for their technical prowess and insights regarding behavioral tracking, quantitative analysis and system development. Thanks also to Gary Atchison, Natalie Snyder, and Madeline Sigrist for their helpful review of this manuscript.

References

1. Drummond, R.A., Russom, C.L., Geiger, D.L. and DeFoe, D.L., Behavioral and morphological changes in fathead minnows, *Pimephales promelas*, as diagnostic endpoints for screening chemicals according to modes of action, in *Aquatic Toxicology and Environmental Fate*, Poston, T.M. and Perdy, R., Eds., American Society for Testing and Materials, Philadelphia, PA, 1986, pp. 415–435.
2. Drummond, R.A. and Russom, C.L., Behavioral toxicity syndromes: a promising tool for assessing toxicity mechanisms in juvenile fathead minnows, *Environ. Toxicol. Chem.*, 9, 37–46, 1990.
3. Diamond, J.M., Parson, M.J. and Gruber D., Rapid detection of sublethal toxicity using fish ventilatory behavior, *Environ. Toxicol. Chem.*, 9, 3–12, 1990.
4. Baldwin, I.G., Harman, M.M.I. and Neville, D.A., Performance characteristics of a fish monitor for detection of toxic substances. 1. Laboratory trials, *Water Res.*, 28 (10), 2191–2199, 1994a.
5. Baldwin, I.G., Harman, M.M.I., Neville, D.A. and George, S.G., Performance characteristics of a fish monitor for detection of toxic substances. 2. Field trials, *Water Res.*, 28 (10), 2201–2208, 1994b.
6. Gruber, D., Frago, C. and Rasnake, W.J., Implementation of a multiple biomonitoring approach to evaluate the potential for impact from an industrial discharge, *J. Aquat. Ecosys. Health*, 3 (4), 259–271, 1994.
7. Balk, L., Larsson, A. and Foerlin, L., Baseline studies of biomarkers in the feral female perch (*Perca fluviatilis*) as tools in biological monitoring of anthropogenic substances, Pollutant Responses in Marine Organisms (Primo 8), *Mar. Environ. Res.*, 42 (1–4), 203–208, 1996.
8. Rice, P.J., Drews, C.D., Klubertany, T.M., Bradbury, S.P. and Coats, J.R., Acute toxicity and behavioral effects of chlorophyrophos, permethrine, phenol, strychnine, and 2,4,-dinitro phenol to 30 day old Japanese medaka (*Oryzias atipes*), *Environ. Toxicol. Chem.*, 16, 696–704, 1997.
9. Biological Monitoring, Inc., Blacksburg, VA, *http://www.biomon.com/biosensor.html*.
10. Rand, G.M., Behavior, in *Fundamentals of Aquatic Toxicology: Methods and Applications*, Rand, G.M. and Petrocelli, S.R., Eds., Hemisphere Publishing, New York, 1985, pp. 221–256.
11. Gray, R.H., Fish behavior and environmental assessment, *Environ. Toxicol. Chem.*, 9, 53–68, 1990.
12. Little, E.E. and Finger, S.E., Swimming behavior as an indicator of sublethal toxicity in fish, *Environ. Toxicol. Chem.*, 9, 13–19, 1990.
13. Døving, K.B., Assessment of animal behavior as a method to indicate environmental toxicity, *Comp. Biochem. Physiol.* C., 100, 247–252, 1991.
14. Weber, D.N. and Spieler, R.E., Behavioral mechanisms of metal toxicity in fishes, in *Aquatic Toxicology: Molecular, Biochemical, and Cellular Perspectives*, Malins, D.C. and Ostrander, G.K., Eds., CRC Press, Boca Raton, FL, 1994, pp. 421–467.
15. Little, E.E. and Brewer, S.K., Neurobehavioral toxicity in fish, in *Target Organ Toxicity in Marine and Freshwater Teleosts New Perspectives: Toxicology and the Environment, Vol. 2. Systems*, Schlenk, D. and Benson, W.H., Eds., Taylor & Francis, London, 2001, pp. 139–174.
16. Natural Resource Damage Assessments (NRDA), Final rule, *Federal Register*, 51, 27674–27753, 1986.
17. Boyd, W., Williams, P. and Brewer, S., Altered behaviour of invertebrates living in polluted environments, in *Behaviour in Ecotoxicology*, Dell' Omo, G., Ed., John Wiley & Sons, New York, 2002, pp. 415–435.
18. Bookhout, C.G., Monroe, R.J., Forward, R.B. and Costlow, J.D. Jr., Effects of soluble fractions on drilling fluids on development of crabs, *Rhithropanopeus harrissi* and *Callinectes sapidus*, *Water Air Soil Pollut.*, 21 (1–4), 183–197, 1984.
19. Kumari, B.N., Yellamma, K. and Mohan, P.M., Toxicological evaluation fenvalerate on the freshwater field crab, *Oziotelphusa senex senex*, *Environ. Ecol.*, 5 (3), 451–455, 1987.

20. Dodson, S.I., Hanazazto, T. and Gorski, P.R., Behavioral responses of *Daphnia pulex* exposed to carbaryl and *Chaoborus* kairomone, *Environ. Toxicol. Chem.*, 14 (1), 43–50, 1995.
21. Ham, K.D. and Peterson, M.J., Effect of fluctuating low-level chlorine concentration on valve movement behavior of the Asiatic clam, *Corbicula fluminea*, in *13th Annual Meeting*, Society of Environmental Toxicology and Chemistry, Pensicola, FL, 1992.
22. McCloskey, J.T. and Newman, M.C., Asiatic clam (*Corbicula fluminea*) and viviparid snail (*Campeloma decisum*) sediment preference as a sublethal response to low level metal contamination, *Arch. Environ. Contam. Toxicol.*, 28, 195–202, 1995.
23. Keenleyside, M.H.A., Diversity and adaptation in fish behavior, in: *Zoophysiology*, vol. XI, Springer-Verlag, Berlin, West Germany, 1979, p. 208.
24. Vogl, C., Grillitsch, B., Wytek, R., Hunrich Spieser, O. and Scholz, W., Qualification of spontaneous undirected locomotor behavior of fish for sublethal toxicity testing. I. Variability of measurement parameters under general test conditions, *Environ. Toxicol. Chem.*, 18 (12), 2736–2742, 1999.
25. Brewer, S.K., Little, E.E., DeLonay, A.J., Beauvais, S.L., Jones, S.B. and Ellersieck, M.R., Behavioral dysfunctions correlate to altered physiology in rainbow trout (*Oncorynchus mykiss*) exposed to cholinesterase-inhibiting chemicals, *Arch. Environ. Contam. Toxicol.*, 40, 70–76, 2001.
26. Bridges, C.M., Tadpole swimming performance and activity affected by acute exposure to sublethal levels of carbaryl, *Environ. Toxicol. Chem.*, 16 (9), 1935–1939, 1997.
27. Little, E.E., Fairchild, J.F. and DeLonay, A.J., Behavioral methods for assessing the impacts of contaminants on early life stage fishes, in *American Fisheries Society Symposium. 14: Water Quality and the Early Life Stages of Fishes*, Fuiman, L., Ed., American Fisheries Society, Bethesda, MD, 1993a, pp. 67–76.
28. Henry, M.G. and Atchison, G.J., Behavioral changes in social groups of bluegills exposed to copper, *Trans. Am. Fish. Soc.*, 115, 590–595, 1986.
29. Alcock, J., *Animal Behavior: An Evolutionary Approach*, Sinauer Associates, Inc., Sunderland, MA, 1997.
30. Kleerekoper, H., Waxman, J.B. and Matis, J., Interaction of temperature and copper ions as orienting stimuli in the locomotor behavior of the goldfish (*Carassius auratus*), *J. Fish. Res. Board Can.*, 30, 725–728, 1973.
31. Giattina, J.D., Cherry D.S., Cairns, J. and Larrick, S.R., Comparison of laboratory and field avoidance behavior of fish in heated chlorinated water, *Trans. Am. Fish. Soc.*, 110, 526–535, 1981.
32. Birtwell, I.K. and Kruzynski, G.M., *In situ* and laboratory studies on the behaviour and survival of Pacific salmon (genus *Oncorhynchus*), environmental bioassay techniques and their application, *Hydrobiologia*, 188–189, 543–560, 1989.
33. Saglio, P. and Trijasse, S., Behavioral responses to atrazine and diuron in goldfish, *Arch. Environ. Contam. Toxicol.*, 35, 484–491, 1998.
34. Zhou, T. and Weis, J.S., Swimming behavior and predator avoidance in three populations of *Fundulus heteroclitus* larvae after embryonic exposure and/or larval exposure to methylmercury, *Aquat. Toxicol.*, 43, 131–148, 1998.
35. Radakov, D.V., *Schooling in the Ecology of Fish*, John Wiley & Sons, New York, 1973.
36. Kane, A.S., Salierno, J., Deamer-Melia, N., Glasgow, H. and Burkholder, J., Unpublished data.
37. Munro, A.L.S. and Hastings, T.S., Furunculosis, in *Bacterial Diseases of Fish*, Inglis, V., Roberts, R.J. and Bromage, N.R., Eds., Blackwell Scientific Publications, Boston, 1993, pp. 122–142.
38. Austin, B. and Austin, D.A., *Bacterial Fish Pathogens: Disease of Farmed and Wild Fish*, 3rd revised ed., Praxis Publishing, Chichester, England, 1999.
39. Noga, E.J., *Fish Disease: Diagnosis and Treatment*, Iowa State University Press, Ames, IA, 1996.
40. Bullock, G.L. and Conroy, D.A., Bacterial gill disease, in *Diseases of Fishes, Book 2A, Bacterial Diseases of Fishes*, Snieszko, S.F. and Axelrod, H.R., Eds., TFH Publications Inc., Jersey City, 1971, pp. 77–87.

41. Ferguson H.W., Ostland V.E., Byrne P. and Lumsden J.S., Experimental production of bacterial gill diseases in trout by horizontal transmission and by bath challenge, *J. Aquat. Anim. Health*, 3, 118–123, 1991.
42. Kane, A.S., Dykstra, M.J., Noga, E.J., Reimschuessel, R., Baya, A., Driscoll, C., Paerl, H.W. and Landsberg, J., Etiologies, observations and reporting of estuarine finfish lesions, *Mar. Environ. Res.*, 50, 473–477, 2000.
43. Mudarris, M. and Austin, B., Systemic diseases in turbot *Scophthalmus maximus* caused by a previously unrecognized *Cytophaga*-like bacterium, *Dis. Aquat. Organisms*, 6, 161–166, 1989.
44. Mudarris, M. and Austin, B., Histopathology of a gill and systemic disease of turbot (*Scophthalmus maximus*) caused by a *Cytophaga*-like bacterium (CLB), *Bull. Eur. Assoc. Fish Pathologists*, 12, 120–123, 1992.
45. Cann, D.C. and Taylor, L.Y., An outbreak of botulism in rainbow trout, *Salmo gairdneri* Richardson, farmed in Britain, *J. Fish Dis.*, 5, 393–399, 1982.
46. Mellergaard, S. and Bloch, B., Herpes virus-like particles in angelfish, *Pterophyllum altum*, *Dis. Aquat. Organisms*, 5, 151–155, 1988.
47. Plumb, J.A., *Edwardsiella* septicaemia, in *Bacterial Diseases of Fish*, Inglis, V., Roberts, R.J. and Bromage, N.R., Eds., Blackwell Scientific Publications, Boston, 1993, pp. 61–79.
48. Bradley, T.M., Medina, D.J., Chang, P.W. and McClain, J., Epizootic epitheliotropic disease of lake trout (*Salvelinus namaycush*): history and viral etiology, *Dis. Aquat. Organisms*, 7, 195–201, 1989.
49. Udey, L.R., Young, R. and Sallman, B., *Eubacterium* spp. ATCC 29255: an anaerobic bacterial pathogen of marine fish, *Fish Health News*, 5, 3–4, 1976.
50. Richards, R.H., Holliman, A. and Helagson, S., *Exophiala salmonis* infection in Atlantic salmon, *Salmo salar* L, *J. Fish Dis.*, 1, 357–368, 1978.
51. Alderman, D.J., Fungal diseases of aquatic animals, in *Microbial Diseases of Fish: Society for General Microbiology*, Special Publication 9, Robert, R.J., Ed., Academic Press, New York, 1982, pp. 189–242.
52. Carson, J., Schmidtke, L.M. and Munday, B.L., *Cytophaga johnsonae* a putative skin pathogen of juvenile farmed barramundi, *Lates calcarifer* Bloch, *J. Fish Dis.*, 16, 209–218, 1993a.
53. Holt, R.A., Rohovec, J.S. and Fryer, J.L., Bacterial cold-water disease, in *Bacterial Diseases of Fish*, Inglis, V., Roberts, R.J. and Bromage, N.R., Eds., Blackwell Scientific Publications, Boston, 1993, pp. 3–22.
54. Teshima, C., Kudo, S., Ohtani, Y. and Saito, A., Kidney pathology from the bacterium *Hafnia alvei*: experimental evidence, *Trans. Am. Fish. Soc.*, 121, 599–607, 1992.
55. Roberts, R.J., Motile *aeromonad septicaemia*, in *Bacterial Diseases of Fish*, Inglis, V., Roberts, R.J. and Bromage, N.R., Eds., Blackwell Scientific Publications, Boston, 1993, pp. 143–155.
56. Wolke, R.E. and Stroud, R.K., Piscine mycobacteriosis, in *Mycobacterial Infections of Zoo Animals*, Montali, R.J., Ed., Smithsonian Institute Press, Washington, D.C., 1978, pp. 269–275.
57. Dulin, M.P., A review of tuberculosis (mycobacteriosis) in fish, *Vet. Med. Small Anim. Clinician*, 74, 731–735, 1979.
58. Giavenni R., Finazzi M., Poli, G. and Grimalki, C.N.R., Tuberculosis in marine tropical fishes in an aquarium, *J. Wildlife Dis.*, 16, 161–168, 1980.
59. van Duijn, C., Tuberculosis in fishes, *J. Small Anim. Pract.*, 22, 391–411, 1981.
60. Hoffman, G.L., Whirling Disease of Trout, Fish disease leaflet No. 47, United States Fish and Wildlife Service, Washington, D.C., 1976.
61. Frerichs, G.N., Mycobacteriosis: nocardiosis, in *Bacterial Diseases of Fish*, Inglis, V., Roberts, R.J. and Bromage, N.R., Eds., Blackwell Scientific Publications, Boston, 1993, pp. 219–233.
62. Burkholder, J.M., Noga, E.J., Hobbs, C.H., Glasgow, H.B. and Smith, S.A., A new 'phantom' dinoflagellate, causative agent of major estuarine fish kills, *Nature*, 358, 407–410, 1992.
63. Burkholder, J.M., Glasgow, H.B. and Hobbs, C.W., Fish kills linked to a toxic ambush-predator dinoflagellate: distribution and environmental conditions, *Mar. Ecol. Prog. Ser.*, 124, 43–61, 1995.

64. Marshall, H.G., Gordon, A.S., Seaborn, D.W., Dyer, B., Dunstan, W.M. and Seaborn, A.M., Comparative culture and toxicity studies between the toxic dinoflagellate *Pfiesteria picicida* and a morphologically similar cryptoperidiniopsoid dinoflagellate, *J. Exp. Mar. Biol. Ecol.*, 255, 51–74, 2000.
65. Noga, E.J., Fungal diseases of marine and estuarine fish, in *Pathobiology of Marine and Estuarine Organisms*, Couch, J.A. and Fournie, J.W., Eds., CRC Press, Boca Raton, FL, 1993, pp. 85–109.
66. Klein, B.U., Kleingeld, D.W. and Bohm, K.H., First isolations of *Plesiomonas shigelloides* from samples of cultured fish in Germany, *Bull. Eur. Assoc. Fish Pathologists*, 14, 165–166, 1993.
67. Malsberger, R.G. and Lautenslager, G., Fish viruses: rhabdovirus isolated from a species of the family Cichlidae, *Fish Health News*, 9, i/ii, 1980.
68. Turnbull, J.F., Epitheliocystis and salmonid rickettsial septicaemia, in *Bacterial Diseases of Fish*, Inglis, V., Roberts, R.J. and Bromage, N.R., Eds., Blackwell Scientific Publications, Boston, 1993, pp. 237–254.
69. Saeed, M.O., Alamoudi, M.M. and Al-Harbi, A.H., A *Pseudomonas* associated with disease in cultured rabbitfish *Siganus rivulatus* in the Red Sea, *Dis. Aquat. Organisms*, 3, 177–180, 1987.
70. Fijan, N., Infectious dropsy in carp: a disease complex, *Symp. Zool. Soc. Lond.*, 30, 39–51, 1972.
71. Kitao, T., Streptococcal infections, in *Bacterial Diseases of Fish*, Inglis, V., Roberts, R.J. and Bromage, N.R., Eds., Blackwell Scientific Publications, Boston, 1993, pp. 196–210.
72. Michel, C., Nougayrede, P., Eldar, A., Sochon, E. and de Kinkelin, P., *Vagococcus salmoninarum*, a bacterium of pathological significance in rainbow trout *Oncorhynchus mykiss* farming, *Dis. Aquat. Organisms*, 30, 199–208, 1997.
73. Colorni, A., Paperna, I. and Gordin, H., Bacterial infections in gilthead sea bream *Sparus aurata* cultured in Elat, *Aquaculture*, 23, 257–267, 1981.
74. Austin, B., Stobie, M., Robertson, P.A.W., Glass, H.G., Stark, J.R. and Mudarris, M., *Vibrio alginolyticus*: the cause of gill disease leading to progressive low-level mortalities among juvenile turbot, *Scophthalmus maximus* L., in a Scottish aquarium, *J. Fish Dis.*, 16, 277–280, 1993.
75. Hjeltnes, B. and Roberts, R.J., Vibriosis, in *Bacterial Diseases of Fish*, Inglis, V., Roberts, R.J. and Bromage, N.R., Eds., Blackwell Scientific Publications, Boston, 1993, pp. 109–121.
76. Carson, J. and Schmidtke, L.M., Opportunistic infection by psychrotrophic bacteria of cold-comprised Atlantic salmon, *Bull. Eur. Assoc. Fish Pathologists*, 13, 49–52, 1993b.
77. Stevenson, R., Flett, D. and Raymond, B.T., Enteric redmouth (ERM) and other enterobacterial infections of fish, in *Bacterial Diseases of Fish*, Inglis, V., Roberts, R.J. and Bromage, N.R., Eds., Blackwell Scientific Publications, Boston, 1993, pp. 80–105.
78. Exley, C., Avoidance of aluminum by rainbow trout, *Environ. Toxicol. Chem.*, 19 (4), 933–939, 2000.
79. Allin, C.J. and Wilson, R.W., Effects of pre-acclimation to aluminum on the physiology and swimming behaviour of juvenile rainbow trout (*Oncorhynchus mykiss*) during a pulsed exposure, *Aquat. Toxicol.*, 51, 213–224, 2000.
80. Sprague, J.B., Avoidance of copper-zinc solutions by young salmon in the laboratory, *J. Water Pollut. Control Fed.*, 36, 990–1004, 1964.
81. Wibe, E., Nordtug, T. and Jenssen, B.M., Effects of *bis*(tributyltin)oxide on antipredator behavior in three spine stickleback, *Gasterosteus aculeatus* L., *Chemosphere*, 44, 475–481, 2001.
82. Weis, P. and Weis, J.S., Schooling behavior of *Menidia medidia* in the presence of the insecticide Sevin (Carbaryl), *Mar. Biol.*, 28, 261–263, 1974.
83. Sullivan, J.F., Atchison, G.J., Kolar, D.J. and McIntosh, A.W., Changes in the predator-prey behavior of fathead minnows (*Pimephales promelas*) and largemouth bass (*Micropterus salmoides*) caused by cadmium, *J. Fish. Res. Board Can.*, 35 (4), 446–451, 1978.
84. Sloman, K.A., Scott, G.R., Diao, Z., Rouleau, C., Wood, C.M. and McDonald, D., Cadmium affects the social behaviour of rainbow trout, *Oncorhynchus mykiss*, *Aquat. Toxicol.*, 65, 171–185, 2003.
85. Grillitsch, B., Vogl, C. and Wytek, R., Qualification of spontaneous undirected locomotor behavior of fish for sublethal toxicity testing. II. Variability of measurement parameters under toxicant-induced stress, *Environ. Toxicol. Chem.*, 18 (12), 2743–2750, 1999.

86. Ellgaard, E.G., Tusa, J.E. and Malizia, A.A. Jr., Locomotor activity of the bluegill *Lepomis macrochirus*: hyperactivity induced by sublethal concentrations of cadmium, chromium and zinc, J. *Fish Biol.*, 12 (1), 19–23, 1978.
87. Summerfelt, R.C. and Lewis, W., Repulsion of green sunfish by certain chemicals, *J. Water Pollut. Control Fed.*, 39, 2030–2038, 1967.
88. Sloman, K.A., Baker, D.W., Wood, C.M. and McDonald, G., Social interactions affect physiological consequences of sublethal copper exposure in rainbow trout, *Oncorhynchus mykiss*, *Environ. Toxicol. Chem.*, 21 (6), 1255–1263, 2001.
89. Folmar, L.C., Overt avoidance reaction of rainbow trout fry to nine herbicides, *Bull. Environ. Contam. Toxicol.*, 15, 509–514, 1976.
90. Rice, S.D., Moles, A., Taylor, T.L. and Karinen, J.F., Sensitivity of 39 Alaskan marine species to Cook Inlet crude oil and No. 2 fuel oil, in *Proceedings 1979 Oil Spill Conference (prevention, behavior, control, cleanup)*, Los Angeles, CA, USA, 19–22 March 1979, American Petroleum Institute, Washington, D.C., 1979, pp. 549–554.
91. Hara, T.J., Law, Y.M.C. and Macdonald, S., Effects of mercury and copper on the olfactory response in rainbow trout, *Salmo gairdneri*, *J. Fish. Res. Board Can.*, 33 (7), 1568–1573, 1976.
92. Kleerekoper, H., Westlake, G.F., Matis, J.H. and Gensler, P.J., Orientation of goldfish (*Carassius auratus*) in response to a shallow gradient of a sublethal concentration of copper in an open field, *J. Fish. Res. Board Can.*, 29, 45–54, 1972.
93. Black, J.A. and Birge, W.J., An Avoidance Response Bioassay for Aquatic Pollutants, Water Resources Research Institute Research Report 123, University of Kentucky, Kentucky, 1980.
94. Sutterlin, A.M. and Gray, R., Chemical basis for homing of Atlantic salmon (*Salmo salar*) to a hatchery, J. *Fish. Res. Board Can.*, 30 (7), 985–989, 1973.
95. Giattina, J.D., Garton, R.R. and Stevens, D.G., Avoidance of copper and nickel by rainbow trout as monitored by a computer-based data acquisition system, *Trans. Am. Fish. Soc.*, 111 (4), 491–504, 1982.
96. Hansen, J.A., Marr, J.C.A., Lipton, J., Cacela, D. and Bergman, H.L., Differences in neurobehavioral responses of chinook salmon (*Oncorhynchus tshawytscha*) and rainbow trout (*Oncorhynchus mykiss*) exposed to copper and cobalt. 1. Behavioral avoidance, *Environ. Toxicol. Chem.*, 18, 1972–1978, 1999a.
97. Ogilvie, D.M. and Anderson, J.M., Effect of DDT on temperature selection by young Atlantic salmon, *Salmo salar*, *J. Fish. Res. Board Can.*, 22, 503–512, 1965.
98. Ellgaard, E.G., Ochsner, J.C. and Cox, J.K., Locomotor hyperactivity induced in the bluegill sunfish, *Lepomis macrochirus*, by sublethal concentrations of DDT, *Can. J. Zool.*, 55 (7), 1077–1081, 1977.
99. Miller, D.L. and Ogilvie, D.M., Temperature selection in brook trout (*Salvelinus fontinalis*) following exposure to DDT, PCB or phenol, *Bull. Environ. Contam. Toxicol.*, 14, 545–551, 1975.
100. Faulk, C.K., Fuiman, L.A. and Thomas, P., Parental exposure to *ortho-*, *para-*dichlorodiphenyltrichloroethane impairs survival skills of Atlantic croaker (*Micropogonias undulates*) larvae, *Environ. Toxicol. Chem.*, 18, 254–262, 1999.
101. Gardner, D.R., The effect of some DDT and methoxychlor analogs on temperature selection and lethality in brook trout fingerlings, *Pest. Biochem. Physiol.*, 2 (4), 437–446, 1973.
102. Kynard, B., Avoidance behavior of insecticide susceptible and resistant populations of mosquito fish to four insecticides, *Trans. Am. Fish. Soc.*, 3, 557–561, 1974.
103. Smith, G. and Weis, J.S., Predator-prey relationships in mummichogs (*Fundulus heteroclitus* (L.)): effects of living in a polluted environment, *J. Exp. Mar. Biol. Ecol.*, 209, 75–87, 1997.
104. Svecevicius, G., Avoidance response of rainbow trout *Oncorhynchus mykiss* to heavy metal model mixtures: a comparison with acute toxicity tests, *Bull. Environ. Contam. Toxicol.*, 67, 680–687, 2001.
105. Jones, B.F., Waren, C.E., Bond, C.E. and Doudoroff, P., Avoidance reactions of salmonid fishes to pulp mill effluents, *Sewage Industrial Wastes*, 28, 1403–1413, 1956.
106. Smith, W. and Saalfeld, R.W., Studies on the Columbia River smelt, *Thaleichthys pacificus* (Richardson), Washington Department of Fisheries, *Fish. Res. Paper*, 1, 3–26, 1955.

107. Jones, J.R.E., The reactions of *Pygosteus pungitius* L. to toxic solutions, *J. Exp. Biol.*, 24, 110–122, 1947.
108. Domanik, A.M. and Zar, J.H., The effect of malathion on the temperature selection response of the common shiner, *Notropis cornutus* (Mitchill), *Arch. Environ. Contam. Toxicol.*, 7 (2), 193–206, 1978.
109. Kwak, I.-S., Chon, T.-S., Kang, H.-M., Chung, N.-I., Kim, J.-S., Koh, S.C., Lee, S.-K. and Kim, Y.-S., Pattern recognition of the movement tracks of medaka (*Oryzias latipes*) in response to sub-lethal treatments of an insecticide by using artificial neural networks, *Environ. Pollut.*, 120, 671–681, 2002.
110. Rand, G.M., The effect of exposure to a subacute concentration of parathion on the general locomotor behavior of the goldfish, *Bull. Environ. Contam. Toxicol.*, 18 (2), 259–266, 1977.
111. Weis, J.S. and Weis P., Effects of exposure to lead on behavior of mummichog (*Fundulus heteroclitus* L. larvae, J. *Exp. Mar. Biol. Ecol.*, 222, 1–10, 1998.
112. Schneider, M.J., Barraclough, S.A., Genoway, R.G. and Wolford, M.L., Effects of phenol on predation of juvenile rainbow trout Salmo gairdneri, *Environ. Pollut. Ser. A*, 23 (2), 121–130, 1980.
113. Hasler, A.D. and Wisby, W.J, Use of fish for olfactory assay of pollutants (phenols) in water, *Trans. Am. Fish. Soc.*, 79, 64–70, 1950.
114. Wildish, D.J., Akagi, H. and Poole, N.J., Avoidance by Herring of Sulfite Pulp Mill Effluents, International Council for the Exploration of the Sea (ICES), Fisheries Improvement Committee, CM 1976/E, 26, 1976, 8 p.
115. Hogue, C.C., Avoidance responses of rainbow trout and Utah chub to rotenone, *N. Am. J. Fish. Manage.*, 19 (1), 171–179, 1999.
116. Lindahl, P.E. and Marcstrom, A., On the preference of roaches (*Leuciscus rutilis*) for trinitrophenol, studied with the fluvarium technique, *J. Fish. Res. Board Can.*, 15, 685–694, 1958.
117. Pedder, S.C.J. and Maly, E.J., The avoidance response of groups of juvenile brook trout, *Salvelinus fontinalis*, to varying levels of acidity, *Aquat. Toxicol.*, 8 (2), 111–119, 1986.
118. Kitamura, S. and Ikuta, K., Effects of acidification on salmonid spawning behavior, *Water, Air, Soil Pollut.*, 130 (1–4), 875–880, 2001.
119. Israeli-Weinstein, D. and Kimmel, E., Behavioral response of carp (*Cyprinus carpio*) to ammonia stress, *Aquaculture*, 165, 81–93, 1998.
120. Woltering, D.M., Hedtke, J.L. and Weber, L.J., Predator-prey interactions of fishes under the influence of ammonia, *Trans. Am. Fish. Soc.*, 107 (3), 500–504, 1978.
121. Daoust, P.Y. and Ferguson, H.W., Nodular gill disease: a unique form of proliferative gill disease in rainbow trout *Salmo gairdneri* Richardson, *J. Fish Dis.*, 8, 511–522, 1985.
122. Scott, A.L. and Rogers, W.A., Histological effects of prolonged sublethal hypoxia on channel catfish, *Ictalurus punctatus* (Rafinesque), *J. Fish Dis.*, 3, 305–316, 1980.
123. Francis-Floyd, R., Behavioral diagnosis, Veterinary Clinics of North America, *Small Anim. Pract.*, 16, 303–314, 1988.
124. Wannamaker, C.M. and Rice, J.A., Effects of hypoxia on movements and behavior of selected estuarine organisms from the southeastern United States, *J. Exp. Mar. Biol. Ecol.*, 249, 145–163, 2000.
125. Jones, J.R.E., The reactions of fish to water of low oxygen concentration, *J. Exp. Biol.*, 29, 403–415, 1952.
126. Schweldler, T.E., Tucker, C.S., Beleau, C.S. and Beleau, M.H., Non-infectious diseases, in *Channel Catfish Culture*, Tucker, C.S., Ed., Elsevier, Amsterdam, 1985, pp. 497–541.
127. Hartwell, S.I., Cherry D.S. and Cairns, J. Jr., Field validation of avoidance of elevated metals by fathead minnows (*Pimephales promelas*) following in situ acclimation, *Environ. Contam. Toxicol.*, 6, 189–200, 1987.
128. Woodward, D.F., Goldstein, J.N. and Farag, A.M., Cutthroat trout avoidance of metals and conditions characteristic of a mining waste site: Coeur d'Alene River, Idaho, *Trans. Am. Fish. Soc.*, 126, 699–706, 1997.

129. Sprague, J.B., Avoidance reactions of rainbow trout to zinc sulfate solutions, *Water Res.*, 2, 367–372, 1968.
130. Hansen, J.A., Rose, J.D., Jenkins, R.A., Gerow, K.G. and Bergman, H.L., Chinook salmon (*Oncorhynchus tshawytscha*) and rainbow trout (*Oncorhynchus mykiss*) exposed to copper. 2. Neurophysiological and histological effects on the olfactory system, *Environ. Toxicol. Chem.*, 18, 1979–1991, 1999b.
131. Beitinger, T.L., Behavioral reactions for the assessment of stress in fishes, *Great Lakes Res.*, 16, 495–528, 1990.
132. Hara, T.J. and Thompson, B.E., The reaction of whitefish, *Coregonus clupeaformis*, to the anionic detergent sodium lauryl sulphate and its effects on their olfactory responses, *Water Res.*, 12, 893–897, 1978.
133. Timms, A.M., Kleerekoper, H. and Matis, J., Locomotor response of goldfish, channel catfish, and largemouth bass to a "copper-pollluted" mass of water in an open field, *Water Resour. Res.*, 8, 1574–1580, 1972.
134. Lawrence, M. and Scherer, E., Behavioural Responses of Whitefish and Rainbow Trout to Drilling Fluids, Technical Report No. 502, Fisheries and Marine Service, Winnepeg (Manitoba), 47 p., 14 figures, 3 tables, 21 references, 1974.
135. Atema, J., Sublethal effects of petroleum fractions on the behavior of the lobster, *Homarus americanus*, and the mud snail, *Nassaruis obsoletus*, in *Uses, Stresses, and Adaptations to the Estuary*, Wiley, M., Ed., Academic Press, New York, 1976, pp. 302–312.
136. McNicol, R.E. and Scherer, E., Behavioral responses of lake whitefish (*Coregonus clupeaformis*) to cadmium during preference-avoidance testing, *Environ. Toxicol. Chem.*, 10, 225–234, 1991.
137. Cherry, D.S. and Cairns, J. Jr., Biological monitoring, preference and avoidance studies, *Water Res.*, 16, 263–301, 1982.
138. Hara, T.J., Brown S.B. and Evans, R.E., Pollutants and chemoreception in aquatic organisms, in *Aquatic Toxicology*, Nriagu, J.O., Ed., John Wiley & Sons, New York, 1983, 525 p.
139. Brewer, S.K., Little E.E., DeLonay, A.J., Beauvais, S.B. and Jones, S.B., The use of automated monitoring to assess behavioral toxicology in fish, linking behavior and physiology, in *Environmental Toxicology and Risk Assessment: Standardization of Biomarkers For Endocrine Disruption and Environmental Assessment*, vol. 8, ASTM STP 1364, Henshel, D.S., Black, M.C. and Harrass, M.C., Eds., American Society for Testing and Materials, Philadelphia, PA, 1999, pp. 370–386.
140. Little, E.E., Dwyer, F.J., Fairchild, J.F., DeLonay, A.J. and Zajicek, J.L., Survival and behavioral response of bluegill during continuous and pulsed exposures to the pyrethroid insecticide ES-fenvalerate, *Environ. Toxicol. Chem.*, 12, 871–878, 1993b.
141. Cleveland, L., Little, E.E., Ingersoll, C.G., Wiedmeyer, R.H. and Hunn, J.B., Sensitivity of brook trout to low pH, low calcium and elevated aluminum concentrations during laboratory pulse exposures, *Aquat. Toxicol.*, 19, 303–317, 1991.
142. Miller, D.C., Lang, W.H., Graeves, J.O.B. and Wilson, R.S., Investigations in aquatic behavioral toxicology using a computerized video quantification system, in *Aquatic Toxicology and Hazard Assessment: Fifth Conference*, ASTM STP 766, Pearson, J.G., Foster, R.B. and Bishop, W.E., Eds., American Society for Testing and Materials, Philadelphia, PA, 1982, pp. 206–220.
143. Smith, E.H. and Bailey, H.C., Development of a system for continuous biomonitoring of a domestic water source for early warning of contaminants, in *Automated Biomonitoring: Living Sensors as Environmental Monitors*, Gruber, D.S. and Diamond, J.M., Eds., Ellis Horwood Ltd., Chichester, UK, 1988, 208 p.
144. Laurence, G.C., Comparative swimming abilities of fed and starved larval largemouth bass (*Micropterus salmoides*), *J. Fish Biol.*, 4 (1), 73–78, 1972.
145. Steele, C.W., Effects of exposure to sublethal copper on the locomotor behavior of the sea catfish, *Arius felis*, *Aquat. Toxicol.*, 4, 83–93, 1983.
146. Mathers, R.A., Brown, J.A. and Johansen, P.H., The growth and feeding behaviour responses of largemouth bass (*Micropterus salmoides*) exposed to PCP, *Aquat. Toxicol.*, 6 (3), 157–164, 1985.

147. Sandheinrich, M.B. and Atchison, G.J., Sublethal toxicant effects on fish foraging behavior: empirical vs. mechanistic approaches, *Environ. Toxicol. Chem.*, 9, 108–120, 1990.
148. Weis, J.S., Smith, G., Zhou, T., Santiago-Bass, C. and Weis, P., Effects of contaminants on behavior: biochemical mechanisms and ecological consequences, *BioScience*, 5193, 209–217, 2001.
149. Nacci, D.E, Champlin, C.L, McKinney, R. and Jayaraman, S., Predicting the occurrence of genetic adaptation to dioxinlike compounds in populations of the estuarine fish, *Fundulus heteroclitus*, *Environ. Toxicol. Chem.*, 21 (7), 1525–1532, 2002.
150. Shedd, T.R., van der Schalie, W.H., Widder, M.W., Burton, D.T. and Burrows, E.P., Long-term operation of an automated fish biomonitoring system for continuous effluent acute toxicity surveillance, *Bull. Environ. Contam. Toxicol.*, 66, 392–399, 2001.
151. Pavlov, D.S. and Kasumyan, A.O., Patterns and mechanisms of schooling behavior in fish: a review, *J. Ichthyol.*, 40, S163–S231, 2000.
152. Partridge, B.L., The structure and function of fish schools, *Scientific American*, 246, 114–123, 1982.
153. Holcombe, G.W., Fiandt, J.T. and Phipps, G.L., Effects of pH increases and sodium chloride additions on the acute toxicity of 2,4-dichlorophenol to the fathead minnow, *Water Res.*, 14, 1073–1077, 1980.
154. Hoglund, L.B., The reactions of fish in concentration gradients, *Fish Board of Sweden Institute of Freshwater Resources, Drottingholm Republic*, 43, 1–147, 1953.
155. Bishai, H.M., Reactions of larvae and young salmonids to water of low oxygen concentration, *J. du Conseil*, 27, 167–180, 1962a.
156. Bishai, H.M., Reactions of larvae and young salmonids to different hydrogen ion concentrations, *J. du Conseil*, 27, 187–191, 1962b.
157. Sprague, J.B, Elson, P.F. and Saunders, R.L., Sublethal copper-zinc pollution in a salmon river — a field and laboratory study, *Int. J. Air Water Pollut.*, 9, 531–543, 1965.
158. Hill, G., Oxygen preference in the spring cavefish, *Chologaster agassizi*, *Trans. Am. Fish. Soc.*, 97, 448–454, 1968.
159. Hansen, D.J., Avoidance of pesticides by untrained sheepshead minnows, *Trans. Am. Fish. Soc.*, 98, 426–429, 1969.
160. Kleerekoper, H., *Olfaction in Fishes*, Indiana University Press, Bloomington, 1969.
161. Sprague, J.B. and Drury, D.E., Avoidance reactions of salmonid fish to representative pollutants, in *Advances in Water Pollution Research*, Jenkins, S.H., Ed., Pergamon Press, London, 1969, pp. 169–179.
162. Rehnoldt, R. and Bida, G., Fish avoidance reactions, *Bull. Environ. Contam. Toxicol.*, 5, 205–206, 1970.
163. Davy, F.B., Kleerekoper, H. and Gensler, P., Effects of exposure to sublethal DDT on the locomotor behavior of the goldfish (*Carassius auratus*), *J. Fish. Res. Board Can.*, 29, 1333–1336, 1972.
164. Davy, F.B., Kleerekoper, H. and Matis, J.H., Effects of exposure to sublethal DDT on the exploratory behavior of goldfish (*Carassius auratus*), *Water Resour. Res.*, 9, 900–905, 1973.
165. Hansen, D.J., Mathews, E., Nall, S.L. and Dumas, D.P., Avoidance of pesticides by untrained mosquitofish, *Gambusia affinis*, *Bull. Environ. Contam. Toxicol.*, 8, 46–51, 1972.
166. Scherer, E. and Nowak, S., Apparatus for recording avoidance movements of fish, *J. Fish. Res. Board Can.*, 30, 1594–1596, 1973.
167. Westlake, G.F. and Lubinski, K.S., A chamber to monitor the locomotor behavior of free-swimming aquatic organisms exposed to simulated spills, in *Proceedings of the 1976 National Conference on Control of Hazardous Material Spills*, Information Transfer, Inc., Rockville, MD, 1976, pp. 64–69.
168. Spoor, W.A., Neiheisel, T.W. and Drummond, R.A., An electrode chamber for recording respiratory and other movements of free-swimming animals, *Trans. Am. Fish. Soc.*, 100, 22–28, 1971.

169. Spoor, W.A. and Drummond, R.A., An electrode for detecting movement in gradient tanks, *Trans. Am. Fish. Soc.*, 101 (4), 714–715, 1972.
170. Drummond, R.A. and Carlson, R.W., Procedures for Measuring Cough (Gill Purge) Rates of Fish, EPA-600/3-77-133, EPA Environmental Research Laboratory, Duluth, MN, USA, 1977.
171. Waller, W.T. and Cairns, J. Jr., The use of fish movement patterns to monitor zinc in water, *Water Res.*, 6, 257–269, 1972.
172. Kane, A.S., Salierno, J.D., Gipson, G.T., Molteno, T. and Hunter, C., A video-based movement analysis system to quantify behavioral stress responses in fish, Water Res. 38: 3993–4001, 2004.
173. Byers, J.A. and Unkrich, M.A., Electronic light intensity control to simulate dusk and dawn conditions, *Ann. Entomol. Soc. Am.*, 76, 556–558, 1983, http://www.wcrl.ars.usda.gov/cec/papers/aesa83.htm.
174. Alpharma, MS222 (Tricane methane sulphonate), World Wide Web Alpharma Technical Bulletin. Animal Health Ltd., Fordingbridge, Hampshire, UK, May 2001, www.alpharmaanimalhealth.co.uk/VPDF/MS%20222.pdf.
175. Westlake, G.F., Kleerekoper, G.H. and Matis, J., The locomotor response of goldfish to a steep gradient of copper ions, *Water Resour. Res.*, 10, 103–105, 1974.
176. Hayward, J., Oxford, J.D., and Walley, W.B. 1989. Three implications of fractal analysis of particle outlines. *Comput. Geosci.* 15: 199–207.

chapter thirty-three

Measuring metals and metalloids in water, sediment, and biological tissues

Michael C. Newman and Yuan Zhao
Virginia Institute of Marine Science

Contents

1-56670-664-5/05/$0.00+$1.50
© 2005 by CRC Press

Introduction

Outbreaks of cadmium (Itai–Itai disease) and mercury (Minamata disease) poisoning during the 1950s made us acutely aware of the adverse consequences of high metal concentrations in our environment. Measurement of metals and metalloids became an integral component of our efforts to monitor and correct effects of anthropogenic emissions. The widespread introduction of commercial atomic absorption spectrophotometers (AAS) in the early 1960s[1,2] contributed enormously to the rapid increase in essential data. Today flame and flameless capabilities are incorporated together in most AAS units, allowing convenient measurement of elements present in "milligram per gram" to "microgram per kilogram" concentrations. Well-established preconcentration procedures are used to remove analytes from interfering matrices, as well as to concentrate them in small volumes. Flame and furnace chemistries are now sufficiently well understood to allow effective matrix modification for most elements.

Inductively coupled plasma-mass spectrometry (ICP-MS), a more recent technology, can also be used for rapid ultratrace multielement analysis. It consists of an ICP ion source, a quadrupole or magnetic sector mass filter, and an ion detection system.[3] The detection sensitivity of ICP-MS is generally better than the graphite furnace AAS. One important feature is that it can detect and quantify small variations on isotopic compositions in geological and environmental samples. (Relative isotope abundances of the metals are shown in Appendix I.) Only the most fundamental techniques for measuring metals dissolved in waters or present in solid samples are described herein. More up-to-date information about measurement of metals with ICP-MS can be found in the literature, e.g., References 4–6.

Materials required

General

Chemical safety

The analyst should not prepare or use any of the following reagents without fully understanding steps necessary for their safe use. Eye protection, protective clothing, and plastic gloves should be worn when handling strong acids. Always handle strong acids in a fume hood. The methods described here avoid the use of perchloric acid as it can react violently with organic materials. Nitric acid is a safer and spectroscopically cleaner; however, digestion of some samples with nitric acid may be unacceptable.

High purity acids

High purity nitric and hydrochloric acids may be purchased or produced. Commercially available acids include Ultrex® (Baker, Phillipsburg, NJ), Suprapur® (Merck, Darmstadt, Germany), and Aristar® (BDH Chemicals, Poole, England). Sub-boiling distillation with a Teflon® still, such as that sold by Berghof/America, Inc. (Raymond, NH), may also be used to generate these acids. Store distilled acids in acid-cleaned Teflon containers.

Acid cleaning of laboratory ware

No one procedure ensures both optimal allocation of effort and contamination-free analyses for all situations. Initially, the analyst should use procedural blanks to gain an understanding of the effort required for efficient, contamination-free analyses. Contamination control is one of the most crucial aspects in analysis of many trace metals.

Although expensive, it is preferable to use Teflon containers and laboratory ware. Linear polyethylene, polycarbonate plasticware and, perhaps, Pyrex are less costly alternatives. For example, some analysts may find a Pyrex filtration apparatus more convenient than the suggested polycarbonate unit in the dissolved metals procedure below. Pyrex volumetric flasks are a necessary compromise for careful volumetric measurement, although liquids may be weighed to avoid potential contamination from volumetric glassware. Rubber stoppers or seals, soft glass, Bakelite® caps, and caps made with contaminating seals, such as paper or aluminum, should not be used.

Clean all plasticware and glassware with a noncationic detergent (e.g., Acationox®, American Scientific Products, Boston, MA, or Citranox®, Alconox, White Plains, NY, 10603) before acid cleaning. Acid soak all materials that will touch the sample at least 24 h in 50% (v/v) concentrated nitric acid in deionized water. Rinse items 7–10 times with deionized water and place them in a contamination-free area to dry (e.g., under a Class 100 laminar flow clean hood). Adjust the duration of soaking, rinsing, and level of contamination control during drying in balance with the concentration of analyte expected. Items for the analysis of lead in water may require extreme attention to restricting contact with dust particles during drying, yet those used for the analysis of calcium in sediment will require much less attention. Use of procedural blanks aids in adjusting procedures to the appropriate level of rigor. The results from procedural blanks should dictate the specific details of washing and preparation, not rote adherence to any prescribed cleaning steps.

Many problems are eliminated if laboratory wares used for trace element measurement are used only for those analyses. Such dedicated laboratory wares should be stored in sealed plastic bags, e.g., Ziploc bags. Acid-cleaned pipette tips can be conveniently stored in an acid-cleaned, Teflon or linear polyethylene bottle with a polyethylene cap. Use these items as soon as reasonable after acid cleaning to minimize contamination during storage.

Soak membrane filters for dissolved metals analysis in dilute, high purity hydrochloric acid at least for 4 h. A 0.5 *N* HCl soaking solution can be made for this purpose by carefully adding 4 ml of concentrated, high purity hydrochloric acid to 92 ml of deionized water. Soak the filters for longer duration if blanks indicate contamination. Soaking can be done in a covered, acid-cleaned Teflon beaker (100 ml). Rinse filters thoroughly with metal-free deionized water prior to use. Acid-washed plastic or Teflon-coated tweezers should be used carefully to handle the filters.

If dissection is required in preparation of biological samples, metal-cutting tools can easily introduce metals, such as iron, zinc, and chromium. The authors use plastic, disposable utensils, such as those sold in grocery stores for picnics, as dissecting utensils. Plastic knives may be sharpened with nonmetallic abrasives, but they should be carefully cleaned and checked for contamination before use. These plastic utensils may be destroyed by prolonged soaking in strong acid. A dilute (10%, v/v) concentrated nitric acid soak for 12 h is probably adequate. However, the researcher should test one utensil in any cleaning procedure prior to general implementation, as some are destroyed by more concentrated acid solutions.

Standard materials

Standard reference materials are the most effective means of troubleshooting analytical problems during method development and of documenting accuracy during sample analysis. Sample spikes are useful in the absence of a standard material or to augment method troubleshooting. The U.S. National Institute of Standards and Technology (NIST) has standard materials for water, sediment, and animal tissue. The author also uses materials supplied by the National Resource Council of Canada. Veillon[7] and Van Loon[8] list other sources of standard materials for biological tissues. At present, convenient suppliers include:

Standard reference materials

- National Institute of Standards and Technology, 100 Bureau Drive, Stop 2322, Gaithersburg, MD 20899-2322, USA
- U.S. Geological Survey, Crustal Imaging & Characterization Team, Box 25046, MS 973, Denver, CO 80225, USA
- Brammer Standard Co., Inc., 14603 Benfer Rd, Houston, TX 77069-2895, USA
- National Research Council of Canada, Institute for National Measurement Standards, 1200 Montreal Road, Bldg. M-36, Ottawa, Ontario, Canada K1A 0R6
- BSI, Customer Services, 389 Chiswick High Road, London, W4 4AL

Standard metal solutions

Stock solutions of 1000 μg/ml for preparing AAS standards can be purchased from several sources, including Fisher Scientific (Pittsburgh, PA), Sigma (St. Louis, MO), and Perkin-Elmer (Norwalk, CT). Standards have a finite shelf life and should not be used beyond their expiration dates.

Matrix modifiers

Atomic absorption spectrophotometry measures the absorption of light at a specific wavelength (λ) by the ground state atom (M^0) during excitation: $M^0 + E_\lambda \rightarrow M^*$. The absorption is proportional to the number of ground state atoms in the light path. In turn, the number of ground state atoms is a function of the concentration of the element in the sample and physicochemical reactions occurring in the flame or graphite furnace (Figure 33.1). The reactions occurring in the flame or furnace can be manipulated with matrix modifiers to optimize production of ground state atoms and, thus, enhance absorption (sensitivity).

For flame AAS, the sample is aspirated to form small droplets in a premix chamber. These droplets are swept into the flame and form dry "clotlets" of salts and other solids. The clotlets melt and vaporize in the flame. Simple metal compounds, such as the metal monoxides illustrated here, undergo a sequence of transitions that, at equilibrium, determine the measured population of ground state atoms (M^0):

$$MO + E_{\text{heat}} \leftrightarrow M^0 + O^0 \tag{33.1}$$

$$M^0 + E_\lambda \leftrightarrow M^* \tag{33.2}$$

$$M^0 + E_{\text{heat}} \leftrightarrow M^+ + e^- \tag{33.3}$$

Selection of a hot flame (e.g., nitrous oxide–acetylene, 2900°C) instead of a cool flame (e.g., air–acetylene, 2300°C) for elements with metal monoxide dissociation constants greater than approximately 5 eV assures that reaction (33.1) will produce ample amounts of M^0.[2] (Pertinent dissociation constants are listed in Appendix I.) A reagent (e.g., La) may also be added, which reacts with components in a droplet that would otherwise combine with the analyte to make it resistant to dissociation. If the flame is too hot, ionization in reaction (33.3) shifts the equilibrium to decrease the amount of M^0 present (Figure 33.2).

Elements with low ionization energies (see Appendix I), such as the alkali and alkaline-earth metals, are most prone to ionization.[9] Rare earth elements can also have ionization problems with a hot (nitrous oxide–acetylene) flame.[9] Adding excess amounts

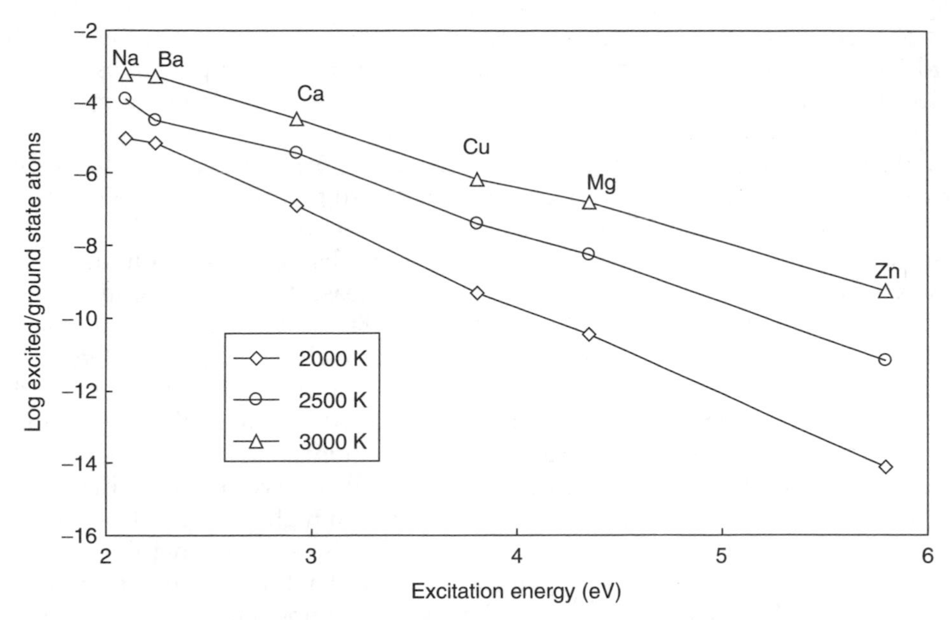

Figure 33.1 The relative amount of atoms excited and at ground state for flames with three different temperatures (Na, Ba, Ca, Cu, Mg, and Zn). Both temperature of the flame and the excitation energy determine the proportion of atoms that are excited. Regardless of excitation energy, the log (number of excited atoms/number of ground state atoms) increases with flame temperature. For any given temperature, the proportion of atoms excited decreases with increasing excitation energies. (Data from Table VIII-5 of Reference 60.)

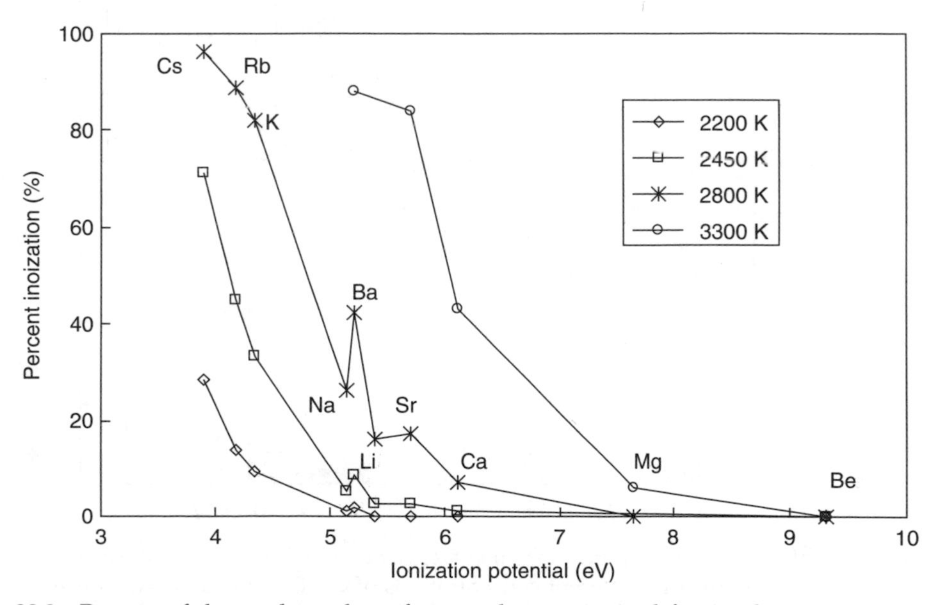

Figure 33.2 Percent of the total number of atoms that are ionized for 10 elements aspirated into different temperature flames. Note that some (e.g., Cs) have high proportions of atoms ionized even in a lower temperature (2450 K, oxygen–hydrogen) flame. Others (e.g., Mg) have minimal ionization even in the hottest (3300 K, nitrous oxide–acetylene) flame. (Data from Table 12-3 of Reference 61.)

of any element that ionizes readily (i.e., an ionization buffer) can shift the equilibrium back to favor M^0 (Figure 33.3). As is evident from their ionization potentials (E_i), cesium ($E_i = 3.89$), potassium ($E_i = 4.34$), sodium ($E_i = 5.14$) or, even, lanthanum ($E_i = 5.58$) can be used as ionization buffers.

For flameless AAS, a matrix modifier might be added for several reasons. In the furnace, a sample is dried onto a graphite surface and then heated to drive off any compounds that would interfere during subsequent measurement of analyte. After such pyrolysis, the sample is finally heated to a temperature sufficient to dissociate compounds containing the analyte. The amount of ground state atoms is measured at this atomization stage. Obviously, any material remaining after the pyrolysis stage can interfere with measurement during atomization. The presence of several salts of the analyte with their different dissociation constants may also change the shape (broaden or skew) of the absorbance curve during atomization and compromise quantification. Matrix modifiers (e.g., the Pd–Mg modifier described in the following or thiourea addition for cadmium analysis[10]) can be added to the sample that increase the temperature at which the analyte dissociates. In this way, more potentially interfering compounds can be driven off with higher temperatures during the pyrolysis stage without significant loss of analyte. This is especially helpful for elements with low boiling temperatures (see Appendix I) or those that form volatile compounds.

A matrix modifier can also be added to make components of the sample more volatile. Ediger[11] recommended additions of ammonium nitrate to samples containing large amounts of sodium chloride because the resulting sodium nitrate and ammonium chloride salts could be driven off at low pyrolysis temperatures.

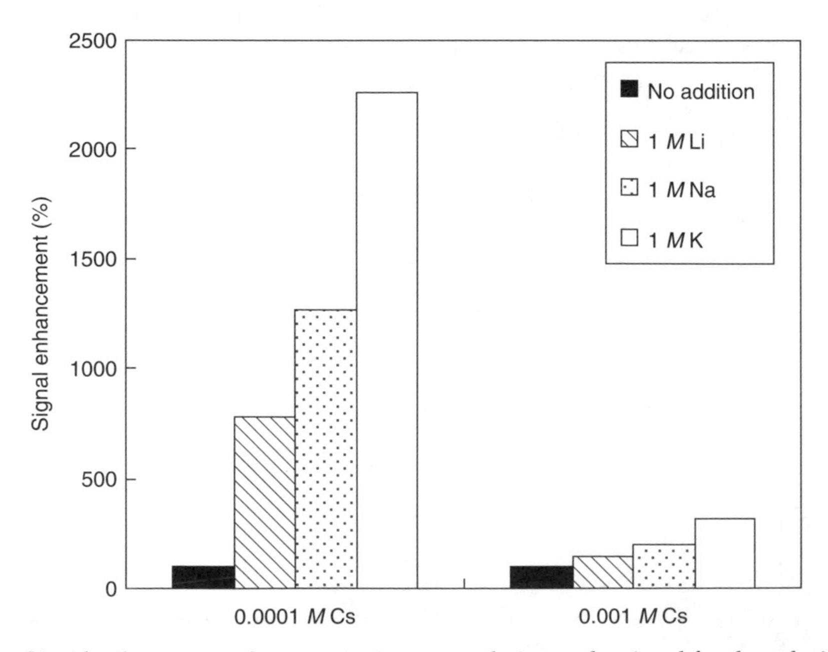

Figure 33.3 Signal enhancement (percentage increase relative to the signal for the solution with no addition of Li, Na, or K) for two different concentrations of Cs. Cesium, an easily ionized element (Figure 33.2), has a greatly enhanced signal if an excess of another easily ionized element is added. Elements with the lowest ionization potentials (see Figure 33.2) produce the best enhancement. The enhancement increases as more ionization buffer is added relative to the amount of Cs, e.g., 1 *M* ion buffer to 0.0001 *M* Cs versus 1 *M* ionic buffer to 0.001 *M* Cs. (Air–acetylene flame data from Table 12-4 of Reference 61.)

The excess of matrix modifier can also result in the preponderance of only one salt to dissociate during atomization. For example, the addition of excess H_3PO_4 to lead samples greatly improves the shape of the atomization curve by favoring formation of a single lead phosphate salt prior to atomization.[12] Similarly, Ediger[11] suggested that cadmium can be stabilized during furnace AAS by adding ammonium phosphate.

In general, the reagents recommended by Perkin-Elmer[13,14] and Schlemmer and Welz[15] are used here. Additional discussion of matrix modifiers can be obtained from References 1 and 12. The reagents described in Appendix I for analysis of each element are produced and used as described in the following. Preparation of $Pd(NO_3)_2$ and $Mg(NO_3)_2$ solutions are taken from Schlemmer and Welz.[15] The $Pd(NO_3)_2$ and $Mg(NO_3)_2$ reagents may also be purchased from Perkin-Elmer.

$Pd(NO_3)_2$ *stock solution* (1% w/v Pd): A minimum volume of Ultrex® or equivalent concentrated nitric acid is used to dissolve 1.0 g of palladium metal powder (Aldrich Chemical Co., Milwaukee, WI), and the dissolved metal is brought to 100 ml with deionized water. A small amount of metal may be dissolved first with mild heating and additional small amounts added until all is dissolved. $Pd(NO_3)_2$ can also be purchased as a salt (Sigma or Aldrich) to make this solution.

$Mg(NO_3)_2$ *stock solution* (1% w/v $Mg(NO_3)_2$): Dissolve 1.729 g of $Mg(NO_3)_2 \cdot 6H_2O$ in 100 ml of deionized water.

$NH_4H_2PO_4$ *stock solution* (1% w/v $NH_4H_2PO_4$): Dissolve 1.000 g of ammonium phosphate dibasic in 100 ml of deionized water.

5 μg Pd + 3 μg $Mg(NO_3)_2$ *reagent*: Use Eppendorf or similar calibrated pipettes with acid-cleaned tips to add the following volumes of stock solutions to a 25-ml, acid-cleaned volumetric flask: 1.250 ml of 1% Pd stock solution and 0.750 ml of the 1% $Mg(NO_3)_2$ stock solution. Bring to volume with deionized water. Add 10 μl per furnace injection for matrix modification as indicated in Appendix I.

5 μg $Mg(NO_3)_2$ *reagent*: Use Eppendorf or similar calibrated pipettes with acid-cleaned tips to add 1.250 ml of the 1% $Mg(NO_3)_2$ stock solution to a 25-ml, acid-cleaned volumetric flask. Bring to volume with deionized water. Add 10 μl per furnace injection for matrix modification as indicated in Appendix I.

15 μg $Mg(NO_3)_2$ *reagent*: Use Eppendorf or similar calibrated pipettes with acid-cleaned tips to add 3.750 ml of the 1% $Mg(NO_3)_2$ stock solution to a 25-ml, acid-cleaned volumetric flask. Bring to volume with deionized water. Add 10 μl per furnace injection for matrix modification as indicated in Appendix I.

50 μg $NH_4H_2PO_4$ + 3 μg $Mg(NO_3)_2$ *reagent*: Use Eppendorf or similar calibrated pipettes with acid-cleaned tips and acid-cleaned volumetric pipettes to add the following volumes of solutions to a 25-ml, acid-cleaned volumetric flask: 12,500 ml of 1% $NH_4H_2PO_4$ stock solution and 0.750 ml of 1% $Mg(NO_3)_2$ stock solution. Bring to volume with deionized water. Add 10 μl per furnace injection for matrix modification as indicated in Appendix I.

$LaCl_3$ *reagent* (50 g/l La): (Caution: the following solution reacts vigorously during preparation.) Dissolve very small amounts of lanthanum oxide (La_2O_3) at a time and, after complete dissolution of each small amount, add a small amount more. Cautiously continue this process until all is dissolved. Carefully and slowly dissolve a total of 29.3 g of La_2O_3 in 250 ml of concentrated hydrochloric acid. Then cautiously dilute to 500 ml with deionized water. Add 1 ml of this $LaCl_3$ solution to 10 ml of each sample, blank and standard prior to analysis. Alternatively, 133.6750 g of lanthanum chloride heptahydrate ($LaCl_3 \cdot 7H_2O$) can be added to 1 l of deionized water to make this 50 g/l La solution.

KCl *reagent* (50 g/l K): Dissolve 95 g of KCl in 1 l of deionized water. Add 200 μl of this reagent to 10 ml of each sample, blank and standard prior to analysis.

Dissolved metals in water

- Acid-cleaned 500-ml Teflon or linear polyethylene sample bottles
- Acid-cleaned 60-ml, wide mouth, Teflon or linear polyethylene bottles for filtered samples
- Acid-cleaned 47-mm, polycarbonate filtration funnel and receiving flask
- Acid-soaked 0.45-μm membrane filter (e.g., Millipore HAWP 047 or AAWP 047 filters)
- Acid-cleaned plastic or Teflon-coated tweezers for handling the filters
- Acid-cleaned 500-ml squirt bottle containing deionized water
- Source of ultrapure deionized water with consistently low concentrations of the elements of interest (e.g., ASTM Type I reagent grade water)
- Vacuum pump with a liquid trap
- Class 100 laminar flow hoods if required
- Plastic gloves (without talc powder)
- Calibrated fixed-volume pipettes (e.g., Eppendorf pipettes) of appropriate volumes
- Acid-cleaned pipette tips of appropriate sizes
- An appropriate standard material
- Insulated chest containing ice for field chilling of samples
- Plastic Ziploc® bags

Metals in biological tissue or sediments

- Acid-cleaned Teflon beakers with a spout to allow effective pouring of the digest (25, 50, or 100 ml volume depending on the weight of sample to be digested, one per sample, including blanks and standard materials)
- Acid-cleaned Teflon watch glasses adequate to completely cover the Teflon beakers or loose-fitting beaker lids, such as those sold by Berghof/American, Inc. (Raymond, NH)
- Acid-cleaned Teflon funnels for digest transfer from beakers to volumetric flasks
- Acid-cleaned 60-ml, Teflon or linear polyethylene bottles for sample digests
- Acid-cleaned, plastic dissecting utensils if required
- Acid-cleaned, Class A Pyrex 25-ml volumetric flasks with stoppers
- Acid-cleaned, Class A Pyrex 5-ml volumetric pipette
- Acid-resistant pipette bulb
- Calibrated fixed-volume pipettes (e.g., Eppendorf pipettes) of appropriate volumes
- Acid-cleaned pipette tips of appropriate sizes
- Acid-cleaned 500-ml squirt bottle of deionized water
- Source of ultrapure deionized water with consistently low concentrations of the elements of interest (e.g., ASTM Type I reagent grade water)
- Plastic gloves (without talc powder)
- 0° to 100°C thermometer (preferably traceable to an NIST thermometer)
- Hot plate capable of holding a constant temperature of 80°C
- Pyrex pan to serve as a water bath into which the Teflon beakers may be placed
- Appropriate standard materials

Procedures

Dissolved metals in water

Dissolved metals are procedurally defined as those passing through a 0.45-μm filter.

Sampling

Sufficient numbers of acid-cleaned sample bottles should be carried to the field inside sealed plastic bags. Bring enough to take replicate samples at all or a subset of sample sites. Travel blanks may be produced by pouring a deionized water "sample" into acid-cleaned containers in the field. Travel blanks should be handled like all other samples. Field spikes or field processing of solutions of known metal concentrations are also helpful in tracking accuracy and precision of the entire sampling and measurement process.

Rinse each sample bottle (including the cap) with sample at least four times prior to filling with the final sample. Do not allow the sample to touch your hand or any other contaminating materials prior to entering the sample bottle. Do not agitate sediments or dislodge materials from vegetation during sampling. Do not take a sample near the water surface as some metals are concentrated in the surface microlayer. Seal the sample bottle in a Ziploc® plastic bag and place it into an ice-filled cooler chest. Process the sample as soon as reasonable, usually within 24 h of sampling. Ideally, the sample should be filtered immediately. Adjust the above sampling instructions based on results from spiked samples and travel blanks.

Filtration

Ideally, one acid-cleaned filtration apparatus should be used for each sample. If this is unreasonable, samples could be filtered in sequence through one or a few filtration apparatus. If information is available about the relative concentrations expected in the samples, samples suspected to have low concentration should be filtered first so as to minimize potential contamination among samples. If there are extreme differences in metal concentrations among samples, samples can be grouped and filtered through separate apparatus. Regardless, results from filtration blanks should be used in any final judgment of the method adequacy.

With deionized water, thoroughly rinse all surfaces of the filtration apparatus that will contact the sample. Rinse the filter with deionized water and place it into the apparatus. Filter 50–100 ml of deionized water through the unit and discard the filtrate. Repeat this process at least three times. Use caution when disconnecting the vacuum tubing from the filtration flask because contaminating particulates can be sucked into your sample during the abrupt equalization of pressure. Initially use blanks and known metal solutions to assess the adequacy of these procedures for your specific needs and adjust as required.

Place 50 ml of water sample into this thoroughly rinsed apparatus. Allow at least 25 ml to filter into the receiving flask. Rinse the receiving flask with this filtered sample and discard the rinse. Repeat this sample rinsing process. Place more samples into the filtration funnel and collect approximately 60 ml of filtered sample. Transfer approximately 10 ml of this filtered sample into an acid-cleaned, Teflon or polyethylene bottle. Use this sample aliquot to completely rinse the bottle and then discard the rinse. Fill the bottle with the remaining 50 ml of sample. Add 100 μl of concentrated, ultraclean nitric acid to the sample using a pipette with an acid-cleaned tip. This 0.2% (v/v) nitric acid preserved sample should have a pH less than 2. (Do not place the pH probe into the sample to check the pH as this could result in gross contamination. Instead use another aliquot of filtered sample for assessing the adequacy of acidification.) Add more acid if required for adequate adjustment of your sample type.

Other methods exist for sampling dissolved metals. Windom et al.[16] pumped samples through cartridge filters in the field. The filtrate flowed directly into a polypropylene sample bottle and was acidified under a portable clean bench. This procedure requires considerable field effort and equipment but, if feasible, eliminates many ambiguities regarding changes in unfiltered water during transport and potential contamination upon contact with various laboratory wares.

Often, dissolved element concentrations are below the detection limit or, in the case of marine waters, present in a high salt matrix. The metals must then be removed from their original matrix and concentrated before analysis using one of several methods. They may be removed with solid resins (e.g., References 17 and 18), chelated and extracted with a solvent (e.g., References 19–25), or chelated and co-precipitated (e.g., Reference 26). If the chelation-extraction procedure (e.g., Reference 27) is used, it is important to keep in mind that pH of the sample is critical to effective extraction[25] that humic acids can interfere,[24] and that the potential for contamination increases with preconcentration.

Storage and analysis

Minimal storage time of dissolved metal samples is recommended. Although regulatory guidelines for maximum storage times range from 28 days to 6 months, spiked samples and sample blanks should be used to assess the best storage time for your particular needs. Store samples at 4°C, away from any potential source of contamination. They may be stored in the dark to avoid precipitation if silver concentrations are to be measured. Evaporation during storage can increase sample concentrations and should be avoided. A mark indicating the sample volume on the outside of each container can help in monitoring any potential volume changes.

Allow the sample to come to room temperature before analysis. To avoid contamination, never sample directly from the sample bottle. Instead, pour an aliquot from the bottle into an acid-cleaned container and discard the aliquot after use. Analyze the sample according to Appendix I and the AAS manufacturer's instructions.

Calculation of free metal concentration

The method described here applies to the measurement of dissolved metal, which includes free metal ions, metal–ligand complexes, metal monoxides, and other species. Notionally, the most bioavailable and toxic form is the free metal ion.[28–30] The free ion concentrations can be estimated with chemical equilibrium calculations, either manually or with computer software (e.g., Visual MINTEQ[31]). Figure 33.4 illustrates the fraction of free copper or silver to the dissolved copper or silver at different pH values, in both seawater and river water.

Metals in biological tissue or sediments

Sampling, dissection, and storage

It is especially important with solid samples, such as sediments or tissues, that a large enough sample be taken so as to be representative. Too often treated as homogeneous

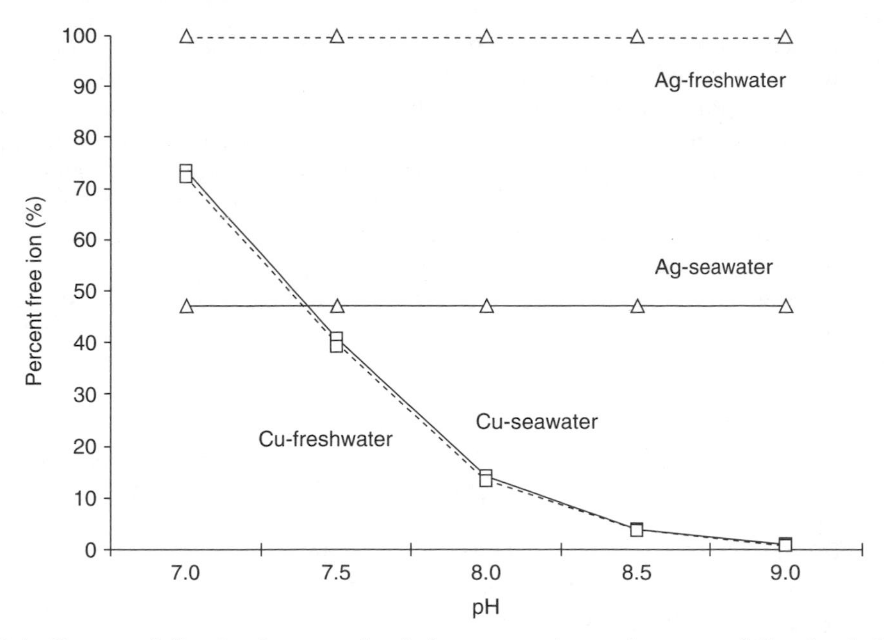

Figure 33.4 Percent of dissolved copper that is free copper ion and percent of dissolved silver that is free silver ion in normal seawater and river water at different pH conditions (calculated with Visual MINTEQ). The percent of free Cu^{2+} to dissolved Cu varies with pH of the water. In contrast, that of dissolved Ag is much less variable with the same water component.

materials, these materials range from poorly mixed to well-mixed, heterogeneous samples. Methods for calculating representative weights are provided by Ingamells and Switzer,[32] Ingamells,[33] Wallace and Kratochvi1,[34] and Newman.[35]

Sediments should be taken from a specified depth as metal concentrations and species can change rapidly with depth. If possible, nonmetallic or samplers with nonmetallic coatings (e.g., a Teflon-coated dredge) should be used to minimize contamination. The sediment may be mixed, put into an acid-cleaned, Teflon or plastic bottle, and stored on ice until return to the laboratory. It should be stored at 4°C, away from any potential source of contamination until processed.

Biological samples could be tissues, organs, or entire organisms depending on the goals of the study. If the entire animal or the gut alone is to be analyzed, the organism must be allowed ample time to clear its gut of contaminating food. Tissues and organs may be carefully dissected from the animal using acid-cleaned, plastic utensils. Whole animals or dissected tissues should be placed into acid-cleaned, wide mouth containers (Teflon or plastic) and stored frozen until freeze-drying can be performed. The 20-ml polypropylene scintillation vials with linerless caps are ideal for small samples. Placing sample vials in a Ziploc® bag along with several ice cubes can reduce sublimation from samples during prolonged storage. Appropriate standard reference materials should be processed in tandem with samples from this step onward. They will reflect the quality of the process from storage through measurement.

Freeze-drying

Loosening or removing caps from storage vials allows freeze-drying of samples in their storage containers. Freeze-dry until no more weight (water) is lost from the sample. Tighten caps onto vials prior to storage under desiccation. (Do not use desiccant that sheds small particulates that could contaminate the sample.) Digest the samples, including the standard reference materials, as soon as reasonable after freeze-drying. Note that volatile metals, including those forming volatile organic compounds, such as mercury, can be lost during freeze-drying.[36] Wet tissue can be analyzed for mercury and expressed later as dry weight using a wet-to-dry weight ratio. Relative to volatile elements, it is important to realize that most standard reference materials are freeze-dried materials that may have lost volatile compounds of analytes before they were analyzed for certified concentrations. Consequently, the lack of any apparent loss during freeze-drying of a standard material in tandem with samples may not accurately reflect loss from sample. Splits of materials analyzed before and after freeze-drying would better reflect such loss.

Digestion

Wear proper eye protection, protective clothing, and plastic gloves while working with strong acids. Digestion should be done in a hood capable of efficiently removing the generated acid fumes.

Weigh 0.1000–0.5000 g of freeze-dried tissue or sediment and record the exact weight for later calculations. (The optimum weight needed to be representative can be estimated as described by Newman.[35]) It is best to use the same weight for all samples. Place the sample into an acid-cleaned Teflon beaker and cover immediately with an acid-cleaned Teflon watch glass or lid. Approximately 5–10% of all samples to be digested should be reference materials and an additional 5–10% should be procedural blanks (beakers without sediment or tissue added). Whether 1 in 5 or 10 samples is a blank or standard material will depend in the element of interest and its concentration in the materials being analyzed. Low concentrations of easily contaminated analytes require numerous blanks and standard materials.

After all samples have been placed into beakers, slowly add 5 ml of concentrated, ultraclean nitric acid to each. More acid may be required to completely wet, and cover some types of samples, e.g., light powders. Make certain that all of the material contacts acid and no internal pockets or edges remain dry. Allow the covered beakers containing sample and acid to sit for 1 h at room temperature. This predigestion reduces the risk of vigorous foaming and loss of sample upon heating. Next, carefully place the covered beakers onto the hot plate and allow them to reflux at 80°C for 4 h. (A beaker filled with water and placed among the beakers containing samples can be used to monitor digestion temperature.) If there are wide deviations in temperature on the surface of the hot plate, a Pyrex pan partially filled with water can be placed onto the hot plate to provide more uniform heating of samples. Rotation of samples on the hot plate may also ensure that heating is less variable among samples. Periodically check to ensure that the entire bottom of each beaker remains wet and that the partially digested samples remain covered with acid. Add more acid if required to keep the entire sample wet with hot acid.

After 4 h, carefully remove the watch glass or lid from each beaker. (Some samples may require additional time to digest adequately. Use the digests of standard materials to assess the completeness of digestion or loss of analyte by excessive digestion.) Gently and carefully rinse the droplets of acid from the watch glass or lid back into the beaker using deionized water in a squirt bottle. Place the samples back onto the hot plate and allow the digest volume to evaporate to approximately 2.5 ml. After cooling, quantitatively transfer the digest to a 25-ml, acid-cleaned Class A volumetric flask using successive rinses with deionized water. Allow each digest to cool in the volumetric flask and bring the volume to nearly 25 ml with deionized water. Gently mix the digest in the flask and allow it to cool again. Carefully bring the digest to 25 ml volume. This digest has a 10% (v/v) nitric acid matrix. If concentrations allow further dilution, a 50-ml volumetric flask can be used instead to produce a 5% (v/v) nitric acid matrix. Carefully transfer the digest to an acid-cleaned Teflon, polypropylene, or polyethylene storage container.

This procedure has been used successfully by Newman and Mitz[37] and Newman et al.[38] for Pb and Zn analysis in biological tissues. Standard materials were used to indicate the adequacy of the digestion for Pb[37] (National Research Council Canada TORT-l standard) and Zn[38] (U.S. EPA Quality Control Samples, Metals in Fish). However, it may not be adequate for samples with high concentrations of lipids. Incompletely digested lipids produce a turbid digest with a surface film. An inadequate digestion will be indicated by spurious values upon repeated analysis of digest aliquots or unacceptably low concentrations for standard reference materials. More rigorous digestion is required in this case. Clegg et al.[39] and Sinex et al.[40] compare alternative digestion procedures for biological tissue and sediments, respectively. Siemer and Brinkley[41] and Chung and Tsai[42] describe simple refluxing systems for acid digests, and Uhrberg[43] describes an acid digestion bomb. Very effective methods are described that use conventional microwave ovens[44,45] or microwave ovens specifically designed for sample digestion (see the following).

Microwave ovens specifically designed for sample digestion (e.g., CEM MARS 5 Microwave Digester, Matthews, NC) are widely used at this time.[46–48] This method can provide more complete digestion than the method introduced above. The associated procedures for the CEM MARS 5 Microwave Digester are briefly introduced here. After freeze-drying, sample aliquots are weighed on an analytical balance and the exact weight recorded for use in later calculations. Transfer the entire sample to an acid-cleaned, microwave digestion vessel. Add 5 ml of concentrated, ultraclean nitric acid to the vessel and rinse the vessel sides with the acid to sweep the entire sample to the bottom. Place 5 ml of concentrated, ultraclean nitric acid as a blank in another vessel and cover it with a special top with glass tubing for the temperature probe. Make sure the rupture membrane is inserted into each blue Teflon vent cover. Insert the vessel into the support module and fasten it with a torque wrench. Thread the temperature probe into the blank vessel lid and attach the pressure sensor to the top of the vented lid. Insert the vessel modules into the turntable, make sure the weight is evenly balanced, and place the turntable into the microwave oven. Close the door firmly after attaching the temperature probe and pressure sensor to the microwave oven. Select the appropriate method and begin the digestion. During the first few minutes of the digestion, observe the vessels to ensure that no steam is venting. Venting indicates missing or damaged rupture membranes and the digestion should be stopped. Caution should be taken when removing things from the microwave oven because the vessels may be hot and pressured. After the programmed digestion is finished, wait until the temperature has dropped to room temperature. Disconnect the pressure and temperature probes. Place the turntable under an acid fume hood. When removing the module, point the vent away from yourself and slowly open the vent.

Carefully decant the sample from the vessel into a 25-ml volumetric flask. Rinse the vessel sides thrice with small volumes of ultrapure deionized water and transfer that water into the 25-ml volumetric flask. Do not exceed the 25 ml volume. Use the ultrapure deionized water to bring the sample volume to 25 ml, mix well, and store in a clean sample container. These samples are ready to be analyzed.

The nitric acid digest of sediments might not accurately reflect the concentration of bioavailable metal in many cases. Digestive enzyme, gut amino acids, and other factors can affect the bioavailability. Chemical procedures have also been developed for determining metal concentrations in various sediment fractions. Concentrations in the various fractions may be related to solid phase species and bioavailability. Tessier et al.[49,50] describe one such technique in current use. If applied, it is important to acknowledge that the resulting fractions are procedurally defined ones and artifacts might be present.[51,52] Pai et al.[53] provide specific recommendations for graphite furnace AAS analysis of the complex sediment extract matrices generated with this procedure. Digestive solubilization, which is a major route of exposure for many metals in sediments (e.g., Reference 54), can be qualified with biomimetic techniques. The digestive fluids extracted from the gut lumens of deposit feeders are added to sediments. The mixture is incubated for a certain period of time before the dissolved metals in the digestive fluids are measured by AAS or ICP-MS.[55]

Storage and analysis

Digests should be analyzed as soon as reasonable. Storage under refrigeration will reduce change in volume due to evaporation. Store the digests away from any contaminating materials.

Allow the sample to come to room temperature before analysis. To avoid contamination, never sample directly from the sample bottle. Instead, pour an aliquot from the bottle into an acid-cleaned container. Discard the aliquot after use; do not return it to the sample bottle. Analyze the sample according to Appendix I and the AAS or ICP-MS manufacturer's instructions. Any dilutions of sample digests should be performed in acid-cleaned containers with special attention given to changes in the matrix, for example, a significant change in viscosity associated with dilution. Matrix matching is often required for many elements and sample dilutions. The method of standard addition can be used to assess the amount of matrix matching required.

Results and discussion

Concentrations are estimated using aqueous or solid standards matching the sample. The method of standard addition can be used if there are matrix effects. Particularly with graphite furnace AAS and samples requiring various dilutions, it is prudent to assume that a matrix effect is present unless shown to be otherwise.

Although method detection limit is more commonly reported, measurement relative uncertainty also provides the end user with sufficient information to assess data quality.[35] If the data include "below detection limit" observations, they must be treated as censored data in subsequent statistical analyses.[56,57] It is also essential that results of standard reference materials or, minimally, spiked samples recoveries, be reported. Expected reference material values and acceptable limits should also be reported. Comment must also be made regarding results from procedural blank analysis. Without this information, the end user has no means of determining the quality of measurement.

Appendix: Analytical details for specific elements

Extracted information from the works of Dean,[58] Lide,[59] Van Loon,[8] Varma,[1] Perkin-Elmer,[13,14] and Price.[2] Sensitivities are expressed in milligram per liter to produce a 0.0044 Abs unit (1% absorption) for flame AAS. For furnace AAS, characteristic masses are picogram to produce a 0.0044 Abs unit for a Perkin-Elmer transverse heated graphite atomizer.[14] Because values for sensitivity and characteristic mass vary among instruments, those given reflect only general ranges and relative values among wavelengths: AW, atomic weight; D_0, dissociation energy of the metal monoxide in eV; BP, boiling point in °C; and E_i, ionization energy in eV. For those using ICP-MS to determine isotopic compositions, published relative abundances of each elements' isotopes are listed as percentages of total element abundance, e.g., Ag-107 51.84% below indicates that 51.84% of Ag in a sample will be the Ag-107 isotope.

Ag (silver, AW 107.868, D_0 2.2, E_i 7.57, BP 2163, Ag-107 51.84%, Ag-109 48.16%)

General: Silver solutions should be stored in an amber or opaque container as light could cause silver to precipitate.[13] Hydrochloric acid should not be used because silver will precipitate with chloride.

Flame technique: Use a lean air–acetylene flame to obtain sensitivities of 0.1 mg/l (λ = 328.1 nm) and 0.2 mg/l (λ = 338.3 nm).[2] Large amounts of aluminum can cause signal suppression.[2] Bromide, chromate, iodate, iodide, permanganate, tungstate, and chloride may precipitate silver from solution.[13]

Flameless technique: Perkin-Elmer[14] recommends addition of 5 μg Pd and 3 μg of $Mg(NO_3)_2$ to the sample. (Reagent described above.) The pyrolysis and atomization temperatures should be in the ranges of 800°C and 1500°C, respectively. The characteristic mass is approximately 4.5 pg at 328.1 nm.

ICP-MS technique: With a Perkin-Elmer SCIEX ELAN 6000 ICP-MS, the detection limit is less than 1 ng/l.

Al (aluminum, AW 26.98154, D_0 5.3, E_i 5.99, BP 2520, Al-27 100%)

General: During preparation of working standards, aluminum can quickly adsorb to glass volumetric flasks. Always add aliquots of the stock solution to acidified (nitric acid) deionized water at the bottom of each volumetric flask to avoid adsorption.

Flame technique: Respectively, the 308.2, 309.3, and 396.2 nm wavelengths have sensitivities of approximately 1.5, 1.1, and 1.1 mg/l[13] with the use of a rich nitrous oxide–acetylene flame and the presence of excess La or K to suppress ionization. ($LaCl_3$ and KCl solutions described above.) Price[2] reported slight signal suppression by calcium, silicon, and perchloric and hydrochloric acids. Acetic acid, titanium, and iron may enhance signal.[13]

Flameless technique: Perkin-Elmer[14] recommends addition of 15 μg of $Mg(NO_3)_2$ to the sample. (Reagent described above.) The pyrolysis and atomization temperatures should be in the ranges of 1200°C and 2300°C, respectively. The characteristic mass is approximately 31 pg at 309.3 nm.

ICP-MS technique: With a Perkin-Elmer SCIEX ELAN 6000 ICP-MS, the detection limit is 1–10 ng/l.

As (arsenic, AW 74.9216, D_0 5.0, E_i 9.81, BP 615 (hydride AsH_3-55, As-75 100%)

Flame technique: Generally, a hydride generation method is used. However, a rich air–acetylene flame will produce sensitivities of 0.78, 1.0, and 2.0 mg/l at the 189.0, 193.7, and 197.2 nm wavelengths. Background absorption is high at these wavelengths.

Flameless technique: Perkin-Elmer[10] recommends addition of 5 μg Pd and 3 μg of $Mg(NO_3)_2$ to the sample. (Reagent described above.) The pyrolysis and atomization temperatures should be in the ranges of 1200°C and 2000°C, respectively. The characteristic mass is approximately 40.0 pg at 193.7 nm.

ICP-MS technique: With a Perkin-Elmer SCIEX ELAN 6000 ICP-MS, the detection limit is 1–10 ng/l.

Ba (barium, AW 137.33, D_0 5.8, E_i 5.21, BP 1805, Ba-130 0.11%, Ba-132 0.10%, Ba-134 2.42%, Ba-135 6.59%, Ba-136 7.85, Ba-137 11.23%, Ba-138 71.7%)

Flame technique: Price[2] and Varma[1] suggest a rich nitrous oxide–acetylene flame with an ionization buffer (KCl solution described above) at 553.6 nm to obtain a sensitivity of approximately 0.4 mg/l. Price[2] recommends analysis without an ionization buffer if the less sensitive 455.4-nm wavelength is used (sensitivity = 2.0 mg/l).

Flameless technique: The pyrolysis and atomization temperatures should be in the ranges of 1200°C and 2300°C, respectively. The characteristic mass is approximately 15.0 pg at 553.6 nm.

ICP-MS technique: With a Perkin-Elmer SCIEX ELAN 6000 ICP-MS, the detection limit is less than 1 ng/l.

Be (beryllium, AW 9.01218, D_0 4.6, E_i 9.32, BP 2472, Be-9 100%)

General: This toxic element should be handled with appropriate caution.

Flame technique: A rich nitrous oxide–acetylene flame is recommended. Price[2] reports a sensitivity of 0.05 mg/l at 234.9 nm. Absorption may be enhanced by nitric and sulfuric acids.[2] High concentrations of aluminum, sodium, silicon, and magnesium can depress sensitivity.[1,2,13]

Flameless technique: Perkin-Elmer[14] recommends addition of 15 μg of $Mg(NO_3)_2$ to the sample. (Reagent described above.) The pyrolysis and atomization temperatures should be in the ranges of 1500°C and 2300°C, respectively. The characteristic mass is approximately 2.5 pg at 234.9 nm.

ICP-MS technique: With a Perkin-Elmer SCIEX ELAN 6000 ICP-MS, the detection limit is 1–10 ng/l.

Bi (bismuth, AW 208.9804, D_0 3.6, E_i 7.29, BP 1564, Bi-209 100%)

Flame technique: Price[2] reports the following approximate sensitivities: 0.8 (223.1 nm), 1.5 (222.8 nm), 10 (227.7 nm), and 2.2 (306.8 nm) mg/l with a lean air–acetylene flame.

Flameless technique: Perkin-Elmer[14] recommends addition of 5 μg Pd and 3 μg of $Mg(NO_3)_2$ to the sample. (Reagent described above.) The pyrolysis and atomization temperatures should be in the ranges of 1100°C and 1700°C, respectively. The characteristic mass is approximately 60.0 pg at 223.1 nm.

ICP-MS technique: With a Perkin-Elmer SCIEX ELAN 6000 ICP-MS, the detection limit is less than 1 ng/l.

Ca (calcium, AW 40.08, D_0 4.8, E_i 6.11, BP 1494, Ca-40 96.95%, Ca-42 0.65%, Ca-43 0.14%, Ca-44 2.86%, Ca-46 0.004%, Ca-48 0.19%)

Flame technique: With a lean nitrous oxide–acetylene flame and La (or alkali salt) additions, sensitivity is approximately 0.03 mg/l but increases twofold with use of a stoichiometric or lean air–acetylene flame at 422.7 nm.[2] Sensitivity is 20 mg/l at 239.9 nm (air–acetylene flame). [KCl and $LaCl_3$ solutions described above can be used to reduce ionization (KCl) or chemical interferences ($LaCl_3$).] With the lean air–acetylene flame, high concentrations of aluminum, beryllium, phosphorus, silicon, titanium, vanadium, or zirconium can interfere with analysis.[1,2,13]

Flameless technique: The pyrolysis and atomization temperatures should be in the ranges of 1100°C and 2500°C, respectively. The characteristic mass is approximately 1.0 pg at 422.7 nm.

ICP-MS technique: With a Perkin-Elmer SCIEX ELAN 6000 ICP-MS, the detection limit is 10–100 ng/l.

Cd (cadmium, AW 112.41, D_0 1.5, E_i 8.99, BP 767, Cd-106 1.25%, Cd-108 0.89%, Cd-110 12.49%, Cd-111 12.80%, Cd-112 24.13%, Cd-113 12.22%, Cd-114 28.73%, Cd-116 7.49%)

General: With graphite furnace AAS, contamination can be a significant problem. Use special care in acid cleaning and sample preparation. This element is relatively toxic and should be handled appropriately.

Flame technique: With a lean air–acetylene flame, sensitivity is 0.015 mg/l at 228.8 nm and 20 mg/l at 326 nm.[2] High concentrations of silicate interfere with analyses.[13]

Flameless technique: Perkin-Elmer[10] recommends addition of 50 μg $NH_4H_2PO_4$ and 3 μg of $Mg(NO_3)_2$ to the sample. (Reagent described above.) The pyrolysis and atomization temperatures should be in the ranges of 700°C and 1400°C, respectively. The characteristic mass is approximately 1.3 pg at 228.8 nm.

ICP-MS technique: With a Perkin-Elmer SCIEX ELAN 6000 ICP-MS, the detection limit is 1–10 ng/l.

Co (cobalt, AW 58.9332, D_0 3.8, E_i 7.88, BP 2928, Co-59 100%)

Flame technique: Sensitivities using a rich air–acetylene flame for the 240.7, 242.5, 252.1, and 341.3 nm wavelengths are 0.08, 0.2, 0.5, and 4.0 mg/l, respectively.[2] Perkin-Elmer[13] and Varma[1] note that large amounts of transition and heavy metals depress signal.

Flameless technique: Perkin-Elmer[14] recommends addition of 15 μg of $Mg(NO_3)_2$ to the sample. (Reagent described above.) The pyrolysis and atomization temperatures should be in the ranges of 1400°C and 2400°C, respectively. The characteristic mass is approximately 17.0 pg at 242.5 nm.

ICP-MS technique: With a Perkin-Elmer SCIEX ELAN 6000 ICP-MS, the detection limit is 1–10 ng/l.

Cr (chromium, AW 51.996, D_0 4.4, E_i 6.77, BP 2672, Cr-50 4.35%, Cr-52 83.79%, Cr-53 9.50%, Cr-54 2.36%)

Flame technique: Sensitivities using a rich air–acetylene flame for the 357.9, 359.3, 360.5, and 425.4 nm wavelengths are approximately 0.05, 0.15, 0.2, and 0.5 mg/l, respectively.[2] Phosphate, nickel, and iron may interfere with an air–acetylene flame.[1,2,13] Perkin-Elmer[13] and Varma[1] indicate that the addition of calcium can eliminate the effect of phosphates and the addition of 2% (w/v) ammonium chloride can reduce the effect of iron.

Flameless technique: Perkin-Elmer[14] recommends addition of 15 μg of $Mg(NO_3)_2$ to the sample. (Reagent described above.) The pyrolysis and atomization temperatures should be in the ranges of 1500°C and 2300°C, respectively. The characteristic mass is approximately 7.0 pg at 357.9 nm.

ICP-MS technique: With a Perkin-Elmer SCIEX ELAN 6000 ICP-MS, the detection limit is 1–10 ng/l.

Cu (copper, AW 63.546, D_0 3.6, E_i 7.73, BP 2563, Cu-63 69.17%, Cu-65 30.83%)

Flame technique: Using a lean air–acetylene flame, Price[2] reports the following sensitivities for 217.9, 222.6, 244.2, 249.2, 324.8, and 327.4 nm wavelengths are approximately 0.6, 2.0, 40, 10, 0.04, and 0.1 mg/l, respectively.

Flameless technique: Perkin-Elmer[14] recommends addition of 5 μg Pd and 3 μg of $Mg(NO_3)_2$ to the sample. (Reagent described above.) The pyrolysis and atomization temperatures should be in the ranges of 1200°C and 1900°C, respectively. The characteristic mass is approximately 17.0 pg at 324.8 nm.

ICP-MS technique: With a Perkin-Elmer SCIEX ELAN 6000 ICP-MS, the detection limit is 1–10 ng/l.

Fe (iron, AW 55.847, D_0 4.2, E_i 7.90, BP 2862, Fe-54 9.59%, Fe-56 91.72%, Fe-57 2.20%, Fe-58 0.28%)

Flame technique: Price[2] lists the following sensitivities for the 248.3, 252.3, 271.9, 296.7, 302.1, 344.1, 372.0, 382.4, and 386.0 nm wavelengths: 0.08, 0.3, 0.5, 1.0, 0.7, 5.0, 1.0, 30, and 2.0 mg/l, respectively, using a lean air–acetylene flame. Cobalt, copper, nickel, or nitric acid can depress sensitivity, but their effects are much reduced by adjusting the flame to a very lean flame.[1,13] Organic acid, such as citric acid, may depress signal, but this effect may be minimized by using phosphoric acid.[1] Silicon effects can be reduced by the addition of 0.2% (w/v) calcium chloride.[1,13]

Flameless technique: Perkin-Elmer[14] recommends addition of 15 μg of $Mg(NO_3)_2$ to the sample. (Reagent described above.) The pyrolysis and atomization temperatures should be in the ranges of 1400°C and 2100°C, respectively. The characteristic mass is approximately 12.0 pg at 248.3 nm.

ICP-MS technique: With a Perkin-Elmer SCIEX ELAN 6000 ICP-MS, the detection limit is 1–10 ng/l.

K (potassium, AW 39.0983, D_0 2.5, E_i 4.34, BP 759, K-39 93.2%, K-40 0.012%, K-41 6.73%)

Flame technique: Price[2] lists the following sensitivities for 404.4, 766.5, and 769.9 nm wavelengths with a lean air–acetylene flame: 10.0, 0.2, and 0.1 mg/l, respectively. Ionization is overcome by the addition of La, Na, or Cs.[1,2,13] The $LaCl_3$ solution described above may be used for this purpose.

Flameless technique: The pyrolysis and atomization temperatures should be in the ranges of 900°C and 1500°C, respectively. The characteristic mass is approximately 2.0 pg at 766.5 nm.

ICP-MS technique: With a Perkin-Elmer SCIEX ELAN 6000 ICP-MS, the detection limit is 10–100 ng/l.

Mg (magnesium, AW 24.305, D_0 4.1, E_i 7.65, BP 1090, Mg-24 78.9%, Mg-25 10.00%, Mg-26 11.10%)

Flame technique: Use a lean nitrous oxide–acetylene flame to get approximate sensitivities of 25, 0.2, and 0.005 mg/l for the 202.5, 279.6, and 285.2 nm wavelengths.[2] La ($LaCl_3$ solution described above) or K (KCl solution described above) additions may be required, as well as the hot nitrous oxide–acetylene flame. Perkin-Elmer[13] recommends a lean air–acetylene flame. The $LaCl_3$ solution can reduce chemical interference from aluminum, silicon, titanium, zirconium, and phosphorus.[1] The KCl solution reduces ionization.

Flameless technique: The pyrolysis and atomization temperatures should be in the ranges of 800°C and 1900°C, respectively. The characteristic mass is approximately 0.4 pg at 285.2 nm.

ICP-MS technique: With a Perkin-Elmer SCIEX ELAN 6000 ICP-MS, the detection limit is 1–10 ng/l.

Mn (manganese, AW 54.9380, D_0 4.2, E_i 7.43, BP 2062, Mn-55 100%)

Flame technique: Price[2] reports sensitivities of 1, 0.025, 100, and 0.5 mg/l using a stoichiometric to lean air–acetylene flame for the 222.2, 279.5, 321.7, and 403.1 nm wavelengths. Perkin-Elmer[13] suggests that interference by silicon may be overcome by the addition of 0.2% (w/v) calcium chloride. Varma[1] notes interferences from phosphate, perchlorate, iron, nickel, and cobalt that are reduced or eliminated using a lean air–acetylene or nitrous oxide–acetylene flame. Varma[1] suggests addition of K (KCl solution described above) to eliminate any ionization.

Flameless technique: Perkin-Elmer[14] recommends addition of 5 μg Pd and 3 μg of $Mg(NO_3)_2$ to the sample. (Reagent described above.) The pyrolysis and atomization temperatures should be in the ranges of 1300°C and 1900°C, respectively. The characteristic mass is approximately 6.3 pg at 279.5 nm.

ICP-MS technique: With a Perkin-Elmer SCIEX ELAN 6000 ICP-MS, the detection limit is 1–10 ng/l.

Mo (molybdenum, AW 95.94, D_0 6.3, E_i 7.09, BP 4639, Mo-92 14.84%, Mo-94 9.25%, Mo-95 15.92%, Mo-96 16.68%, Mo-97 9.55%, Mo-98 24.13%, Mo-100 9.63)

Flame technique: Price[2] reports sensitivities of 0.5, 1.5, 10.0, and 2.0 mg/l for the 313.3, 317.0, 320.8, and 379.8 nm wavelengths using a rich air–acetylene flame. A rich nitrous oxide–acetylene flame has a sensitivity of 0.2 mg/l for the 313.3 nm wavelength. Interference from calcium, chromium, manganese, nickel, strontium, sulfate, and iron can be reduced by using a nitrous oxide–acetylene flame.[1,2,13] Varma[1] suggests the addition of either 0.5% (w/v) aluminum, 2% (w/v) ammonium chloride, or 0.1% (w/v) sodium sulfate to reduce interferences.

Flameless technique: Perkin-Elmer[14] recommends addition of 5 μg Pd and 3 μg of $Mg(NO_3)_2$ to the sample. (Reagent described above.) The pyrolysis and atomization temperatures should be in the ranges of 1500°C and 2400°C, respectively. The characteristic mass is approximately 12.0 pg at 313.3 nm.

ICP-MS technique: With a Perkin-Elmer SCIEX ELAN 6000 ICP-MS, the detection limit is less than 1 ng/l.

Na (sodium, AW 22.98977, D_0 2.7, E_i 5.14, BP 883, Na-23 100%)

Flame technique: Sensitivities of 5.0 and 0.02 mg/l are reported by Price[2] for 330.2 and 589.0 nm wavelengths using a lean air–acetylene flame. The addition of an alkali salt (KCl solution described above) is recommended.[1] Noise produced by calcium can interfere at the 589.0 nm wavelength.[1]

Flameless technique: The pyrolysis and atomization temperatures should be in the ranges of 900°C and 1500°C, respectively. The characteristic mass is approximately 1.2 pg at 589.0 nm.

ICP-MS technique: With a Perkin-Elmer SCIEX ELAN 6000 ICP-MS, the detection limit is 1–10 ng/l.

Ni (nickel, AW 58.69, D_0 4.0, E_i 7.64, BP 2914, Ni-58 68.27%, Ni-60 26.1%, Ni-61 1.13%, Ni-62 3.59%, Ni-64 0.91%)

Flame technique: For the wavelengths 232.0, 234.6, 247.7, 339.1, 341.5, 346.2, and 352.5 nm, Price[2] reports sensitivities of 0.1, 0.5, 50, 5, 0.2, 1.0, and 2.5 mg/l, respectively with a lean air–acetylene flame. Perkin-Elmer[13] and Varma[1] reports that high concentrations of iron, cobalt, and chromium can depress the signal. A nitrous oxide–acetylene flame can be used to reduce this interference.

Flameless technique: The pyrolysis and atomization temperatures should be in the ranges of 1100°C and 2300°C, respectively. The characteristic mass is approximately 20.0 pg at 232.0 nm.

ICP-MS technique: With a Perkin-Elmer SCIEX ELAN 6000 ICP-MS, the detection limit is 1–10 ng/l.

Pb (lead, AW 207.2, D_0 3.9, E_i 7.42, BP 1750, Pb-204 1.40%, Pb-206 24.10%, Pb-207 22.10%, Pb-208 52.40%)

General: This element is toxic and should be handled appropriately. Low concentration work requires extreme care to avoid contamination.

Flame technique: Price[2] reports sensitivities of 0.12, 6.0, and 0.2 mg/l for the 217.0, 261.4, and 283.3 nm wavelengths with a lean air–acetylene flame. Note that, although the 217.0 nm wavelength is more sensitive than the 283.3 nm wavelength, the associated noise is higher than that of the 283.3 nm wavelength. The lower noise and only marginally lower sensitivity of the 283.3 nm wavelength make it slightly more useful.

Flameless technique: Perkin-Elmer[14] recommends addition of 50 μg $NH_4H_2PO_4$ and 3 μg of $Mg(NO_3)_2$ to the sample. (Reagent described above.) The pyrolysis and atomization temperatures should be in the ranges of 850°C and 1500°C, respectively. The characteristic mass is approximately 30.0 pg at 283.3 nm.

ICP-MS technique: With a Perkin-Elmer SCIEX ELAN 6000 ICP-MS, the detection limit is less than 1 ng/l.

Pt (platinum, AW 195.08, D_0 3.6, E_i 9.0, BP 3827, Pt-190 0.01%, Pt-192 0.79%, Pt-194 32.90%, Pt-195 33.80%, Pt-196 25.30%, Pt-198 7.20%)

Flame technique: Using a lean air–acetylene flame, Price[2] reports respective sensitivities for the 217.5, 265.9, 299.8, and 306.5 nm wavelengths of 10, 2.5,15, and 5 mg/l. Interferences from high concentrations of many elements, ammonium ion, sulfuric acid, perchloric acid, and phosphoric acid can be reduced by adding 0.2% (w/v) La in 1% (v/v) hydrochloric acid.[1,13] Perkin-Elmer[13] also note that a nitrous oxide–acetylene flame will reduce these interferences.

Flameless technique: The pyrolysis and atomization temperatures should be in the ranges of 1300°C and 2200°C, respectively. The characteristic mass is approximately 220.0 pg at 265.9 nm.

ICP-MS technique: With a Perkin-Elmer SCIEX ELAN 6000 ICP-MS, the detection limit is less than 1 ng/l.

Sb (antimony, AW 121.75, D_0 3.9, E_i 8.64, BP 1587 (hydride SbH_3-17), Sb-121 57.3%, Sb-123 42.7%)

Flame technique: For the 206.8, 217.6, and 231.2 nm wavelengths and a lean to stoichiometric air–acetylene flame, Price[2] reports sensitivities of 0.5, 1.1, and 1.9 mg/l, respectively. Although the 206.8 nm wavelength is more sensitive, it is noisier than the 217.6 nm wavelength.

Flameless technique: Perkin-Elmer[14] recommends addition of 5 μg Pd and 3 μg of $Mg(NO_3)_2$ to the sample. (Reagent described above.) The pyrolysis and atomization temperatures should be in the ranges of 1300°C and 1900°C, respectively. The characteristic mass is approximately 55.0 pg at 217.6 nm.

ICP-MS technique: With a Perkin-Elmer SCIEX ELAN 6000 ICP-MS, the detection limit is less than 1 ng/l.

Se (selenium, AW 78.96, D_0 4.4, E_i 9.75, BP 685, Se-74 0.90%, Se-76 9.00%, Se-77 7.60%, Se-78 23.50%, Se-80 49.60%, Se-82 9.40%)

Flame technique: Price[2] reports a sensitivity of 0.8 mg/l for a "just luminous" lean air–acetylene flame with the 196.1 nm wavelength. At this wavelength, there is much light scatter from the flame and background correction is essential.[13]

Flameless technique: Perkin-Elmer[14] recommends addition of 5 μg Pd and 3 μg of $Mg(NO_3)_2$ to the sample. (Reagent described above.) The pyrolysis and atomization temperatures should be in the ranges of 1300°C and 1900°C, respectively. The characteristic mass is approximately 45.0 pg at 196.1 nm.

ICP-MS technique: With a Perkin-Elmer SCIEX ELAN 6000 ICP-MS, the detection limit is 10–100 ng/l.

Sn (tin, AW 118.69, D_0 5.7, E_i 7.34, BP 2603 (hydride SnH_4-52), Sn-112 0.97%, Sn-114 0.65%, Sn-115 0.36%, Sn-116 14.70%, Sn-117 7.70%, Sn-118 24.30%, Sn-119 8.60%, Sn-120, 32.40%, Sn-122 4.60%, Sn-124 5.60%)

Flame technique: Sensitivities at 224.6 and 286.3 nm using a rich nitrous oxide–acetylene flame are 0.8 and 2.5 mg/l, respectively.[2] Price[2] reports sensitivities for the 270.6, 286.3, and 303.4 nm wavelengths to be 25, 10, and 50 mg/l with a rich air–acetylene flame, respectively. A rich nitrous oxide–acetylene flame and 328.1 nm wavelength seem the best combination because of the slightly lower noise at this wavelength relative to the slightly more sensitive 224.6 nm wavelength.

Flameless technique: Perkin-Elmer[13,14] recommends addition of 5 μg Pd and 3 μg of $Mg(NO_3)_2$ to the sample. (Reagent described above.) The pyrolysis and atomization temperatures should be in the ranges of 1400°C and 2200°C, respectively. The characteristic mass is approximately 90.0 pg at 286.3 nm.

ICP-MS technique: With a Perkin-Elmer SCIEX ELAN 6000 ICP-MS, the detection limit is 1–10 ng/l.

Ti (titanium, AW 47.88, D_0 6.9, E_i 6.83, BP 3289, Ti-46 8.00%, Ti-47 7.30%, Ti-48 73.80%, Ti-49 5.50%, Ti-50 5.40%)

Flame technique: Price[2] recommends a rich nitrous oxide–acetylene flame for all three wavelengths (364.3, 365.4, and 337.2 nm). At 364.3, he gives a sensitivity of 1.5 mg/l. Addition of an excess of alkali salt (KCl solution described above) is recommended to reduce ionization.[1,2] The presence of hydrofluoric acid, iron, and many other elements will enhance the signal for this element.[1,2] Sulfuric acid will greatly reduce sensitivity.[1]

Flameless technique: The pyrolysis and atomization temperatures should be in the ranges of 1500°C and 2500°C, respectively. The characteristic mass is approximately 70.0 pg at 364.3 nm.

ICP-MS technique: With a Perkin-Elmer SCIEX ELAN 6000 ICP-MS, the detection limit is 1–10 ng/l.

V (vanadium, AW 50.9415, D_0 6.7, E_i 6.75, BP 3409, V-50 0.25%, V-51 99.75%)

Flame technique: Price[2] gives a sensitivity of 1.0 mg/l for the doublet (318.39 and 318.34) using a stoichiometric nitrous oxide–acetylene flame. High concentrations of aluminum, titanium, iron, and phosphoric acid can enhance the signal.[1,2] An ionization buffer (KCl solution described above) is required.

Flameless technique: The pyrolysis and atomization temperatures should be in the ranges of 1200°C and 2400°C, respectively. The characteristic mass is approximately 42.0 pg at 318.4 nm.

ICP-MS technique: With a Perkin-Elmer SCIEX ELAN 6000 ICP-MS, the detection limit is 1–10 ng/l.

Zn (zinc, AW 65.38, D_0 2.9, E_i 9.39, BP 907, Zn-64 48.60%, Zn-66 27.90, Zn-67 4.10%, Zn-68 18.80%, Zn-70 0.60%)

General: Zinc contamination can be a problem when working with low concentrations.

Flame technique: Using a stoichiometric to lean air–acetylene flame, Price[2] reports sensitivities of 0.012 and 150 mg/l for the 213.9 and 307.6 nm wavelengths, respectively.

Flameless technique: Perkin-Elmer[13,14] recommends addition of 5 μg of $Mg(NO_3)_2$ to the sample. (Reagent described above.) The pyrolysis and atomization temperatures should be in the ranges of 700°C and 1800°C, respectively. The characteristic mass is approximately 1.0 pg at 213.9 nm.

ICP-MS technique: With a Perkin-Elmer SCIEX ELAN 6000 ICP-MS, the detection limit is 1–10 ng/l.

References

1. Varma, A., *CRC Handbook of Atomic Absorption Analysis, vol.* 1, CRC Press, Boca Raton, FL, 1984, 510 pp.
2. Price, W.J., *Analytical Atomic Absorption Spectrometry*, Heyden & Son Ltd., London, 1972, 239 pp.
3. Clesceri, L.S., Greenberg, A.E. and Eaton, A.D., Eds., *Standard Methods for the Examination of Water and Wastewater*, 20th ed., American Public Health Association, Washington, D.C., 1999.
4. Becker, J.S., State-of-the-art and progress in precise and accurate isotope ratio measurements by ICP-MS and LA-ICP-MS, *J. Anal. At. Spectrom.*, 17, 1172–1185, 2002.
5. Mota, J.P.V., Encinar, J.R., Fernández de la Campa, M.R., Alonso, J.I.G. and Sanz-Medel, A., Determination of cadmium in environmental and biological reference materials using isotope dilution analysis with a double focusing ICP-MS: a comparison with quadrupole ICP-MS, *J. Anal. At. Spectrom.*, 14, 1467–1473, 1999.
6. Taylor, H.E., *Inductively Coupled Plasma-Mass Spectrometry: Practices and Techniques*, Academic Press, San Diego, 2001, 293 pp.
7. Veillon, C., Trace element analysis of biological samples, *Anal. Chem.*, 58, 851A–866A, 1986.
8. Van Loon, J.C., *Selected Methods of Trace Metal Analysis: Biological and Environmental Samples*, John Wiley & Sons, New York, 1985, 357 pp.
9. Pinta, M., *Modern Methods for Trace Element Analysis*, Ann Arbor Science Publishers, Ann Arbor, MI, 1978, 492 pp.
10. Suzuki, M. and Ohta, K., Reduction of interferences with thiourea in the determination of cadmium by electrothermal atomic absorption spectrometry, *Anal. Chem.*, 54, 1686–1689, 1982.
11. Ediger, R.D., Atomic absorption analysis with the graphite furnace using matrix modification, *Atom. Abs. Newslett.*, 14, 127–130, 1975.
12. Carnrick, G., Schlemmer, G. and Slavin, W., Matrix modifiers: their role and history for furnace AAS, *Am. Lab. (FAIRFIELD CONN)* Feb. 1991, 118–131, 1991.
13. Perkin-Elmer, *Analytical Methods for Atomic Absorption Spectrophotometry*, Perkin-Elmer Corp., Norwalk, CT, 1982.
14. Perkin-Elmer, *The THGA Graphite Furnace: Techniques and Recommended Conditions*, Perkin-Elmer Corp., Norwalk, CT, 1992.
15. Schlemmer, G. and Welz, B., Palladium and magnesium nitrates, a more universal modifier for graphite furnace atomic absorption spectrometry, *Spectrochim. Acta*, 418, 1157–1165, 1986.
16. Windom, H.L., Byrd, J.T., Smith, R.G., Jr. and Huan, F., Inadequacy of NASQAN data for assessing metal trends in the nation's rivers, *Environ. Sci. Technol.*, 25, 1137–1142, 1991.
17. Colella, M.B., Siggia, S. and Barnes, R.M., Poly (acrylamidoxime) resin for determination of trace metals in natural waters, *Anal. Chem.*, 52, 2347–2350, 1980.
18. Koide, M., Lee, D.S. and Stallard, M.O., Concentration and separation of trace metals from seawater using a single anion exchange bead, *Anal. Chem.*, 56, 1956–1959, 1984.
19. Brooks, R.R, Presley, B.J. and Kaplan, I.R., APDC-MIBK extraction system for the determination of trace elements in saline waters by atomic-absorption spectrophotometry, *Talanta*, 14, 809–816, 1967.
20. Bruland, K.W., Franks, R.P., Knauer, G.A. and Martin, J.H., Sampling and analytical methods for the determination of copper, cadmium, zinc and nickel at the nanogram per liter level in seawater, *Anal. Chim. Acta*, 105, 233–245, 1979.
21. Danielsson, L.-G., Magnusson, B. and Westerlund, S., An improved metal extraction procedure for the determination of trace metals in seawater by atomic absorption spectrometry with electrothermal atomization, *Anal. Chim. Acta*, 98, 47–57, 1978.

22. Jan, T.K. and Young, D.R., Determination of microgram amounts of some transition metals in seawater by methyl isobutyl ketone-nitric acid successive extraction and flameless atomic absorption spectrophotometry, *Anal. Chem.*, 50, 1250–1253, 1978.
23. Kinrade, J.D. and Van Loon, J.C., Solvent extraction for use with flame atomic absorption spectrometry, *Anal. Chem.*, 46, 1894–1898, 1974.
24. Pakalns, P. and Farrar, Y.J., The effect of surfactant on the extraction-atomic absorption spectrophotometric determination of copper, iron, manganese, lead, nickel, zinc, cadmium and cobalt, *Water Res.*, 11, 145–151, 1977.
25. Koirtyohann, S.R. and Wen, J.W., Critical study of the APDC-MIBK extraction system for atomic absorption, *Anal. Chem.*, 45, 1986–1989, 1973.
26. Boyle, E.A. and Edmond, J.M., Determination of trace elements in aqueous solution by APDC chelate co-precipitation, in *Analytical Methods in Oceanography*, Gibb, T.R.P., Jr., Ed., vol. 6, American Chemical Society, Washington, D.C., 1975, pp. 44–55 (238 pp.).
27. APHA, AWWA, WPCF, *Standard Methods for the Examination of Water and Wastewater*, vol. 3, American Public Health Association, Washington, D.C., 1989, pp. 1–163.
28. Allen, H.E., Hall, R.H. and Brisbin, T.D., Metal speciation: effects on aquatic toxicity, *Environ. Sci. Technol.*, 14, 441–442, 1980.
29. Borgmann, U., Metal speciation and toxicity of free ions to aquatic biota, in *Aquatic Toxicology*, Nriagu, J.O., Ed., John Wiley & Sons, New York, 1983.
30. Campbell, P.G.C. and Tessier, A., Ecotoxicology of metals in the aquatic environment: geochemical aspects, in *Ecotoxicology: A Hierarchical Treatment*, Newman, M.C. and Jagoe, C.H., Eds., CRC Press, Boca Raton, FL, 1996.
31. Visual MINTEQ version 2.22, KTH, Department of Land and Water Resources Engineering, Stockholm, Sweden, 2003.
32. Ingamells, C.O. and Switzer, P., A proposed sampling constant for use in geochemical analysis, *Talanta*, 20, 547–568, 1973.
33. Ingamells, C.O., New approaches to geochemical analysis and sampling, *Talanta*, 21, 141–155, 1974.
34. Wallace, D. and Kratochvil, B., Visman equations in the design of sampling plans for chemical analysis of segregated bulk materials, *Anal. Chem.*, 59, 226–232, 1987.
35. Newman, M.C., *Quantitative Methods in Aquatic Ecotoxicology*, CRC/Lewis Publishers, Boca Raton, FL, 1995, 426 pp.
36. Sivasankara Pillay, K.K., Thomas, C.C. Jr., Sondel, J.A. and Hyche, C.M., Determination of mercury in biological and environmental samples by neutron activation analysis, *Anal. Chem.*, 43, 1419–1425, 1971.
37. Newman, M.C. and Mitz, S.V., Size dependence of zinc elimination and uptake from water by mosquitofish *Gambusia affinis* (Baird and Girard), *Aquat. Toxicol.*, 12, 17–32, 1988.
38. Newman, M.C., Mulvey, M., Beeby, A., Hurst, R.W. and Richmond, L., Snail (*Helix aspersa*) exposure history and possible adaptation to lead as reflected in shell composition, *Arch. Environ. Contam. Toxicol.*, 27, 346–351, 1994.
39. Clegg, M.S., Keen, C.L., Lönnerdal, B. and Hurley, L.S., Influence of ashing techniques on the analysis of trace elements in animal tissue. I. Wet ashing, *Biol. Trace Element Res.*, 3, 107–115, 1981.
40. Sinex, S.A., Cantillo, A.Y. and Helz, G.R., Accuracy of acid extraction methods for trace metals in sediments, *Anal. Chem.*, 52, 2342–2346, 1980.
41. Siemer, D.D. and Brinkley, H.G., Erlenmeyer flask-reflux cap for acid sample decomposition, *Anal. Chem.*, 53, 750–751, 1981.
42. Chung, S.-W. and Tsai, W.-C., Atomic absorption spectrometric determination of heavy metals in foodstuffs using a simple digester, *At. Spectrosc.*, 13, 185–189, 1992.
43. Uhrberg, R., Acid digestion bomb for biological samples, *Anal. Chem.*, 54, 1906–1908, 1982.
44. Hewitt, A.D. and Reynolds, C.M., Dissolution of metals from soils and sediments with a microwave-nitric acid digestion technique, *At. Spectrosc.*, 11, 187–192, 1990.

45. Kojima, I., Kato, A. and Iida, C., Microwave digestion of biological samples with acid mixture in a closed double PTFE vessel for metal determination by "one-drop" flame atomic absorption spectrometry, *Anal. Chim. Acta*, 264, 101–106, 1992.
46. Lajunen, L.H.J., Piispanen, J. and Saari, E., Microwave dissolution of plant samples for AAS analysis, *At. Spectrosc.*, 13, 127–131, 1992.
47. Mincey, D.W., Williams, R.C., Giglio, J.J., Graves, G.A. and Pacella, A.J., Temperature controlled microwave oven digestion system, *Anal. Chim. Acta*, 264, 97–100, 1992.
48. Lan, W.G., Wong, M.K and Sin, Y.M., Comparison of four microwave digestion methods for the determination of selenium in fish tissue by using hydride generation atomic absorption spectrometry, *Talanta*, 41, 195–200, 1994.
49. Tessier, A., Campbell, P.G.C. and Bisson, M., Sequential extraction procedure for the speciation of particulate trace metals, *Anal. Chem.*, 51, 844–851, 1979.
50. Tessier, A. and Campbell, P.G.C., Partitioning of trace metals in sediments, in *Metal Speciation. Theory, Analysis and Application*, Kramer, J.R. and Allen, H.E., Eds., vol. 9, Lewis Publishers, Chelsea, MI, 1988, pp. 183–217 (357 pp.).
51. Rendell, P.S., Batley, G.E. and Cameron, A.J., Adsorption as a control of metal concentrations in sediment extracts, *Environ. Sci. Technol.*, 14, 314–318, 1980.
52. Tipping, E., Hetherington, N.B., Hilton, J., Thompson, D.W., Bowles, E. and Hamilton-Taylor, J., Artifacts in the use of selective chemical extraction to distributions of metals between oxides of manganese and iron, *Anal. Chem.*, 57, 1944–1946, 1985.
53. Pai, S., Lin, F., Tseng, C. and Sheu, D., Optimization of heating programs of GFAAS for the determination of Cd, Cu, Ni and Pb in sediments using sequential extraction technique, *Int. J. Environ. Anal. Chem.*, 50, 193–205, 1993.
54. Landrum, P.F., Bioavailability and toxicokinetics of polycyclic aromatic hydrocarbons sorbed to sediments for the amphipod *Pontoporeia hoyi*, *Environ. Sci. Tech.*, 23, 588–595, 1989.
55. Mayer, L., Chen, Z., Findlay, R., Fang, J., Sampson, S., Self, L., Jumars, P., Quetél, C. and Donard, O., Bioavailability of sedimentary contaminants subject to deposit-feeder digestion, *Environ. Sci. Tech.*, 30, 2641–2645, 1996.
56. Newman, M.C., Dixon, P.M., Looney, B.B. and Pinder, J.E. III, Estimating mean and variance for environmental samples with below detection limit observations, *Water Res. Bull.*, 25, 905–916, 1989.
57. Newman, M.C. and Dixon, P.M., UNCENSOR: a program to estimate means and standard deviations for data sets with below detection limit observations, *Am. Env. Lab. (FAIRFIELD CONN)*, April 1990, 1990, 27–30 (http://www.vims.edu/env/research/software/vims_software.html for software download).
58. Dean, J.A., *Lange's Handbook of Chemistry*, vol. 4, McGraw-Hill, New York, 1992, pp. 23–35.
59. Lide, D.R., Ed., *CRC Handbook of Chemistry and Physics*, 73rd ed., CRC Press, Boca Raton, FL, 1992.
60. Reynolds, R.J., Aldous, K. and Thompson, K.C., *Atomic Absorption Spectroscopy. A Practical Guide*, Barnes & Noble Inc., New York, 1970.
61. Willard, H.H., Merritt, L.L., Jr. and Dean, J.A., *Instrumental Methods of Analysis*, D. Van Nostrand Co., New York, 1974.

chapter thirty-four

Estimation of inorganic species aquatic toxicity

James P. Hickey
U.S. Geological Survey

Contents

Introduction

Background

Researchers, manufacturers, and regulating agencies must evaluate properties for chemicals that are either present in or could be released into the environment. The routine use of over 70,000 synthetic chemicals stresses the need for this information. However, minimal physical and toxicity data are available for only about 20% of these compounds. The cost of testing all of these chemicals is prohibitive, so researchers and managers increasingly rely on predictive models [i.e., quantitative structure–activity relationships (QSARs)] for chemical property estimation, hazard evaluation, and information to direct research and set priorities. While property estimation is routine for organic species, QSARs usable for inorganic materials are few.

The linear solvation energy relationship (LSER) developed by Kamlet and coworkers (Reference 1 contains a compilation of the Kamlet et al. LSER development references) for neutral organic species is suitable for estimation of environmental properties. The form of the LSER equation used here relates many chemical solution properties to the solute's LSER variables (Table 34.3):

$$\log(\text{property}) = mV_\text{i}/100 + s\pi^* + b\beta_m + a\alpha_m \qquad (34.1)$$

1-56670-664-5/05/$0.00+$1.50
© 2005 by CRC Press

where V_i is the intrinsic (van der Waals) molecular volume, π^* is the solute ability to stabilize a neighboring charge or dipole by nonspecific dielectric interactions, and β_m and α_m are the solute ability to accept or donate a hydrogen in a hydrogen bond. The coefficients m, s, b, and a are constants for a particular set of conditions, determined by multiple linear regression of the LSER variable values for a series of chemicals with the measured value for a particular chemical property.

Inorganic LSER variable value development

Since LSER has proven to be a useful QSAR for organics,[2] LSER variable values were devised for the remaining periodic table elements and ions.[3] For the steric parameter $V_i/100$, atomic parachors[4] were transformed through characteristic molecular volumes V_x to intrinsic molecular volumes, V_i.[5] The use of a calculated volume was preferable, and any coordination sphere of water was considered to be included in the LSER theory (see References 1, 2, 6, and 7, also references therein). For the other variables (dipolar π^* and hydrogen bonding β and α), reasonably complete sets of appropriate elemental property values (*vide infra*) were not available to serve as a proportional reference for value calculation for the remaining periodic table elements. A heuristic development process was used, in which descriptive solution chemistry[8,9] served as a frame of reference to correlate trends from measured physical property data with existing LSER parameter values in order to suggest parameter values for unassigned elements. The property data used were: (1) for π^*, dipole moments, polarizabilities, and electronegativity; (2) for β and α, ionization potentials (basicity), electron affinities (acidity), and pK values.

Inorganic structures were taken as likely solution species.[8,9] The LSER values for the neutral inorganic molecule or salt were calculated as the sums of the neutral molecule or salt component contributions (see Table 34.1 for component values, Table 34.2 for whole salt/molecule values, and the section "Method" for a calculation example). The volumes of a salt's component ions were summed with no compensation for bonds between

Table 34.1 Inorganic component species LSER values

Ion	$V_i/100$[a]	π^*	β	α
Li^{+1}	0.158	0.05	0.00	0.10
Na^{+1}	0.229	0.00	0.05	0.00
K^{+1}	0.360	0.10	0.06	0.00
Rb^{+1}	0.417	0.17	0.06	0.03
Cs^{+1}	0.533	0.20	0.10	0.03
Be^{+2}	0.144	0.00	0.00	0.00
Mg^{+2}	0.216	0.00	0.00	0.10
Ca^{+2}	0.349	0.00	0.00	0.10
Sr^{+2}	0.406	0.00	0.00	0.10
Ba^{+2}	0.529	0.00	0.00	0.10
B^{+3}	0.131	0.03	0.00	0.40
Al^{+3}	0.202	0.06	0.00	0.20
V^{+5}	0.316	0.05	0.10	0.10
$Cr^{+3/+6}$	0.305	0.05	0.05	0.20
Mo	0.362	0.10	0.05	0.10
Mn^{+2}	0.294	0.10	0.10	0.00
$Fe^{+2/+3}$	0.283	0.05	0.10	0.05

Co^{+2}	0.272	0.10	0.05	0.35
Ni^{+2}	0.261	0.10	0.05	0.30
Cu^{+2}	0.251	0.10	0.05	0.35
Ag^{+1}	0.307	0.15	0.05	0.25
Zn^{+2}	0.240	0.15	0.00	0.25
Cd^{+2}	0.296	0.20	0.00	0.25
Hg^{+2}	0.319	0.19	0.00	0.55
Tl^{+1}	0.311	0.03	0.00	0.15
Pt^{+2}	0.337	0.15	0.05	0.25
As^{+3}	0.207	0.23	0.13	0.05
Sb^{+3}	0.263	0.35	0.09	0.07
Bi^{+3}	0.294	0.18	0.10	0.08
Sn^{+2}	0.274	0.05	0.03	0.00
Pb^{+2}	0.302	0.00	0.00	0.00
Ce^{+3}	0.507	0.00	0.00	0.20
Yb^{+3}	0.405	0.04	0.03	0.00
Th^{+4}	0.496	0.02	0.02	0.10
U^{+6}	0.479	0.03	0.03	0.10
F^{-1} covalent	0.077	0.08	0.19	0.06
F^{-1} ionic	0.077	0.18	0.29	0.06
Cl^{-} covalent	0.149	0.35	0.15	0.06
Cl^{-1} ionic	0.149	0.60	0.40	0.06
Br^{-} covalent	0.185	0.43	0.17	0.05
Br^{-1} ionic	0.185	0.68	0.32	0.05
I^{-} covalent	0.242	0.45	0.18	0.04
I^{-1} ionic	0.242	0.70	0.33	0.04
$ClO_3{}^{-1}$	0.269	0.50	0.40	0.30
$ClO_4{}^{-1}$	0.309	0.00	0.40	0.42
$BrO_3{}^{-1}$	0.458	0.60	0.30	0.41
$BrO_4{}^{-1}$	0.549	0.20	0.57	0.53
$IO_3{}^{-1}$	0.402	0.50	0.45	0.40
$IO_4{}^{-1}$	0.515	0.25	0.50	0.57
−OH ionic	0.105	0.45	0.50	0.00
−OH covalent	0.105	0.40	0.47	0.33
−SH ionic	0.176	0.25	0.20	0.00
−SH covalent	0.176	0.35	0.16	0.03
=O	0.091	0.34	0.10	0.12
=S	0.162	0.24	0.05	0.05
−O−	0.091	0.27	0.45	0.00
O^{-2} ionic	0.091	0.10	0.15	0.00
S^{-2} covalent	0.162	0.50	0.23	0.00
S^{-2} ionic	0.162	0.10	0.00	0.00
Se^{-2}	0.196	0.20	0.00	0.00
$SO_3{}^{-2}$	0.282	0.65	0.82	0.36
$SO_4{}^{-2}$	0.322	0.65	0.82	0.00
$S_2O_3{}^{-2}$	0.597	1.62	1.28	0.17
−OS(=O)2OH	0.336	1.00	0.80	0.75
$PO_4{}^{-3}$	0.336	0.45	0.87	0.00
$HPO_4{}^{-2}$	0.350	0.95	0.80	0.75
$H_2PO_4{}^{-1}$	0.364	0.95	0.75	0.75
$_2$(−O)(H)P(=O)	0.310	0.75	0.75	0.00

Continued

Table 34.1 Inorganic component species LSER values

Ion	$V_i/100$[a]	π^*.	β	α
(−O)(HO)(H)P(=O)	0.325	0.68	0.47	0.33
$P_2O_7^{-4}$	0.581	0.90	1.74	0.00
VO_3^{-1}	0.436	1.07	0.40	0.46
CrO_4^{-2}	0.465	1.41	0.45	0.68
$HCrO_4^{-1}$	0.505	1.52	0.85	0.56
$Cr_2O_7^{-2}$	0.839	2.41	1.15	1.12
MnO_4^{-1}	0.454	1.46	0.50	0.48
WO_4^{-2}	0.531	0.67	1.00	0.36
SeO_4^{-2}	0.356	1.14	0.84	0.24
SeO_3	0.316	0.80	0.74	0.12
AsO_3	0.327	1.25	0.43	0.41
AsO_4	0.367	1.59	0.53	0.53
$HAsO_4$	0.375	1.65	0.90	0.74
$B_4O_7^{-2}$	0.498	1.04	1.54	1.60
NO_2^{-1}	0.184	0.53	0.49	0.00
NO_3^{-1}	0.224	0.50	0.49	0.00
$[S{=}C{=}N]^{-1}$	0.282	0.63	0.22	0.00
$[S{=}N{=}C]^{-1}$	0.282	0.85	0.42	0.00
−OC(=O)H	0.212	0.62	0.37	0.00
$-OC(=O)CH_3$	0.308	0.65	0.80	0.06
CO_3^{-2}	0.230	0.44	0.55	0.00
−OC(=O)OH	0.252	0.55	0.48	0.55
$[-O_2CCO_2-]^{-2}$	0.345	1.10	0.45	0.24
−C#N covalent	0.171	0.45	0.11	0.22
$C\#N^{-1}$ ionic	0.171	0.70	0.30	0.22
H_2O	0.119	0.45	0.45	0.45
Coordinated water	0.068	0.25	0.00	0.55
NH_3	0.146	0.15	0.65	0.00
NH_4^{+1}	0.160	0.00	0.00	0.05

[a] Volumes computed from V_x as explained in the section "Inorganic LSER variable value development" and used in Table 34.2.

cationic and anionic species. These suggested values were processed through a solubility estimation equation[7] [Equation (34.2), Table 34.3] and compared with the large table of solubility values (aqueous solubility at 20°C) from the 44th Edition of the *Handbook of Chemistry and Physics*.[10] Overall agreement within approximately an order of magnitude between observed and estimated solubilities for the range of compounds containing the element(s) in question was the criterion for acceptance of the parameter value. This criterion was adopted since no error limits were reported for the solubility listing, which was compiled from numerous sources. Use of this large series of test compounds ensured numerous cation/anion permutations. Variable values for elements with the simplest solution chemistry (e.g., alkali and alkaline earth metals as halides), as well as for those elements with solution chemistry analogous to the 11 traditional LSER elements, were developed first. These parameter values were then used to develop values for more complex elements, in more complex salts or molecules. The process was repeated until all elements and common atom aggregates (e.g., anions, such as sulfates, metal oxides, etc.) were assigned LSER values.

Table 34.2 Inorganic compound LSER values and organism acute toxicities

Salt	LSER values				Golden Orfe, log LC_{50} (mM/l)				*Daphnia magna*, log EC_{50} (mM/l)				Microtox, log EC_{50} (μM/l)			
	$V_i/100$	π	β	α	%[a]	Pred.[b]	P – O[c]	Obs.[d]	%	Pred.[e]	P – O	Obs.[f]	%	Pred.[g]	P – O	Obs.[h]
$Na_2B_4O_7$	0.956	1.04	1.64	1.60	0.92	1.47	0.87	0.60	1.00	–1.14	–0.37	–0.77				
$Na_2Cr_2O_7$	1.297	1.49	1.25	0.64	0.91	–1.13	–1.00	–0.12	1.00	–3.19	–0.86	–2.33				
$K_2Cr_2O_7$	1.559	1.69	1.27	0.64					1.00	–4.73	–2.97	–1.76				
Na_2WO_4	0.989	0.67	1.10	0.36					1.00	–0.32	–0.01	–0.30				
$KClO_3$	0.629	0.60	0.46	0.30	0.75	0.29	–1.17	1.46	1.00	0.60	–0.25	0.86				
$KClO_4$	0.669	0.10	0.46	0.42	0.75	0.35	–0.96	1.31	1.00	1.14	0.31	0.83				
NaC#N	0.400	0.70	0.35	0.22	0.92	1.12	3.97	–2.85								
KC#N	0.531	0.80	0.36	0.22					0.50	0.33	2.42	–2.09				
$NaNO_2$	0.413	0.53	0.54	0.00	0.92	2.22	1.30	0.91	1.00	2.14	1.74	0.40				
$LiNO_3$	0.382	0.55	0.49	0.10									0.97	5.39	0.08	5.30
$NaNO_3$	0.453	0.50	0.54	0.00									0.92	5.01	–0.14	5.15
KNO_3	0.584	0.60	0.55	0.00									0.75	3.29	–1.06	4.35
$CsNO_3$	0.757	0.70	0.59	0.03									0.75	2.30	–1.91	4.20
$Be(NO_3)_2$	0.592	1.00	0.98	0.00	0.95	2.73	2.80	–0.07	0.80	0.93	1.94	–1.02				
$Mg(NO_3)_2$	0.664	1.00	0.98	0.10									0.62	2.89	–0.17	3.06
$Ca(NO_3)_2$	0.797	1.00	0.98	0.10									0.60	2.21	–0.78	2.99
$Sr(NO_3)_2$	0.854	1.00	0.98	0.10									0.39	1.27	–0.82	2.10
$Ba(NO_3)_2$	0.977	1.00	0.98	0.10									0.40	0.94	–1.05	1.99
$Cr(NO_3)_3$	1.247	1.55	1.52	0.20									0.91	1.30	–0.76	2.06
$Mn(NO_3)_2$	1.012	1.10	1.08	0.00									0.94	2.34	–0.66	3.00
$Fe(NO_3)_3$	1.225	1.55	1.57	0.05									0.22	0.45	0.00	0.44
$Cu(NO_3)_2$	0.969	1.10	1.03	0.35									0.76	1.55	–0.13	1.68
$Co(NO_3)_2$	0.990	1.10	1.03	0.35									0.87	1.64	–0.92	2.56
$Ni(NO_3)_2$	0.979	1.10	1.03	0.30									0.77	1.58	–0.54	2.12
$AgNO_3$	0.801	0.65	0.54	0.25	0.98	–0.23	2.40	–2.63	1.00	–0.13	4.57	–4.70	0.98	2.21	2.24	–0.03
$Zn(NO_3)_2$	0.958	1.15	0.98	0.25									0.80	1.63	0.39	1.24
$Cd(NO_3)_2$	1.014	1.20	0.98	0.25	0.40	–0.02	0.63	–0.64	1.00	–1.40	1.31	–2.71	0.71	1.10	0.09	1.02
$Hg(NO_3)_2$	1.037	1.19	0.98	0.55												
$La(NO_3)_3$	0.981	1.53	1.52	0.00									0.19	0.71	0.10	2.99

Continues

Table 34.2 Continued

Salt	LSER values				Golden Orfe, log LC_{50} (mM/l)				*Daphnia magna*, log EC_{50} (mM/l)				Microtox, log EC_{50} (μM/l)			
	$V_i/100$	π	β	α	%[a]	Pred.[b]	P – O[c]	Obs.[d]	%	Pred.[e]	P – O	Obs.[f]	%	Pred.[g]	P – O	Obs.[h]
$TlNO_3$	0.535	0.53	0.49	0.15	0.40	0.52	0.69	−0.17	1.00	1.35	3.57	−2.22				
$Pb(NO_3)_2$	0.750	1.00	0.98	0.00									0.40	1.68	1.65	0.02
$BeSO_4$	0.466	0.65	0.82	0.00					1.00	2.10	2.61	−0.51				
$MgSO_4$	0.538	0.65	0.82	0.10					1.00	1.67	0.87	0.80				
$BaSO_4$	0.851	0.65	0.82	0.10					1.00	0.19	1.05	−0.86				
$MnSO_4$	0.886	0.75	0.92	0.00					1.00	0.10	1.36	−1.26				
$FeSO_4$	0.88	0.70	0.92	0.05					0.40	0.07	0.96	−0.89				
$CuSO_4$	0.843	0.75	0.87	0.35	0.76	0.57	2.87	−2.30	0.80	−0.08	3.22	−3.30				
$ZnSO_4$	0.832	0.80	0.82	0.25					0.80	−0.09	1.98	−2.07				
$CdSO_4$	0.888	0.85	0.82	0.25					1.00	−0.46	1.32	−1.78				
NaF	0.306	0.18	0.34	0.06	0.92	2.20	1.01	1.20	0.80	2.30	1.40	0.91				
NaCl	0.378	0.60	0.45	0.06					1.00	2.00	0.76	1.24				
KCl	0.509	0.70	0.46	0.06					1.00	1.23	0.67	0.56				
$MgCl_2$	0.514	1.20	0.80	0.22									1.00	4.63	−0.69	5.32
$CaCl_2$	0.647	1.20	0.80	0.22					1.00	0.10	−0.89	0.98	1.00	3.65	−1.71	5.36
$SrCl_2$	0.704	1.20	0.80	0.22					0.40	−0.07	−0.10	0.03				
$BaCl_2$	0.827	1.20	0.80	0.22	0.40	0.12	−0.50	0.62								
$MnCl_2$	0.772	0.80	0.40	0.12									0.73	1.43	−0.85	2.28
$CoCl_2$	0.750	0.80	0.35	0.47					1.00	−0.62	0.97	−1.60				
$NiCl_2$	0.739	0.80	0.35	0.42	0.83	−0.88	−1.53	0.64	0.83	−0.43	0.64	−1.07	0.83	1.27	−0.83	2.10
$CuCl_2$	0.729	0.80	0.35	0.47									0.87	1.31	0.93	0.39
$ZnCl_2$	0.718	0.85	0.30	0.37	0.86	−0.98	−0.17	−0.81	0.86	−0.46	0.53	−0.99	0.86	1.29	0.21	1.08
$CdCl_2$	0.774	0.90	0.30	0.37									0.41	0.42	−1.48	1.90
$HgCl_2$	0.797	0.89	0.30	0.67	0.10	−0.20	2.53	−2.73	0.11	−0.13	4.19	−4.32	0.11	0.04	0.08	−0.04
$SnCl_2$	0.572	0.75	0.33	0.12					0.40	0.24	0.98	−0.74				
$PbCl_2$	0.600	0.70	0.30	0.12									0.74	2.21	2.26	−0.05
NaBr	0.414	0.68	0.37	0.05	0.92	1.34	−0.64	1.99	1.00	1.59	−0.25	1.83				
NaI	0.471	0.70	0.38	0.04	0.92	1.08	−0.75	1.82	1.00	1.31	2.37	−1.06				
Na_2S	0.620	0.50	0.33	0.00	0.92	0.31	0.81	−0.49	1.00	0.90	1.94	−1.04				
Na_2SeO_3	0.774	0.80	0.84	0.12	0.92	1.15	1.37	−0.22	1.00	0.31	1.35	−1.03				

NaH_2PO_4	0.579	0.95	0.85	0.75	0.92	1.08	−0.06	1.14				
Na_2HAsO_4	0.833	1.44	1.01	0.05	0.92	−0.87	0.12	−0.26	1.00	−0.79	0.47	−1.26
Na_3AsO_3	1.014	1.25	0.58	0.41					1.00	−2.23	−1.15	−1.08
$Pb(OAc)_2$	0.918	1.30	1.60	0.12	0.40	1.31	1.09	0.22	0.40	−0.03	2.15	−2.18
NaOAc	0.537	0.65	0.85	0.06					0.50	0.88	−1.06	1.94

[a] Percentage, as (free + MCl_x), the bioavailable metal fraction, from References 11–13. For metal nitrates, $[MCL_x] = 0$.

[b] Estimated using Equation (34.5) (Table 34.3), from Reference 6.

[c] P – O = (predicted – observed), the prediction accuracy.

[d] Observed values from Reference 19.

[e] Estimated using Equation (34.4) (Table 34.3), from Reference 2.

[f] Observed values from References 15–18.

[g] Estimated using Equation (34.3), from Reference 2.

[h] Observed values from References 11, 12, and 14.

Acute aquatic toxicity estimation development

With solubility a fundamental factor in aquatic acute toxicity, the utility of these inorganic species LSER values were proven using published toxicity equations. Lastly, bioavailable metal fraction values for the various inorganic solution species adapted from the works of McCloskey et al.,[11] Newman and McCloskey,[12] and Tatara et al.,[13] were applied to the organism toxicity estimations. These refined toxicities were then compared with literature endpoint data to determine the prediction accuracy.

Equations (34.3)–(34.5) in Table 34.3 were used to demonstrate the estimation of aquatic baseline toxicities (narcoses) for three organism types spanning eukaryotic, invertebrate, and vertebrate species: the phosphorescent algae Microtox (*Vibrio fischeri*, Beijerinck 1889),[2] the water flea *Daphnia magna*,[2] and the decorative goldfish Golden Orfe (*Leuciscus idus melanotus*).[6] For the Microtox test, data from References 11 and 12 were supplemented with data from Reference 14. *D. magna* data reported by Bringmann and Kuhn[15,16] were supplemented with data from the works of Khangarot and Ray[17] and LeBlanc.[18] Golden Orfe data were obtained from the work of Juhnke and Ludemann.[19]

In this chapter, these new LSER variable values are made available to estimate acute aquatic toxicities for a range of inorganic salts and molecules for aquatic species ranging from algae through fish using published LSER equations. These inorganic LSER variable values will enhance the application and utility of LSER (e.g., for estimation of their environmental behavior and accurate inorganic compound hazard screening).

Materials required

A PC or Mac, EXCEL or other version of a spreadsheet, and any storage media the user is familiar with. No laboratory work is involved, as this is purely an estimation method. For the calculation(s), the user will need the table of LSER values presented in this chapter (Table 34.1), the two-step example with comments provided under the section "Method," and any published (or unpublished) LSER equation(s) for aquatic toxicity the user may wish to use. The user is also referred to earlier works[1,2] for calculating properties for organic compounds using LSER.

Table 34.3 LSER equations

1. General LSER equation (see References 1, 2, 6, and 7, also references therein):
   $\log(\text{property}) = mV_i / 100 + s\,\pi^* + b\,\beta_m + a\,\alpha_m$ (34.1)
2. Solubility estimation[7]: (M/L, 20°C)
   $\log(S_w) = 0.05 - 5.85V_i/100 + 1.09\pi^* + 5.23\beta$; $r^2 = 0.9889$, $n = 115$, SD $= 0.153$ (34.2)
3. Microtox (*Vibrio fischeri* Beijerinck 1889) (uM/l, 15 min, 20°C)[2]:
   $\log(EC_{50}) = 7.49 - 7.39V_i/100 - 1.38\pi^* + 3.70\beta - 1.66\alpha$; $r^2 = 0.97$, $n = 40$, SD $= 0.319$ (34.3)
4. *Daphnia magna* (mM/l, 48 h, 20°C)[2]:
   $\log(EC_{50}) = 4.18 - 4.73V_i/100 - 1.67\pi^* + 1.48\beta - 0.93\alpha$; $r^2 = 0.95$, $n = 53$, SD $= 0.221$ (34.4)
5. Golden Orfe (*Leuciscus idus melanotus*) (M/L, 48 h, 20°C)[6]:
   $\log(LC_{50}) = 2.90 - 5.71V_i/100 - 0.92\pi^* + 4.36\beta - 1.27\,\alpha$; $r^2 = 0.94$, $n = 32$, SD $= 0.246$ (34.5)

Method

Inorganic LSER variable value determination

The calculation of the LSER values for any inorganic species, using the values for various components in Table 34.1 is a simple sum of the parts. In some cases, principally the halogens and chalcogens, the best professional judgment must be used to determine whether some components are covalently bonded in solution, or exist as anions or cations in solution. But the process is straightforward as shown by the following example.

Example: Calculation of LSER values for $BaCl_2$, using Table 34.1.

Ion	$V_i/100$	π^*	β	α
Ba^{+2}	0.529	0.00	0.00	0.10
Cl^{-1} (ionic)	0.149	0.60	0.40	0.06
Cl^{-1} (ionic)	0.149	0.60	0.40	0.06
$\Sigma(BaCl_2)$	0.827	1.20	0.80	0.22

To estimate the acute toxicity of an inorganic species to an organism, process the LSER values determined above through predetermined QSAR equations from the literature that correlate the LSER structural and electronic parameters to overall toxicity (such as those presented in Table 34.3), and then apply a metal species bioavailability factor to the result. The following example demonstrates the relative simplicity of the calculation.

Example: Calculation of the acute toxicity of $BaCl_2$ to the Golden Orfe (European goldfish) as shown in Table 34.2.

To estimate the acute toxicity of $BaCl_2$ to the Golden Orfe (mM/l, 48 h, 20°C),[6] use the LSER parameter values derived above, and Equation (34.5)[6] in Table 34.3:

$$\begin{aligned}\log(LC_{50}) &= 2.90 - 5.71V_i/100 - 0.92\pi^* + 4.36\beta - 1.27\alpha \\ &= 2.90 - 5.71(0.827) - 0.92(1.20) + 4.36(0.80) - 1.27(0.22) \\ &= 2.90 - 4.722 - 1.10 + 3.49 - 0.28 \\ &= 0.288\text{, if all of the salt were bioavailable}\end{aligned}$$

As presented in Table 34.2, the barium bioavailability, as percentage or (free metal + MCl_x), is 0.40. So,

$$\log(LC_{50}) = 0.228 \times 0.40\text{, or } 0.12$$

The observed toxicity as $\log(LC_{50})$ from Table 34.2 is reported to be 0.62. Prediction accuracy ($P - O$) is 0.50, within half of an order of magnitude. For systems with several prominent solution components (e.g., phosphate salts), the above procedure is followed with the LSER values for each component determined, and the bioavailability factor applied. The toxicity for each component is then multiplied by its relative proportion in solution, and the results are summed for an overall toxicity. The major components, their relative proportions, and bioavailability factors for solution equilibria can be determined for the solution conditions with available software, such as MINEQL+.[11–13]

Results and discussion

To demonstrate the utility of the process described in this chapter, toxicities, as baseline narcosis [$\log(EC_{50})$ and $\log(LC_{50})$] estimated by Equations (34.3)–(34.5) in Table 34.3 to three organisms for numerous inorganic species, are tabulated in Table 34.2 as bioavailable fractions (%), predicted values (Pred.), prediction accuracy (P – O), and observed values (Obs.). Prediction accuracy is presented graphically in Figures 34.1–34.3, plotted

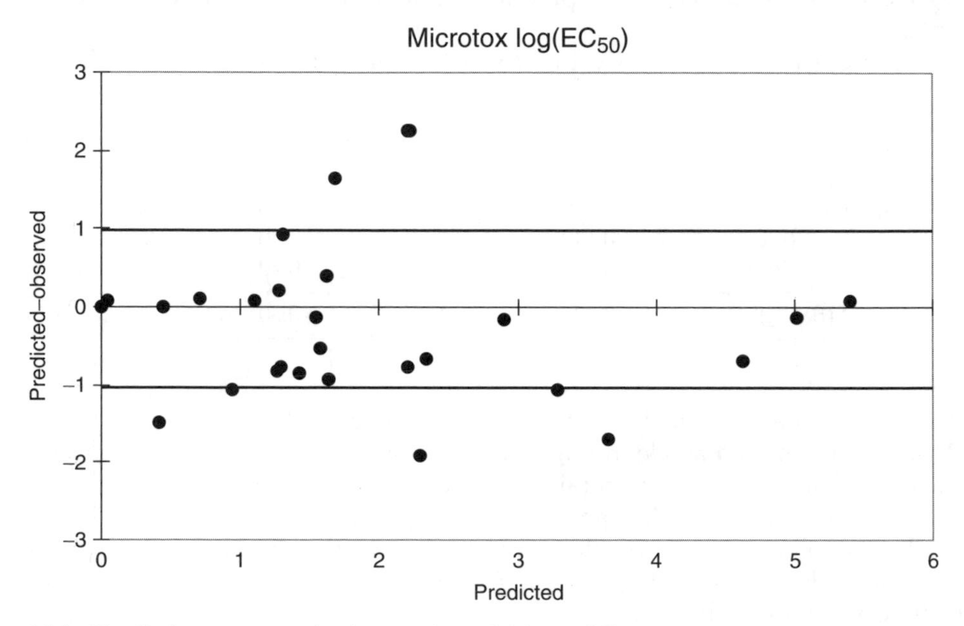

Figure 34.1 Prediction accuracy for inorganic toxicities to Microtox.

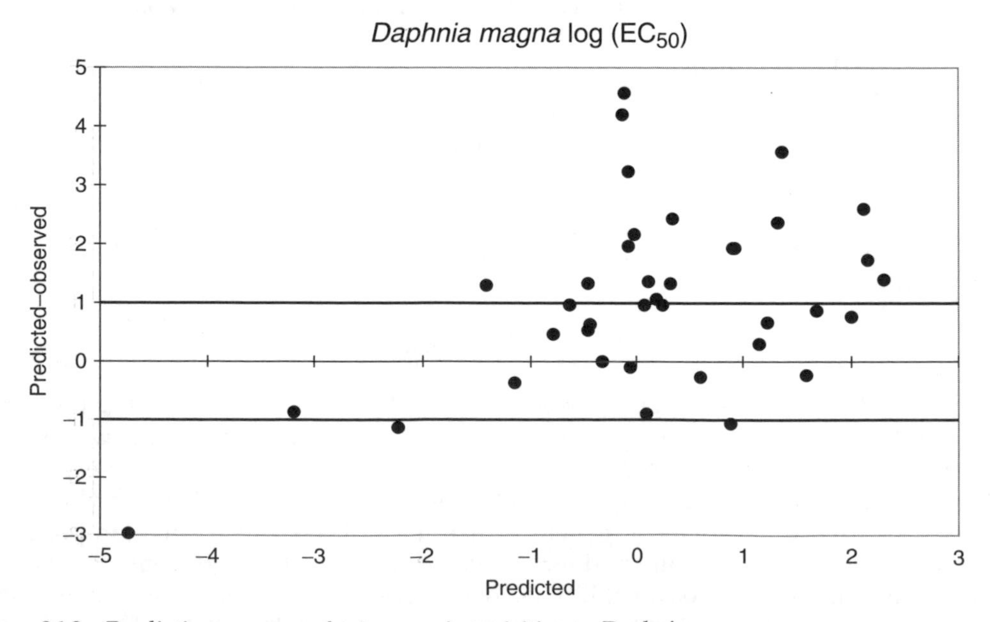

Figure 34.2 Prediction accuracy for inorganic toxicities to *Daphnia magna*.

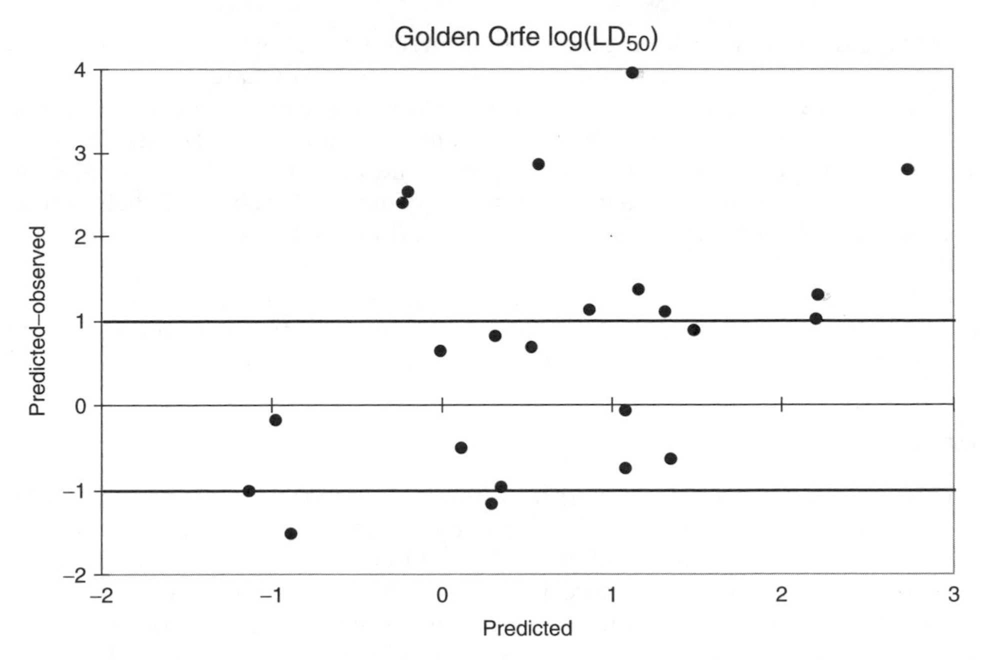

Figure 34.3 Prediction accuracy for inorganic toxicities to Golden Orfe.

against the predicted value. The tighter the data point cluster around the 0-line or x-axis, the closer the agreement between prediction and observation, and the better the LSER values serve to estimate the observed property with the particular equation used.

Estimated toxicities for many of the compounds were at or within an order of magnitude of the measured value. For the Golden Orfe and *D. magna*, two-thirds (16 in 23 and 26 in 39, respectively) of the estimated acute toxicities were within 1.3 log-units of the measured value; for the Microtox test, 80% were within 1.3 log-units, or 33 in 39 tested. Outliers in common for all organisms that were more toxic than the narcosis toxicity estimated for their tested species included cyanides, Pb^{+2} and Be^{+2} salts, $AgNO_3$, $CuSO_4$, and $HgCl_2$. Each organism also had a small number of other outliers more or less toxic than estimated.

LSER can be used as a much-needed, simple although unorthodox tool for estimating aquatic baseline toxicities (narcosis) for inorganic and organometal compounds when used with LSER equations that represent a wide range of organism types. Since the toxicity equations were developed for nonionic organic compounds, the whole salt as a neutral material was used to address this criterion and to indirectly address the question of elemental valence. As the reference data here were gathered from a number of sources for each organism, with each source having an anticipated variation in data measurement accuracy, an arbitrary outer difference limit of $\pm$ 1.3 log-units was used to define a reasonable fit. Use of a factor to account for the bioavailable metal fraction[11–13] improved the narcosis prediction accuracy, often markedly. Where multiple solution species are possible, the dominant species is not necessarily the toxic species. The toxic species will likely resemble the species successfully transported across the cell membrane, the bioavailable fraction.[20] A better definition of both the toxic solution species structure(s) and the bioavailable fraction could yield a more accurate toxicity estimation.

Compounds with a positive prediction accuracy (P – O) greater than 1.3 log-units likely have a dominant specific mode of action other than narcosis. A considerable deviation from the 0-line is considered to demonstrate the action of an enhanced toxicity over narcosis. As LSER equations are developed for the more specialized modes of action, the inorganic compounds that act accordingly should be modeled better by those equations. The different responses to a given inorganic compound (Table 34.2) between such a wide range of organism types are expected[18,21] and are reflected in the coefficients for toxicity equations (Table 34.3). For the outliers that were less toxic than estimated, the reasons for each were not understood. The deviations may reflect the error limits unavailable from the data sources, the error limits of the equations, improper species modeling, including bioavailability, or the predominance of another mode of action.

Conclusions

The acute aquatic toxicities (narcoses) for a range of organism types may be estimated with LSER for a large number of inorganic and organometal species, many with an accuracy at or within an order of magnitude. Optimum estimations make use of a bioavailable metal fraction and a more accurate structure(s) for the toxic solution species. The estimated toxicities for a number of salts were quite different from the observed values, likely due to a dominant specific toxicity mechanism other than baseline narcosis. The tool still requires fine-tuning.

THIS IS CONTRIBUTION # 1068 FROM THE USGS/GREAT LAKES SCIENCE CENTER, ANN ARBOR, MI 48105

References

1. Hickey, J.P., Linear solvation energy relationships (LSER): "Rules of Thumb" for $V_i/100$, π^*, β_m, α_m estimation and use in aquatic toxicology, in *Techniques in Aquatic Toxicology*, Ostrander, G., Ed., CRC Press/Lewis Publishers, Boca Raton, FL, 1996a, Chapter 23 (most of Kamlet et al. LSER development references are compiled within).
2. Hickey, J.P., Aldridge, A.J., Passino, D.R.M. and Frank, A.M., Expert systems for environmental applications, in *ACS Symposium Series 431: Environmental Expert Systems*, Hushon, J., Ed., American Chemical Society, Washington D.C., 1990, pp. 90–107.
3. Hickey, J.P., Modeling Main Group and Heavy Metal Environmental Behavior Using Linear Solvation Energy Relationships (LSER), Abstract 570, QSAR Symposium, 17th National SETAC Meeting, Washington, D.C., November 17–21, 1996, 1996b; Hickey, J.P., Modeling Main Group and Heavy Metal Environmental Behavior Using Linear Solvation Energy Relationships (LSER). II. Abstract 570, QSAR Symposium, 18th National SETAC Meeting, San Francisco, CA, November 16–20, 1997; Hickey, J.P., LSER Estimation of Main Group and Heavy Metal Baseline Toxicity, Poster Abstract, Computational Methods in Toxicology Workshop, Dayton, OH, April 20–22, 1998, 1998a; Hickey, J.P., LSER Estimation of Main Group and Heavy Metal Solution Behavior, Platform Abstract and Presentation, QSAR98 Workshop, Baltimore, MD, May 16–20, 1998, 1998b.
4. McGowan, J.C., Molecular volumes and structural chemistry (Recueil des travaux chimiques des Pays-Bas.), *Rec. Trav. Chim.*, 75, 193–208, 1956.
5. Abraham, M.H. and McGowan, J.C., The use of characteristic volumes to measure cavity terms in reversed phase liquid chromatography, *Chromatographia*, 23, 243–246, 1987.
6. Kamlet, M.J., Doherty, R.M., Abraham, M.H. and Taft, R.W., Solubility properties in biological media. 12. Regarding the mechanism of nonspecific toxicity or narcosis by organic nonelectrolytes, *Quant. Struct.-Act. Relat.*, 7, 71–78, 1988.

7. Kamlet, M.J., Doherty, R.M., Abraham, M.H., Carr, P.W., Doherty, R.F. and Taft, R.W., Linear solvation energy relationships. 41. Important differences between aqueous solubility relationships for aliphatic and aromatic solutes, *J. Chem. Phys.*, 91, 1996–2004, 1987.
8. Cotton, F.A. and Wilkinson, G., *Advanced Inorganic Chemistry: A Comprehensive Text*, 2nd ed., Wiley Interscience, New York, 1962 (this edition contains the best treatment of inorganic solution chemistry in the series).
9. Greenwood, N.N. and Earnshaw, A., *Chemistry of the Elements*, Pergamon Press, New York, 1984.
10. Weast, R.C., et al., Solubilities of inorganic compounds in water, in *Handbook of Chemistry and Physics*, 44th ed., Hodgman, C.D., Ed., Chemical Rubber Company, Cleveland, 1961, pp. 1694–1709.
11. McCloskey, J.T., Newman, M.C. and Clark, S.B., Predicting the relative toxicity of metal ions using ion characteristics: Microtox bioluminescence assay, *Environ. Toxicol. Chem.*, 15, 1730–1737, 1996.
12. Newman, M.C. and McCloskey, J.T., Predicting relative toxicity and interactions of divalent metal ions: Microtox bioluminescence assay, *Environ. Toxicol. Chem.*, 15, 275–281, 1996.
13. Tatara, C.P., Newman, M.C., McCloskey, J.T. and Williams, P.L., Predicting relative metal toxicity with ion characteristics: *Caenorhabditis elegans* LC_{50}, *Aquat. Toxicol.*, 39, 279–290, 1997.
14. deZwart, D. and Sloof, W., The Microtox as an alternative assay in the acute toxicity assessment of water pollutants, *Aquat. Toxicol.*, 4, 129–138, 1983.
15. Bringmann, V.-G. and Kuhn, R., Befunde der Schadwirkung wassergefardender Stoffe gegen Daphnia magna, *Z. Wasser Abwasser-Forschung*, 10, 161–166, 1977.
16. Bringmann, V.-G. and Kuhn, R., Ergebnisse der Schadwirkung wassergefahrdender Stoffe gegen *Daphnia magna* in einem weiterenwickelten standardisierten Testverfahren, *Z. Wasser Abwasser-Forschung* 15, 1–6, 1982.
17. Khangarot, B.S. and Ray, P.K., Investigation of correlation between physicochemical properties of metals and their toxicity to the water flea *Daphnia magna* Strauss, *Ecotoxico. Environ. Safety*, 18, 109–120, 1989.
18. LeBlanc, G., Interspecies relationships in acute toxicity of chemicals to aquatic organisms, *Environ. Toxicol. Chem.*, 3, 47–60, 1984.
19. Juhnke, I. and Ludemann, D., Ergibnesse der Untersuchung von 200 chemischen Verbindungen auf akute Fischtoxizitat mit dem Goldorfentest, *Z. Wasser Abwasser-Forschung*, 11, 161–164, 1978.
20. Newman, M.C. and Jagoe, C.H., Ligands and the bioavailability of metals in aquatic environments, in *Bioavailability: Physical, Chemical, and Biological Interactions*, Hamelink, J.L., Landrum, P.F., Bergman, H.R. and Benson, W.H., Eds., CRC Press, Boca Raton, FL, 1994, pp. 39–60.
21. Gover, R.A., Toxic effects of metals, in *Casarett and Doull's Toxicology: The Basic Science of Poisons*, 3rd ed., Klaasen, C.D., AmDur, M.O. and Doull, J., Eds., Macmillan, Dordrecht, 1986, pp. 582–635.

chapter thirty-five

Determining aromatic hydrocarbons and chlorinated hydrocarbons in sediments and tissues using accelerated solvent extraction and gas chromatography/mass spectrometry

C.A. Sloan, D.W. Brown, R.W. Pearce, R.H. Boyer, J.L. Bolton, D.G. Burrows, D.P. Herman, and M.M. Krahn
National Oceanic and Atmospheric Administration

Contents

1-56670-664-5/05/$0.00+$1.50
© 2005 by CRC Press

Introduction

Numerous and varied environmental studies have involved analyses of marine sediments and tissues (e.g., marine mammal blubber, shellfish, fish muscle or liver, etc.) for toxic contaminants, such as chlorinated pesticides, polychlorinated biphenyls, and aromatic hydrocarbons (AHs). High quality data with documented quality assurance (QA) are needed from such studies in order for valid conclusions to be drawn and appropriate decisions reached, for instance, in damage assessment situations, for policy making, and to facilitate comparisons of the results with those of other studies. New techniques are continually being sought and evaluated to further optimize analytical accuracy, precision, sensitivity, robustness, efficiency, and safety, as well as to minimize contamination from laboratory materials and from sample-to-sample carry-over.

Recent improvements assessed and adopted by this laboratory for determining selected chlorinated hydrocarbons (CHs, Table 35.1) and AHs (Table 35.2) at "nanogram per gram" concentrations are presented. They are among the latest revisions[1] to the methods of Sloan et al.,[2] and Krahn et al.,[3] which updated the method of MacLeod et al.[4] Here, we provide detailed descriptions of the improved procedures for sample extraction, cleanup by gravity-flow silica/alumina columns and by size-exclusion high-performance liquid chromatography (HPLC), and quantitation of CHs and AHs by on-column injection gas chromatography/mass spectrometry (GC/MS). One of the recent procedural modifications is the extraction of samples using accelerated solvent extraction (ASE), which reduced time, labor, solvent use, hazardous waste, and potential exposure of analysts to extraction solvent (in this case, dichloromethane). This ASE method provides an exhaustive extraction of organic compounds from various matrices, while excluding water. Thus, both AHs and CHs are quantitatively recovered in a single extract, and the percent of nonvolatile extractable material, as well as lipid classes, can be determined from a portion of the ASE extract. Another major modification is the use of GC/MS selected-ion monitoring (SIM)[5] in place of GC/electron-capture detection for CH analyses. Thus, the concentrated cleaned-up extracts are analyzed for both AHs and CHs (separately) using the same GC/MS system and configuration, as well as similar operating conditions. As MS instrumentation has improved, the sensitivity of GC/MS SIM for CH detection has approached that of electron-capture detection while being more selective, thus reducing interferences. Alterations made to the GC/MS system that allow greater instrument stability and accuracy include replacing the electron ionization filaments with chemical ionization filaments in order to use a higher source temperature, using a cool on-column injection system in the GC in place of a splitless injection system, and adding a guard column before the analytical column. Quantitation has also been optimized by using point-to-point calibration (also known as pointwise linear calibration), which provides a better data fit over the entire range of GC/MS calibration standards than would be given by a single equation, such as linear, weighted linear, or quadratic regression.

The method described here incorporates QA measures (including a method blank, a sample of a reference material that has certified concentrations for many analytes, and, as

Table 35.1 Chlorinated hydrocarbon analytes including polychlorinated biphenyl (PCB) congeners (by IUPAC number)

PCB 17	2,4′-DDD
PCB 18	4,4′-DDD
PCB 28	2,4′-DDE
PCB 31	4,4′-DDE
PCB 33	2,4′-DDT
PCB 44	4,4′-DDT
PCB 49	Hexachlorobenzene
PCB 52	α-Hexachlorocyclohexane
PCB 66	β-Hexachlorocyclohexane
PCB 70	γ-Hexachlorocyclohexane
PCB 74	Aldrin
PCB 82	Dieldrin
PCB 87	Endosulfan I
PCB 95	Endosulfan II
PCB 99	*trans*-Nonachlor
PCB 101 + PCB 90[a]	*cis*-Nonachlor
PCB 105	Heptachlor
PCB 110	Heptachlor epoxide
PCB 118	Oxychlordane
PCB 128	*trans*-Chlordane
PCB 138 + PCB 163 + PCB 164[a]	*cis*-Chlordane
PCB 149	Mirex
PCB 151	
PCB 153 + PCB 132[a]	
PCB 156	
PCB 158	
PCB 170 + PCB 190[a]	
PCB 171	
PCB 177	
PCB 180	
PCB 183	
PCB 187 + PCB 159 + PCB 182[a]	
PCB 191	
PCB 19	
PCB 195	
PCB 199	
PCB 205	
PCB 206	
PCB 208	
PCB 209	

[a] These analytes are quantitated and reported as the sum of their concentrations because they co-elute on GC by this method.

needed, replicate samples) into each batch of samples. Internal standards (also known as surrogate standards) are added to the samples before extraction to monitor and account for any losses during sample preparation. Additional internal standards are added

Table 35.2 Aromatic hydrocarbon analytes

Naphthalene
1-Methylnaphthalene
2-Methylnaphthalene
Biphenyl
2,6-Dimethylnaphthalene
Acenaphthylene
Acenaphthene
2,3,5-Trimethylnaphthalene
Fluorene
Dibenzothiophene
Phenanthrene
Anthracene
1-Methylphenanthrene
Fluoranthene
Pyrene
Benz[*a*]anthracene
Chrysene + triphenylene[a]
Benzo[*b*]fluoranthene
Benzo[*j*]fluoranthene + benzo[*k*]fluoranthene[a]
Benzo[*e*]pyrene
Benzo[*a*]pyrene
Perylene
Indeno[1,2,3-*cd*]pyrene
Dibenz[*a,h*]anthracene + dibenz[*a,c*]anthracene[a]
Benzo[*ghi*]perylene

[a] These analytes are quantitated and reported as the sum of their concentrations because they co-elute on GC by this method.

before sample cleanup to account for the amount of total extract fractionated by HPLC, and before GC/MS to measure the recovery of the extraction and HPLC internal standards. Analytes are measured based on multiple concentration levels of GC/MS calibration standards. These procedures can be applied to a variety of sediment and tissue matrices across a wide range of contaminant concentrations and with any moisture or lipid content. Low limits of quantitation (on the order of a nanogram per gram or lower) can be achieved even when only small sample amounts (2 g or less) are available, such as when analyzing individual, small fish tissues or marine mammal biopsy samples.

Materials required

Reference to a company or product does not imply endorsement by the U.S. Department of Commerce to the exclusion of others that may be suitable.

Equipment

Samples are extracted using an Accelerated Solvent Extractor® (ASE 200, Dionex, Salt Lake City, UT) with 33-cc extraction cells (048764, Dionex). Nitrogen (Grade 4.8, PraxAir, Danbury, CT) is used for ASE pneumatics and cell purging. The HPLC system used for sample cleanup consists of several components: a Waters 515 HPLC pump (Waters, Milford, MA), a Waters 717 Plus autosampler (Waters), an in-line filter (2-μm particle size, 7302, Rheodyne, Inc., Cotati, CA) after the autosampler, a six-port valve (7030, Rheodyne, Inc.), an Envirosep ABC® size-exclusion guard column (60 mm × 21.2 mm, Phenomenex, Inc., Torrance, CA), an Envirosep ABC size-exclusion preparatory column (350 mm × 21.2 mm, Phenomenex, Inc.), a Spectra 100 UV/VIS detector (Spectra-Physics, San Jose, CA), a Gilson FC 204 fraction collector (Gilson Co., Middleton, WI), and a computer system with Rainin Dynamax® DA software (Version 1.2B4, Rainin Instrument Co., Inc., Woburn, MA). It also includes a ventilated solvent reservoir, in which the solvent is degassed by helium (Grade 5.0, 99.999%, PraxAir). The helium is purified by an in-line hydrocarbon trap (Big Supelco Carb HC®, 24564, Supelco, Bellefonte, PA) and an in-line oxygen trap (Supelco Pure O, 2-2449, Supelco, Bellefonte, PA). The GC/MS system consists of an Agilent 7683 autosampler with an on-column injection syringe and needle guide (Agilent Technologies, Wilmington, DE), an Agilent 6890N gas chromatograph with an cool on-column injection port (Agilent Technologies), a deactivated fused-silica guard column (10-m × 0.53-mm, 460-2535-10, Agilent Technologies), a DB-5® (J & W Scientific, Folsom, CA) GC column (60-m × 0.25-mm, 0.25-μm film thickness, 122-5062, Agilent Technologies), an Agilent 5973N Mass Selective Detector® (Agilent Technologies), and a computer system with ChemStation® software (Version DA, Agilent Technologies). The helium carrier gas (grade 5, ultra-high purity, 99.999%, PraxAir) is filtered through an indicating moisture trap (GMT-4-HP, Agilent Technologies), a disposable oxygen trap (803088, Supelco, Inc.) and an indicating oxygen trap (4004, Alltech Associates, Inc., Deerfield, IL). The guard column is connected to the GC column by a glass, universal column union (705-0825, Agilent Technologies) using polyimide-sealing resin (500-1200, Agilent Technologies). Other necessary equipment includes a gas leak detector (Leak Detective II, 20413, Restek Corporation, Bellefonte, PA), an analytical balance, a muffle furnace, drying ovens at 120°C and 170°C, a steam table, tube heaters (720000-0000, Kontes, Vineland, NJ, with glass-cylinder shroud, wire tube holder, and aluminum inserts bored out to accept 50-ml conical tubes), and a vial heater (we use a HPLC column heater, SP 8792, Spectra-Physics) modified with a rack to hold 2-ml GC vials, which allows the solvent level in the vials to be visually monitored).

Reagents

Only high purity solvents, i.e., dichloromethane (High Purity, 300-4, Burdick & Jackson, Muskegon, MI), isooctane (Optima, 0301-4, Fischer Scientific, Fair Lawn, NJ), and methanol (Optima, A454-4, Fischer Scientific) are used in sample analysis. The following reagents are heated in a muffle furnace at 700°C for 18 h then stored at 120°C: sodium sulfate (reagent grade, anhydrous, granular, S1461, Spectrum, Gardena, CA), magnesium sulfate (anhydrous, M65-3, Fischer Scientific), and Davisil® (WR Grace & Co.) silica (100–200 mesh, grade 634, S734-1, Fischer Scientific). Immediately prior to use, alumina (80–200 mesh, O537-01, JT Baker, Phillipsburg, NJ) is heated at 120°C for 2 h. Sand (Ottawa, kiln-dried, 30–40 mesh, SX0075-3, EM Science, Gibbstown, NJ) is soaked overnight in a 1:3 v/v mixture of concentrated nitric acid (reagent grade, A200-500, Fischer Scientific), concentrated hydrochloric acid (reagent grade, A144-212, Fischer Scientific), then it is washed three times each with water, methanol, and dichloromethane, in that order, then dried, and stored at 120°C. All heated reagents are allowed to cool to room temperature in a desiccator immediately prior to use. Glass wool (Pyrex®, Corning, Inc., Corning, NY) is heated in a muffle furnace at 400°C for 18 h then cooled and stored at room temperature. Copper (reagent grade, granular, 20–30 mesh, 1720-05, JT Baker) is activated before use by submersing it in concentrated hydrochloric acid, stirring it with a glass rod, and allowing it to stand for 5 min. The copper is then washed three times each with water, methanol, and dichloromethane, in that order, and covered with dichloromethane until used to avoid contact with air.

Disposable labware

The following items are low-cost, one-use only. ASE collection vials (60-ml with cap and septa, 048784, Dionex) are purchased pre-cleaned. Boston round bottles (250-ml, 28-mm i.d. mouth) are rinsed with acetone. The following items are heated in a muffle furnace at 400°C for 18 h then stored at room temperature before use: ASE glass fiber filters (047-017, Dionex), conical centrifuge tubes (50-ml, 18-mm i.d. mouth), chromatography columns (plain, custom, 22-mm i.d., 25-cm length, DJ's Glass Factory, San Jose, CA), jars (10-oz, 2.5-in. i.d.), GC vials (2-ml, C4000-1W, National Scientific Co., Duluth, GA), HPLC vials (4-ml, 99300-A, Sun International, Wilmington, NC), low-volume vial inserts for the GC vials (C4000-54R, National Scientific Co.), and low-volume vial inserts for the HPLC vials (200756, Sun International). The bottles, jars, and tubes are fitted with caps lined with Teflon® (E.I. Du Pont de Nemours & Company, Inc., Wilmington, DE), and the vials are assembled with caps and Teflon-lined septa. Teflon boiling chips (AW0919120, All-World Scientific, Lynnwood, WA) are rinsed three times with dichloromethane, allowed to dry, and stored at room temperature.

Standard solutions

All solutions are prepared in-house using isooctane as the solvent, except for the HPLC retention-time standard, which is prepared using dichloromethane. Concentrations given here are approximate; actual concentrations vary slightly from batch to batch of the solutions. The GC calibration standards and the internal standard solutions, including GC and HPLC internal standards, are prepared such that the concentrations of the internal standards in final sample extracts analyzed by GC/MS are approximately equal to the concentrations of the internal standards in the GC calibration standards. The internal standard for quantitating CH analytes (CH I-Std) contains PCB 103 and 4,4′-dibromooctafluorobiphenyl (DOB) at 1 ng/μl each compound. The HPLC internal standard for CHs (CH HPLC I-Std) contains tetrachloro-*m*-xylene (TCMX) at 1 ng/μl. The GC internal standard for CHs (CH GC I-Std) contains tetrachloro-*o*-xylene (TCOX) at 1 ng/μl. The internal standard for quantitating AH analytes (AH I-Std) contains naphthalene-d_8, acenaphthene-d_{10}, and benzo[*a*]pyrene-d_{12} at 1.7 ng/μl each. The HPLC internal standard for AHs (AH HPLC I-Std) contains phenanthrene-d_{10} at 1.6 ng/μl. The GC internal standard for AHs (AH GC I-Std) contains hexamethylbenzene (HMB) at 10 ng/μl. The HPLC retention-time standard contains DOB and perylene at 1.5 ng/μl each compound. The CH GC calibration standards contain the CHs as in Table 35.1. They are prepared at the following concentrations: Level 1 at 0.001 ng/μl each compound, Level 2 at 0.003 ng/μl each compound, Level 3 at 0.01 ng/μl each compound, Level 4 at 0.03 ng/μl each compound, Level 5 at 0.1 ng/μl each compound, Level 6 at 0.3 ng/μl each compound, Level 7 at 1.0 ng/μl each compound, Level 8 at 4.0 ng/μl each compound, Level 9 at 10 ng/μl each compound, Level 10 at 20 ng/μl each compound, Level 11 at

100 ng/μl each compound; plus all levels contain PCB 103, DOB, TCMX, and TCOX at 0.3 ng/μl each compound. The AH GC calibration standards contain the AHs in Table 35.2 at the following concentrations: Level 1 at 0.0011 ng/μl each compound, Level 2 at 0.0044 ng/μl each compound, Level 3 at 0.015 ng/μl each compound, Level 4 at 0.044 ng/μl each compound, Level 5 at 0.11 ng/μl each compound, Level 6 at 0.33 ng/μl each compound, Level 7 at 1.13 ng/μl each compound, Level 8 at 3.3 ng/μl each compound, and Level 9 at 10.5 ng/μl each compound; plus all levels contain naphthalene-d_8, acenaphthene-d_{10}, benzo[*a*]pyrene-d_{12}, and phenanthrene-d_{10} at 0.5 ng/μl each compound in all levels, and HMB at 3.2 ng/μl.

Reference materials

Standard Reference Materials® (SRM) are purchased from the National Institute of Standards and Technology (NIST), Gaithersburg, MD. SRM 1941b "Organics in Marine Sediment" is analyzed with sediment samples, SRM 1945 "Organics in Whale Blubber" is analyzed with marine mammal blubber samples, and SRM 1974b "Organics in Mussel Tissue (*Mytilus edulis*)" is analyzed with other tissue samples (e.g., shellfish, fish muscle, or liver, etc.).

Procedures

Sample extraction

Sediments or tissues are customarily analyzed in batches of 10–14 samples. Each batch includes a method blank, a sample of an appropriate SRM and, as needed, one or more replicate sediment or tissue samples. If only one of the two classes of analytes (CHs or AHs) is to be quantitated in the samples, all the internal standards for the other class of analytes are omitted. Prior to removing a portion of sample for analysis, tissues are homogenized to the extent possible. Standing water is decanted from sediment containers, and all pebbles, biota, and other extraneous material are discarded, and then the sediment is stirred. A specified amount of sediment or tissue sample is transferred to a 10-oz jar and weighed to the nearest 0.001 g. Typically, the amount of sediment or tissue analyzed is approximately 2 g of wet sample, except the amount of blubber analyzed is approximately 0.5 g. To absorb the water from the sample, allowing greater extraction efficiency of the organic compounds, 15 cc of sodium sulfate is added to the sample in the jar and mixed thoroughly with the sample. To avoid clumping and hardening of the sodium sulfate, the sample is mixed immediately after adding the sodium sulfate to the jar. The sample continues to be mixed until it appears dry, then 15 cc of magnesium sulfate is added to the sample in the jar and mixed thoroughly for increased absorption of water from the sample. Two glass-fiber filters are placed at the bottom of an ASE cell, then the sample/drying agent mixture is transferred to the ASE cell. The bottom of the cell is tapped firmly but carefully on the countertop during the transfer to completely settle the cell contents. The remaining cell volume is filled with sodium sulfate. Again, the bottom of the cell is tapped firmly but carefully on the counter top to completely settle the cell contents, leaving approximately 3 mm void depth at the top of the cell. One glass-fiber filter is placed on top of the sodium sulfate in the cell without overlapping the rim of the cell. AH I-Std solution (75 μl) and CH I-Std solution (75 μl) are added onto the top filter in the cell, then approximately 1 ml of dichloromethane is added onto the filter to rinse the internal standards into the cell. The cell threads are cleared of sample/drying agent as necessary, and the cell is capped firmly but not forcefully. The ASE is prepared for extracting the batch of samples by loading the sample cells and collection vials into their respective carousels, and filling the ASE solvent reservoir with dichloromethane. The ASE system rinse function is activated three times to flush the solvent lines prior to extracting a batch of samples. The samples are extracted using an ASE schedule with the rinse parameter set to "ON," and a method with a 0-min preheat, a 5-min heat, two 5-min static cycles at 2000 psi and 100°C, a 115%-volume flush and a 180-s purge. After the extraction is complete, AH HPLC I-Std solution (75 μl) and CH HPLC I-Std solution (75 μl) are added to the extract in the ASE collection vial, and the extract is thoroughly mixed.

Sample cleanup

Silica/alumina column chromatography

A gravity-flow silica/alumina chromatography column is used for removing extraneous polar compounds from the sample extract. The column is prepared by positioning an approximately 1-cm high plug of glass wool at the bottom of a chromatography column, then adding 10 cc of alumina, followed by 20 cc of silica, and 5 cc of sand. To slightly deactivate the column packing, which improves the recovery of high molecular weight AHs, 35 ml of 10% methanol in dichloromethane is slowly added to the column and allowed to drain into a waste container. Then, to flush the methanol from the column, 35 ml of dichloromethane is slowly added to the column and allowed to drain into the waste container. All subsequent eluant is collected in a 250-ml bottle. The sample extract in the ASE collection vial is transferred into the column without disturbing the packing. To elute the analytes, 35 ml of dichloromethane is added to the column. After the eluant is collected, several boiling chips are added to the bottle, and the sample volume is reduced to approximately 20 ml using a steam table. The sample is transferred to a 50-ml centrifuge tube, a boiling chip is added to the tube, and the sample volume is reduced to approximately 1 ml using a tube heater. For all samples in a batch of sediments, activated copper is added to the tube a few grains at a time until no further discoloring of the copper occurs, then the tube is capped and stored overnight. The sample is transferred to a HPLC vial with a low-volume insert (leaving the copper in the tube in the case of sediment samples), and the sample volume is brought to 1 ml as necessary.

Size-exclusion high-performance liquid chromatography

Size-exclusion HPLC is used for removing extraneous high molecular weight compounds (e.g., biogenic material) from the sample extracts by collecting a fraction of the eluant containing the analytes, which elute after the high molecular weight compounds. The system is operated using a mobile phase of 100% dichloromethane at 5 ml/min and ambient temperature, while the UV detector monitors the wavelength 254 nm. The HPLC system is calibrated when any component of the system (e.g., column or length of tubing) is changed, to determine the relationship between the fraction collection times and the retention times of the components in the HPLC retention-time standard.[1] In this standard, DOB is the first peak to elute, at approximately 14 min, and perylene is the last peak, at approximately 20 min. The beginning of the elution of DOB, which is determined during the system calibration, denotes the beginning of the fraction containing AHs and CHs; the ending of the elution of perylene, also determined during the system calibration, denotes the ending of the fraction. The HPLC retention-time standard is chromatographed immediately prior to chromatographing a batch of samples in order to adjust the fraction collection times as necessary using the relationship determined from the system calibration. This standard is also chromatographed in the middle and at the end of the batch of samples to monitor the retention-time stability and adjust the fraction collection times if necessary. Samples are chromatographed by injecting 500 μl, and the "AH/CH" fraction of each sample is collected in a 50-ml tube. The remaining extract in the HPLC vial is reserved in a freezer in case of need. The size-exclusion guard column is backflushed for 2–3 min after the sequence of samples and standards has finished. A boiling chip is added to the 50-ml tube containing the AH/CH fraction, and the fraction volume is reduced to 1 ml using a tube heater. AH GC I-Std solution (30 μl), CH GC I-Std solution (30 μl), and isooctane (70 μl) are added to the tube, and the contents are mixed thoroughly. The "AH/CH" fraction is transferred from the tube to a GC vial, a small boiling chip is added to the vial, and the fraction volume is reduced to 100 μl using a vial heater. The concentrated fraction is transferred to a GC vial with a low-volume insert.

Gas chromatography/mass spectrometry

GC/MS is used to measure the concentrations of the analytes and the internal standards in the samples. Operating conditions for the GC/MS are listed in Table 35.3. The GC/MS is checked for air leaks, both spectrally and with a handheld electronic leak detector, and the MS tune is performed or checked for stability prior to analyzing a sequence of samples.

GC/MS for quantitating AHs

For AHs, the GC oven temperature is programmed for an initial temperature of 80°C held for 1 min, a first ramp to 200°C at 10°C/min, a second ramp to 300°C at 4°C/min, and a final temperature of 300°C held for 15 min. Sediment samples are analyzed using scan mode, scanning all masses from 60 to 300 amu. Tissue samples are analyzed using SIM mode, scanning only the quantitation ions during the specified time window as shown in Table 35.4. Scan mode provides greater ability to confirm analyte identification in complex sediment samples, whereas SIM mode provides greater sensitivity for tissue samples, which generally contain lower concentrations of contaminants. The quantitation ions for AHs in tissues shown in Table 35.4 are also the ions used for quantitating AHs in sediments. The time windows for SIM mode are adjusted as necessary when the GC column is changed or the guard column is trimmed. Prior to analyzing samples, the sensitivity of the GC/MS is checked by analyzing the lowest level of AH GC calibration standard that will be used (Level 1 for tissues; Level 3 for sediments). One or more batches of sediment samples are analyzed in a sequence that includes one each of AH GC calibration standard Levels 3–6, 8, and 9, alternating with the samples, plus one Level 7 before the first sample and replicate Level 7 analyses at the middle and end of each batch. One or more batches of tissue samples are analyzed in a sequence that includes one each of AH GC calibration standard Levels 1–3 and Levels, 5–7, alternating with the samples in the sequence, plus one Level 4 before the first sample and replicate Level 4 analyses at the middle and end of each batch. Included at the beginning of each sequence is an injection of a sample in the batch, which is not quantitated but serves as a GC/MS system conditioner. Alternating the standards and samples in the sequence improves the GC/MS stability, which is monitored using the replicate Level 7 or Level 4 analyses.

Table 35.3 Gas chromatograph/mass spectrometer operating conditions

Injection volume	2 μl
Injection technique	Cool on-column
Inlet temperature program	Oven temperature plus 3°C
Transfer-line temperature	300°C
Carrier gas	Helium
Carrier gas linear velocity	30 cm/s
Carrier gas flow	1.3 ml/min constant flow
Column	
Material	Fused-silica capillary tubing
Length	60 m
Internal diameter	0.25 mm
Stationary phase	DB-5® (J & W Scientific)
Phase composition	5% phenyl, 95% methylpolysiloxane
Film thickness	0.25 μm
Guard column	
Material	Deactivated fused-silica capillary tubing
Length	10 m
Internal diameter	0.53 mm
Quadrupole temperature	150°C
Ion-source pressure	$\leq 5 \times 10^{-5}$ Torr
Ion-source temperature	300°C
Ionization mode	Electron ionization (70 eV)
Filament	Chemical ionization filament
Tune mode	AHs: autotune CHs: autotune, then the multiplier voltage is increased for the desired sensitivity
Emission current maximum	35 μA
Scan rate	Approximately 10–20 scans per peak

GC/MS for quantitating CHs

For CHs, the GC oven temperature is programmed for an initial temperature of 80°C held for 1 min, a first ramp to 150°C at 10°C/min, a second ramp to 195°C at 0.5°C/min, a third ramp to 315°C at 3°C/min, and a final temperature of 315°C held for 15 min. Sediment and tissue samples are analyzed using SIM mode for optimal sensitivity, scanning only the quantitation ions during the specified time window as shown in Table 35.5. The time windows for SIM mode are adjusted as necessary when the GC column is changed or the guard column is trimmed. Prior to analyzing samples, the sensitivity of the GC/MS is checked by analyzing the lowest level of CH GC calibration standard that will be used (Level 2 for blubber; Level 1 for other tissues and sediments). One or more batches of blubber samples are analyzed in a sequence that includes one of each CH GC calibration standard Levels 2–4 and Levels 6–11 alternating with the samples, plus one Level 5 before the first sample and replicate Level 5 analyses at the middle and end of each batch. One or more batches of sediment samples or tissue samples, other than blubber, are analyzed in a sequence that includes one of each CH GC calibration standard Levels 1–4 and 6 alternating with the samples, plus one Level 5 before the first sample and replicate Level 5 analyses at the middle and end of each batch. The replicate Level 5 analyses are monitored for GC/MS stability. Included at the beginning of each sequence is an injection of a sample in the batch, which is not quantitated but serves as a GC/MS system conditioner.

Table 35.4 Selected-ion monitoring for quantitating aromatic hydrocarbons in tissue samples

Approximate time window (min)	Compounds	Quantitation ion (atomic mass units)
7–12	Naphthalene-d_8	136
	Naphthalene	128
	2-Methylnaphthalene	142
	1-Methylnaphthalene	142
12–13	Biphenyl	154
	2,6-Dimethylnaphthalene	156
13–14	Hexamethylbenzene	147
	Acenaphthylene	152
	Acenaphthene-d_{10}	164
	Acenaphthene	154
14–17	2,3,5-Trimethylnaphthalene	170
	Fluorene	166
17–20	Dibenzothiophene	184
	Phenanthrene-d_{10}	188
	Phenanthrene	178
	Anthracene	178
20–29	1-Methylphenanthrene	192
	Fluoranthene	202
	Pyrene	202
29–42	Benz[*a*]anthracene	228
	Chrysene + triphenylene	228
	Benzo[*j*]fluoranthene + benzo[*k*]fluoranthene	252
	Benzo[*e*]pyrene	252
	Benzo[*a*]pyrene-d_{12}	264
	Benzo[*a*]pyrene	252
	Perylene	252
42–53	Indeno[1,2,3-*cd*]pyrene	276
	Dibenz[*a,h*]anthracene + dibenz[*a,c*]anthracene	278
	Benzo[*ghi*]perylene	276

Calculations of results

The acquired GC/MS data for the samples and calibration standards are processed using the Agilent ChemStation software to determine the areas of the MS response for the analytes and internal standards. These areas are then used by a BASIC program written in-house to compute the analyte amounts in the samples (nanogram analyte per sample) based on point-to-point calibration. The point-to-point calibration method plots the relative response factors of the calibration standards versus their analyte areas, then the linear equations generated between each pair of consecutive points are used for the calibration,[1] i.e., for each analyte in each sample, a relative response factor is computed by using the analyte area and interpolating between the relative response factors of the analyte in the two consecutive calibration standards whose areas for that analyte bracket the area of the analyte in the sample. When a particular analyte in a given sample has an MS area that is larger than its area in the highest-level calibration standard, the analyte amount is calculated using the relative response factor of that analyte in the highest level of calibration standard used. The resulting concentration is footnoted as exceeding the

Table 35.5 Selected-ion monitoring for quantitating chlorinated hydrocarbons in tissue and sediment samples

Approximate time window (min)	Compounds	Quantitation ion (atomic mass units)
15–29	Tetrachloro-*o*-xylene	207
	Tetrachloro-*m*-xylene	207
29–35	α-Hexachlorocyclohexane	181
	Hexachlorobenzene	284
	4,4′-Dibromooctafluorobiphenyl	456
35–39	β-Hexachlorocyclohexane	219
	γ-Hexachlorocyclohexane	181
39–45	PCB 18	256
	PCB 17	256
45–55	PCB 31	256
	PCB 28	256
	PCB 33	256
	Heptachlor	272
55–65	PCB 52	292
	PCB 49	292
	Aldrin	263
	PCB 44	292
65–69	PCB 103	326
69–76	Heptachlor epoxide	353
	Oxychlordane	115
	PCB 74	292
	PCB 70	292
	PCB 66	292
	PCB 95	326
76–79	*trans*-Chlordane	373
79–86	2,4′-DDE	246
	Endosulfan I	241
	PCB 101 + PCB 90	326
	cis-Chlordane	373
	PCB 99	326
	trans-Nonachlor	409
86–93	Dieldrin	79
	PCB 87	326
	4,4′-DDE	318
	PCB 110	326
	2,4′-DDD	235
93–98	PCB 82	326
	PCB 151	360
	Endosulfan II	241
98–103	PCB 149	360
	PCB 118	326
	cis-Nonachlor	409
	4,4′-DDD	165
	2,4′-DDT	235
103–106	PCB 153 + PCB 132	360
	PCB 105	326
106–110	4,4′-DDT	165
	PCB 138 + PCB 163 +PCB 164	360
	PCB 158	360

Table 35.5 Continued

Approximate time window (min)	Compounds	Quantitation ion (atomic mass units)
110–113	PCB 187 + PCB 159 + PCB 182	394
	PCB 183	394
	PCB 128	360
113–115	PCB 177	394
	PCB 171	394
	PCB 156	360
115–121	PCB 180	394
	PCB 191	394
	Mirex	272
	PCB 170 + PCB 190	394
	PCB 198	430
	PCB 199	430
121–153	PCB 208	464
	PCB 195	430
	PCB 194	430
	PCB 205	430
	PCB 206	464
	PCB 209	498

calibration range and is therefore an estimate. When an analyte is not detected in a sample or has an area that is smaller than that analyte's area in the lowest level of GC calibration standard used, the concentration of the analyte in that sample is reported to be less than the value of its lower limit of quantitation. The lower limit of quantitation for a particular analyte in a given sample is the concentration that would be calculated if the analyte had an MS area equivalent to its area in the lowest-level calibration standard used in the calibration. Percent recoveries of internal standards are calculated using single-point calibration. The CH internal standard is PCB 103 for all analytes in Table 35.1 except when samples contain extremely high concentrations of PCBs, including significant amounts of PCB 103, in which case DOB is used as the CH internal standard. The AH internal standards are naphthalene-d_8 for the analytes naphthalene through 2,6-dimethylnaphthalene, acenaphthene-d_{10} for analytes acenaphthene through pyrene, and benzo[*a*]pyrene-d_{12} for analytes benz[*a*]anthracene through benzo[*ghi*]perylene in Table 35.2. For CH analyses, the percent recovery of each sample's CH internal standard is computed based on the CH GC internal standard and then adjusted for the CH HPLC internal standard recovery (to account for the amount injected on HPLC), which is also computed based on the CH GC internal standard. For these calculations, the Level 5 GC calibration standard is used for the relative response factors of the internal standards. Similarly, for AH analyses, the percent recovery of each sample's AH internal standards are computed based on the AH GC internal standard and then adjusted for the AH HPLC internal standard recovery, which is also computed based on the AH GC internal standard. For these calculations, the Level 4 GC calibration standard in the case of SIM analyses, or the Level 7 GC calibration standard in the case of scan analyses, is used for the relative response factors of the internal standards.

Quality assurance

Quality assurance measures are incorporated into the analyses of each batch of samples. Specific QA criteria for batches and individual samples, as well as the percent of the total results that must meet the criteria, depend on the type of project for which the samples are analyzed and the purpose of the data. Typical criteria are presented here. When QA criteria are not met, the data may be footnoted as such or the sample(s) reanalyzed. The stability of the GC/MS is evaluated using the repetitions of the mid-level GC calibration standard analyzed periodically in the sequence of samples and other levels of GC calibration standards. The GC/MS is considered stable if the area of an analyte relative to the area of its internal standard in a given repetition is within $\pm$ 15% of the respective average for the repetitions. The percent recoveries of internal standards in samples are considered acceptable if they are 60–130%. When three or more replicate samples are analyzed, the precision of the results can be evaluated based on the relative standard deviation (RSD) of the concentrations of each analyte. The precision is considered acceptable if the RSD is <15% for at least 90% of the analytes. For duplicate samples, this translates to a relative percent difference of <30% for at least 90% of the analytes. At least one sample of an appropriate SRM from the National Institute of Standards and Technology (NIST) is analyzed in each batch of samples to indicate the accuracy of the data for the entire batch. NIST provides certified concentrations and uncertainty values (for approximately 95% confidence interval) for many of the analytes in Tables 35.1 and 35.2. For the analytes having NIST-certified concentrations, upper and lower control limits are set as is done by NIST in the Intercomparison Exercise Program for Organics in the Marine Environment, such that the control limits are 30% beyond each end of the 95% confidence interval. The accuracy of an SRM result is considered acceptable if the determined concentration falls within the upper and lower control limits. This criterion does not apply to analytes with concentrations below their limit of quantitation. The analysis of the SRM is acceptable if at least 70% of the analytes having NIST-certified concentrations are within their control limits. Analytes which do not have NIST-certified concentrations, as a result, do not have QA criteria for accuracy, but can be monitored and compared to NIST reference concentrations where available or to previous in-house results.

Results and discussion

The analytical procedures described here resulted in better data quality, improved safety, and increased cost-effectiveness as new techniques were adopted. Specifically, the ASE has provided quantitative extraction of the analytes, while decreasing (1) solvent use, (2) hazardous waste, (3) potential exposure of analysts to solvent vapors, and (4) labor for sample extraction and cleaning equipment. This ASE method uses optimal amounts of drying agents and sample sizes to robustly and reproducibly extract samples with better quantitative results than previous methods (i.e., using a homogenizing probe for tissues and a ball-mill tumbler for sediments[2]). GC/MS has been another focal point for improving analyses. GC/MS SIM eliminated many interferences with measuring CHs that were present in analyses by a comparable method using GC/electron-capture detection. Cool on-column GC injection avoids the molecular weight loading discrimination, thermal degradation, and adsorption of analytes to active sites that occurred during splitless injection. In conjunction with the cool on-column injector, a 0.53-mm i.d. guard column attached to the analytical column has maintained good GC resolution for a longer period of time by reducing build-up of nonvolatile compounds on the analytical column. The guard column provides the added benefit of focusing the analytes, which improves peak shape and resolution for better quantitation. Setting the GC carrier gas to constant flow instead of constant pressure also improved peak shape and resolution. Increasing the MS ion-source temperature (from 230°C to 300°C) greatly reduced inconsistencies in the MS response caused by varying amounts of extraneous compounds that are present in the samples but not in the GC calibration standards. The higher source temperature was made possible by replacing the standard electron-ionization filaments with chemical-ionization filaments, which have ceramic insulators that can withstand higher temperatures. Alternating the standards and samples in the sequence, rather than analyzing all the standards before or after all the samples, further contributed to improved GC/MS stability. The mode of processing data has also had a significant impact on data quality. Quantitation using point-to-point calibration was found to be more accurate than using a single regression equation for the expansive range of analyte concentrations that can be determined by this method. This range is defined by the lowest-level calibration standard used, which is near the instrument detection limit, and the highest-level calibration standard used, which for AHs in sediment samples or CHs in blubber samples is near the level of overloading the GC column. Relatively large errors in calculating concentrations have occurred when a single quadratic regression equation, with or without weighting, was used, particularly near the extremes of the calibration range.

Before method modifications were employed, they were validated in this laboratory by analyzing and applying QA criteria to replicate samples of a variety of NIST SRMs, such as SRM 1974b "Organics in Mussel Tissue (*M. edulis*)," NIST SRM 1941b "Organics in Marine Sediment," and NIST SRM 1945 "Organics in Whale Blubber." During the application of this method in analyses of environmental sediment or tissue samples, QA measures, including the analysis of an appropriate (by matrix type) SRM, are continuously monitored and documented with the analysis of every batch of samples. The proficiency of the method has also been demonstrated and compared with other laboratories' results through participation in the annual NIST Intercomparison Exercise Program for Organics in the Marine Environment, which involves replicate analyses of both an SRM and an unknown sample of a similar matrix.

SRM 1974b is used as the reference material in batches of tissue samples other than blubber. This material is analytically challenging because of its high moisture content (89.87%) and low CH and AH concentrations. SRM 1974b has certified concentrations for 35 of our CH analytes and 17 of our AH analytes (co-eluting compounds are counted as one analyte, as noted in Tables 35.1 and 35.2, and their certified concentrations are summed for comparison). These certified concentrations range from 0.269 to 14.73 ng/g wet weight for the CHs and from 0.494 to 18.04 ng/g wet weight for the AHs. Tables 35.6 and 35.7 show representative examples of our results for these CH and AH analytes, respectively, quantitated in replicate samples of SRM 1974b as part of a recent NIST intercomparison exercise for the analysis of mussel tissue. The NIST-certified concentrations and the control limits are also shown for comparison. In these analyses, the measured concentrations of 31 CHs and 15 AHs were within the control limits, and 2 CHs were below their limit of quantitation (2,4′-DDE at <0.397 ng/g and PCB 170 at <0.385 ng/g). The measured concentrations of PCB 31, 4,4′-DDD and benz[*a*]anthracene were greater than their upper control limit, whereas that for naphthalene was less than its lower control limit. The RSDs of all measured analyte concentrations ranged from 0.4% to

Table 35.6 Measured and certified concentrations of chlorinated hydrocarbons in National Institute of Standards and Technology (NIST) Standard Reference Material® 1974b "Organics in Mussel Tissue (*Mytilus edulis*)" analyzed as part of the NIST Intercomparison Exercise Program for Organics in the Marine Environment, 2003

Analyte	Measured average (ng/g wet weight) ($n = 3$)	RSD[a] (%)	Certified concentration[b] (ng/g wet weight)	LCL[c]	UCL[d]
trans-Chlordane	0.98	1.7	1.14 ± 0.17	0.68	1.70
cis-Chlordane	1.12	1.6	1.36 ± 0.10	0.88	1.90
trans-Nonachlor	1.07	2.3	1.30 ± 0.14	0.81	1.87
2,4′-DDE	<0.397	–	0.336 ± 0.044	0.204	0.494
4,4′-DDE	4.27	1.0	4.15 ± 0.38	2.64	5.89
2,4′-DDD	0.87	7.2	1.09 ± 0.16	0.65	1.63
4,4′-DDD	5.07	2.4	3.34 ± 0.22	2.18	4.63
PCB 18	1.07	2.8	0.84 ± 0.13	0.50	1.26
PCB 28	2.55	5.3	3.43 ± 0.25	2.23	4.78
PCB 31	5.04	1.4	2.88 ± 0.23	1.86	4.04
PCB 44	3.75	0.5	3.85 ± 0.20	2.56	5.27
PCB 49	4.98	0.5	5.66 ± 0.23	3.80	7.66
PCB 52	6.51	0.4	6.26 ± 0.37	4.12	8.62
PCB 66	7.15	2.1	6.37 ± 0.37	4.20	8.76
PCB 70	7.45	2.0	6.01 ± 0.22	4.05	8.10
PCB 74	4.28	0.6	3.55 ± 0.23	2.32	4.91
PCB 82	1.17	1.2	1.16 ± 0.14	0.71	1.69
PCB 87	4.61	1.0	4.33 ± 0.36	2.78	6.10
PCB 95	5.85	0.5	6.04 ± 0.36	3.98	8.32
PCB 99	6.15	1.2	5.92 ± 0.27	3.96	8.05
PCB 101	11.1	0.7	10.7 ± 1.1	6.7	15.3
PCB 105	4.73	1.2	4.00 ± 0.18	2.67	5.43
PCB 110	10.3	0.7	10.0 ± 0.7	6.5	13.9
PCB 118	10.6	1.0	10.3 ± 0.4	6.9	13.9
PCB 128	1.90	1.8	1.79 ± 0.12	1.17	2.48
PCB 138 + PCB 163	13.4	1.8	11.2 ± 1.5[e]	6.8	16.5
PCB 149	7.08	1.5	7.01 ± 0.28	4.71	9.48
PCB 151	1.75	0.7	1.86 ± 0.16	1.19	2.63
PCB 153 + PCB 132	15.6	1.2	14.7 ± 1.1[f]	9.5	20.5
PCB 156	0.792	3.4	0.718 ± 0.080	0.447	1.037
PCB 158	1.112	2.4	0.999 ± 0.096	0.632	1.424
PCB 170	<0.385	–	0.269 ± 0.034	0.165	0.394
PCB 180	1.15	1.0	1.17 ± 0.10	0.75	1.65
PCB 183	1.23	1.2	1.25 ± 0.03	0.85	1.66
PCB 187	2.76	0.6	2.94 ± 0.15	1.95	4.02

[a] Relative standard deviation of the measured concentrations.

[b] NIST-certified concentration with uncertainty value.

[c] Lower control limit (ng/g wet weight) = 0.7 × (certified concentration – uncertainty value).

[d] Upper control limit (ng/g wet weight) = 1.3 × (certified concentration + uncertainty value).

[e] Certified concentration and uncertainty value of PCB 138 are summed with NIST reference concentration and uncertainty value of PCB 163, respectively.

[f] Certified concentrations and uncertainty values of PCB 153 and PCB 132 are summed, respectively.

Table 35.7 Measured and certified concentrations of aromatic hydrocarbons in National Institute of Standards and Technology (NIST) Standard Reference Material® 1974b "Organics in Mussel Tissue (Mytilus edulis)" analyzed as part of the NIST Intercomparison Exercise Program for Organics in the Marine Environment, 2003

Analyte	Measured average (ng/g wet weight) ($n = 3$)	RSD[a] (%)	Certified concentration[b] (ng/g wet weight)	LCL[c]	UCL[d]
Naphthalene	1.43	11.3	2.43 ± 0.12	1.62	3.32
Fluorene	0.466	2.1	0.494 ± 0.036	0.321	0.689
Phenanthrene	2.65	3.2	2.58 ± 0.11	1.73	3.50
Anthracene	0.551	1.0	0.527 ± 0.071	0.319	0.777
1-Methylphenanthrene	0.82	1.9	0.98 ± 0.13	0.60	1.44
Fluoranthene	19.6	1.3	17.1 ± 0.7	11.5	23.1
Pyrene	20.15	1.1	18.04 ± 0.6	12.21	24.23
Benz[*a*]anthracene	7.09	3.7	4.74 ± 0.53	2.95	6.85
Chrysene + triphenylene	13.9	2.3	10.6 ± 1.7[e]	6.2	16.0
Benzo[*b*]fluoranthene	7.19	3.1	6.46 ± 0.59	4.11	9.17
Benzo[*j* + *k*]fluoranthenes[f]	5.79	2.7	6.15 ± 0.47[e]	3.98	8.61
Benzo[*e*]pyrene	10.9	2.7	10.3 ± 1.1	6.4	14.8
Benzo[*a*]pyrene	2.81	1.0	2.80 ± 0.38	1.69	4.13
Perylene	1.15	1.4	0.99 ± 0.14	0.60	1.47
Indeno[1,2,3-*cd*]pyrene	2.68	1.0	2.14 ± 0.11	1.42	2.93
Dibenz[*a,h* + *a,c*]anthracenes[g]	0.656	0.8	0.539 ± 0.044[h]	0.347	0.758
Benzo[*ghi*]perylene	3.80	1.7	3.12 ± 0.33	1.95	4.49

[a] Relative standard deviation of the measured concentrations.

[b] NIST-certified concentration with uncertainty value.

[c] Lower control limit (ng/g wet weight) = 0.7 × (certified concentration − uncertainty value).

[d] Upper control limit (ng/g wet weight) = 1.3 × (certified concentration + uncertainty value).

[e] Certified concentrations and uncertainty values of coeluting compounds are summed, respectively.

[f] Benzo[*j*]fluoranthene + benzo[*k*]fluoranthene.

[g] Dibenz[*a,h*]anthracene + dibenz[*a,c*]anthracene.

[h] Certified concentration and uncertainty value of dibenz[*a,h*]anthracene are summed with NIST reference concentration and uncertainty value of dibenz[*a,c*]anthracene, respectively.

7.2% for CHs and from 0.8% to 11.3% for AHs. The percent recoveries of internal standards ranged from 102% to 107% for CHs and from 84% to 95% for AHs.

The reference material used in batches of sediments is SRM 1941b, which has certified concentrations for 34 of our CH analytes and 17 of our AH analytes. These certified concentrations range from 0.378 to 6.75 ng/g dry weight for the CHs and from 73.2 to 848 ng/g dry weight for the AHs. Measuring CHs in this material is particularly challenging because of the low concentrations of CHs in the presence of high concentrations of AHs and other contaminants. As part of a recent NIST intercomparison exercise for the analysis of marine sediment, our analyses of replicate samples of SRM 1941b (Tables 35.8 and 35.9) resulted in 25 CHs and 17 AHs being within the control limits, and 3 CHs being below their limit of quantitation (*t*-nonachlor at <0.413 ng/g, *c*-nonachlor at <0.421 ng/g and PCB 195 at <0.417 ng/g). The measured concentrations of PCB 31, PCB 105, PCB 110, 4,4′-DDD, and hexachlorobenzene were greater than their upper control limit, whereas that for *cis*-chlordane was less than its lower control limit. The RSDs of all measured analyte concentrations ranged from 0.3% to 6.8% for CHs and from 0.8% to 2.9% for AHs.

Table 35.8 Measured and certified concentrations of chlorinated hydrocarbons in National Institute of Standards and Technology (NIST) Standard Reference Material® 1941b "Organics in Marine Sediment" analyzed as part of the NIST Intercomparison Exercise Program for Organics in the Marine Environment, 2003

Analyte	Measured average (ng/g dry weight) (*n* = 3)	RSD[a] (%)	Certified concentration[b] (ng/g dry weight)	LCL[c]	UCL[d]
Hexachloro-benzene	8.53	0.6	5.83 ± 0.38	3.82	8.07
trans-Chlordane	0.611	2.5	0.566 ± 0.093	0.331	0.857
cis-Chlordane	0.51	4.1	0.85 ± 0.11	0.52	1.25
trans-Nonachlor	<0.413	–	0.438 ± 0.073	0.256	0.664
cis-Nonachlor	<0.421	–	0.378 ± 0.053	0.228	0.560
4,4′-DDE	4.25	4.0	3.22 ± 0.28	2.06	4.55
4,4′-DDD	8.88	2.5	4.66 ± 0.46	2.94	6.66
PCB 18	2.99	1.2	2.39 ± 0.29	1.47	3.48
PCB 28	5.31	2.9	4.52 ± 0.57	2.77	6.62
PCB 31	5.05	1.9	3.18 ± 0.41	1.94	4.67
PCB 44	4.23	1.2	3.85 ± 0.20	2.56	5.27
PCB 49	4.14	0.8	4.34 ± 0.28	2.84	6.01
PCB 52	5.93	1.4	5.24 ± 0.28	3.47	7.18
PCB 66	6.24	0.9	4.96 ± 0.53	3.10	7.14
PCB 87	1.48	6.8	1.14 ± 0.16	0.69	1.69
PCB 95	4.29	3.1	3.93 ± 0.62	2.32	5.92
PCB 99	3.24	0.6	2.90 ± 0.36	1.78	4.24
PCB 101	6.25	2.5	5.11 ± 0.34	3.34	7.09
PCB 105	2.11	2.4	1.43 ± 0.10	0.93	1.99
PCB 110	6.68	1.8	4.62 ± 0.36	2.98	6.47
PCB 118	5.21	2.6	4.23 ± 0.19	2.83	5.75
PCB 128	0.854	2.0	0.696 ± 0.044	0.456	0.962
PCB 138 + PCB 163	6.35	2.0	4.88 ± 0.34[e]	3.18	6.79
PCB 149	5.03	1.5	4.35 ± 0.26	2.86	5.99
PCB 153 + PCB 132	7.78	1.3	6.75 ± 0.59[f]	4.31	9.54
PCB 156	0.600	1.8	0.507 ± 0.090	0.292	0.776
PCB 170	1.47	4.2	1.35 ± 0.090	0.88	1.87
PCB 180	3.40	0.7	3.24 ± 0.51	1.91	4.88
PCB 183	0.860	0.5	0.979 ± 0.087	0.624	1.386
PCB 187	2.22	0.7	2.17 ± 0.22	1.37	3.11
PCB 194	1.30	5.6	1.04 ± 0.06	0.69	1.43
PCB 195	<0.417	–	0.645 ± 0.060	0.410	0.917
PCB 206	2.69	5.9	2.42 ± 0.19	1.56	3.39
PCB 209	4.71	0.3	4.86 ± 0.45	3.09	6.90

[a] Relative standard deviation of the measured concentrations.

[b] NIST-certified concentration with uncertainty value.

[c] Lower control limit (ng/g dry weight) = 0.7 × (certified concentration - uncertainty value).

[d] Upper control limit (ng/g dry weight) = 1.3 × (certified concentration + uncertainty value).

[e] Certified concentration and uncertainty value of PCB 138 are summed with NIST reference concentration and uncertainty value of PCB 163, respectively.

[f] Certified concentration and uncertainty value of PCB 153 are summed with NIST reference concentration and uncertainty value of PCB 132, respectively.

Table 35.9 Measured and certified concentrations of aromatic hydrocarbons in National Institute of Standards and Technology (NIST) Standard Reference Material® 1941b "Organics in Marine Sediment" analyzed as part of the NIST Intercomparison Exercise Program for Organics in the Marine Environment, 2003

Analyte	Measured average (ng/g dry weight) (n = 3)	RSD[a] (%)	Certified concentration[b] (ng/g dry weight)	LCL[c]	UCL[d]
Naphthalene	896	1.9	848 ± 95	527	1230
Fluorene	76	1.1	85 ± 15	49	130
Phenanthrene	453	1.5	406 ± 44	253	585
Anthracene	203	0.8	184 ± 18	116	263
1-Methyl-phenanthrene	78.5	1.3	73.2 ± 5.9	47.1	102.8
Fluoranthene	712	1.4	651 ± 50	421	911
Pyrene	604	1.4	581 ± 39	379	806
Benz[*a*]-anthracene	390	2.1	335 ± 25	217	468
Chrysene + triphenylene	490	2.9	399 ± 36[e]	254	566
Benzo[*b*]fluoranthene	520	2.3	453 ± 21	302	616
Benzo[*j* + *k*]fluoranthenes[f]	494	1.5	442 ± 23[g]	293	605
Benzo[*e*]pyrene	398	1.5	325 ± 25	210	455
Benzo[*a*]pyrene	360	2.5	358 ± 17	239	488
Perylene	492	1.0	397 ± 45	246	575
Indeno[1,2,3-*cd*]pyrene	380	1.1	341 ± 57	199	517
Dibenz[*a,h* + *a,c*] anthracenes[h]	83	1.8	90 ± 15[e]	52	137
Benzo[*ghi*]perylene	304	2.3	307 ± 45	183	458

[a] Relative standard deviation of the measured concentrations.

[b] NIST-certified concentration with uncertainty value.

[c] Lower control limit (ng/g dry weight) = 0.7 × (certified concentration − uncertainty value).

[d] Upper control limit (ng/g dry weight) = 1.3 × (certified concentration + uncertainty value).

[e] Certified concentrations and uncertainty values of coeluting compounds are summed, respectively.

[f] Benzo[*j*]fluoranthene + benzo[*k*]fluoranthene.

[g] Certified concentration and uncertainty value of benzo[*j*]fluoranthene are summed with reference concentration and uncertainty value of benzo[*k*]fluoranthene, respectively.

[h] Dibenz[*a,h*]anthracene + dibenz[*a,c*]anthracene.

The percent recoveries of internal standards ranged from 103% to 107% for CHs and from 81% to 90% for AHs.

SRM 1945 is used as the reference material in batches of blubber samples. This SRM has certified concentrations for 41 of our CH analytes, ranging from 3.30 to 445 ng/g wet weight. Measuring CHs across this range of concentrations in this material is also challenging because of the high percent of nonvolatile extractable material (74.29%) and high concentrations of other contaminants in this SRM. In another recent NIST intercomparison exercise, which was for the analysis of marine mammal blubber, our analyses of replicate samples of SRM 1945 (Table 35.10) resulted in 36 CHs being within the control limits and none being below its limit of quantitation. The measured concentrations of PCB 87, PCB 138 + PCB 163 + PCB 164, PCB 194, PCB 206, and PCB 209 were greater than their upper control limit. The RSDs of all measured analyte concentrations ranged from 0.3% to 4.5%, and the percent recoveries of the internal standard ranged from 107% to 109%.

The results for the three different SRMs summarized above met the targeted QA criteria for overall accuracy, precision, and internal standard recoveries, and demonstrate

Table 35.10 Measured and certified concentrations of chlorinated hydrocarbons in National Institute of Standards and Technology (NIST) Standard Reference Material® 1945 "Organics in Whale Blubber" analyzed as part of the NIST Intercomparison Exercise Program for Organics in the Marine Environment, 2003

Analyte	Measured average (ng/g wet weight) (n = 3)	RSD[a] (%)	Certified concentration[b] (ng/g wet weight)	LCL[c]	UCL[d]
Hexachloro-benzene	27.2	0.8	32.9 ± 1.7	21.8	45.0
α-Hexachloro-cyclohexane	16.2	1.4	16.2 ± 3.4	8.96	25.5
γ-Hexachloro-cyclohexane	3.12	4.5	3.30 ± 0.81	1.74	5.34
Mirex	38.3	0.4	28.9 ± 2.8	18.3	41.2
Heptachlor epoxide	12.4	0.4	10.8 ± 1.3	6.7	15.7
Oxychlordane	23.0	3.3	19.8 ± 1.9	12.5	28.2
cis-Chlordane	56.9	0.8	46.9 ± 2.8	30.9	64.6
trans-Nonachlor	182	0.3	231 ± 11	154	315
cis-Nonachlor	58.4	0.4	48.7 ± 7.6	28.8	73.2
2,4′-DDE	14.66	1.6	12.28 ± 0.87	7.99	17.10
4,4′-DDE	488	0.5	445 ± 37	286	627
2,4′-DDD	21.7	1.0	18.1 ± 2.8	10.7	27.2
4,4′-DDD	133	0.8	133 ± 10	86	186
2,4′-DDT	89	0.3	106 ± 14	64	156
4,4′-DDT	257	0.4	245 ± 15	161	338
PCB 18	3.08	1.3	4.48 ± 0.88	2.52	6.97
PCB 44	11.7	0.5	12.2 ± 1.4	7.6	17.7
PCB 49	16.9	0.6	20.8 ± 2.8	12.6	30.7
PCB 52	38.1	0.3	43.6 ± 2.5	28.8	59.9
PCB 66	22.3	1.8	23.6 ± 1.6	15.4	32.8
PCB 87	25.3	0.8	16.7 ± 1.4	10.7	23.5
PCB 95	39.9	0.3	33.8 ± 1.7	22.5	46.2
PCB 99	54.8	0.5	45.4 ± 5.4	28.0	66.0
PCB 101 + PCB 90	77.5	0.6	65.2 ± 5.6[e]	41.7	92.0
PCB 105	27.1	0.4	30.1 ± 2.3	19.5	42.1
PCB 110	34.8	0.4	23.3 ± 4.0	13.5	35.5
PCB 118	83.8	0.4	74.6 ± 5.1	48.7	104
PCB 128	24.5	0.3	23.7 ± 1.7	15.4	33.0
PCB 138 + PCB 163 + PCB 164	183.5	0.3	131.5 ± 7.4[e]	86.9	180.6
PCB 149	85.2	0.6	106.6 ± 8.4	68.7	149.5
PCB 151	27.5	0.8	28.7 ± 5.2	16.5	44.1
PCB 153 + PCB 132	243	0.4	213 ± 19[f]	136	302
PCB 156	12.6	0.6	10.3 ± 1.1	6.4	14.8
PCB 170 + PCB 190	50.3	0.6	40.6 ± 2.6[e]	26.6	56.2
PCB 180	143.7	0.4	106.7 ± 5.3	71.0	145.6
PCB 183	39.6	0.5	36.6 ± 4.1	22.8	52.9
PCB 187	114.2	0.5	105.1 ± 9.1	67.2	148.5
PCB 194	60.8	0.4	39.6 ± 2.5	26.0	54.7
PCB 195	11.6	0.4	17.7 ± 4.3	9.4	28.6
PCB 206	50.6	0.4	31.1 ± 2.7	19.9	43.9
PCB 209	18.4	0.6	10.6 ± 1.1	6.7	15.2

[a] Relative standard deviation of the measured concentrations.

[b] NIST-certified concentration with uncertainty value.

[c] Lower control limit (ng/g wet weight) = 0.7 × (certified concentration - uncertainty value).

[d] Upper control limit (ng/g wet weight) = 1.3 × (certified concentration + uncertainty value).

[e] Certified concentration and uncertainty value are for the co-eluting compounds combined as stated by NIST.

[f] Certified concentration and uncertainty value are for PCB 153 only; NIST does not provide a certified or reference value for PCB 132.

the applicability of this method. In each exercise, over 70% of the analytes having certified concentrations above their limit of quantitation were within the control limits (94% of CHs and 88% of AHs in SRM 1974b, 81% of CHs and 100% of AHs in SRM 1941b, and 88% of CHs in SRM 1945). Also, the RSDs for all analytes were less than the criteria of 15% maximum, and the percent recoveries of the internal standards in all samples were well within the criteria of 60% to 130%. This level of data quality has been achieved or surpassed routinely, using the techniques incorporated in the method given here. The QA measures also show where difficulties in the analyses occur, e.g., high or low biases in some analyte results. Because it is desirable to meet the SRM QA criteria for 100% of the certified analytes as an indication of attaining that level of accuracy for all sample results, further improvements in all parts of the method continue to be investigated for their impact on data quality.

Acknowledgments

We are pleased to acknowledge the support given by John Stein and Tracy Collier, Northwest Fisheries Science Center, and we thank the past and present Environmental Assessment Program chemists for their expert contributions. We also thank William Reichert and Jon Buzitis for manuscript review.

References

1. Sloan, C.A., Brown, D.W., Pearce, R.W., Boyer, R.H., Bolton, J.L., Burrows, D.G., Herman, D.P. and Krahn, M.M., *Northwest Fisheries Science Center Procedures for Extraction, Cleanup and Gas Chromatography/Mass Spectrometry Analysis of Sediments and Tissues for Organic Contaminants*, U.S. Department of Commerce, NOAA Tech. Memo. NMFS-NWFSC-59, 2004, 47 pp.
2. Sloan, C.A., Adams, N.G., Pearce, R.W., Brown, D.W. and Chan, S.-L., Northwest fisheries science center organic analytical procedures, in *Sampling and Analytical Methods of the National Status and Trends Program, National Benthic Surveillance and Mussel Watch Projects 1984–1992, Volume IV, Comprehensive Descriptions of Trace Organic Analytical Methods*, Lauenstein, G.G. and Cantillo, A.Y., Eds., U.S. Department of Commerce, NOAA Tech. Memo. NOS ORCA 71 (Chapter 2).
3. Krahn, M.M., Wigren, C.A., Pearce, R.W., Moore, L.K., Bogar, R.G., MacLeod, W.D. Jr., Chan, S.-L. and Brown, D.W., *Standard Analytical Procedures of the NOAA National Analytical Facility, 1988: New HPLC Cleanup and Revised Extraction Procedures for Organic Contaminants*, U.S. Department of Commerce, NOAA Tech. Memo. NMFS F/NWC-153, 1988, 52 pp.
4. MacLeod, W.D., Brown, D.W., Friedman, A.J., Burrows, D.G., Maynes, O., Pearce, R.W., Wigren, C.A. and Bogar, R.G., *Standard Analytical Procedures of the NOAA National Analytical Facility, 1985–1986: Extractable Toxic Organic Compounds*, U.S. Department of Commerce, NOAA Tech. Memo. NMFS F/NWC-92, 1985, 121 pp.
5. Burrows, D.G., Brown, D.W. and MacLeod, W.D., A twenty-five fold increase in GC/MS sensitivity attained by switching through a sequence of ten MID descriptors during capillary GC analysis, presented at the 38th Annual ASMS Conference on Mass Spectrometry and Allied Topics, Tucson, AZ, June 3–8, 1990, 16 pp.

chapter thirty-six

Histological preparation of invertebrates for evaluating contaminant effects

Esther C. Peters
Tetra Tech, Inc.
Kathy L. Price
Cooperative Oxford Laboratory
Doranne J. Borsay Horowitz
U.S. Environmental Protection Agency

Contents

1-56670-664-5/05/$0.00+$1.50
© 2005 by CRC Press

Introduction

Although many studies in toxicologic pathology evaluate the effects of toxicants on fishes because of their similarities with other vertebrates, invertebrates can also provide insights into toxicant impacts on ecosystems. Invertebrates not only serve as food resources (e.g., worms, clams, shrimp, insects), but also as agents of physical habitat change (e.g., reef-building corals, bioeroding sponges, bioturbating worms), among other roles. Adverse effects on individuals and populations of any species due to toxicant exposures from water, sediment, or food can ultimately alter community relationships and have direct and indirect impacts on members of the community.

Many invertebrates are sedentary or have a limited range of mobility compared to fishes. Because of this, their health may reflect local environmental conditions better than fishes. Therefore, invertebrates can be used to study chronic or acute exposures to single toxicants or mixtures in the field and laboratory[1,2] (see also Chapters 6 and 7 in this volume). However, many such studies of invertebrate species limit their observations to gross descriptions, percent of organisms killed by the toxicant, or reductions in growth and reproduction. Knowing that an animal died or did not reproduce, however, may not be as important as knowing why. Toxicant effects can be measured using many techniques and tools, but the only techniques that can provide visual images of the internal condition of the cells and tissues of an organism are observations made using light and electron microscopy. This field is known as histology. Structure and composition of cells and tissues reflect their function and overall functioning of the organism. Histotechniques provide data that can be used in conjunction with procedures from other fields (e.g., physiology, biochemistry, molecular biology, ecology) to improve our understanding of invertebrate susceptibilities to contaminants, mechanisms of toxicant damage to target cells and organs, and impact of the exposure on the individual, population, and community.[3–8]

This chapter will discuss the preparation of invertebrate tissues for light microscopical examination using paraffin-embedding, a common procedure in many hospitals and veterinary medical diagnostic operations. The resulting product is a section of tissue typically 2–10 μm thick mounted on a glass microscope slide (histoslide). Dyes or other compounds can be used to enhance particular features of tissues for study with bright-field illumination or epifluorescence. Although basic steps for tissue fixation, dehydration, clearing, embedding, sectioning, and staining have remained similar since the late 1800s, recent developments in equipment, techniques, and the ability to identify specific molecules in tissues could require different approaches to improve the microscopic appearance of tissues or to address specific research questions.

Three kinds of data can be generated from histoslides to test hypotheses: descriptive, qualitative, and quantitative data.[8] Study objectives must be established prior to collecting and processing organisms to identify the most appropriate methodology for producing the required information (i.e., the correct type, quantity, and quality of histological data). The hypothesis(es) being tested and the endpoints (data) that will be collected using microscopic examinations of the stained tissue sections should be framed. Qualitative (ranked) data can be compared using nonparametric statistics, and quantitative data can be compared using parametric statistics. Adequate numbers of control (unexposed) animals need to be included in every study for comparison with exposed animals. The number of replicates needed from exposed and "pristine" sites, or from laboratory exposures and controls, should be statistically determined to ensure valid data. In the case of endangered or threatened species, such as stony corals, a minimum of three samples from different colonies from each exposure concentration should be examined to provide an indication of individual variability and normal tissue conditions within that field site or experiment. You should plan to collect the invertebrates from the cleanest water first, then sample from the least to most contaminated, rinsing equipment between samples. Also consider limitations on resources versus data needs, i.e., how many histoslides of which portions of tissue will you need to examine to test the hypothesis(es), versus what will you be able to examine based on time, expertise, or funding constraints.

Since the ability to obtain the data is dependent on the quality of the histoslide produced, this procedure focuses on two steps: *fixation* to preserve the tissues without postmortem autolysis, and *staining* to enable the observer to see the structure and composition of the cells and tissues so that comparisons can be made with established criteria for cell damage, or with control specimens. Interpreting structure and composition seen in the tissue section under the microscope to derive these data requires special training, which cannot be covered in this chapter. Please contact any of the authors for assistance in locating comparative histopathologists and courses or workshops for such training. One resource for information, histoslides, and other materials on invertebrates exposed to contaminants or affected by other diseases is the Registry of Tumors in Lower Animals in Sterling, Virginia, VA, USA, a project funded by the National Cancer Institute (see www.pathology-registry.org, for more information).

Materials required

The equipment, supplies, and reagents needed to prepare invertebrates for histologic examination are presented in three separate lists, below. These materials are similar to those used in hospital and research histology laboratories for processing human or other vertebrate tissues. This procedure is based on the same histotechniques used in these facilities and more detailed information is presented in many books and journals (e.g., References 9–13, *Journal of Histotechnology*). You can use an internet search engine to locate resources for new and used equipment. Processing, embedding, sectioning, staining, and coverslipping can be performed manually or with the use of automated equipment. If possible, using an automated tissue processor and embedding center, at a minimum, will reduce time and artifacts, and improve the consistency of the histoslides. Be sure to keep equipment clean and in working order by following manufacturers' instructions.

Equipment

- Fume hood, providing ventilation and sufficient air exchange to protect worker health
- Tissue processor for paraffin-embedding (e.g., the Technicon Autotechnicon Mono Tissue Processor is a basic model, more features in Leica TP1020, Shandon Hypercenter, or many other models on the market)
- Embedding center (e.g., Tissue Tek III Embedding Console System)
- Microtome (e.g., American Optical, Leitz 1512, Leica RM2125T Manual Rotary Microtome, Microm HM 310 Manual Rotary Microtome)
- Blade holder for microtome disposable blades
- Temperature-controlled water bath (e.g., Fisher Model 134 Tissue Prep Flotation Bath)
- Slide warmer or oven (many models available)
- Balance (for weighing out chemicals and stains, to $\pm$ 0.01 g)
- Magnetic stirrer/heater and coated stir bars (for mixing solutions)

Supplies

- Personal protective equipment (disposable latex or nitrile gloves, goggles, lab coat, closed-toed shoes)
- Dissecting tools (e.g., Clauss poultry shears; stainless steel scissors; scalpels with disposable surgical blades, razor blades or knives; needle probes; 12 and 6-in., and fine curved and pointed dissecting forceps)
- Weighing papers or pans
- Paper towels
- Cork or plastic trimming board or wax-based dissecting pan
- Plastic ruler (or rule marked on trimming board) or calipers (for measuring animal size)
- Pencils
- Waterproof paper for labels
- Plastic or glass jars or containers with tight-fitting lids
- Sturdy resealable plastic bags
- Stainless steel or plastic cassettes (type may depend on tissue processor used)
- Stainless steel or disposable molds for embedding
- Mold-release spray (for stainless steel molds)
- Narrow- and wide-blade plastic paint scrapers (for removing paraffin from equipment, counters, floor)
- Thermometer(s) (0–100°C in 1°C gradations, for checking temperature of molten paraffin, water bath, oven)
- Embedding rings (optional, depends on type of cassette used)
- Disposable knife blades for the microtome (be sure to get proper size for holder).

Note: Traditional one-piece steel blades, which require skillful sharpening, may be used; however, due to potential sand and grit in tissues, the more expensive disposable blades save time

- Ice tray or pan filled with water and frozen
- Needle probes or small brushes

- Flat-bladed forceps
- Slide racks to fit in staining dishes
- Staining dishes (glass or plastic, 300–500 ml size, disposable plastic food storage dishes with lids that fit slide racks can be used for water rinses)
- Well-cleaned microscope slides, frosted end (use Plus (+)™ charged slides for immunohistochemistry (IHC) or stains which tend to loosen tissues from slide)
- Glass coverslips (24 × 40, 50, or 60 mm to cover the tissue section on the unfrosted area of the slide, thickness No. 1 or 1.5)
- Mounting medium, xylene-compatible (Permount® or synthetic)
- Absorbent tissues (e.g., Scotties® or Kleenex®)
- Lint-free tissues (e.g., Kimwipes®)
- Cardboard or metal flat slide holders or trays
- Microscope slide boxes (to hold 25 or 100 histoslides)

Reagents

Note: Many of these reagents are hazardous chemicals and should be handled with appropriate care and caution; review all MSDS before using.[14]

- Formaldehyde (37–40%)
- Other chemicals as noted in the selected fixative formulation(s)
- Undenatured ethanol solutions (70%, 80%, 95%, and 100%)
- Xylenes (or xylene substitute, clearing agent)
- Paraffin
- Certified stains and/or staining solutions (some can be purchased premade)
- Ammonium hydroxide (reagent-grade, in solution)
- Sodium hydroxide (reagent-grade, pellets)
- Hydrochloric acid (reagent-grade, concentrated, 12 *N*)
- Ethylenediaminetetraacetic acid (EDTA), disodium salt dihydrate, reagent-grade ($Na_2ClOH_{14}O_8N_2 \cdot 2H_2O$, CAS No. 6381-92-6)
- Deionized or distilled water

Procedure

An overview of the paraffin-embedding procedure is presented here with the steps in chronological order, as it applies to a variety of invertebrates. Each laboratory should develop written standard operating procedures (SOPs). SOPs are used to guide processing for a particular invertebrate to be studied, inform technician(s) on how the laboratory operates, how equipment should be used, tips to obtain the best results with available equipment, and describe safe handling of chemicals and emergency procedures.

1. Select and prepare fixative solution

Fixation is the most important step in the overall procedure. Chemicals are used to precipitate and denature proteins, cross-link molecules, and preserve membranes in tissue samples. The fixative solution stops all metabolic reactions, prevents autolysis by intracellular enzymes, maintains the spatial relationships of cellular components and extracellular substances, and brings out differences in refractive indexes among tissue elements for better visibility, as well as enhancing the staining reactions.[11] Each of the chemicals used in the fixation process has advantages and drawbacks and will produce different results, both in terms of tissue preservation and in the ability to further demonstrate metabolic processes, presence of microorganisms, or immunoreactivity through the application of special stains. However, all fixatives are not compatible with all staining procedures.

(a) Use Table 36.1 to select a fixative for a particular invertebrate.
   - Consider hypothesis(es) being tested and endpoints (data) that will be collected from stained tissue sections; which stains will be used and why they are necessary.
   - Note special fixation and handling requirements for the chemicals.
   - Determine the most appropriate fixation times for the animals and whether they can remain in fixative or require immediate rinsing and storage in another solution.
   - Note the origin of the animal. Marine organisms, especially those that are osmoconformers, cannot be fixed in a low-osmolality (or osmolarity) fixative, but require the osmolality (= salt content) of the fixative to be approximately the same or slightly less than the salinity or osmolality of the water from which they were collected to prevent swelling and bursting of cells. Fixing organisms from fresh or brackish estuarine waters in a high-osmolality fixative solution can result in shrinkage of cells and tissues.
   - Calculate volume of fixative needed, approximately 10–20 times the tissue volume.

(b) Prepare the fixative.
   - Wear personal protective equipment.
   - Weigh chemicals carefully and check calculations of proportions.
   - Follow mixing instructions.
   - Store the fixative in a plastic or glass jug or carboy.

2. *Inventory samples*

Planning collection efforts well in advance of a study can maximize tissue preservation and minimize unnecessary artifacts and cellular changes, which can result from improper handling. Each animal must be uniquely identified with a label, in pencil or waterproof ink. A written record of the field collection data, such as date, source or site, and other pertinent information, must accompany the animals back to the laboratory. Because postmortem autolysis increases with time after death, animals should be fresh (alive, apparently healthy or morbid), at ambient or chilled temperature. Freezing causes extensive cellular disruption and should be avoided. The species and sample collection information must be recorded in an accessioning log prior to processing. Additional notes should be kept in laboratory notebooks.

(a) Assign each specimen a unique number or code to facilitate identification and sample tracking during processing.

(b) The accessioning log should include, for example:
   - Identifying number or code
   - Collection location and exposure conditions
   - Experimental exposure data

Table 36.1 Recommended fixatives for selected invertebrates for histologic examination using light microscopy

Fixative and source	Invertebrate group	Anticipated staining uses			Comments
		Routine	Special	Immunohisto-chemistry	
Bouin's solution (Reference 11, p. 20; can be ordered premade)	Marine, freshwater, or terrestrial invertebrates	Yes	Yes, good for stains requiring mordant	No	Excellent preservation. Do not leave in this fixative longer than 6–12 h. Wash thoroughly in 70% undenatured ethanol, changing frequently to remove the picric acid. Store in 70% ethanol
Davidson's or Dietrich's solutions (see text)	Marine or freshwater mollusks, crustaceans	Yes	Yes, but will need mordant solution for some stains	No	Good preservation in some organisms, in others more vacuolation of cells observed; intracellular granules not as well preserved. Fix at least 48 h. May be stored in this fixative up to a few months, then transfer to 70% undenatured ethanol
Helly's solution (see text)	Marine sponges, cnidarians, annelids, mollusks, echinoderms	Yes	Yes, good for stains requiring mordant	No	Excellent preservation. Fix only for 8–16 h. Tissues must be washed in water for 24 h and then stored in 70% undenatured ethanol
10% Neutral buffered formalin (Reference 10, p. 28; Reference 11, p. 13; can be ordered premade)	Freshwater or terrestrial invertebrates	Yes	Yes, but will need mordant solution for some stains	Yes, extensive time in formalin might require antigen-retrieval procedures	Fix for at least 24 h (small animals) or for IHC to 72 h. Can remain in this solution after fixation
10% Seawater-formalin solution (see text)	All marine invertebrates	Yes	Yes, but will need mordant solution for some stains	Yes, extensive time in formalin might require antigen-retrieval procedures	Preservation of tissues not optimal, use only if another fixative is not available. Fix for at least 48 h. Can remain in this solution after fixation
Z-Fix (order concentrate from Anatech, Ltd.)	All invertebrates	Yes	Yes	Yes	Dilute concentrate with freshwater for freshwater or terrestrial invertebrates and with seawater for marine invertebrates. Fix for at least 24 h. Store in this solution after fixation

Figure 36.1 Trimming tissues to fit into cassettes. (Photo by Jeffrey Greenberg, U.S. EPA, Atlantic Ecology Division.)

- Sacrifice date
- Species
- Project title or other information
- Researcher's name

(c) As each animal is inventoried, examine it carefully and record gross observations (e.g., changes in tissue color or morphology, ulcerations or abrasions, masses or cysts, degeneration or sloughing of tissue, internal or external parasites). It is better to have more information and field or experimental data, than less.

3. *Trim and fix tissue*

After recording gross observations (including size and weight), each animal must be trimmed into 2- to 3-mm thick portions so the fixative will penetrate evenly into tissues (Figure 36.1). Animals can be anesthetized[15] prior to necropsy. They can either be trimmed and fixed immediately, or partially fixed until firm enough to trim, then returned to fixative. Small specimens can be placed directly into processing cassettes for fixation. For large specimens, organs can be dissected out and sliced to permit fixative penetration. Alternatively, the animal's body cavities can be injected with fixative (use blunt end of syringe or wide-bore needle syringe) in addition to immersion. Begin working with specimens from the cleanest location first, to most contaminated, rinsing dissecting board and implements between animals to avoid cross-contamination of tissues.

(a) Wear gloves and handle animals with care.
- Two or more layers of gloves, heavy rubber, or chain-mail gloves might be needed when working with sponges, spiny crustaceans, or sea urchins.
- Long-spined urchins can be held with tongs while trimming their spines to about 1.5 cm length with scissors for easier handling.
- Gloves not only protect you from physical injury but also from exposure to potentially harmful chemicals and pathogenic microorganisms.

(b) Place animal on a dissecting board, keep moist (cover with a damp paper towel if needed).
(c) Trim, using scalpel or razor blade, according to Table 36.2 and the study plan.
(d) Mark particular locations or lesions with colored surgical marking ink (applied to blotted-almost-dry tissue) to facilitate later identification (this step can also be done after fixation).
(e) Select container large enough to hold organism(s) and add fixative (10–20 times the tissue volume).
(f) Add trimmed organism(s) to fixative solution:
- If animal sections are small enough to fit into processing cassettes without further trimming, then place sections in cassette and add a waterproof label marked with a solvent-resistant histological marking pen (not just waterproof) or a pencil recording the inventory or accessioning number or code before closing cassette.
- Multiple specimens can be fixed in a single large container of fixative (when individual identification is not needed), by placing separate specimen samples in different cassettes, or by placing larger individuals in mesh or plastic bags with holes cut in the sides to allow penetration of solutions (the latter options permit keeping a label with each specimen).
- Smaller animals can be fixed in containers with tight-fitting lids, either individually or in groups. Extremely small organisms (e.g., micromollusks, copepods, or worms) or pieces of tissue can be placed in fine mesh mini-cassettes or cassettes with biopsy sponges to prevent loss through the cassette holes during processing.
- Fixed material can also be enrobed in agar or HistoGel™ (see next step), and the resulting block inserted in cassette for fixation.

(g) Unused tissues can be stored in the fixative depending on the solution used (see Table 36.1), in 70% ethanol, or discarded following the protocol for your facility.
(h) To avoid contamination of tissues, rinse or wipe off dissecting board before trimming the next organism.

4. *Remove excess fixative, decalcify tissue (optional), and place in cassettes*

After fixation, tissues are ready to be rinsed and undergo a final trimming, if necessary, to fit into cassettes and embedding molds. Calcium carbonate endoskeletons or exoskeletons of invertebrates (e.g., stony corals, sclerites of gorgonians, tests and Aristotle's lantern of echinoderms, chitinous shells and stomachs of arthropods, small mollusks fixed in their shells) must be decalcified before final trimming to permit sectioning. Not all of these materials will be readily removed during decalcification, but it should help.

Acidic decalcification solutions can affect end results. Routine or special histochemical stains are not affected by this treatment, provided they are washed thoroughly to remove the acid. For delicate tissues, acid-based solutions can be diluted 1:1 with deionized water. Acid-based decalicifiers work quickly; however, preserving antigens for study using IHC requires slow decalcification of small pieces of tissue, using EDTA at neutral pH and room temperature. The formula for this solution is provided in Appendix. If you are uncertain of which decalcification method to use and have enough material, divide the specimen so more than one technique can be used.

Table 36.2 Trimming and fixation of invertebrate tissues for histological examination

Invertebrate group	Small (<5-mm thick)	Large (>5-mm thick)
Sponges *Note*: Sclero-sponges are extremely hard and cannot be sectioned with razor knives. Embedding in plastic and grinding to prepare thin sections (as is done for bone) might be useful with this group	Immerse whole in fixative	Trim into slices with disposable razor blade to 5-mm thickness, then immerse in fixative. (*Note*: Siliceous spicules in the sponges will dull the blade quickly, change frequently.) Boring sponges in shells or calcareous substratum can be fixed in entirety, but try to keep portions relatively thin to ensure penetration of the fixative
Cnidarians or coelenterates	Relax using an anesthetic, if desired.[15] (*Example*: Add a 2% solution of magnesium sulfate to the collection water of the animals. Gradually increase the concentration of the magnesium sulfate in the collection water until it reaches about 30%. Carefully drain off overlying water, then add fixative to this container. After about 30 min, change solution to fresh fixative.) If not relaxing, immerse whole animal in fixative Minimize time in Bouin's and Helly's solutions	Relax using an anesthetic, if desired[15] Stony corals: Immerse whole or portion in fixative. Tissue is thin and will easily be penetrated by the fixative. The carbonate skeleton is removed after fixation Gorgonians: Trim into 5-mm thick portions. Trim branch length to fit the size of the fixation container Sea anemones: Immerse in fixative for approximately 10 min to firm tissue, then slice open sagittally to permit fixation in the gastric cavity. Or inject fixative into the mouth, slice as needed, and immerse in fixative Jellyfish or comb jellies: Immerse whole animal in fixative. Slice larger animals at intervals to permit fixative penetration or dissect out portions needed for study and immerse in fixative All of the above: Minimize time in Bouin's and Helly's solutions
Annelids	Relax worms using an anesthetic, if desired[15] Immerse whole worms less than 3 cm long in the fixative Final trim: Cross-sections, sagittal, and longitudinal sections should be made	Relax worms using an anesthetic, if desired[15] Immerse larger worms in fixative for 15 min, then cut longitudinally to 5-mm thick sections Very large worms will require dissection of desired parts Final trim: Cross-sections, sagittal, and longitudinal sections should be made
Mollusks	Bivalves: Carefully separate shell halves slightly with a scalpel to allow fixative to penetrate. Immerse whole animal and shell in fixative Gastropods: Crack the shells of smaller snails using pliers or a hammer (gently!) to allow preservative infiltration. Immerse whole animal with shell in fixative	Bivalves: Insert a shellfish knife between the valves being careful not to damage the tissues. Separate the valves by turning the knife sideways. Gently slide the knife along the dorsal valve to detach the mantle and adductor muscle. Remove the shell. Detach the adductor muscle from the ventral valve using the same technique (see Reference 37) Univalves (e.g., limpets, slipper shells): Immerse whole animal and shell in the fixative

		Gastropods: Gastropods can be quite difficult to remove from their shells. The tip of the spire can be removed, then the shell is cracked and slowly chipped away from the opening and up along the spiral using bone shears, and carefully the animal is removed from the remaining shell fragments. Immerse whole animal in fixative
		Cephalopods: Immerse whole animal in fixative until firm
		All of the above: After the whole animal has been in fixative for approximately 30 min, the body can be sagittally or parasagittally sectioned at 5-mm intervals, leaving the sections attached at one end. Use a firm pressure and slow sawing motion with razor blade or scalpel. Return the tissue to the fixative for the specified amount of time
Crustaceans	Copepods, amphipods, crabs, hermit crabs, shrimp: Immerse whole animals in fixative[38]	Immerse in fixative for 30 min or inject fixative into the animal's hepatopancreas, stomach, and midgut
	Zooplankton samples (can contain an assortment of crustaceans and other phyla): Add Dietrich's fixative to the collecting medium. Decant solution off after the animals die, and refill the container with fresh fixative. The animals can be stored indefinitely in Dietrich's	Barnacles: Crack shells with poultry shears and place in fixative
		Crabs: Cut abdomen away from carapace, pull body apart, and return to fixative
		Hermit crabs: Pull from their shells. Cut into several sections and immerse in fixative
		Shrimp: Bisect or trisect longitudinally but leave the tail intact. Return to fixative
		Lobsters, large shrimp: Loosen the carapace from the thorax and cut it off behind eyes. Remove the intestine, reproductive tract, heart, and hepatopancreas. Cut the body through midline from posterior to anterior. Ventral nerve fibers and ganglions lie between the muscle bundles and are easily teased out after fixation
Echinoderms	Immerse in fixative	Inject fixative through the test (exoskeleton) in several places
		If necessary, insert scalpel and cut around peristome to permit fixative to infiltrate the organs
		Test can also be split open to permit fixative penetration. Immerse in fixative

Figure 36.2 Self-siphoning water bath. Continuous filling and draining with tap water thoroughly rinses decalcified tissues (here in cassettes) for 24 h. (Photo by Jeffrey Greenberg, U.S. EPA, Atlantic Ecology Division.)

(a) Rinse fixed tissues briefly in freshwater before final trimming, unless they have been stored in 70% undenatured ethanol.
  - As noted in Table 36.1, Bouin's and Helly's solutions require extensive washing to remove excess fixative. As appropriate to the fixative, either water or 70% ethanol can be used for this procedure, until the wash solution is no longer colored.
  - For Helly's solution, a self-siphoning water bath is preferred (Figure 36.2); alternatively, use constant flow from a tap into a water dish containing the tissue or several changes of static water.

(b) Using fresh scalpel or razor blades, trim fixed tissue into 2- to 3-mm thick sections that fit easily into labeled cassettes.
  - Use firm pressure on blade and a slow sawing motion.
  - Switch to a new blade as the old one dulls to ensure a smooth cut with minimum disruption or tearing of tissue.
  - Do not force sections to fit into cassettes, as this will yield incomplete penetration of reagents during subsequent processing.
  - If tissue cannot be cut easily, it should be decalcified first [see Steps 4(e)–4(i) below].

(c) Place the side of the tissue that will be viewed microscopically facing down in cassette for appropriate orientation in embedding molds following processing.
  - More than one piece of small tissue can be placed in a single cassette.
  - Trimming must be determined based on study objectives. One study might need only a cross-section of a clam; another might need multiple cassettes to hold portions of gills, gonad, heart, intestines, hepatopancreas, and muscles from a lobster.
  - Ganglia and nerve fibers from large organisms must be processed for shorter times during subsequent tissue processing. They should be placed in separate cassettes and marked to indicate this during necropsy and fixation.

(d) After final trimming of undecalcified or decalcified (and washed) invertebrate samples, place cassettes in a large container and cover them with 70% ethanol, cap tightly. They can be kept in this solution indefinitely.

(e) If the tissue will not cut easily, decalcification should be performed. Check your invertebrate carefully for possible calcareous material. For example, echinoderms that feed by scraping algae off calcareous surfaces, such as the long-spined sea urchin, may form rounded balls of calcareous material in their intestines, and the balls should be carefully removed or decalcified prior to processing.
  - Copepods and amphipods often require no further decalcification after several days of fixation in Dietrich's solution or Davidson's solution as the weakly acidic nature of these fixatives aids in decalcification.
  - Delicate tissues of some specimens could benefit from enrobing the specimen in low melting point (35°C) agar (dissolve 1.5 g in 100 ml at 90°C with stirring) or commercially available HistoGel (which comes with instructions) prior to embedding. This enrobing will help to maintain the architectural relationships of tissue elements after decalcification has removed the calcium carbonate. Blot the specimen dry with a paper towel first so the cooled agar (about 60°C) will adhere, place the specimen in a disposable mold and cover with the agar. Place the mold in a vacuum oven set to 50°C and apply two cycles of vacuum, while the tissue is in the molten agar to help the agar coat the specimen and remove air bubbles. Turn off the oven to let the agar cool and gel. Scrape the agar or HistoGel gently away from an area of calcified tissue to permit the decalcifying solution to penetrate.

(f) Place tissue in a glass or plastic jar filled with appropriate room temperature decalcifying solution.
  - The amount of reagent needed will vary with amount of calcified tissue.
  - For acid-based solution, add a tissue or paper towel to top of the solution and place container under a fume hood.
  - For neutral pH EDTA solution, suspend tissues off bottom and keep them above the layer of chelated calcium that will form on the bottom as decalcification progresses.
  - For either solution, loosely cap the container. Change decalcifying solution in container once or twice daily until calcium carbonate is no longer present. For some organisms, this is easily seen when an acid decalcifier stops bubbling, or when an attempt to trim the tissue results in easy cutting. Other procedures for checking the decalcification endpoint can be used, such as chemical tests or X-radiography.

(g) When decalcification is confirmed, trim tissue and place in numbered processing cassettes [see Step 4(c)].

(h) Place the cassettes of final trimmed, decalcified tissues in running tap water for 24 h to remove all acid (Figure 36.2). Rinse EDTA-decalcified tissues in tap water for about 30 min. (If not rinsed, adverse staining reactions will occur.)

(i) Store the tissues in 70% undenatured ethanol as described in Step 4(d).

4. *Process and embed tissue*

Although the steps of dehydration, clearing, and infiltration with paraffin can be performed manually, using glass jars and an oven (for melting the paraffin), automation provides more controlled timing, heating, and cooling options and greatly improves the speed and consistency of tissue processing (Figure 36.3). We have used both "simple"

Figure 36.3 A basic tissue processor can speed up the dehydration, clearing, and infiltration steps and provide consistent results. (Photo by E.C. Peters.)

(e.g., graduated series of ethanol and xylene) and special formulations (e.g., S-29 and UC-670; 2-2-DMP) for dehydration and clearing of invertebrate tissues and did not find appreciable differences for the many species processed.

The length of time in dehydrating and clearing solutions does, however, depend on the thickness and density of tissues being processed and can greatly affect results. For example, bivalves require up to 1 h in each solution, including 3 changes of molten paraffin, whereas delicate coral tissue should be restricted to approximately 15–30 min in each change of dehydrating and clearing agent, with no more than 15 min in each change of molten paraffin. A reduced-time processing run, such as those used for biopsies in clinical settings, should also be used for marine cell-blocks and fine micromollusk samples, with no more than 10–15 min per solution. A low melting point (56–58°C) paraffin-based embedding medium that will produce a moderately firm block is preferred for most tissues. If available, gentle vacuum should be applied throughout processing, or minimally for the final change of paraffin. Table 36.3 summarizes suggested processing times for the solutions typically used. Do not allow the cassettes to dry out at any point of the process.

(a) *Automated processing*: Set up the processor and follow operating instructions to program the times for each solution. Fill processor chamber or basket with cassettes directly from storage solution and immediately immerse in the first solution. Gentle agitation or stirring helps the solutions infiltrate the tissues.

(b) *Manual hand processing*: Carefully transfer tissues or cassettes into each solution. Gently swirl or agitate containers every 10 min to aid infiltration of solutions into the tissues.
- Due to their small size, copepods or other small invertebrates can be easily transferred from one solution to the next by using a Pasteur pipette under a dissecting microscope.
- A dye (e.g., phloxine or eosin) added to the last 95% ethanol station will stain the tissue and aid in providing color contrast in paraffin during embedding and sectioning. The dye will wash out as the microscope slide goes through the staining process.

Table 36.3 Suggested times in solutions for processing invertebrates. Times are guidelines only and should be altered if needed. RT, room temperature, approximately 21°C

Station	Solution	Tissue <2 mm all dimensions[a]		Tissue <2 mm thick, delicate[b]		Tissue 2–5 mm thick, dense	
		Time (min)	Temperature	Time (min)	Temperature	Time (h)	Temperature
1	70% ethanol	10	RT	30	RT	1	RT
2	80% ethanol	10	RT	30	RT	1	RT
3	95% ethanol	10	RT	15	RT	1	RT
4	100% ethanol	10	RT	15	RT	1	RT
5	100% ethanol	10	RT	15	RT	1	RT
6	100% ethanol	10	RT	15	RT	1	RT
7	Xylene	10	RT	15	RT	1	RT
8	Xylene	10	RT	15	RT	1	RT
9	Xylene	10	RT	15	RT	1	RT
10	Paraplast	10	58°C	30	58°C	1	58°C
11	Paraplast	10	58°C	15	58°C	1	58°C
12	Paraplast with vacuum, if available	Not needed		15 min, if available	58°C	1, if available	58°C

[a] Such as tissue biopsies, zooplankton, eggs, nerve fibers, and ganglions.

[b] Such as corals.

(c) While tissues are processing, set up the embedding center (Figure 36.4). Be sure the paraffin has melted before starting to process tissues, as this can take several hours or overnight.

(d) When tissues have completed their last paraffin, transfer them to clean molten paraffin in the embedding center or another container in an oven.

(e) Using heated forceps, remove one cassette from container, open it, and remove paper label and set it aside. If using embedding rings with base molds, write inventory number from label on ring using pencil.

- *Embedding center*: Keep forceps in heated wells when not in use.

Figure 36.4 Basic embedding center consists of molten paraffin dispenser (top center), paraffin bath for tissues (back right), mold warming tray (front right), forceps warmers (front center right), hot plate (front center), and cold plate for hardening the blocks (left). (Photo by Jeffrey Greenberg, U.S. EPA, Atlantic Ecology Division.)

- *Manual embedding*: A small alcohol lamp may be used to warm forceps. Warning — overheated forceps can burn tissues.

(f) Half-fill a pre-warmed clean base mold with molten paraffin and quickly transfer the tissue piece(s). Keep orientation of tissue the same as was trimmed into the cassette, so the area of interest is placed on bottom of the mold.
- Keep forceps warm between handling of specimens. Paraffin congeals on cold forceps causing specimen to stick.
- Should this occur, put forceps and tissue back into molten paraffin until congealed paraffin remelts and specimen is free.

(g) Cool base mold on cold plate or an ice tray, gently holding tissue against bottom with forceps until trapped in hardening paraffin. Do not let surface paraffin become hardened.

(h) Quickly place numbered cassette or embedding ring on top of base mold and fill with molten paraffin.

(i) Place paper label on top and keep filled mold on cold plate or ice tray until completely hardened.

(j) When completely hardened, remove block from mold. Do not force out of mold. Blocks can be stored at room temperature indefinitely.

5. *Section and mount tissue*

Set the thickness setting on microtome (Figure 36.5) at 4–6 μm. Disposable microtome blades are preferred when sectioning most invertebrates, especially for tissues that might contain spicules, debris, sand particles, or calcified areas.

A pan of ice, ice tray, or cold pack should be placed on one side of the microtome and a water bath is placed on the other side (Figure 36.5). A water bath filled with deionized water and heated to approximately 45°C should be warm enough to soften thin paraffin sections, but not hot enough to cause rapid melting and distortion of tissues (about 10–15°C below the melting point of the paraffin). Remove dust particles and debris from surface of water bath at regular intervals to keep debris from getting on or under tissue sections. This is accomplished by dragging an unfolded paper towel or Kimwipe on

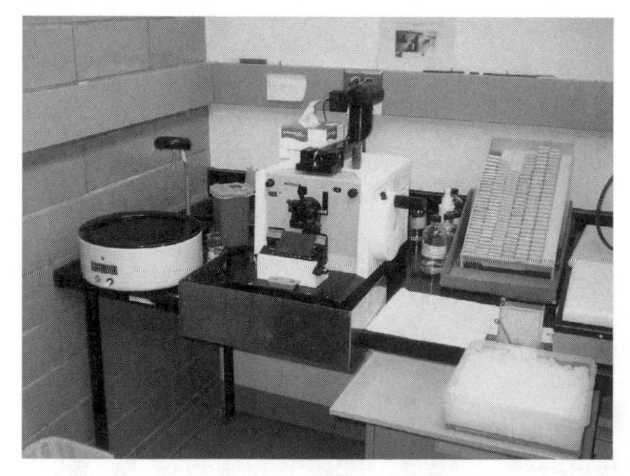

Figure 36.5 Basic microtome set up. A water bath is placed to the left of the microtome. (Photo by Jeffrey Greenberg, U.S. EPA, Atlantic Ecology Division.)

surface to pick up debris. Place blocks to be chilled tissue side down on ice. Microscope slides can be labeled with the inventory number before or after sections are mounted. The number of slides made for each block will depend on the number of stains to be applied to the sections.

(a) Place cooled block in microtome chuck and be sure all adjustments are finger-tight on block, blade, and holder.
(b) Trim block face into tissue by taking about 10 thick sections, then form a ribbon of about 10 sections using the automatic feed.[10,11]
(c) Use fingers and a brush or dissecting needle to lift ribbon and float it on the warm water bath.
(d) Select suitable section(s) and gently separate them. Hold a clean microscope slide at a 45° angle and insert into water bath. Use the needle to bring the section(s) to the slide. Holding the section(s) over the slide, slowly draw the slide out of the water with the section(s) attached.
   - In general, no adhesive is necessary for attaching tissues to slides; however, for long or harsh staining procedures use adhesives, such as gelatin, albumin, and poly-L-lysine, or Plus (+) charged slides.
   - Mounting two or more sections on one slide ensures that differences seen in one section can be confirmed and traced in adjacent sections.
   - Change blade as frequently as necessary to achieve good sections (with long blades, they can often be moved two or three times in the holder to use all portions of the blade surface).
   - Nicks in the blade will tear sections. With difficult sections, enough tissue can often be recovered by careful and slow placement of partial ribbons on the slide to obtain the necessary tissue sections for diagnosis and observation.
   - Soaking the block with a water-dampened wipe between sections can often compress tears and yield a suitable ribbon.
   - If the tissue has not been completely decalcified, the block can be melted down, and tissues run through a reverse process so further decalcification can be accomplished, or the cut block can be placed in decal solution for several hours to several days to decalcify it enough to permit sectioning.
   - Another aid to manipulating difficult tissues is the use of an intermediate room temperature 5% ethanol solution before the warm water bath. Float sections on the surface of the room temperature ethanol solution and tease small folds and wrinkles out prior to transferring with a clean glass slide onto the warm water bath.
   - Dry serial sectioning is another useful technique. The paraffin ribbon containing the tissue sections is not floated on a water bath but carefully placed in a shallow cardboard box until all required sections are cut. (Air currents can be treacherous for this method.) Place a microscope slide on a slide warmer and flood with water. Cut the paraffin ribbon into appropriately sized lengths and float them on the slide. Multiple ribbons can be placed on the same slide in this manner and, if done skillfully, can even provide a means of measuring structures within the organism. The slide must remain flat on warming table until water has completely dried and sections have adhered to slide.
   - Arthropods present challenges in sectioning, because chitin, a dense nitrogen-containing polysaccharide related to cellulose, forms the exoskeleton of these organisms, and it can easily destroy microtome blades. Face the block,

then apply either a water-dampened tissue or a thin film of a depilatory cream for a few minutes, wipe off, then try sectioning with a new blade. Alternatively, try using 4% phenol, as recommended for insects.[10] These tissues often require more care and attention when sectioning, but good sections can be obtained with patience.

(e) Drain each slide vertically for a few minutes, then place it on the slide warmer at 45–50°C for up to 4 h, or place it in a slide rack and put it in an oven at 60°C for up to 4 h.

(f) After they are thoroughly dried, store the mounted tissue sections in slide racks or slide boxes at room temperature until ready to be stained.

(g) Once an adequate number of slides have been made, blocks can be sealed by dipping in molten paraffin. Should additional sections become necessary at a later date, carefully realign the face of the block with the microtome blade.

6. *Stain tissue*

Excellent texts are available that explain the chemistry involved in applying dyes to highlight specific features of the cells and tissues. Staining enables changes in structure and composition, whether natural or induced by toxicants or pathogenic microorganisms, to be visualized with microscopy.[9–13] This section discusses the basic steps and provides details on routine and special stains that have proven helpful in aquatic toxicology studies of invertebrates, as well as one of the fastest growing areas of specialization, immunohistochemical protocols, which identify the location of specific molecules within cells and tissues. The term "biomarkers" includes measurable molecular, biochemical, and cellular changes caused by chemical contaminants or other stressors that might help identify causal mechanisms underlying observed effects at the population or community levels. Some biochemical biomarkers under investigation are molecules that indicate the structural integrity or functional integrity of enzyme pathways or that are induced during exposure to detoxify and remove contaminants. These biochemical biomarkers hold promise in identifying the most subtle irreversible changes occurring in cells and tissues before they are evident during histopathological examinations with light microscopy.

All staining processes share the same basic steps, whether performed manually (Figure 36.6a) or by using an automated stainer (Figure 36.6b):

(a) Place slides in racks.

(b) Move the racks through various solutions contained in glass or plastic slide dishes.
   - Paraffin is removed with several changes of xylene and the sections are rehydrated by passing them through graded ethanol solutions (100%, 95%, and 80%) to deionized or distilled water.
   - The time spent in each staining solution might need to be adjusted for some invertebrates, depending on the fixative used and whether the tissue was decalcified, as well as how recently the staining solution was prepared.
   - If the protocol indicates the staining reaction should be checked under a microscope, do this and adjust intensity of the staining reaction by increasing

Figure 36.6 Slide staining options: (a) Performing the routine H&E procedure in glass staining dishes. Dishes containing xylene are under the fume hood to prevent human exposure. (Photo by E.C. Peters.) (b) Automated slide stainer. (Photo by Jeffrey Greenberg, U.S. EPA, Atlantic Ecology Division.)

time in the stain or in the differentiation solution (which removes excess stain).

(c) Following staining, dehydrate the sections again through graded ethanols and 3 changes of xylene.

(d) Apply a coverslip over the stained section(s) using an appropriate mounting medium, such as Permount or one of the newer acrylic compounds.
- Most people develop their own coverslipping techniques.
- The finished slide should not have any air bubbles under the coverslip, excess mounting medium on the sides, or streaks of medium on the surface.
- Clean the surface of the coverslipped slide with a folded absorbent tissue dipped in xylene or xylene substitute.

Routine staining

General tissue staining procedures use one to three dyes to differentiate the nucleus from the cytoplasm and to distinguish types of tissue. Most staining procedures call for one stain to demonstrate an important cell component's structure, such as the nucleus, and one or more counterstains to act as a contrast color on the rest of the tissue. Hematoxylin and eosin (H&E) is known as a "routine" staining protocol, and should always be performed on one section from each block to interpret the condition of the nuclei and cytoplasm.

- Hematoxylin is a basic dye that binds to nucleic acids and other basophilic substances, such as mucopolysaccharides, tinting them various shades of blue or purple.
- A slide is properly stained with hematoxylin when cytoplasm and connective tissue are clear or light tan, and nuclei are distinctly purplish, but chromatin inside the nuclei is visible as a darker purple dust or streaks.
- Note that basophilic secretions, such as those of mucous secretory cells, may or may not stain purplish and should not be used as an indicator.
- The tissue is typically counterstained with eosin, an acidic dye, which stains acidophilic components, such as cytoplasm and connective tissue, with shades of red to pink (i.e., it should not be a single bright pink color).
- A slide is properly stained with eosin when cytoplasm and connective tissue are a clear and strong pink, but not heavy, i.e., cytoplasmic inclusions and pigments should be distinguishable; and different shades of pink are seen in connective tissue and muscle bundles.

Various H&E protocols can be found in reference books,[9–11] and some premade solutions are now available commercially. Two types of hematoxylin staining procedures have been developed: *regressive*, in which the tissue is first overstained with hematoxylin, then differentiated by removing excess stain with acid alcohol; and *progressive*, which relies on ionic bonding to only stain the acidic nuclear material. One popular regressive formulation is Harris's hematoxylin (Figure 36.6a) and results in crisp nuclear staining with distinctive chromatin. It works well with tissues that have been decalcified. Progressive hematoxylins, such as Mayer's or Gill's, should only stain the basophilic components and do not require the differentiation step. A number of formulations for eosin also exist, some of which are also commercially available, and it is often combined with phloxine to increase the intensity.

Special histochemical staining

Additional staining protocols can be performed on replicate tissue sections from each block to highlight certain structural features, confirm composition of secretions, or aid in identifying parasites and pathogens.[9–13] We provide two protocols here that have been helpful in toxicologic pathology studies. Modified Cason's is a quick trichrome stain primarily used to distinguish connective tissue from muscle fibers. It can also assist in examining the condition of symbiotic algae in corals, gonad development, and fine ducts. Some toxicants are irritants, provoking mucus secretions as a protective mechanism in invertebrates. Modified Movat's pentachrome protocol distinguishes changes in the mucopolysaccharides of mucous secretory cells. Changes in molecular structure, composition, and pH are reflected in differential binding of the alcian blue and saffron solutions, resulting in blue, yellow, or green.

Modified Cason's trichrome for connective tissue stain

Adapted from Reference 16. Prepare Cason's trichrome solution (see Appendix). Use a control tissue slide (a section known to have connective tissue present).

(a) Deparaffinize and hydrate to water (3 min each in 3 changes of xylenes and 3 changes of 100% undenatured ethanol, 1 change each 95% and 80% ethanol, and distilled or deionized water).
(b) If tissue sections were not fixed in Bouin's solution, Helly's solution, Z-Fix, or Zenker's-type, put the slides into a staining dish overnight in a solution of Helly's without the formaldehyde. The picric acid or metal ions in the Helly's will bind to the connective tissue and enhance the binding of the aniline blue.
(c) Stain 5 min in Cason's trichrome solution.
(d) Wash 3–5 s in running tap water.
(e) Dehydrate rapidly through 95% ethanol (1 min) then 3 changes of 100% ethanol (2 min each). Water removes the Cason's trichrome solution from the tissues, so be careful that it does not wash out too much.
(f) Clear in 2 changes of xylene (3 min each).
(g) Mount in Permount or other synthetic resin.

Results:

- Collagen — blue
- Nuclei, muscle fibers, yolk proteins — red

Modified Movat's pentachrome stain

This modification of Movat's procedure, which was developed by P.P. Yevich and C.A. Barszcz at the EPA's Narragansett, Rhode Island, laboratory, does not include the stains and steps for elastin found in mammalian tissues and is more appropriate for marine invertebrates. The formulas for the necessary solutions are in Appendix.

(a) Deparaffinize and move through 3 changes of 100% ethanol.
(b) Place slides in alcian blue solution for 30 min.
(c) Wash slides in running tap water for 3 min.
(d) Place slides in alkaline alcohol for 120 min.
(e) Transfer slides to rinse in distilled water for 2 min.
(f) Place slides in Weigert's iron hematoxylin for 1 min.
(g) Wash slides in running tap water, 15 min.
(h) Rinse slides in deionized water, 2 min.
(i) Stain slides with woodstain scarlet solution, 1 min.
(j) Rinse in 0.5% aqueous glacial acetic acid, 2 min.
(k) Differentiate in 5% aqueous phosphotungstic acid solution, 7–10 min. Check slides under the microscope.

(l) Rinse in 0.5% aqueous glacial acetic acid, 2 min.
(m) Place slides in 3 changes of 100% undenatured ethanol, 2 min each.
(n) Transfer slides to alcoholic saffron solution, for 7–10 min. Cover the staining dish and seal edges with tape after inserting slides to prevent rehydration with moisture in the air. After staining, return solution quickly to bottle and cap tightly.
(o) Rinse slides in 100% ethanol, 3 changes, 2 min each,
(p) 3 changes of xylene, 2 min each. Coverslip.

Results (different fixatives and decalcification solutions may result in color variations in your sections):

- Nuclei — black
- Connective tissue — pale blue-green to yellow
- Mucous secretory cells — green, blue, yellow
- Muscle fibers — red

Price's modified Twort's procedure for Gram-positive and Gram-negative bacteria
Tissue damage can be caused by pathogenic microorganisms. A toxicant can damage the animal's immune system or cause tissue necrosis, which then attracts microorganisms that feed on or use released tissue compounds as substrates. This stain will help identify Gram-positive and Gram-negative bacteria. Formulas for the solutions in this staining procedure are in Appendix.

(a) Deparaffinize and rehydrate tissue sections to water.
(b) Stain with filtered crystal violet solution, 2 min.
(c) Rinse in tap water, then drain.
(d) Place in iodine solution, 2 min.
(e) Rinse in tap water, then drain.
(f) Do a half-second dip in acetone, then without pausing to drain, plunge immediately into running tap water for 2–3 min.
(g) Counterstain in Twort's stain for 3–5 min.
(h) Wipe off the back of the slide, then carefully blot front of slide by placing tissue side down on bibulous paper or a lint-free Kimwipe. (Use a rolling motion to avoid damaging tissue section.)
(i) After blotting, place the slides in a clean, dry rack to avoid transferring any excess Twort's stain into the alcohol.
(j) Dehydrate rapidly: about 15 gentle dips in 100% alcohol, then 2 changes of fresh 100% alcohol for 2 min each.
(k) Clear in 2 or 3 changes of xylene, 3 min each, and mount coverslip with permanent media.

Results:

- Gram-positive organisms — blue
- Gram-negative organisms — pink to red
- Nuclei — red
- Most cytoplasmic structures — green

Immunohistochemical staining

Histological dyes cannot uniquely stain particular cells, cell structures, or tissue components. IHC developed from the discovery of antigens that induce a detectable immune response in a vertebrate by producing a specific antibody capable of binding to the epitope of the antigen (proteins, polysaccharides, nucleic acids, and polymers). An antigen present in a tissue section on a microscope slide can be demonstrated by applying a solution containing the appropriate antibody, which then binds to the epitope(s) of the antigen. The antibody is visualized by attaching either a fluorochrome or an enzymatic chromogen to the antibody. A fluorochrome is a substance that absorbs light not visible to the human eye (short wavelengths, such as UV) and emits light that is visible (longer wavelengths). An example of a fluorochrome is fluoroscein isothiocyanate (FITC), and epifluorescence microscopy must be used to see where the antibody is binding. Enzyme IHC is used for light microscopic detection. An enzyme is used that, in the presence of substrate and chromogen, produces visible color. For example, horseradish peroxidase is used with hydrogen peroxide and 3-amino-9-ethylcarbazole (AEC) to form a colored compound. Mayer's hematoxylin is used as a counterstain in this procedure because it does not contain alcohol that would decolorize the section.

While IHC has become an extremely valuable tool in the diagnostic process of human disease, it is unclear how far down the evolutionary chain this antigen–antibody complex is conserved. IHC is typically species-specific and most antibodies are made against human or other mammalian (and a few fish) antigens, although many have been successfully used in lower animals. Some investigators are now producing invertebrate-specific antibodies. Antibodies should always be fresh (note expiration dates) and stored as directed by the manufacturer.

It is beyond the scope of this chapter to cover more than the basics of IHC, and there are many excellent texts[17] journal articles [e.g., *Journal of Histotechnology*, Special Issue on Diagnostic Immunohistochemistry, Vol. 25 (4), 2002] available. When performing a preliminary investigation, the primary decision will be whether the direct method or indirect method of staining is most appropriate. In either case, fixation is important to anchor the water-soluble antigens in place but can mask the epitopes. A variety of fixatives have been developed that limit masking reactions and procedures have also evolved to unmask antigens by using heat from a microwave oven or by proteolytic digestion, but these are not always successful.

Several different approaches may need to be undertaken to determine the most effective protocol. The methods used, how the antibody is prepared (monoclonal — more specific but less sensitive; polyclonal — more sensitive but less specific), the need for unmasking and how it should be done are variables that need to be addressed. From a technical standpoint, the procedure is fairly straightforward but interpretation of staining results requires skill and experience.

When evaluating the usefulness of an antibody, the use of several different control slides is essential as numerous factors can impact results. Control slides serve to verify both positive and negative staining.

- *Positive control*: a slide containing tissue that has the antigen of interest. This slide verifies the viability of the antibody. If the positive control fails to stain, then the antibody solution may be suspect. The tissue used as a control should have undergone the same fixation and processing protocol as the tissue under investigation.

- *Procedural control*: a slide of the tissue being investigated treated with a ubiquitous antibody, such as vimentin or ubiquitin. This will verify that the tissue being tested is immunoreactive, and the procedure being followed is conducive to achieving a positive result.
- *Negative control*: two slides — one of the verifiable positive control tissue, and one from the investigative tissue. Both slides follow through all the same procedural steps as the test slides, but plain buffer is substituted for the primary antibody. This will ensure that any positive reaction obtained is due to the application of the primary antibody.

Dilution factors for the antibody must also be determined. Optimal staining is achieved when background staining is at a minimum and specific staining is of medium to high intensity. For direct immunostaining, only the primary antiserum needs to be tested at different dilutions. Use of the indirect method will require a checkerboard titration to determine the optimal concentrations of the antibody solutions.[17] Seventeen test slides will need to be used: sixteen for the dilutions plus one that receives neither the primary nor the secondary antiserum. If the optimal set of dilutions seems to fall between tested examples, set up a second checkerboard titration within the indicated range. Include control slides in the dilution tests. Once staining has started, do not allow the tissue sections to dry out.

Secondary antiserum dilution	Primary antiserum dilution			
	1:50	1:100	1:500	1:1000
1:50	1	2	3	4
1:100	5	6	7	8
1:500	9	10	11	12
1:1000	13	14	15	16

(a) Cut sections at 3–4 μm and use poly-L-lysine-coated or positively charged slides to help tissue sections adhere to the slides. Dry at 58°C for 2 h.
(b) Deparaffinize (3 changes xylene, 5 min each) and rehydrate slides (2 changes 100% ethanol, 5 min each; 2 changes 95% ethanol 5 min each) to distilled water.
(c) Circle the tissue sections with a wax pencil or immuno-marker to act as a dam, to prevent the antibody solution from running off the tissue. This will also reduce the amount of antibody solution needed for the procedure.
(d) Place two glass rods in a shallow dish with lid or special IHC tray that contains a paper towel moistened with IHC buffer. Lay the slides horizontally on top of the two parallel glass rods.
(e) Inhibit endogenous peroxidase activity by applying 3% hydrogen peroxide solution for 5–10 min.
(f) Wash slides in distilled water for 1 min.
(g) Wash slides in buffer for 5 min.
(h) Incubate with primary antibody for 20 min.
(i) Rinse with 3 changes of buffer, 5 min each. Discard buffer after each rinse.
(j) Apply and incubate with the link antibody (if used) for 20 min.

(k) Rinse with 3 changes of buffer, 5 min each. Discard buffer after each rinse.
(l) Apply and incubate chromogen substrate or fluorochrome solution for 10 min.
(m) Rinse with 3 changes of buffer, 5 min each. Discard buffer after each rinse.
(n) Counterstain, if desired, and coverslip with water-based media.
(o) Examine the tissue sections with brightfield or epifluorescence microscopy, as appropriate, and record the intensity of the immunostaining to the background on a scoring sheet for quality control purposes. If the staining is too intense or shows too much background staining, increase the dilution of the antibody.
(p) Seal edges of histoslides with glue, cement, or fingernail polish for longer storage when using aqueous mounting methods to prevent drying out.

Results and discussion

Knowledge of invertebrate histology is critical to understanding their physiology, reproduction, biochemistry, systematics, molecular biology/genetics, immunology, embryology, and ecology. Examination of tissues by light microscopy can be used to evaluate the health of an animal following exposures to contaminated materials or infectious agents, or to determine the presence or absence of pathological changes. These changes can be identified by morphological alterations, changes in the rate of occurrence of normal features (e.g., mitotic figures), reactions to special histochemical or immunohistochemical stains, or through variations in normal staining characteristics, which might be indicative of changes in biochemical composition or the existence of degenerative processes.[3,4,7] The study of such changes in cells and tissues — histopathology — is important in the invertebrates, as well as in the vertebrates. The normal range of appearance of cells and tissues must be understood first, because cellular appearance can change due to metabolic activity, diet and feeding times, daily light–dark cycles, or the reproductive cycle.

Study of vertebrate[3] and invertebrate pathology resources[18,19] is required, as well as consulting experts on different organisms. Although there are few atlases on the anatomy and histology of invertebrates, they need to be consulted.[20–25] Two papers have reviewed toxicant effects on invertebrates.[4,7,26] The mechanisms by which changing environmental conditions, toxicants, biotoxins, or pathogens cause disease appear varied, and will also differ with the species and individual (genetic polymorphisms), their defense mechanisms, and immune systems. Interpretation also involves converting the 2-dimensional images seen in light and transmission electron microscopy to their original 3-dimensional structures.

A study of a small burrowing clam, *Macoma balthica,* collected in the Gulf of Riga, provides one instructive example of the application of invertebrate histopathology in aquatic toxicology and illustrates some of the decisions that must be made regarding processing tissues, data collection efforts, and interpretation of multiple parameters.[27,28] This study, one of six subprojects to look at pollution in this Gulf, was undertaken in collaboration with Dr. Kari Lehtonen of the Finnish Institute of Marine Research, as part of the multinational Nordic Environmental Research Programme financed mainly by the Nordic Council of Ministers.[29] The Gulf of Riga is a semi-enclosed estuary under a strong influence of the Daugava River, with water exchange with the Baltic Sea. The eastern and southern (city of Riga) watersheds are heavily industrialized, contributing nutrients, polycyclic aromatic hydrocarbons (PAHs), polychlorinated biphenyls (PCBs), and metals to the water and sediments of the Gulf. The clam larvae settled at a particular location and were chronically exposed to potentially toxic organic and inorganic chemicals. Contact with and ingestion of water and sediment, and ingestion of food particles as deposit/detritus feeders[30] resulted in a bioaccumulation of these chemicals in the tissues. The plan to collect and examine clams from sites throughout the Gulf was not coordinated with sediment evaluation or clam tissue collection for analyses of chemical contaminants, which was unfortunate, based on the preliminary histopathology results.

The fixative selected was Helly's, which nicely preserves the intracytoplasmic vesicles and provides crisp staining. However, these clams live in very low salinity water (average of 5.7 practical salinity units) and the high-osmolarity of the fixative caused vacuolation and distortion of oocytes and other changes that were consistently observed in the tissues. Otherwise, cells and tissues were well fixed. The small clams (*ca.* 15 mm average maximum diameter) were carefully removed from their shells using a scalpel and fixed for 12–14 h, rinsed in water for 24 h, and stored in 70% ethanol. Clams were trimmed by slicing sagittally through the mid-line and both halves embedded (one cut side up, the other cut side down) in Paraplast. Because of their small size, two to four clams were included in each cassette and paraffin block. On sectioning the clams, it was apparent that particulate material in the stomach caused microtome blade damage, indicating that depuration of gut contents did not occur as expected when the animals were held in buckets of clean ambient water for 24 h before fixation. Sections were stained with hematoxylin and eosin, and on some, Cason's, periodic acid-Schiff (with and without diastase digestion to demonstrate glycogen), and modified Movat's pentachrome procedures were used on additional sections from the blocks. Despite some difficulties with the fixation and sectioning, sections of sufficient quality were obtained for histopathologic examination.

Collection of data involved examining each of the 558 clams and recording lesions and other observations using a lesion coding system[31] modified for bivalves. The first challenge of the study occurred during processing and embedding when the clam halves became mixed up. This resulted in confusion about which lesion belonged to which of the two to four clams in the block, so to solve this, each histoslide was placed on a light table and the locations of the tissue sections traced onto the data sheet. Microscopic examination then permitted the halves to be assigned to one clam, based on the minor differences in tissue structure and composition that can be used to distinguish among individuals. Each clam was then assigned a subnumber for the block; and all observations from each clam could then be recorded on the data sheets (Figure 36.7). The lesion coding system also permitted recording which tissues and organs were present on the sections for each clam. Not all organs were present for each clam on the histoslide; thus, deeper sections needed to be cut from those blocks before statistical comparisons of the intensity/severity, prevalence, and distribution of some of the organ-specific lesions could be performed (e.g., kidney concretions, because the kidney was not sectioned for all clams).

Although the lesion coding system provided a concise way to record all histopathological information for each clam, the amount of data generated for 558 clams was daunting. Both qualitative (presence/absence, descriptions) and semiquantitative data (extent: focal, multifocal, diffuse; severity: minimal = 1 to severe = 5) were recorded. Image analysis, to provide count and measurement data, was not undertaken, although possible proliferation of blood vessels with hypertrophy of endothelial cells and variable size of lysosomes in digestive tubule cells could be examined this way.

As noted previously, the amount of time available for data collection and analysis must be factored into the study. In this case, time ran out and a complete multivariate analysis and comparison with data from other studies in the Gulf has yet to be performed. The limited analyses conducted revealed that neoplasms (of gill and gonad), blood vessel proliferation, thickened connective tissue, and extracellular kidney concretions with

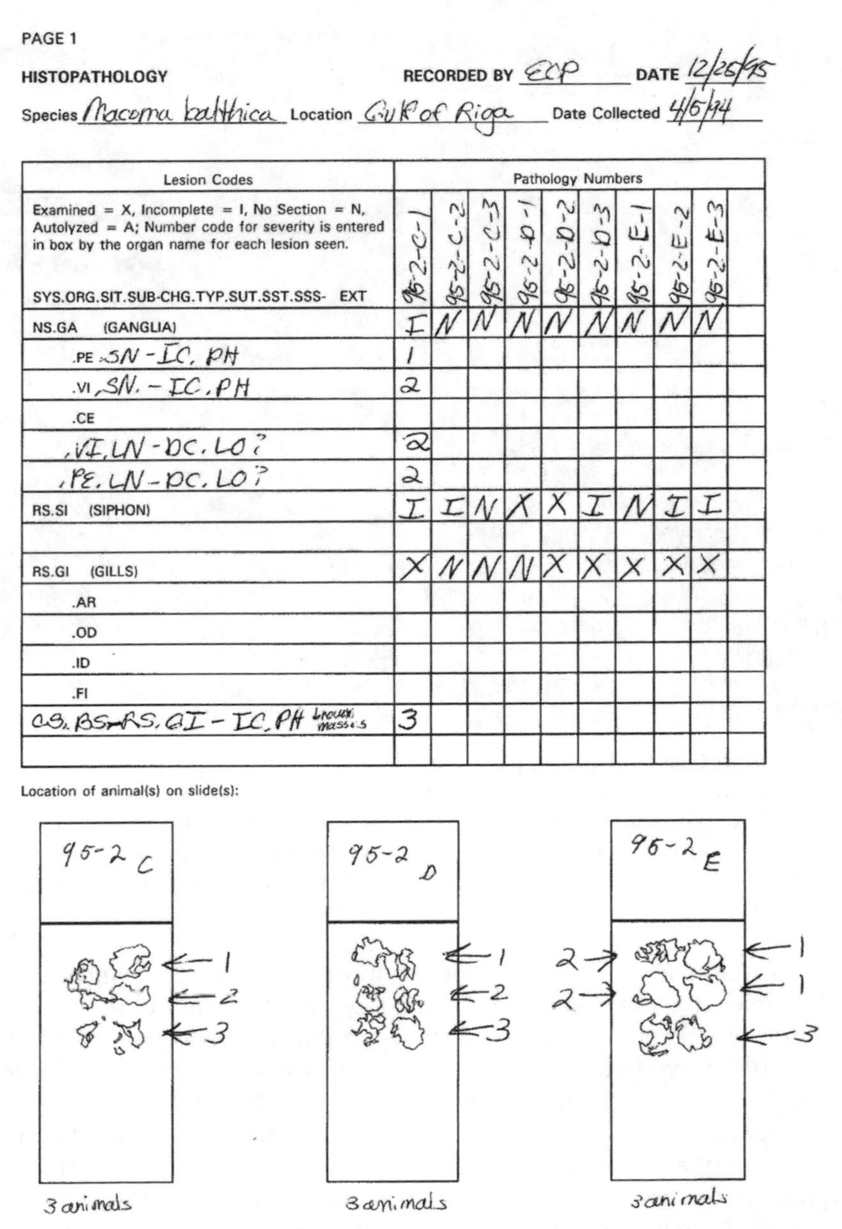

PAGE 1

HISTOPATHOLOGY RECORDED BY ECP DATE 12/26/95

Species Macoma balthica Location Gulf of Riga Date Collected 4/6/94

Lesion Codes	Pathology Numbers									
Examined = X, Incomplete = I, No Section = N, Autolyzed = A; Number code for severity is entered in box by the organ name for each lesion seen. SYS.ORG.SIT.SUB-CHG.TYP.SUT.SST.SSS- EXT	95-2-C-1	95-2-C-2	95-2-C-3	95-2-D-1	95-2-D-2	95-2-D-3	95-2-E-1	95-2-E-2	95-2-E-3	
NS.GA (GANGLIA)	I	N	N	N	N	N	N	N	N	
.PE SN - IC, PH	1									
.VI SN. - IC.PH	2									
.CE										
.VI.LN - DC.LO?	2									
.PE.LN - DC.LO?	2									
RS.SI (SIPHON)	I	I	N	X	X	I	N	I	I	
RS.GI (GILLS)	X	N	N	N	X	X	X	X	X	
.AR										
.OD										
.ID										
.FI										
CS.BS-RS.GI - IC.PH brown masses	3									

Location of animal(s) on slide(s):

Figure 36.7 First page of three-page data sheet for recording lesion codes for each clam, identified by the pathology numbers at the top of the columns and on the sketches of each histoslide at the bottom of the page. Numbers indicate the severity of the lesion; an "X" means the tissue was present on the histoslide, an "N" means it was not present, and a "I" means that the amount of tissue was insufficient to score. (Data sheet designed by E.C. Peters.)

atrophy and necrosis of the renal epithelium were found at stations in the industrialized area of the Gulf (Figure 36.8a–d). The relationship of bivalve neoplasms and other lesions to chemical contaminants is unclear but appears to be likely, especially for PCBs and metals at the mouth of the Daugava River.[29] However, sediment analyses indicated that the Gulf overall was not significantly polluted compared to the Baltic Sea.

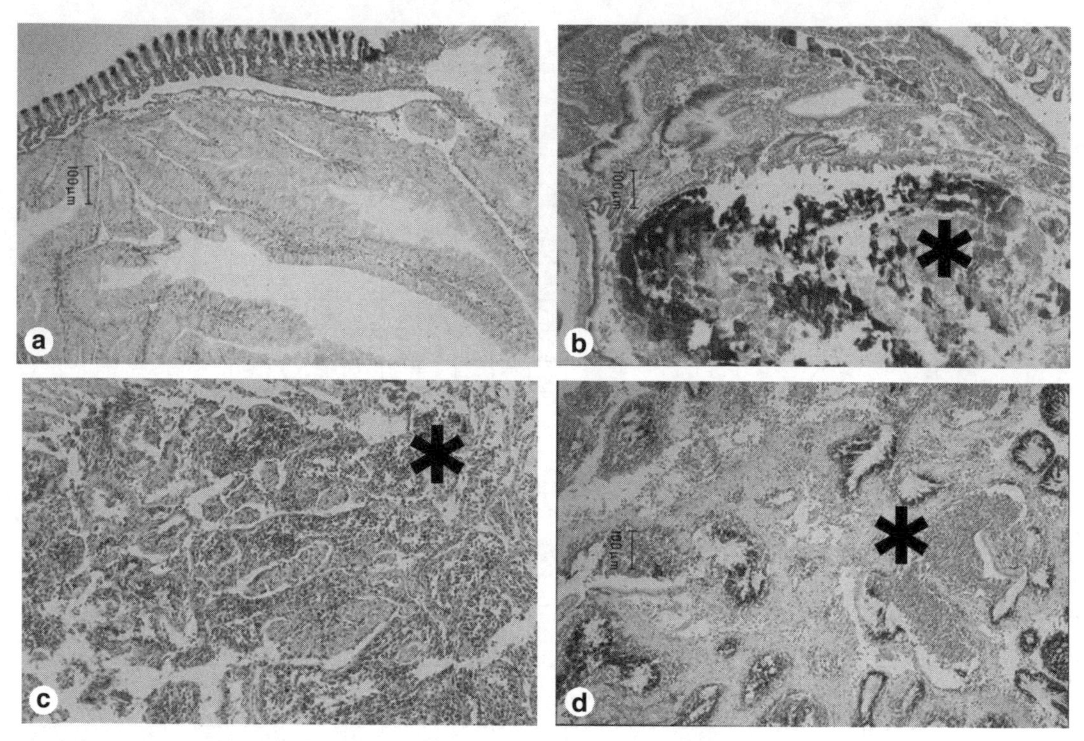

Figure 36.8 **(see color insert following page 464)** Example lesions affecting the health of individual clams and population status in the Gulf of Riga: (a) normal kidney, gill at top; (b) kidney concretion at "*"; (c) germinoma, undifferentiated cells filling gonadal follicles to left of "*"; (d) abnormal thickening of connective tissue among gonadal follicles in male clam to left of "*", infiltration of hemocytes in follicles to right of "*". (Photomicrographs by E.C. Peters.)

Although preliminary research on induction of stress proteins began using tissues from other samples of these clams, the use of Helly's fixative did not permit examination of the induction of stress biomarker proteins using immunohistochemical procedures. A study of *M. balthica* exposed to contaminated sediments from Sydney Harbour, Cape Breton, Nova Scotia, Canada, for 28 days in the laboratory used sediment chemistry analyses, histopathological examinations, and immunohistochemical staining for the enzymes acid phosphatase, adenosine triphosphatase (ATP), salmon CYP1A (P450), and γ-glutamyl transpeptidase (GGT) in the digestive gland.[32] The enzymes generally showed increased activity in sediment-exposed animals compared to controls. Some of the differences were dependent on the condition of the tissues, as revealed by the histopathological examinations. The responses of these clams tested using the U.S. EPA bioaccumulation test method[33] showed early indication of stress due to chronic biologic exposure to organic and inorganic contaminants.

Despite the issues raised in this discussion, it should be noted that the large number of animals in the Gulf of Riga study from diverse stations provided insights into individual variability and also revealed new information on the life histories of these clams. For example, no juveniles were found at the easternmost station where neoplasms were present in the clams, but juveniles dominated at the westernmost station nearest the Baltic Sea. Gonadal follicles might have been damaged by gymnophallid trematode sporocysts or metacercaria at most stations except the most industrialized southeastern location in the Gulf, and all clams, no matter when collected during the year, had mature gonads (*M. balthica* in the Baltic Sea has a pronounced seasonal reproductive cycle). The "Swiss cheese" appearance of digestive diverticular tubule cells might have been related to the unusually high lipid content of the clams in some areas of the Gulf [disturbed lipid metabolism, neutral lipid/lipofuscin accumulation(?)]. However, a special fixative would have to be used when clams were collected to verify that the clear circular vacuoles had been filled with lipid, since lipid is removed during tissue processing. Different nutritional conditions in different parts of the Gulf were suggested in the biochemistry studies and aid in the interpretation of the histopathology and analytical chemistry data.[27,32,34,35]

Microscopic observations provide additional clues to verify whether gross observations of invertebrate mortality or morbidity, reduced growth and reproduction, or other effects were actually due to the toxicant, or were the result of parasites or pathogenic microorganism infestations or infections, starvation, genetics, or other factors. Sometimes toxicant-exposed animals might survive better than controls, because the toxicant preferentially kills parasites or pathogens that damage the host. Impacts on invertebrates can produce delayed, chronic, and indirect effects on the long-term survival and reproduction of their predators, influencing recovery from contamination events.[36] Research correlating body burdens of contaminants with histopathologic biomarkers of cellular alteration (contaminant-specific or nonspecific lesions) is needed to link exposures with mechanisms of action (e.g., necrosis, neoplasia, metaplasia, oocyte atresia, skeletal deformities). Laboratory exposure and field verification of the lesions that develop in invertebrates, as well as measurements of contaminant levels, will provide the necessary information for ecological risk assessments.

Appendix

The following fixatives are either not readily available in vertebrate-oriented histotechnique manuals or are modifications of similar fixative solutions and are presented here. Each formula makes 1 l of fixative. Each volume (in cm^3) of tissue will require a minimum of 10–20 times the volume of fixative.

Davidson's solution

Davidson's solution[12,23] is similar to Dietrich's,[7,38] although it has higher concentrations of the formaldehyde and glacial acetic acid, but both of these can be used for the same organisms. Davidson's I is recommended for crustaceans and Davidson's II has been used for bivalves.[37,40] Both decalcify, as well as fix, the tissues; however, secretions and granules might not be well preserved. Mix the ingredients in order:

	Davidson's I (ml)	Davidson's II (ml)	Dietrich's (ml)
Glycerin		100	
Formaldehyde (37–40%) (CAS No. 50–000)	225	200	100
95% undenatured ethanol (CAS No. 64-17-5)	310	300	290
Tap or distilled water	345	300[a]	590
Glacial acetic acid (CAS No. 64-19-7)	120	100	20

[a] Filtered ambient seawater for Davidson's II will appear cloudy when made with high salinity seawater.

Helly's fixative

Replacing the original mercuric chloride with zinc chloride reduced the risks of mercury exposure and no precipitate forms in the tissues[39]; however, zinc chloride must be handled carefully as it is hygroscopic and can cause burns.

Zinc chloride (CAS No. 7646-85-7)	50 g
Potassium dichromate (CAS No. 7778-50-9)	25 g
Water	1000 ml (distilled, deionized, or seawater)

Add 50 ml of 37–40% formaldehyde (CAS No. 000-050-000) just before placing animals in this solution. It will not work 24 h after the formaldehyde is added, and specimens should not be left in the solution more than 16–20 h or they will become brittle.

10% seawater formalin

For field use, this is the simplest fixative solution for marine organisms:

Formaldehyde (37–40%) (CAS No. 50-000)	100 ml
Seawater	900 ml (filter to remove debris, if possible)

This EDTA-based decalcifying solution is useful for fragile tissues or when IHC is to be performed on the tissue sections:

10% neutral EDTA

Disodium EDTA ($Na_2ClOH_{14}O_8N_2 \cdot 2H_2O$)	300 g
Distilled or deionized water	3000 ml

Place a large glass flask or beaker on a magnetic stirrer and add a large stir bar. Turn on the stirrer and add the EDTA to the water while stirring. Then add solid NaOH pellets to the solution while titrating to bring the pH to 7.0 (approximately 29 g). The NaOH will speed up the dissolution of the EDTA.

The following staining solutions are useful when working with invertebrates:

Cason's trichrome solution

Dissolve each chemical in order:

Distilled or deionized water	200 ml
Phosphotungstic acid (CAS No. 12067-99-1)	1.0 g
Orange G (CI 16230) (CAS No. 1936-15-8)	2.0 g
Aniline blue, WS (CI 42780) (CAS No. 28631-66-5)	0.5[a] g
Acid fuchsin (CI 42685) (CAS No. 3244-88-0)	6.0[b] g

[a] Adapted from Reference 16, modified amount from original procedure of 1.0 g.

[b] Modified amount from original procedure of 3.0 g.

Alcian blue solution

Combine:

Alcian blue	1 g
Deionized water	100 ml
Glacial acetic acid	1 ml

Solution can be saved and reused many times.

Alkaline alcohol solution

Combine:

Ammonium hydroxide	10 ml
95% undenatured ethanol	100 ml

Mix and store in amber glass bottle.

Wiegert's iron hematoxylin

Prepare two solutions (A and B) and mix them together just prior to staining. Discard mixed solution after staining each rack of slides.

Solution A	
Hematoxylin crystals	1 g
95% undenatured ethanol	100 ml
Solution B	
29% aqueous ferric chloride	4 ml
Deionized water	95 ml
Concentrated hydrochloric acid	1 ml
Woodstain scarlet solution: Combine	
Woodstain scarlet N.S. conc.	0.1 g
Deionized water	95.5 ml
Glacial acetic acid	0.5 ml

Should be saved and reused. If you cannot find woodstain scarlet, prepare crocein scarlet-acid fuchsin working solution.[10] Discard this solution after each use.

Aqueous glacial acetic acid solution

Mix:

Glacial acetic acid	0.5 ml
Deionized water	99.5 ml

This can be made up ahead of time, but do not reuse after staining.

5% aqueous phosphotungstic acid solution

Combine in a 100-ml volumetric flask:

Phosphotungstic acid 5 g
Deionized water to fill the flask to the 100-ml mark

This can be made up ahead of time, but do not reuse after staining.

Alcoholic saffron solution

Combine as directed:

Saffron (stain)	6 g
100% (absolute) ethanol	100 ml

Place in an airtight, stoppered glass container to prevent dehydration. Extract the saffron by placing the solution in an oven at 58°C for 48 h. May be reused many times, store in a tightly sealed bottle (saffron is expensive!).

Crystal violet

Crystal violet	1 g
25% ethanol	200 ml

Gram's iodine

Place in small clean glass vial with stopper:

Iodine	1 g
Potassium iodide	2 g
Distilled water	10 ml

Shake or grind until dissolved. Transfer solution to a 300-ml volumetric flask and fill to the mark with distilled water. Stopper and shake to mix.

Twort's stain

For stock solution, mix the following together and store in a screw-capped bottle. It is stable for at least a year:

0.36 g neutral red (CI 50040) dissolved in 180 ml 95% ethanol
0.02 g fast green FCF (CI 42053) dissolved in 20 ml of 95% ethanol

Working solution

Mix this immediately prior to use.

Stock solution	48 ml
Distilled water	144 ml

References

1. Kramer, K.J.M., *Biomonitoring of Coastal Waters and Estuaries*, CRC Press, Boca Raton, FL, 1994.

2. Rand, G.M., *Fundamentals of Aquatic Toxicology: Effects, Environmental Fate, and Risk Assessment*, 2nd ed., Taylor & Francis, Washington, D.C., 1995.
3. Cotran, R.S., Kumar, V. and Collins, T., *Robbins Pathologic Basis of Disease*, 6th ed., W.B. Saunders, Philadelphia, PA, 1999.
4. Hinton, D.E., Baumann, P.C., Gardner, G.R., Hawkins, W.E., Hendricks, J.D., Murchelano, R.A. and Okihiro, M.S., Histopathologic biomarkers, in *Biomarkers: Biochemical, Physiological, and Histological Markers of Anthropogenic Stress*, Huggett, R.J., Kimerle, R.A., Mehrle, P.M. Jr. and Bergman, H.L., Eds., Lewis Publishers, Boca Raton, FL, 1992, pp. 155–209.
5. Couch, J.A. and Fournie, J.W., Eds., *Pathobiology of Marine and Estuarine Organisms*, CRC Press, Boca Raton, FL, 1993.
6. Hoffman, D.J., Rattner, B.A., Burton, G.A, Jr. and Cairns, J. Jr., *Handbook of Ecotoxicology*, Lewis Publishers, Boca Raton, FL, 1995.
7. Yevich, P.P. and Yevich, C.A., Use of histopathology in biomonitoring marine invertebrates, in *Biomonitoring of Coastal Waters and Estuaries*, Kramer, K.J.M., Ed., CRC Press, Boca Raton, FL, 1994, pp. 179–204.
8. Jagoe, C.H., Responses at the tissue level: quantitative methods in histopathology applied to ecotoxicology, in *Ecotoxicology: A Hierarchical Treatment*, Newman, M.C. and Jagoe, C.H., Eds., Lewis Publishers, Boca Raton, FL, 1996, pp. 163–196.
9. Sheehan, D.C. and Hrapchak, B.B., *Theory and Practice of Histotechnology*, The C.V. Mosby Company, St. Louis, MO, 1980.
10. Prophet, E.B., Mills, B., Harrington, J.B. and Sobin, L.H., *AFIP Laboratory Methods in Histotechnology*, Armed Forces Institute of Pathology, American Registry of Pathology, Washington, D.C., 1992.
11. Carson, F.L., *Histotechnology: A Self Instructional Text*, ASCP Press, Chicago, IL, 1997.
12. Presnell, J.K. and Schreibman, M.P., *Humason's Animal Tissue Techniques*, 5th ed., The John's Hopkin's University Press, Baltimore, 1997.
13. Callis, G. and Sterchi, D., Eds., *Animal Processing Manual*, Veterinary, Industry and Research Committee, National Society for Histotechnology, Bowie, MD, 2002.
14. Dapson, J.C. and Dapson, R.W., *Hazardous Materials in the Histopathology Laboratory: Regulations, Risks, Handling, and Disposal*, 3rd ed., Anatech Ltd., Battle Creek, MI, 1995.
15. Ross, L.G. and Ross, B., *Anaesthetic and Sedative Techniques for Aquatic Animals*, Blackwell Scientific, Oxford, 1999.
16. Cason, J.E., A rapid one-step Mallory–Heidenhain stain for connective tissue, *Stain Technol.*, 25, 225–226, 1950.
17. Wordinger, R.J., Miller, G.W. and Nicodemus, D.S., *Manual of Immunoperoxidase Techniques*, 2nd ed., American Society of Clinical Pathologists (ASCP) Press, Chicago, IL, 1987.
18. Sparks, A.K., *Invertebrate Pathology: Noncommunicable Diseases*, Academic Press, New York, 1972.
19. Sparks, A.K., *Synopsis of Invertebrate Pathology Exclusive of Insects*, Elsevier Science Publishers, New York, 1985.
20. Anderson, D.T., *Atlas of Invertebrate Anatomy*, University of New South Wales Press, Sydney, Australia, 1996.
21. Harrison, F.W., et al., Eds., *Microscopic Anatomy of Invertebrates* (Vol. I. Protozoa, Vol. 2. Placozoa, Porifera, Cnidaria, and Ctenophora, Vol. 3. Platyhelminthes and Nemertina, Vol. 4. Aschelminthes, Vol. 5. Mollusca I, Vol. 6. Mollusca II, Vol. 7. Annelida, Vol. 8. Chelicerate Arthropoda, Vol. 9. Crustacea, Vol. 10. Decapod Crustacea, Vol. 11. Insecta (Parts A and B), Vol. 12. Onychophora, Chilopoda, and Less Protostomata, Vol. 13. Lophophorates and Entoproca, Vol. 14. Echinodermata, Vol. 15. Hemichordata, Chaetognatha, and the Invertebrate Chordates), Wiley-Liss, New York, 1991–1998.
22. Kennedy, V.S., Newell, R.I.E. and Eble, A.F., Eds., *The Eastern Oyster*, Crassostrea virginica, Maryland Sea Grant College, University of Maryland System, College Park, MD, 1996.
23. Bell, T.A. and Lightner, D.V., *A Handbook of Normal Penaeid Shrimp Histology*, World Aquaculture Society, Baton Rouge, LA, 1988.

24. Stachowitsch, M., *The Invertebrates: An Illustrated Glossary*, Wiley-Liss, New York, NY, 1982.
25. Yevich, P.P., Yevich, C. and Pesch, G., *Effects of Black Rock Harbor Dredged Material on the Histopathology of the Blue Mussel* Mytilus edulis *and Polychaete Worm* Nephtys incisa *After Laboratory and Field Exposures*, Technical Report D-87-8. U.S. Army Engineer Waterways Experiment Station, Vicksburg, MS, 1987.
26. Gardner, G.R., Chemically induced histopathology in aquatic invertebrates, in *Pathobiology of Marine and Estuarine Organisms*, Couch, J.A. and Fournie, J.W., Eds., CRC Press, Boca Raton, FL, 1993, pp. 359–391.
27. Lehtonen, K.K., *Gulf of Riga Project: Physiology of Macrofauna*, Internal report to the Finnish Institute of Marine Research, Helsinki, Finland, 1998.
28. Peters, E.C. and Lehtonen, K.K., Chemical and other stressors in the Gulf of Riga: interpreting multiple lesions in the clam *Macoma balthica*, in *Abstracts of the National Shellfisheries Association Annual Meeting*, 20–24 April 1997, Fort Walton Beach, FL, 1997, p. 351.
29. NCM, Nordic Environmental Research Programme for 1993–1997. Final Report and Self-Evaluation. TemaNord 1999:548. Nordic Council of Ministers, Copenhagen. ISBN 92-2893-0332-9, ISSN 0908-6692, 1999.
30. Luoma, S.N., The developing framework of marine ecotoxicology: pollutants as a variable in marine ecosystems? J. *Exp. Mar. Biol. Ecol.*, 200, 29–55, 1996.
31. Reimschussel, R., Bennet, R.O. and Lipsky, M.M., A classification system for histological lesions, J. *Aquat. Animal Health*, 4, 135–143, 1992.
32. Tay, K.-L., Teh, S.J., Doe, K., Lee, K. and Jackman, P., Histopathologic and histochemical biomarker responses of Baltic clam, *Macoma balthica*, to contaminated Sydney Harbour sediment, Nova Scotia, Canada, *Environ. Health Perspect.*, 111 (3), 273–280, 2003.
33. USEPA, *Guidance Manual: Bedded Sediment Bioaccumulation Tests*. EPA/600/R-93/183. U.S. Environmental Protection Agency, Office of Research and Development, Washington, D.C., 1993.
34. Teh, S.J., Clark, S.L., Brown, C.L., Luoma, S.N. and Hinton, D.E., Enzymatic and histopathologic biomarkers as indicators of contaminant exposure and effect in Asian clam (*Potamocorbula amurensis*), *Biomarkers*, 4 (6), 497–509, 1999.
35. Bright, D.A. and Ellis, D.V., Aspects of histology in *Macoma carlottensis* (Bivalvia: Tellinidae) and in situ histopathology related to mine-tailings discharge, *Mar. Biol. Assoc. UK*, 69, 447–464, 1989.
36. Peterson, C.H., Rice, S.D., Short, J.W., Esler, D., Bodkin, J.L., Ballachey, B.E. and Irons, D.B., Long-term ecosystem response to the Exxon Valdez oil spill, *Science*, 302, 2082–2086, 2003.
37. Howard, D.W., Lewis, E.J., Jane Keller, B. and Smith, C.S., In press. Histological Techniques for Marine Bivalve Mollusks and Crustaceans. NOAA Technical Memorandum NOS NCCOS 6. U.S. Department of Commerce, National Oceanic and Atmospheric Administration, Woods Hole, MA.
38. Barszcz, C.A. and Yevich, P.P., Preparation of copepods for histopathological examination, *Trans. Amer. Micros. Soc.*, 95, 104–108, 1976.
39. Barszcz, C.A. and Yevich, P.P., The use of Helly's fixative for marine invertebrate histopathology, *Comp. Pathol. Bull.*, 7 (3), 4, 1975.
40. Bancroft, J.D. and Stevens, A., Eds., *Theory and Practice of Histological Techniques*, 4th ed., Churchill Livingstone, London, 1996.

chapter thirty-seven

Isolation of genes in aquatic animals using reverse transcription-polymerase chain reaction and rapid amplification of cDNA ends

Jeanette M. Rotchell
University of Sussex

Contents

1-56670-664-5/05/$0.00+$1.50
© 2005 by CRC Press

Introduction

Aquatic environments are receiving areas for significant levels of chemical contaminants. Such contaminants have been implicated in specific disease endpoints in aquatic organisms. Our laboratory has been using reverse transcription-polymerase chain reaction (RT-PCR) and rapid amplification of cDNA ends (RACE) in studies to identify and characterize the genes involved in the responses of aquatic organisms to toxicants. We have been particularly interested in the role of specific cancer genes, oncogenes, and tumor suppressor genes, in contaminant-induced tumor development in aquaria and feral populations of fish. For example, we have characterized the normal retinoblastoma (Rb) cDNA sequence from medaka (*Oryzias latipes*)[1] and proceeded to analyze it for the presence of chemically induced damage in hepatic tumor tissue using SSCP mutation detection techniques.[2,3] We have developed a protocol that works well with many species of fish, as well as invertebrates, for a variety of cancer genes provided care is taken in the primer design stage. The approach can be adapted to investigate virtually any gene provided some knowledge regarding sequence homology is available in the literature. This protocol is divided into five sections:

1. Isolation of RNA
2. Treatment of RNA with DNase I to remove genomic DNA contamination
3. Production of first strand cDNAs by reverse transcription
4. Isolation of an internal fragment of the targeted gene using PCR and degenerate primers, sub-cloning and sequence characterization
5. Isolation of the 5′ and 3″ cDNA ends using gene-specific primers (GSPs)

Materials required

Chemicals from specific suppliers that have worked well in our laboratory are noted by the accompanying catalog number. All other chemicals can be obtained from any reputable supplier, such as Invitrogen/Gibco.

Total RNA isolation

Equipment required:

- Tissue homogenizer, if preparations are from small (<100 mg) tissue samples

Reagents required:

- RNeasy Total RNA isolation kit (Qiagen, #74104), which includes mini spin columns, RNase-free Microfuge tubes, lysis buffer (RLT), two wash buffers RW1 and RPE, and RNase-free dH_2O

- 70% ethanol
- β-Mercaptoethanol
- RNaseAWAY (Invitrogen, #10328-011)

DNase I treatment

Equipment required:

- 37°C water bath or thermocycler

Reagents required:

- DNase I — RNase-free, 1 unit/μl (Promega, #M6101)
- 10× DNase buffer (400 m*M* Tris–HCl [pH 8.0 at 25°C], 100 m*M* $MgSO_4$, 10 m*M* $CaCl_2$). Supplied with enzyme
- Stop buffer (20 m*M* EGTA, pH 8.0 at 25°C). Supplied with enzyme
- Ethanol
- 3 *M* sodium acetate, pH 5.2
- Diethylpyrocarbonate (depc)-dH_2O

First strand synthesis

Equipment required:

- Thermocycler
- 37°C, 42°C, 65°C, and 70°C water baths or heat blocks

Reagents required:

- Superscript™ First-strand synthesis system for RT-PCR (Invitrogen, #11904-018), which includes: 10× buffer (200 m*M* Tris–HCl [pH 8.4], 500 m*M* KCl), Superscript II reverse transcriptase (50 units/μl), oligo (dT)20 primer (0.5 μg/μl), 25 m*M* $MgCl_2$, 0.1 *M* dithiothreitol (DTT), 10 m*M* deoxyribonucleotides (dNTPs), RNase-OUT (40 units/μl), *E. coli* RNase H (2 units/μl), depc-dH_2O.

PCR of internal fragment, sub-cloning and sequence characterization

Equipment required:

- Thermocycler
- Water bath at 42°C
- Shaking and static incubators at 37°C
- Standard horizontal agarose gel electrophoresis apparatus (e.g., BioRad, #170-4487)

Reagents required:

- Taq DNA polymerase, such as Platinum™ Taq DNA polymerase (Invitrogen, #10966-018) (without proofreading capacity, i.e., *not* Pfu, Pfx, or Vent). Supplied with 10× buffer and 50 m*M* $MgSO_4$.

- 10 m*M* dNTP mix.
- Two (degenerate) amplification primers (forward and reverse) specific for your target mRNA.
- Sub-cloning reagents (Invitrogen, #K2040-01), which includes: T4 DNA ligase, 10 × ligation buffer (60 m*M* Tris–HCl [pH 7.5], 60 m*M* $MgCl_2$, 50 m*M* NaCl, 1 mg/ml bovine serum albumin, 1 m*M* ATP, 20 m*M* DTT, 10 m*M* spermidine), cloning vector, SOC medium (2% tryptone, 0.5% yeast extract, 10 m*M* NaCl, 2.5 m*M* KCl, 10 m*M* $MgCl_2$, 10 m*M* $MgSO_4$, 20 m*M* glucose), and competent cells.
- Plasmid preparation reagents (Promega, #A7100), which includes: cell resuspension solution (50 m*M* Tris–HCl [pH 7.5], 10 m*M* EDTA, 100 μg/ml RNase A), cell lysis solution (0.2 *M* NaOH, 1% SDS), neutralization solution (1.32 *M* potassium acetate), Wizard® Minipreps DNA purification resin, column wash solution (80 m*M* potassium acetate, 8.3 m*M* Tris–HCl [pH 7.5], 40 μ*M* EDTA), and syringe barrels. A vacuum manifold device (Promega, #A7231) simplifies the procedure but is not compulsory.

RACE

Equipment required:

- Thermocycler
- Standard horizontal agarose gel electrophoresis apparatus (e.g. BioRad, #170-4487)
- Water bath at 42°C
- Shaking and static incubators at 37°C

Reagents required:

- SMART™ RACE cDNA amplification reagents (BD Biosciences, #K1811-1), which includes: SMART oligonucleotide (10 μ*M*), 3′-RACE and 5′-RACE cDNA synthesis (CDS) primers (10 μ*M*) Powerscript™ reverse transcriptase, 5× buffer (250 m*M* Tris–HCl [pH 8.3], 375 m*M* KCl, 30 m*M* $MgCl_2$), 20 m*M* DTT, 10× universal primer mix (UPM): 0.4 μ*M* of long primer and 0.2 μ*M* of short primer, 10 m*M* dNTPs, tricine–EDTA buffer (10 m*M* tricine–KOH [pH 8.5], 1 m*M* EDTA).
- Advantage 2 PCR reagents for PCR reactions (BD Biosciences, #K1910-y, free if bought with SMART reagents), which includes: Taq DNA polymerase, 10× buffer, 10 m*M* dNTPs, depc-dH_2O.
- Two GSPs.
- Sub-cloning reagents (Invitrogen, # K2040-01) as above.

Optional: SSCP-PCR analysis

Equipment required:

- Thermocycler
- PhastSystem gel electrophoresis and gel development apparatus (Amersham Biosciences, #18-1018-24)

Reagents required:

- Taq DNA polymerase, such as Platinum Taq (Invitrogen, #10966-018). Supplied with 10× buffer and 50 m*M* $MgSO_4$
- 10 m*M* dNTP mix
- Gene-specific amplification primers (forward and reverse) that span your target cDNA
- Phastgel homogeneous 12.5 (Amersham Biosciences, #17-0623-01)
- Phastgel DNA silver staining reagents (Amersham Biosciences, #17-1596-01)
- DNA buffer strips (Amersham Biosciences, #17-1599-01)
- Stop solution/gel loading dye (95% formamide, 10 m*M* NaOH, 0.25% bromophenol blue, 0.25% xylene cyanol)

Procedures

Figure 37.1 is a schematic overview of the procedures to be described, as well as potential applications for the normal cDNA sequence derived from RT-PCR/RACE. The specific protocols detailed herein are total RNA isolation, DNase I treatment, first strand synthesis, PCR of an internal target cDNA fragment, and RACE.

Total RNA isolation

Isolation of total RNA can be conducted using a variety of techniques, but those based on the method of Chomcynski and Saachi[4] have been successfully employed for many tissues types taken from a range of species. Commercial products, such as the RNeasy reagents from Qiagen, exploit silica-gel-based membrane binding and spin column technology to speed up the method still further. The latter technique avoids the use of toxic (phenol) reagents, and also works well with a variety of tissues. The resulting total RNA is suitable for RT-PCR, RACE, and Northern hybridization.

Total RNA isolation protocol

The tissue sample should be fresh or snap frozen in liquid nitrogen and stored at −70°C. The homogenizer should be wiped with RNaseAWAY to reduce possible contamination with RNases. Ethanol must be added to wash buffer RPE before first use. β-Mercaptoethanol should be added to the lysis buffer (RLT) just before use. The general protocol is as recommended by the reagent's supplier with the following changes that have been optimized for small fish (eye and liver) samples.

- Drop ~20 mg of tissue into 350 μl of RLT inside a Microfuge tube and quickly homogenize using three return (up and down) strokes. Check for large tissue pieces and repeat if necessary. Keep the homogenate on ice.
- Spin for 3 min at high speed in a microcentrifuge and extract supernatant to a clean tube.
- Add 350 μl of 70% ethanol and mix by pipetting up/down.
- Add 700 μl of sample to a minicolumn within a collection tube, spin for 15 s at 10,000 rpm. Repeat until the sample is loaded onto the column.
- Add 700 μl of wash buffer RW1, and spin for 15 s at 10,000 rpm.

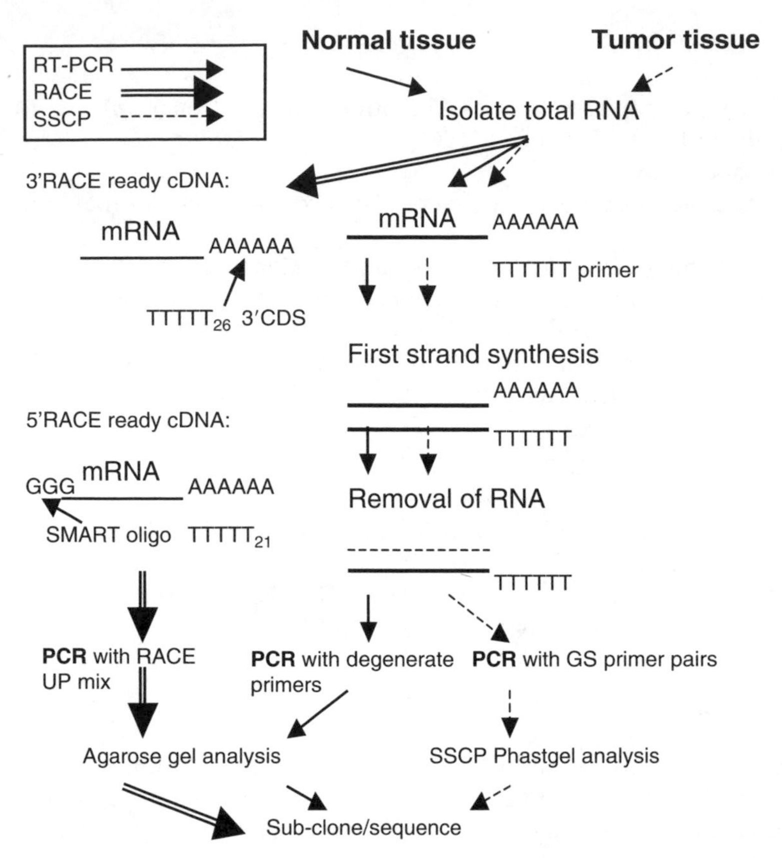

Figure 37.1 A schematic overview of the RT-PCR and RACE procedure. The first step, RT-PCR to isolate an internal fragment of the desired cDNA, and its methodological pathway, is shown by a regular arrow. Subsequently, RACE analysis and its methodological pathway are represented by an arrow with double lines. The last step, using the sequence derived for the normal cDNA to screen for DNA damage and its methodological pathway, is represented by an arrow with a gapped line. Abbreviations: GS, gene-specific; UP, universal primer.

- Place the spin column in a clean collection tube, add 500 μl of buffer RPE, and spin for 15 s at 10,000 rpm. Empty the collection tube.
- Add another 500 μl of buffer RPE, and spin for 2 min at maximum speed. Empty the collection tube.
- Place the spin column in a clean Microfuge tube and add 30 μl of RNase-free dH_2O. Wait for 5 min at room temperature and spin for 1 min at 10,000 rpm.
- Measure absorbance at 260 nm to estimate concentration of nucleic acid.
- Store the resulting total RNA preparation on ice if proceeding to the DNase treatment or at −20°C.

DNase I treatment

It is very likely that the total RNA preparation contains some contaminating genomic DNA that may serve as an inappropriate template during PCR.

DNase I treatment protocol

- In a Microfuge tube, combine approximately 20 μg of total RNA, 10 μl of 10× DNase buffer, 10 μl (10 units) of DNase, and depc-treated dH_2O to a final volume of 100 μl.
- Incubate for 1 h at 37°C.
- Repeat RNA isolation protocol above using 350 μl of lysis buffer RLT in the first step.
- Measure absorbance at 260 nm to estimate concentration of RNA.
- Store the RNA at −20°C or −70°C for the longer term.

First strand synthesis

In this step, the RNA is used as a template to produce a cDNA strand. The enzyme used is reverse transcriptase that uses RNA (or single strand DNA) as a template, provided a suitable primer is present. A first strand cDNA synthesis reaction can be primed using three different methods that exploit different primers: random hexamers, oligo (dT), and GSPs. The protocol described uses oligo (dT) primers, which hybridize to the 3′ poly (A) tails that are located on most eukaryotic mRNAs. The resulting product is an RNA–cDNA hybrid that is subsequently digested with RNase H to remove the RNA template.

First strand synthesis protocol

- The total reaction volume is 20 μl.
- Prepare two tubes as follows:

	Sample (label RT+)	Control (label RT−)
RNA (~2.5 μg)	X μl	X μl
10 m*M* dNTP mix	1 μl	1 μl
Oligo (dT) (0.5 μg/μl)	1 μl	1 μl
Depc-treated dH_2O	To 10 μl	To 10 μl

- Incubate each tube at 65°C for 5 min, and then place on ice for at least 1 min.
- Prepare the following reaction mixture (for *n* samples plus 1 RT-control, prepare reaction mix for *n* + 3 reactions):

	Per each reaction (μl)
10× RT buffer	2
25 m*M* $MgCl_2$	4
0.1 *M* DTT	2
RNaseOUT	1

- Add 9 μl of reaction mixture to each RNA/primer mixture, mix and spin gently.
- Incubate at 42°C for 2 min.

- Add 1 μl of reverse transcriptase to each tube (except the "RT–" tube), mix and incubate at 42°C for 50 min.
- Stop the reactions by incubating at 70°C for 15 min. Place on ice.
- Briefly spin the tubes before adding 1 μl of RNase H to each and incubate at 37°C for 20 min.
- Place on ice if proceeding to the next step, or at −20°C.

PCR of internal fragment, sub-cloning and sequence characterization

The first strand cDNAs can be used to amplify the target cDNA directly using PCR. Any *Taq* DNA polymerase can be used but, if the resulting products are to be sub-cloned, a *Taq* that adds adenine overhangs is desirable. The design of the PCR primers is critical in obtaining a target cDNA. The primers degeneracy, length, GC content, and subsequent melting temperature will affect the results achieved. Those primers with high degeneracy, short length, and low melting temperature will give rise to the production of many nonspecific PCR products. For guidance on how to design degenerate primers, the papers of Rotchell et al.[1,2] and Rotchell and Ostrander,[3] which include primer designs based on homology alignments for the fish *ras* genes and vitellin/aromatase genes, respectively, may be of help.

Once a PCR product of the expected size has been obtained, it is then possible, exploiting the adenine overhangs, to sub-clone into a cloning vector that possesses complimentary thymine overhangs. It is possible to sequence PCR products directly, but in our experience it is rare to amplify a single PCR product of the expected size using degenerate primers, and this creates sequencing results that are hard to interpret due to the presence of extra templates. The preferred technique in our lab is to excise the desired PCR product from an agarose gel, purify it and then ligate into a vector using a 1:1 ratio (vector/insert). The size of the insert is important in determining the amount of vector to use (and it is important to refer to the kit instructions for further guidance). The vector is then transformed into competent cells and plated to obtain single colonies. The LB plates used would have been prepared the preceding day. The overnight culture derived from the single colony can then be used for a plasmid preparation that is subsequently sequenced to characterize the PCR product obtained.

PCR protocol

The following PCR protocol serves as a starting point only, it is usually necessary to optimize several of the parameters (e.g., the Mg concentration, the cDNA concentration, the use of bovine serum albumin, the cycling conditions, and the number of cycles) before obtaining the desired PCR product.

- The final reaction volume is 50 μl.
- Add the following to a 0.2-ml thin walled PCR tube (the following is for one reaction):

10× PCR buffer	5 μl
50 m*M* $MgCl_2$	3 μl
10 m*M* dNTP mix	1 μl
10 μ*M* forward primer	1 μl

10 μ*M* Reverse primer	1 μl
Taq DNA polymerase	0.5 (1.25 units)
First strand cDNA	3 μl
depc-treated dH_2O	To a final volume of 50 μl

- Mix the components and spin briefly.
- Add mineral oil if your thermocycler does not have a heated lid.
- Perform 20–40 cycles of PCR with optimized conditions for your target cDNA. The following is a starting point for a 1-kb target cDNA — 1 cycle at 94°C for 4 min, followed by 30 cycles of 94°C for 45 s (denaturation), 50°C for 50 s (annealing), 72°C for 1 min (extension).
- Analyze 10 μl of amplified products using agarose gel electrophoresis.

Sub-cloning and sequence characterization protocol

PCR products generated with *Taq* DNA polymerase have adenine overhangs at the 3′ end, and this can be exploited using the TA cloning technique. Cloning is achieved in two steps: ligation and transformation.

- Combine the following ligation reagents in a 200-μl PCR tube on ice:

Fresh PCR product	X μl to give a 1:1 (vector/insert) ratio
10× ligation buffer	1 μl
Vector (25 ng/μl)	2 μl
Sterile dH_2O to a final volume	9 μl
T4 DNA ligase (4 Weiss units)	1 μl

- Incubate the reaction at 14°C for at least 4 h.
- Keep on ice or store at −20°C until ready to transform.
- Prepare a water bath at 42°C.
- Warm SOC medium to room temperature, prepare LB plates (the day before use) containing 50 μg/ml of either kanamycin or ampicillin and warm to room temperature. Spread each plate with 40 μl of 40 mg/ml X-Gal.
- Thaw on ice one tube containing a 50-μl aliquot of competent cells per ligation reaction.
- Add 2 μl of ligation reaction and stir gently using the pipette tip.
- Incubate on ice for 30 min.
- Heat shock at 42°C for exactly 30 s and return to ice.
- Add 250 μl of SOC medium and shake tubes at 37°C for 1 h at 225 rpm.
- Spread 10–200 μl from each transformation tube on a separate LB agar plate that contains the X-Gal and antibiotic. Plating two different volumes should ensure that you obtain at least one plate from which it is possible to pick a single colony later.
- Incubate at 37°C overnight. In the morning, place the plates at 4°C for 3 h to allow color development.
- To determine the presence of the desired insert, pick at least 10 colonies for plasmid preparation and restriction digest analysis. Pick each colony into 10 ml of LB broth containing 50 μg/ml of either kanamycin or ampicillin and grow overnight at 37°C in a shaking incubator.

Plasmid preparation:

- Prepare a 3-ml pellet from the overnight culture by spinning cells (in two 1.5-ml batches, emptying the supernatant after each spin) for 2 min at low speed (~10,000*g*) in a Microfuge. Alternatively, prepare a 3-ml pellet using a larger collection tube and a tabletop centrifuge.
- Resuspend cells using 200 μl of cell resuspension solution.
- Add 200 μl of cell lysis solution and invert tube several times to mix.
- Add 200 μl of neutralization solution and invert to mix.
- Spin at low speed in a Microfuge for 5 min.
- Decant the clear supernatant into 1 ml of resin and load both into a barrel of the minicolumn/syringe assembly.
- Apply a vacuum to pass the sample/resin through the minicolumn.
- Wash the minicolumn with 2 ml of wash solution using the vacuum.
- Disassemble the minicolumn/syringe, placing the minicolumn in a Microfuge tube and then spin at low speed to remove residual wash solution.
- Place the minicolumn in a clean Microfuge tube and add 50 μl of depc-dH_2O. Wait for 1 min and then elute the DNA from the column by spinning for 20 s at low speed. Elution using TE buffer is not recommended for several methods of sequencing.
- Store at −20°C.

Sequencing: We have found that sending 1 μg of plasmid DNA to commercial sequencing companies (such as MGWBiotech — www.mwg-biotech.com) is faster and more cost-effective than carrying out the reaction in your own laboratory, especially if you only have a small number of samples to process. Alternatively, use an onsite sequencing core facility if available.

RACE

Once an internal fragment of your target cDNA has been sequenced, it is then possible to design GSPs and use the RACE technique to isolate the entire cDNA. It is possible to obtain full-length cDNAs without having to construct or screen a cDNA library. If the gene of interest is remarkably well conserved across species and phyla, it may, in theory, be possible to use degenerate primers, omitting the previous internal fragment PCR step, to isolate the target cDNA. In practice, we have only had success with GSPs.

The reverse transcriptase employed has terminal transferase activity, adding cytosine residues to the 3′ end of the first strand cDNAs. This, in turn, provides a terminal stretch to which the SMART oligo provided can anneal, serving as an extended template for reverse transcription. A complete cDNA copy of the mRNA molecule is synthesized, which contains a SMART sequence at the end, and can then be used in 5′ and 3′ PCR reactions.

RACE protocol

For the RACE technique, at least two GSPs are required. In terms of the primer design, the two GSPs should be 23–28 nucleotides in length, 50–70% GC content, and have a melting temperature of at least 60°C (70°C is better). The location of the primers can also affect results, those that will produce a PCR product of 2 kb or less is optimal.

First strand synthesis:

- The final reaction volume is 10 μl.
- To prepare the first strand cDNAs, combine the following in separate 200-μl tubes.

(a) For the preparation of 5′-RACE ready cDNA:

X μl of total RNA (~1 μg)
1 μl of 5′-CDS primer
1 μl of SMART oligo
X μl of depc-dH_2O to a final volume of 5 μl

(b) For the preparation of 3′-RACE ready cDNA:

X μl of total RNA (~1 μg)
1 μl of 3″-CDS primer A
X μl of depc-dH_2O to a final volume of 5 μl

- Mix the contents, spin briefly, and incubate at 70°C for 2 min.
- Cool the tubes on ice for 2 min and then spin briefly to collect the contents at the bottom of the tube.
- Add the following components to each reaction tube:

5× first strand buffer	2 μl
DTT (20 m*M*)	1 μl
dNTP mix (10 m*M*)	1 μl
Powerscript reverse transcriptase	1 μl

- Mix by pipetting up and down, spin briefly and incubate at 42°C for 1.5 h.
- Dilute with 250 μl of tricine–EDTA buffer.
- Incubate at 70°C for 7 min.
- Store samples at −20°C.

Before performing PCR it is advisable to carry out the control reactions supplied by the manufacturer to ensure that all components are working correctly. Once confirmed, the 5′ and 3′ PCR reactions can be performed as follows (using the advantage reagents listed above).

- Prepare enough PCR master mix for each reaction plus one extra by combining the following (1 reaction mix) —

Depc-dH_2O	34.5 μl
10× PCR buffer	5 μl
dNTP (10 m*M*)	1 μl
50× polymerase mix	1 μl
Total reaction volume	50 μl

- Mix and spin briefly, keep on ice.
- Combine the following in separate tubes:

For the 5′-RACE PCR:

Tube number	1	2	3	4
5′-RACE-ready cDNA (μl)	2.5	2.5	2.5	2.5
Universal primer mix (μl)	5	0	5	0
GSP 1 (10 μ*M*) (μl)	1	1	0	1
GSP 2 (10 μ*M*) (μl)	0	1	0	0
Depc-dH_2O (μl)	0	4	1	5
Master mix (μl)	41.5	41.5	41.5	41.5
Total reaction volume (μl)	50	50	50	50

For the 3′-RACE PCR

Tube number	5	6	7	8
3′-RACE-ready cDNA (μl)	2.5	2.5	2.5	2.5
Universal primer mix (μl)	5	0	5	0
GSP 1 (10 μ*M*) (μl)	0	1	0	0
GSP 2 (10 μ*M*) (μl)	1	1	0	1
Depc-dH_2O (μl)	0	4	1	5
Master mix (μl)	41.5	41.5	41.5	41.5
Total reaction volume (μl)	50	50	50	50

- Mix all tubes and spin briefly.
- Amplify using the following program:

5 cycles	94°C	5 s
	72°C	3 min
5 cycles	94°C	5 s
	70°C	10 s
	72°C	3 min
25 cycles	94°C	5 s
	68°C	10 s
	72°C	3 min

- Analyze 10 μl of amplified products using agarose gel electrophoresis.
- Sub-clone PCR product of desired length and sequence.

Optional application: SSCP analysis for the detection of mutations in tumor tissue samples

Having determined the normal sequence of a cDNA, such as that for a proto-oncogene or tumor suppressor gene, it is then a relatively simple task to screen tumor tissue samples for the presence of mutational damage that may have been an early cause of that condition. In a number of studies, investigators have also built mutational profiles that relate specific chemical exposure to mutational "hot spots" of DNA damage within critical genes, such as the *ras* gene.[2]

The technique that we routinely use to screen for the presence of mutations is SSCP analysis of PCR products, followed by sequencing to characterize the actual nature of any DNA damage detected. The screening procedure can be completed in a day using the PhastSystem detailed in the section "Materials required." Alternatively, it is possible to use sequencing gel electrophoresis apparatus (a less expensive equipment option) to conduct SSCP analysis, but this method usually requires long gel running times (overnight), radioisotope labeling, and autoradiography (overnight).

SSCP protocol:

- Design *overlapping* GSPs that span your target cDNA and generate PCR products of approximately 200 bp in size (see Figure 37.2 as an example).
- Using the protocols detailed earlier, isolate total RNAs and construct first strand cDNAs from your normal (control) and tumor tissue samples.
- Add the following to a 0.2-ml thin walled PCR tube. (The following mix is for one reaction only. Set up one reaction for each primer pair that spans your target cDNA, for each normal/tumor sample to be analyzed. For instance, if your target cDNA is divided into 10 overlapping PCR primer pairs and you wish to analyze two normal and two tumor tissue samples, you would set up 10 × 4 reactions.)

10× PCR buffer	5 μl
50 m*M* $MgCl_2$	3 μl
10 m*M* dNTP mix	1 μl
10 μ*M* forward primer	1 μl
10 μ*M* reverse primer	1 μl
Taq DNA polymerase	0.5 (1.25 units)
First strand cDNA	3 μl
Depc-treated dH_2O	To a final volume of 50 μl

- Mix the components and spin briefly.
- Add mineral oil if your thermocycler does not have a heated lid.
- Perform 20–40 cycles of PCR with optimized conditions for your target cDNA. The following is a starting point for a 200-bp target cDNA – 1 cycle at 94°C for 4 min, followed by 30 cycles of 94°C for 45 s (denaturation), 60°C for 50 s (annealing), 72°C for 30 s (extension).

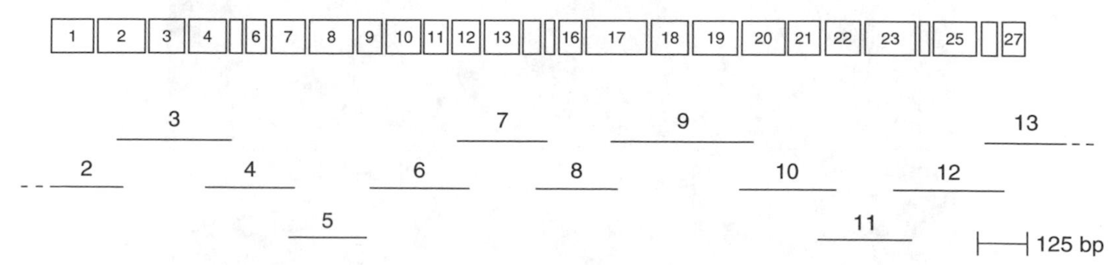

Figure 37.2 A map of the Rb cDNA and the location of overlapping PCR products obtained using the primer pairs (PPs) designed. To facilitate SSCP analysis, the PCR primer pairs are designed to produce amplification products that are approximately 200 bp in length.

- Mix 5 μl of PCR product with 5 μl of stop solution, denature at 98°C for 3 min and cool on ice.
- Pre-run gels at 400 V, 5 mA, 1 W for 10-volt hours (Vh).
- Analyze 1 μl of "PCR products/stop solution" mix using Phastgel electrophoresis.
- Stain the gel using the protocol supplied with the silver stain reagents.

Results and discussion

Anticipated results

The desired outcome of RT-PCR and RACE is the isolation and subsequent characterization of the desired target cDNA. The protocol detailed here will allow the investigator to isolate, firstly, an internal fragment of the target cDNA to serve as a starting point for the amplification of the two remaining cDNA ends. By sequencing the internal fragment of this target cDNA, it is possible to design GSPs that can be used in parallel with RACE primers and RACE-ready first strand template. Following amplification with such primers, it is possible that more than one 3′-RACE product, in particular, may be obtained. Additional products may represent truncated versions of the desired cDNA or closely related members of that gene family. In any case, it is important to sequence the product to confirm its identity as the desired target cDNA. An example from our work, involving the isolation of the Rb cDNA from medaka, can help to illustrate some of the typical features of a RACE result. Figure 37.3 displays an agarose gel of the 5′-RACE and 3′-RACE results using medaka total RNA and GSPs for the Rb cDNA. In the 5′ reaction, there is a single amplified fragment of approximately 2.8 kb, which was the expected size in this instance. For the 3′ reaction, at least three amplified products of approximately 1.9 kb, 1.6 kb, and 750 bp were visible. The expected size in this instance

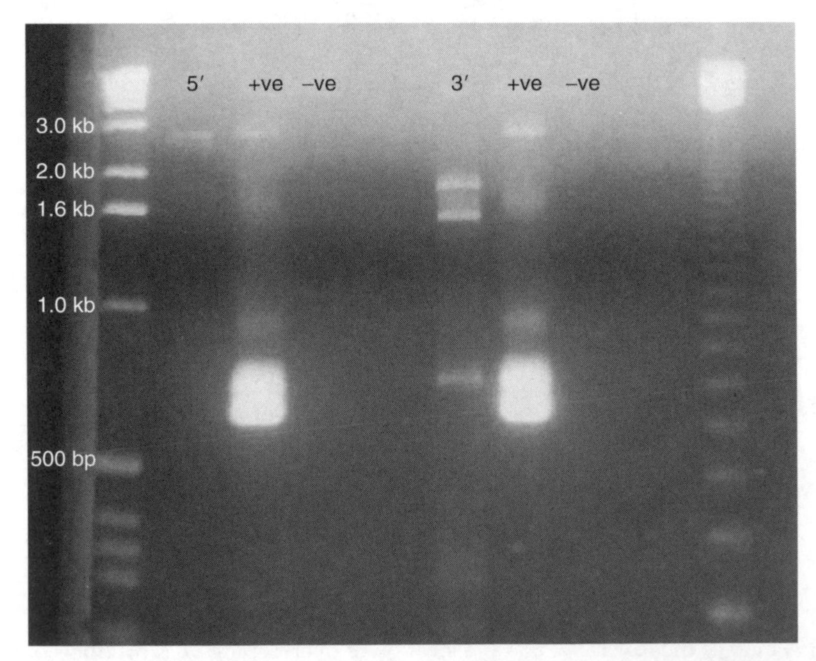

Figure 37.3 A representative agarose gel result of the 5′-RACE and 3′-RACE reactions from the medaka Rb cDNA.

was 1.7 kb, so the two larger fragments were purified from the agarose, sub-cloned, and sequenced. Once the sequence had been characterized, the smaller of the two fragments aligned well with the Rb cDNA sequences already published in the literature. This figure demonstrates a commonly observed feature of RACE results in that more than one amplified product is obtained and these may represent the presence of truncated forms or related family members of the cDNA of interest or simply be artifacts.

A similar RACE strategy has been adopted by investigators in the characterization of a variety of genes involved in aquatic toxicology applications. Recent examples include the characterization of fish alpha estrogen receptors,[5,6] an invertebrate vitellogenin cDNA,[7] and a fish cytochrome P450 2X1 detoxification enzyme cDNA.[8]

SSCP application: anticipated results

Laboratory studies have demonstrated that a variety of tumors, such as hepatocellular tumors, can be induced experimentally in aquaria fish using a variety of compounds.[9,10] The histopathology of the induced tumors in fish often corresponds broadly with the human conditions, and it is anticipated that the molecular etiology may correspond similarly. Furthermore, understanding the underlying cause of experimentally induced tumors should help elucidate the cause of environmentally induced tumors.

In our laboratory we have been interested in medaka fish tumors in particular, and exploiting the unique opportunity that existed to examine the role of the Rb gene in malignancy using a fish model. Using the RACE protocol and SSCP application described herein, the mutational screening of the Rb cDNA from normal liver and eye, as well as chemically induced medaka eye and liver tumor samples, were conducted.[1,11] Figure 37.4

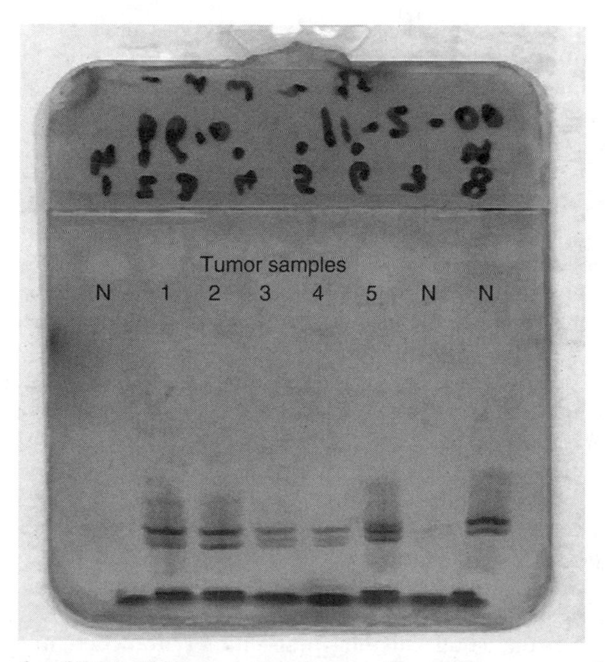

Figure 37.4 An example of PCR-SSCP analysis of normal medaka eye samples (N) and eye tumor samples (T1–T5) with Rb-specific primers PP10 (spanning exons 21–23). Abnormal electrophoretic patterns are observed using primers PP10 with samples T1, T3, T4, and T5.

displays a typical SSCP gel result of, in this instance, an analysis of eye normal and chemically induced tumor samples. In this investigation, abnormal electrophoretic patterns were observed in four samples (Figure 37.4: samples T1, T3, T4, and T5). Sequence analysis of these PCR products revealed point mutations and two deletions within the Rb cDNA sequence. No mutations were detected in the normal eye samples. However, it is possible to detect changes using normal/control samples in the sequence of the Rb cDNA that represent polymorphic variation.

Summary

In summary, RT-PCR combined with RACE is an effective method to isolate and characterize genes that are involved in the responses of aquatic organisms to toxicants. Having a limited amount of prior sequence information from related species, usually obtainable from the Genbank database, is sufficient to enable an investigator to design degenerate primers for their targeted gene. Inversely, having too limited sequence information available can be the major drawback of this technique. In such instances, the alternative approach of constructing and screening a cDNA library for the desired gene (though more time consuming) may be more successful. As more relevant cDNA sequences (of critical growth controlling proteins and detoxification enzymes, for instance) become available for a greater variety of aquatic organisms, it will become a much simpler task to isolate such genes using cross-species degenerate primers in a chosen species of interest using the RT-PCR and RACE methods.

References

1. Rotchell, J.M., Shim, J., Blair, J.B., Hawkins, W.E. and Ostrander, G.K., Cloning of the Retinoblastoma cDNA from the Japanese medaka (*Oryzias latipes*) and preliminary evidence of mutational alterations in chemically-induced retinoblastomas, *Gene*, 263, 231–237, 2001.
2. Rotchell, J.M., Lee, J.-S., Chipman, J.K. and Ostrander, G.K., Structure, expression and activation of fish ras genes, *Aquat. Toxicol.*, 55, 1–21, 2001.
3. Rotchell, J.M. and Ostrander, G.K., Molecular markers of endocrine disruption in aquatic organisms, *J. Toxicol. Environ. Health (B)*, 6, 453–495, 2003.
4. Chomcynski, P. and Saachi, N., Single-step method of RNA isolation by acid guanidinium thiocyanate-phenol-chloroform extraction, *Anal. Biochem.*, 162, 156–159, 1987.
5. Choi, C.Y. and Habibi, H.R., Molecular cloning of estrogen receptor alpha and expression pattern of estrogen receptor subtypes in male and female goldfish, *Mol. Cell Endocrinol.*, 204, 169–177, 2003.
6. Karels, A.A. and Brouwer, M., Cloning, sequencing and phylogenetic/phenetic classification of an estrogen receptor alpha (alpha) subtype of sheepshead minnow (*Cyprinodon variegatus*), *Comp. Biochem. Physiol. (B) Biochem. Mol. Biol.*, 135, 263–272, 2003.
7. Matsumoto, T., Nakamura, A.M., Mori, K. and Kayano, T., Molecular characterization of a cDNA encoding putative vitellogenin from the Pacific oyster *Crassostrea gigas*, *Zool. Sci.*, 20, 37–42, 2003.
8. Schlenk, D., Furnes, B., Zhou, X. and Debusk, B.C., Cloning and sequencing of cytochrome P450 2X1 from channel catfish (*Ictalurus punctatus*), *Mar. Environ. Res.*, 54, 391–394, 2002.
9. Hawkins, W.E., Fournie, J.W., Overstreet, R.M. and Walker, W.W., Development of aquarium fish models for environmental carcinogenesis: tumor induction in seven species, *J. Appl. Toxicol.*, 5, 261–264, 1985.

10. Hawkins, W.E., Walker, W.W., Overstreet, R.M., Lytle, J.S. and Lytle, T.F., Carcinogenic effects of some polycyclic aromatic hydrocarbons on the Japanese medaka and guppy in water borne exposures, *Sci. Total Environ.*, 94, 155–167, 1990.
11. Rotchell, J.M., Unal, E., Van Beneden, R.J. and Ostrander, G.K., Induction of retinoblastoma tumor suppressor gene mutations in chemically-induced liver tumors in medaka (*Oryzias latipes*), *Mar. Biotechnol.*, 3, S44–S49, 2001.

chapter thirty-eight

Analysis of mutations in λ transgenic medaka using the cII *mutation assay*

Richard N. Winn and Michelle B. Norris
University of Georgia

Contents

1-56670-664-5/05/$0.00+$1.50
© 2005 by CRC Press

Introduction

The ability to identify induced mutations and to understand their roles in a variety of disease processes within whole animals pose significant challenges. The basis of these challenges is primarily a sampling problem. Mutations in genes, even those induced following exposure to potent chemical or physical mutagens, typically occur at low frequencies (e.g. ~1 spontaneous mutation/10^{-5} to 10^{-7} loci). Consequently, to distinguish mutant genes among a very large number of nonmutant genes, the gene carrying the mutation must be identified accurately and recovered efficiently. Until recently, routine *in vivo* mutation analyses have been hampered by the lack of methods capable of meeting these requirements. For example, although a variety of *in vivo* genotoxicity assays provide information related to the ability of the chemical to alter DNA or the structure of a chromosome by using such endpoints as the induction of DNA adducts, DNA repair, DNA strand breaks, and chromosomal damage, these endpoints are limited in that they are not direct measures of induced gene mutations. Other assays using endogenous genes are either insensitive, or are limited to specific tissues or developmental stages,[1–3] or unavailable in such organisms as fish.

Mutation assays based on transgenic animals are meeting many of the fundamental challenges confronting studies of *in vivo* mutagenesis. Developed originally in rodent models,[4–8] and now in fish,[9–11] these transgenic mutation assays share a similar general approach. The transgenic animals have been produced that carry prokaryotic vectors that harbor a specific gene that serves as a mutational target for quantifying spontaneous and induced mutations. After treatment with a mutagen, and allowing sufficient time for the mutations to manifest, genomic DNA is isolated from target tissues. The vectors are then separated from the animal's genomic DNA and shuttled into specialized indicator bacteria that facilitate distinguishing mutant and nonmutant genes.

In addition to a common general approach, transgenic mutation assays share numerous benefits for *in vivo* mutation analyses. These assays permit the screening of a large number of copies of a locus rapidly, thereby providing statistically reliable information on the frequency of mutations in virtually any tissue in the animal, while reducing the need for large numbers of animals. Mutations are quantified directly at the level of single genes, the endpoint of DNA damage and repair. The mutation target transgenes are genetically neutral affording the persistence and accumulation of mutations over time without being subjected to selection in the animal.[12–14] In addition, the recovered mutant gene can be sequenced to characterize the mutation at the molecular level and to assist in identifying the mutagen's possible mechanism of action.

In this chapter, we describe features of, and procedures for, conducting mutation analyses using the *cII* mutation assay based on the λ transgenic medaka.[10] The λ transgenic medaka carry multiple copies of the bacteriophage λLIZ vector, which harbors the *lacI* and *cII* bacterial genes that serve as mutational targets. The most widely used transgenic mutation assay in rodents is also based on this bacteriophage vector,[5] providing ample opportunities for studies of comparative mutagenesis in identical mutation target genes carried by different species. Mutation analyses are feasible in either the *lacI* or *cII* target genes by using different assay procedures. Here, we focus on a description of the *cII* mutation assay, a positive-selection assay that uses the *cII* target gene as a logistically simple and cost-effective alternative to the *lacI* mutation assay.[15] Following the descriptions of procedures, we summarize specific examples of analyses of induced mutations detected in the *cII* target gene recovered from λ transgenic medaka exposed to four chemical mutagens. These examples highlight many features of the λ transgenic medaka and illustrate some of the more important factors to be considered when designing and conducting *in vivo* mutation studies.

Overview of procedures for mutation analyses using λ transgenic medaka

1. Treat animals with chemical or physical mutagens
2. Isolate genomic DNA
3. Prepare G1250 plating culture
4. Perform packaging reaction and plate the packaged DNA samples
5. Examine titer plates and mutant screening plates
6. Verify putative λ *cII*– mutants
7. Characterize λ *cII*– mutants by sequencing

Materials required

Animals

The λ transgenic medaka, designated lineage λ310, were generated using the orange-red strain (Himedaka), strain that originated from the Gulf Coast Research Laboratories (Gulf Coast, MS). The λ310 lineage carries ~75 copies per haploid genome of the bacteriophage λLIZ vector (~45 kb) harboring the *lacI* and *cII* bacterial genes that are used as mutational targets.[10] The animals used in these studies were obtained from in-house stocks maintained at the Aquatic Biotechnology and Environmental Laboratory (ABEL) at the University of Georgia. Medaka was selected for this application because it is widely used in environmental toxicology and is the fish species of choice in carcinogenesis bioassays and germ-cell mutagenesis studies. The small size, sensitivity, well-characterized histopathology, short generation time, and cost-effective husbandry contribute to the utility of this species in routine toxicity testing.

Tissue dissection and DNA isolation

Equipment

Equipment required for tissue dissection and DNA isolation is listed in Table 38.1.

Supplies and reagents

Supplies and reagents required for tissue dissection and DNA isolation are listed in Table 38.1.

Solutions:

- Phenol: chloroform (0.5 ml/sample)

In a sterile 50 ml polypropylene centrifuge tube, mix equal volumes of saturated phenol with chloroform. Prepare on day of use and equilibrate at room temperature.

- 20 mg/ml proteinase K (15 μl/sample)

Prepare on day of use.

- 1× SSC stock solution (1 l)
  - 8.77 g NaCl
  - 4.41 g sodium citrate

Dissolve salts in 800 ml of water. Adjust the pH to 7.0. Bring the volume up to 1000 ml. Sterilize by autoclaving at 121°C for 30 min.

Table 38.1 Materials required for tissue dissection and DNA extraction

Refrigerated Microfuge w/rotor for 1.5-ml tubes
37°C incubator
Rotating platform
Pipetters 20–100 μl
1.5-ml sterile Microfuge tubes
50-ml sterile polypropylene centrifuge tubes
2-ml glass Pasteur pipettes
Wide bore pipette tips 1–1000 μl
Pipette tips for pipetters
Saturated phenol (Fisher biotech grade)/0.5 ml
Chloroform (reagent grade)/1 ml
100% ethanol/1 ml
8 *M* potassium acetate (pH not adjusted)/50 μl
Proteinase K (Invitrogen, #25530-03)/10 mg
Sodium chloride, 100 g
Sodium citrate, 100 g
Sodium dodecyl sulfate (99% pure)/20 g

- 20% SDS (w/v) stock solution (100 ml)
  20.0 g SDS

Dissolve 20 g of SDS in water to obtain a final volume of 100 ml.

- Homogenization buffer (10 ml)
  10 ml 1× SSC
  625 μl 20% SDS

Add SDS solution to the 1× SSC in a sterile tube. Prepare for each sample to be extracted.

- TE, pH 7.5
  10 m*M* Tris, pH 7.5
  1 m*M* EDTA, pH 8.0

λ cII mutation assay

Equipment

Equipment required for the λ *cII* mutation assay is listed in Table 38.2.

Supplies and reagents

Supplies and reagents required for the λ *cII* mutation assay are listed in Table 38.3. Table 38.4 lists the amounts necessary for the analysis of one DNA sample.

In vitro packaging extracts:

λ bacteriophage *in vitro* packaging extracts (see Appendix I for protocol for preparation, or use commercially available reagents from Stratagene, La Jolla, CA).

Table 38.2 Equipment required for the λ *cII* mutation assay

37°C incubator
24°C incubator (± 0.5°C)
30°C shaking incubator
30°C water bath
55°C water bath
Refrigerated centrifuge (50-ml centrifuge tubes)
Temperature data-logger, accurate to 0.1°C
Autoclave
UV/Vis spectrophotometer
Microwave
Light box
Vortex
Electric pipetters
Pipetters 20–100 μl

Table 38.3 Supplies and reagents required for the λ *cII* mutation assay

Glass culture tubes and lids (12 × 75 mm)	Casein digest (Difco #211610/0116-17)
Petri dishes (100 × 15 mm plastic)	Thiamine hydrochloride (Invitrogen #13540-018)
Latex gloves	Gelatin (Sigma #G-2500)
10 ml sterile disposable serological pipettes	Magnesium sulfate 7-hydrate crystal
1.5 ml sterile Microfuge tubes	Kanamycin powder
Pipetters 20–100 μl	Maltose (Difco #216830/0168-17)
Wide bore pipette tips for 0–200 μl	Trizma base (Sigma #T1503)
Sterile loop for cell culture	Hydrochloric acid
Agar (Invitrogen Select Agar #30391-049)	

Table 38.4 Quantities of reagents used for analysis of one DNA sample using the λ *cII* mutation assay

Reagents	Quantity
TB-1 agar plates	13
10 m*M* $MgSO_4$	100 ml
SM buffer	100 ml
20% maltose-1 *M* $MgSo_4$	100 μl
TB-1 liquid media	10 ml
TB-1 Top agar	40 ml
TB-1/Kan+ plates	1

Solutions:

- 20% (w/v) maltose–1 *M* $MgSO_4$ (100 ml)
  20 g maltose
  1.6 g $MgSO_4 \cdot 7H_2O$

Dissolve in water and bring to a final volume of 100 ml. Filter sterilize and store at 4°C for 6 months.

- 10 m*M* $MgSO_4$ (1 l)
  2.46 g $MgSO_4 \cdot 7H_2O$

Dissolve in water and bring to a final volume of 1000 ml. Sterilize in an autoclave for 30 min at 121°C and store at room temperature for up to 1 year.

- 1 *M* Tris buffer, pH 7.5 (1 l)
  121.1 g Trizma® base

Dissolve in 800 ml of water and add approximately 65 ml of hydrochloric acid to adjust the pH to 7.5. It is important that the pH is measured at room temperature using a pH probe that will accurately measure the pH of Tris buffers. Adjust the final volume to 1000 ml. Sterilize in an autoclave for 30 min at 121°C and store at room temperature for up to 1 year.

- SM buffer (1 l)
  - 5.8 g NaCl
  - 2 g $MgSO_4 \cdot 7H_2O$
  - 50 ml 1 *M* Tris buffer, pH 7.5
  - 5 ml 2% (w/v) gelatin

Dissolve in water and bring the final volume to 1000 ml. Sterilize in an autoclave for 30 min at 121°C and store at room temperature for up to 1 year.

Media preparation:

All media should be prepared at least 24 h in advance of conducting the assay. The selection of casein digest and agar used to prepare the media is important to the success of the mutation assay. It is critically important that Difco brand casein digest is used. Other brands have been shown to alter mutant selection or quality of the mutant screening plates. Select agar obtained from Invitrogen is also recommended for preparation of the media.

- TB-1/Kan+ plates (250 ml = 6 plates)
  - 1.25 g NaCl
  - 2.5 g casein digest
  - 3.0 g agar
  - 250 μl 0.1% thiamine HCl

Dissolve in approximately 200 ml of water and adjust the pH to 7.0. Adjust the final volume to 250 ml. Sterilize in autoclave for 30 min at 121°C. Cool to 55°C.

12.5 mg kanamycin

Add the kanamycin after cooling media. Mix on stir plate, then pour 40 ml into each Petri dish. Store plates in plastic sleeves at 4°C for up to 2 weeks.

- TB-1 agar plates (1 l = 25 plates)
  - 5 g NaCl
  - 10 g casein digest
  - 12 g agar
  - 1 ml 0.1% thiamine HCl

Dissolve in approximately 800 ml of water. Adjust the pH to 7.0. Adjust the final volume to 1000 ml. Sterilize in an autoclave for 30 min at 121°C. Cool to 55°C and pour approximately 40 ml/plate. Store plates upside down in plastic sleeves at 4°C for up to 2 weeks.

- TB-1 liquid media (1 l)
  - 5 g NaCl
  - 10 g casein digest
  - 1 ml 0.1% thiamine HCl

Dissolve in approximately 800 ml of water. Adjust pH to 7.0. Adjust the final volume to 1000 ml. Prepare 100-ml aliquots into small glass bottles. Sterilize in an autoclave for 30 min at 121°C. Store at room temperature for up to 3 months.

- TB-1 Top agar (1 l = 25 packaged DNA samples)
  - 5 g NaCl
  - 10 g casein digest
  - 7 g agar
  - 1 ml 0.1% thiamine HCl

Dissolve in approximately 800 ml of water. Adjust the pH to 7.0 before heating. Adjust the final volume to 1000 ml and heat on a stir plate to form a homogenous suspension. Prepare 100 ml aliquots using small glass bottles. Sterilize in an autoclave for 30 min at 121°C. Melt completely in microwave before using. Each DNA sample uses 40 ml of top agar (~2.5 ml/plate). Store at room temperature for up to 3 months.

Sequencing of λ cII– *mutants*

Equipment

- Thermocycler
- Gel electrophoresis system
- Microfuge

Supplies and reagents

Supplies and reagents required for sequencing are listed in Table 38.5.

Procedures

Overview of λ cII *mutation assay procedures*

The *cII* mutation assay is based on the role the *cII* protein plays in the commitment of bacteriophage λ to the lysogenic cycle in *Escherichia coli* host cells (Figure 38.1). To select the mutant λ *cII–*, a specialized strain of *E. coli* (*hfl–*) is used to extend the longevity of the *cII* product. After isolation of fish genomic DNA, *in vitro* packaging procedures simultaneously excise and package the vector into viable phage particles. To determine the total number of packaged phage, a sub-sample of a package of DNA solution is mixed with *E. coli* cells, mixed with top agar and incubated on titer plates at 37°C overnight. To select *cII* mutants, the remaining packaged phage are mixed with *E. coli* host cells, plated, and incubated at 24°C for 40 h. The phage with wild-type *cII* produce lysogens and are indistinguishable in the *E. coli* lawn, whereas phage that carry a mutation in *cII* form plaques on the bacterial lawn when incubated at 24°C. Mutant frequencies are calculated by dividing the total number of *cII* mutant plaque forming units (PFUs) on the selective mutant screening plates by the estimated total λ+ and *cII* assayed on the titer plates.

Table 38.5 Supplies and reagents required for sequencing of λ *cII* mutants

Wide bore pipette tips	TB-1/Kan+ plates, 1
Glass culture tubes and lids (12 × 75 mm)	TB-1 liquid media, 100 ml
PCR nucleotide mix (Roche #1814362)	Dimethyl sulfoxide (DMSO)
Vent DNA polymerase (New England Biolabs)	SM buffer
TB-1 plates, 1 per sequence	

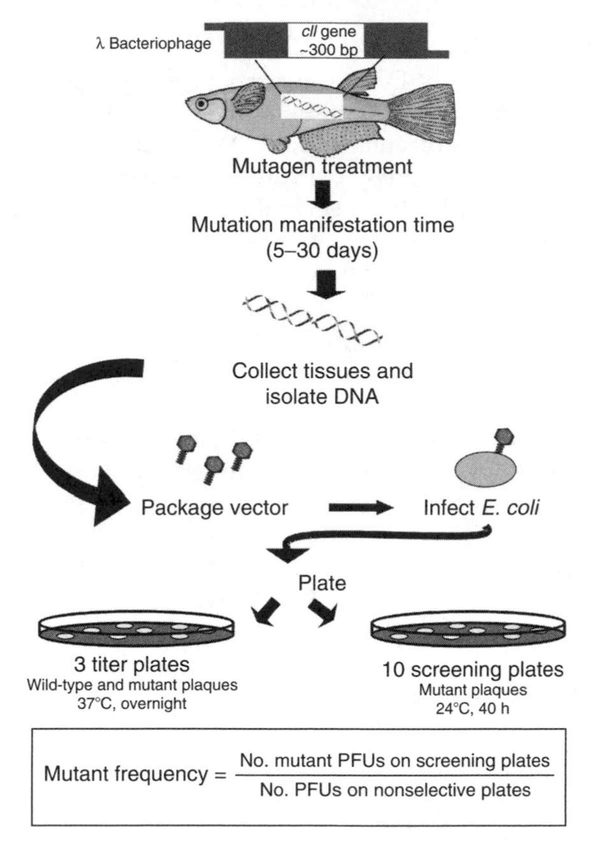

Figure 38.1 Overview of the bacteriophage λ-based transgenic medaka mutation assay using the *cII* gene (~300 bp) as a mutational target. After mutagen treatment and allowing sufficient time for mutations to manifest (~5–30 days), tissues are collected, and genomic DNA is isolated. The λ vector (~45 kb) is excised and packaged using *in vitro* packaging procedures. Individually packaged phage infect the G1250 *E. coli* host extending the longevity of the *cII* product to facilitate selection of mutant *cII* incubated at 24°C. λ Phage with a mutation in *cII* are selected by forming plaques while phage containing wild-type *cII* produce lysogens and are indistinguishable in the *E. coli* lawn. To determine the total number of plaques screened, a dilution of the infected *E. coli* is incubated at 37°C. The *cII*–mutant frequency is the ratio of the number of mutant λ *cII*– plaques selected to the total number of plaques screened.

Mutagen treatment protocols

A variety of methods, including static, single-pulse exposures, multiple-pulse exposures, and sophisticated chronic exposures using flow-through exposure systems used commonly for testing carcinogenicity in small fish species, may be appropriate for exposing λ transgenic medaka to chemical mutagens. See the section "Results and discussion" for additional comments about this subject.

Dissection of tissues

Following euthanization, standard procedures commonly used to dissect tissues from fish may be used. However, care should be taken to reduce potential DNA degradation by quickly removing tissues/organs and immediately placing samples on ice for immediate DNA isolation, or preferably, flash freezing in liquid nitrogen.

Isolation of high molecular weight DNA from dissected tissues

Isolation of high molecular weight DNA is essential for efficient *in vitro* packaging of the bacteriophage, and the ultimate success of the mutation assay. The following protocol has been optimized to isolate high molecular weight DNA from various fish tissues, including liver, eyes, skin, testes, and fins, thereby permitting mutation analyses from single tissues without pooling multiple samples from several individuals. Other DNA extraction procedures may not provide DNA of sufficient quality and quantity for this application. The authors may be contacted for procedures required to prepare DNA from whole fish.

1. Add tissue to a 1.5-ml Microfuge tube containing 500 μl of homogenization buffer, then add 15 μl of cold proteinase K solution to sample. Vortex briefly to mix.
2. Using small scissors finely mince each tissue into several pieces. Alternatively, use sterile pestles for 1.5-ml tubes to homogenize the tissue.
3. Immediately place samples into 37°C incubator and digest on rotating platform for 15–30 min or until tissue has digested.
4. Add 500 μl phenol/chloroform to digested tissue and invert gently 5 times.
5. Centrifuge samples at 1300 *g* at 4°C for 10 min.
6. Remove supernatant using a wide bore pipette tip and place it into a clean Microfuge tube.
7. Add 50 μl of 8 *M* potassium acetate to supernatant. Gently mix by inversion, then add an equal volume of chloroform and gently invert 5 times to mix.
8. Centrifuge samples at 1300 *g* at 4°C for 10 min.
9. Remove supernatant using a wide bore tip and place it into a clean Microfuge tube. Add 1 ml 100% ethanol.
10. Invert gently approximately 10 times or until the DNA has precipitated completely.
11. If the DNA is visible, collect by spooling with a flame-sealed Pasteur pipette. Carefully remove excess ethanol from the pipette using a paper tissue (e.g., Kimwipe) without touching the DNA.
12. Resuspend DNA in 25–50 μl TE (pH 7.5) in a 1.5-ml Microfuge tube. Resuspension volume should be sufficiently low as to maintain DNA in a highly viscous solution.
13. Alternatively, if the DNA is not visible, spin sample at 1300 *g* at 4°C for 10 min. Decant ethanol, dry tube around pellet. Do not over dry the pellet. Resuspend in 10 μl TE and, if necessary, make only minor adjustments to resuspension volumes to avoid over diluting the DNA.
14. Samples may remain on bench-top overnight to resuspend. Do not pipette the DNA for at least 12 h. Mix the samples the next day by pipetting up and down several times with a wide bore tip. Standard methods used to quantify the DNA concentration are not necessary, as they do not typically provide reliable predictors of potential success in the efficiency of packaging the DNA for this assay. Store the DNA at 4°C until use. Do not freeze the DNA.

λ cII mutation assay procedures

Schematic overview of procedures

- Day 1: streak TB-1/Kan+ plate with G1250 *E. coli* host strain

- Day 2: prepare G1250 liquid culture
- Day 3: package DNA samples and plate
- Day 4: count plaques on titer plates; calculate recoveries
- Day 5: count plaques on mutant screening plates; calculate mutant frequencies

It is recommended that 2–14 samples be analyzed at a time. An assay should ideally consist of a blocked design for each analysis, in which at least a sample from each treatment is analyzed simultaneously to account for potential variability in assay procedures (i.e., samples from all controls or all treatments should not be analyzed separately).

Day 1, preparation of the G1250 streak plate:

1. Using a sterile loop, streak a TB-1/Kan+ plate with frozen (−80°C) G1250 stock cells.
2. Incubate the streak plate upside down at 30°C for 24 h. Following incubation, wrap the G1250 streak plate in parafilm and store upside down in refrigerator. If properly stored, a streaked plate can be used for up to 1 week.

Day 2, preparation of G1250 liquid culture:

1. Add 10 ml TB-1 liquid media and 100 μl 20% maltose–1 *M* $MgSO_4$ to a sterile 50-ml conical tube.
2. Transfer 3 G1250 colonies from streak plate into the tube using a sterile pipette tip (be sure lids on the cultures are loose, so that air exchange will occur).
3. Shake culture at 265 rpm in a 30°C incubator for 16–21 h.

Prepare G1250 liquid culture 16–21 h before plating (e.g., 5:00 p.m.). Prepare one 10-ml culture for every four DNA samples (e.g., if plating 12 DNA samples, make 3 10-ml cultures).

Day 3, packaging and plating the DNA sample: Procedures are described in the following for using packaging extracts that are prepared in the laboratory. See Appendix I for procedures. Alternatively, follow the manufacturer's recommendations for using commercially available extracts.

1. Adjust water bath temperature to 30°C. Monitor the water bath to ensure it has stabilized at 30°C prior to the following steps.
2. Prior to performing the first packaging reaction, remove 13 TB-1 plates per sample from the refrigerator, shake out excess water, and lay out (upside down) to warm and dry.
3. Turn on 55°C water bath.
4. Label and lay out plates consisting of 3 titer plates (e.g., T-20 titer plates) and 10 mutant screening plates per sample (Table 38.6).
5. Set up a rack with 13 sterile glass culture tubes per sample, and label the same as the 13 plates for each sample.

Part 1, packaging reaction:

1. Using a wide bore tip, mix DNA samples to be plated by pipetting up and down 5–6 times. Aliquot 10 μl of DNA from each sample into a new Microfuge tube.

Table 38.6 Summary of materials and conditions for plating a packaged DNA sample

Tube labels	G1250 plating culture (μl)	1:100 dilution of packaged DNA	Undiluted packaged DNA (μl)	Corresponding TB-1 plates	Incubation temperature (°C)	Incubation time
T-20 A	200	20 μl of rep. A	—	T-20 A	37	O/n
T-20 B	200	20 μl of rep. B	—	T-20 B	37	O/n
T-20 C	200	20 μl of rep. C	—	T-20 C	37	O/n
Screening 1–10	200	—	100	Screening 1–10	24	40 h

2. At approximately 12:00 noon (13–18 of the G1250 liquid culture), begin the packaging reaction by removing the packaging extracts from the −80°C freezer. 1 Tube A and 1 Tube B will be needed for every 2 samples. Thaw the tubes quickly, centrifuge briefly to collect contents and, using a wide bore pipette tip, combine contents of Tube A (40 μl) and Tube B (60 μl). Mix by pipetting up and down 6 times. One tube of combined packaging extracts will package 2 DNA samples.
3. Using a wide bore pipette tip, pipette 50 μl of the combined packaging extracts into each DNA sample and mix by pipetting up and down 6 times. It is very important not to introduce bubbles into the mixtures, as these will reduce packaging efficiency.
4. Incubate at 30°C for 1.5 h.
5. Following the 1.5-h incubation period, continue the packaging reaction by repeating steps 2 and 3 above except using a standard pipette tip instead of a wide bore tip.
6. Melt top agar in microwave oven. Place melted top agar in 55°C water bath. It is very important that the agar be completely melted. Swirl and inspect to verify no unmelted pieces of agar are present.
7. 1 h following step 5 (above) begin preparation of the G1250 plating culture by centrifuging the tubes of G1250 overnight culture at 3200 *g* for 10 min at 4°C.
8. Decant supernatant and resuspend pellet in 10 ml of 10 m*M* $MgSO_4$.
9. Using a spectrophotometer, measure absorbance ($OD_{600\,nm}$) by adding 100 μl of cells to 900 μl of 10 m*M* $MgSO_4$. Multiply absorbance by 10 for OD value.
10. Dilute cell suspension to OD = 0.5.
11. Place cells on ice until ready for use. This is the G1250 plating culture.
12. Following the second 1.5-h incubation, add 990 μl sterile SM buffer to the samples.
13. Vortex for 10 s.
14. Place samples on ice. These are referred to as the packaged DNA samples.

Part 2, plating of λ cII:

1. For the first DNA sample, add 200 μl of the G1250 plating culture into each of the 13 sterile glass culture tubes for the packaged DNA sample.
2. Label 3 Microfuge Tubes A, B, and C.
3. Add 990 μl SM buffer to each one.
4. Add 10 μl of packaged DNA to Tubes A, B, and C and vortex to mix.
5. Add 20 μl from Tube A to 1 of the 3 culture tubes representing the T-20 titer plates. Repeat with Tubes B and C.

4. Add 100 μl of the undiluted packaged DNA sample to each of the 10 remaining culture tubes for the 10 mutant screening plates.
5. Incubate at room temperature for 30 min.
6. Repeat steps 1–5 above for each of the remaining DNA samples. Set timers such that all samples are incubated for exactly 30 min. It is important to finish these steps for all DNA samples before the 30-min incubation is completed for the first sample.
7. Following the 30-min incubation, add 2.5 ml of top agar to each culture tube, mix, and immediately pour onto an appropriately labeled TB-1 plate while swirling gently to evenly distribute the top agar.
8. Repeat for remaining samples. Allow at least 5 min for the agar to solidify on each plate.
9. Transfer plates to appropriate incubators within 30 min of plating. Incubate the titer plates upside down at 37°C overnight. Incubate the mutant screening plates upside down at 24°C for 40 h. Monitor the temperature in this incubator using a data-logger. Internal incubator temperature should vary no more than ± 0.5°C (Table 38.6).

Day 4, count titer plates and calculate recovery:

1. Following the overnight 37°C incubation, count PFUs on titer plates to determine the total number of λ bacteriophage recovered from the fish genomic DNA. A light box partially covered with black paper is helpful to assist in the counting. PFUs are easiest to see in the contrast between the light and black background. Use the following to determine the recovery for each sample:

$$\text{recovery} = \frac{\text{(mean No. plaques/titer plate)}}{\text{(No. } \mu\text{l of dilution/titer plate)}} \times \text{ dilution} \times (100\ \mu\text{l/plate}) \times 10 \text{ plates}$$

   Example: Using a mean number of 161 PFUs counted per plate in the T-20 titer plate group, and 20 μl of a 100-fold dilution plated on each T-20 titer plate, the total number of PFUs recovered is 161/20 μl of diluted phage per titer plate × 100 (the dilution factor) × 100 μl of undiluted phage per mutant screening plate × 10 mutant screening plates = 805,000 total plaques screened.

2. After counting PFUs, plates may be discarded. Sterilize plates (original TB-1/Kan+ plate, titer plates, and unwanted mutant screening plates) in an autoclave at 121°C for 30 min prior to discarding.

Day 5, count PFUs on mutant screening plates and calculate cII– mutant frequency:

1. Count PFUs on mutant screening plates at 40 h. This can be best achieved by holding the plates next to a lamp. Individual PFU may vary in size. Be careful not to count plating artifacts (e.g., bubbles). The *cII* mutant frequency is calculated as follows:

$$\text{mutation frequency} = \frac{\text{total No. PFUs on screening plates}}{\text{total No. PFUs recovered}}$$

Example: Using a total of 23 PFUs counted on the 10 mutant screening plates, and a total recovery of 805,000 PFUs for the sample, the mutation frequency is $23/805{,}000 = 2.8 \times 10^{-5}$.

2. Mutant screening plates should be saved for later coring, mutant verification, and sequencing procedures if desired. Store plates by placing upside down in Petri dish bags at 4°C.

Sequencing of λ cII– mutants

Schematic overview of procedures

1. Core mutant plaques and elute phage
2. Re-plate eluted phage and incubate plates at 24°C for 40 h
3. Perform PCR amplification of re-plated mutants
4. Sequence samples following the appropriate protocol for available sequencing system

Day 1:

1. Determine percentage of plaques to be cored from each plate.
2. Add 500 μl sterile SM buffer into a sterile Microfuge tube for each core.
3. Core an individual plaque from a plate using a 1000-μl wide bore pipette tip by stabbing the tip into the plate.
4. Place entire agar plug containing the plaque into a microcentrifuge tube containing SM buffer.
5. Repeat for remaining samples.
6. Elute phage at 4°C overnight (for long-term storage, transfer an aliquot of the eluted phage to a clean microcentrifuge tube, add DMSO to a final concentration of 7% [v/v], and store at −80°C).
7. Prepare one 10-ml G1250 liquid culture for every 50 plaques re-plated (see Day 2 of the λ *cII* mutation assay).

Day 2:

1. Prepare a G1250 plating culture (see steps 7–11 on Day 3 of the λ *cII* mutation assay).
2. Add 200 μl of the G1250 plating culture to each culture tube.
3. Add 1 μl of eluted phage solution to culture tube, mix, and incubate for 30 min at room temperature.
4. Following the 30-min incubation, add 2.5 ml top agar to each culture tube, pour onto plate, and swirl to distribute evenly.
5. Incubate the plates upside down at 24°C for 40 h.

Phage isolation and PCR amplification for sequencing:

1. Determine number of samples to be sequenced.
2. Pick an individual plaque from a plate containing the re-plated phage (verification plate) using a 200-μl wide bore tip and place into 25 μl sterile water in a

screw-top tube. It is very important to scrape only the top layer of the plate containing the plaque. Avoid taking any agar, as it will inhibit the PCR amplification.

3. Repeat for remaining samples to be sequenced.
4. Prepare a master mix for the PCR reaction to obtain the following concentration of reagents in a final 25-μl volume:
   - 1× PCR buffer
   - 200 μ*M* dNTPs
   - 0.5 μ*M* primers (*cII* primer 1: 5′ to 3′ AAA AAG GGC ATC AAA TTA AAC C; *cII* primer 2: 5′ to 3′ CCG AAG TTG AGT ATT TTT GCT GT)
   - 1 unit of DNA polymerase
5. Place tubes containing the scraped phage into boiling water for 5 min.
6. Centrifuge tubes for 3 min at maximum speed.
7. Immediately pipette 2.5 μl of supernatant into the PCR reaction.
8. Perform the PCR reaction using the following amplification conditions:
   - 95° for 3 min
   - 30 cycles of:
   - 95° for 1 min
   - 55° for 1 min
   - 72° for 1 min
   - followed by 72° for 10 min
   - 4°C storage

Results and discussion

Considerations for conducting mutation analyses

A number of factors that may influence the induction of mutations in animals must be considered when designing and conducting an assessment of mutations using the λ transgenic medaka (Figure 38.2). Genotoxicity will be a function of uptake, distribution, detoxification, metabolic activation, DNA repair, and the types of DNA adducts formed. These factors may be modified by, for example, age of animals at the time of exposure, concentration of chemical at exposure, duration of exposure, and the time of sampling after exposure. Whereas some of the variables that influence the induction of mutation are specific to detecting mutations using the λ *cII* mutation assay, others (e.g., routes of exposure, duration of exposure, chemical concentrations, and characteristics of the chemical) are similarly considered when designing and conducting toxicity assessments using different endpoints. The following describes some of the more important factors related to performing mutation analysis using the λ transgenic medaka. Many of the features of detecting mutations in the λ transgenic medaka are shared with the transgenic rodent mutation assays. Consequently, additional guidance to addressing these mutation analyses in fish may be obtained from the transgenic rodent literature.[16]

Mutagen treatment regimen

Small laboratory fish species, including the λ transgenic medaka, provide significant flexibility in their use in toxicity assessments. Depending on the characteristics of the test compound (e.g., potent mutagens) and research aims (e.g., screening potential mutagenicity of a chemical), short-term chemical treatments conducted over one to several hours may be sufficient to achieve significant inductions of mutations. However, some

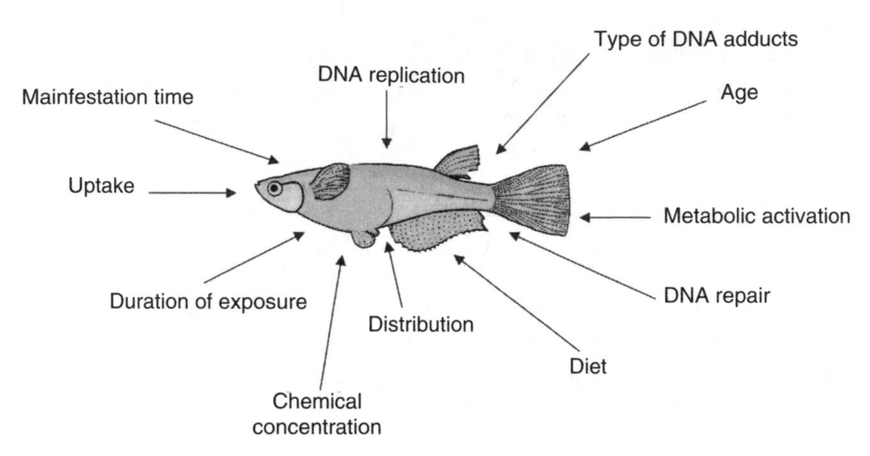

Figure 38.2 Schematic overview of important factors that affect or modify the induction of mutations to be considered in the design and conduct of *in vivo* mutation analyses.

chemicals may, in addition to producing DNA adducts, induce cell proliferation. Mutations in transgenic mutation assays are detected using genetically neutral targets, thereby avoiding the potential selective pressures on the mutant frequency *in vivo* and allowing the accumulation and persistence of mutations.[12–14] As a consequence of the accumulation of mutations in these loci over time, the use of repeated or chronic chemical treatments will increase the sensitivity of the mutation assay. A chemical exposure regimen consisting of repeated treatments will facilitate the administration of the chemical during those periods, in which cell proliferation may have been induced by an earlier exposure. Consequently, chemically induced mutations should be studied ideally under conditions of sub-chronic or chronic administration if resources are available. Alternatively, short-term multiple doses protocols, for example, at a minimum of 3 days, and up to 2–3 weeks, to allow chemical administration in the presence of chemically induced cell turnover, could be considered. Thus far, data from studies on the transgenic medaka indicate that mutations accumulate linearly using daily chemical treatments, including exposures extending greater than 9 months (unpublished data).

The selection of the concentrations of the chemicals to be administered should be determined from standard range finding tests of toxicity. However, in considering the potential adverse toxic effects of compounds administered at high concentrations, and the possible need to increase number of treatments when exposing at low chemical concentrations, a compromise between the number of treatments and the exposure concentration may have to be made.

Numbers of animals per treatment

The efficient recovery of the λ bacteriophage from the λ transgenic medaka genomic DNA (typically >300,000 PFUs/sample) and, the low variability in mutational responses observed among animals within treatments, contribute to the ability to use relatively small numbers of animals per treatment. Statistical analysis revealed that as few as 6–7 animals may be required to detect a 50% induction of mutations above the spontaneous *cII* mutant frequency (e.g., $2–5 \times 10^{-5}$ for most tissues), and 2–3 animals may be required to detect a 100% induction of mutations.[10] As a general rule, 10–15 fish per treatment should provide sufficient material for analyses. Additional tissues can be held frozen for further analyses as needed.

Mutation manifestation time

The interval between mutagen treatment and sample processing, termed mutation manifestation time, is an important factor to consider in the design and interpretation of mutation studies.[17–19] The mutation manifestation time is important because DNA replication is necessary for genetic damage to become fixed as a mutation. Consequently, mutation manifestation time will directly influence the frequency of mutations observed. The time for mutations to manifest is affected by several variables, including tissue/cell type (e.g., rate of cell turnover), mutagen (e.g., potent or weak), and mutagen treatment regimen (e.g., number of treatments). For example, a tissue with a rapid cell turnover, such as intestinal epithelium, sampled within several days of a short-term exposure to a potent mutagen would likely exhibit a significant induction of mutations. In contrast, when using a weakly mutagenic compound and/or when using a target tissue with slower cell turnover, such as liver, a manifestation time greater than 30 days may be required. Sampling times may have to be determined on a case-by-case basis depending on the variables. As a guide, when collecting multiple tissues, all tissues should be sampled at the time appropriate for the tissue with the slowest cell turnover. Thus far, results indicate that a mutation manifestation time of 15 days may be sufficient time to detect a significant fold induction of mutations over a spontaneous frequency in most fish tissues, although a mutation manifestation time greater than 30 days may be required to detect induction by weak mutagens, and a 5-day sampling time is probably suitable only for the most potent mutagens.

Target tissues/organs

As with other toxicity assessments, the choice of target tissue/organ to be examined in the *cII* mutation assay will be dictated by the experimental questions. The liver is among the more widely studied target organs in a variety of toxicological assessments. Analyses of mutations in the liver of λ transgenic medaka have proven valuable in demonstrating that many aspects of *in vivo* mutagenesis, including the modes of mutagen action of several chemicals, are shared with transgenic rodent models.[10,11]

The efficiency with which the λ bacteriophage is recovered from the fish genomic DNA is one of the most important factors that can influence the ability to perform the *cII* mutation assay. DNA extraction procedures and other protocols described here have been optimized to assist in isolating high molecular weight DNA required for λ bacteriophage *in vitro* packaging from virtually any medaka tissue, including liver, testes, eye, fin, blood, skin, embryos, and fry, without the need to pool DNA from multiple samples.

Number of PFUs needed to determine mutant frequency

As the number of PFUs recovered from the fish genomic DNA approaches ~300,000 PFUs/sample, the average spontaneous mean frequency will become less variable. Consequently, based on experience, at least 300,000 PFUs from each tissue should be analyzed to calculate the spontaneous or induced mutant frequency.

Spontaneous mutant frequency

The measure of the frequency of spontaneous *cII*– mutants provides the basis for comparing mutations induced after mutagen exposure. The sensitivity of a mutation assay is defined by the magnitude of the induced mutational response compared to the frequency of spontaneous mutations. Spontaneous mutant frequencies will vary among different tissues with the lowest mean mutant frequencies exhibited by testes ($1–2 \times 10^{-5}$), followed by whole fish ($3–4 \times 10^{-5}$), and then liver ($4–6 \times 10^{-5}$).

Statistics

Application of statistics to data obtained from transgenic mutation assays is consistent with statistical approaches used commonly in toxicity studies. Nonparametric tests, such as the generalized Cochran–Armitage test, are generally used with data from these assays.[16] Typically, a positive response will be indicated if the mean mutant frequency of a treatment group exceeds at least 2 times that of an untreated and/or historical control.

Sequencing of mutants

One of the strengths of the *cII* mutation assay is the ability to obtain information about the types of mutations induced. The small size of the *cII* target gene (~300 base pairs, bp) facilitates efficient characterization of the mutations by direct sequencing of the entire gene. Mutations detected in λ-based systems consist primarily of point mutations, base pair substitutions, with some frameshifts, and small insertions and deletions. Although sequencing data may not be required in all cases, sequencing can be very informative. In addition to the importance of using sequencing to verify a λ *cII*– mutant, sequencing may provide important information on the spectra of mutations that may be highly characteristic of specific compounds, as well as may disclose possible mechanisms of mutagen action. When only marginal increases of mutations may be observed, sequencing may disclose actual mutation induction. In this instance, an induction of a specific class may be too small in absolute numbers to be apparent in the overall mutant frequency, but sequencing may reveal it. Similarly, sequencing may be particularly useful in examining whether small increases in mutant frequencies after exposure to chemicals at low environmental concentrations are accompanied by shifts in mutational spectra. Sequencing will also aid in reducing the variability in mutant frequencies by identifying and removing clusters or "jackpot" mutants that have arisen from a single mutation.

Examples of studies of chemical mutagenesis

The following are examples of studies of induced mutagenesis in the λ transgenic medaka using four chemicals, ethylnitrosourea (ENU), dimethylnitrosamine (DMN), diethylnitrosamine (DEN), and benzo(*a*)pyrene (BaP). These studies illustrate many of the features of the *cII* mutation assay based on medaka, and highlight some of the important issues to be considered when conducting studies of *in vivo* mutagenesis.

Ethylnitrosourea

The feasibility of the *cII* mutation assay in detecting induced mutations in the λ transgenic medaka was first demonstrated by using the model mutagen, ENU.[10] ENU is a well-characterized, and widely used direct-acting alkylating agent, and potent germ-cell mutagen. Using a static exposure regimen, fish (6–10 hemizygous, 2- to 6-month old transgenic fish/treatment) were immersed 1 h in water containing 0, 60, or 120 mg/l of ENU. Fish were rinsed, transferred to clean water, and held for 5, 15, 20, or 30 days until euthanized with an overdose of MS222. Fish treated similarly without mutagen exposure served as controls. Whole fish, or dissected tissues (liver, testes), were flash frozen in liquid nitrogen and stored at −80°C until processed for DNA isolation.

The *cII* target gene was highly responsive to ENU treatment, exhibiting responses that were dependent on chemical exposure concentration. In whole fish, sampled 15 days after exposure, *cII* mutant frequencies were induced significantly, 2.7-fold ($8.0 \pm 1.0 \times 10^{-5}$), and 4-fold ($12.0 \pm 1.9 \times 10^{-5}$), over untreated fish ($3.0 \pm 0.3 \times 10^{-5}$) at 60 and 120 mg/l, respectively (Table 38.7). The time between mutagen exposure and tissue sampling, or mutation manifestation time, was shown to have a significant influence on the mutant frequencies, and the influence on mutant frequencies was tissue-specific. Mutant frequencies in liver of treated fish did not increase significantly above that of untreated fish ($4.4 \pm 0.7 \times 10^{-5}$) at 5 days ($5.1 \times 10^{-5}$), but increased significantly 3.5-fold (15.6×10^{-5}) at 15 days, 5.7-fold (25.1×10^{-5}) at 20 days, and 6.7-fold by 30 days after mutagen treatment. In testes, the mutant frequencies were elevated 5.2-fold above that of the untreated fish within 5 days after mutagen exposure, reaching a peak or threshold 10-fold induction at 15 days (Table 38.8). The relatively higher magnitude of mutation inductions and shorter mutation manifestation time observed in testes compared with liver reflect apparent differences in cell proliferation rates and/or mutagen action in these tissues. The different frequencies of mutations exhibited by these tissues at equivalent mutagen treatments illustrate the significant utility of this mutation assay for examining tissue-specific mutagenesis.

To characterize the spectra of spontaneous and mutagen-induced mutations, the phenotype of individual *cII* mutant phage obtained from the mutant screening plates was first verified using selective plating conditions. A 446-bp product, including the entire *cII* gene (~300 bp), was then sequenced for each of the isolated mutants. Sequence analyses revealed that single base substitutions comprised the majority of spontaneous and ENU-induced mutations in fish, with the large percentage of G:C to A:T mutations at CpG sites (Table 38.9). ENU is a direct acting mutagenic agent producing 0^6-ethylguanine, 0^4-ethylthymidine, and 0^2-ethylthymidine in DNA, promoting G:C to A:T and A:T to G:C transitions, as well as A:T to T:A transversions. The proportion of mutations at A:T base

Table 38.7 Frequencies of mutants recovered from whole fish treated with ENU[10]

Treatment	Mean mutant frequency $\times 10^{-5} \pm$ SEM (n = 6–9)
Control	3.0 ± 0.3
60 mg/l	8.0 ± 1.0
120 mg/l	12.0 ± 1.9

Table 38.8 Frequencies of mutants from medaka sampled at different times after exposure to 120 mg/l ENU[10]

	Mutant frequency $\times 10^{-5} \pm$ SEM			
	5 days (n)	15 days (n)	20 days (n)	30 days (n)
Control liver	3.8 ± 0.7 (9)	5.2 ± 0.6 (6)	4.8 ± 1.9 (4)	3.6 ± 0.5 (5)
Exposed liver	5.1 ± 0.8 (9)	15.6 ± 1.8 (6)	25.1 ± 2.3 (6)	29.8 ± 2.0 (5)
Control testes	2.1 ± 0.5 (5)	2.2 ± 0.3 (6)	2.9 ± 0.8 (5)	1.3 ± 0.1 (6)
Exposed testes	10.3 ± 1.1 (8)	19.6 ± 1.2 (8)	17.5 ± 1.6 (7)	16.4 ± 2.1 (9)

Table 38.9 Comparison of mutational spectra from untreated and mutagen-treated λ transgenic medaka

	Spontaneous	ENU 120 mg/l	DMN 600 mg/l	BaP 50 μg/l
Total mutations	89	39	44	98
Mutations outside of *cII*	5	3	6	5
Independent mutations[a]	74	36	34	81
	Percentage (*n*)	Percentage (*n*)	Percentage (*n*)	Percentage (*n*)
Transitions				
G:C to A:T	20 (15)	28 (10)	73 (25)	6 (5)
At CpG sites[b]	47 (7)	30 (3)	12 (3)	40 (2)
A:T to C:G	12 (9)	25 (9)	3 (1)	1 (1)
Transversions				
G:C to T:A	20 (15)	6 (2)	0 (0)	59 (48)
G:C to C:G	11 (8)	6 (2)	6 (2)	22 (18)
A:T to T:A	4 (3)	28 (10)	3 (1)	1 (1)
A:T to C:G	7 (5)	3 (1)	0 (0)	1 (1)
Frameshift				
+1	14 (10)	3 (1)	3 (1)	3 (2)
−1	11 (8)	0 (0)	9 (3)	3 (2)
Other	1 (1)	3 (1)	3 (1)	4 (3)

[a] Clonal mutations were subtracted.

[b] Percent occurring at transitions [e.g., 7 of the 15 G:C to A:T transitions (47%) were at CpG sites for the spontaneous mutations].

pairs increased from 31% in untreated fish to 59% in ENU-treated livers, with the bulk of the increase being A:T to T:A transversions. This shift in the frequency of mutations is characteristic of the greater mutagenic effect of ENU at A:T base pairs.[20,21] Also consistent with the known mode of action of ENU was the reduction in the frequency of frameshift mutations observed in ENU-exposed fish compared to that of the untreated animals. ENU has been shown to not induce high numbers of frameshift mutations.[22]

Dimethylnitrosamine

Induction of mutations in the λ transgenic medaka was examined after exposure to DMN, a potent liver carcinogen used extensively as a model mutagen in carcinogenesis studies.[11] DMN is among a class of nitrosamines that induced hepatic carcinogenesis in fish with progressive stages similar to those characterized in rodent hepatic neoplasia.[23] Studies in rodents demonstrated that cell proliferation is a requisite for DMN-produced methyl DNA adducts to become fixed as mutations.[24] The results demonstrate the potent mutagenicity of DMN in medaka and illustrate the importance of considering cell proliferation in the design of *in vivo* mutagenesis studies. Adult fish were exposed to DMN at 0, 300, or 600 mg/l administered in a 96-h static renewal exposure regimen. *cII* mutants were induced 7.1-fold ($16.7 \pm 2.8 \times 10^{-5}$), and more than 16-fold ($38.4 \pm 6.6 \times 10^{-5}$) above untreated fish ($2.1 \pm 0.25 \times 10^{-5}$) 15 days after exposure to DMN at 300 and 600 mg/l, respectively (Table 38.10).

Sequencing disclosed a different spectra of DMN-induced mutations compared to that induced by ENU, revealing different modes of action of these mutagens. Most notably, single-base substitutions were the most frequent mutations in liver of treated

Table 38.10 Frequencies of mutants recovered from medaka liver treated with DMN[10]

Treatment	Mean mutant frequency $\times 10^{-5} \pm$ SEM (n = 6–9)
Control	2.1 ± 0.25
300 mg/l	16.7 ± 2.8
600 mg/l	38.4 ± 6.6

fish, with a large percentage of G:C to A:T transitions. The higher frequency of G:C to A:T transitions compared to transversions is characteristic of DMN exposure (Table 38.9). These mutations are attributed to the mispairing of 0^6-methylguanine with thymine, leading to G:C to A:T transitions during replication.[25]

Diethylnitrosamine

DEN is a potent genotoxic carcinogen, and is one of the most widely used compounds to investigate the mechanisms of carcinogenesis in medaka having been shown to be a strong inducer of hepatic neoplasms.[26,27] To examine the potential mutagenicity of this compound, adult fish (3 months) were held in water containing 100 mg/l DEN for 48 h and then transferred to toxicant-free water for a 30-day mutation manifestation time. The potency of DEN as a mutagen was shown clearly in the livers of the treated fish, which exhibited a mean mutant frequency induced 65-fold ($136.4 \pm 18.24 \times 10^{-5}$) above that of untreated fish ($2.1 \pm 0.33 \times 10^{-5}$) (Table 38.11). DEN has been shown to require only a brief and relatively low concentration to initiate carcinogenesis in the medaka.[28] Considering the proven utility of medaka in carcinogenesis research, and the extensive use of DEN as a model compound for inducing hepatic neoplasms, the mutation data presented here illustrate the potential value of combining the use of λ transgenic medaka and *cII* mutation assay with the assessment of histopathological, and other endpoints in the fish,

Table 38.11 Mutant frequencies in liver of medaka exposed to DEN (48 h, 30-day mutation manifestation)

Treatment	Concentration (mg/l)	Total PFU	Mutant PFU	MF $\times 10^{-5}$	Group mean MF $\times 10^{-5} \pm$ SEM
Control	0	1,225,000	29	2.4	$x = 2.1 \pm 0.33$
		2,560,000	77	3.0	
		2,160,000	48	2.2	
		1,090,000	9	0.8	
		4,505,000	70	1.6	
		4,085,000	35	0.9	
		2,180,000	73	3.4	
		820,000	19	2.3	
DEN	100	3,775,000	6,249	165.5	$x = 136.4 \pm 18.24$
		3,065,000	3,930	128.2	
		2,480,000	2,022	81.5	
		1,490,000	2,543	170.7	
		2,810,000	5216	185.6	
		1,520,000	1319	86.8	

to enhance understanding of the initiation, promotion, and progression of hepatic carcinogenesis.

Benzo(a)pyrene

The responsiveness of the *cII* target to mutagen exposure was further examined using a mutagen that requires metabolic activation, BaP, a persistent contaminant in aquatic systems, and a widely used model compound in laboratory fish carcinogenesis studies[29] and transgenic rodent assays.[21,30] BaP induces the formation of DNA adducts following the metabolism by CYP1A to its ultimate carcinogenic metabolite, BaP 7,8-dihydrodiol-9, 10-epoxide (BPDE). In a simple static-renewal exposure regimen, medaka adults were treated with water containing acetone as a vehicle and 0 or 50 μg/l BaP over 96 h (6 h/day). The fish were then allowed a 15-day mutation manifestation time. Mean mutant frequencies in the liver of fish treated with BaP were induced nearly 5-fold ($14.1 \pm 4.69 \times 10^{-5}$) above that of the controls ($2.97 \pm 0.30 \times 10^{-5}$) (Table 38.12).

Sequence analysis revealed that single base substitutions comprise the majority of spontaneous and BaP-induced mutations in fish (Table 38.9). The majority of mutations at G:C base pairs in BaP-treated fish were G:C to T:A transversions (59%), a spectral shift consistent with the greater mutagenic effect of BaP at G:C base pairs.[21] The BaP metabolite, BPDE, forms covalent bonds at the N2 position of guanine, thereby resulting in point mutations at G:C base pairs, which can lead to oncogene activation. Similar significant increases in G:C to T:A transversions in BaP-treated transgenic mice compared controls have also been observed.[31] Also consistent with previous reports in rodent models, a large proportion of the mutations in the fish were located at CpG sites.

Summary

Studies to date support the continued use of the λ transgenic medaka and the associated *cII* mutation assay for assessing spontaneous and induced *in vivo* mutagenesis. The λ transgenic medaka exhibits chemical concentration-dependent, and tissue-specific

Table 38.12 Mutant frequencies in liver of medaka exposed to BaP (96 h, 15-day mutation manifestation)

Treatment	Concentration (μg/l)	Total PFU	Mutant PFU	MF $\times 10^{-5}$	Group mean MF $\times 10^{-5} \pm$ SEM
Control	0	380,000	12	3.1	$x = 2.9 \pm 0.3$
		1,560,000	61	3.9	
		1,930,000	57	2.9	
		1,518,000	59	3.8	
		936,000	23	2.4	
		1,603,000	28	1.7	
BaP	50	1,238,000	275	22.2	$x = 14.1 \pm 2.1$
		2,460,000	336	13.6	
		1,030,000	105	10.2	
		545,000	66	12.1	
		323,000	40	12.4	

responses to mutagen exposures reflecting known modes of mutagen action. The mutation analyses, when used in combination with measures of other toxicological endpoints, promise to be useful in more fully characterizing health risks associated with chemical exposures in the environment. In addition, the practical benefits afforded by this mutation assay and the medaka as an animal model should facilitate in-depth investigations into the mechanisms of mutagenesis and carcinogenesis.

Acknowledgment

This research was supported by grants from the National Center for Research Resources (NCRR), National Institutes of Health.

Appendix I. Preparation of λ packaging extracts

The *in vitro* packaging materials are essential to the simultaneous excision of the λ bacteriophage from the transgenic fish genomic DNA, and packaging of individual λ bacteriophage. The following protocol consists of a series of steps that may be completed over 5 days. Note that two different cell extracts are prepared in these procedures. Dr. Peter Glazer (Yale University) kindly provided the cell stocks and the original protocols adapted here for preparing these packaging extracts. Dr. Glazer should be contacted to obtain these materials.

Day 1: Prepare LB plates.
Day 2: Streak LB plates with BHB 2688 and NM 759 stock cells
Prepare solutions and NZY media
Autoclave glassware and centrifuge bottles
Day 3: BHB extracts (Tube A), inoculate NZY media with BHB colony.
Day 4: Perform BHB freeze–thaw preparation.
Day 5: Perform ultracentrifugation of frozen BHB cell mixture from Day 4
Freeze BHB cell extracts
Day 6: NM extracts (Tube B), inoculate NZY media with NM colony.
Day 7: Make and freeze NM cell extracts.

Materials required

Equipment

Equipment required for preparation of λ packaging extracts is listed in Table 38.A1.

Supplies and reagents

Supplies and reagents required for preparation of λ packaging extracts are listed in Table 38.A2. Table 38.A3 lists the amounts necessary for preparation of approximately 70 extracts.

Table 38.A1 Equipment required for preparation of λ packaging extracts

30°C incubator
42°C incubator
30°C shaking incubator
32°C shaking incubator
Sonicator (Fisher Scientific Sonic Dismembrator Model 100, setting 3)
Refrigerated centrifuge with rotor capable of spinning 250 ml bottles at 3200 *g*
Ultracentrifuge with Beckman 75 TI rotor or equivalent
65°C water bath
45°C water bath
Vortex
Spectrophotometer

Solutions:

- 1 *M* TrisCl (250 ml)
  30.27 g Tris-base

Dissolve in 100 ml water. Adjust pH to 8.0 using a pH probe that will accurately measure Tris solutions. Adjust the final volume to 250 ml. Sterilize in an autoclave for 30 min at 121°C and store at room temperature for up to 1 year.

- 1 *M* $MgCl_2$ (250 ml)
  50.8 g $MgCl_2$
  Filter-sterilize
  Store at room temperature for 1 year

- 0.1 *M* ATP (1 ml)
  0.06 g ATP
  1 ml H_2O
  Filter-sterilize

- 10% Sucrose solution (10 ml)
  1 g sucrose
  500 μl 1 *M* TrisCl, pH 8.0
  Filter-sterilize; store at 4°C for 3 months

- Lysozyme solution (10 ml)
  0.02 g lysozyme
  100 μl 1 *M* TrisCl, pH 8.0
  Filter-sterilize; store at 4°C; make fresh every time

- Sonication buffer (10 ml)
  200 μl 1 *M* TrisCl, pH 8.0
  20 μl 0.5 *M* EDTA
  3.5 μl B-mercaptoethanol
  Filter-sterilize; store at 4°C for 3 months

Table 38.A2 Supplies and reagents required for preparation of λ packaging extracts

Sterile 1.5-ml centrifuge tubes
Sterile 50-ml centrifuge tubes
Sterile 250-ml centrifuge bottles (6–8)
250-ml Erlenmeyer flasks (2)
1-l Erlenmeyer flasks (4)
5-ml serological pipettes
25-ml serological pipettes
Sterile foam stoppers
Chloroform
Timers
Thermometer
Liquid nitrogen
Loop for cell culture
16 × 76 mm tubes for ultracentrifuge (Nalgene #3425-1613)
Casein digest (Difco #211610/0116-17)
Agar (Invitrogen Select Agar #30391-049)
Yeast extract (Invitrogen #30393-029)
NaCl
$MgCl_2 \cdot 6H_2O$
ATP (Roche #519987)
β-mercaptoethanol
Sucrose
Lysozyme
Spermadine
Putrescine

- Packaging buffer (2.5 ml)
  - 15 μl 1 *M* TrisCl, pH 8.0
  - 0.032 g spermadine
  - 0.02 g putrescine
  - 50 μl 1 *M* $MgCl_2$
  - 5.25 μl B-mercaptoethanol
  - 750 μl 0.1 *M* ATP, pH 7.0
  - Store at 4°C for 3 months

Media preparation: It has been found that the media used for growing the cells is a very important factor in the efficiency of the extracts. It is recommended to stick to the exact brands of agar, casein digest, and yeast to optimize the production of the extracts. Once optimized, any changes should be tested systematically.

- LB plates (0.5 l)
  - 5.0 g casein digest
  - 2.5 g yeast extract
  - 5.0 g NaCl
  - 7.5 g agar

Autoclave for 30 min at 121°C; mix well and cool to 55°C; pour into sterile Petri plates (~40 ml/plate).

Table 38.A3 Quantities[a] of reagents used for preparation of λ packaging extracts

Reagents	Quantity	Reagents	Quantity
LB plates	2/cell line	0.1 *M* ATP (−70°C)	750 μl
NZY media	2 l	β-mercaptoethanol	9 μl
10% sucrose solution (4°C)	3 ml	1 *M* $MgCl_2$	100 μl
Packaging buffer (4°C)	3 ml	Spermadine (−20°C)	0.032 g
Sonication buffer (4°C)	2 ml	Putrescine	0.02 g
1 *M* Tris buffer	900 μl		
Lysozyme solution (4°C) (make fresh on day of use)	300 μl		

[a] These quantities will produce BHB (Tube A) and NM (Tube B) extracts for ~ 70 samples.

- NZY media (1 l)
  - 10 g casein digest
  - 5 g yeast
  - 5 g NaCl
  - 2.01 g $MgCl_2 \cdot 6\ H_2O$
  - pH 7.0, autoclave for 30 min at 121°C

Procedures

Packaging extract A (BHB 2688 cell stock)

Day 1:

1. Using a sterile loop, streak 2-LB plates with frozen BHB 2688 stock cells (−80°C).
2. Incubate the plates overnight, one at 30°C and the other at 42°C.

Use the plate grown at 30°C for inoculation. This plate may be stored at 4°C and used for 1 week. Check the plate grown at 42°C to ensure no growth has occurred. If it has, discard stock cells and re-streak plates with new stock cells.

Day 2:

1. Inoculate 80 ml of NZY broth with a colony of BHB 2688 (30°C plate).
2. Grow overnight in a shaking incubator at 30°C and 265 rpm.

Day 3:

1. Inoculate three 1-l flasks (each containing 500 ml sterile NZY media) with 25 ml each of the BHB 2688 overnight culture. Cap flasks with sterile foam stoppers.
2. Grow the culture in a shaking incubator at 32°C and 175 rpm until the optical density at wavelength 600 nm (OD_{600}) reaches 0.6 (~2.5 h). It is important not to overgrow the culture.
3. Transfer the flasks to a 65°C water bath. Swirl the flasks and monitor the temperature in flasks until they reach 45°C (~5 min).
4. Transfer the flasks to a 45°C water bath for 15 min, swirling the flasks every 5 min.

5. Return the flasks to the shaking incubator now set at 38–39°C, and shake for 2–3 h (do not exceed 40°C).
6. After approximately 2.5 h of incubation time, remove 2 ml of the cell suspension from a flask and aliquot 1 ml into each of 2 culture tubes. Add 3–4 drops of chloroform to one and see if it clears (compared to other tube) when incubated at 37°C for 2–3 min.
7. When the chloroform/cell suspension clears, aliquot flask contents into six 250-ml centrifuge bottles and centrifuge at 3200 *g* at 4°C for 15 min.
8. Immediately decant the supernatant, place the bottles on ice, and dry the inside of the bottles with a lint-free towel. Avoid touching the pellet at the bottom of the bottle.
9. Vortex each bottle for a few seconds.
10. Pipette 3 ml of 10% sucrose solution into one bottle and vortex to resuspend the pellet.
11. Using a 5-ml serological pipette, transfer the resuspended cells/sucrose solution to the next bottle and vortex again. Repeat this process until all six pellets are resuspended together. It is important to avoid introducing air into the solution at this stage.
12. Aliquot 500 μl of the solution into Microfuge tubes (~10 tubes).
13. Add 25 μl of lysozyme solution. Gently mix.
14. Flash-freeze the tubes in liquid nitrogen and store at −80°C.

Day 4:

1. Thaw tubes from the day before on ice for at least 1 h.
2. Add 25 μl of packaging buffer to each tube.
3. Use a small spatula to scoop the material into a centrifuge tube. The material will be highly viscous.
4. Spin in ultracentrifuge at 45,000 rpm for 90 min at 4°C.
5. Remove supernatant using a serological pipette and add it into a chilled 50-ml tube.
6. Aliquot 45 μl into each Microfuge tube (~70 tubes) and freeze in liquid nitrogen. Store at −80°C.

Packaging extract B (NM 759 cell stock)

Day 1:

1. Using a sterile loop, streak 2-LB plates with frozen NM 759 stock cells (−80°C).
2. Incubate the plates overnight, one at 30°C and the other at 42°C.

Use the plate grown at 30°C for inoculation. This plate may be stored at 4°C and used for up to 1 week.

Check the plate grown at 42°C to ensure no growth has occurred. If it has, discard stock cells and re-streak plates with new stock cells.

Day 2:

1. Inoculate 30 ml of NZY media with a colony of NM 759 (30°C plate).
2. Grow overnight in a shaking incubator at 30°C and 265 rpm.

Day 3:

1. Inoculate one 1-l flask containing 500 ml of NZY media with 25 ml of the NM 759 overnight culture. Cap flasks with sterile foam stoppers.
2. Grow the culture in a shaking incubator at 32°C and 175 rpm until the optical density at wavelength 600 nm (OD_{600}) reaches 0.3 (~2.5–3 h).
3. Transfer the flask to a 65°C water bath. Swirl the flask and monitor the temperature in the flask until it reaches 45°C (~5 min).
4. Transfer the flask to a 45°C water bath for 15 min, swirling every 5 min.
5. Return the flask to the shaking incubator now set at 38–39°C, and shake for 2–3 h (do not exceed 40°C).
6. After approximately 2.5 h of incubation time, remove 2 ml of the cell suspension from the flask and aliquot 1 ml into each of 2 culture tubes. Add 3–4 drops of chloroform to one and see if it clears (compared to other tube) when incubated at 37°C for 2–3 min.
7. When the chloroform/cell suspension clears, aliquot flask contents into two 250-ml centrifuge bottles and centrifuge 3200 *g* at 4°C for 15 min.
8. Immediately decant supernatant, place bottles on ice, and dry inside of bottles with a lint-free towel. Avoid touching the pellet at the bottom of the bottle.
9. Vortex each bottle for a few seconds.
10. Pipette 3.6 ml sonication buffer into one bottle and vortex to resuspend pellet.
11. Using a 5-ml serological pipette, transfer the resuspended cells/sonication solution to the second bottle and vortex again.
12. Aliquot mixture into sterile Microfuge tubes (~4 tubes).
13. Adjust sonic dismembrator to setting 3 and sonicate each tube 8 times in 2-s bursts. Keep tubes on ice between bursts for at least 30 s. Samples should decrease in viscosity and foam. Air bubbles are acceptable at this stage.
14. Centrifuge tubes at 8000 *g* for 10 min at 4°C.
15. Transfer supernatant using a wide bore tip to a cooled 15-ml tube.
16. Add 1/2 volume of sonication buffer to supernatant.
17. Add 1/6 volume (original) of packaging buffer to supernatant.
18. Invert tube to mix and aliquot 60 μl of mixture into sterile Microfuge tubes (~90 tubes) using a wide bore tip.
19. Flash-freeze tubes in liquid nitrogen and store at −80°C.

References

1. Jones, I.M., Burkhart, S. and Carrano, A.V., A method to quantify spontaneous and *in vivo* individual thioguanine-resistant mouse lymphocytes, *Mutat. Res.*, 147, 97–105, 1985.
2. Aidoo, A., Lyn-Cook, L., Mittelstaedt, R.A., Heflich, R.H. and Casciano, D.A., Induction of G-thioguanine-resistant lymphocytes on Fischer 344 rats following *in vivo* exposure to *N*-ethyl-*N*-nitrosourea and cyclophosphamide, *Environ. Mol. Mutagen.*, 17, 141–151, 1991.
3. Winton, D.J., Blount, M.A. and Ponder, B.A.J., A clonal marker induced by mutation in mouse intestinal epithelium, *Nature*, 333, 463–466, 1988.
4. Gossen, J.A., De Leeuw, W.J.F., Tan, C.H.T., Zwarthoff, E.C., Berends, F., Lohman, P.H.M., Knook, D.L. and Vijg, J., Efficient rescue of integrated shuttle vectors from transgenic mice: a model for studying mutations *in vivo, Proc. Natl. Acad. Sci. USA*, 86, 7971–7975, 1989.

5. Kohler, S.W., Provost, G.S., Kretz, P.L., Fieck, A., Bullock, W.O., Sorge, J.A., Putman, D.L. and Short, J.M., Analysis of spontaneous and induced mutations in transgenic mice using a lambda ZAP®/*lacI* shuttle vector, *Environ. Mol. Mutagen*, 18, 316–321, 1991.
6. Burkhart, J.G., Winn, R.N., Van Beneden, R.J. and Malling, H.V., Spontaneous and induced mutagenesis in transgenic animals containing ΦX174, *Environ. Mol. Mutagen*, 21 (Suppl. 22), 9, 1993.
7. Boerrigter, M., Dolle, M., Martus, H., Gossen, J.A. and Vijg, J., Plasmid based transgenic mouse model for studying *in vivo* mutations, *Nature*, 377, 657–659, 1995.
8. Manjanatha, M.G., Chen, J.B., Shaddock, J.G.J., Harris, A.J., Shelton, S.D. and Casciano, D.A., Molecular analysis of *lacI* mutations in Rat2™ cells exposed to 7,12-dimethylbenz[*a*]anthracene: evidence for DNA sequence and DNA strand biases for mutation, *Mutat. Res.*, 372, 53–64, 1996.
9. Amanuma, K., Takeda, H., Amanuma, H. and Aoki, Y., Transgenic zebrafish for detecting mutations caused by compounds in aquatic environments, *Nat. Biotechnol.*, 18, 62–65, 2000.
10. Winn, R.N., Norris, M.B., Brayer, K.J., Torres, C. and Muller, S.L., Detection of mutations in transgenic fish carrying a bacteriophage lambda *cII* transgene target, *Proc. Natl. Acad. Sci.*, 97 (23), 12655–12660, 2000.
11. Winn, R.N., Norris, M., Muller, S., Torres, C. and Brayer, K., Bacteriophage lambda and plasmid pUR288 transgenic fish models for detecting *in vivo* mutations, *Mar. Biotechnol.*, 3, S185–S195, 2001.
12. Tao, K.S., Urlando, C. and Heddle, J.A., Comparison of somatic mutation in a transgenic versus host locus, *Proc. Natl. Acad. Sci. USA*, 90, 10681–10685, 1993.
13. Cosentino, L. and Heddle, J.A., A test for neutrality of mutations of the *lacZ* transgene, *Environ. Mol. Mutagen.*, 28, 313–316, 1996.
14. Swiger, R.R., Cosentino, L., Shima, N., Bielas, J.H., Cruz-Munoz, W. and Heddle, A., The *cII* locus in the Muta™Mouse System, *Environ. Mol. Mutagen.*, 34 (2/3), 201–207, 1999.
15. Jakubczak, J.L., Merlina, G., French, J.E., Muller, W.J., Paul, B., Adhya, S. and Garges, S., Analysis of genetic instability during mammary tumor progression using a novel selection-based assay for *in vivo* mutations in a bacteriophage λ transgene target, *Proc. Natl. Acad. Sci. USA*, 93, 9073–9078, 1996.
16. Heddle, J.A., Dean, S., Nohmi, T., Boerrigter, M., Casciano, D., Douglas, G.R., Glickman, B.W., Gorelick, N.J., Mirsalis, J.C., Martus, H., Skopek, T.R., Thybuad, V., Tindall, K.R. and Yajima, N., *In vivo* transgenic mutation assays, *Environ. Mol. Mutagen.*, 35, 253–259, 2000.
17. Hara, T., Sui, H., Kawakami, K., Shimada, Y. and Shibuya, T., Partial hepatectomy strongly increased in the mutagenicity of *N*-ethyl-*N*-nitrosourea in Muta™ Mouse liver, *Environ. Mol. Mutagen.*, 34, 121–123, 1999.
18. Sun, B., Shima, N. and Heddle, J.A., Somatic mutation in the mammary gland: influence of time and estrus, *Mutat. Res.*, 427 (1), 11–19, 1999.
19. Walker, V.E., Jones, I.M., Crippen, T.L., Meng, Q., Walker, D.M., Bauer, M.J., Reilly, A.A., Tates, A.D., Nakamura, J., Upton, P.B. and Skopek, T.R., Relationships between exposure, cell loss and proliferation, and manifestation of *Hprt* mutant T cells following treatment of pre-weaning, weaning, and adult male mice with *N*-ethyl-*N*-nitrosourea, *Mutat. Res.*, 431, 371–388, 1999.
20. Walter, C.A., Intano, G.W., McCarrey, J.R., McMahan, C.A. and Walter, R.B., Mutation frequency declines during spermatogenesis in young mice but increases in old mice, *Proc. Natl. Acad. Sci. USA*, 95, 10015–10019, 1998.
21. Shane, B.S., Lockhart, A.-M.C., Winston, G.W. and Tindall, K.R., Mutant frequency of *lacI* in transgenic mice following benzo[*a*]pyrene treatment and partial hepatectomy, *Mutat. Res.*, 377, 1–11, 1997.
22. Shelby, M.D. and Tindall, K.R., Mammalian germ cell mutagenicity of ENU, IPMS, and MMS, chemicals selected for a transgenic mouse collaboratory study, *Mutat. Res.*, 388, 99–109, 1997.

23. Hawkins, W.E., Walker, W.W. and Overstreet, R.M., Carcinogenicity tests using aquarium fish, in *Fundamentals of Aquatic Toxicology: Effects, Environmental Fate, and Risk Assessment*, Rand, G.M., Ed., Taylor & Francis, Washington, D.C., 1995, pp. 421–446.
24. Mirsalis, J.C., Monforte, J.A. and Winegar, R.A., Transgenic animal models of measuring mutations *in vivo, Crit. Rev. Toxicol.*, 24 (3), 255–280, 1994.
25. Wang, X., Suzuki, T., Itoh, T., Honma, M., Nishikawa, A., Furukawa, F., Takahashi, M., Hayashi, M., Kato, T. and Sofuni, T., Specific mutational spectrum of dimethylnitrosamine in the *lacI* transgene of Big Blue C57BL/6 mice, *Mutagenesis*, 13 (6), 625–630, 1998.
26. Okihiro, M.S. and Hinton, D.E., Progression of hepatic neoplasia in medaka (*Oryzias latipes*) exposed to diethylnitrosamine, *Carcinogenesis*, 20 (6), 933–940, 1999.
27. Bunton, T.E., Use of non-mammalian species in bioassays for carcinogenicity, in *Data on Genetic Effects in Carcinogenic Hazard Evaluation*, McGregor, D.B., Rice, J.M. and Venitt, S., Eds., IARC Scientific Publications, Lyon, 1999, pp. 151–184.
28. Brown-Peterson, N.J., Krol, R.M., Zhu, Y., and Hawkins, W.E., N-Nitrosodiethylamine initation of carcinogenesis in Japanese medaka (*Oryzias latipes*): hepatocelluar proliferation, toxicity, and neoplastic lesions resulting from short-term, low level exposure, *Toxicol. Sci.*, 50, 186–194, 1999.
29. Hawkins, W.E., Walker, W.W., Overstreet, R.M., Lytle, T.M. and Lytle, J.S., Dose related carcinogenic effects of waterborne benzo[*a*]pyrene on livers of two small fish species, *J. Ecotox. Environ. Safety*, 16, 219–236, 1988.
30. Boerrigter, M.E.T.I., High sensitivity for color mutants in *lacZ* plasmid-based transgenic mice, as detected by positive selection, *Environ. Mol. Mutagen.*, 32, 148–154, 1998.
31. Monroe, J.J., Kort, K.L., Miller, J.E., Morino, D.R. and Skopek, T.R., A comparative study of *in vivo* mutation assays: analysis of hprt, *lacI*, and *cII/cI* as mutational targets for *N*-nitroso-*N*-methylurea and benzo[*a*]pyrene in Big Blue™ mice, *Mutat. Res.*, 421, 121–136, 1998.

chapter thirty-nine

Improved methods of conducting microalgal bioassays using flow cytometry

Natasha M. Franklin
McMaster University

Jennifer L. Stauber and Merrin S. Adams
CSIRO Energy Technology

Contents

1-56670-664-5/05/$0.00+$1.50
© 2005 by CRC Press

Introduction

Microalgal toxicity tests have been widely used to evaluate the potential impact of contaminants and nutrient inputs in marine and freshwater ecosystems. Standard tests use population growth and measure chronic toxicity, i.e., the inhibition of cell division rate (growth rate) or cell yield over 48–96 h. To avoid losses of contaminants to the test containers or high cell numbers that may occur over this period, shorter tests that measure acute toxicity, e.g., enzyme inhibition, have also been developed. Tests based on inhibition of esterases, enzymes essential for phospholipid turnover in cell membranes, have been shown to be good surrogates for metabolic activity and cell viability.[1,2–4] However, most conventional techniques used to count and analyze cells in growth rate and enzyme inhibition tests suffer from several limitations.[5–7]

1. They use environmentally unrealistic high cell densities in order to obtain a measurable response, which may alter contaminant speciation, bioavailability, and toxicity.
2. They measure only one effect parameter (e.g., growth or enzyme activity at a time).
3. It is difficult to distinguish between live and dead cells.
4. Cells cannot be counted in the presence of particulates (i.e., the samples require filtering which may remove toxicants).
5. They are single species tests, lacking in environmental realism.

One technique that may be used to overcome these limitations, resulting in more environmentally relevant tests, is flow cytometry. Flow cytometry is a rapid method for the quantitative measurement of individual cells in a moving fluid. Thousands of cells are passed through a light source (usually a laser) and measurements of cell density, light scatter, and fluorescence are collected simultaneously. Although this technique has been widely applied to biomedical and oceanographic studies, flow cytometry has only recently been applied to ecotoxicology.[6–8] Microalgae are ideally suited to flow cytometric analysis as they contain photosynthetic pigments, such as chlorophyll *a*, which autofluoresce when exited by blue light. In addition, specific fluorescent dyes can be used and detected by flow cytometry to provide information about the physiological status of algal cells in response to toxicants.

Tests with both marine and freshwater microalgae have been developed using flow cytometry and applied to testing wastewaters, chemicals, and sediments.[8,9–13] In this chapter, we describe a chronic growth rate inhibition test using flow cytometry to count low algal cell densities and an acute enzyme (esterase) inhibition test.

Materials required

Test species

Any nonchain forming marine or freshwater microalga can be used as the test species. In this chapter, we describe two freshwater tests with the chlorophytes (green algae)

Pseudokirchneriella subcapitata (Korschikov) Hindák, 1990 (previously called *Selenastrum capricornutum*) and a tropical *Chlorella* sp. *P. subcapitata* was obtained from the American Type Culture Collection (ATCC-22662), Maryland, USA. *Chlorella* sp. was isolated from Lake Aesake, Strickland River, Papua New Guinea. This species is available from the Centre for Environmental Contaminants Research, CSIRO Energy Technology, Australia (contact: Jenny.Stauber@csiro.au).

Tests are also described for two marine algae (both pennate diatoms): a planktonic species *Phaeodactylum tricornutum* Bohlin and a benthic species *Entomoneis* cf. *punctulata* Osada and Kobayashi. *P. tricornutum* was obtained from the National Research Centre Istituto di Biofisica, Pisa, Italy, while *E.* cf. *punctulata* was isolated from Little Swanport, Tasmania, Australia, and provided by CSIRO Marine Research (strain No. CS-426).

Equipment:

- Flow cytometer (e.g., Becton Dickinson FACSCalibur TM, BD BioSciences, San Jose, CA, USA)
- Environmental chamber/incubator with built-in light and temperature controls and mechanical shaking platform
- Milli-Q® (Millipore) water purification system or equivalent
- Centrifuge (benchtop), 4 × 30 ml capacity with swing out buckets
- Refrigerator/freezer for storing stock solutions and reagents
- Biohazard cupboard or laminar flow cabinet for aseptic algal culturing
- Analytical balance
- Autoclave
- Adjustable pipettes (5 μl to 5 ml)
- Vortex mixer
- pH meter and buffers
- Light meter
- Conductivity meter
- Thermometer
- Filter apparatus, 47 mm filter holder, 1-l flask, vacuum pump, and tubing
- Magnetic stirrer

Supplies:

- Microalgae
- Fluorescent calibration beads for flow cytometer (e.g., CaliBRITE™ beads, BD BioSciences)
- TruCount™ tubes or equivalent flow cytometer counting beads
- Glass tissue homogenizer (hand-held, 15 ml) with Teflon pestle
- Glass centrifuge tubes, 30 ml capacity
- Glass Erlenmeyer flasks (200 or 250 ml) with loose-fitting glass caps
- Glass scintillation minivials (20–30 ml capacity) with plastic screw-on lids
- Glass volumetric flasks
- Glass beakers
- Glass graduated measuring cylinders
- Centrifuge tube rack
- Weighing trays and spatula
- Magnetic stirrer bars

- Polyethylene wash bottles and storage containers (1–10 l)
- Graduated glass (or sterile disposable serological pipettes) 2 ml
- Disposable glass Pasteur pipettes
- Disposable pipette tips
- Disposable plastic counting cups
- Membrane filter papers, 0.45-μm pore size
- Parafilm or equivalent laboratory sealing film
- Coatasil (AJAX) for silanizing glassware
- Disposable plastic counting tubes (12 × 72 mm Falcon tube or equivalent)
- Polycarbonate vials (50 ml capacity) used for chemical analysis
- Gloves
- Timer
- Seawater (filtered through an acid-washed Millipore HA 0.45-μm membrane filter and stored in polyethylene or glass containers at 4°C)

Chemicals and reagents:

- Fluorescein diacetate (FDA) (Sigma F-7378, dissolved in acetone)
- Propidium iodide (PI) (Sigma P-4170, dissolved in Milli-Q water)
- Salts for algal culture media and synthetic water (AR grade) (Tables 39.1 and 39.2)
- Acids and bases for pH adjustment (HCl, NaOH)
- Acetone (AR grade)
- $CuSO_4 \cdot 5H_2O$ (AJAX Chemicals)
- Nonphosphate detergent and HNO_3 for glassware washing

Procedures

All glassware used exclusively for toxicity testing was soaked in 10% (v/v) nitric acid (HNO_3) overnight and was thoroughly rinsed with Milli-Q water. All polycarbonate vials used for chemical analysis were acid-washed and rinsed in the same manner. General laboratory glassware was washed in a laboratory dishwasher (Lab999, Gallay) with a phosphate free detergent, followed by nitric acid (1%) and six rinses with Milli-Q water.

Maintenance of algal cultures

Freshwater algae

P. subcapitata stock cultures were maintained in clean 250-ml glass Erlenmeyer flasks in 100 ml USEPA medium with EDTA (Table 39.1).[14] Medium was prepared using Milli-Q water as the base and sterilized by filtering through a 0.45-μm sterile membrane filter. Stock cultures were maintained in a controlled environmental chamber on a 24-h light cycle (65 ± 5 μmol photons/m^2/s, cool white fluorescent lighting) at 24°C. Each week, fresh medium was inoculated with about 1 ml of algal stock culture under aseptic conditions in a biohazard cupboard. These cells were in late exponential phase growth and resulted in an algal cell density of the new stock of approximately $2–4 \times 10^4$ cells/ml for all species.

Chlorella sp. was cultured in one-fifth strength modified Jaworki's medium[15] (Table 39.1). The medium (in a Milli-Q water base) was sterilized by autoclaving (50 ml in 250-ml Erlenmeyer flasks, with loose-fitting glass lids). Stock cultures were maintained on a 12:12 h light/dark cycle (75 ± 5 μmol photons/m^2/s, cool white fluorescent lighting) at 27°C. Algae were subcultured weekly using the same procedure as for *P. subcapitata*.

Table 39.1 Algal culture medium for marine and freshwater toxicity tests

Chemical	EPA[a]	JM/5[b]	f/2[c]
Macronutrients (mg/l)			
$NaNO_3$	25.5	18	75
$MgCl_2 \cdot 6H_2O$	12.2	–	–
$CaCl_2 \cdot 2H_2O$	4.41	–	–
$Ca(NO_3)_2 \cdot 4H_2O$	–	2.8	–
$MgSO_4 \cdot 7H_2O$	14.7	10	–
KH_2PO_4	1.04	2.5	–
K_2HPO_4	–	3.5	–
$NaHCO_3$	15.0	3.2	–
$NaH_2PO_4 \cdot 2H_2O$	–	–	5
$Na_2SiO_3 \cdot 5H_2O$	–	–	13
$C_6H_5O_7Fe \cdot 5H_2O$[d]	–	–	2.3
$C_6H_8O_7 \cdot H_2O$[e]	–	–	2.3
Micronutrients (μg/l)			
H_3BO_3	185	496	–
$MnCl_2 \cdot 4H_2O$	416	278	90
$ZnCl_2$	3.27	–	–
$CoCl_2 \cdot 6H_2O$	1.43	–	5
$CuCl_2 \cdot 2H_2O$	0.012	–	–
$Na_2MoO_4 \cdot 2H_2O$	7.26	–	–
$(NH_4)6MoO_4 \cdot 4H_2O$	–	200	–
$FeCl_3 \cdot 6H_2O$	160	–	–
$Na_2EDTA \cdot 2H_2O$	300	450	–
$FeNaEDTA \cdot 2H_2O$	–	450	–
$CuSO_4 \cdot 5H_2O$	–	–	
$ZnSO_4 \cdot 7H_2O$	–	–	1
Thiamine HCl	–	8	100
Biotin	–	8	0.5
Vitamin B_{12}	–	8	0.5
pH	7.5 ± 0.1	7.3 ± 0.1	8.0 ± 0.2

[a] USEPA medium for *Pseudokirchneriella subcapitata.*

[b] One-fifth strength modified Jaworki's medium for *Chlorella* sp.

[c] Half-strength f media for *Entomoneis* cf. *punctulata* and *Phaeodactylum tricornutum.* Natural filtered seawater was used as the test medium base. Concentrations of ambient phosphate and nitrate in seawater <5 μg PO_4/l and <100 μg NO_3/l.

[d] Ferric citrate.

[e] Citric acid.

Marine algae

Both species were maintained in half-strength f medium (f/2).[16] Medium was prepared by adding all stock solutions, except phosphate, to 100 ml of 0.45 μm filtered clean seawater and sterilized by autoclaving. After cooling overnight, phosphate was added to give the final concentration shown in Table 39.1. A 1-ml algal inoculum was added to 100 ml of medium under aseptic conditions. The algae were subcultured weekly and maintained on a 12:12 h light/dark cycle (daylight, 65 ± 5 and 40 ± 5 μmol photons/m^2/s for *P. tricornutum* and *E.* cf. *punctulata*, respectively) at 21°C.

Table 39.2 Final concentration of inorganic components in *Chlorella* sp. synthetic softwater (80–90 mg $CaCO_3$/l) used for toxicity testing (U.S. EPA)[14]

Element	Concentration (mg/l)
Na	26
Ca	14
S	25
Mg	11
K	2.1
Cl	1.9
C	14

Preparation of algae for toxicity tests

Immediately prior to the test, the algal inoculum was prepared and used within 2 h. An exponentially growing stock culture (usually 4–5 days old) was decanted into two glass centrifuge tubes (about 25 ml in each) and centrifuged at low speed (~2500 rpm) for 7 min. The supernatant in each tube was poured off and the algal cell pellet gently resuspended in about 25 ml of Milli-Q water (freshwater algae) or filtered seawater (marine algae). The suspensions were mixed with a vortex mixer for several seconds and then centrifuged again. The centrifuging and rinsing process was repeated three times, resulting in a concentrated algal suspension that was diluted in about 15 ml of Milli-Q water or seawater, ready for counting and inoculating into the toxicity test containers.

Test sample preparation

Although toxicity tests described in this chapter were conducted with individual chemicals, some guidance on the procedure for handling and testing effluents, natural waters, groundwaters, and leachates is also given in the following.

Chemicals

Copper stock solutions were prepared by dissolving the metal salt in Milli-Q water and acidified to pH <2 by the addition of 10 ml HCl (Suprapur grade, Merck) per liter to enable storage and to reduce metal losses to the container walls. Neutralization was carried out by the addition of appropriate amounts of dilute NaOH when the pH of the test medium changed by more than 0.1 unit with the addition of the metals. Subsamples (5 ml) were immediately taken from each flask and replicates combined to give a total volume of 15 ml, which was acidified with 30 fl concentrated HNO_3 (Normatom). Samples were analyzed for total dissolved metals using appropriate analytical methods [e.g., graphite furnace atomic absorption spectrometry (GFAAS) and inductively coupled plasma atomic emission spectrometry (ICPAES)]. Measured, rather than nominal, metal concentrations were used to calculate toxicity indices.

For other chemicals, particularly those that degrade rapidly in aqueous solutions, stock solutions should be prepared fresh on the day of testing. For poorly water-soluble chemicals, mild heating, stirring, or ultrasonic dispersion may be sufficient. If carrier solvents are necessary to dissolve chemicals that are poorly soluble in water, several additional steps are required. Firstly, the prepared stock solution must be diluted in solvent to give working stock solutions at each test concentration. Exactly the same volumes can then be taken from each of the working stocks and spiked into the appropriate test vessels, ensuring that the final concentration of solvent in the test solutions is less than the no-observable-effect concentration (NOEC) for that particular solvent and test species (e.g., 0.3% for ethanol for *P. subcapitata*). Secondly, a set of solvent controls must be included in the bioassay to account for any toxicity caused by the solvent alone. These are prepared by adding the same volume of solvent to diluent water as was added to the test solutions.

Effluents and natural waters

Complex effluents, natural waters, groundwaters, and leachates should be collected and/or extracted according to standard methods to ensure that the sample is representative. For complex effluents, either grab samples or 24-h composite samples are usually collected into prerinsed cleaned containers, preferably glass or polyethylene, filled with no headspace and transported at 4°C (not frozen). About 1 l of sample is sufficient for either the algal growth or esterase inhibition toxicity test. Effluents should be tested within 3 days of sample collection to avoid chemical and biological changes associated with effluent aging and storage.

On receipt of the sample, physicochemical measurements, including pH, conductivity/salinity, temperature, and dissolved oxygen, should be taken immediately. If the sample pH is outside the optimal pH range for the particular algal test species, it can be adjusted or left unadjusted, depending on the purposes of the test. If the conductivity of the sample is greater than 2000 μS and freshwater test species such as *P. subcapitata* are to be used, then the sample may need to be diluted with diluent (control) water. Alternatively, an additional set of controls, adjusted to the same conductivity/salinity as the sample, can be prepared and tested alongside the standard controls.

The sample may be filtered through either a GF/F filter (approximate pore size 0.7 μm) or a 0.45 μm cellulose acetate filter. Once filtered, the physicochemical parameters of the sample should be measured again. However, the advantage of the flow cytometry-based toxicity tests are that the samples do not have to be filtered prior to testing, Testing of unfiltered samples allows detection of toxicants associated with particulate material. However, while bacteria and other algae in the sample may be distinguished from the algal test species, they may alter the growth of the test algae and affect the toxic response. Careful interpretation of results from unfiltered samples is therefore required.

Preparation of test solution

All toxicity tests were conducted in 250-ml borosilicate glass Erlenmeyer flasks that had been precoated with a silanizing solution (Coatasil, AJAX) to reduce adsorption of contaminants, particularly metals, to the vessel walls. Alternatively, tests may be carried out in 30-ml glass minivials (scintillation vials) with plastic screw-on lids if only small contaminant volumes are available or due to space limitations (see the section "Modification to standard flask bioassays using minivials" in the following).

All marine toxicity tests were conducted in filtered seawater at a pH of 8.0 $\pm$ 0.2 and salinity of 34%. Fifty milliliters of seawater were added to each flask and supplemented with 0.5 ml of 26 m*M* sodium nitrate (15 mg NO_3^-/l) and 0.5 ml of 1.3 m*M* potassium dihydrogen phosphate (1.5 mg PO_4^{3-}/l) in order to maintain exponential growth over 72 h.

P. subcapitata test medium (50 ml) consisted of the standard USEPA media (Table 39.1) without EDTA in order to reduce the possibility of metal complexation. However, without EDTA, iron in the medium was found to precipitate, leading to a minor decrease in dissolved copper in the test medium (NB: negligible losses at copper concentrations >20 μg/l). This medium had an alkalinity of 9 mg $CaCO_3$/l and water hardness of 15 mg $CaCO_3$/l.

For *Chlorella* sp., tests were conducted in a synthetic softwater[14] having a water hardness of 80–90 mg $CaCO_3$/l and an alkalinity of 54 mg $CaCO_3$/l. The synthetic softwater was prepared in 5-l volumes by weighing 0.30 g of $CaSO_4 \cdot 2H_2O$, 0.48 g of $NaHCO_3$, 0.30 g $MgSO_4 \cdot H_2O$, and 0.02 g of KCl into a beaker of Milli-Q water. The solution was left overnight and then made up volumetrically to 5 l before filtering through a Millipore HA 0.45 μm membrane filter. The pH was adjusted to 7.5 ± 0.1 with 0.02 *M* NaOH or HCl. An aged and acid-washed 5-l plastic container was used to store the water at 4°C for no longer than 4 weeks before use. The final concentrations of inorganic components in the synthetic softwater are shown in Table 39.2. Fifty milliliters of synthetic softwater were supplemented with 0.5 ml of 26 m*M* sodium nitrate (15 mg NO_3^-/l) and 0.05 ml of 1.3 m*M* potassium dihydrogen phosphate (0.15 mg PO_4^{3-}/l).

All toxicity tests consisted of triplicate controls, together with at least five toxicant concentrations (also in triplicate). Test concentrations were chosen with the aim of encompassing a range of responses from 0% to 90–100% growth inhibition (or esterase activity) and were in a geometric series with a dilution factor of 2 (i.e., 2.5, 5, 10, 20, and 40 μg/l). Range-finding tests were initially performed for all species using widely separated concentrations of each metal to broadly estimate the IC_{50} for later definitive tests. All definitive tests were repeated at least three times.

Modification to standard flask bioassays using minivials

In our laboratory, we have conducted toxicity tests in both flasks and glass minivials (20–30 ml capacity). Minivials have the advantage that only small sample volumes are required, making it an ideal method for testing complex environmental samples (e.g., industrial effluents) that often need a time-consuming filtering step prior to testing. Due to the reduced size of the test vessel, the minivial bioassay also requires much less incubator space, allowing multiple tests to be conducted at the same time.

Test solutions for the minivial bioassay are prepared as mentioned above but with the following modifications:

- The test volume is reduced to 6 ml.
- Additional minivials are prepared (controls and diluent water) for physicochemical measurements at the beginning and end of the test and for chemical analysis, e.g., number of replicate vials/concentration = 6 vials (3 vials for daily cell counts, 2 vials for pH measurement, 1 vial for chemical analysis).

For chemical testing, the chemical is spiked directly into each test vial, ensuring that the spike volume is less than 60 μl.

Flow cytometric analysis

A variety of flow cytometers are currently on the market and suitable for analyzing microalgae. In this chapter, we describe the use of the BD-FACSCalibur (Becton Dickinson

BioSciences, San Jose, CA, USA) flow cytometer. It should be noted that between different flow cytometers the instruments settings and methods of analysis may vary.

The BD-FACSCalibur flow cytometer is a four-color, dual-laser benchtop instrument capable of both cell analysis and cell sorting. It is equipped with an air-cooled argon-ion laser providing 15 mW at an excitation wavelength of 488 nm (blue light) and with standard filter set up. Dual excitation is possible as it also has a diode capable of excitation in the red region of the spectrum.

The instrument has two light-scatter detectors, which serve to identify the morphology of the cell. The forward angle light scatter (FSC < 15°) detector provides information on cell size, while the side angle light scatter (SSC, 15–85°) detector provides information on internal cell complexity/granularity. Fluorescence is collected at a range of wavelengths by photomultiplier tubes (PMTs) with different fluorescence emission filters. Chlorophyll *a* or autofluorescence (present in all algae) is detected as red fluorescence in greater than 600-nm-long pass filter band (FL3). Green fluorescence from cells stained with FDA is collected in the FL1 channel (530 ± 15 nm) and orange fluorescence from cells stained with PI is collected in the FL2 channel. Cells are presented to the flow cytometer and are hydrodynamically focused in a sheath fluid as they pass through the fluorescence and light-scatter detectors. Sheath fluid was high purity Milli-Q water.

Data were displayed and analyzed using the flow cytometric software package CellQuest™ Pro in one-dimensional histograms (256 channels) and two-dimensional dot plots (comprising 64×64 channels) based on a combination of fluorescence and light-scatter signals. Although flow cytometry is able to count cells at very low cell densities, a minimum of 1000 events (cells) per sample should typically be analyzed to achieve a CV of less than 10%.[17] For bioassays using cell inocula of 1×10^4 cells/ml, a preset acquisition time of 120 s was used to obtain sufficient cell numbers. However, at recommended low initial cell densities of 10^2 to 10^3 cells/ml, longer acquisition times (e.g., 300 s) were required. To avoid unnecessarily long counting times when cell numbers have increased over the course of the bioassay, a feature of this flow cytometer allowed data acquisition to be stopped when the number of cells (events) in a specified region (e.g., *Chlorella* sp. cells) reaches a value greater than 1000. All parameters were collected as logarithmic signals and analysis performed at a high flow rate (60 μl/min).

As a quality control procedure, the flow cytometer was calibrated daily using flow cytometer standardization CaliBRITE™ 3 beads (BD BioSciences) or equivalent, to verify instrument performance. Because CaliBRITE beads set up the cytometer for human cell analysis, it was necessary to optimize (adjust) the instrument settings for analyzing microalgae. Typical instrument settings used for analyzing the marine and freshwater microalgae presented in this chapter are listed in Table 39.3. A detailed description of the initial set-up, calibration, and analysis of cells using flow cytometry can be found in the FACSCalibur instrument manual and CellQuest Pro and FACSComp™ software manuals (BD Bioscience).

Protocol 1: Chronic growth rate inhibition test

The growth rate inhibition test measures the decrease in algal growth rate over 72 h. Exponentially growing cells are exposed to various concentrations of the toxicant over several generations under defined conditions. A comparison of growth rates in the controls and the test-exposed algae enables typical toxicity test endpoints, such as the IC_{50}, NOEC, and lowest-observable-effect concentration (LOEC), to be determined. One unique feature

Table 39.3 Flow cytometry instrument settings for the analysis of microalgae

Parameter		Setting		
Threshold				
Primary parameter	LS2	Value	200[a]	
Secondary parameter	None	Value	—	
Compensation				
FL1	0.0% of FL2			
FL2	0.0% of FL1			
FL2	0.0% of FL3			
FL3	0.0% of FL2			
Detectors/amps		Voltage	AmpGain	Mode
P1	FSC	E-1	3.93	Log
P2	SSC	320[b]	1.00	Log
P3	FL1	470[b]	1.00	Log
P4	FL2	470[b]	1.00	Log
P5	FL3	370[b]	1.00	Log
P6	FL1-A	—	1.00	Lin
P7	FL1-W	—	1.00	Lin
Four color	OFF			

[a] The threshold must remain on the left of the distribution of algal cells to ensure that all algal cells are captured for analysis.

[b] These values are a guide only. The operator can adjust these values to alter the position of the algal distribution along the FSC, SSC, FL1, FL2, and FL3 axis. Analyzing healthy cells harvested from a stock culture is the best way to gauge this movement. Generally, algal populations are positioned in the center of the FSC, SSC, FL2, and FL3 axis so that shifts (both increase and decrease in intensity) can be observed. For the FL1 axis it is necessary to position the algal distribution (unstained cells) in the first decade of the logarithmic scale so that high fluorescence intensities can be observed when cells are stained with FDA.

of this bioassay is that the initial inoculum can be lowered to 100 or 1000 cells/ml using the technique of flow cytometry, which is more representative of cell numbers in aquatic systems. This prevents changes in toxicant speciation and subsequent bioavailability, which can lead to an underestimation of toxicity in standard bioassays.

Day 0 (i.e., starting day of test $t = 0\,\text{h}$):

1. The concentrated algal inoculum was prepared as described earlier.
2. For those algal species prone to clumping (e.g., *P. tricornutum* and *E.* cf. *punctulata*), a hand-held glass tissue grinder with Teflon pestle was used to separate clumps in the resuspended cells. Preliminary studies have shown that this does not cause any rupture or damage to these species (checked microscopically). The cell density of the algal suspension was determined by adding a known volume (x μl) to a flask containing 50 ml of the appropriate test medium and counted by flow cytometry (described in the following). From this "counting flask," the volume of algal inoculum required to give an initial cell density of 10,000 cells/ml (standard test) or 1000 cells/ml in low cell density test was determined.
3. Test solutions were prepared as described earlier and left at room temperature.
4. Test flasks were inoculated with the appropriate amount of prewashed algal cells (see the previous section "Preparation of algae for toxicity testing") ($2\text{–}4 \times 10^3$ cells/ml recommended for low cell density test). To ensure that the suspension remained homogenous, the concentrated algal suspension was vortexed between

every 3 and 4 inoculations. The volume of test inoculum added to each vessel did not exceed 0.5% of the total volume in the vessel.

5. The flasks were placed randomly in an environmental cabinet at the specified test conditions for each species (Table 39.4).

NB: For the minivial growth inhibition bioassay, minivials are placed in racks on an electronic shaker in an environmental cabinet set at the conditions defined in Table 39.4 and left shaking for 72 h at 100 rpm.

Days 1–3:

1. Each flask was gently agitated by hand twice daily throughout the test to avoid gas limitation. This was done by swirling the solution approximately six times in the clockwise direction and six times in the anticlockwise direction.
2. A subsample (1.0 ml) from each flask was taken for counting by flow cytometry at 24, 48, and 72 h after test commencement. Cell counts recorded at the end of the 48-h period were designated as Day 2 and at the end of the 72-h period as Day 3 observations. The pH, temperature, and conductivity of one replicate flask at each test concentration were measured and recorded at the end of the test (Day 3).

Determination of cell counts by flow cytometry

For most commercially available flow cytometers, absolute cell counts are obtained by adding a known amount of reference beads into the sample (ratiometric counting). By comparing the algal cell count with the bead count, the cell concentration can be calculated. This approach has been shown to be accurate, particularly when using primary reference bead solutions (i.e., Becton Dickinson TruCount tubes, described in the following) and also provides an internal standard that can be used to assess the performance of

Table 39.4 Culture conditions for algal stock culture maintenance and toxicity tests

Alga	Temperature (°C)	Light[a] (μmol photons/m²/s)	Light/dark cycle (h)	Culture medium
Stock maintenance				
Chlorella sp.	27	75	12:12	JM_5
P. subcapitata	24	65	24:0	U.S. EPA
P. tricornutum	21	65	12:12	f_2
E. cf. *punctulata*	21	40	12:12	f_2
Toxicity tests[b]				
Chlorella sp.	27	140	12:12	Synthetic softwater + 15 mg/l NO_3 + 0.15 mg/l PO_4
P. subcapitata	24	65	24:0	U.S. EPA medium (without EDTA)
P. tricornutum	21	140	12:12	Filtered seawater + 15 mg/l NO_3 + 1.5 mg/l PO_4
E. cf. *punctulata*	21	140[c]	12:12	Filtered seawater + 15 mg/l NO_3 + 1.5 mg/l PO_4

[a] Philips TL 40 W cool white fluorescent lighting.

[b] Suitable pH ranges for growth inhibition bioassays with each species are: pH 6.0–8.0 for *Chlorella* sp; pH 5.3–8.7 for *P. subcapitata*; pH 7.5–8.5 for *E.* cf. *punctulata*; pH 7.0–9.0 for *P. tricornutum*.

[c] Acute esterase inhibition bioassay conducted at a reduced light intensity of 1–4 μmol photons/m²/s.

the instrument (e.g., standardization of light scatter and fluorescence). Unfortunately, this counting method adds additional cost in consumables (e.g., fluorescent beads) and requires extra manipulations/calculations. In our laboratory, we have also used a Bio-Rad Bryte HS flow cytometer, which is particularly simple to operate and inexpensive to count as the instrument takes a known sample volume, so direct algal counts are possible without the need for internal calibration with beads. Unfortunately, this instrument has been discontinued, so protocols with this instrument were not described in this chapter. Alternatively, the flow rate of the instrument could be calibrated daily; however, this approach would require a very stable instrument and the accuracy of this method would need to be verified.

Before daily counting, the test flasks were well mixed. For bioassays with *P. tricornutum* and *E. punctulata*, a disposable pipette tip was used to scratch the bottom of the flasks to remove cells that had adhered to the test vessel. An aliquot of cells (1 ml) was immediately taken and placed directly into a TruCount tube.* The tube was mixed well and checked for the presence of air bubbles under the metal retainer, which can interfere with the analysis (air bubbles are removed by gentle tapping on the tube). To ensure that accurate counts were obtained, all pipettes used for dispensing solutions were calibrated prior to use.

The sample was acquired on the flow cytometer using the appropriate instrument settings for each species (see Table 39.3). The data were analyzed by drawing a gate around the TruCount bead population (see Figure 39.1A) from a dot plot of FL1 versus FL2. This population was then removed (i.e., gated out) from a new plot of FL3 versus SSC to obtain the algal population alone (Figure 39.1B). The region statistics (Figure 39.1) were used to determine the number of bead and algal events within each region.

The absolute number of algal cells in the sample was calculated using the following equation:

$$\frac{\text{No. events in region containing cell population}}{\text{No. events in bead region}} \times \frac{\text{No. beads per test}}{\text{test volume}} = \text{concentration of algal population} \qquad (39.1)$$

(The number of beads in each pellet (beads per test) varies among lots and is printed on the foil pouch.)

Modification of the counting method using a diluted TruCount tube bead stock

To help reduce costs associated with running the bioassay, we have included a modification to the method of counting cells based on diluting TruCount tubes, e.g., rather than using a new TruCount tube for every sample, the TruCount tube becomes a bead stock solution, which is then added to the algal sample to be counted.

The fluorescent bead stock solution is prepared by pipetting 1 ml of Milli-Q water or filtered seawater directly into a TruCount tube containing a known quantity of fluorescent beads. The tube is then mixed thoroughly and 200 μl of bead stock added to a sample tube (12 × 72 mm Falcon tube or equivalent) containing an appropriate aliquot of algal sample (e.g., 0.5 ml). It is recommended that the pipette tip be rinsed with the sample

* Becton Dickinson TruCount Absolute Count Tubes contain a lyophilized pellet of 4.2-μm fluorescent-dyed beads. The pellet is restrained in the bottom of the tube by a stainless-steel retainer. Store tubes in the foil pouch at room temperature and use within 1 h after removal from the foil pouch. Reseal foil pouches immediately after each use. Once the pouch has been opened, the tubes are stable for 30 days.

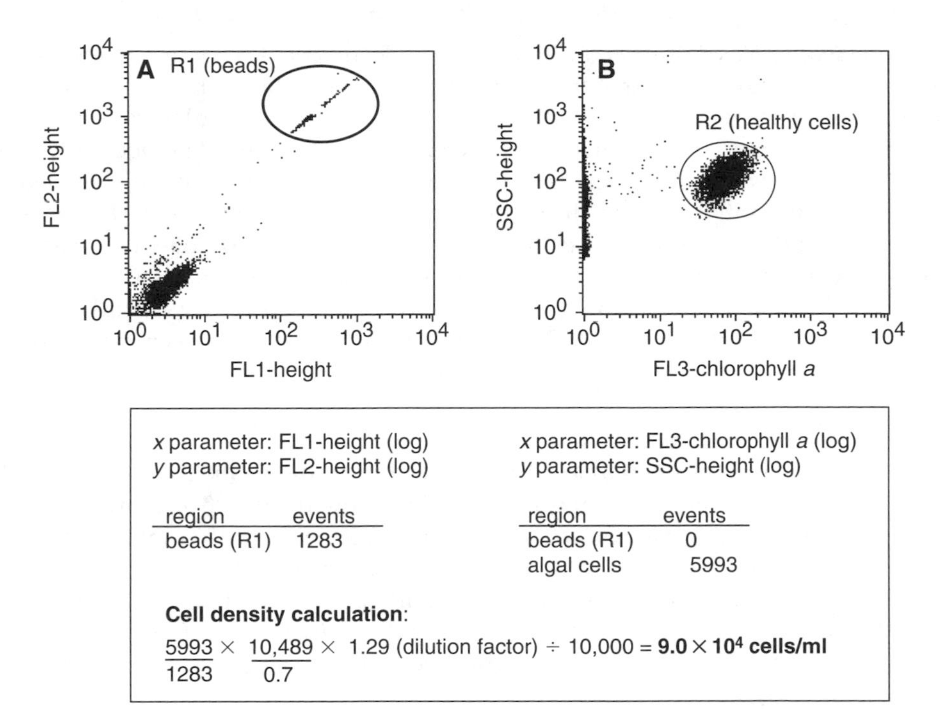

Figure 39.1 Determination of absolute cell counts of *Chlorella* sp. using diluted TruCount tubes (200 μl spike). (A) Dot plot of FL1 versus FL2 fluorescence identifying bead population (R1). (B) Dot plot of FL3 (chlorophyll *a*) versus SSC (side-angle light scatter/cell complexity) identifying healthy algal population.

solution by taking up and dispensing the solution several times. It should be noted that the high precision and accuracy of this counting method is limited only to the microalgae and bead pipetting step. Absolute cell counts are determined using Equation (39.1), with the exception that the number of beads per test is now the number of beads in 200 μl of the bead stock solution (e.g., 52,445 beads in 1 ml = 10,489 beads in 200 μl), and the test volume is increased to include the total volume of the sample (e.g., 0.5 ml algae + 0.2 ml bead spike). A dilution factor is also required to account for the bead volume. At least 1000 bead events must be acquired to ensure the accuracy of this counting method.

Statistical analysis of test data

The growth rate (cell division rate) of algae in each flask over 72 h was calculated using linear regression analysis. A regression line was fitted to a plot of the $\log_{10}$ cell density for each replicate versus time (h). The slope of the regression line is equivalent to the cell division rate per hour (μ) for each treatment. Daily doubling times were calculated by multiplying this value $\mu \times 24 \times 3.32$ (constant). Growth rates of the treated flasks were presented as a percentage of the control growth rate. A concentration–response curve was obtained by plotting the percentage control growth rates versus the measured toxicant concentrations.

The endpoints of the algal growth inhibition test were the 72 h IC_{50}, NOEC, and LOEC. The 72 h IC_{50} is the inhibitory toxicant concentration that gives 50% reduction in algal growth rate over 72 h compared to the controls. IC_{50} values were calculated using a computer program, NYHOLM-3 (modified by Yuri Tsvetnenko from Nyholm et al.[18]).

Calculations were carried out assuming that the concentration–response can be described by the probit function and were performed using weighted linear regression analysis on probit transformed data. Alternatively, the IC_{50} value can be calculated by linear interpolation using commercially available software, such as ToxCalc (Tidepool Software).

A hypothesis-testing approach was used to determine which metal-exposed treatments were significantly different from the controls. Data were initially tested for normality and homogeneity of variance, and Dunnett's Multiple Comparison test (ToxCalc Version 5.0.14, Tidepool Software) was used to estimate the NOEC and LOEC values. The NOEC is the highest tested metal concentration that yielded no statistically significant deviation from the control. The LOEC is the lowest tested metal concentration to yield a statistically significant deviation from the control.

Acceptability of the test

The data were considered acceptable if the growth rate of the control algae increased by a factor of 16 after 72 h, corresponding to a specific growth rate of 0.9/day. In addition, variability in the growth rate of controls should not exceed 10%, and the recorded temperature and water quality parameters throughout the test must be within acceptable limits, with a pH change of not more than 1.5 pH units. For metal toxicants, pH changes of less than 0.5 pH units are desirable. Each test should also include a reference toxicant, e.g., copper, to ensure that the algae are responding in a reproducible way to a known toxicant (i.e., IC_{50} within 2 SD of the mean IC_{50}).

Protocol 2: Acute esterase inhibition tests with P. subcapitata *and* E. *cf.* punctulata

This acute enzyme toxicity test uses flow cytometry to measure inhibition of the enzyme esterase after a 3- or 24-h exposure to toxicant. After this time, the substrate FDA is added, and the cells incubated for 5 min, followed by flow cytometric analysis. Healthy control cells take up FDA, which is cleaved by esterases, releasing a fluorescent product fluorescein that is retained in the cells. This is measured as an increase in algal cellular fluorescence in the green region of the spectrum. Toxicants decrease FDA cleavage by esterases and subsequent green fluorescence, and this is measured as a shift in fluorescence from the healthy control region towards the unhealthy (dead) cell region. This shift, detectable by flow cytometry, is quantified and the percentage of shift out of the control region is calculated and used for determination of EC_{50}, NOEC, and LOEC values.

The test set up and exposure conditions for the acute algal esterase tests was similar to the growth rate bioassay protocol with two exceptions. For toxicity tests with *P. subcapitata*, the starting inoculum used was 1×10^4 cells/ml ($\pm$ 10%), and the test media was adjusted to a pH of 7.8–8.5 prior to use. Acute tests using the benthic marine alga *E.* cf. *punctulata* had a starting inoculum of $2–4 \times 10^4$ cells/ml and were conducted at a low light intensity of 1–4 μmol photons/m^2/s throughout the 3- and 24-h exposure period.

Analysis of cellular esterase activity

A 1-m*M* stock solution was prepared daily by weighing 0.0104 g of FDA into a small glass weighboat and rinsed into a 25-ml volumetric flask with acetone. The dissolved FDA solution was made up to volume with acetone and stored at 4°C.

After a 3- and 24-h exposure, a subsample (4.88 ml) from each flask (control and test concentrations) was transferred into 20-ml glass scintillation vials. An aliquot of the FDA stock solution (125 μl) was added to each vial to give a final FDA concentration of 25 μM and incubated for 5 min prior to flow cytometric analysis for 120 s.

Flow cytometric analysis was similar to that used for the algal growth rate bioassay. After identifying and capturing the algal population on the FL3 versus SSC cytogram, shifts in algal esterase activity were determined by using the histogram of cell number versus FL1 fluorescence gated on the algal population. For each toxicant concentration, the data were expressed as the difference in the percentage of cells falling into three defined metabolic activity states (S1, S2, S3), obtained by merging reference histograms of FDA-stained control and heat-treated samples (see the section "Negative controls" in the following). Region S2 was defined manually around the normal distribution of control healthy cells (stained with FDA) and incorporated greater than 90% of the cells. The percentage of cells falling into regions S1 (decreased FL1 fluorescence) and S3 (enhanced FL1 fluorescence) were recorded and expressed as a percentage decrease in S2 compared to a control according to the following equation:

$$(100 - \%S1_t) \div (100 - \%S1_c) \times 100 = \% \text{ control} \tag{39.2}$$

where $\%S1_t$ is the percentage of treated cells in S1 and $\%S1_c$ is the percentage of control (untreated) cells in S1.

Negative controls

Healthy unstained cells and FDA-stained deactivated cells were included as negative controls. The esterase enzyme in algal cells was deactivated by heat treatment (10 min in a water bath at 100°C) or by the addition of formalin (final concentration of 4% v/v). The FL1 (FDA) fluorescence intensity of the negative controls should fall in the S1 region of the histogram (ideally in the first decade of the log scale). To ensure that shifts in FL1 fluorescence intensity and hence changes in cellular esterase activity could be observed, the algal distribution of the FDA-stained controls and negative controls should not overlap by more than 10%.

Statistical analysis of test data

The percentage of control data was used for all calculations. The EC_{50} value (concentration to cause a 50% inhibition in esterase activity), NOEC, and LOEC values were calculated using the same statistical methods used for the growth rate inhibition test.

Modification of esterase test set up for a minivial assay

Each test concentration was prepared by dispensing 100 ml of test media (Table 39.4) into 120-ml polycarbonate vials and spiked with the copper stock solution to give test concentrations of 5–300 μg Cu/l. The pH of each test concentration was measured and adjusted, if necessary, with NaOH or HCl to within the pH range specified in Table 39.5. Ten milliliters of each solution were dispensed into 20-ml glass scintillation vials coated with a silanizing solution, in triplicate. Controls consisting of test media were also included in triplicate. Two vials were prepared for each replicate, one vial for 3-h FDA analysis and one vial for 24-h analysis. One additional vial per test concentration was prepared for pH measurements throughout the test. Each vial was inoculated with a washed algal inoculum to give a final cell density of 1×10^4 cells/ml ($\pm$ 10%) and $2\text{–}4 \times 10^4$ cells/ml for *P. subcapitata* and *E.* cf. *punctulata*, respectively. The vials were loosely capped and incubated under standard conditions specified in Table 39.5.

Table 39.5 Test conditions for esterase inhibition bioassays with *Pseudokirchneriella subcapitata* and *Entomoneis* cf. *punctulata*

Parameter	*P. subcapitata*	*E.* cf. *punctulata*
Test media	USEPA media (without EDTA)	Filtered seawater + 15 mg/l NO_3 + 1.5 mg/l PO_4
Acceptable pH range	7.8–8.5	6.5–8.5
Acceptable salinity range	NA[a]	15–35%
Initial cell density	1×10^4 cells/ml $\pm$ 10%	$2–4 \times 10^4$ cells/ml
Exposure conditions	24°C, cool-white continuous lighting 65 μmol photons/m²/s	21°C, daylight fluorescence lighting, 1–4 μmol photons/m²/s 12 h light:12 h dark

[a] Not applicable.

Propidium iodide staining

To ensure that a decrease in FDA fluorescence was due to an effect on esterases within the cell, and not due to membrane disruption (i.e., reduced uptake of the dye/reduced retention inside the cell), the nucleotide-binding stain PI was used to assess membrane integrity.[10] A fully intact membrane is impermeable to PI, and DNA will only be stained in cells that are dead or that have compromised membranes. After a 3- to 24-h exposure, a separate sample of toxicant-exposed cells was therefore analyzed for changes in membrane permeability using PI and flow cytometry. Dual staining (FDA and PI in the same sample at the same time) was not performed in this bioassay but is possible when fluorescence compensation is used to eliminate spectral overlap in the dyes.

A stock solution of 100 μ*M* PI (Sigma P-4170 in Milli-Q water) was added to 5 ml aliquots of cells, to give final concentrations 7.5 and 120 μ*M* PI, suitable for *P. subcapitata* and *E.* cf. *punctulata*, respectively. After a 5-min incubation time, the sample was analyzed by flow cytometry, with PI detected as orange fluorescence emission (FL2). Cells with compromised membranes have a higher orange fluorescence than healthy intact cells, and this is seen as a shift to the right on a histogram plot of count versus FL2. If no cells have higher FL2 fluorescence, then any decrease in FDA fluorescence in FL1 is due to true inhibition of esterase activity. Cells killed by heat treatment (100°C for 10 min) or formaldehyde treatment (4% for 24 h) should be included in each experiment as negative controls.

Results and discussion

We have used flow cytometry-based protocols described in this chapter to examine the toxicity of copper to four species of microalgae. Through the ability to gate on chlorophyll *a* fluorescence (FL3), dead cells could easily be distinguished from live cells, so that only healthy cells were included in cell counts used to determine inhibition of cell division rate. This technique is therefore more accurate than automatic particle counters (e.g., Coulter counters) routinely used for algal growth tests, which often include dead cells and particulate matter in cell counts.

A major advantage of flow cytometry for growth inhibition tests is the ability to analyze and count low algal cell densities, more typical of algal concentrations in natural aquatic systems. Most standard test protocols use high initial cell densities of 10^4 to 10^5 cells/ml in order to obtain a measurable algal response. However, at these cell densities, significant adsorption of the test substance to the rapidly growing algal biomass may occur.[19,20] Algal metabolism at high cell densities can also cause drifts in pH due to CO_2 depletion and subsequent chemical alteration of the test substance.[21] Increased exudate production from an increasing number of cells may also influence chemical speciation through the formation of nontoxic metal–exudate complexes.[22] With the move towards miniscale tests to reduce costs,[23,24] surface adsorption losses and use of high cell densities may be even more problematic.

The effect of increasing initial cell density from 10^2 to 10^5 cells/ml on the toxicity of copper to *Chlorella* sp. and *P. subcapitata* is shown in Figure 39.2. As the initial cell density increased, the toxicity of copper to each species decreased (e.g., the curve shifts to the right). This corresponded to a significant decrease ($p < 0.05$) in the 72-h IC_{50} values for copper at initial cell densities of 10^4 and 10^5 cells/ml, compared to those at 10^2 and 10^3 cells/ml (Table 39.6). For *Chlorella* sp., the NOEC values at high initial cell densities (10^4 and 10^5 cells/ml) (4.7 and 9.0 μg Cu/l, respectively) were higher than the 72-h IC_{50} values obtained at the lower initial cell densities (4.4 and 4.6 μg Cu/l for 10^2 and 10^3 cells/ml, respectively).

Similar effects of reduced toxicity with increasing algal cell densities have been reported for a number of contaminants, including copper, cadmium, and lead.[25,26] Measured concentrations of extracellular and intracellular copper at 10^3, 10^4, and 10^5 cells/ml were determined by washing cells in dilute EDTA to remove extracellular copper.[9,27] As cell density increased, less copper was bound to the cells, leading to less copper uptake into the cell and consequently less disruption of cell division. Decreased copper toxicity at higher cell densities was primarily due to greater copper adsorption by algal cells and exudates, resulting in depletion of dissolved copper in solution.[9]

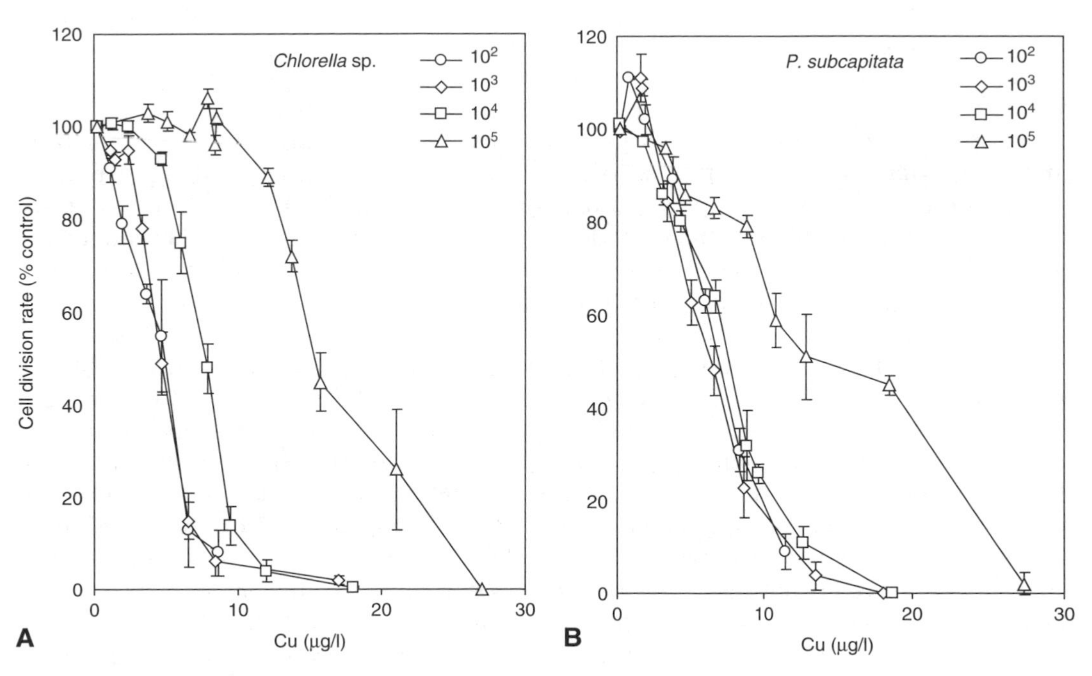

Figure 39.2 Growth inhibition of *Chlorella* sp. and *Pseudokirchneriella subcapitata* exposed to copper for 72 h at initial cell densities of 10^2, 10^3, 10^4, and 10^5 cells/ml. (From Franklin, N.M., Stauber, J.L., Apte, S.C. and Lim, R.P., *Environ. Toxicol. Chem.*, 21, 742–751, 2002.)

Table 39.6 Effect of initial cell density on the toxicity of copper to *Chlorella* sp. and *Pseudokirchneriella subcapitata* after a 72-h exposure. (Modified from Franklin, N.M., Stauber, J.L., Apte, S.C. and Lim, R.P., *Environ. Toxicol. Chem.*, 21, 742–751, 2002.)

Initial cell density	*Chlorella* sp.			*P. subcapitata*		
(cells/ml)	NOEC (μg/l)	LOEC (μg/l)	IC_{50} (μg/l)	NOEC (μg/l)	LOEC (μg/l)	IC_{50} (μg/l)
10^2	1.1	2.0	4.6^a (3.5–6.0)	1.9	3.8	6.6^{de} (5.9–7.3)
10^3	2.4	3.3	4.4^a (3.9–5.0)	3.4	4.8	6.2^d (5.5–6.9)
10^4	4.7	6.0	7.3^b (6.7–8.0)	1.8	3.2	7.5^e (6.8–8.2)
10^5	9.0	12	16^c (14–18)	4.6	6.6	17^f (14–20)

Superscript letters "a–f" denote whether IC_{50} values are significantly ($p < 0.05$) different from each other, i.e., same superscript means no significant ($p > 0.05$) difference.

The finding of decreased toxicity with increasing initial cell densities has important ramifications for static laboratory bioassays with microalgae, suggesting that the algal biomass itself may alter the concentration and speciation of copper in the test solution, leading to an underestimation of copper toxicity. While our research has focused on copper, the same limitations may apply to algal bioassays with other metals. It is therefore a key recommendation that initial cell densities used in standard growth bioassays do not exceed 2 to 4 $\times$ 10^3 cells/ml. In this way, metal bioavailability determined in laboratory algal bioassays should more closely estimate metal bioavailability in natural aquatic systems.

In addition to counting algal cells, the ability of flow cytometry to analyze algal light scatter and fluorescence properties can be exploited as alternative endpoints for acute and chronic toxicity tests.[6,10] Although chronic test endpoints are preferred due to their sensitivity and direct relevance to ecosystems, acute tests are often useful due to their rapid nature and low cost. In particular, chronic algal bioassays can suffer from changing contaminant concentrations over time (as described earlier) making them difficult to interpret.[28,29] By limiting exposure to only a few hours in short-term toxicity tests, such as the enzyme inhibition bioassay described in this chapter, these test limitations can be largely overcome. The esterases involved in the FDA bioassay turn over on a time frame of several hours,[30] making the technique suitable for monitoring short-term algal responses to environmental contaminants.[31]

Esterase activity, measured using FDA and flow cytometry, was a sensitive sublethal indicator of copper toxicity in *P. subcapitata* and *E.* cf. *punctulata* (Table 39.7). As copper concentrations increased, esterase activity decreased in a concentration-dependent manner, and this was seen as a shift in cells from S2 (control region) to S1 on a histogram of log fluorescein fluorescence (Figure 39.3). For *E. punctulata*, greater than 50% of cells had shifted into the lower esterase state (S1) after 24-h at a copper concentration of 85 μg/l, corresponding to an EC_{50} value of 63 $\pm$ 30 μg/l (Figure 39.3; Table 39.7). As little as 40 μg Cu/l caused a significant reduction in FDA fluorescence over the same exposure time (i.e., NOEC = 21 μg/l). The freshwater alga, *P. subcapitata*, showed similar sensitivity to copper with a 24-h EC_{50} value of 51 μg/l. PI staining for both species showed that high concentrations of copper did not significantly alter membrane integrity over 3- to 24-h copper exposures, indicating that the decreased FDA fluorescence was due to true inhibition of intracellular esterases.

Table 39.7 Inhibitory effect of copper on cell division rate and esterase activity (FDA fluorescence) in marine and freshwater algae using flow cytometry. (The data in this table is a compilation of published data from Franklin, N.M., Stauber, J.L. and Lim, R.P., *Environ. Toxicol. Chem.*, 20, 160–170, 2001a; Franklin, N.M., Adams, M.S., Stauber, J.L. and Lim, R.P., *Arch. Environ. Contam. Toxicol.*, 40, 469–480, 2001b; except the esterase data for *Entomoneis* cf. *punctulata*, which are unpublished.)

Test endpoint	Exposure time (h)	I/EC$_{50}$[a] (μg/l)			
		Chorella sp.	*P. subcapitata*	*E. punctulata*	*P. tricornutum*
Cell division rate[b]	48	6 ± 2[c]	6 ± 2	17 (15–20)	9 ± 3
	72	8 ± 2	8 ± 2	10 ± 4	10 ± 4
Esterase activity	3	—	112 (88–143)[d]	173 ± 80	—
(FDA fluorescence inhibition)	24	—	51 (38–70)	63 ± 30	—

[a] Inhibitory or effective concentration giving 50% effect or affecting 50% of cells.

[b] Tests conducted with initial cell density of 2 to 4 × 10^4 cells/ml.

[c] Mean ± SD (n = 3–13).

[d] 95% confidence limit (n = 6–30).

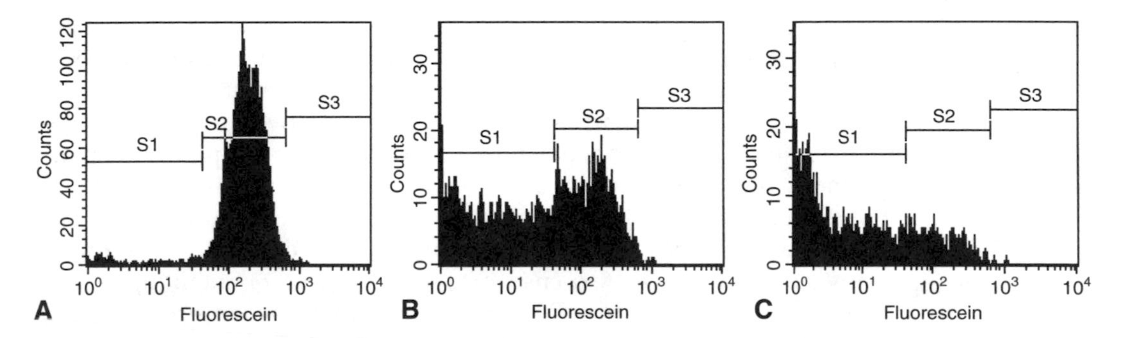

Figure 39.3 Flow cytometric histogram showing shifts in esterase activity (FL1 fluorescence) of *E.* cf. *punctulata* at copper concentrations of 0, 85, and 250 μg/l after a 24-h exposure.

For *E.* cf. *punctulata* and *P. subcapitata*, the FDA inhibition bioassay was less sensitive (~2-fold) after only a 3-h exposure and considerably less sensitive than chronic cell division rate over 48–72 h (Table 39.7). Gala and Giesy,[2] in assessing the toxicity of the surfactant SDS, also found that the 24-h EC$_{50}$ value for cell viability (using FDA) was significantly greater than the 96-h EC$_{50}$ value for specific growth rate. This is not unexpected as this is comparing an acute effect with a chronic effect, and EC$_{50}$ values typically decrease with increasing exposure time. One area where acute enzyme endpoints are particularly useful is assessing the sensitivity of contaminants in sediments. In this situation, growth is often stimulated by ammonia release from sediments, which can mask the toxic effects of contaminants. By contrast, FDA fluorescence is rarely stimulated and is therefore a more suitable endpoint for whole sediment and porewater tests than growth rate.[8]

Analysis of fluorescein fluorescence by flow cytometry, rather than fluorescence microscopy or fluorimetry, has improved the FDA method by providing a rapid and highly precise means of detecting the fluorescence of individual cells.[32] The cell-level perspective provided by flow cytometry offers more detailed information on cell physiology, which may be obscured if only the average characteristics of samples are measured (population-based endpoints assume that each cell behaves in an identical manner). With flow cytometry, cell functions are determined at conditions close to the *in vivo* state, without prolonged exposure to unusual light levels, and without the need for extractions of fluorescent products from cells in solution. In contrast to fluorescence microscopy, flow cytometry can not only distinguish labeled from nonlabeled cells but also estimate the fraction of the population with a positive reaction to the substrate. It can also provide a quantitative measure, in terms of relative fluorescence, of the rate at which the cells process the substrate (i.e., kinetics of reactions).[31] These advantages also apply to measurement of chlorophyll *a* or other fluorescence properties, not just FDA fluorescence.

Although not described in this chapter, another example of how flow cytometry is being used to improve the environmental relevance of laboratory toxicity testing is the development of multispecies algal bioassays. In aquatic systems, algae rarely occur in single species populations (except in algal blooms) but as mixtures of different species. In our laboratory, multispecies tests have recently been developed using mixtures of three freshwater (*Microcystis aeruginosa*, *P. subcapitata*, and *Trachelomonas* sp.) and three marine (*Micromonas pusilla*, *P. tricornutum*, and *Heterocapsa niei*) microalgae. Flow cytometry enabled separation of each algal population on the basis of their characteristic size and pigment fluorescence, so that the effects of copper in single species, versus multispecies tests could be assessed. Single species freshwater bioassays were shown to underestimate the toxicity of copper, whereas the marine single species tests overestimated copper toxicity.[7] Such multispecies bioassays open a new approach to water quality management, and flow cytometry offers great potential for further development in this area. The ability of some flow cytometers to physically separate different cell populations (cell sorting) will also enable wider application of flow cytometry in ecotoxicology in future.

Despite its many advantages, the expense of acquiring and maintaining a flow cytometer may be a potential factor currently limiting its routine use in ecotoxicology. Flow cytometers require considerable investment in equipment, together with skilled operators. However, once the individual test protocols are established, it is relatively easy to train operators in their routine use. Development of flow cytometric methods has predominantly been directed towards medical research and diagnostics involving the analysis of mammalian cells, particularly red and white blood cells. However, with increasing interest in the ecotoxicological applications of flow cytometry, particularly in the area of microbial research, costs will continue to be reduced and instruments will increasingly be tailored to these purposes. Use of flow cytometers in a multiuser central facility or university laboratory may be a suitable option to reduce costs.

Conclusions

The application of flow cytometry to ecotoxicology is enabling the development of more environmentally relevant aquatic toxicity tests with microalgae. Multiparameter, multispecies toxicity tests at low cell densities are now available to better assess the bioavailability of contaminants in aquatic systems. These test protocols are suitable for assessing the toxicity of effluents, sediments, and marine and freshwaters. Such toxicity tests are an important part of an integrated approach to water quality management, combining chemical measurement techniques, geochemical speciation modeling, bioassays, and ecological monitoring, to better understand ecological processes, metal bioavailability, and mechanisms of toxicity to microorganisms.

References

1. Dorsey, J., Yentsch, C., Mayo, S. and McKenna, C., Rapid analytical technique for the assessment of cell metabolic activity in marine microalgae, *Cytometry*, 10, 622–628, 1989.
2. Gala, W.R. and Giesy, J.P., Flow cytometric techniques to assess toxicity to algae, in *Aquatic Toxicology and Risk Assessment*, Landis, W.G. and Van der Schalie, W.H., Eds., vol. 13, American Society for Testing and Materials, Philadelphia, PA, 1990, pp. 237–246.
3. Snell, T.W., Mitchell, J.L. and Burbank, S.E., Rapid toxicity assessment with microalgae using *in vivo* esterase inhibition, in *Techniques in Aquatic Toxicology*, Ostrander, G.K., Ed., Lewis Publishers, 1996.
4. Regel, R.H., Ferris, J.M., Ganf, G.G. and Brookes, J.D., Algal esterase activity as a biomeasure of environmental degradation in a freshwater creek, *Aquat. Toxicol.*, 59, 209–223, 2002.
5. Stauber, J.L. and Davies, C.M., Use and limitations of microbial bioassays for assessing copper bioavailability in the aquatic environment, *Environ. Rev.*, 8, 255–301, 2000.
6. Franklin, N.M., Stauber, J.L. and Lim, R.P., Development of flow cytometry-based algal bioassays for assessing toxicity of copper in natural waters, *Environ. Toxicol. Chem.*, 20, 160–170, 2001a.
7. Franklin, N.M., Stauber, J.L. and Lim, R.P., Development of multispecies algal bioassays using flow cytometry, *Environ. Toxicol. Chem.* 23, 1452–1462, 2004.
8. Adams, M.S. and Stauber, J.L., Development of a whole-sediment toxicity test using a benthic marine microalga, *Environ. Toxicol. Chem.* 23, 1957–1968, 2004.
9. Franklin, N.M., Stauber, J.L., Apte, S.C. and Lim, R.P., The effect of initial cell density on the bioavailability and toxicity of copper in microalgal bioassays, *Environ. Toxicol. Chem.*, 21, 742–751, 2002.
10. Franklin, N.M., Adams, M.S., Stauber, J.L. and Lim, R.P., Development of an improved rapid enzyme inhibition bioassay with marine and freshwater microalgae using flow cytometry, *Arch. Environ. Contam. Toxicol.*, 40, 469–480, 2001b.
11. Stauber, J.L., Franklin, N.M. and Adams, M.S., Applications of flow cytometry to ecotoxicity testing using microalgae, *Trends Biotechnol.*, 20, 141–143, 2002.
12. Blaise, C. and Ménard, L., A micro-algal solid phase test to assess the toxic potential of freshwater sediments, *Water Qual. Res. J. Can.*, 33, 133–151, 1998.
13. Hall, J. and Cumming, A., Flow cytometry in aquatic science, *Water Atmos.*, 11, 24–25, 2003.
14. U.S. Environmental Protection Agency, *Short-term Methods for Estimating the Chronic Toxicity of Effluents and Receiving Waters to Freshwater Organisms*, 4th ed., EPA-821-R-02-013, Washington, D.C., USA, 2002, p. 27.
15. Thompson, A.S., Rhodes, J.C. and Pettman, I., *Culture Collection of Algae and Protozoa: Catalogue of Strains*, Natural Environmental Research Council, Swindon, UK, 1988.
16. Guillard, R.R.L. and Ryther, J.H., Studies on marine planktonic diatoms. I. *Cyclotella nana* Hustedt and *Detonula confervacea* (Cleve) Gran, *Can. J. Microbiol.*, 8, 229–239, 1962.
17. Li, W.K.W., Particles in "particle-free" seawater: growth of ultraphytoplankton and implications for dilution experiments, *Can. J. Fish. Aquat. Sci.*, 47, 1258–1268, 1990.
18. Nyholm, N., Sorensen, P.S., Kusk, K.O. and Christensen, E.R., Statistical treatment of data from microbial toxicity tests, *Environ. Toxicol. Chem.*, 11, 157–167, 1992.
19. Hörnström, E., Toxicity test with algae — a discussion of the batch method, *Ecotoxicol. Environ. Safety*, 20, 343–353, 1990.
20. Mayer, P., Nyholm, N., Verbruggen, E.M.J., Hermens, J.L.M. and Tolls, J., Algal growth inhibition test in filled, closed bottles for volatile and sorptive materials, *Environ. Toxicol. Chem.*, 19, 2551–2556, 2000.

21. Nyholm, N. and Källqvist, T., Methods for growth inhibition toxicity tests with freshwater algae, *Environ. Toxicol. Chem.*, 8, 689–703, 1989.
22. González-Dávila, M., Santana-Casiano, J.M., Pérez-Peña, J. and Millero, F.J., Binding of Cu(II) to the surface and exudates of the alga *Dunaliella tertiolecta* in seawater, *Environ. Sci. Technol.*, 29, 289–301, 1995.
23. Arensberg, P., Hemmingsen, V.H. and Nyholm, N., A miniscale algal toxicity test, *Chemosphere*, 30, 2103–2115, 1995.
24. Ismail, M., Phang, S.-M., Tong, S.-L. and Brown, M.T., A modified toxicity testing method using tropical marine microalgae, *Environ. Monit. Assess.*, 75, 145–154, 2002.
25. Vasseur, P., Pandard, P. and Burnel, D., Influence of some experimental factors on metal toxicity to *Selenastrum capricornutum*, *Toxicol. Assess.*, 3, 331–343, 1988.
26. Moreno-Garrido, I., Lubián, L.M. and Soares, A.M.V.M., Influence of cellular density on determination of EC50 in microalgal growth inhibition tests, *Ecotoxicol. Environ. Safety*, 47, 112–116, 2000.
27. Stauber, J.L. and Florence, T.M., The effect of culture medium on metal toxicity to the marine diatom *Nitzschia closterium* and the freshwater green alga *Chlorella pyrenoidosa*, *Water Res.*, 23, 907–911, 1989.
28. Péry, A.R.R., Bedaux, J.J.M., Zonneveld, C. and Kooijman, S.A.L.M., Analysis of bioassays with time-varying concentrations, *Water Res.*, 35, 3825–3832, 2001.
29. Simpson, S.L., Roland, M.G., Stauber, J.L. and Batley, G.E., Effect of declining toxicant concentrations on algal bioassay endpoints, *Environ. Toxicol. Chem.*, 22, 2073–2079, 2003.
30. Yentsch, C.M., Cucci, T.L. and Mague, F.C., Profiting from the visible spectrum, *Biol. Oceanogr.*, 6, 477–492, 1988.
31. Jochem, F.J., Probing the physiological state of phytoplankton at the single-cell level, *Sci. Mar.*, 64, 183–195, 2000.
32. Shapiro, H.M., *Practical Flow Cytometry*, 3rd ed., Wiley-Liss, New York, 1995.

Index

LUDWIG WITTGENSTEIN

LUDWIG WITTGENSTEIN

The Duty of Genius

RAY MONK

THE FREE PRESS
A Division of Macmillan, Inc.
NEW YORK

Maxwell Macmillan International
NEW YORK OXFORD SINGAPORE SYDNEY

Copyright © 1990 by Ray Monk

All rights reserved. No part of this book may be reproduced or transmitted in any form or by any means, electronic or mechanical, including photocopying, recording, or by any information storage and retrieval system, without permission in writing from the Publisher.

The Free Press
A Division of Macmillan, Inc.
866 Third Avenue, New York, N.Y. 10022

Collier Macmillan Canada, Inc.
1200 Eglinton Avenue East
Suite 200
Don Mills, Ontario M3C 3N1

First American Edition 1990

Printed in the United States of America

printing number

3 4 5 6 7 8 9 10

Library of Congress Cataloging-in-Publication Data

Monk, Ray.
Ludwig Wittgenstein : the duty of genius / Ray Monk.—1st American ed.
p. cm.
Includes bibliographical references and index.
ISBN 0-02-921670-2
1. Wittgenstein, Ludwig, 1889-1951. 2. Philosophers—Austria—Biography. 3. Philosophers—Great Britain—Biography. I. Title
B3376.W564M59 1990
192—dc20
[B] 90-37619
CIP

Excerpt from "The Strayed Poet" by Ludwig Wittgenstein from *Internal Colloquies,* copyright © 1971 by I.A. Richards, reprinted by permission of Harcourt Brace Jovanovich, Inc.

Logic and ethics are fundamentally the same,
they are no more than duty to oneself.

Otto Weininger, *Sex and Character*

CONTENTS

ILLUSTRATIONS

The author and publishers would like to thank the following for permission to use photographs: Dr Milo Keynes (13); Anne Keynes (15); Dr Norman Malcolm (46); Michael Nedo (1–7, 9–12, 19–34, 36, 39–43, 47, 50–54); Neue Pinakothek, Munich (photo: Artothek: 7); Gilbert Pattisson (37–8, 41–2, 44–5); Ferry Radax (16–18, 48), Technische Hochschule, Berlin (now the Technical University of Berlin: 8); Trinity College, Cambridge (14, 35).

ACKNOWLEDGEMENTS

My first thanks must go to Monica Furlong, without whose support this book would never have been started. It was she who persuaded David Godwin (then Editorial Director of Heinemann) to consider the possibility of financing the project. No less essential has been the undying enthusiasm and kind encouragement of David Godwin himself and the equally unstinting support given me by my American publisher, Erwin Glikes of the Free Press.

It was initially feared that the project would founder for lack of co-operation from Wittgenstein's literary heirs. I am happy to report that the exact opposite has been the case. Wittgenstein's three literary executors, Professor Georg Henrik von Wright, Professor G. E. M. Anscombe and the late Mr Rush Rhees have all been exceptionally kind, co-operative and helpful. In addition to granting me permission to quote from Wittgenstein's unpublished manuscripts, they have been assiduous in replying to my many questions and very generous in providing me with information that I would not otherwise have found out.

To Professor von Wright I am particularly grateful for his very patient and detailed replies to my (initially rather primitive) speculations concerning the composition of *Philosophical Investigations*. His articles on the origins of Wittgenstein's two great works and his meticulous cataloguing of Wittgenstein's papers have been indispensable. Professor Anscombe agreed to meet me on a number of occasions to speak to me about her own memories of Wittgenstein and to answer my enquiries. To her I am especially grateful, for allowing me access to Francis Skinner's letters to Wittgenstein.

The kindness shown to me by Mr Rhees was far above and beyond the call of duty. Despite his advancing years and frail health he devoted many hours to discussions with me; during which he revealed his incomparable knowledge of Wittgenstein's work and his many insights into both Wittgenstein's personality and his philosophy. He further showed me many documents, the existence of which I would not otherwise have known. So concerned was he to impart to me as much of what he knew as possible that on one occasion he insisted on paying for me to stay in a hotel in Swansea, so that our discussions would not have to be curtailed by my having to return to London. News of his death came just as I finished work on the book. He will be sorely missed.

Sadly, a number of other friends of Wittgenstein's died while I was researching this book. Roy Fouracre had been ill for a long time, but his wife was kind enough to see me and to supply me with copies of the letters from Wittgenstein to her husband. Similarly kind was Katherine Thomson, whose late husband, Professor George Thomson, expressed a wish to meet me shortly before he died, in order to discuss Wittgenstein's visit to the Soviet Union. Mrs Thomson also showed me some letters and discussed with me her own memories of Wittgenstein. Dr Edward Bevan I met a year or so before his death. His recollections, and those of his widow, Joan Bevan, form the basis of Chapter 27. Tommy Mulkerrins, who provided Wittgenstein with indispensable help during his stay on the west coast of Ireland, was an invalided but exceptionally alert octogenarian when I met him in his cottage in the spring of 1986. His reminiscences have been incorporated into Chapter 25. He too, alas, is no longer with us.

Other friends, happily, are alive and well. Mr Gilbert Pattisson, a close friend of Wittgenstein's from 1929 to 1940, met with me a number of times and provided me with the letters quoted in Chapter 11. Mr Rowland Hutt, a friend of both Wittgenstein and Francis Skinner, took a lively and helpful interest in my work and provided me with the letters quoted in Chapter 23. I am grateful also to Mr William Barrington Pink, Sir Desmond Lee, Professor Basil Reeve, Dr Ben Richards, Dr Casimir Lewy, Mr Keith Kirk, Mrs A. Clement, Mrs Polly Smythies, Professor Wolfe Mays, Mrs Frances Partridge and Madame Marguerite de Chambrier, all of whom took the trouble to speak to me – in some cases over the course of several meetings – about their recollections of Wittgenstein. To Professor Georg Kreisel,

Professor F. A. von Hayek, Mr John King, Professor Wasif A. Hijab, Professor John Wisdom, the late Professor Sir Alfred Ayer and Father Conrad Pepler, I am grateful for replies by letter to my enquiries.

The account of Wittgenstein's work at Guy's Hospital and the Royal Infirmary at Newcastle could not have been written without the help of Wittgenstein's colleagues: Mr T. Lewis, Dr Humphrey Osmond, Dr R. T. Grant, Miss Helen Andrews, Dr W. Tillman, Miss Naomi Wilkinson, Dr R. L. Waterfeld, Dr Erasmus Barlow and Professor Basil Reeve. To Dr John Henderson I am grateful for help in contacting many of these colleagues. Dr Anthony Ryle kindly showed me the letter from his father quoted in Chapter 21, and allowed me to quote from the diary he kept as a child. To him and to Professor Reeve I am also grateful for reading and commenting on an earlier draft of this chapter.

To Mr Oscar Wood, Sir Isaiah Berlin and Dame Mary Warnock I am grateful for their recollections of the meeting of the Jowett Society described in Chapter 24, the only occasion on which Wittgenstein took part in a philosophical meeting at Oxford.

Many people who did not meet Wittgenstein also provided me with valuable help, and in this context I am pleased to acknowledge my gratitude to Professor W. W. Bartley III, Professor Quentin Bell, Mrs Margaret Sloan, Mr Michael Straight, Mr Colin Wilson and Professor Konrad Wünsche, who all replied helpfully to my letters, and to Mrs Anne Keynes, Dr Andrew Hodges and Professor George Steiner, who were kind enough to arrange to meet me to discuss matters arising from my research. Mrs Keynes also kindly provided me with a dissertation on philosophy written by her uncle, David Pinsent.

My research took me far and wide, but two trips in particular should be mentioned; those to Ireland and to Austria. In Ireland I was chauffeured around Dublin, County Wicklow and County Galway by my friend Jonathan Culley, who showed inexhaustible patience and supplied a much-needed (but otherwise missing) sense of urgency and punctuality. In Dublin I was helped by Mr Paul Drury, in Wicklow by the Kingston family and in Connemara by Tommy Mulkerrins. Mr and Mrs Hugh Price, Mrs R. Willoughby and Mr Sean Kent gave much appreciated assistance along the way. My trip to Austria was made pleasant and comfortable by the kindness shown to me by my friend Wolfgang Grüber and by the hospitality of his brother, Heimo. In Vienna I had the pleasure of meeting Mrs Katrina Eisenburger, a granddaughter of Helene Wittgenstein, and another member of the family, Dr Elizabeth Wieser. I also received the kind help of Professor Hermann Hänsel. On my visit to the Wechsel mountains, where Wittgenstein taught at the schools of Trattenbach and Otterthal, I was helped enormously by Dr Adolf Hübner, who not only escorted me around the area and supplied me with copies of the fascinating material he has collected for the Documentation Centre in Kirchberg, but also – with quite extraordinary kindness – took the trouble to re-shoot a series of photographs that I had taken after it had been discovered that my own pictures had been ruined.

To Dr T. Hobbs of the Wren Library, Trinity College, Cambridge; Dr A. Baster of the Wills Library, Guy's Hospital; Miss M. Nicholson of the Medical Research Council Archives; and the staff of the British Library, the Bodleian Library, Oxford, and the Cambridge University Library, I extend my gratitude for their undying courtesy and helpful assistance. To my friend Mr Wolf Salinger I am grateful for the trouble he took on my behalf to uncover whatever records exist at the Technical University of Berlin of the time that Wittgenstein spent as a student of engineering there (when it was the Technische Hochschule). I am grateful also to the staff of the university library for the help they gave Mr Salinger.

One of the most important collections of letters used in the book is that held by the Brenner Archive at the University of Innsbruck. This is a collection of several hundred letters to Wittgenstein (including those from Bertrand Russell and Gottlob Frege used in chapters 6, 7, 8 and 9) that has only recently been made available. I am grateful to Dr P. M. S. Hacker of St John's College, Oxford, for drawing my attention to the existence of this collection, and to Dr Walter Methlagl and Professor Allan Janik of the Brenner Archive for their kindness in allowing me access to it, and for giving up their time to discuss the contents of the letters with me. For permission to quote these and other letters from Bertrand Russell my thanks go to Kenneth Blackwell of the Russell Archive, McMaster University, Hamilton, Ontario.

I owe a special debt of gratitude to Dr Michael Nedo of Trinity College, Cambridge, whose knowledge of Wittgenstein's manuscripts is unrivalled, and who has, over the years, collected photographs, documents and copies of documents connected with Wittgenstein that constitute an immensely useful archive. He not only gave me completely free access to this material, but also devoted much of his time to the discussion of many and various aspects of my research. I am also greatly indebted to him for providing me with copies of his careful transcriptions of the coded remarks in Wittgenstein's manuscripts.

Dr Paul Wijdeveld has been enormously helpful in a similarly varied fashion. He has given me the benefit of the meticulous research he has conducted in connection with his book on the house that Wittgenstein designed, alerted me to the existence of published sources that I would not otherwise have known about, and provided me with copies of drafts of his own work and of the many documents he has discovered relating to Wittgenstein's relationship with Paul Engelmann.

For reading and passing comments on earlier drafts of parts of this book, I am grateful to Dr G. P. Baker of St John's College, Oxford, and Professor Sir Peter Strawson of Magdalen College, Oxford. Dr Baker and his colleague Dr P. M. S. Hacker were also kind enough to make available to me work that they are currently engaged upon. Professor Stephen Toulmin kindly read through the whole manuscript, and made a number of helpful suggestions and constructive criticisms. My editors, David Godwin and Erwin Glikes, have read many earlier drafts, and have also made a great number of useful suggestions. In preparing the manuscript for publication, Alison Mansbridge pointed out many errors that I would not otherwise have noticed, and I am greatly indebted to her for the enthusiasm and meticulousness with which she undertook her difficult task. Dr David McLintock kindly checked the accuracy of my translations of Frege's letters and Wittgenstein's diary entries. He made many important corrections and drew my attention to a number of interesting nuances and allusions I would otherwise have missed. Any errors that remain are, of course, entirely my responsibility.

Without the help of my agent, Mrs Gill Coleridge, I could not have survived the last four years. To Jenny, I owe my most heartfelt thanks for having survived them with me.

London
December 1989 RAY MONK

INTRODUCTION

The figure of Ludwig Wittgenstein exerts a very special fascination that is not wholly explained by the enormous influence he has had on the development of philosophy this century. Even those quite unconcerned with analytical philosophy find him compelling. Poems have been written about him, paintings inspired by him, his work has been set to music, and he has been made the central character in a successful novel that is little more than a fictionalized biography (*The World as I Found It*, by Bruce Duffy). In addition, there have been at least five television programmes made about him and countless memoirs of him written, often by people who knew him only very slightly. (F. R. Leavis, for example, who met him on perhaps four or five occasions, has made his 'Memories of Wittgenstein' the subject of a sixteen-page article.) Recollections of Wittgenstein have been published by the lady who taught him Russian, the man who delivered peat to his cottage in Ireland and the man who, though he did not know him very well, happened to take the last photographs of him.

All this is, in a way, quite separate from the ongoing industry of producing commentaries on Wittgenstein's philosophy. This industry too, however, continues apace. A recent bibliography of secondary sources lists no fewer than 5,868 articles and books about his work. Very few of these would be of interest (or even intelligible) to anyone outside academia, and equally few of them would concern themselves with the aspects of Wittgenstein's life and personality that have inspired the work mentioned in the previous paragraph.

It seems, then, that interest in Wittgenstein, great though it is, suffers from an unfortunate polarity between those who study his work in isolation from his life and those who find his life fascinating but his work unintelligible. It is a common experience, I think, for someone to read, say, Norman Malcolm's *Memoir*, to find themselves captivated by the figure described therein, and then be inspired to read Wittgenstein's work for themselves, only to find that they cannot understand a word of it. There are, it has to be said, many excellent introductory books on Wittgenstein's work that would explain what his main philosophical themes are, and how he deals with them. What they do not explain is what his work has to do with *him* – what the connections are between the spiritual and ethical preoccupations that dominate his life, and the seemingly rather remote philosophical questions that dominate his work.

The aim of this book is to bridge that gap. By describing the life and the work in the one narrative, I hope to make it clear how this work came from this man, to show – what many who read Wittgenstein's work instinctively feel – the unity of his philosophical concerns with his emotional and spiritual life.

I

1889–1919

I

THE LABORATORY FOR SELF-DESTRUCTION

'Why should one tell the truth if it's to one's advantage to tell a lie?'

Such was the subject of Ludwig Wittgenstein's earliest recorded philosophical reflections. Aged about eight or nine, he paused in a doorway to consider the question. Finding no satisfactory answer, he concluded that there was, after all, nothing wrong with lying under such circumstances. In later life, he described the event as, 'an experience which if not decisive for my future way of life was at any rate characteristic of my nature at that time'.

In one respect the episode is characteristic of his entire life. Unlike, say, Bertrand Russell, who turned to philosophy with hope of finding certainty where previously he had felt only doubt, Wittgenstein was drawn to it by a compulsive tendency to be struck by such questions. Philosophy, one might say, came to him, not he to philosophy. Its dilemmas were experienced by him as unwelcome intrusions, enigmas, which forced themselves upon him and held him captive, unable to get on with everyday life until he could dispel them with a satisfactory solution.

Yet Wittgenstein's youthful answer to this particular problem is, in another sense, deeply uncharacteristic. Its easy acceptance of dishonesty is fundamentally incompatible with the relentless truthfulness for which Wittgenstein was both admired and feared as an adult. It is incompatible also, perhaps, with his very sense of being a philosopher. 'Call me a truth-seeker', he once wrote to his sister (who had, in a letter to him, called him a great philosopher), 'and I will be satisfied.'

This points not to a change of opinion, but to a change of character –

the first of many in a life that is marked by a series of such transformations, undertaken at moments of crisis and pursued with a conviction that the source of the crisis was himself. It is as though his life was an ongoing battle with his own nature. In so far as he achieved anything, it was usually with the sense of its being in spite of his nature. The ultimate achievement, in this sense, would be the complete overcoming of himself – a transformation that would make philosophy itself unnecessary.

In later life, when someone once remarked to him that the childlike innocence of G. E. Moore was to his credit, Wittgenstein demurred. 'I can't understand that', he said, 'unless it's also to a *child*'s credit. For you aren't talking of the innocence a man has fought for, but of an innocence which comes from a natural absence of a temptation.'

The remark hints at a self-assessment. Wittgenstein's own character – the compelling, uncompromising, dominating personality recalled in the many memoirs of him written by his friends and students – *was* something he had had to fight for. As a child he had a sweet and compliant disposition – eager to please, willing to conform, and, as we have seen, prepared to compromise the truth. The story of the first eighteen years of his life is, above all, the story of this struggle, of the forces within him and outside him that impelled such transformation.

He was born – Ludwig Josef Johann Wittgenstein – on 26 April 1889, the eighth and youngest child of one of the wealthiest families in Habsburg Vienna. The family's name and their wealth have led some to suppose that he was related to a German aristocratic family, the Seyn-Wittgensteins. This is not so. The family had been Wittgensteins for only three generations. The name was adopted by Ludwig's paternal great-grandfather, Moses Maier, who worked as a land-agent for the princely family, and who, after the Napoleonic decree of 1808 which demanded that Jews adopt a surname, took on the name of his employers.

Within the family a legend grew up that Moses Maier's son, Hermann Christian Wittgenstein, was the illegitimate offspring of a prince (whether of the house of Wittgenstein, Waldeck or Esterházy depends on the version of the story), but there are no solid grounds for believing this. The truth of the story seems all the more doubtful, since it appears to date from a time when the family was attempting (successfully, as we shall see later) to have itself reclassified under the Nuremberg Laws.

The story would no doubt have suited Hermann Wittgenstein himself, who adopted the middle name 'Christian' in a deliberate attempt to dissociate himself from his Jewish background. He cut himself off entirely from the Jewish community into which he was born and left his birthplace of Korbach to live in Leipzig, where he pursued a successful career as a wool-merchant, buying from Hungary and Poland and selling to England and Holland. He chose as his wife the daughter of an eminent Viennese Jewish family, Fanny Figdor, but before their wedding in 1838 she too had converted to Protestantism.

By the time they moved to Vienna in the 1850s the Wittgensteins probably no longer regarded themselves as Jewish. Hermann Christian, indeed, acquired something of a reputation as an anti-Semite, and firmly forbade his offspring to marry Jews. The family was large – eight daughters and three sons – and on the whole they heeded their father's advice and married into the ranks of the Viennese Protestant professional classes. Thus was established a network of judges, lawyers, professors and clergymen which the Wittgensteins could rely on if they needed the services of any of the traditional professions. So complete was the family's assimilation that one of Hermann's daughters had to ask her brother Louis if the rumours she had heard about their Jewish origins were true. '*Pur sang*, Milly', he replied, '*pur sang*.'

The situation was not unlike that of many other notable Viennese families: no matter how integrated they were into the Viennese middle class, and no matter how divorced from their origins, they yet remained – in some mysterious way – Jewish 'through and through'.

The Wittgensteins (unlike, say, the Freuds) were in no way part of a Jewish community – except in the elusive but important sense in which the whole of Vienna could be so described; nor did Judaism play any part in their upbringing. Their culture was entirely Germanic. Fanny Wittgenstein came from a merchant family which had close connections with the cultural life of Austria. They were friends of the poet Franz Grillparzer and known to the artists of Austria as enthusiastic and discriminating collectors. One of her cousins was the famous violin virtuoso, Joseph Joachim, in whose development she and Hermann played a decisive role. He was adopted by them at the age of twelve and sent to study with Felix Mendelssohn. When the composer asked what he should teach the boy, Hermann Wittgenstein replied: 'Just let him breathe the air you breathe!'

Through Joachim, the family was introduced to Johannes Brahms, whose friendship they prized above any other. Brahms gave piano lessons to the daughters of Hermann and Fanny, and was later a regular attender at the musical evenings given by the Wittgensteins. At least one of his major works – the Clarinet Quintet – received its first performance at the Wittgenstein home.

Such was the air the Wittgensteins breathed – an atmosphere of cultural attainment and comfortable respectability, tainted only by the bad odour of anti-Semitism, the merest sniff of which was sufficient to keep them forever reminded of their 'non-Aryan' origins.

His grandfather's remark to Mendelssohn was to be echoed many years later by Ludwig Wittgenstein when he urged one of his students at Cambridge, Maurice Drury, to leave the university. 'There is', he told him, 'no oxygen in Cambridge for you.' Drury, he thought, would be better off getting a job among the working class, where the air was healthier. With regard to himself – to his own decision to stay at Cambridge – the metaphor received an interesting twist: 'It doesn't matter for me', he told Drury. 'I manufacture my own oxygen.'

His father, Karl Wittgenstein, had shown a similar independence from the atmosphere in which he was brought up, and the same determination to manufacture his own. Karl was the exception among the children of Hermann and Fanny – the only one whose life was not determined by their aspirations. He was a difficult child, who from an early age rebelled against the formality and authoritarianism of his parents and resisted their attempts to provide him with the kind of classical education appropriate to a member of the Viennese bourgeoisie.

At the age of eleven he tried to run away from home. At seventeen he got himself expelled from school by writing an essay denying the immortality of the soul. Hermann persevered. He tried to continue Karl's education at home by employing private tutors to see him through his exams. But Karl ran off again, and this time succeeded in getting away. After hiding out in the centre of Vienna for a couple of months, he fled to New York, arriving there penniless and carrying little more than his violin. He managed nevertheless to maintain himself for over two years by working as a waiter, a saloon musician, a bartender and a teacher (of the violin, the horn, mathematics, German and anything else he could think of). The adventure served to establish that he was his own master, and when he returned to Vienna in 1867 he was allowed – indeed, encouraged – to pursue his practical and technical bent, and to study engineering rather than follow his father and his brothers into estate management.

After a year at the technical high school in Vienna and an apprenticeship consisting of a series of jobs with various engineering firms, Karl was offered the post of draughtsman on the construction of a rolling mill in Bohemia by Paul Kupelwieser, the brother of his brother-in-law. This was Karl's great opportunity. His subsequent rise within the company was so astonishingly swift that within five years he had succeeded Kupelwieser as managing director. In the ten years that followed he showed himself to be perhaps the most astute industrialist in the Austro-Hungarian Empire. The fortunes of his company – and, of course, his own personal fortune – increased manifold, so that by the last decade of the nineteenth century he had become one of the wealthiest men in the empire, and the leading figure in its iron and steel industry. As such, he became, for critics of the excesses of capitalism, one of the archetypes of the aggressively acquisitive industrialist. Through him the Wittgensteins became the Austrian equivalent of the Krupps, the Carnegies, or the Rothschilds.

In 1898, having amassed a huge personal fortune which to this day provides comfortably for his descendants, Karl Wittgenstein suddenly retired from business, resigning from the boards of all the steel companies he had presided over and transferring his investments to foreign – principally US – equities. (This last act proved to be remarkably prescient, securing the family fortune against the inflation that crippled Austria after the First World War.) He was by this time the father of eight extraordinarily talented children.

The mother of Karl Wittgenstein's children was Leopoldine Kalmus, whom Karl had married in 1873, at the beginning of his dramatic rise through the Kupelwieser company. In choosing her, Karl was once again proving to be the exception in his family, for Leopoldine was the only partly Jewish spouse of any of the children of Hermann Christian. However, although her father, Jakob Kalmus, was descended from a prominent Jewish family, he himself had been brought up a Catholic; her mother, Marie Stallner, was entirely 'Aryan' – the daughter of an established (Catholic) Austrian land-owning family. In fact, then (until the Nuremberg Laws were applied in Austria, at least), Karl had not married a Jewess, but a Catholic, and had thus taken a further step in the assimilation of the Wittgenstein family into the Viennese establishment.

The eight children of Karl and Leopoldine Wittgenstein were baptized into the Catholic faith and raised as accepted and proud members of the Austrian high-bourgeoisie. Karl Wittgenstein was even given the chance of joining the ranks of the nobility, but declined the invitation to add the aristocratic 'von' to his name, feeling that such a gesture would be seen as the mark of the parvenu.

His immense wealth nevertheless enabled the family to live in the style of the aristocracy. Their home in Vienna, in the 'Alleegasse' (now Argentinergasse), was known outside the family as the Palais Wittgenstein, and was indeed palatial, having been built for a count earlier in the century. In addition to this, the family owned another house, in the Neuwaldeggergasse, on the outskirts of Vienna, and a large estate in the country, the Hochreit, to which they retired during the summer.

Leopoldine (or 'Poldy' as she was known to the family) was, even when judged by the very highest standards, exceptionally musical. For her, music came second only to the well-being of her husband, as the most important thing in her life. It was owing to her that the Alleegasse house became a centre of musical excellence. Musical evenings there were attended by, among others, Brahms, Mahler and Bruno Walter, who has described 'the all-pervading atmosphere of humanity and culture' which prevailed. The blind organist and composer Josef Labor owed his career largely to the patronage of the Wittgenstein family, who held him in enormously high regard. In later life Ludwig Wittgenstein was fond of saying that there had been just six *great* composers: Haydn, Mozart, Beethoven, Schubert, Brahms – and Labor.

After his retirement from industry Karl Wittgenstein became known also as a great patron of the visual arts. Aided by his eldest daughter, Hermine – herself a gifted painter – he assembled a noteworthy collection of valuable paintings and sculptures, including works by Klimt, Moser and Rodin. Klimt called him his 'Minister of Fine Art', in gratitude for his financing of both the Secession Building (at which the works of Klimt, Schiele and Kokoschka were exhibited), and Klimt's own mural, *Philosophie*, which had been rejected by the University of Vienna. When Ludwig's sister, Margarete Wittgenstein, married in 1905, Klimt was commissioned to paint her wedding portrait.

The Wittgensteins were thus at the centre of Viennese cultural life during what was, if not its most glorious era, at least its most dynamic. The period of cultural history in Vienna from the late nineteenth century to the outbreak of the First World War has, quite justifiably, been the centre of much interest in recent years. It has been described as a time of 'nervous splendour', a phrase which might also be used to characterize the environment in which the children of Karl and Poldy were raised. For just as in the city at large, so within the family, beneath the 'all-pervading atmosphere of culture and humanity', lay doubt, tension and conflict.

The fascination of *fin de siècle* Vienna for the present-day lies in the fact that its tensions prefigure those that have dominated the history of Europe during the twentieth century. From those tensions sprang many of the intellectual and cultural movements that have shaped that history. It was, in an oft-quoted phrase of Karl Kraus, the 'research laboratory for world destruction' – the birthplace of both Zionism and Nazism, the place where Freud developed psychoanalysis, where Klimt, Schiele and Kokoschka inaugurated the *Jugendstil* movement in art, where Schoenberg developed atonal music and Adolf Loos introduced the starkly functional, unadorned style of architecture that characterizes the buildings of the modern age. In almost every field of human thought and activity, the new was emerging from the old, the twentieth century from the nineteenth.

That this should happen in Vienna is especially remarkable, since it was the centre of an empire that had, in many ways, not yet emerged from the eighteenth century. The anachronistic nature of this empire was symbolized by its aged ruler. Franz Josef, Emperor of Austria since 1848 and King of Hungary since 1867, was to remain both *kaiserlich* and *königlich* until 1916, after which the ramshackle conglomeration of kingdoms and principalities that had formed the Habsburg Empire soon collapsed, its territory to be divided between the nation states of Austria, Hungary, Poland, Czechoslovakia, Yugoslavia and Italy. The nineteenth-century movements of nationalism and democracy had made its collapse inevitable a long time before that, and for the last fifty or so years of its existence the empire survived by teetering from one crisis to the next, its continuing survival believed in only by those who turned a blind eye to the oncoming tides. For those who wished it to survive, the political situation was always 'desperate, but not serious'.

The emergence of radical innovation in such a state is not, perhaps, such a paradox: where the old is in such transparent decay, the new *has* to emerge. The empire *was* a home for genius, after all, as Robert Musil once famously observed: 'and that, probably, was the ruin of it'.

What divided the intellectuals of *Jung Wien* from their forebears was their recognition of the decay around them, a refusal to pretend that things could go on as they always had. Schoenberg's atonal system was founded on the conviction that the old system of composition had run its course; Adolf Loos's rejection of ornament on the recognition that the baroque adornments to buildings had become an empty shell, signifying nothing; Freud's postulation of unconscious forces on the perception that beneath the conventions and mores of society something very real and important was being repressed and denied.

In the Wittgenstein family this generational difference was played out in a way that only partly mirrors the wider dissonance. Karl Wittgenstein, after all, was not a representative of the Habsburg old order. Indeed, he represented a force that had curiously little impact on the life of Austria-Hungary – that of the metaphysically materialistic, politically liberal and aggressively capitalistic entrepreneur. In England, Germany or – especially, perhaps – in America, he would have been seen as a man of his times. In Austria he remained outside the mainstream. After his retirement from business he published a series of articles in the *Neue Freie Presse* extolling the virtues of American free enterprise, but in doing so he was addressing an issue that had only a marginal place in Austrian politics.

The absence of an effective liberal tradition in Austria was one of the chief factors that set its political history apart from that of other European nations. Its politics were dominated – and continued to be so until the rise of Hitler – by the struggle between the Catholicism of the Christian Socialists and the socialism of the Social Democrats. A side-show to this main conflict was provided by the opposition to both parties – who each in their different ways wished to maintain the supra-national character of the empire – of the pan-German movement led by Georg von Schoenerer, which espoused the kind of anti-Semitic, *Volkisch*, nationalism that the Nazis would later make their own.

Being neither members of the old guard, nor socialists – and certainly not pan-German nationalists – the Wittgensteins had little to contribute to the politics of their country. And yet the values that had made Karl Wittgenstein a successful industrialist were, in another way, the focus of a generational conflict that resonates with the wider tensions of the age. As a successful industrialist, Karl was content to *acquire* culture; his children, and especially his sons, were intent on contributing to it.

Fifteen years separated Karl's eldest child, Hermine, from his youngest, Ludwig, and his eight children might be divided into two distinct generations: Hermine, Hans, Kurt and Rudolf as the older; Margarete, Helene, Paul and Ludwig the younger. By the time the two youngest boys reached adolescence, the conflict between Karl and his first generation of children had dictated that Paul and Ludwig grew up under quite a different régime.

The régime within which Karl's eldest sons were raised was shaped by Karl's determination to see them continue his business. They were not to be sent to schools (where they would acquire the bad habits of mind of the Austrian establishment), but were to be educated privately in a way designed to train their minds for the intellectual rigours of commerce. They were then to be sent to some part of the Wittgenstein business empire, where they would acquire the technical and commercial expertise necessary for success in industry.

With only one of his sons did this have anything like the desired effect. Kurt, by common consent the least gifted of the children, acquiesced in his father's wishes and became in time a company director. His suicide, unlike that of his brothers, was not obviously related to the parental pressure exerted by his father. It came much later, at the end of the First World War, when he shot himself after the troops under his command had refused to obey orders.

On Hans and Rudolf, the effect of Karl's pressure was disastrous. Neither had the slightest inclination to become captains of industry. With encouragement and support, Hans might have become a great composer, or at the very least a successful concert musician. Even by the Wittgenstein family – most of whom had considerable musical ability – he was regarded as exceptionally gifted. He was a musical prodigy of Mozartian talents – a genius. While still in infancy he mastered the violin and piano, and at the age of four he began composing his own work. Music for him was not an interest but an all-consuming passion; it had to be at the centre, not the periphery, of his life. Faced with his father's insistence that he pursue a career in industry, he did what his father had done before him and ran away to America. His intention was to seek a life as a musician. What exactly happened to him nobody knows. In 1903 the family were informed that a year earlier he had disappeared from a boat in Chesapeake Bay, and had not been seen since. The obvious conclusion to draw was that he had committed suicide.

Would Hans have lived a happy life had he been free to devote himself to a musical career? Would he have been better prepared to face life outside the rarefied atmosphere of the Wittgenstein home if he had attended school? Obviously, nobody knows. But Karl was sufficiently shaken by the news to change his methods for his two youngest boys, Paul and Ludwig, who were sent to school and allowed to pursue their own bent.

For Rudolf, the change came too late. He was already in his twenties when Hans went missing, and had himself embarked upon a similar course. He, too, had rebelled against his father's wishes, and by 1903 was living in Berlin, where he had gone to seek a career in the theatre. His suicide in 1904 was reported in a local newspaper. One evening in May, according to the report, Rudolf had walked into a pub in Berlin and ordered two drinks. After sitting by himself for a while, he ordered a drink for the piano player and asked him to play his favourite song, 'I am lost'. As the music played, Rudi took cyanide and collapsed. In a farewell letter to his family he said that he had killed himself because a friend of his had died. In another farewell letter he said it was because he had 'doubts about his perverted disposition'. Some time before his death he had approached 'The Scientific-Humanitarian Committee' (which campaigned for the emancipation of homosexuals) for help, but, says the yearbook of the organization, 'our influence did not reach far enough to turn him away from the fate of self-destruction'.

Until the suicides of his two brothers, Ludwig showed none of the self-destructiveness epidemic among the Wittgensteins of his generation. For much of his childhood, he was considered one of the dullest of this extraordinary brood. He exhibited no precocious musical, artistic or literary talent, and, indeed, did not even start speaking until he was four years old. Lacking the rebelliousness and wilfulness that marked the other male members of his family, he dedicated himself from an early age to the kind of practical skills and technical interests his father had tried unsuccessfully to inculcate into his elder brothers. One of the earliest photographs of him to survive shows a rather earnest young boy, working with apparent relish at his own lathe. If he revealed no particular genius, he at least showed application and some considerable manual dexterity. At the age of ten, for example, he constructed a working model of a sewing machine out of bits of wood and wire.

Until he was fourteen, he was content to feel himself surrounded by genius, rather than possessed of it. A story he told in later life concerned an occasion when he was woken at three in the morning by the sound of a piano. He went downstairs to find Hans performing one of his own compositions. Hans's concentration was manic. He was sweating, totally absorbed, and completely oblivious of Ludwig's presence. The image remained for Ludwig a paradigm of what it was like to be possessed of genius.

The extent to which the Wittgensteins venerated music is perhaps hard for us to appreciate today. Certainly there is no modern equivalent of the form this veneration took, so intimately connected was it with the Viennese classical tradition. Ludwig's own musical tastes – which were, as far as we can judge, typical of his family – struck many of his later Cambridge contemporaries as deeply reactionary. He would tolerate nothing later than Brahms, and even in Brahms, he once said, 'I can begin to hear the sound of machinery.' The true 'sons of God' were Mozart and Beethoven.

The standards of musicality that prevailed in the family were truly extraordinary. Paul, the brother closest in age to Ludwig, was to become a very successful and well-known concert pianist. In the First World War he lost his right arm, but, with remarkable determination, taught himself to play using only his left hand, and attained such proficiency that he was able to continue his concert career. It was for him that Ravel, in 1931, wrote his famous Concerto for the Left Hand. And yet, though admired throughout the world, Paul's playing was not admired within his own family. It lacked taste, they thought; it was too full of extravagant gestures. More to their taste was the refined, classically understated playing of Ludwig's sister Helene. Their mother, Poldy, was an especially stern critic. Gretl, probably the least musical of the family, once gamely attempted a duet with her, but before they had got very far Poldy suddenly broke off. '*Du hast aber kein Rhythmus!*' ('You have no sense of rhythm at all!') she shrieked.

This intolerance of second-rate playing possibly deterred the nervous Ludwig from even attempting to master a musical instrument until he was in his thirties, when he learnt to play the clarinet as part of his training to be a teacher. As a child he made himself admired and loved in other ways – through his unerring politeness, his sensitivity to others, and his willingness to oblige. He was, in any case, secure in the knowledge that, so long as he showed an interest in engineering, he could always rely on the encouragement and approval of his father.

Though he later emphasized the unhappiness of his childhood, he gave the impression to the rest of his family of being a contented, cheerful boy. This discrepancy surely forms the crux of his boyhood reflections on honesty quoted earlier. The dishonesty he had in mind was not the petty kind that, say, allows one to steal something and then deny it, but the more subtle kind that consists in, for example, saying something because it is expected rather than because it is true. It was in part his willingness to succumb to this form of dishonesty that distinguished him from his siblings. So, at least, he thought in later life. An example that remained in his memory was of his brother Paul, ill in bed. On being asked whether he would like to get up or to stay longer in bed, Paul replied calmly that he would rather stay in bed. 'Whereas I in the same circumstances', Ludwig recalled, 'said what was untrue (that I wanted to get up) because I was afraid of the bad opinion of those around me.'

Sensitivity to the bad opinion of others lies at the heart of another example that stayed in his memory. He and Paul had wanted to belong to a Viennese gymnastic club, but discovered that (like most such clubs at that time) it was restricted to those of 'Aryan' origin. He was prepared to lie about their Jewish background in order to gain acceptance; Paul was not.

Fundamentally, the question was not whether one should, on all occasions, tell the truth, but rather whether one had an overriding obligation to *be* true – whether, despite the pressures to do otherwise, one should insist on being oneself. In Paul's case, this problem was made easier following Karl's change of heart after the death of Hans. He was sent to grammar school, and spent the rest of his life pursuing the musical career that was his natural bent. In Ludwig's case, the situation was more complicated. The pressures on him to conform to the wishes of others had become as much internal as external. Under the weight of these pressures, he allowed people to think that his natural bent was in the technical subjects that would train him for his father's preferred occupation. Privately he regarded himself as having 'neither taste nor talent' for engineering; quite reasonably, under the circumstances, his family considered him to have both.

Accordingly, Ludwig was sent, not to the grammar school in Vienna that Paul attended, but to the more technical and less academic Realschule in Linz. It was, it is true, feared that he would not pass the rigorous entrance examinations set by a grammar school, but the primary consideration was the feeling that a more technical education would suit his interests better.

The Realschule at Linz, however, has not gone down in history as a promising training ground for future engineers and industrialists. If it is famous for anything, it is for being the seedbed of Adolf Hitler's *Weltanschauung*. Hitler was, in fact, a contemporary of Wittgenstein's there, and (if *Mein Kampf* can be believed) it was the history teacher at the school, Leopold Pötsch, who first taught him to see the Habsburg Empire as a 'degenerate dynasty' and to distinguish the hopeless dynastic patriotism of those loyal to the Habsburgs from the (to Hitler) more appealing *Völkisch* nationalism of the pan-German movement. Hitler, though almost exactly the same age as Wittgenstein, was two years behind at school. They overlapped at the school for only the year 1904–5, before Hitler was forced to leave because of his poor record. There is no evidence that they had anything to do with one another.

Wittgenstein spent three years at the school, from 1903 to 1906. His school reports survive, and show him to have been, on the whole, a fairly poor student. If one translates the five subject-grades used in the school into a scale from A to E, then he achieved an A only twice in his school career – both times in religious studies. In most subjects he was graded C or D, rising to a B every now and again in English and natural history, and sinking to an E on one occasion in chemistry. If there is a pattern to his results, it is that he was, if anything, weaker in the scientific and technical subjects than in the humanities.

His poor results may in part be due to his unhappiness at school. It was the first time in his life he had lived away from the privileged environment of his family home, and he did not find it easy to find friends among his predominantly working-class fellow pupils. On first setting eyes on them he was shocked by their uncouth behaviour. '*Mist!*' ('Muck!') was his initial impression. To them he seemed (as one of them later told his sister Hermine) like a being from another world. He insisted on using the polite form 'Sie' to address them, which served only to alienate him further. They ridiculed him by chanting an alliterative jingle that made play of his unhappiness and of the distance between him and the rest of the school: '*Wittgenstein wandelt wehmütig widriger Winde wegen Wienwärts*' ('Wittgenstein wends his woeful windy way towards Vienna'). In his efforts to make friends, he felt, he later said, 'betrayed and sold' by his schoolmates.

His one close friend at Linz was a boy called Pepi, the son of the Strigl family, with whom he lodged. Throughout his three years at the school, he experienced with Pepi the love and hurt, the breaks and reconciliations, typical of adolescent attachment.

The effect of this relationship, and of his difficulties with his classmates, seems to have been to intensify the questioning and doubting nature implicit in his earlier reflections. His high marks in religious knowledge are a reflection, not only of the comparative leniency of priests compared with school teachers, but also of his own growing preoccupation with fundamental questions. His intellectual development during his time at Linz owed far more to the impetus of these doubts than to anything he may have been taught at school.

The biggest intellectual influence on him at this time was not that of any of his teachers, but that of his elder sister Margarete ('Gretl'). Gretl was acknowledged as the intellectual of the family, the one who kept abreast of contemporary developments in the arts and sciences, and the one most prepared to embrace new ideas and to challenge the views of her elders. She was an early defender of Freud, and was herself psychoanalysed by him. She later became a close friend of his and had a hand in his (perilously late) escape from the Nazis after the *Anschluss*.

It was no doubt through Gretl that Wittgenstein first became aware of the work of Karl Kraus. Kraus's satirical journal *Die Fackel* ('The Torch') first appeared in 1899, and from the very beginning was a huge success among the intellectually disaffected in Vienna. It was read by everyone with any pretence to understanding the political and cultural trends of the time, and exerted an enormous influence on practically all the major figures mentioned previously, from Adolf

Loos to Oskar Kokoschka. From the first, Gretl was an enthusiastic reader of Kraus's journal and a strong sympathizer of almost everything he represented. (Given the protean nature of Kraus's views, it was more or less impossible to sympathize with quite everything he said.)

Before founding *Die Fackel*, Kraus was known chiefly as the author of an anti-Zionist tract entitled *Eine Krone für Zion* ('A Crown for Zion'), which mocked the views of Theodor Herzl for being reactionary and divisive. Freedom for the Jews, Kraus maintained, could come only from their complete assimilation.

Kraus was a member of the Social Democrat Party, and for the first few years of its publication (until about 1904) his journal was regarded as a mouthpiece for socialist ideas. The targets of his satire were, to a large extent, those a socialist might like to see hit. He attacked the hypocrisy of the Austrian government in its treatment of the Balkan peoples, the nationalism of the pan-German movement, the *laissez-faire* economic policies advocated by the *Neue Freie Presse* (in, for example, Karl Wittgenstein's articles for the paper), and the corruption of the Viennese Press in its willingness to serve the interests of the government and of big business. He led an especially passionate campaign against the sexual hypocrisy of the Austrian establishment, as manifested in the legal persecution of prostitutes and the social condemnation of homosexuals. 'A trial involving sexual morality', he said, 'is a deliberate step from individual to general immorality.'

From 1904 onward, the nature of his attacks became less political than moral. Behind his satire was a concern with spiritual values that was alien to the ideology of the Austro-Marxists. He was concerned to uncover hypocrisy and injustice, not primarily from a desire to protect the interests of the proletariat, but rather from the point of view of one who sought to protect the integrity of an essentially aristocratic ideal of the nobility of truth. He was criticized for this by his friends on the Left, one of whom, Robert Scheu, told him bluntly that the choice before him was that of supporting the decaying old order or supporting the Left. 'If I must choose the lesser of two evils', came Kraus's lofty reply, 'I will choose neither.' Politics, he said, 'is what a man does in order to conceal what he is and what he himself does not know'.

The phrase encapsulates one of the many ways in which the outlook of the adult Wittgenstein corresponds to that of Kraus. 'Just improve yourself', Wittgenstein would later say to many of his friends, 'that is all you can do to improve the world.' Political questions, for him, would always be secondary to questions of personal integrity. The question he had asked himself at the age of eight was answered by a kind of Kantian categorical imperative: one *should* be truthful, and that is that; the question 'Why?' is inappropriate and cannot be answered. Rather, all other questions must be asked and answered within this fixed point – the inviolable duty to be true to oneself.

The determination not to conceal 'what one is' became central to Wittgenstein's whole outlook. It was the driving force that impelled the series of confessions he was to make later in life of the times when he had failed to be honest. During his time at school in Linz, he made the first of these attempts to come clean about himself, when he made some confessions to his eldest sister, Hermine ('Mining'). What formed the subject of these confessions, we do not know; we know only that he was later disparaging about them. He described them as confessions 'in which I manage to appear to be an excellent human being'.

Wittgenstein's loss of religious faith, which, he later said, occurred while he was a schoolboy at Linz, was, one supposes, a consequence of this spirit of stark truthfulness. In other words, it was not so much that he lost his faith as that he now felt obliged to acknowledge that he had none, to confess that he could not believe the things a Christian was supposed to believe. This may have been one of the things he confessed to Mining. Certainly he discussed it with Gretl, who, to help him in the philosophical reflection consequent on a loss of faith, directed him to the work of Schopenhauer.

Schopenhauer's transcendental idealism, as expressed in his classic, *The World as Will and Representation*, formed the basis of Wittgenstein's earliest philosophy. The book is, in many ways, one that is bound to appeal to an adolescent who has lost his religious faith and is looking for something to replace it. For while Schopenhauer recognizes 'man's need for metaphysics', he insists that it is neither necessary nor possible for an intelligent honest person to believe in the literal truth of religious doctrines. To expect that he should, Schopenhauer says, would be like asking a giant to put on the shoes of a dwarf.

Schopenhauer's own metaphysics is a peculiar adaptation of Kant's. Like Kant, he regards the everyday world, the world of the senses, as mere appearance, but unlike Kant (who insists that noumenal reality is unknowable), he identifies as the only true reality the world of the ethical will. It is a theory that provides a metaphysical counterpart to the attitude of Karl Kraus mentioned earlier – a philosophical justification of the view that what happens in the 'outside' world is less important than the existential, 'internal' question of 'what one is'. Schopenhauer's idealism was abandoned by Wittgenstein only when he began to study logic and was persuaded to adopt Frege's conceptual realism. Even after that, however, he returned to Schopenhauer at a critical stage in the composition of the *Tractatus*, when he believed that he had reached a point where idealism and realism coincide.*

Taken to its extreme, the view that the 'internal' has priority over the 'external' becomes solipsism, the denial that there is any reality *outside* oneself. Much of Wittgenstein's later philosophical thinking about the self is an attempt once and for all to put to rest the ghost of this view. Among the books that he read as a schoolboy which influenced his later development, this doctrine finds its most startling expression in *Sex and Character*, by Otto Weininger.

It was during Wittgenstein's first term at Linz that Weininger became a cult figure in Vienna. On 4 October 1903, his dying body was found on the floor in the house in Schwarzspanierstrasse where Beethoven had died. At the age of twenty-three, in an act of self-consciously symbolic significance, he had shot himself in the home of the man whom he considered to be the very greatest of geniuses. *Sex and Character* had been published the previous spring, and had received, on the whole, fairly bad reviews. Had it not been for the sensational circumstances of its author's death, it would probably have had no great impact. As it was, on 17 October, a letter from August Strindberg appeared in *Die Fackel*, describing the work as: 'an awe-inspiring book, which has probably solved the most difficult of all problems'. Thus was the Weininger cult born.

Weininger's suicide seemed to many to be the logical outcome of the argument of his book, and it is primarily this that made him such a *cause célèbre* in pre-war Vienna. His taking of his own life was seen, not as a cowardly escape from suffering, but as an ethical deed, a brave acceptance of a tragic conclusion. It was, according to Oswald Spengler, 'a spiritual struggle', which offered 'one of the noblest

---

* See p. 144.

spectacles ever presented by a late religiousness'. As such, it inspired a number of imitative suicides. Indeed, Wittgenstein himself began to feel ashamed that he had not dared kill himself, that he had ignored a hint that he was *de trop* in this world. This feeling lasted for nine years, and was overcome only after he had convinced Bertrand Russell that he possessed philosophic genius. His brother Rudolf's suicide came just six months after Weininger's, and was, as we have seen, executed in an equally theatrical manner.

Wittgenstein's acknowledgement of Weininger's influence, more than that of any other, ties his life and work to the environment in which he was raised. Weininger is a quintessentially Viennese figure. The themes of his book, together with the manner of his death, form a potent symbol of the social, intellectual and moral tensions of the *fin de siècle* Vienna in which Wittgenstein grew up.

Running throughout the book is a very Viennese preoccupation with the decay of modern times. Like Kraus, Weininger attributes this decay to the rise of science and business and the decline of art and music, and characterizes it, in an essentially aristocratic manner, as the triumph of pettiness over greatness. In a passage reminiscent of the prefaces Wittgenstein would write in the 1930s, for his own philosophical work, Weininger denounces the modern age as:

> . . . a time when art is content with daubs and seeks its inspiration in the sports of animals; the time of a superficial anarchy, with no feeling for Justice and the State; a time of communistic ethics, of the most foolish of historical views, the materialistic interpretation of history; a time of capitalism and of Marxism; a time when history, life and science are no more than political economy and technical instruction; a time when genius is supposed to be a form of madness; a time with no great artists and no great philosophers; a time without originality and yet with the most foolish craving for originality.

Like Kraus, too, Weininger was inclined to identify those aspects of modern civilization that he most disliked as Jewish, and to describe the social and cultural trends of the age in terms of the sexual polarity between the masculine and the feminine. Unlike Kraus, however, Weininger emphasizes these two themes to an obsessive, almost lunatic, extent.

Sex and Character is dominated by an elaborately worked out theory intended to justify Weininger's misogyny and anti-Semitism. The central point of the book, he says in the preface, is to 'refer to a single principle the whole contrast between men and women'.

The book is divided into two parts: the 'biological-psychological', and the 'logical-philosophical'. In the first he seeks to establish that all human beings are biologically bisexual, a mixture of male and female. Only the proportions differ, which is how he explains the existence of homosexuals: they are either womanly men or masculine women. The 'scientific' part of the book ends with a chapter on 'Emancipated Women', in which he uses this theory of bisexuality to oppose the women's movement. 'A woman's demand for emancipation and her qualification for it', he claims, 'are in direct proportion to the amount of maleness in her.' Such women are, therefore, generally lesbians, and as such are on a higher level than most women. These masculine women should be given their freedom, but it would be a grave mistake to let the majority of women imitate them.

The second, and much larger, part of the book discusses Man and Woman, not as biological categories, but as 'psychological types', conceived of as something like Platonic ideas. As actual men and women are all mixtures of masculinity and femininity, Man and Woman do not exist *except* as Platonic forms. Nevertheless, we are all psychologically either a man or a woman. Curiously, Weininger thinks that, whereas it is possible for a person to be biologically male but psychologically female, the reverse is impossible. Thus, even emancipated, lesbian, women are psychologically female. It follows that everything he says about 'Woman' applies to all women and also to some men.

The essence of Woman, he says, is her absorption in sex. She is nothing but sexuality; she is sexuality itself. Whereas men possess sexual organs, 'her sexual organs possess women'. The female is completely preoccupied with sexual matters, whereas the male is interested in much else, such as war, sport, social affairs, philosophy and science, business and politics, religion and art. Weininger has a peculiar epistemological theory to explain this, based on his notion of a 'henid'. A henid is a piece of psychical data before it becomes an idea. Woman thinks in henids, which is why, for her, thinking and feeling are the same thing. She looks to man, who thinks in clear and articulated ideas, to clarify her data, to interpret her henids. That is why women fall in love only with men cleverer than themselves. Thus, the essential difference between man and woman is that, whereas 'the male lives consciously, the female lives unconsciously'.

Weininger draws from this analysis alarmingly far-reaching ethical implications. Without the ability to clarify her own henids, woman is incapable of forming clear judgements, so the distinction between true and false means nothing to her. Thus women are naturally, inescapably, untruthful. Not that they are, on this account, immoral; they do not enter the moral realm at all. Woman simply has no standard of right or wrong. And, as she knows no moral or logical imperative, she cannot be said to have a soul, and this means she lacks free will. From this it follows that women have no ego, no individuality, and no character. Ethically, women are a lost cause.

Turning from epistemology and ethics to psychology, Weininger analyses women in terms of two further Platonic types: the mother and the prostitute. Each individual woman is a combination of the two, but is predominantly one or the other. There is no moral difference between the two: the mother's love for her child is as unthinking and indiscriminate as the prostitute's desire to make love to every man she sees. (Weininger will have nothing to do with any explanation of prostitution based on social and economic conditions. Women are prostitutes, he says, because of the 'disposition for and inclination to prostitution' which is 'deep in the nature of women'.) The chief difference between the two types is the form that their obsession with sex takes: whereas the mother is obsessed with the object of sex, the prostitute is obsessed with the act itself.

All women, whether mothers or prostitutes, share a single characteristic – 'a characteristic which is really and exclusively feminine' – and that is the instinct of match-making. It is the ever present desire of all women to see man and woman united. To be sure, woman is interested first and foremost in her own sexual life, but that is really a special case of her 'only vital interest' – 'the interest that sexual unions shall take place; the wish that as much of it as possible shall occur, in all cases, places and times'.

As an adjunct to his psychological investigation of woman, Weininger has a chapter on Judaism. Again, the Jew is a Platonic idea, a psychological type, which is a possibility (or a danger) for all mankind, 'but which has become actual in the most conspicuous fashion only amongst the Jews'. The Jew is 'saturated with femininity' – 'the most manly Jew is more feminine than the least manly Aryan'. Like the woman, the Jew has a strong instinct for pairing. He has a poor sense of individuality and a correspondingly strong instinct to preserve the race. The Jew has no sense of good and evil, and no soul. He is non-philosophical and profoundly irreligious (the Jewish religion being 'a mere historical tradition'). Judaism and Christianity are opposites: the latter is 'the highest expression of the highest faith'; the former is 'the extreme of cowardliness'. Christ was the greatest of all men because he: 'conquered in himself Judaism, the greatest negation, and created Christianity, the strongest affirmation and the most direct opposite of Judaism'.

Weininger himself was both Jewish and homosexual (and therefore possibly of the psychologically female type), and the idea that his suicide was, in some way, a 'solution' could therefore be easily assimilated within the most vulgar anti-Semitic or misogynist outlook. Hitler, for example, is reported as having once remarked: 'Dietrich Eckhart told me that in all his life he had known just one good Jew: Otto Weininger who killed himself on the day when he realised that the Jew lives upon the decay of peoples.' And the fact that fear of the emancipation of women, and, particularly, of Jews, was a widespread preoccupation in Vienna at the turn of the century no doubt accounts to some extent for the book's enormous popularity. It would later provide convenient material for Nazi propaganda broadcasts.

But why did Wittgenstein admire the book so much? What did he learn from it? Indeed, given that its claims to scientific biology are transparently spurious, its epistemology obvious nonsense, its psychology primitive, and its ethical prescriptions odious, what could he *possibly* have learnt from it?

To see this, we have, I think, to turn aside from Weininger's – entirely negative – psychology of Woman, and look instead at his psychology of Man. Only there do we find something in the book other than bigotry and self-contempt, something that resonates with themes we know to have been central to Wittgenstein's thoughts as a teenager (and, indeed, for the rest of his life), and that provides at least some hint as to what Wittgenstein may have found to admire in it.

Unlike Woman, Man, according to Weininger, has a choice: he can, and must, choose between the masculine and the feminine, between consciousness and unconsciousness, will and impulse, love and sexuality. It is every man's ethical duty to choose the first of each of these pairs, and the extent to which he is able to do this is the extent to which he approximates to the very highest type of man: the genius.

The consciousness of the genius is the furthest removed from the henid stage; 'it has the greatest, most limpid clearness and distinctness'. The genius has the best developed memory, the greatest ability to form clear judgements, and, therefore, the most refined sense of the distinctions between true and false, right and wrong. Logic and ethics are fundamentally the same; 'they are no more than duty to oneself'. Genius 'is the highest morality, and, therefore, it is everyone's duty'.

Man is not born with a soul, but with the potential for one. To realize this potential he has to find his real higher self, to escape the limitations of his (unreal) empirical self. One route to this self-discovery is love, through which 'many men first come to know of their own real nature, and to be convinced that they possess a soul':

> In love, man is only loving himself. Not his empirical self, not the weaknesses and vulgarities, not the failings and smallnesses which he outwardly exhibits; but all that he wants to be, all that he ought to be, his truest, deepest, intelligible nature, free from all fetters of necessity, from all taint of earth.

Naturally, Weininger is here talking of Platonic love. Indeed, for him, there *is* only Platonic love, because: 'any other so-called love belongs to the kingdom of the senses'. Love and sexual desire are not only not the same thing; they are in opposition to one another. That is why the idea of love after marriage is make-believe. As sexual attraction increases with physical proximity, love is strongest in absence of the loved one. Indeed, love *needs* separation, a certain distance, to preserve it: 'what all the travels in the world could not achieve, what time could not accomplish, may be brought about by accidental, unintentional, physical contact with the beloved object, in which the sexual impulse is awakened, and which suffices to kill love on the spot'.

The love of a woman, though it can arouse in a man some hint of his higher nature, is, in the end, doomed either to unhappiness (if the truth about the woman's unworthiness is discovered) or to immorality (if the lie about her perfection is kept up). The only love that is of enduring worth is that 'attached to the absolute, to the idea of God'.

Man should love, not woman, but his own soul, the divine in himself, the 'God who in my bosom dwells'. Thus he must resist the pairing instinct of the woman and, despite pressure from women, free himself from sex. To the objection that this suggestion, if adopted universally, would be the death of the human race, Weininger replies that it would be merely the death of the *physical* life, and would put in its place the 'full development of the spiritual life'. Besides, he says: 'no one who is honest with himself feels bound to provide for the continuity of the human race':

> That the human race should persist is of no interest whatever to reason; he who would perpetuate humanity would perpetuate the problem and the guilt, the only problem and the only guilt.

The choice that Weininger's theory offers is a bleak and terrible one indeed: genius or death. If one can only live as a 'Woman' or as a 'Jew' – if, that is, one cannot free oneself from sensuality and earthly desires – then one has no right to live at all. The *only* life worth living is the spiritual life.

In its strict separation of love from sexual desire, its uncompromising view of the worthlessness of everything save the products of genius, and its conviction that sexuality is incompatible with the honesty that genius demands, there is much in Weininger's work that chimes with attitudes we find Wittgenstein expressing time and again throughout his life. So much so, that there is reason to believe that of all the books he read in adolescence, Weininger's is the one that had the greatest and most lasting impact on his outlook.

Of particular importance, perhaps, is the peculiar twist that Weininger gives to Kant's Moral Law, which, on this account, not only imposes an inviolable duty to be honest, but, in so doing, provides the route for all men to discover in themselves whatever genius they possess. To acquire genius, on this view, is not merely a noble ambition; it is a Categorical Imperative. Wittgenstein's recurring thoughts of suicide between 1903 and 1912, and the fact that these thoughts abated only after Russell's recognition of his genius, suggest that he accepted this imperative in all its terrifying severity.

So much for Wittgenstein's intellectual development as a schoolboy, which, we have seen, was inspired primarily by philosophical reflection and (under Gretl's guidance) fuelled by the reading of philosophers and cultural critics. But what of his development in the technical subjects – his advancement in the skills and knowledge required to succeed in his chosen profession?

Of this we hear surprisingly little. The works by scientists which he read as a teenager – Heinrich Hertz's *Principles of Mechanics*, and Ludwig Boltzmann's *Populäre Schriften* – suggest an interest, not in mechanical engineering, nor even, especially, in theoretical physics, but rather in the philosophy of science.

Both (like the works previously discussed) espouse a fundamentally Kantian view of the nature and method of philosophy. In *Principles of Mechanics* Hertz addresses the problem of how to understand the mysterious concept of 'force' as it is used in Newtonian physics. Hertz proposes that, instead of giving a direct answer to the question: 'What is force?', the problem should be dealt with by restating Newtonian physics without using 'force' as a basic concept. 'When these painful contradictions are removed', he writes, 'the question as to the nature of force will not have been answered; but our minds, no longer vexed, will cease to ask illegitimate questions.'

This passage of Hertz was known by Wittgenstein virtually word for word, and was frequently invoked by him to describe his own conception of philosophical problems and the correct way to solve them. As we have seen, philosophical thinking *began* for him with 'painful contradictions' (and not with the Russellian desire for *certain* knowledge); its aim was always to resolve those contradictions and to replace confusion with clarity.

He may well have been led to Hertz by reading Boltzmann's *Populäre Schriften*, a collection of Boltzmann's more popular lectures, published in 1905. The lectures present a similarly Kantian view of science, in which our models of reality are taken *to* our experience of the world, and not (as the empiricist tradition would have it) derived from it. So ingrained in Wittgenstein's philosophical thinking was this view that he found the empiricist view difficult even to conceive.

Boltzmann was Professor of Physics at the University of Vienna, and there was some talk of Wittgenstein's studying with him after he had left school. In 1906, however, the year that Wittgenstein left Linz, Boltzmann committed suicide, despairing of ever being taken seriously by the scientific world.

Independently of Boltzmann's suicide, it seems, it had been decided that Wittgenstein's further education should advance his technical knowledge rather than develop his interest in philosophy and theoretical science. Accordingly, after leaving Linz he was sent – no doubt urged by his father – to study mechanical engineering at the Technische Hochschule (now the Technical University) in Charlottenburg, Berlin.

Of Wittgenstein's two years in Berlin very little is known. The college records show that he matriculated on 23 October 1906, that he attended lectures for three semesters and that, having completed his diploma course satisfactorily, he was awarded his certificate on 5 May 1908. Photographs of the time show him to be a handsome, immaculately dressed young man, who might well have been – as he is reported to have been a year later in Manchester – a 'favourite with the ladies'.

He lodged with the family of one of his professors, Dr Jolles, who adopted him as their own 'little Wittgenstein'. Much later, after the First World War had produced in him a transformation comparable to, and perhaps even deeper than, the change he had undergone in 1903–4, Wittgenstein was embarrassed by the intimacy he had shared with the Jolles family, and replied to the friendly and affectionate letters he received from Mrs Jolles with stiff politeness. But while he was in Berlin, and for a number of years after he left, he was grateful enough for the warm concern they showed him.

It was a time of competing interests and obligations. Wittgenstein's sense of duty towards his father impelled him to stick at his engineering studies, and he developed an interest in the then very young science of aeronautics. But, increasingly, he found himself gripped, almost against his will, by philosophical questions. Inspired by the diaries of Gottfried Keller, he began to write down his philosophical reflections in the form of dated notebook entries.

In the short term, his father's wishes prevailed, and on leaving Berlin he went to Manchester to further his studies in aeronautics. But in the long term, it was probably already clear to him that the only life he could consider worthwhile was one spent in fulfilment of the greater duty he owed to himself – to his own genius.

2
MANCHESTER

Suppressing his growing preoccupation with philosophical questions, Wittgenstein, in the spring of 1908, at the age of nineteen, went to Manchester to pursue research in aeronautics. It was his apparent intention to construct, and eventually to fly, an aeroplane of his own design.

These were the very pioneering days of aeronautics, when the subject was in the hands of rival groups of amateurs, enthusiasts and eccentrics from America and the various European nations. Orville and Wilbur Wright had not yet amazed the world by staying in the air for a full two and a half hours. Though no substantial successes had been made, and the subject was treated with some amusement and derision by the Press and public, scientists and governments alike were aware of the potential importance of the research. It was a field in which successful innovation could expect handsome rewards, and Wittgenstein's project no doubt had the complete support of his father.

He began his research by experimenting on the design and construction of kites. For this purpose he went to work at the Kite Flying Upper Atmosphere Station, a meteorological observation centre near Glossop, where observations were made using box-kites carrying various instruments. The centre had been set up by the recently retired Professor of Physics Arthur Schuster, who continued to maintain an active interest in its work. The head of the centre was J. E. Petavel, lecturer in meteorology at Manchester, who developed a keen interest in aeronautics and later became one of the leading authorities on the subject.

While working at the observatory Wittgenstein lived at the Grouse Inn, an isolated roadside pub on the Derbyshire Moors. From here, on 17 May, he wrote to Hermine, describing the conditions under which he worked, exulting in the glorious isolation of the Grouse Inn, but complaining of the incessant rain and the rustic standards of the food and the toilet facilities: 'I'm having a few problems growing accustomed to it all; but I am already beginning to enjoy it.'

His job, he said, 'is the most delightful I could wish for':

> I have to provide the observatory with kites – which formerly were always ordered from outside – and to ascertain through trial and error their best design; the materials for this are ordered for me by request from the observatory. To begin with, of course, I had to help with the observations, in order to get to know the demands which would be made of such a kite. The day before yesterday, however, I was told that I can now begin making independent experiments . . . Yesterday I began to build my first kite and hope to finish it by the middle of next week.

He goes on to describe his physical and emotional isolation, and his deep need for a close companion. At the inn he was the only guest apart from 'a certain Mr Rimmer who makes meteorological observations', and at the observatory the only time he had company was on Saturdays, when Petavel arrived with some of his students:

> Because I am so cut off I naturally have an *extraordinarily strong* desire for a friend and when the students arrive on Saturday I always think it will be one of them.

He was too reticent to approach the students, but shortly after this letter a friend came to him. William Eccles, an engineer four years older than himself, came to the observatory to conduct meteorological research. On his arrival at the Grouse Inn, Eccles walked into the common living room to find Wittgenstein surrounded with books and papers which lay scattered over the table and floor. As it was impossible to move without disturbing them, he immediately set to and tidied them up – much to Wittgenstein's amusement and appreciation. The two quickly became close friends, and remained so, with interruptions, until the Second World War.

In the autumn of 1908 Wittgenstein registered as a research student in the Engineering Department of Manchester University. In those days Manchester had very few research students, and the arrangements made for them were somewhat casual. No formal course of study was organized, nor was a supervisor provided to oversee the research. It was not expected that Wittgenstein should work for a degree. Instead, it was understood that he would pursue his own line of research, having at his disposal the laboratory facilities of the university and the interested attention, should he require it, of its professors.

Among these professors was the mathematician Horace Lamb, who held a seminar for research students to which they could bring their problems for his consideration. Wittgenstein seems to have availed himself of this provision. In a letter to Hermine of October he describes a conversation he had had with Lamb, who, he says:

> . . . will try to solve the equations that I came up with and which I showed him. He said he didn't know for certain whether they are altogether solvable with today's methods and so I am eagerly awaiting the outcome of his attempts.

His interest in the solution of this problem was evidently not confined to its aeronautical applications. He began to develop an interest in pure mathematics, and to attend J. E. Littlewood's lectures on the theory of mathematical analysis, and one evening a week he met with two other research students to discuss questions in mathematics. These discussions led to a consideration of the problems of providing mathematics with logical foundations, and Wittgenstein was introduced by one of his fellow students to Bertrand Russell's book on the subject, *The Principles of Mathematics*, which had been published five years previously.

Reading Russell's book was to prove a decisive event in Wittgenstein's life. Though he continued for another two years with research in aeronautics, he became increasingly obsessed with the problems discussed by Russell, and his engineering work was pursued with an ever-growing disenchantment. He had found a subject in which he could become as absorbed as his brother Hans had been in playing the piano, a subject in which he could hope to make, not just a worthwhile contribution, but a *great* one.

The central theme of *The Principles of Mathematics* is that, contrary to the opinion of Kant and most other philosophers, the whole of pure mathematics could be derived from a small number of fundamental, *logical*, principles. Mathematics and logic, in other words, were one and the same. It was Russell's intention to provide a strictly mathematical demonstration of this by actually making all the derivations required to prove each theorem of mathematical analysis from a few trivial, self-evident axioms. This was to be his second volume. In fact, it grew into the monumental three-volume work *Principia Mathematica*. In this, his 'first volume', he lays the philosophical foundations of this bold enterprise, taking issue principally with Kant's at that time widely influential view that mathematics was quite distinct from logic and was founded on the 'structure of appearance', our basic 'intuitions' of space and time. For Russell, the importance of the issue lay in the difference between regarding mathematics as a body of certain, *objective*, knowledge, and regarding it as a fundamentally *subjective* construction of the human mind.

Russell did not become aware until *The Principles of Mathematics* was being printed that he had been anticipated in the main lines of his enterprise by the German mathematician Gottlob Frege, who in his *Grundgesetze der Arithmetik* (the first volume of which was published in 1893) had attempted precisely the task that Russell had set himself. He made a hurried study of Frege's work and appended to his book an essay on 'The Logical and Arithmetical Doctrines of Frege', praising the *Grundgesetze*.

Until then, the *Grundgesetze* had been a much neglected work. Few had bothered to read it, and even fewer had understood it. Russell was perhaps the first to appreciate its importance. In his quick study of Frege's work, however, he had noticed a difficulty that Frege had overlooked. The problem it raised seemed at first a minor one, but its solution was soon to become the cardinal problem in the foundations of mathematics.

To provide a logical definition of number, Frege had made use of the notion of a class, which he defined as the extension of a concept. Thus, to the concept 'man' corresponds the class of men, to the concept 'table', the class of tables, and so on. It was an axiom of his system that to every meaningful concept there corresponds an object, a class, that is its extension. Russell discovered that, by a certain chain of reasoning, this led to a contradiction. For on this assumption there will be some classes that belong to themselves, and some that do not: the class of all classes is itself a class, and therefore belongs to itself; the class of men is not itself a man, and therefore does not. On this basis we form 'the class of all classes which do not belong to themselves'. Now we ask: is *this* class a member of itself or not? The answer either that it is or that it is not leads to a contradiction. And, clearly, if a contradiction can be derived from Frege's axioms, his system of logic is an inadequate foundation upon which to build the whole of mathematics.

Before publishing his discovery, Russell wrote to Frege at the University of Jena to inform him of it. Frege was then preparing the second volume of his *Grundgesetze*, and though he included in it a hasty and unsatisfactory reaction to the paradox, he realized it showed his entire system to be fundamentally flawed. Russell himself proposed to avoid the contradiction by a strategy he called the 'Theory of Types', an outline of which formed the second appendix to the *Principles*. This postulates a hierarchy of *types* of objects, collections of which can legitimately be grouped together to form sets: thus, the first type is individuals, the second, classes of individuals, the third, classes of classes of individuals, and so on. Sets must be collections of objects of the same type; thus, there is no such thing as a set which is a member of itself.

The Theory of Types does indeed avoid the contradiction, but at the expense of introducing into the system a somewhat *ad hoc* measure. It may be true that there are different types of things; it may also be true that there is no such thing as a set which is a member of itself – but these are hardly the sort of trivial, self-evident truths of logic from which Russell had originally intended to proceed. Russell himself was dissatisfied with it, and his book ends with a challenge:

> What the complete solution of the difficulty may be, I have not succeeded in discovering; but as it affects the very foundations of reasoning, I earnestly commend the study of it to the attention of all students of logic.

It was precisely the bait needed to hook Wittgenstein, and, as Russell recommends, he dedicated himself earnestly to the solution of the paradox. He devoted much of his first two terms at Manchester to a close study of both Russell's *Principles* and Frege's *Grundgesetze*, and, some time before April 1909, formulated his first attempt at a solution, which he sent to Russell's friend, the mathematician and historian of mathematics Philip E. B. Jourdain.

That Wittgenstein sent his solution to Jourdain rather than to Russell or Frege perhaps indicates some degree of tentativeness. He presumably came across Jourdain's name in a 1905 issue of the *Philosophical Magazine*, which contains an article by Jourdain on the foundations of mathematics and an article by his professor at Manchester, Horace Lamb. An entry in Jourdain's correspondence book, dated 20 April, shows that he replied to Wittgenstein's attempted solution after first discussing it with Russell. Neither, it seems, were inclined to accept it:

> Russell said that the views I gave in reply to Wittgenstein (who had 'solved' Russell's contradiction) agree with his own.

According to his sister Hermine, Wittgenstein's obsession with the philosophy of mathematics at this time caused him to suffer terribly from the feeling of being torn between conflicting vocations. It may have been Jourdain's dismissal of his 'solution' that persuaded him, for the time being, to stick to aeronautics. Not for another two years did he return to the fray, when he finally contacted both Frege and Russell directly to present to them a more considered philosophical position. Though he had taste enough for philosophical problems, he was yet to be persuaded he had talent for them.

Though still convinced he had neither talent nor taste for aeronautical engineering, Wittgenstein persevered with his attempts to design and construct an aircraft engine. Plans of his proposed engine survive, and show that his idea was to rotate the propeller by means of high-speed gases rushing from a combustion chamber (rather in the way that the pressure of water from a hose is used to turn a rotating lawn sprinkler). The idea was fundamentally flawed, and quite impractical for propelling an aeroplane. It was, however, successfully adopted during the Second World War in the design of certain helicopters.

Wittgenstein had a combustion chamber built specially for him by a local firm, and much of his research consisted of experimenting on this with various discharge nozzles. He was helped by a laboratory assistant called Jim Bamber, whom he later described as: 'one of the very few people with whom I got on well during my Manchester period'. His general irritability at having to dedicate himself to engineering work was compounded by the fiddly nature of the task, and, recalls Bamber, 'his nervous temperament made him the last person to tackle such research':

> . . . for when things went wrong, which often occurred, he would throw his arms about, stamp around and swear volubly in German.

According to Bamber, Wittgenstein would ignore the midday meal break and carry on until the evening, when he would relax either by sitting in a very hot bath ('He used to brag about the temperature of the water') or by going to a concert given by the Hallé Orchestra, occasionally accompanied by Bamber, who has described how 'he used to sit through the concert without speaking a word, completely absorbed'.

Other diversions included outings with Eccles, who by this time had left the university to take up an engineering job in Manchester. One Sunday afternoon remained in Eccles's memory. Wittgenstein had decided he would like to go to the seaside, to Blackpool. On finding that there was no suitable train, he did not seek an alternative but suggested they hire a special train, just for the two of them. He was eventually dissuaded from this by Eccles, and induced to adopt the less expensive (though still, to Eccles's mind, extravagant) option of taking a taxi to Liverpool, where they had a trip on the Mersey ferry.

During his second year at Manchester, Wittgenstein abandoned his attempt to design and construct a jet engine and concentrated instead on the design of the propeller. His work on this was taken sufficiently seriously by the university for him to be elected to a research studentship for what was to be his last year there, 1910–11. He himself was confident enough about the importance and originality of his work to patent his design. His application, together with a provisional specification of his design for 'Improvements in Propellers applicable for Aerial Machines', is dated 22 November 1910. On 21 June 1911 he left a complete specification, and the patent was accepted on 17 August of that year.

By this time, however, Wittgenstein's obsession with philosophical problems had got the better of his resolve to pursue a vocation in engineering. Though his studentship was renewed for the following year, and he is still listed as a student of Manchester University in October 1911, his days as an aeronaut were finished during the summer vacation of that year, when, 'in a constant, indescribable, almost pathological state of agitation', he drew up a plan for a proposed book on philosophy.

3
RUSSELL'S PROTÉGÉ

At the end of the summer vacation of 1911 Wittgenstein, having drawn up a plan of his projected book on philosophy, travelled to Jena to discuss it with Frege – presumably with a view to finding out whether it was worth going on with, or whether he should instead continue with his work in aeronautical research. Hermine Wittgenstein knew Frege to be an old man and was anxious about the visit, fearing that he would not have the patience to deal with the situation, or the understanding to realize the momentous importance the meeting had for her brother. In the event, so Wittgenstein later told friends, Frege 'wiped the floor' with him – one reason, perhaps, why nothing of this proposed work has survived. Frege was, however, sufficiently encouraging to recommend to Wittgenstein that he go to Cambridge to study under Bertrand Russell.

The advice was more propitious than Frege could have known, and was not only to lead to a decisive turning point in Wittgenstein's life, but also to have a tremendous influence on Russell's. For at the very time when Wittgenstein needed a mentor, Russell needed a protégé.

The year 1911 was something of a watershed in Russell's life. He had, in the previous year, finished *Principia Mathematica*, the product of ten years of exhausting labour. 'My intellect never quite recovered from the strain', he writes in his *Autobiography*. 'I have been ever since definitely less capable of dealing with difficult abstractions than I was before.' With the completion of *Principia* Russell's life, both personally and philosophically, entered a new phase. In the spring of 1911 he fell in love with Ottoline Morrell, the aristocratic wife of the Liberal MP Phillip Morrell, and began an affair that was to last until 1916. During the height of his passion he wrote Ottoline as many as three letters a day. These letters contain an almost daily record of Russell's reactions to Wittgenstein – a record which provides a useful corrective to some of the anecdotes he told about Wittgenstein in his later years, when his love of a good story frequently got the better of his concern for accuracy.

Partly under the influence of Ottoline, and partly from the debilitating effect of finishing *Principia*, Russell's work in philosophy began to change. His first work after *Principia* was *The Problems of Philosophy*, his 'shilling shocker', the first of many popular works and the book that first revealed his remarkable gift for the lucid expression of difficult ideas. At the same time he took up the post of lecturer in mathematical logic at Trinity College. His teaching and his work on a book popularizing his thought – together with the fact that *Principia* had left him exhausted – combined to persuade him that from now on his chief task in the development of the ideas of *Principia* lay in encouraging others to continue from where he had left off. At the end of 1911 he wrote to Ottoline: 'I did think the technical philosophy that remains for me to do very important indeed.' But now:

> I have an uneasiness about philosophy altogether; what remains for *me* to do in philosophy (I mean in *technical* philosophy) does not seem of first-rate importance. The shilling shocker really seems to me better worth doing . . . I think really the important thing is to make the ideas I have intelligible.

The influence of Ottoline during this time is most clearly seen in Russell's plans for a book on religion, to be called *Prisons*, which he began while he was still finishing *The Problems of Philosophy*, and which he abandoned some time in 1912. The title of the work comes from a line in *Hamlet* – 'The world's a prison and Denmark one of the worst wards' – and its central idea was that 'the religion of contemplation' could provide the means of escape out of the prisons into which we enclose human life. By 'the religion of contemplation' Russell did not mean a belief in God or immortality – even his infatuation with the deeply religious Ottoline could not persuade him to believe in those. He meant, rather, a mystical union with the universe in which our finite selves are overcome and we become at one with infinity.

For, as he told Ottoline (with dubious accuracy), 'What you call God is very much what I call infinity.'

The project might plausibly be seen as an attempt by Russell to reconcile his own sceptical agnosticism with Ottoline's devout faith. The central conceit of the book reappears in a letter to Ottoline describing the liberating effect of her love for him:

> . . . now there is no prison for me. I reach out to the stars, & through the ages, & everywhere the radiance of your love lights the world for me.

The Russell Wittgenstein met in 1911, then, was far from being the strident rationalist, the offender of the faith, he later became. He was a man in the grip of romance, more appreciative than he had been before, or was to become, of the irrational and emotional side of human character – even to the extent of adopting a kind of transcendental mysticism. Perhaps more important, he was a man who, having decided that his contribution to technical philosophy was finished, was looking for someone with the youth, vitality and ability to build upon the work which he had begun.

There are some indications that Wittgenstein was initially inclined to ignore Frege's advice and to continue with his work at Manchester. Thus we find him still listed as a research student in engineering for the start of the autumn term, his studentship having been renewed for the coming year. It may be that, having been floored by Frege in argument, he had resolved to overcome his obsession with the philosophy of mathematics and to persist with his vocation as an engineer.

He had not, apparently, made any prior arrangement with Russell, when, on 18 October – about two weeks into the Michaelmas term – he suddenly appeared at Russell's rooms in Trinity College to introduce himself.

Russell was having tea with C. K. Ogden (later to become the first translator of *Tractatus Logico-Philosophicus*) when:

> . . . an unknown German appeared, speaking very little English but refusing to speak German. He turned out to be a man who had learned engineering at Charlottenburg, but during his course had

> acquired, by himself, a passion for the philosophy of mathematics & has now come to Cambridge on purpose to hear me.

Two omissions in Wittgenstein's self-introduction are immediately striking. The first is that he does not mention that he had been advised to come to Russell by Frege. The second is that he omits to tell Russell he had studied (indeed, officially, *was studying*) engineering at Manchester. These omissions, though strange, are perhaps indicative of nothing more than Wittgenstein's extreme nervousness; if Russell had the impression that he spoke very little English, he must indeed have been in quite a state.

From what we know of the weeks that followed, it seems to have been Wittgenstein's intention not simply to hear Russell lecture, but to impress himself upon him, with a view to finding out, once and for all – from the horse's mouth, as it were – whether he had any genuine talent for philosophy, and therefore whether he would be justified in abandoning his aeronautical research.

Russell's lectures on mathematical logic attracted very few students, and he often lectured to just three people: C. D. Broad, E. H. Neville and H. T. J. Norton. He therefore had reason to feel pleased when, on the day he first met Wittgenstein, he found him 'duly established' at his lecture. 'I am much interested in my German', he wrote to Ottoline, '& shall hope to see a lot of him.' As it turned out, he saw a lot more of him than he had bargained for. For four weeks Wittgenstein plagued Russell – dominating the discussions during the lectures, and following him back to his rooms after them, still arguing his position. Russell reacted to this with a mixture of appreciative interest and impatient exasperation:

> My German friend threatens to be an infliction, he came back with me after my lecture & argued till dinner-time – obstinate & perverse, but I think not stupid. [19.10.11]

> My German engineer very argumentative & tiresome. He wouldn't admit that it was certain that there was not a rhinoceros in the room . . . [He] came back and argued all the time I was dressing. [1.11.11]

> My German engineer, I think, is a fool. He thinks nothing empirical is knowable – I asked him to admit that there was not a rhinoceros in the room, but he wouldn't. [2.11.11]

[Wittgenstein] was refusing to admit the existence of anything except asserted propositions. [7.11.11]

My lecture went off all right. My German ex-engineer, as usual, maintained his thesis that there is nothing in the world except asserted propositions, but at last I told him it was too large a theme. [13.11.11]

My ferocious German came and argued at me after my lecture. He is armour-plated against all assaults of reasoning. It is really rather a waste of time talking with him. [16.11.11]

In later life Russell made great play of these discussions and claimed he had looked under all the tables and chairs in the lecture room in an effort to convince Wittgenstein that there was no rhinoceros present. But it is clear that for Wittgenstein the issue was metaphysical rather than empirical, to do with what kind of things make up the world rather than the presence or otherwise of a rhinoceros. In fact, the view that he is here so tenaciously advancing prefigures that expressed in the famous first proposition of the *Tractatus*: 'The world is the totality of facts, not of things.'

It will be seen from the above extracts that Russell had, as yet, no great conviction of Wittgenstein's philosophical ability. And yet the responsibility for Wittgenstein's future was soon to rest with him. On 27 November, at the end of the Michaelmas term, Wittgenstein came to Russell to seek his opinion on the question that mattered to him above all else, the answer to which would determine his choice of career and finally settle the conflict of interests with which he had struggled for over two years:

My German is hesitating between philosophy and aviation; he asked me today whether I thought he was utterly hopeless at philosophy, and I told him I didn't know but I thought not. I asked him to bring me something written to help me judge. He has money, and is quite passionately interested in philosophy, but he feels he ought not to give his life to it unless he is some good. I feel the responsibility rather as I really don't know what to think of his ability. [27.11.11]

Before he left Cambridge, Wittgenstein met Russell socially and, for once, was relaxed enough in Russell's company to reveal something of himself apart from his all-consuming involvement with philosophical problems. Russell finally discovered that he was Austrian rather than German, and also that he was 'literary, very musical, pleasant-mannered . . . and, I *think* really intelligent'. The consequence of which was: 'I am getting to like him.'

The real turning point, however, came when Wittgenstein returned to Cambridge in January 1912 with a manuscript he had written during the vacation. On reading it, Russell's attitude towards him changed immediately. It was, he told Ottoline, 'very good, much better than my English pupils do', adding: 'I shall certainly encourage him. Perhaps he will do great things.' Wittgenstein later told David Pinsent that Russell's encouragement had proved his salvation, and had ended nine years of loneliness and suffering, during which he had continually thought of suicide. It enabled him finally to give up engineering, and to brush aside 'a hint that he was *de trop* in this world' – a hint that had previously made him feel ashamed he had *not* killed himself. The implication is that, in encouraging him to pursue philosophy and in justifying his inclination to abandon engineering, Russell had, quite literally, saved Wittgenstein's life.

Over the next term Wittgenstein pursued his studies in mathematical logic with such vigour that, by the end of it, Russell was to say that he had learnt all he had to teach, and, indeed, had gone further. 'Yes', he declared to Ottoline, 'Wittgenstein has been a great event in my life – whatever may come of it':

> I love him & feel he will solve the problems I am too old to solve – all kinds of problems that are raised by my work, but want a fresh mind and the vigour of youth. He is *the* young man one hopes for.

After supervising him for just one term, Russell had identified in Wittgenstein the protégé he was looking for.

What philosophical work Wittgenstein actually did during the three months of this term, we do not know. Russell's letters to Ottoline contain only the most tantalizing hints. On 26 January Wittgenstein proposed: 'a definition of logical *form* as opposed to logical *matter*'. A month later he: 'brought a very good original suggestion, which I think is right, on an important point of logic'. These hints, though, are enough to suggest that Wittgenstein's work was, from the beginning, not directed to the problem: 'What is mathematics?' but to the still more fundamental question: 'What is logic?' This, Russell himself felt, was the most important question left unanswered by *Principia*.

On 1 February 1912 Wittgenstein was admitted as a member of Trinity College, with Russell as his supervisor. Knowing that he had never received any formal tuition in logic, and feeling that he might benefit from it, Russell arranged for him to be 'coached' by the eminent logician and Fellow of King's College, W. E. Johnson. The arrangement lasted only a few weeks. Wittgenstein later told F. R. Leavis: 'I found in the first hour that he had nothing to teach me.' Leavis was also told by Johnson: 'At our first meeting he was teaching me.' The difference is that Johnson's remark was sardonic, Wittgenstein's completely in earnest. It was actually Johnson who put an end to the arrangement, thus providing Russell with the first of many occasions on which he had to use all his tact and sensitivity to point out Wittgenstein's faults to him without upsetting him:

> While I was preparing my speech Wittgenstein appeared in a great state of excitement because Johnson (with whom I had advised him to coach) wrote and said he wouldn't take him any more, practically saying he argued too much instead of learning his lesson like a good boy. He came to me to know what truth there was in Johnson's feeling. Now he is terribly persistent, hardly lets one get a word in, and is generally considered a bore. As I really like him very much, I was able to hint these things to him without offending him.

Wittgenstein made quite a different impression on G. E. Moore, whose lectures he began to attend during this term. 'Moore thinks enormously highly of Wittgenstein's brains', Russell told Ottoline; '– says he always feels W. *must* be right when they disagree. He says during his lectures W. always looks frightfully puzzled, but nobody else does. I am glad to be confirmed in my high opinion of W. – the young men don't think much of him, or if they do it is only because Moore and I praise him.' For Wittgenstein's part, he 'said how much he loves Moore, how he likes and dislikes people for the way they think – Moore has one of the most beautiful smiles I know, and it had struck him'.

Wittgenstein's friendship with Moore was to develop later. But, with Russell, an affectionate bond quickly developed. Russell's admiration knew no bounds. He saw in Wittgenstein the 'ideal pupil', one who 'gives passionate admiration with vehement and very intelligent dissent'. As opposed to Broad, who was the most *reliable* pupil he had had – 'practically certain to do a good deal of useful but not brilliant work' – Wittgenstein was 'full of boiling passion which may drive him anywhere'.

Russell came increasingly to identify with Wittgenstein, to see in him a fellow-spirit, one who brought all his force and passion to bear on *theoretical* questions. 'It is a rare passion and one is glad to find it.' Indeed: 'he has more passion about philosophy than I have; his avalanches make mine seem mere snowballs'. Again and again, one comes across the word 'passion' in Russell's descriptions: 'a pure intellectual passion' that Wittgenstein (like Russell himself) had 'in the highest degree'; 'it makes me love him'. It was almost as though he saw in Wittgenstein a mirror image of himself – or, perhaps more apposite, as though he saw him as his own offspring:

> His disposition is that of an artist, intuitive and moody. He says every morning he begins his work with hope, and every evening he ends in despair – he has just the sort of rage when he can't understand things that I have. [16.3.12]

> I have the most perfect intellectual sympathy with him – the same passion and vehemence, the same feeling that one must understand or die, the sudden jokes breaking down the frightful tension of thought. [17.3.12]

> . . . he even has the same similes as I have – a wall parting him from the truth which he must pull down somehow. After our last discussion, he said 'Well, there's a bit of wall pulled down.'
>
> His attitude justifies all I have hoped about my work. [22.3.12]

Russell noted with approval that Wittgenstein had excellent manners, but, even more approvingly, that: 'in argument he forgets about manners & simply says what he thinks':

> No one could be more sincere than Wittgenstein, or more destitute of the false politeness that interferes with truth; but he lets his feelings and affections appear, and it warms one's heart. [10.3.12]

When, for example, Wittgenstein met a student who happened to be a monk, Russell could report gleefully to Ottoline that he was 'far more terrible with Christians than I am':

> He had liked F., the undergraduate monk, and was horrified to learn that he is a monk. F. came to tea with him and W. at once attacked him – as I imagine, with absolute fury. Yesterday he returned to the charge, not arguing but only preaching honesty. He abominates ethics and morals generally; he is deliberately a creature of impulse and thinks one should be. [17.3.12]

'I wouldn't answer for his technical morals', Russell concluded.

The remark jars. It shows that he had misunderstood the thrust of Wittgenstein's argument. For, if Wittgenstein was preaching honesty, he was obviously not abominating ethics in the sense of arguing for a licence for immorality. He was arguing for a morality based on integrity, on being true to oneself, one's impulses – a morality that came from inside one's self rather than one imposed from outside by rules, principles and duties.

It was a question upon which, for Wittgenstein, a great deal would have hung. In abandoning engineering for philosophy, had he not forsaken what might have been seen as his duty in favour of pursuing something that burned *within* him? And yet, as we have seen – and as Russell was originally told – such a decision needed the justification that, in doing so, he was not simply pursuing a whim, but following a course in which he might conceivably make an important contribution.

Russell's misunderstanding of the point is a hint of things to come, a suggestion that his 'theoretical passion' and Wittgenstein's were not, after all, as similar as he had supposed. By the end of term their relationship was such that Wittgenstein felt able to tell Russell what he liked *and disliked* about his work. He spoke with great feeling about the beauty of *Principia*, and said – what was probably the highest praise he could give it – that it was like music. The popular work, however, he disliked strongly – particularly 'The Free Man's Worship' and the last chapter of *The Problems of Philosophy*, on 'The Value of Philosophy'. He disliked the very idea of saying philosophy had *value*:

> . . . he says people who like philosophy will pursue it, and others won't, and there is an end of it. *His* strongest impulse is philosophy. [17.3.12]

It is hard to believe that Wittgenstein's attitude was quite as straightforward as Russell suggests. After all, for years before becoming Russell's pupil, he had suffered deeply from the conflict of duty and impulse engendered by the fact that philosophy was his strongest impulse. He indeed believed that one *should* be – as his father had been, and as his brother Hans had been, and as all geniuses are – a creature of impulse. But he also had an almost overbearing sense of duty, and was prone to periodically crippling self-doubts. Russell's encouragement had been necessary precisely because it enabled him to overcome these doubts, and to follow his strongest impulse *happily*. His family had been struck by the immediate change that came over him after he had been encouraged to work on philosophy by Russell. And he himself, at the end of this term, told Russell that the happiest hours of his life had been spent in his rooms. But this happiness was caused not simply by his being allowed to follow his impulses, but also by the conviction that – as he had an unusual talent for philosophy – he had the *right* to do so.

It was important to Wittgenstein that Russell should understand him on this point, and on the day he returned to Cambridge for the following term, the theme was renewed. Russell found that he 'wears well . . . quite as good as I thought. I find him strangely exciting', and was still inclined to see no fundamental difference in their temperamental outlook: 'He lives in the same intense excitement as I do, hardly able to sit still or read a book.' Wittgenstein talked about Beethoven:

> . . . how a friend described going to Beethoven's door and hearing him 'cursing, howling and singing' over his new fugue; after a whole hour Beethoven at last came to the door, looking as if he had been fighting the devil, and having eaten nothing for 36 hours because his cook and parlour-maid had been away from his rage. That's the sort of man to be.

But, again, this is not just *anybody* 'cursing, howling and singing'. Would Wittgenstein have felt this to be 'the sort of man to be' if all this fierce absorption had produced only mediocre works? What is implied is that, if one's strongest impulse is to write music, and if, by surrendering completely to that impulse one is able to write sublime music, then one not only has the right to behave impulsively; one has a duty to do so.

Similarly, Russell gave Wittgenstein the licence to behave in the same way, because he recognized in him the quality of genius. He later described Wittgenstein as:

> . . . perhaps the most perfect example I have ever known of genius as traditionally conceived, passionate, profound, intense, and dominating.

He had already come to see in Wittgenstein these qualities at the beginning of this summer term. In his letter to Ottoline of 23 April, he told her: 'I don't feel the subject neglected by my abandoning it, as long as he takes it up', adding, as if in illustration of the qualities needed for the task: 'I thought he would have smashed all the furniture in my room today, he got so excited.'

Wittgenstein asked him how he and Whitehead were going to end *Principia*. Russell replied that they would have no conclusion; the book would just end 'with whatever formula happened to come last':

> He seemed surprised at first, and then saw that was right. It seems to me the beauty of the book would be spoilt if it contained a single word that could possibly be spared.

This appeal to the beauty of the work would no doubt have been sympathetically received by Wittgenstein, who was to take to new heights, with the sparse prose of the *Tractatus*, the austere aesthetic here advanced by Russell.

Already, by the beginning of the summer term, the relationship between the two was beginning to change. Although he was still formally Wittgenstein's supervisor, Russell became increasingly anxious for his approval. Over the Easter vacation he had begun work on a paper on 'Matter', to be delivered to the Philosophical Society at Cardiff University. It was, he hoped, to be a work which demonstrated a renewed vigour – 'a model of cold passionate analysis, setting forth the most painful conclusions with utter disregard of human feelings'. Cold *and* passionate? Russell explained:

> I haven't had enough courage hitherto about matter. I haven't been sceptical enough. I want to write a paper which my enemies will call 'the bankruptcy of realism'. There is nothing to compare to passion for giving one cold insight. Most of my best work has been done in the inspiration of remorse, but any passion will do if it is strong. Philosophy is a reluctant mistress – one can only reach her heart with the cold steel in the hand of passion.

'Cold steel in the hand of passion' – the phrase conjures up perfectly Wittgenstein's own combination of a rigorously logical mind and an impulsive and obsessional nature. He was the very personification of Russell's philosophical ideal.

Russell was, however, to be disappointed with Wittgenstein's reaction to the project. He dismissed the whole subject as 'a trivial problem':

> He admits that if there is no Matter then no one exists but himself, but he says it doesn't hurt, since physics and astronomy and all the other sciences could still be interpreted so as to be true.

A few days later, when Wittgenstein actually read parts of the paper, Russell was relieved to note a change of mind: Wittgenstein was delighted by its radicalism. Russell began his paper by stating baldly that all the arguments advanced hitherto by philosophers intending to prove the existence of matter were, quite simply, fallacious. This, declared Wittgenstein, was the best thing Russell had done. When he saw the rest of the paper he changed his mind again and told Russell that he did not like it after all, 'but only', Russell told Ottoline, clutching at straws, 'because of disagreement, not because of its being badly done'. The paper, for which Russell originally had such high hopes, remained unpublished.

The extraordinarily high opinion which Russell had of Wittgenstein was bound to arouse the curiosity of his friends at Cambridge, particularly among the Apostles, the self-consciously élite conversation society (of which Russell himself was a member) which was at the time dominated by John Maynard Keynes and Lytton Strachey. Wittgenstein became what, in Apostolic jargon, was known as an 'embryo' – a person under consideration for membership. Strachey (who lived in London) came to tea with Wittgenstein at Russell's rooms to inspect this potential Apostle for himself. Wittgenstein had recently read Strachey's *Landmarks in French Literature*, but had not liked it. He told Russell that it made an impression of effort, like the gasps of an asthmatic person. He nevertheless took the trouble to shine at tea sufficiently to impress Strachey. 'Everybody has just begun to discover him', Russell told Ottoline afterwards; 'they all now realize he has genius.'

As to whether Wittgenstein would want to join the Apostles, Russell had doubts:

> Somebody had been telling them about Wittgenstein and they wanted to hear what I thought of him. They were thinking of electing him to the Society. I told them I didn't think he would like the Society. I am quite sure he wouldn't really. It would seem to him stuffy, as indeed it has become, owing to their practice of being in love with each other, which didn't exist in my day – I think it is mainly due to Lytton.

Whether or not he was right in supposing that Wittgenstein would object to the 'stuffy' atmosphere of the homosexual affairs which dominated the Society at that time, he was, as it turned out, right to suggest that Wittgenstein would not like the Apostles.

Strachey's impressions of Wittgenstein, meanwhile, were somewhat mixed. On 5 May he invited him for lunch, but on this second meeting he was not impressed. 'Herr Sinckel-Winckel lunches with me', he wrote to Keynes; 'quiet little man'. Two weeks later the two met again in the rooms of Strachey's brother, James. This time the impression gained was one of exhausting brilliance:

> Herr Sinckel-Winckel hard at it on universals and particulars. The latter oh! so bright – but *quelle souffrance!* Oh God! God! 'If A loves B' – 'There may be a common property' – 'Not analysable in that way at all, but the complexes have certain qualities.' How shall I manage to slink off to bed?

There, for the moment, Wittgenstein's connections with the Apostles rested, until the following October when, after meeting Keynes, 'Herr Sinckel-Winckel' became, briefly and disastrously, 'Brother Wittgenstein'.

From being 'generally considered a bore' by the young men at Cambridge, Wittgenstein was now considered 'interesting and pleasant, though his sense of humour is heavy'. That, at least, was the judgement of one of them, David Pinsent, who met him at one of Russell's 'squashes' (social evenings) early in the summer term. Pinsent was then in his second year as a mathematics undergraduate. The previous year he had himself been an Apostolic 'embryo', but had not been elected. This perhaps indicates how he was seen by the fashionable intellectual élite at Cambridge – interesting but not fascinating, bright but not possessed of genius.

To Wittgenstein, however, Pinsent's musical sensitivity and his equable temperament made him the ideal companion. He seems to have recognized this very swiftly, within a month of knowing Pinsent surprising him by inviting him on a holiday in Iceland, all expenses to be met by Wittgenstein's father. 'I really don't know what to think', Pinsent wrote in his diary:

> . . . it would certainly be fun, and I could not afford it myself, and Vittgenstein [sic] seems very anxious for me to come. I deferred my decision and wrote home for advice: Iceland seems rather attractive: I gather that all inland travelling has to be done on horse back, which would be supreme fun! The whole idea attracts and surprises me: I have known Vittgenstein only for three weeks or so – but we seem to get on well together: he is very musical with the same tastes as I. He is an Austrian – but speaks English fluently. I should say about my age.

Until then their acquaintance had been confined to Pinsent's acting as a subject in some experiments that Wittgenstein was conducting in the psychological laboratory. It seems to have been his intention to investigate scientifically the role of rhythm in musical appreciation. For this he presumably needed a subject with some understanding of music. In his diary Pinsent does not describe what the experiments involved, recording only that taking part in them was 'not bad fun'.

Wittgenstein was helped in this work by the psychologist C. S. Myers, who took the experiments seriously enough to introduce a demonstration of them to the British Society of Psychology. The chief result obtained from them was that, in some circumstances, subjects heard an accent on certain notes that was not in fact there.

Apart from taking part in these experiments two or three times a week, Pinsent's only contact with Wittgenstein before being invited to spend a holiday with him had been at Russell's Thursday evening 'squashes'. After one of these, on 30 May, he reports that he found Wittgenstein 'very amusing':

> . . . he is reading philosophy up here, but has only just started systematic reading: and he expresses the most naive surprise that all the philosophers he once worshipped in ignorance are after all stupid and dishonest and make disgusting mistakes!

But it was not until after Wittgenstein's unexpected invitation that a close friendship began to develop. The following day the two attended a concert together, after which they went to Wittgenstein's rooms, where they stayed talking until 11.30. Wittgenstein was 'very communicative and told me lots about himself'. It was then that he told Pinsent that Russell's encouragement to pursue philosophy had been his salvation, after nine years of suicidal loneliness and suffering. Pinsent added:

> Russell, I know, has a high opinion of him: and has been corrected by him and convinced that he (Russell) was wrong in one or two points of Philosophy: and Russell is not the only philosophical don up here that Vittgenstein has convinced of error. Vittgenstein has few hobbies, which rather accounts for his loneliness. One can't thrive entirely on big and important pursuits like Triposes. But he is quite interesting and pleasant: I fancy he has quite got over his morbidness now.

After this Wittgenstein and Pinsent saw much of each other, going to concerts at the Cambridge University Musical Club, dining together at the Union and meeting for tea in each other's rooms. Wittgenstein even attended a service at the college chapel especially to hear Pinsent read the lesson.

Despite his earlier having been described by Russell as 'terrible' with Christians, this may not have been quite so out of character as it appears. At about the same time, in fact, he surprised Russell by suddenly saying how much he admired the text: 'What shall it profit a man if he gain the whole world and lose his own soul':

> [He] then went on to say how few there are who don't lose their soul. I said it depended on having a large purpose that one is true to. He said he thought it depended more on suffering and the power to endure it. I was surprised – I hadn't expected that kind of thing from him.

The stoicism expressed here seems related to something Wittgenstein later told Norman Malcolm. During one of his holidays at home in Vienna, his previously contemptuous attitude to religion had been changed by seeing a play, *Die Kreuzelscheiber* by the Austrian dramatist and novelist Ludwig Anzengruber.* It was mediocre drama, but in it one of the characters expressed the thought that no matter what happened in the world, nothing bad could happen to *him*. *He* was independent of fate and circumstances. This stoic thought struck Wittgenstein powerfully, and, so he told Malcolm, he saw, for the first time, the possibility of religion.

For the rest of his life he continued to regard the feeling of being 'absolutely safe' as paradigmatic of religious experience. A few months after the conversation with Russell quoted above, we find him reading William James's *Varieties of Religious Experience*, and telling Russell:

> This book does me a *lot* of good. I don't mean to say that I will be a saint soon, but I am not sure that it does not improve me a little in a way in which I would like to improve *very much*: namely I think that it helps me to get rid of the *Sorge* [worry, anxiety] (in the sense in which Goethe used the word in the 2nd part of Faust).

---

* He told Malcolm he was about twenty-one years old when this event took place, which would date it to 1910 or early 1911. Because of the change in Wittgenstein's attitude to religion noted by Russell in the summer of 1912, however, it is tempting to date the episode to the Easter vacation of that year.

Two days after the discussion on losing and retaining one's soul, Russell and Wittgenstein had another conversation which revealed something of the deep differences between their respective ethical views. It arose out of a discussion of Dickens's *David Copperfield*. Wittgenstein maintained that it was wrong for Copperfield to have quarrelled with Steerforth for running away with Little Emily. Russell replied that, in the same circumstances, he would have done the same. Wittgenstein was 'much pained, and refused to believe it; thought one could and should always be loyal to friends and go on loving them'.

Russell then asked him how he would feel if he were married to a woman and she ran away with another man:

> [Wittgenstein] said (and I believe him) that he would feel no rage or hate, only utter misery. His nature is good through and through; that is why he doesn't see the need of morals. I was utterly wrong at first; he might do all kinds of things in passion, but he would not practise any cold-blooded immorality. His outlook is very free; principles and such things seem to him nonsense, because his impulses are strong and never shameful.

'I think he is passionately devoted to me', Russell added. 'Any difference of feeling causes him great pain. My feeling towards him is passionate, but of course my absorption in you makes it less important to me than his feeling is to him.'

Russell seemed slow to appreciate that their differences of feeling were important to Wittgenstein because they touched upon issues of fundamental importance to him. He was slow, too, to understand that Wittgenstein's emphasis on personal integrity (and, in the above case, loyalty) was not something *opposed* to morality, but something which constituted a different morality. It is typical of their fundamentally opposed attitudes, for example, that Russell, even in this, perhaps his most introspective period, should think that keeping one's soul depended upon a 'large purpose that one was true to' – that he was inclined to look outside himself for something to sustain him. It was typical, too, for Wittgenstein to insist that the possibility of remaining uncorrupted rested entirely on one's self – on the qualities one found within. If one's soul was pure (and disloyalty to a friend was one thing that would make it impure), then no matter what happened to one 'externally' – even if one's wife ran away with another man –

nothing could happen to one's *self*. Thus, it is not external matters that should be of the greatest concern, but one's self. The *Sorge* that prevents one facing the world with equanimity is thus a matter of more immediate concern than any misfortune that may befall one through the actions of others.

When attitudes of the most fundamental kind clash, there can be no question of agreement or disagreement, for everything one says or does is interpreted from *within* those attitudes. It is therefore not surprising that there should be frustration and incomprehension on both sides. What is surprising is Russell's rather naïve assumption that he was faced, not with a different set of ideals to his own, but simply with a rather peculiar person, a person whose 'impulses are strong and never shameful'. It is as though he could make sense of Wittgenstein's views only by appealing to some *fact* about Wittgenstein that would make his holding such views intelligible. Finding Wittgenstein's outlook alien and incomprehensible, he could seek only to *explain* it, not to *make sense* of it. He was incapable. as it were, of getting *inside* it.

Again and again, in Russell's letters to Ottoline, one has the feeling that the spirit of Wittgenstein's 'theoretical passion' eluded him. The centrality of the notion of personal integrity in Wittgenstein's outlook was interpreted by him at various times as the rejection of conventional morality, the sign of a pure, uncorrupted nature – and even, on at least one occasion, as a joke. At one of Russell's 'quashes', Wittgenstein defended the view that a study of mathematics would improve a person's taste: 'since good taste is genuine taste and therefore is fostered by whatever makes people think truthfully'. From Russell's account to Ottoline, one gathers he found this argument impossible to take seriously. He describes Wittgenstein's view as a 'paradox', and says 'we were all against him'. And yet there is every reason to suppose that Wittgenstein was speaking in complete earnest: honesty and good taste were, for him, closely intertwined notions.

Wittgenstein was not one to debate his most fundamental convictions. Dialogue with him was possible only if one shared those convictions. (Thus, dialogue with Russell on ethical questions was soon to become impossible.) To one who did not share his fundamental outlook, his utterances – whether on logic or on ethics – would, as likely as not, remain unintelligible. It was a tendency that began to worry Russell. 'I am seriously afraid', he told Ottoline, 'that no one will see the point of what he writes, because he won't recommend it by arguments addressed to a different point of view.' When Russell told him he ought not simply to state what he thought, but should also provide arguments for it, he replied that arguments would spoil its beauty. He would feel as if he were dirtying a flower with muddy hands:

> I told him I hadn't the heart to say anything against that, and that he had better acquire a slave to state the arguments.

Russell had good reason to worry about whether Wittgenstein would be understood, for he increasingly came to feel that the future of his own work in logic should be placed in Wittgenstein's hands. He even felt that when his five-year lectureship at Trinity expired he should give it up and let Wittgenstein take his place. 'It is really amazing how the world of learning has grown unreal to me', he wrote. 'Mathematics has quite faded out of my thoughts, except when proofs bring it back with a jerk. Philosophy doesn't often come into my mind, and I have no *impulse* to work at it.' Despite what he had written in the last chapter of *Problems of Philosophy*, he had lost faith in the value of philosophy:

> I did seriously mean to go back to it, but I found I really couldn't think it was very valuable. This is partly due to Wittgenstein, who has made me more of a sceptic; partly it is the result of a process which has been going on ever since I found you.

The 'process' he mentions was his growing interest, inspired by Ottoline, in non-philosophical work. First, there was *Prisons*, his book on religion; then an autobiography (which he abandoned and, apparently, destroyed); and finally an autobiographical novella called *The Perplexities of John Forstice*, in which, no doubt using some of the material he had written for his autobiography, and quoting extensively from his own letters to Ottoline, he attempted to describe imaginatively his own intellectual pilgrimage from isolation, through moral and political confusion, to clarity and grace. Russell was not at his best with this sort of writing, and none of the works mentioned saw the light of day during his lifetime. 'I do *wish* I were more creative', he lamented to Ottoline. 'A man like Mozart makes one feel such a worm.' In later life he agreed to the posthumous publication of *Forstice*, but with serious reservations:

> . . . the second part represented my opinions during only a very short period. My view in the second part were very sentimental, much too mild, and much too favourable to religion. In all this I was unduly influenced by Lady Ottoline Morrell.

For better or for worse, it was during this 'very short period' that Wittgenstein was making his phenomenal progress in the analysis of logic. And perhaps his acceptance as a philosophical genius owes something to the influence exerted by Ottoline over Russell. If Russell had not been going through such a sentimental phase, he may not have taken to Wittgenstein in the way that he did: 'Wittgenstein brought me the most lovely roses today. He is a treasure' (23.4.12); 'I love him as if he were my son' (22.8.12). And, perhaps, if he had not lost faith and interest in his own contribution to mathematical logic, he might not have been quite so prepared to hand the subject over to Wittgenstein.

As it was, by the end of his first year at Cambridge, Wittgenstein was being primed as Russell's successor. At the end of the summer term, when Hermine visited Cambridge and was taken to meet Russell, she was amazed to hear him say: 'We expect the next big step in philosophy to be taken by your brother.'

At the beginning of the summer vacation Wittgenstein was offered G. E. Moore's old college rooms. He had, until then, been living in lodgings in Rose Crescent, and he accepted Moore's offer gratefully. The rooms were perfectly situated for him, at the top of K staircase in Whewell's Court, with a splendid view of Trinity College. He liked being at the top of the tower, and kept the same set of rooms for the rest of his time at Cambridge, even when he returned in later life, when as a Fellow and, later, as a Professor he would have been entitled to larger and grander rooms.

Wittgenstein chose the furniture for his rooms with great care. He was helped in this by Pinsent:

> I went out and helped him interview a lot of furniture at various shops: he is moving into college next term. It was rather amusing: he is terribly fastidious and we led the shopman a frightful dance, Vittgenstein ejaculating 'No – Beastly!' to 90% of what he shewed us!

Russell, too, was drawn into Wittgenstein's deliberations on the matter, and found it rather exasperating. 'He is *very* fussy', he told Ottoline, 'and bought nothing at all yesterday. He gave me a lecture on how furniture should be made – he dislikes all ornamentation that is not part of the construction, and can never find anything simple enough.' In the end Wittgenstein had his furniture specially made for him. When it arrived it was judged by Pinsent to be 'rather quaint but not bad'.

Neither Pinsent nor Russell was well placed to understand Wittgenstein's fastidiousness in the matter. To appreciate his concern for design and craftsmanship, one would have to have had some experience of construction. Thus, a few years later, we find Eccles, his engineer friend from Manchester, sending him some furniture designs of his own for Wittgenstein's comments, and receiving in reply a carefully considered, and gratefully accepted, verdict.

And to understand the strength of Wittgenstein's feeling against superfluous ornamentation – to appreciate the *ethical* importance it had for him – one would have to have been Viennese; one would have to have felt, like Karl Kraus and Adolf Loos, that the once noble culture of Vienna, which from Haydn to Schubert had surpassed anything else in the world, had, since the latter half of the nineteenth century, atrophied into, in Paul Engelmann's words, an 'arrogated base culture – a culture turned into its opposite, misused as ornament and mask'.

On 15 July Wittgenstein returned to Vienna, having arranged with Pinsent (whose parents had given the proposed holiday in Iceland their blessing) to meet in London during the first week of September. Home life in Vienna was not easy. His father had cancer and had been operated on several times; Gretl was having a baby and experiencing a difficult birth; and he himself was operated on for a hernia, discovered during an examination for military service. This latter he kept secret from his mother, who was in a distraught state looking after the invalid father.

From Vienna he wrote to Russell telling him: 'I am quite well again and philosophizing for all I am worth.' His thought progressed from thinking about the meaning of logical constants (the Russellian signs 'v', '$\sim$', '$\supset$' etc.) to deciding that: 'our problems can be traced down to the *atomic* prop[osition]s'. But in his letters to Russell he gave only hints as to what theory of logical symbolism this progression would result in.

'I am glad you read the lives of Mozart and Beethoven', he told Russell. 'These are the actual sons of God.' He told Russell of his delight on reading Tolstoy's *Hadji Murat*: 'Have you ever read it? If not, you ought to, for it is wonderful.'

When he arrived in London on 4 September he stayed as a guest of Russell's in his new flat in Bury Street. Russell found him a refreshing change from Bloomsbury – 'a great contrast to the Stephens and Stracheys and such would-be geniuses':

> We very soon plunged into logic and have had great arguments. He has a very great power of seeing what are really important problems.
>
> . . . He gives me such a delightful lazy feeling that I can leave a whole department of difficult thought to him, which used to depend on me alone. It makes it easier for me to give up technical work. Only his health seems to me very precarious – he gives one the feeling of a person whose life is very insecure. And I think he is growing deaf.

Perhaps the reference to Wittgenstein's hearing problem is ironic; in any case, it was not true that he couldn't hear, simply that he wouldn't listen – especially when Russell offered some 'sage advice' not to put off writing until he had solved *all* the problems of philosophy. That day, Russell told him, would never come:

> This produced a wild outburst – he has the artist's feeling that he will produce the perfect thing or nothing – I explained how he wouldn't get a degree or be able to teach unless he learnt to write imperfect things – this all made him more and more furious – at last he begged me not to give him up even if he disappointed me.

Pinsent arrived in London the next day, to be met by Wittgenstein, who insisted on taking him by taxi to the Grand Hotel, Trafalgar Square. Pinsent tried in vain to suggest some less luxurious hotel, but Wittgenstein would not hear of it. There was evidently, as Pinsent noted in his diary, to be no sparing of expense on this trip. Once at the hotel Pinsent was told the financial arrangements:

> Wittgenstein, or rather his father, insists on paying for both of us: I had expected him to be pretty liberal – but he surpassed all my expectations: Wittgenstein handed me over £145 in notes, and kept the same amount in notes himself. He also has a letter of Credit for about £200!

From London they went by train to Cambridge ('I need hardly say we travelled first class!'), where Wittgenstein had to attend to some business connected with his new college rooms, and then on to Edinburgh, where they stayed overnight before the boat journey. At Edinburgh, Wittgenstein took Pinsent on a shopping trip; he had not, he insisted, brought enough clothes:

> . . . he is being very fussy about taking enough clothing: he himself has three bags of luggage, and is much perturbed by my single box. He made me buy a second rug in Cambridge and several other odds and ends in Edinborough* this morning: I have resisted a good deal – especially as it is not my money I am thus spending. I got my own back on him, however, by inducing him to buy oilskins, which he has not got.

On 7 September they set off from Leith on the *Sterling*, which to Wittgenstein's disgust looked much like an ordinary cross-Channel steamer – he had expected something grander. He calmed down when they discovered a piano on board and Pinsent, who had brought a collection of Schubert's songs, sat down to play, heartily encouraged by the other passengers. It was a five-day boat journey across fairly rough sea, and both Pinsent and Wittgenstein suffered from it, although Pinsent notes with curiosity that Wittgenstein, though he spent a great deal of time in his cabin lying down, was never actually sick.

They reached Reykjavik on 12 September, and as soon as they had booked into the hotel, they hired a guide to take them on a journey inland, starting the following day. At the hotel they had their first argument – about public schools. The argument got quite heated, until, so Pinsent records, they found that they had misunderstood each other: 'He has an enormous horror of what he calls a "Philistine"

---

* This, as are other eccentric spellings in Pinsert's letters, is authentic.

attitude towards cruelty and suffering – any callous attitude – and accuses Kipling of such: and he got the idea that I sympathised with it'.

A week later, the subject of 'Philistine' attitudes was renewed:

> Wittgenstein has been talking a lot, at different times, about 'Philistines' – a name he gives to all people he dislikes! (*vide supra* – Thursday Sept: 12th) I think some of the views I have expressed have struck him as a bit philistine (views – that is – on practical things (not on philosophy) – for instance on the advantage of this age over past ages and so forth), and he is rather puzzled because he does not consider me really a Philistine – and I don't think he dislikes me! He satisfies himself by saying that I shall think differently as soon as I am a bit older!

It is tempting to see in these arguments a contrast between the pessimism of Viennese *Angst* and the optimism of British stolidity (at least, as it existed before the First World War had weakened even the British faith in 'the advantage of this age over past ages'). But if so, then it was exactly the qualities in Pinsent that made it impossible for him to share Wittgenstein's cultural pessimism that also made him the ideal companion.

Even Pinsent's cheerful equanimity, however, could sometimes be strained by Wittgenstein's nervousness – his 'fussiness', as Pinsent called it. On their second day in Reykjavik they went to the office of the steamship company to secure their berths home. There was some trouble in making themselves understood, but the matter was eventually resolved, at least to Pinsent's satisfaction:

> Wittgenstein however got terribly fussy and talked about our not getting home at all, and I got quite irritated with him: eventually he went off alone and got a man from the bank to act as interpreter and go through the whole business again at the steamship office.

Such temporary lapses in Pinsent's good humour, though infrequent, disturbed Wittgenstein greatly. On 21 September we read:

> Wittgenstein was a bit sulky all the evening: he is very sensitive if I get, momentarily, a bit irritated over some trifle – as I did tonight – I forget what about: the result being that he was depressed and silent

> for the rest of the evening. He is always imploring me not to be irritable: and I do my best, and really, I think, I have not been so often on this trip!

Ten days of the holiday were taken up with a journey inland on ponies. Again, no expense was spared. The cortège consisted of Wittgenstein, Pinsent and their guide, each on a pony, while in front they drove two pack ponies and three spare ponies. During the day they rode and explored the countryside, and in the evenings Wittgenstein taught Pinsent mathematical logic, which Pinsent found 'excessively interesting' – 'Wittgenstein makes a very good teacher'.

Occasionally they explored the countryside on foot, and even once attempted some rock-climbing, at which neither was very proficient. It made Wittgenstein 'terribly nervous':

> His fussyness comes out here again – he is always begging me not to risk my life! It is funny that he should be like that – for otherwise he is quite a good travelling companion.

On their walks the talk would mostly be of logic, Wittgenstein continuing to educate Pinsent on the subject: 'I am learning a lot from him. He is really remarkably clever.'

> I have never yet been able to find the smallest fault in his reasoning: and yet he has made me reconstruct entirely my ideas on several subjects.

When, after their excursion through the Icelandic countryside, they were back at the Reykjavik hotel, Pinsent took the opportunity of having some lighter conversation with 'a very splendid bounder' who had just arrived. It prompted a long discussion about 'such people': 'he simply won't speak to them, but really, I think, they are rather amusing'. The following day: 'Wittgenstein made an awful fuss.' He took such a violent dislike to Pinsent's 'splendid bounder' that he refused to contemplate eating at the same table. To ensure that this didn't happen, he gave orders that their meal was to be served an hour earlier than table d'hôte in all cases. At lunch they forgot, and rather than take the chance, Wittgenstein took Pinsent out to see if they could find anything in Reykjavik. They couldn't. So Wittgenstein ate a few biscuits in his room, and Pinsent went to table d'hôte. In the evening Pinsent found Wittgenstein 'still fairly sulky about the lunch business', but they got their supper an hour early as arranged, and had champagne with it, 'which cheered him up a bit, finally leaving him quite normal'.

Pinsent stayed receptive and cheerful throughout. On the boat home Wittgenstein took him into the engine room and explained to him how the engines worked. He also described the research he was doing in logic. 'I really believe he has discovered something good', Pinsent comments – without, unfortunately, mentioning what it was.

On the return journey, Pinsent persuaded Wittgenstein to spend a night with his family in Birmingham – he was keen to show him off to his parents. The inducement was a concert at the town hall, the programme for which was Brahms's *Requiem*, Strauss's *Salome*, Beethoven's Seventh Symphony and a Bach motet, 'Be not afraid'. Wittgenstein enjoyed the Brahms, refused to go in for the Strauss and left the hall as soon as the Beethoven had finished. At supper, Pinsent's father was suitably impressed when Pinsent got Wittgenstein to explain to him some of the logic that he had taught him during the holiday. 'I think father was interested', he writes, and – less hesitantly – 'certainly he agreed with me afterwards that Wittgenstein is really very clever and acute.'

For Pinsent it had been: 'the most glorious holiday I have ever spent!'

> The novelty of the country – of being free of all considerations about economising – the excitement and everything – all combine to make it the most wonderful experience I have ever had. It leaves almost a mystic-romantic impression on me: for the greatest romance consists in novel sensations – novel surroundings – and so forth, whatever they be provided they are novel.

Not so for Wittgenstein. What remained in his memory were their differences and disagreements – perhaps the very occasions mentioned in Pinsent's diary – Pinsent's occasional irritability, hints of his 'Philistinism', and the incident with the 'bounder'. He later told Pinsent that he had enjoyed it, 'as much as it is possible for two people to do who are nothing to each other'.

4
RUSSELL'S MASTER

> If we now turn to the gifted men, we shall see that in their case love frequently begins with self-mortification, humiliation, and restraint. A moral change sets in, a process of purification seems to emanate from the object loved.
>
> Weininger, *Sex and Character*

Wittgenstein returned to Cambridge from his holiday with Pinsent in an agitated and irritable state. Within days, he had his first major disagreement with Russell. In Wittgenstein's absence, Russell had published a paper on 'The Essence of Religion' in the *Hibbert Journal*. It was taken from his abandoned book, *Prisons*, and was an Ottoline-inspired attempt to present a 'Religion of Contemplation' centred on the notion of 'the infinite part of our life', which 'does not see the world from one point of view: it shines impartially, like the diffused light on a cloudy sea':

> Unlike the finite life, it is impartial; its impartiality leads to truth in thought, justice in action, and universal love in feeling.

In many ways the article anticipates the mystical doctrines that Wittgenstein himself was to advance in the *Tractatus*, particularly in its advocation of a Spinozistic 'freedom from the finite self' (what in the *Tractatus* is called contemplating the world *sub specie aeterni*) and in its repudiation of what Russell calls: 'the insistent demand that our ideals shall be already realized in the world' (compare *TLP* 6.41). Nevertheless, unlike the *Tractatus*, Russell's paper shows no hesitation in articulating this mysticism, and in, for example, using words like 'finite' and 'infinite' in ways that are, strictly speaking, meaningless. In any case, Wittgenstein hated the article and, within days of being back at Cambridge, stormed into Russell's rooms to make his feelings known. He happened to interrupt a letter to Ottoline:

> Here is Wittgenstein just arrived, frightfully pained by my Hibbert article which he evidently *detests*. I must stop because of him.

A few days later Russell spelt out the reasons for Wittgenstein's outburst: 'He felt I had been a traitor to the gospel of exactness; also that such things are too intimate for print.' 'I minded very much', he added, 'because I half agree with him.' He continued to brood on the attack for a few more days:

> Wittgenstein's criticisms disturbed me profoundly. He was so unhappy, so gentle, so wounded in his wish to think well of me.

He minded all the more because of his increasing tendency to look upon Wittgenstein as his natural successor. His own endeavours in the analysis of logic were becoming ever more half-hearted. Having prepared the first draft of a paper entitled 'What is Logic?' he found he couldn't get on with it and felt 'very much inclined to leave it to Wittgenstein'.

Moore, too, felt the force of Wittgenstein's forthright criticism during these first few weeks of October. Wittgenstein began the term by attending Moore's lectures on psychology. 'He was very displeased with them', writes Moore, 'because I was spending a great deal of time in discussing Ward's view that psychology did not differ from the Natural Sciences in subject-matter but only in point of view':

> He told me these lectures were very bad – that what I ought to do was to say what *I* thought, not to discuss what other people thought; and he came no more to my lectures.

Moore adds: 'In this year both he and I were still attending Russell's Lectures on the Foundations of Mathematics; but W. used also to go for hours to Russell's rooms in the evening to discuss Logic with him.'

In fact, Wittgenstein – apparently undergoing the process of self-mortification and moral change described by Weininger – would spend these hours discussing himself as much as logic. He would, according to Russell, 'pace up and down my room like a wild beast for three hours in agitated silence'. Once, Russell asked: 'Are you thinking about logic or your sins?' 'Both', Wittgenstein replied, and continued his pacing.

Russell considered him to be on the verge of a nervous breakdown – 'not far removed from suicide, feeling himself a miserable creature, full of sin' – and was inclined to attribute this nervous fatigue to the fact that: 'he strains his mind to the utmost constantly, at things which are discouraging by their difficulty'. He was supported in this view by a doctor who, so concerned was Wittgenstein about his fits of dizziness and his inability to work, was called and who pronounced: 'It is all nerves.' Despite Wittgenstein's earnest desire to be treated *morally*, therefore, Russell insisted on treating him physically, advising him to eat better and to go out riding. Ottoline did her bit by sending some cocoa. 'I will remember the directions', Russell promised her, '& try to get W. to use it – but I'm sure he won't.'

Wittgenstein did, however, take Russell's advice to go riding. Once or twice a week for the rest of term he and Pinsent would hire horses and take them for what Pinsent regarded as 'tame' rides (i.e. rides that involved no jumping) along the river tow-path to Clay-hithe, or along the Trumpington road to Grantchester. If this produced any effect on Wittgenstein's temperament at all, it did not make him any the less inclined to sudden outbursts of wrath against the moral failings of himself and others.

On 9 November, Russell had arranged to take a walk with Wittgenstein. On the same day, however, he felt obliged to watch Whitehead's son, North, compete in a rowing race on the river. He therefore took Wittgenstein to the river, where they both watched North get beaten in his race. This led to, as Russell put it, a 'passionate afternoon'. He himself found the 'excitement and conventional importance' of the race painful, the more so because North 'minded being beaten horribly'. But Wittgenstein found the whole affair *disgusting*:

> . . . said we might as well have looked on at a bull fight (I had that feeling myself), and that *all* was of the devil, and so on. I was cross North had been beaten, so I explained the necessity of competition

> with patient lucidity. At last we got on to other topics, and I thought it was all right, but he suddenly stood still and explained that the way we had spent the afternoon was so vile that we ought not to live, or at least he ought not, that nothing is tolerable except producing great works or enjoying those of others, that he has accomplished nothing and never will, etc. – all this with a force that nearly knocks one down. He makes me feel like a bleating lambkin.

A few days later, Russell had had enough: 'I told Wittgenstein yesterday that he thinks too much about himself, and if he begins again I shall refuse to listen unless I think he is quite desperate. He has talked it out now as much as is good for him.'

And yet, at the end of November, we find him drawn once more into a discussion with Wittgenstein about Wittgenstein:

> I got into talking about his faults – he is worried by his unpopularity and asked me why it was. It was a long and difficult and passionate (on his side) conversation lasting till 1.30, so I am rather short of sleep. He is a great task but quite worth it. He is a little too simple, yet I am afraid of spoiling some fine quality if I say too much to make him less so.

Something of what Russell means by Wittgenstein's being 'a little too simple' (and also, perhaps, of the source of his supposed unpopularity) we can gather from an entry in Pinsent's diary. On the evening following Wittgenstein's 'passionate afternoon' on the river, he and Pinsent attended a concert at the Cambridge University Music Club, and afterwards went to Wittgenstein's rooms. Farmer, the undergraduate monk mentioned earlier by Russell, appeared. He was, says Pinsent, 'a man Wittgenstein dislikes and believes to be dishonest minded':

> . . . [Wittgenstein] got very excited trying to induce him to read some good book on some exact science, and see what honest thought is. Which would obviously be good for Farmer – as indeed for any one – : but Wittgenstein was very overbearing and let Farmer know exactly what he thought of him, and altogether talked as if he was his Director of Studies! Farmer took it very well – obviously convinced that Wittgenstein is a lunatic.

Wittgenstein's conviction that he was unpopular needs some qualification. During this term, at the very peak of his nervous irritability, he did succeed in making some new and important friendships. In particular, he gained the respect and affection of John Maynard Keynes, who was to be a valuable and supportive friend for the greater part of Wittgenstein's life. Russell first brought the two together on 31 October – 'but it was a failure', he reports, 'Wittgenstein was too ill to argue properly'. But by 12 November we find Keynes writing to Duncan Grant: 'Wittgenstein is a most wonderful character – what I said about him when I saw you last is quite untrue – and extraordinarily nice. I like enormously to be with him.'

Keynes's advocacy was powerful enough to overcome any doubts Lytton Strachey may still have had about Wittgenstein's suitability for membership of the Apostles, and, after his pronouncement of Wittgenstein's genius, the issue was decided. The only remaining doubt was whether Wittgenstein would wish to become a member – whether he really considered it worth his while to meet regularly for discussions with the other members. This, from an Apostolic point of view, was quite extraordinary. 'Have you heard', wrote Keynes in astonishment to Strachey, 'how our new brother's only objection to the Society is that it doesn't happen to be apostolic?'

Russell, with some misgivings, did his best to further the cause. 'Obviously', he wrote to Keynes:

> from [Wittgenstein's] point of view the Society is a mere waste of time. But perhaps from a philanthropic point of view he might be made to feel it worth going on with.

Thus, 'philanthropically', he did what he could to present the Society in a favourable light. He explained to Wittgenstein that, though there was nothing to be got out of the Society in its present state, in former days it had been good and might be again if he were prepared to stick to it. As we have seen, Russell's own objections to the Society centred on their predilection for homosexual 'intrigues'. Wittgenstein's doubts, however, were to do with the fact that, though he liked the 'angels' (graduates) in the Society (Moore, Russell and Keynes, in particular), he had taken a fierce dislike to his fellow 'brothers' – the undergraduate members – and was uncertain whether he could stand the prospect of regular discussions with them.

He objected to their immaturity, telling Keynes that to watch them at Apostolic meetings was to see those who had not yet made their toilets – a process which, though necessary, was indecent to observe.

The 'brothers' in question were Frank Bliss, who had come from Rugby to read classics at King's, and Ferenc Békássy, an aristocratic Hungarian who had been at Bedales before coming to King's. Both were involved in the intrigues that Russell objected to, particularly Békássy, who, it is reported by James Strachey, filled Keynes and Gerald Shove with such lust at his first meeting with the Apostles that they wanted to 'take him' right then and there on the ritual hearthrug. It can hardly have been their involvement in such affairs that made Wittgenstein object to them: otherwise, it would be quite inexplicable that he took no exception to Keynes. His dislike for Békássy may perhaps have had something in it of Austro-Hungarian rivalry. But it was principally Bliss that he objected to – 'He can't stand him', Russell told Ottoline.

With great hesitation and doubt, therefore, Wittgenstein accepted membership and attended his first Saturday meeting on 16 November. At the meeting Moore read a paper on religious conversion and Wittgenstein contributed to the discussion his view that, so far as he knew, religious experience consisted in getting rid of worry (the *Sorge* that he had mentioned to Russell) and had the consequence of giving one the courage not to care what might happen (for, to someone with faith, nothing *could* happen). After the meeting Lytton Strachey was optimistic about the future of the Society, finding the prospect of conflict and bitchiness offered by the new members 'particularly exhilarating':

> Our brothers B[liss] and Wittgenstein are so nasty and our brother Békássy is so nice that the Society ought to rush forward now into the most progressive waters. I looked in on B[liss] on Sunday night and he seemed quite as nasty as Rupert [Brooke] ever was.

On the same day he wrote at length to Sydney Saxon Turner about Russell's objections to Wittgenstein's membership:

> The poor man is in a sad state. He looks about 96 – with long snow-white hair and an infinitely haggard countenance. The election of Wittgenstein has been a great blow to him. He dearly hoped

> to keep him all to himself, and indeed succeeded wonderfully, until Keynes insisted on meeting him, and saw at once that he was a genius and that it was essential to elect him. The other people (after a slight wobble from Békássy) also became violently in favour. Their decision was suddenly announced to Bertie, who nearly swooned. Of course he could produce no reason against the election – except the remarkable one that the Society was so degraded that his Austrian would certainly refuse to belong to it. He worked himself up into such a frenzy over this that no doubt he got himself into a state of believing it – but it wasn't any good. Wittgenstein shows no signs of objecting to the Society, though he detests Bliss, who in return loathes him. I think on the whole the prospects are of the brightest. Békássy is such a pleasant fellow that, while he is in love with Bliss, he yet manages to love Wittgenstein. The three of them ought to manage very well, I think. Bertie is really a tragic figure, and I am very sorry for him; but he is most deluded too.

Strachey was wrong on several counts. Russell had no desire to 'keep Wittgenstein all to himself'; he would have been only too glad to be spared the evening-long examinations of Wittgenstein's 'sins', to which he had been subjected all term. His doubts about the wisdom of electing Wittgenstein to the Apostles – apart from his own disapproval of their homosexuality – had to do chiefly with his feeling that it would 'lead to some kind of disaster'. And in this he was not, as Strachey thought, deluded.

At the beginning of December, Strachey was told by his brother James that: 'The Witter-Gitter man is trembling on the verge of resignation.' Prompted by Moore, Strachey came up to Cambridge to try and persuade Wittgenstein to stay on, but, even after several meetings with both Wittgenstein and Moore, he was unable to do so. At the end of term Russell reported to Ottoline:

> Wittgenstein has left the Society. I think he is right, tho' loyalty to the Society wd. not have led me to say so beforehand.

He added, in terms which suggest he was very far indeed from wanting to keep Wittgenstein all to himself:

> I have had to cope with him a good deal. It is really a relief to think of not seeing him for some time, tho' I feel it is horrid of me to feel that.

To 'Goldie' Lowes Dickenson, Russell repeated his opinion that Wittgenstein had been right to leave, and added that he had tried to dissuade him: 'He is much the most apostolic and the ablest person I have come across since Moore.'

Evidence for the nature of Wittgenstein's work during this Michaelmas term is scanty. On 25 October Pinsent records a visit from Wittgenstein during which he announces a new solution to a problem – 'In the most fundamental Symbolic Logic' – which had puzzled him greatly in Iceland, and to which he had then made only a makeshift solution:

> His latest is quite different and covers more ground, and if sound should revolutionise lots of Symbolic Logic: Russell, he says, thinks it sound, but says nobody will understand it: I think I comprehend it myself however (!). If Wittgenstein's solution works, he will be the first to solve a problem which has puzzled Russell and Frege for some years: it is the most masterly and convincing solution too.

From this we can reconstruct neither the problem nor the solution, although it seems quite likely that it had to do with Wittgenstein's remark to Russell during the summer that: 'our problems can be traced down to the *atomic* prop[osition]s'. Towards the end of term Wittgenstein gave a paper to the Moral Science Club, the philosophy society at Cambridge, which can perhaps be seen as an expansion of this remark. Wittgenstein had played a great part in the discussions of the club during this term, and with Moore's help had persuaded them to adopt a new set of rules, requiring a Chairman to be appointed with the duty of preventing the discussion from becoming futile, and stipulating that no paper should last longer than seven minutes. Wittgenstein's own paper was one of the first to be given under these new rules. On 29 November the minutes record:

> Mr Wittgenstein read a paper entitled 'What is Philosophy?' The paper lasted only about 4 minutes, thus cutting the previous record established by Mr Tye by nearly two minutes. Philosophy was defined as all those primitive propositions which are assumed as true without proof by the various sciences. This defn. was much discussed, but there was no general disposition to adopt it. The discussion kept very well to the point, and the Chairman did not find it necessary to intervene much.

After term, on his way back to Vienna, Wittgenstein called on Frege at Jena, and had a long discussion with him, so he told Russell, 'about our Theory of Symbolism of which, I think, he understood the general outline'. His letters to Russell during January show him to be concerned with 'the Complex Problem' – the question of what corresponds to an atomic proposition if it is true. Suppose, for example, 'Socrates is mortal' is such a proposition: is the fact which corresponds to it a 'complex' made up of the two 'things', Socrates and mortality? This view would require the Platonic assumption of the objective existence of forms – the assumption that there exists, not only individuals, but also abstract entities such as mortality. Such an assumption is, of course, made by Russell in his Theory of Types, with which Wittgenstein became increasingly dissatisfied.

During the vacation this dissatisfaction led to his announcement of one of the central conceptions of his new logic. 'I think that there cannot be different Types of things!' he wrote to Russell:

> . . . every theory of types must be rendered superfluous by a proper theory of symbolism: For instance if I analyse the prop[osition] Socrates is mortal into Socrates, Mortality and (Ex, y)el(x, y) I want a theory of types to tell me that 'Mortality is Socrates' is nonsensical, because if I treat 'Mortality' as a proper name (as I did) there is nothing to prevent me to make the substitution the wrong way round. *But* if I analyse [it] (as I do now) into Socrates and (Ex) x is mortal or generally into x and (Ex) (x) it becomes impossible to substitute the wrong way round, because the two symbols are now of a different *kind* themselves.

He told Russell that he was not quite certain that his present way of analysing 'Socrates is mortal' was correct. But on one point he was most certain: 'all theory of types must be done away with by a theory of symbolism showing that what seem to be *different kinds of things* are symbolised by different kinds of symbols which *cannot* possibly be substituted in one another's places'.

In the face of such a sweeping dismissal of his theory, Russell might have been expected to present a spirited defence of his position – or at least some tough questions as to how his logicist foundations of mathematics might avoid contradiction *without* a theory of types. But he had by this time abandoned logic almost entirely. He spent the vacation working on quite a different subject – the existence of matter. He had, in November, delivered a paper to the Moral Science Club on the subject, in which he reiterated the view expressed in Cardiff earlier in the year that: 'No good argument for or against the existence of matter has yet been brought forward', and posed the question: 'Can we, therefore, know an object satisfying the hypotheses of physics from our private sense-data?' During the vacation he sketched an outline of the way in which he proposed to treat the problem:

> Physics exhibits sensations as functions of physical objects.
>
> But epistemology demands that physical objects should be exhibited as functions of sensations.
>
> Thus we have to solve the equations giving sensations in terms of physical objects, so as to make them give physical objects in terms of sensations.
>
> That is all.

'I am sure I have hit upon a real thing', he told Ottoline, 'which is very likely to occupy me for years to come.' It would require 'a combination of physics, psychology, & mathematical logic', and even the creation of 'a whole new science'. In the letter of January 1913, Wittgenstein was faintly dismissive of the whole project: 'I cannot imagine your way of working from sense-data forward.'

By the beginning of 1913, then, we see Russell and Wittgenstein working on very different projects – Russell on the creation of his 'new science', and Wittgenstein on the analysis of logic. Russell was now fully prepared to accept the latter as Wittgenstein's field rather than his own.

The new basis of their relationship was detected by Pinsent, who near the beginning of term describes an occasion when he and Wittgenstein were together in his rooms:

> Then Russell appeared – to inform me of some alterations he is making in the hours of his lectures – and he and Wittgenstein got talking – the latter explaining one of his latest discoveries in the Fundamentals of Logic – a discovery which, I gather, only occurred to him this morning, and which appears to be quite important and was very interesting. Russell acquiesced in what he said without a murmur.

A couple of weeks after this, upon being convicted by Wittgenstein that some of the early proofs in *Principia* were very inexact, Russell commented to Ottoline: 'fortunately it is his business to put them right, not mine'.

The co-operation between the two had come to an end. In the field of logic, Wittgenstein, far from being Russell's student, had become his teacher.

Wittgenstein was late returning for the start of term owing to his father's long-expected death from the cancer he had suffered with for over two years. The end, when it came, was something of a relief. On 21 January he wrote to Russell:

> My dear father died yesterday in the afternoon. He had the most beautiful death that I can imagine; without the slightest pains and falling asleep like a child! I did not feel sad for a single moment during all the last hours, but most joyful and I think that this death was worth a whole life.

He finally arrived in Cambridge on 27 January, going straight to Pinsent's rooms. About a week later Pinsent records an argument that shows yet another facet of the differences between Russell and Wittgenstein. In 1907 Russell had stood as a parliamentary candidate for the Women's Suffrage Party. Perhaps prompted by this fact (they had just returned from one of Russell's lectures), Wittgenstein and Pinsent got into an argument about women's suffrage. Wittgenstein was 'very much against it':

> . . . for no particular reason except that 'all the women he knows are such idiots'. He said that at Manchester University the girl students spend all their time flirting with the professors. Which disgusts him very much – as he dislikes half measures of all sorts, and disapproves of anything not deadly in earnest.

Wittgenstein's work on logic did nothing, apparently, to improve the rigour of his thought on political questions.

It is perhaps this inability – or, more likely, unwillingness – to bring his analytical powers to bear on questions of public concern that prompted Russell to criticize Wittgenstein for being 'in danger of becoming narrow and uncivilised'. Russell suggested as a corrective some French prose – a suggestion that involved him in a 'terrific contest':

> He raged & stormed, & I irritated him more & more by merely smiling. We made it up in the end, but he remained quite unconvinced. The things I say to him are just the things you would say to me if you were not afraid of the avalanche they wd. produce – & his avalanche is just what mine wd. be! I feel his lack of civilisation & suffer from it – it is odd how little music does to civilise people – it is too apart, too passionate & too remote from words. He has not a sufficiently wide curiosity or a sufficient wish for a broad survey of the world. It won't spoil his work on logic, but it will make him always a very narrow specialist, & rather too much the champion of a party – that is, when judged by the highest standards.

As the comparisons to his own situation with Ottoline indicate, Russell was bemused to find himself in the position of one advocating synthesis rather than analysis. But it should be remembered that even his philosophical preoccupations at this time were moving in that direction – away from the 'narrowness' of logical analysis and towards a broader synthesis of physics, psychology and mathematics. In consequence, his discussions with Wittgenstein became, for him, frustratingly one-sided:

> I find I no longer talk to him about *my* work, but only about his. When there are no clear arguments, but only inconclusive considerations to be balanced, or unsatisfactory points of view to be set

against each other, he is no good; and he treats infant theories with a ferocity which they can only endure when they are grown up. The result is that I become completely reserved, even about work.

As the wearer of Russell's mantle in logic (it is hard to remember that Wittgenstein was still only twenty-four, and officially an undergraduate reading for a BA), Wittgenstein was asked to review a textbook on logic – *The Science of Logic* by P. Coffey – for the *Cambridge Review*. This is the only book review he ever published, and the first published record of his philosophical opinions. In it he presents a Russellian dismissal of the Aristotelian logic advanced by Coffey, but expresses himself with a stridency that exceeds even Russell's, and borders on the vitriolic:

In no branch of learning can an author disregard the results of honest research with so much impunity as he can in Philosophy and Logic. To this circumstance we owe the publication of such a book as Mr Coffey's 'Science of Logic': and only as a typical example of the work of many logicians of to-day does this book deserve consideration. The author's Logic is that of the scholastic philosophers and he makes all their mistakes – of course with the usual references to Aristotle. (Aristotle, whose name is so much taken in vain by our logicians, would turn in his grave if he knew that so many Logicians know no more about Logic to-day than he did 2,000 years ago.) The author has not taken the slightest notice of the great work of the modern mathematical logicians – work which has brought about an advance in Logic comparable only to that which made Astronomy out of Astrology, and Chemistry out of Alchemy.

Mr Coffey, like many logicians, draws a great advantage from an unclear way of expressing himself; for if you cannot tell whether he means to say 'Yes' or 'No', it is difficult to argue against him. However, even through his foggy expression, many grave mistakes can be recognised clearly enough; and I propose to give a list of some of the most striking ones, and would advise the student of Logic to trace these mistakes and their consequences in other books on Logic also.

There follows a list of such mistakes, which are, for the most part, the weaknesses of traditional (Aristotelian) logic customarily pointed out by adherents of Russellian mathematical logic – for instance, that it assumes all propositions to be of the subject-predicate form, that it confuses the copula 'is' (as in 'Socrates is mortal') with the 'is' of identity ('Twice two is four'), and so on. 'The worst of such books as this', the review concludes, 'is that they prejudice sensible people against the study of Logic.'

By 'sensible people' Wittgenstein presumably meant people with some sort of training in mathematics and the sciences, as opposed to the classical training we can probably assume Mr Coffey (along with most traditional logicians) to have had. In this he was echoing a view held by Russell and expressed to Ottoline the previous December:

> I believe a certain sort of mathematicians have far more philosophical capacity than most people who take up philosophy. Hitherto the people attracted to philosophy have been mostly those who loved the big generalizations, which were all wrong, so that few people with exact minds have taken up the subject. It has long been one of my dreams to found a great school of mathematically-minded philosophers, but I don't know whether I shall ever get it accomplished. I had hopes of Norton, but he has not the physique, Broad is all right, but has no fundamental originality. Wittgenstein of course is exactly my dream.

As we have seen, during the Lent term Russell somewhat modified this view: Wittgenstein was exact, but narrow. He had *too* little 'wish for a broad survey of the world', insisted on too great a degree of exactness from infant theories, showed too little patience with 'inconclusive considerations' and 'unsatisfactory points of view'. Perhaps, faced with Wittgenstein's single-mindedness, Russell came to think that a love of the big generalizations was not such a bad thing after all.

For Wittgenstein, his absorption in logical problems was complete. They were not a part of his life, but the whole of it. Thus, when during the Easter vacation he found himself temporarily bereft of inspiration, he was plunged into despair. On 25 March he wrote to Russell describing himself as 'perfectly sterile' and doubting whether he would ever gain new ideas:

> Whenever I try to think about Logic, my thoughts are so vague that nothing ever can crystallize out. What I feel is the curse of all those who have only half a talent; it is like a man who leads you along a dark corridor with a light and just when you are in the middle of it the light goes out and you are left alone.

'Poor wretch!' Russell commented to Ottoline. 'I know his feelings so well. It is an awful curse to have the creative impulse unless you have a talent that can always be relied on, like Shakespeare's or Mozart's.'

The responsibility Wittgenstein had been given by Russell – for the 'next great advance in philosophy' – was a source of both pride and suffering. He assumed it with complete and utter seriousness. He assumed also the role of a kind of custodian in the field of Russellian mathematical logic. Thus, when Frege wrote to Jourdain telling him about his plans to work at the theory of irrational numbers, we find Jourdain ticking him off in Wittgenstein's name:

> Do you mean that you are writing a third volume of the *Grundgesetze der Arithmetik*? Wittgenstein and I were rather disturbed to think that you might be doing so, because the theory of irrational numbers – unless you have got quite a new theory of them – would seem to require that the contradiction has been previously avoided; and the part dealing with irrational numbers on the new basis has been splendidly worked out by Russell and Whitehead in their *Principia Mathematica*.

Wittgenstein returned from the Easter vacation in, according to Russell, a 'shocking state – always gloomy, pacing up and down, waking out of a dream when one speaks to him'. He told Russell that logic was driving him mad. Russell was inclined to agree: 'I think there is a danger of it, so I urged him to let it alone for a bit and do other work.'

There is no record of Wittgenstein's doing any other work during this period – only of his taking up, albeit briefly, an unexpected pastime. On 29 April Pinsent notes: 'I played tennis with Wittgenstein: he has never played the game before, and I am trying to teach him: so it was a rather slow game!' However, a week later: 'I had tea chez: Wittgenstein, and at 5.0, we went up to the "New Field" and played tennis. He was off his game today and eventually got sick of it and stopped in the middle of a game.' And that is the last we hear of tennis.

Wittgenstein came to think that what he needed was not a diversion, but greater powers of concentration. To this end he was prepared to try anything, even hypnosis, and had himself mesmerized by a Dr Rogers. 'The idea is this', writes Pinsent in his diary: 'It is, I believe, true that people are capable of special muscular effort while under hypnotic trance: then why not also special mental effort?'

> So when he is under trance, Rogers is to ask him certain questions about points of Logic, about which Wittgenstein is not yet clear, (– certain uncertainties which no one has yet succeeded in clearing up): and Witt: hopes he will then be able to see clearly. It sounds a wild scheme! Witt: has been twice to be hypnotised – but not until the end of the second interview did Rogers succeed in sending him to sleep: when he did, however, he did it so thoroughly that it took ½ hour to wake him up again completely. Witt: says he was conscious all the time – could hear Rogers talk – but absolutely without will or strength: could not comprehend what was said to him – could exert no muscular effort – felt exactly as if he were under anaesthetic. He felt very drousy for an hour after he left Rogers. It is altogether a wonderful business.

Wonderful it may have been; useful it was not.

Russell, apparently, knew nothing about the scheme (it would surely have made too good a story to omit from his many recorded reminiscences of Wittgenstein if he *had* known of it); by this time, Pinsent was the more trusted confidant. At one of Russell's 'squashes' they are described as 'talking to each other and ignoring the rest of the world'. Pinsent was, perhaps, the *only* person with whom Wittgenstein could relax, and, temporarily at least, take his mind off logic. With Pinsent, Wittgenstein found himself able to enjoy some of the recreations customary for Cambridge undergraduates – riding, playing tennis, and even, on occasion, 'messing about on the river':

> . . . went on the river with Wittgenstein in a canoo. We went up to 'the Orchard' at Grantchester, where we had lunch. Wittgenstein was in one of his sulky moods at first, but he woke up suddenly (as

> always happens with him) after lunch. Then we went on above Byron's pool and there bathed. We had no towels or bathing draws, but it was great fun.

Their strongest common bond, however, was music. Pinsent's diary records innumerable concerts at the Cambridge University Music Club, and also times when they would make music together, Wittgenstein whistling the vocal parts of Schubert's songs while Pinsent accompanied him on the piano. They had the same taste in music – Beethoven, Brahms, Mozart, and above all Schubert. Wittgenstein also seems to have tried to arouse interest in Labor, and Pinsent tells of an occasion when Wittgenstein tried to get a Labor quintet performed at Cambridge. They shared, too, a distaste for what Pinsent describes as 'modern music'. Thus:

> . . . we went to the C.U.M.C., and found Lindley there . . . he and Wittgenstein got arguing about modern music, which was rather amusing. Lindley used not to like modern stuff, but he has been corrupted! These performers always do in the end. [30.11.12]

> Wittgenstein and Lindley came to tea: there was a lot of animated discussion about modern music – Lindley defending it against us two. [28.2.13]

> I came with him to his rooms. Soon afterwards one Mac'Clure turned up – a musical undergraduate – and there was a wild discussion on modern music – Mac'Clure against Witt: and myself. [24.5.13]

And so on. Music did not have to be *very* modern to be deprecated by Wittgenstein, and these entries are just as likely to refer to conversations about, say, Mahler, as Schönberg. Labor aside, neither Wittgenstein nor Pinsent are on record as admiring anything after Brahms.

Wittgenstein asked Pinsent to accompany him on another holiday, this time in Spain, again to be paid for by Wittgenstein, an offer which Pinsent's mother told him was 'too good to refuse'. No doubt intrigued by the munificence of their son's friend, Pinsent's parents were taken to tea at Wittgenstein's rooms. It was an occasion when his exceptionally good manners could be used to good effect. The tea was served in chemical beakers ('because ordinary crockery is too ugly for him!'), and 'except that he was somewhat preoccupied by his duties as host, [Wittgenstein] was in very good form'.

When Pinsent's parents had left, Wittgenstein proceeded to lecture his friend about his character. Pinsent was, he said, 'ideal in all respects':

> . . . except that he [Wittgenstein] feared that with others except himself I was lacking in generous instincts. He specially said – not with himself – but he feared I did not treat my other friends so generously. By 'generously' he did not imply the ordinary crude meaning – but meant feelings of sympathy etc.

Pinsent took all this very well. 'He was very nice about it all and spoke in no way that one could resent.' He was nevertheless inclined to demur at Wittgenstein's judgement. After all, Wittgenstein knew very little of his other friends and his relations with them. He conceded, however, that it might be true that he treated Wittgenstein differently – after all, Wittgenstein was so different from other people ('he is if anything a bit mad') that one *had* to treat him differently.

As Wittgenstein's friendship with Pinsent grew warmer, his relationship with Russell became increasingly strained. Russell was, more and more, inclined to see in Wittgenstein his own faults writ large – to feel that, faced with Wittgenstein, he knew how other people felt when faced with himself. 'He affects me just as I affect you', he told Ottoline:

> I get to know every turn and twist of the ways in which I irritate and depress you from watching how he irritates and depresses me; and at the same time I love and admire him. Also I affect him just as you affect me when you are cold. The parallelism is curiously close altogether. He differs from me just as I differ from you. He is clearer, more creative, more passionate; I am broader, more sympathetic, more sane. I have overstated the parallel for the sake of symmetry, but there is something in it.

Emphasis on this parallel may have misled Russell. He was inclined to see Wittgenstein's faults as those 'characteristic of logicians': 'His faults are exactly mine – always analysing, pulling things up by the roots, trying to get the exact truth of what one feels towards him. I see it is very tiring and deadening to one's affections.' But the story he tells in illustration of this could point to a different moral – not that Wittgenstein was too analytical, but that he himself was too remote:

> I had an awful time with Wittgenstein yesterday between tea and dinner. He came analysing all that goes wrong between him and me and I told him I thought it was only nerves on both sides and everything was all right at bottom. Then he said he never knew whether I was speaking the truth or being polite, so I got vexed and refused to say another word. He went on and on and on. I sat down at my table and took up my pen and began to look through a book, but he still went on. At last I said sharply 'All you want is a little self-control.' Then at last he went away with an air of high tragedy. He had asked me to a concert in the evening, but he didn't come, so I began to fear suicide. However, I found him in his room late (I left the concert, but didn't find him at first), and told him I was sorry I had been cross, and then talked quietly about how he could improve.

Perhaps he had to keep himself aloof to stop himself being submerged. But though Russell could turn a deaf ear to Wittgenstein's personal haranguing, he could not withstand the power of his philosophical onslaughts. During this summer Wittgenstein had a decisive influence on Russell's development as a philosopher – chiefly by undermining his faith in his own judgement. Looking back on it three years later, Russell described it as: 'an event of first-rate importance in my life', which had 'affected everything I have done since':

> Do you remember that at the time when you were seeing Vittoz [Ottoline's doctor] I wrote a lot of stuff about Theory of Knowledge, which Wittgenstein criticised with the greatest severity? . . . I saw he was right, and I saw that I could not hope ever again to do fundamental work in philosophy. My impulse was shattered, like a wave dashed to pieces against a breakwater. I became filled with

> utter despair . . . I *had* to produce lectures for America, but I took a metaphysical subject although I was and am convinced that all fundamental work in philosophy is logical. My reason was that Wittgenstein persuaded me that what wanted doing in logic was too difficult for me. So there was no really vital satisfaction of my philosophical impulse in that work, and philosophy lost its hold on me. That was due to Wittgenstein more than to the war.

The 'stuff about Theory of Knowledge' that Russell mentions was the beginning of what he hoped would be a major work. It grew out of his work on matter, and was stimulated in part by an invitation to lecture in America. He had already written the first chapter of it before he even mentioned it to Wittgenstein. 'It all flows out', he wrote euphorically to Ottoline on 8 May. 'It is all in my head ready to be written out as fast as my pen will go. I feel as happy as a king.' His euphoria lasted only so long as he kept his writing secret from Wittgenstein. The fact that he did so seems to indicate that he was never as convinced of its worth as his letters to Ottoline would suggest. He seemed to know instinctively what Wittgenstein's reaction would be to a work that was metaphysical rather than logical in character. Sure enough: Wittgenstein disliked the very idea of it. 'He thinks it will be like the shilling shocker, which he hates. *He* is a tyrant, if you like.'

Russell pressed on regardless, and before the end of May had written six chapters of what was clearly becoming a sizeable volume. Then came the blow that was to shatter his impulse and convince him that he was no longer capable of fundamental work in philosophy. In discussing the work, Wittgenstein made what at first seemed a relatively unimportant objection to Russell's Theory of Judgement. Initially, Russell was confident that it could be overcome. 'He was right, but I think the correction required is not very serious', he told Ottoline. Just a week later, however, it seemed to him that the very basis of his work had been undermined:

> We were both cross from the heat – I showed him a crucial part of what I had been writing. He said it was all wrong, not realizing the difficulties – that he had tried my view and knew it couldn't work. I couldn't understand his objection – in fact he was very inarticulate – but I feel in my bones that he must be right, and that he has

seen something I have missed. If I could see it too I shouldn't mind, but, as it is, it is worrying, and has rather destroyed the pleasure in my writing – I can only go on with what I see, and yet I feel it is probably all wrong, and that Wittgenstein will think me a dishonest scoundrel for going on with it. Well, well – it is the younger generation knocking at the door – I must make room for him when I can, or I shall become an incubus. But at the moment I was rather cross.

It is a measure of Russell's lack of confidence that, even though he did not understand Wittgenstein's objections, he *felt* they must be justified. 'But even if they are', he wrote with unconvincing equanimity, 'they won't destroy the value of the book. His criticisms have to do with the problems I want to leave to him.' In other words, Wittgenstein's criticisms were logical rather than metaphysical. But if, as Russell believed, the problems of philosophy were *fundamentally* logical, how could it not affect the value of the book? How could the book be sound if its foundations were not? When Wittgenstein was finally able to put his objections in writing, Russell admitted defeat, unreservedly. 'I am very sorry to hear that my objection to your theory of judgment paralyses you', Wittgenstein wrote. 'I think it can only be removed by a correct theory of propositions.' Such a theory was one of the things that Russell had wanted to leave to Wittgenstein. Being convinced that it was at once necessary and beyond his own capabilities, he came to think that he was no longer able to contribute to philosophy of the most fundamental kind.

The conviction produced in him an almost suicidal depression. The huge work on the Theory of Knowledge, begun with such vigour and optimism, was now abandoned. But as he was contractually obliged to deliver the series of lectures in America, he had to continue with his preparations for them, even though he was now convinced that the material he had written for them was fundamentally in error. 'I must be much sunk', he told Ottoline, 'it is the first time in my life that I have failed in honesty over work. Yesterday I felt ready for suicide.' Four months previously he had written: 'Ten years ago I could have written a book with the store of ideas I have already, but now I have a higher standard of exactness.' That standard had been set by Wittgenstein, and it was one he now felt unable to live up to. He did not recover faith in his work until Wittgenstein was out of the way – and even then he felt it necessary to reassure himself in his absence that: 'Wittgenstein would like the work I have done lately.'

It says much for the generosity of Russell's spirit that, although devastated by Wittgenstein's criticisms of his own work, he could yet rejoice when he heard from Wittgenstein – during the late summer of 1913 – that *his* work was going well. He wrote to Ottoline: 'You can hardly believe what a load this lifts off my spirits – it makes me feel almost young and gay.'

Wittgenstein felt himself to have made a substantial breakthrough. When, at the end of August, he met Pinsent in London, he gave him an almost ecstatic account of his 'latest discoveries', which were, according to Pinsent, 'truly amazing and have solved all the problems on which he has been working unsatisfactorily for the last year'. They constituted a system which was: 'wonderfully simple and ingenious and seems to clear up everything':

> Of course he has upset a lot of Russell's work (Russell's work on the fundamental concepts of Logic that is: on his purely Mathematical work – for instance most of his 'Principia' – it has no bearing. Wittgenstein's chief interest is in the very fundamental part of the subject) – but Russell would be the last to resent that, and really the greatness of his work suffers little thereby – as it is obvious that Wittgenstein is one of Russell's disciples and owes enormously to him. But Wittgenstein's work is really amazing – and I really believe that the mucky morass of Philosophy is at last crystallising about a rigid theory of Logic – the only portion of Philosophy about which there is any possibility of man knowing anything – Metaphysics etc: are hampered by total lack of data. (Really Logic is all Philosophy. All else that is loosely so termed is either Metaphysics – which is hopeless, there being no data – or Natural Science e.g.: Psychology.

And yet, frustratingly – despite the fact that he had, apparently, already developed a system of logic that completely transformed philosophy – there is still no written record of Wittgenstein's work. Whether the overstatements that the system 'cleared up everything' and 'solved all the problems' are due to Wittgenstein or to Pinsent is impossible to say. But a few weeks later we learn from a letter of

Wittgenstein's to Russell that: 'There are still some *very* difficult problems (and very fundamental ones too) to be solved and I won't begin to write until I have got some sort of a solution for them.'

Pinsent had arranged to meet Wittgenstein under the impression that he would then be taken on a holiday to Spain. When they met, however, he was told there had been a change of plan. Spain had (for some unspecified reason) given way to three other alternatives: Andorra, the Azores or Bergen, in Norway. Pinsent was to choose – 'He was very anxious to shew no preference for any particular scheme and that I should choose unbiased' – but it was quite obvious that Wittgenstein's choice was Norway, so Pinsent opted for that. (Actually, he would have preferred the Azores, but Wittgenstein feared they would meet crowds of American tourists on the boat, 'which he can't stand!')

> So we are going to Norway and not Spain after all! Why Wittgenstein should have suddenly changed his mind at the last moment I can't think! But I expect it will be great fun in Norway all the same.

Before they left, Wittgenstein travelled to Cambridge to explain his new work to Russell and Whitehead. Both, according to Pinsent, were enthusiastic, and agreed that the first volume of *Principia* would now have to be rewritten in the light of it (if this is so, Whitehead must have had a change of heart later on) with Wittgenstein himself perhaps rewriting the first eleven chapters: 'That is a splendid triumph for him!'

As he assumed (or appeared to assume) more and more responsibility for the future of Russellian mathematical logic, Wittgenstein became ever more nervously susceptible. As they set sail from Hull to Christiania (now Oslo), he revealed himself to be in an extraordinarily fraught frame of mind:

> Soon after we had sailed Wittgenstein suddenly appeared in an awful panic – saying that his portmanteau, with all his manuscripts inside, had been left behind at Hull . . . Wittgenstein was in an awful state about it. Then, just as I was thinking of sending a wireless message about it, – it was found in the corridor outside someone else's cabin!

They reached Christiania, where they stayed overnight, before taking the train to Bergen, on 1 September. At the hotel Wittgenstein, apparently thinking of their occasional differences in Iceland the previous year, remarked to Pinsent: 'We have got on splendidly so far, haven't we?' Pinsent responded with typically English reserve. 'I always find it exceedingly hard to respond to his fervent outbursts, and I suppose this time I instinctively tried to turn it off flippantly – I am horribly shy of enthusiasm about that sort of thing.' His reticence deeply offended Wittgenstein, who said not a word to him for the rest of the evening.

The following morning he was still 'absolutely sulky and snappish'. On the train they had to change their seats at the last moment because Wittgenstein insisted on being away from other tourists:

> Then a very genial Englishman came along and talked to me and finally insisted on our coming into his carriage to smoke – as ours was a non-smoker. Witt: refused to move, and of course I had to go for a short time at least – it would have been violently rude to refuse. I came back as soon as I could and found him in an awful state. I made some remark about the Englishman being a weird person – whereat he turned and said 'I could travel the whole way with him if I pleased.' And then I had it all out with him and finally brought him round to a normal and genial frame of mind.

'I have to be frightfully careful and tolerant when he gets these sulky fits', Pinsent adds. 'He is – in his acute sensitiveness – very like Levin in "Anna Karenina", and thinks the most awful things of me when he is sulky – but is very contrite afterwards':

> I am afraid he is in an even more sensitive neurotic state just now than usual, and it will be very hard to avoid friction altogether. We can always avoid it at Cambridge, when we don't see so much of each other: but he never will understand that it becomes infinitely harder when we are together so much as now: and it puzzles him frightfully.

The quarrel on the train seemed to mark some kind of turning point in their relationship. Wittgenstein is now referred to as 'Ludwig' for the remainder of Pinsent's diary.

On reaching Bergen they went to a tourist office to make enquiries as to where they could find the sort of place that Wittgenstein wanted: a small hotel, somewhere on a fjord, set in pleasant country and entirely away from tourists. A perfect place, in other words, for Wittgenstein to work on logic undisturbed. (It must by now have been obvious that this was the reason for the last-minute change in plan.) He had already begun his work in the hotel in Bergen. 'When he is working', Pinsent noted, 'he mutters to himself (in a mixture of German and English) and strides up and down the room all the while.'

The tourist office found them a place satisfying all their conditions – a small hotel in a tiny village called Öistesjo, on the Hardanger fjord, at which they would be the only foreign tourists, the other ten guests being Norwegian. Once there they went for a short walk, Pinsent, ever the keen photographer, taking his camera, 'which was the cause of another scene with Ludwig':

> We were getting on perfectly amicably – when I left him for a moment to take a photo: And when I overtook him again he was silent and sulky. I walked on with him in silence for half an hour, and then asked him what was the matter. It seemed, my keenness to take that photo: had disgusted him – 'like a man who can think of nothing – when walking – but how the country would do for a golf course'. I had a long talk with him about it, and eventually we made it up again. He is really in an awful neurotic state: this evening he blamed himself violently and expressed the most piteous disgust with himself.

In an ironically apt comparison, Pinsent remarks: 'at present it is no exaggeration to say he is as bad – (in that nervous sensibility) – as people like Beethoven were'. Perhaps he had not been told that Wittgenstein regarded Beethoven as *exactly* 'the sort of man to be'.

Henceforth, Pinsent took great care not to offend or irritate Wittgenstein, and the rest of the holiday passed without another scene. They very quickly settled into a routine that suited Wittgenstein perfectly: they worked in the morning, went walking or sailing in the early afternoon, worked in the late afternoon, and played dominoes in the evening. For Pinsent it was all rather dull – 'just enough to do to keep one from being bored'. There was none of the novelty and romance of a pony cortège through the Icelandic countryside, and in his diary he is forced to linger on what little excitement there was to be found in an empty hotel (the other guests having left soon after Pinsent and Wittgenstein arrived) in an isolated part of Norway – returning again and again, for example, to their attempts to rid the hotel of a wasps' nest they had found in the roof.

For Wittgenstein, however, it was perfect. He was able to write to Russell in a mood of great contentment:

> I am sitting here in a little place inside a beautiful fiord and thinking about the beastly theory of types . . . Pinsent is an enormous comfort to me here. We have hired a little sailing boat and go about with it on the fiord, or rather Pinsent is doing all the sailing and I sit in the boat and work.

One question nagged at him:

> Shall I get anything out??! It would be awful if I did not and all my work would be lost. However I am not losing courage and go on thinking . . . I very often now have the indescribable feeling as though my work was all sure to be lost entirely in some way or other. But I still hope that this won't come true.

Wittgenstein's mood – as always – fluctuated with his ability to work. And it fell to Pinsent to cheer him up when he was depressed about his progress. On 17 September, for example, we read:

> During all the morning and most of the afternoon Ludwig was very gloomy and unapproachable – and worked at Logic all the time . . . I somehow succeeded in cheering him up – back to his normal frame of mind – and after tea we went for a stroll together (it was a fine sunny day). We got talking and it appeared that it had been some very serious difficulty with the 'Theory of Types' that had depressed him all today. He is morbidly afraid he may die before he has put the Theory of Types to rights, and before he has written out all his other work in such a way as shall be intelligable to the world and of some use to the science of Logic. He has written a lot already – and Russell has promised to publish his work if he were to die – but he is sure that what he has already written is not sufficiently well put, so as absolutely to make plain his real methods of

> thought etc: – which of course are of more value than his definite results. He is always saying that he is certain he will die within four years – but today it was two months.

Wittgenstein's feeling that he might die before being able to publish his work intensified during his last week in Norway, and prompted him to write to Russell asking if Russell would be prepared to meet him '*as soon as possible* and give me time enough to give you a survey of the whole field of what I have done up to now and if possible to let me make notes for you *in your presence*'. It is to this that we owe the existence of *Notes on Logic*, the earliest surviving exposition of Wittgenstein's thought.

In his anxiety, this feeling that he *might* die soon became an unchangeable conviction that he was *bound* to do so. Everything he said or did became based on this assumption. He was not *afraid* to die, he told Pinsent – 'but yet frightfully worried not to let the few remaining moments of his life be wasted':

> It all hangs on his absolutely morbid and mad conviction that he is going to die soon – there is no obvious reason that I can see why he should not live yet for a long time. But it is no use trying to dispel that conviction, or his worries about it, by reason: the conviction and the worry he can't help – for he is mad.

Another, related, anxiety was that his work on logic might perhaps, after all, be of no real use: 'and then his nervous temperament had caused him a life of misery and others considerable inconvenience – all for nothing'.

Pinsent seems to have done a marvellous job of keeping Wittgenstein's spirits up throughout these attacks of crippling anxiety – encouraging him, reassuring him, playing dominoes with him, taking him out sailing and, above all, perhaps, playing music with him. During the holiday they put together a repertoire of some forty Schubert songs, Wittgenstein whistling the air and Pinsent playing the accompaniment.

It is perhaps not surprising to find that their perceptions of the holiday differed sharply. Wittgenstein said he had never enjoyed a holiday so much. Pinsent was less enthusiastic: 'I am enjoying myself pretty fairly . . . But living with Ludwig alone in his present neurotic state is trying at times.' On his return, on 2 October, he swore that he would never go away with Wittgenstein again.

At the end of the holiday Wittgenstein 'suddenly announced a scheme of the most alarming nature':

> To wit: that he should exile himself and live for some years right away from everybody he knows – say in Norway. That he should live entirely alone and by himself – a hermits life – and do nothing but work in Logic. His reasons for this are very queer to me – but no doubt they are very real for him: firstly he thinks he will do infinitely more and better work in such circumstances, than at Cambridge, where, he says, his constant liability to interruption and distractions (such as concerts) is an awful hindrance. Secondly he feels that he has no right to live in an antipathetic world (and of course to him very few people are sympathetic) – a world where he perpetually finds himself feeling contempt for others, and irritating others by his nervous temperament – without some justification for that contempt etc:, such as being a really great man and having done really great work.

Part of this argument is familiar: if he is to behave like Beethoven, he ought, like Beethoven, to produce really great work. What is new is the conviction that this is impossible at Cambridge.

Wittgenstein's mind was not definitely made up on the scheme, however, and he continued with preparations for a course of lectures on philosophy that he had agreed to give at the Working Men's College in London. The issue was finally decided when they reached Newcastle on the way home. There Wittgenstein received a letter from Gretl telling him that she and her American husband, Jerome Stonborough, were coming to live in London. That appeared to settle the matter. He could not, he told Pinsent, stand to live in England if he were perpetually liable to visits from the Stonboroughs.

He even convinced Pinsent – who at first was inclined to think the idea absurd – that he *should*, after all, go to Norway to work on logic. For: 'He has settled many difficulties, but there are still others unsolved.' And: 'The great difficulty about his particular kind of work is that – unless he absolutely settles all the foundations of Logic – his work will be of little value to the world.' So: 'There is nothing between doing really great work & doing practically nothing.'

Pinsent appears to have accepted the force of this reasoning, even though it has nothing whatever to do with the fact of the Stonboroughs being in England, does not explain why Wittgenstein has to be alone, and is in sharp contrast to the view he had accepted just a week before (that Wittgenstein's *method* was the important thing, not his *results*). The argument appears, in fact, to be a restatement of the terrible dichotomy raised by Weininger – greatness or nothingness. But in order to make it an intelligible reason for living away from Cambridge, one would perhaps have to add to it two other Weiningerian themes: that love is conducive to greatness, sexual desire inimical to it; and that 'sexual desire increases with physical proximity; love is strongest in the absence of the loved one; it needs a separation, a certain distance to preserve it'.

The possibility of greatness, therefore, demands a separation from the loved one.

5
NORWAY

As might be expected, Russell thought Wittgenstein's plan to live alone in Norway for two years a wild and lunatic one. He tried to talk him out of it by presenting various objections, all of which were brushed aside:

> I said it would be dark, & he said he hated daylight. I said it would be lonely, & he said he prostituted his mind talking to intelligent people. I said he was mad & he said God preserve him from sanity. (God certainly will.)

Before Wittgenstein embarked again for Bergen, it was important for both him and Russell that a written record of his work should be made: for Wittgenstein, because of his conviction that he had only a few more years (or even months) to live; and for Russell, because he hoped to use Wittgenstein's ideas for his forthcoming series of lectures in America – and because he, too, thought it was now or never (he strongly suspected that Wittgenstein would go completely mad and/ or commit suicide during his solitary sojourn in Norway).

The difficulty was that Wittgenstein's 'artistic conscience' (as Russell called it) made him extremely reluctant to write his ideas out in an imperfect form, and – as he had not yet reached a perfect formulation of them – therefore loathe to write anything at all. He wanted simply to explain his ideas to Russell, orally. Russell, who considered Wittgenstein's work 'as good as anything that has ever been done on logic', did his best to follow Wittgenstein's explanations, but in the end, finding the ideas so subtle that he kept forgetting them, begged him to write them out:

> After much groaning he said he couldn't. I abused him roundly and we had a fine row. Then he said he wd. talk & write down any of his remarks that I thought worth it, so we did that, & it answered fairly well. But we both got utterly exhausted, & it was slow.

He was prevented from giving up only by his resolute determination to 'drag W's thoughts out of him with pincers, however he may scream with the pain'.

Eventually he managed to get some written record of Wittgenstein's thoughts by asking the secretary of Philip Jourdain (who had come into Russell's room to borrow a book) to take shorthand notes while Wittgenstein talked and Russell asked questions. These notes were supplemented by a typescript which Wittgenstein dictated a few days later while he was in Birmingham saying his goodbyes to Pinsent. Together, the dictation and typescript constitute *Notes on Logic* – Wittgenstein's first philosophical work.

The work might be seen as an expansion of his remark to Russell earlier in the summer, that the Theory of Types 'must be rendered superfluous by a proper theory of symbolism', and as a preliminary attempt to provide such a theory. In its details, and in its criticisms of Russell, it is indeed very subtle. But its fundamental thought is quite staggeringly simple. It is that: '"A" is the same letter as "A"' (a remark which prompted the shorthand writer to comment, 'Well, that's true anyway'). This apparently trivial truism was to lead to the distinction between showing and saying which lies at the heart of the *Tractatus*. The thought – here in an embryonic form – is that what the Theory of Types *says* cannot be said, and must be *shown* by the symbolism (by our *seeing* that 'A' is the same letter as 'A', the same *type* of letter as 'B', and a different type from 'x', 'y' and 'z').

In addition to this embryonic Theory of Symbolism, *Notes on Logic* contains a series of remarks on philosophy which state unequivocally Wittgenstein's conception of the subject, a conception that remained – in most of these respects at least – unchanged for the rest of his life:

> In philosophy there are no deductions: *it* is purely descriptive.
>
> Philosophy gives no pictures of reality.

Philosophy can neither confirm nor confute scientific investigation.

Philosophy consists of logic and metaphysics: logic is its basis.

Epistemology is the philosophy of psychology.

Distrust of grammar is the first requisite for philosophizing.

After bidding farewell to Pinsent, Wittgenstein left Birmingham on 8 October. 'It was sad parting from him', wrote Pinsent:

> . . . but it is possible he may pay a short visit to England next summer (remaining in Norway till then & going back thither afterwards) when I may see him again. Our acquaintance has been chaotic but I have been very thankful for it: I am sure he has also.

The outbreak of war the following summer meant that this was the last time the two saw each other.

What Wittgenstein needed (or felt he needed) in 1913 was solitude. He found the ideal place: a village called Skjolden, by the side of the Sogne fjord, north of Bergen. There he lodged with the local postmaster, Hans Klingenberg. 'As I hardly meet a soul in this place', he wrote to Russell, 'the progress of my Norwegian is exceedingly slow.' Neither statement is entirely true. In fact, he made friends with a number of the villagers. As well as the Klingenbergs, there was Halvard Draegni, the owner of a local crate-factory, Anna Rebni, a farmer, and Arne Bolstad, then a thirteen-year-old schoolboy. And his progress in Norwegian was so swift that within a year he was able to exchange letters with these friends in their native language. Admittedly, the language in these letters was not excessively complicated or sophisticated. But this was due less to the limitations of his Norwegian than to the nature of the friendships. They were, in fact, the kind of simple, direct and brief letters he liked best: 'Dear Ludwig, how are you? We often think of you' might be a typical example.

He was not, then, entirely divorced from human contact. But he was – and perhaps this is most important – away from *society*, free from the kind of obligations and expectations imposed by bourgeois life, whether that of Cambridge or that of Vienna. His horror of the bourgeois life was based in part on the superficial nature of the relationships it imposed on people, but partly also on the fact that his own nature imposed upon *him* an almost insufferable conflict when faced with it – the conflict between needing to withstand it, and needing to conform to it.

In Skjolden he could be free from such conflicts; he could be himself without the strain of upsetting or offending people. It was a tremendous liberation. He could devote himself entirely to himself – or, rather, to what he felt to be practically the same thing, to his logic. That, and the beauty of the countryside – ideal for the long, solitary walks he needed as both a relaxation and a meditation – produced in him a kind of euphoria. Together they created the perfect conditions in which to think. It was perhaps the only time in his life when he had no doubts that he was in the right place, doing the right thing, and the year he spent in Skjolden was possibly the most productive of his life. Years later he used to look back on it as the one time that he had had some thoughts that were entirely his own, when he had even 'brought to life new movements in thinking'. '*Then* my mind was on fire!' he used to say.

Within just a few weeks he was able to write to Russell with an announcement of important new ideas, the apparently startling consequence of which would be 'that the whole of Logic follows from one P.P. [primitive proposition] only!!'

Russell, meanwhile, was doing his utmost to digest the *Notes on Logic* in readiness for his Harvard lectures. In the preface to the published version of those lectures, he states:

> In pure logic, which, however, will be very briefly discussed in these lectures, I have had the benefit of vitally important discoveries, not yet published, by my friend Mr Ludwig Wittgenstein.

But there were points about which he was still unclear, and he sent Wittgenstein a series of questions hoping for elucidation. Wittgenstein's answers were brief and, for the most part, helpful. But he was too full of new ideas to find the process of going over old ground congenial: 'An account of general indefinables? Oh Lord! It is *too* boring!!! Some other time!'

> Honestly – I *will* write to you about it some time, if by that time you have not found out all about it. (Because it is all quite clear in the manuscript, I think.) But just now I am SO troubled with Identity

> that I really cannot write any long jaw. All sorts of new logical stuff seems to be growing in me, but I can't yet write about it.

In the excitement of this peak of intellectual creativity he found it particularly irksome to give explanations of points he felt were already clear and well established. In a letter of November he tried to explain why he thought the whole of logic had to follow from a single primitive proposition. But when Russell still did not get it, his patience was exhausted:

> I beg you to think about these matters for yourself, it is INTOLERABLE for me, to repeat a written explanation which even the first time I gave only with the *utmost repugnance*.

Nevertheless, he did make an effort to clarify the point. It hinged on his conviction that, given the correct method of displaying the truth possibilities of a proposition, a *logical* proposition can be shown to be either true or false without knowing the truth or falsity of its constituent parts. Thus: 'It is either raining or not raining' will be true whether 'It is raining' is true or false. Similarly, we need know nothing about the weather to know that the statement: 'It is both raining and not raining' is certainly false. Such statements are *logical* propositions: the first is a tautology (which is always true), and the second a contradiction (always false). Now, if we had a method for determining whether or not any given proposition is a tautology, a contradiction or neither, then we would have a single rule for determining *all* the propositions of logic. Express this rule in a proposition, and the whole of logic has been shown to follow from a single (primitive) proposition.

This argument works only if we accept that all true logical propositions are tautologies. That is why Wittgenstein begins his letter to Russell with the following oracular pronouncement:

> All the propositions of logic are generalizations of tautologies and all generalizations of tautologies are propositions of logic. There are no other logical propositions. (I regard this as definitive.)

'The big question now', he told Russell, is: 'how must a system of signs be constituted in order to make every tautology recognizable as such IN ONE AND THE SAME WAY? This is the fundamental problem of logic!'

He was later to tackle this problem using the so-called Truth-Table Method (familiar to all present-day undergraduate students of logic). But, for the moment, the peak of the *crescendo* had passed. As Christmas approached, exhilaration gave way to gloom and Wittgenstein returned to the morbid conviction that he had not long to live, and that he would therefore never publish anything in his lifetime. 'After my death', he insisted to Russell, 'you must see to the printing of the volume of my journal with the whole story in it.'

The letter ends: 'I often think I am going mad.' The insanity was two-edged, the mania of the previous few months turning into depression as Christmas approached. For, at Christmas: 'I must UNFORTUNATELY go to Vienna.' There was no way out of it:

> The fact is, my mother very much wants me to, so much so that she would be grievously offended if I did not come; and she has such bad memories of just this time last year that I have not the heart to stay away.

And yet: 'the thought of going home appals me'. The one consolation was that his visit would be brief and he would be back in Skjolden before long: 'Being alone here does me no end of good and I do not think I could now bear life among people.'

The week before he left, he wrote: 'My day passes between logic, whistling, going for walks, and being depressed':

> I wish to God that I were more intelligent and everything would finally become clear to me – or else that I needn't live much longer!

Complete clarity, or death – there was no middle way. If he could not solve: 'the question [that] is fundamental to the *whole* of logic', he had no right – or, at any rate, no desire – to live. There was to be no compromise.

In agreeing to join his family for Christmas, Wittgenstein was compromising – going against his own impulses – in order to satisfy the duty he felt to his mother. Once there, further compromise would be inevitable. The energy he had successfully channelled into his logic would have once more to be dissipated in the strain of personal relationships. His real preoccupations would have to be driven underground, while, for the sake of his mother and the rest of his family, he adopted the persona of the dutiful son. And, worst of all, he did not have the strength or clarity of purpose to do anything different: he *could* not bring himself to do anything that would risk grievously offending his mother. The experience threw him into a state of paralysing confusion. He was forced to realize that, however close he might be to complete, uncompromising clarity in the field of logic, he was as far away from it as ever in his personal life – in himself. He alternated between resistance and resignation, between ferment and apathy. 'But', he told Russell:

> . . . deep inside me there's a perpetual seething, like the bottom of a geyser, and I keep hoping that things will come to an eruption once and for all, so that I can turn into a different person.

In this state he was, of course, unable to do any work on logic. But, in his torment, was he not squaring up to an equally important, even related, set of problems? 'Logic and ethics', Weininger had written, 'are fundamentally the same, they are no more than duty to oneself.' It was a view Wittgenstein echoed in his letter to Russell, who, as Wittgenstein knew from their discussions in Cambridge, was hardly likely to see it in the same light:

> Perhaps you regard this thinking about myself as a waste of time – but how can I be a logician before I'm a human being! *Far* the most important thing is to settle accounts with myself!

Like his logic, this work on himself could best be done in solitude, and he returned to Norway as soon as possible. 'It's VERY sad', he wrote to Russell, 'but I've once again no logical news for you':

> The reason is that things have gone terribly badly for me in the last weeks. (A result of my 'holidays' in Vienna.) Every day I was tormented by a frightful *Angst* and by depression in turns and even in the intervals I was so exhausted that I wasn't able to think of doing a bit of work. It's terrifying beyond all description the kinds of mental torment that there can be! It wasn't until two days ago

> that I could hear the voice of reason over the howls of the damned and I began to work again. And *perhaps* I'll get better now and be able to produce something decent. But I *never* knew what it meant to feel only *one* step away from madness. – Let's hope for the best!

He had returned with a determination to rid himself once and for all of the sordid compromises in his life. And – though it was a bit like kicking the dog to get even with the boss – he began with his relationship with Russell. The opening salvo was mild enough – a gentle, and disguised, ticking off about Russell's own tendency to compromise:

> All best wishes for your lecture-course in America! Perhaps it will give you at any rate a more favourable opportunity than usual to tell them your *thoughts* and not *just* cut and dried results. THAT is what would be of the greatest imaginable value for your audience – to get to know the value of *thought* and not that of a cut and dried result.

This can hardly have prepared Russell for what was to follow. He replied, as he told Ottoline, in 'too sharp' a manner. What he actually said, we do not know, though it is a reasonable conjecture that he showed some impatience with Wittgenstein's pointed remarks about his forthcoming lectures, that he criticized Wittgenstein's perfectionism (as he had done in the past), and that he justified his own willingness to publish imperfect work.

Whatever it was, it was sufficient – in Wittgenstein's present mood – to convince him that the time had come to break off all relations with Russell. In what was clearly intended to be the last letter he ever wrote to Russell, he explained that he had thought a lot about their relationship, and had 'come to the conclusion that we really don't suit one another':

> This is NOT MEANT AS A REPROACH! either for you or for me. But it is a fact. We've often had uncomfortable conversations with one another when certain subjects came up. And the uncomfortableness was not a consequence of ill humour on one side or the other but of enormous differences in our natures. I beg you most earnestly not to think I want to reproach you in any way or to

> preach you a sermon. I only want to put our relationship in clear terms *in order to draw a conclusion*. – Our latest quarrel, too, was certainly not simply a result of your sensitiveness or my inconsiderateness. It came from deeper – from the fact that my letter must have shown you how totally different our ideas are, E.G. of the value of a scientific work. It was, of course, stupid of me to have written to you at such length about this matter: I ought to have told myself that such fundamental differences cannot be resolved by a letter. And this is just ONE instance out of *many*.

Russell's value-judgements, he conceded, were just as good and as deep-seated as his own, but – for that very reason – there could not be any *real* relation of friendship between them:

> *I shall be grateful to you and devoted to you* WITH ALL MY HEART *for the whole of my life, but I shall not write to you again and you will not see me again either*. Now that I am once again reconciled with you I want to part from you *in peace* so that we shan't sometime get annoyed with one another again and then perhaps part as enemies. I wish you everything of the best and I beg you not to forget me and to think of me often *with friendly feelings*. Goodbye!
>
> Yours *ever*
>
> LUDWIG WITTGENSTEIN

'I dare say his mood will change after a while', Russell told Ottoline after showing her the letter; 'I find I don't care on his account, but only for the sake of logic.' And yet: 'I do really care too much to look at it. It is my fault – I have been too sharp with him.'

He managed to reply in a way that melted Wittgenstein's resolve never to write to him again. On 3 March Wittgenstein wrote saying Russell's letter had been: '*so* full of kindness and friendship that I don't think I have the *right* to leave it unanswered'. Wittgenstein remained, however, resolute on the central point: 'our quarrels don't arise *just* from external reasons such as nervousness or over-tiredness but are – at any rate on *my* side – *very* deep-rooted':

> You may be right in saying that *we ourselves* are not *so very* different, but *our ideals* could not be more so. And that's why we haven't been able and we shan't *ever* be able to talk about anything involving our

> value-judgements without either becoming hypocritical or falling out. *I think this is incontestable*; I had noticed it a long time ago; and it was frightful for me, because it tainted our relations with one another; we seemed to be side by side in a marsh.

If they were to continue with any sort of relationship at all, it would have to be on a different basis, one in which, 'each can be completely frank without hurting the other'. And, as their ideals were fundamentally irreconcilable, these would have to be excluded. They could avoid hypocrisy or strife *only* by 'restricting our relationship to the communication of facts capable of being established objectively, with perhaps also some mention of our friendly feelings for one another':

> Now perhaps you'll say, 'Things have more or less worked, up to the present. Why not go on in the same way?' But I'm *too* tired of this constant sordid compromise. My life has been one nasty mess so far – but need that go on indefinitely?

He therefore presented a proposal that would, he thought, allow their relationship to continue on a 'more genuine basis':

> Let's write to each other about our work, our health, and the like, but let's avoid in our communications any kind of value-judgement.

It was a plan to which he adhered for the remainder of his correspondence with Russell. He continued to sign himself 'your devoted friend'; he wrote of his work and he described his health. But the intimacy which had previously enabled them to talk of 'music, morals and a host of things besides logic' was lost. And what intellectual sympathy survived *this* break was to disappear entirely as a result of the changes brought about in both of them by the First World War – changes that emphasized and heightened the differences in their natures.

As Wittgenstein repeatedly emphasized in his letters, his friendship with Russell had been strained by their differences for over a year – despite Russell's delusion that it was their *similarity* that caused the trouble. Even their philosophical discussions had, long before Wittgenstein left for Norway, lost their co-operative character. In fact, during his last year at Cambridge he had not really discussed his ideas with Russell at all; he had simply reported them – given him, as it were, logical bulletins. As early as the previous November, when he had written to Moore urging him to come to Norway to discuss his work, he had expressed the view that there was no one at Cambridge with whom this was possible – no one 'who is not yet stale and is *really* interested in the subject':

> Even Russell – who is of course most extraordinarily fresh for his age – is no more pliable for *this* purpose.

As his relationship with Russell was first severed, and then placed on a less intimate footing, Wittgenstein's overtures to Moore became ever more insistent. Moore was dragging his feet somewhat about the proposed visit, and was probably regretting ever having promised to undertake it. But Wittgenstein's demands would brook no refusal: '*You must come as soon as Term ends*', he wrote on 18 February:

> I am looking forward to your coming more than I can say! I am bored to death with Logik and other things. But I hope I shan't die before you come for in that case we couldn't discuss *much*.

'Logik' is probably a reference to a work which Wittgenstein was then in the process of writing, and which he planned to show Moore with the intention of its being submitted for the BA degree. In March he wrote: 'I *think*, now, that Logic must be very nearly done if it is not already.' And although Moore, meanwhile, had come up with a new excuse – he needed to stay at Cambridge to work on a paper – Wittgenstein would have none of it:

> Why on earth won't you do your paper *here*? You shall have a *sittingroom* with a splendid view ALL BY YOURSELF and I shall leave you alone as much as you like (*in fact the whole day, if necessary*). On the other hand we *could* see one another whenever both of us should like to. And we *could* even talk over your business (which *might* be fun). Or do you want *so* many books? You see – I've PLENTY to do myself, so I shan't disturb you a bit. *Do* take the

> Boat that leaves Newcastle on the 17th arriving in Bergen on the 19th and do your work here (I might even have a good influence upon it by preventing too many repetitions).

Finally, Moore overcame his reluctance to face the rigours of the journey – and the even more daunting prospect of being alone with Wittgenstein – and agreed to come. He left for Bergen on 24 March and was met there by Wittgenstein two days later. His visit lasted a fortnight, every evening of which was taken up with 'discussions', which consisted of Wittgenstein talking and Moore listening ('*he* discusses', Moore complained in his diary).

On 1 April Wittgenstein began to dictate to Moore a series of notes on logic. Whether these constitute the whole of the work referred to above as '*Logik*', or simply a selection, we can at least assume they contain the most important parts of it. Their central point is the emphatic insistence on the distinction between *saying* and *showing*, which was only implicit in the notes dictated to Russell the previous year. The notes begin:

> Logical so-called propositions *show* the logical properties of language and therefore of the Universe, but *say* nothing.

The notes outline how this distinction allows us to achieve what he had earlier told Russell must be achieved: a Theory of Symbolism which shows the Theory of Types to be superfluous. That there are different types of things (objects, facts, relations etc.) cannot be said, but is *shown* by there being different types of symbols, the difference being one which can immediately be *seen*.

Wittgenstein regarded the work as a considerable advance on the notes he had earlier dictated to Russell, and it was, for the moment at least, his last word on the subject. He wrote to Russell urging him to read Moore's notes. 'I have now relapsed into a state of exhaustion and can neither do any work nor explain what I did earlier':

> However, I explained it *in detail* to Moore when he was with me and he made various notes. So you can best find it all out from him. Many things in it are new. – The best way to understand it all would be if you read Moore's notes for yourself. It will probably be some time before I produce anything further.

On his return to Cambridge, Moore – as he had been instructed by Wittgenstein – made enquiries as to whether '*Logik*' might serve as a BA thesis. In this, he sought the advice of W. M. Fletcher (Wittgenstein's tutor at Trinity), and was told that, according to the regulations governing such theses, Wittgenstein's work would not be eligible as it stood. It was required that the dissertation should contain a preface and notes indicating the sources from which information had been derived and specifying which parts of the dissertation were claimed to be original and which parts were based on the work of others.

Moore, accordingly, wrote to Wittgenstein to explain the situation. Wittgenstein was outraged. His work – 'the next big advance in philosophy' – not entitled to receive a BA!? And just because it was not surrounded with the usual paraphernalia of undergraduate scholarship! This was the limit. It was bad enough to have to offer pearls to swine; to have them rejected was intolerable. On 7 May he gave vent to his feelings in a fiercely sarcastic letter to Moore, which, for the time being, put an end both to his friendship with Moore and to his hopes of obtaining a Cambridge degree:

> Dear Moore,
> Your letter annoyed me. *When I wrote Logik I didn't consult the Regulations*, and therefore I think it would only be fair if you gave me my degree without consulting them so much either! As to a Preface and Notes; I think my examiners will easily see how much I have cribbed from Bosanquet. – If I am not worth your making an exception for me *even in some* STUPID *details* then I may as well go to HELL directly; and if I *am* worth it and you don't do it then – by God – *you* might go there.
> The whole business is too stupid and too beastly to go on writing about it so –
>
> L.W.

The attack on Moore was unjustifiable: he had not made the regulations, nor was it his job to enforce them – he was simply letting Wittgenstein know where he stood with regard to them. He was, moreover, not used to being addressed in such a manner, and was profoundly disturbed by the tone of the letter. The unjustness shocked him deeply, and the fierceness of it made him feel physically sick. His diary for 11–15 May shows him to be still reeling from the blow days after receiving it. He did not reply.

Nor did he reply when, nearly two months later, on 3 July, he received a somewhat friendlier, almost contrite letter, written after Wittgenstein had left Norway to spend the summer in Vienna.

> Dear Moore,
> Upon clearing up some papers before leaving Skjolden I popped upon your letter which had made me so wild. And upon reading it over again I found that I had probably no sufficient reason to write to you as I did. (Not that I like your letter a bit *now*.) But at any rate my wrath has cooled down and I'd rather be friends with you again than otherwise. I consider I have strained myself enough now for I would *not* have written this to many people and if you don't answer this I shan't write to you again.

'Think I won't answer it', Moore wrote in his diary, 'because I really don't want to see him again.' His resolve almost weakened several times during the next few years. Wittgenstein's name would come up in conversation with Russell or with Desmond MacCarthy, and whenever it did he would wonder whether he was right not to have replied. Even though Wittgenstein (indirectly, through Pinsent) pleaded with him to get in touch, however, he did not do so, and the breach in their friendship was not healed until, when Wittgenstein returned to Cambridge in 1929, they happened to meet each other on the train. But throughout these years, thoughts of Wittgenstein haunted him so much that he contemplated writing a diary devoted to: 'what I feel about Wittgenstein'.

After Moore's visit Wittgenstein relapsed, as we have seen, into a state of exhaustion. Incapable, for the moment, of doing further work in logic, he devoted himself instead to building a small house on the side of the Sogne fjord, about a mile away from the village. It was intended to be a more or less permanent residence – or at least a place to live until he had finally solved all the fundamental problems in logic. But work on it was not finished when, in July, he returned to Vienna to escape the tourist season in Norway. He intended to spend only the summer away, partly in Austria with his family and partly on holiday with Pinsent. But he was not to return to Norway until the summer of 1921, by which time the fundamental problems of logic had – temporarily at least – been solved.

6
BEHIND THE LINES

Wittgenstein arrived back at the Hochreit towards the end of June 1914. His intention was to spend the early part of the summer there, before taking a two-week holiday with Pinsent starting at the end of August, and, finally, visiting old friends in England (e.g. Eccles) before returning to Norway in the autumn, there to live in his new house and to complete his book.

Throughout July, as the crisis following the assassination of Archduke Franz Ferdinand worsened and the European powers prepared for war, Wittgenstein and Pinsent exchanged letters discussing their proposed holiday. Should they go to Spain as planned, or to somewhere more remote? Eventually they agreed to meet at the Grand Hotel in Trafalgar Square on 24 August, and to decide then where they should go. In his reply to a letter from Eccles dated 28 June (the very day of the assassination), in which Eccles had told Wittgenstein of his new house and of the baby – 'the little stranger' – that his wife was expecting some time in August, we find Wittgenstein promising, with complete confidence, to visit Eccles in Manchester around 10 September, after he and Pinsent had returned from wherever they had decided to go. 'I hope the little stranger keeps well', Wittgenstein replied, 'and I hope he'll turn out to be a boy.'

Eccles had written asking Wittgenstein's advice on a proposed suite of bedroom furniture – wardrobe, medicine chest and dressing table – which Eccles had designed and was proposing to have manufactured. Such faith did he have in Wittgenstein's judgement on these matters that his new drawing room was a copy of Wittgenstein's room in Cambridge: blue carpet, black paint, yellow walls. 'The effect,' he told Wittgenstein, 'is greatly admired by everyone.'

Eccles's own criteria of good design are outlined in the letter: the greatest utility, the easiest method of construction, and absolute simplicity were, he said, the only things he had considered. They were criteria which Wittgenstein could readily approve. 'Splendid' was his verdict on Eccles's designs, suggesting only some alterations to the wardrobe, based on purely functional considerations. 'I can't see any drawing of a bed', he added:

> . . . or do you wish to take the one which the furniture manufacturers submitted? If so *do* insist that they cut off all those measly fancy ends. And why should the bed stand on rollers? You're not going to travel about with it in your house!? By all means have the other things made after *your* design.

Though Wittgenstein and Eccles were at one in their preference for functional design, stripped of any sort of ornamentation, we can, I think, assume that for Wittgenstein the issue had a cultural, even an ethical, importance that it did not for Eccles. To the intellectuals of *Jung Wien*, the abhorrence of unnecessary ornamentation was at the centre of a more general revolt against what they saw as the empty posturing that characterized the decaying culture of the Habsburg Empire. Karl Kraus's campaign against the *feuilleton*, and Adolf Loos's notoriously unadorned building on the Mischlerplatz, were but two aspects of the same struggle. That Wittgenstein, to some extent at least, identified with this struggle is evident from his admiration for the work of its two chief protagonists.

While in Norway Wittgenstein had had Kraus's *Die Fackel* sent to him, and came across an article written by Kraus about Ludwig von Ficker, a writer who was an admirer of Kraus, and who was himself the editor of a Krausian journal published in Innsbruck, called *Der Brenner* ('The Burner'). On 14 July Wittgenstein wrote to Ficker offering to transfer to him the sum of 100,000 crowns, with the request that he distribute the money 'among Austrian artists who are without means'. 'I am turning to you in this matter', he explained, 'since I assume that you are acquainted with many of our best talents and know which of them are most in need of support.'

Ficker was, quite naturally, dumbfounded by the letter. He had neither met Wittgenstein nor heard of him, and this offer to place at his disposal such a large sum of money (100,000 crowns was the equivalent of £4000 in 1914, and therefore of perhaps £40,000–£50,000 of today's currency) needed, he thought, to be checked. He replied asking whether he could really take it that the offer was meant in all seriousness, that it was not a joke. 'In order to convince you that I am sincere in my offer', Wittgenstein answered, 'I can probably do nothing better than actually transfer the sum of money to you; and this will happen the next time I come to Vienna.' He explained that upon the death of his father he had come into a large fortune, and: 'It is a custom in such cases to donate a sum to charitable causes.' He had chosen Ficker 'because of what Kraus wrote about you and your journal in the *Fackel*, and because of what you wrote about Kraus'.*

Having received this letter, and having arranged to meet Wittgenstein at Neuwaldeggergasse on 26–7 July, Ficker endeavoured to find out something about him from his Viennese friends. From the painter Max von Esterle he learnt that Wittgenstein's father had been one of the richest *Kohlen-Juden* in the empire, and a generous patron of the visual arts. Reassured as to the sincerity of Wittgenstein's offer, Ficker travelled to Vienna to meet him in person and to discuss the allocation of the money. He stayed two days in the Wittgenstein house in Neuwaldeggergasse. Wittgenstein (he said in a memoir published in 1954) reminded him of such figures as Aloysha in *The Brothers Karamazov* and Myshkin in *The Idiot*: 'a picture of stirring loneliness at the first glance'.

Somewhat to Ficker's surprise, very little of their weekend together

---

*Kraus had written of Ficker's journal: 'That Austria's only honest review is published in Innsbruck should be known, if not in Austria, at least in Germany, whose only honest review is published in Innsbruck also.' *Der Brenner* had been started in 1910. Its name echoes that of Kraus's journal ('The Torch'), and announces its intention of extending the work of Kraus. Where Kraus satirized the shoddy thinking and writing prevalent in Austria, Ficker attempted to publish work that was free from shoddiness. His greatest success, and perhaps his greatest claim to fame, was that he was the first to recognize the genius of the poet Georg Trakl. From October 1912 to July 1914, no issue of *Der Brenner* appeared without Trakl's work. He also published the work of Hermann Broch, Else Lasker-Schüler, Carl Dallago and Theodor Haecker, and *Der Brenner* had, by the time Wittgenstein wrote to Ficker, established a reputation as one of the leading literary journals of the German avant-garde.

was taken up with a discussion of the matter in hand. Indeed, the subject of the allocation was broached only on the second day of his visit. Wittgenstein, initially, seemed more anxious to tell Ficker a little about himself. He described his work on logic, and its relation to the work of Frege and Russell. He told also of his hut in Norway, and of how he now lived among Norwegian peasants, and of his intention to return to Norway to continue his work. It is difficult to resist the conclusion that Wittgenstein's offer to Ficker was motivated not only by philanthropy, but also by a desire to establish some contact with the intellectual life of Austria. After all, he had severed communication with his Cambridge friends, Russell and Moore, despairing of their ever understanding his ideals and sensitivities. Perhaps among Austrians he might be better understood.

While in Vienna Ficker introduced Wittgenstein to Adolf Loos. Ficker's introduction was, for Wittgenstein, the highlight of his visit. 'It makes me *very* happy to have been able to meet him', he wrote to Ficker on 1 August. Indeed, so close were their concerns and their attitudes at this time, that Loos himself is reported to have exclaimed on meeting Wittgenstein: 'You are me!'

When they eventually got around to discussing the disbursement of the money, Wittgenstein's one condition was that 10,000 crowns should go to the *Brenner* itself; the rest was to be left to Ficker himself to distribute.

Ficker had already decided who his three main beneficiaries should be: Rainer Maria Rilke, Georg Trakl and Carl Dallago. Each would receive 20,000 crowns. Rilke is one of the few modern poets that Wittgenstein is known to have admired, and he welcomed Ficker's suggestion. Trakl's name, too, he accepted readily. About Dallago he made no comment. Dallago was a bohemian figure, well known at the time as a writer and philosopher. A regular contributor to the *Brenner*, he espoused an anti-materialistic, anti-scientific outlook that embraced Eastern mysticism and a celebration of the emotional, 'feminine', side of human nature.

Of the remaining 30,000 crowns, 5000 each went to the writer Karl Hauer (a friend of Trakl's and an erstwhile contributor to *Die Fackel*) and the painter Oskar Kokoschka; 4000 to Else Lasker-Schüler (a poet and a regular contributor to *Der Brenner*); 2000 each to Adolf Loos and the writers Theodor Haecker, Theodor Däubler, Ludwig Erik Tesar, Richard Weiss and Franz Kranewitter; and 1000 each to

Hermann Wagner, Josef Oberkofler, Karl Heinrich and Hugo Neugebauer.

Another contributor to *Der Brenner*, the expressionist writer Albert Ehrenstein, also seems to have benefited from Ficker's allocation of the money. So, at least, Wittgenstein thought. 'Once I helped him financially without really meaning to', he later told Paul Engelmann. In gratitude, Ehrenstein had sent him two of his books, *Tubutsch* and *Man Screams*, which Wittgenstein declared were: 'just muck if I'm not mistaken'.

It is very doubtful whether he knew the work of most of the artists whom he helped, and still more doubtful that he would have admired it if he had. From his responses to the letters of gratitude that were passed on to him by Ficker, there is no sign at all of any admiration for most of these artists, and, indeed, his reactions reveal a certain disdain for the whole business. The first such letter he received was from Dallago. Wittgenstein sent it straight back to Ficker: 'I do not know whether you have any use for it, but I am returning it anyway.' And when he was later sent a collection of such letters, he returned them all, saying that he did not need them as documents, and: 'as thanks they were – to be frank – for the most part highly distasteful to me. A certain degrading, *almost* swindling tone – etc.'

Something of this distance from the 'needy' artists he had patronized was felt by at least one of the beneficiaries, Theodor Haecker, whose German translations of Kierkegaard were published in *Der Brenner*, and did much to stimulate the interest of Austrian intellectuals in the Danish philosopher before the First World War. Haecker was at first inclined to refuse the money. The condition specified in Wittgenstein's instructions to Ficker was that the money should go to artists *in need*, and this was, he argued, a condition he could not meet. It would be different if a rich person had been so taken by his translations of Kierkegaard that he wished to pay for them: 'but a gift which the sponsor has expressly tied to the condition of the neediness of the recipient, I cannot and will not accept'. In his reply Ficker urged that it was appropriate, and in accordance with the wishes of the benefactor, that Haecker should be granted a portion of the donation. Haecker was reassured, and accepted the money, but there is no sign that Wittgenstein took any more pride in having helped Haecker than in having helped Ehrenstein.

Of only three of the beneficiaries can one say with any certainty that

Wittgenstein both knew of their work and admired it: Loos, Rilke and Trakl. And even here we must add the provisos that, though he admired the tone of Trakl's work, he professed himself incapable of understanding it; that he came to dislike Rilke's later poetry; and that, after the war, he denounced Loos as a charlatan.

Nevertheless, Rilke's letter of thanks he described as 'kind' and 'noble':

> [It] both moved and deeply gladdened me. The affection of any noble human being is a support in the unsteady balance of my life. I am totally unworthy of the splendid present which I carry over my heart as a sign and remembrance of this affection. If you could only convey my deepest thanks and my faithful *devotion* to Rilke.

Of Trakl's poems, he probably knew nothing until he was sent a collection of them by Ficker. He replied: 'I do not understand them, but their *tone* makes me happy. It is the tone of true genius.'

The weekend during which Wittgenstein and Ficker discussed the allocation of money to the artists of the Austro-Hungarian Empire was the weekend that sealed the fate of that empire. The Austro-Hungarian ultimatum to Serbia had been presented on 23 July, and the deadline for their acceptance of its terms was Saturday 25 July, at 6 p.m. No acceptance was received, and accordingly, on 28 July, Austria declared war on Serbia.

Even at this late stage – and within a week, the whole of Europe would be at war – it was not generally realized that this would have any effect on relations between Austria-Hungary and Britain. Public opinion in Britain – in so far as it concerned itself with such things – was sympathetic to the Habsburgs and antagonistic towards the Serbians. The British newspapers were almost as passionate in their denunciation of the murder of the Archduke as were their Austrian counterparts.

Perhaps, then, it is not surprising to read, in a letter to Wittgenstein dated 29 July, Pinsent's confident confirmation of their arrangement to meet at the Grand Hotel on 24 August. The only doubt expressed by him concerns their destination. Should it be Andorra or the Faeroe Islands? Or perhaps somewhere else? 'I suppose Madeira wouldn't suit you', he suggested optimistically. 'Of course', he wrote, without any

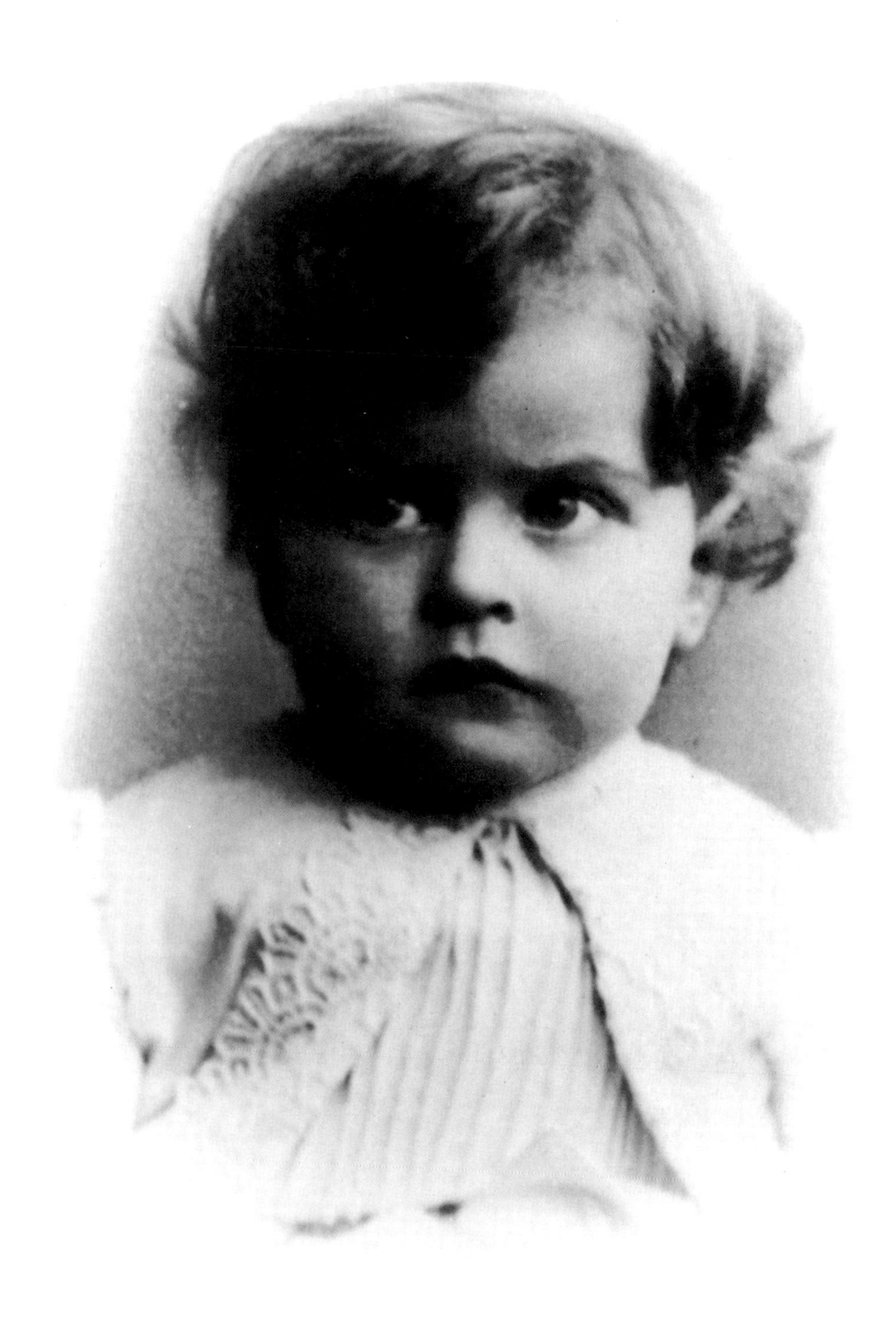

1 The infant Ludwig Wittgenstein

2 The eleven sons and daughters of Hermann and Fanny Wittgenstein. From left to right: Fine, Karl (Ludwig's father), Anna, Milly, Lydia, Louis, Clothilde, Clara, Marie, Bertha and Paul

3 Ludwig, *c.* 1891

4 The grand staircase in the Alleegasse

5 A family photograph taken on Karl and Leopoldine's silver wedding anniversary, 1898, at the family home in Neuwaldeggasse in Vienna. Ludwig is in the front row on the left

6 Ludwig, aged nine

discernible enthusiasm, 'there are out of the way places in the British Isles.' But: 'I don't think we had better go to Ireland, as there will almost certainly be riots & civil war of a sort there soon!' Scotland might do (this had evidently been Wittgenstein's suggestion) – say, Orkney or Shetland, or the Hebrides. And, indeed, this might, in one respect, be preferable to a holiday on the Continent. For:

> Perhaps in view of this European War business we had better not go to Andorra – it might be difficult to get back.

By the absurd logic of what A. J. P. Taylor had called 'war by time-table', 'this European war business' had led, within a few days of Wittgenstein's receiving this letter, to his country and Pinsent's being on different sides in the First World War.

Wittgenstein's first reaction seems to have been to try and get out of Austria, perhaps to England or to Norway. When that failed, and he was told he could not leave, he joined the Austrian army as a volunteer, the rupture he had suffered the previous year having exempted him from compulsory service. 'I think it is magnificent of him to have enlisted', wrote Pinsent in his diary, 'but extremely sad and tragic.'

Although a patriot, Wittgenstein's motives for enlisting in the army were more complicated than a desire to defend his country. His sister Hermine thought it had to do with: 'an intense desire to take something difficult upon himself and to do something other than purely intellectual work'. It was linked to the desire he had felt so intensely since January, to 'turn into a different person'.

The metaphor he had then used to describe his emotional state serves equally to describe the feeling that pervaded Europe during the summer of 1914 – the sense of perpetual seething, and the hope that 'things will come to an eruption once and for all'. Hence the scenes of joy and celebration that greeted the declaration of war in each of the belligerent nations. The whole world, it seems, shared Wittgenstein's madness of 1914. In his autobiography, Russell describes how, walking through the cheering crowds in Trafalgar Square, he was amazed to discover that 'average men and women were delighted at the prospect of war'. Even some of his best friends, such as George Trevelyan and Alfred North Whitehead, were caught up in the enthusiasm and became 'savagely warlike'.

We should not imagine Wittgenstein greeting the news of war against Russia with unfettered delight, or succumbing to the hysterical xenophobia that gripped the European nations at this time. None the less, that he in some sense *welcomed* the war seems indisputable, even though this was primarily for personal rather than nationalistic reasons. Like many of his generation (including, for example, some of his contemporaries at Cambridge, such as Rupert Brooke, Frank Bliss and Ferenc Békássy), Wittgenstein felt that the experience of facing death would, in some way or other, *improve* him. He went to war, one could say, not for the sake of his country, but for the sake of himself.

The spiritual value of facing death heroically is touched upon by William James in *Varieties of Religious Experience* – a book which, as he had told Russell in 1912, Wittgenstein thought might improve him in a way in which he very much wanted to improve. 'No matter what a man's frailties otherwise may be', writes James:

> if he be willing to risk death, and still more if he suffer it heroically, in the service he has chosen, the fact consecrates him forever.

In the diaries Wittgenstein kept during the war (the personal parts of which are written in a very simple code) there are signs that he wished for precisely this kind of consecration. 'Now I have the chance to be a decent human being', he wrote on the occasion of his first glimpse of the enemy, 'for I'm standing eye to eye with death.' It was two years into the war before he was actually brought into the firing line, and his immediate thought was of the spiritual value it would bring. 'Perhaps', he wrote, 'the nearness of death will bring light into life. God enlighten me.' What Wittgenstein wanted from the war, then, was a transformation of his whole personality, a 'variety of religious experience' that would change his life irrevocably. In this sense, the war came for him just at the right time, at the moment when his desire to 'turn into a different person' was stronger even than his desire to solve the fundamental problems of logic.

He enlisted on 7 August, the day after the Austrian declaration of war against Russia, and was assigned to an artillery regiment serving at Kraków on the Eastern Front. He was immediately encouraged by the kindness of the military authorities in Vienna. 'People to whom thousands come each day to ask advice replied kindly and in detail', he commented. It was a good sign; it reminded him of the English way of doing things. He arrived in Kraków on 9 August in a state of excited anticipation: 'Will I be able to work now??! I am anxious to know what lies ahead.'

Wittgenstein's regiment was assigned to the Austrian First Army, and was therefore involved in one of the most absurdly incompetent campaigns of the early months of the war. Both the Russian and the Austrian commands pursued a strategy based on illusion: the Russians thought the mass of Austrian troops would be centred on Lemberg (now Lwów); the Austrians expected to find the bulk of Russian forces further north, around Lublin. Thus, while the Austrian army made easy headway into Russian Poland, the Russians advanced on Lemberg, the largest city in Austrian Galicia, both forces surprised at how little resistance they faced. By the time the Austrian commander, Conrad, realized what had happened, Lemberg had fallen and his First Army was in grave danger of being cut off from its supply lines by Russian troops to the south. He was therefore forced to order a retreat. What had begun as a bold offensive into Russian territory ended in an ignominious withdrawal to a line 140 miles *inside* Austria-Hungary. Had the Austrian army not withdrawn, however, it would have been annihilated by the numerically stronger Russian forces. As it was, 350,000 men out of the 900,000 under Conrad's control died in the confused and fruitless Galician campaign.

Wittgenstein spent most of this campaign on board a ship on the Vistula river – the *Goplana*, captured from the Russians during the initial advance. If he saw any active fighting during these first few months, there is no record of it in his diary. We read instead of great battles heard but not seen, and of rumours that 'the Russians are at our heels'. It is perhaps typical of Wittgenstein's pessimism (justified in this case) that he was all too ready to believe the stories coming in that the Russians had taken Lemberg, but quick to disbelieve the rumour that the Germans had taken Paris. From both stories he drew the same conclusion: 'Now I know that we are lost!' The rumour about Paris, particularly, prompted him, on 25 October, to reflect gloomily on the situation of the Central Powers:

> Such incredible news is always a bad sign. If something really had gone well for us then *that* would be reported and people would not succumb to such absurdities. Because of this, today, more than

> ever, I feel the terrible sadness of our – the German race's – situation. The English – the best race in the world – *cannot* lose. We, however, can lose, and will lose, if not this year then the next. The thought that our race will be defeated depresses me tremendously, because I am German through and through.

The fact that he was inclined to see the war in racial terms perhaps goes some way towards explaining why he found it so difficult to get on with the majority of his fellow crew members. The Austro-Hungarian army was the most multi-racial of all the European armies. Although most of its officers were Germans or Magyars, the bulk of its ordinary soldiers came from the various subject Slavic nationalities of the empire. Wittgenstein's officers he found 'kind, and at times very fine', but as soon as he met his crew mates he pronounced them 'a bunch of delinquents': 'No enthusiasm for anything, unbelievably crude, stupid and malicious.' He could barely see them as human beings:

> When we hear a Chinese talk we tend to take his speech for inarticulate gurgling. Someone who understands Chinese will recognize *language* in what he hears. Similarly I often cannot discern the *humanity* in a man.

Surrounded as he was by alien beings – and appearing to them equally as alien – Wittgenstein found the situation rather like that he had faced at school in Linz. On 10 August, the day after he collected his uniform, the analogy had struck him in terms which suggest a repressed anxiety brought suddenly to the surface: 'Today, when I woke up I felt as though I were in one of those dreams in which, unexpectedly and absurdly, one finds oneself back at school.' And on the *Goplana*, after having been jeered at by the crew, he wrote: 'It was terrible. If there is one thing I have found out it is this: in the whole crew there is not one decent person':

> There is an enormously difficult time ahead of me, for I am sold and betrayed now just as I was long ago at school in Linz.

His sense of isolation was made complete by the knowledge that the people in his life who had helped him to overcome the feeling of loneliness he had had since a pupil at Linz – Russell, Keynes, Pinsent –

were 'on the other side'. 'The last few days I have often thought of Russell', he wrote on 5 October. 'Does he still think of me?' He received a letter from Keynes, but it was of a purely businesslike nature, asking him what should happen after the war to the money that he had arranged to give to Johnson.* 'It hurts to receive a letter of business from one with whom one was once in confidence, and especially in these times.' But it was, above all, to Pinsent that his thoughts turned: 'No news from David. I am completely abandoned. I think of suicide.'

To his few German and Austrian friends Wittgenstein sent greetings in the form of military postcards, and received in reply letters of encouragement and support. The Jolles family in Berlin, in particular, were frequent and enthusiastic correspondents. Elderly and patriotic, they took a vicarious pleasure in reading the news from the Front of their 'little Wittgenstein', and, throughout the war, badgered him to provide more detailed accounts of his exploits. 'I have never thought of you so often and with such joy in my heart as now', wrote Stanislaus Jolles on 25 October. 'Let us hear from you often and soon.' They 'did their bit' by sending him regular parcels of chocolate, bread and cigarettes.

From Frege, too, he received patriotic best wishes. 'That you have enlisted as a volunteer', Frege wrote on 11 October:

> I read with particular satisfaction, and am amazed that you can still dedicate yourself to scientific work. May it be granted to me to see you return from the war in good health and to resume discussions with you. Undoubtedly, we will then, in the end, come closer to each other and understand one another better.

What saved him from suicide, however, was not the encouragement he received from Jolles and Frege, but exactly the kind of personal transformation, the religious conversion, he had gone to war to find. He was, as it were, saved by the word. During his first month in Galicia, he entered a bookshop, where he could find only one book: Tolstoy's *Gospel in Brief*. The book captivated him. It became for him

* Before the war Wittgenstein had arranged with Keynes to donate £200 a year to a research fund administered by King's College to help Johnson continue in his work on logic.

a kind of talisman: he carried it wherever he went, and read it so often that he came to know whole passages of it by heart. He became known to his comrades as 'the man with the gospels'. For a time he – who before the war had struck Russell as being 'more terrible with Christians' than Russell himself – became not only a believer, but an evangelist, recommending Tolstoy's *Gospel* to anyone in distress. 'If you are not acquainted with it', he later told Ficker, 'then you cannot imagine what an effect it can have upon a person.'

His logic and his thinking about himself being but two aspects of the single 'duty to oneself', this fervently held faith was bound to have an influence on his work. And eventually it did – transforming it from an analysis of logical symbolism in the spirit of Frege and Russell into the curiously hybrid work which we know today, combining as it does logical theory with religious mysticism.

But such an influence does not become apparent until a few years later. During the first few months of the war, the spiritual sustenance that Wittgenstein derived from reading Tolstoy's *Gospel* 'kept him alive', in the sense that it allowed him, as he put it, to lighten his external appearance, 'so as to leave undisturbed my inner being'.

It allowed him, that is to say, to put into practice the thought that had struck him while watching *Die Kreuzelscheiber* two (or three) years previously – the idea that, whatever happened 'externally', nothing could happen to *him*, to his innermost being. Thus we find in his diary repeated exhortations to God to help him not to 'lose himself'. This, for him, was far more important than staying alive. What happened to his body was – or, he felt, should have been – a matter of indifference. 'If I should reach my end now', he wrote on 13 September (one of the days on which it was reported that the Russians were advancing towards them), 'may I die a good death, attending myself. May I never lose myself.'

For Wittgenstein, the body belonged only to the 'external world' – the world to which also belonged the 'crude, stupid and malicious' delinquents among whom he now lived. His *soul*, however, must inhabit an entirely different realm. In November he told himself:

> Don't be dependent on the external world and then you have no fear of what happens in it . . . It is x times easier to be independent of things than to be independent of people. But one must be capable of that as well.

His job on the ship was to man the searchlight at night. The loneliness of the task made it much easier to achieve the independence from people he considered necessary to endure the conditions on the boat. 'Through it', he wrote, 'I succeed in escaping the wickedness of the comrades.' Perhaps also his intense desire to distance himself from his external surroundings made it easier for him to resume work on logic. On 21 August he had wondered whether he would ever be able to work again:

> All the concepts of my work have become 'foreign' to me. I do not SEE anything at all!!!

But during the following two weeks – a period spent working at night on the searchlight, and the time during which he first began to read, and find solace in, Tolstoy's *Gospel* – he wrote a great deal. At the end of these two weeks he remarked: 'I am on the path to a great discovery. But will I reach it?'

And yet, the separation between body and spirit was not complete. How could it be? He could distance himself from his surroundings, even from his fellow men, but he could not separate himself from his own body. In fact, coincident with his renewed ability to work on logic, was a revitalized sensuality. The almost jubilant remark quoted above is followed by: 'I feel more sensual than before. Today I masturbated again.' Two days previously he had recorded that he had masturbated for the first time in three weeks, having until then felt almost no sexual desire at all. The occasions on which he masturbated – though clearly not an object of pride – are not recorded with any self-admonition; they are simply noted, in a quite matter-of-fact way, as one might record one's state of health. What appears to emerge from his diary is that his desire to masturbate and his ability to work were complementary signs that he was, in a full sense, *alive*. One might almost say that, for him, sensuality and philosophical thought were inextricably linked – the physical and mental manifestations of passionate arousal.

There are no coded remarks in Wittgenstein's notebooks for the second half of September, the time of the Austrian retreat. It was during this time, however, that he made the great discovery he had felt was imminent. This consisted of what is now known as the 'Picture

Theory of language' – the idea that propositions are a picture of the reality they describe. The story of how this idea occurred to him was told by Wittgenstein in later life to his friend G. H. von Wright, and has since been retold many times. While serving on the Eastern Front, the story goes, Wittgenstein read in a magazine a report of a lawsuit in Paris concerning a car accident, in which a model of the accident was presented before the court. It occurred to him that the model could represent the accident because of the correspondence between the parts of the model (the miniature houses, cars, people) and the real things (houses, cars, people). It further occurred to him that, on this analogy, one might say a *proposition* serves as a model, or *picture*, of a state of affairs, by virtue of a similar correspondence between *its* parts and the world. The way in which the parts of the proposition are combined – the *structure* of the proposition – depicts a possible combination of elements in reality, a possible state of affairs.

From Wittgenstein's notebooks we can date this genesis of the Picture Theory to somewhere around 29 September. On that day he wrote:

> In the proposition a world is as it were put together experimentally. (As when in the law-court in Paris a motor-car accident is represented by means of dolls, etc.)

Throughout October, Wittgenstein developed the consequences of this idea, which he called his 'Theory of Logical Portrayal'. Just as a drawing or a painting portrays pictorally, so, he came to think, a proposition portrays *logically*. That is to say, there is – and must be – a logical structure in common between a proposition ('The grass is green') and a state of affairs (the grass being green), and it is this commonality of structure which enables language to represent reality:

> We can say straight away: Instead of: this proposition has such and such a sense: this proposition represents such and such a situation. It portrays it logically.
>
> Only in this way can *the proposition* be true or false: It can only agree or disagree with reality by being a *picture* of a situation.

Wittgenstein regarded this idea as an important breakthrough. It was, so to speak, a valuable fortress that had to be taken if logic were to be conquered. 'Worked the whole day', he wrote on 31 October:

> Stormed the problem in vain! But I would pour my blood before this fortress rather than march off empty-handed. The greatest difficulty lies in making secure fortresses already conquered. And as long as the whole *city* has not fallen one cannot feel completely secure in one of its fortifications.

But while he himself was on the offensive, the Austrian army was in chaotic and disorderly retreat. The *Goplana* was making its way back towards Kraków, deep inside Austrian territory, where the army was to be quartered for the winter. Before they reached Kraków, Wittgenstein received a note from the poet Georg Trakl, who was at the military hospital there as a psychiatric patient. He had earlier been told of Trakl's situation by Ficker, who had been to Kraków to visit Trakl, and from there had written to Wittgenstein asking him to pay the poet a visit. Trakl felt extremely lonely, Ficker had written, and knew nobody in Kraków who would visit him. 'I would be greatly obliged', Trakl himself wrote, 'if you would do me the honour of paying a visit . . . I will possibly be able to leave the hospital in the next few days to return to the field. Before a decision is reached, I would greatly like to speak with you.' Especially in his present company, Wittgenstein was delighted by the invitation: 'How happy I would be to get to know him! When I arrive at Kraków I hope to meet him! He could be a great stimulus to me.' On 5 November, the day that the *Goplana* finally arrived at Kraków, he was: 'thrilled with the anticipation and hope of meeting Trakl':

> I miss greatly having people with whom I can communicate a little . . . it would invigorate me a great deal . . . In Krakau. It is already too late to visit Trakl today.

This final sentence is charged with the most terrible, unwitting, irony. For as Wittgenstein found out the next morning when he rushed to the hospital, it was indeed too late: Trakl had killed himself with an overdose of cocaine on 3 November 1914, just two days before Wittgenstein's arrival. Wittgenstein was devastated: '*Wie traurig, wie traurig*!!!' ('What unhappiness, what unhappiness!!!') was all he could find to say on the matter.

For the next few days Wittgenstein's diary is full of the misery of his life, the brutality of his surroundings and the failure of his attempts to find a decent person to help him survive. Robbed of such a person in Trakl, he turned his thoughts to Pinsent: 'How often I think of him! And whether he thinks of me half as much.' He discovered that he could get mail through to England via Switzerland, and immediately sent a letter to 'the *beloved* David'. During the weeks that followed, he waited impatiently for a reply. When a letter from Pinsent finally did arrive, on 1 December, it came as such a relief that he kissed it.

In the letter Pinsent told Wittgenstein that he had tried to join the British army but had failed the medical test for a private ('I am too thin') and could not get a commission as an officer. So he was still, reluctantly, reading for his examinations in law. 'When this war is over', he wrote, 'we will meet each other again. Let's hope it will be soon!' 'I think it was *splendid* of you to volunteer for the Army', he added, 'though it is terribly tragic that it should be necessary.'

Wittgenstein answered straight away, and then waited with growing impatience for the reply. And so it went on. '*Keine Nachricht von David*' ('No news from David') and '*Lieben Brief von David*' ('Lovely letter from David') are phrases that appear repeatedly, throughout the winter, in his diary.

Wittgenstein's greatest worry about spending winter in Kraków was not the cold (although he complains of that frequently), but the thought that he would have to share sleeping quarters with the other men – 'from which', he prayed, 'may God release me'. In the event, his prayers were answered: he was, to his enormous relief, promised a room of his own. Even better, in December he was given an entirely new post, a chance to rid himself once and for all of the 'mass of scoundrels' whose company he had endured for four long months. He had wanted to join a balloon section, but when it was discovered that he had a mathematical training he was offered instead a job with an artillery workshop.

Actually, the task Wittgenstein was given at the workshop was mundane clerical work, requiring no mathematical expertise and consisting of compiling a list of all the vehicles in the barracks. For a while all he had to report in his diary was '*Ganzer Tag Kanzlei*' ('At the office all day'), which, so often did it recur, he began to abbreviate as '*G.T.K.*'. The job had its compensations, however, not the least of which was a decent room to himself: 'For the first time in 4 months I am alone in a proper room!! A luxury I *savour.*' He was also, more important, among people he could like and respect, and with whom he could communicate. With his immediate superior, Oberleutnant Gürth, particularly, he formed the nearest thing to a friendship he had yet experienced in the army.

Perhaps because he now had people with whom he could converse, his diary entries at this point become shorter and more formulaic. Apart from '*G.T.K.*', the other constant refrain is: '*Nicht gearbeitet*' ('Have not worked'). Paradoxically (but, upon reflection, perhaps not surprisingly), focussing his mind on logic after a long day at the office among congenial colleagues was more difficult than it had been while he was facing death manning a searchlight amid heavy fighting, and living among people whom he detested. At the workshop he had neither the opportunity nor the desire for the complete solitude he needed for absorption in philosophical problems.

He managed, however, to do some reading. In November he had begun to read Emerson's *Essays*. 'Perhaps', he thought, 'they will have a good influence on me.' Whether they did, he does not say, and Emerson is not mentioned again in his diary. Certainly there is no trace of Emerson's influence in the work that he wrote at this (or indeed at any other) time.

More stimulating was a writer whose view could not have been more antithetical to the Tolstoyan Christianity that Wittgenstein had come to embrace: Friedrich Nietzsche. Wittgenstein had bought in Kraków the eighth volume of Nietzsche's collected works, the one that includes *The Anti-Christ*, Nietzsche's blistering attack upon Christianity. In it, Nietzsche rails against the Christian faith as a decadent, corrupt religion, 'a form of mortal hostility to reality as yet unsurpassed'. Christianity has its origins, according to him, in the weakest and basest aspects of human psychology, and is at root no more than a cowardly retreat from a hostile world:

> We recognize a condition of morbid susceptibility of the *sense of touch* which makes it shrink back in horror from every contact, every grasping of a firm object. Translate such a psychological *habitus* into its ultimate logic – as instinctive hatred of *every* reality, as flight into the 'ungraspable', into the 'inconceivable', as antipathy towards every form, every spatial and temporal concept, towards everything firm . . . as being at home in a world undisturbed by reality of any kind, a merely 'inner world', a 'real' world, an 'eternal' world . . . 'The kingdom of God *is within you*' . . .

This hatred of reality, and the idea to which it gives rise, of the need for redemption through the love of God, are, in Nietzsche's view, the consequence of: 'an extreme capacity for suffering and irritation which no longer wants to be "touched" at all because it feels every contact too deeply . . . The fear of pain, even of the infinitely small in pain – *cannot* end otherwise than in a *religion of love.*'

Though 'strongly affected' by Nietzsche's hostility towards Christianity, and though he felt obliged to admit some truth in Nietzsche's analysis, Wittgenstein was unshaken in his belief that: 'Christianity is indeed the only *sure* way to happiness':

> . . . but what if someone spurned this happiness? Might it not be better to perish unhappily in the hopeless struggle against the external world? But such a life is senseless. But why not lead a senseless life? Is it unworthy?

Even from this it can be seen how close Wittgenstein was, despite his faith, to accepting Nietzsche's view. He is content to discuss the issue in Nietzsche's psychological terms; he does not see it as a question of whether Christianity is *true*, but of whether it offers some help in dealing with an otherwise unbearable and meaningless existence. In William James's terms, the question is whether it helps to heal the 'sick soul'. And the 'it' here is not a *belief* but a practice, a way of living. This is a point that Nietzsche puts well:

> It is false to the point of absurdity to see in a 'belief', perchance the belief in redemption through Christ, the distinguishing characteristic of the Christian: only Christian *practice*, a life such as he who died on the Cross *lived*, is Christian . . . Even today *such* a life is possible, for *certain* men even necessary: genuine, primitive Christianity will be possible at all times . . . *Not* a belief but a doing, above all a *not*-doing of many things, a different *being* . . . States of consciousness, beliefs of any kind, holding something to be true for example – every psychologist knows this – are a matter of complete indifference and of the fifth rank compared with the value of the instincts . . . To reduce being a Christian, Christianness, to a holding of something to be true, to a mere phenomenality of consciousness, means to negate Christianness.

This, we may feel sure, was one of the passages in *The Anti-Christ* that persuaded Wittgenstein that there was some truth in Nietzsche's work. The idea that the essence of religion lay in feelings (or, as Nietzsche would have it, *instincts*) and practices rather than in beliefs remained a constant theme in Wittgenstein's thought on the subject for the rest of his life. Christianity was for him (at this time) 'the only *sure* way to happiness' – not because it promised an after-life, but because, in the words and the figure of Christ, it provided an example, an attitude, to follow, that made suffering bearable.

In the winter months of 1914–15 we read little more in Wittgenstein's diary about his faith. There are no more calls to God to give him strength, no more entries that end: 'Thy will be done.' To bear life in the workshop, it seems, required no divine assistance. Apart from the fact that he had very little time to himself to work on philosophy, life was almost pleasant, at least by comparison with the previous four months.

In any case, it was preferable to life in Vienna. The fact that he had no leave at Christmas to visit his family troubled him not a bit. On Christmas Eve he was promoted to *Militärbeamter* ('Military Official'); on Christmas Day he was invited to a meal in the officers' mess; and on Boxing Day he went out in the evening to a café with a young man he had got to know and like, who had been at college in Lemberg. So passed his Yuletide – quietly, and apparently without any longing to be at home with his family. Through the military post he received Christmas greetings from Jolles (complete, of course, with a chocolate parcel), from the Klingenberg family in Norway, and from Frege. ('Let us hope', wrote Frege, 'for the victory of our warriors and for a lasting peace in the coming year.')

On New Year's Eve, however, Wittgenstein was suddenly told that he had to accompany his superior officer, Oberleutnant Gürth, on a visit to Vienna, where Gürth had some official business to attend to. Wittgenstein's mother was, naturally, delighted by the surprise visit. From his diary, one gathers that Wittgenstein himself retained a cold aloofness. On being reunited with his family, he remarks only that, as New Year's Day was taken up entirely with being with them, he had managed to do no work. He adds, with dispassion (and apparent irrelevance): 'I wish to note that my moral standing is much lower now than it was, say, at Easter.' Of the ten days that he was in Vienna, two of them were spent with Labor, the by now elderly composer, and much of the rest of the time with Gürth. When he arrived back in Kraków his only comment on the visit was that he had 'spent many pleasant hours with Gürth'.

This coolness towards his family suggests a determination not to let them intrude on his inner life, and a fear, perhaps, that to do so would risk undoing the advances he had made in self-discovery and self-possession during his experience of the war. But it seems also to be part of a more general lethargy. He remarks on his exhaustion frequently during this period, especially in relation to his work. On 13 January, for example, he comments that he is not working with any great energy:

> My thoughts are tired. I am not seeing things freshly, but rather in a pedestrian, lifeless way. It is as if a flame had gone out and I must wait until it starts to burn again by itself.

He was dependent, he thought, on an external source of inspiration: '*Only* through a miracle can my work succeed. Only if the veil before my eyes is lifted from the outside. I have to surrender myself completely to my fate. As it has been destined for me, so will it be. I am in the hands of fate.'

His thoughts turned once more to his English friends. He wrote to Pinsent again, and waited impatiently for a reply. 'When will I hear something from David?!' his diary pleads on 19 January. He received a letter from Keynes, but pronounced it 'not very good'. It was, in fact, a very friendly note, but perhaps its tone was too flippant to be of any real comfort. 'I hope you have been safely taken prisoner by now', Keynes wrote:

> Russell and I have given up philosophy for the present – I to give my services to the govt. for financial business, he to agitate for peace. But Moore and Johnson go on just as usual. Russell, by the way, brought out a nice book at about the beginning of the war.
>
> Pinsent had not joined the army by the middle of October but I have not heard since.
>
> Your dear friend Békássy is in your army and your very dear friend Bliss is a *private* in ours.
>
> It must be much pleasanter to be at war than to think about propositions in Norway. But I hope you will stop such self-indulgence soon.

Finally, on 6 February, Wittgenstein was able to announce: 'Lovely letter from David!' The letter had been written on 14 January; in it Pinsent says he has little to say, 'except that I hope to God we shall see each other again after the War'. In contrast to the affectionate, but none the less distancing, 'cleverness' of Keynes's letter, this straightforward expression of friendship was exactly what was craved and needed.

More to Wittgenstein's taste also were probably the short notes he received from the villagers in Skjolden: Halvard Draegni, Arne Bolstad and the Klingenberg family. 'Thanks for your card. We are all healthy. Speak of you often', runs a typical card from Draegni. Wittgenstein's replies were no doubt just as brief, and just as warmly received. The news from Norway was that work on his hut had been completed. 'We all hope', wrote Klingenberg, 'that you will soon be able to return to your new house, which is now finished.' Wittgenstein paid the workers via Draegni, who was surprised to be sent the money; he had not expected Wittgenstein to pay, he wrote, until his return. Draegni was apologetic about the cost: 'If one wants to build as solidly as you have had it made', he explained, 'it will always be more expensive than one initially reckons.'

At the beginning of February Wittgenstein was put in charge of the forge in the workshop, and this added responsibility made it even harder for him to concentrate on philosophy. Apart from having to spend more time in the forge, his supervisory role imposed on him more trouble with his workmates. He was presumably chosen for this task because of his superior engineering skills, but even so it was difficult for him to assume the role of foreman. He reports many difficulties with the men whose work he was supervising, some of which led to great unpleasantness. On one occasion he came close to a duel with a young officer who, one supposes, disliked being told what to do by someone of inferior rank. The effort of trying to impose his will on an intransigent workforce, which neither respected his rank nor were willing to accept the authority of his superior knowledge, drained him of all his energy and strained his nerves almost to breaking point. After only a month of the job – a month in which he had written almost no philosophy at all – Wittgenstein was suicidal, despairing of ever being able to work again.

'One cannot go on like this', he wrote on 17 February. It was clear that something had to change: he had either to be promoted, or transferred to another post. He began to petition Gürth for a change in his situation, but, whether through inefficiency or neglect, nothing was done for a very long time. To add to the constant refrain of '*Nicht gearbeitet*', a new phrase finds its way into the diary at this point: '*Lage unverändert*' ('Situation unchanged'). It must be this period that Hermine had in mind when, in speaking of Wittgenstein's war experiences, she wrote of his repeated efforts to be sent to the Front, and of the 'comical misunderstandings which resulted from the fact that the military authorities with which he had to deal always assumed that he was trying to obtain an easier posting for himself when in fact what he wanted was to be given a more dangerous one'.

It is possible, I think, that Wittgenstein's requests to join the infantry were not so much misunderstood as ignored, and that he was perceived to be of more use to the army as a skilled engineer in charge of a repair depot than as an ordinary foot-soldier. Throughout March, despite repeated entreaties to Gürth, his situation remained unchanged.

Philosophically, the first three months of 1915 were almost completely barren. In other respects, too, Wittgenstein felt dead, unresponsive. (It puzzled him, however, that at such a time, when nothing else moved in him, he could feel sensual and have the desire to masturbate.) When, in February, Ficker sent a posthumously published edition of Trakl's works, his only comment was strikingly dull: 'probably very good'. Thanking Ficker for the volume, he explained that he was now in a sterile period and had 'no desire to assimilate foreign thoughts'. There was, however, something to be hoped for, even from his very unresponsiveness:

> I have this only during a decline of productivity, not when it has *completely* ceased. However – UNFORTUNATELY I now feel completely burnt out. One just has to be patient.

He had, he thought, just to wait for God, for the spirit, to help and inspire him.

Having, in the meantime, nothing to say, he fell silent. From Adele Jolles he received a letter gently berating him for the brevity of his communications from the field. One thing is certain, she told him – he would not make a satisfactory war correspondent or telegraphist.

Couldn't he, for once, send a decent letter, so that one could know where he was and how he was and what he was doing? What did he think of the Italians? Weren't they a pack of scoundrels, deserting the Triple Alliance as they had? 'If I were to write what I think of them', she said, 'my letter would most likely not get through the censors.' She kept up the supply of bread, chocolate and fruit cake, evidently taking pride in the part her 'little Wittgenstein' was playing in the struggle. 'That you willingly volunteered', she told him, 'pleases me ever anew.'

Her husband took pride in the fact that Wittgenstein was at last in a position to put his technical knowledge to some use. 'In any case', he wrote, 'with your skills you are at the right place and, what with the awful Galician roads there must surely be autos aplenty to repair!' Wittgenstein evidently replied that he would rather be with the infantry at the Front than repairing cars behind the lines. Jolles was surprised: 'Don't you think that you could use your technical talents more in the workshop?' His wife, too, despite her patriotic fervour, was concerned: 'Hopefully, your wish to go to the Front will not be fulfilled', she wrote, with motherly anxiety; 'there you are one of very many, and not one of the strongest, here you can play your part more safely.'

Such concern, though no doubt welcome, and perhaps even necessary, was not sufficient. And it was not until after he had received a letter from Pinsent that Wittgenstein was able to shake off his lethargy. On 16 March he was again able to write in his diary: 'Lovely letter from David.' And: 'Answered David. *Very* sensual.' A draft of a reply has been preserved. It reads:

> My *dear* Davy,
> Got today your letter dated January 27th. This is about the limit. I'm now beginning to be more fertile again.

Wittgenstein had asked Pinsent to pass on a message to Moore, and to explain to him how to get a letter through. This Pinsent had done, and 'I hope he will write to you.' It was a forlorn hope. 'I am so sorry if Moore won't behave like a Christian', Pinsent wrote in April; 'as a matter of fact he never acknowledged my letter.'

Not that Moore could keep Wittgenstein out of his thoughts entirely. On 12 October 1915 he recorded in his diary: 'Dream of Wittgenstein':

> . . . he looks at me as if to ask if it is all right, and I can't help smiling as if it was, though I know it isn't; then he is swimming in the sea; finally he is trying to escape arrest as an enemy alien.

On 22 April Wittgenstein was put in charge of the whole workshop, but this, he reports, merely presented him with more unpleasantness to deal with. To help ease the situation he was allowed by Gürth to wear the uniform of an engineer and was provisionally given that title.*

On 30 April Wittgenstein records another 'lovely letter from David', which contained a piece of perhaps surprising news. 'I have been writing a paper on Philosophy', Pinsent told him, 'probably absolute rot!' It was, he said, an attempt to explain 'what Logic as a whole is *about* and what "Truth" is & "Knowledge"'. Though its subject-matter is identical to Wittgenstein's, the resultant work (which still survives) bears little resemblance either to the *Tractatus* or to the earlier *Notes on Logic*. Logic is defined by Pinsent using the notion of 'consistency' rather than that of 'tautology', and the general tenor of its thought owes more to the British empirical tradition (particularly to Moore and Russell) than to Wittgenstein. Nevertheless, Pinsent himself clearly thought of it as a contribution to the issues with which Wittgenstein was concerned. 'I wish you were here & could talk it over with me', he wrote. His letter ends:

> I wish to God this horrible tragedy would end, & I am longing to see you again.

Whether inspired by Pinsent's letter or not, it is remarkable that during his last few months at Kraków – at a time when he was desperately unhappy and intensely frustrated at not being able to obtain another position – Wittgenstein found himself able to work again with renewed vigour. Throughout the months of May and June he was prolific. A large part (approximately a third) of the remarks published as *Notebooks: 1914–1916* were written during this period.

The problem with which he was principally concerned during this time was that of *how* language pictures the world – what features of

---

*Wittgenstein's status had still not been made official when, following the breakthrough of the Central Powers forces on the Eastern Front in the late summer of 1915, the whole depot was moved further north to Sokal, north of Lemberg.

both language and the world make it possible for this picturing to take place:

> The great problem round which everything I write turns is: Is there an order in the world *a priori*, and if so what does it consist in?

Almost against his will, he was forced to the conclusion that there was such an order: the world, as he had insisted to Russell, consists of *facts*, not of things – that is, it consists of things (objects) standing in certain relations to one another. These facts – the relations that exist between objects – are mirrored, pictured, by the relations between the symbols of a proposition. But if language is analysable into *atomic* propositions (as he had earlier insisted), then it looks as though there must be atomic *facts* corresponding to those atomic propositions. And just as atomic propositions are those that are incapable of being analysed any further, atomic facts are relations between *simple* rather than complex objects. Wittgenstein could produce no examples of either an atomic proposition or an atomic fact, nor could he say what a 'simple object' was, but he felt that the very possibility of analysis demanded that there be such things, providing the structure of both language and the world, which allowed the one to mirror the other.

> It does not go against our feeling, that *we* cannot analyse PROPOSITIONS so far as to mention the elements by name: no, we feel that the world must consist of elements. And it appears as if that were identical with the proposition that the world must be what it is, it must be definite.

We can be indefinite and uncertain, but surely the world cannot be: 'The world has a fixed structure.' And this allows the possibility of language having a definite meaning: 'The demand for simple things *is* the demand for definiteness of sense.'

In the middle of this philosophically fruitful period, Wittgenstein received a letter from Russell, written in German and dated 10 May. Russell had, he told Wittgenstein, seen the notes dictated to Moore in Norway but had found them difficult to understand. 'I hope', he wrote, 'with all my heart that after the war you will explain everything to me orally.' 'Since the war began', he added, 'it has been impossible for me to think of philosophy.'

'I'm extremely sorry that you weren't able to understand Moore's notes', Wittgenstein replied:

> I feel that they're very hard to understand without further explanation, but I regard them essentially as definitive. And now I'm afraid that what I've written recently will be still more incomprehensible, and if I don't live to see the end of this war I must be prepared for all my work to go to nothing. – In that case you must get my manuscript printed whether anyone understands it or not.

'The problems are becoming more and more lapidary and general', he told Russell, 'and the method has changed drastically.' The book was to undergo a far more drastic change over the next two years, strangely enough in a way that was prefigured by the development of Pinsent's treatise. In a letter dated 6 April (which Wittgenstein would probably have received some time in May) Pinsent writes that his paper on philosophy has expanded from logic to 'ethics and philosophy in general'. The following year Wittgenstein's own work was to move in a similar direction.

The revitalization of Wittgenstein's work on logic coincided with a dramatic improvement in the position of the Central Powers on the Eastern Front. In March the plight of the Austro-Hungarians had looked desperate. The Russians were forcing them further back into the Carpathian mountains and threatening an incursion into Hungary itself. On 22 March the fortress town of Przemyśl fell, and it was clear that, if disaster were to be avoided, the Austrians would need the help of the superior strength and efficiency of their German allies. Throughout April preparations were thus made for a massive combined German and Austrian assault in Galicia, which on 1 May was launched under the leadership of the German General, von Mackensen. The place chosen to launch the offensive was an area between the localities of Gorlice and Tarnów. The success of the attack surprised even its planners, and a decisive breakthrough was achieved. Throughout the summer months of 1915 the German and Austrian forces swept through the Russian defences with remarkable ease, eventually advancing their position by some 300 miles. Przemyśl and Lemberg were retaken, and Lublin, Warsaw and Brest Litovsk were captured.

If Wittgenstein took any pleasure in the Gorlice–Tarnów breakthrough, there is no indication of it in his diary. Throughout the advance he remained at the workshop in Kraków, growing increasingly resentful of the fact. In Jolles, however, he had a correspondent who could always be relied upon to enthuse over a military success. On 25 March Jolles had written to lament the fall ('after brave resistence') of Przemyśl, and to hope that poor Galicia would be freed from the Russians in the spring. Throughout the campaign Jolles's letters read like a patriotic commentary on the news from the Eastern Front. 'It looks as if the Russian offensive in the Carpathians has ground to a halt', he wrote on 16 April; 'perhaps the occupied part of Galicia can now be successfully liberated!' On 4 May he wrote to say that he had heard a great victory was expected as a result of Mackensen's success: 'May poor Galicia soon be set free from the Russians!'

In the light of Mackensen's breakthrough, he wrote on 17 May, he could understand only too well Wittgenstein's urge to go to the Front. His wife was more concerned about Wittgenstein's safety and whether he was getting enough to eat. 'I write rarely', she explained on 8 April, 'because you yourself write so rarely and so stereotypically, always using the same few words – one has the feeling that you are hardly interested in what one has to write.' 'I am pleased', she added, 'that you are not going to the Front and will instead stay where you are.' In every letter she asked whether food was scarce and whether Wittgenstein needed anything. In his replies Wittgenstein spoke vaguely of the 'unpleasantness' with which he had to deal. 'What kind of unpleasantness?' Adele Jolles asked. 'We are sorry to hear that you have much to cope with, but that you bear it so courageously is splendid and pleases me most sincerely.'

In July he received a letter from Ficker, who was by this time himself in the Austrian army, serving in an Alpine regiment stationed at Brixen. The conditions under which he lived were frightful, Ficker complained: thirty-six men to a room, no chance to be alone at any time of the day or night – and it was likely to stay like that until September. He complained of sleeplessness and spiritual exhaustion; he was so weary he could hardly read or write. 'Sometimes, dear friend, it is as though my whole being were exhausted . . . So thoroughly have these circumstances already undermined my resistance.'

The tone sounded familiar. Wittgenstein replied with a piece of advice based on his own experience of a similar despair. 'I understand your sad news all too well', he wrote:

> You are living, as it were, in the dark and have not found the saving word. And if I, who am essentially so different from you, should offer some advice, it might seem asinine. However, I am going to venture it anyway. Are you acquainted with Tolstoi's *The Gospel in Brief*? At its time, this book virtually kept me alive. Would you buy the book and read it?! If you are not acquainted with it, then you cannot imagine what an effect it can have upon a person.

Perhaps surprisingly, this advice was enthusiastically received. 'God protect you!' Ficker replied. Yes, Wittgenstein was right, he *was* living in the dark: 'for no one had given me the word'. And Wittgenstein had not only given him the word, but also in such a way that he would never forget it: 'God protect you!'

Wittgenstein's letter to Ficker was written in hospital. As a result of an explosion at the workshop he had suffered a nervous shock and a few light injuries. After about a week in hospital he took a much needed three-week leave in Vienna. 'Three weeks' holiday', clucked Adele Jolles, 'after over a year's service and then after an injury and illness is really rather little.' It was, however, probably more than enough from his point of view.

By the time he returned to the repair unit, it had moved from Kraków. In the wake of the Gorlice–Tarnów breakthrough, it had been relocated to Sokal, north of Lemberg, and was housed in an artillery workshop train at the railway station there.

No notebooks have survived from Wittgenstein's time at Sokal, but there is reason to think that it was for him a comparatively happy period. He had at least one fairly close friend, a Dr Max Bieler, who was in charge of a Red Cross hospital train which stood next to the workshop train. Bieler first met Wittgenstein when he was invited to take his meals with the officers at the workshop. He recalls:

> Already during the first meal strikes my eye among the participants who were all officers, a lean and quick man, without military rank, of about twenty-five years. He ate little, drank little and did not smoke, while the rest of the table companions stuffed themselves

> and were very noisy. I asked my table neighbour and he informed me that his name was Ludwig Wittgenstein. I was glad to find among the very young empty-headed career officers a man with a university culture and so sympathetic a man besides. I had the impression that he does not belong in this atmosphere; he was there because he must. I think the sympathy was mutual, because after the meal he invited me to visit his enclosure in the train. And so began our friendship, that lasted several months (almost a year) with daily hour-long conversations, without whisky or cigarettes. After a few days he proposed me to 'thou' him.

During the autumn of 1915 and throughout the following winter, when almost everything was in short supply and conditions at the Front were extremely harsh, the friendship between Bieler and Wittgenstein was of enormous comfort to them both. They had long and animated conversations on philosophical and metaphysical subjects, although, perhaps not surprisingly, these conversations were conducted on terms that were not quite equal. Wittgenstein once told Bieler that he would make a good disciple but that he was no prophet. 'I could say about him', writes Bieler, that: 'he had all the characteristics of a prophet, but none of a disciple.'

Militarily it was a quiet time, with the Russians having to regroup after the disaster of the previous summer, and the Central Powers content to hold their position while they concentrated on the Western Front. It was, evidently, a quiet time too for the repair unit. Wittgenstein, pleased with the results of his recent work on logic, was able to make a preliminary attempt to work it into a book. This, the first version of the *Tractatus*, has unfortunately not survived. We learn of its existence only in a letter to Russell dated 22 October 1915, in which he tells Russell that he is now in the process of writing the results of his work down in the form of a treatise. 'Whatever happens', he told Russell, 'I won't publish anything until you have seen it.' That, of course, could not happen until after the war:

> But who knows whether I shall survive until then? If I don't survive, get my people to send you all my manuscripts: among them you'll find the final summary written in pencil on loose sheets of paper. It will perhaps cost you some trouble to understand it all, but don't let yourself be put off by that.

Russell's reply is dated 25 November. 'I am enormously pleased', he wrote, 'that you are writing a treatise which you want to publish.' He was impatient to see it, and told Wittgenstein that it was hardly necessary to wait until the end of the war. Wittgenstein could send it to America, to Ralph Perry at Harvard, who had, through Russell, already learnt of Wittgenstein's earlier theories of logic. Perry could then send it on to Russell who would publish it. 'How nice it will be when we finally see each other again!' Russell ended.

Frege, too, was told of Wittgenstein's treatise. On 28 November he wrote, in similar vein to Russell: 'I am pleased that you still have time and energy left for scientific work.' If Wittgenstein had followed Russell's suggestion, the work that would have been published in 1916 would have been, in many ways, similar to the work we now know as the *Tractatus*. It would, that is, have contained the Picture Theory of meaning, the metaphysics of 'logical atomism', the analysis of logic in terms of the twin notions of tautology and contradiction, the distinction between saying and showing (invoked to make the Theory of Types superfluous), and the method of Truth-Tables (used to show a logical proposition to be either a tautology or a contradiction). In other words, it would have contained almost everything the *Tractatus* now contains – *except* the remarks at the end of the book on ethics, aesthetics, the soul, and the meaning of life.

In a way, therefore, it would have been a completely different work.

The years during which the book underwent its final – and most important – transformation were years in which Wittgenstein and Russell were not in touch with one another. After the letter of 22 October 1915, Russell heard no more from Wittgenstein until February 1919, after Wittgenstein had been taken prisoner by the Italians. In his *Introduction to Mathematical Philosophy*, written during the final year of the war (while he himself was in prison, serving a sentence for having allegedly jeopardized Britain's relations with the United States), Russell raises the question of how 'tautology' should be defined, and appends the following footnote:

> The importance of 'tautology' for a definition of mathematics was pointed out to me by my former pupil Ludwig Wittgenstein, who was working on the problem. I do not know whether he has solved it, or even whether he is alive or dead.

Communication with Pinsent also stopped for the last two years of the war. On 2 September 1915 Pinsent wrote to say that he had 'given up the study of that damned Law', and was now working for the government. During 1916 Pinsent managed to get three letters through – all written in German – the first of which emphasizes that: 'the war cannot change our personal relationships, it has nothing to do with them'. In these letters Pinsent tells Wittgenstein that he has now received some mechanical training and is employed as an engineer. The last letter Wittgenstein received from him is dated 14 September 1916.

The change in the conception of the book – and the accompanying transformation of Wittgenstein himself – came, then, at a time when he was cut off from his English friends. It is therefore not surprising that, after the war, he was doubtful whether his English friends would be able to understand him. What did they know – what could they know – of the circumstances that had produced the change in him?

An anticipation of the nature of this change can perhaps be seen in the discussions he had in Sokal with Bieler – discussions that, Bieler says, 'sometimes absorbed us so completely that we lost sight of place and time':

> I remember one comical incident. It was New Year's Eve 1915. The local Commandant had invited us all to the Officers' Mess for the New Year's celebrations. When supper was over, getting on for 10 o'clock, the two of us retired to Wittgenstein's room in order to resume yesterday's theme. At about 11 o'clock the officers from the train let us know that it was time to set off in order to arrive at the party in good time. Wittgenstein conveyed to them that they should simply go and we would follow immediately. We quickly forgot about the invitation and the time and continued our discussion until loud voices became audible outside. It was our comrades, returning merrily at 4 a.m. – and we thought it was not yet midnight. The next day we had to make our excuses to the local Commandant and pay him our New Year's compliments belatedly.

Such intensity suggests a wholehearted commitment on Wittgenstein's part. And yet the subject of these discussions was not logic: Wittgenstein did not attempt to teach Bieler, as he had earlier attempted to teach Pinsent, the results of his work. They talked instead of Tolstoy's *Gospel* and of Dostoevsky's *Brothers Karamazov*. The latter Wittgenstein read so often he knew whole passages of it by heart, particularly the speeches of the elder Zossima, who represented for him a powerful Christian ideal, a holy man who could 'see directly into the souls of other people'.

The period during which Wittgenstein and Bieler were together was one of the quietest times on the Eastern Front. It was, for Wittgenstein, a time of relative comfort. Although not an officer, he was in many ways treated like one. He was even provided with a servant – a young Russian boy from a nearby prisoner-of-war camp called Constantin. Bieler recalls: 'Constantin was a good boy and took care of Wittgenstein with great zeal. Wittgenstein treated him very well and in a short time the lean, frail, and dirty prisoner of war was transformed into the most fleshy and cleanest soldier of the whole garrison.'

This period of comparative serenity ended in March 1916, when the Russians, to relieve the pressure on France, launched an attack on the Baltic flank. At the same time, after over a year, a decision was reached by the Austrian authorities on Wittgenstein's status. It was decided that he could not retain the title or the uniform of an *Ingenieur*, but that he could be granted his long-expressed wish to be posted to the Front as an ordinary soldier. The decision came, says Bieler, as 'a heavy blow to us both'. Wittgenstein parted from him in the manner of one who did not expect to return alive:

> He took with him only what was absolutely necessary, leaving everything else behind and asking me to divide it among the troops. On this occasion he told me that he had had a house built beside a Norwegian fjord where he would sometimes take refuge in order to have peace for his work. He now wanted to make me a present of this house. I refused it and took in its place a Waterman's fountain pen.

One of the few personal possessions Wittgenstein packed was a copy of *The Brothers Karamazov*.

If he thought that he would not return from the Front alive, he knew with certainty that he could not return unchanged. In this sense the war really began for him in March 1916.

7
AT THE FRONT

> Undoubtedly, it is the knowledge of death, and therewith the consideration of the suffering and misery of life, that give the strongest impulse to philosophical reflection and metaphysical explanations of the world.
>
> Schopenhauer, *The World as Will and Representation*

If Wittgenstein had spent the entire war behind the lines, the *Tractatus* would have remained what it almost certainly was in its first inception of 1915: a treatise on the nature of logic. The remarks in it about ethics, aesthetics, the soul and the meaning of life have their origin in precisely the 'impulse to philosophical reflection' that Schopenhauer describes, an impulse that has as its stimulus a knowledge of death, suffering and misery.

Towards the end of March 1916 Wittgenstein was posted, as he had long wished, to a fighting unit on the Russian Front. He was assigned to an artillery regiment attached to the Austrian Seventh Army, stationed at the southernmost point of the Eastern Front, near the Romanian border. In the few weeks that elapsed before his regiment was moved up to the front line, he endeavoured to prepare himself, psychologically and spiritually, to face death. 'God enlighten me. God enlighten me. God enlighten my soul', he wrote on 29 March. On the next day: 'Do your best. You cannot do more: and be cheerful':

> Help yourself and help others with all your strength. And at the same time be cheerful! But how much strength should one need for oneself and how much for the others? It is hard to live well!! But it is good to live well. However, not my, but Thy will be done.

When the long-awaited moment came, however, he fell ill and was told by his commanding officer that he may have to be left behind. 'If that happens', he wrote, 'I will kill myself.' When, on 15 April, he was told that he would, after all, be allowed to accompany his regiment, he prayed: 'If only I may be allowed to risk my life in some difficult assignment.' He counted the days until he would at last be in the line of fire, and, when the time came, prayed to God for courage. He noted that since he had been at the Front he had become completely asexual.

Once at the front line he asked to be assigned to that most dangerous of places, the observation post. This guaranteed that he would be the target of enemy fire. 'Was shot at', he recorded on 29 April. 'Thought of God. Thy will be done. God be with me.' The experience, he thought, brought him nearer to enlightenment. On 4 May he was told that he was to go on night-duty at the observation post. As shelling was heaviest at night, this was the most dangerous posting he could have been given. 'Only then', he wrote, 'will the war really begin for me':

> And – maybe – even life. Perhaps the nearness of death will bring me the light of life. May God enlighten me. I am a worm, but through God I become a man. God be with me. Amen

During the following day in the observation post, he waited for the night's shelling with great anticipation. He felt 'like the prince in the enchanted castle'.

> Now, during the day, everything is quiet, but in the night it must be *frightful*. Will I endure it?? Tonight will show. God be with me!!

The next day he reported that he had been in constant danger of his life, but through the grace of God he had survived. 'From time to time I was afraid. That is the fault of a false view of life.' Almost every night at his post he expected to die and prayed to God not to abandon him, to give him courage to look death squarely in the eye without fear. Only then could he be sure that he was living decently: 'Only death gives life its meaning.'

As on the *Goplana* Wittgenstein preferred being in a solitary and dangerous position to being in the company of his comrades. He needed as much, or more, strength from God to face them as he needed to face the enemy. They were 'a company of drunkards, a company of vile and stupid people':

> The men, with few exceptions, hate me because I am a volunteer. So I am nearly always surrounded by people that hate me. And this is the one thing that I still cannot bear. The people here are malicious and heartless. It is almost impossible to find a trace of humanity in them.

The struggle to stop himself from hating these people was, like the struggle against fear in the face of death, a test of his faith: 'The heart of a true believer understands everything.' So, he urged himself: 'Whenever you feel like hating them, try instead to understand them.' He tried, but it was obviously an effort:

> The people around me are not so much mean as *appallingly* limited. This makes it almost impossible to work with them, because they forever misunderstand. These people are not stupid, but limited. Within their circle they are clever enough. But they lack character and thereby breadth.

Finally, he decided that he did not *hate* them – but they disgusted him all the same.

Throughout these first few months at the Front, from March to May, Wittgenstein was able to write a little on logic. He continued with his theme of the nature of functions and propositions and the need to postulate the existence of simple objects. But he added this isolated and interesting remark about the 'modern conception of the world', which found its way unchanged into the *Tractatus* (6.371 and 6.372):

> The whole modern conception of the world is founded on the illusion that the so-called laws of nature are the explanations of natural phenomena.

> Thus people today stop at the laws of nature, treating them as something inviolable, just as God and Fate were treated in past ages.
>
> And in fact both are right and both wrong: though the view of the ancients is clearer in so far as they have a clear and acknowledged terminus, while the modern system tries to make it look as if *everything* were explained.

From Frege he received a postcard encouraging him to keep up his work on logic. 'Your desire not to allow your intellectual work to be abandoned', Frege wrote, 'I find very understandable.' He thanked Wittgenstein for the invitation to come to Vienna to discuss his work, but thought it doubtful whether he would be able to make it. He hoped, nevertheless, to be able to continue their scientific discussions in some way or other. Wittgenstein, however, was to write little on logic for the remainder of the war. And when Frege finally had a chance to read the *Tractatus*, he was unable, in Wittgenstein's eyes, to understand a word of it.

The fighting on the Eastern Front during the months of April and May was light, but in June Russia launched its long-expected major assault, known, after the general who planned and led it, as the 'Brusilov Offensive'. Thus began some of the heaviest fighting of the entire war. The Austrian Eleventh Army, to which Wittgenstein's regiment was attached, faced the brunt of the attack and suffered enormous casualties. It was at precisely this time that the nature of Wittgenstein's work changed.

On 11 June his reflections on the foundations of logic are interrupted with the question: 'What do I know about God and the purpose of life?' He answers with a list:

> I know that this world exists.
>
> That I am placed in it like my eye in its visual field.
>
> That something about it is problematic, which we call its meaning.
>
> That this meaning does not lie in it but outside it.
>
> That life is the world.
>
> That my will penetrates the world.
>
> That my will is good or evil.
>
> Therefore that good and evil are somehow connected with the meaning of the world.

> The meaning of life, i.e. the meaning of the world, we can call God.
>
> And connect with this the comparison of God to a father.
>
> To pray is to think about the meaning of life.
>
> I cannot bend the happenings of the world to my will: I am completely powerless.
>
> I can only make myself independent of the world – and so in a certain sense master it – by renouncing any influence on happenings.

These remarks are not written in code, but presented as if they somehow belonged to the logical work that precedes them. And from this time on reflections of this sort dominate the notebook. It is as if the personal and the philosophical had become fused; ethics and logic – the two aspects of the 'duty to oneself' – had finally come together, not merely as two aspects of the same personal task, but as two parts of the same philosophical work.

In a notebook entry of 8 July, for example, we find: 'Fear in the face of death is the best sign of a false, i.e. a bad life' – not, this time, as a statement of a personal credo but as a contribution to philosophical thought.

At the beginning of the war, after he had received news that his brother Paul had been seriously wounded and assumed that he had lost his profession as a concert pianist, he wrote: 'How terrible! What philosophy will ever assist one to overcome a fact of this sort?' Now, it seems, having experienced the full horrors of the war for himself, he needed, not only a religious faith, but also a philosophy.

That is to say, he needed not only to *believe* in God – to pray to Him for strength and for enlightenment; he needed to *understand* what it was that he was believing in. When he prayed to God, what was he doing? To whom was he addressing his prayers? Himself? The world? Fate? His answer seems to be: all three:

> To believe in a God means to understand the meaning of life.
>
> To believe in God means to see that the facts of the world are not the end of the matter.
>
> To believe in God means to see that life has a meaning.
>
> The world is *given* me, i.e. my will enters the world completely from the outside as into something that is already there.

> (As for what my will is, I don't know yet.)
> *However this may be*, at any rate we *are* in a certain sense dependent, and what we are dependent on we can call God.
> In this sense God would simply be fate, or, what is the same thing: The world – which is independent of our will.
> I can make myself independent of fate.
> There are two godheads: the world and my independent I.
> . . . When my conscience upsets my equilibrium, then I am not in agreement with Something. But what is this? Is it *the world*?
> Certainly it is correct to say: Conscience is the voice of God.

A little later on we read: 'How things stand, is God. God is, how things stand.' By 'how things stand' here, he means both how they stand *in the world* and how they stand in *oneself*. For the self is, as Weininger and Schopenhauer had said, a microcosm of the world.

These thoughts seem to have forced themselves upon him – almost to have taken him by surprise. On 7 July he recorded: 'Colossal exertions in the last month. Have thought a great deal on every possible subject. But curiously I cannot establish the connection with my mathematical modes of thought.' And on 2 August he remarked about his work – as though it had a life of its own – that it had 'broadened out from the foundations of logic to the essence of the world'.

The connection between Wittgenstein's thought on logic and his reflections on the meaning of life was to be found in the distinction he had made earlier between *saying* and *showing*. Logical form, he had said, cannot be expressed *within* language, for it is the form of language itself; it makes itself manifest in language – it has to be *shown*. Similarly, ethical and religious truths, though inexpressible, manifest themselves in life:

> The solution to the problem of life is to be seen in the disappearance of the problem.
> Isn't this the reason why men to whom the meaning of life had become clear after long doubting could not say what this meaning consisted in?

Thus: 'Ethics does not treat of the world. Ethics must be a condition of the world, like logic.' Just as to understand logical form one must see language as a whole, so, to understand ethics, one must see the world as a whole. When one tries to describe what one sees from such a view, one inevitably talks nonsense (as Wittgenstein wrote about his own attempts to do so: 'I am aware of the complete unclarity of all these sentences'), but that such a view is attainable is undeniable: 'There are, indeed, things that cannot be put into words. They *make themselves manifest*. They are what is mystical.'

In discussing this view of the world (the view that sees it as a limited whole), Wittgenstein adopts the Latin phrase used by Spinoza: *sub specie aeternitatis* ('under the form of eternity'). It is the view, not only of ethics, but also of aesthetics:

> The work of art is the object seen *sub specie aeternitatis*; and the good life is the world seen *sub specie aeternitatis*. This is the connection between art and ethics.
>
> The usual way of looking at things sees objects as it were from the midst of them, the view *sub specie aeternitatis* from outside.
>
> In such a way that they have the whole world as background.

These remarks show the unmistakable influence of Schopenhauer. In *The World as Will and Representation*, Schopenhauer discusses, in a remarkably similar way, a form of contemplation in which we relinquish 'the ordinary way of considering things', and 'no longer consider the where, the when, the why, and the whither in things, but simply the *what*':

> Further we do not let abstract thought, the concepts of reason, take possession of our consciousness, but, instead of all this, devote the whole power of our mind to perception, sink ourselves completely therein, and let our whole consciousness be filled by the calm contemplation of the natural object actually present, whether it be a landscape, a tree, a rock, a crag, a building, or anything else. We *lose* ourselves entirely in this object, to use a pregnant expression . . .
>
> It was this that was in Spinoza's mind when he wrote: *Mens aeterna est quatenus res sub specie aeternitatis* ['The mind is eternal in so far as it conceives things from the standpoint of eternity'].

Whether Wittgenstein was rereading Schopenhauer in 1916, or whether he was remembering the passages that had impressed him in his youth, there is no doubt that the remarks he wrote in that year have a distinctly Schopenhauerian feel. He even adopts Schopenhauer's jargon of *Wille* ('will') and *Vorstellung* ('representation' or, sometimes, 'idea'), as in:

> As my idea is the world, in the same way my will is the world-will.

Wittgenstein's remarks on the will and the self are, in many ways, simply a restatement of Schopenhauer's 'Transcendental Idealism', with its dichotomy between the 'world as idea', the world of space and time, and the 'world as will', the *noumenal*, timeless, world of the self. The doctrine might be seen as the philosophical equivalent of the religious state of mind derided by Nietzsche, the morbid sensitivity to suffering which takes flight from reality into 'a merely "inner" world, a "real" world, an "eternal" world'. When this state of mind is made the basis of a philosophy it becomes solipsism, the view that *the* world and *my* world are one and the same thing. Thus we find Wittgenstein saying:

> It is true: Man *is* the microcosm:
> I am my world.

What distinguishes Wittgenstein's statement of the doctrine from Schopenhauer's is that in Wittgenstein's case it is accompanied by the proviso that, when put into words, the doctrine is, strictly speaking, nonsense: 'what the solipsist *means* is quite correct; only it cannot be *said*, but makes itself manifest'.

He had, he thought, reached a point where Schopenhauerian solipsism and Fregean realism were combined in the same point of view:

> This is the way I have travelled: Idealism singles men out from the world as unique, solipsism singles me alone out, and at last I see that I too belong with the rest of the world, and so on the one side *nothing* is left over, and on the other side, as unique, *the world*. In this way idealism leads to realism if it is strictly thought out.

Frege, the thinker Wittgenstein credited with freeing him from his earlier Schopenhauerian idealism, was not, apparently, told of his relapse back into it. A card from Frege dated 24 June remarks again how pleased he is that Wittgenstein is capable of scientific work. 'I can hardly say the same for myself', he writes. His mind is occupied by the war and by the suffering of the people he knows who are engaged in it, one of whom had recently been injured for the second time, while another had been killed in Poland. Of the Brusilov Offensive he says nothing, but remarks how pleased he was at the recapture of Lemberg. In the next card, dated 2 July, he sympathizes with Wittgenstein for the latter's inability to work. He, too, he says, has been incapable of scientific work, but he hopes that after the war he and Wittgenstein can take up again their work on logical questions. On 29 July he again remarks on the low spirits evident from Wittgenstein's recent communications, and hopes to receive a card written in a better spirit soon, but: 'I am always pleased to receive a sign of life from you.'

Nothing in these cards indicates that he was made aware of the fundamental changes taking place in Wittgenstein's thought at this time – that he knew of the widening of Wittgenstein's concerns from the foundations of logic to the essence of the world, or of Wittgenstein's conviction that he had found the point at which solipsism and realism coincide.

Thoughts of Pinsent had never been far from Wittgenstein's mind during the writing of his book. On 26 July he recorded another letter from him. It was written in German and told Wittgenstein of the death of Pinsent's brother, who had been killed in France. 'The war cannot change our personal relationships', Pinsent insisted; 'it has nothing to do with them.' 'This kind, lovely letter', Wittgenstein wrote, 'has opened my eyes to the way in which I live in *exile* here. It may be a salutary exile, but I feel it now as an exile.'

By this time the Austrian forces had been driven back into the Carpathian mountains, pursued by the victorious Russians. The conditions were harsh – 'icy cold, rain and fog', Wittgenstein records. It was: 'a life full of torment':

> Terribly difficult not to lose oneself. For I am a weak person. But the spirit will help me. The best thing would be if I were already ill, then at least I would have a bit of peace.

But to avoid capture or death he had to keep on the move, pursued by the fire of the advancing Russians. 'Was shot at', he wrote on 24 July, 'and at every shot I winced with my whole being. I want so much to go on living.'

Under these circumstances, the question of the identity of the 'philosophical I', the self which is the bearer of moral values, was given a peculiar intensity. On the retreat through the Carpathian mountains Wittgenstein discovered, probably for the first time in his life, what it was like to lose sight of that self and to be overtaken by an instinctual, animal, will to stay alive, a state in which moral values were irrelevant:

> Yesterday I was shot at. I was scared! I was afraid of death. I now have such a desire to live. And it is difficult to give up life when one enjoys it. This is precisely what 'sin' is, the unreasoning life, a false view of life. From time to time I become an *animal*. Then I can think of nothing but eating, drinking and sleeping. Terrible! And then I suffer like an animal too, without the possibility of internal salvation. I am then at the mercy of my appetites and aversions. Then an authentic life is unthinkable.

For the next three weeks, his diary shows him remonstrating with himself about this tendency to sink into a life of sin. 'You know what you have to do to live happily', he told himself on 12 August. 'Why don't you do it? Because you are unreasonable. A bad life is an unreasonable life.' He prayed to God for strength in the struggle against his own weak nature.

Despite these self-admonitions, he in fact showed remarkable courage throughout the campaign. During the first few days of the Brusilov Offensive he was recommended for a decoration in recognition of his bravery in keeping to his post, despite several times being told to take cover. 'By this distinctive behaviour', the report states, 'he exercised a very calming effect on his comrades.' He was quickly promoted, first to *Vormeister* (a non-commissioned artillery rank similar to the British Lance-Bombardier) and then to *Korporal*. Finally, towards the end of August, when the Russian advance had ground to a halt, he was sent away to the regiment's headquarters in Olmütz (Olomouc), Moravia, to be trained as an officer.

Before going to Olmütz Wittgenstein had a period of leave in Vienna. There, he wrote in his diary, he felt depressed and lonely, the one cheerful piece of news being the fact that Loos was still alive. From Loos he received the name and address of a contact in Olmütz: an ex-student of Loos was, at the time, convalescing at his family home there after having been discharged from the army suffering from tuberculosis.

On 28 August Wittgenstein received a letter from Frege suggesting that they enter into a correspondence on logic. When Wittgenstein had enough time, Frege suggested, couldn't he write his thoughts on paper and send them to him? He would then attempt to respond to Wittgenstein's thoughts by letter. 'In this way', wrote Frege, 'perhaps a scientific communication between us could take place, and so at least provide some sort of substitute for a face-to-face discussion.' Wittgenstein appears not to have responded to this suggestion until after he had completed his book. Perhaps the offer came too late; for in the autumn of 1916 he found the discussion partner he needed to work out the new direction of his thoughts.

The student Loos mentioned was Paul Engelmann, a member of a group of young people which formed a self-consciously cultured oasis in what was otherwise a rather culturally barren outpost of the Austro-Hungarian empire. There was Fritz Zweig, a gifted pianist who later became first conductor at the Berlin State Opera House, his cousin Max Zweig, a law student and playwright, and Heinrich ('Heini') Groag, also a law student and later a successful barrister. Groag was, says Engelmann, 'one of the wittiest men I ever met'. Engelmann's brother, too, was a man of sharp wit – later to become famous in Vienna as the cartoonist 'Peter Eng' – although at this time he and Wittgenstein shared a mutual antipathy for one another. Engelmann himself was a disciple of both Adolf Loos and Karl Kraus. After his discharge from the army he devoted himself to assisting Kraus in his campaign against the war, helping to collect the newspaper cuttings that formed the material for Kraus's satirical anti-war propaganda.

Wittgenstein arrived in Olmütz some time in October 1916, and stayed there until shortly before Christmas. He wanted at first to lodge in the tower of Olmütz Town Hall, but upon being told by the watchman that it was not to let, he settled for a room in a tenement block on the outskirts of town. Shortly after moving in, he fell ill with enteritis, and was nursed back to health by Engelmann, with the help of Engelmann's mother, who cooked Wittgenstein light meals, which Engelmann would then deliver to the invalid. On the first occasion that he performed this act of kindness Engelmann spilt some soup on his way up to Wittgenstein's room. On his entering, Wittgenstein exclaimed: 'My dear friend, you are showering me with kindness', to which Engelmann, his coat bespattered, replied: 'I am afraid I have been showering myself.' It was exactly the sort of simple kindness and simple humour that Wittgenstein appreciated, and the scene stayed in his mind. When he was back at the Front he wrote to Engelmann: 'I often think of you . . . and of the time you brought me some soup. But that was your mother's fault as well as yours! And I shall never forget her either.'

Thanks to Engelmann's group of friends, Wittgenstein's time at Olmütz was a happy one. He joined in with their performances of Molière's *Malade imaginaire*, listened appreciatively to Fritz Zweig's piano recitals and, above all, he joined in with their conversations – about literature, music and religion. With Engelmann particularly he was able to discuss with a sympathetic and like-minded listener all the ideas that had come to him during the last six months at the Front. Sometimes, Engelmann recalls, these conversations would be conducted while he was accompanying Wittgenstein on his way back from Engelmann's house to his room on the outskirts of town. If they were still engrossed in discussion by the time they reached the tenement block, they would turn round and continue the conversation while Wittgenstein accompanied Engelmann back.

Engelmann was the closest friend Wittgenstein had had since leaving England. The friendship owed much to the fact that the two met each other at a time when both were experiencing a religious awakening which they each interpreted and analysed in a similar way. Engelmann puts it well when he says that it was his own spiritual predicament that:

> . . . enabled me to understand, from within as it were, his utterances that mystified everyone else. And it was this understanding on my part that made me indispensable to him at that time.

Wittgenstein himself used to say: 'If I can't manage to bring forth a proposition, along comes Engelmann with his forceps and pulls it out of me.'

The image brings to mind Russell's remark about dragging Wittgenstein's thoughts out of him with pincers. And, indeed, it is hard to resist comparing Engelmann and Russell with respect to the roles they played in Wittgenstein's life during the development of the *Tractatus*. Engelmann himself seems to have had the comparison in mind when he wrote that:

> In me Wittgenstein unexpectedly met a person, who, like many members of the younger generation, suffered acutely under the discrepancy between the world as it is and as it ought to be according to his lights, but who tended also to seek the source of that discrepancy within, rather than outside himself. This was an attitude which he had not encountered elsewhere and which, at the same time, was vital for any true understanding or meaningful discussion of his spiritual condition.

And of Russell's introduction to the book, he says:

> [It] may be considered one of the main reasons why the book, though recognized to this day as an event of decisive importance in the field of logic, has failed to make itself understood as a philosophical work in the wider sense. Wittgenstein must have been deeply hurt to see that even such outstanding men, who were also helpful friends of his, were incapable of understanding his purpose in writing the *Tractatus*.

To a certain extent, this is anachronistic. It shows, too, little awareness of the fact that the Wittgenstein Engelmann met in 1916 was not the same as the Wittgenstein Russell had met in 1911. Nor was his purpose in writing the *Tractatus* the same. Russell was not in touch with Wittgenstein at a time when his work 'broadened out from the foundations of logic to the essence of the world'; so far as Russell knew, his purpose in writing the book was to shed light on the nature of logic. Engelmann, one might say, would have been little use to Wittgenstein's development as a philosopher in 1911, when his preoccupations centred on the issues raised by Russell's Paradox.

Nevertheless, it remains true that in 1916 – just as in 1911 – Wittgenstein was fortunate to be in a situation in which he could have daily conversations with, and the almost undivided attention of, a kindred spirit.

It is noteworthy that there are no coded remarks in Wittgenstein's notebook for this time; Engelmann's presence made them unnecessary. There are, however, a number of philosophical remarks. In the main these are a continuation of the Schopenhauerian line of thought begun at the Front. It is likely, I think, that his protracted conversations with Engelmann helped Wittgenstein to formulate the connections between the mystical and the logical parts of the book. Certainly he discussed the book in depth with Engelmann, and from the latter's 'Observations on the *Tractatus*' included in his memoir it is clear that it had been firmly impressed upon him that: 'logic and mysticism have here sprung from the same root'. The central thread that links the logic and the mysticism – the idea of the unutterable truth that makes itself manifest – was an idea that came naturally to Engelmann. Indeed, he later supplied Wittgenstein with what both considered to be an excellent example: a poem by Uhland called 'Count Eberhard's Hawthorn'.

After spending Christmas in Vienna Wittgenstein returned to the Russia Front in January 1917, now an artillery officer attached to a division of the Austrian Third Army stationed just north of the Carpathian mountains. By now the Russians were in disarray and the Front was relatively quiet. He wrote to Engelmann that he was once again capable of work (unfortunately, the manuscripts from this period have not survived). In all probability the work he wrote at this time was concerned with the inexpressibility of ethical and aesthetic truths. In a letter dated 4 April 1917, Engelmann enclosed 'Count Eberhard's Hawthorn', Uhland's poem recounting the story of a soldier who, while on crusade, cuts a spray from a hawthorn bush; when he returns home he plants the sprig in his grounds, and in old age he sits beneath the shade of the fully grown hawthorn tree, which serves as a poignant reminder of his youth. The tale is told very simply, without adornment and without drawing any moral. And yet, as Engelmann says, 'the poem as a whole gives in 28 lines the picture of a life'. It is, he told Wittgenstein, 'a wonder of objectivity':

> Almost all other poems (including the good ones) attempt to express the inexpressible, here that is not attempted, and precisely because of that it is achieved.

Wittgenstein agreed. The poem, he wrote to Engelmann, is indeed 'really magnificent':

> And this is how it is: if only you do not try to utter what is unutterable then *nothing* gets lost. But the unutterable will be – unutterably – *contained* in what has been uttered!

There was, at this time, some reason to think that the war might soon come to an end through a victory for the Central Powers. The government in Russia had collapsed; there had been a breakthrough for the Germans against the French on the Western Front; and the U-Boat campaign against the British seemed to be succeeding. So, at least, Frege thought. 'Let's hope for the best!' he wrote to Wittgenstein on 26 April, listing all these reasons for doing so.

During the quiet time that followed the Russian Revolution, Wittgenstein was given some leave in Vienna. There Frege wrote to him, apologizing for having to refuse his invitation to come to Vienna to discuss his work. 'The journey to Vienna and back again', he explained, 'is, under my present circumstances, too strenuous.' It was clear that if Wittgenstein wanted to discuss his work with Frege, he would have to go to Jena.

In the event, the collapse of the Tsarist government was to lead, initially, to renewed activity on the Eastern Front. The new Minister of War (and, from July, the new Prime Minister), Alexander Kerensky, was determined to continue the struggle, and in July the Russians launched the ill-fated offensive named after him. The will to prolong the war on the part of the ordinary soldiers had, however, by this time been dissipated, and the Russian advance soon ground to a halt. Wittgenstein was awarded the Silver Medal for Valour for his part in the stand made by the Austro-Hungarian forces in defence of their positions at Ldziany. In the counter-offensive that followed he took part in the advance along the line of the river Pruth which led, in August, to the capture of the city of Czernowitz (Chernovtsy) in the Ukraine.

The Russian war effort had by this time completely collapsed, and with it the Kerensky government. The war in the East had been won by the Central Powers. After coming to power with a slogan of 'Bread and Peace', it remained only for the new Bolshevik government to salvage whatever they could from their inevitable surrender.

Throughout the long drawn out negotiations that followed, Wittgenstein remained stationed in the Ukraine, and it was not until 3 March 1918, when Lenin and Trotsky finally gave their signatures to the draconian terms of the Treaty of Brest Litovsk, that he, along with the bulk of the Austro-Hungarian forces, was transferred to the Italian Front.

During these six months of effectively non-combatant service, he seems to have begun the work of arranging his philosophical remarks into something like the form they finally took in the *Tractatus*. The manuscript of an early version of the book (published as *Prototractatus*) appears to date from this time, and we have it from Engelmann that a typed copy of the book existed *before* Wittgenstein left for Italy. This cannot have been the final version, but it is clear that during the winter of 1917–18 the work was beginning to take its final shape.

During this time Wittgenstein was in communication with both Frege and Engelmann. Frege wrote cards expressing the by now customary wish that he and Wittgenstein would be able to meet to discuss logic after the war. Engelmann, who was at this time employed by the Wittgenstein family to make alterations to their house in the Neuwaldeggergasse, wrote on more personal matters. On 8 January 1918 he was bold enough to make an observation on Wittgenstein's spiritual condition. It was, he said, something he had wanted to say when the two met in Vienna during the Christmas vacation, but had neglected to do so. 'If in saying it, I do you an injustice, forgive me':

> It seemed to me as if you – in contrast to the time you spent in Olmütz, where I had not thought so – had no faith. I write this not in an attempt to influence you. But I ask you to consider what I say, and wish for you, that you do what is in your *true* best interests.

Wittgenstein's reply to this is remarkably forbearing. 'It is true', he wrote, 'that there is a difference between myself as I am now and as I was when we met in Olmütz. And, as far as I know, the difference is that I am now *slightly* more decent. By this I only mean that I am slightly clearer in my own mind about my lack of decency':

> If you tell me now I have no faith, you are *perfectly right*, only I did not have it before either. It is plain, isn't it, that when a man wants,

> as it were, to invent a machine for becoming decent, such a man has no faith. But what am I to do? *I am clear about one thing*: I am far too bad to be able to theorize about myself; in fact I shall either remain a swine or else I shall improve, and that's that! Only let's cut out the transcendental twaddle when the whole thing is as plain as a sock on the jaw.

'I am sure you are quite right in all you say', the letter ends. Engelmann had, it seems, at one and the same time spoken twaddle and spoken the truth. It was a combination that Wittgenstein was also to attribute to his own words in the *Tractatus*, but which Russell, as a logician, was to find deeply unsatisfying.

On 1 February 1918 Wittgenstein was promoted *Leutnant*, and on 10 March he transferred to a mountain artillery regiment fighting on the Italian Front. His book now almost complete, he wrote to Frege on 25 March acknowledging the great debt his work owed to the elderly, and still much neglected, logician. Frege replied that he was astonished to read such an effusive acknowledgement:

> Each of us, I think, has taken something from others in our intellectual work. If I have furthered your endeavours more than I thought I had, then I am very pleased to have done so.

In the preface to the final version of the book, Wittgenstein repeats that he is indebted 'to Frege's great works and to the writings of my friend Mr Bertrand Russell for much of the stimulation of my thoughts'.

Within a month of arriving in Italy Wittgenstein fell ill with the enteritis that had troubled him in Olmütz, and sent away to Engelmann for the medicine he had been given then – 'the *only one* that has ever helped me'. Engelmann was slow to reply, and when on 28 May he finally put pen to paper, it was to ask Wittgenstein whether he knew of any remedy for a weakness of the will! His letter crossed with a parcel of books sent by Wittgenstein: 'which you don't deserve as you are too lazy even to answer an urgent enquiry'.

In the meantime, Wittgenstein had spent some time in a military hospital in Bolzano, where he was presumably able to continue work on his book. A letter from Frege of 1 June remarks how pleased he is that Wittgenstein's work is coming to a conclusion, and that he hopes it will all soon be down on paper 'so that it doesn't get lost'.

On the same day Adele Jolles wrote to him in a slightly wounded tone, apologizing for bothering him with another letter when he was so contemptuous of letters and so reluctant to engage in a superficial exchange. The Jolles family were perhaps the first, but by no means the last, of Wittgenstein's friends to fall victim to the changes in him wrought by his experience of war.

By the time of the Austrian offensive of 15 June Wittgenstein was fit enough to take part, and was employed as an observer with the artillery attacking French, British and Italian troops in the Trentino mountains. Once again he was cited for his bravery. 'His exceptionally courageous behaviour, calmness, sang-froid, and heroism', ran the report, 'won the total admiration of the troops.' He was recommended for the Gold Medal for Valour, the Austrian equivalent of the Victoria Cross, but was awarded instead the Band of the Military Service Medal with Swords, it being decided that his action, though brave, had been insufficiently consequential to merit the top honour. The attack, which was to be the last in which Wittgenstein took part, and indeed the last of which the Austrian army was capable, was quickly beaten back. In July, after the retreat, he was given a long period of leave that lasted until the end of September.

It was not in Vienna, but at Wittgenstein's Uncle Paul's house in Hallein, near Salzburg, that what we now know as *Tractatus Logico-Philosophicus* received its final form. One day in the summer of 1918 Paul Wittgenstein came across his nephew unexpectedly at a railway station. He found him desperately unhappy and intent on committing suicide, but managed to persuade him to come to Hallein. There Wittgenstein finished his book.

The most likely cause of this suicidal wish is a letter from Mrs Fanny Pinsent, dated 6 July, written to inform Wittgenstein of the death of her son, David, who had been killed in an aeroplane accident on 8 May. He had been engaged in research on aerodynamics, and had died investigating the cause of a previous accident. 'I want to tell you', she wrote, 'how much he loved you and valued your friendship up to the last.' It was to David's memory that Wittgenstein dedicated the completed book. David was, he wrote to Mrs Pinsent, 'my first and my only friend':

> I have indeed known many young men of my own age and have been on good terms with some, but only in him did I find a real friend, the hours I have spent with him have been the best in my life, he was to me a brother and a friend. Daily I have thought of him and have longed to see him again. God will bless him. If I live to see the end of the war I will come and see you and we will talk of David.

'One thing more,' he added. 'I have just finished the philosophical work on which I was already at work at Cambridge':

> I had always hoped to be able to show it to him some time, and it will always be connected with him in my mind. I will dedicate it to David's memory. For he always took great interest in it, and it is to him I owe far the most part of the happy moods which made it possible for me to work.

This last, as we have seen, refers not only to the time they spent together in Cambridge, Iceland and Norway, but also to the letters that Pinsent wrote during the war, which were, at times, the only things capable of reviving Wittgenstein's spirits sufficiently to enable him to concentrate on philosophy.

Now, having finished the book – having solved the problems he had set out to solve – what struck him most forcibly was the relative unimportance of the task that he had achieved. 'The *truth* of the thoughts that are here communicated', he wrote in the preface, 'seems to me unassailable and definitive'; and he believed himself to have found, 'on all essential points', the solution to the problems of philosophy. But:

> . . . if I am not mistaken in this belief, then the second thing in which the value of this work consists is that it shows how little is achieved when these problems are solved.

He chose as a motto for the book a quotation from Kürnberger: '. . . and anything a man knows, anything he has not merely heard rumbling and roaring, can be said in three words'. The quotation had been used before by Karl Kraus, and it is possible that Wittgenstein took it from Kraus, but it is equally likely he got it straight from Kürnberger (books by Kürnberger were among those sent by

Wittgenstein to Engelmann). In any case, it is extremely apt. The whole meaning of his book, he says in the preface, 'can be summed up as follows: What can be said at all can be said clearly; and whereof one cannot speak thereof one must be silent.'

In its final form, the book is a formidably compressed distillation of the work Wittgenstein had written since he first came to Cambridge in 1911. The remarks in it, selected from a series of perhaps seven manuscript volumes, are numbered to establish a hierarchy in which, say, remark 2.151 is an elaboration of 2.15, which in turn elaborates the point made in remark 2.1, and so on. Very few of the remarks are justified with an argument; each proposition is put forward, as Russell once put it, 'as if it were a Czar's ukase'. The Theory of Logic worked out in Norway before the war, the Picture Theory of Propositions developed during the first few months of the war, and the quasi-Schopenhauerian mysticism embraced during the second half of the war, are all allotted a place within the crystalline structure, and are each stated with the kind of finality that suggests they are all part of the same incontrovertible truth.

Central to the book in all its aspects is the distinction between showing and saying: it is at once the key to understanding the superfluity of the Theory of Types in logic and to realizing the inexpressibility of ethical truths. What the Theory of Types attempts to say can be shown only by a correct symbolism, and what one wants to say about ethics can be shown only by contemplating the world *sub specie aeternitatis*. Thus: 'There is indeed the inexpressible. This *shows* itself; it is the mystical.'

The famous last sentence of the book – 'Whereof one cannot speak, thereof one must be silent' – expresses both a logico-philosophical truth and an ethical precept.

To this extent, as Engelmann has pointed out, the book's central message is allied to the campaign of Karl Kraus to preserve the purity of language by exposing to ridicule the confused thought that stems from its misuse. The nonsense that results from trying to say what can only be shown is not only logically untenable, but ethically undesirable.

On completion of the book Wittgenstein evidently considered these ethical implications to be as important, if not more so, as its implications for logical theory. He wanted it published alongside the work of Kraus. And as soon as it was finished, he sent it to Kraus's publisher,

Jahoda, seemingly expecting the relevance of it to Kraus's work to be readily apparent. At the same time he wrote to Frege offering to send him a copy. In a letter of 12 September Frege says that he would indeed be pleased to see it. He could understand, he wrote, Wittgenstein's feeling that the work might prove fruitless: when one has forged a path up a steep mountain that no one has climbed before, there must be a doubt as to whether anyone else will have the desire to follow one up it. He knew that doubt himself. And yet he had confidence that his work had not all been in vain. In a later letter (15 October) he writes: 'May it be granted to you to see your work in print, and to me to read it!'

Engelmann, too, was promised a copy. Towards the end of September, immediately before he returned to Italy, Wittgenstein travelled to Olmütz, and it was then that Engelmann read the book for the first time. In a letter to Wittgenstein of 7 November he mentions that he often reads in it: 'and it gives me more and more joy, the more I understand it'.

At the end of September Wittgenstein returned to the Italian Front, and for the next month waited impatiently to hear from Jahoda. 'Still no reply from the publisher!' he wrote to Engelmann on 22 October:

> And I feel an insuperable repugnance against writing to him with a query. The devil knows what he is doing with my manuscript. Please be so *very* kind and look up the damned blighter some day when you are in Vienna, and then let me know the result!

A few days later he was informed that Jahoda could not publish the work, 'for technical reasons'. 'I would dearly like to know what Kraus said about it', he told Engelmann. 'If there were an opportunity for you to find out, I should be very glad. Perhaps Loos knows something about it.'

By the time Wittgenstein returned to Italy, the Austro-Hungarian Empire was beginning to break up. The allegiance of the Czechs, the Poles, the Croats and the Hungarians that made up the bulk of its army was no longer given to the Habsburg Empire (in so far as it ever had been), but to the various national states, the creation of which had been promised to them not only by the Allies, but by the Habsburg Emperor himself. After the final Allied breakthrough of 30 October, before any armistice had been signed, large numbers of the men formed themselves into groups of compatriots and simply turned their back on the war, making their way home instead to help found their new nations. Austrian officers frequently found that they had no control whatsoever over the troops that were still nominally under their command. One casualty of this situation was Wittgenstein's brother, Kurt, who in October or November shot himself when the men under him refused to obey his orders.

The Austrians could do nothing but sue for peace, a process which the Italians, presented with a golden opportunity to collect booty and win back territory, were in no mind to hurry. On 29 October an Austrian delegation, carrying a flag of truce, approached the Italians but was sent back because it lacked the proper credentials. It was not until five days later that an armistice was finally signed. In the meantime the Italians had taken about 7000 guns and about 500,000 prisoners – Wittgenstein among them.

Upon capture, he was taken to a prisoner-of-war camp in Como. There he met two fellow officers, who were to remain valuable friends in the years to come: the sculptor Michael Drobil and the teacher Ludwig Hänsel. A story told by Hermine Wittgenstein relates that Drobil had assumed from his ragged and unassuming appearance that Wittgenstein was from a humble background. One day their conversation turned to a portrait by Klimt of a certain Fräulein Wittgenstein. To Drobil's amazement, Wittgenstein referred to the painting as 'the portrait of my sister'. He stared in disbelief: 'Then you're a Wittgenstein are you?'

Wittgenstein got to know Hänsel after attending a class on logic that Hänsel was giving to prisoners who hoped upon release to train as teachers. This led to regular discussions between them, during which Wittgenstein led Hänsel through the elements of symbolic logic and explained the ideas of the *Tractatus* to him. They also read Kant's *Critique of Pure Reason* together.

In January 1919 Wittgenstein (together with Hänsel and Drobil) was transferred to another camp, in Cassino. There they were to remain, as bargaining material for the Italians, until August.

It was while he was held at Cassino that Wittgenstein made the decision that, on return to Vienna, he would train as an elementary school teacher. According to the writer Franz Parak, however, with whom Wittgenstein enjoyed a brief friendship at the prisoner-of-war camp, Wittgenstein would most have liked to have become a priest 'and to have read the Bible with the children'.*

In February Wittgenstein was able to write a card to Russell. 'I am prisoner in Italy since November', he told him, 'and hope I may communicate with you after a three years interruption. I have done lots of logical work which I am dying to let you know before publishing it.'

The card somehow managed to reach Russell at Garsington Manor, where he was staying as a guest of Ottoline Morrell and endeavouring to finish *The Analysis of Mind*, which had been started in Brixton Prison the previous year.

Russell had had, in his way, almost as difficult a time as Wittgenstein. At forty-two, he was too old to fight, but given his implacable opposition to the war he would not have volunteered anyway. His opposition to the war had caused him to be sacked from his lectureship by Trinity College, and had brought him briefly into an uneasy and emotionally fraught collaboration with D. H. Lawrence, which, when it ended, left him with a much firmer distaste for the irrational and impulsive sides of human nature than he had had hitherto.

He had campaigned tirelessly against conscription and published numerous political essays, one of which brought against him the charge of prejudicing relations between Britain and the United States. For this he was imprisoned for six months. To the public he was now better known as a political campaigner than as a philosopher/mathematician. *Principles of Social Reconstruction* and *Roads to Freedom* enjoyed a far wider readership than had *Principles of Mathematics* and *Principia Mathematica*. In prison, however, he returned to philosophical work, writing the *Introduction to Mathematical Philosophy* and beginning *The Analysis of Mind*. Now, in temporary retirement from public controversy and taking advantage of the peaceful surroundings offered at Garsington to reorientate himself back into philosophical thought,

*From Parak's point of view, the friendship was all too brief. Parak, who was seven years younger than Wittgenstein, formed a respect for Wittgenstein that bordered on worship. He hung on to Wittgenstein's every word, hoping, as he says in his memoir, to drink in as much as possible of Wittgenstein's superior knowledge and wisdom. After a time Wittgenstein tired of this and began to withdraw, 'like a mimosa', from Parak's attachment. Parak, he said, reminded him of his mother.

he was only too pleased to re-establish communication with Wittgenstein. He dashed off two cards on successive days:

> Most thankful to hear you are still alive. Please write on Logic, when possible. I hope it will not be long now before a talk will be possible. I too have very much to say about philosophy, etc. [2.3.19]

> Very glad to hear from you – had been anxious for a long time. I shall be interested to learn what you have done in Logic. I hope before long it may be possible to hear all about it. Shall be glad of further news – about your health etc. [3.3.19]

'You can't imagine how glad I was to receive your cards!' Wittgenstein replied, adding that, unless Russell were prepared to travel to Cassino, there was no hope of their meeting 'before long'. He could not write on logic, as he was allowed only two cards a week, but he explained the essential point: 'I've written a book which will be published as soon as I get home. I think I have solved our problems finally.' A few days later he was able, after all, to expand on this, when, thanks to a student who was on his way back to Austria, he had the opportunity to post a full-length letter. 'I've written a book called "Logisch-Philosophische Abhandlung" containing all my work of the last six years', he explained:

> I believe I've solved our problems finally. This may sound arrogant but I can't help believing it. I finished the book in August 1918 and two months after was made Prigioniere. I've got the manuscript here with me. I wish I could copy it out for you; but it's pretty long and I would have no safe way of sending it to you. In fact you would not understand it without a previous explanation as it's written in quite short remarks. (This of course means that *nobody* will understand it; although I believe, it's all as clear as crystal. But it upsets all our theory of truth, of classes, of numbers and all the rest.) I will publish it as soon as I get home.

He repeated that he expected to be in the camp for some time. But, he asked speculatively: 'I suppose it would be impossible for you to come and see me here?'

> . . . or perhaps you think it's colossal cheek of me even to think of such a thing. But if you were on the other end of the world and I *could* come to you I would do it.

In fact, it was impossible for Russell to visit him at Cassino, although, as it turned out, Wittgenstein himself was given an opportunity to leave the camp. Through a relative with connections in the Vatican, strings were pulled to get him released by the Italians. He was to be examined by a doctor and declared medically unfit to stand prolonged confinement. Wittgenstein, however, rejected such privileged treatment, and at the examination insisted vehemently that he was in perfect health.

Russell also pulled strings, and through Keynes (who was at this time with the British delegation at the Versailles Peace Conference) managed to get permission for Wittgenstein to receive books and to be excepted from the rule allowing him only two postcards a week, so that he could engage in learned correspondence. These privileges Wittgenstein did not refuse. They made it possible for him both to send his manuscript to Russell and to receive Russell's newly published book, *Introduction to Mathematical Philosophy*, which Russell regarded as having been influenced by his reading of Wittgenstein's *Notes on Logic.*★

To Wittgenstein, however, the book was confirmation of his suspicion that Russell would not be able to understand his latest work. 'I should never have believed', he wrote after reading it, 'that the stuff I dictated to Moore in Norway six years ago would have passed over you so completely without trace':

> In short, I'm now afraid that it might be very difficult for me to reach any understanding with you. And the small remaining hope that my manuscript might mean something to you has completely vanished . . . *Now* more than ever I'm burning to see it in print. It's galling to have to lug the completed work round in captivity and to see how nonsense has a clear field outside! And it's equally galling to think that no one will understand it even if it does get printed!

Russell's reply is remarkably conciliatory. 'It is true', he wrote, 'that

---

★For the footnote acknowledging his debt, see p. 134.

what you dictated to Moore was not intelligible to me, and he would give me no help.' About his book he explained:

> Throughout the war I did not think about philosophy, until, last summer, I found myself in prison, and beguiled my leisure by writing a popular text-book, which was all I could do under the circumstances. Now I am back at philosophy, and more in the mood to understand.

'Don't be discouraged', he urged; 'you will be understood in the end.'

By the summer of 1919 the three people Wittgenstein most hoped and expected to understand his work – Engelmann, Russell and Frege – had each received a copy. Even on the assumption (confirmed in a later letter to Russell) that this left Wittgenstein himself without a copy, it is something of a mystery how he managed to make three copies of the book.

In a letter of 6 April, Engelmann paid his own tribute to the book with a friendly parody of its system of numbering:

> No writing between the lines!
> 1. Dear Mr Wittgenstein, I am very pleased to hear,
> 2. through your family, that you are well. I
> 3. hope that you do not take it badly that I have
> 4. not written to you for so long, but I had so
> 5. much to write that I preferred to leave it to
> 6. a reunion that I hope will be soon. But I must
> 7. now thank you with all my heart for your
> 8. manuscript, a copy of which I received some time
> 9. ago from your sister. I think I now, on the
> 10. whole, understand it, at least with me you have
> 11. entirely fulfilled your purpose of providing
> 12. somebody some enjoyment through the book; I am
> 13. certain of the truth of your thoughts and
> 14. discern their meaning. Best wishes,
> 15. yours sincerely, Paul Engelmann

Engelmann evidently enjoyed this, so much so that he repeated the format in his next letter of 15 August, written to explain to Wittgenstein why he had so far been unable to obtain a copy of Frege's

Grundgesetze der Arithmetik, which Wittgenstein had asked him to send.

There are some indications that it was Frege's response to the book that Wittgenstein most eagerly awaited. If so, the disappointment must have been all the more great when he received Frege's reactions.

Frege's first impressions are contained in a letter written on 28 June. He begins by apologizing for the late response, and for the fact that, as he had had much to do, he had had little time to read Wittgenstein's manuscript and could therefore offer no firm judgement on it. Almost the whole of his letter is concerned with doubts about the precision of Wittgenstein's language:

> Right at the beginning I come across the expressions 'is the case' and 'fact' and I suspect that *is the case* and *is a fact* are the same. The world is everything that is the case and the world is the collection of facts. Is not every fact the case and is not that which is the case a fact? Is it not the same if I say, A is a fact, as if I say, A is the case? Why then this double expression? . . . Now comes a third expression: 'What is the case, a fact, is the existence of *Sachverhalte*.' I take this to mean that every fact is the existence of a *Sachverhalt*, so that another fact is the existence of another *Sachverhalt*. Couldn't one delete the words 'existence of' and say 'Every fact is a *Sachverhalt*, every other fact is another *Sachverhalt*'. Could one perhaps also say 'Every *Sachverhalt* is the existence of a fact?'

'You see', Frege wrote, 'from the very beginning I find myself entangled in doubt as to what you want to say, and so make no proper headway.' He was unsure what Wittgenstein meant by the terms *Tatsache*, *Sachverhalt* and *Sachlage*, and would need, he said, examples to clarify the terminology. Are there *Sachverhalte* that do not exist? Is every collection of objects a *Sachverhalt*?* Frege's letter must have been a bitter disappointment to Wittgenstein. There is nothing in it to indicate that Frege got past the first page; his questions all relate to the

---

*I have kept the German words here, because for the English reader Frege's confusion is compounded by differences of translation. Ogden translates *Sachverhalt* as 'atomic fact', and *Sachlage* as 'state of affairs'; Pears and McGuinness have 'state of affairs' for *Sachverhalt* and 'situation' for *Sachlage*. Ogden's translation has the merit, at least, of making it clear – as Wittgenstein had to explain to both Frege and Russell – that *Sachverhalte* are what correspond to (true) atomic propositions, and are therefore the constituent parts of *Tatsachen* (facts).

first ten or so propositions in the book, and are all concerned with terminology rather than substance. Of Wittgenstein's Theory of Symbolism and its implications for the understanding of logic, Frege obviously had no grasp at all; still less could he be expected to understand the ethical implications of the book.

Despondently, Wittgenstein pinned his hopes on Russell. In a letter of 19 August he told Russell of Frege's response to the book: 'I gather he doesn't understand a word of it':

> So my only hope is to see *you* soon and explain all to you, for it is VERY hard not to be understood by a single soul!

There were indeed reasons for hoping that Russell might, after all, be brought to an understanding of the book. His initial reactions to it were both more comprehending and more favourable than those of Frege. He had at least managed to read the whole book – 'twice carefully' so he told Wittgenstein. And he had, moreover, formed *some* idea (even if mistaken) of what it was about. 'I am convinced', he wrote on 13 August, 'you are right in your main contention, that logical props are tautologies, which are not true in the sense that substantial props are true.'

In fact, this was not the main contention of the book – at least as Wittgenstein understood it. But, nevertheless, it showed that Russell had understood what Wittgenstein was trying to say about logic. This was, however, Wittgenstein explained in his letter of 19 August, only a 'corollary' to his main contention:

> The main point is the theory of what can be expressed (gesagt) by props – i.e. by language – (and, which comes to the same, what can be *thought*) and what can not be expressed by props, but only shown (gezeigt); which, I believe, is the cardinal problem of philosophy.

This links, I believe, with what Wittgenstein had meant earlier when he said that *Introduction to Mathematical Philosophy* showed that the notes he had dictated to Moore had completely 'passed over' Russell. For though Russell had borrowed Wittgenstein's notion of tautology, he had made no use in the book of the distinction between saying and showing, the distinction that the notes to Moore had introduced. It was not that Russell did not understand the distinction, but rather that he thought it obscure and unnecessary. He later called it 'a curious kind of logical mysticism', and thought that, in logic at least, it could be dispensed with by introducing a higher level language (a 'meta-language') to say the things that could not be said using the original 'object-language'.

To his letter Russell appended a list of questions and doubts about the book. Like Frege, he wanted to know what the difference was between *Tatsache* and *Sachverhalt*. Wittgenstein gave the same answer he had given Frege:

> Sachverhalt is, what corresponds to an Elementarsatz (elementary proposition) if it is true. Tatsache is what corresponds to the logical product of elementary props when this product is true.

Most of the other queries raised by Russell in some way or other arise from his reluctance to accept the idea that some things – logical form, for instance – cannot be expressed in language but have to be shown. Russell objected, for example, to Wittgenstein's summary dismissal of the Theory of Types in proposition 3.331: 'The theory of types, in my view', he told Wittgenstein, 'is a theory of correct symbolism: (a) a simple symbol must not be used to express anything complex; (b) more generally, a symbol must have the same structure as its meaning.' 'That's exactly what one can't say', Wittgenstein replied:

> You cannot prescribe to a symbol what it *may* be used to express. All that a symbol CAN express, it MAY express. This is a short answer but it is true!

In two other replies to points raised by Russell Wittgenstein hammered home the same message:

> . . . Just think that, what you want to *say* by the apparent prop 'there are two things' is *shown* by there being two names which have different meanings.

> . . . 'It is necessary to be given the prop that all elementary props are given.' This is not necessary, because it is even *impossible*. There is no such prop! That all elementary props are given is SHOWN by there being none having an elementary sense which is not given.

Although these questions and answers relate to specific points of logical theory, not very far behind them lies a more general and more important difference. It is no coincidence that Russell's insistence on the applicability of meta-languages abolishes the sphere of the mystical, while Wittgenstein's insistence on the impossibility of saying what can only be shown preserves it.

What was perhaps Russell's gravest doubt, however, remained unanswered. This concerned Wittgenstein's brief discussion of mathematics, and particularly his abrupt dismissal of Set Theory. 'The theory of classes', he writes in proposition 6.031, 'is completely superfluous in mathematics.' As this cuts at the root of everything Russell had achieved in mathematics, he quite naturally found it perturbing:

> If you said classes were superfluous in *logic* I would imagine that I understood you, by supposing a distinction between logic and mathematics; but when you say they are unnecessary in *mathematics* I am puzzled.

To this Wittgenstein said only that it would require a lengthy answer, and: 'you know how difficult it is for me to write on logic'.

Of the final sections of the book, Russell had little to comment: 'I agree with what you say about induction, causality, etc.; at least I can find no ground for disagreeing.' On the remarks on ethics, aesthetics, the soul and the meaning of life, he said nothing.

'*I am sure you are right in thinking the book of first-class importance*', he concluded. 'But in places it is obscure through brevity':

> I have a most intense desire to see you, to talk it over, as well as simply because I want to see you. But I can't get abroad as yet. Probably you will be free to come to England before I am free to go abroad. – I will send back your MS when I know where to send it, but I am hoping you will soon be at liberty.

The letter was sufficiently encouraging to prompt Wittgenstein to seek a meeting as soon as possible. 'I should like to come to England', he wrote, 'but you can imagine it's rather awkward for a German to travel to England now.' The best thing would be to meet in some neutral country – say, Holland or Switzerland. And *soon*. 'The day after tomorrow', he told Russell, 'we shall probably leave the Campo Concentramento and go home. Thank God!'

He was released two days later, on 21 August 1919.

II

1919–28

8
THE UNPRINTABLE TRUTH

Like many war veterans before and since, Wittgenstein found it almost insuperably difficult to adjust to peace-time conditions. He had been a soldier for five years, and the experience had left an indelible stamp upon his personality. He continued to wear his uniform for many years after the war, as though it had become a part of his identity, an essential part, without which he would be lost. It was also perhaps a symbol of his feeling – which persisted for the rest of his life – that he belonged to a past age. For it was the uniform of a force that no longer existed. Austria-Hungary was no more, and the country he returned to in the summer of 1919 was itself undergoing a painful process of adjustment. Vienna, once the imperial centre of a dynasty controlling the lives of fifty million subjects of mixed race, was now the capital of a small, impoverished and insignificant Alpine republic of little more than six million, mostly German, inhabitants.

The parts of the empire in which Wittgenstein himself had fought to defend what had been his homeland were now absorbed into foreign states. Lemberg and Kraków were now in the new state of Poland; the area around the Trentino mountains had been claimed by Italy; and Olmütz, that last outpost of Austro-Hungarian culture, was now in Czechoslovakia – itself a hybrid creation of 'self-determination' – of which Paul Engelmann had become a reluctant citizen. (The problems encountered by Engelmann in obtaining a Czechoslovakian passport kept him from visiting Wittgenstein in Vienna for many months.) To many Austrians, the whole *raison d'être* of their separate identity had been destroyed, and in 1919 the majority voted for *Anschluss* with Germany. If they were to be nothing more than a German state, it was felt, then surely they had better be part of the fatherland. That option was denied them by the Allies, who also, through the war reparations demanded by the Treaties of Versailles and St Germain, ensured that the German people of both German states would remain poor, resentful and revengeful throughout the inter-war period.

Wittgenstein had entered the war hoping it would change him, and this it had. He had undergone four years of active service and a year of incarceration; he had faced death, experienced a religious awakening, taken responsibility for the lives of others, and endured long periods of close confinement in the company of the sort of people he would not previously have shared a railway carriage with. All this had made him a different person – had given him a new identity. In a sense, he was not returning to anything in 1919: everything had changed, and he could no more slip back into the life he had left in 1914 than he could revert to being the 'little Wittgenstein' that the Jolles had known in Berlin. He was faced with the task of re-creating himself – of finding a new role for the person that had been forged by the experiences of the last five years.

His family were dismayed by the changes they saw in him. They could not understand why he wanted to train to become a teacher in elementary schools. Hadn't Bertrand Russell himself acknowledged his philosophic genius, and stated that the next big step in philosophy would come from him? Why did he now want to waste that genius on the uneducated poor? It was, his sister Hermine remarked, like somebody wanting to use a precision instrument to open crates. To this, Wittgenstein replied:

> You remind me of somebody who is looking out through a closed window and cannot explain to himself the strange movements of a passer-by. He cannot tell what sort of storm is raging out there or that this person might only be managing with difficulty to stay on his feet.

Of course, it might be thought that the most natural step for the person in Wittgenstein's analogy to take would be to come in out of the storm. But this Wittgenstein could not do. The hardship suffered during the war was not experienced by him as something from which he sought refuge, but as the very thing that gave his life meaning. To shelter from the storm in the comfort and security which his family's wealth and his own education could provide would be to sacrifice everything he had gained from struggling with adversity. It would be to give up climbing mountains in order to live on a plateau.

It was essential to Wittgenstein, not only that he should not use the privileges of his inherited wealth, but that he could not do so. On his arrival home from the war he was one of the wealthiest men in Europe, owing to his father's financial astuteness in transferring the family's wealth, before the war, into American bonds. But within a month of returning, he had disposed of his entire estate. To the concern of his family, and the astonishment of the family accountant, he insisted that his entire inheritance should be made over to his sisters, Helene and Hermine, and his brother Paul (Gretl, it was decided, was already too wealthy to be included). Other members of the family, among them his Uncle Paul Wittgenstein, could not understand how they could have accepted the money. Couldn't they at least have secretly put some of it aside in case he should later come to regret his decision? These people, writes Hermine, could not know that it was precisely that possibility that troubled him:

> A hundred times he wanted to assure himself that there was no possibility of any money still belonging to him in any shape or form. To the despair of the notary carrying out the transfer, he returned to this point again and again.

Eventually the notary was persuaded to execute Wittgenstein's wishes to the letter. 'So', he sighed, 'you want to commit financial suicide!'

In September 1919, after ridding himself of his wealth and having enrolled at the Lehrerbildungsanhalt in the Kundmanngasse, Wittgenstein took another step towards independence from his privileged background, moving out of the family home in the Neuwaldeggergasse and taking lodgings in Untere Viaduktgasse, a street in Vienna's Third District within easy walking distance of the college.*

This period was one of great suffering for Wittgenstein, and on more than one occasion during these months he contemplated taking

---

*He was in these lodgings scarcely more than a month, but his time there has become the subject of a heated controversy following claims made by the writer William Warren Bartley III. See pp. 581–6.

his own life. He was exhausted and disorientated. 'I'm not quite normal yet', he wrote to Russell soon after his return; and to Engelmann: 'I am not very well (i.e. as far as my state of mind is concerned).' He asked both Russell and Engelmann to come and see him as soon as they could, but neither could manage the trip. Engelmann was having problems obtaining a Czechoslovakian passport, and Russell was engaged on a course of lectures at the London School of Economics (the material of which formed the basis for *The Analysis of Mind*), which would keep him in England until the Christmas vacation. Besides, there was a real possibility that Russell would not be given permission to leave the country – 'for as you may know', he wrote to Wittgenstein, 'I have fallen out with the Government'. He nevertheless suggested that they try to meet at The Hague at Christmas: 'I could manage a week, if the government will let me go.'

The frustration of not being able to reunite with either Engelmann or Russell undoubtedly added to the emotional strain Wittgenstein was suffering. He had the feeling of losing all his old friends and being unable to make any new ones. The meeting he had most looked forward to during the last five years had been denied him by the death of 'dear David' (so he wrote to Mrs Pinsent), and other eagerly anticipated meetings were either frustrated or turned out to be a bitter disappointment. He looked up Adolf Loos, but was, he told Engelmann, 'horrified and nauseated':

> He has become infected with the most virulent bogus intellectualism! He gave me a pamphlet about a proposed 'fine arts office', in which he speaks about a sin against the Holy Ghost. This surely is the limit! I was already a bit depressed when I went to see Loos, but that was the last straw!

Nor was he – a thirty-year-old war veteran – likely to make many friends among the teenagers with whom he attended lectures at the teacher training college. 'I can no longer behave like a grammar-school boy', he wrote to Engelmann, 'and – funny as it sounds – the humiliation is *so* great for me that often I think I can hardly bear it!' He complained in similar spirit to Russell:

> The benches are full of boys of 17 and 18 and I've reached 30. That leads to some very funny situations – and many *very* unpleasant ones too. I often feel miserable!

Though he was embarking on a new career and a new life, and in many ways was deliberately severing the ties that bound him to his family background, he needed to establish some continuity between the person he was before the war and the person he had become. Before he started his course at the Lehrerbildungsanhalt, he spent about ten days living at the Hochreit, in order, as he put it to Engelmann, 'to find something of myself again if I can'.

His family connections, and the ambivalence with which he regarded them, provoked one of the unpleasant situations at college he mentions to Russell. His teacher asked him whether he was related to *the* Wittgensteins, the rich Wittgensteins. He replied that he was. Was he *closely* related? the teacher persisted. To this, Wittgenstein felt compelled to lie: 'Not very.'

The defeat and impoverishment of his home country, the death of his most beloved friend, the frustration at not being able to re-establish old friendships and the strain of putting his whole life on a new footing might themselves be sufficient to account for Wittgenstein's suicidal state during the autumn of 1919. But perhaps the most important cause of his depression was his failure to find a publisher for the *Tractatus* – or even a single person who understood it.

He had, he thought, completed a book that provided a definitive and unassailably true solution to the problems of philosophy. How, then, could he have anticipated such difficulty in finding someone willing to publish it? Even after it had been rejected by Jahoda, Wittgenstein could write confidently from the prison camp at Cassino: 'My book will be published as soon as I get home.'

Within a few days of his return, he took the book to the Viennese offices of Wilhelm Braumüller, the publishers of Otto Weininger's *Sex and Character*. Braumüller, he told Russell, 'naturally neither knows my name nor understands anything about philosophy [and] requires the judgment of some expert in order to be sure that the book is really worth printing':

> For this purpose he wanted to apply to one of the people he relies on here (probably a professor of philosophy). So I told him that no one here would be able to form a judgment on the book, but that *you* would perhaps be kind enough to write him a brief assessment of the value of the work, and if this happened to be favourable that

would be enough to induce him to publish it. The publisher's address is: Wilhelm Braumüller, XI Servitengasse 5, Vienna. Now please write him a few words – as much as your conscience will allow you to.

After receiving Russell's testimonial, Braumüller offered to publish the book on condition that Wittgenstein himself paid for the printing and paper. By the time the offer was made he had no money to pay such costs, but even if he had he would still have refused. 'I consider it indecent', he said, 'to force a work upon the world – to which the publisher belongs – in this way. The writing was *my* affair; but the world must accept it in the normal manner.'

While waiting for Braumüller's decision, he received a letter from Frege – a late reply to Wittgenstein's last letter from Cassino, and to a further letter which Wittgenstein had written since his return to Vienna. Frege was still far from satisfied with the clarity of Wittgenstein's use of the word *Sachverhalt*:

You now write: 'What corresponds to an elementary proposition, if it is true, is the existence of a *Sachverhalt*.' Here you do not explain the expression '*Sachverhalt*', but the whole expression 'the existence of a *Sachverhalt*'.

He was further perturbed by what Wittgenstein had written about the purpose of the book. 'This book will perhaps only be understood by those who have themselves already thought the thoughts which are expressed in it', Wittgenstein had written in the preface (he must also have written something similar to Frege). 'It is therefore not a textbook. Its object would be attained if it afforded pleasure to one who read it with understanding.' This Frege found strange:

The pleasure of reading your book can therefore no longer be aroused by the content which is already known, but only by the peculiar form given to it by the author. The book thereby becomes an artistic rather than a scientific achievement; what is said in it takes second place to the way in which it is said.

He was encouraged, however, by one sentence in Wittgenstein's letter. Responding to Frege's remarks about the identical meanings of his propositions, 'The world is everything that is the case', and 'The world is the totality of facts', Wittgenstein wrote: 'The sense of both propositions is one and the same, but not the ideas that *I* associated with them when I wrote them.' Here Frege was (or thought he was) on home ground, and agreed wholeheartedly with Wittgenstein's point, the more so because it touched on a thought that was dear to him at this time. In order to make Wittgenstein's point, he argued, it was necessary to distinguish a proposition from its sense, thus opening up the possibility that two propositions could have the same sense and yet differ in the ideas associated with them. 'The actual sense of a proposition', he wrote to Wittgenstein, 'is the same for everybody; but the ideas which a person associates with the proposition belong to him alone . . . No one can have another's ideas.'

It was a theme Frege had dealt with in an article he had recently published, a copy of which he enclosed with his letter to Wittgenstein. The article was called '*Der Gedanke*' ('The Thought'), and was published in the journal *Beiträgen zur Philosophie des Deutschen Idealismus*. Though exasperated by Frege's laboured attempts to clarify the meaning of his book ('He doesn't understand a single word of my work', he wrote to Russell after receiving Frege's letter, 'and I'm thoroughly exhausted from giving what are purely and simply explanations'), Wittgenstein seized the opportunity to offer his work to another potentially sympathetic publisher. After rejecting Braumüller's offer to publish it if he would pay for the printing, he asked Frege to investigate the possibility of having it published in the same journal that had published Frege's article.

Frege's reply was not greatly encouraging. He could, he told Wittgenstein, write to the editor of the journal and tell him: 'that I have learnt to know you as a thinker to be taken thoroughly seriously'. But: 'Of the treatise itself I can offer no judgement, not because I am not in agreement with its contents, but rather because the content is too unclear to me.' He could ask the editor if he wished to see Wittgenstein's book, but: 'I hardly think that this would lead to anything.' The book would take up about fifty printed pages, nearly the whole journal, and: 'There seems to me not a chance that the editor would give up a whole edition to a single, still unknown writer.'

If, however, Wittgenstein were prepared to split the book into sections, its publication in a periodical would be more feasible (and, one gathers, would receive more support from Frege himself):

> You write in your preface that the truth of the thoughts communicated seems to you unassailable and definitive. Now could not one of these thoughts, in which the solution to a philosophical problem is contained, itself provide the subject for an article and thus the whole book be divided into as many parts as the number of philosophical problems with which it deals?

This would have the merit, argued Frege, of not frightening the reader away from the book because of its length. And furthermore: 'If the first article, which would have to lay the foundations, met with approval, it would be easier to find a place in the periodical for the rest of the treatise.'

It would also, he thought, help to make Wittgenstein's work clearer. After reading the preface, he told Wittgenstein, one did not really know what to make of the first proposition. One expected to see a question, to have a problem outlined, to which the book would address itself. Instead, one came across a bald assertion, without being given the grounds for it. Wouldn't it be better to make clear to which problems the book was supposed to provide a definitive solution?

'Don't take these remarks badly', Frege ended; 'they are made with good intentions.'

Wittgenstein would have nothing to do with Frege's suggestion. To divide the book up in the manner recommended would be, in his opinion, to 'mutilate it from beginning to end and, in a word, make another work out of it'. As Frege had remarked earlier, the form in which Wittgenstein's thoughts were expressed was essential to the nature of the work. After receiving Frege's letter he abandoned the attempt to have it published in *Beiträgen zur Philosophie des Deutschen Idealismus*.

Reasoning, perhaps, that if the book was too literary for a philosophical journal he would try a literary journal instead, Wittgenstein next thought of von Ficker and *Der Brenner*. By coincidence, the day on which he was about to go to see Loos to get Ficker's address, a letter from Ficker arrived telling him that *Der Brenner* would indeed continue to be published and asking him if he would like to be sent a copy. Immediately, Wittgenstein wrote a long letter to Ficker, explaining the history of his book. 'About a year ago', he wrote, 'I finished a philosophical work on which I had worked for the previous seven years':

> It is quite strictly speaking the presentation of a system. And this presentation is *extremely* compressed since I have only retained in it that which really occurred to me – and how it occurred to me.

Immediately after finishing the work, he continued, he wanted to find a publisher: 'And therein lies a great difficulty':

> The work is very small, only about sixty pages long. Who writes sixty-page brochures about philosophical matters? . . . [only] those certain, totally hopeless hacks who have neither the mind of the great, nor the erudition of the professors, and yet would like to have something printed at any price. Therefore such products are usually printed privately. But I simply cannot mix my life's work – for that is what it is – among these writings.

He then told Ficker of the unsatisfactory responses he had so far had from the publishers of, respectively, Kraus, Weininger and Frege. Finally, he got to the point: 'it occurred to me whether *you* might be inclined to take the poor thing into your protection'. If Ficker thought its publication in *Der Brenner* conceivable, Wittgenstein would send him the manuscript. 'Until then I should only like to say this about it':

> The work is strictly philosophical and at the same time literary, but there is no babbling in it.

Ficker replied with a mixture of encouragement and caution. 'Why hadn't you thought of me immediately?' he asked. 'For you could well imagine that I would take a completely different, i.e. a deeper, interest in your work than a publisher who had only his commercial interests in mind.' Strangely, his letter then dwelt at length on the need for him to keep *his* commercial interests in mind. He had, he said, previously published *Der Brenner* for love, not money. But this could not continue; times were hard, he had a wife and children to support, and printing costs were prohibitively high. In the difficult financial climate that prevailed in Austria after the war, publishing was a risky business, and he had to ensure that he did not take more chances than were necessary. Nevertheless, with the proviso that 'strictly scientific works are not really our field' (and with the awareness that he was still somewhat in Wittgenstein's debt for the benefactions of 1914), he asked to see Wittgenstein's manuscript: 'Rest assured, dear Mr Wittgenstein, that I will do my best to meet your wishes.'

Wittgenstein was sufficiently encouraged by this to send Ficker the manuscript. 'I am pinning my hopes on you', he wrote in the accompanying letter, which also provides one of the most direct statements we have of how he wished his book to be understood. He needed to say *something* about it, he told Ficker: 'For you won't – I really believe – get too much out of reading it. Because you won't understand it; the content will be strange to you':

> In reality, it isn't strange to you, for the point of the book is ethical. I once wanted to give a few words in the foreword which now actually are not in it, which, however, I'll write to you now because they might be a key for you: I wanted to write that my work consists of two parts: of the one which is here, and of everything which I have *not* written. And precisely this second part is the important one. For the Ethical is delimited from within, as it were, by my book; and I'm convinced that, *strictly* speaking, it can ONLY be delimited in this way. In brief, I think: All of that which *many* are *babbling* today, I have defined in my book by remaining silent about it. Therefore the book will, unless I'm quite wrong, have much to say which you want to say yourself, but perhaps you won't notice that it is said in it. For the time being, I'd recommend that you read the *foreword* and the *conclusion* since these express the point most directly.

If this was intended to convince Ficker that the message of the *Tractatus* was, despite appearances, consonant with the aims of *Der Brenner*, it was misjudged. Wittgenstein was asking Ficker to accept that what he wanted to say about ethics had best be said by remaining silent – and, by implication, that much of what Ficker published in *Der Brenner* was mere 'babbling'. His letter was hardly calculated, either, to reassure Ficker's financial worries. A book in which the most important part has been left out could not be expected to be a very attractive proposition to a publisher with a concerned eye to his own solvency.

Ficker's response was cool. He could not give a definite answer, he wrote on 18 November, but there was a possibility that he would not be able to publish Wittgenstein's work. It was, at that moment, in the hands of his friend and colleague, who, as he had explained in the previous letter, was responsible for the financial affairs of the publishing house. The opinion of this colleague was that the work was too specialized to appear in *Der Brenner* – although that was not necessarily his last word on the subject. Nevertheless, Ficker had approached Rilke for advice on where an alternative publisher might be found. Finally, could he show the book to a philosophy professor? He knew someone at Innsbruck University who was familiar with the work of Russell and who was interested to read what Wittgenstein had written. Who knows, he might even be able to help find a publisher for it.

The letter threw Wittgenstein into a state of despondency. 'Do you remember', he wrote to Russell, 'how you were always pressing me to publish something? And now when I should like to, it can't be managed. The devil take it!' To Ficker he replied: 'Your letter, naturally, wasn't pleasant for me, although I wasn't really surprised by your answer. I don't know where I can get my work accepted either. If I were only somewhere else than in this lousy world!' Yes, Ficker could show the book to a professor if he liked, but showing a philosophical work to a professor of philosophy would be like casting pearls before swine – 'At any rate he won't understand a word of it':

> And now, only *one* more request: Make it short and sweet with me. Tell me 'no' quickly, rather then too slowly; that is Austrian delicacy which my nerves are not strong enough to withstand, at the moment.

Alarmed by this note of despair, Ficker wired a telegram: 'Don't worry. Treatise will appear whatever the circumstances. Letter follows.' Much relieved, Wittgenstein replied that he would rather Ficker accepted the book because he considered it worth publishing than because he wanted to do a favour. He nevertheless seemed inclined to accept the offer: 'I think I can say that if you print Dallago, Haecker, etc., *then* you can also print *my* book.' The next letter he received, however, reinforced any doubts he may still have had. Ficker wrote that he was still hoping something would come of Rilke's attempt to find a publisher. But if not, so moved was he by the bitterness and distress evident in Wittgenstein's previous letter, he had decided – even if it meant risking everything he had – to see to the publication of Wittgenstein's work himself. Rather that than disappoint the trust Wittgenstein had placed in him. (By the way, he added, if it did come to that was it absolutely necessary to include the decimal numbers?)

This, obviously, would not do. 'I couldn't accept the responsibility', Wittgenstein wrote to him, 'of a person's (whoever's) livelihood being placed in jeopardy by publishing my book.' Ficker hadn't betrayed his trust:

> . . . for my trust, or rather, simply my hope, was only directed to your perspicacity that the treatise is not junk – unless I am deceiving myself – but not to the fact that you would accept it, without thinking something of it, just *out of kindness toward me and against your interests*.

And, yes, the decimals were absolutely necessary: 'because they alone give the book lucidity and clarity and it would be an incomprehensible jumble without them'. The book had to be published as it was, and for the reason that it was perceived to be worth publishing. Nothing else would do. If Rilke could somehow arrange that, he would be very pleased, but: 'if that isn't possible, we can just forget about it'.

It is difficult to know how much trouble Rilke went to on Wittgenstein's behalf. In a letter from Berne, dated 12 November 1919, he asks Ficker whether his own publisher, Insel-Verlag, might be suitable, and further suggests Otto Reichl, the publisher of Count Keyserling. Nothing came of either suggestion, and no further correspondence on the subject survives.

By this time Wittgenstein was sick to death of the whole business. 'Is there a *Krampus* who fetches evil publishers?' he asked Ficker; and on 16 November he wrote to Engelmann:

> Just how far I have gone downhill you can see from the fact that I have on several occasions contemplated taking my own life. Not from my despair about my own badness but for purely external reasons.

Wittgenstein's despair was alleviated to some extent when, in November, he left his lodgings in Untere Viaduktgasse and moved in with the Sjögren family at their home in St Veitgasse, in Vienna's

Thirteenth District. The Sjögrens were lifelong friends of the Wittgenstein family; the father, Arvid Sjögren, had been a director of a steelworks belonging to the Wittgenstein group, and the mother, Mima, now widowed, was a particularly close friend of Wittgenstein's sister Hermine. Mima was having problems bringing up her three sons alone, and it was thought by the Wittgenstein family that Wittgenstein, by acting as the man of the house, might be able to help her. If he refused to enjoy the benefits of living with his own family, perhaps he could be induced to share the responsibilities of caring for another. This, it was thought, might have a calming effect on him.

To a certain extent, it worked. Wittgenstein's time with the Sjögrens was, in the context of perhaps the most desperately unhappy year of his life, comparatively pleasant. 'Normal human beings are a balm to me', he wrote to Engelmann, 'and a torment at the same time.' With the middle son, Arvid, in particular, he formed a close friendship, and indeed became a sort of father figure to him. Arvid Sjögren was a big, clumsy, gruff boy – a 'bear of a man', he was later called – who continued to look to Wittgenstein for moral guidance throughout most of his life. Under Wittgenstein's influence he abandoned all thought of studying at university and trained instead as a mechanic. In this sense he was, perhaps, Wittgenstein's first disciple, the forerunner of the bright young undergraduates at Cambridge of the 1930s and 1940s who similarly chose to learn an honest trade rather than pursuing the kind of careers for which their education and privileged backgrounds had prepared them.

Throughout November Wittgenstein and Russell exchanged letters in connection with their proposed meeting at The Hague in December; there were dates to arrange, bureaucratic hurdles to overcome, and, in Wittgenstein's case at least, money to raise to finance the trip. 'It is terrible to think of your having to earn a living', Russell wrote to him, after hearing that he had given all his money away, 'but I am not surprised by your action. I am much poorer too. They say Holland is very expensive but I suppose we can endure a week of it without going bankrupt.' To pay Wittgenstein's expenses, Russell bought some furniture and books that Wittgenstein had left behind in a dealer's shop in Cambridge before his trip to Norway. It included the furniture he had chosen so painstakingly in the autumn of 1912. Russell paid £100; it was, he says in his autobiography, the best bargain he ever made.

Russell arrived at The Hague on 10 December, accompanied by his new lover and future wife, Dora Black. They booked into the Hotal Twee Steden. 'Come here as quick as you can after your arrival in The Hague', Russell wrote to Wittgenstein; 'I am impatient to see you – We will find some way to get your book published – in England if necessary.' Wittgenstein arrived a few days later, accompanied by Arvid Sjögren (remembered by Dora Russell as: 'a vague, shadowy figure who spoke little, even at mealtimes'). For Russell and Wittgenstein, the week was taken up with intense discussion of Wittgenstein's book. Wittgenstein was, Russell wrote to Colette on 12 December, 'so full of logic that I can hardly get him to talk about anything personal'. Wittgenstein didn't want to waste a moment of their time together. He would rise early and hammer at Russell's door until he woke, and then discuss logic without interruption for hours on end. They went through the book line by line. The discussions were fruitful: Russell came to think even more highly of the book than he had before, while Wittgenstein had the euphoric feeling that, at last, somebody understood it.

Not that Russell agreed with it entirely. In particular, he refused to accept Wittgenstein's view that any assertion about the world as a whole was meaningless. To Russell, the proposition: 'There are at least three things in the world' was both meaningful and true. During discussion of this point Russell took a sheet of white paper and made three blobs of ink on it: 'I besought him to admit that, since there were these three blobs, there must be at least three things in the world; but he refused resolutely':

> He would admit there were three blobs on the page, because that was a finite assertion, but he would not admit that anything at all could be said about the world as a whole.

'This part of his doctrine', Russell insisted, 'is to my mind definitely mistaken.'

Related to this was Russell's refusal to accept what Wittgenstein had earlier told him was the 'main contention' of the book: the doctrine that what cannot be said by propositions *can* be shown. To Russell this remained an unappealingly mystical notion. He was surprised, he wrote to Ottoline, to find that Wittgenstein had become a complete mystic. 'He has penetrated deep into mystical ways of thought and feeling, but I think (though he wouldn't agree) that what he likes best in mysticism is its power to make him stop thinking.'

He was nonetheless sufficiently impressed by the Theory of Logic in the book to offer to write an introduction, based on their conversation at The Hague, which would attempt to explain the most difficult parts of the book. With an introduction by Russell, now a best-selling author, the publication of the book was almost guaranteed. Wittgenstein returned to Vienna in jubilant mood. 'I enjoyed our time together *very* much', he wrote to Russell on 8 January 1920, 'and I have the feeling (haven't you too?) that we did a great deal of real work during that week.' To Ficker, he wrote: 'The book is now a much smaller risk for a publisher, or perhaps even none at all, since Russell's name is very well known and ensures a quite special group of readers for the book':

> By this I naturally don't mean that it will thus come into the right hands; but at any rate, favourable circumstances are less excluded.

Ficker, who did not reply for over two weeks, was evidently still not convinced that the book would be anything other than a financial liability. 'With or without Russell', he wrote on 16 January, 'the publication of your treatise is, under the present circumstances, a risk that *no* publisher in Austria today can afford to take.' He advised Wittgenstein to have the book published in English first and then – if the opportunity arose – in German.

Anticipating he would have no success with Ficker, Wittgenstein had already made appoaches to another publisher. Through Engelmann, he obtained a recommendation from a Dr Heller to the Leipzig publishing house, Reclam, who, after learning of Russell's introduction, were only too willing to consider the book.

Wittgenstein at once took the manuscript from Ficker and sent it to Reclam, and throughout February and March waited impatiently for Russell's introduction to arrive. When it did, he was immediately disappointed. 'There's so much of it that I'm not in agreement with', he told Russell, 'both where you're critical of me and also where you're simply trying to elucidate my point of view.' He nevertheless had it translated into German, in preparation for its printing, but this only made matters worse: 'All the refinement of your English style', he wrote to Russell, 'was, obviously, lost in the translation and what remained was superficiality and misunderstanding.' He sent the introduction to Reclam but told them it was not for publication; it was to serve only for the publisher's own orientation with regard to the work. As a consequence Reclam, as Wittgenstein had anticipated, rejected the book. He comforted himself with the following argument, which, he told Russell, 'seems to me unanswerable':

> Either my piece is a work of the highest rank, or it is not a work of the highest rank. In the latter (and more probable) case I myself am in favour of it not being printed. And in the former case it's a matter of indifference whether it's printed twenty or a hundred years sooner or later. After all, who asks whether the Critique of Pure Reason, for example, was written in 17x or y.

Russell was at this time visiting Soviet Russia with a Labour Party delegation, and did not see Wittgenstein's letter until his return in June. He reacted with remarkable generosity. 'I don't care twopence about the introduction, but I shall be really sorry if your book isn't printed. May I try, in that case, to have it printed in England?' Yes, Wittgenstein replied, '*you can do what you like with it*'. He himself had given up trying: 'But if you feel like getting it printed, it is entirely at your disposal.'

The comforting argument he had earlier offered Russell did not prevent Wittgenstein from sinking into a deep depression after Reclam's rejection. At the end of May he wrote to Engelmann: 'I have continually thought of taking my own life, and the idea still haunts me sometimes. *I have sunk to the lowest point.* May you never be in that position! Shall I ever be able to raise myself up again? Well, we shall see.'

He was, by this time, living on his own again. At the beginning of April he moved out of the Sjögrens' home and into lodgings once more, this time in Rasumofskygasse, which, like his earlier lodgings, was in Vienna's Third District. 'This change of home was accompanied by operations which I can never remember without a sinking feeling', he told Engelmann. In fact, he fled the house after it became apparent that Mrs Sjögren was in love with him.*

---

* So it was believed, anyway, by certain members of the Sjögren and Wittgenstein families, who (according to Brian McGuinness, op. cit. p. 285) thereafter avoided inviting Mima and Wittgenstein to the same occasions.

Wittgenstein's letters to Russell, and especially to Engelmann, during this period, show him to be desperately, suicidally, depressed. The severity of self-accusation contained in them is extreme even for Wittgenstein, who was always harsh on himself. He attributes his miserableness to his own 'baseness and rottenness', and talks of being afraid that: 'the devil will come and take me one day'.*

For both Wittgenstein and Engelmann, religion was inseparable from an awareness of one's own failings. Indeed, for Engelmann, such awareness was central to the religious outlook:

> If I am unhappy and know that my unhappiness reflects a gross discrepancy between myself and life as it is, I solved nothing; I shall be on the wrong track and I shall never find a way out of the chaos of my emotions and thoughts so long as I have not achieved the supreme and crucial insight that that discrepancy is not the fault of life as it is, but of myself as I am . . .
>
> The person who has achieved this insight and holds on to it, and who will try again and again to live up to it, is religious.

On this view, to be unhappy *is* to find fault with oneself: one's misery can only be the consequence of one's own 'baseness and rottenness'; to be religious is to recognize one's own unworthiness and to take responsibility for correcting it.

This was a theme that dominated the conversations and the correspondence between Wittgenstein and Engelmann, as, for example, in the set of remarks on religion that Engelmann sent Wittgenstein in January:

> Before Christ, people experienced God (or Gods) as something outside themselves.
>
> Since Christ, people (not all, but those who have learnt to see through him) see God as something in themselves. So that one can say that, through Christ, God has been drawn into mankind . . .

---

* Since the publication of Bartley's book it has become natural to interpret these self-admonitions as being in some way connected with the alleged 'Prater episodes'. If there is any connection, however, Engelmann himself was unaware of it. In a diary entry written after Wittgenstein's death, he remarks that he is often asked about Wittgenstein's homosexuality, but can say nothing about it – he and Wittgenstein did not discuss such things.

> . . . Through Christ God has become man.
> Lucifer *wanted* to become God and was not.
> Christ *became* God without wanting to.
> So the wicked thing is to *want* pleasure without deserving it.
> If, however, one *does* right, without wanting pleasure, so joy comes of its own accord.

When Wittgenstein came to comment on these remarks, he did not dispute the truth of them, but only the adequacy of their expression. 'They are still not clear enough', he wrote. 'It must be possible, I believe, to say all these things much more adequately. (Or, not at all, which is even more likely).' Even if their most perfect expression should turn out to be silence, then, they are nonetheless true.

Wittgenstein regarded Engelmann as 'someone who understands man'. When, after the attempt to be published by Reclam had come to nothing, he was feeling emotionally and spiritually demoralized, he felt an urgent need to talk with him. And when, at the end of May, he reached his 'lowest point' and continually thought of suicide, it was to Engelmann he turned for support. He received it in the form of a long letter about Engelmann's own experience. Engelmann wrote that he had recently been worried about his motives for his own work – whether they were decent and honest motives. He had taken some time off to be alone in the countryside, to think about it. The first few days were unsatisfactory:

> But then I did something about which I can tell *you*, because you know me well enough not to regard it as a piece of stupidity. That is, I took down a kind of 'confession', in which I tried to recall the series of events in my life, in as much detail as is possible in the space of an hour. With each event I tried to make clear to myself how I should have behaved. By means of such a general over-view [*übersicht*] the confused picture was much simplified.
> The next day, on the basis of this newly-gained insight, I revised my plans and intentions for the future.

'I don't know at all', he wrote, 'whether something similar would be good or necessary for you now; but perhaps my telling you this would help you now to find something.'

'Concerning what you write about thoughts of suicide,' Engelmann added, 'my thoughts are as follows':

> Behind such thoughts, just as in others, there can probably lie something of a noble motive. But that this motive shows itself in *this* way, that it takes the form of a contemplation of suicide, is certainly wrong. Suicide is certainly a mistake. So long as a person lives, he is never completely lost. What drives a man to suicide is, however, the fear that he is completely lost. This fear is, in view of what has already been said, ungrounded. In this fear a person does the worst thing he can do, he deprives himself of the time in which it would be possible for him to escape being lost.

'You undoubtedly know all this better than I', wrote Engelmann, excusing himself for appearing to have something to teach Wittgenstein, 'but one sometimes forgets what one knows.'

Wittgenstein himself was later to use, more than once, the technique of preparing a confession in order to clarify his own life. On this occasion, however, it was not the advice that did him good, but simply reading about Engelmann's own efforts. 'Many thanks for your kind letter', he wrote on 21 June, 'which has given me much pleasure and thereby perhaps helped me a little, although as far as the merits of my case are concerned I am beyond any outside help':

> In fact I am in a state of mind that is terrible to me. I have been through it several times before: it is the state of *not being able to get over a particular fact*. It is a pitiable state, I know. But there is only one remedy that I can see, and that is of course to come to terms with that fact. But this is just like what happens when a man who can't swim has fallen into the water and flails about with his hands and feet and feels that he *cannot* keep his head above water. That is the position I am in now. I know that to kill oneself is always a dirty thing to do. Surely one *cannot* will one's own destruction, and anybody who has visualized what is in practice involved in the act of suicide knows that suicide is always a *rushing of one's own defences*. But nothing is worse than to be forced to take oneself by surprise.
>
> Of course it all boils down to the fact that I have no faith!

Unfortunately, there is no possible way of knowing what fact he is here talking about. Certainly, it is some fact about himself, and something for which he felt the only remedy to be religious faith.

Without such faith, his life was unendurable. He was in the position of wishing himself dead, but unable to bring himself to suicide. As he put it to Russell: 'The best for me, perhaps, would be if I could lie down one evening and not wake up again.'

'But perhaps there is something better left for me', he added parenthetically. The letter was written on 7 July, the day he received his teaching certificate: perhaps in teaching, it is implied, he would find something worth living for.

Wittgenstein had completed his course at the Lehrerbildungsanhalt satisfactorily, but not without misgivings. The best thing about it, he told Engelmann, was that on teaching practice he was able to read fairy-tales to the children: 'It pleases them and relieves the strain on me.' It was the 'one good thing in my life just now'.

He received help and encouragement from his friend from the prisoner-of-war camp, Ludwig Hänsel, himself a teacher, and a figure well known in Viennese educational circles. On at least one occasion he considered giving the course up, because, he told Hänsel, of the bad relations between himself and his fellow men. Hänsel, perceptively, attributed this to Wittgenstein's chronic sensitivity. 'There is no wall between you and your fellow men', he wrote. 'I have a thicker crust around me.'

At the Lehrerbildungsanhalt Wittgenstein would have been taught in accordance with the principles of the School Reform Movement, which was, under the leadership of the education minister, Otto Glöckel, attempting to reshape the education of the new, post-war, republic of Austria. It was a movement fired with secular, republican and socialist ideals, and it had attracted the good will, and even the participation, of a good number of well known Austrian intellectuals. It was not, however, a movement with which Wittgenstein himself could readily identify. It was not the idea of fitting pupils to live in a democracy that had inspired him to become a teacher, such social and political motives being alien to the fundamentally religious morality that he shared with Engelmann.

Hänsel, too, was a religious man, and for that very reason at odds with the School Reform Movement. He was to become a leading light in a Conservative-Catholic organization called *Der Bund Neuland*, which sought to reform education while maintaining, and indeed increasing, the influence of the Catholic Church. Wittgenstein no more identified himself with this movement, however, than he did with the Glöckel programme. In the struggle between clericals and socialists that dominated the public life of post-war Austria, Wittgenstein occupied an ambivalent position. He shared with the socialists a dislike of the Catholic establishment and a general egalitarianism, while firmly rejecting their secularism and their faith in social and political change. In the politically turbulent and increasingly polarized world of the 1920s, however, such ambivalence and aloofness would always be liable to be misunderstood: to conservative clericals, his contempt for convention was sufficient to establish him as a socialist; to socialists, on the other hand, his individualism and fundamentally religious outlook identified him as a clerical reactionary.

Wittgenstein, then, was trained within the Glöckel programme while distancing himself from some of its objectives. He felt sufficiently uncertain of his standing at the college to ask Hänsel what he had heard said of him by the lecturers there. The entire faculty, Hänsel reported, were united in their praise of him; he was regarded as a serious, capable student-teacher, who knew just what he was doing. The teachers of all his classes – educational theory, natural history, handwriting and music – were all pleased with his work. 'The professor of psychology said with great self-satisfaction that he was very pleased with the noble Lord Wittgenstein.'

Throughout his year as a student-teacher Wittgenstein saw Hänsel regularly, sometimes in the company of their fellow prisoner of war, Michael Drobil. With Hänsel he discussed not only educational matters, but also philosophy. As a learned *Hofrat Direktor*, Hänsel maintained a keen interest in the subject, and in his lifetime published some twenty articles on philosophical subjects (mostly ethics). In a letter of 23 May we find him providing Wittgenstein with a summary of the three kinds of object (actual, ideal and real) distinguished by the 'Critical Realist' O. Külpe, in his book *Die Realisierung*. What, precisely, Wittgenstein's interest in this might have been must remain a mystery, since Külpe is nowhere referred to again. However, further evidence of Wittgenstein's preoccupation at this time with the competing metaphysics of idealism and realism is provided by a letter from Frege – the last Frege is known to have written to Wittgenstein – dated 3 April.

Frege was evidently responding to criticisms Wittgenstein had made of his essay 'The Thought', in which Wittgenstein had spoken of 'deep grounds' for idealism. 'Of course I don't take exception to your frankness', Frege began:

> But I would like to know what deep grounds for idealism you think I have not grasped. I take it that you yourself do not hold the idealist theory of knowledge to be true. So, I think, you recognise that there can, after all, be no deep grounds for this idealism. The grounds for it can then only be apparent grounds, not logical ones.

The rest of this long letter is taken up with an analysis by Frege of the lack of clarity of the *Tractatus*. This time he concentrates solely on the first proposition: 'The world is everything that is the case.' Assuming, he argues, that the 'is' in this statement is the 'is of identity', and further assuming that it is meant to convey information and not simply to provide a definition of 'the world', then, in order for it to mean anything, there must be some way of identifying the sense of 'the world' and that of the phrase 'everything that is the case' *independently* of the statement of their identity. How is this to be done? 'I would be glad', he wrote, 'if you, by answering my questions, could facilitate my understanding of the results of your thinking.'

This is the last preserved communication between the two. Frege died four years later, presumably no nearer to understanding a word of the famous book inspired by his own work. The 'deep grounds' for idealism which Wittgenstein perceived are undoubtedly connected with the account of the world which he gives in propositions 5.6–5.641 of the *Tractatus*. 'The world is *my* world', 'I am my world. (The microcosm.)', and yet I am not *in* my world: 'The subject does not belong to the world; rather it is a limit of the world.' Thus, solipsism, 'when its implications are followed out strictly', coincides with pure realism: 'The self of solipsism shrinks to a point without extension, and there remains the reality co-ordinated with it.' The realism of Frege is thus seen to coincide with the idealism of Schopenhauer and the solipsism of Weininger.

It is a view that gives a philosophical underpinning to the religious individualism adopted by Wittgenstein and Engelmann. I *am* my world, so if I am unhappy about the world, the *only* way in which I can do anything decisive about it is to change myself. 'The world of the happy man is a different one from that of the unhappy man.'

Nevertheless, in a sense Frege was right to find the metaphysics of this view unintelligible. On Wittgenstein's own theory, its expression in words can lead only to nonsense. And yet, though he was unable to explain it to Frege, unable to convince Russell of its truth, and unable to find a publisher for its expression as the outcome of a Theory of Logical Symbolism, Wittgenstein remained firmly convinced of its unassailability. Though he had suffered greatly from 'external' causes in the last year – the death of Pinsent, the defeat of the Habsburg Empire, the problems of publishing his book – he looked only to an 'internal' solution. What, in the final analysis, did it matter if his book remained unpublished? By far the most important thing was to 'settle accounts with himself'.

Thus in the summer, after completing his training as a teacher and after abandoning his book to Russell, he concentrated on what was, to him, the most immediate task: the struggle to overcome his own unhappiness, to combat the 'devils within' that pulled him away from the 'world of the happy man'. To this end he spent the summer working as a gardener at the Klosterneuburg Monastery, just outside Vienna. Working solidly the whole day through seemed to act as a kind of therapy. 'In the evening when the work is done,' he told Engelmann, 'I am tired, and then I do not feel unhappy.' It was a job to which he could bring his customary competence with practical, manual tasks. One day the Abbot of the monastery passed him while he was at work and commented: 'So I see that intelligence counts for something in gardening too.'

However, the therapy was only partially successful. 'External' causes of suffering continued to confine Wittgenstein to the 'world of the unhappy man'. 'Every day I think of Pinsent', he wrote to Russell in August. 'He took half my life away with him. The devil will take the other half.' As the summer vacation drew to an end, and his new life as a primary school teacher beckoned, he had, he told Engelmann, 'grim forebodings' about his future life:

> For unless all the devils in hell pull the other way, my life is bound to become very sad if not impossible.

9
'AN ENTIRELY RURAL AFFAIR'

Though not inspired with the reforming zeal of the adherents to Glöckel's programme, Wittgenstein entered the teaching profession with a still more idealistic set of intentions, and a rather romantic, Tolstoyan conception of what it would be like to live and work among the rural poor.

In keeping with his general ethical *Weltanschauung*, he sought, not to improve their external conditions, but to better them 'internally'. He wanted to develop their intellects by teaching them mathematics, to extend their cultural awareness by introducing them to the great classics of the German language, and to improve their souls by reading the Bible with them. It was not his aim to take them away from their poverty; nor did he see education as a means to equip them for a 'better' life in the city. He wanted, rather, to impress upon them the value of intellectual attainment for its own sake – just as, conversely, he would later impress upon Cambridge undergraduates the inherent value of manual work.

The ideal that emerges from his teaching, whether in the Austrian countryside or at Cambridge University, is a Ruskinian one of honest toil combined with a refined intelligence, a deep cultural appreciation and a devout seriousness; a meagre income, but a rich inner life.

It was important to him to work in an area of rural poverty. However, as was customary for graduates of the Lehrerbildungsanhalt, he was sent to do his probationary teaching year at a school in Maria Schultz am Semmering, a small, pleasant and relatively prosperous town, famous as a pilgrim centre, in the countryside south of Vienna. After a brief inspection of the place, he decided it would not do. He explained to the astonished headmaster that he had noticed the town had a park with a fountain: 'That is not for me, I want an entirely rural affair.' In that case, the headmaster suggested, he should go to Trattenbach, a village the other side of the neighbouring hills. Wittgenstein at once set off on the ninety-minute hike and found, much to his delight, exactly the sort of place he had in mind.

Trattenbach was small and poor. Those of its villagers who had jobs were employed either at the local textile factory or on the neighbouring farms. Life for these villagers was difficult, especially in the deprived years of the 1920s. Wittgenstein was, however (initially, at any rate), enchanted with the place. Soon after his arrival he wrote to Russell, who was then in China at the start of his year's visiting lectureship at the University of Peking, giving his address proudly as 'LW Schoolmaster Trattenbach', and revelling in the obscurity of his new position:

> I am to be an elementary-school teacher in a tiny village called Trattenbach. It's in the mountains, about four hours' journey south of Vienna. It must be the first time that the schoolmaster at Trattenbach has ever corresponded with a professor in Peking.

To Engelmann, a month later, he was even more enthusiastic. He described Trattenbach as 'a beautiful and tiny place' and reported himself to be 'happy in my work at school'. But, he added darkly, 'I do need it badly, or else all the devils in hell break loose inside me.'

His letters to Hänsel during these first few months are written in a similarly cheerful spirit. He relied upon Hänsel to supply him with reading books for his pupils, and would send him requests to order multiple copies of, for example, Grimm's stories, *Gulliver's Travels*, Lessing's fables and Tolstoy's legends. Hänsel visited him regularly at the weekends, as did Arvid Sjögren, Moritz Nähe (the Wittgenstein family photographer) and Michael Drobil. These visits tended, however, to emphasize the already obvious differences between Wittgenstein and the villagers, including his own colleagues, and it was not long before he became the subject of rumour and speculation. One of his colleagues, Georg Berger, once came across Wittgenstein and Hänsel sitting together in the school office. Wittgenstein immediately demanded to know what was being said about him in the village.

Berger hesitated, but on being pressed told Wittgenstein: 'the villagers take you to be a rich baron'.

Berger omitted to use the word, but it was certainly as an *eccentric* aristocrat that Wittgenstein was regarded. '*Fremd*' (strange) was the word most often used by the villagers to describe him. Why, they asked, should a man of such wealth and culture choose to live among the poor, especially when he showed such little sympathy for their way of life and clearly preferred the company of his refined Viennese friends? Why should he live such a meagre existence?

At first Wittgenstein had lodged in a small room in the local guest-house, 'Zum braunen Hirschen', but he quickly found the noise of the dance music coming from below too much for him, and left. He then made a bed for himself in the school kitchen. There, according to Berger (who was, one suspects, one of the chief sources of the stories told by the villagers about Wittgenstein), he would sit for hours by the kitchen window, watching the stars.

He soon established himself as an energetic, enthusiastic but rather strict schoolmaster. In many ways, as his sister Hermine writes, he was a born teacher:

> He is interested in everything himself and he knows how to pick the most important aspects of anything and make them clear to others. I myself had the opportunity of watching Ludwig teach on a number of occasions, as he devoted some afternoons to the boys in my occupational school. It was a marvellous treat for all of us. He did not simply lecture, but tried to lead the boys to the correct solution by means of questions. On one occasion he had them inventing a steam engine, on another designing a tower on the blackboard, and on yet another depicting moving human figures. The interest which he aroused was enormous. Even the ungifted and usually inattentive among the boys came up with astonishingly good answers, and they were positively climbing over each other in their eagerness to be given a chance to answer or to demonstrate a point.

Despite his misgivings about the School Reform Movement, it was among the reformers, such as Putre and Wilhelm Kundt, the District School Superintendent, that Wittgenstein found most encouragement and support during his career as a teacher. His teaching methods shared some of the basic principles of the Reform Movement, the most important of which was that a child should not be taught simply to repeat what it has been told, but should instead be encouraged to think through problems for itself. Thus practical exercises played a large part in his teaching. The children were taught anatomy by assembling the skeleton of a cat, astronomy by gazing at the sky at night, botany by identifying plants on walks in the countryside, architecture by identifying building styles during an excursion to Vienna. And so on. With everything he taught, Wittgenstein attempted to arouse in the children the same curiosity and questioning spirit that he himself brought to everything in which he took an interest.

This naturally worked better with some children than with others. Wittgenstein achieved especially good results with some of the boys that he taught, and with a select group of his favourite pupils, mainly boys, he gave extra tuition outside school hours. To these children, he became a sort of father figure.

However, to those children who were not gifted, or whose interest failed to be aroused by his enthusiasm, he became not a figure of fatherly kindness, but a tyrant. The emphasis he placed on the teaching of mathematics led him to devote the first two hours of each morning to the subject. He believed that it was never too early to begin algebra, and taught mathematics at a far higher level than was expected of his age group. For some of his pupils, the girls especially, the first two hours of the day were remembered with horror for years afterwards. One of them, Anna Brenner, recalls:

> During the arithmetic lesson we that had algebra had to sit in the first row. My friend Anna Völkerer and I one day decided not to give any answers. Wittgenstein asked: 'What do you have?' To the question what is three times six Anna said: 'I don't know.' He asked me how many metres there were in a kilometre. I said nothing and received a box on my ears. Later Wittgenstein said: 'If you don't know I'll take a child from the youngest class in the school who will know.' After the lesson Wittgenstein took me into the office and asked: 'Is it that you don't want to [do arithmetic] or is it that you can't?' I said 'Yes, I want to.' Wittgenstein said to me: 'You are a good student, but as for arithmetic . . . Or are you ill? Do you have a headache?' Then I lied, 'Yes!' 'Then', said Wittgenstein, 'please,

> please Brenner, can you forgive me?' While he said this he held up his hands in prayer. I immediately felt my lie to be a great disgrace.

As this account illustrates, one respect in which Wittgenstein's methods differed sharply from those recommended by Glöckel's reforms was in his use of corporal punishment. Another girl who was weak at mathematics remembers that one day Wittgenstein pulled her hair so hard that when she later combed it a lot of it fell out. The reminiscences of his former pupils abound with stories of the '*Ohrfeige*' (ear-boxing) and '*Haareziehen*' (hair-pulling) they received at his hands.

As news of this brutality reached the children's parents it contributed to a growing feeling against him. It was not that the villagers disapproved of corporal punishment, nor that such methods of discipline were at all unusual, despite Glöckel's recommendations. However, though it was accepted that an unruly boy should have his ears boxed if he misbehaved, it was not expected that a girl who could not grasp algebra should receive the same treatment. Indeed, it was not expected that she *should* grasp algebra.

The villagers (including some of his own colleagues) were, in any case, disposed to take a dislike to this aristocratic and eccentric stranger, whose odd behaviour sometimes amused and sometimes alarmed them. Anecdotes about his *fremdheit* were told and retold, until he became a kind of village legend. There is the story, for example, of how he once got together with two of his colleagues to play a Mozart trio – himself on clarinet, Georg Berger playing the viola part on a violin and the headmaster, Rupert Köllner, playing the piano part. Berger recalls:

> Again and again we had to start from the beginning, Wittgenstein not tiring at all. Finally we were given a break! The headmaster, Rupert Köllner, and I were then so unintentionally inconsiderate as to play by heart some dance tune. Wittgenstein reacted angrily: '*Krautsalat! Krautsalat!*' he cried. He then packed up and went.

Another story concerns the time he attended a catechism at the local Catholic Church. He listened carefully to the questions put to the children by the priest, with the Dean in attendance, and then said suddenly, and very audibly: 'Nonsense!'

But the greatest wonder – and the story for which he was most remembered by the village – concerns the time he repaired the steam engine in the local factory, using an apparently miraculous method. The story is told here by Frau Bichlmayer, the wife of one of Wittgenstein's colleagues, who herself worked at the factory:

> I was in the office when the engine went dead and the factory had to stand idle. In those days we were dependent on steam. And then a lot of engineers came, who couldn't get it to go. Back at home I told my husband what had happened and my husband then told the story in the school office and the teacher Wittgenstein said to him: 'Could I see it, could you obtain permission for me to take a look at it?' Then my husband spoke to the director who said yes, he could come straight away . . . so then he came with my husband and went straight down into the engine room and walked around, saying nothing, just looking around. And then he said: 'Can I have four men?' The director said: yes, and four came, two locksmiths and two others. Each had to take a hammer and then Wittgenstein gave each of the men a number and a different place. As I called they had to hammer their particular spot in sequence: one, four, three, two . . .
>
> In this way they cured the machine of its fault.

For this 'miracle' Wittgenstein was rewarded with some linen, which he at first refused, and then accepted on behalf of the poorer children in his school.

The villagers' gratitude for this miracle, however, did not outweigh their growing mistrust of his *fremdheit*, and throughout the autumn term relations between him and them gradually deteriorated. During this term his sister Hermine kept a watchful and motherly eye on the progress of his new career. She had to do this indirectly, through Hänsel, because, while Wittgenstein welcomed visits from his Viennese friends, his family were under strict instructions not to see him or to offer him any help. Food parcels were returned unopened, and letters left unanswered.

Hänsel was able to reassure Hermine that, though under some strain, Wittgenstein had got through the first term reasonably well. On 13 December, she wrote to him with obvious relief:

> I am indeed very grateful to you for your kind letter. Firstly it reassured me about the anguish Ludwig has endured through the Trattenbachers and their curiosity; his letters of that time give a very encouraging impression and with his laconic way of writing they are doubly reassuring. Secondly, I greatly appreciate everything you say about my brother, although it is, in fact, nothing other than what I myself think. Of course, it is true what you say, though it is not easy having a saint for a brother, and after the English expression: 'I had rather be a live dog than a dead philosopher' I would like to add: I would (often) rather have a happy *person* for a brother than an unhappy *saint*.

Ironically, just a few weeks after this letter, on 2 January 1921, Wittgenstein wrote to Engelmann berating himself for not having chosen the heavenly course:

> I was sorry not to have seen you at Christmas. It struck me as rather funny that you should want to hide from *me*, for the following reason: I have been morally dead for more than a year! From that you can judge for yourself whether I am fine or not. I am one of those cases which perhaps are not all that rare today: I had a task, did not do it, and now the failure is wrecking my life. I ought to have done something positive with my life, to have become a star in the sky. Instead of which I remained stuck on earth, and now I am gradually fading out. My life has really become meaningless and so it consists only of futile episodes. The people around me do not notice this and would not understand; but I know that I have a fundamental deficiency. Be glad of it, if you don't understand what I am writing here.

In the event, Engelmann did not understand. If Wittgenstein felt he had an unfinished task to accomplish, he replied, why did he not do it now – or at least at some future time when he was ready to do so? Furthermore, it was surely wrong for him to talk of a *fundamental* deficiency; as they had discussed before, no one is so lost that their position is irrevocable. This time, however, Engelmann's letter struck the wrong note. 'I cannot at present analyse my state in a letter', Wittgenstein wrote to him. 'I don't think – by the way – that you quite understand it . . . a visit from you would not suit me in the near future. Just now we would hardly know what to do with one another.'

For the time being, at least, Engelmann's place as the person to whom Wittgenstein turned for an understanding of his inner life had been taken by Hänsel. In his memoir of Wittgenstein, Hänsel writes: 'One night, while he was a teacher, he had the feeling that he had been called but had refused.' This perhaps explains Wittgenstein's mention to Engelmann of a task, the fulfilment of which would have brought him to the heavens, but the neglect of which condemned him to remain earth-bound.*

---

*This connects also with a dream quoted by Bartley (from where we do not know), which he says came to Wittgenstein 'possibly in early December 1920'. The dream is as follows:

> I was a priest. In the front hall of my house there was an altar; to the right of the altar a stairway led off. It was a grand stairway carpeted in red, rather like that at the Alleegasse. At the foot of the altar, and partly covering it, was an oriental carpet. And certain other religious objects and regalia were placed on and beside the altar. One of these was a rod of precious metal.
>
> But a theft had occurred. A thief entered from the left and stole the rod. This had to be reported to the police, who sent a representative who wanted a description of the rod. For instance, of what sort of metal was it made? I could not say; I could not even say whether it was of silver or of gold. The police officer questioned whether the rod had ever existed in the first place. I then began to examine the other parts and fittings of the altar and noticed that the carpet was a prayer rug. My eyes began to focus on the border of the rug. The border was lighter in colour than the beautiful centre. In a curious way it seemed to be faded. It was, nonetheless, still strong and firm.

This is the part of Bartley's book that most strongly indicates that in writing it he had access to a manuscript of Wittgenstein's. Bartley not only quotes the dream as though it were described by Wittgenstein himself; he also gives interpretations of it suggested by both Wittgenstein and 'some other party, possibly Hänsel'. Furthermore, unlike the 'Prater episodes', Bartley's information – the content of the dream, its timing, and even the interpretations given of it by Hänsel and Wittgenstein – connects plausibly with information from other sources. Bartley even gives us Wittgenstein's reaction to Hänsel's interpretation (which connects the symbolism of the dream to images taken from the Old Testament):

> It puzzled Wittgenstein to think that *if* such an interpretation were to be attached to the dream, it would be *his* dream.

This reaction, too, is very plausible. According to Bartley, Wittgenstein himself was inclined to interpret the dream in alchemical terms. The rod is at once a phallic symbol (his 'base self') and a symbol of alchemical transformation (the base metal

Or, to be more specific, stuck in Trattenbach. During the spring and summer terms of 1921 Wittgenstein's earlier delight with Trattenbach gradually turned to disgust, as his attempts to educate the village children above the customary expectations met with increasing misunderstanding and resistance from the parents, from the children themselves (those of them who felt unable to meet Wittgenstein's high expectations) and from his own colleagues.

In March he received from Russell a reply to his enthusiastic letter of September. 'I wonder how you like being an elementary school-teacher', Russell wrote, 'and how you get on with the boys':

> It is honest work, perhaps as honest as there is, and everybody now a-days is engaged in some form of humbug, which you escape from.

Russell himself was in good humour, enjoying Peking, and revelling in the occasional affront he caused to (British) conventional morality by openly living 'in sin' with Dora Black. 'I like China and the Chinese', he told Wittgenstein:

> They are lazy, good-natured, fond of laughter, very like nice children – They are very kind and nice to me – All the nations set upon them and say they mustn't be allowed to enjoy life in their own way – They will be forced to develop an army and navy, to dig up their coal and smelt their iron, whereas what they want to do is to make verses and paint pictures (very beautiful) and make strange music, exquisite but almost inaudible, on many stringed instruments with green tassels. Miss Black and I live in a Chinese house, built around a courtyard, I send you a picture of me at the

---

changing into gold or silver), a transformation of which Wittgenstein is unable to convince his conscience, represented by the doubting police.

Thus, if we may be allowed to conflate Wittgenstein's letter to Engelmann, Hänsel's reminiscence and the dream quoted by Bartley, we arrive at a convincing account of the profound change in his temperamental state that is evident during the Christmas holiday period of 1921. Because he could not convince himself that the transformation in himself that he so desired could actually take place, he refused to follow what he regarded as a call to become a priest. The refusal could be explained only by a 'fundamental deficiency', for otherwise the longed-for transformation would surely be possible. He really *was* base metal; he *had* to remain stuck on earth.

> door of my study. My students are all Bolsheviks, because that is the fashion. They are amazed with me for not being more of a Bolshevik myself. They are not advanced enough for mathematical logic. I lecture to them on Psychology, Philosophy, Politics and Einstein. Once in a while I have them to an evening party and they set off fire-works in the courtyard. They like this better than lectures.

Wittgenstein at once let Russell know that his earlier enchantment with Trattenbach had given way to disgust for its inhabitants. 'I am sorry you find the people in your neighbourhood so disagreeable', Russell replied. 'I don't think average human nature is up to much anywhere, and I dare say wherever you were you would find your neighbours equally obnoxious.' No, insisted Wittgenstein, 'here they are much more good-for-nothing and irresponsible than elsewhere'. Russell remained unconvinced:

> I am very sorry you find the people of Trattenbach so trying. But I refuse to believe they are worse than the rest of the human race: my logical instinct revolts against the notion.

'You are right', Wittgenstein at last conceded; 'the Trattenbachers are not uniquely worse than the rest of the human race':

> But Trattenbach is a particularly insignificant place in Austria and the *Austrians* have sunk so miserably low since the war that it's too dismal to talk about. That's what it is.

Russell had told Wittgenstein that he had left the manuscript of the *Tractatus* in England with Dorothy Wrinch, a friend of his, 'a good mathematician and a student of mathematical logic', with instructions to try and get it printed. 'I am determined to get your manuscript published', he affirmed, 'and if it has not been achieved during my absence, I will take the matter in hand as soon as I return.'

Apart from this encouraging news, the one bright spot in Wittgenstein's life during the summer term of 1921 was his relationship with one of his pupils, a boy from one of the poorest families in the village, called Karl Gruber. Gruber was a gifted boy who responded well to Wittgenstein's methods. Like many of Wittgenstein's pupils, he initially found algebra difficult. 'I could not grasp', he recalled later, 'how one could calculate using letters of the alphabet.' However, after receiving from Wittgenstein a box on the ears, he began to knuckle down: 'Soon I was the best at algebra in the class.' At the end of the summer term, he was due to leave the school and start work at the local factory. Wittgenstein was determined to do all he could to continue the boy's education. On 5 July he wrote to Hänsel explaining Gruber's position and asking for advice. Given that his parents could not afford to send him to a boarding school, what could be done? Might a free or a cheap place be found for him in one of the middle schools in Vienna? 'It would in my opinion', he wrote, 'be a great pity for the lad if he could not develop himself further.' Hänsel replied suggesting the possibility of the Calasanzverein, a Catholic establishment in Vienna which took on poor students. In the meantime, however, it was decided that Wittgenstein himself should continue to give the boy lessons, even after he had left the school, and that Hänsel should act as his occasional examiner, testing him to see that he reached the standard required to enter one of the Gymnasiums in Vienna.

In the summer vacation Wittgenstein travelled to Norway with Arvid Sjögren. It was the first time he had been there since 1914, and during the visit he was finally able to see the house that had been built for him in his absence. They left with very little money, and had to spend a night *en route* in a Salvation Army hostel in Hamburg. It was, as he explained in a letter to Hänsel, a working holiday: 'I work from early in the morning until the evening in a kind of carpentry workshop and together with Arvid make crates. In that way I earn myself a heap of money.' As ever, though, the reward he sought for his hard work was peace of mind. 'I think it is very good that I made this journey', he told Hänsel.

Shortly after his return to Trattenbach, Wittgenstein learnt from Russell that his book, finally, was to be published. Russell had returned from China with Dora Black in August, the latter six months pregnant, and his first two months back in England were taken up with arrangements to secure the legitimacy of his child. In China he had been in boat-burning mood, writing to Trinity to resign from the lectureship he had been offered ('because', he later said, 'I was living in open sin') and arranging a divorce from his wife, Alys. But the imminent arrival of a possible heir to the earldom impelled him to take steps towards respectability. He received his decree absolute from Alys on 21 September, married Dora six days later, and the baby, John Conrad, the future 4th Earl Russell, was born on 16 November.

Having taken the necessary steps to ensure that his son would inherit his title, Russell was able to turn his attention to arranging the publication of Wittgenstein's book. Through his friend C. K. Ogden he secured its publication in English in a series of monographs produced by Kegan Paul called The International Library of Psychology, Philosophy and Scientific Method, of which Ogden had recently been made editor. The book was still perceived as a financial liability, but a tolerable one. 'As they can't drop less than £50 on doing it I think it very satisfactory to have got it accepted', Ogden wrote to Russell on 5 November, 'though of course if they did a second edition soon and the price of printing went suddenly down they might get their costs back.'

Independently of these negotiations, Russell's friend Dorothy Wrinch had, while Russell was still in China, secured the book's acceptance in a German periodical called *Annalen der Naturphilosophie*, edited by Wilhelm Ostwald. Russell, knowing how Wittgenstein felt about the piece in its German translation, had left his introduction to the book with Miss Wrinch on the assumption that she would try English publishers. However, after having it rejected by Cambridge University Press, Miss Wrinch – considering, no doubt correctly, that this was her only chance of success – had approached the editors of three German periodicals. Only from Ostwald had she received a positive reply, and then only because of Russell's introduction. 'In any other case I should have declined to accept the article', Ostwald wrote to her on 21 February:

> But I have such an extremely high regard for Mr Bertrand Russell, both for his researches and for his personality, that I will gladly publish Mr Wittgenstein's article in my *Annalen der Naturphilosophie*: Mr Bertrand Russell's Introduction will be particularly welcome.

On 5 November, having received the proofs from Ostwald and a promise from Ogden that it would appear in the Kegan Paul series, Russell wrote to Wittgenstein to let him know what was happening.

He told him Ostwald would publish his introduction: 'I am sorry, as I am afraid you won't like that, but as you will see from his letter, it can't be helped.'

In a phrase which possibly shocked Wittgenstein, Russell told him: 'As for me, I am now married to Miss Black, and expecting a child in a few days':

> We have bought this house [31 Sydney Street, London], and got your furniture from Cambridge, which we like very much. The child will probably be born in your bed.

He urged Wittgenstein to come to England, offering to pay his expenses as further recompense for the furniture: 'Your things are worth much more than I paid for them, and I will pay you more whenever you like. I didn't know when I bought them how much I was getting.' In a later letter he calculated that he owed Wittgenstein a further £200: 'I don't see why I should swindle you because Jolley understated the value of your things.'

Wittgenstein replied on 28 November: 'I must admit I am pleased my stuff is going to be printed', he wrote. 'Even though Ostwald is an utter charlatan':

> As long as he doesn't tamper with it! Are you going to read the proofs? If so, please take care that he prints it exactly as I have it. He is quite capable of altering the work to suit his own taste – putting it into his idiotic spelling, for example. What pleases me most is that the whole thing is going to appear in England.

Russell evidently had little time to read the proofs carefully, and in any case the book had already gone to print before he received them. The proofs were therefore left uncorrected. Far from altering the work to suit his own taste, Ostwald – without, apparently, any interest in or concern for the meaning of the work he was publishing – simply had it printed exactly as it was in typescript. Thus one finds, for example – besides many more ordinary misprints – typewriter symbols where one would expect to find symbols of Russellian logic: '!' for the Sheffer stroke; '/' for the negation sign (and occasionally also for the Sheffer stroke); and the capital letter C for material implication.

Wittgenstein was not consulted by Ostwald at any stage in the publication; nor was he sent any offprints. On being told by Russell that it was finally in print, he had to write to Hänsel to ask him to search for a copy of *Annalen der Naturphilosophie* in the Viennese bookshops. The search was unsuccessful, and it was not until April of the following year, when he was sent a copy by Ogden, that Wittgenstein finally saw how his work had been printed. He was horrified. He regarded it, so he told Engelmann, as a 'pirated edition', and it was not until the English edition appeared, in 1922, that he considered his work to have been properly published.

The wheels for the English edition were set in motion by Russell when, on 6 December, he wrote to Ogden again, sending him Wittgenstein's letter of 28 November:

> Enclosed from Wittgenstein gives all the authority needed for going ahead, so you can tell the publishers it is all right . . . I am much relieved that W. takes the whole affair sanely.

During the winter months of 1921–2, using an offprint of Ostwald's edition, the book was translated into English by Frank Ramsey, then an eighteen-year-old undergraduate at King's, who was a friend of Ogden's and was already recognized as a mathematician of outstanding promise.

Wittgenstein received Ramsey's translation towards the end of March, together with a questionnaire asking for his opinion on particular points that had puzzled both Ogden and Ramsey. In some cases, these puzzles were the outcome of Ostwald's careless printing of the German text; in others, they were due to a faulty understanding of Wittgenstein's intended meaning. Which was which was impossible for Wittgenstein to tell, as he had still not seen a copy of Ostwald's edition. Indeed, he was, by now, doubtful that Ostwald had even printed it – or that he would.

The task of correcting the translation was therefore long and difficult, but by 23 April Wittgenstein had completed a detailed list of comments and suggestions, which he sent to Ogden. In the main his suggestions were motivated by a desire to make the English as natural as possible, and to relax the literalness of Ramsey's translation. Not only was he forced to define particular German words and phrases; he also had to explain what *he* had meant by them and then find an English expression that captured the same meaning and tone. Thus, to a certain extent, the English version is not simply a translation from the German, but a reformulation of Wittgenstein's ideas.

The first question Ogden had raised concerned the title. Ostwald had published it under Wittgenstein's German title, *Logisch-Philosophische Abhandlung*, which, when translated literally, produces the rather awkward. 'Logico-Philosophical Treatise'. Russell had suggested 'Philosophical Logic' as an alternative, while Moore – in a conscious echo of Spinoza's *Tractatus Theologico-Politicus* – had put forward 'Tractatus Logico-Philosophicus' as 'obvious and ideal'. It was not, of course, a title that would reassure the public of the book's accessibility, and Ogden felt slightly uneasy about it. 'As a selling title, he told Russell, '*Philosophical Logic* is better, if it conveys the right impression.'

The matter was settled by Wittgenstein. 'I think the Latin one is better than the present title', he told Ogden:

> For although 'Tractatus logico-philosophicus' isn't *ideal* still it has something like the right meaning, whereas 'Philosophic logic' is wrong. In fact I don't know what it means! There is no such thing as philosophic logic. (Unless one says that as the whole book is nonsense the title might as well be nonsense too.)

The suggestions and comments made by Wittgenstein were given careful consideration by Ogden (who, in his correspondence with Wittgenstein, emerges as the most scrupulous and accommodating editor an author could wish for), and the text was altered in the light of them. By May, work on the English text was more or less complete.

One problem remained. At the time of preparing the typescript Wittgenstein had written a series of supplementary remarks, which were, with one exception, not included in the final text. These supplementary remarks were numbered, and the exception was No. 72, which was intended to be proposition 4.0141, an elaboration of the preceding remark comparing the pictorial relation between language and the world with the relation between a musical thought, a gramophone record and a musical score. However, in Ostwald's edition proposition 4.0141 reads, rather bizarrely: '*(Siehe Ergänzung Nr. 72)*'. He had evidently either lost or never received the supplementary list, and presumably found this no more unintelligible than the other propositions in the book. It was left for Ogden to query Ramsey's translation: '(See Supplement No. 72)'. 'What is this?' asked Ogden. 'There is presumably some mistake.'

In his reply, Wittgenstein explained about the supplements and provided Ogden with a translation of the one he had intended to include in the book. This raised in Ogden's mind the intriguing possibility that there might be more supplements to elucidate and expand what was, after all, a rather difficult – and short – book.

Wittgenstein refused to send any more. 'There can be no thought of printing them', he told Ogden. 'The supplements are exactly what must *not* be printed. Besides THEY REALLY CONTAIN NO ELUCIDATIONS AT ALL, but are still less clear than the rest of my props':

> As to the shortness of the book I am *awfully sorry for it; but what can I do?* If you were to squeeze me like a lemon you would get nothing more out of me. To let you print the Ergänzungen would be no remedy. It would be just as if you had gone to a joiner and ordered a table and he had made the table too short and now would sell you the shavings and sawdust and other rubbish along with the table to make up for its shortness. (Rather than print the Ergänzungen to make the book fatter leave a dozen white sheets for the reader to swear into when he has purchased the book and can't understand it.)

In June, when the book was ready to print, Wittgenstein was sent by Ogden a declaration to sign giving Kegan Paul all publication rights of the book, 'In consideration of their issuing it in German and English in the "International Library of Psychology & Philosophy" under the title *Tractatus Logico-Philosophicus*'. Under the terms of this contract, Wittgenstein was paid nothing for the rights of the book and entitled to no royalties from its sales. When, in 1933, a reprint was planned, he attempted to persuade Kegan Paul to pay him a royalty, but they did not respond, which is why he took his later work to a different publisher. At the time, however, he was less concerned with payment than with ensuring that Fanny Pinsent, David's mother, should be sent a complimentary copy. With every letter he wrote to Ogden during the final stages of publication, he asked him to trace Mrs Pinsent and to make sure she received a copy.

The proofs were ready in July, and Wittgenstein returned them, duly corrected, in the first week of August. The publishers seem to have wanted to print some details of Wittgenstein's biography and the peculiar circumstances in which the book was written, mentioning the prison camp at Monte Cassino, and so on. To this Wittgenstein responded with scathing contempt. 'As to your note about the Italian monastery etc. etc.', he wrote to Ogden on 4 August, 'do as you please':

> . . . only I can't for my life see the point of it. Why should the general reviewer know my age? Is it as much as to say: You can't expect more of a young chap especially when he writes a book in such a noise as must have been on the Austrian front? If I knew that the general reviewer believed in astrology I would suggest to print the date and hour of my birth in front of the book that he might set *the horoscope* for me. (26/IV 1889, 6 p.m.)

By the time the book was published, Wittgenstein had left Trattenbach. He had hinted to Russell as early as 23 October that this was to be his last year there, 'because I don't get on here even with the other teachers', and from then on his life in Trattenbach became progressively more difficult. He was determined to raise the sights of at least the more able of his students, and the private lessons he was giving Karl Gruber were extended to include also some of his better pupils from his new class. These included Emmerich Koderhold and Oskar Fuchs. From the parents of all three he encountered resistance. When he wanted to take Fuchs to Vienna to see a play, he was refused, Fuchs's mother not wishing to entrust her boy to 'that crazy fellow'. When he suggested to Koderhold's father that his son had the ability to attend a grammar school in Vienna, and should do so, he was told that it was out of the question; the boy was needed to help run the farm. His biggest disappointment, however, was with Karl Gruber, the most talented of his students. Every day after school, from four o'clock until half past seven, Wittgenstein conducted Gruber through an intensive study, concentrating on Latin, mathematics, geography and history. From time to time Gruber's progress was examined by Hänsel, especially in Latin, the subject Wittgenstein felt least qualified to teach. The plan was to see Gruber through a grammar school in Vienna. While attending school Gruber was to live with Hermine, and herein lay a difficulty: 'I would have felt it as a humiliation', Gruber later explained:

> I didn't want to beg for alms and would have felt myself to be receiving charity. I would have come there as a 'poor chap' and would have had to say thank you for every bit of bread.

Perhaps for this reason, or perhaps simply because he had been worn down by the effort of studying for three and a half hours every day while working at the local factory, receiving nothing but discouragement from his family, Gruber told Wittgenstein he did not want to continue with his lessons. On 16 February 1921 Wittgenstein wrote to Hänsel: 'Today I had a conversation with Gruber who came to me to bring some books back. It turns out that he has no enthusiasm to go on with his studies . . . Of course he has no conception of where he is now heading. i.e. he does not know how bad a step he is taking. But how should he know. Sad! Sad!'

'I wish you didn't have to work so hard at elementary teaching', Russell wrote to him on 7 February; 'it must be very dreary.' Wittgenstein replied that he had indeed felt very depressed lately, but not because he found teaching in an elementary school repugnant: 'On the contrary!'

> But it is *hard* to have to be a teacher in this country where the people are so completely and utterly hopeless. In this place I do not have a soul with whom I can exchange a single reasonable word. God knows how I will be able to stand it for much longer!

Russell had written of how he 'liked China much better than Europe': 'the people are more civilized – I keep wishing I were back there'. Yes, Wittgenstein replied, 'I can well believe that you found it more pleasant in China than in England, although in England it is doubtless a thousand times better than here.'

In his correspondence with Ogden, also, there are some indications that he was already beginning to look towards England, in order to be with at least a few people to whom he could talk. In his letters he frequently asks after, and asks to be remembered to, his old friends at Cambridge – Johnson and Keynes in particular.

Throughout the summer term he looked forward with great anticipation and pleasure to a proposed meeting with Russell, who was planning to visit the Continent to stay with his brother and wife at their home in Switzerland. Originally, the plan was for Wittgenstein to join the Russells there, but this was changed in favour of meeting for an over-night stay in Innsbruck. The tone of the letters that were exchanged to make this arrangement is warm and friendly, and gives no hint of the differences between the two that were to emerge. They exchanged comments on the baleful situation of Europe, told each other how much they were looking forward to their meeting, and Wittgenstein asked affectionately after Russell's wife and baby ('The little boy is lovely', Russell replied. 'At first he looked exactly like Kant, but now he looks more like a baby.')

And yet the meeting proved to be a great disappointment on both sides, and was, in fact, the last time the two met as friends. According to Dora Russell, it was the 'circumstances of the time' that made it a 'troubled meeting'. Inflation in Austria was then at its height, and: 'The whole place was full of ghouls and vultures, tourists profiting by the cheap currency to have a good time at the Austrians' expense':

> We all tramped the streets trying to find rooms in which to stay; Wittgenstein was in an agony of wounded pride at the state of his country and his inability to show some sort of hospitality.

Eventually, they took a single room, the Russells occupying the bed while Wittgenstein slept on the couch. 'But the hotel had a terrace, where it was pleasant enough to sit while Bertie discussed how to get Wittgenstein to England.' She strenuously denies that they quarrelled on this occasion: 'Wittgenstein was never easy, but I think any differences must have been over their philosophical ideas.'

Russell himself, however, remembers the differences as religious. Wittgenstein was, he said, 'much pained by the fact of my not being a Christian', and was at the time: 'at the height of his mystic ardour'. He: 'assured me with great earnestness that it is better to be good than clever', but was, nevertheless (Russell seems to see an amusing paradox here), 'terrified of wasps and, because of bugs, unable to stay another night in lodgings we had found in Innsbruck'.

In later life Russell gave the impression that, after their meeting at Innsbruck, Wittgenstein considered him too wicked to associate with, and so abandoned all contact. Russell enjoyed being thought wicked, and this is no doubt the aspect of the meeting that stayed freshest in his memory. Wittgenstein did, indeed, disapprove of his sexual mores, and had before their meeting in Innsbruck attempted to steer him in the direction of religious contemplation by suggesting he read Lessing's *Religiösen Streitschriften* (a suggestion Russell did not take up). But it is not true that Wittgenstein avoided all contact with Russell after they had met at Innsbruck; he wrote at least two letters to him in the months following the meeting, each of which begins: 'I have heard nothing from you for a long time.'

The indications are, then, that it was Russell who broke off communication. Perhaps the truth is that he found Wittgenstein's religious earnestness too tiresome to tolerate. For, if it is true that Wittgenstein was at the 'height of his mystic ardour', it is equally true that Russell was at the height of his atheist acerbity. Gone was the Ottoline-inspired transcendentalism of 'The Essence of Religion' and 'Mysticism and Logic'; in its place was a fierce anti-Christianity, which, in his now familiar role as a public speaker and a popular writer, he never lost an opportunity to express.

There is, too, the related, and perhaps still deeper, difference upon which Engelmann places so much emphasis: the difference between trying to improve the world, and seeking only to improve oneself. And, again, it is not just that Wittgenstein had become more introspective and individualistic, but that Russell had become much less so. The war had made him a socialist, and had convinced him of the urgent need to change the way the world was governed; questions of personal morality were subordinated by him to the overriding public concern to make the world a safer place. There is a story told by Engelmann which illustrates this difference in its starkest form, and which must surely refer to the meeting at Innsbruck:

> When, in the 'twenties, Russell wanted to establish, or join, a 'World Organization for Peace and Freedom' or something similar, Wittgenstein rebuked him so severely, that Russell said to him: 'Well, I suppose *you* would rather establish a World Organization for War and Slavery', to which Wittgenstein passionately assented: 'Yes, rather that, rather that!'

If this is true, then it may well have been Russell who regarded Wittgenstein as too wicked to associate with. For there can be no more complete repudiation of the ethical view upon which he based the rest of his life's activities.

In any case, Russell made no further attempt to communicate with

Wittgenstein or to persuade him to come to England. If Wittgenstein was to escape the 'odiousness and baseness' of the Austrian peasantry, it would not be through his old teacher at Cambridge.

That Wittgenstein's spell as an elementary school teacher in Trattenbach had not been a success was due in large measure to his very devotion to the task. His high expectations and his stern means of enforcing them had baffled and frightened all but a minority of his pupils; he had aroused the hostility of their parents and had failed to get on even with his own colleagues. And, as he had been forced to admit by Russell, there was nothing uniquely evil about the people at Trattenbach – he was very likely to encounter the same reaction elsewhere.

There are some indications that, if he could have found something better to do, he would have left school-teaching altogether. As well as talking to Russell about returning to England, he also discussed the possibility with Engelmann of a 'flight to Russia'. What he would have done in either place, he did not know. Certainly not philosophy – he had said all he had to say about *that* in his book.

In the event, September 1922 saw him starting at a new school in the same area as Trattenbach, this time a secondary school in a village called Hassbach. He did so without any great hopes. Before he started there he reported to Engelmann that he had formed 'a very disagreeable impression of the new environment there (teachers, parish priest, etc.)'. These people, he said, 'are not human *at all* but loathsome worms'. It had perhaps been thought that he would find it easier to get on with secondary school teachers, but in fact he found their pretence to 'specialized learning' utterly unbearable, and soon wished to return to an elementary school. He stayed barely a month.

In November he started at a primary school in Puchberg, a pleasant village in the Schneeberg mountains, now a popular skiing resort. Again, he found it difficult to discern any humanity in the people around him; in fact, he told Russell, they were not really people at all, but one-quarter animal and three-quarters human.

He had not been at Puchberg long before he at last received finished copies of the *Tractatus*. He wrote to Ogden on 15 November: 'They really look nice. I wish their contents were half as good as their external appearance.' He wondered whether Johnson – the first two volumes of whose three-volume work on logic had also recently been published – would buy it: 'I should like to know what *he* thinks about it. If you see him please give him my love.'

There was, naturally, no one at Puchberg with whom he could discuss philosophy, but he did, at least, find someone with whom he could share his passion for music in Rudolf Koder, a very talented pianist, who taught music at the school. On hearing Koder playing the 'Moonlight' Sonata, Wittgenstein walked into the music room and introduced himself. From then on the two would meet almost every afternoon to play duets for clarinet and piano – the clarinet sonatas by Brahms and Labor, and arrangements of the clarinet quintets of Brahms and Mozart.

Later, they were joined in these musical sessions by a local coal-miner called Heinrich Postl, a member of the village choir. Postl, who became a good friend and a kind of protégé of Wittgenstein's, was later employed as a porter and caretaker by the Wittgenstein family. Wittgenstein gave him copies of some of his favourite books – Tolstoy's *Gospel in Brief* and Hebel's *Schatzkästlein* – and sought to impress upon him his own moral teaching. Thus, when Postl once remarked that he wished to improve the world, Wittgenstein replied: 'Just improve yourself; that is the only thing you *can* do to better the world.'

Aside from Koder and Postl, Wittgenstein made few friends among the staff and villagers at Puchberg. As at Trattenbach, his teaching inspired a few of his pupils to heights they otherwise would not have attained, and antagonized the parents because of the disruption caused to their work at home.

While Wittgenstein was struggling to teach primary school children, the *Tractatus* was becoming the subject of much attention within the academic community. At Vienna University the mathematician Hans Hahn gave a seminar on the book in 1922, and it later attracted the attention also of a group of philosophers led by Moritz Schlick – the group that evolved into the famous Vienna Circle of Logical Positivists. In Cambridge, too, the *Tractatus* became the centre of discussion for a small but influential group of dons and students. The first public discussion of the book in Cambridge was probably in January 1923, when Richard Braithwaite addressed the Moral Science Club on the subject of 'Wittgenstein's logic as expounded in his *Tractatus Logico-Philosophicus*'.

For a time, Wittgenstein's only contact at Cambridge remained

Ogden, who in March sent Wittgenstein his recently published book *The Meaning of Meaning*, written jointly with the poet and literary critic I. A. Richards. Ogden regarded the book as providing a causal solution to the problem of meaning addressed by Wittgenstein in the *Tractatus*. Wittgenstein regarded it as an irrelevance. 'I think I ought to confess to you frankly', he wrote, 'that I believe you have not *caught the problems* which – for instance – I was at in my book (whether or not I have given the correct solution).' In a letter to Russell of 7 April, he went further:

> A short time ago I received 'The Meaning of Meaning'. It has surely also been sent to you. Is it not a miserable book?! Philosophy is not as easy as that! From this one sees how easy it is to write a thick book. The worst thing is the introduction of Professor Postgate Litt.D.F.B.A. etc. etc. I have seldom read anything so foolish.

It was the second letter Wittgenstein had written to Russell since their ill-starred meeting at Innsbruck, and he was impatient for a reply. 'Write to me sometime', he pleaded, 'how everything's going with you and what your baby's up to; whether he is already studying logic fluently.'

Russell appears not to have replied. Wittgenstein's categoric dismissal of Ogden's work possibly irritated him, since he himself saw little to criticize in the book. It was, in many ways, simply a restatement of what he himself had already said in *The Analysis of Mind*. Shortly afterwards, Wittgenstein was shocked to read in *The Nation* a favourable review by Russell of the book, describing it as 'undoubtedly important'. From Frank Ramsey he learnt that Russell 'does not really think *The Meaning of Meaning* important, but he wants to help Ogden by encouraging the sale of it' – an explanation certain to have increased Wittgenstein's disapproval, and to have confirmed him in his growing belief that Russell was no longer *serious*. In the 1930s Wittgenstein once or twice attempted (unsuccessfully) to interest Russell in the philosophical work that he was then doing, but he never again addressed Russell warmly, as a friend.

Increasingly isolated as he was ('To my great shame', he wrote to Engelmann, 'I must confess that the number of people to whom I can talk is constantly diminishing'), Wittgenstein *needed* friends. When, through Ogden, he was sent Keynes's 'Reconstruction in Europe' –

published as a special supplement to the *Manchester Guardian* – he tried writing directly to Keynes to thank him. 'I should have preferred to have got a line from you personally', he told him, 'saying how you are getting on, etc':

> Or, are you too busy to write letters? I don't suppose you are. Do you ever see Johnson? If so, please give him my love. I should so much like to hear from him too (*not* about my book but about himself).
>
> So do write to me sometime, if you will condescend to do such a thing.

It took Keynes over a year to reply. 'Did Keynes write to me?' Wittgenstein asked Ogden on 27 March 1923. 'If so, please tell him it hasn't reached me.' He even gave Ogden his Puchberg address again – despite having given it him twice before – just in case Keynes's letter had been misdirected.

It was Keynes who could (and eventually did) persuade Wittgenstein to return to England. In the meantime, contact with Cambridge was kept up through a friend of Keynes's, a fellow Apostle and member of King's College: Frank Ramsey.

Of the people at Cambridge who studied the *Tractatus* in its first year of publication, Ramsey was undoubtedly the most perceptive. Though still an undergraduate (in 1923 he was still just nineteen years old), he was commissioned to write a review of Wittgenstein's work for the philosophical journal, *Mind*. The review remains to this day one of the most reliable expositions, and one of the most penetrating criticisms, of the work. It begins in Russellian vein:

> This is a most important book containing original ideas on a large range of topics, forming a coherent system, which whether or not it be, as the author claims, in essentials the final solution of the problems dealt with, is of extraordinary interest and deserves the attention of all philosophers.

But Ramsey then goes on to take issue with some of the misunderstandings contained in Russell's introduction – for example, Russell's misconception that Wittgenstein was concerned with the possibility of a 'logically perfect language' – and to give a fuller and more reliable exposition of the main lines of the book.

When Wittgenstein heard from Ogden that Ramsey intended to visit Vienna in the summer vacation of 1923, he wrote to Ramsey himself, inviting him to Puchberg. Ramsey gratefully accepted, and arrived on 17 September, not quite knowing what to expect. He stayed about two weeks, during which time Wittgenstein devoted about five hours a day – from when he finished school at two o'clock in the afternoon until seven in the evening – to going through the *Tractatus* line by line with him. 'It is most illuminating', Ramsey wrote to Ogden; 'he seems to enjoy this and we get on about a page an hour':

> He is very interested in it, although he says that his mind is no longer flexible and he can never write another book. He teaches in the village school from 8 to 12 or 1. He is very poor and seems to lead a dreary life having only one friend here, and being regarded by most of his colleagues as a little mad.

In going through the book in such detail, Wittgenstein made some corrections and changes to the text which were incorporated in later editions. For both Wittgenstein and Ramsey, it was important that Ramsey should understand the book thoroughly, in every last detail. Wittgenstein was concerned lest Ramsey should forget everything when he returned to England – as Moore had appeared to have done when he came to Norway in 1914. 'It's terrible', Ramsey wrote to his mother, 'when he says "Is that clear" and I say "no" and he says "Damn it's *horrid* to go through that again."'

Ramsey intended to make Wittgenstein's work the basis for a theory of higher mathematics. When they had finished going through the book, he wrote, 'I shall try to pump him for ideas for its further development which I shall attempt':

> He says he himself will do nothing more, not because he is bored, but because his mind is no longer flexible. He says no one can do more than 5 or 10 years work at philosophy. (His book took 7.) And he is sure Russell will do nothing more important.

Wittgenstein seemed to support Ramsey's plan, at least to the extent of agreeing that *something* should take the place of Russell's *Principia*

Mathematica. He struck Ramsey as 'a little annoyed' that Russell was planning a new edition of *Principia*: 'because he thought that he had shown R that it was so wrong that a new edition would be futile. It must be done altogether afresh.'

As for Wittgenstein's present living conditions, Ramsey was somewhat dismayed:

> He is very poor, at least he lives very economically. He has one *tiny* room whitewashed, containing a bed, washstand, small table and one hard chair and that is all there is room for. His evening meal which I shared last night is rather unpleasant coarse bread, butter and cocoa.

He was, however, impressed by Wittgenstein's youthful appearance and his athletic vigour. 'In explaining his philosophy he is excited and makes vigorous gestures but relieves the tension by a charming laugh.' He was inclined to think that Wittgenstein 'exaggerates his own verbal inspiration', but of his genius he had no doubt:

> He is great. I used to think Moore a great man but beside W!

From Wittgenstein's point of view, the discussions with Ramsey provided a stimulating and pleasant – if strenuous – change from his normal routine, and also a welcome link with Cambridge. He told Ramsey that he was likely to leave Puchberg at the end of the school year, but had no firm idea of what he would do after that – perhaps find a job as a gardener, or perhaps come to England to look for work. He asked Ramsey to investigate whether he was entitled to receive a BA degree from Cambridge on the basis of the six terms he had spent with Russell before the war, with perhaps the *Tractatus* being accepted as a BA thesis.

On Ramsey's return to Cambridge for the Michaelmas term, he and Wittgenstein entered into a warm and friendly correspondence. In one of his first letters Ramsey explained (what he had found out from Keynes) that the regulations governing the eligibility of Cambridge degrees had changed. It was no longer possible to obtain a BA degree by keeping six terms' residence and submitting a thesis. If Wittgenstein wanted a degree he would have to come back to Cambridge for at least another year and then submit a thesis. In this way he could hope to obtain a Ph.D.

Through Ramsey, Keynes tried to entice Wittgenstein to England by offering him £50 to pay his expenses. He at first tried to make this offer anonymously, but, on being asked directly, Ramsey had to admit: 'the £50 belong to Keynes':

> He asked me not to say so straight away because he was afraid you might be less likely to take it from him than from an unknown source, as he has never written to you. I can't understand why he hasn't written, nor can he explain, he says he must have some 'complex' about it. He *speaks of you with warm affection and very much wants to see you again*.

Ramsey even wrote to Wittgenstein's nephew, Thomas Stonborough (whom he had got to know at Cambridge), to convince him of the same point: 'Keynes very much wants to see L.W. again and his offer of £50 is really better evidence of that, than his failure to answer letters is of the contrary. He speaks of L.W. with considerable affection.'

This marks the beginning of a long campaign to persuade Wittgenstein, first to visit England for a summer holiday, and then to abandon teaching and resume philosophical work at Cambridge. Ramsey did his best to allay Wittgenstein's fears about entering Cambridge society after such a long absence – an absence in which he had changed a great deal and had lived, to a large extent, away from any kind of society. On 20 December he wrote that he could quite understand this fear, 'but you mustn't give it any weight':

> I could get lodgings in Cambridge and you need not see more of people than you like or feel able to. I can see that staying with people might be difficult as you would inevitably be with them a lot, *but if you lived by yourself you could come into society gradually*.
>
> I don't want you to take this as endorsing your fear of boring or annoying people, for I know *I myself want to see you awfully*, but I just want to say that if you have such a fear surely it would be all right if you were not staying with anyone but lived alone first.

As Ramsey realized later, this line of attack was fruitless – the last thing Wittgenstein wanted was to live *alone* in England. But in any case, by February 1924 he gave up trying to persuade Wittgenstein to come to England for the summer and instead told him of his plan to go to Vienna.

Ramsey had for some time been interested in the prospect of being psychoanalysed. Originally this had been because of the emotional turmoil occasioned by an 'unhappy passion' for a married woman. In the Lent term of 1924 he returned to the idea after suffering from depression. This, together with a desire to have a break from Cambridge before he started his intended academic career, culminated in the decision to spend six months in Vienna. His choice of Vienna was not simply dictated by his desire to undergo psychoanalysis, but was also influenced by the fact that, while there, he could see Wittgenstein regularly to discuss his work.

In connection with his own work, he had recently been to see Russell to help him with the new edition of *Principia Mathematica*. Russell gave him the manuscript of the revisions that he intended to include in the new edition, in order for him to comment on them. What criticisms Ramsey made are unrecorded. The new introduction states merely that 'the authors' (referring to Russell and Whitehead, although in fact Russell alone was responsible for the changes) were 'under great obligations' to Ramsey.

To Wittgenstein, however, Ramsey was quite scathing of the project:

> You are quite right that it is of no importance; all it really amounts to is a clever proof of mathematical induction without using the axiom of reducibility. There are no fundamental changes, identity as it used to be. I felt he was too old: he seemed to understand and say 'yes' to each separate thing, but it made no impression so that 3 minutes afterwards he talked on his old lines. Of all your work he seems now to accept only this: that it is nonsense to put an adjective where a substantive ought to be which helps in his theory of types.

The new edition, indeed, seemed to please no one. While Wittgenstein and Ramsey thought it paid too little attention to Wittgenstein's criticisms, Whitehead considered it too Wittgensteinian and published a paper dissenting from the new ideas that Russell had included.

*

Ramsey went to Vienna in March. He travelled with Thomas Stonborough, and on the journey was briefed by him on some salient facts about the Wittgenstein family – that three of Wittgenstein's brothers had committed suicide, and that three sisters and a fourth brother remained, all of whom lived in Vienna. Having met Thomas Stonborough, it must have become apparent to Ramsey that his assessment of Wittgenstein as 'very poor' would have to be slightly amended. In Paris he was introduced to Jerome Stonborough, Thomas's father, who, he told his mother, 'looked just a prosperous American'.

In Vienna Ramsey saw for himself the scale of the Wittgenstein family wealth when he made the acquaintance of Margarete, who was at this time living in the Schönbrunn Palace: 'She must be collossally wealthy.' He was invited to a dinner party at the palace the following Saturday: 'As far as I could make out the party consisted of Wittgensteins, mostly female, professors, and friends of Tommy, the son, mostly male. So there was a good majority of males.' Music was provided by a professional string quartet, who played first Haydn and then Beethoven. Ramsey preferred the Haydn, but was told that this gave him away – 'what I didn't mind as I couldn't avoid it sooner or later'. After dinner he talked to Paul Wittgenstein – 'a brother, who is a celebrated pianist who lost an arm in the war and now plays with one hand. Lionel had heard of him without connecting him with Ludwig' – and was invited out to lunch by Paul and Hermine.

Having met the family, Ramsey had a better understanding of the completely self-inflicted nature of Wittgenstein's situation. He wrote to Keynes to explain that it was probably no good 'trying to get him to live any pleasanter life, or stop the ridiculous waste of his energy and brain':

> I only see this clearly now because I have got to know one of his sisters and met the rest of the family. They are very rich and extremely anxious to give him money or do anything for him in any way, and he rejects all their advances; even Christmas presents or presents of invalid's food, when he is ill, he sends back. And this is not because they aren't on good terms but because he won't have any money he hasn't earned except for some very specific purpose like to come and see you again. I think he teaches to earn money and would only stop teaching if he had some other way of earning

> which was preferable. And it would have to be really earning, he wouldn't accept any job which seemed in the least to be wangled for him. It is an awful pity.

He even put forward the basis of a psychological explanation: 'it seems to be the result of a terribly strict upbringing. Three of his brothers committed suicide – they were made to work so hard by their father: at one time the eight children had twenty-six private tutors; and their mother took no interest in them.'

At the end of his first week in Vienna, Ramsey travelled to Puchberg to spend a day with Wittgenstein. His mind was chiefly on his psychoanalysis, and he had not intended to talk to Wittgenstein about his work on the foundations of mathematics. It appears, however, that he made some effort to do so, but found Wittgenstein's response disappointing. 'Wittgenstein seemed to me tired', he wrote to his mother, 'though not ill; but it isn't really any good talking to him about work, he won't listen. If you suggest a question he won't listen to your answer but starts thinking of one for himself. And it is such hard work for him like pushing something too heavy uphill.'

After his visit to Puchberg, Ramsey wrote to Keynes underlining the importance of getting Wittgenstein out of the hostile environment in which he had placed himself:

> . . . if he were got away from his surroundings and were not so tired, and had me to stimulate him, he might do some more very good work; and he might conceivably have come to England with that in view. But while he is teaching here I don't think he will do anything, his thinking is so obviously frightfully uphill work as if he were worn out. If I am here during his summer holiday I might try to stimulate him then.

It appears Wittgenstein had asked Ramsey to write to Keynes explaining his attitude to visiting England, convinced that he could not express the matter adequately in English, and that Keynes wouldn't understand it if he wrote in German. Wittgenstein, Ramsey explained, had severe misgivings about coming to England to renew old acquaintances. He felt he could no longer talk to Russell, and the quarrel with Moore had remained unhealed; there remained only Keynes and Hardy. He wanted very much to get to know Keynes again, but only if he could renew their old intimacy; he did not want to come to England and see Keynes only occasionally and establish only a superficial acquaintanceship. He had changed so much since the war, he felt, that unless he spent a lot of time with Keynes, Keynes would never understand him.

He would therefore be prepared to come to England *if* Keynes were prepared to invite him to stay as a guest in his country home, and were willing to spend a great deal of time getting to know him again.

Ramsey ended his explanation of this with a warning:

> I'm afraid I think you would find it difficult and exhausting. Though I like him very much I doubt if I could enjoy him for more than a day or two, unless I had my great interest in his work, which provides the mainstay of our conversation.

But, he added, 'I should be pleased if you did get him to come and see you, as it might possibly get him out of this groove.'

For the time being, Keynes did not respond to the suggestion that he should invite Wittgenstein to spend the summer with him in the country; he possibly considered the demands involved too great. He had, however – on 29 March, apparently before seeing Ramsey's letter – finally replied to Wittgenstein's letter of the previous year. He explained the long delay as being caused by his desire to understand the *Tractatus* before he wrote: 'yet my mind is now so far from fundamental questions that it is impossible for me to get clear about such matters':

> I still do not know what to say about your book, except that I feel certain it is a work of extraordinary importance and genius. Right or wrong, it dominates all fundamental discussions at Cambridge since it was written.

He sent Wittgenstein some of his recent books, including *The Economic Consequences of the Peace*, and urged him to come to England, stressing: 'I would do anything in my power which could make it easier for you to do further work.'

This last statement, for the moment at least, struck the wrong note. It was not that Wittgenstein wanted to resume philosophical work, but that he wanted badly to re-establish old friendships. He replied only in July, writing half in English, half in German, and insisting there was nothing that could be done to enable him to return to philosophy:

> . . . because I myself no longer have any strong inner drive towards that sort of activity. Everything that I really *had* to say, I have said, and so the spring has run dry. That sounds queer, but it's how things are.

On the other hand, he told Keynes, if he had any work to do in England, even sweeping the streets or cleaning boots, 'I would come over with great pleasure'. Without such a job, the only thing that would make it worthwhile for him to come would be if Keynes were prepared to see him on something more than a casual basis. It would be nice, he said, to see Keynes again, but: 'staying in rooms and having tea with you every other day or so would not be *nice enough*'. It would be necessary, for the reasons that Ramsey had already outlined, for them to work hard at establishing an intimate relationship:

> We haven't met since 11 years. I don't know if you have changed during that time, but *I* certainly have tremendously. I am sorry to say I am no better than I was, but I am *different*. And therefore if we shall meet you may find that the man who has come to see you isn't really the one you meant to invite. There is no doubt that, even if we *can* make ourselves understood to one another, a chat or two will *not* be sufficient for the purpose, and that the result of our meeting will be disappointment and disgust on your side and disgust and despair on mine.

As it was, no such complications arose, because no such invitation was forthcoming. Wittgenstein spent the summer in Vienna.

He had already decided that the summer term of 1924 would be his last at Puchberg, although he appears to have been relatively happy at this time. When Ramsey visited him in May he reported to his mother that Wittgenstein seemed more cheerful: 'he has spent weeks preparing the skeleton of a cat for his children, which he seemed to enjoy'. 'But', he wrote, 'he is no good for my work.'

Ramsey's respect for Wittgenstein had not diminished in any way. Later, he was to write:

> We really live in a great time for thinking, with Einstein, Freud and Wittgenstein all alive (and all living in Germany or Austria, those foes of civilisation!).

But although he stayed in Austria throughout the summer, he made little effort to see much of Wittgenstein. When Ogden wrote to him asking for the corrections to the text of the *Tractatus* made during the discussions of the previous year, he replied that he wouldn't be seeing Wittgenstein again until September, shortly before his return to England. Ogden apparently wanted the material in case a new edition should be published, but at the time this looked an unlikely prospect. Ramsey's letter ends: 'I'm sorry so few have sold.'

Ramsey spent the summer completing his course in psychoanalysis and working on his dissertation. While still in Vienna he received the news that, at the extraordinarily early age of twenty-one, he was to become a Fellow of King's College upon his return to Cambridge. Before he left he made only one more visit to Wittgenstein. He told him beforehand: 'I don't much want to talk about mathematics as I haven't been doing much lately.'

This was, in all probability, a polite way of saying that, so long as Wittgenstein continued 'this ridiculous waste of his energy and brain', he was likely to remain 'no good' for Ramsey's work.

In what was to be his last attempt to raise the sights of the children of rural Austria, and to withstand the hostility of their parents and of his fellow teachers, Wittgenstein, in September 1924, started at yet another village school, this time at Otterthal, a neighbouring village of Trattenbach.

Given his experience of Trattenbach, it is perhaps surprising that he should have chosen to return to the Wechsel mountains. But there was some hope that he would enjoy better relations with his colleagues. So, at least, Hermine thought. Almost as soon as Wittgenstein moved to Otterthal, she wrote to Hänsel asking if he had any plans to visit her brother. 'I would, naturally,' she said, 'be very happy if someone could tell me how Ludwig is getting on there, I mean how relations with the school are':

> It *cannot*, I think, be completely without friction, since his teaching programme is so different from that of the other teachers, but at

least one might hope that the friction should not result in his being *ground to dust*.

The head of the school in Otterthal was Josef Putre, whom Wittgenstein had befriended while at Trattenbach. Putre was a socialist and an enthusiastic proponent of Glöckel's School Reform Movement, and in his first two years of teaching Wittgenstein had often turned to him for advice.

There were, of course, differences of opinion between himself and Putre, particularly concerning the role of religion in education. While Putre discouraged praying in schools, Wittgenstein prayed with his pupils every day. When Putre once remarked that he was against paying lip-service to the Catholic faith, and considered it meaningless, Wittgenstein replied: 'People kiss each other; that too is done with the lips.'

Despite his friendship with Putre, Wittgenstein knew within a month that he would find it no easier at Otterthal than he had at Trattenbach. 'It's not going well here', he wrote to Hänsel in October, 'and perhaps now my teaching career is coming to an end':

> It is too difficult for me. Not one but a dozen forces are against me, and what am I?

It was while at Otterthal, however, that Wittgenstein produced what is arguably his most lasting contribution to educational reform in Austria – a contribution that is, furthermore, fully in line with the principles of Glöckel's programme. That is his *Wörterbuch für Volksschulen*, a spelling dictionary for use in elementary schools. The origin of his desire to publish such a book seems to lie in his asking Hänsel to enquire into the cost of dictionaries for use in schools. In the letter to Hänsel quoted above, he says:

> I had never thought the dictionaries would be so frightfully expensive. I think, if I live long enough, I will produce a small dictionary for elementary schools. It appears to me to be an urgent need.

The need for such a dictionary was well recognized by the authorities. There were at the time only two dictionaries available, both of them designed for the task of teaching students to spell. One was too big and too expensive to be used by children at the kind of rural schools at which Wittgenstein taught. The other was too small and badly put together, containing many foreign words which the children were unlikely ever to use, and omitting many words commonly mis-spelt by children. At Puchberg Wittgenstein had overcome this difficulty by getting his pupils to produce their own dictionaries. During German lessons and PE lessons when the weather prevented them from going outside, Wittgenstein wrote words on the blackboard and had the children copy them into their own vocabulary books. These vocabulary books were then sewn together and bound with cardboard covers to produce the finished dictionary.

In discussing this solution to the problem in the preface to the published dictionary, Wittgenstein remarks:

> He who works at the practical level is able to understand the difficulties of this work. Because the result should be that each student receives a clean and, if at all possible, correct copy of the dictionary, and in order to reach that goal the teacher has to control almost every word each student has written. (It is not enough to take samples. I do not even want to talk about the demands on discipline.)

Although he comments on the astonishing improvement of spelling which resulted ('The orthographic conscience had been awakened!'), he clearly had no desire to repeat what had obviously been an arduous and trying task. The *Wörterbuch* was envisaged as a more practical solution to the problem, both for himself and for other teachers in the same position.

In contrast to the *Tractatus*, the publication of the dictionary was achieved quickly and without any great problems. In November 1924, Wittgenstein contacted his former principal at the Lehrerbildungsanhalt, Dr Latzke, to inform him of the plan. Latzke contacted the Viennese publishing house of Hölder-Pichler-Tempsky, who on 13 November wrote to Wittgenstein to say that they would be willing to publish the dictionary. The manuscript was delivered during the Christmas vacation of 1924, and Wittgenstein was sent the proofs the following February.

Wittgenstein's preface is dated 22 April 1925. In it he explains the need for such a dictionary and the considerations that dictated the selection of words and their arrangement. He makes clear that these considerations are based on his own experience as a teacher. 'No word is too common to be entered,' he says, 'since I have experienced that *wo* has been written with the "h" that indicates a long vowel, and *was* with "ss".' It is clear from the preface that Wittgenstein intended his dictionary specifically to meet the needs of elementary schools in rural Austria. Thus, while some perfectly good German words have been omitted because they are not used in Austria, some Austrian dialectical expressions have been included. Dialect is also used to explain distinctions which Wittgenstein's experience had shown were often confused, such as the difference between *das* and *dass*, and the distinction between the accusative *ihn* and the dative *ihm*.

Before the publishers could go ahead with printing the book, they needed assurance that it would be recommended for use in the schools for which it was intended. They therefore submitted it for approval to the provincial board of education for Lower Austria. The report for the board was written by District School Inspector Eduard Buxbaum. In his report, dated 15 May, Buxbaum agrees with Wittgenstein about the need for such a dictionary, and goes so far as to describe this need as 'the most pressing question at the present time'. He also agrees with Wittgenstein's emphasis on words which belong to 'the common everyday vocabulary'. He finds fault, however, with Wittgenstein's selection of words, criticizing him for omitting such common words as *Bibliothek* (library), *Brücke* (bridge), *Buche* (beech tree) etc., and also takes exception to Wittgenstein's preface. Dictating a dictionary to students is, Buxbaum remarks, a strange way to control their spelling. It would have been better, he felt, to have dictated the correct spelling of words only after the children had used those words for themselves. He also finds fault with Wittgenstein's own use of the German language: 'By no means should the mistake of writing "*eine mehrmonatliche Arbeit*" instead of saying "*eine Arbeit von viele Monaten*" ["a work of several months"] creep into the German language, not even into the preface.'

Buxbaum concludes:

> One can express the opinion that the dictionary will be a somewhat useful educational tool for the upper classes of elementary schools and of the Bürgerschulen after the cited shortcomings have been removed. It is the opinion of the undersigned that no

> board of education will find the dictionary in its present form recommendable.

After Wittgenstein's preface was omitted and the words which Buxbaum had mentioned were included, the book received its required official approval. In November, a contract between Wittgenstein and the publisher was drawn up, under the terms of which Wittgenstein received 10 per cent of the wholesale price for each copy sold and ten free copies. The book was published in 1926 and enjoyed a limited success. (It was not reprinted, however, until 1977, by which time its interest was confined to Wittgensteinian scholarship.)

As we have seen, soon after he arrived in Otterthal, Wittgenstein became convinced that he would not for very much longer be able to withstand the pressures of trying to teach in a hostile environment. In February 1925 he wrote to Engelmann:

> I suffer much from the human, or rather inhuman, beings with whom I live – in short it is all as usual!

As before, Wittgenstein found an enthusiastic response from a small group of boys who became his favourites. These formed a special group who stayed behind after school for extra tuition and who were known to Wittgenstein by their Christian names. They were taken by Wittgenstein on outings to Vienna and on walks through the local countryside, and were educated to a standard far beyond what was expected in the kind of rural elementary school that they attended. And, as before, their commitment to their education, and Wittgenstein's commitment to them, aroused the hostility of the parents, who turned down Wittgenstein's suggestions that their children should continue their education at a grammar school. Again, the girls proved more resistant to Wittgenstein's methods, and resented having their hair pulled and their ears boxed because they were unable or unwilling to meet Wittgenstein's unrealistically high expectations, especially in mathematics.

In short, indeed, it was all as usual.

Engelmann, too, was finding life in post-war Europe difficult. Like Wittgenstein, he felt himself to belong to an earlier epoch, but unlike Wittgenstein, he characterized that epoch as essentially *Jewish*. In his memoir, he talks of the 'Austrian-Jewish spirit' and of the 'Viennese-Jewish' culture, which was the inheritance of both himself and Wittgenstein. Wittgenstein, as we shall see, saw it differently. But for both, in their different ways, an awareness of their Jewishness was heightened as the epidemic of European anti-Semitism became ever more virulent. In Engelmann's case, this resulted in his becoming a Zionist and looking to the creation of Israel for a new homeland to replace the one destroyed by the First World War. Though never at any time attracted to Zionism (the religious associations of Palestine would always, for him, have had more to do with the New Testament than the Old), Wittgenstein found something to cheer in Engelmann's desire to settle in the Holy Land. 'That you want to go to Palestine', he wrote, 'is the one piece of news that makes your letter cheering and hopeful for me':

> This may be the right thing to do and may have a spiritual effect. I might want to join you. Would you take me with you?*

Shortly after his letter to Engelmann, Wittgenstein received, completely unexpectedly, a letter from Eccles, his friend in Manchester, from whom he had heard nothing since the war (unlike Pinsent, Russell and Keynes, Eccles was not the sort to exchange friendly letters with a member of an enemy army). Eccles's letter was to provide the catalyst Wittgenstein needed to persuade him to visit England. On 10 March he replied with obvious pleasure at the renewal in contact:

> Dear Eccles,
> I was more than pleased to hear from you, for some reason or other I was convinced that you either were killed in the war, or if alive, that you would hate Germans and Austrians so much that you would have no more intercourse with me.
> . . . I wish I could see you again before long, but when and where we can meet God knows. Perhaps we might manage to meet during the summer vacation, but I haven't got much time and *no* money to

* Engelmann eventually left Europe for Tel Aviv in 1934 and remained there (after 1948 as an Israeli citizen) until his death in 1963. Nothing more was mentioned of the idea that Wittgenstein should join him.

> come to England as I have given *all* my money away, about 6 years ago. Last summer I should have come to England to see a friend of mine Mr Keynes (whose name you might know) in Cambridge. He would have paid my expenses, but I resolved after all not to come, because I was so much afraid that the long time and the great events (external and internal) that lie between us would prevent us from understanding one another. However now – or at least *to-day* I feel as if I might still be able to make myself understood by my old friends and if I get any opportunity I might – w.w.p. come and see you at Manchester.

In a later letter, of 7 May, he accepts Eccles's invitation to stay at his home in Manchester, while stressing the point that had prevented him from staying with Keynes the previous summer (the fact that Keynes had not actually invited him seems to have been dismissed by Wittgenstein as irrelevant):

> England may not have changed since 1913 but *I* have. However, it is no use writing to you about that as I couldn't explain to you the exact nature of the change (though I perfectly understand it). You will see it for yourself when I get there. I should like to come about the end of August.

In July, Wittgenstein wrote to Keynes about his projected visit to England, saying that his mind was not quite made up whether to go or not, and hinting that the final decision depended on Keynes: 'I should rather like to, if I could also see *you* during my stay (about the middle of August). Now please let me know FRANKLY if you have the slightest wish to see me.' Keynes evidently replied encouragingly, and even sent Wittgenstein £10 for the journey. Before he left Wittgenstein wrote: 'I'm awfully curious how we are going to get on with one another. It will be exactly like a dream.'

Wittgenstein arrived in England on 18 August, and stayed with Keynes in his country home in Lewes, Sussex before travelling to Manchester to see Eccles. Despite his earlier insistence to Russell that it was better to be good than to be clever, the experience of exchanging the company of rural peasants for that of some of the finest minds in Europe was delightful to him. From Lewes he wrote to Engelmann:

> I know that brilliance – the riches of the spirit – is not the ultimate good, and yet I wish now that I could die in a moment of brilliance.

When he went to Manchester, both Eccles and his wife were surprised at the great change in him. They went to the railway station to meet him, and found in the place of the immaculately dressed young man, the 'favourite of the ladies' they had known before the war, a rather shabby figure dressed in what appeared to them to be a Boy Scout uniform. The appearance of eccentricity was compounded by Wittgenstein's giving Eccles the (false) impression that he had not yet seen a copy of the *Tractatus*. He asked Mrs Eccles to obtain a copy, and after she had tried in vain to buy one from the booksellers in Manchester, Eccles borrowed one from the university library. 'It was during this period', Eccles states confidently, but mistakenly, in his memoir, 'that he obtained his first copy of the English edition of his *Tractatus*.' Evidently, Wittgenstein very much wanted Eccles to see the book, but was too embarrassed to admit that as the reason for their determined search.

At the end of his stay in England Wittgenstein went to Cambridge, where he was finally reunited with Johnson. 'Tell Wittgenstein', Johnson wrote to Keynes on 24 August, 'that I shall be very pleased to see him once more; but I must bargain that we don't talk on the foundations of Logic, as I am no longer equal to having my roots dug up.' He also met with Ramsey, with whom, however, it appears he quarrelled so fiercely that the two did not resume communication until two years later.

Despite his argument with Ramsey, Wittgenstein's trip was a success. It had served the useful purpose of re-establishing contacts with old friends – contacts he intended to make use of in the likely event that life at Otterthal would become unbearable. 'In case of need I shall probably go to England', as he put it to Engelmann. In letters to both Engelmann and Eccles at the beginning of the new school term in September, he speaks of *trying once again* with his 'old job', as though this coming year were to be his very last attempt at teaching in rural schools. 'However', he told Eccles, 'I don't feel so miserable now, as I have decided to come to you if the worst came to worst, which it certainly will sooner or later.' In October he wrote to Keynes in similar vein, saying he would remain a teacher, 'as long as I feel that the troubles into which I get that way, may do me any good':

> If one has toothache it is good to put a hot-water bottle on your face, but it will only be effective, as long as the heat of the bottle gives you some pain. I will chuck the bottle when I find that it no longer gives me the particular kind of pain which will do my character any good. That is, if people here don't turn me out before that time.

'If I leave off teaching', he added, 'I will probably come to England and look for a job there, because I am convinced that I cannot find anything at all possible in *this* country. In this case I will want your help.'

In the event, the worst did come to the worst, and Wittgenstein had to chuck the hot-water bottle perhaps even sooner than he had anticipated. He left Otterthal and gave up teaching altogether very suddenly, in April 1926. The event which precipitated this was much talked about at the time, and was known to the villagers in Otterthal and the surrounding area as '*Der Vorfall Haidbauer*' ('The Haidbauer Case').

Josef Haidbauer was an eleven-year-old pupil of Wittgenstein's whose father had died and whose mother worked as a live-in maid for a local farmer named Piribauer. Haidbauer was a pale, sickly child who was to die of leukaemia at the age of fourteen. He was not the rebellious type, but possibly rather slow and reticent in giving answers in class. One day, Wittgenstein's impatience got the better of him, and he struck Haidbauer two or three times on the head, causing the boy to collapse. On the question of whether Wittgenstein struck the boy with undue force – whether he ill-treated the child – a fellow pupil, August Riegler, has (with dubious logic) commented:

> It cannot be said that Wittgenstein ill-treated the child. If Haidbauer's punishment was ill-treatment, then 80 per cent of Wittgenstein's punishments were ill-treatments.

On seeing the boy collapse, Wittgenstein panicked. He sent his class home, carried the boy to the headmaster's room to await attention from the local doctor (who was based in nearby Kirchberg) and then hurriedly left the school.

On his way out he had the misfortune to run into Herr Piribauer, who, it seems, had been sent for by one of the children. Piribauer is remembered in the village as a quarrelsome man who harboured a deep-seated grudge against Wittgenstein. His own daughter, Hermine, had often been on the wrong side of Wittgenstein's temper, and had once been hit so hard that she bled behind the ears. Piribauer recalls that when he met Wittgenstein in the corridor, he had worked himself up into a fierce rage: 'I called him all the names under the sun. I told him he wasn't a teacher, he was an animal-trainer! And that I was going to fetch the police right away!' Piribauer hurried to the police station to have Wittgenstein arrested, but was frustrated to find that the single officer who manned the station was away. The following day he renewed the attempt, but was informed by the headmaster that Wittgenstein had disappeared in the night.

On 28 April 1926 Wittgenstein handed in his resignation to Wilhelm Kundt, one of the District School Inspectors. Kundt had, naturally, been told of the 'Haidbauer Case', but reassured Wittgenstein that nothing much would come of it. Kundt placed great value on Wittgenstein's ability as a teacher, and did not want to lose him. He advised him to take a holiday to calm his nerves, and then to decide whether he really wished to give up teaching. Wittgenstein, however, was resolute. Nothing would persuade him to stay on. At the hearing which followed, he was, as Kundt had anticipated, cleared of misconduct. But he had, by then, despaired of accomplishing anything more as a teacher in the Austrian countryside.

The Haidbauer incident was not, of course, the cause of this despair, but simply the event that finally triggered its inevitable culmination in Wittgenstein's resignation. The despair itself had deeper roots. Shortly before the incident, Wittgenstein had met August Wolf, an applicant for the post of headmaster at Otterthal, and had told him:

> I can only advise you to withdraw your application. The people here are so narrow-minded that nothing can be achieved.

10
OUT OF THE WILDERNESS

The most natural thing for Wittgenstein to have done in 1926, after things had come to a head in Otterthal, might have been to avail himself of Keynes's hospitality and return to England. In fact, it was over a year before he once more got in touch with Keynes. He had, he then explained, postponed writing until he had got over the great troubles he had experienced.

Although he had expected to leave Otterthal, and to abandon his career as a teacher, the manner in which he did so left Wittgenstein completely devastated. The trial had been a great humiliation, the more so because, in defending himself against charges of brutality, he had felt the need to lie about the extent of corporal punishment he had administered in the classroom. The sense of moral failure this left him with haunted him for over a decade, and led eventually, as we shall see, to his taking drastic steps to purge himself of the burden of guilt.

In this state he could not contemplate returning to England. Nor, for the moment, did he feel able to return to Vienna. He considered, instead, a complete retreat from worldly troubles. Shortly after his retirement from teaching, he called at a monastery to enquire about the possibility of his becoming a monk. It was an idea that occurred to him at various times in his life, often during periods of great despair. On this occasion he was told by an obviously perceptive Father Superior that he would not find what he expected, and that he was, in any case, led by motives which the order could not welcome. As an alternative, he found work as a gardener with the monk-hospitallers in Hütteldorf, just outside Vienna, camping for three months in the tool-shed of their garden. As it had six years earlier, gardening proved an effective therapy, and at the end of the summer he felt able to return to Vienna to face society.

On 3 June 1926, while he was still working as a gardener, his mother, who had been ill for some time, died at the family home in the Alleegasse, leaving Hermine as the acknowledged head of the family. Whether or not this made it easier for Wittgenstein to return to Vienna, or whether his mother's death influenced him in any way, is impossible to say. But it is striking that from this time on there is a profound change in his attitude to his family. The family Christmas celebrations, which in 1914 had filled him with such dread and produced in him such confusion, were now looked forward to by him with delight. Every Christmas from now until the *Anschluss* of 1938 made it impossible for him to leave England, we find him taking part in the proceedings with enthusiasm – distributing gifts to his nieces and nephews, and joining in with the festive singing and dining with no hint that this compromised his integrity.

Wittgenstein's return to Vienna in the summer of 1926, then, appears to mark the end of an estrangement from his family that goes back at least to 1913, when his father died. On his return, he was offered a kind of work-therapy which, unlike his work as a gardener, would impose upon him the obligation to work with others, and help to bring him back into society. It would, furthermore, allow him an opportunity to put into practice his strongly held views on architectural aesthetics. He was asked by his sister Gretl and by Paul Engelmann to become Engelmann's partner in the design and construction of Gretl's new house.

Engelmann had already carried out some work for the Wittgenstein family. He had worked on the renovation of the family house in Neuwaldeggergasse, and had built for Paul Wittgenstein a room in the Alleegasse for the exhibition of his collection of porcelain. Towards the end of 1925 he was approached by Gretl to be the architect for a new town-house to be built on a plot of land she had bought in one of Vienna's least fashionable areas, on Kundmanngasse, in Vienna's Third District (next to the teacher training college Wittgenstein had attended). Wittgenstein's interest in the project was quickly aroused, and during his last year at Otterthal, whenever he returned to Vienna, he would discuss it with Gretl and Engelmann with great intensity and concern, so that it seemed to Engelmann that Wittgenstein understood Gretl's wishes better than he did himself.

The early plans were drawn up by Engelmann during Wittgenstein's last term of teaching, but after he had left Otterthal, it seemed natural to invite him to join him as a partner in the project. From then on, says Engelmann: 'he and not I was the architect, and although the ground plans were ready before he joined the project, I consider the result to be his and not my achievement'.

The final plan is dated 13 November 1926 and is stamped: 'P. Engelmann & L. Wittgenstein Architects'. Though he never had any architectural training, and was involved only in this one architectural job, there are signs that Wittgenstein began to take this designation seriously, and to see in architecture a new vocation, a new way of re-creating himself. For years he was listed in the Vienna city directory as a professional architect, and his letters of the time are written on notepaper headed: 'Paul Engelmann & Ludwig Wittgenstein Architects, Wien III. Parkgasse 18'. Perhaps, though, this is no more than another statement of his personal independence – an insistence on his status as a freelance professional and a denial that his architectural work for his sister was a mere sinecure.

His role in the design of the house was concerned chiefly with the design of the windows, doors, window-locks and radiators. This is not as marginal as it may at first appear, for it is precisely these details that lend what is otherwise a rather plain, even ugly, house its distinctive beauty. The complete lack of any external decoration gives a stark appearance, which is alleviated only by the graceful proportion and meticulous execution of the features designed by Wittgenstein.

The details are thus everything, and Wittgenstein supervised their construction with an almost fanatical exactitude. When a locksmith asked: 'Tell me, Herr Ingenieur, does a millimetre here or there really matter so much to you?' Wittgenstein roared 'Yes!' before the man had finished speaking. During discussions with the engineering firm responsible for the high glass doors which Wittgenstein had designed, the engineer handling the negotiations broke down in tears, despairing of ever executing the commission in accordance with Wittgenstein's standards. The apparently simple radiators took a year to deliver because no one in Austria could build the sort of thing Wittgenstein had in mind. Castings of individual parts were obtained from abroad, and even then whole batches were rejected as unusable. But, as Hermine Wittgenstein recalls:

> Perhaps the most telling proof of Ludwig's relentlessness when it came to getting the proportions exactly right is the fact that he had the ceiling of one of the rooms, which was almost big enough to be a hall, raised by three centimetres, just when it was almost time to start cleaning the complete house.

Gretl was able to move into the house at the end of 1928. According to Hermine, it fitted her like a glove; the house was an extension of Gretl's personality, 'just as from childhood onwards everything surrounding her had to be original and grand'. For herself, however, Hermine had reservations:

> . . . even though I admired the house very much, I always knew that I neither wanted to, nor could, live in it myself. It seemed indeed to be much more a dwelling for the gods than for a small mortal like me, and at first I even had to overcome a faint inner opposition to this 'house embodied logic' as I called it, to this perfection and monumentality.

It is easy to understand this slight abhorrence. The house was designed with little regard to the comforts of ordinary mortals. The qualities of clarity, rigour and precision which characterize it are indeed those one looks for in a system of logic rather than in a dwelling place. In designing the interior Wittgenstein made extraordinarily few concessions to domestic comfort. Carpets, chandeliers and curtains were strictly rejected. The floors were of dark polished stone, the walls and ceilings painted a light ochre, the metal of the windows, the door handles and the radiators was left unpainted, and the rooms were lit with naked light-bulbs.

In part because of this stark monumentality, and also in part because of the sad fate of Austria itself, the house – which had taken so much time, energy and money to build – has had an unfortunate history. Less then a year after Gretl moved in, the Great Crash of 1929 (though it did not by any means leave her destitute) forced her to lay off many of the staff she needed to run the house as it had been intended to be run, and she took to entertaining, not in the hall, but in the kitchen. Nine years later, after the *Anschluss*, she fled from the Nazis to live in New York, leaving the house empty and in the care of the sole remaining servant. In 1945, after the Russians occupied Vienna, the house was used as a barracks for Russian soldiers and as a stable for their horses. Gretl moved back in 1947 and lived there until her death in 1958, when the house became the property of her son, Thomas Stonborough. Sharing Hermine's reservations about its suitability as a home, Stonborough left it empty for many years before finally, in 1971, selling it to a developer for demolition. It was saved from this fate only by a campaign to have it declared a national monument by the Vienna Landmark Commission, and now survives as a home for the Cultural Department of the Bulgarian Embassy in Vienna, though its interior has been extensively altered to suit its new purpose. Were Wittgenstein to see it in its present state – room dividers removed to form L-shaped rooms, walls and radiators painted white, the hall carpeted and wood-panelled, and so on – it is quite possible he would have preferred it to have been demolished.

Through working for Gretl, Wittgenstein was brought back into Viennese society and, eventually, back into philosophy. While the Kundmanngasse house was being built, Gretl and her family continued to occupy the first floor of the Schönbrunn Palace. Her eldest son, Thomas, had recently returned from Cambridge and was now reading for a Ph.D. at the University of Vienna. At Cambridge he had met a Swiss girl by the name of Marguerite Respinger and had invited her to Vienna. With her, Wittgenstein began a relationship which he at least came to regard as a preliminary to marriage, and which was to last until 1931. She was, as far as anybody knows, the only woman with whom he fell in love.

Marguerite was a lively, artistic young lady from a wealthy background, with no interest in philosophy and little of the devout seriousness that Wittgenstein usually made a prerequisite for friendship. Her relationship with Wittgenstein was, presumably, encouraged by Gretl, although some of his other friends and relations were bemused and rather less than pleased by it. She first met Wittgenstein when, after an accident at the building site, he had hurt his foot and was staying with Gretl's family to convalesce. She was part of a group of young people – which included Thomas Stonborough and the Sjögren brothers, Talle and Arvid – which gathered round his bed to listen to him read. He read something from the Swiss writer Johann Peter Hebel, and, she reports: 'I felt again at home and moved by

7 Margarete Wittgenstein, painted by Gustav Klimt, on the occasion of her wedding to Jerome Stonborough, 1905

8 The Technische Hochschule in Charlottenburg, Berlin

9 The family at the dinner table in thc Hochreit. From left to right: the housemaid Rosalie Hermann, Hermine, Grandmother Kalmus, Paul, Margarete, Ludwig

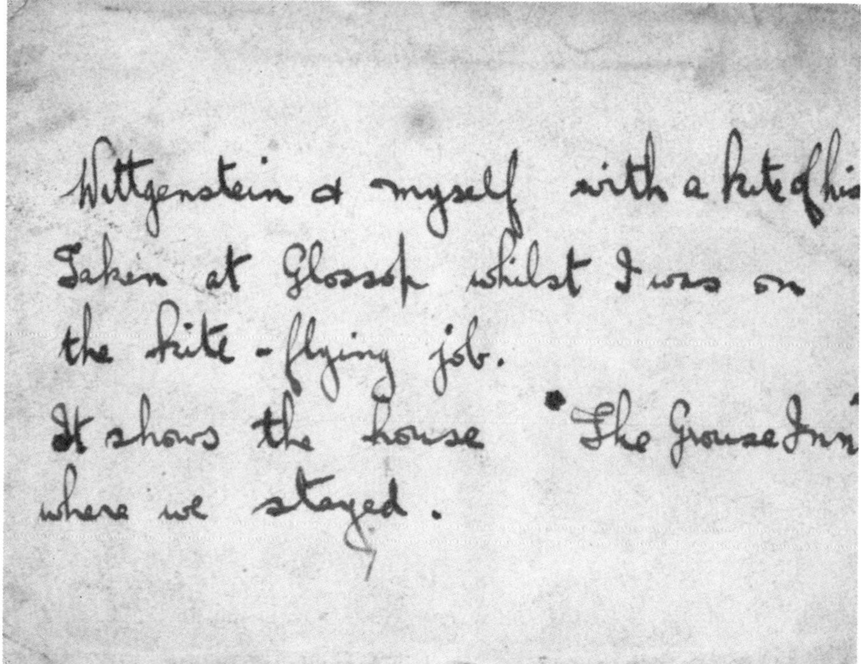

Wittgenstein & myself with a kite of hi
Taken at Glossop whilst I was on
the kite-flying job.
It shows the house "The Grouse Inn"
where we stayed.

10 Ludwig Wittgenstein, aged about eighteen

11 & 12 With Eccles at the Kite-Flying Station in Glossop

13 On the river in Cambridge. On the left, with his arms resting on the top of the seat, is John Maynard Keynes; standing on the right, with the flowered hat and long scarf, is Virginia Woolf; to the left of her, dressed in blazer, white trousers and cravat, is Rupert Brooke

14 Members of the Moral Science Club, Cambridge, *c.* 1913. In the front row, third from left, is James Ward; to the right of him, Bertrand Russell; next to Russell is W. E. Johnson; in the second row, on the far right, is McTaggart; and third from the right, G. E. Moore

15 David Pinsent

353
Bergen.

Bergensbanen
Langs Finsevand

Dear Eccles,
Enclosed please
find a Check
for 23 Norwegian
kroner = ~ 23
& a note from Goodalls
you would
highly oblige
me by paying
the amount to
Goodalls.
Yours ever L.W.
Remember me to Mrs Eccles

16–18 *Left* Postcards from Norway, 1913

19 & 20 *Above* Postcard to Eccles

21–3 Photographs showing Wittgenstein's house in Norway, with sketch-map indicating the position of the house, sent by Wittgenstein to Moore, October 1936

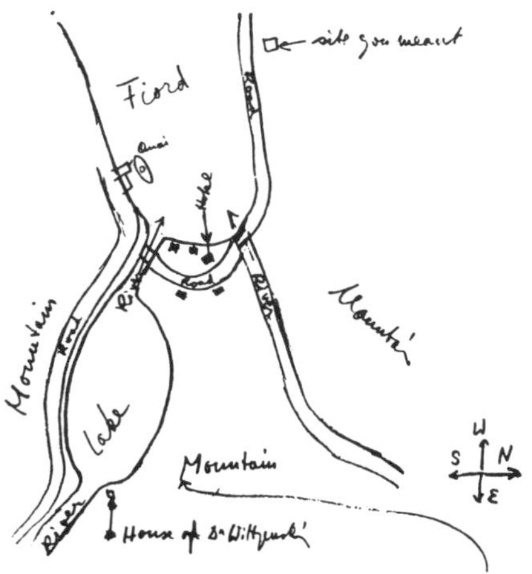

hearing it read with such deep understanding.' Much to the displeasure – and, perhaps, jealousy – of Arvid Sjögren, Wittgenstein's attention was drawn to her. On a similar occasion he asked his audience what they would like him to read, directing his question in particular to Marguerite. 'It doesn't matter what you read', Arvid commented sourly, 'she won't understand it.'

Despite Sjögren's disapproval, Wittgenstein and Marguerite began to see each other almost daily. While she was in Vienna, Marguerite attended the art school, and after her lessons would go to the Kundmanngasse building site to meet Wittgenstein. They would then go together to the cinema to see a Western, and eat together at a café a simple meal consisting of eggs, bread and butter and a glass of milk. It was not quite the style to which she was accustomed. And it required a certain degree of courage for a respectable and fashionable young lady like herself to be seen out with a man dressed, as Wittgenstein invariably was, in a jacket worn at the elbows, an open-neck shirt, baggy trousers and heavy boots. He was, moreover, nearly twice her age. She would on occasion prefer the company of younger, more fashionable, men like Thomas Stonborough and Talle Sjögren. This both puzzled and angered Wittgenstein. 'Why', he would demand, 'do you want to go out with a young thing like Thomas Stonborough?'

To their respective friends a much more puzzling question was why Wittgenstein and Marguerite would want to go out with one another. Arvid Sjögren was not the only close friend of Wittgenstein's who could not get on with her. Another was Paul Engelmann, whom Marguerite disliked in her turn. He was, she says, 'the sort of Jew one didn't like'. 'One' could presumably put up with the Wittgensteins because of their immense wealth, their integration into Viennese society and because they were neither religiously nor fully 'racially' Jewish. But Engelmann was simply *too* Jewish. It may or may not be a coincidence that Wittgenstein's friendship with Engelmann deteriorated at the time that his relationship with Marguerite developed, and that during the time he was in love with her Wittgenstein's attitude to his own Jewishness underwent a profound change.

The relationship was apparently encouraged by Gretl because she considered that Marguerite's company would have a calming and 'normalizing' influence on her brother. This may have been true, and indeed it may have been Marguerite's very lack of intellectual depth that enabled her to exert this influence. Wittgenstein explicitly asked her *not* to try and penetrate his inner world of thought – a request with which she was more than happy to oblige.

Marguerite was used as the model for a bust which Wittgenstein sculpted at this time. The bust, executed in the studios of Michael Drobil, is not exactly a portrait of Marguerite, for although Wittgenstein's interest was primarily in the attitude and expression of the face, it was not her actual expression that he was attempting to capture, but one that he himself was interested in creating. One is reminded – as so often when describing Wittgenstein in love – of what Weininger says in *Sex and Character*:

> Love of a woman is possible only when it does not consider her real qualities, and so is able to replace the actual physical reality by a different and quite imaginary reality.

When the bust was finished it was given to Gretl and displayed in the Kundmanngasse house – an appropriate home for it, for aesthetically it is of a piece with the house. Wittgenstein said of his excursion into architecture:

> . . . the house I built for Gretl is the product of a decidedly sensitive ear and *good* manners, an expression of great *understanding* (of a culture, etc.). But *primordial* life, wild life striving to erupt into the open – that is lacking. And so you could say it isn't *healthy*.

One could say of his sculpture, too, that it lacks 'primordial life'. It thus falls short, on Wittgenstein's own terms, of being a great work of art. For: 'Within all great art there is a WILD animal: *tamed*.' Wittgenstein himself considered the bust as no more than a clarification of Drobil's work.

Even in music, the art for which Wittgenstein had the greatest feeling, he showed above all a great understanding, rather than manifesting 'wild life striving to erupt into the open'. When he played music with others, as he did frequently during this time in Vienna, his interest was in getting it right, in using his acutely sensitive ear to impose upon his fellow musicians an extraordinary exactitude of expression. One could even say that he was not interested in creating music, but in re-creating it. When he played, he was not expressing himself, his own primordial life, but the thoughts, the life, of others.

To this extent, he was probably right to regard himself not as creative, but as reproductive.

Despite Wittgenstein's interest in, and sensitivity to, the other arts, it was only in philosophy that his creativity could really be awakened. Only then, as Russell had long ago noticed, does one see in him 'wild life striving to erupt into the open'.

It was while he was working on Gretl's house that Wittgenstein was brought back to the activity in which he could best express his peculiar genius. Again, Gretl acted as social catalyst, when she brought Wittgenstein into contact with Moritz Schlick, Professor of Philosophy at the University of Vienna.

In bringing these two together Gretl succeeded where Schlick himself had, on more than one occasion over a number of years, failed. He had arrived in Vienna in 1922, the year that the *Tractatus* was published, and was one of the first people in Vienna to read it and to understand its value. In the summer of 1924, after meeting Frank Ramsey at Gretl's house, he wrote to Wittgenstein, addressing his letter to Puchberg:

> As an admirer of your *Tractatus Logico-Philosophicus* I have long intended to get in touch with you. My professorial and other duties are responsible for the fact that I have again and again put off carrying out my intention, though nearly five semesters have passed since I was called to Vienna. Every winter semester I have regular meetings with colleagues and gifted students who are interested in the foundations of logic and mathematics and your name has often been mentioned in this group, particularly since my mathematical colleague Professor Reidemaster reported on your work in a lecture which made a great impression on us all. So there are a number of people here – I am one myself – who are convinced of the importance and correctness of your fundamental ideas and who feel a strong desire to play some part in making your views more widely known.

In the letter Schlick suggested that he should like to visit Wittgenstein in Puchberg. Wittgenstein had, in fact, by this time moved to Otterthal, but the letter eventually found him there, and in his reply he welcomed the possibility of a visit from Schlick. Schlick wrote back quickly, announcing again his intention of coming, but it was not until April 1926, fifteen months later, that he, accompanied by a few chosen pupils, finally made the trip to Otterthal. Schlick's wife has described the spirit in which her husband undertook the journey: 'It was as if he were preparing to go on holy pilgrimage, while he explained to me, almost with awesome reverence, that W. was one of the greatest geniuses on earth.' On arriving in Otterthal the pilgrims were deeply disappointed to be told that Wittgenstein had resigned his post and had left teaching.

Thus Schlick was overjoyed when, in February 1927, he received Gretl's letter inviting him to dinner to meet Wittgenstein. 'Again', says Mrs Schlick, 'I observed with interest the reverential attitude of the pilgrim.' Schlick had, in the meantime, sent Wittgenstein some of his work and had proposed that Wittgenstein join him and some others in discussions on logical problems. In her letter of invitation Gretl responded to this proposal on Wittgenstein's behalf. She told Schlick:

> He asks me to give you his warmest regards and to make his excuses to you, since he feels quite unable to concentrate on logical problems as well as doing his present work, which demands all his energies. He could certainly not have a meeting with a number of people. He feels that if it were with you alone, dear Professor Schlick, he might be able to discuss such matters. It would then become apparent, he thinks, whether he is at present at all capable of being of use to you in this connection.

After meeting Wittgenstein, his wife recalls, Schlick 'returned in an ecstatic state, saying little, and I felt I should not ask questions'. The next day Wittgenstein told Engelmann: 'Each of us thought the other must be mad.' Soon after this, Wittgenstein and Schlick began to meet regularly for discussions. According to Engelmann: 'Wittgenstein found Schlick a distinguished and understanding partner in discussion, all the more so because he appreciated Schlick's highly cultured personality.' But Wittgenstein could not be persuaded to attend meetings of Schlick's 'Circle', a group of philosophers and mathematicians, united in their positivist approach to philosophical problems and their scientific *Weltanschauung*, who met on Thursday evenings to discuss the foundations of mathematics and science, and who later evolved into the Vienna Circle. Wittgenstein told Schlick that he could talk only with somebody who 'holds his hand'.

Nevertheless, by the summer of 1927 Wittgenstein was meeting regularly with a group which met on Monday evenings and which included, in addition to himself and Schlick, a few carefully chosen members of Schlick's Circle. These included Friedrich Waismann, Rudolf Carnap and Herbert Feigl. The success of these meetings depended upon Schlick's sensitive handling of the situation. Carnap recalls that:

> Before the first meeting Schlick admonished us urgently not to start a discussion of the kind to which we were accustomed in the Circle, because Wittgenstein did not want such a thing under any circumstances. We should even be cautious in asking questions, because Wittgenstein was very sensitive and easily disturbed by a direct question. The best approach, Schlick said, would be to let Wittgenstein talk and then ask only very cautiously for the necessary elucidations.

To persuade Wittgenstein to attend these meetings Schlick had to assure him that the discussion would not have to be philosophical; he could discuss whatever he liked. Sometimes, to the surprise of his audience, Wittgenstein would turn his back on them and read poetry. In particular – as if to emphasize to them, as he had earlier explained to von Ficker, that what he had *not* said in the *Tractatus* was more important than what he had – he read them the poems of Rabindranath Tagore, an Indian poet much in vogue in Vienna at that time, whose poems express a mystical outlook diametrically opposed to that of the members of Schlick's Circle. It soon became apparent to Carnap, Feigl and Waismann that the author of *Tractatus Logico-Philosophicus* was not the positivist they had expected. 'Earlier', writes Carnap:

> when we were reading Wittgenstein's book in the Circle, I had erroneously believed that his attitude toward metaphysics was similar to ours. I had not paid sufficient attention to the statements in his book about the mystical, because his feelings and thoughts in this area were too divergent from mine. Only personal contact with him helped me to see more clearly his attitude at this point.

To the positivists, clarity went hand in hand with the scientific method, and, to Carnap in particular, it was a shock to realize that the author of the book they regarded as the very paradigm of philosophical precision and clarity was so determinedly unscientific in both temperament and method:

> His point of view and his attitude toward people and problems, even theoretical problems, were much more similar to those of a creative artist than to those of a scientist; one might almost say, similar to those of a religious prophet or a seer. When he started to formulate his view on some specific philosophical problem, we often felt the internal struggle that occurred in him at that very moment, a struggle by which he tried to penetrate from darkness to light under an intense and painful strain, which was even visible on his most expressive face. When finally, sometimes after a prolonged arduous effort, his answer came forth, his statement stood before us like a newly created piece of art or a divine revelation. Not that he asserted his views dogmatically . . . But the impression he made on us was as if insight came to him as through a divine inspiration, so that we could not help feeling that any sober rational comment or analysis of it would be a profanation.

In contrast to the members of the Circle, who considered the discussions of doubts and objections the best way of testing an idea, Wittgenstein, Carnap recalls, 'tolerated no critical examination by others, once the insight had been gained by an act of inspiration':

> I sometimes had the impression that the deliberately rational and unemotional attitude of the scientist and likewise any ideas which had the flavour of 'enlightenment' were repugnant to Wittgenstein.

Despite these differences in temperament and concerns, Wittgenstein and the members of Schlick's Circle were able to have a number of profitable discussions on philosophical issues, one focus of interest being provided by a recent paper of Frank Ramsey's, 'The Foundations of Mathematics', which Ramsey had delivered as a lecture to the London Mathematical Society in November 1925, and which had been published in the Society's *Proceedings*.

This marked the beginning of Ramsey's campaign to use the work of Wittgenstein on logic to restore the credibility of Frege and Russell's logicist approach to the foundations of mathematics. Until his untimely death in 1930, at the age of twenty-six, it was Ramsey's overriding and abiding aim to repair the theoretical holes in Russell's *Principia* and thus to re-establish the dominance of the logicist school of thought and to nip in the bud the more radical alternative proposed by the increasingly influential intuitionist school led by the Dutch mathematician L. E. J. Brouwer. Broadly speaking, the difference is that, whereas Russell wanted to show that all mathematics could be reduced to logic and thus provide a rigorous logical foundation for all the theorems accepted by pure mathematicians, Brouwer – starting from a fundamentally different conception of both mathematics and logic – wanted to *reconstruct* mathematics in such a way that only those theorems provable from within his system were to be accepted. The rest, which included a good number of well-established theorems, would have to be abandoned as unproven.

Ramsey wanted to use the *Tractatus* theory of propositions to show that mathematics consists of tautologies (in Wittgenstein's sense), and thus that the propositions of mathematics are simply logical propositions. This is not Wittgenstein's own view. In the *Tractatus* he distinguishes between logical and mathematical propositions: only the former are tautologies; the latter are 'equations' (*TLP* 6.22).

Ramsey's aim was thus to show that equations *are* tautologies. At the centre of this attempt was a Definition of Identity which, using a specially defined logical function Q(x, y) as a substitute for the expression x = y, tries, in effect, to assert that x = y is either a tautology (if x and y have the same value) or a contradiction (if x and y have different values). Upon this definition was built a Theory of Functions which Ramsey hoped to use to demonstrate the tautologous nature of mathematics. 'Only so', he thought, 'can we preserve it [mathematics] from the Bolshevik menace of Brouwer and Weyl.'

The paper came to Wittgenstein's attention through Schlick, who had been sent a copy by Ramsey. (Ramsey had not sent it to Wittgenstein himself, because of their quarrel in the summer of 1925.) Wittgenstein evidently read the paper very thoroughly. On 2 July 1927 he wrote to Ramsey criticizing his Definition of Identity at length, and expressing the view that all such theories (those that claim expressions of identity to be either tautologous or contradictory) would not do.

Wittgenstein himself – as Russell had discovered to his consternation in 1919 – had no stake at all in the enterprise of founding mathematics on logic. Indeed, he considered the enterprise misguided. 'The way out of all these troubles', he told Ramsey, 'is to see that neither "Q(x, y)", although it is a very interesting function, nor any propositional function whatever, can be substituted for "x = y".'

Ramsey replied to Wittgenstein's objections twice – once through Schlick, and again directly to Wittgenstein. The gist of his defence was that he had not intended to provide a *Definition* of Identity, but merely a substitute function which was defined in such a way that it did the job of identity statements within his theory and gave him the logical result he wanted.

The exchange is interesting as an illustration of the differences between Wittgenstein and Ramsey, and of what Wittgenstein may have meant when he described Ramsey as a 'bourgeois' thinker. For whereas Wittgenstein's objection attempts to go straight to the heart of the matter, and to demonstrate that Ramsey's whole enterprise of reconstructing Russellian foundations of mathematics was *philosophically* misguided, Ramsey's reply is concerned only with the logical and mathematical question of whether his function will do the task for which it was designed. Thus, according to Wittgenstein, Ramsey was 'bourgeois' in the sense that:

> . . . he thought with the aim of clearing up the affairs of some particular community. He did not reflect on the essence of the state – or at least he did not like doing so – but on how *this* state might reasonably be organized. The idea that this state might not be the only possible one in part disquieted him and in part bored him. He wanted to get down as quickly as possible to reflecting on the foundations – of *this* state. This was what he was good at and what really interested him; whereas real philosophical reflection disturbed him until he put its result (if it had one) to one side and declared it trivial.

The political metaphor, of course, alludes to Ramsey's remark about the 'Bolshevik menace' of Brouwer, and it might be thought that, in his use of this metaphor, Wittgenstein is equating 'real philosophical reflection' with Bolshevism. This is not so. Wittgenstein was not interested in organizing the affairs of *this* state (Russellian logicism), but neither was he interested in replacing it with another (Brouwer's intuitionism). 'The philosopher is not a citizen of *any* community of ideas', he wrote. 'That is what makes him into a philosopher.'

It was possibly this exchange with Ramsey that prompted Wittgenstein, at last, to write to Keynes. It was the first time he had written since he left teaching ('I couldn't stand the hot-water bottle any longer', he explained). He wrote to thank Keynes for his book, *A Short View of Russia*, and to tell him that he expected the house on which he was working to be finished in November of that year (1927), and that he would then like to visit England, 'if anybody there should care to see me'.

'About your book', Wittgenstein wrote, 'I forgot to say that I liked it. It shows that you know that there are more things between heaven and earth etc.'

This strange reason for liking a survey of Soviet Russia is explained by the fact that in it Keynes emphasizes that it is as a new religion, not as an economic innovation, that Soviet Marxism is to be admired. The economic aspects of Leninism he dismisses as: 'a doctrine which sets up as its bible, above and beyond criticism, an obsolete economic textbook, which I know to be not only scientifically erroneous but without interest or application for the modern world'. But the religious fervour accompanying this doctrine he found impressive:

> . . . many, in this age without religion, are bound to feel a strong emotional curiosity towards any religion which is really new and not merely a recrudescence of the old ones and has proved its motive force; and all the more when this new thing comes out of Russia, the beautiful and foolish youngest son of the European family, with hair on his head, nearer both to the earth and to heaven than his bald brothers in the West – who, having been born two centuries later, has been able to pick up the middle-aged disillusionment of the rest of the family before he had lost the genius of youth or become addicted to comfort and to habits. I sympathise with those who seek for something good in Soviet Russia.

The Soviet faith is characterized by Keynes as having in common with Christianity an exalted attitude towards the common man. But, in contrast to Christianity, there is something in it:

> . . . which may, in a changed form and a new setting, contribute something to the true religion of the future, if there be any true religion – *Leninism is absolutely, defiantly non-supernatural, and its emotional and ethical essence centres about the individual's and the community's attitude towards the Love of Money*.

It is not difficult to see how such passages would have earned Wittgenstein's approval, nor how the faith which Keynes describes would have earned his respect and, potentially, his allegiance. Keynes's book, which was written after a brief visit to the Soviet Union, contrasts sharply with Russell's *The Practice and Theory of Bolshevism*, which was published after his own visit in 1920. Russell's book expressed nothing but loathing for the Soviet régime. He, too, draws a parallel with Christianity, but uses precisely that parallel to express his contempt:

> One who believes as I do, that free intellect is the chief engine of human progress, cannot but be fundamentally opposed to Bolshevism as much as to the Church of Rome. The hopes which inspire communism are, in the main, as admirable as those instilled by the Sermon on the Mount, but they are held as fanatically and are as likely to do as much harm.

Wittgenstein's own interest in Soviet Russia dates from soon after the publication of Russell's book – almost as though he thought that if Russell hated it so much there must be something good about it. Since 1922 (when he wrote to Paul Engelmann about 'the idea of a possible flight to Russia which we talked about'), Wittgenstein had been one of those who, in Keynes's words, 'seek for something good in Soviet Russia', and he continued to be attracted to the idea of living and working in the Soviet Union until 1937, when political circumstances made it impossible for him to do so.

Although Keynes proclaims himself a non-believer, in presenting Soviet Marxism as a faith in which there are fervently held religious *attitudes* (towards, for example, the value of the common man and the evils of money-love) but no supernatural *beliefs*, he has, I think, provided an important clue as to what Wittgenstein hoped to find in Soviet Russia.

Wittgenstein's suggestion to Keynes that the Kundmanngasse house would be finished by November 1927 was, for reasons already explained, hopelessly optimistic, and it was not until a year later that he was able to consider the proposed trip to England.

In the meantime he had an opportunity to see and hear for himself the 'Bolshevik menace' that had so disturbed Ramsey. In March 1928 Brouwer came to Vienna to deliver a lecture entitled 'Mathematics, Science and Language', which Wittgenstein attended, together with Waismann and Feigl. After it the three spent a few hours together in a café, and, reports Feigl:

> . . . it was fascinating to behold the change that had come over Wittgenstein that evening . . . he became extremely voluble and began sketching ideas that were the beginnings of his later writings . . . that evening marked the return of Wittgenstein to strong philosophical interest and activities.

It would be wrong to infer from Feigl's report that Wittgenstein underwent a sudden conversion to Brouwerian intuitionism – although there can be no doubt that hearing Brouwer was a tremendous stimulus to him and may well have planted a seed that developed during the following years. There is no evidence in Wittgenstein's earlier work that he was at all aware of Brouwer's ideas, and it may be that Ramsey's reference to him in his 1925 paper was the first he had heard of Brouwer. But references to Brouwer *do* crop up from 1929 onwards – so much so, that when Russell was called upon to report on Wittgenstein's work in 1930 he detected what he obviously considered to be an unhealthy influence:

> . . . he has a lot of stuff about infinity, which is always in danger of becoming what Brouwer has said, and has to be pulled up short whenever this danger becomes apparent.

It is likely, however, that Wittgenstein's excitement after the lecture had as much to do with his disagreements with Brouwer as with his agreement. There is much in the lecture that conflicts with Wittgenstein's own views, both in his early and in his later work. In particular, the Kantian notion of a 'basic mathematical intuition', which forms the philosophical foundation of intuitionism, was something with which Wittgenstein never had any sympathy, at any time in his life. In fact, if anything, his opposition to it strengthened as time went on, until in his 1939 lectures on the foundations of mathematics he told his audience bluntly: 'Intuitionism is all bosh – entirely.'

Nevertheless, there are in Brouwer's outlook certain elements that would have chimed with Wittgenstein's own, especially in his disagreements with the view propounded by Russell and Ramsey. These go deeper than the particular point noted by Russell – that Wittgenstein appeared to accept Brouwer's rejection of the notion of an infinite series in extension – and constitute a philosophical *attitude* that is fundamentally at variance with the 'bourgeois' mentality of Russell and Ramsey. On a general level one could say that Brouwer's philosophical position belongs to the tradition of continental anti-rationalist thought which one associates, for example, with Schopenhauer, and for which Wittgenstein – as Carnap discovered to his surprise – had a great deal of sympathy. (During this period Wittgenstein surprised Carnap by defending Schopenhauer against the criticism of Schlick.) The Vienna Circle, like Russell and Ramsey, placed themselves in a position that would have nothing to do with this anti-rationalist tradition.

More particularly, there are elements in Brouwer's disagreements with Russell's logicism that would have struck a sympathetic chord in Wittgenstein. Brouwer rejected the idea that mathematics either could or needed to be grounded in logic. He further rejected the notion that consistency proofs were essential in mathematics. He also rejected the 'objectivity' of mathematics in the sense that it is usually understood – i.e. for Brouwer, there is no mind-independent mathematical reality about which mathematicians make discoveries. The mathematician, on Brouwer's view, is not a discoverer but a creator: mathematics is not a body of facts but a construction of the human mind.

With all these points Wittgenstein was in agreement, and his later work can be seen as a development of these thoughts into an area that took him far away from the logical atomism of the *Tractatus*. If this development brought him no nearer to intuitionism, it perhaps helped to crystallize his many disagreements, in general and in detail, with the logicist approach to mathematics propounded by Russell and Ramsey – an approach which had guided, even if it had not dictated, the view he had expounded in the *Tractatus*.

Brouwer's lecture may not have persuaded Wittgenstein that the

Tractatus was mistaken, but it may have convinced him that his book was not, after all, the final word on the subject. There might, indeed, be more to be said.

Thus, in the autumn of 1928, when the house was finished and his thoughts turned once more to visiting England, he could, after all, contemplate returning to philosophical work. Not that such an intention is evident from the letters he wrote Keynes. In November he sent Keynes photographs of the house – 'à la Corbusier', as Keynes inaccurately described it in a letter to his wife, Lydia Lopokova – and announced his desire to visit England in December, implying a short, holiday, visit. He 'wants to stay with me here in about a fortnight', Keynes wrote. 'Am I strong enough? Perhaps if I do not work between now and then, I shall be.'

In the event, illness kept Wittgenstein in Vienna throughout December, and when, at the beginning of January, he was finally able to come to England, it was (as Keynes discovered with no apparent great surprise) not to enjoy a holiday in Lewes, nor to look for work sweeping the streets, but to return to Cambridge and work with Ramsey on philosophy.

III

1929–41

11
THE SECOND COMING

'Well, God has arrived. I met him on the 5.15 train.'

Thus was Wittgenstein's return to Cambridge announced by Keynes in a letter to Lydia Lopokova, dated 18 January 1929. Wittgenstein had been back in England just a few hours, and had already informed Keynes of his plan 'to stay in Cambridge permanently':

> Meanwhile we have had tea and now I retire to my study to write to you. I see that the fatigue is going to be crushing. But I must not let him talk to me for more than two or three hours a day.

For Wittgenstein, the experience of returning to a university that had remained largely unchanged throughout the years that had brought such a fundamental transformation in himself – and, moreover, of being greeted by some of the very people he had left in 1913 – was strange, almost eerie. It was, he wrote in his diary, 'as though time had gone backwards'. 'I do not know what awaits me', but whatever it would turn out to be: 'It will prove something! If time doesn't run out':

> At the moment I am wandering about with great restlessness, but around what point of equilibrium I do not know.

Upon his arrival there was an attempt, orchestrated by Keynes, to welcome Wittgenstein back into the Apostolic fold. On Wittgenstein's second day back in England, Keynes held a special supper meeting of the Apostles to celebrate his return. Among those attending were Richard Braithwaite, Frank Ramsey, George Rylands, George Thomson, Alister Watson, Anthony Blunt and Julian Bell – the cream of the current generation of Cambridge intelligentsia. At the meeting Wittgenstein was elected an honorary member (in Apostolic language: an 'Angel'), a gesture of forgiveness by the Society for his attitude towards them in 1912. At a subsequent meeting he was formally 'declared to be absolved from his excommunication at the appropriate time'.

The reason for this unprecedented humility on the part of the Society was that, in his absence, Wittgenstein had become an almost legendary figure among the Cambridge élite, and the *Tractatus* the centre of fashionable intellectual discussion.

But if the Apostles had hoped to claim this 'God' for their own, they were to be disappointed. Wittgenstein attended a few of their meetings, and at dinner parties at Keynes's house in Gordon Square he came into contact with a few members of what might be regarded as their London branch: the Bloomsbury group. But there was little common ground between the peculiarly English, self-consciously 'civilized', aestheticism of Bloomsbury and the Apostles, and Wittgenstein's rigorously ascetic sensibility and occasionally ruthless honesty. There was shock on both sides. Leonard Woolf recalls that he was once appalled by Wittgenstein's 'brutally rude' treatment of Lydia Keynes at lunch. At another lunch, Wittgenstein walked out, shocked at the frank discussion of sex in the presence of ladies. Clearly, the atmosphere of Bloomsbury was not one in which he felt at home. Frances Partridge describes how, in contrast to the Bells, Stracheys and Stephens with whom she mixed, Wittgenstein seemed unable or unwilling to discuss serious matters with members of the opposite sex: 'in mixed company his conversation was often trivial in the extreme, and larded with feeble jokes accompanied by a wintry smile'.

It is possible that at one of Keynes's parties Wittgenstein and Virginia Woolf may have met; if so, neither seems to have made much impression on the other. After Virginia Woolf's death Wittgenstein spoke to Rush Rhees about the effects of her background. She grew up, he said, in a family in which the measure of a person's worth was his distinction in some form of writing or in art, music, science or politics, and she had consequently never asked herself whether there might be other 'achievements'. This could be based on personal acquaintance, but it could equally well be based on hearsay. There are no references to Wittgenstein in Virginia Woolf's diaries, and only a few incidental mentions of him in her letters. In one of these, in a letter to Clive Bell written a few months after Wittgenstein's arrival in Cambridge, she mentions him in connection with Bell's son, Julian:

> . . . Julian, Maynard says, is undoubtedly the most important undergraduate at Kings, and may even get a Fellowship, and Maynard seems highly impressed with him altogether, and his poetry – Julian by the way says he tackled Maynard about Wittgenstein but was worsted.

The reference is interesting only for the fact that it was Julian Bell who was to provide, in a lengthy Drydenesque satire published in Anthony Blunt's student magazine, *The Venture*, a kind of Bloomsbury riposte to what some began to regard as the uncivilized savagery of Wittgenstein's domineering, argumentative style.

In the poem Bell seeks to defend the Bloomsbury creed that 'value is known and found in states of mind' against the *Tractatus* view that such statements are nonsense. Surely, Bell argues, Wittgenstein breaks his own rules:

> For he talks nonsense, numerous statements makes,
> Forever his own vow of silence breaks:
> Ethics, aesthetics, talks of day and night,
> And calls things good or bad, and wrong or right.

Not only does Wittgenstein talk of these things, about which he insists one must be silent; he dominates *all* talk of them:

> . . . who, on any issue, ever saw
> Ludwig refrain from laying down the law?
> In every company he shouts us down,
> And stops our sentence stuttering his own;
> Unceasing argues, harsh, irate and loud,
> Sure that he's right, and of his rightness proud,
> Such faults are common, shared by all in part,
> But Wittgenstein pontificates on Art.

The poem was written as an epistle to a fellow Apostle, Richard Braithwaite, and expressed the view of many of the young Apostolic aesthetes – 'these Julian Bells', as Wittgenstein contemptuously called them – among whom it was greatly enjoyed. When it was published, Fania Pascal says: 'the kindest people enjoyed a laugh; it released accumulated tension, resentment, even fear. For no one could ever turn the tables on Wittgenstein and pay him back in kind.'

If Wittgenstein did not turn his back on the Apostles altogether, this was chiefly because among its members was Frank Ramsey.

During his first year back at Cambridge, Ramsey was not only Wittgenstein's most valued partner in philosophical discussion, but also his closest friend. For the first two weeks after his arrival he lived with the Ramseys at their home in Mortimer Road. Ramsey's wife, Lettice, soon became a close friend and confidante – a woman who, 'at last has succeeded in soothing the fierceness of the savage hunter', as Keynes put it. She had the kind of robust sense of humour and earthy honesty that could make him relax, and gain his trust. With her alone he felt able to discuss his love for Marguerite, although, from a letter by Frances Partridge to her husband Ralph, it appears the confidence was not strictly kept:

> We have seen a lot of Wittgenstein; he confides in Lettice that he is in love with a Viennese lady, but he feels marriage to be sacred, and can't speak of it lightly.

What is surprising here is, not that he could not speak lightly of marriage, but that he could speak of it at all. He was at this time writing regularly and frequently, sometimes daily, to Marguerite, but it was not until about two years later that she realized he intended to make her his wife, and when she did, she beat a hasty retreat. Though flattered by his attention, and over-awed by the strength of his personality, Marguerite did not see in Wittgenstein the qualities she wished for in a husband. He was too austere, too demanding (and, one suspects, just a little too Jewish). Besides, when he made clear his intentions, he also made clear that he had a Platonic, childless, marriage in mind – and that was not for her.

During his first two terms at Cambridge, Wittgenstein's official status was that of an 'Advanced Student' reading for a Ph.D., with

Ramsey, seventeen years his junior, as his supervisor. In practice, he and Ramsey met as equals working on similar, or related, problems, and looking to each other for criticism, guidance and inspiration. Several times a week they would meet for many hours at a time to discuss the foundations of mathematics and the nature of logic. These meetings were described by Wittgenstein in his diary as 'delightful discussions': 'There is something playful about them and they are, I believe, pursued in a good spirit.' There was, he wrote, something almost erotic about them:

> There is nothing more pleasant to me than when someone takes my thoughts out of my mouth, and then, so to speak, spreads them out in the open.

'I don't like taking walks through the fields of science alone', he added.

Ramsey's role in these discussions was akin to that of any other supervisor: to raise objections to what Wittgenstein said. In the preface to the *Investigations*, Wittgenstein says that he was helped by Ramsey's criticism – 'to a degree which I am hardly able to estimate' – to realize the mistakes of the *Tractatus*. In a diary entry of the time, however, he took a less generous view:

> A good objection helps one forward, a shallow objection, even if it is valid, is wearisome. Ramsey's objections are of this kind. The objection does not seize the matter by its root, where the life is, but so far outside that nothing can be rectified, even if it is wrong. A good objection helps directly towards a solution, a shallow one must first be overcome and can, from then on, be left to one side. Just as a tree bends at a knot in the trunk in order to grow on.

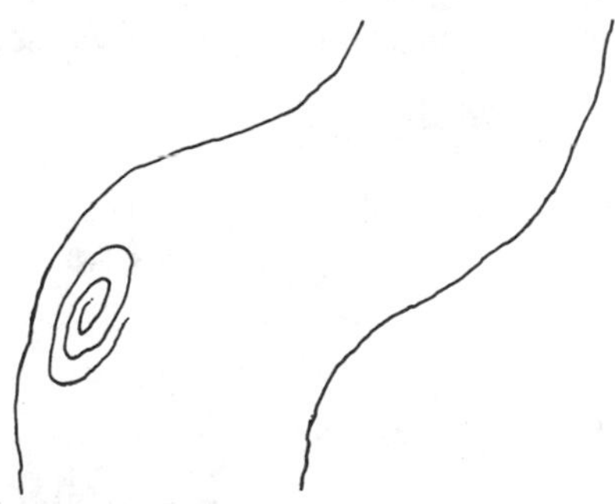

Despite their enormous respect for each other, there were great differences, intellectual and temperamental, between Ramsey and Wittgenstein. Ramsey was a mathematician, dissatisfied with the logical foundations of his subject, who wanted to reconstruct mathematics on sound principles. Wittgenstein was not interested in reconstructing mathematics; his interest lay in extracting the philosophical root from which confusion about mathematics grew. Thus, while Ramsey could look to Wittgenstein for inspiration and Wittgenstein in Ramsey for criticism, frustrations between the two were inevitable. Ramsey once told Wittgenstein bluntly: 'I don't like your method of arguing', while Wittgenstein wrote of Ramsey, in a remark already quoted, that he was a 'bourgeois thinker' who was disturbed by *real* philosophical reflection 'until he put its result (if it had one) to one side and declared it trivial'.

A 'non-bourgeois' thinker whose profound influence on Wittgenstein's development dates from this first year back at Cambridge was Piero Sraffa. Sraffa was a brilliant Italian economist (of a broadly Marxist persuasion), and a close friend of Antonio Gramsci, the imprisoned Italian Communist leader. After jeopardizing his career in his home country by publishing an attack on Mussolini's policies, Sraffa was invited by Keynes to come to King's to pursue his work, and a lectureship in economics at Cambridge was created specially for him. Upon being introduced by Keynes, he and Wittgenstein became close friends, and Wittgenstein would arrange to meet him at least once a week for discussions. These meetings he came to value even more than those with Ramsey. In the preface to the *Investigations* he says of Sraffa's criticism: 'I am indebted to *this* stimulus for the most consequential ideas of this book.'

This is a large claim, and – considering their widely differing intellectual preoccupations – a puzzling one. But it is precisely because Sraffa's criticisms did not concern details (because, one might say, he was not a philosopher or a mathematician) that they could be so consequential. Unlike Ramsey, Sraffa had the power to force Wittgenstein to revise, not this or that point, but his whole perspective. One anecdote that illustrates this was told by Wittgenstein to both Malcolm and von Wright, and has since been retold many times. It concerns a conversation in which Wittgenstein insisted that a proposition and that which it describes must have the same 'logical form' (or 'grammar', depending on the version of the story). To this idea. Sraffa made a Neapolitan gesture of brushing his chin with his fingertips, asking: 'What is the logical form of *that*?' This, according to the story, broke the hold on Wittgenstein of the Tractarian idea that a proposition must be a 'picture' of the reality it describes.

The importance of this anecdote is not that it explains why Wittgenstein abandoned the Picture Theory of meaning (for it does not), but that it is a good example of the way in which Sraffa could make Wittgenstein see things anew, from a fresh perpective. Wittgenstein told many of his friends that his discussions with Sraffa made him feel like a tree from which all branches had been cut. The metaphor is carefully chosen: cutting dead branches away allows new, more vigorous ones to grow (whereas Ramsey's objections left the dead wood in place, forcing the tree to distort itself around it).

Wittgenstein once remarked to Rush Rhees that the most important thing he gained from talking to Sraffa was an 'anthropological' way of looking at philosophical problems. This remark goes some way to explain why Sraffa is credited as having had such an important influence. One of the most striking ways in which Wittgenstein's later work differs from the *Tractatus* is in its 'anthropological' approach. That is, whereas the *Tractatus* deals with language in isolation from the circumstances in which it is used, the *Investigations* repeatedly emphasizes the importance of the 'stream of life' which gives linguistic utterances their meaning: a 'language-game' cannot be described without mentioning their activities and the way of life of the 'tribe' that plays it. If this change of perspective derives from Sraffa, then his influence on the later work is indeed of the most fundamental importance. But in this case, it must have taken a few years for that influence to bear fruit, for this 'anthropological' feature of Wittgenstein's philosophical method does not begin to emerge until about 1932.

Apart from Ramsey and Sraffa, Wittgenstein had little to do with the college dons at Cambridge. After the first few weeks his relations with Keynes were confined largely to business matters, and although Keynes became an invaluable ally whenever Wittgenstein needed anything sorted out with the authorities, he was not a close friend. This, one gathers, was a role that Keynes was quite happy to fit into;

being Wittgenstein's *friend* demanded more time and energy than he was able, or prepared, to give.

G. E. Moore happened to be on the same train from London as Wittgenstein when Wittgenstein first arrived, and immediately their friendship, broken since Wittgenstein's wild letter to Moore in 1914, was resumed. Moore, who was by this time Professor of Philosophy at Cambridge, took responsibility for arranging the grants that enabled Wittgenstein to continue his work; other than this, however, their friendship was personal, rather than philosophical. Though he admired Moore's exactitude of expression, and would occasionally make use of it to find the precise word he wanted to make a particular point, Wittgenstein had little respect for him as an original philosopher. 'Moore?' – he once said – 'he shows you how far a man can go who has absolutely no intelligence whatever.'

Similarly with the, by now elderly, logician W. E. Johnson – another figure from his earlier Cambridge period – Wittgenstein maintained an affectionate friendship, despite the intellectual distance that existed between the two. Wittgenstein admired Johnson as a pianist more than as a logician, and would regularly attend his Sunday afternoon 'at homes' to listen to him play. For his part, though he liked and admired Wittgenstein, Johnson considered his return a 'disaster for Cambridge'. Wittgenstein was, he said, 'a man who is quite incapable of carrying on a discussion'.

Though he was approaching his fortieth birthday, Wittgenstein drew his circle of friends largely from the younger generation at Cambridge – from the undergraduates (of the non-Apostolic kind) who attended the Moral Science Club. It was in the 'sons of the English middle class' who made up this student philosophy society, according to Fania Pascal, that Wittgenstein found the two features he required in a disciple: childlike innocence and first-class brains. This may be so, but it is also true, I think, that Wittgenstein simply found he had more in common with the younger generation. He was, in a sense, very young himself. He even looked young, and at forty was frequently mistaken for an undergraduate himself. But more than this, he had the intellectual freshness and suppleness of youth. 'The mind', he told Drury, 'gets stiff long before the body does'; and in this sense he was still an adolescent. There was, that is, very little in his mental outlook that had become inflexible. He had returned to Cambridge prepared to overhaul all the conclusions he had reached up to now – prepared not only to consider new ways of thinking, but even new ways of life. Thus he was still as unformed, as unsettled into any particular pattern of life, as any undergraduate.

Many who had heard of Wittgenstein as the author of *Tractatus Logico-Philosophicus* imagined him to be an old and dignified German academic, and were unprepared for the youthfully aggressive and animated figure they encountered at meetings of the Moral Science Club. S. K. Bose, for example, who subsequently became one of the circle of Wittgenstein's friends and admirers, recalls:

> My first encounter with Wittgenstein was at a meeting of the Moral Science Club at which I read a paper on 'The nature of moral judgement'. It was a rather largely attended meeting and some people were squatting on the carpet. Among them was a stranger to all of us (except, of course, Professor Moore and one other senior member possibly present). After I had read the paper, the stranger raised some questions and objections in that downright fashion (but never unkind way) which one learned later to associate with Wittgenstein. I have never been able to live down the shame I felt when I learnt, some time later, who my interlocutor had been, and realised how supercilious I had been in dealing with the questions and objections he raised.

Wittgenstein came to dominate the discussions of the Moral Science Club so completely that C. D. Broad, the Professor of Moral Philosophy, stopped attending. He was not prepared, he later said, 'to spend hours every week in a thick atmosphere of cigarette-smoke, while Wittgenstein punctually went through his hoops and the faithful as punctually "wondered with a foolish look of praise"'.

Desmond Lee, another member of Wittgenstein's undergraduate circle of friends, has likened Wittgenstein, in his preference for discussions with younger men, and in the often numbing effect he had on them, to Socrates. Both, he points out, had an almost hypnotic influence on those who fell under their spell. Lee himself was freed from this spell when he left Cambridge, and, though deeply influenced by Wittgenstein, cannot correctly be described as a disciple. His contemporary, Maurice Drury, however, became the first, and perhaps the most perfect, example of the young disciples described by Fania Pascal.

After first meeting Wittgenstein in 1929, almost every major decision in Drury's life was made under his influence. He had originally intended, upon leaving Cambridge, to be ordained as an Anglican priest. 'Don't think I ridicule this for one minute', Wittgenstein remarked upon being told of the plan, 'but I can't approve; no, I can't approve. I would be afraid that one day that collar would choke you.' This was on the second, or possibly the third, occasion on which they had met. On the next, Wittgenstein returned to the theme: 'Just think, Drury, what it would mean to have to preach a sermon every week; you couldn't do it.' After a year at theological college, Drury agreed, and, prompted by Wittgenstein, took a job instead among 'ordinary people'. He worked on projects to help the unemployed, first in Newcastle and then in South Wales, after which, again prompted by Wittgenstein, he trained as a doctor. After the war he specialized in psychiatry (a branch of medicine suggested by Wittgenstein), and from 1947 until his death in 1976 worked at St Patrick's Hospital, Dublin, first as Resident Psychiatrist and then as Senior Consultant Psychiatrist. His collection of essays on philosophical problems in psychiatry, *The Danger of Words*, was published in 1973; though much neglected, it is perhaps, in its tone and its concerns, the most truly Wittgensteinian work published by any of Wittgenstein's students. 'Why do I now bring these papers together?' he asks in the preface, and answers:

> For one reason only. The author of these writings was at one time a pupil of Ludwig Wittgenstein. Now it is well known that Wittgenstein encouraged his pupils (those at least whom he considered had no great originality in philosophical ability) to turn from academic philosophy to the active study and practice of some particular avocation. In my own case he urged me to turn to the study of medicine, not that I should make no use of what he had taught me, but rather that on no account should I 'give up thinking'. I therefore hesitantly put these essays forward as an illustration of the influence that Wittgenstein had on the thought of one who was confronted by problems which had both an immediate practical difficulty to contend with, as well as a deeper philosophical perplexity to ponder over.

Similarly, shortly before his death, Drury published his notes of conversations with Wittgenstein to counteract the effect of 'well-meaning commentators', who 'make it appear that his writings were now easily assimilable into the very intellectual milieu they were largely a warning against'. These notes provide – perhaps more than any other secondary source – information on the spiritual and moral attitudes that informed Wittgenstein's life and work. Drury is the first, but by no means the last, disciple to illustrate that there is an important aspect of Wittgenstein's influence that is not, and cannot be, covered in the large body of academic literature which Wittgenstein's work has inspired. The line of apostolic succession, one might say, extends far beyond the confines of academic philosophy.

One of Wittgenstein's closest undergraduate friends, indeed, was a man who had no interest in philosophy whatever. Gilbert Pattisson met Wittgenstein on the train coming back from Vienna after the Easter break of 1929, and for over ten years the two enjoyed an affectionate, and strictly non-philosophical, friendship, which came to an end during the troubled years of the Second World War, when Wittgenstein began to suspect Pattisson of taking a jingoistic attitude towards the war. Pattisson was (indeed is) a genial, witty and rather worldly character, quite unlike the innocent and over-shy disciples described by Pascal. On completing his studies at Cambridge (with a minimum of academic effort and commitment), he became a chartered accountant in the City of London and led the kind of comfortable life for which his class, upbringing and education had prepared him. With him Wittgenstein could indulge the taste for what Frances Partridge had described as trivia and feeble humour, but which Wittgenstein himself called simply 'nonsense'. To have someone with whom he could 'talk nonsense to by the yard' was, he said, a deep-seated need.

At Cambridge, Pattisson and Wittgenstein would read together magazines like the *Tatler*, delighting in their rich supply of 'nonsense', and enjoying particularly the ludicrous advertisements that used to appear in such journals. They were avid readers, too, of the 'Letters from a satisfied customer' which used to be displayed in the windows of Burton's, 'The Tailor of Taste', and to which Pattisson and Wittgenstein would give exaggerated attention during their shopping trips to buy Wittgenstein's clothes. (It may have seemed to most people that Wittgenstein always wore the same things – an open-neck shirt, grey flannel trousers and heavy shoes; in fact, these items were chosen with meticulous care.)

After Pattisson left Cambridge, he and Wittgenstein would meet whenever Wittgenstein passed through London (as he did frequently on his way to and from Vienna) to go through what Wittgenstein described as their 'ritual'. This consisted of tea at Lyons followed by a film at one of the big cinemas in Leicester Square. Before arriving in London, Wittgenstein would send Pattisson a card letting him know when he was arriving, so that Pattisson could make the necessary arrangements – i.e. search the *Evening Standard* for a cinema that was showing a 'good' film. In Wittgenstein's sense this meant an American film, preferably a Western, or, later, a musical or a romantic comedy, but always one without any artistic or intellectual pretensions. It was understood that Pattisson's work in the City would take second place to this ritual: 'I hope you won't be busybodying in your office', Wittgenstein wrote once, after Pattisson had pleaded pressure of work. 'Remember, even Bismarck could be replaced.'

Wittgenstein's correspondence with Pattisson consists almost entirely of 'nonsense'. In nearly every letter he makes some use of the English adjective 'bloody', which, for some reason, he found inexhaustibly funny. He would begin his letters 'Dear Old Blood' and end them 'Yours bloodily' or 'Yours in bloodiness'. Pattisson would send him photographs cut out from magazines, which he called his 'paintings', and to which Wittgenstein would respond with exaggeratedly solemn appreciation: 'I would have known it to be a Pattisson immediately without the signature. There is that bloodiness in it which has never before been expressed by the brush.' In reply, Wittgenstein would send 'portraits', photographs of distinguished looking middle-aged men, ripped out of newspaper advertisements for self-improvement courses. 'My latest photo', he announced, enclosing one such picture. 'The previous one expressed fatherly kindness only; this one expresses triumph.'

Throughout the correspondence there is a gentle ridicule of the language of the advertiser, the absurdity of the style being invoked simply by using it as though it were the normal way for two friends to write to each other. Sending Wittgenstein a (genuine) photograph of himself, Pattisson writes on the back: 'On the other side is pictured one of our 47/6 suits.' 'Somehow or other', Wittgenstein writes at the end of one letter, 'one instinctively feels that Two Steeples No. 83

Quality Sock is a real man's sock. It's a sock of taste – dressy, fashionable, comfortable.' In a postscript to another, he writes:

> You may through my generosity one of these days get a free sample of Glostora the famous hair oil, may your hair always retain that gloss which is so characteristic for well groomed gentlemen.

Some of the jokes contained in Wittgenstein's letters to Pattisson are, indeed, astonishingly feeble. Enclosing an address that ends 'W.C.1', he draws an arrow to the 'W.C.' and writes: 'This doesn't mean "Lavatory".' And on the back of a postcard of Christ Church Cathedral, Dublin, he writes: 'If I remember rightly this Cathedral was built, partly at least, by the Normans. Of course, it's a long time ago & my memory isn't what it was then.'

Within a few months of being at Cambridge, then, Wittgenstein established a fairly wide circle of friendships, which, to some extent, showed his fears of entering back into society to have been misplaced. And yet he continued to feel a foreigner at Cambridge, to feel the lack of someone like Paul Engelmann or Ludwig Hänsel – someone with whom he could discuss his innermost thoughts and feelings in his own language, and with the certainty that he would be understood. Perhaps for this reason, as soon as he returned to Cambridge, he reverted to a practice he had not kept since the *Tractatus* had been published: he began to make personal, diary-like entries in his notebooks. As before, these were separated from his philosophical remarks by being written in the code he had used as a child. In one of the earliest entries, he remarks how strange it was: 'That for so many years I have never felt the slightest need to make notebook entries', and reflects on the genesis of the habit. In Berlin, when he began to write down thoughts about himself, it had arisen out of a need to preserve something of himself. It had been an important step, and though there was in it something of vanity and imitation (of Keller and Pepys), yet it still fulfilled a genuine requirement; it provided a substitute for a person in whom he could confide.

Wittgenstein could not fully confide in the people at Cambridge because, given the linguistic and cultural differences of which he was far more acutely conscious than they perhaps realized, he could not feel entirely sure that he would be understood. Whenever a misunderstanding arose he was inclined to attribute it to those differences. 'What a statement seems to imply to me it doesn't to you', he wrote to Ramsey after one such misunderstanding. 'If you should ever live amongst foreign people for any length of time & be dependent upon them you will understand my difficulty.'

His feeling that he was dependent upon people to whom he could not make himself understood caused him intense suffering, particularly where money was involved. In May 1929 he wrote a long letter to Keynes attempting to explain these anxieties. 'Please try to understand it before you criticize it', he pleaded, adding: 'To write in a foreign language makes it more difficult.' He had become convinced (with, as we have seen, some justification) that Keynes had grown tired of his conversation. '*Now please don't think that I mind that!*' he wrote. 'Why shouldn't you be tired of me, I don't believe for a moment that I can be entertaining or interesting to you.' What pained him was the fear that Keynes might think he cultivated his friendship in order to receive financial assistance; in his anxiety about this, and about being misunderstood when he spoke in English, he invented a completely fictitious confirmation of this fear:

> In the beginning of this term I came to see you and wanted to return you some money you had lent me. And in my clumsy way of speaking I prefaced the act of returning it by saying 'Oh first I want money' meaning 'first I want to settle the money business' or some such phrase. But you naturally misunderstood me and consequently made a face in which I could read a whole story. And what followed this, I mean our conversation about the society [the Apostles], showed me what amount of negative feeling you had accumulated in you against me.

He was, however, probably right in thinking that Keynes saw himself as his benefactor rather than his friend. But, he insisted, 'I don't accept benefactions except from my friends. (That's why I accepted your help three years ago in Sussex.)' He ended: 'Please don't answer this letter unless you can write a short and kind answer. I did not write it to get explanations from you but to inform you about how I think. So if you can't give me a kind answer in three lines, no answer will please me best.' Keynes's reply to this is a masterpiece of tact and sensitivity:

Dear Ludwig,
What a maniac you are! Of course there is not a particle of truth in anything you say about money. It never crossed my mind at the beginning of this term that you wanted anything from me except to cash a cheque or something of that kind. I have never supposed it possible that you could want any money from me except in circumstances in which I should feel it appropriate to give it. When I mentioned your finances in my note the other day, it was because I had heard that you were bothered with heavy unexpected fees and I wanted, if this was so, to examine a possibility which I think I suggested to you when you first came up, namely that some help might conceivably be got out of Trinity. I had considered whether it could be a good thing for me to do anything myself, and had decided on the whole better not.

No – it was not 'an undertone of grudge' that made me speak rather crossly when we last met; it was just fatigue or impatience with the difficulty, almost impossibility, when one has a conversation about something affecting you personally, of being successful in conveying true impressions into your mind and keeping false ones out. And then you go away and invent an explanation so remote from anything then in my consciousness that it never occurred to me to guard against it!

The truth is that I alternate between loving and enjoying you and your conversation and having my nerves worn to death by it. It's no new thing! I always have – any time these twenty years. But 'grudge' 'unkindness' – if only you could look into my heart, you'd see something quite different.

Without committing himself to the strains of a more intimate friendship with Wittgenstein, Keynes managed to smooth things over to the extent of allowing Wittgenstein to accept his help in good conscience – of becoming a *friendly* benefactor, whose help was offered, and therefore accepted, in the right spirit.

Without some sort of financial assistance, Wittgenstein would not have been able to continue his philosophical work. By the end of his second term, whatever savings he had (presumably from his earnings as an architect) were insufficient to pay his college fees and leave anything for him to live on. Keynes's suggestion that he apply for a research grant from Trinity was taken up, but there were, inevitably, complications. These arose from the fact that the college found it hard to understand why someone from as wealthy a background as Wittgenstein should need a grant of this kind. Did he have any other source of money? he was asked by Sir James Butler, the Tutor in Trinity. He answered, no. Didn't he have any relations who could help? He answered, yes. 'Now as it somehow appears as if I tried to conceal something', he wrote to Moore after this interview, 'will you please accept my written declaration that: not only I have a number of wealthy relations, but also they would give me money if I asked them to, BUT THAT I WILL NOT ASK THEM FOR A PENNY.' His attitude, as he explained in another letter to Moore, was this:

> I propose to do some work, and I have a vague idea, that the College in some cases encourages such work by means of research grants, fellowships, etc. That's to say, I turn out some sort of goods and *if* the College has any use for these goods, I would like the College to enable me to produce them, as long as it *has* a use for them, and as long as I *can* produce them.

His application for a grant was given fulsome support by Frank Ramsey, who in his role as Wittgenstein's supervisor wrote to Moore urging the necessity of such assistance. 'In my opinion', he wrote, 'Mr Wittgenstein is a philosophic genius of a different order from anyone else I know':

> This is partly owing to his great gift for seeing what is essential in a problem and partly to his overwhelming intellectual vigour, to the intensity of thought with which he pursues a question to the bottom and never rests content with mere possible hypothesis. From his work more than that of any other man I hope for a solution of the difficulties that perplex me both in philosophy generally and in the foundations of Mathematics in particular. It seems to be, therefore, peculiarly fortunate that he should have returned to research.

Ramsey's report on the 'goods' that Wittgenstein had produced so far is, however, tantalizingly brief:

> During the last two terms I have been in close touch with his work and he seems to me to have made remarkable progress. He began

> with certain questions in the analysis of propositions which have now led him to problems about infinity which lie at the root of current controversies on the foundation of Mathematics. At first I was afraid that lack of mathematical knowledge and facility would prove a serious handicap to his working in this field. But the progress he had made has already convinced me that this is not so, and that here too he will probably do work of the first importance.

'He is now working very hard', Ramsey adds, 'and, so far as I can judge he is getting on well. For him to be interrupted by lack of money would be a great misfortune for philosophy.'

Perhaps in order to further convince the authorities, Wittgenstein was hurriedly awarded a Ph.D. for his 'thesis', the *Tractatus*, a work that had been in print for seven years and was already regarded by many as a philosophical classic. The examiners were Moore and Russell, the latter having to be somewhat reluctantly dragged up to Cambridge from his school in Sussex. He had had no contact with Wittgenstein since their meeting in Innsbruck in 1922, and was naturally apprehensive. 'I think', he wrote to Moore, 'that unless Wittgenstein has changed his opinion of me, he will not much like to have me as an Examiner. The last time we met he was so much pained by the fact of my not being a Christian that he has avoided me ever since; I do not know whether pain on this account has grown less, but he must still dislike me, as he has never communicated with me since. I do not want him to run out of the room in the middle of the Viva, which I feel is the sort of thing he might do.'

The Viva was set for 18 June 1929, and was conducted with an air of farcical ritual. As Russell walked into the examination room with Moore, he smiled and said: 'I have never known anything so absurd in my life.' The examination began with a chat between old friends. Then Russell, relishing the absurdity of the situation, said to Moore: 'Go on, you've got to ask him some questions – you're the professor.' There followed a short discussion in which Russell advanced his view that Wittgenstein was inconsistent in claiming to have expressed unassailable truths by means of meaningless propositions. He was, of course, unable to convince Wittgenstein, who brought the proceedings to an end by clapping each of his examiners on the shoulder and remarking consolingly: 'Don't worry, I know you'll never understand it.'

In his examiner's report, Moore stated: 'It is my personal opinion that Mr Wittgenstein's thesis is a work of genius; but, be that as it may, it is certainly well up to the standard required for the Cambridge degree of Doctor of Philosophy.'

The day after he received his Ph.D., Wittgenstein was awarded a grant of £100 by Trinity College – £50 for the summer, and £50 for the following Michaelmas term.

Wittgenstein spent the early part of the summer vacation in Cambridge, living as a lodger with Maurice Dobb and his wife at Frostlake Cottage, Malting House Lane. To this period belongs a brief and uneasy friendship with the renowned literary critic F. R. Leavis. They met at one of Johnson's 'at homes', and would occasionally take long walks together. Wittgenstein admired Leavis's personality more than his work; indeed, one might almost say he liked Leavis in spite of his work. He once greeted Leavis with the words: 'Give up literary criticism!' – a piece of advice in which Leavis, with striking misjudgement, saw only the bad influence of Bloomsbury, supposing Wittgenstein to have accepted 'Keynes, his friends and their protégés; as 'the cultural élite they took themselves to be'.

Wittgenstein was, Leavis recalls, working desperately hard at this time, and was chronically short of sleep. On one occasion, when they were out walking together until after midnight, Wittgenstein was so exhausted that on their way back to Malting House Lane he could hardly walk without the support of Leavis's arm. When they finally reached Frostlake Cottage Leavis implored him to go to bed at once. 'You don't understand', Wittgenstein replied. 'When I'm engaged on a piece of work I'm always afraid I shall die before I've finished it. So I make a fair copy of the day's work, and give it to Frank Ramsey for safe-keeping. I haven't made today's copy.'

The work that he was then engaged in writing was the paper entitled 'Some Remarks on Logical Form', which has the distinction of being the only piece of philosophical writing he published after the *Tractatus*. It was printed in the conference proceedings of the 1929 Annual Joint Session of the Aristotelian Society and the Mind Association, the most important of the British conferences of professional philosophers, which that year was held in Nottingham between 12 and 15 July. It is a mark of how quickly his thought was developing at this time, however, that almost as soon as he had sent it off to be printed he disowned it as worthless, and, at the meeting of which it supposedly forms part of the proceedings, read something quite different – a paper on the concept of infinity in mathematics, which has, consequently, been lost to posterity.

'Some Remarks on Logical Form' is nonetheless interesting as a record of a transitory phase in the development of Wittgenstein's philosophy – a phase in which the logical edifice of the *Tractatus*, though crumbling, had not yet been demolished altogether. The paper can be seen as an attempt to answer criticisms made by Frank Ramsey of Wittgenstein's discussion of colour-exclusion in the *Tractatus*. Ramsey's objections were first raised in his review of the *Tractatus*; no doubt they had been explored further in discussions between the two in the first two terms of 1929.

In proposition 6.375 of the *Tractatus* Wittgenstein had insisted: 'Just as the only necessity that exists is *logical* necessity, so too the only impossibility that exists is *logical* impossibility', and had gone on in the following proposition to apply this to the impossibility of something's being, say, both red and blue:

> . . . the simultaneous presence of two colours at the same place in the visual field is impossible, in fact, logically impossible, since it is ruled out by the logical structure of colour.

The problem here is that, if this is so, then the statement 'This is red' cannot be an atomic proposition. In the *Tractatus* it is claimed that atomic propositions are logically independent of one another, with 'This is red' quite clearly *not* being independent of 'This is blue': the truth of one implies the falsehood of the other. Thus, ascriptions of colour have to be complex, susceptible to further analysis. In the *Tractatus* Wittgenstein had appealed to the analysis of colour in terms of the velocities of particles as a way out of this difficulty. Thus, the impossibility of something's being both red and blue appears as the following contradiction: 'a particle cannot have two velocities at the same time; that is to say, it cannot be in two places at the same time'. But, as Ramsey insisted, even at this level of analysis the problem reappears:

> . . . even supposing that the physicist thus provides an analysis of what we mean by 'red', Mr Wittgenstein is only reducing the

> difficulty to that of the *necessary* properties of space, time, and matter or the ether. He explicitly makes it depend on the *impossibility* of a particle being in two places at the same time.

And it is still hard to see, says Ramsey, how this can be a matter of logic rather than of physics.

Ramsey's remarks thus presented Wittgenstein with a challenge: he must either show how the properties of space, time and matter can appear as *logical* necessities, or provide an alternative account of colour-exclusion. In 'Some Remarks on Logical Form', Wittgenstein chose the latter.

He now abandons the claim that atomic propositions are independent; the truth of one can indeed imply the falsity of another, and 'This is both red and blue' is, therefore, 'ruled out'. But if this is so, then there is something seriously amiss with the analysis of the rules of logical form that was offered in the *Tractatus*. For, by the *Tractatus* rules, such constructions are ruled out only if they can be analysed into forms such as 'p and not-p', which can be shown to be contradictory by the Truth-Table Method. The paper therefore ends on a problematic note:

> It is, of course, a deficiency of our notation that it does not prevent the formation of such nonsensical constructions, and a perfect notation will have to exclude such structures by definite rules of syntax . . . Such rules, however, cannot be laid down until we have actually reached the ultimate analysis of the phenomena in question. This, we all know, has not yet been achieved.

In the work written during the following year Wittgenstein made some attempt to provide 'the ultimate analysis of the phenomena in question', and for this short period his work became, as he described it, a kind of phenomenology. Prompted by his discussions with Sraffa, however, he soon gave up the attempt to repair the structure of the *Tractatus*, and abandoned altogether the idea that there *had* to be a commonality of structure between the world and language. Indeed, the point at which he abandoned it is perhaps the point at which he decided he could not read this paper before the conference. For the paper does not present the solution to the problem raised by Ramsey so much as an admission that, within the terms of the *Tractatus*, Wittgenstein had no solution.

Having decided to speak instead on the concept of infinity in mathematics, he wrote to Russell asking him to attend – 'as your presence would improve the discussion immensely and perhaps would be the only thing making it worth while at all'. It was the first and only time in his career that Wittgenstein attended such a conference, and, as he explained to Russell, he had no great hopes for it: 'I fear that whatever one says to them will either fall flat or arouse irrelevant troubles in their minds.' What he had to say about infinity would, he feared, 'be all Chinese to them'.

The Oxford philosopher, John Mabbott, recalls that when he arrived in Nottingham to attend the conference he met at the student hostel a youngish man with a rucksack, shorts and open-neck shirt. Never having seen Wittgenstein before, he assumed that this was a student on vacation who did not know his hostel had been given over to those attending the conference. 'I'm afraid there is a gathering of philosophers going on in here', he said kindly. Wittgenstein replied darkly: 'I too.'

In the event, Russell did not attend, and the conference served only to confirm Wittgenstein's contempt for such gatherings. One positive consequence of the meeting, however, was that he struck up a friendship with Gilbert Ryle, who had, as Ryle writes in his autobiographical notes, 'for some time been a mystified admirer'. According to Wittgenstein it was the serious and interested expression on Ryle's face during Wittgenstein's paper that attracted his attention, and compelled him to make Ryle's acquaintance. Later, Ryle became convinced that Wittgenstein's influence on his students was detrimental, and Wittgenstein that Ryle was not, after all, *serious*. But throughout the 1930s the two enjoyed a cordial relationship, and occasionally accompanied each other on walking holidays. Their conversation on these walks was as likely to touch on films as on philosophy, Ryle stoutly resisting Wittgenstein's contention that, not only had a good British film never been made, but such a thing was an impossibility – almost, one might say (subject to further analysis), a *logical* impossibility.

Wittgenstein's conviction that his paper on infinity would be 'all Chinese' to the philosophers gathered at Nottingham is a typical expression of a recurrent feeling that whatever he said would be liable to be misunderstood. He was, he felt, surrounded by people unable to understand him. Even Ramsey was unable to follow him in his radical departures from the theory of the *Tractatus*. In September we find him complaining in his diary about Ramsey's lack of originality, his inability to see things afresh, as though he had come across the problems for the first time. On 6 October, at the beginning of the Michaelmas term, he recorded a dream that is a kind of allegory of his situation, or at least of how he felt about his position:

> This morning I dreamt: I had a long time ago commissioned someone to make me a water-wheel and now I no longer wanted it but he was still working on it. The wheel lay there and it was bad; it was notched all around, perhaps in order to put the blades in (as in the motor of a steam turbine). He explained to me what a tiresome task it was, and I thought: I had ordered a straightforward paddle-wheel, which would have been simple to make. The thought tormented me that the man was too stupid to explain to him or to make a better wheel, and that I could do nothing but leave him to it. I thought: I have to live with people to whom I cannot make myself understood. – That is a thought that I actually do have often. At the same time with the feeling that it is my own fault.

'The situation of the man, who so senselessly and badly worked on the waterwheel', he adds, 'was my own in Manchester when I made what were, with hindsight, fruitless attempts at the construction of a gas turbine.' But more than that, the dream is a picture of his present intellectual situation now that the *Tractatus* had been proved inadequate. There it lay: ineptly constructed and inadequate for the task, and *still* the man (himself or Ramsey?) was tinkering with it, performing the tiresome and pointless feat of making it still more elaborate, when what was really needed was a completely different, and simpler, kind of wheel.

In November Wittgenstein accepted an invitation from C. K. Ogden, the translator of the *Tractatus*, to deliver a paper to 'The Heretics', a society similar to the Apostles but less élitist and more concerned with science. The society had previously been addressed by such luminaries as H. G. Wells, Bertrand Russell and Virginia Woolf (*Mr Bennett and Mrs Brown* is based on Virginia Woolf's address to the Heretics). This time he chose not to speak 'Chinese', but rather to use the opportunity to try and correct the most prevalent and serious misunderstanding of the *Tractatus*: the idea that it is a work written in a positivist, anti-metaphysical spirit.

In what was to be the only 'popular' lecture he ever gave in his life, Wittgenstein chose to speak on ethics. In it he reiterated the view of the *Tractatus* that any attempt to say anything about the subject-matter of ethics would lead to nonsense, but tried to make clearer the fact that his own attitude to this was radically different from that of a positivist anti-metaphysician:

> My whole tendency and I believe the tendency of all men who ever tried to write or talk on Ethics or Religion was to run against the boundaries of language. This running against the walls of our cage is perfectly, absolutely hopeless. Ethics so far as it springs from the desire to say something about the meaning of life, the absolute good, the absolute valuable, can be no science. What it says does not add to our knowledge in any sense. But it is a document of a tendency in the human mind which I personally cannot help respecting deeply and I would not for my life ridicule it.

He also gave some examples from his own experience of this tendency to 'run against the walls of our cage':

> I will describe this experience in order, if possible, to make you recall the same or similar experiences, so that we may have a common ground for our investigation. I believe the best way of describing it is to say that when I have it I wonder at the existence of the world. And I am then inclined to use such phrases as 'how extraordinary that anything should exist' or 'how extraordinary that the world should exist'. I will mention another experience straight away which I also know and which others of you might be acquainted with: it is, what one might call, the experience of feeling absolutely safe. I mean the state of mind in which one is inclined to say 'I am safe, nothing can injure me whatever happens.'

He went on to show that the things one is inclined to say after such experiences are a misuse of language – they mean nothing. And yet the experiences themselves 'seem to those who have had them, for instance to me, to have in some sense an intrinsic, absolute value'. They cannot be captured by factual language precisely because their value lies beyond the world of facts. In a notebook of the time

Wittgenstein wrote a sentence which he did not include in the lecture, but which crystallizes his attitude perfectly: 'What is good is also divine. Queer as it sounds, that sums up my ethics.'

What is perhaps most striking about this lecture, however, is that it is not about ethics at all, as the term is usually understood. That is to say, there is no mention in it of moral problems, or of how those problems are to be analysed and understood. For Wittgenstein's thoughts on ethics in that sense, we have to turn to his diaries and the records of his conversations.

There is no doubt that, though he regarded ethics as a realm in which nothing was sayable, Wittgenstein did indeed think and say a great deal about moral problems. In fact, his life might be said to have been dominated by a moral struggle – the struggle to be *anständig* (decent), which for him meant, above all, overcoming the temptations presented by his pride and vanity to be dishonest.

It is not true, as some of his friends have insisted, that Wittgenstein was so honest that he was incapable of telling a lie. Nor is it true that he had no trace of the vanity of which he was always accusing himself. Of course, to say this is not to claim that he was, by ordinary standards, either dishonest or vain. He most certainly was not. But there were, equally certainly, occasions on which his concern to impress people overcame his concern to speak the strict truth. In his diary he says of himself:

> What others think of me always occupies me to an extraordinary extent. I am often concerned to make a good impression. I.e. I very frequently think about the impression I make on others and it is pleasant if I think that it is good, and unpleasant if not.

And though, in stating this, he is only remarking on something that is platitudinously true of all of us, yet he is also drawing attention to what he felt to be the biggest barrier between himself and *anständigkeit* – namely, his vanity.

An impression Wittgenstein quite often made, and which no doubt appealed to his vanity, was of being aristocratic. F. R. Leavis, for example, once overheard him remark: 'In my father's house there were seven grand pianos', and immediately wondered whether he was related to the Princess Wittgenstein who figures in the annals of music. It was, in fact, widely believed in Cambridge that he was of the princely German family, the Sayn-Wittgensteins. Though Wittgenstein did not positively encourage this misapprehension, remarks such as the one quoted by Leavis (which is, incidentally, of doubtful truth, there being only three or four grand pianos at the house in the Alleegasse) would have done nothing to correct it. Opinions vary as to what degree of concealment there was about his true background.* Perhaps the most important fact is that Wittgenstein himself felt that he was hiding something – felt that he was allowing people to think of him as an aristocrat when in fact he was a Jew. In December he reported a complex dream, which might be seen as an expression of this anxiety:

> A strange dream:
>
> I see in an illustrated newspaper a photograph of Vertsagt, who is a much talked about hero of the day. The picture shows him in his car. People talk about his disgraceful deeds; Hänsel is standing next to me and also someone else, someone who resembles my brother Kurt. The latter says that Vertsag [sic] is a Jew but has enjoyed the upbringing of a rich Scottish Lord. Now he is a workers' leader (*Arbeiterführer*). He has not changed his name because it is not the custom there. It is new to me that Vertsagt, which I pronounce with the stress on the first syllable, is a Jew, and I see that his name is simply *verzagt* [German for 'faint-hearted']. It doesn't strike me that it is written with 'ts' which I see printed a little bolder than the other letters. I think: must there be a Jew behind every indecency? Now Hänsel and I are on the terrace of a house, perhaps the big log-cabin on the Hochreit, and along the street comes Vertsag in his motor-car; he has an angry face, slightly reddish fair hair and a similarly coloured moustache (he does not look Jewish). He opens fire with a machine-gun at a cyclist behind him who writhes with pain and is mercilessly gunned to the ground with several shots. Vertsag has driven past, and now comes a young, poor-looking girl on a cycle and she too is shot at by Vertsag as he drives on. And these shots, when they hit her breast make a bubbling sound like an almost-empty kettle over a flame. I felt sorry for the girl and thought that it could only happen in Austria that this girl would find no help and compassion; that the people would look on as she suffers and is

---

* Bartley claims that Wittgenstein once pleaded with a cousin who was living in England not to reveal his partly Jewish descent, while most of his friends insist he did nothing at all to hide the truth about his origins.

killed. I myself am afraid to help her because I am afraid of being shot by Vertsag. I go towards her, but try and hide behind a board.

Then I wake up. I must add that in the conversation with Hänsel, first in the presence of the other person and then after he had left us, I am embarrassed and do not want to say that I myself am descended from Jews or that the case of Vertsag is my own case too.

Wittgenstein's waking reflections on the dream have to do mostly with the name of its central character. He thought, strangely, that it was spelt *pferzagt* (which means nothing) and also that it was Hungarian: 'The name had for me something evil, spiteful and very masculine about it.'

But more pertinent perhaps is his very first thought: that the case of Vertsagt is also his own – that of a man who is regarded as a hero and has the looks and upbringing of an aristocrat, but is actually a Jew and a scoundrel. And, what is worse, that he felt too embarrassed, too *verzagt*, to confess it. This feeling of cowardice haunted him for many years, and led him eventually, seven years after this dream, to make a formal confession of the extent of his Jewish background.

What is most disturbing about the dream, however, is Wittgenstein's use of Nazi slogans to express his own internal anxieties. Is a Jew behind every indecency? The question might have come out of *Mein Kampf*, so redolent is it of the Nazi picture of the deceitful, parasitic Jew hiding his real intentions and his real nature as he spreads his poison among the German peoples. Thankfully, the period during which Wittgenstein was inclined to adopt this image (or something not so very different from it) to describe and analyse his own *unanständigkeit* is mercifully brief. It reaches its climax in a series of remarks about Jewishness written in 1931, and after that comes abruptly to an end.

One question which naturally arises out of the dream is not discussed by Wittgenstein: is Vertsagt's shooting of the innocent girl a symbol of his own corrupting influence on Marguerite? There is, of course, no way of answering the question, but there are, I think, reasons for thinking that his plans to marry Marguerite prompted him to make yet deeper and more strenuous efforts to cleanse himself of his own impurities, to unearth all the unpleasant and dishonest sides to his nature that he preferred to keep hidden, in readiness for his commitment to the 'sacred' act he mentioned to Lettice Ramsey.

12
THE 'VERIFICATIONIST PHASE'

Towards the end of 1929 Wittgenstein might have gained some hint of Marguerite's ambivalence about their relationship, and of her doubts about marrying him, when, shortly after he arrived in Vienna to spend Christmas with her and with his family, she announced that she no longer wished to kiss him. Her feelings for him, she explained, were not of the appropriate kind. Wittgenstein did not take the hint. In his diary notes he does not pause to reflect on *her* feelings, but dwells, rather, on his own. He found it painful, he admitted, but at the same time he was not unhappy about it. For, really, everything depended on his spiritual state rather than on the satisfaction of his sensual desires. 'For if the spirit does not abandon me, then nothing that happens is dirty and petty.' 'I will, however', he added, 'have to stand on tiptoe a great deal if I do not want to go under.' The problem, as he saw it, was not to win her over but to conquer his own desires. 'I am a beast and am still not unhappy about it', he wrote on Christmas Day. 'I am in danger of becoming still more superficial. May God prevent it!'

As a technique for avoiding this tendency, or perhaps in order to reveal it, he conceived the idea of writing an autobiography. And here again, *everything* depended on the spirit. On 28 December he wrote:

> The spirit in which one can write the truth about oneself can take the most varied forms; from the most decent to the most indecent. And accordingly it is very desirable or very wrong for it to be written. Indeed, among the true autobiographies that one might write there are all the gradations from the highest to the lowest. I for instance

> cannot write my biography on a higher plane than I exist on. And by the very fact of writing it I do not *necessarily* enhance myself; I *may* thereby even make myself dirtier than I was in the first place. Something inside me speaks in favour of my writing my biography, and in fact I would like some time to spread out my life clearly, in order to have it clearly in front of me, and for others too. Not so much to put it on trial as to produce, in any case, clarity and truth.

Nothing came of this plan, although for the next two or three years he continued to make notes that attempted to expose the 'naked truth' about himself and to reflect upon the nature of a worthwhile autobiography.

Any autobiography he might have written would almost certainly have had more in common with St Augustine's *Confessions* than with, say, Bertrand Russell's *Autobiography*. The writing of it would, that is, have been fundamentally a spiritual act. He considered *Confessions* to be possibly 'the most serious book ever written'. He was particularly fond of quoting a passage from Book I, which reads: 'Yet woe betide those who are silent about you! For even those who are most gifted with speech cannot find words to describe you', but which Wittgenstein, in discussing it with Drury, preferred to render: 'And woe to those who say nothing concerning thee just because the chatterboxes talk a lot of nonsense.'

In conversation with Waismann and Schlick the text was translated even more freely: 'What, you swine, you want not to talk nonsense! Go ahead and talk nonsense, it does not matter!' These free translations, even if they fail to capture Augustine's intended meaning, certainly capture Wittgenstein's view. One should put a stop to the nonsense of chatterboxes, but that does not mean that one should refuse to talk nonsense oneself. Everything, as always, depends on the spirit in which one does it.

To Waismann and Schlick he repeated the general lines of his lecture on ethics: ethics is an attempt to say something that cannot be said, a running up against the limits of language. 'I think it is definitely important to put an end to all the claptrap about ethics – whether intuitive knowledge exists, whether values exist, whether the good is definable.' On the other hand, it is equally important to see that *something* was indicated by the inclination to talk nonsense. He could imagine, he said, what Heidegger, for example, means by anxiety and being (in such statements as: 'That in the face of which one has anxiety is Being-in-the-world as such'), and he sympathized too with Kierkegaard's talk of 'this unknown something with which the Reason collides when inspired by its paradoxical passion'.

St Augustine, Heidegger, Kierkegaard – these are not names one expects to hear mentioned in conversations with the Vienna Circle – except as targets of abuse. Heidegger's work, for example, was used frequently by logical positivists to provide examples of the sort of thing they meant by metaphysical nonsense – the sort of thing they intended to condemn to the philosophical scrapheap.

While Wittgenstein had been in Cambridge, the Circle had formed itself into a self-consciously cohesive group and had made the anti-metaphysical stance that united them the basis for a kind of manifesto, which was published under the title, *Die Wissenschaftliche Weltauffassung: Der Wiener Kreis* ('*The Scientific View of the World: The Vienna Circle*'). The book was prepared and published as a gesture of gratitude to Schlick, who was acknowledged as the leader of the group and who had that year rejected an offer to go to Berlin in order to stay with his friends and colleagues in Vienna. On hearing of the project, Wittgenstein wrote to Waismann to express his disapproval:

> Just because Schlick is no ordinary man, people owe it to him to take care not to let their 'good intentions' make him and the Vienna school which he leads ridiculous by boastfulness. When I say 'boastfulness' I mean any kind of self-satisfied posturing. 'Renunciation of metaphysics!' As if that were something new! What the Vienna school has achieved, it ought to show not say . . . The master should be known by his work.

Apart from an outline of the central tenets of the Circle's doctrine, their manifesto also contained an announcement of a forthcoming book by Waismann entitled *Logik, Sprache, Philosophie*, which was then described as an introduction to the ideas of the *Tractatus*. Despite his misgivings about the manifesto, Wittgenstein agreed to co-operate with the book and to meet regularly with Waismann to explain his ideas.

The discussions were held at Schlick's house. Waismann took fairly complete notes of what Wittgenstein said, partly to use for his projected book and partly in order to keep the other members of the Vienna Circle (whom Wittgenstein refused to meet) informed about Wittgenstein's latest thoughts. These members then quoted Wittgenstein's ideas in their own papers at philosophical conferences etc. In this way Wittgenstein established a reputation as an influential but somewhat shadowy contributor to Austrian philosophical debate. Among some Austrian philosophers there was even speculation that this 'Dr Wittgenstein', about whom they heard a great deal but of whom they saw nothing, was no more than a figment of Schlick's imagination, a mythological character invented as a figurehead for the Circle.

What neither Schlick nor Waismann – still less the other members of the Circle – appreciated in 1929 was how quickly and radically Wittgenstein's ideas were moving away from those of the *Tractatus*. In the ensuing years the conception of Waismann's book was forced to undergo fundamental changes: from its intended beginnings as an exposition of the ideas of the *Tractatus*, it became first a summary of Wittgenstein's modifications of those ideas, and then, finally, a statement of Wittgenstein's entirely new thoughts. After it had reached this final manifestation, Wittgenstein withdrew his co-operation, and the book was never published.*

In his discussions with Schlick and Waismann during the Christmas vacation, Wittgenstein outlined some of the ways in which his views had changed since he had written the *Tractatus*. He explained to them his conviction that the *Tractatus* account of elementary propositions was mistaken, and had to be abandoned – and, with it, his earlier view of logical inference:

> . . . at that time I thought that all inference was based on tautological form. At that time I had not seen that an inference can also have the form: This man is 2m tall, therefore he is not 3m tall.

'What was wrong about my conception', he told them, 'was that I believed that the syntax of logical constants could be laid down

* At least, it was never published in either Waismann's or Wittgenstein's lifetime. In 1965 it appeared in English as *The Principles of Linguistic Philosophy*, but by then the posthumous publication of Wittgenstein's own work had made it more or less obsolete.

without paying attention to the inner connection of propositions.' He now realized, though, that the rules for the logical constants form only a part of 'a more comprehensive syntax about which I did not yet know anything at the time'. His philosophical task now lay in describing this more complicated syntax, and in making clear the role of 'internal connections' in inference.

His thoughts on *how* he was to accomplish this task were, at this time, in a state of flux, changing from one week to the next, and even from one day to the next. A feature of these conversations is how often Wittgenstein begins his remarks with comments like 'I used to believe . . .', 'I have to correct my account . . .', 'I was wrong when I presented the matter in this way . . .', referring, not to positions he had taken in the *Tractatus*, but to views he had expressed earlier in the year, or, perhaps, earlier in the week.

As an example of what he meant by 'syntax' and of the internal connections it established, he imagined someone's saying: 'There is a circle. Its length is 3cm and its width is 2cm.' To this, he says, we could only reply: 'Indeed! What do you mean by a circle then?' In other words, the possibility of a circle that is longer than it is wide is ruled out by what we mean by the word 'circle'. These rules are provided by the syntax, or, as Wittgenstein also says, the 'grammar' of our language, which in this case establishes an 'internal connection' between something's being a circle and its having only *one* radius.

The syntax of geometrical terms prohibits, *a priori*, the existence of such circles, just as the syntax of our colour words rules out the possibility of a thing's being both red and blue. The internal connections set up by these different grammars allow the kind of inferences that had eluded analysis in terms of the tautologies of the *Tractatus*, because each of them forms a *system*:

> Once I wrote [*TLP* 2.1512], 'A proposition is laid against reality like a ruler . . .' I now prefer to say that a *system of propositions* is laid against reality like a ruler. What I mean by this is the following. If I lay a ruler against a spatial object, I lay *all the graduating lines* against it at the same time.

If we measure an object to be ten inches, we can also infer immediately that it is *not* eleven inches etc.

In describing the syntax of these systems of propositions, Wittgenstein was coming close to, as Ramsey had put it, outlining certain '*necessary* properties of space, time, and matter'. Was he, then, in some sense, doing physics? No, he replies, physics is concerned with determining the *truth* or *falsity* of states of affairs; he was concerned with distinguishing *sense* from *nonsense*. 'This circle is 3 cm long and 2 cm wide' is not false, but nonsensical. The properties of space, time and matter that he was concerned with were not the subject of a physical investigation, but, as he was inclined to put it at this time, a *phenomenological* analysis. 'Physics', he said, 'does not yield a description of the structure of phenomenological states of affairs. In phenomenology it is always a matter of possibility, i.e. of sense, not of truth and falsity.'

This way of putting things had, for Schlick, an uncomfortably Kantian ring. It almost sounded as if Wittgenstein were attempting, à la *The Critique of Pure Reason*, to describe the general and necessary features of the 'structure of appearance', and was being led along the road that led to Husserl. With Husserl's phenomenology in mind, he asked Wittgenstein: 'What answer can one give to a philosopher who believes that the statements of phenomenology are synthetic *a priori* judgements?' To this Wittgenstein replied enigmatically: 'I would reply that it is indeed possible to make up words, but I cannot associate a thought with them.' In a remark written at about this time he is more explicit: his view that there are indeed grammatical rules that are not replaceable by tautologies (e.g. arithmetical equations) 'explains – I believe – what Kant means when he insists that 7 + 5 = 12 is not an analytic proposition, but synthetic *a priori*'. In other words, his answer is the familiar one that his investigations *show* what Kant and the Kantians have tried to *say*.

Though disturbed by this pseudo-Kantian strain in Wittgenstein's new reflections, Schlick and (therefore) the other members of the Vienna Circle took comparatively little notice of it. More congenial to the empiricist tenor of their thinking was another point expressed by Wittgenstein in the course of these conversations. This was that, if a proposition is to have a meaning, if it is to say something, we must have some idea of what would be the case if it were true. And therefore we must have some means of establishing its truth or falsity. This became known to the Vienna Circle as 'Wittgenstein's Principle of Verification', and so enthusiastically was it adopted by the members of the Circle that it has been regarded ever since as the very essence of logical positivism. In English it received its best-known and most strident statement in A. J. Ayer's *Language, Truth and Logic* (a title inspired – if that is the word – by Waismann's *Logik, Sprache, Philosophie*), which was published in 1936 and was written after Ayer had spent some time in Vienna sitting in on meetings of the Circle.

The principle is expressed in the slogan: The sense of a proposition is its means of verification; and it was explained by Wittgenstein to Schlick and Waismann as follows:

> If I say, for example, 'Up there on the cupboard there is a book', how do I set about verifying it? Is it sufficient if I glance at it, or if I look at it from different sides, or if I take it into my hands, touch it, open it, turn over its leaves, and so forth? There are two conceptions here. One of them says that however I set about it, I shall never be able to verify the proposition completely. A proposition always keeps a back-door open, as it were. Whatever we do, we are never sure that we are not mistaken.
>
> The other conception, the one I want to hold, says, 'No, if I can never verify the sense of a proposition completely, then I cannot have meant anything by the proposition either. Then the proposition signifies nothing whatsoever.'
>
> In order to determine the sense of a proposition, I should have to know a very specific procedure for when to count the proposition as verified.

Later, Wittgenstein denied that he had ever intended this principle to be the foundation of a Theory of Meaning, and distanced himself from the dogmatic application of it by the logical positivists. He told a meeting of the Moral Science Club in Cambridge:

> I used at one time to say that, in order to get clear how a sentence is used, it was a good idea to ask oneself the question: 'How would one try to verify such an assertion?' But that's just one way among others of getting clear about the use of a word or sentence. For example, another question which it is often very useful to ask oneself is: 'How is this word learned?' 'How would one set about teaching a child to use this word?' But some people have turned this

suggestion about asking for the verification into a dogma – as if I'd been advancing a theory about meaning.

When, in the early 1930s, he was asked by G. F. Stout about his views on verification, Wittgenstein told the following parable, the point of which seems to be that discovering that a sentence has no means of verification is to understand something important about it, but not to discover that there is nothing in it to understand:

> Imagine that there is a town in which the policemen are required to obtain information from each inhabitant, e.g. his age, where he came from, and what work he does. A record is kept of this information and some use is made of it. Occasionally when a policeman questions an inhabitant he discovers that the latter does not do any work. The policeman enters this fact on the record, because this too is a useful piece of information about the man!

And yet, despite these later disavowals, throughout 1930 – in his conversations with Schlick and Waismann, in a list of 'Theses' dictated to Waismann, and in his own notebooks – we find the principle expressed by Wittgenstein in formulations that sound every bit as dogmatic as those of the Vienna Circle and of Ayer: 'The sense of a proposition is the way it is verified', 'How a proposition is verified is what it says . . . The verification is not *one* token of the truth, it is *the* sense of the proposition', and so on. We can, it seems, talk of a 'Verificationist Phase' of Wittgenstein's thought. But only if we distance the verification principle from the logical empiricism of Schlick, Carnap, Ayer etc, and place it within the more Kantian framework of Wittgenstein's 'phenomenological', or 'grammatical', investigations.

In the new year of 1930 Wittgenstein returned to Cambridge to find that Frank Ramsey was seriously ill. He had suffered a spell of severe jaundice and had been admitted to Guy's Hospital for an operation to discover the cause. After the operation his condition became critical, and it became apparent that he was dying. Frances Partridge, a close friend of the Ramseys, has described how, the evening before Frank Ramsey's death, she visited his ward and was surprised to find

Wittgenstein sitting in a small room that opened off the ward a few feet from Frank's bed:

> Wittgenstein's kindness, and also his personal grief, were somehow apparent beneath a light, almost jocose tone which I myself found off-putting. Frank had had another operation from which he had not yet come round properly, and Lettice had had no supper, so the three of us set off to search for some, and eventually found sausage rolls and sherry in the station buffet. Then Wittgenstein went off and Lettice and I returned to our furnace.

Ramsey died at three o'clock the following morning, on 19 January. He was twenty-six years old.

The following day Wittgenstein gave his first lecture. He had been invited to give a course of lectures at the end of the previous term by Richard Braithwaite, on behalf of the Moral Science Faculty. Braithwaite asked him under what title the course should be announced. After a long silence Wittgenstein replied: 'The subject of the lectures would be philosophy. What else can be the title of the lectures but Philosophy.' And, under this uniquely general title, they were so listed for the rest of Wittgenstein's lecturing career.

During the Lent term of 1930 he gave an hour's lecture in the Arts School lecture room each week, followed later in the week by a two-hour discussion held in a room in Clare College lent by R. E. Priestley (later Sir Raymond Priestley), the explorer. Subsequently he abandoned the formality of the lecture room altogether and held both lecture and discussion in Priestley's rooms, until 1931, when he acquired a set of rooms of his own in Trinity.

His lecture style has often been described, and seems to have been quite different from that of any other university lecturer: he lectured without notes, and often appeared to be simply standing in front of his audience, thinking aloud. Occasionally he would stop, saying, 'Just a minute, let me think!' and sit down for a few minutes, staring at his upturned hand. Sometimes the lecture would restart in response to a question from a particularly brave member of the class. Often he would curse his own stupidity, saying: 'What a damn fool I am!' or exclaim vehemently: 'This is as difficult as hell!' Attending the lectures were about fifteen people, mostly undergraduates but including also a few dons, most notably G. E. Moore, who sat in the only armchair available (the others sat on deckchairs) smoking his pipe and taking copious notes. Wittgenstein's impassioned and syncopated performances left a memorable impression on all who heard him, and are vividly described by I. A. Richards (the co-author, with C. K. Ogden, of *The Meaning of Meaning*) in his poem, 'The Strayed Poet':

Your voice and his I heard in those Non-Lectures
– Hammock chairs sprawled skew-wise all about;
Moore in the armchair bent on writing it all out –
Each soul agog for any word of yours.

Few could long withstand your haggard beauty,
Disdainful lips, wide eyes bright-lit with scorn,
Furrowed brow, square smile, sorrow-born
World-abandoning devotion to your duty.

Such the torment felt, the spell-bound listeners
Watched and waited for the words to come,
Held and bit their breath while you were dumb,
Anguished, helpless, for the hidden prisoners.

Poke the fire again! Open the window!
Shut it! – patient pacing unavailing,
Barren the revelations on the ceiling –
Dash back again to agitate a cinder.

'O it's so clear! It's absolutely clear!'
Tense nerves crisp tenser then throughout the school;
Pencils are poised: 'Oh, I'm a bloody fool!
A damn'd fool!' – So: however it appear.

Not that the Master isn't pedagogic:
Thought-free brows grow pearly as they gaze
Hearts bleed with him. But – should you want a blaze,
Try prompting! Who is the next will drop a brick?

Window re-opened, fire attack't again,
(Leave, but leave what's out, long since, alone!)
Great calm; A sentence started; then the groan
Arrests the pencil leads. Round back to the refrain.

Richards's title is apt; Wittgenstein's lecturing style, and indeed his writing style, was curiously at odds with his subject-matter, as though a poet had somehow strayed into the analysis of the foundations of mathematics and the Theory of Meaning. He himself once wrote: 'I think I summed up my attitude to philosophy when I said: philosophy ought really to be written as a *poetic composition*.'

In these lectures Wittgenstein outlined his conception of philosophy as 'the attempt to be rid of a particular kind of puzzlement', i.e. 'puzzles of *language*'. The method it employs is that of spelling out the features of the grammar of our language: grammar tells us what makes sense and what does not – it 'lets us do some things with language and not others; it fixes the degree of freedom'. The colour octahedron is an example of grammar, in this sense, because it tells us that, though we can speak of a greenish blue, we cannot speak of a greenish red. It therefore concerns, not truth, but possibility. Geometry is also in this sense a part of grammar. 'Grammar is a mirror of reality.'

In explaining his view of the 'internal relations' established by grammar, Wittgenstein explicitly contrasts it with the causal view of meaning adopted by Ogden and Richards in *The Meaning of Meaning* and by Russell in *The Analysis of Mind*. A causal relation is *external*. In Russell's view, for example, words are used with the intention of causing certain sensations and/or images, and a word is used correctly 'when the average hearer will be affected by it in the way intended'. To Wittgenstein, this talk of cause and effect misses the point. In his notes he ridiculed Russell's account by the following analogy: 'If I wanted to eat an apple, and someone punched me in the stomach, taking away my appetite, then it was this punch that I originally wanted.'

At the end of term the question again arose of how to provide Wittgenstein with funds necessary for him to pursue his work. The grant made by Trinity the previous summer was all but spent, and the college council apparently had doubts as to whether it was worthwhile renewing it. On 9 March, therefore, Moore wrote to Russell at his school in Petersfield to ask him if he would be prepared to look at the work which Wittgenstein was doing, and report to the college on its value:

> . . . for there seems to be no other way of ensuring him sufficient income to continue his work, unless the Council do make him a

> grant; and I am afraid there is very little chance that they will do so, unless they can get favourable reports from experts in the subject; and you are, of course, by far the most competent person to make one.

As Moore had anticipated, Russell was not very enthusiastic. 'I do not see how I can refuse', he replied:

> At the same time, since it involves arguing with him, you are right that it will require a great deal of work. I do not know anything more fatiguing than disagreeing with him in an argument.

The following weekend Wittgenstein visited Russell at Beacon Hill School and tried to explain the work that he had been doing. 'Of course we couldn't get very far in two days', he wrote to Moore, 'but he seemed to understand a little bit of it.' He arranged to see Russell again after the Easter vacation, in order to give him a synopsis of the work that he had done since returning to Cambridge. Thus Wittgenstein's Easter vacation in Vienna was taken up with the task of dictating selected remarks from his manuscripts to a typist. 'It is a terrible bit of work and I feel wretched doing it', he complained to Moore.

The result of this work was the typescript that has now been published as *Philosophical Remarks*. It is usually referred to as a 'transitional' work – transitional, that is, between the *Tractatus* and *Philosophical Investigations* – and it is perhaps the only work that can be so-called without confusion. It does indeed represent a very transitory phase in Wittgenstein's philosophical development, a phase in which he sought to replace the Theory of Meaning in the *Tractatus* with the pseudo-Kantian project of 'phenomenological analysis' outlined in his discussions with Schlick and Waismann. This project, as we shall see, was soon abandoned – and with it the insistence on the Verification Principle as the criterion for meaningfulness. As it stands, *Philosophical Remarks* is the most verificationist, and at the same time the most phenomenological, of all his writings. It uses the tools adopted by the Vienna Circle for a task diametrically opposed to their own.

Upon his return from Vienna towards the end of April, Wittgenstein visited Russell at his home in Cornwall to show him the manuscript. From Russell's point of view it was not a convenient time. His wife Dora was seven months pregnant with the child of another man (Griffen Barry, the American journalist); his daughter Kate was ill with chickenpox; and his son John had gone down with measles. His marriage was falling apart amid mutual infidelities, and he was working desperately hard, writing the popular journalism, the lectures and potboiling books that paid for his financially draining experiment in educational reform. The pressures on him at this time were such that colleagues at Beacon Hill School seriously considered that he was going insane.

In these troubled circumstances Wittgenstein stayed for a day and a half, after which the beleaguered Russell made a rather tired attempt to summarize Wittgenstein's work in a letter to Moore:

> Unfortunately I have been ill and have therefore been unable to get on with it as fast as I hoped. I think, however, that in the course of conversation with him I got a fairly good idea of what he is at. He uses the words 'space' and 'grammar' in peculiar senses, which are more or less connected with each other. He holds that if it is significant to say 'This is red', it cannot be significant to say 'This is loud'. There is one 'space' of colours and another 'space' of sounds. These 'spaces' are apparently given a priori in the Kantian sense, or at least not perhaps exactly that, but something not so very different. Mistakes of grammar result from confusing 'spaces'. Then he has a lot of stuff about infinity, which is always in danger of becoming what Brouwer has said, and has to be pulled up short whenever this danger becomes apparent. His theories are certainly important and certainly very original. Whether they are true, I do not know; I devoutly hope not, as they make mathematics and logic almost incredibly difficult.

'Would you mind telling me whether this letter could possibly suffice for the Council?' he pleaded with Moore. 'The reason I ask is that I have at the moment so much to do that the effort involved in reading Wittgenstein's stuff thoroughly is almost more than I can face. I will, however, push on with it if you think it is really necessary.' Moore did not consider it necessary, although, unfortunately for Russell, he did not think the letter would suffice as a report to the Council. Russell accordingly rewrote his letter in, as he put it, 'grander language, which the Council will be able to understand', and this was then accepted as a report on Wittgenstein's work, and Wittgenstein was duly awarded a grant for £100. 'I find I can only understand Wittgenstein when I am in good health', Russell explained to Moore, 'which I am not at the present moment.'

Given the litany of Russell's troubles at this time, it is surprising that he coped as well as he did with the rigours of examining Wittgenstein's work. For his part, Wittgenstein was a harsh critic of Russell's predicament. He loathed Russell's popular works: *The Conquest of Happiness* was a 'vomative'; *What I Believe* was 'absolutely not a "harmless thing"'. And when, during a discussion at Cambridge, someone was inclined to defend Russell's views on marriage, sex and 'free love' (expressed in *Marriage and Morals*), Wittgenstein replied:

> If a person tells me he has been to the worst places I have no right to judge him, but if he tells me it was his superior wisdom that enabled him to go there, then I know he is a fraud.

On his arrival back in Cambridge on 25 April, Wittgenstein had reported in his diary the state of progress in his own, more restrained, love life:

> Arrived back in Cambridge after the Easter vacation. In Vienna often with Marguerite. Easter Sunday with her in Neuwaldegg. For three hours we kissed each other a great deal and it was very nice.

After the Easter term Wittgenstein returned to Vienna to spend the summer with his family and Marguerite. He lived on the family estate, the Hochreit, but not in the large house, preferring the woodman's cottage, where he had the peace, quiet and unencumbered surroundings that he needed for his work. He received the £50 grant from Trinity College designed to see him through the summer, but, as he wrote to Moore: 'My life is now very economical, in fact as long as I'm here there is no possibility of spending any money.' One of the few breaks he allowed himself from this work was to write nonsense to Gilbert Pattisson:

> Dear Gil (old beast),
> You have a goal of ambition; of course you have; otherwise you'd be a mere drifter, with the spirit of a mouse rather than of man. You are not content to stay where you are. You want something more

out of life. You deserve a better position & larger earnings for the benefit of yourself & those who are (or will be) dependent on you.

How, you may ask, can I lift myself out of the ranks of the ill paid?? To think about these and other problems I have retired to the above address which is a country place about 3 hours journey from Vienna. I have purchased a new big writing book of which I enclose the label & am doing a good deal of work. I also enclose my photo which has been taken recently. The top of my head has been removed as I don't want it for philosophising. I have found pelmanism the most useful method for the organisation of thought. The little gray books have made it possible to 'card-index' my mind.

Early in the summer Wittgenstein met with Schlick and Waismann at Schlick's house in Vienna, primarily to prepare a lecture that Waismann would deliver at the coming conference on the Theory of Knowledge in the exact sciences to be held in Königsberg in September. Waismann's lecture, 'The Nature of Mathematics: Wittgenstein's Standpoint', would be the fourth in a series covering the main schools of thought on the subject of the foundations of mathematics (the others in the series were: Carnap on logicism, Heyting on intuitionism and von Neumann on formalism). The central point of this lecture was the application of the Verification Principle to mathematics to form the basic rule: 'The meaning of a mathematical concept is the mode of its use and the sense of a mathematical proposition is the method of its verification.' In the event, Waismann's lecture, and all other contributions to the conference, were overshadowed by the announcement there of Gödel's famous Incompleteness Proof.*

---

*Gödel's first and second Incompleteness Theorems state: (1) that within any consistent formal system, there will be a sentence that can neither be proved true nor proved false; and (2) that the consistency of a formal system of arithmetic cannot be proved *within* that system. The first (often known simply as Gödel's Theorem) is widely believed to show that Russell's ambition in *Principia Mathematica*, of deriving all mathematics from within a single system of logic is, in principle, unrealizable. Whether Wittgenstein accepted this interpretation of Gödel's result is a moot point. His comments on Gödel's Proof (see *Remarks on the Foundations of Mathematics*, Appendix to Part I) appear at first sight, to one trained in mathematical logic, quite amazingly primitive. The best, and most sympathetic, discussion of these remarks that I know of is S. G. Shanker's 'Wittgenstein's Remarks on the Significance of Gödel's Theorem', in *Gödel's Theorem in Focus*, ed. S. G. Shanker (Croom Helm, 1988), pp. 155–256.

During the summer, Wittgenstein also dictated to Waismann a list of 'Theses', presumably as a preliminary to the proposed joint book. These theses are for the most part a restatement of the doctrines of the *Tractatus*, but they include also a number of 'elucidations' on the subject of verification. Here the Verification Principle is stated in its most general and direct form: 'The sense of a proposition is the way it is verified', and is elucidated in the following way:

> A proposition cannot say more than is established by means of the method of its verification. If I say 'My friend is angry' and establish this in virtue of his displaying a certain perceptible behaviour, I only mean that he displays that behaviour. And if I mean more by it, I cannot specify what that extra consists in. A proposition says only what it does say and nothing that goes beyond that.

Almost as soon as these theses were written, Wittgenstein became dissatisfied with their formulation, which he came to regard as sharing the mistaken dogmatism of the *Tractatus*. Indeed, Wittgenstein was developing a conception of philosophy without any theses at all. This, in fact, is implied by the remarks about philosophy in the *Tractatus*, especially proposition 6.53:

> The correct method in philosophy would really be the following: to say nothing except what can be said, i.e. propositions of natural science – i.e. something that has nothing to do with philosophy – and then, whenever someone else wanted to say something metaphysical, to demonstrate to him that he had failed to give a meaning to certain signs in his propositions. Although it would not be satisfying to the other person – he would not have the feeling that we were teaching him philosophy – this method would be the only strictly correct one.

However, the *Tractatus* itself, with its numbered propositions, notoriously fails to adhere to this method. Insisting that these propositions are not really propositions at all, but 'pseudo-propositions' or 'elucidations', is an obviously unsatisfactory evasion of this central difficulty. And, clearly, a similar difficulty would attend the Theses being compiled by Waismann. Philosophical clarity must be elucidated in some other way than by the assertion of doctrines. In 1930, at the very time that Waismann was preparing his presentation of Wittgenstein's 'Theses', Wittgenstein wrote: 'If one tried to advance theses in philosophy, it would never be possible to debate them, because everyone would agree to them.'

Instead of teaching doctrines and developing theories, Wittgenstein came to think, a philosopher should demonstrate a technique, a method of achieving clarity. The crystallization of this realization and its implications brought him to, as he put it to Drury, 'a real resting place'. 'I know that my method is right', he told Drury. 'My father was a business man, and I am a business man: I want my philosophy to be businesslike, to get something done, to get something settled.' The 'transitional phase' in Wittgenstein's philosophy comes to an end with this.

13
THE FOG CLEARS

By the time he returned to Cambridge in the autumn of 1930, Wittgenstein had reached the resting place he had mentioned to Drury. He had, that is, arrived at a clear conception of the correct *method* in philosophy. His lectures for the Michaelmas term began on an apocalyptic note: 'The nimbus of philosophy has been lost', he announced:

> For we now have a method of doing philosophy, and can speak of skilful philosophers. Compare the difference between alchemy and chemistry: chemistry has a method and we can speak of skilful chemists.

The analogy of the transition from alchemy to chemistry is, in part, misleading. It is not that Wittgenstein thought he had replaced a mystical pseudo-science with a genuine science, but rather that he had penetrated beyond the cloudiness and the mystique of philosophy (its 'nimbus') and discovered that behind it lies *nothing*. Philosophy cannot be transformed into a science, because it has nothing to find out. Its puzzles are the consequence of a misuse, a misunderstanding, of grammar, and require, not solution, but dissolution. And the method of dissolving these problems consists not in constructing new theories, but in assembling reminders of things we already know:

> What we find out in philosophy is trivial; it does not teach us new facts, only science does that. But the proper synopsis of these trivialities is enormously difficult, and has immense importance. Philosophy is in fact the synopsis of trivialities.

In philosophy we are not, like the scientist, building a house. Nor are we even laying the foundations of a house. We are merely 'tidying up a room'.

This humbling of the 'Queen of the Sciences' is an occasion for both triumph and despair; it signals the loss of innocence that is a symptom of a more general cultural decay:

> . . . once a method has been found the opportunities for the expression of personality are correspondingly restricted. The tendency of our age is to restrict such opportunities; this is characteristic of an age of declining culture or without culture. A great man need be no less great in such periods, but philosophy is now being reduced to a matter of skill and the philosopher's nimbus is disappearing.

This remark, like much else that Wittgenstein said and wrote at this time, shows the influence of Oswald Spengler's *Decline of the West* (1918; English edition 1926). Spengler believed that a civilization was an atrophied culture. When a culture declines, what was a living organism rigidifies into a dead, mechanical, structure. Thus a period in which the arts flourish is overtaken by one in which physics, mathematics and mechanics dominate. This general view, especially as applied to the decline of Western European culture during the late nineteenth and early twentieth centuries, chimed perfectly with Wittgenstein's own cultural pessimism. One day, arriving at Drury's rooms looking terribly distressed, he explained that he had seen what amounts to a pictorial representation of Spengler's theory:

> I was walking about in Cambridge and passed a bookshop, and in the window were portraits of Russell, Freud and Einstein. A little further on, in a music shop, I saw portraits of Beethoven, Schubert and Chopin. Comparing these portraits I felt intensely the terrible degeneration that had come over the human spirit in the course of only a hundred years.

In an age in which the scientists have taken over, the great personality – Weininger's 'Genius' – can have no place in the mainstream of life; he is forced into solitude. He can only potter about tidying up his room, and distance himself from all the house-building going on around him.

*

During the Michaelmas term of 1930 Wittgenstein wrote several drafts of a foreword for a book – not the book he was working on with Waismann, but the typescript he had shown Russell earlier in the year. In each draft he tried to make explicit the spirit in which he was writing, and to distance his work from that of scientists and scientific philosophers: to make clear, as it were, that he was working from within the confines of his own tidy little room.

But here he came up against a familiar dilemma: to whom was he thus explaining his attitude? Those who understood it would surely see it reflected in his work, while those who did not would not understand his explanation of it either. It was a dilemma he discussed with himself in his notebooks: 'Telling someone something he does not understand is pointless, even if you add that he will not understand it. (That so often happens with someone you love.)':

> If you have a room which you do not want certain people to get into, put a lock on it for which they do not have the key. But there is no point in talking to them about it, unless of course you want them to admire the room from outside!
>
> The honourable thing to do is to put a lock on the door which will be noticed only by those who can open it, not by the rest.

'But', he added, 'it's proper to say that I think the book has nothing to do with the progressive civilization of Europe and America. And that while its spirit may be possible only in the surroundings of this civilization, they have different objectives.' In an early draft of the foreword he talks explicitly about his own work in relation to that of Western scientists:

> It is all one to me whether or not the typical western scientist understands or appreciates my work, since he will not in any case understand the spirit in which I write. Our civilization is characterised by the word 'progress'. Progress is its form rather than making progress one of its features. Typically it constructs. It is occupied with building an ever more complicated structure. And even clarity is sought only as a means to this end, not as an end in itself. For me on the contrary clarity, perspicuity are valuable in themselves.
>
> I am not interested in constructing a building, so much as in

> having a perspicuous view of the foundations of possible buildings.
> So I am not aiming at the same target as the scientists and my way of thinking is different from theirs.

In the final draft there is no mention of science or scientists. Wittgenstein talks instead of the spirit 'which informs the vast stream of European and American civilization in which all of us stand', and insists that the spirit of his work is different. But the same effect is achieved by striking a religious note:

> I would like to say 'This book is written to the glory of God', but nowadays that would be chicanery, that is, it would not be rightly understood. It means the book is written in good will, and in so far as it is not so written, but out of vanity, etc., the author would wish to see it condemned. He cannot free it of these impurities further than he himself is free of them.

Again and again in his lectures Wittgenstein tried to explain that he was not offering any philosophical *theory*; he was offering only the means to escape any *need* of such a theory. The syntax, the grammar, of our thought could not be, as he had earlier thought, delineated or revealed by analysis – phenomenological or otherwise. 'Philosophical analysis', he said, 'does not tell us anything new about thought (and if it did it would not interest us).' The rules of grammar could not be justified, nor even described, by philosophy. Philosophy could not consist, for example, of a list of 'fundamental' rules of the sort that determine the 'depth-grammar' (to use Chomsky's term) of our language:

> We never arrive at fundamental propositions in the course of our investigation; we get to the boundary of language which stops us from asking further questions. We don't get to the bottom of things, but reach a point where we can go no further, where we cannot ask further questions.

The 'internal relations' which are established by grammar cannot be further examined or justified; we can only give examples of where rules are used correctly and where they are used incorrectly, and say:

'Look – don't you see the rule?' For example, the relation between a musical score and a performance cannot be grasped causally (as though we find, mysteriously, that a certain score *causes* us to play in a certain way), nor can the rules that connect the two be exhaustively described – for, given a certain interpretation, *any* playing can be made to accord with a score. Eventually, we just have to '*see* the rule in the relations between playing and score'. If we cannot see it, no amount of explanation is going to make it comprehensible; if we can, then there comes a point at which explanations are superfluous – we do not need any kind of 'fundamental' explanation.

Wittgenstein's insistence on this point marks the turning point between his 'transitional' phase and his mature later philosophy. The later developments of his method, for example his use of 'language-games', are of less decisive importance. These developments are of an heuristic nature: they reflect the different ways in which Wittgenstein tried to make people see certain connections and differences – to see their way out of philosophical dilemmas. But the really decisive moment came when he began to take literally the idea of the *Tractatus* that the philosopher has nothing to *say*, but only something to *show*, and applied that idea with complete rigour, abandoning altogether the attempt to say something with 'pseudo-propositions'.

This emphasis on *seeing* connections links Wittgenstein's later philosophy with Spengler's *Decline of the West*, and at the same time provides the key to understanding the connection between his cultural pessimism and the themes of his later work. In *The Decline of the West* Spengler distinguishes between the Principle of Form (*Gestalt*) and the Principle of Law: with the former went history, poetry and life; with the latter went physics, mathematics and death. And on the basis of this distinction he announces a general methodological principle: 'The means whereby to identify dead forms is Mathematical Law. The means whereby to understand living forms is Analogy.' Thus Spengler was concerned to understand history, not on the basis of a series of laws, but rather through seeing analogies between different cultural epochs. What he was concerned above all to combat was a conception of history as 'natural science in disguise' – the 'taking of spiritual-political events, as they become visible day by day on the surface at their face value, and arranging them on a scheme of "causes" and "effects"'. He argued for a conception of history that saw the historian's job, not as gathering facts and providing explanations, but as perceiving the significance of events by seeing the morphological (or, as Spengler preferred to say, physiognomic) relations between them.

Spengler's notion of a physiognomic method of history was, as he acknowledges, inspired by Goethe's notion of a morphological study of nature, as exemplified in Goethe's poem *Die Metamorphose der Pflanze*, which follows the development of the plant-form from the leaf through a series of intermediate forms. Just as Goethe studied 'the Destiny in Nature and not the Causality', Spengler says, 'so here we shall develop the form-language of human history'. Goethe's morphology had as its motivation a disgust with the mechanism of Newtonian science; he wanted to replace this dead, mechanical, study with a discipline that sought to 'recognize living forms *as such*, to see in context their visible and tangible parts, to perceive them as *manifestations* of something within'.

Wittgenstein's philosophical method, which replaces theory with 'the synopsis of trivialities', is in this same tradition. 'What I give', he once said in a lecture, 'is the morphology of the use of an expression.' In *Logik, Sprache, Philosophie*, the work on which he collaborated with Waismann, the connection is made explicit:

> Our thought here marches with certain views of Goethe's which he expressed in the *Metamorphosis of Plants*. We are in the habit, whenever we perceive similarities, of seeking some common origin for them. The urge to follow such phenomena back to their origin in the past expresses itself in a certain style of thinking. This recognizes, so to speak, only a single scheme for such similarities, namely the arrangement as a series in time. (And that is presumably bound up with the uniqueness of the causal schema). But Goethe's view shows that this is not the only possible form of conception. His conception of the original plant implies no hypothesis about the temporal development of the vegetable kingdom such as that of Darwin. What then *is* the problem solved by this idea? It is the problem of synoptic presentation. Goethe's aphorism 'All the organs of plants are leaves transformed' offers us a plan in which we may group the organs of plants according to their similarities as if around some natural centre. We see the original form of the leaf changing into similar and cognate forms, into the leaves of the calyx, the leaves of the petal, into organs that are half petals, half stamens, and so on. We follow this sensuous

> transformation of the type by linking up the leaf through intermediate forms with the other organs of the plant.
>
> That is precisely what we are doing here. We are collating one form of language with its environment, or transforming it in imagination so as to gain a view of the whole of space in which the structure of our language has its being.

Explicit statements of what Wittgenstein is trying to accomplish in his philosophical work are rare, and it is perhaps not surprising that, as Drury has put it, 'well-meaning commentators' have made it appear that Wittgenstein's writings 'were now easily assimilable into the very intellectual milieu they were largely a warning against'. But, after all, when we see somebody tidying a room, we do not usually hear them keeping up a commentary all the while explaining what they are doing and why they are doing it – they simply get on with the job. And it was, on the whole, with this strictly 'business-like' attitude that Wittgenstein pursued his own work.

At the end of the Michaelmas term of 1930 Wittgenstein was awarded a five-year fellowship of Trinity College, the typescript that he had shown to Russell earlier in the year (published after his death as *Philosophical Remarks*) being accepted as a fellowship dissertation, with Russell and Hardy acting as examiners. The award put an end, for the time being, to the problem of funding his philosophical work, and gave him the opportunity of working out the consequences of his new method in the secure knowledge that there was indeed a demand for the 'goods' he intended to provide. Replying to congratulations sent by Keynes, he wrote: 'Yes, this fellowship business is very gratifying. Let's hope that my brains will be fertile for some time yet. God knows if they will!'

Wittgenstein's attack on theory dominates his discussions with Schlick and Waismann during the Christmas vacation of 1930. '*For me*', he told them, 'a theory is without value. A theory gives me nothing.' In understanding ethics, aesthetics, religion, mathematics and philosophy, theories were of no use. Schlick had, that year, published a book on ethics in which, in discussing theological ethics, he had distinguished two conceptions of the essence of the good: according to the first, the good is good because it is what God wants; according to the second, God wants the good because it is good. The second, Schlick said, was the more profound. On the contrary, Wittgenstein insisted, the first is: 'For it cuts off the way to any explanation "why" it is good, while the second is the shallow, rationalist one, which proceeds "as if" you could give reasons for what is good':

> The first conception says clearly that the essence of the good has nothing to do with facts and hence cannot be explained by any proposition. If there is any proposition expressing precisely what I think, it is the proposition 'What God commands, that is good'.

Similarly, the way to any explanation of aesthetic value must be cut off. What is valuable in a Beethoven sonata? The sequence of notes? The feelings Beethoven had when he was composing it? The state of mind produced by listening to it? 'I would reply', said Wittgenstein, 'that whatever I was told, I would reject, and that not because the explanation was false but because it was an *explanation*':

> If I were told anything that was a *theory*, I would say, No, no! That does not interest me – it would not be the exact thing I was looking for.

Likewise the truth, the value, of religion can have nothing to do with the *words* used. There need, in fact, be no words at all. 'Is talking essential to religion?' he asked:

> I can well imagine a religion in which there are no doctrinal propositions, in which there is thus no talking. Obviously the essence of religion cannot have anything to do with the fact that there is talking, or rather: when people talk, then this itself is part of a religious act and not a theory. Thus it also does not matter at all if the words used are true or false or nonsense.
>
> In religion talking is not *metaphorical* either; for otherwise it would have to be possible to say the same things in prose.

'If you and I are to live religious lives, it mustn't be that we talk a lot about religion', he had earlier told Drury, 'but that our manner of life is different.' After he had abandoned any possibility of constructing a philosophical theory, this remark points to the central theme of his later work. Goethe's phrase from *Faust*, '*Am Anfang war die Tat*' ('In the beginning was the deed'), might, as he suggested, serve as a motto for the whole of his later philosophy.

The deed, the activity, is primary, and does not receive its rationale or its justification from any theory we may have of it. This is as true with regard to language and mathematics as it is with regard to ethics, aesthetics and religion. 'As long as I can play the game, I can play it, and everything is all right', he told Waismann and Schlick:

> The following is a question I constantly discuss with Moore: Can only logical analysis explain what we mean by the propositions of ordinary language? Moore is inclined to think so. Are people therefore ignorant of what they mean when they say 'Today the sky is clearer than yesterday'? Do we have to wait for logical analysis here? What a hellish idea!

Of course we do not have to wait: 'I must, of course, be able to understand a proposition without knowing its analysis.'

The greater part of his discussions with Waismann and Schlick that vacation was taken up with an explanation of how this principle applies to the philosophy of mathematics. So long as we can use mathematical symbols correctly – so long as we can apply the rules – no 'theory' of mathematics is necessary; a final, fundamental, justification of those rules is neither possible nor desirable. This means that the whole debate about the 'foundations' of mathematics rests on a misconception. It might be wondered why, given his Spenglerian conviction of the superiority of music and the arts over mathematics and the sciences, Wittgenstein troubled himself so much over this particular branch of philosophy. But it should be remembered that it was precisely this philosophical fog that drew him into philosophy in the first place, and that to dispel it remained for much of his life the primary aim of his philosophical work.

It was the contradictions in Frege's logic discovered by Russell that had first excited Wittgenstein's philosophical enthusiasm, and to resolve those contradictions had seemed, in 1911, the fundamental task of philosophy. He now wanted to declare such contradictions trivial, to declare that, once the fog had cleared and these sorts of problems had lost their nimbus, it could be seen that the real problem was not the contradictions themselves, but the imperfect vision that made them look like important and interesting dilemmas. You set up a game and discover that two rules can, in certain cases, contradict one another. So what? 'What do we do in such a case? Very simple – we introduce a new rule and the conflict is resolved.'

They had seemed interesting and important because it had been assumed that Frege and Russell were not just setting up a game, but revealing the foundations of mathematics; if their systems of logic were contradictory, then it looked as if the whole of mathematics were resting on an insecure base and needed to be steadied. But this, Wittgenstein insists, is a mistaken view of the matter. We no more need Frege's and Russell's logic to use mathematics with confidence than we need Moore's analysis to be able to use our ordinary language.

Thus the 'metamathematics' developed by the formalist mathematician David Hilbert is unnecessary.* Hilbert endeavoured to construct a 'meta-theory' of mathematics, seeking to lay a provably consistent foundation for arithmetic. But the theory he has constructed, said Wittgenstein, is not metamathematics, but mathematics: 'It is another calculus, just like any other one.' It offers a series of rules and proofs, when what is needed is a clear *view*. 'A proof cannot dispel the fog':

> If I am unclear about the nature of mathematics, no proof can help me. And if I am unclear about the nature of mathematics, then the question about its consistency cannot arise at all.

The moral here, as always, is: '*You cannot gain a fundamental understanding of mathematics by waiting for a theory.*' The understanding of one game cannot depend upon the construction of another. The analogy with games that is invoked so frequently in these discussions prefigures the later development of the 'language-game' technique, and replaces the earlier talk of 'systems of propositions'. The point of the analogy is that it is obvious that there can be no question of a *justification* for a game: if one can play it, one understands it. And

---

* Hilbert's formalist approach to the foundations of mathematics was announced in a lecture entitled 'On the Foundations of Logic and Arithmetic', delivered to the Third International Congress of Mathematicians in Heidelberg in 1904, and developed in a series of papers published in the 1920s. Two of the most important are reprinted in English translations in Jean von Heijenoort, ed., *From Frege to Gödel: A Source Book in Mathematical Logic* (Harvard, 1967).

similarly for grammar, or syntax: 'A rule of syntax corresponds to a configuration of a game . . . Syntax cannot be justified.'

But, asked Waismann, couldn't there be a theory of a game? There is, for example, a theory of chess, which tells us whether a certain series of moves is possible or not – whether, for instance, one can checkmate the king in eight moves from a given position. 'If, then, there is a theory of chess', he added, 'I do not see why there should not be a theory of the game of arithmetic, either, and why we should not use the propositions of this theory to learn something substantial about the possibilities of this game. This theory is Hilbert's meta-mathematics.'

No, replies Wittgenstein, the so-called 'theory of chess' is itself a calculus, a game. The fact that it uses words and symbols instead of actual chess pieces should not mislead us: 'the demonstration that I can get there in eight moves consists in my actually getting there in the symbolism, hence in doing with signs what, on a chess-board, I do with chessmen . . . and we agree, don't we?, that pushing little pieces of wood across a board is something inessential'. The fact that in algebra we use letters to calculate, rather than actual numbers, does not make algebra the theory of arithmetic; it is simply another calculus.

After the fog had cleared there could be, for Wittgenstein, no question of meta-theories, of theories of games. There were only games and their players, rules and their applications: 'We cannot lay down a rule for the application of another rule.' To connect two things we do not always need a third: 'Things must connect directly, without a rope, i.e. they must already stand in a connection with one another, like the links of a chain.' The connection between a word and its meaning is to be found, not in a theory, but in a practice, in the *use* of the word. And the direct connection between a rule and its application, between the word and the deed, cannot be elucidated with another rule; it must be *seen*: 'Here *seeing* matters essentially: as long as you do not see the new system, you have not got it.' Wittgenstein's abandonment of theory was not, as Russell thought, a rejection of serious thinking, of the attempt to understand, but the adoption of a different notion of what it is to understand – a notion that, like that of Spengler and Goethe before him, stresses the importance and the necessity of 'the understanding that consists in seeing connections'.

14
A NEW BEGINNING

For Wittgenstein, everything depended on the spirit. This is as true of his philosophy as it is of his personal relationships. What distinguished his rejection of metaphysics from that of the logical positivists, for example, was, above all else, the spirit in which it was done. In the forewords he had written in the Michaelmas term of 1930 he had tried to make the spirit of his work explicit. In 1931, he considered another possibility, a way of *showing* what he had previously tried to say. 'I think now', he wrote, 'that the right thing would be to begin my book with remarks about metaphysics as a kind of magic':

> But in doing this I must neither speak in defence of magic nor ridicule it.
>
> What it is that is deep about magic would be kept. –
>
> In this context, in fact, keeping magic out has itself the character of magic.
>
> For when I began in my earlier book to talk about the '*world*' (and not about this tree or this table), was I trying to do anything except conjure up something of a higher order by my words?

He was dissatisfied by these remarks, and wrote '*S*' (for '*schlecht*' = 'bad') by the side of them. But they are nonetheless revealing of his intentions. Given that he could not now, as he had in the *Tractatus*, attempt to 'conjure up' something of a higher order with words, with a theory, he wanted, as it were, to *point* to it. Just as speech is not essential in religion, so words cannot be essential to revealing what is true, or deep, in metaphysics.

Indeed, what is deep in metaphysics, as in magic, is its expression of a fundamentally religious feeling – the desire to run up against the limits of our language, which Wittgenstein had spoken of in connection with ethics, the desire to transcend the boundaries of reason and to take Kierkegaard's 'leap of faith'. This desire in all its manifestations was something for which Wittgenstein had the deepest respect, whether in the philosophies of Kierkegaard and Heidegger, the *Confessions* of St Augustine, the prayers of Dr Johnson, or the devotion of monastic orders. Nor was his respect confined to its Christian forms. *All* religions are wonderful, he told Drury: 'even those of the most primitive tribes. The ways in which people express their religious feelings differ enormously.'

What Wittgenstein felt was 'deep' about magic was precisely that it is a primitive expression of religious feeling. In connection with this, he had long wanted to read *The Golden Bough*, Sir James Frazer's monumental account of primitive ritual and magic, and in 1931 Drury borrowed the first volume from the Cambridge Union library. There are thirteen volumes in all, but Wittgenstein and Drury, though they read it together for some weeks, never got beyond a little way into the first volume, so frequent were Wittgenstein's interruptions to explain his disagreements with Frazer's approach. Nothing could have been more calculated to arouse his ire than Frazer's treatment of magical rituals as though they were early forms of science. The savage who sticks a pin in an effigy of his enemy does so, according to Frazer, because he has formed the mistaken scientific hypothesis that this will injure his opponent. This, on Wittgenstein's view, was to 'explain' something deep by reducing it to something incomparably more shallow. 'What narrowness of life we find in Frazer!' he exclaimed. 'And as a result: how impossible for him to understand a different way of life from the English one of his time!'

> Frazer cannot imagine a priest who is not basically an English parson of our times with all his stupidity and feebleness . . .
>
> Frazer is much more savage than most of his savages, for these savages will not be so far from any understanding of spiritual matters as an Englishman of the twentieth century. His explanations of the observances are much cruder than the sense of the observances themselves.

The wealth of facts which Frazer had collected about these rituals would, Wittgenstein thought, be more instructive if they were presented without any kind of theoretical gloss and arranged in such a way that their relationships with each other – and with our own rituals – could be *shown*. We might then say, as Goethe had said of the plant-forms he had described in *Metamorphose der Pflanze*: '*Und so deutet das Chor auf ein geheimnes Gesetz*' ('And all this points to some unknown law'):

> I can set out this law in an hypothesis of evolution, or again, in analogy with the schema of a plant I can give it the schema of a religious ceremony, but I can also do it just by arranging the factual material so that we can easily pass from one part to another and have a clear view of it – showing it in a perspicuous way.
>
> For us the conception of a perspicuous presentation is fundamental. It indicates the form in which we write of things, the way in which we see things. (A kind of Weltanschauung that seems to be typical of our time. Spengler.)
>
> This perspicuous presentation makes possible that understanding which consists just in the fact that we 'see the connections'.

A *morphology* of magic rituals would, then, preserve what was deep about them, without either ridiculing or defending them. It would, in this way, have 'the character of magic'. Similarly, Wittgenstein hoped, his new method of philosophy would preserve what was to be respected in the old metaphysical theories, and would itself have the character of metaphysics, without attempting the conjuring tricks of the *Tractatus*.

There is an analogy here, too, with Wittgenstein's projected autobiography. This, too, he intended, would reveal his essential nature without any kind of explanation, justification or defence. He took it for granted that what would be revealed would be an 'unheroic', perhaps even an 'ugly', nature. But he was concerned above all that, in laying his real character bare, he should not deny it, make light of it, or, in some perverse way, take pride in it:

> If I may explain in a simile: If a street loafer were to write his biography, the danger would be that he would either
>
> (a) deny that his nature was what it is,

> or (b) would find some reason to be proud of it,
>
> or (c) present the matter as though this – that he has such a nature – were of no consequence.
>
> In the first case he lies, in the second he mimics a trait of the natural aristocrat, that pride which is a vitium splendidum and which he cannot really have any more than a crippled body can have natural grace. In the third case he makes as it were the gesture of social democracy, placing culture above the bodily qualities – but this is deception as well. He is what he is, and this is important and means something but is no reason for pride, on the other hand it is always the object of his self-respect. And I can accept the other's aristocratic pride and his contempt for my nature, for in this I am only taking account of what my nature is and of the other man as part of the environment of my nature – the world with this perhaps ugly object, my person, as its centre.

As Rush Rhees has pointed out, there is something Weiningerian about Wittgenstein's conception of writing an autobiography, a conception that sees it almost as a spiritual duty. 'Putting together a complete autobiography', Weininger writes in *Sex and Character*, 'when the need to do this originates in the man himself, is always the sign of a superior human being':

> For in the really faithful memory the root of piety lies. A man of real character, faced with the proposal or demand that he abandon his past for some material advantage or his health, would reject it, even if the prospect were of the greatest treasures in the world or of happiness itself.

It is in 1931, the year in which his planned autobiography received its greatest attention, that references to Weininger and Weiningerian reflections abound in Wittgenstein's notebooks and conversations. He recommended *Sex and Character* to his undergraduate friends, Lee and Drury, and to Moore. Their response was understandably cool. The work that had excited the imagination of pre-war Vienna looked, in the cold light of post-war Cambridge, simply bizarre. Wittgenstein was forced to explain. 'I can quite imagine that you don't admire Weininger very much', he wrote to Moore on 28 August, 'what with that beastly translation and the fact that W. must feel very foreign to you':

> It is true that he is fantastic but he is great and fantastic. It isn't necessary or rather not possible to agree with him but the greatness lies in that with which we disagree. It is his enormous mistake which is great. I.e. roughly speaking if you add a ~ to the whole book it says an important truth.

What he meant by this elliptical remark remains obscure. On the subject of Weininger's central theme that women and femininity were the sources of all evil, Wittgenstein admitted to Drury: 'How wrong he was, my God he was wrong.' But this hardly reveals the important truth obtained by negating the whole book. The negation of an absurdity is not an important truth, but a platitude ('Women are *not* the source of all evil'). Perhaps he meant that Weininger had captured the essential characteristics of Man and Woman but charged the wrong suspect. In his dream about 'Vertsagt', after all, it is the woman who is the victim, while the perpetrator of the crime is a man whose very name has something disagreeably 'masculine' about it.

Certainly in his autobiographical notes there is nothing to suggest that he considered his 'unheroic', 'ugly' nature to be attributable to any supposed feminine traits.

There are, however, several remarks that indicate that he was inclined to accept a Weiningerian conception of Jewishness, and that he considered at least some of his less heroic characteristics to have something to do with his Jewish ancestry. Like Weininger, Wittgenstein was prepared to extend the concept of Jewishness beyond the confines of such ancestry. Rousseau's character, for example, he thought 'has something Jewish about it'. And, like Weininger, he saw some affinity between the characteristics of a Jew and those of an Englishman. Thus: 'Mendelssohn is not a peak, but a plateau. His Englishness'; 'Tragedy is something un-Jewish. Mendelssohn is, I suppose, the most untragic of composers.'

But – and in this he is also following Weininger – it is clear that for most of the time when he talks of 'Jews' he is thinking of a particular racial group. Indeed, what is most shocking about Wittgenstein's remarks on Jewishness is his use of the language – indeed, the slogans – of racial anti-Semitism. The echo that really disturbs is not that of *Sex and Character*, but that of *Mein Kampf*. Many of Hitler's most outrageous suggestions – his characterization of the Jew as a parasite 'who like a noxious bacillus keeps spreading as soon as a favourable medium invites him', his claim that the Jews' contribution to culture has been entirely derivative, that 'the Jew lacks those qualities which distinguish the races that are creative and hence culturally blessed', and, furthermore, that this contribution has been restricted to an *intellectual* refinement of another's culture ('since the Jew . . . was never in possession of a culture of his own, the foundations of his intellectual work were always provided by others') – this whole litany of lamentable nonsense finds a parallel in Wittgenstein's remarks of 1931.

Were they not written by Wittgenstein, many of his pronouncements on the nature of Jews would be understood as nothing more than the rantings of a fascist anti-Semite. 'It has sometimes been said', begins one such remark, 'that the Jews' secretive and cunning nature is a result of their long persecution':

> That is certainly untrue; on the other hand it is certain that they continue to exist despite this persecution only because they have an inclination towards such secretiveness. As we may say that this or that animal has escaped extinction only because of its capacity or ability to conceal itself. Of course I do not mean this as a reason for commending such a capacity, not by any means.

'They' escape extinction only because they avoid detection? And therefore they are, of necessity, secretive and cunning? This is anti-Semitic paranoia in its most undiluted form – the fear of, and distaste for, the devious 'Jew in our midst'. So is Wittgenstein's adoption of the metaphor of illness. 'Look on this tumour as a perfectly normal part of your body!' he imagines somebody suggesting, and counters with the question: 'Can one do that, to order? Do I have the power to decide at will to have, or not to have, an ideal conception of my body?' He goes on to relate this Hitlerian metaphor to the position of European Jews:

> Within the history of the peoples of Europe the history of the Jews is not treated as their intervention in European affairs would actually merit, because within this history they are experienced as a sort of disease, and anomaly, and no one wants to put a disease on the same level as normal life [and no one wants to speak of a disease as if it had the same rights as healthy bodily processes (even painful ones)].

> We may say: people can only regard this tumour as a natural part of their body if their whole feeling for the body changes (if the whole national feeling for the body changes). Otherwise the best they can do is *put up with* it.
>
> You can expect an individual man to display this sort of tolerance, or else to disregard such things; but you cannot expect this of a nation, because it is precisely not disregarding such things that makes it a nation. I.e. there is a contradiction in expecting someone *both* to retain his former aesthetic feeling for the body and *also* to make the tumour welcome.

Those who seek to drive out the 'noxious bacillus' in their midst, he comes close to suggesting, are right to do so. Or, at least, one cannot expect them – as a nation – to do otherwise.

It goes without saying that this metaphor makes no sense without a racial notion of Jewishness. The Jew, however 'assimilated', will never be a German or an Austrian, because he is not of the same 'body': he is experienced by that body as a growth, a disease. The metaphor is particularly apt to describe the fears of Austrian anti-Semites, because it implies that the more assimilated the Jews become, the more dangerous becomes the disease they represent to the otherwise healthy Aryan nation. Thus it is quite wrong to equate the anti-Semitism implied by Wittgenstein's remarks with the 'Jewish self-hatred' of Karl Kraus. The traits which Kraus disliked, and which he took to be Jewish (acquisitiveness etc.), he attributed not to any racial inheritance but to the social and religious isolation of the Jews. What he attacked primarily was the 'ghetto-mentality' of the Jews; far from wanting to keep Jew and non-Jew separate, and regarding the Jew as a 'tumour' on the body of the German people, he campaigned tirelessly for the complete assimilation of Jews: 'Through dissolution to salvation!'

From this perspective Kraus was far better placed than Wittgenstein to understand the horror of Nazi propaganda – and, one might add, more perceptive in recognizing its intellectual precedents. Wittgenstein, of course, could see that the Nazis were a barbarous 'set of gangsters', as he once described them to Drury, but at the time he was recommending Spengler's *Decline of the West* to Drury as a book that might teach him something about the age in which they were living, Kraus was drawing attention to the affinities between Spengler and the

Nazis, commenting that Spengler understood the *Untergangsters* of the West – and that they understood him.

Though alarming, Wittgenstein's use of the slogans of racist anti-Semitism does not, of course, establish any affinity between himself and the Nazis. His remarks on Jewishness were fundamentally introspective. They represent a turning inwards of the sense of cultural decay and the desire for a New Order (which is the path that leads from Spengler to Hitler) to his own internal state. It is as though, for a brief time (after 1931 there are, thankfully, no more remarks about Jewishness in his notebooks), he was attracted to using the then current language of anti-Semitism as a kind of metaphor for himself (just as, in the dream of Vertsagt, the image of the Jew that was propagated by the Nazis – an image of a cunning and deceptive scoundrel who hides behind a cloak of respectability while committing the most dreadful crimes – found a ready response in his fears about his own 'real' nature). And just as many Europeans, particularly Germans, felt a need for a New Order to replace their 'rotten culture', so Wittgenstein strived for a new beginning in his life. His autobiographical notes were essentially confessional, and 'a confession', he wrote in 1931, 'has to be a part of your new life'. Before he could begin anew, he had to take stock of the old.

What is perhaps most ironic is that, just as Wittgenstein was beginning to develop an entirely new method for tackling philosophical problems – a method that has no precedent in the entire tradition of Western philosophy (unless one finds a place for Goethe and Spengler in that tradition) – he should be inclined to assess his own philosophical contribution within the framework of the absurd charge that the Jew was incapable of original thought. 'It is typical for a Jewish mind', he wrote, 'to understand someone else's work better than [that person] understands it himself.' His own work, for example, was essentially a clarification of other people's ideas:

> Amongst Jews 'genius' is found only in the holy man. Even the greatest of Jewish thinkers is no more than talented. (Myself for instance.) I think there is some truth in my idea that I really only think reproductively. I don't believe I have ever *invented* a line of thinking. I have always taken one over from someone else. I have simply straightaway seized on it with enthusiasm for my work of clarification. That is how Boltzmann, Hertz, Schopenhauer, Frege,

> Russell, Kraus, Loos, Weininger, Spengler, Sraffa have influenced me. Can one take the case of Breuer and Freud as an example of Jewish reproductiveness? – What I invent are new *similes*.

This belittling of his own achievement may have been a way of guarding himself from his own pride – from believing that he really was, as he once lightheartedly described himself in a letter to Pattisson, 'the greatest philosopher that ever lived'. He was acutely aware of the dangers of false pride. 'Often, when I have had a picture well framed or have hung it in the right surroundings', he wrote, 'I have caught myself feeling as proud as if I had painted it myself.' And it was against the background of such pride that he felt forced to remind himself of his limitations, of his 'Jewishness':

> The Jew must see to it that, in a literal sense, 'all things are as nothing to him'. But this is particularly hard for him, since in a sense he has nothing that is particularly his. It is much harder to accept poverty willingly when you *have* to be poor than when you might also be rich.
>
> It might be said (rightly or wrongly) that the Jewish mind does not have the power to produce even the tiniest flower or blade of grass that has grown in the soil of another's mind and to put it into a comprehensive picture. We aren't pointing to a fault when we say this and everything is all right as long as what is being done is quite clear. It is only when the nature of a Jewish work is confused with that of a non-Jewish work that there is any danger, especially when the author of the Jewish work falls into the confusion himself, as he so easily may. (Doesn't he look as proud as though he had produced the milk himself?)

So long as he lived, Wittgenstein never ceased to struggle against his own pride, and to express doubts about his philosophical achievement and his own moral decency. After 1931, however, he dropped the language of anti-Semitism as a means of expressing those doubts.

Wittgenstein's remarks on Jewishness, like his projected autobiography, were essentially confessional, and both seem in some way linked to the 'sacred' union he had planned for himself and

Marguerite. They coincide with the year in which his intention to marry Marguerite was pursued with its greatest earnestness.

Early in the summer he invited Marguerite to Norway, to prepare, as he thought, for their future life together. He intended, however, that they should spend their time separately, each taking advantage of the isolation to engage in serious contemplation, so that they would be spiritually ready for the new life that was to come.

Accordingly, while he stayed in his own house, he arranged lodgings for Marguerite at the farmhouse of Anna Rebni, a tough seventy-year-old woman who lived with her hundred-year-old mother. During the two weeks she spent there, Marguerite saw very little of Wittgenstein. When she arrived at the farmhouse, she unpacked her bags to find a Bible which Wittgenstein had slipped in there, together with a letter, significantly tucked into Corinthians I 13 – St Paul's discourse on the nature and virtue of love. It was a heavy hint that she did not take. Instead of meditating, praying and reading the Bible – which is how Wittgenstein spent much of his time – she did what Pinsent had done in 1913, and made the best she could of what little Skjolden had to offer in the way of entertainment. She took walks around the farm, went swimming in the fjord, got to know the villagers and learnt a little Norwegian. After two weeks she left for Rome to attend her sister's wedding, determined that the one man she was *not* going to marry was Ludwig Wittgenstein. Not only did she feel she could never rise to meet the demands a life with Wittgenstein would present; also, and equally important, she knew Wittgenstein would never be able to give her the sort of life she wanted. He had made it clear, for example, that he had absolutely no intention of having children, thinking that to do so would simply be to bring another person into a life of misery.

For some of his time in Norway, Wittgenstein was joined by Gilbert Pattisson, whose visit overlapped by about a week that of Marguerite, and who no doubt did something to lighten Wittgenstein's mood during the three weeks that he was there – although, as usual, Pattisson found it necessary to get away from Wittgenstein from time to time, taking himself to Oslo for a night to 'paint the town red'.

The visit to Norway may have put an end to any idea there might have been of Wittgenstein's marrying Marguerite, but it did not (or not immediately) result in a breaking off of friendship. For three weeks during the late summer of 1931 they saw each other almost every day at the Hochreit, where Wittgenstein stayed, as before, at the woodman's cottage on the edge of the estate, while Marguerite was a guest of Gretl's at the family residence. In a volume of reminiscences written for her grandchildren, she comments, in a phrase that recalls the role of David Pinsent: 'My presence brought him the peace which he needed while he was nurturing his ideas.'

At the Hochreit Wittgenstein worked on the completion of his book, which at this time had the working title of *Philosophical Grammar*, a title he admitted might have the smell of a textbook, 'but that doesn't matter, for behind it there is the book'.

Wittgenstein had a peculiarly laborious method of editing his work. He began by writing remarks into small notebooks. He then selected what he considered to be the best of these remarks and wrote them out, perhaps in a different order, into large manuscript volumes. From these he made a further selection, which he dictated to a typist. The resultant typescript was then used as the basis for a further selection, sometimes by cutting it up and rearranging it – and then the whole process was started again. Though this process continued for more than twenty years, it never culminated in an arrangement with which Wittgenstein was fully satisfied, and so his literary executors have had to publish either what they consider to be the most satisfactory of the various manuscripts and typescripts (*Philosophical Remarks*, *Philosophical Investigations*, *Remarks on the Philosophy of Psychology*), or a selection from them (or rearrangement of them) made by the executors themselves (*Philosophical Grammar*, *Remarks on the Foundations of Mathematics*, *Culture and Value*, *Zettel*). These we now know as the works of the later Wittgenstein, though in truth not one of them can be regarded as a completed work.

For this frustrating circumstance we can blame the fastidiousness about publication that had so infuriated Russell in 1913, and was soon to exasperate even further the unfortunate Friedrich Waismann. For in 1931, as Wittgenstein was just beginning to formulate some sort of satisfactory presentation of his new thought, Waismann was under the impression that his own presentation of Wittgenstein's ideas, the book that had been announced in 1929 under the title of *Logik, Sprache, Philosophie*, was nearing completion. On 10 September Schlick wrote to Waismann from California, commenting that he assumed the book would be ready for publication soon and that it would appear in print by the time he returned to Vienna the following Easter.

Waismann, however, had seen little of Wittgenstein that summer. Shortly before the end of the vacation, Wittgenstein met him in Vienna to present him with the latest typescripts to have been culled from his recent work. They discussed the changes in the proposed book that would have to be made in the light of this new work, and, on the basis of these discussions, Waismann rewrote his 'Theses' and sent the new version to Schlick. Wittgenstein, meanwhile, was becoming increasingly concerned that Waismann might misrepresent his new thoughts. In November he wrote to Schlick about 'this Waismann thing', and apologized for keeping him waiting for a final edition. He stressed that he wanted to honour his obligations to Schlick, but: 'For the thing itself I have no enthusiasm. I am convinced that Waismann would present *many* things in a form *completely* different to what I regard as correct.'

The central problem was that the book as originally conceived was now redundant. Wittgenstein's ideas had changed so fundamentally that he could no longer present them in a form that was essentially an updated version of the *Tractatus*. 'There are', he told Schlick, '*very*, *very* many statements in the book with which I now disagree!' The *Tractatus* talk about 'elementary propositions' and 'objects' had, he said, been shown to be erroneous, and there was no point in publishing a work that simply repeated the old mistakes. The *Tractatus* analysis of the proposition must be replaced by a 'perspicuous representation' of grammar which would throw overboard 'all the dogmatic things that I said about "objects", "elementary propositions" etc.'

Wittgenstein next met Waismann during the Christmas vacation of 1931, and it was then that he made clear to him his view that the whole conception of the book would have to be changed. He explained the implications of his new thinking for the status of philosophical theses:

> If there were theses in philosophy, they would have to be such that they do not give rise to disputes. For they would have to be put in such a way that everyone would say, Oh yes, that is of course obvious. As long as there is a possibility of having different opinions and disputing about a question, this indicates that things have not yet been expressed clearly enough. Once a perfectly clear

> formulation – ultimate clarity – has been reached, there can be no second thoughts or reluctance any more, for these always arise from the feeling that something has now been asserted, and I do not yet know whether I should admit it or not. If, however, you make the grammar clear to yourself, if you proceed by very short steps in such a way that every single step becomes perfectly obvious and natural, no dispute whatsoever can arise. Controversy always arises through leaving out or failing to state clearly certain steps, so that the impression is given that a claim has been made that could be disputed.

About the *Tractatus*, he told Waismann that in it he had 'still proceeded dogmatically . . . I saw something from far away and in a very indefinite manner, and I wanted to elicit from it as much as possible.' 'But', he added firmly, 'a rehash of such theses is no longer justified.' He insisted that notes taken by Waismann of this discussion should be sent to Schlick in California, and that Waismann should inform Schlick of the change in plan and explain the reasons for it.

When Wittgenstein returned to Cambridge in the new year of 1932 he wrote to Schlick asking whether he had received Waismann's notes, and whether he could 'make head or tail of it'. Schlick evidently thought he could, for he persisted in encouraging Waismann to continue with the project. Like Wittgenstein, Waismann did so for Schlick's sake. For 'the thing' itself, we can assume that he had no more enthusiasm than Wittgenstein. The following Easter, his already unenviable position was made even more difficult when Wittgenstein proposed a new procedure: instead of Waismann receiving material for the book directly from Wittgenstein, he was to be dependent on Schlick for typescripts that Wittgenstein would send to him. Wittgenstein had, in other words, completely lost faith in Waismann as a communicator of his ideas; Waismann was no longer, for example, given the responsibility of presenting Wittgenstein's new ideas to the members of the Vienna Circle.

Almost all of Wittgenstein's energies were by now devoted to producing his own presentation of his new thoughts. He experimented with many different formulations – numbered remarks, numbered paragraphs, an annotated table of contents etc. In his lectures, as though to orientate himself within the Western tradition, he went through C. D.

Broad's taxonomy of philosophical styles and theories, given in Broad's own series of undergraduate lectures, 'Elements of Philosophy'. The method of Hume and Descartes, he rejected, but said of Kant's critical method: 'This is the right sort of approach.' With regard to the distinction between the deductive and dialectical methods of speculative philosophy – the first represented by Descartes, the second by Hegel – he came down, with reservations, on the side of Hegel:

> . . . the dialectical method is very sound and a way in which we do work. But it should not try to find, from two propositions, a and b, a further more complex proposition, as Broad's description implied. Its object should be to find out where the ambiguities in our language are.

Of Broad's three 'theories of truth' – the Correspondence Theory, the Coherence Theory and the Pragmatic Theory – he was dismissive: 'Philosophy is not a choice between different "theories"':

> We can say that the word ['truth'] has at least three different meanings; but it is mistaken to assume that any one of these theories can give the whole grammar of how we use the word, or endeavour to fit into a single theory cases which do not seem to agree with it.

What replaces theory is *grammar*. During this series of lectures Moore made a spirited attempt to insist that Wittgenstein was using the word 'grammar' in a rather odd sense. He presented to Wittgenstein's class a paper distinguishing what he took to be the usual meaning from its Wittgensteinian use. Thus, he argued, the sentence: 'Three men was working' is incontrovertibly a misuse of grammar, but it is not clear that: 'Different colours cannot be in the same place in a visual field at the same time' commits a similar transgression. If this latter is also called a misuse of grammar, then 'grammar' must mean something different in each case. No, replied Wittgenstein. 'The right expression is "It does not have sense to say . . ."' Both kinds of rules were rules in the same sense. 'It is just that some have been the subject of philosophical discussion and some have not':

> Grammatical rules are all of the same kind, but it is not the same mistake if a man breaks one as if he breaks another. If he uses 'was' instead of 'were' it causes no confusion; but in the other example the analogy with physical space (c.f. two people in the *same* chair) does cause confusion. When we say we can't think of two colours in the same place, we make the mistake of thinking that this is a proposition, though it is not; and we would never try to say it if we were not misled by an analogy. It is misleading to use the word 'can't' because it suggests a wrong analogy. We should say, 'It has no sense to say . . .'

The grammatical mistakes of philosophers, then, differed from the ordinary mistakes mentioned by Moore only in being more pernicious. To study these mistakes was, therefore, pointless – indeed, it was worse than that, and could only do harm; the point was not to study them, but to free oneself from them. Thus, to one of his students, Karl Britton, Wittgenstein insisted that he *could* not take philosophy seriously so long as he was reading for a degree in the subject. He urged him to give up the degree and do something else. When Britton refused, Wittgenstein only hoped that it would not kill his interest in philosophy.

Similarly, he urged Britton, as he urged most of his students, to avoid becoming a teacher of philosophy. There was only one thing worse, and that was becoming a journalist. Britton should do a real job, and work with ordinary people. Academic life was detestable. When he returned from London, he told Britton, he would overhear one undergraduate talking to another, saying, 'Oh, really!' and know that he was back in Cambridge. The gossip of his bedmaker in college was much preferable to the insincere cleverness of high table.

Maurice Drury had already taken Wittgenstein's advice, and was working with a group of unemployed shipbuilders in Newcastle. As the project neared its completion, however, he was tempted into applying for a post as a lecturer in philosophy at Armstrong College, Newcastle. In the event, the post was given to Dorothy Emmett, and Drury went to South Wales to help run a communal market garden for unemployed miners. 'You owe a great debt to Miss Emmett', Wittgenstein insisted; 'she saved you from becoming a professional philosopher.'

Despite this contempt for the profession, Wittgenstein kept a jealous and watchful eye on the use to which his ideas were put by academic philosophers, and in the summer of 1932 was involved in what amounts to a *Prioritätstreit* with Rudolf Carnap. It was occasioned by an article by Carnap entitled '*Die physikalische Sprache als Universalsprache der Wissenschaft*', which was published in *Erkenntnis*, the journal of the Vienna Circle (it was later published in English, as *The Unity of Science*). The article is an argument for 'physicalism' – the view that *all* statements, in so far as they are worthy of inclusion into a scientific study, are ultimately reducible to the language of physics, whether the science concerned treats of physical, biological, psychological or social phenomena. It is indebted, as Carnap acknowledges, to the views of Otto Neurath, the most rigorously positivist of the Vienna Circle philosophers.

Wittgenstein, however, was convinced that Carnap had used the ideas he himself had expressed in conversations with the Vienna Circle, and had done so without proper acknowledgement. In August 1932, in two letters to Schlick and again in a letter to Carnap himself, Wittgenstein insists that his annoyance at Carnap's article is a purely ethical and personal matter, and is in no way to do with a concern to establish his authorship of the thoughts published by Carnap, or with his anxiety about his reputation within the academic community. On 8 August he wrote to Schlick:

> . . . from the bottom of my heart it is all the same to me what the professional philosophers of today think of me; for it is not for them that I am writing.

And yet his point was that the ideas published under Carnap's name – about, for example, ostensive definition and the nature of hypotheses – were properly speaking *his* ideas. Carnap took them, he alleged, from records of his conversations with Waismann. When Carnap replied that his central argument was concerned with *physicalism*, about which Wittgenstein had said nothing, Wittgenstein objected that the basic idea was to be found in the *Tractatus*: 'That I had not dealt with the question of "physicalism" is untrue (only not under that – horrible – name), and [I did so] with the brevity with which the whole of the *Tractatus* is written.'

With the publication of Carnap's article Wittgenstein's philosophical conversations with Waismann finally came to an end. Their last recorded discussion, indeed, is taken up with an attempt by Wittgenstein to refute Carnap's suggestion that his (Carnap's) conception of a hypothesis was taken from Poincaré, rather than from Wittgenstein himself. After that, Waismann was not to be trusted with privileged access to Wittgenstein's new ideas.

Wittgenstein's growing distrust of Waismann, and his pique at what he saw as Carnap's impertinence, coincided with renewed efforts on his part to compose a publishable presentation of his work.

During his sojourn at the Hochreit during the summer of 1932, he dictated to a typist a large selection of remarks from the eight manuscript volumes he had written during the previous two years. (In the letter to Schlick of 8 August he mentions that he is spending up to seven hours a day in dictation.) The result was what has become known to Wittgenstein scholars as 'the Big Typescript'. This, more than any of the other typescripts left by Wittgenstein, presents the appearance of a finished book, complete with chapter headings and a table of contents, and forms the basis of what has been published as *Philosophical Grammar*. It is, however, by no means identical with the published text.

In particular, an interesting chapter entitled 'Philosophy' has been omitted from the published version. 'All that philosophy can do', he says there, 'is to destroy idols.' 'And', he adds, in a swipe at the Vienna Circle, 'that means not making any new ones – say out of "the absence of idols".' It is not in practical life that we encounter philosophical problems, he stresses, but, rather, when we are misled by certain analogies in language to ask things like 'What is time?', 'What is a number?' etc. These questions are insoluble, not because of their depth and profundity, but rather because they are nonsensical – a misuse of language. Thus:

> The real discovery is the one that makes me capable of stopping doing philosophy when I want to – the one that gives philosophy peace, so that it is no longer tormented by questions which bring *itself* in question. – Instead, we now demonstrate a method, by examples; and the series of examples can be broken off. – Problems are solved (difficulties eliminated), not a *single* problem . . . 'But then we will never come to the end of our job!' Of course not, because it has no end.

This conception of philosophy, which sees it as a task of clarification that has no end, and only an arbitrary beginning, makes it almost impossible to imagine how a satisfactory book on philosophy *can* be written. It is no wonder that Wittgenstein used to quote with approval Schopenhauer's dictum that a book on philosophy, with a beginning and an end, is a sort of contradiction. And it comes as no surprise that, almost as soon as he had finished dictating the Big Typescript, he began to make extensive revisions to it. The section he revised least, however, was that on the philosophy of mathematics (hence the complete reproduction of those chapters in *Philosophical Grammar*). It is unfortunate that his work in that area has not received the same attention as have his remarks on language.

Not only did Wittgenstein himself regard his work on mathematics as his most important contribution to philosophy; it is also in this work that it is most apparent how radically his philosophical perspective differs from that of the professional philosophy of the twentieth century. It is here that we can see most clearly the truth of his conviction that he was working against the stream of modern civilization. For the target at which his remarks are aimed is not a particular view of mathematics that has been held by this or that philosopher; it is, rather, a conception of the subject that is held almost universally among working mathematicians, and that has, moreover, been dominant throughout our entire culture for more than a century – the view, that is, that sees mathematics as a *science*.

'Confusions in these matters', he writes in the Big Typescript, 'are entirely the result of treating mathematics as a kind of natural science':

> And this is connected with the fact that mathematics has detached itself from natural science; for, as long as it is done in immediate connection with physics, it is clear that *it* isn't a natural science. (Similarly, you can't mistake a broom for part of the furnishing of a room as long as you use it to clean the furniture.)

Wittgenstein's philosophy of mathematics is not a contribution to the debate on the foundations of the subject that was fought during the first half of this century by the opposing camps of logicists (led by Frege and Russell), formalists (led by Hilbert) and intuitionists (led by Brouwer and Weyl). It is, instead, an attempt to undermine the whole basis of this debate – to undermine the idea that mathematics *needs* foundations. All the branches of mathematics that were inspired by this search for 'foundations' – Set Theory, Proof Theory, Quantificational Logic, Recursive Function Theory, etc. – he regarded as based on a philosophical confusion. Thus:

> Philosophical clarity will have the same effect on the growth of mathematics as sunlight has on the growth of potato shoots. (In a dark cellar they grow yards long.)

Wittgenstein knew, of course, that with regard to mathematics, if not with regard to his entire philosophical enterprise, he was tilting at windmills. 'Nothing seems to me less likely', he wrote, 'than that a scientist or mathematician who reads me should be seriously influenced in the way he works.' If, as he repeatedly emphasized, he was not writing for professional philosophers, still less was he writing for professional mathematicians.

15
FRANCIS

Wittgenstein's quixotic assault on the status of pure mathematics reached a peak during the academic year of 1932–3. During this year he gave two sets of lectures, one entitled 'Philosophy', and the other 'Philosophy for Mathematicians'. In the second of these he attempted to combat what he regarded as the baleful influence on undergraduate mathematicians of the textbooks that were used to teach them. He would read out extracts from Hardy's *Pure Mathematics* (the standard university text at that time) and use them to illustrate the philosophical fog that he believed surrounded the whole discipline of pure mathematics – a fog he thought could be dispelled only by uprooting the many commonly held assumptions about mathematics that are so deeply embedded as to be very rarely examined.

The first of these is that mathematics stands upon the logical foundations given to it by Cantor, Frege and Russell, among others. He began his lectures with a straightforward statement of his position on this question. 'Is there a substratum on which mathematics rests?' he asked rhetorically:

> Is logic the foundation of mathematics? In my view mathematical logic is simply part of mathematics. Russell's calculus is not fundamental; it is just another calculus. There is nothing wrong with a science before the foundations are laid.

Another of these assumptions is the idea that mathematics is concerned with the discovery of *facts* that are in some way objectively true (about something or other). *What* they are true of, and in what this objectivity consists, has, of course, been the subject-matter of the philosophy of mathematics since the time of Plato, and philosophers have traditionally been divided between those who say that mathematical statements are true about the *physical* world (empiricists) and those who, feeling this view does not do justice to the inexorability of mathematics, claim that they are true of the *mathematical* world – Plato's eternal world of ideas or forms (hence, Platonists). To this division Kant added a third view, which is that mathematical statements are true of the 'form of our intuition', and this was roughly the view of Brouwer and the intuitionist school. But, for Wittgenstein, the whole idea that mathematics is concerned with the discovery of truths is a mistake that has arisen with the growth of pure mathematics and the separation of mathematics from physical science (the unused broom being mistaken for part of the furniture). If, says Wittgenstein, we looked on mathematics as a series of *techniques* (for calculating, measuring etc.), then the question of what it was *about* simply would not arise.

The view of mathematics that Wittgenstein is attacking is stated very succinctly in a lecture given by Hardy, which was published in *Mind* in 1929 under the title 'Mathematical Proof'. Hardy – who appears to have regarded his excursion into philosophy as a kind of light relief from the serious business of his work as a mathematician – states unequivocably:

> . . . no philosophy can possibly be sympathetic to a mathematician which does not admit, in one manner or another, the immutable and unconditional validity of mathematical truth. Mathematical theorems are true or false; their truth or falsity is absolute and independent of our knowledge of them. In *some* sense, mathematical truth is part of objective reality . . . [mathematical propositions] are in one sense or another, however elusive and sophisticated that sense may be, theorems concerning reality . . . They are not creations of our minds.

Both the tone and the content of this lecture infuriated Wittgenstein. He told his class:

> The talk of mathematicians becomes absurd when they leave mathematics, for example, Hardy's description of mathematics as not being a creation of our minds. He conceived philosophy as a decoration, an atmosphere, around the hard realities of mathematics

> and science. These disciplines, on the one hand, and philosophy on the other, are thought of as being like the necessities and decoration of a room. Hardy is thinking of philosophical opinions. I conceive of philosophy as an activity of clearing up thought.

In relation to mathematics, Wittgenstein had, at this time, a fairly clear idea of the way in which he wanted to present this activity of clarification; it was in the presentation of his more general philosophical position that he was still feeling his way towards a satisfactory formulation. Philosophy was, for him, like mathematics, a series of techniques. But whereas the mathematical techniques already existed, and his role consisted in persuading his audience to see them as techniques (and not as true or false propositions), the philosophical techniques he wished to advance were of his own creation, and were still in their infancy.

In the series of lectures entitled 'Philosophy', Wittgenstein introduced a technique that was to become increasingly central to his philosophical method: the technique of inventing what he called 'language-games'. This is the method of inventing imaginary situations in which language is used for some tightly defined practical purpose. It may be a few words or phrases from our own language or an entirely fictitious language, but what is essential is that, in picturing the situation, the language cannot be described without mentioning the *use* to which it is put. The technique is a kind of therapy, the purpose of which is to free ourselves from the philosophical confusions that result from considering language in isolation from its place in the 'stream of life'.

As examples of the sort of thinking he was attempting to liberate his audience from, Wittgenstein mentioned his own earlier work and that of Russell. Both, he said, had been misled, by concentrating on *one* type of language, the assertoric sentence, into trying to analyse the whole of language as though it consisted of nothing but that type, or as though the other uses of language could be analysed as variations on that basic theme. Thus, they had arrived at an unworkable notion – the 'atomic proposition':

> Russell and I both expected to find the first elements, or 'individuals', and thus the possible atomic propositions, by logical analysis . . . And we were both at fault for giving no examples of atomic propositions or of individuals. We both in different ways

pushed the question of examples aside. We should not have said 'We can't give them because analysis has not gone far enough, but we'll get there in time.'

He and Russell had had too rigid a notion of proposition, and the purpose of the language-game method was, so to speak, to loosen such notions. For example, he asked his audience to consider the language-game of teaching a child language by pointing to things and pronouncing the words for them. Where in this game, he asked, does the use of a proposition start? If we say to a child: 'Book', and he brings us a book, has he learnt a proposition? Or has he learnt propositions only when there is a question of truth and falsity? But then, still, one word – for example, the word: 'Six' in answer to the question: 'How many chairs?' – might be true or false. And is it, therefore, a proposition? It does not matter, Wittgenstein implies, how we answer these questions; what matters is that we see how arbitrary *any* answers to them would be, and thus how 'fluid' our concepts are – too fluid to be forced into the kind of analysis once advocated by Russell and himself:

> I have wanted to show by means of language-games the vague way in which we use 'language', 'proposition', 'sentence'. There are many things, such as orders, which we may or may not call propositions and not only one game can be called language. Language-games are a clue to the understanding of logic. Since what we call a proposition is more or less arbitrary, what we call logic plays a different role from that which Russell and Frege supposed.

Among those attending these lectures was a twenty-year-old undergraduate student of mathematics, then in his third year at Trinity, who was soon to become the most important person in Wittgenstein's life – his constant companion, his trusted confidant, and, even, his most valued collaborator in philosophical work.

Francis Skinner had come up to Cambridge from St Paul's in 1930, and was recognized as one of the most promising mathematicians of his year. By his second year at Cambridge, however, his mathematical work had begun to take second place to his interest in Wittgenstein. He became utterly, uncritically and almost obsessively devoted to Wittgenstein. What it was about him that attracted Wittgenstein, we can only guess. He is remembered by all who knew him as shy, unassuming, good-looking and, above all, extraordinarily gentle. But attracted Wittgenstein certainly was. As with Pinsent and Marguerite, Skinner's mere presence seemed to provide Wittgenstein with the peace he needed to conduct his work. In 1932 Wittgenstein left a note concerning the work he was then endeavouring to finish which suggests that he himself regarded Skinner's relation to that work as parallel to Pinsent's to the *Tractatus*:

> In the event of my death before the completion or publication of this book my notes should be published as fragments under the title 'Philosophical Remarks' and with the dedication: 'To Francis Skinner'.

Skinner's letters were kept by Wittgenstein, and were found among his possessions after his death, and from them we can reconstruct something of how the relationship developed. (Wittgenstein's letters to Skinner were retrieved by Wittgenstein after Skinner's death and were, presumably, burnt.) The first letter to survive is dated 26 December 1932, and is written to thank Wittgenstein for a Christmas tree he had given. Two days later Skinner writes: 'I am glad to read that you think about me. I think about you a lot.'

But it was not until the Easter break of 1933 that they became 'Francis' and 'Ludwig' to each other, and that Skinner began to express himself in terms that suggest he was writing – albeit nervously and self-consciously – to a beloved. On 25 March, while on holiday in Guernsey, he wrote:

> Dear Ludwig,
> I have thought about you a lot since we left last Saturday. I hope I think about you in the right way. When we were talking about the case which your sister gave you, I smiled several times, and you said you could see it wasn't a kind smile. Sometimes when I am thinking about you, I have smiled in the same sort of way. I always knew it was wrong to smile, because immediately afterwards I tried to put it out of my mind, but I didn't know how unkind it was.
>
> . . . I am staying for a few days on an island in the Channel, where some of the people speak French. I remember when I once asked if you could speak French, you told me you were taught when you were young by a lady who stayed in your house who was very nice indeed. When I thought about this this morning, I hoped that you

would be pleased if you knew how much I enjoyed remembering things like this you have told me.

Francis

We must assume that Wittgenstein found the childlike simplicity – one might almost say simple-mindedness – revealed in this letter endearing. Certainly Skinner's letters show nothing of the 'cleverness' that Wittgenstein so disliked in many of the students and dons at Cambridge. He was not the type to be overheard saying 'Oh, really!' Neither do his letters show any trace of egotism. In his devotion to Wittgenstein (which he maintained for the rest of his tragically short life), Skinner surrendered his own will almost entirely. Everything else took second place. His sister has recalled that when she and her mother came to Trinity to meet Francis, they would be met by him rushing down the stairs and hushing them with: 'I'm busy. I've got Dr Wittgenstein here. We're working. Come back later.'

Skinner is the most perfect example of the childlike innocence and first-class brains that Fania Pascal has described as the prerequisites of Wittgensteinian disciplehood. He came from a family steeped in the value of academic achievement. His father was a physicist at Chelsea Polytechnic, and his two elder sisters had both been at Cambridge before him, the first to study classics, the second, mathematics. It was expected – indeed, regarded as inevitable – that Francis would pursue an academic career. Had it not been for Wittgenstein's intervention, he would almost certainly have done so.

So complete was Skinner's absorption in Wittgenstein during his final year as an undergraduate that when, in the summer of 1933, he graduated with a first-class degree in mathematics and was awarded a postgraduate scholarship, his family had the impression that this was to allow him to continue working with Wittgenstein. In fact, the award was given by Trinity with the intention that he should use it to pursue mathematical research.

It had by this time become difficult for Skinner to endure the long summer vacations that Wittgenstein spent away from Cambridge. At the end of the summer he wrote: 'I feel much further away from you, and am longing to be nearer you again.' He sent Wittgenstein a series of picture postcards depicting scenes of his home town of Letchworth in Hertfordshire. On the front of the cards he scribbled remarks which were ostensibly designed to explain a little about the town, but which in fact are much more revealing of Skinner's own frame of mind, showing that, with Wittgenstein hundreds of miles away, Letchworth was the very last place on earth he wished to be.

On a card showing Howard Corner, he explained that the 'Garden City' of Letchworth had been founded by Sir Ebenezer Howard, who had wanted everyone to have the chance of living in the country. 'The result', he wrote, 'is something incredibly depressing and bloody (for me, at any rate).' On a card showing the Broadway: 'This is the road to the town & station. There is a row of houses one side. They always make me feel very miserable.' On a photograph of the Spirella Works: 'This is the largest factory in Letchworth . . . the garden looks to me quite uninteresting and dead always.' The last two cards show Leys Avenue – 'a very dull and depressing street. All the people are disagreeably dressed and have such mean expressions on their faces' – and East Cheap – 'an absurd name . . . When I'm in these streets I feel surrounded by gossip.'

His relationship with Wittgenstein was to provide him with some sort of escape from this 'dead' and 'dull' existence, and eventually – much to the consternation of his family – a release from their expectations. It was also to provide him with a new set of expectations, to which he zealously conformed. For the three years of his postgraduate scholarship he worked assiduously with Wittgenstein on the preparation of his work for publication, and when the time came he abandoned academic life altogether for a job that Wittgenstein considered more suitable for him.

Wittgenstein's advice to his friends and students to leave academia was based on his conviction that its atmosphere was too rarefied to sustain proper life. There is no oxygen in Cambridge, he told Drury. It didn't matter for him – he manufactured his own. But for people dependent on the air around them, it was important to get away, into a healthier environment. His ideal was a job in the medical profession. He had already nudged Marguerite in this direction, and she was at this time training to become a nurse in Berne, a project in which Wittgenstein took a great personal interest. Their relationship had lost any trace of romantic involvement, and Marguerite had fallen in love with Talle Sjögren, but Wittgenstein would still occasionally travel to Berne to see how Marguerite was getting on in her training.

Now, in the summer of 1933, after the completion of his project working with unemployed miners in South Wales, Drury decided he, too, wanted to train as a nurse. He was told, however, that with his education he would be more useful if he trained as a doctor. Upon being told of this, Wittgenstein immediately took the matter into his own hands. He arranged for Keynes and Gilbert Pattisson to lend Drury the necessary funds, and sent a telegram to Drury urging him: 'Come to Cambridge at once.' Drury was hardly out of the train before Wittgenstein announced: 'Now there is to be no more argument about this: it has all been settled already, you are to start work as a medical student at once.' He was later to say that, of all his students, it was in his influence on Drury's career that he could take most pride and satisfaction.

On more than one occasion Wittgenstein himself thought seriously of training to be a doctor, and of escaping the 'deadness' of academic philosophy. He might be able to generate oxygen – but what was the point of providing lungs for a corpse? He knew, of course, that a great many philosophers wanted to know his latest thoughts, for by 1933 it was widely known, particularly in Cambridge and Vienna, that he had radically changed his position since the publication of the *Tractatus*. He resolutely refused to accept that it was for *them* – for the 'philosophic journalists' – that he was preparing his new work, but still he could not bear to see his oxygen being recycled by them. In March 1933, he had been pained to see an article by Richard Braithwaite in a collection entitled *Cambridge University Studies*, in which Braithwaite outlined the impression that various philosophers, including Wittgenstein, had made upon him. The fact that Braithwaite might have been regarded as presenting Wittgenstein's present views prompted Wittgenstein to write a letter to *Mind* disclaiming all responsibility for the views that had been attributed to him: 'Part of [Braithwaite's] statements can be taken to be inaccurate representations of my views', he wrote; 'others again clearly contradict them.' He ended:

> That which is retarding the publication of my work, the difficulty of presenting it in a clear and coherent form, *a fortior* prevents me from stating my views within the space of a letter. So the reader must suspend his judgement about them.

The same edition of *Mind* carries a contrite apology from Braithwaite, which ends, however, with a sting in its tail: 'The extent to which I have misrepresented Dr Wittgenstein cannot be judged until the appearance of the book which we are all eagerly awaiting.'

16

LANGUAGE-GAMES: *The Blue and Brown Books*

After Wittgenstein's return to Cambridge for the academic year of 1933–4, Wittgenstein and Skinner were rarely to be seen apart: they both had rooms in college; they walked together, talked together, and whatever social life they had (chiefly going to the cinema to watch Westerns and musicals) was shared. Above all, perhaps, they worked together.

Wittgenstein began the term, as he had the previous year, by giving two sets of lectures, one entitled 'Philosophy', and the other 'Philosophy for Mathematicians'. The second set, much to his dismay, proved particularly popular, between thirty and forty people turning up – far too many for the kind of informal lectures he wanted to deliver. After three or four weeks he amazed his audience by telling them that he could no longer continue to lecture in this way, and that he proposed instead to dictate his lectures to a small group of students, so that they could be copied and handed out to the others. The idea, as he later put it to Russell, was that the students would then 'have something to take home with them, in their hands if not in their brains'. The select group included his five favourite students – Skinner, Louis Goodstein, H. M. S. Coxeter, Margaret Masterman and Alice Ambrose. The duplicated set of notes was bound in blue paper covers and has been known ever since as 'The Blue Book'.

This was the first publication in any form of Wittgenstein's new method of philosophy, and as such it created great interest. Further copies were made and distributed, and the book reached a far wider audience than Wittgenstein had expected – much wider, indeed, than he would have wished. By the late 1930s, for example, it had been distributed among many members of the philosophy faculty of Oxford. The *Blue Book* was thus responsible for introducing into philosophic discourse the notion of a 'language-game' and the technique, based upon it, for dissolving philosophical confusion.

In many ways the *Blue Book* can be regarded as an early prototype for subsequent presentations of Wittgenstein's later philosophy. Like all future attempts to arrange his work in coherent form (including the *Brown Book* and *Philosophical Investigations*), it begins with 'one of the great sources of philosophical bewilderment' – i.e. the tendency to be misled by substantives to look for something that corresponds to them. Thus, we ask: 'What is time?', 'What is meaning?', 'What is knowledge?', 'What is a thought?', 'What are numbers?' etc., and expect to be able to answer these questions by naming some *thing*. The technique of language-games was designed to break the hold of this tendency:

> I shall in the future again and again draw your attention to what I shall call language games. These are ways of using signs simpler than those in which we use the signs of our highly complicated everyday language. Language games are the forms of language with which a child begins to make use of words. The study of language games is the study of primitive forms of language or primitive languages. If we want to study the problems of truth and falsehood, of the agreement and disagreement of propositions with reality, of the nature of assertion, assumption, and question, we shall with great advantage look at primitive forms of language in which these forms of thinking appear without the confusing background of highly complicated processes of thought. When we look at such simple forms of language the mental mist which seems to enshroud our ordinary use of language disappears. We see activities, reactions, which are clear-cut and transparent.

Connected with the inclination to look for a substance corresponding to a substantive is the idea that, for any given concept, there is an 'essence' – something that is common to all the things subsumed under a general term. Thus, for example, in the Platonic dialogues, Socrates seeks to answer philosophical questions such as: 'What is knowledge?' by looking for something that all examples of knowledge have in common. (In connection with this, Wittgenstein once said that his method could be summed up by saying that it was the exact opposite of that of Socrates.) In the *Blue Book* Wittgenstein seeks to replace this notion of *essence* with the more flexible idea of *family resemblances*:

> We are inclined to think that there must be something in common to all games, say, and that this common property is the justification for applying the general term 'game' to the various games; whereas games form a *family* the members of which have family likenesses. Some of them have the same nose, others the same eyebrows and others again the same way of walking; and these likenesses overlap.

The search for essences is, Wittgenstein states, an example of 'the craving for generality' that springs from our preoccupation with the method of science:

> Philosophers constantly see the method of science before their eyes, and are irresistibly tempted to ask and answer questions in the way science does. This tendency is the real source of metaphysics, and leads the philosopher into complete darkness.

Wittgenstein's avoidance of this tendency – his complete refusal to announce any general conclusions – is perhaps the main feature that makes his work difficult to understand, for without having the moral pointed out, so to speak, it is often difficult to see the point of his remarks. As he himself once explained at the beginning of a series of lectures: 'What we say will be easy, but to know why we say it will be very difficult.'

During the Christmas vacation of 1933 Skinner wrote to Wittgenstein every few days, telling him how much he missed him, how often he thought about him and how much he was longing to see him again. Every last moment he had spent with Wittgenstein was recalled with fond affection:

> After I stopped waving my handkerchief to you I walked through Folkestone and took the train at 8.28 back to London. I thought about you and how wonderful it had been when we said goodbye

. . . I loved very much seeing you off. I miss you a great deal and think about you a lot.

With love,
Francis

At the family Christmas at the Alleegasse, Marguerite (who continued to spend Christmas in Vienna as the guest of Gretl) caused something of a sensation by announcing her engagement to Talle Sjögren. Encouraged by Gretl, but in the face of disapproval from her father, Marguerite decided on an extremely short engagement, and she and Talle were married on New Year's Eve. Her father, at least, was a safe distance away in Switzerland. Wittgenstein was not. Recalling her wedding day, she writes:

> My despair reached its zenith when Ludwig came to see me on the Sunday morning, an hour before my wedding. 'You are taking a boat, the sea will be rough, remain always attached to me so that you don't capsize', he said to me. Until that moment I hadn't realised his deep attachment nor perhaps his great deception. For years I had been like soft putty in his hands which he had worked to shape into a better being. He had been like a Samaritan who gives new life to someone who is failing.

It is hard to believe that she had not appreciated until that day how deep was Wittgenstein's attachment to her. It is characteristic, however, of many of his friendships that she should have felt his involvement in her life to have had a fundamentally *ethical* purpose. 'He conjured up a vision of a better you', as Fania Pascal has put it. It was, after all, partly because she did not want to live with this kind of moral pressure that Marguerite had chosen to marry someone else.

For most of 1934 Wittgenstein continued to work on three different, but related, projects, which attempted to solve the problem he had described in his letter to *Mind* – that of presenting his philosophical method 'in a clear and coherent form'. At Cambridge, as well as dictating the *Blue Book*, he also made copious revisions to the Big Typescript – 'pottering about with it', as he put it to Russell. (The results of this 'pottering about' have been incorporated into the first part of *Philosophical Grammar*). In Vienna he continued to co-operate (albeit with increasing reluctance and with ever-growing misgivings) with the scheme to publish a book with Waismann. In the Easter vacation of 1934, this scheme took a new turn: it was now proposed that Waismann and Wittgenstein should be co-authors, with Wittgenstein providing the raw material and taking control of the form and structure, and Waismann being responsible for writing it up in a clear and coherent way. Waismann, that is to say, was given what Wittgenstein himself regarded as the most difficult part of the job.

With each new arrangement Waismann's position seemed to get worse. By August he was complaining to Schlick about the difficulties of writing a book with Wittgenstein:

> He has the great gift of always seeing things as if for the first time. But it shows, I think, how difficult collaborative work with him is, since he is always following up the inspiration of the moment and demolishing what he has previously sketched out . . . all one sees is that the structure is being demolished bit by bit and that everything is gradually taking on an entirely different appearance, so that one almost gets the feeling that it doesn't matter at all how the thoughts are put together since in the end nothing is left as it was.

Wittgenstein's habit of following up the inspiration of the moment applied not only to his work, but also to his life. In 1934, despite the fact that he was then involved in two projects to prepare a book for publication (*Logik, Sprache, Philosophie* in Vienna, and *Philosophical Grammar* in England), he conceived the idea of giving up academic life altogether and going, with Skinner, to live in Russia, where they would both seek work as manual labourers. Skinner's family was naturally apprehensive about the idea, but for Skinner himself it had the inestimable advantage that he would be with Wittgenstein all the time. He had begun to regard being with Wittgenstein almost as a necessity; away from Wittgenstein nothing looked the same or felt the same. 'If I am with you', he wrote during the Easter vacation, 'I can feel everything deeply.' It was a constant theme of his letters:

> I thought of you a lot. I longed to have you with me. The night was very wonderful and the stars looked particularly beautiful. I longed to be able to feel everything in the way I would feel it if I was with you. [25.3.34]

> I long to be with you in any open space. I think of you a lot and how wonderful our walks have been. I look forward *enormously* to our tour next week. Yesterday I got your Easter card which was very lovely. I thought the houses in the street on the other card looked very beautiful. I should like to have looked at them with you. [4.4.34]

Skinner also stressed in his letters the *moral* necessity of having Wittgenstein by his side, as though without Wittgenstein's guidance he would fall into the hands of the devil. The most remarkable instance of this occurs in a letter written on 24 July 1934, the day after Skinner had waved goodbye to Wittgenstein at Boulogne. The letter begins with the by now customary remarks about how 'wonderful and sweet' it had been waving goodbye; he then goes on to describe how sinful he became as soon as he had been left alone in Boulogne. He had visited a casino, lost ten francs and then, despite his best resolutions, had been tempted to return, this time winning fifty francs. Disgusted with himself, he vowed to return to England on the afternoon boat, but when the time came to leave he was drawn once more into the casino. By this time, he was a lost soul:

> I started off again by playing very carefully and keeping myself very restrained. Then I began to lose slightly and I suddenly lost my restraint and care and played more and more recklessly. I became quite feverishly excited and unable to control myself. Altogether I lost about 150 francs. I first lost all the French money which I had with me, about 80 francs, and then I changed a 10/- note into French money and lost it all and then I changed all the loose English silver I had and lost it all. I then left the Casino at about 5 o'clock. When I got out into the fresh air I suddenly felt in what a horrible unnatural and loathesome way I had been acting since I had begun to gamble. It seemed dreadful that I should have felt such eagerness to win money. I suddenly realised in what a mean filthy state of degradation I had fallen. I felt physically stirred up and excited in my body. I walked about the streets for a time in a wretched state of mind. I felt I understood why gamblers often committed suicide for the feeling of degradation is so awful. I felt a most terrible philistine. I felt I had been destroying myself. I then returned to the Hotel and washed myself all over.

Skinner is no Dostoevsky, and his depiction of his own moral depravity has a curiously unconvincing ring, but the effect he is trying to achieve is surely something like that of the Russian novels he knew Wittgenstein to have admired. His tale, with its evocation of desperate suicidal guilt, seems to point inexorably towards the necessity of redemptive religion. Indeed, he goes on to describe how, after washing his hands, he sought out the church in Boulogne that he had visited with Wittgenstein. Inside the church: 'I thought of you a lot. I felt comforted by the Church though I was hardly able to look at it at all.' He adds: 'I felt I would be a terrible rascal and be utterly unworthy of your love if I wrote without saying anything.'

The religious theme is renewed a few weeks later, on 11 August, when Skinner writes quoting the passage from *Anna Karenina* where the almost suicidal Levin says: 'I cannot live without knowing what I am.' The passage ends: 'But Levin did not hang himself, or shoot himself, but lived and struggled on.' 'When I read this last sentence', Skinner tells Wittgenstein, in a phrase that echoes many of Wittgenstein's own, 'I suddenly realised I was reading something terrific':

> I suddenly seemed to understand what all I had been reading meant. I went on reading the following chapters and it all seemed written with enormous truth. I felt as if I was reading the chapters of the Bible. I didn't understand it all but I felt it was religion. I wanted very much to tell you this.

Skinner and Wittgenstein had, by this time, begun to take Russian lessons together in preparation for their impending visit to the Soviet Union. Their teacher was Fania Pascal, the wife of the Marxist intellectual and Communist Party member Roy Pascal. In discussing Wittgenstein's motives for wanting to go to Russia, Mrs Pascal remarks: 'To my mind, his feeling for Russia would have had at all times more to do with Tolstoy's moral teachings, with Dostoyevsky's spiritual insights, than with any political or social matters.' The tone and content of Skinner's letters seems to confirm this. And yet it was not the Russia of Tolstoy and Dostoevsky that Wittgenstein and Skinner wanted to visit, and in which they planned to find work; it was the Russia of Stalin's Five Year Plans. And neither of them could possibly have been so politically naïve or so ill-informed as not to recognize any difference between the two.

Wittgenstein probably struck Pascal as 'an old-time conservative' because of his hostility to Marxism. But many of Wittgenstein's other friends received a very different impression. George Thomson, for example, who knew Wittgenstein well during the 1930s, speaks of Wittgenstein's 'growing political awareness' during those years, and says that, although he did not discuss politics very often with Wittgenstein, he did so 'enough to show that he kept himself informed about current events. He was alive to the evils of unemployment and fascism and the growing danger of war.' Thomson adds, in relation to Wittgenstein's attitude to Marxism: 'He was opposed to it in theory, but supported it in practice.' This chimes with a remark Wittgenstein once made to Rowland Hutt (a close friend of Skinner's who came to know Wittgenstein in 1934): 'I am a communist, *at heart*.' It should be remembered, too, that many of Wittgenstein's friends of this period, and particularly the friends on whom he relied for information about the Soviet Union, were Marxists. In addition to George Thomson, there were Piero Sraffa, whose opinion Wittgenstein valued above all others on questions of politics, Nicholas Bachtin and Maurice Dobb. There is no doubt that during the political upheavals of the mid-1930s Wittgenstein's sympathies were with the working class and the unemployed, and that his allegiance, broadly speaking, was with the Left.

However, it remains true that Russia's attraction for Wittgenstein had little or nothing to do with Marxism as a political and economic theory, and much to do with the sort of life he believed was being led in the Soviet Union. This emerged during a conversation which Wittgenstein and Skinner had with Maurice Drury in the summer of 1934, when they spent a summer holiday at Drury's brother's cottage in Connemara, on the west coast of Ireland. On their arrival Drury prepared for them a rather elaborate meal consisting of roast chicken followed by suet pudding and treacle. Wittgenstein expressed his disapproval, insisting that while they were staying at Connemara they should eat nothing but porridge for breakfast, vegetables for lunch and a boiled egg in the evening. When the subject of Russia came up, Skinner announced that he wanted to do something 'fiery', a way of thinking that struck Wittgenstein as dangerous. 'I think', said Drury, 'Francis means that he doesn't want to take the treacle with him.' Wittgenstein was delighted. 'Oh, that is an excellent expression: I understand what that means entirely. No, we don't want to take the treacle with us.'

Presumably, for Wittgenstein, the life of a manual labourer in Russia was the epitome of a life without treacle. During the following year, to give Skinner some taste of what that would be like, he arranged for Skinner, together with Rowland Hutt, to spend six weeks working on a farm during the winter months. Wittgenstein himself came out one morning at six o'clock on a cold February day to help with the work.

During the year 1934–5 Wittgenstein dictated what is now known as the *Brown Book*. This, unlike the *Blue Book*, was not a substitute for a series of lectures, but rather an attempt by Wittgenstein to formulate the results of his own work for his own sake. It was dictated to Skinner and Alice Ambrose, who sat with Wittgenstein for between two and four hours a day for four days a week. The *Brown Book* is divided into two parts, corresponding, roughly, to the method and its application. Part I, introducing the method of language-games, reads almost like a textbook. After an introductory paragraph describing St Augustine's account of 'How, as a boy, he learned to talk', it consists of seventy-two numbered 'exercises', many of which invite the reader to, for example:

> Imagine a people in whose language there is no such form of sentence as 'the book is in the drawer' or 'water is in the glass', but wherever we should use these forms they say, 'The book can be taken out of the drawer', 'The water can be taken out of the glass'. [p. 100]

> Imagine a tribe in whose language there is an expression corresponding to our 'He has done so and so', and another expression corresponding to our 'He can do so and so', this latter expression, however, being used only where its use is justified by the same fact which would also justify the former expression. [p. 103]

> Imagine that human beings or animals were used as reading machines; assume that in order to become reading machines they need a particular training. [p. 120]

The book is difficult to read because the *point* of imagining these various situations is very rarely spelt out. Wittgenstein simply leads the reader through a series of progressively more complicated language-games, occasionally pausing to remark on various features of the games he is describing. When he does make the point of these remarks explicit, he claims it is to ward off thoughts that may give rise to philosophical puzzlement. It is as though the book was intended to serve as a text in a course designed to nip in the bud any latent philosophizing. Thus we are first introduced to a language which contains just four nouns – 'cube', 'brick', 'slab' and 'column' – and which is used in a building 'game' (one builder shouts, 'Brick!' and another brings him a brick). In subsequent games this proto-language is supplemented by the addition first of numerals, and then of proper nouns, the words 'this' and 'there', questions and answers, and finally colour words. So far, no philosophical moral has been drawn other than that, in understanding how these various languages are used, it is not necessary to postulate the existence of mental images; all the games could be played with or without such images. The unspoken point of this is to loosen the hold of the idea that mental images are an essential concomitant to any meaningful use of language.

It is not until we have been led through another series of language-games, which introduce, first, the notion of an infinite series, and then the notions of 'past', 'present' and 'future', that Wittgenstein explicitly mentions the relevance of all this to philosophical problems. After describing a series of language-games with more or less primitive means of distinguishing one time of day from another, he contrasts these with our own language, which permits the construction of such questions as: 'Where does the present go when it becomes past, and where is the past?' 'Here', he says, 'is one of the most fertile sources of philosophic puzzlement.' To a reader studying the *Brown Book* as a work of philosophy, this statement, the only mention of philosophy in the first thirty pages of the book, comes as something of a relief. Such questions, he states, arise because we are misled by our symbolism into certain analogies (in this case, the analogy between a past event and a *thing*, the analogy between our saying 'Something has happened' and 'Something came towards me'). Similarly: 'We are inclined to say that both "now" and "six o'clock" refer to points in time. This use of words produces a puzzlement which one might express in the question: "What is the 'now'?" – for it is a moment of time and yet it can't be said to be either the "moment at which I speak" or "the moment at which the clock strikes", etc., etc.' Here, in relation to what is essentially St Augustine's problem of time, Wittgenstein finally spells out the point of his procedure:

> Our answer is: the function of the word 'now' is entirely different from that of a specification of time – This can easily be seen if we look at the role this word really plays in our usage of language, but it is obscured when instead of looking at the *whole language game*, we only look at the contexts, the phrases of language in which the word is used.

There is no indication that Wittgenstein considered publishing the *Brown Book*. On 31 July 1935 he wrote to Schlick describing it as a document which shows 'the way in which I think the whole stuff should be handled'. Perhaps, as he was then planning to leave philosophy altogether to take up manual work in Russia, it represents an attempt to present the results of his seven years' work in philosophy in a way that would enable someone else (perhaps Waismann) to make use of them.

It is unlikely, however, that he would ever have been satisfied with the attempts of another to represent his thought faithfully. Again and again attempts were made by others to present his ideas, and again and again he reacted angrily, accusing whoever used his ideas of plagiarism if they did not acknowledge their debt, or misrepresentation if they did. During the dictation of the *Brown Book*, it was the turn of Alice Ambrose to encounter his wrath on this point. She planned to publish in *Mind* an article entitled 'Finitism in Mathematics', in which she would present what she took to be Wittgenstein's view on the matter. The article annoyed Wittgenstein intensely, and he tried hard to persuade her not to publish it. When she and G. E. Moore, who was then editor of the journal, refused to succumb to this pressure, he abruptly ended any association with her. In the letter to Schlick mentioned above, however, he blames not her, but the academics who encouraged her to go ahead with the article. The fault lay primarily, he thought, with the curiosity of academic philosophers to know what his new work was all about before he felt able to publish his results himself. Reluctant as he was to cast pearls before swine, he was nonetheless determined they should not be offered counterfeits.

17
JOINING THE RANKS

In his letter to Schlick of 31 July 1935, Wittgenstein wrote that he would probably not be coming to Austria that summer:

> At the beginning of September I want to travel to Russia & will either stay there or, after about two weeks, return to England. What, in that case, I will do in England is still completely uncertain, but I will probably not continue in philosophy.

Throughout the summer of 1935 he made preparations for his impending visit to Russia. He met regularly with those of his friends, many of them members of the Communist Party, who had been to Russia or who might be able to inform him about conditions there. Possibly he hoped, too, that they could put him in contact with people who could help to find work there for himself and Skinner. Among these friends were Maurice Dobb, Nicholas Bachtin, Piero Sraffa and George Thomson. The impression they received was that Wittgenstein wanted to settle in Russia as a manual worker, or possibly to take up medicine, but in any case to abandon philosophy. At a meeting with George Thomson in the Fellows' garden at Trinity he explained that, since he was giving up philosophical work, he had to decide what to do with his notebooks. Should he leave them somewhere or should he destroy them? He talked at length to Thomson about his philosophy, expressing doubts about its worth. It was only after urgent appeals by Thomson that he agreed not to destroy his notebooks, but to deposit them instead in the college library.

Wittgenstein was not the only one at Cambridge then seeking in

Soviet Russia an alternative to the countries of Western Europe, menaced as those countries were by the growth of fascism and the problems of mass unemployment. The summer of 1935 was the time when Marxism became, for the undergraduates at Cambridge, the most important intellectual force in the university, and when many students and dons visited the Soviet Union in the spirit of pilgrimage. It was then that Anthony Blunt and Michael Straight made their celebrated journey to Russia, which led to the formation of the so-called 'Cambridge Spy Ring', and that the Cambridge Communist Cell, founded a few years earlier by Maurice Dobb, David Hayden-Guest and John Cornford, expanded to include most of the intellectual élite at Cambridge, including many of the younger members of the Apostles.

Despite the fact that Wittgenstein was never at any time a Marxist, he was perceived as a sympathetic figure by the students who formed the core of the Cambridge Communist Party, many of whom (Hayden-Guest, Cornford, Maurice Cornforth etc.) attended his lectures. But Wittgenstein's reasons for wanting to visit Russia were very different. His perception of the decline of the countries of Western Europe was always more Spenglerian than Marxian, and, as we have remarked earlier, it is likely that he was extremely attracted to the portrait of life in the Soviet Union drawn by Keynes in his *Short View of Russia* – a portrait which, while deprecating Marxism as an economic theory, applauded its practice in Russia as a new religion, in which there were no supernatural beliefs but, rather, deeply held religious attitudes.

Perhaps because of this, Wittgenstein felt he might be understood by Keynes. 'I am sure that you partly understand my reasons for wanting to go to Russia', he wrote to him on 6 July, 'and I admit that they are partly bad and even childish reasons but it is true also that behind all that there are deep and even good reasons.' Keynes, in fact, disapproved of Wittgenstein's plan, but despite this did all he could to help Wittgenstein overcome the suspicions of the Soviet authorities. Wittgenstein had had a meeting at the Russian Embassy with an official called Vinogradoff, who, he told Keynes, was 'exceedingly careful in our conversation . . . He of course knew as well as anyone that recommendations might help me but it was quite clear that he wasn't going to help me get any.' Characteristically, Keynes went straight to the top, and provided Wittgenstein with an introduction to

Ivan Maisky, the Russian Ambassador in London: 'May I venture to introduce to you Dr Ludwig Wittgenstein . . . who is a distinguished philosopher [and] a very old and intimate friend of mine . . . I should be extremely grateful for anything you could do for him.' He added: 'I must leave it to him to tell you his reasons for wanting to go to Russia. He is not a member of the Communist Party, but has strong sympathies with the way of life which he believes the new regime in Russia stands for.'

At his meeting with Maisky, Wittgenstein took great pains to appear both respectable and respectful. Keynes had warned him that, while Maisky was a Communist, that did not mean he would not wish to be addressed as 'Your Excellency', nor that he had any less respect than any other high bourgeois official for standards of formality and politeness. Wittgenstein took the advice to heart. The meeting was one of the few occasions in his life when he wore a tie, and he used the phrase 'Your Excellency' as often as he could. Indeed, as he later told Gilbert Pattisson, he was so anxious to show respect to the ambassador that he made a great show of wiping his shoes well on the mat – on the way *out* of the room. After the meeting Wittgenstein reported to Keynes that Maisky was: 'definitely nice and in the end promised to send me some addresses of people in Russia of whom I might get useful information. He did not think that it was utterly hopeless for me to try to get permission to settle in Russia though he too didn't think it was likely.'

Apart from these – not very encouraging – meetings at the Russian Embassy, Wittgenstein also tried to make contacts through the Society for Cultural Relations with the Soviet Union (SCR). The SCR was founded in 1924 and was (indeed, is) an organization dedicated to improving cultural links between Britain and the Soviet Union. It organizes lectures, discussions and exhibitions, and publishes its own magazine, the *Anglo-Soviet Journal*, each issue of which in the 1930s carried an advertisement for the tours of Russia organized by *Intourist*, the Soviet travel company ('For the experience of a lifetime visit USSR' etc.). Because (unlike its companion organization, the Society of Friends of Soviet Russia) its aims were cultural rather than political, the SCR numbered among its members many non-Communists such as Charles Trevelyan and, indeed, Keynes himself. By 1935, however, it was dominated by much the same people (Hayden-Guest, Pat Sloan etc.) as the Society of Friends. On 19 August Wittgenstein went to the offices of the SCR to meet with Miss Hilda Browning, its Vice-Chairman. The following day he reported to Gilbert Pattisson:

> My interview with Miss B. went off better than I had expected. At least I got one useful piece of information: – that my only chance of getting a permission to settle in R. is to go there as a tourist & talk to officials; & that all I can do at this end is to try to get letters of introduction. Also Miss B. told me that she would give me such letters to two places. This is, on the whole, better than nothing. It doesn't however settle anything & I'm as much in the dark as ever, not only about what they'll let me do but also about what I want to do. It is shameful, but I change my mind about it every two hours. I see what a perfect ass I am, at bottom & feel rather rotten.

The two places to which he had secured an introduction were the Institute of the North and the Institute of National Minorities. These were both educational institutes dedicated to improving the level of literacy among the ethnic minorities in the Soviet Union. Though considering this 'better than nothing', Wittgenstein did not particularly want a teaching job. But, as he had been told by Keynes, he was likely to get permission to settle in the Soviet Union only if he received an invitation from a Soviet organization: 'If you were a qualified technician of a sort likely to be useful to them', Keynes wrote to him, 'that might not be difficult. But, without some such qualification, which might very well be a medical qualification, it would be difficult.' Wittgenstein, who throughout his life harboured a desire to become a doctor, considered the possibility of studying medicine in England with the intention of practising in Russia, and even secured from Keynes a promise to finance his medical training. What he really wanted, however, was to be allowed to settle in Russia as a manual worker. But, as became increasingly clear to him, it was extremely unlikely that he would receive an invitation from a Soviet organization to do that. The one thing that was not in short supply in Soviet Russia was unskilled labour.

By the time he left for Leningrad, on 7 September, all Wittgenstein had managed to obtain were the introductions from Hilda Browning and a few names and addresses of people living in Moscow. He was seen off at Hay's Wharf in London by Gilbert Pattisson, Francis being too ill to make the trip. It was understood, however, that he was looking for work on Francis's behalf as well as his own. On the same boat was Dr George Sacks, who has recalled that he and his wife sat opposite Wittgenstein at meal-times. Next to Wittgenstein sat an American Greek Orthodox priest. Wittgenstein, who seemed depressed and preoccupied, sat staring into space, not speaking to anyone, until one day he introduced himself to the priest by sticking up his hand and exclaiming: 'Wittgenstein!' to which the priest replied by saying his own name. For the rest of the journey he was silent.

He arrived in Leningrad on 12 September, and for the next two weeks his pocket diary is full of the names and addresses of the many people he contacted in an effort to secure an offer of employment. In Leningrad he visited, as well as the Institute of the North, the university professor of philosophy, Mrs Tatiana Gornstein, and offered to give a philosophy course at Leningrad University. At Moscow he met Sophia Janovskaya, the professor of mathematical logic, with whom he struck up a friendship that lasted through correspondence long after he had returned to England. He was attracted to her by the forthrightness of her speech. Upon meeting him for the first time, she exclaimed: 'What, not the great Wittgenstein?' and during a conversation on philosophy she told him quite simply: 'You ought to read more Hegel.' From their discussions of philosophy, Professor Janovskaya received the (surely false) impression that Wittgenstein was interested in dialectical materialism and the development of Soviet philosophic thought. It was apparently through Janovskaya that Wittgenstein was offered first a chair in philosophy at Kazan University, and then a teaching post in philosophy at the University of Moscow.

In Moscow Wittgenstein also met two or three times with Pat Sloan, the British Communist who was then working as a Soviet trade union organizer (a period of his life recalled in the book *Russia Without Illusions*, 1938). It seems quite likely that these meetings centred on Wittgenstein's continuing hopes to work in some manual capacity. If so, they were apparently unsuccessful. George Sacks recalls that in Moscow: 'we [he and his wife] heard that Wittgenstein wanted to work on a collective farm, but the Russians told him his own work was a useful contribution and he ought to go back to Cambridge'.

On 17 September, while still in Moscow, Wittgenstein received a letter from Francis urging him to stay as long as it took to find work. 'I wish I could be with you and see things with you', he wrote. 'But I feel it is as though I was with you.' From this letter it appears that Wittgenstein and Skinner were planning to spend the following academic year preparing the *Brown Book* for publication, prior, presumably, to their settling in the Soviet Union. This makes sense in that the coming academic year, 1935–6, was to be the final year of both Skinner's three-year postgraduate scholarship and Wittgenstein's five-year fellowship at Trinity. 'I think a lot about the work which we are going to do next year', Francis told him. 'I feel that the spirit of the method which you used last year is so good':

> Everything, I feel, is absolutely simple and yet it's all full of light. I feel it will be very good to go on with it and get it ready for publication. I feel that the method is so valuable. I hope very much we shall be able to get on with it. We will do our best.

'I'd like to say again', he added, 'that I hope you will stay longer in Moscow than the time you arranged for if you feel there is any chance that you might learn more. It would be valuable for both of us.'

Wittgenstein evidently saw no reason to extend his stay. His visit had served only to confirm what he had been told before he left: that he was welcome to enter the Soviet Union as a teacher, but unwelcome as a worker on a collective farm. On the Sunday before he left he wrote a postcard to Pattisson asking to be met in London:

> My dear Gilbert!
> Tomorrow evening I shall leave Moskow (I am staying in the rooms which Napoleon had in 1812). The day after tomorrow my boat sails from Leningrad & I can only hope that Neptune will have a heart when he sees me. My boat is due in London on Sunday 29th [September]. Could you either meet the boat or leave a message for me at my Palace (usually called 'Strand Palace')? I'm certainly looking forward to seeing your old and bloody face again. Ever in blood.
>
> Ludwig
>
> P.S. If the censor reads this it serves him right!

After his return to England Wittgenstein very rarely discussed his trip to Russia. He sent Francis to deliver a report to Fania Pascal, which told her of his meeting with Mrs Janovskaya and his offer of an academic job at Kazan, and concluded with a statement that: 'He had taken no decisions about his future.' What the report did not provide were any of Wittgenstein's impressions of Soviet Russia – any hint of whether he liked or disliked what he had seen. On this he remained, apart from one or two isolated comments, completely silent. The reason he gave friends for this silence was that he did not wish his name to be used, as Russell had allowed *his* name to be used (after the publication of *Theory and Practice of Bolshevism*), to support anti-Soviet propaganda.

This suggests that, had he been open about his impression of the Soviet Union, he would have drawn an unflattering picture. An important clue to his attitude perhaps lies in his remark to Gilbert Pattisson that living in Russia was rather like being a private in the army. It was difficult, he told Pattisson, for 'people of our upbringing' to live there because of the degree of petty dishonesty that was necessary even to survive. If Wittgenstein thought of life in Russia as comparable to his experience on the *Goplana* during the First World War, it is perhaps not surprising that he showed so little inclination to settle there after he returned from his brief visit.

He nevertheless repeatedly expressed his sympathy for the Soviet régime and his belief that, as material conditions for ordinary Soviet citizens were improving, the régime was strong and unlikely to collapse. He spoke admiringly of the educational system in Russia, remarking that he had never seen people so eager to learn and so attentive to what they were being told. But perhaps his most important reason for sympathizing with Stalin's régime was that there was in Russia so little unemployment. 'The important thing', he once said to Rush Rhees, 'is that the people have *work*.' When the regimentation of life in Russia was mentioned – when it was pointed out that, although they were employed, the workers there had no freedom to leave or change their jobs – Wittgenstein was not impressed. 'Tyranny', he told Rhees with a shrug of his shoulders, 'doesn't make me feel indignant.' The suggestion that 'rule by bureaucracy' was bringing class distinctions in Russia, however, did arouse his indignation: 'If anything could destroy my sympathy with the Russian regime, it would be the growth of class distinctions.'

For two years after his return from Russia Wittgenstein toyed with the idea of taking up the teaching post in Moscow that he had been offered. During this time he continued to correspond with Sophia

Janovskaya, and when he went away to Norway he arranged with Fania Pascal for Janovskaya to be sent insulin for her diabetes. As late as June 1937 he remarked in a letter to Engelmann: 'perhaps I shall go to Russia'. Shortly after this, however, the offer of a post was withdrawn, because (according to Piero Sraffa) by this time all Germans (including Austrians) had become suspect in Russia.

Nevertheless, even after the show trials of 1936, the worsening of relations between Russia and the West and the Nazi-Soviet Pact of 1939, Wittgenstein continued to express his sympathy with the Soviet régime – so much so that he was taken by some of his students at Cambridge to be a 'Stalinist'. This label is, of course, nonsense. But at a time when most people saw only the tyranny of Stalin's rule, Wittgenstein emphasized the problems with which Stalin had to deal and the scale of his achievement in dealing with them. On the eve of the Second World War he asserted to Drury that England and France between them would not be able to defeat Hitler's Germany; they would need the support of Russia. He told Drury: 'People have accused Stalin of having betrayed the Russian Revolution. But they have no idea of the problems that Stalin had to deal with; and the dangers he saw threatening Russia.' He immediately added, as though it were somehow relevant: 'I was looking at a picture of the British Cabinet and I thought to myself, "a lot of wealthy old men".' This remark recalls Keynes's characterization of Russia as 'the beautiful and foolish youngest son of the European family, with hair on his head, nearer to both the earth and to heaven than his bald brothers in the West'. Wittgenstein's reasons for wanting to live in Russia, both the 'bad and even childish' reasons and the 'deep and even good' ones, had much to do, I think, with his desire to dissociate himself from the old men of the West, and from the disintegrating and decaying culture of Western Europe.

It is also, of course, one more manifestation of his perennial desire to join the ranks. The Soviet authorities knew, just as the Austrian authorities had in 1915, that he would be more use to them as an officer than as a private; and Wittgenstein himself realized that he could not really tolerate life among the 'petty dishonesty' of the ordinary soldiers. Yet he continued to wish it could be otherwise.

When, in the autumn of 1935, Wittgenstein began the final year of his fellowship at Trinity, he still had little idea of what he would do after it had expired. Perhaps he would go to Russia – perhaps, like Rowland Hutt, get a job among 'ordinary people'; or perhaps, as Skinner had wanted, he would concentrate on preparing the *Brown Book* for publication. One thing seemed sure: he would not continue to lecture at Cambridge.

His lectures during this last year centred on the theme of 'Sense Data and Private Experience'. In these lectures he tried to combat the philosopher's temptation to think that, when we experience something (when we see something, feel pain etc.), there is some thing, a sense datum, that is the primary content of our experience. He took his examples, however, not from philosophers but from ordinary speech. And when he quoted from literature, it was not from the great philosophical works, nor from the philosophical journal *Mind*, but from Street & Smith's *Detective Story Magazine*.

He began one lecture by reading a passage from Street & Smith in which the narrator, a detective, is alone on the deck of a ship in the middle of the night, with no sound except the ticking of the ship's clock. The detective muses to himself: 'A clock is a bewildering instrument at best: measuring a fragment of infinity: measuring something which does not exist perhaps.' Wittgenstein told his class that it is much more revealing and important when you find this sort of confusion in something said 'in a silly detective story' than it is when you find it in something said 'by a silly philosopher':

> Here you might say 'obviously a clock is not a bewildering instrument at all'. – If in some situation it strikes you as a bewildering instrument, and you can then bring yourself round to saying that of course it is not bewildering – then this is the way to solve a philosophical problem.
>
> The clock becomes a bewildering instrument here because he says about it 'it measures a fragment of infinity, measuring something which does not exist perhaps'. What makes the clock bewildering is that he introduces a sort of entity which he then can't see, and it seems like a ghost.
>
> The connection between this and what we were saying about sense data: What is bewildering is the introduction of something we might call 'intangible'. It seems as though there is nothing intangible about the chair or the table, but there is about the fleeting personal experience.

A recurrent theme of Wittgenstein's lectures for that year was his concern to uphold, against philosophers, our ordinary perception of the world. When a philosopher raises doubts, about time or about mental states, that do not occur to the ordinary man, this is not because the philosopher has more insight than the ordinary man, but because, in a way, he has less; he is subject to temptations to misunderstand that do not occur to the non-philosopher:

> We have the feeling that the ordinary man, if he talks of 'good', of 'number' etc., does not really understand what he is talking about. I see something queer about perception and he talks about it as if it were not queer at all. Should we say he knows what he is talking about or not?
>
> You can say both. Suppose people are playing chess. I see queer problems when I look into the rules and scrutinise them. But Smith and Brown play chess with no difficulty. Do they understand the game? Well, they play it.

The passage is redolent of Wittgenstein's own doubts about his status as a philosopher, his weariness of 'seeing queer problems' and his desire to start playing the game rather than scrutinizing its rules. His thoughts turned again to the idea of training as a doctor. Drury was at this time preparing for his first MB examination at Dublin, and Wittgenstein wrote asking him to make enquiries about the possibility of his entering the medical school there, his training presumably to be paid for by Keynes. He suggested to Drury that the two of them might practise together as psychiatrists. Wittgenstein felt that he might have a special talent for this branch of medicine, and was particularly interested in Freudian psychoanalysis. That year he sent Drury, as a birthday present, Freud's *Interpretation of Dreams*, telling Drury that when he first read it he said to himself: 'Here at last is a psychologist who has something to say.'

Wittgenstein's feeling that he would have made a good psychiatrist seems to rest on a belief that his style of philosophizing and Freudian psychoanalysis required a similar gift. Not, of course, that they are the same technique. Wittgenstein reacted angrily when his philosophical method was dubbed 'therapeutic positivism' and compared with psychoanalysis. When, for example, A. J. Ayer drew the comparison in an article in the *Listener*, he received from Wittgenstein a strongly worded letter of rebuke. However, Wittgenstein was inclined to see some sort of connection between his work and Freud's. He once described himself to Rhees as a 'disciple of Freud', and at various times summed up the achievements of both himself and Freud in strikingly similar phrases. 'It's all excellent similes', he said in a lecture of Freud's work; and of his own contribution to philosophy: 'What I invent are new *similes.*' This ability to form a synoptic view by constructing illuminating similes and metaphors was, it appears, what he wished to contribute to psychiatric medicine.

As the year wore on, however, Wittgenstein's interest in training as a doctor or in getting any other sort of job declined in favour of the idea of finishing his book. At the end of the year, as his fellowship was coming to an end, Wittgenstein discussed the possibilities open to him with a number of his favourite students. The latest of these was the postgraduate student Rush Rhees. Rhees had arrived in Cambridge in September 1935 to study under G. E. Moore, after having studied philosophy previously at Edinburgh, Göttingen and Innsbruck. He had, at first, been put off attending Wittgenstein's lectures by the mannerisms of Wittgenstein's students, but in February 1936 he overcame these misgivings and attended all the remaining lectures of that year. He became one of Wittgenstein's closest friends, and remained so until Wittgenstein's death. In June 1936 Wittgenstein invited Rhees to tea and discussed with him the question of whether he should try to get a job of some sort, or go somewhere by himself and spend his time working on his book. He told Rhees: 'I still have a little money. And I could live and work by myself as long as that lasts.'

This latter idea held sway, and when Wittgenstein and Skinner visited Drury in Dublin later that June, the issue of training as a psychiatrist was not raised. What perhaps clinched the decision was news of the death of Moritz Schlick. Wittgenstein was in Dublin when he heard that Schlick had been murdered – shot on the steps of Vienna University by a mentally deranged student. The fact that the student later became a member of the Nazi Party gave rise to rumours that the killing had a political motive, although the evidence suggests that the student had a more personal grudge against Schlick, who had rejected his doctoral thesis. Upon hearing the news, Wittgenstein immediately wrote to Friedrich Waismann:

Dear Mr Waismann,
The death of Schlick is indeed a great misfortune. Both you and I have lost much. I do not know how I should express my sympathy, which as you know, I really feel, to his wife and children. If it is possible for you, you would do me a great kindness if you contact Mrs Schlick or one of the children and tell them that I think of them with warm sympathy but I do not know what I should write to them. Should it be impossible for you (externally or internally) to convey this message, please let me know.

With kind sympathy and regards
Yours
Ludwig Wittgenstein

Schlick's death finally put an end to any idea there might still have been to fulfil the plans made in 1929 for Waismann and Wittgenstein to co-operate together on a book. With Waismann's exasperation at Wittgenstein's constant changes of mind, and Wittgenstein's distrust of Waismann's understanding of him, only their mutual respect for Schlick, and Schlick's encouragement to persist with the project, had provided even a slender hope of completion. After Schlick's death Waismann decided to work without Wittgenstein, and signed a contract to finish the book himself and have it published under his own name. The book reached the galley-proof stage in 1939, but was then withdrawn.

Wittgenstein, meanwhile, decided to do as he had done in 1913 – to go to Norway, where he could live alone without any distractions and bring his work to completion. It is possible that his decision was made in the wake of Schlick's death, but possible too that it was prompted by the more personal reason of needing to get away from the 'distraction' of his relationship with Francis, whose three-year post-graduate studentship ended at the same time as Wittgenstein's fellowship.

Until the summer of 1936 it seems to have been understood that whatever Wittgenstein and Francis would do – train as doctors, go to Russia, work with 'ordinary' people or work on Wittgenstein's book – they would do together. So, at least, it was understood by Francis. It is doubtful, however, whether Wittgenstein ever took Francis seriously as a philosophical collaborator; he was useful to *dictate* ideas to, especially when, as in the case of the *Blue* and *Brown Books*, the dictation was done in English. But for the *discussion* of ideas, the clarification of thoughts, Francis was no use; his awed respect for Wittgenstein paralysed him and got in the way of his making any useful contribution. 'Sometimes', Wittgenstein told Drury, 'his silence infuriates me and I shout at him, "Say something, Francis!"' 'But', he added, 'Francis isn't a thinker. You know Rodin's statue called *The Thinker*; it struck me the other day that I couldn't imagine Francis in that attitude.'

For similar reasons Wittgenstein discouraged Francis from going on with academic work. 'He would never be happy in academic life', he decided, and Francis, as always, accepted his decision. It was not, however, the view of Francis's family, nor that of many of his friends. Louis Goodstein, for example, who was Francis's contemporary at both St Paul's and Cambridge and who later became Professor of Mathematical Logic at Leicester University, thought that Francis might have had a promising career as a professional mathematician. He was one of the first people to be told by Francis of his decision to abandon mathematics, and he disapproved strongly, seeing in the decision only the unfortunate influence of Wittgenstein's own dislike of academic life. So, too, did Francis's family. His mother particularly came to dislike deeply the influence that Wittgenstein was exerting on her son. She reacted with great consternation both to the plan to settle in Russia and to the idea that Francis should abandon his potentially brilliant academic career. His sister, Priscilla Truscott, was equally incredulous. 'Why?' she demanded. '*Why?*'

For Francis, however, the only person whose opinion mattered was Wittgenstein, and he resolutely adhered to Wittgenstein's decision, even when that meant living away from Wittgenstein himself and working in a job which made little use of his talents and in which he felt exploited. It was not to train as a doctor, but as a factory mechanic, that Skinner left university, and not alongside Wittgenstein, but on his own. The idea that he might train as a doctor was impractical: his parents could not afford to see him through medical studies, and Keynes's promise to finance Wittgenstein's medical training did not extend to Francis. Francis volunteered to fight in the Spanish Civil War alongside the International Brigade, but was turned down because of his physical disability. (Francis, whose health was always precarious, was lame in one leg as a result of the osteomyelitis from which he suffered as a young boy and from which he was always subject to renewed attacks.)

Wittgenstein's (and therefore Skinner's) second choice of career after medicine was that of a mechanic. And so in the summer of 1936 Francis was taken on as a two-year apprentice mechanic at the Cambridge Instrument Company. For most of the time he was employed in making mainscrews, a repetitive and tiring task that he neither enjoyed nor found at all interesting; it was simply a drudge which he put up with for Wittgenstein's sake. Fania Pascal, however, believes that Skinner was happier among working people than among people of his own class. The working people were, she says, kinder and less self-conscious. This is perhaps true, although for the first few years at the factory Francis spent little time socializing with his colleagues. His evenings were spent either alone or with friends from the university – the Bachtins, Rowland Hutt and Pascal herself. What he wanted more than anything else was to live and work together with Wittgenstein, and this he had been denied by Wittgenstein himself.

Francis did not have a Weiningerian conception of love; he did not believe that love needed a separation, a certain distance, to preserve it. Wittgenstein, on the other hand, probably shared Weininger's view. While in Norway he recorded in his diary that he realized how unique Francis was – that he really appreciated him – only when he was away from him. And thus it was, perhaps, precisely to get away from him that he decided to go to Norway.

Before he left for Norway Wittgenstein took a holiday in France with Gilbert Pattisson, the two touring the Bordeaux region together in a car. Pattisson was one of the relatively few people with whom Wittgenstein could relax and enjoy himself. For Pattisson's part, however, Wittgenstein's company could be a little too heavy. He accordingly insisted, as he had before in 1931, on spending at least a few nights of the holiday away from Wittgenstein at a fashionable resort, where he could indulge himself in unfettered luxury – wining, dining and gambling. On the one occasion when Wittgenstein accompanied Pattisson in the pleasures of gambling, he showed himself to be a novice in the art of wasting money. They went together to the Casino Royan, where they played roulette, a game obviously new to Wittgenstein. He studied the game carefully before remarking to Pattisson, incredulously: 'I don't see how you *can* win!' Sometimes, it seems, there is more point in scrutinizing the rules than in playing the game.

18
CONFESSIONS

Wittgenstein's leaving for Norway in August 1936 is strongly reminiscent of his earlier departure in October 1913. In both cases he was leaving for an indefinite period to accomplish a definite task – the preparation of a final formulation of his philosophical remarks. In both cases, too, he was leaving behind someone he loved.

The difference is that in 1913 Pinsent had had no wish to accompany him. It is doubtful whether Pinsent ever realized how much Wittgenstein was in love with him, and almost certain that he did not return that love. He was 'thankful' for his 'acquaintance' with Wittgenstein, but not in any way dependent upon it. In October 1913 Pinsent's training as a lawyer figured much larger in his concerns than his friendship with Wittgenstein, a break from whom possibly came to him as something of a relief.

For Francis, however, his relationship with Wittgenstein was the very centre of his life: if asked, he would have dropped everything to go and live with him in Norway. 'When I got your letter', he wrote just a few weeks after they were separated, 'I wished I could come and help you clean your room.' His life in Cambridge without Wittgenstein was lonely and dreary. He no longer got on well with his family, he could no longer participate in Wittgenstein's work, and, though he persevered with it for Wittgenstein's sake, he disliked his job at the factory. As he had no doubt been asked by Wittgenstein to do, he gave regular reports of his work. They sound far from enthusiastic: 'My work goes on all right. I am working at mainscrews' (21.8.36); 'My work is going on all right. I am almost reaching the end of making the mainscrews. Last week I had to do some hand tool work on them, which was difficult at first. I am now polishing them up ready to be nickelled' (1.9.36); 'I have an order for 200 draught and pressure gauges. I wish there weren't going to be so many' (14.10.36). Eventually, after a discussion with Rowland Hutt about his position at the factory, even the mild and compliant Francis was moved to express his dissatisfaction:

> I don't feel clear about my relation to the firm. I don't feel clear whether I am getting work which makes full use of me. It seems to me (and Hutt agrees with me) that there is a line to be drawn between being favoured exceptionally and letting them do anything with you. For instance the foreman said to me that if I had been going to be there five years he could have advanced me very quickly, but as I was only going to be there two years and the firm knew anyhow I would not ever be much use to them, it was very different.

He tried, he said, to remember what Wittgenstein had told him about being 'hopeful, grateful & thoughtful', but in such circumstances it was not easy. He does not say so, but one can imagine him thinking that in such a position he had little to hope for, nothing to be grateful for and nothing to occupy his thoughts except that he would rather be with Wittgenstein. His conversation with Hutt, he told Wittgenstein, 'made me feel how much I wished I could have you here to talk with you'. Time and again in his letters he emphasizes that: 'I think of you a lot with great love.' Wittgenstein's side of this correspondence has not survived, but the form of these declarations of love sometimes suggests that they were written to reassure doubts that Wittgenstein might have expressed: 'My feelings for you haven't changed at all. This is the honest truth. I think of you a lot and with great love.'

It is likely that the advice to stay 'hopeful, grateful & thoughtful' was all that Francis received in the way of sympathetic understanding from Wittgenstein. In Norway, Wittgenstein gave more thought to himself and to his work – the two, as ever, being inextricably linked – than he did to Francis. And, as in 1913–14, and again in 1931, being alone in Norway proved conducive to thinking seriously about both logic and his sins.

'I do believe that it was the right thing for me to come here thank God', he wrote to Moore in October. 'I can't imagine that I could have worked anywhere as I do here. It's the quiet and, perhaps, the wonderful scenery; I mean, its quiet seriousness.' To the news that both Moore and Rhees were finding it difficult to write anything, Wittgenstein replied that this was a good sign: 'One can't drink wine while it ferments, but that it's fermenting shows that it isn't dish-water'. 'You see', he added, 'I still make beautiful similes.'

Wittgenstein sent Moore a map, showing his hut in relation to the fjord, the neighbouring mountains and the nearest village. The point was to illustrate that it was impossible for him to get to the village without rowing. In clement weather this was not too bad, but by October it was wet and cold. He wrote to Pattisson: 'The weather has changed from marvellous to rotten. It rains like hell now. Two days ago we had our first snow.' Pattisson responded by sending Wittgenstein a sou'wester, with which Wittgenstein was very pleased. Recalling the 'Letters from a Satisfied Customer', he wrote: '"Both fit and style are perfect" as they always write to Mr Burton, the tailor of taste.'

He had taken with him a copy of the *Brown Book*, with the intention of using it as the basic material out of which to construct a final version of his book. For over a month, he worked on a revision of it, translating it from English into German and rewriting it as he went along. At the beginning of November he gave this up, writing, in heavy strokes: '*Dieser ganze "Versuch einer Umarbeitung" vom (Anfang) bis hierher ist nichts wert*' ('This whole attempt at a revision, from the beginning right up to here, is worthless'). He explained in a letter to Moore that when he read through what he had written so far he found it all, 'or nearly all, boring and artificial':

> For having the English version before me had cramped my thinking. I therefore decided to start all over again and not to let my thoughts be guided by anything but themselves. – I found it difficult the first day or two but then it became easy. And so I'm writing now a new version and I hope I'm not wrong in saying that it's somewhat better than the last.

This new version became the final formulation of the opening of Wittgenstein's book. It forms, roughly, paragraphs 1–188 of the published text of *Philosophical Investigations* (about a quarter of the book), and is the only section of Wittgenstein's later work with which he was fully satisfied – the only part he never later attempted to revise or rearrange, or indicate that he would *wish* to revise if he had time.

To a large extent it follows the arrangement of the *Brown Book*, beginning with St Augustine's account of how he learnt to talk, using that to introduce the notion of a language-game and then proceeding to a discussion of following a rule. In this final version, however, the passage from Augustine's *Confessions* is actually quoted, and the point of beginning with that passage is spelt out more clearly:

> These words, it seems to me, give us a particular picture of the essence of human language. It is this: the individual words in language name objects – sentences are combinations of such names. – In this picture of language we find the roots of the following idea: Every word has a meaning. This meaning is correlated with the word. It is the object for which the word stands.

The rest of the book was to examine the implications of this idea and the traps into which it led philosophers, and to suggest routes out of those traps. These routes all begin by dislodging the (pre-philosophical) picture of language expressed by Augustine, which gives rise to the philosophical idea mentioned above. In this way, Wittgenstein hoped to dig out philosophical confusion by its pre-philosophical roots.

The quotation from St Augustine is not given, as is sometimes thought, to present a *theory* of language, which Wittgenstein will then show to be false. *Confessions*, after all, is not (primarily, at least) a philosophical work, but a religious autobiography, and in the passage quoted Augustine is not *theorizing*, but is describing how he learnt to talk. And this is precisely why it is appropriate for presenting the target of Wittgenstein's philosophical enterprise. Though it expresses no theory, what is contained in Augustine's account is a *picture*. And, for Wittgenstein, *all* philosophical theories are rooted in just such a picture, and must be uprooted by the introduction of a new picture, a new metaphor:

> A simile that has been absorbed into the forms of our language produces a false appearance and this disquiets us.

> A *picture* held us captive. And we could not get outside it, for it lay in our language and language seemed to repeat it to us inexorably.

The final version of the beginning of Wittgenstein's book differs from the *Brown Book* in that, rather than simply leading the reader through a series of language-games without any explanation, he pauses now and again to elucidate his procedure and to guard against some possible misunderstandings of it:

> Our clear and simple language-games are not preparatory studies of a future regularization of language – as it were first approximations, ignoring friction and air-resistance. The language-games are set up as *objects of comparison* which are meant to throw light on the facts of our language by way not only of similarities, but also of dissimilarities.

> It is not our aim to refine or complete the system of rules for the use of our words in unheard-of ways.
>
> For the clarity that we are aiming at is indeed *complete* clarity. But this simply means that the philosophical problems should *completely* disappear. The real discovery is the one that makes me capable of stopping doing philosophy when I want to. – The one that gives philosophy peace, so that it is no longer tormented by questions which bring *itself* in question. – Instead we now demonstrate a method, by examples; and the series of examples can be broken off. – Problems are solved (difficulties eliminated), not a *single* problem.

Anticipating a natural reaction to his conception of philosophy and its method, he asks: 'Where does our investigation get its importance from, since it seems only to destroy everything interesting, that is, all that is great and important? (As it were all the buildings, leaving behind only bits of stone and rubble.)' He answers: 'What we are destroying is nothing but houses of cards and we are clearing up the ground of language on which they stand.' And, changing the metaphor, but keeping to the same point:

> The results of philosophy are the uncovering of one or other piece of plain nonsense and of bumps that the understanding has got by

running its head up against the limits of language. These bumps make us see the value of the discovery.

Whether such explanations would mean anything to people who have not themselves experienced such 'bumps' remains doubtful. But then, the method was not developed for such people, just as Freudian analysis was not developed for the psychologically unconcerned. *Philosophical Investigations* – more, perhaps, than any other philosophical classic – makes demands, not just on the reader's intelligence, but on his *involvement*. Other great philosophical works – Schopenhauer's *World and Representation*, say – can be read with interest and entertainment by someone who 'wants to know what Schopenhauer said'. But if *Philosophical Investigations* is read in this spirit it will very quickly become boring, and a chore to read, not because it is intellectually difficult, but because it will be practically impossible to gather what Wittgenstein is 'saying'. For, in truth, he is not *saying* anything; he is presenting a technique for the unravelling of confusions. Unless these are *your* confusions, the book will be of very little interest.

Connected with the degree of personal involvement required to make sense of it, there is another reason why it seems appropriate to begin the book with a quotation from St Augustine's *Confessions*. And that is that, for Wittgenstein, *all* philosophy, in so far as it is pursued honestly and decently, begins with a confession. He often remarked that the problem of writing good philosophy and of thinking well about philosophical problems was one of the will more than of the intellect – the will to resist the temptation to misunderstand, the will to resist superficiality. What gets in the way of genuine understanding is often not one's lack of intelligence, but the presence of one's pride. Thus: 'The edifice of your pride has to be dismantled. And that is terribly hard work.' The self-scrutiny demanded by such a dismantling of one's pride is necessary, not only to be a decent person, but also to write decent philosophy. 'If anyone is unwilling to descend into himself, because this is too painful, he will remain superficial in his writing':

> Lying to oneself about oneself, deceiving yourself about the pretence in your own state of will, must have a harmful influence on [one's] style; for the result will be that you cannot tell what is genuine in the style and what is false . . .

24 With the family at the Hochreit

25 Ludwig, Helene and Paul at the Hochreit

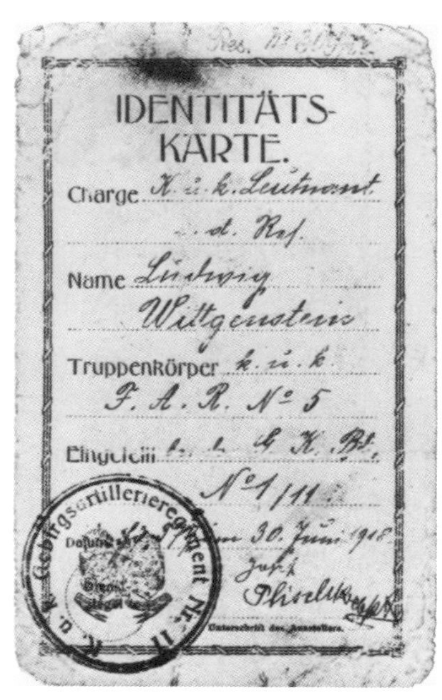

IDENTITÄTS-
KARTE.

Charge K. u. k. Leutnant
d. Res.

Name Ludwig
Wittgenstein

Truppenkörper k. u. k.
F. A. R. No 5

Eingeteilt
No 1/11

Datum am 30. Juni 1918

Unterschrift des Ausstellers.

K. u. k. Gebirgsartillerieregiment Nr. 11

K. u. k. Gebirgsartillerieregiment

Datum am 30. Juni 1918

Ludwig Wittgenstein

Unterschrift des Besitzers

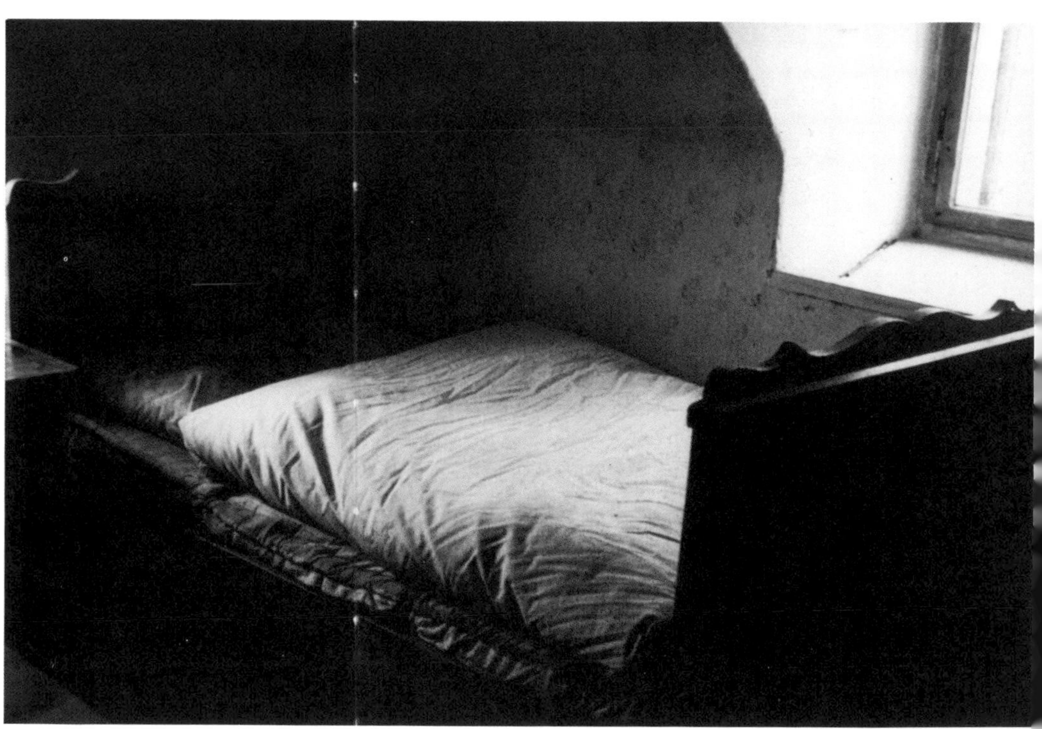

26 & 27 Wittgenstein's military identity card during the First World War

28 Wittgenstein's room in the guest-house 'Zum Braunen Hirschen', Trattenbach

29 Wittgenstein with his pupils in Puchberg am Schneeberg

30 Frank Ramsey

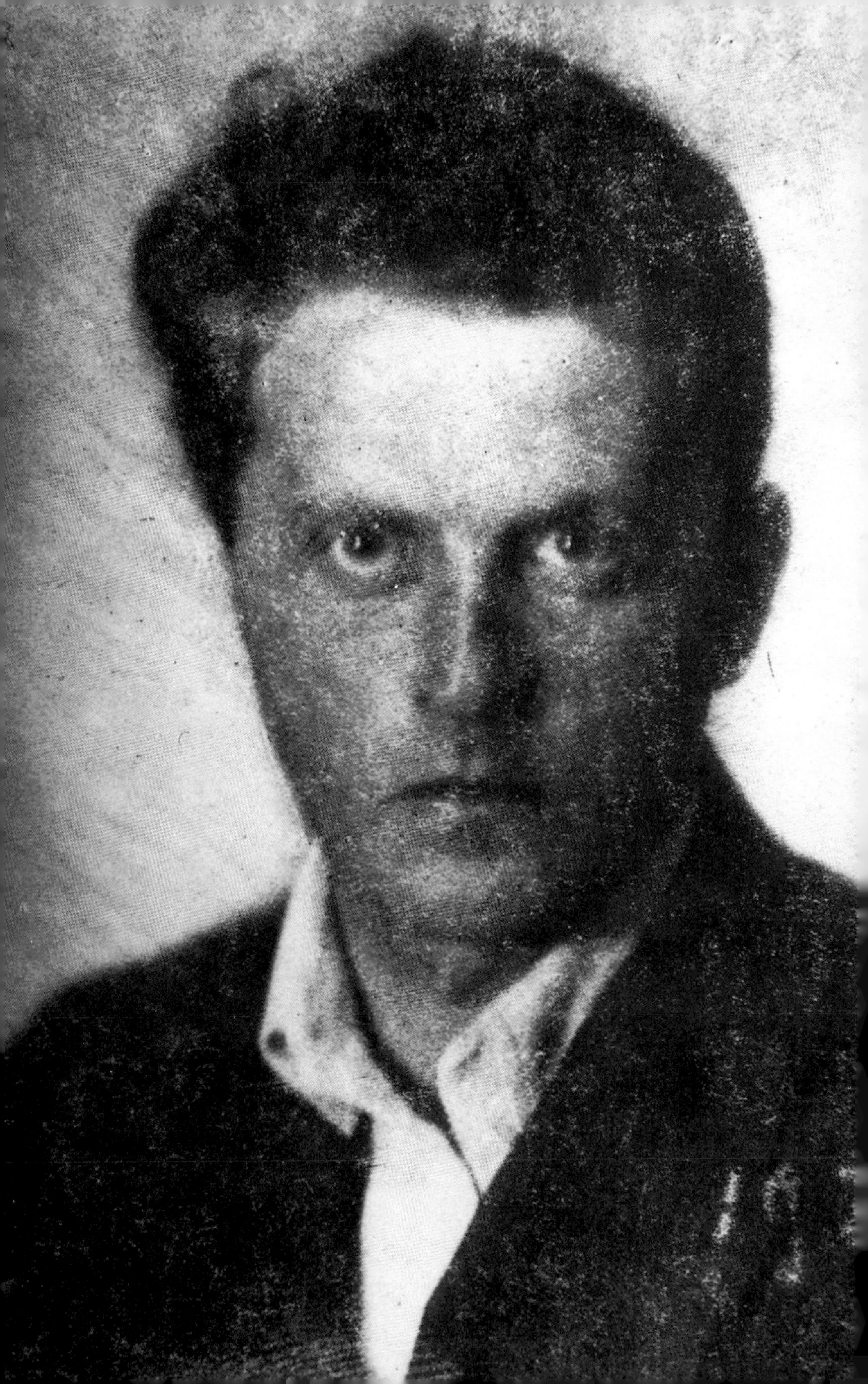

31 Wittgenstein, 1925

32 & 33 Examples of a window-catch and door-handle designed by Wittgenstein for his sister's house in the Kundmanngasse

34 The Kundmanngasse house

35 & 36 Portraits of Wittgenstein: *above* on being awarded a scholarship from Trinity College, 1929; *right* the Fellowship Portrait, 1930

Dear old Blood, I'm sure you'll
be interested to see me
as I walk with a sister & a
friend of mine in an exhibitio
of bloody modern houses. Don't
I look enterprising?! You can
gather from this picture that
we had very hot weather but
not that there was a terrific
thunderstorm half an hour after this
was taken. I am, old god, yours
in bloodyness Ludwig

37 & 38 Postcard to Gilbert Pattisson from Vienna: 'Dear old Blood, I'm sure you'll be interested to see me as I walk with a sister & a friend of mine [Margarete Stonborough and Arvid Sjögren] in an exhibition of bloody modern houses. Don't I look enterprising?! You can gather from this picture that we had very hot weather but not that there was a terrific thunderstorm half an hour after this was taken. I am, old god, yours in bloodyness Ludwig'

> If I perform to myself, then it's this that the style expresses. And then the style cannot be my own. If you are unwilling to know what you are, your writing is a form of deceit.

It is no coincidence that Wittgenstein wrote the set of remarks with which he remained most satisfied at a time when he was most ruthlessly honest about himself – when he made the most intense efforts to 'descend into himself' and admit to those occasions on which his pride had forced him to be deceitful.

During the months in which he prepared the final formulation of the beginning of his book, Wittgenstein also prepared a confession, describing the times in his life when he had been weak and dishonest. His intention was to read the confession out to members of his family and to a number of his closest friends. He presumably felt that to admit the deceit to himself was not enough; properly to 'dismantle the pride' that had given rise to his weakness would involve confession to other people. It was, for him, a matter of the utmost importance, and accordingly, in November 1936, he wrote to, among others, Maurice Drury, G. E. Moore, Paul Engelmann, Fania Pascal and, of course, Francis Skinner, telling them that he *had* to meet them some time during the Christmas period. The only one of these letters to survive is the one to Moore, although we can guess that the others were broadly similar. He told Moore that, besides his work, 'all sorts of things have been happening inside me (I mean in my mind)':

> I won't write about them now, but when I come to Cambridge, as I intend to do for a few days about the New Year, I hope to God I shall be able to talk to you about them; and I shall then want your advice and your help in some very difficult and serious matters.

To Francis he must have been a little more direct, telling him that what he had in mind was to make a confession. In a letter of 6 December, we find Francis promising: 'Whatever you say to me can't make any difference to my love for you. I am terribly rotten in every way myself.' What was of more importance for Francis was that he would, at last, be seeing Wittgenstein again: 'I think of you a lot and of our love for each other. This keeps me going and gives me cheerfulness and helps me to get over despondency.' Three days later he repeated the promise: 'Whatever you have to tell me about yourself can't make any difference to my love for you . . . There won't be any question of my forgiving you as I am a much worse person than you are. I think of you a lot and love you always.'

Wittgenstein spent Christmas in Vienna, and delivered the confession to Engelmann, some members of his family and, probably, to some other friends as well (surely, one would think, Hänsel must have been included). None of these people left any record of what the confession contained. When Engelmann published his letters from Wittgenstein he omitted the one that mentioned the confession; in all probability, he destroyed it. In the New Year Wittgenstein visited Cambridge and made his confession to G. E. Moore, Maurice Drury, Fania Pascal, Rowland Hutt and Francis.

Moore, Drury and Francis died without revealing the secret of what the confession contained, and we therefore have only the recollections of Pascal and Hutt to depend on. How the others reacted to the confession we do not know, although Pascal most likely captures the spirit of Drury's and Moore's responses when she remarks that, without being told, she knows they: 'listened patiently, said little, but showed friendly participation, implying by manner and look that there was no need for him to make this confession, but if he thought he should, well and good, so be it'. According to Drury, however, he did not *listen* to the confession, but read it. Drury adds that Moore had already read it, and, according to Wittgenstein, had seemed very distressed at having to do so. Other than this, Drury says nothing in his memoir about the confession. As for Francis, Pascal is no doubt correct in speculating: 'He would have sat transfixed, profoundly affected, his eyes undeviatingly on Wittgenstein.'

For both Rowland Hutt and Fania Pascal, listening to the confession was an uncomfortable experience. In Hutt's case the discomfort was simply embarrassment at having to sit in a Lyons café while opposite him sat Wittgenstein reciting his sins in a loud and clear voice. Fania Pascal, on the other hand, was exasperated by the whole thing. Wittgenstein had phoned at an inconvenient moment to ask whether he could come and see her. When she asked if it was urgent she was told firmly that it was, and could not wait. 'If ever a thing could wait', she thought, facing him across the table, 'it is a confession of this kind and made in this manner.' The stiff and remote way in which he delivered his confession made it impossible for her to react with sympathy. At one point she cried out: 'What is it? You want to be perfect?' '*Of course* I want to be perfect', he thundered.

Fania Pascal remembers two of the 'sins' confessed to by Wittgenstein. As well as these, there were a number of more minor sins, which elude her memory. Some of these have been remembered by Rowland Hutt. One concerned the death of an American acquaintance of Wittgenstein's. Upon being told by a mutual friend of this death, Wittgenstein reacted in a way that was appropriate to hearing sorrowful news. This was disingenuous of him, because, in fact, it was not news to him at all; he had already heard of the death. Another concerned an incident in the First World War. Wittgenstein had been told by his commanding officer to carry some bombs across an unsteady plank which bridged a stream. He had, at first, been too afraid to do it. He eventually overcame his fear, but his initial cowardice had haunted him ever since. Yet another concerned the fact that, although most people would have taken him to be a virgin, he was not so: as a young man he had had sexual relations with a woman. Wittgenstein did not use the words 'virgin' or 'sexual relations', but Hutt is in no doubt this is what he meant. The actual words used by Wittgenstein have eluded his memory. They were, he thinks, something like: 'Most people would think that I have had no relationship with women, but I have.'

The first of the 'sins' that Fania Pascal does remember was that Wittgenstein had allowed most people who knew him to believe that he was three-quarters Aryan and one-quarter Jewish, whereas, in fact, the reverse was the case. That is to say, of Wittgenstein's grandparents, three were of Jewish descent. Under the Nuremberg Laws, this made Wittgenstein himself a Jew, and Pascal is surely right in linking this confession with the existence of Nazi Germany. What Wittgenstein did not tell her, but what she subsequently discovered, was that not one of his 'Jewish' grandparents was actually a Jew. Two were baptized as Protestants, and the third as a Roman Catholic. 'Some Jew', she remarks.

So far, all these 'crimes' are sins of omission: they concern only cases in which Wittgenstein *failed* to do something, or declined to correct a misleading impression. The final, and most painful, sin concerns an actual untruth told by Wittgenstein. At this stage in the confession, Pascal recalls, 'he had to keep a firmer control on himself, telling in a clipped way of the cowardly and shameful manner in which he had behaved'. Her account of this confession, however, gives a strangely distorted impression of the incident she describes:

> During the short period when he was teaching at a village school in Austria, he hit a little girl in his class and hurt her (my memory is, without details, of a physically violent act). When she ran to the headmaster to complain, Wittgenstein denied he had done it. The event stood out as a crisis of his early manhood. It may have been this that made him give up teaching, perhaps made him realise that he ought to live as a solitary.

This is distorted in a number of ways. First, Wittgenstein was in his late thirties when the incident at Otterthal occurred, which is surely a little old to be described as 'early manhood'. More important, Pascal seems to have had no idea that physically violent acts were, by all accounts, a not infrequent occurrence in Wittgenstein's classes, nor that Wittgenstein actually stood trial before a court to answer charges of violence. It is possible that Wittgenstein did not tell her these things – that he used the isolated incident as a symbol of his misdemeanours at Otterthal. But it is also possible – and I think not unlikely – that Pascal's memory is at fault. She was, after all, in no mood to listen to Wittgenstein's confession, and was further alienated by his manner while delivering it. Rowland Hutt remembers the confession, not as concerning a denial made to a headmaster about an isolated incident, but rather as an admission to having told lies in a court case. Put like this, it both squares better with the accounts given by the Otterthal villagers and explains better why the deception so haunted Wittgenstein.

There is no doubt that of all the deceptions confessed to by Wittgenstein, his behaviour at Otterthal was felt to be the greatest burden, and he went to much greater lengths than Pascal and Hutt could have known to relieve himself of it. In the same year in which he made his confessions, Wittgenstein astounded the villagers of Otterthal by appearing at their doorsteps to apologize personally to the children whom he had physically hurt. He visited at least four of these children (and possibly more), begging their pardon for his ill-conduct towards them. Some of them responded generously, as the Otterthal villager Georg Stangel recalls:

> I myself was not a pupil of Wittgenstein's, but I was present when shortly before the war Wittgenstein visited my father's house to apologise to my brother and my father. Wittgenstein came at midday, at about 1 o'clock, into the kitchen and asked me where Ignaz is. I called my brother, my father was also present. Wittgenstein said that he wanted to apologise if he had done him an injustice. Ignaz said that he had no need to apologise, he had learnt well from Wittgenstein. Wittgenstein stayed for about half an hour and mentioned that he also wanted to go to Gansterer and Goldberg to beg their pardon in a similar way.

But at the home of Mr Piribauer, who had instigated the action against Wittgenstein, he received a less generous response. There he made his apologies to Piribauer's daughter Hermine, who bore a deep-seated grudge against him for the times he had pulled her by the ears and by the hair in such a violent fashion that, on occasion, her ears had bled and her hair had come out. To Wittgenstein's plea for pardon, the girl responded only with a disdainful, 'Ja, ja.'

One can imagine how humiliating this must have been for Wittgenstein. And it might almost seem that the point of humbling himself in this way was precisely that: to punish himself. But this, I think, would be to misunderstand the purpose of his confessions and apologies. The point was not to *hurt* his pride, as a form of punishment; it was to *dismantle* it – to remove a barrier, as it were, that stood in the way of honest and decent thought. If he felt he had wronged the children of Otterthal, then he ought to apologize to them. The thought might have occurred to anyone, but most people would entertain the idea and then dismiss it for various reasons: it happened a long time ago; the villagers would not understand such an apology, and would think it very strange; the journey to Otterthal is difficult in winter; it would be painful and humiliating to offer such an apology and, given the other reasons, not worth the trouble; and so on. But to find these reasons compelling, as, I think, most of us would, is in the end to submit to cowardice. And this, above all else, is what Wittgenstein was steadfastly determined not to do. He did not, that is, go to Otterthal to *seek* pain and humiliation, but rather with the determination to go through with his apology despite it.

In reflecting upon the effects of his confession he wrote:

> Last year with God's help I pulled myself together and made a confession. This brought me into more settled waters, into a better relation with people, and to a greater seriousness. But now it is as though I had spent all that, and I am not far from where I was before. I am cowardly beyond measure. If I do not correct this, I shall again drift entirely into those waters through which I was moving then.

Wittgenstein regarded his confessions as a kind of surgery, an operation to remove cowardice. Characteristically, he regarded the infection as malignant and in need of continued treatment. It was characteristic, too, for him to regard a mere physical injury as trivial by comparison. Soon after he returned to Norway in the New Year of 1937 he suffered an accident and broke one of his ribs. Whereas his moral condition had been a matter of urgency, this was simply brushed aside with a joke. He told Pattisson: 'I thought of having it removed & of having a wife made of it, but they tell me that the art of making women out of ribs has been lost.'

If the confessions had any effect on Francis, it was possibly that of emboldening him to speak his mind a bit more freely – to reveal some of the things *he* had kept hidden. 'I feel it is wrong to hide things from you', he wrote in March 1937, 'even though I do so because I am ashamed of myself.' But in his case, it was not past deeds that he revealed, but present feelings, and in particular the feeling that he did not want to be in Cambridge working at the factory, but with Wittgenstein, and preferably working with him: 'I sometimes wish that we could do some work together, any sort of work. I feel you are part of my life.' What worried him was not his own moral state (and certainly not Wittgenstein's), but their relationship – his fear that they were growing apart or that they might be forced apart by circumstances:

> I think a lot about our relation. Are we going to act independently of each other, will I be able to act independently of you? What will happen if there is a war? Or if we are permanently separated? I am so terribly deficient in courage. I long for you often. I do feel you are near me in whatever state of mind I am in, and would feel so even if I did something very bad. I am always your old heart. I love to think about you.

It was painful for Francis to think that he was no longer involved in Wittgenstein's work – to recognize that he was no longer in any sense Wittgenstein's collaborator. In May he wrote: 'I don't think I have ever understood your present work thoroughly, and I think it would be good for me to try and understand it better.' The letter contains a report of a meeting with Sraffa, from which, he said, he had 'learnt a lot and it had done me good'. Sraffa had: 'talked in a very nice way about working men'. But, as a working man himself, Francis was beginning to find, to his great consternation, that the problems of philosophy seemed rather remote to him now:

> I have been trying to think lately about what use philosophy is to me now. I don't want to lose my intellectual conscience. I want to make some use of all those years I spent learning philosophy. I don't want now just to have been made a cleverer person. I want to keep in mind the importance of trying to use words correctly . . . I think also I ought not to forget that philosophical problems are really important problems to me.

This letter, dated 27 May, was written to Wittgenstein in Vienna. His work in Norway during the spring of 1937 had gone badly – 'partly', he told Moore, 'because I've been troubled about myself a lot', and he spent the summer, first with his family, and then in East Road with Francis. At Cambridge he undertook work that, presumably, Francis *could* help him with: he dictated a typescript of the remarks, written the previous winter, that now form the first 188 paragraphs of *Philosophical Investigations.* On 10 August he left again for Norway.

That Wittgenstein returned to Norway full of trepidation is apparent from his diary entries of this period. On the ship to Skjolden he notes that he has managed to write a little, but that his mind is not 'wholeheartedly' on his work. A few days later he describes himself as: 'vain, thoughtless, anxious' – anxious, that is, about living alone. 'Am afraid I will be depressed and not able to work':

> I would now like to live with somebody. To see a human face in the morning. – On the other hand, I have now become so *soft*, that it

> would perhaps be good for me to have to live alone. Am now extraordinarily contemptible.

'I have the feeling', he writes, 'that I would not be completely without ideas, but that the solitude will depress me, that I will not be able to work. I am afraid that in my house all my thoughts will be killed off, that there a spirit of despondency will take complete possession of me.' But where else could he work? The idea of living in Skjolden but not in his house disturbed him, and in Cambridge, 'I could *teach*, but not write as well.' The following day he was: 'unhappy, helpless and thoughtless', and it occurred to him: 'how unique and irreplaceable Francis is. And yet how little I realise this when I am with him':

> Am completely ensnared in pettiness. Am irritable, think only of myself, that my life is wretched, and at the same time I have no idea of how wretched it is.

He could not face moving back into his house. His room, which before had seemed charming, now struck him as alien and unfriendly. He took lodgings instead with Anna Rebni, but in so doing he had to wrestle with his conscience. It struck him as 'weird' (*unheimlich*) that he should live with her and leave his own house standing empty: 'I am ashamed to have this house and not to live in it. It is, however, strange that this shame should be such a powerful feeling.' After a night at Rebni's house, he wrote that he felt strange being there: 'I do not know whether I have either a right or any good reason to live here. I have neither any real need for solitude nor any overpowering urge to work.' He felt weak at the knees. 'Is it the climate?? – It is terrible how easily I am overcome by anxiety [*die Sorge*].' He thought of moving back to his own house, 'but I am frightened of the sadness that can overcome me there'. It is difficult, he wrote, to walk uphill, and one does it only reluctantly. He himself felt too weak to make the effort. He was, for a day or two, inclined to think that the trouble was physical rather than psychological. 'Am now really ill', he wrote on 22 August, 'abdominal pains and temperature.' The following evening, however, he recorded that his temperature was normal, but that he felt as tired as ever. It was not until 26 August that he recorded the first sign of recovery: he was able once more to look at the Norwegian scenery with pleasure. He had that day received two letters ('showered with gifts' as he put it) – one from Francis and another from Drury, '*both stirringly lovely*'. On the same day he finally – a year after he had first gone to live in Norway – wrote to Francis to invite him to join him. 'May it go well. And may it be given to me to be halfway decent.'

Francis accepted the invitation with alacrity. On 23 August he had written: 'You said in one of your letters "I wish I had you here". Would I be of any help to you if I could come to see you? You know I would come and would love to come.' Now: 'I would love very much to come and see you. I definitely think it would do me good. I'm quite sure of it.' Owing, however, to a blister on his leg that had to be operated on, it was not until the third week of September that he was able to travel.

During this time Wittgenstein gradually recovered his mental stability and his ability to work, and was able to move back into his house. 'The way to solve the problem you see in life', he wrote on 27 August, 'is to live in a way that will make what is problematic disappear':

> The fact that life is problematic shows that the shape of your life does not fit into life's mould. So you must change the way you live and, once your life does fit into the mould, what is problematic will disappear.
>
> But don't we have the feeling that someone who sees no problem in life is blind to something important, even to the most important thing of all? Don't I feel like saying that a man like that is just living aimlessly – blindly, like a mole, and that if only he could see, he would see the problem?
>
> Or shouldn't I say rather: a man who lives rightly won't experience the problem as *sorrow*, so for him it will be a bright halo round his life, not a dubious background.

In these terms Wittgenstein saw himself as neither blind, nor one who lives rightly. He felt the problem of life as a problem, as sorrow. Inevitably, he identified the problem as himself: 'I conduct myself badly and have mean and shabby feelings and thoughts' (26.8.37); 'I am a coward, that I notice again and again, on all sorts of occasions' (2.9.37); 'Am irreligious, but with *Angst*' (7.9.37). The 'but' in the last sentence seems to have been some reassurance, as though if he felt his lack of faith with anxiety, that at least proved that he was not living blindly – it gave him at least the possibility of living 'with a bright halo round his life'. On 4 September he wrote:

> Christianity is not a doctrine, not, I mean, a theory about what has happened and will happen to the human soul, but a description of something that actually takes place in human life. For 'consciousness of sin' is a real event and so are despair and salvation through faith. Those who speak of such things (Bunyan for instance) are simply describing what has happened to them, whatever gloss anyone may want to put on it.

It was, as ever, the God within himself that he sought – the transformation of his own despair into faith. He chastized himself when, during the violent storms of the following few days, he found himself tempted to curse God. It was, he told himself, 'just wicked and superstitious'.

By 11 September Wittgenstein's ability to work had revived sufficiently for him to start writing into one of his large manuscript volumes (rather than his notebooks), but he was frightened, he said, that he would write 'in a stilted and bad style'. He found he could just about manage to work, but he could find no pleasure in so doing: 'It's as though my work had been drained of its juice', he wrote on 17 September.

The following day he travelled to Bergen to meet Francis. He felt, he wrote, very sensual: in the night, when he could not sleep, he had sensual fantasies. A year ago he had been much more decent – more *serious*. After Francis had arrived at the house, Wittgenstein was 'sensual, susceptible, indecent' with him: 'Lay with him two or three times. Always at first with the feeling that there was nothing wrong in it, *then* with shame. Have also been unjust, edgy and insincere towards him, and also cruel.' Whether this was the only occasion on which he and Francis were sexually intimate, we do not know. It is certainly the only occasion mentioned in his coded remarks. What is striking is the juxtaposition of his account of their lying together with observations of his lovelessness towards Francis. Or perhaps what he is expressing is his *fear* of becoming loveless, as though he expected to find that Weininger was right when he wrote: 'physical contact with the beloved object, in which the sexual impulse is awakened . . . suffices to kill love on the spot'.

During the ten days or so that Francis stayed at Wittgenstein's house there is only one coded remark: 'Am very impatient!' (25.9.37) On 1 October, the day that Francis left, however, he wrote:

> The last 5 days were nice: he settled into the life here and did everything with love and kindness, and I was, thank God, not impatient, and truly I had no reason to be, except for my own rotten nature. Yesterday I accompanied him as far as Sogndal; returned today to my hut. Somewhat depressed, also tired.

For Francis, of course, the sensuality and the intimacy of their first night together in Wittgenstein's house had no Weiningerian connotations. He could give himself over to his 'susceptibility' to Wittgenstein without any fear of losing his love. In an undated letter, for example, he writes: 'I often remember all the things we have done together in the past and also the things we did here in Cambridge. This makes me long for you sometimes very violently'; and his letters immediately after his visit to Norway are repetitious in their affirmation of how 'wonderful' the visit had been:

> I think *constantly* of you and of the wonderful time I have had with you. It was wonderful that it was possible. It was so lovely being with you and living in the house with you. It was a wonderful gift to us. I hope it will do me a lot of good. [undated]

> I am often thinking now how good I felt when I was with you and how wonderful it was being with you and looking at the landscape with you. You were most wonderfully good to me. It has done me a lot of good being with you . . . It was wonderful being with you. [14.10.37]

During his stay, Francis had helped Wittgenstein clean his room, just as he had longed to do a year earlier. Wittgenstein's horror of uncleanliness impelled him to adopt a particularly rigorous method of cleaning his floor: he would throw wet tea-leaves over it, to soak up the dirt, and then sweep them up. He performed this task frequently wherever he stayed, and resolutely refused to have a carpet in any room that he lived in for any length of time. When Francis returned to his flat in East Road he adopted this fastidiousness as a sort of memento of his stay:

> I'm thinking of you a lot. I also think often how lovely it was cleaning your room with you. When I got back I decided I wouldn't put my carpet down even though it had been beaten because I know I can't keep it properly clean. I now have to sweep my room. I like to do it because it reminds me of when I was with you. I'm glad I learnt how to do it properly then.

Francis also put on Wittgenstein's mantle when he attended a meeting of the Moral Science Club. In his report of the meeting he eschews the modest and mild tone of almost all his other recorded utterances, and displays an uncharacteristic fierceness, borrowed, one suspects, from Wittgenstein:

> Prof. Moore wasn't present and Braithwaite took the chair. The paper was on ethics. I must say I thought Braithwaite was most revolting in the discussion. He took away all the seriousness from it. He never talked as if he had any responsibility for the discussion, or as if the discussion had a serious purpose. There was constant laughter throughout the discussion, a good deal of it provoked by him. I wouldn't have minded if what he had said had just been bad but I hated his lack of seriousness. This prevents anything useful and valuable coming from the discussion.

In his diary Wittgenstein describes this as a 'lovely letter from Fr.':

> He writes . . . how miserably bad the discussion was under Braithwaite's chairmanship. It is dreadful. But I wouldn't know what to do about it, for the other people are not serious enough either. Also, I would be *too cowardly* to do anything decisive.

In another letter, Francis mentions, in a similarly disapproving manner, Fania Pascal's lectures on 'Modern Europe', a course on current events that she had agreed to give to the Workers' Educational Association. Here Wittgenstein did attempt a decisive intervention: he wrote Pascal what she describes as a 'harsh and hectoring' letter, which 'caused the greatest explosion of fury on my part, a fury that rankled the more since I would not dare to express it to him'. Wittgenstein wrote that she must on no account give the course – that it was wrong for her, it was evil and damaging. *Why* he thought this, and what exactly the letter said, we will never know; Pascal tore it up in a fit of anger.

The first of Francis's letters did not reach Wittgenstein until about two weeks after Francis had left Skjolden. This, though not an extraordinarily long delay, was enough to confirm his fears. On 16 October he wrote: 'Have not heard from Francis for about 12 days and am rather worried, because he has not yet written from England. God, how much misery and wretchedness there is in this world.' The next day he received the first letter: 'Am relieved and gladdened. God may help us.'

In the meantime, he had received a short visit from Ludwig Hänsel's son, Hermann: 'He made a *good* impression. I have no *very* close relationship with him, because he is rough-grained [*grobkörnig*] and I am not entirely suited to rough-grained people.' But, though the grain was rough, the wood was good, Hänsel, 'who is much more decent than I', showed him what a shabby person he was: 'how worried I am that something will corrupt me; how *annoyed* if the slightest thing is ruined'. He worried that he might lose his energy for work, his imagination. Images of decay impressed themselves upon him:

> [I] just took some apples out of a paper bag where they had been lying for a long time. I had to cut half off many of them and throw it away. Afterwards when I was copying out a sentence I had written, the second half of which was bad, I at once saw it as a half-rotten apple.

And was there not, he asked himself, something feminine about this way of thinking, whereby: 'Everything that comes my way becomes a picture for me of what I am thinking about at the time.' It was as though, in Weininger's terms, he had relapsed into thinking in henids rather than in concepts.

Throughout November and December, his last two months in Norway, Wittgenstein's diary is full of the fears, anxieties and unpleasant thoughts that assailed him. He thought about illness and death – his own, his friends' and his family's. He worried that something would happen to him before he could leave. He fretted about his relationship with Anna Rebni, and about what he would do after he had left Norway. Would his book be finished by then? Would he be able to work on his own again, or should he go somewhere where he could be with someone – perhaps to Dublin, to Drury?

He worried, too, about his sensuality and his ability to love. He recorded the occasions on which he masturbated, sometimes with shame, and sometimes with bewildered doubt: 'How bad is it? I don't know. I guess it is bad, but I have no reason to think so.' Was his ability to love, with a clean and pure heart, threatened by the sexual desire manifested in his urge to masturbate?

> Think of my earlier love, or infatuation, for Marguerite and of my love for Francis. It is a bad sign for me that my feelings for M could go so completely cold. To be sure, there is a difference here; but still *my coldheartedness* remains. May I be forgiven; i.e., may it be possible for me, to be sincere and loving. [1.12.37]

> Masturbated last night. Pangs of conscience. But also the conviction that I am too weak to withstand the urge and the temptation if they and the images which accompany them offer themselves to me without my being able to take refuge in others. Yet only *yesterday evening* I was reflecting on the need to lead a pure life. (I was thinking of Marguerite and Francis.) [2.12.37]

Throughout all these worries, anxieties and fears, he tried to work on his book. During these months he wrote most of the remarks that now form Part I of *Remarks on the Foundations of Mathematics*, although at the time of writing he intended them to form the second half of the work he had written the previous year. The remarks are an application of the method described in the earlier work to the problems in the philosophy of mathematics, trying to show that these problems arise out of 'the bewitchment of our intelligence by means of language'. In particular, he uses his 'anthropological' method in an attempt to dissolve the way of thinking that gave rise to the logicism of Frege and Russell. By imagining tribes with conventions or ways of reasoning different to our own, and by constructing metaphors different to ones commonly employed, he tries to weaken the hold of certain analogies, certain 'similes that have been absorbed into the forms of our language'. He attacks, for example, the Platonism that regards logical propositions as analogous to factual propositions. 'Isn't there a truth corresponding to logical inference?' he makes his interlocutor ask. 'Isn't it *true* that this follows from that?' Well, replies Wittgenstein, what would happen if we made a different inference? How would we get into conflict with the truth?

> How should we get into conflict with truth, if our footrules were made of very soft rubber instead of wood and steel? – 'Well, we shouldn't get to know the correct measurement of the table.' – You mean: we should not get, or could not be sure of getting, *that* measurement which we get with our rigid rulers.

The point here is that the criteria for correct or incorrect reasoning are not provided by some external realm of Platonic truths, but, rather, by ourselves, by 'a *convention*, or a *use*, and perhaps our practical requirements'. The convention of using rigid rulers rather than floppy ones is not *truer*; it is simply more useful.

Wittgenstein also attacks the simile that lies at the heart of logicism: the analogy between a mathematical proof and a logical argument. In a logical argument, connections are made between various (empirical) propositions, with the intention of establishing the truth of a conclusion: All men are mortal; Socrates is a man; *therefore* Socrates is mortal. The result of a mathematical proof, on the other hand, is never the truth of an empirical proposition, but the establishing of a generally applicable *rule*. In this particular assault, Wittgenstein had to show the disanalogy between mathematical and empirical propositions, but his remarks on this point are less than completely satisfying. Occasionally his own dissatisfaction is acknowledged in the body of the text: 'I am merely – in an unskilful fashion – pointing to the *fundamental* difference, together with an apparent similarity, between the roles of an arithmetical proposition and an empirical proposition.' He was never happy with his presentation of this point, or with his treatment of other issues in the philosophy of mathematics, and over the next six years was to attempt to improve on it again and again.

Wittgenstein was dissatisfied with this work as he was writing it. In his diary he criticizes it often, and severely. The style, he says repeatedly, is bad, too uncertain, and he has constantly to cross out and change what he has written: 'I am nervous when writing and all my thoughts are short of breath. And I feel constantly that I cannot *completely* justify the expression. That it tastes bad.' This is indicative of his own nervousness, and of the fact that he got so little sleep and had been so long without seeing the sun. The weather was upsetting him; it was too cold. The fjord was already completely frozen, and the lake was beginning to ice over as well. He could no longer row, but had to walk across the ice, and this, too, worried him. He began to count the days until he could leave to go to Vienna for Christmas. Of course, he could leave at any time, but would this be right?

> I would like to flee, but that would be wrong and I simply can't do it. On the other hand, perhaps I could – I could pack *tomorrow* and leave the next day. But would I want to? Would it be right? Is it not right to stick it out here? Sure. I would leave tomorrow with a *bad* feeling. 'Stick it out', a voice says to me. There is also some vanity in this desire to stick it out, but something better as well. – The only cogent reason to leave here any earlier or at once would be that now I could perhaps work better somewhere else. For it is a fact that the pressure I am under at the moment makes it almost impossible for me to work and perhaps in a few more days really impossible.

When, during the following few days, he was able to work again, he thanked God for a gift he did not deserve. He always felt, he wrote, what a truly devout person never feels – that God was responsible for what he was: 'It is the opposite of piety. Again and again I want to say: "God, if you do not help me, what can I do?"' And although this attitude accords with what the Bible teaches, it is not that of a truly devout man, for such a one would assume responsibility for himself. 'You must *strive*', he urged himself; 'never mind God.'

Despite these urgings, he remained 'sensual, weak and mean', and subject to all the old anxieties – that he would not be able to leave because something would happen to him, that he would become ill or have an accident on the way home. He was troubled, too, by all the problems of spending the winter in Norway that Russell had pointed out in 1913: 'The ever-changing, difficult weather, cold, snow, sheet ice, etc., and the darkness and my exhaustion make everything very difficult.' He was sent, of course, encouragement and loving concern from Francis:

> I'm sorry you are having storms. Please be very careful about going across the lake. I will be thinking about you a great deal. I love to remember our time together in Norway. It does me good to think about it.

But when his last night in Norway came, on 10 December, he greeted it with some relief, writing that it was perfectly possible that he would never return.

On the ship to Bergen Wittgenstein wrote of Christ's Resurrection and of what inclined even him to believe in it. If Christ did not rise from the dead, he reasoned, then he decomposed in the grave like any other man. '*He is dead and decomposed.*' He had to repeat and underline the thought to appreciate its awfulness. For if that were the case, then Christ was a teacher like any other, 'and can no longer *help*; and once more we are orphaned and alone. So we have to content ourselves with wisdom and speculation.' And if that is all we have, then: 'We are in a sort of hell where we can do nothing but dream, roofed in, as it were, and cut off from heaven.' If he wanted to be saved, to be redeemed, then wisdom was not enough; he needed faith:

> And faith is faith in what is needed by my *heart*, my *soul*, not my speculative intelligence. For it is my soul with its passions, as it were with its flesh and blood, that has to be saved, not my abstract mind. Perhaps we can say: Only *love* can believe the Resurrection. Or: it is *love* that believes the Resurrection. We might say: Redeeming love believes even in the Resurrection; holds fast even to the Resurrection.

Ultimately, then, perhaps what he needed to do to escape the hell of being alone was to love; if he could do this, then he could overcome his doubts, believe in the Resurrection, and thus be saved. Or, perhaps, it was that he needed first to *be* loved, by God:

> What combats doubts is, as it were, *redemption*. Holding fast to *this* must be holding fast to that belief. So what that means is: first you must be redeemed and hold on to your redemption – then you will see that you are holding fast to this belief.

First, you must be redeemed: 'Then *everything* will be different and it will be "no wonder" if you can do things that you cannot do now.' Such as: believing in the Resurrection. So, it seems, belief in the Resurrection is necessary for salvation, but salvation is needed to believe in the Resurrection. Who is to break the vicious circle: himself or God?

As he escaped the hell of being alone in Norway, Wittgenstein seemed to be saying that his escape from the wider hell, from the greater aloneness, was God's responsibility.

He could confess his sins, but it was not for him to forgive them.

19
FINIS AUSTRIAE

In December 1937, just as in July 1914, Wittgenstein arrived back in Austria from Norway at a critical moment in his country's history. As the earlier crisis had led to the end of the Habsburg Empire, so the present crisis was to lead to the end of Austria itself.

That Hitler had both the intention and the means to incorporate Austria into his German *Reich* should, in December 1937, have been no surprise to anyone who cared to think about it. *Mein Kampf* had been in print since 1925, and on its very first page Hitler declares: 'German-Austria must return to the great German mother country . . . One blood demands one Reich.' And, a few pages later: 'in my earliest youth I came to the basic insight which has never left me, but only became more profound: *That Germanism could be safeguarded only by the destruction of Austria.*' After the failure of the attempted Nazi *Putsch* of 1934, Hitler had been pursuing this policy of the destruction of Austria by 'legal' means, and in the treaty for the 'Normalization of Relations between Austria and Germany' of July 1936, Austria had acknowledged itself to be a 'German State', and the Austrian Chancellor, Schuschnigg, had had to admit into his cabinet two Nazi members of the 'Nationalist Opposition'. Hitler's subsequent repudiation of the Versailles Treaty, his rearmament campaign, and the unwillingness of Britain, France, Russia and Italy to intervene made it inevitable that this Nazi opposition would one day govern Austria, not as an independent country, but as part of Nazi Germany.

With very few exceptions, the large Jewish population of Vienna were slow to realize – or, perhaps, reluctant to admit to themselves – the likely consequences of the impending *Anschluss*. Even those who admitted its inevitability could not bring themselves to believe its possible repercussions. Surely, it was urged, the Nuremberg Laws couldn't be enforced in Austria. The Jewish population was too well assimilated into the mainstream of Austrian life: there were too many Jews in high places, too many marriages between Jew and non-Jew, too many loyal Austrian citizens who just happened to have Jewish forebears. How could these laws be applied in a country in which the distinction between Aryan and non-Aryan had become so blurred?

So, at least, it was reasoned by Hermine Wittgenstein. Writing her memoirs, in 1945, she found it inconceivable that she could have been so naïve – 'but', she adds, 'cleverer people than I also regarded the threatening political events with the same obtuseness'. No doubt in the light of what was to follow, the Christmas of 1937 is remembered by her in particularly rosy terms. She describes how delighted she was that all four of her brothers and sisters, together with members of their own families, were present (by this time she and Ludwig were the only childless members of the family, while Helene was the head of a large extended family of her own, the mother of four and the grandmother of eight), and how they, together with pupils and ex-pupils from the school at which she taught, sang carols, talked of old times, laughed and joked, and, most ironically, gathered around the Christmas tree to sing the Austrian national anthem. 'When, at midnight, the feast had to come to an end, we were all of one mind that it had been the most lovely Christmas ever, and we were already talking of the Christmas we would have next year.'

In Wittgenstein's diary-notes of the time there is none of this sentimental *gemütlichkeit*. But neither is there any mention of political events. It is impossible to believe, however, that he was as naïve about the situation as his sister. Admittedly, his only source of information during his stay in Norway was the *Illustrated London News*, which he had sent to him by Fania Pascal; on the other hand, it should be remembered that he had been in Cambridge twice during the last year, and there would have had the inestimable benefit of Piero Sraffa's informed political analysis and judgement. His confession of his non-Aryan origins in January was made, I believe, with awareness both of the terms of the Nuremberg Laws and of the possibility of their future application to Austrian citizens.

Nevertheless, in his diary he does not discuss politics. He writes, instead, of himself – of his mental and physical exhaustion after his trying time in Norway, and of how difficult he finds it to converse with the people he is with, how he can hardly speak to them, his mind is so befogged, and of how *unnecessary* he feels himself to be there. He writes also of Freud:

> Freud's idea: In madness the lock is not destroyed, only altered; the old key can no longer unlock it, but it could be opened by a differently constructed key.

Perhaps here, too, he is writing about himself, and a feeling that if only he could find a new key, he could unlock the doors to his own prison, and then '*everything* will be different'.

In the first week of January he confined himself to bed, suffering from gallbladder trouble, although scarcely believing that that was really why he felt so tired and weak. In bed he reflected on his sensuality and dwelt on his feelings for Francis. It was often the case, he wrote, that when he was unwell he was open to sensual thoughts and susceptible to sensual desires. He thought of Francis with sensual desires, 'and that is bad, but that's what it's like now'. He worried that it had been so long since he had heard from Francis, and, as always, was inclined to think the worst – to consider, for example, the possibility that Francis had died: 'Thought: it would be good and right if he had died, and thereby taken my "folly" away.' This dark, solipsistic, thought is immediately retracted, although only in part: 'Although, there again, I only *half* mean that.'

This qualification is, if anything, still more shocking. After giving the matter a second thought, was he really still even *half-way* inclined to think that the death of Francis would be a good thing?

'I am cold and wrapped up in myself', he wrote, reflecting on the fact that he did not feel himself to be in any loving relationship with anyone in Vienna. He was inclined to think that the comfortable life at the Alleegasse was bad for him, but where else was he to go? The solitude of his house in Norway had proved unendurable, and he had no wish to reenter academic life at Cambridge. Again, Dublin presented itself as an attractive alternative. There he could be with Drury, and perhaps even join Drury in training to be a psychiatrist. Everything was in flux; he did not know what he would do any more than he knew where he wanted to live. Of one thing, however, he was sure: that he needed to be with someone with whom he could talk.*

Wittgenstein arrived in Dublin on 8 February, and moved into Drury's old flat in Chelmsford Road. On his second day there he described himself as: 'irreligious, ill-tempered, gloomy'. He was in the 'hateful situation' of being incapable of work, of not knowing what he should do, of having to simply vegetate and wait. He was, he said, still inclined to *lie*: 'Again and again I see that I cannot resolve to speak the truth about myself. Or that I admit it to myself only for a moment and then forget about it.' His vanity, his cowardice, his fear of the truth, prompted him to keep hidden things about himself that he did not want to acknowledge: 'until I am no longer *clever* enough to find them'. Two days later he was beginning to regret that he had come to Dublin, where he could apparently do nothing; 'on the other hand I will have to wait, because nothing is still very clear'. During these first few weeks in Dublin he wrote very little philosophy; his philosophical thoughts had, so to speak, been lulled to sleep: 'It is entirely as though my talent lies in a kind of half-slumber.'

While his philosophical thoughts slept, his ideas of becoming a psychiatrist awoke. He asked Drury to arrange for him to visit St Patrick's Hospital so that he could meet patients who were seriously mentally ill. This was, he told Drury, a matter of great interest to him. After the visit he wrote (in English): 'See the sane man in the maniac! (and the mad man in yourself.)', and for the next few weeks he went two or three times a week to visit some of the long-stay patients. He remained, however, uncertain as to what, if anything, it might lead to.

Drury was at this time in his final year of training as a doctor, and was spending his period of residence in the City of Dublin Hospital. He told Wittgenstein that, when working in the casualty department,

---

*The fact that Wittgenstein chose to be with Drury rather than Francis demands explanation. Unfortunately, however, we are forced to speculate on the point; in his diary he does not even consider the idea that he should go to Cambridge to be with Francis. Perhaps it was not Francis, but Cambridge, that he was avoiding, or perhaps he was attracted to Dublin by the possibility that there he might train as a doctor. But in the light of the remarks already quoted there must also be a possibility that his desire for Francis, and Francis's almost overwhelming desire for him, made Cambridge unattractive – that he considered the sensuality between himself and Francis incompatible with the change he wished to see in himself.

he was disturbed by his clumsiness, and wondered whether he had made a mistake in taking up medicine. Wittgenstein, whatever ambivalence he felt about his own plans to take up medicine, was quick to quash Drury's doubts. The next day Drury received a letter from him which stated emphatically: 'You didn't make a mistake because there was nothing at the time you knew or ought to have known that you overlooked.' He urged Drury: 'Don't think about yourself, but think about others':

> Look at people's sufferings, physical and mental, you have them close at hand, and this ought to be a good remedy for your troubles. Another way is to take a rest whenever you ought to take one and collect yourself. (Not with me because I wouldn't rest you.) . . . Look at your patients more closely as human beings in trouble and enjoy more the opportunity you have to say 'good night' to so many people. This alone is a gift from heaven which many people would envy you. And this sort of thing ought to heal your frayed soul, I believe. It won't rest it; but when you are healthily tired you can just take a rest. I think in some sense you don't look at people's faces closely enough.

The letter ends: 'I wish you good thoughts but chiefly good feelings.'

The first mention in Wittgenstein's diary of the crisis facing Austria during the early months of 1938 occurs on 16 February. 'Can't work', he then wrote:

> Think a great deal of a possible change of my nationality. Read in today's paper that a further compulsory rapprochement between Austria and Germany has taken place – But I don't know what I should really do.

It was on that day that the Nazi leader of the 'Nationalist Opposition', Dr Arthur Seyss-Inquart, was appointed the Austrian Minister of the Interior, and the significance of the Berchtesgaden meeting between Hitler and Schuschnigg became apparent to the world.

This meeting had been held on 12 February, and had initially been celebrated in Austria as a sign of more cordial relations between the two countries. It was only later realized that in this 'friendly discussion' Hitler had demanded of Schuschnigg that Nazi ministers be appointed in charge of the Austrian police, army and financial affairs, and had threatened: 'You will either fulfill my demands in three days, or I will order the march into Austria.' On 15 February *The Times* reported:

> If Herr Hitler's suggestion that Dr von Seyss-Inquart should be made Austrian Minister of the Interior with control of the Austrian police were granted, it would in the general view of anti-Nazis in Austria mean that before long the words 'finis Austriae' would be written across the map of Europe.

The following day the paper commented drily on the fact that, immediately after taking his ministerial oath, Seyss-Inquart left Vienna for Berlin: 'That the first act of the Minister of the Interior is to pay a visit to a foreign country is a fair indication of the unusual situation in which Austria finds itself after the Hitler–Schuschnigg meeting.'

During the following few weeks Wittgenstein kept a close watch on developments. Every evening he asked Drury: 'Any news?', to which, presumably, Drury responded by telling Wittgenstein what had been reported that day. Reading Drury's recollections, however, one wonders which newspapers he read. His account of the days leading up to the *Anschluss* is, to say the least, somewhat strange. He writes that on the evening of 10 March he told Wittgenstein that all the papers reported that Hitler was poised to invade Austria. Wittgenstein replied, with quite breath-taking naïvety: 'That is a ridiculous rumour. Hitler doesn't want Austria. Austria would be no use to him at all.' The next evening, according to Drury, he had to tell him that Hitler had indeed taken over Austria. He asked Wittgenstein if his sisters would be in any danger. Again Wittgenstein replied with quite extraordinary insouciance: 'They are too much respected, no one would dare to touch them.'

From this account one might think that Wittgenstein had forgotten what he had read in the papers on 16 February – that he knew nothing about the threat to Austria, that he was entirely ignorant about the nature of the Nazi régime, and that he was unconcerned about the safety of his family. All this is quite assuredly false, and one can only think that he gave this misleading impression to Drury because he did not wish to add to Drury's burdens. That Drury was inclined to accept Wittgenstein's responses at face value perhaps says much about his unquestioning attitude towards Wittgenstein, and about his own political naïvety. It is also possible, I think, that Wittgenstein, who tended to compartmentalize his friendships, did not think it worth discussing these matters with Drury. With Drury he discussed religious questions; it was Keynes, Sraffa and Pattisson upon whom he relied for discussion of political and worldly affairs.

However, even on its own account – without, that is, bringing to it anything else we might know or suspect about Wittgenstein's awareness of events – Drury's story is a little puzzling. For if he told Wittgenstein the news every evening, he would, for example, have told him on 9 March about Schuschnigg's announcement that he was to hold a plebiscite asking the Austrian people to vote for or against an independent Austria. It was this announcement that prompted Hitler, on the following day, to move his forces up to the Austrian border in readiness for an invasion. Now if Wittgenstein reacted to this latter piece of news by denying that Hitler wanted Austria, what did he (or Drury for that matter) think was the point of Schuschnigg's plebiscite? Why should Austria's independence have needed reaffirming? Independence from whom?

Furthermore, the day after the troops had gathered on the border was not the day on which Hitler took over Austria, but rather the day on which Schuschnigg resigned and Seyss-Inquart became Chancellor. Hitler and the German troops did not cross the border until the day after that, 12 March, when they were invited by the new Chancellor, and it was then that the *Anschluss* was formally effected. This might be seen as a quibble, but the events of these three days are etched clearly into the minds of all who lived through them, and for Wittgenstein, if not for Drury, the change in the state of affairs on each of these days would have been of momentous importance. On 10 March Austria was an independent state under Schuschnigg; on the 11th it was an independent state under Nazi rule; and on the 12th it was part of Nazi Germany. To an Austrian family of Jewish descent the difference between the second and third days was decisive: it marked the difference between being an Austrian citizen and a German Jew.

On the day of the *Anschluss* Wittgenstein wrote in his diary: 'What I hear about Austria disturbs me. Am unclear what I should do, whether to go to Vienna or not. Think chiefly of Francis and that I do not want to leave him.' Despite his assurances to Drury, Wittgenstein was extremely concerned about the safety of his family. His first reaction was to go immediately to Vienna to be with them; what stopped him was his fear that, if he did so, he would never see Francis again. He nevertheless wrote to his family offering to come to Vienna if they needed him.

The only letter that survives from the correspondence between Wittgenstein and Sraffa is a long analysis of Wittgenstein's position following the *Anschluss*, which was written by Sraffa on 14 March, the day of Hitler's triumphant procession through Vienna. It demonstrates clearly the calibre of informed political opinion and advice that was open to Wittgenstein through Sraffa, and shows that Wittgenstein must have written to him straight away for advice on the possible consequences of his leaving for Vienna.

The letter begins:

> Before trying to discuss, probably in a confused way, I want to give a clear answer to your question. If, as you say, it is of 'vital importance' for you to be able to leave Austria and return to England, there is no doubt – *you must not go to Vienna*.

Sraffa pointed out that the Austrian frontier would be closed to the exit of Austrians, and that, though these restrictions might soon be lifted, there was every chance that, were he to go to Vienna, Wittgenstein would not be allowed out for a long time. 'You are aware no doubt that now you are a German citizen', Sraffa continued:

> Your Austrian passport will certainly be withdrawn as soon as you enter Austria: and then you will have to apply for a German passport, which may be granted if and when the Gestapo is satisfied that you deserve it . . .
>
> As to the possibility of war, I do not know: it may happen at any moment, or we may have one or two more years of 'peace'. I really have no idea. But I should not gamble on the likelihood of 6 months' peace.

Wittgenstein must also have asked Sraffa whether it would help his situation if he were to become a lecturer at Cambridge, for he goes on:

> If however you decided in spite of all to go back to Vienna, I think: a) it would certainly increase your chance of being allowed out of Austria if you were a lecturer in Cambridge; b) there would be no difficulty in your entering England, once you are let out of Austria (of Germany, I should say); c) *before* leaving Ireland or England you should have your passport changed with a German one, at a German Consulate: I suppose they will begin to do so in a very short time; and you are more likely to get the exchange effected here than in Vienna; and, if you go with a German passport, you are more likely (though not at all certain) to be let out again.

'You must be careful', Sraffa warned, 'about various things':

> 1) if you go to Austria, you must have made up your mind not to say that you are of Jewish descent, or they are sure to refuse you a passport;
> 2) you must not say that you have money in England, for when you are there they could compel you to hand it over to the Reichsbank;
> 3) if you are approached, in Dublin or Cambridge, by the German Consulate, for registration, or change of passport, be careful how you answer, for a rash word might prevent your ever going back to Vienna;
> 4) take great care how you write home, stick to purely personal affairs, for letters are certainly censored.

With regard to the question of a change of nationality, Sraffa advised that, if Wittgenstein had made up his mind to apply for Irish citizenship, then he should do so before his Austrian passport was taken away from him, for it would be easier as an Austrian than as a German. On the other hand:

> In the present circumstances I should not have qualms about British nationality if that is the only one which you can acquire without waiting for another ten years' residence: also you have friends in England who could help you to get it: and certainly a Cambridge job would enable you to get it quickly.

Sraffa, who was leaving for Italy the following Friday, invited Wittgenstein to come to Cambridge to discuss the matter if he could make it before then, but warned: 'afterwards letters will be forwarded to me in Italy, so take care what you say, that you may be writing to the Italian censor'. He ends: 'Excuse this confused letter', forcing one to wonder what levels of clarity and precision he reached in the rest of his correspondence.

'You are aware no doubt that now you are a German citizen.' On the day that Sraffa wrote these dreaded words, Wittgenstein's diary shows him to be wrestling with precisely this awareness:

> I am now in an extraordinarily difficult situation. Through the incorporation of Austria into the German Reich I have become a German citizen. That is for me a frightful circumstance, for I am now subject to a power that I do not in any sense recognise.

Two days later, he was, 'in my *head* and with my mouth', decided upon the loss of his Austrian nationality and resigned to the thought of emigrating for several years: 'it can't be any different. But the thought of leaving my people alone is dreadful.'

Upon receiving Sraffa's letter Wittgenstein immediately left Dublin for Cambridge to discuss the situation with him. On 18 March he reported in his diary:

> Sraffa advised me yesterday that I should, for the time being, not go to Vienna under any circumstances, for I could not now help my people and in all probability would not be allowed to leave Austria. I am not *fully* clear what I should do, but, for the time being, I think Sraffa is right.

Following this conversation with Sraffa, Wittgenstein decided upon a course of action. First, he would secure an academic job at Cambridge, and then apply for British citizenship. In connection with both aims he immediately wrote to Keynes for help. He began by explaining to Keynes the situation – that by the annexation of Austria he had become a German citizen, and, by the Nuremberg Laws, a German Jew: 'The same, of course, applies to my brother and sisters (not to their children, *they* count as Aryans).' 'I must say', he added, 'that the idea of becoming (or being) a German citizen, even apart from all the nasty consequences, is APPALLING to me. (This may be foolish, but it just is so.)' He outlined Sraffa's arguments against his going to Vienna – that his Austrian passport would be taken away from him, that, as a Jew, he would not be issued with a new passport, and that, therefore, he would be unable to leave Austria or ever again get a job. Presented with the choice of being a German Jew or a British university lecturer, he was forced, with some reluctance, to choose the latter:

> The thought of acquiring British citizenship had *occurred* to me before; but I have always rejected it on the ground: that I do not wish to become a sham-english-man (I think you will understand what I mean). The situation has however entirely changed for me now. For now I have to choose between two new nationalities, one of which deprives me of *everything*, while the other, at least, would allow me to work in a country in which I have spent on and off the greater part of my adult life, have made my greatest friends and have done my best work.
>
> . . . As to getting a job at Cambridge you may remember that I was an 'assistant faculty lecturer' for 5 years . . . Now it is for *this* that I shall apply, for there is no other job vacant. I had, in fact, thought of doing so anyway; though not now, but perhaps next autumn. But it would be important now for me to get a job *as quickly as possible*; for a) it would help me in becoming naturalised and b) if I failed in this and *had* to become a 'German' I would have more chance to be allowed out of Austria again on visiting my people if I had a JOB in England.

On Sraffa's advice Wittgenstein asked Keynes for an introduction to a solicitor – 'one who is an expert in this kind of thing' – to help with his application for naturalization. 'I want to add that I'm in no sort of financial difficulties. I shall have about 300 or 400 £ and can therefore easily hold out for another year or so.'

Keynes's reply to this letter has not survived, but it is clear that he did what he could to secure Wittgenstein a place at the university, and to help him in his application for British citizenship. Wittgenstein, however, was typically anxious that Keynes might have misunderstood his situation, and sent Keynes's letter to Pattisson, asking him to 'smell' it. Above all, he was concerned that Keynes might present him to the university authorities and to the Home Office as a member of that most wretched of species, the impoverished refugee. He was therefore suspicious of Keynes's suggestion that he might be eligible for a grant from the Academic Assistance Council. This, he told Pattisson, 'is a body who helps people, say refugees, who have no money, & accepting there [*sic*] assistance would not only not be quite fair of me but also place me in an *entirely* wrong category'. So nervous was he on this point that he was inclined to doubt whether he should use Keynes's introduction to a solicitor:

> I have a *vague* fear that this introduction, if slightly wrongly worded, might make things more awkward for me; it may, for instance, represent me as a *refugee* of sorts & stress a wrong aspect of the business.

His anxieties proved to be unfounded. The university responded quickly, and he was given a lecturing post starting from the beginning of the following term.

Of great concern to Wittgenstein throughout the long wait for his British passport was the situation of his family. It was difficult for him to know how much danger they were in, and he was not reassured when, soon after the *Anschluss*, he received this note (written in English):

> My dear Ludwig,
> Not a day passes but that Mining and I talk about you; our loving thoughts are always with you. Please do not worry about us, we are quite well really and in best spirits and ever so happy to be here. To see you again will be our greatest joy.
>
> Lovingly yours,
> Helene

In his diary Wittgenstein dismissed this (no doubt correctly) as: 'reassuring *sounding* news from Vienna. Obviously written for the censor.'

In fact, both Helene and Hermine were slow to appreciate the danger they were in, and when realization finally came they panicked. Hermine recalls how, one morning soon after the *Anschluss*, Paul announced, with terror in his voice: 'We count as Jews!' Hermine could understand why this should strike such fear into Paul's heart.

His career as a concert pianist mattered a great deal to him, and as a Jew he would not be allowed to perform; besides, he liked to go for long walks in the country, and all those signs proclaiming '*Juden verboten*' must make his walks a good deal less pleasant. But for herself, the fact that she counted as a Jew under German law seemed to mean very little. She spent most of her life inside her own four walls, and apart from the fact that a few people who used to greet her in public might no longer do so, her life would surely go on much the same.

At first Paul tried to get reassurances that the family would be treated as Aryans, on the grounds that they had always been loyal and patriotic citizens, and had accomplished much for their country. To this end he and Gretl (who, as an American citizen, was in no personal danger) travelled to Berlin to negotiate with the Nazi authorities. Their appeal came to nothing. Unless they could produce evidence of a second Aryan grandparent, they were told, they would remain Jews.

Another branch of the family, the descendants of Wittgenstein's Aunt Milly, made efforts to establish the Aryan credentials of Hermann Christian Wittgenstein. In the Berlin archives a report survives written by Brigitte Zwiauer, Milly's granddaughter, pleading Hermann Christian's case. It is addressed to the *Reichstelle für Sippenforschung* ('Department for Genealogical Research', the Nazi ministry responsible for determining who was and who was not Aryan), and states that Hermann Christian is known in the family to be the illegitimate son of a member of the princely family of Waldeck. Zwiauer admits that there is no direct proof of this, but stresses that there is also no proof to the contrary; that, although Hermann Christian was brought up in the Jewish community, there is no evidence that he was actually the son of a Jew. As indirect proof of his Aryan origins, she encloses a photograph of the eleven children of Hermann and Fanny. 'That these children descend from two fully Jewish parents', she argues, 'appears to us biologically impossible.' The report points out that Hermann chose the middle name 'Christian' for himself, and was known as an anti-Semite who, in adult life, avoided association with the Jewish community and did not allow his offspring to marry Jews. The report is dated 29 September 1938, but its plea was ignored until nearly a year later, when the Nazis saw some advantage to themselves in accepting it.

Hermine, Gretl and Helene would have nothing to do with this report. As far as they were concerned, Hermann Christian was the son of Moses Maier, and if that meant that, under German law, they were regarded as Jews, then so be it. Paul would probably have gone along with whatever steps were necessary to escape the consequences of being a Jew in the German *Reich*. As it was, he saw no hope of being reclassified and therefore sought only to leave Greater Germany as soon as he could. He urged Hermine and Helene to do the same – to leave everything behind and go to Switzerland. When a house is burning down, he argued, the sensible thing to do is to jump out of the window and forget about one's possessions inside. Hermine, however, could not bring herself to leave her friends, her family and her beloved Hochreit, and neither could Helene face leaving her children and grandchildren. Both refused to go. In July 1938, after many harsh words had been exchanged, Paul left his sisters in Austria and went to Switzerland alone.

Helene and Hermine left Vienna to spend the summer at the Hochreit, still convinced that their status as Jews would not place them in any danger. They were shaken out of this conviction by Gretl, who in September came to the Hochreit and told them that outside Germany it was widely believed by those who were well informed that war would break out at any moment (this being the time of the Czechoslovakia crisis), and that it was also known that the Jews in Germany would be rounded up and placed in concentration camps, where they would be insufficiently fed and treated very badly. Gretl urged Hermine and Helene to leave Austria.

By this time, however, it was no longer possible for German Jews to enter Switzerland, and some other plan had to be devised. Upon Gretl's suggestion, Hermine agreed to buy Yugoslav passports for herself and Helene from a Jewish lawyer in Vienna. She apparently believed that this was the way in which the Yugoslav government conferred nationality, for she says she had no idea that what they were buying were false passports until Arvid Sjögren, who travelled to Yugoslavia on their behalf to collect them, reported that they had been produced in a workshop specializing in forged documents.

Nevertheless, Hermine went ahead with the plan and travelled to Munich herself to obtain visas for Switzerland using the false passports. Soon afterwards, the police began to investigate this particular source of forgeries, and before they could begin their flight to Switzerland Hermine and Helene were arrested, together with Gretl and Arvid. They each spent two nights in prison, except Gretl, who was detained a further night. At the subsequent trial Gretl did everything she could to present herself as solely responsible for the whole fiasco, a claim that was accepted by the magistrate, although according to Hermine their best defence was their appearance and their manner of speech. Appearing before the court was not a collection of the grubby, smelly, kaftan-wearing Jews described in *Mein Kampf*, but proud members of a famous and wealthy high-bourgeois Austrian family. All four were cleared of the charges brought against them.

How much Wittgenstein himself knew of this story is impossible to say. Enough, anyway, for him to be sick with worry about his sisters' situation. In a letter to Moore of October 1938 he speaks of 'the great nervous strain of the last month or two', and attributes it to the fact that: 'My people in Vienna are in great trouble.' The wait for his British passport became almost unendurable, as he longed to be able to use it to travel to Vienna to do whatever he could to help his sisters. In the midst of all this anxiety, the sight of Neville Chamberlain returning from Munich proclaiming 'Peace in our time' was too much to bear. He sent Gilbert Pattisson one of the postcards printed to celebrate Chamberlain's 'success'. Beneath a picture of Chamberlain and his wife the legend reads: 'The Pilgrim of Peace. Bravo! Mr Chamberlain.' On the back Wittgenstein wrote: 'In case you want an Emetic, there it is.'

In the winter of 1938–9 the Reichsbank began to make enquiries about the huge amounts of foreign currency held by the Wittgenstein family. Under Nazi law the Reichsbank was empowered to compel the family to hand this money over to them. Owing to the complicated arrangement under which the wealth was held, however, it was difficult for them to get their hands on it. This circumstance suggested to Gretl another possibility of securing the safety of her sisters: they would agree to hand over the foreign currency in return for a written declaration that Hermine and Helene would be treated as Aryans.

So began a long series of negotiations between the authorities in Berlin and the Wittgensteins, which culminated in the Nazis agreeing to accept the report prepared by Brigitte Zwiauer the previous year in return for the transfer of the Wittgensteins' foreign currency. The negotiations were complicated by the disagreement between Paul and the rest of the family. Paul, who by this time had left Switzerland and was living in America, was against doing a deal with the Nazis in order to satisfy his sisters' perverse desire to stay in Austria. It would be wrong, he argued, to help the Nazis by placing in their hands such a large fortune. (Hermine attributes this latter argument to Paul's advisors, who, she points out, were, without exception, Jews – as though only a Jew would think such a consideration relevant.)

These wrangles continued throughout the spring of 1939, with Gretl travelling between New York, Berlin and Vienna, trying to come to an agreement that suited all parties, and the matter was still unsettled when Wittgenstein finally received his British passport on 2 June 1939. Barely a month later, he used it to travel to Berlin, Vienna and New York, with the aim of helping Gretl to reach a settlement. It was not, as Hermine says, the sort of thing to which her brother was suited, either by experience or by temperament. Furthermore (though she does not point this out), the irony of bribing the Nazis to accept a lie about the very thing to which he had confessed just two years previously can hardly have escaped him. He nevertheless entered into the negotiations with all the considerable precision and tenacity at his disposal. 'And if', Hermine adds, 'he did not achieve in New York what he had had in mind, then the blame was really not his.' It was, she implies elsewhere, Paul's.

Despite Paul's objections, the result of these negotiations was that a great deal of the family wealth was transferred from Switzerland to the Reichsbank, and the *Reichstelle für Sippenforschung* issued a formal declaration to its Viennese office that Hermann Christian Wittgenstein was, without qualification, *deutschblütig*. Consequently, in August 1939, Hermine and Helene, and all the other grandchildren of Hermann Christian, received certificates stating that they were *Mischlinge* (of mixed Jewish blood) rather than Jews. Later, in February 1940, the Berlin authorities went further, and issued a proclamation that the regulations covering *Mischlinge* were not applicable to the descendants of Hermann Christian Wittgenstein, and that 'their racial classification under the Reich Citizenship Law [the Nuremberg Laws] presents no further difficulties'. Hermine and Helene were thus able to survive the war relatively untroubled.

20
THE RELUCTANT PROFESSOR

Whether Wittgenstein would ever have returned to Cambridge had it not been for the *Anschluss*, one cannot say. His attempts to find a niche in life outside academia had, however, been at best inconclusive. Though he sometimes talked of finding a job among 'ordinary' people, as he had encouraged Skinner and Hutt to do, he seems to have made little effort to do so. His plans to work in Russia and/or to train as a doctor, though pursued with greater purpose, never crystallized into a firm and unequivocable intention. He would, perhaps, have continued to try and find the peace of mind and concentration he required to finish his book, perhaps in Dublin with Drury or in Norway by himself. But his savings, of £300 or £400, would not have lasted a lifetime. Eventually, he would have had to have found some paid employment. That is, as he had put it to Moore in 1930, he would have had to have found someone who had a use for the sort of goods he produced. And the place where these goods were in most demand was, inevitably, in academic life, and particularly in Cambridge. It is, therefore, perfectly possible that at some time or other he would have applied for a lectureship. What one can say with certainty, however, is that if it had not been for the *Anschluss*, this would not have been as early as April 1938.

This is not only because Wittgenstein was then not eager to return to teaching, but also because he was apprehensive about his relationship with Francis. As indicated by his diary entries in the New Year, he was deeply concerned about the sensuality that existed between himself and Francis, and anxious whether, on his part at least, such sensual desires were compatible with true love. He would have preferred to have loved Francis at a distance, away from the temptation of his sensual 'susceptibility'. And yet, his fear of losing Francis altogether had now brought him back to Cambridge, and more firmly than ever into the sphere of that temptation.

Upon his return he moved into Francis's lodgings above the grocer's shop on East Road, and for over a year they lived, as Francis had always wished them to live, as a couple. Their period as collaborators on Wittgenstein's work had long come to an end. While Wittgenstein lectured and continued to work on his book, Francis worked at the factory. There are no letters from Francis during this period, and no relevant remarks among Wittgenstein's coded diary-notes, so we do not know how or why their relationship deteriorated during this year. All we know is that by 1939 it had deteriorated, and that for the following two years it was only Francis's undyingly faithful, perhaps even clinging, love for Wittgenstein that kept it going. Wittgenstein's love for Francis, it seems, did not – perhaps could not – survive the physical closeness that he at once craved and feared.

Among his students at this time, Wittgenstein found a new generation of disciples. In order to keep his class down to a size with which he felt comfortable, he did not announce his lectures in the usual way in the *Cambridge University Recorder*. Instead, John Wisdom, Moore and Braithwaite were asked to tell those students they thought would be interested about the classes. No more than about ten students attended. Among this select band were Rush Rhees, Yorick Smythies, James Taylor, Casimir Lewy and Theodore Redpath. The class was small enough for them all to get to know Wittgenstein fairly well, although it was Rhees, Taylor and Smythies that became particularly close friends during this period.

The classes were held in Taylor's rooms. Taylor, about whom almost nothing is mentioned in the published memoirs, was a Canadian, a graduate of Toronto University, who had come to Cambridge to study under G. E. Moore and, through Moore, became a friend of Wittgenstein's. After the war he was offered a lectureship in philosophy at an Australian university, but died in a pub brawl in Brisbane while on his way to take up the post. Smythies is one of those mysterious figures who are referred to repeatedly in the published texts, but about whom little is ever said. He was a devoted disciple of Wittgenstein's and a truly Wittgensteinian character in the sense that, though he never became a professional philosopher, he never ceased to think seriously and deeply about philosophical problems. He remained a close friend of Wittgenstein's for the rest of Wittgenstein's life. When he left Cambridge he became a librarian at Oxford. In later life he suffered from paranoid schizophrenia and became a patient of Maurice Drury. He died in tragic circumstances in 1981. Regarding such people, one is reminded of the fact that those whom Wittgenstein influenced most strongly, particularly in the 1930s (one thinks of Drury, Skinner and Hutt, as well as Smythies), did not enter academic life. There is therefore a large and important aspect of Wittgenstein's influence that is not, and cannot be, reflected in the large body of academic literature that Wittgenstein's work has inspired. The only one of these to have published anything is Maurice Drury, whose collection of essays on philosophical and psychological issues, *The Danger of Words*, though it has been almost completely ignored in the secondary literature, is, in its attitudes and concerns, more truly Wittgensteinian than almost any other secondary text.

On vacation from his final year's training as a doctor, Drury managed to attend one of the lectures in Wittgenstein's new course. During this lecture Wittgenstein told one of the students to stop making notes:

> If you write these spontaneous remarks down, some day someone may publish them as my considered opinions. I don't want that done. For I am talking now freely as my ideas come, but all this will need a lot more thought and better expression.

Fortunately, this request was ignored, and notes from these lectures have indeed been published.*

These lectures are unique among Wittgenstein's corpus. Their subject-matter alone would make them so, for they are concerned, not with mathematics or philosophy generally, but with aesthetics and religious belief. This difference is less radical than it might appear, for Wittgenstein brings to his discussion of these subjects many of the same examples that he used in other contexts – Cantor's Diagonal Proof, Freud's confusion between cause and reason and so on – so that

* See *Lectures and Conversations on Aesthetics, Psychology and Religious Belief*, ed. Cyril Barrett (Blackwell, 1978).

his discussion of aesthetics, for example, looks not so very different from his discussions of the philosophy of mathematics or the philosophy of psychology. What distinguishes these lectures is their tone. Precisely because he was speaking in a spontaneous and unguarded manner, they provide one of the most unambiguous statements of his purpose in philosophy, and of how this purpose connects with his personal *Weltanschauung*. Through them, it is made even clearer that his target was not merely, as he had put it in the *Blue Book*, the damage that is done when philosophers 'see the method of science before their eyes and are irresistibly tempted to ask and answer questions in the way that science does'; it was, more generally, the wretched effect that the worship of science and the scientific method has had upon our whole culture. Aesthetics and religious belief are two examples – for Wittgenstein, of course, crucially important examples – of areas of thought and life in which the scientific method is not appropriate, and in which efforts to make it so lead to distortion, superficiality and confusion.

Wittgenstein told his audience that what he was doing was 'persuading people to change their style of thinking'. He was, he said, 'making propaganda' for one style of thinking as opposed to another. 'I am honestly disgusted with the other', he added. The 'other' he identified as the worship of science, and he therefore spent some time in these lectures execrating what he considered to be powerful and damaging forms of evangelism for this worship – the popular scientific works of the time, such as Jeans's *The Mysterious Universe*:

> Jeans has written a book called *The Mysterious Universe* and I loathe it and call it misleading. Take the title . . . I might say the title *The Mysterious Universe* includes a kind of idol worship, the idol being Science and the Scientist.

In discussing aesthetics, Wittgenstein was not attempting to contribute to the philosophical discipline that goes by that name. The very idea that there could be such a discipline was a consequence, or perhaps a symptom, of the 'other'. He was, instead, trying to rescue questions of artistic appreciation from that discipline, particularly from the idea that there could be a kind of science of aesthetics:

> You might think Aesthetics is a science telling us what's beautiful – almost too ridiculous for words. I suppose it ought to include also what sort of coffee tastes well.

When Rhees asked Wittgenstein about his 'theory' of deterioration (referring to one of Wittgenstein's examples, which was the deterioration of the German musical tradition), Wittgenstein reacted with horror to the word: 'Do you think I have a theory? Do you think I'm saying what deterioration is? What I do is describe different things called deterioration.'

Rather than trying to answer the traditional questions of aesthetics ('What is beauty?' etc.), Wittgenstein gives a succession of examples to show that artistic appreciation does not consist (as one might think from reading some philosophical discussions of aesthetics) in standing before a painting and saying: 'That is beautiful.' Appreciation takes a bewildering variety of forms, which differ from culture to culture, and quite often will not consist in *saying* anything. Appreciation will be *shown*, by actions as often as by words, by certain gestures of disgust or satisfaction, by the way we read a work of poetry or play a piece of music, by how often we read or listen to the same piece, and how we do so. These different forms of appreciation do not have any one thing in common that one can isolate in answer to the question: 'What is artistic appreciation?' They are, rather, linked by a complicated series of 'family resemblances'. Thus:

> It is not only difficult to describe what appreciation consists in, but impossible. To describe what it consists in we would have to describe the whole environment.

Above all, in seeking to answer the why and how of aesthetic understanding, we are not looking for a *causal* explanation. There is no science of aesthetics, and neither can the results of some other science, such as physics, or some pseudo-science, such as psychology, be brought to bear on these questions. Wittgenstein quotes two kinds of explanation from the work of Freud, which illustrate, respectively, the kind of reductive account that he thought should be avoided at all costs, and the other 'style of thinking' that he was trying to promote.

The first comes from *The Interpretation of Dreams*, and concerns Freud's 'explanation' of what his patient had described to him as a pretty dream. In retelling the dream Freud puts certain words in capitals to indicate – with a nod and a wink, as it were – the sexual allusions:

> She was descending from a height . . . She was holding a BIG BRANCH in her hand; actually it was like a tree, covered over with RED BLOSSOMS . . . Then she saw, after she had got down, a manservant who was combing a similar tree, that is to say he was using a PIECE OF WOOD to drag out some THICK TUFTS OF HAIR that were hanging down from it like moss.

And so on. Later in the dream the woman comes across people taking branches and throwing them into the road where they lay (LAY) about. She asks whether she might also take one – that is, explains Freud, whether she might pull one down, i.e. masturbate (in German, the phrase 'to pull one down' is equivalent to the English 'to toss oneself off'). Freud adds: 'The dreamer quite lost her liking for this pretty dream after it had been interpreted.'

Wittgenstein's response to this is to say that Freud has cheated his patient: 'I would say to the patient: "Do these associations make the dream not beautiful? It was beautiful. Why shouldn't it be?"' Freud's reduction of the pretty elements of the dream to bawdy innuendo has a certain charm, a certain fascination, but it is wrong to say that Freud has shown what the dream is *really* about. Wittgenstein compared it to the statement: 'If we boil Redpath at 200 degrees C. all that is left when the water vapour is gone is some ashes, etc. This is all Redpath really is.' Saying this, he says, might have a certain charm, 'but it would be misleading to say the least'.

The sort of Freudian explanations that Wittgenstein mentioned with approval are those contained in *Jokes and their Relation to the Unconscious*. Wittgenstein does not give any examples, but perhaps a simple one will suffice. In the early part of the book, Freud discusses a joke in Heine's *Reisebilder*. One of Heine's characters, a humble lottery-agent, in boasting about his relations with Baron Rothschild, remarks: 'He treated me quite as his equal – quite familionairely'. The reason this makes us laugh, Freud claims, is not only that it is a clever abbreviation of the thought that Rothschild treated the man as his equal, quite familiarly, so far as a millionaire can, but also because it brings out a suppressed subsidiary thought: that there is actually something rather unpleasant about being treated with a rich man's condescension.

If we are inclined to accept this sort of explanation, Wittgenstein asks, on what grounds do we do so?

> 'If it is not causal how do you know it's correct?' You say: 'Yes, that's right.' Freud transforms the joke into a different form which is recognised by us as an expression of the chain of ideas which led us from one end to another of a joke. An entirely new account of a correct explanation. Not one agreeing with experience but one accepted.

It was essential to this form of explanation, he stressed, that: 'You have to give the explanation that is accepted. This is the whole point of the explanation.' And this is precisely the sort of explanation one wants in aesthetics: not one that establishes a *cause* for something's being beautiful, or for our regarding something as beautiful, but rather one that, by showing connections we had not thought of previously, *shows* what is beautiful about it – shows why, for example, a certain piece of music or a certain play, poem etc. is correctly regarded as a great work.

In his lectures Wittgenstein gave some examples from his own experience of what happens when one begins to understand the greatness of an artistic work. He had, he said, read the work of Friedrich Klopstock, the eighteenth-century poet, and initially failed to see anything in it. Then he realized that the way to read him was to stress his metre abnormally:

> When I read his poems in this new way, I said: 'Ah-ha, now I know why he did this.' What had happened? I had read this kind of stuff and had been moderately bored, but when I read it in this particular way, intensely, I smiled, said: 'This is *grand*, etc. But I might not have said anything. The important fact was that I read it again and again. When I read these poems I made gestures and facial expressions which were what would be called gestures of approval. But the important thing was that I read the poems entirely differently, more intensely, and said to others: 'Look! This is how they should be read.'

Another example he might have given was *The King of the Dark Chamber* by the Indian poet, Rabindranath Tagore. Wittgenstein had first read this play, in a German translation (it was originally written in Bengali), in 1921, when Tagore was at the height of his fame and enormously popular in Europe, particularly in Germany and Austria. He had then written to Engelmann that, despite its great wisdom, the play had failed to make a deep impression on him. He was not *moved*:

> It seems to me as if all that wisdom has come out of the ice box; I should not be surprised to learn that he got it all second-hand by reading and listening (exactly as so many among us acquire their knowledge of Christian wisdom) rather than from his own genuine *feeling*. Perhaps I don't understand his tone; to me it does not ring like the tone of a man possessed by the truth. (*Like for instance Ibsen's tone.*) It is possible, however, that here the translation leaves a chasm which I cannot bridge. I read *with interest* throughout, but without being gripped. That does not seem to be a good sign. For this is a subject that could have gripped me – or have I become so deadened that nothing will touch me any longer? A possibility, no doubt.
>
> – Again, I do not feel for a single moment that here is a drama taking place. I merely understand the allegory in an abstract way.

Just a few months after this, he wrote to Hänsel saying that he had been rereading Tagore, 'and this time with much *more* pleasure'. 'I now believe', he told Hänsel, 'that there is *indeed* something grand here.' *The King of the Dark Chamber* subsequently became one of his favourite books, one of those he habitually gave or lent to his friends. And at about the time of his lectures on aesthetics he reread the play together with Yorick Smythies, this time in an English translation made by Tagore himself. Again, it seems, the translation left a chasm, and in order to overcome this – in order to, as it were, defrost the text – Smythies and Wittgenstein prepared their own translation. Among Smythies's papers was found a typed copy of their version of Act II of the play, headed:

> THE KING OF THE DARK CHAMBER, by Rabindranath Tagor [*sic*] *translated from* the English of Rabindranath Tagor *into* English used by L. Wittgenstein and Yorick Smythies, *by* L. Wittgenstein and Yorick Smythies.

Almost all the changes introduced by Smythies and Wittgenstein involve substituting modern, idiomatic words and phrases for Tagore's old-fashioned 'poetic' diction. Thus, where Tagore has 'chamber', they have 'room' (except in the title), and where Tagore wrote: 'He has no dearth of rooms', they write: 'He's not short of rooms', and so on.

The play is an allegory of religious awakening, and it echoes many of Wittgenstein's own thoughts on the subject. The King of the title is never seen by his subjects, some of whom doubt his existence, while others believe that he is so ugly he dare not reveal himself. Others, such as the maidservant Surangama, are so devoted to the King and so worshipful that they do not ask to *see* him; they *know* him to be a being without comparison to other mortals. Only these people, who have completely overcome their own pride in subjection to their master, have a sense of when the King is approaching and when he is present. The play concerns the awakening – or, one might say, the humbling, the subjugation – of the King's wife, Sudarshana. She is shown first as a proud Queen, bemoaning the cruelty of her husband, whom she can meet only in a room that is kept forever dark. She longs to see him, to know whether he is handsome, and out of that longing falls in love with another king, whom she meets in the world outside and mistakes for her husband. Only when she has been brought by this mistake to complete despair, when she feels utterly humiliated and degraded and has cast away her pride, can she be reconciled with her real husband, before whom she now bows with total servility. Only, that is, when Queen Sudarshana is brought down to the level of the servant Surangama can she become enlightened. The play ends with her realization that *everything* of any real value is conferred upon her by the King, who can now say to her: 'Come, come with me now, come outside – *into the light!*'

The part of the play translated by Wittgenstein and Smythies is a conversation between Surangama and Sudarshana in which the servant tries to explain to the Queen how she became so completely devoted to the King, though she has never seen him, and though he caused her to suffer greatly when he banished her father from the kingdom. When the King exiled her father, the Queen asks, didn't Surangama feel bitterly oppressed? 'It made me furious', the servant replies:

> I was on the road to ruin and destruction: when that road was closed to me, I seemed left without any support, without help or shelter. I raved and raged like a wild animal in a cage – I wanted, in my powerless anger, to tear everyone to pieces.

'But how did you become devoted to the King who had done all this?' Sudarshana asks. When did this change of feeling take place? 'I couldn't tell you', comes the reply:

> I don't know myself. A day came when all the rebel in me knew itself beaten, and then my whole nature bowed down in humble resignation in the dust. And then I saw . . . I saw that he was as incomparable in beauty as he was in terror. I was saved, I was rescued.

Wittgenstein's translation of Tagore might fruitfully be read in conjunction with his lectures on religious belief, for in those passages he translated, Tagore expresses Wittgenstein's own religious ideal. That is, like Surangama, Wittgenstein did not wish to see God or to find reasons for His existence. He thought that if he could overcome *himself* – if a day came when his whole nature 'bowed down in humble resignation in the dust' – then God would, as it were, come to him; he would then be saved.

In his lectures on religious belief he concentrates only on the first part of this conviction – the denial of the necessity to have reasons for religious beliefs. In their rejection of the relevance of the scientific mode of thought, these lectures are of a piece with those on aesthetics. They might also be seen as an elaboration of his remark to Drury: 'Russell and the parsons between them have done infinite harm, infinite harm.' Why pair Russell and the parsons in the one condemnation? Because both have encouraged the idea that a philosophical justification for religious beliefs is necessary for those beliefs to be given any credence. Both the atheist, who scorns religion because he has found no *evidence* for its tenets, and the believer, who attempts to *prove* the existence of God, have fallen victim to the 'other' – to the idol-worship of the scientific style of thinking. Religious beliefs are not analogous to scientific theories, and should not be accepted or rejected using the same evidential criteria.

The kind of experience that can make a man religious, Wittgenstein insists, is not at all like the experience of drawing a conclusion from an experiment, or of extrapolating from a collection of data. He takes as an example someone who dreams of the Last Judgement, and now says he knows what it will be like:

> Suppose someone said: 'This is poor evidence.' I would say: 'If you want to compare it with the evidence for it's raining to-morrow it is no evidence at all.' He may make it sound as if by stretching the point you may call it evidence. But it may be more than ridiculous as evidence. But now, would I be prepared to say: 'You are basing your belief on extremely slender evidence, to put it mildly?' Why should I regard this dream as evidence – measuring its validity as though I were measuring the validity of the evidence for meteorological events?
>
> If you compare it with anything in Science which we call evidence, you can't credit that anyone could soberly argue: 'Well, I had this dream . . . therefore . . . Last Judgement.' You might say: 'For a blunder, that's too big.' If you suddenly wrote numbers down on the blackboard, and then said: 'Now, I'm going to add', and then said: '2 and 21 is 13', etc. I'd say: 'This is no blunder.'

On the question of how we *are* to accept or reject religious beliefs, and of what we are to believe about such things as the existence of God, the Last Judgement, the immortality of the soul etc., Wittgenstein is, in these lectures, non-committal:

> Suppose someone said: 'What do you believe, Wittgenstein? Are you a sceptic? Do you know whether you will survive death?' I would really, this is a fact, say 'I can't say. I don't know', because I haven't any clear idea what I'm saying when I'm saying 'I don't cease to exist', etc.

It is clear from remarks he wrote elsewhere, however (for example in his remarks written on the boat to Bergen and quoted earlier), that he thought that if he could come to believe in God and the Resurrection – if he could even come to attach some *meaning* to the expression of those beliefs – then it would not be because he had found any evidence, but rather because he had been redeemed.

Still there is a persistent and nagging doubt about how Wittgenstein expected, or hoped, this redemption to come about – whether, so to speak, it was in his hands or God's.

On this central question *The King of the Dark Chamber* is, like Wittgenstein, ambiguous. After Sudarshana has been saved, she remarks to the King: 'You are not beautiful, my lord – you stand beyond all comparisons!' To which the King replies: 'That which can be comparable with me lies within yourself.' 'If this be so', says Sudarshana, 'then that too is beyond comparison':

> Your love lives in me – you are mirrored in that love, and you see your face reflected in me: nothing of this is mine, it is all yours.

And yet elsewhere in the play, it is the King who holds up the mirror. Those who think he is ugly, we are told, do so because they fashion the King after the image of themselves they see reflected there. And so, one wants to ask, is 'that which lies beyond all comparisons' within us or not? What do we need to do in order to see it – polish up the mirror that is our self so that *it* can be reflected, or look with open eyes at the mirror and see it reflected *in ourselves*? Perhaps here we run up against the limits of meaningful language, and go beyond the applicability of the Law of Excluded Middle and the Law of Contradiction.* 'It', perhaps, both is and is not within us, and to find it we must both search within ourselves and recognize our dependence on something, some power, outside ourselves.

Perhaps the difference between allowing 'it' to be reflected in us and finding it in the reflection of us is not as great as it seems. In both cases we must remove the dirt that obscures the reflection. In this respect Wittgenstein laboured hard, polishing off the slightest speck, determined not to let himself get away with the smallest misdemeanour. In October 1938, for example, he wrote to George Thomson's mother-in-law earnestly apologizing for a transgression of quite extraordinary inconsequence:

> Dear Mrs Stewart,
> I must apologise for an untruth I told you today in Miss Pate's

* The Law of Excluded Middle states that either a proposition or its denial must be true; the Law of Contradiction states that *both* cannot be true.

office. I said that I had seen Mrs Thomson recently in Birmingham; & only when I came home this evening it occurred to me that this wasn't true at all. I stayed with the Bachtin's a few weeks ago in Birmingham & I *tried* to see Mrs Thomson & we had a talk on the phone; but I wasn't able to see her. When I talked to you this afternoon what was in my head was that I had seen Mrs Thomson at your house before she went to Birmingham. Please forgive my stupidity.

Yours sincerely,
L. Wittgenstein

In the context of his search for redemption through the dismantling of his pride, Wittgenstein's philosophical work occupies a curiously ambivalent place. On the one hand it is undoubtedly informed by the same attitudes that directed that search. On the other hand, it was itself the greatest source of his pride. Though he tried repeatedly to exclude from his work any question of pride, and to write, as he put it, 'to the glory of God' rather than out of vanity, yet we find again and again that it was to his philosophical work more than to anything else that he brought what Russell called his 'pride of Lucifer'.

In the summer of 1938 he prepared for publication a typescript based on the work he had written in Norway. This typescript constitutes the very earliest version of *Philosophical Investigations*. 'For more than one reason', he wrote in the preface:

> what I publish here will have points of contact with what other people are writing today. – If my remarks do not bear a stamp which marks them as mine, – I do not wish to lay any further claim to them as my property.

And yet, that they were his property was enormously important to him, and that Carnap, Braithwaite, Waismann, Ambrose and others had published ideas derivative of them was precisely the reason he was now prepared to go into print. In a later preface he admitted as much:

> I was obliged to learn that my results (which I had communicated in lectures, typescripts and discussions), variously misunderstood, more or less angled or watered down, were in circulation. This stung my vanity and I had difficulty in quieting it.

But if pride gave birth to his desire to publish, it also prevented his doing so. In September the book was offered to Cambridge University Press, who agreed to publish the German original with a parallel English translation. About a month later, however, the Press was told that Wittgenstein was now uncertain about the publication of his book, and the project had been, for the time being, shelved.

There were two reasons for Wittgenstein's doubts. One, the most important, was that he became increasingly dissatisfied with the second half of the book, which dealt with the philosophy of mathematics. The other was concerned with the problems in translating his work.

On Moore's recommendation, Wittgenstein asked Rush Rhees to undertake the translation. It was a formidable task – not because Wittgenstein's German is difficult (in the way that, for example, Kant's German is difficult), but rather because Wittgenstein's language has the singularly rare quality of being both colloquial and painstakingly precise.

Rhees laboured on the translation throughout the Michaelmas term of 1938. During this time he met with Wittgenstein regularly to discuss the problems that arose. In January 1939 he had to leave Cambridge to visit the United States, and so left a typescript of his work with Wittgenstein. Wittgenstein, who was never easily pleased with any attempt by others to represent his thoughts, was horrified at what he saw.

By this time the question of having a decent English edition of his work had acquired an importance outside his plans to publish. He had by then decided to apply for the post of Professor of Philosophy, which had become vacant on G. E. Moore's resignation, and he wanted to submit the translated portion of his book in support of his application. He was, in any case, convinced that he would not be elected, partly because one of the other applicants was John Wisdom, whom he felt sure would get it, and partly because one of the electors was R. G. Collingwood of Oxford, a man who was sure to disapprove of Wittgenstein's work. More than compensating for these two disadvantages, however, was the fact that also among the electors was John Maynard Keynes. Wittgenstein hurriedly attempted to improve on Rhees's translation in time for Keynes to read through the English version. 'I needn't say the whole thing is absurd', he wrote to Moore, 'as he couldn't make head or tail of it if it were translated very well.'

Wittgenstein would probably have been awarded the chair with or without Keynes's support, and regardless of the quality of the translation. By 1939 he was recognized as the foremost philosophical genius of his time. 'To refuse the chair to Wittgenstein', said C. D. Broad, 'would be like refusing Einstein a chair of physics.' Broad himself was no great admirer of Wittgenstein's work; he was simply stating a fact.

On 11 February Wittgenstein was duly elected professor. It was, inevitably, an occasion for both expressing pride and condemning it. 'Having got the professorship is very flattering & all that', he wrote to Eccles, 'but it might have been very much better for me to have got a job opening and closing crossing gates. I don't get any kick out of my position (except what my vanity & stupidity sometimes gets).' This in turn helped with his application for British citizenship, and on 2 June 1939 he received his British passport. No matter how illiberal their policy on the admission of Austrian Jews, the British government could hardly refuse citizenship to the Professor of Philosophy at the University of Cambridge.

More serious than the problems with translation, as far as the publication of Wittgenstein's remarks was concerned, was his dissatisfaction with what he had written on the philosophy of mathematics. In the three terms of 1939 he devoted a series of lectures to the subject. They are, to a certain extent, similar in theme to the previous year's lectures on aesthetics and religious belief, only now it is Russell and the *logicians* who between them have done infinite harm, and mathematics that has to be rescued, from the clutches of philosophical theorists. The strategy of these lectures was, indeed, announced in the earlier lectures on aesthetics, when, in discussing Cantor's Diagonal Proof, he expressed his loathing of it and his view that it was only the 'charm' of such proofs (he presumably meant by this the fascination it exerts to be told that it is provable that there exists an infinite number of different infinite cardinalities) that gave them their interest. 'I would', he said, 'do my utmost to show the effects of this charm, and of the associations of "Mathematics"':

> Being Mathematics . . . it looks incontrovertible and this gives it a still greater charm. If we explain the surrounding of the expression we see that the thing could have been expressed in an entirely

> different way. I can put it in a way in which it will lose its charm for a great number of people and certainly will lose its charm for me.

The aim, then, was to reinterpret mathematics – to redescribe it in such a way that the mathematical realm that appeared to have been opened up by Cantor's Proof was presented, not as a fascinating world awaiting the discovery of mathematicians, but as a swamp, a quagmire of philosophical confusions. The mathematician Hilbert had once said: 'No one is going to turn us out of the paradise which Cantor has created.' 'I would say', Wittgenstein told his class: 'I wouldn't dream of trying to drive anyone out of this paradise':

> I would do something quite different: I would try to show you that it is not a paradise – so that you'll leave of your own accord. I would say, 'You're welcome to this; just look about you.'

The lectures on mathematics form part of Wittgenstein's general attack on the idol-worship of science. Indeed, he perceived this particular campaign as the most important part of that struggle. 'There is no religious denomination', he once wrote, 'in which the misuse of metaphysical expressions has been responsible for so much sin as it has in mathematics.' The 'charm' exerted by the metaphysics of mathematics was even more potent than that exerted by such books as Jeans's *The Mysterious Universe*, and an even more powerful influence in the idolization of science, which was, Wittgenstein thought, the most significant symptom, and perhaps even a contributory cause, of the decay of our culture.

It was thus his task to destroy that metaphysics. A feature of these lectures is that, in attempting to accomplish that task, he does not, as he had earlier, discuss mathematics itself with any degree of technical sophistication. He does not, for example, as he had in 1932–3, read out extracts from Hardy's textbook *A Course of Pure Mathematics*; nor does he, as he had in *Philosophical Grammar*, subject particular proofs (such as Skolem's Proof of the Associative Law) to rigorous and detailed analysis. Technical details are eschewed altogether. When he discusses Russell's Paradox, for instance, he does so in a way that is, from a mathematical point of view, quite extraordinarily primitive:

> Take Russell's contradiction. There are concepts which we call predicates – 'man', 'chair', and 'wolf' are predicates, but 'Jack' and

> 'John' are not. Some predicates apply to themselves and others don't. For instance 'chair' is not a chair, 'wolf' is not a wolf, but 'predicate' is a predicate. You might say this is bosh. And in a sense it is.

This lack of sophistication has, I think, a propagandist purpose. Wittgenstein's use of casual, everyday, language in discussion of problems in mathematical logic, and his simple dismissal as 'bosh' of the terms in which those problems have been raised, serves as an antidote to the seriousness and earnestness with which they have been discussed by those who have fallen for their 'charm' (including, for example, himself, in 1911). But also, for the problems he wished to raise, technical details were irrelevant. 'All the puzzles I will discuss', he said in his first lecture, 'can be exemplified by the most elementary mathematics – in calculations which we learn from ages six to fifteen, or in what we easily might have learned, for example, Cantor's proof.'

This series of lectures was remarkable in having among its audience a man who was one of the ablest exponents of the view that Wittgenstein was attacking, and also one of the greatest mathematicians of the century: Alan Turing. During the Easter term of 1939, Turing too gave classes under the title, 'Foundations of Mathematics'. They could not have been more different from Wittgenstein's. Turing's course was an introduction to the discipline of mathematical logic, in which he took his students through the technique of proving mathematical theorems from within a strictly axiomatic system of logic. Lest it be thought that *his* lectures had anything to do with the 'foundations of mathematics' in this sense, Wittgenstein announced:

> Another idea might be that I was going to lecture on a particular branch of mathematics called 'the foundations of mathematics'. There is such a branch, dealt with in *Principia Mathematica*, etc. I am not going to lecture on this. I know nothing about it – I practically only know the first volume of *Principia Mathematica*.

That it was at one time thought (by both himself and Russell) that he would be responsible for rewriting sections of the *Principia*, he does not mention. His present series of lectures was relevant to that branch of mathematics only in the sense of trying to undermine the rationale for its existence – of trying to show that: 'The *mathematical* problems of what is called foundations are no more the foundations of mathematics for us than the painted rock is the support of the painted tower.'

The lectures often developed into a dialogue between Wittgenstein and Turing, with the former attacking and the latter defending the importance of mathematical logic. Indeed, the presence of Turing became so essential to the theme of the discussion that when he announced he would not be attending a certain lecture, Wittgenstein told the class that, therefore, that lecture would have to be 'somewhat parenthetical'.

Wittgenstein's technique was not to reinterpret certain particular proofs, but, rather, to redescribe the whole of mathematics in such a way that mathematical logic would appear as the philosophical aberration he believed it to be, and in a way that dissolved entirely the picture of mathematics as a science which discovers facts about mathematical objects (numbers, sets etc.). 'I shall try again and again', he said, 'to show that what is called a mathematical discovery had much better be called a mathematical invention.' There was, on his view, nothing for the mathematician to discover. A proof in mathematics does not establish the truth of a conclusion; it fixes, rather, the *meaning* of certain signs. The 'inexorability' of mathematics, therefore, does not consist in *certain knowledge* of mathematical truths, but in the fact that mathematical propositions are *grammatical*. To deny, for example, that two plus two equals four is not to disagree with a widely held view about a matter of fact; it is to show ignorance of the meanings of the terms involved. Wittgenstein presumably thought that if he could persuade Turing to see mathematics in this light, he could persuade anybody.

But Turing was not to be persuaded. For him, as for Russell and for most professional mathematicians, the beauty of mathematics, its very 'charm', lay precisely in its power to provide, in an otherwise uncertain world, unassailable truths. ('Irrefragability, thy name is mathematics!' as W. V. Quine once put it.) Asked at one point whether he understood what Wittgenstein was saying, Turing replied: 'I understand but I don't agree that it is simply a question of giving new meanings to words.' To this, Wittgenstein – somewhat bizarrely – commented:

> Turing doesn't object to anything I say. He agrees with every word. He objects to the idea he thinks underlies it. He thinks we're

> undermining mathematics, introducing Bolshevism into mathematics. But not at all.

It was important to Wittgenstein's conception of his philosophical method that there could be no disagreements of opinion between himself and Turing. In his philosophy he was not advancing any theses, so how could there possibly be anything to disagree with? When Turing once used the phrase: 'I see your point', Wittgenstein reacted forcefully: 'I have no point.' If Turing was inclined to object to what Wittgenstein was saying, it could only be because he was using words in a different way to Wittgenstein – it *could* only be a question of giving meanings to words. Or, rather, it could only be a question of Turing's not understanding Wittgenstein's use of certain words. For example, Turing was inclined to say that there could be experiments in mathematics – that is, that we could pursue a mathematical investigation in the same spirit in which we might conduct an experiment in physics: 'We don't know how this might turn out, but let's see . . .' To Wittgenstein, this was quite impossible; the whole analogy between mathematics and physics was completely mistaken, and one of the most important sources of the confusions he was trying to unravel. But how was he to make this clear without opposing Turing's view with a view of his own? He had to: (a) get Turing to admit that they were both using the word 'experiment' in the same sense; and (b) get him to see that, in that sense, mathematicians do not make experiments.

> Turing thinks that he and I are using the word 'experiment' in two different ways. But I want to show that this is wrong. That is to say, I think that if I could make myself clear, then Turing would give up saying that in mathematics we make experiments. If I could arrange in their proper order certain well-known facts, then it would become clear that Turing and I are not using the word 'experiment' differently.
>
> You might say: 'How is it possible that there should be a misunderstanding so very hard to remove?'
>
> It can be explained partly by a difference of education.

It might also be explained by the fact that Turing refused to leave his mathematician's paradise, or that he suspected Wittgenstein of

Bolshevism. What could not account for it, on Wittgenstein's view, was that there was here a substantive difference of opinion. 'Obviously', he told his class, 'the whole point is that I must not have an opinion.'

Wittgenstein, however, quite clearly did have very strong opinions – opinions that were, moreover, at variance with the conception of their subject held by most professional mathematicians. His suggestion that Turing suspected him of 'introducing Bolshevism into mathematics' is an allusion to Frank Ramsey's 1925 essay 'The Foundations of Mathematics', in which he had talked of rescuing mathematics from the 'Bolshevik menace' of Brouwer and Weyl, who, in their rejection of the Law of Excluded Middle, had regarded as illegitimate certain standard proofs in conventional analysis. To Turing, however, it must have seemed that Wittgenstein's Bolshevism was of a far more extreme sort. It was not, after all, the Law of Excluded Middle that Wittgenstein challenged, but the Principle of Contradiction.

All the conventional schools of thought on the foundations of mathematics – logicism, formalism and intuitionism – agree that if a system has a hidden contradiction in it, then it is to be rejected on the grounds of being inconsistent. Indeed, the whole point of providing mathematics with sound logical foundations was that Calculus, as traditionally understood, is manifestly inconsistent.

In his lectures, Wittgenstein ridiculed this concern for 'hidden contradictions', and it was to this that Turing voiced his most dogged and spirited dissent. Take the case of the Liar Paradox, Wittgenstein suggested:

> It is very queer in a way that this should have puzzled anyone – much more extraordinary than you might think: that this should be the thing to worry human beings. Because the thing works like this: if a man says 'I am lying' we say that it follows that he is not lying, from which it follows that he is lying and so on. Well, so what? You can go on like that until you are black in the face. Why not? It doesn't matter.

What was puzzling about this sort of paradox, Turing tried to explain, is 'that one usually uses a contradiction as a criterion for having done something wrong. But in this case one cannot find anything done wrong.' Yes, replied Wittgenstein, because nothing *has* been done wrong: 'One may say, "This can only be explained by a theory of types." But what is there which needs to be explained?'

Turing clearly needed to explain, not only why it was puzzling, but also why it *mattered*. The real harm of a system that contains a contradiction, he suggested, 'will not come in unless there is an application, in which case a bridge may fall down or something of the sort'. In the following lecture, he returned to the fray, and almost the entire lecture was taken up with a debate between the two on the importance of discovering 'hidden contradictions':

> *Turing*: You cannot be confident about applying your calculus until you know that there is no hidden contradiction in it.
> *Wittgenstein*: There seems to me to be an enormous mistake there. For your calculus gives certain results, and you want the bridge not to break down. I'd say things can go wrong in only two ways: either the bridge breaks down or you have made a mistake in your calculation – for example you multiplied wrongly. But you seem to think there may be a third thing wrong: the calculus is wrong.
> *Turing*: No. What I object to is the bridge falling down.
> *Wittgenstein*: But how do you know that it will fall down? Isn't that a question of physics? It may be that if one throws dice in order to calculate the bridge it will never fall down.
> *Turing*: If one takes Frege's symbolism and gives someone the technique of multiplying in it, then by using a Russell paradox he could get a wrong multiplication.
> *Wittgenstein*: This would come to doing something which we would not call multiplying. You give him a rule for multiplying and when he gets to a certain point he can go in either of two ways, one of which leads him all wrong.

'You seem to be saying', suggested Turing, 'that if one uses a little common sense, one will not get into trouble.' 'No', thundered Wittgenstein, 'that is NOT what I mean at all.' His point was rather that a contradiction cannot lead one astray because it leads nowhere at all. One cannot calculate wrongly with a contradiction, because one simply cannot use it to calculate. One can do nothing with contradictions, except waste time puzzling over them.

After two more lectures Turing stopped attending, convinced, no doubt, that if Wittgenstein would not admit a contradiction to be a fatal flaw in a system of mathematics, then there could be no common ground between them. It must indeed have taken a certain amount of courage to attend the classes as the single representative of all that Wittgenstein was attacking, surrounded by Wittgenstein's acolytes and having to discuss the issues in a way that was unfamiliar to him. Andrew Hodges, in his excellent biography of Turing, expresses surprise at what he sees as Turing's diffidence in these discussions, and gives as an example the fact that, despite long discussions about the nature of a 'rule' in mathematics, Turing never offered a definition in terms of Turing machines. But, surely, Turing realized that Wittgenstein would have dismissed such a definition as irrelevant; the discussion was conducted at a more fundamental level. Wittgenstein was attacking, not this or that definition, but the very motivation for providing such definitions.

With the certain exception of Alister Watson, and the possible exception of others, it is likely that many of those who attended these lectures did not fully grasp what was at stake in the arguments between Wittgenstein and Turing, nor fully understand how radically Wittgenstein's views broke with anything that had previously been said or written on the philosophy of mathematics. They were, on the whole, more interested in Wittgenstein than in mathematics. Norman Malcolm, for one, has said that, though he was aware that 'Wittgenstein was doing something important', he 'understood almost nothing of the lectures' until he restudied his notes ten years later.

Malcolm was then a doctoral student at Harvard, who had arrived in Cambridge in the Michaelmas term of 1938 to study with Moore, and quickly fell under the spell of Wittgenstein's personality. It is in his memoir that that personality is most memorably and (in the opinion of many who knew Wittgenstein) accurately described. Wittgenstein warmed to Malcolm's kindness and his human understanding, and during Malcolm's brief time at Cambridge the two became close friends. Upon Malcolm's return to the United States, he became, as well as a cherished correspondent, an invaluable supplier of Wittgenstein's favourite journal, Street & Smith's *Detective Story Magazine*, at a time when American magazines became unavailable in England.

Why Wittgenstein insisted – and insist he did; when Malcolm sent some other brand, Wittgenstein gently admonished him, asking why he had tried to be original instead of sticking to the 'good, old, tried out stuff' – on Street & Smith is a mystery: it was at this time practically indistinguishable from its more famous rival, *Black Mask*. Both published 'hard-boiled' detective stories written largely by the same group of writers, the most famous of which are: Carroll John Daly, Norbert Davis, Cornell Woolrich and Erle Stanley Gardner. Raymond Chandler published but one story in Street & Smith, a lesser known piece called 'No Crime in the Mountains', and Dashiell Hammett had, by this time, given up writing for the 'pulps' altogether.

In one respect, at least, the ethos of the hard-boiled detective coincides with Wittgenstein's own: they both, in their different ways, decry the importance of the 'science of logic', exemplified in the one case by *Principia Mathematica* and in the other by Sherlock Holmes. 'I am not the deducting, deducing book type of detective', explains Race Williams in a typical Street & Smith story:

> I'm a hard working, plugging sort of guy who can recognize a break when I see it and act at the same minute, at the same second or even split second if guns are brought into it.

This fast-acting, fast-shooting, honest sort of a guy bears an obvious similarity to movie cowboys, and it is probably no coincidence that the Western was Wittgenstein's favourite genre. By the late 1930s, however, his taste had broadened to include musicals. His favourite actresses, he told Malcolm, were Carmen Miranda and Betty Hutton. Exhausted and disgusted by his lectures, he would invariably go to see a 'flick' after them, accompanied by Malcolm, Smythies or one of his other friends from the class. He would always sit at the front of the cinema, where he could be totally immersed in the picture. He described the experience to Malcolm as 'like a shower bath', washing away his thoughts of the lecture.

It was the custom at that time to play the national anthem at the end of the film, at which point the audience was expected to rise to their feet and stand respectfully still. This was a ceremony that Wittgenstein could not abide, and he would dash out of the cinema before it could begin. He also found the movie newsreels, which used to be shown between films, unbearable. As war with Germany approached, and the newsreels became more and more patriotic and jingoistic, Wittgenstein's anger increased. Among his papers there is a draft of a letter addressed to their makers, accusing them of being 'master pupils of Goebbels'. It was at this time that his friendship with Gilbert Pattisson, of ten years' standing, came to an end, when he perceived in Pattisson's attitude to the war something he took to be jingoism. His friendship with Norman Malcolm was threatened by a similar issue. Passing a newspaper vendor's sign which announced the German government's accusation that the British had attempted to assassinate Hitler, Wittgenstein commented that he would not be surprised if it were true. Malcolm demurred. Such an act was, he said, incompatible with the British 'national character', Wittgenstein reacted angrily to this 'primitive' remark:

> . . . what is the use of studying philosophy if all that it does for you is to enable you to talk with some plausibility about some abstruse questions of logic, etc., & if it does not improve your thinking about the important questions of everyday life, if it does not make you more conscientious than any . . . journalist in the use of the DANGEROUS phrases such people use for their own ends.

The rift healed before Malcolm's return to the United States in February 1940, but for a time Wittgenstein ceased his habit of taking a walk with Malcolm before his lectures.

Wittgenstein had reason to be wary of the nationalist sentiment and anti-German feeling that was being whipped up in readiness for the coming war. On the day that war was declared, 3 September 1939, he and Skinner were in Wales, visiting Drury and staying in an hotel in Pontypridd. The following morning he was told to report to the local police station, his German name having aroused the suspicions of the hotel manageress. By this time he was a British national, and he had no trouble establishing the fact, but, as he said to Skinner and Drury, he would in future have to be very careful.

For the first two years of the war Wittgenstein was forced to remain a lecturer at Cambridge, despite strenuous efforts to find alternative work related to the war effort, such as joining the ambulance brigade. In September 1937, when things had been going badly with his work, he had urged himself to do something else. But: 'how should I find the strength to do something different now' he had asked, 'unless I were being forced, as in a war?' When war did come he found that, far from forcing him to do something else, it was preventing him from doing so. The doors to his doing something 'useful' were closed by his German name and Austrian background. While he continued to lecture and to work on the second half of his book, he longed to get away from Cambridge and to be, in some way or other, engaged in the struggle. 'I feel I will die slowly if I stay there', he told John Ryle. 'I would rather take a chance of dying quickly.'

He tried, unsuccessfully, to dissuade Malcolm from an academic career (Smythies, he felt sure, would never be offered an academic appointment anyway – he was 'too serious'). Couldn't Malcolm do some manual work instead? On a ranch or on a farm, say? Malcolm declined. He returned to Harvard, collected his Ph.D. and took up a teaching post at Princeton. In his letters Wittgenstein repeated his warnings. Congratulating Malcolm on his doctorate, he urged him to make good use of it and not to cheat himself or his students: 'Because, unless I'm very much mistaken, *that's* what will be expected from you.' Wishing him luck with his academic appointment, he stressed again that the temptation for Malcolm to cheat himself would be overwhelming: '*Only by a miracle* will you be able to do decent work in teaching philosophy.'

By the time war broke out, Skinner's period as an apprentice at the Cambridge Instrument Company had come to an end, and he seems to have made an attempt to return to theoretical work. In a letter dated 11 October 1939, written from Leeds, he mentions collaborating on a book (one assumes a mathematics textbook) with his old mathematics tutor, Ursell. The project was presumably abandoned (at least, I have been able to find no trace of such a book being published). In the letter Skinner says how difficult he now finds this sort of work, and mentions that he may soon return to Cambridge to look for a job. He also alludes to some sort of break between himself and Wittgenstein, a problem in their relationship, for which, characteristically, he assumes full blame:

> I feel very unhappy that I should have given you cause to write that you feel I'm away from you. It's a terrible thing that I have acted in a way that might loosen what is between us. It would be a catastrophe for me if anything happened to our relation. Please forgive me for what I have done.

What he had done he does not say, and no doubt did not know; he knew only that he was losing Wittgenstein's love. After his return to Cambridge, he and Wittgenstein lived separately – he in East Road, and Wittgenstein in his favoured set of rooms in Whewell's Court.

After Skinner's death Wittgenstein repeatedly chastized himself for having been unfaithful to him in the last two years of his life. It is a reasonable conjecture that this guilt is connected with Wittgenstein's feelings for a young working-class colleague of Skinner's called Keith Kirk. In 1939, Kirk, who was then nineteen, worked as an apprentice alongside Skinner, and they became friends when he began to ask Skinner questions about the mathematics and mechanics of the instruments they were using. Skinner, who was too reticent to be an adequate teacher, introduced Kirk to Wittgenstein, and from then on Wittgenstein gave Kirk regular lessons on physics, mathematics and mechanics to help him with the City and Guilds professional examinations for which he was then preparing.

To Kirk these lessons from a Cambridge professor were nothing but an unexpected and extremely welcome source of help, and a remarkable opportunity. From Wittgenstein's diary, however, it appears that he thought rather more of the relationship than one would expect:

> See K once or twice a week; but am doubtful whether the relationship is the right one. May it be genuinely good [13 June 1940]

> Occupied myself the *whole day* with thoughts of my relations with Kirk. For the most part, *very* insincere and fruitless. If I wrote these thoughts down, one would see how low and dishonest – how *indecent* they were [7 October 1940]

Throughout 1940 and the first half of 1941 Kirk came regularly to Wittgenstein's rooms in Trinity for his unpaid lessons. Wittgenstein taught without a textbook; instead, he would ask Kirk a series of questions that forced him to think through the problems from first principles. Thus, a lesson might begin with Wittgenstein asking Kirk what happens when water boils – What are bubbles? Why do they rise to the surface? and so on. The amount that Kirk learnt from these lessons therefore depended to a large extent on his own ability to think, and, as in Wittgenstein's philosophy classes, there would frequently be long silences. However, according to Kirk, what he learnt in these lessons has stayed with him ever since, and the style of thinking imparted to him by Wittgenstein has been of lasting benefit to him.

Kirk never had the slightest idea that Wittgenstein's feelings for him were anything other than those of a helpful teacher. After the lesson he would occasionally accompany Skinner and Wittgenstein to see a Western at the local cinema, but apart from that he saw little of Wittgenstein outside his periods of instruction.

The lessons came to an end in 1941, when Kirk was sent by the Ministry of War to work in Bournemouth on Air Ministry Research. The move put an end to his City and Guilds studies, but not immediately to his friendship with Wittgenstein. Wittgenstein did what he could to stay in touch. He once went down to Bournemouth to see how Kirk was getting on, and when Kirk came back to Cambridge Wittgenstein would invariably arrange to meet him.

It was during one of these latter visits that Wittgenstein called on Kirk in an extremely distraught state to tell him that Francis had been taken seriously ill with polio and had been admitted into hospital. A few days later, on 11 October 1941, Francis died.

Wittgenstein's initial reaction was one of delicate restraint. In letters to friends telling them of Francis's death, he managed a tone of quiet dignity. To Hutt, for example, he wrote:

> My dear Ro[w]land,
> I have to give you very terrible news.
> Francis fell ill four days ago with poliomyelitis & died yesterday morning. He died without *any* pain or struggle *entirely* peacefully. I was with him. I think he has had one of the happiest lives I've known anyone to have, & also the most peaceful death.
> I wish you good and kind thoughts.
> As always,
> Ludwig

By the time of the funeral, however, his restraint had gone. He has been described by Skinner's sister as behaving like a 'frightened wild animal' at the ceremony, and after it, she recalls, he refused to go to the house but was seen walking round Letchworth with Dr Burnaby, the tutor of Trinity, looking 'quite wild'. He would not, in any case, have been unreservedly welcome at the Skinner home. Skinner's family were always mistrustful of the influence he had exerted on their delicate boy, and his mother, who believed that his job at the Cambridge Instrument Company had hastened Francis's death, refused to speak to Wittgenstein at the funeral.

But Wittgenstein's guilt over Francis was entirely unconnected with the way in which he had influenced him. It had to do with more internal matters – with how Wittgenstein himself had felt towards Francis during the last few years of his life. On 28 December 1941, he wrote:

> Think a lot about Francis, but always only with remorse over my lovelessness; not with gratitude. His life and death seem only to accuse me, for I was in the last 2 years of his life very often loveless and, in my heart, unfaithful to him. If he had not been so boundlessly gentle and true, I would have become *totally* loveless towards him.

Immediately after this passage he goes on to discuss his feelings for Kirk: 'I see Keith often, and what this really means I don't know. Deserved disappointment, anxiety, worry, inability to settle into a pattern of life.' About seven years later, in July 1948, he wrote: 'Think a great deal about the last time I was with Francis; about my odiousness towards him . . . I cannot see how I can ever in my life be freed from this guilt.'

Wittgenstein's infatuation with Kirk – entirely unspoken, unacknowledged and unreciprocated as it was – exemplifies in its purest form a feature that had characterized his earlier loves for Pinsent and for Marguerite; namely, a certain indifference to the feelings of the other person. That neither Pinsent nor Marguerite – and certainly not Kirk – were in love with him seemed not to affect his love for them. Indeed, it perhaps made his love easier to give, for the relationship could be conducted safely, in the splendid isolation of his own feelings. The philosophical solipsism to which he had at one time been attracted, and against which much of his later work is addressed (he characterized his later work as an attempt to show the fly the way out of the fly-bottle), has its parallel in the emotional solipsism in which his romantic attachments were conducted. With Francis that isolation was threatened, and, in the face of that threat, Wittgenstein had withdrawn, like the porcupines of Schopenhauer's fable, behind his spiky exterior.

IV

1941–51

21
WAR WORK

For the first two years of war, a recurrent theme of conversation with Wittgenstein was his frustration at not being able to find work outside academic life. He found it intolerable to be teaching philosophy while a war was being fought, and wanted more than anything else to be able to contribute to the war effort. His chance to do so came through his friendship with the Oxford philosopher Gilbert Ryle. Gilbert's brother, John Ryle, was Regius Professor of Physics at Cambridge, but in 1940 he had returned to Guy's Hospital to help them prepare for the Blitz. In September 1941 Wittgenstein wrote to John Ryle asking to meet him at Guy's. Ryle invited him to lunch, and was immediately impressed. 'He is one of the world's famousest philosophers', he wrote to his wife. 'He wears an open green shirt and has a rather attractive face':

> I was so interested that after years as a Trinity don, so far from getting tarred with the same brush as the others, he is overcome by the deadness of the place. He said to me 'I feel I will die slowly if I stay there. I would rather take a chance of dying quickly.' And so he wants to work at some humble manual job in a hospital as his war-work and will resign his chair if necessary, but doesn't want it talked about at all. And he wants the job to be in a blitzed area. The works department are prepared to take him as an odd job man under the older workmen who do all the running repairs all over the hospital. I think he realises that his mind works so differently to most people's that it would be stupid to try for any kind of war-work based on intelligence. I have written to him tonight

to tell him about this job but am not trying to persuade him unduly.

Someday I must bring him and also one or two of the Canadians down to see you.

Wittgenstein clearly needed no undue persuasion, for a week or so after this letter was written he started work at Guy's. Not, however, as an odd-job man but as a dispensary porter.

John Ryle respected Wittgenstein's wish that his change of job from Professor of Philosophy at Cambridge to dispensary porter at Guy's Hospital should not be talked about, and does not seem to have mentioned to any of the staff at Guy's that the new porter was 'one of the world's famousest philosophers'. One indication of his discretion is that Humphrey Osmond, a good friend of Ryle's and the editor during the war of the in-house journal *Guy's Gazette* (and therefore always on the look-out for an interesting story), did not find out that Wittgenstein had been at Guy's until after the publication of Norman Malcolm's memoir in 1958. It is fortunate that Ryle kept his silence, for if the *Gazette* had run a 'Famous philosopher at Guy's' piece, there is no doubt that Wittgenstein would have reacted with the utmost rage.

While he was at Guy's, Wittgenstein lived and dined with the medical staff at Nuffield House. (This in itself would have been enough to distinguish him from the other porters in the hospital, because the non-medical staff usually lived outside the hospital grounds and dined separately from the doctors.) Shortly after his arrival at Nuffield House, he was enthusiastically greeted at dinner by the hospital haematologist, Dr R. L. Waterfield. Waterfield had been at Cambridge and had attended meetings of the Moral Science Club. Upon being recognized, Wittgenstein turned as white as a sheet and said: 'Good God, don't tell anybody who I am!' But whether through Waterfield or some other source – and despite the fact that *Guy's Gazette* never got hold of the story – many of the staff at Guy's knew perfectly well who Wittgenstein was. Everyone there who knew him at all knew him as 'Professor Wittgenstein'.

Wittgenstein's job as a porter was to deliver medicines from the dispensary to the wards, where, according to John Ryle's wife, Miriam, he advised the patients not to take them. His boss at the pharmacy was Mr S. F. Izzard. When asked later if he remembered Wittgenstein as a porter, Izzard replied: 'Yes, very well. He came and worked here and after working here three weeks he came and explained how we should be running the place. You see, he was a man who was used to thinking.' After a short while, he was switched to the job of pharmacy technician in the manufacturing laboratory, where one of his duties was to prepare Lassar's ointment for the dermatological department. When Drury visited Wittgenstein at Guy's, he was told by a member of staff that no one before had produced Lassar's ointment of such high quality.

Wittgenstein arrived at Guy's needing a friend. After the death of Francis, and with the departure of Kirk for Bournemouth, he was desperately lonely. He needed some sort of emotional contact. '*One word* that comes from your heart', he had written to Rowland Hutt on 20 August 1941, 'would mean more to me than 3 pages out of your head!' And on 27 November: 'I can't write about Francis, and what you write about him, though in a sense true, somehow doesn't click with *my* thoughts of him.' He told Hutt of his job at the dispensary, how he earned twenty-eight shillings a week and of how hard the work was. 'I hope my body'll be able to stick it. My soul is *very* tired, & isn't at all in a good state; I mean, not at all as it should be.' 'Perhaps', he added, 'when we see each other again it will help us in some way.'

It was important for Wittgenstein that if he and Hutt did meet they should do so for long enough for it to mean something. In subsequent letters he stressed the importance of meeting on a Sunday, the only day on which he was not working at the hospital:

> If, however, you *can't* come on a Sunday, a week day would have to do. In this case to come even half an hour too late would be unwise; for under the circs, it's quite easy for us to make a mess of things, but ever so undesirable!

'It is on the whole', he explained in another letter, 'not a good plan *for people like us* to see each other *hurriedly*. If possible we should be together leisurely.' After Hutt had shown some hesitation about the proposed meeting, he was told by Wittgenstein to wait three months before they tried to see each other:

> As long as you find it difficult to say that you want to see me, as you write, why should you see me? I want to see people who want to see *me*; & if the time should come (& perhaps it will come soon) when nobody'll want to see me, I think I'll see nobody.

The fear that his body would not be able to cope with the work of a dispensary porter was a real one. He was, by now, fifty-two, and was beginning to look (and feel) *old*. 'When I finish work about 5', he told Hutt, 'I'm so tired I often can hardly move.' But if his body was weak, his spirit was, in the wake of Francis's death, almost broken. He spent Christmas with the Barbrooke family, who owned the grocery on East Road below Francis's flat. It was a melancholy occasion. On New Year's Eve, he wrote to Hutt:

> I feel, on the whole, lonely & am afraid of the months & years to come! . . . I hope that you have some happiness & that you appreciate whatever you have of it more than I did.

In the New Year of 1942, John Ryle fulfilled the promise he had made to his wife and took Wittgenstein back to their home in Sussex, to meet her. Happily, the weekend is recorded in the diary of their son, Anthony, who was then fourteen years old. His first impressions were not entirely favourable:

> Daddy and another Austrian (?) professor called Winkenstein (spelling?) arrived at 7.30. Daddy rather tired. Wink is awful strange – not a very good english speaker, keeps on saying 'I mean' and 'its "tolerable"' meaning intolerable.

By the end of the following day, although making a closer approximation to the spelling of Wittgenstein's name, Anthony was still far from being won over by his father's new friend:

> In the morning Daddy, Margaret, goats, Tinker & I went for a walk. Frosty but sunny. Witkinstein spent the morning with the evacuees. He thinks we're terribly cruel to them.
>
> We spent the afternoon argueing – he's an impossible person everytime you say anything he says 'No No, that's not the point.' It probably isn't his point, but it is ours. A tiring person to listen to.

> After tea I showed him round the grounds and he entreated me to be kind to the miserable little children – he goes far too much to the other extreme – Mommy wants them to be good citizens, he wants them to be happy.

The Ryles rented a farmhouse in Sussex, and the 'miserable little children' were evacuees – two working-class boys from Portsmouth whom Mrs Ryle had taken on as a political statement. They joined a group of children she organized to knit gloves for the Russian Red Cross. Although she looked after these children well, she evidently maintained strict discipline among them. On occasions when John Ryle was at home, or when they had visitors, the Ryle family, to some extent, kept a certain distance from the evacuees – dining, for example, in separate rooms. While Wittgenstein was staying there, he insisted on showing his support and sympathy for the children by dining with them.

It is easy to see why Wittgenstein should have liked and respected John Ryle. Like Wittgenstein, Ryle fitted uneasily into academic life at Cambridge, and it is clear that he shared Wittgenstein's preference for the dangers of working in a blitzed hospital over the 'deadness' of Cambridge. While at Cambridge he had been politically active, and had stood in the 1940 election as a left-wing independent candidate. From 1938 onwards he had been active in getting Jewish doctors out of Austria and Germany (which is presumably why Wittgenstein is described by Anthony Ryle as 'another Austrian professor').

Ryle's kindness is remembered with warmth and gratitude by many of the staff who served at Guy's during the Blitz. Many of them were young and, unlike Ryle, who had served in the First World War, had had no experience of war. Humphrey Osmond's memories of the dangers of working at Guy's during the heavy bombing – and Ryle's inspiration in helping the staff to cope with those dangers – are typical:

> The hospital had scores of firebombs dropped on it & at least a dozen exploded or unexploded bombs on its premises . . . Under the pressure of bombing & taking in many casualties the small staff left at Guy's itself knew each other pretty well . . . I used to firewatch on the roof at Guy's . . . We spent a lot of time gossiping & drinking tea . . . We used to camp out in the basement of Nuffield

> House. Ryle was a wise & intelligent man, whose serenity, which had been tempered in the trenches during WW1, was a great support to those like myself who disliked being bombed.

In April Wittgenstein underwent an operation at Guy's to remove the gall-stone that had been troubling him for a number of years. His mistrust of English doctors (he was inclined to believe that both Ramsey's death and Skinner's might have been avoided if they had received proper medical care) impelled him to insist on remaining conscious during the operation. Refusing a general anaesthetic, he had mirrors placed in the operating theatre so that he could watch what was happening. To help him through what was undoubtedly a painful ordeal, John Ryle sat with him throughout the operation, holding his hand.

Apart from Ryle, Wittgenstein's few friends at Guy's tended to be technicians rather than doctors. One of these was Naomi Wilkinson, a radiographer and a cousin of Ryle's. Miss Wilkinson used to organize gramophone recitals at the hospital, at which Wittgenstein was a regular attender. He took an intense interest in the choice of records, and was often very critical of the selection. As a result of this common interest in music, he and Miss Wilkinson became friends, and, as with many of his friends, she was invited to join him for tea at Lyons. At one of these teas she asked him how many people he thought understood his philosophy. He pondered the question for a long time before he replied: 'Two – and one of them is Gilbert Ryle.' He did not, unfortunately, say who the second one was. And perhaps his choice of Gilbert Ryle indicates nothing more than that, even in his fifties, he had not entirely lost his childhood good manners – his inclination to say things he thought would please the other person.

Naomi Wilkinson's gramophone recitals perhaps provided one of the elements in a dream that Wittgenstein recorded while he was working at Guy's:

> Tonight I dreamt: my sister Gretl gave Louise Politzer a present: a bag. I saw the bag in the dream, or rather just its steel lock, which was very big and square and very finely constructed. It looked like one of those complicated old padlocks one sometimes sees in museums. In this lock there was, among other things, a mechanism by which the words 'From your Gretl', or something similar, were

> spoken through the key-hole. I thought about how intricate the mechanism of this device must be and whether it was a kind of gramophone and of what sort of material the records could be made, whether they were possibly made of steel.

Wittgenstein himself offers no interpretation of this dream, but given his preoccupation at this time with Freud's work, his previous use of the metaphor of a lock to describe Freud's central idea, and the fact that Gretl was the member of his family most closely associated with Freud, it is possible, I think, to regard this dream as being *about* the interpretation of dreams. Dreams seem to say something, and a skilful use of Freud's work will enable us to hear what they say (through, as it were, the key-hole of Freud's theories), but the mechanism that lies behind their saying something and the material from which dream symbols are constructed (the unconscious) is intricate and complicated, too complicated to be understood in terms of Freud's rather crude analogy with nineteenth-century mechanics.

Such, anyway, were the central themes of the discussions Wittgenstein had with Rhees in the summer of 1942. He went to Swansea to stay with Rhees, partly in order to recuperate from his gall-stone operation, and the two of them would take walks along the South Wales coastline, of which Wittgenstein was extremely fond. Rhees was at that time one of the very few people left alive whom Wittgenstein valued as a partner in philosophical discussion, but it is striking that, at a time when his philosophical work centred mainly on the philosophy of mathematics, his conversations with Rhees dealt with the nature of Freud's explanations in psychology.

There *is* a sense, he stressed, in which the images in a dream might be regarded as symbols, a sense in which we can speak of a dream language, even if the symbols are not understood by the dreamer. This can emerge when we discuss the dream with an interpreter and accept his interpretation. Similarly, when we draw apparently meaningless doodles and an analyst asks us questions and traces associations, we might come to an explanation of why we drew what we did: 'we may then refer to the doodling as a kind of writing, as using a kind of language, although it was not understood by anyone'. But it was important to Wittgenstein to dissociate this kind of explanation from those given in science. Explanations of dreams or doodles do not proceed by the application of laws, 'and to me the fact that there *aren't* actually any such laws seems important'. Freud's explanations have more in common with a mythology than with science; for example, Freud produces no evidence for his view that anxiety is always a repetition of the anxiety we felt at birth, and yet 'it is an idea which has a marked attraction':

> It has the attraction which mythological explanations have, explanations which say that this is all a repetition of something that has happened before. And when people do accept or adopt this, then certain things seem clearer and easier for them.

Freud's explanations, then, are akin to the elucidations offered by Wittgenstein's own work. They provide, not a causal, mechanical theory, but:

> . . . something which people are inclined to accept and which makes it easier for them to go certain ways: it makes certain ways of behaving and thinking natural for them. They have given up one way of thinking and adopted another.

It was in this respect that Wittgenstein described himself to Rhees at this time as a 'disciple' or 'follower' of Freud.

Wittgenstein's philosophical preoccupations throughout most of the Second World War centred on the philosophy of mathematics. Most of what he wrote at this time is an attempt to improve on the remarks written during his last months in Norway, and thus improve the section of the *Investigations* that was based on them. During the time he worked at Guy's he filled three notebooks with remarks on mathematics. They, and the manuscript volume compiled from them, have now been published and form Parts IV, V, VI and VII of *Remarks on the Foundations of Mathematics.*

Though in its general outlines it is of a piece with his earlier work on the subject, the assault on mathematical logic is expressed in more caustic terms. It is, perhaps, his most polemical work.

In his essay 'Mathematics and the Metaphysicians', Russell provides the most perfect summary of the target of Wittgenstein's polemic. 'One of the chief triumphs of modern mathematics', Russell writes, 'consists in having discovered what mathematics really is':

> All pure mathematics – Arithmetic, Analysis, and Geometry – is built up by combinations of the primitive ideas of logic, and its propositions are deduced from the general axioms of logic, such as the syllogism and the other rules of inference . . . The subject of formal logic has thus shown itself to be identical with mathematics.

He goes on to discuss the problems of the infinitesimal, the infinite and continuity:

> In our time, three men – Weierstrass, Dedekind, and Cantor – have not merely advanced the three problems, but have completely solved them. The solutions, for those acquainted with mathematics, are so clear as to leave no longer the slightest doubt or difficulty. This achievement is probably the greatest of which our age has to boast.

Wittgenstein's work is an attack on both the conception of mathematics that is outlined here, and also the attitude towards it that is revealed. 'Why do I want to take the trouble to work out what mathematics is?' he asks in one of the notebooks kept during his period at Guy's:

> Because we have a mathematics, and a special conception of it, as it were an ideal of its position and function, – and this needs to be clearly worked out.
>
> It is my task, not to attack Russell's logic from within, but from without.
>
> That is to say: not to attack it mathematically – otherwise I should be doing mathematics – but its position, its office.

For Wittgenstein, formal logic had not shown itself to be identical with mathematics; to say it had: 'is almost as if one tried to say that cabinet-making consisted in glueing'. Nor has mathematical logic finally shown us what mathematics is. It has rather: 'completely deformed the thinking of mathematicians and of philosophers'. And the work of Weierstrass, Dedekind and Cantor, far from being the greatest achievement of our age, was, in relation to the rest of mathematics: 'a cancerous growth, seeming to have grown out of the normal body aimlessly and senselessly'.

In order to show that logic and mathematics are different techniques, and that the results in mathematical logic do not have the importance attributed to them by Russell (in understanding the concepts of infinity, continuity and the infinitesimal), Wittgenstein tries a number of tacks, including, for example, trying to show that the notions of infinity, continuity and the infinitesimal as actually used in mathematics and everyday life have not been clarified by the definitions of them given by Cantor, Dedekind and Weierstrass, but, rather, distorted.

But the centre of his attack consists in trying to show that the methods of proof that typify mathematics are disanalogous to those used in logic. A proof in logic consists in a series of propositions intended to establish the truth of a conclusion. What Wittgenstein wants to show is that a proof in mathematics consists rather in a series of *pictures* intended to establish the usefulness of a technique.

For example, he sees no reason why this picture:

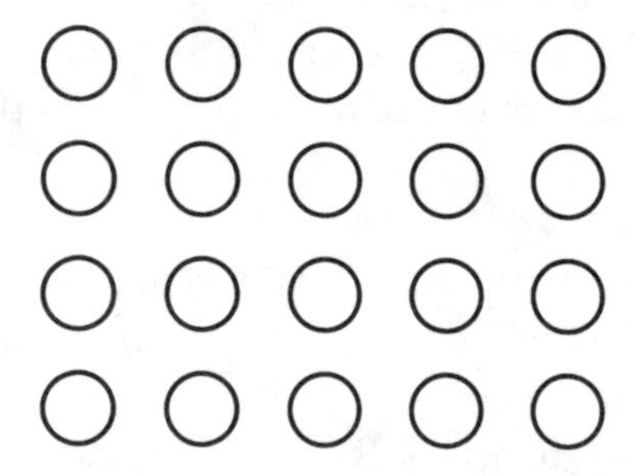

should not be regarded as a proof of the Commutative Law for multiplication – i.e.: $(a \times b) = (b \times a)$. For someone might, by looking at the picture first one way and then the other, see that (5×4) is equal to (4×5) and then come to use the principle of commutation in all other cases.

Here there is no question of propositions or conclusions, and consequently the question of what the commutative law, if true, is true *about* does not arise. And if this sort of picture, rather than the axiomatic systems of logic, were regarded as paradigmatic, then there would be no reason at all to think that mathematical logicians had, as Russell would have it, 'discovered what mathematics really is'. In their work on the 'foundations of mathematics' they had simply drawn a different sort of picture, and invented a different sort of technique.

But the point of emphasizing the role of pictures in mathematics is not simply to destroy a particular conception of the subject. It is also to substitute for it a conception of mathematical reasoning which stresses the role of 'seeing connections'. In order to grasp the principle of commutation from the picture above we need to see this:

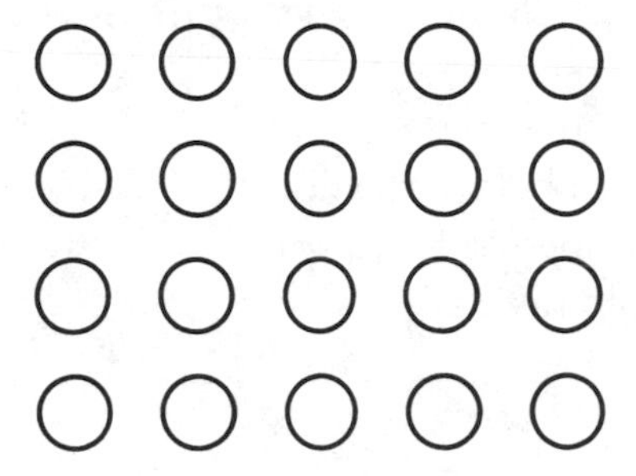

as the same as this:

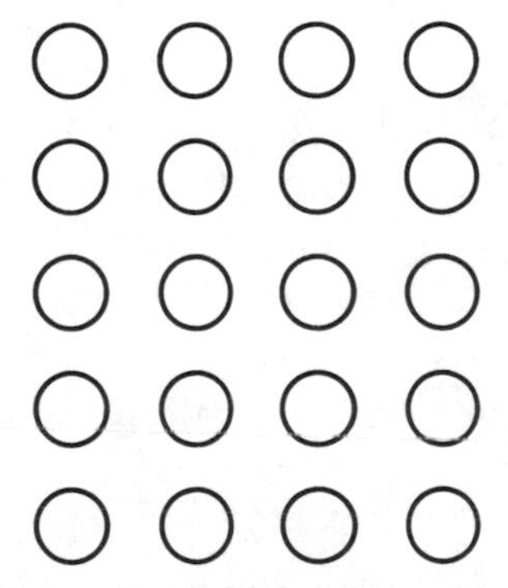

If we could not 'see the connection', the proof would not convince us of anything. The understanding of the proof, then, is a good, if rudimentary, example of the kind of understanding that forms the basis of Wittgenstein's *Weltanschauung*. Mathematical proofs, like his own philosophical remarks, should be seen as 'perspicuous representations', the point of which is to produce 'just that understanding which consists in seeing connections'.

In this way, curious though it sounds, proofs in pure mathematics are analogous to the explanations offered in Freudian psychoanalysis. And perhaps the clue to Wittgenstein's shift in concerns, from mathematics to psychology, lies in his finding Freud's 'patterns' more interesting than the 'pictures' of mathematicians.

It would, one suspects, have been something of relief for Wittgenstein to have been able to place the events of his own life into some kind of pattern.

'I no longer feel any hope for the future of my life', he wrote on 1 April 1942:

> It is as though I had before me nothing more than a long stretch of living death. I cannot imagine any future for me other than a ghastly one. Friendless and joyless.

A few days later:

> I suffer greatly from the fear of the complete isolation that threatens me now. I cannot see how I can bear this life. I see it as a life in which every day I have to fear the evening that brings me only dull sadness.

At Guy's he felt that he had to keep himself busy. 'If you can't find happiness in stillness', he told himself, 'find it in running!'

> But what if I am too tired to run? 'Don't talk of a collapse until you break down.'
>
> Like a cyclist I have to keep pedalling, to keep moving, in order not to fall down.

'My unhappiness is so complex', he wrote in May, 'that it is difficult to describe. But probably the main thing is still *loneliness.*'

After Skinner's death Kirk had returned to Bournemouth, and, as he had with Skinner, Wittgenstein began to fret about not receiving letters from him. On 27 May he noted:

> For ten days I've heard nothing more from K, even though I pressed him a week ago for news. I think that he has perhaps broken with me. A *tragic* thought!

In fact, Kirk married in Bournemouth, made a successful career for himself in mechanical engineering and never saw Wittgenstein again. But for Kirk there had been nothing to 'break' from. It had never occurred to him that Wittgenstein might be, in any sense, homosexual, or that their relationship was anything other than that between a teacher and a pupil.

As if acknowledging this, Wittgenstein wrote, in the same diary entry: 'I have suffered much, but I am apparently incapable of *learning* from my life. I suffer still *just* as I did many years ago. I have not become any stronger or wiser.'

Some comfort – some relief from this desperate loneliness – came from a friendship with a young colleague of his in the dispensary at Guy's, Roy Fouracre. It was, one gathers, principally Fouracre's warmth and jovial good humour that endeared him to Wittgenstein. Sometimes, Wittgenstein told Drury, he would be rushed or agitated and Roy would say to him, 'Steady, Prof.' This he liked.

Fouracre would visit Wittgenstein in his room on the third floor of Nuffield House. The room, like his rooms at Cambridge, was completely bare, and Fouracre was surprised to see no philosophy books at all, but only neat piles of detective magazines. Fouracre was at the time studying for a correspondence course in modern languages, and would often sit in Wittgenstein's room reading while Wittgenstein sat quite still and silent. During these times Wittgenstein would be preparing for the lectures that he gave in Cambridge every alternate weekend. On the other weekends Wittgenstein and Fouracre would go on outings, perhaps to the zoo or to Victoria Park in Hackney, where they would go rowing on the lake.

Like many who knew Wittgenstein well, Fouracre remembers his virtuoso whistling. He recalls Wittgenstein's ability to whistle whole movements of symphonies, his showpiece being Brahms's *St Anthony Variations*, and that when other people whistled something wrong, Wittgenstein would stop them and firmly tell them how it should go – something that didn't endear him to his fellow workers in the dispensary.

Fouracre's background could hardly have been more different to that of Skinner. Whereas Skinner had been brought up in a middle-class home in Letchworth and had been educated at public school and Cambridge, Fouracre lived in a council house in Hackney, East

London, and had started work at the age of fifteen. But their personal qualities were, in many ways, similar. Fania Pascal's description of Skinner:

> He could be cheerful and liked the company of others. Without guile of any kind, he was incapable of thinking evil of anyone. He could and did learn to be more practical, though alas, he would always be too unselfish, too self-effacing.

could serve equally as a description of Fouracre. Like Skinner, Fouracre was considerably younger than Wittgenstein – in his early twenties to Wittgenstein's fifty-two. And while it would be wrong to imply that Wittgenstein's friendship with Fouracre provided anything like a substitute for his love for Skinner, it is true that for the eighteen months they worked together Fouracre filled a role in Wittgenstein's life similar to that which Skinner had filled at Cambridge. That is, he provided Wittgenstein with some sort of *human* contact: he was, like Francis, someone whose mere presence had a reassuring effect.

In many of his later letters to Fouracre, Wittgenstein mentions Guy's with warmth and with some element, perhaps, of nostalgia:

> I'm sorry to hear the atmosphere at Guy's is getting worse. It's difficult to imagine. [8.6.49]

> I wonder what the job news are you write about. I suppose it isn't that they are errecting a huge statue of me in front of Nuffield House. Or is it? Of course, no monument of stone could really show what a wonderful person I am. [15.12.50]

Fouracre obviously replied to the last suggestion telling Wittgenstein that all the statues to him at Guy's had been pulled down. 'I'm glad to hear [it]', Wittgenstein wrote in his next letter, 'as long as it wasn't done in any disrespectful way!'

Of the medical staff at Guy's, the only person to have gained Wittgenstein's trust and friendship, apart from John Ryle, seems to have been Basil Reeve, a young doctor (then in his early thirties) with an interest in philosophy. When he heard from Reg Waterfield that the new member at the dining table (whom he had previously thought 'looked interesting and rather lost amongst the hospital physicians') was Ludwig Wittgenstein, he decided to try to make his acquaintance. He therefore started sitting next to Wittgenstein at dinner, and eventually a friendship developed between the two. The topic of conversation, however, hardly ever turned to philosophy, but would centre instead on art or architecture or music, or people Wittgenstein had known, or even Freudian interpretations of some of the medical conversation that went on at the dinner table. Later, it became centred on Reeve's own work, in which Wittgenstein began to take a keen interest.

Reeve was at Guy's working in the Medical Research Council's Clinical Research Unit with a colleague, Dr Grant. Early in the Blitz the laboratories of this unit were destroyed by bombing, and, unable to pursue their original studies, Grant and Reeve started studying the plentiful air-raid casualties that were being admitted to Guy's at this time. Their object was to try and familiarize themselves with the condition of 'wound shock', which would occur not only in battle casualties but in any condition of acute traumatic injury.

Grant and Reeve's initial problem was that, in spite of a detailed study of the scientific literature, there seemed to be no satisfactory way of defining clinically the condition of 'wound shock'. Some authors identified the condition on the basis of the presence of haemoconcentration (the development of an abnormally high concentration of red cells in the blood thought to be due to the leakage of plasma from the blood into the tissues), while others recognized it as a syndrome of low blood pressure, skin pallor and rapid pulse. Quite early on in the research, therefore, Grant recommended that the very concept of 'wound shock' should be abandoned and that detailed observations of casualties should be made without using the term. In January 1941 – ten months before Wittgenstein came to Guy's – Grant produced a memorandum on the observations required in cases of wound shock, outlining his objections to the concept:

> Recent experience of air raid casualties shows that in spite of all the work already done, specially in the last war, but little is known about the nature and treatment of traumatic or wound shock. In the first place there is in practice a wide variation in the application of the diagnosis of 'shock'. We cannot yet foretell and we are often in doubt about treatment. Moreover, the lack of a common basis of

diagnosis renders it impossible to assess the efficacy of the various methods of treatment adopted.

There is good ground, therefore, for the view that it would be better to avoid the diagnosis of 'shock' and to replace it by an accurate and complete record of a patient's state and progress together with the treatment given.

It is clear, I think, why Wittgenstein should find this radical approach to the problem interesting and important. Grant's way of dealing with the problem of 'shock' has an obvious parallel with Heinrich Hertz's way of dealing with the problems of 'force' in physics. In *The Principles of Mechanics* Hertz had proposed that, instead of giving a direct answer to the question: 'What is force?' the problem should be dealt with by restating Newtonian physics without using 'force' as a basic concept. Throughout his life, Wittgenstein regarded Hertz's solution to the problem as a perfect model of how philosophical confusion should be dispelled, and frequently cited – as a statement of his own aim in philosophy – the following sentence from Hertz's introduction to *The Principles of Mechanics*:

> When these painful contradictions are removed, the question as to the nature of force will not have been answered; but our minds, no longer vexed, will cease to ask illegitimate questions.

In a conscious echo of this sentence, Wittgenstein wrote:

> In my way of doing philosophy, its whole aim is to give an expression such a form that certain disquietudes disappear. (Hertz).

And we might say of Grant's proposal to avoid the diagnosis of 'shock' that its whole aim was to: 'give an expression such a form that certain disquietudes disappear'.

Grant's approach was not, however, universally well received – especially by the army. Colonel Whitby of the Army Blood Transfusion Service responded to Grant's report in the following letter to the Medical Research Council:

> Quite a lot of the preamble, and some of the discussion was devoted to a diatribe against the word 'Shock'. I do not feel that this point needs to be so heavily emphasised.

> It is not justifiable to throw over the findings of the last war. These men were not fools . . . they at least established the fundamental fact that lowered blood pressure was a sign very constantly observed. Grant would throw over the whole of the valuable MRC literature of the last war because their records do not attain his standard of detail.

As Wittgenstein realized when he discussed the project with Reeve, the problem with the theories of wound shock formulated during the First World War was not primarily that their standard of detail was inadequate, but that they were operating with an unusable concept. It was precisely the 'diatribe against the word "Shock"' that most interested him. (Reeve remembers that when they came to write an annual report, Wittgenstein suggested printing the word 'shock' upside-down to emphasize its unusability.)

Because of the interest he showed in the project, Wittgenstein was introduced by Reeve to Dr Grant, who was immediately impressed by the acuteness and relevance of the many questions and suggestions he made concerning the investigations. During 1942, the heavy bombing of London, which had provided Grant's team with a steady supply of research material, began to peter out. The unit therefore began to search elsewhere for casualties suitable for their observations. On two occasions, stations of Bomber Command were visited and a number of instances of injury to air-crew, incurred in bombing raids, were observed. But to make progress in the research a more constant supply of injured people was needed, so arrangements were made for the unit to move to the Royal Victoria Infirmary, Newcastle, which admitted a large number of heavy industrial and road injuries to its wards. At the time this move was planned Wittgenstein told Reeve he would like to go with the Unit to Newcastle.

The Unit moved to Newcastle in November 1942. Grant's technician, however, refused to go, and remembering Wittgenstein's interest in the project, Grant offered him the post. By the spring of 1943, Fouracre had left Guy's to join the army, and there was, presumably, little else to keep Wittgenstein there. In April 1943 Grant wrote to Dr Herrald at the Medical Research Council office in London saying:

> Professor Ludwig Wittgenstein, of whom I told you, joined the unit as Laboratory Assistant on the 29th April for a probationary

> period of one month. As I arranged with you, he will be paid at the rate of £4 a week.

As he was earning only twenty-eight shillings a week as a dispensary porter, this would have constituted a significant increase in income. When the probationary period of a month was up Grant wrote again to Herrald confirming the arrangement, adding: 'He is proving very useful.'

The change from a manual job to the more cerebral task of assisting the research unit was no doubt welcome to Wittgenstein, and not only because he was finding the physical demands of the work as a porter a little difficult. Just before he left Guy's, on 17 March, he wrote to Hutt about the value of thinking. 'I *imagine*', he wrote, 'that thinking a little more than you probably do would do you good. – I hope your family isn't keeping you from thinking!? If anyone does they act *very* foolishly.' Hutt had, by now, left Woolworth's and was to join the army. He had written to Wittgenstein about some trouble he had with his superiors. 'I imagine', Wittgenstein replied, 'that it's partly external & partly internal':

> I mean, they mayn't be as decent with you as you deserve, – but you *tend* to be unreliable. I.e. you tend to be alternatively cold, warm, & lukewarm; & you mustn't be surprised if people sometimes disregard your periods of warmth & treat you as if you could be cold & lukewarm only.

Before he left for Newcastle, Wittgenstein spent some time in Swansea with Rush Rhees. There he resumed the conversations he had the previous summer about Freud. Again, it was the idea that dream symbols form a kind of language that interested him – the fact that we naturally think that dreams *mean* something, even if we do not know *what* they mean. Analogously, he told Rhees of the five spires of Moscow Cathedral: 'On each of these there is a different sort of curving configuration. One gets the strong impression that these different shapes and arrangements must mean something.' The question at issue was the extent to which Freud's work is useful in enabling us to interpret dreams. He stressed that what was wanted was not an explanation but an interpretation. Thus a scientific theory of dreams – which might, for example, enable us to predict that a dreamer could be brought to recall certain memories after providing us with a description of a dream – would not even touch the problem. Freud's work is interesting precisely because it does not provide such a scientific treatment. What puzzles us about a dream is not its causality but its *significance*. We want the kind of explanation which 'changes the aspect' under which we see the images of a dream, so that they now make sense. Freud's idea that dreams are wish fulfilments is important because it 'points to the sort of interpretation that is wanted', but it is too general. Some dreams obviously are wish fulfilments – 'the sexual dreams of adults, for instance'. But it is strange that these are precisely the kind of dreams ignored by Freud:

> Freud very commonly gives what we might call a sexual interpretation. But it is interesting that among all the reports of dreams which he gives, there is not a single example of a straightforward sexual dream. Yet these are as common as rain.

This again is connected with Freud's determination to provide a *single* pattern for all dreams: all dreams must be, for him, expressions of longing, rather than, for example, expressions of fear. Freud, like philosophical theorists, had been seduced by the method of science and the 'craving for generality'. There is not one type of dream, and neither is there one way to interpret the symbols in a dream. Dream symbols do mean something – 'Obviously there are certain similarities with language' – but to understand them requires not some general theory of dreams, but the kind of multi-faceted skill that is involved, say, in the understanding of a piece of music.

In April Wittgenstein left Swansea to join Grant's research unit in Newcastle. The members of the unit, Basil Reeve, Dr Grant and Miss Helen Andrews, Grant's secretary, all lodged in the same house in Brandling Park, an area within walking distance of the hospital. The house belonged to a Mrs Moffat. Miss Andrews remembers the arrival of Wittgenstein:

> There was a vacant room at Mrs Moffat's so he joined us there. By this time we had settled in & adapted to our unusual surroundings, but Prof. W. did not easily fit in. He came down to breakfast in a bright & chatty mood, while we all shared a Manchester Guardian &

did not talk much. In the evening when we relaxed, he would not join us at dinner, but preferred to eat in his bedroom. Mrs Moffat, grumbling, put his meal on a tray & he came downstairs & fetched it. (I thought this was rude to Dr Grant.)

We had a sitting room with a good coal fire & he never spent an evening there with us. He went to a cinema almost every evening, but could not remember anything about the films when asked about them the following day. He just went to relax.

Not long after Wittgenstein's arrival, the members of the unit had to leave the house in Brandling Park, because of Mrs Moffat's ill health. They each found separate digs, but, recalls Miss Andrews, 'Prof. W. had difficulty in finding anywhere to live because as he had a foreign accent, looked a bit shabby & said he was a professor, most landladies were quite naturally suspicious.'

That Wittgenstein went every evening to see a film is an indication of how hard he worked at Newcastle, and how seriously he took the work. It is reminiscent of his remark to Drury:

You think philosophy is difficult enough but I can tell you it is nothing to the difficulty of being a good architect. When I was building the house for my sister in Vienna I was so completely exhausted at the end of the day that all I could do was go to a 'flick' every night.

Another indication of this is that while at Newcastle he wrote no philosophy at all, whereas while he was at Guy's he filled three notebooks with remarks on the philosophy of mathematics. He did not confine himself to fulfilling his duties as a technician, but took an intense and active interest in the thinking behind the research. Although both Grant and Reeve benefited from discussing their ideas with Wittgenstein, and encouraged his interest in their work, they sometimes found his absorption in the research a little too intense. Miss Andrews remembers that, because the unit worked so hard, Grant would sometimes suggest that they all take a day off and go for a walk together along Hadrian's Wall. She noticed that Wittgenstein was never invited to accompany them on these communal walks, and asked Grant why he was left out. She was told that if he came with them it would defeat the purpose of the walk, because he: 'talked shop all the time'.

Although he was not invited on these 'rest-day' walks, both Grant and Reeve remember accompanying Wittgenstein on many other occasions on walks along the Roman wall. Usually the conversation centred on their research, but with Reeve in particular Wittgenstein would often discuss more personal matters. He talked to Reeve, for example, about his early childhood, mentioning that he did not talk until he was four years old. He told Reeve a childhood memory that he had also related to Drury, and which obviously had a great significance for him. In the lavatory of his home, he said, some plaster had fallen from the wall, and he always saw this pattern as a duck, but it frightened him: it had the appearance for him of those monsters that Bosch painted in his *Temptations of St Anthony*.

Reeve would at times ask Wittgenstein about philosophy, but Wittgenstein characteristically discouraged his interest in the subject. He emphasized to Reeve that, unlike his own subject of medicine, philosophy was absolutely useless, and that unless you were compelled to do it, there was no point in pursuing it. 'You do *decent* work in medicine', he told Reeve; 'be content with that.' 'In any case', he would add mischievously, 'you're too stupid.' It is interesting, however, that forty years later Reeve should say that he had been influenced in his thinking by Wittgenstein in two important ways: first, to keep in mind that things are as they are; and secondly, to seek illuminating comparisons to get an understanding of how they are.

Both these ideas are central to Wittgenstein's later philosophy. Wittgenstein, indeed, thought of using Bishop Butler's phrase: 'Everything is what it is, and not another thing' as a motto for *Philosophical Investigations*. And the importance of illuminating comparisons not only lies at the heart of Wittgenstein's central notion of: 'the understanding which consists in seeing connections', but was also regarded by Wittgenstein as characterizing his whole contribution to philosophy. Wittgenstein's conversations with Reeve, like his work in helping Grant and Reeve to clarify their ideas about 'shock', show that there are more ways of having a philosophical influence than by discussing philosophy. Wittgenstein imparted a way of thinking and understanding, not by saying what was distinctive about it, but by showing how it can be used to clarify one's ideas.

Both Grant and Reeve recall that Wittgenstein's influence played an important part in the thinking embodied in the introduction to the final report of the unit, which, significantly, did not use the word 'shock' in its main title, but was called, rather, *Observations on the General Effects of Injury in Man*. The main thread of argument is the same as that in Grant's original memorandum of January 1941, but the 'diatribe against the word "Shock"' is expressed in even stronger terms:

> In practice we found that the diagnosis of shock seemed to depend on the personal views of the individual making it rather than on generally accepted criteria. Unless we were acquainted with these views we did not know what to expect when called to the bedside. The label alone did not indicate what signs and symptoms the patient displayed, how ill he was or what treatment he required. The only common ground for diagnosis that we could detect was that the patient seemed *ill*. We were led, therefore, to discard the word 'shock' in its varying definitions. We have not since found it to be of any value in the study of injury; it has rather been a hindrance to unbiased observation and a cause of misunderstanding.

Whether this was written by Wittgenstein or not, it had an effect that he hoped his philosophical work would have – namely, it put an end to many misguided lines of research. The Report of the Medical Research Council for 1939–45 says of the work done under Grant:

> [It] threw grave doubt upon the value of attacking the 'shock' problem as if wound 'shock' were a single clinical and pathological entity. In consequence, several lines of investigation started for the Committee at the beginning of the war were abandoned.

It had the effect, in fact, that Wittgenstein had hoped for in his later work on the philosophy of mathematics – the effect that sunlight has on the growth of potato shoots.

The purpose of the research conducted by Grant and Reeve was not primarily to campaign against the use of the word 'shock' in diagnosing the effects of injury, but to uncover other, more fruitful, diagnoses and treatments than those derived from the research done in the First World War. For this they needed detailed observations of the effects of injuries. Wittgenstein's part in this practical aspect of the work was to cut frozen sections of tissue and stain them in order to detect the presence of, for example, fat. He apparently did this very well.

As well as this histological work, Wittgenstein was asked by Grant to assist in his investigations into Pulsus Paradoxus, a pulse pressure that varies with respiration, which often occurred in badly injured patients. In this he seems to have introduced a technological innovation by inventing a better apparatus for recording pulse pressure than the one they had. Both Grant and Reeve remember that this apparatus was innovatory, but neither could remember the details of it. The only description we have of this apparatus, therefore, is the one Drury gives when he describes the time he used his leave from the army to visit Wittgenstein in Newcastle:

> After the end of the campaign in North Africa I was posted back to England to prepare for the Normandy landing. Having a period of disembarkation leave, I travelled up to Newcastle to spend a few days with Wittgenstein . . . He took me to his room in the Research Department and showed me the apparatus which he himself had designed for his investigation. Dr Grant had asked him to investigate the relationship between breathing (depth and rate) and pulse (volume and rate). Wittgenstein had so arranged things that he could act as his own subject and obtain the necessary tracings on a revolving drum. He had made several improvements in the original apparatus, so much so that Dr Grant had said he wished Wittgenstein had been a physiologist and not a philosopher. In describing to me his results so far he made a characteristic remark: 'It is all very much more complicated than you would imagine at first sight.'

Drury's visit to Newcastle also provides us with a revealing conversation, showing an interesting change in Wittgenstein's attitude to sex. By 1943, it seems, far from accepting Weininger's view that sex and spirituality were incompatible, Wittgenstein was sympathetic to a view of the sexual act which saw it as an object of religious reverence. Drury relates that while he was staying in Newcastle with Wittgenstein, the two of them caught a train to Durham and took a walk by the river there. As they walked Drury talked to Wittgenstein about his experiences in Egypt, and especially about seeing the temples at Luxor. He told Wittgenstein that, although seeing the temples was a wonderful experience, he had been surprised and shocked to find on the wall of one of the temples a bas-relief of the god Horus, with an erect phallus, in the act of ejaculation and collecting the semen in a bowl. Wittgenstein responded to this story with an encouraging rejection of Drury's implied disapproval:

> Why in the world shouldn't they have regarded with awe and reverence that act by which the human race is perpetuated? Not every religion has to have St Augustine's attitude to sex.

When Wittgenstein had first moved to Newcastle, he had responded with an even more dismissive frankness to another remark of Drury's. Drury had written to him wishing him luck in his new work, and added that he hoped Wittgenstein would make lots of friends. Wittgenstein replied:

> It is obvious to me that you are becoming thoughtless and stupid. How could you imagine I would ever have 'lots of friends'?

Though harshly expressed, this was doubtless true. Wittgenstein's only friend at Newcastle seems to have been Basil Reeve. He got on well with Grant, and they shared an interest in music (Grant can remember Wittgenstein's enthusiastic agreement when he once mentioned that he disliked the opening of Beethoven's 'Emperor' Concerto), but there was little of the warmth of feeling – the simple human contact – that Wittgenstein had shared at Guy's with Roy Fouracre. Grant was too absorbed in his own work for that. Wittgenstein had complained to Fouracre about the lack of human contact his philosophical work at Cambridge provided, but at Newcastle, as his letters to Norman Malcolm show, he began to miss his Cambridge friends:

> I haven't heard from Smythies for many months. I know he is in Oxford but he doesn't write to me. – [Casimir] Lewy is still at Cambridge . . . Rhees is still lecturing at Swansea . . . I hope you'll see Moore & find him in good health. [11.9.43]

> I am feeling rather lonely here & may try to get to some place where I have someone to talk to. E.g. to Swansea where Rhees is a lecturer in philosophy. [7.12.43]

More important, perhaps, he began to miss the opportunity to pursue his own work. Absorption in Grant and Reeve's work was no longer enough:

> I too regret that for external & internal reasons I can't do philosophy, for that's the only work that's given me real satisfaction. No other work really bucks me up. I'm extremely busy now & my mind is kept occupied the whole time but at the end of the day I just feel tired & sad.

The 'internal reasons' were Wittgenstein's doubts about whether he was still capable of good work in philosophy. He used to say to Reeve, time and again: 'my brain has gone', and would often talk about his days in Norway in 1913 with a wistful longing: '*Then* my mind was on fire . . . but now it's gone.' The 'external reasons' were the demands of the job that he was doing in Newcastle, and rather unsatisfactory lodgings. He was also finding continued exposure to hospital registrars and house officers, with their outspoken and often ribald comments about their patients, more and more difficult to accept, and he required more and more help from Reeve in understanding the responses of young doctors to the stresses of their profession.

Perhaps as a consequence of these frustrations, together with the extra limitations on Reeve's time caused by the arrival in Newcastle of his wife and young baby, even Wittgenstein's relationship with Reeve began to deteriorate. Wittgenstein – ever a possessive friend – began to demand more of Reeve's time and attention while the demands of Reeve's work and home life limited the amount he could give. They eventually parted on bad terms. Wittgenstein's words of goodbye to Reeve were: 'You're not such a nice person as I had thought.' For his part, Reeve felt relieved that he would not have to go on giving Wittgenstein the emotional support he demanded.

Given Wittgenstein's frustrated yearning to return to philosophical work, and his worsening relationship with Reeve, it was probably a relief for him to hear that Grant and Reeve were leaving Newcastle. Their research had begun to emphasize the need for further study of the effects of blood loss and tissue damage, and this necessitated access to injuries of a more severe degree than were available in civil life. They therefore needed to conduct their research on the battlefield, and so, towards the end of 1943, arrangements were made to transfer them to Italy.

Grant's successor was Dr E. G. Bywaters, who, like Grant and Reeve, had previously carried out observations on air-raid casualties in London. Before he left, Grant wrote to Dr A. Landsborough Thomson at the Medical Research Council Head Office in London, telling him:

> Wittgenstein has agreed to continue as laboratory assistant in the meantime, but the duration of his stay will depend on how he gets on with Bywaters.

He urged him to keep Wittgenstein on as an employee of the Council, even if he decided to leave Newcastle:

> Wittgenstein has taken up laboratory work as a contribution to the war effort, as I told you he is Professor of Philosophy at Cambridge. If he decides that he cannot continue here after Reeve and I go, it seems a pity not to use him further . . . He has a first class brain and a surprising knowledge of physiology. He is an excellent man to discuss problems with. On the practical side, he has served us well as a laboratory assistant and in addition has by himself made new observations on man on the respiratory variations of blood pressure devising his own apparatus and experiments. He is not an easy man to deal with, but given suitable conditions, can be a helpful and stimulating colleague. I presume that after the war, he will return to his Chair at Cambridge.

Grant and Reeve finally left for Italy at the end of January 1944. Wittgenstein was, as we have seen, feeling lonely and depressed before Bywaters arrived, and although he continued to fulfil his duties as a technician with care and attention, he did not care to socialize. Bywaters remembers:

> He was reserved & rather withdrawn: when in conversation at coffee or teatime, philosophical subjects came up, he refused to be drawn. I was disappointed in this, but pleased with his meticulous & conscientious approach to the frozen sections of lung & other organs that he prepared for me. I remember him as an enigmatic,

> non-communicating, perhaps rather depressed person who preferred the deck chair in his room to any social encounters.

Bywaters had been there only three weeks when he had to write to his head office asking for help in finding a new technician:

> Professor Wittgenstein has been doing the histological work here for Dr Grant . . . He has now received a letter from Cambridge requesting him to spend the next three months or longer writing a treatise on his own subject (philosophy).

A week later, on 16 February, he wrote:

> Professor Wittgenstein left us today: he has been called back to his Cambridge Chair to write a treatise on Philosophy, which has been in the air for the last year or so, but they now want it on paper.

So, on 16 February 1944, Wittgenstein left Newcastle and returned to Cambridge. The reference in Bywaters's letter to a treatise which has 'been in the air for the last year or so' but which 'they now want on paper' makes it plausible to assume that 'they' refers not to Cambridge University, but to Cambridge University Press.

In September 1943 Wittgenstein had approached the Press with a suggestion that they publish his new book, *Philosophical Investigations*, alongside his old book, the *Tractatus*. This idea had occurred to him earlier in the year, when he and Nicholas Bachtin had been reading the *Tractatus* together. He also mentioned the idea to Reeve, saying that he liked the idea of publishing a refutation of ideas in the *Tractatus* alongside the *Tractatus*. Cambridge University Press confirmed their acceptance of this offer on 14 January 1944, which ties in with Bywaters's first letter saying: 'He has now received a letter from Cambridge . . .' This plan, however, like the previous plan accepted by the University Press in 1938, was never carried out.

22
SWANSEA

Despite Bywaters's impression that Wittgenstein had had to leave the research unit at Newcastle because he had been 'called back to his Cambridge Chair', Wittgenstein was determined to avoid going back to Cambridge if at all possible. He wanted to finish his book before he resumed his duties as a professor, and for that purpose he considered Swansea a much better place. The thought of going to Swansea had occurred to him the previous December, after he had been told that Grant and Reeve would have to leave Newcastle in the new year. He wanted, as he had said in a letter to Malcolm, to be with someone with whom he could discuss philosophy, and Rhees was the obvious choice. 'I don't know if you remember Rhees', he wrote. 'You saw him in my lectures I believe. He was a pupil of Moore's & is an excellent man & has a real talent for philosophy, too.'

A week before he left Newcastle, however, it suddenly occurred to him that he might not be able to spend a long period working in Swansea. As he explained to Rhees, he was on leave of absence from his duties as a professor because he was doing 'important' war work:

> If, e.g., I leave here & try to find another job, say in a hospital, I have to let the general Board know & they have to approve the new job. Now if I come to Cambridge next week they'll want to know what I'm doing & I'll have to tell them that I want to do some philosophy for a couple of months. And in this case they can say: if you want to do philosophy you're not doing a war job & have to do your philosophy in Cambridge.

> ... I'm almost sure that I couldn't work at Cambridge now! I hope I'll be able to come to Swansea.

Wittgenstein's fears proved to be unfounded, and after spending a few weeks in Cambridge he was granted leave of absence to go to Swansea to work on his book. He left Cambridge in March 1944, and did not have to return until the following autumn.

The prospect of daily discussions with Rhees was not the only attraction of Swansea. Wittgenstein loved the Welsh coastline, and, perhaps more important, found the people in Swansea more congenial than those in Cambridge. 'The weather's foul', he was to tell Malcolm in 1945, 'but I enjoy not being in Cambridge':

> I know quite a number of people here whom I like. I seem to find it more easy to get along with them here than in England. I feel much more often like smiling, e.g. when I walk in the street, or when I see children, etc.

Through an advertisement in a newspaper, Rhees found lodgings for him in the home of a Mrs Mann, who lived by the coast at Langland Bay. So ideal was this position that when Mrs Mann wrote to Wittgenstein explaining that she had changed her mind and could not, after all, take him in as a lodger, he refused to accept it and insisted on moving in regardless. He stayed with her throughout the spring of 1944, and she indeed proved to be a good landlady, taking good care of him throughout his periodic bouts of ill health.

Soon after he moved in with Mrs Mann he entered into a correspondence with Rowland Hutt which illustrates something of what Fania Pascal may have had in mind when she wrote that, if you had committed a murder, or if you were about to change your faith, Wittgenstein would be the best man to consult, but that for more ordinary anxieties and fears he could be dangerous: 'his remedies would be all too drastic, surgical. He would treat you for original sin.'

Hutt was at this time serving in the Royal Army Medical Corps and was dissatisfied with his position. He hoped to be given a commission and to be able to work in a laboratory or an operating theatre. Deeply depressed, he wrote to Wittgenstein complaining of his situation. Though he always encouraged any desire to take on a medical position, Wittgenstein treated Hutt's problem as one of the soul rather than one concerning his career. 'I hadn't a good impression from your letter', he wrote to Hutt on 17 March. 'Though it's very difficult for me to say what's wrong':

> I feel as though you were getting more and more slovenly. I'm not *blaming* you for it, & I'ld have no right to do so. But I've been thinking what's to be done about it. Seeing a psychologist – unless he is a very extraordinary *man* – won't get you far.

He was, in any case, inclined to doubt whether Hutt would be particularly good in an operating theatre: 'You've got to be rather quick there & resourceful, & I just don't know if you are.' But: 'One thing seems to me fairly clear: that you mustn't go on in a humiliating or demoralising position.' The central problem, as Wittgenstein saw it, was that of preserving Hutt's self-respect. If he couldn't get a commission, and if he was not prepared to make a good job of whatever he was doing, then he should apply to be sent, in whatever capacity he could, to a unit near the Front. There, Wittgenstein told him, 'you'll at least live some sort of *life*':

> I have extremely little courage myself, much less than you; but I have found that whenever, after a long struggle, I have screwed my courage up to do something I always felt *much* freer & happier after it.

'I know that you have a family', he wrote, anticipating the most obvious objection to the wisdom of this advice, 'but you won't be any use to your family if you're no use to yourself.' And if Lotte, Hutt's wife, did not see this now, 'she'll see it one day'.

This piece of advice, like others that Wittgenstein gave friends during the Second World War, is quite obviously based on his own experience of the Great War. For example, before he left to embark for 'D-Day' Maurice Drury came to Swansea to say goodbye to Wittgenstein, who left him with these words:

> If it ever happens that you get mixed up in hand-to-hand fighting, you must just stand aside and let yourself be massacred.

'I felt', writes Drury, 'that this advice was one that he had had to give himself in the previous war.' Similarly, when Norman Malcolm enlisted in the US Navy he was sent by Wittgenstein a 'foul copy' (presumably second-hand and a bit scruffy) of Gottfried Keller's novel *Hadlaub*. The advantage of its not being in pristine condition, Wittgenstein wrote, 'is that you can read it down in the engine-room without making it *more* dirty'. He obviously pictured Malcolm working in some manual capacity on a steam vessel similar to the *Goplana*. It is as though the war gave him the opportunity of vicariously reliving, through his younger friends, the intense and transforming events of 1914–18.

If he had been in Hutt's position, one feels, he would have had no hesitation in applying to be sent to the Front – just as he had in 1915.

But his advice to Hutt was also based on a more general attitude. 'I think', he told him, 'you must stop creeping & start *walking* again':

> When I talked of courage, by the way, I didn't mean, make a row with your superiors; particularly not when it's entirely useless & just shooting off your mouth. I meant: take a burden & try to *carry* it. I know that I've not any right to say this. I'm not much good at carrying burdens myself. But still it's all I've got to say; until perhaps I can see you.

Hutt wrote back showing no inclination to accept the advice and reporting that he had recently seen a psychologist. 'I wish I knew more about military matters', Wittgenstein replied with impatient irony:

> I can't understand, e.g., what a psychologist has to do with your medical category in the Army. Surely there's nothing wrong with you *mentally*! (Or if there is the psychologist won't know it.)

He repeated the outlines of his previous advice. If he couldn't get a commission, then the only thing for Hutt to do was: 'To do the job *which you've got really well*; so well that you don't lose your self-respect doing it':

> I don't know if you understand me. It is an intelligent thing to do to use whatever means one has to get a better, or more suitable job. *But* if those means fail then there comes a moment when it no longer

> makes sense always to complain & to kick but when you've got to *settle down*. You're like a man who moves into a room & says 'Oh this is only temporary' & doesn't unpack his trunks. Now this is all right – for a time. But if he *can't* find a better place, or *can't* make up his mind to risk moving perhaps into another town altogether, then the thing to do is to unpack his trunks & settle down whether the room's good or not. For *anything's* better than to live in a state of waiting.

'This war *will* end', he insisted, '& the most important thing is what sort of person *you'll* be when it's over. That's to say: when it's over you ought to be a *man*. And you won't be if you don't train yourself now':

> The first thing to do is: stop kicking when it's fruitless. It seems to me you ought either apply to be posted somewhere nearer the front & just *risk* it, or, if you don't want to do that, *sit down* where you are, don't think of moving but only of doing that work well which you've got *now*.

'I'll be quite frank with you', he added, with another suggestion that indicates he was perhaps projecting his own history on to Hutt's situation, '& say that I think it *might* be better for you not to be within reach of your family':

> Your family is, of course, soothing, but it may also have a softening effect. And for certain sores you want to make the skin *harder*, not softer. I mean, I have an idea (perhaps it's as wrong as hell) that your family makes it more difficult, or impossible, for you to settle down & get down to your work without looking right or left. Also you should perhaps look a little more inside you, & this also is perhaps impossible with your family around. In case you show this letter to Lotte & she wildly disagrees with me I'll say this: maybe she wouldn't be a good wife if she didn't disagree, but that doesn't mean that what I tell you mayn't be *true*!

Still hankering after a commission, Hutt wrote telling Wittgenstein he had seen his commander and might do so again soon. 'It seems to me', Wittgenstein replied, 'that you're alternatively living in unfounded hopes & in despair . . . To pester your Commander about a commission now seems to me silly. *Nothing* has changed since you've been refused!'

> You write: 'All these steps have made or will make my position here satisfactory, or at least a lot better.' This is all nonsense, & it makes me really sick to read it. The one step which can make things more satisfactory for you is a step that has to be taken inside you. (Though I won't say that getting away from your family mightn't help.)

The exchange on this subject ended in June, with Wittgenstein having the last word. 'I wish you *luck* and *patience*!' he wrote in conclusion, '& no more dealings with psychologists.'

By this time, Wittgenstein had moved out of Mrs Mann's house and into the home of a Methodist minister, the Reverend Wynford Morgan. On his first visit to the house Mrs Morgan, acting the solicitous hostess, asked him whether he would like some tea, and also whether he would like this and that other thing. Her husband called out to her from another room: 'Do not ask; *give*.' It was a remark that impressed Wittgenstein enormously, and he repeated it to his friends on a number of occasions.

But in other respects Wittgenstein was less favourably impressed with his host. He made fun of him for having his walls lined with books that he never read, accusing him of having them there simply to impress his flock. When Morgan asked Wittgenstein whether he believed in God, he replied: 'Yes I do, but the difference between what you believe and what I believe may be infinite.'

This remark does not, of course, refer to the difference between Methodism and other forms of Christianity. Wittgenstein was no more a Catholic than he was a Methodist. Of his friends who had converted to Catholicism, he once remarked: 'I could not possibly bring myself to believe all the things that they believe.' One of these was Yorick Smythies, who wrote to Wittgenstein announcing his conversion at the time that Wittgenstein was lodging with Reverend Morgan. Wittgenstein was very concerned, the more so as he thought he might have unwittingly been partly responsible for the conversion by encouraging Smythies to read Kierkegaard. His reply to Smythies was oblique: 'If someone tells me he has bought the outfit of a tightrope walker I am not impressed until I see what is done with it.'

The point of this analogy is clarified in one of his notebooks:

> An honest religious thinker is like a tightrope walker. He almost looks as though he were walking on nothing but air. His support is the slenderest imaginable. And yet it really is possible to walk on it.

Though he had the greatest admiration for those who could achieve this balancing act, Wittgenstein did not regard himself as one of them. He could not, for example, bring himself to believe in the literal truth of reported miracles:

> A miracle is, as it were, a gesture which God makes. As a man sits quietly and then makes an impressive gesture, God lets the world run on smoothly and then accompanies the words of a saint by a symbolic occurrence, a gesture of nature. It would be an instance if, when a saint has spoken, the trees around him bowed, as if in reverence. – Now, do I believe this happens? I don't.
>
> The only way for me to believe in a miracle in this sense would be to be impressed by an occurrence in this particular way. So that I should say e.g.: 'It was impossible to see these trees and not to feel that they were responding to the words.' Just as I might say 'It is impossible to see the face of this dog and not to see that he is alert and full of attention to what his master is doing.' And I can imagine that the mere report of the words and life of a saint can make someone believe the reports that the tree bowed. But I am not so impressed.

The belief in God which he acknowledged to Morgan did not take the form of subscribing to the truth of any particular doctrine, but rather that of adopting a religious attitude to life. As he once put it to Drury: 'I am not a religious man but I cannot help seeing every problem from a religious point of view.'

Next door to Morgan lived the Clement family, with whom Wittgenstein quickly made friends – a good example of his remark to Malcolm that he found it easier to get along with people in Swansea than in England. He particularly liked Mrs Clement, who invited him to have Sunday lunch with her family every week. 'Isn't she an angel?' he said of her to her husband one Sunday lunch-time. 'Is she?' replied Mr Clement. 'Damn it all, man! of course she is!' roared Wittgenstein.

In fact, he was so impressed by Mrs Clement that he wished to lodge in her house rather than with Morgan. The Clements had not until then taken lodgers, and had no great desire to do so, but, on being pressed, they assented to Wittgenstein's moving in. Wittgenstein's association with the Clements lasted for the next three years, and during his last years at Cambridge he would spend his vacations as a guest in their home.

The Clements had two daughters, Joan, aged eleven, and Barbara, aged nine, and while he was staying there Wittgenstein was treated almost as part of the family. Finding the name 'Wittgenstein' something of a mouthful, they all called him 'Vicky', although it was made clear that they were the only people allowed to do so. Unusually, Wittgenstein took his meals with the family while he was staying with the Clements. He also joined in with other aspects of family life. In particular, he enjoyed playing *Ludo* and *Snakes and Ladders* with the girls – so much so that on one occasion, when a game of *Snakes and Ladders* in which Wittgenstein was particularly absorbed had gone on for over two hours, the girls had to plead with him against his will to leave the game unresolved.

He took an active interest, too, in the education of the two girls. The elder, Joan, was at the time taking her scholarship examinations for the local grammar school. On the day that the results were announced, Wittgenstein came home to find her in tears. She had been told that she had failed. Wittgenstein was emphatic that this could not be so. 'Damn it all!' he said. 'We'll see about that!' With Joan and her mother following anxiously, Wittgenstein marched down to Joan's school to confront the teacher who had told her that she had failed. 'I am stunned that you say she failed', he told the teacher, 'and I can tell you on authority that she *must* have passed.' The somewhat intimidated teacher checked the records and discovered, much to everybody's relief, that there had indeed been a mistake and that Joan had gained enough marks to pass the examination. The teacher was denounced as an 'incompetent fool' by Wittgenstein, but although both his judgement and Joan's ability had been vindicated, Mrs Clement was ashamed to show her face again at the school.

Apart from his self-imposed family responsibilities and his almost daily walks with Rhees, Wittgenstein's time at Swansea was taken up largely with writing. He had taken with him the typescript of the 1938 version of the *Investigations*, and the notebooks and hard-back ledger volumes that he had written while he had been working at Guy's, and he set to work on a revision of the book which he hoped would be ready for the publishers by the time he had to return to Cambridge the following autumn.

For the first two months that he was in Swansea, the focus of his work was on the philosophy of mathematics. He resumed work on the notebook he had kept at Guy's and to which he had given the title 'Mathematics and Logic'. His chief preoccupation in this notebook is with the notion of following a rule. Part 1 of the 1938 version had ended with remarks concerning the confusions associated with this notion, and Part 2 began with an attempt to unravel these confusions as a preliminary to a discussion of issues in the philosophy of mathematics. In the reworked version of the *Investigations* which was published after his death, however, the discussion of following a rule is used instead as a preliminary to a discussion of issues in the philosophy of psychology. This change was effected in Swansea in the spring and summer months of 1944.

How quickly and radically Wittgenstein's interests shifted while he was at Swansea is illustrated by two incidents that are separated only by a few months. The first occurred soon after he had moved there, and is connected with a short biographical paragraph that John Wisdom was writing of Wittgenstein for inclusion in a biographical dictionary. Before publication, Wisdom sent the piece to Wittgenstein for his comments. Wittgenstein made only one change; he added a final sentence to the paragraph, which read: 'Wittgenstein's chief contribution has been in the philosophy of mathematics.' Two or three months later, when Wittgenstein was at work on the set of remarks which have since become known as the 'Private Language Argument', Rhees asked him: 'What about your work on mathematics?' Wittgenstein answered with a wave of his hand: 'Oh, someone else can do that.'

Of course, switching from the philosophy of mathematics to the philosophy of psychology and back again, using the problems in one area as analogies to illustrate points in the other, was something that Wittgenstein had done in his lectures, notebooks and conversations since the early 1930s. Neither was his interest in combating the idea that a private language is possible new in 1944: he had discussed it in his lectures as early as 1932. What is significant about the shift in 1944 is that it was permanent: Wittgenstein never again returned to the attempt to arrange his remarks on mathematics in a publishable form, and he spent the rest of his life arranging, rearranging and revising his thoughts on the philosophy of psychology. Moreover, this seemingly permanent shift came at a time when he seemed most anxious to complete the part of his book dedicated to the philosophy of mathematics.

The clue to this shift, I think, lies in Wittgenstein's changing conception of his book, and in particular in his recognition that his remarks on following a rule should serve, not as a preliminary to his discussion of mathematics, but rather as an overture to his investigations of both mathematical and psychological concepts. Despite his remark to Rhees that 'someone else can do that', and despite the fact that he never returned to his work on mathematics, Wittgenstein continued to regard his remarks on mathematics as belonging to his *Philosophical Investigations*. Thus the preface to the book, written in 1945, still lists 'the foundations of mathematics' as one of the subjects about which the book is concerned, and as late as 1949 he wrote in one of his notebooks:

> I want to call the enquiries into mathematics that belong to my Philosophical Investigations 'Beginnings of Mathematics'.

So the change should be considered first and foremost as a change in Wittgenstein's conception of the force of his remarks on following a rule. They now led, not in one direction, but in two, and after recognizing this fact Wittgenstein was more inclined to follow the line that led to the investigation of psychological concepts. Although he did not live long enough to retrace his steps and continue along the other branch of the forked road, he did not give up his idea that it was there to be followed. Thus the final remark of the *Investigations* – 'An investigation is possible in connexion with mathematics which is entirely analogous to our investigation of psychology' – ties in with his remark to Rhees. Although *he* had not drawn out all the implications of the first part of his book, it was still possible that it could be done by someone else.

In conversation with Rhees Wittgenstein once remarked that he could feel really active only when he changed his philosophical position and went on to develop something new. He gave as an example of this something that he considered to be an important change in his philosophical logic, concerning his view of the relation between 'grammatical' and 'material' propositions. Previously, he said, he had regarded this distinction to be fixed. But now he thought that the boundary between the two was fluid and susceptible to change. In truth, this appears as a change of emphasis rather than of opinion, for even in the 1938 version of the *Investigations* he had not treated the distinction as fixed. But neither had he particularly emphasized its fluidity. And it is this emphasis that dictated the course of his work in the summer of 1944.

The distinction between the two types of proposition lies at the heart of Wittgenstein's entire philosophy: in his thinking about psychology, mathematics, aesthetics and even religion, his central criticism of those with whom he disagrees is that they have confused a grammatical proposition with a material one, and have presented as a discovery something that should properly be seen as a grammatical (in Wittgenstein's rather odd sense of the word) innovation.

Thus, in his view, Freud did not discover the unconscious; rather, he introduced terms like 'unconscious thoughts' and 'unconscious motives' into our grammar of psychological description. Similarly, Georg Cantor did not discover the existence of an infinite number of infinite sets; he introduced a new meaning of the word 'infinite' such that it now makes sense to talk of a hierarchy of different infinities. The question to ask of such innovations is not whether these 'newly discovered' entities exist or not, but whether the additions they have made to our vocabulary and the changes they have introduced to our grammar are useful or not. (Wittgenstein's own view was that Freud's were and that Cantor's were not.)

Wittgenstein had many ways of characterizing grammatical propositions – 'self-evident propositions', 'concept-forming propositions', etc. – but one of the most important was in describing them as *rules*. In emphasizing the fluidity of the grammatical/material distinction, he was drawing attention to the fact that concept-formation – and thus the establishing of rules for what it does and does not make sense to say – is not something fixed by immutable laws of logical form (as he had thought in the *Tractatus*) but is something that is always linked with a custom, a practice. Thus, different customs or practices would presuppose different concepts from the ones *we* find useful. And this in turn would involve the acceptance of different rules (to determine what does and does not make sense) to the ones we, in fact, have adopted.

The concern with grammatical propositions was central to Wittgenstein's philosophy of mathematics because he wanted to show that the 'inexorability' of mathematics does not consist in certain knowledge of mathematical truths, but rather in the fact that mathematical propositions are grammatical. The certainty of '2 + 2 = 4' consists in the fact that we do not use it as a description but as a rule.

In his last writings on the philosophy of mathematics – as in his conversation with Rhees – Wittgenstein was increasingly concerned with the connection between rule-following and customs:

> The application of the concept of 'following a rule' presupposes a custom. Hence it would be nonsense to say: just once in the world someone followed a rule (or a signpost; played a game, uttered a sentence, or understood one; and so on).

This is such a general point that in the notebook from which the remark comes – written in 1944 – it is not obvious that Wittgenstein has mathematics in mind at all. And the connection between this point and Wittgenstein's argument against the possibility of a private language is obvious:

> I may give a new rule today, which has never been applied, and yet is understood. But would it be possible, if no rule had ever actually been applied?
>
> And if it is now said: 'Isn't it enough for there to be an imaginary application?' the answer is: No.

In this way it seemed perfectly natural to restructure the book so that the section on following a rule led not into the philosophy of mathematics but into the argument against the possibility of a private language. In the work that he did during this summer Wittgenstein extended Part I of the 1938 version of *Investigations* to about double its previous length, adding to it what are now considered to be the central parts of the book: the section on rule-following (paragraphs 189–242 of the published version), and the section on the 'privacy of experience' (the so-called 'Private Language Argument' in paragraphs 243–421).

In August he began to make what looks like an attempt at a final arrangement of the book, which he intended to have finished before he left Swansea in the autumn. Then, he told Hutt, 'I'll probably take on a war-job again.' In a later letter, of 3 September, he writes: 'What I'll do when I have to leave at the beginning of Oct. I don't yet know & I hope events will take the decision out of my hands.' With the Allies making swift progress through France, and the Russians advancing on Poland, it was by now clear that the war would end soon with a defeat for Germany. In this Wittgenstein saw no grounds for rejoicing. 'I'm pretty sure', he told Hutt, 'that the peace after this war will be more horrible than the war itself.'

Whether because he could not find a suitable war job, or because his leave of absence could not be extended, when the time came to leave Swansea, Wittgenstein was forced to return to Cambridge. He did so grudgingly, not least because his book remained unfinished. Before he left Swansea he had a typescript prepared of the parts he regarded as publishable. (This more or less corresponds to the final version up to paragraph 421.) Having given up hope of arranging to his satisfaction the section of the book he had previously considered the most important (that on the philosophy of mathematics), his one remaining hope was to complete his 'first volume': his analysis of psychological concepts.

23
THE DARKNESS OF THIS TIME

In October 1944 Wittgenstein returned to Cambridge, frustrated at not having finished his book and not at all enthusiastic at the prospect of resuming his lecturing responsibilities.

Russell was also back in Cambridge, having spent the last six years living and working in America. There his life had become unbearable, owing to the hysteria and outrage whipped up against him by the more conservative elements in American society in response to his widely-publicized views on marriage, morals and religion, and he had gratefully accepted the invitation to a five-year lectureship in the quieter and calmer environment of Trinity College. He arrived, however, to find himself out of fashion with English academic philosophers, among whom Moore and Wittgenstein were now far more influential than Russell himself. He brought back with him the manuscript of his *History of Western Philosophy*, which, although it enjoyed a huge commercial success (it was for many years the main source of Russell's income), did not improve his reputation as a philosopher.

Although he maintained his admiration for the keenness of Russell's intellect, Wittgenstein detested the popular work that Russell had published since the 1920s. 'Russell's books should be bound in two colours', he once said to Drury:

> . . . those dealing with mathematical logic in red – and all students of philosophy should read them; those dealing with ethics and politics in blue – and no one should be allowed to read them.

Russell, Wittgenstein thought, had achieved all he was ever likely to achieve. 'Russell isn't going to kill himself doing philosophy now', he told Malcolm, with a smile. And yet Malcolm recalls that, during the 1940s, on the rare occasions that Russell and Wittgenstein both attended the Moral Science Club, 'Wittgenstein was deferential to Russell in the discussion as I never knew him to be with anyone else.'

Russell, for his part, could see no merit whatever in Wittgenstein's later work. 'The earlier Wittgenstein', he said, 'was a man addicted to passionately intense thinking, profoundly aware of difficult problems of which I, like him, felt the importance, and possessed (or at least so I thought) of true philosophical genius':

> The later Wittgenstein, on the contrary, seems to have grown tired of serious thinking and to have invented a doctrine which would make such an activity unnecessary.

Not surprisingly, therefore, when the two remet in the autumn of 1944 (after a break of about fourteen years), there was little warmth between them. 'I've seen Russell', Wittgenstein wrote to Rhees after being back about a week; he 'somehow gave me a *bad* impression'. And after that he had little or nothing to do with his former teacher.

Russell's contempt for Wittgenstein's later work was undoubtedly heightened by (but not entirely attributable to) his personal pique at being left philosophically isolated. The philosophical problems with which he was chiefly concerned were no longer regarded as fundamental. Partly under Wittgenstein's influence, the Theory of Knowledge had been subordinated to the analysis of meaning. Thus, when *Human Knowledge: Its Scope and Limits* – a work which Russell conceived as a major statement of his philosophical position – was published in 1948, it was greeted with cool indifference. Russell's greatest contempt, therefore, was reserved for Wittgenstein's disciples:

> It is not an altogether pleasant experience to find oneself regarded as antiquated after having been, for a time, in the fashion. It is difficult to accept this experience gracefully. When Leibniz, in old age, heard the praises of Berkeley, he remarked: 'The young man in Ireland who disputes the reality of bodies seems neither to explain himself sufficiently nor to produce adequate arguments. I suspect him of

> wishing to be known for his paradoxes.' I could not say quite the same of Wittgenstein, by whom I was superseded in the opinion of many British philosophers. It was not by paradoxes that he wished to be known, but by a suave evasion of paradoxes. He was a very singular man, and I doubt whether his disciples knew what manner of man he was.

Moore had not suffered in the same way, but although relations between him and Wittgenstein were still friendly, by 1944 he was too old and infirm wholeheartedly to welcome the arduous prospect of frequent and prolonged philosophical discussions with Wittgenstein. His wife therefore limited Wittgenstein's visits to one and a half hours, much to Wittgenstein's displeasure. 'Moore is as nice as always', he told Rhees:

> I couldn't see him for long as we were interrupted by Mrs Moore. She told me later that Moore wasn't really as well as he seemed & that he mustn't have long conversations. I have good reason for believing that this, on the whole, is baloney. Moore has queer blackouts at times but then he's an oldish man. For his age he is obviously fit. Mrs Moore, however, doesn't like the idea of his seeing me. Perhaps she's afraid that I might criticise the book which was written about him &, generally, have a bad effect on his morale.

The book Wittgenstein mentions is *The Philosophy of G. E. Moore*, a collection of articles by a number of distinguished philosophers on various aspects of Moore's philosophy, edited by P. A. Schilpp and published in 1942. Moore agreed to the book's production, and wrote a short autobiographical piece specially for it. Wittgenstein strongly disapproved. 'I fear', he wrote to Moore after hearing about the book, 'that you may now be walking at the edge of that cliff at the bottom of which I see lots of scientists and philosophers lying dead, Russell amongst others.' When the book came out Moore was in the USA, and their meeting in the autumn of 1944 was thus the first opportunity Wittgenstein would have had since the book's publication to renew his criticism of it. Dorothy Moore's anxieties, therefore, were probably entirely justified.

In fact, Wittgenstein was wrong to blame her solely for the time regulation imposed on his meetings with Moore. Moore had suffered a stroke while in America, and his wife was acting on instructions from his doctor to forbid any kind of excitement or fatigue. She therefore limited his discussions with all his philosophical friends to one and a half hours. Wittgenstein, she says, was the only one of them who resented this: 'He did not realise how exhausting he could be, so much so that at least on one occasion Moore said to me beforehand "Don't let him stay too long."'

Wittgenstein, however, continued in his belief that Moore was being forced by his wife to cut short his conversations with him. Two years later he told Malcolm that he considered it unseemly that Moore, 'with his great love for truth', should be forced to break off a discussion before it had reached its proper end. He should discuss as long as he liked, and if he became very excited or tired and had a stroke and died – well, that would be a decent way to die: 'with his boots on'.

Nothing should come between a philosopher and his search for the truth. 'Thinking is sometimes easy, often difficult but at the same time thrilling', he wrote to Rhees:

> But when it's most important it's just disagreeable, that is when it threatens to rob one of one's pet notions & to leave one all bewildered & with a feeling of worthlessness. In these cases I & others shrink from thinking or can only get ourselves to think after a long sort of struggle. I believe that you too know this situation & I wish you *lots of courage*! though I haven't got it myself. We are all *sick* people.

His mind turned to the argument he had had with Malcolm at the beginning of the war, when Malcolm had talked about the British 'national character'. This was a case in point – an example of when, precisely because it is disagreeable, thinking clearly is most important. 'I then thought', he wrote to Malcolm:

> . . . what is the use of studying philosophy if all that it does for you is enable you to talk with some plausibility about some abstruse questions of logic, etc., & if it does not improve your thinking about the important questions of everyday life, if it does not make you more conscientious than any . . . journalist in the use of DANGEROUS phrases such people use for their own ends.

> You see, I know that it's difficult to think well about 'certainty', 'probability', 'perception', etc. But it is, if possible, still more difficult to think, or try to think, really honestly about your life & other people's lives. And the trouble is that thinking about these things is not thrilling, but often downright nasty. And when it's nasty then it's most important.

Malcolm had not written for some time, and – perhaps thinking of his break with Russell in 1914 – Wittgenstein began to think that this was because Malcolm feared they would clash if they talked on serious non-philosophical subjects. 'Perhaps I was quite wrong', he wrote:

> But anyway, if we live to see each other again let's not shirk digging. You can't think decently if you don't want to hurt yourself. I know all about it because I am a shirker.

In fact, the lapse in Malcolm's correspondence had nothing to do with the quarrel recalled by Wittgenstein, nor with his supposed feeling that they would not 'see eye to eye in very serious matters'. It had more to do with the demands of his job as an officer with the US Navy, which prevented his replying to Wittgenstein's letter until May 1945, when he wrote acknowledging that his remark about 'national character' had been foolish. Unfortunately for him, he arrived in Britain before his reply reached Wittgenstein. When his ship came into Southampton, Malcolm obtained leave to visit Wittgenstein in Cambridge. Wittgenstein had evidently interpreted his failure to reply as a sign that he was indeed a 'shirker', unwilling to dig deep. When Malcolm arrived at Whewell's Court Wittgenstein did not even greet him, but nodded grimly and invited him to sit down to a supper of powdered eggs. 'We sat in silence for a long time', recalls Malcolm. 'He was cold and severe the whole time. We were not in touch with one another at all.'

The day after this meeting Wittgenstein received Malcolm's letter and immediately wrote him a warm, conciliatory reply: 'Had I had it before I saw you it would have made getting into contact with you rather easier.' He suggested that they should, from now on, call each other by their Christian names. But it seems perfectly possible that, had Wittgenstein not received this acknowledgement from Malcolm of the foolishness of his remarks on 'national character', and of the need to 'dig deep', their friendship would have come to a close.

During the last year of the war, both in his attempts to finish his book and in his efforts to present his thoughts to an uncomprehending audience in his lectures, Wittgenstein felt himself to be struggling against superficiality and stupidity – his own as well as other people's – and everything else in his life was subordinated to that struggle. 'This war', he wrote to Hutt, 'I believe, has a *bad* effect on *all* of us. (It also seems to be gradually killing me, although I'm in good health.)'

One of the few people he regarded as an ally in this struggle was Rhees. When Rhees wrote to Wittgenstein about his own frustrations with teaching logic to uninterested students in Swansea, Wittgenstein responded with empathy and encouragement:

> I'm sorry to hear about the depressing circumstances under which you are working. Please don't give in or despair! I know how immensely depressing things can look; &, of course, I'm the first man to think of running away, but I hope you'll pull yourself together. I wonder what lines for a logic course I recommended. Anyhow, there is nothing more difficult than to teach logic with any success when your students are all half asleep. (I've heard Braithwaite snore in my lectures.) Please go the *bloody rough* way! – I wish you *one moderately* intelligent & awake pupil to sweeten your labour!
>
> . . . I repeat; Please go the bloody rough way! Complain, swear but go on. The students are stupid but they get something out of it.

He was dissatisfied with his own students. 'My class is exceedingly poor', he wrote to Rhees. 'I have so far 6 people, none of whom is really good.'

But a much bigger source of dissatisfaction was the fact that his book was still so far from completion. He told Rhees: 'I have no hope whatever to finish my book in the near future.' This engendered in him a feeling of worthlessness, exacerbated by the reading of other people's books:

> I have recently been reading a fair amount; a history of the Mormons & two books of Newman's. The chief effect of this

> reading is to make me feel a little more my worthlessness. Though I'm aware of it only as a slumbering man is aware of certain noises going on around him which, however, don't wake him up.

His lectures dealt with the problems in the philosophy of psychology that he had concentrated on in Swansea the previous summer. He had thought of using as a text William James's *Principles of Psychology* – primarily to illustrate the conceptual confusions that he was concerned to combat – but, as he told Rhees, 'you were right; I didn't take James as my text but just talked out of my own head (or through my own hat)'. In fact, what he was doing in these lectures was thinking through the problems that he was concerned with in the section of the *Investigations* that he was then writing.

Those problems centred on the issue between those who assert and those who deny the existence of mental processes. Wittgenstein wanted to do neither; he wanted to show that both sides of the issue rest on a mistaken analogy:

> How does the philosophical problem about mental processes and states and about behaviourism arise? – The first step is the one that altogether escapes notice. We talk of processes and states and leave their nature undecided. Sometime perhaps we shall know more about them – we think. But that is just what commits us to a particular way of looking at the matter. For we have a definite concept of what it means to learn to know a process better. (The decisive movement in the conjuring trick has been made, and it was the very one that we thought quite innocent.) – And now the analogy which was to make us understand our thoughts falls to pieces. So we have to deny the yet uncomprehended process in the yet unexplored medium. And now it looks as if we had denied mental processes. And naturally we don't want to deny them.

'What is your aim in philosophy?' he asks himself immediately after this passage, and answers: 'To show the fly the way out of the fly-bottle.' William James's textbook was used to provide examples of the sort of things that people are led to say when they are caught in this particular fly-bottle.

For example, when discussing the concept of 'the Self' James describes what happens when he tries to glance introspectively at his own 'Self of selves'. He records that what he is mostly conscious of during these attempts at introspection are motions of the head. So he concludes:

> . . . The 'Self of selves', when carefully examined, is found to consist mainly of the collection of these peculiar motions of the head or between the head and throat.

What this shows, according to Wittgenstein, is: 'not the meaning of the word "self" (so far as it means something like "person", "human being", "he himself", "I myself"), nor any analysis of such a thing, but the state of a philosopher's attention when he says the word "self" to himself and tries to analyse its meaning.' And, he adds, 'a good deal could be learned from this'.

Like his use of St Augustine to illustrate the confused picture of language that he wanted to combat, and his use of Russell to illustrate confusions in the philosophy of mathematics, Wittgenstein's use of James to provide examples of confusion in the philosophy of psychology implies no lack of respect. Just as he told Malcolm that he used the quotation from Augustine to begin the *Investigations* because: 'the conception must be important if so great a mind held it', so he cited James in his remarks on psychology precisely because he held him in high regard. One of the few books that he insisted Drury should read was James's *Varieties of Religious Experience*. Drury told him that he had already read it: 'I always enjoy reading anything of William James. He is such a human person.' Yes, Wittgenstein replied: 'That is what makes him a good philosopher; he was a real human being.'

During the Christmas vacation in Swansea, the prospect of finishing the book soon appeared brighter, and Wittgenstein returned to Trinity confident that the time for publication was near. The final version of the preface to the book is dated 'Cambridge, January 1945'.

In the preface he describes the book as 'the precipitate of philosophical investigations which have occupied me for the last sixteen years' (i.e. since his return to Cambridge in 1929), and says about his remarks:

> I make them public with doubtful misgivings. It is not impossible that it should fall to the lot of this work, in its poverty and in the

> darkness of this time, to bring light into one brain or another – but, of course, it is not likely.

Evidently the 'doubtful misgivings' triumphed over the inclination to publish. Wittgenstein did not deliver his typescript to the publishers, but instead spent the rest of the year working on an extension to it that considerably expands his investigations into psychological concepts.

For this extension he selected remarks from the manuscript volumes he had written since 1931. He worked on this throughout the Lent and Easter terms of 1945, and by the summer he was ready to dictate the selection to a typist. On 13 June he wrote to Rhees:

> The Term's over & my thoughts travel in the direction of Swansea. I've been working fairly well since Easter. I am now dictating some stuff, remarks, some of which I want to embody in my first volume (if there'll ever be one). This business of dictating will take roughly another month or 6 weeks. After that I could leave Cambridge.

Two weeks later he was in more frustrated mood. His work, he told Malcolm, was 'going damn slowly. I wish I could get a volume ready for publishing by next autumn; but I probably shan't. I'm a bloody bad worker!'

In fact the work of dictating his remarks kept him in Cambridge until August. The typescript he had produced was not conceived by him as a final version of the book, but rather as something from which – together with the typescript he had produced in Swansea the previous year – a final version could be compiled. Nevertheless, he was now confident that a publishable version was in sight. 'I might publish by Christmas', he told Malcolm:

> Not that what I've produced is good, but it is now about as good as I can make it. I think when it'll be finished I ought to come into the open with it.

During the months of preparing this typescript, he became increasingly oppressed by the 'darkness of this time'. The final stages of the Second World War were accompanied by scenes of savagery and inhumanity on a scale previously unimaginable. In February the bombardment of Dresden by British and American air forces left the city almost completely devastated, and killed 130,000 civilians. In April Berlin fell to the Allies and Vienna to the Russians, with appalling casualties on both sides. Shortly before the German surrender of 7 May, pictures were released of the heaps of rotting corpses discovered by the Allies at the concentration camps at Belsen and Buchenwald. On 14 May Wittgenstein wrote to Hutt: 'The last 6 months have been more nauseating than what went before. I wish I could leave this country for a while & be alone somewhere as I was in Norway.' Cambridge, he said, 'gets on my nerves!'

In the British elections of July he voted for the Labour Party, and strongly urged his friends to do the same. It was important, he felt, to get rid of Churchill. He was convinced, as he put it to Malcolm, 'that this peace is only a truce':

> And the pretence that the complete stamping out of the 'aggressors' of this war will make this world a better place to live in, as a future war could, of course, only be started by them, stinks to high heaven &, in fact, promises a horrid future.

Thus, when the Japanese finally surrendered in August, the celebrations in the streets of Swansea did nothing to lift his spirits. 'We've had two VJ days', he wrote to Malcolm, '& I think there was much more noise than real joy.' In the aftermath of the war he could see only gloom. When Hutt was demobbed, Wittgenstein wrote to him wishing him 'lots of luck' – 'I really mean: strength to bear whatever comes.' He had, he told Hutt, been feeling unwell lately: 'partly because I'm having trouble with one of my kidneys, partly because whatever I read of the beastliness of the Allies in Germany & Japan makes me feel sick'.

In the context of reports of the chronic food shortages in Germany and Austria, and of the policy of the British army not to 'fraternize' with their conquered enemy, and – in the midst of this – of calls in the Press to *punish* the German people for the war, Wittgenstein was gratified to read in the *News Chronicle* an article by Victor Gollancz, calling for an end to 'self-righteousness in international affairs' and a determination to feed the German people: 'not because if we don't we ourselves may suffer, but simply because it is right to feed our starving neighbours'. After remarking on Gollancz's article to Rhees, he was lent Gollancz's earlier pamphlet, *What Buchenwald Really Means*. Writing as 'a Jew who believes in Christian ethics', Gollancz attacked the British Press for their reaction to the horrors of Buchenwald, and pointed out that it was wrong to hold *all* Germans responsible. He further attacked the whole concept of 'collective guilt' as a throwback to the Old Testament, from which the example of Christ ought to have liberated us.

Wittgenstein was powerfully struck by both the strengths and the weaknesses revealed by Gollancz in his call for a humane attitude towards the Germans. On 4 September he wrote to him praising him for his *News Chronicle* article. He was, he wrote, 'glad to see that someone, publicly and in a conspicuous place, called a devilry a devilry'. About the Buchenwald pamphlet, he told Gollancz:

> I am deeply in sympathy with your severe criticism (and it cannot be too severe) of the cruelty, meanness and vulgarity of the daily press and of the BBC. (Our cinema news reels are, if possible, more poisonous still.) It is because I strongly sympathise with your attitude to these evils that I think I ought to make what seems to me a serious criticism of your polemic against them.

Gollancz, he said, had weakened the impact of his criticism by embellishing it with subsidiary points, 'which, even if they are not weak and dubious, draw the reader's attention from the main issue, and make the polemic ineffectual'. If Gollancz wanted to be heard 'above the shouting of the daily press and the radio', he would do well to stick to the point:

> If you really want people to remove the dirt, don't talk to them about the philosophical issues of the value of life and happiness. This, if it does anything, will start academic chat.
>
> In writing about the wrong attitude of people towards the Buchenwald horrors, e.g., did you wish to convince only those who agree with you about the Old and New Testament? Even if they do, your lengthy quotations serve to sidetrack their attention from the one main point. If they don't – and an enormous number who might be seriously shaken by your argument do not – they will feel that all this rigmarole makes the whole article smell fishy. All the more so as they will not gladly give up their former views.

I will stop now. – If you ask me why, instead of criticising you, I don't write articles myself, I should answer that I lack the knowledge, the facility of expression and the time necessary for any decent and effective journalism. In fact, writing this letter of criticism to a man of your views and of your ability is the nearest approach to what is denied me, i.e., to write a good article myself.

The letter reveals Wittgenstein's sound appreciation of the art of polemic. The general outlines of his advice to Gollancz were repeated about a year later to Rush Rhees. Rhees had written an article in which he attacked Gilbert Ryle for the latter's enthusiastic review of Karl Popper's *The Open Society and its Enemies*, in which Popper tars Plato, Hegel and Marx with the same brush, accusing them all of being advocates of totalitarianism. Wittgenstein told Rhees he agreed with the tendency of his article, but criticized him for making too many gestures and not landing enough square blows:

Polemic, or the art of throwing eggs, is, as you well know, as highly skilled a job as, say, boxing . . . I'd love you to throw eggs at Ryle – but keep your face straight and throw them well! The difficulty is: not to make superfluous noises or gestures, which don't harm the other man but only yourself.

Gollancz, however, received Wittgenstein's advice with disdainful indifference. His reply (addressed to 'L. Wiltgenstein, Esq.') is brief and dismissive: 'Thank you for your letter, which I'm sure was very well meant.' Wittgenstein took the rebuff good-humouredly. 'Well, that's rich!' he said to Rhees with a smile, and threw Gollancz's note in the fire.

Despite his fears for the future of Europe, and his conviction that there would soon be another, even more horrible, war, Wittgenstein was able to spend the late summer of 1945 enjoying a holiday in Swansea. Or, at least, as he put it to Malcolm, 'enjoying my absence from Cambridge'.

'My book is gradually nearing its final form', he told Malcolm at the end of the summer:

> . . . & if you're a good boy & come to Cambridge I'll let you read it. It'll probably disappoint you. And the truth is: it's pretty lousy. (Not that I could improve on it essentially if I tried for another 100 years.) This, however, doesn't worry me.

These last two sentences were not true; it did worry him that his book was not as good as he thought it ought to be, and he did think he could improve it. And it is for precisely these two reasons that it remained unpublished at the time of his death.

He dreaded going back to Cambridge to resume his professorial duties, and implored Malcolm to come to England soon, 'before I make up my mind to resign the absurd job of a prof. of philosophy. It is a kind of living death.'

The final version of what is now *Philosophical Investigations*, Part I, was prepared during the Michaelmas and Lent terms of 1945–6. From the typescript he had dictated during the summer he selected about 400 remarks to add to the work he had done in Swansea in 1944, and, after some rearrangement and renumbering, this produced the 693 numbered paragraphs of which the work now consists.

Thus, roughly speaking, the development of the book falls into three identifiable stages: paragraphs 1–188 constitute Part I of the 1938 version; paragraphs 189–421 were added in 1944; and paragraphs 421–693 form the extension added in 1945–6, which was in turn compiled from manuscripts dating from 1931–45.

This complicated patchwork is well described by Wittgenstein in his preface:

> After several unsuccessful attempts to weld my results together into a whole, I realised that I should never succeed. The best that I could write would never be more than philosophical remarks; my thoughts were soon crippled if I tried to force them on in any single direction against their natural inclination. – And this was, of course, connected with the very nature of the investigation. For this compels us to travel over a wide field of thought criss-cross in every direction. – The philosophical remarks in this book are, as it were, a number of sketches of landscapes which were made in the course of these long and involved journeyings.
>
> The same or almost the same points were always being approached afresh from different directions, and new sketches

> made. Very many of these were badly drawn or uncharacteristic, marked by all the defects of a weak draughtsman. And when they were rejected a number of tolerable ones were left, which now had to be arranged and sometimes cut down, so that if you looked at them you could get a picture of the landscape. Thus this book is really only an album.

Even now, there were sketches in this album that he was dissatisfied with, and he made no attempt to publish this final rearrangement. However, for the rest of his life he spoke of this typescript as 'my book', and went through it paragraph by paragraph with a number of his most trusted friends and students, so that when he died there would be at least a few people to whom his book was not totally unintelligible.

He was convinced his book would be fundamentally misunderstood – *especially* by academic philosophers – and this was undoubtedly another reason why the book remained unpublished in his lifetime. In a rewritten version of the preface he states: 'It is not without reluctance that I deliver this book to the public':

> It will fall into hands which are not for the most part those in which I like to imagine it. May it soon – this is what I wish for it – be completely forgotten by the philosophical journalists, and so be preserved for a better sort of reader.

Wittgenstein's hostility towards professional philosophy and his dislike of Cambridge remained constant throughout his academic career, but in the years of 'reconstruction in Europe' that followed the Second World War, they seemed to become fused with a kind of apocalyptic vision of the end of humanity. During the Easter vacation of 1946, he renewed the acquaintance of Karl Britton, previously a student of his, and now a lecturer in philosophy at Swansea University. One afternoon, during a long walk along the coast, Wittgenstein told Britton that he had become convinced that a new war was being planned, and that atomic weapons would put an end to everything: 'They mean to do it, they mean to do it.'

What links this apocalyptic anxiety with his hostility to academic philosophy is his detestation of the power of science in our age, which on the one hand encouraged the philosopher's 'craving for generality', and on the other produced the atomic bomb. In a curious sense, he even welcomed the bomb, if only the fear of it could do something to diminish the reverence with which society regarded scientific progress. At about the same time as his conversation with Britton, he wrote:

> The hysterical fear over the atom bomb now being experienced, or at any rate expressed, by the public almost suggests that at last something really salutary has been invented. The fright at least gives the impression of a really effective bitter medicine. I can't help thinking: if this didn't have something good about it the philistines wouldn't be making an outcry. But perhaps this too is a childish idea. Because really all I can mean is that the bomb offers a prospect of the end, the destruction, of an evil, – our disgusting soapy water science. And certainly that's not an unpleasant thought.

'The truly apocalyptic view of the world', he wrote, 'is that things do not repeat themselves.' The end might indeed come:

> It isn't absurd, e.g., to believe that the age of science and technology is the beginning of the end for humanity; that the idea of great progress is a delusion, along with the idea that the truth will ultimately be known; that there is nothing good or desirable about scientific knowledge and that mankind, in seeking it, is falling into a trap. It is by no means obvious that this is not how things are.

In either case, scientific progress would come to an end. But the most pessimistic view, for him, was one which foresaw the triumph of science and technology:

> Science and industry, and their progress, might turn out to be the most enduring thing in the modern world. Perhaps any speculation about a coming collapse of science and industry is, for the present and for a long time to come, nothing but a dream; perhaps science and industry, having caused infinite misery in the process, will unite the world – I mean condense it into a single unit, though one in which peace is the last thing that will find a home.
>
> Because science and industry do decide wars, or so it seems.

'The darkness of this time', therefore, is directly attributable to the worship of the false idol of science against which his own work had been directed since the early 1930s. Thus, his 'dream' of the coming collapse of science and industry was an anticipation of an age in which his type of thinking would be more generally accepted and understood. It is linked with his remark to Drury: 'My type of thinking is not wanted in this present age, I have to swim so strongly against the tide. Perhaps in a hundred years people will really want what I am writing.' And yet, if 'they' mean to do it, and the apocalyptic view is not absurd, then that time might never come. There never would be an age in which his type of thinking was wanted.

As Wittgenstein's political forebodings drew him closer to the Left, his identification of the worship of science as the greatest evil kept him at some distance from Marxism. Looking through Max Eastman's *Marxism: Is it Science?* (which he took from Rhees's bookshelves), he commented on Eastman's view that Marxism has to be made more scientific if it is to help revolution:

> In fact, nothing is more *conservative* than science. Science lays down railway tracks. And for scientists it is important that their work should move along those tracks.

He shared with the Communists a fierce dislike of the complacency of the British establishment, and he wanted to see some sort of revolution. But he wanted that revolution to be a rejection of the scientific *Weltanschauung* of our age, not an endorsement of it.

In any case, the extent to which he could identify himself with a party was limited by his conception of himself as a philosopher – one who in the ruthless search for truth would willingly abandon whatever 'pet notions' he had formed. At this time Rhees felt he ought to join the (Trotskyist) Revolutionary Communist Party, because, as he put it to Wittgenstein, 'I find more and more that I am in agreement with the chief points in their analysis and criticism of present society and with their objectives.' Wittgenstein was sympathetic, but tried to dissuade him on the grounds that his duties as a loyal party member would be incompatible with his duties as a philosopher. In doing philosophy, he insisted, you have got to be ready constantly to change the direction in which you are moving, and if you are thinking as a philosopher you cannot treat the ideas of Communism differently from others.

Ironically, at a time when his interest in political affairs was at its strongest, and his sympathies with the Left at their peak, he lost the opportunity to have discussions with the Marxist intellectual for whom he had the greatest respect. In May 1946 Piero Sraffa decided he no longer wished to have conversations with Wittgenstein, saying that he could no longer give his time and attention to the matters Wittgenstein wished to discuss. This came as a great blow to Wittgenstein. He pleaded with Sraffa to continue their weekly conversations, even if it meant staying away from philosophical subjects. 'I'll talk about anything', he told him. 'Yes', Sraffa replied, 'but in *your* way.'

Whether this was a contributory factor or not, throughout the summer term of 1946 Wittgenstein thought increasingly often of resigning his Chair and of leaving Cambridge. When he returned to Swansea for the summer vacation his dislike of both Cambridge and academic philosophy was at its peak. In Rhees's absence, Karl Britton had to face the fury of this distaste:

> One day in July . . . Wittgenstein rang me up and explained that his friend was away and that he wished me to take him out. However, he seemed, on the whole, very hostile. The journal *Mind* had just published two papers on 'Therapeutic Positivism' and (as I afterwards found out) this had much annoyed and upset him. With me he was angry too for going to the Joint Session of the Mind Association and the Aristotelian Society, the annual jamboree of philosophers: he took it as a sign of frivolity and of ulterior interests. He railed against professional philosophers, mourned the state of philosophy in England and asked: 'What can one man do alone?' When I told him that the next jamboree was to be held at Cambridge in 1947 and that I was to read a paper, he said: 'Very well, to me it is just as if you had told me that there will be bubonic plague in Cambridge next summer. I am very glad to know and I shall make sure to be in London.' (And so he was.)

Later that day Wittgenstein had tea at Britton's home. He was in a more genial mood, talking (in contrast to his dislike of London and Cambridge) of his liking for Swansea. He told Britton that he liked the North of England, too, recalling an occasion in Newcastle when he had asked the bus conductor where to get off for a certain cinema. The conductor at once told him that that particular cinema was showing a bad film, and he ought to go to another. This prompted a heated argument in the bus as to which film Wittgenstein ought to see and why. He liked that, he told Britton; it was the sort of thing that would have happened in Austria.

The final comparison is revealing, and perhaps partly accounts for the vehemence with which he, at this time, attacked what he called 'the disintegrating and putrefying English civilization'. Put simply: he was missing Vienna. He had not been there since before the *Anschluss*, and had had very little contact since then with his family and friends in Austria.

To be a professor was bad enough, but to be an English professor became, in the end, unbearable.

24
A CHANGE OF ASPECT

Wittgenstein's pessimism about the fate of humanity was not caused by the catastrophic events that brought the Second World War to a close – as we have seen, it has a much longer history; but those events seemed to reinforce in him the certitude of a long-held conviction that mankind was headed for disaster. The mechanical means of killing people that had been employed, and the fearsome displays of technological might that had been witnessed – the fire-bombs at Dresden, the gas-ovens of the concentration camps, the atomic bombs unleashed on Japan – established powerfully and finally that 'science and industry do decide wars'. And this seemed further to convince him in his apocalyptic view that the end of mankind was the consequence of replacing the spirit with the machine, of turning away from God and placing our trust in scientific 'progress'.

His notebooks of the post-war years abound with reflections of this sort. A picture that intruded upon him, he wrote, was of our civilization, 'cheaply wrapped in cellophane, and isolated from everything great, from God, as it were'. The houses, cars and other trappings of our environment struck him as 'separating man from his origins, from what is lofty and eternal, etc'. It was as though life itself was coming to an end, suffocated by the trappings of our industrial age. And, of course, it was futile to expect to alter this course by pointing it out. Is this journey really necessary? One might ask the question, but it was hardly likely that, in response, mankind would say: 'On second thoughts, no.' Yet Wittgenstein continued in his work of undermining the way of thinking that, he thought, lay at the root of the whole disaster. And in his disciples he had people who could continue this work after his death. Not that he had any wish to be found a school, or anything of that sort. 'I am by no means sure', he wrote, 'that I should prefer a continuation of my work by others to a change in the way people live which would make all these questions superfluous.'

The problem could have only an existential, never a theoretical, solution. What was required was a change of spirit: 'Wisdom is cold and to that extent stupid. (Faith on the other hand is a passion.)' To breathe again, it was no use merely thinking correctly; one had to act – to, as it were, rip the cellophane away and reveal the living world behind it. As he put it: '"Wisdom is grey". Life on the other hand and religion are full of colour.' The passion of religious faith was the only thing capable of overcoming the deadness of theory:

> I believe that one of the things Christianity says is that sound doctrines are all useless. That you have to change your *life*. (Or the *direction* of your life.)
>
> It says that wisdom is all cold; and that you can no more use it for setting your life to rights than you can forge iron when it is *cold*.
>
> The point is that a sound doctrine need not *take hold* of you; you can follow it as you would a doctor's prescription. – But here you need something to move you and turn you in a new direction. – (I.e. this is how I understand it.) Once you have been turned round, you must *stay* turned round.
>
> Wisdom is passionless. But faith by contrast is what Kierkegaard calls a *passion*.

What Russell had long ago mistakenly identified with his own theoretical passion was, in fact, the very repudiation of it: Wittgenstein's was a devoutly anti-theoretical passion. Russell's later remark that Wittgenstein liked mysticism for its power to stop him thinking, and his jibe that Wittgenstein had adopted a doctrine to make serious thinking unnecessary, are in fact much nearer the mark, if we equate 'serious thinking' with the attempts to formulate a true theory.

Wittgenstein's ideal of '*primordial* life, wild life striving to erupt into the open' – even though he rarely felt himself to live up to it – is a key to understanding both the purpose of his work and the direction of his life. In so far as he felt himself to be too theoretical, too 'wise', he felt deadened. The need for passion, for religion, was not just something he saw in the world around him; it was something he felt in himself. He felt himself to share exactly the faults characteristic of our age, and to need the same remedy: faith and love. And just as our age finds belief in God impossible, so he too found that he could not pray: 'it's as though my knees were stiff. I am afraid of dissolution (of my own dissolution), should I become soft.'

In love, too, though he felt a deep need for it, he often felt himself incapable, frightened. And, of course, frightened of its being taken away from him, all too conscious of its possible impermanence and of its uncertainty. In 1946 – and it probably came as some relief to find that he was, after all, still capable of loving someone – he fell in love with Ben Richards, an undergraduate student of medicine at Cambridge. Richards had what one by now perceives as the qualities that warmed Wittgenstein's heart: he was extraordinarily gentle, a little timid, perhaps even docile, but extremely kind, considerate and sensitive.

In his mood of deep despair after the Second World War, Wittgenstein found at least some solace in his love for Ben – even if, at times, it seemed that it offered this only to provide him with something else to worry about. 'I am very sad, very often sad', he wrote on 8 August 1946. 'I feel as though my life is now coming to an end':

> The *only* thing that my love for B. has done for me is this: it has driven the other small worries associated with my position and my work into the background.

The anxieties of being in love were perhaps the hardest to bear. And Ben was very young – nearly forty years younger than Wittgenstein. Wasn't it easy to imagine, he wrote on 12 August, that Ben would completely grow out of his love for him, 'just as a boy no longer remembers what he had felt as a young child'? And so, a few days later, when he was waiting impatiently for a letter from Ben, nothing struck him as more probable, indeed more natural, than that Ben had forsaken him. And yet, every morning, when he again found no letter from Ben, it seemed strange to him: 'I feel as though there was something I have not yet *realised*; as though I had to find some standpoint from which to see the truth more clearly.'

These accounts of the almost unbearable anguish Wittgenstein felt while waiting for a letter from his loved one strike a familiar chord. It was the same with Pinsent, and with Skinner, and even with Kirk. Yet in his love for Ben, there is a new note, a break with the solipsism of the past. On 14 August, he wrote – as though it had struck him for the first time:

> It is the mark of a *true* love that one thinks of what the *other* person suffers. For he suffers too, is also a poor devil.

Perhaps the fly had at last found its way out of the fly-bottle. And, furthermore, discovered that life was not necessarily any better outside it. To expose oneself to the elements could even be dangerous. 'I feel', he wrote on 18 August, 'that my mental health is hanging on a thin thread':

> It is of course the worry and anxiety about B. that have so worn me out. And yet that would not be the case if I were not so easily set aflame, 'highly inflammable'.

In former times, he reflected, people went into monasteries: 'Were they stupid or insensitive people? – Well, if people like that found they needed to take such measures in order to be able to go on living, the problem cannot have been an easy one!'

But if love, whether human or divine, was the solution to the problem, it was not one that could be taken; it had to be bestowed as a gift. Thus, to combat his anxieties about other philosophers publishing ideas derived from him, he would remind himself that his work was worthwhile only 'if a light shines on it from above':

> And if that happens – why should I concern myself that the fruits of my labours should be stolen? If what I am writing really has some value, how could anyone steal the value from me? And if the light from above is lacking, I can't in any case be more than clever.

And in connection with his love for Ben, he wrote:

> 'For our desires conceal from us even what we desire. Blessings come from above in their own guises etc.' I say that to myself whenever I receive the love of B. For I well know that it is a great and rare gift; I know too that it is a rare jewel – and also that it is not entirely of the sort of which I had dreamed.

Of course, there were other reasons to get out of Cambridge. On the very day that he arrived back there from Swansea, on 30 September, Wittgenstein wrote:

> Everything about the place repels me. The stiffness, the artificiality, the self-satisfaction of the people. The university atmosphere nauseates me.

To Fouracre he wrote: 'What I miss most is someone I can talk nonsense to by the yard.' Fouracre was the only person at Guy's with whom he had maintained contact. In 1943, shortly after he had married, Fouracre had joined the army, and was sent to the Far East. He did not return home until February 1947. While he was away he was deeply missed by Wittgenstein, who wrote to him with remarkable frequency, urging him to: 'come home from that bl . . . Sumatra or wherever you are'. Not all these letters have survived, but the affection which Wittgenstein held for Fouracre is evident from those that have, including a series of six written in as many months – from August to December 1946 – every one of which ends with the exclamation: 'God bless you!' and includes a request for Fouracre to come home quickly.

The first of these six is dated August 1946, and mentions some heather which Wittgenstein picked for Fouracre and sent to him in the Far East. It describes the 'lousy' situation in Europe and ends: 'So when you'll come back you won't find anything marvellous. But I hope you'll come soon nevertheless. It'll save me a lot of trouble picking flowers & posting them to Sumatra!'

In their lightness of tone, and the preponderance in them of the kind of 'nonsense' that Wittgenstein enjoyed, these letters are reminiscent of those to Pattisson. There is hardly one of them that does not contain a joke or a playful remark:

> Sorry you don't get post regularly, & particularly my letters which are full of content. I mean, paper, ink, & air. – The mosquitos don't bite you because you're so nice – because you aren't – but because you're so bl . . . awful & its the blood they want. – I hope that the Dutch will soon take over for food & send you back! [7.10.46]

> Why in hell you don't get my letters beats me! Do you think the Censor keeps them as souvenirs because they are so marvellous? I

> shouldn't be surprised! – Well, for God's sake finish your tours to South Sumatra & to Central Sumatra & take a plane (I don't mean the sort a joiner uses) & get home. [21.10.46]

> I'm feeling far better now than I did at the beginning of term. I then felt very lousy & had queer attacks of exhaustion. Finally in despair I went to see a physician here in Cambridge . . . Well, he advised me this & that & in the end he mentioned that I might try a Vitamin B preparation . . . So, I took Vit. B. tablets without the slightest hope that they'd help, & to my great surprise they did help. I'm taking them regularly now & I haven't had any more attacks of exhaustion. In fact, when I'm all tanked up with Vit. B. I get so witty that the jokes get jammed & can't come out. Isn't that terrible? [9.11.46]

The kind of simple, uncomplicated, relationship he had with Fouracre remained for him a model of what was possible outside academic life. In his letter of 21 October he wrote:

> I'm thinking every day of retiring from my job & taking on something else which might bring me into a more human contact with my fellow men. But what I'll do God Knows! for I'm already a pretty old codger.

The letter ends with the familiar refrain: 'I hope you'll come back from that bl . . . Sumatra.'

'Should I carry on teaching?' he asked himself at the beginning of November, after a meeting of the Moral Science Club, disgusted by the vanity and stupidity of his own performance there. The 'atmosphere', he wrote, was 'wretched'.

His dominance of these meetings was noted with disapproval by other philosophers at Cambridge (Broad and Russell, in particular), and by many of the visiting lecturers. On 26 October a clash took place that has since become famous, when Karl Popper addressed the club on the question: 'Are there Philosophical Problems?' Popper's chosen subject, and his manner of addressing it, was deliberately designed to provoke Wittgenstein (whom Popper thought denied the existence of philosophical problems). And provoke him it did, although exactly in what way has become lost in the mists of legend. Stories have been told that have Popper and Wittgenstein coming to blows with one

39 Wittgenstein and Francis Skinner in Cambridge

40 Wittgenstein with his niece Marie Stockert (a daughter of Helene)

41 *Left* Postcard sent to Pattisson by Wittgenstein while on holiday in Tours, France, 1949

42 *Right* A typical bit of 'nonsense' sent by Wittgenstein to Gilbert Pattisson: 'My latest photo. The previous one expressed fatherly kindness only; this one expresses triumph'

43 Wittgenstein on holiday in France with Gilbert Pattisson, July 1936

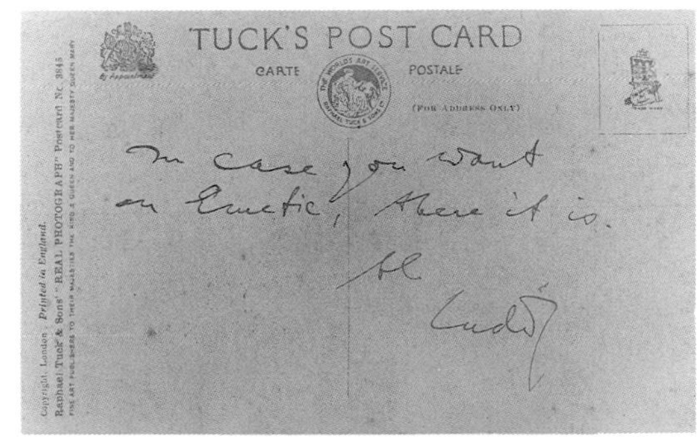
TUCK'S POST CARD
CARTE POSTALE
(FOR ADDRESS ONLY)

In case you want
an Emetic, there it is.
L
Ludwig

44 & 45 Wittgenstein's acerbic reaction to Chamberlain's diplomacy at Munich: 'In case you want an Emetic, there it is'

46 Wittgenstein in the Fellows' Garden at Trinity, 1939 (taken by Norman Malcolm)

47 From Wittgenstein's photo-album: Francis Skinner; and family and friends at Christmas in Vienna

48 The Kingstons' farmhouse in Co. Wicklow, Ireland, where Wittgenstein stayed in 1948

49 Tommy Mulkerrins outside his tiny, crowded cottage in Connemara, together with his mother, aunt and sisters

50 Wittgenstein in Swansea (taken by Ben Richards)

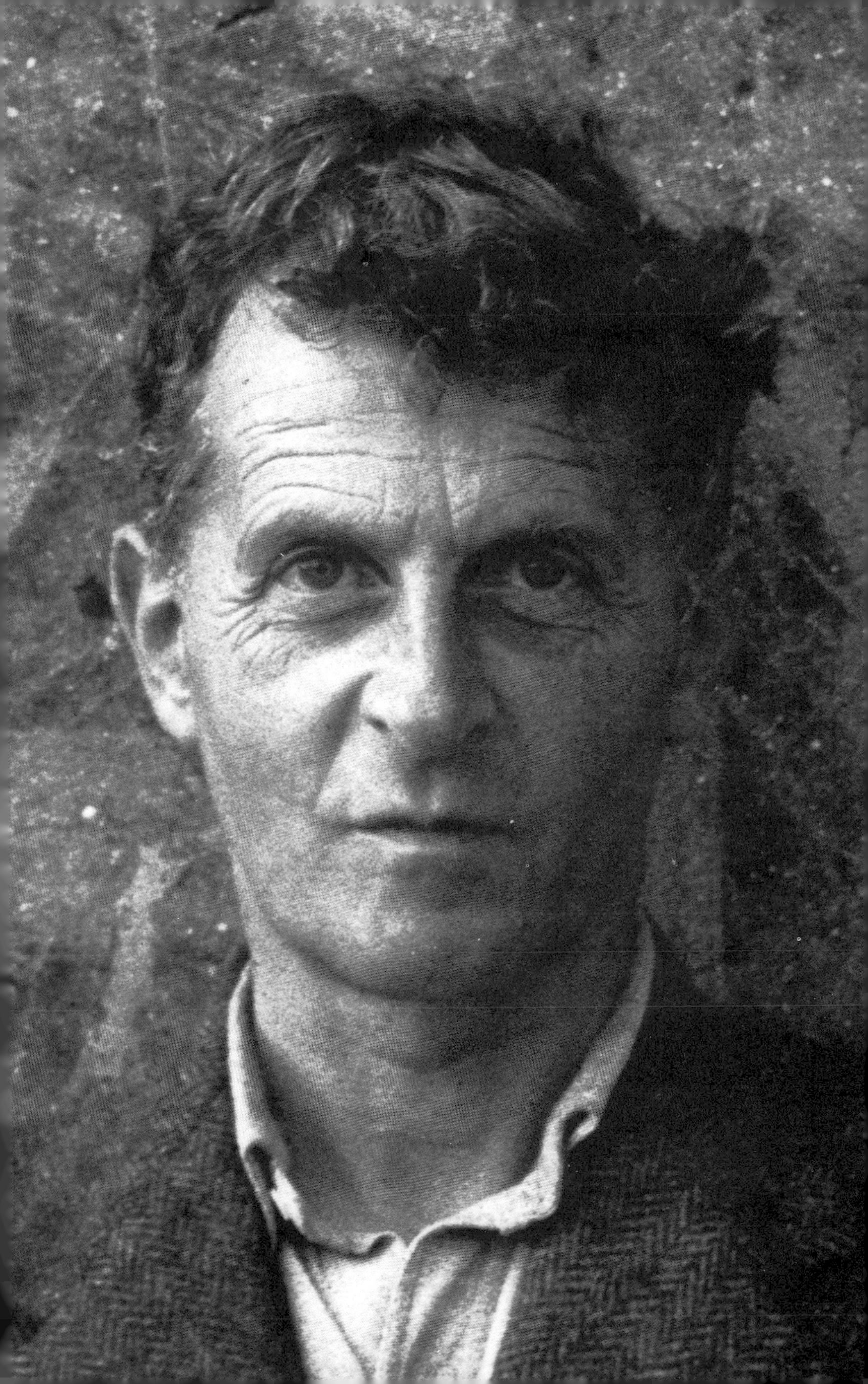

51 Wittgenstein and Ben Richards in London

52 One of the last photographs taken of Wittgenstein, in the garden at the von Wrights' home in Cambridge; Wittgenstein had taken the sheet from his bed and draped it behind him

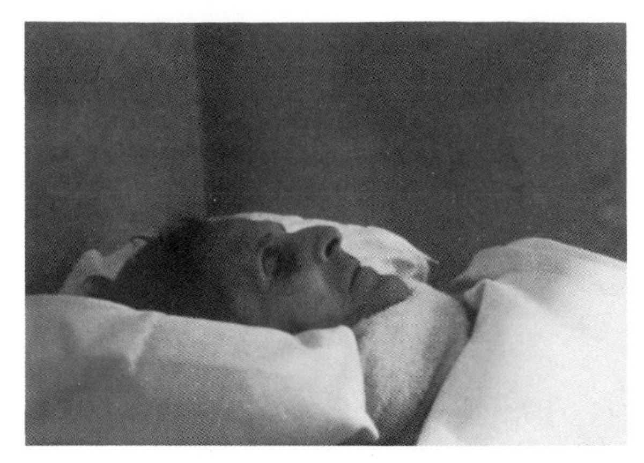

53 Wittgenstein on his death-bed

54 Wittgenstein's grave, St Giles, Cambridge

another, each armed with a poker. In his autobiography Popper scotches this rumour, only to replace it with another tale, the details of which have in turn been challenged by some of those who were present at the time. According to Popper, he and Wittgenstein engaged in an animated exchange on the existence or otherwise of philosophical problems, and he gave as an example the question of the validity of moral rules. Wittgenstein, who had all the while been playing with a poker, then stood up, poker in hand, and demanded an example of a moral rule. 'Not to threaten visiting lecturers with pokers', Popper replied, whereupon Wittgenstein stormed out of the room. Russell was present at the meeting, and made it known that his sympathies were with Popper. An alternative account of the argument has Popper and Wittgenstein each accusing the other of confusing the issue, until Wittgenstein exasperatedly stormed out, with Russell calling after him: 'Wittgenstein, it is you who are creating all the confusion.'

Whatever happened, it did nothing to affect the fervent allegiance to Wittgenstein that was given by most of the young Cambridge philosophers at this time. Gilbert Ryle writes that on his occasional visits to the Moral Science Club he was disturbed to find that: 'veneration for Wittgenstein was so incontinent that mentions, for example, my mentions, of any other philosopher were greeted with jeers':

> This contempt for thoughts other than Wittgenstein's seemed to me pedagogically disastrous for the students and unhealthy for Wittgenstein himself. It made me resolve, not indeed to be a philosophical polyglot, but to avoid being a monoglot; and most of all to avoid being one monoglot's echo, even though he was a genius and a friend.

Wittgenstein, Ryle thought, 'not only properly distinguished philosophical from exegetic problems but also, less properly, gave the impressions:

> . . . first that he himself was proud not to have studied other philosophers – which he had done, though not much – and second, that he thought that people who did study them were academic and therefore unauthentic philosophers.

To a certain extent Ryle is writing here as an Oxford man (his criticisms are given in the context of extolling the virtues of the Oxford tutorial system), but what he says about Wittgenstein's attitude towards reading the great works of the past is perfectly true. 'As little philosophy as I have read', Wittgenstein wrote, 'I have certainly not read too little, *rather too much*. I see that whenever I read a philosophical book: it doesn't improve my thoughts at all, it makes them worse.'

This attitude would never have been tolerated at Oxford, where respect for things past is in general much stronger than at Cambridge, and where a training in philosophy is inseparable from a reading of the great works in the subject. It is almost inconceivable that a man who claimed proudly never to have read a word of Aristotle would have been given any tutorial responsibilities at all at Oxford, let alone be allowed to preside over the affairs of the department. From Wittgenstein's point of view, Oxford was a 'philosophical desert'.

The only time he is known to have addressed an audience of Oxford philosophers was in May 1947, when he accepted an invitation to speak to the Jowett Society. He was to reply to a paper presented by Oscar Wood, the undergraduate secretary of the society, on Descartes' '*Cogito, ergo sum*'. The meeting was held in Magdalen College and was unusually well attended. Mary Warnock, a contemporary of Wood's, noted in her diary: 'Practically every philosopher I'd ever seen was there.' Among the more notable philosophers present were Gilbert Ryle, J. O. Urmson, Isaiah Berlin and Joseph Pritchard. In his reply to Wood's paper Wittgenstein ignored altogether the question of whether Descartes' argument was valid, and concentrated instead on bringing his own philosophical method to bear on the problem raised. To established Oxford orthodoxy, in the person of Joseph Pritchard, this was an unwelcome novelty:

> *Wittgenstein*: If a man says to me, looking at the sky, 'I think it will rain, therefore I exist,' I do not understand him.
> *Pritchard*: That's all very fine; what we want to know is: is the *cogito* valid or not?

Pritchard – described by Mary Warnock in her diary as: 'extremely old and deaf with a terrible cough. Totally tactless' – several times interrupted Wittgenstein in an effort to get him to address the question of whether Descartes' *cogito* was a valid inference or not. And every time he did so, Wittgenstein avoided the question, implying that it was unimportant. What Descartes was concerned with, Pritchard retorted, was far more important than any problem that Wittgenstein had discussed that evening. He then, in Mary Warnock's words, 'shuffled out in disgust'. He died about a week later.

Though the majority feeling at the meeting was that Pritchard had been unbearably rude, there was also a certain degree of sympathy with his objections, and a feeling, too, that Wittgenstein had treated Wood with unjustifiable disdain in not addressing his reply to the substance of Wood's paper. Wittgenstein's ahistorical, existential method of philosophizing could, in the context of respect for the great philosophers engendered at Oxford, readily be taken for arrogance.

The person indirectly responsible, as Wood's intermediary, for bringing Wittgenstein to Oxford on this occasion was Elizabeth Anscombe. Anscombe had been an undergraduate at St Hugh's, Oxford, and had come to Cambridge as a postgraduate student in 1942, when she began attending Wittgenstein's lectures. When Wittgenstein resumed lecturing in 1944, she was one of his most enthusiastic students. For her, Wittgenstein's therapeutic method was felt as a tremendous liberation, a 'medicine' that succeeded, where more theoretical methods had failed, in freeing her from philosophical confusion. 'For years', she writes, 'I would spend time, in cafés, for example, staring at objects saying to myself: "I see a packet. But what do I really see? How can I say that I see here anything more than a yellow expanse?"':

> I always hated phenomenalism and felt trapped by it. I couldn't see my way out of it but I didn't believe it. It was no good pointing to difficulties about it, things which Russell found wrong with it, for example. The strength, the central nerve of it remained alive and raged achingly. It was only in Wittgenstein's classes in 1944 that I saw the nerve being extracted, the central thought 'I have got this, and I define "yellow" (say) as this' being effectively attacked.

In 1946–7 she was again in Oxford, having taken up a research fellowship at Somerville College, but she continued to go to Cambridge once a week to attend tutorials with Wittgenstein in the company of another student, W. A. Hijab. These tutorials, at the request of both Hijab and Anscombe, dealt with issues in the philosophy of religion. By the end of the year she had become one of Wittgenstein's closest friends and one of his most trusted students, an exception to his general dislike of academic women and especially of female philosophers. She became, in fact, an honorary male, addressed by him affectionately as 'old man'. 'Thank God we've got rid of the women!' he once said to her at a lecture, on finding, to his delight, that no (other) female students were in attendance.

Anscombe was, at this time, an enthusiastic admirer of Kafka, and in an effort to share her enthusiasm she lent Wittgenstein some of his novels to read. 'This man', said Wittgenstein, returning them, 'gives himself a great deal of trouble not writing about his trouble.' He recommended, by comparison, Weininger's *The Four Last Things* and *Sex and Character*. Weininger, he said, whatever his faults, was a man who really did write about his trouble.

This directness – this determination to strip away all inessentials and all pretence, to 'extract the root' – could be unsettling as well as inspiring, and Anscombe was fairly rare in finding it liberating. Iris Murdoch, who attended a few of Wittgenstein's last series of lectures, found both him and his setting 'very unnerving':

> His extraordinary directness of approach and the absence of any sort of paraphernalia were the things that unnerved people . . . with most people, you meet them in a framework, and there are certain conventions about how you talk to them and so on. There isn't a naked confrontation of personalities. But Wittgenstein always imposed this confrontation on all his relationships. I met him only twice and I didn't know him well and perhaps that's why I always thought of him, as a person, with awe and alarm.

The student for whom Wittgenstein had the greatest respect during this period was Georg Kreisel. Originally from Graz, Kreisel came to Trinity in 1942 as an undergraduate mathematician, and attended the lectures on the philosophy of mathematics that Wittgenstein gave during the war. In 1944 – when Kreisel was still only twenty-one – Wittgenstein shocked Rhees by declaring Kreisel to be the most able philosopher he had ever met who was also a mathematician. 'More able than Ramsey?' Rhees asked. 'Ramsey?!' replied Wittgenstein. 'Ramsey was a mathematician!'

Although he had not written on the philosophy of mathematics for over two years, during 1946 and 1947 Wittgenstein had regular discussions with Kreisel on the subject. Unusually, the tenor of these discussions was set by Kreisel rather than Wittgenstein, and when Wittgenstein's remarks on mathematics were published after his death, Kreisel expressed astonishment at their tendency. After reading *Remarks on the Foundations of Mathematics*, Kreisel wrote, he realized that the topics raised in the discussions he had had with Wittgenstein: 'were far from the centre of his interest though he never let me suspect it'.

Stimulated by his discussions with Kreisel, Wittgenstein, in his last year at Cambridge, added regular seminars on the philosophy of mathematics to his weekly classes on the philosophy of psychology. Kreisel, however, remembers the discussions as being of more value than the seminars. Wittgenstein's public performances, he said, he found 'tense and often incoherent'.

Kreisel was not the stuff of which disciples are made, and after leaving Cambridge he studied with Kurt Gödel and became a leading figure in the very branch of mathematics against which Wittgenstein's work is an assault: the 'cancerous growth' of mathematical logic. 'Wittgenstein's views on mathematical logic are not worth much', he later wrote, 'because he knew very little and what he knew was confined to the Frege–Russell line of goods.' When the *Blue and Brown Books* were published, his dismissal was couched in still stronger, perhaps even bitter, terms. 'As an introduction to the significant problems of traditional philosophy', he wrote in his review, 'the books are deplorable':

> This is largely based on a personal reaction. I believe that early contact with Wittgenstein's outlook has hindered rather than helped me to establish a fruitful perspective on philosophy as a discipline in its own right.

Wittgenstein himself often felt that he had a bad influence on his students. 'The only seed I am likely to sow', he said, 'is a certain jargon.' People imitated his gestures, adopted his expressions, even wrote philosophy in a way that made use of his techniques – all, it seems, without understanding the point of his work.

He tried again and again to make this point clear. His last series of lectures began with an emphatic and unambiguous statement of their purpose: that of resolving the confusions to which the notion of psychology as the 'science of mental phenomena' gives rise:

> These lectures are on the philosophy of psychology. And it may seem odd that we should be going to discuss matters arising out of, and occurring in a science, seeing that we are not going to do the science of Psychology and we have no particular information about the sort of things that are found when the science is done. But there are questions, puzzles that naturally suggest themselves when we look at what psychologists may say, and what non-psychologists (and we) say.
>
> Psychology is often defined as the science of Mental Phenomena. This is a little queer, as we shall see: contrast it with physics as the science of physical phenomena. It is the word 'phenomena' which may be troublesome. We get the idea: on the one hand you have phenomena of one kind which do certain things, on the other, phenomena of another kind which do other things: so how do the two sorts of things compare? But perhaps it makes no sense to say that both do the sort of things the other does. 'The science of mental phenomena' – by this we mean what everybody means, namely, the science that deals with thinking, deciding, wishing, desiring, wondering . . . And an old puzzle comes up. The psychologist when he finds his correlations finds them by watching people doing things like screwing up their noses, getting rises in blood pressure, looking anxious, accepting this after S seconds, reflecting that after S plus 3 seconds, writing down 'No' on a piece of paper, and so on. So where is the science of mental phenomena? Answer: You observe your own mental happenings. How? By introspection. But if you observe, i.e., if you go about to observe your own mental happenings you alter them and create new ones: and the whole point of observing is that you should not do this – observing is supposed to be just the thing that avoids this. Then the science of mental phenomena has this puzzle: I can't observe the mental phenomena of others and I can't observe my own, in the proper sense of 'observe'. So where are we?

His answer to this last question is: in a fog, a set of confusions that cannot be resolved by the accumulation of more data – either by introspection or by behavioural analysis. Nor can they be resolved by a *theory* of thinking. The only thing capable of clearing the fog is a conceptual investigation, an analysis of the use of words like 'intention', 'willing', 'hope' etc., which shows that these words gain their meaning from a form of life, a 'language-game', quite different from that of describing and explaining physical phenomena.

The lectures for the first two terms covered roughly the same ground as that covered in the last third of *Investigations*, Part I: the question 'What is thinking?', the analysis of 'mental phenomena' and the investigation of particular psychological concepts such as 'intention', 'willing', 'understanding' and 'meaning'.

Wittgenstein had by now a good appreciation of the ways in which his approach to philosophical problems was liable to be misunderstood, and he devoted much time in these lectures to an attempt to describe his philosophical method. In addition, he gave a talk to the Moral Science Club on: 'what I believe philosophy is, or what the method of philosophy is' (as he put it in a letter to Moore asking him to attend). One common cause of confusion was that, though he began with a question ostensibly about a *phenomenon* ('What is thinking?'), he ended up with an investigation into the way we use *words* (like 'thinking').

In his second lecture he summed up the unease that many people feel about this procedure when he summarized what had been said in the previous session:

> Now let us go back to last day. You must remember I suggested (i) we want analysis. This wouldn't do unless it meant (ii) we want the definition of thinking. And then I made a fishy step. I suggested: Perhaps we really want the use of 'thinking'. 'But', you say, 'clearly, we don't want to know about the "use of words". And, in a sense, we clearly don't.

This is, we don't want to know about the use of words for its own sake. The point of describing the (real and imagined) use of words was to loosen the hold of the confused way of looking at things that is the product of the philosopher's 'impoverished diet' of examples:

> What I give is the morphology of the use of an expression. I show that it has kinds of uses of which you had not dreamed. In

> philosophy one feels forced to look at a concept in a certain way. What I do is suggest, or even invent, other ways of looking at it. I suggest possibilities of which you had not previously thought. You thought that there was one possibility, or only two at most. But I made you think of others. Furthermore, I made you see that it was absurd to expect the concept to conform to those narrow possibilities. Thus your mental cramp is relieved, and you are free to look around the field of use of the expression and to describe the different kinds of uses of it.

Another problem with this method was that, in providing a richer diet of examples, Wittgenstein was in danger of leading his students through the trees without giving them a glimpse of the wood. As two of these students, D. A. T. Gasking and A. C. Jackson, recall, the difficulty in following the lectures: 'arose from the fact that it was hard to see where all this rather repetitive concrete detailed talk was leading to – how the examples were inter-connected and how all this bore on the problems which one was accustomed to put oneself in abstract terms'.

This problem, too, Wittgenstein was aware of. 'I am showing my pupils details of an immense landscape', he wrote, 'which they cannot possibly know their way around.' In his lectures he elaborated on the analogy:

> In teaching you philosophy I'm like a guide showing you how to find your way round London. I have to take you through the city from north to south, from east to west, from Euston to the embankment and from Piccadilly to the Marble Arch. After I have taken you many journeys through the city, in all sorts of directions, we shall have passed through any given street a number of times – each time traversing the street as part of a different journey. At the end of this you will know London; you will be able to find your way about like a born Londoner. Of course, a good guide will take you through the more important streets more often than he takes you down side streets; a bad guide will do the opposite. In philosophy I'm a rather bad guide.

In his writing, too, Wittgenstein was concerned that he might be spending too much time traversing side streets. He was, he said, a long way from knowing 'what I do and don't need to discuss' in the book:

> I still keep getting entangled in details without knowing whether I ought to be talking about such things at all; and I have the impression that I may be inspecting a large area only eventually to exclude it from consideration.

Although he referred to the typescript he had prepared the previous year as 'my book', he was deeply dissatisfied with it, particularly with the last third of it – the analysis of psychological concepts that had largely been drawn from earlier manuscripts. Nevertheless, one afternoon a week he met with Norman Malcolm (who was in Cambridge during Wittgenstein's last year, on a Guggenheim Fellowship), to discuss the book. He lent Malcolm a copy of the typescript with the idea that they should read it through together, paragraph by paragraph. The procedure, as Malcolm recalls it, was this:

> Starting at the beginning of the work, Wittgenstein first read a sentence aloud in German, then translated it into English, then made some remarks to me about the meaning of it. He then went on to the next sentence; and so on. At the following meeting he started at the place where we had last stopped.

'The reason I am doing this', Wittgenstein explained, 'is so there will be at least one person who will understand my book when it is published.' This is slightly odd, in the sense that he had at this time no intention of publishing this typescript, and was already working on a reformulation of its final section. Contemporaneous with his discussions with Malcolm is a series of manuscript volumes from which he hoped to produce a more satisfactory presentation of his investigations into psychological concepts. Not that any time was wasted, for the style of their meetings had changed before they reached the last section of the book. Wittgenstein's form of 'discussion' proved to be too rigidly exegetical for Malcolm's taste; he wanted to discuss philosophical problems that were currently puzzling *him*. And so Wittgenstein gradually relaxed the procedure.

During the Michaelmas term of 1946, Wittgenstein's love for Ben Richards provided him with moments of happiness and prolonged periods of torment. 'All is happiness', he wrote on 8 October. 'I could not write like this now if I had not spent the last 2 weeks with B. And I could not have spent them as I did if illness or some accident had intervened.'

But the happiness was fragile – or, at least, he felt it to be so. 'In love I have too little *faith* and too little *courage*', he wrote on 22 October:

> But I am easily hurt and afraid of being hurt, and to protect oneself in *this* way is the death of all love. For real love one needs *courage.* But this means one must also have the courage to make the break and renounce [one's love], in other words to endure a mortal wound. But I can only hope to be spared the worst.

'I do not have the courage or the strength and clarity to look the facts of my life straight in the face', he wrote a few days later. One of these facts, he thought, was that: 'B. has for me a *pre*-love [in German there is a pun here: *Vorliebe* means a liking, a preference], something that can not last':

> How it will fade I don't know of course. Nor do I know how some part of it might be preserved, alive, not pressed between the leaves of a book as a memento.

He felt sure that he would lose Ben, and that conviction made it painful to carry on in the affair. It presented a 'frightful difficulty of my life': 'I do not know whether and how I can bear to continue *this* relationship with *this* prospect.'

Neither, however, could he bear the thought of ending the relationship: 'Whenever I imagine myself having made the break, the loneliness terrifies me.' And in any case, was it not a great and wonderful gift from the heavens, which it would be almost blasphemous to throw away? The pain and suffering of either continuing or ending the affair seemed more than he could possibly endure.

But, he insisted the following day: 'love is a *joy*. Perhaps a joy mixed with pain, but a joy nevertheless.' And if it wasn't a joy, then it wasn't love. 'In love I have to be able to rest secure.' As it was, his doubts would give him no rest. That Ben was warm-hearted, he did not doubt. 'But can you reject a warm heart?' The question immediately prompted the central doubt: 'is it a heart that beats warmly for *me*?' He quotes in English (and, therefore, presumably from Ben) the saying: 'I'll rather do anything than to hurt the soul of friendship', and continues in English (but this time surely in his own words): 'I must know: he won't hurt *our friendship*.' Having fallen for Ben, he demanded, not just friendship, not just fondness, but love:

> A person cannot come out of his skin. I cannot give up a demand that is anchored deep inside me, in my whole life. For *love* is bound up with nature; and if I became unnatural, the love would have to end. – Can I say: 'I will be reasonable and no longer demand it?' . . . I can say: Let him do as he pleases, – it will be different some day – *Love*, that is the pearl of great price that one holds to one's heart, that one would exchange for *nothing*, that one prizes above all else.* In fact it *shows* – if one has it – what great value *is*. One learns what it *means* to single out a precious metal from all others.

'The frightening thing is the uncertainty.' And in this uncertainty Wittgenstein's imagination tormented him with all manner of frightful possibilities. 'Trust in God', he told himself. But the whole point was that he was unable to trust in anything:

> From where I am to a trust in God is a *long* way. Joyous hope and fear are close cousins. I can't have the one without its bordering on the other.

And there was, too, a doubt as to whether he had any right to fall in love. In doing so, was he not being unfaithful to the memory of Francis? 'Ask yourself this question', he wrote on 10 November:

> . . . when you die who will mourn for you; and *how deep* will their mourning be? Who mourns for F., how deeply do I – who have more reason to mourn than anyone – mourn for him? Did he not deserve that someone should mourn him their whole life long? If anyone did, it was he.

Francis, however, was in God's hands: 'Here one would like to say: God will take care of him and give him what a bad person denies him.'

---

* A reference to Matthew 13:45–6: 'the Kingdom of heaven is like unto a merchant man, seeking goodly pearls, who, when he had found one pearl of great price went and sold all that he had, and bought it.' I'm grateful to Dr David McLintock for drawing my attention to this allusion.

His own life, though, was entirely in his own hands. An isolated phrase two days later simply notes: 'The fundamental insecurity of life.' The foundations might at any moment give way. 'Don't be too cowardly to put a person's friendship to the test', he urged. He had to know whether his relationship with Ben would stand up to the pressure put upon it: 'The walking-stick that looks pretty so long as one carries it, but bends as soon as you rest your weight upon it, is worth nothing.'

Better, surely, to walk without a stick than use one that could not be relied upon:

> Can you not be cheerful even without his love? Do you *have* to sink into despondency without this love? Can you not live without this prop? For that is the question: can you *not* walk upright without leaning on this staff? Or is it only that you cannot *resolve* to give it up. Or is it both? – You *mustn't* go on expecting letters that don't arrive.

In so far as it was used as a prop, the relationship was not worthy: 'It is not love that draws me to this prop, but the fact that I cannot stand securely on my own two feet alone.'

Without Ben, certainly, his life would be more lonely and miserable. But why not suffer? After all: 'Some men are ill their whole lives and the only happiness they know is that which comes from a few painless hours following a long period of intense suffering (a blessed sigh of relief)':

> Is it so unheard of for a person to suffer, e.g., that an elderly person is tired and lonely – yes even, that he becomes half-crazy?

Exhaustion, loneliness, madness – these were his lot, and he had to accept them: 'Only nothing theatrical. Of that you must guard against.'

The hardest feat was to love with hope, and not to despair if those hopes were not fulfilled: 'The belief in a benevolent father is actually the expression of just this life.'

To live like that would be a real solution, an achievement against which his philosophical work would pale into insignificance: 'What good does all my talent do me, if, at heart, I am unhappy? What help is it to me to solve philosophical problems, if I cannot settle the chief, most important thing?' And what real use were his lectures?

> My lectures are going well, they will never go better. But what effect do they leave behind? Am I helping anyone? Certainly no *more* than if I were a great actor playing out tragic roles for them. What they learn is not worth learning; and the personal impression I make does not serve them with anything. That's true for all of them, with, perhaps, one or two exceptions.

During the summer term of 1947 Wittgenstein resolved to give up lecturing. He told Georg von Wright that he would resign his professorship, and that, when he did, he would like to see von Wright as his successor.

The lectures that Wittgenstein gave in his last term are of particular interest, because they introduce the issues that were to preoccupy him for the next two years, and which found their final expression in the typescript that now forms Part II of *Philosophical Investigations*. It was in these lectures that he first introduced the famous ambiguous figure of the duck–rabbit:

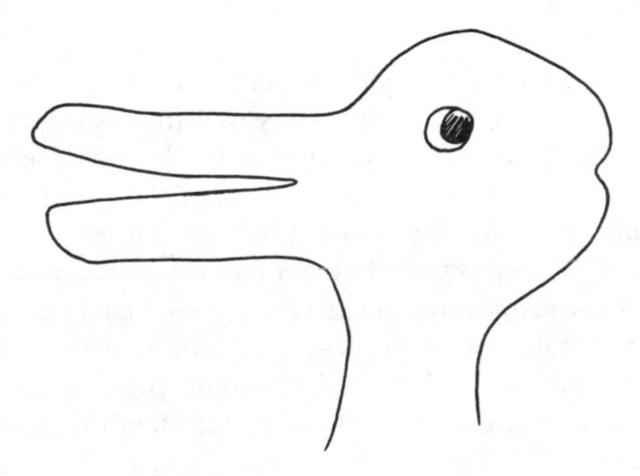

> Suppose I show it to a child. It says 'It's a duck' and then suddenly 'Oh, it's a rabbit.' So it recognises it as a rabbit. – This is an experience of recognition. So if you see me in the street and say 'Ah, Wittgenstein.' But you haven't an experience of recognition all the time. – The experience only comes at the moment of change from

duck to rabbit and back. In between, the aspect is as it were dispositional.

The point about the figure is that it can be seen under more than one aspect: the same drawing can be seen as a duck and as a rabbit. And it is this phenomenon of *seeing-as* that interested Wittgenstein. In describing this sort of phenomenon, there is a great temptation to talk of psychological states as if they were objects of some kind. For example, we might say that when we see it now as a duck, now as a rabbit, the external figure – the drawing – has not changed; what has changed is our internal picture – our sense-datum. And if this idea were generalized, it would lead to the very theory of sensory experience that is the target of Wittgenstein's philosophy of psychology – the phenomenalist notion that the objects of our immediate experience are the private, shadowy entities that empiricists call sense-data. It is for fear of this kind of generalization that one of the first points Wittgenstein makes about aspect-seeing – in the lecture quoted above and in the *Investigations* – is that it is not typical; we do not see everything as something:

> It would have made as little sense for me to say 'Now I am seeing it as . . .' as to say at the sight of a knife and fork 'Now I am seeing this as a knife and fork.'

But although the experience of seeing-as is not typical of all perception, it is of particular importance to Wittgenstein, and not only because of the dangers of phenomenalism. It could be said of his philosophical method that its aim is to change the aspect under which certain things are seen – for example, to see a mathematical proof not as a sequence of propositions but as a picture, to see a mathematical formula not as a proposition but as a rule, to see first-person reports of psychological states ('I am in pain' etc.) not as descriptions but as expressions, and so on. The 'understanding that consists in seeing connections', one might say, is the understanding that results from a change of aspect.

As he acknowledges in the *Investigations*, Wittgenstein took the duck–rabbit figure from Joseph Jastrow's *Fact and Fable in Psychology* (1900), but his discussion of aspect-seeing owes far more to Wolfgang Köhler than it does to Jastrow. It is Köhler's *Gestalt Psychology* (1929), and especially the chapter on 'Sensory Organization', that Wittgenstein has in mind in much of his discussion. Many of the lectures began with Wittgenstein reading a short passage from the book.

To understand Wittgenstein's interest in Köhler, we have, I think, to understand their common inheritance from Goethe. For both Köhler and Wittgenstein the word '*Gestalt*' had connotations of a way of understanding that had its origins in Goethe's morphological studies (of colour, plants and animals). And both, in very different ways, used this Goethean conception as a central plank in their thinking.

The German word '*Gestalt*' usually means 'shape' or 'form'. Köhler, however, following Goethe, used it to mean something quite different:

> In the German language – at least since the time of Goethe, and especially in his own papers on natural science – the noun 'gestalt' has two meanings: besides the connotation of 'shape' or 'form' as a property of things, it has the meaning of a concrete individual and characteristic entity, existing as something detached and having a shape or form as one of its attributes. Following this tradition, in gestalttheorie the word 'gestalt' means any segregated whole.

This notion of a 'segregated whole', or, as Köhler often puts it, an 'organized whole', forms the basis of Köhler's anti-behaviourist psychology. As against the behaviourist's mechanical model of stimulus-response, Köhler uses what he calls a 'dynamic' model of human behaviour which emphasizes the active role of organization in perception. Our perceptions, says Köhler, are not of discrete stimuli but of organized *Gestalten*: we do not, for example, see three dots on a page; we form them into a triangle and see them as a whole, a *Gestalt*.

Köhler's programme for a 'dynamic' understanding of human psychology has a close parallel with Goethe's programme for a 'dynamic' understanding of nature. Just as Köhler is opposed to the mechanism implicit in behaviourism, so Goethe began his scientific studies in response to a desire to see an alternative to the mechanistic Newtonian science of his day.

Goethe's first venture in the morphological understanding of natural forms was his study of plants. His idea – developed on his 'Italian Journey' – was that plant-life could be studied systematically (but non-mechanically) if all plants could be seen under the aspect of a single *Gestalt*. For each type of natural phenomenon – for example, plants and animals – there was to be a single form, the *Urphänomen*, of which all instances of that type could be seen as metamorphoses. In the case of plants, this *Urphänomen* would be the *Urpflanze* (the 'original plant').

In Goethe's work, however, there is some confusion about the nature of this *Urpflanze*; at one time he regarded it as an actual plant that might one day be discovered:

> Here [in Italy] where, instead of being grown in pots or under glass as they are with us, plants are allowed to grow freely in the open fresh air and fulfil their natural destiny, they become more intelligible. Seeing such a variety of new and renewed forms, my old fancy suddenly came back to mind: Among this multitude might I not discover the *Urpflanze*? There certainly must be one. Otherwise, how could I reognize that this or that form was a plant if all were not built upon the same basic model?

A month later, however, he conceived it, not as something discoverable in nature, but as something created by himself and brought to nature as a measure of possibilities:

> The *Urpflanze* is going to be the strangest creature in the world, which Nature herself shall envy me. With this model and the key to it, it will be possible to go on for ever inventing plants and know that their existence is logical; that is to say, if they do not actually exist, they could.

The difference between these two conceptions is of fundamental importance. The first makes Goethe's morphology look like a spurious kind of pseudo-evolutionary theory – as though the task were a Darwinian one of finding a plant from which all others are (causally) derived. The second makes clear that the *Urpflanze* cannot be used to make any causal inferences; the task of morphology is not to discover empirical laws (of evolution etc.) but to present us with an *übersicht* (a 'synoptic view') of the whole field of plant-life. It is this second conception that forms the connection between Goethe's work and Wittgenstein's.

Goethe's morphology provided Wittgenstein with an example of a study that seeks to clarify without explaining the phenomena with which it deals. This type of study consists in seeing analogies. It is crucial, however, to Wittgenstein's understanding of this morphological technique, that the *Gestalten* that are used as *Urphänomene* (the *Urpflanze* etc.) are not themselves objects, any more than ideas or concepts are objects. We see or recognize a *Gestalt*, not in the sense that we see a physical object, but in the sense that we see or recognize a likeness. This distinction is centrally important, but because *Gestalt*, *Urphänomen*, *Urpflanze* etc. are all nouns, and we can talk of seeing and recognizing them, is easily lost sight of. Thus, in his discussion of aspect-seeing in the *Investigations*, Wittgenstein begins with a masterfully clear statement of the distinction:

> Two uses of the word 'see'.
>
> The one: 'What do you see there?' – 'I see this' (and then a description, a drawing, a copy). The other: 'I see a likeness between these two faces' – let the man I tell this to be seeing the faces as clearly as I do myself.
>
> The importance of this is the difference of category between the two 'objects' of sight.

This ambiguity of the word 'see' lies at the root of a disagreement that Goethe had with Schiller about the *Urpflanze* when Goethe tried to explain his conception:

> I explained to him with great vivacity the *Metamorphosis of Plants* and, with a few characteristic strokes of the pen, conjured up before his eyes a symbolical plant.

Schiller refused to regard this 'symbolical plant' as an object of vision:

> . . . when I had ended, he shook his head saying: This has nothing to do with *experience*, it is an idea.

But Goethe was unmoved, and insisted that what he was talking about was something he had seen:

> Well, so much the better; it means that I have ideas without knowing it, and even *see them with my eyes* . . . if he takes for an idea what to me is an experience then there must, after all, prevail some mediation, some relationship between the two.

On Wittgenstein's view, both Goethe and Schiller could be said to be right: Schiller is right to insist that the *Urpflanze* belongs in the same category as ideas (rather than that of physical objects), and Goethe is right to insist that, in some sense, he sees it with his own eyes. The philosophical task is to explain how this can be so – to describe the phenomenon of seeing-as in such a way that it does not appear paradoxical that a *Gestalt* (an 'aspect', an 'organized whole') is at one and the same time an idea and an 'object' of vision.

The issues raised by Köhler's *Gestalt Psychology*, then, were central to Wittgenstein's concerns. Köhler's treatment of them, however, falls foul of the very conceptual confusions that Wittgenstein had been trying to dispel in his 'Private Language Argument'. These confusions begin with Köhler's description of a *Gestalt* as 'a concrete individual and characteristic entity, existing as something detached and *having* a shape or form as one of its attributes'. This already makes it sound as if what was being described was an object, a private object. And this is exactly the sort of object that Köhler needs for his theory of perception, because he wants to say that the 'organization' of an object of perception is as much a part of it as its colour and shape. This blurs the distinction between a physical object and a mental construct (an idea etc.), and results in a rather confused notion of a somewhat shadowy *thing*:

> If you put the 'organisation' of a visual impression on a level with colours and shapes, you are proceeding from the idea of the visual impression as an inner object. Of course this makes this object into a chimera; a queerly shifting construction.

Wittgenstein was likewise unhappy about Köhler's use of the phrase 'visual reality' to describe what it is that changes when we 'organize' a perception in different ways. For example, we would not ordinarily see the number 4 in the following figure until it was pointed out to us:

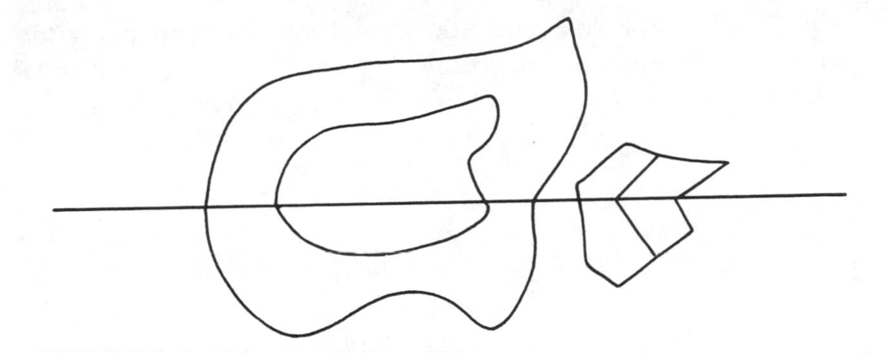

Köhler says about this:

> When I tell the reader that the number 4 is before him in the field, he will undoubtedly find it [see below]; but if he is not influenced by theoretical prejudices, he will confess that the form of the 4 did not exist as a visual reality at first and that, if it began to exist later on, that meant a transformation of visual reality.

In his lectures Wittgenstein ridiculed this passage as follows:

> Now Köhler said: 'you see two visual realities'. As opposed to what? To interpreting, presumably. How does he do this? [i.e., how is this established?] It won't do to ask people. Köhler never says it will; but he says 'If you're not blinded by theory you'll admit there are two visual realities.' But of course, he can't mean only that

---

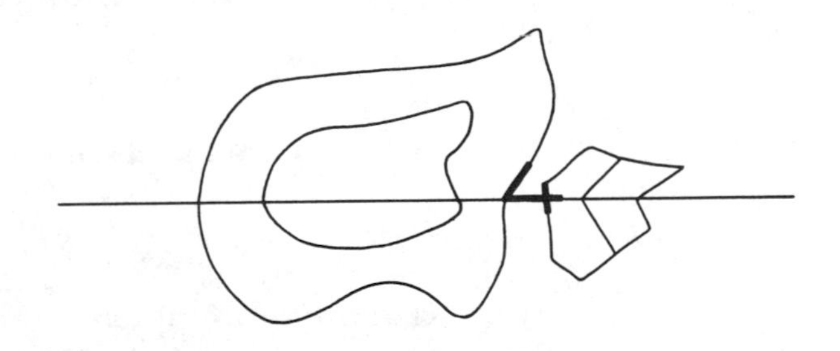

> those who don't hold a certain theory will say 'There are two visual realities.' He must intend to say that whether or not you're (1) blinded by theory, or (2) whether or not you do say one thing or the other, you must, to be right, say 'there are two visual realities'.

But in these cases of ambiguous figures (where we first see a duck and then a rabbit, first see two unknown forms with a horizontal line and then see the number 4 hidden in the figure), if we are not to say that our visual reality has changed, or that the organization of the figure has changed, then what are we to say? What *has* changed? Typically, Wittgenstein wants to describe the process in such a way that this question does not arise. Like all cases of philosophical confusion, it is the question itself that misleads. 'It makes no sense to ask: "What has changed"', Wittgenstein told his class. 'And the answer: "the organisation has changed" makes no sense either.'

He did not, however, find it very easy to formulate a felicitous description of aspect-seeing that removed the confusions inherent in Köhler's description of it. Two years after these lectures he showed Drury the duck–rabbit picture and said to him: 'Now you try and say what is involved in seeing something as something. It is not easy. These thoughts I am now working on are as hard as granite.'

This strain perhaps shows in the paradoxical, and even contradictory, descriptions that were finally published in the *Investigations*:

> The expression of a change of aspect is the expression of a new perception and at the same time of the perception's being unchanged.

> 'Seeing as . . .' is not part of perception. And for that reason it is like seeing and again not like.

On one thing he *was* clear: however it is described, it must not be by recourse to a 'private object':

> . . . above all do not say 'After all my visual impression isn't the drawing: it is this which I can't show to anyone.' – Of course it is

> not the drawing, but neither is it anything of the same category, which I carry within myself.

He was also emphatic that the question to ask about changes of aspect was not: '*What* changes?' but: 'What *difference* does the change make?' Thus in his discussion of Köhler's example of the hidden 4, Wittgenstein replaces talk of 'a transformation of visual reality' with talk of the consequences of seeing the figure differently:

> Köhler says that very few people would of their own accord see the figure 4 in the drawing
>
> 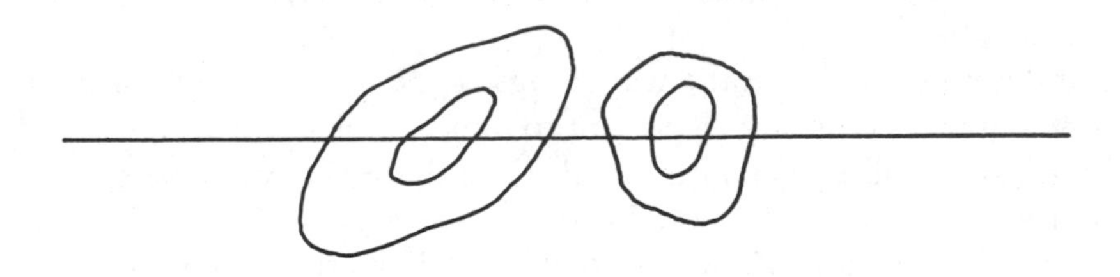
>
> and that is certainly true. Now if some man deviates radically from the norm in his description of flat figures or when he copies them, what difference does it make between him and normal humans that he uses different 'units' in copying and describing? That is to say, how will such a one go on to differ from normal humans in yet other things?

In the case of a drawing, the consequence of seeing it differently might be that it is copied differently (someone might, for example, with the drawing above, start with the figure 4); in the case of a piece of music, hearing it differently might result in its being sung, played or whistled differently; in the case of a poem, it might be read differently. From these examples we can perhaps see that Wittgenstein's dictum: 'An "inner process" stands in need of outward criteria' (*PI*, I, 580) could have (and did have) a vastly different motivation from the superficially similar slogans of the behaviourists.

But this is especially clear when we consider that in the case of a philosophical *Weltanschauung* the consequence of a 'change of aspect' might be a change of *life*. In Wittgenstein's case, the consequence – the 'outward criteria' – that he earnestly hoped for was a culture which treated music, poetry, art and religion with the same respect and seriousness with which our present society treats science.

Was there any point in urging such a change of aspect?

> A philosopher says 'Look at things like this!' – but in the first place that doesn't ensure that people will look at things like that, and in the second place his admonition may come altogether too late; it's possible, moreover, that such an admonition can achieve nothing in any case and that the impetus for such a change in the way things are perceived has to originate somewhere else entirely.

But that this 'change in the way things are perceived' should happen somehow was crucially important to him. It was not true, as he and Engelmann had earlier insisted, that the discrepancy between how things are and how they ought to be always pointed to an internal change. It was impossible not to allow the external to encroach, to have an effect. And, somehow, one had to try and change things.

Or, at least, to change one's external surroundings. Wittgenstein was now convinced he had to get out of England. 'In this country', he wrote on 13 April, 'there is no more obvious reaction for people like me than misanthropy.' The fact that one could not possibly imagine a revolution taking place in England made it all the more depressing: 'It is as though one could say of this country: it has a damp, cold *spiritual* climate.' Ten days later:

> Cambridge grows more and more hateful to me. The disintegrating and putrefying English civilization. A country in which politics alternates between an evil purpose and *no* purpose.

'[I] feel myself to be an alien in the world', he wrote in July. 'If you have no ties to either mankind or to God, then you *are* an alien.'

As soon as the term was over he travelled to Swansea, where he was joined for two weeks by Ben. Though he had not yet formally resigned his Chair, he had resolved to leave England and to live alone.

His thoughts turned first to Norway, and then to Ireland. In August he went to Dublin to visit Drury, who had recently been appointed as a psychiatrist at St Patrick's Hospital in Dublin. Wittgenstein was intensely interested in the new post: 'I wouldn't be altogether surprised', he told Drury, 'if this work in psychiatry turned out to be the right thing for you. You at least know that "There are more things in heaven and earth" etc.' Drury lent him the book which was the basis of the treatment given at St Patrick's – Sargant and Slater's *Physical Methods of Treatment in Psychiatry* – to which Wittgenstein responded with a characteristic combination of an enthusiastic appreciation of the value of sound scientific technique together with an urgent reminder of its limitations:

> This is an excellent book. I like the spirit in which it is written. I am going to get Ben to read this book. I can quite understand that you would adopt the attitude 'Let's see now what these methods of treatment will accomplish.'
>
> I don't want for one moment to underestimate the importance of the work you are doing; but don't ever let yourself think that all human problems can be solved in this way.

At the end of August he returned to Cambridge, resolved to resign his Chair but still undecided whether to go to Norway or to Ireland. His plan was to spend about a month in Vienna and then, as he put it to von Wright:

> . . . go somewhere where I can be alone for a longish time and, if possible, to finish a part of my book . . . I haven't told the Cambridge authorities anything about it so far, as it's not absolutely certain. (Though just now I can't see how it can be avoided, I mean, my leaving Cambridge.)

'My mind just now is in *great* disorder', he told von Wright:

> It's partly due to this: that I dread seeing Vienna again after all that's happened and, in a way, I also dread chucking my job at Cambridge. But I'll get over it.

The thought of returning to a Vienna that he knew would be much altered for the worse was dreadful. And in this case the reality was possibly even worse than the expectation. The city was still occupied by the Russian army, who had for a time used the house Wittgenstein had built for Gretl as a barracks and stables. The occupying army was despised by the Austrians, and tales of brutality, rape and pillage were common. Gretl's servant, who had loyally done her best to protect the Kundmanngasse house, had herself received rough treatment from the Russians. The whole situation would have been bleak and depressing. Friedrich von Hayek, a distant cousin of Wittgenstein's, remembers meeting him on a train on the way back from the visit. According to Hayek: 'he was reacting to having encountered the Russians at Vienna (as an army of occupation) in a manner which suggested that he had met them in the flesh for the first time and that this had shattered all his illusions'. Though entirely wrong in thinking that this was Wittgenstein's first encounter with Russian people, Hayek's impression of his anger and disillusionment is no doubt correct. Indeed, it is difficult to imagine any other reaction.

As soon as he returned from Vienna, Wittgenstein handed in his resignation. He was told that he could take the Michaelmas term as sabbatical leave. So, though he did not formally cease to be a professor until the end of 1947, he was relieved of the burden both of giving lectures and of living in Cambridge.

Before he left he spent a month preparing a typescript of his recent work on the philosophy of psychology. This has now been published as *Remarks on the Philosophy of Psychology*, Volume I. Wittgenstein had it typed, however, not as a separate, publishable, work, but as material to use in his attempt to revise the last third of *Philosophical Investigations*. 'It's mostly bad', he told von Wright, 'but I've got to have it in a handy form, i.e. typewritten, because it may possibly give rise to better thoughts when I read it'. He added:

> I am in no way optimistic about my future, but as soon as I had resigned I felt that it was the only natural thing to have done.

It is hard not to see, in his flight to Ireland and solitude, an attempt to escape, not only Cambridge, lecturing and the English people, but, even more painful, the torments of being close to his beloved. His ostensible reason for being alone was to finish his book, but though he wrote a good deal in the years that he spent in Ireland, it is difficult to see in this work a concerted effort to bring it to completion. In this work he was pursuing an entirely new line of thought, and the strongest impression one gets from it is of Wittgenstein 'philosophising for all he is worth' – of doing the 'only work that really bucks me up'.

25
IRELAND

Wittgenstein spent his first two weeks in Ireland living at Ross's Hotel in Dublin. Whenever Drury was free from hospital duties he and Wittgenstein went together to look at possible lodgings in or around Dublin. None could offer the solitariness and peace that was required, but the problem was temporarily solved by a friend of Drury's at St Patrick's, Robert McCullough. McCullough had been in the habit of spending his holidays at a farmhouse at Red Cross, in County Wicklow, which belonged to Richard and Jenny Kingston, and they had told him they were prepared to take in a permanent guest. This information was passed on to Wittgenstein, who immediately set off from Dublin to 'case the joint' (by this time, his vocabulary contained a liberal sprinkling of phrases borrowed from American detective fiction). He was enchanted by Wicklow. 'On my journey down in the bus', he told Drury on his return, 'I kept remarking to myself what a really beautiful country this is.'

Soon after moving into the Kingstons' farmhouse, however, he wrote to Rhees saying he felt 'cold and uncomfortable' there: 'I may, in a couple of months, move to a much more isolated place in the West of Ireland.' But after a few weeks he became much more acclimatized, and on Drury's first visit to Red Cross all seemed to be going well. Wittgenstein told him: 'Sometimes my ideas come so quickly that I feel as if my pen was being guided. I now see clearly that it was the right thing for me to give up the professorship. I could never have got this work done while I was in Cambridge.'

Being away from the 'disintegrating and putrefying English civilization' that Cambridge represented was undoubtedly one of the chief attractions of living in Ireland. When von Wright wrote to him about his hesitation in applying for the Cambridge Chair in Philosophy, Wittgenstein replied that he understood perfectly, and in fact had assumed von Wright would not apply, for: 'the prospect of becoming English, or a refugee in England, seemed to me anything but attractive in our time'.

When von Wright finally did apply, Wittgenstein's encouragement was tempered by a fearful warning:

> Cambridge is a dangerous place. Will you become superficial? smooth? If you don't you will have to suffer terribly. – The passage in your letter which makes me feel particularly uneasy is the one about your feeling enthusiasm at the thought of teaching at Cambridge. It seems to me: if you go to Cambridge you must go as a SOBER man. May my fears have no foundation, and may you not be tempted beyond your powers!

Apart from not being in Cambridge, the chief attraction of living at Red Cross was the beauty of the Wicklow countryside. The winter was mild, and Wittgenstein could go for a walk almost every day. 'There is nothing like the Welsh coast here', he wrote to Rhees, 'but the colours are most wonderful & make up for everything.' And to his sister, Helene, he wrote:

> The country here would not have so many attractions for me if the colours were not often so wonderful. I think it must be to do with the atmosphere, for not only the grass, but also the sky, the sea and even everything that is brown are all magnificent. – I feel a good deal better here than in Cambridge.

He took his notebook with him on his walks around Red Cross, and would often work outdoors. A neighbour of the Kingstons', who often saw Wittgenstein out on his favourite walk, reports that he once passed him sitting in a ditch, writing furiously, oblivious of anything going on around him. This is presumably one of those occasions when, as he told Drury, his ideas came so quickly that he felt as if his pen were being guided. He was, however, cautious in attributing too much importance to these moods of inspiration:

> In a letter (to Goethe I think) Schiller writes of a 'poetic mood'. I think I know what he means, I believe I am familiar with it myself. It is a mood of receptivity to nature in which one's thoughts seem as vivid as nature itself. But it is strange that Schiller did not produce anything better (or so it seems to me) and so I am not entirely convinced that what I produce in such a mood is really worth anything. It may be that what gives my thoughts their lustre on these occasions is a light shining on them from behind. That they do not themselves glow.

He had taken with him both the typescript that is now *Philosophical Investigations*, Part I, and that which is now *Remarks on the Philosophy of Psychology*, Volume I. From these – together with the remarks he was writing at Red Cross – he hoped to put together a final version of the first part of his book. (The second part – that dealing with the philosophy of mathematics – could by now be considered an abandoned project.) He reported to all his friends that the work was going fairly well. There is, however, some indication that he was already inclined to leave the task of publishing to his literary executors. 'Heaven knows if I'll ever publish this work', he wrote to von Wright, 'but I should like you to look at it after my death if you survive me. There is a good deal of hard thinking in it.'

Wittgenstein was prevented from working as hard as he would have liked by ill health. Despite claiming to Rhees, on 5 February 1948, that: 'I am in very good bodily health', he was in fact suffering from painful attacks of indigestion. To combat this he would keep by his side as he worked a tin of 'Scragg's' charcoal biscuits. He had so much faith in this remedy (the Kingstons' children, Maude and Ken, remember him eating very little else) that he would frequently have to walk to Arklow to replenish his stock. The biscuits, however, did not appear to solve the problem: 'My work's going moderately well', he wrote to Malcolm in January, '& I think it might even go very well if I weren't suffering from some kind of indigestion which I don't seem to be able to shake off.'

Far worse (but perhaps in some way linked to his poor digestion) was his deteriorating nervous condition. On 3 February he wrote:

> Feel unwell. Not physically, but mentally. Am frightened of the onset of insanity. God alone knows whether I am in danger.

If the nearness of Ben had been the cause of his mental instability during his last year at Cambridge, absence from him made him no more sane. On 5 February he reported to Malcolm: 'occasionally queer states of nervous instability about which I'll only say that they're rotten while they last & teach one to pray'. And on the same day he wrote to Rhees:

> My nerves, I'm afraid, often misbehave. Of course they're tired and old nerves. – My work, on the whole, goes fairly well. Again it's the work of an old man: for, though I am not really old, I have, somehow, an old soul. May it be granted to me that my body doesn't survive my soul!

'I often believe that I am on the straight road to insanity', he told von Wright a month later. 'It is difficult for me to imagine that my brain should stand the strain very long.'

He spent the following two weeks in a state of acute depression, unable to work and increasingly dissatisfied with his accommodation. He had originally been pleased with his hosts. 'They are very quiet', he had written to von Wright in December. 'I have my meals in my room and am very little disturbed.' But in March, Ken, the youngest of the family (he was then eleven years old), had a friend to stay. The two of them shared a bedroom, and would be up late at night talking and laughing together. When Wittgenstein banged furiously on the wall to tell them to be quiet, they took it as a joke. But Wittgenstein was truly at his wits' end. He sent a telegram to Drury in Dublin asking him to book a room at Ross's Hotel and to come and see him there, as a matter of urgency. Drury recalls: 'As soon as he had arrived I went down to see him. He looked distressed and agitated':

> WITTGENSTEIN: It has come.
> DRURY: I don't understand; what has happened?
> WITTGENSTEIN: What I have always dreaded: that I would no longer be able to work. I have done no work at all for the past two weeks. And I can't sleep at nights. The people under my room sit up late talking and the continual murmur of voices is driving me crazy.

Drury prescribed some tablets to help Wittgenstein sleep, and told him that his brother's cottage on the west coast of Ireland was now empty, and that Wittgenstein was welcome to make use of it. There, at least, he would find peace and solitude.

Relieved, Wittgenstein returned to Red Cross to think it over. He spent Easter with the Kingstons, but was still unable to work, and therefore resolved to take up Drury's offer. His state of mind, however, improved considerably – as did his relations with the Kingstons – and shortly before he left he presented the children with a big, bright green Easter egg full of chocolates; on the day he left for the west coast, 28 April, he signed the visitors' book with the remark: 'Thank you for a very good time.'

There is no reason to see in the remark any irony or insincerity – it was no doubt a genuine expression of gratitude to the Kingstons. But the last two months, at least, of his stay at Red Cross could hardly be called 'a very good time', as is shown by the letter he wrote Rhees a week before he left:

> I often thought of you these days &, although this may sound awful, often thought; thank God I wrote to you not to come to see me at Easter. For the last 6 or 8 weeks have been a bad time for me. First I suffered terrible depressions, then I had a bad flu, & all the time I didn't know where to go from here. I am now gradually getting better & I intend leaving here next week & so go to Rosro in the West. This has great disadvantages (it's a 10 hours journey from Dublin) but there's nothing else I can do, so far as I can see. So, had you come, you'd have found me in a state of great tribulation. Wish me some strength, some courage, & luck! During the last month my work has hardly progressed at all, & only in the last few days I've been able to think a little (I mean about philosophy; for my brain, though dull, wasn't inactive, I wish it had been!).

Wittgenstein knew Rosro, the cottage in Connemara, from 1934, when he had spent a holiday there in the company of Francis Skinner and Maurice Drury. It stands facing the sea at the mouth of the Killary harbour, and the countryside around it is dominated by a range of mountains with peaks of extraordinary angularity, known as 'The Twelve Pins'. The cottage was built as a coastguard station, but fell into disuse after the First World War. During the early 1920s it remained unoccupied, used only by the IRA as a place to hide prisoners until, in 1927, it was bought by Maurice Drury's brother,

Miles, as a holiday cottage. There are a few neighbouring cottages, but it is many miles away from any shop, post office or any other village or town amenity. This isolation, though it had, as Wittgenstein anticipated, 'great disadvantages', was necessary if he were to enjoy the freedom from interruption he thought essential for his work.

On his arrival there Wittgenstein was met by Thomas Mulkerrins ('Tommy', as Wittgenstein, like everybody else in Killary, learnt to call him), an employee of the Drury family, who lived in a tiny cottage about half a mile away from Rosro and who was paid £3 a week to look after the Drurys' holiday home. (He supplemented this inconsiderable wage by gathering peat and fishing for mackerel.) Tommy had been told by Drury that Wittgenstein had suffered a nervous breakdown, and had been asked to help in whatever way he could. So every morning he walked down to Rosro to deliver milk and peat, and to check that Wittgenstein was all right. Wittgenstein found him (so he told Malcolm): 'quite nice & certainly better company than the people I stayed with in Co Wicklow'.

In conversation with Rhees, later, he was more critical, describing the whole Mulkerrins family as people ill disposed to do any work. He was shocked to see that Tommy's mother, though an excellent sempstress, went about in rags, and that though Tommy himself was a competent carpenter every chair in their cottage had a broken leg. In his diary, Tommy – 'the man upon whom I am totally dependent here' – is described simply as 'unreliable'.

Unreliable or not, Tommy was all he had. His immediate neighbours, the Mortimer family, considered him completely mad, and would have nothing to do with him. They even forbade him to walk on their land, on the grounds that he would frighten their sheep. He therefore had to take a long and circuitous route along the road if he wanted to walk on the hills behind Rosro. On one of these walks, the Mortimers saw him stop suddenly and, using his walking-stick as an implement, draw an outline figure (a duck–rabbit?) in the dirt on the track, which he stood staring at in complete absorption for a long time before resuming his walk. This confirmed their original assessment. So, too, did Wittgenstein's vehement outburst one night when the barking of the Mortimers' dog disturbed his concentration. He appeared to the Mortimers, in fact, in much the same way that he had earlier appeared to the villagers of rural Austria.

Tommy, too, thought Wittgenstein a little odd. But partly because of his loyalty to the Drury family (Miles Drury had once dived off a boat to save Tommy from drowning), and partly because he came to enjoy the company of 'the Professor', he was prepared to do what he could to make Wittgenstein's stay at Rosro as comfortable and as enjoyable as possible. He did his best, for example, to satisfy Wittgenstein's strict standards of cleanliness and hygiene. At Wittgenstein's suggestion, he brought with him every morning, not only the milk and peat, but also his own used tea-leaves. Every morning the leaves were sprinkled on the kitchen floor to absorb the dirt, and were then swept up. Tommy was also called upon to rid the cottage of 'slaters' (wood-lice). This he did by showering the whole cottage with a suffocating amount of disinfectant powder. Wittgenstein, who throughout his life had a dread of bugs of any kind, was pleased with the result, preferring the threat of suffocation to the sight of wood-lice.

Rosro cottage had two rooms, a bedroom and a kitchen, and it was in the latter that Wittgenstein spent most of his time. He did not, however, use the kitchen to prepare his meals. While at Rosro he lived almost entirely on tinned food ordered from a grocer's shop in Galway. Tommy was concerned about this diet. 'Tinned food will be the death of you', he once said. 'People live too long anyway', came the grim reply. Instead, Wittgenstein used the kitchen as a study, and when Tommy came in the morning he would often find him sitting at the kitchen table writing on to loose sheets of paper clipped together. Almost every day there would be a pile of rejected sheets, which it was Tommy's job to burn.

One morning when Tommy arrived at Rosro, he heard Wittgenstein's voice and, on entering the cottage, was surprised to find 'the Professor' alone. 'I thought you had company', he said. 'I did', Wittgenstein answered. 'I was talking to a very dear friend of mine – myself.' The remark is echoed in one of his notebooks of the period:

> Nearly all my writings are private conversations with myself. Things that I say to myself tête-à-tête.

Apart from the time he spent with Tommy, Wittgenstein's solitude at Rosro was interrupted only by a brief visit from Ben Richards, who spent a couple of weeks there in the summer of 1948. Together they went on Wittgenstein's favourite walks up the hills and along the coast, admiring the magnificently varied flora and fauna of the area.

Wittgenstein had taken a particular interest in the different kinds of birds that are to be seen at Killary. (Northern Divers, Cormorants, Curlews, Oyster Catchers, Puffins and Terns are all fairly common along that part of the west Ireland coast.) At first he used to ask Tommy to identify the birds for him. He would describe a bird he had seen, and Tommy would do his best to name it, although, as he freely admits: 'maybe it wasn't always the right name I gave him'. Having caught him out a few times, Wittgenstein relied instead on the illustrated handbooks sent to him by Drury.

In order to gain a better view of the sea-birds, Wittgenstein wanted to build a hut on one of the small islands off the Killary coast. He was eventually dissuaded from this by Tommy (whose job it would have been to construct it) on the grounds that a small wooden hut would not be strong enough to withstand the exposed conditions on the island. Instead, Tommy took Wittgenstein out in a rowing-boat; while Tommy rowed, Wittgenstein would either look out for sea-birds or sit silently in contemplation. Occasionally, while out in the boat, they would chat, Wittgenstein reminiscing about his time in Norway, when he would have to row across the fjord to fetch his supplies, and Tommy answering Wittgenstein's questions about the history of Killary.

Wittgenstein took an interest, too, in the more domestic birds, the robins and chaffinches, that used to come to the cottage in search of crumbs. He would encourage them by leaving food out for them, and eventually they grew so tame that they would come to him at the kitchen window, and eat off his hand. When he left Rosro he gave Tommy some money with which to buy food to provide for the birds, who had now come to expect a daily feed. By the time Tommy next visited the cottage, however, he found the birds' tameness had been their undoing. While waiting by the window to be fed, they had fallen easy prey to the local cats.

The way of life at Rosro, though strenuous, seems to have provided the necessary conditions for an improvement in Wittgenstein's mental and physical well-being. He had, as we have seen, arrived there in poor shape. 'I've had a bad time lately: soul, mind & body', he wrote to Malcolm on 30 April, a few days after he arrived. 'I felt exceedingly depressed for many weeks, then felt ill & now I'm weak & completely dull. I haven't done any work for 5–6 weeks.' But within a month, the solitude of the cottage, the beauty of the coastal scenery, the company of the birds and Tommy Mulkerrins's good-humoured (if not entirely dependable) support had effected a change for the better. Wittgenstein found himself once again able to work.

His biggest complaint about the way of life there was that he had to do all his own housework. This he found an infuriating nuisance, but, as he wrote to Malcolm's wife, Lee, 'it's undoubtedly a great blessing, too, because it keeps me sane, it forces me to live a regular life & is in general good for me although I curse it every day'.

The remoteness of Rosro was a problem only in so far as the scarcity of American pulp fiction was concerned. The nearest village was ten miles away, and the selection of books to be had there was so poor that, in the periods between his regular parcels of 'mags' from Norman Malcolm, Wittgenstein was forced to resort to reading Dorothy Sayers. This, he told Malcolm, 'was so bl . . . foul that it depressed me'. Malcolm's supply of the 'real thing' came as a relief: 'when I opened one of your mags it was like getting out of a stuffy room into the fresh air'.

By chance, however, he did manage to find at the village shop a paperback copy of his favourite detective novel, *Rendezvous with Fear* by Norbert Davis. He had read Davis's book during his last year at Cambridge, and had enjoyed it so much that he lent it to both Moore and Smythies (he later gave a copy to Ben Richards as well). Seeing it again, he couldn't resist the temptation to buy it and reread it, and having done so his regard for it increased even further: 'For though, as you know', he wrote to Malcolm, 'I've read hundreds of stories that amused me & that I liked reading, I think I've only read two perhaps that I'd call good stuff, & Davis's is one of them.' He asked Malcolm to try to find out more about Davis:

> It may sound crazy, but when I recently re-read the story I liked it again so much that I thought I'd really like to write to the author and thank him. If this is nuts don't be surprised, for so am I.

Unfortunately, Malcolm reports: 'As I recall, I was unable to obtain any information about this author.' This is a pity, because by 1948 Norbert Davis was, in fact, in sore need of encouragement. He was, along with Dashiell Hammett and other *Black Mask* writers, one of the pioneers of the American 'hard-boiled' detective story. In the early 1930s he had given up a career as a lawyer to write detective stories, and then enjoyed ten years as a successful author. By the late 1940s, however, he had fallen on hard times. Shortly after Wittgenstein's letter to Malcolm, Davis wrote to Raymond Chandler saying that fourteen of his last fifteen stories had been rejected for publication, and asking Chandler to lend him $200. He died in poverty the following year, unaware of his rare (probably unique) distinction in having written a book which Wittgenstein liked sufficiently to want to write a letter of thanks to its author.

Wittgenstein's gratitude is, no doubt, partly accounted for by the scarcity of detective fiction in Connemara. But why should he have rated *Rendezvous with Fear* above all the other detective stories he had read (of which there was a great number)?

The answer perhaps lies in the humour of the novel, which is in fact its most striking feature. The detective in the story, Doan, is distinguished from such figures as Sam Spade and Philip Marlowe by his rather comically unprepossessing appearance: he is a short, fat man who is followed everywhere by an enormous, well-trained Great Dane called Carstairs. The feature of Davis's style that particularly impressed Raymond Chandler was the casual way in which he killed off his characters, and this is particularly evident in *Rendezvous with Fear*. For example, after setting the scene by describing the tourists at the Azteca, a hotel in South America, Davis introduces 'Garcia':

> All this was very boring to a man who, for the time being, was named Garcia. He sat and drank beer the general colour and consistency of warm vinegar, and glowered. He had a thin, yellowish face and a straggling black moustache, and he was cross-eyed. He should really have been more interested in the tourists coming from the Hotel Azteca, because in a short time one of them was going to shoot him dead. However, he didn't know that, and had you told him he would have laughed. He was a bad man.

When Doan shoots Bautiste Bonofile, another 'bad man', the romantic but naïve heroine, Jane, asks with concern: 'Is he hurt?' 'Not a bit', says Doan, 'he's just dead.'

'Humour is not a mood but a way of looking at the world',

Wittgenstein wrote while he was at Rosro. 'So if it is correct to say that humour was stamped out in Nazi Germany, that does not mean that people were not in good spirits, or anything of that sort, but something much deeper and more important.' To understand what that 'something' is, it would perhaps be instructive to look at humour as something strange and incomprehensible:

> Two people are laughing together, say at a joke. One of them has used certain somewhat unusual words and now they both break out into a sort of bleating. That might appear very extraordinary to a visitor coming from quite a different environment. Whereas we find it completely reasonable. (I recently witnessed this scene on a bus and was able to think myself into the position of someone to whom this would be unfamiliar. From that point of view it struck me as quite irrational, like the responses of an outlandish animal.)

Understanding humour, like understanding music, provides an analogy for Wittgenstein's conception of philosophical understanding. What is required for understanding here is not the discovery of facts, nor the drawing of logically valid inferences from accepted premises – nor, still less, the construction of *theories* – but, rather, the right point of view (from which to 'see' the joke, to hear the expression in the music or to see your way out of the philosophical fog). But how do we explain or teach what is meant by the 'right point of view'?

> So how do we explain to someone what 'understanding music' means? By specifying the images, kinaesthetic sensations, etc., experienced by someone who understands? *More likely*, by drawing attention to his expressive movements. – And we really ought to ask what function explanation has here. And what it means to speak of: understanding what it means to understand music. For some would say: to understand that means: to understand music itself. And in that case we should have to ask 'Well, can someone be taught to understand music?', for that is the only sort of teaching that could be called explaining music.
>
> There is a certain *expression* proper to the appreciation of music, in listening, playing, and at other times too. Sometimes gestures form part of this expression, but sometimes it will just be a matter of how a man plays, or hums, the piece, now and again of the

comparisons he draws and the images with which he as it were illustrates the music. Someone who understands music will listen differently (e.g. with a different expression on his face), he will talk differently, from one who does not. But he will show that he understands a particular theme not just in manifestations that accompany his hearing or playing that theme but in his understanding for music in general.

Appreciating music is a manifestation of the life of mankind. How should we describe it to someone? Well, I suppose we should first have to describe *music*. Then we could describe how human beings react to it. But is that all we need do, or must we also teach him to understand it for himself? Well, getting him to understand and giving him an explanation that does not achieve this will be 'teaching him what understanding is' in *different* senses of that phrase. And again, teaching him to understand poetry or painting may contribute to teaching him what is involved in understanding music.

These remarks on understanding music – like those on humour quoted earlier – have been published in a collection of remarks 'which do not belong directly with his philosophical works although they are scattered amongst the philosophical texts' (editor's preface to *Culture and Value*). But their connection with Wittgenstein's philosophical work is more direct than this would suggest. One of his chief philosophical preoccupations while he was at Rosro was the problem of aspect-seeing. To discuss this problem he would often imagine people who were 'aspect-blind' (or, as he sometimes put it, 'gestalt-blind') – people who were unable to see something as something. His remarks on what it is like to be unable to see a joke, or unable to appreciate music, are not distinct from this philosophical preoccupation; they are part of it.

'What would a person who is blind towards these aspects be lacking?' Wittgenstein asks, and replies: 'It is not absurd to answer: the power of imagination.' But the imagination of individuals, though necessary, is not sufficient. What is further required for people to be alive to 'aspects' (and, therefore, for humour, music, poetry and painting to mean something) is a culture. The connection between Wittgenstein's philosophical concern with aspect-seeing and his cultural concerns is therefore simple and direct. It is made clear in the following sequence of remarks (written at Rosro, and, it should be added, occurring in Wittgenstein's *bona fide* philosophical work):

> What is lacking to anyone who doesn't understand the question which way the letter F is facing, where, for example, to paint a nose on? Or to anyone who doesn't find that a word loses something when it is repeated several times, namely, its meaning; or to someone who doesn't find that it then becomes a mere sound?
> *We* say: 'At first something like an image was there.'
>
> Is it that such a person is unable to appreciate a sentence, judge it, the way those who understand it can? Is it that for him the sentence is not alive (with all that implies)? Is it that the word does not have an aroma of meaning? And that therefore he will often react differently to a word than we do? – It *might* be that way.
>
> But if I hear a tune with understanding, doesn't something special go on in me – which does not go on if I hear it without understanding? And what? – No answer comes; or anything that occurs to me is insipid. I may indeed say: 'Now I've understood it', and perhaps talk about it, play it, compare it with others etc. *Signs* of understanding may accompany hearing.
>
> It is wrong to call understanding a process that accompanies hearing. (Of course its manifestation, expressive playing, cannot be called an accompaniment of hearing either.)
>
> For how can it be explained what 'expressive playing' is? Certainly not by anything that accompanies the playing. – What is needed for the explanation? One might say: a culture. – If someone is brought up in a particular culture – and then reacts to music in such-and-such a way, you can teach him the use of the phrase 'expressive playing'.

Seeing aspects, understanding music, poetry, painting and humour, are reactions that belong to, and can only survive within, a culture, a form of life:

> What is it like for people not to have the same sense of humour? They do not react properly to each other. It's as though there were a custom amongst certain people for one person to throw another a ball which he is supposed to catch and throw back; but

some people, instead of throwing it back, put it in their pocket.

Thus, if it is true that humour was stamped out in Nazi Germany, this would mean, not just that people were not in good spirits, but that the Nazis had been successful in destroying a whole way of life, a way of looking at the world and the set of reactions and customs that go with it. (It would mean that the Nazis had, so to speak, pocketed the ball.)

The philosophical difficulty about aspect-seeing arises from the *prima facie*, and puzzling, fact that, though the aspect changes, the thing that is seen does not; the same drawing is now a duck, and now a rabbit. Likewise, it is the same joke, poem, painting or piece of music that is now just an extraordinary, outlandish piece of behaviour, words on a page, splashes on canvas or an incoherent noise, and now (when it is understood) funny, moving, beautiful or wonderfully expressive: 'What is incomprehensible is that *nothing*, and yet *everything*, has changed.'

Wittgenstein's remark about philosophy – that it 'leaves everything as it is' – is often quoted. But it is less often realized that, in seeking to change nothing but the way we look at things, Wittgenstein was attempting to change *everything*. His pessimism about the effectiveness of his work is related to his conviction that the way we look at things is determined, not by our philosophical beliefs, but by our culture, by the way we are brought up. And in the face of this, as he once said to Karl Britton: 'What can one man do alone?'

> Tradition is not something a man can learn; not a thread he can pick up when he feels like it; any more than a man can choose his own ancestors.
>
> Someone lacking a tradition who would like to have one is like a man unhappily in love.

Wittgenstein had a tradition – one that he loved dearly: the German/Austrian literature, art and (especially) music of the nineteenth century. But he was acutely aware that, for the greater part of his life, this tradition was no longer alive. In this sense, he was not so much unhappily in love, as desperately bereaved. The physical isolation at Connemara that he felt necessary to pursue his work matches the sense of cultural isolation that pervades it.

★

Wittgenstein stayed at Rosro throughout the summer of 1948, from May to August. During these four months he wrote a great deal. But the demands of the life-style and his uncertain health combined to make him feel too weak to accomplish what he had set out to achieve. 'I get tired, bodily and mentally, very easily', he told von Wright. He was, he wrote in his diary, 'too soft, too weak, and so too lazy to achieve anything significant':

> The industry of great men is, among other things, a sign of their strength, quite apart from their inner wealth.

He was, in addition, plagued by attacks of melancholy that he was inclined to personify, as though haunted by a ghost. 'Don't let grief vex you', he wrote on 29 June:

> You should let it into your heart. Nor should you be afraid of madness. It comes to you perhaps as a friend and not as an enemy, and the only thing that is bad is your resistance. Let grief into your heart. Don't lock the door on it. Standing outside the door, in the mind, it is frightening, but in the *heart* it is not.

A little later, on 11 July, the ghost is identified:

> Think a great deal of the last time with Francis, of my odiousness towards him. I was at that time very unhappy; *but with a wicked heart*. I cannot see how I will ever in my life be freed from this guilt.

The psychological and physical strains of living alone in Rosro could not, he thought, be withstood for much longer. He found it almost inconceivable that he could stand to spend the winter there. 'But', he wrote on 17 July, 'I have decided to *try* and do so':

> I pray a good deal. But whether in the right spirit I do not know. – Without the blessings of C and B [Con Drury and Ben] I could not live here.

He asked Tommy whether he would consider having him as a guest for the winter. Tommy declined. His tiny, two-room, cottage was already over-crowded with himself, his mother and his sister. Wittgenstein also approached Mrs Phillips, the proprietor of nearby Kylemore House (now Kylemore Hotel), but was told that she took guests only during the summer. If he was to stay in Connemara, the only option open to him was to live alone at Rosro.

He left Connemara in August, travelling first to Dublin to visit Drury, and then to Uxbridge to stay with Ben at his family home. In September he left for Vienna to visit Hermine, who was seriously ill with cancer.

On his return he spent a couple of weeks at Cambridge dictating a typescript compiled from the work he had written in Ireland. This has now been published as *Remarks on the Philosophy of Psychology*, Volume II. But, like its predecessor, it was not conceived as a separate work, its intended – or, perhaps, ostensible – purpose being to provide, in a convenient form, a set of remarks to use in the revision of the *Investigations*.

By 16 October the work was completed, and Wittgenstein returned to Dublin, intending initially to go on to Rosro. From Vienna he had written to Tommy, asking him to have the cottage ready for his return. He had, however, as we have seen, serious misgivings about returning. And as Wittgenstein's doctor, Drury, too, was concerned about his spending the winter in a place where, if he fell ill, there would be no one to look after him, and no way of getting him medical attention. Moreover, Wittgenstein found that in the warm, comfortable, and above all quiet room that he occupied at the top of the Dublin hotel in which he was staying, he could work quite well. In the event, therefore, he spent the winter as a guest at Ross's Hotel.

Ross's Hotel was, in 1948, a large but not especially luxurious hotel on Parkgate Street, close to Phoenix Park. (It still stands, but is now extensively altered and has been renamed the Ashling Hotel.) It was known locally as a 'Protestant' hotel: many of the permanent guests there were Protestants, and it was used by Protestant clergymen when they came to Dublin to attend conferences and meetings. 'When I look at the faces of the clergy here in Dublin', Wittgenstein remarked to Drury, 'it seems to me that the Protestant ministers look less smug than the Roman priests. I suppose it is because they know they are such a small minority.'

Of more importance to him, however, was the fact that it was only a short walk away from the Zoological Gardens in Phoenix Park.

Through Drury he became a member of the Royal Zoological Society, which allowed him free access to the gardens and the right to have his meals in the members' room. While he was in Dublin he saw Drury almost every day: they would meet for lunch either at the members' room of the Zoological Gardens or at Bewley's Café in Grafton Street, where the waitress quickly became accustomed to Wittgenstein's unvarying diet, and would bring him an omelette and coffee without his having to order it. Drury also introduced him to the Botanical Gardens at Glasnevin, where the heated Palm House provided a warm and congenial place in which to work during the winter.

During the winter months in Dublin Wittgenstein worked with great intensity. 'I'm anxious to make hay during the very short period when the sun shines in my brain', he told Malcolm on 6 November. On one occasion, when he and Drury had planned to have lunch together, Drury arrived at the hotel to be told: 'Just wait a minute until I finish this.' Wittgenstein then continued to write for two hours without saying a word. When he finally finished he seemed quite unaware that it was now long past their lunch-time.

The work he wrote in Dublin has been published under the title *Last Writings on the Philosophy of Psychology*. The title has misled many into thinking that it is Wittgenstein's last work. It is not; it predates, for example, Part II of *Philosophical Investigations*, *On Certainty* and *Remarks on Colour*. It is, however, the last of the series of manuscript volumes, begun in Cambridge in 1946, in which he attempts to provide a better, more perspicuous, analysis of psychological concepts than that given in Part I of the *Investigations*. It is a continuation of his attempts to exhibit the multiplicity and complexity of psychological concepts (such as 'fear', 'hope', 'belief' etc.) in a way that exposes the barrenness and confusion of 'the philosopher's search for generality'. The work is full of fine distinctions, designed to show, among other things, the danger of assuming that all sentences in the indicative mode can be regarded as *descriptions*:

> I hear the words 'I am afraid'. I ask: 'In what connection did you say that? Was it a sigh from the bottom of your heart, was it a confession, was it self-observation . . . ?'

On one of their walks in Phoenix Park, Drury mentioned Hegel. 'Hegel seems to me to be always wanting to say that things which look different are really the same', Wittgenstein told him. 'Whereas my interest is in showing that things which look the same are really different.' He was thinking of using as a motto for his book the Earl of Kent's phrase from *King Lear* (Act I, scene iv): 'I'll teach you differences.'

His concern was to stress life's irreducible variety. The pleasure he derived from walking in the Zoological Gardens had much to do with his admiration for the immense variety of flowers, shrubs and trees and the multitude of different species of birds, reptiles and animals. A theory which attempted to impose a single scheme upon all this diversity was, predictably, anathema to him. Darwin had to be wrong: his theory 'hasn't the necessary multiplicity'.

The concepts with which Wittgenstein is particularly concerned in these 'Last Writings' are those of 'thinking' and 'seeing'. More particularly, his concern is with the relation between the two. Of central importance to his whole later work is the idea that there is a kind of seeing that is also a kind of thinking (or, at least, a kind of *understanding*): the seeing of connections. We *see* a connection in the same sense that we see an aspect, or a *Gestalt* To distinguish this sense of 'see' from that in which we see a physical object, and to describe the connections and the differences between this sense of 'see' and the concepts of 'thinking' and 'understanding', is the central task of the work written at Ross's Hotel.

'Now you try and say what is involved in seeing something as something', Wittgenstein challenged Drury; 'it is not easy. These thoughts I am now working on are as hard as granite.' In reply Drury quoted James Ward: '*Denken ist schwer*' ('Thinking is difficult'), a response which perhaps prompted the following notebook entry:

> 'Denken ist schwer' (Ward). What does this really mean? Why is it difficult? – It is almost like saying 'Looking is difficult.' Because looking intently is difficult. And it's possible to look intently without seeing anything, or to keep thinking you see something without being able to see clearly. Looking can tire you even when you don't see anything.

It was on the same day that Wittgenstein remarked to Drury: 'It is impossible for me to say one word in my book about all that music has meant in my life. How then can I hope to be understood?' The work that he was writing at the time, however, does contain a strong hint of this. For in drawing attention to the sense of 'seeing' (or 'hearing'), in the act of which we understand something, the paradigmatic example of music is never far from his thoughts:

> We say that someone has the 'eye of a painter' or 'the ear of a musician', but anyone lacking these qualities hardly suffers from a kind of blindness or deafness.

> We say that someone doesn't have a 'musical ear', and 'aspect-blindness' is (in a way) comparable to this inability to hear.

The example of understanding music was important to him, not only because of the immense importance of music in his own life, but also because it is clear that the meaning of a piece of music cannot be described by naming anything that the music 'stands for'. And in this way: 'Understanding a sentence is much more akin to understanding a theme in music than one may think.'

'I would like it if some day you were able to read what I am writing now', Wittgenstein told Drury. But the demands of Drury's job at St Patrick's, and his relative lack of acquaintance with the specific philosophical problems with which Wittgenstein was concerned, forbade any detailed discussion of Wittgenstein's work. Indeed, Drury recalls that it was Wittgenstein's express decision not to discuss philosophy with him: 'I think he felt that his own thinking was so much more developed than mine that there was a danger of swamping me and of my becoming nothing but a pale echo of himself.' Nor did Wittgenstein read through his current work with Ben Richards, who came to stay with him at Ross's Hotel for a week or two in November.

In December, however, Wittgenstein had a chance to discuss his work in detail, when he was joined at the hotel first by Elizabeth Anscombe, who stayed for the first two weeks of December, and then by Rush Rhees, who came to Dublin immediately after Anscombe to spend Christmas with Wittgenstein. Wittgenstein had already decided that Rhees should be the executor of his will, and also, perhaps, that Anscombe and Rhees should be two of his literary executors. In any case, with both of them he read through the work that he had written in the previous two months, and discussed his attempts to revise *Philosophical Investigations*, using this new material together with some of the remarks contained in the two typescripts he had prepared in the previous two years.

Rhees left Dublin on the first day of the new year, while Wittgenstein stayed on at Ross's Hotel hoping to continue his good run of work. Early in January, however, he fell ill with a complaint similar to that which had dogged him during the previous year. He described it to Malcolm as: 'some sort of infection of the intestines'. 'Of course it hasn't done my work any good', he added. 'I had to interrupt it completely for a week & after that it just crawled along, as I do when I take a walk these days.'

He felt tired, ill and old. He wondered whether this would be his last illness. He also felt isolated. 'Drury, I think, is growing more and more unfaithful', he wrote on 29 January. 'He has found friends with whom he can live more easily.' His doctor diagnosed him as suffering from nothing more serious than gastro-enteritis, but he was inclined to distrust the doctor, and ignored the prescribed treatment. On 11 February he reported 'great weakness and pain'. Mining, he had heard, was dying – 'a *great* loss for me and for everyone'. She had many and varied talents, he wrote, that were not exposed to the day, but hidden: 'like human innards *should* be'.

Throughout February he was still able to work, but not with anything like the intensity and industry of which he had been capable before Christmas. By the end of March even this limited capacity for work had deserted him, and for a few months after that he wrote nothing at all. During this fallow period he read a fair amount. Drury was a member of the Royal Dublin Society's library, and used to borrow books from there on Wittgenstein's behalf. He recalls that what Wittgenstein generally wanted to read was history – Macaulay's *Essays Critical and Historical*, Livy's account of the second Punic War, Morley's *Life of Cromwell*, Ségur's *L'Histoire de Napoléon* and Bismarck's *Gedanken und Erinnerungen*. They were, for the most part, books Wittgenstein had read before. In 1937, for example, he had written of the Macaulay essays:

> [They] contain many excellent things; but his value judgements about people are tiresome and superfluous. One feels like saying to him: stop gesticulating! and just say what you have to say.

And in 1942 he had written to Rhees that he was reading Livy's account of Hannibal's invasion of Italy: 'this interests me immensely'. A favourite passage (so he told Drury) concerned the incident when, after the battle of Cannae, Hannibal has the battle-field searched for the bodies of the two consuls, in order that he could show his respect for them.

In his present state, he wrote in his diary, he would not try to work unless it came easily to him: 'for otherwise, even if I strive, nothing will come out'. At the beginning of March he was again joined at the hotel by Ben, who stayed for ten days: 'Nice time. Always loving.' But even while he enjoyed being with Ben he was conscious of the fact that he was unwell. He slept badly and was troubled by thoughts of the future: '*Do not know how it will turn out.*' A few days after Ben left, he wrote: 'Often it is as though my soul were dead.'

His conversations with Drury turned with increasing frequency to religious subjects. He contrasted Drury's 'Greek' religious ideas with his own thoughts, which were, he said, 'one hundred per cent Hebraic'. Drury had admired Origen's vision of a final restitution of all things, a restoration to their former glory of even Satan and the fallen angels, and commented sadly on its condemnation as a heresy. 'Of course it was rejected', Wittgenstein insisted:

> It would make nonsense of everything else. If what we do now is to make no difference in the end, then all the seriousness of life is done away with.

Wittgenstein's 'Hebraic' conception of religion was, Drury suggested, based on the sense of awe which one feels throughout the Bible. In illustration of this he quoted from Malachi: 'But who may abide the day of his coming and who shall stand when he appeareth?' (Mal. 3:2). This stopped Wittgenstein in his tracks: 'I think you have just said something very important. Much more important than you realize.'

Central to Wittgenstein's 'Hebraic' conception of religion (like that of his favourite English poet, Blake) is the strict separation of philosophy from religion: 'If Christianity is the truth then all the philosophy that is written about it is false.' In conversation with Drury he sharply distinguished the more philosophical St John's Gospel from the others: 'I can't understand the Fourth Gospel. When I read those long discourses, it seems to me as if a different person is speaking than in the synoptic Gospels.'

But what of St Paul? In 1937 he had written: 'The spring which flows gently and limpidly in the Gospels seems to have froth on it in Paul's Epistles.' He had then seen in St Paul, in contrast to the humility of the Gospels, 'something like pride or anger'. In the Gospels you find huts; in Paul a church: 'There all men are equal and God himself is a man; in Paul there is already something like a hierarchy; honours and official positions.' But now, he told Drury, he saw that he had been wrong: 'It is one and the same religion in both the Gospels and the Epistles.'

And yet, within his fundamentally ethical conception of religious faith, he still found the Pauline doctrine of predestination hard to embrace. For, like Origen's teaching, it seems to have the consequence that: 'what we do now is to make no difference in the end'. And if this is so, how can the seriousness of life be upheld?

In 1937 Wittgenstein had characterized the Pauline doctrine as one that could have arisen only from the most dreadful suffering: 'It's less a theory than a sigh, or a cry.' At his own 'level of devoutness' it could appear only as 'ugly nonsense, irreligiousness':

> If it is a good and godly picture, then it is so for someone at a quite different level, who might use it in his life in a way completely different from anything that would be possible for me.

In 1949, he could no longer speak of it as 'irreligious'. But neither could he quite see how it could be used as a 'good and godly picture':

> Suppose someone were taught: there is a being who, if you do such and such or live thus and thus, will take you to a place of everlasting torment after you die; most people end up there, a few get to a place of everlasting happiness. – This being has selected in advance those who are to go to the good place and, since only those who have lived a certain sort of life go to the place of torment, he has also arranged in advance for the rest to live like that.
>
> What might be the effect of such a doctrine? Well, it does not mention punishment, but rather a sort of natural necessity. And if you were to present things to anyone in this light, he could only react with despair or incredulity to such a doctrine.

> Teaching it could not constitute an ethical upbringing. If you wanted to bring someone up ethically while yet teaching him such a doctrine, you would have to teach it to him after having educated him ethically, representing it as a sort of incomprehensible mystery.

Though Wittgenstein had, as yet, no medical grounds for thinking so, he felt that his death might come soon. When Malcolm wrote asking about his financial situation, he replied that he had enough to live on for two years: 'What happens after that time I don't yet know. Maybe I won't live that long anyway.'

In April he left for Vienna to visit Mining on her death-bed. He stayed for three or four weeks, returning to Dublin on 16 May. From there he wrote to Malcolm that Mining was still alive, but that there was no hope of her recovering: 'While I was in Vienna I was hardly able to write at all. I felt so rotten myself.'

Soon after arriving back in Dublin, he went, on Drury's advice, to see the Professor of Medicine at Trinity College for a diagnosis of the intestine trouble and the general feeling of exhaustion that had dogged him since the beginning of the year. It was suspected that he might have a growth in his stomach, but after being admitted to hospital for a full investigation he was told that no such growth showed up on the X-ray, and the only findings made were that he had an atypical and unexplained anaemia. He was put on a treatment of iron and liver extract, and although he still found himself unable to concentrate on philosophy, his condition gradually improved.

He was anxious to overcome his anaemia quickly for two reasons. First, because he had at last decided to accept a long-standing invitation from Norman Malcolm to spend the summer at the Malcolms' home in Ithaca, USA (after first insisting in jest that, if he came, Malcolm was to introduce him to his favourite film star, Betty Hutton). He had booked a passage on the *Queen Mary*, sailing on 21 July. The second reason was that, before he left for America, he wanted to spend a few weeks in Cambridge preparing a final, polished typescript of the work with which he had been occupied since 1946.

During his period of recovery he remained in Dublin, and it was presumably during this time that he prepared the fair manuscript copy of what is now *Philosophical Investigations*, Part II. As a restful diversion from this work, Drury proposed giving him a record-player and some records of his choice. Wittgenstein declined. It would never do, he said; it would be like giving him a box of chocolates: 'I wouldn't know when to stop eating.' On the other hand, Drury himself, he said, ought to listen to music when he was tired after his work. And so the next morning he had a radio set delivered to Drury's rooms. Shortly after this Drury remarked on the great improvement of recording techniques evident in the records he heard on the radio. This elicited from Wittgenstein a typically Spenglerian reflection:

> It is so characteristic that, just when the mechanics of reproduction are so vastly improved, there are fewer and fewer people who know how the music should be played.

On 13 June, Drury and Wittgenstein listened together to a radio discussion between A. J. Ayer and Father Copleston on 'The Existence of God'. Ayer, Wittgenstein said, 'has something to say, but he is incredibly shallow'. Copleston, on the other hand, 'contributed nothing at all to the discussion'. To attempt to justify the beliefs of Christianity with philosophical arguments was entirely to miss the point.

A week later he left Dublin. One senses that, in packing his large pile of notebooks, manuscripts and typescripts, he was not only winding up his affairs in Dublin, but also bringing to a close his entire contribution to philosophy. He told Drury of a letter he had received from Ludwig Hänsel in which Hänsel had expressed the hope that Wittgenstein's work would go well, if it should be God's will. 'Now that is all I want', he said, 'if it should be God's will':

> Bach wrote on the title page of his *Orgelbüchlein*, 'To the glory of the most high God, and that my neighbour may be benefited thereby.' That is what I would have liked to say about my work.

The use of the past tense here is telling; it indicates that he now considered his work to be all but over.

He spent the month before his journey to the United States staying alternately with von Wright in Cambridge and with Ben Richards in Uxbridge. Von Wright had just finished his first year as Wittgenstein's successor as Professor of Philosophy of Cambridge, and was living in a rented house ('Strathaird') in Lady Margaret Road. During his stay there Wittgenstein occupied a separate apartment of two rooms and had his meals with the family (von Wright, his wife and their two children). 'There is one thing that I'm afraid of', he wrote to von Wright before he came to stay with him: 'I may not be able to discuss philosophy. Of course it's possible that things will have changed by then, but at present I'm quite incapable of even thinking of philosophical problems. My head is *completely* dull.'

His chief concern during these few weeks in Cambridge was to dictate to a typist the manuscript containing his final selection of the remarks written over the last three years, which now forms Part II of *Philosophical Investigations*. This is the last typescript that Wittgenstein is known to have prepared, and as such it represents the culmination of his attempts to arrange his remarks on psychological concepts into a publishable form.

It does not represent, however, the completion of that task: as he had told Elizabeth Anscombe in Dublin, he regarded this new selection as material to use in the revision of *Philosophical Investigations*, Part I. As he never carried out this work of revision himself, the book as we now have it has the rather unsatisfying two-part structure whereby the second 'part' is no more than material to be used in the revision of the first. Moreover, the work that was originally conceived of as the 'second part' – Wittgenstein's analysis of mathematical concepts – does not appear in the book at all. Wittgenstein's painstaking fastidiousness over the structure of his book has had the ironic result that his work has been published in a form very far removed from his original conception.

The longest section of this new typescript is that concerned with the problems of aspect-seeing, and is a distillation of the work, already discussed, that he had written on that topic over the previous three years. This section constitutes roughly half (thirty-six pages in the printed version) of the entire typescript. He told Rhees, however, that the section he was particularly satisfied with was that concerned with 'Moore's Paradox' (Section X). He was pleased, he said, to have condensed his many remarks on this paradox into such a relatively short section (three printed pages).

'Moore's Paradox' is the name Wittgenstein gave to the absurdity of stating a proposition and then saying that one does not believe it – for example: 'There is a fire in this room and I don't believe there is.' The title 'Moore's Paradox' is perhaps a misnomer: Wittgenstein believed, probably erroneously, that Moore had discovered this type of absurdity. (Indeed, he once remarked to Malcolm that its discovery was the only work of Moore's that had greatly impressed him.) The interest Wittgenstein took in the Paradox arises from the fact that, although anybody who uttered such a statement would ordinarily be taken to be contradicting themselves, it is not formally a contradiction. That is, the two statements: 'There is a fire in this room' and: 'RM does not believe there is a fire in this room' do not contradict one another.

Wittgenstein first came across the Paradox in a paper by Moore given to the Moral Science Club in October 1944. He immediately wrote to Moore urging him to publish his 'discovery' and explaining why he considered it so important:

> You have said something about the *logic* of assertion. Viz: It makes sense to say 'Let's suppose: p is the case and I don't believe that p is the case', whereas it makes *no* sense to assert 'I-p is the case and I don't believe that p is the case.' This *assertion* has to be ruled out and *is* ruled out by 'common sense', just as a contradiction is. And this just shows that logic isn't as simple as logicians think it is. In particular: that contradiction isn't the *unique* thing people think it is. It isn't the *only* logically inadmissible form and it is, under certain circumstances, admissible. And to show that seems to me the chief merit of your paper.

This was not how Moore himself saw it. He was inclined to say that, as the Paradox did not issue in formal contradiction, it was an absurdity for psychological, rather than logical, reasons. This Wittgenstein vigorously rejected:

> If I ask someone 'Is there a fire in the next room?' and he answers 'I believe there is' I can't say: 'Don't be irrelevant. I asked you about the fire not about your state of mind!'

Any investigation into what it does and does not make sense to assert was, for Wittgenstein, a part of logic, and to point out that, in this sense: 'logic isn't as simple as logicians think it is' was one of the chief concerns of his own investigations. It was an aspect of Wittgenstein's later work that was noted early by Bertrand Russell, who in his report to the Council of Trinity College of 1930 remarked that Wittgenstein's theories were 'novel, very original, and indubitably important'. But: 'Whether they are true, I do not know. As a logician, who likes simplicity, I should wish to think that they are not.'

'Moore's Paradox' interested Wittgenstein as an illustration that, contrary to the logician's desire for simplicity, the forms of our language cannot be squeezed without distortion into the pigeon-holes created for them by the categories of formal logic. The statement: 'I believe there is a fire in the next room' is used to assert, albeit hesitantly, that there is a fire in the next room; it is not used to assert a state of mind. ('Don't regard a hesitant assertion as an assertion of hesitancy.') This distinguishes it from the statements: 'I believed then that there was a fire in the next room'; and: 'He believes there is a fire in the next room' – both of which would ordinarily be taken to be about people's beliefs, rather than about fires. This feature of the logic of our language forbids us from constructing the convenient form: 'x believes/believed p' and thinking that the form remains unchanged whatever values are given to x and p: 'I believe there is a fire in the next room' is not the same kind of assertion as 'I believed there was a fire in the next room':

> 'But surely "I believed" must tell of just the same thing in the past as "I believe" in the present!' – Surely $\sqrt{-1}$ must mean just the same in relation to -1, as $\sqrt{1}$ means in relation to 1! This means nothing at all.

If we regard the form $\sqrt{x}$ as having a single meaning, whatever the value of x, we get into a hopeless tangle when we consider $\sqrt{-1}$. For, given our ordinary rules of multiplication, the square root of minus one can be neither a positive nor a negative number, and within the realm of 'real numbers' there is nothing left for it to be. And yet $\sqrt{-1}$ has a use: it is an essential notion in many important branches of pure and applied mathematics. But to give it a meaning it has been found necessary to construct different meanings of 'multiplication', 'square root' and even 'number' such that the square root of minus one is said to be, not a real number, but i, an 'imaginary number' (or, as it is sometimes called, an 'operator'). Given this revised framework, $i^2 = -1$, and the notion of the square root of minus one is not only unproblematic but is made the basis of a whole Theory of 'Complex

Numbers'. Wittgenstein was interested in the square root of minus one for exactly the same reason that he was interested in 'Moore's Paradox': it illustrates the fact that superficial similarities of form can disguise very important differences of meaning.

This latter idea is one of the main themes of the book, justifying Wittgenstein's suggestion to Drury that he might use the Earl of Kent's phrase: 'I'll teach you differences' as a motto, and it is particularly evident in the analysis of psychological concepts in *Philosophical Investigations,* Part II. Just as he wished to show that logic isn't as simple as logicians think it is, so he wished to show that psychological concepts and the sentences in which they are used, are not as uniform as philosophers and psychologists would wish them to be. In both cases the aim is to disencourage the 'craving for generality' – to encourage people to look before they think.

For example, the question: 'What does the sentence "I am afraid" mean?' does not have a single answer that would be adequate to cover all the occasions on which the sentence might be used. For, as in the case of the square roots of one and minus one, the differences between the various uses might be just as important as the similarities:

> We can imagine all sorts of things here, for example:
>
> 'No, no! I am afraid!'
> 'I am afraid. I am sorry to have to confess it.'
> 'I am still a bit afraid, but no longer as much as before.'
> 'At bottom I am afraid, though I won't confess it to myself.'
> 'I torment myself with all sorts of fears.'
> 'Now, just when I should be fearless, I am afraid!'
>
> To each of these sentences a special tone of voice is appropriate, and a different context. It would be possible to imagine people who as it were thought more definitely than we, and used different words where we use only one.

To understand what 'I am afraid' means on a particular occasion one might have to take into account the tone of voice and the context in which it is uttered. There is no reason to think that a general theory of fear would be much help here (still less a general theory of language). Far more to the point would be an alert and observant sensitivity to people's faces, voices and situations. This kind of sensitivity can be gained only by experience – by attentive looking and listening to the people around us. Once, when Wittgenstein and Drury were walking together in the west of Ireland, they came across a five-year-old girl sitting outside a cottage. 'Drury, just look at the expression on that child's face', Wittgenstein implored, adding: 'You don't take enough notice of people's faces; it is a fault you ought to try to correct.' It is a piece of advice that is implicitly embodied in his philosophy of psychology: 'An inner process stands in need of outward criteria.' But those outward criteria stand in need of careful attention.

What is 'internal' is not hidden from us. To observe someone's outward behaviour – if we understand them – is to observe their state of mind. The understanding required can be more or less refined. At a basic level: 'If I see someone writhing in pain with evident cause I do not think: all the same, his feelings are hidden from me.' But at a deeper level some people, and even whole cultures, will always be an enigma to us:

> It is important for our view of things that someone may feel concerning certain people that their inner life will always be a mystery to him. That he will never understand them. (Englishwomen in the eyes of Europeans.)

This is because the commonality of experience required to interpret the 'imponderable evidence', the 'subtleties of glance, gesture and tone', will be missing. This idea is summed up in one of Wittgenstein's most striking aphorisms: 'If a lion could talk, we could not understand him.'

The abstractions and generalities, the laws and principles, that result from theorizing can, on Wittgenstein's view, only hinder our attempts towards a better understanding of this 'imponderable evidence'. But how, in the absence of theory, are we to better our understanding, to deepen our insight?

Take, for example, one of the hardest and one of the most important distinctions to make concerning our understanding of people: the distinction between a genuine and an affected expression of feeling:

> Is there such a thing as 'expert judgment' about the genuineness of expressions of feeling? – Even here there are those whose judgment is 'better' and those whose judgment is 'worse'.

Correcter prognoses will generally issue from the judgments of those with better knowledge of mankind.

Can one learn this knowledge? Yes; some can. Not, however, by taking a course in it, but through '*experience*'. – Can someone else be a man's teacher in this? Certainly. From time to time he gives him the right *tip*. – This is what 'learning' and 'teaching' are like here. – What one acquires here is not a technique; one learns correct judgments. There are also rules, but they do not form a system, and only experienced people can apply them right. Unlike calculating rules.

An example of such a teacher might be the figure of Father Zossima in Dostoevsky's *The Brothers Karamazov*:

> It was said by many people about the elder Zossima that, by permitting everyone for so many years to come to bare their hearts and beg his advice and healing words, he had absorbed so many secrets, sorrows, and avowals into his soul that in the end he had acquired so fine a perception that he could tell at the first glance from the face of a stranger what he had come for, what he wanted and what kind of torment racked his conscience.

In describing Father Zossima, Dostoevsky is here describing Wittgenstein's ideal of psychological insight. When, after being persuaded by Wittgenstein to read *The Brothers Karamazov*, Drury reported that he had found the figure of Zossima very impressive, Wittgenstein replied: 'Yes, there really have been people like that, who could see directly into the souls of other people and advise them.'

Such people, Wittgenstein suggests, have more to teach us about understanding ourselves and other people than the experimental methods of the modern-day 'science' of psychology. This is not because the science is undeveloped but because the methods it employs are inappropriate to its task:

> The confusion and barrenness of psychology is not to be explained by calling it a 'young science'; its state is not comparable with that of physics, for instance, in its beginnings. (Rather with that of certain branches of mathematics. Set Theory.) For in psychology there are experimental methods and *conceptual confusion*. (As in the other case

> conceptual confusion and methods of proof.) The existence of the experimental method makes us think we have the means of solving the problems which trouble us; though problem and method pass one another by.

Philosophical Investigations, Part II, ends with a suggestion of what the second volume of Wittgenstein's book might have contained:

> An investigation is possible in connexion with mathematics which is entirely analogous to our investigation of psychology. It is just as little a mathematical investigation as the other is a psychological one. It will not contain calculations, so it is not for example logistic. It might deserve the name of an investigation of the 'foundations of mathematics'.

By 12 July the work of dictating this typescript was finished, and Wittgenstein left Cambridge to spend the week remaining before his trip to the United States with Ben Richards in Uxbridge. During the remaining two years of his life, although he continued to write philosophy, he made no further attempt to restructure his book in the way that he had intended. *Philosophical Investigations* has therefore reached us in the somewhat transitory state in which it was left in the summer of 1949.

26

A CITIZEN OF NO COMMUNITY

The last two years of Wittgenstein's life have about them something of the nature of an epilogue. The task of arranging his work for publication, though not complete, was now over – at least for him. He had by now accepted that his book – the work that had been the centre of his life for nearly twenty years – would not appear in his lifetime. The job of editing it and seeing to its posthumous publication was in the hands of others. And in other ways, too, he was dependent on other people in a way that he had not been since before the First World War. He had no income, no home of his own, and little taste for the solitariness and fierce independence that before he had craved. His last two years were spent living as a guest of his friends and disciples – with Malcolm in Ithaca, von Wright in Cambridge, and Elizabeth Anscombe in Oxford.

But his motives for living with others were not primarily financial. Indeed, there was actually no financial need for him to do so: as he had earlier told Malcolm, he had enough money saved from his salary at Cambridge to last another two years. The need to live with others was partly emotional, partly physical (he was increasingly ill and in need of attention), and also partly intellectual. So long as he lived, he wanted to live as a philosopher, and though he now felt, for the most part, unable to live alone and write, he did feel able to discuss philosophy. Thus we find, to a much greater extent than hitherto, the stimulus for his philosophical thinking being provided by the thoughts and problems of others. The work he wrote in his last two years, though naturally in many ways of a piece with the *Investigations*, is in another respect quite distinct from it; it is much more directed to the solution of other people's problems. It has the character that he himself had earlier attributed to all his work – that of clarifying the work of others – and it is written much more consciously than his other work with a view to being useful. It is as though he wished to reward the hospitality of his hosts by availing them of his most prized possession: his philosophical talent.

In his exchange of letters with Malcolm before he left for the States, Wittgenstein returned again and again to the question of whether, in coming to Ithaca, he would be of any use to Malcolm philosophically. 'My mind is tired & stale', he wrote in April. 'I think I could discuss philosophy if I had someone here to discuss with, but alone I can't concentrate on it.' Two months later he wrote: 'I *know* you'd extend your hospitality to me even if I were *completely* dull & stupid, but *I* wouldn't want to be a mere dead weight in your house. I want to feel that I can at least give a *little* for so much kindness.'

He set sail on 21 July 1949, travelling on the *Queen Mary* across the Atlantic. 'My anaemia is as good as cured', he wrote before he left, insisting that it was unnecessary for Malcolm to meet him from the boat: 'Maybe, like in the films, I'll find a beautiful girl whom I meet on the boat & who will help me.' Malcolm was, nevertheless, there to meet him, and was surprised to find him apparently fit and strong, 'striding down the ramp with a pack on his back, a heavy suitcase in one hand, cane in the other'.

In some respects, at least, the Malcolms found him an undemanding guest. He insisted on eating bread and cheese at all meals, declaring that he did not care what he ate so long as it was always the same.

The Malcolms lived on the edge of the settled area around Ithaca, just outside the boundaries of the Cayuga Heights, and Wittgenstein often took long walks in the nearby countryside. He took a great interest in the unfamiliar flora of the area. Stuart Brown, a colleague of Malcolm's at Cornell University, remembers that, on one occasion at least, this unfamiliarity resulted in astonished disbelief:

> Ordinarily, he would refuse the offer of a ride. But one afternoon, after it had begun to rain, I stopped to offer him a ride back to the Malcolms. He accepted gratefully, and once in the car asked me to identify for him the seed pods of a plant which he picked. 'Milk-

weed', I told him and pointed out the white sap for which the weed is named. He then asked me to describe the flowers of the plant. I failed so miserably that I at length stopped the car by a grown up field, walked out and picked him more plants, some with flowers and some with seeds. He looked in awe from flowers to seed pods and from seed pods to flowers. Suddenly he crumpled them up, threw them down on the floor of the car, and trampled them. 'Impossible!' he said.

Brown was one of a group of philosophers at Cornell with whom Wittgenstein held discussions. The others included Max Black, Willis Doney, John Nelson and Oets Bouwsma. 'I'm doing my old work here' was how he put it in a letter to Roy Fouracre.* He had, he said, often thought of Fouracre – 'particularly because I often thought I might take up my old work at Guy's or some such place, & now I'm such an old cripple that I couldn't possibly do the work I used to in the dispensary lab'. To Malcolm, too, he expressed an anxiety about what he was to do with the remainder of his life: 'When a person has only one thing in the world – namely a certain talent – what is he to do when he begins to lose that talent?'

In the meantime, there was plenty of demand and appreciation at Cornell for that talent. Together with Malcolm he attended a surprisingly large number of seminars and discussions. There were regular meetings with Brown, Bouwsma and Black, in which they discussed a variety of philosophical topics, seminars with Doney to read the *Tractatus*, and meetings with Bouwsma to discuss Frege's 'On Sense and Reference'. He also met with Nelson and Doney to discuss a problem about memory, an occasion remembered by Nelson as 'probably the most philosophically strenuous two hours I have ever spent':

Under the relentless probing and pushing of his enquiry my head felt almost as if it were ready to burst . . . There was no quarter

---

* Fouracre had finally returned from Sumatra in February 1947, and from then until Wittgenstein's death the two continued their friendship, meeting regularly both in London and in Cambridge. Whenever Wittgenstein was away he would write to Fouracre with the same regularity as he had when Fouracre was in the army. Letters survive written from Dublin, Vienna and Oxford, as well as Ithaca.

given – no sliding off the topic when it became difficult. I was absolutely exhausted when we concluded the discussion.

Nelson's reactions were typical. Though the topics of these meetings were usually suggested by others, the discussions were invariably dominated by Wittgenstein, who demanded from the participants a degree of absorption and attentive rigour to which they were unaccustomed. After one such discussion, Bouwsma asked Wittgenstein whether such evenings robbed him of his sleep. He said that they did not. 'But then', recalls Bouwsma:

> . . . he added in all seriousness and with the kind of smile Dostoyevsky would suggest in such a circumstance: 'No but do you know, I think I may go nuts.'

Apart from Malcolm, Bouwsma was the person with whom Wittgenstein spent most time. He seemed to see in him the quality of seriousness that he considered essential in a discussion partner. Unlike the others, Bouwsma was the same age as Wittgenstein. He had been Malcolm's tutor at the University of Nebraska, and Malcolm was one of several students whom Bouwsma had encouraged to go to Cambridge to study with G. E. Moore. Bouwsma himself had been deeply influenced by Moore's work, having abandoned his earlier Hegelianism under the impact of Moore's refutations of idealism. Later, through another of his students, Alice Ambrose, he came across Wittgenstein's *Blue Book*, and made a close study of it.

After a few meetings at Malcolm's house in the company of others, Wittgenstein arranged to meet Bouwsma alone. It was then that the conversation quoted above took place. Wittgenstein had come to see Bouwsma primarily to ask him whether he thought their discussion had been 'any good' – did Bouwsma get anything out of it? 'I am a very vain person', he told him. 'The talk wasn't good. Intellectually, it may have been, but that isn't the point . . . My vanity, my vanity.' He discussed with Bouwsma his reasons for resigning his position at Cambridge:

> First I wanted to finish my book . . . Second, why should I teach? What good is it for X to listen to me? Only the man who thinks gets any good out of it.

He made an exception of a few students, 'who had a certain obsession and were serious'. But most of them came to him because he was clever, 'and I am clever, but it's not important'.

What was important was that his teaching should have a good effect, and in this respect the students with whom he was most satisfied were those who had not become professional philosophers – Drury, for example, and Smythies, and those that had become mathematicians. Within professional philosophy he thought his teaching had done more harm than good. He compared it to Freud's teachings, which, like wine, had made people drunk. They did not know how to use the teaching soberly. 'Do you understand?' he asked. 'Oh yes', Bouwsma replied. 'They had found a formula.' 'Exactly.'

That evening Bouwsma took Wittgenstein in his car up to the top of the hill overlooking the town. The moon was up. 'If I had planned it', said Wittgenstein, 'I should never have made the sun at all':

> See! How beautiful! The sun is too bright and too hot . . . And if there were only the moon there would be no reading and writing.

In addition to the other meetings mentioned, Wittgenstein had a number of private discussions with Malcolm. These are of special interest, because they provided the primary stimulus for the work which Wittgenstein produced in the last eighteen months of his life.

He had taken with him to Ithaca copies of both parts of *Philosophical Investigations*, so that he and Malcolm could read it through together. He told Malcolm that, though the book was not in a completely finished state, he did not now think that he could give the final polish to it in his lifetime. And, although he did not want to take it to a publisher in its unfinished state, he did want his friends to read and understand it. He therefore considered having it mimeographed and distributed among his friends with expressions of dissatisfaction, like: 'This is not quite right' or: 'This is fishy', written in parentheses after remarks that needed revision. Malcolm did not like the plan, and advised Wittgenstein against it; mimeograph copies, he thought, were an unfitting form of publication for a work of such importance.

A more time-consuming alternative would have been for Wittgenstein to go through the book paragraph by paragraph with each of his friends separately. This he seems to have made some attempt to do. Soon after arriving at Ithaca, he suggested that he and Malcolm should read through the book in this way – just as they had begun to do in Cambridge in 1946. Again, however, Malcolm found the procedure too confining, and after a few meetings the project was once more abandoned. Instead, they began a series of discussions on a philosophical issue that had more immediate relevance to Malcolm's own work.

The topic of these discussions was Moore's attempt to refute philosophical scepticism in his articles 'Proof of an External World' and 'A Defence of Common Sense'. Scepticism asserts that nothing about the external world can be known with certainty, not even that it is external. Moore's 'Proof of an External World' begins with an attempt to prove that some external objects at least can be shown with certainty to exist, his famous example being the existence of his own hands:

> I can prove now, for instance, that two human hands exist. How? By holding up my two hands, and saying, as I make a certain gesture with the right hand, 'Here is one hand', and adding, as I make a certain gesture with the left, 'and here is another.'

In 'A Defence of Common Sense' Moore provides a list of common-sense beliefs, all of which he claims to know with certainty to be true. These include: that a body exists which is Moore's body; that this body has been, throughout its existence, not far from the surface of the earth; that the earth had existed for many years before Moore was born; etc.

Not long before Wittgenstein's visit Malcolm had published an article criticizing Moore for incorrectly using the verb 'to know' in these claims to knowledge. Holding up a hand and saying: 'I know that this is a hand', or pointing to a tree and saying: 'I know for certain that this is a tree' is, Malcolm maintained, a senseless use of 'know'. Moore had written a spirited defence of his use of 'know' in a letter to Malcolm, and now Malcolm had the opportunity of finding out Wittgenstein's view of the matter; he was determined not to waste it.

In conversation with Malcolm, Wittgenstein insisted that: 'An expression has meaning only in the stream of life.' So whether or not Moore's statements were senseless depends on whether we can imagine an occasion on which they could sensibly be used: 'To understand a sentence is to be prepared for one of its uses. If we can't think of any use for it at all, then we don't understand it at all.' In this way:

> Instead of saying that Moore's statement 'I know that this is a tree' is a misuse of language, it is better to say that it has no clear meaning, and that Moore himself doesn't know how he is using it . . . It isn't even clear to him that he is not giving it an ordinary usage.

We could, Wittgenstein thought, imagine ordinary uses for some of Moore's statements more easily than for others: 'It isn't difficult to think of usages for "I know that this is a hand"; it is more difficult for "I know that the earth has existed for many years".'

Moore, of course, was not using his statements in an 'ordinary' way; he was using them to make a philosophical point. He was not informing his readers that he had two hands; he was attempting to refute philosophical scepticism. On this point Wittgenstein was quite clear that Moore had failed:

> When the sceptical philosophers say 'You don't know' and Moore replies 'I do know', his reply is quite useless, unless it is to assure them that he, Moore, doesn't feel any doubt whatever. But that is not what is at issue.

Wittgenstein's own view of scepticism remained that succinctly expressed in the *Tractatus*: 'Scepticism is *not* irrefutable, but obviously nonsensical, when it tries to raise doubts where no questions can be asked.' And it is in connection with this view of scepticism that he found something philosophically interesting about Moore's 'common-sense propositions'. They do not give examples of 'certain knowledge', but, rather, examples of cases in which doubt is nonsensical. If we could seriously doubt that Moore was holding up two hands, there would be no reason not to doubt anything else, including the trustworthiness of our senses. And in that case the whole framework in which we raise doubts and answer them would collapse: 'Certain propositions belong to my "frame of reference". If I had to give *them* up, I shouldn't be able to judge *anything*.' One such proposition might be the statement: 'That's a tree', said while standing in front of a tree:

> If I walked over to the tree and could touch nothing I might lose confidence in everything my senses told me . . . Moore said 'I know that there's a tree', partly because of the feeling that if it turned out *not* to be a tree, he would have to 'give up'.

The idea that there are certain judgements (among them, some of Moore's statements of common sense) that belong to our frame of reference, and as such cannot sensibly be doubted, was developed by Wittgenstein in the work written during the eighteen months left of his life following his visit to the United States.*

At the beginning of the autumn term Malcolm took Wittgenstein along to a meeting of the graduate students of philosophy at Cornell University. His presence there, as John Nelson has recalled, had a tremendous impact. 'Just before the meeting was to get underway', Nelson writes, 'Malcolm appeared approaching down the corridor':

> On his arm leaned a slight, older man, dressed in windjacket and old army trousers. If it had not been for his face, alight with intelligence, one might have taken him for some vagabond Malcolm had found along the road and decided to bring out of the cold.
>
> . . . I leaned over to Gass and whispered, 'That's Wittgenstein.' Gass thought I was making a joke and said something like, 'Stop pulling my leg.' And then Malcolm and Wittgenstein entered. [Gregory] Vlastos was introduced and gave his paper and finished. Black, who was conducting this particular meeting, stood up and turned to his right and it became clear, to everyone's surprise . . . that he was about to address the shabby older man Malcolm had brought to the meeting. Then came the startling words; said Black, 'I wonder if you would be so kind, Professor Wittgenstein . . .' Well, when Black said 'Wittgenstein' a loud and instantaneous gasp went up from the assembled students. You must remember: 'Wittgenstein' was a mysterious and awesome name in the philosophical world of 1949, at Cornell in particular. The gasp that went up was just the gasp that would have gone up if Black had said 'I wonder if you would be so kind, Plato . . .'

* This work has now been published as *On Certainty*.

Soon after this meeting Wittgenstein fell ill and was admitted to hospital for an examination. He had already booked his return passage to England in October, and was extremely frightened that he would have to remain in the United States as a result of the operation. He feared that, like Mining, he would be found to have cancer, and would be bed-ridden for the rest of his life. On the day before he went into hospital he said to Malcolm in a frenzy:

> I don't want to die in America. I am a European – I want to die in Europe . . . What a fool I was to come.

The examination, however, did not find anything seriously wrong with him, and he recovered sufficiently well over the next two weeks to return to England as planned, arriving in London at the end of October. His original plan was to spend a few days in Cambridge with von Wright and then return to Ross's Hotel in Dublin. Soon after arriving in London, however, he fell ill once more, and it was not until 9 November that he was able to go to Cambridge, still too ill to contemplate the journey to Dublin.

Drury had once told Wittgenstein that if he ever needed to see a doctor when in Cambridge he should consult Dr Edward Bevan. Drury had got to know Bevan during the war, when they had been in the same unit in the army, and had been impressed by his ability. By a coincidence, Bevan was also von Wright's family doctor. Soon after his arrival at Cambridge, therefore, Wittgenstein was examined by Dr Bevan. The final diagnosis was given on 25 November: cancer of the prostate.

Wittgenstein was in no way shocked to learn that he had cancer. He was, however, shocked to hear that something could be done about it. Cancer of the prostate often responds well to hormone therapy, and Wittgenstein was therefore immediately prescribed oestrogen. He was told that, with the help of this hormone, he could reasonably hope to live for another six years. 'I am sorry that my life should be prolonged in this way', he wrote to Rhees: 'six months of this half-life would be plenty.'

A few days after he learnt that he had cancer he wrote to Helene, asking if it would be convenient for him to come to Vienna to stay in the family home in the Alleegasse. 'My health is very bad', he told her, 'and I can therefore do no work. In Vienna I hope to find peace . . . If I could have my old room in the Alleegasse (with the overhead light), that would be good.'

Having warned her that she would find him in bad health, and that he would have to spend part of each day in bed, he told her nothing more about the nature of his illness. He was determined that his family should not know that he had cancer. Before he left for Vienna he wrote to Malcolm imploring him not to let anyone know about his illness: 'This is of the greatest importance for me as I plan to go to Vienna for Christmas & not to let my family know about the real disease.'

He flew to Vienna on 24 December, and moved into his old room at the Alleegasse. With Hermine lying in bed dying of cancer, and with Wittgenstein himself having the unmistakable pallor of a man stricken with the same disease, it is unlikely that his family did not guess the real nature of his illness. Wittgenstein, however, continued with his attempts at concealment, sending a telegram to von Wright with the euphemistic message: 'ARRIVED VIENNA EXCELLENT HEALTH AND SPIRITS. NOTIFY FRIENDS.'

For the first month that he was there, Wittgenstein did no writing at all. He allowed himself the luxury of leading the kind of comfortable life for which the family home was so admirably equipped. At the Alleegasse he was well fed and well looked after, and even entertained. 'I haven't been to a concert so far', he wrote to von Wright:

> but I hear a fair amount of music. A friend of mine [Rudolf Koder] plays the piano to me (very beautifully) and one of my sisters and he play piano duets. The other day they played two string quartets by Schumann and a sonata written for 4 hands by Mozart.

'I'm very happy & being treated *very* well', he wrote to Dr Bevan. Bevan had written to him about Ben, and had obviously commented on Ben's timidity. 'He isn't so much *timid*', Wittgenstein explained, 'as *very* shy & *very* repressed, particularly before he knows someone *well*':

> I wish I knew how important it really is that he should get a job at *Barts*. He seems to regard it as important. But I wish he could get out of London! I have an idea Barts isn't good for him. I *don't* mean by this that he is in any danger of becoming superficial, or snobbish, or anything like that. There is no danger there. But I wish he could be with more simple & more kindly people with whom he could open up, or he will get more & more withdrawn.

To his doctor, Wittgenstein was prepared to dwell a little on the state of his health. It was, as one would expect (especially in an Austrian winter – 'we had −15 C', he told Bevan), rather precarious:

> I had a rather nasty cold lately, accompanied by stomach troubles, I'm afraid I was on the verge of seeing a doctor & very worried about *that*, but it cleared up by itself & I'm almost as good as new again.

Of course, he could only really feel 'as good as new' in so far as he was capable of philosophizing. 'Colours spur us to philosophize. Perhaps that explains Goethe's passion for the theory of colours', he had written in 1948, and in January 1950 it was exactly with the intention of spurring himself to philosophize that he began reading Goethe's *Farbenlehre* ('Theory of Colour'). 'It's partly boring and repelling, but in some ways also very instructive and philosophically interesting', he told von Wright. Its chief merit, as he put it to Malcolm, was that it 'stimulates me to think'.

Eventually, it also stimulated him to write. A short series of twenty remarks inspired by reading Goethe's *Farbenlehre* have survived, presumably written during what was to be Wittgenstein's last visit to Vienna. They are now published as Part II of *On Colour*.

In them Wittgenstein connects Goethe's Theory of Colour, as he had earlier connected Goethe's other scientific work, with his own philosophical investigations. Contrary to Goethe himself, who regarded his theory as a successful refutation of Newton's Theory of Optics, Wittgenstein was clear that whatever interest the theory had, it was not as a contribution to physics. It was, rather, a *conceptual* investigation. From Wittgenstein's point of view this made it more, not less, interesting:

> I may find scientific questions interesting, but they never really grip me. Only conceptual and aesthetic questions do that. At bottom I am indifferent to the solution of scientific problems; but not of the other sort.

To be sure, Goethe's studies, like scientific investigations, were based on careful observations, but these observations do not enable us to construct explanatory laws. They do, however, enable us to clarify certain concepts. Take, for example, the proposition: 'Blending in white removes the colouredness from the colour; but blending in yellow does not.' What kind of proposition is this?

> As I mean it, it can't be a proposition of physics. Here the temptation to believe in a phenomenology, something midway between science and logic, is very great.

It can't be a proposition of physics, because its contrary is not false but meaningless: 'If someone didn't find it to be this way, it wouldn't be that he had experienced the contrary, but rather that we wouldn't understand him.' Therefore, to analyse this proposition (and others like it) is not to clarify a matter of fact, whether physical or phenomenological; it is to clarify certain concepts ('colour', 'colouredness', 'white' etc.). Thus:

> Phenomenological analysis (as e.g. Goethe would have it) is analysis of concepts and can neither agree with nor contradict physics.

On 11 February Hermine died. 'We had expected her end hourly for the last 3 days', Wittgenstein wrote to von Wright the next day. 'It wasn't a shock.'

His own health, meanwhile, continued to improve, and he was able to meet with Elizabeth Anscombe (who was in Vienna to improve her German in preparation for translating Wittgenstein's work) two or three times a week. Anscombe acted as a further stimulus to his attempts to recover his ability for philosophical work. At a meeting of the Austrian College Society in Alpbach, she had met Paul Feyerabend, who was then a student at the University of Vienna. She gave Feyerabend manuscripts of Wittgenstein's work and discussed them with him. Feyerabend was then a member of the Kraft Circle, an informal philosophy club founded by students at the university who were dissatisfied with the official course of lectures. It was exactly the kind of informal gathering at which Wittgenstein felt able to discuss philosophy publicly, and eventually he was persuaded to attend. Feyerabend recalls:

> Wittgenstein who took a long time to make up his mind and then appeared over an hour late gave a spirited performance and seemed

to prefer our disrespectful attitude to the fawning admiration he encountered elsewhere

This meeting of the Kraft Circle is probably the only public meeting of philosophers that Wittgenstein attended while he was in Vienna. His regular meetings with Anscombe, however, probably helped to 'spur him to philosophize'. As well as the twenty remarks on Goethe's theory of colour, there is also a series of sixty-five remarks which continue the topic of his conversations with Malcolm. These have now been published as the first sixty-five remarks of *On Certainty*. In them Wittgenstein insists that, like 'Moore's Paradox', Moore's 'Defence of Common Sense' is a contribution to logic. For: 'everything descriptive of a language-game is part of logic'.

The line of thought here is strikingly reminiscent of the *Tractatus* (as Wittgenstein himself later acknowledges: *On Certainty*, p. 321). The idea is that, if the contrary of a proposition makes sense, then that proposition can be regarded as an empirical hypothesis, its truth or falsity being dependent on the way things stand in the world. But if the contrary of a proposition does not make sense, then the proposition is not descriptive of the world but of our conceptual framework; it is then a part of logic.

Thus: 'Physical objects exist' is not an empirical proposition, for its contrary is not false but incomprehensible. Similarly, if Moore holds up two hands and our reaction is to say: 'Moore's hands don't exist', our statement could not be regarded as false but as unintelligible. But if this is so, then these 'framework propositions' do not describe a body of knowledge; they describe the way in which we understand the world. In this case, it makes no sense to claim, as Moore does, that you know them with certainty to be true:

> If 'I know etc', is conceived as a grammatical proposition, of course the 'I' cannot be important. And it properly means 'There is no such thing as a doubt in this case' or 'The expression "I do not know" makes no sense in this case.' And of course it follows from this that 'I know' makes no sense either.

There is an important parallel between these remarks on Moore and the remarks on Goethe. In both cases Wittgenstein's concern is to point out that what look like experiential propositions should actually be seen as grammatical propositions, descriptive not of our experience but of the framework within which our experience can be described. In some ways his discussion of both is an application of a general truth enunciated in the *Investigations*:

> If language is to be a means of communication there must be agreement not only in definitions but also (queer as this may sound) in judgments. This seems to abolish logic but does not do so.

Such statements as: 'Blending in white removes the colouredness from the colour' and: 'The earth has existed for a long time' are examples of such judgements. Recognizing them as such does not abolish logic, but it does considerably expand and complicate it, in such a way as to include within its domain discussions, for example, of Goethe's Theory of Colour and of Moore's 'Defence of Common Sense'.

The work written in Vienna is no more than the beginning of these discussions. Compared to the work written in Dublin the previous year, it is dull stuff. It has none of the aphoristic compression of that work, nor does it contain any of the kind of startlingly imaginative metaphors that characterize Wittgenstein's best work. It does show, however, that, as Wittgenstein's health gradually recovered, so too did his capacity for philosophical writing.

Wittgenstein left Vienna on 23 March and returned to London, where he stayed for a week at the home of Rush Rhees's wife, Jean, in Goldhurst Terrace. Being back in England was, he wrote, 'woeful'. The order of the place was 'loathsome'. The people seemed dead; every spark of life was extinguished.

On 4 April he moved back to von Wright's house in Cambridge, where he found waiting for him an invitation from Oxford University to deliver the John Locke lectures for 1950. These are an annual series of prestigious and relatively well-paid lectures which are traditionally delivered by a distinguished visiting philosopher. Despite the financial rewards (he was offered £200 to give them), he was not tempted. He was told that there would be a large audience of over 200 students, and that there was not to be any discussion during the lectures. No two conditions could be more likely to put him off. He told Malcolm: 'I don't think I can give formal lectures to a large audience that would be any good.'

Concerned that Wittgenstein's money would soon run out, Malcolm approached the Rockefeller Foundation on his behalf. He told Wittgenstein that he had managed to interest a director of the foundation, Chadbourne Gilpatrick, in the possibility of awarding him a research grant. Wittgenstein's gratitude for this was tempered by a remarkable piece of fiercely honest self-assessment. There were, of course, reasons to accept the grant:

> The thought of being able to live where I like, of not having to be a burden or a nuisance to others, of doing philosophy when my nature inclines me to do it is, of course, pleasant for me.

But, he told Malcolm, he could not take the money unless the Rockefeller Foundation 'knew the complete truth about me':

> The truth is this. a) I have not been able to do any sustained good work since the beginning of March 1949. b) Even before that date I could not work well for more than 6 or 7 months a year. c) As I'm getting older my thoughts become markedly less forceful & crystallize more rarely & I get tired very much more easily. d) My health is in a somewhat labile state owing to a constant slight anaemia which inclines me to catch infections. This further diminishes the chance of my doing really good work. e) Though it's impossible for me to make any definite predictions, it seems to me likely that my mind will never again work as vigorously as it did, say, 14 months ago. f) I cannot promise to publish anything during my lifetime.

He asked Malcolm to show this letter to the foundation's directors. 'It is obviously impossible to accept a grant under false pretences, & you *may* unintentionally have presented my case in too rosy a light.' 'I believe', he added, 'that as long as I live & as often as the state of my mind permits it I will think about philosophical problems & try to write about them':

> I also believe that much of what I wrote in the past 15 or 20 years may be of interest to people when it's published. But it is, nevertheless, perfectly possible that all that I'm going to produce will be flat, uninspired & uninteresting.

When, eight months later, he was visited by Gilpatrick, Wittgenstein told him: 'in my present state of health & intellectual dullness I couldn't accept a grant'.

He attributed his 'intellectual dullness' in part to the oestrogens that he was taking to alleviate the symptoms of his cancer. While taking them he found the intense concentration required to write philosophy difficult to achieve. 'I'm doing some work', he told Malcolm, on 17 April, 'but I get stuck over simple things & almost all I write is pretty dull.'

The work in question forms Part III of *On Colour*, and is a continuation of the remarks on Goethe's *Farbenlehre* that he wrote in Vienna. In some ways it bears out Wittgenstein's own assessment of it: it is a repetitive and rather laboured attempt to clarify the 'logic of colour concepts', in particular the concepts of 'primary colour', 'transparency' and 'luminosity'. Wittgenstein's dissatisfaction with it is manifest: 'That which I am writing about so tediously, may be obvious to someone whose mind is less decrepit.' It does, however, contain a marvellously succinct dismissal of Goethe's remarks on the general characteristics of various colours:

> One and the same musical theme has a different character in the minor than in the major, but it is completely wrong to speak of the character of the minor mode in general. (In Schubert the major often sounds more sorrowful than the minor.)
>
> And in this way I think that it is worthless and of no use whatsoever for the understanding of painting to speak of the characteristics of the individual colours. When we do it, we are really thinking of special uses. That green as the colour of a tablecloth has this, red that effect, does not allow us to draw any conclusions as to their effect in a picture.

> Imagine someone pointing to a spot in the iris in a face by Rembrandt and saying 'the wall in my room should be painted this colour'.

It was while he was staying with the von Wrights in April 1950 that the last photographs of Wittgenstein were taken. They show Wittgenstein and von Wright sitting together in folding chairs in front of a bed sheet. This odd arrangement was, as K. E. Tranøj (who took the photographs) remembers, Wittgenstein's own idea:

> In the late spring of 1950 we had tea with the von Wrights in the garden. It was a sunny day and I asked Wittgenstein if I could take a photograph of him. He said, yes, I could do that, if I would let him sit with his back to the lens. I had no objections and went to get my camera. In the meantime Wittgenstein had changed his mind. He now decided I was to take the picture in the style of a passport photograph, and von Wright was to sit next to him. Again I agreed, and Wittgenstein now walked off to get the sheet from his bed; he would not accept Elizabeth von Wright's offer of a fresh sheet from her closets. Wittgenstein draped the sheet, hanging it in front of the verandah, and pulled up two chairs.

On 25 April Wittgenstein left Cambridge to move into Elizabeth Anscombe's house in St John Street, Oxford. 'I like to stay with the von Wrights', he told Malcolm, 'but the two children are noisy & I need quiet.' At Anscombe's house he occupied a room on the second floor, the ground floor being occupied by Frank and Gillian Goodrich, and the first by Barry Pink. Soon after he moved in he told von Wright: 'The house isn't very noisy but not very quiet, either. I don't know yet how I shall get on. The lodgers seem all to be rather nice, and one of them even very nice.'

The 'very nice' one was Barry Pink, who was then attending art college. His interests were many and varied: 'Pink wants to sit on six stools at once', Wittgenstein once remarked, 'but he only has one arse.' Pink had long been a friend of Yorick Smythies, and, like Smythies and Anscombe, was a convert to Catholicism. He found in Wittgenstein someone willing and able to converse with him about the whole range of his interests – art, sculpture, stone-masonry, machine-construction and so on.

The two took walks together around Oxford, and for a time Pink became a confidant. They were able to discuss their thoughts, feelings and lives with a fair degree of frankness. They discussed, for example, the tendency to hide one's real nature. In connection with this, Pink asked Wittgenstein whether he thought his work as a philosopher, even his being a philosopher, had anything to do with his homosexuality. What was implied was that Wittgenstein's work as a philosopher may in some way have been a device to hide from his homosexuality. Wittgenstein dismissed the question with anger in his voice: 'Certainly not!'

Wittgenstein planned to spend the summer in Norway with Ben, who was then in his final year as a medical student at Bart's in London. In July, however, Ben failed his final qualifying exams, and so had to stay in London during the summer to study for the 'retake' in September. Their holiday was therefore postponed until the autumn, and Wittgenstein stayed in Oxford throughout the summer, devoting himself to a continuation of the remarks on colour that he had written in Cambridge.

In the same manuscript notebook as the remarks on colour is a series of remarks on Shakespeare, which have been published in *Culture and Value* (pp. 84–6). Wittgenstein had long been troubled by his inability to appreciate the greatness of Shakespeare. In 1946, for example, he had written:

> It is remarkable how hard we find it to believe something that we do not see the truth of for ourselves. When, for instance, I hear the expression of admiration for Shakespeare by distinguished men in the course of several centuries, I can never rid myself of the suspicion that praising him has been the conventional thing to do; though I have to tell myself that this is not how it is. It takes the authority of a Milton really to convince me. I take it for granted that he was incorruptible. – But I don't of course mean by this that I don't believe an enormous amount of praise to have been, and still to be, lavished on Shakespeare without understanding and for the wrong reasons by a thousand professors of literature.

One of the difficulties he had in accepting Shakespeare as a great poet was that he disliked many of Shakespeare's metaphors and similes: 'Shakespeare's similes are, *in the ordinary sense*, bad. So if they are all the same good – and I don't know whether they are or not – they must be a law to themselves.' An example he discussed with Ben was the use of a portcullis as a metaphor for teeth in Mowbray's speech in *Richard II*: 'Within my mouth you have engaol'd my tongue/ Doubly portcullis'd with my teeth and lips.'

A more fundamental difficulty was Wittgenstein's dislike of English culture in general: 'I believe that if one is to enjoy a writer one has to like the culture he belongs to as well. If one finds it indifferent or distasteful, one's admiration cools off.' This did not prevent Wittgenstein from admiring Blake or Dickens. The difference is that, in Shakespeare, Wittgenstein could not see a writer whom he could admire as a great human being:

> I could only stare in wonder at Shakespeare; never do anything with him . . .
>
> 'Beethoven's great heart' – nobody could speak of 'Shakespeare's great heart' . . .
>
> I do not think that Shakespeare would have been able to reflect on the 'lot of the poet'.
>
> Nor could he regard himself as a prophet or as a teacher of mankind.
>
> People stare at him in wonderment, almost as at a spectacular natural phenomenon. They do not have the feeling that this brings them into contact with a great *human being*. Rather with a phenomenon.

In Dickens, on the other hand, Wittgenstein did find an English writer whom he could respect for his 'good universal art' – in the Tolstoyan sense of art that is intelligible to everyone, and that espouses Christian virtues. He gave a pocket edition of *A Christmas Carol*, bound in green leather and plastered with cheerful 'Merry Xmas' stickers, to Fouracre as a belated Christmas present when Fouracre returned from Sumatra. The choice of book, of course, is significant. F. R. Leavis recalls that Wittgenstein knew *A Christmas Carol* practically by heart, and the book is, in fact, placed by Tolstoy in his treatise *What is Art?* in the very highest category of art 'flowing from the love of God'. It is thus a highly appropriate gift to make within a friendship that stands out in Wittgenstein's life as a rare example of his Tolstoyan respect for 'the common man', exemplified in a simple and straightforward affection for an ordinary working man.

In the late summer of 1950 Wittgenstein resumed his remarks on the philosophical significance of Moore's 'common-sense propositions'. This work now constitutes remarks 65–299 of *On Certainty*. In it Wittgenstein elaborates on the idea that Moore's statements have the peculiarity that their denial is not just false, but incomprehensible:

> If Moore were to pronounce the opposite of those propositions which he declares certain, we should not just not share his opinion: we should regard him as demented.

Thus: 'if I make certain false statements, it becomes uncertain whether I understand them.' Moore gives us examples of such statements. Another example might be knowing where one lives:

> For months I have lived at address A, I have read the name of the street and the number of the house countless times, have received countless letters here and have given countless people the address. If I am wrong about it, the mistake is hardly less than if I were (wrongly) to believe I was writing Chinese and not German.
>
> If my friend were to imagine one day that he had been living for a long time past in such and such a place, etc. etc., I should not call this a mistake, but rather a mental disturbance, perhaps a transient one.

A mistake becomes, for us, a mental disturbance when it contradicts, not just this or that proposition which we believe to be true, but the whole framework within which we can give grounds for our beliefs. The only occasion Wittgenstein can think of in which it would be appropriate for Moore to assert: 'I know that I have not left the surface of the earth', would be one in which he is faced by people operating within a vastly different framework:

> I could imagine Moore being captured by a wild tribe, and their expressing the suspicion that he has come from somewhere between the earth and the moon. Moore tells them that he knows etc. but he can't give them the grounds for his certainty, because they have fantastic ideas of human ability to fly and know nothing about physics. This would be an occasion for making that statement.

But this example indicates that a different framework is not necessarily evidence of insanity. In 1950 it was absurd to suppose that someone might have been in outer space and returned to earth. We have now grown accustomed to the idea. Frameworks change, both between different cultures and within a culture between different times.

This, though, is not a point against Wittgenstein. On the contrary, he stresses that a framework itself cannot be justified or proven correct; it provides the limits within which justification and proof take place:

> Everything that I have seen or heard gives me the conviction that no man has ever been far from the earth. Nothing in my picture of the world speaks in favour of the opposite.
>
> But I did not get my picture of the world by satisfying myself of its correctness; nor do I have it because I am satisfied of its correctness. No: it is the inherited background against which I distinguish between true and false.

Frameworks change: what was once dismissed as absurd may now be accepted; hard and fast certainties become dislodged and abandoned. Nevertheless, we cannot make sense of anything without some sort of framework, and with any particular framework there has to be a distinction between propositions that, using that framework, describe the world, and those that describe the framework itself, though this distinction is not fixed at the same place for ever:

> . . . the river-bed of thoughts may shift. But I distinguish between the movements of the waters on the river-bed and the shift of the bed itself; though there is not a sharp division of the one from the other.

We do not, however, need to consider imaginary wild tribes to find examples of people with a world picture fundamentally different to our own:

> I believe that every human being has two parents; but Catholics believe that Jesus only had a human mother. And other people might believe that there are human beings with no parents, and give no credence to all the contrary evidence. Catholics believe as well that in certain circumstances a wafer completely changes its nature, and at the same time that all evidence proves the contrary. And so if Moore said 'I know that this is wine and not blood', Catholics would contradict him.

This remark was possibly prompted by a conversation about Transsubstantiation that Wittgenstein had with Anscombe about this time. He was, it seems, surprised to hear from Anscombe that it really was Catholic belief that 'in certain circumstances a wafer completely changes its nature'. It is presumably an example of what he had in mind when he remarked to Malcolm about Anscombe and Smythies: 'I could not possibly bring myself to believe all the things that they believe.' Such beliefs could find no place in his own world picture. His respect for Catholicism, however, prevented him from regarding them as mistakes or 'transient mental disturbances'.

'I have a world picture. Is it true or false? Above all it is the substratum of all my enquiring and asserting.' There is no reason why a religious faith should not provide this substratum, and why religious beliefs should not be part of 'the inherited background against which I distinguish between true and false'. But for this a thoroughgoing religious education and instruction may be necessary: 'Perhaps one could "convince someone that God exists" by means of a certain upbringing, by shaping his life in such and such a way.'

But what intelligibility can religious belief have in the absence of such an upbringing? Wittgenstein seems to have thought the concept of God can, in some cases (his own for example), be forced upon one by life:

> Life can educate one to a belief in God. And experiences too are what bring this about; but I don't mean visions and other forms of experience which show us the 'existence of this being', but, e.g., sufferings of various sorts. These neither show us an object, nor do they give rise to conjectures about him. Experiences, thoughts, – life can force this concept on us.
>
> So perhaps it is similar to the concept of 'object'.

Of course, in this case, the form that this faith takes is unlikely to be acceptance, or even comprehension, of the Catholic doctrines of the Virgin Conception and Transsubstantiation. What is forced upon one is, rather, a certain attitude:

> The attitude that's in question is that of taking a certain matter seriously and then, beyond a certain point, no longer regarding it

> as serious, but maintaining that something else is even more important.
>
> Someone may for instance say it's a grave matter that such and such a man should have died before he could complete a certain piece of work; and yet, in another sense, this is not what matters. At this point one uses the words 'in a deeper sense'. Actually I should like to say that in this case too the *words* you utter or what you think as you utter them are not what matters, so much as the difference they make at various points in your life. How do I know that two people mean the same when each says he believes in God? And the same goes for belief in the Trinity. A theology which insists on the use of *certain particular* words and phrases, and outlaws others, does not make anything clearer (Karl Barth). It gesticulates with words, as one might say, because it wants to say something and does not know how to express it. *Practice* gives the words their sense.

Wittgenstein's example in the second paragraph is not, of course, arbitrary. But if, as it implies, the completion of *Philosophical Investigations* before his death is not what matters, then what is this 'something else' that, 'in a deeper sense', is even more important?

The answer seems to be: his reconciliation with God. In the autumn Wittgenstein asked Anscombe if she could put him touch with a 'non-philosophical' priest. He did not want to discuss the finer points of Catholic doctrine; he wanted to be introduced to someone to whose life religious belief had made a practical difference. She introduced him to Father Conrad, the Dominican priest who had instructed Yorick Smythies during his conversion to Catholicism. Conrad came to Anscombe's house twice to talk to Wittgenstein. 'He wanted', Conrad recalls, 'to talk to a priest as a priest and did not wish to discuss philosophical problems':

> He knew he was very ill and wanted to talk about God, I think with a view to coming back fully to his religion, but in fact we only had, I think, two conversations on God and the soul in rather general terms.

Anscombe, however, doubts that Wittgenstein wanted to see Conrad 'with a view to coming back fully to his religion', if by that Conrad means that Wittgenstein wanted to return to the Catholic

Church. And, given Wittgenstein's explicit statements that he could not believe certain doctrines of the Catholic Church, it seems reasonable to accept her doubt.

In September Ben successfully retook his final qualifying exams, leaving him free to join Wittgenstein on their postponed visit to Norway. In the first week of October, therefore, they set off on the long and difficult journey to Wittgenstein's remote hut by the side of the Sogne fjord.

To travel so far north at that time of the year might have been considered a foolhardy risk for Wittgenstein to take in his precarious state of health. It was, however, Ben's health that suffered from the cold. After they had been in Norway a short time he developed bronchitis and had to be moved from Wittgenstein's hut into a nursing home further up the fjord. They then moved into Anna Rebni's farmhouse, where they spent the remainder of their holiday.

Ben had taken with him a copy of J. L. Austin's recently published translation of Frege's *Foundations of Arithmetic*, and while in Norway he and Wittgenstein spent much time reading and discussing Frege's work. Wittgenstein began to think that he might, after all, be able to live alone once again in Norway and work on philosophy.

On his return to Oxford, he wrote to von Wright that, in spite of Ben's illness, 'we enjoyed our stay enormously':

> We had excellent weather the whole time and were surrounded by the greatest kindness. I decided then and there that I'd return to Norway to work there. I get no real quiet here. If all goes well I shall sail on Dec. 30th and go to Skjolden again. I don't think I'll be able to stay in my hut because the physical work I've got to do there is too heavy for me, but an old friend told me that she'd let me stay at her farmhouse. Of course I don't know whether I'm able any more to do decent work, but at least I'm giving myself a real chance. If I can't work there I can't work anywhere.

He even booked a passage on a steamer from Newcastle to Bergen for 30 December. Shortly before Christmas, however, he was told by Anna Rebni that she could not put him up after all. Wittgenstein was, in any case, in no fit state to make the journey. While visiting Dr Bevan to have an examination before his proposed trip to Norway, he had fallen ill in Bevan's house, and had to remain there for Christmas. None of this deterred him from his plan, however, and after Christmas he wrote to Arne Bolstad, another friend in Norway, asking him if he knew of a suitable place for him to live and work in isolation. This, too, came to nothing.

His plans to go to Norway thwarted, Wittgenstein tried another favoured refuge: a monastery. Father Conrad made arrangements on Wittgenstein's behalf for him to stay at a Blackfriars Priory in the Midlands, where he could live the life of a Brother, doing the house chores such as washing up and, most important, where he could be alone.

By January 1951, however, Wittgenstein's health made all these plans impractical. He needed constant medical attention. As his condition worsened, he had to travel to Cambridge with increasing frequency to see Dr Bevan. In addition to the hormones he was taking, he was also given X-ray treatment at Addenbrooke's Hospital.

He had a deep horror of the idea of dying in an English hospital, but Bevan promised him that, if necessary, he could spend his last days being looked after at Bevan's own home. At the beginning of February, Wittgenstein decided to accept his offer, and so moved to Cambridge, there to die in Bevan's home: 'Storeys End'.

27
STOREYS END

Wittgenstein arrived at the Bevans' home resigned to the fact that he would do no more work. He had written nothing since his visit to Norway, and now that he had been forced to give up the idea of living and working by the side of the Sogne fjord, his only wish was that these unproductive last months of his life should be few in number. 'I can't even *think* of work at present', he wrote to Malcolm, '& it doesn't matter, if only I don't live too long!'

Mrs Bevan was initially somewhat frightened of Wittgenstein, especially after their first meeting, which was something of an ordeal for her. Before Wittgenstein moved into their house, Dr Bevan had invited him for supper to introduce him to his wife. She had been warned by her husband that Wittgenstein was not one for small talk and that she should be careful not to say anything thoughtless. Playing it safe, she remained silent throughout most of the evening. But when Wittgenstein mentioned his visit to Ithaca, she chipped in cheerfully: 'How lucky for you to go to America!' She realized at once she had said the wrong thing. Wittgenstein fixed her with an intent stare: 'What do you mean, *lucky*?'

After Wittgenstein had been there a few days, however, she began to relax in his company, and eventually they became close friends. Not that he was a particularly easy guest:

> He was very demanding and exacting although his tastes were very simple. It was *understood* that his bath would be ready, his meals on time and that the events of the day would run to a regular pattern.

It was also understood that Wittgenstein should pay for nothing while he was there – not even for the items on the shopping lists that he would leave lying on the table for Mrs Bevan to collect when she went out. These items would include food and books, and, of course, every month, Street & Smith's *Detective Story Magazine*.

When they had become friends, Wittgenstein and Mrs Bevan would, as part of the regular pattern of each day, walk to the local pub at six o'clock in the evening. Mrs Bevan remembers: 'We always ordered two ports, one I drank and the other one he poured with great amusement into the Aspidistra plant – this was the only dishonest act I ever knew him to do.' Conversation with Wittgenstein was, in spite of her first experience of him, surprisingly easy: 'It was remarkable that he never discussed or tried to discuss with me, subjects which I did not understand, so that in our relationship I never felt inferior or ignorant.' This is not to say, however, that the significance of all his remarks was always transparently clear; perhaps the most gnomic was his comment on Peter Geach, Elizabeth Anscombe's husband. When Mrs Bevan asked Wittgenstein what Geach was like, he replied solemnly: 'He reads Somerset Maugham.'

In February Wittgenstein wrote to Fouracre:

> I have been ill for some time, about 6 weeks, & have to spend part of the day in bed. I don't know when I'll come to London again. If there's no chance of it I'll let you know & you might be able to visit me here on a Sunday sometime.

He does not tell Fouracre that he has cancer, and as his condition deteriorated sharply soon after this letter was written, it is unlikely that Wittgenstein met Fouracre again in London. But the fact that it should even have been suggested at this stage illustrates how important to him his meetings with his ex-colleague at Guy's had become.

At the end of February it was decided that there was no further point in continuing Wittgenstein's hormone and X-ray treatment. This, even when accompanied by the information that he could not expect to live more than a few months, came as an enormous relief to him. He told Mrs Bevan: 'I am going to work now as I have never worked before.' Remarkably, he was right. During the two months left of his life Wittgenstein wrote over half (numbered paragraphs 300–676) of the remarks which now constitute *On Certainty*, and in doing so produced what many people regard as the most lucid writing to be found in any of his work.

The work picks up the threads of Wittgenstein's earlier discussions of Moore's 'Defence of Common Sense', but explores the issues in much greater depth, and expresses the ideas with much greater clarity and succinctness than hitherto. Even when he is chiding himself for his own lack of concentration, he does so with an amusingly apt simile: 'I do philosophy now like an old woman who is always mislaying something and having to look for it again: now her spectacles, now her keys.' Despite this self-deprecation, he was in no doubt that the work he was now writing would be of interest: 'I believe it might interest a philosopher, one who can think himself, to read my notes. For even if I have hit the mark only rarely, he would recognize what target I had been ceaselessly aiming at.'

The target he was aiming at was the point at which doubt becomes senseless – the target at which he believed Moore to have made an inaccurate shot. We cannot doubt everything, and this is true not only for practical reasons, like insufficient time or having better things to do; it is true for the intrinsic, logical reason that: 'A doubt without an end is not even a doubt.' But we do not reach that end with statements that begin: 'I know . . .' Such statements have a use only in the 'stream of life'; outside of it, they appear absurd:

> I am sitting with a philosopher in the garden; he says again and again 'I know that that's a tree', pointing to a tree that is near us. Someone else arrives and hears this, and I tell them: 'This fellow isn't insane. We are only doing philosophy.'

We reach the end of doubt, rather, in practice: 'Children do not learn that books exist, that armchairs exist, etc., etc., – they learn to fetch books, sit in armchairs, etc. etc.' Doubting is a rather special sort of practice, which can be learnt only after a lot of non-doubting behaviour has been acquired: 'Doubting and non-doubting behaviour. There is a first only if there is a second.' The thrust of Wittgenstein's remarks is to focus the attention of philosophers away from words, from sentences, and on to the occasions in which we use them, the contexts which give them their sense:

> Am I not getting closer and closer to saying that in the end logic cannot be described? You must look at the practice of language, then you will see it.

His attitude is summed up by Goethe's line in Faust: '*Im Anfang war die Tat*' ('In the beginning was the deed'), which he quotes with approval, and which might, with some justification, be regarded as the motto of *On Certainty* – and, indeed, of the whole of Wittgenstein's later philosophy.

The last remark of *On Certainty* was written on 27 April, the day before Wittgenstein finally lost consciousness. The day before that was his sixty-second birthday. He knew it would be his last. When Mrs Bevan presented him with an electric blanket, saying as she gave it to him: 'Many happy returns', he stared hard at her and replied: 'There will be no returns.' He was taken violently ill the next night, after he and Mrs Bevan had returned from their nightly stroll to the pub. When told by Dr Bevan that he would live only a few more days, he exclaimed 'Good!' Mrs Bevan stayed with him the night of the 28th, and told him that his close friends in England would be coming the next day. Before losing consciousness he said to her: 'Tell them I've had a wonderful life.'

The next day Ben, Anscombe, Smythies and Drury were gathered at the Bevans' home to be with Wittgenstein at his death. Smythies had brought with him Father Conrad, but no one would decide whether Conrad should say the usual office for the dying and give conditional absolution, until Drury recollected Wittgenstein's remark that he hoped his Catholic friends prayed for him. This decided the matter, and they all went up to Wittgenstein's room and kneeled down while Conrad recited the proper prayers. Shortly after this, Dr Bevan pronounced him dead.

The next morning he was given a Catholic burial at St Giles's Church, Cambridge. The decision to do this was again prompted by a recollection of Drury's. He told the others:

> I remember that Wittgenstein once told me of an incident in Tolstoy's life. When Tolstoy's brother died, Tolstoy, who was then a stern critic of the Russian Orthodox Church, sent for the parish priest and had his brother interred according to the

> Orthodox rite. 'Now', said Wittgenstein, 'that is exactly what I should have done in a similar case.'

When Drury mentioned this, everyone agreed that all the usual Roman Catholic prayers should be said by a priest at the graveside, although Drury admits: 'I have been troubled ever since as to whether what we did then was right.' Drury does not expand on this, but the trouble perhaps stems from doubt as to whether the story about Tolstoy quite fits the occasion. For the point of the story is that, although not himself an adherent of the Orthodox Church, Tolstoy had the sensitivity to respect his brother's faith. But in Wittgenstein's case the position is reversed: it was Anscombe and Smythies, and not he, who adhered to the Catholic faith.

Wittgenstein was not a Catholic. He said on a number of occasions, both in conversation and in his writings, that he could not bring himself to believe the things that Catholics believe. Nor, more important, did he practise Catholicism. And yet there seems to be something appropriate in his funeral being attended by a religious ceremony. For, in a way that is centrally important but difficult to define, he had lived a devoutly religious life.

A few days before his death Wittgenstein was visited in Cambridge by Drury, and remarked to him: 'Isn't it curious that, although I know I have not long to live, I never find myself thinking about a "future life". All my interest is still on this life and the writing I am still able to do.' But if Wittgenstein did not think of a future life, he did think of how he might be judged. Shortly before his death he wrote:

> God may say to me: 'I am judging you out of your own mouth. Your own actions have made you shudder with disgust when you have seen other people do them.'

The reconciliation with God that Wittgenstein sought was not that of being accepted back into the arms of the Catholic Church; it was a state of ethical seriousness and integrity that would survive the scrutiny of even that most stern of judges, his own conscience: 'the God who in my bosom dwells'.

APPENDIX: BARTLEY'S WITTGENSTEIN AND THE CODED REMARKS

One of the books that has done most to stimulate interest in Wittgenstein's life in recent years has been W. W. Bartley III's short study, *Wittgenstein*. This is an account of Wittgenstein's 'lost years', from 1919 to 1929, during which he abandoned philosophy and worked as an elementary school teacher in rural Austria. Bartley's chief interest in writing the book seems to have been to emphasize the philosophical relevance of this part of Wittgenstein's life, and, in particular, the influence on Wittgenstein's later philosophy of the educational theories of the Austrian School Reform Movement (the movement that shaped educational policy in Austria after the First World War).

Interest in Bartley's book, however, has tended to focus, not on his main themes, but almost exclusively on the sensational claims he makes towards the beginning of it about Wittgenstein's sexuality. The interest generated by these assertions is, in my opinion, disproportionate, but I feel obliged to say something about them. The question I was asked most during the writing of this book was: 'What are you going to do about Bartley?' – meaning: what response was I going to give in my book to Bartley's claims about Wittgenstein's homosexual promiscuity?

What are these claims? According to Bartley, while Wittgenstein was training as a school teacher and living on his own in lodgings in Vienna, he discovered an area in the nearby *Prater* (a large park in Vienna, analogous, perhaps, to Richmond Park in London), where 'rough young men were ready to cater to him sexually'. Once Wittgenstein had discovered this place, Bartley maintains:

> [he] found to his horror that he could scarcely keep away from it. Several nights each week he would break away from his rooms and make the quick walk to the Prater, possessed, as he put it to friends, by a demon he could

> barely control. Wittgenstein found he much preferred the sort of rough blunt homosexual youth he could find strolling the paths and alleys of the Prater to those ostensibly more refined young men who frequented the Sirk Ecke in the Kärtnerstrasse and the neighbouring bars at the edge of the city. [*Wittgenstein*, p. 40]

In an 'Afterword', written in 1985 and published in a revised edition of his book, Bartley clears up one widespread misinterpretation of this passage. He had not, it seems, meant to imply that the 'rough young men' in question were prostitutes. But, cleared of this misunderstanding, he stands by the truth of what he says.

He does not, however, clear up the mystery of how he knows it to be true. Neither in this revised edition of the book nor in the original does he give any source for these claims. He says merely that his information is based on 'confidential reports from his [Wittgenstein's] friends'.

Ever since it appeared, this passage has been the subject of a heated and apparently unresolvable controversy. Many who knew Wittgenstein well felt outraged, and vented their anger in reviews and letters to periodicals pouring scorn on Bartley's book and swearing that his claims about Wittgenstein's sexuality were false – that they *had* to be false, since the Wittgenstein they knew could not have done such things.

On the other hand, many who did not know Wittgenstein but who had read his published correspondence and the memoirs of him written by his friends and students felt inclined to believe what Bartley said – felt, in fact, that Bartley had provided the key to Wittgenstein's tormented personality. A few (though not Bartley himself) even thought that these sexual encounters provided the key to understanding Wittgenstein's philosophy. Colin Wilson, for example, in his book *The Misfits: A Study of Sexual Outsiders* (the theme of which is the connection between genius and sexual perversion), states that it was only after he had read Bartley's book that he felt he understood Wittgenstein's work.

Many people, it seems, find it so natural to think of Wittgenstein as guiltily and promiscuously homosexual that they are inclined to accept Bartley's claims without any evidence. It somehow 'fits' with their image of Wittgenstein – so much so that the picture of Wittgenstein guiltily wandering the paths of the Prater in search of 'rough young homosexual youths' has become an indelible part of the public image of him. Wittgenstein is, I was once assured, the 'Joe Orton of philosophy'.

Another reason, I think, why Bartley's claims have been generally accepted is the widespread feeling that Wittgenstein's friends, and especially his literary executors, would not admit the truth of such things, even if they knew them to be true. There is, it is felt, a cover-up going on. One of

Wittgenstein's executors, Professor Elizabeth Anscombe, gave ammunition to that view when, in a letter to Paul Engelmann (published in the introduction to Engelmann's *Letters from Ludwig Wittgenstein with a Memoir*), she stated:

> If by pressing a button it could have been secured that people would not concern themselves with his personal life I should have pressed the button.

Further ammunition has been provided by the executors' attitude to the personal remarks that Wittgenstein wrote in his philosophical manuscripts – the so-called coded diaries.

These remarks were separated by Wittgenstein from his philosophical remarks by being written in a very simple code that he had learnt as a child (whereby a = z, b = y, c = x etc.). The simplicity of the code, and the fact that Wittgenstein used it to write instructions concerning the publication of his work, suggest that he used it, not to disguise what he was saying from posterity, but, rather, to disguise it from someone who, say, happened to lean over his shoulder or who happened to see his manuscript volume lying on a table.

A collection of these remarks, the less personal of them, has been published under the title *Culture and Value*. The more personal remarks have remained unpublished. In the microfilm edition of Wittgenstein's complete manuscripts, these more personal remarks have been covered up with bits of paper.

All this has (*a*) increased people's curiosity about what the coded remarks contain; and (*b*) confirmed their view that the executors are hiding something. And this, in turn, has helped to create a climate of opinion favourable to the acceptance of Bartley's otherwise extraordinary allegations. 'Aha!' people have thought. 'So that's what Anscombe has been covering up all these years!'

Bartley himself has made use of this climate of opinion to defend himself against accusations of peddling falsehoods. In the aforementioned 'Afterword' (which is subtitled 'A Polemical Reply to my critics'), he alleges that the executors were bluffing when they expressed their outrage at his book. For all the time they:

> . . . had coded notebooks, in Wittgenstein's own hand, written in a very simple cipher and long since decoded and transcribed, corroborating my statements about his homosexuality.

Now this is actually not true. In the coded remarks Wittgenstein *does* discuss his love for, first David Pinsent, then Francis Skinner, and finally Ben Richards (this is over a period of some thirty years or so), and in that sense they do 'corroborate' his homosexuality. But they do nothing to corroborate *Bartley*'s statements about Wittgenstein's homosexuality. That is, there is not a word in them about going to the Prater in search of 'rough young men', nor is there anything to suggest that Wittgenstein engaged in promiscuous behaviour at any time in his life. Reading them, one would rather get the impression that he was incapable of such promiscuity, so troubled does he seem by even the slightest manifestation of sexual desire (homosexual or heterosexual).

Not many people would have been able to point this out, because few have ever seen the coded remarks. Indeed, the way Bartley himself talks of 'coded notebooks' suggests that his information, too, is second-hand – that he has not actually seen the sources he mentions. There simply *are* no coded notebooks. The coded remarks are not gathered together in two volumes (as Bartley appears to think), but are scattered throughout the eighty or so notebooks that constitute Wittgenstein's literary and philosophical *Nachlass*. This alleged 'corroboration', then, is entirely spurious.

Of the many attempts to refute Bartley, the most often quoted are those by Rush Rhees and J. J. Stonborough in *The Human World* (No. 14, Feb. 1972). They are, in my opinion, unsuccessful. Rhees, indeed, does not even attempt to refute Bartley in the ordinary sense of showing what Bartley says to be false. The gist of his argument is that, even if what Bartley says is true, it was 'foul' of him to repeat it. Stonborough's piece, stripped of its bombast, its heavy-handed irony and its moral indignation, contains only one, rather flimsy, argument: that if Wittgenstein had behaved as Bartley suggests, he would have been blackmailed. This argument is quite easily dealt with by Bartley in his 'Afterword'. By concentrating on the morality of Bartley's book, rather than on the veracity of his information, Rhees and Stonborough have, I believe, merely clouded the issue, and, inadvertently, let Bartley off the hook.

The only way in which Bartley's statements could be effectively refuted is by showing, either that the information he received was false, or that he had misinterpreted it. And before that can even be attempted, it is necessary to know what that information was. This, Bartley has resolutely refused to reveal.

Elsewhere in Bartley's book there are signs that, in writing it, he had access to a manuscript of Wittgenstein's dating from the years 1919–20. The most striking indication of this occurs on page 29 (of the revised edition), when he quotes a dream-report of Wittgenstein's and comments on Wittgenstein's own interpretation of the dream. I find it impossible to imagine how Bartley's information here could have come from any other source than a document written by Wittgenstein himself. If it strains credulity to think that Wittgenstein's friends supplied Bartley with accounts of his trips to the

Prater, it positively defies belief that they gave him reports of Wittgenstein's dreams, told in the first person.

Interestingly, in the aforementioned coded remarks, Wittgenstein does occasionally record and comment on his dreams (three examples can be found in this book, on pages 276, 279 and 436). And the discussion of the dream that Bartley quotes, though more elaborate than anything else that has been preserved, is entirely consistent with the interest that Wittgenstein showed at various times in Freud's techniques of interpreting dreams.

Thus, there is every reason to believe that the dream-reports that Bartley gives are real, and, therefore, a *prima facie* reason to think that Bartley had access to a manuscript, the existence of which is unknown to Wittgenstein's literary executors (indeed, which has been kept from them). The executors have no manuscripts belonging to the years 1919 and 1920, even though it is quite likely that there were some.

If this (admittedly highly speculative) hypothesis is correct, then this manuscript might also be the source for the alleged 'Prater episodes'. In correspondence with Bartley, I asked him directly whether there was such a manuscript or not. He neither confirmed nor rejected the suggestion; he said only that to reveal his source of information would be to betray a confidence, and that he was not prepared to be so dishonourable. I therefore regard the hypothesis as still awaiting falsification.

In writing this book, I have had unrestricted access to all the coded remarks in possession of the literary executors, and permission to quote any of them that I wish. I have chosen to quote virtually all the remarks that are in any way revealing of Wittgenstein's emotional, spiritual and sexual life. (Discretion, as Lytton Strachey once said, is not the better part of biography.) I have left nothing out that would lend support to the popular notion that Wittgenstein was tormented by his homosexuality, although I myself believe this to be a simplification that seriously misrepresents the truth.

What the coded remarks reveal is that Wittgenstein was uneasy, not about homosexuality, but about sexuality itself. Love, whether of a man or a woman, was something he treasured. He regarded it as a gift, almost as a divine gift. But, together with Weininger (whose *Sex and Character* spells out, I believe, many attitudes towards love and sex that are implicit in much that Wittgenstein said, wrote and did), he sharply differentiated love from sex. Sexual arousal, both homo- and heterosexual, troubled him enormously. He seemed to regard it as incompatible with the sort of person he wanted to be.

What the coded remarks also reveal is the extraordinary extent to which Wittgenstein's love life and his sexual life went on only in his imagination. This is most striking in the case of Keith Kirk (for whom Wittgenstein formed a brief obsession that he regarded as 'unfaithful' to his love for Francis

Skinner; see pages 426 to 428), but it is also evident in almost all of Wittgenstein's intimate relationships. Wittgenstein's perception of a relationship would often bear no relation at all to the perception of it held by the other person. If I had not met Keith Kirk, I would have been almost certain, from what I had read in the coded remarks, that he and Wittgenstein had had some kind of 'affair'. Having met Kirk, I am certain that whatever affair there was existed only in Wittgenstein's mind.

If I may be allowed a final twist to my speculations concerning Bartley: I believe it to be possible that his information came from coded remarks contained in a manuscript written between 1919 and 1920, but that he has been too hasty in inferring from those remarks that Wittgenstein engaged in sexually promiscuous behaviour. It would be entirely in keeping with what else we know about Wittgenstein that he did indeed find the 'rough, young, homosexual youths' that he discovered in the Prater fascinating, that he returned again and again to the spot from where he could see them, and that he recorded his fascination in diary form in his notebooks. But it would also be entirely in keeping with what we know that the youths themselves knew nothing at all about his fascination, and indeed were unaware of his existence. If Wittgenstein was 'sexually promiscuous' with street youths, it was, I believe, in the same sense that he was 'unfaithful' to Francis Skinner.

REFERENCES

Wittgenstein's manuscripts are kept in the Wren Library, Trinity College, Cambridge. They are cited here according to the numbers assigned them by Professor G. H. von Wright in his article 'The Wittgenstein Papers' (see *Wittgenstein*, Blackwell, 1982). In the text, the extracts have been given in English. Where this follows a previously published translation, I cite here only the published reference. Where it is my translation of an extract previously published (or quoted) only in German, I give the manuscript reference, together with a reference to the original publication. In cases where I have translated a previously unpublished extract, I give the manuscript reference, together with the original German text. I have followed this procedure also with some of the more important letters from Frege to Wittgenstein.

Wittgenstein's letters to Bertrand Russell, G. E. Moore, J. M. Keynes, W. Eccles, Paul Engelmann, Ludwig von Ficker, C. K. Ogden and G. H. von Wright have been published in the various editions of letters listed in the Bibliography. His letters to Ludwig Hänsel are published as an appendix in *Der Volksschullehrer Ludwig Wittgenstein*, by Konrad Wünsche. His letters to his sisters Hermine and Helene, and to his friends Roy Fouracre, Rowland Hutt, Gilbert Pattisson, Rush Rhees, Moritz Schlick and Friedrich Waismann are hitherto unpublished unless otherwise indicated, and remain in private hands.

The letters to Wittgenstein from Engelmann, Eccles, Gottlob Frege, von Ficker, Hänsel, Adele and Stanislav Jolles, Ogden, David Pinsent and Russell are held by the Brenner Archive, Innsbruck. Those from Francis Skinner are in private hands.

The letters from Russell to Lady Ottoline Morrell are held by the Humanities Research Center, University of Texas, and have been extensively quoted in at least three previous publications: *The Life of Bertrand Russell*, by Ronald W. Clark; *Wittgenstein: A Life*, by Brian McGuinness; and 'The Early

Wittgenstein and the Middle Russell', by Kenneth Blackwell (in *Perspectives on the Philosophy of Wittgenstein*, ed. Irving Block).

The letters between Drs Grant, Bywater, Herrald and Landsborough Thomson quoted in Chapter 21 are held by the Medical Research Council (MRC) Library, London. Records of the interviews with Adolf Hübner quoted in Chapter 9 are held by the Wittgenstein Documentation Centre in Kirchberg, Lower Austria.

In the notes that follow, *Recollections of Wittgenstein*, ed. Rush Rhees, is abbreviated as *Recollections*; Ludwig Wittgenstein as 'LW'; and his co-correspondents as follows:

PE	Paul Engelmann
WE	W. Eccles
GF	Gottlob Frege
LF	Ludwig von Ficker
RF	Roy Fouracre
LH	Ludwig Hänsel
RH	Rowland Hutt
AJ	Adele Jolles
SJ	Stanislav Jolles
PEJ	P. E. Jourdain
JMK	John Maynard Keynes
LL	Lydia Lopokova
GEM	G. E. Moore
NM	Norman Malcolm
OM	Lady Ottoline Morrell
CKO	C. K. Ogden
DP	David Pinsent
FP	Fanny Pinsent
GP	Gilbert Pattisson
BR	Bertrand Russell
FR	Frank Ramsey
RR	Rush Rhees
FS	Francis Skinner
LS	Lytton Strachey
MS	Moritz Schlick
FW	Friedrich Waismann
GHvW	G. H. von Wright

1 THE LABORATORY FOR SELF-DESTRUCTION

p. 3 'Why should one tell the truth': Wittgenstein's recollection of this episode is contained in a document found among Wittgenstein's papers; quoted by Brian McGuinness in *Wittgenstein: A Life*, pp. 47–8.

p. 3 'Call me a truth-seeker': LW to Helene Salzer (née Wittgenstein); quoted in Michael Nedo and Michele Ranchetti, *Wittgenstein: Sein Leben in Bildern und Texten*, p. 292.

p. 4 'I can't understand that': Malcolm, *Memoir*, p. 116.

p. 6 'There is no oxygen in Cambridge': *Recollections*, p. 121.

p. 12 'our influence did not reach far enough': *Jahrbuch für sexuelle Zwischenstufen*, VI, p. 724; quoted by W. W. Bartley in *Wittgenstein*, p. 36.

p. 13 woken at three: this account was given by Wittgenstein to Rush Rhees, who mentioned it to the author in conversation.

p. 13 'I can begin to hear the sound of machinery': *Recollections*, p. 112.

p. 14 '*Du hast aber kein Rhythmus!*': quoted by Rush Rhees, in conversation with the author.

p. 14 'Whereas I in the same circumstances': from the document referred to on p. 3.

p. 16 '*Wittgenstein wandelt wehmütig*': recalled in a letter (12.4.76) from a fellow pupil of Wittgenstein's at Linz, J. H. Stiegler, to Adolf Hübner; quoted by Konrad Wüsche in *Der Volksschullehrer Ludwig Wittgenstein*, p. 35. I am indebted to Paul Wijdeveld for the translation.

p. 17 'A trial involving sexual morality': quoted by Frank Field in *Karl Kraus and his Vienna*, p. 56.

p. 17 'If I must choose': ibid., p. 51.

p. 17 Politics 'is what a man does': ibid., p. 75.

p. 20 'a time when art is content': Weininger, *Sex and Character*, pp. 329–30.

p. 21 'her sexual organs possess women': ibid., p. 92.

p. 22 'the male lives consciously': ibid., p. 102.

p. 22 'disposition for and inclination to prostitution': ibid., p. 217.

p. 22 'a characteristic which is really and exclusively feminine': ibid., p. 255.

p. 22 'the interest that sexual unions shall take place': ibid., p. 258.

p. 22 'but which has become actual': ibid., p. 303.

p. 23 'the most manly Jew': ibid., p. 306.

p. 23 'the extreme of cowardliness': ibid., p. 325.

p. 23 'conquered in himself Judaism': ibid., pp. 327–8.

p. 24 'it has the greatest, most limpid clearness and distinctness': ibid., p. 111.

p. 24 'they are no more than duty to oneself': ibid., p. 159.

p. 24 'Genius is the highest morality': ibid., p. 183.

p. 24 'many men first come to know of their own real nature': ibid., p. 244.
p. 24 'In love, man is only loving himself': ibid., p. 243.
p. 24 'what all the travels in the world': ibid., p. 239.
p. 24 'attached to the absolute': ibid., p. 247.
p. 25 'no one who is honest with himself': ibid., p. 346.
p. 26 'When these painful contradictions are removed': Hertz, *Principles of Mechanics*, p. 9.

2 MANCHESTER

p. 29 'I'm having a few problems': LW to Hermine Wittgenstein, 17.5.08.
p. 29 'Because I am so cut off': ibid.
p. 30 '. . . will try to solve': LW to Hermine Wittgenstein, Oct. 1908.
p. 32 'What the complete solution': Russell, *Principles of Mathematics*, p. 528.
p. 33 'Russell said': Jourdain, correspondence book, 20.4.09.
p. 34 'one of the very few people': LW to WE, 30.10.31.
p. 34 'his nervous temperament': J. Bamber to C. M. Mason, 8.3.54; printed as an appendix to Wolfe Mays, 'Wittgenstein in Manchester'.
p. 34 'He used to brag': ibid.
p. 34 'he used to sit through the concert': ibid.
p. 35 'in a constant, indescribable, almost pathological state': *Recollections*, p. 2.

3 RUSSELL'S PROTÉGÉ

p. 36 There exists some disagreement about whether Wittgenstein met Frege or Russell first. The account I give here agrees with that of Brian McGuinness in *Wittgenstein: A Life*, and follows that given by Hermine Wittgenstein in her memoir, 'My Brother Ludwig', *Recollections*, pp. 1–11. It is also supported by G. H. von Wright, who recounts in his 'Biographical Sketch' the story told to him by Wittgenstein that he went first to see Frege in Jena and then (on Frege's advice) to Cambridge to see Russell. Russell, however, was of the opinion that Wittgenstein had not met Frege before he came to Cambridge, and this opinion is shared by some of Wittgenstein's friends, including Elizabeth Anscombe and Rush Rhees (who expresses it in his editorial notes to Hermine's recollections). Professor Anscombe has suggested to me that, as Hermine Wittgenstein was an elderly woman by the time she wrote her reminiscences, her memory might have been at fault. However, in the absence of any conclusive reason for thinking this, I have trusted Hermine's account.

p. 36 'My intellect never recovered': Russell, *Autobiography*, p. 155.
p. 37 'I did think': BR to OM, 13.12.11.
p. 38 'What you call God': BR to OM, 29.12.11.
p. 38 '. . . now there is no prison': BR to OM, July 1911.
p. 38 '. . . an unknown German': BR to OM, 18.10.11.
p. 39 'I am much interested': ibid.
p. 39 'My German friend': BR to OM, 19.10.11.
p. 39 'My German engineer very argumentative': BR to OM, 1.11.11.
p. 39 'My German engineer, I think, is a fool': BR to OM, 2.11.11.
p. 40 '. . . was refusing to admit': BR to OM, 7.11.11.
p. 40 'My lecture went off all right': BR to OM, 13.11.11.
p. 40 'My ferocious German': BR to OM, 16.11.11.
p. 40 'My German is hesitating': BR to OM, 27.11.11.
p. 41 'literary, very musical, pleasant-mannered': BR to OM, 29.11.11.
p. 41 'very good, much better than my English pupils': BR to OM, 23.1.12.
p. 41 'Wittgenstein has been a great event in my life': BR to OM, 22.3.12.
p. 41 'a definition of logical *form*': BR to OM, 26.1.12.
p. 42 'brought a very good original suggestion': BR to OM, 27.2.12.
p. 42 'I found in the first hour': *Recollections*, p. 61.
p. 42 'At our first meeting': ibid.
p. 42 'While I was preparing my speech': BR to OM, 2.3.12.
p. 42 'Moore thinks enormously highly': BR to OM, 5.3.12.
p. 43 'ideal pupil': BR to OM, 17.3.12.
p. 43 'practically certain to do a good deal': BR to OM, 15.3.12.
p. 43 'full of boiling passion': ibid.
p. 43 'It is a rare passion': BR to OM, 8.3.12.
p. 43 'he has more passion': BR to OM, 16.3.12.
p. 43 'His disposition is that of an artist': BR to OM, 16.3.12.
p. 43 'I have the most perfect intellectual sympathy with him': BR to OM, 17.3.12.
p. 43 '. . . he even has the same similes': BR to OM, 22.3.12.
p. 43 'in argument he forgets': BR to OM, 10.3.12.
p. 44 'far more terrible with Christians': BR to OM, 17.3.12.
p. 45 'he says people who like philosophy': BR to OM, 17.3.12.
p. 45 'wears well': BR to OM, 23.4.12.
p. 46 'perhaps the most perfect example': Russell, *Autobiography*, p. 329.
p. 46 'I don't feel the subject neglected': BR to OM, 23.4.12.
p. 46 'He seemed surprised': ibid.
p. 47 'a model of cold passionate analysis': BR to OM, 24.4.12.

p. 47 'a trivial problem': BR to OM, 23.4.12
p. 47 'but only because of disagreement': BR to OM, 26.5.12.
p. 48 'Everybody has just begun to discover him': BR to OM, 2.5.12.
p. 48 'Somebody had been telling them': ibid.
p. 48 'Herr Sinckel-Winckel lunches with me': LS to JMK, 5.5.12.
p. 48 'Herr Sinckel-Winkel hard at it': LS to JMK, 17.5.12.
p. 49 'interesting and pleasant': Pinsent, Diary, 13.5.12.
p. 49 'I really don't know what to think': Pinsent, Diary, 31.5.12.
p. 50 '. . . he is reading philosophy up here': Pinsent, Diary, 30.5.12.
p. 50 'very communicative': Pinsent, Diary, 1.6.12.
p. 51 '. . . then went on to say': BR to OM, 30.5.12.
p. 51 'This book does me a lot of good': LW to BR, 22.6.12.
p. 52 'much pained, and refused to believe it': BR to OM, 1.6.12.
p. 52 '[Wittgenstein] said (and I believe him)': ibid.
p. 53 'since good taste is genuine taste': BR to OM, 17.5.12.
p. 53 'I am seriously afraid': BR to OM, 27.5.12.
p. 54 'I told him': ibid.
p. 54 'It is really amazing': BR to OM, 24.7.12.
p. 54 'I did seriously mean to go back to it': BR to OM, 21.5.12.
p. 54 'I do *wish* I were more creative': BR to OM, 7.9.12.
p. 55 '. . . the second part represented my opinions': BR to Anton Felton, 6.4.68.
p. 55 'Wittgenstein brought me the most lovely roses': BR to OM, 23.4.12.
p. 55 'I love him as if he were my son': BR to OM, 22.8.12.
p. 55 'We expect the next big step': *Recollections*, p. 2.
p. 55 'I went out': Pinsent, Diary, 12.7.12.
p. 56 'He is *very* fussy': BR to OM, 5.9.12.
p. 56 'rather quaint but not bad': Pinsent, Diary, 14.10.12.
p. 56 'arrogated base culture': Engelmann, *Memoir*, p. 130.
p. 56 'I am quite well again': LW to BR, summer 1912.
p. 57 'I am glad you read the lives of Mozart and Beethoven': LW to BR, 16.8.12.
p. 57 'a great contrast': BR to OM, 4.9.12.
p. 57 'This produced a wild outburst': BR to OM, 5.9.12.
p. 58 'Wittgenstein, or rather his father': Pinsent, Diary, 5.9.12.
p. 58 '. . . he is being very fussy': Pinsent, Diary, 7.9.12.
p. 58 'He has an enormous horror': Pinsent, Diary, 12.9.12.
p. 59 'Wittgenstein has been talking a lot': Pinsent, Diary, 19.9.12.
p. 59 'Wittgenstein however got terribly fussy': Pinsent, Diary, 13.9.12.
p. 59 'Wittgenstein was a bit sulky all the evening': Pinsent, Diary, 21.9.12.

p. 60 'His fussyness comes out': Pinsent, Diary, 15.9.12.
p. 60 'I am learning a lot': Pinsent, Diary, 18.9.12.
p. 60 'he simply won't speak to them': Pinsent, Diary, 24.9.12.
p. 61 'still fairly sulky': Pinsent, Diary, 25.9.12.
p. 61 'I really believe': Pinsent, Diary, 29.9.12.
p. 61 'I think father was interested': Pinsent, Diary, 4.10.12.
p. 61 'the most glorious holiday': Pinsent, Diary, 5.10.12.

4 RUSSELL'S MASTER

p. 62 'the infinite part of our life': Russell, 'The Essence of Religion', *Hibbert Journal*, XI (Oct. 1912), pp. 42–62.
p. 63 'Here is Wittgenstein': BR to OM, early Oct. 1912.
p. 63 'He felt I had been a traitor': BR to OM, 11.10.12.
p. 63 'Wittgenstein's criticisms': BR to OM, 13.10.12.
p. 63 'very much inclined': BR to OM, 14.10.12.
p. 63 'He was very displeased with them': Moore, undated letter to Hayek; quoted in Nedo, op. cit., p. 79.
p. 64 'pace up and down': Russell, *Autobiography*, p. 330.
p. 64 'not far removed from suicide': BR to OM, 31.10.12.
p. 64 'he strains his mind': BR to OM, 5.11.12.
p. 64 'I will remember the directions': BR to OM, 4.11.12.
p. 64 'passionate afternoon': BR to OM, 9.11.12.
p. 65 'I told Wittgenstein': BR to OM, 12.11.12.
p. 65 'I got into talking about his faults': BR to OM, 30.11.12.
p. 65 'a man Wittgenstein dislikes': Pinsent, Diary, 9.11.12.
p. 66 'but it was a failure': BR to OM, 31.10.12.
p. 66 'Wittgenstein is a most wonderful character': JMK to Duncan Grant, 12.11.12.
p. 66 'Have you heard': JMK to LS, 13.11.12.
p. 66 'Obviously from his point of view': BR to JMK, 11.11.12.
p. 67 'take him': James Strachey to Rupert Brooke, 29.1.12; see Paul Delany, *The Neo-pagans*, p. 142.
p. 67 'He can't stand him': BR to OM, 10.11.12.
p. 67 'Our brothers B and Wittgenstein': LS to JMK, 20.11.12.
p. 67 'The poor man is in a sad state': LS to Sydney Saxon Turner, 20.11.12.
p. 68 'The Witter-Gitter man': JS to LS, early Dec. 1912.
p. 68 'Wittgenstein has left the Society': BR to OM, undated, but either 6 or 13 Dec. 1912.
p. 69 'He is much the most apostolic': BR to 'Goldie' Lowes Dickenson, 13.2.13.
p. 69 'His latest': Pinsent, Diary, 25.10.12.

p. 70 'Mr Wittgenstein read a paper': minutes of the Moral Science Club, 29.11.12.
p. 70 'about our Theory of Symbolism': LW to BR, 26.12.12.
p. 70 'I think that there cannot be different Types of things!': LW to BR, Jan. 1913.
p. 71 'No good argument': see Ronald W. Clark, *The Life of Bertrand Russell*, p. 241.
p. 71 'Physics exhibits sensations': Russell, 'Matter', unpublished MS; quoted by Kenneth Blackwell in 'The Early Wittgenstein and the Middle Russell', in *Perspectives on the Philosophy of Wittgenstein*, ed. Irving Block.
p. 71 'I am sure I have hit upon a real thing': BR to OM, 9.11.12.
p. 72 'Then Russell appeared': Pinsent, Diary, 4.2.13.
p. 72 'fortunately it is his business': BR to OM, 23.2.13.
p. 72 'My dear father died': LW to BR, 21.1.13.
p. 72 'very much against it': Pinsent, Diary, 7.2.13.
p. 73 'terrific contest': BR to OM, 6.3.13.
p. 73 'I find I no longer talk to him about *my* work': BR to OM, 23.4.13.
p. 75 'I believe a certain sort of mathematicians': BR to OM, 29.12.12.
p. 76 'Whenever I try to think about Logic': LW to BR, 25.3.13.
p. 76 'Poor wretch!': BR to OM, 29.3.13.
p. 76 'Do you mean': PEJ to GF, 29.3.13.
p. 76 'shocking state': BR to OM, 2.5.13.
p. 76 'I played tennis with Wittgenstein': Pinsent, Diary, 29.4.13.
p. 76 'I had tea chez: Wittgenstein': Pinsent, Diary, 5.5.13.
p. 77 'The idea is this': Pinsent, Diary, 15.5.13.
p. 77 'talking to each other': BR to OM, 16.5.13.
p. 77 'went on the river': Pinsent, Diary, 4.6.13.
p. 78 '. . . we went to the C.U.M.C.': Pinsent, Diary, 30.11.12.
p. 78 'Wittgenstein and Lindley came to tea': Pinsent, Diary, 28.2.13.
p. 78 'I came with him': Pinsent, Diary, 24.5.13.
p. 79 'because ordinary crockery is too ugly': Pinsent, Diary, 16.6.13.
p. 79 'He affects me': BR to OM, 1.6.13.
p. 80 'His faults are exactly mine': BR to OM, 5.6.13.
p. 80 'an event of first-rate importance': BR to OM, 1916; the letter is reproduced in Russell, *Autobiography*, pp. 281–2.
p. 81 'It all flows out': BR to OM, 8.5.13.
p. 81 'He thinks it will be like the shilling shocker': BR to OM, 13.5.13.
p. 81 'He was right': BR to OM, 21.5.13.
p. 81 'We were both cross from the heat': BR to OM, 27.5.13.
p. 82 'But even if they are': BR to OM, undated.
p. 82 'I am very sorry': LW to BR, 22.7.13.

p. 82 'I must be much sunk': BR to OM, 20.6.13.
p. 82 'Ten years ago': BR to OM, 23.2.13.
p. 83 'Wittgenstein would like the work': BR to OM, 18.1.14.
p. 83 'You can hardly believe': BR to OM, 29.8.13.
p. 83 'latest discoveries': Pinsent, Diary, 25.8.13.
p. 84 'There are still some *very* difficult problems': LW to BR, 5.9.13.
p. 84 'He was very anxious': Pinsent, Diary, 25.8.13.
p. 84 'That is a splendid triumph': Pinsent, Diary, 29.8.13.
p. 84 'Soon after we had sailed': Pinsent, Diary, 30.8.13.
p. 85 'We have got on splendidly': Pinsent, Diary, 2.9.13.
p. 85 'absolutely sulky': ibid.
p. 86 'When he is working': Pinsent, Diary, 3.9.13.
p. 86 'which was the cause of another scene': Pinsent, Diary, 4.9.13.
p. 86 'just enough to do to keep one from being bored': Pinsent, Diary, 23.9.13.
p. 87 'I am sitting here': LW to BR, 5.9.13.
p. 87 'During all the morning': Pinsent, Diary, 17.9.13.
p. 88 '*as soon as possible*': LW to BR, 20.9.13.
p. 88 'but yet frightfully worried': Pinsent, Diary, 20.9.13.
p. 88 'I am enjoying myself pretty fairly': Pinsent, Diary, 23.9.13.
p. 89 'suddenly announced a scheme': Pinsent, Diary, 24.9.13.
p. 89 'He has settled many difficulties': Pinsent, Diary, 1.10.13.
p. 90 'sexual desire increases with physical proximity': Weininger, *Sex and Character*, p. 239.

5 NORWAY

p. 91 'I said it would be dark': BR to Lucy Donnelly, 19.10.13.
p. 92 'After much groaning': BR to OM, 9.10.13.
p. 92 'In philosophy there are no deductions': *Notes on Logic*; printed as Appendix I in *Notebooks 1914–16*, pp. 93–107.
p. 93 'It was sad': Pinsent, Diary, 8.10.13.
p. 93 'As I hardly meet a soul': LW to BR, 29.10.13.
p. 94 '*Then* my mind was on fire!': quoted by Basil Reeve in conversation with the author.
p. 94 'that the whole of Logic': LW to BR, 29.10.13.
p. 94 'In pure logic': Russell, *Our Knowledge of the External World*, preface, pp. 8–9.
p. 94 'An account of general indefinables?': LW to BR, Nov. 1913.
p. 95 'I beg you': LW to BR, Nov. or Dec. 1913.
p. 95 'All the propositions of logic': ibid.
p. 96 'My day passes': LW to BR, 15.12.13.
p. 97 'But deep inside me': LW to BR, Dec. 1913 or Jan. 1914. In *Letters to*

Russell, Keynes and Moore this letter is dated June/July 1914, but, as Brian McGuinness argues in *Wittgenstein: A Life*, p. 192, it seems more plausible to assume that it was written during the Christmas period of 1913.

p. 97 'It's VERY sad': LW to BR, Jan. 1914.
p. 98 'come to the conclusion': LW to BR, Jan. or Feb. 1914.
p. 99 'I dare say': BR to OM, 19.2.14.
p. 99 '*so* full of kindness and friendship': LW to BR, 3.3.14.
p. 101 'who is not yet stale': LW to GEM, 19.11.13.
p. 101 '*You must come*': LW to GEM, 18.2.14.
p. 101 'I *think*, now': LW to GEM, March 1914.
p. 102 'Logical so-called propositions': 'Notes Dictated to G. E. Moore in Norway'; printed as Appendix II in *Notebooks 1914–16*.
p. 102 'I have now relapsed': LW to BR, early summer 1914.
p. 103 'Your letter annoyed me': LW to GEM, 7.5.14.
p. 104 'Upon clearing up some papers': LW to GEM, 3.7.14
p. 104 'Think I won't answer it': Moore, Diary, 13.7.14; quoted on p. 273 of Paul Levy, *G. E. Moore and the Cambridge Apostles.*

6 BEHIND THE LINES

p. 105 'I hope the little stranger': LW to WE, July 1914.
p. 106 'The effect is greatly admired': WE to LW, 28.6.14.
p. 106 'I am turning to you in this matter': LW to LF, 14.7.14.
p. 107 'In order to convince you': LW to LF, 19.7.14.
p. 107 'That Austria's only honest review': quoted in the notes to *Briefe an Ludwig von Ficker*, and translated by Allan Janik in 'Wittgenstein, Ficker and "*Der Brenner*"', C. G. Luckhardt, *Wittgenstein: Sources and Perspectives*, pp. 161–89.
p. 107 From the painter Max von Esterle: see Walter Methlagl, 'Erläuerungen zur Beziehung zwischen Ludwig Wittgenstein und Ludwig von Ficker', *Briefe an Ludwig von Ficker*, pp. 45–69.
p. 107 'a picture of stirring loneliness': quoted by Janik, op cit., p. 166. The quotation comes from Ficker, 'Rilke und der Unbekannte Freund', first published in *Der Brenner*, 1954, and reprinted in *Denkzettel und Danksagungen*, 1967.
p. 108 'It makes me *very* happy': LW to LF, 1.8.14.
p. 108 'You are me!': reported by Engelmann, op. cit., p. 127.
p. 109 'Once I helped him': LW to PE, 31.3.17.
p. 109 'I do not know': LW to LF, 1.8.14.
p. 109 'as thanks they were': LW to LF, 13.2.15.
p. 109 'but a gift': see Methlagl, op. cit., p. 57.
p. 110 '[It] both moved and deeply gladdened me': LW to LF, 13.2.15.

p. 110 'I suppose Madeira wouldn't suit you': DP to LW, 29.7.14.
p. 111 'an intense desire': *Recollections*, p. 3.
p. 111 'average men and women': Russell, *Autobiography*, p. 239.
p. 112 'No matter what a man's frailties': James, *Varieties of Religious Experience*, p. 364.
p. 112 'Now I have the chance': this and the following extract from Wittgenstein's diaries are quoted by Rush Rhees in his 'Postscript' to *Recollections*, pp. 172–209.
p. 112 'People to whom thousands come': *'Leute, die von Tausenden täglich um Rat gefragt werden, gaben freundliche und ausführliche Antworten'*: MS 101, 9.8.14.
p. 113 'Will I be able to work now??!': *'Werde ich jetzt arbeite können??! Bin gespannt auf mein kommendes Leben!'*: ibid.
p. 113 'Such incredible news': *'Solche unmögliche Nachrichten sind immer ein sehr schlechtes Zeichen. Wenn wirklich etwas für uns Günstiges vorfällt, dann wird* das *berichtet und niemand verfällt auf solche Absurditäten. Fühle darum heute mehr als je die furchtbare Traurigkeit unserer – der deutschen Rasse – Lage. Denn dass wir gegen England nicht aufkommen können, scheint mir so gut wie gewiss. Die Engländer – die beste Rasse der Welt –* können *nicht verlieren. Wir aber können verlieren und werden verlieren, wenn nicht in diesem Jahr so im nächsten. Der Gedanke, dass undere Rasse geschlagen werden soll, deprimiert mich furchtbar, denn ich bin ganz und gar deutsch'*: MS 101, 25.10.14.
p. 114 'a bunch of delinquents': quoted by Rhees, op. cit., p. 196.
p. 114 'When we hear a Chinese talk': MS 101, 21.8.14; this translation is taken from *Culture and Value*, p. 1.
p. 114 'Today, when I woke up': MS 101, 10.8.14; quoted in Nedo, op. cit., p. 161.
p. 114 'It was terrible': MS 101, 25.8.14; quoted ibid., p. 70.
p. 114 'There is an enormously difficult time': ibid.; quoted in Rhees, op. cit., p. 196.
p. 115 'No news from David': *'Keine Nachricht von David. Bin ganz verlassen. Denke an Selbstmord'*: MS 102, 26.2.15.
p. 115 'I have never thought of you': SJ to LW, 25.10.14.
p. 115 'That you have enlisted': GF to LW, 11.10.14.
p. 116 'If you are not acquainted with it': LW to LF, 24.7.15.
p. 116 'If I should reach my end now': MS 101, 13.9.14; quoted (but translated slightly differently) in Rhees, op. cit., p. 194.
p. 116 'Don't be dependent on the external world': MS 102, Nov. 1914; quoted in Rhees, op. cit., p. 196.
p. 117 'All the concepts of my work': *'Ich bin mit allen den Begriffen meiner*

Arbeit ganz und gar "unfamiliar". Ich SEHE gar nichts!!!': MS 101, 21.8.14.

p. 117 'I am on the path': MS 101, 5.9.14; quoted in Nedo, op. cit., p. 168.

p. 117 'I feel more sensual': *'Bin sinnlicher als früher. Heute wieder o . . .'*: MS 101, 5.9.14.

p. 118 told by Wittgenstein to G. H. von Wright: see *Biographical Sketch*, p. 8.

p. 118 'In the proposition': *Notebooks*, p. 7.

p. 118 'We can say': *Notebooks*, p. 8.

p. 118 'Worked the whole day': *'Den ganzen Tag gearbeitet. Habe das Problem* verzeifelt *gestürmt! Aber ich will eher mein Blut von dieser Festung lassen, ehe ich unverrichteter Dinge abziehe. Die grösste Schwierigkeit ist, die einmal eroberten Forts zu halten bis man ruhig in ihnen sitzen kann. Und bis nicht die* stadt *gefallen ist, kann man* nicht *für immer ruhig in einem der Forts sitzen'*: MS 102, 31.10.14.

p. 119 'I would be greatly obliged': Trakl to LW, Nov. 1914.

p. 119 'How happy I would be': *'Wie gerne möchte ich ihn kennen lernen! Höffentlich treffe ich ihn, wenn ich nach Krakau komme! Vielleicht wäre es mir eine grosse Stärkung'*: MS 102, 1.11.14.

p. 119 'I miss greatly': *'Ich vermisse sehr einen Menschen, mit dem ich mich ein wenig ausreden kann . . . es würde mich sehr stärken . . . In Krakau. Es ist schon zu spät, Trakl heute noch zu besuchen'*: MS 102, 5.11.14.

p. 120 'How often I think of him!': MS 102, 11.11.14.

p. 120 'When this war is over': DP to LW, 1.12.14.

p. 120 'For the first time in 4 months': MS 102, 10.12.14.

p. 121 'Perhaps they will have a good influence': MS 102, 15.11.14.

p. 121 'We recognize a condition of morbid susceptibility': Nietzsche, *The Anti-Christ*, p. 141.

p. 122 'An extreme capacity for suffering': ibid., p. 142.

p. 122 'Christianity is indeed the only *sure* way to happiness': *'Gewiss, das Christentum ist der einzige* sichere *Weg zum Glück; aber wie wenn einer dies Glück verschähte?! Könnte es nicht besser sein, unglücklich im hoffnungslosen Kampf gegen die äussere Welt zu Grunde zu gehen? Aber ein solches Leben ist sinnlos. Aber warum nicht ein sinnloses Leben führen? Ist es unwürdig?'*: MS 102, 8.12.14.

p. 122 'It is false to the point of absurdity': Nietzsche, op. cit., p. 151.

p. 123 'Let us hope': GF to LW, 23.12.14.

p. 123 'I wish to note': *'Notieren will ich mir, dass mein moralischer Stand jetzt viel tiefer ist als etwa zu Ostern'*: MS 102, 2.1.15.

p. 124 'spent many pleasant hours': MS 102, 10.1.15.

p. 124 'My thoughts are tired': *'Meine Gedanken sind müde. Ich sehe die Sachen nicht frisch, sondern alltäglich, ohne Leben. Es ist als ob eine*

Flamme erloschen wäre und ich muss warten, bis sie von selbst wieder zu brennen anfängt': MS 102, 13.1.15.

p. 124 '*Only* through a miracle': 'Nur *durch Wunder kann sie gelingen. Nur dadurch, indem von ausserhalb mir der Schleier von meinen Augen weggenommen wird. Ich muss mich ganz in mein Schicksal ergeben. Wie es über mich verhängt ist, so wird es werden. Ich lebe in der Hand des Schicksals'*: MS 102, 25.1.15.

p. 124 'When will I hear something from David?!': MS 102, 19.1.15.

p. 124 'I hope you have been safely taken prisoner': JMK to LW, 10.1.15.

p. 125 'Lovely letter from David!': MS 102, 6.2.15.

p. 125 'except that I hope to God': DP to LW, 14.1.15.

p. 125 'We all hope': Klingenberg to LW, 26.2.15.

p. 125 'If one wants to build as solidly': Halvard Draegni to LW, 4.2.15.

p. 126 'comical misunderstandings': *Recollections*, p. 3.

p. 126 'probably very good': MS 102, 8.2.15.

p. 126 'no desire to assimilate foreign thoughts': LW to LF, 9.2.15.

p. 127 'If I were to write what I think': AJ to LW, 12.2.15.

p. 127 'In any case': SJ to LW, 20.2.15.

p. 127 A draft of a reply: this draft (written in English) is now in the Brenner Archive, Innsbruck.

p. 127 'I hope he will write to you': DP to LW, 27.1.15.

p. 127 'I am so sorry': DP to LW, 6.4.15.

p. 127 'Dream of Wittgenstein': quoted in Levy, op. cit., p. 274.

p. 128 'I have been writing a paper on Philosophy': DP to LW, 2.3.15.

p. 129 'The great problem': 1.6.15; *Notebooks*, p. 53.

p. 129 'It does not go against our feeling': 17.6.15; ibid., p. 62.

p. 129 'The demand for simple things': ibid., p. 63.

p. 129 'I hope with all my heart': BR to LW, 10.5.15.

p. 130 'I'm extremely sorry': LW to BR, 22.5.15.

p. 131 'It looks as if the Russian offensive': SJ to LW, 16.4.15.

p. 131 'May poor Galicia': SJ to LW, 4.5.15.

p. 131 'I write rarely': AJ to LW, 8.4.15.

p. 131 'What kind of unpleasantness?': ibid.

p. 131 'Sometimes, dear friend': LF to LW, 11.7.15.

p. 132 'I understand your sad news': LW to LF, 24.7.15.

p. 132 'God protect you': LF to LW, 14.11.15.

p. 132 'Three weeks' holiday': AJ to LW, 12.8.15.

p. 132 'Already during the first meal': Dr Max Bieler, letter to G. Pitcher, quoted by Sister Mary Elwyn McHale in her MA dissertation: 'Ludwig Wittgenstein: A Survey of Source Material for a Philosophical Biography', p. 48. I have here preserved Dr Bieler's own English.

p. 133 'Whatever happens': LW to BR, 22.10.15.
p. 134 'I am enormously pleased': BR to LW, 25.11.15.
p. 134 'I am pleased that you still have time': GF to LW, 28.11.15.
p. 134 'The importance of "tautology"': Russell, *Introduction to Mathematical Philosophy*, p. 205.
p. 135 'sometimes absorbed us so completely': Dr Max Bieler, letter to G. Pitcher, 30.9.61; quoted in McGuinness, op. cit., pp. 234–5. The translation, one assumes, is either by McGuinness or Pitcher – it is certainly not Dr Bieler's English (compare previous extract).
p. 136 'Constantin was a good boy': Bieler, op. cit.
p. 136 'The decision came as a heavy blow': ibid.

7 AT THE FRONT

p. 137 'God enlighten me': MS 103, 29.3.16.
p. 137 'Do your best. You cannot do more': *'Tu du dein bestes. Mehr kannst du nicht tun: und sei heiter. Lass dir an dir selbst genügen. Denn andere werden dich nicht stützen oder doch nur für kurze Zeit. (Dann wirst du diesen lästig werden.) Hilf dir selbst und hilf andern mit deiner ganzen Kraft. Und dabei sei heiter! Aber wieviel Kraft soll man für sich und wieviel für die anderen brauchen? Schwer ist es gut zu leben!! Aber das gute Leben ist schön. Aber nicht mein, sondern dein Wille geschehe'*: MS 103, 30.3.16.
p. 138 'If that happens': MS 103, 2.4.16.
p. 138 'If only I may be allowed': MS 103, 15.4.16.
p. 138 'Was shot at': MS 103, 29.4.16.
p. 138 'Only then': *'Dann wird für mich erst der Krieg anfangen. Und-kann sein – auch das Leben. Vielleicht bringt mir die Nähe des Todes das Licht des Lebens. Möchte Gott mich erleuchten. Ich bin ein Wurm, aber durch Gott werde ich zum Menschen. Gott stehe mir bei. Amen'*: MS 103, 4.5.16.
p. 138 'like the prince': *'Bin wie der Prinz im verwünschten Schloss auf dem Aufklärerstand. Jetzt bei Tag ist alles ruhig aber in der Nacht da muss es* fürchterlich *zugehen! Ob ich es aushalten werde?? Die heutige Nacht wird es zeigen. Gott stehe mir bei!!'*: MS 103, 5.5.16.
p. 138 'From time to time': MS 103, 6.5.16.
p. 139 'Only death gives life its meaning': MS 103, 9.5.16.
p. 139 'The men, with few exceptions': MS 103, 27.4.16; quoted in Rhees, op. cit., p. 197.
p. 139 'The heart of a true believer': MS 103, 8.5.16.
p. 139 'Whenever you feel like hating them': MS 103, 6.5.16; quoted in Rhees, op. cit., p. 198.

p. 139 'The people around me': MS 103, 8.5.16; quoted in Rhees, op. cit., p. 198.

p. 139 'The whole modern conception of the world': see *Notebooks*, p. 72. I have here adopted the translation published in the Pear/McGuinness edition of the *Tractatus*.

p. 140 'Your desire not to allow': GF to LW, 21.4.16.

p. 140 'What do I know about God and the purpose of life?': *Notebooks*, p. 72.

p. 141 'Fear in the face of death': ibid., p. 75.

p. 141 'How terrible!': MS 101, 28.10. 14.

p. 141 'To believe in a God': *Notebooks*, p. 74.

p. 142 'How things stand, is God': ibid., p. 79.

p. 142 'Colossal exertions': MS 103, 6.7.16.

p. 142 'broadened out': MS 103, 2.8.16.

p. 142 'The solution to the problem of life': 6 and 7.7.16; *Notebooks*, p. 74; see *Tractatus*, 6.521.

p. 142 'Ethics does not treat of the world': 24.7.16; *Notebooks*, p. 77.

p. 143 'I am aware of the complete unclarity': ibid., p. 79.

p. 143 'There are, indeed': *Tractatus*, 6.522.

p. 143 'The work of art': 7.10.16; *Notebooks*, p. 83.

p. 143 'no longer consider the where': Schopenhauer, *The World as Will and Representation*, I, p. 179.

p. 144 'As my idea is the world': 17.10.16; *Notebooks*, p. 85.

p. 144 'a merely "inner" world': Nietzsche, *The Anti-Christ*, p. 141.

p. 144 'It is true': 12.10.16; *Notebooks*, p. 84.

p. 144 'What the solipsist *means*': *Tractatus*, 5.62.

p. 144 'This is the way I have travelled': 15.10.16; *Notebooks*, p. 85.

p. 145 'I can hardly say the same': GF to LW, 24.6.16.

p. 145 'I am always pleased': GF to LW, 29.7.16.

p. 145 'The war cannot change our personal relationships': DP to LW, 31.5.16.

p. 145 'This kind, lovely letter': MS 103, 26.7.16.

p. 145 'icy cold, rain and fog': *'eisige Kälte, Regen und Nebel. Qualvolles Leben. Furchtbar schwierig sich nicht zu verlieren. Denn ich bin ja ein schwacher Mensch. Aber der Geist hilft mir. Am besten wärs ich wäre schon krank, dann hätte ich wenigstens ein bisschen Ruhe'*: MS 103, 16.7.16.

p. 146 'Was shot at': MS 103, 24.7.16.

p. 146 'Yesterday I was shot at': *'Wurde gestern beschossen. War verzagt! Ich hatte Angst vor dem Tode. Solch einen Wunsch habe ich jetzt zu leben. Und es ist schwer, auf das Leben zu verzichten, wenn man es einmal gern hat. Das ist eben "Sünde", unvernünftiges Leben, falsche*

Lebensauffassung. Ich werde von Zeit zu Zeit zum Tier. *Dann kann ich an nichts denken als an essen, trinken, schlafen. Furchtbar! Und dann leide ich auch wie ein Tier, ohne die Möglichkeit innerer Rettung. Ich bin dann meinen Gelüsten und Abneigungen preisgegeben. Dann ist an ein wahres Leben nicht zu denken'*: MS 103, 29.7.16.

p. 146 'You know what you have to do': MS 103, 12.8.16.
p. 146 'By this distinctive behaviour': quoted McGuinness, op. cit., p. 242.
p. 147 'In this way': GF to LW, 28.8.16.
p. 147 'one of the wittiest men': Engelmann, op. cit., p. 65.
p. 148 'My dear friend': ibid., p. 68.
p. 148 'I often think of you': LW to PE, 31.3.17.
p. 148 '. . . enabled me to understand': Engelmann, op. cit., p. 72.
p. 148 'If I can't manage': ibid., p. 94.
p. 149 'In me': ibid., p. 74.
p. 149 '[it] may be considered': ibid., p. 117.
p. 150 'the poem as a whole': PE to LW, 4.4.17.
p. 151 'really magnificent': LW to PE, 9.4.17.
p. 151 'Let's hope for the best': GF to LW, 26.4.17.
p. 151 'The journey to Vienna': GF to LW, 30.6.17.
p. 152 'If in saying it': PE to LW, 8.1.18.
p. 152 'It is true': LW to PE, 16.1.18.
p. 153 'Each of us': GF to LW, 9.4.18.
p. 153 'which you don't deserve': LW to PE, 1.6.18.
p. 154 'so that it doesn't get lost': GF to LW, 1.6.18.
p. 154 'His exceptionally courageous behaviour': quoted in McGuinness, op. cit., p. 263.
p. 154 'I want to tell you': FP to LW, 6.7.18.
p. 154 'My first and my only friend': LW to FP; quoted in Nedo, op. cit.
p. 156 'as if it were a Czar's ukase': Russell, *My Philosophical Development*, p. 88.
p. 156 'There is indeed the inexpressible': *TLP* 6.522.
p. 157 'May it be granted': GF to LW, 15.10.18.
p. 157 'and it gives me more and more joy': PE to LW, 7.11.18.
p. 157 'Still no reply': LW to PE, 22.10.18.
p. 157 'for technical reasons': LW to PE, 25.10.18.
p. 158 'the portrait of my sister': see *Recollections*, p. 9.
p. 159 'and to have read': Parak, *Am anderen Ufer*.
p. 159 'I am prisoner in Italy': LW to BR, 9.2.19.
p. 160 'Most thankful to hear': BR to LW, 2.3.19.
p. 160 'Very glad to hear from you': BR to LW, 3.3.19.
p. 160 'You can't imagine': LW to BR, 10.3.19.
p. 160 'I've written a book': LW to BR, 13.3.19.

p. 161 'I should never have believed': LW to BR, 12.6.19.
p. 161 'It is true': BR to LW, 21.6.19.
p. 162 'No writing between the lines!': PE to LW, 6.4.19.
p. 163 'Right at the beginning': '*Gleich zu Anfang treffe ich die Ausdrücke "der Fall sein" und "Tatsache" und ich vermute, dass* der Fall sein *und* eine Tatsache sein *dasselbe ist. Die Welt ist alles, was der Fall ist und die Welt ist die Gesamtheit der Tatsachen. Ist nicht jede Tatsache der Fall und ist nicht, was der Fall ist, eine Tatsache? Ist nicht dasselbe, wenn ich sage, A sei eine Tatsache wie wenn ich sage, A sei der Fall? Wozu dieser doppelte Ausdruck? . . . Nun kommt aber noch ein dritter Ausdruck: "Was der Fall ist, die Tatsache, ist das Bestehen von Sachverhalten". Ich verstehe das so, dass jede Tatsache das Bestehen eines Sachverhaltes ist, so dass eine andre Tatsache das Bestehen eines andern Sachverhaltes ist. Könnte man nun nicht die Worte "das Bestehen" streichen und sagen: "Jede Tatsache ist ein Sachverhalt, jede andre Tatsache ist ein anderer Sachverhalt. Könnte man vielleicht auch sagen "Jeder Sachverhalt ist das Bestehen einer Tatsache"? Sie sehen: ich verfange mich gleich anfangs in Zweifel über das, was Sie sagen wollen, und komme so nicht recht vorwärts*': GF to LW, 28.6.19.
p. 164 'I gather he doesn't understand a word': LW to BR, 19.8.19.
p. 164 'I am convinced': BR to LW, 13.8.19.
p. 164 'The main point': LW to BR, 19.8.19.
p. 165 'a curious kind of logical mysticism': Russell, *My Philosophical Development*, pp. 84–5.
p. 165 'Sachverhalt is': LW to BR, 19.8.19.
p. 165 'The theory of types': BR to LW, 13.8.19.
p. 165 'That's exactly what one can't say': LW to BR, 19.8.19.
p. 165 '. . . Just think': ibid.
p. 166 'If you said': BR to LW, 13.8.19.
p. 166 'you know how difficult': LW to BR, 19.8.19.
p. 166 'I agree with what you say': BR to LW, 13.8.19.
p. 166 'I should like to come to England': LW to BR, 19.8.19.

8 THE UNPRINTABLE TRUTH

p. 170 'You remind me of somebody': *Recollections*, p. 4.
p. 171 'A hundred times': ibid.
p. 171 'So you want to commit financial suicide': quoted by Rush Rhees, *Recollections*, p. 215.
p. 172 'I'm not quite normal yet': LW to BR, 30.8.19.
p. 172 'I am not very well': LW to PE, 25.8.19.
p. 172 'for as you may know': BR to LW, 8.9.19.
p. 172 'horrified and nauseated': LW to PE, 2.9.19.

p. 172 'I can no longer behave like a grammar-school boy': LW to PE, 25.9.19.
p. 172 'The benches are full of boys': LW to BR, 6.10.19.
p. 173 'to find something of myself': LW to PE, 2.9.19.
p. 173 'Not very': LW to LH, Sept. 1919.
p. 173 'My book will be published': LW to FP, 24.3.19.
p. 173 'naturally neither knows my name': LW to BR, 30.8.19.
p. 174 'I consider it indecent': LW to LF, undated, but probably Nov. 1919.
p. 174 'You now write': *'Sie schreiben nun: "Was einem Elementarsatze entspricht, wenn er wahr ist, ist das Bestehen eines Sachverhaltes". Hiermit erklären Sie nicht den Ausdruck "Sachverhalt", sondern den ganzen Ausdruck, "das Bestehen eines Sachverhaltes" . . . Die Freude beim Lesen Ihres Buches kann also nicht mehr durch den schon bekannten Inhalt, sondern nur durch die Form erregt werden, in der sich etwa die Eigenart des Verfassers ausprägt. Dadurch wird das Buch eher eine künstlerische als eine wissenschaftliche Leistung; das, was darin gesagt wird, tritt zurück hinter das, wie es gesagt wird'*: GF to LW, 16.9.19.
p. 175 'The sense of both propositions': quoted Frege, ibid.
p. 175 'The actual sense of a proposition': *'Der eigentliche Sinn des Satzes ist für alle derselbe; die Vorstellungen aber, die jemand mit dem Satze verbindet, gehören ihm allein an; er ist ihr Träger. Niemand kann die Vorstellungen eines Andern haben'*: ibid.
p. 175 'He doesn't understand': LW to BR, 6.10.19.
p. 175 'that I have learnt to know you': GF to LW, 30.9.19.
p. 176 'mutilate it from beginning to end': LW to LF, op. cit.
p. 176 'About a year ago': ibid.
p. 177 'Why hadn't you thought of me': LF to LW, 19.10.19.
p. 178 'I am pinning my hopes on you': LW to LF, undated, but almost certainly Nov. 1919.
p. 179 'Do you remember': LW to BR, 27.11.19.
p. 179 'Your letter, naturally, wasn't pleasant': LW to LF, 22.11.19.
p. 179 'Don't worry': LF to LW, 28.11.19.
p. 179 'I think I can say': LW to LF, 4.12.19.
p. 179 Ficker wrote that he was still hoping: LF to LW, 29.11.19.
p. 180 'I couldn't accept': LW to LF, 5.12.19.
p. 180 Rilke's letter to Ficker: reproduced in full in Ficker, op. cit., pp. 212–14.
p. 180 'Is there a *Krampus*': LW to LF, 5.12.19.
p. 180 'Just how far': LW to PE, 16.11.19.
p. 181 'Normal human beings': ibid.
p. 181 'It is terrible': BR to LW, 14.10.19.

p. 182 'Come here as quick as you can': BR to LW, undated, but certainly Dec. 1919.
p. 182 'a vague, shadowy figure': Dora Russell, *The Tamarisk Tree*, I, p. 79.
p. 182 'so full of logic': BR to Colette, 12.12.19.
p. 182 'I besought him to admit': Russell, *My Philosophical Development*, p. 86.
p. 182 'He has penetrated deep into mystical ways': BR to OM, 20.12.19.
p. 183 'I enjoyed our time together': LW to BR, 8.1.20.
p. 183 'The book is now a much smaller risk': LW to LF, 28.12.19.
p. 183 'With or without Russell': LF to LW, 16.1.20.
p. 183 'There's so much of it': LW to BR, 9.4.20.
p. 183 'All the refinement': LW to BR, 6.5.20.
p. 184 'I don't care twopence': BR to LW, 1.7.20.
p. 184 '*you can do what you like*': LW to BR, 7.7.20.
p. 184 'I have continually thought': LW to PE, 30.5.20.
p. 184 'This change of home': LW to PE, 24.4.20.
p. 185 'If I am unhappy': Engelmann, op. cit., pp. 76–7.
p. 185 'Before Christ': PE to LW, 31.12.19.
p. 186 'They are still not clear': LW to PE, 9.1.20.
p. 186 'But then I did something': PE to LW, 19.6.20.
p. 187 'Many thanks for your kind letter': LW to PE, 21.6.20.
p. 188 'The best for me': LW to BR, 7.7.20.
p. 188 'It pleases them': LW to PE, 19.2.20.
p. 188 'There is no wall': LH to LW, 17.1.20.
p. 188 'The professor of psychology': LH to LW, 5.3.20.
p. 190 'Of course I don't take exception': *'Natürlich nehme ich Ihnen Ihre Offenheit nicht übel. Aber ich möchte gerne wissen, welche tiefen Gründe des Idealismus Sie meinen, die ich nicht erfasst hätte. Ich glaube verstanden zu haben, dass Sie selbst den erkenntnistheoretischen Idealismus nicht für wahr halten. Damit erkennen Sie, meine ich, an, dass es tiefere Gründe für diesen Idealismus überhaupt nicht gibt. Die Gründe dafür können dann nur Scheingründe sein, nicht logische'*: GF to LW, 3.4.20.
p. 191 'In the evening': LW to PE, 20.8.20.
p. 191 'So I see that intelligence counts': see *Recollections*, p. 123.
p. 191 'He took half my life': LW to BR, 6.8.20.
p. 191 'For unless all the devils in hell': LW to LF, 20.8.20.

9 'AN ENTIRELY RURAL AFFAIR'

p. 193 'That is not for me': recalled by Leopold Baumrucker in an interview with Adolf Hübner, 18.4.75.
p. 193 'I am to be an elementary-school teacher': LW to BR, 20.9.20.

p. 193 'a beautiful and tiny place': LW to PE, 11.10.20.
p. 194 'the villagers take you': see Luise Hausmann, 'Wittgenstein als Volkschullehrer', *Club Voltaire*, IV, pp. 391–6.
p. 194 'He is interested in everything himself': *Recollections*, p. 5.
p. 195 'During the arithmetic lesson': Anna Brenner, interview with Adolf Hübner, 23.1.75.
p. 196 'Again and again': Berger in Hausmann, op. cit., p. 393.
p. 197 'I was in the office': Frau Bichlmayer, quoted in Nedo, op. cit., p. 164–5
p. 198 'I am indeed very grateful': Hermine to LH, 13.12.20.
p. 198 'I was sorry': LW to PE, 2.1.21.
p. 198 In the event, Engelmann did not understand: what follows is a summary of a letter from Engelmann to Wittgenstein, undated, but almost certainly Jan. 1921.
p. 198 'I cannot at present analyse my state in a letter': LW to PE, 7.2.21.
p. 199 'One night': Hänsel, 'Ludwig Wittgenstein (1889–1951)', *Wissenschaft und Weltbild*, Oct. 1951, pp. 272–8.
p. 199 'I was a priest': see Bartley, *Wittgenstein*, p. 29.
p. 199 'It puzzled Wittgenstein': ibid., p. 30.
p. 200 'I wonder how you like being an elementary school-teacher': BR to LW, 11.2.21.
p. 201 'I am sorry you find': BR to LW, 3.6.21.
p. 201 'here they are much more good-for-nothing': LW to BR, 23.10.21.
p. 201 'I am very sorry': BR to LW, 5.11.21.
p. 201 'You are right': LW to BR, 28.11.21.
p. 201 'a good mathematician': BR to LW, 3.6.21.
p. 202 'I could not grasp': Karl Gruber, interview with Adolf Hübner, 16.1.75; quoted in Wünsche, *Der Volkschullehrer Ludwig Wittgenstein*, p. 150.
p. 202 'It would in my opinion': LW to LH, 5.7.21.
p. 202 'I work from early in the morning': LW to LH, 23.8.21.
p. 202 'because I was living in open sin': Russell, note on a letter from Littlewood, 30.1.21; quoted in Clark, op. cit., p. 485.
p. 203 'As they can't drop less than £50': CKO to BR, 5.11.21; full text of the letter reproduced in Russell, *Autobiography*, pp. 353–4.
p. 203 'In any other case': Ostwald to Dorothy Wrinch, 21.2.21; quoted by G. H. von Wright in 'The Origin of the "*Tractatus*", *Wittgenstein*, pp. 63–109.
p. 204 'I am sorry, as I am afraid you won't like that': BR to LW, 5.11.21.
p. 204 'I must admit': LW to BR, 28.11.21.
p. 205 'Enclosed from Wittgenstein': BR to CKO, 28.11.21.
p. 206 'As a selling title': CKO to BR, 5.11.21.

p. 206 'I think the Latin one is better': LW to CKO, 23.4.22.
p. 207 'What is this?': CKO to LW, 20.3.22.
p. 207 'There can be no thought': LW to CKO, 5.5.22.
p. 207 'in consideration of their issuing it': the declaration was enclosed by Ogden in a letter to Wittgenstein dated 18.6.22.
p. 208 'As to your note': LW to CKO, 4.8.22.
p. 208 'because I don't get on': LW to BR, 23.10.21.
p. 208 'I would have felt it as a humiliation': Gruber, op. cit.
p. 209 'Today I had a conversation': LW to LH, 16.2.22.
p. 209 'I wish': BR to LW. 7.2.22.
p. 209 'On the contrary!': LW to BR, undated, but no doubt Feb. 1922. The letter is not included in *Letters to Russell, Keynes and Moore*, but will, I believe, be published in a forthcoming edition of Wittgenstein's correspondence. It is among the collection held by the Brenner Archive.
p. 210 'The little boy is lovely': BR to LW, 9.5.22.
p. 210 'circumstances of the time': Dora Russell, op. cit., p. 160.
p. 210 'much pained by the fact': BR to GEM, 27.5.29.
p. 210 'at the height of his mystic ardour': Russell, *Autobiography*, p. 332.
p. 210 'assured me with great earnestness': ibid.
p. 211 wrote at least two letters: these are now in the Brenner Archive.
p. 211 'When, in the twenties': Engelmann, quoted in Nedo, op. cit.
p. 212 'a very disagreeable impression': LW to PE, 14.9.22.
p. 212 he told Russell: in a letter now in the possession of the Brenner Archive; undated, but probably Nov. or Dec. 1922.
p. 212 'They really look nice': LW to CKO, 15.11.22.
p. 213 'Just improve yourself': quoted by Postl in an interview with Adolf Hübner, 10.4.75.
p. 214 'I think I ought to confess': LW to CKO, March 1923.
p. 214 'A short time ago': LW to BR, 7.4.23; letter now in the possession of the Brenner Archive.
p. 214 'does not really think': FR to LW, 20.2.24; reproduced in *Letters to C. K. Ogden*, pp. 83–5.
p. 214 'To my great shame': LW to PE, 10.8.22.
p. 215 'I should have preferred': LW to JMK, 1923.
p. 215 'Did Keynes write to me?': LW to CKO, 27.3.23.
p. 215 'This is a most important book': Ramsey, 'Critical Notice of L. Wittgenstein's "Tractatus Logico-Philosophicus",' *Mind*, Oct. 1923, pp. 456–78.
p. 216 'It is most illuminating': FR to CKO, undated.
p. 216 'It's terrible': FR to his mother, 20.9.23.
p. 216 'I shall try': ibid.

p. 217 'He is very poor': ibid.
p. 218 'the £50 belong to Keynes': FR to LW, 20.12.23.
p. 218 'Keynes very much wants to see L. W.': FR to Thomas Stonborough, Nov. or Dec. 1923; letter now in the Brenner Archive.
p. 218 'but you mustn't give it any weight': FR to LW, 20.12.23.
p. 219 'You are quite right': FR to LW, 20.2.24.
p. 220 'looked just a prosperous American': FR to his mother, dated March 1924 ('In the train Innsbruck–Vienna Sunday').
p. 220 'As far as I could make out': FR to his mother, dated simply 'Sunday', but certainly written from Vienna, March 1924.
p. 220 'trying to get him': FR to JMK, 24.3.24; quoted in Nedo, op. cit., p. 191.
p. 221 'Wittgenstein seemed to me tired': FR to his mother, 30.3.24.
p. 221 '. . . if he were got away': FR to JMK, 24.3.24.
p. 222 'I'm afraid I think': ibid.
p. 222 'yet my mind': JMK to LW, 29.3.24.
p. 223 '. . . because I myself no longer': LW to JMK, 4.7.24.
p. 223 'he has spent weeks': FR to his mother; quoted in *Letters to C. K. Ogden*, p. 85.
p. 224 'We really live in a great time for thinking': FR to his mother, 22.7.24; quoted in Nedo, op. cit., p. 188.
p. 224 'I'm sorry so few have sold': FR to CKO, 2.7.24.
p. 224 'I don't much want to talk about mathematics': FR to LW, 15.9.24.
p. 224 'I would, naturally': Hermine to LH, autumn 1924.
p. 225 'People kiss each other': quoted in Josef Putre, 'Meine Erinnerungen an den Philosophen Ludwig Wittgenstein', 7 May 1953.
p. 225 'It's not going well': LW to LH, Oct. 1924.
p. 226 'He who works': Wittgenstein, preface to the *Wörterbuch für Volksschulen*, trans. into English in the edition prepared by Adolf Hübner, together with Werner and Elizabeth Leinfellner, Hölder-Pichler-Tempsky, 1977.
p. 227 'No word is too common': ibid., p. xxxiii.
p. 227 'the most pressing question': Buxbaum's report is quoted in full by Adolf Hübner in his editor's introduction to the *Wörterbuch*.
p. 228 'I suffer much': LW to PE, 24.2.25.
p. 229 'That you want to go to Palestine': ibid.
p. 229 'I was more than pleased': LW to WE, 10.3.25.
p. 230 'England may not have changed': LW to WE, 7.5.25.
p. 230 'I should rather like to': LW to JMK, 8.7.25.
p. 230 'I'm awfully curious': LW to JMK, July or Aug. 1925.
p. 231 'I know that brilliance': LW to PE, 19.8.25.

p. 231 'It was during this period': Eccles, op. cit., p. 63.
p. 231 'Tell Wittgenstein': W. E. Johnson to JMK, 24.8.25.
p. 231 'In case of need': LW to PE, 9.9.25.
p. 231 'as long as I feel': LW to JMK, 18.10.25.
p. 232 'It cannot be said': August Riegler, interview with Adolf Hübner, 3.6.76.
p. 233 'I called him all the names under the sun': Franz Piribauer, interview with Adolf Hübner, 20.4.75.
p. 233 'I can only advise you': quoted by August Wolf in an interview with Adolf Hübner, 10.4.75.

10 OUT OF THE WILDERNESS

p. 236 'he and not I was the architect': Engelmann in a letter to F. A. von Hayek; quoted in Nedo, op. cit., p. 206.
p. 236 'Tell me, Herr Ingenieur': *Recollections*, pp. 6–7.
p. 237 'Perhaps the most telling proof': ibid., p. 8.
p. 237 '. . . even though I admired the house': quoted in Leitner, *The Architecture of Ludwig Wittgenstein*, p. 23.
p. 238 'I felt again at home': Marguerite Sjögren (*née* Respinger, now de Chambrier), *Granny et son temps*, p. 101.
p. 239 'It doesn't matter': quoted by Marguerite de Chambrier in conversation with the author.
p. 239 'Why do you want': ibid.
p. 239 'the sort of Jew one didn't like': ibid.
p. 240 'Love of a woman': *Sex and Character*, p. 249.
p. 240 '. . . the house I built for Gretl': *Culture and Value*, p. 38.
p. 240 'Within all great art': ibid., p. 37.
p. 241 'As an admirer': MS to LW, 25.12.24; quoted in *Ludwig Wittgenstein and the Vienna Circle*, p. 13.
p. 242 'It was as if': Mrs Blanche Schlick to F. A. von Hayek, quoted in Nedo, op. cit.
p. 242 'Again': ibid.
p. 242 'He asks me': Gretl to MS, 19.2.27; quoted in *Ludwig Wittgenstein and the Vienna Circle*, p. 14.
p. 242 'returned in an ecstatic state': Mrs Schlick, ibid.
p. 242 'Wittgenstein found Schlick': Engelmann, op. cit., p. 118.
p. 243 'Before the first meeting': Carnap's recollections of Wittgenstein appeared first in his 'Autobiography' in Paul Schlipp, ed., *The Philosophy of Rudolf Carnap*, and are reprinted in K. T. Fann, *Ludwig Wittgenstein: The Man and His Philosophy*, pp. 33–9.
p. 244 'His point of view': ibid.
p. 245 'Only so can we preserve it': Ramsey, 'The Foundations of

Mathematics', reprinted in *Essays in Philosophy, Logic, Mathematics and Economics*, pp. 152–212.

p. 246 'The way out of all these troubles': LW to FR, 2.7.27.

p. 246 '. . . he thought with the aim': *Culture and Value*, p. 17.

p. 247 'The philosopher is not a citizen': *Zettel*, p. 455.

p. 247 'I couldn't stand the hot-water bottle': LW to JMK, summer 1927.

p. 247 'a doctrine which sets up as its bible': Keynes, *A Short View of Russia*, p. 14.

p. 247 '. . . many, in this age without religion': ibid., p. 13.

p. 248 '. . . which may, in a changed form': ibid., p. 15.

p. 248 'One who believes as I do': Russell, *Practice and Theory of Bolshevism*, p. 18.

p. 249 '. . . it was fascinating': Feigl; quoted in Nedo, op. cit., p. 223.

p. 249 '. . . he has a lot of stuff about infinity': BR to GEM, 5.5.30.

p. 250 'Intuitionism is all bosh': *Lectures on the Foundations of Mathematics*, p. 237.

p. 251 'à la Corbusier': JMK to his wife, 18.11.28; quoted in Nedo, op. cit., p. 222.

II THE SECOND COMING

p. 255 'Well, God has arrived': JMK to LL, 18.1.29.

p. 255 'as though time had gone backwards': MS 105; quoted in Nedo, op. cit., p. 225.

p. 256 'brutally rude': Leonard Woolf, *An Autobiography*, II: *1911–1969*, p. 406.

p. 256 'in mixed company': Frances Partridge, *Memories*, p. 160.

p. 257 '. . . Julian, Maynard says': Virginia Woolf, *A Reflection of the Other Person: Letters 1929–31*, p. 51.

p. 257 'For he talks nonsense': Julian Bell, 'An Epistle On the Subject of the Ethical and Aesthetic Beliefs of Herr Ludwig Wittgenstein', first published in *The Venture*, No. 5, Feb. 1930, pp. 208–15; reprinted in Irving M. Copi and Robert W. Beard, ed., *Essays on Wittgenstein's Tractatus*.

p. 258 'the kindest people': *Recollections*, p. 17.

p. 258 'at last has succeeded': JMK to LL, 25.2.29.

p. 258 'We have seen a lot of Wittgenstein': Partridge, op. cit., p. 159.

p. 259 'delightful discussions': MS 105; quoted in Nedo, op. cit., p. 225.

p. 259 'A good objection': '*Ein guter Einwand hilft vorwärts, ein flacher Einwand, selbst wenn er recht hat, wirkt ermattend. Ramseys Einwände sind von dieser Art. Der Einwand fasst die Sache nicht an ihrer Wurzel, wo das Leben ist, sondern schon so weit aussen wo sich nichts mehr rectifizieren* *lässt, selbst wenn es falsch ist. Ein guter Einwand hilft unmittelbar zur Lösung, ein flacher muss erst überwunden werden und kann dann von weiter unten herauf (wie eine überwundene abgestorbene Stelle) zur Seite liegengelassen werden. Wie wenn sich der Baum an der vernarbten Stelle vorbei krümmt um weiter zu wachsen*': MS 107, p. 81.

p. 260 'I don't like your method': quoted in Moore, 'Wittgenstein's Lectures in 1930–33', *Philosophical Papers*, pp. 252–324.

p. 261 'What is the logical form of *that*?': see Malcolm, *Memoir*, p. 58.

p. 262 'Moore? he shows you': *Recollections*, p. 51.

p. 262 'a disaster for Cambridge': ibid., p. 103.

p. 262 'The mind gets stiff': ibid., p. 105.

p. 263 'My first encounter with Wittgenstein': this recollection is contained in a letter from S. K. Bose to John King, 5.4.78, a copy of which Mr King very kindly sent me.

p. 264 'Don't think I ridicule this': *Recollections*, p. 101.

p. 264 'well-meaning commentators': ibid., p. xi.

p. 266 'I would have known': LW to GP, summer 1931.

p. 266 'Somehow or other': LW to GP, summer 1930.

p. 267 'You may through my generosity': LW to GP, Oct. 1931.

p. 267 'If I remember rightly': LW to GP, 16.2.38.

p. 267 'that for so many years': MS 107, pp. 74–5.

p. 268 'What a statement seems to imply': LW to FR, undated; see *Briefe*, p. 261.

p. 268 'Please try to understand': LW to JMK, May 1929.

p. 269 'What a maniac you are!': JMK to LW, 26.5.29.

p. 270 'Now as it somehow appears': LW to GEM, 18.6.29.

p. 270 'I propose to do some work': LW to GEM, 15.6.29.

p. 270 'In my opinion': FR to GEM, 14.7.29; quoted in Nedo, op. cit., p. 227.

p. 271 'I think that unless Wittgenstein': BR to GEM, 27.5.29.

p. 271 'I have never known anything so absurd': quoted in Rhees; see Nedo, op. cit., p. 227.

p. 271 Then Russell: the account of the viva that follows is based on that given by Alan Wood in his biography of Russell, *The Passionate Sceptic*, p. 156.

p. 272 'Give up literary criticism!': *Recollections*, p. 59.

p. 272 'You don't understand': ibid., p. 61.

p. 273 'even supposing': Ramsey, review of *Tractatus*, Copi and Beard, op. cit., p. 18.

p. 274 'It is, of course': 'Some Remarks on Logical Form', reprinted in Copi and Beard, op. cit., pp. 31–7.

p. 275 'as your presence': LW to BR, July 1929.

p. 275 'I'm afraid there is a gathering': John Mabbott, *Oxford Memories*, p. 79.

p. 275 'for some time': Ryle, 'Autobiographical', in Oscar P. Wood and George Pitcher, ed., *Ryle*.

p. 276 'This morning I dreamt': MS 107, p. 153.

p. 277 'My whole tendency': the 'Lecture on Ethics' is published in *Philosophical Review*, Jan. 1965, pp. 3–26.

p. 278 'What is good is also divine': *Culture and Value*, p. 3.

p. 278 'What others think of me': *'Was die anderen von mir halten beschäftigt mich immer ausserordentlich viel. Es ist mir sehr oft darum zu tun, einen guten Eindruck zu machen. D.h. ich denke sehr häufig über den Eindruck den ich auf andere mache und es ist mir angenehm, wenn ich denke, dass er gut ist und unangenehm im anderen Fall'*: MS 107, p. 76.

p. 278 'In my father's house': *Recollections*, p. 54.

p. 279 'A strange dream': *'Ein seltsamer Traum:*

*Ich sehe in einer Illustrierten Zeitschrift eine Photographie von Evighztg (*Vertsagt*), der ein viel besprochener Tagesheld ist. Das Bild stellt ihn in seinem Auto dar. Es ist von seinen Schandtaten die Rede; Hänsel steht bei mir und noch jemand anderer ähnlich meinem Bruder Kurt. Dieser sagt, dass Vertsag ein Jude sei aber die Erziehung eines reichen schottischen Lords genossen habe. Jetzt ist er Arbeiterführer. Seinen Namen habe er nicht geändert weil das dort nicht Sitte sei. Es ist mir neu dass Vertsagt den ich mit der Betonung auf der ersten Silbe ausspreche, ein Jude ist, und ich erkenne dass ja sein Name einfach verzagt heisst. Es fällt mir nicht auf, dass es mit "ts" geschrieben ist was ich ein wenig fetter als das übrige gedruckt sehe. Ich denke: muss denn hinter jeder Unanständigkeit ein Jude stecken. Nun bin ich und Hänsel auf der Terrasse eines Hauses etwa des grossen Blockhauses auf der Hochreit und auf der Strasse kommt in seinem Automobil Vertsag; er hat ein böses Gesicht ein wenig rötlich blondes Haar und einen solchen Schnauzbart (er sieht nicht jüdisch aus). Er feuert nach rückwärts mit einem Maschinengewehr auf einen Radfahrer, der hinter ihm fährt und sich vor Schmerzen krümmt und der unbarmherzig durch viele Schüsse zu Boden geschossen wird. Vertsag ist vorbei und nun kommt ein junges Mädchen ärmlich aussehend auf einem Rade daher und auch sie empfängt die Schüsse von dem weiterfahrenden Vertsag. Und diese Schüsse die ihre Brust treffen machen ein brodelndes Geräusch wie ein Kessel in dem sehr wenig Wasser ist über einer Flamme. Ich hatte Mitleid mit dem Mädchen und dachte nur in Österreich kann es geschehen dass dieses Mädchen kein hilfreiches Mitleid findet und die Leute zusehen wie sie leidet und umgebracht wird. Ich selbst fürchte mich auch davor ihr zu helfen weil ich die Schüsse Vertsags fürchte. Ich nähere mich ihr, suche aber Deckung hinter einer Planke. Dann erwache ich. Ich muss nachtragen, dass in dem*

Gespräch ob Hänsel erst in Anwesenheit des anderen dann nachdem er uns verlassen hat ich mich geniere und nicht sagen will dass ich ja selbst von Juden abstamme oder dass der Fall Vertsags ja auch mein Fall ist. Nach dem Erwachen komme ich darauf dass ja verzagt nicht mit "ts" geschrieben wird, glaube aber sonderbarerweise dass es mit "pf" geschrieben wird "pferzagt". Ich habe den Traum gleich nach dem Erwachen notiert. Die Gegend die in dem Traum etwa der Gegend hinter der Hochreiter Kapelle entspricht (die Seite gegen den Windhut) stelle ich mir im Traum als einen steilen bewaldeten Abhang und eine Strasse im Tal vor wie ich es in einem anderen Traum gesehen habe. Ähnlich einem Stück der Strasse von Gloggnitz nach Schlagl. Als ich das arme Mädchen bedauere sehe ich undeutlich ein altes Weib, welches sie bedauert aber sie nicht zu sich nimmt und ihr hilft. Das Blockhaus auf der Hochreit ist auch nicht deutlich, wohl aber die Strasse und was auf ihr vorgeht. Ich glaube ich hatte eine Idee dass der Name wie ich ihn im Traume ausspreche "Vért-sagt" ungarisch ist. Der Name hatte für mich etwas böses, *boshaftes, und sehr männliches'*: MS 107, p. 219, 1.12.29.

12 THE 'VERIFICATIONIST PHASE'

p. 281 'For if the spirit': MS 108, p. 24, 19.12.29.
p. 281 'I am a beast': MS 108, p. 38, 25.12.29.
p. 281 'The spirit in which one can write the truth': *'Die Wahrheit über sich selbst kann man in dem verschiedensten Geiste schreiben. Im anständigsten und unanständigsten. Und danach ist es sehr wünschenswert oder sehr unrichtig, dass sie geschrieben werde. Ja, es gibt unter den wahrhaften Autobiographien die man schreiben könnte, alle Stufen vom Höchsten zum Niedrigsten. Ich zum Beispiel kann meine Biographie nicht höher schreiben als ich bin. Und durch die blosse Tatsache, dass ich sie schreibe, hebe ich mich nicht* notwendigerweise, *ich* kann *mich dadurch sogar schmutziger machen als ich schon war. Etwas in mir spricht dafür, meine Biographie zu schreiben und zwar möchte (m)ich mein Leben einmal klar ausbreiten, um es klar vor mir zu haben und auch für andere. Nicht so sehr, um darüber Gericht zu halten, als um jedenfalls Klarheit und Wahrheit zu schaffen'*: MS 108, pp. 46–7, 28.12.29.
p. 282 'the most serious book': *Recollections*, p. 90.
p. 282 'And woe to those': ibid.
p. 282 'What, you swine': *Ludwig Wittgenstein and the Vienna Circle*, p. 69.
p. 282 'I think it is definitely important': ibid.
p. 283 'Just because Schlick': quoted ibid., p. 18.
p. 284 '. . . at that time': ibid., p. 64.
p. 285 'Indeed!': ibid., p. 78.
p. 285 'Once I wrote': ibid., pp. 63–4.

p. 286 'Physics does not yield': ibid., p. 63.
p. 286 'I would reply': ibid., p. 68.
p. 286 'explains – I believe': *Philosophical Remarks*, p. 129.
p. 287 'If I say, for example': *Ludwig Wittgenstein and the Vienna Circle*, p. 47.
p. 287 'I used at one time to say': quoted in Gasking and Jackson, 'Wittgenstein as a Teacher'; see Fann, op. cit., pp. 49–55.
p. 288 'Imagine that there is a town': quoted Malcolm, op. cit., p. 55.
p. 289 'Wittgenstein's kindness': Partridge, op. cit., p. 170.
p. 289 'The subject of the lectures': recalled by S. K. Bose in the letter to John King of 5.4.78.
p. 290 'Your voice and his': I. A. Richards, 'The Strayed Poet', in *Internal Colloquies*, Routledge, 1972, pp. 183–6.
p. 291 'I think I summed up': *Culture and Value*, p. 24.
p. 291 'the attempt to be rid': *Lectures 1930–1932*, p. 1.
p. 291 'when the average hearer': Russell, *The Analysis of Mind*, p. 198.
p. 291 'If I wanted to eat an apple': *Philosophical Remarks*, p. 64.
p. 291 'for there seems': GEM to BR, 9.3.30.
p. 292 'I do not see how I can refuse': BR to GEM, 11.3.30.
p. 292 'Of course, we couldn't get very far': LW to GEM, March or April 1930.
p. 293 'Unfortunately I have been ill': BR to GEM, 5.5.30.
p. 294 'I find I can only understand Wittgenstein': BR to GEM, 8.5.30.
p. 294 'If a person tells me': *Recollections*, p. 112.
p. 294 'Arrived back in Cambridge': *'Nach den Osterferien wieder in Cambridge angekommen. In Wien oft mit Marguerite. Ostersonntag mit ihr in Neuwaldegg. Wir haben uns viel geküsst drei Stunden lang und es war sehr schön'*: MS 108, p. 133, 25.4.30.
p. 294 'My life is now very economical': LW to GEM, 26.7.30.
p. 294 'Dear Gil (old beast)': LW to GP, summer 1930.
p. 296 'A proposition cannot say more': see *Ludwig Wittgenstein and the Vienna Circle*, p. 244.
p. 296 'If one tried to advance': *Philosophical Investigations*, I, 128.
p. 297 'I know that my method is right': *Recollections*, p. 110.

13 THE FOG CLEARS

p. 298 'The nimbus of philosophy': *Lectures: 1930–1932*, p. 21.
p. 298 'What we find out': ibid., p. 26.
p. 299 '. . . once a method has been found': ibid., p. 21.
p. 299 'I was walking about': *Recollections*, p. 112.
p. 300 'Telling someone something': *Culture and Value*, p. 7.
p. 300 'It is all one to me': ibid.

p. 301 'I would like to say': *Philosophical Remarks*, preface.
p. 301 'Philosophical analysis': *Lectures: 1930–1932*, p. 35.
p. 301 'We never arrive': ibid., p. 34.
p. 302 'The means whereby': Oswald Spengler, *The Decline of the West*, p. 4.
p. 302 'taking of spiritual-political events': ibid., p. 6.
p. 303 'so here we shall develop': ibid., p. 26.
p. 303 'recognize living forms': quoted by Erich Heller in 'Goethe and the Scientific Truth', *The Disinherited Mind*, pp. 4–34.
p. 303 'What I give': see Malcolm, op. cit., p. 43.
p. 303 'Our thought here': Waismann, *Principles of Linguistic Philosophy*, pp. 80–81.
p. 304 'Yes, this fellowship business': LW to JMK, Dec. 1930.
p. 304 '*For me*': *Ludwig Wittgenstein and the Vienna Circle*, p. 117.
p. 305 'For it cuts off': ibid., p. 115.
p. 305 'I would reply': ibid., p. 116.
p. 305 'I can well imagine a religion': ibid., p. 117.
p. 305 'If you and I': *Recollections*, p. 114.
p. 306 'As long as I can play the game': *Ludwig Wittgenstein and the Vienna Circle*, p. 120.
p. 306 'The following is a question': ibid., pp. 129–30.
p. 307 'What do we do': ibid., p. 120.
p. 307 'It is another calculus': ibid., pp. 121–2.
p. 307 '*You cannot*': ibid., p. 129.
p. 308 'A rule of syntax': ibid., p. 126.
p. 308 'If, then': ibid., p. 133.
p. 308 'the demonstration': ibid.
p. 308 'Things must connect directly': ibid., p. 155.
p. 308 'Here *seeing* matters': ibid., p. 123.

14 A NEW BEGINNING

p. 309 'I think now': *Remarks on Frazer's Golden Bough*, p. vi.
p. 310 'even those': *Recollections*, p. 102.
p. 310 'What narrowness': *Remarks on Frazer*, p. 5.
p. 311 'I can set out this law': ibid., pp. 8–9.
p. 311 'If I may explain': '*Wenn ich es durch einen Vergleich klar machen darf: Wenn ein "Strassenköter" seine Biographie schriebe, so bestünde die Gefahr a) dass er entweder seine Natur verleugnen, order b) einen Grund ausfindig machen würde, auf sie stolz zu sein, oder c) die Sache so darstellen, als sei diese seine Natur eine nebensächliche Angelegenheit. Im ersten Falle lügt er, im zweiten ahmt er eine nur dem Naturadel natürliche Eigenschaft, den Stolz, nach, der ein vitium splendidum ist, das er*

ebensowenig wirklich besitzen kann, wie ein krüppelhafter Körper natürliche Gracie. Im dritten Fall macht er gleichsam die sozialdemokratische Geste, die die Bildung über die rohen Eigenschaften des Körpers stellt, aber auch das ist ein Betrug. Er ist was er ist und das ist zugleich wichtig und bedeutsam, aber kein Grund zum Stolz, andererseits immer Gegenstand der Selbstachtung. Ja ich kann den Adelsstolz des Andern und seine Verachtung meiner Natur anerkennen, den ich erkenne ja dadurch nur meine Natur an und den andern der zur Umgebung meiner Natur, die Welt, deren Mittelpunkt dieser vielleicht hässliche Gegenstand, meine Person, ist': MS 110. pp. 252–3, 1.7.31.

p. 312 'Putting together a complete autobiography': quoted in Rhees, *Recollections*, p. 182.

p. 312 'I can quite imagine': LW to GEM, 23.8.31.

p. 313 'How wrong he was': *Recollections*, p. 91.

p. 313 'has something Jewish': *Culture and Value*, p. 20.

p. 313 'Mendelssohn is not a peak': ibid., p. 2.

p. 313 'Tragedy is something un-Jewish': ibid., p. 1.

p. 313 'who like a noxious bacillus': Hitler, *Mein Kampf*, p. 277.

p. 314 'the Jew lacks those qualities': ibid., p. 275.

p. 314 'since the Jew': ibid., p. 273.

p. 314 'It has sometimes been said': *Culture and Value*, p. 22.

p. 314 'Look on this tumour': ibid., p. 20.

p. 316 *Untergangsters*: quoted in Field, op. cit., p. 207.

p. 316 'a confession': *Culture and Value*, p. 18.

p. 316 'It is typical': ibid., p. 19.

p. 316 'Amongst Jews': ibid., pp. 18–19.

p. 317 'Often, when I have': ibid., p. 19.

p. 317 'The Jew must see to it': ibid.

p. 319 'My presence': Sjögren, op. cit., p. 122.

p. 319 'but that doesn't matter': see *Philosophical Grammar*, p. 487.

p. 320 'For the thing itself': LW to MS, 20.11.31.

p. 320 'There are *very, very* many statements': ibid.

p. 320 'If there were theses': *Ludwig Wittgenstein and the Vienna Circle*, p. 183.

p. 321 'still proceeded dogmatically': ibid., p. 184.

p. 321 'make head or tail': LW to MS, 4.3.32.

p. 322 'This is the right sort of approach': *Lectures: 1930–1932*, p. 73.

p. 322 '. . . the dialectical method': ibid., p. 74.

p. 322 'Philosophy is not a choice': ibid., p. 75.

p. 322 'The right expression is': ibid., p. 97.

p. 323 'Grammatical rules': ibid., p. 98.

p. 323 'You owe a great debt': see *Recollections*, p. 123.

p. 324 '. . . from the bottom of my heart': LW to MS, 8.8.32.
p. 324 'That I had not dealt': ibid. (I am indebted to Dr P.M.S. Hacker for this translation.)
p. 325 'All that philosophy can do': MS 213, p. 413; quoted by Anthony Kenny in 'Wittgenstein on the Nature of Philosophy', *The Legacy of Wittgenstein*, pp. 38–60. See also, in the same collection of essays, 'From the Big Typescript to the "*Philosophical Grammar*"', pp. 24–37.
p. 326 'Confusions in these matters': *Philosophical Grammar*, p. 375.
p. 327 'Philosophical clarity': ibid., p. 381.
p. 327 'Nothing seems to me less likely': *Culture and Value*, p. 62.

15 FRANCIS

p. 328 'Is there a substratum': *Wittgenstein's Lectures Cambridge 1932–1935*, p. 205.
p. 329 'no philosophy can possibly': Hardy, 'Mathematical Proof', *Mind*, Jan. 1929, pp. 1–25.
p. 329 'The talk of mathematicians': *Lectures: 1932–5*, p. 225.
p. 330 'Russell and I': ibid., p. 11.
p. 331 'I have wanted to show': ibid., pp. 12–13.
p. 332 'In the event of my death': MS 114.
p. 332 'I am glad': FS to LW, 28.12.32.
p. 332 'Dear Ludwig': FS to LW, 25.3.33.
p. 333 'I'm busy': quoted Pascal, *Recollections*, p. 23.
p. 333 'I feel much further': FS to LW, 2.10.33.
p. 335 'Come to Cambridge at once': *Recollections*, p. 124.
p. 335 'Part of [Braithwaite's] statements', *Mind*, 42 (1933), pp. 415–16.
p. 335 'The extent to which': ibid.

16 LANGUAGE-GAMES: *THE BLUE AND BROWN BOOKS*

p. 337 'I shall in the future': *Blue Book*, p. 17.
p. 338 'We are inclined': ibid.
p. 338 'Philosophers constantly see': ibid., p. 18.
p. 338 'What we say will be easy': *Lectures: 1932–5*, p. 77.
p. 338 'After I stopped': FS to LW, 17.12.33.
p. 339 'My despair': Sjögren, op. cit., p. 137.
p. 340 'He has the great gift': FW to MS, 9.8.34.
p. 340 'If I am with you': FS to LW, 25.3.34.
p. 340 'I thought of you a lot': ibid.
p. 341 'I long to be with you': FS to LW, 4.4.34.
p. 341 'I started off again': FS to LW, 24.7.34.

p. 342 'When I read this': FS to LW, 11.8.34.
p. 343 'growing political awareness': George Thomson, 'Wittgenstein: Some Personal Recollections', *Revolutionary World*, XXXVII, no. 9, pp. 86–8.
p. 343 'I am a communist': quoted by Rowland Hutt in conversation with the author.
p. 343 'I think Francis means': *Recollections*, pp. 125–6.
p. 344 'Imagine a people': *Brown Book*, p. 100.
p. 344 'Imagine a tribe': ibid., p. 103.
p. 344 'Imagine that human beings': ibid., p. 120.
p. 345 'Here is one of the most fertile sources': ibid., p. 108.
p. 345 'We are inclined to say': ibid.
p. 346 'Our answer is': ibid.
p. 346 'the way in which': LW to MS, 31.7.35.

17 JOINING THE RANKS

p. 347 'At the beginning of September': LW to MS, 31.7.35.
p. 348 'I am sure that you partly understand': LW to JMK, 6.7.35.
p. 348 'exceedingly careful': ibid.
p. 349 'May I venture': Keynes's introduction is reproduced in full in *Letters to Russell, Keynes and Moore*, pp. 135–6.
p. 349 'definitely nice': LW to JMK, July 1935.
p. 350 'My interview with Miss B': LW to GP, dated simply 'Tuesday', which is consistent with its being 20.8.35.
p. 350 'If you were a qualified technician': JKM to LW, 10.7.35.
p. 351 'Wittgenstein!': this story is told by George Sacks in *A Thinking Man as Hero*, a television play by Hugh Whitemore, first broadcast on BBC 2, April 1973. I am very grateful to Mr Whitemore for drawing my attention to this source.
p. 351 'we heard that Wittgenstein': ibid.
p. 351 'I wish I could be with you': FS to LW, 17.9.35.
p. 352 'My dear Gilbert!': LW to GP, Sept. 1935.
p. 353 'He had taken no decisions': *Recollections*, p. 29.
p. 353 'The important thing': ibid., p. 205.
p. 353 'Tyranny': ibid.
p. 353 'If anything': ibid.
p. 354 'perhaps I shall go to Russia': LW to PE, 21.6.37.
p. 354 according to Piero Sraffa: see John Moran, 'Wittgenstein and Russia', *New Left Review*, LXXIII (May–June 1972), pp. 83–96.
p. 354 'People have accused Stalin': *Recollections*, p. 144.
p. 355 'in a silly detective story': 'The Language of Sense Data and Private Experience – I (Notes taken by Rush Rhees of Wittgenstein's

Lectures, 1936)', *Philosophical Investigations*, VII, no.1 (Jan. 1984), pp. 1–45.

p. 356 'We have the feeling': ibid., no.2 (April 1984), p. 139.

p. 356 'Here at last': *Recollections*, p. 136.

p. 357 'It's all excellent similes': Moore, 'Wittgenstein's Lectures', op. cit., p. 316.

p. 357 'What I invent': *Culture and Value*, p. 19.

p. 357 'I still have a little money': *Recollections*, p. 209.

p. 359 'Sometimes his silence infuriates me': *Recollections*, p. 127.

p. 360 'I don't see how you *can* win!': quoted by Gilbert Pattisson in conversation with the author.

18 CONFESSIONS

p. 361 'When I got your letter': FS to LW, 6.9.36.

p. 362 'I don't feel clear': FS to LW, 1.11.36.

p. 362 'My feelings for you': FS to LW, 26.10.36.

p. 363 'I do believe': LW to GEM, Oct. 1936.

p. 363 'The weather has changed': LW to GP, Oct. 1936.

p. 363 '"Both fit and style are perfect"': LW to GP, 2.2.37.

p. 363 'or nearly all': LW to GEM, 20.11.36.

p. 364 'These words': *Philosophical Investigations*, I, 1.

p. 364 'A simile that has been': ibid., 112.

p. 365 'A *picture* held us': ibid., 115.

p. 365 'Our clear and simple': ibid., 130.

p. 365 'It is not our aim': ibid., 133.

p. 365 'The results of philosophy': ibid., 119

p. 366 'The edifice of your pride': *Culture and Value*, p. 26.

p. 366 'If anyone is unwilling': quoted in Rhees, *Recollections*, p. 174.

p. 367 'all sorts of things': LW to GEM, 20.11.36.

p. 367 'Whatever you say': FS to LW, 6.12.36.

p. 368 'Whatever you have to tell me': FS to LW, 9.12.36.

p. 368 'listened patiently': *Recollections*, p. 38.

p. 368 'He would have sat transfixed': ibid.

p. 368 'If ever a thing could wait': ibid., p. 35.

p. 369 remembered by Rowland Hutt: and told to me in the course of several conversations.

p. 369 'he had to keep a firmer control': *Recollections*, p. 37.

p. 371 'I myself was not a pupil': Georg Stangel, interview with Adolf Hübner, 19.2.75.

p. 371 'Ja, ja': Leopold Piribauer, interview with Adolf Hübner, 3.12.74.

p. 372 'Last year': quoted in Rhees, *Recollections*, p. 173.

p. 372 'I thought of having it removed': LW to GB, 18.11.37.

p. 372 'I feel it is wrong': FS to LW, 1.3.37.
p. 373 'I don't think I have ever understood': FS to LW, 27.5.37.
p. 373 'partly because I've been troubled': LW to GEM, 4.3.37.
p. 373 'vain, thoughtless, frightened': *'Eitel, gedankenlos, ängstlich . . . Ich möchte jetzt bei jemandem wohnen. In der Früh ein menschliches Gesicht sehen. Anderseits bin ich jetzt wieder so verweichlicht, dass es vielleicht gut wäre allein sein zu müssen. Bin jetzt ausserordentlich verächtlich . . . Ich habe das Gefühl, dass ich jetzt nicht ganz ohne Ideen wäre, aber dass mich die Einsamkeit bedrücken wird, dass ich nicht arbeiten werde können. Ich fürchte mich, dass in meinem Haus alle meine Gedanken werden getötet werden. Dass dort ein Geist der Niedergeschlagenheit von mir ganz Besitz ergreifen wird'*: MS 118, 16.8.37.
p. 374 'unhappy, helpless and thoughtless': *'Unglücklich, ratlos und gedankenlos . . . Und da kam mir wieder zum Bewusstsein, wie einzig Francis ist und unersetzlich. Und wie wenig ich doch das weiss, wenn ich mit ihm bin.*

Bin ganz in Kleinlichkeit verstrickt. Bin irritiert, denke nur an mich und fühle, dass mein Leben elend ist, und dabei habe ich auch gar keine Ahnung, wie elend es ist': MS 118, 17.8.37.

p. 374 'I am ashamed': MS 118, 18.8.37.
p. 374 'I do not know whether I have either a right': MS 118, 19.8.37.
p. 374 'Am now really ill': MS 118, 22.8.37.
p. 375 'showered with gifts': MS 118, 26.8.37.
p. 375 'You said in one of your letters': FS to LW, 23.8.37.
p. 375 'I would love very much to come': FS to LW, 30.8.37.
p. 375 'The way to solve the problem': MS 118, 27.8.37.
p. 375 'I conduct myself badly': MS 118, 26.8.37.
p. 375 'I am a coward': MS 118, 2.9.37.
p. 375 'Am irreligious': MS 118, 7.9.37.
p. 376 'Christianity is not a doctrine': MS 118, 4.9.37; see *Culture and Value*, p. 28.
p. 376 'just wicked and superstitious': MS 118, 7.9.37.
p. 376 'It's as though my work had been drained': MS 118, 17.9.37.
p. 376 'sensual, susceptible, indecent': MS 118, 22.9.37.
p. 376 'Lay with him two or three times': *'Zwei oder dreimal mit ihm gelegen. Immer zuerst mit dem Gefühl, es sei nichts Schlechtes,* dann *mit Scham. Bin auch ungerecht, auffahrend und auch falsch gegen ihn gewesen und quälerisch'*: ibid.
p. 377 'Am very impatient!': MS 119, 25.9.37.
p. 377 'The last 5 days': *'Die letzten 5 Tage waren schön: er hatte sich in das Leben hier hineingefunden und tat alles mit Liebe und Güte, und ich war, Gott sei Dank, nicht ungeduldig, und hatte auch wahrhaftig keinen Grund,*

ausser meine eigene böse Natur. Begleitete ihn gestern bis Sogndal; heute in meine Hütte zurück. Etwas bedrückt, auch müde': MS 119, 1.10.37.

p. 377 'I often remember': FS to LW, undated.

p. 377 'I think *constantly* of you': FS to LW, undated.

p. 377 'I am often thinking': FS to LW, 14.10.37.

p. 378 'I'm thinking of you a lot': FS to LW, 26.10.37.

p. 378 'Prof. Moore wasn't present': FS to LW, 22.10.37.

p. 378 'lovely letter from Fr.': *'Lieben Brief von Fr., er schreibt über eine Sitzung des Mor.Sc.Cl. und wie elend schlecht die Diskussion unter Braithwaits Vorsitz war. Est ist scheusslich. Aber ich wüsste nicht, was dagegen zu machen wäre, denn die andern Leute sind auch zu wenig ernst. Ich wäre auch* zu feig, *etwas Entscheidendes zu tun'*: MS 119, 27.10.37.

p. 378 'harsh and hectoring' letter: see *Recollections*, p. 32.

p. 379 'Have not heard from Francis': MS 119, 16.10.37.

p. 379 'Am relieved': MS 119, 17.10.37.

p. 379 'He made a *good* impression': MS 119, 10.10.37.

p. 379 '[I] just took some apples': *Culture and Value*, p. 31.

p. 380 'How bad is it?': *'Heute Nacht onaniert. Wie schlecht ist es? Ich weiss es nicht. Ich denke mir, es ist schlecht, aber habe keinen Grund'*: MS 120, 21.11.37.

p. 380 'Think of my earlier love': *'Denke an meine frühere Liebe, oder Verliebtheit, in Marguerite und an meine Liebe für Francis. Es ist ein schlimmes Zeichen für mich, dass meine Gefühle für M. so gänzlich erkalten konnten! Freilich ist hier ein Unterschied; aber* meine Herzenskälte *besteht.*

Möge mir vergeben werden; d.h. aber: möge es mir möglich sein, aufrichtig und liebevoll zu sein': MS 120, 1.12.37.

p. 380 'Masturbated tonight': *'Heute nacht onaniert. Gewissensbisse, aber auch die Überzeugung, dass ich zu schwach bin, dem Drang und der Versuchung zu widerstehen, wenn die und die Vorstellungen sich mir darbieten, ohne dass ich mich in andere* flüchten *kann.* Gestern abend *noch hatte ich Gedanken über die Notwendigkeit der Reinheit meines Wandels. (Ich dachte an Marguerite und Francis.)'*: MS 120, 2.12.37.

p. 380 'the bewitchment of our intelligence': *Philosophical Investigations*, I, 109.

p. 381 'Isn't there a truth': *Remarks on the Foundations of Mathematics*, I, p. 5.

p. 381 'I am merely': ibid., I, p. 110.

p. 382 'I am nervous': *'Ich bin beim Schreiben nervös und alle meine Gedanken kurz von Atem. Und ich fühle immer, dass ich den Ausdruck nicht* ganz *verteidigen kann. Dass er schlecht schmeckt'*: MS 119, 11.11.37.

p. 382 'I would like to run away': *'Ich möchte fliehen, aber es wäre unrecht und ich kann es gar nicht. Vielleicht aber könnte ich es auch – ich könnte* morgen *packen und den nächsten Tag abfahren. Aber möchte ich es? Wäre es richtig? Ist es nicht richtig, hier noch auszuhalten? Gewiss. Ich würde mit einem* schlechten *Gefühl morgen abfahren. "Halt es aus", sagt mir eine Stimme. Es ist aber auch Eitelkeit dabei, dass ich aushalten will; und auch etwas Besseres. – Der einzige triftige Grund hier früher oder gleich abzureisen wäre, dass ich anderswo jetzt vielleicht besser arbeiten könnte. Denn es ist Tatsache, dass der Druck, der jetzt auf mir liegt, mir das Arbeiten beinahe unmöglich macht und vielleicht in einigen Tagen wirklich unmöglich'*: MS 120, 22.11.37.

p. 382 'It is the opposite of piety': MS 120, 28.11.37.
p. 382 'You must *strive*': ibid.
p. 382 'The ever-changing, difficult weather': MS 120, 30.11.37.
p. 383 'I'm sorry you are having storms': FS to LW, 1.11.37.
p. 383 '*He is dead and decomposed*': MS 120, 12.12.37. This long diary entry is given in full in *Culture and Value*, p. 33.

19 *FINIS AUSTRIAE*

p. 385 'German-Austria must return': Hitler, *Mein Kampf*, p. 3.
p. 385 'in my earliest youth': ibid., p. 15.
p. 386 'but cleverer people than I': Hermine Wittgenstein, *Familienerinnerungen*, p. 148.
p. 386 'When, at midnight': ibid.
p. 387 'Freud's idea': MS 120, 2.1.38, *Culture and Value*, p. 33.
p. 387 'and that is bad': MS 120, 5.1.38.
p. 387 'Thought: it would be good': MS 120, 4.1.38.
p. 387 'I am cold': ibid.
p. 388 'irreligious': MS 120, 10.2.38.
p. 388 'on the other hand': MS 120, 14.2.38.
p. 388 'It is entirely as though': MS 120, 15.2.38.
p. 389 'You didn't make a mistake': the letter is reproduced in full in *Recollections*, pp. 95–6.
p. 389 'Can't work': *'Kann nicht arbeiten. Denke viel über einen eventuellen Wechsel meiner Nationalität nach. Lese in der heutigen Zeitung, dass eine weitere zwangsweise Annäherung Österreichs an Deutschland erfolgt ist. – Aber ich weiss nicht, was ich eigentlich machen soll'*: MS 120, 16.2.38.
p. 390 'You will either fulfil my demands': quoted in Roger Manvell and Heinrich Fraenkel, *Hitler: The Man and the Myth*, p. 142.
p. 390 'If Herr Hitler's suggestion': quoted in George Clare, *Last Waltz in Vienna*, p. 166.
p. 390 'That the first act': quoted ibid.
p. 390 'That is a ridiculous rumour': *Recollections*, p. 139.

p. 391 'What I hear about Austria': *'Was ich von Österreich höre, beunruhigt mich. Bin im Unklaren darüber, was ich tun soll, nach Wien fahren oder nicht. Denke hauptsächlich an Francis und dass ich ihn nicht verlassen will'*: MS 120, 12.3.38.

p. 392 'Before trying to discuss': Piero Sraffa to LW, 14.3.38.

p. 394 'I am now in an extraordinarily difficult situation': MS 120, 14.3.38; quoted in Nedo, op. cit., p. 296.

p. 394 'In my *head*': MS 120, 16.3.38; quoted ibid.

p. 394 'Sraffa advised me': *'In Cambridge: Sraffa riet mir gestern vorläufig auf keinen Fall nach Wien zu gehen, da ich meinen Leuten jetzt nicht helfen könnte und* aller *Wahrscheinlichkeit nach nicht mehr aus Österreich herausgelassen würde. Ich bin nicht* völlig *klar darüber, was ich tun soll, aber ich glaube vorläufig, Sraffa hat recht'*: MS 120, 18.3.38.

p. 394 'The same, of course': LW to JMK, 18.3.38. The letter is given in full in *Briefe*, pp. 278–9.

p. 396 'is a body who helps people': LW to GP, 26.3.38.

p. 396 'My dear Ludwig': reproduced in Nedo, op. cit., p. 300.

p. 396 'reassuring *sounding* news': MS 120, 25.3.38.

p. 396 Hermine recalls: see Hermine Wittgenstein, op. cit., pp. 154–81.

p. 397 'That these children': Brigitte Zwiauer, 'An die Reichstelle für Sippenforschung', 29 Sept. 1938.

p. 399 'the great nervous strain': LW to GEM, 19.10.38.

p. 399 'In case you want an Emetic': LW to GP, Sept. 1938.

p. 400 'And if': Hermine Wittgenstein, op. cit., p. 120.

p. 400 received certificates: these were issued from Berlin, and still survive. Hermine's is dated 30.8.39, and states: 'Hermine Maria Franziska Wittgenstein of 16 Argentinierstrasse, Vienna IV, born in Eichwald, Teplich on 1.12.1874, is of mixed Jewish blood, having two racially Jewish grandparents as defined by the first Reich Citizenship Law of 14 Nov. 1935'.

p. 400 'their racial classification': this document, dated 10 Feb. 1940, is reproduced in Nedo, op. cit., p. 303.

20 THE RELUCTANT PROFESSOR

p. 403 'If you write': *Recollections*, p. 141.

p. 404 'persuading people': *Lectures and Conversations on Aesthetics, Psychology & Religious Belief*, p. 28.

p. 404 'Jeans has written a book': ibid., p. 27.

p. 405 'You might think': ibid., p. 11.

p. 405 'Do you think I have a theory?': ibid., p. 10.

p. 405 'It is not only difficult': ibid., p. 7.

p. 406 'She was descending from a height': Freud, *The Interpretation of Dreams*, pp. 463–5.
p. 406 'I would say': *Lectures and Conversations*, p. 24.
p. 406 'he treated me quite as his equal': see Freud, *Jokes and their Relation to the Unconscious*, pp. 47–52.
p. 407 'If it is not causal': *Lectures and Conversations*, p. 18.
p. 407 'When I read his poems': ibid, p. 4.
p. 408 'It seems to me': LW to PE, 23.10.21.
p. 408 'I now believe': LW to LH, Nov. 1921.
p. 408 'THE KING OF THE DARK CHAMBER': This fragment is now in the possession of Mrs Peg Rhees; it was shown to me by Rush Rhees, who very kindly supplied me with a copy of it.
p. 410 'Russell and the parsons': *Recollections*, p. 102.
p. 411 'Suppose someone said: "This is poor evidence"': *Lectures and Conversations*, p. 61.
p. 411 'Suppose someone said: "What do you believe"': ibid., p. 70.
p. 412 'You are not beautiful': Tagore, *King of the Dark Chamber*, p. 199.
p. 412 'Dear Mrs Stewart': LW to Mrs Stewart, 28.10.38, now in the possession of Mrs Katherine Thomson.
p. 414 'I needn't say': LW to GEM, 2.2.39.
p. 415 'To refuse the chair': quoted in *Recollections*, p. 141.
p. 415 'Having got the professorship': LW to WE, 27.3.39.
p. 415 'I would do my utmost': *Lectures and Conversations*, p. 28.
p. 416 'I would say': *Wittgenstein's Lectures on the Foundations of Mathematics: Cambridge, 1939*, p. 103.
p. 416 'There is no religious denomination': *Culture and Value*, p. 1.
p. 416 'Take Russell's contradiction': *Lectures on the Foundations of Mathematics*, p. 222.
p. 417 'All the puzzles': ibid., p. 14.
p. 417 'Another idea': ibid.
p. 417 'The *mathematical* problems': *Remarks on the Foundations of Mathematics*, VII, p. 16.
p. 418 'somewhat parenthetical': *Lectures on the Foundations of Mathematics*, p. 67.
p. 418 'I shall try': ibid., p. 22.
p. 418 'I understand, but I don't agree': *Lectures on the Foundations of Mathematics*, p. 67.
p. 419 'I see your point': ibid., p. 95.
p. 419 'Turing thinks': ibid., p. 102.
p. 420 'Obviously': ibid., p. 55.
p. 420 'introducing Bolshevism': ibid., p. 67.
p. 420 'It is very queer': ibid., pp. 206–7.

p. 421 'will not come in': ibid., p. 211.
p. 421 *Turing*: 'You cannot be confident': ibid., pp. 217–18.
p. 421 'You seem to be saying': ibid., p. 219.
p. 422 Andrew Hodges: see *The Enigma of Intelligence*, note (3.39), pp. 547–8.
p. 422 'Wittgenstein was doing something important': Malcolm, *Memoir*, p. 23.
p. 423 'I am not the deducting, deducing book type': Street & Smith's *Detective Story Magazine*, Jan. 1945. Race Williams was the creation of Carroll John Daly, the writer credited with having invented the 'hard-boiled' detective story.
p. 424 'what is the use of studying philosophy': this is Wittgenstein's reaction as it was remembered by him in a later letter, LW to NM, 16.11.44. See *Memoir*, pp. 93–4.
p. 424 'how should I find the strength': MS 118, 12.9.37; quoted in Baker and Hacker, *An Analytical Commentary*, p. 11.
p. 425 'I feel I will die slowly': quoted in a letter from John Ryle to his wife, Miriam. Letter now in the possession of Dr Anthony Ryle.
p. 425 'Because, unless I'm very much mistaken': LW to NM, 22.6.40.
p. 425 '*Only by a miracle*': LW to NM, 3.10.40.
p. 425 'I feel very unhappy': FS to LW, 11.10.39.
p. 426 'See K once or twice a week': '*Sehe K ein – bis zweimal die Woche; bin aber zweifelhaft darüber, inwieweit das Verhältnis das richtige ist. Möge es wirklich gut sein*': MS 117, 13.6.40.
p. 426 'Occupied myself the *whole* day': '*Habe den* ganzen *Tag mich mit Gedanken über mein Verhältnis zu Kirk beschäftigt. Grösstenteils* sehr *falsch und fruchtlos. Wenn ich diese Gedanken aufschriebe, so sähe man wie tiefstehend und ungerade //schlüpferig// meine Gedanken sind*': MS 123, 7.10.40.
p. 427 'My dear Ro[w]land': LW to RH, 12.10.41.
p. 427 'frightened wild animal': quoted in Pascal, *Recollections*, p. 26.
p. 428 'Think a lot about Francis': '*Denke viel an Francis, aber immer nur mit Reue wegen meiner Lieblosigkeit; nicht mit Dankbarkeit. Sein Leben und Tod scheint mich nur anzuklagen, denn ich war in den letzten 2 Jahren seines Lebens sehr oft lieblos und im Herzen untreu gegen ihn. Wäre er nicht so unendlich sanftmütig und treu gewesen, so wäre ich* gänzlich *lieblos gegen ihn geworden . . . Keit(h?) sehe ich oft, und was das eigentlich heisst, weiss ich nicht. Verdiente Enttäuschung, Bangen, Sorge, Unfähigkeit mich in eine Lebensweise niederzulassen*': MS 125, 28.12.41.
p. 428 'Think a great deal about the last time I was with Francis': '*Denke viel an die letzte Zeit mit Francis: an meine Abscheulichkeit mit ihm . . . Ich kann nicht sehen, wie ich je im Leben von dieser Schuld befreit werden kann*': MS 137, 11.7.48.

21 WAR WORK

p. 431 'He is one of the world's famousest philosophers': John Ryle, letter to Miriam Ryle, 29.9.41.

p. 432 'Good God': story told to me by Dr R. L. Waterfield.

p. 433 'Yes, very well': quoted by Ronald MacKeith in a letter to *Guy's Hospital Gazette*, XC (1976), p. 215.

p. 433 '*One word* that comes from your heart': LW to RH, 20.8.41.

p. 433 'I can't write about Francis': LW to RH, 27.11.41.

p. 433 'If, however': LW to RH, dated 'Sunday'.

p. 433 'It is on the whole': LW to RH, 'Wednesday'.

p. 434 'As long as you find it difficult': LW to RH, 26.1.42.

p. 434 'I feel, on the whole, lonely': LW to RH, 31.12.41.

p. 434 'Daddy and another Austrian': This diary is still in the possession of its author, Dr Anthony Ryle.

p. 435 'The hospital had scores of firebombs': letter from Dr H. Osmond to the author, 4.2.86.

p. 436 'Two – and one of them is Gilbert Ryle': told to the author by Miss Wilkinson.

p. 436 'Tonight I dreamt': MS 125, 16.10.42; quoted in Nedo, op. cit., p. 305.

p. 437 'we may then refer': *Lectures and Conversations*, p. 44.

p. 437 'and to me': ibid., p. 42.

p. 438 'it is an idea': ibid., p. 43.

p. 438 '. . . something which people are inclined to accept': ibid., p. 44.

p. 438 'One of the chief triumphs': Russell, 'Mathematics and the Metaphysicians', *Mysticism and Logic*, pp. 59–74.

p. 439 'Why do I want to take': *Remarks on the Foundations of Mathematics*, VII, p. 19.

p. 439 'is almost as if': ibid., V, p. 25.

p. 439 'completely deformed the thinking': ibid., VII, p. 11.

p. 442 'I no longer feel any hope': *'Ich fühle keine Hoffnung mehr für die Zukunft in meinem Leben. Es ist als hätte ich nur mehr eine lange Strecke lebendigen Todes vor mir. Ich kann mir für mich keine Zukunft als eine grässliche vorstellen. Freundlos und freudlos'*: MS 125, 1.4.42.

p. 442 'I suffer greatly': *'Ich leide sehr unter Furcht vor der gänzlichen Vereinsamung, die mir jetzt droht. Ich kann nicht sehen, wie ich dieses Leben ertragen kann. Ich sehe es als ein Leben, in dem ich mich jeden Tag werde vor dem Abend fürchten müssen, der mir nur dumpfe Traurigkeit bringt'*: MS 125, 9.4.42.

p. 442 'If you can't find happiness': *'"Wenn du das Glück nicht in der Ruhe finden kannst, finde es im Laufen!" Wenn ich aber müde werde, zu laufen? "Sprich nicht vom Zusammenbrechen, ehe du zusammenbrichst."*

Wie ein Radfahrer muss ich nun beständig treten, mich beständig bewegen, um nicht umzufallen': MS 125, 9.4.42.

p. 442 'My unhappiness is so complex': *'Mein Unglück ist so komplex, dass es schwer zu beschreiben ist. Aber wahrscheinlich ist doch* Vereinsamung *die Hauptsache'*: MS 125, 26.5.42.

p. 442 'For ten days': *'Höre seit 10 Tagen nichts mehr von K, obwohl ich ihn vor einer Woche um dringende Nachricht gebeten habe. Ich denke, dass er vielleicht mit mir gebrochen hat. Ein* tragischer *Gedanke!'*: MS 125, 27.5.42.

p. 443 'I have suffered much': *'Ich habe viel gelitten, aber ich bin scheinbar unfähig aus meinem Leben zu* lernen. *Ich leide noch immer so wie vor vielen Jahren. Ich bin nicht stärker und nicht weiser geworden'*: *ibid.*

p. 444 'He could be cheerful': *Recollections*, p. 24.

p. 444 'I'm sorry to hear': LW to RF, 8.6.49.

p. 444 'I wonder': LW to RF, 15.12.50.

p. 444 'I'm glad to hear': LW to RF, 1.2.51.

p. 445 'Recent experience': Dr R. T. Grant, 'Memorandum on the Observations Required in Cases of Wound Shock', MRC Archives.

p. 446 'In my way of doing philosophy': MS 213 (The 'Big Typescript'), p. 421.

p. 446 'Quite a lot of the preamble': Colonel Whitby to Dr Landsborough Thomson, 5.7.41, MRC Archives.

p. 447 'Professor Ludwig Wittgenstein': Dr Grant to Dr Herrald, 30.4.43, MRC Archives.

p. 448 'He is proving very useful': Dr Grant to Dr Herrald, 1.6.43.

p. 448 'I *imagine*': LW to RH, 17.3.43.

p. 448 'On each of these': *Lectures and Conversations*, p. 45.

p. 449 'points to the sort of explanation': ibid., p. 47.

p. 449 'Freud very commonly': ibid.

p. 449 'Obviously': ibid., p. 48.

p. 449 'There was a vacant room': Miss Helen Andrews to the author, 12.11.85.

p. 450 'You think philosophy is difficult': *Recollections*, p. 106.

p. 451 'You do *decent* work': quoted by Dr Basil Reeve in conversation with the author.

p. 452 'In practice': introduction to *Observations on the General Effects of Injury in Man* (HMSO, 1951).

p. 452 '[It] threw grave doubt': *Medical Research in War*, Report of the MRC for the years 1939–45, p. 53.

p. 453 'After the end': *Recollections*, p. 147.

p. 454 'Why in the world': quoted in Drury, *Recollections*, p. 148.

p. 454 'It is obvious to me': quoted ibid., p. 147.

p. 454 'I haven't heard from Smythies': LW to NM, 11.9.43.
p. 454 'I am feeling lonely': LW to NM, 7.12.43.
p. 455 'I too regret': LW to NM, 11.9.43.
p. 456 'Wittgenstein has agreed': Dr Grant to Dr Landsborough Thomson, 13.12.43.
p. 456 'He was reserved': Dr E. G. Bywaters to the author, 9.11.85.
p. 457 'Professor Wittgenstein has been doing': Bywaters to Cuthbertson, 8.2.44.
p. 457 'Professor Wittgenstein left us today': Bywaters to Herrald, 16.2.44.

22 SWANSEA

p. 458 'I don't know if you remember Rhees': NM to LW, 7.12.43.
p. 458 'If, e.g., I leave here': LW to RR, 9.2.44.
p. 459 'The weather's foul': LW to NM, 15.12.45.
p. 459 'his remedies would be all too drastic': *Recollections*, p. 32.
p. 460 'I hadn't a good impression': LW to RH, 17.3.44.
p. 460 'If it ever happens': *Recollections*, p. 149.
p. 461 'is that you can read it': LW to NM, 24.11.42.
p. 461 'I think you must stop creeping': LW to RH, 17.3.44.
p. 461 'I wish I knew more': LW to RH, 24.3.44.
p. 462 'It seems to me': LW to RH, 20.4.44.
p. 463 'I wish you *luck*': LW to RH, 8.6.44.
p. 463 'Yes I do': quoted by Rush Rhees in conversation with the author.
p. 464 'If someone tells me': quoted in Drury, *Recollections*, p. 88.
p. 464 'An honest religious thinker': *Culture and Value*, p. 73.
p. 464 'A miracle is, as it were': ibid., p. 45.
p. 464 'I am not a religious man': *Recollections*, p. 79.
p. 464 'Isn't she an angel': quoted by Mrs Clement in conversation with the author.
p. 466 'Wittgenstein's chief contribution': quoted by Rush Rhees in conversation with the author. (Rhees had no doubt that his memory was correct, but it must be pointed out that Professor John Wisdom had, when I asked him, no recollection of the episode.)
p. 466 'What about your work on mathematics?': quoted by Rush Rhees in conversation with the author.
p. 467 'I want to call': MS 169, p. 37.
p. 469 'The application of the concept': *Remarks on the Foundations of Mathematics*, VI, p. 21.
p. 469 'I may give a new rule': ibid., p. 32.
p. 470 'I'll probably take on a war-job': LW to RH, 3.8.44.
p. 470 'What I'll do': LW to RH, 3.9.44.

23 THE DARKNESS OF THIS TIME

p. 471 'Russell's books': quoted in *Recollections*, p. 112.
p. 472 'Russell isn't going to kill himself': see Malcolm, op. cit., p. 57.
p. 472 'The earlier Wittgenstein': Russell, *My Philosophical Development*, p. 161.
p. 472 'I've seen Russell': LW to RR, 17.10.44.
p. 472 'It is not an altogether pleasant experience': Russell, op. cit., p. 159.
p. 473 'Moore is as nice as always': LW to RR, 17.10.44.
p. 473 'I fear': LW to GEM, 7.3.41; see *Briefe*, p. 254.
p. 474 'He did not realise': quoted in Sister Mary Elwyn McHale, op. cit., p. 77.
p. 474 'with his great love for truth': Malcolm, op. cit., p. 56.
p. 474 'Thinking is sometimes easy': LW to RR, 17.10.44.
p. 474 'I then thought': LW to NM, 16.11.44.
p. 475 'We sat in silence': Malcolm, op. cit., p. 36.
p. 475 'Had I had it before': LW to NM, 22.5.45.
p. 476 'This war': LW to RH, dated only 'Thursday', but probably autumn 1944.
p. 476 'I'm sorry to hear': LW to RR, 28.11.44.
p. 476 'you were right': ibid.
p. 477 'How does the philosophical problem': *Philosophical Investigations*, I, 308.
p. 478 'The "Self of selves"': William James, *Principles of Psychology*, I, 301.
p. 478 'not the meaning of the word': *Philosophical Investigations*, I, 410.
p. 478 'I always enjoy reading anything of William James': *Recollections*, p. 106.
p. 479 'The Term's over': LW to RR, 13.6.45.
p. 479 'going damn slowly': LW to NM, 26.6.45.
p. 479 'I might publish by Christmas': LW to NM, 17.8.45.
p. 480 'The last 6 months': LW to RH, 14.5.45.
p. 480 'this peace is only a truce': LW to NM, undated.
p. 480 'lots of luck': LW to RH, 8.9.45.
p. 480 'self-righteousness in international affairs': quoted in Ruth Dudley Edwards, *Victor Gollancz: A Biography*, p. 406.
p. 481 'glad to see that someone': LW to Victor Gollancz, 4.9.45; reproduced in full, ibid., pp. 406–7.
p. 482 'Polemic, or the art of throwing eggs': *Recollections*, p. 203.
p. 482 'L. Wiltgenstein, Esq.': quoted in Edwards, op. cit., p. 408. Wittgenstein's reaction was told to me by Rush Rhees.
p. 482 'enjoying my absence from Cambridge': LW to NM, 8.9.45.
p. 482 'My book is gradually': LW to NM, 20.9.45.
p. 484 'It is not without reluctance': *Culture and Value*, p. 66.

p. 484 'They mean to do it': see Britton, op. cit., p. 62.
p. 485 'The hysterical fear': *Culture and Value*, p. 49.
p. 485 'The truly apocalyptic view': ibid., p. 56.
p. 485 'Science and industry': ibid., p. 63.
p. 486 'My type of thinking': quoted Drury, *Recollections*, p. 160.
p. 486 'In fact, nothing is more *conservative*': quoted Rhees, *Recollections*, p. 202.
p. 486 'I find more and more': see *Recollections*, pp. 207–8.
p. 487 'I'll talk about anything': told to the author by Rush Rhees.
p. 487 'One day in July': Britton, op. cit., p. 62.
p. 488 'the disintegrating and putrefying English civilization': MS 134; quoted in Nedo, op. cit., p. 321.

24 A CHANGE OF ASPECT

p. 489 'cheaply wrapped': *Culture and Value*, p. 50.
p. 490 'I am by no means sure': ibid., p. 61.
p. 490 'Wisdom is cold': ibid., p. 56.
p. 490 '"Wisdom is grey"': ibid., p. 62.
p. 490 'I believe': ibid., p. 53.
p. 491 'it's as though my knees were stiff': ibid., p. 56.
p. 491 'I am very sad': *'Ich bin sehr traurig, sehr oft traurig. Ich fühle mich so, als sei das jetzt das Ende meines Lebens . . . Das* eine, *was die Liebe zu B. für mich getan hat ist: sie hat die übrigen kleinlichen Sorgen meine Stellung und Arbeit betreffend in den Hintergrund gejagt'*: MS 130, p. 144, 8.8.46.
p. 491 'just as a boy': MS 131, 12.8.46.
p. 491 'I feel as though': MS 131, 14.8.46.
p. 492 'It is the mark of a *true* love': MS 131, 14.8.46; quoted in Nedo, op. cit., p. 325.
p. 492 'I feel that my mental health': *'Ich fühle, meine geistige Gesundheit hängt an einem dünnen Faden. Es ist natürlich die Sorge und Angst wegen B., die mich so abgenützt hat. Und doch könnte auch das nicht geschehen, wenn ich nicht eben leicht entzündbar wäre, "highly inflammable"'*: MS 131, 18.8.46.
p. 492 'Were they stupid': MS 131, 20.8.46; *Culture and Value*, p. 49.
p. 492 'if a light shines on it': *Culture and Value*, pp. 57–8.
p. 492 '"For our desires shield us"': *'"Denn die Wünsche verhüllen uns selbst das Gewünschte. Die Gaben kommen herunter in ihren eignen Gestalten etc." Das sage ich mir wenn ich die Liebe B's empfange. Denn dass sie das grosse, seltene Geschenk ist, weiss ich wohl; dass sie ein seltener Edelstein ist, weiss ich wohl, – und auch, dass sie nicht ganz von der Art ist, von der ich geträumt hatte'*: MS 132, 29.9.46.

p. 493 'Everything about the place': *'Alles an dem Ort stösst mich ab. Das Steife, Künstliche, Selbstgefällige der Leute. Die Universitätsatmosphäre ist mir ekelhaft'*: MS 132, 30.9.46.

p. 493 'What I miss most': LW to RF, 9.11.46.

p. 493 'So when you'll come back': LW to RF, Aug. 1946.

p. 493 'Sorry you don't get post': LW to RF, 7.10.46.

p. 493 'Why in hell': LW to RF, 21.10.46.

p. 494 'I'm feeling far better': LW to RF, 9.11.46.

p. 494 'I'm thinking every day': LW to RF, 21.10.46.

p. 495 'Not to threaten visiting lecturers': Popper's account is given in *Unended Quest: An Intellectual Autobiography*, pp. 122–3.

p. 495 'veneration for Wittgenstein': Ryle in Wood & Pitcher, op. cit., p. 11.

p. 496 'As little philosophy as I have read': MS 135, 27.7.47.

p. 496 'Practically every philosopher': Mary Warnock, diary, 14.5.47. I am very grateful to Lady Warnock, and to Oscar Wood and Sir Isaiah Berlin for their recollections of this meeting. The part of the exchange between Wittgenstein and Pritchard that is quoted comes from *A Wittgenstein Workbook*, p. 6.

p. 497 'For years': Anscombe, *Metaphysics and the Philosophy of Mind*, pp. vii–ix.

p. 498 'This man': quoted by Professor Anscombe in conversation with the author.

p. 498 'very unnerving': Iris Murdoch, quoted by Ved Mehta in *The Fly and the Fly Bottle*, p. 55.

p. 498 'More able than Ramsey?': story told to me by Rush Rhees.

p. 499 'were far from the centre': Kreisel, 'Wittgenstein's "*Remarks on the Foundations of Mathematics*"', *British Journal for the Philosophy of Science*, IX (1958), pp. 135–58.

p. 499 'tense and often incoherent': Kreisel, 'Critical Notice: "*Lectures on the Foundations of Mathematics*"', in *Ludwig Wittgenstein: Critical Assessments*, pp. 98–110.

p. 499 'Wittgenstein's views on mathematical logic': Kreisel, *British Journal for the Philosophy of Science*, op. cit., pp. 143–4.

p. 499 'As an introduction': Kreisel, 'Wittgenstein's Theory and Practice of Philosophy', *British Journal for the Philosophy of Science*, XI (1960), pp. 238–52.

p. 500 'These lectures are on the philosophy of psychology': from the notes taken by A. C. Jackson. These notes (together with those of the same lectures taken by P. T. Geach and K. J. Shah) have now been published as *Wittgenstein's Lectures on Philosophical Psychology: 1946–7*, but at the time of writing this I was dependent on privately circulated copies. There may be some slight variations between the notes as I give them and as they have been published.

p. 501 'Now let us go back to last day': ibid.

p. 501 'What I give is the morphology': Malcolm, op. cit., p. 43.

p. 502 'arose from the fact': Gasking and Jackson, 'Wittgenstein as a Teacher', Fann, op. cit., pp. 49–55.

p. 502 'I am showing my pupils': *Culture and Value*, p. 56.

p. 502 'In teaching you philosophy': Gasking and Jackson, op. cit., p. 51.

p. 503 'I still keep getting entangled': *Culture and Value*, p. 65.

p. 503 'Starting at the beginning': Malcolm, op. cit., p. 44.

p. 503 'All is happiness': *'Alles ist Glück. Ich könnte jetzt so nicht schreiben, wenn ich nicht die letzten 2 Wochen mit B. verbracht hätte. Und ich hätte sie nicht so verbringen können, wenn Krankheit oder irgend ein Unfall dazwischen gekommen wäre'*: MS 132, 8.10.46.

p. 504 'In love': *'Ich bin in der Liebe zu wenig* gläubig *und zu wenig* mutig . . . *ich bin leicht verletzt und fürchte mich davor, verletzt zu werden, und sich in* dieser *Weise selbst schonen ist der Tod aller Liebe. Zur wirklichen Liebe braucht es* Mut. *Das heisst aber doch, man muss auch den Mut haben, abzubrechen, und zu entsagen, also den Mut eine Todeswunde zu ertragen. Ich aber kann nur hoffen, dass mir das Fürchterlichste erspart bleibt'*: MS 132, 22.10.46.

p. 504 'I do not have the courage': *'Ich habe nicht den Mut und nicht die Kraft und Klarheit den Tatsachen meines Lebens gerade in's Gesicht zu schauen. – B. hat zu mir eine* Vor-*Liebe. Etwas, was nicht halten kann.* Wie *diese verwelken wird, weiss ich natürlich nicht. Wie* etwas *von ihr zu erhalten wäre, lebendig, nicht gepresst in einem Buch als Andenken, weiss ich auch nicht . . . Ich weiss nicht, ob und wie ich es aushalten werde,* dies *Verhältnis mit* dieser *Aussicht weiterzunähren . . . Wenn ich mir vorstelle, dass ich es abgebrochen hätte, so fürchte ich mich vor der Einsamkeit'*: MS 133, 25.10.46.

p. 504 'love is a *joy*': *'Die Liebe ist ein* Glück. *Vielleicht ein Glück mit Schmerzen, aber ein Glück . . . In der Liebe muss ich sicher* ruhen *können. – Aber kannst du ein warmes Herz zurückweisen; Ist es ein Herz, das warm für* mich *schlägt;* – I'll rather do anything than to hurt the soul of friendship. – I must know: he won't hurt *our friendship. Der Mensch kann aus seiner Haut nicht heraus. Ich kann nicht eine Forderung, die tief in mir, meinem ganzen Leben verankert, liegt, aufgeben. Denn die* Liebe *ist mit der Natur verbunden; und würde ich unnatürlich, so würde //müsste// die Liebe aufhören. – Kann ich sagen: "Ich werde vernünftig sein, und das nicht mehr verlangen."? . . . Ich kann sagen: Lass ihn gewähren, – es wird einmal anders werden. –* Die Liebe, *die ist die Perle von grossem Wert, die man am Herzen hält, für die man* nichts *eintauschen will, die man als das wertvollste schätzt. Sie* zeigt *einem überhaupt – wenn man sie hat – was grosser Wert* ist. *Man lernt, was es* heisst*: ein Edelmetall von allen andern aussondern . . . Das Furchtbare ist die Ungewissheit . . . "Auf Gott vertrauen" . . . Von da, wo ich bin, zum Gottvertrauen ist ein* weiter *Weg. Freudevolle Hoffnung und Furcht sind einander verschwistert. Ich kann die eine nicht haben ohne dass sie an die andre grenzt'*: MS 133, 26.10.46.

p. 505 'Trust in God': ibid.

p. 505 'Ask yourself these questions: *'Frag dich diese Frage: Wenn du stirbst, wer wird dir nachtrauern; und* wie tief *wird die Trauer sein? Wer trauert um F.; wie tief trauere ich um ihn, der mehr Grund zur Trauer hat als irgend jemand? Hat er* nicht *verdient, dass jemand sein ganzes Leben lang um ihn trauert? Wenn jemand so er. Da möchte man sagen: Gott wird ihn aufheben und ihm* geben, *was ein schlechter Mensch ihm versagt'*: MS 133, 10.11.46.

p. 506 'The fundamental insecurity': MS 133, 12.11.46.

p. 506 'Don't be too cowardly': MS 133, 15.11.46.

p. 506 'Can you not cheer up': *'Kannst du nicht auch ohne seine Liebe fröhlich sein?* Musst *du ohne diese Liebe in Gram versinken? Kannst du ohne diese Stütze nicht leben? Denn das ist die Frage: kannst du* nicht *aufrecht gehn, ohne dich auf diesen Stab zu lehnen? Oder kannst du dich nur nicht* entschliessen *ihn aufzugeben? Oder ist es beides? – Du* darfst *nicht immer Briefe erwarten, die nicht kommen . . . Es ist nicht Liebe, was mich zu dieser Stütze zieht, sondern, dass ich auf meinen zwei Beinen allein nicht sicher stehen kann'*: MS 133, 27.11.46.

p. 506 'Some men': *'Mancher Mensch ist im ganzen Leben krank und kennt nur das Glück, das der fühlt, der nach langen heftigen Schmerzen ein paar schmerzlose Stunden hat. (Es ist ein seeliges Aufatmen.)'*: MS 133, 23.11.46.

p. 506 'Is it so unheard of': *'Ist es so unerhört, dass ein Mensch leidet, dass z.B. ein ältlicher Mensch müde und einsam ist, ja selbst, dass er halb verrückt wird?'*: MS 133, 2.12.46.

p. 506 'Only nothing theatrical': MS 133, 12.2.46.

p. 506 'The belief in a benevolent father': MS 134, 4.4.47.

p. 506 'What good does all my talent do me': *'Wozu dient mir all meine Geschicklichkeit, wenn ich im Herzen unglücklich bin? Was hilft es mir, philosophische Probleme zu lösen, wenn ich mit der Hauptsache nicht ins Reine kommen kann?'*: MS 134, 13.4.47.

p. 507 'My lectures are going well': *'Meine Vorlesungen gehen gut, sie werden nie besser gehen. Aber welche Wirkung lassen sie zurück? Helfe ich irgendjemand? Gewiss nicht* mehr, *als wenn ich ein grosser Schauspieler wäre, der ihnen Tragödien vorspielt. Was sie lernen, ist nicht wert gelernt* *zu werden; und der persönliche Eindruck nützt ihnen nichts. Das gilt für Alle, mit vielleicht einer, oder zwei Ausnahmen'*: MS 133, 19.11.46.

p. 507 'Suppose I show it to a child': notes taken by P. T. Geach; see *Wittgenstein's Lectures on Philosophical Psychology*, p. 104 (but see also reference for p. 500).

p. 508 'It would have made': *Philosophical Investigations*, II, p. 195.

p. 509 'In the German language': Wolfgang Köhler, *Gestalt Psychology*, p. 148.

p. 510 'Here where': Goethe, *Italian Journey*, pp. 258–9.

p. 510 'The *Urpflanze*': ibid., p. 310.

p. 511 'Two uses': *Philosophical Investigations*, II, p. 193.

p. 511 'I explained to him': quoted Heller, op. cit., p. 6.

p. 512 'Well, so much the better': ibid.

p. 513 'When I tell the reader': Köhler, op. cit., p. 153.

p. 513 'Now Köhler said': see *Wittgenstein's Lectures on Philosophical Psychology*, pp. 329–30.

p. 514 'It makes no sense to ask': see ibid., p. 104.

p. 514 'Now you try': *Recollections*, p. 159.

p. 514 'The expression of a change': *Philosophical Investigations*, II, pp. 196–7.

p. 515 'Köhler says': *Remarks on the Philosophy of Psychology*, I, p. 982.

p. 516 'A philosopher says': *Culture and Value*, p. 61.

p. 516 'In this country': *'Für Leute wie mich liegt in diesem Lande nichts näher als Menschenhass. Gerade dass man sich in all dieser Solidität auch keine Revolution denken kann macht die Lage noch viel hoffnungsloser. Es ist als hätte diese ganze grenzenlose Öde "come to stay". Es ist als könnte man von diesem Land sagen, es habe ein nasskaltes geistiges Klima'*: MS 134, 13.4.47.

p. 516 'Cambridge grows more and more hateful': *'Cambridge wird mir mehr und mehr verhasst.* The disintegrating and putrefying English civilization. *Ein Land, in dem die Politik zwischen einem bösen Zweck und* keinem *Zweck schwankt'*: MS 134, 23.4.47.

p. 516 '[I] feel myself to be an alien': *'fühle mich fremd //als Fremdling// in der Welt. Wenn dich kein Band an Menschen und kein Band an Gott bindet, so* bist *du ein Fremdling'*: MS 135, 28.7.47.

p. 517 'I wouldn't be altogether surprised': *Recollections*, p. 152.

p. 517 'This is an excellent book': ibid.

p. 517 '. . . go somewhere': LW to GHvW, 27.8.47.

p. 518 'he was reacting': quoted by John Moran, op. cit., p. 92.

p. 518 'It's mostly bad': LW to GHvW, 6.11.47.

25 IRELAND

p. 520 'cold and uncomfortable': LW to RR, 9.12.47.
p. 520 'Sometimes my ideas': *Recollections*, pp. 153–4.
p. 521 'the prospect of becoming English': LW to GHvW, 22.12.47.
p. 521 'Cambridge is a dangerous place': LW to GHvW, 23.2.48.
p. 521 'There is nothing like the Welsh coast': LW to RR, 5.2.48.
p. 521 'The country here': LW to Helene, 10.1.48; quoted in Nedo, op. cit., p. 326.
p. 522 'In a letter': *Culture and Value*, pp. 65–6.
p. 522 'Heaven knows if I'll ever publish': LW to GHvW, 22.12.47.
p. 522 'I am in very good bodily health': LW to RR, 5.2.48.
p. 522 'My work's going moderately well': LW to NM, 4.1.48.
p. 522 'Feel unwell': MS 137, 3.2.48.
p. 523 'occasionally queer states': LW to NM, 5.2.48.
p. 523 'My nerves, I'm afraid': LW to RR, 5.2.48.
p. 523 'I often believe': LW to GHvW, 17.3.48.
p. 523 'They are very quiet': LW to GHvW, 22.12.47.
p. 523 'As soon as he had arrived': *Recollections*, pp. 154–5.
p. 524 'I often thought of you': LW to RR, 15.4.48.
p. 525 'quite nice': LW to NM, 5.6.48.
p. 525 'the man upon whom': MS 137, 17.7.48; quoted in Nedo, op. cit., p. 326.
p. 526 'Tinned food': story told to the author by Thomas Mulkerrins.
p. 526 'I thought you had company': ibid.
p. 526 'Nearly all my writings': *Culture and Value*, p. 77.
p. 527 'I've had a bad time': LW to NM, 30.4.48.
p. 528 'it's undoubtedly a great blessing': LW to Lee Malcolm, 5.6.48.
p. 528 'For though as you know': LW to NM, 4.6.48.
p. 528 'As I recall': see footnote to above letter.
p. 529 Davis wrote to Raymond Chandler: see Frank MacShane, ed., *Selected Letters of Raymond Chandler* (Cape, 1981), p. 167.
p. 529 'All this was very boring': Norbert Davis, *Rendezvous with Fear*, p. 9.
p. 529 'Is he hurt?': ibid., p. 86.
p. 529 'Humour is not a mood': *Culture and Value*, p. 78.
p. 530 'So how do we explain': ibid., p. 70.
p. 531 'What would a person be lacking?': *Remarks on the Philosophy of Psychology*, II, p. 508.
p. 532 'What is lacking': ibid., pp. 464–8.
p. 532 'What is it like': *Culture and Value*, p. 83.
p. 533 'What is incomprehensible': *Remarks on the Philosophy of Psychology*, II, p. 474.

p. 533 'leaves everything as it is': *Philosophical Investigations*, I, 124.
p. 533 'Tradition is not something a man can learn': *Culture and Value*, p. 76.
p. 534 'I get tired': LW to GHvW, 26.5.48.
p. 534 'too soft, too weak': *Culture and Value*, p. 72.
p. 534 'Don't let the grief vex you!': *'Lass dich die Trauer nicht verdriessen! Du solltest sie ins Herz einlassen und auch den Wahnsinn nicht fürchten! Er kommt vielleicht als Freund und nicht als Feind zu dir und nur dein* Wehren *ist das Übel. Lass die Trauer ins Herz ein, verschliess ihr nicht die Tür. Draussen vor der Tür im Verstand stehend ist sie furchtbar, aber im* Herzen *ist sie's nicht'*: MS 137, 29.6.48.
p. 534 'Think a great deal': MS 137, 11.7.48.
p. 534 'But I have decided to *try*': MS 137, 17.7.48.
p. 535 'When I look at the faces': *Recollections*, p. 166.
p. 536 'I'm anxious to make hay': LW to NM, 6.11.48.
p. 536 'Just wait a minute': *Recollections*, p. 156.
p. 536 'I hear the words': *Last Writings*, I, p. 47.
p. 536 'Hegel seems to me': *Recollections*, p. 157.
p. 537 'hasn't the necessary multiplicity': ibid., p. 160.
p. 537 'Now you try': ibid., p. 159.
p. 537 '"Denken ist schwer"': *Culture and Value*, p. 74.
p. 537 'It is impossible for me': *Recollections*, p. 160.
p. 538 'We say that someone has the "eyes of a painter"': *Last Writings*, I, p. 782.
p. 538 'We say that someone doesn't have a musical ear': ibid., p. 783; cf. *Philosophical Investigations*, II, xi, p. 214.
p. 538 'I would like it': *Recollections*, p. 160.
p. 538 'I think he felt': ibid., p. 97.
p. 539 'some sort of infection': LW to NM, 28.1.49.
p. 539 'Drury, I think': MS 138, 29.1.49.
p. 539 'great weakness and pain': MS 138, 11.2.49.
p. 539 '[They] contain many excellent things': *Culture and Value*, p. 27.
p. 540 'for otherwise': MS 138, 2.3.49.
p. 540 'Nice time': MS 138, 15.3.49.
p. 540 'Often it is as though my soul were dead': MS 138, 17.3.49.
p. 540 'Of course it was rejected': *Recollections*, p. 161.
p. 540 'If Christianity is the truth': *Culture and Value*, p. 83.
p. 540 'I can't understand the Fourth Gospel': *Recollections*, p. 164.
p. 541 'The spring which flows gently': *Culture and Value*, p. 30.
p. 541 'It is one and the same': *Recollections*, p. 165.
p. 541 'If it is a good and godly picture': *Culture and Value*, p. 32.
p. 541 'Suppose someone were taught': ibid., p. 81.

p. 542 'What happens': LW to NM, 18.2.49.
p. 542 'While I was in Vienna': LW to NM, 17.5.49.
p. 543 'It is so characteristic': *Recollections*, p. 163.
p. 543 'has something to say': ibid., p. 159.
p. 543 'Now that is all I want': ibid., p. 168.
p. 544 'There is one thing that I'm afraid of': LW to GHvW, 1.6.49.
p. 545 'You have said something': LW to GEM, Oct. 1944.
p. 546 'Don't regard': *Philosophical Investigations*, II, p. 192.
p. 546 'But surely': ibid., p. 190.
p. 547 'We can imagine': ibid., p. 188.
p. 548 'Drury, just look at the expression': *Recollections*, p. 126.
p. 548 'If I see someone': *Philosophical Investigations*, II, p. 223.
p. 548 'It is important': *Culture and Value*, p. 74.
p. 548 'If a lion could talk': *Philosophical Investigations*, II, p. 223.
p. 548 'Is there such a thing': *ibid.*, p. 227.
p. 549 'It was said by many people': Dostoevsky, *The Brothers Karamazov*, p. 30.
p. 549 'Yes, there really have been people like that': *Recollections*, p. 108.
p. 549 'The confusion and barrenness': *Philosophical Investigations*, II, p. 232.

26 A CITIZEN OF NO COMMUNITY

p. 552 'My mind is tired & stale': LW to NM, 1.4.49.
p. 552 'I *know* you'd extend your hospitality': LW to NM, 4.6.49.
p. 552 'My anaemia': LW to NM, 7.7.49.
p. 552 'Ordinarily': Stuart Brown; quoted in McHale, op. cit., p. 78.
p. 553 'particularly because': LW to RF, 28.7.49.
p. 553 'probably the most philosophically strenuous': quoted ibid., p. 80.
p. 554 'But then, he added': Bouwsma's notes of his conversations with Wittgenstein have now been published; see *Wittgenstein Conversations 1949–1951*, p. 9.
p. 554 'I am a very vain person': ibid.
p. 555 'If I had planned it': ibid., p. 12.
p. 556 'I can prove now': Moore, 'Proof of an External World', *Philosophical Papers*, p. 146.
p. 556 'To understand a sentence': Malcolm, op. cit., p. 73.
p. 557 'Instead of saying': ibid., p. 72.
p. 557 'When the sceptical philosophers': ibid., p. 73.
p. 557 'Scepticism is *not* irrefutable': *Tractatus*, 6.51.
p. 557 'Certain propositions': Malcolm, op. cit., p. 74.
p. 558 'If I walked over to that tree': ibid., p. 71.
p. 558 'Just before the meeting': quoted in McHale, op. cit., pp. 79–80.

p. 559 'I don't want to die in America': Malcolm, op. cit., p. 77.
p. 559 'I am sorry that my life should be prolonged': LW to RR, 4.12.49.
p. 559 'My health is very bad': LW to Helene, 28.11.49; quoted in Nedo, op. cit., p. 337.
p. 560 'This is of the greatest importance': LW to NM, 11.12.49.
p. 560 'ARRIVED': LW to GHvW, 26.12.49.
p. 560 'I haven't been to a concert': LW to GHvW, 19.1.50.
p. 560 'I'm very happy': LW to Dr Bevan, 7.2.50.
p. 561 'Colours spur us to philosophise': *Culture and Value*, p. 66.
p. 561 'It's partly boring': LW to GHvW, 19.1.50.
p. 561 'I may find': *Culture and Value*, p. 79.
p. 562 'As I mean it': *On Colour*, II, p. 3.
p. 562 'If someone didn't find': ibid., p. 10.
p. 562 'Phenomenological analysis': ibid., p. 16.
p. 562 'We had expected her end': LW to GHvW, 12.2.50.
p. 562 'Wittgenstein, who took a long time': Paul Feyerabend, *Science in a Free Society*, p. 109.
p. 563 'everything descriptive': *On Certainty*, p. 56.
p. 563 'If "I know etc."': ibid., p. 58.
p. 564 'If language': *Philosophical Investigations*, I, 242.
p. 564 'woeful': see MS 173, 24.3.50.
p. 564 'I don't think I can give formal lectures': LW to NM, 5.4.50.
p. 565 'The thought': LW to NM, 17.4.50.
p. 566 'in my present state': LW to NM, 12.1.51.
p. 566 'I'm doing some work': LW to NM, 17.4.50.
p. 566 'That which I am writing about': *On Colour*, III, p. 295.
p. 566 'One and the same': ibid', p. 213.
p. 566 'Imagine someone': ibid., p. 263.
p. 567 'In the late spring': K. E. Tranøj, 'Wittgenstein in Cambridge, 1949–51', *Acta Philosophica Fennica*, XXVII, 1976; quoted in Nedo, op. cit., p. 335.
p. 567 'I like to stay': LW to NM, 17.4.50.
p. 567 'The house isn't very noisy': LW to GHvW, 28.4.50.
p. 567 'Pink wants to sit on six stools': quoted by Barry Pink in conversation with the author (he was perfectly sure that 'arse' was the word Wittgenstein used).
p. 568 'Certainly not!': quoted by Barry Pink in conversation with the author.
p. 568 'It is remarkable': *Culture and Value*, p. 48.
p. 568 'Shakespeare's similes': ibid., p. 49.
p. 568 'I believe': ibid., p. 85.
p. 569 'I could only stare': ibid., pp. 84–5.

p. 570 'If Moore were to pronounce': *On Certainty*, p. 155.
p. 570 'For months I have lived': ibid., pp. 70–71.
p. 570 'I could imagine': ibid., p. 264.
p. 571 'Everything that I have seen': ibid., p. 94.
p. 571 'The river-bed': ibid., p 97.
p. 571 'I believe': ibid., p. 239.
p. 572 'I have a world picture': ibid., p. 162.
p. 572 'Perhaps one could': *Culture and Value*, p. 85.
p. 572 'Life can educate one': ibid. p. 86.
p. 572 'The attitude that's in question': ibid., p. 85.
p. 573 'He wanted': Father Conrad to the author, 30.8.86.
p. 574 'we enjoyed our stay': LW to GHvW, 29.1.51.

27 STOREYS END

p. 576 'I can't even *think* of work': LW to NM, undated.
p. 576 'How lucky': this, and all other remarks of Wittgenstein's to Mrs Bevan quoted in this chapter, are given as they were told to me by Mrs Bevan in a series of conversations I had with her during 1985–7.
p. 576 'He was very demanding': extract from a written statement of her memories of Wittgenstein by Mrs Bevan.
p. 577 'I have been ill': LW to RF, 1.2.51.
p. 578 'I do philosophy': *On Certainty*, p. 532.
p. 578 'I believe': ibid., p. 387.
p. 578 'A doubt without an end': ibid., p. 625.
p. 578 'I am sitting': ibid., p. 467.
p. 578 'Doubting and non-doubting': ibid., p. 354.
p. 579 'Am I not': ibid., p. 501.
p. 579 'I remember': *Recollections*, p. 171.
p. 580 'Isn't it curious': ibid., p. 169.
p. 580 'God may say to me': *Culture and Value*, p. 87.

SELECT BIBLIOGRAPHY

Below are listed the principal printed sources used in the writing of this biography. For an exhaustive bibliography of works by and about Wittgenstein, see V. A. and S. G. Shanker, ed., *Ludwig Wittgenstein: Critical Assessments, V: A Wittgenstein Bibliography* (Croom Helm, 1986).

Anscombe, G. E. M. *Metaphysics and the Philosophy of Mind*, Collected Philosophical Papers, II (Blackwell, 1981)

Augustine, Saint *Confessions* (Penguin, 1961)

Ayer, A. J. *Wittgenstein* (Weidenfeld & Nicolson, 1985)

—— *Part of My Life* (Collins, 1977)

—— *More of My Life* (Collins, 1984)

Baker, G. P. *Wittgenstein, Frege and the Vienna Circle* (Blackwell, 1988)

Baker, G. P. and Hacker, P. M. S. *Wittgenstein: Meaning and Understanding* (Blackwell, 1983)

—— *An Analytical Commentary on Wittgenstein's Philosophical Investigations*, I (Blackwell, 1983)

—— *Wittgenstein: Rules, Grammar and Necessity: An Analytical Commentary on the Philosophical Investigations*, II (Blackwell, 1985)

—— *Scepticism, Rules and Language* (Blackwell, 1984)

Bartley, W. W. *Wittgenstein* (Open Court, rev. 2/1985)

Bernhard, Thomas *Wittgenstein's Nephew* (Quartet, 1986)

Block, Irving, ed., *Perspectives on the Philosophy of Wittgenstein* (Blackwell, 1981)

Bouwsma, O. K. *Philosophical Essays* (University of Nebraska Press, 1965)

—— *Wittgenstein: Conversations 1949–1951*, ed. J. L. Craft and Ronald E. Hustwit (Hackett, 1986)

Clare, George *Last Waltz in Vienna* (Pan, 1982)

Clark, Ronald W. *The Life of Bertrand Russell* (Jonathan Cape and Weidenfeld & Nicolson, 1975)

Coope, Christopher, et al. *A Wittgenstein Workbook* (Blackwell, 1971)
Copi, Irving M. and Beard, Robert W., ed., *Essays on Wittgenstein's Tractatus* (Routledge, 1966)
Dawidowicz, Lucy S. *The War Against the Jews 1933–45* (Weidenfeld & Nicolson, 1975)
Deacon, Richard *The Cambridge Apostles: A History of Cambridge University's Elite Intellectual Secret Society* (Robert Royce, 1985)
Delany, Paul *The Neo-pagans: Rupert Brooke and the Ordeal of Youth* (The Free Press, 1987)
Dostoevsky, Fyodor *The Brothers Karamazov* (Penguin, 1982)
Drury, M. O'C. *The Danger of Words* (Routledge, 1973)
Duffy, Bruce *The World As I Found It* (Ticknor & Fields, 1987)
Eagleton, Terry 'Wittgenstein's Friends', *New Left Review*, CXXXV (September–October 1982); reprinted in *Against the Grain* (Verso, 1986)
Fann, K. T., ed., *Ludwig Wittgenstein: The Man and His Philosophy* (Harvester, 1967)
Feyerabend, Paul *Science in a Free Society* (Verso, 1978)
Ficker, Ludwig von *Denkzettel und Danksagungen* (Kösel, 1967)
Field, Frank *The Last Days of Mankind: Karl Kraus and His Vienna* (Macmillan, 1967)
Frege, Gottlob *The Foundations of Arithmetic* (Blackwell, 1950)
—— *Philosophical Writings* (Blackwell, 1952)
—— *Philosophical and Mathematical Correspondence* (Blackwell, 1980)
—— *The Basic Laws of Arithmetic* (University of California Press, 1967)
Freud, Sigmund *The Interpretation of Dreams* (Penguin, 1976)
—— *Jokes and Their Relation to the Unconscious* (Penguin, 1976)
Gay, Peter *Freud: A Life for Our Time* (Dent, 1988)
Goethe, J. W. *Italian Journey* (Penguin, 1970)
—— *Selected Verse* (Penguin, 1964)
Grant, R. T. and Reeve, E. B. *Observations on the General Effects of Injury in Man* (HMSO, 1951)
Hacker, P. M. S. *Insight and Illusion: Themes in the Philosophy of Wittgenstein* (Oxford, rev. 2/1986)
Hänsel, Ludwig 'Ludwig Wittgenstein (1889–1951)', *Wissenschaft und Weltbild* (October 1951), p. 272–8
Haller, Rudolf *Questions on Wittgenstein* (Routledge, 1988)
Hayek, F. A. von 'Ludwig Wittgenstein' (unpublished, 1953)
Heller, Erich *The Disinherited Mind: Essays in Modern German Literature and Thought* (Bowes & Bowes, 1975)
Henderson, J. R. 'Ludwig Wittgenstein and Guy's Hospital', *Guy's Hospital Reports*, CXXII (1973), pp. 185–93

Hertz, Heinrich *The Principles of Mechanics* (Macmillan, 1899)
Hilmy, S. Stephen *The Later Wittgenstein: The Emergence of a New Philosophical Method* (Blackwell, 1987)
Hitler, Adolf *Mein Kampf* (Hutchinson, 1969)
Hodges, Andrew *Alan Turing: The Enigma of Intelligence* (Burnett, 1983)
Iggers, Wilma Abeles *Karl Kraus: A Viennese Critic of the Twentieth Century* (Nijhoff, 1967)
James, William *The Varieties of Religious Experience* (Penguin, 1982)
—— *The Principles of Psychology*, 2 vols (Dover, 1950)
Janik, Allan and Toulmin, Stephen *Wittgenstein's Vienna* (Simon & Schuster, 1973)
Jones, Ernest *The Life and Work of Sigmund Freud* (Hogarth, 1962)
Kapfinger, Otto *Haus Wittgenstein: Eine Dokumentation* (The Cultural Department of the People's Republic of Bulgaria, 1984)
Kenny, Anthony *Wittgenstein* (Allen Lane, 1973)
—— *The Legacy of Wittgenstein* (Blackwell, 1984)
Keynes, J. M. *A Short View of Russia* (Hogarth, 1925)
Köhler, Wolfgang *Gestalt Psychology* (G. Bell & Sons, 1930)
Kraus, Karl *Die Letzten Tage der Menschenheit*, 2 vols (Deutscher Taschenbuch, 1964)
—— *No Compromise: Selected Writings*, ed. Frederick Ungar (Ungar Publishing, 1984)
—— *In These Great Times: A Karl Kraus Reader*, ed. Harry Zohn (Carcanet, 1984)
Kreisel, G. 'Wittgenstein's "*Remarks on the Foundations of Mathematics*"', *British Journal for the Philosophy of Science*, IX (1958), pp. 135–58
—— 'Wittgenstein's Theory and Practice of Philosophy', *British Journal for the Philosophy of Science*, XI (1960), pp. 238–52
—— 'Critical Notice: "*Lectures on the Foundations of Mathematics*"', in *Ludwig Wittgenstein: Critical Assessments*, ed. S. G. Shanker (Croom Helm, 1986), pp. 98–110
Leitner, Bernhard *The Architecture of Ludwig Wittgenstein: A Documentation* (Studio International, 1973)
Levy, Paul *G. E. Moore and the Cambridge Apostles* (Oxford, 1981)
Luckhardt, C. G. *Wittgenstein: Sources and Perspectives* (Harvester, 1979)
Mabbott, John *Oxford Memories* (Thornton's, 1986)
McGuinness, Brian *Wittgenstein: A Life. Young Ludwig 1889–1921* (Duckworth, 1988)
—— , ed., *Wittgenstein and His Times* (Blackwell, 1982)
McHale, Sister Mary Elwyn *Ludwig Wittgenstein: A Survey of Source Material for a Philosophical Biography* (MA thesis for the Catholic University of America, 1966)

Malcolm, Norman *Ludwig Wittgenstein: A Memoir* (with a Biographical Sketch by G. H. von Wright) (Oxford, rev. 2/1984)
Manvell, Roger and Fraenkel, Heinrich *Hitler: The Man and the Myth* (Grafton, 1978)
Mays, W. 'Wittgenstein's Manchester Period', *Guardian* (24 March 1961)
—— 'Wittgenstein in Manchester', in '*Language, Logic, and Philosophy': Proceedings of the 4th International Wittgenstein Symposium* (1979), pp. 171–8
Mehta, Ved *The Fly and the Fly-Bottle* (Weidenfeld & Nicolson, 1963)
Moore, G. E. *Philosophical Papers* (Unwin, 1959)
Moran, John 'Wittgenstein and Russia', *New Left Review*, LXXIII (May–June 1972)
Morton, Frederic *A Nervous Splendour* (Weidenfeld & Nicolson, 1979)
Nedo, Michael and Ranchetti, Michele *Wittgenstein: Sein Leben in Bildern und Texten* (Suhrkamp, 1983)
Nietzsche, Friedrich *Twilight of the Idols* and *The Anti-Christ* (Penguin, 1968)
Ogden, C. K. and Richards, I. A. *The Meaning of Meaning* (Kegan Paul, 1923)
Parak, Franz *Am anderen Ufer* (Europäischer Verlag, 1969)
Partridge, Frances *Memories* (Robin Clark, 1982)
Popper, Karl *Unended Quest: An Intellectual Autobiography* (Fontana, 1976)
Ramsey, F. P. 'Critical Notice of L. Wittgenstein's "Tractatus Logico-Philosophicus"', *Mind*, XXXII, no. 128 (October 1923), pp. 465–78
—— *Foundations: Essays in Philosophy, Logic, Mathematics and Economics* (Routledge, 1978)
Rhees, Rush 'Wittgenstein' [review of Bartley, op. cit.] *The Human Word*, XIV (February 1974)
—— *Discussions of Wittgenstein* (Routledge, 1970)
—— *Without Answers* (Routledge, 1969)
—— , ed., *Recollections of Wittgenstein* (Oxford, 1984)
Russell, Bertrand *The Principles of Mathematics* (Unwin, 1903)
—— *The Problems of Philosophy* (Home University Library, 1912)
—— *Our Knowledge of the External World* (Unwin, 1914)
—— *Mysticism and Logic* (Unwin, 1918)
—— *Introduction to Mathematical Philosophy* (Unwin, 1919)
—— *The Analysis of Mind* (Unwin, 1921)
—— *The Practice and Theory of Bolshevism* (Unwin, 1920)
—— *Marriage and Morals* (Unwin, 1929)
—— *The Conquest of Happiness* (Unwin, 1930)
—— *In Praise of Idleness* (Unwin, 1935)
—— *An Inquiry into Meaning and Truth* (Unwin, 1940)
—— *History of Western Philosophy* (Unwin, 1945)
—— *Human Knowledge: Its Scope and Limits* (Unwin, 1948)

Russell, Bertrand *Logic and Knowledge*, ed. R. C. March (Unwin, 1956)
—— *My Philosophical Development* (Unwin, 1959)
—— *Autobiography* (Unwin, 1975)
Russell, Dora *The Tamarisk Tree*, I: *My Quest for Liberty and Love* (Virago, 1977)
Ryan, Alan *Bertrand Russell: A Political Life* (Allen Lane, 1988)
Schopenhauer, Arthur *Essays and Aphorisms* (Penguin, 1970)
—— *The World as Will and Representation*, 2 vols (Dover, 1969)
Shanker, S. G. *Wittgenstein and the Turning Point in the Philosophy of Mathematics* (Croom Helm, 1987)
Sjögren, Marguerite *Granny et son temps* (privately printed in Switzerland, 1982)
Skidelsky, Robert *John Maynard Keynes*, I: *Hopes Betrayed 1883–1920* (Macmillan, 1983)
Spengler, Oswald *The Decline of the West* (Unwin, 1928)
Sraffa, Piero *Production of Commodities By Means of Commodities* (Cambridge, 1960)
Steiner, G. *A Reading Against Shakespeare*, W. P. Ker Lecture for 1986 (University of Glasgow, 1986)
Tagore, Rabindranath *The King of the Dark Chamber* (Macmillan, 1918)
Thomson, George 'Wittgenstein: Some Personal Recollections', *The Revolutionary World*, XXXVII–IX (1979), pp. 87–8
Tolstoy, Leo *A Confession and Other Religious Writings* (Penguin, 1987)
—— *Master and Man and Other Stories* (Penguin, 1977)
—— *The Kreutzer Sonata and Other Stories* (Penguin, 1985)
—— *The Raid and Other Stories* (Oxford, 1982)
Waismann, F. *The Principles of Linguistic Philosophy*, ed. R. Harré (Macmillan, 1965)
Walter, Bruno *Theme and Variations: An Autobiography* (Hamish Hamilton, 1947)
Weininger, Otto *Sex and Character* (Heinemann, 1906)
Wittgenstein, Hermine, *Familienerinnerungen* (unpublished)
Wood, Oscar P. and Pitcher, George, ed., *Ryle* (Macmillan, 1971)
Wright, G. H. von *Wittgenstein* (Blackwell, 1982)
—— 'Ludwig Wittgenstein, A Biographical Sketch', in Malcolm, op. cit.
Wuchterl, Kurt and Hübner, Adolf *Ludwig Wittgenstein im Selbstzeugnissen und Bilddokumenten* (Rowohlt, 1979)
Wünsche, Konrad, *Der Volksschullehrer Ludwig Wittgenstein* (Suhrkamp, 1985)

TEXTS

Review of P. Coffey *The Science of Logic*, *The Cambridge Review*, XXXIV (1913), p. 351

'Notes on Logic', in *Notebooks 1914–16*, pp. 93–107
'Notes Dictated to G. E. Moore in Norway', in *Notebooks 1914–16*, pp. 108–19
Notebooks 1914–16, ed. G. E. M. Anscombe and G. H. von Wright (Blackwell, 1961)
Prototractatus – An Early Version of Tractatus Logico-Philosophicus, ed. B. F. McGuinness, T. Nyberg and G. H. von Wright (Routledge, 1971)
Tractatus Logico-Philosophicus, trans. C. K. Ogden and F. P. Ramsey (Routledge, 1922)
Tractatus Logico-Philosophicus, trans. D. F. Pears and B. F. McGuinness (Routledge, 1961)
Wörterbuch für Volksschulen, ed. Werner and Elizabeth Leinfelner and Adolf Hübner (Hölder-Pichler-Tempsky, 1977)
'Some Remarks on Logical Form', *Proceedings of the Aristotelian Society*, IX (1929), pp. 162–71; reprinted in *Essays on Wittgenstein's Tractatus*, ed. I. M. Copi and R. W. Beard (Routledge, 1966)
'A Lecture on Ethics', *Philosophical Review*, LXXIV, no. 1 (1968), pp. 4–14
Philosophical Remarks, ed. Rush Rhees (Blackwell, 1975)
Philosophical Grammar, ed. Rush Rhees (Blackwell, 1974)
Remarks on Frazer's Golden Bough, ed. Rush Rhees (Brynmill, 1979)
The Blue and Brown Books (Blackwell, 1975)
'Notes for Lectures on "Private Experience" and "Sense Data"', ed. Rush Rhees, *Philosophical Review*, LXXVII, no. 3 (1968), pp. 275–320; reprinted in *The Private Language Argument*, ed. O. R. Jones (Macmillan, 1971), pp. 232–75
'Cause and Effect: Intuitive Awareness', ed. Rush Rhees, *Philosophia*, VI, nos. 3–4 (1976)
Remarks on the Foundations of Mathematics, ed. R. Rhees, G. H. von Wright and G. E. M. Anscombe (Blackwell, 1967)
Philosophical Investigations, ed. G. E. M. Anscombe and R. Rhees (Blackwell, 1953)
Zettel, ed. G. E. M. Anscombe and G. H. von Wright (Blackwell, 1981)
Remarks on the Philosophy of Psychology, I, ed. G. E. M Anscombe and G. H. von Wright (Blackwell, 1980)
Remarks on the Philosophy of Psychology, II, ed. G. H. von Wright and Heikki Nyman (Blackwell, 1980)
Last Writings on the Philosophy of Psychology, I: *Preliminary Studies for Part II of Philosophical Investigations*, ed. G. H. von Wright and Heikki Nyman (Blackwell, 1982)
Remarks on Colour, ed. G. E. M. Anscombe (Blackwell, 1977)
On Certainty, ed. G. E. M. Anscombe and G. H. von Wright (Blackwell, 1969)

Culture and Value, ed. G. H. von Wright in collaboration with Heikki Nyman (Blackwell, 1980)

NOTES OF LECTURES AND CONVERSATIONS

Ludwig Wittgenstein and the Vienna Circle: Conversations Recorded by Friedrich Waismann ed. B. F. McGuinness (Blackwell, 1979)

'Wittgenstein's Lectures in 1930–33', in G. E. Moore, *Philosophical Papers* (Unwin, 1959), pp. 252–324

Wittgenstein's Lectures: Cambridge, 1930–1932, ed. Desmond Lee (Blackwell, 1980)

Wittgenstein's Lectures: Cambridge, 1932–1935, ed. Alice Ambrose (Blackwell, 1979)

'The Language of Sense Data and Private Experience – Notes taken by Rush Rhees of Wittgenstein's Lectures, 1936', *Philosophical Investigations*, VII, no. 1 (1984), pp. 1–45; continued in *Philosophical Investigations*, VII, no. 2 (1984), pp. 101–40

Lectures and Conversations on Aesthetics, Psychology and Religious Belief, ed. Cyril Barrett (Blackwell, 1978)

Wittgenstein's Lectures on the Foundations of Mathematics: Cambridge, 1939, ed. Cora Diamond (Harvester, 1976)

Wittgenstein's Lectures on Philosophical Psychology 1946–47, ed. P. T. Geach (Harvester, 1988)

CORRESPONDENCE

Briefe, Briefwechsel mit B. Russell, G. E. Moore. J. M. Keynes, F. P. Ramsey, W. Eccles, P. Engelmann und L. von Ficker, ed. B. F. McGuinness and G. H. von Wright (Suhrkamp, 1980)

Letters to Russell, Keynes and Moore, ed. G. H. von Wright assisted by B. F. McGuinness (Blackwell, 1974)

Letters to C. K. Ogden with Comments on the English Translation of the Tractatus Logico-Philosophicus, ed. G. H. von Wright (Blackwell/Routledge, 1973)

Letters from Ludwig Wittgenstein with a Memoir by Paul Engelmann, ed. B. F. McGuinness (Blackwell, 1967)

Briefe an Ludwig von Ficker, ed. G. H. von Wright with Walter Methlagl (Otto Müller, 1969)

'Letters to Ludwig von Ficker', ed. Allan Janik, in *Wittgenstein: Sources and Perspectives*, ed. C. G. Luckhardt (Harvester, 1979), pp. 82–98

'Some Letters of Ludwig Wittgenstein', in W. Eccles, *Hermathena*, XCVII (1963), pp. 57–65

Letter to the Editor, *Mind*, XLII, no. 167 (1933), pp. 415–16

'Some Hitherto Unpublished Letters from Ludwig Wittgenstein to Georg Henrik von Wright', *The Cambridge Review* (28 February 1983)

INDEX

Throughout the index, Ludwig Wittgenstein *is noted as* L.W.